PIPING HANDBOOK

Other Books of Interest

Avallone & Baumeister • MARKS' STANDARD HANDBOOK FOR MECHANICAL ENGINEERS
Bhushan & Gupta • HANDBOOK OF TRIBOLOGY
Brady & Clauser • MATERIALS HANDBOOK
Bralla • HANDBOOK OF PRODUCT DESIGN FOR MANUFACTURING
Brunner • HANDBOOK OF INCINERATION SYSTEMS
Corbitt • STANDARD HANDBOOK OF ENVIRONMENTAL ENGINEERING
Ehrich • HANDBOOK OF ROTORDYNAMICS
Elliot • STANDARD HANDBOOK OF POWERPLANT ENGINEERING
Freeman • STANDARD HANDBOOK OF HAZARDOUS WASTE TREATMENT AND DISPOSAL
Ganić & Hicks • THE McGRAW-HILL HANDBOOK OF ESSENTIAL ENGINEERING INFORMATION AND DATA
Gieck • ENGINEERING FORMULAS
Grimm & Rosaler • HANDBOOK OF HVAC DESIGN
Harris • HANDBOOK OF ACOUSTICAL MEASUREMENTS AND NOISE CONTROL
Harris & Crede • SHOCK AND VIBRATION HANDBOOK
Hicks • STANDARD HANDBOOK OF ENGINEERING CALCULATIONS
Hodson • MAYNARD'S INDUSTRIAL ENGINEERING HANDBOOK
Jones • DIESEL PLANT OPERATIONS HANDBOOK
Juran & Gryna • JURAN'S QUALITY CONTROL HANDBOOK
Kurtz • HANDBOOK OF APPLIED MATHEMATICS FOR ENGINEERS AND SCIENTISTS
Karassik et al. • PUMP HANDBOOK
Parmley • STANDARD HANDBOOK OF FASTENING AND JOINING
Rohsenow et al. • HANDBOOK OF HEAT TRANSFER FUNDAMENTALS
Rohsenow et al. • HANDBOOK OF HEAT TRANSFER APPLICATIONS
Rosaler & Rice • STANDARD HANDBOOK OF PLANT ENGINEERING
Rothbart • MECHANICAL DESIGN AND SYSTEMS HANDBOOK
Schwartz • COMPOSITE MATERIALS HANDBOOK
Schwartz • HANDBOOK OF STRUCTURAL CERAMICS
Shigley & Mischke • STANDARD HANDBOOK OF MACHINE DESIGN
Townsend • DUDLEY'S GEAR HANDBOOK
Tuma • HANDBOOK OF NUMERICAL CALCULATIONS IN ENGINEERING
Tuma • ENGINEERING MATHEMATICS HANDBOOK
Wadsworth • HANDBOOK OF STATISTICAL METHODS FOR ENGINEERS AND SCIENTISTS
Young • ROARK'S FORMULAS FOR STRESS AND STRAIN

PIPING HANDBOOK

Mohinder L. Nayyar, P.E. Editor in Chief

Engineering Specialist
Bechtel Corporation

The fifth edition of this Handbook was edited by
Reno C. King, B.M.E., M.M.E., D.Sc., P.E.

Professor of Mechanical Engineering and Assistant Dean,
School of Engineering and Science, New York University
Registered Professional Engineer

The first four editions of this Handbook were edited by
Sabin Crocker, M.E.

Fellow, ASME: Registered Professional Engineer

Sixth Edition

McGraw-Hill, Inc.
New York St. Louis San Francisco Auckland Bogotá
Caracas Lisbon London Madrid Mexico Milan
Montreal New Delhi Paris San Juan São Paulo
Singapore Sydney Tokyo Toronto

Library of Congress Cataloging-in-Publication Data

Nayyar, Mohinder L.
Piping handbook / [edited by] Mohinder L. Nayyar.—6th ed.
p. cm.
ISBN 0-07-046881-8
1. Pipe—Handbooks, manuals, etc. 2. Pipe-fitting—Handbooks, manuals, etc. I. Nayyar, Mohinder L.
TJ415.P54 1992
621.8′672—dc20 91-42836
CIP

10 11 12 13 14 15 DOC/DOC 9 9 8

ISBN 0-07-046881-8

The sponsoring editor for this book was Robert W. Hauserman, the editing supervisor was Nancy Young, and the production supervisor was Donald F. Schmidt. This book was set in Times Roman. It was composed by McGraw-Hill's Professional Book Group composition unit.

Printed and bound by R. R. Donnelley & Sons Company.

This book is printed on acid-free paper.

CONTENTS

HONORS LIST

CONTRIBUTORS

Charles L. Arnold *Principal Pipeline Consultant, Bechtel Corporation, P.O. Box 193965, 50 Beale Street, San Francisco, CA 94119* (CHAP. C5)

Daniel L. Arnold, P.E. *Engineering Manager, Rolf Jensen & Associates, Inc., 3390 Peachtree Road, N.E., Suite 1000, Atlanta, GA 30326* (CHAP. C2)

Dr. Chakrapani Basavaraju *Engineering Specialist, Bechtel Corporation, 9801 Washington Boulevard, Gaithersburg, MD, 20878-5356* (APP. E2)

Paul A. Bourquin *Senior Vice President, Wolff & Munier, Inc., 50 Broadway, Hawthorne, NY 10532* (CHAP. C4)

Charles A. Bullinger *Engineering Specialist, Bechtel Corporation, 9801 Washingtonian Boulevard, MD 20878-5356* (CHAP. B3)

John W. Burr *Development Engineer, Union Carbide Industrial Gases, Linde Division, P.O. Box 44, Tonawanda, NY 14150* (CHAP. C8)

Joseph H. Casiglia, P.E. *Principal Engineer, Piping, Detroit Edison, 2000 Second Ave., Detroit, MI 48226* (CHAP. B2)

R. C. Cipolla *Cryogenic Equipment Engineer, Union Carbide Industrial Gases, Linde Division, P.O. Box 44, Tonawanda, NY 14150* (CHAP. C8)

Alton B. Cleveland, Jr. *President, Jacobus Technology, Inc., 7901 Beech Craft Ave., Gaithersburg, MD 20879* (CHAP. B3)

G. C. Crim *Engineering Specialist, Bechtel Corporation, 9801 Washingtonian Boulevard, Gaithersburg, MD 20878-5356* (CHAP. B4)

Sabin Crocker, Jr., P.E. *307 Claggett Drive, Rockville, MD 20851. Formerly Project Engineer, Bechtel Power Corporation, Gaithersburg, MD 20878* (CHAP. B1)

C. J. Erickson *Engineering Consultant, E. I. DuPont De NeMours & Co., P.O. Box 6090, Newark, DE 19714-6090* (CHAP. B6)

Peter H. O. Fischer *Manager, Pipeline Operations, Bechtel Corporation, 12440 East Imperial Highway, Norwalk, CA 90650* (CHAP. C6)

Theodore F. Fisher *Process Engineer, Union Carbide Industrial Gases, Linde Division, P.O. Box 44, Tonawanda, NY 14150* (CHAP. C8)

Michael Frankel, CIPE *Chief Plumbing Engineer, Jacobs Engineering Group, Inc., 270 Davidson Ave., Somerset, NJ 08875* (CHAP. C13)

Michael G. Gagliardi *Manager, Mechanical Nuclear Engineering, Ebasco Services Inc., Two World Trade Center, New York, NY 10048* (CHAP. C1, APP. E4)

Ramesh L. Gandhi *Chief Slurry Engineer, Bechtel Corporation, P.O. Box 193965, 50 Beale Street, San Francisco, CA 94119* (CHAP. C11)

Lucy A. Gebhart *Pipeline Engineer, Bechtel Corporation, P.O. Box 193965, 50 Beale Street, San Francisco, CA 94119* (CHAP. C5)

Ervin L. Geiger, P.E. *Engineering Supervisor, Bechtel Corporation, 9801 Washingtonian Boulevard, Gaithersburg, MD 20878-5356* (CHAP. A2, APPS. E1, E3)

Edward F. Gerwin *Life Fellow ASME, 1515 Grampian Boulevard, Williamsport, PA 17701* (CHAP. A6)

Richard C. Getz, P.E. *Chief Piping Engineer, United Engineers & Constructors, Inc., 30 South 17th Street, P.O. Box 8223, Philadelphia, PA 19101* (CHAPS. C7, C10)

Frank E. Grunwald *Engineering Supervisor, Bechtel Corporation, 9801 Washingtonian Boulevard, Gaithersburg, MD 20878-5356* (CHAP. B5)

J. K. Howell *Cold Box Engineer, Union Carbide Industrial Gases, Linde Division, P.O. Box 44, Tonawanda, NY 14150* (CHAP. C8)

Thomas C. LaCroix, P.E. *Engineering Specialist, Bechtel Corporation, 9801 Washingtonian Boulevard, Gaithersburg, MD 20878-5356* (CHAP. B5)

Dr. Ashok L. Lagvankar *Vice President and Chief Engineer, Harza Environmental Services, Inc., Sears Towers, 233 South Wacker Drive, Chicago, IL 60606* (CHAP. C12)

Louis J. Liberatore *Staff Engineer, Ebasco Services Inc., Two World Trade Center, New York, NY 10048* (CHAP. C1)

Alfred Lohmeier *Materials Engineer, Formerly Vice President, Sumitomo Corporation of America, 345 Park Ave., New York, NY 10154* (CHAP. A5)

Joseph T. Lonsdale *Manager, International Sales, Chemlex Division, Raychem Corporation, 2415 Bay Road, Redwood City, CA 94063* (CHAP. B6)

German R. Mayer *Manager, Pipeline Projects, Bechtel Corporation, P.O. Box 193965, 50 Beale Street, San Francisco, CA 94119* (CHAP. C6)

M. F. Melchioris *Piping Design Engineers, Union Carbide Industrial Gases, Linde Division, P.O. Box 44, Tonawanda, NY 14150* (CHAP. C8)

Stanley A. Mruk *Executive Director, Plastics Pipe Institute, Wayne Interchange Plaza II, 155 Route 46 West, Wayne, NJ 07470* (CHAP. D1)

Mohinder L. Nayyar, P.E. *Engineering Specialist, Bechtel Corporation, 9801 Washingtonian Boulevard, Gaithersburg, MD 20878-5356* (CHAP. A1, A4)

Charles J. Obst, P.E. *Engineering Specialist, Bechtel Corporation, 9801 Washingtonian Boulevard, Gaithersburg, MD 20878-5356* (CHAP. B5)

Kenneth J. Oswald *Technical Services Manager, Smith Fiberglass Products, Inc. 2700 West 65th St., Little Rock, AR 72209* (CHAP. D2)

Akhil Prakash, P.E. *Supervisor Engineer, WCNOC, 8200 East 32nd Street North, Wichita, KS 67226* (TABLE A3.13)

Sam P. Soling *Consultant, St. Onge Ruff and Associates, Inc., 617 W. Market Street, Box M42, York, PA 17405* (CHAP. C9)

Dr. William S. Sun *Engineering Specialist, Bechtel Corporation, 9801 Washingtonian Boulevard, Gaithersburg, MD 20878-5356* (CHAP. B4)

Dr. Tadeusz J. Swierzawski *Consultant, Mechanical Division, Stone & Webster Engineering Corporation, 245 Summer Street, Boston, MA 02107* (CHAPS. B8, C3)

James M. Tanzosh *Supervisor, Materials Engineering, Babcock & Wilcox Co., 20 S. Van Buren Ave., Barberton, OH 44203* (CHAP. A3)

N. P. Theophilos *Design Engineer, Union Carbide Industrial Gases, Linde Division, P.O. Box 44, Tonawanda, NY 14150* (CHAP. C8)

Daniel A. Van Duyne *Assistant Chief Engineer, Mechanical Division, Stone & Webster Engineering Corporation, 245 Summer Street, Boston, MA 02107* (CHAPS. B8, C3)

John P. Velon *Vice President, Harza Environmental Services, Inc., Sears Towers, 233 South Wacker Drive, Chicago, IL 60606* (CHAP. C12)

James Waldo *General Supervisor, Production, Pittsburgh Corning Corporation, 2700 W. 16th Street, Sedalia, MO 65301* (CHAP. B7)

Norman H. White *Applications Engineer, Union Carbide Industrial Gases, Linde Division, P.O. Box 44, Tonawanda, NY 14150* (CHAP. C8)

Robert Zawierucha *Materials Engineer, Union Carbide Industrial Gases, Linde Division,P.O. Box 44, Tonawanda, NY 14150* (CHAP. C8)

REVIEWERS

S. S. Arora, P.E. *Engineering Supervisor, Bechtel Corporation, 9801 Washingtonian Boulevard, Gaithersburg, MD 20878-5356*

Dr. C. Basavarajru *Engineering Specialist, Bechtel Corporation, 9801 Washingtonian Boulevard, Gaithersburg, MD 20878-5356*

M. C. Bode *Rohm and Haas Company, Box 584, Bristol, PA 19007*

E. B. Branch *Mechanical Design Director, Sargent & Lundy Engineers, 55 Monroe Street, Chicago, IL 60603*

A. J. Breugelmans, P.E. *Chairman ASME B31.1, 705 3rd Ave., Lyndhurst, NJ 07071*

Richard E. Chambers *Principal, Simpson, Gumpertz & Hager, Inc., 297 Broadway, Arlington, MA 02174*

L. F. Clynch, P.E. *Conoco Mid-Continental Division, Transportation Department, P.O. Box 1267, Ponca City, OK 74603*

Sabin Crocker, Jr., P.E. *307 Claggett Drive, Rockville, MD 20878. Formerly Project Engineer, Bechtel Power Corporation, Gaithersburg, MD 20878*

A. R. Faulkner, P.E. *Engineering Specialist, Bechtel Corporation, 9801 Washingtonian Boulevard, Gaithersburg, MD 20878-5356*

Russel P. Fleming, P.E. *Vice President of Engineering, National Fire Sprinkler Association, Inc., Robin Hill Corporate Park, Route 22, P.O. Box 1000, Patterson, NY 12563*

E. L. Geiger, P.E. *Engineering Supervisor, Bechtel Corporation, 9801 Washingtonian Boulevard, Gaithersburg, MD 20878-5356*

E. Glynn Jones, P.E. *Fellow ASME, Manager Pipeline Technology, Bechtel Corporation, P.O. Box 193965, 50 Beale Street, San Franciso, CA 94119*

G. T. Kitz *Sargent & Lundy Engineers, 55 Monroe Street, Chicago, IL 60603*

Stanley A. Mruk *Executive Director, Plastics Pipe Institute, Wayne Interchange Plaza II, 155 Route 46 West, Wayne, NJ 07470*

Frank Murphy, CIPE *Vice President, Glassman & Associates, P.C., 1600 Spring Hill Road, Vienna, VA 22182*

Mohinder L. Nayyar, P.E. *Engineering Specialist, Bechtel Corporation, 9801 Washingtonian Boulevard, Gaithersburg, MD 20878-5356*

Kenneth J. Oswald *Technical Services Manager, Smith Fiberglass Product, Inc., 2700 West 65th Street, Little Rock, AR 72209*

J. K. Parikh *Principal Engineer, Florida Power and Light Company, 759 South Federal Highway, Stuart, FL 34994*

C. Podczerwinski *Sargent & Lundy Engineers, 55 Monroe Street, Chicago, IL 60603*

David C. Putnam *President, TBV, Inc., 103 Worcester/Providence Road, P.O. Box 72, Sutton, MA 01527*

B. C. Seam, P.E. *Engineering Supervisor, Bechtel Corporation, 9801 Washingtonian Boulevard, Gaithersburg, MD 20878-5356*

Gursharan Singh *Engineering Supervisor, Bechtel Corporation, 9801 Washingtonian Boulevard, Gaithersburg, MD 20878-5356*

A. K. Vij, P.E. *Engineering Specialist, Bechtel Corporation, 9801 Washingtonian Boulevard, Gaithersburg, MD 20878-5356*

Dr. Jagdish K. Virmani *Engineering Specialist, Bechtel Corporation, 9801 Washingtonian Boulevard, Gaithersburg, MD 20878-5356*

Austin G. Walther, P.E. *Walther Engineering Consultants, 2341 Morningside Circle, Santa Rosa, CA 95405*

Eberhard V. Welker *Principal Pipeline Engineer, Bechtel Corporation, San Francisco, CA 94119*

Horace E. Wetzell, Jr. *Vice President, The Smith & Oby Company, 6107 Carnegie Avenue, Cleveland, OH 44103*

Royce L. Williams *Analytical Engineer II, Duke Power Company, 500 S. Church Street, P.O. Box 1006, Charlotte, NC 28201*

Wartan J. Wartan *Consultant, Sargent & Lundy Engineers, 55 East Monroe Street, Chicago, Il 60603.*

Mohammed N. Vohra *Consulting Engineer, 9314 Northgate Road, Laurel, Md 20723*

Dr. Chalres Kung *14736 Main Cove Terrace, Gaithersubrg, Md 20878*

TECHNICAL AND ADMINISTRATIVE SUPPORT

Joy C. Ballinger *Administrative Assistant, Mechanical Engineering Department, Bechtel Corporation, 9801 Washington Boulevard, Gaithersburg, MD 20878-5356*

Darya Nabavian *Mechanical Engineer, Bechtel Corporation, 9801 Washington Boulevard, Gaithersburg, MD 20878-5356*

Angela T. Taylor *Senior Engineering Technician, Bechtel Corporation, 9801 Washingtonian Boulevard, Gaithersburg, MD 20878-5356*

PREFACE

While undertaking the preparation of the sixth edition of the *Piping Handbook*, we planned to reorganize, reformat, and modernize the contents to try to meet the needs of the present and future users of this handbook. We applied our energies toward this main objective and strived to enhance the utility of this handbook by focusing on approaches dealing with practical aspects of piping in industrial and commercial applications.

This sixth edition of the *Piping Handbook* is divided into five parts, A through E. Part A deals with piping fundamentals, Part B delineates generic design considerations irrespective of the contents of the piping, Part C covers specific piping systems generally encountered in industrial and commercial facilities, Part D provides an insight into thermoplastics and fiberglass piping, and Part E contains technical data, conversion tables, and reference material. We believe that the user will find the organization of the contents easy to use. Depending upon the need, level of piping knowledge, and requirements, the user may find it convenient to focus on a specific part of the handbook.

All contributors and reviewers of this handbook are engaged in the field of piping and have established themselves as authorities in their fields of expertise. They have offered their best. They have included real life examples to explain theoretical and practical concepts, including the requirements of applicable industry codes, standards, and regulations.

The scope of the book has been expanded to cover thermoplastics and fiberglass piping. The present day advancements in the manufacturing, design, fabrication, installation, examination, testing, operation, and maintenance of piping systems have been included, where possible. We know we have not achieved perfection; however, it remains our goal.

Credit for meeting the expectations of the user for any need belongs to the distinguished individuals included in the Honors List. Any time a user finds the contents inadequate or deficient in any respect, the responsibility is solely mine. I will welcome any and all constructive criticism of the handbook. You, the users, can make the difference. So please, do not hold back. Let your views be known and honored in the form of improvements in subsequent printings of this edition.

I feel humbled and privileged in offering my heartfelt gratitude to each and everyone listed in the Honors List for their perseverance, devotion, dedication, and hard work to make this team project a reality. Especially, I wish to express my thanks to Ervin L. Geiger and Sabin Crocker, Jr., for their continuous support from the beginning until the finish line as Associate Editors.

Finally, I must recognize the active and inactive support of my wife, Prabha; daughters, Mukta and Mahak; and my son, Manav, for letting me devote their share of my life to this handbook.

Mohinder L. Nayyar

HOW TO USE THIS HANDBOOK

As with any handbook, the user of this handbook can seek the topic covered either with the help of the table of contents or the index. However, an understanding of the organization and the format of this handbook will enhance its utility.

The contents of this handbook are organized in five parts:

Part A. Piping Fundamentals: There are six chapters in Part A, numbered Al through A6, dealing with commonly used terminology associated with piping, units, piping components, materials, piping codes and standards, manufacturing of piping, and fabrication and installation of piping. Each chapter is a self-contained unit. Figures, and tables are numbered sequentially preceded by the chapter number. For example, in the case of Chapter A1, the figures are numbered as Fig. A1.1, Fig. A1.2, and so on, and tables are numbered as Table A1.1, Table A1.2, and so on. Pages are numbered sequentially throughout each part, starting with A.1.

Part B. Generic Design Considerations: The Part B consists of eight chapters. The topics covered deal with generic design considerations which may be applicable to any piping system irrespective of the fluid or the mixture carried by the piping. The topics are design documents, design bases, piping layout, stress analysis, piping supports, heat tracing, thermal insulation, and flow of fluids. The chapter, page, figure, and table numbering scheme is similar to that describe for Part A.

Part C. Piping Systems: There are 13 chapters in Part C, each dealing with a specific type of piping system or systems involving application specific considerations. The piping systems covered include water, fire protection, steam, building services, oil, gas, chemical and refinery, cryogenic, refrigeration, toxic and hazardous wastes, slurry and sludge, stormwater and wastewater, and plumbing. The numbering approach for Part C is similar to Part A.

Part D. Nonmetallic Piping: Part D has two chapters, D1 and D2. Chapter D1 addresses thermoplastics piping, and Chapter D2 covers fiberglass piping systems. The numbering scheme for pages, figures, and tables is similar to the one followed for Part A.

Part E. Appendixes: Part E of the handbook contains reference material. It consists of 4 appendixes, E1 through E4. They include conversion tables, pipe and tube properties, and pressure drop tables.

Depending upon the need, level of piping knowledge, and requirements, the user of this handbook may find it very convenient to locate the desired information by focusing on a specific part of the handbook.

P • A • R • T • A

PIPING FUNDAMENTALS

CHAPTER A1
DEFINITIONS, ABBREVIATIONS, AND UNITS

Mohinder L. Nayyar, P. E.
Engineering Specialist
Bechtel Corporation
Gaithersburg, MD

GENERAL DEFINITIONS

Absolute Viscosity. Absolute viscosity or the coefficient of absolute viscosity is a measure of the internal resistance. In the centimeter, gram, second (CGS) or metric system, the unit of absolute viscosity is the poise, which is equal to 100 centipoise. The English units used to measure or express viscosity are slugs per foot second or pound force seconds per square foot. Sometimes, the English units are also expressed as pound mass per foot second or poundal seconds per square foot. Refer to Chapter B8 of this handbook.

Adhesive Joint. A joint made in plastic piping by the use of an adhesive substance which forms a continuous bond between the mating surfaces without dissolving either one of them. Refer to Part D of this handbook.

Air-Hardened Steel. A steel that hardens during cooling in air from a temperature above its transformation range.[1]

Alloy Steel. A steel which owes its distinctive properties to elements other than carbon. Steel is considered to be alloy steel when the maximum of the range given for the content of alloying elements exceeds one or more of the following limits[2]:

Manganese	1.65 percent
Silicon	0.60 percent
Copper	0.60 percent

or in which a definite range or a definite minimum quantity of any of the following elements is specified or required within the limits of the recognized field of constructional alloy steels:

Aluminum	Nickel
Boron	Titanium
Chromium (up to 3.99 percent)	Tungsten
Cobalt	Vanadium
Columbium	Zirconium
Molybdenum	

or any other alloying element added to obtain a desired alloying effect.

Small quantities of certain elements are unavoidably present in alloy steels. In many applications, these are not considered to be important and are not specified or required. When not specified or required, they should not exceed the following amounts:

Copper	0.35 percent
Chromium	0.20 percent
Nickel	0.25 percent
Molybdenum	0.06 percent

Ambient Temperature. The temperature of the surrounding medium, usually used to refer to the temperature of the air in which a structure is situated or a device operates.

Anchor. A rigid restraint providing substantially full fixation, permitting neither translatory nor rotational displacement of the pipe.

Annealing. Heating a metal to a temperature above a critical temperature and holding above that range for a proper period of time, followed by cooling at a suitable rate to below that range for such purposes as reducing hardness, improving machinability, facilitating cold working, producing a desired microstructure, or obtaining desired mechanical, physical, or other properties.[3] (A softening treatment is often carried out just below the critical range which is referred to as a *subcritical* annealing.)

Arc Cutting. A group of cutting processes wherein the severing or removing of metals is effected by melting with the heat of an arc between an electrode and the base metal. (Includes carbon, metal, gas metal, gas tungsten, plasma, and air carbon arc cutting.) *See also* Oxygen Cutting.

Arc Welding. A group of welding processes wherein coalescence is produced by heating with an electric arc or arcs, with or without the application of pressure and with or without the use of filler metal.[3,4]

Assembly. The joining together of two or more piping components by bolting, welding, caulking, brazing, soldering, cementing, or threading into their installed location as specified by the engineering design.

Automatic Welding. Welding with equipment which performs the entire welding operation without constant observation and adjustment of the controls by an operator. The equipment may or may not perform the loading and unloading of the work.[3,5]

Backing Ring. Backing in the form of a ring that can be used in the welding of piping to prevent weld spatter from entering a pipe and to assure full penetration of the weld to the inside of the pipe wall.

Ball Joint. A component which permits universal rotational movement in a piping system.[5]

Base Metal. The metal to be welded, brazed, soldered, or cut. It is also referred to as *parent* metal.

Bell-Welded Pipe. Furnace-welded pipe produced in individual lengths from cut-length skelp, having its longitudinal butt joint forge welded by the mechanical pressure developed in drawing the furnace-heated skelp through a cone-shaped die (commonly known as a *welding bell*), which serves as a combined forming and welding die.

Bevel. A type of edge or end preparation.

Bevel Angle. The angle formed between the prepared edge of a member and a plane perpendicular to the surface of the member. See Fig. A1.1.

Blank Flange. A flange that is not drilled but is otherwise complete.

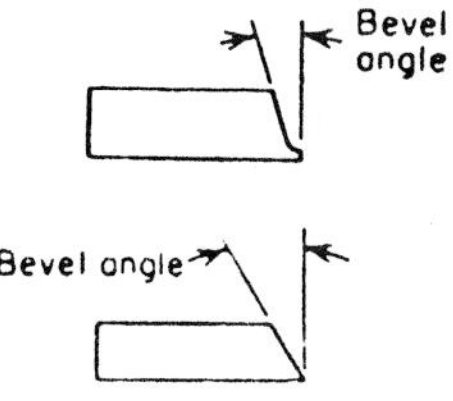

FIGURE A1.1 Bevel angle.

Blind Flange. A flange used to close the end of a pipe. It produces a blind end which is also known as a *dead end.*

Bond. The junction of the weld metal and the base metal or the junction of the base metal parts when weld metal is not present. See Fig. A1.2.

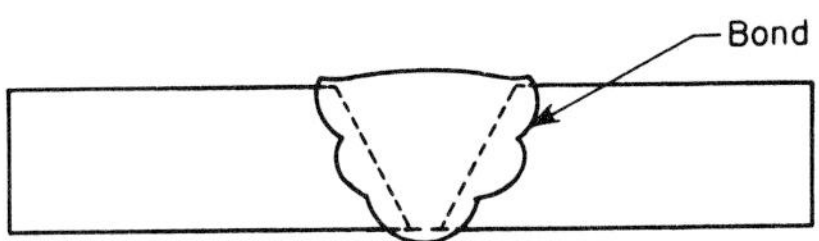

FIGURE A1.2 Bond between base metal and weld metal.

Branch Connection. The attachment of a branch pipe to the run of a main pipe with or without the use of fittings.

Braze Welding. A method of welding whereby a groove, fillet, plug, or slot weld is made using a nonferrous filler metal having a melting point below that of the base metals, but above 800°F. The filler metal is not distributed in the joint by capillary action.[5] (*Bronze* welding, the term formerly used, is a misnomer.)

Brazing. A metal joining process wherein coalescence is produced by use of a nonferrous filler metal having a melting point above 800°F but lower than that of the base metals joined. The filler metal is distributed between the closely fitted surfaces of the joint by capillary action.[5]

Butt Joint. A joint between two members lying approximately in the same plane.[5]

Butt Weld. Welded along a seam that is butted edge to edge. See Fig. A1.3.

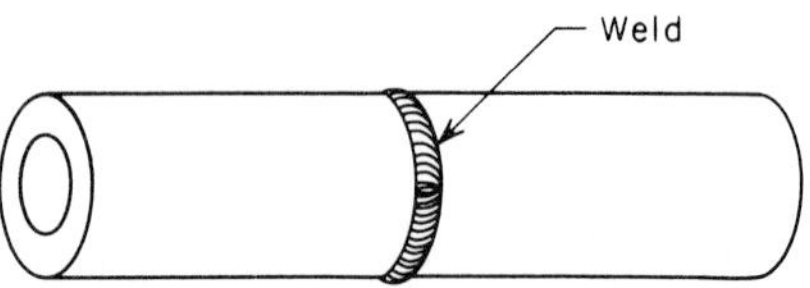

FIGURE A1.3 A circumferential butt-welded joint.

Bypass. A small passage around a large valve for warming up a line. An emergency connection around a reducing valve, trap, etc., to use in case they are out of commission.

Carbon Steel. A steel which owes its distinctive properties chiefly to the carbon (as distinguished from the other elements) which it contains. Steel is considered to be carbon steel when no minimum content is specified or required for aluminum, boron, chromium, cobalt, columbium, molybdenum, nickel, titanium, tungsten, vanadium, or zirconium or for any other element added to obtain a desired alloying effect; when the specified minimum for copper does not exceed 0.40 percent; or when the maximum content specified for any of the following elements does not exceed the percentages noted; manganese 1.65 percent, silicon 0.60 percent, copper 0.60 percent.[2]

Cast Iron. A generic term for the family of high carbon-silicon-iron casting alloys including gray, white, malleable, and ductile iron.

Centrifugally Cast Pipe. Pipe formed from the solidification of molten metal in a rotating mold. Both metal and sand molds are used. After casting, the pipe is machined, to sound metal, on the internal and external diameters to the surface roughness and dimensional requirements of the applicable material specification.

Certificate of Compliance. A written statement that the materials, equipment, or services are in accordance with the specified requirements. It may have to be supported by documented evidence.[6]

Certified Material Test Report (CMTR). A document attesting that the material is in accordance with specified requirements, including the actual results of all required chemical analyses, tests, and examinations.[6]

Chamfering. The preparation of a contour, other than for a square groove weld, on the edge of a member for welding.

Cold Bending. The bending of pipe to a predetermined radius at any temperature below some specified phase change or transformation temperature but especially at or near room temperature. Frequently, pipe is bent to a radius of 5 times the nominal pipe diameter.

Cold Working. Deformation of a metal plastically. Although ordinarily done at room temperature, cold working may be done at a temperature and rate at which strain hardening occurs. Bending of steel piping at 1300°F would be considered a cold-working operation.

Companion Flange. A pipe flange suited to connect with another flange or with a flanged valve or fitting. A loose flange which is attached to a pipe by threading, van stoning, welding, or similar method as distinguished from a flange which is cast integrally with a fitting or pipe.

Consumable Insert. Preplaced filler metal which is completely fused into the root of the joint and becomes part of the weld.[1] See Fig. A1.4.

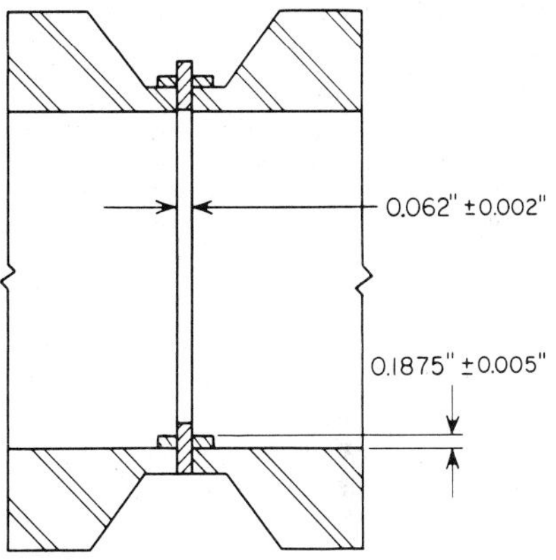

FIGURE A1.4 Consumable insert ring inserted in pipe joint eccentrically for welding in horizontal position.

Continuous-Welded Pipe. Furnace-welded pipe produced in continuous lengths from coiled skelp and subsequently cut into individual lengths, having its longitudinal butt joint forge-welded by the mechanical pressure developed in rolling the hot-formed skelp through a set of round pass welding rolls.[3]

Contractor. The entity responsible for furnishing materials and services for fabrication and installation of piping and associated equipment.

Control Piping. All piping, valves, and fittings used to interconnect air, gas, or hydraulically operated control apparatus or instrument transmitters and receivers.[2]

Controlled Cooling. A process of cooling from an elevated temperature in a predetermined manner to avoid hardening, cracking, or internal damage or to produce a desired metallurgical microstructure. This cooling usually follows the final hot-forming or postheating operation.

Corner Joint. A joint between two members located approximately at right angles to each other in the form of an *L*. See Fig. A1.5.

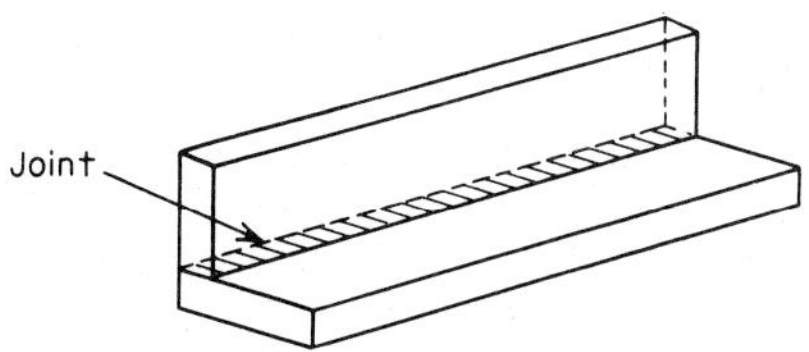

FIGURE A1.5 Corner joint.

Coupling. A threaded sleeve used to connect two pipes. Commercial couplings have internal threads to fit external threads on pipe.

Covered Electrode. A filler metal electrode, used in arc welding, consisting of a metal core wire with a relatively thick covering which provides protection for the molten metal from the atmosphere, improves the properties of the weld metal, and stabilizes the arc. Covered electrodes are extensively used in shop fabrication and field erection of piping of carbon, alloy, and stainless steels.

Crack. A fracture-type imperfection characterized by a sharp tip and high ratio of length and depth to opening displacement.

Creep or Plastic Flow of Metals. At sufficiently high temperatures all metals flow under stress. The higher the temperature and stress, the greater the tendency to plastic flow for any given metal.

Cutting Torch. A device used in oxygen, air, or powder cutting for controlling and directing the gases used for preheating and the oxygen or powder used for cutting the metal.

Defect. A flaw or an imperfection of such size, shape, orientation, location, or properties as to be rejectable per the applicable minimum acceptance standards.[7]

Density. The density of substance is defined as the mass of the substance per unit volume. It may be expressed in a variety of units.

Deposited Metal. Filler metal that has been added during a welding operation.[8]

Depth of Fusion. The distance that fusion extends into the base metal from the surface melted during welding. See Fig. A1.6.

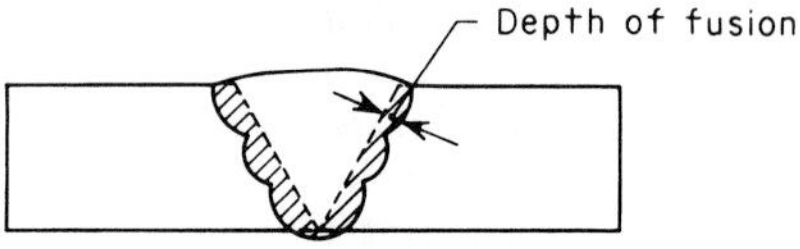

FIGURE A1.6 Depth of fusion.

Designer. Responsible for assuring that the engineering design of piping complies with the requirements of the applicable code and standard and any additional requirements established by the owner.

Dew Point. The temperature at which the vapor condenses when it is cooled at constant pressure.

Dilatant Liquid. If the viscosity of a liquid increases as agitation is increased at constant temperature, the liquid is termed *dilatant.* Examples are clay slurries and candy compounds.

Discontinuity. A lack of continuity or cohesion; an interruption in the normal physical structure of material or a product.[7]

Double Submerged Arc-Welded Pipe. Pipe having a longitudinal butt joint produced by at least two passes, one of which is on the inside of the pipe. Coalescence is produced by heating with an electric arc or arcs between the bare metal electrode or electrodes and the work. The welding is shielded by a blanket of granular, fusible material on the work. Pressure is not used and filler metal for the inside and outside welds is obtained from the electrode or electrodes.

Ductile Iron. A cast ferrous material in which the free graphite is in a spheroidal form rather than a flake form. The desirable properties of ductile iron are achieved by means of chemistry and a ferritizing heat treatment of the castings.

Eddy Current Testing. This is a nondestructive testing method in which eddy-current flow is induced in the test object. Changes in the flow caused by variations in the object are reflected into a nearby coil or coils for subsequent analysis by suitable instrumentation and techniques.

Edge Joint. A joint between the edges of two or more parallel or nearly parallel members.

Edge Preparation. The contour prepared on the edge of a member for welding. See Fig. A1.7.

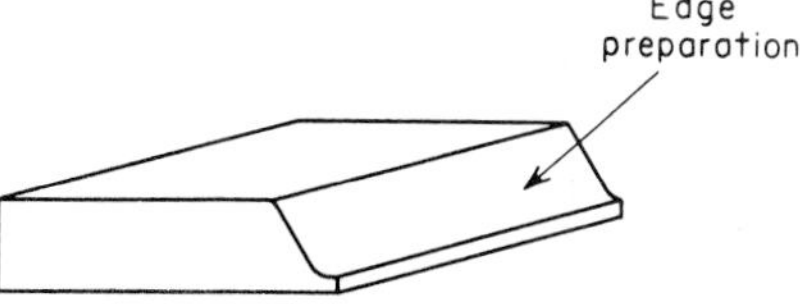

FIGURE A1.7 Edge preparation.

Electric-Flash-Welded Pipe. Pipe having a longitudinal butt joint, wherein coalescence is produced simultaneously over the entire area of abutting surfaces by the heat obtained from resistance to the flow of electric current between the two surfaces and by the application of pressure after heating is substantially completed. Flashing and upsetting are accompanied by expulsion of metal from the joint.[4]

Electric-Fusion-Welded Pipe. Pipe having a longitudinal or spiral butt joint wherein coalescence is produced in the preformed tube by manual or automatic electric-arc welding. The weld may be single or double and may be made with or without the use of filler metal.[4]

Electric-Resistance-Welded Pipe. Pipe produced in individual lengths or in continuous lengths from coiled skelp and subsequently cut into individual lengths having a longitudinal butt joint wherein coalescence is produced by the heat obtained from resistance of the pipe to the flow of electric current in a circuit of which the pipe is a part and by the application of pressure.[3]

Electrode. *See Covered Electrode.*

End Preparation. The contour prepared on the end of a pipe, fitting, or a nozzle for welding. The particular preparation is prescribed by the governing code. Refer to Chapter A6 of this handbook.

Engineering Design. The detailed design developed from process requirements and conforming to an established design criteria, including all necessary drawings and specifications, governing a piping installation.[5]

Equipment Connection. An integral part of such equipment as pressure vessels, heat exchangers, pumps, etc., designed for attachment of pipe or piping components.[8]

Erection. The complete installation of a piping system, including any field assembly, fabrication, testing, and inspection of the system.[5]

Erosion. Destruction of materials by the abrasive action of moving fluids, usually accelerated by the presence of solid particles.[9]

Examination. Denotes the procedures for all visual observation and nondestructive testing.[5]

Expansion Joint. A flexible piping component which absorbs thermal and/or terminal movement.[5]

Extruded Nozzles. The forming of nozzle (tee) outlets in pipe by pulling hemispherically or conically shaped dies through a circular hole from the inside of the pipe. Although some cold extruding is done, it is generally performed on steel after the area to be shaped has been heated to temperatures between 2000 and 1600°F.

Extruded Pipe. Pipe produced from hollow or solid round forgings, usually in a hydraulic extrusion press. In this process the forging is contained in a cylindrical die. Initially a punch at the end of the extrusion plunger pierces the forging. The extrusion plunger then forces the contained billet between the cylindrical die and the punch to form the pipe, the latter acting as a mandrel.

One variation of this process utilizes autofrettage (hydraulic expansion) and heat treatment, above the recrystallization temperature of the material, to produce a wrought structure.

Fabrication. Primarily, the joining of piping components into integral pieces ready for assembly. It includes bending, forming, threading, welding, or other operations upon these components, if not part of assembly. It may be done in a shop or in the field.[5]

Face of Weld. The exposed surface of a weld on the side from which the welding was done.[5,8]

Filler Metal. Metal to be added in welding, soldering, brazing, or braze welding.[8]

Fillet Weld. A weld of an approximately triangular cross section joining two surfaces approximately at right angles to each other in a lap joint, tee joint, corner joint, or socket weld.[5] See Fig. A1.8.

Fire Hazard. Situation in which a material of more than average combustibility or explosibility exists in the presence of a potential ignition source.[5]

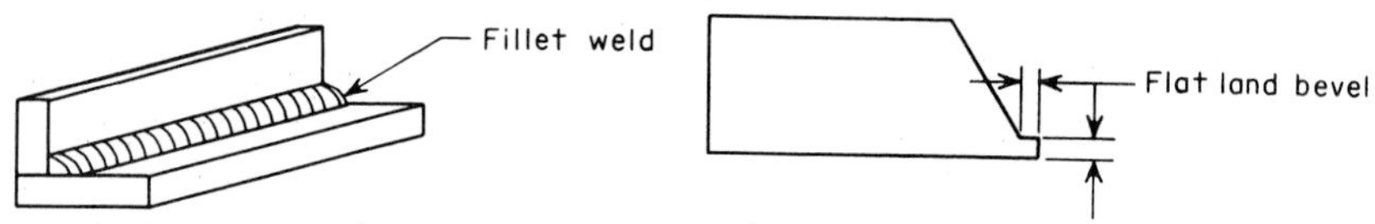

FIGURE A1.8 Fillet weld.

FIGURE A1.9 Flat land bevel.

Flat Land Bevel. A square extended root face preparation extensively used in inert-gas, root-pass welding of piping. See Fig. A1.9.

Flat Position. The position of welding wherein it is performed from the upper side of the joint and the face of the weld is approximately horizontal. See Fig. A1.10.

Flaw. An imperfection or unintentional discontinuity which is detectable by a nondestructive examination.[7]

Flux. Material used to dissolve, prevent accumulation of, or facilitate removal of oxides and other undesirable substances during welding, brazing, or soldering.

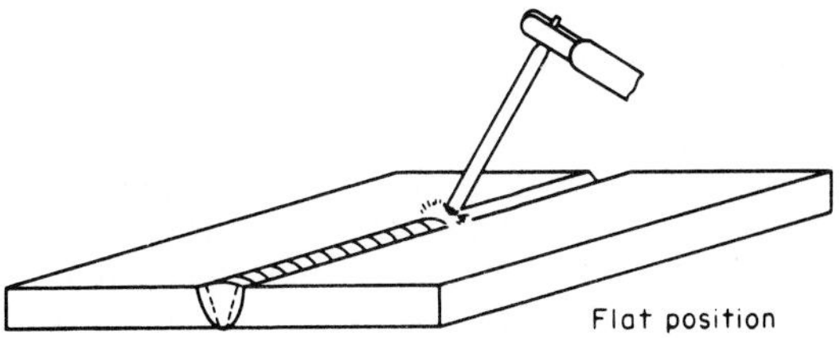

FIGURE A1.10 Welding in the flat position.

Flux-Cored Arc Welding (FCAW). An arc welding process that employs a continuous tubular filler metal (consumable) electrode having a core of flux for shielding. Added shielding may or may not be obtained from an externally supplied gas or gas mixture.

Forge Weld. A method of manufacture similar to hammer welding. The term *forge welded* is applied more particularly to headers and large drums, while *hammer welded* usually refers to pipe.

Forged and Bored Pipe. Pipe produced by boring or trepanning of a forged billet.

Full Fillet Weld. A fillet weld whose size is equal to the thickness of the thinner member joined.[8]

Fusion. The melting together of filler and base metal, or of base metal only, which results in coalescence.[8]

Fusion Zone. The area of base metal melted as determined on the cross section of a weld. See Fig. A1.11.

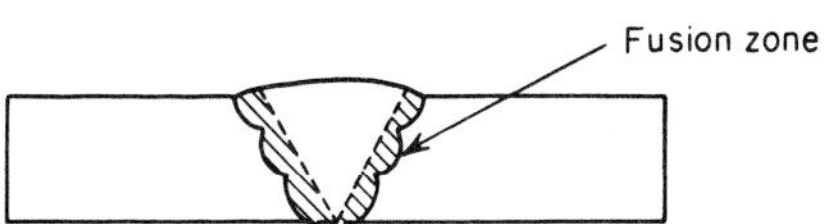

FIGURE A1.11 Fusion zone, the section of the parent metal which melts during the welding process.

Galvanizing. A process by which the surface of iron or steel is covered with a layer of zinc.

Gas Metal Arc Welding (GMAW). An arc welding process that employs a continuous solid filler metal (consumable) electrode. Shielding is obtained entirely from an externally supplied gas or gas mixture.[4,8] (Some methods of this process have been called *MIG* or CO_2 welding.)

Gas Tungsten Arc Welding (GTAW). An arc welding process that employs a tungsten (nonconsumable) electrode. Shielding is obtained from a gas or gas mixture. Pressure may or may not be used and filler metal may or may not be used. (This process has sometimes been called *TIG* welding.) When shielding is obtained by the use of an inert gas such as helium or argon, this process is called *inert-gas tungsten arc* welding.[8]

Gas Welding. A group of welding processes wherein coalescence is produced by heating with a gas flame or flames, with or without the application of pressure, and with or without the use of filler metal.[4]

Groove. The opening provided for a groove weld.

Groove Angle. The total included angle of the groove between parts to be joined by a groove weld. See Fig. A1.12.

Groove Face. That surface of a member included in the groove. See Fig. A1.13.

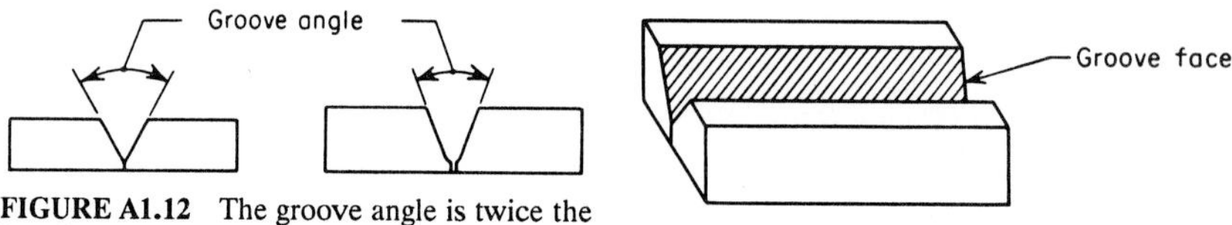

FIGURE A1.12 The groove angle is twice the bevel angle.

FIGURE A1.13 A groove face.

Groove Radius. The radius of a J or U groove. See Fig. A1.14.

Groove Weld. A weld made in the groove between two members to be joined. The standard types of groove welds are square, single V, single-bevel, single U, single J, double V, double U, double-bevel, double J, and flat-land single and double V groove welds. See Fig. A1.15 for a typical groove weld.

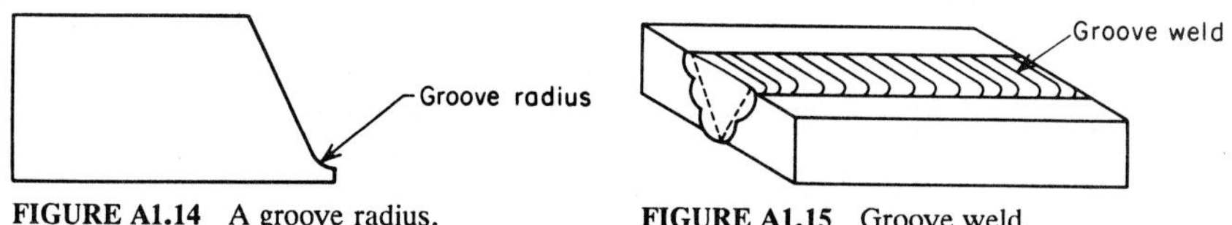

FIGURE A1.14 A groove radius.

FIGURE A1.15 Groove weld.

Hammer Weld. Method of manufacturing large pipe (usually 20 in and larger) by bending a plate into circular form, heating the overlapped edges to a welding temperature, and welding the longitudinal seam with a power hammer applied to the outside of the weld while the inner side is supported on an overhung anvil.

Hangers and Supports. Hangers and supports include elements which transfer the load from the pipe or structural attachment to the supporting structure or equipment. They include hanging-type fixtures such as hanger rods, spring hangers, sway braces, counterweights, turnbuckles, struts, chains, guides, and anchors and bearing-type fixtures such as saddles, bases, rollers, brackets, and sliding supports.[5] Refer to Chap. B5 of this handbook.

Header. A pipe or fitting to which a number of branch pipes are connected.

Heat-Affected Zone. That portion of the base metal which has not been melted but whose mechanical properties or microstructure have been altered by the heat of welding or cutting.[8] See Fig. A1.16.

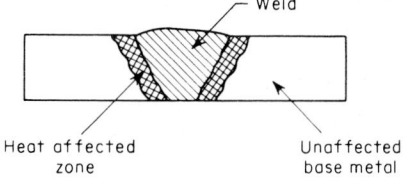

FIGURE A1.16 Welding zones.

Heat Fusion Joint. A joint made in thermoplastic piping by heating the parts sufficiently to permit fusion of the materials when the parts are pressed together.

Horizontal Fixed Position. In pipe welding, the position of a pipe joint in which the axis of the pipe is approximately horizontal and the pipe is not rotated during the operation.

Horizontal-Position Fillet Weld. The position of welding in which welding is performed on the upper side of an approximately horizontal surface and against an approximately vertical surface. See Fig. A1.17.

Horizontal-Position Groove Weld. The position of welding in which the weld axis lies in an approximately horizontal plane and the face of the weld lies in an approximately vertical plane. See Fig. A1.18.

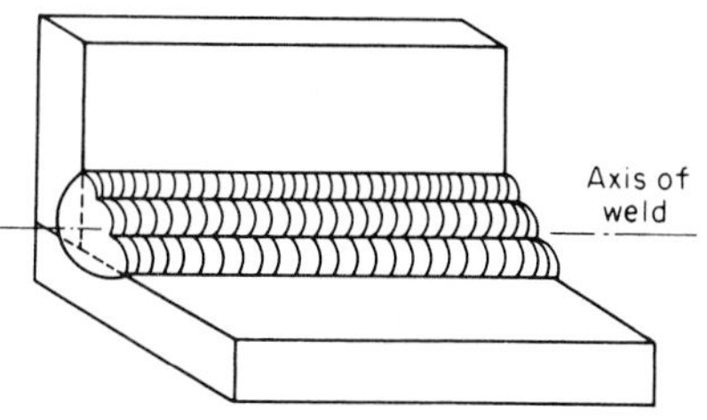

FIGURE A1.17 Horizontal position fillet weld.

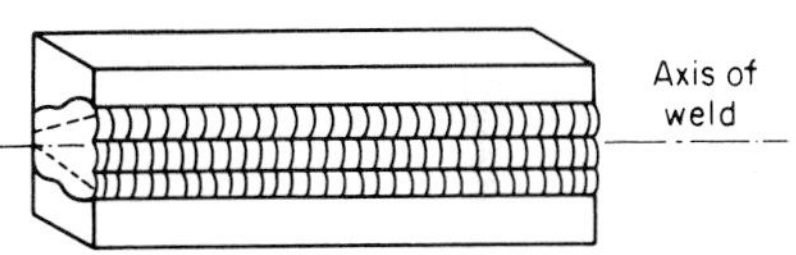

FIGURE A1.18 Horizontal position groove weld.

Horizontal Rolled Position. The position of a pipe joint in which welding is performed in the flat position by rotating the pipe. See Fig. A1.19.

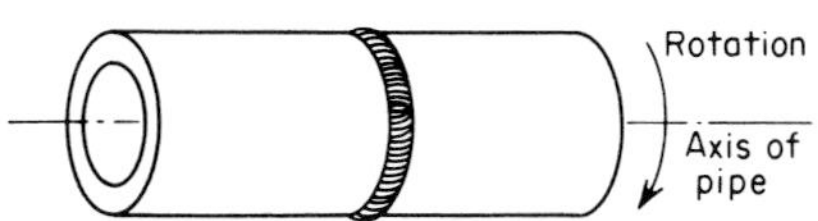

FIGURE A1.19 Horizontal rolled position.

Hot Bending. Bending of piping to a predetermined radius after heating to a suitably high temperature for hot working. On many pipe sizes, the pipe is firmly packed with sand to avoid wrinkling and excessive out of roundness.

Hot Taps. Branch piping connections made to operating pipelines, mains, or other facilities while they are in operation.

Hot Working. The plastic deformation of metal at such a temperature and rate that strain hardening does not occur. Extruding or swaging of chrome-moly piping at temperatures between 2000 and 1600°F would be considered hot-forming or hot-working operations.

Hydraulic Radius. The ratio of area of flowing fluid divided by the wetted perimeter.

$$\text{Hydraulic radius} = \frac{\text{area of flowing fluid}}{\text{wetted perimeter}}$$

Impact Test. A test to determine the behavior of materials when subjected to high rates of loading, usually in bending, tension, or torsion. The quantity measured is the energy absorbed in breaking the specimen by a single blow, as in Charpy or Izod tests.

Imperfection. A condition of being imperfect; a departure of a quality characteristic from its intended condition.[5]

Incomplete Fusion. Fusion which is less than complete and which does not result in melting completely through the thickness of the joint.

Indication. The response or evidence from the application of a nondestructive examination.[5]

Induction Heating. Heat treatment of completed welds in piping by means of placing induction coils around the piping. This type of heating is usually performed during field erection in those cases where stress relief of carbon- and alloy-steel field welds is required by the applicable code.

Inspection. Activities performed by an authorized inspector to verify whether an item or activity conforms to specified requirements.

Instrument Piping. All piping, valves, and fittings used to connect instruments to main piping, to other instruments and apparatus, or to measuring equipment.[2]

Interpass Temperature. In a multiple-pass weld, the minimum or maximum temperature of the deposited weld metal before the next pass is started.

Interrupted Welding. Interruption of welding and preheat by allowing the weld area to cool to room temperature as generally permitted on carbon-steel and on chrome-moly alloy-steel piping after sufficient weld passes equal to at least one-third of the pipe wall thickness or two weld layers, whichever is greater, have been deposited.

Joint. A connection between two lengths of pipe or between a length of pipe and a fitting.

Joint Penetration. The minimum depth a groove weld extends from its face into a joint, exclusive of reinforcement.[5] See Fig. A1.20.

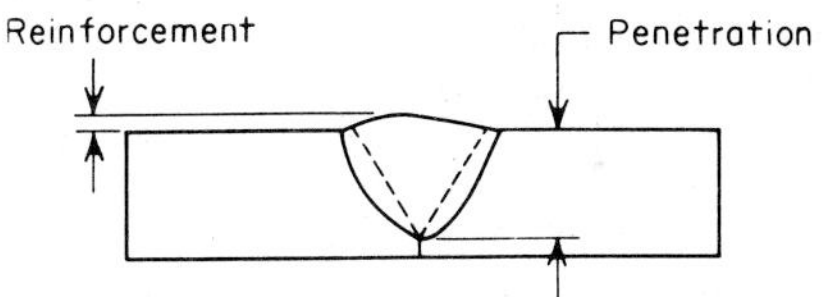

FIGURE A1.20 Weld joint penetration.

Kinematic Viscosity. The ratio of the absolute viscosity to the mass density. In the metric system, kinematic viscosity is measured in stokes or square centimeter per second. Refer to Chap. B8 of this handbook.

Laminar Flow. Fluid flow in a pipe is usually considered laminar if the Reynolds number is less than 2000. Depending upon many possible varying conditions, the flow may be laminar at a Reynolds number as low as 1200 or as high as 40,000; however, such conditions are not experienced in normal practice.

Lap Weld. Welded along a longitudinal seam in which one part is overlapped by the other. A term used to designate pipe made by this process.

Lapped Joint. A type of pipe joint made by using loose flanges on lengths of pipe whose ends are lapped over to give a bearing surface for a gasket or metal-to-metal joint.

Liquid Penetrant Examination or Inspection. This is a nondestructive examination method for finding discontinuities that are open to the surface of solid and essentially nonporous materials. This method is based on capillary action or capillary attraction by which the surface of a liquid in contact with a solid is elevated or depressed. A liquid penetrant, usually a red dye, is applied to the clean surface of the specimen. Time is allowed for the penetrant to seep into the opening. The excess penetrant is removed from the surface. A developer, normally white in color, is applied to aid in drawing the penetrant up or out to the surface. The red penetrant is drawn out of the discontinuity, which is located by the contrast and distinct appearance of the red penetrant against the white background of the developer.

Local Preheating. Preheating a specific portion of a structure.

Local Stress-relief Heat Treatment. Stress-relief heat treatment of a specific portion of a weldment. This is done extensively with induction coils, resistance coils, or propane torches in the field erection of steel piping.

Machine Welding. Welding with equipment which performs the welding operation under the observation and control of an operator. The equipment may or may not perform the loading and unloading of the work.

Magnetic Particle Examination or Inspection. This is a nondestructive examination method to locate surface and subsurface discontinuities in ferromagnetic materials. The presence of discontinuities is detected by the use of finely divided ferromagnetic particles applied over the surface. Some of these magnetic particles are gathered and held by the magnetic leakage field created by the discontinuity. The particles gathered at the surface form an outline of the discontinuity and generally indicate its location, size, shape, and extent.

Malleable Iron. Cast iron which has been heat-treated in an oven to relieve its brittleness. The process somewhat improves the tensile strength and enables the material to stretch to a limited extent without breaking.

Manual Welding. Welding wherein the entire welding operation is performed and controlled by hand.[5]

Mean Velocity of Flow. Under steady state of flow, the mean velocity of flow at a given cross section of pipe is equal to the rate of flow Q divided by the area of cross section A. It is expressed in feet per second.

$$v = \frac{Q}{A}$$

where v = mean velocity of flow, in feet per second

Q = rate of flow, in cubic feet per second

A = area of cross section, in square feet

Mechanical Joint. A joint for the purpose of mechanical strength or leak resistance or both, where the mechanical strength is developed by threaded, grooved, rolled, flared, or flanged pipe ends or by bolts, pins, and compounds, gaskets, rolled ends, caulking or machined and mated surfaces. These joints have particular application where ease of disassembly is desired.[5]

Mill Length. Also known as *random length.* The usual run-of-mill pipe is 16 to 20 ft in length. Line pipe and pipe for power plant use are sometimes made in double lengths of 30 to 35 ft.

Miter. Two or more straight sections of pipe matched and joined on a line bisecting the angle of junction so as to produce a change in direction.[4]

Newtonian Liquid. A liquid is called *newtonian* if its viscosity is unaffected by the kind and magnitude of motion or agitation to which it may be subjected as long as the temperature remains constant. Water and mineral oil are examples of a newtonian liquid.

Nipple. A piece of pipe less than 12 in long that may be threaded on both ends or one end and provided with ends suitable for welding or a mechanical joint. Pipe over 12 in long is regarded as cut pipe. Common types of nipples are close nipple, about twice the length of a standard pipe thread and without any shoulder; shoulder nipple, of any length and having a shoulder between the pipe threads; short nipple, a shoulder nipple slightly longer than a close nipple and of a definite length for each pipe size which conforms to manufacturer's standard; long nipple, a shoulder nipple longer than a short nipple which is cut to specific length.

Nominal Pipe Size (NPS). A dimensionless designator of pipe. It indicates standard pipe size when followed by the specific size designation number without an inch symbol (e.g., NPS 1 1/2, NPS 12).[2]

Nominal Thickness. The thickness given in the product material specification or standard to which manufacturing tolerances are applied.[5]

Nondestructive Examination or Inspection. Inspection by methods that do not destroy the item, part, or component to determine its suitability for use.

Normalizing. A process in which a ferrous metal is heated to a suitable temperature above the transformation range and is subsequently cooled in still air at room temperature.[5]

Nozzle. As applied to piping, this term usually refers to a flanged connection on a boiler, tank, or manifold consisting of a pipe flange, a short neck, and a welded attachment to the boiler or other vessel. A short length of pipe, one end of which is welded to the vessel with the other end chamfered for butt welding is also referred to as a *welding* nozzle.

Overhead Position. The position of welding in which it is performed from the underside of the joint.

Oxidizing Flame. An oxyfuel gas flame having an oxidizing effect caused by excess oxygen.

Oxyacetylene Cutting. An oxygen-cutting process in which metals are severed by the chemical reaction of oxygen with the base metal at elevated temperatures. The necessary temperature is maintained by means of gas flames obtained from the combustion of acetylene with oxygen.

Oxyacetylene Welding. A gas welding process in which coalescence is produced by heating with a gas flame or flames obtained from the combustion of acetylene with oxygen and with or without the addition of filler metal.

Oxyfuel Gas Welding (OFW). A group of welding processes in which coalescence is produced by heating with a flame or flames obtained from the combustion of fuel gas with oxygen, with or without the application of pressure and with or without the use of filler metal.

Oxygen Cutting (OC). A group of cutting processes used to sever or remove metals by means of the reaction of oxygen with the base metal at elevated temperatures. In the case of oxidation-resistant metals the reaction is facilitated by use of a chemical flux or metal powder.[8]

Oxygen Gouging. An application of oxygen cutting wherein a chamfer or groove is formed.

Pass. A single progression of a welding or surfacing operation along a joint, weld deposit, or substrate. The result of a pass is a weld bead, layer, or spray deposit.[8]

Peel Test. A destructive method of examination that mechanically separates a lap joint by peeling.[8]

Peening. The mechanical working of metals by means of hammer blows.

Pickle. The chemical or electrochemical removal of surface oxides. Following welding operations, piping is frequently *pickled* in order to remove mill scale, oxides formed during storage, and the weld discolorations.

Pipe. A tube with a round cross section conforming to the dimensional requirements for nominal pipe size as tabulated in ANSI B36.10 and ANSI B36.19. For special pipe having diameter not listed in above mentioned standards, the nominal diameter corresponds with the outside diameter.[5]

Pipe Alignment Guide. A restraint in the form of a sleeve or frame that permits the pipeline to move freely only along the axis of the pipe.[8]

Pipe Supporting Fixtures. Elements that transfer the load from the pipe or structural attachment to the support structure or equipment.[8]

Pipeline or Transmission Line. A pipe installed for the purpose of transmitting gas, liquids, slurries, etc., from a source or sources of supply to one or more distribution centers or to one or more large volume customers; a pipe installed to interconnect source or sources of supply to one or more distribution centers or to one or more large volume customers; or a pipe installed to interconnect sources of supply.[2]

Piping System. Interconnected piping subject to the same set or sets of design conditions.[1]

Plasma Cutting. A group of cutting processes wherein the severing or removing of metals is effected by melting with a stream of hot ionized gas.[1]

Plastic. A material which contains as an essential ingredient an organic substance of high to ultrahigh molecular weight, is solid in its finished state, and at some stage of its manufacture or processing can be shaped by flow. The two general types of plastic are thermoplastic and thermosetting.

Polarity. The direction of flow of current with respect to the welding electrode and workpiece.

Porosity. Presence of gas pockets or voids in metal.

Positioned Weld. A weld made in a joint which has been so placed as to facilitate the making of the weld.

Postheating. The application of heat to a fabricated or welded section subsequent to a fabrication, welding, or cutting operation. Postheating may be done locally, as by induction heating, or the entire assembly may be postheated in a furnace.

Postweld Heat Treatment. Any heat treatment subsequent to welding.[5]

Preheating. The application of heat to a base metal immediately prior to a welding or cutting operation.[5]

Pressure. The force per unit that is acting on a real or imaginary surface within a fluid is the pressure or intensity of pressure. It is expressed in pounds per square inch:

$$p = 144 \cdot w \cdot h + p_a$$

where p = absolute pressure at a point, in pounds per square inch
w = specific weight, in pounds per cubic foot
h = height of fluid column above the point, in feet
p_a = atmospheric pressure, in pounds per square inch

The gauge pressure at a point is obtained by designating atmospheric pressure as zero:

$$p = 144 \cdot w \cdot h$$

where p = gauge pressure

To obtain absolute pressure from gauge pressure, add the atmospheric pressure to the gauge pressure.

Pressure Head. From the definition of pressure, expression p/w is the pressure head. It can be defined as the height of the fluid above a point, and it is normally measured in feet.

Purging. The displacement during welding, by an inert or neutral gas, of the air inside the piping underneath the weld area in order to avoid oxidation or contamination of the underside of the weld. Gases most commonly used are argon, helium, and nitrogen (the last principally limited to austenitic stainless steel). Purging can be done within a complete pipe section or by means of purging fixtures of a small area underneath the pipe weld.

Quenching. Rapid cooling of a heated metal.

Radiographic Examination or Inspection. Radiography is a nondestructive test method which makes use of short wave length radiations, such as *x*-rays or gamma rays, to penetrate objects for detecting the presence and nature of macroscopic defects or other structural discontinuities. The shadow image of defects or discontinuities is recorded either on a fluorescent screen or on photographic film.

Reinforcement. In branch connections, reinforcement is material around a branch opening that serves to strengthen it. The material is either integral in the branch components or added in the form of weld metal, a pad, a saddle, or a sleeve. In welding, reinforcement is weld metal in excess of the specified weld size.

Reinforcement Weld. Weld metal on the face of a groove weld in excess of the metal necessary for the specified weld size.[5]

Repair. The process of physically restoring a nonconformance to a condition such that an item complies with the applicable requirements, including the code requirements.[6]

Resistance Weld. Method of manufacturing pipe by bending a plate into circular form and passing electric current through the material to obtain a welding temperature.

Restraint. A structural attachment, device, or mechanism that limits movement of the pipe in one or more directions.[8]

Reverse Polarity. The arrangement of direct current arc welding leads with the work as the negative pole and the electrode as the positive pole of the welding arc; a synonym for direct current electrode positive.[8]

Reynolds Number. A dimensionless number. It is defined as the ratio of the dynamic forces of mass flow to the shear stress due to viscosity. It is expressed as

$$R = \frac{Dv\rho}{\mu}$$

where R = Reynolds number
v = mean velocity of flow, in feet per second
ρ = weight density of fluid, pounds per cubic foot
D = internal diameter of pipe, in feet
μ = absolute viscosity, in pound mass per foot second poundal seconds per square foot

Rolled Pipe. Pipe produced from a forged billet which is pierced by a conical mandrel between two diametrically opposed rolls. The pierced shell is subsequently rolled and expanded over mandrels of increasingly larger diameter. Where closer dimensional tolerances are desired, the rolled pipe is cold- or hot-drawn through dies, and machined. One variation of this process produces the hollow shell by extrusion of the forged billet over a mandrel in a vertical, hydraulic piercing press.

Root Edge. A root face of zero width.

Root Face. That portion of the groove face adjacent to the root of the joint. This portion is also referred to as the *root land.* See Fig. A1.21.

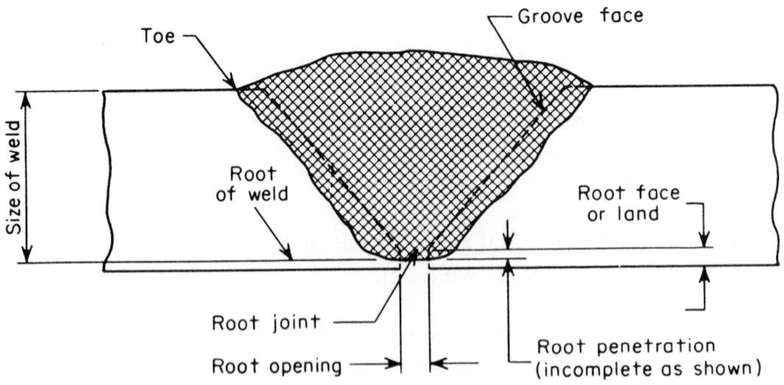

FIGURE A1.21 Nomenclature at joint of groove weld.

Root of Joint. That portion of a joint to be welded where the members to be joined approach nearest to each other. In cross section, the root of a joint may be a point, a line, or an area. See Fig. A1.21.

Root Opening. The separation, between the members to be joined, at the root of the joint.[5] See Fig. A1.21.

Root Penetration. The depth which a groove weld extends into the root of a joint as measured on the centerline of the root cross section. Sometimes welds are considered unacceptable if they show incomplete penetration. See Fig. A1.21.

Root Reinforcement. Weld reinforcement at the side other than that from which welding was done.

Root Surface. The exposed surface of a weld on the side other than that from which welding was done.

Run. The portion of a fitting having its end in line, or nearly so, as distinguished from branch connections, side outlets, etc.

Saddle Flange. Also known as *tank flange* or *boiler flange.* A curved flange shaped to fit a boiler, tank, or other vessel and receive a threaded pipe. A saddle flange is usually riveted or welded to the vessel.

Sample Piping. All piping, valves, and fittings used for the collection of samples of gas, steam, water, oil, etc.[2]

Sargol. A special type of joint in which a lip is provided for welding to make the joint fluid tight, while mechanical strength is provided by bolted flanges. The Sargol joint is used with both Van Stone pipe and fittings.

Sarlun. An improved type of Sargol joint.

Schedule Numbers. They indicate approximate values of the expression $1000 \times P/S$, where P is the service pressure and S is the allowable stress, both expressed in pounds per square inch.

Seal Weld. A fillet weld used on a pipe joint primarily to obtain fluid tightness as opposed to mechanical strength; usually used in conjunction with a threaded joint.[8]

Seamless Pipe. A wrought tubular product made without a welded seam. It is manufactured by hot-working steel or, if necessary, by subsequently cold-finishing the hot-worked tubular product to produce the desired shape, dimensions, and properties.

Semiautomatic Arc Welding. Arc welding with equipment which controls only the filler metal feed. The advance of the welding is manually controlled.[3]

Semisteel. A high grade of cast iron made by the addition of steel scrap to pip iron in a cupola or electric furnace. More correctly described as *high-strength gray iron.*

Service Fitting. A street ell or street tee having a male thread at one end.

Shielded Metal Arc Welding (SMAW). An arc welding process in which coalescence is produced by heating with an electric arc between a covered metal electrode and the work. Shielding is obtained from decomposition of the electrode covering. Pressure is not used and filler metal is obtained from the electrode.[8]

Shot Blasting. Mechanical removal of surface oxides and scale on the pipe inner and outer surfaces by the abrasive impingement of small steel pellets.

Single-Bevel-, Single-J-, Single-U-, Single-V-Groove Welds. All are specific types of groove welds and are illustrated in Fig. A1.22.

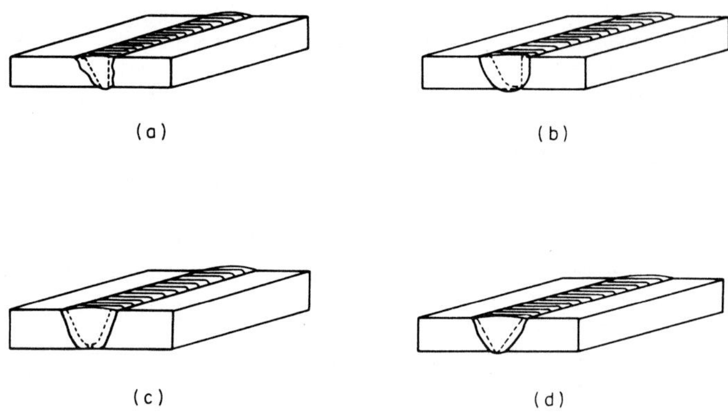

FIGURE A1.22 Groove welds. (*a*) Single bevel; (*b*) single J; (*c*) double U; (*d*) double V.

Single-Welded Butt Joint. A butt joint welded from one side only.[8]

Size of Weld. For a groove weld, the joint penetration which is the depth of chamfering plus the root penetration. See Fig. A1.21. For fillet welds, the leg length of the largest isosceles right triangle which can be inscribed within the fillet-weld cross section. See Fig. A1.23.

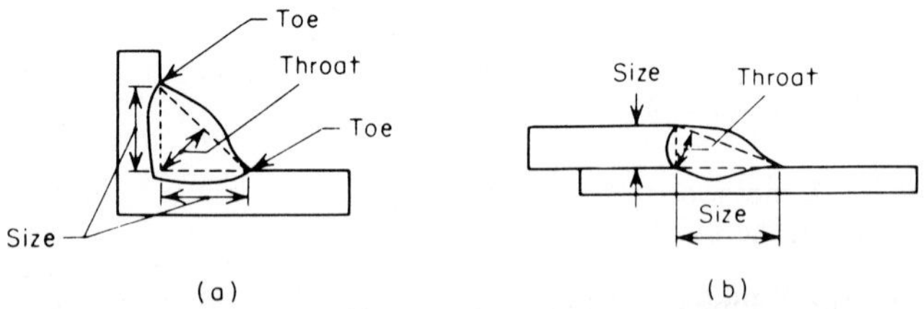

FIGURE A1.23 Size of weld (*a*) in fillet weld of equal legs, (*b*) in fillet weld of unequal legs.

Skelp. A piece of plate prepared by forming and bending, ready for welding into pipe. Flat plates when used for butt-welded pipe are called *skelp*.

Slag Inclusion. Nonmetallic solid material entrapped in weld metal or between weld metal.[8]

Slurry. A two-phase mixture of solid particles in an aqueous phase.[9]

Socket Weld. Fillet-type seal weld used to join pipe to valves and fittings or to other sections of pipe. Generally used for piping whose nominal diameter is 2 in or smaller.

Soldering. A metal-joining process wherein coalescence is produced by heating to a suitable temperature and by using a nonferrous alloy fusible at temperatures below that of the base metals being joined. The filler metal is distributed between closely fitted surfaces of the joint by capillary action.[5]

Solution Heat Treatment. Heating an alloy to a suitable temperature, holding at that temperature long enough to allow one or more constituents to enter into solid solution, and then cooling rapidly enough to hold the constituents in solution.

Solvent Cement Joint. A joint made in thermoplastic piping by the use of a solvent or solvent cement which forms a continuous bond between the mating surfaces.

Source Nipple. A short length of heavy-walled pipe between high-pressure mains and the first valve of bypass, drain, or instrument connections.

Spatter. In arc and gas welding, the metal particles expelled during welding that do not form part of the weld.[8]

Spatter Loss. Difference in weight between the amount of electrode consumed and the amount of electrode deposited.

Specific Gravity. The specific gravity of a substance is the ratio of its weight to the weight of an equal volume of water at standard conditions.

Specific Volume. The volume of a unit mass of a fluid is its specific volume, and it is measured in cubic feet per pound mass.

Specific Weight. The weight of a unit volume of a fluid is its specific weight. In English units, it is expressed in pounds per cubic foot.

Spiral Riveted. A method of manufacturing pipe by coiling a plate into a helix and riveting together the overlapped edges.

Spiral Welded. A method of manufacturing pipe by coiling a plate into a helix and fusion-welding the overlapped or abutted edges.

Spiral-Welded Pipe. Pipe made by the electric-fusion-welded process with either a butt joint, a lap joint, or a lock-seam joint.

Square-Groove Weld. A groove weld in which the pipe ends are not chamfered. Square-groove welds are generally used on piping and tubing of wall thickness no greater than 1/8 in.

Stainless Steel. An alloy steel having unusual corrosion-resisting properties, usually imparted by nickel and chromium.

Standard Dimension Ratio (SDR). The ratio of outside pipe diameter to wall thickness of thermoplastic pipe. It is calculated by dividing the specified outside diameter of the pipe by the specified wall thickness in inches.

Statically Cast Pipe. Pipe formed by the solidification of molten metal in a sand mold.

Straight Polarity. The arrangement of direct current arc welding leads in which the work is the positive pole and the electrode is the negative pole of the welding arc; a synonym for direct current electrode negative.

Stress Relieving. Uniform heating of a structure or portion thereof to a sufficient temperature to relieve the major portion of the residual stresses, followed by uniform cooling.[5]

Stringer Bead. A type of weld bead made by moving the electrode in a direction essentially parallel to the axis of the bead. There is no appreciable transverse oscillation of the electrode. The deposition of a number of string beads is known as *string beading* and is used extensively in the welding of austenitic stainless-steel materials. *See also* Weave Bead.

Structural Attachments. Brackets, clips, lugs, or other elements welded, bolted, or clamped to the pipe support structures, such as stanchions, towers, building frames, and foundation. Equipment such as vessels, exchangers, and pumps are not considered pipe-supporting elements.

Submerged Arc Welding (SAW). An arc welding process that produces coalescence of metals by heating them with an arc or arcs drawn between a bare metal electrode or electrodes and the base metals. The arc is shielded by a blanket of granular fusible material. Pressure is not used and filler metal is obtained from the electrode and sometimes from a supplementary welding rod, flux, or metal granules.

Supplemental Steel. Structural members that frame between existing building framing steel members and are significantly smaller in size than the existing steel.[8]

Swaging. Reducing the ends of pipe and tube sections with rotating dies which are pressed intermittently against the pipe or tube end.

Swivel Joint. A joint which permits single-plane rotational movement in a piping system.

Tack Weld. A small weld made to hold parts of a weldment in proper alignment until the final welds are made.

Tee Joint. A welded joint between two members located approximately at right angles to each other in the form of a T.

Tempering. A process of heating a normalized or quench-hardened steel to a temperature below the transformation range and, from there, cooling at any rate desired. This operation is also frequently called *stress relieving.*

Testing. An element of verification for the determination of the capability of an item to meet specified requirements by subjecting the item to a set of physical, chemical, environmental, or operating conditions.[6]

Thermoplastic. A plastic which is capable of being repeatedly softened by increase of temperature and hardened by decrease of temperature.[2] Refer to Chapter D1 of this handbook.

Thermosetting Plastic. Plastic which is capable of being changed into a substantially infusible or insoluble product when cured under application of heat or chemical means.[2] Refer to Chapter D2 of this handbook.

Thixotropic Liquid. If the viscosity of a liquid decreases as agitation is increased at constant temperature, the liquid is called *thixotropic.* Examples include glues, greases, paints, etc.

Throat of a Weld. A term applied to fillet welds. It is the perpendicular distance from the beginning of the root of a joint to the hypotenuse of the largest right triangle that can be inscribed within the fillet-weld cross section. See Fig. A1.23.

Toe of Weld. The junction between the face of a weld and the base metal.[8] See Fig. A1.23.

Transformation Range. A temperature range in which a phase change is initiated and completed.

Transformation Temperature. A temperature at which a phase change occurs.

Trepanning The removal by destructive means of a small section of piping (usually containing a weld) for an evaluation of weld and base-metal soundness. The operation is frequently performed with a hole saw.

Tube. A hollow product of round or any other cross section having a continuous periphery. Round tube size may be specified with respect to any two, but not all three, of the following: outside diameter, inside diameter, wall thickness. Dimensions and permissible variations (tolerances) are specified in the appropriate ASTM or ASME specifications.

Turbinizing. Mechanical removal of scale from the inside of the pipe by means of air-driven centrifugal rotating cleaners. The operation is performed on steel pipe bends after hot bending to remove loose scale and sand.

Turbulent Flow. Fluid flow in a pipe is usually considered turbulent if the Reynolds number is greater than 4000. Fluid flow with a Reynolds number between 2000 and 4000 is considered to be in "transition."

Ultrasonic Examination or Inspection. A nondestructive method in which beams of high-frequency sound waves that are introduced into the material being inspected are used to detect surface and subsurface flaws. The sound waves travel through the material with some attendant loss of energy and are reflected at interfaces. The reflected beam is detected and analyzed to define the presence and location of flaws.

Underbead Crack. A crack in heat-affected zone or in previously deposited weld metal paralleling the underside contour of the deposited weld bead and usually not extending to the surface.

Undercut. A groove melted into the base material adjacent to the toe or root of a weld and left unfilled by weld material.[8]

Van Stoning. Hot upsetting of lapping pipe ends to form integral lap flanges, the lap generally being of the same diameter as that of the raised face of standard flanges.

Vapor Pressure. The pressure exerted by the gaseous form, or vapor, of liquid. When the pressure above a liquid equals its vapor pressure, boiling occurs. If the pressure at any point in the flow of a liquid falls below the vapor pressure or becomes equal to the vapor pressure, the liquid flashes into vapor. This is called *cavitation.* The vapor thus formed travels with the liquid and collapses where the pressure is greater than vapor pressure. This could cause damage to piping and other components.

Vertical Position. With respect to pipe welding, the position in which the axis of the pipe is vertical, with the welding being performed in the horizontal position. The pipe may or may not be rotated.

Viscosity. In flowing liquids, the internal friction or the internal resistance to relative motion of the fluid particles with respect to each other.

Weave Bead. A type of weld bead made with oscillation of the electrode transverse to the axis of the weld. Contrast to string bead.

Weld. A localized coalescence of material wherein coalescence is produced either by heating to suitable temperatures, with or without the application of pressure, or by application of pressure alone, with or without the use of filler material.

Weld Bead A weld deposit resulting from a pass.

Weld Metal. That portion of a weld which has been melted during welding. The portion may be the filler metal or base metal or both.

Weld Metal Area. The area of the weld metal as measured on the cross section of a weld.

Weld Penetration. *See* Joint Penetration and Root Penetration.

Weld-Prober Sawing. Removal of a boat-shaped sample from a pipe weld for examination of the weld and its adjacent base-metal area. This operation is usually performed in graphitization studies.

Weld Reinforcement. Weld material in excess of the specified weld size.

Weldability. The ability of a metal to be welded under the fabrication conditions imposed into a specific, suitably designed structure and to perform satisfactorily in the intended service.

Welded Joint. A localized union of two or more members produced by the application of a welding process.

Welder. One who is capable of performing a manual or semiautomatic welding operation.[8]

Welder Performance Qualification. Demonstration of a welder's ability to produce welds in a manner described in a welding procedure specification that meets prescribed standards.

Welding Current. The current which flows through the electrical welding circuit during the making of a weld.

Welding Fittings. Wrought- or forged-steel elbows, tees, reducers, and similar pieces for connection by welding to each other or to pipe. In small sizes, these fittings are available with counterbored ends for connection to pipe by fillet welding and are known as *socket-weld* fittings. In larger sizes, the fittings are supplied with ends chamfered for connection to pipe by means of butt welding and are known as *butt-welding* fittings.

Welding Generator. The electrical generator used for supplying welding current.

Welding Machine. Equipment used to perform the welding operation.

Welding Operator. One who operates a welding machine or automatic welding equipment.[8]

Welding Procedure. The detailed methods and practices involved in the production of a weldment.[1]

Welding Procedure Qualification Record. Record of welding data and test results of the welding procedure qualifications, including essential variables of the process and the test results.

Welding Procedure Specification (WPS). The document which lists the parameters to be used in construction of weldments in accordance with the applicable code requirements.[1]

Welding Rod. Filler metal, in wire or rod form, used in gas-welding and brazing procedures and those arc-welding processes where the electrode does not furnish the filler metal.

Welding Sequence. The order of making the welds in a weldment.

Weldment. An assembly whose component parts are to be joined by welding.[5]

Wrought Iron. Iron refined in a plastic state in a puddling furnace. It is characterized by the presence of about 3 percent of slag irregularly mixed with pure iron and about 0.5 percent carbon and other elements in solution.

Wrought Pipe. The term *wrought pipe* refers to both wrought steel and wrought iron. Wrought in this sense means "worked" as in the process of forming furnace-welded pipe from skelp or seamless pipe from plates or billets. The expression *wrought pipe* is thus used as a distinction from cast pipe. Wrought pipe in this sense should not be confused with *wrought-iron pipe,* which is only one variety of wrought pipe. When wrought-iron pipe is referred to, it should be designated by its complete name.

FORCES, MOMENTS, AND EQUILIBRIUM

Simple Forces. When two or more forces act upon a body at one point, they may be single or combined into a resultant force. Conversely, any force may be resolved into component forces. In Fig. A1.24, let the vectors F_1 and F_2 represent two forces acting on a point O. The resultant force F is represented in direction and magnitude by the diagonal of the parallelogram of which F_1 and F_2 are the sides. Conversely, any force F may be resolved into component forces by a reverse of the above operation.

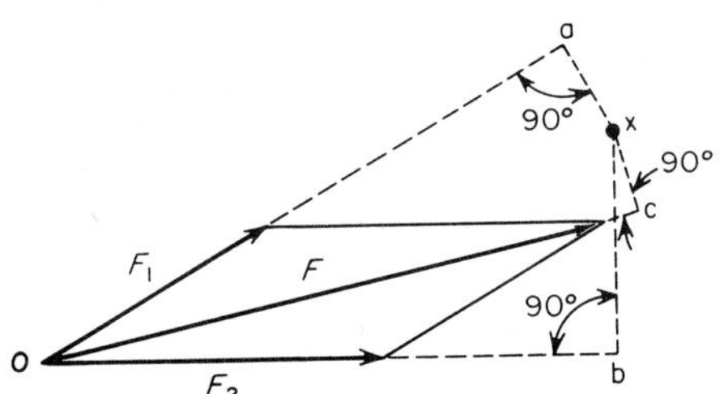

FIGURE A1.24 Vectors and moments.

Moments. The moment of a force with respect to a given point is the tendency of that force to produce rotation around it. The magnitude of the moment is represented by the product of the force and the perpendicular distance from its line of action to the point or center of moment. In the English system of weights and measures, moments are expressed as the product of the force in pounds and the length of the moment arm in feet or inches, the unit of the moment being termed the pound foot or the pound inch. Moments acting in a clockwise direction are designated as positive, and those acting in a counterclockwise direction are negative. They may be added and subtracted algebraically, as moments, regardless of the direction of the forces themselves.

With respect to Fig. A1.24, moments about an arbitrary point x are calculated as follows: Extend the line of action of F_1 until its extension intersects the perpendicular ax drawn from point x. Draw bx from x perpendicular to F_2. The sum of moments about point x due to the two forces is then

$$\Sigma M_x = F_1 \times ax - F_2 \times bx$$

Alternatively, since F_1 and F_2 have been shown to be the vector equivalent of the resultant F, the moments about x can be calculated as

$$\Sigma M_x = F \times cx$$

Couples. Two parallel forces of equal magnitude acting in opposite directions constitute a couple. The moment of the couple is the product of one of the forces and the perpendicular distance between the two. A couple has no single resultant and can be balanced only by another couple of equal moment of opposite sign.

Law of Equilibrium. When a body is at rest, the external forces acting upon it must be in equilibrium and there must be a zero net moment on the body. This means that (1) the algebraic sums of the components of all forces with reference to any three axes of reference at right angles with each other must each be zero and (2) the algebraic sum of all moments with reference to any three such axes must be zero. When the forces all lie in the same plane, the algebraic sums of their components with respect to any two axes must be equal to zero and the algebraic sum of all moments with respect to any point in the plane must be zero.

WORK, POWER, AND ENERGY

Work. When a body is moved against a resistance, work must be done upon it. The amount of work done is the product of the force and the distance through which it acts. The unit of work in the English system is the foot pound, which is the amount of work done by a force of 1 lb acting through a distance of 1 ft. The following symbols are used in this section in defining the interrelation of work, power, and energy:

A = area, in^2 or ft^2 as noted
F = force, lbf
g = local acceleration of gravity, ft/s^2
g_c = conversion constant, $ft \cdot lbf/(lbm \cdot s^2)$
h = vertical distance, ft
H = enthalpy, Btu
hp = horsepower
kw = kilowatts
KE = kinetic energy, ft · lbf
PE = potential energy, ft · lbf
p = pressure, psi
l = distance, ft
T = time, s
v = velocity, ft/s
V = volume, ft^3

w = weight, lb
W = work, ft · lb

According to the above definition of work, the following expressions may be written to represent work:

$$W = \int F\, dl = \int \frac{w}{g_c} g\, dh = \int pA\, dl = \int p\, dV$$

If the force is independent of distance, if the process takes place at sea level, if pressure and area are independent of distance, and if pressure is independent of volume, respectively, the above expressions reduce to

$$W = F(l_2 - l_1) = w(h_2 - h_1) = pA(l_2 - l_1) = p(V_2 - V_1)$$

where the subscripts 2 and 1 refer to final and initial states, respectively. The above expressions contain no term involving time, since the measure of work is independent of the time interval during which it is performed.

Power. Power is the time rate of performing work. The English unit of power is the horsepower, which is defined at 33,000 ft · lb/min or 550 ft · lb/s. Electrical power is commonly expressed in watts or kilowatts, 1 kW being equivalent to 1.34 hp and 1 hp to 0.746 kW. The expressions for horsepower corresponding to those given above for work are

$$\text{hp} = \frac{W}{550T} = \frac{Fl}{550T} \quad \text{etc.}$$

Electrical power is the product of volts × amperes, i.e.,

$$\text{kW} = \frac{\text{volts} \times \text{amperes}}{1000}$$

The above expression for the determination of electrical power is strictly true for direct current and for alternating current with a zero power factor. For the latter case, if the power factor is different from zero, the expression becomes

$$\text{kW} = \frac{\text{volts} \times \text{amperes}}{1000} \times \text{power factor}$$

Energy. Energy is the capacity for doing work possessed by a system through virtue of work having previously been done upon it. Whenever work has been done upon a system in producing a *change* in either its *motion,* its *position,* or its *molecular condition,* the system has acquired the capacity for doing work. Energy may be that due to motion, termed *kinetic energy*; that due to position, termed *potential energy*; or that due to molecular activity or configuration and is manifest as a change in its internal or stored energy. These three forms of energy are mutually convertible. In the English system, the units of energy are the foot pound and the Btu, which are related by the fact that 1 Btu is equivalent to 778 ft · lb. Some of the more common expressions for energy as as follows:

1. The potential energy of a body of weight w lb mass which has been raised h ft against gravity is PE = $(wg/g_c)h$.

2. The kinetic energy possessed by a body of weight w lb mass moving at a velocity v fps is KE $= wv^2/2g_c$.
3. If the body of 1, initially at rest, were to fall freely through the distance h, its potential energy would be converted to kinetic energy and it would acquire a velocity v determined as follows:

$$\text{PE} = wh = \text{KE} = wv^2/2g_c \qquad \text{hence} \quad wh = wv^2/2h \qquad \text{and} \qquad v = \sqrt{2g_c\, h}$$

4. The energy, resulting from its temperature, of a gas in motion is measured by its specific enthalpy h with units of Btu per pound mass. This energy is available for conversion to kinetic energy, as given by

$$\Delta H = \Delta \text{KE}$$

$$w\Delta h = \frac{w}{778 \times 2g_c}(v_2^2 - v_1^2)$$

If the initial velocity v_1 is negligible, there is obtained

$$v_2 = 223.7\sqrt{\Delta h}$$

5. Energy is measured in the English system in horsepower hours, kilowatthours, Btu, and foot pounds. The relation among these units is as follows:

$$1 \text{ hp} \cdot \text{h} = 0.746 \text{ kWh} = 2546 \text{ Btu} = 1980{,}788 \text{ ft} \cdot \text{lb}$$

$$1 \text{ kWh} = 1.34 \text{ hp} \cdot \text{h} = 3412 \text{ Btu} = 2654{,}536 \text{ ft} \cdot \text{lb}$$

HEAT AND TEMPERATURE

Units of Heat. The unit of heat commonly used in the English system is the British thermal unit, or Btu, and is approximately equal to the quantity of heat that must be transferred to one pound of water in order that its temperature be raised one degree Fahrenheit. In laboratory work and throughout much of the world, the calorie is the common unit of heat. A gram calorie is the approximate quantity of heat that must be transferred to 1 gram of water in order to raise its temperature by 1°C. The kilocalorie, sometimes called the kilogram calorie, is equal to 1000 gram calories.

The definitions above are indicated as being approximate because, over the temperature range from freezing to boiling points of water, different quantities of heat are required to produce a unit temperature change. For this reason, the calorie and the Btu have been defined in international units as

$$1 \text{ IT calorie} = 1/860 \text{ international watthour}$$

$$1 \text{ Btu} = 251.996 \text{ IT calories}$$

In most engineering work, it is sufficiently accurate to use the relations that 1 kg · cal = 3.968 Btu and that 1 Btu = 0.252 kg · cal.

Units of Temperature. The relative "hotness" or "coldness" of a body is denoted by the term *temperature.* The temperature of a substance is measured by noting its effect upon a thermometer or pyrometer whose thermal properties are known. The mercury thermometer is suitable for measuring temperatures from −39 to about 600°F. This limit may be extended to 1000°F if the capillary tube above the mercury is filled with nitrogen or carbon dioxide under pressure. High temperatures must be measured with thermocouples or optical pyrometers. The most commonly used thermometer scales are the Fahrenheit and the Celsius. Thermometer scales have as their bases the melting and boiling points of water, both measured at atmospheric pressure. The relation of the Fahrenheit and Celsius scales is as follows:

	Absolute zero	Freezing point of water	Boiling point of water
Degrees Fahrenheit	−459.6	32	212
Degrees Celsius............	−273	0	100

The relation between the two scales is

$$°C = \frac{5}{9}(°F - 32) \qquad \text{and} \qquad °F = \frac{9}{5}°C + 32$$

in which C is the reading on the Celsius scale and F is the reading on the Fahrenheit scale.

In certain calculations, it is necessary to express the temperature in "absolute" units. The absolute temperature associated with the Fahrenheit scale is called the Rankine temperature, and that associated with the Celsius scale is termed the Kelvin temperature. The relationships among these scales are as follows:

$$R = °F + 459.6$$

$$K = °C + 273$$

$$R = 1.8\ K$$

$$K = \frac{5}{9} R$$

where R and K designate absolute temperatures on the Rankine and Kelvin scales, respectively.

Specific Heat. The specific heat of a substance is the quantity of heat required to produce a unit temperature change in a unit mass of that substance. Typical units are calories per gram per degree Celsius and Btu per pound per degree Fahrenheit. The numerical value of the specific heat is a function of the process by which the unit temperature change is effected; if a gas expands at constant pressure owing to the addition of heat, work is done by the walls of the containing vessel on the surrounding atmosphere and the heat addition must be greater than would have been required to cause the same temperature change at constant volume. The two most frequently used specific heats are those at constant volume and constant pressure, and they are represented symbolically as c_v and c_p, respectively.

The definition of specific heat given in the preceding paragraph is convenient for engineering applications. By thermodynamic analysis, it can be shown that the two specific heats referred to are given by

$$c_v = \left(\frac{\partial u}{\partial T}\right)_v$$

$$c_p = \left(\frac{\partial h}{\partial T}\right)_p$$

where u and h represent internal energy and enthalpy, respectively, and v and p indicate that volume or pressure remains constant during the measurement of the corresponding specific heat.

The specific heats of most substances vary with temperature. For a general functional relationship, the mean value of specific heat over a temperature range from T_1 to T_2 is given by

$$c_{\text{mean}} = \frac{\int_{T_1}^{T_2} c(T)\, dT}{T_2 - T_1}$$

If the algebraic relationship between specific heat and temperature is not known but the relation is available in the form of a graph or table, it is usually sufficiently accurate to evaluate the average or mean specific heat at the average of temperature over the temperature range in question.

LENGTHS, AREAS, SURFACES, AND VOLUMES

List of Symbols

A = angle, deg*
C = length of chord
d = diameter of circle or sphere = $2r$
h = height of segment, altitude of cone, etc., as explained in context
π = ratio of circumference to diameter of circle = 3.1416
θ = angle in radian measure*
S = length of arc, slant height, etc., as explained in context
r = radius of circle or sphere = $d/2$
R = mean radius of curvature for pipe bends

Areas are expressed in square units and volumes in cubical units of the same system in which lengths are measured.

Triangle. Area = one-half base × altitude.

Circle. (See Fig. A1.25):

Circumference = $\pi d = 2\pi r$.
Area = $\pi r^2 = \pi d^2/4$.

*Degrees can be converted to radian measure by multiplying by 0.0175, since 2π radians = 360°. Hence, $\theta = 0.0175A$.

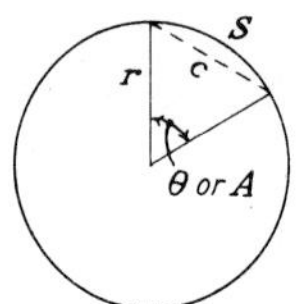

FIGURE A1.25 Length of arc and chord.

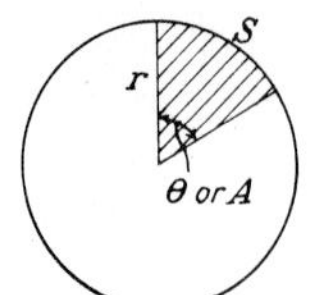

FIGURE A1.26 Area of sector.

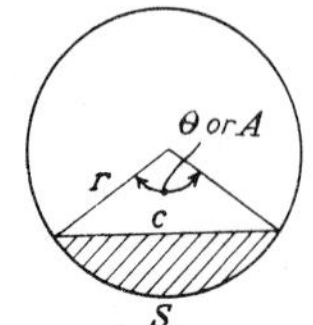

FIGURE A1.27 Area of segment, method 1.

Length of arc, $S = \theta r = 0.0175Ar$.

Length of chord, $C = 2r \sin \theta/2 = 2r \sin A/2$.

Area of Sector. (See Fig. A1.26):

$$\text{Area} = \tfrac{1}{2}rS = \tfrac{1}{2}r^2\theta = \pi r^2A/360 = 0.008727r^2A$$

Area of Segment. (Method 1 Fig. A1.27). Find area of sector having same arc and area of triangle formed by chord and radii of sector. The area of the segment equals the sum of these two areas if the segment is greater than a semicircle, and it equals their difference if the segment is less than a semicircle.

$$\text{Area} = \tfrac{1}{2}r^2(\theta \pm \sin \theta)$$

$$= \tfrac{1}{2}r^2(0.0175A + \sin A)$$

Area of Segment. [Method 2 (approximate)]:*

When $h = 0$ to $\tfrac{1}{4}d$, area $= h\sqrt{1.766dh - h^2}$.

When $h = \tfrac{1}{4}d$ to $\tfrac{1}{2}d$, area $= h\sqrt{0.017d^2 + 1.7dh - h^2}$.

When $h = \tfrac{1}{2}d$ to d, subtract area of empty sector from area of entire circle.

Offset Bends. (See Fig. A1.28.) The relation of D, R, H, and L is determined by geometry for the general case shown in Fig. A1.28*a* and A1.28*b* as follows: Consider the diagonal line joining the centers of curvature of the two arcs in either figure as forming the hypotenuse of three right-angled triangles and write an equation between the squares of the two other sides, thus,

$$2\sqrt{(D/2)^2 + R^2} = \sqrt{(2R - H)^2 + L^2}$$

squaring both sides and solving for each term in turn,

$$D = \sqrt{H^2 - 4HR + L^2}$$

$$L = \sqrt{D^2 + 4HR - H^2}$$

$$H = 2R - \sqrt{4R^2 - L^2 + D^2}$$

*For a sketch and table of volumes in partly full horizontal tanks, see Table A1.4. The greatest error possible by this method is 0.23 percent.

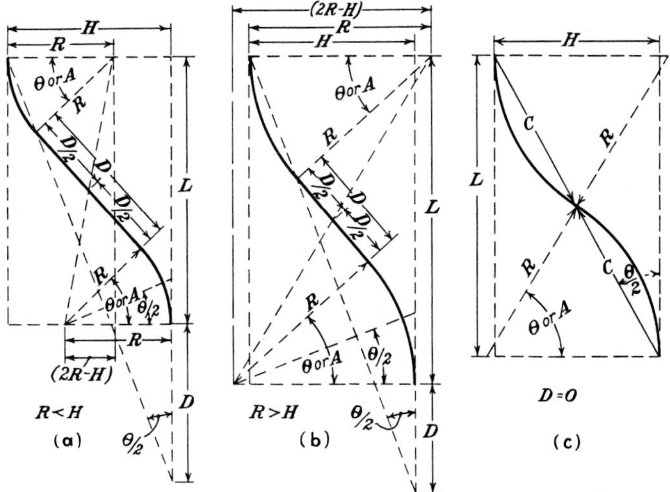

FIGURE A1.28 Offset bends.

$$R = \frac{L^2 + H^2 - D^2}{4H}$$

$\theta = 2 \tan^{-1}(HL + D)$ (from similarity of triangles, see Fig. A1.28a. When $D = 0$ (Fig. A1.28c),

$$L = \sqrt{(4R - H)H}$$

$$R = \frac{L^2 + H^2}{4H}$$

$$C = \sqrt{RH}$$

Length of pipe in offset:

$$L = 2R\theta + D$$

$$= 0.035RA + D$$

$$= 4R \tan^{-1}\left(\frac{H}{L + D}\right) + D$$

where the angle is expressed in radians.

Cylinder

$$\text{Area} = 2\pi rh + 2\pi r^2 = 2\pi r(h + r)$$

where r = radius of base
h = height

$$\text{Volume} = \pi r^2 h$$

Pyramid. Right pyramid (i.e., vertex directly above center of base):

Lateral area = one-half slant height × perimeter of base

Volume = one-third altitude × area of base

Cone

Volume = one-third area of base × perpendicular distance from vertex to plane of base

Right circular cone:

$$\text{Lateral area} = \pi rs$$

$$\text{Volume} = \tfrac{1}{3}\pi r^2 h$$

where s = slant height
r = radius of base
h = perpendicular distance form vertex to plane of base

Frustum of right circular cone:

$$\text{Lateral area} = \pi s(r + r')s = \sqrt{(r - r')^2 + h^2}$$

where r = radius of lower base
r' = radius of upper base
h = height of frustum
s = slant height of frustum

$$\text{Volume} = \tfrac{1}{3}\pi h(r^2 + rr' + r'^2)$$

Sphere

$$\text{Area} = 4\pi r^2 = \pi d^2$$

$$\text{Volume} = \tfrac{4}{3}\pi r^2 = \tfrac{1}{6}\pi d^3$$

ACRONYMS AND ABBREVIATIONS

Listed below are some abbreviations and acronyms which are associated with activities related to piping.

AAE	American Association of Engineers
ACI	American Concrete Institute
ACRI	Air Conditioning and Refrigeration Institute
ACSI	Association of Consulting Structural Engineers
A-E	Architect-engineer
AEC	American Engineering Council
AESC	American Engineering Standards Committee
AFFFA	American Forged Fitting and Flange Association
AIDD	American Institute of Design and Drafting

AIME	American Institute of Mechanical Engineers
AISC	American Institute of Steel Construction
AISE	Association of Iron and Steel Engineers
AISI	American Iron and Steel Institute
AMAA	Adhesives Manufacturers Association of America
AMCA	Air Moving and Conditioning Association
AMFIE	Association of Mutual Fire Insurance Engineers
AMICE	Associate Member of Institute of Civil Engineers
ANS	American Nuclear Society
ANSI	American National Standards Institute
API	American Petroleum Institute
ARI	Air Conditioning and Refrigeration Institute
ASBC	American Standard Building Code
ASCE	American Society of Civil Engineers
ASCEA	American Society of Civil Engineers and Architects
ASCHE	American Society of Chemical Engineers
ASE	Amalgamated Society of Engineers
ASEA	American Society of Engineers and Architects
ASEE	American Society of Engineering Education
ASHACE	American Society of Heating and Air-Conditioning Engineers
ASHRAE	American Society of Heating, Refrigerating and Air-Conditioning Engineers
ASME	American Society of Mechanical Engineers
ASNE	American Society of Naval Engineers
ASRE	American Society of Refrigeration Engineers
ASSE	American Society of Safety Engineers; American Society of Sanitary Engineers
ASTM	American Society for Testing and Materials
ASWG	American steel and wire gauge
AWG	American wire gauge
AWS	American Welding Society
AWWA	American Water Works Association
BHN	Brinell hardness number
CAD	Computer-aided design
CADD	Computer-aided design drafting
db	Dry bulb
DCCP	Design Change Control Program
DCEA	Directory of Civil Engineering Abbreviations
DCN	Design/Drawing Change Notice
DIN	Deutsches Institu für Normung; German Standards Institute
DIS	Ductile Iron Society

EJMA	Expansion Joint Manufacturers Association
FCI	Fluid Controls Institute
FMA	Forging Manufacturers Association
GPM	Gallons per minute
GPS	Gallons per second
HI	Hydraulic Institute
HVAC	Heating, ventilating, and air conditioning
IAEE	International Association of Earthquake Engineers
IFHTM	International Federation for the Heat Treatment of Materials
IFI	Industrial Fasteners Institute
IGSCC	Intergranular Stress Corrosion Cracking
ISA	International Standards Association; Instrument Society of America
NACE	National Association of Corrosion Engineers
NAPE	National Association of Power Engineers
NASE	National Association of Stationary Engineers
NASPD	National Association of Steel Pipe Distributors
NCPI	National Clay Pipe Institute
NFPA	National Fire Protection Association
NIMA	National Insulation Manufacturers Association
NPT	National Pipe Thread
NSPE	National Society of Professional Engineers
NSSS	Nuclear Steam System Supplier
OBE	Operating Basis Earthquake
OSHA	Occupational Safety and Health Act or Administration
PACE	Professional Association of Consulting Engineers
PFI	Pipe Fabrication Institute
PJA	Pipe Jacking Association
PLCA	Pipe Line Contractors Association
ppb	Parts per billion
PFF	Plumbers and Pipe Fitters Union
PPI	Plastic Pipe Institute
ppm	Parts per million
PPMS	Plastic Pipe Manufacturers' Society
PRI	Plastics and Rubber Institute
RWMA	Resistance Welding Manufacturers' Association
SAE	Society of Automotive Engineers
SCC	Stress corrosion cracking
SMA	Solder Markers' Association; Steel Manufacturers' Association
VMA	Valve Manufacturers' Association

USEFUL TABLES

Following are tables of units and measures associated with piping. For convenience of calculation, Table A1.1 provides decimal equivalents of eights, sixteenths, thirty-seconds, and sixty-fourths of an inch. Table A1.2 provides diameters, and thicknesses of wire and sheet-metal gauges. Table A1.3 provides volume of contents, in cubic feet and U.S. gallons, of cylindrical tanks of various diameters and 1 ft in length, when completely filled. Table A1.4 lists the contents of pipes and cylindrical tanks per foot of length for any depth of liquid.

TABLE A1.1 Decimal Equivalents of Eighths, Sixteenths, Thirty-seconds, and Sixty-Fourths of an Inch

Eighths	9/32 = 0.28125	19/64 = 0.296875
1/8 = 0.125	11/32 = 0.34375	21/64 = 0.328125
1/4 = 0.250	13/32 = 0.40625	23/64 = 0.359375
3/8 = 0.375	15/32 = 0.46875	25/64 = 0.390625
1/2 = 0.500	17/32 = 0.53125	27/64 = 0.421875
5/8 = 0.625	19/32 = 0.59375	29/64 = 0.453125
3/4 = 0.750	21/32 = 0.65625	31/64 = 0.484375
7/8 = 0.875	23/32 = 0.71875	33/64 = 0.515625
Sixteenths	25/32 = 0.78125	35/64 = 0.546875
1/16 = 0.0625	27/32 = 0.84375	37/64 = 0.578125
3/16 = 0.1875	29/32 = 0.90625	39/64 = 0.609375
5/16 = 0.3125	31/32 = 0.96875	41/64 = 0.640625
7/16 = 0.4375	Sixty-fourths	43/64 = 0.671875
9/16 = 0.5625	1/64 = 0.015625	45/64 = 0.703125
11/16 = 0.6875	3/64 = 0.046875	47/64 = 0.734375
13/16 = 0.8125	5/64 = 0.078125	49/64 = 0.765625
15/16 = 0.9375	7/64 = 0.109375	51/64 = 0.796875
Thirty-seconds	9/64 = 0.140625	53/64 = 0.828125
1/32 = 0.03125	11/64 = 0.171875	55/64 = 0.859375
3/32 = 0.09375	13/64 = 0.203125	57/64 = 0.890625
5/32 = 0.15625	15/64 = 0.234375	59/64 = 0.921875
7/32 = 0.21875	17/64 = 0.265625	61/64 = 0.953125
		63/64 = 0.984375

TABLE A1.2 Wire and Sheet-metal Gauges*

Gauge No.	American wire gauge, or Brown and Sharpe (for copper wire)	Steel wire gauge, or Washburn and Moen or Roebling (for steel wire)	Birmingham wire gauge (B.W.G.) or Stubs' iron wire (for steel wire or sheets)	Stubs steel wire gauge	British Imperial standard wire gauge (S.W.G.)	U. S. standard gauge for sheet metal (iron and steel) 480 lb per cu ft	AISI inch equivalent for U. S. steel sheet thickness	British standard for iron and steel, sheets and hoops 1914 (B.G.)
0000000	......	0.4900	.....	.....	0.500	0.500	......	0.6666
000000	......	0.4615	.....	.....	0.464	0.469	......	0.625
00000	......	0.4305	.....	.....	0.432	0.438	......	0.5883
0000	0.460	0.3938	0.454	.....	0.400	0.406	......	0.5416
000	0.410	0.3625	0.425	.....	0.372	0.375	......	0.5000
00	0.365	0.3310	0.380	.....	0.348	0.344	......	0.4452
0	0.325	0.3065	0.340	.....	0.324	0.312	......	0.3964
1	0.289	0.2830	0.300	0.227	0.300	0.281	......	0.3532
2	0.258	0.2625	0.284	0.219	0.276	0.266	......	0.3147
3	0.229	0.2437	0.259	0.212	0.252	0.250	0.2391	0.2804
4	0.204	0.2253	0.238	0.207	0.232	0.234	0.2242	0.2500
5	0.182	0.2070	0.220	0.204	0.212	0.219	0.2092	0.2225
6	0.162	0.1920	0.203	0.201	0.192	0.203	0.1943	0.1981
7	0.144	0.1770	0.180	0.199	0.176	0.188	0.1793	0.1764
8	0.128	0.1620	0.165	0.197	0.160	0.172	0.1644	0.1570
9	0.114	0.1483	0.148	0.194	0.144	0.156	0.1495	0.1398
10	0.102	0.1350	0.134	0.191	0.128	0.141	0.1345	0.1250
11	0.091	0.1205	0.120	0.188	0.116	0.125	0.1196	0.1113
12	0.081	0.1055	0.109	0.185	0.104	0.109	0.1046	0.0991
13	0.072	0.0915	0.095	0.182	0.092	0.094	0.0897	0.0882
14	0.064	0.0800	0.083	0.180	0.080	0.078	0.0747	0.0785
15	0.057	0.0720	0.072	0.178	0.072	0.070	0.0673	0.0699
16	0.051	0.0625	0.065	0.175	0.064	0.062	0.0598	0.0625
17	0.045	0.0540	0.058	0.172	0.056	0.056	0.0538	0.0556
18	0.040	0.0475	0.049	0.168	0.048	0.050	0.0478	0.0495
19	0.036	0.0410	0.042	0.164	0.040	0.0438	0.0418	0.0440
20	0.032	0.0348	0.035	0.161	0.036	0.0375	0.0359	0.0392
21	0.0285	0.0317	0.032	0.157	0.032	0.0344	0.0329	0.0349
22	0.0253	0.0286	0.028	0.155	0.028	0.0312	0.0299	0.0313
23	0.0226	0.0258	0.025	0.153	0.024	0.0281	0.0269	0.0278
24	0.0201	0.0230	0.022	0.151	0.022	0.0250	0.0239	0.0248
25	0.0179	0.0204	0.020	0.148	0.020	0.0219	0.0209	0.0220
26	0.0159	0.0181	0.018	0.146	0.018	0.0188	0.0179	0.0196
27	0.0142	0.0173	0.016	0.143	0.0164	0.0172	0.1064	0.0175
28	0.0126	0.0162	0.014	0.139	0.0148	0.0156	0.0149	0.0156
29	0.0113	0.0150	0.013	0.134	0.0136	0.0141	0.0135	0.0139
30	0.0100	0.0140	0.012	0.127	0.0124	0.0125	0.0120	0.0123
31	0.0089	0.0132	0.010	0.120	0.0116	0.0109	0.0105	0.0110
32	0.0080	0.0128	0.009	0.115	0.0108	0.0102	0.0097	0.0098
33	0.0071	0.0118	0.008	0.112	0.0100	0.0094	0.0090	0.0087
34	0.0063	0.0104	0.007	0.110	0.0092	0.0086	0.0082	0.0077
35	0.0056	0.0095	0.005	0.108	0.0084	0.0078	0.0075	0.0069
36	0.0050	0.0090	0.004	0.106	0.0076	0.0070	0.0067	0.0061
37	0.0045	0.0085	.....	0.103	0.0068	0.0066	0.0064	0.0054
38	0.0040	0.0080	.....	0.101	0.0060	0.0062	0.0060	0.0048
39	0.0035	0.0075	.....	0.099	0.0052	......	......	0.0043
40	0.0031	0.0070	.....	0.097	0.0048	......	......	0.0039
41	......	0.0066	.....	0.095	0.0044	......	......	0.0034
42	......	0.0062	.....	0.092	0.0040	......	......	0.0031
43	......	0.0060	.....	0.088	0.0036	......	......	0.0027
44	......	0.0058	.....	0.085	0.0032	......	......	0.0024
45	......	0.0055	.....	0.081	0.0028	......	......	0.0022
46	......	0.0052	.....	0.079	0.0024	......	......	0.0019
47	......	0.0050	.....	0.077	0.0020	......	......	0.0017
48	......	0.0048	.....	0.075	0.0016	......	......	0.0015
49	......	0.0046	.....	0.072	0.0012	......	......	0.0014
50	......	0.0044	.....	0.069	0.0010	......	......	0.0012

*Diameters and thicknesses in decimal parts of an inch.

TABLE A1.3 Contents, in Cubic Feet and U.S. Gallons, of Cylindrical Tanks of Various Diameters and 1 Ft in Length, When Completely Filled*

Diameter, in inches	For 1 ft. in length		Length, in inches, of cylinder of 1 cu. ft. capacity	Diameter, in inches	For 1 ft. in length		Length, in inches, of cylinder of 1 cu. ft. capacity
	Cubic feet; also, area in square feet	U. S. gal., 231 cu. in.			Cubic feet; also, area in square feet	U. S. gal., 231 cu. in.	
12½	0.8522	6.375	14.080	21¼	2.463	18.42	4.872
12⅝	0.8693	6.503	13.800	21½	2.521	18.86	4.760
12¾	0.8866	6.632	13.530	21¾	2.580	19.30	4.651
12⅞	0.9041	6.763	13.270	22	2.640	19.75	4.545
13	0.9218	6.895	13.020	22¼	2.700	20.20	4.445
13⅛	0.9395	7.028	12.780	22½	2.761	20.66	4.347
13¼	0.9575	7.163	12.530	22¾	2.823	21.12	4.251
13⅜	0.9757	7.299	12.300	23	2.885	21.58	4.160
13½	0.994	7.436	12.070	23¼	2.948	22.05	4.070
13⅝	1.013	7.578	11.850	23½	3.012	22.53	3.990
13¾	1.031	7.712	11.640	23¾	3.076	23.01	3.901
13⅞	1.051	7.855	11.420	24	3.142	23.50	3.819
14	1.069	7.997	11.230	25	3.409	25.50	3.520
14⅛	1.088	8.139	11.030	26	3.678	27.58	3.263
14¼	1.107	8.281	10.840	27	3.976	29.74	3.018
14⅜	1.127	8.431	10.650	28	4.276	31.99	2.806
14½	1.147	8.578	10.460	29	4.587	34.31	2.616
14⅝	1.167	8.730	10.280	30	4.909	36.72	2.444
14¾	1.187	8.879	10.110	31	5.241	39.21	2.290
14⅞	1.207	9.029	9.940	32	5.585	41.78	2.149
15	1.227	9.180	9.780	33	5.940	44.43	2.020
15⅛	1.248	9.336	9.620	34	6.305	47.16	1.903
15¼	1.268	9.485	9.460	35	6.681	49.98	1.796
15⅜	1.289	9.642	9.310	36	7.069	52.88	1.698
15½	1.310	9.801	9.160	37	7.467	55.86	1.607
15⅝	1.332	9.964	9.010	38	7.876	58.92	1.527
15¾	1.353	10.121	8.870	39	8.296	62.06	1.446
15⅞	1.374	10.278	8.730	40	8.727	65.28	1.375
16	1.396	10.440	8.600	41	9.168	68.58	1.309
16¼	1.440	10.772	8.330	42	9.621	71.91	1.247
16½	1.485	11.11	8.081	43	10.085	75.44	1.190
16¾	1.530	11.45	7.843	44	10.559	78.99	1.136
17	1.576	11.79	7.511	45	11.045	82.62	1.087
17¼	1.623	12.14	7.394	46	11.541	86.33	1.040
17½	1.670	12.49	7.186	47	12.048	90.13	0.996
17¾	1.718	12.85	6.985	48	12.566	94.00	0.955
18	1.768	13.22	6.787	49	13.095	97.96	0.916
18¼	1.817	13.59	6.604	50	13.635	102.00	0.880
18½	1.867	13.96	6.427	51	14.186	106.12	0.846
18¾	1.917	14.34	6.259	52	14.748	110.32	0.814
19	1.969	14.73	6.094	53	15.320	114.60	0.783
19¼	2.021	15.12	5.938	54	15.904	118.97	0.755
19½	2.074	15.51	5.786	55	16.499	122.82	0.727
19¾	2.128	15.92	5.639	56	17.104	127.95	0.702
20	2.182	16.32	5.500	57	17.720	132.55	0.677
20¼	2.237	16.73	5.365	58	18.347	137.24	0.654
20½	2.292	17.15	5.236	59	18.985	142.02	0.632
20¾	2.348	17.56	5.110	60	19.637	146.89	0.611
21	2.405	17.99	4.989				

*To find the capacity of pipes greater than the largest given in the table, look in the table for a pipe one-half the given size and multiply its capacity by 4, or one of one-third its size, and multiply its capacity by 9, etc.

1 gal = 231 in^3; 1 ft^3 = 7.4805 gal.

TABLE A1.4 Contents of Pipes and Cylindrical Tanks—Axis Horizontal—Flat Ends—per Foot of Length for any Depth of Liquid

h = Depth of liquid inches	d = diameter of tank, inches									
	12		18		24		30		36	
	Gal.	Cu. ft.	Gal.	Cu. ft.	Gal.	Cu. ft.	Gal.	Cu. ft.	Gal.	Cu. ft.
2	0.64	0.0860	0.80	0.1072	0.93	0.1244	1.05	0.1400	1.15	0.154
4	1.73	0.2317	2.18	0.2920	2.57	0.3440	2.90	0.3878	3.21	0.429
6	2.94	0.3927	3.85	0.5149	4.59	0.6140	5.23	0.6988	5.80	0.775
8	4.14	0.5537	5.67	0.7578	6.85	0.9152	7.85	1.049	8.75	1.17
10	5.23	0.6994	7.55	1.009	9.26	1.238	10.72	1.432	12.0	1.60
12	5.87	0.7854	9.38	1.252	11.75	1.571	13.72	1.833	15.4	2.03
14	.....	.....	11.04	1.476	14.24	1.903	16.82	2.248	19.0	2.54
16	.....	.....	12.43	1.659	16.65	2.226	19.90	2.660	22.6	3.02
18	.....	.....	13.22	1.767	18.91	2.527	23.00	3.075	26.4	3.53
20	.....	.....	.....	.....	20.93	2.797	26.00	3.476	29.6	3.95
22	.....	.....	.....	.....	22.57	3.017	28.85	3.859	33.4	4.46
24	.....	.....	.....	.....	23.50	3.1416	31.49	4.209	37.4	5.00
26	.....	.....	.....	.....	.....	.....	33.82	4.521	40.4	5.40
28	.....	.....	.....	.....	.....	.....	35.67	4.768	43.7	5.84
30	.....	.....	.....	.....	.....	.....	36.72	4.908	46.6	6.23
32	.....	.....	.....	.....	.....	.....	.....	.....	49.1	6.55
34	.....	.....	.....	.....	.....	.....	.....	.....	51.2	6.85
36	.....	.....	.....	.....	.....	.....	.....	.....	52.9	7.07
38										
40										
42										
44										
46										
48										
50										
52										
54										
56										
58										
60										
64										
68										
72										
76										
80										
84										

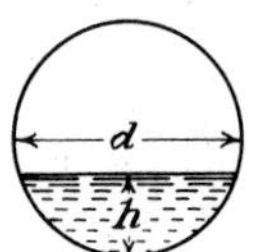

Formulas for determination of approximate capacity of horizontal cylindrical tanks for any depth. Given:—diameter of tank d and height of segment h.

To find area of segment

$$\text{when } h = 0 \text{ to } \tfrac{1}{4}d;\ \text{area} = h\sqrt{1.766dh - h^2}$$

$$\text{when } h = \tfrac{1}{4}d \text{ to } \tfrac{1}{2}d;\ \text{area} = h\sqrt{0.017d^2 + 1.7dh - h^2}$$

1 cu ft = 7.4805 U. S. gal

TABLE A1.4 Contents of Pipes and Cylindrical Tanks—Axis Horizontal—Flat Ends—per Foot of Length for any Depth of Liquid *(Continued)*

h = depth of liquid inches	d = diameter of tank, inches							
	42		48		54		60	
	Gal.	Cu. ft.	Gal.	Cu. ft.	Gal.	Cu. ft.	Gal.	Cu. ft
2	1.25	0.167	1.36	0.182	1.43	0.191	1.47	0.197
4	3.49	0.465	3.72	0.496	3.98	0.531	4.19	0.560
6	6.31	0.843	6.90	0.921	7.25	0.967	7.48	1.00
8	9.57	1.28	10.3	1.37	11.0	1.47	11.6	1.55
10	13.3	1.77	14.2	1.89	15.2	2.02	16.2	2.16
12	16.9	2.26	18.6	2.48	19.7	2.63	21.0	2.81
14	21.0	2.80	22.8	3.04	24.2	3.23	26.3	3.52
16	25.2	3.36	27.4	3.66	29.4	3.92	31.4	4.19
18	29.4	3.92	32.3	4.31	34.8	4.64	36.9	4.93
20	33.8	4.51	37.0	4.94	40.1	5.35	42.8	5.72
22	38.2	5.10	42.0	5.61	45.6	6.08	48.8	6.53
24	42.5	5.67	47.0	6.27	51.0	6.80	54.7	7.30
26	46.8	6.25	52.0	6.94	56.7	7.56	61.0	8.15
28	50.9	6.80	57.0	7.61	62.3	8.33	66.9	8.94
30	55.0	7.34	61.7	8.23	67.8	9.05	73.4	9.81
32	58.8	7.86	66.6	8.89	73.4	9.79	79.7	10.7
34	62.4	8.19	71.3	9.52	78.8	10.5	85.9	11.5
36	65.6	8.75	75.7	10.1	84.2	11.2	92.6	12.4
38	68.4	9.13	79.9	10.7	89.4	11.9	98.0	13.1
40	70.7	9.44	83.7	11.2	94.4	12.6	105	14.0
42	72.0	9.61	87.4	11.7	99.5	13.3	110	14.7
44	.....	.....	90.3	12.1	104	13.9	115	15.4
46	.....	.....	92.7	12.4	108	14.4	121	16.2
48	.....	.....	94.0	12.6	112	14.9	126	16.8
50	.....	.....	.....	.....	115	15.4	131	17.5
52	.....	.....	.....	.....	117	15.6	135	18.0
54	.....	.....	.....	.....	119	15.9	139	18.6
56	.....	.....	.....	.....	.....	.....	143	19.1
58	.....	.....	.....	.....	.....	.....	145	19.4
60	.....	.....	.....	.....	.....	.....	147	19.6
64								
68								
72								
76								
80								
84								

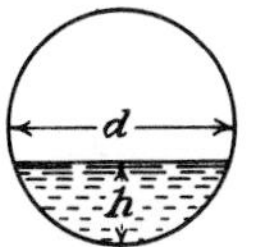

Formulas for determination of approximate capacity of horizontal cylindrical tanks for any depth. Given:—diameter of tank d and height of segment h.

To find area of segment

$$\text{when } h = 0 \text{ to } \tfrac{1}{4}d;\ \text{area} = h\sqrt{1.766dh - h^2}$$

$$\text{when } h = \tfrac{1}{4}d \text{ to } \tfrac{1}{2}d;\ \text{area} = h\sqrt{0.017d^2 + 1.7dh - h^2}$$

1 cu ft = 7.4805 U. S. gal

TABLE A1.4 Contents of Pipes and Cylindrical Tanks—Axis Horizontal—Flat Ends—per Foot of Length for any Depth of Liquid (*Continued*)

h = Depth of liquid inches	d = diameter of tank, inches							
	66		72		73		84	
	Gal.	Cu. ft.	Gal.	Cu. ft.	Gal.	Cu. ft.	Gal.	Cu. ft.
2	1.57	0.210	1.65	0.220	1.73	0.229	1.77	0.236
4	4.42	0.580	4.64	0.618	4.81	0.641	4.95	0.661
6	8.04	1.07	8.10	1.16	8.78	1.17	9.13	1.22
8	12.2	1.63	12.8	1.71	13.4	1.78	13.9	1.86
10	17.0	2.27	17.7	2.36	18.6	2.48	19.7	2.63
12	22.1	2.96	23.6	3.14	24.2	3.22	25.7	3.43
14	27.6	3.68	28.9	3.85	30.2	4.03	31.5	4.21
16	33.3	4.45	35.0	4.66	36.1	4.81	38.2	5.10
18	39.3	5.25	41.7	5.55	43.3	5.78	45.8	6.12
20	45.5	6.08	48.0	6.40	50.3	6.72	52.5	7.01
22	51.8	6.93	54.7	7.29	57.6	7.69	60.0	8.02
24	58.3	7.79	61.9	8.25	64.8	8.65	68.8	9.19
26	65.0	8.68	68.7	9.15	72.1	9.63	75.7	10.1
28	71.8	9.59	76.0	10.2	80.0	10.7	84.1	11.2
30	78.6	10.5	83.5	11.1	88.0	11.8	91.6	12.3
32	85.4	11.4	90.7	12.1	96.0	12.8	101	13.5
34	92.3	12.3	98.2	13.1	104	13.9	109	14.5
36	99.1	13.2	106	14.1	112	14.9	117	15.6
38	106	14.2	113	15.1	120	16.0	126	16.8
40	113	15.1	121	16.1	128	17.1	135	18.0
42	119	15.9	128	17.1	136	18.2	144	19.2
44	126	16.8	136	18.2	144	19.2	153	20.4
46	132	17.6	143	19.1	152	20.3	162	21.6
48	138	18.4	150	20.0	160	21.4	171	22.8
50	144	19.2	157	21.0	168	22.4	179	23.9
52	150	20.0	164	21.9	176	23.5	187	25.0
54	156	20.8	170	22.7	183	24.4	196	26.2
56	161	21.5	176	23.5	191	25.5	204	27.2
58	165	22.0	182	24.3	198	26.4	212	28.3
60	169	22.6	188	25.1	205	27.4	219	29.2
64	176	23.5	198	26.4	218	29.1	235	31.4
68	.....	.....	207	27.6	230	30.7	250	33.4
72	.....	.....	211	28.2	239	31.9	262	35.0
76	.....	.....	.....	.....	246	32.8	274	36.6
80	.....	.....	.....	.....	.....	.....	283	37.8
84	.....	.....	.....	.....	.....	.....	288	38.5

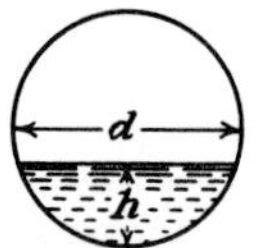

Formulas for determination of approximate capacity of horizontal cylindrical tanks for any depth. Given:—diameter of tank d and height of segment h.

To find area of segment

when $h = 0$ to $\frac{1}{4}d$; area $= h\sqrt{1.766dh - h^2}$

when $h = \frac{1}{4}d$ to $\frac{1}{2}d$; area $= h\sqrt{0.017d^2 + 1.7dh - h^2}$

1 cu ft = 7.4805 U. S. gal

UNITS AND CONVERSION TABLES

The units and conversion factors for commonly used quantities associated with piping are given in tables contained in Appendix E1, of this handbook.

REFERENCES

1. ASME B31, Code for Pressure Piping, Section B31.3, Chemical Plant and Petroleum Refinery Piping, American Society of Mechanical Engineers, New York, 1990.
2. ASME B31, Code for Pressure Piping, Section B31.8, Gas Transmission and Distribution Piping Systems, American Society of Mechanical Engineers, New York, 1989 ed. with 1990 addendum.
3. ASME B31, Code for Pressure Piping, Section B31.5, Refrigeration Piping, American Society of Mechanical Engineers, New York, 1987 ed. with 1989 addendum.
4. ASME B31, Code for Pressure Piping, Section B31.4, Liquid Transportation Systems for Hydrocarbons, Liquid Petroleum Gas, Anhydrous Ammonia, and Alcohols, American Society of Mechanical Engineers, New York, 1989.
5. ASME B31, Code for Pressure Piping, Section B31.1, Power Piping, American Society of Mechanical Engineers, New York, 1989 ed. with 1990 addendum.
6. ASME Boiler and Pressure Vessel Code, Section III, Nuclear Power Plant Components, American Society of Mechanical Engineers, New York, 1989 ed. with 1990 addendum.
7. ASME Boiler and Pressure Vessel Code, Section XI, Rules for Inservice Inspection of Nuclear Power Plant Components, American Society of Mechanical Engineers, New York, 1989 ed. with 1990 addendum.
8. ASME B31, Code for Pressure Piping, Section B31.9, Building Services Piping, American Society of Mechanical Engineers, New York, 1988.
9. ASME B31, Code for Pressure Piping, Section B31.11, Slurry Transportation Piping Systems, American Society of Mechanical Engineers, New York, 1989.

CHAPTER A2

PIPING COMPONENTS

Ervin L. Geiger, P.E.
Engineering Supervisor
Bechtel Corporation
Gaithersburg, Maryland

The term *piping* refers to the overall network of pipes, fitting, flanges, valves, and other components that comprise a conduit system used to convey fluids. Whether a piping system is used to simply convey fluids from one point to another or to process and condition the fluid, piping components serve an important role in the erection and operation of the system. A system used solely to convey fluids may consist of only a relatively few components, such as valves and fittings, where as a complex chemical processing system may consist of a variety of components used to measure, control, condition, and convey the fluids. In the following sections, the characteristics and functions of the various piping components are described.

PIPE AND TUBE PRODUCTS

Commercial pipe and tube products are grouped into various classifications generally based on the application or use and not on the manufacturing method. Most tubular products fall into one of three very broad classifications: (1) pipe, (2) pressure tubes, and (3) mechanical tubes. Each classification falls into various subgroupings, which may be defined and standardized differently by the different trade or user groups. Moreover, the same standard materials specifications may apply to several of the (user) classifications. For example, ASTM A53 pipe may be used for applications representing refrigeration, pressure, and nipple service. Cost considerations enter also into the selection of specific piping materials. In some sizes, prices of pipe made to different material specifications may vary, whereas in other sizes, they may be identical.

Within the broad use classifications listed above, the method classifications described in the preceding section are often referred to. These primarily involve (1) seamless wrought, (2) seamless cast, and (3) seam-welded tubular products. The large variety of single and combination of pipe- or tube-forming methods can produce different characteristics and properties. In addition, the final finishing can involve *hot-finished* and *cold-finished products.* Cold-finishing be accomplished by *reducing* or by *expanding* during the sizing process. Table A2.1 lists the standard pipe sizes and wall thicknesses commercially available.

Piping

On the basis of *user* classification, the more commonly used types of pipe are tabulated in Table A2.2. This listing ignores method of manufacture, size range, wall thickness, and finish, for which the different user groups may have developed different standard requirements.

Standard Pipe. *Mechanical service pipe* is produced in three classes of wall thickness—standard weight, extra strong, and double extra strong. It is available as welded or seamless pipe of ordinary finish and dimensional tolerances, produced in sizes up to 12 in nominal outside diameter (OD) inclusive. This pipe is used for structural and mechanical purposes. Certain applications have other requirements for size, surface finish, or straightness.

Refrigeration pipe (also known as ice-machine or ammonia pipe) may be butt-welded, lap-welded, electric-resistance-welded, or seamless and is intended for use as a conveyor of refrigerants. This pipe is suitable for coiling, bending, and welding. The sizes commonly used are ¾ to 2 in inclusive, produced in random and double random lengths in standard line pipe sizes and weights. Double random lengths are used as ice-rink pipe. It can be produced with plain ends, with threaded ends only, or with threaded ends and line pipe couplings, as desired.

Dry-kiln pipe is butt-welded, electric-resistance-welded, or seamless pipe, for use in the lumber industry. It is produced in standard-weight pipe sizes of ¾, 1, and 1¼ in. Joints are designed to permit subsequent "makeup" after expansion has occurred. Dry-kiln pipe is commonly produced with threaded ends and couplings and in random lengths.

Pressure Pipe. Pressure pipe is used for conveying fluids or gases at normal, subzero, or elevated temperatures or pressures. It generally is not subjected to external heat application. The range of sizes is ⅛-in nominal size to 80-in actual OD in various wall thicknesses. Pressure piping is furnished in random lengths, with threaded or plain ends, as required. Jointers are not customarily produced. Pressure pipe generally receives a hydrostatic test by the mill.

Line Pipe. Line pipe is welded or seamless pipe produced in sizes from ⅛-in nominal OD to 36-in actual OD, inclusive. It is used principally for conveying gas, oil, or water. Line pipe is produced with ends plain, threaded, beveled, grooved, flanged, or expanded, as required for various types of mechanical couplers or for welded joints. When threaded ends and couplings are required, recessed couplings are normally supplied.

Water-well Pipe. Water-well pipe is welded or seamless steel pipe used for conveying water for municipal and industrial applications. Pipelines for such purposes involve flow mains, transmission mains, force mains, water mains, or distribution mains. The mains are generally laid underground. Sizes range from ⅛ to 96 in in a variety of wall thicknesses. Pipe is produced with ends suitably prepared for mechanical couplers, with plain ends beveled for welding, with ends fitted with butt straps for field welding, or with bell-and-spigot joints with rubber gaskets for field joining. Pipe is produced in double random lengths of about 40 ft,

TABLE A2.1 Standard Commercial Pipe Sizes Commonly Used* (Per ASME/ANSI B36.10-1985 and B36.19-1985)

Nominal pipe size	Outside diameter	Nominal wall thickness														
		Sch. 5S†	Sch. 10S†	Sch. 10	Sch. 20	Sch. 30	Stand-ard‡	Sch. 40	Sch. 60	Extra-strong§	Sch. 80	Sch. 100	Sch. 120	Sch. 140	Sch. 160	XX strong
⅛	**0.405**	...	**0.049**	...	...	...	*0.068*	**0.068**	...	*0.095*	**0.095**					
¼	**0.540**	...	**0.065**	...	...	...	*0.088*	**0.088**	...	*0.119*	**0.119**					
⅜	**0.675**	...	**0.065**	...	...	...	*0.091*	**0.091**	...	*0.126*	**0.126**					
½	**0.840**	**0.065**	**0.083**	...	...	...	*0.109*	**0.109**	...	*0.147*	**0.147**	...	...	...	**0.188**	**0.294**
¾	**1.050**	**0.065**	**0.083**	...	...	...	*0.113*	**0.113**	...	*0.154*	**0.154**	...	...	...	**0.219**	**0.308**
1	**1.315**	**0.065**	**0.109**	...	...	...	*0.133*	**0.133**	...	*0.179*	**0.179**	...	...	...	**0.250**	**0.358**
1¼	**1.660**	**0.065**	**0.109**	...	...	...	*0.140*	**0.140**	...	*0.191*	**0.191**	...	...	...	**0.250**	**0.382**
1½	**1.900**	**0.065**	**0.109**	...	...	...	*0.145*	**0.145**	...	*0.200*	**0.200**	...	...	...	**0.281**	**0.400**
2	**2.375**	**0.065**	**0.109**	...	...	...	*0.154*	**0.154**	...	*0.218*	**0.218**	...	...	...	**0.344**	**0.436**
2½	**2.875**	**0.083**	**0.120**	...	...	...	*0.203*	**0.203**	...	*0.276*	**0.276**	...	...	...	**0.375**	**0.552**
3	**3.5**	**0.083**	**0.120**	...	...	...	*0.216*	**0.216**	...	*0.300*	**0.300**	...	...	...	**0.438**	**0.600**
3½	**4.0**	**0.083**	**0.120**	...	...	...	*0.226*	**0.226**	...	*0.318*	**0.318**					
4	**4.5**	**0.083**	**0.120**	...	...	...	*0.237*	**0.237**	...	*0.337*	**0.337**	...	**0.438**	...	**0.531**	**0.674**
5	**5.563**	**0.109**	**0.134**	...	...	...	*0.258*	**0.258**	...	*0.375*	**0.375**	...	**0.500**	...	**0.625**	**0.750**
6	**6.625**	**0.109**	**0.134**	...	...	...	*0.280*	**0.280**	...	*0.432*	**0.432**	...	**0.562**	...	**0.719**	**0.864**
8	**8.625**	**0.109**	**0.148**	...	**0.250**	**0.277**	*0.322*	**0.322**	**0.406**	*0.500*	**0.500**	**0.594**	**0.719**	**0.812**	**0.906**	**0.875**
10	**10.750**	**0.134**	**0.165**	...	**0.250**	**0.307**	*0.365*	**0.365**	**0.500**	*0.500*	**0.594**	0.719	**0.844**	1.000	**1.125**	**1.000**
12	**12.750**	**0.156**	**0.180**	...	**0.250**	**0.330**	*0.375*	**0.406**	0.562	*0.500*	**0.688**	**0.844**	**1.000**	1.125	**1.312**	**1.000**

TABLE A2.1 Standard Commercial Pipe Sizes Commonly Used* (Per ASME/ANSI B36.10-1985 and B36.19-1985) (*Continued*)

Nominal pipe size	Outside diameter	Nominal wall thickness														
		Sch. 5S†	Sch. 10S†	Sch. 10	Sch. 20	Sch. 30	Stand-ard‡	Sch. 40	Sch. 60	Extra-strong§	Sch. 80	Sch. 100	Sch. 120	Sch. 140	Sch. 160	XX strong
14	**14.0**	**0.156**	**0.188**	**0.250**	**0.312**	**0.375**	**0.375**	**0.438**	0.594	**0.500**	**0.750**	**0.938**	1.094	1.250	1.406	
16	**16.0**	**0.165**	**0.188**	**0.250**	**0.312**	**0.375**	**0.375**	**0.500**	**0.656**	**0.500**	**0.844**	**1.031**	**1.219**	1.438	1.594	
18	**18.0**	**0.165**	**0.188**	**0.250**	**0.312**	**0.438**	**0.375**	**0.562**	**0.750**	**0.500**	0.938	**1.156**	1.375	1.562	1.781	
20	**20.0**	**0.188**	**0.218**	**0.250**	**0.375**	**0.500**	**0.375**	**0.594**	**0.812**	**0.500**	**1.031**	1.281	1.500	1.750	1.969	
22	**22.0**	**0.188**	**0.218**	**0.250**	**0.375**	**0.500**	**0.375**	...	**0.875**	**0.500**	1.125	1.375	1.625	1.875	2.125	
24	**24.0**	**0.218**	**0.250**	**0.250**	**0.375**	**0.562**	**0.375**	**0.688**	**0.969**	**0.500**	1.218	1.531	1.812	2.062	2.344	
26	**26.0**	...	...	**0.312**	**0.500**	...	**0.375**	...	...	**0.500**						
28	**28.0**	...	...	**0.312**	**0.500**	**0.625**	**0.375**	...	...	**0.500**						
30	**30.0**	**0.250**	**0.312**	**0.312**	**0.500**	**0.625**	**0.375**	...	...	**0.500**						
32	**32.0**	...	...	**0.312**	**0.500**	**0.625**	**0.375**	**0.688**	...	**0.500**						
34	**34.0**	...	...	**0.312**	**0.500**	**0.625**	**0.375**	**0.688**	...	**0.500**						
36	**36.0**	...	...	**0.312**	**0.500**	**0.625**	**0.375**	**0.750**	...	**0.500**						
38	**38.0**	...	...	...	...	...	**0.375**	...	...	**0.500**						
40	**40.0**	...	...	...	...	...	**0.375**	...	...	**0.500**						
42	**42.0**	...	...	...	...	...	**0.375**	...	...	**0.500**						

All dimensions are given in inches.

The decimal thicknesses listed for the respective pipe sizes represent their nominal or average wall dimensions. The actual thicknesses may be as much as 12.5 percent under the nominal thickness because of mill tolerance. Thicknesses shown in lightface for Schedule 60 and heavier pipe are not currently supplied by the mills, unless a certain minimum tonnage is ordered.

*Additional nominal wall thicknesses established as standard sizes are listed in ANSI/ASME Standard B36.10-1985. Also, Schedule 10S is available in carbon steel in sizes 12 in. and smaller.

†Schedules 5S and 10S are available in corrosion-resistant materials and Schedule 10S is also available in carbon steel.

‡Thicknesses shown in italics are available also in stainless steel, under the designation Schedule 40S.

§Thicknesses shown in italics are available also in stainless steel, under the designation Schedule 80S.

TABLE A2.2 Major Pipe Classifications and Examples of Applications

Type of pipe	Uses
Standard	Mechanical (structural) service pipe, low-pressure service pipe, refrigeration (ice machine) pipe, ice-rink pipe, dry-kiln pipe
Pressure	Liquid, gas or vapor service pipe, service for elevated temperature or pressure, or both
Line	Threaded or plain ends, gas, oil, and steam pipe
Water well	Reamed and drifted, water-well casing, drive pipe, driven well pipe, pump pipe, turbine pump pipe
Oil country tubular goods	Casing, well tubing, drill pipe
Other pipe	Conduit, piles, nipple pipe, sprinkler pipe, bedstead tubing

single random lengths of about 20 ft, or in definite cut lengths, as specified. Wall thicknesses vary from 0.068 in for ⅛-in nominal OD to 1.00 in for 96-in actual OD.

When required, water-well pipe is produced with a specified coating or lining or both. For example, cement-mortar coatings are extensively used.

Oil Country Goods. *Casing* is used as a structural retainer for the walls of oil or gas wells and is also used to exclude undesirable fluids and to confine and conduct oil or gas from productive subsurface strata to the ground level. Casing is produced in sizes 4½- to 20-in OD, inclusive. Size designations refer to actual outside diameter and weight per foot. Ends are commonly threaded and furnished with couplings. When required, the ends are prepared to accommodate other types of joints.

Drill pipe is used to transmit power by rotary motion from ground level to a rotary drilling tool below the surface and also to convey flushing mediums to the cutting face of the tool. Drill pipe is produced in sizes 2⅜- to 6⅝-in OD, inclusive. Size designations refer to actual outside diameter and weight per foot. Drill pipe is generally upset, either internally or externally or both, and is furnished with threaded ends and couplings, threaded only, or prepared to accommodate other types of joints.

Tubing is used with the casing of oil wells to conduct oil to ground level. It is produced in sizes 1.050- to 4.500-in OD, inclusive, in several weights per foot. Ends are threaded and fitted with couplings and may or may not be upset externally.

Other Classifications. *Rigid conduit pipe* is welded or seamless pipe intended especially for the protection of electrical wiring systems. Conduit pipe is not subjected to hydrostatic tests unless so specified. It is furnished in standard-weight pipe sizes from ¼ to 6 in, in 10-ft lengths, with plain ends or with threaded ends and couplings, as specified.

Piling pipe is welded or seamless pipe for use as piles, where the cylinder section acts as a permanent load-carrying member or where it acts as a shell to form cast-in-place concrete piles. Specifications provide for the choice of three grades by minimum tensile strength in which the sizes listed are 8⅝- to 24-in ODs in a variety of wall thicknesses and in two length ranges. Ends are plain or beveled for welding.

Nipple pipe is standard-weight, extra-strong, or double-extra-strong welded or seamless pipe produced for the manufacture of pipe nipples. Standard-weight pipe with threaded ends is also used in sprinkler systems. Nipple pipe is commonly produced in random lengths with plain ends in nominal sizes ⅛ to 12 in. Close OD tolerances, sound welds, good threading properties, and surface cleanliness are essential in this product. It is commonly coated with oil or zinc and well protected in shipment. Nipple pipe is generally made to ASTM A53.

Standard Sizes. Standard pressure, line, and other pipe with plain ends for welding or with threaded ends are standardized in accordance with two scales. Diameters of 12 in and less have a *nominal* size which represents approximately that of the inside diameter of standard-weight pipe. The nominal outside diameter is standard, regardless of weight. The increase in wall thickness results in a decrease of the inside diameter.

The standardization of pipe sizes over 12 in is based on the actual outside diameter, the wall thickness, and the weight per foot.

The principal dimensions of commercial piping materials are summarized in App. E2.

Standard Weights. The weight of standard piping materials is tabulated in App. E2. The weights of butt-welding fittings are given in Tables A2.21 through A2.26. The weights of reducing fittings are approximately the same as for full-size fittings.

The weights of welding reducers are for one size reduction and are thus only approximately correct for other reductions.

For special materials, the equations listed below for weights of tubes and weights of contents of tubes may be helpful.

$$\text{Weight of tube, lb/ft} = F \times 10.68 \times (T \times D - T^2)$$

where T = wall thickness, in
D = outside diameter, in
F = relative weight factor

The weight of tube calculation is based on low-carbon steel weighing 0.2833 lb/in^3 and is extended to other materials through the factor F.

Relative weight factor F	
Aluminum	0.35
Brass	1.12
Cast iron	0.91

Copper	1.14
Ferritic stainless steel	0.95
Austenitic stainless steel	1.02
Steel	1.00
Wrought iron	0.98

$$\text{Weight of contents of a tube, lb/ft} = G \times 0.3405 \times (D - 2T)^2$$

where G = specific gravity of contents
T = tube wall thickness, in.
D = tube outside diameter, in.

The weight per foot of steel pipe is subject to the tolerances listed in Table A2.3.

TABLE A2.3 Weight Tolerances of Steel Piping

Specification	Sizes (NPS)	Tolerance	
ASTM A53	All	+10%, −10%	(1)
ASTM A106	All	+10%, −3.5%	(2)
ASTM A369, A333, A335	12 & under	+10%, −3.5%	
	Over 12	+10%, −5%	(3)
ASTM A671, A672, A691	All	Governed by plate spec.	
		ASTM A285, A515, A516	(4)
ASTM A312, A376, A430	12 & under	+10%, −3.5%	
	Over 12	+10%, −5%	(3)
API 5L	All	+10%, −3.5%	

(1) As specified in ASTM A53-90a.
(2) As specified in ASTM A106-90.
(3) As specified in ASTM A530-90.
(4) As specified in ASTM A671-89a, A672-89b, and A691-89a, respectively.

Pressure Tubing

Pressure-tube applications commonly involve external heat applications, as in boilers or superheaters.

Pressure tubing is produced to the actual outside diameter and minimum or average wall thickness specified by the purchaser. Pressure tubing may be hot- or cold-finished.

The wall thickness is normally given in decimal parts of an inch rather than as a fraction or gauge number. When gauge numbers are given without reference to a system, Birmingham Wire Gauge (BWG) is implied. Weights of commercial tubing are given in App. E3, Table E3.1.

Pressure tubing is usually made from steel produced by the open-hearth, basic oxygen, or electric-furnace processes.

Seamless pressure tubing may be either hot-finished or cold-drawn. Cold-drawn steel tubing is frequently process-annealed at temperatures above 1200°F. To assure quality, maximum hardness values are frequently specified.

Hot-finished or cold-drawn seamless low-alloy steel tubes generally are process-annealed at temperatures between 1200 and 1350°F.

Austenitic stainless-steel tubes are usually annealed at temperatures between 1800 and 2100°F, with specific temperatures varying somewhat with each grade. This is generally followed by pickling, unless bright-annealing was done.

Mechanical Tubing

Unlike pipe and pressure tubes, mechanical tubing is generally classified by the method of manufacture and the degree of finish. Examples of classifications are seamless hot-finished, cold-drawn welded, or flash-in-grade, etc.

Seamless Tubes. Seamless tubes are available as either *hot-* or *cold-*finished. They are normally made in sizes from 0.187-in OD to 10.750-in OD.

Dimensions for hot-finished mechanical tubes are given in App. E3, Table E3.2. Dimensions for cold-finished tubes are listed in App. E3, Table E3.3.

Welded Tubes. Welded tubes generally are produced by electric resistance methods. Where required, the welding flash is removed with a cutting tool. Industry practice normally recognizes a number of finish conditions which are summarized in Table A2.4.

Flash-in-type tubing is generally limited to applications where nothing is inserted in the tube.

Flash-controlled tubing is used where moderate control of the inside diameter is required. Generally the outside and inside diameters are specified.

The designation *sink-draw tubes* is specified where close control over the outer diameter is required with normal tolerance applying to the wall thickness. Smoothness of the inside surface is not controlled, except that the flash is generally controlled to a height of 0.005 or 0.010 in maximum.

Mandrel-drawn tubes usually are normalized after welding by passing the tubes through a continuous atmosphere-controlled furnace. After descaling, the tubes are cold-drawn through a die with a mandrel on the inside of the tube. These tubes provide maximum control over surface finish, outside or inside diameters, and wall thickness. The normalizing heat treatment removes the effects of welding and provides a uniform microstructure around the tube circumference.

The different finish classifications may result in substantial differences in the mechanical properties of the steel material.

Typical examples for low-carbon steel material are given in Table A2.5. Differences in carbon content and other chemistry, heat treatment, etc., may significantly change these typical values.

Other Tubing Types. Among other tube classifications are sanitary tubing usually made of 18 Cr-8 Mo stainless steel and available as seamless or welded. This tubing is used extensively in the dairy, beverage, and food industries. Sanitary tubing is generally available in sizes from 1- to 4-in OD. It may be either hot- or cold-finished. It is normally heat-treated at temperatures above 1900°F.

Some welded tube is also produced by fusion-welding methods of either the

TABLE A2.4 Finish Classifications Normally Recognized as Welded Mechanical Tubing

	Type	Condition
A	Flash-in, hot-rolled	Made by longitudinally forming and welding hot-rolled strip; the interior welding flash remains in place
B	Flash-in, cold-rolled	Made by longitudinally forming and welding cold-rolled strip; the interior flash remains in place
C	Flash controlled—0.010 max, hot-rolled	Same as (A), except that the interior flash is partially removed so that the height remaining does not exceed 0.010 in.
D	Flash controlled—0.010 max, cold-rolled	Same as (C), except that cold-rolled strip has been used
E	Flash controlled—0.005 max, hot-rolled	Same as (C), except that flash height on tube inside does not exceed 0.005″
F	Flash controlled—0.005 max, cold-rolled	Same as (E), except that cold-rolled strip has been used
G	Sink-drawn, hot-rolled	Tubing with flash controlled to 0.010″ max, which has been descaled and cold-drawn through a die to cold-finish the exterior surface
H	Sink-drawn, cold-rolled	Tubing with flash controlled to 0.010″ max, which has been cold-drawn through a die to cold-finish the exterior surface
I	Mandrel-drawn	Cold-rolled tubing with the interior flash removed and drawn through a die and over a mandrel to cold-finish the exterior and interior surfaces

inert-gas tungsten-arc-welding or gas shielded consumable metal-arc-welding types. This tubing is generally more expensive than the resistance-welded types.

The butt-welded cold-finished tubes are made from hot- or cold-rolled strip and fusion welded. This tubing is usually furnished as sink- or mandrel-drawn.

Butt-welded tubing is made in heavier wall thicknesses than the resistance-welded tube.

TABLE A2.5 Typical Mechanical Properties of Resistance-welded Mechanical Carbon-Steel Tubing

Type	Yield strength, psi	Tensile strength, psi	Elongation, %	Hardness, Rockwell B
Flash-in or flash controlled, hot-rolled	48,000	58,000	29	68
Flash-in or flash controlled, cold-rolled	68,000	76,000	17	84
Normalized	34,000	52,000	39	61
Sink-drawn	73,000	76,000	20	84
Mandrel-drawn	80,000	83,000	15	86

Several tubing materials used in the automobile industry are covered by specifications of the Society of Automotive Engineers, SAE Handbook.

Brazed Steel Tubing. In the automotive, refrigeration, and stove industries, extensive use is made of copper-brazed steel tubes. Applications include fuel and brake lines, oil lines, heating and cooling lines, and similar services.

The tubing is produced from copper-coated steel strip. A single or double strip is formed into tubing which is heated in a reducing atmosphere to effect a brazed joint between the mating surfaces.

ASTM Specification A254, Specification for Copper Brazed Steel Tubing, recognizes two classifications, each consisting of the two types shown in Fig. A2.1.

This tubing is available in sizes from 3⁄16-in OD to 5⁄8-in OD. Wall thicknesses range from 0.025 in. to 0.035 in. To improve surface finish and tolerances, this tubing is sometimes sink-drawn.

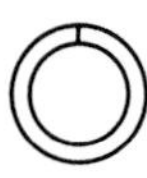

FIGURE A2.1 Construction of copper-brazed steel tubing, as specified in ASTM A254.

PIPE FITTINGS

The major piping materials are also produced in the form of standard fittings. Among the more widely used materials are cast-iron, malleable-iron, brass, copper, cast-steel, and wrought-steel. Other major nonferrous piping materials are also produced in the form of cast and wrought fittings.

Cast-iron fittings are made by conventional founding methods for a variety of joints including bell-and-spigot, flanged, and mechanical (gland-type) or other proprietary designs.

Cast-Iron Fittings

Cast-iron fittings are covered by a number of ASME/ANSI and Federal Standards. They are:

ANSI/ASME B16.1	Cast Iron Pipe Flanges and Flanged Fittings, Class 25, 125, 250, and 800 (The Standard also includes bolt, nut, and gasket data.)
ANSI/ASME B16.4	Cast Iron Threaded Fittings, Class 125 and 250
ANSI/ASME B16.12	Cast Iron Threaded Drainage Fittings
ANSI/AWWA A21.10/C110	Ductile Iron and Gray Iron Fittings 3 Inch through 48 Inch for Water and Other Liquids

Cast-Iron Threaded Fittings

Cast-iron threaded fittings are covered by ANSI/ASME Standard B16.4.

The standard specifies the following for Class 125 and Class 250 tees, crosses, 45° and 90° elbows, reducing tees, caps, couplings, and reducing couplings in sizes ranging from 1/4 in through 12 in, inclusive. However, in Class 250, the Standard only covers 45° and 90° elbows, straight tees, and straight crosses.

- Pressure-temperature ratings
- Size and method of designating openings of reducing fittings
- Marking
- Minimum requirements for materials
- Dimensions and tolerances
- Threading
- Coatings

Tables A2.6 and A2.7 give the American National Standard dimensions for cast-iron threaded fittings for Classes 125 and 250. The dimensions conform to ANSI/ASME B16.4-1985.

The pressure-temperature ratings of the class 125 and class 250 fittings are listed in Table A2.8. The ratings are independent of the contained fluid and are the maximum nonshock pressures at the listed temperatures. Minimum material must conform to Class A of ASTM A126. Body thickness at any point is to be no less than 90 percent of the specified minimum metal thickness. All fittings are threaded with ANSI/ASME B1.20.1 pipe threads. Cast-iron threaded fittings customarily are furnished in black finish only.

TABLE A2.6 Dimensions of Class 125 Cast-Iron Threaded 90° and 45° Elbows, Tees, and Crosses (Straight Sizes) (ANSI/ASME B16.4-1985)

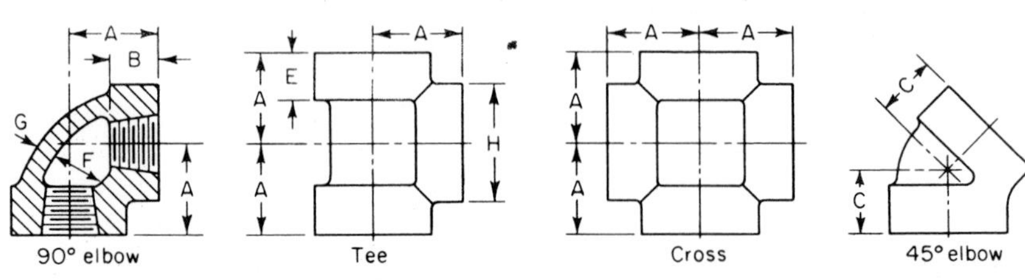

Nominal pipe size	Center to end, elbows, tees, and crosses *A*	Center to end, 45° elbows *C*	Length of thread, min. *B*	Width of band, min. *E*	Inside diameter of fitting *F*		Metal thickness *G*	Outside diameter of band, min. *H*
					Max.	Min.		
¼	0.81	0.73	0.32	0.38	0.58	0.54	0.11	0.93
⅜	0.95	0.80	0.36	0.44	0.72	0.67	0.12	1.12
½	1.12	0.88	0.43	0.50	0.90	0.84	0.13	1.34
¾	1.31	0.98	0.50	0.56	1.11	1.05	0.15	1.63
1	1.50	1.12	0.58	0.62	1.38	1.31	0.17	1.95
1¼	1.75	1.29	0.67	0.69	1.73	1.66	0.18	2.39
2½	1.94	1.43	0.70	0.75	1.97	1.90	0.20	2.68
2	2.25	1.68	0.75	0.84	2.44	2.37	0.22	3.28
2½	2.70	1.95	0.92	0.94	2.97	2.87	0.24	3.86
3	3.08	2.17	0.98	1.00	3.60	3.50	0.26	4.62
3½	3.42	2.39	1.03	1.06	4.10	4.00	0.28	5.20
4	3.79	2.61	1.08	1.12	4.60	4.50	0.31	5.79
5	4.50	3.05	1.18	1.18	5.66	5.56	0.38	7.05
6	5.13	3.46	1.28	1.28	6.72	6.62	0.43	8.28
8	6.56	4.28	1.47	1.47	8.72	8.62	0.55	10.63
10	8.08*	5.16	1.68	1.68	10.85	10.75	0.69	13.12
12	9.50*	5.97	1.88	1.88	12.85	12.75	0.80	15.47

All dimensions given in inches.
*This applies to elbows and tees only.

Malleable-Iron Threaded Fittings

Malleable-iron fittings are also extensively produced. They are generally made with threaded joints. Malleable-iron threaded fittings for Classes 150 and 300 are standardized in ANSI/ASME B16.3. The Standard specifies the same categories of parameters for Classes 150 and 300 fittings as discussed under ANSI/ASME B16.4 for cast-iron fittings. The fittings are available in a variety of configurations from NPS 1/8 through NPS 6. The pressure-temperature ratings of these fittings are listed in Table A2.9. As with cast-iron fittings, the ratings are independent of the contained fluid and are maximum nonshock pressures at the listed temperatures. Malleable-iron fittings are furnished either black, galvanized, or as otherwise ordered by the buyer. Standard dimensions of elbows, tees, and crosses are shown in Tables A2.10 and A2.11. The galvanized threaded fittings commonly

TABLE A2.7 Dimensions of Class 250 Cast-Iron Threaded 90° and 45° Elbows, Tees, and Crosses (Straight Sizes) (ANSI/ASME B16.34-1985)

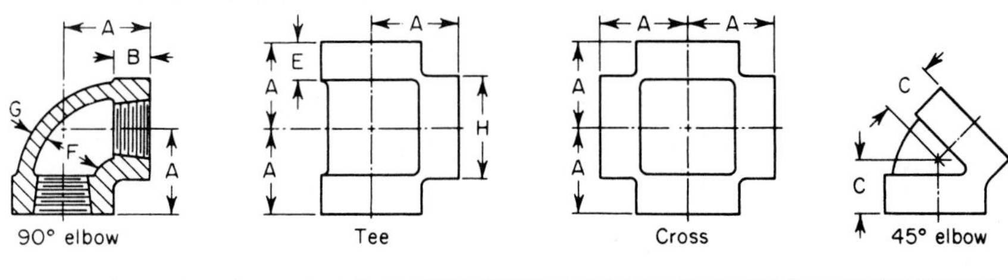

Nominal pipe size	Center to end, elbows, tees, and crosses A	Center to end, 45° elbows C	Length of thread, min. B	Width of band, min. E	Inside diameter of fitting F Max.	Inside diameter of fitting F Min.	Metal thickness G	Outside diameter of band, min. H
¼	0.94	0.81	0.43	0.49	0.58	0.54	0.18	1.17
⅜	1.06	0.88	0.47	0.55	0.72	0.67	0.18	1.36
½	1.25	1.00	0.57	0.60	0.90	0.84	0.20	1.59
¾	1.44	1.13	0.64	0.68	1.11	1.05	0.23	1.88
1	1.63	1.31	0.75	0.76	1.38	1.31	0.28	2.24
1¼	1.94	1.50	0.84	0.88	1.73	1.66	0.33	2.73
1½	2.13	1.69	0.87	0.97	1.97	1.90	0.35	3.07
2	2.50	2.00	1.00	1.12	2.44	2.37	0.39	3.74
2½	2.94	2.25	1.17	1.30	2.97	2.87	0.43	4.60
3	3.38	2.50	1.23	1.40	3.60	3.50	0.48	5.36
3½	3.75	2.63	1.28	1.49	4.10	4.00	0.52	5.98
4	4.13	2.81	1.33	1.57	4.60	4.50	0.56	6.61
5	4.88	3.19	1.43	1.74	5.66	5.56	0.66	7.92
6	5.63	3.50	1.53	1.91	6.72	6.62	0.74	9.24
8	7.00	4.31	1.72	2.24	8.72	8.62	0.90	11.73
10	8.63	5.19	1.93	2.58	10.85	10.75	1.08	14.37
12	10.00	6.00	2.13	2.91	12.85	12.75	1.24	16.84

All dimensions given in inches.

The Class 250 standard for threaded fittings covers only the straight sizes of 90° and 45° elbows, tees, and crosses.

TABLE A2.8 Pressure-Temperature Rating of ANSI/ASME B16.4 Cast-Iron Fittings

Temperature (°F)	Class 125 (psi)	Class 250 (psi)
−20 to 150	175	400
200	165	370
250	150	340
300	140	310
350	125*	300
400	...	250†

*Permissible for service temperature up to 353°F, reflecting the temperature of saturated steam at 125 psig.

†Permissible for service temperature up to 406°F, reflecting the temperature of saturated steam at 250 psig.

TABLE A2.9 Pressure-Temperature Rating of ANSI/ASME B16.3 Malleable-Iron Threaded Fittings

Temperature (°F)	Class 150 (psig)	Class 300 (psig)		
		Sizes ¼–1	Sizes 1¼–2	Sizes 2½–3
−20 to 150	300	2000	1500	1000
200	265	1785	1350	910
250	225	1575	1200	825
300	185	1360	1050	735
350	150*	1150	900	650
400	...	935	750	560
450	...	725	600	475
500	...	510	450	385
550	...	300	300	300

*Permissible for service temperature up to 366°F, reflecting the temperature of saturated steam at 150 psig.

TABLE A2.10 Dimensions of Class 150 Malleable-Iron Threaded 90° and 45° Elbows, Tees, and Crosses (Straight Sizes) (ANSI/ASME B16.3-1985)

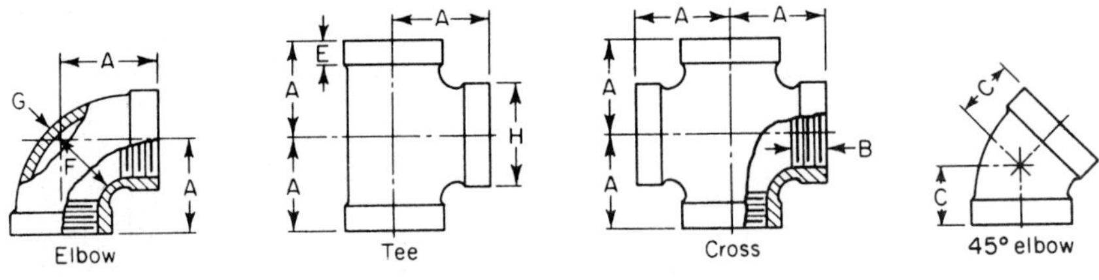

Nominal pipe size	Center to end, elbows, tees, and crosses *A*	Center to end, 45° elbows *C*	Length of thread, min. *B*	Width of band, min. *E*	Inside diameter of fitting *F* Min.	Inside diameter of fitting *F* Max.	Metal thickness *G*	Outside diameter of band, min. *H*
⅛	0.69	...	0.25	0.20	0.40	0.43	0.09	0.69
¼	0.81	0.73	0.32	0.21	0.54	0.58	0.09	0.84
⅜	0.95	0.80	0.36	0.23	0.67	0.72	0.10	1.01
½	1.12	0.88	0.43	0.25	0.84	0.90	0.10	1.20
¾	1.31	0.98	0.50	0.27	1.05	1.11	0.12	1.46
1	1.50	1.12	0.58	0.30	1.31	1.38	0.13	1.77
1¼	1.75	1.29	0.67	0.34	1.66	1.73	0.14	2.15
1½	1.94	1.43	0.70	0.37	1.90	1.97	0.15	2.43
2	2.25	1.68	0.75	0.42	2.37	2.44	0.17	2.96
2½	2.70	1.95	0.92	0.48	2.87	2.97	0.21	3.59
3	3.08	2.17	0.98	0.55	3.50	3.60	0.23	4.28
3½	3.42	2.39	1.03	0.60	4.00	4.10	0.25	4.84
4	3.79	2.61	1.08	0.66	4.50	4.60	0.26	5.40
5	4.50	3.05	1.18	0.78	5.56	5.66	0.30	6.58
6	5.13	3.46	1.28	0.90	6.62	6.72	0.34	7.77

All dimensions in inches.

TABLE A2.11 Dimensions of Class 300 Malleable-Iron Threaded 90° and 45° Elbows, Tees, and Crosses (Straight Sizes) (ANSI/ASME B16.3-1985) (Refer to Table A2.10 for nomenclature)

Nominal pipe size	Center to end, elbows, tees, and crosses *A*	Center to end, 45° elbows *C*	Length of thread, *B*, min.	Width of band, *E*, min.	Inside diameter of fitting *F* Min.	Inside diameter of fitting *F* Max.	Metal thickness *G*	Outside diameter of band, *H* min.
¼	0.94	0.81	0.43	0.38	0.54	0.58	0.14	0.93
⅜	1.06	0.88	0.47	0.44	0.67	0.72	0.15	1.12
½	1.25	1.00	0.57	0.50	0.84	0.90	0.16	1.34
¾	1.44	1.13	0.64	0.56	1.05	1.11	0.18	1.63
1	1.63	1.31	0.75	0.62	1.31	1.38	0.20	1.95
1¼	1.94	1.50	0.84	0.69	1.66	1.73	0.22	2.39
1½	2.13	1.69	0.87	0.75	1.90	1.97	0.24	2.68
2	2.50	2.00	1.00	0.84	2.37	2.44	0.26	3.28
2½	2.94	2.25	1.17	0.94	2.87	2.97	0.31	3.86
3	3.38	2.50	1.23	1.00	3.50	3.60	0.35	4.62

used in water piping for homes are Class 150 malleable iron. Minimum properties of malleable iron are required to meet ASTM A197 "Cupola Malleable Iron" requirements. The fittings are threaded with ANSI/ASME B1.20.1 pipe threads.

Cast-Brass and Cast-Bronze Threaded Fittings

Cast-brass and -bronze threaded fittings are commonly produced for use with brass pipe. The fittings are manufactured in accordance with ANSI/ASME B16.15 in pressure classes 125 and 250. The standard establishes pressure-temperature ratings, size and method of designating openings of reducing fittings, marking, minimum requirements for casting quality and materials. Dimensions of the more commonly used configuration are listed in Table A2.12. Nonshock pressure-temperature ratings, which are independent of the contained fluid, are listed in Table A2.13. The permitted materials for the fittings are:

ASTM B62, alloy C83600
ASTM B584, alloy C83800 and C84400
ASTM B16, alloy C36000 (bar stock)*
ASTM B140, alloy C32000 or C31400 (bar stock)*

Soldered-Joint Fittings

Soldered-joint wrought metal and cast-brass or -bronze fittings for use with copper water tubes are covered by ASTM B88 and ANSI H23.1. The fittings are made in accordance with ANSI B16.22 and ANSI B16.18, respectively. Standard

*Used for manufacture of threaded plugs, bushings, and caps.

TABLE A2.12 Dimensions of Classes 125 and 250 Cast-Bronze or Brass Threaded Elbows, Tees, and Crosses (Straight Sizes) (ANSI/ASME B16.15-1985)*

Class 125 fittings

90° elbow Tee Cross 45° elbow

Nominal pipe size	Center to end, elbows, tees, and crosses *A*	Length of thread, min. *B*	Center to end 45° elbows *C*	Band length, min. *E*	Inside diameter of cast fitting *F* Min.	Inside diameter of cast fitting *F* Max.	Metal thickness† *G*	Band diameter, min. *H*
1/8	0.54	0.25	0.42	0.14	0.41	0.44	0.08	0.67
1/4	0.71	0.32	0.56	0.16	0.54	0.58	0.08	0.81
3/8	0.82	0.36	0.63	0.17	0.68	0.72	0.09	1.00
1/2	1.01	0.43	0.78	0.19	0.84	0.90	0.09	1.17
3/4	1.18	0.50	0.89	0.23	1.05	1.11	0.10	1.42
1	1.43	0.58	1.06	0.27	1.32	1.39	0.11	1.72
1 1/4	1.69	0.67	1.22	0.31	1.66	1.73	0.12	2.10
1 1/2	1.84	0.70	1.30	0.34	1.90	1.97	0.13	2.38
2	2.12	0.75	1.45	0.41	2.38	2.45	0.15	2.92
2 1/2	2.70	0.92	1.95	0.48	2.88	2.98	0.17	3.49
3	3.08	0.98	2.17	0.55	3.50	3.60	0.19	4.20
4	3.79	1.08	2.61	0.66	4.50	4.60	0.22	5.31

Class 250 fittings

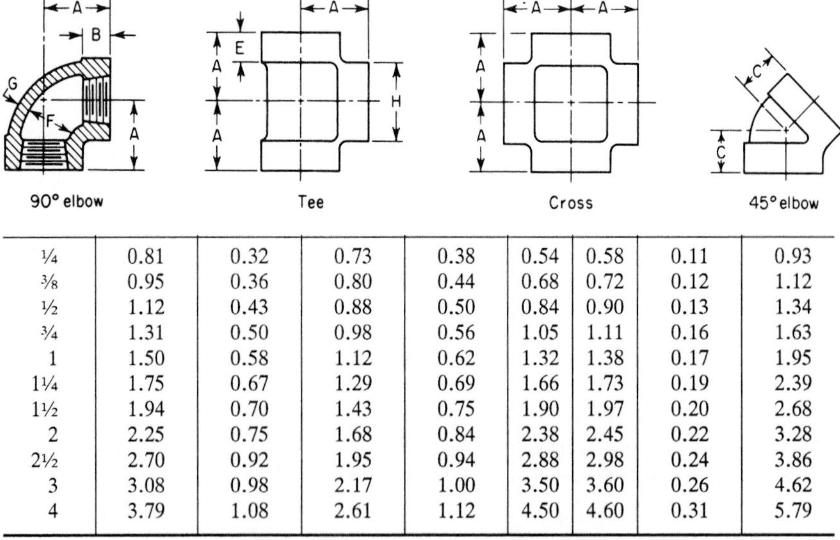

90° elbow Tee Cross 45° elbow

Nominal pipe size	*A*	*B*	*C*	*E*	*F* Min.	*F* Max.	*G*	*H*
1/4	0.81	0.32	0.73	0.38	0.54	0.58	0.11	0.93
3/8	0.95	0.36	0.80	0.44	0.68	0.72	0.12	1.12
1/2	1.12	0.43	0.88	0.50	0.84	0.90	0.13	1.34
3/4	1.31	0.50	0.98	0.56	1.05	1.11	0.16	1.63
1	1.50	0.58	1.12	0.62	1.32	1.38	0.17	1.95
1 1/4	1.75	0.67	1.29	0.69	1.66	1.73	0.19	2.39
1 1/2	1.94	0.70	1.43	0.75	1.90	1.97	0.20	2.68
2	2.25	0.75	1.68	0.84	2.38	2.45	0.22	3.28
2 1/2	2.70	0.92	1.95	0.94	2.88	2.98	0.24	3.86
3	3.08	0.98	2.17	1.00	3.50	3.60	0.26	4.62
4	3.79	1.08	2.61	1.12	4.50	4.60	0.31	5.79

*All dimensions are in inches.

Several other fittings produced to the requirements of ANSI/ASME B16.15 are available from various manufacturers. These types include couplings, reducers, return bends, 4-branches, caps and reducing elbows and tees. Manufacturers should be consulted for availability and dimensions.

†The actual metal thickness must be at least 90 percent of the indicated values.

TABLE A2.13 Pressure-Temperature Rating for Classes 125 and 250 Cast-Bronze Threaded Fittings (ANSI/ASME B16.15-1985)

Temperature (°F)	Class 125 (psi)	Class 250 (psi)
−20 to 150	200	400
200	190	385
250	180	365
300	165	335
350	150	300
400	125	250

cast-brass or -bronze fittings are illustrated in Fig. A2.2. The laying length of cast-brass fittings (center-to-shoulder distance) is included in the specification, but because of the variety of forming methods, the laying length of wrought-metal fittings has not been standardized.

Fittings with this type of joint made with 50-50 tin-lead solder, 95-tin 5-antimony, and solder melting above 1100°F have the pressure-temperature ratings shown in Table A2.14. (Note: lead-bearing solder is not permitted for potable water service.)

Wrought copper fittings normally have a minimum copper content of 83 percent. Cast-brass fittings conform to ASTM B62 and have a nominal composition of 85 percent copper, 5 percent tin, 5 percent lead, and 5 percent zinc.

The minimum requirements for 50-50 tin-lead solder generally used with these fittings are covered in ASTM B32 alloy grade 50A. Metal thickness tolerances and general dimensions of fittings are given in Table A2.15. For laying length dimensions of reducing elbows, tees, crosses, and couplings in both straight and reducing sizes, references should be made to ANSI B16.18. Elbows and coupling adapters with male and female pipe thread ends also are available.

TABLE A2.14 Pressure Ratings of Solder Joints (ANSI/ASME B16.18-1984). Maximum Working Pressure (psi).

Solder used in joints	Working temperatures (°F)	⅛–1 in, incl.*	1¼–2 in, incl.*	2½–4 in, incl.*	5–8 in, incl.*
50-50 tin-lead†	100	200	175	150	135
	150	150	125	100	90
	200	100	90	75	70
	250	85	75	50	45
95-5, tin-antimony	100	500	400	300	270
	150	400	350	275	250
	200	300	250	200	180
	250	200	175	150	135
Solders melting at or above 1100°F	‡	‡	‡	‡	‡

*Standard water tube sizes.
†ASTM B32 Alloy Grade 50A.
‡Rating to be consistent with materials and procedures employed.

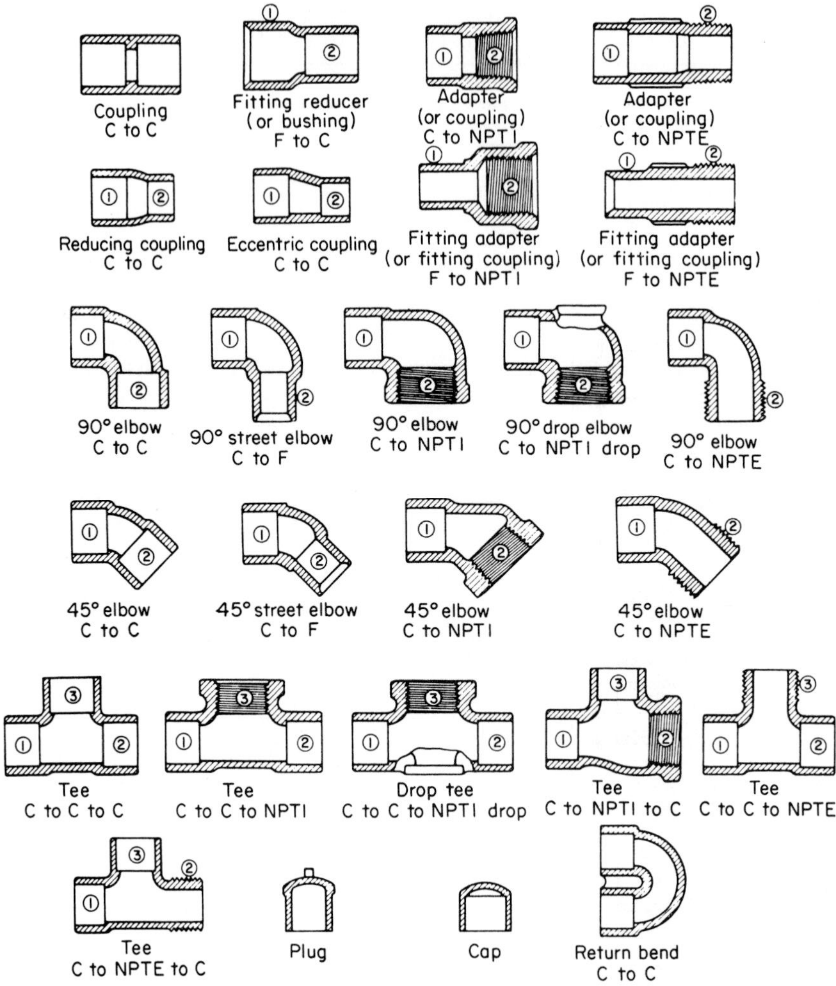

FIGURE A2.2 Standard cast-bronze solder-joint pressure fittings, as covered by American National Standard ANSI B16.18-1984. C, Solder-joint fitting end (female) made to receive copper tube diameter. F, Solder-joint fitting end (male) made to copper tube diameter. NPTI, Internal American Standard taper pipe thread. NPTE, External American Standard taper pipe thread.

Cast-Iron Flanged Fittings

Cast-iron flanged fittings are produced in accordance with American National Standard ANSI/ASME B16.1. The standard specifies pressure-temperature ratings, sizes, marking, minimum requirements for materials, dimensions and tolerances, bolting, gasketing, and testing requirements. The fittings are manufactured in a variety of configurations (tees, elbows, crosses, laterals, etc.) in pressure classes 25, 125, 250, and 800. Not all sizes and styles are available in all ratings. The sizes available in each class are listed below:

TABLE A2.15 Dimensions of Soldered-joint Fittings (ASME/ANSI B16.22-1989), (ANSI/ASME B16.18-1984). All Dimensions in Inches.

All dimensions in inches.

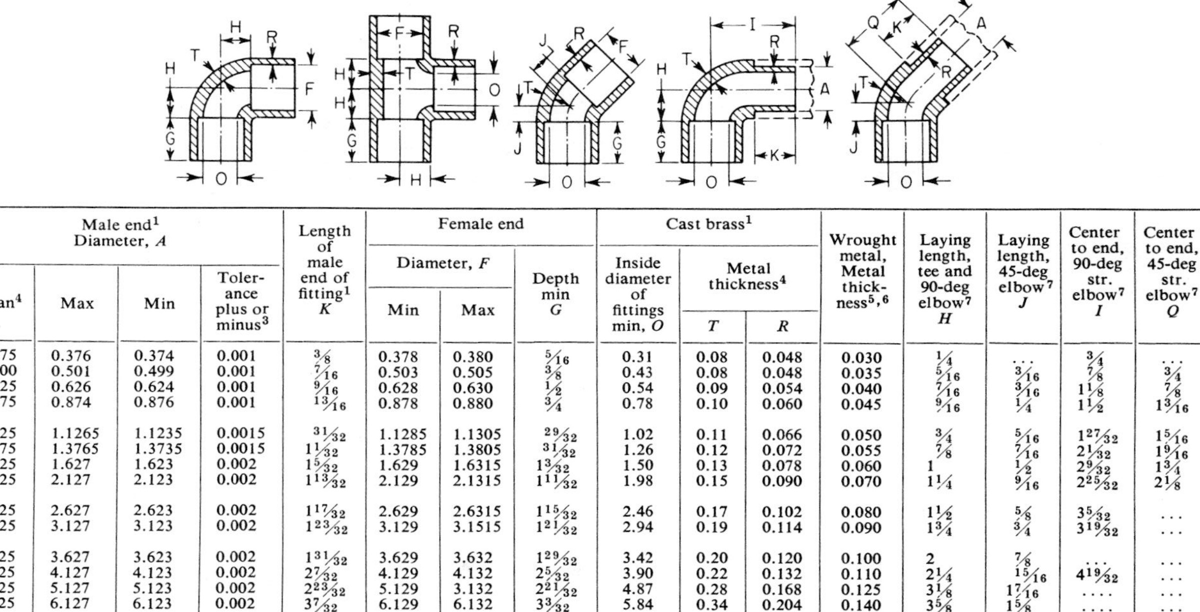

Nominal size[2]	Male end[1] Diameter, A: Mean[4]	Male end Diameter, A: Max	Male end Diameter, A: Min	Male end: Tolerance plus or minus[3]	Length of male end of fitting[1] K	Female end Diameter, F: Min	Female end Diameter, F: Max	Female end: Depth min G	Cast brass[1]: Inside diameter of fittings min, O	Cast brass Metal thickness[4]: T	Cast brass Metal thickness[4]: R	Wrought metal, Metal thickness[5,6]	Laying length, tee and 90-deg elbow[7] H	Laying length, 45-deg elbow[7] J	Center to end, 90-deg str. elbow[7] I	Center to end, 45-deg str. elbow[7] Q	Inspection tolerances, applied to H, I, J, and Q
1/4	0.375	0.376	0.374	0.001	3/8	0.378	0.380	5/16	0.31	0.08	0.048	0.030	1/4	...	3/4	...	±3/64
3/8	0.500	0.501	0.499	0.001	7/16	0.503	0.505	3/8	0.43	0.08	0.048	0.035	5/16	3/16	7/8	3/4	±3/64
1/2	0.625	0.626	0.624	0.001	9/16	0.628	0.630	1/2	0.54	0.09	0.054	0.040	7/16	3/16	1 1/8	7/8	±1/16
3/4	0.875	0.874	0.876	0.001	13/16	0.878	0.880	3/4	0.78	0.10	0.060	0.045	9/16	1/4	1 1/2	1 3/16	±1/16
1	1.125	1.1265	1.1235	0.0015	31/32	1.1285	1.1305	29/32	1.02	0.11	0.066	0.050	3/4	5/16	1 27/32	1 5/16	±5/64
1 1/4	1.375	1.3765	1.3735	0.0015	1 1/32	1.3785	1.3805	31/32	1.26	0.12	0.072	0.055	7/8	7/16	2 1/32	1 9/16	±5/64
1 1/2	1.625	1.627	1.623	0.002	1 5/32	1.629	1.6315	1 3/32	1.50	0.13	0.078	0.060	1	1/2	2 9/32	1 3/4	±5/64
2	2.125	2.127	2.123	0.002	1 13/32	2.129	2.1315	1 11/32	1.98	0.15	0.090	0.070	1 1/4	9/16	2 25/32	2 1/8	±5/64
2 1/2	2.625	2.627	2.623	0.002	1 17/32	2.629	2.6315	1 15/32	2.46	0.17	0.102	0.080	1 1/2	5/8	3 5/32	...	±7/64
3	3.125	3.127	3.123	0.002	1 23/32	3.129	3.1515	1 21/32	2.94	0.19	0.114	0.090	1 3/4	3/4	3 19/32	...	±7/64
3 1/2	3.625	3.627	3.623	0.002	1 31/32	3.629	3.632	1 29/32	3.42	0.20	0.120	0.100	2	7/8	...	...	±7/64
4	4.125	4.127	4.123	0.002	2 7/32	4.129	4.132	2 5/32	3.90	0.22	0.132	0.110	2 1/4	15/16	4 19/32	...	±1/8
5	5.125	5.127	5.123	0.002	2 23/32	5.129	3.132	2 21/32	4.87	0.28	0.168	0.125	3 1/8	1 7/16	....	...	±1/8
6	6.125	6.127	6.123	0.002	3 7/32	6.129	6.132	3 3/32	5.84	0.34	0.204	0.140	3 5/8	1 5/8	....	...	±5/32

[1]These dimensions are used for wrought-metal fittings as well as for cast-brass fittings.

[2]This size is the nominal bore of the tube.

[3]From American Standard Specifications for Copper Water Tube, ANSI H23.1 (ASTM B88).

[4]Patterns shall be designed to produce body thicknesses given in the table. Metal thickness at no point shall be less than 90 percent of the thicknesses given in the table.

[5]This dimension has the same thickness as Class L tubing.

[6]These dimensions are nominal, but in every case the thickness of wrought fittings should be at least as heavy as the tubing with which it is to be used. The dimensions R and T are equal for wrought fittings.

[7]Apply to cast fittings only. Laying lengths have not been standardized for wrought fittings.

NOTE: Wrought fittings, as well as cast fittings, must be provided with a shoulder or stop at the bottom end of the socket.

Pressure class	Size range (in)
25	4 through 72
125	1 through 96
250	1 through 30
800	2 through 12

The nonshock pressure-temperature ratings for the four pressure classes are listed in Table A2.16. The materials of construction are ASTM A 126 class A or B as shown in Table A2.16.

Cast- and Forged-Steel and Nickel-Alloy Flanged Fittings

Flanged fittings of steel and nickel alloys are manufactured in accordance with ASME B16.5. The standard covers ratings, materials, dimensions, tolerances, marking, testing, and methods of designating openings for pipe flanges and flanged fittings in sizes NPS 1/2 through NPS 24 and in rating classes 150, 300, 400, 600, 900, 1500, and 2500. However, not all sizes are available in all pressure classes. Dimensions of more commonly used fittings are given in Table A2.17. The standard also contains recommendations and requirements for bolting and gaskets.

Reducing fittings with raised-face flanges have the same center-to-face dimensions as those of straight-size fittings of the largest opening; face-to-face dimensions for reducers are as listed for the larger opening.

Center-to-face dimensions shown for fittings with ring-joint flanges apply to straight sizes only. For reducing fittings and reducers, the dimensions shown for raised face flanges of the largest opening should be used. On Class 400 and higher, the 1/4-in raised face for each flange should be subtracted. However, the 1/16-in raised face in Classes 150 and 300 should not be subtracted. The height of the ring-joint raised face L applying to each flange should be added.

When calculating the "laying length" of fittings with ring joints, the distance D between flange faces (when the ring is compressed) must be included.

Within each pressure class, the dimensions of the fittings are held constant, irrespective of the materials being used. Since the physical properties of different materials vary, the pressure-temperature ratings within each pressure class vary with the material. As an example, a Class 600 forged carbon steel (A105) flange is rated at 1270 psig at 400°F, whereas a Class 600 forged stainless steel (A182 F304) flange is rated at 940 psig at 400°F. The matrix of materials and pressure classes is too numerous to reproduce here and, therefore, the reader is referred to ASME/ASSI B16.5 for the flanged fitting pressure-temperature ratings.

Forged-Steel Threaded and Socket-Welding Fittings

Forged-steel socket welding and threaded fittings are manufactured in accordance with ANSI B16.11. The standard covers pressure-temperature ratings, dimensions, tolerances, marking, and material requirements for forged carbon and alloy steel fittings in the styles and sizes listed in Tables A2.18 and A2.19. Laying tolerances are listed in Table A2.20. Although these fittings are available in sizes up to NPS-4, size limitations may be imposed by certain codes. Acceptable material forms are forgings, bars, seamless pipe, and seamless tubes which conform

TABLE A2.16 Pressure-Temperature Rating of Cast-Iron Pipe Flanges and Flanged Fittings (ANSI/ASME B16.1-1989).

	Class 25*, ASTM A 126, Class A		Class 125, ASTM A 126				Class 250,* ASTM A 126				Class 800*, ASTM A 126, Class B
			Class A	Class B			Class A	Class B			
Temperature °*F*	NPS 4-36	NPD 42-96	NPS 1-12	NPS 1-12	NPS 14-24	NPS 30-48	NPS 1-12	NPS 1-12	NPS 14-24	NPS 30-48	NPS 2-12
−20 to 150	45	25	175	200	150	150	400	500	300	300	800
200	40	25	165	190	135	115	370	460	280	250	...
225	35	25	155	180	130	100	355	440	270	225	...
250	30	25	150	175	125	85	340	415	260	200	...
275	25	25	145	170	120	65	325	395	250	175	...
300	...	...	140	165	110	50	310	375	240	150	...
325	...	...	130	155	105	...	295	355	230	125	...
353†	...	...	125	150	100	...	280	335	220	100	...
375	...	...	...	145	...	...	265	315	210	...	...
406‡	...	...	...	140	...	...	250	290	200	...	...
425	...	...	...	130	...	...	...	270	...	...	...
450	...	...	...	125	...	...	...	250	...	...	...

Pressure is in lb/in^2 gauge.
NPS is nominal pipe size.
Hydrostatic tests are not required unless specified by user. The test pressure is equal to 1.5 times the 100°F pressure rating.
*Limitations:
(1) Class 25. When Class 25 cast-iron flanges and flanged fittings are used for gaseous service, the maximum pressure shall be limited to 25 psig. Tabulated pressure-temperature ratings above 25 psig for Class 25 cast-iron flanges and flanged fittings are applicable for nonshock hydraulic service only.
(2) Class 250. When used for liquid service, the tabulated pressure-temperature ratings in NPS 14 and larger are applicable to Class 250 flanges only and not to Class 250 fittings.
(3) Class 800. The tabulated rating is not a steam rating and applies to nonshock hydraulic pressure only.
†353°F (max.) to reflect the temperature of saturated steam at 125 psig.
‡406°F (max.) to reflect the temperature of saturated steam at 250 psig.

TABLE A2.17 Dimensions of Typical Commercial Cast-Steel Flanged Fittings (From ASME/ANSI B16.5-1988)

90° elbow — 90° long radius elbow — 45° elbow — Tee — Cross — 45° lateral — Reducer

Nominal pipe size	1/16-in. raised-face						Ring joint							
	AA	*BB*	*CC*	*EE*	*FF*	*GG*	*HH*	*JJ*	*KK*	*LL*	*MM*	*NN*	L^1	D^2
150 lb														
1	3½	5	1¾	5¾	1¾	4½	3¾	5¼	2	6	2		¼	5/32
1¼	3¾	5½	2	6¼	1¾	4½	4	5¾	2¼	6½	2		¼	5/32
1½	4	6	2¼	7	2	4½	4¼	6¼	2½	7¼	2¼		¼	5/32
2	4½	6½	2½	8	2½	5	4¾	6¾	2¾	8¼	2¾		¼	5/32
2½	5	7	3	9½	2½	5½	5¼	7¼	3¼	9¾	2¾		¼	5/32
3	5½	7¾	3	10	3	6	5¾	8	3¼	10¼	3¼		¼	5/32
3½	6	8½	3½	11½	3	6½	6¼	8¾	3¾	11¾	3¼		¼	5/32
4	6½	9	4	12	3	7	6¾	9¼	4¼	12¼	3¼		¼	5/32
5	7½	10¼	4½	13½	3½	8	7¾	10½	4¾	13¾	3¾	See note[3]	¼	5/32
6	8	11½	5	14½	3½	9	8¼	11¾	5¼	14¾	3¾		¼	5/32
8	9	14	5½	17½	4½	11	9¼	14¼	5¾	17¾	4¾		¼	5/32
10	11	16½	6½	20½	5	12	11¼	16¾	6¾	20¾	5¼		¼	5/32
12	12	19	7½	24½	5½	14	12¼	19¼	7¾	24¾	5¾		¼	5/32
14	14	21½	7½	27	6	16	14¼	21¾	7¾	27¼	6¼		¼	⅛
16	15	24	8	30	6½	18	15¼	24¼	8¼	30¼	6¾		¼	⅛
18	16½	26½	8½	32	7	19	16¾	26¾	8¾	32¼	7¼		¼	⅛
20	18	29	9½	35	8	20	18¼	29¼	9¾	35¼	8¼		¼	⅛
24	22	34	11	40½	9	24	22¼	34¼	11¼	40¾	9¼		¼	⅛

Nominal pipe size	1/16-in. raised-face						Ring joint							
	AA	*BB*	*CC*	*EE*	*FF*	*GG*	*HH*	*JJ*	*KK*	*LL*	*MM*	*NN*	L^1	D^2
300 lb														
1	4	5	2 1/4	6 1/2	2	4 1/2	4 1/4	5 1/4	2 1/2	6 3/4	2 1/4		1/4	5/32
1 1/4	4 1/4	5 1/2	2 1/2	7 1/4	2 1/4	4 1/2	4 1/2	5 3/4	2 3/4	7 1/2	2 1/2		1/4	5/32
1 1/2	4 1/2	6	2 3/4	8 1/2	2 1/2	4 1/2	4 3/4	6 1/4	3	8 3/4	2 3/4		1/4	5/32
2	5	6 1/2	3	9	2 1/2	5	5 5/16	6 13/16	3 5/16	9 5/16	2 13/16		5/16	7/32
2 1/2	5 1/2	7	3 1/2	10 1/2	2 1/2	5 1/2	5 13/16	7 5/16	3 13/16	10 13/16	2 13/16		5/16	7/32
3	6	7 3/4	3 1/2	11	3	6	6 5/16	8 1/16	3 13/16	11 5/16	3 5/16		5/16	7/32
3 1/2	6 1/2	8 1/2	4	12 1/2	3	6 1/2	6 13/16	8 13/16	4 5/16	12 13/16	3 5/16		5/16	7/32
4	7	9	4 1/2	13 1/2	3	7	7 5/16	9 5/16	4 13/16	13 13/16	3 5/16		5/16	7/32
5	8	10 1/4	5	15	3 1/2	8	8 5/16	10 9/16	5 5/16	15 5/16	3 13/16		5/16	7/32
6	8 1/2	11 1/2	5 1/2	17 1/2	4	9	8 13/16	11 13/16	5 13/16	17 13/16	4 5/16		5/16	7/32
8	10	14	6	20 1/2	5	11	10 5/16	14 5/16	6 5/16	20 13/16	5 5/16	See note[3]	5/16	7/32
10	11 1/2	16 1/2	7	24	5 1/2	12	11 13/16	16 13/16	7 5/16	24 5/16	5 13/16		5/16	7/32
12	13	19	8	27 1/2	6	14	13 5/16	19 5/16	8 5/16	27 13/16	6 5/16		5/16	7/32
14	15	21 1/2	8 1/2	31	6 1/2	16	15 5/16	21 13/16	8 13/16	31 5/16	6 13/16		5/16	7/32
16	16 1/2	24	9 1/2	34 1/2	7 1/2	18	16 13/16	24 5/16	9 13/16	34 13/16	7 13/16		5/16	7/32
18	18	26 1/2	10	37 1/2	8	19	18 5/16	26 13/16	10 5/16	37 13/16	8 5/16		5/16	7/32
20	19 1/2	29	10 1/2	40 1/2	8 1/2	20	19 7/8	29 3/8	10 7/8	40 7/8	8 7/8		3/8	7/32
24	22 1/2	34	12	47 1/2	10	24	22 15/16	34 7/16	12 7/16	47 15/16	10 7/16		7/16	1/4

TABLE A2.17 Dimensions of Typical Commercial Cast-Steel Flanged Fittings (from ASME/ANSI B16.5-1988). (*Continued*)

Nominal pipe size	¼-in. raised-face					Ring joint						
	AA	*CC*	*EE*	*FF*	*GG*	*HH*	*KK*	*LL*	*MM*	*NN*	L^1	D^2
400 lb (for sizes smaller than 4 in., use 600 lb)												
4	8	5½	16	4½	8¼	8 1/16	5 9/16	16 1/16	4 9/16	See note[3]	5/16	7/32
5	9	6	16¾	5	9¼	9 1/16	6 1/16	16 13/16	5 1/16		5/16	7/32
6	9¾	6¼	18¾	5¼	10	9 13/16	6 5/16	18 13/16	5 5/16		5/16	7/32
8	11¾	6¾	22¼	5¾	12	11 13/16	6 13/16	22 5/16	5 13/16		5/16	7/32
10	13¼	7¾	25¾	6¼	13½	13 5/16	7 13/16	25 13/16	6 5/16		5/16	7/32
12	15	8¾	29¾	6½	15¼	15 1/16	8 13/16	29 13/16	6 9/16		5/16	7/32
14	16¼	9¼	32¾	7	16½	16 5/16	9 5/16	32 13/16	7 1/16		5/16	7/32
16	17¾	10¼	36¼	8	18½	17 13/16	10 5/16	36 5/16	8 1/16		5/16	7/32
18	19¼	10¾	39¼	8½	19½	19 5/16	10 13/16	39 5/16	8 9/16		5/16	7/32
20	20¾	11¼	42¾	9	21	20 7/8	11 3/8	42 7/8	9 1/8		3/8	7/32
24	24¼	12¾	50¼	10½	24½	24 7/16	12 15/16	50 7/16	10 11/16		7/16	1/4
600 lb												
½	3¼	2	5¾	1¾	5	3 7/32	1 31/32	5 23/32	1 23/32	See note[3]	7/32	1/8
¾	3¾	2½	6¾	2	5	3¾	2½	6¾	2		1/4	5/32
1	4¼	2½	7¼	2¼	5	4¼	2½	7¼	2¼		1/4	5/32
1¼	4½	2¾	8	2½	5	4½	2¾	8	2½		1/4	5/32
1½	4¾	3	9	2¾	5	4¾	3	9	2¾		1/4	5/32
2	5¾	4¼	10¼	3½	6	5 13/16	4 5/16	10 5/16	3 9/16		5/16	3/16
2½	6½	4½	11½	3½	6¾	6 9/16	4 9/16	11 9/16	3 9/16		5/16	3/16
3	7	5	12¾	4	7¼	7 1/16	5 1/16	12 13/16	4 1/16		5/16	3/16
3½	7½	5½	14	4½	7¾	7 9/16	5 9/16	14 1/16	4 9/16		5/16	3/16
4	8½	6	16½	4½	8¾	8 9/16	6 1/16	16 9/16	4 9/16		5/16	3/16
5	10	7	19½	6	10¼	10 1/16	7 1/16	19 9/16	6 1/16		5/16	3/16
6	11	7½	21	6½	11¼	11 1/16	7 9/16	21 1/16	6 9/16		5/16	3/16
8	13	8½	24½	7	13¼	13 1/16	8 9/16	24 9/16	7 1/16		5/16	3/16
10	15½	9½	29½	8	15¾	15 9/16	9 9/16	29 9/16	8 1/16		5/16	3/16
12	16½	10	31½	8½	16¾	16 9/16	10 1/16	31 9/16	8 9/16		5/16	3/16
14	17½	10¾	34¼	9	17¾	17 9/16	10 13/16	34 5/16	9 1/16		5/16	3/16
16	19½	11¾	38½	10	19¾	19 9/16	11 13/16	38 9/16	10 1/16		5/16	3/16
18	21½	12¼	42	10½	21¾	21 9/16	12 5/16	42 1/16	10 9/16		5/16	3/16
20	23½	13	45½	11	23¾	23 5/8	13 1/8	45 5/8	11 1/8		3/8	3/16
24	27½	14¾	53	13	27¾	27 11/16	14 15/16	53 3/16	13 3/16		7/16	7/32

Nominal pipe size	1/4-in. raised-face					Ring joint						
	AA	*CC*	*EE*	*FF*	*GG*	*HH*	*KK*	*LL*	*MM*	*NN*	L^1	D^2
900 lb (for sizes smaller than 3 in., use 1,500 lb)												
3	7 1/2	5 1/2	14 1/2	4 1/2	7 3/4	7 9/16	5 9/16	14 9/16	4 9/16	See note[3]	5/16	5/32
4	9	6 1/2	17 1/2	5 1/2	9 1/4	9 1/16	6 9/16	17 9/16	5 9/16		5/16	5/32
5	11	7 1/2	21	6 1/2	11 1/4	11 1/16	7 9/16	21 1/16	6 9/16		5/16	5/32
6	12	8	22 1/2	6 1/2	12 1/4	12 1/16	8 1/16	22 9/16	6 9/16		5/16	5/32
8	14 1/2	9	27 1/2	7 1/2	14 3/4	14 9/16	9 1/16	27 9/16	7 9/16		5/16	5/32
10	16 1/2	10	31 1/2	8 1/2	16 3/4	16 9/16	10 1/16	31 9/16	8 9/16		5/16	5/32
12	19	11	34 1/2	9	17 3/4	19 1/16	11 1/16	34 9/16	9 1/16		5/16	5/32
14	20 1/4	11 1/2	36 1/2	9 1/2	19	20 7/16	11 11/16	36 11/16	9 11/16		7/16	5/32
16	22 1/4	12 1/2	40 3/4	10 1/2	21	22 7/16	12 11/16	40 15/16	10 11/16		7/16	5/32
18	24	13 1/4	45 1/2	12	24 1/2	24 1/4	13 1/2	45 3/4	12 1/4		1/2	3/16
20	26	14 1/2	50 1/4	13	26 1/2	26 1/4	14 3/4	50 1/2	13 1/4		1/2	3/16
24	30 1/2	18	60	15 1/2	30 1/2	30 7/8	18 3/8	60 3/8	15 7/8		5/8	7/32
1,500 lb												
1/2	4 1/4	3	....	....	....	4 1/4	3	.....	.....	See note[3]	1/4	5/32
3/4	4 1/2	3 1/4	....	....	....	4 1/2	3 1/4	.....	.....		1/4	5/32
1	5	3 1/2	9	2 1/2	5	5	3 1/2	9	2 1/2		1/4	5/32
1 1/4	5 1/2	4	10	3	5 3/4	5 1/2	4	10	3		1/4	5/32
1 1/2	6	4 1/4	11	3 1/2	6 1/4	6	4 1/4	11	3 1/2		1/4	5/32
2	7 1/4	4 3/4	13 1/4	4	7 1/4	7 5/16	4 13/16	13 5/16	4 1/16		5/16	1/8
2 1/2	8 1/4	5 1/4	15 1/4	4 1/2	8 1/4	8 5/16	5 5/16	15 5/16	4 9/16		5/16	1/8
3	9 1/4	5 3/4	17 1/4	5	9 1/4	9 5/16	5 13/16	17 5/16	5 1/16		5/16	1/8
4	10 3/4	7 1/4	19 1/4	6	10 3/4	10 13/16	7 5/16	19 5/16	6 1/16		5/16	1/8
5	13 1/4	8 3/4	23 1/4	7 1/2	13 3/4	13 5/16	8 13/16	23 5/16	7 9/16		5/16	1/8
6	13 7/8	9 3/8	24 7/8	8 1/8	14 1/2	14	9 1/2	25	8 1/4		3/8	1/8
8	16 3/8	10 7/8	29 7/8	9 1/8	17	16 9/16	11 1/16	30 1/16	9 5/16		7/16	5/32
10	19 1/2	12	36	10 1/4	20 1/4	19 11/16	12 3/16	36 3/16	10 7/16		7/16	5/32
12	22 1/4	13 1/4	40 3/4	12	23	22 9/16	13 9/16	41 1/16	12 5/16		9/16	3/16
14	24 3/4	14 1/4	44	12 1/2	25 3/4	25 1/8	14 5/8	44 3/8	12 7/8		5/8	7/32
16	27 1/4	16 1/4	48 1/4	14 3/4	28 1/4	27 11/16	16 11/16	48 11/16	15 3/16		11/16	5/16
18	30 1/4	17 3/4	53 1/4	16 1/2	31 1/2	30 11/16	18 3/16	53 11/16	16 15/16		11/16	5/16
20	32 3/4	18 3/4	57 3/4	17 3/4	34	33 3/16	19 3/16	58 3/16	18 3/16		11/16	3/8
24	38 1/4	20 3/4	67 1/4	20 1/2	39 3/4	38 13/16	21 5/16	67 13/16	21 1/16		13/16	7/16

[1]*L* = height of raised face of ring joint flanges.

[2]*D* = approximate distance between flange faces when ring is compressed.

[3]Center-to-face dimensions shown for fittings with ring-joint flanges apply to straight sizes only. For reducing fittings and reducers, use dimensions shown for raised-face flanges of largest opening; 400-lb and higher classes, subtract the 1/4-in raised face for each flange (do not subtract the 1/16-in raised face in 150- and 300-lb classes); add height of ring-joint raised face (*L*) applying to each flange.

For calculating the "laying length" of fittings with ring joints, add the approximate distance (*D*) between flange faces when ring is compressed to the center-to-face dimensions in these tables.

TABLE A2.18 Dimensions of Typical Commercial Forged-Steel Threaded Fittings (ANSI B16.11-1980)

90° elbow · Tee · 45° elbow · Cross · Coupling · Reducer · Half coupling · Pipe cap

	Dimensions, in.											
	1/8	1/4	3/8	1/2	3/4	1	1 1/4	1 1/2	2	2 1/2	3	4
						2,000 lb						
A	0.81	0.81	0.97	1.12	1.31	1.50	1.75	2.00	2.38	3.00	3.38	4.19
B	0.88	0.88	1.00	1.31	1.50	1.81	2.19	2.44	2.97	3.62	4.31	5.756
C	0.69	0.69	0.75	0.88	1.00	1.12	1.31	1.38	1.69	2.06	2.50	3.12
T	0.125	0.125	0.125	0.125	0.125	0.145	0.153	0.158	0.168	0.221	0.236	0.258
						3,000 lb						
A	0.81	0.97	1.12	1.31	1.50	1.75	2.00	2.38	2.50	3.25	3.75	4.50
B	0.88	1.00	1.31	1.50	1.81	2.19	2.44	2.97	3.31	4.00	4.75	6.00
C	0.69	0.75	0.88	1.00	1.12	1.31	1.38	1.69	1.72	2.06	2.50	3.12
T	0.125	0.13	0.138	0.161	0.170	0.196	0.208	0.219	0.281	0.301	0.348	0.440
N	0.68	0.75	0.88	1.12	1.38	1.75	2.25	2.50	3.00	3.62	4.25	4.75
P	1.25	1.38	1.50	1.88	2.00	2.38	2.62	3.12	3.38	3.62	4.25	4.75
R	0.75	1.00	1.00	1.25	1.44	1.62	1.75	1.75	1.88	2.38	2.58	2.69
						6,000 lb						
A	0.97	1.12	1.31	1.50	1.75	2.00	2.38	2.50	3.25	3.75	4.19	4.50
B	1.00	1.31	1.50	1.81	2.19	2.44	2.97	3.31	4.00	4.75	5.75	6.00
C	0.75	0.88	1.00	1.12	1.31	1.38	1.69	1.72	2.06	2.50	3.12	3.12
T	0.250	0.260	0.275	0.321	0.336	0.391	0.417	0.436	0.476	0.602	0.655	0.735
N	0.88	1.00	1.25	1.50	1.75	2.25	2.50	3.00	3.62	4.25	5.00	6.25
P	1.25	1.38	1.50	1.88	2.00	2.38	2.62	3.12	3.38	3.62	4.25	4.75
R	...	1.06	1.06	1.31	1.50	1.69	1.81	1.88	2.00	2.50	2.69	2.94

Manufacturers catalogs should be consulted for dimensions of street elbows and of laterals since these two types of fittings are no longer covered by ANSI Standards.

TABLE A2.19 Dimensions of Typical Commercial Forged-Steel Socket-Welding Fittings (ANSI B16.11-1980)

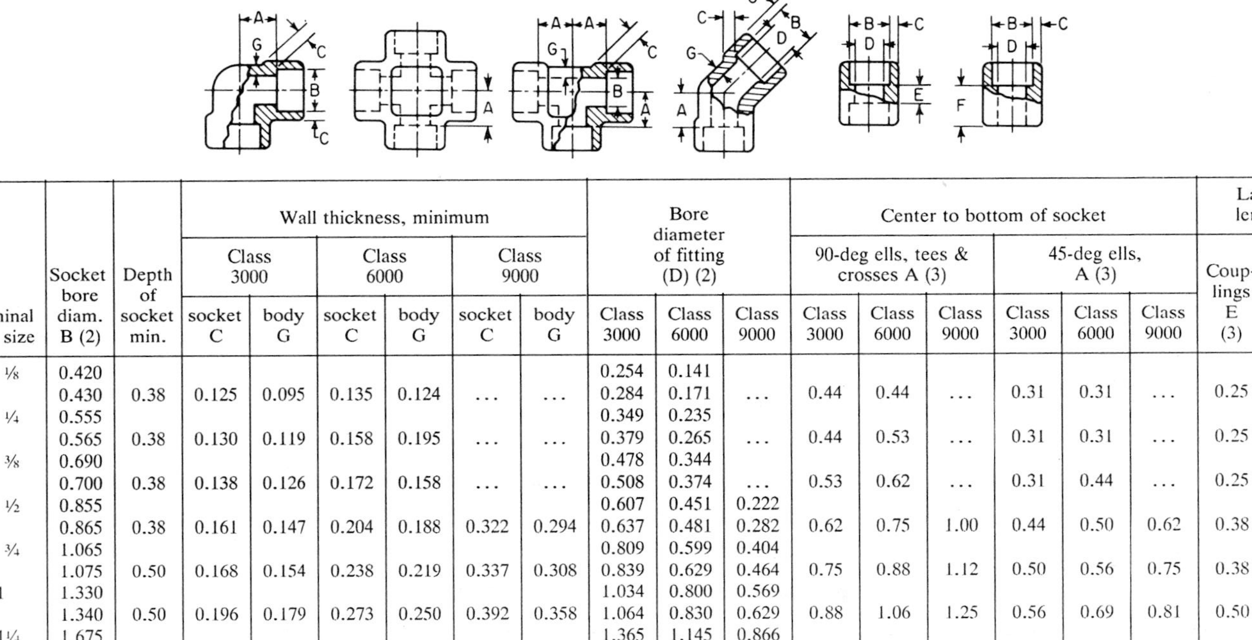

Nominal pipe size	Socket bore diam. B (2)	Depth of socket min.	Wall thickness, minimum: Class 3000 socket C	Class 3000 body G	Class 6000 socket C	Class 6000 body G	Class 9000 socket C	Class 9000 body G	Bore diameter of fitting (D) (2): Class 3000	Class 6000	Class 9000	Center to bottom of socket: 90-deg ells, tees & crosses A (3): Class 3000	Class 6000	Class 9000	45-deg ells, A (3): Class 3000	Class 6000	Class 9000	Laying lengths: Couplings E (3)	Half couplings F (3)
⅛	0.420								0.254	0.141									
	0.430	0.38	0.125	0.095	0.135	0.124	. . .	. . .	0.284	0.171	. . .	0.44	0.44	. . .	0.31	0.31	. . .	0.25	0.62
¼	0.555								0.349	0.235									
	0.565	0.38	0.130	0.119	0.158	0.195	. . .	. . .	0.379	0.265	. . .	0.44	0.53	. . .	0.31	0.31	. . .	0.25	0.62
⅜	0.690								0.478	0.344									
	0.700	0.38	0.138	0.126	0.172	0.158	. . .	. . .	0.508	0.374	. . .	0.53	0.62	. . .	0.31	0.44	. . .	0.25	0.69
½	0.855								0.607	0.451	0.222								
	0.865	0.38	0.161	0.147	0.204	0.188	0.322	0.294	0.637	0.481	0.282	0.62	0.75	1.00	0.44	0.50	0.62	0.38	0.88
¾	1.065								0.809	0.599	0.404								
	1.075	0.50	0.168	0.154	0.238	0.219	0.337	0.308	0.839	0.629	0.464	0.75	0.88	1.12	0.50	0.56	0.75	0.38	0.94
1	1.330								1.034	0.800	0.569								
	1.340	0.50	0.196	0.179	0.273	0.250	0.392	0.358	1.064	0.830	0.629	0.88	1.06	1.25	0.56	0.69	0.81	0.50	1.12
1¼	1.675								1.365	1.145	0.866								
	1.685	0.50	0.208	0.191	0.273	0.250	0.418	0.382	1.395	1.175	0.926	1.06	1.25	1.38	0.69	0.81	0.88	0.50	1.19
1½	1.915								1.595	1.323	1.070								
	1.925	0.50	0.218	0.200	0.307	0.281	0.438	0.400	1.625	1.353	1.130	1.25	1.50	1.50	0.81	1.00	1.00	0.50	1.25

TABLE A2.19 Dimensions of Typical Commercial Forged-Steel Socket-Welding Fittings (ANSI B16.11-1980) (*Continued*)

Nominal pipe size	Socket bore diam. B (2)	Depth of socket min.	Wall thickness, minimum: Class 3000 socket C	Class 3000 body G	Class 6000 socket C	Class 6000 body G	Class 9000 socket C	Class 9000 body G	Bore diameter of fitting (D) (2): Class 3000	Class 6000	Class 9000	Center to bottom of socket: 90-deg ells, tees & crosses A (3): Class 3000	Class 6000	Class 9000	45-deg ells, A (3): Class 3000	Class 6000	Class 9000	Laying lengths: Couplings E (3)	Half couplings F (3)
2	2.406								2.052	1.674	1.473								
	2.416	0.62	0.238	0.218	0.374	0.344	0.477	0.436	2.082	1.704	1.533	1.50	1.62	2.12	1.00	1.12	1.12	0.75	1.62
2½	2.906								2.439										
	2.921	.062	0.301	0.276	...	0.375	...	...	2.499	...	...	1.62	...	...	1.12	...	...	0.75	1.69
3	3.535								3.038										
	3.550	0.62	0.327	0.300	...	0.438	...	...	3.098	...	...	2.25	...	...	1.25	...	...	0.75	1.75
4	4.545								3.996										
	4.560	0.75	0.368	0.337	...	0.531	...	...	4.056	...	...	2.62	...	...	1.62	...	...	0.75	1.88

[1]Dimensions for caps and reducers are not standardized. Refer to manufacturer's literature for dimensions.
[2]Values are lower/upper limits.
[3]For tolerances, refer to Table A2.20.

TABLE A2.20 Center to Bottom and Laying Length Tolerances for Classes 3000, 6000, and 9000 Socket-Welding Fittings (from ANSI B16.11-1980)

NPS Dimensions	Tolerances plus or minus		
	A	*E*	*F*
1/8	0.03	0.06	0.03
1/4	0.03	0.06	0.03
3/8	0.06	0.12	0.06
1/2	0.06	0.12	0.06
3/4	0.06	0.12	0.06
1	0.08	0.16	0.08
1 1/4	0.08	0.16	0.08
1 1/2	0.08	0.16	0.08
2	0.08	0.16	0.08
2 1/2	0.10	0.20	0.10
3	0.10	0.20	0.10
4	0.10	0.20	0.10

Refer to Table A2.19 for nomenclature.

to the chemical compositions, melting processes, and mechanical property requirements of ASTM A105, A182, or A350.

Threaded fittings are available in pressure classes 2000, 3000, and 6000. Socket-welded fittings are available in pressure classes 3000, 6000, and 9000.

Limitations on fitting service conditions are as provided for by the applicable code for the material of construction of the fitting. The maximum allowable pressure of the fitting is equal to that computed for straight seamless pipe of equivalent material, considering manufacturing tolerance, corrosion allowance, and mechanical strength allowance. Also, for socket-welding fittings, the pressure rating must be matched to the pipe wall thickness to assure that the flat of the band can accommodate the size of the fillet-weld required by the applicable code. The recommended fitting pressure class for the various pipe wall thickness is as follows:

Pipe schedule and designation	Threaded	Socket welded
80/XS or less	2000	3000
160	3000	6000
XXS	6000	9000

Internal threads of threaded fittings are in accordance with ANSI/ASME B1.20.1—Pipe Threads, General Purpose (Inch).

Wrought-Steel Butt-Welding Fittings

Wrought-steel welding fittings include elbows, tees, crosses, reducers, laterals, lap-joint stub ends, caps, and saddles.

Wrought-steel fittings are made to the dimensional requirements of ASME/ANSI B16.9 in sizes NPS 1/2 through 48. Also, short-radius elbows and returns are produced in accordance with ASME/ANSI B16.28 in sizes NPS 1/2 through

NPS 24. The wrought fitting materials conform to ASTM A234, A403, *or* A420, the grades of which have chemical and physical properties equivalent to that of the mating pipe. ANSI B16.9 requires that the pressure-temperature rating of the fitting equal or exceed that of the mating pipe of the same or equivalent material, same size, and same nominal wall thickness. The pressure-temperature rating may be established by analysis or by proof testing. Short-radius elbows and returns manufactured under ANSI B16.28 are rated at 80 percent of the rating calculated for seamless straight pipe of the same size and nominal thickness and same or equivalent material. Therefore, both standards require that in lieu of specifying any pressure rating, the pipe wall thickness and pipe material type with which the fittings are intended to be used be identified on the fitting.

Pressure testing of the fittings is not required by either standard. However, the fittings are required to be capable of withstanding, without leakage, a test pressure equal to that prescribed in the specification of the pipe with which the fitting is recommended to be used.

Both ASME/ANSI B16.9 and B16.28 prescribe dimensions and manufacturing tolerances of wrought butt-welded fittings. The standards establish laying dimensions which remain fixed for each size and type of fitting irrespective of the fitting wall thickness. Tables A2.21 through A2.26 list laying dimensions and approximate weights for selected fitting sizes, wall thicknesses, and configurations.

Laterals are not governed by any national standard. However, dimensions of laterals commonly used are given in Table A2.25. Working pressures are rated at 40 percent of the allowable working pressure established for pipe from which laterals are made. Where full allowable pipe pressures must be met, the laterals are generally made from heavier pipe with ends machined to match standard pipe dimensions. Dimensional tolerances of laterals vary not more than $\pm\frac{1}{32}$ in. for sizes up to and including NPS 8 and $\pm\frac{1}{16}$ in. for sizes NPS 10 through 24.

Standard dimensions of caps are given in Table A2.26. Pressure-temperature ratings are identical to those of seamless pipe for the same size, thickness or schedule, and material grades. Caps conform to ASME Boiler and Pressure Vessel Code requirements. Welding caps are formed from steel plate and are stress-relieved after forming. They are ellipsoidal in shape, the minor axis being equal to half the major axis.

Radii R and r closely approximate the actual semiellipsoidal shape.

Forged Branch Fittings

Under the various pressure piping codes, branch connections may be made by welding the branch pipe or a welding outlet fitting to the run pipe, provided sufficient reinforcement is available to compensate for the material removed from the run pipe to create the branch opening. The reinforcement may be in the form of excess material already available in the run and branch pipes or may be added. At the writing of this book, national standards governing the dimensions, tolerances, and manufacture of welding outlet fittings had not been issued. However, MSS-SP-79, 1987, has been developed to cover forged carbon steel 90° branch outlet fittings in butt-welding, socket-welding, and threaded outlet ends. The standard provides essential dimensions, finish, tolerances, and testing requirements. Because of the absence of strict standards, manufacturers produce welding outlet fittings of their own proprietary designs. These fittings must comply with the codes governing the systems in which they are to be installed. The fittings, when installed in accordance with the manufacturers' recommendations, include the required reinforcement. The dimensions of these fittings vary; stan-

TABLE A2.21 Dimensions of Typical Commercial 90° Long-Radius Butt-Welding Elbows (ASME/ANSI B16.9-1986)

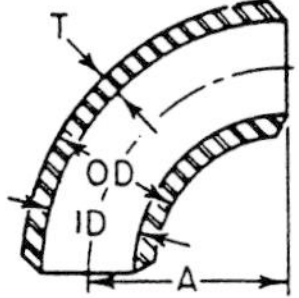

Nominal pipe size	Outside diameter, OD	Inside diameter, ID	Wall thickness T	Center to face, A	Pipe schedule number*	Weight (approx), lb†
			Standard			
½	0.840	0.622	0.109	1½	40	0.2
¾	1.050	0.824	0.113	1⅛	40	0.2
1	1.315	1.049	0.133	1½	40	0.4
1¼	1.660	1.380	0.140	1⅞	40	0.6
1½	1.900	1.610	0.145	2¼	40	0.9
2	2.375	2.067	0.154	3	40	1.4
2½	2.875	2.469	0.203	3¾	40	2.9
3	3.500	3.068	0.216	4½	40	4.5
3½	4.000	3.548	0.226	5¼	40	6.4
4	4.500	4.026	0.237	6	40	8.7
5	5.563	5.047	0.258	7½	40	14.7
6	6.625	6.065	0.280	9	40	22.9
8	8.625	7.981	0.322	12	40	46.0
10	10.750	10.020	0.365	15	40	81
12	12.750	12.000	0.375	18	•‡	119
14	14.000	13.250	0.375	21	30	154
16	16.000	15.250	0.375	24	30	201
18	18.000	17.250	0.375	27	•‡	256
20	20.000	19.250	0.375	30	20	317
22	22.000	21.250	0.375	33	20	385
24	24.000	23.250	0.375	36	20	458
26	26.000	25.250	0.375	39	•‡	539
28	28.000	27.250	0.375	42	•‡	626
30	30.000	29.250	0.375	45	•‡	720
32	32.000	31.250	0.375	48	•‡	818
34	34.000	33.250	0.375	51	•‡	926
36	36.000	35.250	0.375	54	•‡	1040
42	42.000	41.250	0.375	63	•‡	1420

dardized dimensions and properties must be obtained from the manufacturers. Also, designers must consider the appropriate parameters (e.g., stress intensification factors). Figure A2.3 shows several types of welding fittings, which are proprietary; the terminology used varies with the manufacturer. The fittings are produced in carbon and alloy steels under the ASTM specifications for forgings permitted by applicable codes.

TABLE A2.21 Dimensions of Typical Commercial 90° Long-Radius Butt-Welding Elbows (ASME/ANSI B16.9-1986) (*Continued*)

Nominal pipe size	Outside diameter, OD	Inside diameter, ID	Wall thickness *T*	Center to face, *A*	Pipe schedule number*	Weight (approx), lb†
			Extra strong			
½	0.840	0.546	0.147	1½	80	0.3
¾	1.050	0.742	0.154	1⅛	80	0.3
1	1.315	0.957	0.179	1½	80	0.5
1¼	1.660	1.278	0.191	1⅞	80	0.8
1½	1.900	1.500	0.200	2¼	80	1.0
2	2.375	1.939	0.218	3	80	2.0
2½	2.875	2.323	0.276	3¾	80	3.8
3	3.500	2.900	0.300	4½	80	6.1
3½	4.000	3.364	0.318	5¼	80	8.7
4	4.500	3.826	0.337	6	80	11.9
5	5.563	4.813	0.375	7½	80	20.6
6	6.625	5.761	0.432	9	80	34.1
8	8.625	7.625	0.500	12	80	69
10	10.750	9.750	0.500	15	60	109
12	12.750	11.750	0.500	18	•‡	157
14	14.000	13.000	0.500	21	•‡	202
16	16.000	15.000	0.500	24	40	265
18	18.000	17.000	0.500	27	•‡	338
20	20.000	19.000	0.500	30	30	419
22	22.000	21.000	0.500	33	30	508
24	24.000	23.000	0.500	36	•‡	606
26	26.000	25.000	0.500	39	20	713
28	28.000	27.000	0.500	42	20	829
30	30.000	29.000	0.500	45	20	953
32	32.000	31.000	0.500	48	20	1090
34	24.000	33.000	0.500	51	20	1230
36	36.000	35.000	0.500	54	20	1380
42	42.000	41.000	0.500	63	•‡	1880
			Schedule 160†			
1	1.315	0.815	0.250	1½	160	0.6
1¼	1.660	1.160	0.250	1⅞	160	1.0
1½	1.900	1.338	0.281	2¼	160	1.4
2	2.375	1.689	0.343	3	160	2.9
2½	2.875	2.125	0.375	3¾	160	4.9
3	3.500	2.624	0.438	4½	160	8.3
4	4.500	3.438	0.531	6	160	17.6
5	5.563	4.313	0.625	7½	160	32.2
6	6.625	5.189	0.718	9	160	53
8	8.625	6.813	0.906	12	160	117
10	10.750	8.500	1.125	15	160	226
12	12.750	10.126	1.312	18	160	375

TABLE A2.21 Dimensions of Typical Commercial 90° Long-Radius Butt-Welding Elbows (ASME/ANSI B16.9-1986) (*Continued*)

Nominal pipe size	Outside diameter, OD	Inside diameter, ID	Wall thickness *T*	Center to face, *A*	Pipe schedule number*	Weight (approx), lb†
			Double extra strong			
¾	1.050	0.434	0.308	1⅛	•‡	0.4
1	1.315	0.599	0.358	1½	•‡	0.7
1¼	1.660	0.896	0.382	1⅞	•‡	1.2
1½	1.900	1.100	0.400	2¼	•‡	1.8
2	2.375	1.503	0.436	3	•‡	3.4
2½	2.875	1.771	0.552	3¾	•‡	6.5
3	3.500	2.300	0.600	4½	•‡	10.7
3½	4.000	2.728	0.636	5¼	•‡	15.4
4	4.500	3.152	0.674	6	•‡	21.2
5	5.563	4.063	0.750	7½	•‡	37.2
6	6.625	4.897	0.864	9	•‡	61
8	8.625	6.875	0.875	12	•‡	114

*Pipe schedule numbers in accordance with ASME/ANSI B36.10.
†Weights are not tabulated in ASME/ANSI B16.9.
‡This size and thickness does not correspond with any schedule number.

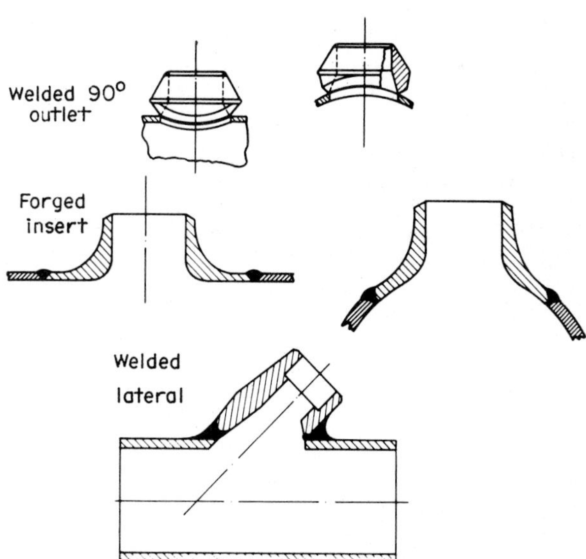

FIGURE A2.3 Typical welding outlet fittings.

TABLE A2.22 Dimensions of Typical Commercial 90° Short-Radius Elbows (ASME/ANSI B16.28-1986)

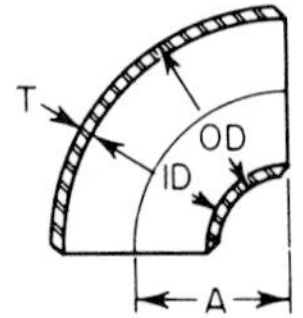

Nominal pipe size	Outside diameter, OD	Inside diameter, ID	Wall thickness, *T*	Center to face, *A*	Pipe schedule number*	Weight (approx), lb†
			Standard			
1	1.315	1.049	0.133	1	40	0.3
1¼	1.660	1.380	0.140	1¼	40	0.4
1½	1.900	1.610	0.145	1½	40	0.6
2	2.375	2.067	0.154	2	40	1.0
2½	2.875	2.469	0.203	2½	40	1.9
3	3.500	3.068	0.216	3	40	3.0
3½	4.000	3.548	0.226	3½	40	4.2
4	4.500	4.026	0.237	4	40	5.7
5	5.563	5.047	0.258	5	40	9.7
6	6.625	6.065	0.280	6	40	15.2
8	8.625	7.981	0.322	8	40	30.5
10	10.750	10.020	0.365	10	40	54
12	12.750	12.000	0.375	12	•‡	79
14	14.000	13.250	0.375	14	30	102
16	16.000	15.250	0.375	16	30	135
18	18.000	17.250	0.375	18	•‡	171
20	20.000	19.250	0.375	20	20	212
22	22.000	21.250	0.375	22	•‡	256
24	24.000	23.250	0.375	24	20	305
26§	26.000	25.250	0.375	26	•‡	359
28	28.000	27.250	0.375	28	•‡	415
30	30.000	29.250	0.375	30	•‡	480
32	32.000	31.250	0.375	32	•‡	546
34	34.000	33.250	0.375	34	•‡	617
36	36.000	35.250	0.375	36	•‡	692
42	42.000	41.250	0.375	48	•‡	1079

TABLE A2.22 Dimensions of Typical Commercial 90° Short-Radius Elbows (ASME/ANSI B16.28-1986) (*Continued*)

Nominal pipe size	Outside diameter, OD	Inside diameter, ID	Wall thickness, T	Center to face, A	Pipe schedule number*	Weight (approx), lb†
			Extra strong			
1½	1.900	1.500	0.200	1½	80	0.7
2	2.375	1.939	0.218	2	80	1.3
2½	2.875	2.323	0.276	2½	80	2.5
3	3.500	2.900	0.300	3	80	4.0
3½	4.000	3.364	0.318	3½	80	5.7
4	4.500	3.826	0.337	4	80	7.8
5	5.563	4.813	0.375	5	80	13.7
6	6.625	5.761	0.432	6	80	22.6
8	8.625	7.625	0.500	8	80	45.6
10	10.750	9.750	0.500	10	60	72
12	12.750	11.750	0.500	12	•‡	104
14	14.000	13.000	0.500	14	•‡	135
16	16.000	15.000	0.500	16	40	177
18	18.000	17.000	0.500	18	•‡	225
20	20.000	19.000	0.500	20	30	278
22	22.000	21.000	0.500	22	30	338
24	24.000	23.000	0.500	24	•‡	404
26§	26.000	25.000	0.500	26	20	474
28	28.000	27.000	0.500	28	20	581
30	30.000	29.000	0.500	30	20	634
32	32.000	31.000	0.500	32	20	722
34	34.000	33.000	0.500	34	20	817
36	36.000	35.000	0.500	36	20	913
42	42.000	41.000	0.500	48	•‡	1,430

*Pipe schedule numbers in accordance with ASME/ANSI B36.10.
†Filling weights are not tabulated in ANSI/ASME B16.28.
‡This size and thickness has no corresponding schedule number.
§Dimensional data for pipe sizes 26 in and larger are not included in ASME/ANSI B16.28.

TABLE A2.23 Dimensions of Typical Commercial Straight Butt-Welding Tees (ASME/ANSI B16.9-1986)

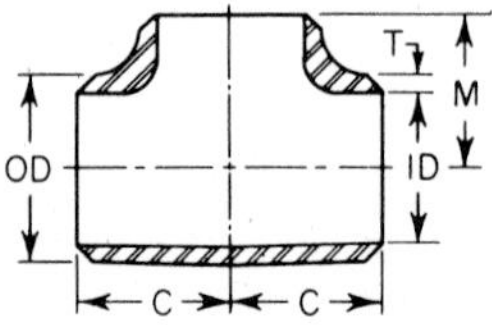

Nominal pipe size	Outside diameter, OD	Inside diameter, ID	Wall thickness, *T*	Center to end, *C*	Center to end, *M*	Pipe schedule number*	Weight (approx), lb†
			Standard				
½	0.840	0.622	0.109	1	1	40	0.3
¾	1.050	0.824	0.113	1⅛	1⅛	40	0.4
1	1.315	1.049	0.133	1½	1½	40	0.8
1¼	1.660	1.380	0.140	1⅞	1⅞	40	1.3
1½	1.900	1.610	0.145	2¼	2¼	40	2.0
2	2.375	2.067	0.154	2½	2½	40	2.9
2½	2.875	2.469	0.203	3	3	40	5.2
3	3.500	3.068	0.216	3⅜	3⅜	40	7.4
3½	4.000	3.548	0.226	3¾	3¾	40	9.8
4	4.500	4.026	0.237	4⅛	4⅛	40	12.6
5	5.563	5.047	0.258	4⅞	4⅞	40	19.8
6	6.625	6.065	0.280	5⅝	5⅝	40	29.3
8	8.625	7.981	0.322	7	7	40	53
10	10.750	10.020	0.365	8½	8½	40	91
12	12.750	12.000	0.375	10	10	•‡	132
14	14.000	13.250	0.375	11	11	30	172
16	16.000	15.250	0.375	12	12	30	219
18	18.000	17.250	0.375	13½	13½	•‡	282
20	20.000	19.250	0.375	15	15	20	354
22	22.000	21.250	0.375	16½	16½	20	437
24	24.000	23.250	0.375	17	17	20	493
26	26.000	25.250	0.375	19½	19½	•‡	634
28	28.000	27.250	0.375	20½	20½	•‡	729
30	30.000	29.250	0.375	22	22	•‡	855
32	32.000	31.250	0.375	23½	23½	•‡	991
34	34.000	33.250	0.375	25	25	•‡	1136
36	36.000	35.250	0.375	26½	26½	•‡	1294

TABLE A2.23 Dimensions of Typical Commercial Straight Butt-Welding Tees (ASME/ANSI B16.9-1986) (*Continued*)

Nominal pipe size	Outside diameter, OD	Inside diameter, ID	Wall thickness, *T*	Center to end, *C*	Center to end, *M*	Pipe schedule number*	Weight (approx), lb†
			Extra strong				
½	0.840	0.546	0.147	1	1	80	0.3
¾	1.050	0.742	0.154	1⅛	1⅛	80	0.5
1	1.315	0.957	0.179	1½	1½	80	0.9
1¼	1.660	1.278	0.191	1⅞	1⅞	80	1.6
1½	1.900	1.500	0.200	2¼	2¼	80	2.4
2	2.375	1.939	0.218	2½	2½	80	3.7
2½	2.875	2.323	0.276	3	3	80	6.4
3	3.500	2.900	0.300	3⅜	3⅜	80	9.4
3½	4.000	3.364	0.318	3¾	3¾	80	12.6
4	4.500	3.826	0.337	4⅛	4⅛	80	16.4
5	5.563	4.813	0.375	4⅞	4⅞	80	26.4
6	6.625	5.761	0.432	5⅝	5⅝	80	42.0
8	8.625	7.625	0.500	7	7	80	76
10	10.750	9.750	0.500	8½	8½	60	118
12	12.750	11.750	0.500	10	10	•‡	167
14	14.000	13.000	0.500	11	11	•‡	203
16	16.000	15.000	0.500	12	12	40	271
18	18.000	17.000	0.500	13½	13½	•‡	351
20	20.000	19.000	0.500	15	15	30	442
22	22.000	21.000	0.500	16½	16½	30	548
24	24.000	23.000	0.500	17	17	20	607
26	26.000	25.000	0.500	19½	19½	20	794
28	28.000	27.000	0.500	20½	20½	20	910
30	30.000	29.000	0.500	22	22	20	1065
32	32.000	31.000	0.500	23½	23½	20	1230
34	34.000	33.000	0.500	25	25	20	1420
36	36.000	35.000	0.500	26½	26½	20	1610
			Schedule 160*				
½	0.840	0.466	0.187	1	1	160	0.4
¾	1.050	0.614	0.218	1⅛	1⅛	160	0.6
1	1.315	0.815	0.250	1½	1½	160	1.1
1¼	1.660	1.160	0.250	1⅞	1⅞	160	1.9
1½	1.900	1.338	0.281	2¼	2¼	160	3.0
2	2.375	1.689	0.343	2½	2½	160	4.9
2½	2.875	2.125	0.375	3	3	160	7.8
3	3.500	2.626	0.438	3⅜	3⅜	160	12.2
4	4.500	3.438	0.531	4⅛	4⅛	160	22.8
5	5.563	4.313	0.625	4⅞	4⅞	160	38.5
6	6.625	5.189	0.718	5⅝	5⅝	160	59
8	8.625	6.813	0.906	7	7	160	120
10	10.750	8.500	1.125	8½	8½	160	222
12	12.750	10.126	1.312	10	10	160	360

TABLE A2.23 Dimensions of Typical Commercial Straight Butt-Welding Tees (ASME/ANSI B16.9-1986) (*Continued*)

Nominal pipe size	Outside diameter, OD	Inside diameter, ID	Wall thickness, *T*	Center to end, *C*	Center to end, *M*	Pipe schedule number*	Weight (approx), lb†
Double extra strong							
½	0.840	0.252	0.294	1	1	•‡	0.4
¾	1.050	0.434	0.308	1⅛	1⅛	•‡	0.6
1	1.315	0.599	0.358	1½	1½	•‡	1.3
1¼	1.660	0.896	0.382	1⅞	1⅞	•‡	2.4
1½	1.900	1.100	0.400	2¼	2¼	•‡	3.7
2	2.375	1.503	0.436	2½	2½	•‡	5.7
2½	2.875	1.771	0.552	3	3	•‡	9.8
3	3.500	2.300	0.600	3⅜	3⅜	•‡	14.8
3½	4.000	2.728	0.636	3¾	3¾	•‡	20.2
4	4.500	3.152	0.674	4⅛	4⅛	•‡	26.6
5	5.563	4.063	0.750	4⅞	4⅞	•‡	43.4
6	6.625	4.897	0.864	5⅝	5⅝	•‡	68
8	8.625	6.875	0.875	7	7	•‡	118

*Pipe schedule numbers in accordance with ASME/ANSI B36.10. Other thicknesses available.
†Fitting weights are not tabulated in ASME/ANSI B16.9.
‡This size and thickness does not correspond with any schedule number.

TABLE A2.24 Dimensions of Typical Commercial Concentric and Eccentric Butt-Welding Reducers (ASME/ANSI B16.9-1986)

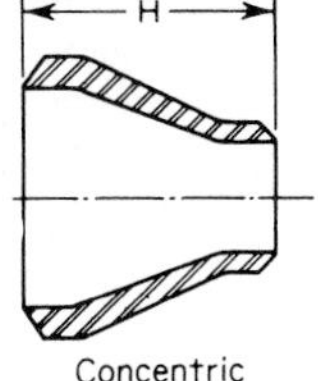

Concentric

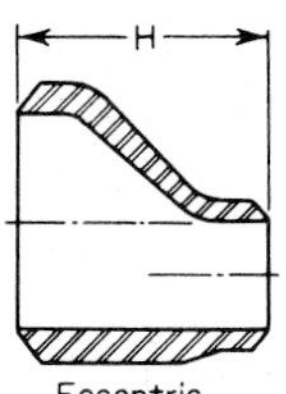

Eccentric

Nominal pipe size	Length, *H*	Weight (approx), lb (concentric or eccentric)			
		Standard	Extra strong	Schedule 160	Double extra strong
¾ × ⅜	1½	0.2	0.3	0.3	. . .
¾ × ½	1½	0.2	0.3	0.3	0.4
1 × ⅜	2	0.3	0.4	0.4	0.4
1 × ½	2	0.3	0.4	0.5	0.5
1 × ¾	2	0.3	0.4	0.5	0.5
1¼ × ½	2	0.5	0.5	0.6	0.7
1¼ × ¾	2	0.5	0.5	0.6	0.7
1¼ × 1	2	0.5	0.6	0.7	0.8
1½ × ½	2½	0.5	0.6	0.8	1.0
1½ × ¾	2½	0.5	0.6	0.9	1.0
1½ × 1	2½	0.6	0.7	0.9	1.0
1½ × 1¼	2½	0.6	0.8	1.0	1.2
2 × ¾	3	0.8	1.0	1.4	1.7
2 × 1	3	0.9	1.0	1.4	1.6
2 × 1¼	3	0.9	1.1	1.4	1.8
2 × 1½	3	0.9	1.2	1.6	1.9
2½ × 1	3½	1.3	1.7	2.3	3.0
2½ × 1¼	3½	1.4	1.7	2.2	3.1
2½ × 1½	3½	1.5	1.8	2.2	3.0
2½ × 2	3½	1.6	2.0	2.7	3.3

TABLE A2.24 Dimensions of Typical Commercial Concentric and Eccentric Butt-Welding Reducers (ASME/ANSI B16.9-1986) (*Continued*)

Nominal pipe size	Length, *H*	Weight (approx), lb (concentric or eccentric)			
		Standard	Extra strong	Schedule 160	Double extra strong
3 × 1¼	3½	1.7	2.2	3.1	4.1
1½	3½	1.8	2.1	3.1	4.0
2	3½	2.0	2.6	3.4	4.0
2½	3½	2.1	2.8	3.7	4.6
3½ × 1¼	4	2.3	3.2	...	5.8
1½	4	2.5	3.1	...	5.8
2	4	2.7	3.5	...	5.7
2½	4	2.9	3.8	...	5.9
3	4	3.0	4.0	...	6.8
4 × 1½	4	2.7	3.8	5.6	6.6
2	4	3.1	3.9	5.6	6.6
2½	4	3.3	4.4	5.5	6.3
3	4	3.5	4.7	6.5	7.7
3½	4	3.6	4.8	...	8.2
5 × 2	5	5.0	6.6	10.6	12.2
2½	5	5.5	7.2	10.2	11.7
3	5	5.7	7.8	10.2	11.1
3½	5	5.8	8.0	...	13.3
4	5	5.9	8.3	12.4	14.2
6 × 2½	5½	7.6	9.9	15.8	18.8
3	5½	8.0	11.1	15.1	18.5
3½	5½	8.1	11.6	...	17.3
4	5½	8.1	12.0	17.2	19.1
5	5½	8.6	12.6	18.8	21.4
8 × 3½	6	12.8	16.1	...	27.9
4	6	13.1	18.6	26.9	25.7
5	6	13.4	19.5	29.6	29.2
6	6	13.9	20.4	32.1	32.7
10 × 4	7	21.1	25.3	50	
5	7	21.8	28.7	48	
6	7	22.3	29.8	50	
8	7	23.2	31.4	58	
12 × 5	8	30.5	39.1	78	
6	8	31.1	40.6	75	
8	8	32.1	37.4	86	
10	8	33.4	43.6	94	
14 × 6	13	55	74		
8	13	57	76		
10	13	60	79		
12	13	63	83		

*Weights in italics are for eccentric reducers in sizes 32, 34, and 36 in. For all other sizes, weights shown are for either concentric or eccentric reducers.

TABLE A2.25 Dimensions of Typical Commercial Butt-Welding Laterals

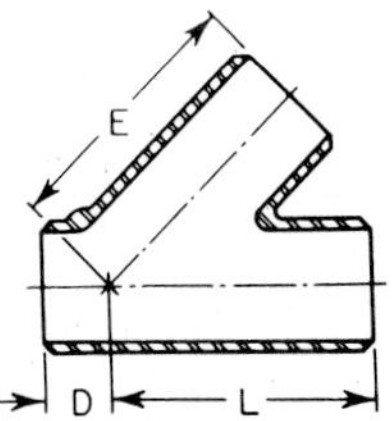

Nominal pipe size	Standard *L* and *E*	Standard *D*	Weight (approx), lb	Extra strong *L* and *E*	Extra strong *D*	Weight (approx), lb
			Straight			
1	5¾	1¾	1.7	6½	2	2.5
1¼	6¼	1¾	2.4	7¼	1¼	3.8
1½	7	2	3.2	8½	2½	5.4
2	8	2½	5.0	9	2½	7.7
2½	9½	2½	9.2	10½	2½	13.5
3	10	3	12.6	11	3	18.8
3½	11½	3	17.2	12½	3	25.6
4	12	3	20.8	13½	3	32.8
5	13½	3½	31.4	15	3½	49.8
6	14½	3½	42.4	17½	4	79
8	17½	4½	76	20½	5	140
10	20½	5	124	24	5½	202
12	24½	5½	180	27½	6	273
14	27	6	218	31	6½	340
16	30	6½	275	34½	7½	433
18	32	7	326	37½	8	526
20	35	8	396	40½	8½	628
24	40½	9	544	47½	10	882

TABLE A2.26 Dimensions of Typical Commercial Butt-Welding Standard Caps (ASME/ANSI B16.9-1986 (Except as Noted))

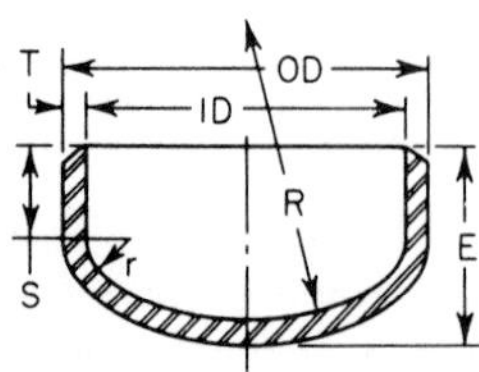

Nominal pipe size	Outside diameter, OD	Inside diameter, ID	Wall thickness, *T*	Length, *E*	Tangent, *S*	Dish radius, *R*	Knuckle radius, *r*	Pipe schedule number[1]	Weight (approx), lb
½	0.840	0.622	0.109	1	0.74	0.54	0.10	40	0.1
¾	1.050	0.824	0.113	1¼	0.93	0.72	0.14	40	0.2
1	1.315	1.049	0.133	1½	1.10	0.92	0.17	40	0.3
1¼	1.660	1.380	0.140	1½	1.02	1.35	0.23	40	0.4
1½	1.900	1.610	0.145	1½	0.95	1.41	0.27	40	0.4
2	2.375	2.067	0.154	1½	0.83	1.81	0.34	40	0.6
2½	2.875	2.469	0.203	1½	0.68	2.15	0.41	40	0.9
3	3.500	3.068	0.216	2	1.02	2.69	0.51	40	1.4
3½	4.000	3.548	0.226	2½	1.39	3.11	0.59	40	2.1
4	4.500	4.026	0.237	2½	1.26	3.52	0.67	40	2.5
5	5.563	5.047	0.258	3	1.48	4.42	0.84	40	4.2
6	6.625	6.065	0.280	3½	1.70	5.31	1.01	40	6.4
8	8.625	7.981	0.322	4	1.68	6.98	1.33	40	11.3
10	10.750	10.020	0.365	5	2.13	8.77	1.67	40	20.0
12	12.750	12.000	0.375	6	2.62	10.50	2.00	▲[2]	29.5
14	14.000	13.250	0.375	6½	2.81	11.60	2.21	30	35.3
16	16.000	15.250	0.375	7	2.81	13.34	2.54	30	44.3
18	18.000	17.250	0.375	8	3.31	15.08	2.88	▲[2]	57
20	20.000	19.250	0.375	9	3.81	16.84	3.21	20	71
22	22.000	21.250	0.375	10	4.31	18.60	3.54	20	86
24	24.000	23.250	0.375	10½	4.31	20.35	3.88	20	102
26	26.000	25.250	0.375	10½	3.81	22.10	4.21	▲[2]	110
28	28.000	27.250	0.375	10½	3.31	23.85	4.54	▲[2]	120
30	30.000	29.250	0.375	10½	2.81	25.60	4.88	▲[2]	125
32	32.000	31.250	0.375	10½	2.31	27.35	5.21	▲[2]	145
34	34.000	33.250	0.375	10½	1.81	29.10	5.54	▲[2]	160
36	36.000	35.250	0.375	10½	1.31	30.85	5.88	▲[2]	175
42	42.000	41.250	0.375	12	1.31	36.10	6.88	▲[2]	230

[1]Pipe schedule numbers in accordance with ANSI/ASME B36.10.
[2]This size and thickness does not correspond with any schedule number.

VALVES

Valves are extensively used in piping systems to interrupt, divert, or regulate the flow of fluids. Based on the valve's function, a change in the valve's state may be manually initiated or may be automatically initiated by a signal from a control device, or the valve may automatically respond to changing system conditions. The various valve types and their applications are described in the following sections.

Valves are manufactured in standard pressure and temperature ratings in accordance with ANSI/ASME B16.1, B16.34, and B16.24 for cast-iron, steel, and bronze materials, respectively. The pressure-temperature rating of ANSI/ASME B16.34 parallels that of ANSI/ASME B16.5 except that B16.34 provides for an increase in pressure rating for welding-end valves that receive additional prescribed nondestructive examinations.

Valve Categories

Stop (Isolation) Valves. As the name implies, stop valves are used to stop flow or isolate a portion of the system until it is desirable to achieve flow downstream of the valve. The basic design requirement of stop valves is to offer minimum resistance to flow in the fully open position and to exhibit tight shut-off characteristics when fully closed. Gate, globe, ball, butterfly, plug, and diaphragm valves satisfy the above requirements in varying degrees and, therefore, are widely used in shut-off service—the actual type of valve selected is dictated by several parameters, including:

- Pressure drop
- Seat leakage
- Fluid properties
- System leakage
- Actuation requirements
- Initial cost
- Maintenance

Regulating Valves. Regulating valves are used extensively in piping systems to regulate the flow of fluid. Whether the desired effect is to control flow, pressure, or temperature, the task is accomplished by increasing or decreasing the flow through the valve in response to a signal from a pressure, flow, or temperature controller.

The primary requirement of a flow control valve is to predictably regulate the flow with respect to its open position and impart the required pressure drop without sustaining damage. Specially designed globe, needle, butterfly, ball, plug, and diaphragm valves are capable of satisfying these requirements in varying degrees. The manufacturer's literature should be consulted for the limitations placed on a particular valve.

Back-Flow Prevention. Check valves are generally used for the prevention of back-flow. The valves are self-actuating—the valve's disk is kept open by the forward flow of fluid and quickly closed by reverse flow. In certain applications, pneumatic actuators may be used to assist in the rapid closure of the valves on reversal of flow.

Pressure-Relief Devices. Pressure-relief devices are used to protect piping and equipment from being subjected to pressures that exceed their design pressures. Generally, the seating of relief valves is accomplished by a compressed spring which exerts a force on the valve disc, pressing it against the valve seat. When the force exerted by the fluid on the valve disc exceeds the spring force, the valve automatically opens to release the excess pressure. Other designs incorporate a pilot valve which uses system pressure to control the movement of the disk. Another type of pressure-relieving device, although not a valve, is a rupture disk (Fig. A2.4). The rupture disk is designed to burst open at a predetermined pressure. A rupture disk cannot be reseated and, therefore, must be replaced once it has performed its relieving function. Rupture disks have the advantage of being leak-tight up to the rupture pressure and of being capable of relieving large rates of flow. The set pressure of rupture disks cannot be adjusted.

Valve Designs

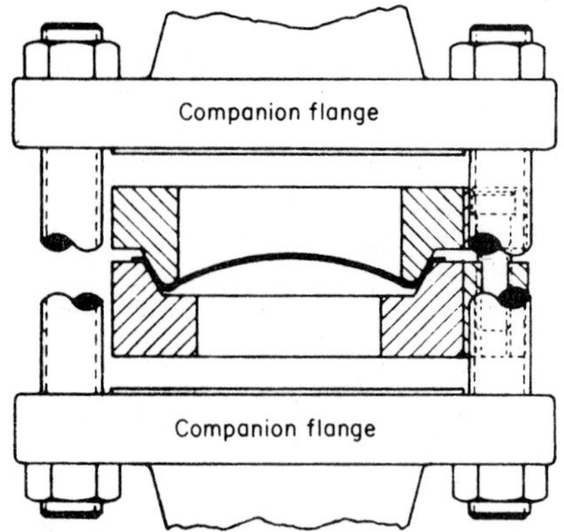

FIGURE A2.4 The rupture disk, a type of relief device.

Gate Valves. Gate valves are primarily designed to serve as isolation valves. In service, these valves generally are either fully open or fully closed. When fully open, the fluid or gas flows through the valve in a straight line with very little resistance. As a result, the pressure loss through the valve is minimal.

Gate valves should not be used in the regulation or throttling of flow because accurate control is not possible. Furthermore, high-flow velocity in partially opened valves may cause erosion of the disks and seating surfaces. Vibration may also result in chattering of the partially opened valve disk.

An exception to the above are specially designed gate valves that are used for low-velocity throttling. An example involves guillotine gate valves for pulp stock.

Gate valves consist of three major components: body, bonnet, and trim. The body is generally connected to the piping by means of flanged, screwed, or welded connections. The bonnet, containing the moving parts, is joined to the body, generally with bolts, to permit cleaning and maintenance. The valve trim consists of the stem, the gate, the wedge, or disk, and the seat rings. Two basic types of gate valves are the manufactured-wedge type and the double disk type, and there are several variations within each of these types.

Wedge Type. In the wedge or disk-wedge types (Figs. A2.5 to A2.7) either a tapered solid or tapered split wedge is used. When the valve is closed, the gate disk is wedged on both sides against the seat. In split-wedge gate valves (Fig. A2.7), the two-piece wedge disk is seated between matching tapered seats in the body. This type is preferred where the body seats might be distorted due to pipeline strain.

In the rising stem valves (Fig. A2.5), the operating threads are out of direct contact with the fluid or gas. The nonrising stem type (Fig. A2.6) is preferred where space is limited and where the fluid passing through the valve will not corrode or erode the threads or leave deposits on the threads. Also, the nonrising stem valve is preferred for buried service.

In the rising-stem type of valve, the upper part of the stem is threaded and a nut is fastened solidly to the handwheel and held in the yoke by thrust collars. As the handwheel is turned, the stem moves up or down. In the nonrising stem valve, the lower end of the stem is threaded and screws into the disk, vertical motion of the stem being restrained by a thrust collar. The rising-stem valve requires a greater amount of space when opened. However, it is generally preferred because the position of the stem indicates at once whether the valve is open or closed. Nonrising stem valves are sometimes provided with an indicator for this purpose.

Double-Disk Type. In the double-disk parallel-seat valves (Figs. A2.8 and A2.9), the disks are forced against the valve seats by a wedging mechanism as the stem is tightened.

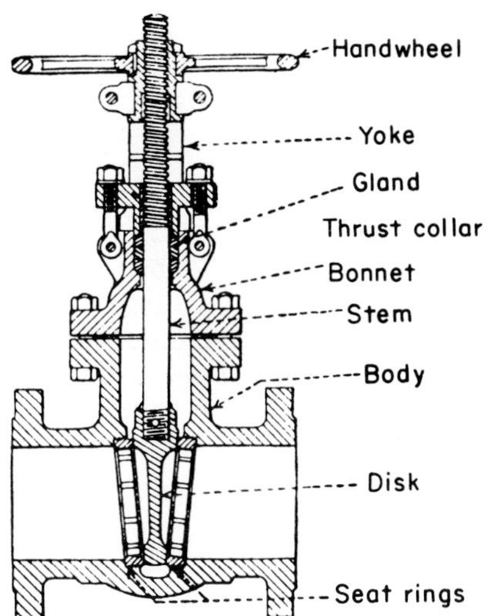

FIGURE A2.5 Rising-stem solid-wedge gate valve for 250-psig steam service.

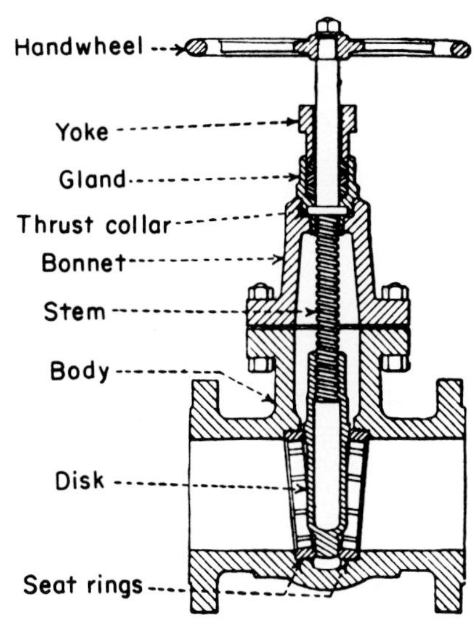

FIGURE A2.6 Non-rising-stem solid-wedge gate valve for 250-psig steam service.

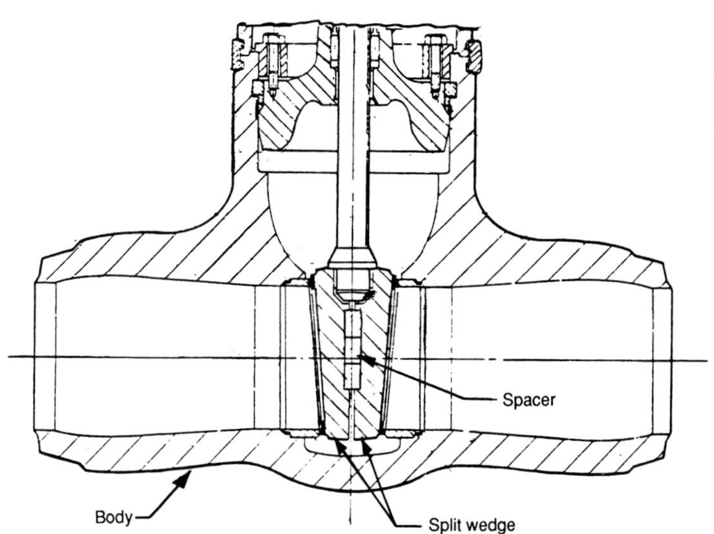

FIGURE A2.7 Split wedge gate valve.

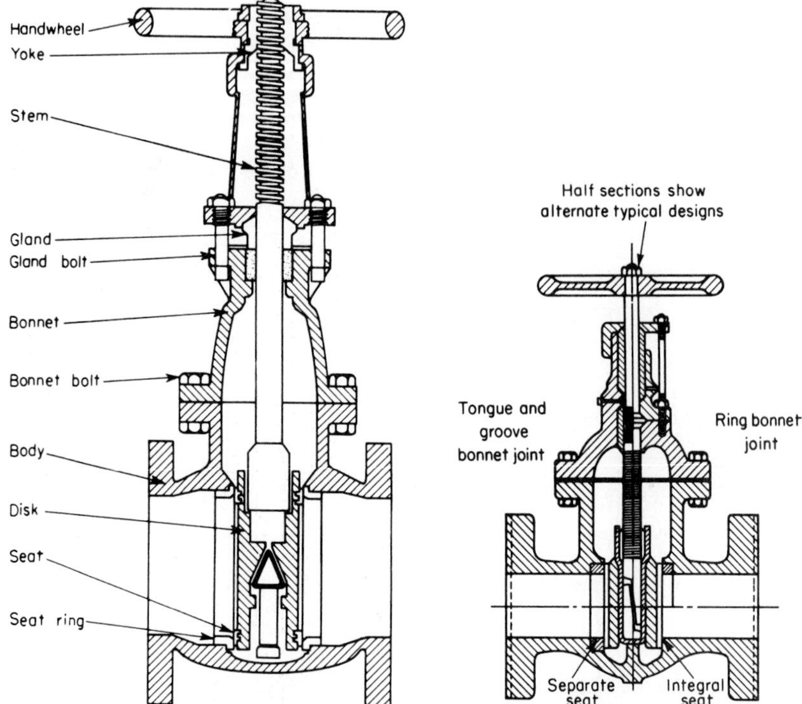

FIGURE A2.8 Double-disk rising-stem flanged-end gate valve for 150-psig service.

FIGURE A2.9 Double-disk non-rising-stem gate valve.

Some double-disk parallel-seat valves employ a design (Fig. A2.10) which depends for its tightness mainly upon the fluid pressure exerted against one or the other side of the disk. The major advantage of this type is that the disk cannot be jammed into the body, which otherwise might make it difficult to open the valve. This is particularly important where motors are used for opening and closing the valve.

Unlike the wedge in a wedge gate valve which only comes into contact with the seat rings when the valve is nearly closed, each disk in the parallel-seat valve slides against its seat while the valve is being opened or closed. Consequently, these components must be made of metals which do not gall or tear when in sliding contact with each other. The double-disk parallel-seat gate valve is often favored for high-temperature steam service because it is less likely to stick in the closed position as a result of change in temperature.

Plug Valves. Plug valves, also called *cocks,* generally are used for the same full-flow service as gate valves, where quick shutoff is required. They are used for steam, water, oil, gas, and chemical liquid service.

Plug valves are not generally designed for the regulation of flow. Nevertheless, in some applications, specially designed plugs are used for this purpose, particularly for gas-flow throttling.

The basic design of plug valves is illustrated in Fig. A2.11. Full flow is ob-

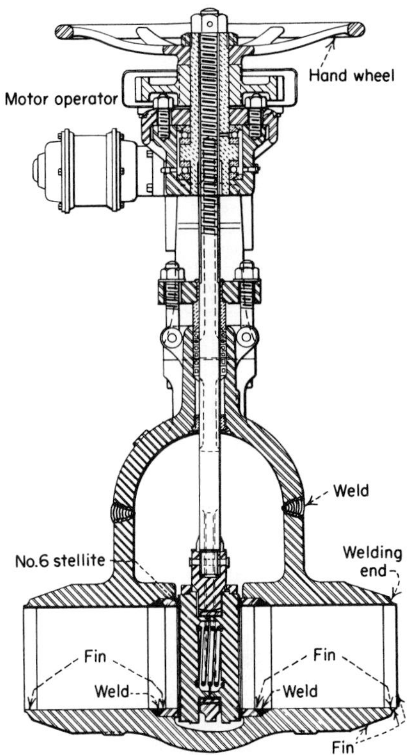

FIGURE A2.10 Parallel seat gate valve showing welded construction for high-temperature service with welded-in seat ring.

tained when the opening in the tapered plug is aligned in the direction of flow. When the plug is rotated a quarter turn, flow is terminated.

The body and tapered plug represent the essential features in plug valves. Careful design of the internal contours of the valve produces maximum flow efficiency. The port in the tapered plug is generally rectangular. However, valves are also available with round ports. Major valve patterns or types are identified as regular, venturi, short, round-port, and multiport.

The regular pattern employs the tapered form of port openings, the area of which is from 70 to 100 percent of the internal pipe area. In some cases, the face-to-face lengths are greater than those of standard gate valves. The venturi pattern provides streamlined flow and thus permits reduction in the port size. The port opening area is approximately 35 percent of the internal pipe area. The round-port full-bore pattern has a circular port through the plug and body equal to or greater than the inside diameter of the pipe or fitting. Operating efficiency is equal to or greater than that of gate valves of the same size.

Use of multiport valves (Fig. A2.12) is advantageous in many installations because it provides simplification of piping and convenience in operation. One three-way or four-way multiport valve may be used in place of two, three, or four straightway valves.

Major types of plug valves involve lubricated and nonlubricated designs. Lubricant-seal plug valves are less subject to seizing or wear and may exhibit somewhat greater resistance to corrosion in some service environments. Nonlubricated plug valves are used where maintenance must be kept to a minimum. Both types of valves provide a bubble-tight closure and are of compact size.

Plug valves generally can be readily repaired or cleaned without necessitating removal of the body from the piping system. They are available for pressure service from vacuum to 10,000 psi and temperatures from −50 to 1500°F. Also, plug valves are available with a wide variety of linings suitable for many chemical service applications.

Lubricant-Seal Valves. In lubricant-seal valves, channels for the admission of the lubricant surround the ports to ensure positive sealing against internal or external leakage. The lubricant pressure developed by a turn of the lubricant screw or injection of lubricant with a pressure gun exerts a powerful hydraulic jacking action on the plug, momentarily lifting it from the seat and making it easy to turn.

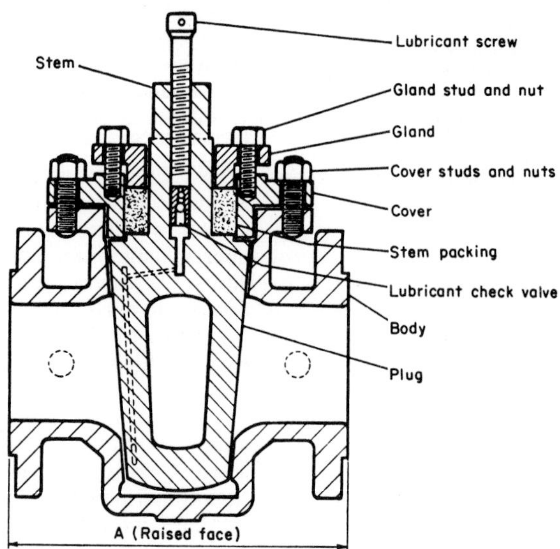

FIGURE A2.11 Plug valve with lubricant system, as specified in API Standard 600.

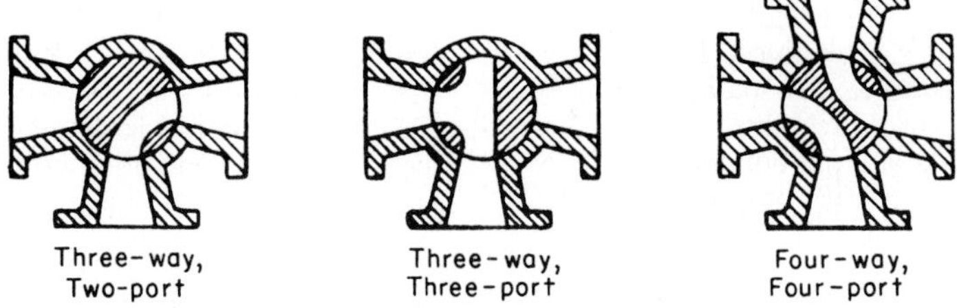

FIGURE A2.12 Multiport valves. (*a*) Three-way, Two-port; (*b*) three-way, Three-port; (*c*) Four-way, Four-port.

Since the lubricant pressure is greater than the line pressure, it is virtually impossible for solids to lodge between the valve body and plug.

The functions of pressure lubrication in plug valves are (1) hydraulic action, keeping the plug in free working condition, (2) maintenance of positive seal against internal and external leakage, (3) free turning even of the large sizes and against heavy differential pressure, and (4) protection of working surfaces from wear and corrosion. This principle of pressure lubrication makes it possible to take full advantage of the inherent simplicity, compactness, and positive rotary action of the tapered plug valve.

The stem or shank used to rotate the plug is sealed by screwed or bolted packing glands.

Lubricants. The word *lubricant* does not precisely define the part this material plays in the efficient functioning of lubricated plug valves. More properly such valves might be called *plastic sealed valves* and the lubricant could better be designated *plastic sealant.* The use of an effective lubricant is most important since, in operation, the valve structure and plastic sealing film are an integral unit, and each component is dependent on the other for ultimate performance.

The lubricant in effect becomes a structural part of the valve, since it provides a flexible and renewable seat. This eliminates the necessity of "force fits" and metal-to-metal "distortable-seat" contacts to effect a seal. For this purpose, the lubricant must exhibit proper elasticity as well as resistance to solvents and chemicals to avoid the destructive action of the line fluid and to form an impervious seal around each body port even under pressure. The film of lubricant also protects the metal surfaces between the plug and body from corrosion. The seal formed by the lubricant transmitted in a system of lubricant grooves circuiting each port aids in maintaining the essential film on the metal closure surfaces.

The lubricant film also eases valve operation. By preventing metal-to-metal contact on internal valve parts, galling and seizing are minimized. The lubricant must provide a high degree of lubrication to the bearing surfaces of the valve over a wide temperature range. The valve manufacturer's literature should be consulted for the proper selection of lubricant.

Diaphragm Valves. Diaphragm valves offer advantages in certain low-pressure applications not possible with other types of valves. Their fluid passages are smooth and streamlined, minimizing pressure drop. They are suitable for moderate throttling applications, and they exhibit excellent leak-tight characteristics, even when conveying liquids containing suspended solids. The fluid stream is isolated from the working parts of the valve, preventing contamination of the fluid and corrosion of the operating mechanism. Since there is no leak path around the valve stem, the valve is virtually leak-tight. This feature makes the valve indispensable where leakage into or out of the system cannot be tolerated.

Diaphragm valves (Fig. A2.13) consist of a rigid body formed with a weir placed in the flow-path, a flexible diaphragm which forms the upper pressure boundary of the valve, a compressor which is used to force the diaphragm against the weir, and the bonnet and handwheel which secure the diaphragm to the body and actuate the compressor.

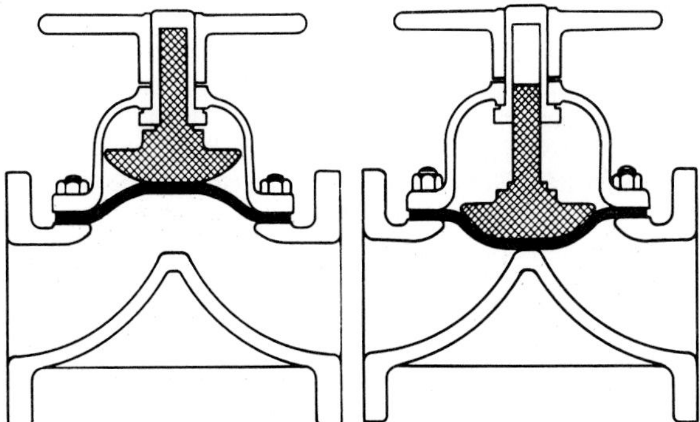

FIGURE A2.13 Sketch of weir-type diaphragm valve in open and closed positions.

The maximum pressure that these valves can be subjected to is a function of the diaphragm material and the service temperature. Also, the rated design life of the valve is influenced by the service conditions. Furthermore, the system hy-

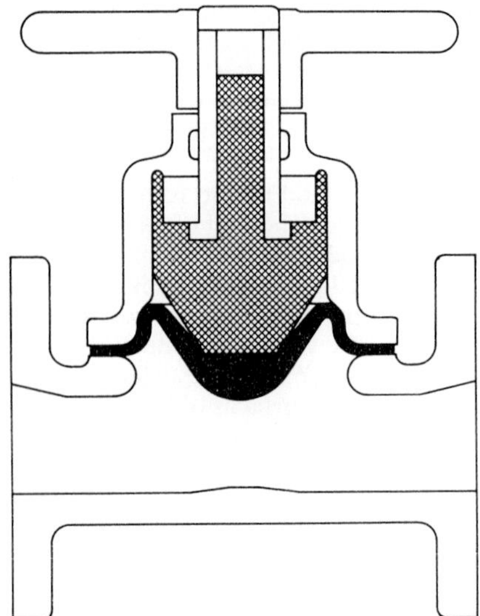

FIGURE A2.14 Straightway-type diaphragm valve.

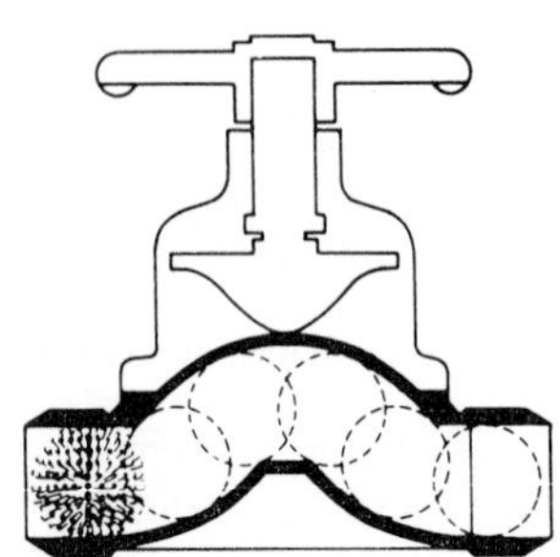

FIGURE A2.15 Full-bore-type diaphragm valve illustrating passage of ball-brush cleaner through valve.

drostatic test pressure must not exceed the maximum pressure rating of the diaphragm.

Variations of the weir diaphragm valve are the straightway (Fig. A2.14) and the full-bore types (Fig. A2.15). When the straightway valve is open, its diaphragm lifts high for full streamline flow in either direction. When the valve is closed, the diaphragm seals tight for positive closure even with gritty or fibrous materials in the line.

The full-bore type of valve is most extensively used in the beverage industry. It permits ball-brush cleaning with either steam or caustic soda, without opening or removing the valve from the line.

Diaphragm valves are available in a wide choice of body, diaphragm, and lining materials that are suitable for service with a wide variety of chemicals. For severe corrosive applications, diaphragm valves are made of stainless steel or PVC plastics, or they are lined with glass, rubber, lead, plastics, titanium, or still other materials. Some of the common materials used for diaphragms are listed in Table A2.27.

Ball Valves. The ball valve (Fig A2.16) is a quarter-turn valve suitable for gas, compressed air, liquid, and slurry service. The use of soft-seat materials such as nylon, delrin, synthetic rubbers, and fluorinated polymers imparts excellent sealing ability. With fluorinated polymer seats, ball valves can be used for service temperatures ranging from −450 to 500°F; with graphite seats, service temperatures as high as 1000°F are possible. Also, with metal-backing seats, the valves can be used in fire-safe services. Ball valves are similar to plug valves in operation. They are nonbinding and provide leak-tight closure. The valves exhibit negligible resistance to flow because of their smooth body and port.

Major components of the ball valve are the body, spherical plug, and seats. Ball valves are made in three general patterns—venturi port, full port, and re-

TABLE A2.27 Typical Materials Used for Diaphragms

Valve type	Service	Material	Temp, F	
			Min	Max
Conventional weir	Abrasive	Soft natural rubber	−30	180
	Water	Natural rubber	−30	180
	Food and beverage	White natural rubber	0	160
	Weak chemical, air, oil	Neoprene	−30	200
	Weak chemical, high vacuum	Reinforced Neoprene	−30	200
	Other chemicals, gases	Black chlorinated butyl	−20	250
	Food and beverage	White chlorinated butyl	−10	225
	Special for hydrogen peroxide	Clear Tygon	0	150
	Oils and gasoline	Hycar (gen. purpose)	10	180
	Oxidizing services	Hypalon	0	225
	Brewery services	Pure gum rubber	−30	160
	Special service on temperature	Silicone	50	350
	Radioactive conditions	G.R.S.	−10	225
	Severe chemicals, solvents	Teflon[1]	−30	325
	Severe chemicals	Kel-F[1]	60	250
	Specific acids	Polyethylene	10	135
Full flow	Cold beer	White rubber	−30	160
	Hot wort and cold beer	White chlorinated butyl	−10	225
	Cold beer	Pure gum rubber	−30	160
Straightway	Water	Natural rubber	−30	180
	Chemical, air, oil	Neoprene	0	180
	Oils and gasoline	Hycar (gen. purpose)	10	180
	Fatty acids	Black chlorinated butyl	0	225
	Oxidizing services	Hypalon	0	200
	Food and beverage	White chlorinated butyl	−10	200

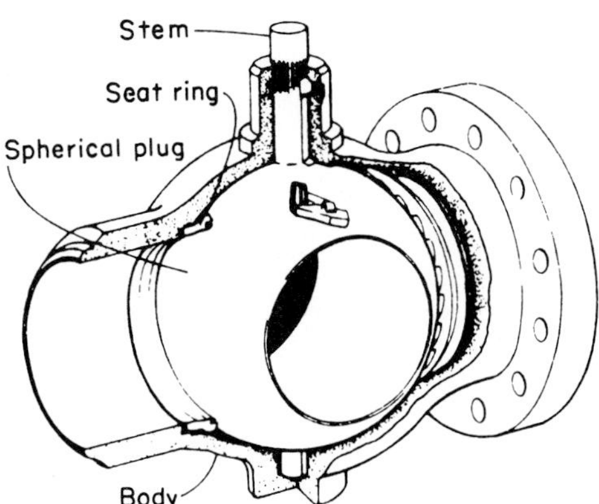

FIGURE A2.16 Ball valve in closed position.

duced port. The full-port valve has an inside diameter equal to the inside diameter of the pipe. In the venturi and reduced port styles, the port is generally one pipe size smaller than the line size. Stem sealing is accomplished by bolted packing glands and O-ring seals. Valves are also available with a lubricant-seal system

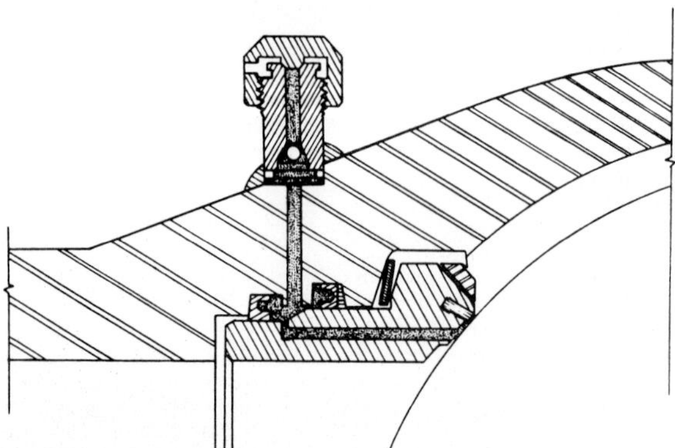

FIGURE A2.17 Lubricant-seal system in a ball valve.

that is similar to that available for plug valves. A typical lubrication system is illustrated in Fig. A2.17.

Globe Valves. Conventional globe valves may be used for isolation service. Although these valves exhibit slightly higher pressure drops than straight-through valves (e.g., gate, plug, ball, etc.), they may be used where the pressure drop through the valve is not a controlling factor. Also, wye-pattern (Fig. A2.18) and angle-pattern (Fig. A2-19) globe valves exhibit improved flow characteristics over the tee-pattern (Fig. A2.20) globe valve. Because the entire system pressure exerted on the disk is transferred to the valve stem, the practical size limit for

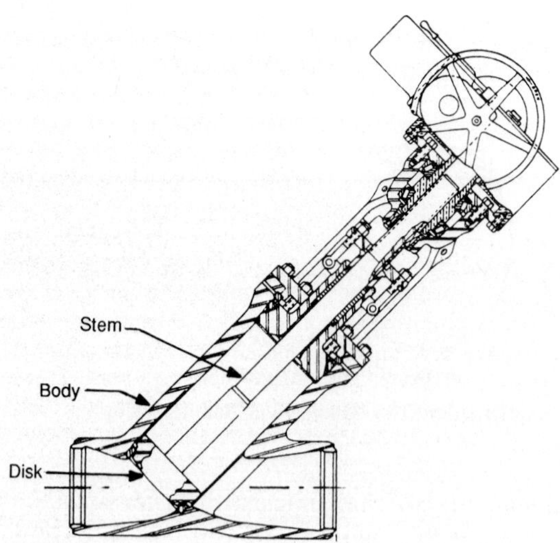

FIGURE A2.18 Large wye-pattern globe valve (with gear actuator).

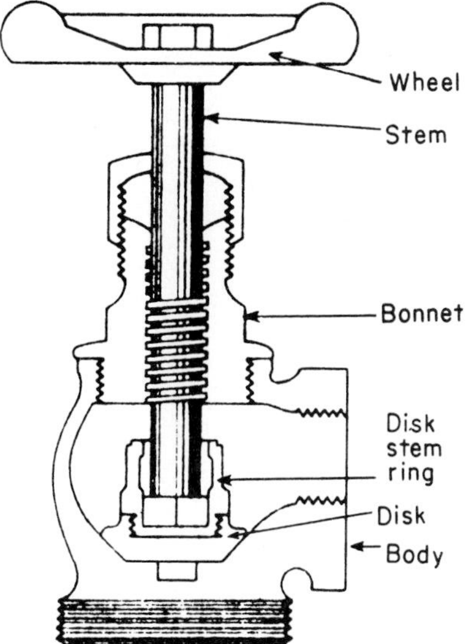

FIGURE A2.19 Angle globe valve with screwed ends.

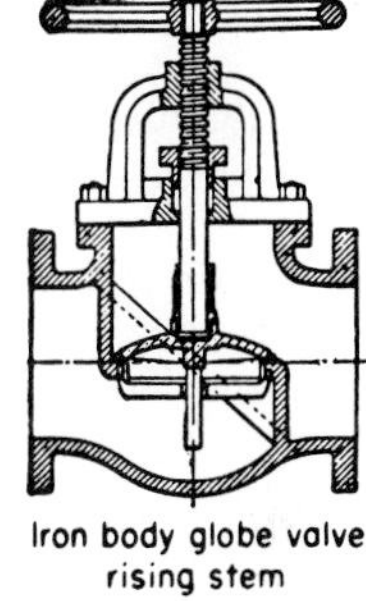

FIGURE A2.20 A typical large globe valve with flanged ends.

these valves is NPS 12. Larger valves would require that enormous forces be exerted on the stem to open or close the valve under pressure.

Globe valves are extensively employed to control flow. The range of flow control, pressure drop, and duty must be considered in the design of the valve to avert premature failure and to assure satisfactory service. Valves subjected to high differential pressure throttling service require a specially designed valve trim.

A typical large globe valve with flanged ends is illustrated in Fig. A2.20 and a large wye-pattern globe is illustrated in Fig. A2-18. Globe valves usually have rising stems, and the larger sizes are of the outside screw-and-yoke construction. Components of the globe valve are similar to those of the gate valve. This type of valve has seats in a plane parallel or inclined to the line of flow.

Maintenance of globe valves is relatively easy since the disks and seats are readily refurbished or replaced. This makes globe valves particularly suitable for services which require frequent valve maintenance. Where valves are operated manually, the shorter disk travel offers advantages in saving operator time, especially if the valves are adjusted frequently.

The principal variation in globe-valve design is in the types of disk employed. Plug-type disks have a long, tapered configuration with a wide bearing surface. This type of seat provides maximum resistance to the erosive action of the fluid stream. In the composition disk, the disk has a flat face that is pressed against the seat opening like a cap. This type of seat arrangement is not as suitable for high differential pressure throttling.

The conventional disk, in contrast to the plug type, provides a thin contact between the taper of the conventional seat and the face of the disk. This narrow contact area tends to break down hard deposits that may form on the seats and thus facilitates pressure-tight closure. This arrangement allows for good seating and moderate throttling.

In cast-iron globe valves, disk and seat rings are regularly furnished of bronze. In steel-globe valves for temperature up to 750°F, the trim is generally furnished of stainless steel and so provides resistance to seizing and galling. The mating faces are normally heat-treated to obtain differential hardness values. Other trim materials, including cobalt-based alloys, are also used.

The seating surface is ground to ensure full bearing surface contact when the valve is closed. For lower pressure classes, alignment is maintained by a long disk lock-nut. For higher pressures, disk guides are cast into the valve body. The disk turns freely on the stem to prevent galling of the disk face and seat ring. The stem bears against a hardened thrust plate, eliminating galling of the stem and disk at the point of contact.

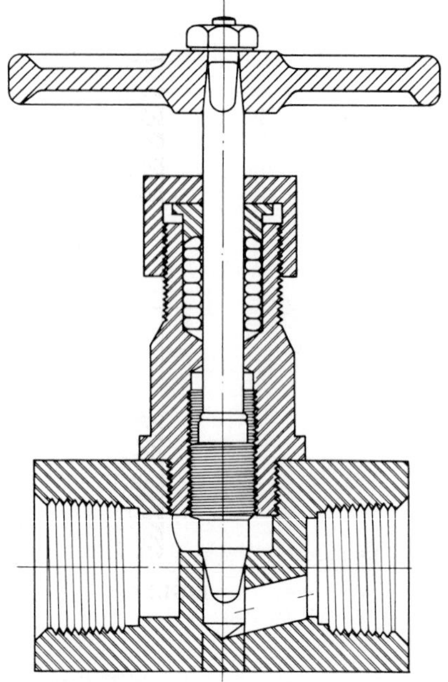

FIGURE A2.21 Needle valve for accurate throttling of flow.

Needle Valves. Needle valves generally are used for instrument, gauge, and meter line service. Very accurate throttling is possible with needle valves and, therefore, they are extensively used in applications that involve high pressures and/or high temperatures. In needle valves (Fig A2.21), the end of the stem is needle-pointed. The needle fits accurately into its seat and thus assures tight closure with the least effort.

Butterfly Valves. Butterfly valves are low-pressure valves of efficient design which are used to control and regulate flow. They are characterized by fast operation and low pressure drop. They require only a quarter-turn from closed to full-open position. A typical flanged butterfly valve is illustrated in Fig. A2.22. Butterfly valves are produced in flanged, wafer, and welding-end styles (Figs. A2.22, A2.23, and A2.24, respectively). The welding-end style is a specially engineered valve for a specific application. Butterfly valves are available with metal-to-metal seats, soft seats, and with fully lined body and disk. The soft seats permit bubble-tight shutoff and the full lining enhances erosion and corrosion resistance. Available seat materials include:

- Buna N
- Neoprene
- Fluorel
- Hypalon
- EPDM

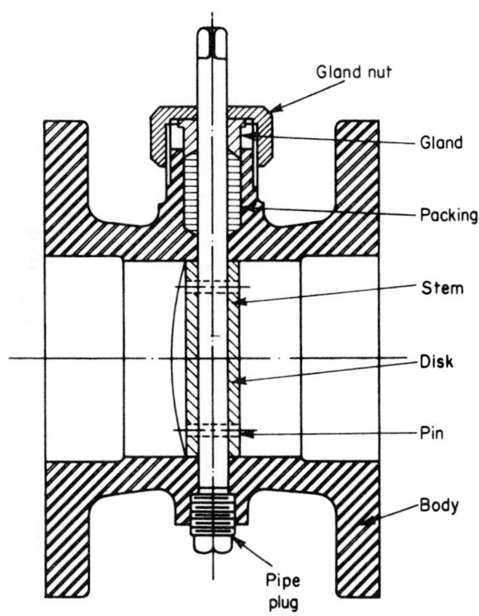

FIGURE A2.22 Typical flanged-end butterfly valve.

FIGURE A2.23 Wafer butterfly valve. (*Courtesy, Keystone International, Inc.*)

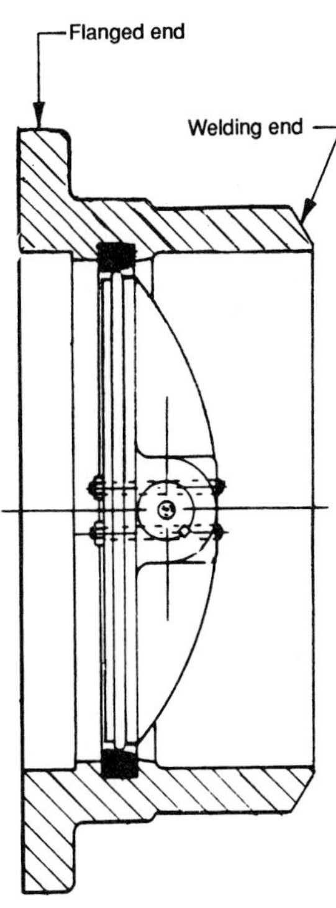

FIGURE A2.24 Butterfly valve; welding end on one end.

Check Valves. Check valves are designed to prevent backflow in lines. The five principal types of check valves used are the tee-pattern lift check, the swing check, the tilting disk check, the wye-pattern lift check, and the ball check, illustrated in Figs. A2.25 to A2.29, respectively. The swing check is the more commonly used.

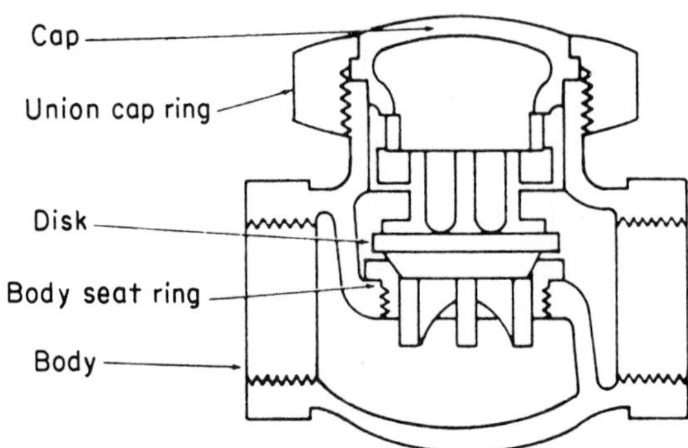

FIGURE A2.25 Lift check valve.

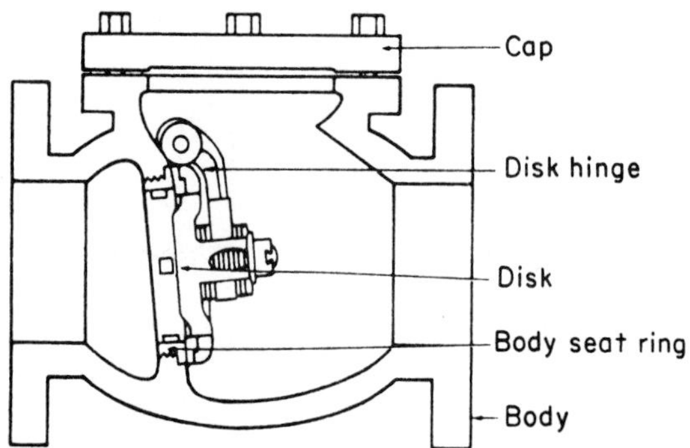

FIGURE A2.26 Swing check valve.

The force of gravity plays an important role in the functioning of a check valve and, therefore, the position of the valve must always be given consideration. Lift and ball-check valves must always be placed so that the direction of lift is vertical. Swing checks must be located to ensure that the disk will always be closed freely and positively by gravity.

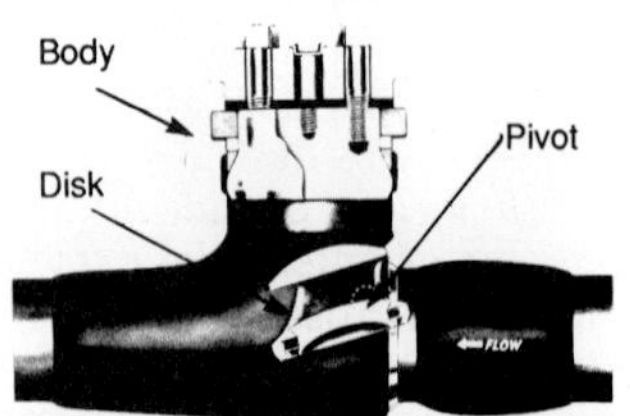

FIGURE A2.27 Tilting disk check valve. (*Courtesy, BTR Inc./Edward Valve.*)

The flow velocity of the fluid through the valve has a significant affect on the life of the check valve. The valve should be sized such that the fluid velocity under normal design conditions is sufficient to keep the disk

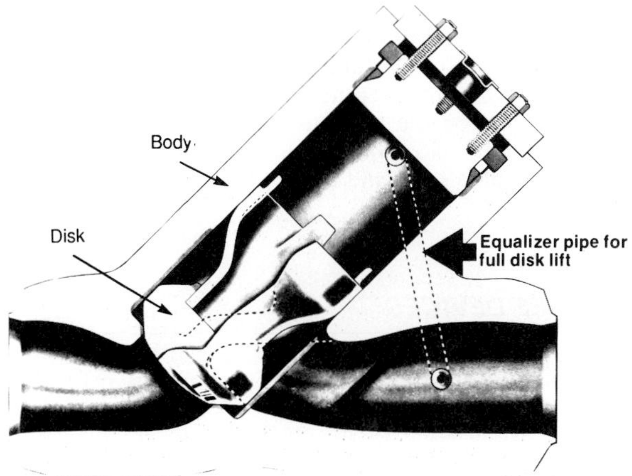

FIGURE A2.28 Wye-pattern lift check valve. (*Courtesy, BTR Inc./Edward Valve*)

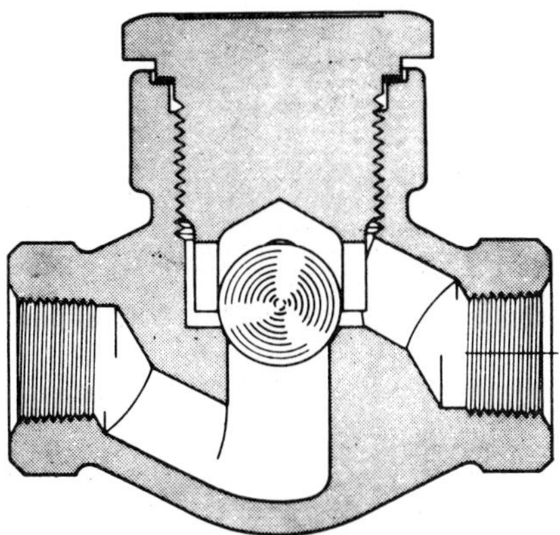

FIGURE A2.29 Ball check valve.

fully open and pressed against the stop. This minimizes disk fluttering, which is the primary cause of valve failure.

Lift check valves are particularly adapted for high-pressure service where velocity of flow is high. In lift check valves, the piston disk is accurately guided by long contact and a close sliding fit with the perfectly centered dash pot. The walls of the piston and dash pot are of approximately equal thickness. Large steam jackets are located outside of the dash pot and inside the piston to eliminate sticking because of differential expansion. The seat ring is of a barrel-type design of heavy uniform cross section. It is normally screwed in and seal-welded. The flow opening is full port size.

In swing check valves, the disks on smaller sizes are of the same material as the trim. On larger valve sizes, the disks are hard faced. The disk is swung from lugs cast integrally with the body. Swing arm-and-disk assembly may, if necessary, be repaired or renewed by removing the bolted valve cover.

Tilting-Disk Check Valves. The tilting-disk check valve is designed to overcome some of the weaknesses inherent in conventional check valves. A combination of design features enable the valve to open fully and remain steady at lower flow velocities and to close quickly upon cessation of forward flow. These attributes prolong the valve's life and reduce flow-induced dynamic loads on the piping system.

PRESSURE-RELIEF DEVICES

Safety Valves and Pressure-Relief Devices

Safety valves and pressure-relief valves are automatic pressure-relieving devices used for overpressure protection of piping and equipment. Safety valves (Fig. A2.30) are generally used in gas or vapor service because their opening and reseating characteristics are commensurate with the properties and potential hazards of compressible fluids. The valves protect the system by releasing excess pressure. Under normal pressure, the valve disk is held against the valve seat by a preloaded spring. As the system pressure increases, the force exerted by the

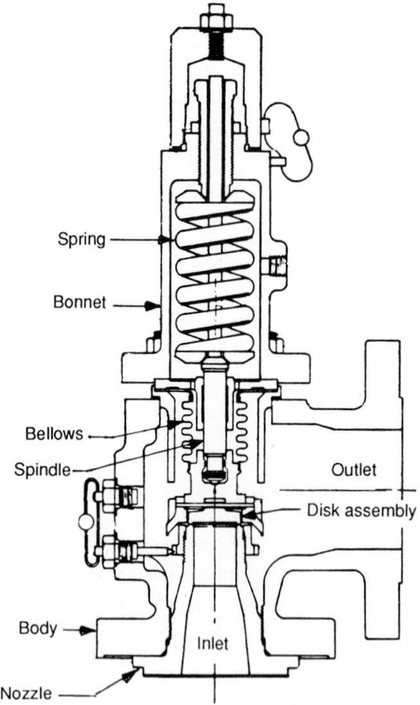

FIGURE A2.30 Safety valve.

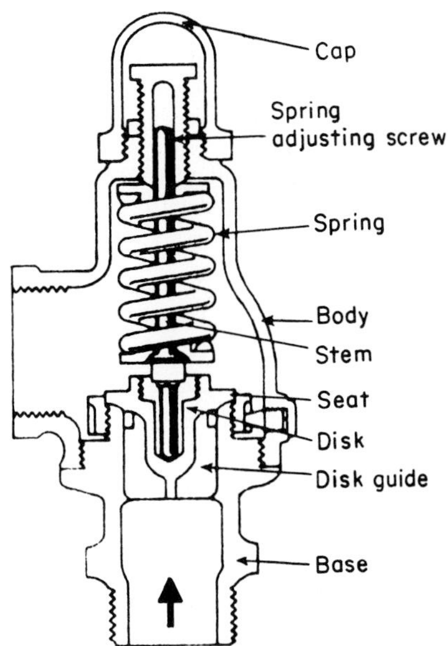

FIGURE A2.31 Relief valve opens when line pressure exceeds preset loading on the spring.

fluid on the disk approaches the spring force. As the forces equalize, fluid begins to flow past the seat. The valve disk is designed in such a way that the escaping fluid exerts a lifting pressure over an increased disk surface area, thereby overcoming the spring force and enabling the valve to rapidly attain near-full lift.

An added benefit to the safety valve disk design is that the pressure at which the valve reseats is below the initial set pressure, thereby reducing the system pressure to a safe level prior to reseating. The ratio of the difference between the set pressure and the reseating pressure to the set pressure is referred to as the *blowdown*.

Pressure-relief valves (Fig. A2.31) are used primarily in liquid service. These valves function in a similar way to safety-relief valves except that since liquids do not expand, there is no additional lifting force on the disk and, therefore, the valve lift is proportional to the system pressure. Also, the valves reseat when the pressure is reduced below the set pressure.

Rupture Disks. A special type of pressure-relief device involves rupture disks (Fig. A2.4). The design generally involves flanges with special machined hold-down seats to prevent slippage of the disks. The disks are designed to rupture automatically at a predetermined pressure. These devices have a particular advantage when large volumes of gas or liquid must be relieved quickly.

Rupture-disk devices may also be used with spring-loaded safety valves. By employing rupture disks for relief at a pressure set approximately 5 to 10 percent above the set pressure of the safety valve, the rupture disk will fail if the regular relief valve does not operate properly. Also, where system leakage cannot be tolerated, rupture disks may be placed between the valve and the component being protected. When the rupture disk's design rating is exceeded, it will burst, and the relief valve will open when its set pressure is exceeded.

Gate or plug valves may be installed ahead of rupture disks. With the rupture disk in place, these valves must be left open to assure that the system is protected. Closing of these valves may be necessary to terminate flow when the disk is replaced after rupture or for maintenance.

Aluminum, copper, and stainless steels are most widely used as disk materials. Metal disks coated with neoprene or other plastics are also used. In highly corrosive service, the precious metals silver, gold, and platinum are also employed.

TRAPS

Steam Traps

The function of a steam trap is to discharge condensate from steam piping or steam heating equipment without permitting live steam to escape. Some principal types of steam traps are:

- Float
- Thermostatic
- Thermodynamic
- Inverted bucket

The float type (Fig. A2.32) consists of a chamber containing a float and arm mechanism which modulates the position of a discharge valve. As the level of condensate in the trap rises, the valve is opened to emit the condensate. This type of valve tends to discharge a steady stream of liquid since the valve position is proportional to the rate of incoming condensate. Because the discharge valve is below the water line, float-type steam traps must employ a venting system to discharge noncondensable gases. This is generally accomplished with a thermostatic element which opens a valve when cooler noncondensable gases are present but closes the valve in the presence of the hotter steam. The thermostatic steam trap (Fig. A2.33) contains a thermostatic element which opens and closes a valve in response to fluid temperature. Condensate collected upstream of the valve is subcooled, cooling the thermostat, which, in turn, exposes the discharge port. When the cooler condensate is discharged and the incoming condensate temperature approaches saturation temperature, the thermostat closes the discharge port. Because of its principal of operations, the thermostatic trap operates intermittently under all but maximum condensate loads. The inverted bucket steam trap (Fig. A2.34) consists of a chamber containing an inverted bucket (the opening at the bottom) which actuates a discharge valve through a linkage. The valve is open when the bucket rests at the bottom of the trap. This allows air to escape during warm-up until the bottom of the bucket is sealed by rising condensate. The valve remains open as long as condensate is flowing, and trapped air bleeds out through a small vent in the top of the bucket. When steam enters the trap, it fills the bucket, causing the bucket to float, so it rises and closes the valve. The steam slowly escapes through the bucket vent and condenses, thus allowing the bucket to sink and reopen the valve for condensate flow. Small amounts of air and noncondensable gases (such as carbon dioxide) that enter the trap during normal operation are also vented through the small opening in the top of the bucket, which prevents the trap from becoming air-bound.

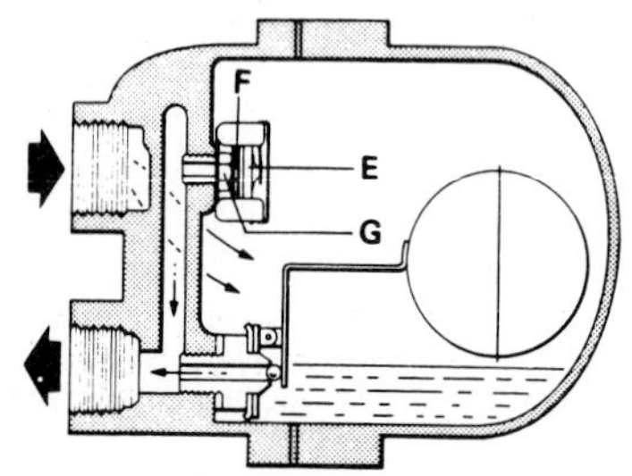

FIGURE A2.32 Float steam trap. (*Spirax Sarco Inc.*)

Another type of trap, the thermostatic impulse trap, is illustrated in Fig. A2.35. In this type, flashing of the hot condensate tends to force a small piston

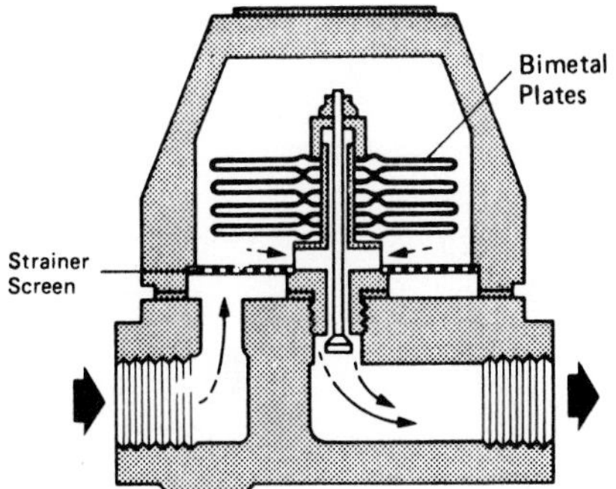

Trap Strainer Unit

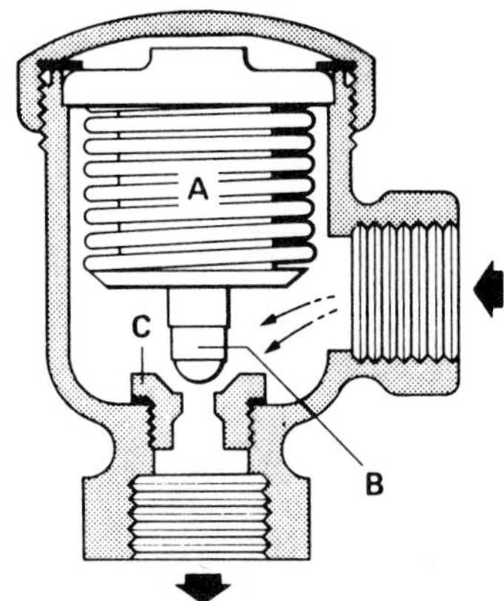

Balanced Pressure Thermostatic Trap

FIGURE A2.33 Thermostatic steam trap. (*a*) Trap strainer unit; (*b*) Balanced pressure thermostatic trap. (*Spirax Sarco Inc.*)

into the discharge opening when the temperature of the condensate approaches within about 30°F of the saturation temperature. As soon as the condensate collected in the drain system cools sufficiently below the flash temperature, the trap opens and discharges the accumulated water until the temperature of the condensate once more approaches the saturation temperature and flashes, thereby closing the trap and again repeating the cycle. A small orifice permits a continuous discharge of steam, flashed vapor, or noncondensable gas when the trap is closed.

Single orifices are sometimes used to remove condensate from high-pressure, high-temperature steam lines. Where the drains are required only in bringing the line up to temperature, the use of orifices, in conjunction with valves, is particularly desirable.

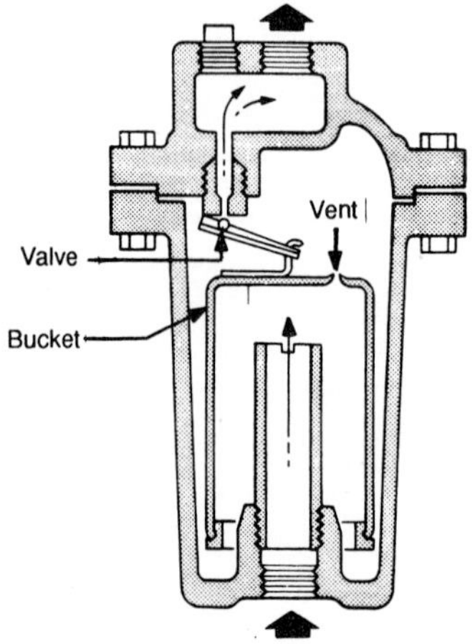

FIGURE A2.34 Inverted bucket steam trap. (*Spirax Sarco Inc.*)

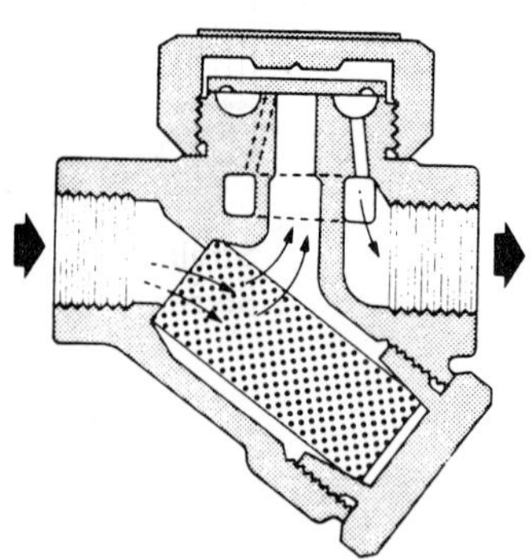

FIGURE A2.35 Thermodynamic steam trap (with integral strainer). (*Spirax Sarco Inc.*)

Air (Drain) Traps

Air (drain) traps are used to discharge condensed liquid from a gas system. The drain trap operates on the same principle as the float steam trap except that the drain trap does not contain a thermostatic element.

STRAINERS

Strainers are used in piping systems to protect equipment sensitive to dirt and other particles that may be carried by the fluid. During system start-up and flushing, strainers may be placed upstream of pumps to protect them from construction debris that may have been left in the pipe. Figure A2.36 depicts a typical start-up strainer. Permanent strainers may be installed upstream of control valves, traps, and instruments to protect them from corrosion products that may become dislodged and carried throughout the piping system.

Strainers are available in a variety of styles, including wye and basket. The wye strainer (Fig. A2.37) is generally used upstream of traps, control valves, and instruments. The wye strainer resembles a lateral branch fitting with the strainer element installed in the branch. The end of the lateral branch is removable to permit servicing of the strainer. Also, a blow-off connection may be provided in the end cap to flush the strainer.

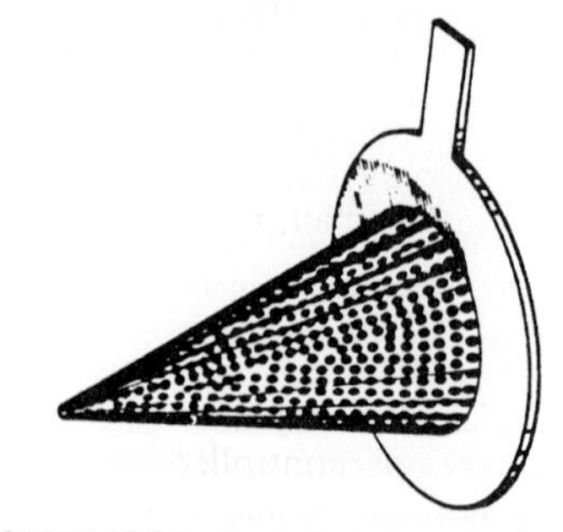

FIGURE A2.36 Conical start-up strainer.

Basket strainers (Fig. A2.38) are generally used where high flow capacity is required. The basket strainer is serviced by removing the cover, which yields access to the basket. Basket strainers are also available in a duplex

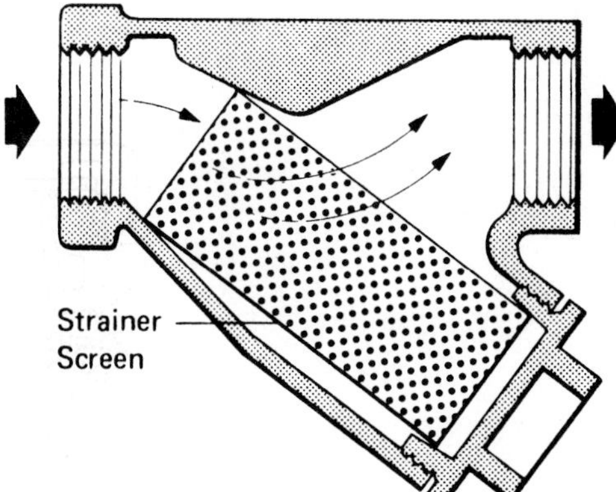

FIGURE A2.37 Wye strainer.

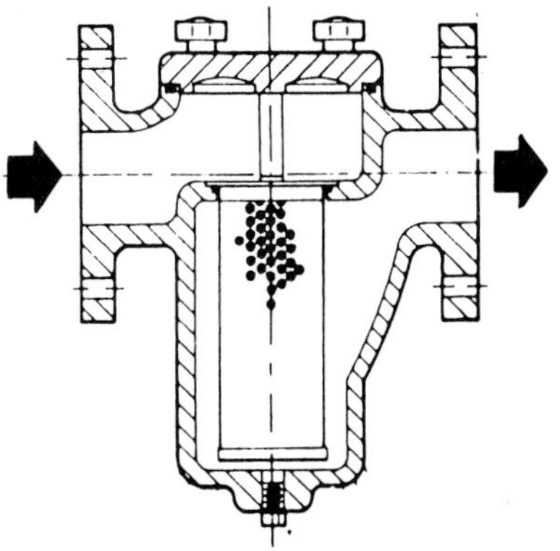

FIGURE A2.38 Basket strainer.

style which consists of two parallel basket strainers and diverting valves which permit diversion of the flow through one of the strainer elements while the other element is being serviced—an essential feature where flow cannot be interrupted.

EXPANSION JOINTS

Expansion joints are used in piping systems to absorb thermal expansion where the use of expansion loops are undesirable or impractical. Expansion joints are available in slip, ball, metal bellows, and rubber bellows configurations.

Slip-Type Expansion Joints

Slip-type expansion joints (Fig. A2.39*a*) have a sleeve that telescopes into the body. Leakage is controlled by packing located between the sleeve and the body. Leakage is minimal and can be near zero in many applications. A completely

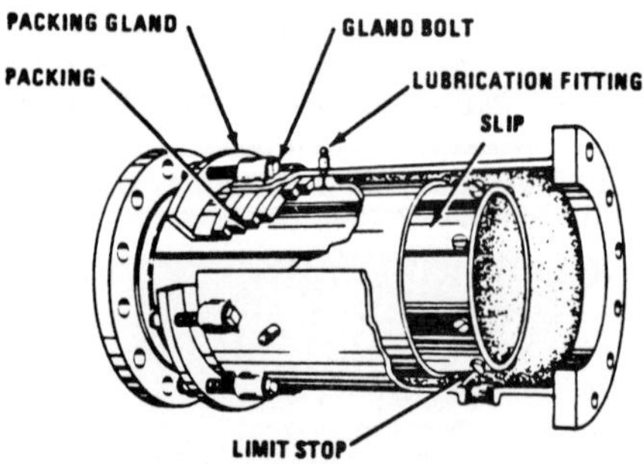

SINGLE-ENDED SLIP-TYPE EXPANSION JOINT WITH ADJUSTABLE PACKING GLAND

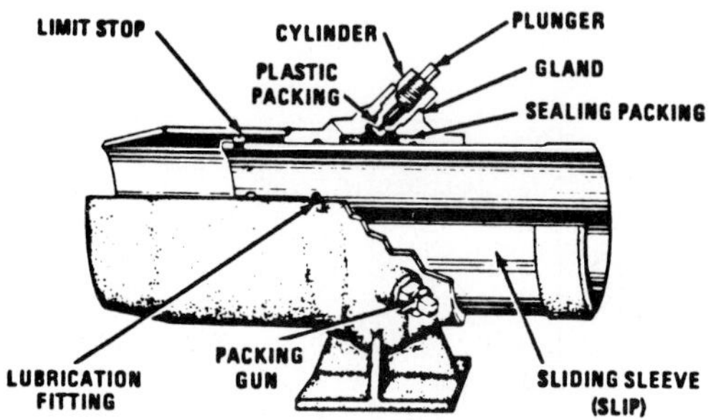

SINGLE-ENDED SLIP-TYPE EXPANSION JOINT WITH GUNS FOR REPLENISHING PACKING

(a)

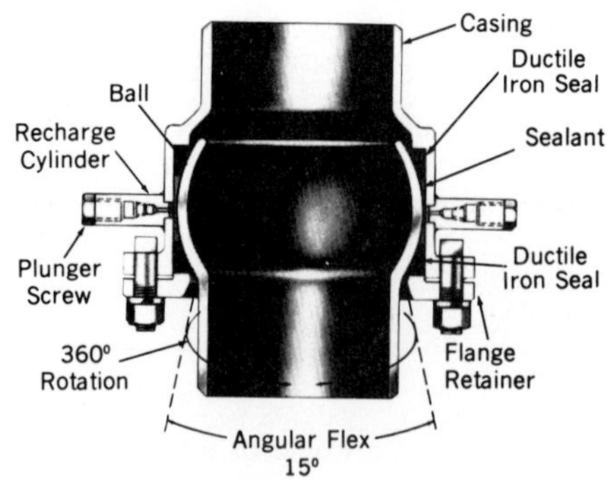

(b)

FIGURE A2.39 (*a*) Slip-type expansion joint. (*Yarway Co.*) (*b*) typical ball expansion joint. (*Barco Co.*)

leakfree seal cannot be assured; thus these expansion joints are ruled out where zero leakage is required. The packing is subject to wear due to cyclic movement of the sleeve when connected piping expands and contracts. Thus, these joints require periodic maintenance, either to compress the packing by tightening a packing gland or to replace or replenish the packing. Replacement of the packing rings is necessary when leakage develops in a joint that has an adjustable packing gland that has been tightened to its limit. Some designs provide for packing replenishment rather than replacement. These are usually called gun-packed or ram-packed slip joints. Since the packing can wear away, some packing material may be picked up in the line fluid. This rules out the use of slip joints in systems where such contamination of fluid cannot be tolerated.

Slip-type expansion joints are particularly suited for lines having straight-line (axial) movements of large magnitude. Slip joints cannot tolerate lateral offset or angular rotation (cocking) since this would cause binding, galling, and possibly leakage due to packing distortion. Therefore, the use of proper pipe alignment guides is essential.

Ball Expansion Joints

Ball expansion joints (Fig. A2.39*b*) consist of a socket and ball with a sealing mechanism placed between them. The seals are of rigid materials, and, in some designs, a pliable sealant may be injected into the cavity located between the ball and socket. The joints are capable of absorbing angular and axial rotation; however, they cannot accommodate movement along the longitudinal axis of the joint. Therefore, an offset must be installed in the line to absorb pipe axial movement.

Bellows Expansion Joints

Bellows-type expansion joints (Fig. A2.40) do not have packing; thus they do not have the potential leakage or fluid contamination problems sometimes associated with slip joints. Likewise, they do not require the periodic maintenance (lubrication and repacking) that is associated with slip joints. Bellow joints absorb expansion and contraction by means of a flexible bellows that is compressed or extended. They can also accommodate direction changes by various combinations of compression on one side and extension on an opposing side. Thus, they can adjust to lateral offset and angular rotation of the connected piping. However, they are not capable of absorbing torsional movement. Typically, the bellows is corrugated metal and is welded to the end pieces. To provide the requisite flexibility, the metal bellows is considerably thinner than the associated piping. Thus these expansion joints are especially susceptible to rupture by overpressure. A bel-

FIGURE A2.40 Metal bellows expansion joint.

lows can also fail because of metal fatigue if the accumulated flexing cycles exceed the designed fatigue life (cyclic life) of the bellows or if the flexing extremes exceed the designed compression and extension limits.

Rubber Expansion Joints

Rubber expansion joints (Fig. A2.41) are similar in design to metal bellows expansion joints except that they are constructed of fabric and wire reinforced elastomers. They are most suitable for use in cold water service where large movements must be absorbed (e.g., condenser circulating water).

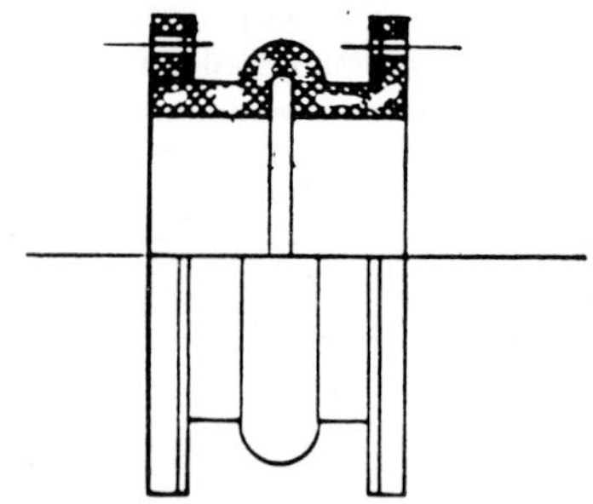

FIGURE A2.41 Rubber expansion joint.

THREADED JOINTS

Threaded joints are normally used in low-pressure small bore, nonflammable service, although threaded iron pipe is commonly used in domestic gas piping and threaded joints up to NPS 12 have been used in low-pressure liquid service.

For quality joints, it is essential to have smooth clean threads. A proper form for a pipe threading die is shown in Fig. A6.35. Because cut-thread surfaces are somewhat imperfect, thread sealants (pipe dope) and lubricants are often used to ensure a leak-tight joint. Lubricants such as linseed oil or a compound containing powdered zinc or nickel are sufficient to produce a leak-tight joint in well-made threads. Imperfect threads may require white lead or plumber's tape to provide a good seal. In high-pressure piping where leakage cannot be tolerated, the threaded joints may be seal welded. Where seal welding is employed, all exposed threads should be covered to prevent cracking in the weld.

Dimensional Standards. Dimensional standards for threads are established in ANSI/ASME Standard B1.20.1. This standard specifies dimensions, tolerances, and gauging for tape and straight pipe threads, including certain special applications. The normal type of pipe joint employs a tapered external and tapered internal thread, but straight pipe threads are used to advantage for certain types of pipe couplings, grease cup, fuel and oil fittings, mechanical joints for fixtures, and conduit and hose couplings.

Pressure-Tight Joints.

Pressure-tight joints for low-pressure service are sometimes made with straight internal threads and the American Standard taper external threads. The ductility of the coupling enables the straight thread to conform to the taper of the pipe thread. In commercial practice straight-tapped couplings are furnished for standard-weight (Schedule 40) pipe 2 in and smaller. If taper-tapped couplings are required for standard-weight pipe sizes 2 in and smaller, line pipe in accordance with API 5L should be ordered. The thread lengths should be in accordance with the American Standard for Pipe Threads, ANSI/ASME B1.20.1. Taper-tapped couplings are furnished on extra-strong (Schedule 80) pipe in all sizes and on standard-weight 2½ in and larger.

Dryseal pipe threads machined in accordance with ANSI/ASME B1.20.3 are also employed for pressure-tight joints, particularly where the presence of a lubricant or sealer would contaminate the flow medium. Threads are similar to the pipe threads covered by ANSI/ASME B1.20.1; the essential difference is that, in dryseal pipe threads, the truncation of the crest and root is controlled to assure metal-to-metal contact coincident with or prior to flank contact, thus eliminating spiral leakage paths. Dryseal pipe threads are used in refrigerant systems and for fuel and hydraulic control lines in aircraft, automotive, and marine service. Thread sizes up to NPS 3 are covered by ANSI/ASME B1.20.3.

Hose Nipples and Couplings. Hose coupling joints are ordinarily used with a gasket and made with straight internal and external loose-fitting threads. There are several standards of hose threads having various diameters and pitches, one of which is based on the American Standard Pipe Thread. With this thread series, it is possible to join small hose sizes ½ to 2 in, inclusive, to ends of standard pipe having American Standard External Taper Pipe Threads, using a gasket to seal the joint.

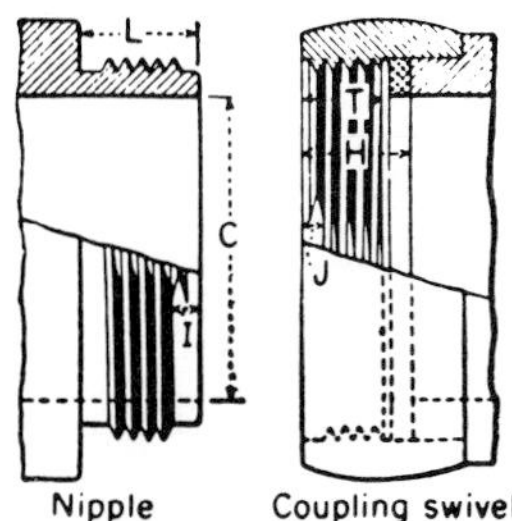

FIGURE A2.42 Typical fire-hose coupling.

ANSI/ASME B1.20.7 applies to the threaded parts of hose couplings, valves, nozzles, and all other fittings used in direct connection with hose intended for fire protection or for domestic and industrial general services. However, fire hose coupling dimensions and threads vary with fire districts and the local fire authority must be consulted.

American Standard ANSI B26 covers the threaded parts of fire-hose couplings, hydrant outlets, standpipe connections, and all other fittings on fire lines where fittings of 2½- through 6-in nominal diameter are used (Fig. A2.42). The threads conform to the American Standard (National) form having an included angle of 60° and are truncated top and bottom.

Bolted Joints

The use of bolted joints is advantageous in the following circumstances:

- The components cannot be serviced in line.
- The components being joined are not capable of being welded.
- Quick field assembly is required.
- The component or pipe section must be frequently removed for service.

Bolted piping components are manufactured in accordance with several national standards. Also, several manufacturers produce proprietary bolted connections which offer cost and time savings over conventional flanged connections. However, proprietary designs must be used within the limitations of the applicable codes.

Cast-Iron Flanges. Cast-iron flanges are produced in accordance with ANSI/ASME B16.1. The standard establishes dimensional requirements, pressure ratings, materials, and bolting requirements. The pressure-temperature ratings and materials requirements for cast-iron flanges are the same as for cast-iron flanged fittings. The pressure-temperature ratings are given in Table A2.16. The dimensions for Class 125 and Class 250 cast-iron flanges are listed in Table A2.28. Dimensions for bolting are listed in Table A2.29. It should be noted that the Class 125 and Class 250 flanges can be mated with ANSI/ASME B16.5 Class 150 and Class 300 steel flanges, respectively. When a Class 150 flange is bolted to a Class 125 cast-iron flange, a flat-faced flange, the steel flange should be flat faced.

TABLE A2.28 Dimensions of Typical Commercial Cast-Iron Companion Flanges Manufactured in Accordance with ANSI/ASME B16.1-1989

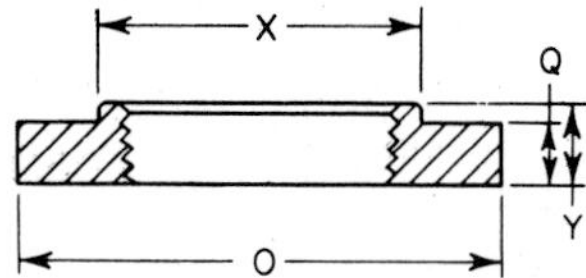

Companion flange, class 125

Size (in)	Diameter of flange *O* (in)	Thickness of flange* (min.) *Q* (in)	Diameter of hub (min.) *X* (in)	Length through hub* (min.) *Y* (in)	Weight (approx) each (lb)	
					Cast iron	Malleable†
1	4¼	7/16	1 15/16	11/16	1.75	
1¼	4⅝	½	2 5/16	13/16	2.00	
1½	5	9/16	2 9/16	⅞	2.25	2.25
2	6	⅝	3 1/16	1	4.00	4.00
2½	7	11/16	3 9/16	1⅛	6.00	6.00
3	7½	¾	4¼	1 3/16	7.63	7.63
3½	8½	13/16	4 13/16	1¼	9.00	9.00
4	9	15/16	5 5/16	1 5/16	11.75	11.75
5	10	15/16	6 7/16	1 7/16	14.00	14.00
6	11	1	7 9/16	1 9/16	16.50	16.50
8	13½	1⅛	9 11/16	1¾	26.00	26.00
10	16	1 3/16	11 15/16	1 15/16	37.75	
12	19	1¼	14 1/16	2 3/16	50.50	
14 OD	21	1⅜	15⅜	2¼	80.00	
16 OD	23½	1 7/16	17½	2½	100.00	
18 OD	25	1 9/16	19⅝	2 11/16	106.00	
20 OD	27½	1 11/16	21¾	2⅞	128.00	
24 OD	32	1⅞	26	3¼	202.00	

TABLE A2.28 Dimensions of Typical Commercial Cast-Iron Companion Flanges Manufactured in Accordance with ANSI/ASME B16.1-1989 (*Continued*)

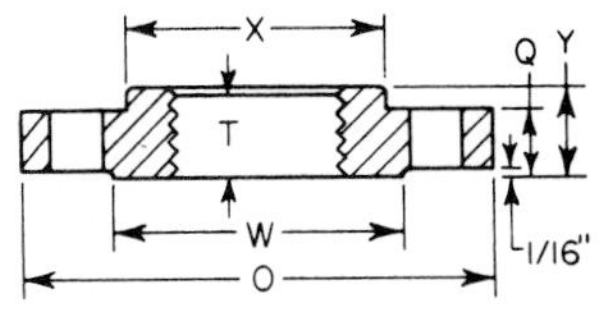

Companion flange, class 250

Size (in)	Diameter of flange *O* (in)	Thickness of flange (min.) *Q* (in)	Diameter of hub (min.) *X* (in)	Length through hub‡ (min.) *Y* (in)	Length of threads (min.) *T* (in)	Diameter of raised face *W* (in)	Weight (approx) each (lb)	
							Cast iron	Malleable†
1½	6⅛	13/16	2¾	1⅛	0.87	3 9/16	5.75	
2	6½	⅞	3 5/16	1¼	1.00	4 3/16	6.50	6.50
2½	7½	1	3 15/16	1 7/16	1.14	4 15/16	9.50	9.50
3	8¼	1⅛	4⅝	1 9/16	1.20	5 11/16	12.33	12.33
3½	9	1 3/16	5¼	1⅝	1.25	6 5/16	16.00	
4	10	1¼	5¾	1¾	1.30	6 15/16	20.00	20.00
5	11	1⅜	7	1⅞	1.41	8 5/16	24.00	24.00
6	12½	1 7/16	8⅛	1 15/16	1.51	9 11/16	32.00	32.00
8	15	1⅝	10¼	2 3/16	1.71	11 15/16	51.00	51.00
10	17½	1⅞	12⅝	2⅜	1.92	14 1/16	77.00	
12	20½	2	14¾	2 9/16	2.12	16 7/16	103.00	

*All 125-lb cast-iron standard flanges have a plain face.

†Dimensional standards have not been established for malleable-iron companion flanges; they are generally produced to the same dimensions as cast-iron flanges of the same class.

‡Minimum thickness of 250-lb flanges includes 1/16-in raised face.

Steel and Nickel-Alloy Flanges. Steel and nickel-alloy flanges up to NPS 24 are produced in accordance with ASME/ANSI B16.5. Steel flanges NPS 26 through NPS 60 are produced in accordance with ASME B16.47. Also, orifice flanges are produced in accordance with ASME/ANSI B16.36. The standards specify materials, dimensions, pressure-temperature ratings, and recommendations for bolting and gasketing. Flanges manufactured to B16.5 and B16.47 may be cast or forged. Also, blind flanges may be fabricated from specific plate materials. The most commonly used materials are forged carbon steel (ASTM A105) and forged low alloy and stainless steel (ASTM A182). The standards cover seven pressure classes (Classes 150, 300, 400, 600, 900, 1500, and 2500) in a variety of styles and materials. Figures A2.43 and A2.44 show typical flange styles. The dimensions of each style within each pressure class are held constant irrespective of the material. Therefore, within each pressure class, the pressure-temperature rating varies with the material properties.

Proprietary Bolted Connections. There are various proprietary bolted pipe joining systems produced that are not formally addressed by any standard. Under the various piping codes, pressure-retaining components not covered by standards

TABLE A2.29 Bolting Dimension for Cast-Iron Flanges

Size (in)	Diameter of bolt circle	Number of bolts	Diameter of bolts	Diameter of bolt holes	Length of bolts
		Class 125			
1	3⅛	4	½	⅝	1¾
1¼	3½	4	½	⅝	2
1½	3⅞	4	½	⅝	2
2	4¾	4	⅝	¾	2¼
2½	5½	4	⅝	¾	2½
3	6	4	⅝	¾	2½
3½	7	8	⅜	¾	2¾
4	7½	8	⅝	¾	3
5	8½	8	¾	⅞	3
6	9½	8	¾	⅞	3¼
8	11¾	8	¾	⅞	3½
10	14¼	12	⅞	1	3¾
12	17	12	⅞	1	3¾
14	18¾	12	1	1⅛	4¼
16	21¼	16	1	1⅛	4½
18	22¾	16	1⅛	1¼	4¾
20	25	20	1⅛	1¼	5
24	29½	20	1¼	1⅝	5½

		Class 250			
Size (in)	**Diameter of bolt circle**	**Diameter of bolt holes**	**Number of bolts**	**Size of bolts**	**Length of bolts**
1½	4½	⅞	4	¾	2¾
2	5	¾	8	⅝	2¾
2½	5⅞	⅞	8	¾	3¼
3	6⅝	⅞	8	¾	3½
3½	7¼	⅞	8	¾	3½
4	7⅞	⅞	8	¾	3¾
5	9¼	⅞	8	¾	4
6	10⅝	⅞	12	¾	4
8	13	1	12	⅞	4½
10	15¼	1⅛	16	1	5¼
12	17¾	1¼	16	1⅛	5½

specifically cited as acceptable for use under the "Code" may be used provided their design is proven by analysis or proof testing or a combination of both.

Flange Types

Flanges differ in method of attachment to the pipe, that is, whether they are screwed, welded, or lapped. Contact

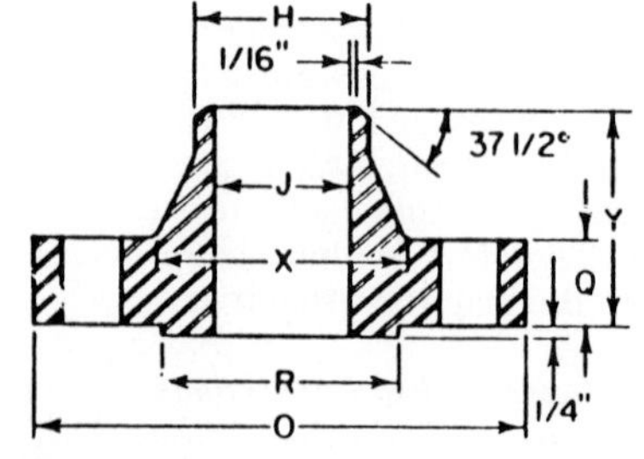

FIGURE A2.43 Typical integral flange (welding neck flange).

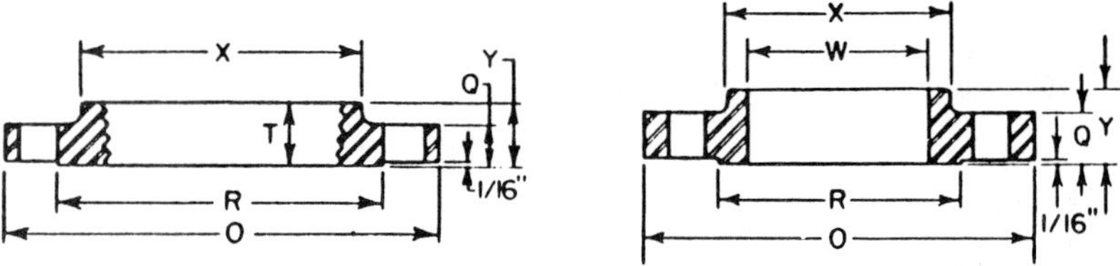

FIGURE A2.44 Typical loose flanges (threaded and slip on).

surface facings may be plain, serrated, grooved for ring joints, seal-welded, or ground and lapped for metal-to-metal contact. Some common types of joints and facings are shown in Fig. A2.45.

In Section VIII, Unfired Pressure Vessels, of the ASME Boiler and Pressure Vessel Code, three types of circular flanges are defined, and these are designated as loose type (Fig. A2.44), integral-type (Fig. A2.43), and optional-type flanges. Under the code, the welds and other details of construction shall satisfy the dimensional requirements stated therein.

Loose-Type Flanges. This (slip-on) type covers those designs in which the flange has no direct connection to the nozzle neck or the vessel or pipe wall and designs where the method of attachment is not considered to give mechanical strength equivalent of integral attachment.

Integral-Type Flanges. This type covers designs in which the flange is cast or forged integrally with the nozzle neck or the vessel or pipe wall, butt-welded thereto, or attached by other forms of arc or gas welding of such a nature that the flange and nozzle neck or vessel or pipe wall is considered to be the equivalent of an integral structure. In welded construction, the nozzle neck or the vessel or pipe wall is considered to act as a hub.

Optional-Type Flanges. This type covers designs where the attachment of the flange to the nozzle neck or the vessel or pipe wall is such that the assembly is considered to act as a unit, which shall be calculated as an integral flange, except that for simplicity the designer may calculate the construction as a loose-type flange, provided that stipulated load values are not exceeded.

It is important in flange design to select materials and to proportion dimensions of bolts, flanges, and gaskets to ensure that the necessary compression will be maintained on the joint faces over the expected life of the equipment.

Several distinct phases of the problem are involved: (1) type of flange facing, (2) finish of contact surfaces, (3) gasket type and proportions, (4) bolt load required to secure and maintain a tight joint, and (5) proportions of flange needed to support the bolt load.

Types of Flange Facing. There are numerous types of contact facings for flanges, the most simple of which is the plain face provided with a "smooth tool finish." Class 125 cast-iron flanged fittings are provided with this type of facing. For steel flanges and fittings, the typical facings (Fig. A2.46) are taken from the American Standard for Steel Pipe Flanges and Flanged Fittings, ASME/ANSI B16.5 and ASME B16.47. The raised face, the lapped, and the large male-and-female facings have the same dimensions, which provide a relatively large contact area.

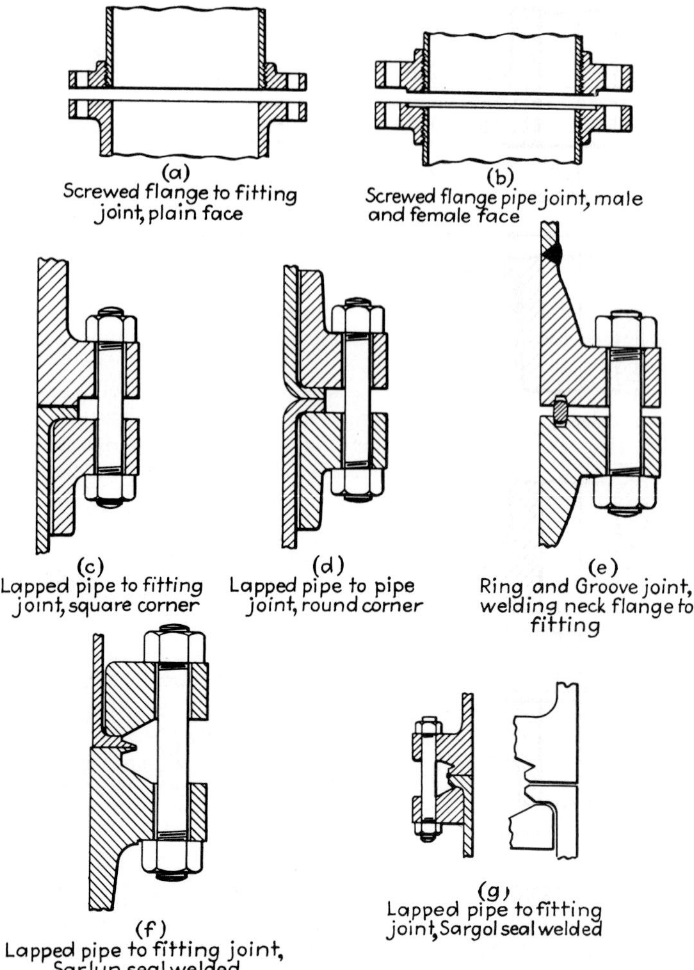

FIGURE A2.45 Commonly used flanged joints. (*a*) Screwed flange to fitting joint, plain face; (*b*) screwed flange pipe joint, male and female face; (*c*) lapped pipe to fitting joint, square corner; (*d*) lapped pipe to pipe joint, round corner; (*e*) ring and groove joint, welding neck flange to fitting; (*f*) lapped pipe to fitting joint, Sarlun seal welded; and (*g*) lapped pipe to fitting joint, Sargol seal welded.

Where metal gaskets are used with these facings, the gasket area should be reduced to increase the gasket compression.

The flange-facing types illustrated in Fig. A2.46 range in size and contact area in the following order: large tongue-and-groove, small tongue-and-groove, small male-and-female, and ring joint. Because of the small gasket contact area, a tight joint may be secured with the ring-type facing using low bolting loads, thereby resulting in lowered flange stresses (ASME/ANSI B16.5). The Sargol and Sarlun facings, which have lips for seal welding, are used frequently for severe service conditions. Seal welding is not always performed since, if it is properly made, a

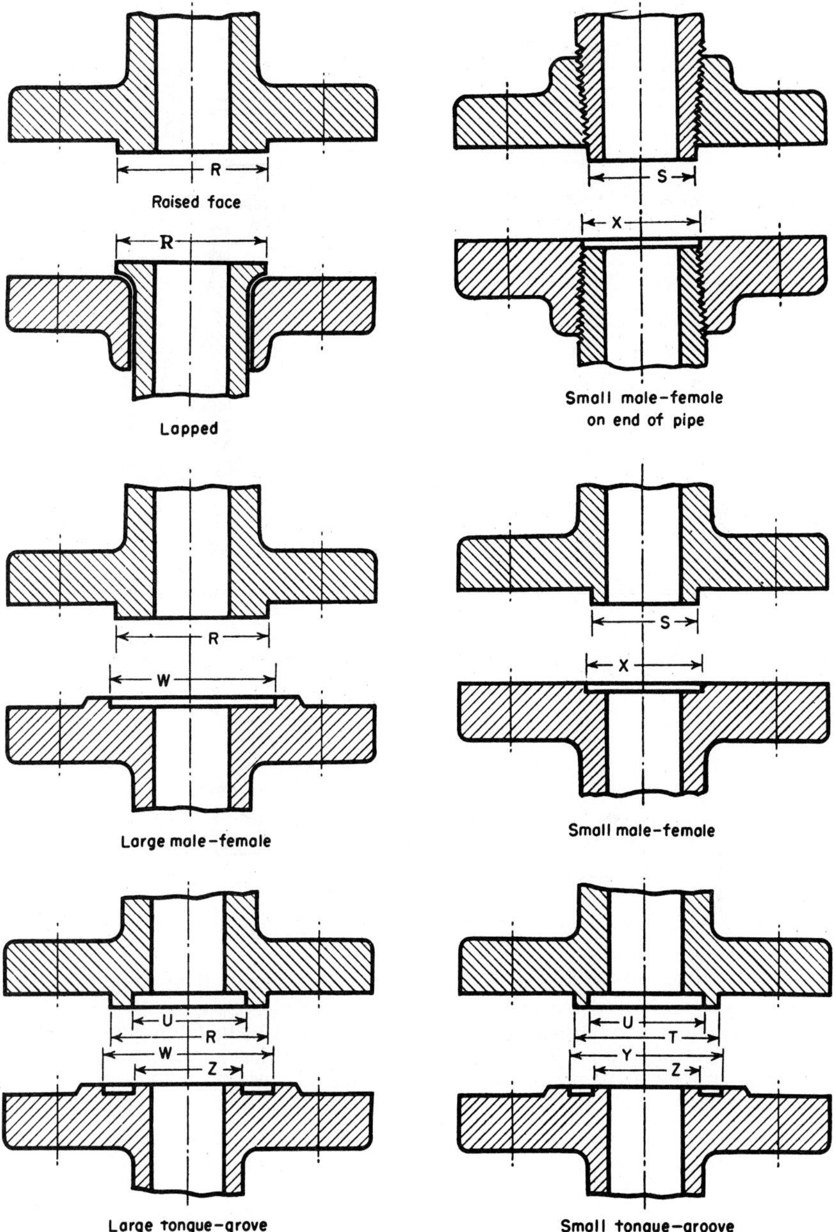

FIGURE A2.46 Typical flange facings (for dimensions, see Table A2.35) (Section I, ASME Boiler and Pressure Vessel Code).

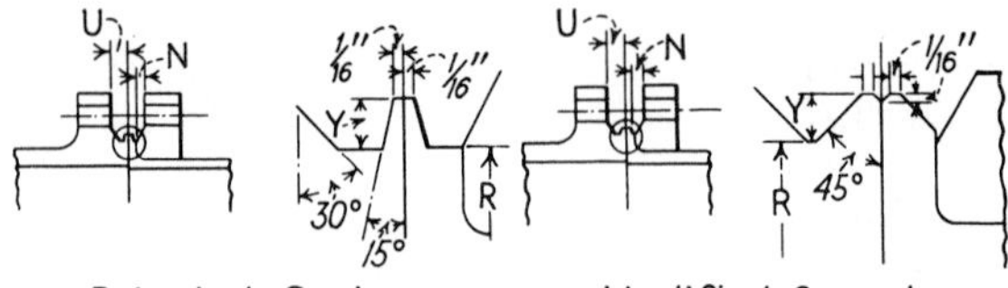

Nominal pipe size	Basic raised-face, outside diameter R	Height of face		Height of front hub		Height of welding projections	
		Sargol[1] U	Sarlun[2] U	Sargol[1] N	Sarlun[2] N	Sargol[1] Y	Sarlun[2] Y
2½	4⅛	½	1 1/16	¼	5/16	⅛	5/16
3	5	⅝	1 1/16	5/16	5/16	3/16	5/16
4	6 3/16	¾	¾	7/16	13/32	5/16	5/16
5	7 5/16	⅞	⅞	½	½	⅜	⅜
6	8½	1	1	9/16	9/16	⅜	⅜
8	10⅝	1⅛	1⅛	⅝	⅝	⅜	⅜
10	12¾	1¼	1¼	⅝	¾	⅜	⅜
12	15	1 5/16	1 5/16	⅝	13/16	⅜	⅜
14	16½	1⅜	1⅜	⅝	⅞	⅜	⅜
16	18½	1⅜	1⅜	⅝	⅞	⅜	⅜

All dimensions in inches.
[1] Dimensions of modified Sargol joint.
[2] Dimensions of Sarlun facings recommended by Sargent and Lundy, Inc.

FIGURE A2.47 Typical facing dimensions for Sargol and Sarlun joints, 150- to 2500-lb flanges (see footnotes).

tight joint often can be maintained without the welded seal, thus facilitating disassembly. Typical facing dimensions for Sarlun and Sargol joints are shown in Fig. A2.47. Special types of facing of individual design intended for a specific service are numerous. Economic considerations generally make it desirable to use a standard facing wherever possible.

Selection of the type of facing depends to a considerable extent on the nature of the service. However, it is not possible to determine exactly which facing should be used. Prior experience is usually relied on as a guide. Plain-face joints with red-rubber gaskets have been found satisfactory for temperatures up to 220°F, whereas serrated raised-face joints with graphite-steel-composition gaskets are commonly used for temperatures up to 750°F. For high temperatures and pressures, faces giving a high contact pressure for a given bolt load are customary, such as the tongue-and-groove and ring joints. However, with high contact pressures, the gasket load must be checked to ensure that the gasket is not overcompressed. An equally successful joint for most types of service can be made by using a profile-serrated metal gasket contacting the flange facing, which may be the plain male-to-male raised-face type.

Contact Surface Finish. The surface finish is an important factor in determining the extent to which a gasket must flow to secure an impervious seal. Bolting that results in adequate gasket flow to form a satisfactory seal with a smooth contact surface may be inadequate to secure a tight joint with a rough surface. The finish may vary from that produced by rough casting surfaces to that produced by grinding and lapping. Less gasket flow will be necessary for the latter than for the former. The finish most frequently provided on cast-iron and steel pipe flanges is the smooth tool finish. A serrated finish frequently is provided for steel flanges, particularly when using a graphite-composition gasket with a wide contact area such as is furnished on raised, lapped, or large tongue-and-groove facings. The serrated finish consists of spiral or concentric grooves, usually about 1/64 in deep with 32 serrations per inch.

Where metal gaskets are used, a smooth surface produced by grinding or lapping is usually provided. The Sargol and Sarlun facings mate metal to metal without a gasket, in which case a mirrorlike finish is necessary. This is usually produced by grinding and lapping. It is evident that the surface finish varies with the type of contact face and gasket used and, therefore, should be specified accordingly.

Gaskets

Since it is expensive to grind and lap joint faces to obtain fluid-tight joints, a gasket of some softer material is usually inserted between contact faces. Tightening the bolts causes the gasket material to flow into the minor machining imperfections, resulting in a fluid-tight seal. A considerable variety of gasket types is in common use. Soft gaskets, such as cork, rubber, vegetable fiber, graphite, or asbestos, are usually plain with a relatively smooth surface. The semimetallic design combines metal and a soft material, the metal to withstand the pressure, temperature, and attack of the confined fluid, and the soft material to impart resilience. Various designs involving corrugations, strip-on-edge, metal jackets, etc., are available. In addition to the plain, solid, and flat-surface metal gaskets, various modified designs and cross-sectional shapes of the profile, corrugated, serrated, and other types are used. The object in general has been to retain the advantage of the metal gasket but to reduce the contact area to secure a seal without excessive bolting load. Effective gasket widths are given in various sections of the ASME Boiler and Pressure Vessel Code.

Gasket Materials. Gaskets are made of materials which are not chemically affected by the fluid in the pipe and which are resistant to deterioration by temperature. Gasket materials may either be metallic or nonmetallic. Metallic ring-joint gasket materials are covered by ANSI Standard B16.20, Ring-joint Gaskets and Grooves for Steel Pipe Flanges. Nonmetallic gaskets are covered in ANSI Standard B16.21, Nonmetallic Gaskets for Pipe Flanges. Typical selections of gasket materials for different services are shown in Table A2.30.

Gasket Compression. In the usual type of high-pressure flange joint, a narrow gasket face or contact surface is provided to obtain higher unit compression on the gasket than is obtainable on full-face gaskets used with low-pressure joints. The compression on this surface and on the gasket if the gasket is used, before internal pressure is applied, depends on the bolt loading used. In the case of standard raised-face joints of the steel-flange standards, these gasket compressions range from 28 to 43 times the rated working pressure in the 150- to 400-lb standards, and from 11 to 28 times in the 600- to 2500-lb standards for an assumed bolt stress of 60,000 psi. For the lower-pressure standards, using composition gaskets, a bolt stress of 30,000 psi usually is adequate. The effect of applying the internal pressure is to decrease the compression on the contact surfaces since part of the bolt tension is used to support the pressure load.

TABLE A2.30 Selections of Gasket Materials for Different Services

Fluid	Application	Gasket material*
Steam (high pressure)	Temp up to 1000°F	Spiral-wound comp. asbestos or graphite
	Temp up to 1000°F	Steel, corrugated, or plain
	Temp up to 1000°F	Monel, corrugated, or plain
	Temp up to 1000°F	Hydrogen-annealed furniture iron
	Temp up to 1000°F	Stainless steel 12 to 14 percent chromium, corrugated
	Temp up to 1000°F	Ingot iron, special ring-type joint
	Temp up to 750°F	Comp. asbestos, spiral-wound
	Temp up to 600°F	Woven asbestos, metal asbestos
	Temp up to 600°F	Copper, corrugated or plain
Steam (low pressure)	Temp up to 220°F	Red rubber, wire inserted
Water	Hot, medium, and high pressures	Black rubber, red rubber, wire inserted
	Hot, low pressures	Brown rubber, cloth inserted
	Hot	Comp. asbestos
Water	Cold	Red rubber, wire inserted
	Cold	Black rubber
	Cold	Soft rubber
	Cold	Asbestos
	Cold	Brown rubber, cloth inserted
Oils (hot)	Temp up to 750°F	Comp. asbestos
	Temp up to 1000°F	Ingot iron, special ring-type joint
Oils (cold)	Temp up to 212°F	Cork or vegetable fiber
	Temp up to 300°F	Neoprene comp. asbestos
Air	Temp up to 750°F	Comp. asbestos
	Temp up to 220°F	Red rubber
	Temp up to 1000°F	Spiral-wound comp. asbestos
Gas	Temp up to 1000°F	Asbestos, metallic
	Temp up to 750°F	Comp. asbestos
	Temp up to 600°F	Woven asbestos
	Temp up to 220°F	Red rubber
Acids	(Varies; see section on corrosion)	Sheet lead or alloy steel
	Hot or cold mineral acids	Comp. blue asbestos
		Woven blue asbestos
Ammonia	Temp up to 1000°F	Asbestos, metallic
	Temp up to 700°F	Comp. asbestos
	Weak solutions	Red rubber
	Hot	Thin asbestos
	Cold	Sheet lead

*Several gasket manufacturers have introduced nonasbestos, nonmetallic gasket materials for use in high-temperature service. These materials are proprietary and, therefore, the manufacturers should be consulted for specific applications.

The initial compression required to force the gasket material into intimate contact with the joint faces depends upon the gasket material and the character of the joint facing. For soft-rubber gaskets, a unit compression stress of 4000 to 6000 psi usually is adequate. Laminated asbestos gaskets in serrated faced joints perform satisfactorily if compressed initially at 12,000 to 18,000 psi. Metal gaskets such as copper, Monel, and soft iron should be given initial compressions considerably in excess of their yield strengths. Unit pressures of 30,000 to 60,000 psi have been used successfully with metal gaskets. Various forms of corrugated and serrated metal gaskets are available which enable high unit compression to be obtained without excessive bolt loads. These are designed to provide a contact area that will flow under initial compression of the bolts so as to make an initially pressure-tight joint, but at the same time the compressive stresses in the body of the gasket are sufficiently low as to be comparable to the long-time load-carrying ability of the bolting and flange material at high temperatures.

The residual compression on the gasket necessary to prevent leakage depends on how effective the initial compression has been in forming intimate contact with the flange joint faces. Tests show that a residual compression on the gasket of only one to two times the internal pressure, with the pressure acting, may be sufficient to prevent leakage where the joint is not subjected to bending or to large and rapid temperature changes. Since joints in piping customarily must withstand both these disturbing influences, minimum residual gasket compressions of 4 to 6 times the working pressure should be provided for in the design of pipe joint.

Relation of Gaskets to Bolting. There is a tendency as indicated in the ASME Rules for Bolted Flanged Connections to assign lower residual contact-pressure ratios ranging from about 1 for soft-rubber gaskets to 6 or 7 for solid-metal gaskets. Whereas these are said to have proved satisfactory service for heat-exchanger and pressure-vessel flanges, the more severe service encountered by pipe flanges due to bending moments and large temperature changes is considered by many to warrant designing on the basis of the larger residual gasket-compression ratios recommended in the previous paragraph. The lack of understanding of the mechanics of gasket action, the variety of gasket materials, shapes, widths, and thicknesses, the variety of facings used, the variation in flange stiffness, and the uncertainties in bolt pull-up are among the factors that render difficult a precise solution to the problem of gasket design.

Rules for bolting and flange design are contained in Sections III and VIII of the ASME Boiler and Pressure Vessel Code.

Bolting

Bolting material for cast-iron flanges is listed in ASME/ANSI B16.1. Generally, ASTM A307, Grade B material is suitable. For steel flanges, acceptable bolting material is listed in ASME/ANSI B16.5.

Threading. Bolts and nuts normally are threaded in accordance with ANSI Standard for Unified Screw Threads B1.1. In diameters 1 in and smaller, Class 2A fits on the bolt or stud and Class 2B on the nut applies with the coarse thread series. In diameters 1⅛ in and larger, the eight-pitch thread series applies with the same fit.

Grade-7 bolts are threaded by roll threading after heat treatment. Roll threading cold works the surface uniformly. The resulting compressive stresses provide substantially increased fatigue strength at the thread root, which is usually the weakest point. The thread root is the weakest point because it is the smallest cross-sectional area in the bolt. The stressed area A_s of a bolt is computed from

$$A_s = 0.7854\left(D - \frac{0.9743}{N}\right)^2$$

where D is the nominal bolt diameter and N represents the threads per inch.

Bolts with fine threads will exhibit a slightly higher proof strength (of about 10 percent) than bolts with coarse threads (as illustrated in Fig. A2.48, provided that the length of engagement with the mating internal thread is sufficient to guarantee a tensile failure through the bolt rather than failure by thread stripping.

In practical bolt assemblies fine threads are considered weaker because of reduced thread height. Fine threads have limited application for threaded assemblies. They should be used for adjustment rather than as a clamping force.

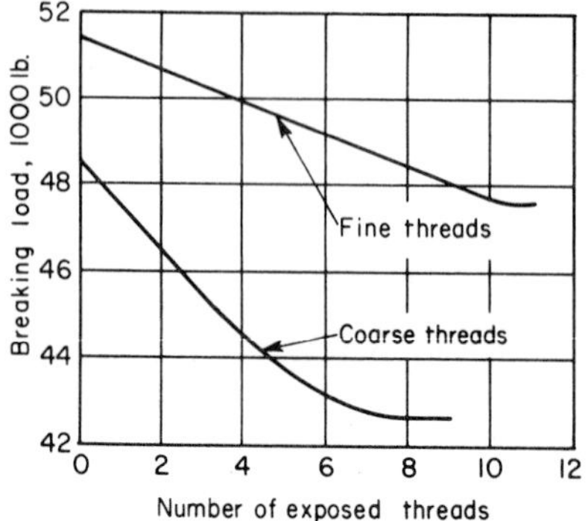

FIGURE A2.48 Comparison of proof strength on fine and coarse threads, SAE Grade 5, ¾-in bolts.

Dimensions

The dimensions applicable to bolting materials are given in ANSI/ASME B16.5, American Standard Pipe Flanges and Flanged Fittings.

Securing and Tightening. For the average low- and medium-pressure installations bolts are made in staggered sequence with wrenches which will usually result in adequately tight joints. For the high pressure and temperature joints, it becomes increasingly more important to make up each stud to a definite tension. Torque wrenches are sometimes used for this purpose.

In exceptional cases where a more positive method is desired, the studs may be tightened until a definite elongation has been attained. For this condition, an initial cold tension of 30,000 to 35,000 psi in each stud is recommended. Since the modulus of elasticity of stud material is about 30×10^6 psi, a tension of 30,000 psi would result in a unit elongation of $30{,}000/(30 \times 10^6)$ or 0.001 in/in of effective length. The effective length is the distance between nut faces plus one nut thickness. Special studs with ground ends are required to make micrometer measurements for this purpose. After the joint has been in service, periodic checks of the actual cold lengths as compared with the tabulated lengths will detect any permanent elongation of the studs. Permanent elongation will indicate overstressing, relaxation, and creep. When these conditions become severe, new studs may be required to maintain the joint properly.

Special thread lubricants are available both for temperatures below 500°F and

TABLE A2.31 Turning Efforts to Tighten Eight-pitch-thread Bolts

Nominal diameter of bolt, in.	Number of threads, per in.	Tensile stress area, A_s	Stress[1]			
			30,000 lb/sq in.		60,000 lb/sq in.	
			Torque, ft-lb	Force per bolt, lb	Torque, ft-lb	Force per bolt, lb
½	13	0.1419	30	4,257	60	8,514
9⁄16	12	0.182	45	4,560	90	10,920
⅝	11	0.226	60	6,780	120	13,560
¾	10	0.334	100	10,020	200	20,040
⅞	9	0.606	160	18,180	320	36,360
1	8	0.462	245	13,860	490	27,720
1⅛	8	0.790	355	23,700	710	47,400
1¼	8	1.000	500	30,000	1,000	60,000
1⅜	8	1.233	680	36,990	1,360	73,980
1½	8	1.492	800	44,760	1,600	89,520
1⅝	8	1.78	1,100	53,400	2,200	106,800
1¾	8	2.08	1,500	62,400	3,000	124,800
1⅞	8	2.41	2,000	72,300	4,000	144,600
2	8	2.77	2,200	83,100	4,400	166,200
2¼	8	3.56	3,180	106,800	6,360	213,600
2½	8	4.44	4,400	133,200	8,800	266,400
2¾	8	5.43	5,920	162,900	11,840	325,800
3	8	6.51	7,720	195,300	15,440	390,600
3¼	8	7.69	.....	230,700	......	461,400
3½	8	8.96	.....	268,800	......	537,600
3¾	8	10.34	.....	310,300	......	620,400
4	8	18.11	.....	354,300	......	708,600

[1]Stress has been calculated on the basis of stressed area A_s where $A_s = 0.7854\,(D - 0.9743/N)^2$ in which D is the nominal bolt diameter and N is threads per inch.

from 500 to 1000°F. Such lubricants not only facilitate initial tightening but also permit easier disassembly after service.

Table A2.31 illustrates the turning effort required for tightening well-lubricated threads and bearing surfaces. Tests with no lubricant on threads and bearing surfaces may increase torque requirements by 75 to 100 percent to secure a given bolt stress.

WELDED AND BRAZED JOINTS

Welded and brazed joints are the most commonly used methods for joining piping components because they are stronger and more leak tight than threaded and flanged joints. Furthermore, they do not add weight to the piping system as flanges do, and they do not require an increase in pipe wall thickness to compensate for threading as threaded joints do.

Pipe-Weld Joint Preparation and Design

Butt Welds. The most common type of joint employed in the fabrication of welded-pipe systems is the circumferential butt joint. It is the most satisfactory joint from the standpoint of stress distribution. Its general field of application is pipe to pipe, pipe to flange, pipe to valve, and pipe to fitting joints. Butt joints may be used for all sizes, but fillet-welded joints can often be used to advantage for pipe NPS 2 in and smaller.

The profile of the weld edge preparations for butt welds may be any configuration the welding organization deems suitable for making an acceptable weld. However, to standardize the weld edge preparation on butt-welded commercial piping components, standard weld-edge preparation profiles have been established in ASME/ANSI B16.25. These weld-edge preparation requirements are also incorporated into the standards governing the specific components (e.g., B16.9, B16.5, B16.34). Figures A2.49, A2.50, and A2.51 illustrate the various standard weld-edge profiles for different wall thickness.

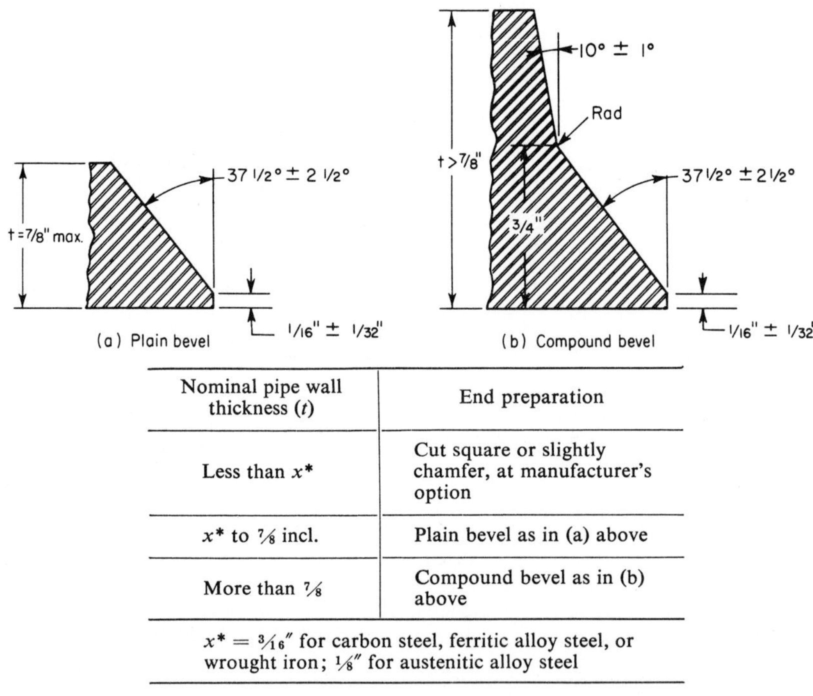

Nominal pipe wall thickness (t)	End preparation
Less than x^*	Cut square or slightly chamfer, at manufacturer's option
x^* to 7/8 incl.	Plain bevel as in (a) above
More than 7/8	Compound bevel as in (b) above

x^* = 3/16" for carbon steel, ferritic alloy steel, or wrought iron; 1/8" for austenitic alloy steel

FIGURE A2.49 Basic welding bevel for all components (without backing ring, or with split ring).

On piping, the end preparation is normally done by machining or grinding. On pipe of heavier wall thicknesses, machining is generally done on post mills. On carbon- and low-alloy steels, oxygen cutting and beveling is also used, particularly on pipe of wall thicknesses below 1/2 in. However, the slag should be removed by grinding prior to welding.

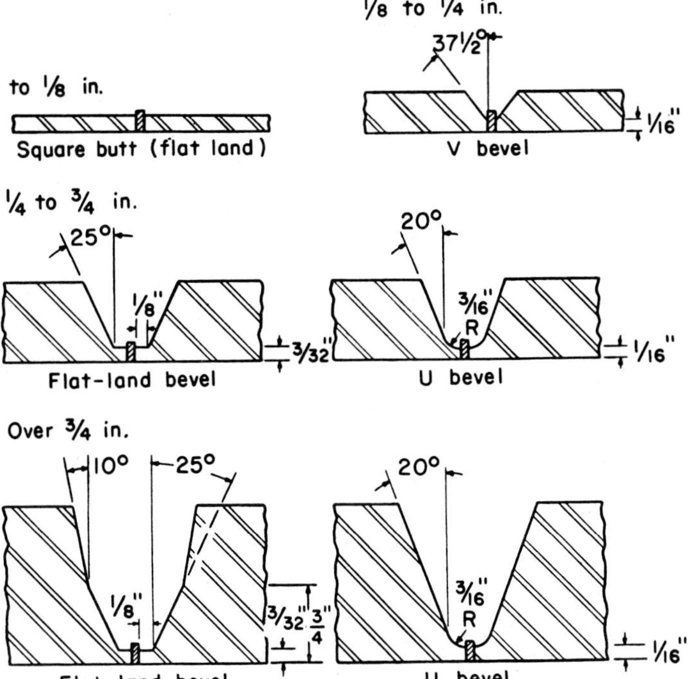

FIGURE A2.50 Typical end preparations for pipe which is to be welded by the inert-gas tungsten-arc welding process.

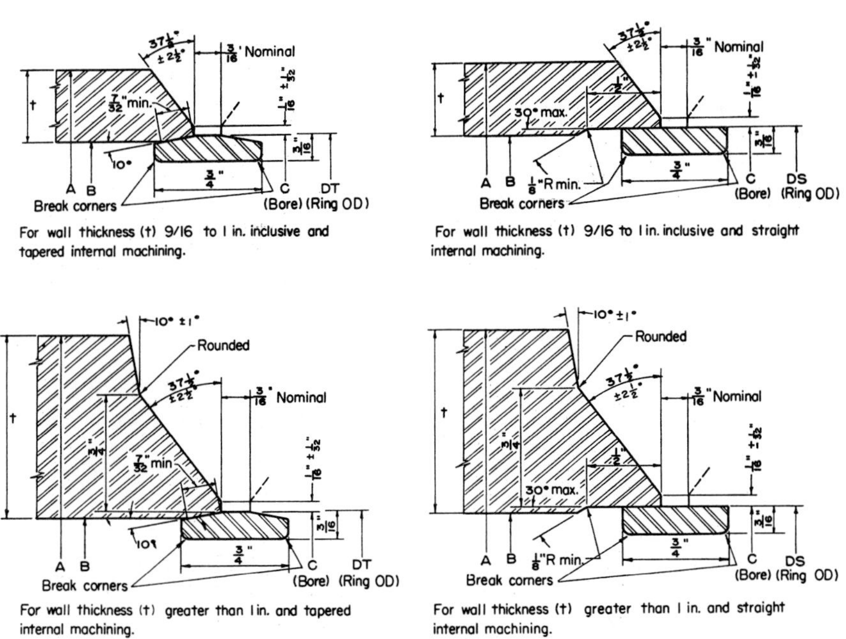

FIGURE A2.51 End-preparation and backing-ring requirements for critical service applications employing flat or taper machined solid backing rings. See Table A2.42 for dimensional data.

Because of fairly broad permissible eccentricity and size tolerances of pipe and fittings, considerable mismatch may be encountered on the inside of the piping. Limitations on fit-up tolerances are included in several piping codes. For severe service applications, internal machining may be required to yield proper fit-up. When machining the inside diameter, care should be taken to ensure that minimum wall requirements are not violated. Table A2.32 lists the counterbore dimensions typically specified.

When piping components of unequal wall thickness are to be welded, care should be taken to provide a smooth taper toward the edge of the thicker member. The length of the taper desirable is normally 3 times the offset between the components, as outlined in ASME Section I, III, and ASME/ANSI B31.1 Codes. The two methods of alignment which are recommended are shown in Fig. A2.52.

The wall thickness of cast-steel fittings and valve bodies is normally greater than that of the pipe to which they are joined. In order to provide a gradual transition between piping and components, the ASME Boiler and Pressure Vessel Code and the ASME Code for Pressure Piping permit the machining of the cylindrical ends of fittings and valve bodies to the nominal wall thickness of the adjoining pipe. However, in no case is the thickness of a valve permitted to be less than 0.77 t_{min} at a distance of 1.33 t_{min} from the weld end where t_{min} is the minimum valve thickness required by ASME/ANSI B16.34. The machined ends may be extended back in any manner, provided that the longitudinal section comes within the maximum slope line indicated in Fig. A6.23. The transition from the pipe to the fitting or valve end at the joint must be such as to avoid sharp reentrant angles and abrupt changes in slope.

End Preparation for Inert-gas Tungsten-Arc Root-Pass Welding. The pipe end bevel preparations shown in Fig. A2.51 are considered adequate for shielded metal-arc welding, but they pose some problems in inert-gas tungsten-arc welding. When this process is used, extended U or flat-land bevel preparations are considered more suitable since the extended land reduces the heat sink, thereby affording better weld penetration. The end preparations apply to inert-gas tungsten-arc welding of carbon- and low-alloy steel piping, stainless-steel piping, and most nonferrous piping materials. On aluminum piping, the flat-land bevel preparations are preferred by some fabricators.

Backing Rings. Backing rings are employed in some piping systems, particularly where pipe joints are welded primarily by the shielded metal-arc welding process with covered electrodes. For example, a significant number of pipe welds for steam power plants and several other applications are made with the use of backing rings. On the other hand, in many applications backing rings are not used, since they may restrict flow, provide crevices for the entrapment of corrosive substances, enhance susceptibility to stress corrosion cracking, or introduce still other objectionable features. Thus, there is little, if any, use made of backing rings in most refinery piping, radioactive service piping, or chemical process piping.

The use of backing rings is primarily confined to carbon- and low-alloy steel and aluminum piping. Carbon-steel backing rings are generally made of a mild carbon steel with a maximum carbon content of 0.20 percent and a maximum sulfur content of 0.05 percent. The latter requirement is especially important since high sulfur in deposited weld metal (which could be created by an excessive sulfur content in such rings) may cause weld cracks. Split backing rings are satisfactory for noncritical piping systems.

TABLE A2.32 Dimensions for Internal Machining and Backing Rings for Heavy-wall Pipe in Critical Applications

Nominal pipe size	Schedule No. or wall	Nominal OD "A"	Nominal ID "B"	Nominal wall thickness t	Machined ID of pipe "C" tolerance $^{+0.010}_{-0.000}$	OD of backing ring: Tapered ring *DT* tolerance $^{+0.010}_{-0.000}$	OD of backing ring: Straight ring *DS* tolerance $^{+0.000}_{-0.010}$
3	XXS	3.500	2.300	0.600	2.409	2.419	2.409
4	XXS	4.500	3.152	0.674	3.279	3.289	3.279
5	160	4.500	4.313	0.625	4.428	4.438	4.428
	XXS	5.563	4.063	0.750	4.209	4.219	4.209
6	120	6.625	5.501	0.562	5.600	5.610	5.600
	160	6.625	5.187	0.719	5.326	5.336	5.326
	XXS	6.625	4.897	0.864	5.072	5.082	5.072
8	100	8.625	7.437	0.594	7.546	7.554	7.544
	120	8.625	7.187	0.719	7.326	7.336	7.326
	140	8.625	7.001	0.812	7.163	7.173	7.163
	XXS	8.625	6.875	0.875	7.053	7.063	7.053
	160	8.625	6.813	0.906	6.998	7.008	6.998
10	80	10.750	9.562	0.594	9.671	9.679	9.669
	100	10.750	9.312	0.719	9.451	9.461	9.451
	120	10.750	9.062	0.844	9.234	9.242	9.232
	140	10.750	8.750	1.000	8.959	8.969	8.959
	160	10.750	8.500	1.125	8.740	8.750	8.740
12	60	12.750	11.626	0.562	11.725	11.735	11.725
	80	12.750	11.374	0.688	11.507	11.515	11.505
	100	12.750	11.062	0.844	11.234	11.242	11.232
	120	12.750	10.750	1.000	10.959	10.969	10.959
	140	12.750	10.500	1.125	10.740	10.750	10.740
	160	12.750	10.126	1.312	10.413	10.423	10.413
14 OD	60	14.000	12.812	0.594	12.921	12.929	12.919
	80	14.000	12.500	0.750	12.646	12.656	12.646
	100	14.000	12.124	0.938	12.319	12.327	12.317
	120	14.000	11.812	1.094	12.046	12.054	12.044
	140	14.000	11.500	1.250	11.771	11.781	11.771
	160	14.000	11.188	1.406	11.498	11.508	11.498
16 OD	60	16.000	14.688	0.656	14.811	14.821	14.811
	80	16.000	14.312	0.844	14.484	14.492	14.482
	100	16.000	13.938	1.031	14.155	14.165	14.155
	120	16.000	13.562	1.219	13.826	13.836	13.826
	140	16.000	13.124	1.438	13.442	13.452	13.442
	160	16.000	12.812	1.594	13.171	13.179	13.169
18 OD	40	18.000	16.876	0.562	16.975	16.985	16.975
	60	18.000	16.500	0.750	16.646	16.656	16.646
	80	18.000	16.124	0.938	16.319	16.312	16.317
	100	18.000	15.688	1.156	15.936	15.946	15.936
	120	18.000	15.250	1.375	15.553	15.563	15.553
	140	18.000	14.876	1.562	15.225	15.235	15.225
	160	18.000	14.438	1.781	14.842	14.852	14.842
20 OD	40	20.000	18.812	0.594	18.921	18.929	18.919
	60	20.000	18.376	0.812	18.538	18.548	18.538
	80	20.000	17.938	1.031	18.155	18.165	18.155
	100	20.000	17.438	1.281	17.717	17.727	17.717
	120	20.000	17.000	1.500	17.334	17.344	17.334
	140	20.000	16.500	1.750	16.896	16.906	16.896
	160	20.000	16.062	1.969	16.515	16.523	16.513
22 OD	...	22.000	20.750	0.625	20.865	20.875	20.865
	60	22.000	20.250	0.875	20.428	20.438	20.428
	80	22.000	19.750	1.125	19.990	20.000	19.990
	100	22.000	19.250	1.375	19.553	19.563	19.553
	120	22.000	18.750	1.625	19.115	19.125	19.115
	140	22.000	18.250	1.875	18.678	18.688	18.678
	160	22.000	17.750	2.125	18.240	18.250	18.240
24 OD	30	24.000	22.876	0.562	22.975	22.985	22.975
	40	24.000	22.624	0.688	22.757	22.765	22.755
	60	24.000	22.062	0.969	22.265	22.273	22.263
	80	24.000	21.562	1.219	21.826	21.836	21.826
	100	24.000	20.938	1.531	21.280	21.290	21.280
	120	24.000	20.376	1.812	20.788	20.798	20.788
	140	24.000	19.876	2.062	20.350	20.360	20.350
	160	24.000	19.312	2.344	19.859	19.867	19.857

[1]Stress has been calculated on basis of stressed area, A_s, where $A_s = 0.7854\,(D - 0.9743/N)^2$ in which D is the nominal bolt diameter and N is threads per inch.

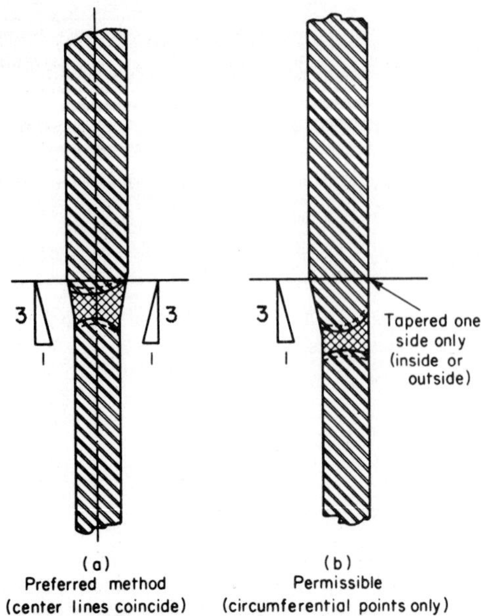

FIGURE A2.52 Recommended welding-end sections for pipe, valves, and fittings of unequal thickness. (*a*) Preferred method (center lines coincide); (*b*) permissible (circumferential points only).

For the more critical service applications involving carbon- and low-alloy steel piping, solid flat or taper-machined backing rings are preferred in accordance with the recommendations shown in Pipe Fabrication Institute Standard ES1 and illustrated in Fig. A2.52 and Table A2.32.

When a machined backing rings is desired, it is a general recommendation that welding ends be machined on the inside diameter in accordance with the Pipe Fabrication Institute standard for the most critical services and then only when pierced seamless pipe is used that complies with the applicable specifications of the American Society for Testing and Materials. Such critical services include high-pressure steam lines between boiler and turbines and high-pressure boiler-feed discharge lines, as encountered in modern steam power plants. It is also recommended that the material of the backing ring be compatible with the chemical composition of the pipe, valve, fitting, or flange with which it is to be used. Where materials of dissimilar composition are being joined, the composition of the backing ring may be that of the lower alloy.

On turned-and-bored and fusion-welded pipe, the design of the backing ring and internal machining, if any, should be a matter of agreement between the customer and the fabricator. Regardless of the type of backing rings used, it is recommended that the general contour of the welding bevel shown in Fig. A2.51 be maintained.

When machining piping for backing rings, the resulting wall thickness should be not less than that required for the service pressure. Wherever internal machining for machined backing rings is required on pipe and welding fittings in smaller sizes and lower schedule numbers than those listed in Table A2.32, weld metal may have to be deposited on the inside of the pipe in the area to be machined. This is to provide satisfactory contact between the machined surface on the pipe inside and the machined backing ring. For such cases, the machining dimension should be a matter of agreement between the fabricator and the purchaser.

Whenever pipe and welding fittings in the sizes and schedule numbers listed in Table A2.32 have plus tolerance on the outside diameter, it also may be necessary to deposit weld metal on the inside of the pipe or welding fitting in the area to be machined. In such cases, sufficient weld metal should be deposited to result in an ID not greater than the nominal ID given in Table A2.32 for the particular pipe size and wall thickness involved.

Experience indicates that machining to dimension C for the pipe size and schedule number listed in Table A2.32 generally will result in a satisfactory seat contact of 7⁄32 in minimum (approximately 75 percent minimum length of contact) between pipe and the 10° backing ring. Occasionally, however, it will be necessary to deposit weld metal on the inside diameter of the pipe or welding fitting in order to provide sufficient material for machining a satisfactory seat.

In welding butt joints with backing rings, care should be exercised to ensure good fusion of the first weld pass into the backing ring in order to avoid lack of weld penetration or other types of stress-raising notches.

Consumable Insert Rings. The chemical composition of a piping base metal is established primarily to provide it with certain mechanical, physical, or corrosion-resisting properties. Weldability characteristics, if considered at all, are of secondary concern. On the other hand, the chemical composition of most welding filler metals is determined with primary emphasis on producing a sound, high-quality weld.

The steel-making process employed in the manufacture of welding filler metals permits closer control of the composition range, which is usually considerably narrower than would be practical for the piping base metal where much larger tonnages of steel are involved.

On some base metals, the welding together by fusion of only the base-metal compositions may lead to such welding difficulties as cracking or porosity. The addition of filler metal tends to improve weld quality. However, in inert-gas tungsten-arc welding, the addition of welding filler metal from a separate wire which the welder feeds with one hand while manipulating the tungsten-arc torch with the other is a cumbersome process and interferes with welding ease. The welder may leave areas with lack of penetration, which generally are considered unacceptable as can be seen, for example, in the rules of the ASME Boiler and Pressure Vessel Code. Since some types of serious weld defects are detected only with difficulty during inspection (if they are detected at all), it is extremely important to provide the easiest welding conditions for the welder to produce quality welds.

One technique to produce high-quality welds is to employ consumable insert rings of proper composition and dimensions. Consumable insert rings which are available commercially are shown in Fig. A2.53. The three primary functions of consumable insert rings are to: (1) provide the easiest welding conditions and thereby minimize the effects of undesirable welding variables caused by the "human" element, (2) give the most favorable weld contour to resist cracking resulting from weld-metal shrinkage and hot shortness, or brittleness, in hot metal, and (3) produce metallurgically the soundest possible weld-metal composition of desirable strength, ductility, and toughness properties.

The best welding conditions are obtained where the flat-land and extended U-bevel preparations are used. These joint preparations are particularly helpful where welding is done in the horizontal fixed pipe position (5G), since they ensure a flat or slightly convex root contour and provide by far the greatest resistance to weld cracking in those alloys particularly susceptible to microfissuring.

The weld-root contour conditions to be expected from different bevel preparations and consumable insert rings are illustrated in Fig. A2.54. Where sink is not acceptable, it is considered obligatory to use consumable insert rings with the special flat-land or extended U-bevel preparation. In horizontal-rolled (1G) and vertical-position (2G) welding, the insert ring should be placed concentrically into the beveled pipe.

In horizontal fixed-position (5G) welding, the insert ring should be placed eccentric to the centerline of the pipe (as shown in Fig. A2.55). In this position, the insert ring compensates for the downward sag of the molten weld metal and aids in obtaining smooth, uniform root contour along the inner diameter and the joint.

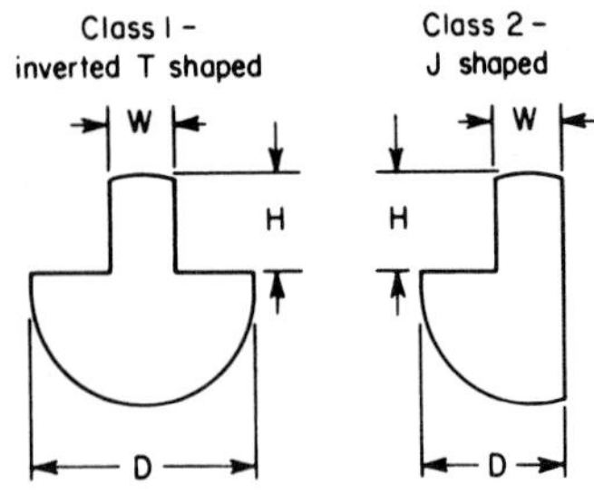

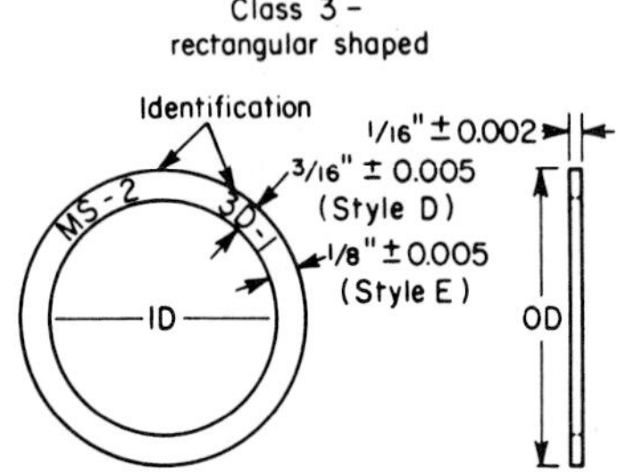

FIGURE A2.53 Commercial consumable insert rings used in pipe welding (MIL-I-23413). Style D: for diameters 2 in and larger. On Schedule 5 for diameters 5 in and larger; Style E: for diameters less than 2 in. On Schedule 5 for diameters less than 5 in.

Fillet Welds

Circumferential fillet-welded joints are generally used for joining pipe to socket joints in sizes NPS 2 in and smaller. Figure A2.56 illustrates three typical fillet-welded joints. These types of welds are subjected to shearing and bending stresses, and adequate penetration of the pieces being joined is essential. This is particularly important with the socket joint, since the danger of washing down the end of the hub may obscure, by reason of fair appearance, the lack of a full and sound fillet weld. This condition is one which cannot be detected in the finished weld by the usual visual inspection. Additionally, a 1/16-in gap (before welding) must be maintained between the pipe end and the base of the fitting to allow for differential expansion of the mating elements.

There are service applications where socket welds are not acceptable. Piping systems involving nuclear or radioactive service or corrosive service with solutions which promote stress corrosion cracking or concentration cell action generally require butt welds in all pipe sizes with complete weld penetration to the inside of the piping.

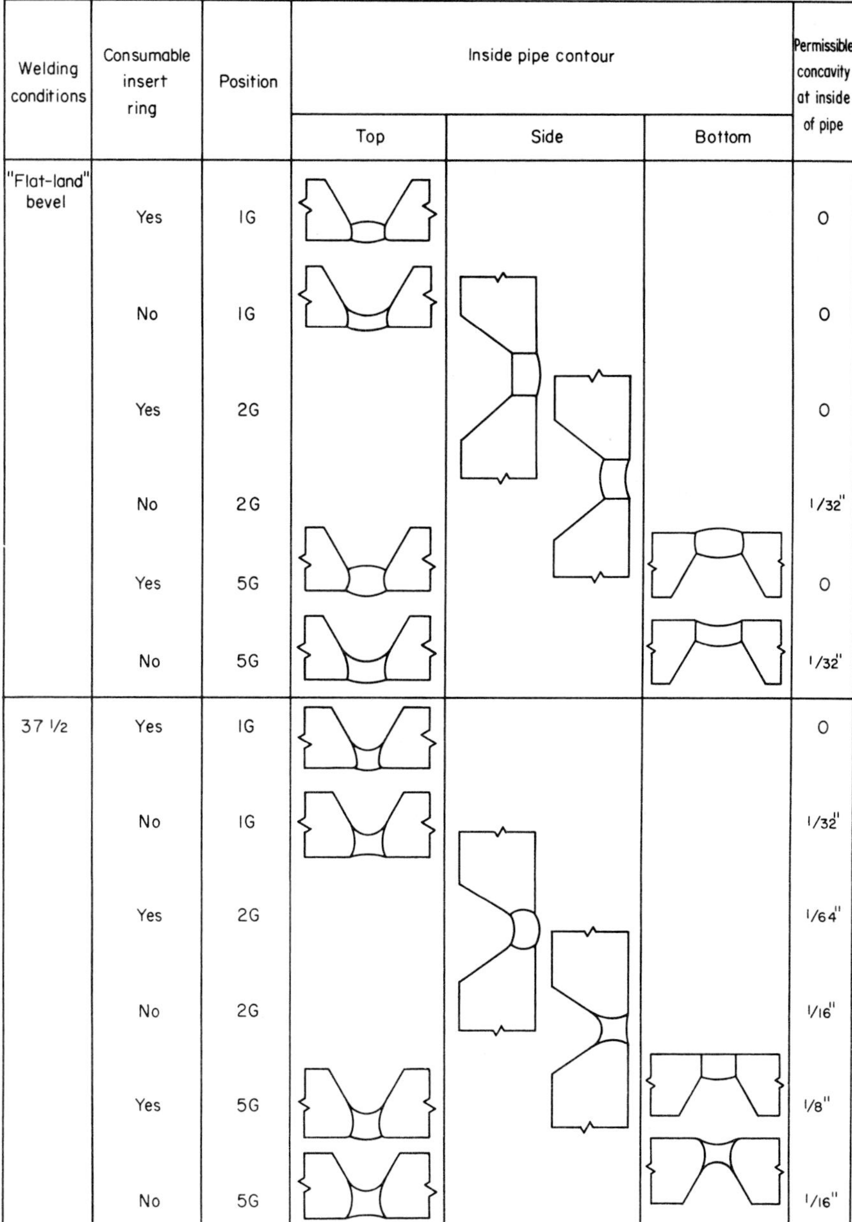

Welding conditions	Consumable insert ring	Position	Inside pipe contour			Permissible concavity at inside of pipe
			Top	Side	Bottom	
"Flat-land" bevel	Yes	1G				0
	No	1G				0
	Yes	2G				0
	No	2G				1/32"
	Yes	5G				0
	No	5G				1/32"
37 1/2	Yes	1G				0
	No	1G				1/32"
	Yes	2G				1/64"
	No	2G				1/16"
	Yes	5G				1/8"
	No	5G				1/16"

FIGURE A2.54 Root-contour conditions which can be expected as the result of normal pipe welding with the gas tungsten-arc-welding process. In 5G (horizontal-fixed) position welding the insert ring is positioned eccentric to the centerline of the pipe, as illustrated in Fig. A2.55.

Brazed Joints

Lap or shear-type joints generally are necessary to provide capillary attraction for brazing of connecting pipe. Square-groove butt joints may be brazed, but the results are unreliable unless the ends of the pipe or tube are accurately prepared, plane and square, and the joint is aligned carefully, as in a jig. High strengths may be obtained with butt joints if they are properly prepared and brazed. However, owing to the brittleness of the brazing alloy, they are not normally applicable.

The alloys generally used in brazing exhibit their greatest strength when the thickness of the alloy in the lap area is minimal. Thin alloy sections also develop the highest ductility. For brazing ferrous and nonferrous piping with silver- and copper-base brazing alloys, the thickness of the brazing alloy in the joint generally should not be over 0.006 in and preferably not over 0.004 in. Thicknesses less than 0.003 in may make assembly difficult, while those greater than 0.006 in tend to produce joints having lowered strength. The brazing of certain aluminum alloys is similar in most respects to the brazing of other materials. However, joint clearances should be greater because of a somewhat more sluggish flow of the brazing alloys. For aluminum, a clearance of 0.005 to 0.010 in will be found satisfactory. Care must be exercised in fitting dissimilar metals, since the joint clearance at brazing temperature is the controlling factor. In these cases, consideration must be given to the relative expansion rates of the materials that are being joined.

The length of lap in a joint, the shear strength of the brazing alloy, and the average percentage of the brazing surface area that normally bonds are the principal factors determining the strength of brazed joints. The shear strength may be calculated by multiplying the width by the length of lap by the percentages of bond area and by taking into consideration the shear

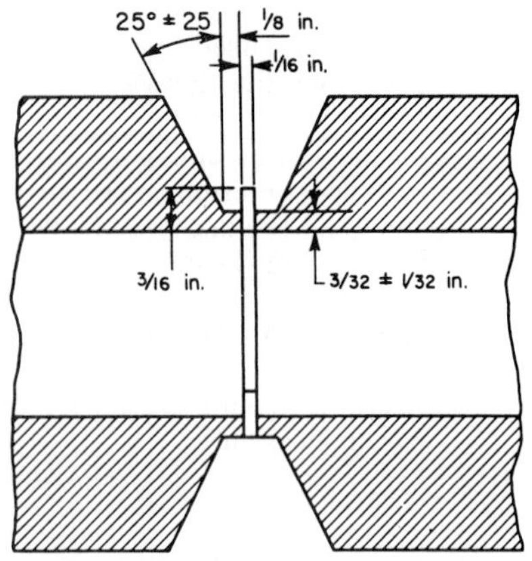

FIGURE A2.55 Eccentric insertion of consumable insert ring in pipe welded in the fixed horizontal pipe position.

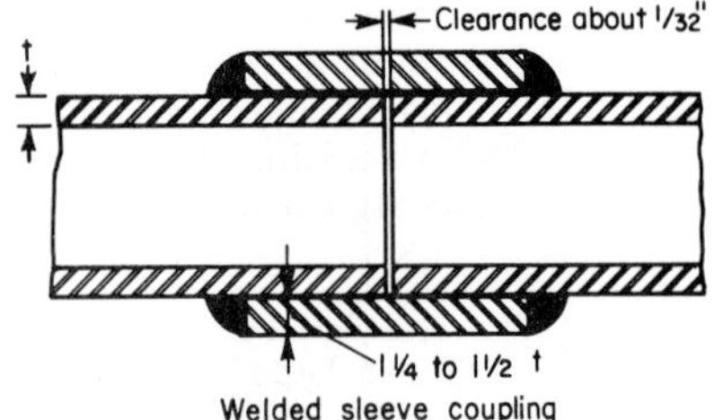

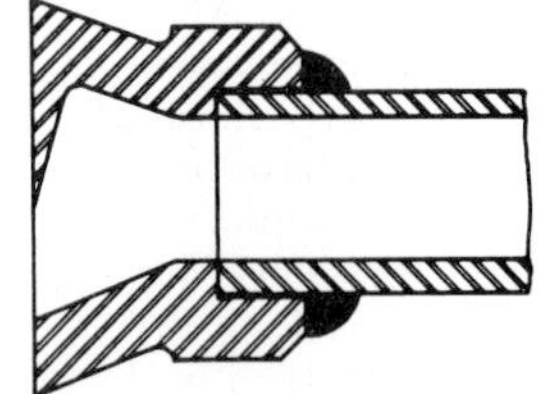

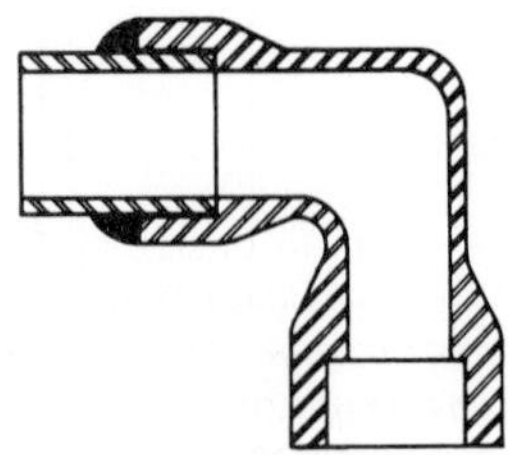

FIGURE A2.56 Examples of typical fillet-welded joints.

strength of the alloy used. An empirical method of determining the lap distance is to take it as twice the thickness of the thinner or weaker member joined. Normally this will provide adequate strength, but in cases of doubt, the fundamental calculations should be employed.

Such detailed determinations are generally unnecessary for brazed piping, since commercial brazing fittings are available in which the length of lap is predetermined at a safe value. For brass and copper pipe, cast- or wrought-bronze and wrought-copper fittings are available. A bore of correct depth to accept the pipe is provided, and midway down this bore may be a groove into which, at the time of manufacture, a ring of brazing alloy is inserted. Since the alloy is preplaced in fittings with such a groove, separate feeding of brazing alloy by hand is generally unnecessary.

JOINING CAST-IRON PIPE

Bell-and-Spigot Joint

This joint for underground cast-iron pipe was developed as far back as 1785. Standard dimensions are shown in Table A2.33.

The joint may be made up with lead and oakum, sulfur compounds, or cement. Lead and oakum constitute the prevailing joint sealer for sanitary systems. Bell-and-spigot joints are usually received for sanitary sewer systems.

Mechanical (Gland-Type) Joint

This modification of the bell-and-spigot joints, as designated in Federal Specification WW-P-421 and ANSI/AWWA C111/A21.11, is illustrated in Table A2.34. This joint is commonly used for low- and intermediate-pressure gas-distribution systems, particularly those conveying natural gas or dry manufactured gas. Mechanical joints are also used for water lines, sewage, and process piping. In the mechanical (gland-type) joint shown in Fig. A2.57 the lead and oakum of the conventional bell-and-spigot joint are supplanted by a stuffing box in which a rubber or composition packing ring, with or without a metal or canvas tip or canvas backing, is compressed by a cast-iron follower ring drawn up with bolts. In addition to making an inherently tight joint even under considerable pressure, this arrangement has the advantage of permitting relatively large lateral deflections (3½° to 7°), as well as longitudinal expansion or contraction.

Tyton Joint

The Tyton joint is designed to contain an elongated grooved gasket. The inside contour of the socket bell provides a seat for the circular rubber in a modified bulb-shaped gasket. An internal ridge in the socket fits into the groove of the gasket. A slight taper on the plain end of the pipe facilitates assembly.

Standard dimensions are given in Table A2.35. The maximum joint deflection angle is 5° for sizes through 12 in, 4° for 14 and 16 in, and 3° for 18, 20, and 24 in. Either all-bell U.S. standardized mechanical joint fittings or bell-and-spigot all-bell fittings with poured or cement joints can be used with Tyton-joint pipe.

TABLE A2.33 Standard Dimensions of Bell-and-Spigot Joints for Pipe Centrifugally Cast in metal Molds

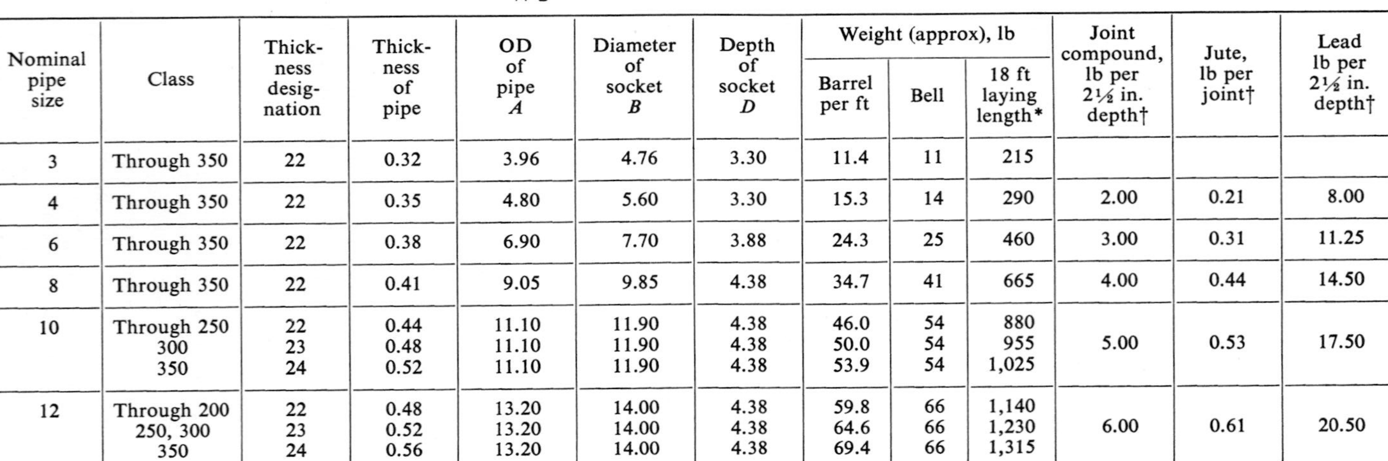

Nominal pipe size	Class	Thickness designation	Thickness of pipe	OD of pipe *A*	Diameter of socket *B*	Depth of socket *D*	Weight (approx), lb			Joint compound, lb per 2½ in. depth†	Jute, lb per joint†	Lead lb per 2½ in. depth†
							Barrel per ft	Bell	18 ft laying length*			
3	Through 350	22	0.32	3.96	4.76	3.30	11.4	11	215			
4	Through 350	22	0.35	4.80	5.60	3.30	15.3	14	290	2.00	0.21	8.00
6	Through 350	22	0.38	6.90	7.70	3.88	24.3	25	460	3.00	0.31	11.25
8	Through 350	22	0.41	9.05	9.85	4.38	34.7	41	665	4.00	0.44	14.50
10	Through 250 300 350	22 23 24	0.44 0.48 0.52	11.10 11.10 11.10	11.90 11.90 11.90	4.38 4.38 4.38	46.0 50.0 53.9	54 54 54	880 955 1,025	5.00	0.53	17.50
12	Through 200 250, 300 350	22 23 24	0.48 0.52 0.56	13.20 13.20 13.20	14.00 14.00 14.00	4.38 4.38 4.38	59.8 64.6 69.4	66 66 66	1,140 1,230 1,315	6.00	0.61	20.50

TABLE A2.33 Standard Dimensions of Bell-and-Spigot Joints for Pipe Centrifugally Cast in metal Molds (*Continued*)

14	50	21	0.48	15.30	16.10	4.50	69.7	78	1,335	7.00	0.81	24.00
	100	22	0.51	15.30	16.10	4.50	73.9	78	1,410			
	150	22	0.51	15.30	16.10	4.50	73.9	78	1,410			
	200	23	0.55	15.30	16.10	4.50	79.5	78	1,510			
	250, 300	24	0.59	15.30	16.10	4.50	85.1	78	1,610			
	350	25	0.64	15.30	16.10	4.50	92.0	78	1,735			
16	50, 100	22	0.54	17.40	18.40	4.50	89.2	96	1,700	8.25	0.94	33.00
	150	22	0.54	17.40	18.40	4.50	89.2	96	1,700			
	200	23	0.58	17.40	18.40	4.50	95.6	96	1,815			
	250	24	0.63	17.40	18.40	4.50	103.6	96	1.960			
	300, 350	25	0.68	17.40	18.40	4.50	111.4	96	2,100			
18	50	21	0.54	19.50	20.50	4.50	100.4	114	1.920	9.25	1.00	36.90
	100	22	0.58	19.50	20.50	4.50	107.6	114	2,050			
	150	22	0.58	19.50	20.50	4.50	107.6	114	2,050			
	200	23	0.63	19.50	20.50	4.50	116.5	114	2,210			
	250	24	0.68	19.50	20.50	4.50	125.4	114	2,370			
	300	25	0.73	19.50	20.50	4.50	134.3	114	2,530			
	350	26	0.79	19.50	20.50	4.50	144.9	114	2,720			
20	50	21	0.57	21.60	22.60	4.50	117.5	133	2,250	10.50	1.25	40.50
	100	22	0.62	21.60	22.60	4.50	127.5	133	2,430			
	150	22	0.62	21.60	22.60	4.50	127.5	133	2,430			
	200	23	0.67	21.60	22.60	4.50	137.5	133	2,610			
	250	24	0.72	21.60	22.60	4.50	147.4	133	2,785			
	300	25	0.78	21.60	22.60	4.50	159.2	133	3,000			
	350	26	0.84	21.60	22.60	4.50	170.9	133	3,210			
24	50	21	0.63	25.80	26.80	4.50	155.4	179	2,975	13.00	1.50	52.50
	100	22	0.68	25.80	26.80	4.50	167.4	179	3,190			
	150	23	0.73	25.80	26.80	4.50	179.4	179	3,410			
	200, 250	24	0.79	25.80	26.80	4.50	193.7	179	3,665			
	300	25	0.85	25.80	26.80	4.50	207.9	179	3,920			
	350	26	0.92	25.80	26.80	4.50	224.4	179	4,220			

*Includes weight of bell.
†Not included in ASA B21.6-1962.

TABLE A2.34 Standard Dimensions of Mechanical (Gland-Type) Joints (ANSI/AWWA C111/A21.11-1985)

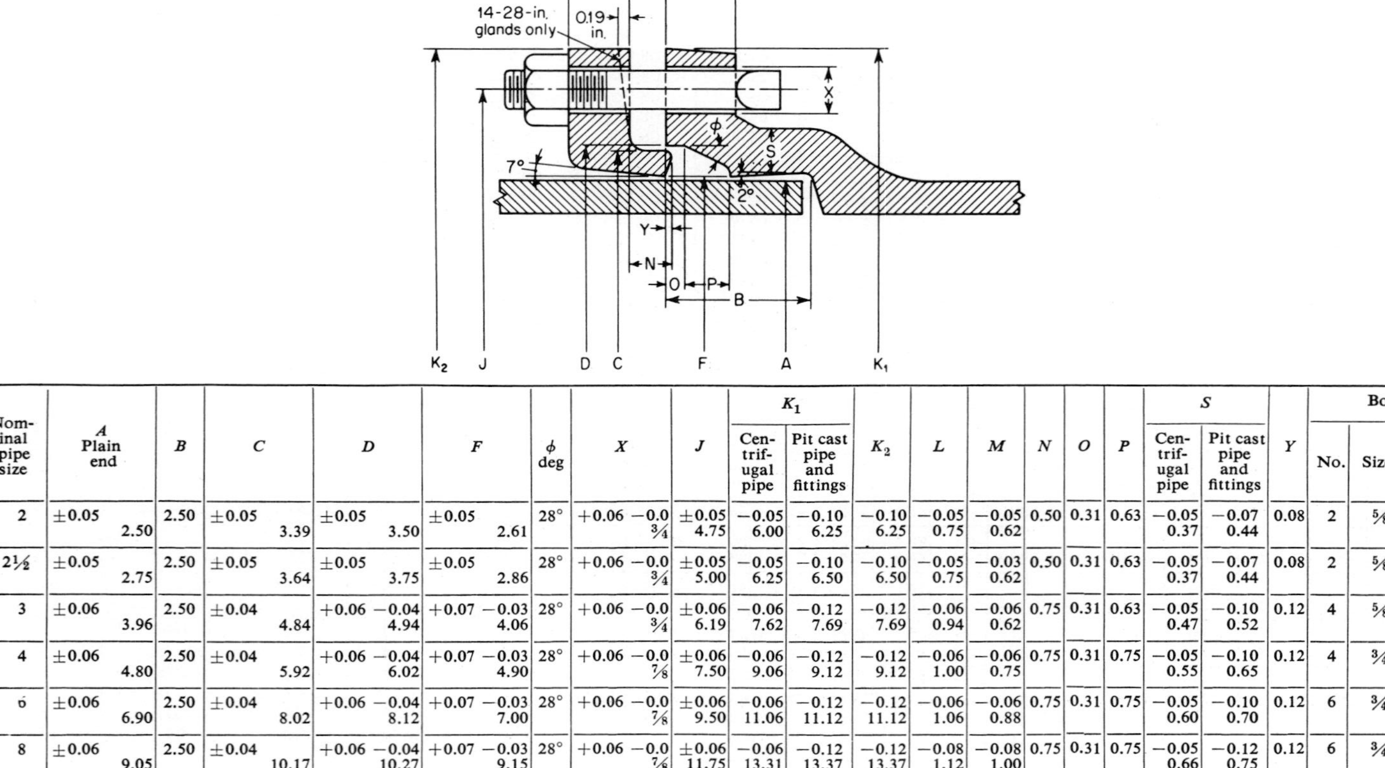

Nominal pipe size	A Plain end	B	C	D	F	φ deg	X	J	K_1 Centrifugal pipe	K_1 Pit cast pipe and fittings	K_2	L	M	N	O	P	S Centrifugal pipe	S Pit cast pipe and fittings	Y	Bolts No.	Bolts Size	Bolts Length
2	±0.05 2.50	2.50	±0.05 3.39	±0.05 3.50	±0.05 2.61	28°	+0.06 −0.0 ¾	±0.05 4.75	−0.05 6.00	−0.10 6.25	−0.10 6.25	−0.05 0.75	−0.05 0.62	0.50	0.31	0.63	−0.05 0.37	−0.07 0.44	0.08	2	⅝	2½
2½	±0.05 2.75	2.50	±0.05 3.64	±0.05 3.75	±0.05 2.86	28°	+0.06 −0.0 ¾	±0.05 5.00	−0.05 6.25	−0.10 6.50	−0.10 6.50	−0.05 0.75	−0.03 0.62	0.50	0.31	0.63	−0.05 0.37	−0.07 0.44	0.08	2	⅝	2½
3	±0.06 3.96	2.50	±0.04 4.84	+0.06 −0.04 4.94	+0.07 −0.03 4.06	28°	+0.06 −0.0 ¾	±0.06 6.19	−0.06 7.62	−0.12 7.69	−0.12 7.69	−0.06 0.94	−0.06 0.62	0.75	0.31	0.63	−0.05 0.47	−0.10 0.52	0.12	4	⅝	3
4	±0.06 4.80	2.50	±0.04 5.92	+0.06 −0.04 6.02	+0.07 −0.03 4.90	28°	+0.06 −0.0 ⅞	±0.06 7.50	−0.06 9.06	−0.12 9.12	−0.12 9.12	−0.06 1.00	−0.06 0.75	0.75	0.31	0.75	−0.05 0.55	−0.10 0.65	0.12	4	¾	3½
6	±0.06 6.90	2.50	±0.04 8.02	+0.06 −0.04 8.12	+0.07 −0.03 7.00	28°	+0.06 −0.0 ⅞	±0.06 9.50	−0.06 11.06	−0.12 11.12	−0.12 11.12	−0.06 1.06	−0.06 0.88	0.75	0.31	0.75	−0.05 0.60	−0.10 0.70	0.12	6	¾	3½
8	±0.06 9.05	2.50	±0.04 10.17	+0.06 −0.04 10.27	+0.07 −0.03 9.15	28°	+0.06 −0.0 ⅞	±0.06 11.75	−0.06 13.31	−0.12 13.37	−0.12 13.37	−0.08 1.12	−0.08 1.00	0.75	0.31	0.75	−0.05 0.66	−0.12 0.75	0.12	6	¾	4

10	±0.06 11.10	2.50	+0.06 −0.04 12.22	+0.06 −0.04 12.34	+0.07 −0.03 11.20	28°	+0.06 −0.0 ⅞	±0.06 14.00	−0.06 15.62	−0.12 15.69	−0.12 15.62	−0.08 1.19	−0.08 1.00	0.75	0.31	0.75	−0.06 0.72	−0.12 0.80	0.12	8	¾	4
12	±0.06 13.20	2.50	+0.06 −0.04 14.32	+0.06 −0.04 14.44	+0.07 −0.03 13.30	28°	+0.06 −0.0 ⅞	±0.06 16.25	−0.06 17.88	−0.12 17.94	−0.12 17.88	−0.08 1.25	−0.08 1.00	0.75	0.31	0.75	−0.06 0.79	−0.12 0.85	0.12	8	¾	4
14	+0.05 −0.08 15.30	3.50	+0.07 −0.05 16.40	+0.07 −0.05 16.54	+0.06 −0.07 15.44	28°	+0.06 −0.0 ⅞	±0.06 18.75	−0.08 20.25	−0.12 20.31	−0.12 20.25	−0.12 1.31	−0.12 1.25	0.75	0.31	0.75	−0.08 0.85	−0.12 0.89	0.12	10	¾	4
16	+0.05 −0.08 17.40	3.50	+0.07 −0.05 18.50	+0.07 −0.05 18.64	+0.06 −0.07 17.54	28°	+0.06 −0.0 ⅞	±0.06 21.00	−0.08 22.50	−0.12 22.56	−0.12 22.50	−0.12 1.38	−0.12 1.31	0.75	0.31	0.75	−0.08 0.91	−0.12 0.97	0.12	12	¾	4½
18	+0.05 −0.08 19.50	3.50	+0.07 −0.05 20.60	+0.07 −0.05 20.74	+0.06 −0.07 19.64	28°	+0.06 −0.0 ⅞	±0.06 23.25	−0.08 24.75	−0.15 24.83	−0.15 24.75	−0.12 1.44	−0.12 1.38	0.75	0.31	0.75	−0.08 0.97	−0.15 1.05	0.12	12	¾	4½
20	+0.05 −0.08 21.60	3.50	+0.07 −0.05 22.70	+0.07 −0.05 22.84	+0.06 −0.07 21.74	28°	+0.06 −0.0 ⅞	±0.06 25.50	−0.08 27.00	−0.15 27.08	−0.15 27.00	−0.12 1.50	−0.12 1.44	0.75	0.31	0.75	−0.08 1.03	−0.15 1.12	0.12	14	¾	4½
24	+0.05 −0.08 25.80	3.50	+0.07 −0.05 26.90	+0.07 −0.05 27.04	+0.06 −0.07 25.94	28°	+0.06 −0.0 ⅞	±0.06 30.00	−0.08 31.50	−0.15 31.58	−0.15 31.50	−0.12 1.62	−0.12 1.56	0.75	0.31	0.75	−0.08 1.08	−0.15 1.22	0.12	16	¾	5
30	+0.08 −0.06 32.00	4.00	+0.08 −0.06 33.29	+0.08 −0.06 33.46	+0.08 −0.06 32.17	20°	+0.06 −0.0 1⅛	±0.06 36.88	−0.12 39.12	−0.18 39.12	−0.18 39.12	−0.12 1.81	−0.12 2.00	0.75	0.38	1.00	−0.10 1.20	−0.15 1.50	0.12	20	1	6
36	+0.08 −0.06 38.30	4.00	+0.08 −0.06 39.59	+0.08 −0.06 39.76	+0.08 −0.06 38.47	20°	+0.06 −0.0 1⅛	±0.06 43.75	−0.12 46.00	−0.18 46.00	−0.18 46.00	−0.12 2.00	−0.12 2.00	0.75	0.38	1.00	−0.10 1.35	−0.15 1.80	0.12	24	1	6
42	+0.08 −0.06 44.50	4.00	+0.08 −0.06 45.79	+0.08 −0.06 45.96	+0.08 −0.06 44.67	20°	+0.06 −0.0 1⅜	±0.06 50.62	−0.12 53.12	−0.18 53.12	−0.18 53.12	−0.12 2.00	−0.12 2.00	0.75	0.38	1.00	−0.10 1.48	−0.15 1.95	0.12	28	1¼	6
48	+0.08 −0.06 50.80	4.00	+0.08 −0.06 52.09	+0.08 −0.06 52.26	+0.08 −0.06 50.97	20°	+0.06 −0.0 1⅜	±0.06 57.50	−0.12 60.00	−0.18 60.00	−0.18 60.00	−0.12 2.00	−0.12 2.00	0.75	0.38	1.00	−0.10 1.61	−0.15 2.20	0.12	32	1¼	6

[1]The thickness of the bell, S, shall in all instances be equal to, and generally exceed by at least 10 percent, the nominal wall thickness of the pipe or fitting of which it is a part.

[2]Cored holes may be tapered an additional 0.06 in in diameter.

[3]In the event of ovalness of the plain end outside diameter, the mean diameter measured by a circumferential tape shall not be less than the minimum diameter shown in the table. The minor axis shall not be less than the above minimum diameter plus an additional minus tolerance of 0.04 in for sizes 8-12 in, 0.07 in for sizes 14–24 in, and 0.10 in for sizes 30–48 in.

[4]K_1 and K_2 are the dimensions across the bolt holes. For sizes 2 and 2¼ in, both flange and gland may be oval shaped. For sizes 3–48 in, the gland may be polygon shaped.

[5]Mechanical joints require the use of specially designed bolts. See ANSI/AWWA C111/A2111-1985.

Mechanical Lock-Type Joint

For installations where the joints may tend to come apart owing to sag or lateral thrust in the pipeline, a mechanical joint having a self-locking feature is used to resist end pull. This joint is similar to the gland-type mechanical joint except that in the locked joint the spigot end of the pipe is grooved or has a recess to grip the gasket. Although only slight expansion or contraction can be accommodated in this type of joint, it does allow the usual 3½° to 7° angular deflection. The lock-type joint finds application aboveground in the process industries and in river crossings on bridges or trestles, as well as in submarine crossings or in unusually loose or marshy soils. Where the locking feature is on the spigot rather than on the bell, this type of pipe can be used with the regular line of mechanical-joint fittings.

FIGURE A2.57 Mechanical (gland-type) joint for cast-iron pipe.

Mechanical Push-On-Type Joint

Where a low-cost mechanical joint is desired, the roll-on type can be used. In this joint, a round rubber gasket is placed over the spigot end and is pulled into the bell by mechanical means, thus pulling the ring into place in the bottom of the bell. Outside the rubber gasket, braided jute is wedged behind a projecting ridge in the bell. This serves to confine the gasket under pressure in the joint. A bituminous compound is used to seal the mouth of the bell and to aid in retaining the hemp and the rubber gasket. In addition, when pipe is laid in cold climates where electrical thawing of mains and services is sometimes necessary, a cold lead strip about ¼ in^2 can be calked in between the hemp and bitumastic to provide an electrical circuit through the joint. Either bell-and-spigot or mechanical (gland-type) fittings are used with this line of pipe.

Mechanical Screw-Gland-Type Joint

This type of mechanical joint for cast-iron pipe makes use of a coarse-threaded screw gland drawn up by means of a spanner wrench to compress a standard rubber or composition packing gasket. The joint allows from 2° to 7° angular deflection, as well as expansion or contraction without danger of leaks. A lead ring, inserted in the bell ahead of the gasket, seals off the contents of the line from the gasket. The ring also provides an electrical circuit through the joint for thawing out frozen underground mains and service lines by the electrical method. The screw-gland joint is used in piping which conveys water, gas, oil, and other fluids at considerable pressure. The gaskets and lead rings are interchangeable with those used in equivalent lines of mechanical joints of the bolted-gland type. A full line of fittings is available for use with screw-gland pipe.

TABLE A2.35 Standard Dimensions of Tyton Joints

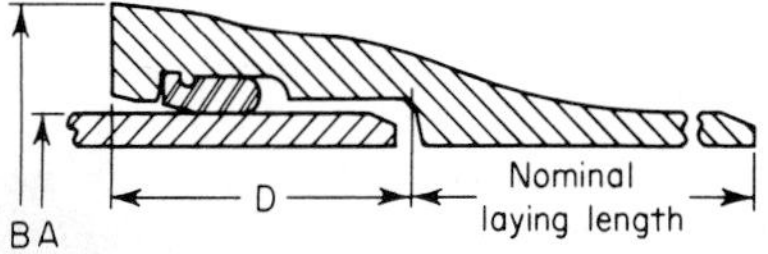

Nominal pipe size	Class	Thickness designation	Thickness of pipe	OD of pipe *A*	OD of bell *B*	Depth of socket *D*	Weight (approx), lb		
							Barrel per ft	Bell	18 ft length*
3	Through 350	22	0.32	3.96	6.08	3.00	11.4	11	215
4	Through 350	22	0.35	4.80	7.22	3.15	15.3	14	290
6	Through 350	22	0.38	6.90	9.47	3.38	24.3	25	460
8	Through 350	22	0.41	9.05	12.00	3.69	34.7	41	665
10	Through 250	22	0.44	11.10	14.20	3.75	46.0	54	880
	300	23	0.48				50.0		955
	350	24	0.52				53.9		1,025
12	Through 200	22	0.48	13.20	16.35	3.75	59.8	66	1,140
	250, 300	23	0.52				64.6		1,230
	350	24	0.56				69.4		1,315
14	50	21	0.48	15.30	19.15	5.00	69.7	78	1,335
	100, 150	22	0.51				73.9		1,410
	200	23	0.55				79.5		1,510
	250, 300	24	0.59				85.1		1,610
	350	25	0.64				92.0		1,735
16	Through 150	22	0.54	17.40	21.36	5.00	89.2	96	1,700
	200	23	0.58				95.6		1,815
	250	24	0.63				103.6		1,960
	300, 350	25	0.68				111.4		2,100
18	50	21	0.54	19.50	23.56	5.00	100.4	114	1,920
	100, 150	22	0.58				107.6		2,050
	200	23	0.63				116.5		2,210
	250	24	0.68				125.4		2,370
	300	25	0.73				134.3		2,530
	350	26	0.79				144.9		2,720
20	50	21	0.57	21.60	25.80	5.00	117.5	133	2,250
	100, 150	22	0.62				127.5		2,430
	200	23	0.67				137.5		2,610
	250	24	0.72				147.4		2,785
	300	25	0.78				159.2		3,000
	350	26	0.84				170.9		3,210
24	50	21	0.63	25.80	30.32	5.00	155.4	179	2,975
	100	22	0.68				167.4		3,190
	150	23	0.73				179.4		3,410
	200, 250	24	0.79				193.7		3,665
	300	25	0.85				207.9		3,920
	350	26	0.92				224.4		4,220

*Includes weight of bell.

Ball-and-Socket Joints

For river crossings, submarine lines, or other places where great flexibility is necessary, cast-iron pipe can be obtained with ball-and-socket joints of the mechanical-gland types as shown in Fig. A2.58. Provision is made for longitudinal expansion and contraction and a positive stop against disengagement of the joint is a feature of the design. As much as 15° angular deflection can be accommodated without leakage. This pipe is heavy enough to remain underwater where laid without requiring river clamps or anchorage devices. The pipe may be pulled across streams with a cable, since the joints are positively locked against separating, or it may be laid direct from a barge, bridge, or pontoons, without the services of a diver. The mechanical ball-and-socket joint is suitable for use with water, sewage, air, gas, oil, and other fluids at considerable pressure.

Either bell-and-spigot or mechanical (gland-type) fittings can be used with this line of pipe, although the integral ball present on the spigot end of some designs has to be cut off before the pipe can be inserted in a regular bell.

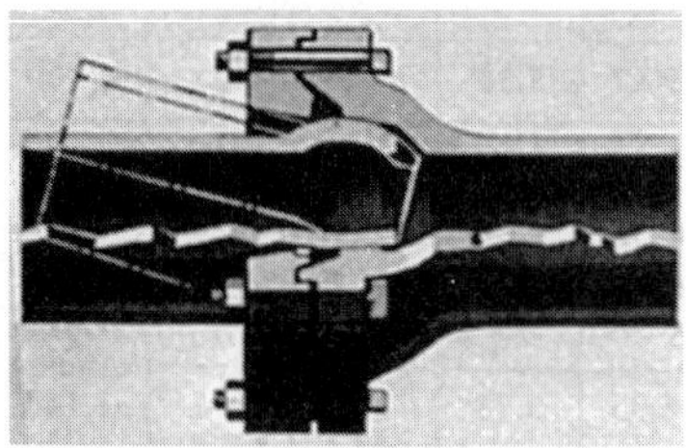

FIGURE A2.58 Ball-and-socket mechanical joint for cast-iron pipe.

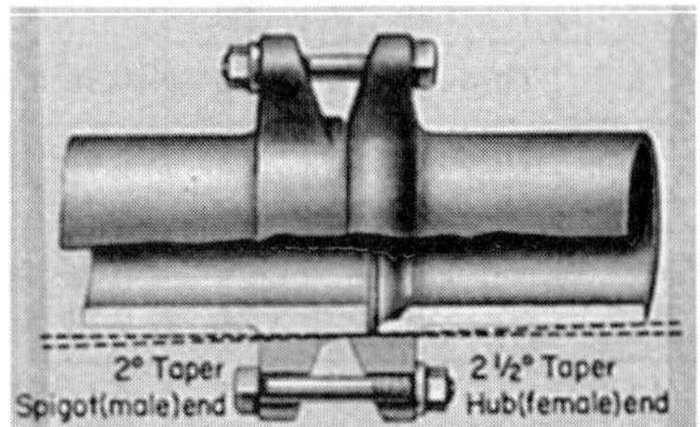

FIGURE A2.59 Universal cast-iron pipe joint.

Universal Pipe Joints

This type of cast-iron pipe joint (shown in Fig. A2.59) has a machined taper seat which obviates the need for calking or for a compression gasket. The joint is pulled up snugly with two bolts, after which the nuts are backed off slightly, thus enabling the lock washers to give enough to avoid overstressing the socket or lugs. Pipe is made in 12- to 20-ft lengths to the usual pressure classes and can be bought as Type III under Federal Specification WW-P-421. Universal-joint fittings are available for use with the pipe. This type of joint is used to some extent in pipe diameters of 4 to 24 in for underground water-supply systems, but it is not considered suitable for gas service, and it does not permit much angular displacement or expansive movement.

Compression-Sleeve Coupling

The type of joint shown in Fig. A2.60 is used with plain-end pipe of either cast iron or steel. It is widely known under the trade names of Dresser coupling and Dayton coupling. Compression-sleeve couplings are used extensively for air, gas, oil, water, and other services above- or underground. With a joint of this type, it is necessary to anchor or brace solidly at dead ends or turns to prevent the line from pulling apart. Compression couplings and fittings with screwed packing glands are available for use with small-size cast-iron or steel pipe. In welded

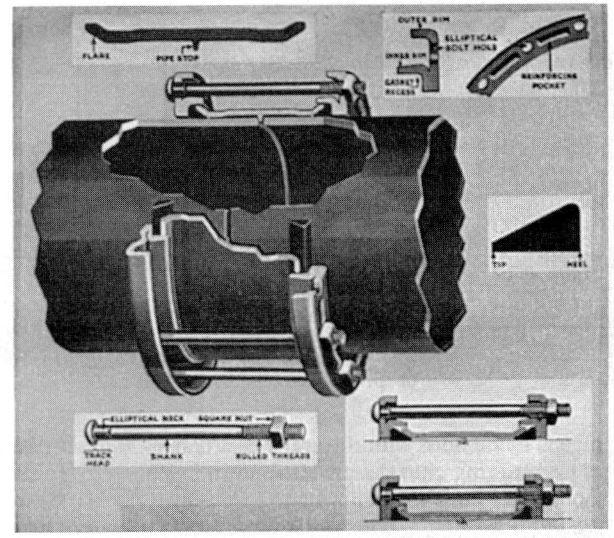

FIGURE A2.60 Compression sleeve (Dresser) coupling for plain-end cast-iron or steel pipe.

transmission lines for oil or gas where any significant change in temperature is expected, a certain percentage of the joints may be made up with compression couplings instead of welding in order to allow for expansion.

Grooved Segmented-Ring Coupling

The type of split coupling shown in Fig. A2.61 is used with either cast-iron or steel pipe that has grooves near the ends which enable the coupling to grip the pipe in order to prevent disengagement of the joint. The couplings are manufactured in a minimum of two segments for smaller pipe sizes and several segments for larger pipe sizes. Grooved-end fittings are available for use with the couplings. With proper choice of gasket material, the joint is suitable for use above- or underground with nearly any fluid or gas. The joint's advantages are:

FIGURE A2.61 Victualic coupling for grooved-end cast-iron or steel pipe.

- Ability to absorb minor angular and axial deflections
- Ability to increase gasket sealing force with increased system pressure
- Simplicity for rapid erection or dismantling for systems requiring frequent disassembly.

The coupling is also available in a style where grooving of the pipe ends is not required. Joint separation is prevented by the use of hardened steel inserts (teeth) which grab the mating pipe ends.

Flanged Joints

Flanged cast-iron pipe is used aboveground for low and intermediate pressures in water-pumping stations, gas works, power and industrial plants, oil refineries, booster stations for water, and gas and oil transmission lines. Flanges usually are faced and drilled according to ANSI B16.1.

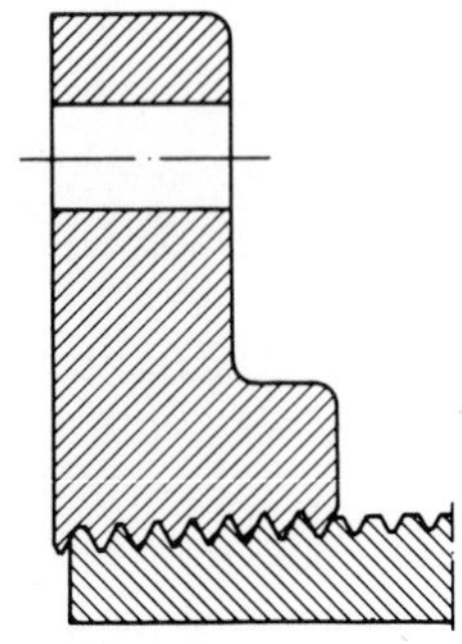

FIGURE A2.62 Screwed-on cast-iron flange.

Cast-iron pipe is made both with integrally cast flanges and with threaded companion flanges for screwing onto the pipe (as shown in Figs. A2.62 and A2.63). In the latter case, the outside diameter of the pipe conforms to iron-pipe-size (IPS) dimensions to allow for the threads provided. It is available in sizes 3 through 24 in and in length to 18 ft. For lengths less than 3 ft, in sizes 3 through 12 in, the flanges may be cast integrally with the pipe, rather than screwed on the pipe, at the manufacturer's option.

Standard dimension of flanged joints for silver brazing are shown in Table A2.36.

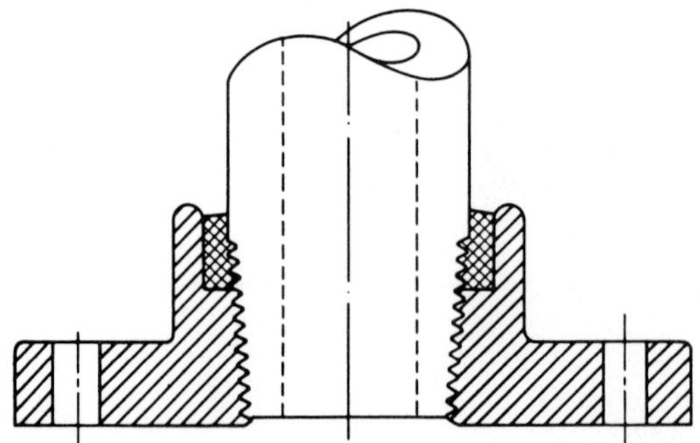

FIGURE A2.63 High-hub cast-iron flanges with bitumastic to protect the exposed threads.

CONCRETE, CEMENT, AND CEMENT-LINED PIPE

Nonreinforced-Concrete Pipe

Nonreinforced-concrete pipe for the conveyance of sewage, industrial waste, and storm water is made in sizes from 4 to 36 in. It is produced in accordance with the following specifications: ASTM C14, Standard Specifications for Concrete Sewer

TABLE A2.36 Standard Dimensions of Class 125 Flanged Joints for Silver Brazing with Centrifugally Cast Pipe

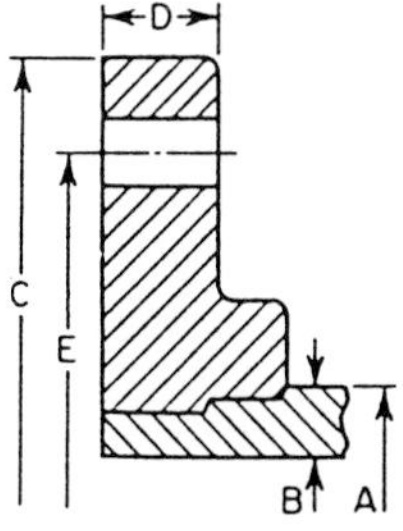

Nominal pipe size	Pipe: Outside diameter *A**	Flanges: Outside diameter *C*	Flanges: Thickness *D*†	Flanges: Bolt circle *E*	Flanges: Weight, lb each	Bolts: Number	Bolts: Diameter
2	2.50	6	¾	4¾	4	4	⅝
3	3.96	7½	1	6	7	4	⅝
4	4.80	9	1⅛	7½	13	8	⅝
6	6.90	11	1¼	9½	17	8	¾
8	9.05	13½	1⅜	11¾	27	8	¾
10	11.10	16	1½	14¼	38	12	⅞
12	13.20	19	1½	17	58	12	⅞

*For centrifugally ductile iron pipe in metal molds in sizes 3 in and larger, see ANSI A 21.51.
†Thickness *D* is slightly heavier than for standard cast-iron flanges in ASME/ANSI B16.1-1989.

Nonreinforced-concrete drain tile is used for land drainage and for subsurface drainage of highways, railroads, airports, and buildings. It is made in sizes from 4 through 36 in in accordance with the following specifications: ASTM C412, Standard Specification for Concrete Drain Tile and AASHO M178, Standard Specification for Concrete Drain Tile. Drain tile is available in the standard-quality, extra-quality, and special-quality classifications.

Perforated concrete pipe used for underdrainage is made in accordance with ASTM Specification C444, Specifications for Perforated Concrete Pipe. This pipe is also made in sizes 4 through 36 in and is available in the standard-strength and extra-strength classification.

Concrete irrigation pipe used for the conveyance of irrigation water under low hydrostatic heads and for land drainage is made in accordance with ASTM Specification C118, Standard Specifications for Concrete Pipe for Irrigation or Drainage. It is made in sizes 4 through 24 in.

Nonreinforced-concrete irrigation pipe for use with rubber-type gasket joints is made for conveyance of irrigation water at water pressures of 30 ft of head or higher depending on diameter. Such pipe is made in sizes 6 through 24 in in accordance with ASTM Specification C505, Specifications for Nonreinforced Concrete Irrigation Pipe with Rubber-type Gasket Joints. Physical and dimensional requirements of standard-strength bell-and-spigot nonreinforced-concrete sewer pipe are tabulated in Table A2.37.

Jointing. Rubber-gasketed joints for C14 and C76 pipe are covered by ASTM Specification C443, Joints for Circular Concrete Sewer and Culvert Pipe, Using

TABLE A2.37 Physical and Dimensional Requirements of Class 1, Bell-and-Spigot Nonreinforced Concrete Sewer Pipe (ASTM C1488; for Class 2 and Class 3 refer to ASTM C14)

Internal diameter (in) (1)	Min thickness of barrel T (in) (2)	Min laying length[1,4] L (ft) (3)	Inside diameter at mouth of socket[2] D_s (in) (4)	Depth of socket L_s (in) (5)	Min. taper of socket HL_s (6)	Min. thickness of socket[3] T_s (7)	Minimum strength (lb/lin ft): Three-edge bearing method (8)	Minimum strength (lb/lin ft): Sand-bearing method[4] (9)	Max. absorption (%) (10)
4	5⁄8	2½	6	1½	1:20	$3T/4$, all sizes	1500	1,500	8
6	5⁄8	2½	8¼	2	1:20		1500	1,650	8
8	¾	2½	10¾	2¼	1:20		1500	1,950	8
10	7⁄8	3	13	2½	1:20		1600	2,100	8
12	1	3	15¼	2½	1:20		1300	2,250	8
15	1¼	3	18¾	2½	1:20		2000	2,620	8
18	1½	3	22¼	2¾	1:20		2200	3,000	8
21	1¾	3	25¾	2¾	1:20		2400	3,300	8
24	2⅛	3	29½	3	1:20		2600	3,600	8

[1]Shorter lengths may be used for closures and specials.

[2]When pipe is furnished having an increase in thickness over that given in column 2, The diameter *at the inside* of the socket shall be increased by an amount equal to twice the increase of the barrel.

[3]This measurement shall be taken ¼ in from the outer end of the socket.

[4]Not included in ASTM Specification C14.

Flexible, Watertight, Rubber-type Joints. Gaskets must meet the following requirements:

Tensile strength, minimum psi	1200
Elongation at break, minimum, %	350
Shore durometer hardness, nominal	
Minimum	35 (see note)
Maximum	65
Compression set, maximum percent of original deflection, 70°C (158°F) for 22 h	25
Accelerated aging and changes in properties after conditioning in a circulating hot-air oven for 96 hr at 70°C (158°F):	
Decrease in tensile strength, maximum, percent of original	15
Decrease in elongation, maximum, percent of original	20
Water absorption by weight, 48 h at 70°C, maximum %	10

Note: Allowable variation ±5 from specified nominal hardness.

Reinforced Concrete Pipe

Reinforced concrete pipe for the conveyance of sewage, industrial wastes, and storm water and for the construction of culverts is made in sizes from 12 to 144 in. Reinforced concrete pipe may or may not be manufactured for use with rubber gaskets to seal the joints. It is usually manufactured in accordance with the following specifications:

ASTM C76—Specifications for Reinforced Concrete Culvert, Storm Drain and Sewer Pipe

AASHO M170—Specifications for Reinforced Concrete Culvert, Storm Drain and Sewer Pipe

Federal SS-P-375—Pipe, Concrete (Reinforced, Sewer)

Reinforced-concrete pipe may be made with either tongue-and-groove or bell-and-spigot joints. When made for use with rubber gaskets, the joints must conform to ASTM Specification C443 or AASHO Specification M198-Specifications for Joints for Circular Concrete Sewer and Culvert Pipe, Using Flexible Watertight, Rubber-type Gaskets.

Concrete pipe is available also in both an arch and an elliptical cross section. These pipes are made in accordance with the following specifications:

ASTM C506—Specifications for Reinforced Concrete Arch Culvert, Storm Drain and Sewer Pipe

ASTM C507—Specifications for Reinforced Concrete Elliptical Culvert, Storm Drain and Sewer Pipe

In each of the standards covering reinforced concrete pipe, five strength classes are defined in terms of minimum three-edge bearing load at a crack width of 0.01 in and at the ultimate strength of the pipe.

The strength class required for a given installation is determined by computing the earth load and live loads which will be transferred to the pipe under the conditions anticipated. This load is then converted to an equivalent three-edge bearing load by dividing it by a bedding factor. The bedding factor depends upon installation conditions and is always greater than 1.0.

Reinforced and Prestressed-Concrete Pressure Pipe

In one form or another, reinforced-concrete pressure pipe has been in use since 1909 for water-supply lines, water-transmission lines, water-intake lines, ocean outfalls, and sewer force mains. In recent years, it has been used in water-distribution systems and, to some extent, in industrial piping. Sizes range from 12 through 180 in for pressure services ranging up to 600 psi, depending on the type and size of pipe.

Four types of reinforced concrete pressure pipe are normally recognized. These are:

1. Reinforced-concrete pressure pipe
2. Reinforced-concrete cylinder pipe
3. Prestressed-concrete cylinder pipe
4. Pretensioned-concrete cylinder pipe

Large footages of reinforced-concrete cylinder pipe are in service, the oldest dating back to about 1920. Prestressed-concrete steel cylinder pipe was first used in 1942, and it now accounts for the major footage of concrete pressure pipe with thousands of miles of it in service. Prestressed-concrete pipe without a steel cylinder is also in use.

Reinforced-concrete pressure pipe is made in accordance with a number of different specifications:

AWWA C300—American Water Works Association Standard for Reinforced Concrete Pressure Pipe, Steel Cylinder Type, Not Prestressed (Table A2.38).

AWWA C301—American Water Works Association Standard for Prestressed Concrete Pressure Pipe, Steel Cylinder Type (Table A2.39).

AWWA C302—American Water Works Association Standard for Reinforced Concrete Pressure Pipe, Noncylinder Type (Table A2.40).

ASTM C361—American Society for Testing and Materials Specification for Reinforced Concrete Low-Head Pressure Pipe

AWWA C303

American Water Works Association Standard for Reinforced Concrete Pressure Pipe, Steel Cylinder Type, Pretensioned

The first two of the above are intended for higher pressures, and they incorporate a light-gauge steel cylinder in a reinforced-concrete or prestressed concrete envelope. The third is a reinforced-concrete pipe, normally with steel bell-and-spigot rings for a rubber O-ring gasket joint, for heads up to 100 ft of water.

Concrete pressure pipe is generally available in 16- or 20-ft lengths. It can be

TABLE A2.38 Requirements for Nonprestressed, Concrete-Lined Water Pipe (AWWA C300-89)

Pipe ID (in)	Min. thickness		Circumferential reinforcement spacing		Minimum thickness of sheet or plate (in)
	Pipe wall (in)	Concrete lining (in)	Min* (in)	Max* (in)	
30	3½	1	1¼	4	.179
36	4	1	1¼	4	.179
42	4½	1	1¾	5	¼
48	5	1¼	1¾	5	¼
54	5½	1¼	1¾	5	5/16
60	6	1¼	1¾	5	5/16
66	6½	1½	2	6	⅜
72	7	1½	2	6	⅜
78	7½	1½	2	6	7/16
84	8	1½	2	6	7/16
90	8	1¾	2¼	6	½
96	8½	1¾	2¼	6	½

NOTE: For pipe larger than 96 in in diameter, dimensions and details of design shall be subject to approval by the purchaser.

*Not specified by AWWA C300-89

TABLE A2.39 Requirements for Prestressed, Concrete-Lined or Coated Pipe (AWWA C301-84)

Pipe ID (in)	Pie with lined cylinder		Pipe with embedded cylinder		Coating thickness (in)	
	Core thickness (in)	Max design pressure (psi)	Core thickness (in)	Max design pressure (psi)	Mortar, min. over the wire	Concrete, nominal over the core
16	1	250	...	...	⅝	1½
18	1⅛	250	...	...	⅝	1½
20	1¼	250	...	...	⅝	1½
24	1½	200	2¼	275	⅝	1½
30	1⅞	200	2¼	240	⅝	1½
36	2¼	200	2¼	210	⅝	1½
42	2⅝	175	2⅝	190	⅝	1½
48	3	150	3	175	⅝	1½
54	...		4	200	⅝	1½
60	...		4½	200	⅝	1½
66	...	...	5	200	⅝	1½
72	...	...	5¼	200	⅝	1½
78	...	...	5¾	200	⅝	1½
84	...	...	6¼	200	⅝	1½
90	...	...	6½	200	⅝	1½
96	...	...	6½	200	⅝	1½

NOTE: For embedded cylinder pipe larger than 96 in in diameter, dimensions and details of design shall be subject to approval by the purchaser.

TABLE A2.40 Requirements for Noncylinder, Nonprestressed Concrete Pipe (AWWA C302-87)

Pipe ID (in)	6000-psi centrifugal concrete		4500-psi poured concrete		Circumferential reinforcement spacing		Min. total steel area per lin ft, (in^2)
	Min pipe wall thickness (in)	Min. no. of cages	Nominal pipe wall thickness (in)	Min. no. of cages	Min. (in)	Max. (in)	
12	2	1	...	...	1¼	4	0.08
15	2	1	...	...	1¼	4	0.11
16	2⅛	1	...	...	1¼	4	0.12
18	2¼	1	...	...	1¼	4	0.14
20	2⅜	1	...	...	1¼	4	0.16
21	2⅜	1	...	...	1¼	4	0.17
24	2½	1	3	1	1¼	4	0.20
27	2⅝	1	3¼	1	1¼	4	0.23
30	2¾	1	3½	1	1¼	4	0.25
36	3	1	4	2	1¼	4	0.30
42	3½	2	4½	2	1¾	5	0.35
48	4	2	5	2	1¾	5	0.40
54	4½	2	5½	2	1¾	5	0.45
60	5	2	6	2	1¾	5	0.52
66	5½	2	6½	2	2¼	6	0.61
72	6	2	7	2	2¼	6	0.71
78	6½	2	7½	2	2½	6	0.81
84	7	2	8	2	2¼	6	0.90
90	7½	2	8	2	2¼	6	1.00
96	8	2	8½	2	2¼	6	1.09

NOTE: For pipe larger than 96 in in diameter, dimensions and details of design shall be subject to approval by the purchaser.

furnished to any pressure class or classes best suited for a specific project. Prestressed-concrete-lined cylinder pipe (Fig. A2.64) is manufactured in sizes 16 through 48 in and for most pressures encountered in water-works systems. Prestressed-concrete-embedded cylinder pipe (Fig. A2.65) is produced in sizes 24 in and larger and for an even greater pressure range than lined cylinder pipe. Prestressed-concrete pipe without a steel cylinder covers about the same diameter and pressure range as embedded cylinder pipe.

Reinforced-concrete pressure pipe operating under relatively high pressures is manufactured with a watertight flexible-expansion joint. The joints are of the bell-and-spigot type (see Figs. A2.64 and A2.65) with the joint surfaces formed by steel rings in the ends of the pipe.

For pipe operating under moderate pressures, a rubber gasket is used to make the joint watertight. Steel joint rings are not used. In the non-cylinder-type pipe, joint rings are joined by the longitudinal reinforcement extending through the pipe. In the cylinder-type pipe, the rings are welded to the ends of the cylinder so as to form a continuous watertight membrane throughout the length of each pipe.

A rectangular groove is provided on the spigot end. Into it is placed a continuous round rubber gasket. As the pipes are pushed together, this gasket is com-

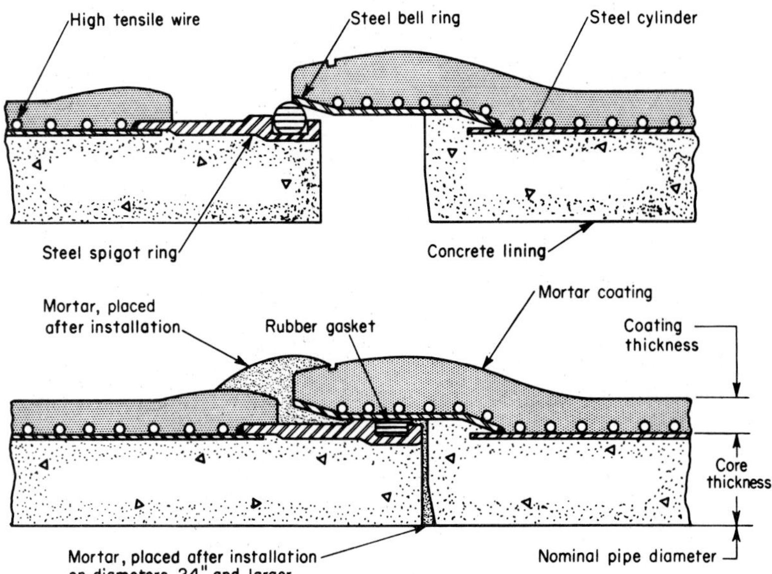

FIGURE A2.64 Prestressed concrete-lined cylinder pipe illustrating joint before closing and completed joint.

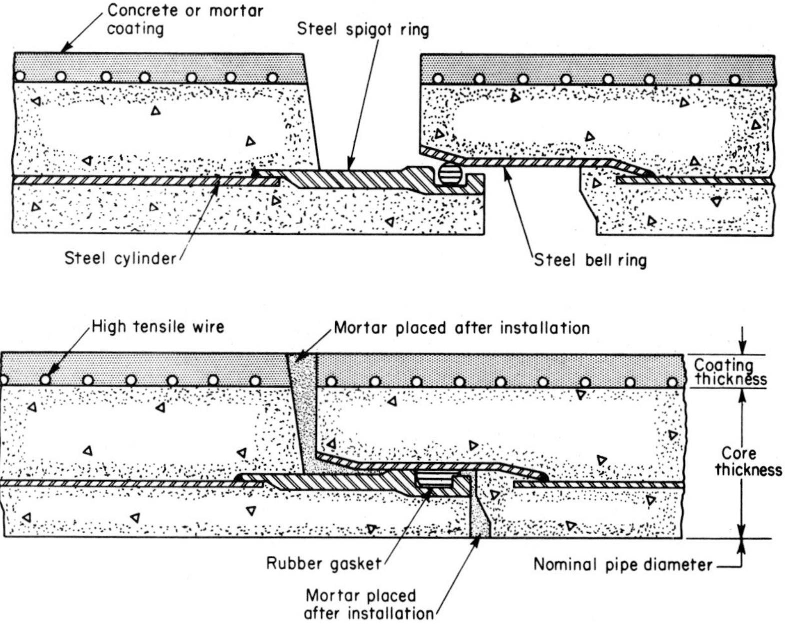

FIGURE A2.65 Prestressed concrete-embedded cylinder pipe, illustrating joint before closing and completed joint.

pressed into the groove by the flared portion of the bell. With the pipes pushed home, the gasket is confined on all four sides. The shape and dimensions of the joint rings or of the concrete surfaces are such as to make the joints self-centering even without the presence of the gasket. The weight of one pipe can be transmitted to the adjoining pipe only by the contact of the concrete surfaces. This prevents undue distortion of the gasket as the pipes move or deflect owing to expansion and contraction or settlement. Cement mortar or premolded asphalt is placed in the spaces between the ends of the pipe, both inside and outside of the joint, merely for the purpose of further protecting the steel joint rings. This mortar or asphalt does not serve to render the joint watertight. Certain jointing compounds are also used in this manner rather than mortar.

The ring forming the bell end of the pipe is covered on its exterior surface with reinforced concrete. The ring forming the spigot end should be lined on its inner surface with concrete. The gaskets sealing the joints are normally made of rubbers compounded to meet specified requirements to assure that a watertight and permanent seal is provided. The gasket should be a continuous ring of a size and cross section that is sufficient to completely fill the groove on the spigot when the pipes are laid. The rubber gasket is the sole element on which watertightness of the joint depends. Cement mortar or plastic materials are used to complete making the joint. However, they should not be depended upon for watertightness.

A *double rubber-gasket joint* is sometimes used with certain types of concrete pressure pipe carrying irrigation water under relatively low pressure. The pipe is made with a spigot on each end. The rubber gasket is placed in the spigot groove, and a double bell ring of steel or reinforced plastic is placed over the end of the pipe. Half the double bell ring extends beyond the end of the pipe. Steel bell rings usually receive an exterior coating of mortar.

All types of concrete pipe can be readily tapped under pressure for either threaded or flanged connections. The highly alkaline environment provided by the concrete and mortar encasement of the steel components prevents galvanic corrosion even in very low resistivity soils. A sustained Haxen-Williams "C" of 140 to 150 has been measured over the years in repeated tests of a number of concrete pressure pipelines.

Long-radius curves can be accomplished with joint openings on straight pipe or with beveled end pipe or bevel adapters. Elbows for short-radius deflections, reducers, tees, wyes, and closures are standard items. Special fittings tailored to suit specific requirements can be supplied.

The steel-cylinder-type pipes have been used principally in municipal and industrial water-supply systems for pressures grater than 40 psi and for sewer force mains. One of the major applications of the reinforced-concrete pipe without a steel cylinder has been steam electric-plant condenser-cooling water intake and discharge lines.

Installation of concrete pressure pipe is simple and rapid. AWWA Manual M9, Installation of Concrete Pipe, provides detailed information on alternate methods and equipment for constructing pipelines of this material. It also contains information on tapping concrete pressure pipe.

Cement-Lined Steel and Cast-Iron Pipe

Cement-lined steel and cast-iron pipe is covered by the following specifications:

AWWA C205—Standard for Cement-mortar Protective Lining and Coating for Steel Water Pipe.

AWWA C104—Cement Mortar Lining for Cast Iron Pipe and Fittings for Water (also ASA Standard A21.4).

Cement-lined pipe is well established for use in cold-water lines. Substantial quantities of cement-lined steel pipe are used for other applications where corrosion is more of a problem. The largest user, by far, is the petroleum industry for use in oil-field flow lines, pipe lines, tubing, and casing. Cement-lined pipe is particularly suitable for these applications because of the presence in the oil fields of salt water, hydrogen sulfide, carbon dioxide, and other corrosive material. Other applications include lines in salt works for handling brine, discharge lines in coal mines for carrying highly corrosive sulfur water, lines in paper and pulp mills for handling diluted acids and corrosive waste liquids, and lines in process plants where water or other liquids must be kept free from iron contamination or rust.

Cement-lined pipe is generally joined with screwed seal rings which prevent the corrosive liquid from coming in contact with steel. Flanged joints are also extensively used. Some prefabrication is done of piping assemblies involving welding of the steel joints. Field joining of the preassembled welded assemblies is then done with flanged ends. Cement of course must not be at the pipe ends being welded. After welding, these are filled with mortar.

CLAY PIPE

Vitrified and unglazed clay pipe is used for the conveyance of sewage, industrial wastes, and storm water in sizes from 4 to 36 in. The following ASTM Specifications have been issued to cover the physical properties, the performance requirements and the methods of installation and testing of vitrified clay pipe:

ASTM C12—Recommended Practice for Installing Vitrified Clay Sewer Pipe

ASTM C13—Specifications for Standard Strength Clay Sewer Pipe

ASTM C200—Specifications for Extra Strength Clay Sewer Pipe

ASTM C211—Specifications for Perforated Clay Pipe

ASTM C301—Methods of Testing Clay Pipe

ASTM C425—Vitrified Clay Pipe Joints Using Materials Having Resilient Properties

Unglazed clay pipe is covered by ASTM Specification C278, Specifications for Extra Strength Unglazed Clay Pipe.

Nearly all vitrified clay pipe is manufactured with a factory-fabricated compression-type joint. Three types of compression joints are normally produced by the clay pipe industry. These are classified in ASTM Specification C425, Vitrified Clay Pipe Joints Using Materials Having Resilient Properties. Features of the three types are outlined below:

Type I includes those joints in which the same resilient material is used both on the spigot end and in the bell of the pipe, except that the material of one of the controlled complementary jointing ends, either in the bell or on the spigot, may vary in hardness from the other. These preformed ends, by controlled complementary design, will form a positive mating pattern of closure.

Type II includes those joints in which different types of materials having resilient characteristics are used in a similar manner as described for Type I, except that the material of one of the controlled complementary jointing ends, either in the bell or on the spigot, may vary in hardness and resiliency from the other.

Type III includes those joints which rely upon a gasket or compression ring of a resilient material having a controlled and calculated shape which will be compressed within the annular space to form a closing seal. True round rings of rigid or flexible material placed on the spigot end and within the bell of the pipe form predictable dimensions within which the gasket will be compressed. The gasket or ring may or may not be attached to one or the other joining faces prior to socketing.

The resilient materials used in the manufacture of these joints are vinyl chloride plastisols, polyurethanes, polyester reins, and natural or synthetic rubber. These jointing materials are selected for their chemical resistance and physical properties. Various joint designs have been developed utilizing these jointing materials, but all designs must meet the rigid performance requirements outlined in the above ASTM specification.

MISCELLANEOUS PIPING

Glass Piping

Glass pipe and pipe fittings are manufactured to ASTM C599. The pipe is used extensively in the chemical, food, beverage, and pharmaceutical industries. Other applications include service in plating plants, paper mills, and textile-finishing plants. A borosilicate glass of very low alkali content and low thermal expansivity is the basic material.

The high heat resistance of borosilicate glass allows operation at temperatures up to 450°F.

A less-expensive type of sanitary glass piping using standard metal fittings is recommended for the dairy industry where pressures do not exceed 25 psi.

Properties. Typical physical properties of borosilicate glass involve the following:

Linear coefficient of expansion	0.0000018 in per degree F between 32 and 572°F or 0.22 in/100 ft/100°F temperature range (0.0000032 per degree C)
Thermal conductivity at 212°F	0.73 Btu/(h·ft·°F)
Specific heat	0.20 Btu/(lb°F)

The low coefficient of thermal expansion (approximately one-third to one-quarter that of steel) provides good resistance to thermal shock. This permits cleaning or sterilization with low-pressure steam. Especially when sanitary conditions are involved, the low-pressure steam is often followed by cold liquids. The maximum sudden (instantaneous) temperature drop which may be applied safely varies from about 200°F for 1- to 3-in-diameter pipe to 160°F for 6-in-diameter pipe.

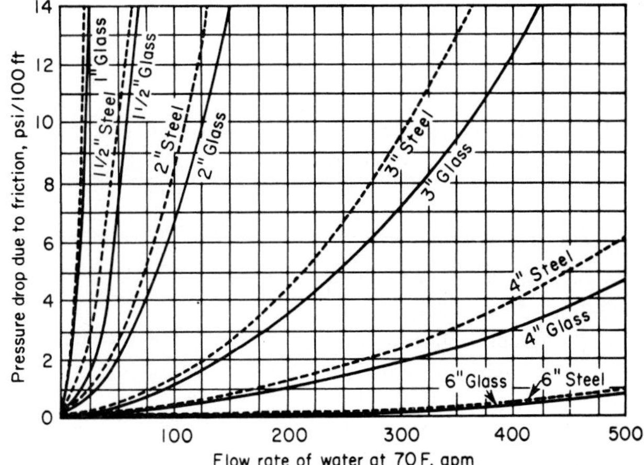

FIGURE A2.66 Calculated pressure drop for water at 70°F in glass and steel pipe (calculated from equations for friction factors by Drew, Koo, and McAdams, *Trans. AICE,* vol. 28, no. 56, 1933; and Wilson, McAdams, and Seltzer, *Ind. Eng. Chem.,* vol. 14, no. 105, 1922).

Pressure-drop data of glass are shown graphically in Fig. A2.66. Data for steel pipe are included for comparison.

The pressure drop through a mitred elbow is approximately twice that through a sweep elbow. But the mitred elbow is usually used because it has a more constant cross section and hence is considered to be stronger.

Permissible deflections and working stresses for borosilicate glass pipe are given in Table A2.41 and in Fig. A2.67. This confirms the importance of gasketed joints in assuring flexibility without inducing excessive stresses. It should be noted that a change of pipe direction increases the maximum recommended deflections. Recommended deflections are additive. For Teflon Type-T gaskets, these deflections should be reduced by approximately 25 percent.

Availability. Pipe is made in lengths up to 10 ft. Nominal dimensions and weights of flanged glass pipe are given in Table A2.42.

TABLE A2.41 Working Stresses and Flexibilities of Borosilicate Glass Pipe

Pipe size, in.	Approximate deflections based on ¼-in.-thick interface gaskets								
	Deflection, in.						Stress, psi	Side thrust, lb	
	A	*B*	C_1	*D*	*E*	*F*		5 ft	10 ft
1	0.22	0.52	0.82	0.21	0.36	0.72	300	0.8	0.4
1¼	0.16	0.36	0.56	0.14	0.24	0.48	300	1.8	0.9
2	0.11	0.26	0.40	0.10	0.17	0.34	275	2.8	1.4
3	0.07	0.14	0.21	0.053	0.09	0.18	250	6.5	3.2
4	0.05	0.11	0.17	0.042	0.072	0.145	225	13	6.5
6	0.030	0.07	0.10	0.025	0.042	0.084	200	30	15

Source: Corning Glass Works.

Pipe of 1-, 1½-, 2-, and 3-in nominal size is designed to operate with 60 psi pressure; the 4-in size is limited to 35 psi; and the 6-in size is limited to 20 psi. A compression type of flanged joint using cast-iron or aluminum flanges is available for sizes up to and including 6 in.

Standard fittings are also readily available (Figs. A2.68 and A2.69). Standard dimensions are tabulated in Tables A2.43 and A2.44. Special fittings involving sweep elbows, U-bends, reducing tees and crosses, and eccentric reducers can also be furnished.

Where fittings are used, the equivalent-length additions are listed in Table A2.45.

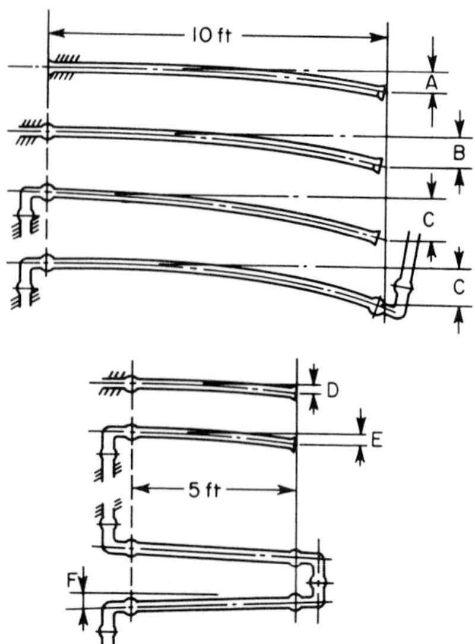

FIGURE A2.67 Permissible deflections for glass pipe, with dimensions given in Table A2.51.

Joints. Joints in glass pipe are normally made with metal flanges which are cushioned from the glass with molded asbestos inserts. An interface gasket of Teflon or other suitable material (depending on the fluid conveyed) is gripped between the pipe ends when the flange bolts are tightened. This prevents contact between the metal flanges and the fluid.

Typical pipe joints between glass pipe and between glass and metal pipe are illustrated in Figs. A2.70 and A2.71, respectively.

Pipe Support. Conventional pipe hangers are used for the support of glass pipe. Rod or guide sleeve hangers with an adjustable clevis are satisfactory. Rigid clamp hangers are used to prevent sidesway of the pipe when the hanger rods are extremely long and long pipe runs are involved.

The pipe hangers are normally padded with rubber or asbestos. Hangers are also supplied with neoprene or rubber coatings applied by dipping.

TABLE A2.42 Nominal Dimensions and Weights of Flanged Glass Pipe

Nominal size, in.	Outside diameter, in., average	Wall thickness, in., average	Weight per ft, lb
1	1.312	0.156	0.6
1½	1.844	0.172	1.0
2	2.344	0.172	1.1
3	3.406	0.203	2.0
4	4.530	0.265	3.4
6	6.656	0.328	6.3

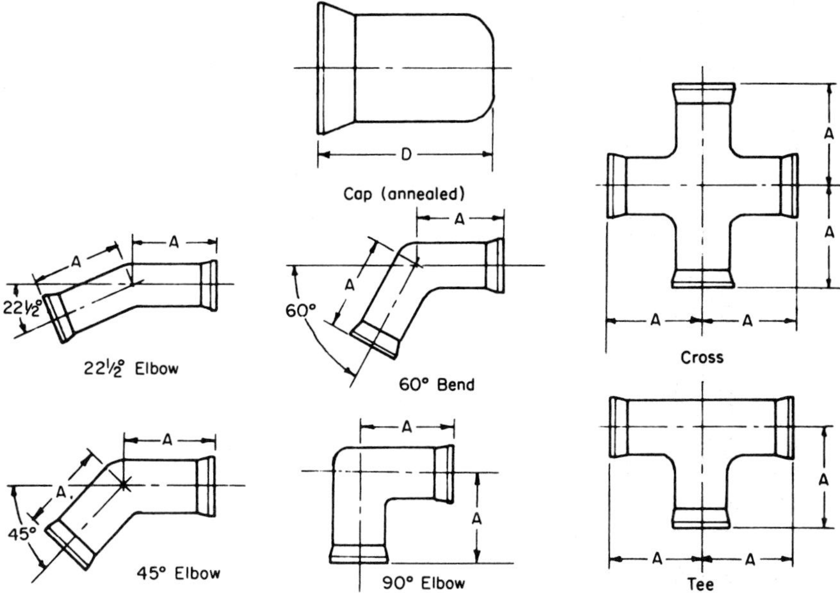

FIGURE A2.68 Standard glass pipe fittings. (*Corning Glass Works*)

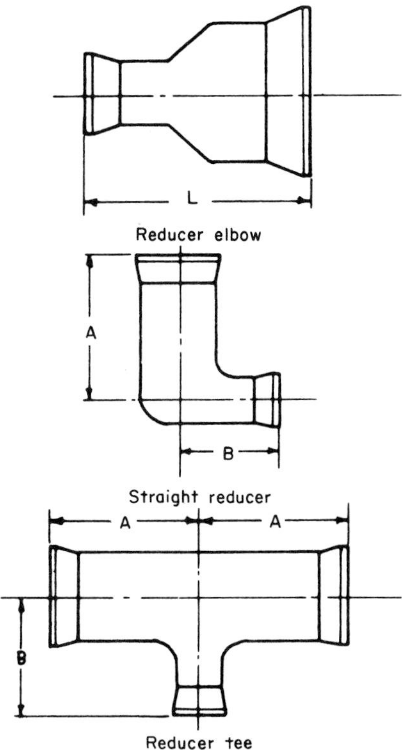

FIGURE A2.69 Standard glass pipe reducing fittings. (*Corning Glass Works*)

Minimum support distances between hanger locations are given in Table A2.46.

Hangers should be fastened to a rigid structure to prevent the hanger from sagging and stressing the glass pipe. Wood beams, especially out of doors, have a tendency to warp and sag. Hangers should always be adjusted to the glass pipe, not the pipe to the hangers.

Vertical glass lines should be supported by plates beneath metal flanges or by padded saddles beneath 90° elbows at the bottom of vertical rises. An anchor point is a rigid support for the glass line tying it into the building structure, to fixed equipment such as tanks and pumps or to independently supported valves. Only one anchor point should be in every straight run of pipe. Two anchor points may be used if there is an expansion joint between them.

Wood-Lined Steel Pipe

Where chemical solutions are to be handled under pressures, steel pipe

TABLE A2.43 Dimensions of Standard Glass Fittings Shown in Fig. A2.68

Pipe size, in.	$A \pm 1/32$ in.	$D \pm 1/32$ in.
1	2¾	2 11/16
1½	3½	3 7/16
2	4	3⅞
3	5	4 11/16
4	7	5 15/16
6	9	6 7/16

TABLE A2.44 Dimensions of Standard Reducing Glass Fittings Shown in Fig. A2.69

Size, in.	$A \pm 1/32$ in.	$B \pm 1/32$ in.	$L \pm 1/32$ in.
1½ × 1	3½	3	4
2 × 1	4	3	4
2 × 1½	4	3½	4
3 × 1	5	3½	5
3 × 1½	5	4	5
3 × 2	5	4½	5
4 × 1	7	4	7
4 × 1½	7	4½	7
4 × 2	7	5	7
4 × 3	7	5½	7
6 × 1	9	5	9
6 × 1½	9	5½	9
6 × 2	9	6	9
6 × 3	9	6½	9
6 × 4	9	8	9

Face-to-centerline dimensions for the 1-in legs are 5 in; 1½-in legs are 5½ in; and 2-in legs are 6in.

TABLE A2.45 Equivalent Length of Glass Fittings

Fitting	Equivalent length, ft					
	1 in.	1½ in.	2 in.	3 in.	4 in.	6 in.
45-deg mitred ell	1	1.3	2	3	4	6
90-deg mitred ell	5	7.5	10	15	20	30
90-deg sweep ell	2.7	4	5.3	8	11	16
U-bend	5	7.5	10	15	20	30
Tee, along run	1.7	2.5	3.3	5	6.7	10
Tee, entering run	5	7.5	10	15	20	30
Tee, entering branch	5.8	8.7	12	18	23	35

Source: Corning Glass Works.

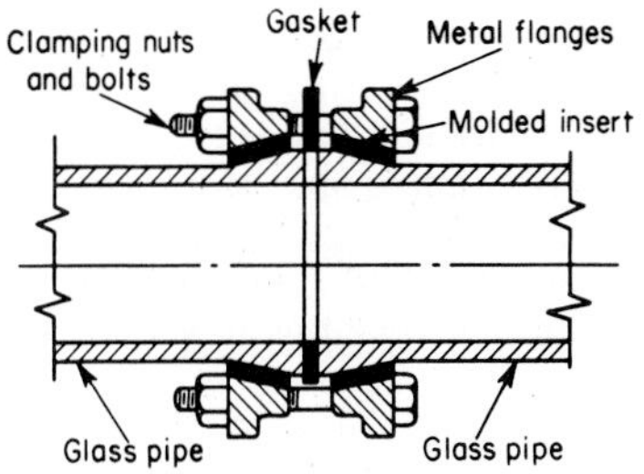

FIGURE A2.70 Flanged joint between glass pipe. (*Corning Glass Works*)

lined with wood staves is successfully used. This type of pipe is also particularly adaptable, even under low pressures, for certain types of pipelines located in the interior of industrial plants where definite alignment in length and fitting, suspension, and other installation features prompt its use. The wood staves that compose the lining offer resistance to the action of the solutions carried, and the metal pipe shell forms the outer casing that serves to carry the stresses incidental to the pressures present.

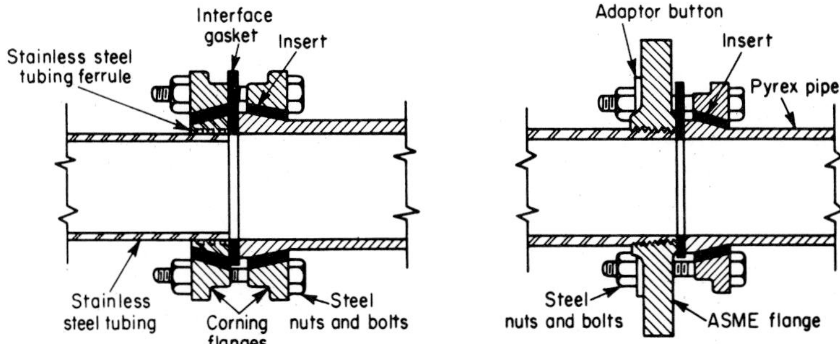

FIGURE A2.71 Typical flanged joints between glass pipe and plain or flanged steel pipe. (*Corning Glass Works*)

This classification of pipe is used in pulp-and-paper mills for conveying sulfite and sulfate stock, groundwood stock, diluted bleach liquor and bleached stock, and water. The wood lining contributes to clean paper making, eliminating the formation of the scale which otherwise would drop into the stock being carried. The accumulation of slime often found in pipelines is practically eliminated. It is also used in mining operations for transporting acid mine waters under high pres-

TABLE A2.46 Minimum Distances between Supports for Glass Pipe (ft)

Pipe size, in.	Fluid in pipe		
	Gas	Specific gravity less than 1.3	Specific gravity, 1.3 and greater
1	8	8	7
1½	9	9	7
2	9	9	8
3	9	9	8
4	10	10	8
6	10	10	8

sures. In chemical, textile, and process industries, wood-lined steel pipe is used in applications carrying acid solutions and gases.

This type of pipe is equipped with standard flanged joints and is easy to install. Field cutting to length and flanging, if necessary, is a simple operation, performed by sawing off the pipe to length required and welding a flange in place.

Fittings for this class of pipe, whether of the standard types of ells, tees, crosses, wyes, or reducers or of special nature are also lined with wood staves.

REFERENCES

1. ASTM Specification A307, Specification for Low-carbon Steel Externally and Internally Threaded Standard Fasteners.
2. ASTM Specification A449, Specification for Quenched and Tempered Steel Bolts and Studs.
3. ASTM Specification A354, Specification for Quenched and Tempered Alloy-steel Bolts and Studs With Suitable Nuts.
4. ASTM Specification A320-63, Specifications for Alloy-steel Bolting Materials for Low-temperature Service.
5. Society of Automotive Engineers, *SAE Handbook,* 1966 ed.

CHAPTER A3
PIPING MATERIALS

James M. Tanzosh
Babcock & Wilcox
Barberton, Ohio

The selection of materials for piping applications is a process that requires consideration of material characteristics appropriate for the required service. Material selected must safely withstand the given operating conditions of temperature and pressure during the intended design life. Mechanical strength must be appropriate for long-term service and capable of withstanding expected operational variables such as thermal or mechanical cycling. Extremes in application temperature can raise issues with material capabilities ranging from brittle fracture toughness at one end of the temperature spectrum to adequacy of creep strength and oxidation resistance at the other end of the temperature spectrum.

In addition, the operating environment surrounding the pipe or piping component must be considered. Degradation of material properties or loss of effective load carrying cross section can occur through corrosion, erosion, or a combination of the two. The substances contained and channeled in the piping are also an important factor.

The fabricability of the material selected must also be considered: The ability of the selected material to be bent or formed, suitability for welding or other methods of joining, the ease of heat treatment, and uniformity and stability of the resultant microstructure and material properties all contribute to the attractiveness and economy of a given piping material. The selection process should lead to the most economical material that meets the requirements of the service conditions and the codes and standards that apply.

Applicable design and construction codes, such as the ASME Boiler and Pressure Vessel Code and pressure piping codes such as ASME B31.1, identify acceptable materials for piping systems within their jurisdiction. These codes specify the design rules and the material design allowable stresses and other properties required to accomplish the design task. However, the information supplied is generally only adequate to assure safe operation under the expected thermal and mechanical conditions developed under steady state and, sometimes as in nuclear construction, cyclic operation. These codes do not directly and explicitly address the many other environmental and material degradation issues that should be considered by the design and materials engineers in arriving at a piping system that is not only safe to operate but will offer long-term, reliable service and function.

Thus, simply selecting materials and designing to "the Code" can sometimes lead to premature end of life of piping system components. This chapter will identify the important metallurgical characteristics of piping materials of construction and describe how they affect or can be effected by operation of all the materials available to the engineer. Carbon and low alloy steels come closest to being the ideal material of construction. The majority of piping applications use iron-based metals, those will be emphasized in this chapter.

MATERIAL PROPERTIES OF PIPING MATERIALS

Piping material behavior can be understood and predicted by studying a number of properties of the material. Appreciation of how a material will perform must extend all the way to the atomic components of the material. Metals are crystalline in structure and composed of atoms in precise locations within a space lattice.

The smallest component of the crystalline structure is called a unit cell; the unit cell is the smallest repeating building block of the material. For example, iron and iron-based alloys exist in two unit cell forms, the body-centered cubic (BCC) and the face-centered cubic (FCC) structure, shown in Fig. A3.1. They are differentiated in the way the atoms are arranged in repeating patterns. The body-centered cubic structure is represented by a cube with atoms at all eight corners and one atom in the center of the cube. The face-centered lattice is represented by atoms at the eight corners of the cube plus one atom at the center of each of the cube's six faces. The crystal structure naturally assumed by a material dictates some of the fundamental properties of the material. For example, FCC materials are generally more ductile than BCC materials. This is basically because FCC crystals are the most tightly packed of metallic structures and, as such, allow for more planes of atoms to slide across one another with the least amount of resistance (the fundamental atomic motion involved in what is called plasticity).

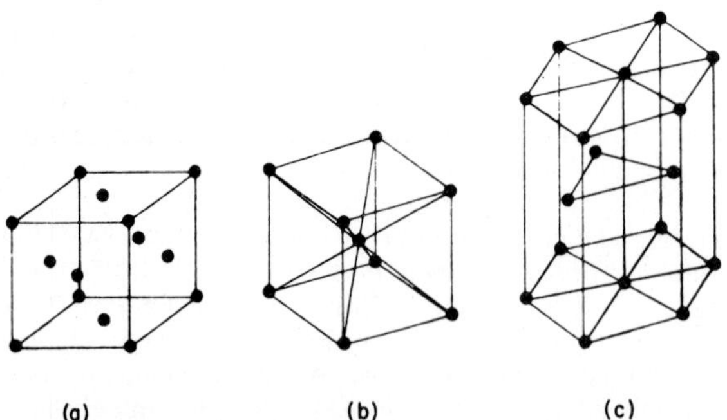

FIGURE A3.1 The three most common crystal structures in metals and alloys. (*a*) Face-centered cubic (FCC); (*b*) body-centered cubic (BCC); (*c*) hexagonal close-packed (HCP).

Metallic materials consist of these and other ordered crystal structures. Some metals, most notably iron, change crystal structure as temperature varies. Structure may also change as certain other elements are added in the form of alloying additions. These changes are used to advantage by metallurgists and are the basis for developing and manipulating important material behavior, such as the heat treatability of carbon and low alloy steels.

Plastics may be defined as synthetic material whose chief component is a resin or resin equivalent. The term *plastic* covers a broad range of materials that contain, as an essential ingredient, one or more organic polymetic substances. Plastics possess large molecular weight formed by the chemical combination of carbon-hydrogen atom chains (monomers to polymers). The atomic structure is thus ordered and predictable but is dissimilar from that of metals. Many plastics have greater strength per unit weight than metal, but they suffer due to lower impact strength, chemical stability, and thermal and aging stability. However, plastics fill an important niche in the piping engineer's repertoire.

Ceramic materials are composed of the oxides of metal arranged in ordered atomic structures similar to that of metals. The atomic constituents are electronically different, resulting in rigid, predictable behavior, but with an inherent lack of plasticity compared to metals.

Glasses form the other extreme of the atomic structure spectrum. Their atomic makeup is essentially that of a liquid; the structure is actually a solid with no ordered arrangement of atoms.

These atomic characteristics (i.e., the natural arrangement of the atoms), as well as the specific elements involved and their electronic characteristics, establish the fundamental properties of engineering materials. The properties that an engineer requires to design and construct a piping system are a manifestation of the longer range effects of atomic structure. These properties fall into three categories—chemical, mechanical, and physical.

Chemical Properties of Metals

Chemical properties are herein defined as those material characteristics dictated by the elemental constituency of the solid. This is usually measured by the relative atomic weight percent of the various elements (metals or nonmetals) or compounds within the material.

Metals are not usually used in their pure form. Rather, secondary elements are purposely added to improve or modify their behavior. This addition of secondary elements is called alloying, and the elements added fall into two categories, based on the relative size of the atoms. Atoms significantly smaller than those of the parent metal matrix fit into spaces between the atoms, in the lattice's interstices, and are called interstitial alloying elements. Carbon added to iron, creating steel, is the most common of these. Larger sized atoms will substitute for parent metal atoms in their matrix locations, thus the name substitutional alloying elements. Examples are zinc substituting for copper atoms in copper, creating brass and tin substituting for copper, creating bronze alloys.

Pure metals possess low strength. Adding an alloying element will increase the strength of the metal's atomic matrix because the atomic lattice is strained locally by the foreign atom, creating a larger impediment for the sliding of planes of atoms across one another during plastic flow. This is true whether the alloying element is interstitial or substitutional; however, the former generally serve as better lattice strengtheners. Strength properties are often improved to the detriment of ductility. Proper alloying, combined with appropriate metal processing and heat treatment, results in optimization of material properties.

Elements are also added to metals to improve or modify corrosion or oxidation performance, improve manufacturability (e.g., machinability), and improve electrical properties, among other effects. However, it is important to note that alloying done to accomplish optimization of one material property may act to the detriment of others.

In carbon steels, the most common construction materials, elements always present in varying amounts are carbon, manganese, phosphorous, sulfur, and silicon. Small amounts of other elements may be found either entering as gases during the steel-making process (hydrogen, oxygen, nitrogen) or introduced through the ores or metal scrap used to make the steel (nickel, copper, molybdenum, chromium, tin, antimony, etc.). The specific effect of each of these elements on steel properties is addressed later in the chapter. Addition of significant quantities of the interstitial element carbon will result in high strength and hardness but to the detriment of formability and weldability. A great amount of research has gone into the development of the principal metals used in piping design and construction, thus the specification limits must be vigorously adhered to in order to assure reliability, predictability, and repeatability of material behavior.

The number of elements alloyed with a parent metal, and the acceptable range of content of each, is identified in the material specification (e.g., ASTM, API, ASME). Tests to determine the elemental constituency of an alloy have been standardized and are also described in ASTM specifications. The material specifications also stipulate whether the chemical analysis of an alloy may be reported by analyzing a sample of the molten metal or taken from a specimen removed from the final product. The former is commonly referred to as a ladle analysis; the latter as a product or check analysis. This "chemistry" of a construction material is reported on a material test report that may be supplied by the material manufacturer upon request.

Mechanical Properties of Metals

Mechanical properties are critically important to the design process. They are defined as the characteristic response of a material to applied force. The method of measurement of these properties is done with standardized test methods described in ASTM specifications.

Properties fall into two general categories—strength and ductility. Some properties, such as material toughness, are dependent on both strength and ductility. The most widely known and used material properties, as defined by ASTM,[1] are given below.

Modulus of Elasticity (Young's Modulus). The modulus of elasticity is the ratio of normal stress to corresponding strain for tensile or compressive stresses. This ratio is linear through a range of stress known as Hooke's law. The material behavior in this range is elastic (i.e., if the applied load is released, the material will return to its original, unstressed shape). The value of the slope in the elastic range is defined as Young's modulus.

The modulus of elasticity is measured using the tension test, the most widely used test applied to engineering materials. The test consists of applying a gradually increasing load in either tension or compression, in a testing machine, to a standardized test specimen (Fig. A3.2). The applied load is continuously monitored, as is the test specimen elongation or contraction under load. These mea-

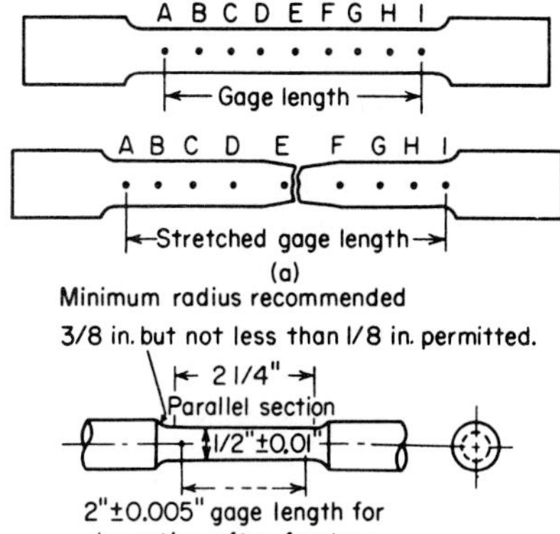

FIGURE A3.2 Tension-test specimens. (*a*) Strip specimen showing measurements which are taken to determine elongation; (*b*) standard round specimen with 2-in gauge length.

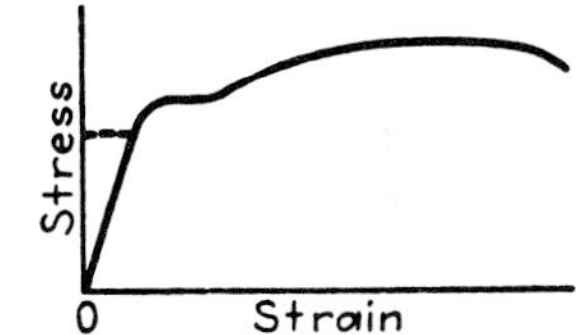

FIGURE A3.3 Stress-strain diagram.

sured quantities are generally represented on a coordinate axis, called a stress-strain curve (Fig. A3.3). The modulus of elasticity, as well as other strength properties, are established from this curve. Values of modulus of elasticity for a number of construction materials are given in Table A3.1.

Yield Strength. Upon loading a specimen beyond the point where elastic behavior can be maintained, the specimen will begin to deform in a plastic manner. This is the yield strength. Most materials do not abruptly transform from purely elastic to purely plastic behavior. Rather, a gradual transition occurs as represented by a curve, or "knee" in the stress-strain curve. Lacking an abrupt and easily definable point representing transition from elastic to plastic behavior, several standardized methods have been defined by ASTM to determine the yield strength used as the engineering property. The most common is termed the 0.2 percent offset method. In this approach a line is drawn parallel to the elastic portion of the curve anchored to a point displaced 0.2 percent along the strain axis (Fig. A3.4). The yield strength corresponds to the calculated value of the load indicated at the intersection point of the drawn line, divided by the original cross-sectional area in the gauge length of the tensile specimen. By convention, this test is performed at a constant rate of strain and is reported as newtons per squared meter or, in English units, as pounds per square inch of cross section.

Ultimate Tensile Strength. Upon further increase of applied load under constant strain rate, the specimen will continue to stretch until the loss of load carrying cross section caused by specimen thinning during the test (due to the Poisson ratio) cannot withstand further load increase. The ultimate tensile strength consti-

TABLE A3.1 Modulus of Elasticity U.S. Units for Metals*

Material	E = Modulus of Elasticity, Msi (Millions of psi), at Temperature, °F																				
	−425	−400	−350	−325	−200	−100	70	200	300	400	500	600	700	800	900	1000	1100	1200	1300	1400	1500
Ferrous Metals																					
Gray cast iron	...	...	...	...	...	...	13.4	13.2	12.9	12.6	12.2	11.7	11.0	10.2	...	...	...	...	...	...	...
Carbon steels, C ≤ 0.3%	31.9	...	...	31.4	30.8	30.2	29.5	28.8	28.3	27.7	27.3	26.7	25.5	24.2	22.4	20.4	18.0	...	...	...	...
Carbon steels, C > 0.3%	31.7	...	...	31.2	30.6	30.0	29.3	28.6	28.1	27.5	27.1	26.5	25.3	24.0	22.2	20.2	17.9	15.4	...	...	...
Carbon–moly steels	31.7	...	...	31.1	30.5	29.9	29.2	28.5	28.0	27.4	27.0	26.4	25.3	23.9	22.2	20.1	17.8	15.3	...	...	...
Nickel steels, Ni 2%–9%	30.1	...	...	29.6	29.1	28.5	27.8	27.1	26.7	26.1	25.7	25.2	24.6	23.0	...	...	...	...	...	...	...
Cr–Mo steels, Cr ½%–2%	32.1	...	...	31.6	31.0	30.4	29.7	29.0	28.5	27.9	27.5	26.9	26.3	25.5	24.8	23.9	23.0	21.8	20.5	18.9	...
Cr–Mo steels, Cr 2¼%–3%	33.1	...	...	32.6	32.0	31.4	30.6	29.8	29.4	28.8	28.3	27.7	27.1	26.3	25.6	24.6	23.7	22.5	21.1	19.4	...
Cr–Mo steels, Cr 5%–9%	33.4	...	...	32.9	32.3	31.7	30.9	30.1	29.7	29.0	28.6	28.0	27.3	26.1	24.7	22.7	20.4	18.2	15.5	12.7	...
Chromium steels, Cr 12%, 17%, 27%	31.8	...	...	31.2	30.7	30.1	29.2	28.5	27.9	27.3	26.7	26.1	25.6	24.7	22.2	21.5	19.1	16.6	...	...	...
Austenitic steels (TP304, 310, 316, 321, 347)	30.8	...	...	30.3	29.7	29.0	28.3	27.6	27.0	26.5	25.8	25.3	24.8	24.1	23.5	22.8	22.1	21.2	20.2	19.2	18.1
36% Ni steel (Invar)	19.2	19.1	19.1	19.2	19.5	20.0	20.8	21.5	22.0	22.5	22.6	22.6	...	...	...	...	...	...	...	...	...
Copper and Copper Alloys																					
Comp. and leaded–Sn bronze (C83600, C92200)	...	...	...	14.8	14.6	14.4	14.0	13.7	13.4	13.2	12.9	12.5	12.0	...	...	...	...	...	...	...	...
Naval brass, Si– & Al–bronze (C46400, C65500, C95200, C95400)	...	...	...	15.9	15.6	15.4	15.0	14.6	14.4	14.1	13.8	13.4	12.8	...	...	...	...	...	...	...	...
Copper (C11000)	...	...	...	16.9	16.6	16.5	16.0	15.6	15.4	15.0	14.7	14.2	13.7	...	...	...	...	...	...	...	...
Copper, red brass, Al–bronze (C10200, C12000, C12200, C12500, C14200, C23000, C61400)	...	...	...	18.0	17.7	17.5	17.0	16.6	16.3	16.0	15.6	15.1	14.5	...	...	...	...	...	...	...	...
90Cu–10Ni (C70600)	...	...	...	19.0	18.7	18.5	18.0	17.6	17.3	16.9	16.6	16.0	15.4	...	...	...	...	...	...	...	...
Leaded Ni–bronze	...	...	...	20.1	19.8	19.6	19.0	18.5	18.2	17.9	17.5	16.9	16.2	...	...	...	...	...	...	...	...
80Cu–20Ni (C71000)	...	...	...	21.2	20.8	20.6	20.0	19.5	19.2	18.8	18.4	17.8	17.1	...	...	...	...	...	...	...	...
70Cu–30Ni (C71500)	...	...	...	23.3	22.9	22.7	22.0	21.5	21.1	20.7	20.2	19.6	18.8	...	...	...	...	...	...	...	...
Nickel and Nickel Alloys																					
Monel 400 (N04400)	28.3	...	...	27.8	27.3	26.8	26.0	25.4	25.0	24.7	24.3	24.1	23.7	23.1	22.6	22.1	21.7	21.2	...	...	...
Alloys G, G1, 20 Mod. (N06007, N08320)	30.3	...	...	29.5	29.2	28.6	27.8	27.1	26.7	26.4	26.0	25.7	25.3	24.7	24.2	23.6	23.2	22.7	...	...	...
Alloys 800, 800H, X (N08800, N08810, N06002)	31.1	...	...	30.5	29.9	29.4	28.5	27.8	27.4	27.1	26.6	26.4	25.9	25.4	24.8	24.2	23.8	23.2	...	...	...
Alloys C-4, C276 (N06455, N10276)	32.5	...	...	31.6	31.3	30.6	29.8	29.1	28.6	28.3	27.9	27.6	27.1	26.5	25.9	25.3	24.9	24.3	...	...	...
Nickel 200, 201, Alloy 625 (N02200, N02201, N06625)	32.7	...	...	32.1	31.5	30.9	30.0	29.3	28.8	28.5	28.1	27.8	27.3	26.7	26.1	25.5	25.1	24.5	...	...	...
Alloy 600 (N06600)	33.8	...	...	33.2	32.6	31.9	31.0	30.2	29.9	29.5	29.0	28.7	28.2	27.6	27.0	26.4	25.9	25.3	...	...	...
Alloy B (N10001)	33.9	...	...	33.3	32.7	32.0	31.1	30.3	29.9	29.5	29.1	28.8	28.3	27.7	27.1	26.4	26.0	25.3	...	...	...
Alloy B-2 (N10665)	34.2	...	...	33.3	33.0	32.3	31.4	30.6	30.1	29.8	29.4	29.0	28.6	27.9	27.3	26.7	26.2	25.6	...	...	...
Unalloyed Titanium																					
Grades 1, 2, 3, and 7	...	...	...	...	...	...	15.5	15.0	14.6	14.0	13.3	12.6	11.9	11.2	...	...	...	...	...	...	...

*These data are for information, and it is not to be implied that materials are suitable for all the temperatures shown. Data are taken from Code for Pressure Piping, ASME B31.3-1990.

TABLE A3.1 Modulus of Elasticity* (*Continued*)

Material	E = Modulus of Elasticity, Msi (Millions of psi), at Temperature, °F									
	−425	−400	−350	−325	−200	−100	70	200	300	400
Aluminum and Aluminum Alloys										
Grades 443, 1060, 1100, 3003, 3004, 6061, 6063 (A24430, A91060, A91100, A93003, A93004, A96061, A96063)	11.4	...	...	11.1	10.8	10.5	10.0	9.6	9.2	8.7
Grades 5052, 5154, 5454, 5652 (A95052, A95154, A95454, A95652)	11.6	...	...	11.3	11.0	10.7	10.2	9.7	9.4	8.9
Grades 356, 5083, 5086, 5456 (A03560, A95083, A95086, A95456)	11.7	...	...	11.4	11.1	10.8	10.3	9.8	9.5	9.0

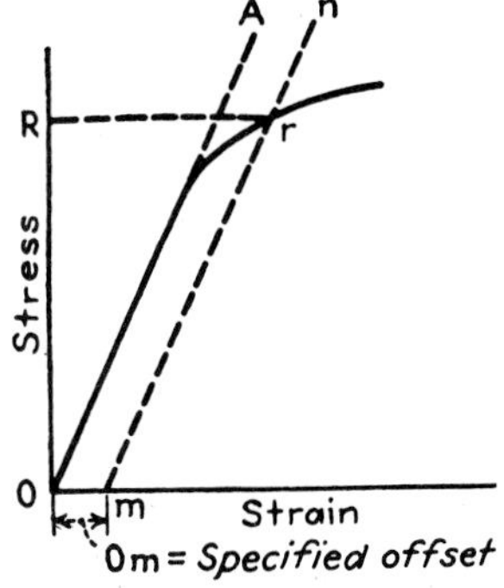

FIGURE A3.4 Offset method of determining yield strength.

tutes the maximum applied load divided by the original specimen cross-sectional area.

Elongation and Reduction of Area. Ductility of the test specimen can be established by measuring the length and minimum diameter before and after testing. Stretch of the specimen is represented as a percent elongation in a given length (usually 2–8 in) and is calculated in the following manner:

$$\text{Percent elongation} = \frac{(\text{final length} - \text{original length})}{\text{original length}} \times 100$$

The diameter of the test specimen will decrease, or neck down, in ductile materials.

Another standard measure of ductility is the reduction of area of the specimen, defined as follows:

$$\text{Percent reduction of area} = \frac{(\text{original cross-sectional area} - \text{final area})}{\text{original area}} \times 100$$

Hardness. Hardness is a measure of the material's ability to resist deformation, usually determined by a standardized test where the surface resistance to indentation is measured. The most common hardness tests are defined by the type and size of the indentor and the amount of load applied. The hardness numbers are a nondimensioned, arbitrary scale, with increasing numbers representing harder surfaces. The two most common hardness test methods are Brinell hardness and Rockwell hardness, each representing a standardized test machine with its own unique hardness scales (Table A3.2). Hardness loosely correlates with ultimate tensile strength in metals (Fig. A3.5).

TABLE A3.2 Conversion Table for Various Hardness Scales

Approximate Equivalent Hardness Numbers for Diamond Pyramid Hardness (Vickers, DPH) for Steel[1]

Diamond pyramid hardness No.	Brinell hardness No., 10-mm ball, 3,000-kg load			Rockwell hardness No.				Rockwell superficial hardness No., superficial brale penetrator			Shore scleroscope hardness No.	Tensile strength (approx.) 1,000 psi
	Standard ball	Hultgren ball	Tungsten carbide ball	A-scale, 60-kg load, brale penetrator	B-scale, 100-kg load, 1/16-in.-diam ball	C-scale, 150-kg load, brale penetrator	D-scale, 100-kg load, brale penetrator	15-N scale, 15-kg load	30-N scale, 30-kg load	45-N scale, 45-kg load		
940	...	...	...	85.6	...	68.0	76.9	93.2	84.4	75.4	97	...
920	...	...	...	85.3	...	67.5	76.5	93.0	84.0	74.8	96	...
900	...	...	...	85.0	...	67.0	76.1	92.9	83.6	74.2	95	...
880	...	...	767	84.7	...	66.4	75.7	92.7	83.1	73.6	93	...
860	...	...	757	84.4	...	65.9	75.3	92.5	82.7	73.1	92	...
840	...	...	745	84.1	...	65.3	74.8	92.3	82.2	72.2	91	...
820	...	...	733	83.8	...	64.7	74.3	92.1	81.7	71.8	90	...
800	...	...	722	83.4	...	64.0	73.8	91.8	81.1	71.0	88	...
780	...	...	710	83.0	...	63.3	73.3	91.5	80.4	70.2	87	...
760	...	...	698	82.6	...	62.5	72.6	91.2	79.7	69.4	86	...
740	...	...	684	82.2	...	61.8	72.1	91.0	79.1	68.6	84	...
720	...	...	670	81.8	...	61.0	71.5	90.7	78.4	67.7	83	...
700	...	615	656	81.3	...	60.1	70.8	90.3	77.6	66.7	81	...
690	...	610	647	81.1	...	59.7	70.5	90.1	77.2	66.2	..	...
680	...	603	638	80.8	...	59.2	70.1	89.8	76.8	65.7	80	329
670	...	597	630	80.6	...	58.8	69.8	89.7	76.4	65.3	..	324
660	...	590	620	80.3	...	58.3	69.4	89.5	75.9	64.7	79	319
650	...	585	611	80.0	...	57.8	69.0	89.2	75.5	64.1	..	314
640	...	578	601	79.8	...	57.3	68.7	89.0	75.1	63.5	77	309
630	...	571	591	79.5	...	56.8	68.3	88.8	74.6	63.0	..	304
620	...	564	582	79.2	...	56.3	67.9	88.5	74.2	62.4	75	299
610	...	557	573	78.9	...	55.7	67.5	88.2	73.6	61.7	..	294
600	...	550	564	78.6	...	55.2	67.0	88.0	73.2	61.2	74	289
590	...	542	554	78.4	...	54.7	66.7	87.8	72.7	60.5	..	284
580	...	535	545	78.0	...	54.1	66.2	87.5	72.1	59.9	72	279
570	...	527	535	77.8	...	53.6	65.8	87.2	71.7	59.3	..	274
560	...	519	525	77.4	...	53.0	65.4	86.9	71.2	58.6	71	269
550	505	512	517	77.0	...	52.3	64.8	86.6	70.5	57.8	..	264
540	496	503	507	76.7	...	51.7	64.4	86.3	70.0	57.0	69	260
530	488	495	497	76.4	...	51.1	63.9	86.0	69.5	56.2	..	254
520	480	487	488	76.1	...	50.5	63.5	85.7	69.0	55.6	67	250
510	473	479	479	75.7	...	49.8	62.9	85.4	68.3	54.7	..	244
500	465	471	471	75.3	...	49.1	62.2	85.0	67.7	53.9	66	240
490	456	460	460	74.9	...	48.4	61.6	84.7	67.1	53.1	..	234
480	448	452	452	74.5	...	47.7	61.3	84.3	66.4	52.2	64	230

470	**441**	**442**	**442**	**74.1**	...	**46.9**	**60.7**	**83.9**	**65.7**	**51.3**	..	224
460	**433**	**433**	**433**	**73.6**	...	**46.1**	**60.1**	**83.6**	**64.9**	**50.4**	62	219
450	**425**	**425**	**425**	**73.3**	...	**45.3**	**59.4**	**83.2**	**64.3**	**49.4**	..	214
440	**415**	**415**	**415**	**72.8**	...	**44.5**	**58.8**	**82.8**	**63.5**	**48.4**	59	210
430	**405**	**405**	**405**	**72.3**	...	**43.6**	**58.2**	**82.3**	**62.7**	**47.4**	..	204
420	**397**	**397**	**397**	**71.8**	...	**42.7**	**57.5**	**81.8**	**61.9**	**46.4**	57	200
410	**388**	**388**	**388**	**71.4**	...	**41.8**	**56.8**	**81.4**	**61.1**	**45.3**	..	195
400	**379**	**379**	**379**	**70.8**	...	**40.8**	**56.0**	**81.0**	**60.2**	**44.1**	55	190
390	**369**	**369**	**369**	**70.3**	...	**39.8**	**55.2**	**80.3**	**59.3**	**42.9**	..	185
380	**360**	**360**	**360**	**69.8**	(110.0)	**38.8**	**54.4**	**79.8**	**58.4**	**41.7**	52	180
370	**350**	**350**	**350**	**69.2**	...	**37.7**	**53.6**	**79.2**	**57.4**	**40.4**	..	175
360	**341**	**341**	**341**	**68.7**	(109.0)	**36.6**	**52.8**	**78.6**	**56.4**	**39.1**	50	170
350	**331**	**331**	**331**	**68.1**	...	**35.5**	**51.9**	**78.0**	**55.4**	**37.8**	..	166
340	**322**	**322**	**322**	**67.6**	(108.0)	**34.4**	**51.1**	**77.4**	**54.4**	**36.5**	47	161
330	**313**	**313**	**313**	**67.0**	...	**33.3**	**50.2**	**76.8**	**53.6**	**35.2**	..	156
320	**303**	**303**	**303**	**66.4**	(107.0)	**32.2**	**49.4**	**76.2**	**52.3**	**33.9**	45	151
310	**294**	**294**	**294**	**65.8**	...	**31.0**	**48.4**	**75.6**	**51.3**	**32.5**	..	146
300	**284**	**284**	**284**	**65.2**	(105.5)	**29.8**	**47.5**	**74.9**	**50.2**	**31.1**	42	141
295	**280**	**280**	**280**	**64.8**	...	**29.2**	**47.1**	**74.6**	**49.7**	**30.4**	..	139
290	**275**	**275**	**275**	**64.5**	(104.5)	**28.5**	**46.5**	**74.2**	**49.0**	**29.5**	41	136
285	**270**	**270**	**270**	**64.2**	...	**27.8**	**46.0**	**73.8**	**48.4**	**28.7**	..	134
280	**265**	**265**	**265**	**63.8**	(103.5)	**27.1**	**45.3**	**73.4**	**47.8**	**27.9**	40	131
275	**261**	**261**	**261**	**63.5**	...	**26.4**	**44.9**	**73.0**	**47.2**	**27.1**	..	129
270	**256**	**256**	**256**	**63.1**	(102.0)	**25.6**	**44.3**	**72.6**	**46.4**	**26.2**	38	126
265	**252**	**252**	**252**	**62.7**	...	**24.8**	**43.7**	**72.1**	**45.7**	**25.2**	..	124
260	**247**	**247**	**247**	**62.4**	(101.0)	**24.0**	**43.1**	**71.6**	**45.0**	**24.3**	37	121
255	**243**	**243**	**243**	**62.0**	...	**23.1**	**42.2**	**71.1**	**44.2**	**23.2**	..	119
250	**238**	**238**	**238**	**61.6**	99.5	**22.2**	**41.7**	**70.6**	**43.4**	**22.2**	36	116
245	**233**	**233**	**233**	**61.2**	...	**21.3**	**41.1**	**70.1**	**42.5**	**21.1**	..	114
240	**228**	**228**	**228**	**60.7**	98.1	**20.3**	**40.3**	**69.6**	**41.7**	**19.9**	34	111
230	219	219	219	...	96.7	(18.0)	...	...	...	...	33	106
220	209	209	209	...	95.0	(15.7)	...	...	...	...	32	101
210	200	200	200	...	93.4	(13.4)	...	...	...	...	30	97
200	190	190	190	...	91.5	(11.0)	...	...	...	...	29	92
190	181	181	181	...	89.5	(8.5)	...	...	...	...	28	88
180	171	171	171	...	87.1	(66.0)	...	...	...	...	26	84
170	162	162	162	...	85.0	(3.0)	...	...	...	...	25	79
160	152	152	152	...	81.7	(0.0)	...	...	...	...	24	75
150	143	143	143	...	78.7	...	...	...	...	...	22	71
140	133	133	133	...	75.0	...	...	...	...	...	21	66
130	124	124	124	...	71.2	...	...	...	...	...	20	62
120	114	114	114	...	66.7	...	...	...	...	...	..	57
110	105	105	105	...	62.3	...	...	...	...	...	..	...
100	95	95	95	...	56.2	...	...	...	...	...	..	...
95	90	90	90	...	**52.0**	...	...	...	...	...	..	...
90	86	86	86	...	**48.0**	...	...	...	...	...	..	...
85	81	81	81	...	41.0	...	...	...	...	...	..	...

[1]The values in **boldface type** correspond to the values in the joint SAE-ASM-ASTM hardness conversions as printed in ASTM, E140 Table 1. Values in parentheses are beyond normal range and are given for information only.

TABLE A3.2 Conversion Table for Various Hardness Scales (*Continued*)

Approximate Equivalent Hardness Numbers for Brinell Hardness Numbers for Steel[1]

Brinell indentation diam, mm	Brinell hardness No.,[2] 10-mm ball, 3,000-kg load			Diamond pyramid hardness No.	Rockwell hardness No.[3]				Rockwell Superficial hardness No., superficial ball penetrator			Shore scleroscope hardness No.	Tensile strength (approx) 1,000 psi
	Standard ball	Hultgren ball	Tungsten carbide ball		A-scale, 60-kg load, brale penetrator	B-scale, 100-kg load, 1/16-in.-diam ball	C-scale, 150-kg load, brale penetrator	D-scale, 100-kg load, brale penetrator	15-N scale, 15-kg load	30-N scale, 30-kg load	45-N scale, 45-kg load		
...	...	...	...	940	85.6	...	68.0	76.9	93.2	84.4	75.4	97	...
...	...	...	...	920	85.3	...	67.5	76.5	93.0	84.0	74.8	96	...
...	...	...	...	900	85.0	...	67.0	76.1	92.9	83.6	74.2	95	...
...	...	...	767	880	84.7	...	66.4	75.7	92.7	83.1	73.6	93	...
...	...	...	757	860	84.4	...	65.9	75.3	92.5	82.7	73.1	92	...
2.25	...	...	745	840	84.1	...	65.3	74.8	92.3	82.2	72.2	91	...
...	...	...	733	820	83.8	...	64.7	74.3	92.1	81.7	71.8	90	...
...	...	...	722	800	83.4	...	64.0	73.8	91.8	81.1	71.0	88	...
2.30	...	...	712	...	...	...	...	...	...	...	...	..	...
...	...	...	710	780	83.0	...	63.3	73.3	91.5	80.4	70.2	87	...
...	...	...	698	760	82.6	...	62.5	72.6	91.2	79.7	69.4	86	...
...	...	...	684	740	82.2	...	61.8	72.1	91.0	79.1	68.6	..	...
2.35	...	...	682	737	82.2	...	61.7	72.0	91.0	79.0	68.5	84	...
...	...	...	670	720	81.8	...	61.0	71.5	90.7	78.4	67.7	83	...
...	...	...	656	700	81 3	...	60.1	70.8	90.3	77.6	66.7	..	...
2.40	...	...	653	697	81.2	...	60.0	70.7	90.2	77.5	66.5	81	...
...	...	...	647	690	81.1	...	59.7	70.5	90.1	77.2	66.2	..	...
...	...	...	638	680	80.8	...	59.2	70.1	89.8	76.8	65.7	80	329
...	...	...	630	670	80.6	...	58.8	69.8	89.7	76.4	65.3	..	324
2.45	...	...	627	667	80.5	...	58.7	69.7	89.6	76.3	65.1	79	323
2.50	...	601	...	677	80.7	...	59.1	70.0	89.8	76.8	65.7	..	328
	...	...	601	640	79.8	...	57.3	68.7	89.0	75.1	63.5	77	309
2.55	...	578	...	640	79.8	...	57.3	68.7	89.0	75.1	63.5	..	309
	...	...	578	615	79.1	...	56.0	67.7	88.4	73.9	62.1	75	297
2.60	...	555	...	607	78.8	...	55.6	67.4	88.1	73.5	61.6	..	293
	...	...	555	591	78.4	...	54.7	66.7	87.8	72.7	60.6	73	285
2.65	...	534	...	579	78.0	...	54.0	66.1	87.5	72.0	59.8	..	279
	...	...	534	569	77.8	...	53.5	65.8	87.2	71.6	59.2	71	274
2.70	...	514	...	553	77.1	...	52.5	65.0	86.7	70.7	58.0	..	266
	...	...	514	547	76.9	...	52.1	64.7	86.5	70.3	57.6	70	263
2.75	495	...	...	539	76.7	...	51.6	64.3	86.3	69.9	56.9	..	259
	...	495	...	530	76.4	...	51.1	63.9	86.0	69.5	56.2	..	254
	...	...	495	528	76.3	...	51.0	63.8	85.9	69.4	56.1	68	253
2.80	477	...	...	516	75.9	...	50.3	63.2	85.6	68.7	55.2	..	247
	...	477	...	508	75.6	...	49.6	62.7	85.3	68.2	54.5	66	243
2.85	461	...	...	495	75.1	...	48.8	61.9	84.9	67.4	53.5	..	237
	...	461	...	491	74.9	...	48.5	61.7	84.7	67.2	53.2	65	235

2.90	444	...	...	474	74.3	...	47.2	61.0	84.1	66.0	51.7	..	226
	...	444	...	472	74.2	...	47.1	60.8	84.0	65.8	51.5	63	225
2.95	429	429	429	455	73.4	...	45.7	59.7	83.4	64.6	49.9	61	217
3.00	415	415	415	440	72.8	...	44.5	58.8	82.8	63.5	48.4	59	210
3.05	401	401	401	425	72.0	...	43.1	57.8	82.0	62.3	46.9	58	202
3.10	388	388	388	410	71.4	...	41.8	56.8	81.4	61.1	45.3	56	195
3.15	375	375	375	396	70.6	...	40.4	55.7	80.6	59.9	43.6	54	188
3.20	363	363	363	383	70.0	...	39.1	54.6	80.0	58.7	42.0	52	182
3.25	352	352	352	372	69.3	(110.0)	37.9	53.8	79.3	57.6	40.5	51	176
3.30	341	341	341	360	68.7	(109.0)	36.6	52.8	78.6	56.4	39.1	50	170
3.35	331	331	331	350	68.1	(108.5)	35.5	51.9	78.0	55.4	37.8	48	166
3.40	321	321	321	339	67.5	(108.0)	34.3	51.0	77.3	54.3	36.4	47	160
3.45	311	311	311	328	66.9	(107.5)	33.1	50.0	76.7	53.3	34.4	46	155
3.50	302	302	302	319	66.3	(107.0)	32.1	49.3	76.1	52.2	33.8	45	150
3.55	293	293	293	309	65.7	(106.0)	30.9	48.3	75.5	51.2	32.4	43	145
3.60	285	285	285	301	65.3	(105.5)	29.9	47.6	75.0	50.3	31.2	..	141
3.65	277	277	277	292	64.6	(104.5)	28.8	46.7	74.4	49.3	29.9	41	137
3.70	269	269	269	284	64.1	(104.0)	27.6	45.9	73.7	48.3	28.5	40	133
3.75	262	262	262	276	63.6	(103.0)	26.6	45.0	73.1	47.3	27.3	39	129
3.80	255	255	255	269	63.0	(102.0)	25.4	44.2	72.5	46.2	26.0	38	126
3.85	248	248	248	261	62.5	(101.0)	24.2	43.2	71.7	45.1	24.5	37	122
3.90	241	241	241	253	61.8	100.0	22.8	42.0	70.9	43.9	22.8	36	118
3.95	235	235	235	247	61.4	99.0	21.7	41 4	70.3	42.9	21.5	35	115
4.00	229	229	229	241	60.8	98.2	20.5	40.5	69.7	41.9	20.1	34	111
4.05	223	223	223	234	...	97.3	(18.8)	...	...	...	...	..	...
4.10	217	217	217	228	...	96.4	(17.5)	...	...	...	...	33	105
4.20	207	207	207	218	...	94.6	(15.2)	...	...	...	...	32	100
4.30	197	197	197	207	...	92.8	(12.7)	...	...	...	...	30	95
4.40	187	187	187	196	...	90.7	(10.0)	...	...	...	...	..	90
4.50	179	179	179	188	...	89.0	(8.0)	...	...	...	...	27	87
4.60	170	170	170	178	...	86.8	(5.4)	...	...	...	...	26	83
4.70	163	163	163	171	...	85.0	(3.3)	...	...	...	...	25	79
4.80	156	156	156	163	...	82.9	(0.9)	...	...	...	...	..	76
4.90	149	149	149	156	...	80.8	...	...	...	...	...	23	73
5.00	143	143	143	150	...	78.7	...	...	...	...	...	22	71
5.10	137	137	137	143	...	76.4	...	...	...	...	...	21	67
5.20	131	131	131	137	...	74.0	...	...	...	...	...	..	65
5.30	126	126	126	132	...	72.0	...	...	...	...	...	20	63
5.40	121	121	121	127	...	69.8	...	...	...	...	...	19	60
5.50	116	116	116	122	...	67.6	...	...	...	...	...	18	58
5.60	111	111	111	117	...	65.7	...	...	...	...	...	15	56

[1]The values in **boldface type** correspond to the values in the joint SAE-ASM-ASTM hardness conversions as printed in ASTM E140, Table 3.

[2]Brinell numbers are based on the diameter of impressed indentation. If the ball distorts (flattens) during test, Brinell numbers will vary in accordance with the degree of such distortion when related to hardness determined with a Vickers diamond pyramid, Rockwell brale, or other penetrator which does not sensibly distort. At high hardnesses, therefore, the relationship between Brinell and Vickers or Rockwell scales is affected by the type of ball used. Steel balls (standard or Hultgren) tend to flatten slightly more than carbide balls, resulting in larger indentation and lower Brinell number than shown by a carbide ball. Thus, on a specimen of 640 Vickers, a Hultgren ball will leave a 2.55-mm impression (578 Bhn), and the carbide ball a 2.50-mm impression (601 Bhn). Conversely identical impression diameters for both types of ball will correspond to different Vickers or Rockwell values. Thus, if both impressions are 2.55 mm (578 Bhn), material tested with a Hultgren ball has a Vickers hardness of 640, while material tested with a carbide ball has a Vickers hardness of 615.

[3]Values in parentheses are beyond normal range.

TABLE A3.2 Conversion Table for Various Hardness Scales (*Continued*)

Approximate Equivalent Hardness Numbers for Rockwell C Hardness Numbers for Steel[1]

Rockwell C-scale hardness No.	Diamond pyramid hardness No.	Brinell hardness No., 10-mm ball, 3,000-kg load			Rockwell hardness No.			Rockwell superficial hardness No., superficial brale penetrator			Shore scleroscope hardness No.	Tensile strength (approx) 1,000 psi
		Standard ball	Hultgren ball	Tungsten carbide ball	A-scale, 60-kg load, brale penetrator	B-scale, 100-kg load, $\frac{1}{16}$-in.-diam ball	D-scale, 100-kg load, brale penetrator	15-N scale, 15-kg load	30-N scale, 30-kg load	45-N scale, 45-kg load		
68	940	...	...	...	85.6	...	76.9	93.2	84.4	75.4	97	...
67	900	...	...	...	85.0	...	76.1	92.9	83.6	74.2	95	...
66	865	...	...	...	84.5	...	75.4	92.5	82.8	73.3	92	...
65	832	...	...	739	83.9	...	74.5	92.2	81.9	72.0	91	...
64	800	...	...	722	83.4	...	73.8	91.8	81.1	71.0	88	...
63	772	...	...	705	82.8	...	73.0	91.4	80.1	69.9	87	...
62	746	...	...	688	82.3	...	72.2	91.1	79.3	68.8	85	...
61	720	...	...	670	81.8	...	71.5	90.7	78.4	67.7	83	...
60	697	...	613	654	81.2	...	70.7	90.2	77.5	66.6	81	...
59	674	...	599	634	80.7	...	69.9	89.8	76.6	65.5	80	326
58	653	...	587	615	80.1	...	69.2	89.3	75.7	64.3	78	315
57	633	...	575	595	79.6	...	68.5	88.9	74.8	63.2	76	305
56	613	...	561	577	79.0	...	67.7	88.3	73.9	62.0	75	295
55	595	...	546	560	78.5	...	66.9	87.9	73.0	60.9	74	287
54	577	...	534	543	78.0	...	66.1	87.4	72.0	59.8	72	278
53	560	...	519	525	77.4	...	65.4	86.9	71.2	58.6	71	269
52	544	500	508	512	76.8	...	64.6	86.4	70.2	57.4	69	262
51	528	487	494	496	76.3	...	63.8	85.9	69.4	56.1	68	253
50	513	475	481	481	75.9	...	63.1	85.5	68.5	55.0	67	245
49	498	464	469	469	75.2	...	62.1	85.0	67.6	53.8	66	239
48	484	451	455	455	74.7	...	61.4	84.5	66.7	52.5	64	232
47	471	442	443	443	74.1	...	60.8	83.9	65.8	51.4	63	225
46	458	432	432	432	73.6	...	60.0	83.5	64.8	50.3	62	219
45	446	421	421	421	73.1	...	59.2	83.0	64.0	49.0	60	212
44	434	409	409	409	72.5	...	58.5	82.5	63.1	47.8	58	206
43	423	400	400	400	72.0	...	57.7	82.0	62.2	46.7	57	201
42	412	390	390	390	71.5	...	56.9	81.5	61.3	45.5	56	196
41	402	381	381	381	70.9	...	56.2	80.9	60.4	44.3	55	191

40	392	371	371	371	70.4	...	55.4	80.4	59.5	43.1	54	186
39	382	362	362	362	69.9	...	54.6	79.9	58.6	41.9	52	181
38	372	353	353	353	69.4	...	53.8	79.4	57.7	40.8	51	176
37	363	344	344	344	68.9	...	53.1	78.8	56.8	39.6	50	172
36	354	336	336	336	68.4	(109.0)	52.3	78.3	55.9	38.4	49	168
35	345	327	327	327	67.9	(108.5)	51.5	77.7	55.0	37.2	48	163
34	336	319	319	319	67.4	(108.0)	50.8	77.2	54.2	36.1	47	159
33	327	311	311	311	66.8	(107.5)	50.0	76.6	53.3	34.9	46	154
32	318	301	301	301	66.3	(107.0)	49.2	76.1	52.1	33.7	44	150
31	310	294	294	294	65.8	(106.0)	48.4	75.6	51.3	32.5	43	146
30	302	286	286	286	65.3	(105.5)	47.7	75.0	50.4	31.3	42	142
29	294	279	279	279	64.7	(104.5)	47.0	74.5	49.5	30.1	41	138
28	286	271	271	271	64.3	(104.0)	46.1	73.9	48.6	28.9	41	134
27	279	264	264	264	63.8	(103.0)	45.2	73.3	47.7	27.8	40	131
26	272	258	258	258	63.3	(102.5)	44.6	72.8	46.8	26.7	38	127
25	266	253	253	253	62.8	(101.5)	43.8	72.2	45.9	25.5	38	124
24	260	247	247	247	62.4	(101.0)	43.1	71.6	45.0	24.3	37	121
23	254	243	243	243	62.0	100.0	42.1	71.0	44.0	23.1	36	118
22	248	237	237	237	61.5	99.0	41.6	70.5	43.2	22.0	35	115
21	243	231	231	231	61.0	98.5	40.9	69.9	42.3	20.7	35	113
20	238	226	226	226	60.5	97.8	40.1	69.4	41.5	19.6	34	110
(18)	230	219	219	219	...	96.7	...	...	...	...	33	106
(16)	222	212	212	212	...	95.5	...	...	...	...	32	102
(14)	213	203	203	203	...	93.9	...	...	...	...	31	98
(12)	204	194	194	194	...	92.3	...	...	...	...	29	94
(10)	196	187	187	187	...	90.7	...	...	...	...	28	90
(8)	188	179	179	179	...	89.5	...	...	...	...	27	87
(6)	180	171	171	171	...	87.1	...	...	...	...	26	84
(4)	173	165	165	165	...	85.5	...	...	...	...	25	80
(2)	166	158	158	158	...	83.5	...	...	...	...	24	77
(0)	160	152	152	152	...	81.7	...	...	...	...	24	75

[1]The values in **boldface type** correspond to the values in the joint SAE-ASM-ASTM hardness conversions as printed in ASTM E140, Table 2. Values in parentheses are beyond normal range and are given for information only.

TABLE A3.2 Conversion Table for Various Hardness Scales (*Continued*)

Brinell Hardness Numbers (10–mm Ball Diameter)

Inden-tation diam, mm	Load, kg						Inden-tation diam, mm	Load, kg						Inden-tation diam, mm	Load, kg					
	500	1,000	1,500	2,000	2,500	3,000		500	1,000	1,500	2,000	2,500	3,000		500	1,000	1,500	2,000	2,500	3,000
2.00	158	316	473	632	788	945	3.50	50.3	101	151	201	252	302	5.00	23.8	47.6	71.5	95.2	119	143
2.05	150	300	450	600	750	899	3.55	48.9	97.8	147	196	244	293	5.05	23.3	46.6	70.0	93.2	117	140
2.10	143	286	428	572	714	856	3.60	47.5	95.0	143	190	238	285	5.10	22.8	45.6	68.5	91.2	114	137
2.15	136	272	409	544	681	817	3.65	46.1	92.2	139	184	231	277	5.15	22.3	44.6	67.0	89.2	112	134
2.20	130	260	390	520	650	780	3.70	44.9	89.8	135	180	225	269	5.20	21.8	43.6	65.5	87.2	109	131
2.25	124	248	373	496	621	745	3.75	43.6	87.2	131	174	218	262	5.25	21.4	42.8	64.0	85.6	107	128
2.30	119	238	356	476	593	712	3.80	42.4	84.8	128	170	212	255	5.30	20.9	41.8	63.0	83.6	105	126
2.35	114	228	341	456	568	682	3.85	41.3	82.6	124	165	207	248	5.35	20.5	41.0	61.5	82.0	103	123
2.40	109	218	327	436	545	653	3.90	40.2	80.4	121	161	201	241	5.40	20.1	40.2	60.5	80.4	101	121
2.45	104	208	314	416	522	627	3.95	39.1	78.2	118	156	196	235	5.45	19.7	39.4	59.0	78.8	98.5	118
2.50	100	200	301	400	500	601	4.00	38.1	76.2	115	152	191	229	5.50	19.3	38.6	58.0	77.2	96.5	116
2.55	96.3	193	289	385	482	578	4.05	37.1	74.2	112	148	186	223	5.55	18.9	37.8	57.0	75.6	95.0	114
2.60	92.6	185	278	370	462	555	4.10	36.2	72.4	109	145	181	217	5.60	18.6	37.2	55.5	74.4	92.5	111
2.65	89.0	178	267	356	445	534	4.15	35.3	70.6	106	141	177	212	5.65	18.2	36.4	54.5	72.8	90.8	109
2.70	85.7	171	257	343	429	514	4.20	34.4	68.8	104	138	172	207	5.70	17.8	35.6	53.5	71.2	89.2	107
2.75	82.6	165	248	330	413	495	4.25	33.6	67.2	101	134	167	201	5.75	17.5	35.0	52.5	70.0	87.5	105
2.80	79.6	159	239	318	398	477	4.30	32.8	65.6	98.5	131	164	197	5.80	17.2	34.4	51.5	68.8	85.8	103
2.85	76.8	154	231	307	384	461	4.35	32.0	64.0	96.0	128	160	192	5.85	16.8	33.6	50.5	67.2	84.2	101
2.90	74.1	148	222	296	371	444	4.40	31.2	62.4	93.5	125	156	187	5.90	16.5	33.0	49.6	66.0	82.5	99.2
2.95	71.5	143	215	286	358	429	4.45	30.5	61.0	91.5	122	153	183	5.95	16.2	32.4	48.7	64.8	81.2	97.3
3.00	69.1	138	208	276	346	415	4.50	29.8	59.6	89.5	119	149	179	6.00	15.9	31.8	47.8	63.6	79.5	95.5
3.05	66.8	134	201	267	334	401	4.55	29.1	58.2	87.0	116	145	174	6.05	15.6	31.2	46.9	62.4	78.0	93.7
3.10	64.6	129	194	258	324	388	4.60	28.4	56.8	85.0	114	142	170	6.10	15.3	30.6	46.0	61.2	76.7	92.0
3.15	62.5	125	188	250	313	375	4.65	27.8	55.6	83.5	111	139	167	6.15	15.1	30.2	45.2	60.4	75.3	90.3
3.20	60.5	121	182	242	303	363	4.70	27.1	54.2	81.5	108	136	163	6.20	14.8	29.6	44.4	59.2	73.8	88.7
3.25	58.6	117	176	234	293	352	4.75	26.5	53.0	79.5	106	133	159	6.25	14.5	29.0	43.6	58.0	72.6	87.1
3.30	56.8	114	171	227	284	341	4.80	25.9	51.8	78.0	104	130	156	6.30	14.2	28.4	42.8	56.8	71.3	85.5
3.35	55.1	110	166	220	276	331	4.85	25.4	50.8	76.0	102	127	152	6.35	14.0	28.0	42.0	56.0	70.0	84.0
3.40	53.4	107	161	214	267	321	4.90	24.8	49.6	74.5	99.2	124	149	6.40	13.7	27.4	41.3	54.8	68.8	82.5
3.45	51.8	104	156	207	259	311	4.95	24.3	48.6	73.0	97.2	122	146	6.45	13.5	27.0	40.5	54.0	67.5	81.0

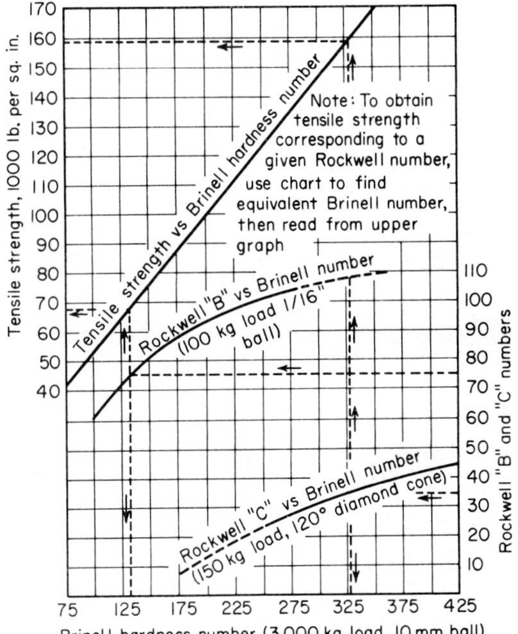

FIGURE A3.5 Conversion chart for Brinell and Rockwell hardness numbers, giving corresponding tensile strength for steel. Based on hardness conversion table. (*SAE Handbook,* 1964.)

Toughness. Sudden fracture, exhibiting little ductility in the vicinity of the break, occurs in certain metals when load is rapidly applied. The capability of a material to resist such a brittle fracture is a measure of its toughness.

Highly ductile materials, for example, those possessing an FCC lattice, exhibit considerable toughness across a full range of temperatures. Other materials, such as BCC-based carbon steels, possess a level of toughness that is dependent on the metal temperature when the load is applied. In these metals, a transition from brittle-to-ductile behavior occurs over a narrow range of temperatures.

The two most common testing methods used to measure metal toughness is the Charpy Impact test, defined in ASTM specification E-23, and the Drop-Weight test, defined in ASTM E-208. The Charpy test uses a small machined specimen with a machined notch that is struck by a pendulum weight (Fig. A3.6). The toughness of the specimen, measured in kilojoules or footpounds of force, is a measure of the energy loss of the pendulum as it passes through and breaks the specimen (Fig. A3.7). Typical impact behavior versus test temperature is shown in Fig. A3.8.

The Drop-Weight test is similar in principal but uses a larger specimen with a brittle, notched weld bead used as the crack starter (Fig. A3.9). A weight is dropped from a height onto the specimen, which had been cooled or heated to the desired test temperature. The test determines the nil-ductility transition temperature (NDTT), defined as the specimen temperature when, upon striking, the crack propagates across the entire specimen width.

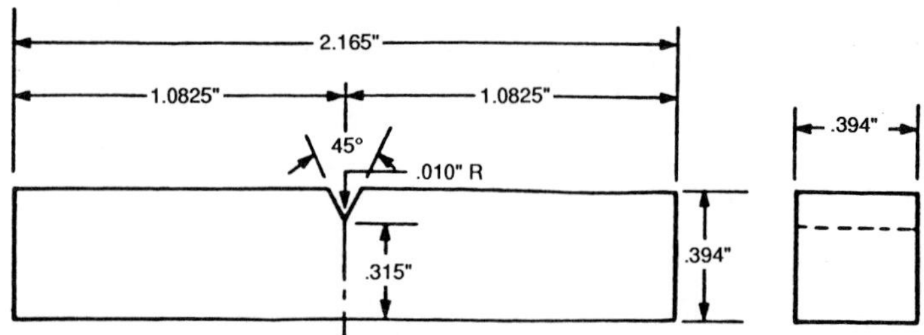

FIGURE A3.6 Charpy V-notch Impact test specimen. (ASTM Specification E23.)

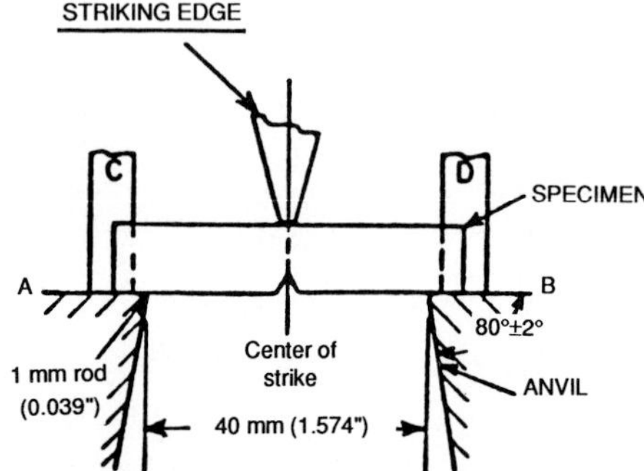

FIGURE A3.7 Charpy V-notch specimen placement during strike by testing anvil. (ASTM Specification E23.)

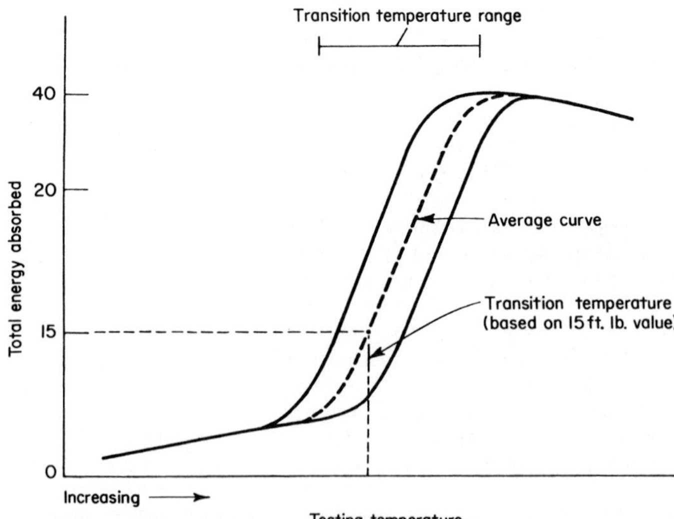

FIGURE A3.8 Transition temperature range and transition temperature in Charpy Impact test.

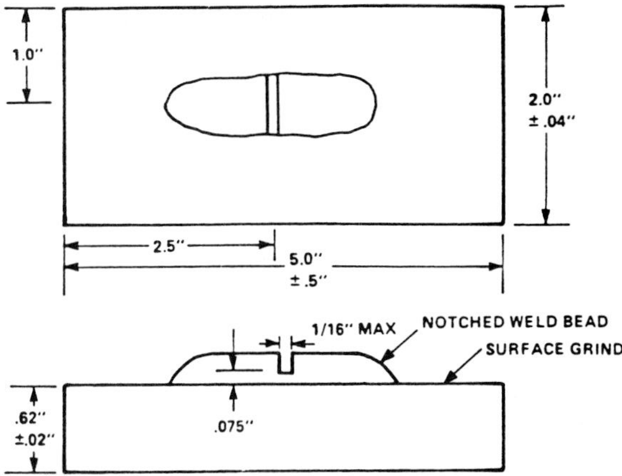

FIGURE A3.9 Drop-Weight test specimen with brittle weld deposit on specimen face; machined notch to act as crack starter. Impact load applied from side opposite weld deposit. (ASTM Specification E208.)

The Charpy brittle transition temperature (sometimes called the Charpy "fix" temperature), and the Drop-Weight NDTT are both important design considerations for those materials that can exhibit poor toughness and may operate in lower temperature regimes. In pressure vessel and piping design codes, limits are placed on the minimum use temperature of the material based on adding an increment of margin over and above the Charpy fix, or NDTT. Operating at or above this elevated temperature is usually sufficient to avoid brittle, catastrophic failure, for example, the case when operating at a temperature on the "upper shelf" of the Charpy V-notch toughness versus temperature curve.

Fatigue Resistance. The ability of a metal to resist crack initiation and further propagation under repeated cyclic loading is a measure of its fatigue resistance. Several standardized test methods have been developed to test metals, machined to particular geometries, where applying a repeating load range. Loads are generally applied through bending, cantilevered, or push-pull load application in suitably outfitted testing machines. Either constant applied stress or strain ranges can be used to determine material response.

The most common representation of fatigue test data is an *S-N* curve, relating stress (*S*) required to cause specimen failure in a given number of cycles (*N*) (Fig. A3.10*a*). These tests are generally performed on smooth specimens, but they can also be run with stress-concentrating mechanisms machined in the specimen surface, such as notches. The effect of stress concentrations on fatigue life cycles can also be estimated from the smooth specimen *S-N* curve by calculating the intensified stress due to the particular geometry and intersecting the curve at that point on the stress axis.

As the applied load range decreases, ferritic steels exhibit a point at which an infinite number of cycles can be absorbed without causing failure. This level of stress is called the endurance limit. Many of the other metals do not exhibit this behavior but rather exhibit an increasing, but finite, number of cycles to failure with decreasing cyclic load (Fig. A3.10*b*).

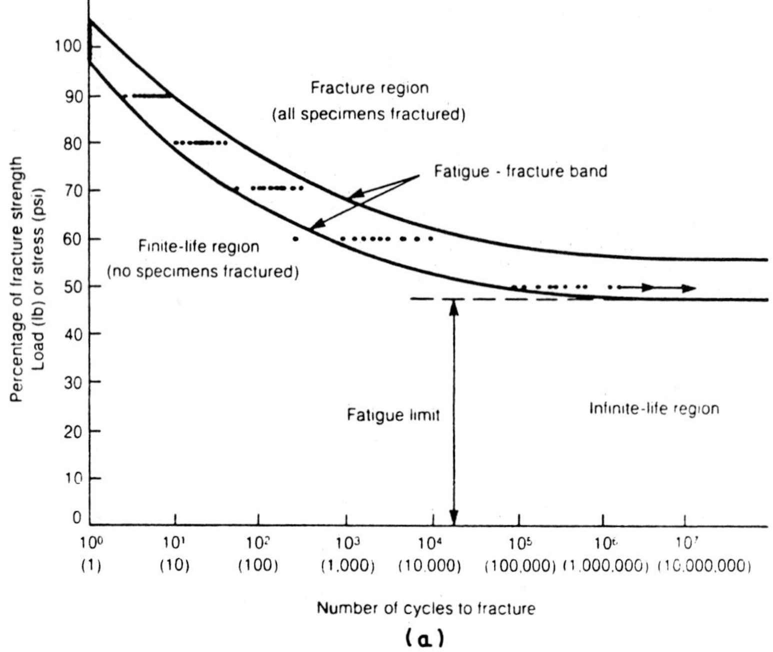

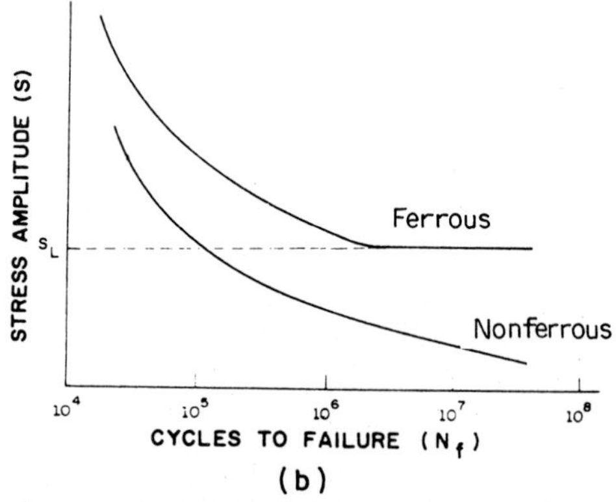

FIGURE A3.10 *S-N* curves that typify fatigue test results (*a*) for testing medium-strength steels and (*b*) showing typical curve shape for ferrous and nonferrous materials. S_L is the endurance limit. (*Atlas of Fatigue Curves,* American Society for Metals, ©1986.)

The fatigue resistance of a material at a given applied stress or strain range is a function of a number of variables, including material strength and ductility. Results may vary significantly for different surface finishes, product forms of the same material (Fig. A3.11), material internal cleanliness, test specimen orientation, and levels of residual stress, among other factors. Test environment can also have a profound effect on test results (Fig. A3.12). Therefore, fatigue test results characteristically exhibit significant scatter.

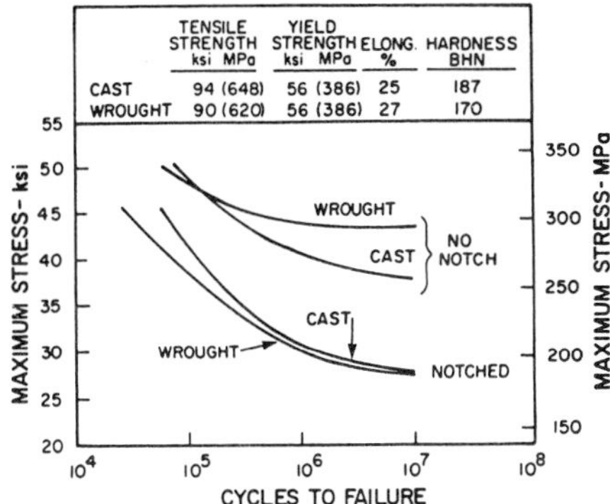

FIGURE A3.11 Fatigue characteristics (*S-N* curve) for cast and wrought 1040 steel in the normalized and tempered condition, both notched and unnotched. R. R. Moore rotating beam tests, K_t = 2.2. (*Atlas of Fatigue Curves,* ASM.)

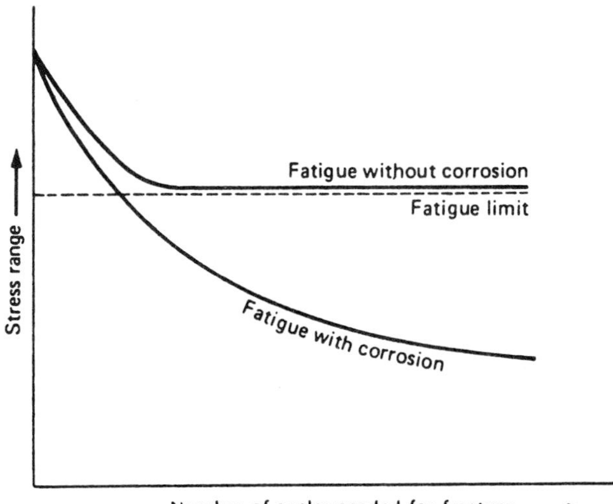

FIGURE A3.12 Effect of alternating stresses with and without corrosion for ferrous material that normally exhibits an endurance limit. (*Atlas of Fatigue Curves,* ASM.)

Fatigue design curves are generated from test data by incorporating large safety margins applied to the average property curve. In U.S. design codes, the fatigue design curve is commonly generated by taking the lesser of 1/20 times the cycles to failure or 1/2 of the stress to cause failure. A new curve is constructed taking the lower bound of these two factored curves.

When considering fatigue of metals in design, a further safety margin is often applied against the cycles to failure at a given stress amplitude. For example, if a component is continuously cycled over the same stress range, a design limit on allowable cycles may correspond to the cycle life multiplied by a factor such as 0.8. This is a common safety margin used in vessel and piping design.

As is normally the case, components may experience a wide variety of cyclic stress ranges at various temperatures over their life. The effect of this array of cyclic parameters on fatigue life can be estimated by an approach referred to as life fraction summation. In this design practice, the percentage of life used up in cycling at a certain stress range is calculated, corresponding to the ratio of the number of actual service duty cycles to the total number of cycles to failure at that stress range. This calculation is performed for all the various stress ranges and duty cycles anticipated. The fractions thereby calculated are summed and compared to the design limit (1.0 with no safety margin or 0.8 or some other value depending on the design safety factor that applies).

Elevated Temperature Tensile and Creep Strength. Tensile tests are performed at elevated temperatures to characterize the material's yield and ultimate tensile properties at potential use temperatures above room temperature. A heating chamber is combined with a conventional tensile testing machine, and special strain-measuring extensometers capable of withstanding the test temperatures are used. Generally, as temperature increases, yield and ultimate strengths decrease and ductility increases.

Creep is defined as the time-dependent deformation of a material that occurs under load at elevated temperatures. The test is performed by holding a specimen, similar in configuration to a tensile specimen, at a uniform temperature, under a constant load (usually using a dead weight), and allowing the specimen gradually to elongate to ultimate failure. The practice is defined in ASTM Specification E-139.

The simplest test method records only the applied stress (based on original test specimen cross section), time to failure, and total elongation at failure. This is called a stress-rupture test. If periodic measurements of strain accumulation versus test duration are also taken, the test is referred to as a creep-rupture test.

A representation of a typical creep strain-versus-time data is shown in Fig. A3.13. Three stages of creep behavior are exhibited. Upon initial loading, instantaneous straining occurs. Almost immediately, the rate of creep strain accumulation (creep rate) is high, but it is continuously decreasing. The test progresses into a phase in which the strain rate slows and becomes fairly constant for a long period of time. Finally, with decreasing load-bearing cross section of the specimen due to specimen stretching and necking, applied stress begins to increase steadily, as does the creep rate, until failure occurs. These three regions are termed the primary, secondary, and tertiary stages of creep. The intent of safe design practice is to avoid the third stage, where strain accumulations are rapid and material behavior less predictable.

After accumulating a number of rupture data points (i.e., time to failure of a metal at various applied stresses), the data are generally represented as a stress-rupture curve (Fig. A3.14). Each curve represents the time to failure at various

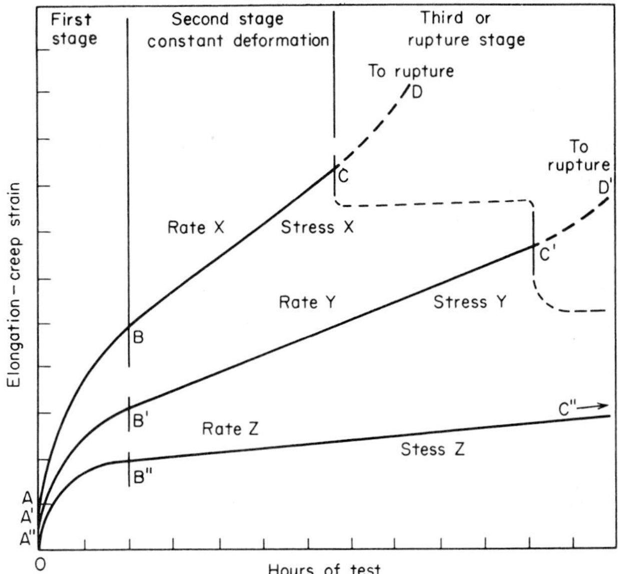

FIGURE A3.13 Creep time versus elongation curves at a given temperature.

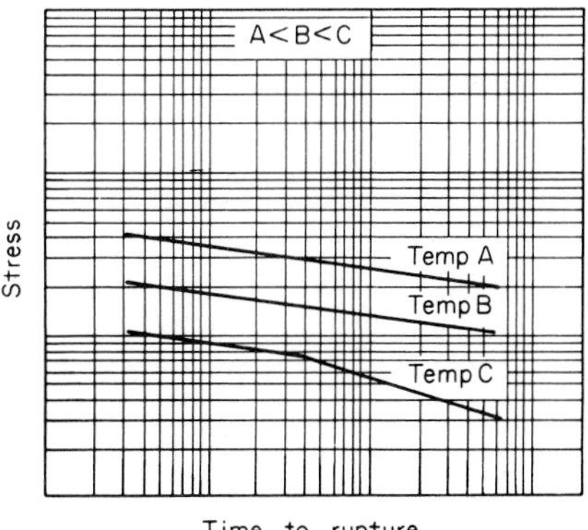

FIGURE A3.14 Typical stress-to-rupture curves.

applied stresses at a given test temperature. Another useful property that can be measured in these tests is the creep rate during the second stage of creep for a given applied stress and temperature. This, along with time to onset of the tertiary creep stage, are useful properties to the design engineer and are used in establishing allowable design tension stresses in design codes.

Metals that experience creep will accumulate a progressively larger amount of microscopic damage to the structure of the material. Damage is first observed microscopically as small cavities, or voids, that begin appearing in the grain boundaries of the metal, particularly at triple points (i.e., where three grains come together). Further progression of damage entails formation of more voids along many of the adjacent grain boundaries, until ultimately they link together to form grain boundary microcracks. With more time, these form larger macrocracks that lead to ultimate failure of the metal component.

The determination of a metal's degree of creep damage and its consequence on the continued safe operation of the component has developed into a sophisticated science referred to as component condition assessment, or estimation of remaining life. This will be addressed in more detail later in this chapter.

A practice essentially identical to cyclic fraction life summation used in fatigue design can be used in material creep analysis to estimate the percentage of creep life expended. Here the individual life fraction corresponds to the amount of time a component spends at a given stress and temperature, compared to the total time to failure given on the stress rupture curve for the same applied stress and temperature. All of these fractions for all the operating conditions are then added together and compared to an appropriate design limit (1.0 or less).

Physical Properties of Metals

Physical properties are those, other than mechanical properties, that pertain to the physics of a material. Physical properties of importance to the materials and design engineer are material density, thermal conductivity, thermal expansion, and specific heat.[2]

Density. Density is the ratio of the mass of a material to its volume. In vessel and piping design, the density of a construction material versus its strength per unit area of cross section is often an important consideration.

Thermal Conductivity. The characteristic ability of a material to transmit energy in the form of heat from a high-temperature source to a point of lower temperature is thermal conductivity. The ability to transmit heat is usually expressed as a coefficient of thermal conductivity (k) whose units are a quantity of heat transmitted through a unit thickness per unit time per unit area per unit difference in temperature. For example:

$$k = \frac{\text{Btu ft}}{\text{h ft}^2\Delta°\text{F}} \text{ or } \frac{\text{cal cm}}{\text{s cm}^2\Delta°\text{C}}$$

The lower the value of k, the more resistant the material is to the flow of thermal energy. Good insulators possess low coefficients of thermal conductivity.

Thermal conductivity is a function of the temperature of the material. For example, the coefficient of thermal conductivity of carbon steel decreases as temperature increases, thereby decreasing its ability to transfer heat energy. With austenitic stainless steels, the k value increases with temperature; however, it remains lower than that for carbon steels in normal piping system temperature ranges.

Thermal Expansion. Thermal expansion, which is expressed as the coefficient of linear expansion, is a ratio of the change in length per degree of temperature to a length at a given standard temperature (such as room temperature or the freezing point of water). The units of the coefficient are length of growth per unit length per degree of temperature. The value of the coefficient varies with temperature.

Specific Heat. Specific heat is a measure of the quantity of heat required to raise a unit weight of a material one degree in temperature.

Values of physical properties of materials of interest are given in Tables A3.3, and A3.4.

Other Metallurgical Properties of Metals

In addition to the properties described, other characteristics of metals can have an important effect on the design process. These may profoundly affect the uni-

TABLE A3.3 Thermal Conductivity and Expansion of Piping Material

Material	Thermal conductivity		Linear thermal expansion average	
	Btu/(hr F $\frac{ft^2}{ft}$)	Temperature, F	Microinch/(in. F)	Temperature range, F
Pure iron	43 28	70 752	6.83 8.97	68–212 932–1112
Gray cast iron	27 23.7	70 752	5.83	32–212
Malleable cast iron ferritic	...	...	6.6	70–750
Malleable cast iron pearlitic	...	...	6.6	70–750
Nodular iron	18 (Pearlitic) 20 (Ferritic)	212 212	6.46 6.97 7.49 7.69	68–212 68–600 68–1000 68–1400
Wrought iron	34 26	212 752		
Wrought carbon steel: 0.06 C, 0.38 Mn	34	32	7.83 7.95 8.02 8.21 8.36	70–800 70–900 70–1000 70–1100 70–1200
0.23 C, 0.635 Mn	30 25 17	70 752 2192	6.50	68–212
0.435 C, 0.69 Mn	...	...	6.44 8.39	68–212 68–1292
1.22 C, 0.35 Mn	26 22 16	70 752 2192	5.89 9.33	122–212 932–1832
Carbon–½ Mo	25.8	212	7.70 7.85 7.95 8.07	68–800 68–1000 68–1100 68–1200
1¼ Cr–½ Mo	17.9	212	7.32 7.44 7.56 7.63 7.74 7.82	70–800 70–900 70–1000 70–1100 70–1200 70–1300
2¼ Cr–1 Mo	16.3	212	7.49 7.65 7.72 7.78 7.84 7.88	70–800 70–900 70–1000 70–1100 70–1200 70–1300

TABLE A3.3 Thermal Conductivity and Expansion of Piping Material (*Continued*)

Material	Thermal conductivity		Linear thermal expansion average	
	$Btu/\left(hr\ F\ \frac{ft^2}{ft}\right)$	Temperature, F	Microinch/(in. F)	Temperature range, F
5 Cr–½ Mo	21.2	212	6.44	0–212
	20.8	392	6.91	70–800
	20.4	572	7.02	70–900
	19.8	752	7.10	70–1000
	19.5	932	7.19	70–1100
			7.31	70–1200
			7.35	70–1300
9 Cr–1 Mo	...	...	6.28	70–300
			6.60	70–800
			6.75	70–900
			6.81	70–1000
			6.95	70–1100
			7.07	70–1200
			7.13	70–1300
3½% Ni steel	21	212		
	14	1472		
Type 304 wrought	9.4	212	9.6	32–212
	10.3	392	9.9	32–600
	11.0	572	10.2	32–1000
	11.8	752	10.4	32–1200
	12.5	932	11.2	32–1800
CF-8 cast	9.2	212		
	12.1	1000		
Type 316 wrought	9.0	212	8.9	32–212
	12.1	932	9.0	32–600
			9.7	32–1000
			10.3	32–1200
			11.1	32–1500
CF-8M cast	9.4	212	8.9	68–212
	12.3	1000	9.7	68–1000
Type 321 wrought	9.3	212	9.3	32–212
	10.2	392	9.5	32–600
	11.1	752	10.3	32–1000
	11.9	932	10.7	32–1200
	12.8		11.2	32–1500
Type 347 wrought	9.3	212	9.3	32–212
	10.2	392	9.5	32–600
	11.1	572	10.3	32–1000
	11.9	752	10.6	32–1200
	12.8	932	11.1	32–1500
CF-8C cast	9.3	212	9.3	68–212
	12.8	1000	10.3	68–1000
405 wrought	...	...	6.0	32–212
			6.4	32–600
			6.7	32–1000
			7.5	32–1200

TABLE A3.3 Thermal Conductivity and Expansion of Piping Material (*Continued*)

Material	Thermal conductivity		Linear thermal expansion average	
	Btu/(hr F $\frac{ft^2}{ft}$)	Temperature, F	Microinch/(in. F)	Temperature range, F
CA15 cast	14.5 16.7	212 1000	5.5 6.4 6.7	68–212 68–1000 68–1300
410 wrought	14.4 16	212 752	6.1 7.2 7.6	32–212 32–1000 32–1832
446 wrought	12.1 14.1	212 932	5.9 6.3 7.6	32–212 32–1000 32–1832
CC50 cast	12.6 17.9 20.3 24.2	212 1000 1500 2000	5.9 6.4	68–212 68–1000
Aluminum 1100	128	70	12.2 13.1 13.7 14.2	−58 to +68 68–212 68–392 68–572
Aluminum 6061	99 (0 temper) 90 (T4 temper) 90 (T6 temper)	70 70 70	12.1 13.0 13.5 14.1	−76 to +68 68–212 68–392 68–572
Aluminum 43	82 (as cast) 94 (annealed)	70 70	12.2 12.8 13.3	68–212 68–392 68–572
Aluminum 356	97 sand cast T51 88 sand cast T6	70 70	11.9 12.8 13.0	68–212 68–392 68–572
Copper (DHP)	196	68	9.8	68–572
Red brass	92	68	10.4 cold rolled	68–572
Yellow brass	67	68	11.3	68–572
Admiralty brass	64	68	11.2	68–572
Manganese bronze	61	68	11.8	68–572
Cupronickel (70–30)	17	68	9.0	68–572
Aluminum bronze (3)	44	68	9.0	68–572
Beryllium copper	33–41 cold worked 48–68 precipitation hardened	68	9.3 9.4 9.9	68–212 68–392 68–572

TABLE A3.3 Thermal Conductivity and Expansion of Piping Material (*Continued*)

Material	Thermal conductivity		Linear thermal expansion average	
	$\mathrm{Btu}\Big/\left(\mathrm{hr\ F}\,\frac{\mathrm{ft}^2}{\mathrm{ft}}\right)$	Temperature, F	Microinch/(in. F)	Temperature range, F
Chemical lead	......	......	16.3 14.7	65–212 −130 to +66
50/40 SnPb solder	27	129	13.0	60–230
Nickel (A) wrought	35	32–212	7.4	77–212
Monel (70 Ni–30 Cu) (wrought)	15	32–212	7.8	32–212
Inconel	8.4	70–212	6.4 8.3	68–212 70–1000
Incoloy	6.8	32–212	8.0	32–212
Hastelloy B	6	......	5.3 7.8	70–200 70–1600
Hastelloy C	5	70	6.6 8.2	70–200 70–1600
Tin	36	32	12.8	32–212
Titanium (99.0%)	9.0–11.5 12.4	68 1500	4.8 5.6 5.7	68–200 68–1200 68–1600
Tantalum	31	68	3.6	

formity, achievable level, or stability of mechanical strength and ductility over long periods of usage.

Grain Size. Upon solidification from the molten state, metals take crystalline form. Rather than a single, large crystal, the material consists of many small crystals that initiated independently and nearly simultaneously from separate nuclei sites. These individual crystals are called grains, and their outer surfaces are called grain boundaries. Grains form initially during the solidification process, but they may also reform, grow, or rearrange while in the solid state.

Some properties of many engineering metals are very dependent on grain size (Fig. A3.15). For example, austenitic stainless steels, such as type 304 (18% Cr–8% Ni-Fe), possess excellent creep strength when the material possesses a coarse grain structure but poor strength when the material has a fine (small) grain structure.

If this same austenitic material is plastically cold worked, these grains will become distorted and possess high levels of lattice strain and residual stress. Subsequent heat treatment can cause the crystal lattice to reform unstrained grains initiating at lattice defects that act as nuclei. The process, called recrystallization, results in an initially very small grain size as the nucleated stressfree grains begin to grow. If heavily strained material is placed into elevated temperature service at

TABLE A3.4 Some Physical Properties of Piping Materials

Material	Density, lb/in.^{3}	Specific heat, mean (temperature, F)	Melting temperature, F
Pure iron	0.2845	0.112 (122–212) 0.170 (1562–1652)	2781–2799
Gray cast iron	0.251–0.265	...	2150–2360
Malleable cast iron ferritic	0.260–0.265	0.11 (at 70) 0.165 (at 800)	2750
Malleable cast iron pearlitic	0.264	0.11 (at 70) 0.165 (at 800)	2750
Nodular iron	0.257	...	2050–3150
Wrought iron	0.28	...	2750
0.06 C, 0.38 Mn	0.2844	0.115 (122–212) 0.264 (1292–1382)	2600
0.23 C, 0.35 Mn	0.2839	0.116 (122–212) 0.342 (1292–1382)	2600
0.43 C, 0.69 Mn	0.2834	0.116 (122–212) 0.227 (1292–1472)	2600
1.22 C, 0.35 Mn	0.2839	0.116 (122–212) 0.499 (1292–1382)	2600
Carbon–½ Mo	0.28	...	2600–2800
1¼ Cr–½ Mo	0.283	0.114 (122–212)	2600–2800
2¼ Cr–1 Mo	0.283	0.11	2600–2800
5 Cr–½ Mo	0.28	0.11	2700–2800
9 Cr–1 Mo	0.28	0.11	2700–2800
3½% Ni steel	0.28	0.115 (212)	...
Type 304 wrought	0.29	0.12 (32–212)	2550–2650
CF-8 cast	0.28	0.12	2600
Type 316 wrought	0.29	0.12 (32–212)	2500–2550
CF-8M cast	0.28	0.12	2550
Type 321 wrought	0.29	0.12 (32–212)	2550–2600
Type 347 wrought	0.29	0.12 (32–212)	2550–2660
CF-8C cast	0.28	0.12	2550–2600
405 wrought	0.28	0.11	2700–2790

TABLE A3.4 Some Physical Properties of Piping Materials (*Continued*)

Material	Density, lb/in.^{3}	Specific heat (mean) temperature, F	Melting temperature, F
CA15 cast	0.275	0.11	2750
410 wrought	0.28	0.11	2700–2790
446 wrought	0.273	0.144	2550–2750
CC50 cast	0.272	0.12	2725
Aluminum 1100	0.098	0.23 (212)	1190–1215
Aluminum 6061	0.098	0.23 (212)	1080–1200
Aluminum 4043	0.097	0.23 (212)	1065–1170
Aluminum 356	0.097	0.23 (212)	1035–1135
Copper (DHP)	0.323	0.092	1981
Red brass (wrought)	0.316	0.09	1810–1880
Yellow brass (wrought)	0.306	0.09	1660–1710
Admiralty brass (wrought)	0.308	0.09	1650–1720
Manganese bronze (wrought)	0.302	0.09	1590–1630
Cupronickel (70–30) (wrought)	0.323	0.09	2140–2260
Aluminum bronze (3) (wrought)	0.281	0.09	1910–1940
Beryllium copper (wrought)	0.297	0.10 (86–212)	1587–1750
Chemical lead	0.4097	0.0309	618
50/50 Sn Pb solder	0.321	0.046	361–421
Nickel (A) (wrought)	0.321	0.13	2615–2635
Monel (70 Ni–30 Cu) (wrought)	0.319	0.127	2370–2460
Inconel (wrought)	0.307	0.11	2540–2600
Incoloy (wrought)	0.290	0.12	2540–2600
Hastelloy B (wrought)	0.334	0.091	2410–2460
Hastelloy C (wrought)	0.323	0.092	2320–2380
Tin	0.26	0.0534	449.4
Titanium (99.0%)	0.163	0.125	3002–3038
Tantalum	0.600	0.034	5425

Source: Reprinted with permission from SAE J 933 ©1989. Society of Automotive Engineers, Inc.

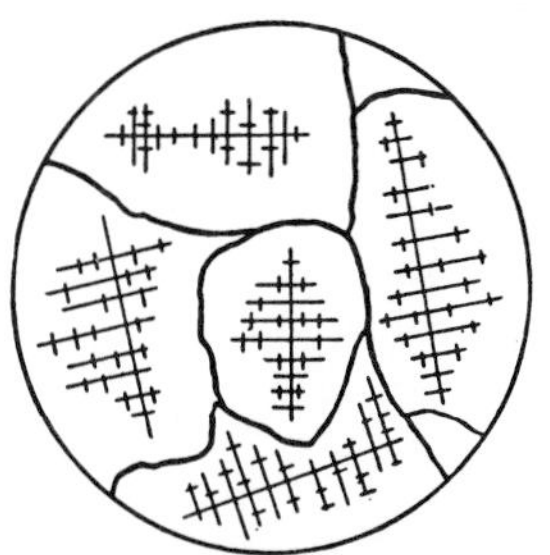

FIGURE A3.15 Sketch illustrating individual grain growth from nuclei and dendrites.

temperatures sufficient to cause recrystallization, initially it will exhibit good creep strength, but when the grains begin to reform, there is poor creep rupture strength. The material would only return to its prestrained creep strength level if additional heat treatment is performed, resulting in further grain growth.

Grain size is a material characteristic that is sometimes directly inspected in the base material testing and certification process. The test entails retrieving a piece of the material, metallographically polishing it, and etching the specimen with a weak acid solution which reveals the grain boundaries under magnification. The test is described in ASTM Specification E-112. Grain size can be measured and reported in a number of ways. The most commonly used method is as an ASTM grain size number, corresponding to the exponent in the following equation:

$$\text{Number of grains per square inch viewed} = 2^n \text{ at 100 magnifications}$$

ASTM has correlated this grain size number, which increases as grain diameter decreases, to a series of photographs representing the grain structure at 100 magnifications. The grain size number can then be estimated by a visual comparative method. Examples of this comparative standard are shown in Fig. A3.16.

Fine-grained carbon and low alloy steels tend to possess better notch toughness and ductility than do coarse-grained steel. As noted earlier, as operating temperature increases into the creep regime, engineering material strength properties are usually enhanced with coarser grains. Although this is an oversimplified, and perhaps overstated rule of thumb, it is important for the engineer to take grain size into account for critical structures.

Hardenability. Hardenability is a property of certain steels that allows them to be strengthened, or hardened, by heat treating. In carbon and alloy steels, for example, hardening is accomplished by heating the material to a temperature above about 1550°F, where the material completely changes its crystal structure from BCC to FCC. When this change is followed by rapid cooling or quenching, usually in water or oil, the result is a crystal structure akin to the original BCC but distorted along one of the unit cell directions. In the case of steels, the result is a martensitic structure possessing a lattice, termed a body-centered tetragonal (BCT), with a larger volume per unit cell than the starting BCC.

The maximum hardness achieved in a quenched structure is primarily a function of the steel's carbon content. The higher the carbon content, the higher the hardness. The depth into the material to which a high hardness is achieved for a given quenching operation is a function of the total alloy content within the steel. The substitutional alloying element nickel has perhaps the strongest effect on increasing the depth to which hardness extends.[3] Other elements creating a similar, if less potent effect, are manganese and boron, a substitutional and interstitial alloying element, respectively.

To test the hardenability of steels, a standard test specimen and procedure, which is a rating of a combination of the highest hardness achievable and the depth to which significant elevation of hardness occurs, has been adopted. The

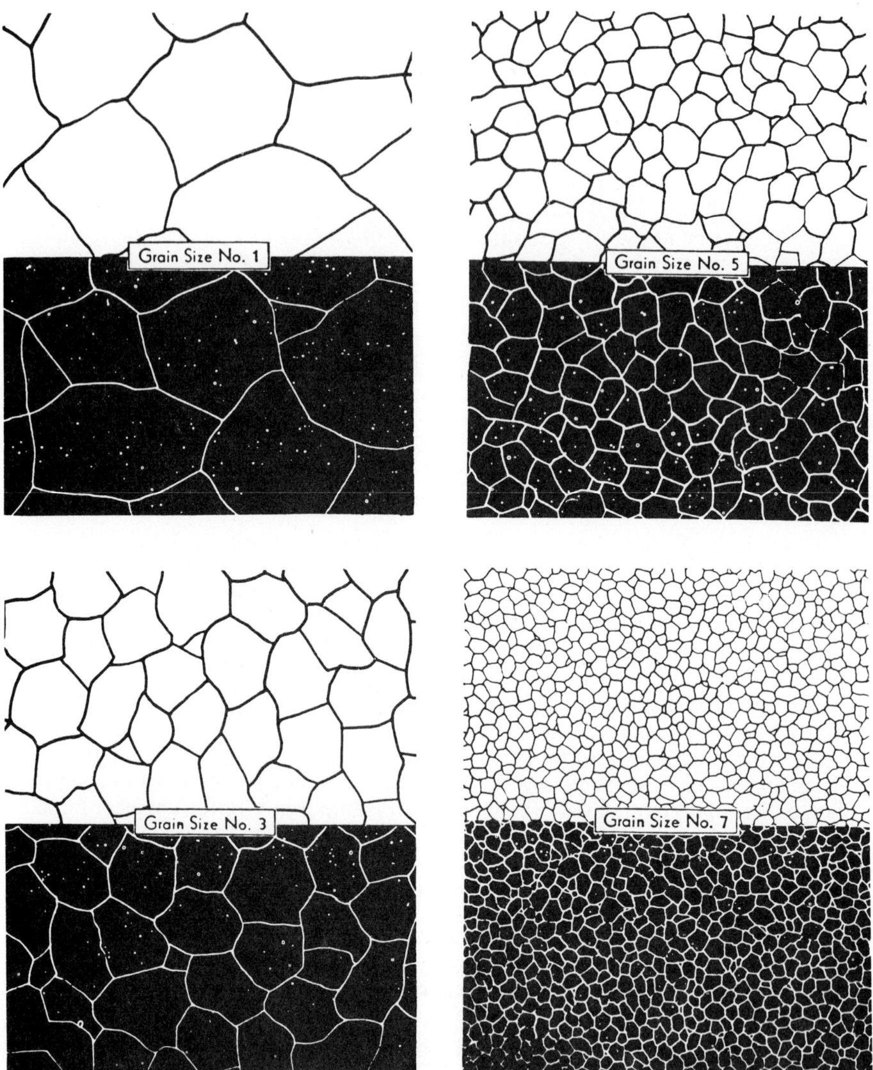

FIGURE A3.16 ASTM grain size charts for classification of steels. 100 magnifications. (*Reproduced by permission of ASTM.*)

test is called the Jominy End-Quench test and entails a 1-in diameter cylindrical specimen machined from the metal in question and heated to a temperature in its austenitic phase (FCC) region. The heated specimen is removed from the heating oven and quickly set in a water-quenching fixture, operating under prescribed conditions of water temperature and flow rate, quenching only the cylinder end face. Upon cooling, the cylinder is parted longitudinally (axially) down the center, and a series of Rockwell hardness readings are taken from the quenched edge. A hardness scan for several alloy steels is shown in Fig. A3.17. The Jominy test procedure is defined in ASTM A-255.

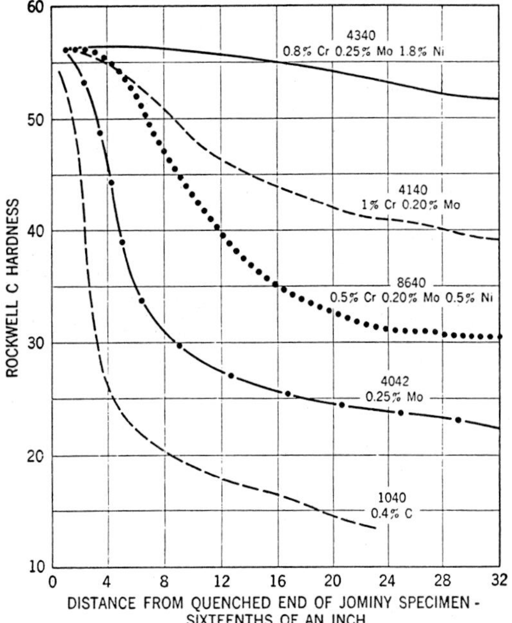

FIGURE A3.17 Jominy end-quench hardenability curves for various 0.40 percent carbon steels. (*Molybdenum Steels Iron Alloys,* Archer et al., Climax Molybdenum Company.)

Many other metal alloys harden or strengthen with special aging or tempering heat treatments. However, this trait is normally not referred to as hardenability. These alloys will be discussed in more detail later in the chapter.

Property Stability. The material mechanical properties that exist at the beginning of life may degrade with service time. In particular, alloys that depend on heat treatment or cold working to develop their strength may weaken if operated for long times at elevated temperatures. The actual exposure to service temperatures acts as a continuation of the heat-treating process, albeit at a significantly reduced rate of effect. In many engineering metals, this effect is actually a property degradation due to overtempering of the material.

A number of thermodynamic relationships exists that relate material strength to time and temperature of exposure for carbon and alloy steels. The most famous and widely used is the Hollomon-Jaffe parameter (HJP), defined as follows:

$$\text{HJP} = (T + 273)(C + \log t) \times 10^{-3}$$

where T is temperature in degrees Celsius, t is time in hours, and C is a constant, usually about 20 for carbon steels.[4] Using this equation and solving for HJP for a given set of time and temperature conditions, the engineer can determine the time at a different temperature of interest that can result in an equivalent metallurgical effect.

A limitation exists on the range of temperatures over which the predictive capability of the Holloman-Jaffe equation can be considered reliable. Phenomenologically, the same metallurgical processes must be in effect over the range of temperatures under consideration. For example, if a phase change occurs or if other important microstructural constituents, such as carbides, are not stable at the two temperatures being compared, the correlation is not valid.

Design codes using allowable design stresses based on creep properties of the metals, by the nature of the long-term rupture tests involved, take these degrading tendencies into account. However, it is not always appreciated that the time-dependent properties, such as ultimate tensile strength and yield strength, can be decreased significantly below the starting property level by the same long-term service. This fact would be important to an engineer concerned with designing a high temperature structure that must tolerate shock loads, such as seismic effects, that can occur near the end of life of the component. Degradation of properties and the mechanisms involved are discussed later in this chapter.

METALLIC MATERIALS

Metals are divided into two types: ferrous, which covers iron and iron-based alloys, and nonferrous, which covers other metals and alloys. Metallurgy deals with the extraction of metals from ores and with the combination, treatment, and processing of metals into useful engineering materials. This section presents the fundamental metallurgical concepts and practices associated with the most common engineering metals and outlines metallurgical considerations appropriate in the selection of metals for piping system construction.

Ferrous Metals

Metallic iron, one of the most common metals, is rarely found in nature in its pure form. It occurs in the form of mineral oxides (Fe_2O_3 or Fe_3O_4) and, as such, comprises about 6 percent of the earth's crust. The first step in the production of iron and steel is the reduction of the ore with coke and limestone in the blast furnace. In this process, the oxygen is removed from the ore, leaving a mixture of iron and carbon and small amounts of other elements as impurities. Coke is the reducing element and source of heat. The limestone ($CaCO_3$) acts as a fluxing agent which combines with impurities of the ore in the molten state and floats them to the top of the molten metal pool where they can be removed as slag. The product removed from the blast furnace is called pig iron, an impure form of iron containing about 4 percent carbon by weight percent. Liquid pig iron cast from the blast furnace is sometimes used directly for metal castings. More often, the iron is remelted in a cupola, or furnace, to refine it further and adjust its composition.[5]

Cast Iron

Pig iron that has been remelted is known as cast iron, a term applicable to iron possessing carbon in excess of 2 weight percent. Compared with steel, cast iron is inferior in malleability, strength, toughness, and ductility. On the other hand, cast iron has better fluidity in the molten state and can be cast satisfactorily into complicated shapes. It is also less costly than steel. The most important types of cast iron are white and gray cast irons.

White cast iron is so known because of the silvery appearance of its fracture surface when broken. In this alloy, the carbon is present in the form of iron carbide (Fe_3C), also known as cementite. This carbide is chiefly responsible for high hardness, brittleness, and poor machinability characteristic of white cast iron. Chilled iron, a form of white cast iron, is cast against metal chills that cause rapid cooling, promoting the formation of cementite. Consequently, a structure is obtained which possesses high wear and abrasion resistance, the principal attribute of the material, but retains white cast iron's characteristic brittleness.

Malleable cast iron is the name given to white cast iron that has been heat treated to change its cementite into nodules of graphite. The iron becomes more malleable because, in this condition, the carbon as carbide no longer exists continuously through the metal matrix.

Gray iron is a widely used type of cast iron. In this alloy, the carbon is predominantly in the form of graphite flakes. The typical appearance of a fracture of this iron is gray since the graphite flakes are exposed. The strength of gray iron depends on the size of the graphite particles and the amount of cementite formed with the graphite. The strength of the iron increases as the graphite crystal size decreases and the amount of cementite increases. This material is easily machined. A wide range of tensile strengths can be achieved by alloying with other elements such as nickel, chromium, and molybdenum.

Another member of the cast iron family is the so-called "ductile" iron. It is a high carbon magnesium-treated product containing graphite in the form of spheroids. Ductile iron is similar to gray cast iron in machinability, but it possesses superior mechanical properties. This alloy is especially suited for pressure castings. By special procedures (casting against the chill) it is possible to obtain a carbide-containing, abrasion-resistant surface with an interior possessing good ductility.[5]

Steel

Steel is defined as an alloy of iron with not more than 2 weight percent carbon. The most common method of producing steel is to refine pig iron by oxidation of impurities and excess carbon, which have a greater affinity for oxygen than iron.

The principal reduction processes used are the basic-oxygen process (BOP) and the electric furnace process, each representing a type of furnace in which the refining takes place. The BOP primarily uses molten pig iron as the initial furnace charge; the electric furnace can use a charge of selected steel scrap. Another process, called the basic open-hearth process, is no longer in use in the United States. Although it constituted the major steel-producing process for decades, it has succumbed to the more advanced and economical BOP and electric furnaces.[6]

The pig iron is reduced to the desired steel composition through use of acid and/or basic reactions with fluxing agents, heat, oxygen, and time. Excess carbon is oxidized and lost as gas; impurities are floated to the surface. Often, desired alloying elements are added to the molten pool. The steel can be further refined by using one of various methods of vacuum degassing. As the name suggests, the molten steel is passed through a vacuum chamber to remove entrained gases such as oxygen, hydrogen, and carbon dioxide. This operation is performed when extra steel cleanliness is desired and results in improved and more uniform properties in the final product.

The molten steel is then cast into ingot molds, which are further reduced by hot working in rolling and drawing operations. Alternately, the molten steel may be directly cast into continuous smaller billet or hollow products. The latter process is called continuous casting and has become the preferred method of making steel since it avoids the costly ingot reduction operations.

Alloying additions are made, if required, to the molten steel either while in the reducing furnace as noted, in the ladle into which the steel is put, or in the ingot into which steel is poured from the ladle.

While molten in the furnace, oxygen is forceably injected into the steel to refine the charge. The oxygen combines with excess carbon and is released as a gas. Excessive oxygen is, however, unavoidably left in the molten steel. This results in the formation of oxide inclusions in the steel or porosity which appear upon solidification. The process of removing the oxygen is known as deoxidizing practice. Deoxidation is achieved by adding silicon, aluminum, or other deoxidizing agents to the molten steel, the amount of which determines the degree of deoxidation and the type of steel. The common names given to these various steel types are killed steel, semikilled steel, and rimmed steel.

Killed steel is deoxidized almost completely; that is, sufficient deoxidizing agent is added to the molten pool to combine with all the excess entrained oxygen. The result is a large number of tiny oxides in the melt. The lack of gas in the molten pool gives the effect of "killing" any visible bubbling activity of the steel, thus the name. Killed steel has a more uniform composition than any of the other types and usually possesses the best formability at room temperature. A fine-grained structure results from this practice because the many oxides formed act as initiation sites of new grains upon solidification and subsequent recrystallization. This fine-grained character offers toughness superior to the other types of steel.

Rimmed steel has no purposeful addition of deoxidizing agents and is characterized by relatively violent bubbling and stirring action in the ingot mold. It exhibits a marked difference in composition across and from top to bottom of the ingot. The outer rim, or outer edge of the solidified ingot, is a relatively pure and ductile material. The amounts of carbon, phosphorus, sulfur, and nonmetallic inclusions in this rim are lower than the average composition of the whole ingot. The amount of these constituents in the inner portion, or core, is higher than the ingot average. This type of steel costs less to make than the other types of steel and is widely used for structural applications, where good surface appearance of the final product is desired.

Semikilled steel is only partially deoxidized with silicon, aluminum, or both, taking advantage of the positive attributes of killed and rimmed steel.

After casting, or "teaming," into the ingot molds, the steel is normally further reduced in size and modified in shape by mechanical working. The majority of the reduction process is done hot. During hot working, sufficient heat is maintained to ameliorate the working effects and maintain a structure that is relatively soft and ductile throughout the reduction process.

The steel in the form of ingot, slab, bar, or billet is first brought to the proper temperature throughout and is then passed through rolls or dies. The flow of metal is continuous and preferentially in one (longitudinal) direction. The cross-sectional area is reduced, and the metal is shaped to the desired form. The internal structure of the steel is also favorably affected. The working reduces the grain size of the material and tends to homogenize the overall structure, compared with cast or unworked steel. Processes used to manufacture pipe and tube are addressed in Chap. 5.

PHYSICAL METALLURGY OF STEEL

Like all metals, iron and steel are crystalline in structure and composed of atoms in a fixed lattice. As noted earlier, iron may exist in one of two cubic forms: body centered (BCC) or face centered (FCC).

At room temperature, pure iron is composed of a body-centered cubic lattice. In this form it is known as alpha iron, also called ferrite, which is soft, ductile, and magnetic. When heated above about 768°C (1415°F), alpha iron loses its magnetism but retains its body-centered crystalline structure. This temperature is called the Fermi temperature. The crystal structure changes to face-centered cubic at about 910°C (1670°F) at which temperature alpha iron is transformed to gamma iron, the FCC form, and remains nonmagnetic. As the temperature rises further, another phase change occurs at 1410°C (2570°F), when delta iron is formed. This phase is again body centered like that of the low temperature alpha iron. It is stable to the melting temperature. When cooling very slowly from the liquid state, the phases appear in reverse order.

The solid-state transformations of atomic structure, which occur in pure iron during heating to and cooling from the melting point, are called allotropic changes. The temperatures at which these changes take place are known as transformation or critical temperatures.

When carbon is added to iron and steel is produced, the same changes in phase occur but a more complex relationship with temperature occurs. The effects of varying amounts of carbon content in iron on phase stability as temperature varies is shown in Fig. A3.18. This figure is called an equilibrium phase diagram, and in this case it is the familiar iron-carbon (Fe-C) phase diagram. With this diagram, one can determine which stable phase the steel will assume at a given composition and temperature. Likewise, the effect of increasing or decreasing the amount of carbon content in iron on these critical temperatures can be predicted.

Phase diagrams are plotted in weight or atomic percent (horizontal axis) versus temperature (vertical axis). A single phase region usually represents an area of high concentration of a single element or an intermetallic single phase stable over a range of composition and temperature. Between these single-phase regions are regions where multiple phases coexist in relative amounts at any given temperature approximated by the proximity of the specific composition to the single-phase regions. On the Fe-C diagram, single-phase regions are represented by those marked as alpha, gamma, and delta and Fe_3C (or cementite) which is a stable intermetallic phase.

The critical transformation temperatures in steel are the A_1, corresponding to about 723°C (1335°F), and A_3, referred to as the lower and upper critical temperatures of steel. The A_3 constitutes the boundary with the gamma phase, and its temperature varies with carbon content. The lower critical temperature, on the other hand, stays constant over the entire range of steel compositions.

These critical temperatures, as well as the entire phase diagram, represent transformations that occur under controlled, very slow cooling and heating conditions (i.e., equilibrium conditions). More rapid heating and cooling rates, like those encountered in normal steel processing, change these critical temperatures

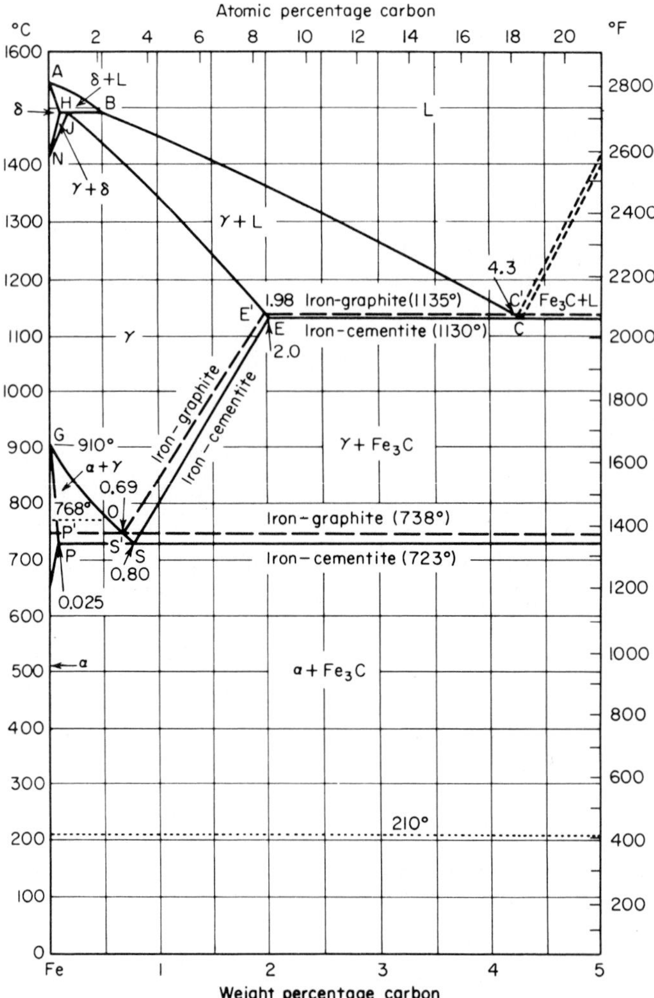

FIGURE A3.18 Iron-carbon equilibrium diagram.

upward and downward, respectively. Additions of other alloying elements also will shift the critical transformation points.

The effective use by the metallurgist of the knowledge of this and similar phase diagrams allows for the manipulation of properties of engineering materials by varying chemistry and heat treatment. For steel, the principal phases that exist and their properties are summarized as follows:

Austenite: Austenite is a single-phase solid solution of carbon in gamma iron (FCC). It exists in ordinary steels only at elevated temperatures; it is also found at room temperature in certain stainless steels (e.g., 18 Cr-8 Ni type), classified as austenitic stainless steels. This structure has high ductility and toughness.

Ferrite: Ferrite is alpha iron (BCC) containing a small amount of carbon (0.04–0.05%) in solid solution. This phase is soft, ductile, and relatively weak.

Cementite: Iron carbide, Fe_3C, is a compound containing 6.67 percent carbon, which is very hard and extremely brittle. Cementite appears as part of most steel structures, the form of which depends on the specifics of the heat treatment which the steel has received (see below).

Pearlite: Pearlite is a mixture of alternating plates of iron carbide (cementite) and ferrite (lamellar structure), which forms on slow cooling from within the gamma range. This condition generally represents a good blend of strength, ductility, and fair machinability. It is the equilibrium structure in steel.

Bainite: Bainite is a structure harder and stronger than pearlite, which is a mixture of ferrite and cementite. It forms by the transformation of austenite in many steels during fairly rapid cooling but not fast enough to cause martensite formation. The structure consists of ferrite and iron carbide, but unlike pearlite, the aggregate is nonlamellar.

Martensite: The hardest constituent achievable by heat treating of steels, martensite is formed by the rapid cooling of austenite to a temperature below the martensite start, or M_s, temperature. Martensite consists of a distorted cubic unit cell (body-centered tetragonal), which contains substantial quantities of carbon in interstitial solution in the lattice. The M_s temperature varies with steel composition.

Bainite and martensite will not be found on the Fe-C phase diagram because they are the direct result of cooling steel at an accelerated rate which prevents atomic diffusion required to maintain equilibrium conditions.

The effects of nonequilibrium cooling of a steel is represented on an isothermal transformation diagram, or a time-temperature transformation (T-T-T) diagram. An example of each is shown in Fig. A3.19. The horizontal axis of the diagram is time, usually log scale; the vertical axis is temperature. A single diagram represents a given steel alloy composition and depicts the various equilibrium and nonequilibrium phases that will be formed and their mix, with a given cooling rate from a starting temperature in the austenitic phase region. The diagram is used by entering it at the alloys temperature at time = 0, represented as a point of the vertical axis. The cooling rate describes the time-temperature path taken by the material from the starting point, through the field of transformation phases, to the final point of sample cooling. The metallurgical phases or constituents in the final state can thus be predicted. The continuous path followed between the two points also has a bearing on final microstructure. The T-T-T diagram is similar to the equilibrium phase diagram in that single and multiple phase fields are depicted. However, it differs from the equilibrium diagram in that it is a dynamic representation of phase formation with time. Thus, quickly cooling to a given temperature above the M_s will result, for example, in coexistence of austenite, ferrite, and cementite (*A, F,* and *C* on the figure). However, as time progresses at that temperature, the austenite continues to decompose into more ferrite and cementite until complete transformation is achieved. Cooling to below the M_s temperature causes transformation to martensite. If the path of cooling intersected the "nose" of the T-T-T curve, some ferrite will form and be combined with the martensite in the final microstructure, since martensite can only be formed by quenching austenite. The ferrite that formed on cooling is stable and unaffected by further cooling.

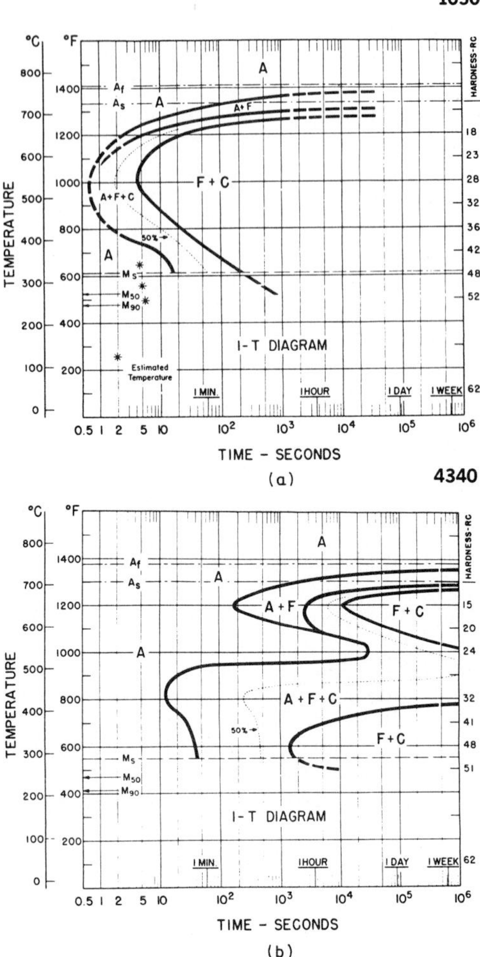

FIGURE A3.19 Isotherm transformation diagrams for AISI 1050 (*a*) and AISI 4340 (*b*). (From I-T Diagrams, *United States Steel,* ©1963.)

ALLOYING OF STEEL

The alloying of carbon steel with other elements in an attempt to obtain a wide range of desired properties is a mature science. The known effect of adding certain elements on the properties of steel are summarized below.[7]

Carbon: In general, an increase in carbon content produces higher ultimate strength and hardness but may lower ductility and toughness of steel alloys. Carbon also increases air-hardening tendencies and weld hardness. In low-alloy steel for high-temperature applications, the carbon content is usually restricted to a maximum of about 0.15 percent in order to assure optimum ductility for welding, expanding, and bending operations. An increasing carbon content lessens the thermal and electrical conductivities of steel.

Phosphorus: High phosphorus content has an undesirable effect on the properties of carbon steel, notably on shock resistance and ductility (see the section on temper embrittlement). Phosphorus, however, is effective in improving machinability. In steels, it is normally controlled to less than 0.04 weight percent.

Silicon: Silicon, which is used as a deoxidizing agent, increases the tensile strength without increasing brittleness when limited to less than about 2 percent. Silicon increases resistance to oxidation, increases electrical resistivity, and decreases hysteresis losses. Thus, it is used for electrical application. Additions of silicon may reduce creep-rupture strength.

Manganese: Manganese is normally present in all commercial steels. The manganese combines with sulfur, thus improving hot working characteristics. In alloy steels, manganese decreases the critical cooling rate to cause a hardened, or martensitic, structure and thus contributes to deep hardening.

Nickel: As an alloying element in alloy steels, nickel is a ferrite strengthener and toughener and is soluble in all proportions. Nickel steels are easily hardened because nickel lowers the critical cooling rate necessary to produce hardening on quenching. In heat-treated steel, nickel increases the strength and toughness. In combination with chromium, nickel produces alloy steels with higher impact and fatigue resistance than are possible with straight carbon steels.

Chromium: As an alloying element in steel, chromium is miscible in iron as a solid solution and forms a complex series of carbide compounds. Chromium is essentially a hardening element and is frequently used with a toughening element, such as nickel, to produce superior mechanical properties. At higher temperatures, chromium contributes increased strength and is ordinarily used in conjunction with molybdenum. Additions of chromium significantly improve the elevated temperature oxidation resistance of steels.

Molybdenum: In steel, molybdenum can form a solid solution with the iron and, depending on the molybdenum and carbon content, can also form a carbide. A deeper hardening steel results. The molybdenum carbide is very stable and is responsible for matrix strengthening in long-term creep service.

Vanadium: Vanadium is one of the strong carbide formers. It dissolves to some degree in ferrite, imparting strength and toughness. Vanadium steels show a much finer grain structure than steels of a similar composition without vanadium.

Boron: Boron is usually added to steel to improve hardenability, that is, to increase the depth of hardening during quenching.

Aluminum: Aluminum is widely used as a deoxidizer in molten steel and to control grain size. When added to steel in controlled amounts, it produces a fine grain size.

Sulfur: Sulfur, which is present to some degree in all steel (less than 0.04 weight percent), forms a nonmetallic inpurity that, in large amounts, results in cracking during forming at high temperatures (hot shortness). Combining it with manganese forms an MnS compound that is relatively harmless.

Copper: Copper dissolves in steel and strengthens the iron as a substitutional element. The use of copper in certain alloys increases the resistance to atmospheric corrosion and increases the yield strength. However, excessive amounts of copper (usually above 0.3 percent) is harmful to elevated temperature performance since the lower melting point element segregates to grain boundaries and locally melts (liquates), causing intergranular separation under applied stress.

In general, when alloying elements are used in combination, they can complement each other and give greater overall benefits than when used singly in much larger quantities.

CLASSIFICATION OF STEELS

There are literally hundreds of wrought grades of steel that range in composition with the variation of the major and minor alloying elements. The simplest of these classes is known as plain carbon steels, with carbon varying between approximately 0.05 and 1.0 weight percent. Within this broad range fall three general groups according to carbon content:

1. Low carbon steels: 0.05–0.25 percent carbon
2. Medium carbon steels: 0.25–0.50 percent carbon
3. High carbon steels: 0.50 percent and greater carbon content

Alloy steels are generally considered to be steels to which one or more alloying elements, other than carbon, have been added to give them special properties different than those of straight carbon steels. From the standpoint of composition, steel is considered to be an alloy steel when amounts of manganese, silicon, or copper exceed the maximum limits for the carbon steels or when purposeful addition of minimum quantities of other alloying elements are added. These could be chromium, molybdenum, nickel, copper, cobalt, niobium, vanadium, or others.

The next higher class of alloyed steel useful to the piping industry is ferritic and martensitic stainless steels. These are steels alloyed with chromium contents above about 12 percent. These materials possess good corrosion resistance due to the chromium. They retain a ferritic (BCC) crystal structure, thus allowing the grades to be hardened by heat treatment.

When sufficient nickel is added to iron-chromium alloys, an austenitic (FCC) structure is retained at room temperature. A family of austenitic stainless steels exists, possessing an excellent combination of strength, ductility, and corrosion resistance. These steels cannot be hardened by quenching, since the austenite does not transform to martensite. A stronger type of stainless steel has been developed which takes advantage of precipitation reactions within the metal matrix made possible by addition of elements such as aluminum, titanium, copper, and nitrogen. These materials are referred to as precipitation—hardenable stainless steels. Both martensitic and austenitic stainless steels can be enhanced in this manner. As annealed, these materials are soft and readily formed. When fully hardened through aging heat treatments, they attain their full strength potential.

STEEL HEAT TREATING PRACTICES

Various heat treatments can be used to manipulate specific properties of steel such as hardness and ductility, to improve machinability, to remove internal stresses, or to obtain high strength levels and impact properties. The heat treatments of steel commonly used—annealing, normalizing, spheroidizing, hardening (quenching), and tempering—are briefly described.[8]

Annealing

Several types of annealing processes are used on carbon and low alloy steel. These are generally referred to as full annealing, process annealing, and spheroidizing annealing.

In full annealing, the steel is heated to just above the upper critical (A_3) temperature, held for a length of time sufficient to fully austenitize the material structure, and then allowed to cool at a slow, controlled rate in the furnace. The microstructure of fully annealed low carbon steel consists of ferrite and pearlite. A full anneal provides a relatively soft, ductile material, free of internal stresses.

The process anneal, sometimes referred to as stress relieving, is carried out at temperatures below the lower critical (A_1) temperature. The process anneal is used to improve the ductility and decrease residual stresses in work-hardened steel.

The usual purpose of spheroidizing is to soften the steel and improve its machinability. Heating steel that possesses a pearlite microstructure for a long time just below the lower critical temperature, followed by very slow cooling will cause spheroidization. This is an agglomeration of the iron carbide, which eventually assumes a spheroidal shape. The properties of this product normally represent the softest condition that can be achieved in the grade of steel being heat treated.

The austenitic stainless steels are annealed differently from carbon steels. First, since they posses a fully austenitic structure, the temperature used is not related to a critical transformation temperature. Rather, the intent of the anneal is to remove residual strain in the lattice, recrystallize the metal grains, and dissolve any iron and chromium carbides that may exist in the matrix material. The temperature selected is usually at or above 1038°C (1900°F). Second, the cooling rate from the annealing temperature is normally as rapid as possible. This suppresses the reformation of carbides at the austenitic grain boundaries during cooling. Formation of grain boundary carbides results in local depletion of chromium in the matrix in the vicinity of the carbides, rendering this thin band of material susceptible to attack in a number of corrosive media. This susceptible condition is referred to as sensitization, and the resultant corrosion is called intergranular attack.

The temperature range in which carbides are most apt to form in austenitic stainless steels is between 454 and 816°C (850–1500°F). Slow cooling through or holding in this zone will sensitize the steel. The degree of sensitization that will occur can be greatly reduced by adding small amounts of elements that possess a stronger tendency to form carbides than the chromium. Two such elements, niobium and titanium, are added to form the so-called stabilized austenitic stainless steels. Alternately, the carbon content can be held as low as possible, thereby resulting in as few carbides as possible. These are termed the *L grade* stainless steels.

Ferritic and martensitic stainless steels will also be adversely affected by slow cooling from annealing temperatures. When slowly cooled, or held in the temperature range of 400–510°C (750–950°F), these materials embrittle (see the discussion on 885°F embrittlement).

Normalizing

Normalizing is similar to the annealing process except that the steel is allowed to cool in air from temperatures above the upper critical temperature. Normalizing relieves the internal stresses caused by previous working. Although it produces sufficient softness and ductility for many purposes, it leaves the steel harder and with higher tensile strength than after annealing. Normalizing is often followed by tempering.

Hardening (Quenching)

When steels of the higher carbon grades are heated to produce austenite then cooled rapidly (quenched), the austenite transforms into martensite. Martensite is formed at temperatures usually below about 210°C (400°F), depending on the carbon content and the type and amount of alloying steel. It is the hardest form of heat-treated steel and has high strength and resistance to abrasion. Martensitic steels have poor impact strength and are difficult to machine.

Tempering

Tempering is a secondary heat treatment performed on some normalized and almost all hardened steel structures. The object of tempering is to remove some of the brittleness by allowing certain solid-state transformations to occur. It involves heating to a predetermined temperature, always below the lower critical temperature, followed by a controlled rate of cooling. In most cases the hardness of the steel is reduced by tempering, its toughness is increased, and residual stresses are eliminated. The higher the tempering temperature used for a given time, the more pronounced is the property change. Some steels may become embrittled on slowly cooling from certain tempering temperatures. Steels so affected are said to be *temper brittle.* To overcome this difficulty, steels of that type are cooled rapidly from the tempering temperature. Temper embrittlement is discussed elsewhere in this chapter.

DEGRADATION OF SERVICE MATERIALS

A number of metallurgically based processes can occur in steels which contribute to loss of engineering strength and even premature failure. Several of these are addressed in the following.

Aging of Properties

Several steels that have accumulated considerable service time are known to have their properties modified, usually to their detriment. This phenomenon has been called *aging,* and it occurs in materials that are heat treated or cold worked to achieve high strength levels and are used at elevated temperatures. These materials are potentially more susceptible to failure later in their life after the condition has developed.

Aging in this case should not be confused with aging used to represent the purposeful heat treatment performed to some types of nonferrous alloys. In the context here, *aging* refers to the normally slowly progressing metallurgical reaction that occurs in a number of alloys while at operating temperatures for extended periods of time. Specific types of this behavior, that is, temper embrittlement and "885" embrittlement, are addressed below.

Components that experience considerable service time contain materials that have "aged" with time. The materials of special interest are those that experience the higher operating temperatures; for example, ferritic steels above 482°C (900°F) and austenitic stainless steels at or above 538°C (1000°F). A recent study, sponsored by the ASME Boiler and Pressure Vessel Code, attempted to identify and quantify these effects. The effort was the result of concerns for near end of life seismic loadings in elevated temperature nuclear boilers.

Data gathered from a number of sources have shown that the room and elevated temperature yield strengths of both ferritic and austenitic steels may be reduced after long exposure times. Ultimate tensile strength is affected, but to a lesser degree. The yield strength reductions can amount to as much as 40 percent in ferritic and 20 percent in austenitic steels.

Creep tests have also been run after long-term (e.g., 10,000 h) static exposure to elevated temperatures. No substantial negative effect on creep properties were noted in these tests.

In the case of ASME Boiler and Pressure Vessel Code and ASME B31.1 Power Piping Code, the degradation of yield strength does not generally violate or invalidate the conservatism built into their design rules, as shown by the allowable design stresses. For example, in Section I of the ASME Boiler and Pressure Vessel Code, addressing design and construction of power boilers, the material design-allowable stresses for wrought materials are established by applying the following factors to base material properties (the lowest calculated value of the following is assumed to be the design-allowable stress at a given temperature):

- 1/4 specified minimum tensile strength at room temperature
- 2/3 specified minimum yield strength at room temperature
- 1/4 tensile strength at the temperature of interest above room temperature
- 2/3 yield strength at the temperature of interest above room temperature
- 67 percent of the average stress to cause rupture in 100,000 h
- 80 percent of the minimum stress to cause rupture in 100,000 h
- 100 percent of the stress to produce 0.01 percent strain in 1000 h

In this manner, short-term properties, stress rupture strength, and creep rate are all taken into account. Typically, at the lower end of the temperature use range, the factored tensile and yield strength controls and at higher temperatures, the creep properties set the allowable stresses.

Since most aging occurs at the higher temperatures, tensile and yield strength degradation does not normally cause concern. However, if large shock loads can occur late in component life, as is possible under seismic conditions, these short-term, time-independent properties can be critical to the components' continued safe operation.

As an illustration of the effect that reduced yield strength has on fatigue, consider that with a lower yield strength, more plastic strain will result from a given high thermal or mechanically induced stress. Since these stresses are usually due to operational transients, stress reversals can occur during continued operation (i.e., temperature stabilization at steady state or, ultimately, component shutdown). The greater the plastic strain cycle, the greater the damage and the sooner the failure.

As components age with increasing service time, they become less resilient to significant operational transients. That is, materials are less likely to withstand these transients than earlier in their life, not only because more cycles have continued to accumulate (toward an end-of-life limit) but also because material strength properties are degrading with time.

Temper Embrittlement

Temper embrittlement occurs in carbon and alloy steels when aged in the temperature range roughly between 350°C (660°F) and 550°C (1020°F). The property most significantly affected is toughness. The time in which this occurs is a function of the steel's chemical composition, heat treatment condition, fabrication history, and service temperature. The most severe degradation occurs in weld regions. Due to Cr-Mo steel's extensive use in the petrochemical and power boiler industries, most of the studies have concentrated on that family of steels.

It has been recognized for some years that a steel's susceptibility to temper embrittlement is due to the existence and amount of the trace elements antimony (Sb), phosphorus (P), tin (Sn), and arsenic (As), with P and Sn having the greatest effect. Other elements that may contribute to reduced toughness are silicon, manganese, and copper. Beneficial effects can be gained by additions of molybdenum and aluminum. The dependency of temper embrittlement severity and chemistry has led to the development of a number of embrittlement "factors"[9]:

Bruscato made the first attempt to combine the effects of various elements into a single factor, known as the embrittlement factor X, which is expressed as follows:

$$X = \frac{1}{100}(10\text{P} + 5\text{Sb} + 4\text{Sn} + \text{As})$$

The concentration of elements is in parts per million (ppm).

Miyano and Adachi arrived at a J factor defined as

$$J = (\%\text{Si}) + (\%\text{Mn}) \times (\%\text{P}) + (\%\text{Sn}) \times 10^4$$

Finally, Katsumata et al. asserted that the following embrittlement factor (EF) was appropriate for 2¼Cr-1Mo and 3Cr-1Mo steels:

$$\text{EF} = (\%\text{Si}) + (\%\text{Mn}) + (\%\text{Cu}) + (\%\text{Ni}) \cdot Y$$

$$Y = \frac{1}{100}(10\text{P} + 5\text{Sn} + \text{Sb} + \text{As})(\text{in ppm})$$

All of these are useful in assessing the relative susceptibility of various steel compositions to temper embrittlement. In all cases, the larger the factor, the more susceptible the particular heat of steel is to embrittlement.

The type of heat treatment applied to the materials may also affect a material's susceptibility to temper embrittle. For example, a number of experimenters have confirmed that 2¼Cr-1Mo alloy steel's susceptibility increases as austenitizing temperature used during its heat treatment increases. Inversely, susceptibility is lowest after an intercritical hold at a temperature between the lower and upper critical temperature (Ac_1 and Ac_3, respectively). This effect is believed to be associated with the grain size achieved during the hold time, a larger grain being more detrimental. Although intercritically treated materials are less susceptible to temper embrittlement, they are weaker in the as-heat-treated condition.

In a parallel fashion, the degree of embrittlement, as measured by loss of toughness or shift of nil ductility temperature, is decreased if the material is more substantially tempered before the embrittling treatment. In this context, *tempered* represents the planned heat treatment that typically follows a normalizing or austenitizing and rapid quenching operation. A longer or higher temperature temper results in a softer, less strong, and more ductile condition, usually accompanied by good fracture toughness.

Luckily, the temper-embrittled condition is reversible. Heat treatment for short periods of time at temperatures well above the upper critical point will result in reestablishing nearly virgin properties in these materials.

Hydrogen Attack

Hydrogen attack is one of the most important problems with materials in ammonia synthesis, oil refining, and coal gasification equipment.[9]

The first major failure of an ammonia convertor attributable to hydrogen damage was in 1933. Since then more failures and untold damage to materials have been accumulated, with the majority of damage occurring in the welds or weld heat-affected zones of these components.

When carbon and low alloy steels are held in hydrogen at high temperature and pressure for an extended period of time, these materials can suffer degrading effects to their tensile and creep-rupture properties. This is accompanied by the formation of intergranular fissures, blisters on the surface, and loss of carbon content (decarburization). The phenomenon is called hydrogen attack and is generally attributed to the formation of methane (CH_4) within the steel.

The microstructural damage occurs when methane bubbles form and grow around precipitates at the grain boundaries within the material. The continued growth of the bubbles causes grains to separate along their boundaries and the bubbles, or voids, to coalesce. The rate of growth of the bubbles is a function of the ease by which the steel carbides give up carbon atoms to the intruding hydrogen atoms to form the methane. The more stable the carbide, the slower this reaction will take place. Thus, it has been long recognized that additions of chromium and molybdenum, both strong carbide stabilizers, improves hydrogen-attack resistance of steels. Addition of other carbide-stabilizing elements such as titanium and tungsten has also assisted in reducing susceptibility.

Weld regions are more susceptible to hydrogen damage because they possess less stable carbides. Also, carbon content of the base material is an important variable in determining the susceptibility of a steel to hydrogen damage. In general, steels used in this service are kept below 0.20 weight percent carbon content. Certain other elements, such as nickel and copper, are also known to have a detrimental effect.

Nelson curves have proven indispensable in the selection of materials in hydrogen service. These curves (Fig. A3.20), which were originally based on experience gathered over several decades, have been revised based on new experiences gained. They identify a "safe" regime in which an alloy will perform acceptably at various temperatures and hydrogen partial pressures. Where these curves have not proved conservative, they have been associated with weld heat-

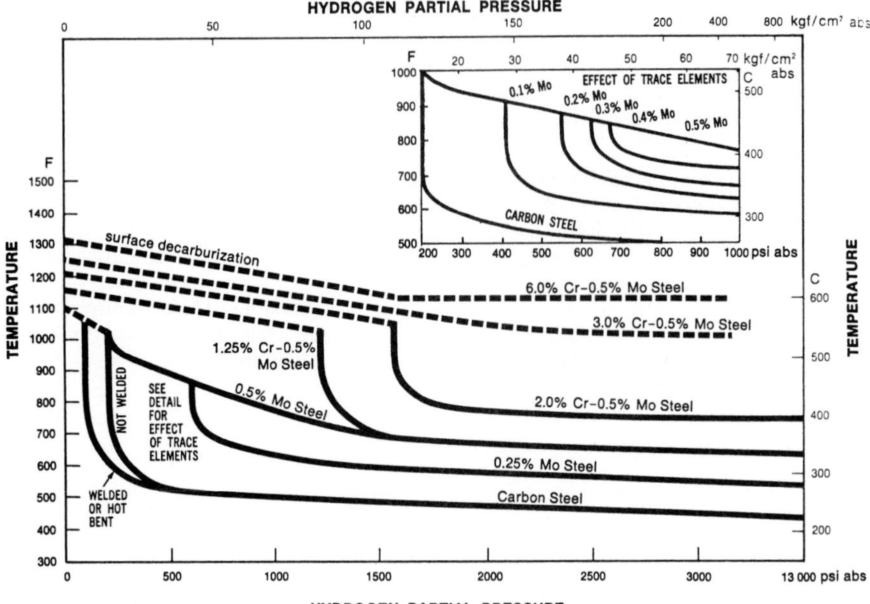

FIGURE A3.20 The classic Nelson diagram indicating the choice of steel warranted to avoid hydrogen attack as a function of operating temperature and partial pressure of hydrogen. Austenitic materials are satisfactory at all temperatures and pressure from hydrogen damage. (Dunn et al., *Molybdenum's Place in the Pressure-Vessel Field,* Climax Molybdenum Company.)

affected zones that had been inadequately postweld heat treated (PWHT). The high residual stresses and high hardness left in the weld region contribute to accelerated damage. For this reason, most specifications for hydrogen service equipment stipulate a maximum hardness in weld regions that will assure adequacy of the PWHT. The limit is usually placed at 210 Brinell hardness, corresponding approximately to a 100,000 psi ultimate tensile strength.

Austenitic stainless steels are essentially immune to hydrogen damage. The numerous sites within the FCC lattice in which the hydrogen atoms can be safely accommodated and the inherent ductility of the lattice give austenitic materials this freedom from hydrogen damage. However, when stainless overlay weld metal has been used over carbon or low alloy vessel steels, hydrogen-induced cracking can occur at the weld fusion line just inside the ferritic material.

885°F Embrittlement

One of the limitations of ferritic stainless steels (those alloys of iron possessing greater than about 14 percent chromium), has been the loss of toughness at room temperature that occurs after these materials are exposed for long times to temperatures in the range of 320 (610) to 538°C (1000°F). This is commonly referred to as 885°F embrittlement, corresponding approximately to the temperature at which many of the alloys degrade the fastest.

The compositional effects in commercial alloys on 885°F embrittlement have not been systematically investigated. However, it is clear that the degree of embrittlement increases as chromium content increases. The effects of other elements are not clear. Of these, most important is carbon, and it has been reported as having from no effect to a retarding effect on embrittlement.

This phenomenon results in increased hardness and strength, with a corresponding decrease in ductility, fracture toughness, and corrosion resistance. Loss of toughness can be particularly severe and, in fact, has tended to relegate the use of this class of alloy to temperature regimes below which significant embrittlement can occur.

Graphitization

Graphitization is a time- and temperature-dependent nucleation and growth process in which iron carbide in the form of pearlite first spheroidizes and later forms graphite nodules. There are two general types:

1. Formation of randomly, relatively uniformly distributed graphite nodules in the steel. This reduces the room temperature mechanical strength somewhat but does not affect the creep-rupture strength at elevated temperature.
2. A concentrated formation of graphite most frequently along the edges of the heat-affected zone of weldments. This is referred to as chain graphite, since a plane of nodules exists paralleling the weld bead contours.

   The formation of these nodules, when aligned through the wall of a pressure part, creates planes of weakness subject to rupture. Fracture characteristically occurs without prior warning.

The first graphitization failure of a low-carbon steam piping material occurred in the early 1940s. The failure occurred after five and a half years of service in a steam line made of aluminum-killed carbon-molybdenum steel. The fracture surface was located approximately 1/16 in from the fusion zone of a butt weld. The failure precipitated numerous and extensive research programs to understand the key variables of the mechanism and to determine the steels which would resist graphitization.

Research has helped in understanding the problem, and has led to restrictions adopted by the various design codes on use of materials subject to graphitization. Carbon steel and carbon-molybdenum grades are the most susceptible, the latter being more so. Relative susceptibility of these two grades is also dependent on the amount of aluminum content; the more aluminum, the greater the susceptibility. Additions of chromium in amounts as low as 0.5 weight percent make the steel essentially immune to graphitization.

The ASME Code permits the use of carbon and carbon-molybdenum steels in Section I boiler applications up to 1000°F. A cautionary note is provided in the allowable stress tables of Section I, indicating the carbon steels and carbon-molybdenum steels may be susceptible to graphitization at temperatures above about 800 and 875°F, respectively. ASME B31.1 has a similar precautionary note specifying limits of 775 and 850°F, respectively. Graphitization is a mechanism dependent on diffusion and is not associated with a precise temperature of initiation (occurs sooner at higher temperatures). Thus, the differences between the design codes only reflects different levels of conservatism in dealing with the failure mode. Many manufacturers extend even more severe restrictions, some prohibiting the use of these steels in piping applications outside the boiler or pressure vessel where rupture creates a serious safety hazard. Substitution of chromium-containing steel grades, such as SA-335 P2 (½Cr-½Mo), P11 (1¼Cr-½Mo), and P22 (2¼Cr-1Mo), is normally recommended for these applications.

Intergranular Attack

When an unstabilized austenitic stainless steel is held at a temperature within the range of 454–816°C (850–1500°F), chromium carbides will quickly and preferentially form at the austenitic grain boundaries. The formation of these carbides deletes the surrounding grain matrix of chromium atoms, rendering the thin zone adjacent to grain boundary susceptible to corrosive attack in aqueous environments. This condition is called sensitization and the resulting corrosion is called intergranular attack (IGA). When also in the presence of local high tension stresses, the result can be intergranular stress corrosion cracking (IGSCC). Avoidance of these failure mechanisms is best achieved by minimizing sensitization (fast cool from anneal; stabilized or L-grade steels) and eliminating local stresses.

The area of piping components most often attacked is weld regions. Sensitization can readily occur in a narrow band of base material in the heat-affected zone, caused by the heat of the weld pool. Corrosion of this area has been called *knife line attack* due to the characteristic appearance of a thin crack along a weld edge.

Sigmatization

A hard, brittle, nonmagnetic phase will form in some Fe-Cr and Fe-Ni-Cr alloys upon prolonged exposure between about 593 and 800°C (1100–1475°F). Those austenitic stainless steels containing higher alloy content, such as type 310 (25%Cr-20%Ni) are susceptible, as well as any grades that possess residual ferrite in their microstructure, a constituent which will transform to sigma, preferentially at grain boundaries.

The most detrimental effect of sigma is reduction of toughness. Charpy V-notch impact toughness can degrade to less than 14 J (10 ft-lb) at room temperature if as much as 10 percent of the volume of material transforms. Toughness is usually not significantly degraded at higher temperatures, above about 538°C (1000°F).

Chemically, sigma is not as resistant to oxidizing media as the austenite, such as in acidic environments; thus, the materials will undergo intergranular attack.

At normal metal operating temperatures in power plants, sigmatization of pressure piping made of these high alloy materials takes long times to form. Once formed, the phase can be redissolved by subjecting the material to an annealing heat treatment.

Creep Damage and Estimation of Remaining Creep Life

The type of damage observed in components operating at high temperatures and high stress typically progresses in stages occurring over a considerable period of time. Elongation or swelling of the component may be observed. Material damage manifests itself in the microstructure in characteristic form at grain boundaries. Voids will form first, which then subsequently link up to form cracks. These cracks increase in size or severity as the end-of-life condition is approached.

Severe damage indications invariably signal the need for near-term corrective action. Such corrective action may entail repair or replacement of the component in question, depending upon the extent of the damage and the feasibility of repair. It is important to note that, except in the most severe cases, damage is not readily detectable by visual examination using the naked eye or even by conventional nondestructive techniques such as ultrasonic, magnetic particle, or liquid penetrant examination methods.

The degree of microstructural damage can be assessed by conventional metallographic procedures that may either take a destructive sampling approach or use nondestructive in-place (in situ) methods. Since the determination of the structural damage allows for a ready estimation of expended creep-rupture life, these inspection methods have recently been adopted to piping and other structural components. The power piping industry, in particular, has seen a wholesale application of metallographic examination to components that have experienced extensive time in elevated temperature service. Several serious steam line ruptures have caused deaths, serious injury, and significant lost operating time of fossil energy power plants. The steam lines that have come under the greatest scrutiny are reheat superheater piping which, based on their relatively large diameter and thin wall, have been made from rolled and welded plates. The failures have been associated with the longitudinal weld regions, which are inherently more susceptible to problems due to danger of latent defects (lack of fusion, slag entrapment, solidification cracks) and the variability in mechanical properties across the welds heat-affected zone.

Destructive sampling of material surfaces of suspected creep-damaged components, to allow for metallographic examination, has evolved to the point at which there can be minimal disturbance to surrounding material. Test samples are either trepanned through thickness, or smaller sliver (boat-shaped) samples are removed by sawing, electrodischarge machining, or another method. However, arc gouging or any other form of heat-producing mechanism must be avoided. It not only can significantly metallurgically alter surrounding material but also can damage the destructive sample, sometimes rendering it unusable for microscopic analysis. The small sample pieces, once properly removed, are metallographically prepared in the standard fashion. These are then examined at high magnification in metallurgical microscopes for evidence of creep damage. The area from which this sample was removed must be weld repaired, using all the required preheat, postweld heat treatment, and weld inspections.

Alternately, an evaluation of microstructure can be performed in place on the component surface in the area of interest using a procedure called replication, which provides, in a manner of speaking, a fingerprint image of the surface. The area to be examined is first carefully polished to a mirrorlike finish using ever-increasing fineness of sandpapers or grinding disks, then polishing compounds. The surface is then etched with an appropriate acid. Next, a thin, softened plastic film is applied to the surface. Upon drying, the film hardens, retaining the microstructure in relief. When properly done by skilled technicians, the resolution of the metal structure at magnifications up to 500× or higher is almost equal to that achieved on an actual metal sample. The disadvantage of the replication method is that only the surface of the material can be examined, leaving any subsurface damage undetected. However, this method has proven useful when applied to weld regions or other high-stressed areas where damage is suspected.

Remaining creep-life determination done in this fashion is not exact; the correlation between the type and degree of damage and expended creep life is only approximate. In most cases, follow-up inspection several years hence is necessary to determine the rate of damage progression. Usually, when a network of microcracks has been generated, it is time to consider repair or replacement.

The science of estimating the expected growth rate of these cracks by creep has evolved rapidly in the 1980s. Armed with sufficient baseline creep data of a given alloy, formulas have been developed that can predict creep crack growth rates reasonably accurately. Analysis can also be made as to whether a pipeline would leak before break; that is, weep fluid for a time before catastrophic rupture. All of these tools are available to the piping designer and operating management; they will not be discussed in any greater detail in this chapter.

Oxide Thickness and Estimation of Remaining Creep Life

Another method for estimating remaining creep life of certain high-temperature tubing and piping components that has been developed considers the amount of metal oxide scale that has formed on the metal's surface. Understandably, this method only applies when the tubular items contain relatively benign substances under oxidizing conditions. It has found its use in steam-carrying piping and components. This method is based on the knowledge that a given thickness of oxide scale on the tube or pipe surface represents growth for a certain time at some temperature. Since oxide growth kinetics of many alloys are well characterized, the "effective" temperature at which the tube was operating for a known time (service life) can be estimated. The combination of effective temperature and time can then be compared to the typical creep life of the alloy at an applied stress or stresses that are known to have acted on the component during its service life.

As noted, the two principal tools needed by the metallurgist to estimate life using the oxide measurement technique are steam oxidation data for the alloy in question and uniaxial creep-rupture data for that alloy across the temperature range of interest. This latter information can be found for many widely used ferrous alloy piping materials in ASTM references. The specific steps followed in this approach are as follows:

1. Oxide thickness is measured either metallographically on a sample or using specialized ultrasonic techniques. Operating time is known.
2. The "effective" operating temperature is determined from the oxidation data. The effective temperature is defined as the constant temperature at which the particular tube metal would have had to have operated at for the known service time to have resulted in the measured oxide thickness. (This is an approximation since the tube or pipe would have operated at various temperatures, perhaps even in upset conditions well above the "design" temperature limit.)
3. The hoop stress is calculated using an appropriate formula knowing the tube or pipe size and operating pressure.
4. The Larsen-Miller parameter (LMP) is calculated for the service time and effective temperature of the subject tube. The LMP is defined as

$$\text{LMP} = T(20 + \log t) \times 10^{-3}$$

where T is temperature in degrees Rankine and t is time in hours. This is a simple factor representing the actual condition of the operating component.

5. Uniaxial creep-rupture data are obtained for the alloy in question. Examples of data for 1¼Cr-½Mo-Si and 2¼Cr-1Mo, taken from creep data sources ASTM DS50 and DS6S2, are shown in Figs. A3.21 and A3.22. These rupture data are normally represented by curves of minimum and average behavior, and lists applied stress versus LMP.
6. The ASTM rupture curve is entered on the stress axis at the level of appropriate calculated operating stress (from step 3). In this manner, the LMP rep-

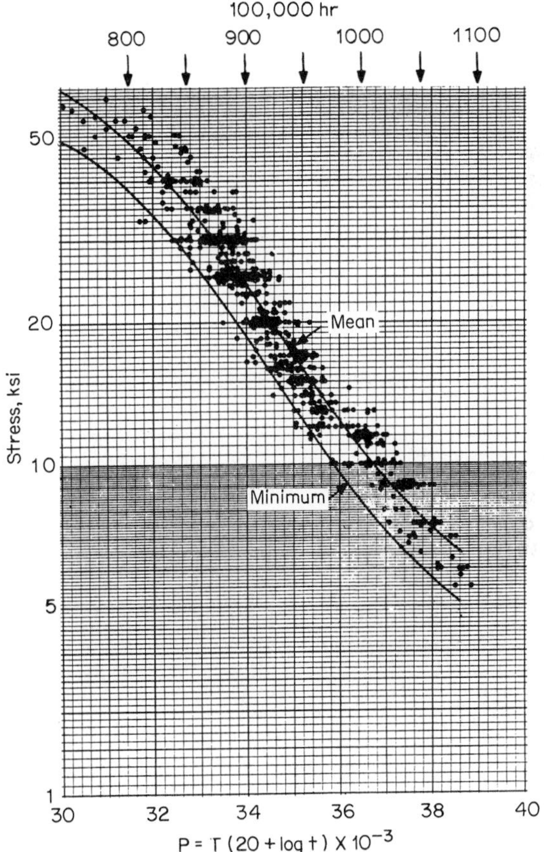

FIGURE A3.21 Variation of Larson–Miller rupture parameter with stress for wrought 1¼ Cr ½ Mo-Si steel. (*Evaluation of the Elevated Temperature Tensile and Creep-Rupture Properties of ½ Cr-½ Mo, 1 Cr-½ Mo, and 1¼ Cr-½ Mo-Si Steels,* ASTM Data Series Publication DS50.)

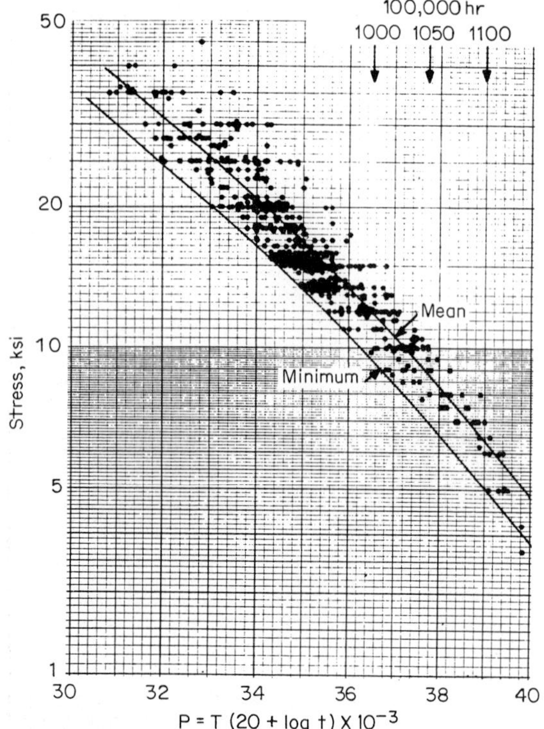

FIGURE A3.22 Variation of Larson–Miller parameter with stress for rupture of annealed 2¼ Cr-1 Mo steel. (*Supplemental Report on the Elevated-Temperature Properties of Chromium-Molybdenum Steels,* ASTM Data Series Publication DS 652.)

resenting the expected minimum and average total creep life at that stress is determined.

7. The operating LMP calculated in step 4 is compared to the LMPs derived in step 6. The differential in time represented by these parameters can be easily calculated from the Larsen-Miller formula, and the percentage of expended life versus minimum and average expected life can be determined by taking a ratio of these values.

This method for estimating remaining creep life has found its greatest use in the fossil power boiler industry, particularly for ferritic alloy steam piping and superheater tubing. Since a great majority of the operating power boilers in the United States is approaching the originally intended lifetime, the method is critical for establishing when major repair or replacement is necessary to restore the unit to safer and more reliable operation.

MATERIAL SPECIFICATIONS

The American Iron and Steel Institute (AISI) and the Society of Automotive Engineers (SAE) have devised a standardized numbering system for the various classes of carbon and alloy steels that has gained widespread acceptance in North America.

This system uses a four-digit number for carbon and low-alloy steels and a three-digit number for stainless steels. Regarding the former, the first two digits represent the major alloying elements of the grade. The last two digits represent the nominal carbon content of each alloy in hundreds of weight percent. For example, 10*XX* represents simple carbon steels, and 41*XX* represents steels with chromium-molybdenum as the major alloying elements. In both classes, a specific grade possessing a nominal carbon content of 0.20 percent would be, respectively, 1020 and 4120. In this fashion, the possible alloy steels can be systematically identified.

Table A3.5 lists the carbon and alloy steel grade categories recognized by AISI and SAE.

The stainless steels are assigned a three-digit code by AISI. Austenitic stainless steels composed of chromium, nickel, and manganese are the 2*XX* series; chromium-nickel austenitic stainless steels are 3*XX*; ferritic and martensitic stainless steels are 4*XX*. In the case of stainless steels, the last two digits represent a unique overall composition rather than the level of carbon.

Because of increasing international technical community involvement and cooperation and because each country possesses its own alloy numbering system, a worldwide universal system of material identification was needed. The Unified Numbering System (UNS) was the result. In this system a letter is followed by a five-digit number which, taken together, uniquely defines each particular composition. Many of the conventions adopted in the AISI/SAE system were incorporated into the UNS numbers, as shown on Fig. A3.5.

The AISI and SAE specifications for alloys control only material composition. Additional control over minimum properties, heat treatment, and other inspections was necessary to assure reproducibility and reliability of the materials for their intended purpose. The ASTM, ASME, and API have generated a series of comprehensive material specifications that extend this control. Table A3.6 lists common ASME specification and grade numbers for the common piping system materials of construction. Table A3.7 gives equivalencies between selected piping material grades in ASME with the UNS.

Copper and Copper Alloys

The use of copper and copper alloys is limited to temperatures below the lower recrystallization temperature for the particular alloy. This is the temperature at which cold-worked specimens begin to soften. This recrystallization is usually accompanied by a marked reduction in tensile strength. Typical classes of wrought copper-based materials are given in Table A3.8.

Brasses containing 70 percent or more of copper may be used successfully at temperatures up to 200°C (400°F), whereas those containing only 60 percent of copper should not be used at temperatures above 150°C (300°F).

The ASME Boiler and Pressure Vessel limits the use of brass and copper pipe and tubing (except for heater tubes) to temperatures not to exceed 208°C (406°F). The ASME B31 Code for Pressure Piping also limits brass and copper pipe and tubing to this temperature for steam, gas, and air piping.

Table A3.9 lists a number of ASME specifications for copper and copper alloy piping and tubing.

TABLE A3.5 Carbon and Alloy Steel Grade Categories: AISI, SAE, UNS

Numerals and digits		
UNS	SAE/AISI	Types of Identifying Elements
		Carbon steels
G10*XX*0	10*XX*	Nonresulfurized, manganese 1.00% maximum
G11*XX*0	11*XX*	Resulfurized
G12*XX*0	12*XX*	Rephosphorized and resulfurized
		Alloy steels
G13*XX*0	13*XX*	Manganese steels
G23*XX*0	23*XX*	Nickel steels
G25*XX*0	25*XX*	Nickel steels
G31*XX*0	31*XX*	Nickel-chromium steels
G32*XX*0	32*XX*	Nickel-chromium steels
G33*XX*0	33*XX*	Nickel-chromium steels
G34*XX*0	34*XX*	Nickel-chromium steels
G40*XX*0	40*XX*	Molybdenum steels
G41*XX*0	41*XX*	Chromium-molybdenum steels
G43*XX*0	43*XX*	Nickel-chromium-molybdenum steels
G44*XX*0	44*XX*	Molybdenum steels
G46*XX*0	46*XX*	Nickel-molybdenum steels
G47*XX*0	47*XX*	Nickel-chromium-molybdenum steels
G48*XX*0	48*XX*	Nickel-molybdenum steels
G50*XX*0	50*XX*	Chromium steels
G51*XX*0	51*XX*	Chromium steels
G50*XX*6	50*XXX*	Chromium steels
G51*XX*6	51*XXX*	Chromium steels
G52*XX*6	52*XXX*	Chromium steels
G61*XX*0	61*XX*	Chromium-vanadium steels
G71*XX*0	71*XXX*	Tungsten-chromium steels
G72*XX*0	72*XX*	Tungsten-chromium steels
G81*XX*0	81*XX*	Nickel-chromium-molybdenum steels
G86*XX*0	86*XX*	Nickel-chromium-molybdenum steels
G87*XX*0	87*XX*	Nickel-chromium-molybdenum steels
G88*XX*0	88*XX*	Nickel-chromium-molybdenum steels
G92*XX*0	92*XX*	Silicon-manganese steels
G93*XX*0	93*XX*	Nickel-chromium-molybdenum steels
G94*XX*0	94*XX*	Nickel-chromium-molybdenum steels
G97*XX*0	97*XX*	Nickel-chromium-molybdenum steels
G98*XX*0	98*XX*	Nickel-chromium-molybdenum steels
		Carbon and alloy steels
G*XXXX*1	*XX*B*XX*	B denotes boron steels
G*XXXX*4	*XX*L*XX*	L denotes leaded steels
		Stainless steels
S2*XXXX*	302*XX*	Chromium-nickel steels
S3*XXXX*	303*XX*	Chromium-nickel steels
S4*XXXX*	514*XX*	Chromium steels
S5*XXXX*	515*XX*	Chromium steels
		Experimental steels
None	Ex...	SAE experimental steels

Source: Reprinted with permssion from SAE J402 ©1988 Society of Automotive Engineers, Inc..

TABLE A3.6 Selected Piping System Materials—ASME Specifications

Metal or Alloy	ASME Specification Number	Grade	Other comments*
	Pipe		
Carbon steel	SA-53	A	48,000 UTS/30,000YS
Carbon steel	SA-106	B	60,000 UTS/35,000YS
Carbon steel	SA-106	C	70,000 UTS/40,000YS
½ Cr-½ Moly	SA-335	P2	55,000 UTS/30,000YS
1 Cr-½ Moly	SA-335	P12	—
1¼ Cr–½ Mo-Si	SA-335	P11	—
2¼ Cr-1 Mo	SA-335	P22	—
5 Cr-1 Mo	SA-335	P5	—
9 Cr-1 Mo	SA-335	P9	60,000 UTS/30,000YS
9 Cr–1 Mo-V	SA-335	P91	85,000 UTS/60,000YS
304H	SA-376	TP304H	0.04% Min carbon
304H	SA-430	FP304H	Forged and bored pipe
316H	SA-376	TP316H	75,000 UTS/30,000YS
	Forgings/fittings		
Carbon steel	SA-105	—	Rolled or forged bar
Carbon steel	SA-181	Cl70	70,000 UTS/36,000YS
Carbon steel	SA-266	Cl2	70,000 UTS/30,000YS
Carbon-Moly	SA-182	F1	0.5% Mo
½ Cr-½ Moly	SA-182	F2	—
1 Cr-½ Moly	SA-182	F12	70,000 UTS/40,000YS
1¼ Cr–½ Mo-Si	SA-182	F11a	75,000 UTS/45,000YS
1¼ Cr–½ Mo-Si	SA-182	F11b	60,000 UTS/30,000YS
2¼ Cr-1 Mo	SA-234	WP12	Fittings
5 Cr-1 Mo	SA-336	F5A	80,000 UTS/50,000YS
9 Cr-1 Mo-V	SA-234	WP91	Fittings
304H	SA-336	F304H	1900°F Min anneal
	Tubing		
Carbon steel	SA-178	A	Electric resistance welded
Carbon steel	SA-210	A1	60,000 UTS/37,000YS
Carbon-Moly	SA-209	T1a	Seamless
½ Cr-½ Moly	SA-213	T2	60,000 UTS/30,000YS
2¼ Cr-1 Moly	SA-213	T22	60,000 UTS/30,000YS
9 Cr–1 Mo-V	SA-213	T91	Normalized and tempered
304H	SA-213	TP304H	75,000UTS/30,000YS

*UTS and YS in psi.

Nickel and Nickel Alloys

Nickel is a tough, malleable metal that offers good resistance to oxidation and corrosion. When combined with copper as the secondary element, the well-known series of Monel alloys are created. Nickel, Monel, and various modifications of these materials are used in piping systems, turbine blading, valves, and miscellaneous power plant accessories handling steam. The presence of even a small amount of sulfur in a reducing environment will result in embrittlement at temperatures of 370–650°C (700–1200°F). By adding Cr, Co, Mo, Ti, Al, or Nb,

TABLE A3.7 Cross-Reference ASME to UNS Selected Pipe and Tubing Specifications

ASME Specification (and grade)	UNS number
SA-53 (E-A)(S-A)	K02504
SA-53 (E-B) (S-B)	K03005
SA-106 (A)	K02501
SA-106 (B)	K03006
SA-106 (C)	K03501
SA-178 (A)	K01200
SA-178 (C)	K03503
SA-209 (T1)	K11522
SA-209 (T1a)	K12023
SA-209 (T1b)	K11422
SA-210 (A1)	K02707
SA-210 (C)	K03501
SA-213 (T2)	K11547
SA-213 (T3b)	K21509
SA-213 (T5)	K41545
SA-213 (T7)	S50300
SA-213 (T9)	S50400
SA-213 (T11)	K11597
SA-213 (T12)	K11562
SA-213 (T21)	K31545
SA-213 (T22)	K21590
SA-213/SA-312 (304)	S30400
SA-213/SA-312 (304H)	S30409
SA-213/SA-312 (304L)	S30403
SA-213/SA-312 (304N)	S30451
SA-213/SA-312 (310)	S31000
SA-213/SA-312 (316)	S31600
SA-213/SA-312 (316H)	S31609
SA-213/SA-312 (316L)	S31603
SA-213/SA-312 (316N)	S31651
SA-213/SA-312 (321)	S32100
SA-213/SA-312 (321H)	S31209
SA-213/SA-312 (347)	S34700
SA-213/SA-312 (347H)	S34709
SA-213/SA-312 (348)	S34800
SA-213/SA-312 (348H)	S34809
SA-335 (P1)	K11522
SA-335 (P2)	K11547
SA-335 (P5)	K41545
SA-335 (P7)	S50300
SA-335 (P9)	S50400
SA-335 (P11)	K11597
SA-335 (P12)	K11562
SA-335 (P21)	K31545
SA-335 (P22)	K21590

TABLE A3.8 Nominal Compositions of Wrought Copper Materials

Alloy	Composition
Coppers	
Electrolytic tough pitch (ETP)	99.90 Cu-0.04 O
Phosphorized, high residual phosphorus (DHP)	99.90 Cu-0.02 P
Phosphorized, low residual phosphorus (DLP)	99.90 Cu-0.005 P
Lake	Cu-8 oz/ton Ag
Silver bearing (10-15)	Cu-10 to 15 oz/ton Ag
Silver bearing (25-30)	Cu-25 to 30 oz/ton Ag
Oxygenfree (OF) (no residual deoxidants)	99.92 Cu (min)
Free cutting	99 Cu-1Pb
Free cutting	99.5 Cu-0.5 Te
Free cutting	99.4 Cu-0.6 Se
Chromium copper (heat treatable)(b)	Cu + Cr and Ag or Zn
Cadmium copper (b)	99 Cu-1Cd
Tellurium nickel copper (heat treatable) (b)	98.4 Cu-1.1 Ni-0.5 Te
Beryllium copper (heat treatable)	Cu-2 Be-0.25 Co or 0.35 Ni
Plain brasses	
Gilding, 95%	95 Cu-5 Zn
Commercial bronze, 90%	90 Cu-10 Zn
Red brass, 85%	85 Cu-15 Zn
Low brass, 80%	80 Cu-20 Zn
Cartridge brass, 70%	70 Cu-30 Zn
Yellow brass, 65%	65 Cu-35 Zn
Muntz metal	60 Cu-40 Zn
Free-cutting brasses	
Leaded commercial bronze (rod)	89 Cu-9.25 Zn-1.75 Pb
Leaded brass strip (B121-3)	65 Cu-34 Zn-1 Pb
Leaded brass strip (B121-5)	65 Cu-33 Zn-2 Pb
Leaded brass tube (B135-3)	66 Cu-33.5 Zn-0.5 Pb
Leaded brass tube (B135-4)	66 Cu-32.4 Zn-1.6 Pb
Medium-leaded brass rod	64.5 Cu-34.5 Zn-1 Pb
High-leaded brass rod	62.5 Cu-35.75 Zn-1.75 Pb
Free-cutting brass rod (B16)	61.5 Cu-35.5 Zn-3 Pb
Forging brass	60 Cu-38 Zn-2 Pb
Architectural bronze	57 Cu-40 Zn-3 Pb
Miscellaneous brasses	
Admiralty (inhibited)	71 Cu-28 Zn-1 Sn
Naval brass	60 Cu-39.25 Zn-0.75 Sn
Leaded naval brass	60 Cu-37.5 Zn-1.75 Pb-0.75 Sn
Aluminum brass (inhibited)	76 Cu-22 Zn-2 Al
Manganese brass	70 Cu-28.7 Zn-1.3 Mn
Manganese bronze rod A (B138)	58.5 Cu-39 Zn-1.4 Fe-1 Sn-0.1 Mn
Manganese bronze rod B (B138)	65.5 Cu-23.3 Zn-4.5 Al-3.7 Mn-3 Fe

TABLE A3.8 Nominal Compositions of Wrought Copper Materials (*Continued*)

Alloy	Composition
Phosphor bronzes	
Grade A	95 Cu-5 Sn
Grade B (rod, B139, alloy B1)	94 Cu-5 Sn-1 Pb
Grade C	92 Cu-8 Sn
Grade D	90 Cu-10 Sn
Grade E	98.75 Cu-1.25 Sn
444 bronze rod (B139, alloy B2)	88 Cu-4 Zn-4 Sn-4 Pb
Miscellaneous bronzes	
Silicon bronze A(c)	Cu-3 Si-1 Mn
Silicon bronze B(c)	Cu-1.75 Si-0.3 Mn
Aluminum bronze, 5%	95 Cu-5 Al
Aluminum bronze, 7%	91 Cu-7 Al-2 Fe
Aluminum bronze, 10%(d)	Cu-9.5 Al
Aluminum-silicon bronze	91 Cu-7 Al-2 Si
Nickel-containing alloys	
Cupro-nickel, 10%	88.5 Cu-10 Ni-1.5 Fe
Cupro-nickel, 30%	69.5 Cu-30 Ni-0.5 Fe
Nickel silver A	65 Cu-17 Zn-18 Ni
Nickel silver B	55 Cu-27 Zn-18 Ni
Leaded nickel silver rod (B151)	62 Cu-19 Zn-18 Ni-1 Pb

Source: *American Society for Metals, Metals Handbook,* vol. 1, 8th ed., p. 961.

the high temperature strength and creep resistance of the nickel-based materials can be substantially increased. However, these alloys possess low ductility values and require special care in forming these materials, even at elevated temperatures. Table A3.10 lists ASME specifications for nickel-based alloy piping and tubing.

Aluminum and Aluminum Alloys

Aluminum and many of its alloys are highly resistant to atmospheric corrosion and to attack by many chemical agents, with the exception of strong alkalis. However, they are subject to galvanic attack if coupled with more noble materials.

Additions of alloying elements increases strength, but to the detriment of thermal and electrical conductivity, and lowers the materials melting point. Alloying with Cu, Mg, and Si creates heat-treatable alloys that are age hardenable. Maximum strength can usually be achieved by heating to about 150–260°C (300–500°F). Effects of working or precipitation hardening can be removed by annealing at temperatures of 315–425°C (600–800°F). A system has been devised to designate alloys of aluminum based on the major alloying constituent; see Table A3.11. Typical classes of wrought aluminum-based materials are given in Table A3.12. The UNS number for each alloy is determined by taking the alloy number in Table A3.12 and preceding it with A9. Thus, for example, the UNS numbers for alloy 6061 is A96061. Table A3.13 provides a list of material specifications acceptable for design and construction of piping systems within the jurisdiction of the ASME Boiler and Pressure Vessel Code and the ASME B31 Code for Pressure Piping.

TABLE A3.9 Copper and Copper-Based Pipe and Tubing Alloy Specifications

ASME specification	UNS grade number	Characteristics
SB-42/SB-68	C10200	99.95 Cu
SB-42/SB-68	C12000	99.90 plus low Phos
SB-42/SB-68	C12200	99.9 plus high Phos
SB-43	C23000	Red brass
SB-75/SB-111	C10200	Oxygen free
SB-75/SB-111	C12000	—
SB-75/SB-111	C12200	—
SB-75/SB-111	C14200	Phosphorized, arsenical
SB-111	C23000	Red brass
SB-111	C28000	Muntz metal
SB-111	C44300	Admiralty metal
SB-111	C44400	Cu-Zn
SB-111	C44500	Cu-Zn
SB-111	C60800	Aluminum bronze
SB-111	C68700	Aluminum brass
SB-111	C70400	95-5 Cu-Ni
SB-111	C70600	90-10 Cu-Ni
SB-111	C71000	80-20 Cu-Ni
SB-111	C71500	70-30 Cu-Ni
SB-315	C65500	High-Si bronze
SB-466	C70600	90-10 Cu-Ni
SB-466	C71500	70-30 Cu-Ni
SB-467	C70600	Welded 90-10
SB-467	C71500	Welded 70-30

TABLE A3.10 Nickel and Nickel-Based Pipe and Tubing Alloy Specifications

ASME specification	UNS grade number	Characteristics
SB-161	N02200	Nickel 200; 99% Ni
SB-161	N02201	Low carbon
SB-163/SB-407	N08800	Alloy 800 tubing (Ni-Fe-Cr)
SB-163/SB-165	N04400	70-30 Ni-Cu Monel
SB-163/SB-167	N06600	Alloy 600 (Ni-Cr-Fe)
SB-163/SB-167	N06690	Alloy 690 (60-30-10)
SB-163/SB-423	N08825	Alloy 825

TABLE A3.11 Designation System for Wrought Aluminum and Aluminum Alloy

Composition	Alloy No.
Aluminum, 99.0% min and greater	1*XXX*
Aluminum alloys grouped by major alloying element	
Copper	2*XXX*
Manganese	3*XXX*
Silicon	4*XXX*
Magnesium	5*XXX*
Magnesium and silicon	6*XXX*
Zinc	7*XXX*
Other elements	8*XXX*
Unused series	9*XXX*

Source: Reprinted with permission from SAE J933 ©1989 Society of Automotive Engineers, Inc.

TABLE A3.12 Chemical Composition Limits for Wrought Aluminum Alloys Where No Range Is Given (Single Number Indicates Maximum Permissible Percentage)

Alloy number	Si	Fe	Cu	Mn	Mg	Cr	Ni	Zn	Ti	Others	
										Each	Total
EC	(Al 99.45 min)										
1100	1.0 Si + Fe		0.20	0.05	—	—	—	0.10	—	0.05(a)	0.15
1060	0.25	0.35	0.05	0.03	0.03	—	—	0.05	0.03	0.03(a)	—
1085	0.10	0.12	0.03	0.02	0.02	—	-	0.03	0.02	0.01(b)	—
1099	(Al 99.99 min)										
2011	0.40	0.7	5.0-6.0	—	—	—	—	0.30	—	0.05(c)	0.15
2014	0.50-1.2	1.0	3.9-5.0	0.40-1.2	0.20-0.8	0.10	—	0.25	0.15	0.05(a)	0.15
2017	0.8	1.0	3.5-4.5	0.40-1.0	0.20-0.8	0.10	—	0.25	—	0.05	0.15
2117	0.8	1.0	2.2-3.0	0.20	0.20-0.50	0.10	—	0.25	—	0.05	0.15
2618	0.25	0.9-1.3	1.9-2.7	—	1.3-1.8	—	0.9-1.2	—	0.04-0.10	0.05	0.15
2219	0.20	0.30	5.8-6.8	0.20-0.40	0.02	—	—	0.10	0.02-0.10	0.05(d)	0.15
X2020	0.40	0.40	4.0-5.0	0.30-0.8	0.03	—	—	0.25	0.10	0.05(e)	0.15
2024	0.50	0.50	3.8-4.9	0.30-0.9	1.2-1.8	0.10	—	0.25	—	0.05	0.15
3003	0.6	0.7	0.20	1.0-1.5	—	—	—	0.10	—	0.05(a)	0.15
3004	0.30	0.7	0.25	1.0-1.5	0.8-1.3	—	—	0.25	—	0.05(a)	0.15
4032	11.0-13.5	1.0	0.50-1.3	—	0.8-1.3	0.10	0.50-1.3	0.25	—	0.05	0.15
4043	4.5-6.0	0.8	0.30	0.05	0.05	—	—	0.10	0.20	0.05(a)	0.15
5005	0.40	0.7	0.20	0.20	0.50-1.1	0.10	—	0.25	—	0.05	0.15
5050	0.40	0.7	0.20	0.10	1.0-1.8	0.10	—	0.25	—	0.05(a)	0.15
5052	0.45 Si + Fe		0.10	0.10	2.2-2.8	0.15-0.35	—	0.10	—	0.05(a)	0.15
5154	0.45 Si + Fe		0.10	0.10	3.1-3.9	0.15-0.35	—	0.20	0.20	0.05(a)	0.15
5155	0.30	0.70	0.25	0.20-0.60	3.5-5.0	0.05-0.25	—	0.25	0.15	0.05	0.15

TABLE A3.12 Chemical Composition Limits for Wrought Aluminum Alloys (*Continued*)
Where No Range Is Given (Single Number Indicates Maximum Permissible Percentage)

Alloy number	Si	Fe	Cu	Mn	Mg	Cr	Ni	Zn	Ti	Others	
										Each	Total
5454	0.40 Si + Fe		0.10	0.50-1.0	2.4-3.0	0.05-0.20	—	0.25	0.20	0.05	0.15
5056	0.30	0.40	0.10	0.05-0.20	4.5-5.6	0.05-0.20	—	0.10	—	0.05(a)	0.15
5456	0.40 Si + Fe		0.10	0.50-1.0	4.7-5.5	0.05-0.20	—	0.25	0.20	0.05	0.15
5357	0.12	0.17	0.07	0.15-0.45	0.8-1.2	—	—	—	—	0.05	0.15
5457	0.08	0.10	0.20	0.15-0.45	0.8-1.2	—	—	—	—	0.03	0.10
5557	0.10	0.12	0.15	0.10-0.40	0.40-0.8	—	—	—	—	0.03	0.10
5083	0.40	0.40	0.10	0.30-1.0	4.0-4.9	0.05-0.25	—	0.25	0.15	0.05	0.15
5086	0.40	0.50	0.10	0.20-0.7	3.5-4.5	0.05-0.25	—	0.25	0.15	0.05	0.15
6151	0.6-1.2	1.0	0.35	0.20	0.45-0.8	0.15-0.35	—	0.25	0.15	0.05	0.15
6351	0.7-1.3	0.6	0.10	0.40-0.8	0.40-0.8	—	—	—	0.20	0.05	0.15
6053	(f)	0.35	0.1	—	1.1-1.4	0.15-0.35	—	0.10	—	0.05	0.15
6061	0.40-0.8	0.7	0.15-0.40	0.15	0.8-1.2	0.15-0.35	—	0.25	0.15	0.05	0.15
6062	0.40-0.8	0.7	0.15-0.40	0.15	0.8-1.2	0.04-0.14	—	0.25	0.15	0.05	0.15
6063	0.20-0.6	0.35	0.10	0.10	0.45-0.9	0.10	—	0.10	0.10	0.05	0.15
6066	0.9-1.8	0.50	0.7-1.2	0.6-1.1	0.8-1.4	0.40	—	0.25	0.20	0.05	0.15
7072(g)	0.7 Si + Fe		0.10	0.10	0.10	—	—	0.8-1.3	—	0.05	0.15
7075	0.50	0.7	1.2-2.0	0.30	2.1-2.9	0.18-0.40	—	5.1-6.1	0.20	0.05	0.15
7277	0.50	0.7	0.8-1.7	—	1.7-2.3	0.18-0.35	—	3.7-4.3	0.10	0.05	0.15
7178	0.50	0.7	1.6-2.4	0.30	2.4-3.1	0.18-0.40	—	6.3-7.3	0.20	0.05	0.15
7079	0.30	0.40	0.40-0.8	0.10-0.30	2.9-3.7	0.10-0.25	—	3.8-4.8	0.10	0.05	0.15
X8001	0.17	0.45-0.7	0.15	—	—	—	0.9-1.3	—	—	0.05(h)	0.15

Source: *American Society for Metals, Metals Handbook,* Vol. 1, 8th Ed., pg. 917.

TABLE A3.13 Material Specifications and Their Acceptance by ASME Codes[a]

Specification[b,c]	Description	B31.1	B31.3	B31.4	B31.5	B31.8	B31.9	B31.11	SEC III	SEC VIII[d]
A 53	Pipe, C. S.	Yes	Yes	Yes	Yes	Yes	Yes	Yes	Yes	Yes
A106	Pipe, C. S.	Yes	Yes	Yes	Yes	Yes	Yes	Yes	Yes	Yes
A120	Pipe, steel black, & galvanized	—	—	—	—	Yes	—	Yes	—	—
A134	Pipe, arc-welded steel	Yes	Yes	Yes	Yes	Yes	—	Yes	Yes	—
A135	Pipe, elec. resis. welded steel	Yes	Yes	Yes	Yes	Yes	Yes	Yes	—	—
A139	Pipe, arc-welded	Yes	Yes	Yes	Yes	Yes	—	Yes	—	—
A211	Pipe, spiral welded	Yes	—	—	Yes	Yes	Yes	Yes	—	—
A312	Pipe, aust. S. S.	Yes	Yes	—	Yes	—	Yes	—	Yes	—
A333	Pipe, steel for low temp. service	Yes	Yes	—	Yes	—	—	—	Yes	Yes
A335	Pipe, ferr. alloy St.	Yes	Yes	—	—	—	—	—	Yes	Yes
A358	Pipe, arc-welded Cr-Ni allo steel	Yes	Yes	—	Yes	—	—	—	Yes	—
A369	Pipe, ferr. alloy St.	Yes	Yes	—	—	—	—	—	Yes	Yes
A376	Pipe, aust. S. S.	Yes	Yes	—	Yes	—	—	—	Yes	Yes
A381	Pipe, metal-arc-welded steel	—	Yes	Yes	—	Yes	—	Yes	—	—
A409	Pipe, aust. S. S.	—	Yes	—	Yes	—	—	—	Yes	Yes
A426	Pipe, ferr. alloys	Yes	Yes	—	—	—	—	—	Yes	—
A430	Pipe, aust. alloys	Yes	Yes	—	—	—	—	—	Yes	Yes
A451	Pipe, aust. alloys	Yes	Yes	—	—	—	—	—	Yes	—
A452	Pipe, aust. cold wrought alloys	Yes	Yes	—	—	—	—	—	Yes	Yes
A524	Pipe, C. S. for low temp.	—	Yes	Yes	—	—	—	Yes	—	Yes
A530	Pipe, special C., & alloy steel	—	—	Yes	—	—	—	Yes	—	—
A587	Pipe, resis. welded C. S.	Yes	Yes	—	Yes	—	—	—	Yes	Yes
A660	Pipe, centrifugal cast	—	—	—	—	—	—	—	Yes	Yes
A671	Pipe, arc-welded S. for L. T. service	Yes	Yes	Yes	—	Yes	—	Yes	Yes	—
A672	Pipe, arc-welded S. for H. P. service	Yes	Yes	Yes	—	Yes	—	Yes	Yes	—
A691	Pipe, arc-welded S. for H. T. service	Yes	Yes	—	—	—	—	Yes	Yes	—
A714	Pipe, high strength, & low alloy steel	Yes	—	—	—	—	—	—	—	—
AP15L	Pipe, line	Yes	Yes	Yes	Yes	Yes	—	Yes	—	—
AP15LU	Pipe, ultra-high test H/T line	—	—	Yes	—	—	—	Yes	—	—

[a]This table was contributed by Akhil Prakash, P.E., Supervisor Engineer, WCNOC, Wichita, KS 67226.

A178	Tube, arc-welded C. S.	Yes	—	—	Yes	—	—	—	Yes	Yes
A179	Tube, low C. S.	Yes	Yes	—	Yes	—	—	—	—	—
A192	Tube, C. S.	Yes	—	—	Yes	—	—	—	—	—
A199	Tube, cold drawn int. alloy steel	Yes	—	—	—	—	—	—	—	—
A210	Tube, med. C. S.	Yes	—	—	Yes	—	—	—	Yes	Yes
A213	Tube, ferr. & aust. alloys	Yes	—	—	Yes	—	—	—	Yes	Yes
A214	Tube, elec. resis. welded C. S.	Yes	—	—	Yes	—	—	—	—	Yes
A226	Tube, C. S. for H. P. service	Yes	—	—	Yes	—	—	—	Yes	Yes
A249	Tube, aust. S. S.	Yes	—	—	Yes	—	—	—	Yes	Yes
A254	Tube, copper brazed steel	Yes	—	—	Yes	—	Yes	—	—	—
A268	Tube, ferr. S. S.	Yes	Yes	—	—	—	—	—	Yes	Yes
A269	Tube, aust. S. S.	—	Yes	—	Yes	—	—	—	—	—
A271	Tube, aust. S. S.	—	—	—	Yes	—	—	—	—	—
A334	Tube, C. S. Ni alloys	—	Yes	—	Yes	—	—	—	Yes	Yes
A450	Tube, carbon, ferr. & aust. alloys	Yes	—	—	—	—	—	—	—	—
A539	Tube, elect. resistant welded	—	—	—	—	—	—	—	Yes	—
A556	Tube, seamless	—	—	—	—	—	—	—	—	Yes
A557	Tube, welded	—	—	—	—	—	—	—	—	Yes
A688	Tube, welded S. S.	—	—	—	—	—	—	—	Yes	Yes
A789	Tube, S. S.	—	Yes	—	—	—	—	—	—	—
A790	Tube, S. S.	—	Yes	—	—	—	—	—	—	—
A105	Forgings, C. S. for H. T. service	Yes	Yes	Yes	Yes	Yes	Yes	Yes	—	—
A181	Forgings, carbon steel	Yes	Yes	Yes	Yes	—	Yes	Yes	Yes	Yes
A182	Forgings, alloy steel for H. T. service	Yes	Yes	Yes	Yes	—	—	Yes	Yes	Yes
A234	Fittings, wrought C., & ferr. alloys	Yes	Yes	Yes	Yes	—	Yes	Yes	—	—
A336	Forgings, steel	—	—	—	—	—	—	—	Yes	Yes
A350	Forgings, C. S. & low alloy steels	—	Yes	Yes	Yes	—	—	Yes	Yes	Yes
A403	Fittings, aust. S. S.	Yes	Yes	—	Yes	—	Yes	—	Yes	Yes
A420	Fittings, wrought C. S. & alloys	Yes	Yes	—	Yes	—	—	Yes	Yes	Yes
A508	Forgings, steel	—	—	—	—	—	—	—	Yes	Yes
A522	Forgings, 8% & 9% nickel alloys	—	—	—	Yes	—	—	—	—	—

TABLE A3.13 Material Specifications and Their Acceptance by ASME Codes[a] (*Continued*)

Specification[b,c]	Description	B31.1	B31.3	B31.4	B31.5	B31.8	B31.9	B31.11	SEC III	SEC VIII[d]
A541	Forgings, steel	—	—	—	—	—	—	—	Yes	Yes
A592	Forgings, fittings	—	—	—	—	—	—	—	Yes	Yes
A694	Forgings, carbon & alloy steels	—	—	Yes	—	—	—	Yes	—	—
A705	Forgings, S. S.	—	—	—	—	—	—	—	Yes	—
A707	Forgings, carbon & alloy steels	—	—	—	—	—	—	Yes	—	—
A723	Forgings, S. S.	—	—	—	—	—	—	—	Yes	Yes
A727	Forgings, S. S.	—	—	—	—	—	—	—	Yes	Yes
API600	Steel gate valves	—	—	Yes	—	—	—	—	—	—
API602	Compact carbon steel gate valves	—	—	Yes	—	—	—	—	—	—
API603	Corrosion resistant gate valves	—	—	Yes	—	—	—	—	—	—
API6A	Wellhead equipment	—	—	Yes	—	Yes	—	—	—	—
API6D	Pipeline valves	—	—	Yes	—	Yes	—	Yes	—	—
A193	Bolts, alloys & S. S. for H. T. service	Yes	Yes	Yes	Yes	Yes	—	Yes	Yes	Yes
A194	Nuts, C. & alloy steel for H. T. service	Yes	Yes	Yes	Yes	Yes	—	Yes	—	—
A307	Fasteners, low carbon steel	Yes	Yes	Yes	Yes	Yes	—	Yes	—	—
A320	Bolts, alloy steel for L. T. service	—	Yes	Yes	Yes	Yes	—	Yes	Yes	Yes
A325	Bolting material, C. S.	—	Yes	Yes	—	—	—	Yes	Yes	Yes
A354	Bolts & studs, quenched & tempered	Yes	Yes	Yes	Yes	Yes	—	Yes	Yes	Yes
A437	Bolting material, S. S.	—	Yes	—	—	—	—	—	Yes	—
A449	Bolts & studs, quenched & tempered	—	—	Yes	—	Yes	—	Yes	Yes	Yes
A453	Bolting material S. S.	Yes	Yes	—	—	—	—	—	Yes	Yes
A540	Bolting material, steel	—	—	—	—	—	—	—	Yes	Yes
A564	Hot rolled & cold finished S. S.	Yes	—	—	—	—	—	—	—	—
A675	Bolting material, C. S.	—	Yes	—	—	—	—	—	—	—

A 47	Castings, malleable iron	—	—	—	Yes	—	Yes	—	—	Yes
A 48	Castings, gray iron	Yes	—	—	Yes	—	Yes	—	—	—
A126	Castings, gray iron	Yes	—	Yes	Yes	—	Yes	Yes	—	—
A197	Castings, cupola malleable iron	—	—	—	Yes	—	Yes	—	—	—
A216	Castings, carbon steel	Yes	Yes	Yes	Yes	—	—	Yes	Yes	Yes
A217	Castings, alloy steels	Yes	Yes	Yes	Yes	—	—	Yes	Yes	Yes
A278	Castings, gray iron	Yes	—	—	Yes	—	Yes	—	—	Yes
A351	Castings, ferr. & aust. steel	Yes	Yes	—	Yes	—	—	—	Yes	Yes
A352	Castings, ferr. & mart. steel	—	Yes	—	Yes	—	—	—	Yes	Yes
A377	Castings, cast & ductile iron	—	—	—	—	—	Yes	Yes	—	—
A389	Castings, ferr. alloy special H/T	—	—	—	—	—	—	—	—	—
A395	Castings, ductile iron	Yes	—	Yes	Yes	Yes	Yes	Yes	—	Yes
A487	Castings, steel	—	—	Yes	—	—	—	Yes	Yes	Yes
A494	Castings	—	Yes	—	—	—	—	—	—	—
A536	Castings, ductile iron	—	—	—	—	—	Yes	Yes	—	—
A743	Castings, corrosion resistant	—	—	—	Yes	—	—	—	—	—
A744	Castings, corrosion resistant	—	—	—	Yes	—	—	—	—	—

[a]The table provides the list of some of the acceptable material specifications approved by the ASME Codes. The User is advised to refer to the latest edition of the Code for specific grades and other limitations.

[b]Add prefix S to each specification number when using ASME Codes Sections III and VIII.

[c]The above material table is based on the following editions: B31.1-1989, including 1989(A); B31.3-1990; B31.4-1986; B31.5-1987; B31.8-1986; B31.9-1988; B31.11-1986, including 1988(A); Section III-1986, including 1988(A); Section VIII-1986, including 1988(A).

[d]Yes indicates acceptance for use by the listed code.

REFERENCES

1. ASTM Committee on Terminology, *Compilation of ASTM Standard Definitions,* Fifth Edition (PCN 03-001082-42), 1982.
2. Charles Mantell, *Engineering Materials Handbook,* First Edition, McGraw-Hill Book Co., New York, 1958.
3. Bain and Paxton, *Alloying Elements in Steel,* American Society for Metals, 1939.
4. Dunn, Whiteley, and Fairhurst, *Molybdenum's Place in the Pressure-Vessel Field,* Climax Molybdenum Company.
5. *Steam, Its Generation and Use,* The Babcock & Wilcox Company, New York 1972.
6. *The Making, Shaping and Treating of Steel,* United States Steel, 9th Edition, 1971.
7. *Alloying Elements and Their Effects.* Hardenability Republic Steel Corporation, 1979.
8. *Metals Handbook,* Ninth Edition, Volume 4, Heat Treating, American Society for Metals, 1981.
9. *Temper Embrittlement and Hydrogen Embrittlement in Pressure Vessel Steels,* JPVRC Report No.2, The Iron and Steel Institute of Japan, 1979.

CHAPTER A4

PIPING CODES AND STANDARDS

Mohinder L. Nayyar, P.E.
Engineering Specialist
Bechtel Corporation
Gaithersburg, Maryland

Codes usually set forth minimum requirements for design, materials, fabrication, erection, test, and inspection of piping systems, whereas standards contain design and construction rules and requirement for individual piping components such as elbows, tees, returns, flanges, valves, and other in-line items. Compliance to Code is generally mandated by regulations imposed by regulatory and enforcement agencies. At times, the insurance carrier for the facility leaves hardly any choice for the owner but to comply with the requirements of a Code or Codes to ensure safety of the workers and the general public. Compliance to standards is normally required by the rules of the applicable Code or the purchaser's specification.

Each Code has limits on its jurisdiction, which are precisely defined in the Code. Similarly, the scope of application for each standard is defined in the standard. Therefore, users must become familiar with limits of application of a Code or standard before invoking their requirements in design and construction documents of a piping system.

The Codes and standards, which relate to piping systems and piping components, are published by various organizations. These organizations have committees made up of representatives from industry associations, manufacturers, professional groups, users, government agencies, insurance companies, and other interest groups. The committees are responsible for maintaining, updating, and revising the Codes and standards in view of technological developments, research, experience feedback, problems, and changes in referenced Codes, standards, specifications, and regulations. The revisions to various Codes and standards are published periodically. Therefore, it is important that the engineers, designers, and other professional and technical personnel stay informed with the latest editions, addenda, or revisions of the Codes and standards affecting their work.

While designing a piping system in accordance with a Code or a standard, the designer must comply with the most restrictive requirements which apply to any of the piping elements.

In regard to applicability of a particular edition, issue, addenda, or revision of a Code or standard, one must be aware of the national, state provincial, and local laws and regulations governing its applicability in addition to the commitments made by the owner and the limitations delineated in the Code or standard.

This chapter covers major Codes and standards related to piping. Some of these Codes and standards are discussed briefly, whereas others are listed for convenience of reference.

AMERICAN SOCIETY OF MECHANICAL ENGINEERS

The American Society of Mechanical Engineers (ASME) is one of the leading organizations in the world which develops and publishes Codes and standards. The ASME established a committee in 1911 to formulate rules for the construction of steam boilers and other pressure vessels. This committee is now known as the ASME Boiler and Pressure Vessel Committee, and it is responsible for the ASME Boiler and Pressure Vessel Code. In addition, the ASME has established other committees which develop many other Codes and standards, such as the ASME B31, Code for Pressure Piping. These committees follow the procedures accredited by the American National Standards Institute (ANSI).

ASME BOILER AND PRESSURE VESSEL CODE

The ASME Boiler and Pressure Vessel Code contains eleven sections:

Section I Power Boilers

Section II Material Specifications

Section III

Division 1 Nuclear Power Plant Components
Division 2 Concrete Reactor Vessel and Containments

Section IV Heating Boilers

Section V Nondestructive Examination

Section VI Recommended Rules for Care and Operation of Heating Boilers

Section VII Recommended Rules for Care of Power Boilers

Section VIII

Division 1 Pressure Vessels
Division 2 Pressure Vessels (Alternative Rules)

Section IX Welding and Brazing Qualifications

Section X Fiberglass Reinforced Plastic Pressure Vessels

Section XI Rules for In Service Inspection of Nuclear Power Plant Components.

Primarily, Sections I, II, III, IV, V, VIII, IX, and XI specify rules and requirements for piping. Sections V and IX are supplementary sections of the Code because they have no jurisdiction of their own unless invoked by reference in the Code of construction, such as Section I or III.

Editions and Addenda

Code editions are published every 3 years incorporating the additions and revisions made to the Code during the preceding three years.

Colored-sheet addenda, which include additions and revisions to individual sections of the Code, are published annually. Before the 1986 edition of the Code, addenda were published semiannually as summer and winter addenda.

Interpretations

ASME issues written replies to inquiries concerning interpretation of technical aspects of the Code. The interpretations for each individual section are published separately as part of the update service to that section. They are issued semiannually up to the publication of the next edition of the Code. Interpretations are not part of the Code edition or the addenda.

Code Cases

The Boiler and Pressure Vessel Committee meets regularly to consider proposed additions and revisions to the Code, to formulate cases to clarify the intent of the existing requirements, and/or to provide, when the need is urgent, rules for materials or construction not covered by existing Code rules. The code cases are published in the appropriate Code case book: (1) *Boiler and Pressure Vessel* and (2) *Nuclear Components*. Supplements are published and issued to the Code holders or buyers up to the publication of the next edition of the Code.

Code case(s) can be reapproved or annulled by the ASME Council. Reapproved Code case(s) can be used after approval by the Council. However, the use of Code case(s) is subject to acceptance by the regulatory and enforcement authorities having jurisdiction. An annulled Code case may become a part of the addenda or edition of the Code or just disappear after its annulment because there may not be any need for it.

ASME SECTION I: POWER BOILERS

Scope

ASME Section I has total administrative jurisdiction and technical responsibility for boiler proper; refer to Fig. A4.1. The piping defined as boiler external piping (BEP) is required to comply with the mandatory certification by Code symbol stamping, ASME data forms, and authorized inspection requirements, called Administrative Jurisdiction, of ASME Section I; however, it must satisfy the technical requirements (design, materials, fabrication, installation, nondestructive examination, etc.) of ASME B31.1, Power Piping Code.[1]

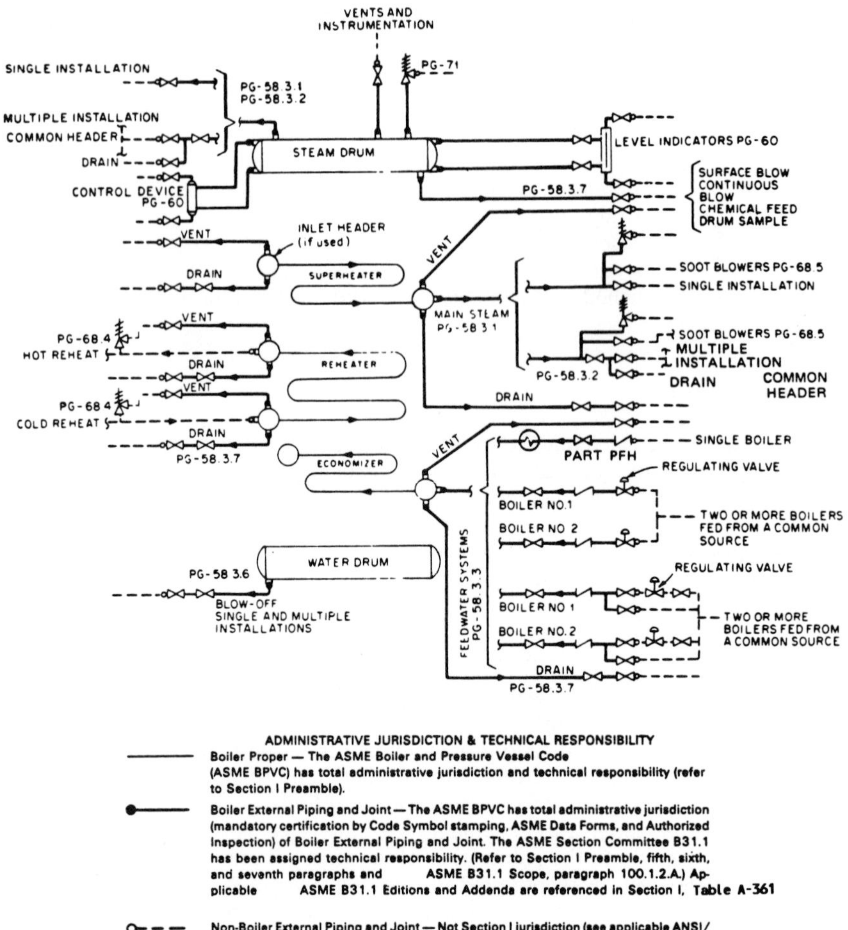

FIGURE A4.1 ASME Section I jurisdictional limits and clarification for jurisdiction over boiler external piping (BEP) and nonboiler external piping (NBEP). (*Figure PG-58. 3.1, ASME Section I.*)

Effective Edition, Addenda, and Code Cases

Code editions are effective and may be used on or after the date of publication printed on the title page.

Code addenda are effective and may be used on or after the date of issue. Revisions become mandatory as minimum requirements six months after such date of issuance, except for boilers (or pressure vessels) contracted for before the end of the six-month period.

Use of revisions and Code cases that are less restrictive than former requirements of the applicable edition and addenda shall not be made without having assurance that they have been accepted by the proper authorities in the jurisdiction in which the power boiler (component) is to be installed.

Use of Code cases is permissible beginning with the ASME council approval date published on the Code case.

ASME SECTION II: MATERIAL SPECIFICATIONS

Scope

ASME Section II contains material specifications, which are invoked for construction of items within the scope of the various sections of the ASME Boiler and Pressure Vessel Code and ASME B31, Code for Pressure Piping.[2] Therefore, ASME Section II is considered a supplementary section of the Code. It consists of three parts.

Part A: Ferrous Materials. Part A contains material specifications for steel pipe, flanges, plates, bolting materials, and castings and wrought, cast, and malleable iron. These specifications are identified by the prefix SA followed by a number such as SA-53 or SA-106.

Part B: Nonferrous Materials. Part B contains material specifications for aluminum, copper, nickel, titanium, zirconium, and their alloys. These specifications are identified by the prefix SB followed by a number such as SB-61 or SB-88.

Part C: Welding Materials. Part C contains material specifications for welding rods, electrodes filler materials, brazing materials, and so on. These specifications are identified by the prefix SFA followed by a number such as SFA-5.1 or SFA-5.27.

Effective Edition, Addenda, and Code Cases

The application of ASME Section II is mandatory only when referenced by other sections of the ASME Boiler and Pressure Vessel Code, ASME B31 and various other industry Codes and standards.

The applicable edition and addenda of ASME Section II shall be corresponding to the edition and addenda of the referencing Code/section.

Use of a later or the latest edition and addenda of ASME Section II is permissible provided it is acceptable to the enforcement authorities having jurisdiction over the site where the component is to be installed.

For items within the scope of ASME Section XI, the effective edition and addenda of ASME Section II shall be in accordance with the requirements of ASME Section XI.

In case of nonnuclear items or applications, the effective edition addenda and Code case shall be determined as described for ASME Section I.

Use of Code cases related to materials for ASME Section III applications may be made in accordance with the recommendations of Regulatory Guide 1.85, Materials Code Case Acceptability, ASME Section III, Division 1. The Code cases, as approved with or without limitations and listed in Regulatory Guide 1.85, may be used. The Code case(s) not listed as approved in Regulatory Guide 1.85 by the Nuclear Regulatory Commission (NRC) may only be used after seeking approval from the NRC.

ASME SECTION III: NUCLEAR POWER PLANT COMPONENTS

Scope

Division 1 of ASME Section III contains requirements for piping classified as ASME Class 1, Class 2, and Class 3. The ASME Section III does not delineate the criteria for classifying piping into Class 1, Class 2, or Class 3; it specifies the requirements for design, materials, fabrication installation, examination, testing, inspection, certification, and stamping of piping systems after they have been classified Class 1, Class 2, or Class 3 based upon the applicable design criteria and Regulatory Guide 1.26, Quality Group Classifications and Standards for Water-Steam, and Radio-Waste-Containing Components of Nuclear Power Plants. Subsections NB, NC, and ND of ASME III specify the construction requirements for Class 1, Class 2, and Class 3 components, including piping, respectively. Subsection NF contains construction requirements for component supports, and Subsection NCA, which is common to Divisions 1 and 2, specifies general requirements for all components within the scope of ASME Section III.[3]

The construction requirements for ASME Class 1, Class 2, and Class 3 piping are based upon their degree of importance of the safety, with Class 1 piping being subjected to the most stringent requirements and Class 2 to lesser and Class 3 to the least stringent requirements. It is noted that a nuclear power plant does have piping systems other than ASME Class 1, Class 2, and Class 3, which are constructed to Codes other than ASME Section III. For example, the fire protection piping systems are constructed to National Fire Protection Association (NFPA) standards, and most of the nonnuclear piping systems are constructed to ASME B31.1, Power Piping Code.

While joining piping systems or components of different classification, the more restrictive requirements shall govern, except that connection between piping and other components, such as vessels, tanks, heat exchangers, and valves, shall be considered part of the piping. For example, a weld between ASME Class 1 valve and ASME Class 2 piping shall be made in compliance with the requirements of Subsection NC, which contains rules for ASME Class 2 components, including piping; refer to Fig. A4.2.

Effective Edition, Addenda, and Code Cases

Selection of effective edition and addenda of ASME Section III shall be based upon the following guidelines:

Only the approved edition(s) and addenda of ASME Section III, incorporated by reference in 10 CFR 50.55a, Paragraph (b) (1) are to be used for construction of items within the scope of ASME Section III.

The latest published edition and addenda of ASME Section III may not be approved by the NRC; therefore, their use can only be made after seeking special permission from the NRC. Refer to 10 CFR 50.55a, Codes and Standards from time to time to find which edition and addenda of ASME Section III have been approved by the NRC.[4]

As per Subsubarticle NCA-1140, in no case shall the Code edition and addenda dates established in the design specifications be earlier than three years prior to the date the nuclear power plant construction permit application is docketed. In addition, the guidelines of preceding paragraphs shall apply.[3]

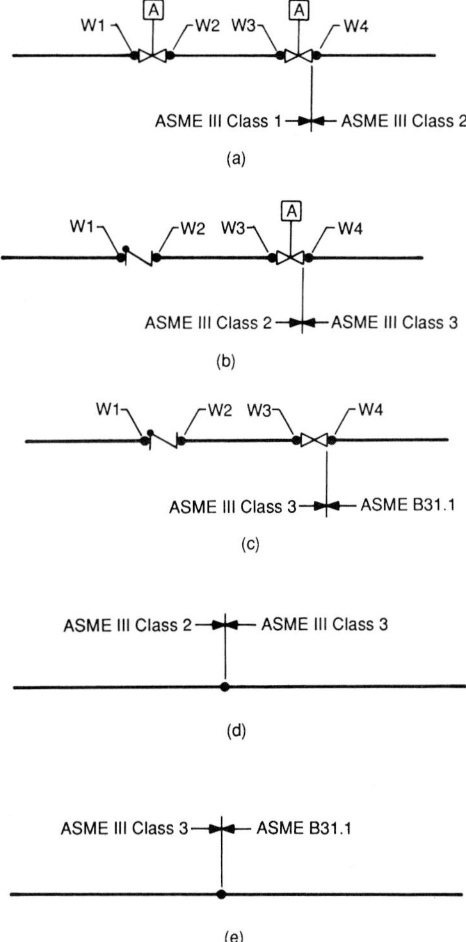

FIGURE A4.2 Code jurisdiction at interface welds between ASME III piping and components, and ASME/ANSI B31.1 piping. (*a*) Welds W1, W2, and W3 are between ASME III Class 1 piping and ASME III Class 1 valves/components. These welds shall comply with the requirements for ASME III Class 1 components. Weld W4 is between ASME III Class 1 valve and ASME III Class 2 piping. This weld shall comply with the requirements for ASME III Class 2 components; (*b*) Welds W1, W2, and W3 are between ASME III Class 2 piping and ASME III Class 2 valves/components. These welds shall comply with the requirements for ASME III Class 2 components. Weld W4 is between ASME Class 2 valve/component and ASME III Class 3 piping. This weld shall comply with the requirements for ASME III Class 3 components; (*c*) Welds W1, W2, and W3 are between ASME III Class 3 piping and ASME III Class 3 valves/components. These welds shall comply with the requirements for ASME III Class 3 components. Weld W4 is between ASME III Class 3 valve/component and ASME B31.1 piping. This weld shall comply with the requirements of ASME B31.1; (*d*) The connecting weld between two different ASME III classes of piping shall comply with more stringent requirements of the connecting classes of piping. In this case, the weld shall meet the requirements for ASME III Class 2 components; (*e*) The connecting weld between ASME III Class 3 and ASME B31.1 piping shall comply with more stringent requirements of ASME III Class 3 piping.

Code edition and addenda later than those established in the design specification and documents per the above delineated approach may be used provided they are approved for use. Also, specific provisions within an edition or addenda later than those established in the design specifications and documents may be used provided all related requirements are met.

All Code items, including piping systems, may be constructed to a single Code edition and addenda, or each item may be constructed to individually specified Code edition and addenda.

The use of Code case(s) is optional. Only the U.S. NRC approved Code cases with or without limitations or additional requirements published in the following regulatory guides may be used without a specific request to the NRC for approval:

- *Regulatory Guide 1.84:* Design and Fabrication Code Case Acceptability ASME Section III, Division 1.

- *Regulatory Guide 1.85:* Materials Code Case Acceptability ASME Section III, Division 1.

The Code cases not listed as approved in Regulatory Guides 1.84 and 1.85 may be used only after seeking permission from NRC for the specific application.[4]

ASME SECTION V: NONDESTRUCTIVE EXAMINATION

Scope

ASME Section V comprises Subsection A, Subsection B, and mandatory and nonmandatory appendixes. Subsection A delineates the methods of nondestructive examination, and Subsection B contains various ASTM standards covering nondestructive examination methods that have been adopted as standards. The standards contained in Subsection B are for information only and are nonmandatory unless specifically referenced in whole or in part in Subsection A or as referenced in other Code sections and other Codes, such as ASME B31, Pressure Piping Code.

The nondestructive examination requirements and methods included in ASME Section V are mandatory to the extent they are invoked by other Codes and standards or by the purchaser's specification.[5] For example, ASME Section III requires radiographic examination of some welds in accordance with Article 2 of ASME Section V.

ASME Section V does not contain acceptance standards for the nondestructive examination methods covered in Subsection A. The acceptance criteria or standards shall be those contained in the referencing Code or standard.

Effective Edition, Addenda, and Code Cases

The applicable edition and addenda of ASME Section V shall correspond to the edition and addenda of the referencing code.

ASME SECTION VIII: PRESSURE VESSELS

Scope

The rules of ASME Section VIII constitute construction requirements for pressure vessels. Division 2 of ASME Section VIII delineates alternative rules of construction to Division 1 requirements. However, there are some differences between the scopes of two divisions.

The rules of ASME Section VIII applies to flanges, bolts, closures, and pressure relieving devices of a piping system when and where required by the Code governing the construction of the piping. For example, ASME B31.1 requires that the safety and relief valves on nonboiler external piping, except for reheat safety valves, shall be in accordance with the requirements of ASME Section VIII, Division 1, UG-126 through UG-133.[6,7]

Effective Edition, Addenda, and Code Cases

Editions are effective and may be used on or after the date of publication printed on the title page.

Addenda are effective and may be used on or after the date of issue.

Addenda and revisions become mandatory as minimum requirements six months after date of issuance, except for pressure vessels contracted for prior to the end of the six-month period.

Code cases may be used beginning with the date of their approval by the ASME.

Use of revisions and addenda and Code cases that are less restrictive than former requirements must not be made without having assurance that they have been accepted by the proper authorities in the jurisdiction where the pressure vessel is to be installed.

ASME SECTION IX: WELDING AND BRAZING QUALIFICATIONS

Scope

ASME Section IX consists of two parts—Part QW and Part QB—which deal with welding and brazing, respectively. In addition, ASME Section IX contains mandatory and nonmandatory appendixes.

ASME Section IX requirements relate to the qualification of welders, welding operators, brazers, and brazing operators and the procedures used in welding and brazing. They establish the basic criteria for welding and brazing observed in the preparation of welding and brazing requirements that affect procedure and performance.[8]

ASME Section IX is a supplemental Code. The requirements of ASME Section IX apply when referenced by the governing Code or specified in purchaser's specification. It is usually referenced in other sections of the ASME Boiler and Pressure Vessel Code and the ASME B31, Pressure Piping Code.

Effective Edition, Addenda, and Code Cases

The applicable edition and addenda of ASME Section IX shall correspond to the edition and addenda of the referencing code. However, the later or the latest edition or addenda of ASME Section IX may be used, provided it is acceptable to the enforcement authorities having jurisdiction.

For safety-related items of an operating nuclear power plant, application of ASME Section IX will be in accordance with the requirements of ASME Section XI, Rules for Inservice Inspection of Nuclear Power Plant Components.

For nonsafety-related items, the following guidelines will apply:

- Editions are effective and may be used on or after the date of publication on the title page.
- Addenda are effective and may be used on or after the date of issue.
- Addenda and revisions become mandatory as minimum requirements 6 months after the date of issue, except for pressure vessels or boilers contracted for prior to the end of the 6-month period.
- Code cases may be used beginning with the date of their approval by the ASME.
- Use of revisions and addenda and Code cases that are less restrictive than former requirements must not be made without having assurance they have been accepted by the proper authorities in the jurisdiction where the item is to be installed.

ASME SECTION XI: RULES FOR INSERVICE INSPECTION OF NUCLEAR POWER PLANT COMPONENTS

Scope

ASME Section XI comprises three divisions, each covering rules for inspection and testing of components of different types of nuclear power plants. These three divisions are as follows:

- *ASME Section XI, Division 1:* Rules for Inspection and Testing of Components of Light-Water-Cooled Plants
- *ASME Section XI, Division 2:* Rules for Inspection and Testing of Components of Gas-Cooled Plants
- *ASME Section XI, Division 3:* Rules for Inspection and Testing of Components of Liquid-Metal-Cooled Plants

Since the publication of the first edition of ASME Section XI in 1971, a significant number of changes and additions has been incorporated, and as such, the organization of the later versions of ASME Section XI, Division 1, is considerably different from the first edition.

ASME Section XI, Division 1, provides the rules and requirements for in-service inspection and in-service testing of light-water-cooled nuclear power plants. The rules and requirements identify, as a minimum, the areas subject to inspection, responsibilities, provisions for accessibility and inspectability, examination methods and procedures, personnel qualifications, frequency of inspection, record keeping and report requirements, procedures for evaluating inspection results and subsequent disposition of results of evaluations, and repair requirements.

Division 1 also provides for the design, fabrication, installation, and inspection of replacements. The jurisdiction of Division 1 of ASME Section XI covers individual components and complete power plants that have met all the requirements of the construction Code, commencing at that time when the construction Code requirements have been met, irrespective of physical location.

When portions of systems or plants are completed at different times, jurisdiction of Division 1 shall cover only those portions on which all of the construction code requirements have been met. Rules of ASME Section XI apply to ASME Classes 1, 2, 3, and MC components and their supports, core support structures, pumps, and valves.

Rules of ASME Section XI, Division 1, apply to modifications made to ASME III components and their supports after all of the original construction Code requirements have been met.[9]

Rules of ASME Section XI, Division 1, apply to systems, portions of systems, components, and their supports not originally constructed to ASME Section III requirements but based upon their importance to safety if they were classified as ASME Classes 1, 2, 3, and MC.

Effective Edition, Addenda, and Code Cases

Section 10 CFR 50.55a, Codes and Standards, of the Code of Federal Regulations requires compliance with ASME Section XI for operating nuclear power plants. In addition, 10 CFR 50.55a, Paragraph (b)(2) delineates the editions and addenda of ASME Section XI that are approved for use. Only the approved editions and addenda of ASME Section XI are to be used. The latest published edition and addenda may not be approved by the U.S. NRC; therefore, they can only be used after seeking special permission from the U.S. NRC.[4]

It is recommended that one refer to 10 CFR 50.55a from time to time to determine which edition and addenda of ASME Section XI have been approved by the U.S. NRC and which edition and addenda may be applicable to a nuclear power plant at a particular time.

The requirements of 10 CFR 50.55a are based on the construction permit (CP) date and the operating license (OL) date of the nuclear plant.

Code edition and addenda later than those established for a particular application in conformance with the requirements of 10 CFR 50.55a may be used provided they are approved and all related requirements of respective editions or addenda are met.

While establishing a particular edition and addenda of ASME Section XI, consider the limitations and modifications to the specific editions and addenda delineated in Paragraph (b) (2) of 10 CFR 50.55a, and ensure compliance to those limitations and modifications, as applicable.

For repairs and replacements, the applicable edition and addenda shall be the one in effect for that inservice inspection (ISI) interval during which the repairs and replacements are to be made. Refer to articles IWA-4000 and IWA-7000 of ASME Section XI.

Applicable Code Cases

Like the Code edition and addenda, Code cases are regularly reviewed by the U.S. NRC. The NRC-approved Code cases with or without limitations or additional requirements are published in the Regulatory Guide 1.147, In Service Inspection Code Case Acceptability of ASME Section XI, Division 1.[4]

Acceptance or endorsement by the NRC staff applies only to those Code cases or Code case revisions with the date of ASME Council approval, as shown in the Regulatory Guide 1.147.

ASME B31: CODE FOR PRESSURE PIPING

Starting with Project B31 in March 1926, the first edition of American Tentative Standard Code for Pressure Piping was published in 1935. In view of continuous industry developments and increase in diversified needs over the years, decisions were made to publish several sections of the Code for Pressure Piping. Since December 1978, the American National Standards Committee B31 was reorganized as the ASME Code for Pressure Piping B31 committee under procedures developed by the ASME and accredited by ANSI.

Presently, the following sections of ASME B31, Code for Pressure Piping are published:

ASME B31.1	Power Piping
ASME B31.3	Chemical Plant and Petroleum Refinery Piping
ASME B31.4	Liquid Transportation Systems for Hydrocarbons, Liquid Petroleum Gas, Anhydrous Ammonia, and Alcohols
ASME/ANSI B31.5	Refrigeration Piping
ASME B31.8	Gas Transmission and Distribution Piping Systems
ASME/ANSI B31.9	Building Services Piping
ANSI/ASME B31.11	Slurry Transportation Piping Systems

ASME B31.1: POWER PIPING CODE

Scope

ASME B31.1, Power Piping Code, covers minimum requirements for the design, material, fabrication, erection, test, and inspection of power and auxiliary service piping systems for electric generation stations, industrial and institutional plants, central and district heating plants, and district heating systems. It does not apply to piping systems covered by other sections of the Code for Pressure Piping and other piping which is specifically excluded from the scope of this Code.[7]

As explained earlier, the BEP is required to meet administrative jurisdictional requirements of ASME Section I; however, pipe connections meeting all other requirements of ASME B31.1 but not exceeding nominal pipe size (NPS) 1/2 may be welded to boiler external pipe or boiler headers without inspection and stamping required by ASME Section I.

Nonboiler external piping is defined as all the piping covered by ASME B31.1 with the exception of BEP. The nonboiler external piping must be constructed in accordance with the requirements of this Code.

In addition to the piping systems covered by other sections of ASME B31, Pressure Piping Code, ASME B31.1 does not cover the following:

- Economizers, heaters, pressure vessels, and component covered by ASME Boiler and Pressure Vessel Code (except the connecting piping not covered by the ASME Boiler and Pressure Vessel Code shall meet the requirements of ASME B31.1).

- Building heating and distribution steam piping designed for 15 psig or less or hot water heating systems piping designed for 30 psig or less.
- Piping for roof and floor drains, plumbing, sewers and sprinkler, and other fire protection systems.
- Piping for hydraulic or pneumatic tools and their components downstream of the first stop valve off the system distribution header.
- Piping for marine or other installations under federal control.
- Piping covered by other sections of ASME B31 and ASME Section III.
- Fuel gas piping within the scope of ANSI Z 223.1, National Fuel Gas Code.
- Pulverized fuel piping within the scope of NFPA.

The requirements of this Code apply to central and district heating systems for distribution of steam and hot water away from the plant, whether underground or elsewhere, and geothermal steam and hot water piping both to and from well heads.

The construction of fuel gas or fuel oil piping brought to plant site from a distribution system inside the plant property line is governed by the requirements of ASME B31.1 when the meter assembly is located outside the plant property line. In case the meter assembly is located within the plant property line, the requirements of this Code shall apply to the fuel gas and fuel oil piping downstream from the outlet of the meter assembly; see Fig. A4.3.

This Code also applies to gas and oil systems piping other than that shown in Fig. A4.3, air systems, hydraulic fluid systems piping, and the steam jet cooling systems piping which are part of the power plant cycle. In addition, building services within the scope of ASME/ANSI B31.9 but outside the limits of Paragraph 900.1.2 of B31.9 are required to be designed in accordance with ASME B31.1.[10]

Effective Edition, Addenda, and Code Cases

Prior to the publication and implementation of ASME Section III for construction of nuclear power plant components, in some nuclear power plants the safety-related piping systems, now classified as ASME Classes 1, 2, and 3, were constructed to earlier versions of ANSI B31.1. Therefore, the repairs and replacements of those safety-related piping systems may be made in accordance with the edition and addenda of ANSI B31.1 used for the original construction or the later edition and addenda of ANSI B31.1. Refer to Article IWA-4000 and Article IWA-7000 of ASME Section XI for requirements related to repairs and replacements, respectively.

For power piping systems other than the nuclear safety-related piping systems constructed and new piping systems to be constructed to ASME B31.1, the following guidelines shall be used to determine the effective edition and addenda of ANSI B31.1:

Editions are effective and may be used on or after the date of publication printed on the title page.

Addenda are effective and may be used on or after the date of publication printed on the title page.

The latest edition and addenda, issued 6 months prior to the original contract date for the first phase of the activity covering a piping system(s) shall be the

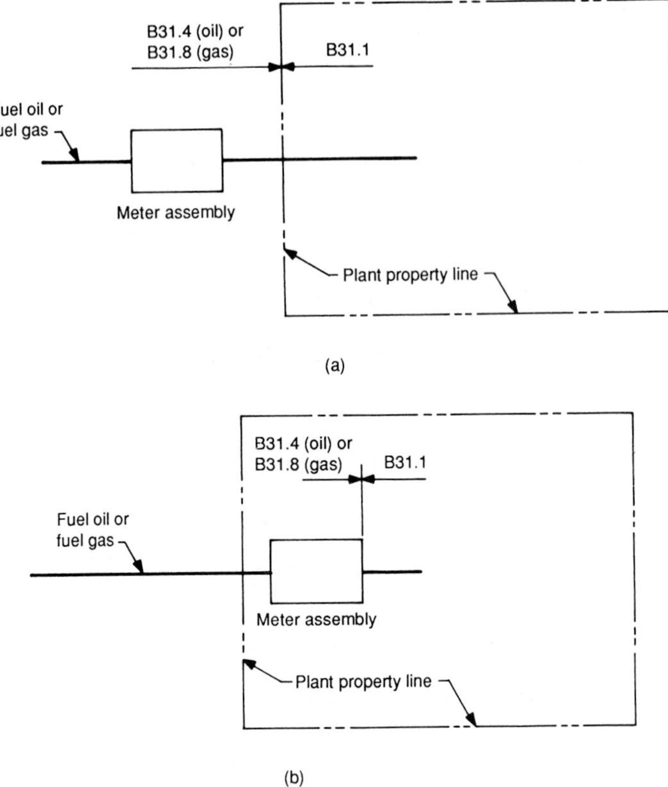

Figure A4.3 Jurisdiction of ASME B31.1, B31.4, and B31.8 over fuel gas and fuel oil piping.

governing document for design, materials, fabrication, erection, examination, and testing activities for the piping system(s) until the completion of the work and initial operation.[7]

Unless agreement is specifically reached between the contracting parties, no Code edition and/or addenda shall be retroactive.

Code cases may be used after they have been approved by the ASME Council. The provisions of a Code case may be used even after its expiration or withdrawal, provided the Code case was effective on the original contract date or was adopted prior to completion of work and the contracting parties agreed to its use.

Do not use revisions and Code cases that are less restrictive than former requirements without having assurance that they have been accepted by the proper authorities in the jurisdictions where the piping is to be installed.

ASME B31.3: CHEMICAL PLANT AND PETROLEUM REFINERY PIPING

Scope

This Code prescribes requirements for the materials, design, fabrication, assembly, erection, examination, inspection, and testing of all piping within the property limits of facilities engaged in the processing or handling of chemical petroleum or related products. Figure A4.4 provides an illustration of the scope of ASME B31.3.[11] The requirements of ASME B31.3 apply to piping for all fluids, including raw, intermediate, and finished chemicals; petroleum products, gas, steam, air, and water; fluidized solids; and refrigerants.

In case of packaged equipment, the interconnecting piping with the exception of refrigeration piping shall be in compliance with the requirements of ASME B31.3. The refrigeration piping may conform to either ASME B31.3 or ASME B31.5.

The requirements of ASME B31.3 do not apply to piping systems designed for internal gauge pressures at or above 0 but less than 15 psig provided the fluid handled is nonflammable, nontoxic, and not damaging to human tissue.

The following piping and equipment are not required to comply with the requirements of ASME B31.3:

- Power boiler and the boiler external piping
- Piping covered by ASME B31.4, B31.8, or B31.11, although located on the company property
- Piping covered by applicable governmental regulations
- Piping for fire protection systems
- Plumbing, sanitary sewers, and storm sewers
- Tubes, tube headers, crossovers, and manifolds of fired heaters, which are internal to the heater enclosures
- Pressure vessels, heat exchangers, pumps, compressors, and other fluid-handling or processing equipment, including internal piping and connections for external piping

Effective Edition, Addenda, and Code Cases

The effective edition, addenda, and Code cases shall be determined similarly to the approach delineated for ASME B31.1 for piping systems other than the nuclear safety-related piping systems.

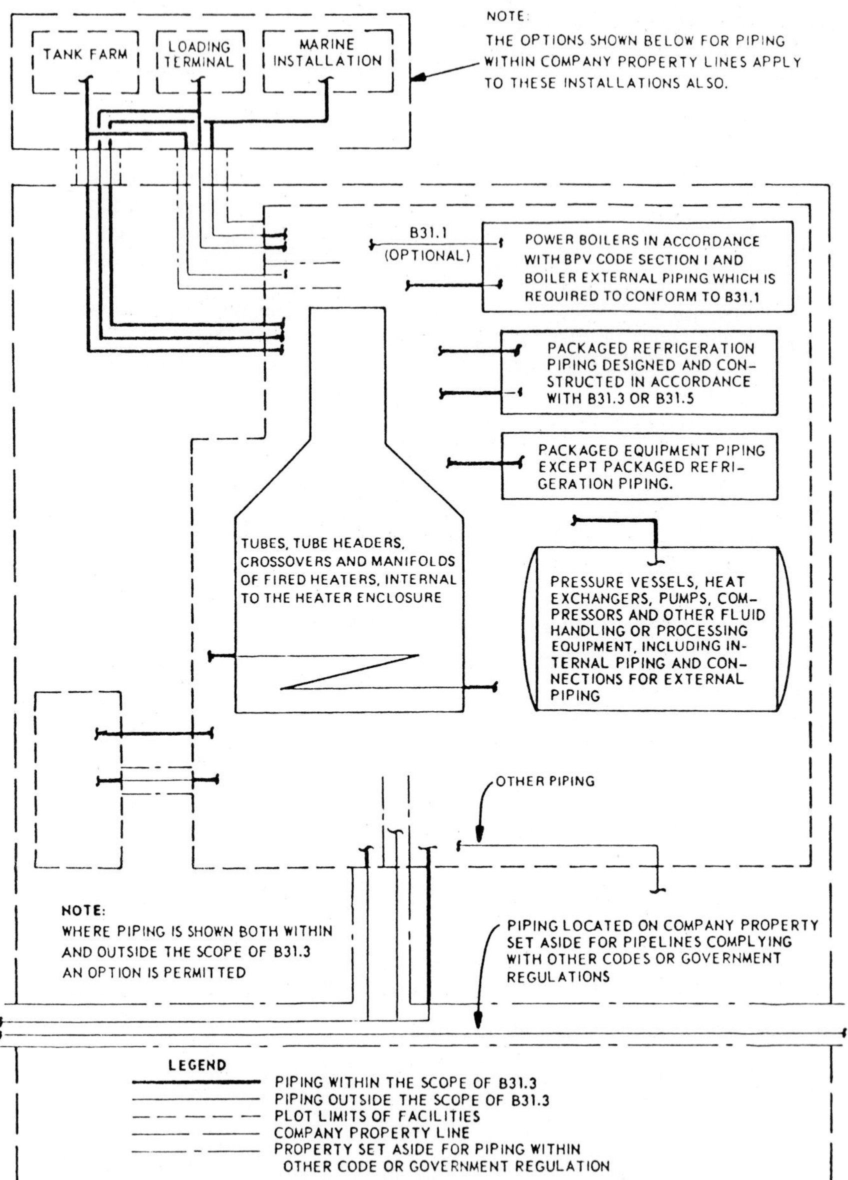

Figure A4.4 B31.3 jurisdictional limits and options. (*ASME B31.3.*)

ASME B31.4: LIQUID TRANSPORTATION SYSTEMS FOR HYDROCARBONS, LIQUID PETROLEUM GAS, ANHYDROUS AMMONIA, AND ALCOHOLS

Scope

Section B31.4 of the ASME Pressure Piping Code specifies minimum requirements for the design, materials construction, assembly, inspection, testing of piping transporting liquids such as crude oil, condensate, natural gasoline, natural gas liquids, liquefied petroleum gas, liquid alcohol, liquid anhydrous ammonia, and liquid petroleum products between producers' lease facilities, tank farms, natural gas-processing plants, refineries, stations, ammonia plants, terminals, and other delivery and receiving points.[12]

The scope of ASME B31.4 also includes the following:

1. Primary and associated auxiliary liquid petroleum and liquid anhydrous ammonia piping at pipeline terminals, tank farms, pump stations, pressure-reducing stations, and metering stations, including scraper traps, strainers, and prover loops
2. Storage and working tanks, including pipetype storage fabricated from pipe and fittings and piping interconnecting these facilities
3. Liquid petroleum and liquid anhydrous ammonia piping located on property which has been set aside for such piping within petroleum refinery, natural gasoline, gas processing, ammonia, and bulk plants
4. Those aspects of operation and maintenance of liquid petroleum and liquid anhydrous ammonia transportation piping systems relating to the safety and protection of the general public, operating company personnel, environment, property, and the piping systems

ASME B31.4 does not apply to

1. Auxiliary piping such as water, air steam, lubricating oil, gas, and fuel
2. Pressure vessels, heat exchangers, pumps, meters, and other such equipment, including internal piping and connections for piping except as limited by Paragraph 423.2.4 (b) of ASME B31.4
3. Piping designed for internal pressures:
   a. At or below 15 psi gauge pressure regardless of temperature
   b. Above 15 psi gauge pressure if design temperature is below minus 20°F or above 250°F
4. Casing, tubing, or pipe used in oil wells, wellhead assemblies, oil and gas separators, crude oil production tanks, other producing facilities, and pipelines interconnecting these facilities
5. Petroleum refinery, natural gasoline, gas processing, ammonia, and bulk plant piping, except as covered within the scope of the Code
6. Gas transmission and distribution piping
7. The design and fabrication of proprietary items of equipment, apparatus, or instruments, except as limited by this Code
8. Ammonia refrigeration piping systems provided for in ASME/ANSI B31.5, Refrigeration Piping Code

The rules of this Code provide protection of the general public, operating company personnel, as well as reasonable protection of the piping system against vandalism and accidental damage by others and reasonable protection of the environment.

Effective Edition, Addenda, and Code Cases

To determine the effective edition, addenda, and Code cases for an application within the jurisdiction of ASME B31.4, follow the requirements delineated for ASME B31.1 for piping systems other than nuclear safety-related piping systems.

ASME/ANSI B31.5: REFRIGERATION PIPING

Scope

This section of ASME B31, Pressure Piping Code, contains minimum requirements for the materials, design, fabrication, assembly, erection, test, and inspection of refrigerant and secondary coolant piping for temperatures as low as −320°F, except when other sections of the Code cover requirements for refrigeration piping.[13]

ASME/ANSI B31.5 does not apply to the following:

- Self-contained or unit systems subject to the requirements of Underwriters' Laboratories (UL) or other nationally recognized testing laboratory
- Water piping
- Piping designed for external or internal gauge pressure not exceeding 15 psig

Effective Edition, Addenda, and Code Cases

To determine the effective edition, addenda, and Code cases for piping systems within the jurisdiction of ASME/ANSI B31.5, follow the guidelines delineated for nonnuclear piping systems within the jurisdiction of ASME B31.1.

ASME B31.8: GAS TRANSMISSION AND DISTRIBUTION PIPING SYSTEMS

Scope

A pipeline or transmission line is defined as that pipe which transmits gas from a source or sources of supply to one or more large volume customers or to a pipe used to interconnect sources of supply. ASME B31.8 prescribes requirements for the design, fabrication, installation, testing, and safety aspects of operation and maintenance of gas transmission and distribution piping systems, including gas pipelines, gas compressor stations, gas metering and regulation stations, gas mains, and service lines up to the outlet of the customer's meter set assembly.[14]

Also included within the scope of ASME B31.8 are gas storage equipment of the closed pipe type, fabricated or forged from pipe or fabricated from pipe and fittings, and gas storage lines.

The requirements of ASME B31.8 also apply to the use of elements of piping systems, including, but not limited to, pipe, valves, fittings, flanges, bolting, gaskets, regulators, pressure vessels, pulsation dampeners, and relief valves.

The requirements of ASME B31.8 are applicable to operating and maintenance procedures of existing installations and to the update of existing installations.

ASME B31.8 does not apply to the following:[14]

1. Design and manufacture of pressure vessels covered by the ASME Boiler and Pressure Vessel Code
2. Piping with metal temperatures above 450°F or below −20°F
3. Piping beyond the outlet of the customer's meter set assembly (refer to ANSI Z223.1)
4. Piping in oil refineries or natural gasoline extraction plants, gas-treating plant piping other than the main gas stream piping in dehydration, and all other processing plants installed as part of a gas transmission system, gas manufacturing plants, industrial plants, or mines. (See other applicable sections of the ASME Code for Pressure Piping, B31.)
5. Vent piping to operate at substantially atmospheric pressures for waste gases of any kind
6. Wellhead assemblies, including control valves, flow lines between wellhead and trap or separator, or casing and tubing in gas or oil wells
7. The design and manufacture of proprietary items of equipment, apparatus, or instruments
8. The design and manufacture of heat exchangers
9. Liquid petroleum transportation piping systems (refer to ANSI/ASME B31.4)

Effective Code Edition, Addenda, and Code Cases

To determine the effective edition, addenda, and Code cases to be invoked for an application or piping systems within the jurisdiction of ASME B31.8, follow the criteria delineated for nonnuclear piping systems within the scope of ASME B31.1:

> No edition and addenda shall be applied retroactively to existing installations insofar as design, fabrication, installation, and testing at the time of construction are concerned. Further, no edition and addenda shall be applied retroactively to established operating pressures of existing installations, except as provided for in Chapter V of ASME B31.8.

ASME/ANSI B31.9: BUILDING SERVICES PIPING

Scope

ASME/ANSI B31.9 applies to the following building services:

- Water for heating and cooling
- Condensing water
- Steam or other condensate
- Steam
- Vacuum
- Compressed air and other nontoxic and nonflammable gases

The requirements of this Code also apply to boiler external piping for steam boilers with 15 psig maximum pressure and for water heating units having 160 psig maximum pressure and 250°F maximum temperature.[10] It is noted that the boiler external piping exceeding the above limits of pressure and temperature fall within the scope of ASME B31.1 and ASME Section I.

This Code places size and thickness limitations on the piping made of different materials. The requirements of this Code shall apply to the piping of up to and including those sizes and thicknesses. These limitations are as follows:

- *Carbon steel:* 20 in O.D. and 0.500 in wall
- *Stainless steel:* NPS 12 and 0.500 in wall
- *Aluminum:* NPS 12
- *Brass and copper:* NPS 12 (12.125 in O.D. for copper tubing)
- *Plastics:* NPS 8
- *Cast iron:* NPS 18

Piping made of other materials permitted by this Code may also be used for building services.

The piping with working pressures or temperatures exceeding the limits shown in Table A4.1 shall be designed and constructed in compliance with the requirements of ASME B31.1.[10]

The piping systems within the jurisdiction of the following are excluded from the scope of this Code:

- ASME B31.3, ASME B31.4, ASME B31.5

TABLE A4.1 Working Pressures and Temperature Limits (B31.9)*

Pressure limit		Temperature limit	
Service	Pressure	Service	Temperature
Steam, air, and nonfuel gases	125 psig	Steam and condensate	355°F
Liquids	300 psig	Other gases and vapors	200°F
		Other nonflammable liquids	250°F
Vacuum	1 atm	Flammable liquids external pressure	Atmospheric boiling temperature

*Building services piping systems beyond the above-described limits shall be designed in accordance with the requirements of ASME B31.1.[10]

- Local plumbing code (hot and cold potable water piping and sanitary and storm drainage piping systems)
- NFPA standards (fire protection systems, liquefied fuel gases, industrial and medical gas systems piping)
- ANSI Z223.1 (fuel gas piping)
- ANSI/NFPA 30 (fuel oil piping)

ASME/ANSI B31.9 does not cover requirements for economizers, heaters, pumps, tanks, heat exchangers, and other equipment within the scope of ASME Boiler and Pressure Vessel Code.

Effective Code Edition, Addenda, and Code Cases

For any specific application, the effective edition, addenda, and Code cases of ASME/ANSI B31.9 shall be determined in accordance with the approach followed for ASME B31.1 for piping systems other than nuclear safety-related piping systems.

ANSI/ASME B31.11: SLURRY TRANSPORTATION PIPING SYSTEMS

Scope

Like ASME B31.4, this section of ASME B31, Pressure Piping Code, specifies minimum requirements for the design, materials, construction, assembly, inspection, testing, operation, and maintenance of piping transporting aqueous slurries of nonhazardous materials, such as oil, mineral ores, and concentrates, between a slurry processing plant or terminal and a receiving plant or terminal.[15]

The requirements of ANSI/ASME B31.11 also apply to the following:

1. Primary and auxiliary slurry piping at storage facilities, pipeline terminals, pump stations, and pressure reducing stations, including piping up to the first valve of attached auxiliary water lines
2. Slurry piping storage facilities and other equipment located on property which has been set aside for the slurry transportation system
3. Those aspects of operation and maintenance of slurry transportation piping systems which relate to the safety and protection of the general public, operating company personnel, environment, property, and the piping systems

Refer to Fig. A4.5 for facilities within the scope of ANSI/ASME B31.11.

Effective Code Edition, Addenda, and Code Cases

The effective edition, addenda, and Code cases applicable for piping within the scope of ANSI/ASME B31.11 shall be determined by following the guidelines de-

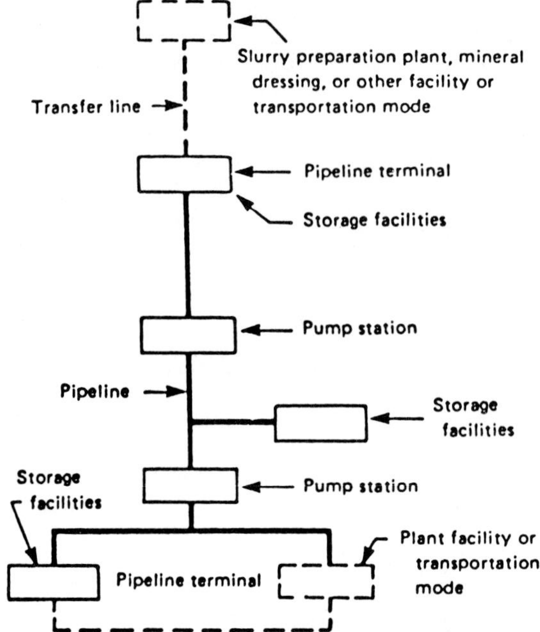

Figure A4.5 Scope of ANSI/ASME B31.11 facilities indicated by solid lines are within the scope of ANSI/ASME B31.11. (*Source: Figure 1100.11, ANSI/ASME B31.11.*)

lineated for ASME B31.1 for piping systems other than the nuclear safety-related piping systems.

ASME PERFORMANCE TEST CODES

The ASME Performance Test Codes (PTC) were originally known as Power Test Codes. These Codes provide standard directions and rules for conducting and reporting tests of specific materials such as fuels, equipment, and processes or functions related to power plants. Listed below are some of the Performance Test Codes which may be of interest in regard to piping:

PTC 1	General Instructions
PTC 2	Definitions and Values
PTC 4.1	Steam Generating Units
PTC 4.3	Air Heaters
PTC 4.4	Gas Turbine Heat Recovery Steam Generators
PTC 6	Steam Turbine
PTC 6A	Appendix A to Test Code for Steam Turbine
PTC 7	Reciprocating Steam-Driven Displacement Pumps
PTC 7.1	Displacement Compressors, Vacuum Pumps, and Blowers

PTC 10	Compressors and Exhaustors
PTC 12.2	Code on Steam Condensing Apparatus
PTC 12.3	Deaerators
PTC 14	Evaporating Apparatus
PTC 16	Gas Producers and Continuous Gas Generators
PTC 18	Hydraulic Prime Movers Including Index Method of Testing
PTC 19.2	Pressure Measurement Instrument and Apparatus
PTC 19.3	Temperature Measurement Instrument and Apparatus
PTC 19.5	Application Part II of Fluid Meters
PTC 19.16	Density Determination of Solids and Liquids
PTC 19.17	Determination of the Viscosity of Liquids Instruments and Apparatus
PTC 22	Gas Turbine Power Plants
PTC 25.3	Safety and Relief Valves
PTC 32.1	Nuclear Steam Supply Systems

AMERICAN NATIONAL STANDARDS INSTITUTE

The American National Standards Institute (ANSI) was earlier known as the American Standards Association (ASA). For a short period of time, from 1967 to 1969, it was called the United States of America Standards Institute (USASI).

ANSI provides a forum for development or obtaining a consensus for approval of standards having national impact and serves as a focal point for distribution of national and other standards, including those developed and issued by the international organization for standardization (ISO) and foreign governments. Development and approval functions are performed by committees representing a cross section of affected interests, such as engineering societies, manufacturers, trade institutes, fabricators, builders, universities, unions, insurance companies, and government agencies. Many of the committees are chaired or sponsored by engineering societies, such as ASME and the Institute of Electrical and Electronics Engineers (IEEE).

Safety is the basic objective of the engineering design and construction requirements contained in standards developed, approved, and distributed by the ANSI. The ANSI standards include prohibition for practices considered unsafe and cautions where advisory warnings, instead of prohibitions, are deemed necessary.

This chapter provides a brief discussion of various sections of ASME B31, Pressure Piping Code, which was earlier known as ANSI B31, Pressure Piping Code. It is envisioned that other ANSI standards may eventually become known as ASME standards; however, they shall be subjected to approval of the ANSI. The following ANSI standards contain provisions related to piping.

ANSI Standards

A13.1-81	Scheme for the Identification of Piping Systems (R 1985)
A112.1.2-42	Air Gaps in Plumbing Systems (Reaffirmation and Redesignation of A40.4-1942); Errata (R 1979)

A112.6.1M-79	Supports for Off-the-Floor Plumbing Fixtures for Public Use (Revision of ANSI A112.6.1-1972)
A112.18.1M-89	Plumbing Fixture Fittings
A112.19.1M-87	Enameled Cast Iron Plumbing Fixtures
A112.19.3M-87	Stainless Steel Plumbing Fixtures (Designed for Residential Use) (Revision of ANSI A112.19.3-1976)
A112.21.1M-80	Floor Drains (Revision of USAS A112.21.1-1968)
A112.21.2M-83	Roof Drains (Revision of ANSI A112.21.2-1971)
A112.26.1M-84	Water Hammer Arresters (Revision of A112.26.1-1969)(R 1975)
A112.36.2M-83	Cleanouts (Revision of ANSI A112.36.2-1975)
B1.20.1-83	Pipe Threads General Purpose (Inch) (Revision and Redesignation of ASME/ANSI B2.1-1968)
B1.20.3-76	Dryseal Pipe Threads (Inch) (Revision and Redesignation of B2.2-1968 (R 1982)
B1.20.4-76	Dryseal Pipe Threads (Metric Translation of B1.20.3-1976) (Partial Revision and Conversion of ANSI B2.2-1968) (R 1982)
B16.1-89	Cast Iron Pipe Flanges and Flanged Fittings
B16.3-85	Malleable Iron Threaded Fittings; Classes 150 and 300
B16.4.85	Cast Iron Threaded Fittings; Classes 125 and 250
B16.5-88	Pipe Flanges and Flanged Fittings
B16.9-86	Factory-Made Wrought Steel Butt-Welding Fittings
B16.10-86	Face-To-Face and End-To-End Dimensions of Valves
B16.11-80	Forged Steel Fittings, Socket-Welding and Threaded
B16.12-83	Cast Iron Threaded Drainage Fittings
B16.14-83	Ferrous Pipe Plugs, Bushings, and Locknuts with Pipe Threads
B16.15-85	Cast Bronze Threaded Fittings; Classes 125 and 250
B16.18-84	Cast Copper Alloy Solder Joint Pressure Fittings
B16.22-89	Wrought Copper and Copper Alloy Solder Joint Pressure Fittings
B16.23-84	Cast Copper Alloy Solder Joint Drainage Fittings
B16.24-79	Bronze Pipe Flanges and Flanged Fittings; Classes 150 and 300
B16.25-86	Butt-Welding Ends
B16.26-88	Cast Copper Alloy Fittings for Flared Copper Tubes
B16.28-86	Wrought Steel Butt-Welding Short Radius Elbows and Returns
B16.29-86	Wrought Copper and Wrought Copper Alloy Solder Joint Drainage Fittings
B16.32-84	Cast Copper Alloy Solder Joint Fittings for Solvent Drainage Systems
B16.33-81	Manually Operated Metallic Gas Valves for Use in Gas Piping Systems up to 125 psig (Sizes 1/2 through 2)
B16.34-88	Valves—Flanged, Threaded, and Welding End
B16.36-88	Orifice Flanges
B16.37-80	Hydrostatic Testing of Control Valves

B16.38-85	Large Metallic Valves for Gas Distribution (Manually Operated, NPS 2 1/2 to 12, 125 psig Maximum)
B16.39-86	Malleable Iron Threaded Pipe Unions; Classes 150, 250, and 300
B16.40-85	Manually Operated Thermoplastic Gas Shutoffs and Valves in Gas Distribution Systems
B16.41-83	Functional Qualification Requirements for Power Operated Active Valve Assemblies for Nuclear Power Plants (R 1989)
B16.42-87	Ductile Iron Pipe Flanges and Flanged Fittings; Classes 150 and 300
B16.43-82	Wrought Copper and Copper Alloy Solder Joint Fittings for Sovent Drainage Systems
B16.45-87	Cast Iron Fittings for Sovent Drainage Systems
B16.47-90	Large Diameter Steel Flanges NPS 26 through NPS 60
B18.2.1-81	Square and Hex Bolts and Screws (Inch Series) Including Hex Cap Screws and Lag Screws; Supplement B18.2.1
B18.2.2-87	Square and Hex Nuts (Inch Series)
B18.2.3.5M-79	Metric Hex Bolts; Errata—May 1981
B18.2.3.6M-79	Metric Heavy Hex Bolts (R 1989)
B18.2.4.1M-79	Metric Hex Nuts, Style 1; Errata—May 1981
B18.2.4.2M-79	Metric Hex Nuts, Style 2
B18.2.4.3M-79	Metric Slotted Hex Nuts
B18.2.4.4M-82	Metric Hex Flange Nuts
B18.2.4.5M-79	Metric Hex Jam Nuts
B18.2.4.6M-79	Metric Heavy Hex Nuts
B18.5-78	Round Head Bolts (Inch Series)
B18.5.2.1M-81	Metric Round Head Short Square Neck Bolts
B18.5.2.2M-82	Metric Round Head Square Neck Bolts
B18.15-85	Forged Eyebolts
B18.18.1M-87	Inspection and Quality Assurance for General Purpose Fasteners
B18.18.3M-87	Inspection and Quality Assurance for Special Purpose Fasteners
B18.18.4M-87	Inspection and Quality Assurance for Fasteners for Highly Specialized Engineered Applications
B18.21.1-72	Lock Washers (Revision and Redesignation of B27.1-1965)
B18.22M-81	Metric Plain Washers
B18.22.1-65	Plain Washers (Reaffirmation and Redesignation of ASA B27.2-1965)
B32.5-77	Preferred Metric Sizes for Tubular Metal Products Other Than Pipe (R 1988)
B32.6M-84	Preferred Metric Equivalents of Inch Sizes for Tubular Products Other Than Pipe (Revision of ANSI B32.6-1877)

B36.10M-85	Welded and Seamless Wrought Steel Pipe (Revision of ANSI B36.10)
B36.19M-85	Stainless Steel Pipe (Revision of ANSI B36.19)
MFC-1M-79	Glossary of Terms Used in the Measurement of Fluid Flow in Pipes (R 1986)
MFC-6M-87	Measurement of Fluid Flow in Pipes Using Vortex Flow Meters
MFC-7M-87	Measurement of Gas Flow by Means of Critical Flow Venturi Nozzles
N45.2.1-80	Cleaning of Fluid Systems and Associated Components for Nuclear Power Plants
N278.1-75	Self-Operated and Power-Operated Safety-Related Valves Functional Specification Standard, Reactor Plants and Their Maintenance (R 1984)
NQA-1-89	Quality Assurance Program Requirements for Nuclear Facilities; Addenda NQA-1A-1989
NQA-2-89	Quality Assurance Requirements for Nuclear Facility Applications; Addenda NQA-2A-1990
TDP-1-85	Recommended Practices for the Prevention of Water Damage to Steam Turbines Used for Electric Power Generation (Fossil)
TDP-2-85	Recommended Practices for the Prevention of Water Damage to Steam Turbines Used for Electric Power Generation (Revision of ASME Standard No. TWDPS-1-1973, Part 2) (Nuclear)

ANS1 Guides/Manuals

1986	Guide for Gas Transmission and Distribution Piping Systems-1986; Addenda 1-1986, Addenda 2-1987, Addenda 3-1987
B31 Guide-77	Corrosion Control for ANSI B31.1, Power Piping Systems
B31 Guide-84	Manual for Determining the Remaining Strength of Corroded Pipelines (A Supplement to ASME B31 Code for Pressure Piping)
1001-88	Performance Requirements for Pipe Applied Atmospheric Type Vacuum Breakers
1003-81	Performance Requirements for Water Pressure Reducing Valves for Domestic Water Supply Systems
1029-81	Performance Requirements for Water Supply Valves; Mixing Valves and Single Control Mixing Valves
1032-80	Performance Requirements for Dual Check Valve Type Backflow Preventers for Carbonated Beverage Dispensers
1037-90	Performance Requirements for Pressurized Flushing Devices (Flushometers) for Plumbing Fixtures
1045-87	Performance Standard and Installation Procedures for Aluminum Drain, Waste and Vent Pipe with End Cap Components

Other ASME/ANSI Publications

The following is a list of additional ASME/ANSI publications which are of interest to people engaged in the piping design, construction, operation, and maintenance activities:

B16.20-73	Ring-Joint Gaskets and Grooves for Steel Pipe Flanges
B16.21-78	Nonmetallic Flat Gaskets for Pipe Flanges
MFC-3M-89	Measurement of Fluid Flow in Pipes Using Orifice, Nozzle, and Venturi

AMERICAN SOCIETY FOR TESTING AND MATERIALS

The American Society for Testing and Materials (ASTM) is a scientific and technical organization that develops and publishes voluntary standards on the characteristics and performance of materials, products, systems, and services. The standards published by the ASTM include test procedures for determining or verifying characteristics, such as chemical composition, and measuring performance, such as tensile strength and bending properties. The standards cover refined materials, such as steel, and basic products, such as machinery and fabricated equipment. The standards are developed by committees drawn from a broad spectrum of professional, industrial, and commercial interests. Many of the standards are made mandatory by references in applicable piping Codes.

The ASTM standards are published in a set of 67 volumes. Each volume is published annually to incorporate new standards and revisions to existing standards and delete obsolete standards. Listed below are the 67 volumes, divided among 16 sections, published by the ASTM.

Section 1: Iron and Steel Products

Volume 01.01	Steel—Piping, Tubing, Fittings
Volume 01.02	Ferrous Castings; Ferroalloys; Shipbuilding
Volume 01.03	Steel—Plate, Sheet, Strip, Wire
Volume 01.04	Steel—Structural, Reinforcing, Pressure Vessel, Railway
Volume 01.05	Steel—Bars, Forgings, Bearing, Chain, Springs
Volume 01.06	Coated Steel Products

Section 2: Nonferrous Metal Products

Volume 02.01	Copper and Copper Alloys
Volume 02.02	Die-Cast Metals, Aluminum and Magnesium Alloys
Volume 02.03	Electrical Conductors

Volume 02.04	Nonferrous Metals-Nickel, Lead, Tin Alloys, Precious, Primary, Reactive Metals
Volume 02.05	Metallic and Inorganic Coatings; Metal Powders, Sintered P/M Structural Parts

Section 3: Metals Test Methods and Analytical Procedures

Volume 03.01	Metals—Mechanical Testing: Elevated and Low-Temperature Tests, Metallography
Volume 03.02	Wear and Erosion, Metal Corrosion
Volume 03.03	Nondestructive Testing
Volume 03.04	Magnetic Properties; Metallic Materials for Thermostats, Electrical Resistance, Heating, Contacts
Volume 03.05	Chemical Analysis of Metals and Metal-Bearing Ores
Volume 03.06	Analytical Atomic Spectroscopy, Surface Analysis

Section 4: Construction

Volume 04.01	Cement, Lime, Gypsum
Volume 04.02	Concrete and Aggregates
Volume 04.03	Road and Paving Materials, Traveled Surface Characteristics
Volume 04.04	Roofing, Waterproofing, and Bituminous Materials
Volume 04.05	Chemical-Resistant Materials; Vitrified Clay, Concrete, Fiber-Cement Products; Mortars; Masonry
Volume 04.06	Thermal Insulation; Environmental Acoustics
Volume 04.07	Building Seals and Sealants; Fire Standards; Building Constructions
Volume 04.08	Soil and Rock; Building Stones; Geotextiles
Volume 04.09	Wood

Section 5: Petroleum Products, Lubricants, and Fossil Fuels

Volume 05.01	Petroleum Products and Lubricants (I): D 56-D 1947
Volume 05.02	Petroleum Products and Lubricants (II): D 1949-D 3601
Volume 05.03	Petroleum Products and Lubricants (III): D 3602-latest; Catalysts
Volume 05.04	Test Methods for Rating Motor, Diesel, and Aviation Fuels
Volume 05.05	Gaseous Fuels; Coal and Coke

Section 6: Paints, Related Coatings, and Aromatic

Volume 06.01	Paint—Tests for Formulated Products and Applied Coatings
Volume 06.02	Paint—Pigments, Resins, and Polymers; Cellulose

Volume 06.03 Paint—Fatty Oils and Acids, Solvents, Miscellaneous; Aromatic Hydrocarbons

Section 7: Textiles

Volume 07.01 Textiles—Yarns, Fabrics, and General Test Methods

Volume 07.02 Textiles—Fibers, Zippers

Section 8: Plastics

Volume 08.01 Plastics (I): C 177-D 1600

Volume 08.02 Plastics (II): D 1601-D 3099

Volume 08.03 Plastics (III): D 3100-latest

Volume 08.04 Plastic Pipe and Building Products

Section 9: Rubber

Volume 09.01 Rubber, Natural, and Synthetic—General Test Methods; Carbon Black

Volume 09.02 Rubber Products, Industrial—Specifications and Related Test Methods; Gaskets; Tires

Section 10: Electrical Insulation and Electronics

Volume 10.01 Electrical Insulation (I) Solids, Composites, and Coatings

Volume 10.02 Electrical Insulation (II) Wire and Cable, Heating and Electrical Tests

Volume 10.03 Electrical Insulating Liquids and Gas; Electrical Protective Equipment

Volume 10.04 Electronics (I)

Volume 10.05 Electronics (II)

Section 11: Water and Environmental Technology

Volume 11.01 Water (I)

Volume 11.02 Water (II)

Volume 11.03 Atmospheric Analysis; Occupational Health and Safety

Volume 11.04 Pesticides; Resource Recovery; Hazardous Substances and Oil Spill Responses; Waste Disposal; Biological Effects

Section 12: Nuclear, Solar, and Geothermal Energy

Volume 12.01 Nuclear Energy (I)

Volume 12.02	Nuclear (II), Solar, and Geothermal Energy

Section 13: Medical Devices

Volume 13.01	Medical Devices

Section 14: General Methods and Instrumentation

Volume 14.01	Analytical Methods—Spectroscopy; Chromatography; Computerized Systems
Volume 14.02	General Test Methods, Nonmetal; Laboratory Apparatus; Statistics Methods; Appearance of Materials; Durability of Nonmetallic Materials
Volume 14.03	Temperature Measurement

Section 15: General Products, Chemical Specialties, and End Use Products

Volume 15.01	Refractures; Carbon and Graphite Products; Activated Carbon
Volume 15.02	Glass; Ceramic Whitewares
Volume 15.03	Space Simulation; Aerospace and Aircraft; High Modulus Fibers and Composites
Volume 15.04	Soap; Polishes; Leather; Resilient Floor Coverings
Volume 15.05	Engine Coolants; Halogenated Organic Solvents; Industrial Chemicals
Volume 15.06	Adhesives
Volume 15.07	End Use Products
Volume 15.08	Fasteners
Volume 15.09	Paper; Packaging; Flexible Barrier Materials; Business Copy Products

Section 00: Index

Volume 00.01	Subject and Alphanumeric Index

AMERICAN GAS ASSOCIATION

The following publications of the American Gas Association (AGA) are of interest to people associated with the design, construction, operation, and maintenance of gas systems piping.

Z223.1-88	National Fuel Gas Code

Z22.3-88 National Fuel Gas Code Handbook

AMERICAN PETROLEUM INSTITUTE

The American Petroleum Institute (API) publishes specifications (Spec.), bulletins (Bull.), recommended practices (RP), standards (Std.), and other publications (Publ.) as an aid to procurement of standardized equipment and materials. These publications are primarily intended for use by the petroleum industry. However, they can be and are used by others in that they are referenced in a code or invoked in the purchased order/specification governing the design and construction of piping systems. For example API Specification 5L and the API Standard 605 are referenced in ASME B31.1, Power Piping Code.

The following documents, which relate to piping, are published by the API.

Specifications (Spec.)

Spec. 2B-90	Specification for the Fabrication of Structural Steel Pipe; Fourth Edition
Spec. 5AR-81	Specification for Reinforced Thermosetting Resin Casing and Tubing
Spec. 5L-90	Specification for Line Pipe; Thirty-Eighth Edition
Spec. 5LC-88	Specification for CRA Line Pipe; First Edition
Spec. 6D-91	Specification for Pipeline Valves (Gate, Plug, Ball, and Check Valves); Twentieth Edition
Spec. 6FA-85	Specification for Fire Test for Valves; First Edition
Spec. 6FC-89	Specification for Fire Test for Valves with Selective Backseats; First Edition
Spec. 7J-85	Specification for Drill Pipe/Casing Protectors
Spec. 15AR-87	Specification for Fiberglass Casing and Tubing; Third Edition
Spec. 15HR-88	Specification for High Pressure Fiberglass Line Pipe; First Edition
Spec. 15LE-87	Specification for Polyethylene Line Pipe (PE); Second Edition
Spec. 15LP-87	Specification for Thermoplastic Line Pipe (PVC and CPVC); Sixth Edition
Spec. 15LR-90	Specification for Low Pressure Fiberglass Line Pipe; Sixth Edition

Bulletins (Bull.)

Bull. 5A2-88	Bulletin on Thread Compounds for Casing, Tubing, and Line Pipe; Sixth Edition
Bull. 5C3-89	Bulletin on Formulas and Calculations for Casing, Tubing, Drill Pipe, and Line Pipe Properties; Fifth Edition

Bull. 6AF-89	Bulletin on Capabilities of API Flanges Under Combinations of Load; First Edition
Bull. 6F2-87	Bulletin on Fire Resistance Improvements for API Flanges

Recommended Practices (RP)

RP 5A5-89	Recommended Practice for Field Inspection of New Casing, Tubing, and Plain-End Drill Pipe; Fourth Edition
RP 5B1-88	Recommended Practice for Gauging and Inspection of Casing, Tubing, and Line Pipe Threads; Third Edition
RP 5L1-90	Recommended Practice for Railroad Transportation of Line Pipe; Fourth Edition
RP 5L2-87	Recommended Practice for Internal Coating of Line Pipe for Non-Corrosive Gas Transmission Service; Third Edition
RP 5L3-78	Recommended Practice for Conducting Drop-Weight Tear Tests on Line Pipe; Second Edition
RP 5L5-75	Recommended Practice for Marine Transportation of Line Pipe; First Edition
RP 5L6-79	Recommended Practice for Transportation of Line Pipe on Inland Waterways; First Edition
RP 5L7-88	Recommended Practices for Unprimed Internal Fusion Bonded Epoxy Coating of Line Pipe; Second Edition
RP 5L8-90	Recommended Practice for Field Inspection of New Line Pipe; First Edition
RP 6G-82	Recommended Practice for Through Flowline (TFL) Pump Down Systems
RP 10E-87	Recommended Practice for Application of Cement Lining to Steel Tubular Goods, Handling, Installation, and Joining
RP 11V7-90	Recommended Practice for Repair, Testing, and Setting Gas Lift Valves; First Edition
RP 15L4-76	Recommended Practice for Care and Use of Reinforced Thermosetting Resin Line Pipe
RP 17B-88	Recommended Practice for Flexible Pipe
RP 37-80	Recommended Practice (Superseded By RP 5C5) for Proof-Test Procedure for Evaluation of High-Pressure Casing and Tubing Connection Designs
RP 520 PT I-90	Recommended Practice for Sizing, Selection, and Installation of Pressure-Relieving Devices in Refineries, Part I—Sizing and Selection; Fifth Edition
RP 520 PT II 88	Recommended Practice Sizing, Selection, and Installation of Pressure-Relieving Devices in Refineries, Part II—Installation; Third Edition
RP 574-90	Inspection of Piping, Tubing, Valves, and Fittings; First Edition (Replaces Guide for Inspection of Refinery Equipment Chapter XI)
RP 1102-81	Recommended Practice for Liquid Petroleum Pipelines Crossing Railroads and Highways

RP 1107-78	Recommended Pipe Line Maintenance Welding Practices
RP 1109-71	Recommended Practice for Marking Liquid Petroleum Pipeline Facilities
RP 1110-81	Recommended Practice for Pressure Testing of Liquid Petroleum Pipelines

Standards (Std.)

Std. 5B-88	Specification for Threading, Gauging, and Thread Inspection of Casing, Tubing, and Line Pipe Threads; Thirteenth Edition
Std. 526-84	Flanged Steel Safety-Relief Valves; Third Edition
Std. 527-78	Commercial Seat Tightness of Safety Relief Valves with Metal-to-Metal Seats; Second Edition
Std. 594-82	Wafer Check Valves; Third Edition
Std. 595-79	Cast-Iron Gate Valves, Flanged Ends
Std. 597-81	Steel Venturi Gate Valves, Flanged and Buttwelding Ends (Withdrawn)
Std; 598-82	Valve Inspection and Test; Fifth Edition
Std. 599-88	Steel and Ductile Iron Plug Valves; Third Edition
Std. 600-81	Steel Gate Valves, Flanged and Butt-Welding Ends; Eighth Edition
Std. 601-88	Metallic Gaskets for Raised-Face Pipe Flanges and Flanged Connections
Std. 602-85	Compact Steel Gate Valves; Fifth Edition
Std. 603-84	Class 150, Cast, Corrosion-Resistant, Flanged-End Gate Valves
Std. 604-81	Ductile Iron Gate Valves, Flanged Ends
Std. 605-88	Large-Diameter Carbon Steel Flanges (Nominal Pipe Sizes 26 through 60; Classes 75, 150, 300, 400, 600, and 900); Fourth Edition
Std. 606-89	Compact Steel Gate Valves—Extended Body; Third Edition
Std. 607-85	Fire Test for Soft-Seated Quarter Turn Valves; Third Edition
Std. 608-89	Metal Ball Valves—Flanged and Butt-Welding Ends; First Edition
Std. 609-83	Butterfly Valves, Lug-Type and Wafer Type; Third Edition
Std. 1104-88	Welding of Pipelines and Related Facilities; Seventeenth Edition

Publications (Publ.)

Publ. 1113-86	Pipeline Supervisory Control Checklist

AMERICAN WATER WORKS ASSOCIATION

The American Water Works Association (AWWA) publishes standards that cover requirements for pipe and piping components used in water treatment and distribution systems, including specialty items such as fire hydrants. They also publish several AWWA manuals relative to design, installation, operation, management, and training. The AWWA standards are used for design, fabrication, and installation of large diameter piping for water systems not covered by ASME Boiler and Pressure Vessel Code, ASME B31, Code for Pressure Piping, and other codes. Conformance to AWWA standards is required either by being referenced in the Codes governing the construction of water systems piping or by the enforcement authorities having jurisdiction over the water systems piping.

Refer to Table C1.2 of Chapter C1, Part C of this handbook for a comprehensive listing of AWWA publications and standards dealing with piping.

AMERICAN WELDING SOCIETY

The American Welding Society (AWS) publishes handbooks, manuals, guides, recommended practices, specifications, and codes. The specifications for filler metals are in the AWS A5 series. The filler metal specifications are usually cited in design documents. The welding procedures are in the D10 series. The AWS handbook is published in five volumes and is intended to be an aid to the user and producer of welded products.

The following is a list of AWS publications directly related to piping. The AWS A5 series filler metal specifications are not included in the list since they can be used for a multitude of items other than piping.

AWS Welding Handbook

Volume 1	Fundamentals of Welding
Volume 2	Welding Processes
Volume 3	Welding Processes
Volume 4	Engineering Applications—Materials
Volume 5	Engineering Applications—Design Brazing Manual, Soldering Manual
AWS A3.0	Welding Terms and Definitions, including Terms for Brazing, Soldering, Thermal Spraying, and Thermal Cutting
AWS A5.01	Filler Metal Procurement Guidelines
AWS D10.4	Recommended Practices for Welding Austenitic Chromium-Nickel Stainless Steel Piping and Tubing
AWS D10.6	Recommended Practices for Gas Tungsten Arc Welding of Titanium Pipe and Tubing
AWS D10.7	Recommended Practices for Gas Shielded Arc Welding of Aluminum and Aluminum Alloy Pipe
AWS D10.8	Recommended Practices for Welding of Chromium—Molybdenum Steel Piping and Tubing
AWS D10.9	Specification for Qualification of Welding Procedures and Welders for Piping and Tubing
AWS D10.10	Recommended Practices for Local Heating of Welds in Piping and Tubing

AWS D10.11 Recommended Practices for Root Pass Welding of Pipe Without Backing

AWS D10.12 Recommended Practices and Procedures for Welding Low Carbon Steel Pipe

AIR-CONDITIONING AND REFRIGERATION INSTITUTE

The Air-Conditioning and Refrigeration Institute (ARI) publishes standards, guidelines, and directories of certification. Some of these standards and guidelines, listed below, may be used for design and construction of refrigeration piping systems.

Standards

720-88 Refrigerant Access Valves and Hose Connectors

760-87 Solenoid Valves for Use with Volatile Refrigerants

770-84 Refrigerant Pressure Regulating Valves

Guidelines

C-88 Guideline for ARI Recommended Dimensions of Steel Solder/Braze Fittings

AMERICAN SOCIETY OF HEATING REFRIGERATING AND AIR-CONDITIONING ENGINEERS

The following standards, guidelines, and handbooks published by the American Society of Heating Refrigeration and Air-Conditioning Engineers (ASHRAE) relate to piping:

Standards

41.3-89 Standard Method for Pressure Measurement

41.4-84 Standard Method for Measurement of Proportion of Oil in Liquid Refrigerant

41.6-82 Standard Method for Measurement of Moist Air Properties

41.7-84 Standard Method for Measurement of Flow of Gas

41.8-89 Standard Methods of Measurement of Flow of Liquids in Pipes Using Orifice Flowmeters

Guidelines

1-89 Guidelines for Commissioning of HVAC Systems

2-86 Guidelines for Engineering Analysis of Experimental Data

Handbooks

1988 ASHRAE Handbook Equipment

1989 ASHRAE Handbook Fundamentals I-P Edition

1989 ASHRAE Handbook Fundamentals SI Edition

1987 ASHRAE Handbook HVAC Systems and Applications

1990 ASHRAE Handbook Refrigeration Systems and Applications I-P Edition

1990 ASHRAE Handbook Refrigeration Systems and Applications SI Edition

AMERICAN SOCIETY OF SANITARY ENGINEERS

The American Society of Sanitary Engineers (ASSE) publishes many standards, some of which are ANSI approved. The following is a list of standards which contain requirements related to sanitary piping.

Standards

1001-88 Performance Requirements for Pipe Applied Atmospheric Type Vacuum Breakers

1003-81 Performance Requirements for Water Pressure Reducing Valves for Domestic Water Supply Systems

1029-81 Performance Requirements for Water Supply Valves; Mixing Valves and Single Control Mixing Valves

1032-80 Performance Requirements for Dual Check Valve Type Backflow Preventers for Carbonated Beverage Dispensers

1037-90 Performance Requirements for Pressurized Flushing Devices (Flushometers) for Plumbing Fixtures

1045-87 Performance Standard and Installation Procedures for Aluminum Drain, Waste, and Vent Pipe with End Cap Components

AMERICAN SOCIETY OF CIVIL ENGINEERS

The following documents, which are published by the American Society of Civil Engineers (ASCE), contain information related to piping. The contents of these documents can be used in design and construction of appropriate piping systems.

Design and Construction of Sanitary and Storm Sewers, 1969 Edition
Design and Operation of Pipeline Control Systems, 1984 Edition
Glossary: Water and Wastewater Control Engineering, 1969 Edition
Pipeline Materials and Design, 1984 Edition
Report on Pipeline Location, 1965 Edition
Guidelines for Seismic Design of Oil and Gas Pipeline Systems, 1984 Edition
Seismic Response of Buried Pipes and Structural Components, 1983 Edition

AMERICAN SOCIETY FOR NONDESTRUCTIVE TESTING

The American Society for Nondestructive Testing (ASNT) publishes recommended practices concerning procedures, equipment, and qualification of personnel for nondestructive testing. The following practice is referred in several codes and standards which contain requirements for piping:

SNT-TC-1A Recommended Practice for Nondestructive Testing Personnel Qualification

AMERICAN IRON AND STEEL INSTITUTE

The following publication of the American Iron and Steel Institute (AISI) provides design guidelines for use of stainless steel in piping systems.

SS910-80 Design Guidelines for Stainless Steel in Piping Systems

AMERICAN NUCLEAR SOCIETY

The following American Nuclear Society (ANS) standards contain requirements for nuclear power plant piping systems:

51.7-76 Single Failure Criteria for PWR Fluid Systems
51.10-79 Auxiliary Feedwater System for Pressurized Water Reactor
56.2-84 Containment Isolation Provisions for Fluid Systems after a LOCA
56.3-77 Overpressure Protection of Low Pressure Systems Connected to the Reactor Coolant Pressure Boundary
56.4-83 Pressure and Temperature Transient Analysis for Light Water Reactor Containment
58.2-88 Design Basis for Protection of Light Water Nuclear Power Plants Against Effects of Postulated Pipe Rupture
59.51-89 Fuel Oil Systems for Emergency Diesel Generators

BUILDING OFFICIALS CONFERENCE OF AMERICA

The Building Officials Conference of America (BOCA) publishes a series of national codes, manuals, training aids, and other documents which contain technical requirements and other information related to piping. The following is a partial list of these publications:

National Codes

National Building Code
National Mechanical Code
National Plumbing Code
National Private Sewage Disposal Code
National Fire Prevention Code
National Energy Conservation Code

Manuals

BOCA National Code Interpretations
Fire Protection Systems Workbook
Plumbing Materials and Sizing Selector

DUCTILE IRON PIPE RESEARCH ASSOCIATION

The following documents published by the Ductile Iron Pipe Research Association contain guidelines, requirements, and other technical information related to ductile iron piping systems:

Pipe Material Comparison Booklet
Ductile Iron Pipe Characteristics/Applications
Design of Ductile Iron Pipe on Supports
Ductile Iron Pipe for Wastewater Applications
Cement-Mortar Linings for Ductile Iron Pipe
Direct Tapping Comparison Study
Ductile Iron Pipe Energy Savings
Ductile Iron Pipe in Deep Trench Installations
Ductile Iron Pipe Installation Guide

Ductile Iron Pipe Subaqueous Crossings
Concentrated Consulting Engineering Programs
Polyethylene Encasements
Corrosion Control Seminars
Ductile Iron Pipe Thrust Restraint Design, 1986 Edition
Inside Diameter/Velocity and Headloss/Pumping Costs

EXPANSION JOINT MANUFACTURERS ASSOCIATION

The Expansion Joint Manufacturers Association (EJMA) publishes a handbook called the *Standards of the Expansion Joint Manufacturers Association.* The book contains manufacturing standard practices as well as comprehensive and detailed engineering data concerning pipe expansion joint types, installation layouts and locations, movements, forces, moments, cycle-life expectancy, effects of corrosion, erosion, and testing.

FLUID CONTROLS INSTITUTE

The Fluid Controls Institute (FCI) publishes voluntary standards that have been developed by consensus of the member companies. The following is a list of the FCI publications:

Steam Trap Classification and Operating Principles, 1987 Edition
Steam Trap Pressure Rating Standards, 1977 Edition
Filled Thermal System Terminology/Definitions, 1970 Edition
Control Valve Seat Leakage Standards, 1982 Edition
Spring-Diaphragm Actuated Control Valve Power Signal Standard
Temperature Regulators Pressure Ratings Standard, 1984 Edition
Pressure-Reducing Regulators Pressure Rating Standard, 1984 Edition
Standard Terminology for Regulators, 1978 Edition
"Y" Type Strainers Pressure Rating Standard, 1973 Edition
Pipeline Strainers Pressure Rating Standard, 1978 Edition
Spring Loaded Lift Disc Check Valve Standard, 1980 Edition
Solenoid Valves Pressure/Flow Rating Process Gas, 1978 Edition
Solenoid Valves Pressure/Flow Rating Process Liquid, 1978 Edition
Solenoid Valves Test Conditions/Procedures, 1979 Edition
Testing Water Hammer Characteristics of Valves, 1982 Edition

Valve Flow Coefficient CV Metric Definition, 1985 Edition
Steam Traps Production/Performance Tests, 1985 Edition
Regulator Terminology, 1986 Edition

FACTORY MUTUAL ENGINEERING & RESEARCH CORPORATION

The Factory Mutual Engineering & Research Corporation is usually referred to as Factory Mutual (FM). FM performs examinations of equipment, materials, and services before listing them as "approved" in the guide *Approval Guide to Equipment, Materials and Services.* The following publications of FM may be of interest to people involved with fire protection systems' piping:

Approval Guide to Equipment, Materials and Services
Automatic Sprinkler Systems
Fixed Extinguishing Systems
Oil Safety Shutoff Valves
Gas Safety Shutoff Valves
Supplemental Data
Property Loss Control Catalog

FLUID SEALING ASSOCIATION

The Fluid Sealing Association publishes various documents related to gaskets and seals used in mechanical joints to maintain leak-tightness of the fluid piping and ducting systems. These documents include the following:

Ducting Systems Technical Handbook
Non-Metallic Gasket Handbook
Compression Packings Handbook
Molded Packings Handbook
Rubber Expansion Joints/Flexible Pipe Connectors
Mechanical Seal Handbook
Glossary of Terms

HEAT EXCHANGE INSTITUTE

The Heat Exchange Institute (HEI) publishes voluntary standards developed to express the consensus of member companies concerned with the fabrication of heat exchangers and similar equipment. The following are typical of the standards published by this institute:

Standards for Steam Jet Ejectors
Standards for Steam Surface Condensers
Standards for Field Testing, Addendum Standards for Steam Jet Ejectors
General Construction Standards for Ejector Components Other Than Ejector Condensers
Construction Standards for Surface Type Condensers for Ejector Service
Code for the Measurement of Sound from Steam Jet Ejectors
Standards for Direct Contact Barometric and Low Level Condensers
Method and Procedure for the Determination of Dissolved Oxygen
Standard for Power Plant Heat Exchangers

HYDRAULIC INSTITUTE

The following is the list of the Hydraulic Institute (HI) publications:

Hydraulic Institute Standards, Fourteenth Edition
Engineering Data Book

The information previously contained in the *Pipe Friction Manual* has been incorporated into the *Engineering Data Book*; the *Pipe Friction Manual* is no longer available.

INSTITUTE OF ELECTRICAL AND ELECTRONICS ENGINEERS

The following standards published by the IEEE are of interest to people involved in the design and construction of nuclear power plant piping systems:

IEEE 323-83 Standard for Qualifying Class 1E Equipment for Nuclear Power Generating Stations

IEEE 336-85 Standard Installation, Inspection, and Testing Requirements for Power, Instrumentation, and Control Equipment at Nuclear Facilities

IEEE 344-87 Recommended Practice for Seismic Qualification of Class 1E Equipment for Nuclear Power Generating Stations

IEEE 352-87 Guide for General Principles of Reliability Analysis of Nuclear Power Generating Station Safety Systems

IEEE 379-88 Standard Application of the Single Failure Criterion to Nuclear Power Generating Station Safety Systems

IEEE 382-85 Standard for Qualification of Actuators for Power Operated

Valve Assemblies with Safety-Related Functions for Nuclear Power Plants

INSTRUMENT SOCIETY OF AMERICA

The Instrument Society of America (ISA) develops and publishes periodicals, books, standards, recommended practices, monographs, references, and training aids pertaining to instruments and automated controls. The following publications of the ISA contain information related to piping:

Recommended Practices (RP)

RP 7.1-56	Recommended Practice for Pneumatic Control Circuit Pressure
RP 7.7-84	Recommended Practice for Producing Quality Instrument Air
RP 16.5-61	Recommended Practice for Installation, Operation, Maintenance Instructions for Glass Tube Variable Area Meters (Rotometers)
RP 31.1-72	Recommended Practice for Specification, Installation, and Calibration of Turbine Flowmeters
RP 42.1-82	Recommended Practice for Nomenclature for Instrument Tube Fittings
RP 60.9-81	Recommended Practice for Piping Guide for Control Centers
RP 75.06-81	Recommended Practice for Control Valve Manifold Designs
RP 75.18-89	Recommended Practice for Control Valve Position Stability
RP 75.21-89	Recommended Practice for Process Data Presentation for Control Valves

Standards

S 5.1-84	Instrumentation Symbols and Identification
S 5.2-76	Binary Logic Diagrams for Process Operations
S 5.3-83	Graphic Symbols for Distributed Control/Shared Display Instrumentation, Logic, and Computer
S 5.4-89	Standard Instrument Loop Diagrams
S 5.5-85	Graphic Symbols for Process Displays
S 7.3-75	Quality Standard for Instrument Air
S 7.4-81	Air Pressures for Pneumatic Controllers, Transmitters, and Transmission Systems

S 12.4-70	Instrument Purging for Reduction of Hazardous Area Classification
S 18.1-79	Annunciator Sequences and Specifications (R 1985)
S 20-81	Specification Forms for Process Measurement and Control Instruments, Primary Elements, and Control Valves
S 26-68	Dynamic Response Testing of Process Control Instrumentation
S 37.3-75	Strain Gauge Pressure Transducers, Specifications, and Tests (R 1982)
S 37.5-75	Strain Gauge Linear Acceleration Transducers, Specifications, and Tests (R 1982)
S 37.6-76	Specifications and Tests of Potentiometric Pressure Transducers (R 1982)
S 37.8-77	Specifications and Tests for Strain Gauge Force Transducers (R 1982)
S 51.1-79	Process Instrumentation Terminology
S 67.01-79	Transducer and Transmitter Installation for Nuclear Safety Applications (R 1987)
S 67.02-80	Nuclear-Safety-Related Instrument Sensing Line Piping and Tubing Standards for Use in Nuclear Power Plants
S 67.03-82	Light Water Reactor Coolant Pressure Boundary Leak Detection
S 67.04-88	Setpoints for Nuclear Safety-Related Instrumentation
S 67.10-86	Sample-Line Piping and Tubing Standard for Use in Nuclear Power Plants
S 75.01-85	Flow Equations for Sizing Control Valves
S 75.02-88	Control Valve Capacity Test Procedure
S 75.03-84	Face-to-Face Dimensions for Flanged Globe-Style Control Valve Bodies (ANSI Classes 125, 150, 250, 300, and 600)
S 75.04-84	Face-to-Face Dimensions for Flangeless Control Valves (ANSI Classes 150, 300, and 600)
S 75.05-83	Control Valve Terminology
S 75.07-87	Laboratory Measurement of Aerodynamic Noise Generated by Control Valves
S 75.08-85	Installed Face-to-Face Dimensions for Flanged Clamp or Pinch Valves
S 75.11-84	Inherent Flow Characteristic and Rangeability of Control Valves
S 75.12-87	Face-to-Face Dimensions for Socket Weld-End and Screwed-End Globe-Style Control Valves (ANSI Classes 150, 300, 600, 900, 1500, and 2500)
S 75.14-84	Face-to-Face Dimensions for Butt-Weld-End Globe Style Control Valves (ANSI Classes 4500)
S 75.15-86	Face-to-Face Dimensions for Butt-Weld-End Globe-Style Control Valves (ANSI Classes 150, 300, 600, 900, 1500, and 2500)

S 75.16-86 Face-to-Face Dimensions for Flanged Globe-Style Control Valve Bodies (ANSI Classes 900, 1500, and 2500)

S 75.17-89 Control Valve Aerodynamic Noise Prediction Standard

S 75.19-89 Hydrostatic Testing of Control Valves (Formerly ASME/ANSI B16.37-80)

Handbook

ISA Handbook of Control Valves

MANUFACTURERS STANDARDIZATION SOCIETY OF THE VALVE AND FITTINGS INDUSTRY

The Manufacturers Standardization Society (MSS) publishes Standard Practices (SP) which provide a basis for common practice by the manufacturers, the user, and the general public. Compliance to the Standard Practices of MSS is required by reference in a code, specification, sales contract, law or regulation. The MSS is also represented on the committees of other standardization groups, such as ANSI and ASME. Many of the ANSI B16 series standards were originally developed as MSS Standard Practices. Once a Standard Practice is adopted as ANSI standard, it is discontinued as an MSS Standard Practice.

The following is a complete list of MSS Standard Practices published and in use at the present time:

Standards Practices (SP)

SP-6-90 Standard Finishes for Contact Faces of Pipe Flanges and Connecting-End Flanges of Valves and Fittings

SP-9-87 Spot Facing for Bronze, Iron, and Steel Flanges

SP-25-78 Standard Marking System for Valves, Fittings, Flanges, and Unions (R 1988)

SP-42-90 Class 150 Corrosion Resistant Gate, Globe, Angle, and Check Valves with Flanged and Butt-Weld Ends

SP-43-82 Wrought Stainless Steel Butt-Welding Fittings, Including Reference to Other Corrosion Resistant Materials (R 1986)

SP-44-90 Steel Pipe Line Flanges

SP-45-82 By-Pass and Drain Connection Standard (R 1987)

SP-51-86 Class 150LW Corrosion Resistant Cast Flanges and Flanged Fittings

SP-53-85 Quality Standard for Steel Castings and Forgings for Valves, Flanges, and Fittings and Other Piping Components Magnetic Particle Examination Method (R 1990)

SP-54-85	Quality Standard for Steel Casting for Valves, Flanges, and Fittings and Other Piping Components, Radiographic Examination Method (R 1990)
SP-55-85	Quality Standard for Steel Castings for Valves, Flanges, and Fittings and Other Piping Components, Visual Method (R 1990)
SP-58-88	Pipe Hangers and Supports—Materials, Design, and Manufacture
SP-60-82	Connecting Flange Joint Between Tapping Sleeves and Tapping Valves (R 1986)
SP-61-85	Pressure Testing of Steel Valves
SP-65-90	High Pressure Chemical Industry Flanges and Threaded Stubs for Use with Lens Gaskets
SP-67-90	Butterfly Valves
SP-68-88	High Pressure-Offset Seat Butterfly Valves
SP-69-83	Pipe Hangers and Supports—Selection and Application
SP-70-84	Cast Iron Gate Valves, Flanged, and Threaded Ends
SP-71-84	Cast Iron Swing Check Valves, Flanged and Threaded Ends
SP-72-87	Ball Valves with Flanged or Butt-Welding Ends for General Service
SP-73-86	Brazing Joints for Wrought and Cast Copper Alloy Solder Joint Pressure Fittings
SP-75-88	Specification for High Test Wrought Butt-Welding Fittings
SP-77-84	Guidelines for Pipe Support Contractual Relationships and Responsibilities of the Pipe Hanger Contractor with the Purchaser's Engineer or the Pipe Fabricator and/or Erector (R 1990)
SP-78-87	Cast Iron Plug Valves, Flanged and Threaded Ends
SP-79-89	Socket-Welding Reducer Inserts
SP-80-87	Bronze Gate, Globe, Angle, and Check Valves
SP-81-81	Stainless Steel, Bonnetless Flanged Knife Gate Valves (R 1986)
SP-82-76	Valve Pressure Testing Methods (R 1986)
SP-83-87	Steel Pipe Unions Socket-Welding and Threaded
SP-84-90	Valves—Socket Welding and Threaded Ends
SP-85-85	Cast Iron Globe and Angle Valves Flanged and Threaded Ends
SP-86-87	Guidelines for Metric Data in Standards for Valves, Flanges, Fittings, and Actuators
SP-87-82	Factory-Made Butt-Welding Fittings for Class 1 Nuclear Piping Applications (R 1986)
SP-88-83	Diaphragm Type Valves (R 1988)
SP-89-85	Pipe Hangers and Supports—Fabrication and Installation Practices
SP-90-86	Guidelines on Terminology for Pipe Hangers and Supports
SP-91-84	Guidelines for Manual Operation of Valves
SP-92-87	Valve User Guide

SP-93-87	Quality Standard for Steel Castings and Forgings for Valves, Flanges, and Fittings and Other Piping Components Liquid Penetrant Examination Method
SP-94-83	Quality Standard for Ferritic and Martensitic Steel Castings for Valves, Flanges, and Fittings, and Other Piping Components Ultrasonic Examination Method (R 1987)
SP-95-86	Swage(d) Nipples and Bull Plugs
SP-96-86	Guidelines on Terminology for Valves and Fittings; Addendum May 1990
SP-97-87	Forged Carbon Steel Branch Outlet Fittings—Socket Welding, Threaded, and Butt-Welding Ends
SP-98-87	Protective Epoxy Coatings for the Interior of Valves and Hydrants
SP-99-89	Instrument Valves
SP-100-88	Qualification Requirements for Elastomer Diaphragms for Nuclear Service Diaphragm Type Valves
SP-101-89	Part-Turn Valve Actuator Attachment Flange and Driving Component Dimensions and Performance Characteristics
SP-102-89	Multi-Turn Valve Actuator Attachment Flange and Driving Component Dimensions and Performance Characteristics
SP-103-90	Wrought Copper and Copper Alloy Insert Fittings for Polybutylene Systems
SP-104-90	Wrought Copper LW Solder Joint Pressure Fittings

NATIONAL FIRE PROTECTION ASSOCIATION

The National Fire Protection Association (NFPA) is a voluntary association of members representing all aspects of fire protection, such as professional societies, educational institutions, public officials, insurance companies, equipment manufacturers, builders and contractors, and transportation groups. The NFPA publishes codes, standards, guides, and recommended practices in a 12-volume set of books called the *National Fire Codes.* Conformance to the *National Fire Codes* may be required by federal, state, and local laws and regulations. Sometimes insurance companies may leave no choice for the owner/user of the facility but to comply with the fire protection and prevention requirements of the applicable National Fire Codes.

Volumes 1 through 8 contain actual text of the National Fire Codes and Standards. The requirements contained in these volumes have been judged suitable for legal adoption and enforcement. Volumes 9 through 11 contain Recommended Practices and Guides considered to be good engineering practices. Volume 12 contains Formal Interpretations, Tentative Interim Amendments, and Errata that relate to the documents in Volumes 1 through 11.

Here is a list of NFPA publications. For specific NFPA Codes and Standards related to fire protection systems piping, refer to Chapter C2, Part C of this handbook.

National Fire Protection Association Publications

Technical Committee Documentation and Reports
Automatic Sprinkler Systems Handbook, Fourth Edition
Flammable and Combustible Liquids Code Handbook, Third Edition
Fire Litigation Handbook
Fire Protection Guide on Hazardous Materials
Fire Protection Handbook, Sixteenth Edition
Liquefied Petroleum Gases Handbook, Second Edition
Life Safety Code Handbook, Fourth Edition
National Electrical Code
National Fuel Gas Code Handbook, First Edition
National Fire Codes and Standards, Volumes 1 through 12

PIPE FABRICATION INSTITUTE

The Pipe Fabrication Institute (PFI) publishes advisory Engineering Standards (ES) and Technical Bulletins (TB) intended to serve the needs of the pipe-fabricating industry at the design level and in actual shop operations. The PFI standards contain minimum requirements; however, the designer or fabricator may consider specifying additional requirements beyond the scope of PFI publications. The use of PFI standards or bulletins is voluntary. The following provides a complete listing of PFI publications.

Engineering Standards (ES)

ES-1-86 Internal Machining and Solid Machined Backing Rings for Circumferential Butt Welds (R 1988)
ES-2-87 Method of Dimensioning Piping Assemblies (R 1990)
ES-3-81 Fabricating Tolerances (R 1990)
ES-4-85 Hydrostatic Testing of Fabricated Piping (R-1988)
ES-5-87 Cleaning of Fabricated Piping
ES-7-80 Minimum Length and Spacing for Welded Nozzles (R 1988)
ES-11-75 Permanent Marking on Piping Materials (R 1990)
ES-16-85 Access Holes, Bosses, and Plugs for Radiographic Inspection of Pipe Welds (R 1988)
ES-20-85 Wall Thickness Measurement by Ultrasonic Examination (R 1988)
ES-21-85 Internal Machining and Fit-Up of GTAW Root Pass Circumferential Butt Welds (R 1989)
ES-22-90 Recommended Practice for Color Coding of Piping Materials

ES-24-84	Pipe Bending Methods, Tolerances, Process, and Material Requirements (R 1990)
ES-25-88	Random Radiography of Pressure Retaining Girth Butt Welds
ES-26-84	Welded Load Bearing Attachments to Pressure Retaining Piping Materials (R 1990)
ES-27-86	"Visual Examination" the Purpose, Meaning, and Limitation of the Term (R 1989)
ES-29-79	Abrasive Blast Cleaning of Ferritic Piping Materials (R 1984)
ES-30-86	Random Ultrasonic Examination of Butt Welds (R 1989)
ES-31-88	Standard for Protection of Ends of Fabricated Piping Assemblies
ES-32-80	Tool Calibration (R 1988)
ES-33-82	Circumferential Butt Welds in the Arc of Pipe Bends (R 1988)
ES-34-83	Painting of Fabricated Piping (R 1989)
ES-35-84	Nonsymmetrical Bevels and Joint Configurations for Butt Welds (R 1990)
ES-36-88	Branch Reinforcement Work Sheets
TB1-88	Pressure-Temperature Ratings of Seamless Pipe Used in Power Plant Piping Systems
TB3-85	Guidelines Clarifying Relationships and Design Engineering Responsibilities Between Purchasers' Engineers and Pipe Fabricator or Pipe Fabricator Erector (R 1988)

PLASTICS PIPE INSTITUTE

People interested in the application of plastics piping systems may find the following Plastics Pipe Institute (PPI) publications of help:

PPI Handbook of Polyethylene Piping
Engineering Basics of Plastics Piping
Plastic Piping Manual

In addition, the PPI publishes technical reports, technical notes, recommendations, and statements dealing with plastics piping.

STEEL STRUCTURES PAINTING COUNCIL

The Steel Structures Painting Council (SSPC) publishes specifications, which include surface preparation, SP; pretreatment, PT; paint application, PA; and paint and paint systems, PS. These specifications specify practical and economical methods of surface preparation and painting steel structures. They are used to clean and paint piping and other steel equipment. With the exception of paint and paint system specifications, the following are the commonly used SSPC specifications:

Surface Preparation Specifications

SSPC-Vis 1	Pictorial Surface Preparation Standard for Painting Steel Surfaces
SSPC-Vis 2	Standard Method of Evaluating Degree of Rusting on Painted Steel Surfaces
SSPC-SP 1	Solvent Cleaning
SSPC-SP 2	Hand Tool Cleaning
SSPC-SP 3	Power Tool Cleaning
SSPC-SP 4	Flame Cleaning of New Steel
SSPC-SP 5	White Metal Blast Cleaning
SSPC-SP 6	Commercial Blast Cleaning
SSPC-SP 7	Brush-Off Blast Cleaning
SSPC-SP 8	Pickling
SSPC-SP 9	Weathering Followed by Blast Cleaning
SSPC-SP 10	Near-White Blast Cleaning

Pretreatment Specifications (PT)

SSPC-PT 1	Wetting Oil Treatment
SSPC-PT 2	Cold Phosphate Surface Treatment
SSPC-PT 3	Basic Zinc Chromate-Vinyl Butyral Washcoat
SSPC-PT 4	Hot Phosphate Surface Treatment

Paint Application (PA) Guides

SSPC-PA 1	Shop, Field, and Maintenance Painting
SSPC-PA 2	Measurement of Dry Paint Thickness with Magnetic Gauges

TUBULAR EXCHANGER MANUFACTURERS ASSOCIATION

The Tabular Exchanger Manufacturers Association (TEMA) publishes standards for use by manufacturers and users. The 1988 edition of these standards is the latest.

UNDERWRITERS LABORATORIES

The UL is a nonprofit organization that develops specifications and standards directed toward assuring the safety of materials, products, and equipment when used in accordance with the conditions for which they were designed. It also tests items for conformance to these and other nationally recognized standards and publishes lists of items approved as a result of the tests. The NFPA Codes require that items to be used in fire protection and prevention systems be approved and listed. The UL published UL Fire Protection Equipment List (such as a listing of fire loop piping material and equipment manufacturers) is one of the publications normally used by those involved in piping associated with the fire protection systems.

FOREIGN CODES AND STANDARDS

The basic principles of piping design and construction may not differ much from one country to another, but the requirements of country specific codes and standards may vary substantially. Therefore, the personnel involved in the engineering design, construction, operation, and maintenance of piping systems must make sure that the requirements of applicable codes and standards are complied with to ensure the safety of the general public and workers associated with the facility.

The user is advised to verify the latest applicable version/edition of the Code and/or standard before invoking their requirements for any application.

REFERENCES

1. *ASME Boiler and Pressure Vessel Code,* Section I, Power Boilers, 1989 Edition with 1990 Addendum, American Society of Mechanical Engineers, New York.
2. *ASME Boiler and Pressure Vessel Code,* Section II, Material Specifications, 1989 Edition with 1990 Addendum, American Society of Mechanical Engineers, New York.
3. *ASME Boiler and Pressure Vessel Code,* Section III, Division 1, Nuclear Power Plant Components, 1989 Edition with 1990 Addendum, American Society of Mechanical Engineers, New York.
4. Code of Federal Regulations, Title 10, Part 50, Section 50.55a, Codes and Standards, January 1, 1991, Office of the Federal Register National Archives and Records Administration, Washington, D.C.
5. *ASME Boiler and Pressure Vessel Code,* Section V, Nondestructive Examination, 1989 Edition with 1990 Addendum, American Society of Mechanical Engineers, New York.
6. *ASME Boiler and Pressure Vessel Code,* Section VIII, Pressure Vessels, 1989 Edition with 1990 Addendum, American Society of Mechanical Engineers, New York.
7. *ASME B31, Code for Pressure Piping,* Section B31.1, Power Piping, 1989 Edition with Addenda B31.1A-1989, American Society of Mechanical Engineers, New York.
8. *ASME Boiler and Pressure Vessel Code,* Section IX, Welding and Brazing Qualifications, 1989 Edition with 1990 Addendum, American Society of Mechanical Engineers, New York.
9. *ASME Boiler and Pressure Vessel Code,* Section XI, Rules for Inservice Inspection of Nuclear Power Plant Components, 1989 Edition with 1990 Addendum, American Society of Mechanical Engineers, New York.
10. *ASME B31, Code for Pressure Piping,* Section B31.9, Building Services Piping, 1988 Edition, American Society of Mechanical Engineers, New York.

11. *ASME B31, Code for Pressure Piping,* Section B31.3, Chemical Plant and Petroleum Refinery Piping, 1990 Edition, American Society of Mechanical Engineers, New York.
12. *ASME B31, Code for Pressure Piping,* Section B31.4, Liquid Transportation Systems for Hydrocarbons, Liquid Petroleum Gas, Anhydrous Ammonia, and Alcohols, 1989 Edition with Addenda B31.1A-1989, American Society of Mechanical Engineers, New York.
13. *ASME B31, Code for Pressure Piping,* Section B31.5, Refrigeration Piping, 1987 Edition with Addenda B31.5A-1989, American Society of Mechanical Engineers, New York.
14. *ASME B31, Code for Pressure Piping,* Section B31.8, Gas Transmission and Distribution Piping Systems, 1989 Edition with Addenda B31.8A-1990, American Society of Mechanical Engineers, New York.
15. *ASME B31, Code for Pressure Piping,* Section B31.11, Slurry Transportation Piping Systems, 1989 Edition, American Society of Mechanical Engineers, New York.

CHAPTER A5
MANUFACTURING OF METALLIC PIPE

Alfred Lohmeier
Materials Engineer
(Formerly Vice President-Technical)
Sumitomo Corporation of America
New York, New York

DEVELOPMENT OF COMMERCIAL PIPE-MAKING

There have been increasing societal demands for modern structures and facilities and concomitant increased emphasis on safety and reliability of equipment under all operating conditions. Piping manufacturing processes have been developed to provide the quality and reliability commensurate with these demands, together with economically feasible production methods. To meet the more stringent reliability goals, the quality control of the piping manufacturing process from the production of the raw material to the finished product is of significant importance. Driving the need for process quality improvement are the social and economic consequences of equipment failure in critical applications such as power generation, chemical and petroleum production, and transportation.

This chapter considers the methods by which different types of metallic pipe are produced. It also considers the various steel-making processes which are important to the ultimate quality of the manufactured pipe. For a discussion of pipes made of thermoplastic and fiberglass, refer to Part D of this handbook.

Historical Background

The history of pipe manufacturing goes back to the use of hollow wooden logs to provide water for medieval cities. The use of cast iron pipes in England and France became prevalent in the early nineteenth century. The first major cast iron water pipeline was obtained for Philadelphia in 1817 and for New York in 1832. Distribution of gas for gaslights was initiated in England, using sheet iron drawn through a die to a cylindrical shape and welding the edges together. In 1887 the first pipe was made of Bethlehem steel in the United States.

Seamless pipe manufacture was attempted in the mid-nineteenth century by various means; the Mannesmann process was developed in Germany in 1885 and operated commercially in England in 1887. The first seamless pipe mill in the United States was built in 1895.

In the early twentieth century, seamless tubes gained wide acceptance as the Industrial Revolution proceeded with the automobiles, oil refineries, oil pipelines, oil wells, and fossil power generation boilers. At that time, the welded tube had not achieved the reliability of present-day electric resistance welded tubes.

The development of pipe and tube production methods, together with the development of steel alloys capable of withstanding the demanding environmental conditions of temperature, chemistry, pressure, and cyclic thermal and pressure load application have enabled pipe and tube to be used reliably in the most critical applications, ranging from Alaskan pipelines to nuclear power generation plants.

World Tubular Product Production Capability

World production and consumption of iron and steel tubular products makes up almost 14 percent of the worldwide crude steel conversion. World production of steel tubular products, according to United Nations figures, amounted to 65.8 million metric tons (Mt) in 1986. At that time, 21.4 million Mt of seamless and 44.4 million Mt of welded tubular products were produced.

Major Tubular Product Producing Countries

In 1986, the three leading producers of iron and steel tubular products were the Union of Soviet Socialist Republics (20.0 Mt), European Economic Community (13.1 Mt), and Japan (10.5 Mt). The production of iron and steel tubular products will vary at these levels, depending on a wide range of worldwide economic factors such as oil exploration, power generation plant construction, and automotive production. For example, in economic climates where oil prices are low, there is less incentive to drill new oil wells. Consequently, the production of steel pipe for oil drilling casings would be reduced. Similar examples of steel pipe production as a function of economic climate can be seen in the power generation and automotive industries. Total world production of pipe is an integration of the effects of the local national economic climates throughout the world.

FERROUS PIPE-MAKING PROCESSES

Iron-making

The making of steel for ferritic piping begins with the smelting of iron ore found in deposits in the crust of the earth throughout the world in forms such as hematite and magnetite. In preparation for the smelting process, the iron ore may be treated by any of several methods to convert it into a suitable form for introduction into the blast furnaces. One method is sintering, which converts ores into a porous mass called clinkers. Another is smelting, which is performed in a blast furnace. The process involves the chemical reaction of iron ore with limestone, coke, and air under heat reducing the iron ore to iron. The "pig" iron obtained from the blast furnace is used as the basic component in the steel-making process.

Steel-making

Steel for piping can be produced in several ways (Fig. A5.1), depending on the facilities available and the desired characteristic of the steel. Generally, steel requires the removal of carbon from the pig iron to a degree required by the carbon steel properties desired. Alloy steel also requires the addition of alloying elements such as chromium, nickel, manganese, and molybdenum to provide the special properties associated with the alloying element.

Bessemer Converter. The Bessemer method of making steel (due to Sir Henry Bessemer in 1856) consisted of blowing a current of cold air through the molten pig iron, thereby using the oxygen in the air to burn carbon and other impurities from the melt. After burning out the carbon in the pig iron, the exact amount of carbon required for the steel is reintroduced into the heat.

Basic-Oxygen Process. The basic-oxygen process (BOP) is essentially the same as the Bessemer process except that it uses pure oxygen (instead of air) together with burned lime converted from limestone. This process burns out the impurities more quickly and completely and provides for more precise control of the steel chemistry.

Open-Hearth Furnace. The open-hearth furnace is used to produce much of the steel in the United States; however, it is being superseded by the basic oxygen process. Its significant advantage is the ability to use scrap steel as well as pig iron as ferrous stock in producing steel. The open-hearth furnace is a large rect-

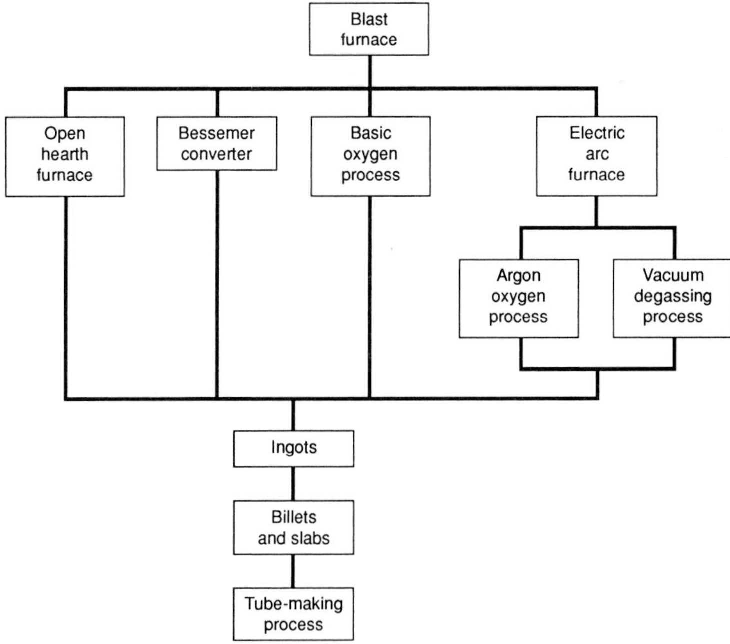

FIGURE A5.1 Piping steel-making processes.

angular brick floor, or hearth, completely covered with a brick structure through which the charge of ferrous stock and limestone is introduced. It is fueled with coke gas, oil, or tar introduced through a burner playing a flame across the hearth while the products of combustion escape through the furnace wall away from the burner. An advantage of the open-hearth process is that testing for carbon content during the heating is possible, allowing adjustments to be made to the feedstock at that time to control the chemistry of the product.

Electric Arc Furnace. The electric arc furnace is a large kettle-shaped chamber lined with fire brick into which a charge of steel scrap with coke is melted by means of heat produced by an electric arc. Since no burning of fuel is required, the oxygen of the steel can be controlled and kept to a minimum. Alloying elements can be added without the fear of oxidation. Because of the control of heat time, temperature, and chemistry, the electric arc furnace is used in the production of high-quality alloy steels.

Argon Oxygen Process. The argon oxygen process (AOD) is used in the production of specialty steels with low carbon and sulfur and high chromium content. A charge of steel of almost the desired properties is introduced into a basic oxygen furnacelike vessel, and controlled amounts of oxygen and argon are introduced into the melt. This reducing process conserves valuable chromium.

Vacuum Degassing Process. When exceptionally high quality steel is required, steel can be "degassed" in a vacuum environment. This vacuum degassing process provides strong reduction in hydrogen, oxygen, nitrogen, inclusions, and contaminants, such as lead, copper, tin, and arsenic.

Ingots, Blooms, and Billets. Ingots, blooms, and billets are the shapes into which the molten metal is solidified before using it in the particular pipe-making (or other) process. An ingot is poured from the molten steel and after solidification goes to the blooming mill to be rolled into square blooms which are further formed onto bar rounds. Alternately, in the case of large pipe, the ingot may be formed into pierced billets to be used in the seamless tube-making process.

Continuous Casting Process. Although the development of the continuous casting process (Fig. A5.2) began in the nineteenth century, it was after World War II that its use became of great commercial interest. In the continuous casting process, molten steel is poured from the melting furnace to a ladle feeding a reservoir called a tundish. The tundish feeds a lubricated mold that has a cooled copper surface, and the solidifying steel is continuously drawn from the mold. In the case of piping steel, the mold is the shape of the billet or slab used in the tube-making process. There are many types of continuous casting processes, ranging from vertical to horizontal, with variations of bent sections in between. This process is now used in more than half the world's steel production. In Japan, 85 percent of the total steel produced is by the continuous casting process.

Pipe- and Tube-Forming Processes

There are basically two types of pipe- and tube-forming processes, namely, seamless and welded. Each process imparts unique properties to the pipe or tube. Seamless pipe or tube does not have the presence of a welded seam along the length of the pipe. This seam has traditionally been believed to be a potential

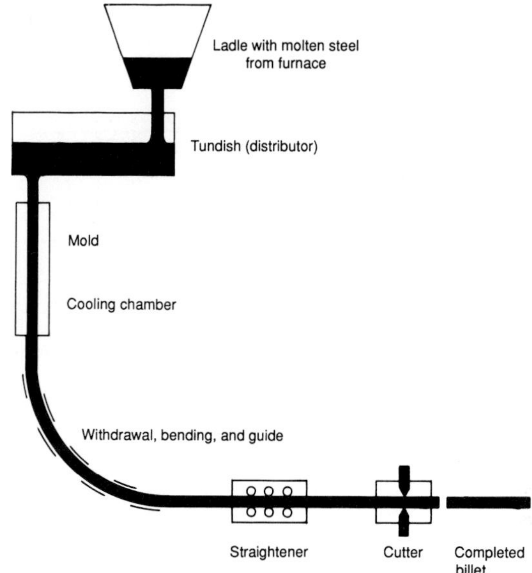

FIGURE A5.2 Continuous casting process.

weakness. The development of automated welding processes and quality control, however, has made this a virtually nonexistent concern. The control of thickness uniformity and concentricity is relatively easy with welded pipe and tube. In general, the seamless pipe is more expensive to produce. The classification of cylindrical tubular products in terms of either pipe or tube is a function of end use. This is discussed further under Tubular Product Classification.

Seamless Pipe. Seamless tube and pipe (Fig. A5.3) are manufactured by first producing a tube hollow which is larger in diameter and thickness than the final tube or pipe. The billet is first pierced by either a rotary (Mannesmann) piercer or by a press piercing method. For tubes of small diameter, the mandril mill process is used. For medium outside diameter tubes of carbon or low alloy steel, the Mannesmann plug mill process is used. Large diameter, heavy wall carbon steel, alloy, and stainless pipe is manufactured by the Erhardt push bench process. High alloy and special shaped pipe are manufactured by the Ugine Sejournet extrusion type process. These processes are performed with the material at hot metal forming temperatures. Further cold processing may or may not be performed to obtain further dimensional accuracy, surface finish, and surface metallurgical structure.

Mandrel (Pilger) Mill Process. In the mandrel (pilger) mill process (Fig. A5.4), a steel billet is heated to forging temperature and placed in between the rolls of a hot rotary piercing mill. A piercing point is placed at the center of the billet, and the rotating rolls are designed to advance the billet over the piercing point, thereby forming a hole through the center of the billet along its entire length as it advances into the tilted roles. A mandrel of outside diameter approximately that of the inside finished pipe diameter is pressed into the pierced hole of the billet. This combination of mandrel and billet is placed between rolls of a pilger-mill

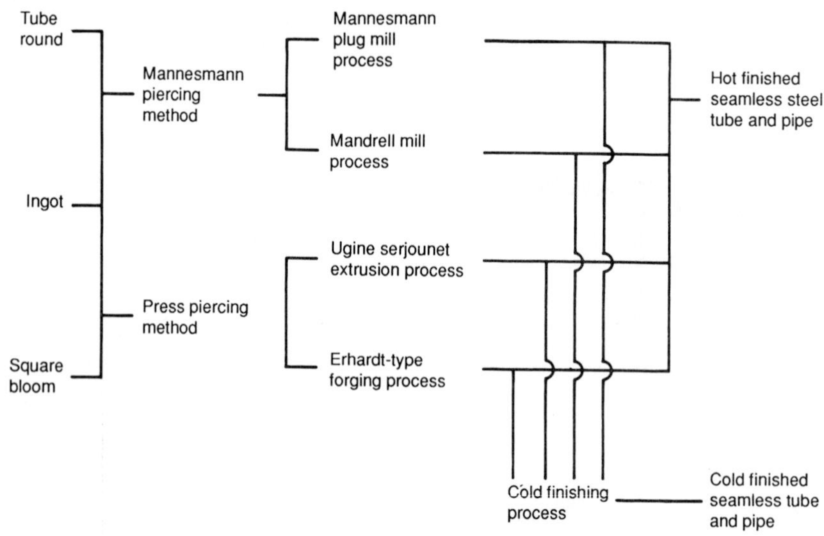

FIGURE A5.3 Seamless tube and pipe manufacturing processes.

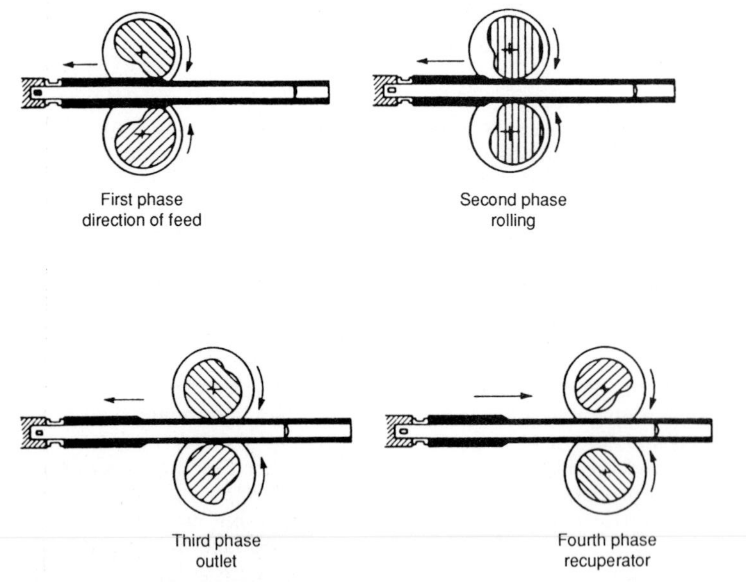

FIGURE A5.4 Mandrel (pilger) mill process.

having a cam-shaped contour revolving counter to the direction in which the billet is being forced by means of a hydraulic and pneumatic ram mechanism.

In the pilger-mill, the rolls first grab the hot billet and after some rotation form a shaft. The pressure of the rolls forces the billet backward, and the resulting tube section is squeezed and smoothed out in the adjacent part of the roll groove. This process is equivalent to forging the billet against the mandrel and driving the billet and mandrel against the ram. After reaching the open portion of the cam shape, the ram mechanism again forces the billet into the rolls. Following the pilger-mill process, the tube is reheated and passed through a reducer or sizer to provide a more uniform diameter. The resulting tube or pipe is called hot finished seamless.

The pilger-mill process is slower than conventional drawing. However, since large reductions in diameter are possible in a single pass, the process is applied to the production of tubes of small diameter such as heat exchangers, fossil fuel boilers, and nuclear steam generators.

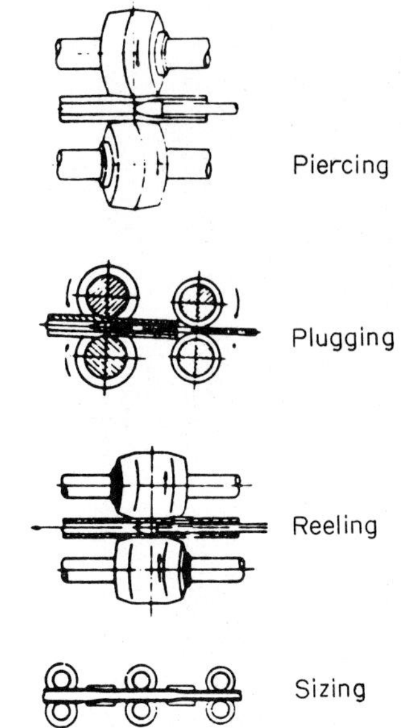

FIGURE A5.5 Mannesmann plug-mill process.

Mannesmann Plug-Mill Process. In the Mannesmann plug-mill process (Fig. A5.5) the billet may be pierced in two hot rotary piercers because of the greater reduction needed for medium size pipe and tube. Following the piercing process, the pierced billet is placed in a plug mill which reduces the diameter by rotating the tube over a mandrel. Following the plug-mill, the tube having some ovality is inserted between the rolls of reelers which provide for dimensional correction and burnish the inside and outside diameters of the tube. Finally, after reheating, the tube reenters a reeler and sizing rollers to provide for greater dimensional uniformity.

The Mannesmann plug-mill process is a standard process for making large quantities of thin wall stainless steel tube or pipe of uniform size and roundness throughout its entire length.

Ugine Sejournet Type Extrusion Processes. The Ugine Sejournet extrusion process (Fig. A5.6) is used for high alloy steel tubes and pipe such as those of stainless steel and specially shaped pipe. A descaled billet, heated to approximately 2300°F, is placed in the vertical press compartment with an extrusion die at its bottom. After applying a hydraulic ram to the billet, a piercing mandrel within the ram punches the billet, producing a cylinder from which the punch piece is ejected through the extrusion die opening. Following this, the ram is activated to apply pressure to the billet, and the billet is extruded through the annulus formed between the piercing mandrel and die cavity. In horizontal presses, piercing is done as a separate operation, or a hollow is used with a mandrel and die. The mandrels and dies are made of high alloy steels containing tungsten, molybdenum, and chromium having Rockwell C hardness values of approximately 46. Powdered glass is the lubricant used in this process.

Forged Seamless Pipe. Forged pipe is used for large diameter (10–30 in) and thickness (1.5–4 in) pipe, where equipment availability and cost for other seamless grades are limiting. There are two processes available for the production of forged seamless pipe, namely, forged and bored pipe and hollow forged pipe.

Forged and Bored Seamless Pipe. In the forged and bored process a billet or ingot heated to forging temperature is elongated by forging in heavy presses or forging hammers to a diameter slightly larger than that of the finished pipe. After

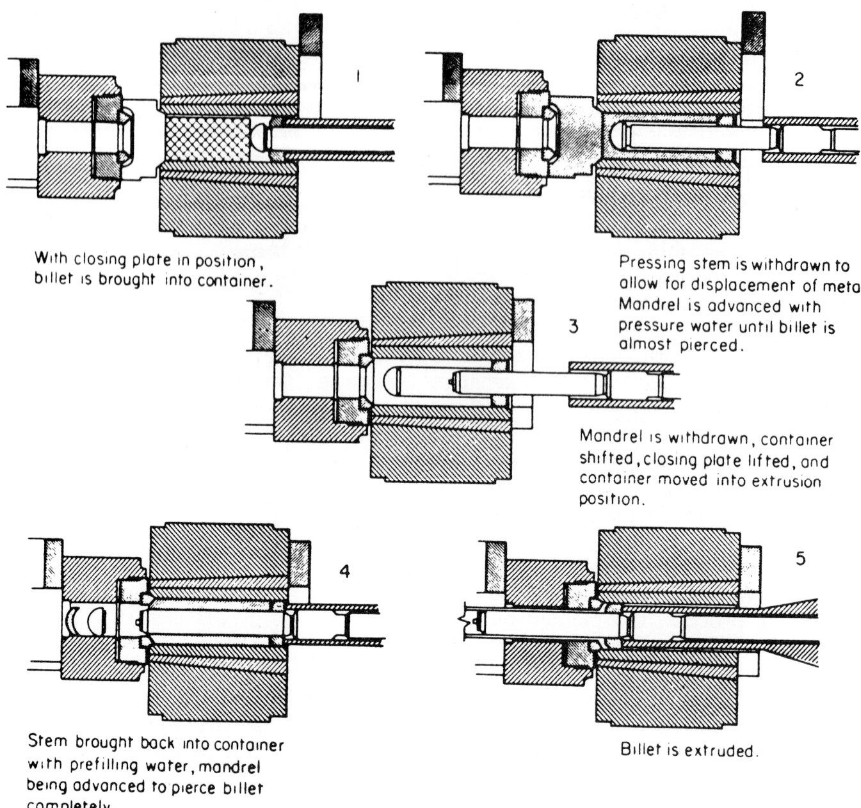

1
2
With closing plate in position, billet is brought into container.
Pressing stem is withdrawn to allow for displacement of metal. Mandrel is advanced with pressure water until billet is almost pierced.
3
Mandrel is withdrawn, container shifted, closing plate lifted, and container moved into extrusion position.
4
5
Stem brought back into container with prefilling water, mandrel being advanced to pierce billet completely.
Billet is extruded.

FIGURE A5.6 Ugine Sejournet extrusion process.

turning in a lathe to the desired outside diameter, the inside diameter is bored to the specified internal diameter dimensions. The resulting pipe can be made to very close tolerances. Sections 50 ft long have been produced by this process.

Hollow Forged Seamless Pipe Erhardt Type Process. The Erhardt process (Fig. A5.7), developed by Heinrich Erhardt in Germany (1891), consists of heating a square ingot to forging temperature, placing it into a circularly hollow die, and incompletely piercing it with a vertical piercing mandrel such that a cup shape is obtained. As a result of the piercing at forging temperature, the square ingot becomes the (circular) shape of the die. After reheating, the cup-shaped shell is mounted on a mandrel and pushed through a series of dies to the desired diameter and wall thickness, after which the cupped end is removed and the inside and outside diameters are machined. This process is used for large diameter and heavy wall seamless pipe for boiler headers and main steam line piping. It can be applied to produce low and medium carbon steel pipe (ASTM A53, A106, A161, A179, A192, A210), stainless steel pipe (TP329, TP304, TP304L, TP321, TP347, TP316), and high nickel alloys (A333, A334).

Cold and Hot Finishing of Seamless Pipe and Tube. Pipe that has been produced by the Mannesmann plug-mill, Mandrel mill, Ugine Sejournet, or Erhardt

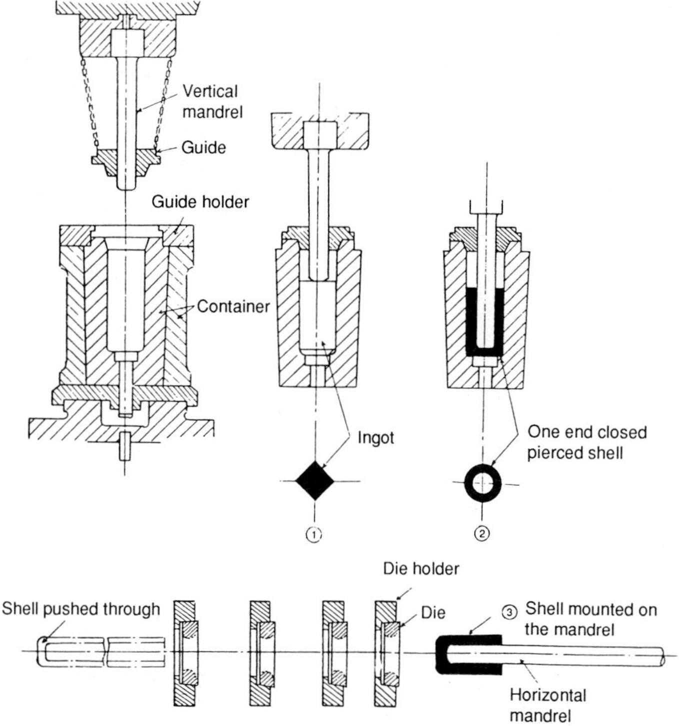

FIGURE A5.7 Hollow-forged seamless pipe-Erhardt-type process.

forging process can be used as hot finished seamless steel pipe or tube if the application does not require further finishing. If further finishing is required, the pipe or tube may be further reduced by a cold reduction process (Fig. A5.8). If the cold reduction processes are used, the reduced tube must be heat treated in a furnace such as a bright annealing furnace or in a continuous barrel furnace. Subsequent to the heat treatment of the cold finished pipe, the pipe must pass through a straightening process which corrects any nonstraight sections caused in the pipe by the heat treatment of cold reduced pipe. The straighteners are either a series of rolls through which the pipes pass cold or a device which bends the pipe at discreet locations along the pipe. The resulting product is called cold finished seamless pipe or tube.

In applications of tube to fossil fuel boilers, cold finishing is sometimes specified. Cold finishing improves the surface finish and dimensional accuracy. Some boiler manufacturers, however, consider the hot finished tube surface satisfactory and specify it as such because of its reduced cost.

Welded Pipe. Welded pipe is produced by forming a cylinder from flat steel sheets coming from a hot strip mill. The strip mill takes the square bloom from the blooming mill and reduces it into plates, skelp, or coils of strip steel to be fed into the particular welding process equipment. Butt-weld pipe is made by furnace heating and forge welding or by fusion welding using electric resistance, flash, submerged-arc welding, inert-gas tungsten-arc welding, or gas-shielded consum-

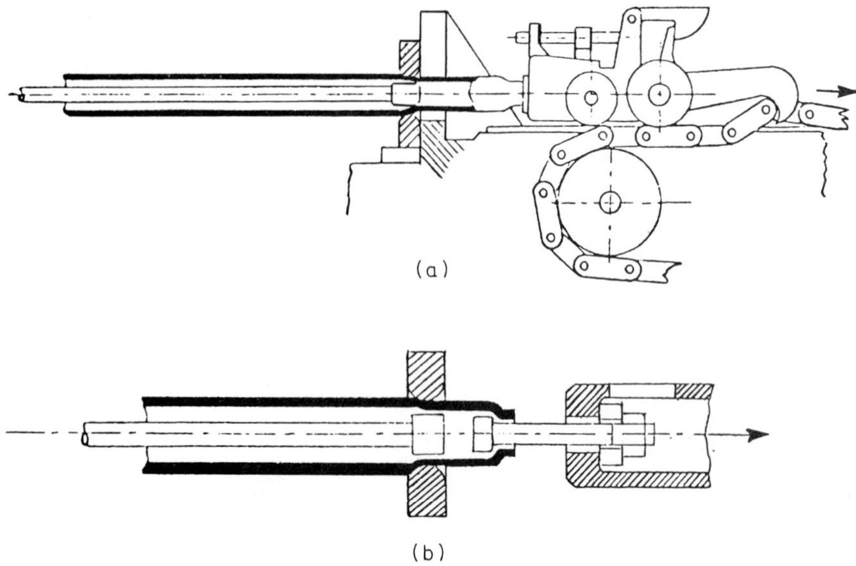

FIGURE A5.8 Cold and hot finishings of seamless pipe and tube.

able metal-arc welding. The welded seam is either parallel to the tube axis or in a spiral direction about the tube center line.

Furnace-Welded (Continuous or Butt-Welded) Pipe. This is a low cost carbon steel pipe below 4 in diameter made of steel from open-hearth or basic oxygen Bessemer steel. In this process, skelp is heated to welding temperature in a continuous furnace and passed through forming and welding rolls, welding the strip edges at the same time the tube is formed. Strips can be consecutively resistance welded to each other to form a continuous pipe.

Fusion-Welded Pipe. Fusion-welded pipe is produced by resistance welding, induction welding, or arc welding.

Electric Resistance Welded Pipe. In the electric resistance welded pipe (ERW) process (Fig. A5.9), upon exiting the forming mill, the longitudinal edges of the cylinder formed are welded by flash welding, low-frequency resistance welding, high-frequency induction welding, or high-frequency resistance welding. All processes begin with the forming of the cylinder with the longitudinal seam butt edges ready to be welded.

In the flash-welding process, the butted cylinder surfaces are placed in contact; a voltage is applied across the contact causing metal flashing along the seam length, raising the steel temperature locally to the metal-forming temperature. After this, the seam edges are pressed together and a pressure fusion weld is formed at a temperature lower than the steel melting point. The upset material along both the inside and outside of the seam is removed with a scarfing tool. This process is used to produce high-strength carbon steel pipe from 4 to 36 in in diameter.

In the low-frequency resistance method, electric current and pressure are simultaneously applied, and the resulting heat causes melting of the edges. The resulting seam is similar to that from the flash-welding process and requires removal of the upset material. Postweld heat treatment may be desirable for stress

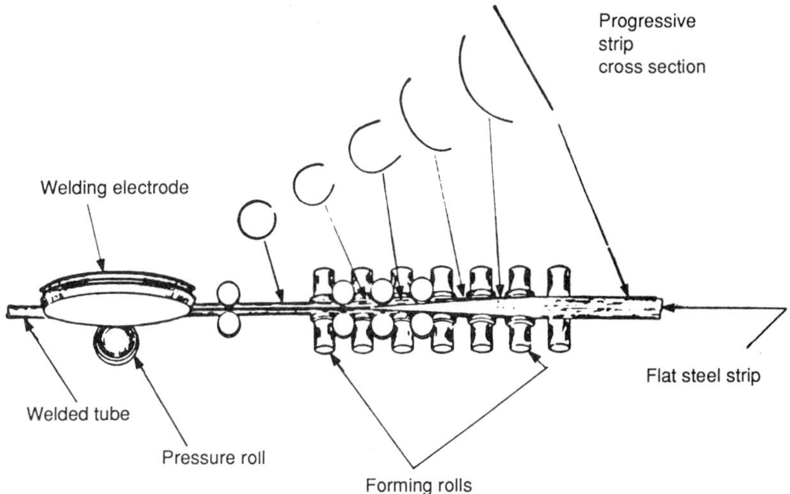

FIGURE A5.9 Electric resistance welded pipe process.

relief, tempering, or recrystallization. This process is applied to pipe of outside diameters up to 22 in.

High-frequency welding process, using an alternating current of more than 400,000 Hz, is similar to the low-frequency resistance welding process. The lower inductance path followed by the current produces a smaller high temperature band that minimizes the amount of upset material. This process is used for pipe up to 42 in in diameter.

For production of small diameter pipe at high rates of production, the high-frequency induction-welding process may be used. In this process, an induction coil raises the seam temperature to welding temperature. The rapid increase in temperature caused by the high-frequency current causes little upsetting because of the resulting control of temperature and fusion.

Several arc-welding processes are used in commercial welded pipe production. These include the submerged-arc-welding process, the inert-gas tungsten-arc-welding process, and the gas-shielded consumable metal-arc-welding process.

In the submerged-arc-welding process, bare wire consumable electrodes are added to the weld metal under a blanket of flux. The melting flux creates a protective atmosphere of inert gas and a slag blanket over the solidifying weld metal. In heavy wall piping, simultaneous submerged arc seam welds on inner and outer sides of the pipe are sometimes used to build up the weld thickness to the desired level. In modern pipe production facilities, automated equipment is used to control all variables in the submerged arc process, including relative movement speed between pipe and welding heads, wire feed rate, welding current, and flux feed rate. Submerged-arc seam-welded pipe is used in critical high-temperature or high-pressure applications in the electrical power generation, process, and chemical industries.

For carbon steel and stainless steel pipe of smaller wall thickness, the inert-gas tungsten-arc-welding process is used. The weld is protected by an inert gas such as argon or helium which forms a blanket over the weld metal. For thicker wall pipe, a filler wire may be fed into the protective gas blanket. For thin wall pipe, no filler is used. A number of variations in pipe forming before welding is used, including molding, pressing, or rolling strip into cylinders.

Spiral Welded Pipe. Lightweight pipe for temporary or light operation duty such as in water systems applications may be made by the spiral-welded process. In this process, narrow strips of steel sheet are helically wound into cylinders. The edges of this strip can either be butting or overlapping and are welded by any of several electric arc-welding processes.

Cast Pipe

Cast Iron Pipe. There are four basic types of cast iron: white iron, gray iron, ductile iron, and malleable iron. White iron is characterized by the prevalence of carbides which impart high compressive strength, hardness, and resistance to wear. Gray cast iron has graphite in the microstructure, giving good machinability and resistance to wear and galling. Ductile iron is gray iron with small amounts of magnesium or cesium which bring about nodularization of the graphite, resulting in both high strength and ductility. Malleable iron is white cast iron which has been heat treated to provide for ductility.

Cast iron pipe is extensively used for underlying water, sewage, and gas distribution systems because of its long life expectancy. Specifications for this pipe can be found under Federal Specification W-W-P-421b-Pipe, Cast Iron, Pressure (for Water and Other Liquids). Cast iron pipe is produced from four processes: vertical pit casting, horizontal casting, centrifugal sand mold casting, and centrifugal metal mold casting.

Vertical Pit Process. The vertical pit process for producing pipe requires a sand mold formed into a pipe pattern of the outer surface of the pipe, into which a separately made core is placed. The molten iron is poured into the vertical annulus between the outer mold and the core. American Standard Specifications for Cast Iron Pit Cast Pipe for Water or Other Liquids are available in ASTM Specification A 377. Pit cast pipe specifications for the gas industry may be found in American Gas Association (AGA) Standards for Cast Iron Pipe and Special Castings. ASTM Designation A142 provides specifications for pit cast culvert pipe.

Horizontal Process. In the horizontal cast iron pipe process, horizontal outer molds are made in halves, with a core formed around a perforated horizontal bar. After the top half is placed on the bottom half, the molten iron is introduced in a manner preventing ladle slag from entering the mold.

Centrifugally Cast Iron Pipe. There are two types of centrifugal casting machines—horizontal and vertical. Pipe is most commonly produced in the horizontal machine. The centrifugal castings are formed after molten metal is poured into a rotating mold. The mold continues its rotation until solidification of the metal is complete, after which the casting is removed. Molds can be made of sand or, for permanent molds, can be made of graphite, carbon, or steel. The centrifugal casting process provides a means of producing high-quality castings which are defectfree due to the absence of shrinking. These castings cool from the outside to the inside, providing a desirable directional solidification which results in cleaner and denser castings than those resulting from static casting methods.

Cast Steel Pipe. Cast steel pipe is produced by either static or centrifugal casting processes. In the horizontal centrifugal casting machine, the molten steel is introduced into the rotating mold of sand, ceramics, or metal. Centrifugally cast pipe can be obtained in diameters up to 54 in. Application of this pipe can be found in paper mill rolls, gun barrels, and high-temperature and pressure service in refineries (temperatures above 1000°F). This process is also used for high nickel and high-nickel alloys.

Cold Wrought Steel Pipe. Centrifugally cast stainless steel pipe can be cold expanded subsequent to casting by internally applied pressure to form cold wrought pipe. The process, called hydroforging, applied to austenitic stainless steels provides for recrystallization and grain refinement of the centrifugally cast material grain structure.

NONFERROUS PIPE-MAKING PROCESSES

Aluminum and Aluminum Alloy Tube and Pipe

Aluminum tubular products include both pipe and tube. They are hollow wrought products produced from a hollowed ingot by either extrusion or by welding flat sheet, or skelp, to a cylindrical form. General applications are available in alloys 1100, 2014, 2024, 3003, 5050, 5086, 6061, 6063, and 7075. For shell and tube heat exchanger applications, alloys 1060, 3003, 5052, 5454, and 6061 are available. Pipe is available only in alloys 3003, 6061, and 6063. The designation numbers indicate the particular alloying element contained in the aluminum alloy, such as copper, manganese, silicon, magnesium, and zinc and the control of the impurities. The numerical designation system consists of four numbers, *abcd,* where *a* designates the major alloying element in the aluminum alloy: 1 for 99% pure aluminum, 2 for copper, 3 for manganese, 4 for silicon, 5 for magnesium, 6 for magnesium and silicon, 7 for zinc, and 8 for other element. Digit *b* designates an alloy modification for groups 2 through 8 and an impurity limits for group 1. Digits *c* and *d* indicate the specific alloy for groups 2 through 8 and the purity of group 1.

Copper and Copper Alloy Tube and Pipe

Copper tube and pipe have a wide range of application throughout the chemical, process, automotive, marine, food and beverage, and construction industries. Unified Numbering System (UNS) designations (CXXXXX) have been established for many alloys of copper. ASTM and ASME specifications have been developed for copper tube and pipe. Seamless pipe and tube are covered by ASTM B466, B315, B188, B42, B302, B75, B135, B68, B360, B11, B395, B280, B306, B251, B372, and B88. ASME specifications include SB466, SB315, SB75, SB135, SB111, SB395, and SB359.

Tubes and pipe of copper and copper alloys are produced by either of two processes—piercing and extrusion or welding skelp formed into cylindrical shape. The seamless pipe or tube produced through the extrusion process is the most common commercial form of copper and copper alloy tubular products.

Hot Piercing Process. In the Mannesmann piercing process, a heated copper billet is first pierced, then rolled over a mandrel which determines the inside diameter of the pipe. Following the piercing operation, the pierced shell is drawn through a die and over a plug to obtain the finished outside and inside diameters.

Extrusion Process. In the extrusion process, the heated copper or copper alloy billets are formed into shells by heavy hydraulic presses. The hollowed out billet is then extruded through a die and over a mandrel to form the outside and inside diameters of the pipe.

Cold Drawing Process. The cold drawing process uses mother pipe which is placed on a draw bench which pulls a cold tube through one or a multiplicity of dies and over a mandrel to reduce the pipe gradually to its finished dimensions.

Other Processes. Other processes of significance are the cup-and-draw process for large diameter pipe and the tube-rolling process which reduces copper tubing by means of cold working over a mandrel with oscillating tapered dies.

Nickel and Nickel Alloy Pipe and Tube

Nickel and nickel alloy pipe and tube, because of their high strength and generally good resistance to oxidation and corrosion, are used in the chemical industry and in steam-generation equipment for nuclear-power-generation plants. Applications of nickel are found in tubes and pipe of pure nickel and binary and tertiary alloys of nickel, such as Ni-Cu (Monel 400 and Monel K-500), Ni-Mo and Ni-Si (Hastelloy B), Ni-Cr-Fe (Inconel 600 and Inconel 800), and Ni-Cr-Mo (Hastelloy C-276 and Inconel 625) alloys. The alloys are used in applications requiring corrosion resistance to water, acids, alkalis, salts, fluorides, chlorides, and hydrogen chloride. The alloy must be carefully selected to provide for resistance to the specific corrosion media found in the environment.

Nickel and nickel alloy pipe and tube are produced by the Ugine-Sejournet extrusion process in which a shell is formed by hydraulic piercing of a billet by a ram and subsequent extrusion. Alternately, the billet may be initially pierced by means of drilling.

Titanium and Titanium Alloy Tube and Pipe

Titanium and its alloys have provided the engineering designer with an important alternative to aluminum. They are lightweight and have high strength at moderately elevated temperatures, good toughness, and excellent corrosion resistance. Their applications have been found in a wide range of industries, including aerospace, heat exchange equipment, chemical plants, and power-generation facilities.

There is a wide range of alloying systems to which titanium may be produced. The alloying elements possibly include aluminum, molybdenum, nickel, tin, manganese, chromium, and vanadium. UNS numbers are used to identify the many available alloys and forms of titanium.

Titanium and alloys of titanium pipe and tube are produced from a melt of raw titanium "sponge" and alloying metals in a vacuum electric arc furnace. An ingot is obtained which is reduced to a billet. The billet provides the stock for the extrusion process, from which the tube or pipe is formed. The process consists of initially piercing the billet, then passing a heated shell through a die and over a mandrel.

COMMERCIAL PIPE AND TUBE SIZES

The standard pipe sizes and other pipe properties are given in App. E2, and the standard tube sizes and other tube properties are given in App. E3.

TUBULAR PRODUCT CLASSIFICATION

Pipe and tubing are considered to be separate products, although geometrically they are quite similar. "Tubular products" infers cylindrical products which are hollow, and the classification of "pipe" or "tube" is determined by the end use.

Piping Classification

Tubular products called pipe include standard pipe, conduit pipe, piling pipe, transmission (line) pipe, water-main pipe, oil country tubular goods (pipe), water-well pipe, and pressure pipe. Standard pipe, available in ERW or seamless, is produced in three weight (wall-thickness) classifications: standard, extra strong, and double extra strong (either seamless or welded). ASTM and American Petroleum Institute (API) provide specifications for the many categories of pipe according to the end use. Other classifications within the end use categorization refer to the method of manufacture of the pipe or tube, such as seamless, cast, and electric resistance welded. Pipe and tube designations may also indicate the method of final finishing, such as hot finished and cold finished.

Tubing Classification

Pressure tubes are differentiated from pressure pipe in that they are used in externally fired applications while carrying pressurized fluid inside the tube.

Structural tubing is used for general structural purposes related to the construction industry. ASTM provides specifications for this type of tubing.

Mechanical tubing is produced to meet particular dimensional, chemical, and mechanical property and finish specifications which are a function of the end use, such as machinery and automotive parts. This category of tubing is available in welded (ERW) and seamless form.

SPECIALTY TUBULAR PRODUCTS

There are many specialty tubular products designed for special applications requiring unique manufacturing methods for production. Examples are the rifled boiler tube, the finned heat exchanger tube, the duplex tube, and the double-wall tube.

The rifled boiler tube (Fig. A5.10) is used to provide an improved heat transfer surface on the inner surface of a boiler tube. The rifling twist, similar to that of a rifle, is produced by specially shaped mandrels over which the tube is drawn.

The finned heat exchanger (Fig. A5.11) tube provides improvement in thermal efficiency by providing an extended surface from the base tube surface. The extended surface is produced by turning the tube through special sets of dies which raise fins from part of the base tube material. These fins can be coarse or fine depending on the equipment developed for producing fins.

Duplex or composite tubes have been developed to provide a different material on the inside and outside of the tube to meet the requirements of a different environment on either side of the tube. One method of producing a composite tube is by providing a bimetal mother tube before the extrusion or drawing pro-

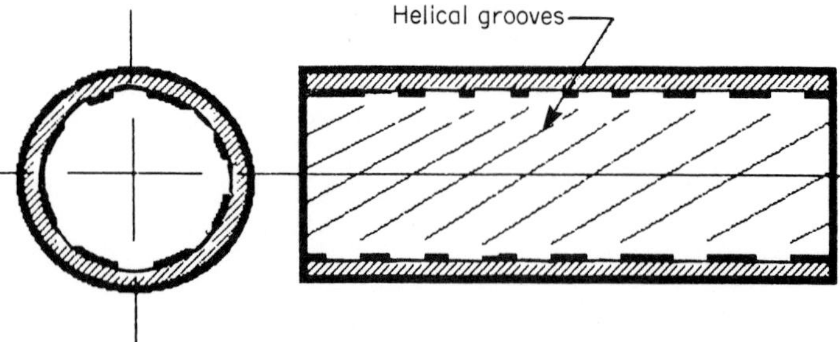

FIGURE A5.10 Single rifled boiler tube.

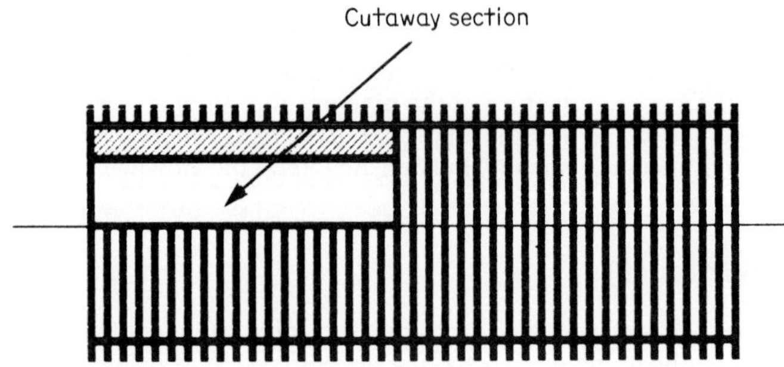

FIGURE A5.11 Finned heat exchanger tube.

cess. Careful development of this process will yield a composite tube with an excellent bond between the two materials.

Double-wall tubes (Fig. A5.12) are used in applications requiring leak detection to avoid a catastrophic mixture of the fluids on either side of the tube. An inert detecting gas can be placed in the annulus between the two tubes to sense very small amounts of leakage from either tube so as to allow careful shutdown of the system. This tube is manufactured by inserting one tube inside the other, then

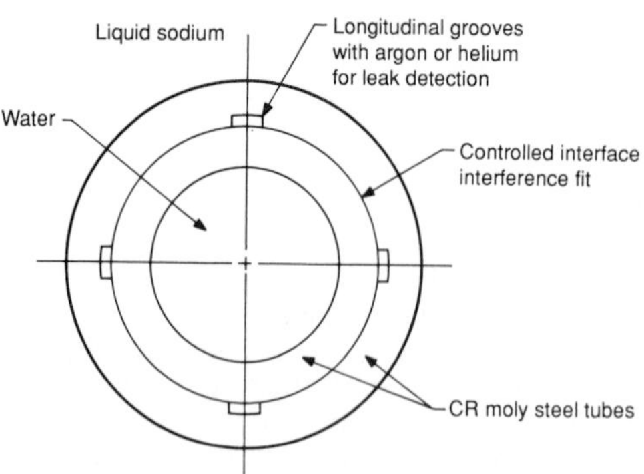

FIGURE A5.12 Double-wall leak-detecting tube.

drawing the combined tube through dies or over mandrels which provide a calibrated prestress between the two tubes. This type of tube was developed for application to a fast breeder reactor sodium-water steam generator.

ENGINEERING SELECTION OF PIPE MANUFACTURING METHODS

The selection of the appropriate pipe manufacturing method by the design engineering specification deserves consideration. For many applications, the codes and standards specified in the procurement contract provide for little room to select an optimal manufacturing method. The safest procedure is to obtain the price and schedule from suppliers before firming the piping specifications. At times the selection of the pipe with the best manufacturing process might be tempered by project cost or delivery considerations. In such cases, much is required of the engineer to consider whether lesser quality will be able to meet the desired reliability standard. It therefore is essential that the engineer is aware of the alternates and their operating history of success and failure before an appropriate alternative decision can be accepted. It must also be recognized that choices based on economic considerations alone may prove to be the more costly in the face of the down-time costs of failure.

CHAPTER A6
FABRICATION AND INSTALLATION OF PIPING SYSTEMS

Edward F. Gerwin, P.E.
ASME Fellow

INTRODUCTION

Background

The term *fabrication* applies to the cutting, bending, forming, and welding of individual pipe components to each other and their subsequent heat treatment and nondestructive examination (NDE) to form a unit (piping subassembly) for installation.

The term *installation* refers to the physical placement of piping subassemblies, valves, and other specialty items in their required final location relative to pumps, heat exchangers, turbines, boilers, and other equipment; assembly thereto by welding or mechanical methods; final NDE; heat treatment; leak testing; and cleaning and flushing of the completed installation.

Depending on the economics of the particular situation, fabrication may be accomplished in a commercial pipe fabrication shop, or a site fabrication shop, where portions of the piping system are fabricated into subassemblies or modules for transfer to the location of the final installation.

Commercial pipe shops have specialized equipment for bending and heat treatment which is not normally available at installation sites. They also have certain types of automatic welding equipment which permits welding to be performed more efficiently and economically than in field locations where fixed position, manual arc welding is most often employed.

As a general rule piping NPS 2½ and larger for nuclear and fossil power plants, chemical plants, refineries, industrial plants, resource recovery, and cogeneration units are most often shop fabricated. Piping NPS 2 and smaller is often shop fabricated where special heat treatment or cleaning practices may be required; otherwise it is field fabricated. Pipe lines and other systems involving long runs of essentially straight pipe sections welded together are usually field assembled.

In recent years, the infusion of new bending technologies, new welding processes, new alloys, fracture toughness limitations, and mandatory quality assurance (QA) programs have made piping fabrication and installation much more complex than in the past. Greater emphasis is being placed on written procedures for QA and quality control (QC) programs, special processes, and qualification and certification of procedures and personnel.

Improper selection of fabrication or installation practices can result in a system which will not function properly or will fail before its expected life span. Accordingly, fabrication and installation contractors must work closely with the designer and be aware of the mandatory requirements of the applicable codes, the unique requirements and limitations of the materials and those of the fabrication, and the installation techniques being applied.

Codes and Standards Considerations

There are a great many codes and standards which apply to piping. These are discussed in detail in Chap. A4.

It is incumbent on the fabricator and/or installer to be familiar with the details of these codes and standards since some codes have the force of law. As an example, the ASME B31.1 Power Piping Code[1] is referenced by ASME Section I Power Boilers[2] for piping classed as Boiler External Piping. The latter, which is law in most states and Canadian provinces contains rules for code stamping, data reports, and third-party inspection. Piping under ASME Section III[3] also has legal standing. Most other piping codes are used for contractual agreements.

Most codes reference ASME Section V[4] for nondestructive examination methodology and ASME Section IX[5] for welding requirements.

Each of the codes covers a different piping application, and each has evolved in a different way over the years. For specific practices, some have mandatory requirements, while others only have recommendations. Heat treatment requirements may vary from one to another. The manner in which the code-writing bodies have perceived the hazardous nature of different applications has led to differing NDE requirements.

Generally, the codes are reasonably similar, but the owner, designer, fabricator, and installer must meet the specifics of the applicable code to ensure a satisfactory installation. It is essential that the designer be very familiar with the code being used and that purchasing specifications for material, fabrication, and installation be very specific. Reference to the code alone is not sufficient. In the design, a particular allowable stress for a specific material, grade, type, product form, and/or heat-treated condition was selected. The specifications issued for material purchase and fabrication must reflect these specifics to assure that the proper materials and fabrication practices are used.

As an example: Type 304 stainless steel has a specified carbon content of 0.08 percent maximum. There is no specified minimum. Footnotes in the B31.1 Code Allowable Stress Tables for Type 304 indicate that for use over 1000°F (538°C), the allowable stresses apply only when the carbon content is 0.04 percent or higher. It is essential that this requirement be put in the purchasing specification if the design temperature exceeds 1000°F (538°C).

Similarly, in the B31.1 Code, low chrome alloy electric fusion welded pipe has differing allowable stresses depending upon whether the plate from which it was made was annealed or normalized and tempered. If this material is to be heated above the lower critical temperature during fabrication by hot bending or forming, the designer should specify a postbending heat treatment appropriate for the allowable stress level used in the design.

It is also incumbent upon the fabricator and/or installer to be very familiar with the applicable code. Each project should be reviewed in detail. "Standard shop practices" may not always produce the desired result. Communication between the designer, fabricator, and installer is essential. All should be familiar with the various standards used in piping design. Most piping systems are composed of items which conform to some dimensional standards such as ASME B36.10[6] and ASME B36.19 for pipe, B16.5[7] for flanges, etc. Other dimensional standards are issued by the Manufacturers Standardization Society (MSS)[8] and the American Petroleum Institute (API).[9]

The Pipe Fabrication Institute (PFI)[10] publishes a series of Engineering Standards which outline suggested practices for various fabrication processes. These standards give excellent guidance for many aspects of piping fabrication not covered by the codes.

The American Welding Society (AWS)[11] publishes a number of recommended practices for welding of pipe in various materials.

Materials Considerations

Piping systems are fabricated from a great variety of metals and nonmetals, material selection being a function of the environment and service conditions. Materials must conform to the standards and specifications outlined in the governing code. Some codes such as ASME Section III impose additional requirements on materials beyond those in the material specifications. All fabrication and installation practices applied to these materials must be conducted so as to assure that the final installation exhibits all of the properties implicit in the design. For example, hot bending of certain austenitic stainless steels in the sensitization range will reduce their corrosion resistance if they are not subsequently heat treated. Accordingly, a heat treatment to restore these properties should be specified.

Consideration must also be given to the various types of piping products, their tolerances, alloys, heat-treated conditions, weldability, and formability. Pipe is made by a variety of processes and depending on the method of manufacture can have differing tolerances.

Most product forms also come in a variety of alloys, and the choice of a fabrication process may be governed by the alloy. ASME Section IX has developed a system of P Numbers and Group Numbers. This system groups material specifications by chemical composition and/or physical properties. Those with like compositions and properties are grouped together to minimize the number of welding procedure qualifications required. This method of grouping can also be applied to other fabrication processes as well.

FABRICATION

Drawings

Installation Drawings. Current industry practice is for the designer to prepare plans and sections or isometric drawings of the required piping system. These, together with line specifications, outline all the requirements needed for the fabrication and installation. Usually the weld bevel requirements for field welds are specified to assure compatibility between all the system components to be field welded. Frequently the shop welding bevels are left to the discretion of the fabricator, provided, of course, the required weld quality is attainable.

Location and numbers of field welds are an economic consideration of available pipe lengths, shipping or heat-treating limitations, and field installation limitations.

Shop Details. A piping system prefabricated at a commercial pipe fabrication shop is usually divided into subassemblies or spools.

The manner in which a system is divided depends on many factors: available lengths of straight pipe, dimensional and weight limitations for shipping and heat treatment, field welding clearance requirements, and sometimes scheduling needs.

Bending, forging, special heat treatment, cleaning, and as much welding as possible are normally performed in the shop. Every attempt is made to minimize the number of field welds, but this must be balanced economically against the added costs of transportation and greater field rigging problems because of larger, heavier, more complex assemblies. Where the site conditions are adverse to normal field erection practices, much of the plant can be fabricated in modules for minimal onsite installation work. Once the number and locations of field welds have been decided, the fabricator will prepare detailed drawings of each subassembly.

Each subassembly drawing will show the required configuration; all necessary dimensions required for fabrication; reference to auxiliary drawings or sketches; size, wall thickness, length, alloy, and identification of the materials required; code and classification; reference to special forming, welding, heat treatment, NDE, and cleaning requirements; need for third-party inspection; weight and piece identification number. See Fig. A6.1.

Tolerances. In order to assure installation of a system within a reasonable degree of accuracy, all the components involved must be fabricated to some set of tolerances on those dimensions which effect the system length. Tolerances on valve dimensions are given in B16.34,[12] those of welding fittings in B16.9,[13] and those for flanges and flanged fittings in B16.5, B16.1,[14] etc. The assembly of these components will result in "tolerance stack-up," which could have a significant impact on the overall dimensions, particularly in a closely coupled system.

Piping subassembly tolerances normally conform to PFI-ES-3 "Fabricating Tolerances."[15] Usually the terminal dimensions are held to ±⅛ in, but can be held more closely upon agreement with the fabricator.

In order to assure that tolerance stack-up is held to a minimum, the manner in which shop details are dimensioned should be carefully studied. As an example, assemblies with multiple nozzles can result in large deviations if these are dimensioned center to center. A better way is to select a base point and dimension all nozzles from this location. This assures that all nozzles are ±⅛ in from the base point. See Fig. A6.2.

For angle bends, terminal dimensions are required since a small variation in angle with long ends can result in serious misalignment. See Fig. A6.3.

Sometimes assemblies which have been fabricated within tolerance may not fit in the field because of tolerance stack-ups on equipment to which they are attached. This will be addressed in the section, "Installation."

Procedures and Travelers. The need to assure better control of fabrication processes has led the use of written procedures for most operations. Fabricators will have a library of written procedures controlling cutting, welding, bending, heat treatment, nondestructive examination, and testing. Welding procedures in most

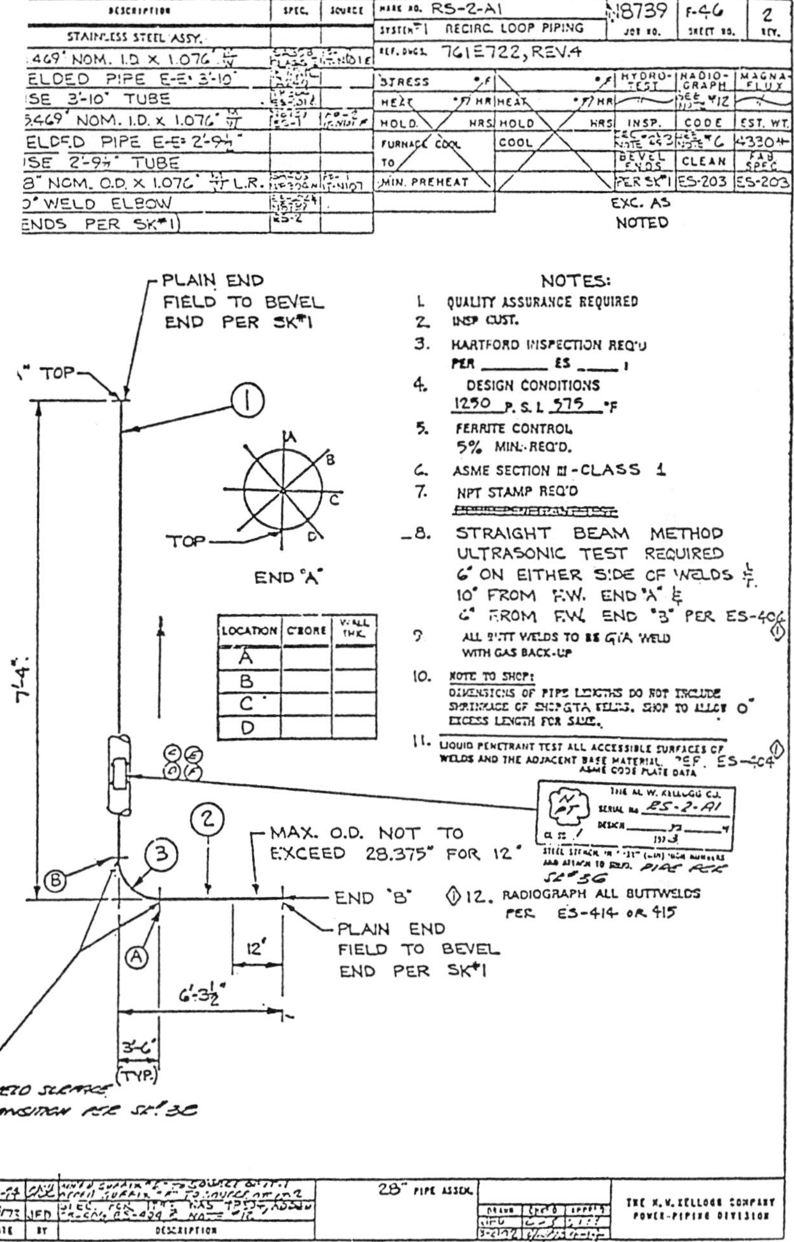

FIGURE A6.1 Shop detail. (*Pullman Power Products Corporation*)

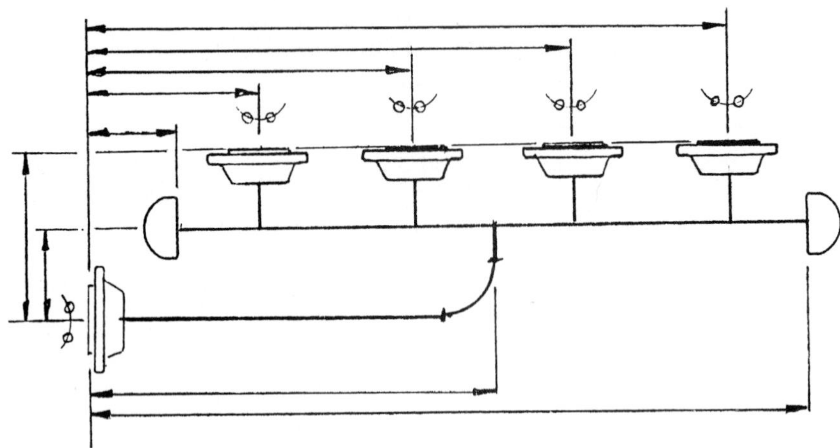

FIGURE A6.2 Dimensioning.

codes are qualified under ASME Section IX, which requires written Welding Procedure Specifications (WPSs) backed up by Procedure Qualification Records (PQRs). Similarly ASME Section V requires NDE to be performed to written procedures.

Frequently, piping fabricators use a system of travelers to control flow through the shop. This practice is well suited to fabrication of piping subassemblies under QA or QC programs, where record keeping is required. It also affords the purchaser and the third-party inspector opportunities for establishing "hold points" where they may wish to witness certain operations or review certain records.

Fabrication Practices

Cutting and Beveling. The methods of cutting plate or pipe to length can be classed as mechanical or thermal.

Mechanical methods involve the use of saws, abrasive discs, lathes, and pipe-cutting machines or tools. See Fig. A6.4.

Thermal methods are oxyfuel gas cutting or electric arc cutting. Oxyfuel gas cutting is a process wherein severing of the metal is effected by the chemical reaction of the base metal with oxygen at an elevated temperature. In the cutting torch, a fuel such as acetylene, propane, or natural gas is used to preheat the base metal to cutting temperature. A high-velocity stream of oxygen is then directed at the heated area resulting in an exothermic reaction and severing of the material. Oxyfuel gas cutting is widely used for cutting carbon steels and low alloys. It does, however, lose its effectiveness with increasing alloy content.

For higher alloy materials, some form of arc cutting is required. Plasma arc cutting is the process most frequently employed. It involves an extremely high temperature (30,000 to 50,000 K), a constricted arc, and a high-velocity gas. The torch generates an arc which is forced to pass through a small-diameter orifice and concentrate its energy on a small area to melt the metal. At the same time a gas such as argon, hydrogen, or a nitrogen-hydrogen mixture is also introduced at

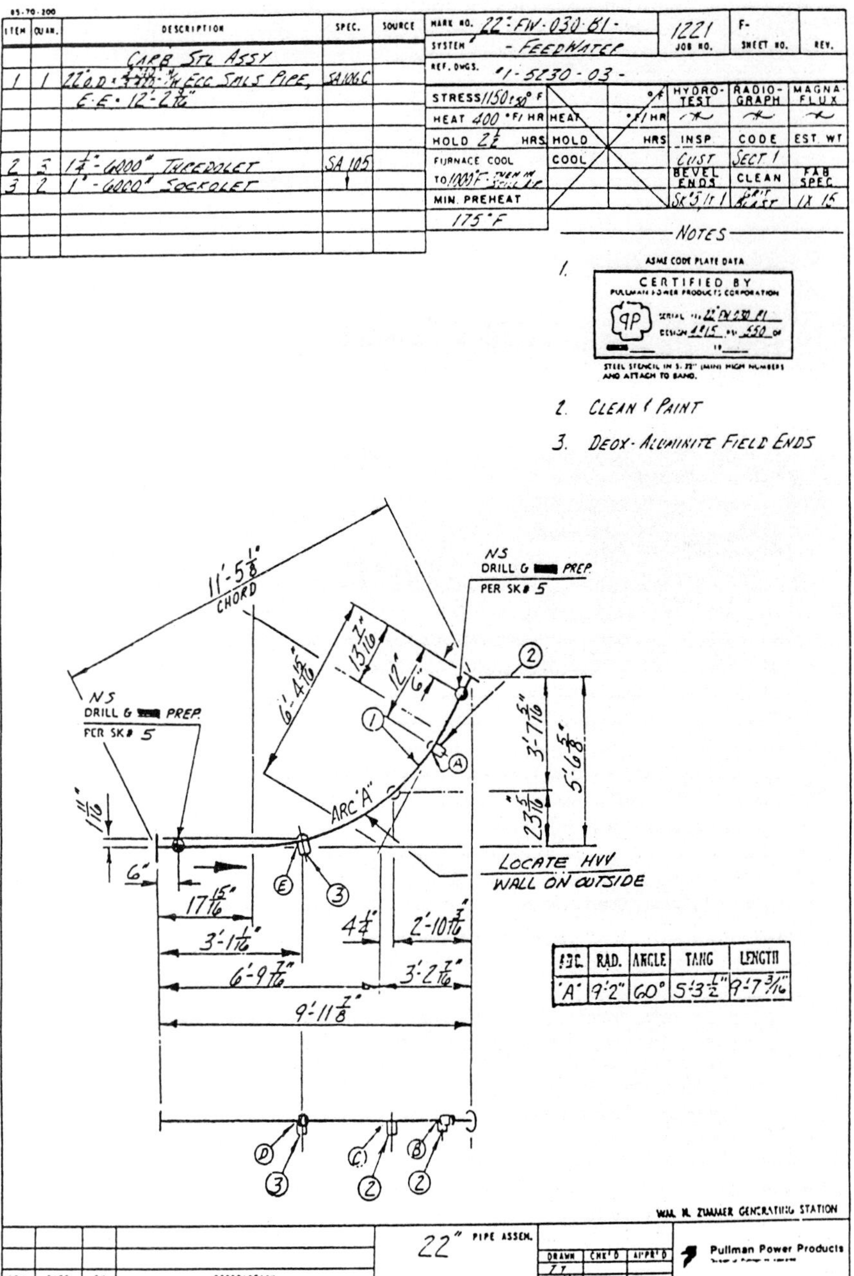

FIGURE A6.3 Dimensioning a bend. (*Pullman Power Products Corporation*)

FIGURE A6.4 Pipe cutting machine. (*Pullman Power Products Corporation*)

the orifice where it expands and is accelerated through the orifice. The melted metal is removed by the jet-like action of the gas stream.

Because oxyfuel gas and arc cutting involve the application of heat, preheating may be advisable in some cases.

A very detailed description of oxyfuel gas and arc cutting is presented in *The Welding Handbook.*[16]

Weld end bevels can also be prepared by the mechanical or thermal methods described above. Both mechanical and thermal methods are used to apply the V bevel, which is used in the vast majority of piping applications. For compound and U bevels or those which may involve a counterboring requirement, horizontal boring mills are most appropriate. Various factors to be considered in selecting a weld end bevel are discussed in the section, "Welding Joint Design."

Forming. The term *forming* as it relates to piping fabrication encompasses bending, extruding, swaging, lapping, and expanding. All of these operations entail the use of equipment normally only available in pipe fabrication shops. Although the availability of welding fittings in the form of elbows, tees, reducers, and lapped-joint stub ends may reduce the need for certain of these operations, economics may dictate their use, especially where special pipe sizes are involved.

Bending

Economics. The use of bends versus welding fittings for changes in direction should be carefully evaluated from an economic viewpoint. Bends whose radii range from 3 to 5 times the nominal pipe diameter will offer the least pressure drop while still affording adequate flexibility to the system. Since each bend eliminates a welding fitting and at least one weld with its attendant examination, bending is very often the economic choice. In the case of special pipe sizes which are frequently used for main steam, reheat, and feedwater lines in large central power generating units, bending may be the only option available.

Limitations. The metal being bent should preferably exhibit good ductility and a low rate of strain hardening. Most metals used in piping systems fulfill these requirements. A successful bend is also a function of its diameter, thickness, and bending radius. As the diameter to thickness ratio increases and the bending radius decreases, there is greater probability of flattening and buckling. Each bending process has differing capabilities, so the selection of a bending process rests on the availability of equipment and/or practices capable of handling the material, diameter, thickness, and bending radius involved.

Accept and Reject Criteria. The codes have certain requirements for the acceptability of finished bends:

1. *Thinning:* In every bending operation the outer portion of the bend (extrados) stretches and the inner portion (intrados) compresses. This results in a thinning of the extrados and a thickening of the intrados. Because of uncertainties introduced by the pipe manufacturing method, by the pipe tolerances, and by those introduced by the pipe bending operation itself, it is not possible to exactly predetermine the degree of thinning. However, it can be approximated by multiplying the thickness before bending by the ratio:

$$\frac{R}{R + r}$$

where r = the radius of the pipe (½ the outside diameter)
R = the radius of the bend

The codes require that the wall thickness at the extrados after bending be at least equal to the minimum wall thickness required for straight pipe. Accordingly, the fabricator must assure that the wall thickness ordered has sufficient margin for this effect.

Although the codes do not comment on the resulting increased thickness of the intrados, this thickness does serve to offset the increased stresses caused by internal pressure which are found at this location. (See *Theory and Design of Modern Pressure Vessels.*[17])

2. *Ovality:* A second acceptance criteria is ovality. During the bending operation, the cross section of the bend arc frequently assumes an oval shape whose major axis is perpendicular to the plane of the bend. See Fig. A6.5. The degree of ovality is determined by the difference between the major and minor axes divided by the nominal diameter of the pipe.

Where the bend is subject to internal pressure, the pressure tries to reround the cross section by creating secondary stresses in the hoop direction. Some codes consider an ovality of 8 percent acceptable in this case. Where the bend is subject to external pressure, the pressure tries to collapse the cross section. The ASME B31.3 Code[18] recommends a 3 percent maximum ovality when one bend is subject to external pressure.

3. *Buckling:* Bending of pipe with large diameter to thickness ratios often results in buckling rather than thickening of the intrados even where internal mandrels or other devices are employed to minimize it. The codes do not address this subject. It is, however, often the subject of "good workmanship" debates. The PFI gives a criteria which has been generally accepted. This appears in PFI ES-24.[19] An acceptable buckle is one where the ratio of the distance be-

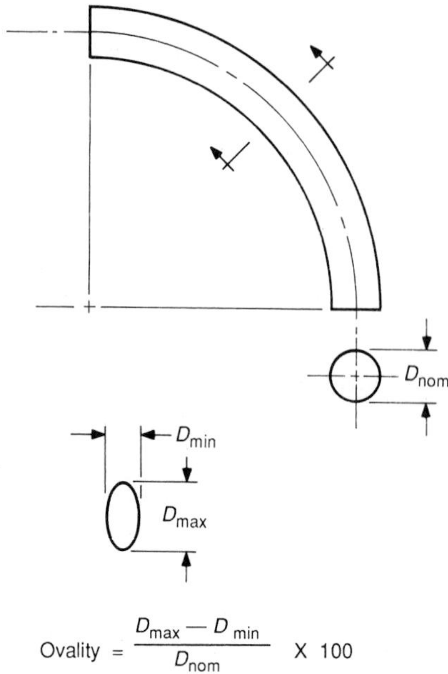

$$\text{Ovality} = \frac{D_{max} - D_{min}}{D_{nom}} \times 100$$

FIGURE A6.5 Bend ovality.

tween two crests divided by the depth of the average crest to valley is equal to or greater than 12. See Fig. A6.6.

Bending Methods. Pipe is bent by a variety of methods using bending tables or bending machines with and without the application of heat. The selection of one method over another is a function of economics, materials properties, pipe size, bending radius, and equipment availability. The arc length of the bend may be heated in order to reduce the yield strength of the material. Higher bending temperatures result in lowering the yield strength ànd reduction of the bending energy required.

Cold bending normally infers bending at ambient temperature, while hot bending infers the application of heat. However, definitions given in B31.1 and ASME Section III create an exception to this for *ferritic* materials. These codes define cold bending of *ferritic* steels as any operation where the bending is performed at a temperature 100°F (55°C) below the lower critical or lower. Ferritic materials undergo a phase change on heating and cooling. On heating, this change starts at a temperature called the *lower critical.* (See the section "Ferritic Steels.")

Ferrous Pipe and Tubes

1. *Cold bending:* Where sufficient quantities of repetitive bends are required, ferrous pipes and tubes up to NPS 10 or 12 with wall thickness of ½ in or less are most often bent at ambient temperature using some type of bending machine.

There are a great variety of cold bending machines available with varying degrees of sophistication from simple manually operated single plane bending devices to numerically controlled hydraulically operated machines capable of multiplane bends.

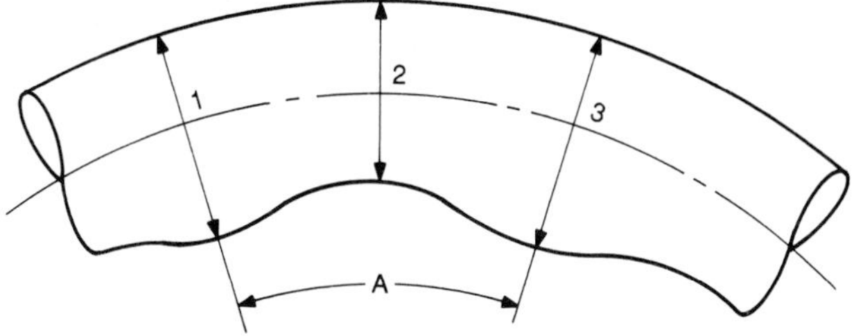

Depth of average crest to valley is the sum of the outside diameters of the two adjoining crests divided by two, minus the outside diameter of the valley.

$$\text{Depth} = \frac{(OD)_1 + (OD)_3}{2} - (OD)_2$$

Ratio of the distance between crests to depth must be equal to or greater than 12.

$$\frac{A}{\text{Depth}} \geq \frac{12}{1}$$

FIGURE A6.6 Suggested pipe buckling tolerance. (*Pipe Fabrication Institute PFI ES-24*)

In ram-type bending, two pressure dies which are free to rotate are mounted in a fixed position on the machine frame. The pipe to be bent is positioned against these dies. A ram then presses a forming die against the pipe and the pressure dies wipe the pipe around the forming die. See Fig. A6.7. Ram bending is usually applied to heavier wall thicknesses.

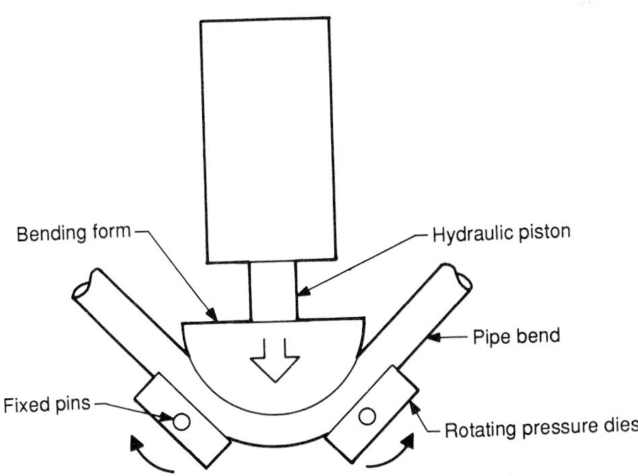

FIGURE A6.7 Ram bender.

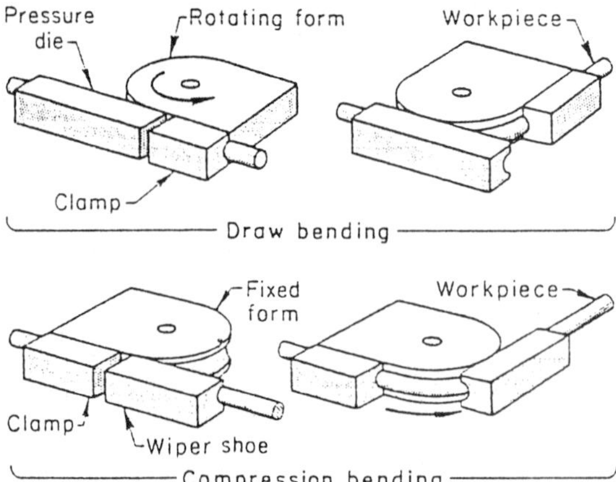

FIGURE A6.8 Comparison of the essential elements of draw bending and compression bending. (*Metals Handbook*[20])

In compression bending the pipe is clamped to a stationary bending die and wiped around it by a follower. As in all bends, the extrados thins and the intrados thickens or compresses. The degree of compression is greater than the thinning in this method. Compression bending is usually limited to heavier walls and larger bending radii. See Fig. A6.8 to compare compression and draw bending.

In rotary draw bending the pipe is clamped to a rotating bending form and drawn past a pressure die which is usually fixed. See Fig. A6.9. The degree of thinning of the extrados is greater than the compression of the intrados. This method permits bending of thinner wall pipe and tubes at smaller bending radii.

FIGURE A6.9 Tooling for a draw bend application. (*Teledyne Pines*)

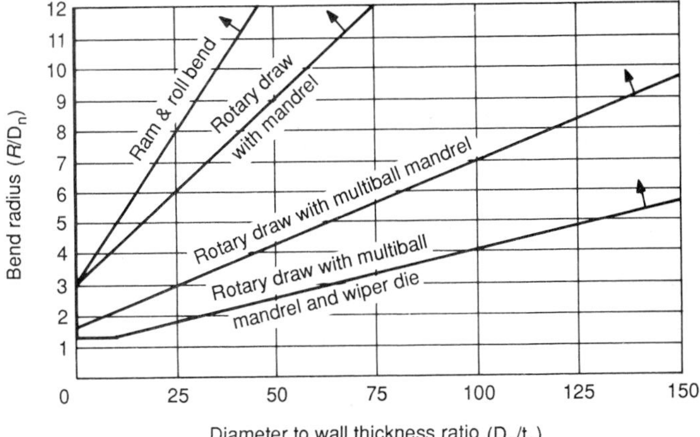

FIGURE A6.10 Cold bending ranges. (*Pipe Fabrication Institute PFI ES-24*)

To accommodate lighter walls and tighter radii it is often advisable to provide internal support to minimize flattening or buckling. Usually this takes the form of an internal mandrel. As the diameter to thickness ratio increases and the bending radius decreases, mandrels using follower balls are employed. See Fig. A6.10.

Roll bending is often used for coiling. One of its great advantages is that the bending radius is not dependent on a fixed radius die, and consequently there is great flexibility in choosing a bending radius. In roll bending three power-driven rolls, usually in pyramid form, are used. The pipe to be bent is placed between the two lower rolls and the upper roll. Bending is accomplished by adjusting the rolls relative to each other as necessary to attain the required diameter. See Fig. A6.11.

Pipe can also be cold bent on a bending table in the manner described for hot bending below, except that for ferritic materials the bending temperature is kept at least 100°F below the lower critical.

2. *Hot bending:* In those cases where suitable cold bending equipment is unavailable, hot bending may be employed. For hot bending of ferrous materials the pipe to be bent is usually heated to temperatures in the range of 1900 to 2050°F (1038 to 1121°C). For austenitic materials these temperatures may introduce sensitization, and for ferritic materials they will exceed the critical temperature where metallurgical phase changes occur. See the section "Heat Treatment" for a discussion of these subjects.

The traditional method of hot bending is performed on a bending table. Depending on the diameter to thickness ratio, the pipe to be bent may be packed with sand to provide more rigidity and thus reduce the tendency for buckling. A rule of thumb is to sand fill if the diameter to thickness ratio is 10 to 1 or greater for 5-diameter bends. However, when the diameter to thick-

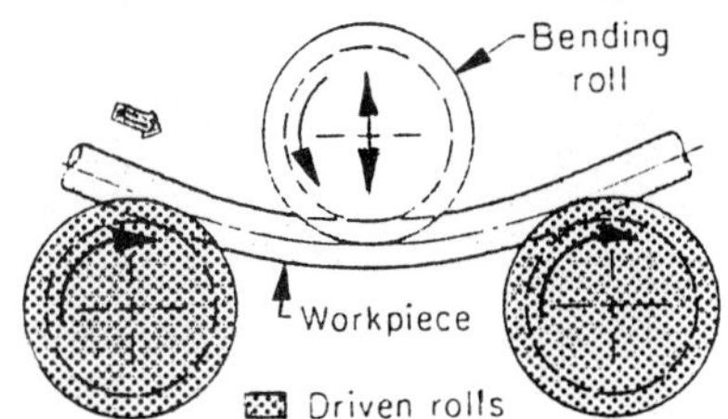

FIGURE A6.11 Operating essentials in one method of three-roll bending. (*Metals Handbook*[20])

ness ratio approaches 30 to 1, sand begins to lose its effectiveness and buckles will appear. As the diameter of the pipe increases, the probability of buckling will increase since the sand fill will not expand in proportion to the pipe, leaving a void between the pipe and packing. It becomes pronounced around NPS 24.

After the pipe has been packed with sand, it is placed in a specially designed bending furnace. The furnace is usually gas fired through ports along its length, placed to direct the flames around the pipe and avoid direct flame impingement. The furnace is controlled by thermocouples or pyrometers to assure that the required bending temperature is attained but not exceeded. Depending on the length of arc to be bent, it may be necessary to make the bend in more than one heat.

After the segment to be bent has attained the required temperature throughout its thickness, the pipe is placed on the bending table. One end is restrained by holding pins and the other is pulled around by block and tackle powered by a winch. As bending progresses, the arc is checked against a bending template. Repositioning of the holding pins may be necessary. See Fig. A6.12. For ferritic steels, it is recommended that the bending be completed above the upper critical temperature of the metal, usually about 1600°F.

There are certain limits as to the combination of diameters, thicknesses, and bending radii which can be accommodated by the hot table bend method. PFI Standard ES-24 contains a chart of suggested limits for bend radius versus diameter to wall thickness ratios. See Fig. A6.13.

To fulfill the need for a bending process beyond the capabilities of hot table bending, the M. W. Kellogg Co. developed the increment bending process which was further refined by Pullman Power Products Corp. In this process one end of the pipe is fixed in an anchor box while a clamp connected to a hydraulic piston is attached to the other. A gas torch ring burner assembly is positioned at one end of the arc to be bent. The burner assembly is sized to heat a length of arc (increment) about 1 to 2 times the pipe wall thickness. The increment length is selected

FIGURE A6.12 Hot bending on a table. (*Pullman Power Products Corporation*)

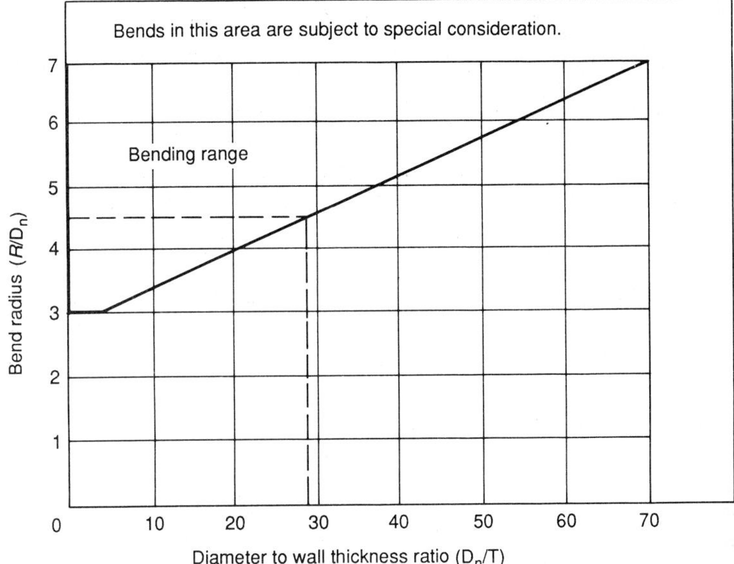

FIGURE A6.13 Limits for hot bending on a table. (*Pipe Fabrication Institute PFI ES-24*)

to be less than the buckling wave length of the pipe. The increment is then heated to bending temperature. Optical pyrometers are used to control the heating to assure that proper temperature is attained but not exceeded. At bending temperature the hydraulic piston pulls the clamped end a fixed amount to bend the heated increment. The increment is then water cooled, the torch ring moved to the next increment, and the process is repeated. As many as 350 increments may be required for a typical NPS 24- × ⅜-in (9.5 mm), 90°, 5-diameter bend.

The process can produce bends in sizes from NPS 8 to 48 with bending radii of 3 pipe diameters and larger in ferrous and nickel-alloy materials. Because the heat is applied from one side only, thicknesses are limited to 2 in (50 mm) and less.

In more recent years a more sophisticated piece of bending equipment has entered the pipe bending field, notably the Induction Bender. In this process the increment to be bent is heated by an induction coil, and the bending operation is continuous. The pipe to be bent is inserted in the machine, and the start of the arc is positioned under the induction coil. The portion of the pipe upstream of the coil is clamped to a rotating arm fixed to the required bend radius. The downstream portion of the pipe is pushed hydraulically through the coil, where it attains bending temperature. Since it is clamped to the rotating arm, a bending moment is imposed on the pipe and it bends as it moves through the coil. As soon as it has been bent, the heated section is quenched to restore its prior rigidity.

The Induction Bender is manufactured in several sizes depending on the expected combinations of pipe size and bending radius. These range from NPS 3½ to 64 and from 8 to 400 in (200 to 10,000 mm) in radius. Since induction is used as the heating method, wall thicknesses as heavy as 4 in (100 mm) can be bent.

3. *Nonferrous pipe and tubes:* Although most of the equipment used to bend ferrous materials is also used for bending nonferrous materials, the details of bending do differ from those for ferrous materials and also vary between the several nonferrous materials themselves. Accordingly, it is wise to obtain specific procedural information from the materials' producers or from other reliable sources such as the latest edition of *The Metals Handbook.*[20] Certain nonferrous materials can be hot bent.

Aluminum and aluminum alloys can be bent cold using the same types of bending equipment used for ferrous materials. Alloys in the annealed condition are easiest to bend, but care is required in selecting tooling because of the low tensile strength and high ductility of these materials. Alloys with higher tempers and heat-treatable alloys require larger bending radii for satisfactory results. It is seldom necessary to heat aluminum for bending: however, non-heat-treated materials can be heated to 375°F (190°C) with minimal loss of properties. Heat-treated alloys require specific time-temperature control. More detailed information is available from the manufacturers of aluminum products.

Copper and copper alloy pipe and tube can be readily bent to relatively small radii. Although copper can be bent hot, the vast majority is done cold. For draw bending an internal mandrel is required and for other methods internal support is recommended. For very tight radii a snug fitting forming block and shoe which practically surround the pipe at the point of bending are needed to preclude buckling.

Hot bending of copper and copper alloys particularly in larger diameters and walls is common. Pipes are usually sand filled and contoured bending dies are recommended. See Table A6.1. More information can be obtained from the Copper Development Association.[21]

Nickel and nickel-alloy pipe can be cold bent with the same type of bending equipment used for ferrous materials. Use of material in the annealed condition is preferred. For bends with radii 6 diameters and less, filler material or internal mandrels are required. Draw bending with internal mandrels is the preferred method for close radius bending. Galling can become a problem and chromium-plated or hard bronze-alloy mandrels should be used.

Nickel and nickel alloys can be hot bent using the same practices as for ferrous steels. Sand filling is appropriate. Care should be taken to assure that the sand and heating fuel are low in sulphur and that any marking paints or crayons or lubricants have been removed. These materials can be bent over a wide temperature range. See Table A6.2. The best bending is usually between 1850 to 2100°F (1010 to 1149°C). Other nickel alloys may exhibit carbide precipitation and should not be worked in the sensitization range. Postbending heat treatment may be required. For more information contact nickel product manufacturers such as Huntington Alloys.[22]

TABLE A6.1 Temperature Range for Hot Bending of Copper and Copper Alloy Pipe and Tube

Material	Spec no.	Alloy	Temperature range (°F)
Copper deoxidized	SB-42, SB-75	C10200	1400–1600
Red brass	SB-43, SB-135	C23000	1450–1650
Copper silicon A	SB-315	C65500	1300–1600
70-30 CuNi	SB-466, SB-467	C71500	1700–2000
80-20 CuNi	SB-466	C71000	1600–1900
90-10 CuNi	SB-466, SB-467	C70600	1400–1800

Source: Adapted from ASME Boiler & Pressure Vessel Code 1989 ed., Section VIII Div. 1 Table NF-4.

TABLE A6-2 Temperature Range for Hot Bending of Nickel and Nickel Alloy Pipe and Tube

Material	Spec. no.	Alloy	Temperature range (°F)
Nickel	SB-161	N02200	1200–2300
Low carbon nickel	SB-161	N02201	1200–2300
Nickel-copper	SB-165	N04400	1700–2150
Ni-Cr-Fe	SB-167, SB-517	N06600, N06690	1850–2300
Ni-Fe-Cr	SB-407, SB-514	N08800, N08810	1850–2200

Source: Adapted from ASME Boiler & Pressure Vessel Code 1989 ed., Section VIII Div. 1 Table NF-4.

Titanium can be bent using draw bending equipment. However, those parts of the equipment which will wipe against the inner and outer surfaces of the pipe should be of aluminum bronze to minimize galling. For better formability, the pipe, the pressure die, and the mandrel should be heated to a temperature between 350 and 400°F (177 and 204°C). Unalloyed titanium can be hot worked in the temperature range of 1000 to 1400°F (538 to 760°C). Titanium alloy grade 12 requires a temperature range of 1400 to 1450°F (760 to 788°C). Heat treatment of titanium is recommended after forming. This is usually a furnace treatment at 1000 to 1100°F (538 to 593°C) for a minimum of ½ hr for the unalloyed grades and 1 h for the alloy (grade 12). Prolonged exposure to temperatures in excess of 1100°F (593°C) will result in heavy scaling and require some type of descaling treatment.

Other Forming Operations. Some additional forming operations which can be performed in a pipe shop are extrusion, swaging, and lapping. Extrusions involve forming outlets in pipe by pulling or pushing a hemispherical or conical die from the inside of the pipe through an opening in the wall. The work may be done hot or cold depending on the characteristics of the material. Ferritic steels, austenitic steels, and nickel alloys are usually formed hot; aluminum and copper are usually formed cold. In order to assure that the outlet will have sufficient reinforcement, it is necessary to increase the wall thickness of the header as a function of the outlet size desired. An increase of 30 percent may be needed for large outlet to header ratios. See Fig. A6.14.

Swaging involves the size reduction of pipe ends by forging, pressing, or rolling operations. The operation is usually used to produce reductions of one to two pipe sizes. Ferritic steels, austenitic steels, and nickel alloys are usually formed hot. Aluminum and copper are formed cold.

In lapped joints, a loose flange is slipped over the end of the pipe which is then heated to forging temperature, upset, and flared at right angles to the pipe axis. After heat treatment and cooling, the lapped section is machined on the face to attain a good gasket surface and on the back for good contact with the flange. The finished thickness of the lapped flange should be equal to or exceed the thickness of the pipe.

Layout, Assembly, and Preparation for Welding. In fabrication shops, piping subassemblies are often assembled on layout tables. A projection of the subassembly is laid out on the table in chalk. This establishes the base line for locating the components and terminal dimensions of the subassembly and the components are assembled relative to the layout. Prior to fit up, it is essential that all weld sur-

FIGURE A6.14 Extruding an outlet nozzle.

faces be properly cleaned of rust, scale, grease, paint, and other foreign substances which might contaminate the weld. If moisture is present, the weld joint should be preheated. For alloy steels the heat affected zone (HAZ) which results from thermal cutting should be removed by grinding or machining.

Depending on the configuration of the subassembly and root opening required by the welding procedure, some allowance may be required for weld shrinkage in the longitudinal direction. Actual shrinkage is difficult to predict and can vary considerably because of the many variables involved. For most open butt and backing ring joints, one-half the root opening is a reasonable allowance. For joints with other root configurations it may be as little as $\frac{1}{16}$ in for the lighter walls increasing to as much as $\frac{5}{32}$ in for walls 4 to 5 in thick.

Each weld joint should be carefully aligned within required tolerances using alignment fixtures, spacers, or jigs if necessary. Poor alignment may result in a poor weld. Once alignment is attained, the joint is usually tack welded to maintain the alignment. The process used for tacking is usually that being used for the root pass weld. Numbers and size of tacks should be kept to a minimum, but if the subassembly is to be moved elsewhere for weld out, their size must be sufficiently large so as not to crack during the moving operation. Temporary lugs or spacer bars may also be used for this purpose provided they are of a compatible material, the temporary welds are removed, and the surface examined after removal to assure sound metal. Tack welds made by the shielded metal arc welding (SMAW) or gas metal arc welding (GMAW) processes at the root of a weld should be removed or ground smooth since they can become a source of lack of fusion. For GTAW root welds, tacks usually fuse the adjacent lands to each other or to the insert, and filler metal is often not used. Tack welds are then fused into the weld during the root pass without further preparation. After tacking, the recommended practice is to complete the root pass and one or more weld out passes before starting to complete the weld by other processes, to avoid burning through the relatively thin root.

Welding. Welding constitutes the bulk of the work involved in fabrication of modern piping systems, so it is essential for all involved to have a good working knowledge of this subject.

Procedure and Personnel Qualifications. All of the ASME Boiler and Pressure Codes and most of the ASME B31 Pressure Piping Codes reference ASME Section IX for the requirements for qualifying welding procedures and welding personnel. The ASME B31.4,[23] B31.8,[24] and B31.11[25] Codes also permit qualification to API-1104,[26] published by the American Petroleum Institute. ASME B31.5[27] permits qualification to AWS D10.9.[28]

The purpose of procedure qualification is to assure that the particular combination of welding process, base metal, filler material, shielding fluxes or gases, electrical characteristics, and subsequent heat treatment is capable of producing a joint with the required chemical and physical characteristics.

The purpose of personnel qualification is to assure that the welder or welding machine operator is capable of performing the operation in accordance with a qualified procedure in the required position.

Procedure Qualification. ASME Section IX requires the preparation of a Welding Procedure Specification (WPS), which lists the various parameters to be used during welding. When each WPS is qualified, the parameters used in the qualification are recorded in a Procedure Qualification Record (PQR).

For each type of welding process, ASME Section IX has established a series of variables. These are base metal, filler metal, position, preheat, postweld heat treatment, shielding gases, joint configuration, electrical characteristics, and technique. Base metal must not only be considered from a chemical and physical properties point of view, but in piping, the diameter and thickness of the test coupon limits the qualification to certain sizes. Differing fluxes, use of solid or gaseous backing, and single- or multipass techniques are some of the other variables which must be considered. Careful study of Section IX, AWS D10.9, or of API 1104 as may be applicable is in order.

The variables for welding are classed as essential, supplementary essential, and nonessential. The manner in which the variables are classed can vary depending on the welding process. That is, what may be classed as an essential variable for one may be a nonessential variable for another. For a given process, each combination of essential variables must be qualified separately. A change in any one of them requires a new qualification.

When welds must meet certain fracture toughness requirements, the supplementary essential variables become essential and the procedure must be requalified for the particular combination of essential and supplementary essential variables.

Nonessential variables do not require requalification but should be referenced in the WPS.

Personnel Qualification. The fabricator and/or installer must qualify each welder or welding operator for the welding processes to be used during production welding. The performance qualification must be in accordance with a qualified WPS. Each performance qualification is also governed by a series of essential variables which are a function of the welding process for which the welder is being qualified.

The welder or welding operator may be qualified by mechanical tests or in some cases by radiographic examination of the test coupon. The record of each performance qualification is kept on a Welder/Welding Operator Performance Qualification (WPQ). Under ASME Section IX rules, a qualified welder who has not welded in a specific process within a specified period of time must be requalified for that process. API 1104 and AWS D10.9 have similar requalification provisions.

Welding Processes. Currently the most commonly used welding processes for fabrication of piping are SMAW, submerged arc welding (SAW), gas tungsten arc welding (GTAW), GMAW, and flux core arc welding (FCAW). Some special applications may involve plasma arc welding (PAW) or electron beam welding (EBW), but their application to piping is still rare. However, any welding process which can be qualified under the requirements of ASME Section IX is acceptable. Detailed descriptions of these various processes and their variations may be found in the *Welding Handbook.*[16] This section will limit discussion to their application to piping.

For shop work, the best efficiency in all welding processes is attained when the pipe axis is horizontal and the piece is rotated so that welding is always done in the flat position. This is referred to as the 1G position. Other positions are 2G (pipe vertical and fixed, weld horizontal), 5G (pipe horizontal and fixed, weld a combination of flat, vertical and overhead), and 6G (pipe inclined at 45° and fixed). See ASME Section IX.

Shielded Metal Arc Welding. SMAW has been the mainstay for pipe welding for many years, but it is rapidly being displaced by newer more efficient processes. It is a process where an arc is manually struck between the work and a flux-coated electrode which is consumed in the weld. The core wire serves as the filler material and the flux coating disintegrates to provide shielding gases for the molten metal, scavengers, and deoxidizers for the weld puddle and a slag blanket to protect the molten metal until it is sufficiently cool to prevent oxidation. It can be used in all positions, for upward or downward progression and for root pass welding depending on the flux composition. Each weld pass is about ⅛ in thick, and before subsequent passes are made the slag must be removed and the surface prepared by removing irregularities which could entrap slag during subsequent passes.

Submerged Arc Welding. Unlike SMAW, SAW is an automatic or semiautomatic process. For circumferential welds in pipe the welding head is fixed for flat welding and the work is rotated under the head (1G position). It is used most efficiently in groove butt welds in heavy wall materials with pipe sizes NPS 6 and larger. The arc is created between the work and a bare solid wire or composite electrode which is consumed during the operation. The electrode comes in coils. Shielding is accomplished by a blanket of granular, fusible material called a *flux* which covers the arc and molten metal by forming a slag blanket. Particular wire-flux combinations are required to assure that the deposited weld has the needed chemical and physical properties. This process has the greatest deposition rate and accordingly is the preferred process wherever possible. Because of the high heat input, care must be taken to assure that the interpass temperature is controlled to minimize sensitization in austenitic stainless steels or loss of notch toughness in ferritic steels. High heat input can also result in excessive penetration, so this process cannot be used effectively for root pass welding unless the root is deposited against a backing ring or sufficient backing is provided by two or more weld passes made by the shielded metal arc or gas shielded arc process.

Gas Shielded Arc Welding. The term *gas shielded arc welding* applies to those welding processes where the arc and molten metal are shielded from oxidation by some type of inert gas rather than by a flux.

1. *Gas tungsten arc welding:* GTAW is a form of gas shielded arc welding where the arc is generated between the work and a tungsten electrode which is not consumed. The filler metal must be added from an external source, usually as bare filler rod or preplaced consumable insert. The filler metal is melted by the heat of the arc and shielding gases are usually argon or helium. Alloying elements are always in the filler material. GTAW is considered to be the most desirable process for making root welds of highest quality. Techniques using added filler metal or preplaced filler metal as inserts are equally effective in manual and automatic applications.

Automatic versions can be used in all positions provided sufficient clearance is available for the equipment. Automatic versions also require tighter fit-up requirements since the equipment is set to specific parameters and will not recognize variations outside of tolerance such as a welder would do in manual applications.

In automatic GTAW, the welding head orbits the weld joint on a guide track placed on the pipe adjacent to the joint to be welded. The welding head contains motors and drive wheels needed to move the head around the track, a torch to create the arc, and a spool of filler wire. Welding current, voltage, travel speed, wire feed rate, and oscillation are controlled from an external source. These parameters may be varied by the operator as the welding head traverses the weld. Oscillation and arc energy can be adjusted to permit greater dwell time and heat input into the side walls. Automatic GTAW welds are usually deposited as a series of stringer beads to minimize the effects of high interpass temperature.

2. *Gas metal arc welding:* GMAW is a type of gas shielded welding generally used in the manual mode but adaptable to automation. The filler wire is the electrode. It is furnished in coils or spools of solid wire which is fed automatically into the joint, melted in the arc, and deposited in the weld groove. Alloying elements are in the wire and shielding gas may be argon, helium, nitrogen, carbon dioxide, or combinations thereof, depending on the application.

Depending on the equipment and the heat input settings, filler metal can be transferred across the arc in several modes. In short-circuiting transfer, the electrode actually touches the work where it short circuits, melts, and restarts the arc. This process has low heat input and accordingly low penetrating power. It can often result in lack of fusion. Because of the low heat input, however, it can be effectively used for open butt root pass welding.

In spray transfer, the heat input parameters are sufficiently high to transfer the molten electrode across the arc as small droplets. Argon or argon-rich gases are used for shielding, resulting in a very stable spatterfree arc. Because of the high arc energy it is normally used in the flat (1G) position. For all position welding, a procedure which superimposes high amplitude pulses of current on a low-level steady state current at regular intervals is often used. This results in a discrete transfer of metal with lower heat input needed for all position welding.

3. *Flux core arc welding:* FCAW is a variation of GMAW where a composite electrode is substituted for the solid wire. The electrode is a tubular wire containing a flux material. Depending on the application, the arc may be self-shielding, or shielding gases may be used. Because of its high deposition rate this process is rapidly being developed for shop and field welding of piping.

Base Metal. Base metal is one of the essential variables for welding qualification. Because there are so many base metals to be welded, ASME Section IX has established a system of P Numbers and Group Numbers. Each base metal is assigned to a specific P Number depending on characteristics such as composition, weldability, and mechanical properties. Each P Number is further subdivided into Group Numbers depending on fracture toughness properties. See Table A6.3. When a procedure is qualified with a base metal within a particular P Number, it is also qualified for all other base metals within that P Number. When fracture toughness is a requirement, qualification is limited to base metals within the same P Number *and* Group Number. For example: A-106 Gr. B pipe is P-No.1 Gr.No.1, while an A-105 flange is P-No.1 Gr.No.2. Since both are P-No.1,

TABLE A6.3 ASME P Numbers and Group Numbers for Some Typical Piping Materials

P no.	Group no.	Nominal composition and strength
1	1	Carbon steel 60 ksi and under
	2	60 to 75 ksi
3	1	C-½ Mo, ½ Cr-½ Mo 65 ksi and under
	2	70 to 75 ksi
4	1	1 Cr-½ Mo, 1¼ Cr-½ Mo
5	1	2¼ Cr-1 Mo, 3 Cr-1 Mo
	2	5 Cr-½ Mo, 7 Cr-½ Mo, 9 Cr-1 Mo
8	1	Type 304, 316 stainless
	2	Type 309, 310 stainless
9B	1	3½ nickel steel

Source: Selected from ASME Boiler & Pressure Vessel Code Section IX, 1989 ed.

qualification on either qualifies both when fracture toughness is not a factor. However, should fracture toughness become a requirement, a separate qualification would be required for each to itself and to each other.

Filler Metals. Electrodes, bare wire, wire-flux combinations, and consumable inserts which form a part of the finished weld are classed as filler materials. Most are covered by AWS and ASME specifications. See ASME Section II Part C.[29]

When the filler material is part of the electric circuit, it is designated as an electrode. If it is fed externally and melted by the heat of the arc, it is designated as a rod. Coated electrodes for SMAW come in straight lengths. Bare rods for GTAW come in straight lengths or spools. Electrode wire for GMAW and SAW are in spools or coils, while composite electrodes for FCAW are in spools.

Each specification incorporates a system of identification so that the filler materials manufactured by different suppliers which have equivalent characteristics are identified by the same number.

For qualification purposes, they are classified in ASME Section IX with F Numbers and A Numbers. Changes in filler metal from one F Number or A Number to another require requalification.

One of the problems associated with coated electrodes for SMAW is the introduction of hydrogen into the arc atmosphere and finished weld resulting in hydrogen-induced cracking. To minimize this problem low-hydrogen-type coatings are used, but these can absorb moisture from the atmosphere. Once a sealed can of electrodes is opened, the electrodes should be stored in an oven at about 250°F or other temperature recommended by the manufacturer. Baking to remove moisture is recommended for electrodes which have been out of the oven for several hours. Refer to the manufacturers' recommendations.

A problem associated with welding of fully austenitic stainless steel is microfissuring. To combat this problem the chemical composition of the filler material is adjusted to produce a weld deposit with small amounts of ferrite. ASME III requires that electrodes and rods used in welding austenitic stainless steels contain a minimum of 5 percent ferrite. Ferrite, however, can be a problem at cryogenic and high temperatures. For cryogenic services the weld metal may not possess the fracture toughness capabilities of the base metal, and the ferrite content should be kept as low as possible. Alternatively, fully austenitic fillers may be required, but these are more crack sensitive. For very high temperatures ferrite in the weld may convert to a brittle phase called *sigma.* For this reason applications over about 800°F usually require a minimum of 3 percent ferrite for weldability but not exceeding 7 percent to minimize sigma formation.

Preheat and Interpass Temperature. Ferritic materials undergo metallurgical phase changes when cooling from welding temperature to ambient. Mild steels which contain no more than 0.20 percent carbon and 1 percent manganese can be welded without preheat when the thickness is 1 in or less. However, as the chemical composition changes by increases of carbon, manganese, and silicon or the addition of chromium, and certain other alloying elements, preheating becomes increasingly important since the higher carbon and chrome molybdenum steels can develop undesirable martensitic, matensitic-bainitic, and other mixed phase structures when cooled rapidly from welding temperatures.

There is also a potential for hydrogen from SMAW electrode coatings or from moisture on the base metal surface to be dissolved in the weld. Also as the weld cools, stresses caused by shrinkage are imposed on the parts and distortion can result, and as thickness increases, thermal shock from the heat of welding can induce cracking more readily.

Preheating prior to welding is a solution to most of these problems. Preheating slows the cooling rate of the weld joint and results in a more ductile metallurgical structure in the weld metal and HAZ. It permits dissolved hydrogen to diffuse more readily and helps to reduce shrinkage, distortion, and possible cracking caused by the resultant residual stresses. It raises the temperature of the material sufficiently high to be above the brittle fracture transition zone for most materials.

The codes vary regarding preheat requirements. Some have mandatory requirements while others give suggested levels. For example, for carbon steel welding, the B31.1 Code *requires* preheating to a temperature of 175°F when the carbon content exceeds 0.30 percent *and* the thickness of the joint exceeds 1 in. B31.3 *recommends* preheating to 175°F when the base metal specified strength exceeds 71 ksi *or* the wall thickness is equal to or greater than 1 in. ASME III Section *suggests* a preheat of 200°F when the maximum carbon content is 0.30 percent or less *and* the wall thickness exceeds 1½ in for P-No.1 Gr.No.1, or 1 in for P-No.1 Gr.No.2. It also suggests a 250°F preheat for materials with carbon in excess of 0.30 percent and wall thicknesses exceeding 1 in. The ASME B31.4 and B31.8 Codes require preheat based on carbon equivalents. When the carbon content (by ladle analysis) exceeds 0.32 percent, or the carbon equivalent (C + ¼ Mn) exceeds 0.65 percent, preheating is required. The reader is advised to consult the specific codes for preheating requirements. See Table A6.4 for some typical preheat requirements.

While it is preferred that preheat be maintained during welding and into the postweld heat treatment cycle without cooling, this may not always be practical. The B31.1 Code permits slow cooling of the weld to room temperature provided the completed weld deposit is a minimum of ⅜ in or 25 percent of the final thickness, whichever is less. For P-No.5 Gr.No.2 materials some type of intermediate stress relief is required.

Too much heat during welding can also be a problem. Where notch toughness is a requirement, prolonged exposure to temperatures exceeding 600°F can temper the base metal. Controlling the interpass temperature is required to minimize this problem. Interpass temperature control means allowing the temperature of the joint to cool below some specified level before the next pass is deposited.

In welding of austenitic stainless steels, sensitization of the base metal HAZ

TABLE A6.4 Typical Preheat Requirements

P no.	Temp. (°F)	Composition and thickness limits
1	175	*Both* a max. specified carbon content >0.30% *and* thickness >1 in
	50	All others
3	175	*Either* a min. specified tensile strength >60 ksi, *or* thickness >½ in
	50	All others
4	250	*Either* a min. specified tensile strength >60 ksi, *or* thickness >½ in
	50	All others
5	400	*Either* a min. specified tensile strength >60 ksi, *or* *both* a min. specified Cr content >6.0% *and* thickness >½ in
	300	All others
6	400	All materials
7	50	All materials
8	50	All materials
9	250	All P-No. 9A materials
	300	All P-No. 9B materials
10E	300	Max. interpass temperature of 450°F

Source: ASME B31.1 1989 ed.

will result from the heat of welding. Here the solution is to weld with as low a heat input as possible, at the highest possible speed to minimize the precipitation of carbides (sensitization). A maximum interpass temperature of 300 to 350°F is usually employed.

Weld Joint Design

Butt Welds. A butt joint is defined as one in which the members being joined are in the same plane. The circumferential butt joint is the most universally used method of joining pipe to itself, fittings, flanges, valves, and other equipment. The type of end preparation may vary depending on the particular preferences of the individual, but in general the bevel shape is governed by a compromise between a root sufficiently wide to assure a full penetration weld but not so wide as to require a great deal of filler metal.

In the shop, the inside surface of large-diameter pipe joints is often accessible. In this case the joint is most often double welded (welded from both sides) and a double V bevel is used. For heavier walls, machined double U bevels can be used. However, the vast majority of piping butt welds must be made from one side only. For this situation the most frequently specified shapes are the V bevel, Compound bevel, and U bevel, all of which can have varying angles, lands, and tolerances. See Fig. A6.15. Recent advances in SAW narrow gap welding as applied to piping butt welds have cut the volume of filler metal significantly in pipe walls that are 2 in and thicker. The 30° or 37½° (60 or 75° included angle) V bevel is most often performed integrally with the cutting operation by machine, oxyfuel gas, or arc cutting. Other bevel shapes such as the compound V, U, J bevels, or combinations thereof require machining in lathes or boring mills.

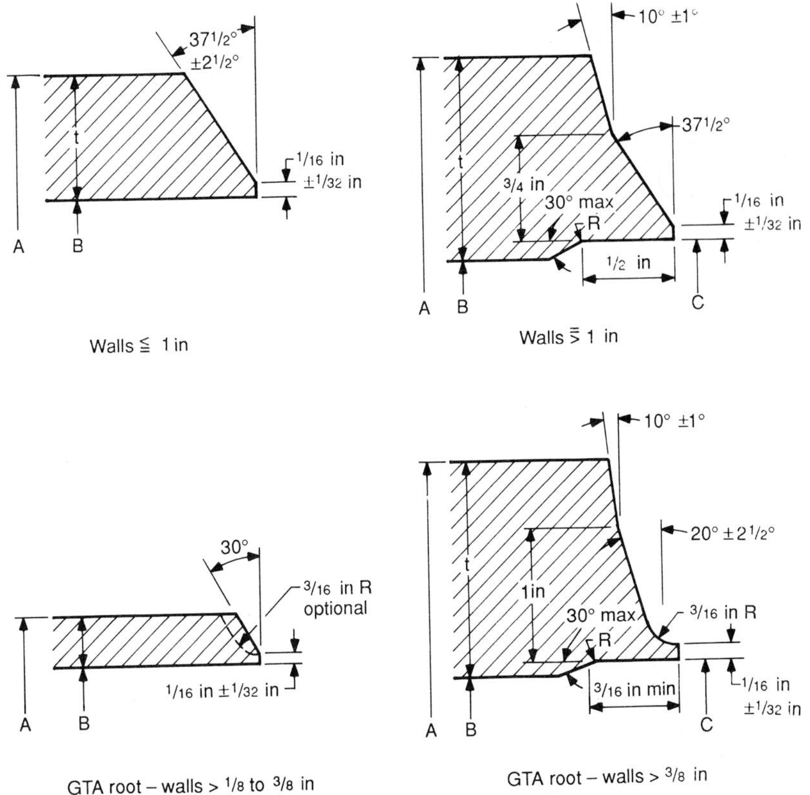

FIGURE A6.15 Typical weld end bevels. (*a*) Walls ≤1 in; (*b*) walls >1 in; (*c*) GTA root walls > ⅛ to ⅜ in; (*d*) GTA root walls > ⅜ in.

1. *Alignment:* Alignment for butt welding can often be a frustrating task since it is influenced by the material; pipe diameter, wall thickness, and out of roundness tolerances; welding process needs; and design requirements.

When a joint can be double welded, the effects of misalignment are minimized since both inner and outer weld surfaces can be blended into the base metal and any remaining offsets can be faired out. ASME Section III gives a table of allowable offsets due to misalignment in double-welded joints. See Table A6.5. All resulting offsets must be faired to a 3:1 taper over the finished weld.

For single-welded joints alignment can be more difficult since the inside surface is not accessible. The degree of misalignment is influenced by many factors and depending on the type of service application may or may not be significant. The various codes impose limits on inside diameter misalignment. This is to assure that the stress intensification resulting from the misalignment is kept within a reasonable value. The B31.1 Code requires that the misalignment between ends to be joined not exceed 1⁄16 in, unless the design specifically permits greater amounts. See Fig. A6.16. The B31.4 and B31.8 Codes do not require special treatment unless the difference in the nominal walls of the adjoining ends exceed 3⁄32 in. ASME Section III on the other hand requires that the inside *diameters* of

TABLE A6.5 Maximum Allowable Offset in Joints Welded from Both Sides

Section thickness (in)	Direction of joints	
	Longitudinal	Circumferential
Up to ½, incl.	¼t	¼t
Over ½ to ¾, incl.	⅛ in	¼t
Over ¾ to 1½, incl.	⅛	3/16 in
Over 1½ to 2, incl.	⅛ in	⅛t
Over 2	Lesser of 1/16t or ⅜ in	Lesser of ⅛t or ¾ in

Source: ASME Boiler & Pressure Vessel Code Section III 1989 ed.

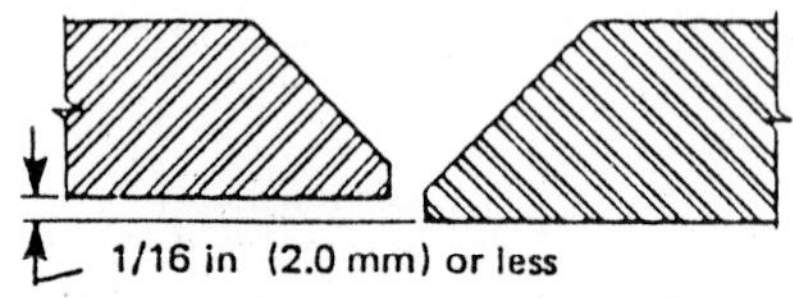

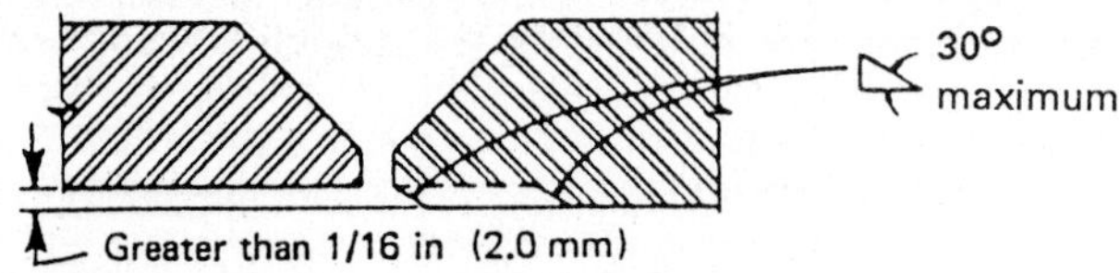

FIGURE A6.16 Butt welding of piping components with internal misalignment. (*ASME B31.1 Power Piping Code, 1989 ed.*)

the adjoining sections match within 1/16 in to assure good alignment. Counterboring is usually required to attain this degree of alignment.

The welding process and NDEs to be employed also bear on misalignment limits. Some welding processes can tolerate fairly large misalignments while others, notably gas tungsten arc root pass welding with and without consumable inserts require closer tolerances. See PFI ES-21.[30] Radiographic or ultrasonic examinations of misaligned areas may show unacceptable indications if the degree of misalignment is too great.

A review of the tolerances permitted in the manufacture of various types of pipe, fittings, and forgings immediately reveals that in many situations the probable inside diameter and wall thickness variations more often than not will produce unacceptable misalignment situations. Out of roundness in lighter wall materials can add to the problem.

When most of the pipe comes from the same rolling and the fittings from the same manufacturing lot, variations in tolerances are minimal and the pipe and fittings can be assembled for most common applications without a great deal of adjustment. Out of round problems in lighter walls are handled with internal or external round-up devices.

To assure that all components will be capable of alignment in the field, it is common practice for the designer to specify that the inside diameters of all matching components be machine counterbored to some specified dimension. This practice is also desirable for shop welding of heavier wall piping subassemblies. PFI ES-21 contains a set of uniform dimensions for counterboring of seamless hot-rolled pipe ordered to A106 or A335 by NPS and schedule number. See Table A6.6.

The C dimension is determined from the following equation:

$$C = A - \tfrac{1}{32}\text{ in} - 2 \times t_m - 0.010\text{ in}$$

where A = pipe outside diameter
$\frac{1}{32}$ in = under tolerance
t_m = the mill minimum wall = 0.875XT nominal
0.010 in = a boring tolerance

For other types of seamless pipe, longitudinally welded pipe, forged and bored pipe, and other specialties, the tolerances on the outside diameter and wall thickness are different. The machining tolerance required for some welding processes may also be different. However, similar logic may be applied in determining C dimensions for these products.

It should be noted from Table A6.6 that the tabulation applies to wall thickness greater than ½ in. While one can calculate a C dimension for lighter walls, the combination of outside diameter tolerance and wall thickness tolerance will usually result in a calculated C which is often smaller than the actual bore of the pipe. The difference is most often relatively small and the existing diameter will usually be suitable for alignment of most welds. In those cases where it is considered essential, the outside diameter at the end can be sized to provide stock for machining, but care is required to assure that the minimum wall is maintained. Where counterboring is used, the machined surface should taper into the existing inside surface at an angle of 30° maximum. See Fig. A6.17.

There are many instances where round-up devices and counterboring are insufficient remedies for misalignment. On occasion it may be necessary to expand the ends where counterboring would violate minimum wall requirements. Most of the codes permit the use of weld metal deposits (weld buildup) both on the inside and outside surfaces of the weld end in order to attain the required alignment. In using this alternative, consideration must be given to other factors such as radial shrinkage, imperfections in the weld buildup which may show on NDE, need for pre- and postweld heat treatment, and possible sensitization of austenitic stainless steels.

2. *Unequal wall thickness:* In most piping systems there are components such as valves, castings, heavier header sections, and equipment nozzles which are welded to the pipe. In such instances the heavier sections are machined to match the lighter pipe wall and the excess thickness tapered both internally and externally to form a transition zone. Limits imposed by the various codes for this transition zone are fairly uniform. The external surface of the heavier component is tapered at an angle of 30° maximum for a minimum length equal to 1½ times the pipe minimum wall thickness and then at 45° for a minimum of ½ times the pipe minimum wall. Internally, either a straight bore followed by a 30° slope or a taper bore at a maximum slope of 1 to 3 for a minimum distance of 2 times the pipe minimum wall are required. See Fig. A6.17. The surface of the weld can also be tapered to accommodate differing thickness. This taper should not exceed 30° although some codes limit the taper to 1 to 4. It may be necessary to deposit weld metal to assure that these limits are not violated.

TABLE A6.6 Internal Machining for Circumferential Butt Welds

Nominal pipe size	Schedule number or wall	Nominal O.D. A (in)	Nominal I.D. B (in)	Nominal wall thickness t (in)	Machined I.D. of Pipe C Tolerance +0.010, −0.040 (in)
3	XXS	3.500	2.300	0.600	2.409
4	XXS	4.500	3.152	0.674	3.279
5	160	4.500	4.313	0.625	4.428
	XXS	5.563	4.063	0.750	4.209
6	120	6.625	5.501	0.562	5.600
	160	6.625	5.187	0.719	5.327
	XXS	6.625	4.897	0.864	5.072
8	100	8.625	7.437	0.594	7.546
	120	8.625	7.187	0.719	7.327
	140	8.625	7.001	0.812	7.163
	XXS	8.625	6.875	0.875	7.053
	160	8.625	6.813	0.906	6.998
10	80	10.750	9.562	0.594	9.671
	100	10.750	9.312	0.719	9.452
	120	10.750	9.062	0.844	9.234
	140	10.750	8.750	1.000	8.959
	160	10.750	8.500	1.125	8.740
12	60	12.750	11.626	0.562	11.725
	80	12.750	11.374	0.688	11.507
	100	12.750	11.062	0.844	11.234
	120	12.750	10.750	1.000	10.959
	140	12.750	10.500	1.125	10.740
	160	12.750	10.126	1.312	10.413
14 O.D.	60	14.000	12.812	0.594	12.921
	80	14.000	12.500	0.750	12.646
	100	14.000	12.124	0.938	12.319
	120	14.000	11.812	1.094	12.046
	140	14.000	11.500	1.250	11.771
	160	14.000	11.188	1.406	11.498
16 O.D.	60	16.000	14.688	0.656	14.811
	80	16.000	14.312	0.844	14.484
	100	16.000	13.938	1.031	14.155
	120	16.000	13.562	1.219	13.827
	140	16.000	13.124	1.438	13.442
	160	16.000	12.812	1.594	13.171
18 O.D.	40	18.000	16.876	0.562	16.975
	60	18.000	16.500	0.750	16.646
	80	18.000	16.124	0.938	16.319
	100	18.000	15.688	1.156	15.936
	120	18.000	15.250	1.375	15.553
	140	18.000	14.876	1.562	15.225
	160	18.000	14.438	1.781	14.842

TABLE A6.6 Internal Machining for Circumferential Butt Welds (*Continued*)

Nominal pipe size	Schedule number or wall	Nominal O.D. *A* (in)	Nominal I.D. *B* (in)	Nominal wall thickness *t* (in)	Machined I.D. of Pipe *C* Tolerance +0.010, −0.040 (in)
20 O.D.	40	20.000	18.812	0.594	18.921
	60	20.000	18.376	0.812	18.538
	80	20.000	17.938	1.031	18.155
	100	20.000	17.438	1.281	17.717
	120	20.000	17.000	1.500	17.334
	140	20.000	16.500	1.750	16.896
	160	20.000	16.062	1.969	16.515
22 O.D.	—	22.000	20.750	0.625	20.865
	60	22.000	20.250	0.875	20.428
	80	22.000	19.750	1.125	19.990
	100	22.000	19.250	1.375	19.553
	120	22.000	18.750	1.625	19.115
	140	22.000	18.250	1.875	18.678
	160	22.000	17.750	2.125	18.240
24 O.D.	30	24.000	22.876	0.562	22.975
	40	24.000	22.624	0.688	22.757
	60	24.000	22.062	0.969	22.265
	80	24.000	21.562	1.219	21.827
	100	24.000	20.938	1.531	21.280
	120	24.000	20.376	1.812	20.788
	140	24.000	19.876	2.062	20.350
	160	24.000	19.312	2.344	19.859

Source: Pipe Fabrication Institute PFI ES-21.

Fillet Welds. Circumferential fillet welds are used in piping systems to join slip-on flanges and socket welding fittings and flanges to pipe. In welding slip-on flanges to pipe, the pipe is inserted into the flange and welded with two fillet welds, one between the outside surface of the pipe and the hub of the flange and the other between the inside surface of the flange and the thickness of the pipe. See Fig. A6.18. Alignment is relatively simple since the pipe fits inside the flange. The B31.1 Code requires that the fillet between the hub and the pipe have a minimum weld leg of 1.09 times the pipe nominal wall or the thickness of the hub, whichever is smaller. The weld leg of the front weld must be equal to the pipe nominal wall or ¼ in, whichever is smaller. The gap between the outside diameter of the pipe and flange inside diameter may increase with size, so the size of the fillet leg should be adjusted to compensate for this situation.

Fillet welds are also used for circumferential welding of pipe to socket fittings. Socket weld fittings and flanges are available in sizes up to NPS 4 but are most frequently used in sizes NPS 2 and smaller. Alignment is not a problem since the pipe fits into the fitting socket. Some codes require that the fillet have uniform leg sizes equal to 1.09 times the pipe nominal wall or be equal to the socket wall, whichever is smaller. In making up socket joints it is recommended that the pipe not be bottomed in the socket before welding. B31.1 and ASME Section III suggest a 1⁄16-in gap. In high-temperature service especially, the pipe inside the socket will expand to a greater degree than the socket itself, and the differential

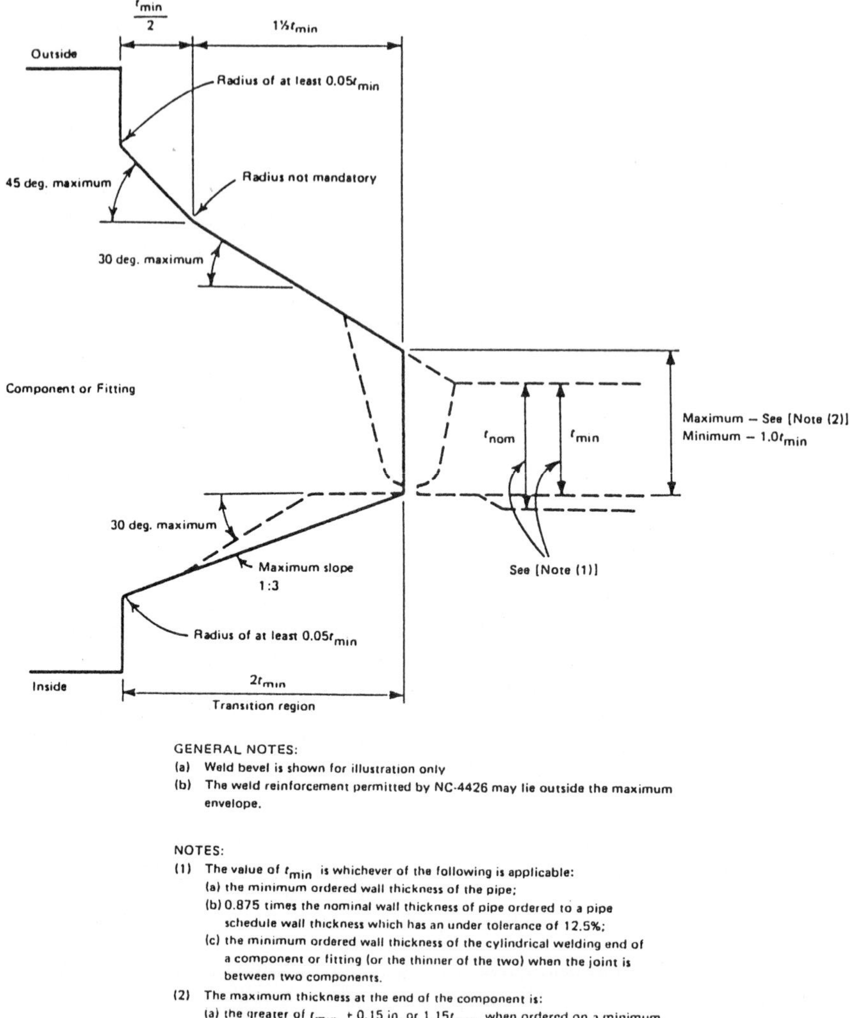
$\frac{t_{min}}{2}$
$1\frac{1}{2}t_{min}$
Outside
Radius of at least $0.05t_{min}$
45 deg. maximum
Radius not mandatory
30 deg. maximum
Component or Fitting
t_{nom}
t_{min}
Maximum – See [Note (2)]
Minimum – $1.0t_{min}$
30 deg. maximum
Maximum slope 1:3
See [Note (1)]
Radius of at least $0.05t_{min}$
Inside
$2t_{min}$
Transition region
GENERAL NOTES:
(a) Weld bevel is shown for illustration only
(b) The weld reinforcement permitted by NC-4426 may lie outside the maximum envelope.
NOTES:
(1) The value of t_{min} is whichever of the following is applicable:
(a) the minimum ordered wall thickness of the pipe;
(b) 0.875 times the nominal wall thickness of pipe ordered to a pipe schedule wall thickness which has an under tolerance of 12.5%;
(c) the minimum ordered wall thickness of the cylindrical welding end of a component or fitting (or the thinner of the two) when the joint is between two components.
(2) The maximum thickness at the end of the component is:
(a) the greater of t_{min} + 0.15 in. or $1.15t_{min}$ when ordered on a minimum wall basis;
(b) the greater of t_{min} + 0.15 in. or $1.0t_{nom}$ when ordered on a nominal wall basis.

FIGURE A6.17 Welding end transitions—maximum envelope. (*ASME Boiler and Pressure Vessel Code, Sec. III, 1989 ed.*)

expansion may result in unwanted shear stress in the fillet and possible cracking during operation.

Intersection-Type Weld Joints. Intersection-type weld joints occur when the longitudinal axes of the two components meet at some angle. Such is the case where nozzle, lateral, and wye intersections are fabricated by welding. Weld joints in these cases may be butt, fillet, or a combination thereof. Nozzles are made either by set-on or set-through construction. In set-on construction, the

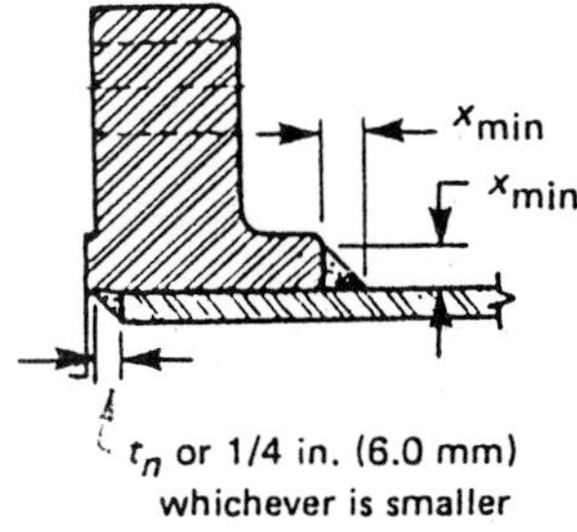

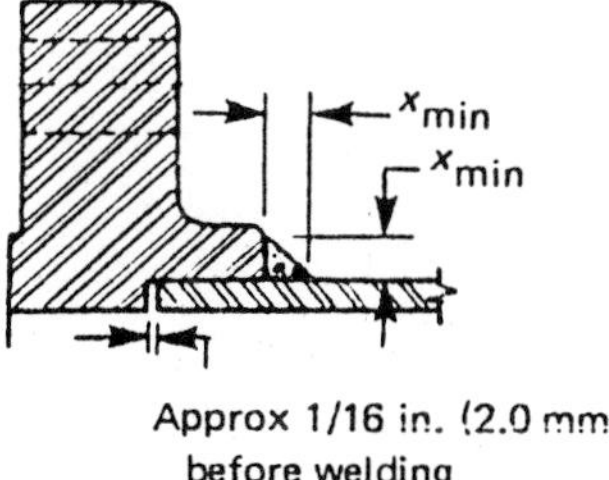

t_n = nominal pipe wall thickness

x_{min} = 1.09t_n or thickness of the hub, whichever is smaller

FIGURE A6.18 Slip-on and socket welding flange welds. (*ASME B31.1, 1989 ed.*)

opening in the header pipe is made equal to the inside diameter of the branch pipe. The branch pipe is contoured to the outside diameter of the header and beveled so that the weld is made between the outside surface of the header and through the thickness of the branch. The through thickness weld is covered by a fillet weld to blend it into the header pipe surface. In set-through construction an opening is cut in the header pipe equal to the outside diameter of the branch pipe and beveled. The branch pipe is contoured to match the inside diameter of the header. See Fig. A6.19. The weld is between the outside surface of the branch and through the thickness of the header and is covered with a fillet weld to blend it into the outside surface of the branch. Either type of construction is acceptable; the usual practice is to set-on since the volume of required weld metal is less. However, when the header is made from a plate product which may contain laminations, set-through construction is preferred.

Small nozzles are frequently made with socket welding or threaded couplings set on the header. In these cases it is difficult to assure complete root penetration and specially designed couplings which permit drilling through the bore to remove the root of the weld are often used. See Fig. A6.19.

Welded nozzle construction cannot be used at the full rating of the pipe involved and suitability for particular pressure temperatures must be verified by component design methods found in Part B of this book. In all cases there must be a through thickness weld of the branch to the header. Where reinforcing pads are used, they should also be joined to the header by a weld through their thickness. See Fig. A6.19 for typical details. In designing headers with multiple outlet nozzles, sufficient clearance is needed between adjacent nozzles to provide accessibility for welding. Nozzles with reinforcing pads or flanges need greater clearance. PFI ES-7[31] gives suggested minimum spacings.

Root Pass Welding. The integrity of any weld rests primarily with the quality of the root pass. In double-welded joints the root pass serves as a backing for passes welded from the first side. Before welding begins from the opposite side, the root area is usually removed to sound metal. In most cases, however, pipe welds must be made from one side only and the inside surface of the root weld is not accessible for conditioning.

Backing Rings. The earliest solution to root pass welding was the use of a backing ring using the SMAW process. This usually assured good penetration

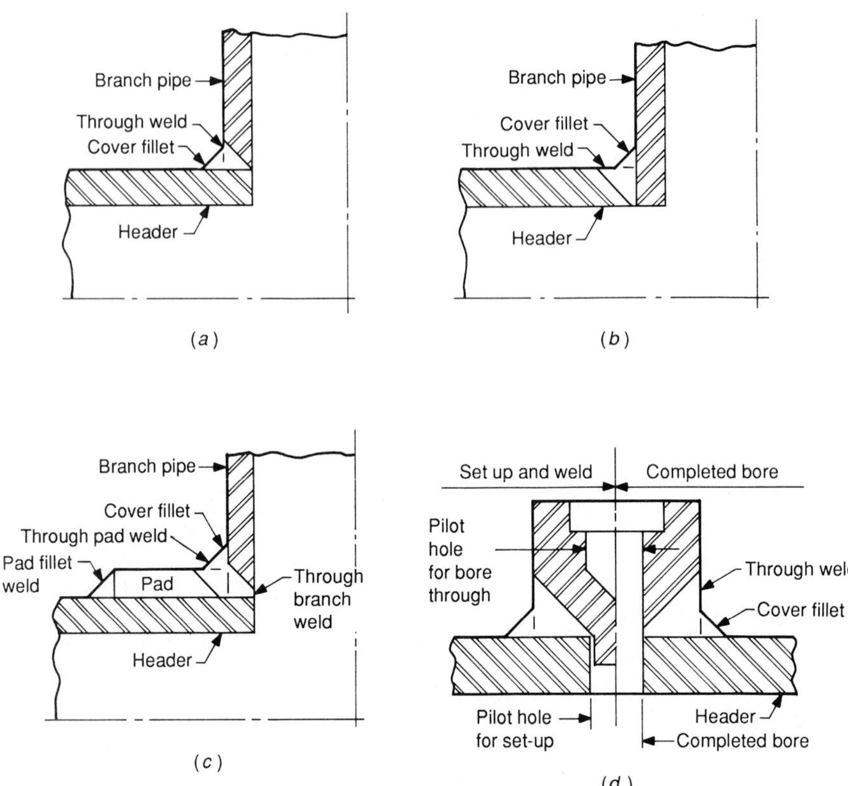

FIGURE A6.19 Types of branch nozzle construction. (*a*) Set-on; (*b*) set-through; (*c*) set-on with reinforcing pad; (*d*) special drill through socket weld coupling.

and is still used for many applications. However, commercial rings used with nominal pipe dimensions may result in unwanted flow restriction, crevices for entrapment of corrosion products, and notch conditions which could result in cracking during service. Prior to the introduction of GTAW root welding, piping systems which required the highest possible quality were welded using counterboring of the pipe to close tolerances and machined backing rings. This reduced problems significantly, but the crack potential still remained. See PFI ES-1.[32]

Open Butt Root Welds. In petrochemical services backing rings often could not be used, and the practice of open butt welding with shielded metal arc electrodes was and still is used. Welders require considerably more skill. Welding is most often performed with E-XX10 electrodes, which are more controllable than the low-hydrogen types but are also more prone to porosity.

GTAW Root Welds. The introduction of GTAW represented a breakthrough in root pass welding. Because of the greater expense involved, its application is usually limited to applications requiring high-quality root welds. The weld end bevels are carefully prepared by machining and counterboring where necessary to meet the close tolerances required. The joint involves butted or open lands and the weld is made with filler metal added or with a preplaced consumable insert.

The latter have a decided advantage in that they eliminate a good deal of the variability introduced by hand feeding of filler wire. Consumable inserts come in a variety of shapes each requiring somewhat differing fit-up tolerances. See PFI ES-21. Some types can be used for root pass welding in lighter wall materials (½ in and less) without the need for counterboring. Depending on the service, the inside surface of the molten weld puddle is often shielded from oxidation by an inert gas inside the pipe contained between dams. See Fig. A6.20. A small, controlled, positive pressure on the backing gas can aid in better controlling the shape of the root inside diameter.

When the root pass is made by the GTAW process, the resulting finished weld is relatively thin. In depositing the second and third passes, the first pass may be remelted. As it resolidifies, it shrinks radially, resulting in a small concave depression on the inside of the weld. This condition is usually considered acceptable provided the resulting thickness through the finished weld is equal to or greater than the required minimum wall, and the concavity blends smoothly into the adjacent base metal.

GMAW Root Welds. Many fabricators and/or installers take advantage of the low penetrating power of GMAW in the short-circuiting mode to use it for open butt root pass welding where the quality level of GTAW root pass welding is not required. The balance of the weld is made by other processes. Care must be taken to assure that unmelted wire does not penetrate the joint and remain.

Welding of Ferrous Piping Materials

Carbon Steels. Carbon steels are classed as P-No.1 by ASME Section IX. See Table A6.3. The vast majority of carbon steel pipe is used for services below 775°F. Joints are most often V bevels with commercial backing rings or open butt roots and are welded out with SMAW, SAW, GMAW, and FCAW. For services which require high quality, GTAW root welds with SMAW, SAW, and FCAW weld-outs are most prevalent. Most carbon steel filler metal is produced to weld 60,000- and 70,000-psi material. More often than not fabricators use the 70,000-psi filler for all carbon steel welding. For SMAW the most popular electrode is E-7018, although for open butt root pass welding using SMAW E-6010 is still the choice. FCAW welding is rapidly replacing SMAW because it can deposit at a much higher rate. Preheating and postweld heat treating are required depending on the carbon content and wall thickness. For typical preheat and postweld heat treatment requirements see Tables A6.4 and A6.7. When working to a specific code, be sure to use the requirements found in that code.

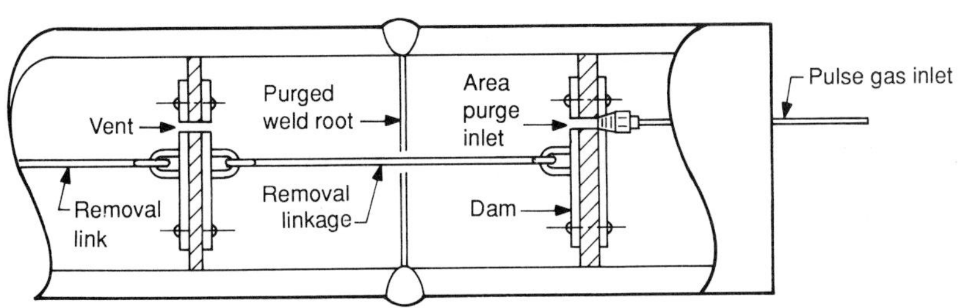

FIGURE A6.20 Typical shop purging arrangement.

TABLE A6.7 Some Typical Time and Temperature Cycles for Heat Treatment

P no.	Heating rate	Holding temperature range*				Minimum holding time at temperature	Cooling program
		SR or T	N	A	CST		
P-1	Above 800°F heat at a rate of 400°F/h divided by the thickness in inches but not faster than 400°F or less than 100°F	1100–1250°F	1600–1700°F	1500–1600°F	N/A	1 h/in of thickness but not less than 30 min or more than 2 h plus 15 min for each additional inch over 2 in	SR or T—Cool at 400°F/h divided by the thickness in inches but not faster than 400°F/h; need not be lower than 100°F/h down to 800°F N—Remove from furnace at normalizing temperature and cool in still air to 800°F; temper if necessary A—Furnace cool to 800°F at a rate of 400°F/h divided by the thickness in inches but not faster than 400°F/h; need not be slower than 100°F/h
P-3	Same as P-1	1100–1250°F	1600–1700°F	1500–1600°F	N/A		
P-4	Same as P-1	1100–1250°F	1725–1775°F	1625–1675°F	N/A	1 h/in of thickness but not less than 30 min or more than 5 h plus 15 min for each additional inch over 5 in	
P-5	Same as P-1	1250–1400°F	1725–1775°F	1625–1675°F	N/A		
P-8	Same as P-1	Not required	N/A	N/A	1900–2000°F	1 h/in of thickness but not less than 30 min or more than 2 h plus 15 min for each additional inch over 2 in	CST—Remove from furnace at holding temperature and quench in water to 300°F within 2 min

*SR = stress relief, T = temper, N = normalize, A = anneal, CST = carbide solution treatment
Source: Pullman Power Products Corporation.

Carbon Molybdenum Steels. Carbon molybdenum steels are classed as P-No.3. Currently this material has very little use because of unfavorable experience with graphitization at temperatures over 800°F.

Chromium Molybdenum Steels. The chromium molybdenum steels are primarily used for service temperatures from 800 to 1050°F. They range from ½ Cr-½ Mo to 9 Cr-1 Mo and are classed by ASME Section IX as P-No.3, P-No.4, and P-No.5. The preponderance of usage is in the 1¼ Cr-½ Mo-Si and 2¼ Cr-1 Mo grades. Welding usually consists of GTAW root welds with filler metal added or preplaced inserts. The balance of the weld is made by SAW for welds which can be performed in the 1G position and SMAW for fixed position welds. FCAW is rapidly overtaking SMAW for these materials also. See Tables A6.4 and A6.7 for typical preheat and postweld heat treatment requirements. Note that in B31.3 hardness limits are imposed to verify the adequacy of any heat treatment, and above critical heat treatment may be necessary to attain the maximum hardness limit.

Martensitic and Ferritic Stainless Steels. The martensitic and ferritic grades of stainless steels are not often encountered in piping systems. They are a group of steels with chromium contents ranging from 11.5 to 30 percent. Martensitic stainless steels are those which are capable of transformation to martensite under most cooling conditions and therefore can be hardened. Ferritic stainless steels on the other hand contain sufficient chromium and other ferrite formers such as aluminum, niobium, molybdenum, and titanium so that they cannot be hardened by heat treatment. ASME Section IX classes martensitic stainless steels as P-No.6 and ferritic stainless steels as P-No.7. The user should consult the *Welding Handbook*[16] for suggested welding processes and the applicable code for specific preheating and postweld heat treatment requirements.

Austenitic Stainless Steels. Austenitic stainless steels are classed as P-No.8. Piping systems of austenitic stainless steels represent a fairly significant proportion of a fabricator's and/or installer's work, since they appear in nuclear power plants, chemical plants, paper mills, food processing, and other applications where cleanliness and corrosion resistance are mandatory and even in fossil power plants where their high-temperature properties are needed. Most root welding is done by the GTAW process and the inside of the root is protected by purging with argon, helium, or nitrogen to prevent formation of hard chromic oxides. GTAW is used for weld-out in lighter walls and combinations of GTAW, SMAW, and SAW are used for heavier sections. Filler metal must contain some ferrite to preclude microfissuring as described in the section "Filler Metals." To minimize the precipitation of carbides (sensitization) during welding, interpass temperatures are usually limited to 300 to 350°F. Heat treatment after welding is not mandatory. For corrosion services heating during fabrication could be detrimental since it would serve to enhance sensitization. The effects of sensitization can be mitigated by a carbide solution heat treatment as described in the section "Heat Treatment." Low-carbon grades of stainless steels welded with L grade electrodes are also used in services where sensitization can be a problem.

Low-Temperature Steels. The term *low-temperature steel* is applied to a variety of steels which exhibit good notch toughness properties at temperatures down to cryogenic levels.

The B31.1 and B31.3 Codes permit the use of most steel down to −20°F. Below this, certain grades of carbon and nickel steel with good toughness and austenitic stainless steels are needed. Welding procedures and welding filler metals must be tested to assure suitability for the intended service. B31.3 gives details of such requirements. Root pass welding using GTAW, with SMAW and

SAW weld-out, is commonly used. Some FCAW is used in the carbon steels and low-nickel steels.

A preheat of 200°F is suggested by B31.3 for the low-nickel steels followed by a postweld heat treatment consisting of a stress relieve at 1100 to 1175°F when the wall exceeds ¾ in. For the 9 percent nickel steel a preheat of 50°F and a stress relieve at 1025 to 1085°F followed by cooling at a rate greater than 300°F/h down to 600°F is required.

Certain nonferrous materials are also suitable for low-temperature service. See the following section.

Welding of Nonferrous Metals

Aluminum. Aluminum and aluminum alloys have high thermal conductivity, high coefficients of thermal expansion, and high fluidity in the molten state. The predominant welding methods used for joining them are GMAW and GTAW, both manually and in automatic modes. Joint designs are much like those used for ferritic metals, except that the included angles are usually 60 to 75° increasing to 90 or 110° for welding overhead. The root pass may be welded against a permanent aluminum backing strip or removable stainless steel backup or with an open butt or consumable insert. Joint cleanliness is very important, so oil, grease, and dirt must be removed. For heavy oxide, wire brushing or chemical cleaning may be required. Preheating is normally not needed but may be required when the mass of the parts is large enough to conduct the heat of welding away from the joint faster than it can be supplied by the arc. Depending on the welding process used, as the weld thickness increases from about ¼ to 1 in, a preheat of 200 to 600°F may be required. Since the properties and tempers of certain alloys may be affected, care should be exercised when preheat is applied. Shielding gases are usually helium or argon. For critical applications and heavier sections a mixture of 75 percent helium, 25 percent argon is recommended. Heat treatment after welding is not required.

It is important to remember that the annealing effect of the heat of welding can reduce the strength level of cold worked and heat-treatable alloys. In this case the allowable stress value for the material in the annealed condition should be used for design. An exception to this can be made in the case of heat-treatable materials when the finished weldment is subjected to the same heat treatment which produced the original temper and both the base metal and weld joint are similarly affected.

Aluminum and aluminum alloys are suitable for service temperatures down to −452°F. See B31.3 for information on this subject.

Copper and Copper Alloys. Although copper and copper alloys can be welded by other processes, GTAW welding is commonly applicable for all position welding of most copper and copper alloys. GMAW with pulsed current can also be used for some alloys. Shielding gases may be argon, helium, or mixtures thereof. Argon is preferred for walls to ⅛ in, but a 75 percent helium, −25 percent argon mixture is most often used for heavier walls and weld positions other than flat (1G).

Like aluminum, the coppers have high thermal conductivity and high coefficient of thermal expansion. Accordingly, preheating is recommended to compensate for heat loss at the joint due to the metal mass and to reduce distortion. Welding current should not be used to compensate for heat loss. The degree of preheat is a function of alloy, welding process, and metal mass. More heat input is needed for the pure coppers with decreasing amounts needed as the alloy content increases. Preheat should increase with wall thickness, from about 200°F for

¼-in wall increasing to 750°F minimum for walls ⅝ in and over. Surface cleanliness is very important and some alloys require a chemical cleaning to remove oxides. Copper-nickel alloys are susceptible to hot cracking if sulfur is present.

The heat of welding will soften the HAZ of cold worked material, and it will be weaker than the base metal. When precipitation hardenable alloys are used, it is recommended that welding be done on base metal in the annealed condition and the entire weldment be given the precipitation hardening heat treatment. For detailed information refer to the *Welding Handbook,*[16] the *Metals Handbook,*[20] or contact the Copper Development Association.

Many coppers are suitable for services down to −325°F. See ASME B31.3.

Nickel and Nickel Alloys. Nickel and its alloys can be welded by SMAW, GTAW, and GMAW. SAW is limited to certain compositions. Welding is similar to austenitic stainless steels except that the molten metal is more sluggish and does not wet as well. Larger groove angles may be required. Preheat is not required, but welding at temperatures below 60°F or in the presence of moisture is not recommended. A low interpass temperature is suggested. For GTAW welding shielding gas is normally argon, but helium or an argon-helium mix may be used. The inside surface of GTAW root welds should be shielded with an inert gas. GMAW in the spray, pulsed, globular, or short-circuiting modes may be used with argon or argon-helium mixtures as shielding. Postweld heat treatment is not usually required. Many nickel and nickel alloys may be used down to −325°F. For more detailed information refer to the *Welding Handbook,*[16] the *Metals Handbook,*[20] and ASME B31.3.

Titanium. Titanium and its alloys are normally welded using the GTAW and GMAW processes. It is vital that the HAZ and molten metal be protected from the atmosphere by a blanket of inert gas during welding. Most welding is done in a protective chamber purged with an inert gas or by using trailing shields. Precleaning is extremely important. Use of degreasers, stainless steel wire brushes, or chemical solutions may be required. Preheating or postweld heat treatment are not normally required. See the *Welding Handbook*[16] and the *Metals Handbook.*[20]

Dissimilar Metals. Until now we have discussed welding where both items being joined are essentially the same material and are joined with a filler metal of similar chemistry and physical properties. Occasions arise where metals of different chemical composition and physical properties must be joined.

In joining dissimilar metals, normal welding techniques may be employed if the two base metals have melting temperatures within about 200°F of each other. Otherwise different joining techniques are required.

In designing a welding procedure for dissimilar metals, a great many factors must be considered. Service conditions such as temperature, corrosion, and the degree of thermal cycling may apply. The effects of dilution of the two base metals by the filler and each other must be evaluated to assure a sound weld with suitable chemical, physical, metallurgical, and corrosion-resistant properties. Similarly, preheat and postweld heat treatment requirements for one base metal may not be suitable for the other.

It is usually necessary to qualify a separate welding procedure for the particular combination of base metals and filler material. ASME Section IX should be consulted for specifics.

As a general rule, when welding within a family such as ferritic to ferritic, austenitic to austenitic, or nickel alloy to nickel alloy, the filler metal may be of the same nominal composition as either of the base metals or of an intermediate composition. The filler metal normally used to weld the lower alloy is most often preferred.

In welding dissimilar materials, selection of preheating and postweld heat treatment requires a great deal of care. What is desirable for one metal may be detrimental to another. Some compromise may be required.

Establishing a welding procedure for welding ferritic to austenitic steels requires careful consideration of the service conditions. For moderate service temperatures (below 800°F), where the thickness of the ferritic side does not require postweld heat treatment, austenitic stainless steel electrodes are often the choice. Some prefer electrodes such as type 309 or 310 because of their higher chrome content. Because of the thickness involved, the ferritic member may require some type of postweld heat treatment. In this case the preferred method is to "butter" the ferritic weld surface with a nickel-chrome-iron (NiCrFe) filler metal such as ERNiCrMo-3 (see ASME Section II Part C SFA-5.14) and postweld heat treat the buttered section as required for the ferritic composition. The buttered section is then prepared for welding, set up with the austenitic side, and the weld between the butter and austenitic base metal is completed with NiCrFe filler metal without subsequent postweld heat treatment. See Fig. A6.21.

For high-temperature service (above 800°F) the buttering procedure described above is also recommended. There is a difference in coefficients of expansion between the ferritic and austenitic metals. This difference will result in expansion stresses above the yield point at the weld juncture while at operating temperature. At higher temperatures there is also greater probability of diffusion of carbon from the ferritic side to the austenitic side. The NiCrFe "butter" minimizes the carbon diffusion problem and has an expansion coefficient which is interme-

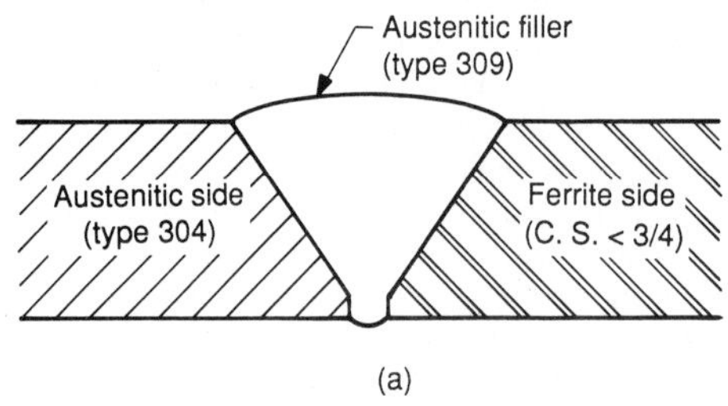

(a)

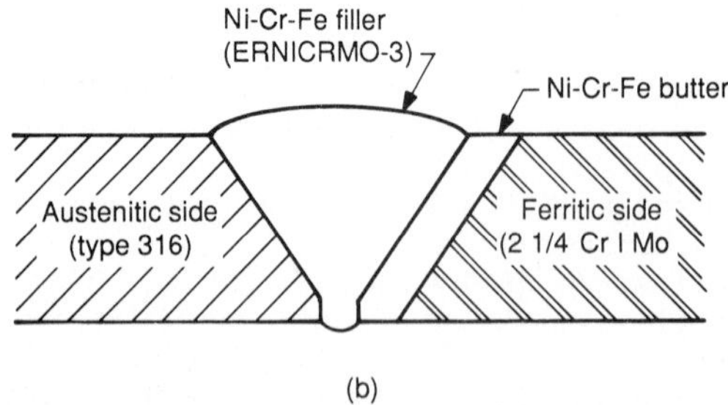

(b)

FIGURE A6.21 Dissimilar metal welds. (*a*) For moderate temperature service; (*b*) for high-temperature services or where stress relief on the ferritic side is required.

diate between the two base metals, thus reducing but not eliminating the thermal stress at the interface. Where a transition from ferritic to austenitic steels is required in high-temperature applications involving cyclic services, a transition piece of a high-nickel alloy such as UNS N06600 with two welds is often used to reduce thermal fatigue damage.

In welding nonferrous metals to ferrous or other nonferrous metals, a filler metal with a melting point comparable to the lower melting point base metal is usually recommended.

Nickel and nickel alloys are invariably welded to ferrous metals with nickel-alloy filler metals. Sulfur embrittlement can be a problem with nickel to ferritic welds just as it is in nickel to nickel welds. Copper-nickel and nickel-copper alloys should not be joined with filler materials containing iron or chromium since hot cracking may result.

Copper and copper alloys can be welded to carbon steel with silicon bronze or aluminum bronze electrodes, but the preferred method is to butter the carbon steel side with nickel and weld the copper to the nickel butter with nickel filler. This will preclude hot cracking of the copper because of iron dilution. The copper side may require preheat. Copper can easily be welded to nickel using nickel, copper-nickel, or nickel-copper filler metal. When welding nickel alloys which contain iron or chromium to copper, the nickel alloy should be buttered with nickel.

Aluminum and titanium generally cannot be welded to ferrous or other nonferrous metal using currently available welding procedures, and special joining procedures must be employed.

Clad, Metal-Coated, and Lined Pipe. There are instances when it is economically desirable to construct a piping system from relatively inexpensive material but with an interior surface having corrosion or erosion resistant properties. Clad pipe may be made by seam welding of clad plate, by weld metal overlay of the inside surface, or by centrifugal casting of a pipe with two metal layers. Lined pipe is made by welding a liner, sometimes as strips, to the inside surface of the pipe. Metal-coated pipe is made by dipping, metal spraying, or plating the entire pipe.

Before choosing construction which requires welding of clad, lined, or metal-coated pipe, such factors as filler metal compatibility, filler metal strength relative to the base metal strength, dilution of base metal into the finished weld, and need for postfabrication heat treatment must be considered. Because it is not possible to cover the great many combinations of base metals and cladding, lining, or metal coatings, some examples of the more common applications will be given.

For corrosion services, a carbon steel base material, clad or lined with austenitic stainless steel, is often used. The cladding is usually about 3/32 to 5/32 in thick. Where the inside of the weld is accessible, the preferred method is to weld the base metal from the outside with carbon steel filler metal, back gouge the root from the inside, and weld the root from the inside with two or more passes of austenitic filler metal to minimize dilution from the base metal. See Fig. A6.22*a*.

Where the inside surface is not accessible, a backing strip of the same composition as the cladding, fillet welded to the cladding on the upstream side may be used. The root weld between the two clad surfaces and the austenitic backing strip is then made with austenitic filler metal. The root weld can also be made with the GTAW process using austenitic filler or preplaced inserts. The carbon steel should be removed for a sufficient distance back to preclude dilution into the root weld. In most instances, the balance of the weld is usually made with austenitic filler metal since it is not good practice to deposit carbon steel or low-alloy steel directly against the stainless steel deposit. See Fig. A6.22*b*. In some cases, nickel-base alloys are used for cladding where high-temperature corrosion

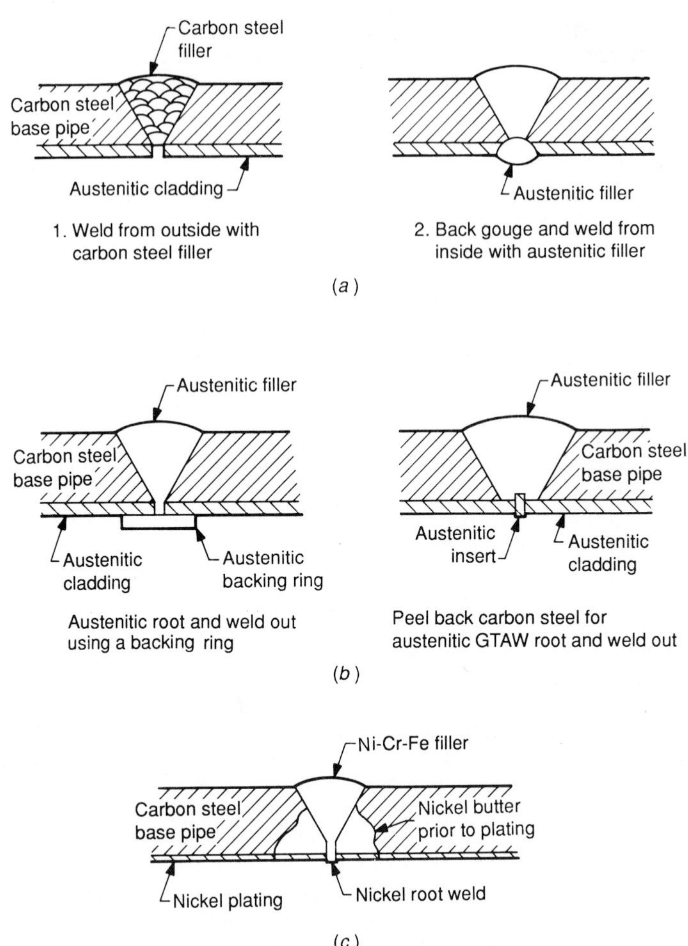

FIGURE A6.22 Examples of welding clad, lined, and plated pipe. (*a*) Clad pipe welded from both sides; (*b*) clad pipe welded from one side only; (*c*) nickel-plated pipe.

is involved. The joints may be treated much like the austenitic cladding, except that appropriate nickel-base filler metals are used.

Some services require the use of carbon steel pipe nickel plated on the inside surface. Since the plating is relatively thin, different approaches are needed. First, as much fabrication as possible should be done prior to plating. For joints to be welded after plating, the ends to be prepared for welding should be buttered with nickel filler metal and machined to the required contour prior to plating. The root weld is made using the GTAW process with nickel filler metal. See Fig. A6.22*c*.

Some occasions require the use of aluminized pipe. Steel pipe is prefabricated and coated with aluminum by immersion in a bath of molten aluminum or by metal spray. Where the inside of the weld will not be accessible for metal spray, one method of joining is to counterbore the ends and use a solid machined backing ring which is fit and welded into one side of the joint prior to coating. After coating, the weld is made using an appropriate base metal process and filler, taking care not to blister the aluminum coating on the underside of the backing ring.

Galvanized steel pipe is often used for external corrosion applications. Since welding of galvanized pipe releases toxic vapors and since the welded area most often cannot be regalvanized, welding of galvanized pipe is not recommended. It is preferable that the assemblies be fabricated with provisions for mechanical joining in the field and then galvanized.

For services involving erosion, carbon steel pipe is often lined with cement or some type of abrasion-resistant material which cannot be welded. In this case the joints are butted together to minimize the gap between the adjacent linings. The weld is then made between the two carbon steel weld bevels, recognizing that full penetration through the carbon steel joint may not be achieved and that additional thickness may be necessary for strength. The gap between the adjacent linings is usually not a problem if only erosion is present.

Brazing and Soldering

Brazing. For services involving the ASME Boiler and Pressure Vessel Code or the B31 Code for Pressure Piping, brazing procedures and brazers must be qualified in accordance with ASME Section IX similar to welding procedures and welders. See the section "Procedure and Personnel Qualification."

There are a great many types of brazing processes. In establishing a brazing procedure, consideration must be given to the ability of the filler metal to produce suitable physical properties, its melting point and wettability, possible base metal and filler metal interactions, loss of base metal properties, increased sensitization to corrosion, increased hardness in the base metal due to brazing temperature, and the need for postbrazing heat treatments.

Since most piping materials can be welded, the use of brazing for joining is rather limited. It is most often used for joining coppers and for combinations of metals which cannot be welded.

Brazing is a process wherein the base metals do not melt, the filler metal has a liquidus above 840°F, and the filler metal wets the base metal and is drawn into the joint by capillary action.

Although butt or scarf joints can be used, a lapped joint with an overlap of 3 times the thickness of the thinner member gives the best joint efficiency and ease of fabrication. It should be noted that typical copper or brass fittings have a depth of socket based on the strength of tin-lead solders. When brazing is used, only a small percentage of that depth is needed. Required clearance between the faying surfaces usually vary from 0.001 to 0.010 in depending on the filler and flux combination used during the operation. The flux melts upon application of heat and is displaced by the molten filler metal. Flux residue should be removed after the operation is complete. Silver, copper-phosphorus, and copper-zinc filler metals are most often used for copper brazing.

Torch brazing is commonly used for fabrication and installation of copper piping systems. For torch brazing, the type of fuel gas selected is a function of the melting temperature required to melt the filler metal. For piping joints NPS 2 and larger, use of a second torch to preheat may be desirable.

In brazing metals with differing coefficients of expansion, it is preferable that the metal with the higher expansion coefficient form the socket and the metal of the lower expansion coefficient form the pipe or tube. Clearance between the parts at room temperature must be adjusted so there will be a suitable clearance at brazing temperature. On cooling, the greater contraction of the socket will put the joint in a compressive stress state.

Soldering. Unlike welding and brazing, ASME Section IX has no requirements for qualification of soldering procedures or personnel. Soldering is much like brazing in that the base metals are not melted, the faying surfaces are wetted by the filler, and the filler is drawn into the joint by capillary action. However, the melting point of the filler metal is lower than 840°F, usually between 450 and 500°F. Since the strength of soldering filler metals is considerably less than that of brazing fillers, a longer overlap is required to develop a joint equal to base metal strength. A clearance of about 0.003 in is preferred for optimum strength.

A good soldered joint depends again on the cleanliness of the faying surfaces. Fluxes are used to assist in the wetting action by removing tarnish films and to prevent oxidation. Rosin fluxes and organic fluxes are used for most materials. Inorganic fluxes may be required for certain other materials that can be soldered, while in some cases precoating of the material with a surface that can be soldered may be required. Most piping applications use tin-lead solders. These range in composition from 5 percent tin, 95 percent lead to 70 percent tin, 30 percent lead, with 50 percent tin, 50 percent lead the most common. Tin-antimony and tin-silver solders are also frequently used. For soldering aluminum, tin-zinc and zinc-aluminum are used.

For additional information refer to the *Welding Handbook*[16] and *The Theory and Technique of Soldering and Brazing of Piping Systems.*[33]

Heat Treatment

Purpose. Heat treatment during piping fabrication is performed for a variety of reasons (i.e., to soften material for working, to relieve fabrication stresses, to restore metallurgical and physical properties, etc.). During fabrication, ferritic steels undergo phase changes during heating and cooling, while the austenitic stainless steels and nonferrous piping materials do not; consequently differing criteria must be applied.

Ferritic Steels. Ferritic steels undergo a phase change on heating and cooling during fabrication operations because their principal component (iron) is allotropic; that is, it undergoes a change in crystalline structure with temperature. At room temperature iron favors a body-centered cubic (BCC) structure called *alpha iron,* but on heating to 1670°F it changes to a face-centered cubic (FCC) structure called *gamma iron* and subsequently at 2534°F it reverts to a BCC called *delta iron.* The addition of carbon to the iron to form steel and additions of other elements such as chromium, manganese, molybdenum, and nickel to form alloys modify the temperatures at which transformation occurs and the manner in which the crystalline structure forms into grains.

As an example, a melt of 0.30 percent carbon steel will first begin to solidify as delta iron and a liquid, then at about 2680°F to an interstitial solid solution of carbon in gamma iron called *austenite.* At about 1500°F this will transform into a mixture of austenite and ferrite which at 1333°F becomes ferrite and pearlite. Ferrite is alpha iron which contains small amounts of carbon (up to a maximum of about 0.02 percent) in solid solution. The excess carbon not in solid solution with the ferrite forms as iron carbide (Fe_3C) or cementite. The cementite forms as thin plates alternating with ferrite. This structure is known as *pearlite.*

The temperatures at which the transformations occur are called critical temperatures or transformation temperatures. The lower critical temperature, usually designated A_1, is that point on heating where the BCC ferrite and pearlite phase begins to transform to FCC austenitic structure, and the upper critical temperature, A_3, is the temperature at which the transformation is complete. Between these two points the structure is a mix of ferrite-pearlite and austenite. These temperatures are of importance in postbending and postwelding heat treatments as well as qualification of welding procedures.

The critical temperatures are a function of chemical composition and as such will vary with alloy. Some approximate methods of calculating critical temperature are found in *Welding Metallurgy*[34] and *The Making, Shaping and Treating of Steel.*[35] Some approximate lower critical temperatures are given in Table A6.8.

Critical temperatures are affected by heating and cooling rates. An increase in heating rate will serve to increase the transformation temperatures, while an increase in cooling rate will tend to depress them. The more rapid the rate of heating or cooling, the greater the variation from the critical temperature at equilibrium conditions. Most sources will indicate the lower and upper critical temperatures on heating as A_{c1} and A_{c3}, respectively, and the upper and lower on cooling as the A_{r3} and A_{r1}, respectively. In the case of our 0.30 percent carbon steel, cooling from the austenite phase through the critical range at a rate of 50°F/h or less will result in the soft, ductile ferrite-pearlite structure. On the other hand, extremely rapid cooling from the austenite phase down to temperatures 600°F or lower can result in an extremely hard structure called *martensite.* This is because the austenite FCC crystals did not have time to transform to BCC ferrite and cementite.

Heat treatments which are applied to ferritic steels are related to the critical temperatures and depending on which is applied will have differing results. These are annealing, normalizing, normalizing and tempering, and stress relieving. See Fig. A6.23.

Annealing is used to reduce hardness, improve machinability, or produce a more uniform microstructure. It involves heating to a temperature above the upper critical or to a point within the critical range, holding for a period of time to assure temperature uniformity, followed by a slow furnace-controlled cooling through the critical range.

Normalizing is used to refine and homogenize the grain structure and to provide more uniform mechanical properties and higher resistance to impact loadings. It involves heating to a temperature above the upper critical temperature, holding for a time to permit complete transformation to austenite, and cooling in still air from the austenitizing temperature.

A normalized structure may be pearlitic, bainitic, or even martensitic depending on the cooling rate. If there is a concern for excessive hardness and attendant

TABLE A6.8 Approximate Lower Critical Temperatures

Material	Approximate lower critical temperature [°F (°C)]
Carbon steel	1340 (725)
Carbon molybdenum steel	1350 (730)
1¼ Cr-½ Mo	1430 (775)
2¼ Cr-1 Mo, 3 Cr-1 Mo	1480 (805)
5 Cr-½ Mo	1505 (820)
7 Cr-½ Mo	1520 (825)
9 Cr-½ Mo	1490 (810)

Source: From ASME B31.1 1989 ed.

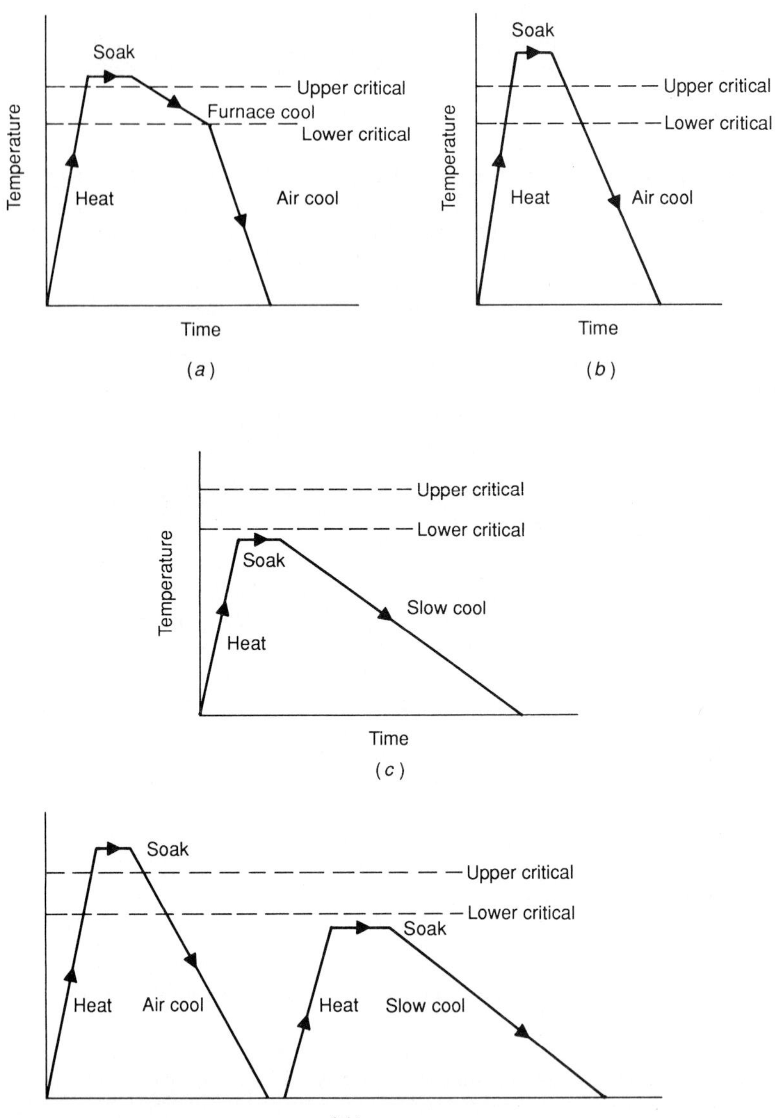

FIGURE A6.23 Heat treatment cycles. (*a*) Anneal; (*b*) normalize; (*c*) stress relief; (*d*) normalize and temper.

low ductility, a tempering treatment may follow the normalizing treatment. Tempering involves heating to a temperature below the lower critical and slowly cooling to room temperature, much like a stress relief. The degree of tempering depends on the tempering temperature selected. The higher the tempering temperature, the greater the degree of softening.

A stress-relieving heat treatment is primarily intended to reduce residual stresses resulting from bending and welding. It involves heating to a temperature below the lower critical; holding for a predetermined time, depending on thickness and material, to permit the residual stresses to creep out; and then slow cooling to room temperature.

Some typical time-temperature cycles are shown in Table A6.7.

Austenitic Stainless Steels. Austenitic stainless steels do not undergo phase changes like the ferritic steels. They remain austenitic at all temperatures and so heat treatments usually do not apply. When austenitic stainless steels are to be used in corrosive services, cold working and heating for bending may significantly lower their corrosion resistance. Cold working may result in residual stresses, and heating operations can result in sensitization. Both factors contribute to intergranular stress corrosion cracking (IGSCC). When austenitic stainless steels are heated in the range of about 800 to 1600°F, carbon in excess of about 0.02 percent will come out of solution and diffuse to the grain boundaries where it will combine with adjacent chromium to form chromium carbide ($Cr_{23}C_6$). This phenomenon is called *sensitization.* These grain boundaries are then preferentially attacked by corrosive media. The heat treatment often applied to cold worked and sensitized stainless steels to restore corrosion resistance is a carbide solution heat treatment. In this procedure, the material is heated to a temperature above the sensitization range, usually about 1950 to 2100°F, and held there sufficiently long to permit the carbides to dissolve and the carbon to go back into solid solution. The material is then removed from the furnace and rapidly cooled through the sensitization range preferably by quenching in water. The rapid cooling does not give the carbon sufficient time to come out of solution and corrosion resistance is restored to the sensitized area.

Obviously carbide solution heat treatment is limited by the furnace size and quenching facilities. It is most frequently applied to bends but is also useful in reducing sensitization and residual stresses in welds.

Nonferrous Materials. Bending and forming of nonferrous materials may result in undesirable work-hardening. Some nickel alloys may be subject to carbide precipitation when hot bent or formed. Materials that can be hardened by precipitation require other considerations. Depending on the final use, it may be desirable to perform some type of postbending or forming heat treatment. Because of the great many new materials being developed and used, it is suggested that the user contact the material manufacturers or material associations for their recommendations on the specific material and service.

Heat Treatment Methods. Shop heat treatments are most often carried out in specifically designed heat treatment furnaces, but local stress relieving of welds may also involve induction, resistance, or torch heating.

Above critical heat treatments, such as annealing, normalizing, and normalizing and tempering for ferritic steel and carbide solution heat treatment for austenitic stainless steels, are performed in large heat treatment furnaces. These same furnaces are also used for stress-relieving heat treatments of ferritic steels. Such furnaces are generally fired with natural gas, propane, or low-sulfur oil. Depending on their design, they may attain temperatures up to 2300°F, which covers the entire spectrum of temperatures commonly encountered in piping applications. Heating and cooling rates and holding temperatures are automatically controlled. Larger furnaces may have two or more zones, each independently controlled. Records of furnace zone temperatures and material temperatures are obtained using recording potentiometers.

When assemblies are too large or furnaces are not available, local stress relieving of individual welds may be accomplished in the shop using electrical induction, electrical resistance, or gas torch heating.

Induction equipment involves alternating current frequencies of the order of 60 to 400 Hz. Induction generates heat within the wall of the pipe. This has the advantage of a more uniform temperature through the thickness with greater uniformity at the lower frequencies. The heat treatment cycle is controlled automatically with thermocouples attached directly on or adjacent to the weld. The weld and thermocouple are covered with insulating material. The induction field is generated in copper cables or solid or water-cooled copper coils external to the insulation. See Fig. A6.24.

Resistance heating involves the use of direct current in suitable lengths of nichrome heating wire. Various configurations and sizes of prefabricated heating elements consisting of heating wires separated by ceramic beads are available commercially. Depending on the size, wall thickness, and desired heating temperature, multiple heating units and combinations of elements may be needed. The weld and heating elements are covered with insulating blankets to retain the heat. Since heating is from one side, a somewhat wider heating band on the outside may be needed to assure that the inside of the pipe attains the required temperature. Thermocouples attached directly to the weld or adjacent to it are used to control heating, holding, and cooling temperatures.

Torch heating can often be used for stress relieving, but where controlled heating and cooling rates are mandated, it may be less than satisfactory. Single torches may be used for pipe up to about NPS 3, but ring burners are needed for larger sizes.

Exothermic heating is often used in the field and is discussed in the section "Installation."

Heat Treatment Considerations

Furnace Heat Treatment. To assure that heat treatments attain the results intended (i.e., correct heating and cooling rates, desired holding temperature in all parts, etc.), it is very important that all controlling and recording instruments be calibrated on a regular basis. The furnace should be inspected and a temperature survey made to assure that all locations within it are capable of attaining

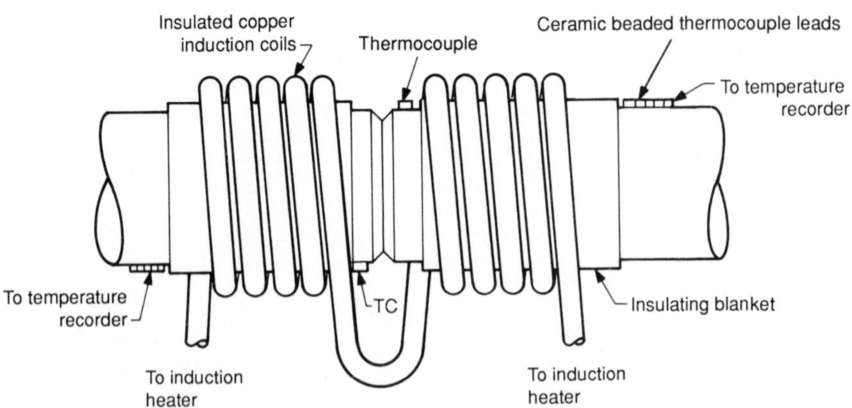

FIGURE A6.24 Setup for preheat, maintenance of preheat during welding, and stress relief using induction heating.

and maintaining specific temperatures within some reasonable tolerance. This is particularly important if the zone temperatures are used as the basis for acceptance of the heat treatment. If there is any concern, it might be advisable to attach thermocouples to the parts being heat treated themselves.

When piping subassemblies are placed in the furnace, they should be supported to permit exposure of the underside to the radiant and convection heating surface. Supports should be located so as to avoid sagging. Care should be taken to avoid any flame impingement directly on surfaces being heat treated.

The ends of assemblies being heat treated should be closed but not sealed to minimize oxidation of the inside surfaces. Occasions may arise where special surface finishes on the pipe inside surface or on flow meter sections could be adversely affected by oxidation caused by heat treatment. In such cases the inside of the assembly can be purged with an inert gas to minimize oxidation.

Assemblies should be so placed as to assure the uniform application of heat. Heating and cooling rates must be selected to assure heating through the full thickness and to minimize distortion caused by uneven heating. The faster the rate of heating or cooling, the more probability of distortion. Assemblies with massive flanges, fittings, or other unusual configurations should be treated more carefully than those with butt welds only. Many of the codes have specified heating and cooling rates which are considered reasonable.

Local Heat Treatment. When an assembly is too large for a furnace to accommodate, it may be fabricated in sections which are individually furnace heat treated and later joined by welding. The final butt welds may then be locally heat treated in the same fashion as field welds. The most common practice is the use of induction or resistance heating. When preheating is an essential part of the welding operation, the induction or resistance equipment can be used for preheating, maintaining preheat during welding, and, finally, stress relieving.

A proper stress-relieving operation will assure that the weld and HAZ through the full thickness will attain the required temperature for the required time. The B31.1 Code requires that the heated band be at least 3 times the thickness of the thickest part being joined. With induction or resistance heating the heating elements themselves often have greater coverage. Depending on the massiveness of the joint being heated, one or more pieces of heating equipment may be needed. Controlling and recording thermocouples are located on or adjacent to the weld. Usually locally heat treated shop welds are in the 5G position (pipe horizontal, weld vertical). For small pipe sizes, a single thermocouple located at the 12 o'clock position may suffice, but for larger diameters and heavier walls at least two and preferably four, located at 90° intervals, should be employed to assure uniformity of heating.

Judicious use of insulating material should be employed to minimize heat loss. When joining parts of differing masses, concentrate more heating effort on the more massive part.

If it is necessary to locally stress relieve a branch connection, not only the branch weld itself but the entire circumference of the header for a distance of at least 2 times the header thickness on either side of the branch should be heated. Heating of the weld alone, while resulting in a satisfactory stress relief, could distort the header significantly.

Heating and cooling during local stress relief of pipe to pipe joints can be more rapid than for furnace applications since there is less chance of distortion unless, of course, the heating is not applied uniformly. Ends of the assembly should be closed but not sealed to reduce heat loss on the inside surface due to air flow. The main concern is assurance that the inside surface of the weld attains the required temperature for the required time.

Local stress relieving with torches or gas ring burners can be effectively employed but must be limited to situations where controlled heating and cooling rates are not a factor.

Code Requirements

Postbending and Postforming Requirements. The designer of the piping system should specify the type of heat treatment required to assure appropriate physical, metallurgical, or corrosion-resistant properties. As an example, a normalize or normalize and temper may be required to assure certain notch toughness properties for nuclear or low-temperature applications, or a carbide solution heat treatment for cold worked austenitic stainless steel may be required to preclude IGSCC. This should be agreed upon well before any fabrication starts.

The codes have certain mandatory heat treatment requirements which must be observed as a minimum, normally a stress-relieving treatment. Such heat treatment is usually in accordance with the postweld heat treatment tables given in the applicable code. Differing requirements apply depending on whether the bending or forming was performed hot or cold. According to B31.3, cold bending is performed at a temperature below the transformation range (below the lower critical), and hot bending is performed at a temperature above the transformation range (above the upper critical). B31.1 and ASME Section III make the break between hot and cold bending at a temperature 100°F below the lower critical.

B31.3 requires heat treatment after cold bending when (1) specified in the engineering design, (2) the calculated elongation will exceed 5 percent for materials requiring notch toughness properties, and (3) the calculated elongation will exceed 50 percent of the specified minimum elongation indicated in the material specification for P-No.1 through P-No.6 materials. For hot bending and forming, heat treatment is required for all thicknesses of P-Nos.3, 4, 5, 6, and 10A materials.

B31.1 and ASME Section III on the other hand require heat treatment after bending or forming in accordance with the postweld heat treatment table of the applicable code for P-No.1 materials with a nominal wall thickness exceeding ¾ in unless the bending or forming was completed above 1650°F. All ferritic alloy materials of NPS 4 or larger or with a nominal wall thickness of ½ in or greater which are hot bent or formed must receive an annealing, normalizing and tempering, or a tempering heat treatment to be specified by the designer, or if cold bent or formed, the heat treatment at the required time and temperature cycle specified in the postweld heat treatment table for the material involved.

The codes have no requirements for postbending or forming heat treatments of austenitic stainless steels or nonferrous materials.

Postwelding Heat Treatment Requirements. Before applying any postwelding heat treatment (PWHT), it should be noted that for work under ASME Section IX, postwelding heat treatment is an essential variable for welding procedure qualification. For P-Nos.1, 3, 4, 5, 6, 9, 10, and 11 materials there are five possible conditions of heat treatment, each requiring separate qualifications. These are:

1. No PWHT
2. PWHT below the lower critical temperature (stress relief)
3. PWHT above the upper critical temperature (normalize or anneal)
4. PWHT above the upper critical temperature followed by heat treatment below the lower critical temperature (normalize and temper)
5. PWHT between the upper and lower critical temperatures

For other materials, two conditions apply: no PWHT or PWHT within a specified temperature range.

Accordingly, for shop work, it may be necessary to qualify welding procedures for several possible heat treatment situations. For field work only the no heat treatment or stress-relieving situations will normally apply.

When required by the codes, heat treatment consists of a stress-relieving operation. Other heat treatments such as annealing, normalizing, or solution heat treatment may be applied but are not mandatory. However, the welding procedure must have been qualified for the heat treatment applied.

Each code has its own definition regarding governing thicknesses, its own exemptions, differing temperature and holding requirements, heating and cooling rates, etc., reflecting the differing concerns and needs of individual industries. The codes are also constantly evolving as the committees obtain and review new data. Accordingly, the reader should refer to the applicable edition of the code of interest for requirements.

At the time of this writing, the following is a comparison of the heat treatment requirements for carbon steel materials.

B31.1 requires heat treatment of P-No.1 Gr.Nos.1, 2, and 3 in the temperature range of 1100 to 1200°F for 1 h/in of thickness for the first 2 in plus 15 min for each additional inch over 2 in, with a 15-min minimum. Exempted are welds with a nominal thickness of ¾ in or less, and a 200°F preheat must be applied when either of the base metals exceed 1 in. The nominal thickness is defined as the lesser of the thickness of the weld or the thicker of the base metals being joined at the weld. The thickness of the weld is further defined as the thicker of the abutting edges in a groove weld, the throat of a fillet weld, the depth of a partial penetration weld, and the depth of the cavity for repair welds. Thickness as it relates to branch welds is a function of the header thickness, the branch thickness, and reinforcing pad thickness.

B31.1 also requires controlled heating and cooling at temperatures above 600°F. The rate shall not exceed 600°F/h or 600°F/h divided by one-half the maximum thickness at the weld in inches, whichever is less.

Section III requires heat treatment of P-No.1 materials in the temperature range of 1100 to 1250°F for 30 min when the thickness is ½ in or less, for 1 h/in of thickness for thickness over ½ to 2 in, and 2 h plus 15 min for each additional inch of thickness over 2 in. In this case the thickness is defined as the lesser of (1) the thickness of the weld, (2) the thinner of the pressure retaining parts being joined, or (3) for structural attachment welds, the thickness of the pressure retaining material.

ASME Section III exempts P-No.1 materials in piping systems from mandatory heat treatment based on thickness and carbon content. When the materials being joined are 1½ in or less, the following exemptions apply: (1) a carbon content of 0.30 percent or less with a nominal thickness of 1¼ in or less, (2) a carbon content of 0.30 percent or less with a nominal wall thickness of 1½ in when a preheat of 200°F is applied, (3) a carbon content over 0.30 percent with a nominal wall thickness of ¾ in or less, and (4) a carbon content over 0.30 percent and a nominal wall of 1½ in when a preheat of 200°F is applied.

ASME Section III also requires controlled heating and cooling. Above 800°F the rate shall not exceed 400°F/h divided by the maximum thickness in inches but not to exceed 400°F. The rate need not be less than 100°F/h. Time and temperature recordings must be made available to the Authorized Nuclear Inspector.

B31.5 requires heat treatment of P-No.1 materials greater than ¾ in in the temperature range of 1100 to 1200°F for 1 h/in of wall thickness with a 1 h minimum.

The governing thickness is the thicker of the abutting edges for butt welds and the throat thickness for fillet socket and seal welds. Controlled heating and cooling rates are specified.

B31.3 has similar requirements except that differing thickness definitions are applied to branch, fillet, and socket welds, and there are no specified heating or cooling rates.

B31.4 and B31.11 both require stress relieving when the wall thickness exceeds 1¼ in, or 1½ in if a 200°F preheat is applied. No specific temperature is specified. B31.8 on the other hand requires stress relief if the carbon content exceeds 0.32 percent, the carbon equivalent (C + ¼ Mn) exceeds 0.65 percent, or the wall thickness exceeds 1¼ in. Carbon steels are to be heat treated at 1100°F or higher as stated in the qualified welding procedure.

Requirements for postweld heat treatment of many different ferrous alloy steels are given in the various codes. As in the case of the carbon steels, there are variations in requirements from code to code.

In the case of dissimilar metals welding, the codes most often specify that the heat treatment which invokes the higher temperature requirement be applied to the weld joint. In applying this criteria many factors should be considered. See the section "Dissimilar Metals" for some options. Another possibility is to take advantage of longer time and lower temperature heat treatments permitted by some codes.

In the end, the best source of information for specific requirements regarding heat treatment is the particular code mandated by law or contract. Where none is invoked, the various codes can be used as guides.

Verification Activities—Inspection, Nondestructive Examination, Testing, and Quality Assurance and Quality Control

Introduction. Activities involved in verifying that fabrication meets the specified quality level may be broadly categorized as inspection, NDE, testing and QA and QC.

The terms *inspection, examination,* and *testing* are still often used interchangeably. The ASME Boiler and Pressure Vessel Codes have begun to establish specific definitions for these terms. The B31 Codes present a mixture of usages, some following the ASME Boiler and Pressure Vessel Code lead while others are less definitive. The reader is directed to the individual codes to see how these terms are used. In general, the ASME Boiler and Pressure Vessel Code practice will be followed in this section.

Inspection relates to those activities performed by the owner, the owner's agent, or a third party. All other activities are usually performed by fabricator personnel.

The term *examination* is applied to nondestructive methods of examination, while *testing* refers to traditional hydrostatic and pneumatic tests for leakage. QA and QC relate to in-plant programs whose function is to control the various activities which affect quality.

Inspection. Inspection, as used in ASME Section I, III and B31.1 for Boiler External Piping, covers those activities which the authorized inspector (AI) or authorized nuclear inspector (ANI) performs in verifying compliance with the applicable code. The AI or ANI is employed by a third party; is independent of the owner, fabricator, or installer; is an employee of a state or municipality in the United States, a Canadian province, or an insurance company authorized to write boiler insurance; and is qualified by written examination as required by state or provincial rules.

In the B31 Piping Codes, inspection is the verification activity performed by the owner or the owner's agent. Specific requirements for qualification of inspectors are outlined in the individual code sections.

The manner in which an inspector verifies compliance is generally left to the discretion of the individual. It may take the form of detailed visual examinations; witnessing of actual operations such as bending, welding, heat treatment, or NDEs; review of records; or combinations thereof. Much relies on the degree of confidence the inspector has in the fabricator's programs and personnel. B31.3 has mandatory sampling requirements for this activity.

Examinations

Types of Examinations. When used in the various codes, examination refers to the verification work performed by employees of the fabricator, much of which falls into the category of NDE. NDEs most often referenced by code and applied to the fabrication and installation of piping components and systems are:

Visual
Radiographic
Ultrasonic
Liquid penetrant
Magnetic particle

Eddy current examination is often used to evaluate the quality of straight lengths of pipe as they are manufactured but is not often used in fabrication activities. Although not referenced by most codes, bubble testing, halogen diode probe testing, or helium mass spectrometer leak testing may be invoked by contract when, in the opinion of the designer, they will contribute to the integrity of the system. While these methods are referred to as leak tests, their methodology is outlined in Article 10 of ASME Section V Nondestructive Examination.

Accept-reject criteria and the extent to which the various NDEs are to be applied are in the applicable code.

The following are brief descriptions of NDEs as they apply to piping. For much more detailed information the reader is referred to various publications of the American Society for Nondestructive Testing (ASNT),[36] particularly the Nondestructive Testing Handbooks.

1. *Visual examination:* Visual examination is probably the oldest and most widely used of all examinations. It is used to ascertain alignment of surfaces, dimensions, surface condition, weld profiles, markings, and evidence of leaks, to name a few. In most instances the manner of conducting a visual examination is left to the discretion of the examiner or inspector, but more recently, written procedures outlining such things as access, lighting, angle of vision, use of direct or remote equipment, and checklists defining the observations required are being used. Visual examination takes place throughout the fabrication cycle along with QA and QC checks. At setup, this would consist of verifying materials, weld procedures, welder qualifications, filler metal, and weld alignment and on completion of fabrication, such things as terminal dimensions, weld profile, surface condition, and cleanliness.

2. *Radiographic examination:* When the need for greater integrity in welding must be demonstrated, the most frequently specified examination is radiography. Since the internal condition of the weld can be evaluated, it is referred to as a volumetric examination.

Radiographic sources used for examination of piping are usually x-rays or gamma rays from radioactive isotopes. While x-ray equipment is often used, it has limitations in that it often requires multiple exposures for a single joint, and special equipment, such as linear accelerators, are needed for heavier thicknesses. Although x-ray machines produce films with better clarity, they are not as practical in the field because of space limitations and portability. In the field, radioactive isotopes are used almost exclusively because of their portability and ease of access. For wall thicknesses up to about 2½ in of steel, the most commonly used isotope is iridium 192. Beyond this cobalt 60 is used for wall thickness up to about 7 in.

Radioactive sources normally used in piping work range in intensity from a few curies up to about 100 curies. Each source decays in intensity in accordance with its particular half-life. As the intensity decays, longer exposure times are required. Iridium 192 has a half-life of 75 days, while cobalt 60 has a 5.3-year half-life.

Radioactive sources have finite dimensions and as a result produce a shadow effect on the film. This is referred to as geometric unsharpness, and it is directly proportional to the source size and inversely proportional to the distance between the source and the film. ASME Section V has established limits for geometric unsharpness.

Ideally for pipe, the source is placed inside the pipe and at the center of the weld being examined, with film on the outside surface of the weld, thus permitting one panoramic exposure. Where geometric unsharpness precludes this practice, the source may be placed on the inside on the opposite wall and a portion of the weld is shot. Several exposures will be needed. The source may also be placed outside the pipe and the exposure made through two walls. Again this requires multiple exposures and longer exposure times. See Fig. A6.25.

A radiograph is considered acceptable if the required essential hole or wire size from the image quality indicator is visible on the film. See ASME Section V for information on this subject.

3. *Ultrasonic examination:* Ultrasonic examination is used in piping for the detection of defects in welds and materials as well as for determining material thickness. A short burst of acoustic energy is transmitted into the piece being examined and echoes reflect from the various boundaries. An analysis of the time and amplitude of the echo provides the examination results.

A clock in the equipment acts to initiate and synchronize the other elements. It actuates a pulsar to send a short-duration electrical signal to a transducer, usually at a frequency of 2.5 MHz. The transducer converts the electrical signal to mechanical vibration. The vibration as ultrasound passes through a couplant (such as glycerine) and through the part at a velocity which is a function of the material. As the sound reflects from various boundaries, it returns to the initiating transducer or sometimes to a second one where it is converted back to an electrical signal which is passed to a receiver amplifier for display on a cathode-ray tube. The horizontal axis of the display relates to time and the vertical axis relates to amplitude. The indication on the extreme left will show the time and amplitude of the signal transmitted from the transducer. Indications to the right will show the time and degree of reflection from various boundaries or internal discontinuities.

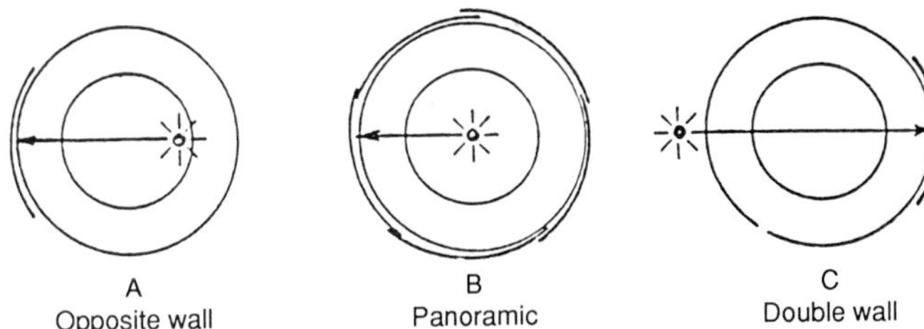

Given:
1. 30-in dia. X 7-in wall C/S pipe
2. 10 cu CO_{60} X 0.051 dia. source
3. 50 cu CO_{60} X 0.125 dia. source
4. 100 cu CO_{60} X 0.181 dia. source
5. Max. allowable geometric unsharpness $Ug = \frac{Ft}{D} = 0.07$ in (ASME Sect. V Art. 2 para. T-251)
6. Use Kodak "AA" Film with a 2.5 film density

Min. SFD = 5.1 in for 10 cu AX 0.051 in dia.
12.5 in for 50 cu X 0.125 in dia.
18.1 for 100 cu X 0.181 in dia.

Using the parameter established in 1 above as well as 5 and 6, it would be possible to:

(*a*) Use a 10-cu source as established in 2 above and shoot the pipe using the panoramic technique, sketch B above, with an exposure time of 3 hrs or

(*b*) Use a 50-cu source as established in 3 above and shoot the pipe using the panoramic technique, sketch B above, with an exposure time of 1 hour 15 min or

(*c*) Use a 100-cu source as established in 4 above and shoot the pipe using the opposite wall technique, sketch A above, with an exposure time of 50 min. As many as six exposures might be required—total 5 hrs.

FIGURE A6.25 Effect of source size on radiographic technique.

The ability of an ultrasonic examination to detect discontinuities depends a great deal on the part geometry and defect orientation. If the plane of the defect is normal to the sound beam, it will act as a reflecting surface. If it is parallel to the sound beam, it may not present a reflecting surface and accordingly may not show on the oscilloscope. Therefore, the search technique must be carefully chosen to assure that it will cover all possible defect orientations.

The most serious defect in a pipe butt weld is that which is oriented in the radial direction. The most commonly used technique for detecting such defects is the shear wave search. In this procedure, the transducer is located to one side of the weld at an angle to the pipe surface. The angle is maintained by a lucite block which transmits the sound from the transducer into the pipe. The sound will travel at an angle through the pipe and weld. Being at an angle, it will reflect from the pipe surfaces until it is attenuated. Any surface which is normal to the beam, however, will reflect a portion of the sound back to the transducer and show as

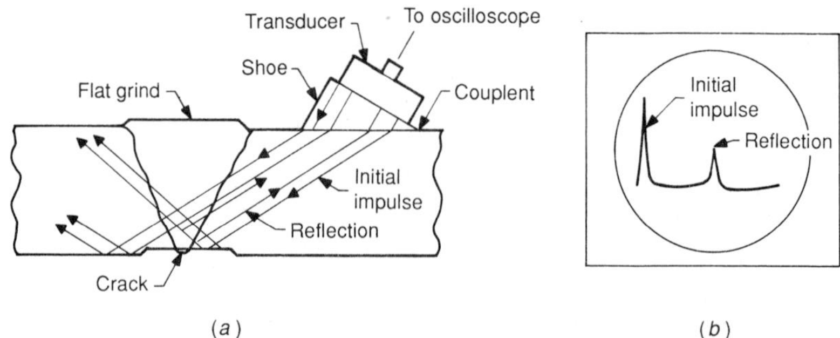

FIGURE A6.26 Ultrasonic shear wave search. (*a*) Search arrangement; (*b*) oscilloscope.

an indication on the oscilloscope. See Fig. A6.26. If the beam angle and the material thickness are known, the reflecting surface can be located and evaluated.

Prior to and periodically during each search, the equipment is calibrated against artificial defects of known size and orientation in a calibration block. The block must be representative of the material being searched (i.e., an acoustically similar material, with appropriate thickness, outside contour, surface finish, and heat-treated condition).

A variation of ultrasonic examination can be used to measure material thickness. If the speed of sound within the material is known, the time it takes for the signal to traverse the thickness and return can be converted to a thickness measurement.

4. *Liquid penetrant examination:* Penetrant-type examinations are suitable for surface examinations only but are very sensitive. They require a fairly smooth surface, since surface irregularities such as grinding mark indications can be confused with defect indications. The surface to be examined is thoroughly cleaned with a solvent and then coated with a penetrating-type fluid. Sufficient time is allowed to permit the fluid to penetrate into surface discontinuities. The excess penetrant is removed by wiping with cloths until all evidence of the penetrant is removed. A developer which acts somewhat like a blotter is then applied to the surface. This draws the penetrant out of the discontinuity, and it will appear on the surface as an indication. Obviously, the success of the examination depends on the visibility of the indication. To enhance this, the penetrant contains colored dyes which can be seen under normal light or fluorescent dyes which are viewed under ultraviolet light. The most common case is a red dye penetrant with a white developer.

5. *Magnetic particle examination:* Magnetic particle examination is essentially a surface-type examination although some imperfections just below the surface are detectable. This type of examination is limited to materials which can be magnetized (paramagnetic materials), since it relies on the lines of force within a magnetic field.

The item to be examined is subjected to a current which will produce magnetic lines of force within the item. The surface is then sprayed with a fine iron powder. The powder will align itself with the lines of force. Any discontinuity normal to the lines of force will produce a leakage field around it and a consequent buildup of powder which will pinpoint the defect. The examination must be repeated at 90° to detect discontinuities which were parallel to the original field.

There are a great many variations of magnetic particle examination depending on the manner in which the field is applied and whether the particles are wet or dry and fluorescent or colored.

Methodology. The ASME Boiler and Pressure Vessel, B31.1 and B31.3, require certain NDEs to be performed in accordance with the methods described in ASME Section V Nondestructive Examination. The pipeline codes, B31.4, B31.8, and B31.11, refer to API-1104 for Radiographic Procedures. In some cases, particularly in visual examination, requirements are given but no specific methodology is stated. In others, alternative parameters or qualification requirements are given. The specific requirements of the individual codes should be consulted.

Qualification Requirements. Qualification of procedures and personnel used in NDEs are required by most codes. When ASME Section V or API-1104 are invoked by the referencing code, a written procedure is required and it must be demonstrated to the satisfaction of the AI, ANI, owner, or owner's agent, whichever is applicable. Similarly personnel who perform NDEs must be trained, qualified, and certified. The most frequently invoked qualification document is SNT-TC-1A[37]; it is also accepted by B31.1 for qualification of personnel performing visual examinations. Some codes permit alternatives, such as AWS-QC-1.[38]

Extent of Examination. The applicable code will define the extent of examination required for piping systems under its coverage. The degree of examination and the examination method and alternatives are a function of the degree of hazard which might be expected to occur in the event of failure. Pressure, temperature, toxicity of the fluid, and release of radioactive substances are some of the considerations. Added layers of examinations may be required as the perceived hazard increases.

Accept-Reject Criteria. The applicable code will also define the items to be examined and the accept-reject criteria to be applied. Table A6.9 shows the acceptance standards applicable to the visual examination of butt welds under B31.1. Other piping codes have similar but not necessarily identical criteria.

Table A6.10 shows acceptance standards for radiographic examination. Indications interpreted as cracks, incomplete penetration, or lack of fusion are not permitted. Porosity and elongated indications are kept within certain limits. The acceptance standards for ultrasonic examination are similar.

Both magnetic particle and liquid penetrant examinations have identical limits. See Table A6.11.

Other types of NDEs, such as acoustic emission, bubble testing, and mass spectrometer testing, are not required by the various codes. They can be invoked by contract and the acceptance standards must be a matter of agreement between the contracting parties.

TABLE A6.9 Acceptance Standards for Visual Examination

The following indications are unacceptable:
1. Cracks on external surfaces
2. Surface undercut greater than $\frac{1}{32}$ in (1.0 mm) deep
3. Weld reinforcement greater than specified in ASME Table 127.4.2
4. Lack of fusion on surface
5. Incomplete penetration (when inside surface is readily accessible)

Source: From ASME B31.1 1989 ed.

TABLE A6.10 Acceptance Standards for Radiography

The following types of discontinuities are unacceptable:

1. Any type of crack or zone of incomplete fusion or penetration
2. Any other elongated indication with a length greater than:
   a. ¼ in (6.0 mm.) for t* up to ¾ in (19.0 mm)
   b. ⅓ t for t from ¾ in (6.0 mm) to 2¼ in (57.0 mm) inclusive
   c. ¾ in (19.0 mm) for t over 2¼ in (57.0 mm) where t is the thickness of the thinner portion of the weld
3. Any group of indications in a line that has an aggregate length greater than t in a length of 12 t, except where the distance between successive indications exceeds 6L where L is the longest indication in the group
4. Porosity in excess of that shown as acceptable in Appendix A-250 of Section I of the Boiler and Pressure Vessel Code.

*t pertains to the thickness of the weld being examined. If a weld joins two members having different thicknesses at the weld, t is the thinner of these thicknesses.
Source: From ASME B31.1 1989 ed.

TABLE A6.11 Acceptance Standards for Magnetic Particle and Liquid Penetrant Examinations

The following relevant indications are unacceptable:

1. Any cracks or linear indications
2. Rounded indications with dimensions greater than 3/16 in (5.0 mm)
3. Four or more rounded indications in a line separated by 1/16 in (2.0 mm) or less edge to edge
4. Ten or more rounded indications in any 6 in^2 (3870 mm^2) of surface with the major dimension of this surface not to exceed 6 in (150 mm) with the area taken in the most unfavorable location relative to the indications being evaluated

Source: From ASME B31.1 1989 ed.

Testing. All of the piping codes outline some type of pressure test to determine leak tightness. Since the completed piping system is usually subjected to some type of test in the field after installation, shop testing of subassemblies is infrequent. In those cases where the assembly cannot be field tested, where welds in the assembly will not be exposed for examination during the field test, and in other special situations, shop testing may be required. Shop testing must meet all of the requirements for field testing. See the section "Installation" for particulars.

Quality Assurance and Quality Control. ASME Section III has very specific requirements for QA programs. ASME Section I has requirements for QC programs. The B31 Piping Codes do not require any formal written program at this time. Refer to these codes for detailed information on this subject.

Cleaning and Packaging. Cleanliness of piping subassemblies is a matter of agreement between the fabricator and purchaser. As a minimum the fabricator will clean the inside of the subassembly of loose scale, weld spatter, machining chips, etc., usually with jets of compressed air. For those systems which require a greater degree of cleanliness several options are available. For specific information refer to PFI Standard ES-5 "Cleaning of Fabricated Pipe."[39] See also the following specifications published by the Steel Structures Painting Council.[40]

SSPC—SP 2 Hand Tool Cleaning

SSPC—SP 3 Power Tool Cleaning

SSPC—SP 6 Commercial Blast Cleaning

SSPC—SP 8 Pickling

SSPC—SP 10 Near-white Blast Cleaning

For ferritic steels the inside surfaces may be cleaned by turbinizing to remove loosely adhering mill scale and heavy rust. Wire brushing and grinding may also be employed for removal of more tightly adhering scale, rust, etc.; however, the most effective method for removal of tight scale is blasting with sand shot or grit. For guidance on blasting methods and degrees of cleanliness refer to PFI Standard ES-29 "Abrasive Blast Cleaning of Ferritic Piping Materials."[41]

Pickling is an equally effective method of cleaning. It is most often used for cleaning large quantities of straight tubes prior to fabrication or small size (about NPS 4) subassemblies where blasting is not as effective. Its application is limited by the availability and size of pickling tanks. A hot solution of sulfuric acid (H_2SO_4) is most commonly used, although cold hydrochloric acid (HCl) is also recommended. See SSPC—SP 8 "Pickling."

Austenitic stainless steels normally do not require cleaning except for a degreasing with solvent saturated cloths to remove traces of greases or cutting oils. Subassemblies which have been heated for bending or which have been given a carbide solution heat treatment will have a tightly adhering chromic oxide scale. Pickling and passivating in a solution of hydrofluoric and nitric acid will remove the scale and passivate the exposed surface. Here again, the equipment for pickling may limit the size of the subassembly. See ASTM A-380 published by the American Society for Testing Materials.[42] Blasting may also be used, but new silica sand or aluminum-oxide grit is required. Sand or grit previously used on ferritic pipe will contaminate the pipe surface with iron particles, and it will subsequently rust. The blasted surface should be treated with a solution of nitric acid to passivate the surface.

For extreme cleanliness, steam degreasing and rinsing with demineralized water may be employed.

The external surfaces of pipe may be left as is, painted, or otherwise preserved. See PFI Standard ES-34 "Painting of Fabricated Piping."[43]

Depending on the need for maintaining rustfree interior surfaces, the pipe inside diameter may be coated with different preservatives, or desiccants may be employed during shipping and storage.

For shipping, the ends of subassemblies are equipped with some type of end protection to preclude damage to weld end bevels or flange faces during shipment and field handling. See PFI Standard ES-31 "Standard for Protection of Ends of Fabricated Piping Assemblies."[44]

During shop operations, it is common practice to move piping assemblies with overhead or floor cranes usually with chain or wire rope slings. For austenitic stainless steels and nonferrous materials which could be damaged or contaminated, use of nylon slings is recommended.

INSTALLATION

Drawings

Drawings used for piping system installation may vary greatly. Often orthographic projections of the building showing several systems or single systems, depending on complexity, are used. In many cases single or multiple isometric drawings of a single system are used. These of course are not to scale but are convenient for planning, progress recording, or record keeping when required by quality programs. In all cases, where prefabricated subassemblies are being erected, these drawings will have been marked up to show the locations and mark numbers of the individual subassemblies, the location and designations of field welds, and the locations and markings of hangers.

Erection Planning

Planning is vitally important in installing a piping system. Many factors must be considered, among them accessibility to the building location, coordination with other work, availability and accessibility of suitable welding and heat treatment equipment, availability and qualification of welders and welding procedures, rigging, scaffolding, and availability of terminal equipment.

Each of the system components should also be carefully checked to assure correctness. Valves and other specialty items in particular should be checked to assure they are marked with flow arrows, that the handwheels or motor operators are properly oriented, and that the material to be welded is compatible with the material of the piping. Special valves for use in carbon steel systems are sometimes furnished as 5 percent chrome material, and thermowells are often not of the same chemical composition as the pipe. This may not be apparent from the drawings. Such a preliminary check will indicate the need for alternate welding procedures and preclude problems later.

The location of the work and accessibility to it should be viewed. It may not be possible to install an overly long subassembly after other equipment or building structure is in place. A common practice in the power field is to have large, heavy assemblies often found in the main steam and reheat lines of large central stations erected with the structure. In other cases, a preliminary review may show interferences from an existing structure, cable trays, ducts, or other piping which are not apparent from the drawings. The locations of the terminal points on equipment should be checked to assure that they are correct. The type, size, rating, or weld preparation of the connection should be checked to assure that it will match the piping. Solutions to any problems can be devised with the designer before work starts.

The ideal way to begin erection is to start at some major piece of equipment or at a header with multiple outlets. Install the permanent hangers if possible. If these are to be welded to the structure, some prudence should be exercised, since the final location of the line may warrant some small relocation to assure that the hanger is properly oriented relative to the piping in its final position. Obviously a certain number of temporary supports will be needed. Welding of temporary supports to the building structure or to the piping itself should be avoided or used only with the approval of the responsible engineers. Variable spring and constant-support-type hangers should normally be installed with locking pins in place, assuring that they function as a rigid support during the erection cycle. Where welded attachments to the pipe are involved, it is preferred that they be installed in the shop as part of the subassembly.

If possible, the major components of the system should be erected in their approximate final position prior to the start of any welding. This will reveal any unusually large discrepancies which may result from equipment mislocation, fabrication error, or tolerance accumulations. Adjustments or corrections can then be decided upon. Long, multiplane systems can absorb considerable tolerance accumulation without the need to modify any part. Short, rigid systems may not be able to accommodate any tolerance accumulation, and it may be necessary to rework one or more parts.

Cold Spring

Both the B31.1 and B31.3 Codes address cold springing in detail. Cold spring is the intentional stressing and elastic deformation of the piping system during the erection cycle to permit the system to attain more favorable reactions and stresses in the operating condition.

The usual procedure is to fabricate the system dimensions short by an amount equal to some percentage of the calculated expansion value in each direction. The system is then erected with a gap at some final closure weld, equal to the "cut shorts" in each direction. Forces and moments are then applied to both ends as necessary to bring the final joint into alignment. Once this is done, it is usually necessary to provide anchors on both sides of the joint to preserve alignment during welding, postweld heat treatment, and final examination. When the weld is completed and the restraining forces are removed, the resulting reactions are absorbed by the terminal points, and the line is in a state of stress. During start-up the line expands as the temperature increases, and the levels of stress and terminal reactions resulting from the initial cold spring will decrease. For the 100 percent cold sprung condition, the reactions and stress will be maximum in the cold condition and theoretically zero in the hot condition. It should be borne in mind that it is very difficult to assure that a perfect cold spring has been attained and for this reason the codes do not permit full credit in the flexibility calculations. Also remember that lines operating in the creep range will ultimately attain the fully relaxed condition. Cold spring merely helps it get there faster. Cold spring was historically applied to high-temperature systems such as main steam and hot reheat lines in central power stations, but this practice is not as prevalent anymore.

For those involved with the repair of lines which have been cold sprung, or which have achieved some degree of creep, caution should be exercised when cutting into such lines since the line will be in a state of stress when cold. The line should be anchored on either side of the proposed cut to prevent a possible accident.

Joint Alignment

In aligning weld joints for field welding it may be necessary to compromise between a perfect weld fit-up and the location of the opposite (downstream) end of the assembly. The weld bevel may not be perfectly square with the longitudinal axis of the assembly. Even a $\frac{1}{32}$-in deviation across the face of the weld bevel can result in an unacceptable deviation from the required downstream location if the joint is aligned as perfectly as possible. Often such a small gap can be tolerated in the welding. If, in order to maintain the downstream location, the gap at the joint is excessive, the joint should be disassembled, and the land filed or ground as needed to attain the required alignment of the weld joint while still maintaining the required downstream position. Flanged connections should be made up hand tight so that advantage can be taken of the bolt hole clearances to translate or rotate the assembly for better alignment of downstream connections.

Weld shrinkage of field welds may or may not be important in field assembly. In long flexible systems, they may be ignored. For more closely coupled systems, particularly those using GTAW root pass welding, this factor should be considered. The degree of longitudinal shrinkage across a weld varies with welding process, heat input, thickness, and weld joint detail. See the section "Layout, Assembly, and Preparation for Welding." In extreme cases closure pieces may be used. Here, the system is completed except for the final piece. A dummy assembly is then fabricated in place and the closure assembly is fabricated to match the dimensions of the dummy assembly with weld shrinkage of the final welds taken into account.

Cutting, Bending, Welding, Heat Treatment, and Examination

Cutting, bending, and welding operations in the field parallel those used in the shop. See the section "Fabrication."

Mechanical and oxyfuel gas cutting are most commonly used in the field. Plasma cutting may occasionally be used.

Bending, if used at all, is limited to small-diameter piping using relatively simple bending equipment at ambient temperatures. Occasionally in order to correct for misalignment, larger diameter ferritic piping is bent at temperatures below the lower critical. Please note that this procedure is limited to ferritic materials. Any application of heat to austenitic materials will result in sensitization and loss of corrosion properties. See the section "Bending." For smaller pipe sizes, torches may be used to supply heat, but for larger heavier wall materials and where better temperature control is warranted, heat may be applied by induction or resistance heating units in the same manner as local stress relieving. See the section "Local Heat Treatment." The heating units are applied to the section of the pipe to be bent. The section of the line upstream of the area to be bent should be anchored to preclude translation or rotation of the installed portion of the line. The anchor should preferably be not more than one or two pipe diameters from the area to be heated. Once the bend area has attained the required temperature, a bending force can be applied on the downstream leg of the pipe until the required bend arc has been obtained. Since most ferritic materials still have reasonably high yield strengths even at lower critical temperatures, care should be exercised. Large bending forces may damage the building structure or crack the line being bent. Apply a reasonable force for the conditions and allow the imposed stress in the bend arc to be relieved by the heat. Then repeat. Progress in this fashion until the required bend is accomplished. Some small amount of overbending may be required to offset the deflection which will occur in the unheated section of pipe between the heated arc and the pulling device. When the bend is completed and allowed to cool, all restraints may then be removed. Little if any force should be needed to align the downstream joint; otherwise additional bending may be needed to further correct the situation. No further heat treatment of the bend arc is needed since the temperatures applied in this bending method are below the lower critical temperature. Corrections to lines with large section modulus or where the required bend arc is large should preferably be made in a shop since better controls can be exercised.

Field welding is more often than not in a fixed position. Welders should be qualified in the 6G position since this qualifies for all positions.

Welding will be done using SMAW, GMAW, FCAW, and GTAW. Some welding processes can be automated using orbital welding techniques. Such practice can result in fewer repairs provided the bevels and alignment are within tolerance and the welding parameters are carefully selected.

Field postweld heat treatment also follows the practices outlined in the section "Heat Treatment" for local stress relieving of ferritic materials. This usually involves induction or resistance heating units with recording devices. For small pipe welds, torch heating using temperature-sensitive crayons to control temperature is sometimes used. Exothermic heating to stress-relieve welds is still used on occasion for outdoor applications where heating rates are not required to be controlled. Exothermic materials are preformed to pipe contour and sized to reflect the wall thickness and desired stress-relieving temperature. They are placed around the weld and ignited, attaining temperature in 5 or 10 min. The actual maximum temperature attained may vary.

NDE in the field will follow the practices outlined in the section "Verification Activities." Radiography is usually limited to radioactive isotopes, although occasionally *x*-ray equipment may find a use. Most surface examination is conducted using liquid penetrant methods, since magnetic particle equipment is not as convenient in the field. Ultrasonics are used for thickness verification and in certain situations as an alternative to radiography of welds when permitted by the governing code.

Mechanical Joints

Threaded joints probably represent the oldest method of joining piping systems. The dimensional standards for taper pipe threads are given in ASME B1.20.1.[45] This document gives all required dimensions including number of threads per inch, pitch diameter, and normal engagement lengths for all pipe diameters. Thread cutting should be regarded as a precise machining operation. A typical threading die is shown on Fig. A6.27. For steel pipe the lip angle should be about

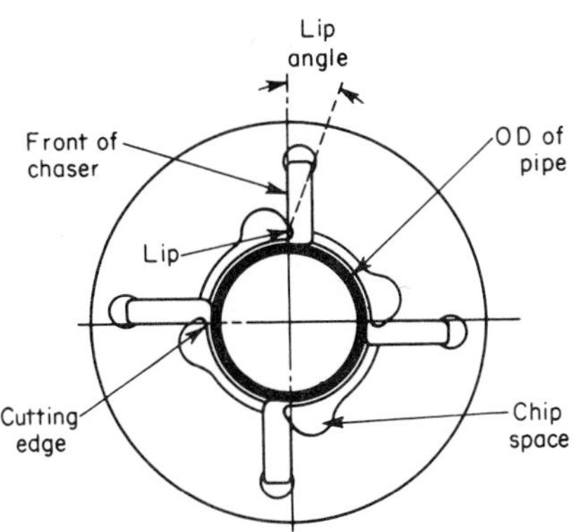

FIGURE A6.27 Theading die.

25°, but for brass it should be much smaller. Improper lip angle results in rough or torn threads. Since pipe threads are not perfect, joint compounds are used to provide leak tightness. The compounds selected, of course, should be compatible with the fluid carried and should be evaluated for possible detrimental effects on system components. Manufacturers' recommendations should be followed.

Where the presence of a joint compound is undesirable, dryseal pipe threads in accordance with ASME B1.20.3[46] may be employed. These are primarily found in hydraulic and pneumatic control lines and instruments.

Flanged joints are most often used where disassembly for maintenance is desired. A great deal of information regarding the selection of flange types, flange tolerances, facings and gasketing, and bolting is found in B16.5. The limitations regarding cast iron to steel flanges, as well as gasket and bolting selection, should be carefully observed. The governing code will usually have further requirements.

Gasket surfaces should be carefully cleaned and inspected prior to making up the joint. Damaged or pitted surfaces may leak. Appropriate gaskets and bolting must be used. The flange contact surfaces should be aligned perfectly parallel to each other. Attempting to correct any angular deviation perpendicular to the flange faces while making up the joint may result in overstressing a portion of the bolts and subsequent leakage. The proper gasket should be inserted making sure that it is centered properly on the contact surfaces. Bolts should be tightened hand tight. If necessary for alignment elsewhere, advantage may be taken of the bolt hole tolerances to translate or rotate in the plane of the flanges. In no case should rotation perpendicular to the flange faces be attempted. When the assembly is in its final location, bolts should be made up wrench tight in a staggered sequence. The bolt loading should exert a compressive force of about twice that generated by the internal pressure to compensate not only for internal pressure but for any bending loads which may be imposed on the flange pair during operation. For a greater guarantee against leakage, torque wrenches may be employed to load each bolt or stud to some predetermined value. Care should be exercised to preclude loading beyond the yield point of the bolting. In other cases, special studs that have had the ends ground to permit micrometer measurement of stud elongation may be used. Flange pairs which are to be insulated should be carefully selected since the effective length of the stud or bolt will expand to a greater degree than the flange thicknesses and leakage will occur. Thread lubricants should be used, particularly in high-temperature service to permit easier assembly and disassembly for maintenance.

There are a great variety of mechanical joints used primarily for buried cast-iron pipelines carrying water or low-pressure gas. They are primarily of the bell and spigot type with variations involving the use of bolted glands, screw-type glands, and various types of gasketing. The reader is referred to Federal Specification WW-P-421,[47] to AWWA Standards C-111,[48] C-150,[49] and C-600,[50] and to catalogs for proprietary types. For reinforced concrete pipe AWWA Standards C-300,[51] C-301,[52] C-302,[53] and Federal Specification SS-P-381[54] should be consulted. Compression-sleeve couplings such as the Dresser coupling (see Fig. A6.28) and the Victualic coupling (see Fig. A6.29) are widely used for above and below ground services, both with cast-iron and steel pipe. Consult the manufacturers' catalogs for more information.

Tubing

Copper, aluminum, steel, and stainless steel tubing are frequently used in hydraulic, pneumatic, and sampling systems. Installation is most often concerned with protec-

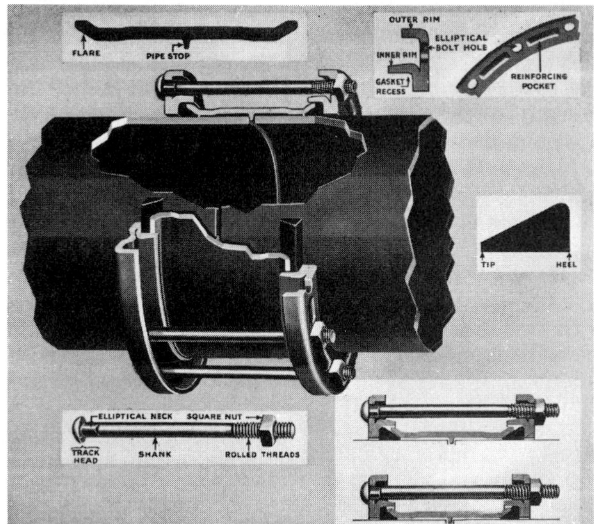

FIGURE A6.28 Compression sleeve (Dresser) coupling for plain end cast-iron or steel pipe.

tion of such materials from damage, since they are often associated with control systems. The manner of protection is left to the designer's judgment.

Lighter wall tubing is often bent using small compression-type benders. Tubing is joined to itself and to pipe size fitting and components with a variety of proprietary tubing fittings which are described in Chap. A2. Some heavier wall stainless steel tubing is welded using specially designed socket welding fittings. GTAW welding with filler metal added is used for such applications.

FIGURE A6.29 Victualic coupling for grooved-end cast-iron or steel pipe.

Pipe Supports

This section offers some thoughts on the installation of piping supports. The design, manufacture, and influence of supports on the system flexibility are outlined in Chaps. B4 and B5 of this book.

As pointed out earlier, economics and efficiency dictate that it is preferable to install the permanent supports for a system as the first step, thus minimizing the need for temporary supports. In so doing considerable judgment should be exercised, since there can be minor variations between the as-designed and as-installed line location. Resilient and constant effort supports should be locked with stops to preclude change in supporting effort as the line is being installed.

Only after the line has been completely welded, tested, and insulated should the stops be removed. Once removed, the resilient and constant effort supports should be carefully adjusted to their "cold" positions. This may take several iterations, since adjustment of any one will change the loading on those adjacent. Systems with multiple constant effort supports can be especially troublesome. Since the support design is most often based on theoretical values of weight of the pipe, insulation, and the fluid, there will be some difference between the actual and calculated supporting effort. Where rigid supports are involved, this variation will be taken up automatically. Where a system is designed with multiple resilient or constant effort supports, every effort should be made to incorporate one or more rigid supports in the design to absorb the variation between actual and theoretical loads. Otherwise it may be necessary, with the approval of the designer, to modify the spring load carrying settings.

As the line goes to operating temperature, it should be carefully observed to assure that there are no unforeseen interferences with its required expansion, particularly at nearby structures, floor sleeves, or adjacent lines or by restrained branch connections. Some modification may be required to assure free expansion of the line. All resilient supports and constant effort supports should be checked during initial start-up to assure that they are functioning properly, and after the line has been at operating temperature for several hours, they should be checked to verify that they are in the required "hot" operating condition. It may be necessary to readjust some units to match the calculated "hot" loading. These settings should be checked on a regular basis for the first few weeks of service, particularly in systems operating in the creep range, since the temperature will begin to relieve locked-in construction stresses and the line may choose a different, more relaxed location. Readjustments may be required. If after some time in service, the resilient and constant effort supports still require significant adjustment (i.e., the system cannot be balanced), a complete review of the flexibility analysis, expansion calculations, weight calculations, hanger design, and installation procedures should be made to determine the cause. Resilient and constant effort support units which are not functioning in the spring range (i.e., they have become "solid" or "loose") may impose undesirably high stresses in the line if they are not corrected, which can lead to premature failure or significantly reduced system life.

Leak Testing

At one time, complex shapes were pressure tested to determine their suitability for the service intended. This involved stressing the component to a point above service stresses, but below bursting stress, and was referred to as a pressure test. Currently most codes require some type of test to determine leak tightness rather than service suitability.

The most common method of leak testing for piping systems is the hydrostatic test. Usually this involves water at ambient temperature as the test medium. B31.1 requires that the system be pressurized to 1.5 times the design pressure, ASME III, to 1.25 times the design pressure, and B31.3 requires a test pressure of 1.5 times the design pressure adjusted by the ratio of the allowable stress at test temperature divided by the allowable stress at operating temperature. In each case, however, the test pressure of unisolated equipment or some function of the yield stress of the line material may be a limiting factor. See the applicable code for particulars. The line must be held at test pressure for at least 10 min, but may be reduced as permitted in the applicable code until the examination for leakage is complete.

Depending on the specific situation, alternative test fluids may be employed. As an example, in a liquid sodium system where water could be very hazardous or in cases where the possibility of freezing exists, a hydrocarbon or other fluid might be used.

In instances where water or other liquids are unacceptable, or where supports may not be adequate to carry the added weight of water, pneumatic tests may be performed. Pneumatic tests are potentially more dangerous than hydrostatic tests and extreme care should be exercised. B31.1 and ASME III require the pneumatic test be performed at not less than 1.2 times the design pressure, while B31.3 limits the test to 1.1 times design. In each case, the limits regarding equipment and yield strength cited above for hydrostatic tests also apply.

Prior to the test a detailed review of the section of the line to be tested should be made with the following in mind:

1. Temporary supports for those sections where the permanent supports were not designed to take the additional weight of the test fluid.
2. Isolation or restraints on expansion joints.
3. Isolation of equipment or valves which may be overstressed at test pressure.
4. Location of test pump and the need for additional test gauges if there is a significant head variation due to elevation differential.
5. Location of vents and drains.
6. Location of a relief valve to preclude excessive overpressure due to possible thermal expansion of the test fluid.
7. Consideration of the probable ambient test temperature relative to the expected brittle fracture toughness of the system materials. Heating the water may be a solution.
8. Alternative test fluid.
9. Accessibility to the weld joints for inspection. Some codes require that the weld joints be left exposed until after the test.
10. Assurance that no part of the system will exceed 90 percent of its yield strength.

It is advisable to prepare a written procedure outlining the scope and boundaries of each test to assure that it is performed in a safe manner. The codes vary a bit on the required test pressures, time at test pressure, pressure during inspection for leakage and whether alternative tests may be performed. It is advisable to look at each one specifically.

REFERENCES

1. ASME B31.1 "Power Piping Code." American Society of Mechanical Engineers, 345 East 47th Street, New York, NY 10017.
2. ASME Section I "Power Boiler Code." ASME.
3. ASME Section III "Nuclear Power Plant Components." ASME.
4. ASME Section V "Nondestructive Examination." ASME.

5. ASME Section IX "Welded and Brazing Qualifications." ASME.
6. ASME/ANSI B36.10 "Welding and Seamless Wrought Steel Pipe." ASME.
7. ASME/ANSI B16.5 "Pipe Flanges and Flanged Fittings." ASME.
8. Manufacturers Standardization Society of the Valve and Fitting Industry, Inc., 127 Park Street, N.E., Vienna, VA 22180.
9. The American Petroleum Institute, 1220 L Street, N.W., Washington, D.C. 20005.
10. The Pipe Fabrication Institute, P.O. Box 173, Springdale, PA 15144.
11. The American Welding Society, 550 N.W. LeJeune Road, P.O. Box 351040, Miami, FL 33135.
12. ASME/ANSI B16.34 "Valves—Flanged and Buttwelding End." ASME.
13. ASME/ANSI B16.9 "Factory-Made Wrought Steel Butt Welding Fittings." ASME.
14. ASME/ANSI B16.1 "Cast Iron Pipe Flanges and Flanged Fittings." ASME.
15. PFI ES-3 "Fabricating Tolerances." PFI.
16. *Welding Handbook,* 7th ed., The American Welding Society, 550 N. W. LeJeune Road, P.O. Box 351040, Miami, FL 33135.
17. Harvey, John F., *Theory and Design of Modern Pressure Vessels,* 2d ed., p 47, Van Nostrand Reinhold Company, 450 West 33d Street, New York, NY 10001.
18. ASME B31.3 "Chemical Plant and Petroleum Refinery Piping." ASME.
19. PFI ES-24 "Pipe Bending Methods, Tolerances, Process and Material Requirements." PFI.
20. *The Metals Handbook,* The American Society for Metals, Metals Park, OH 44073.
21. Copper Development Assn. Inc., Greenwich Office Park 2, P.O. Box 1840, Greenwich, CT 06836.
22. Huntington Alloys, Inc., Guyan River Road, P.O. Box 1958, Huntington, WV 25720.
23. ASME B31.4 "Liquid Transportation Systems for Hydrocarbons, Liquid Petroleum Gas, Anhydrous Ammonia and Alcohols." ASME.
24. ASME B31.8 "Gas Transmission and Distribution Piping Systems." ASME.
25. ASME B31.11 "Slurry Transportation Piping Systems." ASME.
26. API 1104 "Standard for Welding Pipe Lines and Related Facilities." The American Petroleum Institute.
27. ASME B31.5 "Refrigeration Piping." ASME.
28. AWS D10.9 "Qualification of Welding Procedures and Welders for Pipe and Tubing, Specification for." AWS.
29. ASME Section II, Part C, "Welding Rods, Electrodes and Filler Metals." ASME.
30. PFI ES-21 "Internal Machining and Fit-Up of GTAW Root Pass Circumferential Butt Welds." PFI.
31. PFI ES-7 "Minimum Length and Spacing for Welded Nozzles." PFI.
32. PFI ES-1 "Internal Machining and Solid Machined Backing Rings for Circumferential Butt Welds." PFI.
33. Sosnin, H. A., *The Theory and Technique of Soldering and Brazing of Piping Systems,"* NIBCO Inc., Elkhart, IN 46514.
34. Linnert, George E., *Welding Metallurgy,* 3d ed., vol. 2, American Welding Society.
35. United States Steel, *The Making, Shaping and Treating of Steel.* 10th ed./latest technology, Association of Iron and Steel Engineers, Suite 2350, Three Gateway Center, Pittsburgh, PA 15222.
36. The American Society for Nondestructive Testing, 1711 Harlingate Ln, P.O. Box 28518, Columbus, OH 43228-0518.

37. SNT-TC-1A, "Personnel Qualification and Certification in Nondestructive Testing." ASNT.
38. AWS QC-1, "Standards and Guide for Qualification and Certification of Welding Inspectors." AWS.
39. PFI ES-5 "Cleaning of Fabricated Pipe." PFI.
40. Steel Structures Painting Council, 4400 Fifth Ave., Pittsburgh, PA 15213.
41. PFI ES-29 "Abrasive Blast Cleaning of Ferritic Piping Materials." PFI.
42. ASTM A-380 "Standard Practice for Cleaning and Descaling Stainless Steel Parts, Equipment and Systems." American Society for Testing Materials, 1916 Race St. Philadelphia, PA 19103-1187.
43. PFI ES-34 "Painting of Fabricated Pipe." PFI.
44. PFI ES-31 "Standard for Protection of Ends of Fabricated Piping Assemblies." PFI.
45. B1.20.1 "Pipe Threads, General Purpose." ASME.
46. B1.20.3 "Dryseal Pipe Threads." ASME.
47. WW-P-421 "Pipe, Cast Gray and Ductile Iron, Pressure (for Water and Other Liquids)." General Services Administration, Specification Section, Room 6039, 7th and D Streets S. W., Washington, D.C. 20407.
48. C-111/A21.11 "Rubber Gasketed Joint for C.I. Pipe and Fittings." American Water Works Association, 6666 W. Quincy Ave., Denver, CO 80235.
49. C-150/A21.50 "Thickness Design of Ductile Iron Pipe." AWWA.
50. C-600 "Installation of C.I. Water Mains." AWWA.
51. C-300 "Reinforced Concrete Water Pipe—Steel Cylinder Type not Prestressed." AWWA.
52. C-301 "Reinforced Concrete Water Pipe—Steel Cylinder Type Prestressed." AWWA.
53. C-302 "Reinforced Concrete Water Pipe—Non Cylinder Type—not Prestressed." AWWA.
54. SS-P-381 "Pipe Concrete (Pressure, Reinforced Pretensioned Reinforcement Steel Cylinder Type)." by General Services Administration.

P • A • R • T • B

GENERIC DESIGN CONSIDERATIONS

CHAPTER B1
HIERARCHY OF DESIGN DOCUMENTS

Sabin Crocker, Jr., P.E.
Formerly Project Engineer
Bechtel Power Corporation
Gaithersburg, Maryland

In today's atmosphere of complex projects, extended liabilities, tight cost controls, and strict quality standards, it is essential that all phases of a project, from inception to operation, be effectively communicated and correctly executed. To this end, contract documents, design documents, fabrication details, procedures, and specifications are developed to communicate, monitor, and document the design, fabrication, and erection of piping systems precisely. The number and variety of documents to be prepared for a particular piping system are determined not so much by the importance of the piping system as by the complexity of the system and by the interface requirements among the owner, the design organization, the contractor, the material suppliers, and the regulatory agencies. In the following sections, the documentation requirements of a complex project are illustrated to provide a broad overview of the hierarchy of documents. However, since the specific requirements of a project are driven by many variables, such as the owner's requirements, budgets, market conditions, company practice, and licensing requirements, the discussion contained herein should be considered as a guideline only.

In this chapter, the principal design documents normally prepared for a complex piping system are identified, and their roles in the overall project is described. It should be noted, however, that future developments in engineering tools may dictate variations from the documents described here. With the ongoing development of sophisticated computer software, several related design documents may be developed as one database with the capability to extract information as required. For example, pressure drop and pipe stress calculations may be executed from a physical piping drawings database.

The production and management of design documents may be influenced by outside parties. There are many industry and national standards which provide guidance in the preparation and control of design documents. The references listed below contain standards for the preparation and control of design documents. The list is based on current practices in the United States. Where need for clarity exists, the reference is accompanied by a statement of the field of applicability. The list is not all inclusive; engineers responsible for the preparation of design documents must, from time to time, review the current codes and standards in order to take advantage of the changes in the industry which are expected to continue as computerized drafting and preparation of text and recordkeeping improve.

The military (MIL) and Department of Defense (DOD) standards and specifications referenced below provide a number of generally useful concepts and procedures for producing and maintaining design documents. Not all of the information is directly applicable to the field of interest of this handbook; however, the referenced documents should not be ignored by anyone setting up a system of design document production and control because the information is of fundamental importance to an organized approach to the task.

1. ASME Boiler and Pressure Vessel Code, Section III, NCA-3252: Contents of Design Specifications. (The American Society of Mechanical Engineers, 345 East 47th Street, New York, New York 10017.)
2. ANSI/ASME N626.3: Qualifications and Duties of Personnel Engaged in ASME Boiler and Pressure Vessel Code Section III, Divisions 1 and 2 Certifying Activities. (The American National Standards Institute, 1430 Broadway, New York, New York 10018.)
3. U.S. Department of Defense Index of Specifications and Standards (DOD ISS). Naval Publication and Form Center ATT: NPODS, 5801 Tabor Avenue, Philadelphia, Pennsylvania 19120:

   MIL-STD-481A Configuration Control; Engineering Changes, Deviation and Waivers, Short Form
   DOD-STD-480A Configuration Control Engineering Changes, Deviations and Waivers
   MIL-STD-483A Configuration Management Practices for Systems, Equipment, and Computer Programs
   DOD-D-1000B Drawing, Engineering, and Associated List
   ANSI Y14.1; must be ordered from ASME unless the requestor is in the navy

4. ANSI Y14.1, Drawing Sheet Size and Format (For ANSI Y, use ASME address.)
5. ANSI Y14.2, Line Conventions and Lettering. Drawings prepared under this standard usually are adequate for micrographic reproduction. (For ANSI Y, use ASME address.)
6. ANSI Y14.5M, Dimensioning and Tolerancing (For ANSI Y, use ASME address.)
7. ISA-S5.1: Instrument Symbols and Identification. This reference includes a standard for the amount of instrumentation detail the piping designer would show on the piping and instrumentation diagram. A fairly current edition has been adopted by the U.S. Nuclear Regulatory Commission; users should verify that subsequent editions remain acceptable for U.S. nuclear work. (Instrument Society of America, 67 Alexander Drive, P.O. Box 12277, Research Triangle Park, North Carolina 27709.)
8. NMA MS102, National Micrographics Association Drafting Guide for Microfilm. Drawings prepared under this standard usually are adequate for micrographic reproduction. (8728 Colesville Road, Silver Spring, Maryland 20910.)

Also, government regulations may dictate certain documentation requirements. In the United States, federal, state, and local laws apply to various piping systems; in other countries, similar laws may prevail. For example, the U.S. Code of Federal Regulations, Title 10, Part 50 mandates strict requirements for the design, construction, and operation of piping systems in a nuclear production or utilization facility. State and local governments generally invoke the ASME piping and pressure vessel codes (ASME Sections I through XI, ASME B31.1, B31.3, etc.). Laws aimed at protecting the environment may also have an impact on the design of some systems. Therefore, the designer should become familiar with the regulations at the location of the project.

PROJECT EVOLUTION

A typical project evolution can generally be divided into three principal stages. The first stage consists of inception, assignment of responsibilities, preliminary design, and estimating. The second stage consists of detailed design, procurement, and definitive cost estimating. (Depending on the project schedule, production of hardware, site preparations, and construction may also be initiated at this stage.) The third stage consists of completion of engineering, production of equipment, erection, start up, and commercial operation of the systems. Although the demarcation between the various stages can vary with the organization performing the work, the concepts remain unchanged. The following discussion describes the various documents normally prepared in each phase of the project.

FIRST STAGE

Contract Specifications

Contract specifications are normally prepared under the sponsorship of the owner to define the owner's basic requirements of the piping system(s). The contract specification establishes the applicable codes and standards, the owner's requirements, and the obligations of the parties involved in the project. It is expected that the organization charged with the design will develop all design data (to the satisfaction of the owner) not expressly provided by the owner in the contract specification.

Codes and Standards

Although codes and standards are not prepared by any of the parties involved in a project, they are discussed here because of their importance.

New or retrofit construction of almost every type of facility is usually regulated by a government agency at the federal, state, and/or local level. Regulators establish safety standards to protect workers and the general public. These regulations may be documents directly issued by the regulators or industry standards made law by reference. Furthermore, in locales where no laws exist to regulate safety standards, the owner or underwriter may mandate compliance with certain national construction codes or standards. Also, the piping industry regulates the safety of its projects by establishing minimum design requirements. Therefore, before proceeding with the design, the design engineer must establish what the regulatory requirements are. For a detailed discussion of codes and standards, refer to Chap. A4 in Sec. A of this handbook.

Because codes and standards have a significant impact on the project, it is paramount that the codes and standards to which the project must conform be established in the early stages of the project, preferably in the contract specification.

Design Criteria

Before proceeding with the detailed design, it is usually advantageous for the project to prepare a comprehensive set of documents defining the system design criteria. This criteria may be part of an overall project design criteria or may be a separate document prepared solely for the piping design. In either case, the design criteria should reiterate the design requirements delineated in the contract specification and should define the applicable codes and standards, environmental conditions, design parameters, and other pertinent design bases which will govern the work. The design criteria can be updated as the design progresses to reflect any change in the design basis.

Calculations

Calculations are documents prepared to support the establishment of flow rates, system pressures, temperatures, pipe and vessel wall thickness, heat transfer rates, and other design parameters. Calculations are also prepared for the pipe stress and flexibility analysis and pipe support design.

Because calculations form the foundations for the piping system design, their quality is paramount. The documents used for procurement, construction, and licensing must be supported by suitable calculations. Because the design process is an iterative process, it is usual to issue calculations at various stages of completeness based on firmness of the design input. The intended use of the calculation must be commensurate with the reliability of the results. Calculations may be issued "preliminary status" for in-house review, estimating, and bidding. Unless project conditions dictate otherwise, award of contracts, fabrication, and erection should be based only on final calculations derived from firm data. If calculations are not firm, precautions must be taken to minimize the impact of any changes resulting from changes in the calculation's results.

System Descriptions

The many organizations engaged in piping system design may vary the format and content of system descriptions (SDs) from that described herein based on their needs. Systems whose operation may have public safety implications (e.g., nuclear power plant systems and chemical plant systems) may require very detailed system descriptions.

The purpose of the system description is to set forth, specifically in writing, the functions, intent of the design, and major features of the system. Since some systems need mechanical, electrical, and control system discipline inputs, each discipline may prepare its own system description or, alternatively, one document containing all discipline criteria may be prepared. Customarily, the mechanical discipline assumes the lead for preparing system descriptions for piping systems. The four major topics covered in a typical SD are (1) system design bases; (2) operating modes (start up, normal, shutdown, emergency, or as appropriate); (3) descriptions and performance ratings of major equipment; and (4) control concepts.

The design basis stated in the SD is used to develop the system flow diagram (which shows the features necessary to accomplish the design basis) and the piping and instrumentation diagram (which shows the basic controls, interlocks, pipes, valves, vessels, pumps, and miscellaneous equipment comprising the system). Therefore, the system description is important in the development of the documents used in procurement, manufacturing, fabrication, and erection.

Table B1.1 presents topics to be covered in a system description applicable to fossil and nuclear power plants. Systems for other uses may not require all of the topics listed or may require topics not covered in the table. In any event, the SD should include no more or no less than what is required to fulfill the purpose stated above. In this example, no more information is presented than is necessary to obtain the construction and operating permits from the governmental authorities and to provide direction to the users of the document (designers, engineers, operators, vendors, etc.).

TABLE B1.1 Topical Outline: System Description

Function

Design bases
- Safety design bases*
- Contract design bases†
- Codes and standards

Description
- General description
- System operation
- Description of principal components

Unresolved matters‡

References to the sources of design criteria

*Applicable if Code and/or law prescribes safety design bases. Another area subject to law is the environment; for some systems environmental design bases would have to be identified. †Power plant practice names this section Power Generation Design Bases. Other industries use titles appropriate to their conditions. ‡This section is used to track unresolved matters during design development and is closed out with "none" in the final issue of the system description.

System Flow Diagram

The system flow diagram (SFD) illustrates the system description. To realize full benefit from the SFD, it should be issued with the SD. The SFD extends the purpose stated for the SD by schematically showing the operational relationships among the major system components and by stating the design and expected process variables for selected modes of operation. Table B1.2 presents a list of top-

TABLE B1.2 Topics Covered: System Flow Diagram*

Flow diagrams should show the following:
- All major equipment
- Equipment names
- Equipment identification number
- Major bypass and recirculation lines
- Control valves
- Valves required to demonstrate routing for all modes
- Pipe sizes if required by office practice
- Interconnections to other systems
- Equipment ratings or capacities

Flow diagrams should not show the following:
- Pipe class
- Pipe line numbers unless required by office practice
- Minor bypass lines
- Isolation and shutoff valves
- Maintenance vents and drains
- Relief and safety valves unless they serve in a mode
- Instruments
- Code class information
- Seismic class information

*It is impossible to generalize about SFDs. The foregoing do and do not list is subject to as many variations as there are design agents and vendors of packaged systems. For example, pipeline numbers, code class, seismic class, and other information may be found on a flow diagram intended for presentation to another organization having responsibility for the preparation of piping and instrumentation diagrams and other construction documents. In such instances, that flow diagram and system description convey most of the information required by others to develop the system and the rest of the information is defined in the contract or attachments to it.

ics to be included and excluded from the SFD. Figure B1.1 is a portion of a typical system flow diagram based on the concepts described above.

Piping and Instrumentation Diagram

The piping and instrument diagram (P&ID) provides a schematic representation of the piping, process control, and instrumentation which show the functional relationships among the system components. The P&ID also provides important information needed by the constructor and manufacturer to develop the other construction input documents (the isometric drawings or orthographic physical layout drawings). The P&ID provides direct input to the field for the physical design and installation of field-run piping. For clarity, it is usual to use the same general layout of flow paths on the P&ID as used on the flow diagram.

The P&ID ties together the system description, the flow diagram, the electrical control schematic, and the control logic diagram (beyond the scope of this handbook). It accomplishes this by showing all of the piping, equipment, principal instruments, instrument loops, and control interlocks. The P&ID contains a minimum amount of text in the form of notes (the system description minimizes the need for text on the P&ID). The first P&ID in the set for the job should con-

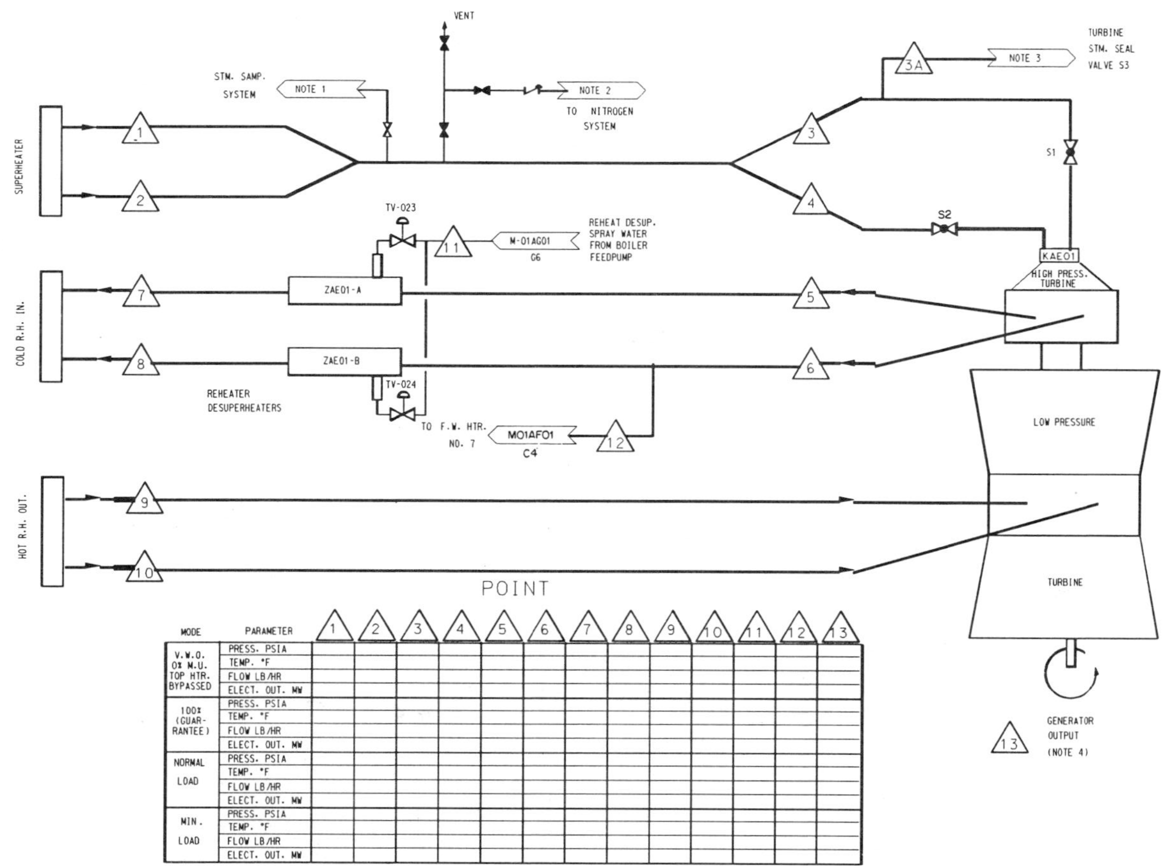

MODE	PARAMETER	1	2	3	4	5	6	7	8	9	10	11	12	13
V.W.O. 0% M.U. TOP HTR. BYPASSED	PRESS. PSIA													
	TEMP. °F													
	FLOW LB/HR													
	ELECT. OUT. MW													
100% (GUAR-RANTEE)	PRESS. PSIA													
	TEMP. °F													
	FLOW LB/HR													
	ELECT. OUT. MW													
NORMAL LOAD	PRESS. PSIA													
	TEMP. °F													
	FLOW LB/HR													
	ELECT. OUT. MW													
MIN. LOAD	PRESS. PSIA													
	TEMP. °F													
	FLOW LB/HR													
	ELECT. OUT. MW													

FIGURE B1.1 Typical system flow diagram.

TABLE B1.3 Topics Covered: Piping and Instrumentation Diagram

The P&ID should show the following:
- Mechanical equipment
- All valves associated with the process piping
- Instruments significant to the process piping, including:
  - Process pipes
  - Vents and drains
  - Special fittings
  - Sampling lines
  - Permanent start-up and flushing lines
- Specific information as applicable to job:
  - Instrument designations
  - Equipment names and numbers
  - Pipeline identification
  - Valve identification
- All size transitions in line:
  - Reducers and increasers, swages, and so on
- Direction of flow
- Interfaces for class changes
- Seismic category
- Quality level
- Interconnection references
- Annunciation inputs
- Plant computer inputs
- Vendor and contractor interfaces
- Identification of components and subsystems by others with reference to a vendor drawing for details if details are not shown on the P&ID
- Intended physical sequence of equipment, branch lines, reducers, and so on

The P&ID should not show the following unless there is a compelling need to do so for clarity. Items below identified by an asterisk (*) may be shown, if necessary.
- Instrument root valves
- Equipment rating or capacity
- Control relays*
- Manual switches*
- Indicating lights*
- Primary instrument tubing and valves
- Pressure, temperature, and flow data
- Vendor package piping which has no interface with engineering or construction*
- Elbows, tees, and similar standard fittings
- Extensive explanatory notes

Remarks: The P&ID for a defined system should be limited to coverage of that system to the maximum practical extent. Other systems that interface with the subject system are shown in phantom if such portions are detailed elsewhere.

Wherever a line is broken off as a matter of drafting convenience, both the break and the continuation are labeled so that one can readily trace the line from both sides of the break. This applies whether the break and continuation are on the same sheet or different sheets of the drawing.

Except for very simple P&IDs, the drawing should have the horizontal and vertical borders marked to permit reference to any small area of the drawing, as by continued at G-7. Care should be taken to ensure that these markings are within the sized field of the drawing so they will always be reproduced with the drawing regardless of the process used.

tain a legend defining all symbols used; if some symbols are defined elsewhere, it may be appropriate only to reference their source. The P&ID is also used by the start-up organizations for preparing flushing, testing, and blow-out procedures for the piping system and by the plant operators to operate the system. The correctness and completeness of the SD, SFD, and P&ID drawings are crucial to the success of the start-up program.

Table B1.3 shows what is included and what is not included in the P&ID. Figure B1.2 shows a portion of a typical P&ID which complies with the intent of ISA-S5.1 (Instrument Symbols and Identification) and with recent electric utility standard practice.

Piping Physical Sketches and Composite Drawings

Isometric or orthographic piping sketches are made, initially of the large bore piping and later of the small bore piping as determined by project requirements. Portions of other piping systems and structural, electrical, control, and HVAC information may be included in these sketches (in which case they are called composite drawings). These piping drawings form the basis for the working physical drawings, such as the system isometrics. The sketches and composites are not used for construction or manufacture; they represent the precomputer version of the present database in a computer-aided design (CAD) system or one of its derivative systems, which can provide the design study more efficiently than manual methods in many instances today. Computer-aided design and drafting (CADD) software packages are available from a number of commercial sources.

SECOND STAGE

With the completion of the initial draft of the calculations and P&IDs, the piping design progresses to the second stage, which consists of the detailed physical design of the piping system, piping stress analysis, and procurement of materials. The types of documents typically issued during this phase are purchase specifications, design specifications, piping erection specifications, piping orthographic drawings, piping isometric drawings, fabrication details, and equipment lists.

Design Specifications

Design specifications are prepared to define the performance requirements for services and materials. The design specifications may be prepared for in-house design work to control the quality of work performed, or the specification may be issued to govern contracted services.

For nuclear power plant systems and components required to meet ASME Section III Code requirements, the design specification takes on added importance. A requirement of the ASME Section III Code is that a design specification be prepared to provide a complete basis for ASME Section III Division 1 Construction. As a minimum, the ASME design specification must define the boundaries of the item covered, the design requirements, environmental conditions,

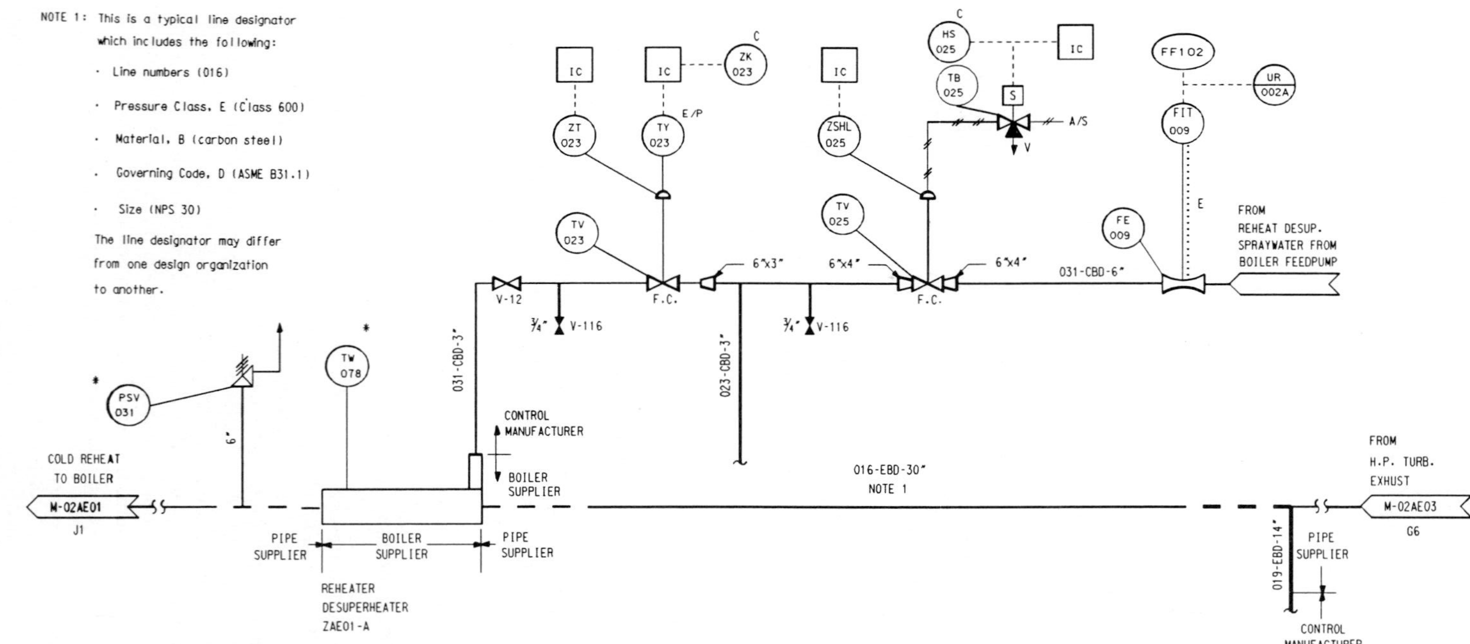

FIGURE B1.2 Typical system piping and instrument diagram.

ASME Code classification, material requirements, operating requirements, and effective edition, addenda and code cases used for construction. Design specifications can stand by themselves or can consist of a cover specification incorporating, by reference, all of the design documents required to perform the work.

Procurement Specifications

The procurement specification for piping fabrication should define the scope of work, applicable codes and standards, environmental conditions, material requirements, fabrication and examination requirements, testing requirements, and documentation requirements. The specification should reference the applicable design documents (e.g., piping isometric drawings) which govern the work. These documents should be made available to the supplier.

Erection Specification

The specification for piping erection should specify the applicable codes and standards governing the work. The specification should also delineate any fabrication, quality, examination, testing, and documentation requirements imposed beyond that required by the code. Also, all design and fabrication documents should be incorporated into the erection specification.

Physical Design Drawings

General practice among engineering organizations is to prepare orthographic piping drawings. With the development of three-dimensional computer aided design (three-dimensional CAD) software, the designer can check for interferences and can generate different views. Once the orthographic drawings are completed, they may be issued for piping fabrication and construction. For complex piping systems, it is common practice to develop separate piping isometric drawings for each pipe run. For pipe stress analysis, fabrication, and installation, the piping isometric drawings are easier to use than the orthographic drawings because all of the information on the isometric drawing pertains to the piping of interest—the drawing is not cluttered with extraneous information. Also, the isometric drawing is more easily visualized. Figure B1.3 depicts a typical isometric drawing.

Whether the scope of the project dictates that piping spools be shop fabricated or field fabricated, isometric drawings are indispensable in relaying the pipe fabrication requirements to the craft.

Stress Analyses

Piping stress analyses is a term applied to calculations which address the static and dynamic loadings resulting from effects of gravity, temperature changes, internal and external pressures, changes in fluid flow rate, seismic activity, fire, and other environmental conditions. Codes and standards establish the minimum scope of stress analyses. Some codes prescribe loading combinations with not-to-exceed stress limits. For specific code requirements, refer to other chapters in

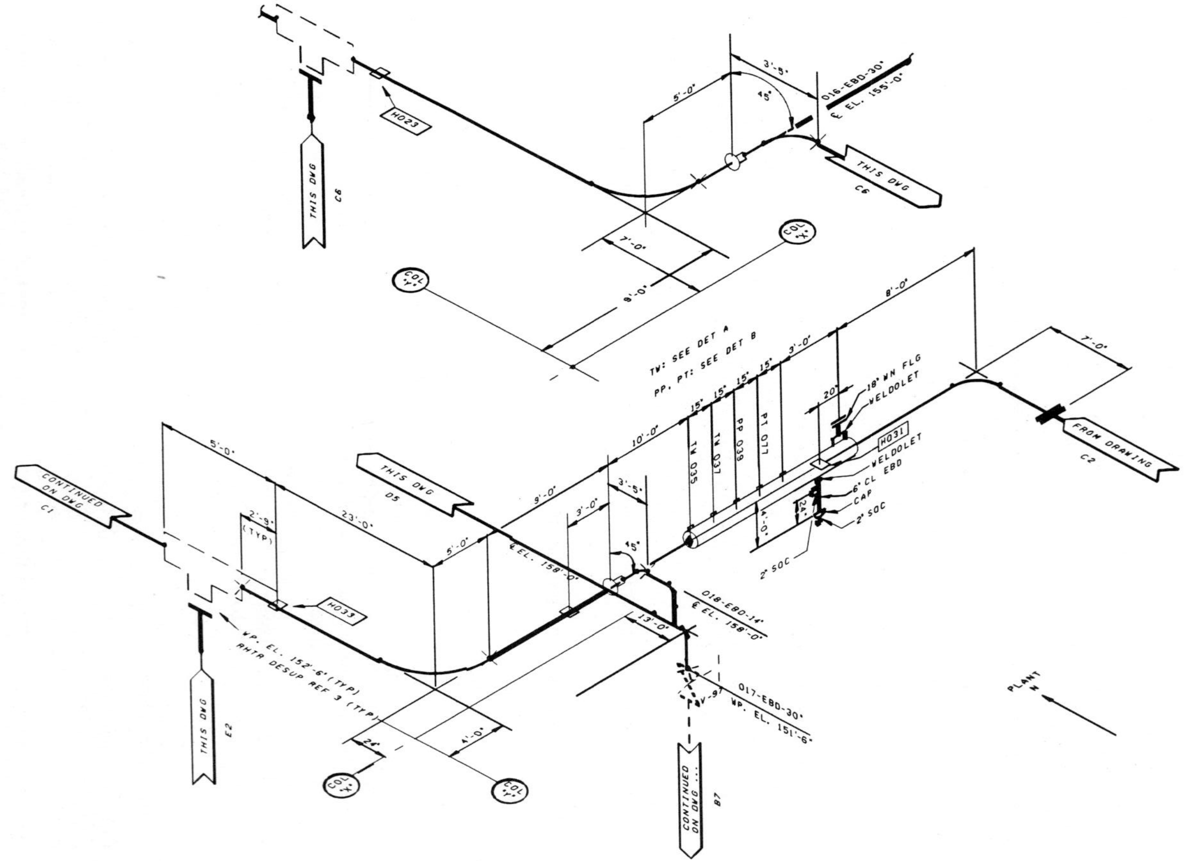

FIGURE B1.3 Typical piping isometric drawing.

this handbook where code requirements relevant to the type of piping system involved are discussed.

Piping Spool Drawings

The piping spool drawing (also referred to as spool sheet) is the shop detail drawing developed by the pipe fabricator from the piping isometric, or the piping orthographic, for each prefabricated section of piping. Generally, a spool sheet covers only one spool, whether the spool consists of only one fitting or of many piping elements (limited in size only by shipping or handling capabilities). The spool sheet should specify all dimensions, materials, fabrication procedures, examination and testing requirements, and code stamping requirements, as applicable. The spool sheet may also be used to document the performance of the required operations. Figure B1.4 is an illustration of a typical spool sheet.

It is customary for the engineering organization to review the spool sheets to ensure that the code, specification, and design documents are properly interpreted by the fabricator.

Equipment Drawings

Certified equipment drawings are generally submitted to the engineer by the equipment supplier. These drawings may be subject to approval by the engineer to ensure that the specification requirements are met. Also, the drawings are used to finalize the facility design—that is, the physical and operating characteristics of the actual procured equipment are incorporated into the design.

The supplier documentation should, as a minimum, contain the following information:

- Component outline dimensions
- Location and size of service connections
- Bill of materials
- Service requirements
- Maintenance procedures
- Instruction manual and list of recommended spare parts
- Diagrams for piping, logic, electrical wiring, and instrumentation

Pipe Support Drawings

The design and procurement of pipe supports are usually accomplished in one of the following ways: The supports can be preengineered by the engineer, or the design can be contracted to a pipe support manufacturer. In the first case, the engineer prepares a detailed pipe support drawing containing the necessary fabrication details and bill of materials. The pipe support fabricator only needs to fabricate the support in accordance with the drawing. In the second case, the engineer specifies the direction of restraint, the type of support (e.g., variable, constant, or rigid), and the support load. The support manufacturer prepares detailed shop fabrication drawings for the shop's use. The support drawings should contain the same level of detail as described for spool sheets. As with spool sheets,

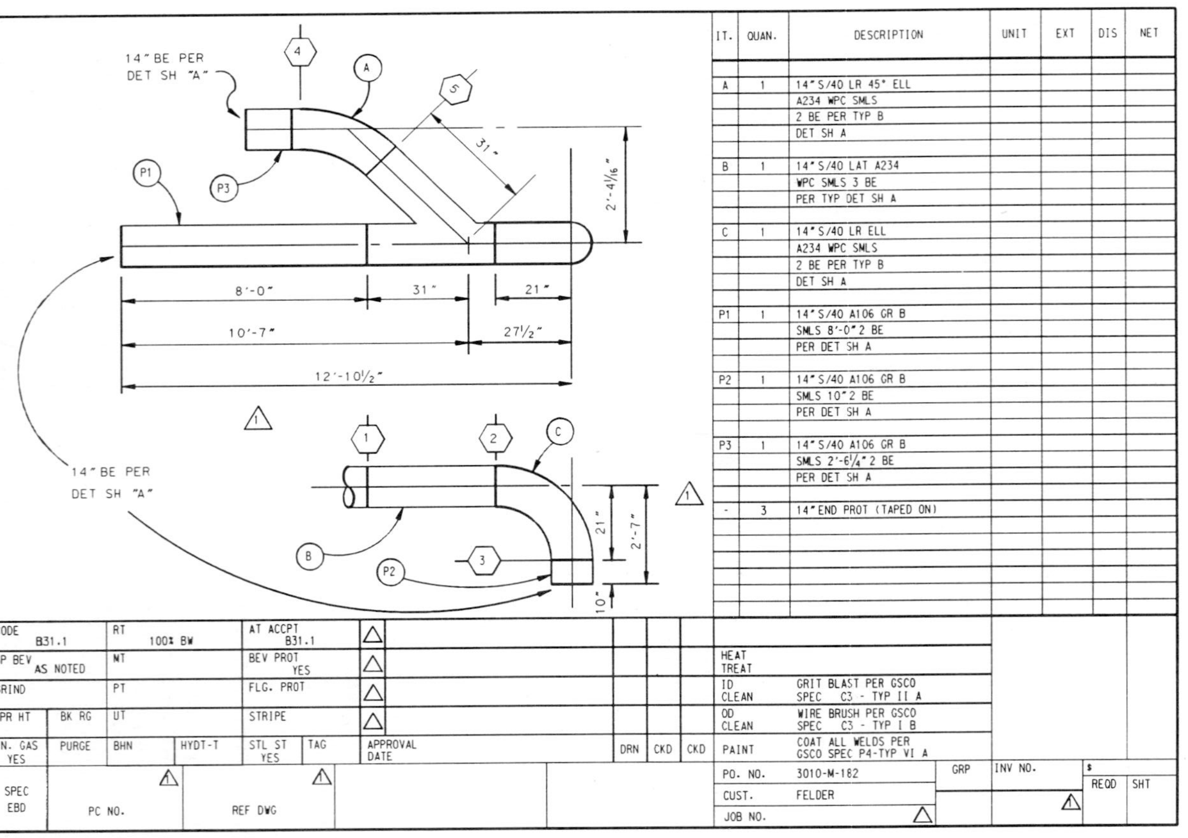

FIGURE B1.4 Typical spool detail sheet.

the engineer reviews the pipe support drawings for compliance with the purchase specification.

THIRD STAGE

As the piping system design reaches completion and materials begin to arrive at the job site, the project enters the third stage. It begins with the installation of the piping system.

During installation of the piping system, it is imperative that deviations from the design precipitated by variations in field conditions be controlled and reconciled by engineering. For this purpose, engineers and constructors have developed several vehicles by which deviations are identified and resolved. The following sections describe some of these vehicles. The titles of these documents may vary from organization to organization; however, their purpose remains the same.

Field Change Request

If a contractor cannot install an item as shown on the design documents (or if he or she identifies an advantageous alternative to the design), the contractor may request a design change by completing a field change request form. The request should identify the existing condition, the reason for the change, and the proposed alternate. Engineering can expeditiously evaluate the condition and prepare a reply, often on the same form.

Nonconformance Report

When an installation does not conform to the design document, a nonconformance report should be generated to document the discrepancy. The report is submitted to the engineer for evaluation and disposition. If the engineer determines that the condition still satisfies the functional requirements of the original design, he or she may accept the nonconformance. If, however, the deviation cannot be accepted, the engineer will require that the nonconforming item be repaired or replaced to meet the original design requirement.

Start-Up Field Report

As each system installation is completed, the contractor turns the system over to the start-up group for system check out and performance testing. Any deficiencies encountered by start up are documented on start-up field reports, which are submitted to the engineer for resolution. The report should describe the deficiency and should, if possible, offer a recommendation for corrective action.

Conclusion

The timely preparation of quality design documents is crucial to the economic and successful installation of a piping system. When prepared in the proper sequence, the engineering documents build on the information developed in the foregoing documents, allow for an organized and methodical development of the project. This way, the number of iterations in the design process is minimized and schedule and costs are controlled.

CHAPTER B2
DESIGN BASES

Joseph H. Casiglia, P.E.
Principal Engineer—Piping
The Detroit Edison Company
Detroit, Michigan

The design of a piping system is a straightforward process. The technology is extensive and diverse. Piping design requires the application of theory from a number of engineering disciplines, including fluid mechanics, statics, dynamics, strength of materials, and physical metallurgy. It also requires the knowledge and application of a number of codes and standards.

This chapter identifies and explores the various facets of piping design and highlights parameters that must be considered in completing the design process. The design bases discussed herein are generic in nature and should be considered during the course of design of any piping system, regardless of its function. In some cases, a number of the parameters discussed may not be applicable; however, they all should be looked at when the initial design concept for the system is formulated and developed.

DEFINITION OF THE TERM DESIGN BASES

Design bases are the physical attributes, loading and service conditions, environmental factors, and materials-related factors which must be considered in the detailed design of a piping system to ensure its pressure integrity over its design life.

Physical Attributes

Physical attributes are those parameters which govern the size, layout, and dimensional limits or proportions of the piping system. Dimensional standards have been established for most piping components such as fittings, flanges, and valves, as well as for the diameter and wall thickness of standard manufactured pipe. Those standards are identified in the section "Use of Codes and Standards in Piping Systems Design." Certain types of piping systems require special design practices for configuration control to assure constructability or in-service performance. For example, high-temperature, high-pressure piping systems are usually designed with weld joints spaced a minimum of one pipe diameter apart to facilitate radiographic examination of the joints. Steam and wet-gas systems are designed to maintain the pipe runs at some minimum pitch to ensure adequate drainage of condensate or other liquids that may separate from the gas stream. Pipelines which are subject to frequent plugging should be designed with adequate clearance to allow for ready disassembly and maintenance.

Loading and Service Conditions

Loading conditions, or *loads,* are forces, moments, pressure changes, temperature changes, thermal gradients, or any other parameters which affect the state of stress of the piping system. Typical examples of loading conditions include internal pressure, piping system dead-weight, steady-state or transient temperatures, wind loads, or snow and ice loads. Loads may be external to the piping system such as environmental temperature changes or wind loads, or they may be internal to the system such as internal fluid pressure or temperature changes.

Service conditions are combinations of loads or load-sets which occur simultaneously and therefore require that the piping system be designed to withstand their combined effects. Occasionally, the service conditions will be specified by the piping design code. Examples are found in the American Society of Mechanical Engineers Boiler and Pressure Vessel Code, Section III, Nuclear Power Plants Components, Paragraph NCA-2142, where service conditions are defined directly, and the American Society of Mechanical Engineers Code for Pressure Piping, ASME B31.1, Power Piping, Paragraph 104.8, where service conditions are specified under the topic of analysis of piping components due to the effects of sustained and occasional loads.

Where service conditions are not specified by a particular code, the designer should review the various loading conditions that the piping system is exposed to and formulate the combinations that must be considered in design. Reference to the most commonly used piping design codes listed in the section "Use of Codes and Standards in Piping Systems Design" will provide the designer with guidance in setting appropriate design stress limits.

Environmental Factors

When used within the context of this chapter, the term *environmental factors* refers to operating conditions that result in progressive physical or chemically induced deterioration of the piping system which ultimately leads to a breach of the pressure boundary or a gross structural failure. Failures that are the result of environmental factors are usually slow to progress and frequently involve localized areas of the piping system. The most common examples of environmental factors include *corrosion, erosion,* and physical damage. While corrosion and erosion mechanisms can act independently, a combined reaction known as *erosion/corrosion* has also been found to occur in wet-steam and water piping systems constructed of plain carbon steels.

Materials-Related Considerations

Materials-related considerations are the specific chemical, metallurgical, and physical properties of a piping system's material constituents which can ultimately determine its suitability for a particular service. Proper materials selection can be the crucial parameter in determining the adequacy of the performance of a piping system where extremes of temperature, chemical attack, or erosion are significant factors in its operation.

Pressure Integrity

Pressure integrity is the maintenance of a leak-tight condition in piping systems' pressure-containing boundary coincident with the control of the rate of stress or strain within predefined criteria limits. Pressure integrity is not synonymous with leakage integrity; the latter being only an assurance of a leak-tight condition without regard for the state of stress or structural stability of the pressure boundary. Maintenance of the pressure integrity of a piping system, within predefined criteria limits, is a major objective of the design process.

USE OF CODES AND STANDARDS IN PIPING SYSTEMS DESIGN

In practice, the assurance that the design and construction of a piping system will meet prescribed pressure-integrity requirements is accomplished through the use of published codes and standards. Numerous codes and standards have been formulated and published by major interest groups of the piping and pressure vessel industry. The most widely used codes and standards for piping systems design are those published by the *American Society of Mechanical Engineers (ASME)* and accredited by the *American National Standards Institute (ANSI).*

Differentiation between Codes and Standards

Codes and standards both provide criteria through which pressure integrity can be assured and simplified design rules to ensure adherence to the criteria. Many designers and engineers think the terms *code* and *standard* are synonymous, or at least somewhat interchangeable, but this understanding is incorrect.

Codes. Piping *codes* provide specific design criteria such as permissible materials of construction, allowable working stresses, and load-sets which must be considered in design. In addition, rules are provided to determine the minimum wall thickness and structural behavior due to the effects of internal pressure, dead weight, seismic loads, live loads, thermal expansion, or other imposed internal or external loads. Piping codes provide design rules for nonstandard fittings and for the reinforcement of openings in the pipe wall. They do not provide design rules for standard in-line components such as valves, flanges, and standard fittings; rather they define the design requirements for these classes of components by reference to industry standards.

The use of specific codes for the design and construction of piping systems is frequently mandated by statute or regulations imposed by regulatory and enforcement agencies.

Typically codes are structured around technology or industry-user lines. For example, ASME B31.1, Power Piping, covers piping systems in power plants, district heating plants, district distribution piping systems, and general industrial piping systems while ASME B31.3, Chemical Plant and Petroleum Refinery Piping, is structured around the petrochemical industry. Any one of the above-named industrial facilities might have a pipeline with similar service requirements; a high-pressure steam main, for example. However, the requirements of the specific code, as influenced by the needs and experience of the user-industry, will dictate its design and construction.

The major piping design and construction codes are listed later in the section "Reference Codes and Standards." The systems and subsystems covered by these codes are defined in their scope sections. The scope sections of all potentially applicable codes should be reviewed in the predesign phase of a project to determine which code, or codes, should be applied to its design and construction. In some cases, multiple codes may be required for the design and construction of the same piping system, depending upon its location. For example, a steam main serving a petrochemical plant from a major utility's district heating system would be designed and constructed to ASME B31.1, up to the petrochemical plant property line. The balance of the piping on the petrochemical plant's property would be designed to ASME B31.3. In the case of a natural gas main serving a utility power house, the outdoor piping is designed and constructed to ASME B31.8 up to and including the meter set, and the in-plant piping is designed and constructed to ASME B31.1.

Sometimes, different piping systems within the same building or facility will be designed and constructed to different codes. For example, most of the piping systems in a utility power plant are designed and constructed to ASME B31.1. However, the building heating and air conditioning piping systems are designed and constructed to ASME B31.9.

Standards. *Standards* provide specific design criteria and rules for individual components or classes of components such as valves, flanges, and fittings. There are two general types of standards: dimensional and pressure integrity.

Dimensional standards provide configuration control information for components. The main purpose of dimensional standards is to assure that similar components manufactured by different suppliers will be physically interchangeable. Conformity to a particular dimensional standard during the manufacture of a product does not imply that all such similarly configured products will provide equal performance. For example, two different styles of NPS 10 Class 150 flanged-end gate valves could be manufactured, in part, to ASME/ANSI Standard B16.10, Face-to-Face and End-to-End Dimensions of Ferrous Valves. The valves would be physically interchangeable between mating flanges in a particular piping system. However, because of completely different seat and disc design, one valve might be capable of meeting far more stringent seat-leakage criteria than the other.

Pressure-integrity standards provide performance criteria. Components designed and manufactured to the same standards will function in an equivalent manner. For example, all NPS 10 Class 150 ASTM A105 flanges, which are constructed in accordance with ASME/ANSI Standard B16.5, Pipe Flanges and Flanged Fittings, have a pressure-temperature rating of 230 psig at 300°F.

Standards are not normally mandated by statute or regulation; rather they are usually invoked by a construction code or purchaser's specification.

The ANSI Pressure Classification System

The ANSI Pressure Classification System meets the needs of industry by providing quantitative performance standards for a wide range of piping components, based upon a manageable number of operational variables. This system defines predetermined pressure-temperature ratings that components are designed to meet.

A number of different ANSI pressure-temperature standards have been devised for piping components. The standards in current use in the piping industry are listed in the section "Reference Codes and Standards." In this section the pressure classification system used in ASME/ANSI Standard B16.5, Pipe Flanges and Flanged Fittings,[1] will be used for illustration. However, the concepts covered are generally applicable to all the ASME/ANSI pressure-integrity standards.

Flanges manufactured in accordance with ASME/ANSI B16.5 are made from materials categorized into 33 material or material alloy groups. There are 10 carbon and low-alloy steel material groups, 7 high-alloy steel material groups, and 16 nonferrous metals groups. Within each of the 33 material groups is a subgrouped listing of ASTM materials specifications for forgings, castings, and plates. In addition, acceptable bolting materials are also specified. Partial listings of the various material groups, subgroups, and bolting materials are shown in Tables B2.1 and B2.2. For the complete list, see ASME/ANSI B16.5.

For any single ANSI flange pressure class, all flanges made from any material in a material group carry the same pressure-temperature rating.

ASME/ANSI B16.5 provides seven individual pressure classes for flanges. They are Classes 150, 300, 400, 600, 900, 1500, and 2500. The pressure-temperature ratings for flanges representing all 33 material groups are organized within seven tables, one for each pressure class. Table B2.3 is adapted from ASME/ANSI Standard B16.5 and is typical of the seven flange rating tables. It provides the pressure-temperature ratings for Class 600 flanges. The table is organized with the material groups listed across the top of the table and the maximum working temperatures along the left-hand border. The body of the table provides the pressure ratings for flanges from each material group, at the given temperature.

In practice, the use of ASME/ANSI B16.5 to determine a flange rating is quite simple. The procedure is outlined below:

1. Determine the maximum operating pressure and temperature for the flange in question.
2. Select a flange material and therefore a material group from one of the 33 listed material groups. Be aware that some of the qualifying notes concerning maximum operating temperatures for various materials may influence the final material selection.
3. Enter the appropriate material group of one of the seven pressure rating tables at the next increment of temperature listed which is higher than the desired maximum operating temperature. Start with the Class 150 table and proceed upward until a pressure rating for the desired temperature which exceeds the required operating pressure is found. The table in which this condition is satisfied dictates the required pressure class and the pressure-temperature rating of the flange.

As an example, assume an ASTM A105 carbon steel flange is required to satisfy a pressure rating of 1060 psig at 650°F. ASTM A105 is a Material Group 1.1 material. Entering Table B2.3 under Material Group 1.1, at a temperature of 650°F, a Class 600 flange is found to have a rating of 1075 psig at 650°F. Therefore, a Class 600 ASTM A105 flange is suitable for the stated conditions. When

TABLE B2.1 Materials Used for ASME B16.5 Flange Construction

Material groups		Product forms								
Material group no.	Nominal designation steel	Forgings			Castings			Plates		
		Spec. no.	Grade		Spec. no.	Grade		Spec. no.	Grade	
1.1	Carbon	A 105	...		A 216	WCB		A 515	70	
		A 350	LF2	...	...			A 516	70	
	C-Mn-Si	...			...			A 537	Cl.1	...
1.2	Carbon	...			A 216	WCC		...		
		...			A 352	LCC	...	...		
	2½Ni	...			A 352	LC2	...	A 203	B	...
	3½Ni	A 350	LF3	...	A 352	LC3	...	A 203	E	...
1.3	Carbon	...			A 352	LCB		A 515	65	
		...			...			A 516	65	
	2½Ni	...			...			A 203	A	...
	3½Ni	...			...			A 203	D	...
1.4	Carbon	A 350	LF1	...	...			A 515	60	
		...			...			A 516	60	...
1.5	C-½Mo	A 182	F1		A 217	WC1		A 204	A	
		...			A 352	LC1	...	A 204	B	

1.7	C-½Mo	. . .			. . .		A 204	C	
	½Cr-½Mo	A 182	F2	. . .	. . .		. . .		
	Ni-Cr-½Mo	. . .			A 217	WC4	. . .		
	Ni-Cr-1Mo	. . .			A 217	WC5	. . .		
1.9	1Cr-½Mo	A 182	F12		. . .		. . .		
	1¼Cr-½Mo	A 182	F11		A 217	WC6	A 387	11 Cl.2	. . .
1.10	2¼Cr-1Mo	A 182	F22	. . .	A 217	WC9	A 387	22 Cl.2	. . .
1.13	5Cr-½Mo	A 182	F5	. . .	A 217	C5	. . .		
		A 182	F5a	. . .	. . .		. . .		
1.14	9Cr-1Mo	A 182	F9	. . .	A 217	C12	. . .		
2.1	18Cr-8Ni	A 182	F304		A 351	CF3	A 240	304	
		A 182	F304H	. . .	A 351	CF8	A 240	304H	. . .
2.2	16Cr-12Ni-2Mo	A 182	F316		. . .		A 240	316	
		A 182	F316H				A 240	316H	
							A 240	317	
				. . .	. . .				. . .
	18Cr-13Ni-3Mo	. . .			. . .				
	18Cr-9Ni-2Mo	. . .			A 351	CF3M	. . .		
					A 351	CF8M	. . .		
		. . .							

Source: Adapted from ASME/ANSI B16.5, *Pipe Flanges and Flanged Fittings,* American Society of Mechanical Engineers, New York, 1988, Table 1A.

TABLE B2.2 Bolting Materials Used with ASME B16.5 Flanges

Bolting materials*									
High strength*			Intermediate strength*			Low strength*		Nickel and special alloy*	
Spec. no.	Grade		Spec. no.	Grade		Spec. no.	Grade	Spec. no.	Grade
A 193	B7	...	A 193	B5	...	A 193	B8 Cl.1	B 164	...
A 193	B16	...	A 193	B6	...	A 193	B8C Cl.1		
			A 193	B6X	...	A 193	B8M Cl.1	B 166	...
A 320	L7		A 193	B7M	...	A 193	B8T Cl.1		
A 320	L7A		A 193	B8 Cl.2		A 193	B8A	B 335	N10665
A 320	L7B		A 193	B8C Cl.2		A 193	B8CA		
A 320	L7C		A 193	B8M Cl.2		A 193	B8MA	B 408	...
A 320	L43		A 193	B8T Cl.2		A 193	B8TA		
								B 473	...
A 354	BC	...	A 320	B8 Cl.2		A 307	B		
A 354	BD	...	A 320	B8C Cl.2				B 574	N10276
			A 320	B8F Cl.2		A 320	B8 Cl.1		
A 540	B21	...	A 320	B8M Cl.2		A 320	B8C Cl.1		
A 540	B22	...	A 320	B8T Cl.2		A 320	B8M Cl.1		
A 540	B23	...				A 320	B8T Cl.1		
A 540	B24	...	A 449	...					
			A 453	651					
			A 453	660					

*For notes refer to ASME/ANSI B 16.5.

Source: Adapted from ASME/ANSI B16.5, *Pipe Flanges and Flanged Fittings,* American Society of Mechanical Engineers, New York, 1988, Table 1B.

TABLE B2.3 Pressure-Temperature Ratings for Class 600 ASME B16.5 Flanges

Material group no.	1.1	1.2	1.3	1.4	1.5	1.7	1.9	1.10	1.13	1.14	2.1	2.2	2.3	2.4	2.5	2.6	2.7
					Alloy steels						Austenitic steels						
Temp., °F	Carbon steel				C-½Mo	½Cr-½Mo, Ni-Cr-Mo	1Cr-½Mo, 1¼Cr-½Mo	2¼Cr-1Mo	5Cr-½Mo	9Cr-1Mo	Type 304	Type 316	Type 304 L / Type 316 L	Type 321	Types 347, 348	Type 309	Type 310
−20 to 100	1480	1500	1390	1235	1390	1500	1500	1500	1500	1500	1440	1440	1200	1440	1440	1345	1345
200	1350	1500	1315	1125	1360	1500	1425	1430	1500	1500	1200	1240	1015	1220	1270	1210	1210
300	1315	1455	1275	1095	1305	1455	1345	1355	1455	1455	1055	1120	910	1090	1175	1140	1140
400	1270	1410	1235	1060	1280	1410	1315	1295	1410	1410	940	1030	825	990	1110	1065	1065
500	1200	1330	1165	995	1245	1330	1285	1280	1330	1330	875	955	765	915	1035	1010	1010
600	1095	1210	1065	915	1210	1210	1210	1210	1210	1210	830	905	720	875	985	955	955
650	1075	1175	1045	895	1175	1175	1175	1175	1175	1175	815	890	700	855	960	930	930
700	1065	1135	1035	895	1135	1135	1135	1135	1135	1135	805	865	685	840	935	910	910
750	1010	1010	945	885	1065	1065	1065	1065	1065	1065	795	845	670	830	920	895	895
800	825	825	780	740	1015	1015	1015	1015	995	1015	790	830	660	825	910	870	870
850	535	535	535	535	975	975	975	975	880	975	780	810	645	815	890	850	850
900	345	345	345	345	900	900	900	900	705	900	770	790	...	810	865	830	830

TABLE B2.3 Pressure-Temperature Ratings for Class 600 ASME B16.5 Flanges (*Continued*)

Material group no.	1.1	1.2	1.3	1.4	1.5	1.7	1.9	1.10	1.13	1.14	2.1	2.2	2.3	2.4	2.5	2.6	2.7
					Alloy steels						Austenitic steels						
Temp., °F	Carbon steel				C-½Mo	½Cr-½Mo, Ni-Cr-Mo	1Cr-½Mo, 1¼Cr-½Mo	2¼Cr-1Mo	5Cr-½Mo	9Cr-1Mo	Type 304	Type 316	Type 304 L / Type 316 L	Type 321	Types 347, 348	Type 309	Type 310
950	205	205	205	205	560	685	755	755	520	740	750	775	...	775	775	775	775
1000	105	105	105	105	330	425	445	535	385	585	645	725	...	715	725	670	700
1050	...					380	275	400	280	380	620	720	...	695	720	585	665
1100	...						190	225	205	225	515	645	...	605	645	445	585
1150	...						105	205	140	150	390	550	...	475	550	345	495
1200	...						70	110	90	105	310	410	...	365	345	260	410
1250	...										220	365	...	280	245	200	325
1300	...										165	275	...	210	185	160	240
1350	...										125	205	...	165	135	115	160
1400	...										90	150	...	125	105	90	110
1450	...										70	115	...	95	80	60	75
1500	...										50	85	...	75	70	50	55

Note: Pressures are in psig.

Source: Adapted from ASME/ANSI B16.5, *Pipe Flanges and Flanged Fittings,* American Society of Mechanical Engineers, New York, 1988, Table 2.

using the tables, linear interpolation between listed temperatures to determine intermediate pressure ratings is permitted.

Reference Codes and Standards

The following listing identifies the codes and standards used for most design work done for modern power and industrial piping systems. It has been prepared as a ready reference.

The ASME Boiler and Pressure Vessel Code. *The ASME Boiler and Pressure Vessel Code* covers a wide variety of pressure-integrity–related design and construction applications. Certain sections of the Code provide rules for the design of specific piping systems. Those sections are:

Section I: Power Boilers

Section III: Rules for Construction of Nuclear Plant Components

Section IV: Heating Boilers

Section VIII: Pressure Vessels

The ASME Piping Codes. These codes are commonly used for the design of commercial power and industrial piping systems:

B31.1: Power Piping

B31.3: Chemical Plant and Petroleum Refinery Piping

B31.4: Liquid Transportation Systems for Hydrocarbons, Liquid Petroleum Gas, Anhydrous Ammonia and Alcohols

B31.5: Refrigeration Piping

B31.8: Gas Transmission and Distribution Piping Systems

B31.9: Building Services Piping

B31.11: Slurry Transportation Piping Systems

The ASME/ANSI Pressure-Integrity Standards. The standards listed below provide the design and manufacturing criteria for many commonly used piping components:

B16.1: Cast Iron Pipe Flanges and Flanged Fittings (Class 25, 125, 250, and 800)

B16.3: Malleable Iron Threaded Fittings (Class 150 and 300)

B16.4: Cast Iron Threaded Fittings Classes 125 and 250

B16.5: Pipe Flanges and Flanged Fittings (Classes 150 through 2500)

B16.9: Factory Made Wrought Steel Buttwelding Fittings

B16.11: Forged Steel Fittings, Socket-Welding and Threaded (Classes 2000 through 9000)

B16.15: Cast Bronze Threaded Fittings (Class 150 and 300)

B16.18: Cast Copper Alloy Solder Joint Pressure Fittings

B16.22: Wrought Copper and Copper Alloy Solder Joint Pressure Fittings

B16.24: Bronze Pipe Flanges and Flanged Fittings (Class 150 and 300)

B16.28: Wrought Steel Buttwelding Short Radius Elbows and Returns

B16.33: Manually Operated Metallic Gas Valves for Use in Gas Piping Systems Up to 125 psig

B16.34: Valves-Flanged, Threaded and Welding End (Classes 150 through 4500)

B16.36: Orifice Flanges (Class 300, 600, 900, 1500, and 2500)

B16.38: Large Metallic Valves for Gas Distribution (Manually Operated, NPS 2-1/2 to 12, 125 psig Maximum)

B16.39: Malleable Iron Threaded Pipe Unions (Classes 150, 250, and 300)

B16.42: Ductile Iron Pipe Flanges and Flanged Fittings (Classes 150 and 300)

B16.47: Large Diameter Steel Flanges

The ASME/ANSI Dimensional Standards. Listed below are the most commonly used piping-related dimensional standards:

B1.20.1: Pipe Threads, General Purpose

B16.10: Face-to-Face and End-to-End Dimensions of Valves

B16.20*: Ring-Joint Gaskets and Grooves for Steel Pipe Flanges

B16.21*: Non-Metallic Flat Gaskets for Pipe Flanges

B16.25: Buttwelding Ends

B36.10: Welded and Seamless Wrought Steel Pipe

B36.19: Stainless Steel Pipe

PIPING JOINTS

Joint design and selection can have a major impact on the initial installed cost, the long-range operating and maintenance costs, and the performance of a piping system. Factors that must be considered in the joint-selection phase of the project design include material cost, installation labor cost, degree of leakage-integrity required, periodic maintenance requirements, and specific performance requirements. In addition, since codes do impose some limitations on joint applications, joint selection must meet the applicable code requirements. In the paragraphs that follow, the above-mentioned considerations will be briefly discussed for a number of common pipe joint configurations. Figures illustrating many of the common piping system joints are shown in Chap. A2.

Buttwelded Joints

Buttwelding is the most common method of joining piping used in large commercial, institutional, and industrial piping systems. Material costs are low, but labor costs are moderate to high due to the need for specialized welders and fitters. Long-term leakage integrity is extremely good, as are structural and mechanical

*Not an American National Standard.

strength. The interior surface of a buttwelded piping system is smooth and continuous which results in low pressure drop. The system can be assembled with internal weld backing rings to reduce fit-up and welding costs, but backing rings create internal crevices which can trap corrosion products. In the case of nuclear piping systems, these crevices can cause the concentration of radioactive solids at the joints which can lead to operating and maintenance problems. Backing ring crevices can also lead to stress-concentration effects which may promote fatigue cracks under vibratory or other cyclic loading conditions. Buttwelded joints made up without backing rings are more expensive to construct, but the absence of interior crevices will effectively minimize "crud" buildup and will also enhance the piping system's resistance to fatigue failures. Most buttwelded piping installations are limited to 2½-in nominal pipe size or larger. There is no practical upper size limit in buttwelded construction. Buttwelding fittings and piping system accessories are available down to ½-in nominal pipe size. Economic penalties associated with pipe end preparation and fit-up, and special weld procedure qualifications normally preclude the use of buttwelded construction in sizes 2-in and under, except for those special cases where interior surface smoothness and elimination of internal crevices are of paramount importance. Smooth external surfaces give buttwelded construction high aesthetic appeal.

Socket-Welded Joints

Socket-welded construction is a good choice wherever the benefits of high leakage integrity and great structural strength are important design considerations. Construction costs are somewhat lower than with buttwelded joints due to the lack of exacting fit-up requirements and elimination of special machining for buttweld end preparation. The internal crevices left in socket-welded systems makes them less suitable for corrosive or radioactive applications where solids buildup at the joints may cause operating or maintenance problems. Fatigue resistance is lower than buttwelded construction due to the use of fillet welds and abrupt fitting geometry, but it is still better than most mechanical joining methods. Aesthetic appeal is good.

Brazed and Soldered Joints

Brazing and soldering are most often used to join copper and copper-alloy piping systems, although brazing of steel and aluminum pipe and tubing is possible. Brazing and soldering both involve the addition of molten filler metal to a close fitting annular joint. The molten metal is drawn into the joint by capillary action and solidifies to fuse the parts together. The parent metal does not melt in brazed or soldered construction. The advantages of these joining methods are high leakage integrity and installation productivity. Brazed and soldered joints can be made up with a minimum of internal deposits. Pipe and tubing used for brazed and soldered construction can be purchased with interior surfaces cleaned and sealed, making this method of construction popular for medical gases and high-purity pneumatic control installations.

Soldered joints are normally limited to near-ambient temperature systems and domestic water supply. Brazed joints can be used at moderately elevated temperatures. Most brazed and soldered installations are constructed using light-wall tubing so mechanical strength is quite low.

Threaded or Screwed Joints

Threaded or screwed piping is commonly used in low-cost, noncritical applications such as domestic water, fire protection, and industrial cooling water systems. Installation productivity is moderately high, and specialized installation skill requirements are not extensive. Leakage integrity is good for low-pressure, low-temperature installations where vibration is not encountered. Rapid temperature changes may lead to leaks due to differential thermal expansion between the pipe and fittings. Vibration can result in fatigue failures of screwed pipe joints due to the high stress intensification effects caused by the sharp notches at the base of the threads. Screwed fittings are normally made of cast gray or malleable iron, cast brass or bronze, or forged alloy and carbon steel. Screwed construction is commonly used with galvanized pipe and fittings for domestic water and drainage applications. While certain types of screwed fittings are available in up to 12-in nominal pipe size, economic considerations normally limit industrial applications to the 3-in size. Screwed piping systems are useful where frequent disassembly and reassembly are necessary to accommodate maintenance needs or process changes. Threaded or screwed joints must be used within the limitations imposed by the rules and requirements of the applicable code.

Grooved Joints

The main advantages of the grooved joints are their ease of assembly, which results in low labor cost, and generally good leakage integrity. They allow a moderate amount of axial movement due to thermal expansion, and they accommodate some axial misalignment. The grooved construction prevents the joints from separating under pressure. Among their disadvantages are the use of an elastomeric seal, which limits their high-temperature service, and their lack of resistance to torsional loading. While typical applications involve machining the groove in standard wall pipe, light wall pipe with rolled-in grooves may also be used. Grooved joints are used extensively for fire protection, general service water, and low-pressure drainage applications such as floor and equipment drain systems and roof drainage conductors. They are a good choice where the piping system must be disassembled and reassembled frequently for maintenance or process changes.

Flanged Joints

Flanged connections are used extensively in modern piping systems due to their ease of assembly and disassembly; however, they are costly. Contributing to the high cost are the material costs of the flanges themselves and the labor costs for attaching the flanges to the pipe, and then bolting the flanges to each other. Flanges are normally attached to the pipe by threading or welding, although in some special cases a flange-type joint known as a "lap-joint" may be made by forging and machining the pipe end. Flanged joints are prone to leakage in services that see rapid temperature fluctuations, which cause high-temperature differentials between the flange body and the bolting, which eventually causes the bolt stress to relax, allowing the joint to open up. Leakage is also a concern in high-temperature installations where bolt stress relaxation due to creep is experienced. These problems can be minimized by periodic retorquing of the bolted connections to reestablish the required seating pressure on the gasket face. Creep-damaged bolts in high-temperature installations must be periodically replaced to reestablish the required gasket seating pressure. Flanged joints are commonly used to join dissimilar materials, for example: steel pipe to cast iron valves and in systems which require frequent maintenance disassembly and reassembly. Flanged construction is also used extensively in lined piping systems.

Compression Joints

Compression sleeve–type joints are used to join plain end pipe without special end preparations. These joints require very little installation labor and as such result in an economic overall installation. Advantages include the ability to absorb a limited amount of thermal expansion and angular misalignment, and the ability to join two dissimilar piping materials, even if their outside diameters are slightly different. Disadvantages include the use of rubber or other elastomeric seals, which limits their high-temperature application, and the need for a separate external thrust resisting system at all turns and dead-ends to keep the line from separating under pressure. Compression joints are frequently used for temporary piping systems or systems that must be dismantled frequently for maintenance. They are suitable for use with air, other gases, water, and oil in both aboveground and underground service.

Small-diameter compression fittings with all metal sleeves may be used at elevated temperatures and pressures, when permitted by the rules and requirements of the applicable code. They are common in instrument and control tubing installations and other applications where high seal integrity and easy assembly and disassembly are desirable attributes.

LOADING CONDITIONS

In an earlier section, "Definition of the Term *Design Bases,*" loading conditions were identified as one of the five principal elements in the definition of the term *design bases.* This section will identify some of the more common loading conditions and discuss the way they are treated in design.

Loading conditions may be classified as either sustained or occasional. *Sustained loads* act on the piping system during all or at least the great majority of its operating time. These loads are time-invariant. Examples of sustained loads include the dead weight of the pipe plus its contents or the pressure load, including the effects of static head. *Occasional loads* are transient in nature and act during relatively small percentages of the system's total operation time. Examples of occasional loads include surges due to pump start-up and shutdown or pressure depressions and/or peaks due to sudden valve actuations.

Design Pressure

The *design pressure* is the maximum sustained pressure that a piping system must contain without exceeding its code-defined allowable stress limits. In single-compartment systems the design pressure is the maximum differential pressure between the interior and exterior portions of the system. In multicompartment systems the design pressure is the maximum differential pressure between any two adjacent compartments. The design pressure is the pressure which results in the heaviest piping wall thickness and/or the highest component pressure rating. The design pressure is not to be exceeded during any normal steady-state operating mode of the piping system.

In formulating the design pressure, the designer must consider all potential pressure sources. Among the more common sources to be considered are:

- The hydrostatic head due to differences in elevation between the high and low points in the system
- Back pressure effects
- Friction losses
- The shutoff head of in-line pumps
- Frequently occurring pressure surges
- Variations in control system performance

Variations in System Pressure. As previously indicated, the system design pressure is at steady state or sustained maximum pressure. Sustained conditions are those that remain constant over the majority of the total operating time. It is reasonable to expect that short-duration transient system pressure excursions in excess of the steady-state design pressure will occur during normal system operation. These transients, or *occasional pressure* excursions, may be tolerated without increasing the basic system design pressure, provided that the pressure increase does not exceed predefined limits and provided that the amount of time that the transients act does not exceed a specified percentage of the total system operating time.

A number, but not all, of the major piping design codes provide design rules to account for overpressure transients. Among the codes which provide design criteria or guidance are:

- The ASME Boiler and Pressure Vessel Code, Section III, Rules for Construction of Nuclear Power Plant Components
- Various sections of the ASME Code for Pressure Piping, including:
  - B31.1: Power Piping
  - B31.3: Chemical Plant and Petroleum Refinery Piping
  - B31.4: Liquid Transportation Systems for Hydrocarbons, Liquid Petroleum Gas, Anhydrous Ammonia and Alcohols
  - B31.11: Slurry Transportation Piping Systems

The methods used to qualify overpressure conditions for service vary from one code to another. The ASME Code, Section III, uses a rather complex approach in which the range of acceptable overpressure transients is related to the nature of the loading combinations being investigated. The loading combinations are known as *service conditions,* and depending upon their severity and frequency of occurrence, pressure transients of up to two times the design pressure may be tolerated. The interested reader is referred to Subsubarticles NB,NC,ND-3600 of Section III for the details. In contrast to the complex methods adopted by Section III, ASME B31.4 and ASME B31.11 allow pressure transients of up to 10 percent over the system design pressure without restricting the amount of time that the transients may act.

ASME B31.1 and ASME B31.3 provide rules that are about midway in relative complexity from the extremes indicated above. As an example, the acceptance criteria for occasional loads specified in Paragraph 102.2.4 of the ASME B31.1 Code for Power Piping are reproduced below:

> Ratings: Allowance for Variation from Normal Operation. The maximum internal pressure and temperature allowed shall include considerations for occasional loads and transients of pressure and temperature.
>
> It is recognized that variations in pressure and temperature inevitably occur, and therefore the piping system except as limited by component standards referred to in Para. 102.2.1 or by manufacturers of components referred to in Para. 102.2.2, shall be considered safe for occasional short operating periods at higher than design pressure or temperature. The calculated stress resulting from such a variation in pressure and/or temperature may exceed the maximum allowable stress from Appendix A by:
>
> (A) 15% if the event duration occurs less than 10% of any 24 hour operating period; or
>
> (B) 20% if the event duration occurs less than 1% of any 24 hour operating period.[2]

Referring to Paragraph 104.1.2 of the ASME B31.1 Code, one finds Equation (4) for the maximum allowable pressure in a straight pipe[3]:

$$P = \frac{2SE\,(t_m - A)}{D_o - 2y\,(t_m - A)} \tag{B2.1}$$

It can be seen from Eq. (B2.1) that the maximum allowable pressure P varies directly with the allowable stress S. Therefore, the net effect of Paragraph 102.2.4 is to allow short-term pressure excursions of from 15 to 20 percent in excess of the design pressure, as long as the respective time criteria are met.

As indicated above, not all piping codes provide rules for accepting transient pressure excursions in excess of the design pressure. Sections of the ASME Code for Pressure Piping which have no such rules include:

- B31.5: Refrigeration Piping
- B31.8: Gas Transmission and Distribution Piping Systems
- B31.9: Building Services Piping

When designing to a code which has no rules for acceptance of overpressure transients, the designer must increase the design pressure to envelope the transient condition. If, however, no specific design code is being used as a basis for design of a project, the designer may make a reasonable engineering judgment concerning the handling of transient overpressure events. In the absence of any other governing criteria, the following may be used. For transient pressure conditions which exceed the design pressure by 10 percent or less and act for no more than 10 percent of the total operating time, the transient may be neglected and the design pressure need not be increased. For transients whose magnitude or duration is greater than 10 percent of the design pressure or operating time, the design pressure should be increased to envelope the transient.

Determination of the Piping Wall Thickness. The determination of the piping wall thickness is one of the most important calculations of the piping system design process. In arriving at the final specification of the piping wall thickness, the designer must consider a number of important factors. They are:

- Pressure integrity
- Allowances for mechanical strength, corrosion, erosion, wear, threading, grooving, or other joining processes
- Manufacturing variations (tolerance) in the wall thickness of commercial pipe
- Wall thickness reduction due to buttwelding end preparation (counterboring)

While a number of different pipe wall thickness design formulas have been proposed over the years,[4] the ASME piping codes have adopted one or the other of the following formulas for pressure-integrity design:

$$t = \frac{PD_o}{2(SE + Py)} \qquad \text{(Ref. 5)} \tag{B2.2}$$

or

$$t = \frac{PD_o}{2S} \quad \text{or} \quad \frac{PD_o}{2SE} \qquad \text{(Ref. 6)} \tag{B2.3}$$

where t = the minimum wall thickness required to assure pressure integrity, in
P = the design pressure, lb/in^2-gauge
D_o = the outside diameter of the pipe, in
S = the allowable stress, lb/in^2
E = the weld joint efficiency factor for seam-welded pipe (some codes also specify a casting quality factor F for cast piping materials)
y = a dimensionless factor which varies with temperature.

For the precise definition of the method by which either equation is used by the codes, the particular code of interest should be consulted.

Most construction codes require the provision of additional wall thickness, over and above that intended to assure pressure integrity. This additional material allowance is provided in accordance with Eq. (B2.4):

$$t_m = t + c \qquad \text{(Ref. 7)} \tag{B2.4}$$

where t_m = the total minimum wall thickness required to satisfy all the design rules of the code, in
t = the wall thickness required to provide pressure integrity [see Eqs. (B2.2) and (B2.3)]
c* = the additional material allowance, in

The additional material allowance c is made up of a number of individual allowances which are provided to address different loads or conditions the piping system will see during fabrication, installation, and operation. Each allowance is figured separately, and their sum added to the pressure-integrity wall thickness to arrive at the final design minimum wall thickness. The major constituents of c include:

- Wall thickness added to account for progressive deterioration or thinning of the pipe wall in service, due to the effects of corrosion, erosion, and wear.

*In some codes, the term A is used in lieu of c, but the intent is the same.

- Wall thickness added to account for material removed to facilitate joining of the various segments of the piping system. Typical joining methods include threading, grooving, and swagging. If a machining tolerance is required as a part of the joint manufacture, this tolerance must be accounted for in the most conservative manner.
- Wall thickness added to provide mechanical strength. This additional strength might be required to resist external operating loads, or loads associated with shipping and handling.

The effects of pressure result in pipe wall stresses in both the longitudinal and circumferential (*hoop stress*) directions. Typically, the circumferential stress is twice the longitudinal stress. Piping wall thickness selections made using hoop stress–type formulas, such as (B2.2) and (B2.3), result in excess material availability in the longitudinal direction. In most cases, this excess material is adequate to resist bending stresses associated with the dead weight of the pipe, its contents, and in-line components such as valves, flanges, and other piping specialties. In relatively rare cases, such as extremely long spacing of pipe hangers or designs required to resist high localized concentrated loads, it may be necessary to increase the wall thickness to accommodate the bending stresses. Refer to Chap. B4.

Once the design minimum wall thickness t_m is determined, the only remaining step is to specify the actual or purchase wall thickness.

Pipe is manufactured to one of two wall thickness dimensioning procedures: minimum wall thickness and nominal wall thickness.

Pipe purchased to a minimum wall thickness specification will be manufactured using special processes to control the wall thickness. These processes may include custom-made dies, extra rolling passes, or final boring of the inside diameter. The use of minimum wall pipe is normally limited to high-pressure–high-temperature applications where the savings in material weight is sufficient to offset the additional manufacturing cost. Most minimum wall pipe is custom-manufactured.

Piping purchased to a nominal wall thickness specification is manufactured in accordance with the dimensional criteria specified in ANSI Standards B36.10 and B36.19. These standards provide predetermined nominal wall thicknesses, or *schedules,* for various outside diameters of commercially manufactured pipe.

The tolerance on the wall thickness of the pipe varies with the particular manufacturing process employed and with the relevant manufacturing specification. Rolled seamless and seam-welded (without filler metal) pipe has a normal wall thickness tolerance of +0, −12½ percent. Forged and bored pipe has a wall thickness tolerance of +⅛-in, −0. Piping manufactured from rolled and welded plate has a wall thickness tolerance of −0.01 in. There is no plus tolerance for this type of pipe. The ASTM (or ASME) specification for the particular piping material should be consulted to determine the wall thickness tolerance for each application.

When piping is to be joined by buttwelding, the pipe ends are frequently counterbored to facilitate fit-up. The counterbore dimensions for standard pipe wall thicknesses are given in ASME/ANSI Standard B16.25. It is important that the net minimum wall thickness resulting from the counterboring process be compared with the code minimum wall thickness t_m to be sure that an underthickness condition does not occur at the joints.

The following example demonstrates how the previously discussed concepts associated with the design pressure may be applied to a typical problem.

Problem Statement. Two motor-driven boiler feed pumps installed on the ground floor of a power house supply 3000 gpm of water at 350°F to a boiler drum which is 260 ft above grade. Each pump discharge is 8-in pipe, and the common discharge header running up to the boiler drum is 12-in pipe. The system is shown diagrammatically in Fig. B2.1. Each pump discharge pipe has a manual valve that can isolate it from the main header. A relief valve is installed upstream of each pump discharge valve to serve as a minimum flow bypass if the discharge valve is closed while the pump is operating. The back pressure at the boiler drum is 2520 psig. The set pressure of the relief valve is 2780 psig, and the shutoff head of the pump is 8700 ft. The piping material is ASTM A106, Grade C, with an allowable working stress of 17,500 psi, over the temperature range of −20 to 650°F. The corrosion allowance is 0.08 in and the design code is ASME B31.1. Find the required nominal pipe wall thickness for the 12-in header (Zone 1) and the 8-in pump discharge pipes upstream of the isolation valve (Zone 2). Also check the adequacy of the pipe wall thickness in Zone 2 assuming the relief valve on one pump does not operate when its discharge valve is closed.

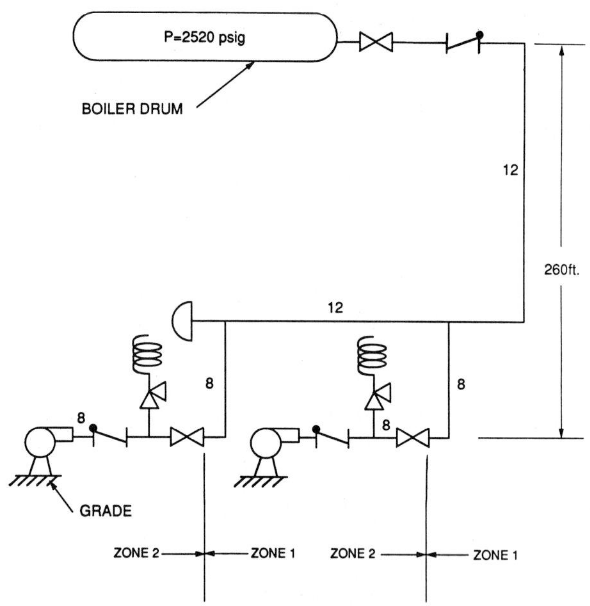

FIGURE B2.1 Simplified power house boiler feed system.

Solution. The specific volume of 350°F saturated water is 0.01799 ft³/lbm. Correcting the specific volume for the effects of compression to 2500 psig results in a revised specific volume of 0.01778 ft³/lbm or a density of 56.24 lb/ft³. The static head P_{st} above the pumps due to the elevation of the boiler drum is:

$$P_{st} = 260 \text{ ft} \times 56.24 \frac{\text{lb}}{\text{ft}^3} \times \frac{1 \text{ ft}^2}{144 \text{ in}^2}$$

$$P_{st} = 102 \frac{\text{lb}}{\text{in}^2}$$

Therefore, the total discharge pressure at the pump exit is:

$$2520 + 102 = 2622 \frac{\text{lb}}{\text{in}^2} - \text{gauge}$$

The design pressure for the piping system in Zone 1 is set just slightly above the maximum operating pressure.. A value of 2650 psig is chosen. The 12-in discharge header wall thickness is determined based upon that pressure.

ASME B31.1 uses the approach provided by Eqs. (B2.2) and (B2.4) to yield a combined minimum wall thickness function represented by Eq. (B2.5):

$$t = \frac{PD_o}{2(SE + Py)} \qquad \text{(B2.2)}$$

$$t_m = t + A \qquad \text{(B2.4)}$$

$$t_m = \frac{PD_o}{2(SE + Py)} + A \qquad \text{(B2.5)}$$

The values of the variables are:

$$P = 2650 \text{ psig}$$
$$D_o = 12.75 \text{ in}$$
$$S = 17{,}500 \text{ psi}$$
$$y = 0.4$$
$$A = 0.08 \text{ in}$$
$$E = 1.0$$

Substituting in Eq. (B2.5) yields the following:

$$t_m = \frac{2650(12.75)}{2(17{,}500 + 2650 \times 0.4)} + 0.08$$

$$t_m = 0.990 \text{ in}$$

The commercial wall thickness tolerance for seamless rolled pipe is +0, −12½ percent; therefore, to determine the nominal wall thickness, t_m must be divided by 0.875:

$$t_{\text{NOM}} = \frac{0.990}{0.875} = 1.13$$

Referring to ANSI B36.10, the nearest commercial 12-in pipe whose wall thickness exceeds 1.13 in is Schedule 160 with a nominal wall thickness of 1.312 in. Therefore 12-in Schedule 160 pipe meeting the requirements of ASTM A106 Grade C is chosen for this application.*

*This calculation does not consider the effects of bending. If bending loads are present, the required wall thickness may increase.

Similarly, the required wall thickness for the 8-in pipe upstream of the isolation valves (Zone 2) is calculated. In this case the set pressure for the relief valves, which is 2780 psig, is used for the design pressure. The outside diameter of 8-in standard pipe is 8.625 in.

Substituting in Eq. (B2.5) yields:

$$t_m = \frac{2780(8.625)}{2(17{,}500 + 2780 \times 0.4)} + 0.08$$

$$= 0.724$$

$$t_{\mathrm{NOM}} = \frac{0.724}{0.875} = 0.828 \text{ in}$$

The required nominal wall thickness is 0.828 in. Referring to ANSI Standard B36.10, XXS pipe with a nominal wall thickness of 0.875 in is selected.

The only remaining step is to check the pipe in Zone 2 to see if the wall thickness is adequate in the event that a relief valve does not open and the pump runs at its shutoff head.

Relief valves are inherently reliable devices, and it is reasonable to assume that a valve failure to open would be a very low probability event. The occasional load criteria of ASME B31.1, Paragraph 102.2.4, will be invoked, and it will be assumed that the relief valve failure-to-open event occurs less than 1 percent of the time. Therefore, the allowable stress is 20 percent higher than the basic code allowable stress of 17,500 psi.

The higher allowable stress is denoted as S':

$$S' = 1.20 \times S$$

$$S' = 1.20 \times 17{,}500 = 21{,}000 \text{ psi}$$

The maximum pressure rating of the 8-in XXS pipe is calculated using Eq. (B2.1):

$$P = \frac{2SE\,(t_m - A)}{D_o - 2y(t_m - A)} \qquad \text{(B2.1)}$$

For this evaluation, the value of S is set to equal to S' and $E = 1.00$ for seamless pipe. t_m is assumed equal to 87½ percent of the nominal wall thickness of the pipe:

$$t_m = 0.875 \times 0.875$$

$$= 0.766 \text{ in}$$

then

$$P = \frac{2 \times 21{,}000(0.776 - 0.08)}{8.625 - 2 \times 0.4(0.776 - 0.08)}$$

$$= 3623 \text{ psig}$$

The shutoff head of the pump was given as 8700 ft. The density of pressurized water at 350°F was previously determined to be 56.24 lb/ft^3.

The pressure equivalent to the shutoff head may be calculated based upon this density:

$$P = 8700 \text{ ft} \times 56.24 \frac{\text{lb}}{\text{ft}^3} \times \frac{1 \text{ ft}^2}{144 \text{ in}^2}$$

$$= 3398 \text{ psig}$$

Since the 3623-psig occasional pressure rating of the 8-in XXS pipe exceeds the 3398-psig shutoff head of the pump, the piping is adequate for the intended service.

The design procedures presented in the forgoing problem are valid for steel or other code-approved wrought materials. They would not be valid for cast iron or ductile iron piping and fittings. For piping design procedures which are suitable for use with cast iron or ductile iron pipe, see ASME B31.1, Paragraph 104.1.2(B).

Design Temperature

The *design temperature* is the temperature at which the allowable stresses for all pressure-retaining parts of the piping system are assigned. The design temperature must be equal to or greater than the maximum sustained temperature that the pressure-retaining components will experience during all normal and expected abnormal modes of operation.

The design temperature of the system's pressure-retaining metal parts is normally assumed equal to the maximum free stream fluid temperature. The affects of any internal or external heat sources such as heat tracing must be considered, as must any temperature excursions occurring as a result of control system response time. The design temperature should be set at or above the peak of these temperature excursions.

While the pressure-integrity design is based upon the design temperature, most other thermally related aspects of the design are based upon the normal operating temperature. The *normal operating temperature* is the temperature achieved by the system fluid while the system is operating at full-load, steady-state, nontransient conditions. It is lower than the design temperature. The normal operating temperature is used as the basis for all thermal design analyses that relate to the structural integrity of the piping system, including the thermal flexibility analysis, the spring hanger sizing and setting calculations, and the thermally induced anchor movement calculations. If a system has more than one "normal" operating mode (i.e., the system runs at different temperatures or has branches that run at different temperatures for different operating modes), then multiple thermal analysis calculations at all normal operating temperatures may be necessary to fully qualify the design.

Dead Weight

The *dead weight* (self-weight) of a piping system consists of the sum of the distributed loads from the weight of the pipe itself, its thermal insulation, and/or other uniformly applied covering materials, plus the sum of the weights of any permanently installed concentrated loads such as valves, strainers, or other in-line appurtenances.

External loads on the piping system such as wind loads, snow and ice loads, and the weight of the fluid contents are considered as live loads. They are distinct from dead weight in that live loads may be variable both in magnitude and/or in the percentage of the total system operating time during which they act. An additional distinction is that the effects of live loads may be removed from the piping system while those of dead weight may not (without dismantling the system, of course).

Both the ASME Boiler and Pressure Vessel Code, Section I, Power Boilers, and the ASME B31.1, Power Piping Code, require that the effects of dead weight and other sustained loads be considered in verifying the pressure integrity of components. Subpart PG-22 of the ASME Boiler and Pressure Vessel Code, Section I, Power Boilers, provides the following generalized rule:

> Stresses due to hydrostatic head shall be taken into account in determining the minimum thickness required unless noted otherwise. Additional stresses imposed by effects other than working pressure or static head which increase the average stress by more than 10% of the allowable working stress shall also be taken into account. These effects include the weight of the component and its contents, and the method of support.[8]

The ASME B31.1, Power Piping, code provides more definitive rules to account for the effects of dead weight. Paragraph 104.8 presents closed-form equations to evaluate the effect of the simultaneous application of the internal pressure, dead weight, and other sustained mechanical loads on the design of a piping system. The equations require the calculation of the piping system stress and the comparison of the stress with set acceptance criteria. Details of this analytical approach are discussed in Chap. B4.

Wind Load

The majority of all piping system installations are indoors where the effects of wind loading can be neglected. However, there are sufficient numbers of outdoor piping installations where wind loading can be a significant design factor. *Wind load,* like dead weight, is a uniformly distributed load which acts along the entire length or portion of the piping system which is exposed to the wind. The difference is that while dead weight loads are oriented in the downward vertical direction, wind loads are usually horizontally oriented and may act in any arbitrary direction. Since wind loads are normally oriented in the horizontal direction, the regular dead weight support system of hangers and anchors will have little or no ability to resist these loads. Consequently, when wind loading is a factor, a separate structural evaluation and wind load support system design will ordinarily be required.

Determination of the magnitude of the wind loadings is based upon empirical procedures developed for the design of buildings and other outdoor structures. Analysis of piping system stresses and support system loads is accomplished by using techniques which are similar to those applied for dead weight design. Details of these procedures are discussed in Chap. B4.

Snow and Ice Loads

Snow and ice loads, like wind loads, need to be considered in the design of piping systems which are installed outdoors, particularly if the installation is made in the northern latitudes. Since snow and ice loads act in the vertical direction, they are treated the same as dead weight loads. In design, they are simply added as additional distributed loads in the dead weight analysis, as discussed earlier in the section "Dead Weight" and in Chap. B4.

Snow Loads. ANSI Standard A58.1, Minimum Design Loads for Buildings and Other Structures,[9] provides recommendations and data for developing design loadings due to snow. The methods used in ANSI A58.1 are generally applicable to sloping or horizontal flat surfaces such as building roofs or grade slabs. While the methods of ANSI A58.1 are completely appropriate for extended flat surfaces, they may be too conservative for application to a smooth round, horizontal pipe which will tend to shed most of the snowfall which may light upon it. The data provided in ANSI Standard A58.1 can, however, be used as part of a rational method to estimate the maximum probable *snow load* on an outdoor piping system and the following procedure may be adopted for piping.

Table B2.4 provides ground snow loading data for 184 locations where that data are recorded. The column marked 2% Annual Probability represents the loading associated with the maximum probable snowfall that is likely to occur in a 50-year period. Based upon the data from Table B2.4, the following relationship for the design snow load for outdoor piping systems may be used:

$$W_s = \tfrac{1}{2} D_o S_{50} \qquad \text{(B2.6)}$$

where W_s = the design snow load to be added to the other distributed loads acting on the pipe, lb/ft
D_o = the outside diameter of the pipe or insulation, ft
S_{50} = the 2 percent probability Snow Loading for the nearest appropriate location from Table B2.4, lb/ft^2

This formula assumes the snow remaining on the pipe will take the shape of an equilateral triangle whose base equals the outside diameter of the pipe.

Ice Loads. Unlike snow loading, no national database exists for ice loading information. Ice storms are sporadic in the frequency of their occurrence and in their intensity. Weather records dating back to the turn of the century for a typical midwestern state relate instances of ice storm deposits of from ⅛ in to 4 in in thickness. The *American Weather Book*[10] cites examples of ice accumulations of up to 8 in in northern Idaho (1961) and 6 in in northwest Texas (1940) and New York State (1942).

The paper "Estimated Glaze Ice and Wind Loads at the Earth's Surface for the Contiguous United States"[11] documents the results of a comprehensive study of ice storm records over the 50-winter period from 1919–20 through 1968–69. A statistical evaluation of the data indicates that 50-year maximum probable values for ice deposits from a single storm vary over the range of 2 to 3 in, with the larger accumulations occurring in the upper Midwest, far West, and Northeast and the smaller accumulations occurring in the lower Midwest and Southeast. Given the relative infrequency of ice storms, this range probably represents a reasonable limit for design. It is suggested that the designer contact local weather or agricultural authorities to determine whether any better regional-specific data exist.

Once the appropriate design thickness is determined, the following formula may be used to estimate the unit loadings on an exposed pipeline due to ice accumulation:

$$\frac{W}{L} = 1.36t(D_o + t) \qquad \text{(B2.7)}$$

where W/L = the unit loading on the pipe, lb/ft
D_o = the outside diameter of the pipe or insulation lagging, in
t = the assumed ice covering thickness, in

TABLE B2.4 Ground Snow Loads at 184 National Weather Service Locations at Which Measurements Are Made

	Ground snow load, lbf/ft^2		
Location	Years of record	Maximum observed	2% annual probability
Alabama			
Huntsville	18	7	7
Arizona			
Flagstaff	28	88	48
Prescott	5	2	3
Winslow	25	12	7
Arkansas			
Fort Smith	22	4	5
Little Rock	22	6	6
California			
Blue Canyon	18	213	255
Mt. Shasta	28	62	69
Colorado			
Alamosa	28	14	15
Colorado Springs	27	16	14
Denver	28	14	15
Grand Junction	28	18	16
Pueblo	26	7	7
Connecticut			
Bridgeport	27	19	23
Hartford	28	23	29
New Haven	17	11	15
Delaware			
Wilmington	27	12	13
Georgia			
Athens	24	5	5
Macon	28	8	8
Idaho			
Boise	26	6	6
Lewiston	24	6	9
Pocatello	28	9	7
Illinois			
Chicago–O'Hare	20	25	18
Chicago	26	37	22
Moline	28	21	17
Peoria	28	27	16
Rockford	14	31	25
Springfield	28	20	23
Indiana			
Evansville	27	11	12
Fort Wayne	28	22	17
Indianapolis	28	19	21
South Bend	28	58	44

TABLE B2.4 Ground Snow Loads at 184 National Weather Service Locations at Which Measurements Are Made (*Continued*)

	Ground snow load, lbf/ft^2		
Location	Years of record	Maximum observed	2% annual probability
Iowa			
Burlington	11	15	17
Des Moines	28	22	22
Dubuque	28	34	38
Sioux City	26	28	33
Waterloo	21	25	36
Kansas			
Concordia	17	12	23
Dodge City	28	10	12
Goodland	27	12	14
Topeka	27	18	19
Wichita	26	8	11
Kentucky			
Covington	28	22	12
Lexington	28	11	12
Louisville	26	11	11
Maine			
Caribou	27	68	100
Portland	28	51	62
Maryland			
Baltimore	28	20	17
Massachusetts			
Boston	27	25	30
Nantucket	16	14	18
Worcester	21	29	39
Michigan			
Alpena	19	34	53
Detroit City	14	6	9
Detroit Airport	22	14	17
Detroit–Willow Run	12	11	21
Flint	25	20	28
Grand Rapids	28	32	37
Houghton Lake	16	33	56
Lansing	23	34	42
Marquette	16	44	53
Muskegon	28	40	43
Sault Ste. Marie	28	68	80
Minnesota			
Duluth	28	55	64
International Falls	28	43	43
Minneapolis–St Paul	28	34	50
Rochester	28	30	50
St. Cloud	28	40	53
Mississippi			
Jackson	27	3	3

TABLE B2.4 Ground Snow Loads at 184 National Weather Service Locations at Which Measurements Are Made (*Continued*)

	Ground snow load, lbf/ft^2		
Location	Years of record	Maximum observed	2% annual probability
Missouri			
Columbia	27	18	21
Kansas City	27	18	18
St. Louis	25	26	16
Springfield	27	9	14
Montana			
Billings	28	21	17
Glasgow	28	18	17
Great Falls	28	22	16
Havre	26	22	24
Helena	28	15	18
Kalispell	17	27	53
Missoula	28	24	23
Nebraska			
Grand Island	27	24	30
Lincoln	8	15	20
Norfolk	28	28	29
North Platte	26	16	15
Omaha	25	23	20
Scottsbluff	28	8	11
Valentine	14	15	22
Nevada			
Elko	12	12	20
Ely	28	9	9
Reno	25	9	11
Winnemucca	24	5	6
New Hampshire			
Concord	28	36	66
New Jersey			
Atlantic City	24	7	11
Newark	27	17	15
New Mexico			
Albuquerque	25	6	4
Clayton	25	8	10
Roswell	22	6	8
New York			
Albany	28	26	25
Binghamton	28	30	35
Buffalo	28	41	42
NYC–Kennedy	7	7	18
NYC–LaGuardia	28	23	18
Rochester	28	33	38
Syracuse	28	32	35

TABLE B2.4 Ground Snow Loads at 184 National Weather Service Locations at Which Measurements Are Made (*Continued*)

	Ground snow load, lbf/ft^2		
Location	Years of record	Maximum observed	2% annual probability
North Carolina			
Asheville	16	7	12
Cape Hatteras	22	5	5
Charlotte	28	8	10
Greensboro	26	14	11
Raleigh-Durham	22	13	10
Wilmington	24	7	9
Winston-Salem	12	14	17
North Dakota			
Bismarck	28	27	25
Fargo	27	24	34
Williston	28	25	25
Ohio			
Akron–Canton	28	16	15
Cleveland	28	27	16
Columbus	27	9	10
Dayton	28	18	11
Manfield	18	31	17
Toledo Express	24	8	8
Youngstown	28	14	12
Oklahoma			
Oklahoma City	24	5	5
Tulsa	21	5	8
Oregon			
Burns City	28	19	24
Eugene	22	22	17
Medford	25	6	8
Pendleton	28	9	11
Portland	25	10	10
Salem	27	5	7
Pennsylvania			
Allentown	28	16	23
Erie	20	20	19
Harrisburg	19	21	23
Philadelphia	27	13	16
Pittsburgh	28	27	22
Scranton	25	13	16
Williamsport	28	18	20
Rhode Island			
Providence	27	22	21
South Carolina			
Columbia	24	9	12
Greenville–Spartanburg	12	4	4

TABLE B2.4 Ground Snow Loads at 184 National Weather Service Locations at Which Measurements Are Made (*Continued*)

Location	Ground snow load, lbf/ft^2		
	Years of record	Maximum observed	2% annual probability
South Dakota			
Aberdeen	16	23	42
Huron	28	41	43
Rapid City	28	14	14
Sioux Falls	28	40	38
Tennessee			
Bristol	27	7	8
Chattanooga	27	5	6
Knoxville	25	10	8
Memphis	27	7	5
Nashville	23	5	8
Texas			
Abilene	23	6	6
Amarillo	26	15	10
Dallas	22	3	3
El Paso	24	5	5
Fort Worth	24	5	6
Lubbock	27	9	10
Midland	25	2	2
San Angelo	22	3	3
Wichita Falls	23	4	5
Utah			
Milford	14	23	16
Salt Lake City	28	9	8
Wendover	13	2	3
Vermont			
Burlington	28	43	37
Virginia			
Dulles Airport	17	15	19
Lynchburg	27	13	16
National Airport	27	16	18
Norfolk	25	9	9
Richmond	28	10	12
Roanoke	27	14	17
Washington			
Olympia	24	23	24
Quillayute	13	21	24
Seattle-Tacoma	28	15	14
Spokane	28	36	41
Stampede Pass	27	483	511
Yakima	27	19	25

TABLE B2.4 Ground Snow Loads at 184 National Weather Service Locations at Which Measurements Are Made (*Continued*)

	Ground snow load, lbf/ft^2		
Location	Years of record	Maximum observed	2% annual probability
West Virginia			
Beckley	8	20	51
Charleston	26	21	20
Elkins	20	22	21
Huntington	18	13	15
Wisconsin			
Green Bay	28	37	36
La Crosse	16	23	32
Madison	28	32	32
Milwaukee	28	34	32
Wyoming			
Casper	28	9	10
Cheyenne	28	18	15
Lander	27	26	20
Sheridan	28	20	25

Source: This material is reproduced with permission from American National Standard A58.1, *Minimum Design Loads for Buildings and Other Structures, Table A10,* Copyright 1982 by the American National Standards Institute. Copies of this standard may be purchased from the American National Standards Institute at 11 West 42d Street, New York, NY 10018.

Table B2.5 provides a tabulation of ice loadings based upon Eq. (B2.7) for piping systems up to 30 in in outside diameter and ice thicknesses up to 3 in.

Seismic (Earthquake) Loads

Under certain circumstances it is necessary or desirable to design a piping system to withstand the effects of an earthquake. Although the applications are not extensive, piping system seismic design technology is well developed and readily accessible. Many of the currently available piping stress analysis computer programs are capable of performing a detailed seismic structural and stress analysis, in addition to the traditional dead weight and thermal flexibility analyses. Some of these programs have been designed to run on desktop microcomputers.

Because of the higher construction costs and design complexities introduced by the application of seismic design criteria, this type of work is normally only done in response to specific regulatory, code, or contractual requirements. An overview of these applications is outlined in the following paragraphs.

Nuclear Power Plants. Title 10, Part 50, of the Code of Federal Regulations requires that safety-related piping systems* in nuclear power plants be designed to withstand the effects of certain severe natural phenomena, including earthquakes.[12]

*A simplified definition of a safety-related system is one whose failure could result in a reduced capacity to mitigate the effects of an accident or which could ultimately result in the uncontrolled release of radioactivity into the environment. See Refs. 13 and 14.

TABLE B2.5 Weight Loadings for Ice Coatings on Horizontal Pipelines

Pipe or lagging outside diameter D_o, in	¼	½	1	1½	2	3
2	0.8	1.7	4.0	7.1	10.9	20.4
2½	0.9	2.0	4.8	8.2	12.2	22.4
3	1.1	2.4	5.4	9.2	13.6	24.5
4	1.5	3.1	6.8	11.2	16.3	28.6
6	2.1	4.4	9.5	15.3	21.8	36.7
8	2.8	5.8	12.2	19.4	27.2	44.9
10	3.5	7.1	15.0	23.5	32.6	53.1
12	4.2	8.5	17.7	27.5	38.1	61.2
14	4.8	9.9	20.4	31.6	43.5	69.4
16	5.5	11.2	23.1	35.7	49.0	77.5
18	6.2	12.6	25.8	39.8	54.4	85.7
20	6.9	13.9	28.6	43.9	59.8	93.8
22	7.6	15.3	31.3	47.9	65.3	102
24	8.3	16.7	34.0	52.0	70.7	110
26	8.9	18.0	36.7	56.1	76.2	118
28	9.6	19.4	39.4	60.2	81.6	127
30	10.3	20.7	42.2	64.3	87.0	135

Note: Ice thickness t: inches. Loadings are in pounds per foot. $W/L = 1.36t(D_o + t)$.

In general, seismic analysis of nuclear power piping is done to demonstrate that the piping system satisfies one of two specific objectives:

Operability: Under this objective, the design of the piping system is such that it will retain its pressure-integrity status and remain capable of performing its design function before, during, and after the occurrence of a postulated seismic event at the plant site. Piping systems designed to meet operability criteria must normally comply with code-specified stress limits during the postulated earthquake.

Structural Integrity: Piping designed to this objective is not required to remain functional or to retain its pressure integrity during or after an earthquake. The only requirement is that the piping system retain its gross structural integrity so that it does not deflect excessively or cause the generation of secondary missiles. Either condition could cause impact and subsequent unacceptable damage to adjacent safety-related structures, systems, or components. Piping designed under this classification is normally allowed to attain stress levels, due to seismic excitation, well in excess of normal code limits.

Of all modern industrial and commercial applications, nuclear power plant piping systems represent the largest single class of seismically qualified piping systems in service. It is estimated that the typical nuclear power unit contains approximately 100,000 ft of seismically qualified piping.[15] It is safe to say that the nuclear piping industry is the largest single contributor to the technology of seismic design of piping systems.

General Building Codes. Many building codes have rules devoted to the seismic design of buildings and other structures. Two examples of such codes are The

Uniform Building Code[16] and The BOCA National Building Code.[17] In addition to these two codes, ANSI Standard A58.1, Minimum Design Loads for Buildings and Other Structures,[9] provides extensive guidance in the area of seismic design.

It should be noted that while both of the above-mentioned codes are intended to be national in scope and applicability, neither is mandatory unless specifically adopted by local statute or ordinance.

Both ANSI A58.1 and the BOCA code are specifically applicable to all piping systems 2½ in and larger in nominal pipe size, with smaller limits of applicability stated for piping systems in boiler rooms (1¼ in and larger) and natural gas piping (1 in and larger). The Uniform Building Code is only applicable to fire protection sprinkler piping systems. Since both The Uniform Building Code and the BOCA code are more specific to the design of buildings and related structures, their application to piping systems requires some interpretation on the part of the designer. It is therefore recommended that the piping designer who intends to apply these codes maintain a close liaison with the local jurisdictional authorities.

Contractual Arrangements. Under certain specific circumstances an agreement might be reached to seismically design a piping system; this agreement would be made between the owner-operator of the system and another organization which has a vested financial interest in it, such as an insurance carrier. An example might involve the construction of a pipeline which carries a hazardous fluid through a seismically active region where no statutory design requirement exists. Under such circumstances the insurance carrier might require the owner to seismically design the line to limit the risk of rupture during a seismic event. Should a piping designer become involved in such an arrangement, every effort should be made to ensure that the design criteria are carefully specified, understood, and agreed to by all parties prior to starting any design work.

Methods of Analysis. There are three methods of analysis in common use for the seismic design of piping systems: *the static coefficient method*; *the response spectra modal analysis method*; and the *time history analysis method.* The static coefficient method is the easiest to apply, but due to simplifying assumptions, it provides a very conservative design. The response spectra modal analysis method is about midway in complexity and provides a lesser degree of conservatism. This is the method used for the vast majority of piping systems analysis and design. The time history analysis method is the least conservative and the most difficult to apply. This approach is used only when the most exacting (and least conservative) results are required. All three methods are discussed, in detail, in Chap. B4.

The Effects of Seismic Analysis on the Overall Design. The design costs associated with the seismic analysis of a piping system go far beyond the simple costs of analyzing the piping system for "just one more load." A piping system is usually a subordinate part of a larger structure, typically a building. Prior to analyzing the piping system to determine its seismic behavior, the analyst must first develop the forcing function. This involves a detailed analysis of the building structure itself to determine its response to the ground motion associated with the postulated earthquake. Development of the postulated ground motion normally requires the consideration of actual earthquake-induced ground motion data taken from geologically similar sites where earthquakes have actually been experienced. Thus it can be seen that the seismic analysis of the piping system is really just the top layer of a multitiered design exercise that requires the consideration of the dynamic interaction of a number of complex structures. Such an exercise requires a considerable expenditure of human resources and computer time.

Seismic qualification of a piping system also leads to greatly increased construction costs. Some of the principal contributors to those costs include:

Higher Loads. If the frequency content of the seismic excitation forces is coincident with the natural frequencies of the piping system, resonant amplification of the forcing function loads will occur. The resulting support system loads will be much higher than corresponding loads caused by dead weight effects alone. These higher loads translate into heavier, and consequently more expensive, supporting structures.

Multiple Load Paths. The multidirectional load characteristics of an earthquake acting on seismically qualified piping systems invariably result in the application of upward and/or lateral reactions on the building structure which would not be present in a typical static design.

Special Supporting Devices. Seismic design of high-temperature piping systems represents an especially challenging exercise. Thermal expansion effects require that the piping system be flexibly supported to allow for free thermal growth. The dynamic aspects of the design usually require that the piping system be rigidly supported during the earthquake to react the seismic loads back into the building structure. The simultaneous consideration of these diametrically opposite requirements results in the need to use a significant number of specialized pipe support devices called *snubbers.* Snubbers lock up and carry load when subjected to the rapidly varying vibratory loads associated with an earthquake yet remain free to permit thermal movement of the piping system during the relatively slow expansion or contraction caused by temperature changes.

Hydraulic Transient Loads

Of all the loading conditions which a piping system may experience in service, hydraulic transients are among the most damaging. The most common form of damage caused by hydraulic transient loads is the failure of pipe supports and supporting structures. However, occasional breaches of pressure integrity are also experienced, particularly where large-diameter thin-walled pipe is involved.

Two common types of hydraulic transient loads are waterhammer and relief valve discharge. Because of their commonality, these two load sets are discussed in further detail below.

Waterhammer. If the velocity of water or other liquid flowing in a pipe is suddenly reduced, a pressure wave results which travels up and down the piping system at the speed of sound in the liquid. Depending upon the initial velocity and physical properties of the liquid and the mechanical properties of the piping system, the peak value of the pressure wave may exceed the steady-state pressure. *Waterhammer* frequently occurs in systems that are subject to rapid changes in fluid flow rate, including systems with rapidly actuated valves, fast-starting pumps, and check valves. Waterhammer is most severe in systems which convey fully condensed liquids; however, it is possible to develop waterhammer-type pressure transients in systems containing two-phase fluids and gases, although the magnitude of the pressure rise for these systems will generally be lower. The techniques used to calculate the magnitude of a waterhammer-induced pressure rise are discussed in Chap. B8.

Waterhammer must be considered in the design of those systems where it is likely to occur. For systems designed to codes which provide higher allowable stress criteria for occasional loads, the waterhammer-induced peak pressure should be evaluated under that loading category. For systems designed to codes which do not provide alternative design criteria for occasional loads, the design pressure may be set high enough to envelope the waterhammer-induced peak pressure. The designer is cautioned that this approach can result in an extremely conservative design, which may be prohibitively expensive. Consequently, consideration of alternatives may be required.

Relief Valve Discharge Loads. Because of their rapid opening characteristics and generally high flow rates, the actuation of relief valves frequently results in the application of significant loads to the attached piping system. These loads are caused by the differential pressures across the valves, differential pressure between the valve discharge and the downstream discharge piping, and differential pressure between the discharge piping and the receiver or atmosphere. In addition, momentum effects caused by velocity changes at high flow rates result in secondary loads which act on the attached (or nonattached) piping. These secondary loads must be reacted by the supporting structures. Actual pipe stress levels in inadequately designed relief valve installations can exceed code allowable values, and failures of such installations are common. Depending upon the pressure and temperature conditions of the system fluid, such failures can represent a personnel safety hazard as well as a costly economic issue.

In order to assist the designer in developing a safe, functional relief valve installation, Appendix II of the ASME B31.1, Power Piping Code, has been issued.[18] This nonmandatory appendix to the code provides an extensive treatment of the relevant factors which must be considered to produce a successful design. Relief valve discharge loadings typically occur during a very small percentage of the total system operating time; consequently, they can be treated as occasional loads. The design for load combinations which include these hydraulic transients can therefore be based upon the higher code-defined stress limits.

Acoustically Induced Vibration

When a piping system is exposed to fluctuating pressure disturbances, or pulsations, it frequently responds by vibrating. The magnitude and nature of the piping system vibration is dependent upon the frequency and energy content of the excitation. Low- to moderate-level periodic excitation, such as the pressure pulsations from positive displacement or constant speed centrifugal pumps, will not ordinarily excite significant levels of response in the piping system as long as the excitation frequencies are well removed from the natural vibrating frequencies of the pipe. If the pulsation frequency of the disturbance coincides with the natural frequencies of the piping system, however, resonant vibration can occur. Resonant response normally results in vibratory amplitudes many times that which would occur if the disturbance did not coincide with the natural frequencies of the piping system. Broad-spectrum or random excitation of the type associated with cavitation, bubble collapse, and rapid pressure reductions also lead to resonant vibration. This type of vibration is known as *self-excited vibration.* The piping system draws energy from the broad-spectrum excitation and responds by vibrating at its own fundamental or harmonic natural frequencies.

Resonant response, whether due to the effects of fixed frequency or random excitation, can lead to unacceptable piping system damage. Cyclic stress reversals associated with resonant vibration can result in short-term fatigue failures, which may occur after only a few hours or days of operation. Reduction in the cyclic stress levels and attendant failures can be accomplished through a number of approaches.

When the excitation is a constant frequency disturbance, "decoupling" the vibrating piping system from the source of excitation can often be accomplished by changing the frequency of the excitation. If the disturbance comes from a positive-displacement pump, changing the running speed will change the disturbance frequency. If a centrifugal pump is involved, a change in running speed or in the number of vanes on the impeller may have a beneficial effect.

Changing the natural response frequencies of the piping system can also mitigate the effects of fixed-frequency vibration. Again, the objective is to "decouple" or "detune" the piping system relative to the disturbance. This can often be accomplished by adding supplemental bracing to the pipe or by breaking the system into smaller segments by introducing flexible elements.

When the excitation is broad spectrum or random in nature, detuning by changing the piping system natural frequencies is usually not effective in solving the vibration problem. The modified piping system will continue to draw energy from the broad-spectrum excitation and vibrate at its new natural frequencies. Mitigation of this class of problems usually requires the reduction of the energy level of the excitation or the "strengthening" of the piping system.

A large number of broad-spectrum vibration problems are the result of high differential pressure reduction systems.[19] Excitation (noise) reduction for these types of systems can often be accomplished by the use of low-noise cage-type pressure-reducing valves or multiple orifices arranged for staged pressure reduction. In other cases, where the excitation is the result of turbulence, geometric changes to the piping system to smooth out the flow or reduce average and local velocities can have a beneficial effect.

The objective of the strengthening process is the reduction of piping system stresses to a level where fatigue failures are substantially eliminated. Much can be accomplished by the elimination of stress concentrations through the removal of geometric discontinuities. Examples of these discontinuities include hanger lugs, insulation supports, and small pipe (vent, drain, test, etc.) connections. Additionally, where changes in section are required, they should be effected by gradual smooth changes in contour and generous fillet radii.

Special attention should be paid to all in-line welding done on pipelines which are subject to fluid-induced vibratory loadings. The use of inert gas root pass welding with filler metal addition is recommended. This technique reduces the potential for the formation of critical root defects, which can lead to crack initiation and propagation. Radiographic and ultrasonic examinations done in excess of the minimum requirements may prove cost effective in identifying stress intensifying volumetric defects. Where such examinations are planned, weld backing rings should not be used since they complicate the job of interpreting the examination results.

Finally, the use of pipe wall thickness in excess of that required for pressure-integrity design alone has been found to be beneficial in mitigating the effects of fluid transient loads.

A recent experience was encountered which required the replacement of the main turbine bypass piping on a large nuclear unit because of multiple short-term acoustically induced vibration failures. The original system was made up of 30- and 24-in OD × ⅜-in wall pipe. The original design of the piping wall thickness was based upon pressure-integrity considerations alone. Several pressure boundary failures were experienced at pipe support lugs, hanger clamps, and vent and drain connections after only a few hours of operation. The replacement material was 30-in OD × 1-in wall and 24-in OD × 1¼-in wall. In addition, special attention was paid to the elimination of nonaxisymmetric discontinuities and minor welded attachments. Lugs were replaced by rings, as shown in Fig. B2.2, and unnecessary small pipe connections were removed. The replacement system has been operated extensively without experiencing any further pressure boundary failures.

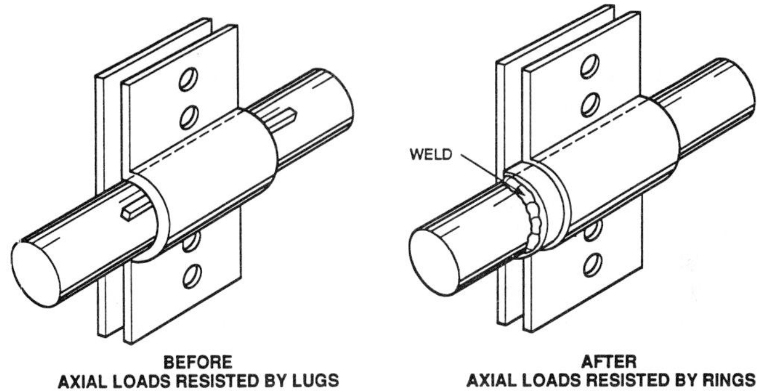

FIGURE B2.2 Use of shear rings instead of lugs to reduce localized pipe wall stress.

Since systems subject to flow-induced vibratory loads usually see those loads over much of their service life, their design should be based upon sustained loading criteria with no increase permitted in allowable stress.

Relative Anchor Movements

Every piping system requires some type of support to function properly. The piping system can be supported from a building or other structure on traditional pipe hangers or from the ground on piers or bents. It even can be supported from another piping system. As long as all of the piping system's support points remain motionless relative to the piping system and relative to each other, the system is unaffected. However, if some of the piping system's supports move relative to the pipe or relative to each other, the piping system will attempt to follow that motion and will experience a change in its state of stress. This condition is called *relative anchor movement.*

Relative movements of a piping system's supports can be caused by a number of phenomena. Some of the more common causes include:

- Thermal expansion–related movement of the connection point on a larger piping system, where the subject system is attached
- Earthquake-induced relative movements of the various points on a building's structure where the subject piping system is supported
- Thermal expansion or mechanically induced movements of a piping connection (nozzle) on a machine, pressure vessel, or heat exchanger

The amount of stress, or more properly stated, the change in stress, that a piping system experiences from relative anchor movements is a function of two variables. They are the magnitude of the anchor movement and the stiffness of the system. As one might expect, larger movements will result in greater changes in stress. Moreover, for a given magnitude of movement, stiffer piping systems will experience greater changes in stress than those that are less stiff. In general, systems that are shorter, have fewer changes in direction, and that are made up of larger-diameter pipe are stiffer than those for which the opposite conditions are true.

Certain phenomena result in a loading case where both the magnitude and direction of the piping system's terminal movements are known, such as movements resulting from the thermal expansion of a pipe. In such a case, the known magnitudes and directions of the anchor movement are input to the piping system stress analysis, and the attendant stress levels are predicted accordingly.

There are other cases, however, where only the magnitude of the anchor movement is known. Examples include earthquake-induced anchor movements or the movements of a building due to wind loading. In this case, the magnitude of the movement is input into analysis as known, but the direction is assumed such that the worst-case change in the state of stress of the piping system under study results. This approach assures that the piping system stress analysis is conducted in the most conservative manner.

ENVIRONMENTAL FACTORS

The generalized definition of the term *stress,* when used in a structural connotation, is force per unit area. The mathematical equation that relates the stress σ to the load F and the load-resisting area A is:

$$\sigma = \frac{F}{A} \tag{B2.8}$$

The breach of a piping system's pressure or structural integrity is invariably the result of its having attained a higher than acceptable state of stress. In the preceding section, it was shown that such an increase in the state of stress of a piping system could result from the application of one or more external (or internal) loads. In terms of Eq. (B2.8), it can be said that loading conditions or loads increase the state of stress of the system by increasing the value of the numerator F of the equation. However, there is a second mechanism through which the state of stress of a piping system may increase, perhaps to the point where a failure occurs. That mechanism is the loss, or deterioration, of the load-resisting area A.

In this chapter the various mechanisms that result in the loss or deterioration of the viable load-resisting base material of a piping system have been generically titled *environmental factors.* An attribute that is common to all environmental factors is that they effectively shorten the useful life of the piping system compared to what it would be if the factors were not present. The following discussion will be limited to four specific environmental factors: corrosion, erosion, physical damage, and erosion-corrosion.

Corrosion

Within the context of this chapter, *corrosion* is the deterioration of the pipe wall due to an electrochemical reaction between the piping material and the process fluid, or the environment.

Corrosion is normally accounted for in design by the provision of additional material in the pipe wall, the use of a suitable coating or lining, or the specification of a corrosion-resistant material. Frequently, the method used to deal with corrosion depends upon the corrosion rate.

For steel corroding in water, the corrosion rate is strongly influenced by the amount of oxygen present and by the temperature. The effect of these variations is shown in Fig. B2.3. Similar data exist in the literature for a wide variety of piping system materials and corrodants. Through the use of these data, the corrosion rate for a given process can often be estimated.

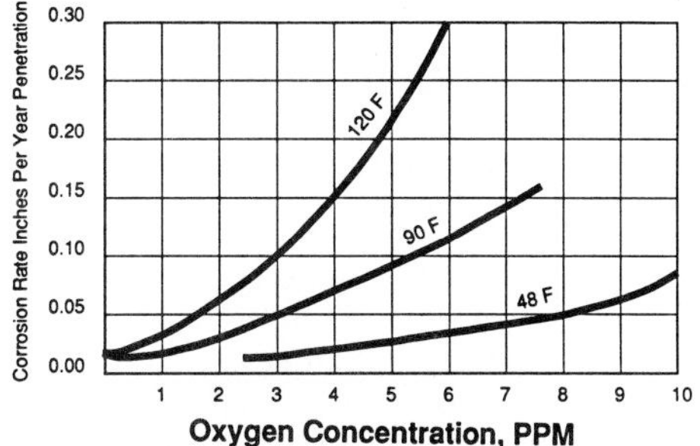

FIGURE B2.3 Effects of oxygen concentration on the corrosion of low carbon steel in tap water at different temperatures. (*Reprinted by permission from* Corrosion Basics: An Introduction, *National Association of Corrosion Engineers, Houston, 1984, Figure 8.1, p. 149.*)

When corrosion is anticipated to occur at a slow, regular rate and this rate can be reliably predicted, it is frequently accommodated by provision of excess material in the pipe wall known as the *corrosion allowance.* This material will be consumed over the design life of the piping system and therefore cannot be counted upon to serve any other purpose such as pressure integrity or mechanical strength. The relationship of the corrosion allowance to the other components of the total pipe wall thickness is discussed earlier in the section "Design Pressure."

In cases where the corrosion rate is prohibitively high or would result in unacceptable contamination of the process fluid, a *pipe lining* or corrosion-resistant piping material may be specified.

Linings tend to be fragile; therefore, their use is limited to applications where abrasion or other physical injury is not likely to occur. Many linings are temperature sensitive and cannot be used in high-temperature service. Most lined pipe is fabricated from plain carbon steel, although fiberglass-reinforced plastic pipe with an integral chemically resistant plastic lining is available. Joining methods are frequently a significant factor in lined pipe design, and the joints must not allow any corrosion-sensitive material to be exposed to the process stream.

Abrasion or other physical attack may damage a lining and expose the corrosion-sensitive substrate. In such cases, a homogeneous material is frequently warranted. If the original corrosion-resistant surface is damaged, the remaining material will continue to resist corrosion. Many corrosion-resistant materials have good high-temperature strength properties and as such are used where linings or plain carbon steels will not survive. Many near ambient temperature, high corrosion rate applications can be successfully and economically accommodated by the use of plastic pipe.

Erosion

Erosion is the wearing away of a surface by abrasion. The abrasion may be the result of particles suspended in the fluid stream, or it may be the result of direct action by the fluid itself.

When the erosion rate is small and consistent, and reliable quantitative data concerning that erosion rate are available, an *erosion allowance* may be provided in the design. The erosion allowance is analogous to the corrosion allowance discussed earlier in the sections "Design Pressure" and "Corrosion." Excess material, over and above that required for pressure integrity and structural and mechanical strength, is provided. This excess material is allowed to waste away over the design life of the piping system.

When the erosion process is not readily quantified, a more qualitative approach to design is normally taken. One approach is to specify special erosion-resistant piping system materials. High-hardness materials are generally effective in resisting erosion. Alternative approaches are to modify the piping system geometry to minimize or eliminate turbulent flow and to direct pipe wall impingement and vortex flow, all of which are known to increase piping system erosion.

Physical Damage

Physical damage or abuse also can be a significant factor in the design of piping systems. This is particularly true of low-pressure, thin-walled piping, which has little resistance to external loadings.

Direct buried pipe is subject to damage from soil pressure and loads from overhead traffic. Uniformly distributed soil pressure loads can normally be estimated with reasonable accuracy and the pipe designed for these loads using methods described in the literature.[20] The effects of heavy concentrated overhead loads cannot be accounted for as easily. Consequently, pipelines which run under heavily trafficked roads or railroad tracks are frequently run through oversized sleeves or conduits which prevent the imposed loads from being directly transmitted to the pipe.

Piping systems of all sizes which carry important services, toxic fluids, or high-pressure, high-temperature fluids should be physically protected from impact from passing motor vehicle traffic, including such vehicles as industrial forklift trucks. The preferred method of protection is to route the piping outside the reach of passing traffic. Where this is not possible, substantial barriers should be erected to protect the piping from impact.

Small-diameter piping takeoffs from large sized headers such as vents, drains, and instrumentation source connections are particularly prone to damage from unspecified external loads. Common design practice is to make the small piping from the header out to the first isolation valve at least one schedule heavier than called for by the pressure design. Similarly, the first isolation or *root valve* is normally made one or two pressure classes heavier than called for by pressure design considerations. These steps normally will make the small lines durable enough to resist random impact or other undefined external loadings that might occur during construction or operation.

Erosion/Corrosion

When iron or steel corrodes in water, a soluble oxide layer called *magnetite* is formed. During steady-state conditions the magnetite attains a constant protective thickness which promotes a uniform corrosion rate. If the magnetite layer is "swept" by a water film deposited by wet steam or by a locally high velocity jet in a liquid stream, the dissolution rate of the magnetite increases. This results in an increase in the localized corrosion rate and an attendant loss of metal from the surface. Since the sweeping away of the oxide layer is an essential part of this corrosion process, it has been named *erosion/corrosion.*

A number of factors have been found to affect the rate of erosion-corrosion in piping systems. In wet-steam systems, percent moisture, material composition, pH and water chemistry, temperature, oxygen level, and flow path geometry have all been found to be significant.[21] In water piping systems, piping material, temperature, pH and oxygen level, and flow path geometry all affect the rate of erosion-corrosion.[22] Among the variables cited above, two that can be readily controlled in design are the piping system materials and the flow path geometry.

Carbon steel is known to be highly susceptible to erosion-corrosion. Both chromium-molybdenum (Cr-Mo) and austenitic stainless steels are significantly less susceptible. The EPRI publication "Erosion/Corrosion in Nuclear Power Plant Steam Piping: Causes and Inspection Program Guidelines"[21] cites a study that indicates the rate of erosion-corrosion in Cr-Mo wet-steam piping is $1/10$ that of carbon steel. Low-alloy Cr-Mo materials can usually be substituted for carbon steel without any other significant design changes. The materials are readily welded to one another, and both have similar physical properties such as tensile and yield strength, density, and thermal expansion coefficient.

The substitution of austenitic stainless steels will normally require some additional engineering. These materials have a thermal expansion rate which averages 50 percent greater than plain carbon steel. Accordingly, increases in terminal reactions and predicted pipe support movements can be expected. These may require that the pipe be rerouted or that different pipe support components be provided.

Piping system geometry plays an important role in mitigating the effects of erosion-corrosion. High localized velocities, vortex flow, jets, and direct stream impingement all increase the rate of magnetite dissolution and therefore increase the rate of erosion-corrosion. Gradual transitions in flow section, pipe size, and geometric changes to smooth out variations in flow velocity and the provision of shallow-angle intersections will all have a beneficial effect. The removal of discontinuities such as weld backing rings and sharp edges at branch connections will also reduce erosion-corrosion. The areas immediately downstream of valves and flow measurement orifices are frequently prone to erosion-corrosion due to vortex formation. These areas will benefit from the addition of flow liners or the substitution of erosion-corrosion–resistant alloys.

MATERIALS-RELATED CONSIDERATIONS

The variety of piping system materials currently in use is extensive and continually growing. The purpose of this section is to provide a brief overview of the common engineering properties of those materials and to describe how those properties influence the design process. For the most part, discussions of specific materials characteristics will be limited to plain carbon and low-alloy steel piping materials. Many of the concepts discussed, however, are applicable to virtually all piping materials.

Strength

Most piping design codes relate the allowable working stresses for materials to their yield strength or ultimate tensile strength at the working temperature. For example, the *allowable working stresses* for materials used for construction in accordance with the ASME B31.1 Power Piping Code are developed using rules defined in the ASME Boiler and Pressure Vessel Code, Section I, Power Boilers, and Section VIII, Pressure Vessels. At any temperature below the creep range, those rules require that the allowable working stress be set at a value no greater than the lowest of the following alternatives[23,24]:

- ¼ of the specified minimum tensile strength at room temperature
- ¼ of the tensile strength at operating temperature
- ⅔ of the specified minimum yield strength at room temperature
- ⅔ of the yield strength at operating temperature

As the temperature of most common pressure-retaining materials increases from ambient, their tensile and yield strengths decrease. Application of the above rules assures that the decreasing strength of piping materials, with increasing temperature, is reflected in the allowable stresses used in design.

At temperatures within the creep range, the allowable working stress is set at a value equal to the lowest of the following[23]:

- 100 percent of the average stress for a creep rate of 0.01 percent/1000 h
- 67 percent of the average stress for rupture at the end of 100,000 h
- 80 percent of the minimum stress for rupture at the end of 100,000 h

When carbon steels are exposed to temperatures greater than 775°F for long periods of time, the carbide phase may convert to graphite. *Graphitization* causes steels to experience brittle fracture at stress levels well below their short-term rupture strength. In recognition of this phenomenon, the ASME B31.1 Power Piping Code provides the following warning statement in the allowable stress tables:

> Upon prolonged exposure to temperatures above 775°F, the carbide phase of carbon steel may be converted to graphite.[25]

For temperatures in excess of 775°F, chromium-molybdenum low-alloy steels or high-alloy stainless steels may be used. These steels offer almost complete freedom from graphitization and enhanced creep-rupture resistance. ASME B31.1 allows the use of these materials at temperatures up to 1200°F.

Toughness

Toughness, or *ductility,* is the ability of a material to resist impact, to withstand repeated reversals of stress, or to absorb energy when stressed beyond the elastic limit. Steel is normally considered to be a ductile material. Contrary to expectation, however, steels sometimes rupture without prior evidence of distress. Under certain conditions, steel may shatter just like glass. In piping, however, this behavior generally occurs only at low temperatures.

The *transition temperature* for any steel is the temperature above which the steel behaves in a predominantly ductile manner and below which it behaves in a predominantly brittle manner. Steel with a high transition temperature is more likely to behave in a brittle manner during fabrication or in service. It follows that a steel with a low transition is more likely to behave in a ductile manner. Therefore, steels with low transition temperatures are generally preferred for service involving severe stress concentrations, impact loading, low operating temperatures, or a combination of all three.

Table B2.6 indicates the low-temperature limitations of various piping materials. Low-alloy steels may be used at temperatures of 0°F when they have Charpy keyhole impact values of at least 15 ft · lb at the lowest design temperature. Austenitic stainless steels with limited carbon content, copper and copper alloys, and aluminum do not experience transitions in impact strength from ductile to

TABLE B2.6 Low-Temperature Operating Limits for Selected Piping Materials

Low-temp. limit, °F	Material and suitable ASTM designation	Comments
Zero	Mild steel (A53, A120, A135)	No requirements other than suitable pressure rating
−20	Mild steel (A53, A135)	Reduce pressure rating 1% for each °F below zero, or Charpy impact test, 15 ft · lb at design temperature
−50	Killed or limited carbon steel (A333, GR-1)	Charpy impact test, 15 ft · lb at design temperature
−150	3½% Ni-steel (A333, GR-3)	Charpy impact test, 15 ft · lb at design temperature
−325	Austenitic stainless steel (Types 304, 316, etc.)	Limited carbon content
No limit	Nonferrous copper, brass, aluminum	Aluminum, copper, brass

brittle fracture and, therefore, may be used for low temperatures without pressure-rating penalties.

Low-temperature piping is generally covered with insulation which, in addition to limiting heat transfer, helps provide protection from external impact. This, however, is not sufficient insurance against the type of damage that could result if a pipe should fracture.

Additional perspectives on materials strength and toughness are provided in Chap. A3.

Corrosion Resistance

Considered as a material property, *corrosion resistance* is a measure of a piping system material's relative inertness to chemical attack from a specific process fluid, at the system's normal operating temperature, or its environment (see earlier section "Corrosion"). The importance of considering the system's operating temperature cannot be overemphasized. It is well known that many chemical reactions are highly temperature dependent. A particular piping system material could be virtually immune to chemical attack by a specific corrodant at one temperature, while it might be prone to excessive attack by the same corrodant at a higher temperature.

Within this context, then, it is clear that there is no such thing as a universally corrosion-resistant material. All common piping system materials react with some process fluids (corrodants) at certain temperatures. Therefore, when pursuing a "corrosion-resistant" material for a specific application, the objective is to identify a material whose corrosion rate in the presence of a specific corrodant is negligible, or at least acceptable, over the design life of the piping system.

It is important also to consider the effect corrosion may have on the process fluid. Under certain conditions, the dissolution of the base metal or the corrosion products into the process stream may require economic or technical considerations that go beyond the piping system's pressure-containing parts. In some cases, the major consideration in choosing a piping system material may be the preservation of the chemical purity of the process fluid. Such is usually the case in choosing piping system materials that handle food products and piping used in many chemical manufacturing operations.

THERMAL INSULATION

Whenever the surface temperature of a piping system differs significantly from that of its surrounding environment, a potential need for an insulation system exists. An insulation system serves three principal purposes. They are:

- The significant reduction in the transfer of thermal (heat) energy to or from the surface of the piping system
- The prevention of moisture formation and collection on the surface of the piping system due to condensation
- The prevention of potentially injurious personnel contact with the surface of the piping system

The reduction in heat transfer to or from the surface of a piping system will minimize the gain or loss in temperature of the process fluid, thus maximizing the capability of the fluid to perform its intended function. Minimizing heat exchange between the piping system and the environment also minimizes the unwanted heating or cooling of the environment. This improves the comfort level for the inhabitants, or improves the operating conditions for adjacent equipment.

Most insulating systems used above ground consist of preformed components which are mechanically attached to the pipe. Low-temperature insulation is frequently made of expanded cellular plastic or rubber foamed materials. Moderate-temperature insulations are frequently made from glass fiber products. High-temperature insulation is usually made of preformed cementations or refractory materials or blankets made from ceramic fibers. Insulation systems used for buried pipe is frequently in loose granular form, so it can be poured loose into the trench to surround the pipe and isolate it from the ground environment.

If the surface temperature of a piping system is below the dew point of the surrounding air, water vapor in the air will condense on the surface of the pipe. This condition can be detrimental. The *condensation* can collect and drip onto surfaces below the pipe, thus doing damage. The condensate can also saturate the piping insulation, thus significantly increasing its thermal conductivity and reducing its insulating capability. In order to prevent condensation on an insulated pipe, the airborne water vapor must be prevented from reaching the pipe surface. This is normally accomplished by providing a *vapor barrier* at the outer surface of the insulation. An adequate vapor barrier may be constructed from a well-fitted metal jacket, an extruded plastic or rubber coating, or a spiral-wrapped impervious tape coating. Whatever the form, the vapor barrier must prevent the airborne water vapor from entering the pores of the insulation and migrating toward the cool pipe surface where condensation can occur.

Extremely hot or cold piping systems can pose a contact safety hazard to personnel in the vicinity. Surface temperatures above 135°F can cause severe burns to unprotected skin, and temperatures below approximately 20°F can cause freeze damage. Thermal insulation can be designed such that the insulation surface temperature is maintained in a safe range. Frequently piping systems which are otherwise uninsulated will have insulation installed in accessible areas to provide personnel protection.

For a complete treatment of the engineering principles involved in designing a thermal insulation system for piping, see Chap. B7.

SIZING OF A PIPING SYSTEM

The term *sizing of a piping system* refers to the completion of two independent design functions: the fluid flow design and the pressure-integrity design. The purpose of the fluid flow design is to determine the minimum acceptable inside diameter of the various segments of the piping system. The purpose of the pressure-integrity design is to determine the minimum acceptable pipe wall thickness and the pressure ratings of the in-line components.

System Fluid Flow Design

The objective of the fluid flow design is to determine the minimum acceptable inside diameter of each segment of the piping system that will accommodate the design flow rate while maintaining the pressure drop and flow velocity within reasonable limits.

Most piping systems use pumps to develop the pressure or head required to maintain the system design flow rates. Piping system pressure drops must be maintained within reasonable values to limit the installed size of the system pumps and their drive motors. Pump and motor size limitations are necessary to control initial system construction costs and continuing system operating costs. The optimum pipe size is based on an economic "trade-off" between the installed capital cost of the piping system and the sum of the capital plus lifetime operating costs of the pumping system.

System flow velocities are limited by design to avoid a number of potential operating problems. These problems have already been discussed in previous sections of this chapter. In the absence of any other formal or more limiting criteria, the flow velocities given for water in Table B2.7 and for steam in Table B2.8 are considered reasonable for normal industrial applications.

TABLE B2.7 Reasonable Design Velocities for Water Flowing through Pipes

Service condition	Reasonable velocity
Boiler feed	8 to 15 ft/s
Pump suction and drain lines	4 to 7 ft/s
General service	4 to 10 ft/s
City	to 7 ft/s

Source: Crane Technical Paper 410, *Flow of Fluids Through Valves, Fittings, and Pipe,* The Crane Company, New York, 1985, pp. 3–6.

The detailed fluid flow design of a piping system requires the consideration of a number of fluid parameters including flow rate, viscosity, density, and pipe wall frictional drag. Further discussions of this aspect of the pipe sizing process are provided in Chap. B8.

Pressure-Integrity Design

The pressure-integrity design of a piping system normally requires the consideration of at least two specific issues. The first is the determination of a minimum or nominal pipe wall thickness, and the second is the determination of the pressure rating of the in-line components such as fittings and valves.

TABLE B2.8 Reasonable Design Velocities for Steam Flowing through Pipes

Condition of steam	Pressure, P, psig	Service	Reasonable velocity, V, ft/min
Saturated	0 to 25	Heating (short lines)	4,000 to 6,000
Saturated	25 and up	Power house equipment, process piping, etc.	6,000 to 10,000
Superheated	200 and up	Boiler and turbine leads, etc.	7,000 to 20,000

Source: Crane Technical Paper 410, *Flow of Fluids Through Valves, Fittings, and Pipe,* The Crane Company, New York, 1985, pp. 3–16.

Determination of the Pipe Wall Thickness. After the fluid design is complete and the minimum inside diameters of the various segments of the piping system are determined, the piping pressure-integrity design may proceed. The major steps in the process are:

1. Using the minimum inside diameter determined from the fluid flow evaluation, select the next larger standard nominal or OD size pipe from the listings provided in ASME/ANSI Standards B36.10 for standard wrought steel pipe or B36.19 for stainless steel pipe (see earlier section "The ANSI Pressure Classification System").
2. Based upon the fluid and service, select a suitable piping material, and if necessary, determine the required corrosion, erosion, joining or mechanical strength allowances.
3. Using equations provided in the design code, calculate the required minimum wall thickness to provide for pressure-integrity and allowances.
4. Refer back to ASME/ANSI Standards B36.10 or B36.19 to select an appropriate nominal wall thickness or schedule.
5. Confirm that the standard manufacturing tolerance will not reduce the nominal wall thickness selected in Step 4 below the minimum required as calculated in Step 3.
6. Confirm that the inside diameter of the pipe selected, based upon the nominal wall thickness selection of Step 4, is compatible with the minimum inside-diameter requirements obtained from the fluid flow evaluation.

The process described above is demonstrated in the following example.

Problem Statement. A carbon steel pipe having a minimum inside diameter of 11.2 in is required to transport water at 700 psig and 90°F. The design code is ASME B31.1, and the design life is 8 years. The water has nominal oxygen content of 1 ppm. Buttwelded construction will be used.

Solution. An economical grade of seam-welded carbon steel pipe (ASTM A53-Grade A) is selected. From ASME B31.1, Appendix A, Table A–1, the allowable working stress at 90°F is determined to be 10,200 psi. From Figure B2.3, the corrosion rate is estimated to be 0.02 inches per year. The pressure integrity design will be based upon ASME B31.1, Paragraph 104.1.2, Equation (3)[26]:

$$t_m = \frac{PD_o}{2(SE + Py)} + A \qquad (B2.9)$$

From ASME/ANSI B36.10, 12 in nominal pipe size (12.75 in OD) is tentatively selected.

Using the stated 8-year design life and 0.02-inches-per-year corrosion rate, the total corrosion allowance of 8 × 0.02 = 0.16 in is calculated. Buttwelded construction is specified; therefore, no additional wall thickness allowance for joining (threading, grooving, etc.) is required.

From ASME B31.1, Table 104.1.2(A), $y = 0.4$ is selected for ferritic steels at temperatures at or below 900°F.

Equation (B2.9) may now be used to calculate the required minimum wall thickness:

$$t_m = \frac{PD_o}{2(SE + Py)} + A \qquad \text{(B2.9)}$$

$$= \frac{700 \times 12.75}{2(10{,}200 + 0.4 \times 700)} + 0.16$$

$$= 0.586 \text{ in}$$

From ASME/ANSI B36.10, under the listings for 12-in nominal pipe size, Schedule 80 pipe with a nominal wall thickness of 0.688 in is tentatively selected.

The wall thickness tolerance for ASTM A53 pipe, which is +0; −12½ percent is checked next:

$$0.688 \times 0.875 = 0.602$$

$$0.602 > 0.586$$

Finally, the nominal inside diameter is checked against the minimum flow diameter:

$$12.75 - 2(0.688) = 11.374 \text{ in}$$

$$11.374 > 11.2$$

The problem requirements are satisfied; 12-in Schedule 80 pipe meeting ASTM Specification A53-Grade A is acceptable.

The previous example did not consider the effects of bending on the pipe wall. In most instances the pressure design will dominate in the determination of pipe wall thickness. However, if the pipe span between supports is unusually long or the pipe has a very heavy in-line component, such as a valve, then the longitudinal bending stress may dominate the design. This facet of the design is considered in the piping stress analysis discussion of Chap. B4.

Determining the Pressure Class for In-Line Components. Two examples are provided here to demonstrate the process used to determine the pressure classification for the in-line components. The first demonstrates the selection process for a standard flange; the second demonstrates the selection process for a special-class valve.

Example B2.1. A 16-in carbon steel pipeline operates at 840 psig and 740°F. Select a standard weld-neck flange for the service.

Solution. Table B2.1 lists various materials of construction for standard pipe flanges. Under Material Group 1.1, ASTM Specification A105 for Forgings, Carbon Steel, for Piping Components, is listed. Next refer to Table B2.3, which lists ANSI pressure-temperature ratings for Class 600 flanges of various material groups. Under Material Group 1.1 it is determined that a Class 600 flange has a pressure-temperature rating of 1010 psig at 750°F. Since this rating exceeds the requirements of 840 psig at 740°F, this flange is acceptable.

Example B2.2. A 12-in buttwelding-end gate valve is required to operate at 2350 psig and 1015°F. The valve material is ASTM A217 Gr WC9. Determine the appropriate ANSI pressure class rating.

Solution. Table B2.9 *a* and *b* lists the pressure-temperature ratings for standard and special class valves of ASTM A217 Grade WC9.

There are two correct answers to this problem. The first and simplest answer is to select a Standard-Class 4500 valve from Table B2.9*a*. This valve has a pressure-temperature rating of 2985 psig at 1050°F and obviously meets the stated requirements. However, this valve may prove to be a very expensive alternative,

TABLE B2.9a Pressure-Temperature Ratings for Standard Class Valves of ASTM A217 Grade WC9

Temperature, °F	Working pressure by classes, psig							
	150	300	400	600	900	1500	2500	4500
−20 to 100	290	750	1,000	1,500	2,250	3,750	6,250	11,250
200	260	715	955	1,430	2,150	3,580	5,965	10,740
300	230	675	905	1,355	2,030	3,385	5,640	10,150
400	200	650	865	1,295	1,945	3,240	5,400	9,720
500	170	640	855	1,280	1,920	3,200	5,330	9,595
600	140	605	805	1,210	1,815	3,025	5,040	9,070
650	125	590	785	1,175	1,765	2,940	4,905	8,825
700	110	570	755	1,135	1,705	2,840	4,730	8,515
750	95	530	710	1,065	1,595	2,660	4,430	7,970
800	80	510	675	1,015	1,525	2,540	4,230	7,610
850	65	485	650	975	1,460	2,435	4,060	7,305
900	50	450	600	900	1,350	2,245	3,745	6,740
950	35	380	505	755	1,130	1,885	3,145	5,660
1000	20	270	355	535	805	1,340	2,230	4,010
1050	20*	200	265	400	595	995	1,660	2,985
1100	20*	115	150	225	340	565	945	1,700
1150	20*	105	140	205	310	515	860	1,545
1200	20*	55	75	110	165	275	460	825

*For welding-end valves only. Flanged-end valves terminate at 1000° F.

TABLE B2.9*b* Pressure-Temperature Ratings for Special Class Valves of ASTM A217 WC9

Temperature, °F	Working pressure by classes, psig							
	150	300	400	600	900	1500	2500	4500
−20 to 100	290	750	1,000	1,500	2,250	3,750	6,250	11,250
200	290	750	1,000	1,500	2,250	3,750	6,250	11,250
300	290	750	1,000	1,500	2,250	3,750	6,250	11,250
400	290	750	1,000	1,500	2,250	3,750	6,250	11,250
500	285	740	985	1,475	2,210	3,685	6,145	11,060
600	285	740	985	1,475	2,210	3,685	6,145	11,060
650	285	740	985	1,475	2,210	3,685	6,145	11,060
700	280	735	980	1,465	2,200	3,665	6,110	10,995
750	280	730	970	1,460	2,185	3,645	6,070	10,930
800	275	720	960	1,440	2,160	3,600	6,000	10,800
850	260	680	905	1,355	2,030	3,385	5,645	10,160
900	230	600	800	1,200	1,800	3,000	5,000	9,000
950	180	470	630	945	1,415	2,360	3,930	7,070
1000	130	335	445	670	1,005	1,670	2,785	5,015
1050	95	250	330	500	745	1,245	2,070	3,730
1100	55	140	190	285	425	710	1,180	2,120
1150	50	130	170	260	385	645	1,070	1,930
1200	25	70	90	140	205	345	570	1,030

Source: Adapted from ASME/ANSI B16.34, *Valves, Flanged, Threaded, and Welding End,* American Society of Mechanical Engineers, New York, 1988, Table 2–1.10.

since Class 4500 valves are massively constructed, and valve prices vary according to the weight of the material used in their construction.

The second alternative is to consider the Special Class 2500 valves whose ratings are provided in Table B2.9*b*. *Special-class valves* undergo mandatory nondestructive examinations and, if necessary, defect repairs to allow them to qualify for higher pressure-temperature ratings. For a more detailed discussion of special-class valves, the reader is referred to ASME/ANSI B16.34.[27] In order to determine whether a Special Class 2500 valve will meet the requirements of Example B2.2, a linear interpolation of the ratings in Table B2.9*b* is required. The process is illustrated below:

Temperature, °F	*Pressure, psig*
1000	2785
1015	*P*
1050	2070

$$\frac{1050 - 1015}{1050 - 1000} = \frac{P - 2070}{2785 - 2070}$$

$$\frac{35}{50} = \frac{P - 2070}{715}$$

$$P = 2070 + \frac{35}{50} \times 715$$

$$= 2571 \text{ psig}$$

Since the interpolated pressure rating of 2571 psig is greater than the specified requirement of 2350 psig, a Special Class 2500 valve will satisfy the requirements of Example B2.2.

REFERENCES

1. ASME/ANSI Standard B16.5, *Pipe Flanges and Flanged Fittings,* American Society of Mechanical Engineers, New York, 1988.
2. ASME B31.1, Code, *Power Piping,* American Society of Mechanical Engineers, New York, 1989, Paragraph 102.2.4.
3. Ibid., Paragraph 104.1.2.
4. Burrows, W. R., Michel, R., and Rankin, A. W., "A Wall Thickness Formula for High-Pressure High-Temperature Piping," American Society of Mechanical Engineers, New York, 1952, Paper No. 52-A-151.
5. ASME B31.3, Code, *Chemical Plant and Petroleum Refinery Piping,* American Society of Mechanical Engineers, New York, 1987, Paragraph 304.1, Equation (3a).
6. Ibid., Equation (3b).
7. Ibid., Equation (2).
8. ASME Boiler and Pressure Vessel Code, Section I, *Rules for Construction of Power Boilers,* American Society of Mechanical Engineers, New York, 1989, Sub-Part PG-22.

9. ANSI Standard A58.1, *Minimum Design Loads for Buildings and Other Structures,* American National Standards Institute, New York, 1982.
10. Ludlum, David W., *The American Weather Book,* Houghton Mifflin, Boston, 1982, p. 268.
11. Tattelman, Paul, and Gringorten, Irving I., "Estimated Glaze Ice and Wind Loads at the Earth's Surface for the Contiguous United States," Air Force Cambridge Research Laboratories, Bedford, Mass., 1973, AFCRL-TR-73-0646.
12. Code of Federal Regulations, *Title 10—Energy, Part 50-Domestic Licensing of Production and Utilization Facilities,* Office of the Federal Register, Washington, D.C., 1987, General Design Criterion No. 2.
13. *Standard Review Plan 3.2.1—Seismic Classification,* Revision 1, U.S. Nuclear Regulatory Commission—Office of Nuclear Reactor Regulation, Washington, D.C., 1981.
14. *Regulatory Guide 1.29—Seismic Design Classification,* Revision 3, U.S. Nuclear Regulatory Commission, Washington, D.C., 1978.
15. *Evaluation of Seismic Designs—A Review of Seismic Design Requirements for Nuclear Power Plant Piping,* United States Nuclear Regulatory Commission Piping Review Committee, Washington, D.C., NUREG-1061, Vol. 2, 1985, p. 2–1.
16. *The Uniform Building Code—1985 Edition,* International Conference of Building Officials, Whittier, Calif., 1985.
17. *The BOCA National Building Code/1987,* 10th edition, Building Officials and Code Administrators International, Inc., Country Club Hills, Ill., 1986.
18. ASME B31.1, Code, *Power Piping,* American Society of Mechanical Engineers, New York, 1989, Appendix II.
19. Carucci, V. A., and Mueller, R. T., "Acoustically Induced Piping Vibration in High Capacity Pressure Reducing Systems," American Society of Mechanical Engineers, 82-WA/PVP-8, 1982.
20. *Steel Pipe—A Guide for Design and Installation,* (M11), 1985, American Water Works Association, Denver, p. 57.
21. Delp, G. A., Robison, J. D., and Sedlack, M. T., "Erosion/Corrosion in Nuclear Power Plant Steam Piping: Causes and Inspection Program Guidelines," Electric Power Research Institute, Palo Alto, Calif., NP-3944, 1985.
22. Jones, R., Chexel, B., Behravesh, M., and Stahlkopf, K., "Single Phase Erosion-Corrosion of Carbon Steel Piping," Electric Power Research Institute, Palo Alto, Calif., 1987.
23. ASME Boiler and Pressure Vessel Code, Section I, *Rules for Construction of Power Boilers,* American Society of Mechanical Engineers, New York, 1989, Appendix A-150, p. 240.
24. ASME Boiler and Pressure Vessel Code, Section VIII, Division I, *Pressure Vessels,* American Society of Mechanical Engineers, New York, 1989, Appendix P, p. 787.
25. ASME B31.1, Code, *Power Piping,* American Society of Mechanical Engineers, New York, 1989, Table A-1, Note (2).
26. ASME B31.1, Code, *Power Piping,* American Society of Mechanical Engineers, New York, 1989, Paragraph 104.1.2.
27. ASME/ANSI Standard B16.34, *Valves—Flanged, Threaded, and Welding End,* American Society of Mechanical Engineers, New York, 1988, Section 8.

CHAPTER B3
PIPING LAYOUT

Charles A. Bullinger
Bechtel Corporation
Gaithersburg, Maryland

A. B. Cleveland, Jr.
President
Jacobus Technology, Inc.
Gaithersburg, Maryland

Piping, historically, has been a major expenditure in the design of any industrial, refinery, petrochemical, or power generating plant, when considering the engineering and design jobhours, the material costs, and the field labor erection costs. Proper planning and execution of the design and routing of pipe in the design office can serve to control these costs and keep them to a minimum.

Piping design is a subject that has been largely overlooked in our colleges and universities until recently, with a few now offering associate in arts (AA) degrees in mechanical engineering technology that include a one-semester course in industrial piping design. However, these courses are limited in that they do not cover the design of high-pressure piping systems, such as main steam or boiler feedwater.

The experienced piping designer has a broad knowledge of plant layout and equipment arrangement and all of the piping systems associated with his or her field of endeavor whether it be refinery, petrochemical, or power. In addition, the designer must have an understanding of the practical application of piping materials, all types of valves, pumps, tanks, pressure vessels, heat exchangers, power boilers, and steam turbine drives.

CODES AND STANDARDS

The various codes and standards applicable to the engineering, design, and fabrication of piping systems are discussed and summarized in Part A, Chap. A4, of this handbook. These codes and standards were written to establish minimum requirements for safe design and construction with very little reference to the physical routing of piping. However, the piping designer should be familiar with them as they apply to his or her work. There are a few specific references to physical piping design of safety relief valve arrangements. These are:

- The ASME Boiler and Pressure Code, Section I, specifies that there is to be no intervening pipe, valves, or fittings between the safety relief valves and the vessel or piping to which they are attached. This means that they must be fitting bound.
- The ASME B31.1, Power Piping Code, Appendix II, Non Mandatory Rules for the Design of Safety Valve Installations, provides guidelines for the physical arrangement of safety valve piping, the most significant being that the distance between the centerline of the valve and the centerline of the discharge elbow must not exceed four times the nominal pipe size of the relief valve outlet.

The Pipe Fabrication Institute has a series of standards covering the fabrication, design, cleaning, and inspection of piping. Three of these standards, listed below, specifically relate to physical piping design:

ES2: Method of Dimensioning Welded Assemblies

ES3: Linear Tolerances, Bending Radii, Minimum Tangents

ES7: Minimum Length and Spacing for Welded Nozzles

For piping and valve drawing symbols, refer to ANSI/ASME Y32.2.3—1949, Graphical Symbols for Pipe Fittings, Valves and Piping.

PIPING LAYOUT CONSIDERATIONS

In order to commence the routing and design of any piping system, the designer will need the following reference information:

- *The system piping and instrumentation diagram (P&ID):* This drawing is the designer's "road map" which shows all equipment, pumps, valves, instrumentation, and other piping specialty items, in sequence, as required for proper system function.
- *The project piping specification:* This document defines the system design and operating pressures and temperatures; piping materials; pipe wall thicknesses or schedules; types of fittings to be used, for example, buttweld, socket weld, or screwed; and the valve and flange pressure rating and insulation requirements. In addition the piping specification defines the fabrication, examination, testing, inspection, and installation requirements, including the requirements for seismic installations, where applicable.
- *Equipment outlines:* These should include their overall dimensions and the pipe size, wall thickness, flange pressure rating, and the locating dimensions of all pipe nozzles and other connections.
- *The project general arrangement or equipment location drawings:* These drawings will indicate the preliminary location of all major pieces of equipment in the plant which the designer will either verify or relocate, as required, to accommodate the physical pipe routing as designed, or redesign the piping to accommodate the particular piece of equipment.

The last of these items, the general arrangement or equipment location drawings, are usually developed by senior-level piping designers during the proposal preparation and are taken over by the project team upon award of the contract. From this point on they are revised and updated as a part of the normal process of design development.

Piping layout then becomes a matter of designing a completely dimensioned routing from one point to another point with all of the branches, valves, piping specialties, and instrumentation as indicated on the P&ID. This statement, however, is an oversimplification of the process since there are many other factors that must be considered, such as interferences, piping flexibility, material costs, pipe supports, operation and maintenance, and safety and construction requirements.

Interferences

One of the most important aspects of piping layout is the avoidance of interferences with other facilities in the plant such as other piping systems, structural steel and concrete, heating, ventilating, and air-conditioning (HVAC) duct work and electric cable tray and conduit. For engineering firms using manual drafting and design, the search for interferences is very tedious and time-consuming since the designer must mentally and visually look for interferences between the system currently being designed and any previously designed systems or facilities, not to mention those systems or facilities in design concurrently. This process is extremely complex at best. Traditionally, this has been accomplished with the use of area composite drawings (see Fig. B3.1), and plastic scale models.

The composite drawings and plastic models show all plant facilities designed to date and are used by the designers to select an interference-free route for the system currently under design; however, the designer still must search out those systems or facilities in design concurrently. Once satisfied that the current system layout is interference free, it will be added to the area composite drawing and the plastic model.

An alternative to composite piping drawings and plastic models for interference detection is the use of computer-aided design (CAD). Specifically, three-dimensional (3D) computer modeling can provide an efficient, accurate, and cost-effective alternative to the traditional manual methods for interference detection. This and other applications of CAD for piping layout are addressed in the section "Application of Computer-Aided Design to Piping Layout."

Piping Flexibility

The effects of the thermal expansion of pipe and fittings as a result of system operating temperature changes cannot be overlooked during the layout and routing of any piping system. The function of piping flexibility or stress analysis has for the most part been delegated to the computer particularly in the case of high-temperature, high-pressure piping systems. The piping stress analyst translates and enters the piping design data into the computer, reviews the output data, and if the system is too rigid, may suggest appropriate corrective redesigns. However, the final responsibility belongs with the piping designer.

In the past, a computer stress analysis, including the development of input data and the interpretation of the output, could be expensive and many thousands

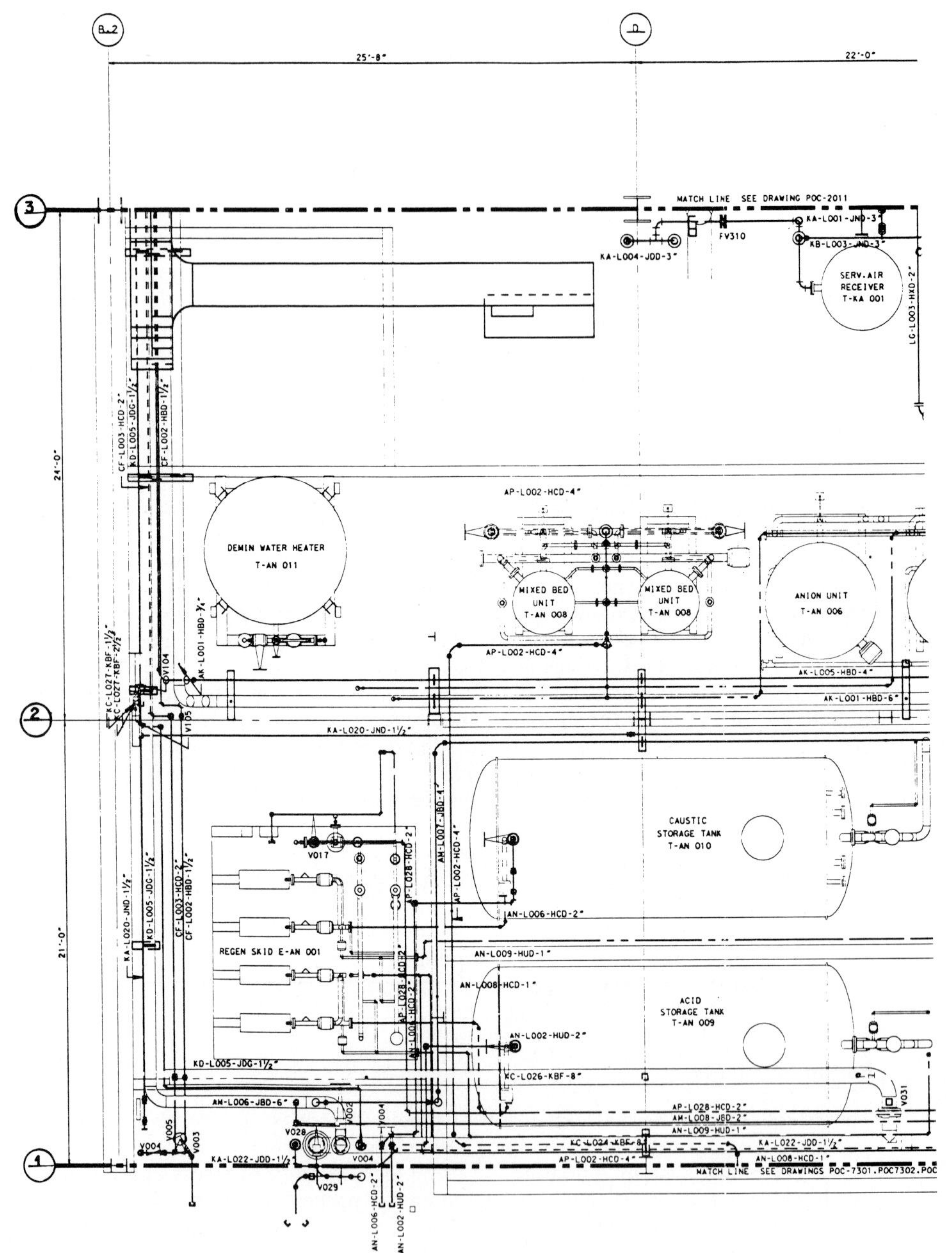

FIGURE B3.1 Area composite drawing.

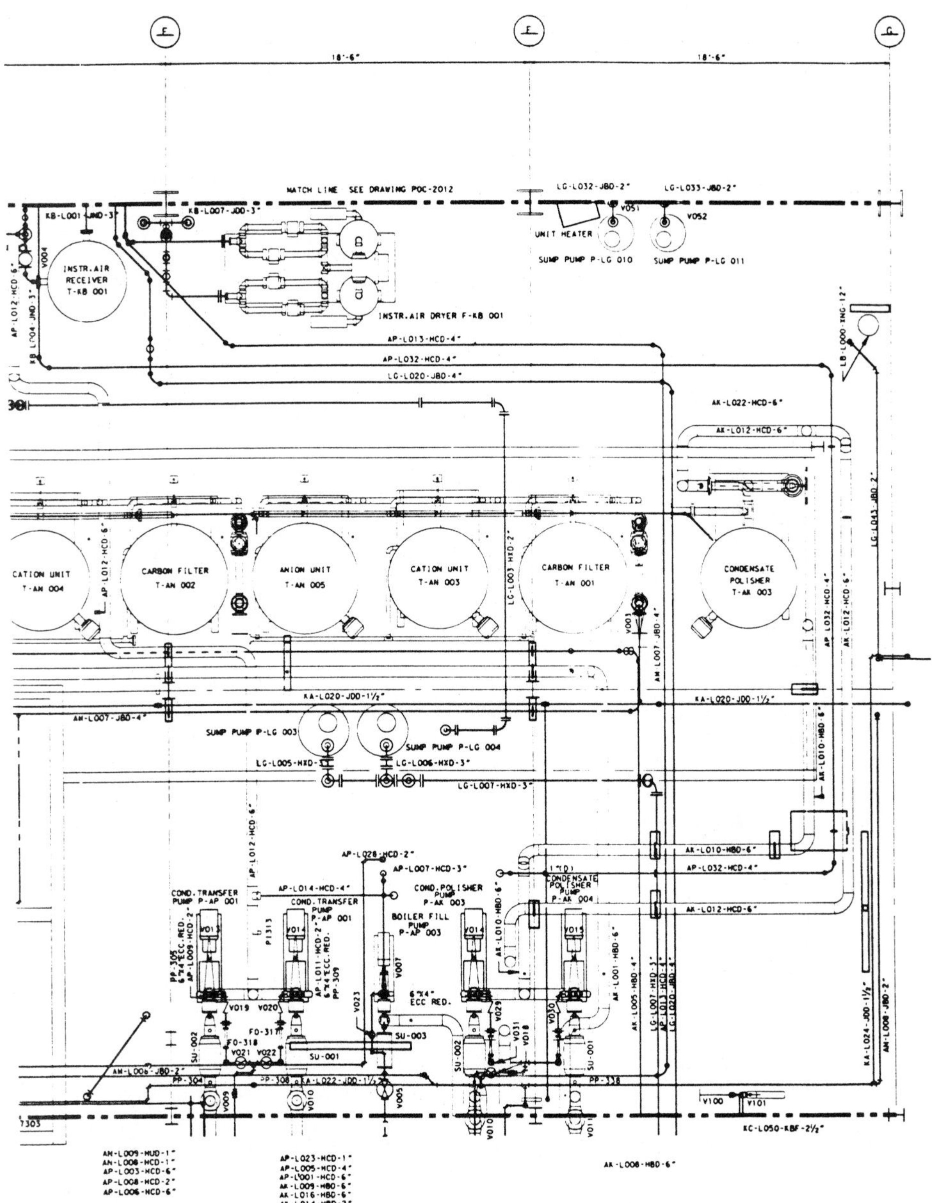

FIGURE B3.1 *Continued*

of dollars spent if numerous iterations of the computer run were needed to arrive at an acceptable system design. The experienced piping designer, with the knowledge and capability of designing piping systems that are inherently flexible, was relied on to keep the number of computer iterations to a minimum. Today, this is much less of a problem with the advent of the personal computer and many computer programs for calculating stresses in piping systems due to thermal expansion and other static and dynamic loads. However, the piping designer must integrate piping flexibility considerations into the piping layout.

The piping designer should route piping with flexibility designed into it using the minimum amount of pipe, pipe fittings, and expansion loops by considering the following:

- Avoid the use of a straight run of pipe between two pieces of equipment or between two anchor points.
- A piping system between two anchor points in a single plane should, as a minimum, be L shaped consisting of two runs of pipe and a single elbow. This type of arrangement should be subjected to a "quick-check" analysis to determine if a formal computer stress analysis is required. A preferred solution in this case may be a series of two or more L-shaped runs of pipe.
- A piping system between two anchor points with the piping in two planes may consist of two L-shaped runs of pipe, for example, one L-shaped run in the horizontal plane and another in the vertical plane. This arrangement should also be subjected to a quick-check analysis.
- A three-plane configuration may consist of a series of L-shaped runs and/or U-shaped expansion loops designed into the normal routing of the system.
- When the expected thermal expansion in any given run of pipe is high, consider the use of an anchor at or near the center of the run, thereby distributing the expansion in two directions.
- For systems consisting of a large-diameter main and numerous smaller branch lines, the designer must ascertain that the branches are flexible enough to withstand the expansion in the main header.
- Systems which are to be purged by steam or hot gas must be reviewed to assure that they will be flexible during the purging operation.
- System or equipment bypass lines may be cold while the main runs are hot, resulting in excessive stresses.
- Temperatures at start-up are often greater than those at operating conditions.
- Closed relief valve and hot blowdown systems should be given special attention.

In addition the piping designer may use a variety of single- and multiplane piping arrangements, such as the L-shaped, the U-shaped, and Z-shaped configurations, in the normal routing of any system as shown in Figs. B3.2 through B3.7.

Chapter B4, Stress Analysis of Piping Systems, discusses in detail pipe stress analysis including quick-check methods that may be used by the piping designer to determine if the system is, or is not, flexible enough and to determine if a more rigorous analysis is required.

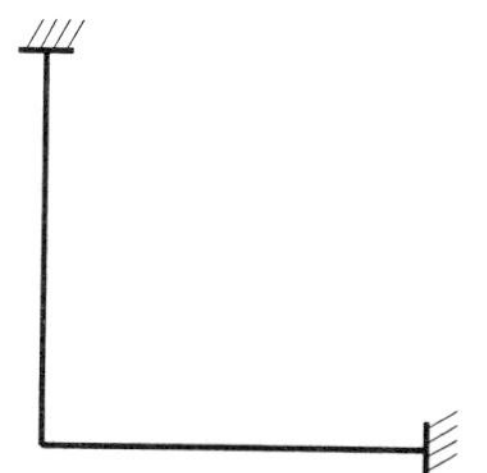

FIGURE B3.2 L-shaped configuration.

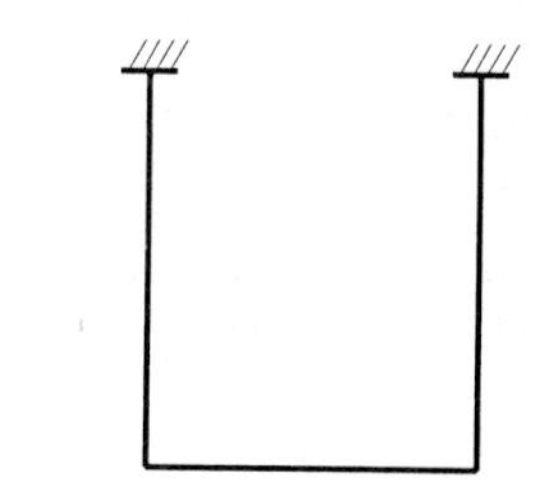

FIGURE B3.3 U-shaped configuration.

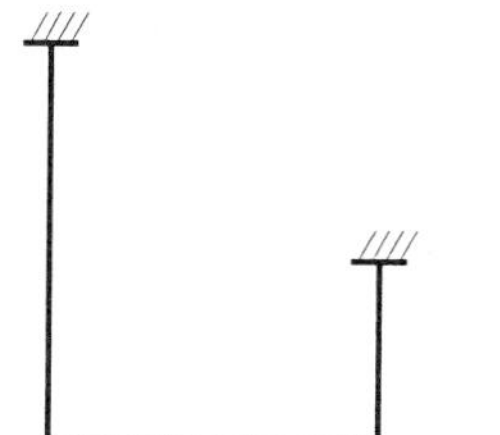

FIGURE B3.4 U-shaped configuration with unequal legs.

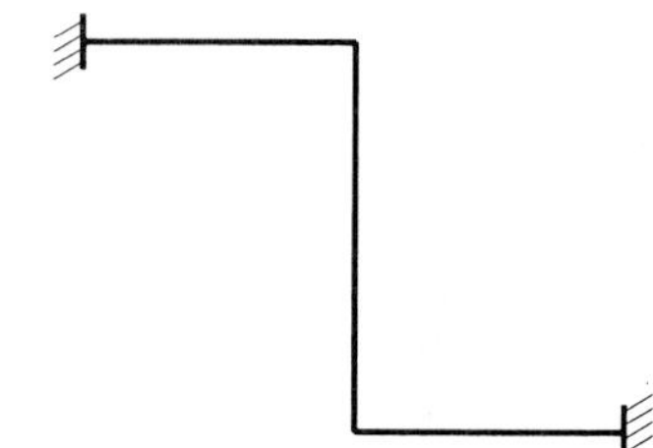

FIGURE B3.5 Z-shaped configuration.

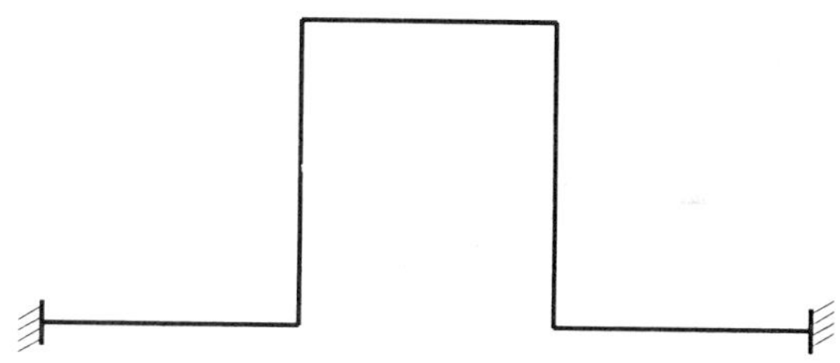

FIGURE B3.6 Expansion loop.

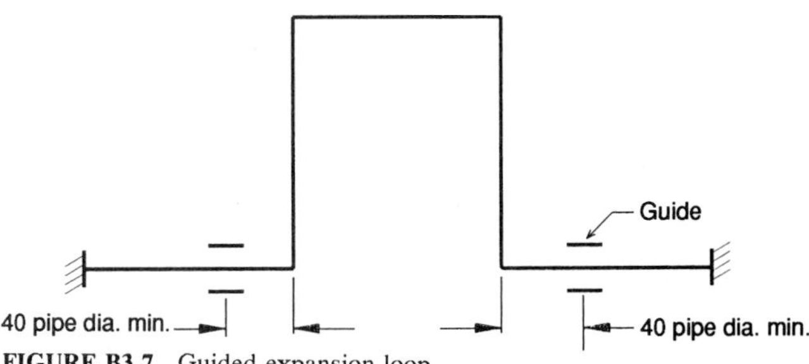

FIGURE B3.7 Guided expansion loop.

Valves

The piping designer must be familiar with the proper application of all types of valves including gate, globe, plug, butterfly, ball, angle, diaphragm, check, pressure relief, and control valves and their methods of operation including manual, chain, gear, air, hydraulic, or motor. The following general guidelines should be applied when locating valves in any piping system:

- Valves should be installed with their stems between the vertically upward and horizontal positions with particular attention given to avoiding head and knee knockers, dripping hazards, and valve stems in the horizontal plane at eye level that may be a safety hazard. Large motor-operated valves should be installed in the vertical upright position where possible to facilitate supporting and maintenance.
- Valves in acid and caustic services should be located below eye level or in such a manner as not to present a safety hazard.
- The location of valves, with consideration for operating accessibility, should be accomplished in the natural routing of the system from point to point, avoiding the use of vertical loops and pockets.
- Valves in overhead piping with their stems in the horizontal position should be located such that the bottom of the handwheel is no more than 6½ ft above the floor or platform. Only infrequently operated valves should be located above this elevation, and then the designer should consider the use of a chain operator or a platform for access.
- Where chain operators are used, the valves should be located such that the chain does not present a safety hazard to the operating personnel.
- A minimum of 4 in of knuckle clearance should be provided around all valve handwheels.
- Valves should not be installed upside down.
- Space should be provided for the removal of all valve internals.

Improper application and placement of valves in the piping system can be detrimental to system function, cause waterhammer, and cause the valves to literally self-destruct. What follows are some specific recommendations and methods of avoiding these problems for some specific types of valves.

Control Valves. All control valve stations should be designed with the valve stem in the vertical upright position and a minimum of three diameters of straight pipe both upstream and downstream of the control valve in order to reduce the turbulence entering and leaving the valve and to provide space for removal of the flange studs or bolts. Where applicable this straight pipe will include the usual reduction in pipe size required to match the control valve size. Space must be provided for flange stud bolt removal where control valve bodies are designed for through-bolt installation.

Butterfly Valves. Butterfly valves should be provided with a minimum of five diameters of straight pipe upstream of the valve, and if this requirement has been met, the valve stem and operator may be oriented in the position best suited for operation and maintenance. When a butterfly valve is preceded by an elbow and this straight-pipe requirement cannot be met, the valve stem must be oriented in the same plane as the elbow, that is, if the elbow is in the vertical plane the valve stem must also be in the vertical plane. This recommendation is based on the fact that the velocity profile of the discharge of an elbow is not symmetrical. The result can be fluid dynamic torques that are twice the magnitude of those found for a valve with a straight run of pipe upstream. The resultant eccentric forces applied to valve disc produces excessive vibration and disc flutter, which eventually may completely destroy the valve.

Check Valves. The preferred installation of any check valve is in a horizontal continuously flooded run of pipe with cap up; however, swing check valves will function properly in vertical runs of pipe with the flow up.

Experience has indicated that check valves are highly susceptible to chattering due to upstream turbulence caused by pipe fittings, and the designer should provide upstream straight pipe in accordance with the valve manufacturer's recommendations. However, where this information is not available, the preliminary design should include a minimum of five diameters of straight pipe upstream of all check valves and in particular tilting disc check valves. In addition the designer should be aware of the fact that this requirement can be as much as 10 diameters of straight pipe depending on the type of valve and the manufacturer.

Safety-Relief Valves. The arrangement for installation of safety and relief valves is very critical and involves the actual location of the valve itself, the design of the vent stack, and the design of any associated drain piping. The designer should adhere to the valve manufacturer's recommendations and the following guidelines; however, these guidelines relate to the power industry and may be used elsewhere, as applicable.

Valve Location

- All relief valves must be in the vertical upright position and fitting bound to the top of a horizontal run of pipe, the pressure source, and must not be located less than one nominal header diameter from any buttweld.
- A safety valve inlet connection in a high-velocity steam line should be located at least 8 to 10 nominal header diameters downstream of any bend in the header to minimize the possibility of acoustically induced vibration. In addition, it should be at least 8 to 10 nominal header diameters either upstream or downstream of any diverging or converging tee or wye fitting.
- No other header branch penetration, for any purpose, should be made in the same circumferential cross section containing the safety valve inlet nozzle.
- Where more than one safety valve or a service branch are to be installed in the same header run, a minimum distance of 24 in or three times the sum of the nozzle inside radii, whichever is greater, shall be provided between the nozzles.
- Where more than two safety valves are located in the same header run, the spacing between valves should be varied such that the distance between two adjacent valves differs by at least an inlet nozzle diameter.

Open Discharge. Open-discharge safety valve installations (see Fig. B3.8) should be in accordance with the following guidelines:

- Open-vent stack diameters shall be the calculated minimum flow diameter required for discharge venting without blowback, except as required to accommodate the movement of the relief valve discharge from the cold to hot position such that the outlet pipe will be centered in the vent stack in the hot position.

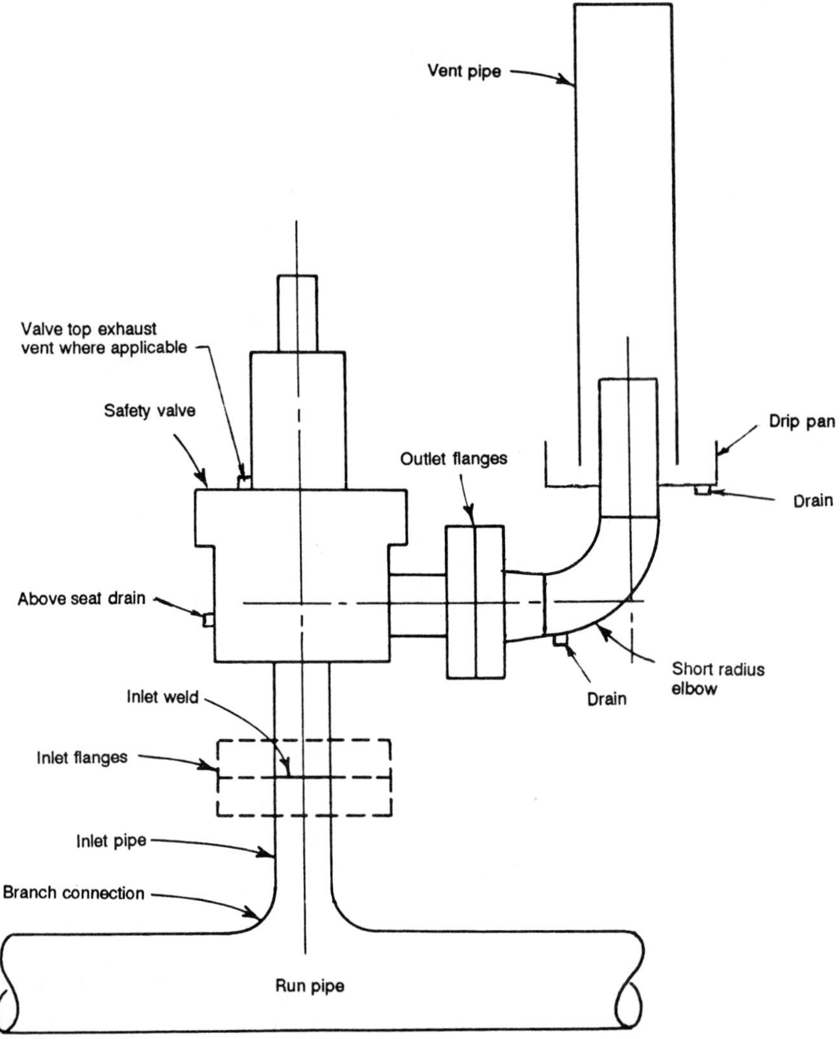

FIGURE B3.8 Safety valve installation (open-discharge system). (*ASME B31.1–1989*)

- The vent stack entry diameter shall be maintained throughout the length of the vent stack; enlarged entry spools, later reduced in size to the calculated minimum flow diameter, are not acceptable.
- The relief valve outlet shall consist of the mating flange and a fitting bound short-radius elbow in order to keep the moment and forces imposed on the valve body to a minimum.
- Vent stacks should be routed, where possible, to provide a straight stack of minimum length. Where offsets or changes in direction are unavoidable, it is desirable to limit the change in direction to 30° or less; however, it could be more. The vent stack should terminate a minimum of 7 ft or higher above the roof level.

Closed Discharge. Closed-discharge piping systems (see Fig. B3.9) are those piped continuously from the valve discharge flange to closed receiver, such as a condenser or blowoff tank. This type of system is required for feedwater heater shell side relief valves to provide protection against the effects of tube rupture, and may be used in other applications. Other than the normal considerations for designing pipe, there are no specific guidelines for the design of closed systems.

Drains. Relief valve and open-vent stack drains are important, in varying degrees, as discussed in the following:

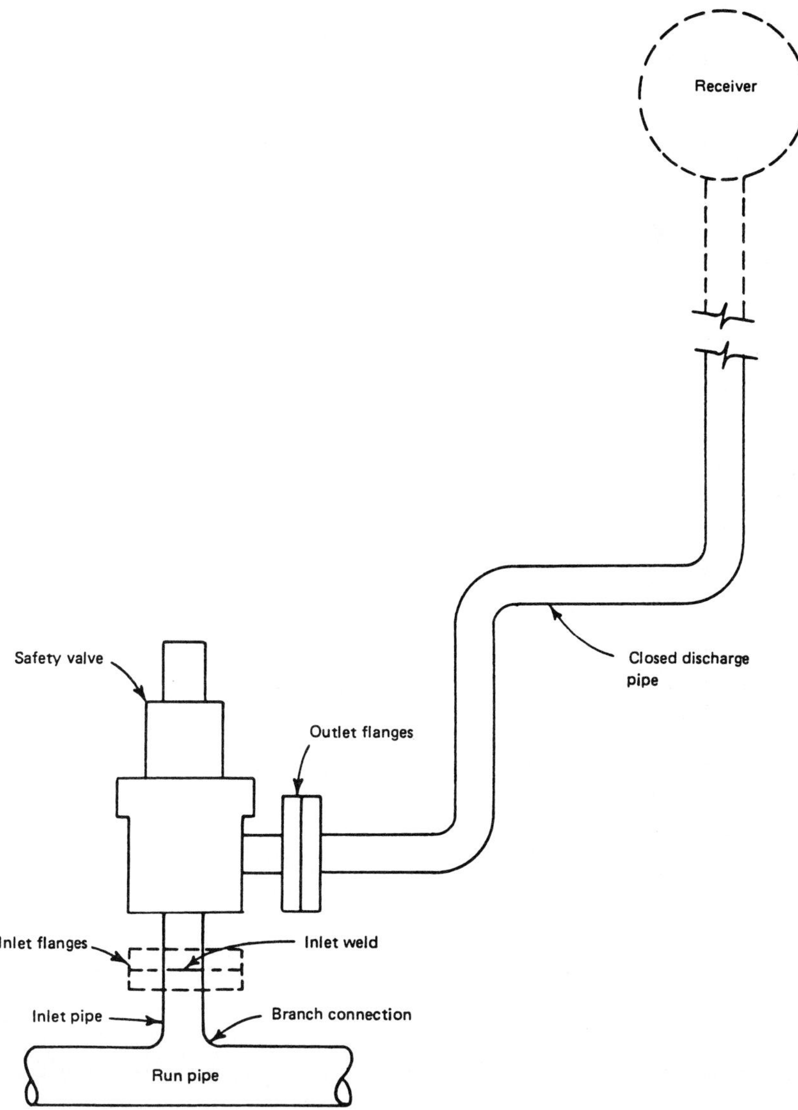

FIGURE B3.9 Safety valve installation (closed-discharge system). (*ASME 31.1–1989*)

- The discharge elbow and above seat body drain points are the most critical for safe valve operation. These drains should be collected into a common, closed drainage system and routed to a point where the drain can safely blow to atmosphere. This system must be sloped continuously downward and stress analyzed to assure that no strain is imposed on the valve body.
- Some relief valves now incorporate a relatively large valve top vent connection that is pressurized when the valve blows. This connection may be piped into the combined discharge elbow and valve body drain system and continued, at full vent pipe size, to the point of the drain discharge.
- The open-discharge vent stack drip pan drain connection is of the least importance and is only intended to carry away any rainfall entering the stack and the residual condensate from the stack following a steam blow.

Piping of Centrifugal Pumps

The piping of centrifugal pumps, particularly the suction piping, can seriously affect the operating efficiency and life expectancy of the pump. Poorly designed suction piping can result in the entrainment of air or vapor into the pump and cause cavitation, which displaces liquid from within the pump casing, results in vibration, and throws the pump out of balance. The cavitation alone can result in severe erosion of the impeller, and the out-of-balance condition may set up a slight eccentric shaft rotation, which will eventually wear out the pump bearings and seals requiring a pump shutdown for overhaul. When routing piping at pumps, the designer should follow the manufacturer's recommendations, the Hydraulic Institute Standards, and the following guidelines:

- The suction and discharge piping must be supported independently of the pump such that very little load is transmitted to the pump casing. The designer may consider the use of expansion joints on either the suction or discharge, or both, as necessary. However, the use of expansion joints should be made only when it is unavoidable.
- The suction of any centrifugal pump must be continuously flooded, and the suction piping shall contain no vertical loops or air pockets.
- When a reduction in pipe size is required at the pump suction, provide an eccentric reducer flat side up.
- The suction side elbow in the piping at horizontal double suction pumps may be fitting bound and in the vertical plane with the flow from either above or below the pump.
- When the suction piping is in the horizontal plane, provide a minimum of three to four diameters of straight pipe between the pump suction connection and the first elbow; the eccentric reducer noted above may be included in this straight section.
- Only long-radius elbows are to be used at or adjacent to any pump suction connection.
- All pump suction lines must be designed to accommodate a conical-type temporary strainer.
- A pipe anchor must be provided between any expansion joint or nonrigid coupling and the pump nozzle it is designed to protect.

- When pump flanges are cast iron flat faced, the mating flanges must also be flat faced and the joint made up with full-faced gaskets and common steel bolts—not high-strength bolts.

Refer to the Hydraulic Institute Standards for arrangement of pump piping.

Vents and Drains

During the course of any system physical routing, the designer should provide high-point vent and low-point drain connections for the following purposes:

- The filling of the piping system with water for hydrostatic testing and operation and the evacuation of entrapped air in the process
- The evacuation of all water used for hydrostatic testing and operation during periods of start-up and maintenance

Systems subject to thermal expansion should be reviewed to assure that they can be properly drained in both the hot and cold positions.

Pipe Supports

The design and engineering of pipe support systems is covered in detail in Chap. B5; however, it is the responsibility of the piping designer to give serious consideration to the means of support during the piping layout, and in doing so, many pipe support problems may be either minimized or avoided altogether. For this reason the piping designer should be familiar with the commercially available pipe support components and their application. Piping should be routed such that the support designer can make use of the surrounding structure to provide logical points of support, anchors, guides, or restraints, with ample space for the appropriate hardware. Banks of parallel pipelines at different elevations should be staggered horizontally and spaced sufficiently apart to permit independent pipe supports for each line. Piping on pipe racks should be routed using bottom-of-pipe (BOP) elevations. The piping designer should work closely with the structural engineer in the spacing of the pipe rack supports and the method of intermediate support in order to prevent pipe sagging.

Insulation

The engineering and selection of thermal insulation materials is covered in Chap. B7, and the piping designer should be familiar with these requirements and specifically with the thickness of insulation for any given system. The location and spacing of piping systems must provide clearance space between the insulation of one pipe and any adjacent pipe and/or other possible interference such as structural steel. The piping designer should also recognize that in some applications insulation may not be required for the prevention of heat loss but will be needed for personnel protection, and the spacing and clearances should be adjusted accordingly.

Heat Tracing

The designer must provide the space and clearances for either electric or steam heat tracing when routing the primary system pipe. The detail design requirements for these systems are covered in Chap. B6.

Operability, Maintenance, Safety, and Accessibility

Operability, maintenance, safety, and accessibility are very dependent upon each other, and certainly if any given piping component is accessible, it would also be assumed to be operable and maintainable; however, maintenance may require additional space for dismantling the component as noted elsewhere in this chapter. It is the responsibility of the piping designer to design a piping arrangement that best satisfies all of these, and other, requirements with the least total cost, that is, the least amount of pipe and fittings and pipe supports.

Operability, from the standpoint of the operating personnel, means being able to perform daily functions in an efficient manner. This is done with consideration for the frequency of operation and the degree of physical effort required to perform it. The designer cannot make every valve and instrument ideally accessible but will concentrate on those requiring the most frequent operation. To do so, he or she must consult with the system engineer and the operating personnel, when possible. In general, an operable valve or instrument is one that can be readily reached when standing at grade or on an elevated floor or platform provided for that purpose. The position of the valve handwheel should be such that the force necessary to operate it can be applied without strain or undue contortions or interference from valves, lines, or other equipment. It is recognized that plant operating personnel will occasionally have to reach for a drain from a kneeling position or a vent valve from a ladder, but these are infrequent operations and as such can be tolerated.

Ease of maintenance actually begins with the development of the plant arrangement and equipment locations by providing sufficient space around each piece of equipment not only for the maintenance of the machinery alone but also space for the pipe and the maintenance of the components in it. These space allocations should include the pull spaces, laydown spaces, and rotor and tube removal spaces for the dismantling of all pieces of equipment. The engineering of the system P&IDs will indicate the need for maintenance facilities in the form of bypasses and block valves that would permit certain pieces of equipment or components to be worked on while the system is operating, or at least with a minimum of downtime. However, it is then up to the designer to design these facilities into the system and to provide the accessibility necessary to accomplish that maintenance including the provision for any lifting gear such as cranes, davits, monorails, and hoists.

There are numerous national, state, and local codes and standards relating to safety, the most notable of which is the Occupational Safety and Health Act of 1970 (OSHA), which became the law of the land on April 28, 1971. Several thousand specific safety and health standards are being enforced under OSHA. These standards have been selected from the key safety standards developed by the American National Standards Institute (ANSI); the American Society for Mechanical Engineers (ASME); and the American Society for Testing and Materials (ASTM); and others, such as the American Water Works Association (AWWA); the American Petroleum Institute (API); and the National Fire Protection Association (NFPA).

Stairs, platforms, ladders, aisles, means of egress aisleways, and minimum headroom allowances designed in accordance with OSHA will provide a safe place to work. From this point on it is up to the piping designer to place equipment, valves, and other piping components in such positions that they do not create hazards. These hazards could include any piping components that presented themselves as "head knockers," "knee knockers," or trippers. The most common cause of these problems are valve stems, and common sense would tell us to place a valve in a horizontal run of pipe with the stem vertical, wherever possible. When this cannot be done, the designer should ascertain that the stem does not project into an access area and become a hazard. The designer should make every effort to keep such projections out of heights of 4½ ft to 6 ft, or specifically at face level. Steam system valves should never be placed at face level in the horizontal position since a packing gland leak may blow steam into the face of an operator; in the event this were a superheated steam leak, the vapor would not be visible. However, this may be too restrictive, and it is not meant to rule out any perfectly safe arrangements of valves at face level if:

- They are outside the limit of a platform.
- They are a part of a manifold of valves all projecting about the same distance and with adequate access space in front of them.
- It is an isolated valve guarded by an adjacent pipe or structural steel.

Accessibility has already been discussed at length in terms of space and the normal platforms and stairways provided in any plant; however, the designer should also review his or her layout and determine if there is a need for any miscellaneous platforms for access to a remotely located valve or component.

Aesthetics

The typical process piping arrangement is not meant to be a thing of beauty but a practical solution to an engineering problem, and the piping designer must avoid increasing the cost of the system purely for aesthetic reasons. He or she should, however, strive for an overall plant piping arrangement that is pleasing to the eye and economical by use of the following guidelines:

- Preplan the overall piping layout such that large concentrations of pipe are routed in neat and orderly pipe chases or on pipe racks. Such piping should be routed parallel to one another with groups of pipes at the same elevation. Branches from these lines should be from either the top or bottom of the main in order to avoid interference with an adjacent parallel line.
- The individual piping arrangements at groups of identical pieces of equipment should be alike not only for appearance but also, when possible, for ease and familiarity in operation.

SPECIFIC SYSTEM CONSIDERATIONS

The power industry, through its many years of experience, has found that piping arrangement and layout can influence the functionality of a piping system. This section will present specific system guidelines and considerations that will enable the piping designer to reduce that influence to a minimum.

Main Steam and Hot and Cold Reheat

In any power plant, be it a base-loaded electric power generation station or an industrial facility power plant, the main steam system is the backbone of the installation since it ties together the two most important and costliest pieces of equipment—the steam generator and the turbine—and is usually the first system designed, giving it the preference in space allocation and routing. Incorporating the recommendations of References 1 and 2, and the following guidelines will enable this system to perform its intended function:

- All piping in this service should be sloped down, in the direction of flow, a minimum of ⅛ in/ft.
- The final design of the main steam and hot reheat lines should be reviewed, with consideration for thermal growth, to determine the location of any necessary low-point drains and to assure that the system can be completely drained in both the hot and cold positions. When these lines are split into more than one branch into the turbine, each branch should be reviewed for low points. Provide a drain connection in each main steam branch just before the turbine stop valve. All drain lines and drain valve ports should have an inside diameter of not less than 1 in to prevent plugging. Main steam piping drains should not be manifolded together with any other drains from the boiler. In addition this review should assure that no condensate can collect in any undrained portion of the system during shutdown.
- Provide a drain pot at the low point of each cold reheat line, which should be fabricated from 6-in, or larger, diameter pipe and be no longer than is required to install the level-sensing devices. Each pot should be provided with a drain line a nominal 2-in minimum size and a full-sized, full-ported automatic power-operated drain valve. Each drain pot should be provided with a minimum of two level-sensing devices.
- Steam lines that are fitted with restricting devices such as orifices or flow nozzles should be adequately drained upstream of the device.
- Valves in all steam services should be installed with the valve stem in the vertical upright position to prevent the entrapment of fluid in the bonnet. Where this is not practical, the stem may be positioned between the vertical and horizontal positions, but in no case below horizontal.
- Main steam safety relief valves should be fitting bound to the main steam headers.
- Sufficient space should be provided around any steam line to allow for insulation, pipe supports and anchors, thermal growth, machine welding, and maintenance repairs and replacements.

Turbine Extraction Steam

Most steam turbines are provided with one or more low- to intermediate-pressure steam extraction points either for boiler feedwater heating or for industrial process service and heating. These systems are extremely critical, particularly from the standpoint of turbine water damage and must be designed in accordance with Refs. 1 and 2 and the following guidelines:

- The routing should be as short and as direct as possible with consideration for thermal growth and piping flexibility.
- Extraction steam piping should be sloped down, in the direction of flow, a minimum of ⅛ in/ft.
- Bleeder trip valves must be located as close to the turbine extraction point as possible, while at the same time keeping the total volume of the system within the turbine manufacturer's recommendations.
- When extraction steam piping is routed through the condenser neck, an expansion joint must be provided in each line and located at the turbine nozzle. The bleeder trip valves in these lines must be located just outside the condenser neck.
- A drain should be located at the low point in the extraction pipe between the turbine and block valve and routed separately to the condenser. A power operated drain valve should be installed in this line that opens automatically upon the closure of the block valve in the extraction pipe.
- There should be no bypasses around the extraction line shutoff or nonreturn valves.
- Unavoidable vertical loops which create low points in the piping downstream of the bleeder trip valves must be provided with continuously drained drip pots.
- Provide a minimum of five diameters of straight pipe downstream of all bleeder trip valves.
- Provide maintenance access to all bleeder trip valves including any miscellaneous platforms, if needed.

Condensate

The condensate collection system from the condenser hotwell presents a unique set of parameters since we are dealing with water at slightly elevated temperatures and at a vacuum pressure. These conditions make the condensate pump suction piping susceptible to flashing and cavitation. The following guidelines apply to the design of condensate pump suction and discharge piping:

- Where two or more condensate pumps are used, the individual runs to each pump must be similar, and if a suction manifold or header is used, the individual pump suction lines from that manifold or header must be similar.
- When the manifold or header is larger than the pump suction size, the manifold or header should be made up of full-sized tees and eccentric reducers, flat side up.
- Each individual pump suction run must be sloped down a minimum of ⅛ in/ft toward the pump and be self-venting back to the condenser.
- Provide a minimum of three to four diameters of straight pipe in the pump suction line, and, in addition, these lines must be fitted with expansion joints and start-up strainers.
- The condensate pump discharge check valve must be located below the hotwell water level and be continuously flooded.

- The discharge header outlet should not be located between the pump discharge connections to the header to avoid a counterflow condition.
- The condensate pump recirculation control valve should be located at the condenser nozzle.

Feedwater

The boiler feedwater pumps normally take a suction from the deaerator storage tank and discharge to the feedwater heaters and then supply the boiler. Here too we have to deal with the possibility of flashing fluid and must assure that the deaerator storage tank is located at an elevation that will provide sufficient net positive suction head (NPSH) at the pump. The following guidelines will apply to the design of this piping:

- The pump suction piping from the deaerator storage tank should drop vertically avoiding any long horizontal runs of pipe. If short horizontal runs are unavoidable, they should be angled vertically down.
- A minimum of three diameters of straight pipe is required at the pump suction. The pump suction strainer may be located in this run of pipe.
- If a reducer is required at the pump suction, it must be eccentric and installed with the flat side up.
- The feed pump discharge swing check valves should be located in horizontal runs of pipe only.
- The feed pump recirculation line control valve should preferably be located at the deaerator storage tank. Horizontal runs are to be avoided in this line at the tank. If the control valve is located in a branch from the pump discharge, the line downstream of the valve must be continuously flooded.

Turbine Drains

This system consists of the turbine casing drains from the turbine to the condenser, a drain collection manifold at the condenser, or other drain vessel as indicated on the system P&ID. References 1 and 2 and the following guidelines should be used in the physical routing of these drains.

- Turbine drain lines and valve ports should be sized for the maximum amount of water to be handled under any operating condition, but in no case may they be less than ¾-in inside diameter.
- Drain lines should be designed for both hot and cold conditions and should slope continuously downward in the direction of flow. Flexibility loops, when required, should be in the plane of the slope or in vertical downward runs.
- Continuous drain orifices, when used, should be located and designed so that they may be cleaned frequently and will not be susceptible to plugging by debris.
- Steam traps are not satisfactory as the only means of draining critical lines; however, they may be used in parallel with automatically operated drain valves.

- No part of any drain line may be below its terminal point at the condenser, drain collection header, or other drain vessel.
- Only drain lines from piping systems of similar pressure may be routed to a common manifold.
- All drain and manifold connections to the condenser must be above the maximum hotwell water level.
- Drainage from other vessels, such as feedwater heaters, steam jet ejectors, and gland steam condensers, that drain water continuously must not be routed to turbine cycle drain manifolds.
- Drain lines should be connected at a 45° angle to the manifold axial centerline with the drain line discharge pointing toward the condenser. Drain line connections at the manifold should be arranged in descending order of pressure, with the highest-pressure source farthest from the manifold opening at the condenser.
- Drain connections to flash tanks must be above the maximum water level in the tank.
- Drains from the upstream and downstream sides of shutoff valves must not be interconnected.
- Drain lines in exposed areas should be protected from freezing.
- All turbine drain drawings must be reviewed and approved by the turbine supplier.

Heater Drains

The heater drains system consists of the feedwater heater drains from one heater to another at a lower pressure, to a drain tank, or the dump line to the condenser. The designer should use extra care in the design and routing of these drain lines particularly where two phase flow conditions may exist. References 1 and 2 and the following guidelines apply to the design and routing of heater drains piping:

- Drains piping from feedwater heaters without an internal drains cooler must immediately drop vertically to provide as much static head as possible upstream of the heater level control valve. Thereafter any horizontal runs must be sloped down a minimum of ¼ in/ft. in the direction of flow.
- Drains piping from feedwater heaters with an internal drains cooler may be routed horizontally without sloping upon leaving the heater.
- Heater level control valves should be located as close as possible to the receiving vessel, with consideration for ease of access and maintenance.
- The heater drains system arrangements must be coordinated with the system engineer for computer analysis to ensure that single-phase water flow is maintained upstream of the heater level control valves and to determine where downstream velocities may require tees and target plates in lieu of elbows for minimizing erosion.
- Heater drain dump lines should enter the condenser at approximately the horizontal centerline of the tube bundle. This location should be coordinated with the condenser manufacturer who will provide the necessary baffle plates to prevent impingement on the condenser tubes.
- Only long-radius elbows should be used in heater drains piping.

- The use of reducers should be avoided, except at the control valves, which are generally smaller than the line size.

Compressed Air

The compressed-air systems provide service air and instrument air throughout the plant. The following guidelines apply to the design and layout of these systems:

- Refer to the compressor manufacturer's instruction manual for the recommended relative lengths of intake and discharge piping versus compressor revolutions per minute (rev/min).
- The compressed-air system equipment arrangement and piping design should be such that the air receiver is the lowest point in the system and any condensate in the system will drain to the air receiver, particularly during periods of shutdown when large amounts of condensate can form. The point to be made here is to preclude any possibility of condensate draining back to the air compressor, where it could cause extensive damage. The compressor discharge piping should be as short and direct as possible through the aftercooler and into the air receiver. The compressed-air system distribution lines and risers should originate from a separate outlet connection on the air receiver and should be sloped back to the air receiver.
- Compressed air line headers, from the vertical risers, should be sloped down $1/16$ in/ft minimum in the direction of flow and drained at their terminations.
- Individual service branches should be taken off of the top of the headers.

Fire Protection

The fire protection system will usually consist of two or more fire pumps taking a suction from the fire water source with the discharge of each pump independently connected to the underground fire main and as widely separated as possible. The underground fire main loop shall completely encircle the plant and may serve multisites if cross-connected between units. The NFPA codes and the following guidelines may be used to design and lay out the yard fire main loop:

- Locate the yard fire main such that all fire hydrants will be a minimum of 50 ft from any building or structure.
- The underground fire main shall be sectionalized in accordance with NFPA code using post indicator valves.
- Post indicator valves shall be provided on each side of any fire pump discharge connection into the fire main loop.
- All fire protection system branches from the yard fire main loop shall be provided with a shutoff valve located not less than 40 ft from the building or structure being served.
- Two-way fire hydrants with individual curb boxes should be provided at every 250- to 300-ft intervals along the yard fire main loop.

Water fire-extinguishing systems within any building may consist of automatic sprinkler systems, spray systems, deluge systems, and hose stations as determined by the project engineering group. The following guidelines shall apply to the design of these systems:

- Large areas, such as below the turbine operating floor, should be divided into sectors each served by an individual branch from the yard fire main loop.
- Each sector should be controlled by an exterior post indicator valve and an alarm check valve or automatic valve located inside the building.
- The maximum area served by any one alarm check valve or automatic sprinkler valve shall not exceed 25,000 ft^2.
- The maximum number of sprinkler heads in any sector shall not exceed 275.
- Provide automatic wet sprinkler systems in the area of the lube oil system below the turbine operating floor and in the ceiling of the clean and dirty lube oil storage tank room.
- Separate water spray systems should be provided in the area of the lube oil system, in addition to the wet sprinkler system noted above, and in the area of the hydrogen seal system.
- Standpipes and hose stations should be provided in accordance with the NFPA code as a complement to the automatic suppression systems noted above.
- The hose stations on any given floor should be fed from above to avoid creating a series of unvented high points.

Cooling Water Systems

There are several types of cooling water systems utilized today in the engineering and design of power generation, petrochem, and industrial plants. The most common system in use for many years in power generation was the direct use of the water from the nearby river, bay, or ocean. In this system a water intake structure would be located along the shoreline and would include as a minimum two or more circulating water pumps, piping, both fixed and traveling intake screens and the necessary crane facilities for the removal, replacement, and maintenance of the pumps and their motors. The intake screens are provided to prevent fish, crabs, and other debris from entering and damaging the pumps. In addition to this main cooling water system, there may be one or more service water systems for the cooling of other miscellaneous equipment throughout the plant. The following guidelines would apply in the design and routing of these systems:

- Where butterfly valves are used follow the guidelines provided for valves.
- Any given heat exchanger inlet and outlet valves should be located close together for ease in balancing the system.
- Avoid unnecessary vertical loops in any closed cooling water system. This type of system will usually include an expansion tank, which should be located at or above the highest point in the system and the outlet from this tank should be piped directly to the pump suction.
- For piping at centrifugal pumps, follow the guidelines provided for piping of centrifugal pumps.
- Consult the Hydraulic Institute Standards and the pump manufacturer's guidelines for layout and arrangement of deep-well type of pumps.
- Since the temperature in these systems is not high and does not vary widely,

the need for piping offsets to accommodate thermal expansion and/or contraction is not of paramount importance.

APPLICATION OF COMPUTER-AIDED DESIGN (CAD) TO PIPING LAYOUT

The piping industry today is very diverse in its use of computer-aided design (CAD). This diversity is shown by the various levels of sophistication of the CAD applications in use by different segments of the industry. Even within the same company, the sophistication of CAD usage can vary widely from discipline to discipline, department to department. This diversity ranges from a surprisingly large portion of the industry in which there is little or no use of CAD to a few who claim to be approaching a "paperless" office. Between these two extremes, most of the industry appears to be using CAD as the initials originally meant: *C*omputer-*A*ided *D*rafting. In this sense, CAD becomes an electronic pencil, not necessarily a design tool.

The meaning of the term *CAD* has evolved as quickly as the technology itself. From its original use as an acronym for computer-aided drafting, it has spawned a whole family of related acronyms: CADD, CAE, CAM, CIM, CAA, and so on. Many of these terms have been applied when describing the design and layout of piping systems. In the minds of many people in our industry, CAD and its related acronyms are still envisioned as simply automated drafting, where CAD is basically the substitution of drawing boards with CAD terminals. While computer-aided drafting represents a significant portion of the application CAD to piping layout, this is changing rapidly. In this section applications beyond simple drafting will be discussed. Therefore the use of the acronym "CAD" will mean *computer-aided design* and will refer to both design and drafting activities related to piping layout.

The entire field of design automation, including CAD, is changing so rapidly that it would be of little value to make recommendations regarding specific hardware and software systems. What may be the best or most cost-effective system today may be out of the picture tomorrow. However, there are some fundamental issues associated with the selection and implementation of a CAD system which should be considered, regardless of the specific supplier of hardware and software.

Computer-Aided Drafting

Currently, as indicated previously, the most significant use of CAD for piping layout is for drafting. Many software systems exist which can function on nearly every type of computer hardware available, including mainframe computers, minicomputers, workstations, and personal computers. Today, the use of CAD for two-dimensional drafting is dominated by CAD software for personal computers. In selecting a system for producing piping drawings, there are several issues which must be considered, regardless of the hardware to be used.

User-Definable Symbols and Menus. Any CAD software, if it is to be of any long-term benefit, must provide the capability to define its own drafting symbols and menus (e.g., tablet, on-screen, etc.) for selecting these symbols. Since piping drawings make extensive use of symbology, defining symbols is of critical importance for significantly increasing drafting productivity. This capability allows the user organization to create and manage libraries of their own symbols, standard details, standard notes, and so on, which can be easily and automatically included in any drawing.

Use of Standard Hardware. Traditionally, many CAD systems were provided by the vendor as a "turnkey system" that included both hardware and software. In these cases, the CAD software was designed to operate specifically on the hardware provided by the vendor. Today, however, many vendors have decoupled the hardware and software, which allows the software to run on a number of hardware platforms. In fact, most of the major providers of CAD software for drafting provide only software, with the users acquiring the hardware independently. This is particularly true for the personal-computer-based CAD systems. By selecting software which can function on a number of types of hardware, the user has the flexibility to more fully take advantage of rapid changes in the hardware market, that is, decreasing prices with increasing performance. If the CAD software can function only on the hardware from one specific vendor, then the user must rely on the hardware vendor to keep pace with the rest of the industry. Typically, the CAD systems today are more often tied to an operating system—for example, DOS, OS/2, Unix—with many vendors providing versions which run with many operating systems.

Availability of Third-Party Software. Certainly not every user can have the luxury of developing his or her own software, particularly beyond the development of symbol libraries and menus. Therefore, before selecting a CAD system, the user should determine how much applications software is available from the vendor or from third parties. For piping layout, the most important applications to look for are those intended for generating orthographic piping drawings and piping isometric drawings. Applications software, specific to piping layout, can significantly increase the productivity of the application of CAD. If little or no applications software exists for the CAD system under consideration, then the user will likely have to develop his or her own applications software or not realize the full value of the CAD system.

Support of User-Developed Software. In cases where no applications software exists, perhaps due to the uniqueness of the user requirements, then the user needs to ensure that applications software can be developed for the specific CAD system. As a minimum, the system should support developing simple "macro" commands which execute a series of commands in response to a single command. Many systems offer macro languages which offer much of the functionality of general-purpose programming languages. For more sophisticated applications, the system should provide interfaces to software written in other programming languages, such as Fortran or C.

Support for Multiple Users. Piping layout does not usually occur in isolation and must reference design information and drawings from other piping designers as well as other disciplines. Therefore, the CAD system must support this type of activity. The CAD system should provide the capability for a designer to have read-only access to the CAD files of other designers for reference, interference checking, or for use as background information for the piping drawings. Systems which have this capability often refer to it as a *reference file* capability. This allows an individual designer to see the file of another designer, as if it were part of his or her file; however, they cannot change any of the data. For personal-computer-based CAD systems, this requires that they be part of some type of local or wide-area network. Without this capability in the CAD software or for personal computers which are not in a network, the data from other designers must be copied and incorporated into the designer's file. This does not allow the designer to see the "active" data of the other designers. In addition, it also greatly increases the storage requirements since many drawings are duplicated, perhaps numerous times. Most importantly, this introduces a more complicated file management problem, making it more difficult to (a) know which file has the most up-to-date information and (b) ensure that everyone references the current data.

Database Capabilities. In order to utilize the CAD system for more than just drafting requires that the system have the ability to create drawings which, in addition to the drawing graphics, contain (or reference in database) other information which can be extracted from the drawing, such as valve numbers and/or line numbers. With this type of capability, bills of material can be generated automatically from the piping drawings. It is even possible to generate the input to the piping stress analysis program from a piping isometric. However, it should be noted that merely having a basic database capability does not mean that it can effectively be used for extracting data from piping drawings. This is the role of applications software developed specifically for piping which automatically generates and manages this information during the creation of the drawing. In the absence of piping applications software, the designer would be required to key in a significant amount of data while generating the drawing. This not only dramatically decreases the productivity of the drawing production but there is also a high possibility of errors.

Training and Implementation. In the past, in implementing the traditional turnkey CAD systems, much of the cost was in the hardware and software. Today, as the cost of hardware and software continues to decline, the majority of the cost is shifting from hardware and software toward training and support. Therefore, the costs associated with the training and implementation of a CAD system, even for two-dimensional drafting, should not be underestimated. In fact, experience has shown that the relative effectiveness of a CAD system is directly related to the amount of training and support the individual users receive.

The precise method of implementing a CAD system is dependent on the company's current organization and method of executing work. Centralized CAD groups working multiple shifts were often the norm with the installation of the large turnkey systems. Now, however, as the cost continues to decrease and the piping design industry in general increases its sophistication in the use of CAD, more effective uses of CAD are being made by placing the workstations right in the piping design groups. Many companies started by training their drafting personnel. But again, experience has shown that even more effective use can be made of the CAD system by training senior-level piping designers. Instead of creating sketches which are then passed onto a drafter, the designer, using the CAD system and piping layout applications software, can create an electronic sketch which is very nearly a finished drawing, leaving very little left to do in the way of drafting. This approach can greatly increase the productivity of the whole design and drafting cycle.

Computer-Aided Design

While the use of CAD for two-dimensional drafting in support of piping layout can provide a number of productivity benefits, there are inherent limitations as to overall benefits to the entire design, fabricate, and construct cycle. While providing benefits in producing the piping drawings (e.g., drafting quality, drafting productivity) and possibly in generating bills of materials, it offers little in the way of improving design productivity. Also, the cost and effort required for interference detection is only marginally improved. Thus two-dimensional drafting, while improving drafting quality and productivity, does little for improving design quality and productivity.

The use of three-dimensional (3D) modeling offers a significant step forward in improving piping design productivity and quality. Systems for 3D piping modeling have existed since the 1970s in a variety of forms. The early systems were geared primarily toward interference detection and materials management and really were not used as design tools per se. Today, a number of systems exist which address the entire piping design cycle. In selecting one of these systems, all of the issues which applied to computer-aided drafting apply to 3D piping design systems. However, there are a number of other issues which must also be considered.

Interactive Design. In order to truly improve piping design productivity, the software should provide the capability to interactively lay out the piping systems directly in the 3D computer model. This allows the piping designer to sit at the graphics workstation, viewing the 3D model and directly add new piping or modify existing piping. Without this capability, the system can provide other benefits, such as in interference detection, but will not necessarily improve the piping design productivity. In fact, without interactive design capabilities, another step is added to the process for entering the data into the 3D model from the 2D design drawings. Many CAD systems provide interactive 3D modeling capabilities, but these are not usually sufficient for 3D piping design. Applications software, specifically aimed at piping design, is required to realize gains in design productivity. Without this type of applications software, 3D modeling is probably only effective for early conceptual design and perhaps detailed modeling of very specific problem areas.

Interference Detection. A major advantage of using 3D computer modeling for piping layout is the ability to automatically check for interferences. This alone can provide a significant improvement in design quality by making it possible to issue a "provable" design, that is, an interference-free design. Many CAD systems, particularly those originally developed for mechanical design, can detect interferences between two 3D objects, but this is not sufficient for checking plant models for interferences in a production environment. As a minimum, the software should provide the following capabilities:

- The software should be able to check interferences for all or part of the plant in a batch mode. This check should include not only piping but all other disciplines as well. The software should have a method of reporting interferences which is easy to interpret and makes it possible to quickly locate the interferences in a large and complex model. Some systems also offer the capability to check for interferences as the piping is being designed. This is especially useful when designing pipe in very congested areas.

- The software should not only check for "hard" interferences, that is, metal to metal, but also for "soft" interferences, such as personnel access areas, equipment removal spaces, insulation, and construction access.
- The software should provide some capability for managing interference resolution over the life of the project. This includes the ability to suppress certain types of interferences and flagging certain specific interferences as acceptable which will not be reported in the future.

Drawing Generation. In order to fully realize the benefits of 3D modeling, the system should provide the capability to automatically or semiautomatically produce the piping drawings, both orthographic and isometric, directly from the 3D piping model. These drawings should be generated in the form of 2D CAD drawings so that they can be managed along with the 2D drawings not generated from the 3D model. For orthographic drawings, the system should be able to represent the piping in the format required by the user, for example, single line; it should be able to automatically remove hidden lines from the model; and it should have some basic capability to automatically place annotation, such as component callouts, into the drawing. For piping isometrics, it is not unreasonable to expect the software to generate the piping isometric completely automatically.

Bills of Material. As a minimum, the software should have the capability to produce a bill of materials for any of the components included in the model. If a user requires stringent control of piping materials, the system should also provide a piping materials control system or an interface to a third-party materials control system.

Interface to Other Systems. Since many disciplines utilize 3D geometry data, the software should have the ability to interface the 3D geometry data with other computer systems. For piping design, this would include the piping stress analysis systems. This could also include interfaces to fabrication equipment, such as numerically controlled pipe bending systems.

Design Review. The use of 3D modeling for piping design impacts the design process in a number of ways. First, the design evolves in the 3D model—not on the drawings, as in the case of 2D design. The drawings are not usually produced until the design is completed. This means that the drawings cannot be used as a means of reviewing the design while the design is in progress. Second, since in some companies the use of 3D design has virtually eliminated the plastic model, the plastic model is also no longer available as a design review tool. Thus the 3D software system should provide, either directly or through an interface, the means of reviewing the 3D computer model on a high-performance graphics terminal. These types of systems provide the capability to "walk through" solid shaded models in real-time for the purposes of design review.

Training and Implementation. Once again, the issues related to computer-aided drafting apply here as well. The primary difference is one of degree. Systems for 3D computer modeling of piping require more training, more support, and a longer learning curve. Also, these types of systems are more pervasive than simple 2D CAD drafting in that they require a higher level of integration between disciplines and departments and thus a higher level of management attention and support. For these systems to be effective, it is imperative that the senior-level design personnel are trained in the use of the system and can use it effectively for piping layout.

Conclusion

Computer-aided drafting and computer-aided design have been used effectively and productively for piping design for a number of years. One of the most important lessons learned from the application of CAD to piping layout, particularly the use of 3D modeling, is that design firms are no longer tied to the same design process and design documentation as when the design was performed manually. The use of 3D piping design provides a number of opportunities for improving the way in which plant design is performed, over and above simply the increase in design productivity. In fact, experience has shown that force fitting 3D piping design into a project organization and design process geared to manual design actually leads to some inefficiencies.

There appear to be several factors which are important to the continued effective application of this technology. Perhaps the most important is the fact that being able to effectively apply this type of software requires training—not only for the individual designers and engineers but also for the supervisors, project engineers, and project managers who control the project work. This type of software opens up new possibilities for improving the way project work is performed. But being able to take advantage of these requires that people at all levels of the project understand the software capabilities as well as its limitations.

REFERENCES

1. ANSI/ASME TDP-1-1985 Recommended Practices for the Prevention of Water Damage to Steam Turbines Used for Electric Power Generation (Fossil), American Society of Mechanical Engineers, New York.
2. ANSI/ASME TDP-2-1985 Recommended Practices for the Prevention of Water Damage to Steam Turbines Used for Electric Power Generation (Nuclear), American Society of Mechanical Engineers, New York.

BIBLIOGRAPHY

ASME B31, Code for Pressure Piping, Section B31.1, Power Piping, 1989 ed., American Society of Mechanical Engineers, New York.

ASME Y32, Code for Graphic Symbols, Section Y32.2.3-1949, Graphic Symbols for Pipe Fittings Valves and Piping, American Society of Mechanical Engineers, New York.

Hydraulic Institute Standards, Hydraulic Institute, Cleveland, OH.

NFPA National Fire Protection Codes, vols. 1–11, National Fire Protection Association, Quincy, MA.

OSHA Regulations, Code of Federal Regulations Title 29 Part 1910.

Pipe Fabrication Institute Standards, Pipe Fabrication Institute, Springdale, PA.

CHAPTER B4
STRESS ANALYSIS OF PIPING SYSTEMS

W. S. Sun
Bechtel Corporation
Gaithersburg, Maryland

G. C. Crim, P.E.
Bechtel Corporation
Gaithersburg, Maryland

Piping stress analysis is a discipline which is highly interrelated with piping layout (Chap. B3) and support design (Chap. B5). The layout of the piping system should be performed with the requirements of piping stress and pipe supports in mind (i.e., sufficient flexibility for thermal expansion; proper pipe routing so that simple and economical pipe supports can be constructed; and piping materials and section properties commensurate with the intended service, temperatures, pressures, and anticipated loadings.). If necessary, layout solutions should be iterated until a satisfactory balance between stresses and layout efficiency is achieved. Once the piping layout is finalized, the piping support system must be determined. Possible support locations and types must be iterated until all stress requirements are satisfied, other piping allowables (e.g., nozzle loads, valve accelerations, and piping movements) are met. The piping supports are then designed (Chap. B5) based on the selected locations and types and the applied loads.

This chapter discusses several aspects of piping stress analysis. The discussion is heavily weighted to the stress analysis of piping systems in nuclear power plants, since this type of piping has the most stringent requirements. However, the discussion is also applicable to the piping systems in ships, aircraft, commercial buildings, equipment packages, refrigeration systems, fire protection piping, petroleum refineries, and so on. Each of the above types of piping must meet the requirements of its applicable code.

FAILURE THEORIES, STRESS CATEGORIES, STRESS LIMITS, AND FATIGUE

Failure Theories

The failure theories most commonly used in describing the strength of piping systems are the maximum principal stress theory and the maximum shear stress theory (also known as the *Tresca criterion*).

The maximum principal stress theory forms the basis for piping systems governed by ANSI/ASME B31 and Subsections NC and ND (Class 2 and 3) of Section III of the ASME Boiler and Pressure Vessel Codes. This theory states that yielding in a piping component occurs when the magnitude of any of the three mutually perpendicular principal stresses exceeds the yield strength of the material.

The maximum shear stress theory is more accurate than the maximum stress theory for predicting both yielding and fatigue failure in ductile metals. This maximum shear stress theory forms the basis for piping of Subsection NB (Class 1) of ASME Section III.[1]

The maximum shear stress at a point τ_{max} is defined as one-half of the algebraic difference between the largest and the smallest of the three principal stresses σ_1, σ_2, and σ_3. If $\sigma_1 > \sigma_2 > \sigma_3$ (algebraically), then $\tau_{max} = (\sigma_1 - \sigma_3)/2$. The maximum shear stress theory states that failure of a piping component occurs when the maximum shear stress exceeds the shear stress at the yield point in a tensile test. In the tensile test, at yield, $\sigma_1 = S_y$ (yield stress), $\sigma_2 = \sigma_3 = 0$. So yielding in the component occurs when

$$\tau_{max} = \frac{\sigma_1 - \sigma_3}{2} = \frac{S_y}{2} \qquad \text{(B4.1)}$$

Equation (B4.1) has an unnecessary operation of dividing both sides by 2 before comparing them. For the sake of simplicity, a stress defined as $2\tau_{max}$ and equal to $\sigma_{max} - \sigma_{min}$ of the three principal stresses has been used for Class 1 piping. This stress is called the *equivalent intensity of combined stresses,* or *stress intensity.* Thus the stress intensity S is directly comparable to the tabulated yield stress values S_y from tensile tests with some factor of safety.

Stress Categories

There are various failure modes which could affect a piping system. The piping engineer can provide protection against some of these failure modes by performing stress analysis according to the piping codes. Protection against other failure modes is provided by methods other than stress analysis. For example, protection against brittle fracture is provided by material selection. The piping codes address the following failure modes: excessive plastic deformation, plastic instability or incremental collapse, and high-strain–low-cycle fatigue. Each of these modes of failure is caused by a different kind of stress and loading. It is necessary to place these stresses into different categories and set limits to them.

The major stress categories are primary, secondary, and peak. The limits of these stresses are related to the various failure modes as follows:

1. The primary stress limits are intended to prevent plastic deformation and bursting.
2. The primary plus secondary stress limits are intended to prevent excessive plastic deformation leading to incremental collapse.
3. The peak stress limit is intended to prevent fatigue failure resulting from cyclic loadings.

Primary stresses which are developed by the imposed loading are necessary to satisfy the equilibrium between external and internal forces and moments of the piping system. Primary stresses are not self-limiting. Therefore, if a primary stress exceeds the yield strength of the material through the entire cross section of the piping, then failure can be prevented only by strain hardening in the material. Thermal stresses are never classified as primary stresses. They are placed in both the secondary and peak stress categories.

Secondary stresses are developed by the constraint of displacements of a structure. These displacements can be caused by either thermal expansion or by outwardly imposed restraint and anchor point movements. Under this loading condition, the piping system must satisfy an imposed strain pattern rather than being in equilibrium with imposed forces. Local yielding and minor distortions of the piping system tend to relieve these stresses. Therefore, secondary stresses are self-limiting. Unlike the loading condition of secondary stresses which cause distortion, peak stresses cause no significant distortion. Peak stresses are the highest stresses in the region under consideration and are responsible for causing fatigue failure. Common types of peak stresses are stress concentrations at a discontinuity and thermal gradients through a pipe wall.

Primary stresses may be further divided into general primary membrane stress, local primary membrane stress, and primary bending stress. The reason for this division is that, as will be discussed in the following paragraph, the limit of a primary bending stress can be higher than the limit of a primary membrane stress.

Basic Stress Intensity Limits

The basic stress intensity limits for the stress categories described above are determined by the application of limit design theory together with suitable safety factors.

The piping is assumed to be elastic and perfectly plastic with no strain hardening. When this pipe is in tension, an applied load producing a general primary membrane stress equal to the yield stress of the material S_y results in piping failure. Failure of piping under bending requires that the entire cross section be at this yield stress. This will not occur until the load is increased above the yield moment of the pipe multiplied by a factor known as the "shape factor" of the cross section. The shape factor for a simple rectangular section in bending is 1.5.

When a pipe is under a combination of bending and axial tension, the limit load depends on the ratio between bending and tension. In Fig. B4.1, the limit stress at the outer fiber of a rectangular bar under combined bending and tension is plotted against the average tensile stress across the section. When the average tensile stress P_m is zero, the failure bending stress is $1.5S_y$. When P_m alone is applied (no bending stress P_b), failure stress is yield stress S_y.

It also can be seen in Fig. B4.1 that a design limit of $(2/3)S_y$ for general primary membrane stress P_m and a design limit of S_y for primary membrane-plus-bending stress $P_m + P_b$ provide adequate safety to prevent yielding failure.

For secondary stresses, the allowable stresses are given in terms of a calculated elastic stress range. This stress range can be as high as twice the yield stress. The reason for this high allowable is that a repetitively applied load which initially stresses the pipe into plastic yielding will, after a few cycles, "shake it down" to elastic action.

This statement can be understood by considering a pipe which is strained in tension to a point ϵ_1 somewhat beyond its yield strain as shown in Fig. B4.2. The calculated elastic stress at this point would be equal to the product of the modu-

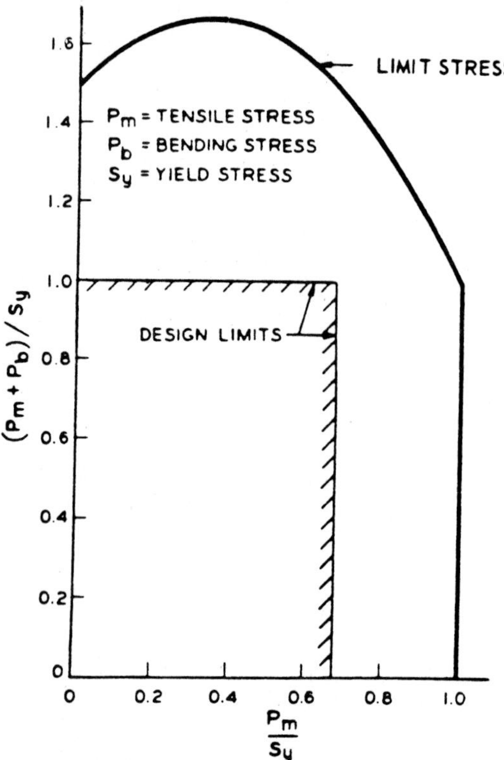

FIGURE B4.1 Limit stress for combined tension and bending (rectangular section). (*ASME, "Criteria."*[1] *Courtesy of ASME.*)

lus of elasticity E and the strain ϵ_1, or $S_1 = E\epsilon_1$. The path $OABC$ is considered as cycling the strain from zero to ϵ_1 (loading) and back to zero (unloading). When the pipe is returned to its original position O, it will retain a residual compressive stress of magnitude $S_1 - S_y$. On each subsequent loading cycle, this residual compression must be overcome before the pipe can go into tension; thus the elastic range has been extended by the value $S_1 - S_y$.

Therefore, the allowable secondary stress range can be as high as $2S_y$ when $S_1 = 2S_y$. When $S_1 > 2S_y$, the pipe yields in compression and all subsequent cycles generate plastic strain EF. For this reason, $2S_y$ is the limiting secondary stress which will shake down to purely elastic action.

Fatigue

As mentioned previously, peak stresses are the highest stresses in a local region and are the source of fatigue failure. The fatigue process may be divided into three stages: crack initiation resulting from the continued cycling of high stress

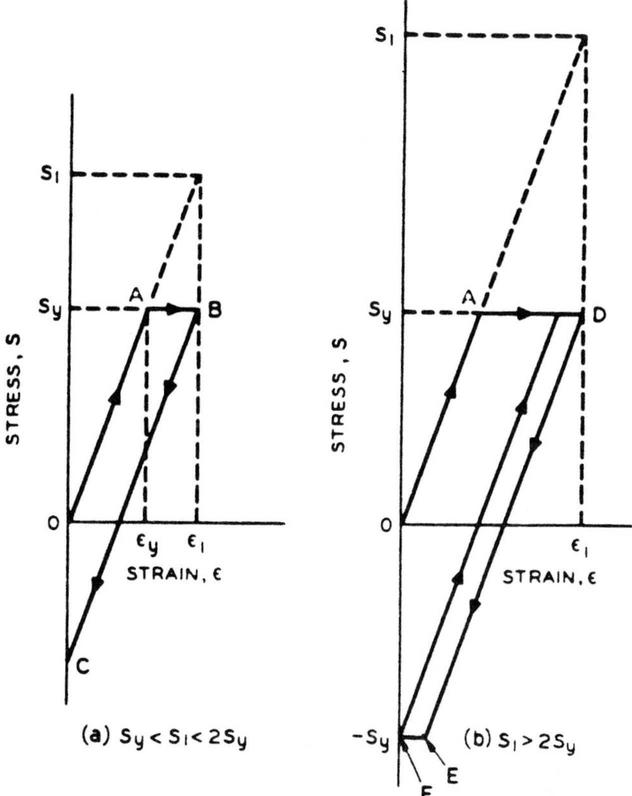

FIGURE B4.2 Strain history beyond yield. (*ASME, "Criteria."*[1] *Courtesy of ASME.*)

concentrations, crack propagation to critical size, and unstable rupture of the remaining section.

Fatigue has long been a major consideration in the design of rotating machinery, where the number of loading cycles is in the millions and can be considered infinite for all practical purposes. This type of fatigue is called *high-cycle fatigue.* High-cycle fatigue involves little or no plastic action. Therefore, it is stress governed. For every material, a fatigue curve, also called the *S–N curve,* can be generated by experimental test[2] which correlates applied stress with the number of cycles-to-failure, as shown in Fig. B4.3. For high-cycle fatigue, the analysis is to determine the endurance limit, which is the stress level which can be applied an infinite number of times without failure.

In piping design, the loading cycles applied seldom exceed 10^5 and are frequently only a few thousand. This type of fatigue is called *low-cycle fatigue.* For low-cycle fatigue, data resulting from experimental tests with stress as the controlled variable are considerably scattered. These undesirable test results are attributable to the fact that in the low-cycle region the applied stress exceeds the

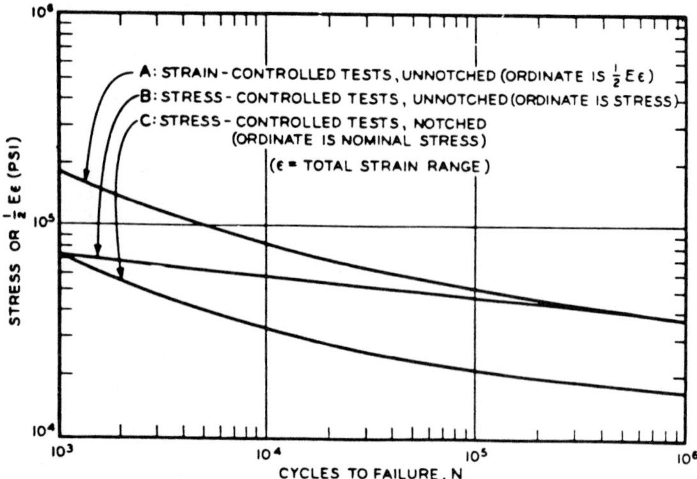

FIGURE B4.3 Typical relationship between stress, strain, and cycles-to-failure. (*ASME, "Criteria."*[1] *Courtesy of ASME.*)

yield strength of the material, thereby causing plastic instability in the test specimen.

However, when strain is used as the controlled variable, the test results in this low-cycle region are consistently reliable and reproducible.

As a matter of convenience, in preparing fatigue curves, the strains in the tests are multiplied by one-half the elastic modulus to give a pseudostress amplitude. This pseudostress is directly comparable to stresses calculated on the assumption of elastic behavior of piping. In piping stress analysis, a stress called the *alternating stress* S_{alt} is defined as one-half of the calculated peak stress. By ensuring that the number of load cycles N which are associated with a specific alternating stress are less than the number allowed in the S–N curve, fatigue failure can be prevented. However, practical service conditions often subject a piping system to alternating stresses of different magnitudes. These changes in magnitude make the direct use of the fatigue curves inapplicable since the curves are based on constant stress amplitude. Therefore, in order to make fatigue curves applicable for piping, it is necessary to take some other approach.

One method of appraising the fatigue failure in piping is to assume that the cumulative damage from fatigue will occur when the cumulative usage factor U equals unity, that is,

$$U = \Sigma U_i = \Sigma \frac{n_i}{N_i} = 1 \qquad (B4.2)$$

where U_i = usage factor at stress level i

n_i = number of cycles operating at stress level i

N_i = number of cycles-to-failure at stress level i as per fatigue curve

CLASSIFICATION OF LOADS, SERVICE LIMITS, AND CODE REQUIREMENTS

Classification of Loads

Primary loads can be divided into two categories based on the duration of loading. The first category is sustained loads. These loads are expected to be present throughout normal plant operation. Typical sustained loads are pressure and weight loads during normal operating conditions. The second category is occasional loads. These loads are present at infrequent intervals during plant operation. Examples of occasional loads are earthquake, wind, and fluid transients such as water hammer and relief valve discharge.

In addition to primary loads, there are expansion loads. Expansion loads are those loads due to displacements of piping. Examples of expansion loads are thermal expansion, seismic anchor movements, thermal anchor movements, and building settlement.

Service Limits

Service levels and their limits are defined for nuclear power plant safety-related piping by the ASME Boiler and Pressure Vessel Code, Section III.[3] They are described below:

1. *Level A service limits:* The piping components or supports must satisfy these sets of limits in the performance of their specified service function. Examples of level A loadings would be operating pressure and weight loadings.
2. *Level B service limits:* The piping component or support must withstand these loadings without damage requiring repair. Examples of level B loadings are fluid transients such as waterhammer and relief valve discharge, and *operating-basis earthquake* (*OBE*), defined as the maximum likely earthquake postulated to occur during plant design life or one-half of the safe shutdown earthquake (see definition below), whichever is higher.
3. *Level C service limits:* The occurrence of stress up to these limits may necessitate the removal of the piping component from service for inspection or repair of damage. An example of level C loading is the combination of fluid transient loads occurring simultaneously with the operating-basis earthquake.
4. *Level D service limits:* These sets of limits permit gross general deformations with some consequent loss of dimensional stability and damage requiring repair, which may require removal of the piping component from service. An example of level D loading would be loading associated with a loss-of-coolant accident or a *safe-shutdown earthquake* (*SSE*), which is defined as the maximum possible earthquake postulated to occur at the site of the plant at any time.

Code Requirements

There are various ASME and ANSI codes which govern the stress analysis of different kinds of pressure piping. These codes contain basic reference data, formulas, and equations necessary for piping design and stress analysis.

Each power plant is committed to a particular edition of a code for different types of piping. For example, the nuclear Class 1, 2, and 3 piping of a power plant may be committed to comply with the ASME Boiler and Pressure Vessel Code, Section III, 1974 Edition, while the nonnuclear piping may be committed to the ANSI B31.1 Power Piping Code, 1973 Edition.

The following sections provide summaries of the ASME and ANSI Codes.

ASME Boiler and Pressure Vessel Code, Section III, Subsection NB.[3] This subsection provides the code requirements of nuclear piping designated as Class 1. The loadings required to be considered for this subsection are the effects of pressure, weight (live and dead loads), thermal expansion and contraction, impact, earthquake, and vibrations. The stress limits which must be met are as follows:

1. *Primary stress intensity:* The primary stress intensity must meet the following requirement:

$$B_1\left(\frac{PD_o}{2t}\right) + B_2\left(\frac{D_o}{2I}\right)M_i \le kS_m \tag{B4.3}$$

where B_1, B_2 = primary stress indices for the specific piping component under investigation
P = design pressure, psi
D_o = outside diameter of pipe, in
t = nominal wall thickness, in
M_i = resultant moment due to a combination of design mechanical loads, in · lb
I = moment of inertia, in^4
kS_m = $1.5S_m$ for service level A; $1.8S_m$ for service level B but not greater than $1.5S_y$; $2.25S_m$ for service level C but not greater than $1.8S_y$; and $3.0S_m$ for service level D but not greater than $2.0S_y$.
S_m = allowable design stress intensity, psi
S_y = yield strength value taken at average fluid temperature under consideration, psi

2. *Primary plus secondary stress intensity range:* The following equations are used to evaluate a stress range as the piping system goes from one service load set (pressure, temperature, and moment) to any other service load set which follows in time. For each specified pair of load sets, the stress range S_n is calculated:

$$S_n = C_1\left(\frac{P_0D_o}{2t}\right) + C_2\left(\frac{D_o}{2I}\right)M_i + C_3E_{ab}|\alpha_aT_a - \alpha_bT_b| \tag{B4.4}$$

where C_1, C_2, C_3 = secondary stress indices for the specific component under consideration
P_0 = range of service pressure, psi
M_i = resultant range of moment, in · lb
E_{ab} = average modulus of elasticity of the two sides of a gross structural discontinuity or material discontinuity at room temperature, psi
α_a, α_b = coefficient of thermal expansion on side a or b of a gross

structural discontinuity or material discontinuity at room temperature, in/(in · °F)

T_a, T_b = range of average temperature on side a or b of gross structural discontinuity or material discontinuity, °F

S_n has the following limit: $S_n \leq 3S_m$.

If this requirement is not met for all pairs of load sets, then the piping component may still be qualified by using the simplified elastic-plastic discontinuity analysis described below; otherwise, the stress analyst may proceed to the fatigue analysis.

3. *Simplified elastic-plastic discontinuity analysis:* If $S_n > 3S_m$ for some pairs of load sets, a simplified elastic-plastic analysis may be performed if thermal stress ratchet is not present. This analysis is required only for the specific load sets which exceeded $3S_m$. The following two equations must be satisfied:

$$S_e = C_2\left(\frac{D_o}{2I}\right)M_i^* \leq 3S_m$$

$$C_1\left(\frac{P_0D_o}{2t}\right) + C_2\left(\frac{D_o}{2I}\right)M_i + C_3'E_{ab}|\alpha_aT_a - \alpha_bT_b| \leq 3S_m \qquad \text{(B4.5)}$$

where S_e = nominal value of expansion stress, psi

M_i^* = resultant range of moments due to thermal expansion and thermal anchor movements, in · lb

M_i = resultant range of moment excluding moments due to thermal expansion and thermal anchor movements, in · lb

C_3' = stress index for the specific component under consideration

For later editions of the code, if $S_n > 3S_m$, the thermal stress ratchet must be evaluated and demonstrated to be satisfactory before a simplified elastic-plastic discontinuity analysis can be done. This ratchet is a function of the $|\Delta T_1|$ (see definition below) range only. The following requirement must be met:

$$|\Delta T_1| \text{ range} \leq \frac{y'S_yC_4}{0.7E\alpha} \qquad \text{(B4.6)}$$

where y' = 3.33, for x = 0.3

= 2.00, for x = 0.5

= 1.20, for x = 0.7

= 0.80, for x = 0.8

$x = PD_o/(2tS_y)$

S_y = yield strength value taken at average fluid temperature under consideration, psi

C_4 = 1.1 for ferritic material and 1.3 for austenitic material

$E\alpha$ = modulus of elasticity E times the mean coefficient of thermal expansion α both at room temperature, psi/°F

P = maximum pressure for the set of conditions under consideration, psi

4. *Peak stress intensity range and fatigue analysis:* For each specified loading condition, peak stress is calculated as follows:

$$S_p = K_1C_1\left(\frac{P_0D_o}{2t}\right) + K_2C_2\left(\frac{D_o}{2I}\right)M_i + \frac{K_3E\alpha|\Delta T_1|}{2(1-\nu)} + K_3C_3E_{ab}|\alpha_aT_a - \alpha_bT_b| + \frac{E\alpha|\Delta T_2|}{1-\nu} \qquad \text{(B4.7)}$$

where K_1, K_2, K_3 = local stress indices for the specific component under consideration

ν = Poisson's ratio of material

$|\Delta T_1|$ = absolute value of range of temperature difference between temperature of outside surface and inside surface of pipe wall, assuming moment generating equivalent linear temperature distribution, °F (see Fig. B4.4)

$|\Delta T_2|$ = absolute value of range of that portion of nonlinear thermal gradient through the wall thickness not included in ΔT_1, °F (see Fig. B4.4)

For each S_p, an alternating stress intensity S_{alt} is determined by

$$S_{alt} = \frac{K_eS_p}{2} \qquad \text{(B4.8)}$$

where $K_e = 1.0$, for $S_n \leq 3S_m$

$= 1.0 + (1 - n)[S_n/(3S_m) - 1]/[n(m - 1)]$, for $3S_m < S_n < 3mS_m$

$= 1/n$, for $S_n \geq 3mS_m$

m, n = material parameters given in Table B4.1

The alternating stress intensities are used to evaluate the cumulative effect of the stress cycles on the piping system. This evaluation is performed as follows:

a. The number of times each stress cycle of type 1, 2, 3, etc., is repeated during the life of the system shall be called n_1, n_2, n_3, and so on. Cycles shall be superimposed such that the maximum possible peak stress ranges are developed.

b. For each type of stress cycle, determine the alternating stress intensity S_{alt}.

c. For each value of S_{alt}, use the applicable design fatigue curve from the code to determine the maximum number of cycles permitted if this were the only cycle occurring. These numbers shall be designated N_1, N_2, N_3, and so on.

d. For each type of stress cycle, calculate the usage factor:

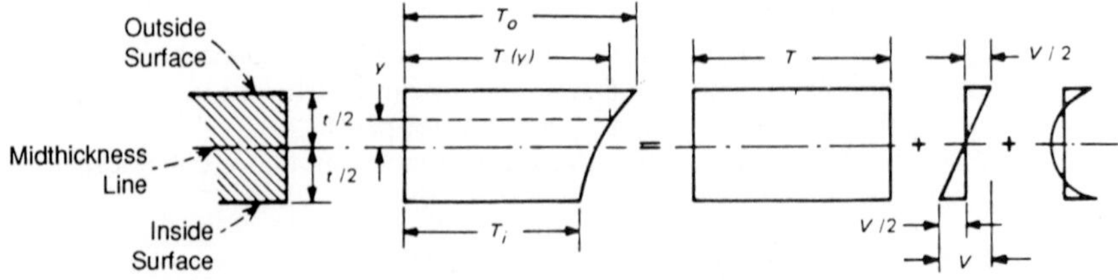

FIGURE B4.4 Decomposition of temperature distribution range. [*Figure NB-3653.2(b)-1, Section III, Division 1, ASME B & PV Code, 1989. Courtesy of ASME.*]

TABLE B4.1 Values of m, n, and T_{max} for Various Classes of Permitted Materials

Materials	m	n	T_{max} °F
Carbon steel	3.0	0.2	700
Low-alloy steel	2.0	0.2	700
Martensitic stainless steel	2.0	0.2	700
Austenitic stainless steel	1.7	0.3	800
Nickel-chromium-iron	1.7	0.3	800
Nickel-copper	1.7	0.3	800

Source: Table NB-3228.5(b)-1, Section III, Division 1, ASME B & PV Code, 1989. (*Courtesy of ASME*)

$$U_1 = \frac{n_1}{N_1}, \; U_2 = \frac{n_2}{N_2}, \; U_3 = \frac{n_3}{N_3}, \; \cdots$$

e. The cumulative usage factor U is the sum of the individual usage factors:

$$U = U_1 + U_2 + U_3 + \cdots$$

ASME Boiler and Pressure Vessel Code, Section III, Subsections NC and ND.[3] These two subsections give the code requirements of nuclear piping designated as Class 2 and Class 3, respectively. The loadings required to be considered for Subsections NC and ND are the effects of pressure, weight, other sustained loads, thermal expansion and contraction, and occasional loads. The stress limits to be met are as follows:

1. *Stresses due to sustained loads:* The calculated stresses due to pressure, weight, and other sustained mechanical loads must meet the allowable $1.5S_h$, that is,

$$\frac{B_1 P D_o}{2t} + \frac{B_2 M_A}{Z} \leq 1.5 S_h \tag{B4.9}$$

where
P = internal design pressure, psi
D_o = outside diameter of pipe, in
t = nominal wall thickness, in
Z = section modulus of pipe, in^3
M_A = resultant moment loading on cross section due to weight and other sustained loads, in · lb
S_h = basic material allowable stress at design temperature, psi

2. *Stresses due to occasional loads:* The calculated stress due to pressure, weight, other sustained loads, and occasional loads must meet the allowables as follows:

$$\frac{B_1 P_{max} D_o}{2t} + \frac{B_2 (M_A + M_B)}{Z} \leq k S_h \tag{B4.10}$$

where M_B = resultant moment loading on cross section due to occasional loads, such as thrusts from relief and safety valves loads from pressure and flow transients, and earthquake, if required. For earthquake, use only one-half the range. Effects of anchor displacement due to earthquake may be excluded if they are included under thermal expansion.

P_{max} = peak pressure, psi

kS_h = $1.8S_h$ for service level B (upset condition) but not greater than $1.5S_y$; $2.25S_h$ for service level C (emergency condition) but not greater than $1.8S_y$; and $3.0S_h$ for service level D (faulted condition) but not greater than $2.0S_y$

S_h = material allowable stress at temperature consistent with the loading under consideration, psi

S_y = material yield strength at temperature consistent with the loading under consideration, psi

3. *Stresses due to thermal expansion:*

a. Thermal expansion stress range must meet the allowable S_A, that is,

$$\frac{iM_C}{Z} \leq S_A \tag{B4.11}$$

where S_A = allowable stress range for expansion stresses = $f(1.25S_c + 0.25S_h)$, psi

f = stress range reduction factor as in Table B4.2

M_C = range of resultant moments due to thermal expansion, in · lb; also include moment effects of anchor displacements due to earthquake if anchor displacement effects were omitted from occasional loadings

S_c = basic material allowable stress at minimum (cold) temperature, psi

S_h = basic material allowable stress at maximum (hot) temperature, psi

i = stress intensification factor

b. If Eq. (B4.11) is not met, the piping may be qualified by meeting the following equation:

$$\frac{PD_o}{4t} + \frac{0.75iM_A}{Z} + \frac{iM_C}{Z} \leq S_h + S_A \tag{B4.12}$$

where $0.75i$ shall not be less than 1.0.

TABLE B4.2 Stress-Range Reduction Factors

Number of equivalent full-temperature cycles, N	f
7,000 and less	1.0
7,000 to 14,000	0.9
14,000 to 22,000	0.8
22,000 to 45,000	0.7
45,000 to 100,000	0.6
100,000 and over	0.5

Source: Table NC-3611.2(e)-1, Section III, Division 1, ASME B & PV Code, 1989. (*Courtesy of ASME*)

4. *Stresses due to nonrepeated anchor movement:* The effect of any single nonrepeated anchor movement (such as building settlement) must meet $3.0S_c$, that is,

$$\frac{iM_D}{Z} \leq 3.0S_c \qquad \text{(B4.13)}$$

where M_D = resultant moment due to any single nonrepeated anchor movement (e.g., predicted building settlement), in · lb

5. The *stress-intensification factor* (*SIF*) is defined as the ratio of the maximum stress intensity to the nominal stress calculated by the ordinary formulas of mechanics. It is used as a safety factor to account for the effect of localized stresses on piping under a repetitive loading. In piping design, this factor is applied to welds, fittings, branch connections, and other piping components where stress concentrations and possible fatigue failure might occur. Usually, experimental methods are used to determine these factors.

   It is recognized that some of the SIFs for the same components are different for different codes. In some cases, different editions of the same code provide different SIFs for a given component. The way that the SIFs are applied to moment loadings is also different for different codes. The B31.1 and ASME Section III codes require that the same SIF be applied to all the three-directional moments while the B31.3, B31.4, B31.5, and B31.8 codes require that different SIFs be applied to the in-plane and out-of-plane moments, with no SIF required for torsion (see Fig. B4.5*a*, and figure note 10).

   Therefore, the stress analyst has to ensure that the appropriate SIFs from the applicable code (i.e., committed code) are used. The formulas for SIFs in the ASME Section III Code (1989 edition) are given in Fig. B4.5*c* for reference.

   Recommended SIFs for some piping components which are not addressed in the code are listed below:

   a. Weldolets or sockolets[4]

   (1) If $r/R > 0.5$,

$$i = \frac{0.9}{h^{2/3}} \qquad \text{where } h = \frac{3.3t}{R} \qquad \text{(B4.14)}$$

where r = mean radius of branch pipe, in
t = wall thickness of run pipe, in
R = mean radius of run pipe, in

   (2) If $r/R \leq 0.5$,

$$i = 1.5\left(\frac{R}{T}\right)^{2/3}\left(\frac{r}{R}\right)^{1/2}\left(\frac{t}{T}\right)\left(\frac{r}{r_p}\right) \qquad \text{(B4.15)}$$

or

$$i = \frac{0.9}{(3.3T/R)^{2/3}} \qquad \text{(B4.16)}$$

whichever is less.

where R = run pipe mean radius, in
T = run pipe wall thickness, in
r = branch pipe mean radius, in
t = branch pipe wall thickness, in
r_p = outer radius of weldolet, in

Item	Description	Flexibility factor n	Stress int. factor i [Notes (9), (10)] In-plane i	Out-of-plane i_o
1	Welding elbow[1,2,3,5,8] or pipe bend	$\frac{1.65}{h}$	$\frac{0.9}{h^{2/3}}$	$\frac{0.75}{h^{2/3}}$
2	Closely spaced miter bend[1,2,3,8] $s < r(1 + \tan\theta)$	$\frac{1.52}{h^{5/6}}$	$\frac{0.9}{h^{2/3}}$	$\frac{0.75}{h^{2/3}}$
3	Widely spaced miter bend,[1,2,4,8] $s \geqq r(1 + \tan\theta)$	$\frac{1.52}{h^{5/6}}$	$\frac{0.9}{h^{2/3}}$	$\frac{0.75}{h^{2/3}}$
4	Welding tee[1,2] per ANSI B16.9	1	$\frac{3}{4}i_o + \frac{1}{4}$	$\frac{0.9}{h^{2/3}}$
5	Reinforced fabricated tee,[1,2,6] with pad or saddle	1	$\frac{3}{4}i_o + \frac{1}{4}$	$\frac{0.9}{h^{2/3}}$
6	Unreinforced fabricated tee[1,2,6]	1	$\frac{3}{4}i_o + \frac{1}{4}$	$\frac{0.9}{h^{2/3}}$
7	Butt-welded joint, reducer, or welding neck flange	1	1.0	
8	Double-welded slip-on flange	1	1.2	
9	Fillet-welded joint, or socket weld flange	1	1.3	
10	Lap-joint flange (with ANSI B16.9 lap-joint stub)	1	1.6	
11	Threaded pipe joint, or threaded flange	1	2.3	
12	Corrugated straight pipe, or corrugated or creased bend[7]	5	2.5	

FIGURE B4.5(*a*) Flexibility factor n and stress-intensification factors i_i and i_o per ANSI/ASME B31.3, B31.4, B31.5, and B31.8 codes.

SIF values for typical weldolet branch connections with $r/R \leq 0.5$ are tabulated in Tables B4.3*a* to B4.3*l*.

b. *Half-couplings (Welding Boss):* For half-couplings with $r/R \leq 0.5$, use the above branch connection Eq. (B4.15) or the unreinforced fabricated tee equation, whichever is less. For half-coupling with $r/R > 0.5$, use the unreinforced fabricated tee formula. Tables B4.4*a* to B4.4*f* give SIFs for commonly used half-coupling configurations. If the half-coupling rating is

Flexibility characteristic h	Sketch
$\frac{tR}{r^2}$	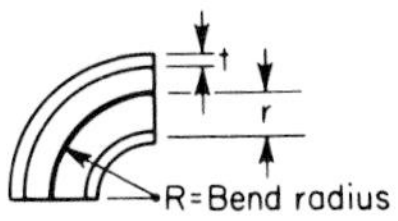 R=Bend radius
$\frac{\cot\theta}{2}\frac{ts}{r^2}$	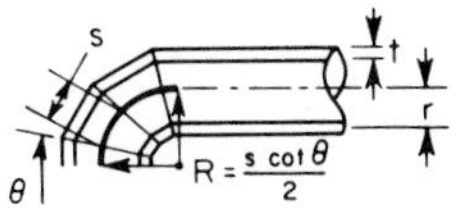 $R=\frac{s\cot\theta}{2}$
$\frac{1+\cot\theta}{2}\frac{t}{r}$	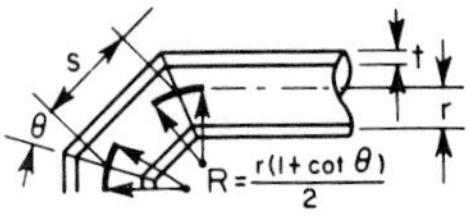 $R=\frac{r(1+\cot\theta)}{2}$
$4.4\frac{t}{r}$	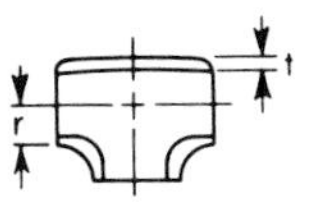
$\frac{(t+\frac{1}{2}T)^{5/2}}{t^{3/2}r}$	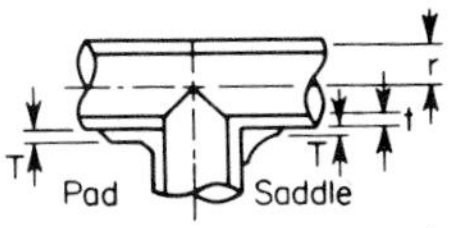 Pad, Saddle
$\frac{t}{r}$	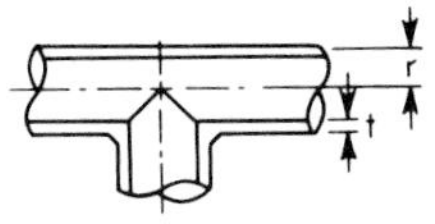

Meaning of symbols:

r = mean radius of matching pipe
t = (for elbows, bends, and miter bends nominal wall thickness of elbow, bend, or miter bend; see Note 5
t = (for tees and nozzles), the nominal wall thickness of the matching pipe
R = bend radius of pipe bend or elbow
θ = one-half angle between adjacent miter axes
s = miter spacing at center line
T = pad or saddle thickness, in.

NOTES:

1. The flexibility factors n and the stress-intensification factors i in the table apply to bending and in no case shall be taken less than unity. Factors for torsion equal unity. Both factors apply over the effective arc length (shown by heavy center lines in the sketches) for curved and miter elbows and to the intersection point for tees.
2. Those flexibility and stress intensification factors which are proportional to a power of the characteristic h may be read from the graph of Fig. B4-5(b)
3. Where flanges are attached at one or both ends, the values of n and i in the table shall be corrected by the multiplicative factors C_f given by the formulas:

$$C_f = h^{1/6} \text{ one end flanged)}$$
$$C_f = h^{1/3} \text{ (both ends flanged)}$$

4. Also includes single-miter joint.
5. Cast butt-welding elbows may have considerably heavier walls than that of the pipe with which they are used. Large errors may be introduced unless the effect of such greater thickness is considered.
6. $h = 4.05t/r$ for $T > 1\frac{1}{2}t$
7. Factors shown apply to bending; flexibility factor for torsion = 0.9.
8. In large-diameter thin-wall elbows and bends internal pressure can significantly decrease both flexibility and stress intensification factors. To correct values obtained from above tabulation so as to account for internal pressure, divide the flexibility factor by the quantity

$$[1 + 6(P/E)(r/t)^{7/3}(R/r)^{1/3}]$$

and divide the stress intensification factor by the quantity

$$[1 + 3.25(P/E)(r/t)^{5/2}(R/r)^{2/3}]$$

where P = internal gage pressure and E = Young's modulus.

9. For items 2 and 3, $i_o = 0.9/h^{2/3}$ per ANSI/ASME B31.3 and B31.8 codes.

10. For ANSI/ASME B31.3 and B31.8 piping, a single intensification factor equal to $0.9/h^{2/3}$ may be used for both i_i and i_o if desired.

FIGURE B4.5(*a*) (*Continued*) Flexibility factor n and stress-intensification factors i_i and i_o per ANSI/ASME B31.3, B31.4, B31.5, and B31.8 codes.

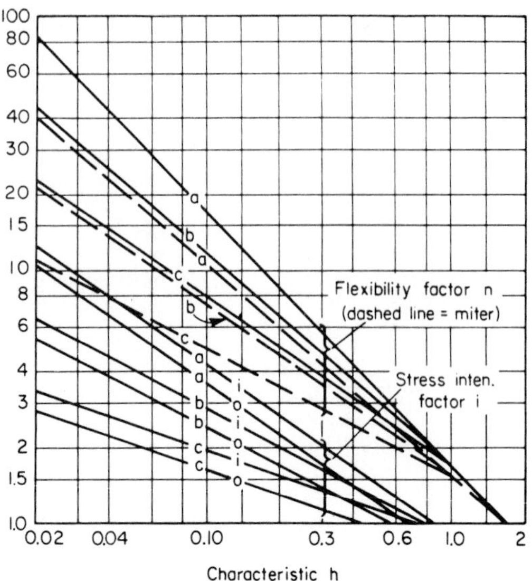

FIGURE B4.5(*b*) Flexibility and stress-intensification factors for those cases where result is a power of the characteristic *h*, defined in Fig. B4.5*a*. Key: a = no flanges; b = one flange; c = two flanges; i = in plane; o = out of plane.

not known, assume a 3000# half-coupling, since this will give the more conservative value.

c. *Sweepolets:* For branch:

$$i = 0.45\left(\frac{R}{T}\right)^{2/3}\left(\frac{r}{R}\right)^{1/2}\left(\frac{t}{T}\right)(F_1)(F_s) \qquad \text{(B4.17)}$$

For run:

$$\text{For } \frac{r}{R} > 0.5, \quad i = 0.40\left(\frac{R}{T}\right)^{2/3}(F_2)(F_s) \qquad \text{(B4.18)}$$

$$\text{For } \frac{r}{R} \leq 0.5, \quad i = 0.8\left(\frac{R}{T}\right)^{2/3}\left(\frac{r}{R}\right)F_s \text{ but not less than 1.5} \qquad \text{(B4.19)}$$

where $F_1 = F_2$ = 1.0 for flush or dressed insert welds
= 1.6 for as-welded insert welds
F_2 = $(0.5 + r/R)$, but not less than 1.0 for as-welded insert welds
F_s = $1 + 0.05(r - 3)$, but not less than 1.0
R = mean radius of run pipe, in
r = mean radius of branch pipe, in
T = nominal wall thickness of run pipe, in
t = nominal wall thickness of branch pipe, in

If a more detailed analysis is desirable, see Ref. 5 for the equations to be used for moment separation.

Description	Flexibility Characteristic h	Flexibility Factor k	Stress Intensification Factor i	Sketch
Welding elbow or pipe bend [Notes (1), (2), (3)]	$\frac{t_n R}{r^2}$	$\frac{1.65}{h}$	$\frac{0.9}{h^{2/3}}$	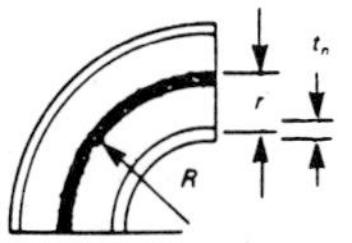 t_n, r, R
Closely spaced miter bend [Notes (1), (2), (3)] $s < r(1 + \tan\theta)$	$\frac{s t_n \cot\theta}{2r^2}$	$\frac{1.52}{h^{5/6}}$	$\frac{0.9}{h^{2/3}}$	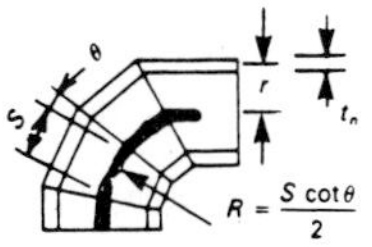 θ, s, r, t_n, $R = \frac{S \cot\theta}{2}$
Widely spaced miter bend [Notes (1), (2), (4)] $s \geq r(1 + \tan\theta)$	$\frac{t_n(1 + \cot\theta)}{2r}$	$\frac{1.52}{h^{5/6}}$	$\frac{0.9}{h^{2/3}}$	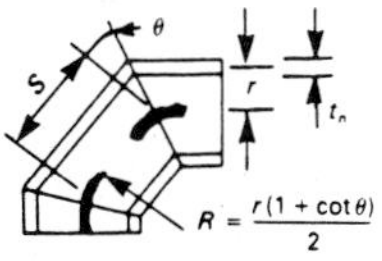 θ, s, r, t_n, $R = \frac{r(1 + \cot\theta)}{2}$
Welding tee per ANSI B16.9 [Notes (1), (2)]	$\frac{4.4\, t_n}{r}$	1	$\frac{0.9}{h^{2/3}}$	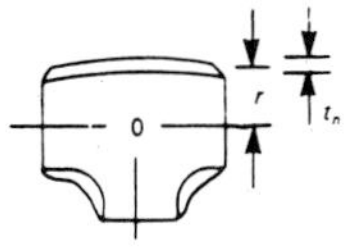 r, t_n, 0
Reinforced fabricated tee [Notes (1), (5), (10)]	$\frac{\left(t_n + \frac{t_e}{2}\right)^{5/2}}{r(t_n)^{3/2}}$	1	$\frac{0.9}{h^{2/3}}$	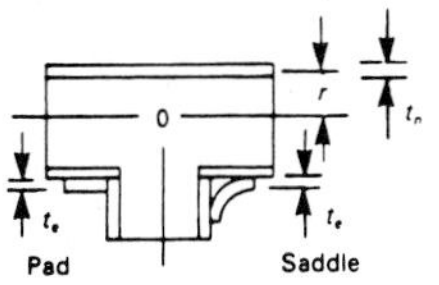 r, t_n, t_e, 0, Pad, Saddle
Unreinforced fabricated tee [Notes (1), (10)]	$\frac{t_n}{r}$	1	$\frac{0.9}{h^{2/3}}$	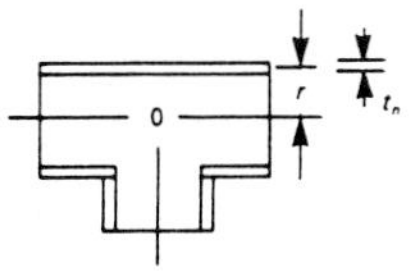 r, t_n, 0

FIGURE B4.5(*c*) Flexibility and stress-intensification factors ($D_o/t_n \leq 100$). (*Figure NC-3673.2(b)-1, Section III, Division 1, ASME B & PV Code, 1989. Courtesy of ASME.*)

Description	Flexibility Factor k	Stress Intensification Factor i	Sketch
Branch connection [Note (6)]	1	For checking branch end $Z = \pi(r'_m)^2 T'_b$ $i = 1.5\left(\frac{R_m}{T_r}\right)^{2/3}\left(\frac{r'_m}{R_m}\right)^{1/2}\left(\frac{T'_b}{T_r}\right)\left(\frac{r'_m}{r_p}\right)$ For checking run ends $Z = \pi(R_m)^2 T_r$ $i = 0.4\left(\frac{R_m}{T_r}\right)^{2/3}\left(\frac{r'_m}{R_m}\right)$ but not less than 1.5	Fig. NC-3673.2(b)-2
Girth butt weld [Note (1)] $t_n \geq 0.237$ in.	1	1.0	
Girth butt weld [Note (1)] $t_n < 0.237$ in.	1	1.9 max. or $0.9(1 + 3\delta/t_n)$ but not less than 1.0	
Circumferential fillet welded or socket welded joints [Note (11)]	1	$2.1/(C_x/t_n)$ but not less than 1.3	Fig. NC-4427-1 sketches (c-1), (c-2), and (c-3)
Brazed joint	1	2.1	Fig. NC-4511-1
30 deg. tapered transition (ANSI B16.25) [Note (1)]	1	1.9 max. or $1.3 + 0.0036\frac{D_o}{t_n} + 3.6\frac{\delta}{t_n}$	
Concentric and eccentric reducers [Note (7)] (ANSI B16.9)	1	2.0 max. or $0.5 + 0.01\alpha\left(\frac{D_2}{t_2}\right)^{1/2}$	t_1 α t_2 D_1 D_2
Threaded pipe joint or threaded flange	1	2.3	
Corrugated straight pipe or corrugated or creased bend [Note (8)]	5	2.5	

FIGURE B4.5(*c*) (*Continued*) Flexibility and stress-intensification factors ($D_o/t_n \leq 100$). (*Figure NC-3673.2(b)-1, Section III, Division 1, ASME B & PV Code, 1989. Courtesy of ASME.*)

d. *Lateral branch connections (45°)*[6,7]*:* For $r_b/r > 0.5$,

$$i = \frac{0.9}{h^{2/3}} \qquad \text{where } h = \frac{1.97t}{r} \tag{B4.20}$$

$$\text{For } \frac{r_b}{r} \leq 0.5, \quad i = 0.537\left(\frac{r}{t}\right)^{2/3}\left(\frac{r_b}{r}\right)^{1/2} \tag{B4.21}$$

(1) The following nomenclature applies:

r = mean radius of pipe, in. (matching pipe for tees and elbows)

t_n = nominal wall thickness of pipe, in. [matching pipe for tees and elbows, see Note (9)]

R = bend radius of elbow or pipe bend, in.

θ = one-half angle between adjacent miter axes, deg.

s = miter spacing at center line, in.

t_e = reinforced thickness, in.

δ = average permissible mismatch at girth butt welds as shown in Fig. NC-4233-1. A value of δ less than $\frac{1}{32}$ in. may be used provided the smaller mismatch is specified for fabrication. For "flush" welds, as defined in Fig. NB-3683.1(c)-1, δ may be taken as zero, i = 1.0, and flush welds need not be ground.

D_o = outside diameter, in.

(2) The flexibility factors k and stress intensification factors i apply to bending in any plane for fittings and shall in no case be taken less than unity. Both factors apply over the effective arc length (shown by heavy center lines in the sketches) for curved and miter elbows, and to the intersection point for tees. The values of k and i can be read directly by entering with the characteristic h computed from the equations given.

(3) Where flanges are attached to one or both ends, the values of k and i shall be corrected by the factor c given below.

(a) One end flanged, $c = h^{1/6}$

(b) Both ends flanged, $c = h^{1/3}$

(4) Also includes single miter joints.

(5) When $t_e > 1.5t_n$, $h = 4.05t_n/r$

(6) The equation applies only if the following conditions are met.

(a) The reinforcement area requirements of NC-3643 are met.

(b) The axis of the branch pipe is normal to the surface of run pipe wall.

(c) For branch connections in a pipe, the arc distance measured between the centers of adjacent branches along the surface of the run pipe is not less than three times the sum of their inside radii in the longitudinal direction or not less than two times the sum of their inside radii along the circumference of the run pipe.

(d) The inside corner radius r_1 [Fig. NC-3673.2(b)-2] for nominal branch pipe size greater than 4 in. shall be between 10% and 50% T_r. The radius r_1 is not required for nominal branch pipe size smaller than 4 in.

(e) The outer radius r_2 is not less than the larger of $T_b/2$, $(T_b' + Y)/2$ [Fig. NC-3673-2(b)-2 sketch (c)] or $T_r/2$.

(f) The outer radius r_3 is not less than the larger of

(1) $0.002\theta\, d_o$

(2) 2 $(\sin \theta)^3$ times the offset for the configurations shown in Fig. NC-3673.2(b)-2 sketches (a) and (b).

(g) $R_m/T_r \leq 50$ and $r'_m/R_m \leq 0.5$.

(h) The outer radius r_2 is not required provided an additional multiplier of 2.0 is included in the equations for branch end and run end stress intensification factors. In this case, the calculated value of i for the branch or run shall not be less than 2.1.

(7) The equation applies only if the following conditions are met.

(a) Cone angle α does not exceed 60 deg.

(b) The larger of D_1/t_1 and D_2/t_2 does not exceed 100.

(c) The wall thickness is not less than t_1 throughout the body of the reducer, except in and immediately adjacent to the cylindrical portion on the small end, where the thickness shall not be less than t_2.

(d) For eccentric reducers, α is the maximum cone angle.

(8) Factors shown apply to bending; flexibility factor for torsion equals 0.9.

(9) The designer is cautioned that cast butt welding elbows may have considerably heavier walls than that of the pipe with which they are used. Large errors may be introduced unless the effect of these greater thicknesses is considered.

(10) The stress intensification factor i shall in no case be taken as less than 2.1.

(11) In Fig. NC-4427-1(c-1) and (c-2), C_x shall be taken as X_{min} and $C_x \geq 1.25\, t_n$. In Fig. NC-4427-1 (c-3), $C_x \geq 0.75\, t_n$. For unequal leg lengths use the smaller leg length for C_x.

FIGURE B4.5(*c*) (*Continued*) Flexibility and stress-intensification factors ($D_o/t_n \leq 100$). (*Figure NC-3673.2(b)-1, Section III, Division 1, ASME B & PV Code, 1989. Courtesy of ASME.*)

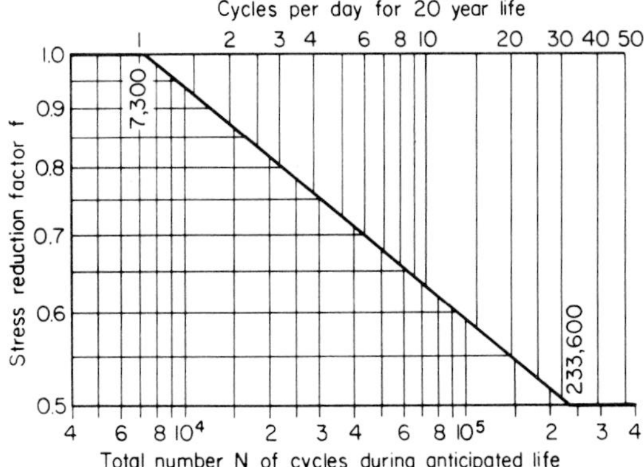

FIGURE B4.5(*d*) Stress range reduction factors. (*Extracted from Refrigeration Piping Code, ASME/ANSI B31.5 1987. Courtesy of ASME/ANSI.*)

TABLE B4.3 (*a*) SIFs for Typical Weldolet Branch Connections (run pipe size: 1½ in)

			Run pipe, 1½ in, nominal			
Branch			10*s*	40	80	160
pipe, in	Sch.	*T*, in	0.109	0.145	0.200	0.281
½	10*s*	0.083	1.665	1.031	1.000	
	40	0.109	2.074	1.285	1.000	1.000
	80	0.147	2.582	1.599	1.000	1.000
	160	0.188			1.086	2.000

TABLE B4.3 (*b*) SIFs for Typical Weldolet Branch Connections (run pipe size: 2 in)

			Run pipe, 2 in, nominal			
Branch			10*s*	40	80	160
pipe, in	Sch.	*T*, in	0.109	0.154	0.218	0.344
½	10*s*	0.083	1.731			
	40	0.109	2.157	1.209	1.000	
	80	0.147			1.000	1.000
	160	0.188			1.000	1.000
¾	10*s*	0.083	1.964			
	40	0.113	2.550	1.429	1.000	
	80	0.154			1.015	1.000
	160	0.219			1.286	1.000

where t = run pipe wall thickness, in
r = run pipe mean radius, in
r_b = branch pipe mean radius, in

These equations are for integrally reinforced branch connections such as latrolets. By the analogy used in Figure NC-3673.2(b)-1 in Section III of the ASME Code, the SIF for unreinforced 45° branch connections (stub-ins) can be obtained by multiplying the factors obtained above by $(4.4)^{2/3} = 2.685$.

e. Pipet:

$$i = \frac{0.9}{h^{2/3}} \tag{B4.22}$$

$$\text{For buttweld pipet, } h = \frac{3.3t}{r}$$

$$\text{For socket-weld pipet, } h = \frac{5.9t}{r}$$

$$\text{For threaded pipet, } h = \frac{4.19t}{r}$$

TABLE B4.3 (*c*) SIFs for Typical Weldolet Branch Connections (run pipe size: 2½ and 3 in)

Branch pipe, in	Sch.	T, in	Run pipe, 2½ in, nominal 10*s* 0.120	40 0.203	80 0.276	160 0.375	Run pipe, 3 in, nominal 10*s* 0.120	40 0.216	80 0.300	160 0.438
½	10*s*	0.083	1.524	1.000			1.576	1.000		
	40	0.109	1.899	1.000	1.000		1.965	1.000	1.000	
	80	0.147		1.000	1.000	1.000		1.000	1.000	1.000
	160	0.188			1.000	1.000			1.000	1.000
¾	10*s*	0.083	1.728	1.000			1.788	1.000		
	40	0.113	2.245	1.000	1.000		2.322	1.000	1.000	
	80	0.154		1.185	1.000	1.000		1.106	1.000	1.000
	160	0.219			1.000	1.000			1.000	1.000
1	10*s*	0.109	2.603	1.078			2.694	1.006		
	40	0.133	3.082	1.277	1.000		3.189	1.192	1.000	
	80	0.179		1.619	1.000	1.000		1.511	1.000	1.000
	160	0.250			1.225	1.000			1.103	1.000

TABLE B4.3 (*d*) SIFs for Typical Weldolet Branch Connections
(run pipe size: 4 and 6 in)

Branch			Run pipe, 4 in, nominal					Run pipe, 6 in, nominal				
			10*s*	40	80	120	160	10*s*	40	80	120	160
pipe, in	Sch.	*T*, in	0.120	0.237	0.337	0.438	0.531	0.134	0.280	0.432	0.562	0.719
½	10*s*	0.083	1.646	1.000				1.462	1.000			
	40	0.109	2.051	1.000	1.000			1.822	1.000	1.000		
	80	0.147		1.000	1.000	1.000			1.000	1.000	1.000	
	160	0.188			1.000	1.000	1.000			1.000	1.000	1.000
¾	10*s*	0.083	1.867	1.000				1.659	1.000			
	40	0.113	2.425	1.000	1.000			2.154	1.000	1.000		
	80	0.154		1.109	1.000	1.000			1.000	1.000	1.000	
	160	0.219			1.000	1.000	1.000			1.000	1.000	1.000
1	10*s*	0.109	2.813	1.000				2.499	1.000			
	40	0.133	3.330	1.066	1.000			2.958	1.000	1.000		
	80	0.179		1.352	1.000	1.000			1.094	1.000	1.000	
	160	0.250			1.009	1.000	1.000			1.000	1.000	1.000
1½	10*s*	0.109	3.762	1.205				3.342	1.000			
	40	0.145	4.855	1.555	1.000			4.313	1.258	1.000		
	80	0.200		2.044	1.132	1.000			1.654	1.000	1.000	
	160	0.281			1.548	1.000	1.000			1.092	1.000	1.000
2	10*s*	0.109						3.907	1.140			
	40	0.154						5.357	1.563	1.000		
	80	0.218							2.117	1.023	1.000	
	160	0.344								1.620	1.041	1.000
2½	10*s*	0.120						4.968	1.449			
	40	0.203						8.027	2.342	1.132		
	80	0.276							3.054	1.477	1.000	
	160	0.375								2.017	1.296	1.000

TABLE B4.3 (*e*) SIFs for Typical Weldolet Branch Connections (run pipe size: 8 in)

			Run pipe, 8 in, nominal				
			10*s*	40	80	120	160
Branch pipe, in	Sch.	*T*, in	0.148	0.322	0.500	0.719	0.906
½	10*s*	0.083	1.295	1.000			
	40	0.109	1.614	1.000	1.000		
	80	0.147		1.000	1.000		
	160	0.188			1.000	1.000	1.000
¾	10*s*	0.083	1.470	1.000			
	40	0.113	1.909	1.000	1.000		
	80	0.154		1.000	1.000		
	160	0.219			1.000	1.000	1.000
1	10*s*	0.109	2.214	1.000			
	40	0.133	2.621	1.000	1.000		
	80	0.179		1.000	1.000		
	160	0.250			1.000	1.000	1.000
1½	10*s*	0.109	2.961	1.000			
	40	0.145	3.821	1.042	1.000		
	80	0.200		1.371	1.000		
	160	0.281			1.000	1.000	1.000
2	10*s*	0.109	3.461	1.000			
	40	0.154	4.746	1.295	1.000		
	80	0.218		1.754	1.000		
	160	0.344			1.329	1.000	1.000
2½	10*s*	0.120	4.401	1.201			
	40	0.203	7.112	1.940	1.000		
	80	0.276		2.530	1.211		
	160	0.375			1.654	1.000	1.000
3	10*s*	0.120	5.049	1.377			
	40	0.216	8.704	2.374	1.136		
	80	0.300		3.172	1.518		
	160				2.102	1.142	1.000

f. Branchlet:

$$i = \frac{0.9}{h^{2/3}} \qquad \text{where } h = \frac{3.8t}{r} \tag{B4.23}$$

g. Reducing elbow:

$$i = \frac{0.9}{(TB/R^2)^{2/3}} \quad \text{or 2.0,} \tag{B4.24}$$

whichever is higher

where T = wall thickness of large end, in
B = actual bend radius, in
R = mean radius of large end, in

TABLE B4.3 (*f*) SIFs for Typical Weldolet Branch Connections (run pipe size: 10 in)

Branch pipe, in	Sch.	T, in	Run pipe, 10 in, nominal: 10*s* 0.165	40 0.365	80 0.594	120 0.844	160 1.125
¾	10*s*	0.083	1.272	1.000			
	40	0.113	1.652	1.000	1.000		
	80	0.154		1.000	1.000		
	160	0.219			1.000	1.000	1.000
1	10*s*	0.109	1.916	1.000			
	40	0.133	2.269	1.000	1.000		
	80	1.179		1.000	1.000		
	160	0.250			1.000	1.000	1.000
1½	10*s*	0.109	2.563	1.000			
	40	0.145	3.308	1.000	1.000		
	80	0.200		1.154	1.000		
	160	0.281			1.000	1.000	1.000
2	10*s*	0.109	2.997	1.000			
	40	0.154	4.108	1.090	1.000		
	80	0.218		1.477	1.000		
	160	0.344			1.035	1.000	1.000
2½	10*s*	0.120	3.810	1.011			
	40	0.203	6.157	1.634	1.000		
	80	0.276		2.131	1.000		
	160	0.375			1.288	1.000	1.000
3	10*s*	0.120	4.371	1.160			
	40	0.216	7.535	2.000	1.000		
	80	0.300		2.672	1.182		
	160	0.438			1.637	1.000	1.000
4	10*s*	0.120	5.172	1.373			
	40	0.237	9.807	2.603	1.152		
	80	0.337		3.572	1.581		
	120	0.438			1.980	1.098	1.000
	160	0.531			2.318	1.286	1.000

ANSI/ASME B31.1 Power Piping Code.[8] This code concerns nonnuclear piping such as that found in the turbine building of a nuclear plant or in a fossil-fueled power plant. Piping services include steam, water, oil, gas, and air. Design requirements of this code cover those for pipe, flanges, bolting, gaskets, valves, relief devices, fittings, and the pressure containing portions of other piping components. It also includes hangers and supports and other equipment items necessary to prevent overstressing the pressure containing components.

The loadings required to be considered are pressure, weight (live, dead, and under test loads); impact (e.g., water hammer); wind; earthquake (where applicable); vibration; and those loadings resulting from thermal expansion and contraction.

The design equations and stress limits are as follows (terms are the same as those for Class 2 and 3 piping except for those defined below):

TABLE B4.3 (*g*) SIFs for Typical Weldolet Branch Connections (run pipe size: 12 in)

			Run pipe, 12 in, nominal					
Branch			10*s*	40*s*	40	80	120	160
pipe, in	Sch.	*T*, in	0.180	0.375	0.406	0.688	1.000	1.312
¾	10*s*	0.083	1.132	1.000	1.000			
	40	0.113	1.471	1.000	1.000			
	80	0.154		1.000	1.000	1.000		
	160	0.219				1.000	1.000	1.000
1	10*s*	0.109	1.706	1.000	1.000			
	40	0.133	2.020	1.000	1.000			
	80	0.179		1.000	1.000	1.000	1.000	1.000
	160	0.250				1.000	1.000	1.000
1½	10*s*	0.109	2.282	1.000	1.000			
	40	0.145	2.944	1.000	1.000			
	80	0.200		1.136	1.000	1.000	1.000	1.000
	160	0.281				1.000	1.000	1.000
2	10*s*	0.109	2.667		1.000			
	40	0.154	3.657	1.073	1.000			
	80	0.218		1.454	1.273	1.000	1.000	1.000
	160	0.344				1.000	1.000	1.000
2½	10*s*	0.120	3.392		1.000			
	40	0.203	5.480	1.609	1.409			
	80	0.276		2.098	1.837	1.000	1.000	1.000
	160	0.375				1.038	1.000	1.000
3	10*s*	0.120	3.890		1.000			
	40	0.216	6.707	1.969	1.724			
	80	0.300		2.630	2.303	1.000	1.000	1.000
	160	0.438				1.319	1.000	1.000
4	10*s*	0.120	4.603		1.183			
	40	0.237	8.730	2.562	2.244			
	80	0.337		3.516	3.079	1.273	1.000	1.000
	120	0.438				1.595	1.000	1.000
	160	0.531				1.868	1.000	1.000

1. *Stress due to sustained loads:* The effects of pressure, weight, and other sustained mechanical loads must meet the requirements of Eq. (B4.25):

$$S_L = \frac{PD_o}{4t} + \frac{0.75iM_A}{Z} \leq 1.0S_h \tag{B4.25}$$

where S_L = sum of the longitudinal stresses due to pressure, weight, and other sustained loads, psi

2. *Stress due to occasional loads:* The effects of pressure, weight, other sustained loads, and occasional loads including earthquake must meet the requirements of Eq. (B4.26):

$$\frac{PD_o}{4t} + \frac{0.75i(M_A + M_B)}{Z} \leq kS_h \tag{B4.26}$$

TABLE B4.3 (*h*) SIFs for Typical Weldolet Branch Connections (run pipe size: 14 in)

Branch pipe, in	Sch.	Run pipe, 14 in, nominal	10	30	40	80	120	160
		T, in	0.250	0.375	0.438	0.750	1.094	1.406
1	10*s*	0.109	1.002	1.000				
	40	0.133	1.186	1.000	1.000			
	80	0.179		1.000	1.000	1.000		
	160	0.250				1.000	1.000	1.000
1½	10*s*	0.109	1.340	1.000				
	40	0.145	1.729	1.000	1.000			
	80	0.200		1.155	1.000	1.000		
	160	0.281				1.000	1.000	1.000
2	10*s*	0.109	1.566	1.000				
	40	0.154	2.147	1.091	1.000			
	80	0.218		1.478	1.140	1.000		
	160	0.344				1.000	1.000	1.000
2½	10*s*	0.120	1.991	1.012				
	40	0.203	3.218	1.635	1.261			
	80	0.276		2.132	1.644	1.000		
	160	0.375				1.000	1.000	1.000
3	10*s*	0.120	2.284	1.160				
	40	0.216	3.938	2.000	1.543			
	80	0.300		2.673	2.062	1.000		
	160	0.438				1.160	1.000	1.000
4	10*s*	0.120	2.703	1.373				
	40	0.237	5.126	2.604	2.009			
	80	0.337		3.573	2.756	1.120		
	120	0.438				1.403	1.000	1.000
	160	0.531				1.643	1.000	1.000
6	10*s*	0.134	3.681	1.870				
	40	0.280	7.434	3.776	2.913			
	80	0.432		5.618	4.334	1.761		
	120	0.562				2.268	1.203	1.000
	160	0.719				2.789	1.480	1.000

where k = 1.15 for occasional loads acting less than 10 percent of operating period; 1.2 for occasional loads acting less than 1 percent of operating period

Note: There is no provision for emergency and faulted conditions.

3. *Thermal expansion stress range:* The effects of thermal expansion must meet the requirements of Eq. (B4.27):

$$\frac{iM_C}{Z} \leq S_A + f(S_h - S_L) \qquad \text{(B4.27)}$$

4. *The requirement for the effects of any single nonrepeated anchor movement is not specified.*

TABLE B4.3 (*i*) SIFs for Typical Weldolet Branch Connections (run pipe size: 16 in)

Branch pipe, in	Sch.	T, in	Run pipe, 16 in, nominal 10	30	40	80	120	160
			0.250	0.375	0.500	0.844	1.219	1.594
1	10s	0.109	1.025	1.000				
	40	0.133	1.213	1.000	1.000			
	80	0.179		1.000	1.000			
	160	0.250				1.000	1.000	1.000
1½	10s	0.109	1.370	1.000				
	40	0.145	1.768	1.000	1.000			
	80	0.200		1.181	1.000	1.000		
	160	0.281				1.000	1.000	1.000
2	10s	0.109	1.602	1.000				
	40	0.154	2.196	1.116	1.000			
	80	0.218		1.512	1.000	1.000		
	160	0.344				1.000	1.000	1.000
2½	10s	0.120	2.037	1.035				
	40	0.203	3.291	1.672	1.034			
	80	0.276		2.181	1.349	1.000		
	160	0.375				1.000	1.000	1.000
3	10s	0.120	2.337	1.187				
	40	0.216	4.028	2.047	1.265			
	80	0.300		2.734	1.691	1.000		
	160	0.438				1.000	1.000	1.000
4	10s	0.120	2.765	1.405				
	40	0.237	5.243	2.664	1.647			
	80	0.337		3.656	2.260	1.000		
	120	0.438				1.179	1.000	1.000
	160	0.531				1.380	1.000	1.000
6	10s	0.134	3.766	1.913				
	40	0.280	7.605	3.864	2.389			
	80	0.432		5.748	3.554	1.480		
	120	0.562				1.905	1.028	1.000
	160	0.719				2.343	1.264	1.000

ANSI/ASME B31.3 Chemical Plant and Petroleum Refinery Piping Code.[9] This code governs all piping within the property limits of facilities engaged in the processing or handling of chemical, petroleum, or related products. Examples are a chemical plant, petroleum refinery, loading terminal, natural gas processing plant, bulk plant, compounding plant, and tank farm. Excluded from the B31.3 code are piping carrying nonhazardous fluid with an internal gauge pressure less than 15 psi and a temperature below 366°F, plumbing, sewers, fire protection systems, boiler external piping per B31.1, as well as pipelines per B31.4 or B31.8.

The loadings required to be considered are pressure, weight (live and dead loads), impact, wind, earthquake-induced horizontal forces, vibration, discharge reactions, thermal expansion and contraction, temperature gradients, and anchor movements.

The governing equations are as follows:

TABLE B4.3 (*j*) SIFs for Typical Weldolet Branch Connections (run pipe size: 18 in)

			Run pipe, 18 in, nominal					
Branch			10	STD	XS	40	80	120
pipe, in	Sch.	*T*, in	0.250	0.375	0.500	0.562	0.938	1.375
1½	10*s*	0.109	1.398	1.000				
	40	0.145	1.804	1.000	1.000			
	80	0.200		1.205	1.000	1.000	1.000	1.000
	160	0.281			1.018	1.000	1.000	1.000
2	10*s*	0.109	1.634	1.000				
	40	0.154	2.241	1.139	1.000			
	80	0.218		1.543	1.000	1.000	1.000	1.000
	160	0.344			1.510	1.242	1.000	1.000
2½	10*s*	0.120	2.078	1.056				
	40	0.203	3.358	1.706	1.055			
	80	0.276		2.225	1.376	1.132	1.000	1.000
	160	0.375			1.880	1.546	1.000	1.000
3	10*s*	0.120	2.384	1.211				
	40	0.216	4.109	2.088	1.291			
	80	0.300		2.790	1.725	1.419	1.000	1.000
	160	0.438			2.388	1.965	1.000	1.000
4	10*s*	0.120	2.821	1.433				
	40	0.237	5.349	2.718	1.681			
	80	0.337		3.730	2.306	1.897	1.000	1.000
	120	0.438			2.889	2.376	1.008	1.000
	160	0.531			3.383	2.782	1.181	1.000
6	10*s*	0.134	3.841	1.952				
	40	0.280	7.758	3.942	2.438			
	80	0.432		5.865	3.627	2.983	1.266	1.000
	120	0.562			4.669	3.840	1.629	1.000
	160	0.719			5.743	4.723	2.004	1.055
8	10*s*	0.148	4.887	2.483				
	40	0.322	8.106	5.237	3.239			
	80	0.500			4.868	4.004	1.699	1.000
	120	0.719			6.907	5.681	2.410	1.269
	160	0.906			8.396	6.906	2.930	1.542

1. *Stresses due to sustained loads:* The sum of the longitudinal stresses S_L due to pressure, weight, and other sustained loads must not exceed S_h (basic allowable stress at maximum metal temperature). The thickness of pipe used in calculating S_L shall be the nominal thickness minus mechanical, corrosion, and erosion allowances.
2. *Stresses due to occasional loads:* The sum of the longitudinal stresses due to pressure, weight, and other sustained loads and of the stresses produced by occasional loads such as earthquake or wind shall not exceed $1.33S_h$. Earthquake and wind loads need not be considered as acting simultaneously.
3. *Stress range due to expansion loads:* The displacement stress range S_E shall not exceed S_A:

TABLE B4.3 (*k*) SIFs for Typical Weldolet Branch Connections (run pipe size: 20 in)

			Run pipe, 20 in, nominal					
			10*s*	20	XS	40	80	120
Branch pipe, in	Sch.	*T*, in	0.218	0.375	0.500	0.594	1.031	1.500
1½	10*s*	0.109	1.789	1.000				
	40	0.145	2.308	1.000	1.000			
	80	0.200		1.227	1.000	1.000	1.000	1.000
	160	0.281		1.675	1.036	1.000	1.000	1.000
2	10*s*	0.109	2.091	1.000				
	40	0.154	2.867	1.159	1.000	1.000		
	80	0.218		1.570	1.000	1.000	1.000	1.000
	160	0.344			1.538	1.156	1.000	1.000
2½	10*s*	0.120	2.659	1.075				
	40	0.203	4.296	1.737	1.074			
	80	0.276		2.266	1.401	1.054	1.000	1.000
	160	0.375			1.914	1.439	1.000	1.000
3	10*s*	0.120	3.050	1.233				
	40	0.216	5.257	2.126	1.315			
	80	0.300		2.840	1.756	1.321	1.000	1.000
	160	0.438			2.432	1.829	1.000	1.000
4	10*s*	0.120	3.608	1.459				
	40	0.237	6.843	2.767	1.711			
	80	0.337		3.797	2.348	1.766	1.000	1.000
	120	0.438			2.942	2.212	1.000	1.000
	160	0.531			3.445	2.590	1.026	1.000
6	10*s*	0.134	4.915	1.987				
	40	0.280	9.925	4.013	2.482			
	80	0.432		5.971	3.693	2.777	1.100	1.000
	120	0.562			4.754	3.575	1.417	1.000
	160	0.719				4.397	1.742	1.000
8	10*s*	0.148	6.252	2.528				
	40	0.322		5.332	3.297			
	80	0.500			4.957	3.727	1.477	1.000
	120	0.719					2.095	1.117
	160	0.906					2.547	1.358

$$S_E \leq S_A \tag{B4.28}$$

where $S_E = (S_b^2 + 4S_t^2)^{1/2}$

S_b = resultant bending stress, psi

$= [(i_i M_i)^2 + (i_o M_o)^2]^{1/2}/Z$

M_i = in-plane bending moment, in · lb

M_o = out-of-plane bending moment, in · lb

i_i = in-plane stress-intensification factor obtained from Fig. B4.5*a* (see also figure note 10)

i_o = out-of-plane stress-intensification factor obtained from Fig. B4.5*a* (see also figure note 10)

TABLE B4.3 (*I*) SIFs for Typical Weldolet Branch Connections (Run pipe size: 24 in)

Branch pipe, in	Sch.	*T*, in	Run pipe, 24 in, nominal 10*s* 0.218	20 0.375	XS 0.500	40 0.594	80 1.031	120 1.500
2	10*s*	0.109	1.716	1.000				
	40	0.154	2.352	1.196	1.000			
	80	0.218		1.620	1.002	1.000	1.000	1.000
	160	0.344			1.586	1.000	1.000	1.000
2½	10*s*	0.120	2.181	1.109				
	40	0.203	3.525	1.792	1.108			
	80	0.276		2.337	1.445	1.000	1.000	1.000
	160	0.375			1.974	1.161	1.000	1.000
3	10*s*	0.120	2.502	1.272				
	40	0.216	4.314	2.193	1.356			
	80	0.300		2.929	1.812	1.066	1.000	1.000
	160	0.438			2.509	1.475	1.000	1.000
4	10*s*	0.120	2.961	1.505				
	40	0.237	5.615	2.854	1.765			
	80	0.337		3.916	2.423	1.425	1.000	1.000
	120	0.438			3.035	1.785	1.006	1.000
	160	0.531			3.553	2.090	1.178	1.000
6	10*s*	0.134	4.033	2.050				
	40	0.280	8.143	4.139	2.560			
	80	0.432		6.158	3.809	2.240	1.262	1.000
	120	0.562			4.904	2.884	1.625	1.106
	160	0.719				3.547	1.999	1.361
8	10*s*	0.148	5.130	2.608				
	40	0.322		5.499	3.402			
	80	0.500			5.113	3.007	1.695	1.153
	120	0.719				4.266	2.404	1.636
	160	0.906				5.186	2.923	1.989
10	10*s*	0.165	7.197	3.659				
	40	0.365		7.865	4.865			
	80	0.594			7.645	4.496	2.534	1.724
	120	0.844				6.375	3.593	2.445
	160	1.125				8.148	4.591	3.125

S_t = torsional stress, psi
$= M_t/(2Z)$
M_t = torsional moment, in · lb
S_A = allowable displacement stress range
$= f(1.25S_c + 0.25S_h)$
$= f[1.25(S_c + S_h) - S_L]$ when $S_h > S_L$
S_c = basic allowable stress at minimum metal temperature, psi
f = stress-range reduction factor per Table B4.2

ANSI/ASME B31.4 Liquid Transportation Systems Piping Code.[10] The scope of ANSI/ASME Code B31.4, Liquid Transportation Systems for Hydrocarbons,

TABLE B4.4 (*a*) SIFs for 3000# Half-Couplings (Branch Pipe Schedule 40) (run pipe size: 1½ to 8 in)

Run pipe			Sch. 40 branch pipe size and thickness					
			½ in	¾ in	1 in	1¼ in	1½ in	2 in
Size, in	Sch.	*T*, in	0.109	0.113	0.133	0.140	0.145	0.154
1½	10*s*	0.109	2.282	3.665	3.665	3.665	3.665	3.665
	40	0.145	1.413	2.989	2.989	2.989	2.989	2.989
	80	0.200	1.000	2.362	2.362	2.362	2.362	2.362
	160	0.281	1.000	1.822	1.822	1.822	1.822	1.822
2	10*s*	0.109	2.373	2.975	4.287	4.287	4.287	4.287
	40	0.154	1.330	1.667	3.359	3.359	3.359	3.359
	80	0.218	1.000	1.000	2.613	2.613	2.613	2.613
	160	0.344	1.000	1.000	1.852	1.852	1.852	1.852
2½	10*s*	0.120	2.089	2.619	3.614	4.580	4.580	4.580
	40	0.203	1.000	1.085	1.497	3.161	3.161	3.161
	80	0.276	1.000	1.000	1.000	2.528	2.528	2.528
	160	0.375	1.000	1.000	1.000	2.008	2.008	2.008
3	10*s*	0.120	2.161	2.710	3.739	4.624	5.249	5.249
	40	0.216	1.000	1.012	1.397	1.728	3.480	3.480
	80	0.300	1.000	1.000	1.000	1.000	2.747	2.747
	160	0.438	1.000	1.000	1.000	1.000	2.073	2.073
4	10*s*	0.120	2.257	2.829	3.904	4.828	5.583	6.239
	40	0.237	1.000	1.000	1.250	1.546	1.788	3.893
	80	0.337	1.000	1.000	1.000	1.000	1.000	3.030
	160	0.531	1.000	1.000	1.000	1.000	1.000	2.168
6	10*s*	0.134	2.005	2.513	3.469	4.289	4.960	6.250
	40	0.280	1.000	1.000	1.012	1.251	1.447	1.823
	80	0.432	1.000	1.000	1.000	1.000	1.000	1.000
	120	0.562	1.000	1.000	1.000	1.000	1.000	1.000
8	10*s*	0.148	1.776	2.227	3.073	3.800	4.394	5.537
	20	0.250	1.000	1.000	1.280	1.583	1.830	2.306
	40	0.322	1.000	1.000	1.000	1.037	1.199	1.510
	80	0.500	1.000	1.000	1.000	1.000	1.000	1.000

Liquid Petroleum Gas, Anhydrous Ammonia, and Alcohols, governs piping transporting liquids such as crude oil, condensate, natural gasoline, natural gas liquids, liquefied petroleum gas, liquid alcohol, liquid anhydrous ammonia, and liquid petroleum products between producers' lease facilities, tank farms, natural gas processing plants, refineries, stations, ammonia plants, terminals and delivery and receiving points. Excluded from B31.4 are auxiliary piping such as water, air, steam, lubricating oil, gas, and fuel; piping with an internal gauge pressure at or below 15 psi regardless of temperature; piping with an internal gauge pressure above 15 psi and a temperature below −20°F or above 250°F; and piping for petroleum refinery, gas transmission and distribution, ammonia refrigeration, and so on, that is covered by other ANSI/ASME B31 Code sections.

The limits of calculated stresses are as follows:

1. *Stresses due to sustained loads:* The sum of the longitudinal stresses due to pressure, weight, and other sustained external loads shall not exceed $0.75S_A$, where $S_A = 0.72S_y$ (specified minimum yield strength).

TABLE B4.4 (*b*) SIFs for 3000# Half-Couplings (Branch Pipe Schedule 40) (run pipe size: 10 to 24 in)

Run pipe			Sch. 40 branch pipe size and thickness					
			½ in	¾ in	1 in	1¼ in	1½ in	2 in
Size, in	Sch.	*T*, in	0.109	0.113	0.133	0.140	0.145	0.154
10	10*s*	0.165	1.538	1.928	2.660	3.290	3.804	4.793
	20	0.250	1.000	1.000	1.329	1.644	1.901	2.395
	40	0.365	1.000	1.000	1.000	1.000	1.010	1.272
	80	0.594	1.000	1.000	1.000	1.000	1.000	1.000
12	10*s*	0.180	1.369	1.716	2.368	2.928	3.386	4.267
	20	0.250	1.000	1.000	1.368	1.692	1.957	2.466
	STD	0.375	1.000	1.000	1.000	1.000	1.000	1.252
	40	0.406	1.000	1.000	1.000	1.000	1.000	1.097
14	10	0.250	1.000	1.007	1.390	1.719	1.988	2.505
	20	0.312	1.000	1.000	1.000	1.187	1.373	1.730
	STD	0.375	1.000	1.000	1.000	1.000	1.010	1.273
	40	0.438	1.000	1.000	1.000	1.000	1.000	1.000
16	10	0.250	1.000	1.030	1.422	1.759	2.034	2.562
	20	0.312	1.000	1.000	1.000	1.215	1.405	1.770
	STD	0.375	1.000	1.000	1.000	1.000	1.033	1.302
	40	0.500	1.000	1.000	1.000	1.000	1.000	1.000
18	10	0.250	1.000	1.051	1.451	1.794	2.075	2.614
	20	0.312	1.000	1.000	1.002	1.239	1.433	1.806
	STD	0.375	1.000	1.000	1.000	1.000	1.054	1.328
	30	0.438	1.000	1.000	1.000	1.000	1.000	1.025
20	10	0.250	1.000	1.070	1.477	1.826	2.112	2.661
	STD	0.375	1.000	1.000	1.000	1.000	1.073	1.352
	30	0.500	1.000	1.000	1.000	1.000	1.000	1.000
	40	0.593	1.000	1.000	1.000	1.000	1.000	1.000
24	10	0.250	1.000	1.103	1.523	1.883	2.178	2.744
	STD	0.375	1.000	1.000	1.000	1.000	1.107	1.395
	XS	0.500	1.000	1.000	1.000	1.000	1.000	1.000
	30	0.562	1.000	1.000	1.000	1.000	1.000	1.000

2. *Stresses due to occasional loads:* The sum of the longitudinal stresses produced by pressure, live and dead loads, and those produced by occasional loads, such as wind or earthquake, shall not exceed $0.8S_y$.
3. *Stresses due to expansion loads*
   a. *Restrained lines:* The net longitudinal compressive stress due to the combined effects of temperature rise and fluid pressure shall be computed from the equation:

$$S_L = E\alpha(T_2 - T_1) - \nu S_H \qquad \text{(B4.29)}$$

where S_L = longitudinal compressive stress, psi
S_H = hoop stress due to fluid pressure, psi
T_1 = temperature at time of installation, °F
T_2 = maximum or minimum operating temperature, °F
E = modulus of elasticity, psi
α = linear coefficient of thermal expansion, in/(in · °F)

TABLE B4.4 (*c*) SIFs for 3000# Half-Couplings (Branch Pipe Schedule 80) (run pipe size: 1½ to 8 in)

Run pipe			Sch. 80 branch pipe size and thickness					
			½ in	¾ in	1 in	1¼ in	1½ in	2 in
Size, in	Sch.	*T*, in	0.147	0.154	0.179	0.191	0.200	0.218
1½	10*s*	0.109	2.841	3.665	3.665	3.665	3.665	3.665
	40	0.145	1.759	2.989	2.989	2.989	2.989	2.989
	80	0.200	1.024	2.362	2.362	2.362	2.362	2.362
	160	0.281	1.000	1.822	1.822	1.822	1.822	1.822
2	10*s*	0.109	2.954	3.792	4.287	4.287	4.287	4.287
	40	0.154	1.655	2.124	3.359	3.359	3.359	3.359
	80	0.218	1.000	1.184	2.613	2.613	2.613	2.613
	160	0.344	1.000	1.000	1.852	1.852	1.852	1.852
2½	10*s*	0.120	2.600	3.337	4.580	4.580	4.580	4.580
	40	0.203	1.077	1.382	1.898	3.161	3.161	3.161
	80	0.276	1.000	1.000	1.132	2.528	2.528	2.528
	160	0.375	1.000	1.000	1.000	2.008	2.008	2.008
3	10*s*	0.120	2.690	3.453	4.742	5.249	5.249	5.249
	40	0.216	1.005	1.290	1.772	2.239	3.480	3.480
	80	0.300	1.000	1.000	1.020	1.290	2.747	2.747
	160	0.438	1.000	1.000	1.000	1.000	2.073	2.073
4	10*s*	0.120	2.809	3.605	4.951	6.239	6.239	6.239
	40	0.237	1.000	1.154	1.585	2.004	2.351	3.893
	80	0.337	1.000	1.000	1.000	1.110	1.302	3.030
	160	0.531	1.000	1.000	1.000	1.000	1.000	2.168
6	10*s*	0.134	2.496	3.203	4.398	5.560	6.523	7.534
	40	0.280	1.000	1.000	1.283	1.622	1.903	2.470
	80	0.432	1.000	1.000	1.000	1.000	1.000	1.194
	120	0.562	1.000	1.000	1.000	1.000	1.000	1.000
8	10*s*	0.148	2.211	2.838	3.897	4.925	5.779	7.502
	20	0.250	1.000	1.182	1.623	2.052	2.407	3.125
	40	0.322	1.000	1.000	1.063	1.344	1.576	2.046
	80	0.500	1.000	1.000	1.000	1.000	1.000	1.000

ν = Poisson's ratio = 0.30 for steel

Then, the equivalent tensile stress is calculated as

$$S_{\text{eqiv}} = S_H + S_L < 0.9S_y \quad \text{(B4.30)}$$

where S_{eqiv} = the equivalent tensile stress, psi

Beam bending stresses shall be included in the longitudinal stress for those portions of the restrained line which are supported above ground.

b. *Unrestrained lines:* Stresses due to expansion for those portions of the piping without substantial axial restraint shall be combined in accordance with the following equation:

$$S_E = (S_b^2 + 4S_t^2)^{1/2} < S_A \quad \text{(B4.31)}$$

where S_E = stress due to expansion, psi

TABLE B4.4 (*d*) SIFs for 3000# Half-Couplings (Branch Pipe Schedule 80) (run pipe size: 10 to 24 in)

Run pipe			Sch. 80 branch pipe size and thickness					
			½ in	¾ in	1 in	1¼ in	1½ in	2 in
Size, in	Sch.	*T*, in	0.147	0.154	0.179	0.191	0.200	0.218
10	10*s*	0.165	1.914	2.457	3.373	4.264	5.003	6.494
	20	0.250	1.000	1.227	1.685	2.130	2.499	3.245
	40	0.365	1.000	1.000	1.000	1.132	1.328	1.724
	80	0.594	1.000	1.000	1.000	1.000	1.000	1.000
12	10*s*	0.180	1.704	2.187	3.003	3.796	4.453	5.781
	STD	0.375	1.000	1.000	1.000	1.114	1.307	1.697
	40	0.406	1.000	1.000	1.000	1.000	1.144	1.486
	80s	0.500	1.000	1.000	1.000	1.000	1.000	1.049
14	10	0.250	1.000	1.284	1.763	2.228	2.614	3.394
	STD	0.375	1.000	1.000	1.000	1.132	1.328	1.724
	40	0.438	1.000	1.000	1.000	1.000	1.024	1.330
	XS	0.500	1.000	1.000	1.000	1.000	1.000	1.066
16	10	0.250	1.023	1.313	1.803	2.279	2.674	3.472
	20	0.312	1.000	1.000	1.246	1.575	1.847	2.398
	STD	0.375	1.000	1.000	1.000	1.158	1.359	1.764
	40	0.500	1.000	1.000	1.000	1.000	1.000	1.091
18	10	0.250	1.044	1.340	1.840	2.325	2.728	3.542
	STD	0.375	1.000	1.000	1.000	1.182	1.386	1.800
	30	0.438	1.000	1.000	1.000	1.000	1.070	1.388
	40	0.562	1.000	1.000	1.000	1.000	1.000	1.000
20	10	0.250	1.062	1.364	1.873	2.367	2.777	3.605
	STD	0.375	1.000	1.000	1.000	1.203	1.411	1.832
	30	0.500	1.000	1.000	1.000	1.000	1.000	1.133
	40	0.593	1.000	1.000	1.000	1.000	1.000	1.000
24	10	0.250	1.096	1.406	1.931	2.441	2.864	3.718
	STD	0.375	1.000	1.000	1.000	1.241	1.456	1.890
	XS	0.500	1.000	1.000	1.000	1.000	1.000	1.169
	30	0.562	1.000	1.000	1.000	1.000	1.000	1.000

$S_b = [(i_i M_i)^2 + (i_o M_o)^2]^{1/2}/Z$
= equivalent bending stress, psi
$S_t = M_t/(2Z)$ = torsional stress, psi
M_i = in-plane bending moment, in · lb
M_o = out-of-plane bending moment, in · lb
M_t = torsional moment, in · lb
i_i = in-plane stress-intensification factor obtained from Fig. B4.5*a*
i_o = out-of-plane stress-intensification factor obtained from Fig. B4.5*a*
Z = section modulus of pipe, in^3

ANSI/ASME B31.5 Refrigeration Piping Code.[11] The scope of this code covers refrigerant and secondary coolant piping for temperatures as low as −320°F. Excluded from this code are piping designed for external or internal gauge pressure not exceeding 15 psi regardless of size; water piping; and any self-contained or

TABLE B4.4 (*e*) SIFs for 6000# Half-Couplings (Branch Pipe Schedule 160) (run pipe size: 1½ to 8 in)

Run pipe			Sch. 160 branch pipe size and thickness					
			½ in	¾ in	1 in	1¼ in	1½ in	2 in
Size, in	Sch.	*T*, in	0.188	0.219	0.250	0.250	0.281	0.344
1½	10*s*	0.109	2.763	3.664	3.665	3.665	3.665	3.665
	40	0.145	1.711	2.459	2.989	2.989	2.989	2.989
	80	0.200	1.000	1.431	2.362	2.362	2.362	2.362
	160	0.281	1.000	1.822	1.822	1.822	1.822	1.822
2	10*s*	0.109	2.873	4.128	4.287	4.287	4.287	4.287
	40	0.154	1.610	2.313	2.979	3.359	3.359	3.359
	80	0.218	1.000	1.290	1.661	2.613	2.613	2.613
	160	0.344	1.000	1.000	1.852	1.852	1.852	1.852
2½	10*s*	0.120	2.529	3.633	4.580	4.580	4.580	4.580
	40	0.203	1.048	1.505	1.939	3.161	3.161	3.161
	80	0.276	1.000	1.000	1.157	2.528	2.528	2.528
	160	0.375	1.000	1.000	1.000	2.008	2.008	2.008
3	10*s*	0.120	2.617	3.759	4.843	5.249	5.249	5.249
	40	0.216	1.000	1.405	1.809	2.481	2.859	3.480
	80	0.300	1.000	1.000	1.042	1.429	2.747	2.747
	160	0.438	1.000	1.000	1.000	1.000	2.073	2.073
4	10*s*	0.120	2.732	3.925	5.056	6.239	6.239	6.239
	40	0.237	1.000	1.257	1.619	2.220	2.558	3.642
	80	0.337	1.000	1.000	1.000	1.230	1.417	2.017
	160	0.531	1.000	1.000	1.000	1.000	1.000	1.168
6	10*s*	0.134	2.427	3.487	4.492	6.159	7.098	7.534
	40	0.280	1.000	1.017	1.310	1.796	2.070	2.947
	80	0.432	1.000	1.000	1.000	1.000	1.001	1.425
	120	0.562	1.000	1.000	1.000	1.000	1.000	1.000
8	10*s*	0.148	2.150	3.089	3.980	5.456	6.288	8.425
	20	0.250	1.000	1.287	1.658	2.273	2.619	3.728
	40	0.322	1.000	1.000	1.086	1.488	1.715	2.442
	80	0.500	1.000	1.000	1.000	1.000	1.000	1.168

unit systems subject to the requirements of Underwriters' Laboratories or other nationally recognized testing laboratory.

The limits of calculated stresses are as follows:

1. *Stresses due to expansion loads:* The expansion stress range S_E shall not exceed the allowable stress range S_A:

$$S_E < S_A = f(1.25S_c + 0.25S_h) \tag{B4.32}$$

where S_E = expansion stress range = $(S_b^2 + 4S_t^2)^{1/2}$, psi
S_b = resultant bending stress
= $[(i_iM_i)^2 + (i_oM_o)^2]^{1/2}/Z$, psi
S_t = torsional stress = $M_t/(2Z)$, psi
M_i = in-plane bending moment, in · lb
M_o = out-of-plane bending moment, in · lb
M_t = torsional moment, in · lb
i_i = in-plane stress-intensification factor obtained from Fig. B4.5*a*

TABLE B4.4 *(f)* SIFs for 6000# Half-Couplings (Branch Pipe Schedule 160) (run pipe size: 10 to 24 in)

Run pipe			Sch. 160 branch pipe size and thickness					
			½ in	¾ in	1 in	1¼ in	1½ in	2 in
Size, in	Sch.	*T*, in	0.188	0.219	0.250	0.250	0.281	0.344
10	10*s*	0.165	1.862	2.674	3.445	4.723	5.444	7.749
	20	0.250	1.000	1.336	1.721	2.360	2.720	3.872
	40	0.365	1.000	1.000	1.000	1.254	1.445	2.057
	80	0.594	1.000	1.000	1.000	1.000	1.000	1.000
12	10*s*	0.180	1.657	2.381	3.067	4.205	4.846	6.898
	STD	0.375	1.000	1.000	1.000	1.234	1.422	2.024
	40	0.406	1.000	1.000	1.000	1.081	1.245	1.773
	80	0.687	1.000	1.000	1.000	1.000	1.000	1.000
14	10	0.250	1.000	1.398	1.800	2.468	2.845	4.050
	STD	0.375	1.000	1.000	1.000	1.254	1.445	2.057
	40	0.438	1.000	1.000	1.000	1.000	1.115	1.587
	60	0.594	1.000	1.000	1.000	1.000	1.000	1.000
16	10	0.250	1.000	1.430	1.842	2.525	2.910	4.142
	STD	0.375	1.000	1.000	1.000	1.283	1.479	2.105
	40	0.500	1.000	1.000	1.000	1.000	1.000	1.301
	60	0.656	1.000	1.000	1.000	1.000	1.000	1.000
18	10	0.250	1.015	1.458	1.879	2.576	2.969	4.226
	STD	0.375	1.000	1.000	1.000	1.309	1.508	2.147
	30	0.438	1.000	1.000	1.000	1.010	1.164	1.657
	40	0.562	1.000	1.000	1.000	1.000	1.000	1.092
20	10	0.250	1.033	1.485	1.912	2.622	3.022	4.302
	STD	0.375	1.000	1.000	1.000	1.333	1.536	2.186
	30	0.500	1.000	1.000	1.000	1.000	1.000	1.352
	40	0.593	1.000	1.000	1.000	1.000	1.000	1.014
24	10	0.250	1.066	1.531	1.972	2.704	3.116	4.436
	STD	0.375	1.000	1.000	1.002	1.374	1.584	2.255
	30	0.562	1.000	1.000	1.000	1.000	1.000	1.147
	40	0.687	1.000	1.000	1.000	1.000	1.000	1.000

i_o = out-of-plane stress-intensification factor obtained from Fig. B4.5*a*

Z = section modulus of pipe, in^3

S_c = basic material allowable stress at minimum (cold) normal temperature, psi

S_h = basic material allowable stress at maximum (hot) normal temperature, psi

f = stress-range reduction factor obtained from Fig. B4.5*d*

2. *Stresses due to sustained loads:* The sum of the longitudinal stresses due to pressure, weight, and other sustained external loading S_L shall not exceed S_h. Where $S_L < S_h$, the difference $(S_h - S_L)$ may be added to the term in parentheses in Eq. (B4.32).

3. *Stresses due to occasional loads:* The sum of the longitudinal stresses produced by pressure, live and dead loads, and those produced by occasional

loads, such as wind or earthquake, may not exceed $1.33S_h$. It is not necessary to consider wind and earthquake as occurring concurrently.

ANSI/ASME B31.8, Gas Transmission and Distribution Piping Code.[12] This code governs most of the pipelines in gas transmission and distribution systems up to the outlet of the customer's meter set assembly. Excluded from this code are piping with metal temperatures above 450°F or below −20°F; piping beyond the outlet of the customer's meter set assembly; piping in oil refineries or natural gas extraction plants, gas treating plants, and so on, which is covered by other ANSI/ASME B31 codes; waste gas vent pipe operating at atmospheric pressures; and liquid petroleum transportation piping. The governing equations are as follows:

1. *Stresses due to pressure and external loads:* The sum of the longitudinal pressure stress and the longitudinal bending stress due to external loads such as weight, wind, and so on, S_L, shall not exceed $0.75S_yFT$:

$$S_L \le 0.75S_yFT \tag{B4.33}$$

where S_y = specified minimum yield strength, psi
T = temperature derating factor obtained from Table B4.5
F = construction-type design factor obtained from Table B4.6. The construction types are associated with the population density of the surrounding area as follows:
Type A: Sparsely populated areas such as deserts, mountains, and farmland
Type B: Fringe areas around cities or towns
Type C: Cities or towns with no buildings over three stories tall
Type D: Areas with taller buildings

2. *Stress range due to expansion loads:* The maximum combined expansion stress range S_E shall not exceed $0.72S_y$:

$$S_E = (S_b^2 + 4S_t^2)^{1/2} < 0.72S_y \tag{B4.34}$$

where S_b = resultant bending stress = iM_b/Z, psi
S_t = torsional stress = $M_t/(2Z)$, psi
M_b = resultant bending moment, in · lb
M_t = torsional moment, in · lb

TABLE B4.5 Temperature Derating Factor T for Steel Pipe

Temperature, °F	Temperature derating factor T
250 or less	1.000
300	0.967
350	0.933
400	0.900
450	0.867

Note: For intermediate temperatures, interpolate for derating factor.
Source: ASME B31.8, 1989, "Gas Transmission and Distribution Piping Systems." (*Courtesy of ASME.*)

TABLE B4.6 Values of Design Factor F

Construction type	Design factor F
A	0.72
B	0.60
C	0.50
D	0.40

Source: ASME B31.8, 1989. (*Courtesy of ASME.*)

i = stress-intensification factor obtained from Fig. B4.5*a* (see also figure note 10)
Z = section modulus of pipe, in^3

3. *Stresses due to pressure, external loads, and expansion loads:* The sum of the longitudinal pressure stress, the longitudinal bending stress due to external loads, and the combined stress due to expansion shall not exceed S_y.

Local Stress. In addition to the general pipe stresses (the pressure stress and the moment stress) as described in the above sections, there are certain local pipe wall stresses produced either by (1) restraint of the pipe radial thermal and internal pressure expansion of pipe-through-structural-steel-type anchors, (2) the transfer of load from the supporting surface to the pipe surface over a contact length along the axis of the pipe, or (3) attachments welded to pipe (e.g., lugs and trunnions).

1. *Local stresses and code requirements:* The local stresses S_L, $S_L{}^*$, $S_L{}^{**}$, and $S_L{}^{***}$ can be expressed as follows:

$$S_L, S_L{}^*, S_L{}^{**}, \text{ or } S_L{}^{***}$$

$$= \text{max. of } [|\sigma_1 + \sigma_1'|, |\sigma_2 + \sigma_2'|, |(\sigma_2 + \sigma_2') - (\sigma_1 + \sigma_1')|] \quad \text{(B4.35)}$$

where S_L = local stress due to dead weight, psi
$S_L{}^*$ = local stress due to dead weight, seismic inertia, and other dynamic loads, psi
$S_L{}^{**}$ = local stress due to thermal expansion and seismic anchor movement, psi
$S_L{}^{***}$ = local stress due to concurrently acting loads, psi
σ_1 = longitudinal membrane stress, psi
σ_2 = circumferential membrane stress, psi
σ_2' = circumferential bending stress, psi
σ_1' = longitudinal bending stress, psi

Strictly speaking, the present piping codes give no specific limits for local stresses. As an industry practice, the calculated local stress is added to the general pipe stress and then compared with the pipe stress allowables specified by the applicable code. As an example, the total (general plus local) pipe stresses for ASME Class 2 and 3 piping shall satisfy the following equations [see Eqs. (B4.9) to (B4.12) for definitions of symbols]:

a. Design loading:

$$\frac{B_1 P D_o}{2t} + \frac{B_2 M_A}{Z} + \left(\frac{2}{3}\right) S_L \le 1.5 S_h \quad \text{(B4.36)}$$

b. Service loadings:

$$\frac{B_1 P_{\max} D_o}{2t} + \frac{B_2(M_A + M_B)}{Z} + \left(\frac{2}{3}\right) S_L{}^*$$

$\leq 1.8S_h$, but not greater than $1.5S_y$ for levels A and B loadings

$\leq 2.25S_h$, but not greater than $1.8S_y$ for level C loadings

$\leq 3.0S_h$, but not greater than $2.0S_y$ for level D loadings (B4.37)

c. Sustained and thermal expansion loadings: Either one of the following equations shall be satisfied:

$$\frac{iM_C}{Z} + \left(\frac{1}{2}\right) S_L{}^{**} \leq S_A \tag{B4.38}$$

or
$$\frac{PD_o}{4t} + \frac{0.75iM_A}{Z} + \frac{iM_C}{Z} + \left(\frac{2}{3}\right) S_L + \left(\frac{1}{2}\right) S_L{}^{**} \leq S_h + S_A \tag{B4.39}$$

d. Local stress limit loading:

$$S_L{}^{***} < 2.5S_y \tag{B4.40}$$

2. *Local stress due to restraint of pipe radial expansion:* The membrane and flexural stresses can be calculated as follows[13]:

σ_1 is negligible

$$\sigma_1{}' = \frac{\pm 12D\lambda^2(\delta_P + \delta_T)}{t^2} \tag{B4.41}$$

$$\sigma_2 = \frac{-(\delta_P + \delta_T)E}{R} + \nu\sigma_1 \tag{B4.42}$$

$$\sigma_2{}' = \nu\sigma_1{}' \tag{B4.43}$$

$$\delta_P = \frac{PR^2}{Et} \tag{B4.44}$$

$$\delta_T = \alpha R \Delta T \tag{B4.45}$$

where P = internal pressure of pipe, psi
R = pipe outside radius, in
E = modulus of elasticity of pipe, psi
t = pipe wall thickness, in
α = coefficient of thermal expansion of pipe, in/(in · °F)
ΔT = range of thermal expansion temperatures, °F
ν = Poisson's ratio
$D = Et^3/[12(1 - \nu^2)]$
$\lambda = [3(1 - \nu^2)/(Rt)^2]^{1/4}$

In the case of the local stress produced by restraint to the pipe radial expansion, $S_L = S_L^* = 0$. For fillet weld, $i = 2.1$ should be used. In addition, the stress check on limit loading is required. Here, $S_L^{***} = S_L^{**}$.

3. *Local stress due to contact:* The membrane and flexural stresses can be calculated as follows[13]:

   a. For line contact:

$$\sigma_1 = -0.52F(R_m)^{1/4}(L)^{-1/2}(t)^{-7/4} \tag{B4.46}$$

(*Note*: σ_1' is considered to be included in this equation.)

$$\sigma_2 = -0.496F(R_m)^{3/4}(L)^{-3/2}(t)^{-5/4} \tag{B4.47}$$

$$\sigma_2' = \pm 2\sigma_1 \tag{B4.48}$$

where F = support load, lb
R_m = mean radius of pipe, in
L = contact length, in
t = pipe wall thickness, in

   b. For point contact: σ_1 and σ_1' are negligible compared to $(\sigma_2 + \sigma_2')$:

$$\sigma_2 = \frac{0.4F}{t^2} \tag{B4.49}$$

$$\sigma_2' = \frac{2.4F}{t^2} \tag{B4.50}$$

   c. In the case of contact stress, the minimum nominal general pipe stress (i.e., the unintensified general pipe stress) may be used in Eqs. (B4.36) to (B4.39) to calculate the total pipe stress. In addition, the stress check on limit loading is not required.

4. *Other types of local stresses:* The above two types of local stresses are commonly encountered by stress analysts. Detailed descriptions and analysis methods for other types of local stresses such as the local stresses at integral welded attachments to pipe (e.g., lugs and trunnions) can be found in technical publications, *Welding Research Council Bulletins 107* and *198* and ASME Code Cases.[14–17]

TYPES OF PIPE LOADING CONDITIONS

1. *Types of loads:* As mentioned previously in the subsection "Classification of Loads," the piping loads are classified into three types: sustained loads, occasional loads, and expansion loads. These three types of loads and the corresponding analysis will be discussed in this section in detail.
2. *Method of analysis:* The piping stress analysis to be performed could be a simplified analysis or a computerized analysis. The choice of the proper analysis depends on the pipe size and the piping code. For small (nominal diameter 2 in and under) pipe except nuclear Class 1 pipe, a cookbook-type, simplified analysis could be performed. For nuclear Class 1 piping, since the require-

ments are more stringent, a computerized analysis is required. A detailed description of a cookbook-type, simplified analysis and a brief description of a computerized analysis are given in the section that follows, "Methods of Analysis." Generally, before computerized analysis is performed, pipe supports may be located using the cookbook method.

Sustained Load: Pressure

Internal pressure in piping usually induces stresses in the pipe wall rather than loads on the pipe supports. This is because pressure forces are balanced by tension in the pipe wall, resulting in zero pipe support loadings. A discussion of unbalanced forces in the pipe created by pressure waves during fluid transients is given in the subsection "Dynamic Loads."

Pressure Stress. The longitudinal stress developed in the pipe due to internal pressure can be calculated as follows:

$$S_{LP} = \frac{PD}{4t}$$

or

$$S_{LP} = \frac{Pd^2}{D^2 - d^2} = P\left(\frac{A_f}{A_m}\right) \tag{B4.51}$$

where S_{LP} = longitudinal stress, psi
P = internal design pressure, psig
D = outside diameter, in
d = inside diameter, in
A_f = flow area, in^2
A_m = metal area, in^2
t = pipe wall thickness, in

The second equation gives pressure stress in terms of the ratio of pipe flow area to metal area. It also provides a more accurate result. Both equations are acceptable to the code.

Expansion Joint. In piping design, elbows, bends, and pipe expansion loops normally provide adequate flexibility for piping thermal expansion and contraction. However, in some cases, this flexibility may not be adequate. As a solution, expansion joints may be used to absorb the expansion and contraction of pipe.

In general, expansion joints are used for the following applications:

1. Where thermal movements would induce excessive stress in normal piping arrangements
2. Where space is inadequate
3. Where reactions transmitted by pipe supports or anchors create large loads on supporting structures
4. Where reactions to equipment terminals are in excess of allowables

When expansion joints are used in piping, the pressure forces can no longer be balanced by tension in the pipe wall, and the pressure forces will be resisted by pipe supports and anchors.

There are many types of expansion joints available, ranging from a piece of rubber hose to metal bellows. The metal bellows expansion joint is most commonly used for power or process piping. Figure B4.6 shows the various components of a bellows expansion joint.

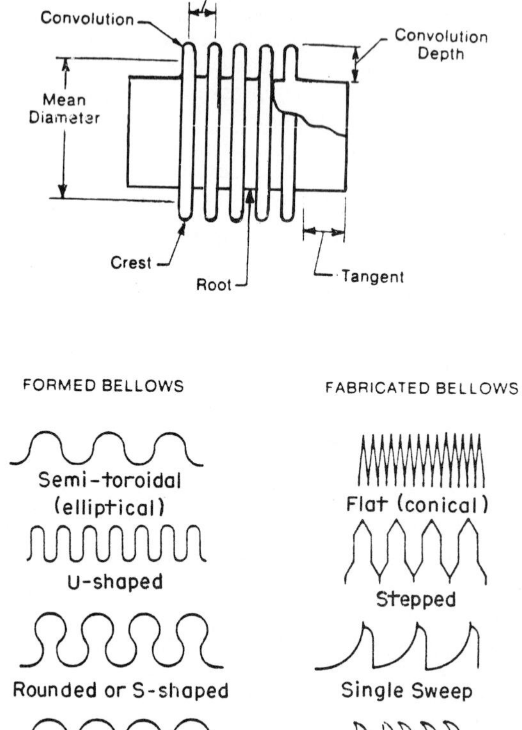

FIGURE B4.6 Bellows expansion joints.

Expansion joints do not have the capability to transmit large pressure forces. Restraints are usually installed on both sides of the expansion joint to prevent the pressure force from pulling apart the joint. The pressure force developed in the expansion joint is equal to the internal pressure times the maximum cross-sectional area over which it is applied. Since an expansion joint increases the flexibility of a piping system, the flexibility (spring rate) of the expansion joint should be incorporated in the piping stress analysis. Typical axial spring rates of bellows can be founded in Ref. 18.

Sustained Load: Weight

The total design weight load of pipe supports includes the weight of the pipe, fittings, insulation, fluid in pipe, piping components such as valves, valve operators, flanges, and so on, and the supports themselves. Supports should be located as specified in Chap. B5.

Hydrotest and Other Occasional Loadings. To assure the integrity and leak tightness of a piping system designed to Section III of ASME Boiler and Pressure Vessel Code or ANSI/ASME B31.1, the codes require that a pressure test be performed prior to placing the system in service. The most commonly used test is the hydrostatic test. When a steam or gas piping system is to be hydrotested, the effects of the weight of the water on the system and its supports must be considered. A hydroweight stress analysis should be performed to assure that the pipe supports, which have been designed for the normal operating condition, are able to withstand the hydrotest loads. If permanent supports cannot withstand these hydrotest loads, temporary supports may be added. Spring supports are available with hydrostatic test stops, which, in effect, transform the units into rigid supports.

Whether or not required by code, other conditions, such as the added weight of a cleaning medium of density greater than that of the process fluid, must be considered in a manner similar to that discussed above. Both dynamic and static loading analyses may be impacted by flushing and blowing-out activities during construction or after major rework.

Thermal Expansion Loads

For weight analysis, the more pipe supports installed, the lower the stress developed in the pipe. However, the opposite is true for the case of piping thermal expansion. When thermal expansion of the piping due to fluid or environmental temperature is restrained at supports, anchors, equipment nozzles, and penetrations, large thermal stresses and loads are caused.

Thermal Modes. Piping systems are generally analyzed for one thermal condition or mode, that is, the maximum operating temperature. However, piping systems that have more than one operating mode with different operating temperatures concurrently in different parts of the piping system should be analyzed for these operating thermal modes.

With the aid of system flow diagrams or *piping and instrumentation drawings* (*P&ID*), the stress analyst can determine the thermal modes required for a particular piping system. For B31.1 piping and ASME Class 2 and 3 piping, the required thermal modes can be determined by using good engineering judgment in selecting the most severe thermal conditions. For ASME Class 1 piping, the required thermal modes can be determined by examining the load histograms specified in the design specification.

Free Thermal Analysis. During the initial stage of piping analysis, an unrestrained (i.e., no intermediate pipe supports) or free thermal analysis may be performed. This analysis is performed for the worst thermal mode and includes only terminal points such as penetrations, anchors, and equipment nozzles. The result of this free thermal analysis usually gives useful information, which can be utilized by the stress analyst in the later stages of the piping analysis. Generally, a resulting thermal expansion stress < 10 ksi means adequate flexibility exists in the piping system. The piping locations with low resulting thermal displacements would be good locations where rigid supports may be installed without adversely affecting the flexibility of the piping system. The resulting equipment nozzle loads could be used to evaluate the capabilities of the equipment for meeting equipment manufacturer's nozzle allowables.

Imposed Thermal Movements. Thermal expansion of equipment causes displacements in the attached piping. Thermal stresses may also be caused due to thermal anchor movements at terminal ends and intermediate restraints. Therefore, appropriate thermal analysis for thermal anchor movements relating to the respective thermal modes should also be performed. Sometimes, it is possible for thermal anchor movements to exist when the piping is cold. In such cases, analysis in the cold condition, with only the thermal anchor movements as input, may be required.

LOCA Thermal Analysis. In nuclear power plants, following a *loss-of-coolant accident* (*LOCA*), the containment (the building structure designed to contain fission products) expands due to the rise in temperature and pressure inside the containment. This containment thermal growth results in large containment penetration anchor movements which affect the connected piping. It is not required to qualify the piping for this faulted condition. Thermal analysis for these LOCA anchor movements is used only for the evaluation of flanges, equipment nozzle loads, and pipe support loads.

Temperature Decay. For piping systems having a portion of the system with stagnant branch lines (dead legs) as shown in Fig. B4.7, it is necessary to consider the temperature decay in the piping. One simple approach to this temperature attenuation problem is as follows:

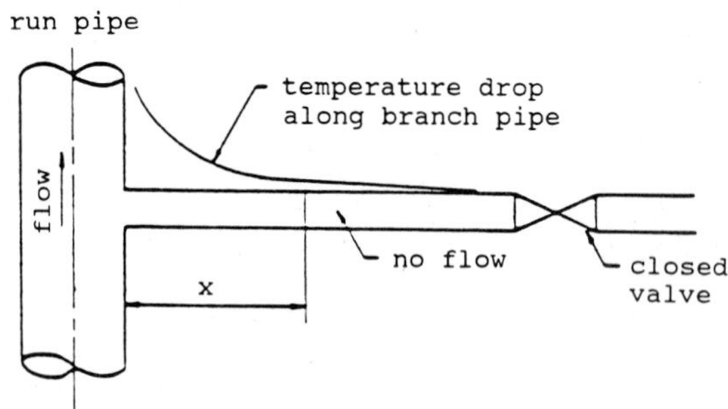

FIGURE B4.7 Temperature decay at dead leg.

1. For a piping system with water, the temperature of the branch pipe is assumed the same as the run pipe up to a length equivalent to 10 times the inside pipe diameter. The remaining portion of the branch pipe may be considered at ambient temperature.

2. For piping system with steam or gas, the temperature of the branch pipe is considered the same as the run pipe up to the closed valve.

For cases such as thermal transient analysis of ASME Class 1 piping, where a more accurate temperature profile along the branch pipe may be required, the approach described in Ref. 19 should be used.

Stress Ranges. The thermal stresses developed in the pipe are in fact "stress ranges," that is, the difference between the unit thermal expansion for the highest operating temperature and for the lowest operating temperature.

For piping systems that do not experience temperatures below ambient temperature, the stress range is the difference between the unit expansion for the maximum thermal mode and that for 70°F. (See later subsection "Seismic Anchor Movement and Building Settlement Analysis.")

For systems with supply from a pool or river which might go below 70°F in the winter, negative coefficients of expansion should be considered in evaluating the stress range.

Occasional Loads: Seismic

The code of Federal Regulation 10CFR Part 50 requires that safety-related piping in nuclear power plants be designed to withstand seismic loadings without loss of capability to perform their function.[20] For nonnuclear piping in regions of high seismic activity, this design requirement should also be considered.

OBE and SSE. Nuclear piping systems and components classified as Seismic Category I are designed to withstand two levels of site-dependent hypothetical earthquakes: the *safe-shutdown earthquake* (*SSE*) and the *operational-basis earthquake* (*OBE*).[21]

For conservatism, the OBE must usually be equal to at least one-half of the SSE. Their magnitudes are expressed in terms of the gravitational acceleration g. Their motions are assumed to occur in three orthogonal directions: one vertical and two horizontal.

Seismic Category I systems are defined as those necessary to assure:

1. The integrity of the reactor coolant pressure boundary
2. The capability to shut down the reactor and maintain it in a safe shutdown condition
3. The capability to prevent or mitigate potential off-site radiation exposure

Types of Seismic Analysis. Generally, piping seismic analysis is performed through one of three methods: time-history analysis, modal response spectrum analysis, or static analysis.

The equation of motion for a piping system subjected to an externally applied loading (seismic excitation) may be expressed as

$$M\ddot{\mathbf{x}} + C\dot{\mathbf{x}} + K\mathbf{x} = \mathbf{f} \qquad \text{(B4.52)}$$

where M = mass matrix of system

C = damping matrix

K = stiffness matrix
$\ddot{\mathbf{x}}$ = acceleration vector
$\mathbf{x}$ = velocity vector
$\mathbf{x}$ = displacement vector
$\mathbf{f}$ = external loading vector, function of time

This equation could be solved by time-history analysis.

Time-History Analysis. Time-history analysis is based on hypothetical earthquake data in the form of ground displacement, velocity, or acceleration versus time. The piping system is represented by lumped masses connected by massless elastic members. The analysis is performed on this mathematical model by the direct numerical integration method.[22,23] At each time step, the piping stresses, displacements, and restraint loads are calculated. Time history simulates the behavior of the piping system during the seismic excitation. The main advantage of time-history analysis is that analytically it is more accurate and less conservative compared to other approaches. The main disadvantages of time-history analysis are the excessive computational time required and the difficulty of obtaining a realistic earthquake input time function.

Modal Response Spectrum Analysis. The seismic response spectrum is a plot of the maximum acceleration response of a number of idealized single-degree-of-freedom oscillators attached to the floor (structure) with certain damping.

These response spectra are based on design response spectra and specified maximum ground accelerations of the plant site. Usually, a series of curves with different damping values for operating and design basis earthquakes for each orthogonal direction are generated, as shown in Fig. B4.8.

In the modal response spectrum analysis, the piping system is idealized as lumped masses connected by massless elastic members. The lumped masses are

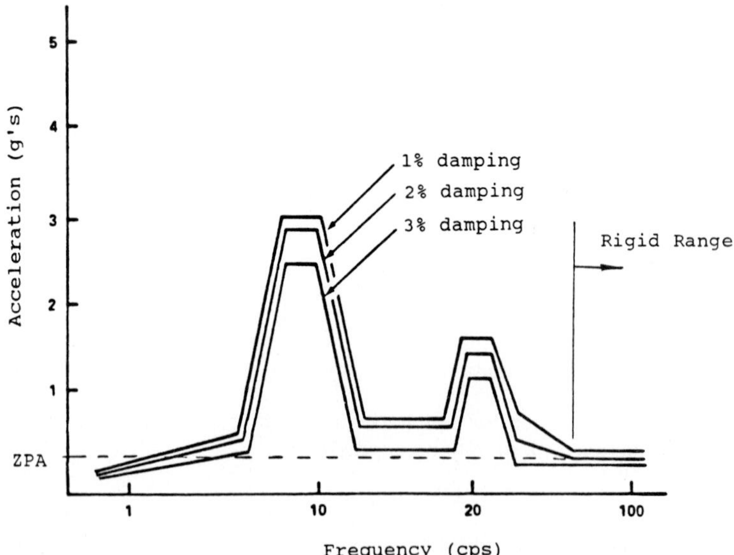

FIGURE B4.8 Response spectrum curves.

carefully located to adequately represent the dynamic properties of the piping system.

After the stiffness and mass matrix of the mathematical model are calculated, the natural frequencies of the piping system and corresponding mode shapes for all significant modes of vibration are also determined using the following equation:

$$(K - W_n^2 M)\phi_n = 0 \tag{B4.53}$$

where K = stiffness matrix
W_n = natural circular frequency for the nth mode
M = mass matrix
ϕ_n = mode shape matrix for the nth mode

The modal spectral acceleration taken from the appropriate response spectrum is then used to find the maximum response of each mode:

$$(Y_n)_{\max} = \frac{\phi_n^t MDSa_n}{W_n^2 M_n} \tag{B4.54}$$

where Sa_n = spectral acceleration value for the nth mode
D = earthquake direction coefficient
$\phi_n{}^t$ = transpose of the nth mode shape
M_n = generalized mass of the nth mode
Y_n = generalized coordinate for the nth mode

Using the maximum generalized coordinate for each mode, the maximum displacements, the effective inertia forces, the effective acceleration, and the internal forces and moments associated with each mode are calculated as follows:

$$X_n = \phi_n(Y_n)_{\max}$$

$$F_n = KX_n$$

$$a_n = M^{-1}F_n$$

$$L_n = bF_n \tag{B4.55}$$

where X_n = displacement matrix due to nth mode
F_n = effective inertia force matrix due to nth mode
a_n = effective acceleration matrix due to nth mode
M^{-1} = the inverse of mass matrix
L_n = internal force and moment matrix due to nth mode
b = force transformation matrix

These modal components are then combined by the appropriate method (see later subsection "Methods for Combining System Responses") to obtain the total displacements, accelerations, forces, and moments for each point in the piping system.

Two types of response spectrum analyses can be performed depending on the pipe routing and attachments to buildings and structures.

Single-Response Spectrum Analysis. This type of analysis is performed using an enveloped response spectrum curve that covers all buildings and elevations to which the piping system is attached.

Multiple-Response Spectrum Analysis. This type of analysis is used where the piping is attached to various buildings or structures that have a wide variation in the amplitude or frequency of accelerations. In such cases, various response spectra curves may be applied at corresponding support and anchor points in the piping system.[24,25]

Static Analysis. Static analysis may be used to evaluate power piping or some piping systems in nuclear power plants. It is performed by analyzing a piping system for the statically applied uniform load equivalent to the site-dependent earthquake accelerations in each of the three orthogonal directions. All rigid restraints and snubbers supporting the pipe in the direction of the earthquake acceleration are included in the analysis. The total seismic effect is obtained by combining the results of the three directions.

The minimum earthquake force for structures described in ANSI A58.1[26] is also one form of static seismic analysis. The code recommends that a lateral seismic force will be assumed to act nonconcurrently in the direction of each of the main axes of the structure in accordance with the formula:

$$V = ZIKCSW \tag{B4.56}$$

where V = lateral seismic force, lb

Z = numerical coefficient, dependent upon the earthquake zone (see Fig. B4.9), 0.1 for Zone 0, 0.25 for Zone 1, 0.50 for Zone 2, and 1.00 for Zone 3

I = occupancy importance factor, usually between 1.0 and 1.5

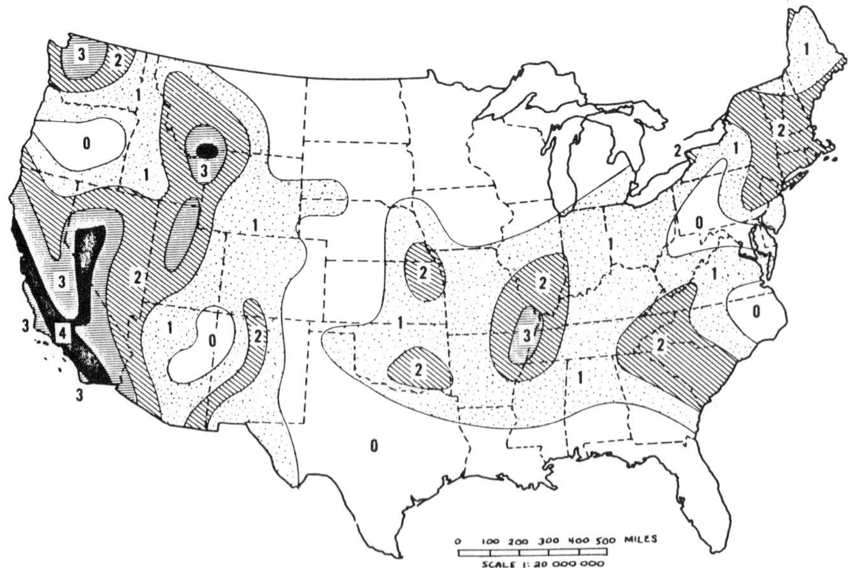

FIGURE B4.9 Map for seismic zones, contiguous 48 states. (*ANSI A58.1, 1982. Courtesy of ANSI.*)

K = horizontal force factor, dependent upon the arrangement of lateral force-resisting elements, usually between 0.67 and 2.50
C = $1/(15T^{1/2})$ but not to exceed 0.12
T = fundamental period of structure, s
S = soil factor, dependent upon the soil profile type, usually between 1.0 to 1.5
W = total dead weight of structure, lb

Damping. Damping is the phenomenon of dissipation of energy in a vibrating system. Each damping value expressed as a percentage of the critical damping is represented in the seismic response spectrum by a separate curve. The higher the damping value, the lower would be the effects of the seismic excitation. The damping values to be used for different levels of the earthquake are given by the NRC (U.S. Nuclear Regulatory Commission) Regulatory Guide 1.61,[27] as shown in Table B4.7.

When a system has both categories of pipe sizes mentioned in the table, dual damping values should be considered in the analysis.

Alternative damping values for response spectrum analysis of ASME Classes 1, 2, and 3 piping are given in ASME Code Case N-411-1,[28,29] as shown in Fig. B4.10. These damping values are applicable to both OBE and SSE. They are also independent of pipe size. As can be seen from Fig. B4.10, the damping values of Code Case N-411-1 are generally higher than the damping values given in Regulatory Guide 1.61. The industry has been applying these higher damping values to existing piping systems to reduce the number of snubbers installed in the plants in order to save snubber maintenance cost. The use of Code Case N-411-1 is acceptable to the NRC subject to the conditions described in the NRC Regulatory Guide 1.84.[30]

Mass Point Spacing. In a seismic analysis, the piping is represented by lumped masses connected by massless elastic members. The locations of these lumped masses are referred to as the *mass points*. In order to accurately represent the piping, the mass points on straight runs of pipe should be no further apart than a

TABLE B4.7 Damping Values
(Percent of critical damping)

Structure or component	Operating-basis earthquake or ½ safe-shutdown earthquake	Safe-shutdown earthquake
Equipment and large-diameter piping systems, pipe diameter greater than 12 in	2	3
Small-diameter piping systems, diameter equal to or less than 12 in	1	2
Welded steel structures	2	4
Bolted steel structures	4	7
Prestressed concrete structures	2	5
Reinforced concrete structures	4	7

Source: U.S. Nuclear Regulatory Commission, Regulatory Guide 1.61.

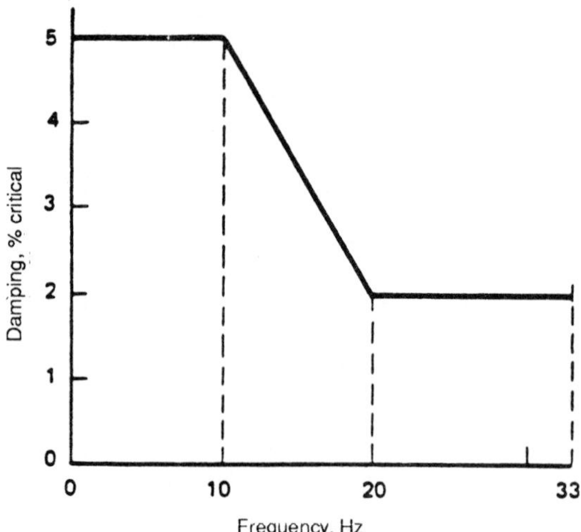

FIGURE B4.10 Code Case N-411 damping values. (*ASME B & PV Code, Case N-411-1, February 1989. Courtesy of ASME.*)

length of pipe which would have a fundamental frequency of 33 Hz (see the later subsection "Cookbook-Type Analysis"). Mass points should also be located at all supports, concentrated weights such as valves, valve operators, flanges, and strainers, and at the end of cantilevered vents and drains. At least two mass points should be placed between supports in the same direction.[31]

Cutoff Frequency, Rigid Range, Zero Period Acceleration, and Missing-Mass Effect. Generally, the piping response spectrum analysis is terminated at a frequency called the *cutoff frequency.* The cutoff frequency is usually specified as the frequency beyond which the spectral acceleration remains constant, and this constant spectral acceleration is known as the *zero period acceleration* (*ZPA*) (see Fig. B4.8).

Supposing a piping system is so designed and supported that the first mode is higher than the cutoff frequency; then as far as the computer program is concerned, this piping system does not receive seismic excitation at all. Consequently, the result of this seismic analysis is invalid because of the artificial constraint specified by the stress analyst.

This phenomenon, known as the *missing-mass effect,*[32,33] could also occur in the following cases:

1. On pipe runs with axial restraint (support, anchor, or nozzle) where the longitudinal frequency could be higher than the cutoff frequency
2. Concentrated masses in a piping system supported in such a manner that the frequency of that portion of piping is high

Most of the computer programs normally used for piping stress analysis have the capability to evaluate the missing-mass effect. These programs usually utilize the acceleration from the spectrum at the cutoff frequency (ZPA) to calculate the missing-mass effect.

Methods for Combining System Responses. In general, there are two approaches for the combination of system responses. One approach, the *absolute sum method* (*ABS*), adds the peak system responses. The second approach, *square root–sum-of-squares method* (*SRSS*), gives a combined response equal to the square root of the sum of the squares of the peak responses. The SRSS method is preferred over the ABS method because not all the peak responses occur simultaneously.

In a response spectrum modal analysis, if the modes are not closely spaced (two consecutive modes are defined as closely spaced if their frequencies differ from each other by less than 10 percent of the lower frequency), responses could be combined by taking the SRSS method. For closely spaced modes, the NRC suggests that the method of combining the responses by the SRSS method may not be conservative. An acceptable method of grouping the closely spaced modes of vibration and combining the responses is described in the NRC Regulatory Guide 1.92.[34,35]

Seismic Anchor Movement and Building Settlement Analysis

A piping system, supported from two seismically independent structures that move out of phase during a seismic event, will experience stresses due to the differential displacement of the supports.

Buried pipe could be considered as supported by the soil. A differential movement during a seismic event between the soil and the building to which the pipe is routed could also cause stresses in the pipe.

Similarly, the differential settlements between two structures or between a building and the adjacent soil will induce stresses in piping which is routed between them.

Seismic Anchor Movement (SAM) Analysis. A seismic anchor movement analysis is required on a piping system where:

1. The piping is supported from two seismically independent structures, or
2. The piping is attached to large equipment having its own modes of vibration (e.g., steam generator, pressurizer, reactor vessel, or reactor coolant pump).

SAM analysis is performed by applying the corresponding seismic displacements of the building and structures at the pipe support and anchor locations. It is usually analyzed by a static method. However, dynamic supports such as snubbers and rigids (including anchors and nozzles) will be active while spring supports remain passive.

SAM displacements from the same building or structure are generally in phase, while those from different buildings or structures are considered out of phase.

When a terminal end of a piping system being analyzed is at a large pipe, the seismic movements from the large pipe analysis should be applied as a SAM displacement in the analysis.

The code allows the consideration of the stress due to SAM as either primary stress [see Eq. (B4.10)] or secondary stress [see Eq. (B4.11)]. However, it will usually be evaluated as secondary stress. Since the stress due to SAM is a cyclic type of stress, it should be combined with other cyclic-type secondary stresses such as thermal expansion stresses.

The total secondary stress range should include the thermal and SAM stress range. If the SAM stress is less than the thermal stress range, the effective secondary stress range is the sum of the SAM stress and the thermal stress range, as shown in Fig. B4.11*a*. If the SAM stress is higher than the thermal stress range, the effective secondary stress range then equals twice the SAM stress, as shown in Fig. B4.11*b*.

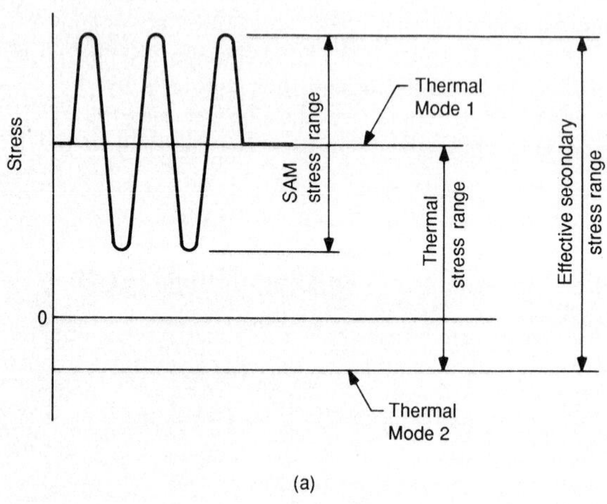

(a)

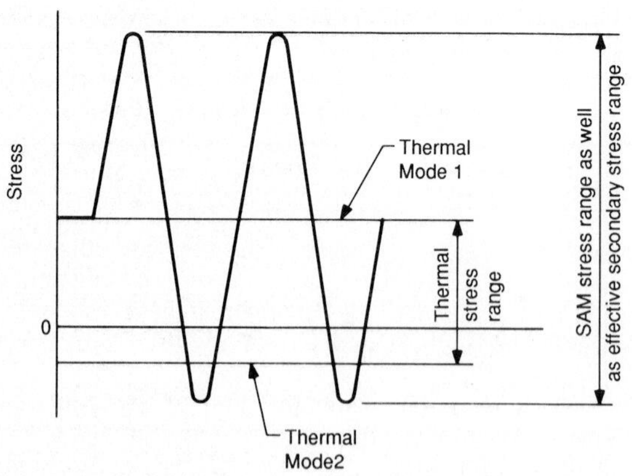

(b)

FIGURE B4.11 Effective secondary stress range.

Building Settlement Analysis. ASME Boiler and Pressure Vessel Code Section III requires that the stresses due to building settlement be evaluated and be considered as secondary stresses. However, the stress due to building settlement is a one-time (single nonrepeated) anchor movement. Therefore, it is not required to combine it with other stresses. From Subsection NC-3653.2(b) of the code, the effects of any single nonrepeated anchor movement shall meet Eq. (B4.13).

Dynamic Loads

The dynamic loads discussed herein are limited to occasional loads (other than seismic loads) frequently encountered in piping stress analysis.

Safety-Relief Valve Discharge Analysis. Safety-relief valves are installed for the purpose of protecting the fluid system from accidental overpressure, or venting the fluid generated in excess of requirement.

The general requirements pertaining to the design of the pressure relief discharge piping are provided in Appendix II of ANSI/ASME B31.1, Subsections NC-3677 and NB-3677 of the ASME Code for different pipe classes.

There are two types of pressure relief valve discharge, namely, open discharge and closed discharge, as shown in the figures of Chap. B3.

Open Discharge. A typical open discharge is the transient due to discharging of steam from a steam line to the atmosphere through relief valves or safety valves. When the steam line pressure reaches the valve set point, the valve opens and decompression waves will travel both upstream and downstream of the valve. This flow transient sets up pressure imbalances along each pipe segment (a straight run of pipe bounded by elbows). The transient forces can be calculated by a computerized method as described in the later subsection "Steam Hammer-Water Hammer Analysis," while the reaction force at the valve exit due to steady-state flow is determined relatively easily.

Closed Discharge. In a closed-discharge system, the fluid is transmitted to its terminal receiver through continuous discharge piping. A typical closed discharge is the transient induced by a sudden opening of the relief and safety valves located on top of the pressurizer in a power plant. A water seal, which is maintained upstream of each valve to minimize leakage, driven by this high discharge pressure, generates a transient thrust force at each pipe segment. The water seal is discharged ahead of the steam as the valve disk lifts. For discharge piping with a water seal, only the first cycle of each event has a transient force based on water in the seal. The remaining cycles would be based on steam occupying the seal piping, and the transient forces would be reduced in magnitude.

Static Analysis. The static method of open discharge described in Appendix II of ANSI/ASME B31.1 can be summarized as follows:

1. The reaction force F due to steady-state flow following the opening of the valve may be computed by

$$F = \frac{WV}{g} + (P - P_a)A \tag{B4.57}$$

where F = reaction force at exit, lbf
W = mass flow rate, lbm/s
V = exit velocity, ft/s

g = gravitational constant,
= 32.2 lbm · ft/lbf · s^2
P = static pressure at exit, psia
P_a = atmospheric pressure, psia
A = exit area, in^2

2. The dynamic load factor (DLF) is used to account for the increased load caused by the sudden application of the discharge load. The DLF value will range between 1.1 and 2.0, depending on the time history of the applied load and the natural frequency of the piping.[36] If the run pipe is rigidly supported and the applied load could be assumed to be a single ramp function, the DLF may be determined in the following manner:

a. Calculate the safety valve installation period *T*:

$$T = 0.1846 \left(\frac{Wh^3}{EI}\right)^{1/2} \tag{B4.58}$$

where T = safety valve installation period, s
W = weight of safety valve, installation piping, flanges, attachments, etc., lb
h = distance from run pipe to centerline of outlet piping, in
E = Young's modulus of inlet pipe, psi, at design temperature
I = moment of inertia of inlet pipe, in^4

b. Calculate the ratio t_o/T where t_o is the time the safety valve takes to go from fully closed to fully open (seconds).

c. For the ratio t_o/T, determine the DLF from data given in Appendix II of ANSI/ASME B31.1, as shown in Fig. B4.12.

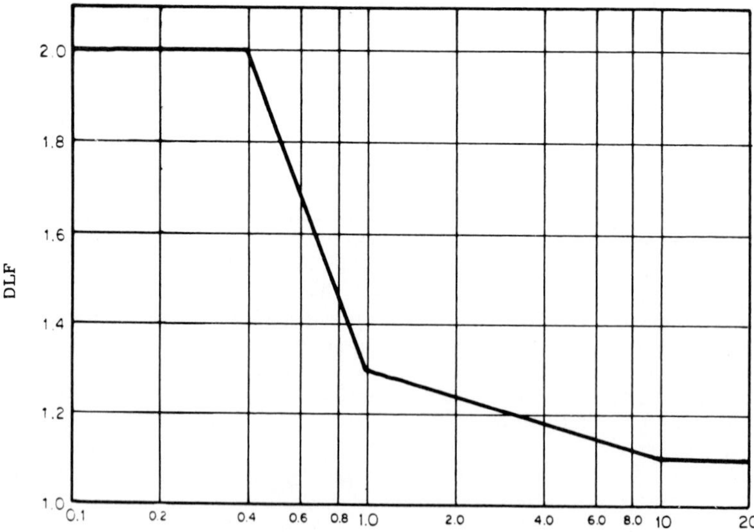

FIGURE B4.12 Hypothetical dynamic load factor (DLF).

3. The moment due to valve reaction force is calculated by simply multiplying the force times the distance from the point in the piping system being analyzed, times a suitable DLF. The stress is then calculated accordingly.

Dynamic Analysis. The reaction force effects are dynamic in nature. A time-history dynamic analysis of the discharge piping is considered to be more accurate. Furthermore, closed-discharge systems do not easily lend themselves to simplified analysis techniques. A time-history analysis (such as the one described in the following subsection "Steam Hammer-Water Hammer Analysis") is required to achieve realistic results.

Steam Hammer-Water Hammer Analysis. The steam hammer-water hammer event is often initiated by intentional actuation of certain flow control devices (main steam stop valve closure, feedwater pump trip, etc.), but in other cases a transient event could be introduced unintentionally as the result of some unforeseen operating condition, component malfunction, or accident (e.g., feedwater line check valve slam following a pipe break upstream of the check valve).[37,38] While these events may produce very complex transient fluid flow, the pipe stress analyst is interested in just the unbalanced force along the pipe segment tending to induce piping vibration.

Calculation of Unbalanced Forces. These time-history unbalanced forces are usually generated through a two-step computerized calculation. The fluid system is modeled as an assemblage of control volumes (e.g., piping volumes or steam generator) interconnected by junctions (e.g., valves, pump, or break). Piping fluid flow data, such as flow area, friction losses, valve closing-opening time, feed pump characteristics, or break characteristics, together with fluid initial conditions (flow rate, pressure, temperature, and mixture quality) are supplied as input to a thermal hydraulic finite difference computer program.[39]

Using this input information and a built-in steam table (fluid thermodynamic state), the first step solves the three equations of conservation (mass, momentum, and energy) at each time step for fluid properties such as pressure, velocity, internal energy, and mixture quality. A typical stop valve closure time history and its associated dynamic pressure time history are shown in Fig. B4.13. The second step utilizes a postprocessor. This postprocessor then accepts the output information from the first step and computes the unbalanced forces in piping segments by applying the momentum theorem.

Static Analysis. Static analysis is simple and saves computer time. It is used when the unbalanced forces are small and the total transient time is long. In the analysis, the peak values of the time-history fluid forcing functions at pipe segments are applied statically to the piping. The piping stress, deflections, and support-nozzle loads are then calculated by the computer program.

To obtain a conservative result for the static analysis, care must be taken in applying a proper dynamic load factor to the unbalanced forces.

Dynamic Analysis. The dynamic analysis generally utilizes either the direct step-by-step integration method (as described in the subsection "Pipe Break Analysis") or the modal-superposition method. In the dynamic analysis, the piping system is idealized as a mathematical model consisting of lumped masses connected by weightless elastic members. These lumped masses are carefully located to adequately represent the dynamic characteristics of the piping system. For computer programs utilizing the modal-superposition method, enough modes (or appropriate cutoff frequency) should be specified in the computer input such that the dynamic response of the piping system can be truly represented. There are no specific guidelines to damping values used in piping fluid transient dynamic anal-

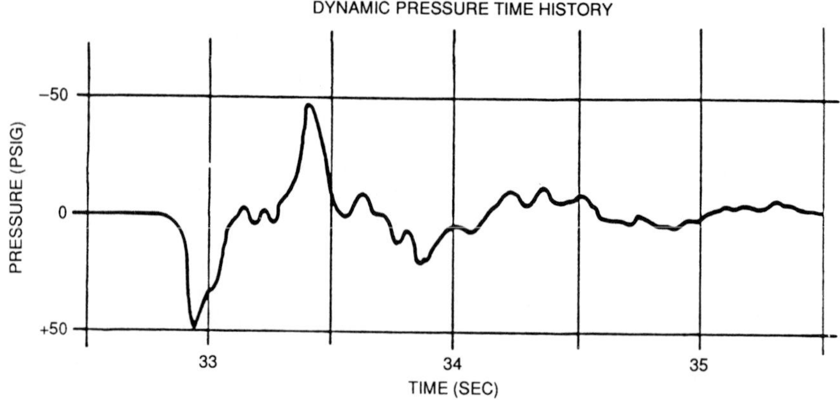

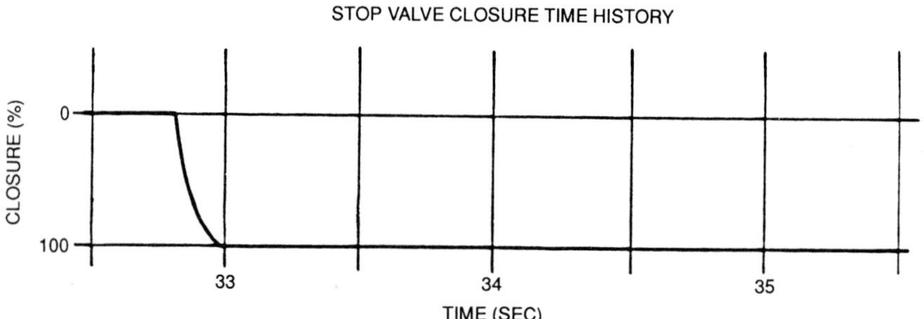

FIGURE B4.13 Steam hammer flow transients.

ysis in the ASME Code or NRC published material. Therefore, it is recommended to use the OBE damping values prescribed in the NRC Regulatory Guide 1.61. Alternative damping values of Code Case N-411-1 are not applicable to the dynamic analysis.

The time-history unbalanced forces are applied to all pipe segments. Snubbers and rigid supports are effective restraints for transient forces. However, axial supports should be avoided in general. An axial support not only requires welded attachments on the pipe but also a pair of supports, which doubles the cost. To support the pipe axially, lateral supports can be used around the elbows. In addition, snubbers should not be located in the immediate vicinity of equipment nozzles. Snubbers located in such areas may not be activated during a fluid transient because of the dead band (built-in manufacturing tolerance) of the snubber hardware.

Stress Allowables. For the steam hammer-water hammer (e.g., feed pump trip) event, the pipe stress from the analysis is combined with stresses due to pressure, dead weight, and OBE in meeting the upset stress allowable. For piping in the turbine building, OBE stress is not included in the stress combination. For some water hammer (e.g., check valve slam) events, the stress from the analysis is combined with stresses due to pressure and dead weight in meeting the faulted stress allowable.

LOCA Analysis. LOCA (loss-of-coolant accident) is a postulated accident that results from the loss of reactor coolant, at a rate in excess of the capability of the reactor coolant makeup system, from breaks in the reactor coolant pressure boundary. Analyses should be performed by the *nuclear steam supply system* (*NSSS*) vendor to confirm the structural design adequacy of the reactor internals and reactor coolant piping (unbroken loop) to withstand the loadings of the most severe LOCA in combination with SSE per the requirements of 10CFR Part 50, Appendix A,[20] and the NRC Standard Review Plan 3.9.2.[40]

The integrity of the secondary system piping (main steam, feedwater, blowdown lines) off the steam generators also has to be assured by the architect-engineer (AE). Additional analyses to demonstrate the structural adequacy of some of the branch piping attached to the broken loop may be required by the NSSS vendor.[41] The information provided herein is limited to the secondary system piping off the steam generators.

Static Analysis. If a substantial separation between the forcing frequencies of the LOCA loading and the natural frequencies of the piping system can be demonstrated, a static analysis may be performed. In the static analysis, the maxima of each of the LOCA displacement components (three deflections and three rotations) are separately applied to the junctions of the *reactor coolant loop* (*RCL*) and the secondary system piping. The results should be combined absolutely and multiplied by an appropriate dynamic load factor.

Dynamic Analysis. The dynamic analysis can be performed in one of the following two ways:

Time-History Analysis. The LOCA displacement time history is applied dynamically to the junctions of the RCL and the secondary system piping. The damping value prescribed in the NRC Regulatory Guide 1.61 for SSE is suggested for this dynamic LOCA analysis. The detailed analysis method is similar to that described in the following subsection "Pipe Break Analysis."[42]

Response Spectrum Analysis. Compared to the time-history analysis, the response spectrum method is favorable for its low computer cost. However, this method may be unnecessarily conservative since the same loading has to be applied to the entire piping system. Because of the nature of the LOCA break and the impacting of the gapped RCL supports, the LOCA motion has much higher frequency content than the seismic excitation. The ZPA of a typical LOCA motion spectrum for a RCL junction is usually higher than that of a typical SSE response spectrum. Therefore, a higher cutoff frequency should be used in the analysis.[43]

Stress Allowables. The resulting stress from the LOCA analysis for the secondary piping system is combined with the stresses due to pressure, dead weight, and SSE in meeting the faulted stress allowables.

Pipe Break Analysis. Although it is extremely improbable that a pipe break will occur as postulated, public safety and the NRC licensing requirements make it necessary that such events must be considered in the design of high-energy piping systems.

A *high-energy piping system* is a piping system that, during normal plant conditions, is maintained at a temperature >200°F, or a pressure >275 psig.

Pipe Break Locations. Pipe breaks are postulated in high-energy piping based on the primary plus secondary stresses and the cumulative usage factor.

1. *ASME Section III, Class 1 Pipe:* Pipe breaks are postulated to occur at terminal ends (the extremities of piping connected to structures, components, or anchors) and at all intermediate locations where:

   a. The primary plus secondary stress intensity range, as calculated by Equation (10) of Subsection NB-3653 [i.e., Eq. (B4.4) of this chapter], exceeds $2.4S_m$ and either Equation (12) or (13) [i.e., Eq. (B4.5) of this chapter] exceeds $2.4S_m$.
   b. Cumulative usage factor exceeds 0.1.
2. *ASME Section III, Class 2 and 3 Pipe:* Pipe breaks are postulated to occur at terminal ends and at all intermediate locations where the primary plus secondary stresses, as calculated by the sum of Equations (9) and (10) of Subsection NC-3653 [i.e., Eqs. (B4.10) and (B4.11) of this chapter], exceed $0.8(1.2S_h + S_A)$.
3. *Nonnuclear piping:* If a rigorous analysis, including seismic loading condition, is done on a high-energy ANSI/ASME B31.1 piping, the requirements of the Class 2 and 3 piping mentioned above will apply. If no analysis is performed, breaks are postulated at the following locations:
   a. Terminal ends
   b. At all fittings, welded attachments, and valves

The detailed pipe break design criteria and guidelines are given in the NRC Standard Review Plan No. 3.6.1 and 3.6.2.[44,45]

No-Break Zone. In the design of nuclear power plants, the region of piping in the containment building penetration areas between the isolation valves requires extra protection so that neither the leak-tight integrity of the containment nor the operability of the containment isolation valves is jeopardized. The extra protection consists of the following:

1. Installing special whip restraints, called *isolation restraints,* to mitigate the effects of the postulated pipe breaks located beyond this region
2. Keeping the primary plus secondary stresses and the cumulative usage factor below certain conservative values
3. Holding the piping stress, the isolation valve acceleration, and the stress at the valve-pipe weld below specified limits during a postulated pipe break outside this region
4. Special construction (welding) requirements and in-service inspection procedures

Because of the stringent design requirements, no pipe breaks are assumed to occur in this region. This area of piping is often referred to as the *no-break zone,* the *break exclusion region,* or the *superpipe area.*

No-Break Zone Piping Analysis. An analysis is required to determine the stresses in no-break zone piping and the accelerations of isolation valves due to a postulated break located beyond this region. During the pipe break event, a portion of the piping and the isolation restraints may enter the inelastic region because of the large pipe break loads imposed on the piping system. A static method, or the energy balance method, is acceptable but usually not used because necessary information on the no-break zone such as isolation valve acceleration is impossible to determine. Therefore, a nonlinear dynamic analysis utilizing the direct step-by-step integration method is necessary for the no-break zone analysis.[46] Computer programs based upon the direct integration method with linear elastic and nonlinear inelastic capabilities are often used for this type of analysis.[47,48]

In the analysis, the piping structural model is similar to that described in the subsection "Steam Hammer-Water Hammer Analysis." The nonlinear effects are accounted for by updating the system stiffness matrix at the end of each time step. The integration time step must be short enough to permit a reliable and stable solution. In addition, suitable system damping values should be used to obtain numerical stability. The time-history pipe break forcing function can be calculated by thermal-hydraulic computer programs as described in the subsection "Steam Hammer-Water Hammer Analysis," or obtained from the acceptable simplified method specified in Appendix B of ANSI/ANS 58.2.[49]

Wind Loads. The wind possesses kinetic energy by virtue of the velocity and mass of the moving air. If an obstacle is placed in the path of the wind so that the moving air is stopped or is deflected, then all or part of the kinetic energy of the wind is transformed into the potential energy of pressure.

A piping system which is located outdoors is usually designed to withstand the maximum wind velocity expected during the system operating life.

Dynamic Pressure. The intensity of wind pressure depends on the shape of the obstacle, the angle of incidence of the wind, and the velocity and density of the air.

For standard air (density of the air = 0.07651 lb/ft^3, temperature = 59°F), the expression for the wind dynamic pressure could be adapted from Bernoulli's equation for fluid flow as follows[26,50]:

$$p = 0.00256V^2C_D \tag{B4.59}$$

where p = dynamic pressure, lb/ft^2
V = basic wind speed, mi/h
C_D = drag coefficient, dimensionless

For the case of piping under wind loading, Eq. (B4.59) can be rewritten as

$$F = 0.000213V^2C_DD \tag{B4.60}$$

where F = linear dynamic pressure loading on projected pipe length, lb/ft
D = pipe diameter, including insulation, in

Basic Wind Speed. The basic wind speed V is the fastest wind speed at 33 ft above the ground in open terrain with scattered obstructions having heights less than 30 ft, as given in Fig. B4.14 for the United States.[26] The basic wind speed used for design shall be at least 70 mi/h.

Drag Coefficient. The drag coefficient C_D is a function of the shape of the structure and a fluid flow factor called the *Reynolds number.* The Reynolds number R is the ratio of the inertial force to the viscous force which a fluid stream exerts on an object. For standard air, the Reynolds number R could be expressed as

$$R = 780VD \tag{B4.61}$$

The drag coefficient C_D for a cylinder (i.e., a pipe) is given versus the Reynolds number in Fig. B4.15.

Wind Loading Analysis. The piping wind loading analysis is usually performed by a static method. In the analysis, the wind loading F is modeled as an uniform load acting over the projected length of the pipe, parallel to the direction of the wind. Two horizontal directions of wind loads (north-south and east-west) are included in the analysis. The design loads are based on the worst case of the two

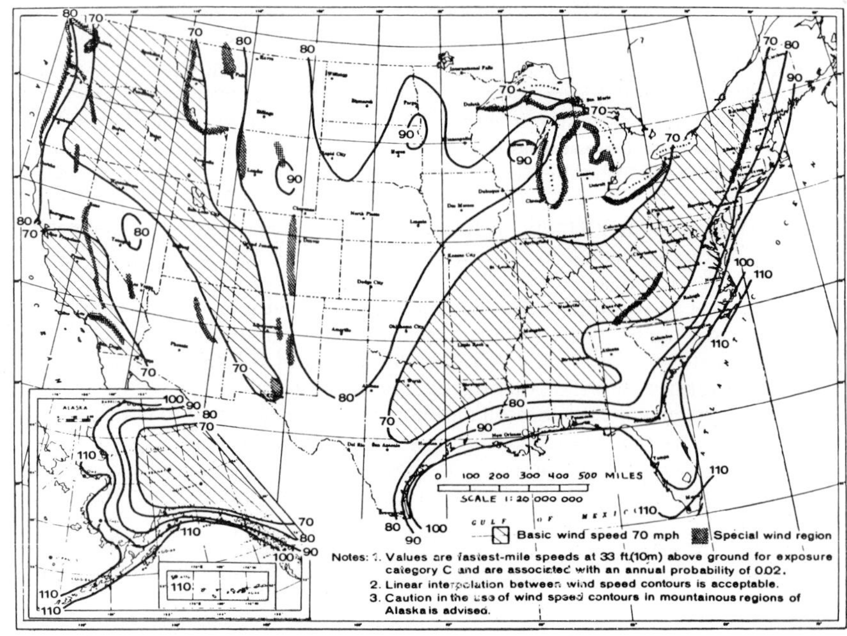

FIGURE B4.14 Basic wind speed (miles per hour). (*ANSI A58.1, 1982. Courtesy of ANSI.*)

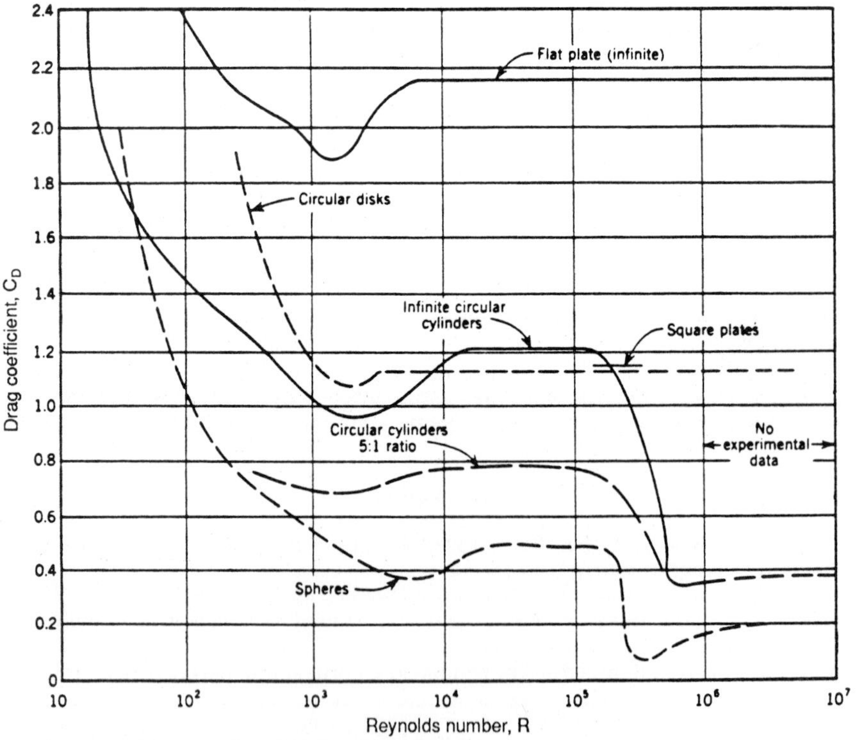

FIGURE B4.15 Drag coefficients for spheres and long cylinders. (*Task Committee on Wind Forces, "Wind Forces."*[50] *Courtesy of ASCE.*)

TABLE B4.8 Gust Response Factor G

Height above ground level, ft	G			
	Exposure A	Exposure B	Exposure C	Exposure D
0–15	2.36	1.65	1.32	1.15
20	2.20	1.59	1.29	1.14
25	2.09	1.54	1.27	1.13
30	2.01	1.51	1.26	1.12
40	1.88	1.46	1.23	1.11
50	1.79	1.42	1.21	1.10
60	1.73	1.39	1.20	1.09
70	1.67	1.36	1.19	1.08
80	1.63	1.34	1.18	1.08
90	1.59	1.32	1.17	1.07
100	1.56	1.31	1.16	1.07
120	1.50	1.28	1.15	1.06
140	1.46	1.26	1.14	1.05
160	1.43	1.24	1.13	1.05
180	1.40	1.23	1.12	1.04
200	1.37	1.21	1.11	1.04
250	1.32	1.19	1.10	1.03
300	1.28	1.16	1.09	1.02
350	1.25	1.15	1.08	1.02
400	1.22	1.13	1.07	1.01
450	1.20	1.12	1.06	1.01
500	1.18	1.11	1.06	1.00

Source: ANSI A58.1, 1982. (*Courtesy of ANSI.*)

directions. Similar to the case of earthquake, the wind loading is considered reversing. For load combination, the wind and the earthquake are assumed to not happen at the same time. A safety factor, the gust response factor G, should also be considered in the analysis. This factor is used to account for the fluctuating nature of wind and its interaction with structures. Its value depends on the exposure categories as shown in Table B4.8, where:

1. *Exposure A:* Large city centers with at least 50 percent of the buildings having a height in excess of 70 ft
2. *Exposure B:* Urban and suburban areas, wooded areas, or other terrain with numerous closely spaced obstructions having the size of single-family dwellings or larger
3. *Exposure C:* Open terrain with scattered obstructions having heights generally less than 30 ft
4. *Exposure D:* Flat, unobstructed coastal areas directly exposed to wind flowing over large bodies of water

METHODS OF ANALYSIS

Cookbook-Type Analysis

The following cookbook-type method is mainly for supporting 2-in and smaller Nuclear Class 2, 3, and B31 piping under gravity, thermal expansion, and seismic loadings. This method is based on standard support span tables. It covers a simplified weight analysis, a simplified thermal analysis, as well as a simplified seismic analysis.

A simplified seismic analysis often requires many pipe supports that are designed to large loads. The cost saving in engineering is offset by increased fabrication and installation cost. The current approach is to analyze the nonseismic piping by simplified methods and all seismic piping by computerized analysis. This greatly reduces the number of required seismic supports and gives an overall cost saving.

Simplified Weight Analysis. A simplified weight analysis is performed by locating the gravity supports based on gravity pipe spans. The maximum gravity pipe spans can be calculated from the following formula:

$$L = \left(\frac{SZ}{1.2W}\right)^{1/2} \qquad \text{(B4.62)}$$

where L = maximum gravity pipe span, ft
Z = section modulus of pipe, in^3
W = distributed weight, lb/ft
S = pipe stress due to gravity, psi

and the corresponding stress S is

$$S = \frac{M}{Z} = \frac{1{\cdot}2WL^2}{Z} \qquad \text{(B4.63)}$$

where M = bending moment, in · lb

Alternatively, the bending stress in empty pipe may be read from Fig. B4.16, and the bending stress in water-filled pipe from Fig. B4.17. The deflection of empty pipe can be read from Fig. B4.18.

The distributed weight of pipe includes the weight of metal, the weight of pipe contents, and the weight of insulation. Pipe material weights are subject to tolerance of applicable manufacturing specifications.

Weights of insulation depend on the composition of insulation material and should be obtained from the insulation manufacturer. Weights of weatherproof protection, if specified, must be added. Insulation thicknesses recommmended by insulation manufacturers do not necessarily agree with insulation specifications for a particular job. Insulation specifications should be reviewed prior to development of final weights of piping.

Weights of insulation should be added to weights of flanges, valves, and fittings. Flange, flanged valve, and flanged fitting weights should include weights of bolts and nuts.

Valve weights vary among particular manufacturers' designs and should include weights of electric-motor operators (if any) or other devices which may be

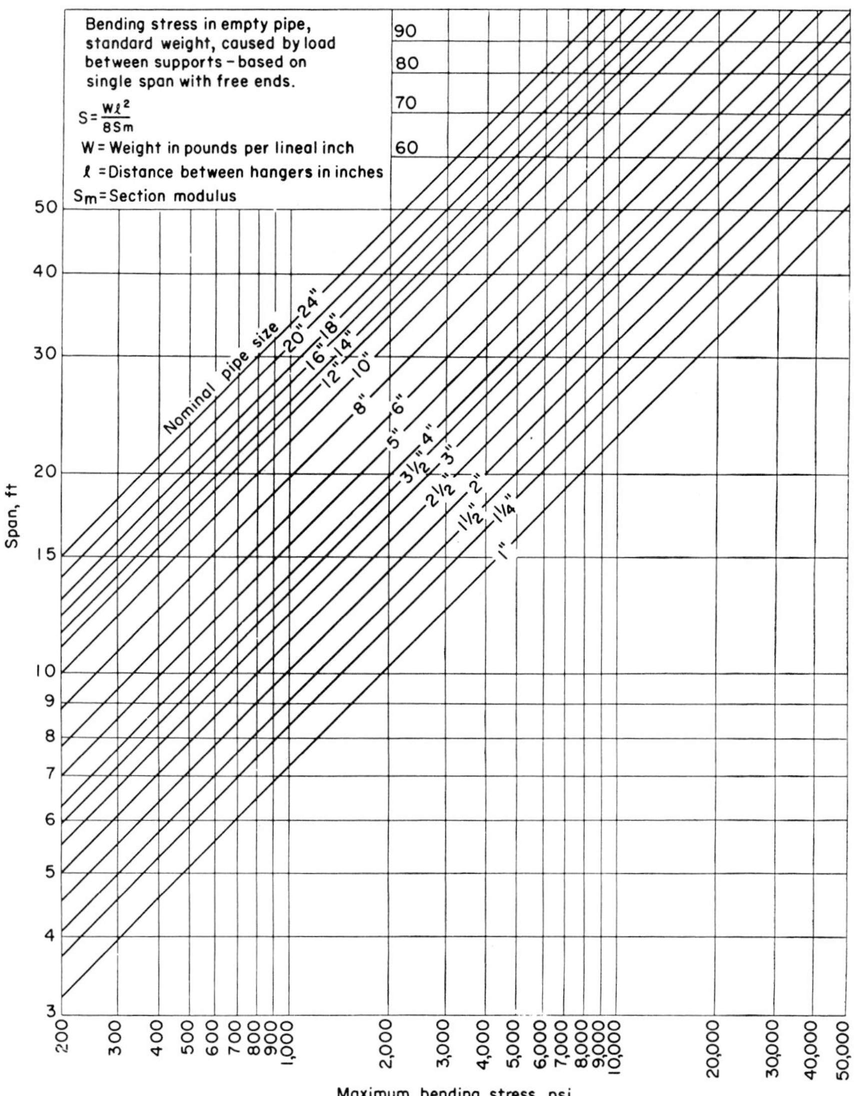

FIGURE B4.16 Bending stress in empty pipe.

specified for particular valves. It is suggested that, wherever possible, valve weights should be obtained from the manufacturer of the particular valves which are to be installed in the piping.

Equation (B4.62) is based on the combination of a simply supported beam model and a fixed-end beam model because the behavior of pipe lies somewhere between these two models.

A table of suggested maximum spans between supports of pipe based on a for-

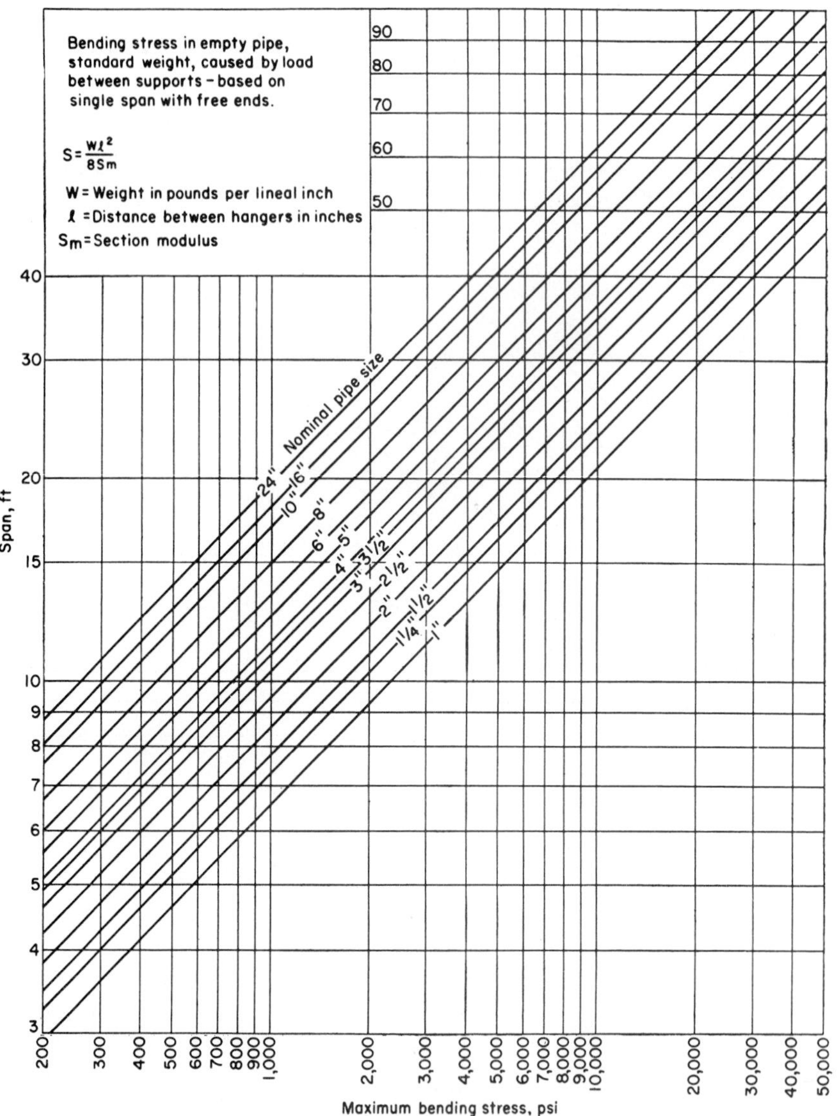

FIGURE B4.17 Bending stress in water-filled pipe.

mula similar to Eq. (B4.62) is given in ASME Codes,[51,52] as shown in Fig. B5.1 of Chap. B5. These spans have been calculated by considering insulated, standard wall thickness and heavier pipe, limited to a maximum stress of 1500 psi and maximum pipe sag of 0.1 in. For small pipe where socket welds are used, Eq. (B4.63) can be rewritten as

$$S = \frac{0.75iM}{Z} = \frac{1.89WL^2}{Z} \tag{B4.64}$$

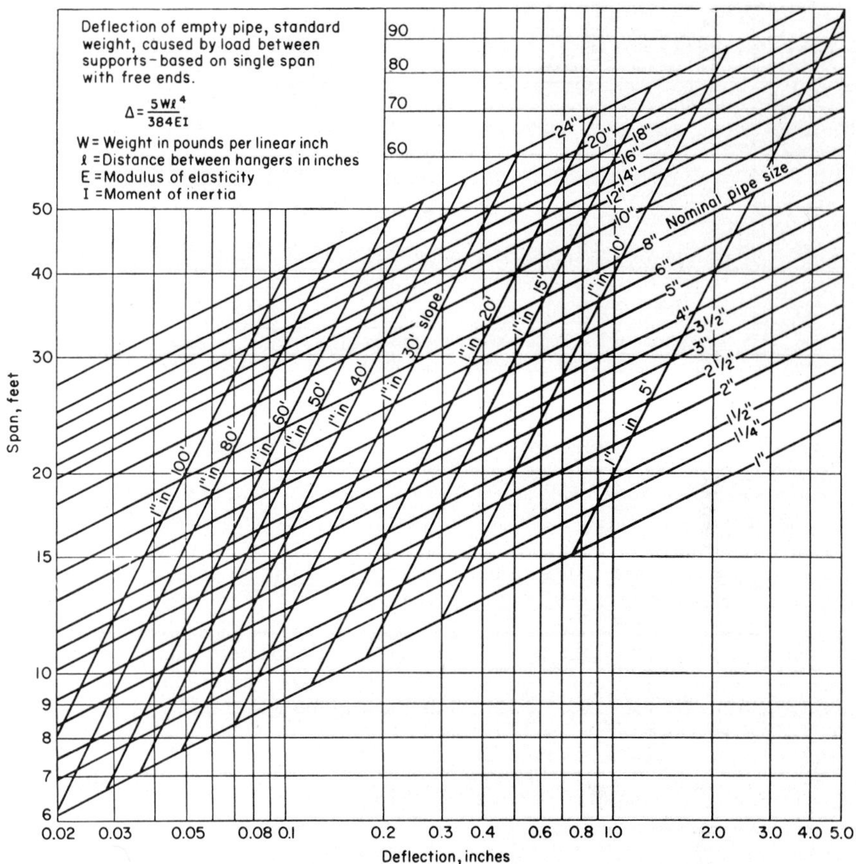

FIGURE B4.18 Deflection of empty pipe.

and

$$L = \left(\frac{SZ}{1.89W}\right)^{1/2} \tag{B4.65}$$

where i = the stress-intensification factor (SIF), 2.1 for socket welds per ANSI/ASME B31.1 and ASME Section III piping codes.

Figures B5.2 and B5.3 of Chap. B5 give the maximum spans for water and steam, air, or gas filled steel pipe, respectively. These tables are based on a pipe stress S of 2000 psi and a socket weld SIF of 2.1. When these suggested weight spans are adhered to, the stress in the piping system due only to gravity load usually need not be explicitly calculated.

Load Calculation by Weight Balance. The following example is used to illustrate a method by which hanger loadings may be determined. The method consists of locating the center of gravity of the specific piping configuration and then, by equating moments, to determine the resultant loads at particular hangers.

A single-plane bend is shown in Fig. B4.19. Hangers are indicated as H-1, H-2, H-3, and H-4. The effects of uniform and concentrated loads are indicated at

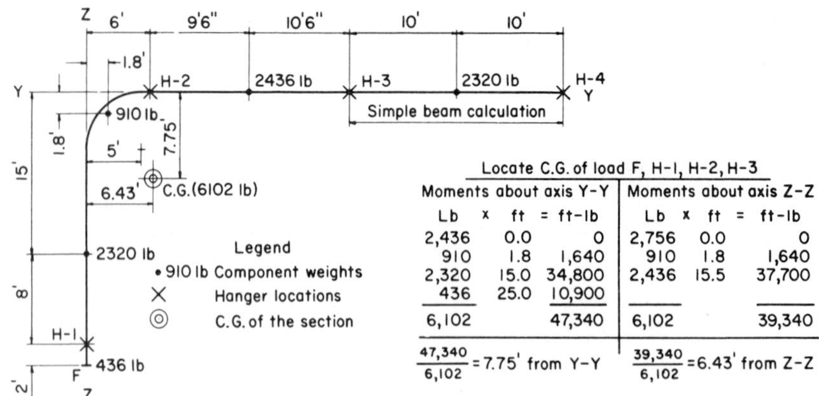

FIGURE B4.19 One-line piping diagram for illustration of load calculation by weight balance.

the points at which these loads act; it is noted that the weight of the 90° bend acts at the centroid of a quarter circle which, in this example, is located 1.8 ft distant from the centerline of the pipe run. The straight pipe length between hangers *H*-3 and *H*-4 is not included in this calculation because it can be analyzed by simple straight-beam theory.

For the piping section which lies between equipment flange *F* and hanger *H*-3, moments are taken about the *Y*–*Y* and *Z*–*Z* axes. As an example, let the center of gravity of this configuration be located *Y* ft from the *Y*–*Y* axis. Then, from equilibrium considerations, the following equation may be written:

$$2436(0) + 910(1.8) + 2320(15) + 436(25) = 6102Y$$

A solution to this equation results in $Y = 7.75$ ft.

Similarly, the distance from the *Z*–*Z* axis to the center of gravity is found to be 6.43 ft.

For convenience, the calculations are made frequently in a tabular fashion as shown on Fig. B4.19.

Let it be now required to determine hanger loadings for the piping configuration of Fig. B4.19 with the stipulation that no load due to weight be imposed on the equipment flange *F*. This is accomplished easily by use of simple geometrical relationships, and the solution is as indicated in Fig. B4.20.

If it were desired to support the piping with two, rather than three, hangers, it would be convenient to eliminate *H*-1 and to relocate *H*-2 to a position at which it would be colinear with the center of gravity and hanger *H*-3. The construction for this arrangement and the associated hanger-load calculations are shown in Fig. B4.21.

In each of the two above cases, one-half of the 2320-lb load between *H*-3 and *H*-4 has been included in the calculations for hanger loading on *H*-3. Thus *H*-4 would be required to support 1160 lb plus, of course, any additional piping load to the right of *H*-4 in Fig. B4.19.

Simplified Thermal Expansion Analysis. This simplified analysis is based on the guided cantilever method. The *guided cantilever* is a cantilever beam restrained in such a way that its free end will not rotate when it is deflected in a direction perpendicular to the longitudinal axis of the beam, as shown in Fig. B4.22.

Construct the triangle H-1, H-2, H-3 superimposing location of C.G. and calculate reactions at H-1, H-2 and H-3

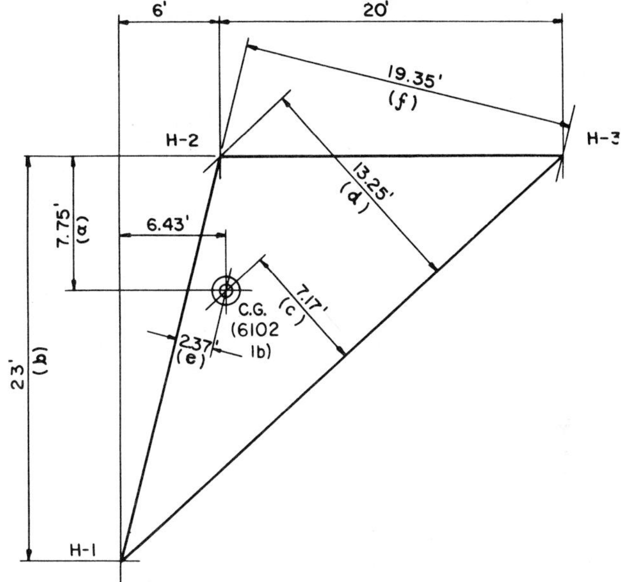

Reaction at $\text{H-1} = \frac{a}{b} \times 6102 = \frac{7.75}{23} \times 6102 = 2055$ lbs

$\text{H-2} = \frac{c}{d} \times 6102 = \frac{7.17}{13.25} \times 6102 = 3301$ lbs

$\text{H-3} = \frac{e}{f} \times 6102 \quad \frac{2.37}{19.35} \times 6102 = 746$ lbs

F = 0 lbs

6102 lbs

Total load on $\text{H-3} = 746 + \frac{2320}{2} = 1906$ lbs

FIGURE B4.20 Hanger load calculations for system of Fig. B4.19. Three hangers with zero reaction at flange *F*.

For piping systems under thermal expansion loads, the behavior of the piping approximates that of a guided cantilever. The thermal growth forces the pipe leg to translate while pipe rotations are restricted by piping continuity. Therefore, this method can be used to check the flexibility of a piping system.

For a guided cantilever, the moment induced by an imposed deflection is

$$M = \frac{6EI\Delta}{L^2} \tag{B4.66}$$

where M = induced moment, in · lb
E = modulus of elasticity, psi
I = moment of inertia, in^4
Δ = deflection, in
L = length of pipe leg perpendicular to deflection, in

The corresponding stress is then

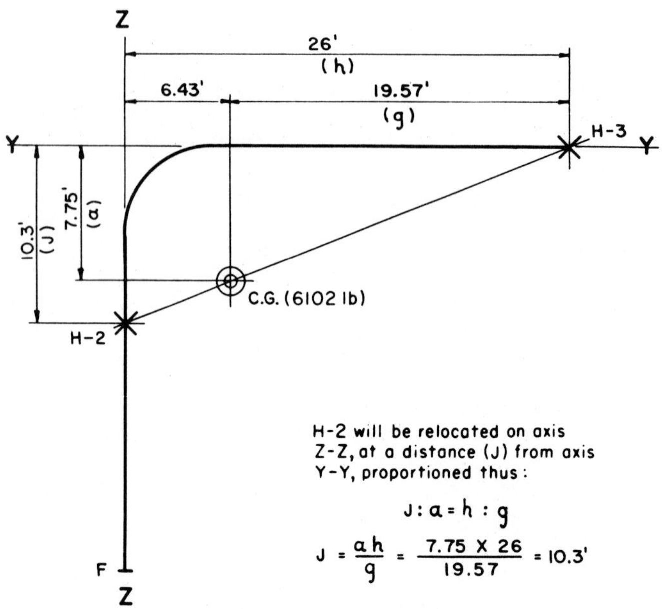

FIGURE B4.21 Hanger load calculations for system of Fig. B4.19 except that one hanger has been eliminated.

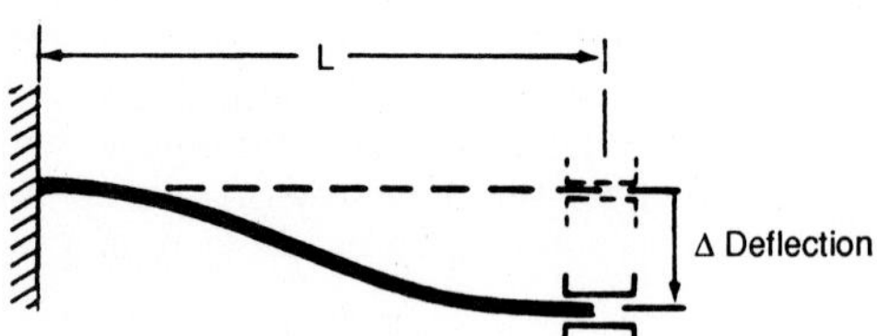

FIGURE B4.22 Guided cantilever.

$$S = \frac{iM}{Z} = \frac{6EI\Delta i}{ZL^2} = \frac{3ED\Delta i}{L^2} \qquad \text{(B4.67)}$$

where S = induced stress, psi
D = outside diameter of pipe, in
Z = section modulus of pipe, in^3

i = stress-intensification factor

Solving the equation for the beam length L gives

$$L = \left(\frac{3ED\Delta i}{S}\right)^{1/2} \tag{B4.68}$$

By determining the proper allowable stress and taking into account the appropriate stress-intensification factor, Eq. (B4.68) gives an estimate of the minimum allowable offset pipe span L required to sustain a piping thermal movement Δ normal to the piping.

Tables B4.9 and B4.10 give the minimum allowable offset span for steel piping ($E = 27.9 \times 10^6$ psi) with socket welds ($i = 2.1$) and without socket welds, respectively. These tables are based on allowable stresses S of 22,500 psi.

Thermal Movement Calculations. The simplified method shown below is one which gives satisfactory approximations of the piping movements. Whenever differences occur between the approximations and actual movements, the approximation of the movement will always be the greater amount.

Step 1. The piping system of Fig. B4.23 is drawn, and on it are shown all known vertical movements of the piping from its cold to hot, or operating, position. These movements will include those supplied by the equipment manufacturers for the terminal point connections. For the illustrated problem, the following vertical movements are known:

TABLE B4.9 Thermal Expansion Minimum Allowable Offset Span (Feet-Inches), Straight Steel Pipe with Socket Welds

Therm.	Pipe size, nominal				
Expan., in	½ in	¾ in	1 in	1½ in	2 in
0.10	2-1¾	2-4¾	2-8	3-2½	3-7
0.15	2-7½	2-11	3-3¼	3-11¼	4-4¾
0.20	3-0¼	3-4½	3-9¼	4-6½	5-1
0.25	3-4½	3-9¼	4-2¾	5-1	5-8
0.30	3-8½	4-1½	4-7½	5-6¾	6-2½
0.35	4-0	4-5½	5-0	6-0	6-8½
0.40	4-3¼	4-9¼	5-4	6-5	7-2¼
0.45	4-6¼	5-0¾	5-8	6-9¾	7-7¼
0.50	4-9¼	5-4	5-11¾	7-2¼	8-0¼
0.60	5-2¾	5-10¼	6-6½	7-10¼	8-9½
0.70	5-7¾	6-3¾	7-0¾	8-6	9-6
0.80	6-0½	6-9	7-6¾	9-1	10-1¾
0.90	6-4¾	7-2	8-0¼	9-7½	10-9¼
1.00	6-9	7-6½	8-5¼	10-1¾	11-4¼
1.10	7-1	7-11	8-10¼	10-7¾	11-10¾
1.20	7-4¾	8-3¼	9-3	11-1½	12-5¼
1.50	8-3¼	9-3	10-4¼	12-5¼	13-10¾
1.75	8-11¼	9-11¾	11-2	13-5¼	15-0¼
1.90	9-3¾	10-4¾	11-7¾	14-0	15-7¾
2.00	9-6½	10-8	11-11¼	14-4¼	16-0¾
2.50	10-8	11-11¼	13-4¼	16-0¾	17-11¼
3.00	11-8¼	13-0¾	14-7¼	17-7	19-8

TABLE B4.10 Thermal Expansion Minimum Allowable Offset Span (Feet-Inches), Straight Steel Pipe, No Socket Weld

Therm. Expan., in	Pipe size, nominal				
	½ in	¾ in	1 in	1½ in	2 in
0.10	1-5¾	1-8	1-10¼	2-2¾	2-5¾
0.15	1-9¾	2-0¼	2-3¼	2-8¾	3-0½
0.20	2-1¼	2-4	2-7¼	3-1¾	3-6¼
0.25	2-4	2-7¼	2-11¼	3-6¼	3-11
0.30	2-6¾	2-10¼	2-2½	3-10¼	4-3½
0.35	2-9¼	3-1	3-5½	4-1¾	4-7¾
0.40	2-11½	3-3¾	3-8¼	4-5¼	4-11¾
0.45	3-1½	3-6	3-11	4-8½	5-3
0.50	3-3¾	3-8¼	4-1¾	4-11¾	5-6½
0.60	3-7½	4-0½	4-6¼	5-5¼	6-1
0.70	3-10¾	4-4½	4-10½	5-10½	6-6¾
0.80	4-2¼	4-8	5-2¾	6-3¼	7-0¼
0.90	4-5	4-11½	5-6½	6-7¾	7-5¼
1.00	4-8	5-1½	5-10	7-0¾	7-10¼
1.10	4-10¾	5-5¾	6-1¾	7-4¼	8-2¾
1.20	5-1¼	5-8½	6-4¾	7-8¼	8-7
1.50	5-8½	6-4¾	7-1¾	8-7	9-7¼
1.75	6-2¼	6-10¾	7-8½	9-3½	10-4½
1.90	6-5¼	7-2¼	8-0½	9-8	10-9¾
2.00	6-7¼	7-4½	8-3	9-11	11-1
2.50	7-4¾	8-3¼	9-3	11-1½	12-5
3.00	8-1¼	9-0¾	10-1¾	12-2¼	13-7½

Point *A*—2 in up, cold to hot
Point *B*—1/16 in up, cold to hot
Point *C*—⅛ in down, cold to hot
H-4—0 in, cold to hot

The operating temperature of the system is given as 1050°F, and the coefficient of expansion for low-chrome steel at 1050°F is 0.0946 in/ft.

The movements at points *D* and *E* are calculated by multiplying the coefficient of expansion by the vertical distance of each point from the position of zero movement on the riser *DE:*

$$55 \text{ ft} \times 0.0946 \text{ in/ft} = 5.20 \text{ in } up \text{ at } D$$

$$20 \text{ ft} \times 0.0946 \text{ in/ft} = 1.89 \text{ in } down \text{ at } E$$

Step 2. A simple drawing is made of the piping between two adjacent points of known movement, extending the piping into a single plane as shown for the portion of the system between *A* and *D*.

The vertical movement at any hanger location will be proportional to its distance from the endpoints:

$$\Delta_1 = \frac{4}{31} \times 3.20$$

$$\Delta_1 = 0.41 \text{ in}$$

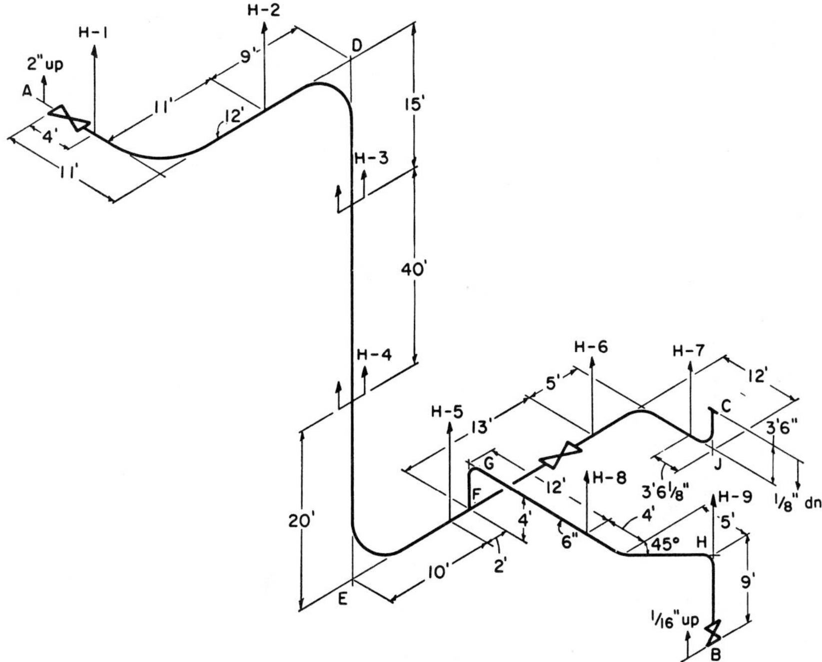

FIGURE B4.23 One-line piping diagram for calculation of hanger movements. Points *A, B,* and *C* are equipment connections. *H*-1, *H*-2, and so on, represent hanger locations.

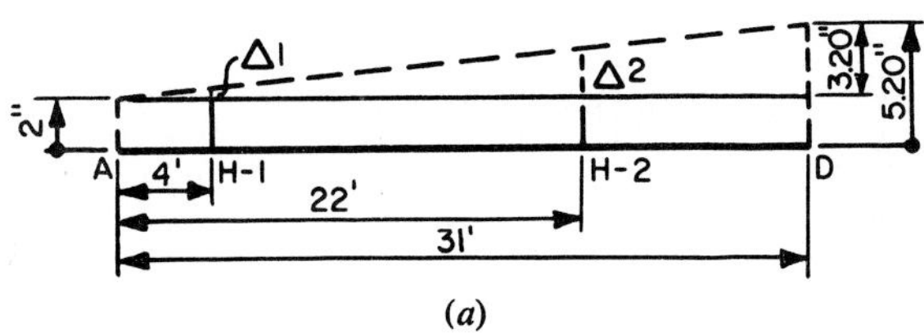

(*a*)

The vertical movement at *H*-1 = 0.41 in + 2 in:

$$\Delta H\text{-}1 = 2.41 \text{ in up}$$

$$\Delta_2 = {}^{22}\!/_{31} \times 3.20$$

$$\Delta_2 = 2.27 \text{ in}$$

The vertical movement at *H*-2 = 2.27 in + 2 in:

$$\Delta H\text{-}2 = 4.27 \text{ in up}$$

Step 3. To calculate the vertical movement at *H*-3, multiply its distance from *H*-4 by the coefficient of expansion:

$$40 \text{ ft} \times 0.0946 \text{ in/ft} = 3.78 \text{ in}$$

$$\Delta H\text{-}3 = 3.78 \text{ in up}$$

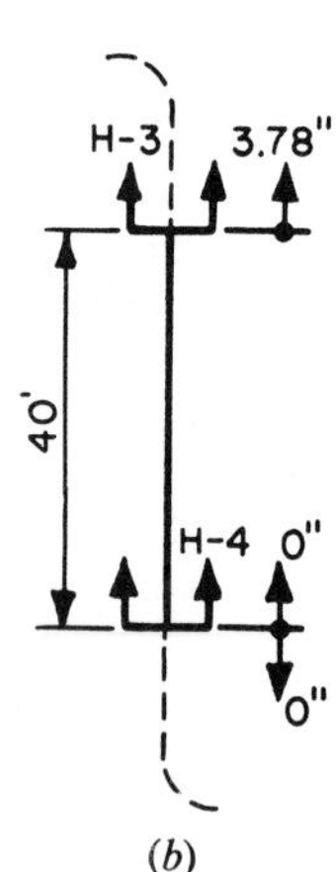

(*b*)

Step 4. The next section of pipe on which there are two points of known movement is the length *E-J*. The movement at *E* was calculated as 1.89 in down:

The movement at *J* is equal to the movement at the terminal point *C* (1/8 in down) plus the amount of expansion of the leg *C-J:*

$$\Delta J = 0.125 \text{ in} + 3.5 \text{ ft} \times 0.0946 \text{ in/ft}$$

$$\Delta J = 0.46 \text{ in down}$$

$$\Delta_7 = 3.5/42 \times 1.43 = 0.12 \text{ in}$$

$$\Delta H\text{-}7 = 0.12 \text{ in} + 0.46 \text{ in}$$

$$\Delta H\text{-}7 = 0.58 \text{ in down}$$

$$\Delta_6 = {}^{17}\!/_{42} \times 1.43 = 0.58 \text{ in}$$

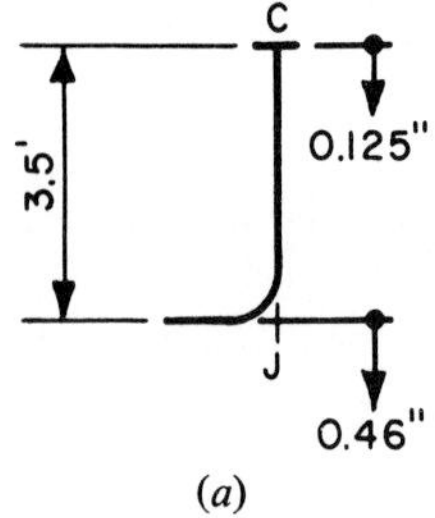

(*a*)

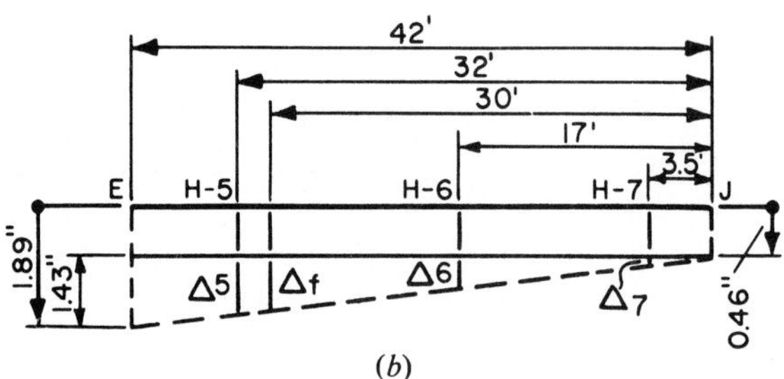

(*b*)

$$\Delta H\text{-}6 = 0.58 + 0.46 \text{ in}$$

$$\Delta H\text{-}6 = 1.04 \text{ in down}$$

$$\Delta_f = {}^{30}\!/_{42} \times 1.43 = 1.02 \text{ in}$$

$$\Delta F = 1.02 + 0.46$$

$$\Delta F = 1.48 \text{ in down}$$

$$\Delta_5 = {}^{32}\!/_{42} \times 1.43 = 1.09 \text{ in}$$

$$\Delta H\text{-}5 = 1.09 + 0.46$$

$$\Delta H\text{-}5 = 1.55 \text{ in down}$$

Step 5. In the section *G-H,* the movement at *G* is equal to the movement at *F minus* the expansion of the leg *GF:*

$$\Delta G = 1.48 \text{ in down} - 4 \text{ ft} \times 0.0946 \text{ in/ft}$$

$$\Delta G = 1.10 \text{ in down}$$

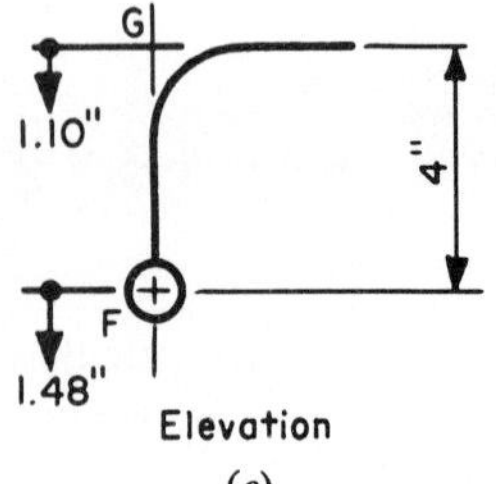

(*c*)

The movement at *H* is equal to the movement of the terminal point *B*(1/16 in up) *plus* the expansion of the leg

B-H:

$$\Delta H = 0.0625 \text{ in up} + 9 \text{ ft} \times 0.0946 \text{ in/ft}$$

$$\Delta H = 0.91 \text{ in up}$$

Since H-9 is located at point H,

$$\Delta H\text{-}9 = \Delta H = 0.91 \text{ in up}$$

$$\Delta_y = \frac{12 \times 2.01}{23.1} = 1.04 \text{ in}$$

$$\Delta H\text{-}8 = 1.10 - 1.04$$

$$\Delta H\text{-}8 = 0.60 \text{ in down}$$

After calculating the movement at each hanger location it is often helpful, for

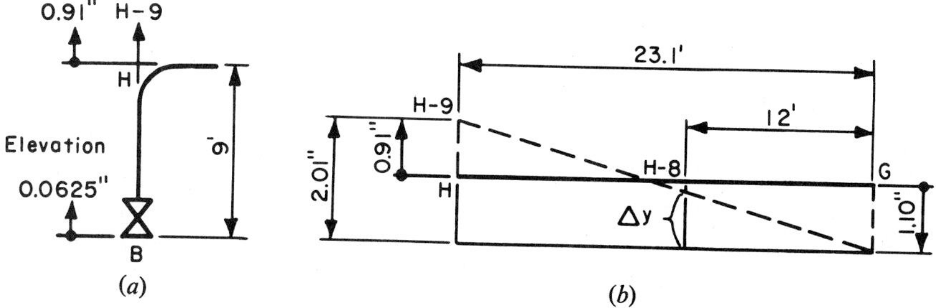

easy reference when selecting the appropriate type hanger, to make a simple table of hanger movements.

Hanger number	Movement, in
H-1	2.41 up
H-2	4.27 up
H-3	3.78 up
H-4	0
H-5	1.55 down
H-6	1.04 down
H-7	0.58 down
H-8	0.06 down
H-9	0.91 up

Calculation of Hanger Loads. A 6-in medium-temperature steam piping system is shown in Fig. B4.24. Terminal movements at equipment flanges A and B are indicated; dimensions of system components and physical data are also given. It is required to determine hanger loadings and also to determine movements at each of the hangers H-1 through H-7.

It is noted that hanger H-3 on the vertical leg has been located 20 ft above the lower horizontal pipe run. Calculations would indicate that the center of gravity of the vertical leg is 16.16 ft above the lower horizontal run. It would not be de-

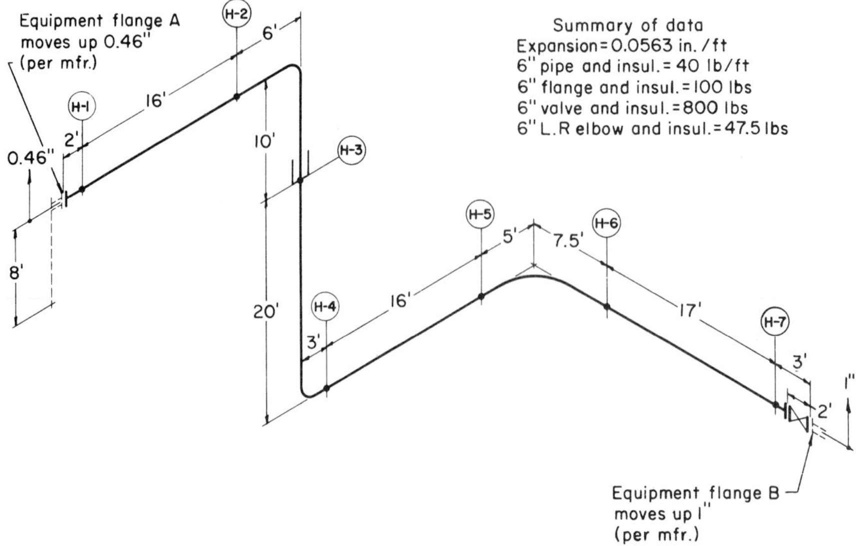

FIGURE B4.24 One-line piping diagram for calculation of hanger loadings and deflections.

sirable to place the hanger at the center of gravity because the hanger would then act as a pivot point and would not resist sway. If the hanger H-3 were placed below the center of gravity, an unstable turnover condition would result. The most desirable location is above the center of gravity; hanger H-3 has thus been placed arbitrarily a distance of 20 ft above the lower horizontal piping run.

Starting with equipment flange A, the system is broken up into component parts between hangers and hanger reactions are calculated. The procedure is indicated in Figs. B4.25*a* to B4.25*g*, and the results are listed in Table B4.11. Hanger deflections, or movements, are determined as shown in Figs. B4.26*a* and B4.26*b*.

Simplified Seismic Analysis. A simplified seismic analysis utilizing simple beam formulas and response spectrum curves is given here. The maximum support spacings are selected from Tables B4.12 to B4.14 so that the fundamental frequency of the span is in the rigid range of the response spectrum.

These seismic spans are based on the following formula:

$$L = \left(\frac{1}{12}\right)\left(\frac{\pi}{2f}\right)^{1/2}\left(\frac{12gEI}{W}\right)^{1/4} \tag{B4.69}$$

where L = maximum seismic spacing, ft
f = desired frequency, cycles/s
g = 386 in/s^2
E = modulus of elasticity, psi
I = moment of inertia, in^4
W = distributed weight, lb/ft

For a system with seismic supports designed in the rigid range, the seismic ac-

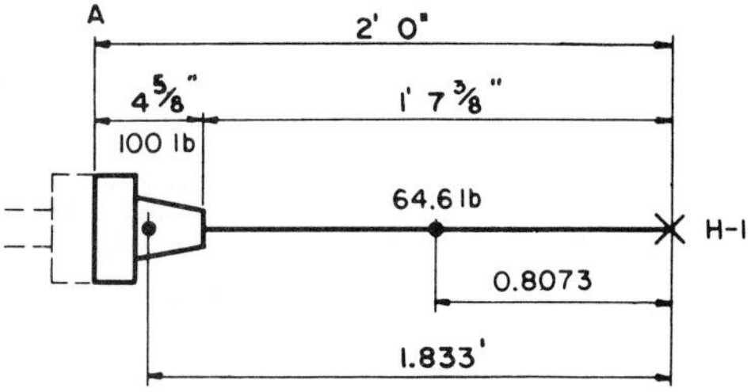

Taking moments about H-1

Ft	X	lb	=	ft-lb
0.8073		64.6	=	52.2
1.833		100.0	=	183.3
		164.6		235.5

Reaction @ flange A = $\frac{235.5}{2.0}$ = 117.8 lbs

Reaction @ H-1 = 164.6 − 117.8 = 46.8 lbs

(a)

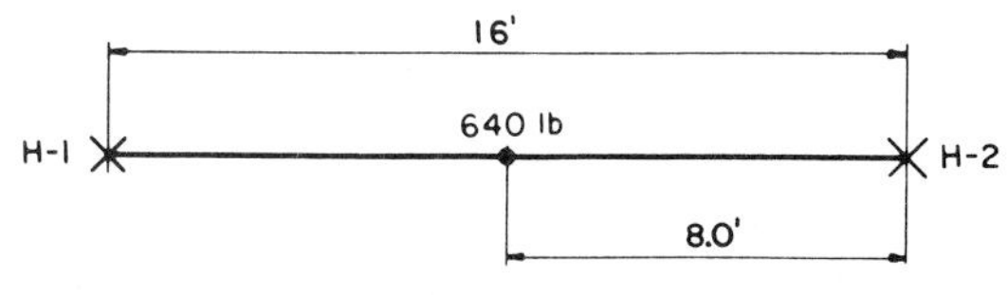

Reactions H-1 and H-2 = $\frac{640}{2}$ = 320 lbs

(b)

FIGURE B4.25 (*a*) Distribution of weight between equipment flange *A* and *H*-1. (*b*) Distribution of weight between *H-1 and H*-2.

celeration of the system is low and consequently the design loads for the system decrease.

The corresponding seismic stress is then

$$S = 0.75i(12)\left(\frac{WL^2}{8Z}\right)(1.5G) \tag{B4.70}$$

where Z = section modulus of pipe, in^3
G = seismic acceleration (OBE or SSE) in gs
i = stress-intensification factor

The number 1.5 in Eq. (B4.70) is a factor to account for the contribution from the higher modes.[40,53]

Computerized Method

Types of Computer Programs. The microcomputer has become the daily tool and workstation of the piping stress analyst. Files which contain data for piping stress

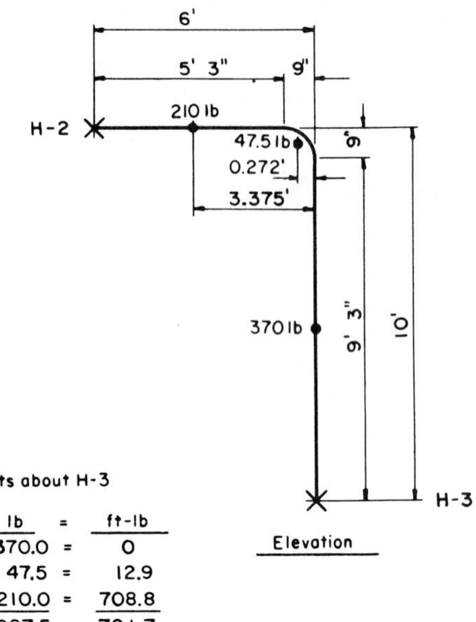

Taking moments about H-3

Ft.	X lb	= ft-lb
0.0	370.0	0
0.272	47.5	12.9
3.375	210.0	708.8
	627.5	721.7

$$\text{Reaction @ H-2} = \frac{721.7}{6.0} = 120.3 \text{ lbs}$$

Reaction @ H-3 = 627.5 − 120.3 = 507.2 lbs

(c)

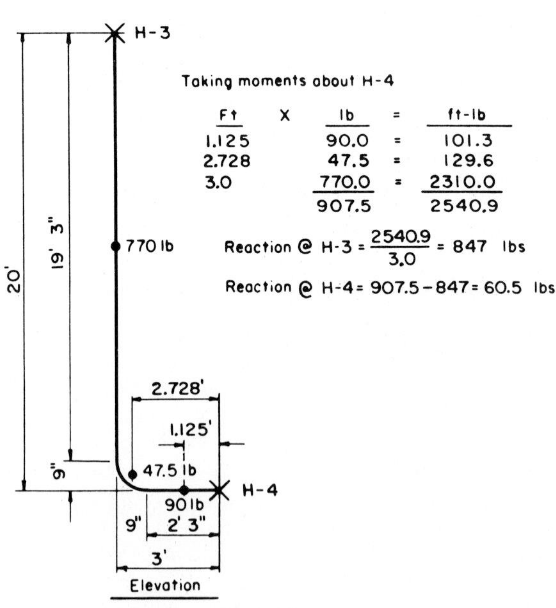

Taking moments about H-4

Ft	X lb	= ft-lb
1.125	90.0	101.3
2.728	47.5	129.6
3.0	770.0	2310.0
	907.5	2540.9

$$\text{Reaction @ H-3} = \frac{2540.9}{3.0} = 847 \text{ lbs}$$

Reaction @ H-4 = 907.5 − 847 = 60.5 lbs

(d)

FIGURE B4.25 (*c*) Distribution of weight between *H*-2 and *H*-3. (*d*) Distribution of weight between *H*-3 and *H*-4.

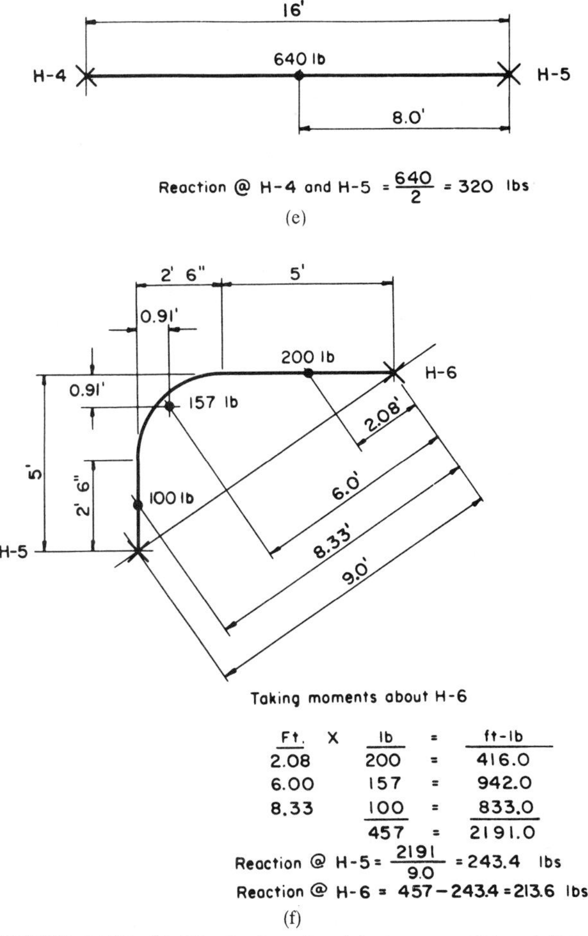

FIGURE B4.25 (*e*) Distribution of weight between *H*-4 and *H*-5. (*f*) Distribution of weight between *H*-5 and *H*-6.

analysis are created, edited, and saved at this workstation. These files are later transferred to the mini- or mainframe computer for the calculation of piping stresses and support loads.

Most of the computer programs for piping stress analysis such as ADLPIPE, NUPIPE, and SUPERPIPE were developed for use on mainframe computers. With the introduction of many powerful microcomputers in the mid-1980s, microcomputer-based programs for piping stress analysis were also developed such as AUTOPIPE and CAESAR II. Some of these new programs are menu driven and user friendly. They help save engineering time and cost. In general, these computer programs may be divided into four classes:

1. Programs that can perform pressure, thermal expansion, dead weight, and external forces (e.g., wind) analyses for ASME Section III, Class 2, 3, ANSI/

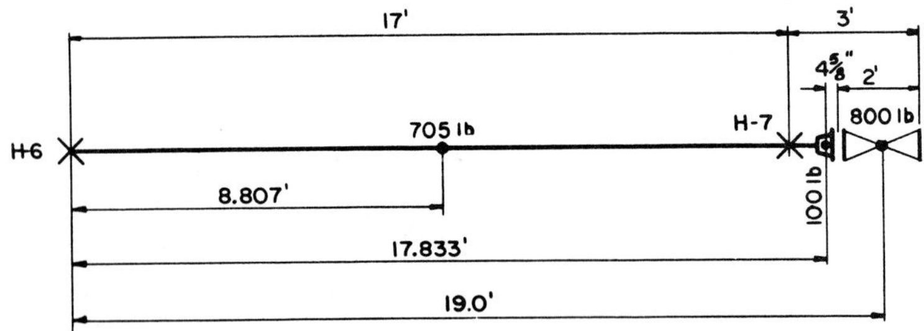

Taking moments about H-6

Ft	X	lb	=	ft-lb
8.807		705	=	6210
17.833		100	=	1783
19.0		800	=	15200
		1605		23193

$$\text{Reaction @ H-7} = \frac{23193}{17.0} = 1364 \text{ lbs}$$

Reaction @ H-6 = 1605 − 1364 = 241 lbs

FIGURE B4.25(*g*) Distribution of weight between *H*-6 and *H*-7 to maintain zero reaction on flange *B*.

TABLE B4.11 Summary of Hanger Loadings

Hanger Mark	Reactions, lb							Hanger load, lb
	A to *H*-1	*H*-1 to *H*-2	*H*-2 to *H*-3	*H*-3 to *H*-4	*H*-4 to *H*-5	*H*-5 to *H*-6	*H*-6 to *H*-7	
Flange *A*	117.8	....	....	....	....	....	....	117.8
H-1	46.8	320.0	....	....	....	....	....	366.8
H-2	....	320.0	120.3	....	....	....	.....	440.3
H-3	....	....	507.2	847.0	....	....	.....	1,354.2
H-4	....	....	....	60.5	320.0	....	.....	380.5
H-5	....	....	....	....	320.0	243.4	.....	563.4
H-6	....	....	....	....	....	213.6	241.0	454.6
H-7	....	....	....	....	....	....	1364.0	1,364.0
Flange *B*	....	....	....	....	....	....	0.0	0.0

ASME B31.1, B31.3, B31.4, B31.5, B31.8, NEMA, API-610, and API-617 piping. Programs such as TRIFLEX, AUTOPIPE, and CAESAR II are in this class. (AUTOPIPE and CAESAR II have response spectrum and SAM analysis capability. However, there is a limit on the number of analyses which can

Flatten out pipe shape into plane and establish movement at top and bottom of vertical leg.
Use method for one vertical leg.

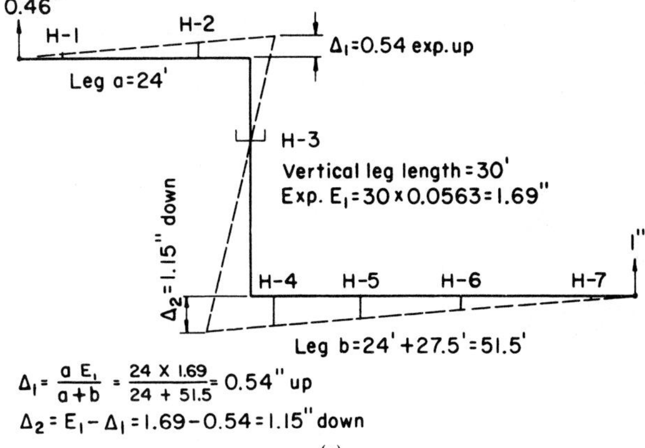

$$\Delta_1 = \frac{a\,E_1}{a+b} = \frac{24 \times 1.69}{24 + 51.5} = 0.54\text{" up}$$

$$\Delta_2 = E_1 - \Delta_1 = 1.69 - 0.54 = 1.15\text{" down}$$

(a)

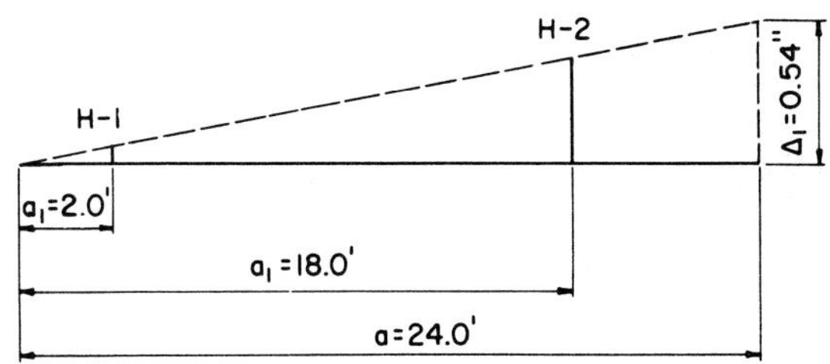

$$\Delta_x = \frac{a_1 \Delta_1}{a}$$

$$\Delta_x @ H\text{-}1 = \frac{2.0 \times 0.54}{24.0} = 0.045\text{"}$$

$$\Delta_x @ H\text{-}2 = \frac{18.0 \times 0.54}{24.0} = 0.405\text{"}$$

(b)

FIGURE B4.26 (*a*) Deflections of vertical leg of Fig. B4.24. (*b*) Determination of deflections at *H*-1 and *H*-2 of Fig. B4.24.

be performed in the same computer run because of the memory capability of microcomputers.)

2. Programs that can perform seismic, independent support motion, thermal transient, and time-history analyses in addition to those mentioned in item 1 for ASME Section III, Class 1, 2, 3, ANSI/ASME B31.1, and B31.3 piping. Programs such as ADLPIPE, ME101, NUPIPE, PIPESD, and SUPERPIPE are in this class.
3. General-purpose programs, such as ANSYS. ANSYS is a general-purpose finite element analysis program which can perform static and dynamic analysis; elastic and plastic analysis; steady-state and transient heat transfer; steady-

TABLE B4.12 Maximum Support Spacing for Seismic Stress for a Frequency of 20 Hz (Feet-Inches), Steel Pipe

Pipe size, in	Pipe sch.	No insul.	Antisweat*		Calcium silicate*					Reflective*
			½ in	1 in	½ in	1 in	1½ in	2 in	2½ in	4 in
½	40	5-4	4-11	4-8	4-9	4-5	4-2	3-10	3-8	3-3
	80	5-4	4-11	4-9	4-10	4-6	4-3	4-0	3-9	3-5
	160	5-3	4-11	4-9	4-10	4-6	4-3	4-0	3-10	3-5
¾	40	6-0	5-7	5-4	5-5	5-1	4-10	4-7	4-4	3-11
	80	6-0	5-7	5-5	5-6	5-3	4-11	4-8	4-5	4-1
	160	5-11	5-7	5-5	5-6	5-3	5-0	4-9	4-7	4-2
1	40	6-9	6-4	6-2	6-3	5-11	5-8	5-4	5-1	4-8
	80	6-9	6-5	6-3	6-4	6-0	5-9	5-6	5-3	4-10
	160	6-8	6-5	6-3	6-4	6-1	5-10	5-7	5-5	5-0
1½	40	8-0	7-8	7-6	7-7	7-4	7-1	6-9	6-7	6-1
	80	8-1	7-10	7-8	7-9	7-6	7-3	7-0	6-9	6-4
	160	8-1	7-10	7-9	7-9	7-7	7-4	7-2	6-11	6-7
2	40	8-11	8-7	8-5	8-6	8-3	8-0	7-9	7-6	7-1
	80	9-1	8-9	8-8	8-8	8-6	8-3	8-0	7-10	7-5
	160	9-1	8-10	8-9	8-9	8-7	8-5	8-3	8-1	7-8

*Insulation type.

TABLE B4.13 Maximum Support Spacing for Seismic Stress for a Frequency of 25 Hz (Feet-Inches), Steel Pipe

Pipe size, in	Pipe sch.	No insul.	Antisweat*		Calcium silicate*					Reflective*
			½ in	1 in	½ in	1 in	1½ in	2 in	2½ in	4 in
½	40	4-10	4-4	4-2	4-3	4-0	3-8	3-6	3-3	2-11
	80	4-9	4-5	4-3	4-4	4-0	3-9	3-7	3-4	3-0
	160	4-8	4-5	4-3	4-4	4-1	3-10	3-7	3-5	3-1
¾	40	5-4	5-0	4-9	4-11	4-7	4-4	4-1	3-10	3-6
	80	5-4	5-0	4-10	4-11	4-8	4-5	4-2	4-0	3-7
	160	5-3	5-0	4-10	4-11	4-8	4-6	4-3	4-1	3-9
1	40	6-0	5-8	5-6	5-7	5-4	5-0	4-10	4-7	4-2
	80	6-0	5-9	5-7	5-8	5-5	5-2	4-11	4-9	4-4
	160	5-11	5-8	5-7	5-8	5-5	5-3	5-0	4-10	4-6
1½	40	7-2	6-10	6-9	6-9	6-6	6-4	6-1	5-10	5-5
	80	7-3	7-0	6-10	6-11	6-8	6-6	6-3	6-1	5-8
	160	7-3	7-0	6-11	6-11	6-9	6-7	6-5	6-2	5-10
2	40	8-0	7-8	7-7	7-7	7-4	7-2	6-11	6-9	6-4
	80	8-1	7-10	7-9	7-9	7-7	7-5	7-2	7-0	6-7
	160	8-1	7-11	7-10	7-10	7-8	7-6	7-4	7-2	6-10

*Insulation type.

TABLE B4.14 Maximum Support Spacing for Seismic Stress for a Frequency of 33 Hz (Feet-Inches), Steel Pipe

Pipe size, in	Pipe sch.	No insul.	Antisweat*		Calcium silicate*					Reflective*
			½ in	1 in.	½ in	1 in	1½ in	2 in	2½ in	4 in
½	40	4-2	3-9	3-7	3-8	3-5	3-2	3-0	2-9	2-6
	80	4-1	3-10	3-8	3-9	3-5	3-3	3-1	2-10	2-7
	160	4-1	3-10	3-8	3-9	3-6	3-4	3-1	2-11	2-8
¾	40	4-7	4-4	4-1	4-3	3-11	3-9	3-6	3-4	3-0
	80	4-7	4-4	4-2	4-3	4-0	3-10	3-7	3-5	3-1
	160	4-6	4-4	4-2	4-3	4-0	3-11	3-8	3-6	3-3
1	40	5-2	4-11	4-9	4-10	4-7	4-4	4-2	3-11	3-7
	80	5-2	5-0	4-10	4-11	4-8	4-5	4-3	4-1	3-9
	160	5-1	4-11	4-10	4-11	4-8	4-6	4-4	4-2	3-11
1½	40	6-2	5-11	5-10	5-10	5-7	5-6	5-3	5-0	4-8
	80	6-3	6-1	5-11	6-0	5-9	5-7	5-5	5-3	4-11
	160	6-3	6-1	6-0	6-0	5-10	5-8	5-7	5-4	5-0
2	40	6-11	6-8	6-7	6-7	6-4	6-2	6-0	5-10	5-6
	80	7-0	6-9	6-8	6-8	6-7	6-5	6-2	6-1	5-8
	160	7-0	6-10	6-9	6-9	6-8	6-6	6-4	6-2	5-11

*Insulation type

state fluid flow analyses; and nonlinear time-history analyses. There are 40 different finite elements available for static and dynamic analysis. Dynamic analyses can be performed either by modal superposition or direct integration.[47]

4. Specialized programs such as PIPERUP. PIPERUP performs nonlinear elastic-plastic analyses of piping systems subjected to concentrated static or dynamic time-history forcing functions. These forces result from fluid jet thrusts at the location of a postulated break in high-energy piping. PIPERUP is an adaptation of the finite element method to the specific requirements of pipe rupture analysis.[48]

Method of Analysis. The piping system is modeled as a series of masses connected by massless springs having the properties of the piping. The mathematical model should include the effects of piping geometry changes, elbow flexibilities, concentrated weights, changes in piping cross sections, and any other parameters affecting the stiffness matrix of the model. Mass point spacing should follow the guidelines specified above. Valves should be modeled as lumped masses at valve body and operator, with appropriate section properties for valve body and valve topworks. Rigid supports, snubbers, springs, and equipment nozzles should be modeled with appropriate spring rates in particular degree of freedom. Stress-intensification factors should be input at the appropriate locations (elbows, tees, branch connections, welds, etc.). Piping distributed weight should include pipe weight, insulation weight, and entrained fluid weight.

Once an accurate model is developed, the loading conditions are applied mathematically:

1. Statically applied loads (dead weight, wind loads, pressure thrust, etc.)
2. Thermal expansion
3. Statically applied boundary condition displacements (seismic anchor movement, LOCA containment displacement, etc.)
4. Response spectrum analysis (seismic, etc.)
5. Dynamically applied boundary condition displacements (LOCA motion, etc.)
6. Dynamically applied forcing functions (steam hammer, etc.)

The results of the analyses should be examined in order to determine if all allowables are met (i.e., piping stress, valve acceleration, nozzle loads, etc.). The loads must be combined using the appropriate load combinations and submitted to structural designers for their analysis.

REFERENCES

1. ASME, "Criteria of the ASME Boiler and Pressure Vessel Code for Design by Analysis in Section III and VIII, Division 2," 1969.
2. Markl, A. R. C., "Fatigue Tests of Piping Components," *Trans. ASME,* 1952.
3. ASME Boiler and Pressure Vessel Code, Section III, Division 1, 1989 edition, including December 31, 1989, addenda.
4. *Bonney Forge Bulletin No. 789,* "Weldolet, Stress Intensification Factors."

5. *Bonney Forge Bulletin No. 775,* "Sweepolet, Stress Intensification Factors and Stress Indices."
6. *Bonney Forge Bulletin No. 785,* "Latrolet, Stress Intensification Factors." Weldolet, Sweepolet, and Latrolet are registered trademarks of Bonney Forge Corp., Allentown, PA.
7. Walsh, D. J., and Woods, G. E., "Determination of Stress Intensification Factors for Integrally Reinforced 45° Latrolet Branch Connections," ASME paper 79-PVP-98, 1979.
8. ASME B31.1, Power Piping Code, 1989 edition, including ASME B31.1a—1989 addenda.
9. ASME B31.3, Chemical Plant and Petroleum Refinery Piping Code, 1990 edition.
10. ASME B31.4, Liquid Transportation Systems for Hydrocarbons, Liquid Petroleum Gas, Anhydrous Ammonia, and Alcohols, 1989 edition.
11. ASME/ANSI B31.5, Refrigeration Piping Code, 1987 edition, including ASME/ANSI B31.5a—1989 addenda.
12. ASME B31.8, Gas Transmission and Distribution Piping Systems, 1989 edition, including ASME B31.8a—1990 addenda.
13. Roark, R. J., and Young, W. C., *Formulas for Stress and Strain,* 5th ed., McGraw-Hill, New York, Tables 30 (p. 463) and 33 (pp. 516 and 517), 1975.
14. Bijlaard, P. P., "Stresses from Local Loadings in Cylindrical Pressure Vessels," *ASME Transactions,* vol. 77, no. 6, August 1955.
15. Wichman, K. R., Hopper, A. G., and Mershon, J. L., "Local Stresses in Spherical and Cylindrical Shells due to External Loadings," *Welding Research Council Bulletin 107,* March 1979 revision of August 1965 edition.
16. Rodabaugh, E. C., Dodge, W. G., and Moore, S. E., "Stress Indices at Lug Supports on Piping Systems," and Dodge, W. G., "Secondary Stress Indices for Integral Structural Attachments to Straight Pipe," *Welding Research Council Bulletin 198,* September 1974.
17. ASME Boiler and Pressure Vessel Code, Code Cases, Nuclear Components, Case N-122-1, 1989; Case N-318-4, 1989; Case N-391-1, 1989; and Case N-392-1, 1989.
18. Standard of the Expansion Joint Manufacturers Association, 1980.
19. Kreith, F., *Principles of Heat Transfer,* 2d ed., International Text Book, Scranton, PA, 1966.
20. 10CFR Part 50, Appendix A, General Design Criterion 2, "Design Bases for Protection Against Natural Phenomena."
21. U.S. Nuclear Regulatory Commission, Regulatory Guide 1.29, "Seismic Design Classification," rev. 3, September 1978.
22. Newmark, N. M., "A Method of Computation for Structural Dynamics," *Journal of Engineering Mechanics Division,* vol. 85, no. EM3, ASCE, July 1959.
23. Bathe, K. J., and Wilson, E. L., "Stability and Accuracy Analysis of Direct Integration Methods," *Earthquake Engineering and Structural Dynamics,* vol. 1, 1973.
24. Wu, R. W., Hussain, F. A., and Liu, L. K., "Seismic Response Analysis of Structural System Subject to Multiple Support Excitation," *Nuclear Engineering and Design,* vol. 47, 1978.
25. Lin, C. W., and Loceff, F., "A New Approach to Compute System Response with Multiple Support Response Spectra Input," *Nuclear Engineering and Design,* vol. 60, 1980.
26. ANSI A58.1, "Minimum Design Loads for Buildings and Other Structures," ASCE 7-88, 1982.
27. U.S. Nuclear Regulatory Commission, Regulatory Guide 1.61, "Damping Values for Seismic Design of Nuclear Power Plants," October 1973.

28. ASME Boiler and Pressure Vessel Code, Case N-411-1, February 20, 1986, reaffirmed on February 20, 1989.

29. *Welding Research Council, Bulletin 300,* "Technical Position on Damping Values for Piping—Interim Summary Report," December 1984.

30. U.S. Nuclear Regulatory Commission, Regulatory Guide 1.84, "Design and Fabrication Acceptability, ASME Section III, Division 1," rev. 26, July 1989.

31. Lin, C. W., "How to Lump the Masses—A Guide to the Piping Seismic Analysis," ASME paper 74-NE-7, June 1974.

32. U.S. Nuclear Regulatory Commission, "Report of the U.S. Nuclear Commission Piping Review Committee—Evaluation of Dynamic Loads and Load Combinations," *NUREG-1061,* vol. 4, December 1984.

33. Dong, M. Y., and Lee, H. M., "Comparative Study of ZPA Effect in Modal Response Spectrum Analysis," ASME Publication PVP, vol. 144, June 1988.

34. U.S. Nuclear Regulatory Commission, Regulatory Guide 1.92, "Combining Modal Responses and Spatial Components in Seismic Response Analysis," rev. 1, February 1976.

35. Singh, A. K., Chu, S. L., and Singh S., "Influence of Closely Spaced Modes in Response Spectrum Method of Analysis," in *Proceedings of the Specialty Conference on Structural Design of Nuclear Power Plant Facilities,* vol. 2, ASCE, December 1973.

36. Biggs, J. M., *Introduction to Structural Dynamics,* McGraw-Hill, New York, 1964.

37. Coccio, C. L., "Steam Hammer in Turbine Piping Systems," ASME 66-WA-FE32, 1966.

38. U.S. Nuclear Regulatory Commission, Report NUREG-0582, "Water-Hammer in Nuclear Power Plants," July 1979.

39. Moore, K. V., and Rettig, W. H., "RELAP 4—A Computer Program for Transient Thermal-Hydraulic Analysis," ANCR-1127, rev. 1, March 1975.

40. U.S. Atomic Energy Commission, Regulatory Standard Review Plan 3.9.2, "Dynamic Testing and Analysis of Systems, Components, and Equipments," November 1974.

41. Patel, M. R., "Auxiliary Line Evaluation for Loss-of-Coolant Accident Using Time-History Plastic Analysis," ASME Publication PVP-PB-022.

42. Sun, W., Lee, R., and Lee, N., "Secondary System Piping Analysis including Seismic and a Loss-of-Coolant Accident," ASME Publication 83 PVP, vol. 73.

43. Kassawara, R. P., Austin, S. C., and Izor, R. C., "The Effect of Reactor Coolant System Rupture Motion on Tributory Piping and Attached Equipment," ASME 80-C2/PVP-24.

44. U.S. Nuclear Regulatory Commission, Standard Review Plan 3.6.1, "Plant Design for Protection against Postulated Piping Failure in Fluid Systems Outside Containment," rev. 1, July 1981.

45. U.S. Nuclear Regulatory Commission, Standard Review Plan 3.6.2, "Determination of Rupture Locations and Dynamic Effects Associated with the Postulated Rupture of Piping," rev. 2, June 1987.

46. Sun, W., and Lee, R., "Pipe Break Isolation Restraint Design for Nuclear Power Plant Containment Penetration Areas," ASME paper 82-PVP-37.

47. DeSalvo, G. J., and Gorman R. W., *ANSYS, Engineering Analysis System, User's Manual,* Swanson Analysis Systems, 1989.

48. *PIPERUP: A Computer Program for Pipe Whip Analysis, User's Manual,* Nuclear Service Corp., 1977.

49. ANSI/ANS 58.2-88, "Design Basis for Protection of Light Water Nuclear Power Plants against Effects of Postulated Pipe Rupture," 1988.

50. Task Committee on Wind Forces, Committee on Loads and Stresses, Structural Divi-

sion, ASCE, "Wind Forces on Structures," *Transactions,* paper no. 3269, vol. 126, part II, 1961.

51. Table NF-3611-1, Subsection NF, Section III, Division 1, ASME Boiler and Pressure Vessel Code, 1989 edition.
52. Table 121.5, ASME B31.1—1989 edition. Support spacing is based on bending stress not exceeding 2300 psi.
53. Stevenson, J. D., and LaPay, W. S., "Amplification Factors to be Used in Simplified Seismic Dynamic Analysis of Piping Systems," ASME paper 74-NE-9, Pressure Vessel and Piping Conference, June 1974.

CHAPTER B5
PIPING SUPPORTS

F. E. Grunwald
Engineering Supervisor
Bechtel Corporation
Gaithersburg, Maryland

T. C. LaCroix, P.E.
Engineering Specialist
Bechtel Corporation
Gaithersburg, Maryland

C. J. Obst, P.E.
Engineering Specialist
Bechtel Corporation
Gaithersburg, Maryland

INTRODUCTION

The correct and economical selection of the supports for any piping system usually presents difficulties of varying degrees, some relatively minor and others of a more critical nature. Proper support selection should be the objective of all phases of design and construction.

Many pipe-support problems may be minimized or avoided if proper attention is given to the means of support during the piping layout design phase. The piping designer's familiarity with support problems, accepted practices, and commercially available pipe support components and their application is extremely important. Good pipe support design begins with good piping design and layout. For example, other considerations being equal, piping should be routed to use the surrounding structure to provide logical and convenient points of support, anchorage, guidance, or restraint, with space available at such points for use of the proper component. Parallel lines, both vertical and horizontal, should be spaced sufficiently apart to allow room for independent pipe attachments for each line.

Pipe-support specifications for individual projects must be written in such a way as to assure proper support under all operating and environmental conditions and to provide for slope, expansion, anchorage, and insulation protection. Familiarity with standard practices, customs of the trade, types and functions of commercial component standard supports and an understanding of their individual advantages and limitations, together with knowledge of existing standards such as MSS SP-58 (Materials and Design of Pipe Supports), MSS SP-69 (Selection and Application of Pipe Supports), MSS SP-89 (Fabrication and Installation of Pipe Supports), and PFI ES-26 (Welded Load Bearing Attachments to Pressure Retaining Boundaries) can be of great help in achieving the desired results.

Unless complete design details are provided by the engineer or piping-system layout designer, the final responsibility for selection and design of pipe supports capable of completely satisfying the system requirements and job specifications rests with the pipe-support engineer and/or designer. Any piping system is inoperable until the pipe hangers and supports have been selected and installed. Experience shows that a high percentage of pipe-support problems cannot be recognized during a cursory examination of the piping drawings but await a detailed analysis by those responsible for the design of the support and the preparation of specific material lists and details. In the interest of properly coordinating support installations with the piping erection schedule, priority should be given to the selection, design, and procurement of pipe-support components.

The dollar value of the support system is generally outweighed many times by the value of pipe, valves, and fittings which are to be supported. Failure to allow sufficient time for the design, procurement, and fabrication of the supports can lead to costly erection delays and the use of temporary supports.

Pipe supports are generally identified by a specification type number, such as a "type 3" double bolt pipe clamp (as in MSS SP-58) or a manufacturer's figure number, a descriptive name, and size. The latter depends upon the type of component; for example, pipe attachments, such as clamps, are prescribed by pipe size, upper attachments are identified by rod size, and variable springs are sized by the calculated load to be supported. In addition, pipe-support components vary in their load-carrying capacity. The load capacity of each type of pipe-support component is given in most manufacturers' catalogs or load capacity data sheets (LCDSs). Various components and their functions will be discussed in greater detail in subsequent sections.

General Design Considerations

Hangers and supports must be designed to meet all static as well as dynamic operational conditions to which the piping and equipment may be subjected. The support system must provide for and control, subject to the requirements of the piping configuration, the movement caused by the thermal expansion and contraction of the piping and connected equipment. Proper design necessitates a thorough knowledge of the complete cyclic behavior of each section of piping to be supported, and an awareness of the proximity of the pipe with respect to building structure, other piping systems, and equipment in the immediate vicinity.

A substantial reduction in the complexity of support design can be realized when piping-support requirements are coordinated properly with the plant and piping design phases. Initial plant design should recognize that the support designer requires access to the piping, sufficient space in which to install the supporting equipment, and adequate structure from which to support the piping. Ideally the support designer should have the opportunity to offer comments on the piping design from the initial design.

Starting with any given set of piping and structural drawings, it is necessary for the support designer to become familiar with the overall design concept and any special requirements that may be called for in the specifications. When dealing with a number of piping systems in any area, it is advisable to superimpose the piping (using a single-line representation) on the structural drawings. A preliminary study very often reveals that consideration has been given to the supporting phase and provisions have been made in the form of steel work patterns, anchor bolts or inserts in concrete, or runs of piping purposely routed near suitable supporting structure. Naturally, full advantage should be taken of such conditions. The piping should also be studied from the standpoint of possibly coordinating supports for one system or pipe run with those of another, thereby providing an orderly supporting arrangement. On the other hand, it may also be necessary, because of loading conditions, to purposely stagger supports in order to distribute the loads on the supporting structures. All of the above should be determined prior to the start of detailed support design for any piping system.

Once a tentative overall pattern of supports for any particular area has been established, it is advisable to begin specific design with the most critical or largest piping systems, thereby reserving the best possible supporting conditions for the most critical lines.

Of prime consideration in design is the determination of support location. Although supports are located ideally to suit the requirements of the piping configuration, some degree of compromise may be required to take the fullest advantage of the available supporting structure. The piping system should first be investigated as a whole. With the use of the allowable support spacing as dictated by code, practice, or special calculation (see Chap. B4), the support points are tentatively located, taking into consideration division of straight runs, concentrated loads, elimination of excessive overhanging sections or bends, and loads on terminal connections.

Next, the tentative locations should then be compared with the available supporting structure, modified as required, and recorded on the superimposed piping previously sketched on the structural drawings. The recording of each support location, along with an indication of any supplementary steel required, serves as a valuable aid in checking clearances and coordinating supports during the course of design, especially when a number of designers are working on the same project.

It should be emphasized that the location of pipe supports is an iterative process in which several support configurations are investigated in succession, with the results of the analysis of one configuration being factored into the selection of the next configuration, until a satisfactory solution is reached.

Once a satisfactory pattern of support locations has been established, the next step is to determine the loading and movement conditions existing at each support point. Here, coverage will be confined to the general considerations regarding only loads and movements. Recommended methods for obtaining or calculating specific loads and movements are covered later under "Determination of Loads and Movements."

Loading Considerations

The magnitude and direction of the design load as determined by the methods of Chap. B4 are used to select and design the proper support or restraint at each selected point along the piping system.

Seismic/Dynamic or Other Loadings

The design of pipe-support components and structural steel must include provisions for seismic/dynamic or other occasional loadings where required. Applicable codes generally permit some increase in allowable stresses for seismic/dynamic loading conditions. Spring supports should be evaluated to determine if additional dynamic movement can be tolerated, and, if not, additional means such as limit stops or dynamic snubbers may be required.

Overloading from Various Causes

ASME B31.1 Power Piping Code allows an occasional short-duration overloading of 15 to 20 percent which can be applied against any form of short-time loading. However, it is appropriate to point out that, where there is a possibility of two or more causes occurring simultaneously, the combined overloading in excess of 20 percent must be added to the design load.

DETERMINATION OF SUPPORT LOCATIONS

Support locations are dependent on many considerations such as pipe size, piping configuration, the location of heavy valves and fittings, and the structure that is available for support. No firm rules or limits exist which will positively fix the location of each support in a piping system. Instead, the engineer must exercise judgment to determine appropriate support locations.

The suggested maximum spans between supports listed in Table B5.1 reflect the practical considerations involved in determining support spacings on straight runs of standard wall piping. The spans are based on a combined bending and shear stress of 1500 psi when the pipe is filled with water, and $\frac{1}{10}$-in sag is allowed between supports. Table spans do not apply where concentrated weights such as valves or heavy fittings exist, or where changes in direction of the piping system occur between support points.

Supports should be placed as close as possible to concentrated loads in order

TABLE B5.1 Suggested Piping-Support Spacing

Nominal pipe size (in)	Suggested maximum span (ft): Water service	Suggested maximum span (ft): Steam, gas, or air service
1	7	9
2	10	13
3	12	15
4	14	17
6	17	21
8	19	24
12	23	30
16	27	35
20	30	39
24	32	42

to keep piping stresses to a minimum. Where changes in piping direction of a system occur between supports, it is a good practice to keep the total length of pipe between the supports equal to or less than 0.75 times the full spans listed in Table B5.1. When practical, a support should be located immediately adjacent to any change in direction of the piping.

For economy in the support of low-pressure, low-temperature systems, support spans may be based on the code allowable total stresses of the pipe and the amount of allowable deflection between supports.

In steam lines with long spans, the deflection caused by piping weight may be large enough to cause an accumulation of condensate at the low points. During plant startup these pockets of condensate can cause flashing and result in undesirable dynamic loads on the system. Natural drainage can be provided by erecting the line with a slope in such a manner that succeeding supports are lower than the points of maximum deflection in preceding spans.

For fluids other than water, the bending stress can be found by using the formulas in Chap. B4. Spans for typical small-diameter piping based on the equations in Chap. B4 are found in Tables B5.2 and B5.3.

TABLE B5.2 Gravity Steel Pipe-Support Spacing (Contents = Water)

		Pipe span (ft and in)			
Pipe size (in)	Pipe sch.	No. insul.	1 in Insul.	1½ in Insul.	2 in Insul.
½	40	6-8	5-4	4-6	3-11
	80	6-7	5-5	4-7	4-0
	160	6-4	5-5	4-9	4-2
¾	40	7-5	6-3	5-6	4-10
	80	7-5	6-4	5-8	5-1
	160	7-3	6-4	5-9	5-2
1	40	8-3	7-0	6-4	5-9
	80	8-3	7-3	6-8	6-1
	160	8-1	7-3	6-9	6-2
1¼	40	8-3	7-0	6-4	5-9
	80	8-3	7-3	6-8	6-1
	160	8-1	7-3	6-9	6-2
1½	40	9-10	8-9	8-1	7-3
	80	9-10	9-1	8-6	7-7
	160	9-11	9-2	8-8	7-11
2	40	10-9	9-10	9-3	8-7
	80	11-1	10-2	9-9	9-2
	160	11-1	10-5	10-0	9-6

DETERMINATION OF LOADS AND MOVEMENTS

The anticipated movement at each support point dictates the basic type of support required. Each type of support selected must be capable of accommodating movements obtained by one of the methods outlined later in this section. It is a

TABLE B5.3 Gravity Steel Pipe-Support Spacing (Pipe Empty, Air, Steam)

Pipe size (in)	Pipe sch.	Pipe span (ft and in) No. insul.	1 in Insul.	1½ in Insul.	2 in Insul.
½	40	7-1	5-6	4-7	4-1
	80	6-10	5-6	4-9	4-2
	160	6-7	5-6	4-9	4-3
¾	40	8-1	6-8	5-9	5-0
	80	7-10	6-8	5-10	5-2
	160	7-5	6-7	5-10	5-4
1	40	9-2	7-6	6-9	6-0
	80	8-10	7-7	6-11	6-3
	160	8-6	7-5	6-11	6-4
1¼	40	9-2	7-6	6-9	6-0
	80	8-10	7-7	6-11	6-3
	160	8-6	7-5	6-11	6-4
1½	40	11-3	9-9	8-11	7-8
	80	11-0	9-10	9-2	8-1
	160	10-6	9-7	9-2	8-3
2	40	12-8	11-2	10-4	9-6
	80	12-5	11-3	10-7	9-11
	160	11-10	11-0	10-6	10-0

good practice to select first the most simple or basic rigid support type and add to the complexity only as conditions warrant. No advantage will be realized in upgrading a support when a simpler, more economical type can be shown to satisfy all the design requirements. Both vertical and horizontal movement must be evaluated.

When piping vertical movement is small, the use of simple rod hangers should be adequate. With small vertical movement and significant horizontal movement, the simple rod hanger will still suffice, provided the overall length is sufficient to keep the angular swing of the rod within reasonable limits—normally accepted as being 4° from the vertical. Consideration should be given to advancing the upper connection some percentage (usually two-thirds) of the total movement as a means for reducing the angularity in the hot position. For piping supported from below, some form of slide must be incorporated to provide for the horizontal movement or, in the case of assured longitudinal movement, a pipe roll may be used. Rollers are usually only used on long runs of piping supported on racks such as found in refinery piping. Suspended hangers with considerable horizontal movement and low headroom will require either single- or double-direction trolleys or rollers. Where both longitudinal and lateral movements are large, consideration may be given to the use of a single-direction trolley oriented on the resultant movement vector.

Piping which is subject to significant vertical movement requires the incorporation of variable springs or constant support springs. Variable supports will normally suffice for spring hangers except for certain systems in power plants, refineries, and some specialized lines in process and chemical plants which may have large vertical movements where the change in load is unacceptable. (In these cases, the use of constant supports is required.) As the name *variable support* implies, the supporting effect varies in relation to the vertical movement of the suspended piping and the resultant compression or extension of the spring coil. The change in supporting force is in proportion to the spring rate and to the amount of movement. When variable supports are used, the loss or gain in supporting force should be considered with respect to added stress on the piping system. For normal installations, various industries have established conventions concerning the use of variable spring supports and constant spring supports which may be regarded as standard practice. For example, the following would be conventional guidelines for power plant piping.

At support locations subject to vertical movement, springs of suitable design to prevent excessive variation in supporting effect should be used to provide for the movement. The amount of variation that can be tolerated should be based on such considerations as pipe-bending effect, control of piping elevation, allowable terminal connection loadings, etc. In general, the deviation in supporting effect should be limited to ±6 percent for systems such as main steam, hot and cold reheat, extraction lines over 750°F discharge in the vicinity of the pumps, and boiler terminal connection. On other systems, the variations in supporting effect should be limited to 25 percent.

For all systems, a greater allowance in percentage of variation is permissible at points of support where the variation in supporting effect is transferred directly to a rigid support or terminal connection specifically designed for the resulting loading condition.

When the load and movement at each selected support location is determined, detailed design of the individual supports can be undertaken. The movement determines the basic type of support (such as spring or rigid, slide or roller); the piping, its temperature, and ambient temperature determine the pipe attachment and remaining support material, respectively; the proximity to supporting structure and available clearance determines the support configuration such as single- or double-rod hanger, stanchion, or bracket-type support; and the load determines the required size for each component of the support.

The design should take full advantage of commercially available load-rated and tested support components to the greatest extent possible. An effort should be made to maintain uniformity and simplicity in design. The support should be functional, provide means for piping elevation adjustment, and be easily installed using normal field labor and equipment.

For economic reasons, the use of commercial catalog items available from piping-support manufacturers eliminates the need for detail design of the numerous types and varying sizes of components required to complete a typical piping-support installation.

SELECTION OF PIPE-SUPPORTING DEVICES

Piping Systems: Temperature Classification. Piping systems, for the purposes of this chapter, are divided into the following temperature categories in order to provide a basis for the selection of supports, anchors, or restraints and guides.

1. Hot systems (120°F and higher).
   a. The temperature range is from 120 to 450°F. Typical examples are low-pressure steam, hot water, and certain process piping.
   b. The temperature range is from 450 to 650°F. Typical examples are boiler plant and industrial steam and hot-water piping systems.
   c. The temperature ranges from 750°F and upward. A typical example is a high-pressure steam power-plant piping system.
2. Ambient systems in which the contents of the line are not heated or cooled by mechanical means. Temperatures range up to 120°F. Plant air and service water are typical systems.
3. Cold systems (usually less than 70°F).
   a. The temperatures range from 32 to 70°F. A typical example is chilled water piping.
   b. The temperature range from 32 to −20°F, as in brine systems.
   c. Temperatures below −20°F, as in cryogenic systems.

Pipe Attachments. Supports for the various systems described above may be selected from Fig. B5.1 in accordance with the following recommendations:

For Type 1*a* systems, MSS-SP-58 support Types 1 and 3 through 12 are used for suspending from above. Rollers, used for supporting from below, should be MSS-SP-58 Types 41 and 43 through 46 with appropriate saddles of MSS-SP-58 Type 39, and sliding supports should be of MSS-SP-58 Types 35 through 38.

For Type 1*b* systems, MSS-SP-58 support Types 1, 3, 4, and 42 are used. Rollers should be of MSS-SP-58 Types 41 and 43 through 46 with appropriate saddles of MSS-SP-58 Type 39.

For Type 1*c* systems, alloy supports are used as required by the line temperature. Supports should be of MSS-SP-58 Types 2, 3, or 42 with saddles of MSS-SP-58 Type 39 and rollers of MSS-SP-58 Types 41 through 47. (In the temperature range 650°F and higher, there is the possibility of metallurgical change if unalloyed carbon steel components are used. It is suggested that those portions of hangers, anchors, and supports in direct contact with piping which operates at or above 650°F be made of materials suitable for these extreme service temperatures.)

For Type 2 systems, hangers can be of MSS-SP-58 Types 1 and 3 through 12 with supports of MSS-SP-58 Types 24, 26, and 35 through 38.

For Type 3 systems, the hanger or support must be outside the insulation and the vapor barrier must be left undisturbed. A MSS-SP-58 Type-40 insulation protection shield (or similar device) must be used to distribute the loading on the insulation. Supports sized for the outside diameter of the insulation can be of MSS-SP-58 Types 1, 4, 6, 7, 9, 10, or 11. For Type 3*c* systems, the welded attachment must be of an alloy material which is compatible with the material of the piping system itself.

Pipe-Covering Protection Saddles. Where insulation is used on the piping system, it is frequently necessary to make some modification at the point of attachment of the support. Since different methods and practices exist for hot lines as contrasted to low-temperature lines (chilled water and brine), they should be treated separately.

Insulation is available in thicknesses shown in Table B5.4. Caution must be taken in the support of insulated piping to assure that a firm attachment is pro-

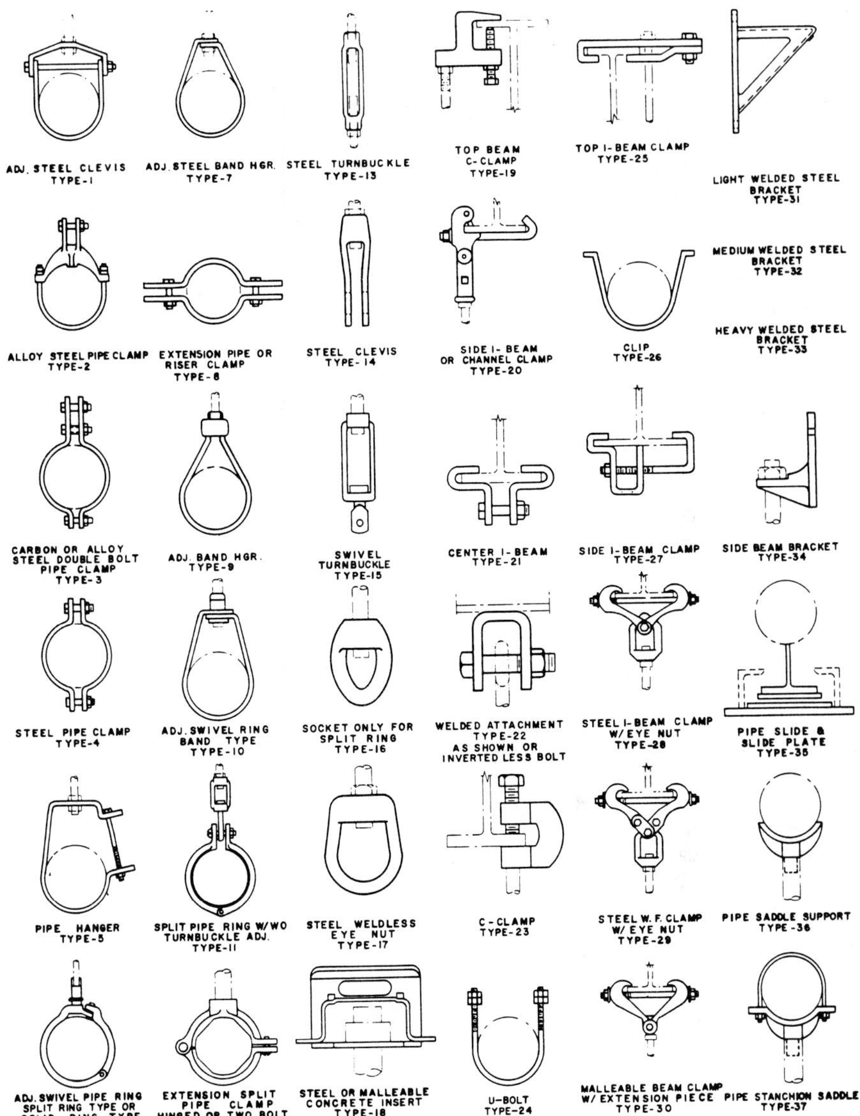

FIGURE B5.1 Types of pipe supports. (*MSS SP-58, 1988 ed.*)

vided. Particularly, with respect to high-temperate lines, provision in the form of pipe-protection saddles is made to assure that the pipe supports do not become overheated. Protection saddles, Type 39 for use with high-temperature insulation should be 12 in long and have approximately 60° of arc. Metal is normally $\frac{1}{8}$ in thick for pipe sizes up to 5 in and $\frac{3}{16}$ in thick for larger pipe sizes. These saddles and matching rollers are available commercially for most sizes and thicknesses listed in Table B5.4.

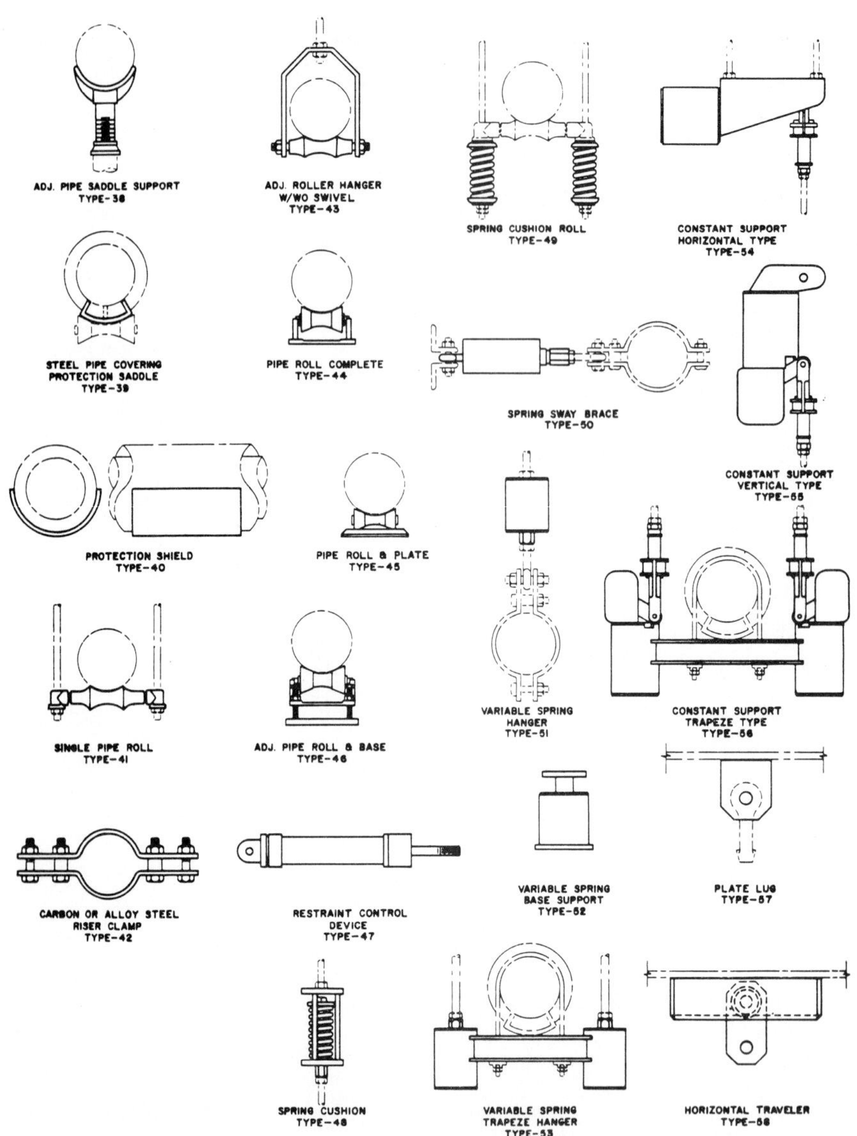

FIGURE B5.1 (*Continued*) Types of pipe supports. (*MSS SP-58, 1988 ed.*)

For support from above, either a conventional type of pipe clamp may be used directly on the pipe and the arms of the clamp (Type 3) extended outside the insulation, or a larger clamp (Type 1 or 4) may be used and lined with saddles or noncrushable insulation material.

Lugs may be welded to the pipe and extended outside the insulation to a steel clevis (Type 14) or, if supported from below, to a sliding shoe or guided roller. The important point is to have the actual attachment or supporting member out-

TABLE B5.4 Simplified Thicknesses for Pipe Insulation

Nominal pipe size, in.	Nominal thickness, in.														
	1		1½		2		2½		3		3½		4	4½	5
	Approximate thickness and outer layer pipe size, in.[1]														
	Thickness	Pipe size	Thickness	Pipe size	Thickness	Pipe size	Thickness	Pipe size	Thickness	Pipe size	Thickness	Pipe size	Thickness	Thickness	Thickness
½	1	2½	1 9/16	3½	2 1/16	4½	2 7/8	6	3 11/32	7	3 7/8	8	4 3/8	4 29/32	5 13/32
¾	29/32	2½	1 15/32	3½	1 31/32	4½	2 3/4	6	3 1/4	7	3 3/4	8	4 9/22	4 13/16	5 5/16
1	1 3/32	3	1 19/32	4	2 1/8	5	2 21/32	6	3 5/32	7	3 21/32	8	4 5/32	4 11/16	5 3/16
1¼	29/32	3	1 21/32	4½	1 15/16	5	2 15/32	6	2 31/32	7	3 15/32	8	3 31/32	4½	5
1½	1 1/32	3½	1 17/32	4½	2 5/16	6	2 27/32	7	3 11/32	8	3 27/32	9	4 3/8	4 7/8	5 3/8
2	1 1/16	4	1 19/32	5	2 1/8	6	2 5/8	7	3 1/8	8	3 5/8	9	4 5/32	4 21/32	5 5/32
2½	1 1/16	4½	1 7/8	6	2 11/32	7	2 7/8	8	3 3/8	9	3 29/32	10	4 13/32	4 29/32	5 9/16
3	1 1/32	5	1 9/16	6	2 1/16	7	2 9/16	8	3 1/16	9	3 19/32	10	4 3/32	4 19/32	5 1/4
3½	1 9/32	6	1 13/16	7	2 9/32	8	2 25/32	9	3 11/32	10	3 27/32	11	4 5/16	....	4 31/32
4	1 1/16	6	1 9/16	7	2 1/16	8	2 9/16	9	3 3/32	10	3 19/32	11	4 3/32	4 3/4	5 3/16
4½	1 9/32	7	1 13/16	8	2 5/16	9	2 13/16	10	3 11/32	11	3 27/32	12	....	4½	4 15/16
5	1 1/32	7	1 17/32	8	2 1/32	9	2 9/16	10	3 1/16	11	3 9/16	12	4 7/32	4 21/32	5 5/32
6	1	8	1½	9	2 1/32	10	2 17/32	11	3 1/32	12	3 11/16	14	4 1/8	4 5/8	5 3/16
7	....	....	1 17/32	10	2 1/32	11	2 17/32	12	3 1/8	14	3 5/8	15	4 1/8	4 5/8	5 1/8
8	....	....	1 17/32	11	2 1/32	12	2 11/16	14	3 1/8	15	3 5/8	16	4 3/16	4 11/16	5 3/16
9	....	....	1 17/32	12	2 5/32	14	2 21/32	15	3 1/8	16	3 21/32	17	4 5/32	4 21/32	5 5/32
10	....	....	1 19/32	14	2 3/32	15	2 19/32	16	3 3/32	17	3 19/32	18	4 3/32	4 19/32	5 3/32
11	....	....	1 19/32	15	2 3/32	16	2 5/8	17	3 3/32	18	3 19/32	19	4 3/32	4 19/32	5 3/32
12	....	....	1 19/32	16	2 3/32	17	2 5/8	18	3 3/32	19	3 19/32	20	4 1/8	4 5/8	5 1/8
14	....	....	1½	17	2	18	2½	19	3	20	3½	21	4	4½	5
16	....	....	1½	19	2	20	2½	21	3	22	3½	23	4	4½	5
18	....	....	1½	21	2	22	2½	23	3	24	3½	25	4	4½	5

[1]Subject to manufacturing tolerances.

side the insulation so that movement of the line will not result in insulation damage.

For low-temperature service, in addition to heat loss or gain, the problem of atmospheric condensation must be considered, and such lines are usually insulated with a material that has an outer covering or seal called a *vapor barrier.* This barrier prevents the insulation from absorbing moisture. For this reason it is not permissible to penetrate the insulation with load-carrying members such as the legs of a conventional high-temperature saddle or a pipe clamp. Since most low-temperature insulations have low compressive strength, it is necessary to provide shields to line the piping insulation and to spread out the bearing area sufficiently to prevent crushing of the insulation. Such shields should fit the outer diameter of the insulation and cover 180° of arc. Metal weights and lengths are given in Table B5.5, taken from MSS SP-58.

TABLE B5.5 Minimum Dimensions: Shields for Insulation Protection

Nominal tubing size (in)	Gauge length (in)	Thickness
½ to 3½	12	18
4	12	16
5	15	16
6	18	16
8	24	14

Attachments to Buildings or Other Structures. When piping is to be hung from steel, Types 20 through 23, 25, and 27 through 30 beam clamps and 57 plate lugs should be used.

When piping is hung from concrete, malleable iron or steel inserts of Type 18 or a continuous strip insert may be used. Embedded anchor bolts may also be used under specific conditions. For wall supports on either concrete or steel brackets, Types 31 through 33 may be used.

In many cases, it is necessary to provide additional structure as a means of upper attachment. Such structure must be designed for the particular load and can be of structural angle, beam, or channel.

Structural steel for pipe supports must meet the allowable stress requirements of the AISC Specification.

Multiple Pipe Runs

Where the bottoms of the various lines are approximately at the same elevation, common supports are advantageous. The supports take the form of trapeze hangers fabricated from structural steel shapes. Multiple pipe runs are also supported on fabricated bents or frames. This is quite common for oil refinery or tunnel work where multiple runs of pipe are relatively near grade, steel or concrete sleepers are used.

On all multiple pipe runs, provisions should be made to keep the lines in their relative positions by the use of either clamps or clips. Lines subject to thermal expansion must be free to slide or roll.

Spring Supports

When there is an appreciable temperature difference between the operating and nonoperating condition of a piping system, there is a resultant expansion or contraction of the pipe caused by the thermal change. When the system consists entirely of horizontal piping, the differential expansion can be taken care of entirely by means of rollers or by swinging rods of sufficient length. When there are vertical portions of the piping system, the thermal change in length causes elevation changes. Movement of terminal points or of the structure to which the supports are attached will also result in elevation changes. These elevation changes must be accommodated by some sort of variable spring or constant support.

Helical-coiled springs are commonly used for such supports. The degree of variation to be provided by the spring support depends on the conditions of the piping system. For systems which operate at temperatures below 750°F, a good rule is that the variation in supporting force be limited to 25 percent of the load.

It is not economical to design custom springs. A more prudent approach is to select spring devices which are available commercially. Commercial spring supports are made in three general types. Spring cushion hangers, MSS-SP-58 Types 48 and 49, are made for light to medium loads and for ¼-in maximum vertical movement. Generally, this type is used in systems where the temperature does not exceed 450°F.

Variable spring supports, MSS-SP-58 Types 51, 52, and 53, are available for a wide range of loads, from about 50 to 30,000 lb. This type of hanger is used on piping systems where the resulting variation in supporting force can be tolerated. The commercial varieties available provide a selection of variability.

Constant load supports, MSS-SP-58 Types 54, 55, and 56, are spring devices in which the varying force of the spring is compensated so that the support variability is within the range of ±6 percent. This type of support is used on systems where there are large thermal movements or critical stress conditions or a combination of both. Such conditions exist on high-pressure, high-temperature steam lines in electric generating stations. These supports are made for loads of approximately 50 to 50,000 lb and for vertical movements up to 16 in. The variability of these supports is not dependent on vertical movement, and the support is adjusted at the factory for specified load and travel. Therefore, extreme care should be taken in determining loads and travel for the selection of hanger size.

Vibration arising from pump pulse and similar conditions can be a problem in piping systems. Where such vibration exists and is in resonance with a spring-supported system, the results can be serious. Such conditions can usually be avoided by judicious use of commercially available spring hangers. Systems that respond to exciting vibrations can be controlled satisfactorily by the use of dampening devices. There are two general types to consider: the coiled spring and the hydraulic vibration dampener.

There are two types of coiled-spring vibration dampeners: the opposed-spring type and the double-acting spring type, MSS-SP-58 Type 50. These devices should be arranged so that the springs are in the neutral position during normal operating conditions of the system.

The hydraulic vibration control is a unit which operates by means of a controlled flow of fluid through an orifice. Resistance to movement increases with the speed of displacement. One distinct advantage of the hydraulic device is that there is a minimum of resistance to thermal movement of the piping.

Both spring and hydraulic cylinder devices may be used to control sway and absorb shocks. These same devices may be used to resist reactions from safety-valve discharges and similar applications. Both the spring and hydraulic types are available commercially in several sizes. Manufacturers' literature gives comprehensive data for selection and performance characteristics.

Hanger Rod

Rod used for pipe-support purposes is usually hot-rolled steel with standard 60° cut threads conforming to the coarse thread series (UNC). Rolled threads to the same standard may be used. It must be pointed out that the length of a rolled thread cannot be increased by running a die over it, since the basic diameter of the rod is less than the size of the threaded portion (smaller than the thread major diameter).

Safe load capacities of threaded rods are based on the area at the root of the thread. A generally accepted standard for such capacities is given by Table B5.6, taken from MSS SP-58.

TABLE B5.6 Load Rating of Threaded Hot-Rolled Steel Conforming to ASTM A36 or A575 Gr. 1020

Nominal rod diameter (in)	Root area of thread (in^2)	Maximum safe load at rod temp of 650°F (lb)
¼	0.027	240
5⁄16	0.046	410
⅜	0.068	610
½	0.126	1,130
⅝	0.202	1,810
¾	0.302	2,710
⅞	0.419	3,770
1	0.552	4,960
1⅛	0.693	6,230
1¼	0.889	8,000
1⅜	1.053	9,470
1½	1.293	11,630
1⅝	1.515	13,630
1¾	1.744	15,690
1⅞	2.048	18,430
2	2.292	20,690
2¼	3.021	27,200
2½	3.716	33,500
2¾	4.619	41,600
3	5.621	50,600
3¼	6.720	60,500
3½	7.918	71,260

TABLE B5.7 Relation between Pipe and Rod Sizes

Pipe size (in)	Rod size (in)
2 and smaller	3/8
2½ to 3½	½
4 and 5	5/8
6	¾
8 to 12	7/8
14 and 16	1
18	1⅛
20	1¼
24	1½

For convenience in ordering and assembling various items, most cataloged pipe-support products have a definite relationship between rod size and pipe size as shown in Table B5.7.

Table B5.7 conforms to Underwriters Laboratories and Factory Mutual requirements in pipe sizes up to 12 in. In general, rod sizes less than 3/8 in should never be used to support piping, regardless of its diameter.

Snubbers

A snubber, or shock arrestor, MSS SP-58 Type 47, is a mechanical or hydraulic type of support that is used to restrain dynamic loads. Such loads can result from a seismic event but are not limited to such. Water hammer, steam hammer, or other sudden impact-type loads are also often restrained by a snubber.

The primary advantage of the snubber is that it can be installed at locations where thermal displacement precludes the use of a "rigid" support. Under the slow and gradual displacement of the pipe due to thermal action, a snubber is designed to slowly telescope (in or out) to accommodate the movement of the pipe. When, however, a sudden impact load acts upon the snubber, internal braking devices engage, thus controlling the movement of the pipe. The snubber is said to "lock up." In this condition, the snubber acts as though it were a rigid restraint.

Snubbers are not capable of supporting gravity loads and, under certain circumstances, the weight of the snubber bears on the pipe and, therefore, added weight should be included in the piping stress analysis.

Riser Supports

A riser or vertical section of a piping system often creates special support problems. The support of service water lines and fire protection systems is governed in the first case by the local building codes and in the latter case by NFPA Codes. These codes dictate the type and frequency of riser supports used on those system; however, piping systems that are subject to thermal movement present more complicated support problems, and each case must be considered separately.

Risers are equivalent to concentrated loads in a piping system; however, in the support of this load, several important points must be considered. These are:

1. Is the support to take the entire riser weight, or is this weight to be distributed among several supports?
2. Are hydrostatic-test conditions more severe than service conditions; that is, will the cold-water-filled condition impose stresses on the support higher than allowable (in cold condition) as compared with the hot operating condition and the imposed stresses? When this decision is made, the system erection sequence should be considered and a determination made whether other supports are effective or ineffective during hydrostatic testing.
3. Is the support to be located at a point of zero vertical movement and hence to be considered a rigid support? If this is the case, the horizontal and vertical movements of riser end points must be analyzed. Pure horizontal movement can be provided for by long support rods which are allowed to swing. However, if vertical movement exists, it may cause tipping, and then it must be assumed that the entire load can transfer to one support rod. In this case, the riser support should be designed for double the calculated load.

The support arms of the pipe attachment or the riser clamp, MSS SP-58 Type 42, must extend out from the pipe far enough to clear any insulation. If springs are a part of the support assembly, the arms must be long enough to provide clearance between the pipe insulation and the springs used.

The final consideration in the design of a riser support is its attachment to the pipe. Since a thermal change occurs between the operating and nonoperating condition of the pipe, the frictional grip of the clamp on the pipe cannot be relied upon and there must be some positive means of engagement between the pipe and the clamp. Sometimes a complete ring is welded to the pipe. The more normal procedure is to use four lugs welded to the pipe, positioned at points which bear on the clamp. These lugs are normally the same material as the pipe and are to be welded on three sides. Suggestions for the selection of welded lugs can be found in PFI ES-26.

SUPPORT REQUIREMENTS FOR SPECIFIC PIPING MATERIALS

Ductile Iron Pressure Pipe

Hanger sizes depend upon the outer diameter of pipe which is to be supported. The diameter of cast-iron pressure pipe exceeds that of nominal-sized steel pipe and varies with its class or intended service. Therefore, selection of supports must consider the actual outside diameter of the piping being installed.

Ductile Iron Soil Pipe

Support sizes correspond to sizes for steel pipe. MSS-SP-58 Types 1, 7, 8, and 11 are normally used.

Asbestos-Cement Pressure Pipe

Support sizes depend upon the class of pipe to be supported. The outer diameter of this pipe exceeds that of steel pipe and varies with its class or service.

For all three categories (i.e., Ductile Iron Pressure Pipe, Ductile Iron Soil Pipe, and Asbestos-Cement Pressure Pipe) support spacing should provide at least one support for each length of pipe, with the support preferably located adjacent to the joint, and the support spacing should be no more than 12 ft. Also, each change of direction or branch connection should be supported.

Buried lines are usually supported by tamped backfill, except that lines buried under building slabs are sometimes supported from the slab with hairpin straps or rods suitable to the soil characteristics.

Special consideration is required in cases where movement occurs either in the terminal points or in the structure to which supports are attached. Spring hangers may be required to allow for deflection.

Copper Tubing

Since supports made for copper tubing are designed to fit the tubing diameter, it is not recommended that supports which have been formed for piping be improvised as tubing supports. Most supports for copper tubing are electroplated with copper for easy identification. MSS-SP-58 Types 1, 8, 9, 10, 11, and 12 are available in tubing sizes.

Where there are indications that electrolysis may occur because the support and tubing are made of dissimilar metals, the support should be lined to prevent such action.

Fire Protection Systems

Sprinkler supports will usually be subject to the consideration and approval of the same insurance agencies that have jurisdiction over the sprinkler-system piping and layout. The supports will be required to conform closely to the standards specified in the National Fire Codes published by the National Fire Protection Association.

Some of the important fire protection codes and standards of concern to the pipe-support engineer follow: NFPA 11—Foam Extinguishing Systems, NFPA 12—Carbon Dioxide Extinguishing Systems, NFPA 12 A/B—Halon Fire Extinguishing Systems, NFPA 13—Sprinkler Systems, NFPA 14—Stand Pipe and Hose Systems, NFPA 15—Water Spray Fixed Systems, NFPA 16—Foam-Water Sprinkler and Spray Systems, NFPA 17—Dry Chemical Extinguishing Systems. If the system is other than a standard building water-sprinkler system, the particular standard should also be consulted; for example, in the design or selection of supports for a foam fire extinguishing system, NFPA 11 should be consulted.

The rules are explicit with respect to support spacing, fasteners, and method of fastening. NFPA also regulates the material used in the construction of supports. Underwriters' Laboratories, Inc., Factory Mutual Engineering Division, and Underwriters' Laboratories, Inc., of Canada have tested and listed or approved all types of supports necessary to meet the various conditions of construction.

The placing of supports along the pipe is an important consideration to assure adequate support for the piping. In addition to the consideration of pipe support, consideration must also be given to sway bracing (especially in deluge systems), riser supports, soundness of the attachment, vibration, and pipe slope for drainage. The stability of the support during a fire, corrosion, and aging is also of prime importance. A sprinkler system may be installed for many years before it is operated, and it must always operate in the manner for which it was designed.

Plastic Pipe

There are many types of plastic pipe, both rigid and semirigid. Under normal conditions, rigid plastic pipe can be supported using conventional supports with the spacing half that used with steel pipe. The support of plastic pipe or tubing should be continuous if, owing to the nature of the plastic, it will become flexible from elevated temperatures or from line contents. The continuous support can be in the form of a light angle or channel into which the plastic pipe is laid.

In some cases wear shoes or pads should be added to plastic pipe where it may rub against steel supports. The use of wear pads will prevent the abrasive action caused by thermal movement, thus preventing damage to the pressure boundary.

It is suggested that recommendations of the manufacturer of the specific plastic pipe also be followed.

Glass Pipe

Glass pipe is used for laboratory service, food processing, and many industrial applications which require the durability or chemical resistance of glass pipe. Because of the nature of glass, special consideration of supports and attachments is necessary.

Standard pipe-support components can be used; however, in all cases the support, even if it is painted or electroplated, must have a padding or cushion to avoid scratching the pipe. The support should fit loosely around the pipe yet contact the pipe in a manner to distribute the load over the largest possible area. Point loading must be avoided. The system of supports must be designed with the least number of rigid anchor points.

Glass pipe will generally require two supports per each 10-ft section. One extra support will be required if there are three or more couplings in a 10-ft section. For maximum protection and accessibility, supports should be placed about 1 ft from each joint or coupling.

Glass-pipe supports and layouts should be designed in accordance with general fundamentals applicable to other piping materials. However, extreme care must be taken to minimize strain and scratching of the glass. The pipe manufacturer and reputable support manufacturers should be consulted in the design of support systems for glass pipe.

DESIGN DETAIL CONSIDERATIONS

ASME Code for Pressure Piping

Specific design requirements for piping support are included in the ASME Code for Pressure Piping in the sections covering Power Piping (B31.1), Chemical Plant and Petroleum Refinery Piping (B31.3), and Refrigeration Piping (B31.5). These requirements must be adhered to on piping installations when piping must conform to these Codes. In most cases, supports conforming to MSS SP-58 are acceptable for these installations.

Government Specifications

The Federal Specification WWH-171, of the latest issue and entitled "Hangers and Supports, Pipe," is the governing specification for all federal agencies. This specification illustrates types of pipe hangers and supports and lists the requirements of certain applications.

Design Temperature

Design temperatures for parts of hangers and supports in direct contact with pipe shall be the temperature of the contained fluid. Parts of hangers and supports not in direct contact with pipe and exterior to any insulation may be designed for one-third fluid temperature or ambient temperature, whichever is greater. Allowable stresses for materials commonly used in the design of pipe hangers and supports are listed in applicable codes and standards.

Welded Fabrication

Welded fabrication shall be accomplished with good engineering practice as prescribed by the American Welding Society or other recognized authorities. Attachments welded directly to the pipe must be of appropriate (compatible) chemical composition, and the process of attachment must conform to the requirements for fabrication of the pipe regarding preheating, welding, and stress relieving.

Cold Spring

The cold-springing procedure for piping systems involves, basically, the cutting short of each segment of piping in an amount equal to some specific percentage of the normal thermal growth of the segment with proper allowance to compensate for possible terminal connection movements. The cumulative effect results in an offset gap between the piping ends at the point of final field closure. The drawing together and alignment of the piping ends by the application of required forces and moments are called cold pull and result in a slight relocation of all points along the entire piping system. This relocation is the only effect cold springing has on supports. The piping is then considered as being in the cold-sprung position, which is equivalent to the cold operating position. The movement at all support points from the erected to the cold-sprung position must be calculated, and provision must be made for this movement in the form of hanger rod adjustment.

Adjustment

It is necessary to provide vertical adjustment to attain the desired elevation of the piping system. On piping supported from above, it may suffice to adjust a MSS-

SP-58 Type 1, 5, 6, 9, or 10 support through the yoke portion by raising or lowering the nut on the rod. For larger ranges of adjustments, it is necessary to provide a turnbuckle, MSS-SP-58 Type 13 or 15, in the hanger rod. It is also practical to select a top attachment whereby adjustment can be made at the top of the hanger rod.

On piping which is supported from below, provision for adjustment can be made with screw thread stanchions, MSS-SP-58 Type 38, or by shims or grout.

Protective Coatings

Protective coatings are normally applied to pipe supports for corrosion resistance. Metallic coatings may be applied by either the electroplating or hot-dip process. Nonmetallic coatings, if selected for specific purposes, should be applied as recommended by the manufacturer of such coating. Consideration must be given to coatings used on threaded parts that are to be assembled after coating.

CHAPTER B6
HEAT-TRACING OF PIPING SYSTEMS

Joseph T. Lonsdale
Raychem Corporation
Redwood City, California

C. J. Erickson
E. I. DuPont De NeMours & Company
Newark, Delaware

The term *heat-tracing* refers to the continuous or intermittent application of heat to a pipeline or vessel in order to replace heat loss to ambient.[1] The major uses of heat-tracing include freeze protection, thawing, maintaining fluids at process temperature (or at pumping viscosities), preventing fluid components from separating, and preventing condensation of gases.

The following examples are typical of the diversity of heat-tracing applications: freeze protection of piped water; transfer of molten process chemicals such as phosphoric acid, sulfur, and p-xylene; low-viscosity maintenance of pumped fluids including petroleum products, vegetable oils and syrups, polymeric and resinous materials, and aqueous concentrates and slurries; avoidance of condensation and subsequent improper burning of fuel gas in refineries; preventing moisture from condensing out of piped natural gas; preventing freezing of control valves and compressor damage; eliminating pipeline corrosion due to wet hydrogen sulfide resulting from condensed moisture.[2]

Heat-tracing may be avoided in situations where heat loss to the environment can be effectively minimized. In cold climates or areas with severe winters, water pipes are often buried below the frost line. Alternatively, they may be kept from freezing by running them through heated buildings.[3]

In cases where flow is intermittent, tracing might be avoided by designing a self-draining system such as those used for steam condensate returns. Pipes may also be cleared after use by means of compressed air, steam, or solvent flushing or "pigging." The self-draining method is suitable only for infrequently used pipes due to the high labor costs involved in cleaning and the potential cost and scope of repair, should a pipe not empty properly.[4]

A third approach in the avoidance of tracing is to design for 100 percent flow. This practice is not recommended since equipment breakdown or process interruption may result in an irreversible drop in the temperature of the piped fluid.

TYPES OF HEAT-TRACING SYSTEMS

Heat-tracing systems can be divided into two broad classes, electric and fluid. Fluid heat-tracing systems utilize heating media at elevated temperatures to transfer heat to a pipeline. The fluid is usually contained in a tube or a small pipe attached to the pipe being traced. If steam is the tracing fluid, the condensate is either returned to the boiler or dumped. If an organic heat-transfer fluid is employed, it is returned to a heat exchanger for reheating and recirculation. In general, heating of tracing fluids can be provided by waste heat from a process stream, burning of fossil fuels, steam, or electricity.

Electric heat-tracing systems convert electric power into heat and transfer it to the pipe and its contained fluid. The majority of commercial electric heat-tracing systems in use today are of the resistive type and take the form of cables placed on the pipe. When current flows through the resistive elements, heat is produced in proportion to the square of the current and the resistance of the elements to current flow I^2R. Other specialized electric tracing systems make use of impedance, induction, and skin conduction effects to generate and transfer heat.

Table B6.1 lists the operating and exposure temperatures and the principal characteristics of the different types of heat-tracing.

FLUID HEAT-TRACING

Steam

A number of desirable features made steam the original heat-tracing system of choice to maintain process temperature and provide freeze protection. Steam's high latent heat from vaporization is ideal for heat-transfer applications. Only a small quantity is required for a large heating load, and it can heat a line quickly, condense at constant temperature, and flow to the point of use without pumping. Steam is universally available and nontoxic.[5]

Today, energy efficiency and the minimizing of expensive labor are priority considerations in selecting an economical heat-tracing system. With the advent of highly reliable electrical heat-tracing, the popularity of steam heat-tracing is declining.

Steam is more expensive to install and maintain than electrical resistance heaters. Periodic leaks and failed steam traps in a steam traced system waste energy and demand additional labor costs for repair and replacement. In addition, a single steam tracer provides 2 to 10 times more heat than most applications require. By contrast, electrical tracing systems provide better temperature control and much more efficient utilization of energy. This means that even though the cost per unit energy is lower for steam, total energy costs for electrical tracing are usually significantly lower.[6]

In most heat-tracing applications, saturated steam is supplied at pressures of 30 to 150 psig (298°F/147°C and 367°F/186°C). The ability to continuously remove condensate via a steam trap assembly allows the steam tracer to provide a constant temperature source of heat.

The overwhelming majority of steam traced piping systems employ external tracing. Straight runs of the steam pipe or tubes are attached to the pipe, and the

TABLE B6.1 Comparison of Heat-Tracing Methods

Heat-tracing method	Max. operational temp.	Max. exposure temp.	Advantages	Disadvantages
			Heat transfer fluids	
Steam	400°F (200°C)	None	Takes advantage of waste steam, explosion environment safe, high heat-transfer rates	Nonuniform heat distribution; expensive to install and maintain; imprecise temperature control; wastes energy
Organics	500–750°F (260–400°C)	None	Moderate temperature control, wide temperature range, low freezing temperatures	Relatively expensive; needs a circulating system; leaks can be hazardous
Glycols	250°F (150°C)	325°F (160°C)	Moderate temperature control; depresses freezing point, providing protection against freezing when not in use; lower operating cost than steam	Relatively expensive (glycols are cheaper than heat process fluids); high installed costs; needs a circulating system; leaks can be hazardous
			Electric	
Self-regulating	150–300°F (65–149°C)	185–420°F (85–215°C)	Will not burn out—most reliable electric heating cable; energy efficient	Limited temperature range
MI cable	1190°F (590°C)	1500°F (800°C)	Rugged; capable of high temperature and high power	Difficult to field cut; a break in the cable causes an open circuit; should not be crossed over itself; can be damaged by moisture penetration
Zone	150–400°F (65–204°C)	250–1000°F (150–538°C)	Can be field cut; if a heating element fails, circuit is maintained	Relatively fragile; can self-destruct from its own heat; can burn out if crossed over itself
Skin effect	400°F (200°C)	450°F (232°C)	Simple components; rugged; needs relatively few energy inputs; can be part of a prefabricated insulated pipe bundle	Impractical for applications less than 5000 ft long; design is complex
Impedance	Up to failure of supply cable and connections	None	High heat-transfer rates and close temp. control; high temp. capability, heating structure; element cannot burn out	Expensive custom design; entire pipeline must be electrically isolated from the support
Inductance	Up to Curie point	None	High-temperature capability; high heat-transfer rates	Very expensive; difficult custom design, not commercially exploited

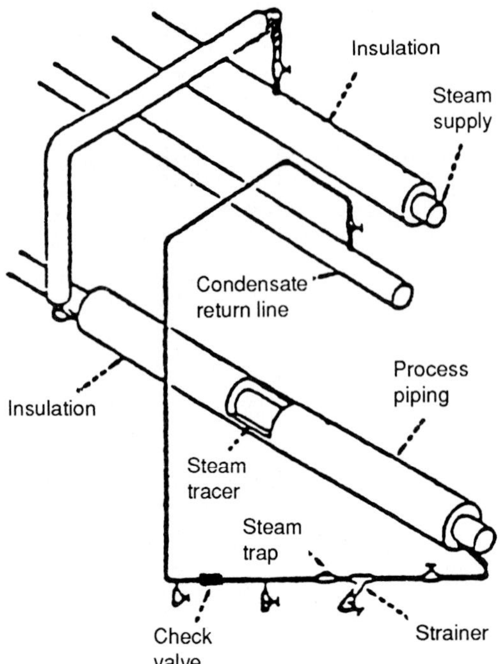

FIGURE B6.1 Typical components of a steam-tracing system. (*I. P. Kohli, "Steam Tracing of Pipelines,"* Chem. Eng., *March 26, 1979, p. 158, Fig. 1.*)

entire assembly covered with preformed sectional insulation (see Fig. B6.1). Valves, fittings, and instruments are *heat sinks* (system components of large surface area and exposed metal surfaces to which system heat will flow and be lost to the environment), and in order to deliver the requisite amount of heat, several loops of the tracing tube are coiled around them before being covered with insulation. This configuration helps reduce *tailing,* that is, the tendency of steam to lose heat and condense along the line with loss of pressure (see Fig. B6.2).

In the majority of applications such as freeze prevention and viscosity maintenance in smaller-diameter pipes, a single tracer provides more than the required heat. However, for processes requiring greater heat input, the heat-transfer characteristics of steam tracers can be significantly improved by placing heat-transfer cement between the trace and the pipeline, greatly increasing the amount of surface for conductive heat transfer.[7] Temperatures of steam tracing systems can vary by as much as +10°F (5½°C) between underground pipelines and +20°F (11°C) for pipelines running above ground. The inability to achieve precise temperature control is attributed to three factors operating in tandem[8]:

1. Saturated steam is delivered at the desired pressure by means of a pressure-reducing valve. As the pressure is reduced, the saturated steam becomes superheated. The excess heat is rapidly dissipated in the system.
2. Uneven contact between the steam tracer and process pipe produces an un-

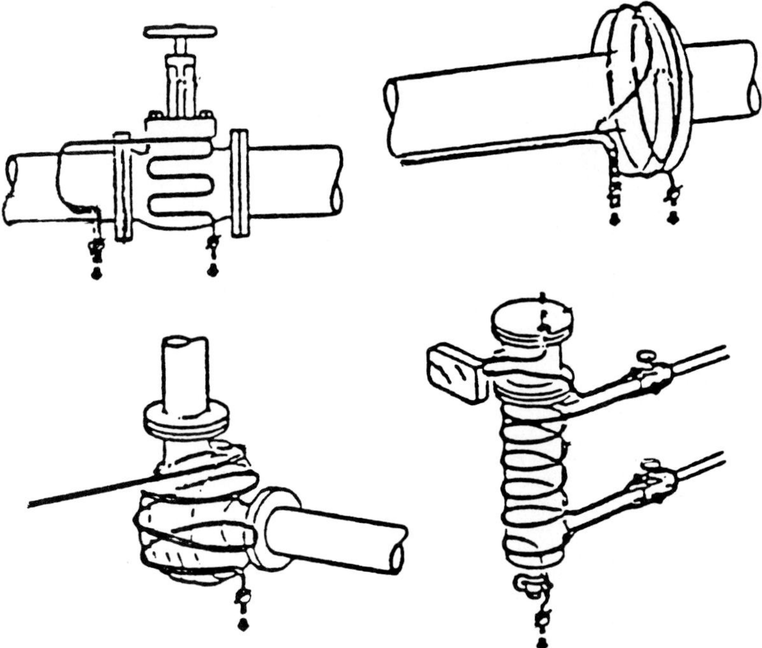

FIGURE B6.2 "Coiling" arrangement for tracing valves, flanges, casings, and instruments. Coils act as expansion joints for steam tracing systems. (*I. P. Kohli, "Steam Tracing of Pipelines,"* Chem. Eng., *March 26, 1979, p. 159, Fig. 3.*)

even distribution of temperature. This effect becomes more significant as the temperature difference between pipe and tracer increases. When the steam becomes superheated, the temperature difference reaches a maximum.

3. Tailing also affects the temperature of the surrounding steam.

A more precise control of steam tracer temperature can be achieved by the use of steam jacketing (see Fig. B6.3) or temperature-sensitive steam valves. However, these methods are rarely used as they provide a level of temperature control inferior to electric heat-tracing, and at a significantly greater cost.

Circulating Media

Circulating media are the most expensive heat-tracing systems and are specified for special-process or ambient conditions (see Table B6.1 and Ref. 9). The virtue of circulating fluids is the ability to provide protection and reasonable control at temperatures above and below those achievable with steam tracing. Circulating media systems can be separated into two classes; oils and organic heat-transfer fluids suitable for high-temperature applications, and glycols with antifreeze properties that make them especially useful in cold climates, where they will not freeze even when used intermittently.

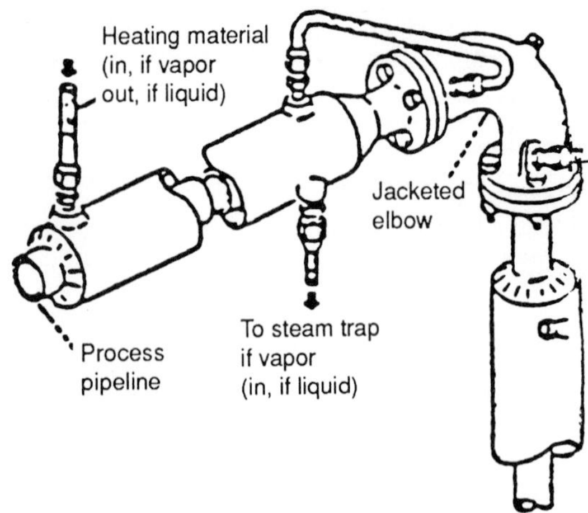

FIGURE B6.3 Steam jacketing is expensive and employed only in special high-heat-demand situations. (*I. P. Kohli, "Steam Tracing of Pipelines,"* Chem. Eng., *March 26, 1979, p. 159, Fig. 2.*)

ELECTRIC RESISTANCE HEAT-TRACING

Introduction

Significant commercial use of electric heat-tracing began to take hold in the 1950s. Electric heat-tracing served as a viable alternative in situations where steam could not be used or was impractical. Typical early applications included the electric tracing of transfer lines for oil, asphalt, and waxes. Electric tracing proved especially useful for long runs of pipe. (Steam tracers are generally limited to runs of 100 to 200 ft. Tracing long or multiple pipe runs with steam can significantly increase both tracing complexity and cost.)

At the outset, hardware had to be adapted from other resistance heating applications. Lead-sheathed soil heating cable was used extensively for waterline freeze protection while longer runs of pipe were traced with mineral-insulated copper-sheathed cable. For higher-temperature service, tubular heaters (normally used for immersion and clamp-on applications) were converted for pipe tracing and controllers were adapted from furnaces and consumer appliances in order to control temperature.

Self-Regulating Heaters

Since their introduction in 1971, self-regulating heaters have become the most popular form of electric heat-tracing and are currently offered by most major vendors of industrial heat-tracing. Self-regulating heat-tracing has an advantage with respect to other heat-tracing products because this technology eliminates the possibility of heater burnout due to an inability to dissipate internally generated heat—the most common cause of heater failure (Fig. B6.4).

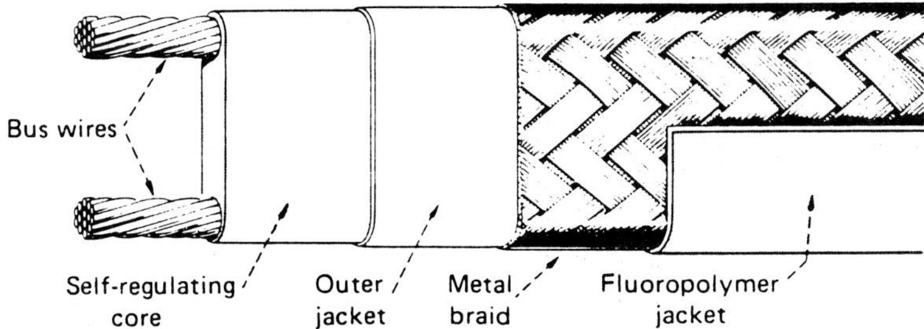

FIGURE B6.4 Components of a self-regulating parallel resistance heat-tracer. (*G. B. Dixon,* Steam versus Electric Process Heat Tracing, *Mississippi Department of Energy and Transportation Annual Conference, April 22–24, 1987, p. 4, Fig. 3.*)

Self-regulating tracers are usually provided in the form of a heater strip consisting of two parallel 20 to 10 American wire gauge (AWG) bus wires embedded in a conductive polymer core which serves as the heating element and over which a polymeric insulator is extruded. The entire assembly may then be covered with a metal braid, an additional polymer jacket or both (see Fig. B6.4). The heater core consists of carbon particles embedded in a polymer matrix. Heat is generated by resistance to current flowing through the conductive polymer heating element. As the temperature of the conductive core increases, so does the electrical resistance. The result is a diminishing output of heat for each successive increment of temperature elevation. Since power output is a function of temperature at any location in the element, the conductive core behaves like a temperature sensitive rheostat guarding against low- as well as high-temperature failure (see Figs. B6.5 and B6.6).

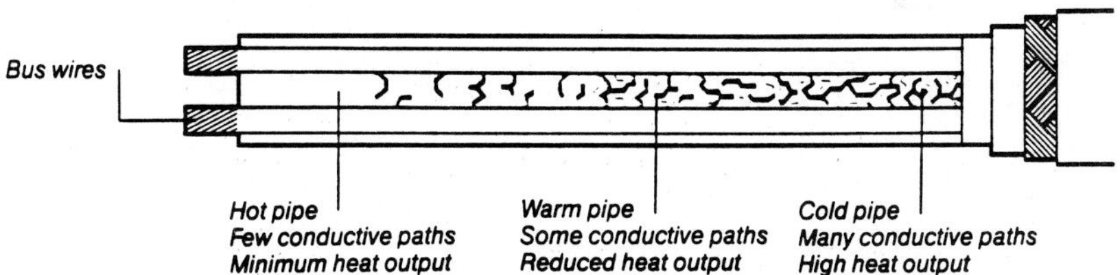

FIGURE B6.5 Relationship of resistive properties to changes in polymer structure with temperature in the conductive core of a self-regulating parallel resistance heat-tracing tape. (*Karen Henry,* Introduction to Heat Tracing, *Cold Regions Research and Engineering Lab, Report No. CRREL-TD-86-1, June 1986, p. 13.*)

Self-regulating tracers can be cut to any desired length and field installed within the limitations of the voltage drop on the hot wires. They have good impact resistance and are routinely handled in the field. The self-regulating feature provides a tremendous boost to operational reliability while cutting installation, maintenance, and energy costs. It also adds a dimension of safety unavailable with any other form of electric resistance tracing product because the heater cannot be destroyed by its own heat output.

The only serious drawback of self-regulating tracers is the upper limit on operating temperatures, 366°F (188°C) for constant exposure and 420°F (215°C) for intermittent exposure.[13] Self-regulating tracers can fail as a result of exposure to excess heat from the piped fluid or from steam cleaning. For this reason, the tracer must be selected to conform with actual process conditions.

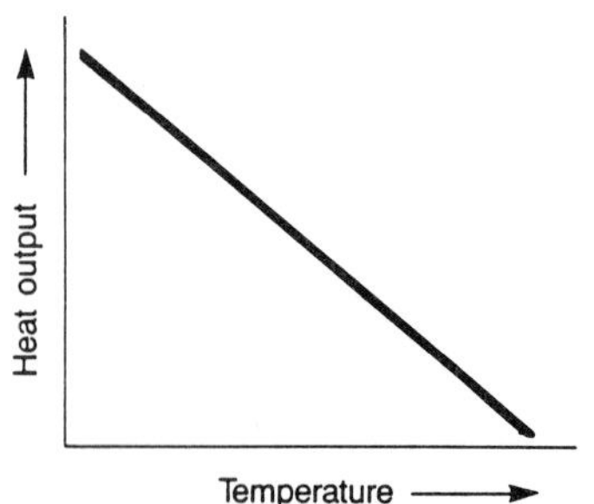

FIGURE B6.6 Graph of resistance versus temperature for a self-regulating parallel resistance heat-tracer. (*G. B. Dixon,* Steam versus Electric Process Heat Tracing, *Mississippi Department of Energy and Transportation Annual Conference, April 22–24, 1987, p. 5, Fig. 4.*)

Zone Heaters

First introduced in 1971, zone heaters were initially the most popular form of parallel resistance heaters and by the late 1970s, were being used in a large percentage of electric heat-tracing applications. Since that time, they have been increasingly replaced by self-regulating heaters.

A typical zone heater consists of two insulated bus wires wrapped with a small-gauge (38 to 41 AWG) nichrome heating wire, covered with polymer insulation, and optionally sheathed in a metallic braid covered with a polymer jacket. The heating wire is connected to alternate bus wires at nodes every 0.3 to 1.2 m, and the distance between connections constitutes a (heating) zone.[14,15,16] Heat is generated by current flowing between the bus wires through the heating wire (see Fig. B6.7).

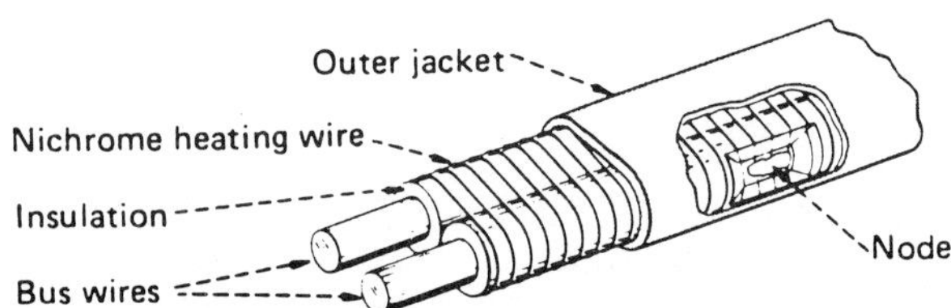

FIGURE B6.7 Components of a zone-type parallel resistance heat-tracer. (*G. B. Dixon,* Steam versus Electric Process Heat Tracing, *Mississippi Department of Energy and Transportation Annual Conference, April 22–24, 1987, p. 5, Fig. 5.*)

The parallel circuit configuration of zone heaters means that output is independent of cable length and systems can be designed and adapted by purchasing cables of a specific wattage which are cut to length in the field. (It is important to remember that the length of cable between the cut and the nearest node will not receive power and should not be depended upon for heater service.) As a result, design and installation costs are significantly reduced.[17] Zone heaters use standard voltages, and their parallel circuitry preserves system function in the event of individual heater element failure (see Fig. B6.8)—an important advantage over series circuits (see Fig. B6.9).[18]

Zone heaters use thinner resistive wires than series heaters and are more susceptible to damage from impact.[19,20,21] Fiberglas-insulated cables are available with an exposure temperature up to 1000°F (538°C), but they are susceptible to moisture. The addition of a fluoropolymer jacket for moisture protection reduces

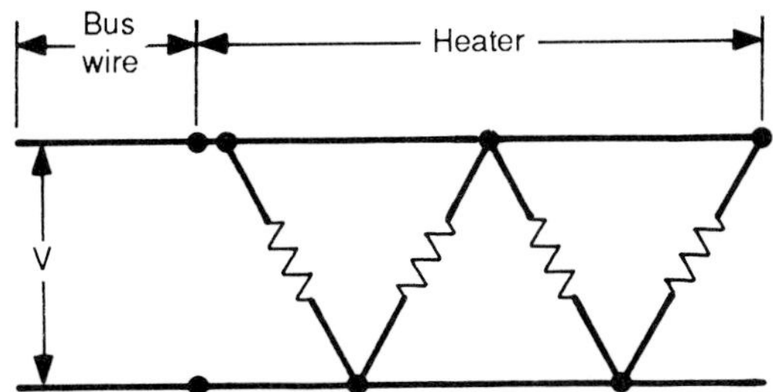

FIGURE B6.8 Simplified circuit diagram for a zone-type parallel resistance heater. (*Karen Henry,* Introduction to Heat Tracing, *Cold Regions Research and Engineering Lab, Report No. CRREL-TD-86-1, June 1986, p. 8, Fig. 3c.)*

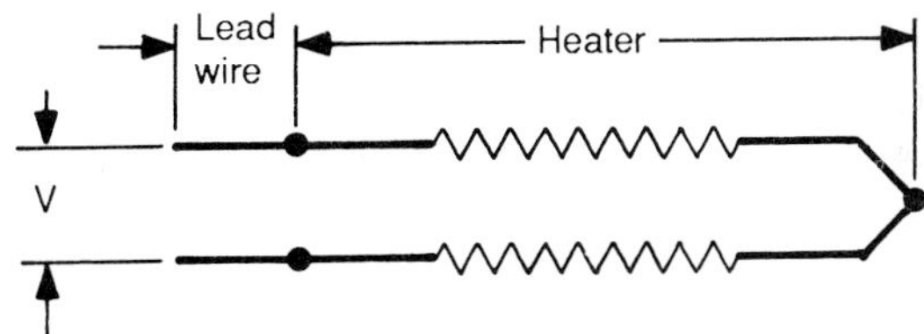

FIGURE B6.9 Simplified circuit diagram for a series-type resistance heater. (*Karen Henry,* Introduction to Heat Tracing, *Cold Regions Research and Engineering Lab, Report No. CRREL-TD-86-1, June 1986, p. 8, Fig. 3a.)*

the exposure temperature rating to 545°F (285°C). Perhaps the biggest drawback of zone heaters is their susceptibility to burnout. With their combination of constant wattage and polymer insulation, zone heaters are vulnerable to destruction from self-generated overheating.[22] As with all constant-wattage heaters, zone heaters to be used in hazardous (classified) areas require factory calculations to determine if the system conforms to the prescribed T rating (see the later section "Area Classification").

Mineral-Insulated Cable (Constant-Wattage Series) Heater

Mineral-insulated (MI) cable was introduced in the early 1950s as an electric-powered alternative to steam and liquid heat-tracing.[23] MI cable is a constant-wattage, series resistance heater in which the entire circuit acts as a continuous heating element.

Heat is generated by current flowing through a nichrome, copper, or other metal conductor, insulated with magnesium oxide and encapsulated in an outer metallic sheath of copper, stainless steel, Inconel, or other suitable metals (see Fig. B6.10).[24]

MI cable is capable of carrying high heating loads. Given the proper conductor and sheath alloys, it can be used in applications up to 1500°F (800°C).[25] Its high impact resistance and general ruggedness allows it to stand up to rough handling in the field. Circuits usually are factory fabricated to length

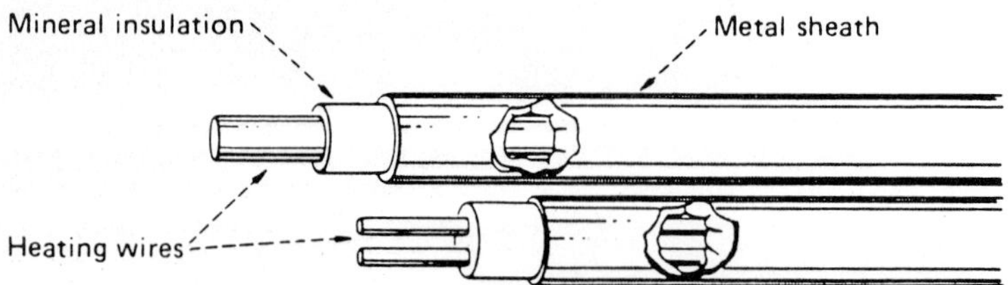

FIGURE B6.10 Components of mineral-insulated heat-tracing cable. (*G. B. Dixon,* Steam versus Electric Process Heat Tracing, *Mississippi Department of Energy and Transportation Annual Conference, April 22–24, 1987, p. 4, Fig. 2.*)

prior to installation, which can be a source of problems when piping changes are made since installations are difficult to field modify. Field fabrication of circuits is sufficiently complex that the training of installation personnel should be supervised by a trained factory technician.[26,27] MI cable system circuits must be individually designed, or variable-voltage controls must be provided to set circuit parameters. Voltage control may also be required for short lengths due to the low resistance.[28]

As with all series circuits, a single break in the cable causes the entire system to fail (breaks in the outer sheathing can cause failure due to absorption of moisture and subsequent loss of insulating properties).[29,30] Another disadvantage of MI cable is the risk of overheating from excessive currents or poor thermal dissipation. Hazardous-area installations must be factory calculated to ensure conformance with the proper T rating (see later section "Area Classification").

Series Resistance Polymer-Insulated Cable

Polymer-insulated series resistance cables can be used with various conductor materials. Nichrome is suitable for short circuits, but the length must either be predetermined to suit the available voltage or field cut and provided with a variable voltage supply. Conductors such as copper offer a measure of self-limiting heater properties, since their resistance increases with temperature. This allows greater latitude of use, and copper conductors with their 600-V limitation and relatively low cost (even with metallic braid and overjacket), are especially favored for long-line applications with this type of heater.

The circuits must be designed and controlled to minimize high temperature, because a failure at one spot disables the complete circuit. The possibility of catastrophic failure (series circuit) due to overheating and melting of the polymer insulation puts this type of tracing in unfavorable competition with parallel resistance heat-tracing systems which dominate in the low- to moderate-temperature application ranges.

SKIN EFFECT TRACING

Skin effect systems are primarily applicable to the tracing of long pipelines. The "skin effect" is based on the tendency of an alternating current to flow in the

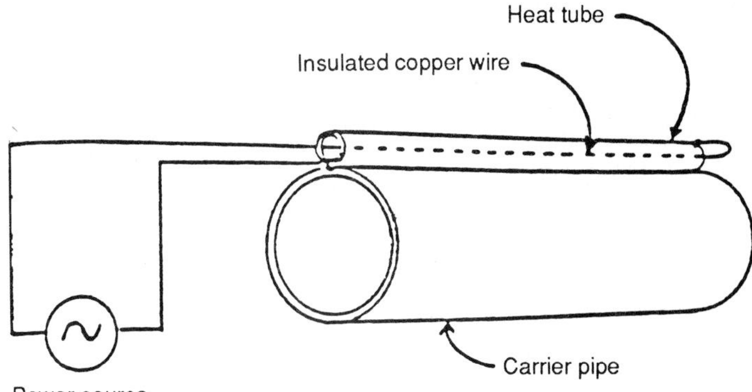

FIGURE B6.11 Components and electrical flow in skin effect heat-tracing. (*N. B. Carson,* A New Method for Heat Tracing Long Pipelines, *ASME, Petroleum Mechanical Engineering Conference, Dallas, Texas, September 1974, p. 1, Fig. 3.*)

layers near the surface (skin) of a current-carrying ferromagnetic conductor (see Fig. B6.11). In a typical skin effect tracer, the heating element is a carbon steel pipe of small diameter welded to the fluid-carrying pipeline to be traced. Running through the heat tube is an insulated, low-resistance copper wire. The alternating magnetic field created by this alternating-current-carrying conductor causes the return current in the small heat tube to be concentrated toward the inner wall of the tube. This phenomenon is called the "proximity effect." Because almost no current flows on the outer surface of the heat tube, there is no measurable potential there and the entire piping system can be grounded at any number of points.[31,32]

The requirement for custom system design makes skin effect systems costly, notwithstanding the ability to make use of ordinary low-cost materials, including prefabricated components, and standard construction techniques. The method maintains a low temperature difference between fluid and tube wall (18°F/10°C),[33] is considered reliable, and is easy to repair. Single-circuit envelopes of up to 7500 m are feasible with supply voltages of 3000 V. Higher supply voltages make even longer length circuits possible.[34] One reference reports a single power station capable of supplying service up to 30 mi (48 km) of pipeline.[35] On the other hand, Carson qualifies this with a practical limitation of 10 mi (16 km), since above the 5 kv supply required for a line of this length, cable and switchgear costs become an increasing consideration.[36] This seems to be confirmed by Ando and Takki, who report the construction of a 68-mi (108-km) skin effect heat-tracing system powered by 12 substations with a transformer voltage of 13,800 V.[37]

Skin effect heat-tracing is generally not cost-effective for pipelines shorter than 1500 m,[38] its upper temperature limit of approximately 400°F (204°C) is set by the maximum exposure temperature of the conducting wire insulation,[39] and the method is not adaptable for complex piping. Installations cannot be modified, and the complete system fails with a single line break.

IMPEDANCE HEAT-TRACING

In impedance heating, the pipe itself becomes the heating element. The generation of heat is produced by resistance to current flow (see Fig. B6.12).[40] Impedance heat-tracing has the ability to reach very high operating temperatures, limited only by the design and contents of the piping system. Since the pipe is the heating element, it is the supply cable and connections which can be vulnerable to burnout.[41] This technique has high heat-transfer rates and uniform heat distribution and provides excellent temperature control at the control point, using any one of several automatic control methods.[42,43]

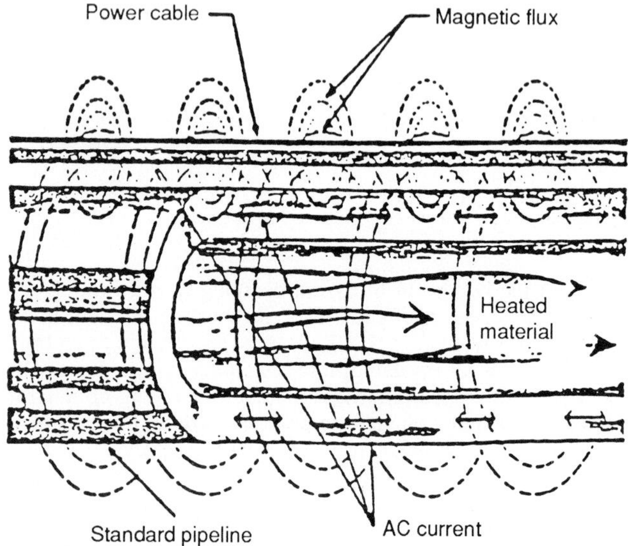

FIGURE B6.12 Components and electrical and magnetic interaction in impedance heat-tracing. (*George Koester II, "Pipe Heat Tracing with Electric Impedance Heating,"* Plant Engineering, *vol. 32, no. 24, November 23, 1978, p. 113, Fig. 1.*)

Impedance tracing is costly and has limited application. Systems are almost exclusively vendor designed, and the engineering can become complicated, especially in attempting to achieve electrical balance in piping systems with multiple branches. Because significant current flows through the pipe, the entire pipeline must be electrically isolated from the support structure and shielded from personnel contact. As a precaution, impedance-traced pipelines are normally operated at 30 V or less.[44] Voltages at 80 V are allowed if ground fault protection is supplied.

INDUCTION HEATING

Induction heat-tracing uses a metallic pipeline as a heating element by placing it in the magnetic field of an alternating current source. Low-resistance wire is wound around a conductive pipeline or vessel, and the alternating current flowing through the coils generates a rapidly changing magnetic field that induces eddy currents and hysteresis losses in the pipeline wall.[45] Induction heating has been most frequently employed for melting metals[46] and would most likely be considered for high-temperature, high-power, heat-tracing applications. The absence of thermal resistance between heat source and pipeline allows very rapid heating. Present systems would involve considerable expense, custom design, and require power inputs at short intervals along the pipeline. Induction methods do not easily lend themselves to the production of uniform heating, and IEEE rates the method as providing only moderate system efficiency.[47]

SELECTION CRITERIA FOR TRACING SYSTEMS

Assuming that methods for avoiding the need for tracing have been considered and rejected, the first step in matching the heat-tracing to the piping system requires an analysis of fundamentals. These include: the type of application, suitability and relative cost of different types of heat-tracing, availability of steam and/or electricity, amount of heat loss which must be made up, requirements for temperature control, and classification of the traced area as a hazardous or ordinary environment due to the presence of flammable substances.

AREA CLASSIFICATION

Areas are classified according to their potential fire hazard as defined by Article 500 of the National Electric Code (NEC).[48] (In industrial applications, verification that electric components meet NEC hazardous-area requirements is issued by an independent testing laboratory.) Under this system, ordinary areas are those not classified as hazardous. Hazardous locations are rated according to class, division, and group. The area class determines the category of combustible atmosphere: flammable gases, vapors, or liquids (Class I); combustible dust (Class II); and combustible fibers (Class III). The division indicates the likelihood of a hazard to be present under different conditions. Hazardous atmospheres with similar combustion properties are listed in the same group.

In order to ensure that the heat-tracing system selected will operate safely in a hazardous environment, it must also be classified according to its NEC temperature identification number, or T rating.[49] This code specifies that the temperature of the exposed surface of the (electrical) equipment not be in excess of 80 percent of the ignition temperature of the combustible atmosphere.

By comparing the T rating to the area classification, an assessment can be made as to heater eligibility for a particular piping system. See the later section "Design Considerations" for a further discussion of T ratings and sheath temperature calculation.

ENVIRONMENTAL CONSIDERATIONS

Environmental factors include whether the area is dry, wet, moderately or severely corrosive or noncorrosive and whether the tracing will experience rough handling or mechanical abuse during installation, operation, or maintenance. These considerations are primarily related to the performance of electrical heat-tracers. Electrical heaters exposed to any of the environmental stresses listed above should be protected accordingly: a metal sheath of a material able to withstand the corrosive agent for MI cable; a braided sheath covered by a polymer jacket formulated for protection from particular classes of corrosives is recommended for polymer-insulated types of heating cables (jacket of modified polyolefin for resistance to moisture and inorganic chemical agents; fluoropolymer for resistance to organic chemicals). The heater must be rated to withstand anticipated maximum exposure temperatures.

HEATER RELIABILITY AND CONSEQUENCE OF FAILURE

In addition to selection of a system with the appropriate T rating, environmental protection, and proper installation, a heat-tracing system must be evaluated in terms of the risk and consequence of failure. Savings in front-end costs may not be justified if the failure of a tracing system incurs the far larger expense of disrupting a process which depends on maintaining acceptable temperature in the traced piping or requires removal and replacement of the thermal insulation and tracing system.

Steam tracing systems have high maintenance costs, but system failure is generally not a concern. The major cause of electric tracing system failure is compromised insulation. Zone heaters and self-regulating heaters are parallel circuited. Failure of a single heating element does not bring down the entire system, and repairs can be made in the field. The series circuitry of MI and polymer-insulated cable allows no such flexibility. A single failure brings the entire system down. MI and polymer-insulated cable are difficult to field repair, requiring the services of a trained technician. See the first section, "Types of Heat-Tracing Systems," for detailed characteristics of different heat-tracing systems.

TRACING OPTIONS FOR DIFFERENT TEMPERATURE RANGES

Table B6.1 classifies heat-tracing technologies according to operating temperature ranges and maximum exposure temperature.

AVAILABILITY OF STEAM AND ELECTRICITY

In process plants, steam is almost universally available and except in regions with substantially developed hydroelectric power, electricity is invariably generated from steam.

When considering the use of electric heat-tracing, especially in some developing countries, the reliability of the power for uninterrupted delivery, available voltages, and the consequences of outages must be evaluated. In practice, heat

TABLE B6.2 Selected Examples of Time versus Temperature Loss in an Insulated Pipe

Pipe fluid	Light fuel oil	Water
Analysis ambient temp.	Static above ground, 0°F	Static above ground, −20°F
Wind speed	20 mi/h	20 mi/h
Pipe	Carbon steel	Carbon steel
Nominal diameter	4 in	2 in
Insulation type and thickness	Fiberglas 2 in	Fiberglas 1 in
Initial temp.	140°F	50°F
Fail temp.	90°F	32°F (frozen solid)
Time, h	7	21 (2 in) 4 (½ in)

Note: In practice, occasional interruption of power will not be a serious concern in the selection of electric heat-tracing. Heat loss from a fluid-filled will is sufficiently slow that intermittent power outages are unlikely to cause a serious lowering of temperature.

Source: Heat Up/Cool Down Analysis Program, Raychem Corp., Process Division, Menlo Park, Calif.

loss from a fluid-filled, insulated pipe is a rather slow process. It will take many times longer than the duration of an intermittent power outage to suffer a serious lowering of temperature.

For example, an above-ground 2-in carbon steel pipe with 1-in of fiberglas insulation containing static water at an initial temperature of 50°F will take 21 h to freeze when the ambient temperature is −20°F (see Table B6.2). Unless power disruption is expected to be of such long duration (an exceedingly rare occurrence), reliability of the power supply is not a determining consideration. One exception to this is the tracing of instrument lines. Because of their small size, instrument lines will freeze much more quickly. In the above example a half-inch instrument line would freeze solid in 4 h (see Table B6.2). While this is much quicker than a pipe, it is still generally less than the duration of the typical short-term power failure.

The most significant factor in a decision between steam and electric tracing is the cost of installing and operating the system, and these costs depend on factors such as the geometry of the piping system, cost of labor and energy, and local tracing practice.[50]

INSTALLED AND OPERATING COSTS

In general, installation of steam tracing is more labor intensive (½-in copper tubing is more difficult to install than the more flexible electric cable). In high-labor-cost areas, such as Europe and North America, the expense of labor can easily offset the higher material cost of electric tracing. In areas where labor costs are considerably lower, steam may have an economic advantage. This is especially true in South America, where steam tracing predominates and the practice is well established.[52]

Since most electricity is produced from steam with about a 40 percent conversion efficiency, raw energy cost will always favor steam by a factor of 2 to 3.

However, steam tracers have inherent disadvantages that usually make electric tracing more cost-effective in overall energy utilization. Maintenance costs also tend to favor electric tracing. As with installation, the maintenance cost differential between electric and steam tracing will tend to be greatest in high-priced labor markets.[53]

The above trade-offs generally apply to both pipe tracing and the tracing of instrument lines. Because they are smaller, instrument lines require less energy, but this only causes a slight decrease in costs for both steam tracing and electric tracing. One major disadvantage steam tracing has in tracing instrument lines is that care must be taken to ensure that the steam tracer does not overheat the line. It is possible for a steam tracer to actually boil a small instrument line. This is generally not a concern with electric tracing as self-regulating heaters can be used.

COMPUTER SELECTION PROGRAMS[54]

Computer programs now exist which can greatly simplify the work of determining the economics of steam versus electric heat-tracing. Many heat-tracing vendors provide user-friendly heat-tracing selection programs for use with their line of products. Input involves supplying answers to a series of fill-in-the-blank questions including size, diameter, and geometry of the piping to be traced, ambient and maintenance temperature, control requirements, labor costs, and exchange rates. The program then calculates worst-case heat loss, determines the components needed to build a steam or electric tracing system to maintain the pipe at the required temperature, and calculates the associated material, installation, and operational costs for both cases.

Cost Comparison of Steam and Electric Tracing

In a study presented at a September 1990 meeting of the IEEE,[55] a major chemical company compared the installed and operating costs of both steam tracing and electric tracing for a freeze protection system and for a process temperature maintenance system.

An existing electric-tracing system was used for the freeze protection analysis. The actual costs of installing and operating this system were compared with detailed engineering estimates of the costs of installing and operating a steam tracing system to do the same task. For the process temperature maintenance study, the costs of an existing steam-tracing system were compared to detailed engineering estimates of the costs of a similar self-regulating electric-tracing system.

The results of this study are summarized in Tables B6.3 and B6.4, and they demonstrate that the electric-tracing system is less costly to install and operate. It is interesting to note that even if the higher energy costs for steam are disregarded, the steam systems still have higher operating costs due to the expensive maintenance required. This suggests that the availability of excess low-pressure steam is not necessarily an adequate justification to select a steam-tracing system instead of an electric-tracing system.

TABLE B6.3 Comparison of Electric and Steam Tracing Freeze Protection Costs

Steam tracing		
Installed costs		$/ft of tracing
Supply piping		$ 6.37
Steam tracer		9.32
Steam trap assemblies		20.33
Condensate return		5.23
Thermal insulation line	$18.65	
Thermal insulation other	12.85	
Thermal insulation total	31.50	31.50
Total steam tracing installation costs		$72.75/ft
Operating costs		$/ft of tracing/yr
Energy tracer		$ 1.85
Energy traps		3.33
Energy supply and return		0.76
Energy total		$ 5.94
Maintenance total		1.46
Total steam tracing operating costs		$ 7.40/ft/yr
Electric tracing		
Installed costs		$/ft of tracing
Heaters		$13.29
Motor control center		0.26
Panels		4.10
Control and distribution		13.53
Thermal insulation total		$19.58
Total costs of electric tracing		$50.76/ft
Operating costs		$/ft of tracing/yr
Energy total		$ 0.11
Maintenance total		1.32
Total costs of electric tracing		$ 1.43/ft/yr

STEAM OR ELECTRIC TRACING: DESIGN CONSIDERATIONS

Figure B6.13 provides a flow diagram designed to assist in choosing between steam or electric tracing. In addition to the distinctions made in the first two sections of this chapter, several additional factors must be considered in determining the type of tracing selected:

- If steam is not available in the vicinity of the pipe being traced, the prohibitive cost of running a long steam supply line to the site usually eliminates the use of steam tracing from further consideration. By the same token, electric power must be available at the site to use electric tracing. When neither steam nor electricity is available, it is usually much cheaper to run electricity than to bring a supply of steam to a remote location.[56]

TABLE B6.4 Comparison of Electric and Steam Tracing Process Maintenance Temperature Costs

	Steam tracing	
Installed costs		$/ft of tracing
Supply piping		$ 27.62
Steam tracer		9.30
Steam trap assemblies		79.38
Condensate return		24.98
Thermal insulation line	$17.72	
Thermal insulation other	55.02	
Thermal insulation total	72.74	72.74
Total cost of steam tracing		$214.02/ft
Operating costs		$/ft of tracing/yr
Energy tracer		$ 3.64
Energy traps		11.02
Energy supply and return		2.12
Energy total		$ 16.78
Maintenance total		6.18
Total cost of steam tracing		$ 22.96/ft/yr
	Electric tracing	
Installed costs		$/ft of tracing
Heaters		$ 30.75
Motor control center		0.44
Panels		6.65
Control and distribution		21.35
Thermal insulation total		19.02
Total costs of electric tracing		$ 78.21/ft
Operating costs		($/ft of tracing/yr
Energy total		$ 0.05
Maintenance total		1.73
Total costs of electric tracing		$ 1.78/ft/yr

- Economics usually favor electric tracing for lines smaller than 12 in in diameter and maintain temperatures below 248°F (120°C).

ELECTRIC SYSTEM DESIGN

Electric heat-tracing is usually marketed through manufacturers' representatives. Major manufacturers of heat-tracing have representation throughout North America and many other parts of the world, and most are capable of designing heat-tracing systems and training plant engineers to design their own systems. Many heat-tracing manufacturers are staffed by professionals who are extremely knowledgeable in heat-tracing practice. Purchasers of heat-tracing products and

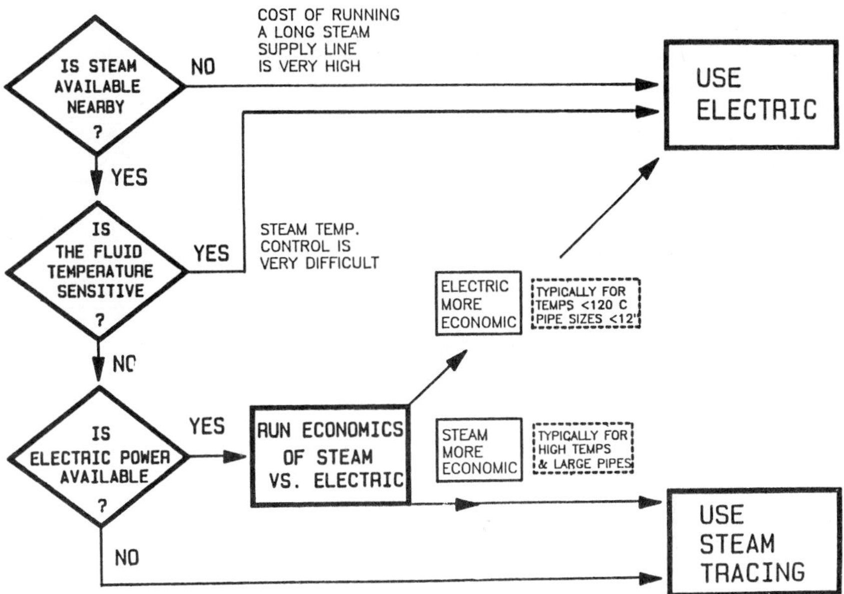

FIGURE B6.13 Steam-electric selection flow diagram. (*Raychem Corp., Menlo Park, Calif.*)

systems can and often do take advantage of the design expertise and related experience these companies make available to their customers.

Computer programs which assist in the development of heat-tracing designs are available from several manufacturers of heat-tracing products. Computer design programs allow for the rapid and comprehensive evaluation of the changes in heat loss resulting from the alteration of system variables. The ease of performing multiple computer evaluations allows more extensive exploration in design optimization and often leads to improvements which might not be accessible if these computations had to be carried out in a less automated fashion. See Tables B6.5 and B6.6.

TABLE B6.5a Input Menu Screen Display for the Computer Program

```
Project ID: LESSON1    Units: English  || DESIGN ID: LESSON1 Name: TEST PROJECT
      Name: TEST PROJECT               ||              Maintain Temp:  48 F
   Done by: YOUR NAME                  ||   Process Operating Temp: 150 F
  Done for: CUSTOMER NAME              ||    Max Htr Exposure Temp: 150 F
Insul Type: FG          Ambient Temp   ||        System Limit Temp: 150 F
   Voltage: 120           Min:   0 F   ||      Area Classification: Ordinary
  Location: Outdoors      Max: 100 F   ||        Chemical Exposure: None
=======================================|| Offscreen Items:
Lines ID: LESSON1Pipe|Insul  Pipe      ||  Heat Loss Safety Factor:  10 %
Ref              Size|Thick Length  #  ||            Pipe Category:  IPS/Steel
 #  Line No      (in)|(in)   (ft)  Vlv ||         Allow Spiralling:  No
== ========== =====|=====  =====   =  ||          Valve HL Factor:   4.30 ft
 1 LINE A        2.00|  1.0  150.0   1 ||        Support HL Factor:   2.00 ft
 2 LINE B        4.00|  1.0  250.0   2 ||  Pipe Length per Support:  20 ft
                     |                 ||      Startup Temperature:    0 F
                     |                 ||          Maximum CB Load:  32 A
                     |                 ||       Allow Uncontrolled:  Yes
```

Source: TraceCalc Program, Raychem Corp., Chemelex Division, Menlo Park, Calif.

TABLE B6.5*b* The Collection of All Data Inputs into the Design Summary Makes It Easier to Optimize the Design by Adjusting Individual Parameters

```
Project Name: TEST PROJECT          Temperatures                     Insulation: GLASS FIBER
    Done For: CUSTOMER NAME              Maintain:  48 F                         (ASTM C547-77)
     Done By: YOUR NAME               Min Ambient:   0 F          K Factor at 50 F: 0.2497 BTU-in/hr-ft2-F
 Design Name: TEST PROJECT            Max Ambient: 100 F   Area Classification Code: Ordinary           Voltage: 120
                                             Mean:  24 F     Chemical Exposure Code: None              Location: Outdoors
  Project ID: LESSON1                Process Oper: 150 F    Heat Loss Safety Factor:  10 %      Maximum CB Load: 32 A
   Design ID: LESSON1                Max Htr Exp: 150 F    Pipe Length per Support: 20  ft        Pipe Category: IPS/Steel
    Lines ID: LESSON1                System Limit: 150 F          Allow Spiralling: No        Support HL Factor:  2.00 ft
                                          Startup:   0 F       Allow Uncontrolled: Yes          Valve HL Factor:  4.30 ft
```

DS Ref #	Pipe Size (in)	Insul Thick (in)	HL Rate (W/ft)	Heater Type	Output Rate (W/ft)	Feet /ft Pipe	of Heater per Valve	per Supp	Pitch (in)	Min T Maintained	Max T	K Factor @ Mean T	Start Load (A/ft)	Oper Load (KW/ft)	Total Pipe (ft)	Total Heater (ft)	Total Comp Sets
1	2.00	1.0	3.1	5BTV1	5.1	1.0	2.6	1.2	0	64	119	0.2398	0.084	0.0058	150	161	1
2	4.00	1.0	5.1	5BTV1	5.1	1.0	4.3	2.0	0	48	113	0.2398	0.084	0.0058	250	283	2

Source: *TraceCalc Program,* Raychem Corp., Chemelex Division, Menlo Park, Calif.

TABLE B6.6*a* The *TraceCalc Program* Line List Reporter Selects a Tracer for Each Line

```
Project Name: TEST PROJECT          Temperatures                     Insulation: GLASS FIBER
    Done For: CUSTOMER NAME              Maintain:  48 F                         (ASTM C547-77)
     Done By: YOUR NAME               Min Ambient:   0 F          K Factor at 50 F: 0.2497 BTU-in/hr-ft2-F
 Design Name: TEST PROJECT            Max Ambient: 100 F   Area Classification Code: Ordinary           Voltage: 120
                                             Mean:  24 F     Chemical Exposure Code: None              Location: Outdoors
  Project ID: LESSON1                Process Oper: 150 F    Heat Loss Safety Factor:  10 %      Maximum CB Load: 32 A
   Design ID: LESSON1                Max Htr Exp: 150 F    Pipe Length per Support: 20  ft        Pipe Category: IPS/Steel
    Lines ID: LESSON1                System Limit: 150 F          Allow Spiralling: No        Support HL Factor:  2.00 ft
                                          Startup:   0 F       Allow Uncontrolled: Yes          Valve HL Factor:  4.30 ft
```

Ref #	Line No	Pipe Size (in)	Insul Thick (in)	Pipe Length (ft)	# Valve	# Support	Heater Type	Trace Ratio	HL Rate (W/ft)	Output Rate (W/ft)	Heater Length (ft)	# Comp Sets	Start Load (Amp)	Oper Load (kVA)	Comment
1	LINE A	2.00	1.0	150.0	1	7	5BTV1	1.0	3.1	5.1	161.1	1	14	0.9	
2	LINE B	4.00	1.0	250.0	2	12	5BTV1	1.0	5.1	5.1	282.7	2	24	1.6	

Source: *TraceCalc Program,* Raychem Corp., Chemelex Division, Menlo Park, Calif.

TABLE B6.6*b* The Bill of Material Report Lists the Types and Quantities of Materials Required to Execute the Design

```
Project Name: TEST PROJECT                        Project ID: LESSON1
    Done For: CUSTOMER NAME                        Design ID: LESSON1
     Done By: YOUR NAME                             Lines ID: LESSON1
 Design Name: TEST PROJECT
```

Quantity	Catalog No	Description
447 ft	5BTV1	Chemelex Auto-Trace Heater
3 Each	PMK-JLP-A	PolyMatrix Power Connection Kit
3 Each	PMK-LE-A	PolyMatrix End Termination Kit
1 Each	AMC-1A	Ambient Sensing Thermostat
40 Each	ETL	Label: "Electric Traced"
12 Rolls	GT66	Glass Tape (66 Ft/Roll)
1 Each	PS-03	Pipe Strap for 1" to 3"
2 Each	PS-10	Pipe Strap for 3" to 10"
150 ft	2.00 X 1.0	Insulation Material
250 ft	4.00 X 1.0	Insulation Material
598 Sq Ft		Cladding Material

Source: *TraceCalc Program,* Raychem Corp., Chemelex Division, Menlo Park, Calif.

The following example demonstrates how a program is used in developing a heat-tracing design. In addition to calculating heat loss, these programs provide a checklist of the information required.

Data Collection

Once the information has been entered, the program will use the data to automatically select the optimum system components. While all the requested information is not always required, the use of a comprehensive inquiry form or a computer program helps ensure that all relevant data is collected.

1. Thermal Data

Temperature at which the pipe is to be maintained:

$$T_m = 40°\text{F}$$

Minimum expected ambient temperature:

$$T_a = 0°\text{F}$$

Comment. T_m and T_a are needed to calculate the heat loss.

Maximum temperature that the heater will be exposed to due to process upsets or steam cleaning:

$$T_e = 140°\text{F}$$

Comment. The design calculation compares T_e with the maximum intermittent exposure temperature of the heater to ensure that adequate safeguards are built in to protect the heater from damage and subsequent failure. This is a serious concern in the case of heat-sensitive polymer-insulated heaters.

Process normal operating temperature:

$$T_p = 50°\text{F}$$

Comment. This temperature is required to ensure that the heater selected can continuously withstand the operating temperature.

System limit temperature (imposed by process fluid, insulation, pipe material, or safety considerations):

$$T_1 = 500°\text{F}$$

Comment. This temperature is required to protect the rest of the system from high temperatures created by the heater. Such temperatures may be a concern where plastic pipes or temperature-sensitive fluids are to be traced. (Most heat-traced fluids are not temperature sensitive and are transferred in steel pipe, so this variable is often of no concern in design considerations.

2. Pipe Data and Insulation Data

Outside pipe diameter:

$$D_1 = 4 \text{ in}$$

Comment. Tables and programs are based on nominal pipe size, actual diameter is required for use in the heat-loss equation.

Outside diameter of insulated pipe:

$$D_2 = 6 \text{ in}$$

Comment. Tables and programs are based on insulation thickness equal to $(D_2 - D_1)/2$.

Thermal conductivity k of the insulation material or type = FG type of insulation.

$$\text{FG} = \text{Fiberglas}$$

The following values are needed to calculate the pipe heat loss:

$$(k = 0.25 \text{ Btu/h} \times {}^\circ\text{F} \times \text{ft}^2/\text{in}),\ D_1,\ D_2, \text{ and } k.$$

The values of k are taken at the mean temperature between the inner and outer surfaces of the insulation. It is noted that the thermal conductivity of the insulation material varies with the mean temperature.

Comment. Computer programs calculate the heat loss based on the insulation type. The thermal conductivity is estimated for a mean insulation temperature. k is temperature dependent, although the variation is small for low and moderate temperatures. Changes in k should be considered for medium- and high-temperature applications.

Pipe length:

$$L_p = 230 \text{ ft}$$

Comment. This value is needed to estimate the length of tracing cable required.

Valves and other heat sinks: Type and number, None, including flanges, valves, hangers, fittings shoes, and anything that could require additional heat-tracing.

*3. **Service Environment Data and Classification.***[57] Areas are classified on the basis of the severity of the fire and/or explosion hazard which may be present. It is essential that the heater selected be used only in areas with classifications for which it has been approved.

Standard Area Classification

- *Ordinary area (nonhazardous):* Areas not having explosive vapors, dust, or fibers are ordinary or nonhazardous areas.
- *Class I, II, III, Division 2:* Areas where explosive concentrations of vapors, dust, or fibers may be present in unusual circumstances. Special heat-tracing strip, usually equipped with a metal braid and an outer jacket, are used in these areas.
- *Class I, II, III, Division 1:* Areas where explosive concentrations of vapors, dust, or fibers may be present in usual circumstances. Measures consider-

ably more stringent than those taken for Division 2 are required in Division 1.

The computation example is for a Class I, Division 2, area, a classification which is typical for the vast majority of hazardous-area designations.

Chemical Exposure

- Dry location (indoors)
- Wet location or limited exposure to aqueous inorganic chemicals
- Exposure to organic chemicals, greases, oils, or solvents

(Some types of heat-tracing can withstand corrosive inorganic chemicals, but not organics.)

Most heaters are designed for use in wet environments. Some heaters will need additional outer jackets if they are to be exposed to organics compounds.[58] The computation example is for an outdoor environment with potential exposure to organic chemicals.

4 Available Voltage. Heaters generally operate on 120 or 240 V. Most 240-V heaters are designed to operate at voltages from 208 to 277 V with minor variations in power output. 240-V heaters are generally preferred in industrial installations because they can support circuit lengths approximately twice as long as 120-V heaters. This allows most jobs to be done with fewer circuits, considerably lowering the overall cost. Both 120- and 240-V systems are available in the computation example.

Heat Loss

The next step in the design of electric heat-tracing is the calculation of the maximum heat loss from the pipe. This is the heat loss from the pipe when the pipe is at its design temperature (maintain temperature), the ambient temperature is at its lowest, and the wind is blowing (comparable to the coldest day of the year with a maximum wind chill). Under these worst-case conditions, the heater must be capable of replacing the heat loss in order to maintain the desired temperature.

In the majority of cases, heat loss in electric tracing applications is usually estimated from tables provided by tracing manufacturers or by means of computer programs.

Table B6.7 is a typical heat-loss table from a heat-tracing manufacturer. Several things should be noted about this table:

- It is based on Fiberglas insulation. Correction factors for other types of insulation are given in the lower left-hand corner.
- A 10 percent safety factor has been included. This is fairly standard practice in the industry.
- The table is based on a 20-mi/h wind. In the absence of wind (indoors), the heat loss will be reduced by about 10 percent. Wind speeds above 20 mi/h will have very little additional effect on the total heat loss; generally less than 1 percent when dealing with an insulated pipe. Although the resistance of the air film can

TABLE B6.7 Pipe Heat-Loss Table*

Insulation thickness (in)	ΔT (°F)	Pipe diameter (IPS) (in): 1/4	1/2	3/4	1	1 1/4	1 1/2	2	2 1/2
	Tubing size (in):		3/4	1	1 1/4	1 1/2	2		
0.5	50	1.9	2.5	2.9	3.5	4.1	4.6	5.5	6.5
	100	3.9	5.2	6.1	7.2	8.6	9.6	11.5	13.5
	150	6.1	8.1	9.5	11.2	13.4	14.9	17.9	21.1
	200	8.5	11.3	13.2	15.6	18.6	20.7	24.9	29.2
1.0	50	1.3	1.6	1.9	2.2	2.5	2.8	3.2	3.8
	100	2.7	3.4	3.9	4.5	5.2	5.8	6.8	7.8
	150	4.2	5.3	6.1	7.0	8.2	9.0	10.6	12.2
	200	5.8	7.4	8.4	9.7	11.3	12.4	14.6	16.9
	250	7.6	9.7	11.0	12.7	14.8	16.3	19.1	22.1
1.5	50	1.1	1.3	1.5	1.7	1.9	2.1	2.4	2.8
	100	2.2	2.8	3.1	3.5	4.0	4.4	5.1	5.8
	150	3.5	4.3	4.8	5.5	6.3	6.9	8.0	9.1
	200	4.8	5.9	6.7	7.6	8.7	9.5	11.0	12.6
	250	6.3	7.8	8.7	9.9	11.4	12.4	14.4	16.5
	300	7.9	9.7	11.0	12.4	14.3	15.6	18.1	20.6
	350	9.6	11.9	13.3	15.1	17.4	19.0	22.0	25.1
2.0	50	0.9	1.1	1.3	1.4	1.6	1.8	2.0	2.3
	100	2.0	2.4	2.7	3.0	3.4	3.7	4.2	4.8
	150	3.1	3.7	4.2	4.7	5.3	5.8	6.6	7.5
	200	4.3	5.2	5.8	6.5	7.4	8.0	9.2	10.4
	250	5.6	6.8	7.5	8.5	9.6	10.4	12.0	13.5
	300	7.0	8.5	9.4	10.6	12.1	13.1	15.0	17.0
	350	8.5	10.3	11.5	12.9	14.7	15.9	18.2	20.6
2.5	50	0.9	1.0	1.2	1.3	1.4	1.6	1.8	2.0
	100	1.8	2.2	2.4	2.7	3.0	3.3	3.7	4.2
	150	2.8	3.4	3.7	4.2	4.7	5.1	5.8	6.5
	200	3.9	4.7	5.2	5.8	6.5	7.0	8.0	9.0
	250	5.1	6.1	6.8	7.6	8.5	9.2	10.5	11.7
	300	6.4	7.7	8.5	9.5	10.7	11.5	13.1	14.7
	350	7.8	9.3	10.3	11.5	13.0	14.0	15.9	17.9
3.0	50	0.8	1.0	1.1	1.2	1.3	1.4	1.6	1.8
	100	1.7	2.0	2.2	2.4	2.7	2.9	3.3	3.7
	150	2.6	3.1	3.4	3.8	4.3	4.6	5.2	5.8
	200	3.6	4.3	4.8	5.3	5.9	6.4	7.2	8.0
	250	4.8	5.7	6.2	6.9	7.8	8.3	9.4	10.5
	300	6.0	7.1	7.8	8.7	9.7	10.4	11.8	13.2
	350	7.3	8.6	9.5	10.5	11.8	12.7	14.3	16.0
4.0	50	0.7	0.9	0.9	1.0	1.1	1.2	1.4	1.5
	100	1.5	1.8	2.0	2.1	2.4	2.5	2.9	3.2
	150	2.4	2.8	3.0	3.4	3.7	4.0	4.4	4.9
	200	3.3	3.9	4.2	4.6	5.2	5.5	6.2	6.8
	250	4.3	5.1	5.5	6.1	6.7	7.2	8.1	8.9
	300	5.4	6.3	6.9	7.6	8.5	9.0	10.1	11.2
	350	6.6	7.7	8.4	9.3	10.3	11.0	12.3	13.6

Insulation factors

Preformed pipe insulation	Insulation factor (f)	Based on K factor @ 50°F mean temp (Btu/h · °F · ft²/in)
Glass fiber (ASTM C547)	1.00	.25
Calcium silicate (ASTM C533)	1.50	.375
Cellular glass (ASTM C552)	1.60	.40
Rigid cellular urethane (ASTM C591)	0.66	.165
Foamed elastomer (ASTM C534)	1.16	.29
Mineral fiber blanket (ASTM C553)	1.20	.30
Expanded perlite (ASTM C610)	1.50	.375

TABLE B6.7 Pipe Heat-Loss Table* (*Continued*)

Pipe diameter (IPS) (in)											
3	3½	4	6	8	10	12	14	16	18	20	24
7.7	8.6	9.6	13.6	17.4	21.4	25.2	27.5	31.3	35.0	38.8	46.2
16.0	18.0	20.0	28.4	36.3	44.6	52.5	57.4	65.2	73.0	80.8	96.3
25.0	28.1	31.2	44.3	56.6	69.6	81.9	89.5	101.7	113.8	126.0	150.2
34.6	39.0	43.3	61.5	78.6	96.6	113.6	124.2	141.1	158.0	174.8	208.5
4.4	4.9	5.4	7.5	9.4	11.5	13.5	14.7	16.6	18.6	20.5	24.4
9.1	10.2	11.2	15.6	19.7	24.0	28.1	30.6	34.7	38.7	42.8	50.9
14.2	15.9	17.5	24.3	30.7	37.4	43.8	47.8	54.1	60.4	66.7	79.4
19.7	22.0	24.2	33.7	42.5	51.9	60.7	66.2	75.0	83.8	92.5	110.0
25.8	28.7	31.7	44.0	55.6	67.9	79.4	86.6	98.1	109.6	121.0	143.9
3.2	3.6	3.9	5.3	6.7	8.1	9.4	10.2	11.5	12.9	14.2	16.8
6.7	7.4	8.1	11.1	13.9	16.8	19.6	21.3	24.0	26.8	29.5	35.0
10.5	11.6	12.7	17.3	21.6	26.2	30.5	33.2	37.5	41.8	46.1	54.6
14.5	16.1	17.6	24.0	30.0	36.3	42.3	46.0	52.0	57.9	63.8	75.7
19.0	21.0	23.0	31.4	39.2	47.5	55.3	60.2	68.0	75.7	83.5	99.0
23.8	26.3	28.8	39.3	49.2	59.6	69.3	75.4	85.1	94.9	104.6	124.0
28.9	32.0	35.0	47.8	59.8	72.4	84.3	91.7	103.5	115.4	127.2	150.8
2.6	2.9	3.1	4.2	5.2	6.3	7.3	7.9	8.9	9.9	10.9	12.9
5.5	6.0	6.6	8.8	10.9	13.1	15.2	16.5	18.6	20.7	22.8	26.9
8.5	9.4	10.2	13.8	17.0	20.5	23.8	25.8	29.0	32.3	35.5	42.0
11.8	13.0	14.2	19.1	23.6	28.4	32.9	35.7	40.2	44.7	49.2	58.2
15.5	17.0	18.5	24.9	30.9	37.2	43.1	46.7	52.6	58.5	64.3	76.1
19.4	21.3	23.2	31.2	38.7	46.6	54.0	58.6	65.9	73.3	80.6	95.3
23.6	25.9	28.3	38.0	47.1	56.6	65.6	71.2	80.2	89.1	98.1	115.9
2.3	2.5	2.7	3.6	4.4	5.2	6.1	6.6	7.4	8.2	9.0	10.6
4.7	5.2	5.6	7.4	9.1	10.9	12.6	13.7	15.3	17.0	18.7	22.0
7.4	8.1	8.7	11.6	14.2	17.0	19.7	21.3	23.9	26.5	29.1	34.3
10.2	11.2	12.1	16.1	19.7	23.6	27.2	29.5	33.1	36.7	40.3	47.5
13.3	14.6	15.8	21.0	25.8	30.9	35.6	38.6	43.3	48.0	52.8	62.2
16.7	18.3	19.8	26.3	32.3	38.7	44.6	48.4	54.3	60.2	66.1	77.9
20.3	22.2	24.1	32.0	39.3	47.1	54.3	58.8	66.0	73.2	80.4	94.7
2.0	2.2	2.4	3.1	3.8	4.5	5.2	5.6	6.3	7.0	7.6	9.0
4.2	4.6	4.9	6.5	7.9	9.4	10.8	11.7	13.1	14.5	15.9	18.7
6.6	7.1	7.7	10.1	12.4	14.7	16.9	18.3	20.5	22.6	24.8	29.2
9.1	9.9	10.7	14.0	17.1	20.4	23.4	25.3	28.3	31.4	34.4	40.4
11.9	12.9	14.0	18.3	22.4	26.6	30.6	33.1	37.1	41.0	45.0	52.8
14.9	16.2	17.5	23.0	28.1	33.4	38.4	41.5	46.5	51.4	56.3	66.2
18.1	19.7	21.3	28.0	34.1	40.6	46.7	50.5	56.5	62.5	68.5	80.5
1.7	1.8	2.0	2.5	3.1	3.6	4.1	4.4	5.0	5.5	6.0	7.0
3.5	3.8	4.1	5.3	6.4	7.5	8.6	9.3	10.3	11.4	12.4	14.5
5.5	6.0	6.4	8.3	10.0	11.8	13.4	14.5	16.1	17.8	19.4	22.7
7.6	8.3	8.9	11.4	13.8	16.3	18.6	20.0	22.3	24.6	26.9	31.4
10.0	10.8	11.6	15.0	18.1	21.3	24.3	26.2	29.2	32.2	35.2	41.1
12.5	13.5	14.6	18.8	22.6	26.7	30.5	32.8	36.6	40.3	44.1	51.5
15.2	16.5	17.7	22.8	27.5	32.4	37.1	39.9	44.5	49.0	53.6	62.6

Valve heat-loss factors

Valve type	Heat-loss factor
Gate	4.3
Butterfly	2.3
Ball	2.6
Globe	3.9

Example:
Heat loss for a 2-in gate valve is 4.3 times the heat loss for one foot of pipe of the same size and insulation.

Pipe heat loss (Q_B) is shown in watts per foot. Heat loss calculations are based on IEEE Std. 515—1983, Equation 1, with the following provisions: pipes insulated with glass fiber in accordance with ASTM C547; pipes located outdoors in a 20 mph wind; no insulating air-space assumed between pipe and insulation; no insulating air-space assumed between the insulation and outer cladding. A 10% safety factor has been included.

Source: *Chemelex Auto-Trace Design Guide,* Raychem Corp., Chemelex Division, Menlo Park, Calif., 1988, pp. 8–9.

change, the total resistance—insulation plus air film—will not be altered significantly since most of that resistance is provided by the insulation.

- The amount of additional heat-tracing needed for valves is indicated in the Valve Heat-Loss Factors table in the lower right-hand corner of Table B6.7. Several additional feet of tracing will be needed for each controller in an insulated enclosure.

The heat loss per foot of pipe is calculated from Table B6.7 by following the procedure indicated in the thermal design guide chart, Fig. B6.14, as follows:

- First, move across the top to locate the 4-in-pipe column.
- Next, moving down the column, stop at the row corresponding to 1 in of insulation (left vertical axis).
- Following the calculation instruction method, a temperature differential ($T_m - T_a$) of 50°F yields a heat loss of 5.4 W/ft.
- Since the actual temperature differential is 40°F (40°F maintain—0°F minimum ambient), interpolation is required, and the final result is: 5.4 W/ft × 40/50 = 4.3 W/ft.

To calculate the heat loss that must be replaced by the heating cable, you need to know:

- ▶ T_M Desired maintenance temperature (°F)
- ▶ T_A Minimum expected ambient temperature (°F)
- ▶ T_E Maximum intermittent exposure temperature (°F)
- ▶ Pipe or tubing size
- ▶ Thermal insulation type and thickness

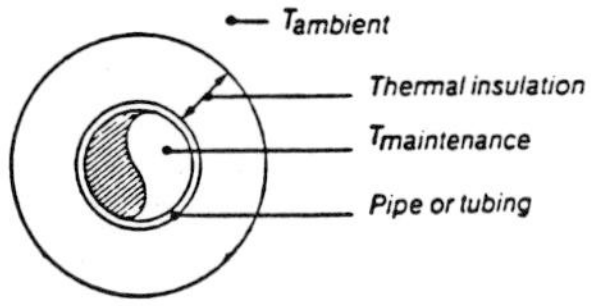

Example:
T_M: 50°F
T_A: −20°F
T_E: 366°F (150 psig steam cleaning)
Pipe size: 6″ steel
Insulation: 2″ calcium silicate

STEP 1	**Calculate temperature differential.** $\Delta T = T_M - T_A$	Calculate $\Delta T = T_M - T_A$ $= 50°F - (-20°F)$ $\Delta T = 70°F$.
STEP 2	**Determine pipe heat loss.** From Table B6.7 (next page), match the pipe size and insulation thickness with the temperature differential (ΔT) to find the base heat loss of the pipe (Q_B). *Note:* *Heat-loss calculations are based on IEEE Std. 515-1983, Equation 1.*	From Table B6.7, 6″ pipe, 2″ insulation and ΔT=70°F, Q_B must be calculated through interpolation: $Q_B = 4.2$ w/ft + 20/50 × (8.8 − 4.2) $= 4.2 + 1.8$ $Q_B = 6.0$ w/ft. @ T_M = 50°F
STEP 3	**Compensate for insulation type.** Multiply the base heat loss of the pipe (Q_B) from Step 2 by the insulation compensation factor (f) from Table B6.7 to get the actual heat loss (Q_T). $Q_T = Q_B \times f$	From Table B6.7, f=1.50 for calcium silicate: $Q_T = Q_B \times f$ $= 6.0$ w/ft. × 1.50 $Q_T = 9.0$ w/ft. @ 50°F

FIGURE B6.14 Thermal design chart. (Chemelex Auto-Trace Design Guide, *Raychem Corp., Chemelex Division, Menlo Park, Calif., 1988, p. 7.)*

This then, is the heat loss, including a 10 percent safety factor, of the pipe with the ambient temperature at 0°F, the pipe at 40°F, and the wind blowing 20 mi/h. The heater selected must be able to provide at least this much heat at a pipe temperature of 40°F.

The equation used to calculate heat loss from IEEE Standard 515-1983 is[59]:

$$q = \frac{T_m - T_a}{R_0 + R_1 + R_i + R_{co}} \tag{B6.1}$$

where q = the heat loss per unit length of the pipe at minimum ambient temperature, Btu/h · ft
T_m = the pipe maintenance temperature, 5°C, 40°F
T_a = the minimum ambient temperature, −18°C, 0°F.
R_0 = resistance to heat flow due to air film around the outside insulation surface
R_1 = resistance of insulation to the heat flow
R_i = resistance of the inside air film between the pipe and the insulation
R_{co} = resistance to heat flow from air film on inside of weather barrier

R_0, R_1, R_i, R_{co}, can be calculated by Eqs. (B6.2), (B6.3), (B6.4), and (B6.5) as follows:

$$R_0 = \frac{1}{\pi D_2 h_0} \tag{B6.2}$$

$$R_1 = \frac{\ln(D_2/D_1)}{2\pi k} \tag{B6.3}$$

$$R_i = \frac{1}{\pi D_1 h_i} \tag{B6.4}$$

$$R_{co} = \frac{1}{\pi D_2 h_{co}} \tag{B6.5}$$

where D_1 = inside diameter of insulation or outside diameter of pipe, ft
D_2 = outside diameter of insulation, ft
h_o = outside air film coefficient from weather barrier to ambient, Btu/h · ft^2 · °F
h_i = the inside air-contact coefficient from pipe to the inner surface of the insulation, Btu/h · ft^2 · °F
h_{co} = the inside air-contact coefficient of weather barrier, Btu/h · ft^2 · °F

Assuming that the insulation is properly installed on a fluid-filled pipe, R_i and R_{co} can be neglected. Under estimates for worst condition, wind speed is assumed to be in excess of 10 m/s, significantly diminishing the contribution of R_0.

With R_i R_{co}, and R_0 eliminated, the heat-loss equation reduces to:

$$q = \frac{T_m - T_a}{[\ln(D_2/D_1)]/2\pi k} \tag{B6.6}$$

The simplified formula with the R_0 term deleted is used only for calculating maximum or worst-condition heat loss. For other calculations, such as maximum uncontrolled equilibrium temperature, this term must be considered. Adjuncts to detailed calculations of heat loss, such as the example above, include formulas for determining the three heat-transfer air film coefficients h_o, h_{co}, and h_i as well as the heat-loss contribution for different types of pipe support.[60]

Once the worst-case heat loss has been determined, refinements can be factored into the calculation in an attempt to improve the overall performance and economics of the design. Iterative calculations may be carried out using different types and thicknesses of insulation, or instead of the lowest ambient temperature, a somewhat higher minimum ambient may be tolerated.

Heater Cable Selection

After the data are collected, the procedure for selecting heating cables is as follows:

1. Select the heater family to be used based on the "maintain" and exposure temperatures. Table B6.8 shows the temperature ratings for a series of commercial heaters. Based on a maintenance temperature (T_m) of 40°F and an exposure temperature T_e of 140°F, the economical choice is the heater family with lowest output capable of sustaining the required maintain temperature requirement and also is capable of withstanding the maximum intermittent exposure temperature.

TABLE B6.8 Heater Cable Temperature Ratings

Family	Maximum maintenance temperature, T_m	Maximum intermittent exposure, T_e
B	150°F (65°C)	185°F (85°C)
Q	225°F (110°C)	225°F (110°C)
X	250°F (121°C)	420°F (215°C)

Select the heating-cable family. Considering the maximum intermittent exposure temperature (T_e) and the desired maintenance temperature (T_m), select the appropriate heating cable family.

Source: *Chemelex Auto-Trace Design Guide,* Raychem Corp., Chemelex Division, Menlo Park, Calif., 1988, p. 10.

Choice: Heater family with T_m = 150°F, T_e = 185°F (B family).

2. Select the power output of the heater based on the heat loss and the desired maintain temperature. In the example computation, a power output of at least 4.3 W/ft at 40°F is required. (Both the power output and the maintain temperature must be specified in order to select the correct heater. A power output of 4.3 W/ft, for example, at 140°F instead of 40°F would lead to the selection of an entirely different heater family.) Figure B6.15 gives the power outputs for various families of heaters. Once again, the most economical selection is the heater with the minimum power output needed to make up the heat loss at a specified T_m of 40°F:

Choice: Heater with power output of 5.7 W/ft at 40°F. (Output specification is based on 1 ft of heater per foot of pipe.)

. In view of the heat replacement requirement, an alternate design solution could have employed a lower-cost heater with an output of only 3.5 W/ft in a spiral configuration of 1.23 ft of heater per foot of pipe. The accepted practice of most experienced users in North America is to avoid spiralling.

Spiralling increases installation costs and time, and in many cases it also increases material costs. For example, analysis of the selection options above demonstrate that although the cost per foot of the 4.3-W/ft heater is about 10 percent greater than the 3.5-W/ft heater, the spiralling requirement increases the overall material cost of the latter by 23 percent. This comparison does not include the additional cost of labor required for a spiral installation. In areas with very low labor costs, the economics of spiralling can be more attractive. Spiralling is most often employed in high-heat-loss configurations where the alternative is the use of multiple strips to generate adequate heat.

In the event that a selection outcome indicates the need for more than one heater, it may be appropriate to review the heater family chosen. For example, while a heater output of 4.3 W/ft at 140°F would require three strips of the heater selection E in Fig. B6.15, that same heat replacement could be more economically supplied by a single heater strip from selection D.

3. Select the voltage classification. 240-V heaters have circuit lengths approximately twice as long as the 120-V heaters. Minimizing the number of circuits is one of the most effective ways to reduce the costs of an electric heat-tracing system.

Choice: 240-V heater.

Most 240-V heaters can be used at voltages ranging from 208 to 277 V, with some power adjustment factors (See Table B6.9).

For self-regulating heaters, it is important to obtain the power adjustment factor from the manufacturer instead of calculating the factor by using the square of the voltage. Ohm's law ($P = V^2/R$) still applies. However, for self-regulating heaters, the electrical resistance R does not remain constant. The result is an adjustment factor smaller than the change calculated from the square of the voltage.

TABLE B6.9 Voltage Adjustment Chart

	240-V Auto-Trace heating cables powered at:				
	208 V		277 V		
Heating cable	Power output, %	Circuit length adjustment factor	Power output, %	Circuit length adjustment factor	Maximum heating-cable circuit length, ft
3BTV2	82	0.99	113	1.03	660
5BTV2	85	0.99	112	1.07	540
8BTV2	89	0.93	108	1.08	420
10BTV2	89	0.93	108	1.05	360
10QTV2	85	0.98	118	1.03	390
15QTV2	91	0.94	109	1.06	340
20QTV2	90	0.92	107	1.06	390
5XTV2	84	0.94	119	0.97	765
10XTV2	83	0.95	119	0.97	540
15XTV2	85	0.94	119	0.97	445
20XTV2	88	0.92	119	0.97	385

Source: Chemelex Auto-Trace Design Guide, Raychem Corp., Chemelex Division, Menlo Park, Calif., 1988, p. 19.

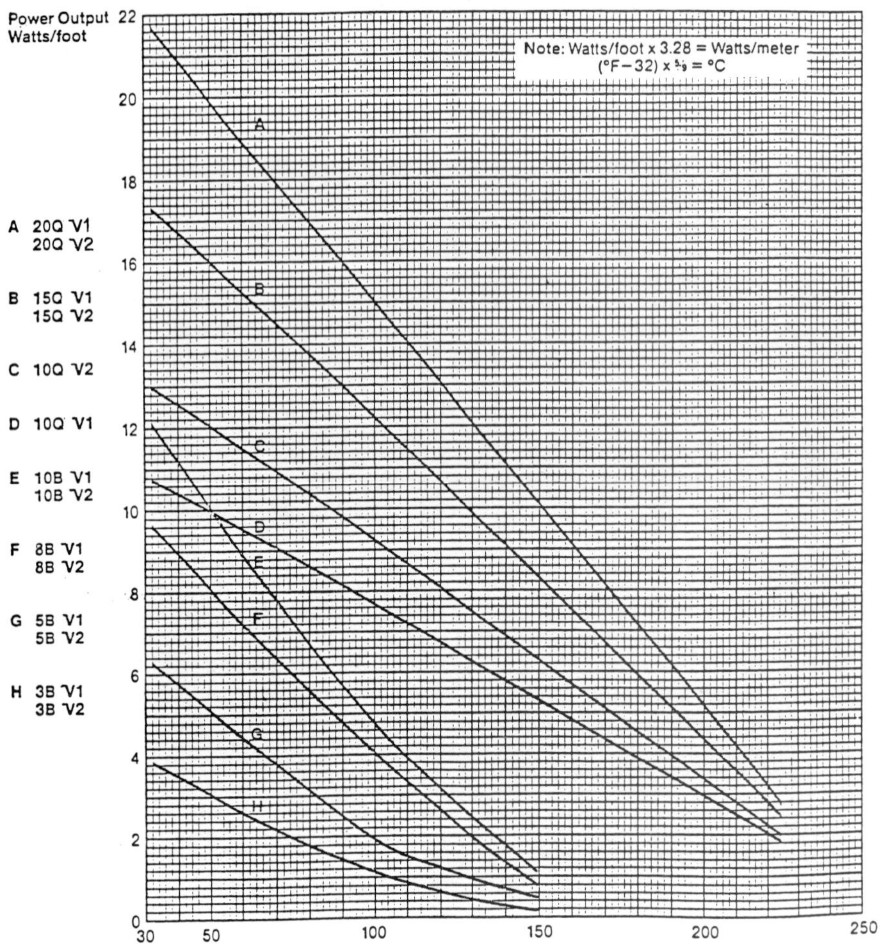

FIGURE B6.15 Thermal output of selected heaters. (Chemelex Auto-Trace Design Guide, *Raychem Corp., Chemelex Division, Menlo Park, Calif., 1988, p. 12.*)

4. Select the heater cable construction. The choices in cable construction are between a base heater with one insulating jacket and no braid or ground path, or a heater equipped with a braid and extra outer jacket. (A braid is required in a hazardous, i.e., classified area.)

The selection criteria for a commercial series of cables is given in Fig. B6.16. Since the sample application is in a classified area, the appropriate type of construction must be used.

Choice: Heater with a tinned braided, tinned copper (braid supplies the ground path required for a hazardous area), and a fluoropolymer outer jacket for chemical resistance.

Electrical Area Classification	Chemical Environment: Dry or Limited Exposure to Mild Chemical Environments			Chemical Environment: Wet or Limited Exposure to Corrosive Chemical Environments		
Ordinary Areas						
Ground path **not required** except as noted.	B base	Q base	X base	B -CT	Q -CT	X -CT
Ground path **recommended** for plastic pipe, stainless steel pipe, and painted surfaces.	***Note:*** *Article 427-22 of the 1987 National Electrical Code* ***requires*** *the use of a ground-path (metal shield) if the heating cable circuit is not equipped with a ground fault protection device. (U.S.A. only)*					
Hazardous (classified) Areas						
Ground path **required** for Division 2, Class I, II & III.	B -CT	Q -CT	X -CT	B -CT	Q -CT	X -CT
	Note: *Heating cables and their component systems are approved for Class I, II, and III, Division 2 areas by Factory Mutual (FM) and the Canadian Standards Association (CSA). For Division 1 area applications consult the manufacturers representative.*					

Heating Cable Constructions:

Base
A conductive polymer core with an insulating jacket that is resistant to mild chemical environments such as aqueous inorganic chemicals.

Outer jacket over braid (-CT)
A braided tinned-copper shield covering the insulating jacket has a fluoropolymer outer jacket that provides superior protection from excessive moisture and chemically corrosive environments such as hydrocarbons, acids, bases, solvents, oils, greases, and fats.

FIGURE B6.16 Area classification and chemical environment selection chart. (Chemelex Auto-Trace Design Guide, *Raychem Corp., Chemelex Division, Menlo Park, Calif., 1988, p. 11.*)

The heater cable selection is now complete. Tables B6.10, B6.11, and B6.12 are the outputs of the computer selection program computed from the data used in the calculation example. As expected, the program arrived at the same heater selection as the manual technique employed in the above discussion. The program output also provides important information that augments the design selection function.

The design summary given in Table B6.10 provides the start-up load in amps per foot as well as the operating load. The Min T maintained and Max T main-

TABLE B6.10 Design Summary Output for Computation Example

Project Name: Freeze Prevention	Temperatures	Insulation: GLASS FIBER			
Done For: Handbook Example	Maintain: 40 F	(ASTM C547-77)			
Done By: Joe Lonsdale	Min Ambient: 0 F	K Factor at 50 F: 0.2497 BTU-in/hr-ft2-F			
Design Name: Freeze Pre	Max Ambient: 95 F	Area Classification Code: Hazardous	Voltage: 240		
	Mean: 20 F	Chemical Exposure Code: Severe	Location: Outdoors		
Project ID: EXAMPLE	Process Oper: 50 F	Heat Loss Safety Factor: 10 %	Maximum CB Load: 40 A		
Design ID: EXAMPLE	Max Htr Exp: 140 F	Pipe Length per Support: 20 ft	Pipe Category: IPS/Steel		
Lines ID: EXAMPLE	System Limit: 500 F	Allow Spiralling: No	Support HL Factor: 0.00 ft		
	Startup: 0 F	Allow Uncontrolled: Yes	Valve HL Factor: 3.28 ft		

DS Ref #	Pipe Size (in)	Insul Thick (in)	HL Rate (W/ft)	Heater Type	Output Rate (W/ft)	Feet of Heater /ft Pipe	Feet of Heater per Valve	Feet of Heater per Supp	Pitch (in)	Min T Maintained	Max T Maintained	K Factor @ Mean T	Start Load (A/ft)	Oper Load (KW/ft)	Total Pipe (ft)	Total Heater (ft)	Total Comp Sets
1	4.00	1.0	4.3	5BTV2-CT	5.6	1.0	2.5	0.0	0	48	110	0.2382	0.042	0.0063	230	232	1

Source: *TraceCalc Program,* Raychem Corp., Chemelex Division, Menlo Park, Calif.

TABLE B6.11 Line List Output for Computation Example

Project Name:	Freeze Prevention	Temperatures		Insulation:	GLASS FIBER		
Done For:	Handbook Example	Maintain:	40 F		(ASTM C547-77)		
Done By:	Joe Lonsdale	Min Ambient:	0 F	K Factor at 50 F:	0.2497 BTU-in/hr-ft2-F		
Design Name:	Freeze Pre	Max Ambient:	95 F	Area Classification Code:	Hazardous	Voltage:	240
		Mean:	20 F	Chemical Exposure Code:	Severe	Location:	Outdoors
Project ID:	EXAMPLE	Process Oper:	50 F	Heat Loss Safety Factor:	10 %	Maximum CB Load:	40 A
Design ID:	EXAMPLE	Max Htr Exp:	140 F	Pipe Length per Support:	20 ft	Pipe Category:	IPS/Steel
Lines ID:	EXAMPLE	System Limit:	500 F	Allow Spiralling:	No	Support HL Factor:	0.00 ft
		Startup:	0 F	Allow Uncontrolled:	Yes	Valve HL Factor:	3.28 ft

Ref #	Line No	Pipe Size (in)	Insul Thick (in)	Pipe Length (ft)	# Valve	# Support	Heater Type	Trace Ratio	HL Rate (W/ft)	Output Rate (W/ft)	Heater Length (ft)	# Comp Sets	Start Load (Amp)	Oper Load (kVA)	Comment
1	101	4.00	1.0	230.0	1	11	5BTV2-CT	1.0	4.3	5.6	232.5	1	10	1.5	

Source: TraceCalc Program, Raychem Corp., Chemelex Division, Menlo Park, Calif.

TABLE B6.12 Bill of Materials for Computation Example

Project Name:	Freeze Prevention	Project ID:	EXAMPLE
Done For:	Handbook Example	Design ID:	EXAMPLE
Done By:	Joe Lonsdale	Lines ID:	EXAMPLE
Design Name:	Freeze Pre		

Quantity	Catalog No	Description
234 ft	5BTV2-CT	Chemelex Auto-Trace Heater
1 Each	PMKG-JLP	PolyMatrix Power Connection Kit
1 Each	PMKG-LE	PolyMatrix End Termination Kit
1 Each	AMC-1H	Ambient-Sensing Thermostat
23 Each	ETL	Label: "Electric Traced"
9 Rolls	GT66	Glass Tape (66 Ft/Roll)
1 Each	PS-10	Pipe Strap for 3" to 10"
230 ft	4.00 X 1.0	Insulation Material
392 Sq Ft		Cladding Material

Source: TraceCalc Program, Raychem Corp., Chemelex Division, Menlo Park, Calif.

tained are the maximum and minimum temperatures the pipe can reach in this application in the absence of thermostatic control.

The Min T maintained (48°F) will be the pipe temperature on the coldest day (0°F) with the wind blowing, and minimum heater output. The Max T maintained will be the pipe temperature on the warmest day (95°F) with no wind, and maximum heater output. These temperature values are useful in determining whether thermostats are needed. (See later section "Control and Monitoring.")

Using the computer heat-tracing program to implement a design change: The actual application for the computation example is a waterline in a process area where many of the process lines are regularly cleaned by purging with 150-psig steam (366°F). It was concluded by the design team that a reasonable possibility existed that the traced line might be inadvertently steam cleaned. What would the consequences be?

The new variable requires that the maximum heater exposure temperature be changed from 140°F to 390°F, and the resulting output is given in Tables B6.13, B6.14, and B6.15. The most significant outcome of the requirement for a higher maximum temperature of exposure is the program's substitution of a heater de-

TABLE B6.13 Design Summary Output for Computation Example, Revised for Steam Exposure

Project Name: Freeze Prevention	Temperatures	Insulation: GLASS FIBER	
Done For: Handbook Example	Maintain: 40 F	(ASTM C547-77)	
Done By: Joe Lonsdale	Min Ambient: 0 F	K Factor at 50 F: 0.2497 BTU-in/hr-ft2-F	
Design Name: Freeze Pre	Max Ambient: 95 F	Area Classification Code: Hazardous	Voltage: 240
	Mean: 20 F	Chemical Exposure Code: Severe	Location: Outdoors
Project ID: EXAMPLE	Process Oper: 50 F	Heat Loss Safety Factor: 10 %	Maximum CB Load: 40 A
Design ID: EXAMPLE	Max Htr Exp: 390 F	Pipe Length per Support: 20 ft	Pipe Category: IPS/Steel
Lines ID: EXAMPLE	System Limit: 500 F	Allow Spiralling: No	Support HL Factor: 0.00 ft
	Startup: 0 F	Allow Uncontrolled: Yes	Valve HL Factor: 3.28 ft

DS Ref #	Pipe Size (in)	Insul Thick (in)	HL Rate (W/ft)	Heater Type	Output Rate (W/ft)	Feet of Heater /ft Pipe	Feet of Heater per Valve	Feet of Heater per Supp	Pitch (in)	Min T Maintained	Max T Maintained	K Factor @ Mean T	Start Load (A/ft)	Oper Load (KW/ft)	Total Pipe (ft)	Total Heater (ft)	Total Comp Sets
1	4.00	1.0	4.3	5XTV2-CT	5.1	1.0	2.7	0.0	0	47	146	0.2382	0.038	0.0067	230	233	1

Source: *TraceCalc Program,* Raychem Corp., Chemelex Division, Menlo Park, Calif.

TABLE B6.14 Line List Output for Computation Example, Revised for Steam Exposure

Project Name: Freeze Prevention	Temperatures	Insulation: GLASS FIBER	
Done For: Handbook Example	Maintain: 40 F	(ASTM C547-77)	
Done By: Joe Lonsdale	Min Ambient: 0 F	K Factor at 50 F: 0.2497 BTU-in/hr-ft2-F	
Design Name: Freeze Pre	Max Ambient: 95 F	Area Classification Code: Hazardous	Voltage: 240
	Mean: 20 F	Chemical Exposure Code: Severe	Location: Outdoors
Project ID: EXAMPLE	Process Oper: 50 F	Heat Loss Safety Factor: 10 %	Maximum CB Load: 40 A
Design ID: EXAMPLE	Max Htr Exp: 390 F	Pipe Length per Support: 20 ft	Pipe Category: IPS/Steel
Lines ID: EXAMPLE	System Limit: 500 F	Allow Spiralling: No	Support HL Factor: 0.00 ft
	Startup: 0 F	Allow Uncontrolled: Yes	Valve HL Factor: 3.28 ft

Ref #	Line No	Pipe Size (in)	Insul Thick (in)	Pipe Length (ft)	# Valve	# Support	Heater Type	Trace Ratio	HL Rate (W/ft)	Output Rate (W/ft)	Heater Length (ft)	# Comp Sets	Start Load (Amp)	Oper Load (kVA)	Comment
1	101	4.00	1.0	230.0	1	11	5XTV2-CT	1.0	4.3	5.1	232.7	1	9	1.6	

Source: *TraceCalc Program,* Raychem Corp., Chemelex Division, Menlo Park, Calif.

TABLE B6.15 Bill of Materials for Computation Example, Revised for Steam Exposure

Project Name: Freeze Prevention	Project ID: EXAMPLE
Done For: Handbook Example	Design ID: EXAMPLE
Done By: Joe Lonsdale	Lines ID: EXAMPLE
Design Name: Freeze Pre	

Quantity	Catalog No	Description
234 ft	5XTV2-CT	Chemelex Auto-Trace Heater
1 Each	PMKG-JLP	PolyMatrix Power Connection Kit
1 Each	PMKG-LE	PolyMatrix End Termination Kit
1 Each	AMC-1H	Ambient-Sensing Thermostat
23 Each	ETL	Label: "Electric Traced"
9 Rolls	GT66	Glass Tape (66 Ft/Roll)
1 Each	PS-10	Pipe Strap for 3" to 10"
230 ft	4.00 X 1.0	Insulation Material
392 Sq Ft		Cladding Material

Source: *TraceCalc Program,* Raychem Corp., Chemelex Division, Menlo Park, Calif.

signed to withstand the temperatures encountered during steam cleaning. The results of the new calculation show that the Min T maintained is relatively unchanged (47°F), but the Max T maintained has increased to 146°F. This increase in the maximum temperature is a function of the original heater's power temperature curve which is considerably steeper when compared with the type selected for the new set of conditions. If a constant-wattage heater had been selected, the maximum temperature would be above 146°F. The design change calculation illustrates two important rules to follow when using self-regulating heaters:

Rule 1. Never risk exposing the heater to temperatures in excess of those for which the heater is rated. If the heater originally specified (T_e = 185°F) had been installed and the pipe were subsequently steam cleaned, the heater would suffer an irreversible resistance increase, that is, it would no longer be functional. The cost of repair—removal of the lagging, removal and possible replacement of the insulation, replacement of the heater, followed by reinstallation of the entire system—would be more than the original price of the system. And this sum is in addition to process-related costs which might result from heater failure such as the replacement of frozen and/or broken pipe, or in lengthy production downtime when a failed tracer shuts down a process.

Because of the potential costs resulting from tracer failure caused by inadvertent exposure to steam, experienced heat-tracing professionals generally use heaters capable of withstanding steam cleaning in any area where steam cleaning is practiced.

Rule 2. Select the heater with the lowest temperature rating capable of withstanding the anticipated maximum exposure temperature.

1. The higher-temperature heaters are generally more expensive because they are made from more costly heat-resistant polymers.
2. The higher temperature generated by high-temperature heaters can cause safety problems. Unnecessary heating also increases the rate of corrosion and wastes energy. Higher-temperature heaters often require a more expensive heater control system.

While the potential savings in purchasing a low-temperature heater are not worth the risk of heater failure from steam cleaning, these additional costs are not justified in applications where there is no risk of steam exposure. Water pipe in a pollution control area where no steam is available is an example.

Component and Accessory Selection

The components necessary to provide power, to terminate, splice, and tee the heat-tracing are provided by the manufacturer. With self-regulating and zone heaters but not for MI cable, heaters, and components, it is important to acquire components designed for the heat-tracing system selected to ensure that the approvals are valid. Each supplier's series of components are unique and generally not interchangeable with those of other manufacturers. In some cases the components from one supplier may work in another's heat-tracing system, but they must be verified to be acceptable for such usage. The components available in this example are shown in Fig. B6.17.

Each manufacturer provides or recommends a complete system of components and accessories to install their heater as part of a total tracing system. Table B6.12 (the design exhibit from the computer solution) shows a complete bill of

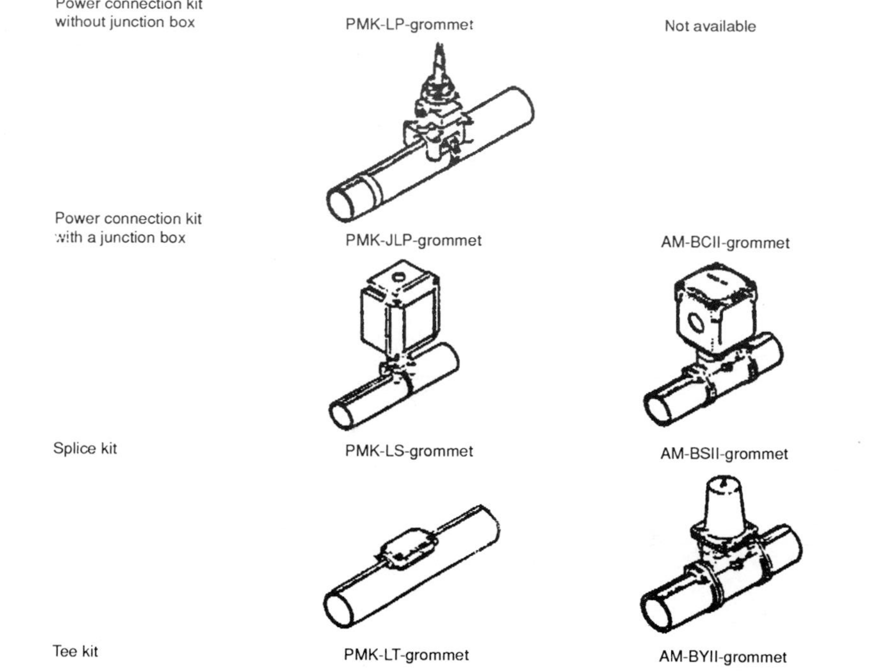

FIGURE B6.17 Component selection guide. (Chemelex Auto-Trace Design Guide, *Raychem Corp., Chemelex Division, Menlo Park, Calif., 1988, pp. 20–21.*)

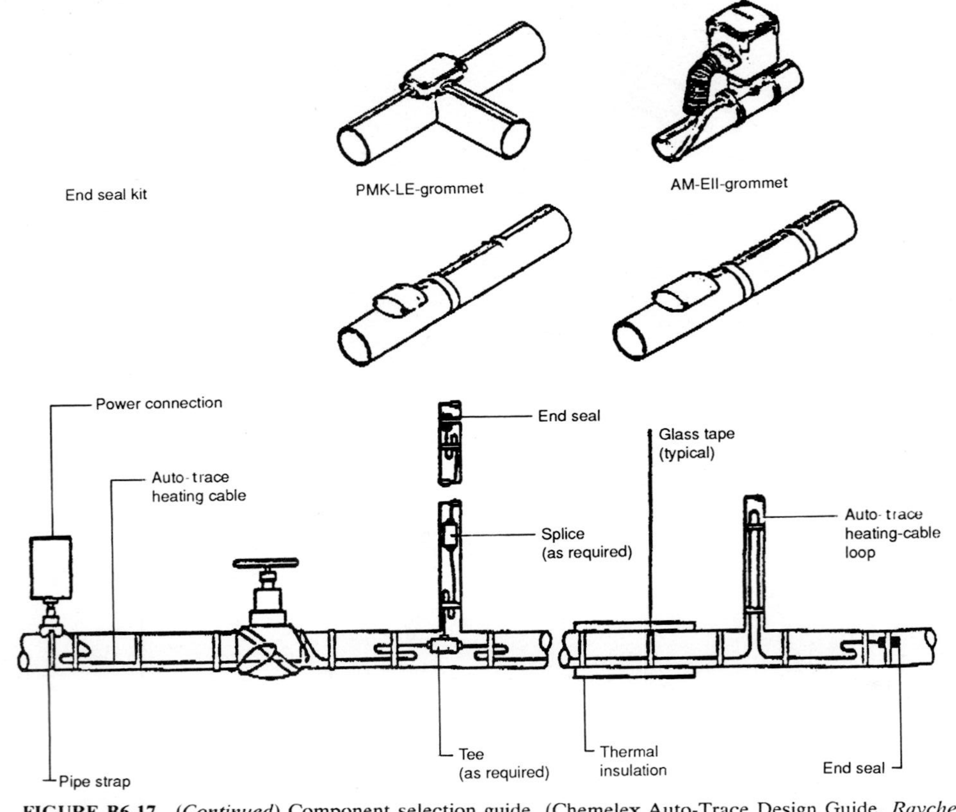

FIGURE B6.17 (*Continued*) Component selection guide. (Chemelex Auto-Trace Design Guide, *Raychem Corp., Chemelex Division, Menlo Park, Calif., 1988, pp. 20–21.*)

material for the design example. In addition to the power connection and the end termination, a splice and/or a tee may be required. The other materials on the list are:

- ETL (electric traced label): The National Electric Code requires that electric traced pipes have signs on the outside of the lagging on alternating sides every 10 ft.
- GT66 glass tape for taping the heater to the pipe.
- PS-10 pipe straps to attach the power connection assembly to the pipe.

Control and Monitoring

Control and monitoring must be considered as part of the system design since these elements can have a significant effect on the circuit layout.

1. Control. Control is the ability to interrupt and restore power to the heat tracer in order to maintain the temperature within a preset range and/or to save energy. Monitoring is any method which provides an ongoing indication of the heating system's operational status. The major control options are:

No Control, or Self-Regulating Control. The heater is constantly supplied with full power, and the self-regulating characteristics of the heater control the pipe temperature. In a limited number of tracing situations, a "no-control" configuration can also be used with constant-wattage heaters. If a circuit without controller is used in a hazardous area, the T rating must be calculated at 120 percent of the rated voltage. (See earlier section "Area Classification.")

Self-regulating control offers the lowest installed cost and the highest reliability. This system uses more energy as the heater is always on. Approximately 8 to 12 percent of the freeze protection systems and 10 to 15 percent of the process temperature systems use self-regulating control.

Ambient-Sensing Control. A thermostat measures the ambient temperature. The heating system is engaged when the ambient temperature drops below a preset level. Ambient sensing offers a degree of control for a very little incremental cost (one thermostat and the required switching device). Seventy to 80 percent of the freeze protection systems, particularly those that use self-regulating heaters, use ambient-sensing control.

Line-Sensing Control. A thermostat measures the temperature of the pipe. Each heater circuit is individually turned on when the temperature indicated by its associated thermostat drops below the design "maintain" temperature (see Fig. B6.18). Line sensing is often used for process control systems. It offers the highest degree of temperature control and the lowest energy use, but it also has the highest installed costs.[61]

Dead Leg Control. A thermostat measures a section of traced pipe that cannot have fluid flow (a "dead leg"). The entire system is turned on when the temperature drops below the design temperature to be maintained. Dead leg sensing requires only one thermostat and eliminates the need for the additional circuits required with line-sensing control. The downside of dead leg sensing is the inability of the system to reduce heat output even when flow conditions within the pipe would dictate such an adjustment. (By definition, the "dead leg" on which the sensor is placed is in a permanent no-flow condition.[62]) Presently, about 5 to 10 percent of the systems installed use dead leg control.

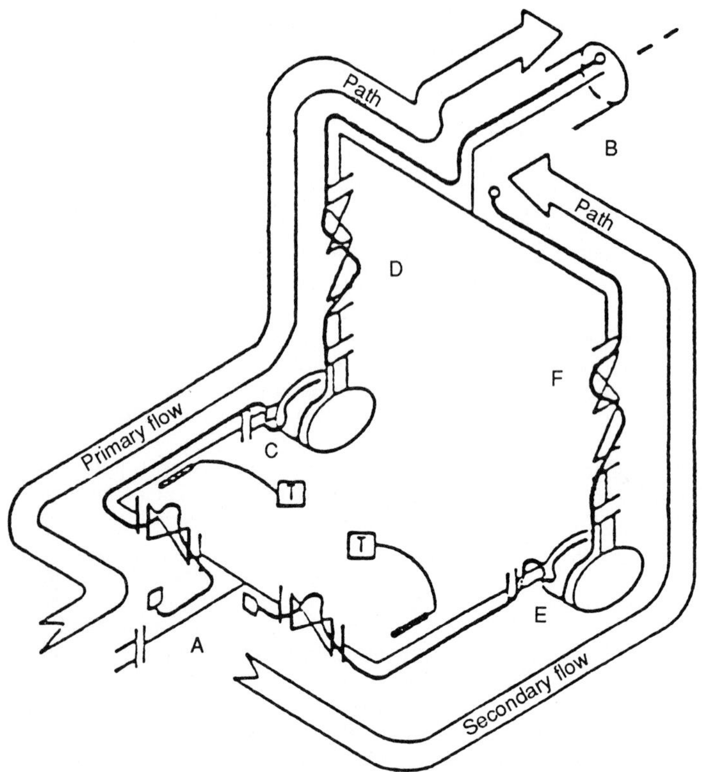

FIGURE B6.18 Line-sensing circuit design for multiple flow paths. (*M. Sarfatti, editor,* Raychem Engineering Manual for Electric Heat-Tracing Systems, *vol. 1, U.S. edition, pp. 2–27, Fig. 3–11.*)

2. ***Monitoring.*** *Monitoring* is the term used to describe any mechanism that provides information about the functional state of the heat-tracing system.[63] Almost all heat-tracing systems employ one of the following monitoring methods:

No Monitoring. The vast majority of heat-tracing systems are not provided with monitoring. Users have found that properly installed heat-tracing systems are very reliable, and under these circumstances, the high cost of monitoring is not generally warranted.

Ground Fault Monitoring Leakage current from the heater strip can be monitored with a ground fault circuit breaker. An annunciator is included to indicate when the breaker has tripped or alternatively, a relay is switched in response to a loss of voltage. This method is also a good way to monitor the heater strip for mechanical abuse since significant mechanical damage to a braided heater will result in leakage current to the braid. Use of ground fault breakers with annunciators or relays provides the highest value of added protection per monitoring dollar invested.

Voltage Monitoring. Voltage at the beginning of the circuit can be sensed by actuating a relay-driven alarm whenever there is no voltage from the circuit breaker. A signal light is the simplest technique for monitoring voltage at the end of the circuit. This system is often used for freeze protection applications with ambient-sensing control and either self-regulating heaters or zone heaters but not with MI cable, as MI cable usually requires line-sensing thermostats.

Although a light indicator is a low-cost system, it requires a visual inspection to detect voltage loss. Thus, there is a good chance that a cut line will lead to a failure before it is detected. The technique does not work well with line-sensing thermostats or dead leg control since there is no way to distinguish if the light is off because of a cut line or a cycling thermostat.

Voltage Sensing with Microprocessor-Based Monitoring and Control Systems. The best of these systems use RTDs (resistance temperature detectors). (See Fig. B6.19.) Since the bus wires of the heaters are used to carry the detector signal, any cut in the line trips the alarm due to signal loss. Microprocessor-based monitoring and control systems are capable of providing very accurate information about the condition of the heat-tracing system and the pipe temperature, but they often cost as much as the heat-tracing system itself. For this reason, they are generally used only for temperature-critical applications.

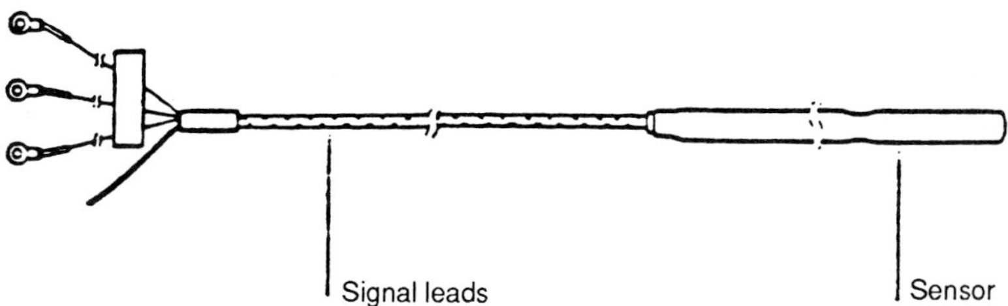

Pipe temperature can be measured using thermocouples or resistance temperature devices (RTDs). These devices are placed along a circuit at locations where temperature readings are desired. The temperatures may be used for control or information only.

FIGURE B6.19 Resistance temperature device. (Chemelex Auto-Trace Installation and Operation Manual, *Raychem Corp., Process Division, Menlo Park, Calif., p. 1.5.3.*)

Temperature Sensing with Thermostatic Monitoring Systems. Thermostats can be used to alarm when the temperature at any monitored point drops below the alarm temperature. This is a good technique when there are only one or two points (such as critical control valves) where the proper temperature must be maintained.

The current draw of the heat tracer can also be monitored with microprocessor-based systems using a simple ammeter indicating the current draw of each circuit. Current monitoring works well with all types of heaters when line-sensing control is used but is a poor choice in applications employing ambient-sensing control and self-regulating heaters.

REFERENCES

1. Henry, K., *Introduction to Heat Tracing,* Cold Regions Research and Engineering Lab, Report No. CRREL-TD-86-1, June 1986, p. 1.

2. Kohli, I. P., "Steam Tracing of Pipelines," *Chem. Eng.,* March 26, 1979, p. 156.
3. Fisch, E., "Winterizing Process Plants," *Chem. Eng.,* August 20, 1984, p. 130.
4. Reference 2.
5. Reference 1, p. 2.
6. Lonsdale, J. T., and Mundy, J. E., "Estimating Pipe Heat-tracing Costs," *Chem. Eng.,* November 29, 1982, pp. 89–93.
7. Reference 2, p. 159.
8. Reference 1, p. 4.
9. Dixon, G. B., *Steam versus Electric Process Heat Tracing,* Mississippi Department of Energy and Transportation Annual Conference, April 22–24, 1987, p. 3.
10. Reference 1, pp. 5–6.
11. Bilbro, J. E., and Levines, J. E., "Electric Heat Tracing—State of the Art," *IEEE Transactions on Industry and General Applications,* vol. IGA-5, no. 4, July/August 1969, p. 476.
12. Lonsdale, J. T., and Mayer, L., *Heat-Tracing Technologies, Energy Use and Temperature Control,* Raychem Corp. Draft Report, 1988, p. 5.
13. *Chemelex Auto-Trace Design Guide,* Raychem Corp., Chemelex Division, Menlo Park, Calif., 1988, p. 7.
14. Reference 6, p. 89.
15. Reference 9, p. 5.
16. Reference 12, pp. 4–5.
17. Reference 1, p. 9.
18. Reference 12, pp. 4–5.
19. Reference 1, p. 9.
20. Reference 6. pp. 89–90.
21. Reference 12, pp. 4–5.
22. Reference 12, pp. 4–5.
23. Reference 12, p. 2.
24. Reference 6, p. 89.
25. Reference 6, p. 89.
26. Reference 12, p. 3.
27. Ohlson, R. N., "A User's Experience With Current Self-Limiting Heat Tracing Cable," IEEE Paper No. PCI-80-25, p. 180.
28. Reference 27, p. 180.
29. Reference 1, pp. 8–9.
30. Reference 12, p. 3.
31. IEEE Recommended Practice for Electrical Impedance, Induction, and Skin Effect Heating of Pipelines and Vessels, ANSI/IEEE Std. 844—1985, pp. 29–30; Erikson, C. J., Chairman, the Electrical Impedance, Induction and Skin Effect Heating Working Group, Standards Subcommittee of the Petroleum and Chemical Industry Committee.
32. Carson, N. B., *A New Method for Heat Tracing Long Pipelines,* ASME, Petroleum Mechanical Engineering Conference, Dallas, Texas, September 15–18, 1974, pp. 2–4.
33. Reference 32, p. 4.
34. Reference 31, p. 30.
35. Reference 1, p. 15.
36. Reference 32, p. 4.

37. Ando, M., and Takki, H., *Application of the SECT Electric Heating System to Long-Distance Pipelines*; Comite Francais Electrothermie, 9th International Congress, October 20–24, 1980, p. 1.

38. Reference 31, p. 30.

39. Reference 1, p. 15.

40. Reference 31, pp. 18–19.

41. Reference 1, p. 15–16.

42. Koester, George II, "Pipe Heat Tracing With Electric Impedance Heating," *Plant Engineering,* vol. 32, no. 24, pp. 113–116, November 23, 1978, p. 115.

43. Reference 1, p. 16.

44. Reference 1, p. 16.

45. Reference 31, p. 23.

46. Reference 1, p. 17.

47. Reference 31, p. 16.

48. Schram, P. J., editor, *National Electrical Code 1987 Handbook,* 4th ed., National Fire Protection Association, Quincy, Mass.

49. Reference 50, Article 500, pp. 659–677.

50. Reference 6, pp. 89–93.

51. *Heat Up/Cool Down Analysis Program,* Raychem Corp., Process Division, Menlo Park, Calif.

52. Reference 12, p. 11.

53. Reference 9, p. 10.

54. *AutoTrace Analysis Program Summary,* Raychem Corp., Menlo Park, Calif.

55. Erikson, C. J., IEEE paper, September 30, 1990.

56. Reference 12, p. 11.

57. Reference 51.

58. Sarfatti, M., editor, *Raychem Engineering Manual for Electric Heat-Tracing Systems,* vol. 1., U.S. edition, pp. 3–33.

59. IEEE Std. 515—1983, Section 6.3.5, p. 14, IEEE Recommended Practice for the Testing, Design, Installation, and Maintenance of Electrical Resistance Heat-Tracing for Industrial Applications, B. C. Johnson, Chairman, Working Group Standards Subcommittee of the Petroleum and Chemical Industry Committee.

60. Reference 59, Appendix A, pp. 27–29.

61. Reference 12, p. 9.

62. Reference 58, pp. 3–19, 3–25.

63. Reference 58, pp. 4–19 through 4–27.

CHAPTER B7

THERMAL INSULATION OF PIPING

James Waldo
General Supervisor, Production
Pittsburgh Corning Corporation
Sedalia, Missouri

Thermal insulation provides many useful purposes in both industrial and commercial piping applications. In the simplest of terms, thermal insulation serves the purpose of reducing heat flow from one surface to another. For hot (above ambient) piping applications thermal insulation reduces heat loss. On cold (below ambient) piping applications the insulation generally serves the purpose of minimizing heat gain.

In some cases the design purpose of the application may seem unrelated to heat loss or heat gain; however, the net result is that heat transfer is retarded. Two illustrations of this are insulation for personnel protection and insulation for condensation control. Insulation for personnel protection involves providing enough insulation to keep the surface temperature below a given design value (usually 140°F). Insulation for condensation control requires enough insulation to keep the surface temperature above the dew point. In both cases the insulation is used to control the surface temperature for a desired effect other than thermal conservation. The effect, however, is that in both cases heat transfer is retarded to maintain the surface temperature at the given design criteria.

Correctly designing and specifying an insulation system is much more involved than just selecting a particular insulation material to be used. This chapter covers some of the practical information necessary to initiate an effective insulation "system" design. The National Insulation Contractor's Association (NICA) in their Insulation Craft Training Program[1] defines insulation as, "those materials or combination of materials which retard the flow of heat." As noted in the NICA definition of insulation, a combination of materials may be used. The emphasis on the word *system* when referring to the purpose of this chapter signifies the importance of considering all of the materials, conditions, and parameters involved in insulation specification and design.

An insulation system is any combination of insulation materials used in conjunction with mastics, adhesives, sealants, coatings, membranes, barriers, and/or other accessory products to provide an efficient assembly for the reduction of heat flow. The engineering of insulation systems can frequently either determine or direct the ultimate performance of the process. Improperly engineered insulation systems are subject to damage and degradation. This degradation will compromise the performance characteristics of the insulation material and, in many cases, the entire process for which the insulation system was specified.

There are many different types of insulation materials available for commercial and industrial piping applications. Each material has its own set of properties and performance characteristics. And, for each insulation material available, a correct application procedure and corresponding accessory materials, or "system," are available.

Before getting into the essence of this chapter, which addresses some of the design parameters, materials, and systems that are commonly incorporated into insulation system design, it is first necessary to review some heat transfer fundamentals.

FUNDAMENTALS OF HEAT TRANSFER

The following definitions, taken from ASTM C168-88a Standard Definitions of Terms Relating to Thermal Insulating Materials[2] and the NICA Insulation Craft Training Program, will be useful in reviewing the basic fundamentals of heat transfer:

Btu: British thermal unit. The amount of energy required to raise 1 lb of water 1°F.

Conduction: The transfer of energy (heat) within a body (material) or between two bodies in physical contact.

Convection: The transfer of heat by movement of parts of a liquid or gas within the liquid or gas because of differences in the density, temperature, etc., of the parts.

Radiation: The transfer of energy (heat) from a higher-temperature body, through space, to another lower-temperature body without warming the space between.

Thermal conductivity (K): The time rate of steady state heat flow through a unit area of a homogeneous material induced by a unit temperature gradient in a direction perpendicular to that unit area. Units are commonly Btu-in/ $(h \cdot ft^2 \cdot °F)$.

Emittance (E): The ratio of the radiant flux emitted by a specimen to that emitted by a black-body at the same temperature and under the same conditions.

Radiance: The rate of radiant emission per unit solid angle and per unit projected area of a source in a stated angular direction from the surface (usually the normal).

Reflectance: The fraction of the incident radiation upon a surface that is reflected from the surface.

Heat flow; heat flow rate (Q): The quantity of heat transferred to or from a system in unit time. Usually measured in Btu/h.

Thermal insulation: A material or assembly of materials used to provide resistance to heat flow.

Thermal insulation system: Applied or installed thermal insulation complete with any accessories, vapor retarder, and facing required.

Heat is transferred by any one or combination of conduction, convection, and/or radiation. Conduction only occurs when there is physical contact. Heat is transferred through most metals very efficiently because metal is a good conductor. A good insulation material is a poor conductor. Convection, with respect to insulation systems, is referred to as the movement of air on or about the surface of an insulated body. And radiation is best described by referring to the warmth you feel when you stand in the sun or by a fire.

Heat transferred through insulation is primarily a function of the resistance of the insulation with respect to its thickness, the operating temperature of the surface being insulated, the surface characteristics of the outer membrane (see emittance above), and the ambient conditions involved.

Thermal conductivity, as defined above, is the rate of heat transfer in one direction (perpendicular to an area) per unit area, per unit temperature differential per unit thickness, per unit time. In the English system of units, typical dimensions are:

$$|K| = \text{Btu}/(\text{h} \cdot {}^\circ\text{F} \cdot \text{ft}^2/\text{ft}) = \text{Btu}/(\text{h} \cdot \text{ft} \cdot {}^\circ\text{F})$$

or

$$|K| = \text{Btu}/(\text{h} \cdot {}^\circ\text{F} \cdot \text{ft}^2/\text{in})$$

or

$$|K| = \text{Btu} \cdot \text{in}/(\text{h} \cdot \text{ft}^2 \cdot {}^\circ\text{F})$$

Heat transferred through flat surface geometry is most commonly represented by Eq. (B7.1):

$$Q = \frac{A(T_i - T_2)}{X/K + 1/f} \tag{B7.1}$$

where Q = the total heat loss in Btu/(h · ft^2)
A = the area of heat flow in square feet
T_i = the inside operating temperature in °F
T_2 = the outside ambient temperature in °F
X = the insulation thickness in inches
K = the thermal conductivity in Btu · in/(h · ft^2 · °F)
1/f = the surface resistance factor

Heat transferred through cylindrical, or pipe insulation, geometry is most commonly represented by Eq. (B7.2) below, and Fig. B7.1:

$$Q = \frac{2(\pi)KL(T_i - T_s)}{\ln (R_o/R_i)} \tag{B7.2}$$

where Q = the total heat loss in Btu/(h · ft^2)
K = the thermal conductivity in Btu · in/(h · ft^2 · °F)
L = the lineal feet involved
T_i = the inside operating temperature in °F
T_s = the outside ambient temperature in °F

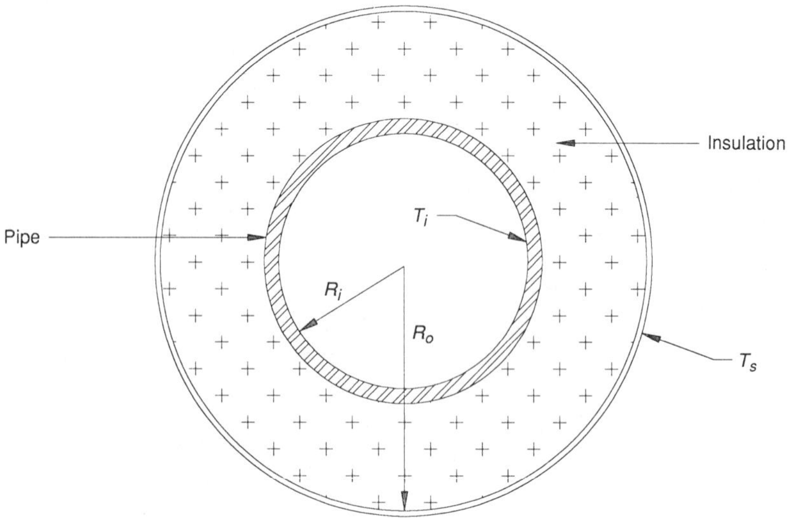

FIGURE B7.1 Cross section of an insulated pipe.

R_i = the bare pipe radius in inches
R_o = the radius to the insulated surface in inches
ln = natural logarithm

Heat transferred through cylindrical, or pipe insulation, geometry with multiple layers of insulation is most commonly represented by Eq. (B7.3) and Fig. B7.2:

$$Q = \frac{(T_i - T_s)}{[R_s \ln (R_1/R_i)/K_1] + [R_s \ln (R_2/R_1)/K_2] + [R_s \ln (R_s/R_2)/K_3] + 1/f} \quad \text{(B7.3)}$$

where Q = the total heat loss in Btu/(h · ft²)
K = the thermal conductivity in Btu · in/(h · ft² · °F)
T_i = the inside operating temperature in °F
T_s = the outside ambient temperature in °F
R_i = the bare pipe radius in inches
R_1 = the radius of the outer surface of the first layer in inches
R_2 = the radius of the outer surface of the second layer in inches
R_s = the outside radius of the outermost layer in inches
ln = natural logarithm
1/f = surface resistance factor

Surface resistance is represented by the inverse of the air film (1/f) factor as seen in Eqs. (B7.1) and (B7.3). When heat flows through a solid material and then out into another atmosphere (usually air), a resistance to heat flow created by the phase change at the interface between the two atmospheres is encountered at the surface separating the solid from the other atmosphere. Therefore, less heat will flow from the surface than if no resistance were offered at this point.

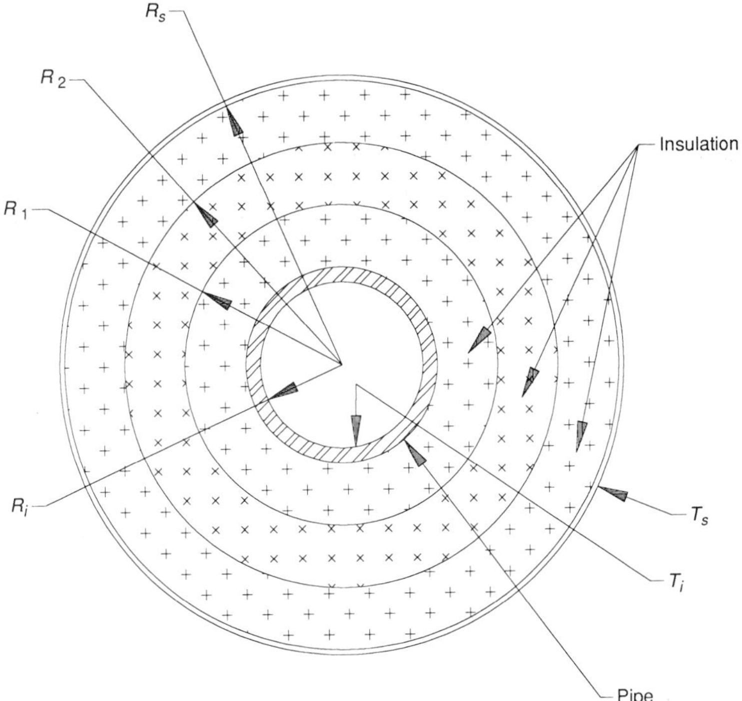

FIGURE B7.2 Cross section of an insulated pipe with multiple layers.

In the case of good conductors of heat, surface resistance is the greater part of the total resistance of heat flow. In connection with efficient insulating materials, however, surface resistance is small compared with the resistance of the materials themselves.

Numerically, surface resistance is the reciprocal of the rate of heat transmission from surface to air. That is, if the rate of heat transmission from surface to air is 2.0 Btu/ft^2 per degree temperature difference per hour, the surface resistance is 0.5 h/ft^2/Btu. A higher rate of heat transmission from the surface indicates a lower surface resistance and vice versa.

Air velocity has a significant impact on surface resistance. Figure B7.3 shows values of surface resistances sufficiently accurate for use in insulation calculations where the surface resistance is usually less than 25 percent and frequently less than 10 percent of the total resistance.

Thermal conductivity is a specific property of a homogeneous material. Its value is not dependent on the area, thickness, or shape of the material. It is a rate and is therefore unaffected by geometry. Heat transmission, however, is dependent upon the geometry of the body to be insulated and it is predominantly governed by the length of the path (thickness).

Thermal conductivity is dependent upon temperature. Standard test methods for determining thermal transmission properties of insulation materials are covered in the following:

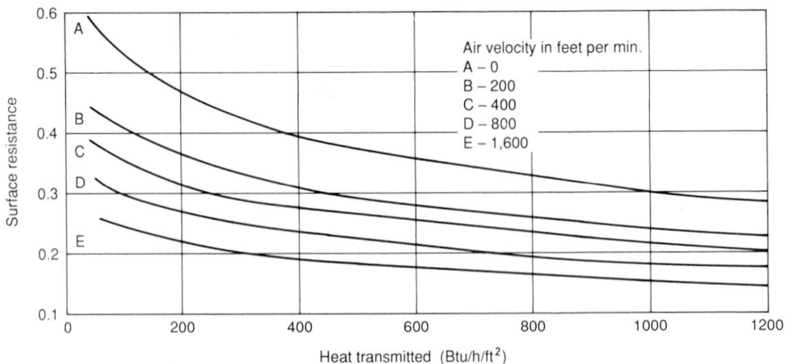

FIGURE B7.3 Surface resistance 1/f at various velocities.

ASTM Standard Test Method C177—Steady-State Thermal Transmission (Properties by Means of the Guarded Hot Plate)

ASTM Standard Test Method C518—Steady-State Thermal Transmission Properties by Means of the Heat Flow Meter

In Table B7.1, conservative thermal conductivity values are shown for several commonly used industrial and commercial insulation materials. The thermal conductivity values are shown as a function of the mean temperature between the inner and outer surfaces of the insulation. This method of expressing thermal conductivities permits their use in the calculation of heat transfer through materials whether used singly or in combination with other materials.

DESIGN PARAMETERS

By providing a medium for retarding heat transfer, thermal insulation offers many useful functions in industrial and commercial piping applications. In specifying an insulation system, it is important to consider the parameters of your process and application needs. These are the whys of insulation system design. Why, or for what purpose, is the pipe going to be insulated? The following are some common design criteria used in insulation system design for piping applications:

- Controlling heat loss on hot piping
- Providing personnel protection
- Providing personnel comfort in commercial buildings
- Reducing heat gain on cold piping
- Limiting or retarding surface condensation
- Providing process control
- Economic optimization or energy conservation
- Providing fire protection

TABLE B7.1 Properties and Limitations of Insulation Materials

Insulation material	Compressive strength (psi)	Maximum temperature (°F)	Minimum temperature (°F)	Permeability (Perm-in)	Thermal conductivity [Btu · in/(h · ft^2 · °F)] (50°F, 200°F)	Flame spread/smoke density index
Calcium silicate	100	1000	250	NA	NA, 0.45	0/0
Cellular glass	100	900	−450	0.005	0.33, 0.45	5/0
Elastomeric foam	NA	200	−40	0.3	0.29, NA	25/50
Fiberglass/ mineral wool	2.5 at 10% 10 at 10%	850 1200	42 42	75 150	0.29, 0.39	25/50
Expanded perlite	70	1000	250	18	NA, 0.55	25/50
Phenolic foam	22 at 10%	300	75	6–7	0.15, 0.25	25/50
Polystyrene foam	25	165	−65	1 to 5	0.23, NA	NA
Polyurethane/polyisocyanurate foams	30	250	−200	2 to 4	0.15, NA	25/50

- Providing freeze protection
- Providing noise control

In many applications these criteria will overlap each other and designing for one condition will benefit by the attainment of another. One example of such overlapping criteria is with controlling heat loss. In designing for a maximum heat loss of a given value, an added benefit may be that the surface temperature is sufficient to provide personnel protection. Another example of this overlapping criteria is condensation control. In humid environments, when insulation is sized according to condensation control parameters, the added benefit will often be an economically optimum design for the reduction of heat gain on the cold pipe.

Environmental, physical, and mechanical conditions play an important role in insulation system design. Indoor applications, for example, generally do not require the complexity of design that outdoor applications require. Similarly, below ambient temperature designs are often more complicated than above ambient temperature designs. The physical abuse and mechanical conditions that an insulation system is subject to are also important to consider in the design process.

The following paragraphs offer a brief explanation of the purpose for considering each of the above items.

Controlling Heat Loss on Hot Piping

The objective behind controlling heat loss on hot piping can be very narrow in scope or very broad, with multifaceted purposes. As already mentioned, insulation in any application serves one primary function, to reduce the heat flow from one surface to another.

Some of the areas that are listed above fall into the category of controlling heat loss on hot piping. These are:

- Providing personnel protection
- Providing personnel comfort in commercial buildings
- Providing process control
- Economic optimization or energy conservation
- Providing freeze protection

One of the most important things to understand about this particular subheading "Controlling Heat Loss on Hot Piping" is the word *control.* Insulation, by itself, does not have the ability to "maintain" or "hold" temperature within a system. Insulation can only provide a means for control. Insulation can limit, retard, reduce, or minimize the rate at which heat flows through or out of the system, but it cannot stop the process. Insulation is merely a resistor to heat flow; it is not a barrier to heat flow.

Insulation *systems*, however, can be designed to maintain a body or a mass at a given temperature. These systems require the use of additional energy input from any of a number of possible sources. Some of the more obvious forms of additional energy input are in the form of heat tracing, pressure, and flow rate. Heat tracing, pressurized systems, and flow-dependent systems will be discussed later in this chapter.

Personnel Protection. When providing for personnel protection, only enough insulation is applied to protect individuals from getting burned by the heat of the pipe surface. Traditionally the upper temperature limit of the surface of the insulation used as a guideline for personnel protection was 140°F. In recent years, however, a more conservative design temperature for personnel protection is 125°F.

The ASTM Standard Guide C1055—Standard Guide for Heated System Surface Conditions That Produce Contact Burn Injuries[2] indicates that the normal metabolism threshold pain level occurs at approximately 111°F. At this point and up to 140°F potential injury is still considered to be reversible. The maximum level of pain occurs at approximately 140°F at which point injury from skin contact may produce irreversible effects. To date there are no mandates or statutes that govern any upper temperature limit for personnel protection; however, many industries have accepted or adopted 125°F as a common practice.

In some design applications where there is clearly no justification for insulation, and insulation could actually be a detriment to the process, fabricated guards are employed to provide personnel protection. In other cases, where guards are impractical and insulation is more appropriate, the insulation is only applied to the piping that is within 7 ft of the ground or platform in high-risk areas. In situations such as this where the insulation does not continue beyond the protected area, it is very important that the insulation system be properly flashed and sealed to prevent water ingress, as shown in Fig. B7.4.

When designing an insulation system for personnel protection, the outer surface conditions become very important. In general, on hot piping, less insulation is required with a painted metal or mastic finish than if a shiny metal finish is used. This is a direct function of the emissivity of the surface material, which is a function of materials' reflectivity and absorptivity. The emittance of a surface material is determined on a scale where a reflective material that is not emitting any infrared energy is rated at 0, and a nonreflective material that is emitting all of its infrared energy is rated at 1. Both of these limits are impractical to attain. Figure B7.5 simplifies the expression of emittance performance as it relates to personnel protection.[3]

In situations where solar loads are high, highly reflective metal jacketing materials reflect much of the radiant heat, thereby creating surfaces which are too hot to touch. Dull textured finishes, such as fabric reinforced mastics, tend to absorb more of the radiant heat (thus it has a higher emittance value), creating a surface condition which is cooler to the touch.

Wind conditions also influence the selection of insulation for personnel protection. In open areas in coastal regions, for example, there is usually a prevailing wind that can be considered in the insulation system design. In this situation, less insulation would be required than if the piping system were in an enclosed space sheltered from the wind, which provides an additional source of heat loss (convection) at the surface. This heat loss cools the surface of the insulation; however, it also decreases the effectiveness of the overall insulation system.

When designing for personnel protection on hot piping, the factors that must be considered are:

1. What are the worst-case ambient temperature and wind conditions the system will be subjected to? Consider the ambient conditions which will create the hottest surface temperature for each application, such as summer weather with no wind and a metal jacketing material. It should be noted that when de-

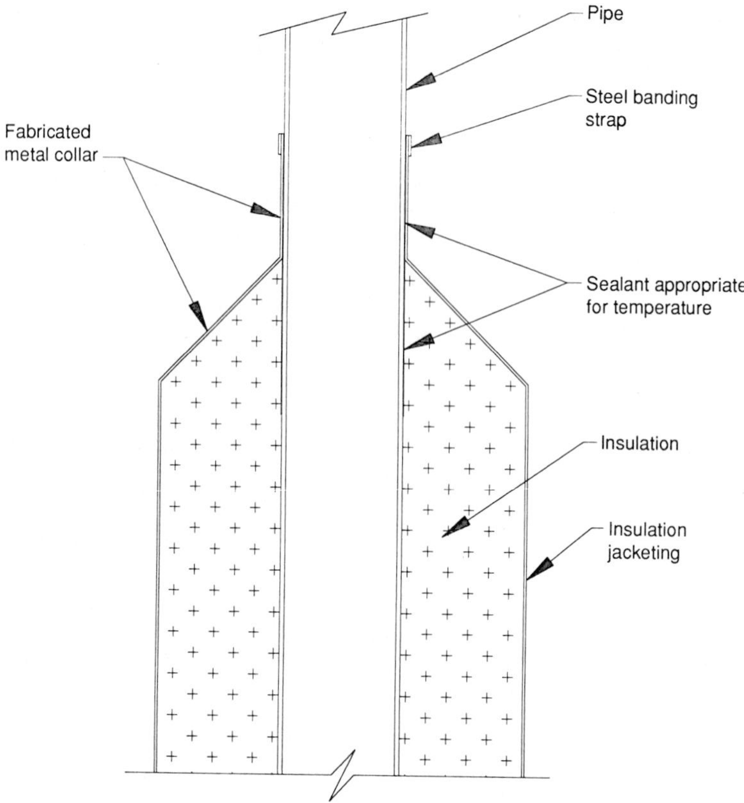

FIGURE B7.4 Flashing and sealing.

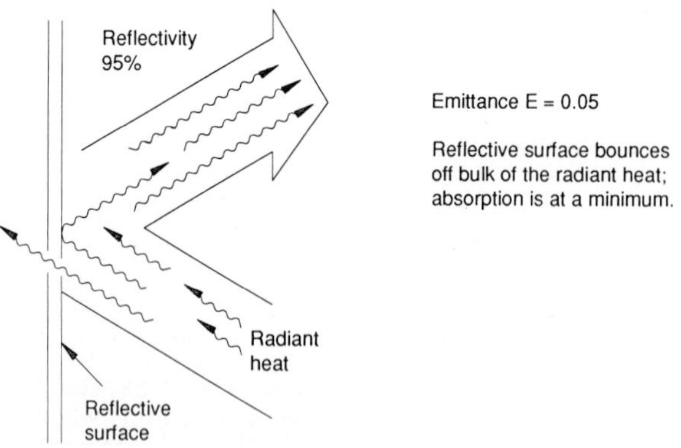

FIGURE B7.5 Emissivity illustration.

signing for worst-case weather conditions, in most applications it is best to use the average worst-case conditions (i.e., average summer weather, average summer wind, etc.). If the absolute worst-case condition is used as the design criteria, the system will usually have an uneconomical and even impractical insulation thickness.

2. What is the risk factor with respect to location of the piping to human contact? Consider the potential of human contact. If there is no opportunity for human contact, insulation for personnel protection may not be necessary. Limited human presence may only require a sign or a fabricated guard. Frequent human presence will require insulation for personnel protection. And frequent human contact will require a thorough investigation into the most efficient insulation system for personnel protection.

Providing Personnel Comfort in Commercial Buildings. In addition to personnel protection, insulation has traditionally been used to provide personnel comfort in enclosed spaces. Heat loss from hot piping in commercial and institutional buildings is primarily an economic consideration. Of course, heat loss into the environment in enclosed spaces can significantly affect the comfort of the personnel occupying the space. However, this heat is usually offset by the climate control systems. For example, in hot summer conditions when air conditioning is required, heat loss into enclosed spaces will significantly affect the load on air compressors and associated climate control equipment. The personnel occupying the space may never actually feel the heat loss from the piping, but it will be evidenced in higher equipment maintenance costs and utility bills.

When designing for personnel comfort in commercial buildings, the factors that must be considered are:

1. What are the worst-case ambient weather conditions the building will be subjected to? Generally when designing mechanical HVAC systems for commercial buildings, the worst-case ambient condition is taken as an average for the seasons. The worst-case in Saskatoon, Saskatchewan, will be an average of the temperature from December through March. Worst case in Miami, Florida, will be an average of the temperature from June through September.
2. What will be the effect on the HVAC systems that will have to offset the heat loss? The additional energy requirement placed on the evaporators, compressors, and water chillers can be substantial. Properly designing and insulating the hot water and chilled water lines in commercial buildings will help to minimize this cost.

Process Control. Process control is a very critical design parameter in many industrial environments. Providing a stable temperature flow throughout a process system is in many cases more important than any other design criteria. For example, in the transport of liquid sulfur through a piping system, it is imperative that the temperature of the sulfur never drop below its freezing point. In this scenario, the time and energy required to get the frozen sulfur into a molten and flowing state again is more expensive than the cost of replacing the transport piping altogether. In another example, providing a uniform temperature heat transfer medium (i.e., steam, synthetic heat transfer fluids, etc.) to a chemical reaction vessel is essential to a proper chemical reaction. Not enough heat or too much heat can completely change or even nullify the chemical reaction. In both of these examples, the cost associated with the improper consideration of process control requirements is significant.

In a paper entitled, "Thermal Insulation Design Concepts,"[4] the following statement relating to process control is made:

> Thermal insulation has a primary function of keeping the process of its equipment and piping hot enough or cold enough to meet operating requirements. This function takes precedence over all others when the following conditions must be controlled:
>
> 1. Process in Elevated Temperature Service:
>   *a.* Excessive reflux condensation in distillation equipment.
>   *b.* Crystallization of solids on equipment or piping walls.
>   *c.* Yield loss or product deterioration from excessive heat input or pressure build-up to compensate for heat loss.

Process control is important in any piping system. However, it is not always the controlling design parameter. For example, a loss of process control in a hot water heating system in a commercial building will probably not cause the building to shut down until the process is stabilized. It may require that the climate control system operate longer and require greater energy input than it would normally have to; however, it could conceivably operate under these conditions indefinitely. As a controlling design parameter, the loss of process control would cause system shutdown, significant product loss, process or product failure, or a significant health and safety hazard.

When designing for process control with respect to heat loss on hot piping, the factors that must be considered are:

1. What is the worst-case ambient temperature the system will be subjected to? For process control purposes, the worst-case ambient temperature is the one which will most adversely effect the process. On hot piping applications, this will be the average winter weather conditions for the geographic location of the facility.
2. What are the temperature limitations of the process being controlled? Evaluate the insulation system based on the amount of temperature change that the process can swing through. This will play an important roll in determining how much insulation is required to control the process.
3. What are the consequences in terms of cost and safety of lost process control? If the system is allowed to extend outside the temperature limitations of the process, what will happen? A good example is steam. When allowed to cool, steam will begin to condense inside the pipe. If allowed to get out of control, this can cause serious process problems. Liquids do not convey as well as gases, the presence of liquids can cause corrosion problems, and it requires more energy to keep the liquid hot than it does to keep the gas hot.

Economic Optimization or Energy Conservation. One of the most common reasons people think of for placing insulation on hot piping is "economic optimization." Actually, however, it is much less common today than in the mid 1970s to see insulation used for this purpose. Prior to the energy crisis of the early 1970s, thermal insulation used on hot piping applications served two primary purposes: personnel protection and process control. These two design parameters are discussed above.

When the cost of energy began to increase in the early 1970s, large energy and oil consumers began to look for means to control their costs. One of the most cost-effective means to control energy costs is to provide an efficient insulation system. An efficient insulation system is theoretically a static system surrounding a body such that the energy requirement to control the heat loss from that body is reduced by a given or predetermined design value.

In order to better understand this concept, it is first useful to examine heat losses from bare surfaces. The rates of heat losses from bare surfaces at temperature differences up to 1000°F are shown in Fig. B7.6. These are average values for still-air conditions, and although there is some variation for different pipe sizes and for different absolute temperatures of surroundings, these variations are small compared to those caused by comparatively low air velocities; therefore, these average values are sufficiently accurate for engineering purposes.

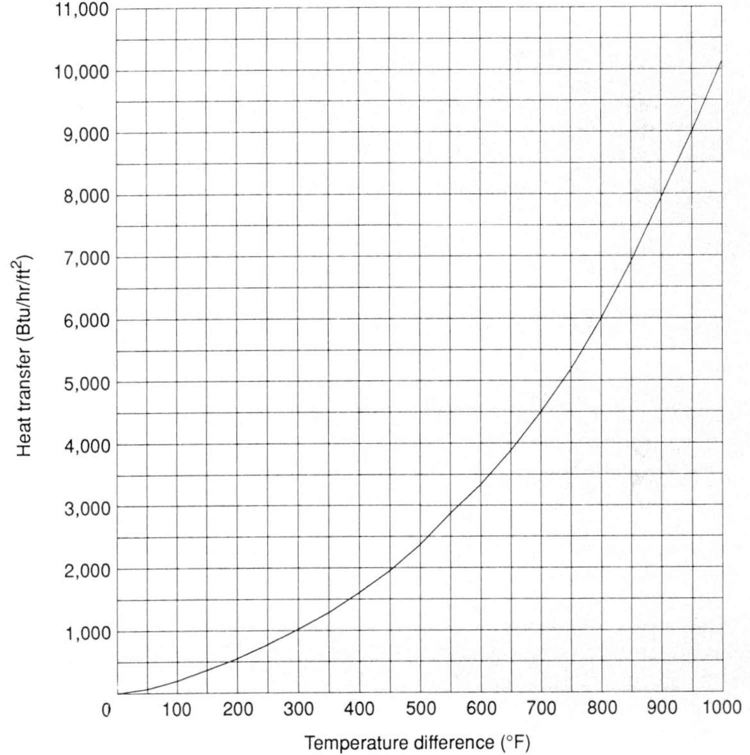

FIGURE B7.6 Heat transfer in still air from bare surfaces at various temperatures.

Heat losses expressed only in Btus are not usually as significant as when expressed in actual costs in dollars. In Fig. B7.7 the equivalent losses in dollars per square foot per year (8760 h) are shown for various temperatures per 1,000,000 Btu (1,000,000 Btu is approximately equivalent to 1000 pounds of steam at 350°F and 150 psig). The value of energy per 1,000,000 Btu is either known or may be

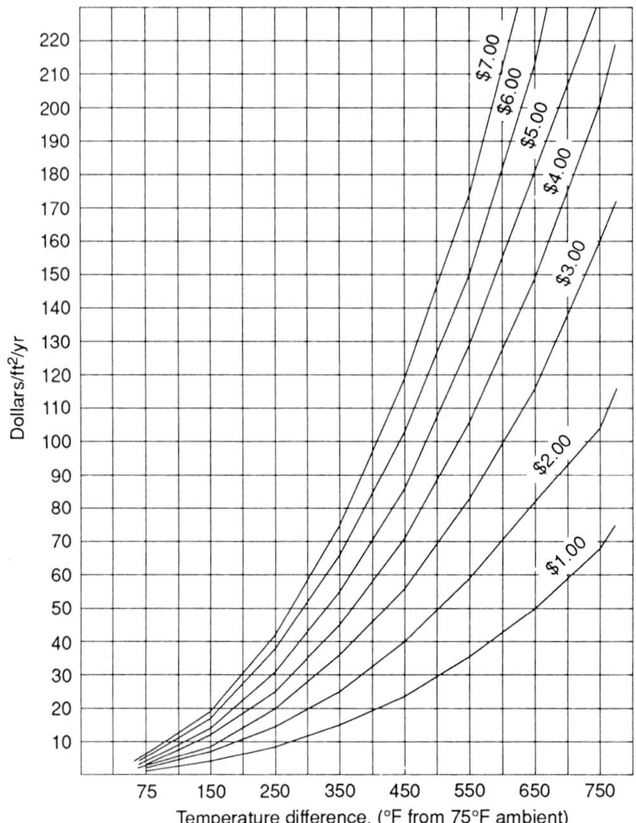

FIGURE B7.7 Equivalent loss in dollars per square foot at various fuel costs per 1,000,000 Btu.

computed readily for a given fuel of known cost and a given efficiency; thus, this procedure renders Fig. B7.7 applicable to a wide range of fuels and conditions.

In order to approximate the most economically optimal insulation thickness, the operating and environmental conditions must first be defined. The operating conditions are the factors that make up the actual process (i.e., pipe size, contents, flow rate, operating temperature). The environmental conditions make up the atmosphere within which the insulated pipe will be operating (i.e., ambient temperatures, wind conditions, annual rainfall, humidity).

Taking as a specific example a square foot of surface area operating 7200 h per year at a temperature of 300°F above that of the surrounding air, the loss per year, at an energy value of $3.50 per 1,000,000 Btu, is $32.00 × 7200/8760 = $26.30. Suitable insulation saving upward of 90 percent of this loss may be applied at a cost considerably less than 1 year's savings when compared to the cost of the energy consumption required with no insulation at all. This illustrates the desirability of insulating such surfaces as boiler drum heads, manways, flanges,

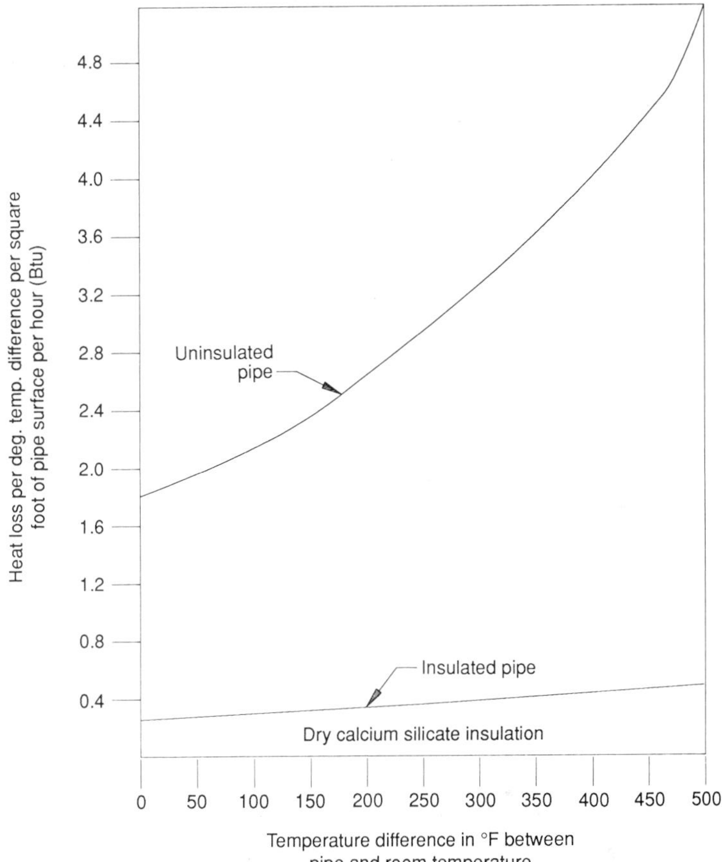

FIGURE B7.8 Heat loss comparison.

and fittings which are frequently left uninsulated, even though adjacent piping is provided with effective insulation.

The magnitude of losses from bare heated surfaces as compared with the relatively small losses from such surfaces when properly insulated is illustrated in Fig. B7.8. In this illustration heat losses per degree temperature difference from bare surfaces are shown by the upper curve and heat losses from insulated surfaces are shown by the lower curve. The area between the two curves represents the savings by insulation.

When designing for economic optimization with respect to heat loss on hot piping, the factors that must be considered are:

1. What are the average ambient temperature conditions that the system will be subjected to? In the case of economic optimization, processes are usually evaluated on an annual basis or on a life-cycle basis if only operated during portions of the year. Taking an average ambient condition over the course of

time during which the process will be in operation will enable the designer to determine the economically efficient insulation thickness.

2. What are the operating temperatures of the process? Each range of temperatures that will be involved in the process should be evaluated on its own criteria. If the process swings through a range of temperatures (cyclical systems), the thickness should be based on the worst-case condition on a time-weighted basis.
3. What is the life expectancy of the system being designed? Life expectancy is a very important variable in the evaluation of economic thickness. A longer life expectancy will justify a more sophisticated insulation system than a quick turnaround project.
4. Will insulation thicknesses specified for economic optimization be a detriment to or an asset to efficient process control? In some cases, the insulation thickness required for economic optimization may not be ideally suited for process control. For example, in some very humid regions, the economic thickness for a chilled water line may not be sufficient to provide adequate condensation control. Another example is hot piping in a quick turnaround project. The economic thickness required to insulate a heat transfer pipe at 650°F for a 2-year life expectancy will be less than what is required to keep the heat transfer fluid at a uniform and stable condition. In both of these examples, the process control condition would prevail over the economic thickness.

Providing Freeze Protection. The final subject, as defined above, with respect to controlling heat loss is insulation system design for the prevention of freezing. Freeze protection can be provided by any of the following, but usually a combination of, three means:

1. Insulation thickness
2. Flow rate
3. Additional heat input

As mentioned earlier, insulation alone cannot "maintain" a temperature. Regardless of the insulation thickness, if a fluid body is stagnant, there is no additional heat input, and the ambient temperature outside the pipe is below the freezing point of the fluid, the fluid will eventually freeze. The thickness of the insulation can slow down the freezing process, but it cannot stop it.

Fluid in motion takes much longer to freeze than stagnant fluid. Therefore, in the design of any freeze protection system the flow rate must be factored in. If there is no flow during significant time intervals, consideration must be given to additional heat input, which is provided by means of various types of heat tracing. Typically these are steam, electric resistance tapes, and heat transfer fluids. Of these three, the most common application for piping purposes is electric resistance tapes. Information on heat input requirements is usually provided as a complimentary service of heat tracing manufacturers and by some, but not all, insulation manufacturers. Refer to Chap. B6 for heat tracing of piping.

When designing to provide for freeze protection, some of the factors that must be considered are:

1. What is the worst-case ambient temperature condition that the system will be subjected to? For freeze protection, the coldest temperature for the longest

period of time must be considered. For example, an underground water pipe that has to traverse a river on the underside of a bridge would have two ambient conditions: the ground temperature at the pipe depth and the winter weather conditions in the geographic location of the bridge. If the bridge is in Houston, Texas, the worst-case condition might be 30°F for 1 day with a 15 mph wind. If the bridge is in Bemidji, Minnesota, the worst-case condition might be −10°F with a 10 mph wind for 20 days.

2. What is the lowest allowable operating temperature of the process? All liquids have different freezing temperatures. For example, water freezes at 32°F, but molten sulfur freezes at 120°F. Both would require freeze protection in Bemidji, but the water line may not require it in Houston.
3. What is the normal operating temperature of the fluid? It is possible in many processes that the normal operating temperature of the fluid is higher than the freezing temperature. In a freeze protection evaluation, the time to reach freezing temperature must be considered in addition to the time to freeze. In the case of the water line in Houston, the normal operating temperature of the water may be 55°F and due to possible drop in temperature, depending on the size of the pipe, it may not even reach 32°F, much less freeze, before the temperature gets back up above 32°F.
4. What are the physical properties (i.e., density, specific heat) of the fluid? It is useful to know the physical properties of the fluid being protected if a detailed analysis is to be performed. For example, a heavy-density fluid will take longer to reach its freezing point than a light-density fluid.
5. What is the flow rate of the fluid? Fluid in motion (with enough pressure behind it) will not freeze. Therefore, in many applications, if the liquid can be kept flowing, the use of insulation for the sole purpose of freeze protection can be eliminated. In most instances, however, it is when the flow is interrupted that the freeze protection is required.
6. What is the maximum downtime that the fluid might remain in a stagnant state? If the maximum downtime that the fluid is stagnant is less than the time required to freeze under the given design conditions, insulation requirements can be reduced or eliminated.

Reducing Heat Gain on Cold Piping

In cold piping applications the main objective of providing insulation is the reduction of heat gain. This process is most often evidenced in terms of providing process control and limiting or retarding surface condensation. The most important controlling factor in the effort to minimize heat gain on cold piping is the prevention of moisture migration, or water intake, into the insulation system. This type of moisture migration will have a dramatic effect on the system performance.

Process Control. As mentioned earlier, in the explanation of process control as it relates to hot piping, process control is often the most important guiding criteria relating to insulation system design. In cold piping applications this statement is even more paramount than it is in hot piping applications. In most cold processes (with the exception of chilled water piping in climate control systems) the maximum allowable heat transfer for process control purposes is 30 to 40 Btu/h/ft^2. The consequences of exceeding this limit are so costly that a safety factor of 4 is frequently employed resulting in a design limitation of 8 to 10 Btu/h/ft^2. By comparison, chilled water piping systems are usually designed around 40 to 50 Btu/h/ft^2 and hot water and steam systems are often designed around 100 Btu/h/ft^2 or more.

In another example, liquefied gases must be kept below their boiling point. This is usually accomplished with a combination of pressure and insulation. If the temperature of the liquid gas is allowed to exceed the process control design parameters, the consequence is either a costly loss of gas through vaporization or a potentially hazardous buildup of pressure.

Cold piping systems are more subject to degradation from the environment than are hot piping systems primarily because of the direction of the vapor driving force in these applications. On hot systems, the water vapor driving force is away from the pipe, and although the ingress of water into the insulation can adversely effect the insulation performance, it is generally considered to be temporary. Conversely, on cold systems, the water vapor driving force is inward toward the pipe. The ingress of water into the insulation system will gradually increase with time, and the moisture will slowly deteriorate the system, eventually destroying it.

For these reasons, it is extremely important that the total insulation system design be very thoroughly thought out to counteract the potential effects of the environment. The use of vapor barrier mastics and low permeability joint sealants is essential to adequate system performance. From a process control standpoint, these materials are equally as important as the insulation itself.

Limiting or Retarding Surface Condensation. Insulation systems can be designed to limit or retard surface condensation, but in most cases, they cannot be designed to "prevent condensation." In some very dry climates, insulation systems can be designed to prevent condensation most of the time; however, even in the driest desert dew settles on the ground in the early morning hours. When dew settles on the surface of an insulation system, it is considered condensation. In humid regions, it is unfeasible to consider designing an insulation system to prevent condensation. In these areas, the insulation thicknesses required of even the most efficient insulations available would be unrealistic from both a financial and a practical standpoint.

Surface condensation can sometimes be a problem, primarily from a maintenance control perspective and, in the long term, from an insulation efficiency standpoint as well. This is particularly a problem where absorptive insulation materials are present. Surface condensation can cause mold and mildew formation, which presents health and safety hazards. When metal jacketing is present, surface condensation can cause pitting and corrosion of the membrane. And, when condensation is allowed to remain for extended periods of time, it can cause membrane degradation, leading to moisture ingress into the insulation.

As with personnel protection, the outer membrane selected for limiting condensation plays an important role in providing good condensation control. The surface temperature of the insulation system is the controlling factor in how often condensation will form and how long it will be present.

When designing to limit or retard surface condensation, the following are some of the factors that should be considered:

1. What are the average summer ambient temperature and wind conditions (not worst case) that the system will be subjected to? For retarding surface con-

densation it is very important to use the average summer conditions and not the worst case. This is because the worst-case ambient weather conditions in the summer months, particularly in the coastal regions, are such that it is unrealistic to try to achieve condensation control. The insulation thicknesses required of even the most efficient insulations systems available are economically unfeasible and impractical for worst-case summer contions.

2. What is the operating temperature of the process? This will have a significant effect on whether or not condensation control needs to be considered. For example, on a cryogenic pipe line the insulation required to provide process control will usually exceed the insulation thickness required to provide condensation control.
3. How important is condensation control in the overall performance of the process? In many applications condensation control is not a particularly important design criteria. Condensation does not present as much of a process problem as it does an aesthetic problem. In indoor or rooftop applications, condensation can be a problem since it presents a building or structural degradation potential.

Providing Fire Protection

As a general rule, insulation materials are better suited for use as thermal insulation than they are as fire protection. However, the National Fire Protection Association (NFPA) and the American Petroleum Institute (API) acknowledge conditions under which some insulation materials may provide "credit" in the design and sizing of pressure relief valves. Accordingly, there are several test methods, some private and some public, which are used to determine the suitability of insulation materials for fire protection applications.

The Flammable and Combustible Liquids Code,[5] NFPA 30, states in section 2-2.5.7 (a) that:

> Insulation systems for which credit is taken shall meet the following performance criteria:
>
> 1. Remain in place under fire exposure conditions.
> 2. Withstand dislodgment when subjected to hose stream impingement during fire exposure. This requirement may be waived where use of solid hose streams is not contemplated or would not be practical.
> 3. Maintain a maximum conductance value of 4.0 Btu's per square foot per degree F (Btu/hr/sq ft/°F) when the outer insulation jacket or cover is at a temperature of 1,660°F (904.4°C) and when the mean temperature of the insulation is 1,000°F (537.8°C).

API Recommended Practice 521[6] requires the same basic performance criteria expectations; however, it includes the following time limitation:

> Section 3.12.2.1 . . . This period of exposure may range from 20 minutes to 1 hour, depending on the adequacy of fire-fighting provisions, the accessibility of equipment, and the degree of skill and training of the fire-fighting group. . . .

In item 1 above, the requirement is that the insulation remain in place under the fire exposure conditions. This is usually accomplished by using a stainless steel jacketing material, of sufficient thickness (generally 0.015 in or greater) to withstand the flame intensity, with stainless steel banding and matching stainless steel clips.

Item 2 above is the hose-stream dislodgment criteria. This criteria is not well defined in any text on this subject; however, API Recommended Practice 521[6] Section 3.12.2.2 states the following:

> The finished installation should ensure that the insulation will not be dislodged when subjected to the fire-water streams used for fire fighting, such as streams from hand lines or monitor nozzles, if installed.

Most insulations used in fire protection applications can withstand this criteria if jacketed and banded in place as described in the explanation of item 1 above.

Item 3 above requires a maximum conductance of 4.0 Btu's per square foot per degree F [Btu/(h · /ft^2 · °F)]. Any insulation material suitable for use as fire protective insulation can meet this criteria.

Prior to specifying any insulation or accessory product material in a fire protection application, it is important to consult with the technical service department of the manufacturer and be advised of any special precautions that may be necessary.

Noise Control

Environmental acoustics is something that can be addressed in thermal insulation system design; however, serious noise problems should be treated as a separate and independent study. For the purpose of preliminary investigation it is useful to know some basic concepts.

Sound absorption is both a process and a property of materials. It is the process of dissipating sound energy and the property possessed by materials, objects, and structures such as rooms of absorbing sound energy. Sound attenuation is the reduction of the intensity of sound as it travels from the source to a receiving location. Refer to ASTM C634-89,[2] Definitions of Terms Relating to Environmental Acoustics.

In thermal insulation design sound attenuation is a natural by-product of the insulation application. Some insulation materials and accessory products provide greater sound attenuation because of their sound absorption characteristics than do others. One of the best thermal insulation materials for sound attenuation is mineral wool. It is available in loose fill, rolls, blankets, boards, and preformed shapes and therefore can be applied in numerous different applications where sound absorption is a desired property of the insulation that is being applied.

The jacketing material used to cover the insulation can play an important role in sound attenuation. For example, a fabric reinforced mastic finish over insulation has better sound absorption properties than does a metal jacket. The environment in which the materials are being used will limit their ability to be considered for sound attenuation characteristics in some cases; however, with composite system design in the planning stage of the process, a combination of materials can often be arranged to provide both thermal and acoustical benefits.

DESIGN CONDITIONS

In addition to defining the purpose for the insulation system, it is also important to define the conditions under which the insulation system will be used. Some conditions to be considered are as follows:

- Indoors or outdoors
- Conditioned or nonconditioned space
- Geographic location (coastal regions, northern climes, southern climes, rainy, dry, etc.)
- Long straight runs or frequent bends
- Personnel traffic area or unaccessible
- Above or below ground

There are numerous types of conditions or combinations of conditions that require consideration in insulation system design. The above list represents just a few of the more obvious ones. Most of these are self-explanatory; however, some of them require some attention to detail. A brief explanation of each item follows.

Indoors or Outdoors

Indoor applications, in general, are much simpler to specify than outdoor applications. The main reason for this difference is that indoor applications are not subjected to rain, snow, or solar loads. Indoor applications are subjected to vapor pressure differential problems as are outdoor ones; however, this problem is not complicated by the other environmental difficulties as are the outdoor applications. Rain, snow, and solar loads are a problem with insulation systems because of moisture migration. The single most detrimental element in any insulation system is the migration of moisture into the insulation system. Water in insulation destroys the insulating value. Approximately 4 percent moisture by volume in an insulation material will increase the thermal conductivity by as much as 70 percent.[7]

Thermal insulations must be dry to function according to design. Water in the insulation causes part or all of the air or gas spaces in the insulation to be filled with water. Its conductivity then approaches that of water instead of the air or gas. The thermal conductivity of water at 70°F is 4.1 Btu · in/(h · ft^2 · °F) as compared to 0.17 Btu · in/(h · ft^2 · °F) for that of air. The heat transmission, therefore, is about 24 times greater for water-saturated insulation than it is for dry insulation.[1]

In addition to this significant drop in thermal performance of the insulation, water in insulation when in contact with metal piping can significantly increase the chance of, and even contribute to, severe corrosion problems. Water and snow, therefore, should be avoided and every precaution should be taken to prevent moisture entry into the system. This is another reason why solar loads present problems in insulation systems that are outdoors.

As mentioned in the section addressing personnel protection, solar loads on the surface of the insulation can significantly increase the surface temperature of the insulation, therefore increasing the risk of injury.

In addition to increasing surface temperature, solar loads on the surface of the insulation cause expansion and contraction of metal jacketing and premature aging of mastic finishes. The expansion and contraction of metal can cause joints to open up or "fishmouth," creating paths for water to enter. It can also cause a gradual loosening up of the bands, leading to a movement of the jacket or loss in high-wind conditions. Premature aging of mastic finishes can lead to cracks or pits in the finish, which also creates a path for water migration into the insulation system.

When designing for indoor or outdoor applications, consideration must be given to the environmental conditions that the system will be subjected to.

Conditioned or Nonconditioned Space

This differs slightly from that of indoor and outdoor in that either space would be considered indoor; however, nonconditioned space is not climate controlled. This provides the advantage of not having to worry about liquid water or solar loads; however, the vapor pressure driving force is still present. This driving force can have a significant detrimental effect on chilled water and other low-temperature piping insulation systems. Water vapor tends to migrate in the direction of the coldest surface. Therefore, any below ambient temperature piping application will have a positive water vapor pressure in the direction from the outside of the insulation to the surface of the pipe. Example B7.1 illustrates how this might occur on a chilled water pipe in the unconditioned space of a commercial building in a relatively temperate climate.

Example B7.1. Chilled Water Pipe Insulation Water Intake. In this illustration, a calculation will be made to show how a typical chilled water pipe insulated with 2 in of polyurethane insulation might become saturated with water simply by the water vapor driving force generated by the ambient conditions around the pipe and the operating conditions of the pipe.

Calculation for the Amount of Condensation per Linear Foot of Piping per Year

Outside diameter (O.D.) of pipe	(NPS 2 assumed) 2.375 in
Service temperature	41°F
Ambient air temperature	70°F
Ambient air relative humidity	70%
Insulation thickness	2-in nominal polyurethane
O.D. of polyurethane insulation	6.63 in (refer to ASTM C588-89[2])
Outside finish	All service vapor retarding jacket (ASJ; formerly vapor barrier jacket): permeance = 0.6 perm
Permeability of polyurethane	4.321 perm · in

Water vapor flow:

$$G = \frac{\Delta P}{R_{\text{vapor}}} \tag{B7.4}$$

where G = water vapor flow
ΔP = water vapor pressure difference, in inches of mercury (inHg)
R_{vapor} = resistance to vapor diffusion for any given material, in 1/perm

$$R_{\text{vapor}} = \frac{\text{thickness}}{\text{permeability}} \; \frac{\text{in} \cdot \text{ft}^2 \cdot \text{h} \cdot \text{inHg}}{\text{grain} \cdot \text{in}}$$

$$R_{\text{vapor}} = \frac{1}{\text{permeance}} \; \frac{\text{ft}^2 \cdot \text{h} \cdot \text{inHg}}{\text{grain}}$$

Calculation for Pressure Drop Across the Insulation System

P_r in ambient air = the real partial pressure of water vapor in air = $R_H \times P_s$

P_s = the partial pressure of water vapor in saturated air

R_H = relative humidity = 0.7

At 70°F, P_s = 0.73964 in Hg[8]

P_r = 0.7 (relative humidity of 70%) × 0.73964 inHg

= 0.517748 inHg

So with a real pressure at 70 percent relative humidity of 0.517748 inHg, the dew point temperature t_s is found to be t_s = 59.8°F

At 41°F,

$$P_s = 0.25765 \text{ inHg}$$

Therefore $$\Delta P = P_r - P_s = 0.517748 \text{ inHg} - 0.25765 \text{ inHg}$$

$$= 0.26010 \text{ inHg}$$

From here, the actual thickness of insulation on the insulated pipe must be converted into an equivalent thickness of insulation in flat plane geometry in order to be used further in the calculation:

$$\text{Equivalent thickness} = \frac{\text{O.D}}{2} \log_e \frac{\text{O.D. insulation}}{\text{O.D. pipe}}$$

$$= \frac{6.63}{2} \log_e \frac{6.63}{2.375}$$

$$= 3.403 \text{ in}$$

Calculation for Total Resistance Against Water Vapor Diffusion. Determining the total resistance against water vapor diffusion is simply a matter of adding up the total resistances from materials in the path of the vapor flow. This is calculated as follows for the vapor retarder:

$$R_{\text{vapor}} = \frac{1}{0.6} = 1.667 \frac{1}{\text{perm}}$$

and for thermal insulation:

$$R_{\text{vapor}} = \frac{3.403}{4.321} = 0.788 \frac{\text{in}}{\text{Perm} \cdot \text{in}}$$

Therefore, the total resistance is

$$R_{\text{vapor}} = R_{\text{vapor}}(\text{vapor retarder}) + R_{\text{vapor}}(\text{thermal insulation})$$

$$R_{\text{vapor}} = 1.667 \frac{1}{\text{perm}} + 0.788 \frac{1}{\text{perm}}$$

$$R_{\text{vapor}} = 2.455 \frac{1}{\text{perm}}$$

Now, knowing ΔP and R_{vapor}, solving for G (water vapor flow) is as follows, using Eq. (B7.4):

$$G = \frac{0.26010}{2.455} = 0.1066 \text{ inHg perm}$$

And since

$$1 \text{ perm} = 1 \frac{\text{grain}}{\text{ft}^2 \cdot \text{h} \cdot \text{inHg}}$$

$$= 0.1066 \frac{\text{grain}}{\text{ft}^2 \cdot \text{h}}$$

At this point the total vapor flow has been calculated and determined to be

$$0.1066 \frac{\text{grain}}{\text{ft}^2 \cdot \text{h}}$$

The next step is to calculate the outside surface area in order to express the results in terms of square feet per lineal foot of insulation.

Calculation for the Outside Surface of the Insulation Expressed in ft² per Lineal Foot

$$6.63 \times (\pi) \times 12 = 249.945 \text{ in}^2\text{/lineal ft}$$

or

$$\frac{249.945}{144} = 1.735 \text{ ft}^2\text{/lineal foot}$$

Calculation for the Migration of Water Vapor that Condenses in the System per Year and per Lineal Foot of Piping

Total water migration = water vapor flow (G) × surface area per lineal foot × hours per day × days per year

$$= 0.1066 \times 1.736 \times 24 \times 365$$

$$= 1611.88 \frac{\text{grain}}{\text{year/lineal foot}}$$

Since

$$1 \text{ grain} = 1 \text{ lb}/7000$$

and

$$1 \text{ lb} = 16 \text{ oz}$$

Therefore,

$$\frac{1611.88 \times 16}{7000} = 3.685 \frac{\text{grain lb oz}}{\text{year/lineal ft grains/lb}}$$

Or, the total amount of water that can pass through the existing water vapor retarder and the thermal insulation system in this illustration equals

$$3.685 \frac{\text{oz}}{\text{year/lineal foot}}$$

As represented in this example, a large amount of moisture can be accumulated in an insulation medium over a relatively short period of time if the insulation system is not designed to prevent the ingress of water vapor. Even with the best application procedures, incorrect materials selection for the environment or service can have very dramatic results in terms of performance losses and maintenance costs.

Geographic Location

It is very important to factor the geographic location of the system being insulated into the design process. The National Weather Service has information available for climates all over the world. Seasonal averages, worst-case conditions, annual rainfall, average and worst-case humidities, and any other climatic data that is necessary should be fully evaluated during the specification process.

Long Straight Runs or Frequent Bends

The layout of the piping with respect to bends and straight runs needs to be considered for the purpose of expansion and contraction control. Depending on the temperature range of the process piping being insulated, expansion and contraction joints in the insulation system may need to be employed.

The differential expansion between the metal substrate and the insulation can be of critical significance, depending on the length of the pipe run. Table B7.2 lists different metals and their linear expansion and contraction rates for various temperature ranges. It is clear that on long runs of piping provision must be made for differential expansion between the insulation and the pipe. This is accomplished through the use of control or expansion joints. There are many different types of control joints that are used in insulation specifications. One system for controlling contraction on cold piping is illustrated in Fig. B7.9. It is advisable to consult the insulation manufacturer for more specific application-dependent recommendations.

Personnel Traffic Area

High traffic areas require insulation that can take abuse. Although it is standard practice to install "Do Not Walk on Pipe" signs anywhere that this might occur, the real world situation is that people walk on the insulated pipes. It is important,

TABLE B7.2 Thermal Expansion and Contraction Properties

Operating temperature (°F)	Inches/100 ft						
	Steel	Stainless steel	Copper	Cellular glass	Polyurethane insulation	Calcium silicate	Mineral wool
−200	−1.62	−2.51	−2.44	−1.50	−16.20	*	**
−100	−1.12	−1.76	−1.70	−0.90	−10.20	*	**
0	−0.50	−0.77	−0.73	−0.40	−4.20	*	**
200	0.99	1.46	1.51	0.70	7.80	*	NIA
400	2.70	3.80	3.89	1.90	*	NIA	NIA
600	4.60	6.24	6.40	3.00	*	Shrinks[1]	Shrinks[2]
800	6.70	8.80	*	4.00	*	Shrinks	Shrinks
1000	8.89	11.48	*	*	*	Shrinks	Shrinks

*Material not recommended for use in this temperature range.

**Material requires special precautions for use in this temperature range. Consult manufacturer for recommendations.

NIA no information available.

[1]Calcium silicate experiences a maximum of 2 percent shrinkage at 1200°F and less than that at lower temperatures. Calcium silicate manufacturers recommend that their materials be double layered for operating temperatures greater than 500°F due to shrinkage cracks that form.

[2]ASTM C547 Standard Specification for Mineral Fiber Preformed Pipe Insulation indicates linear shrinkage as below 2 percent in accordance with the recommended temperature limits. Vitrification and compaction of binder materials may occur at elevated temperatures.

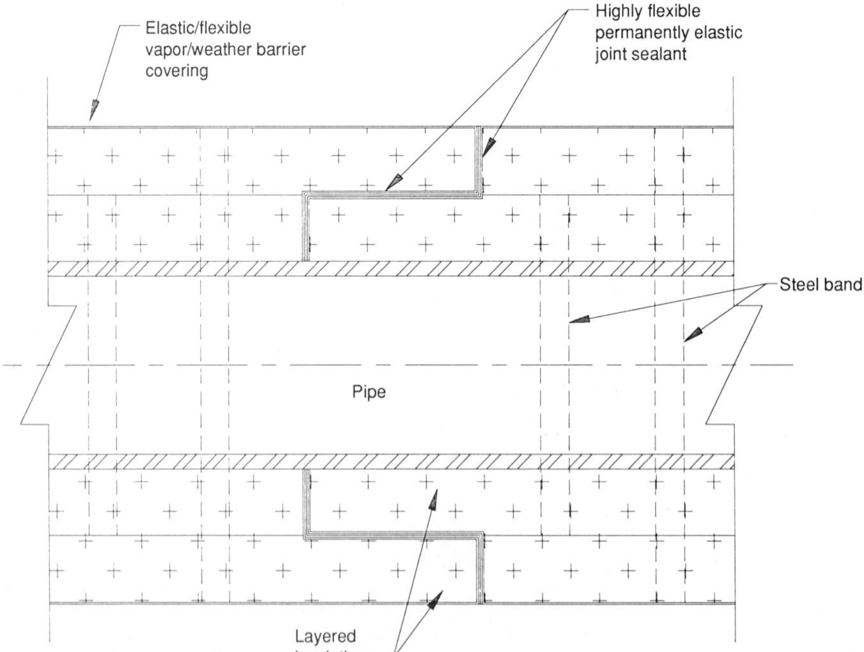

FIGURE B7.9 Cross section of a contraction control joint.

therefore, to consider the use of high compressive strength insulations and/or heavy duty jacketing or membrane covering in these areas.

Above or Below Ground

Above ground insulation applications can be complicated and require thorough design evaluation and analysis for proper specification. Below ground applications are very complicated and mistakes can be extremely costly. It is absolutely essential that every possible condition be considered and a thorough analysis of all variables be accounted for. The two most common means for accommodating underground piping applications that require thermal insulation are:

- Precast trenches
- Direct burial

Precast Trenches. Precast trench systems provide a tunnel for the piping to be channeled through. In this type of application, the insulation system is generally designed exactly as it would be for an above ground application. No special precautions are made for expansion and contraction of the pipe because it is allowed to move freely within the trench as any above ground piping system would. The most important factor to keep in mind when designing an underground precast trench system is ground water.

Excessive ground water infiltration into the trench can cause the piping system to become completely surrounded by water for extended periods of time. This can cause irreparable damage to protective membranes to the point where the insulation becomes totally saturated and literally falls off of the pipe. Also, the condition can lead to serious corrosion problems. Consideration should be given to the use of an impermeable type of insulation material with sealant, adhesive, and covering materials appropriate for the service temperature of the piping being insulated.

Direct Burial. The four most common types of systems employed in direct burial thermal insulation applications are:

1. *Preinsulated:* Preinsulated piping systems for underground direct burial applications generally consist of either plastic or steel piping which is covered with a foamed-in-place polyurethane insulation and covered with a high impact resistant formed plastic covering.

This type of application would look in cross section very much like the cross section shown in Fig. B7.1. These types of systems are most commonly employed in hot water and chilled water piping services. They are not recommended for use on steam lines due to the temperature limitations of the insulation. They can be used on condensate return lines. However, caution must be exercised to avoid overheating. Some condensate return lines are used to evacuate steam which might be released from the primary steam supply line if there is a pressure-relief requirement. If this occurs one time on a preinsulated piping system, it can permanently affect the performance of the insulation and its protective coverings.

2. *Preinsulated conduit with annulus air space:* Preinsulated conduit with annulus air space (conduit systems) has for many years been the primary system used by the U.S. government for steam lines. Thorough analysis of underground direct buried piping systems during the late 1950s and the 1960s revealed that al-

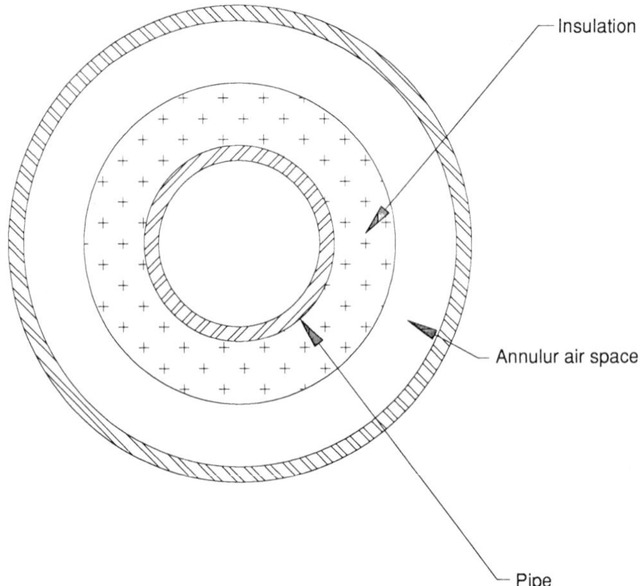

FIGURE B7.10 Cross section of a conduit system.

most no insulation system performed well. The reason was that ground water infiltrated the insulation materials and vaporized on contact with the hot piping, deteriorating the insulation. This led to the development of a system that provided an air space which theoretically allowed any moisture which might enter the system to vent out of the system, allowing the insulation to dry out after saturation. Figure B7.10 shows a cross section of a conduit system.

In actual application, it has been found that the annulus air space is relatively ineffective in providing good moisture venting. However, these systems are still employed in many areas where high-temperature piping must be run underground and the chance of ground water penetration is a concern.

3. *Field installed:* Field installed insulation systems are one of the most cost effective ways to insulate high-temperature lines which must be run underground. The most common material used for field installation is cellular glass, which can be used at operating temperatures up to 900°F. It has a closed cell glass structure which resists penetration from ground water. The cellular glass is usually covered with an asphaltic membrane and sealed with sealants appropriate for the temperature range of the service piping.

4. *Pour-in loose fill:* The most common type of pour-in loose fill insulation for underground applications is calcium carbonate powder. This type of loose fill insulation is poured directly into the trench after the pipe is in position. The pipe is propped up off the ground with wooden blocks to allow the calcium carbonate powder to fill in underneath. The powder is then tamped into place with a mechanical tamper. Plastic bag insulation containers are placed on top of the calcium carbonate prior to backfilling to help prevent the migration of water into the insulation.

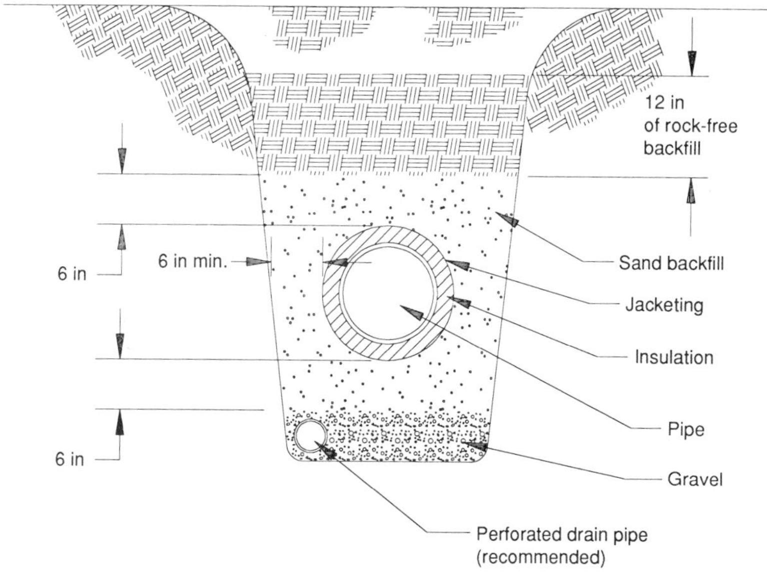

FIGURE B7.11 Backfill detail.

With any of the first three direct burial piping systems, it is very important to consider the location of all pipe anchors, in-line guides, and expansion loops. A thorough pipe-stress analysis must be performed to avoid uncontrolled pipe movement which will destroy the underground thermal insulation system. It is also extremely important to provide good drainage for the piping to try to minimize the potential of ground water infiltration. Figure B7.11 illustrates this type of drainage recommendation. It is best to consult individual manufacturers for their recommendations with respect to application and design.

Heat Flow Calculations

Having the above design parameter information available will allow the specifying engineer to make an educated decision about the thickness for the insulation system used. There are a number of useful sources for information on calculating insulation thickness requirements for various applications. A few of the more common sources for this type of information are:

- *The ASHRAE Fundamentals Handbook*[8]
- *Annual Book of ASTM Standards Volume 4.06 Thermal Insulation and Environmental Acoustics*[2]
- *The Thermal Insulation Handbook*[9]

Following are some examples of heat flow calculations for insulated piping. Example B7.2 shows how the heat gain would be calculated on a typical chilled water pipe in an enclosed space (representing commercial or institutional type construction).

Example B7.2. Typical Chilled Water Pipe Calculation

Pipe size	NPS 2 (2.375-in actual O.D.)
	2.375-in actual O.D.
Operating temperature	42°F
Ambient temperature	95°F
Insulation thickness	2 in nominal
	2.11 in actual
Insulation type	Cellular glass
Length of pipe	120 lineal feet

Using Eq. (B7.2),

$$Q = \frac{2\pi KL(T_i - T_s)}{\ln(R_o/R_i}$$

where $K = 0.33$ Btu · in/(h · ft² · °F) = 0.0275 Btu/(h · ft · °F) (at 50°F mean temperature)

$T_i = 42°F$
$T_s = 95°F$
$R_i = 2.375/2$ in
$R_o = 6.62/2$ in
$L = 120$ ft

$$Q = \frac{2\pi[0.0275 \text{ Btu/(h} \cdot \text{ft} \cdot {}^\circ\text{F})](120 \text{ ft})(42^\circ\text{F} - 95^\circ\text{F})}{\log_e[(6.62/2) \text{ in}/(2.375/2) \text{ in}]}$$

$$Q = \frac{6.2832(0.0275 \text{ Btu/h})(120)(-53)}{\log_e\left(\dfrac{3.31}{1.1875}\right)}$$

$$Q = \frac{-1098.93 \text{ Btu/h}}{1.025} = -1072.13 \text{ Btu/h}$$

Next, to determine the relative heat transfer with respect to area, the following calculation is used:

$$\text{Circumference} = (\pi)d = 3.14159(6.62 \text{ in}) = 20.80 \text{ in}$$

$$\text{Area} = \text{circumference} \times \text{length} = 20.80 \text{ in} \times (120 \text{ ft} \times 12 \text{ in/ft})$$

$$= 29{,}952 \text{ in}^2$$

$$\frac{29{,}952 \text{ in}^2}{144 \text{ in}^2/\text{ft}^2} = 208 \text{ ft}^2$$

So

$$\frac{-1072.13 \text{ Btu/h}}{208 \text{ ft}^2} = -5.15 \text{ Btu/h} \cdot \text{ft}^2$$

In this calculation, the minus sign in the final result indicates that the heat flow is in the direction of the pipe rather than away from the pipe. Therefore, there is a 5.15-Btu/h · ft^2 heat gain in this illustration. This is much greater than what would typically be employed for an application of this type because chilled water piping cannot usually create a serious process upset condition. The thickness selected, therefore, is greater than is necessary and further exploration of thickness requirements is necessary.

The minimum thickness for cellular glass on a NPS 2 pipe is 1 in (specified by cellular glass insulation manufacturers as a limiting factor for handling and durability). The calculation performed again using 1 in of cellular glass instead of 2 in reveals the following:

$$R_i = \frac{2.375}{2 \text{ in}}$$

$$R_o = \frac{4.50}{2 \text{ in}}$$

$$Q = \frac{2\pi[0.0275 \text{ Btu/(h} \cdot \text{ft} \cdot {}^\circ\text{F)}](120 \text{ ft})(42^\circ\text{F} - 95^\circ\text{F})}{\ln [(4.50/2) \text{ in}/(2.375/2) \text{ in}]}$$

$$Q = \frac{6.2832(0.0275 \text{ Btu/h})(120)(-53)}{\ln [(2.25)/(1.1875)]}$$

$$Q = \frac{-1098.93 \text{ Btu/h}}{0.63908} = -1719.55 \text{ Btu/h}$$

Next, to determine the relative heat transfer with respect to area, the following calculation is used:

$$\text{Circumference} = (\pi)d = 3.14159(4.50 \text{ in}) = 14.14 \text{ in}$$

$$\text{Area} = \text{circumference} \times \text{length} = 14.14 \text{ in} \times (120 \text{ ft} \times 12 \text{ in/ft})$$

$$= 20{,}362 \text{ in}^2$$

$$\frac{20{,}362 \text{ in}^2}{144 \text{ in}^2/\text{ft}^2} = 141.4 \text{ ft}^2$$

So

$$\frac{-1719.55 \text{ Btu/h}}{141.4 \text{ ft}^2} = -12.16 \text{ Btu/h} \cdot \text{ft}^2$$

A heat gain on chilled water piping of 12.16 Btu/h · ft^2 is still conservative with respect to process control. The minimum thickness for this application therefore equals 1 in of cellular glass insulation.

In Example B7.3 a calculation of heat loss on a typical steam piping application is shown. In this illustration the following hypothetical conditions will be used:

Example B7.3. Typical Steam Pipe Calculation

Pipe size	NPS 6 6.625-in actual O.D.

Operating temperature	400°F
Ambient temperature	75°F
Insulation thickness	2 in nominal
	2.11 in actual
Insulation type	Calcium silicate
Length of pipe	75 lineal feet

Using Eq. (B7.2),

$$Q = \frac{2(\pi)KL(T_i - T_s)}{\ln(R_o/R_i)}$$

where K = 0.501 Btu · in/(h · ft^2 · °F) (at 165°F mean temperature from Table B7.11)
T_i = 400°F
T_s = 75°F
R_i = 6.625/2 in
R_o = 10.75/2 in

$$Q = \frac{6.29\,[0.501\text{ Btu/(h}\cdot\text{ft}\cdot{}^\circ\text{F)}](75\text{ ft})(400^\circ\text{F} - 75^\circ\text{F})}{\{\ln[(10.75/2)\text{ in}/(6.625/2)\text{ in}]\}12}$$

$$Q = \frac{76812.69\text{ Btu/h}\cdot\text{ft}}{0.484} = 13204.3\text{ Btu/h}$$

Therefore,

$$\frac{13204.3\text{ Btu/h}}{75\text{ ft}} = 176.06\text{ Btu/h}\cdot\text{ft}$$

$$\text{Circumference} = (\pi)d = 3.14159(10.75\text{ in}) = 33.77\text{ in}$$

Therefore, the total surface area per lineal foot of insulation is equal to

$$\text{Circumference} \times 12\text{ in/ft} = 405.27\text{ in}^2\text{/lineal foot}$$

and

$$\frac{405.27\text{ in}^2}{144\text{ in}^2\text{/ft}} = 33.72\text{ ft}^2\text{/ft}$$

$$\text{Heat loss per square foot} = (176.06\text{ Btu/h}\cdot\text{ft})/(2.81\text{ ft}^2\text{/ft})$$

$$= 62.65\text{ Btu/h}\cdot\text{ft}^2$$

For most hot processes, both from a process control perspective and accounting for personnel protection, a desirable heat loss range is between 100 Btu/h · ft^2 and 150 Btu/h · ft^2. Since 62.65 Btu/h · ft^2 is well below this rough guideline, it becomes necessary to perform the calculation again. Solving for the same parameters with 1.5 in of calcium silicate instead of 2 in reveals the following:

$$Q = \frac{6.29\ [0.501\ \text{Btu/(h} \cdot \text{ft} \cdot {}^\circ\text{F)}](75\ \text{ft})(400^\circ\text{F} - 75^\circ\text{F})}{\{\log_e[(9.62/2)\ \text{in}/(6.625/2)\ \text{in}]\}12}$$

$$Q = \frac{6390.88\ \text{Btu/h} \cdot \text{ft}}{0.373} = 17134.01\ \text{Btu/h}$$

Therefore

$$\frac{17134.01\ \text{Btu/h}}{75\ \text{ft}} = 228.45\ \text{Btu/h} \cdot \text{ft}$$

Solving for the surface area per lineal foot of insulation,

$$\text{Circumference} = (\pi)d = 3.14159(9.62\ \text{in}) = 30.22\ \text{in}$$

$$\text{Area per linear foot} = \text{circumference} \times 12\ \text{in/ft} = 362.66\ \text{in}^2/\text{lineal foot}$$

$$= \frac{362.66\ \text{in}^2}{144\ \text{in}^2/\text{ft}} = 2.51\ \text{ft}^2/\text{ft}$$

Therefore it is shown that the heat loss per square foot of surface area with the minimum amount of calcium silicate insulation = (228.45 Btu/h · ft)/(2.51 ft^2/ft) = 90.71 Btu/h · ft^2.

SERVICE CONSIDERATIONS

One other area that needs to be considered when properly designing an insulation system is the service that the piping is providing. This is very important in designing insulation systems because of the different physical properties of the contents of the pipe. The following are some, but not all, generalized service types that are common to industrial and commercial construction:

- Hot water and chilled water
- Steam and condensate return
- Heat transfer fluids
- Hot oils
- Liquefied gas (cryogenic service)
- Sanitary and sewerage water

Hot Water and Chilled Water

Hot water and chilled water lines are generally employed in commercial and institutional facilities as a means of providing climate control. The heated or cooled water is transported through a pipe loop system from the mechanical facilities room of the building or buildings and is used as a heat transfer medium to provide either heating or cooling.

Geographic location is a very important consideration for this type of service. Insulation systems for chilled water piping in Florida are very susceptible to problems with moisture infiltration. Insulation system design in this environment is of critical importance. The correct application of closed cell impermeable insulation materials and highly flexible joint sealants with a vapor retarding membrane is necessary to the success of the entire mechanical system operation. Incorrect application and specification in an environment like this can cost thousands of dollars in mechanical system maintenance and repair, water damage from soggy dripping insulation, and high utility costs from overworking the mechanical systems. In very dry, or arid, regions of the world critical attention to the materials is not as paramount. Water vapor is not as great a threat.

Steam and Condensate Return

Steam and condensate return lines will operate without insulation. Therefore, the chance of a process failure or process upset condition caused by a poorly insulated steam line is rather unlikely. However, the savings that can be achieved by properly insulating steam lines are substantial. This is graphically illustrated in Fig. B7.8 in a heat loss comparison of an uninsulated surface to an insulated surface.

Before the energy crisis of the mid-1970s most steam lines that were insulated were done so without consideration of energy costs, the main purpose being to protect personnel from the extreme temperatures of the surface of a bare steam line. These personnel protection thicknesses, many of which are still in use today, are not adequate for providing optimum energy conservation. One technique used somewhat commonly today to increase energy conservation is often referred to as retrofitting the existing insulation system. A retrofit procedure involves the use of composite insulation systems and/or materials. To retrofit in this sense of the word is to take the existing insulation system that may have been underdesigned at the time and add an extra layer of insulation.

There are several benefits that can be obtained by retrofitting insulation in this fashion. First, the additional insulation will reduce energy consumption and, therefore, operating costs will go down. Second, the increased insulation thickness will provide better personnel protection because of a significantly lowered surface temperature. And finally, but most importantly, by creating a retrofit system, there is an opportunity to take advantage of the optimum performance benefits of more than one material. An illustration of this ability to optimize efficiencies is shown below.

Figure B7.12 shows the cross-over point between thermal conductivity and mean operating temperature of two common industrial thermal insulation materials, cellular glass and calcium silicate. Figure B7.13 shows a cross-sectional view of what a typical retrofitted insulation system might look like and where the interface temperature will occur.

In this illustration, the cross-over point of the two curves shows the ideal temperature at which each material's thermal performance characteristics can be maximized. The two curves cross over at a mean temperature of about 225°F. This shows that the cellular glass insulation has better thermal performance for mean operating temperatures at or below 225°F and the calcium silicate has better thermal performance for mean operating temperatures above 225°F. To maximize the performance of both materials, they can be used together in a composite retrofit system.

In order to use these two materials together and maximize their performance, the optimum interface temperature must be determined first. The interface temperature is the point where the two layers of insulation meet. Knowing that the

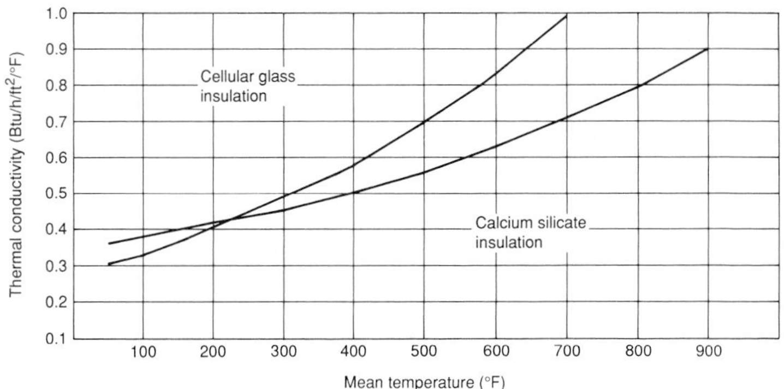

FIGURE B7.12 K-value cross over.

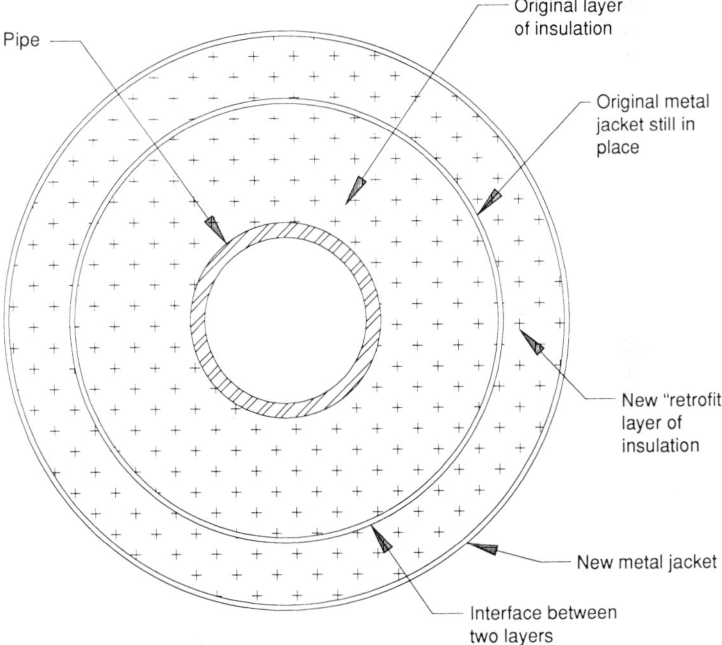

FIGURE B7.13 Cross section of a retrofit system.

cross-over point between the two materials is at 225°F, the interface temperature can be determined by the following calculation shown in Example B7.4:

Example B7.4. Desired Interface Temperature Calculation

Mean temperature = average temperature across the insulation layer

In this example,

$$\text{Mean temperature} = \frac{\text{interface temperature} + \text{ambient temperature}}{2}$$

where mean temperature = 225°F
ambient temperature = 70°F (assumed for this illustration)

Therefore,

$$225°\text{F} = \frac{\text{interface temperature} + 70°\text{F}}{2}$$

$$2(225°\text{F}) = \text{interface temperature} + 70°\text{F}$$

$$2(225°\text{F}) - 70°\text{F} = \text{interface temperature}$$

$$\text{Desired interface temperature} = 380°\text{F}$$

Determining the appropriate thickness for the new retrofitted layer of insulation, in this case cellular glass, is an iterative process. Knowing that the interface temperature should be approximately 380°F provides a target point to work toward. The following example shows the calculation process by which the correct thickness can be determined.

Example B7.5. Retrofit Layer Thickness Calculation. In this example the following theoretical conditions will be employed:

Pipe size	NPS 8 (8.625 actual O.D.)
Operating temperature	600°F
Ambient temperature	70°F
Calcium silicate thickness	2 in nominal; 2.02 in actual

Equation (B7.5) will be used to find the actual interface temperature with the initial assumed insulation thickness of 2 in for the first iteration.

$$T_1 = T_i - Q\frac{R_s \ln (R_1/R_i)}{K_1}$$

where Q = heat loss, Btu/h · ft^2
T_i = temperature of inner surface, 600°F
T_1 = temperature of interface between layers, °F
T_s = temperature of ambient air, 70°F
R_i = inner radius of first layer, 8.625 in/2 = 4.3125 in
R_1 = outer radius of first layer and inner radius of second layer, 12.75 in/2.0 = 6.375 in
R_s = outer radius of second layer, 17.0 in/2 = 8.50 in
K_1 = thermal conductivity of first layer = .52 Btu · in/(h · ft^2 · °F)
ln = natural logarithm
1/f = surface resistance factor = 0.53 (approximated from Fig. B7.3)

In solving for t_2, the heat loss Q of the total retrofitted system must first be calculated. Using Eq. (B7.3), the heat loss for the first trial, assuming that the initial iteration for the second layer with a nominal insulation thickness of 2 in, is as follows:

$$Q = \frac{T_i - T_s}{\{[R_s \ln (R_1/R_i)]/K_1\} + \{[R_s \ln (R_s/R_1)]/K_2\} + 1/\text{f}}$$

where K_2 = the thermal conductivity of the second layer
= 0.42 Btu · in(h · ft^2 · °F)

$$Q = \frac{600°\text{F} - 70°\text{F}}{\dfrac{[8.50 \ln (6.375/4.3125)]}{0.52 \text{ Btu} \cdot \text{in/h} \cdot \text{ft}^2 \cdot °\text{F}} + \dfrac{[8.50 \ln (8.50/6.375)]}{0.42 \text{ Btu} \cdot \text{in/h} \cdot \text{ft}^2 \cdot °\text{F}} + .53}$$

$$Q = \frac{530°\text{F}}{12.74 \text{ Btu/h} \cdot \text{ft}^2 \cdot °\text{F}} = 41.60 \text{ Btu/(h} \cdot \text{ft}^2)$$

Therefore,

$$t_2 = 600°\text{F} - \left(41.60 \frac{8.50 \text{ in} \ln \left(\dfrac{6.375 \text{ in}}{4.3125 \text{ in}}\right)}{0.52 \text{ Btu} \cdot \text{in/(h} \cdot \text{ft}^2 \cdot °\text{F})} \right)$$

$$t_2 = 334.20°\text{F}$$

Since this is lower than the desired interface temperature of 380°F, the iteration process can end at this step. In order to truly maximize the efficiency of the cellular glass and the calcium silicate, the next step would be to perform a second or third calculation with increasing thicknesses until an interface temperature at or around 380°F is achieved. For this trial, however, the thickness of cellular glass used, 2 in, was sufficient to obtain a heat loss of 41.60 Btu/h · ft^2, which is well below what would normally be considered acceptable by today's standards for heat loss. If the interface temperature had come out to be greater than what was tried on the first iteration, the next step would have been to try again with a lesser thickness.

With an interface temperature of 334.2°F, the mean temperature can be recalculated to determine how close to the thermal conductivity cross-over point the 2-in thickness provided. As shown above,

$$\text{Mean temperature} = \frac{\text{interface temperature + ambient temperature}}{2}$$

where interface temperature = 334.2°F
ambient temperature = 70°F (selected for this illustration)

Therefore,

$$\text{Mean temperature} = \frac{334.2°\text{F} + 70°\text{F}}{2} = 202.1°\text{F}$$

This shows that the mean temperature across the outer layer of insulation is 202.1°F, which is only slightly below the cross-over point.

Many old steam lines are insulated with a minimal amount of calcium silicate insulation. As mentioned earlier, this was done before there was a heightened sense of insulation awareness. For these types of applications, this retrofitting option is a very efficient way to revitalize old insulation systems.

Heat Transfer Fluids

Heat transfer fluids are liquids that are used as a means of providing process control. They are generally thermally stable fluids and can be heated or cooled to a given design temperature and transported to the desired process control point while maintaining temperature stability. Insulation system design for heat transfer fluids must take into account the temperature range that the fluid will be operating at and whether it is a constant or a cyclical temperature.

Many catalyst reaction reactor vessels operate at temperatures that swing from as low as −60°F to as high as 450°F. This type of cyclical service presents a complicated insulation system design dilemma. Generally, insulation systems are designed to withstand the ridged environmental and process temperature conditions at either hot or cold operating conditions. When the operating conditions are both hot and cold, a thorough study of all possible combinations of ambient conditions and operating variabilities must be conducted.

In cold process applications the insulation materials specified are usually either plastic foam or cellular glass insulation. In hot process applications the insulation materials specified are usually either mineral fiber, calcium silicate, or cellular glass insulation. In most cases where there is a severe temperature swing as part of the process, a composite system is required. A combination of mineral fiber insulation and cellular glass insulation can be used to provide the protection required. In this scenario, the mineral fiber insulation provides a buffer between the cycling metal substrate and the cellular glass, and the cellular glass insulation provides the protection from damage that can be caused by the vapor driving forces when the system is in its cold temperature stage. Careful consideration of appropriate sealants, adhesives, and accessory materials for the process must be made. It is best to consult with the material manufacturer's technical services departments whenever this type of process is being planned.

Another important consideration when specifying insulation systems for heat transfer fluids is fire safety. Many of the common heat transfer fluids, both organic and inorganic, can present a very serious fire hazard if they are absorbed into a permeable insulation material. It has been shown that some of these fluids, although thermally very stable at their peak operating temperatures, become much less so when absorbed into insulation materials. Spontaneous heating is the problem. Another term for the condition is auto-ignition. Monsanto Industrial Chemicals Company manufactures a product called Therminol which is a very useful and highly effective heat transfer fluid. In the Therminol technical data sheet a description of the problem is offered:[10]

> Organic heat transfer fluids such as Therminol exhibit a slow oxidation reaction with air in the presence of insulating materials when system temperatures are above 500°F (260°C). Porous insulation materials such as calcium silicate offers a large reaction surface in the face of poor heat dissipation conditions, and this, along with possible catalysis from the insulation material can cause a temperature build-up. This temperature build-up can result in ignition of the fluid when the saturated insulation is exposed to air (i.e., should the insulation be opened for repair, etc.).
>
> This phenomenon is not fully understood, but appears not to occur with cellular

> glass, possibly because of its closed cell structure. Cellular glass should be used in all areas where leakage is a possibility and the system temperature is greater than 400°F (200°C).

There are numerous other references which relate similar, if not identical, information about this problem and the precautions that need to be taken in the design and application process. The references for this chapter include a number of publications on this subject. As in the case of cyclical service, it is important to contact the manufacturer before writing the insulation system specification for high-temperature heat transfer fluid applications. The manufacturer of the heat transfer fluid should also be consulted for their own precautions and recommendations.

Hot Oils

Hot oil piping and equipment applications require the same attention to detail as do heat transfer fluids with respect to insulation system design. Although hot oils do not frequently operate at the temperature extremes as do heat transfer fluids, auto-ignition can be just as serious. Fires have occurred in oils at temperatures as low as 176 to 302°F (80 to 150°C), in coal tar distillates at temperatures as low as 212 to 482°F (100 to 250°C), and in mineral oils at temperatures as low as 392 to 572°F (200 to 300°C).[11]

Liquefied Gas (Cryogenic Service)

Cryogenic temperatures range from −40°F (−40°C) down to absolute zero (−459°F). In this temperature range, the primary concern is the prevention of water vapor migration toward the pipe surface. If water vapor is allowed to enter the insulation system, it will rapidly destroy the thermal performance of the insulation.

Cryogenic piping is usually insulated in multiple layers. The best vapor barrier joint sealants available only remain flexible down to a temperature of about −80°F; therefore, the inner layer of insulation is usually left unsealed. The outer layer or layers are completely sealed with vapor barrier joint sealants and the final layer of insulation is usually covered with a vapor barrier mastic or membrane. Plastic foam insulation must be covered with a complete vapor barrier envelope. Cellular glass insulation is sometimes just vapor sealed over the joints and then covered with a weather barrier.

At extreme cryogenic temperatures, there is also the possibility of condensation of oxygen. This can occur anywhere below −290°F. The National Fire Protection Association states:[12]

> Liquid oxygen is the most concentrated common source of oxygen. Contamination of liquid oxygen with most organic substances often renders the mixture subject to violent explosion.
>
> An inadvertent oxygen enriched atmosphere can be created within insulation on piping and equipment containing materials at temperatures below the condensation temperature of oxygen (e.g., liquid hydrogen or nitrogen), if the oxygen in the atmospheric air is condensed within the insulation.

Therefore, there is a risk of liquid oxygen explosion with organic insulation materials and accessory products. Organic foam insulations, if used in the potential presence of condensing oxygen, require that the foam must be protected by a gas-impermeable membrane.[13]

Sanitary and Sewerage Water

Water and sewerage lines are generally only insulated to prevent freezing. All of the considerations appropriate for freeze protection requirements discussed earlier in this chapter should prevail.

MATERIALS

There are many different types of insulation materials available for both commercial and industrial piping applications. There are in fact too many to discuss in detail here. For the purposes of this chapter, a few of the more common commercial and industrial piping insulation types, or classifications, will be described. The following list, sorted alphabetically, contains the material classifications most common to the industrial and commercial piping industry:

- Calcium silicate insulation
- Cellular glass insulation
- Elastomeric foam insulation
- Fiber glass and mineral wool insulations
- Perlite insulation
- Phenolic foam insulation
- Polystyrene foam insulation
- Polyurethane and polyisocyanurate foam insulations

Table B7.1 illustrates some of the properties of insulation materials that are commonly referred to in the insulation material selection process. Tables B7.3 through B7.10 provide thickness guidelines for the materials referenced in Table B7.1. Tables B7.11 through B7.16 provide thermal conductivity data for each of these materials. It is important to remember that the properties of the material alone should not control the specification process; the combined properties of the insulation material and the corresponding accessory products that make up the total system should control the process.

Calcium Silicate Insulation

Calcium silicate is a very rigid heavy-density material used exclusively for applications above 250°F. This insulation material has been a standard for high-temperature applications for many years. Its compressive strengths are very good, and it is noncombustible. It is suitable for temperatures from 250°F up to 1000°F. Calcium silicate is manufactured from a slurry that is poured into molds to make various pipe covering shapes. It is generally available in half-sections,

TABLE B7.3 Calcium Silicate Insulation Thickness Table
Recommended thickness table
Surface emittance = 0.40 Average wind velocity = 5.00 miles/hour
Hot pipe worse case ambient temperature of 90.0°F for a surface temperature of 140.0°F or less

Temperature (°F)	Nominal pipe diameter (inches)															
	0.5	1.0	1.5	2.0	2.5	3.0	4.0	5.0	6.0	8.0	10.0	12.0	14.0	18.0	24.0	Flat
	Calsil thickness (inches)															
200.0	1.0	1.0	1.0	1.0	1.0	1.0	1.0	1.0	1.0	1.5	1.5	1.5	1.5	1.5	1.5	1.5
300.0	1.0	1.0	1.0	1.0	1.0	1.0	1.0	1.0	1.0	1.5	1.5	1.5	1.5	1.5	1.5	1.5
400.0	1.0	1.0	1.0	1.0	1.0	1.0	1.0	1.0	1.0	1.5	1.5	1.5	1.5	1.5	1.5	1.5
500.0	1.0	1.0	1.0	1.0	1.5	1.5	1.5	1.5	1.5	1.5	1.5	1.5	2.0	2.0	2.0	2.0
600.0	1.5	1.5	1.5	1.5	1.5	1.5	2.0	2.0	2.0	2.0	2.0	2.0	2.5	2.5	2.5	2.5
700.0	1.5	1.5	2.0	2.0	1.5	2.0	2.0	2.5	2.5	2.5	2.5	2.5	3.0	3.0	3.0	3.5
800.0	1.5	2.0	2.0	2.0	2.0	2.5	2.5	3.0	3.0	3.0	3.0	3.5	3.5	3.5	4.0	4.5
900.0	2.0	2.0	2.0	2.5	2.5	3.0	3.0	3.5	3.5	3.5	4.0	4.0	4.0	4.5	4.5	5.0

Maximum heat flow = 118.2 Btu/h ft^2 (highest heat flow of any in the table).

TABLE B7.4 Cellular Glass Insulation Thickness Table (Hot)
Recommended thickness table
Surface emittance = 0.40 Average wind velocity = 5.00 miles/hour
Hot pipe worse case ambient temperature of 90.0°F for a surface temperature of 140.0°F or less

Temperature (°F)	Nominal pipe diameter (inches)															
	0.5	1.0	1.5	2.0	2.5	3.0	4.0	5.0	6.0	8.0	10.0	12.0	14.0	18.0	24.0	Flat
	FOAMGLAS* thickness (inches)															
200.0	1.0	1.0	1.0	1.0	1.0	1.0	1.0	1.0	1.0	1.5	1.5	1.5	1.5	1.5	1.5	1.5
300.0	1.0	1.0	1.0	1.0	1.0	1.0	1.0	1.0	1.0	1.5	1.5	1.5	1.5	1.5	1.5	1.5
400.0	1.0	1.0	1.0	1.0	1.0	1.0	1.0	1.0	1.5	1.5	1.5	1.5	1.5	1.5	1.5	1.5
500.0	1.0	1.0	1.5	1.5	1.5	1.5	1.5	1.5	1.5	2.0	2.0	2.0	2.0	2.0	2.0	2.5
600.0	1.5	1.5	1.5	1.5	1.5	2.0	2.0	2.0	2.0	2.5	2.5	2.5	2.5	3.0	3.0	3.0
700.0	1.5	2.0	2.0	2.0	2.0	2.5	2.5	2.5	3.0	3.0	3.0	3.0	3.5	3.5	3.5	4.0
800.0	2.0	2.5	2.5	2.5	2.5	3.0	3.0	3.5	3.5	3.5	4.0	4.0	4.5	4.5	4.5	5.5
900.0	2.5	3.0	3.0	3.5	3.0	3.5	4.0	4.0	4.0	4.5	5.0	5.0	5.5	5.5	5.5	7.0

Maximum heat flow = 114.0 Btu/h ft^2 (highest heat flow of any in the table).
*FOAMGLAS® is a registered trademark of Pittsburgh Corning Corporation.

TABLE B7.5 Cellular Glass Insulation Thickness Table (Cold)
Recommended lower pipe temperature limits for given insulation thickness (°F)
Insulation material—FOAMGLAS*
Ambient temperature = 90.0°F Heat flow limit = −10.0 ±0.1 Btu/(h · ft^2)
Wind velocity = 5.0 miles/hour Emittance = 0.90

	Insulation thickness (inches)																				
NPS	1.0	1.5	2.0	2.5	3.0	3.5	4.0	4.5	5.0	5.5	6.0	6.5	7.0	7.5	8.0	8.5	9.0	9.5	10.0	11.0	12.0
0.50	31	−15	−65	−97	−166	−244															
0.75	41	0	−43	−72	−133	−282	−379	−379													
1.00	32	−2	−47	−101	−160	−229	−311	−423													
1.50	40	9	−10	−52	−98	−152	−213	−295	−382												
2.00	41	10	−25	−64	−109	−161	−229	−300	−383												
2.50	43	28	−2	−37	−76	−120	−178	−237	−306	−408											
3.00	46	19	−10	−43	−81	−130	−179	−236	−319	−397											
4.00	46	22	−4	−35	−74	−113	−157	−220	−279	−346	−424										
5.00	48	26	1	−30	−62	−98	−149	−195	−247	−306	−374										
6.00	50	29	2	−24	−54	−96	−134	−177	−224	−277	−338	−406									
8.00		28	6	−25	−53	−84	−119	−156	−198	−245	−298	−356	−423								
10.00		27	5	−18	−44	−72	−104	−138	−176	−217	−264	−315	−373	−437							
12.00		28	7	−15	−40	−67	−97	−129	−164	−203	−246	−293	−346	−404							
14.00		34	13	−8	−32	−58	−86	−116	−150	−186	−226	−270	−319	−373	−434						
16.00		34	14	−6	−29	−54	−81	−111	−143	−177	−215	−257	−303	−353	−409						
18.00		35	15	−5	−28	−52	−78	−106	−137	−170	−207	−246	−290	−337	−390	−449					
20.00		35	16	−4	−26	−50	−75	−103	−132	−165	−199	−237	−279	−325	−375	−430					
24.00		35	17	−2	−24	−47	−71	−97	−125	−156	−189	−224	−263	−305	−351	−402					
28.00		36	17	−1	−22	−44	−68	−93	−120	−149	−181	−214	−251	−291	−334	−381	−433				
30.00		36	18	−1	−21	−43	−67	−92	−118	−147	−177	−210	−246	−285	−327	−373	−423				
36.00		36	18	0	−20	−41	−64	−88	−113	−141	−170	−201	−235	−271	−311	−353	−400				
42.00		37	19	0	−19	−39	−62	−85	−110	−136	−164	−194	−227	−261	−299	−339	−383	−430			
48.00		37	19	1	−18	−38	−60	−83	−107	−133	−160	−189	−220	−254	−290	−328	−370	−415			
60.00		37	20	2	−16	−37	−58	−80	−103	−128	−154	−182	−211	−243	−277	−313	−352	−394	−439		
72.00		37	20	2	−16	−35	−56	−78	−100	−125	−150	−177	−205	−236	−268	−303	−340	−380	−422		
96.00		38	21	3	−15	−34	−54	−75	−97	−120	−145	−171	−198	−227	−257	−290	−325	−362	−402		
120.00		38	21	3	−14	−33	−53	−73	−95	−118	−142	−167	−193	−221	−251	−282	−315	−351	−389		
168.00		38	21	4	−13	−32	−51	−72	−93	−115	−138	−162	−188	−215	−243	−273	−305	−339	−374		
Flat		38	22	5	−11	−29	−48	−67	−87	−108	−129	−151	−175	−199	−224	−251	−279	−308	−338	−405	

*FOAMGLAS® is a registered trademark of Pittsburgh Corning Corporation.

TABLE B7.6 Fiberglass Insulation Thickness Table

Recommended thickness table

Surface emittance = 0.40 Average wind velocity = 5.00 miles/hour

Hot pipe worse case ambient temperature of 90.0°F for a surface temperature of 140.0°F or less

Temperature (°F)	Nominal pipe diameter (inches)															
	0.5	1.0	1.5	2.0	2.5	3.0	4.0	5.0	6.0	8.0	10.0	12.0	14.0	18.0	24.0	Flat
	Fiberglass thickness (inches)															
200.0	1.0	1.0	1.0	1.0	1.0	1.0	1.0	1.0	1.0	1.5	1.5	1.5	1.5	1.5	1.5	1.5
300.0	1.0	1.0	1.0	1.0	1.0	1.0	1.0	1.0	1.0	1.5	1.5	1.5	1.5	1.5	1.5	1.5
400.0	1.0	1.0	1.0	1.0	1.0	1.0	1.0	1.0	1.0	1.5	1.5	1.5	1.5	1.5	1.5	1.5
500.0	1.0	1.0	1.0	1.0	1.0	1.0	1.0	1.0	1.0	1.5	1.5	1.5	1.5	1.5	1.5	1.5
600.0	1.0	1.0	1.0	1.0	1.5	1.5	1.5	1.5	1.5	1.5	1.5	1.5	2.0	2.0	2.0	2.0
700.0	1.0	1.0	1.5	1.5	1.5	1.5	1.5	2.0	2.0	2.0	2.0	2.0	2.0	2.5	2.5	2.5
800.0	1.5	1.5	1.5	1.5	1.5	2.0	2.0	2.0	2.0	2.5	2.5	2.5	2.5	3.0	3.0	3.0
850.0	1.5	1.5	2.0	2.0	1.5	2.0	2.0	2.5	2.5	2.5	2.5	2.5	3.0	3.0	3.0	3.5

Maximum heat flow = 121.9 Btu/(h · ft^2) (highest heat flow of any in the table).

TABLE B7.7 Mineral Wool Insulation Thickness Table
Recommended thickness table
Surface emittance = 0.40 Average wind velocity = 5.00 miles/hour
Hot pipe worse case ambient temperature of 90.0°F for a surface temperature of 140.0°F or less

Temperature (°F)	Nominal pipe diameter (inches)															
	0.5	1.0	1.5	2.0	2.5	3.0	4.0	5.0	6.0	8.0	10.0	12.0	14.0	18.0	24.0	Flat
	Mineral wool thickness (inches)															
200.0	1.0	1.0	1.0	1.0	1.0	1.0	1.0	1.0	1.0	1.5	1.5	1.5	1.5	1.5	1.5	1.5
300.0	1.0	1.0	1.0	1.0	1.0	1.0	1.0	1.0	1.0	1.5	1.5	1.5	1.5	1.5	1.5	1.5
400.0	1.0	1.0	1.0	1.0	1.0	1.0	1.0	1.0	1.0	1.5	1.5	1.5	1.5	1.5	1.5	1.5
500.0	1.0	1.0	1.0	1.0	1.0	1.5	1.5	1.5	1.5	1.5	1.5	1.5	1.5	1.5	2.0	2.0
600.0	1.0	1.5	1.5	1.5	1.5	1.5	1.5	2.0	2.0	2.0	2.0	2.0	2.0	2.5	2.5	2.5
700.0	1.5	1.5	2.0	2.0	1.5	2.0	2.0	2.5	2.5	2.5	2.5	2.5	3.0	3.0	3.0	3.5
800.0	1.5	2.0	2.0	2.0	2.0	2.5	2.5	3.0	3.0	3.0	3.0	3.5	3.5	3.5	4.0	4.5
900.0	2.0	2.5	2.5	2.5	2.5	3.0	3.0	3.5	3.5	3.5	4.0	4.0	4.5	4.5	4.5	5.5
1000.0	2.5	2.5	3.0	3.0	3.0	3.5	4.0	4.0	4.0	4.5	4.5	5.0	5.0	5.5	5.5	6.5
1100.0	2.5	3.0	3.5	4.0	3.5	4.0	4.5	4.5	5.0	5.0	5.5	6.0	6.0	6.5	6.5	8.0
1200.0	3.0	3.5	4.0	4.5	4.5	5.0	5.0	5.5	5.5	6.0	6.5	7.0	7.0	7.5	7.5	9.5

Maximum heat flow = 119.5 Btu/(h · ft^2) (highest heat flow of any in the table).

TABLE B7.8 Perlite Insulation Thickness Table
Recommended thickness table
Surface emittance = 0.40 Average wind velocity = 5.00 miles/hour
Hot pipe worse case ambient temperature of 90.0°F for a surface temperature of 140.0°F or less

Temperature (°F)	Nominal pipe diameter (inches)															
	0.5	1.0	1.5	2.0	2.5	3.0	4.0	5.0	6.0	8.0	10.0	12.0	14.0	18.0	24.0	Flat
	Perlite thickness (inches)															
200.0	1.0	1.0	1.0	1.0	1.0	1.0	1.0	1.0	1.0	1.5	1.5	1.5	1.5	1.5	1.5	1.5
300.0	1.0	1.0	1.0	1.0	1.0	1.0	1.0	1.0	1.0	1.5	1.5	1.5	1.5	1.5	1.5	1.5
400.0	1.0	1.0	1.0	1.0	1.0	1.0	1.0	1.5	1.5	1.5	1.5	1.5	1.5	1.5	1.5	2.0
500.0	1.0	1.0	1.5	1.5	1.5	1.5	1.5	2.0	2.0	2.0	2.0	2.0	2.0	2.5	2.5	2.5
600.0	1.5	1.5	1.5	2.0	1.5	2.0	2.0	2.5	2.5	2.5	2.5	2.5	3.0	3.0	3.0	3.5
700.0	1.5	2.0	2.0	2.0	2.0	2.5	2.5	2.5	3.0	3.0	3.0	3.0	3.5	3.5	3.5	4.0
800.0	2.0	2.0	2.0	2.5	2.5	3.0	3.0	3.0	3.5	3.5	4.0	4.0	4.0	4.5	4.5	5.0
900.0	2.5	2.5	2.5	3.0	3.0	3.5	3.5	4.0	4.0	4.0	4.5	4.5	5.0	5.0	5.5	6.0
1000.0	2.5	3.0	3.0	3.5	3.5	4.0	4.0	4.5	4.5	5.0	5.0	5.5	5.5	6.0	6.0	7.5

Maximum heat flow = 120.0 Btu/(h · ft^2) (highest heat flow of any in the table).

TABLE B7.9 Polyurethane and Polyisocyanurate Foam Insulation Thickness Table (Hot)
Recommended thickness table
Surface emittance = 0.40 Average wind velocity = 5.00 miles/hour
Hot pipe worse case ambient temperature of 90.0°F for a surface temperature of 140.0°F or less

Temperature (°F)	Nominal pipe diameter (inches)															
	0.5	1.0	1.5	2.0	2.5	3.0	4.0	5.0	6.0	8.0	10.0	12.0	14.0	18.0	24.0	Flat
	Polyurethane/polyisocyanurate thickness (inches)															
200.0	1.0	1.0	1.0	1.0	1.0	1.0	1.0	1.0	1.0	1.5	1.5	1.5	1.5	1.5	1.5	1.5
250.0	1.0	1.0	1.0	1.0	1.0	1.0	1.0	1.0	1.0	1.5	1.5	1.5	1.5	1.5	1.5	1.5
300.0	1.0	1.0	1.0	1.0	1.0	1.0	1.0	1.0	1.0	1.5	1.5	1.5	1.5	1.5	1.5	1.5

Maximum heat flow = 42.9 Btu/(h · ft^2) (maximum heat flow of any in the table).

TABLE B7.10 Polyurethane or Polyisocyanurate Foam Insulation Thickness Table (Cold)
Recommended lower pipe temperature limits for given insulation thickness (°F)
Insulation material—urethane
Ambient temperature = 90.0°F Heat flow limit = − 10.0 ± 0.1 Btu/(h · ft²)
Wind velocity = 5.0 miles/hour Emittance = 0.90

	Insulation thickness (inches)																				
NPS	1.0	1.5	2.0	2.5	3.0	3.5	4.0	4.5	5.0	5.5	6.0	6.5	7.0	7.5	8.0	8.5	9.0	9.5	10.0	11.0	12.0
0.50	−18	−94	−171	−359																	
0.75	0	−69	−138	−289	−425																
1.00	−15	−74	−144	−230	−345																
1.50	−2	−55	−152	−226	−328	−443															
2.00	0	−52	−109	−170	−245	−349															
2.50	3	−74	−128	−189	−265	−384															
3.00	8	−38	−85	−137	−197	−283	−386														
4.00	9	−32	−76	−124	−185	−251	−339														
5.00	13	−25	−67	−118	−167	−225	−322	−416													
6.00	17	−20	−65	−108	−154	−222	−291	−381													
8.00		−21	−59	−109	−153	−202	−261	−338	−421												
10.00		−23	−60	−98	−138	−183	−234	−299	−379	−448											
12.00		−21	−57	−93	−132	−174	−222	−281	−354	−428											
14.00		−12	−46	−82	−119	−159	−204	−257	−323	−399											
16.00		−11	−45	−79	−116	−154	−197	−247	−308	−382	−445										
18.00		−10	−44	−78	−113	−150	−192	−239	−297	−367	−433										
20.00		−10	−42	−76	−111	−147	−187	−233	−288	−355	−422										
24.00		−9	−41	−73	−107	−142	−180	−223	−274	−336	−404										
28.00		−8	−40	−72	−104	−139	−176	−217	−264	−323	−389	−444									
30.00		−8	−39	−71	−103	−137	−174	−214	−260	−317	−382	−439									
36.00		−7	−38	−69	−101	−134	−169	−207	−251	−304	−366	−425									
42.00		−7	−37	−68	−99	−131	−166	−203	−245	−295	−354	−414									
48.00		−6	−36	−67	−98	−129	−163	−199	−240	−288	−345	−405									
60.00		−6	−36	−65	−96	−127	−159	−194	−233	−279	−332	−391	−440								
72.00		−5	−35	−65	−94	−125	−157	−191	−229	−272	−324	−381	−432								
96.00		−5	−34	−63	−93	−123	−154	−187	−223	−264	−313	−368	−420								
120.00		−5	−34	−63	−92	−121	−152	−184	−220	−260	−306	−360	−412								
168.00		−4	−33	−62	−91	−120	−150	−181	−216	−254	−299	−350	−402	−444							
Flat		−4	−32	−60	−88	−116	−144	−174	−206	−241	−281	−327	−377	−421							

TABLE B7.11 Calcium Silicate Insulation K-Value Table
Thermal conductivity as a function of temperature for calcium silicate (calsil)

English units		Non-SI metric units		SI metric units	
Calsil insulation at uniform temperature (°F)	Calsil insulation thermal conductivity [Btu · in/(h · ft^2 · °F)]	Calsil insulation at uniform temperature (°C)	Calsil insulation thermal conductivity [kcal/(h · m · °C)]	Calsil insulation at uniform temperature (K)	Calsil insulation thermal conductivity [W/(m · K)]
0.0	0.369	−17.8	0.0458	255.4	0.0532
50.0	0.378	10.0	0.0469	283.1	0.0545
100.0	0.389	37.8	0.0482	310.9	0.0561
150.0	0.402	65.6	0.0499	338.7	0.0579
200.0	0.417	93.3	0.0517	366.5	0.0601
250.0	0.434	121.1	0.0539	394.3	0.0626
300.0	0.454	148.9	0.0563	422.0	0.0655
350.0	0.476	176.7	0.0591	449.8	0.0687
400.0	0.501	204.4	0.0621	477.6	0.0722
450.0	0.528	232.2	0.0655	505.4	0.0761
500.0	0.557	260.0	0.0692	533.2	0.0804
550.0	0.590	287.8	0.0732	560.9	0.0851
600.0	0.625	315.6	0.0775	588.7	0.0901
650.0	0.663	343.3	0.0822	616.5	0.0956
700.0	0.704	371.1	0.0873	644.3	0.1015
750.0	0.747	398.9	0.0927	672.0	0.1078
800.0	0.794	426.7	0.0985	699.8	0.1145
850.0	0.844	454.4	0.1047	727.6	0.1217
900.0	0.897	482.2	0.1113	755.4	0.1293
950.0	0.953	510.0	0.1183	783.2	0.1374
1000.0	1.012	537.8	0.1256	810.9	0.1460

The values of the thermal conductivity of the insulation shown above were determined by evaluating a polynomial at the insulation temperature. This polynomial, giving the thermal conductivity in (Btu · in)/(h · ft^2 · °F) as a function of temperature in °F, is $K(T) = 0.369 + 0.000158T + 0.000000392T^2 + 0.0000000000094T^3$.

Coefficients are rounded to three significant digits. This curve is based upon an insulation density of 16.00 lb/ft^3 and may be subject to decreasing reliability outside the temperature range from 250.0 to 1000.0°F.

TABLE B7.12 Cellular Glass Insulation K-Value Table
Thermal conductivity as a function of temperature for FOAMGLAS*

English units		Non-SI metric units		SI metric units	
FOAMGLAS* insulation at uniform temperature (°F)	FOAMGLAS* insulation thermal conductivity [Btu · in/(h · ft^2 · °F)]	FOAMGLAS* insulation at uniform temperature (°C)	FOAMGLAS* insulation thermal conductivity [kcal/(h · m · °C)]	FOAMGLAS* insulation at uniform temperature (K)	FOAMGLAS* insulation thermal conductivity [W/(m · K)]
−200.0	0.202	−128.9	0.0251	144.3	0.0292
−150.0	0.221	−101.1	0.0274	172.0	0.0319
−100.0	0.242	−73.3	0.0300	199.8	0.0349
−50.0	0.264	−45.6	0.0328	227.6	0.0381
0.0	0.289	−17.8	0.0359	255.4	0.0417
50.0	0.316	10.0	0.0392	283.1	0.0455
100.0	0.345	37.8	0.0428	310.9	0.0497
150.0	0.377	65.6	0.0467	338.7	0.0543
200.0	0.411	93.3	0.0511	366.5	0.0593
250.0	0.449	121.1	0.0558	394.3	0.0648
300.0	0.491	148.9	0.0609	422.0	0.0708
350.0	0.536	176.7	0.0665	449.8	0.0773
400.0	0.586	204.4	0.0727	477.6	0.0845
450.0	0.640	232.2	0.0795	505.4	0.0924
500.0	0.700	260.0	0.0869	533.2	0.1010
550.0	0.766	287.8	0.0951	560.9	0.1105
600.0	0.839	315.6	0.1041	588.7	0.1210
650.0	0.919	343.3	0.1140	616.5	0.1325
700.0	1.006	371.1	0.1249	644.3	0.1451
750.0	1.102	398.9	0.1368	672.0	0.1590
800.0	1.208	426.7	0.1499	699.8	0.1742
850.0	1.324	454.5	0.1643	727.6	0.1909
900.0	1.451	482.2	0.1800	755.4	0.2092

The values of the thermal conductivity of the insulation shown above were determined by evaluating a polynomial at the insulation temperature. This polynomial, giving the thermal conductivity in Btu · in/(h · ft^2 · °F) as a function of temperature in °F, is $K(T) = 0.289 + 0.000514T + 0.000000436T^2 + 0.000000000227T^3 + 0.000000000000276\,T^4$.

Coefficients are rounded to three significant digits. This curve is based upon an insulation density of 8.00 lb/ft^3 and may be subject to decreasing reliability outside the temperature range from −250.0 to 800.0°F.

This curve for FOAMGLAS cellular glass insulation has been determined from a regression analysis of data obtained by impartial commercial laboratories outside Pittsburgh Corning Corporation.

TABLE B7.13 Fiberglass Insulation K-Value Table
Thermal conductivity as a function of temperature for fiberglass

English units		Non-SI metric units		SI metric units	
Fiberglass insulation at uniform temperature (°F)	Fiberglass insulation thermal conductivity [Btu · in/(h · ft^2 · °F)]	Fiberglass insulation at uniform temperature (°C)	Fiberglass insulation thermal conductivity [kcal/(h · m · °C)]	Fiberglass insulation at uniform temperature (K)	Fiberglass insulation thermal conductivity [W/(m · K)]
0.0	0.195	−17.8	0.0241	255.4	0.0281
50.0	0.216	10.0	0.0268	283.1	0.0311
100.0	0.237	37.8	0.0294	310.9	0.0342
150.0	0.258	65.6	0.0320	338.7	0.0372
200.0	0.280	93.3	0.0347	366.5	0.0403
250.0	0.301	121.1	0.0373	394.3	0.0434
300.0	0.322	148.9	0.0400	422.0	0.0464
350.0	0.343	176.7	0.426	449.8	0.0495
400.0	0.364	204.4	0.0452	477.6	0.0526
450.0	0.386	232.2	0.0479	505.4	0.0556
500.0	0.407	260.0	0.0505	533.2	0.0587
550.0	0.428	287.8	0.0531	560.9	0.0618
600.0	0.449	315.6	0.0558	588.7	0.0648
650.0	0.471	343.3	0.0584	616.5	0.0679
700.0	0.492	371.1	0.0610	644.3	0.0710
750.0	0.513	398.9	0.0637	672.0	0.0740
800.0	0.534	426.7	0.0663	699.8	0.0771
850.0	0.556	454.4	0.0690	727.6	0.0801

The values of the thermal conductivity of the insulation shown above were determined by evaluating a polynomial at the insulation temperature. This polynomial, giving the thermal conductivity in Btu · in/(h · ft^2 · °F) as a function of temperature in °F, is $K(T) = 0.195 + 0.000425T + 0.000000000T^2$.

Coefficients are rounded to three significant digits. This curve is based upon an insulation density of 5.63 lb/ft^3 and may be subject to decreasing reliability outside the temperature range from 42.0 to 850.0°F.

TABLE B7.14 Mineral Wool Insulation K-Value Table
Thermal conductivity as a function of temperature for mineral wool

English units		Non-SI metric units		SI metric units	
Mineral wool insulation at uniform temperature (°F)	Mineral wool insulation thermal conductivity [Btu · in/(h · ft² · °F)]	Mineral wool insulation at uniform temperature (°C)	Mineral wool insulation thermal conductivity [kcal/(h · m · °C)]	Mineral wool insulation at uniform temperature (K)	Mineral wool insulation thermal conductivity [W/(m · K)]
0.0	0.228	−17.8	0.0283	255.4	0.0329
50.0	0.248	10.0	0.0308	283.1	0.0358
100.0	0.271	37.8	0.0336	310.9	0.0391
150.0	0.297	65.6	0.0369	338.7	0.0429
200.0	0.326	93.3	0.0405	366.5	0.0470
250.0	0.358	121.1	0.0445	394.3	0.0517
300.0	0.393	148.9	0.0488	422.0	0.0567
350.0	0.432	176.7	0.0535	449.8	0.0622
400.0	0.473	204.4	0.0586	477.6	0.0682
450.0	0.517	232.2	0.0641	505.4	0.0745
500.0	0.564	260.0	0.0700	533.2	0.0813
550.0	0.614	287.8	0.0762	560.9	0.0885
600.0	0.667	315.6	0.0828	588.7	0.0962
650.0	0.723	343.3	0.0897	616.5	0.1043
700.0	0.782	371.1	0.0971	644.3	0.1128
750.0	0.844	398.9	0.1048	672.0	0.1218
800.0	0.909	426.7	0.1129	699.8	0.1312
850.0	0.978	454.4	0.1213	727.6	0.1410
900.0	1.049	482.2	0.1301	755.4	0.1512
950.0	1.123	510.0	0.1393	783.2	0.1619
1000.0	1.200	537.8	0.1489	810.9	0.1730
1050.0	1.280	565.6	0.1588	838.7	0.1846
1100.0	1.363	593.3	0.1691	866.5	0.1966
1150.0	1.449	621.1	0.1798	894.3	0.2090
1200.0	1.538	648.9	0.1909	922.0	0.2219

The values of the thermal conductivity of the insulation shown above were determined by evaluating a polynomial at the insulation temperature. This polynomial, giving the thermal conductivity in Btu · in/(h · ft² · °F) as a function of temperature in °F, is $K(T) = 0.228 + 0.000372T + 0.000000600T^2$.

Coefficients are rounded to three significant digits. This curve is based upon an insulation density of 11.70 lb/ft³ and may be subject to decreasing reliability outside the temperature range from 42.0 to 1200.0°F.

TABLE B7.15 Perlite Insulation K-Value Table

Thermal conductivity as a function of temperature for perlite

English units		Non-SI metric units		SI metric units	
Perlite insulation at uniform temperature (°F)	Perlite insulation thermal conductivity [Btu · in/(h · ft² · °F)]	Perlite insulation at uniform temperature (°C)	Perlite insulation thermal conductivity [kcal/(h · m · °C)]	Perlite insulation at uniform temperature (K)	Perlite insulation thermal conductivity [W/(m · K)]
0.0	0.388	−17.8	0.0481	255.4	0.0559
50.0	0.412	10.0	0.0511	283.1	0.0594
100.0	0.438	37.8	0.0543	310.9	0.0631
150.0	0.465	65.6	0.0577	338.7	0.0671
200.0	0.494	93.3	0.0613	366.5	0.0712
250.0	0.524	121.1	0.0650	394.3	0.0755
300.0	0.555	148.9	0.0689	422.0	0.0800
350.0	0.587	176.7	0.0729	449.8	0.0847
400.0	0.621	204.4	0.0770	477.6	0.0895
450.0	0.655	232.2	0.0813	505.4	0.0945
500.0	0.691	260.0	0.0857	533.2	0.0996
550.0	0.727	287.8	0.0902	560.9	0.1049
600.0	0.764	315.6	0.0948	588.7	0.1102
650.0	0.802	343.3	0.0996	616.5	0.1157
700.0	0.841	371.1	0.1044	644.3	0.1213
750.0	0.881	398.9	0.1093	672.0	0.1270
800.0	0.921	426.7	0.1143	699.8	0.1328
850.0	0.962	454.4	0.1193	727.6	0.1387
900.0	1.003	482.2	0.1245	755.4	0.1446
950.0	1.045	510.0	0.1296	783.2	0.1507
1000.0	1.087	537.8	0.1349	810.9	0.1567

The values of the thermal conductivity of the insulation shown above were determined by evaluating a polynomial at the insulation temperature. This polynomial, giving the thermal conductivity in Btu · in/(h · ft² · °F) as a function of temperature in °F, is $K(T) = 0.388 + 0.000473T + 0.000000306T^2 - 0.0000000000807T^3$.

Coefficients are rounded to three significant digits. This curve is based upon an insulation density of 12.00 lb/ft³ and may be subject to decreasing reliability outside the temperature range from 250.0 to 1000.0°F.

TABLE B7.16 Polyurethane or Polisocyanurate Foam Insulation K-Value Table
Thermal conductivity as a function of temperature for urethane

English units		Non-SI metric units		SI metric units	
Urethane insulation at uniform temperature (°F)	Urethane insulation thermal conductivity [Btu · in/(h · ft² · °F)]	Urethane insulation at uniform temperature (°C)	Urethane insulation thermal conductivity [kcal/(h · m · °C)]	Urethane insulation at uniform temperature (K)	Urethane insulation thermal conductivity [W/(m · K)]
−100.0	0.179	−73.3	0.0222	199.8	0.0258
−50.0	0.179	−45.6	0.0223	227.6	0.0259
0.0	0.174	−17.8	0.0215	255.4	0.0250
50.0	0.166	10.0	0.0206	283.1	0.0240
100.0	0.165	37.8	0.0205	310.9	0.0238
150.0	0.180	65.6	0.0224	338.7	0.0260
200.0	0.225	93.3	0.0279	366.5	0.0325
250.0	0.316	121.1	0.0392	394.3	0.0455
300.0	0.470	148.9	0.0583	422.0	0.0678
350.0	0.710	176.7	0.0881	449.8	0.1024
400.0	1.059	204.4	0.1314	477.6	0.1528
450.0	1.544	232.2	0.1917	505.4	0.2228
500.0	2.195	260.0	0.2724	533.2	0.3166

The values of the thermal conductivity of the insulation shown above were determined by evaluating a polynomial at the insulation temperature. This polynomial, giving the thermal conductivity in Btu · in/(h · ft² · °F) as a function of temperature in °F, is $K(T) = 0.174 - 0.000155T - 0.000000339T^2 + 0.000000008377T^3 + 0.0000000000181907T^4$.

Coefficients are rounded to three significant digits. This curve is based upon an insulation density of 2.00 lb/ft³ and may be subject to decreasing reliability outside the temperature range from −50.0 to 250.0°F.

quad-sections, and flat blocks. Calcium silicate is applied to the piping with metal bands and generally covered with a metal jacket.

Cellular Glass Insulation

Cellular glass insulation is a high-strength versatile insulation used in temperature services that range from −450°F up to 900°F. Cellular glass insulation is all closed cell glass with no organic binders or fillers. The closed cell glass structure renders it impervious to liquid water and the driving force of water vapor pressure. It is manufactured in flat blocks which are then fabricated into any shape specified. Fabrication techniques are governed by ASTM Standard Recommended Practice for Inner and Outer Diameters of Rigid Thermal Insulation for Nominal Sizes of Pipe and Tubing. Cellular glass is applied to cold piping in single or double layers and usually used with a joint sealant and then covered with either a fabric-reinforced mastic or a metal jacket. On hot piping applications cellular glass is applied to the pipe in single or double layers with metal bands. No sealants are used on hot applications. It is then covered with a metal jacket.

Elastomeric Foam Insulation

Elastomeric foams are used almost exclusively in commercial, institutional, and residential facilities. They are used primarily on hot water and chilled water lines or for water and sewer lines for freeze protection. Elastomeric foams are extruded into pipe dimensions and generally available in ½-in, ¾-in, and 1-in thicknesses. They are also available in sheet form for equipment. These foams are usually taped, wired, or glued in place.

Fiberglass and Mineral Wool Insulations

Fiberglass and mineral wool are actually two separate and distinct types of insulations; however, many of their applications and physical properties are similar. These products are generally used in hot applications but with some restrictions can be used in cold applications as well. Fiberglass is often used for chilled water piping, and other piping, up to a maximum of 850°F. Mineral wool has a peak temperature limit of 1200°F. Fiberglass is made from glass fibers bonded together with resin binders. Mineral wool is made from rock slag fibers and bonded together with resin or clay binders. These materials are generally applied with metal bands, wire, or tape and covered with a metal or nonmetallic flexible jacket. On indoor applications they are frequently covered with an all-service jacket.

Perlite Insulation

Perlite insulation is generally used in the same types of applications as calcium silicate. It is somewhat lighter in density and lower in compressive strength than calcium silicate; however, it is usually treated with a water inhibitor which tends to keep it drier than calcium silicate. Perlite insulation is also made in molds to fit the range of pipe-covering shapes required by industry. It is usually applied with metal bands and covered with a metal jacket.

Phenolic Foam Insulation

Phenolic foam is a very low thermal conductivity organic foam insulation used primarily for plastic piping in freeze protection applications. Phenolic foam insulation is made in a catalyst reaction bun and cut in a fabrication process to the sizes needed for the applications. It is generally applied with tape or wire and covered with an all-service or metal jacket, depending on the ambient condition and the geography.

Polystyrene Insulation

Polystyrene is a very inexpensive efficient thermal insulation used almost exclusively in residential and food processing applications. It comes in expanded boards and extruded buns. The extruded buns are sometimes used to fabricate pipe covering for chilled water lines or water and sewer lines. In residences it is used in the wall panels. In food processing it is used in the walls and on the roofs. It has a low permeability rating and is easy to work with. It is applied with bands, tape, wire, or glue depending on the application.

Polyurethane and Polyisocyanurate Foam Insulations

Polyurethane and polyisocyanurate foams are two chemically different insulation materials; however, their cell structure and physical properties are so similar that they are usually lumped into a common category. This is probably not fair to polyisocyanurate foam insulation because it generally tends to be of higher quality. Both insulations have very good thermal properties. They are used from about −200°F up to 300°F both indoors and outdoors. On cold applications they require multiple layers because of their contraction characteristics. These insulations are manufactured in batch bun processing and then sold to fabricators where they are cut into various shapes and sizes depending on the applications. These insulation materials are usually applied with tape or wire and covered with either a fabric-reinforced mastic or a metal jacket.

ACCESSORY MATERIALS

The accessory materials referenced in the above paragraphs and throughout this chapter are a necessary part of the insulation system. There are many manufacturers and suppliers of these materials and the quality can vary dramatically from one to another. The following are a few of the more common accessory materials used in industrial and commercial insulation system specifications with a brief description of each. See also Table B7.17.

Acrylic Latex Mastic

Acrylic latex mastic is a heavy-bodied weather barrier coating used primarily to cover rigid insulations such as cellular glass and polyurethane. It is generally ap-

TABLE B7.17 ASTM Specification Reference

Material	ASTM specification
Calcium silicate	C533-85
Cellular glass	C552-91
Elastomeric foam	C534-88
Fiberglass and mineral wool	C553-70 and C547-77
Expanded perlite	C610-85
Phenolic foam	C1126-89
Polystyrene foam	C578-87a
Polyurethane/polyisocyanurate foam	C591-85
Acrylic latex mastic	C647-89
Aluminum banding materials	C921-89
Aluminum jacketing	C921-89
ASJ jacketing	C921-89
Asphalt cutback mastic	C647-89
FRP jacketing	C921-89
Hypalon mastic	C647-89
Stainless steel banding materials	C921-89
Stainless steel jacketing	C921-89
6 × 6 polyester mesh fabric	NA
10 × 10 glass scrim	D1668-86

plied in two coats with a 6 × 6 polyester reinforcing fabric for impact and tear resistance. The finished coating thickness is typically about 1⁄16 in thick. This material does not provide vapor protection.

Aluminum Banding Materials

Aluminum bands are used as a securement for many types of insulation materials. The most common sizes specified range between ½, ¾, and 1 in in width and are 0.016 in thick. These bands are secured in place with metal band clips or seals of common dimensions. Aluminum bands should not be used in applications where the insulation may be used for fire protection.

Aluminum Jacketing

Aluminum jacketing materials come in many different sizes and configurations. In piping applications, either smooth or slightly embossed jacketing of 0.016-in thickness is most common. When specifying aluminum jacketing for use with permeable insulation materials and calcium silicate in particular, it is important to specify aluminum jacketing with a factory-applied moisture barrier liner.

ASJ Jacketing

ASJ jacketing stands for all service jacket. This material is a craft paper/foil/scrim laminate material used exclusively on indoor commercial applications. ASJ jacketing is usually factory applied and serves the primary function of providing moisture protection. These are not vapor barriers, but they can provide a water vapor retarding effect when properly applied.

Asphalt Cutback Mastic

Asphalt cutback mastics are heavy-bodied asphalts that are cut with mineral spirits so it can be put on with a trowel or sprayed on. When applied, the mineral spirits dissipates and leaves behind a hard asphaltic vapor barrier finish. It is generally applied in two coats with a 6 × 6 polyester reinforcing fabric for impact and tear resistance. The finished coating thickness is typically about 1⁄16 in. Once applied, asphalt cutback mastics are usually covered with a metal jacket for aesthetic purposes.

FRP Jacketing

FRP stands for fiber resin plastic or fiber reinforced plastic. By either definition it is a hard plastic membrane that is reinforced with glass fibers. FRP jacketing can be used in many of the same applications where aluminum is used. Also, due to its excellent chemical resistance, FRP jacketing is often used where aluminum cannot be used. FRP jacketing materials come in sheet or rolled form and the laps are sealed with a resin sealant recommended by the manufacturer.

Hypalon Mastic

Hypalon is a trade name for a highly flexible and extremely durable vapor barrier mastic material. There are numerous products on the market today that use this product in their compositions to form what are referred to as elastomeric membranes. These elastomeric membranes are also referred to generically as hypalons. Hypalons are usually reinforced with a 10 × 10 or a 10 × 20 fiberglass fabric to provide impact and tear resistance.

Stainless Steel Banding Materials

Stainless steel bands are used to support or secure insulation materials to piping. They are used in many of the same applications as aluminum bands. Stainless steel bands are generally used in either ½ or ¾-in widths and are 0.010 or 0.015 in thick. Stainless steel bands are secured with matching seals or clips. Stainless steel bands should be used whenever the insulation is being used in fire protection applications or where other insulation securement materials are not appropriate.

Stainless Steel Jacketing

Stainless steel jacketing is used to cover insulation materials of all types in various applications. Due to the cost, stainless steel is generally used where it is required for its chemical or fire resistance. Stainless steel jacketing usually has a smooth finish and is 0.010 in thick. On some large-diameter applications 0.015-in-thick stainless steel may be used.

6 × 6 Polyester Mesh Fabric

6 × 6 mesh refers to the number of strands of the primary fiber in 1 in^2 of fabric. A 6 × 6 mesh will have six primary strands going in one direction and six primary strands going perpendicular to that. In a polyester mesh fabric, the primary strands are woven together in a unique way to make a fabric which does not fray, pucker, or fishmouth. This fabric is ideally suited to heavy-bodied mastics such as asphalt cutback and acrylic latex.

Glass Scrim

Glass scrims come in many different configurations. The most common scrims used in industrial and commercial piping applications are 10 × 10 and 10 × 20. As mentioned above, the numbers refer to the number of primary strands in 1 in^2 of fabric. Glass scrims are best suited to light-bodied mastics, paints, and elastomeric membranes.

REFERENCES

1. National Insulation Contractors Association Insulation Craft Training Program, *Theory of Heat Transfer and Moisture Effects on Insulation.*
2. American Society for Testing and Materials Annual Book of Standards, *Thermal Insulation; Environmental Acoustics,* vol. 04.06.
3. Charles M. Pelanne, *Heat Flow Principles,* Johns Manville Refractory Products Research and Development Division, August 1976.
4. Charles W. Sisler, *Thermal Insulation Design Concepts,* Monsanto, Inc.
5. NFPA 30, *The Flammable and Combustible Liquids Code,* 1990 ed.
6. API Recommended Practice 521-82, *Guide for Pressure Relieving and Depressurizing Systems.*
7. Ludwig Adams, *Thermal Conductivity of Wet Insulations,* ASHRAE Journal, October 1974.
8. *ASHRAE Fundamentals Handbook,* Table 2, "Thermodynamic Properties of Water in Saturation," American Society of Heating and Refrigerating Engineering, Inc., 1989.
9. William C. Turner and John F. Malloy, *Thermal Insulation Handbook,* Robert E. Krieger/McGraw-Hill, 1981.
10. *Therminol Heat Transfer Fluid, Data Sheet,* Monsanto Industrial Chemicals Company, MIC-4-081, pp. 1–24.
11. P. C. Bowes, *Fires in Oil Soaked Lagging,* Building Research Establishment Current Paper, CP 35/74, February 1974, pp. 1–11.
12. *Fire Hazards in Oxygen Enriched Atmosphere,* National Fire Protection Association, NFPA 53M, copyright ©1990, NFPA, Quincy, MA 02269.
13. *Precautions for the Proper Usage of Polyurethanes, Polyisocyanurates and Related Materials,* Technical Bulletin, The Upjohn Chemical Division, 107, May 1980, p. 44.

CHAPTER B8
FLOW OF FLUIDS

Tadeusz J. Swierzawski
Consultant, Mechanical Division
Stone & Webster Engineering Corporation
Boston, Massachusetts

Daniel A. Van Duyne
Assistant Chief Engineer, Mechanical Division
Stone & Webster Engineering Corporation
Boston, Massachusetts

The primary objective of this chapter is to show the user the most important logical milestones and the general background of equations and formulas recommended for specific practical applications of fluid flow in pipes, nozzles, and orifices. For details, Refs. 1 through 4 or other equivalent textbooks should be consulted. A nomenclature section is provided to minimize repeating of variable definitions.

NOMENCLATURE

a	Acceleration, ft/s^2
A	Cross-sectional area, ft^2
A_f	Flow area occupied by liquid phase, ft^2
A_g	Flow area occupied by gaseous phase, ft^2
c	Velocity of sound, ft/s
C	Flow coefficient defined by Eq. (B8.39), dimensionless
C_A	Allen flow constant defined by Eq. (B8.87)
C_c	Coefficient of contraction, dimensionless
C_d	Discharge coefficient, dimensionless
C_j	Effective jet discharge coefficient, dimensionless
c_p	Specific heat at constant pressure, Btu/(lb$_m$ · °F)
c_v	Specific heat at constant volume, Btu/(lb$_m$ · °F)
C_v	Velocity coefficient, dimensionless
d	Nozzle or orifice diameter

D	Pipe inside diameter, ft
d_g	Flow height for gaseous phase, ft
E	Energy, Btu
$\dot{E}$	Energy flow rate, Btu/s
f	Darcy-Weisbach friction factor, dimensionless
F	Force, lb_f
F_a	Area factor for thermal expansion of primary elements (see Fig. B8.21)
Fr	Froude number, dimensionless
F_{va}	Approach velocity factor defined by Eq. (B8.36), dimensionless
g	Acceleration of gravity, ft/s^2
g_c	32.174 $(lb_m \cdot ft)/(lb_f \cdot s^2)$ (dimensional conversion factor)
g_s	32.174 ft/s^2 (standard acceleration of gravity)
$\dot{G} = \dot{m}/A$	Mass flux, $lb_m/(ft^2 \cdot s)$
$\dot{G}_c$	Maximum (critical) mass flux, $lb_m/(ft^2 \cdot s)$
$\dot{G}_{fs}$	Superficial mass flux of liquid phase, $lb/(ft^2 \cdot s)$
$\dot{G}_{gs}$	Superficial mass flux of gaseous phase, $lb/(ft^2 \cdot s)$
h	Enthalpy (static), Btu/lb_m
H_f	Friction head, ft
h_{fg}	Latent heat of vaporization $(h_g - h_f)$, Btu/lb
h_o	Stagnation enthalpy, Btu/lb_m
J	778.169 $(ft \cdot lb_f)/Btu$ (dimensional conversion factor)
$j_f = \dot{G}_{fs}v_f$	Superficial velocity (volumetric flux) of liquid phase, ft/s
$j_g = \dot{G}_{gs}v_g$	Superficial velocity (volumetric flux) of gaseous phase, ft/s
k	c_p/c_v (specific heat ratio)
K	Flow resistance coefficient (K factor), dimensionless
L	Length, ft
m	Mass, lb_m
$\dot{m}$	Mass flow rate, lb_m/s
M	Mass, slugs
Ma	Mach number, dimensionless
p	Pressure, lb_f/ft^2
p_o	Stagnation pressure, lb_f/ft^2
Q	Heat, Btu
$\dot{Q}$	Heat flow rate, Btu/s
R	Gas constant, $(ft \cdot lb_f)/(lb_m \cdot °R)$
Re	Reynolds number, dimensionless
r_h	Hydraulic radius, ft
s	Entropy, $Btu/(lb_m \cdot °R)$
s_{fg}	$s_g - s_f$, $Btu/(lb_m \cdot °R)$

s_o	Stagnation entropy, Btu/($lb_m \cdot °R$)
S	Slip ratio, v_g/v_f
t	Time, s
T	Temperature, °R (or °F)
T_o	Stagnation temperature, °R
u	Internal energy, Btu/lb_m
U	Internal energy, Btu
v	Velocity, ft/s
V	Volume, ft^3
w	Work, ($ft \cdot lb_f$)/lb_m
W	Work, $ft \cdot lb_f$
$\dot{W}_t$	Technical work rate (power), $ft \cdot lb_f/s$
x	Mass quality of steam, dimensionless
y	Distance to fixed surface, ft
z	Elevation, ft

Greek Symbols

α	Void fraction, dimensionless
α_s	Quantity defined by Eq. (B8.54)
β	d/D, dimensionless
β_s	Critical pressure ratio, dimensionless
γ	Specific weight, lb_f/ft^3
μ	Dynamic (absolute) viscosity, $(lb_f \cdot s)/ft^2$
μ_f	Viscosity of liquid phase, $(lb_f \cdot s)/ft^2$
μ_g	Viscosity of gaseous phase, $(lb_f \cdot s)/ft^2$
ν	Kinematic viscosity, ft^2/s
ν_f	Kinematic viscosity of liquid phase, ft^2/s
ν_g	Kinematic viscosity of gaseous phase, ft^2/s
π	Pi, approximately 3.14159, dimensionless
ρ	Density, lb_m/ft^3
ρ_f	Density of liquid phase, lb_m/ft^3
ρ_g	Density of gaseous phase, lb_m/ft^3
ψ_s	Dimensionless quantity defined by Eq. (B8.45)
τ	Shearing stress, lb_f/ft^2
υ	Specific volume, ft^3/lb_m
υ_f	Specific volume of saturated liquid, ft^3/lb_m
υ_g	Specific volume of saturated steam, ft^3/lb_m
υ_{fg}	$\upsilon_g - \upsilon_f$, ft^3/lb_m
ϕ_{lo}^2	Two-phase pressure drop multiplier, dimensionless

Subscripts

f Liquid or water phase, saturated liquid
g Steam phase, saturated steam

BASIC FLUID PROPERTIES

A *fluid* is a substance which deforms continuously under the action of shearing forces. Fluids offer no resistance to distortion of form; they yield continuously to tangential forces, no matter how small. Ordinarily, fluids are classified as being liquids or gases. Some classifications also include the vapor form among the group of fluids.

Liquids change volume very slightly with considerable variation in pressure, and when the pressure is removed, they do not dilate significantly.

A *gas* is a fluid which tends to expand to fill completely any vessel in which it is contained. A change in pressure is accompanied by a considerable change in volume.

A *perfect gas* satisfies two conditions: (1) It obeys the gas law $pV = mRT$, and (2) its specific heats are constant. In (1) above, p, V, and T are pressure, volume, and absolute temperature, respectively, and m and R represent the mass of gas and the gas constant, all in proper units. From (2) it follows also that the specific-heat ratio $k = c_p/c_v$ is also constant.

A real gas, at a pressure low with respect to sea-level barometric pressure and at a temperature high with respect to its critical temperature, obeys with reasonable accuracy the perfect-gas law. If higher degrees of accuracy are required, textbooks in thermodynamics list corrections (as, for example, van der Waals' equation) which take into account deviations from ideal conformance.

SURVEY OF DIMENSIONS AND UNITS

Dimensions, although discussed in other sections of this handbook, are discussed briefly to review some fundamentals and to clarify the weight and mass densities commonly used in fluid mechanics analyses.

A dimensional system that will completely describe the events that occur in the science of mechanics can be constructed from three fundamental dimensions: length, mass, and time. The English units in this system are: the pound of mass (lb_m), the foot (ft), and the second (s). The SI (Systeme International) units in this system are: kilogram of mass (kg), meter (m), and second (s).[5]

An alternative system (engineering system) of dimensions uses force instead of mass. The English units are: the pound force (lb_f), the foot, and the second. The dimensions of mass in this system are derived from Newton's relationship that force equals mass times acceleration.

$$F = ma \tag{B8.1}$$

Then, the unit of mass in this system is $(lb_f \cdot s^2)/ft = 1$ slug. A slug is the mass that can be accelerated at the rate of 1 ft/s^2 by a force of 1 standard pound force. By definition, 1 standard pound force will also accelerate 1 pound mass at the

1 slug = 32.174 lb_m

FIGURE B8.1 Slug versus pound mass.

rate of 32.174 ft/s^2. It follows that the slug is the mass of size 32.174 times that of the pound mass (Fig. B8.1), and g_c = 32.174 lb_m/slug = 32.174 $(lb_m \cdot ft)/(lb_f \cdot s^2)$ is the dimensional conversion factor between the two systems of units.

The density ρ is the mass per unit volume of the fluid and is not to be confused with specific weight γ, the weight per unit volume. The interrelation of density and specific weight is $\gamma = \rho g/g_c$. When mass units are lb_m, the g_c factor must always be used with density as ρ/g_c to account for dimensional conversions.

Because of the increasing international commitment of U.S. Engineering and U.S. Industry and Commerce, the need for an accelerated growth in acceptance of the capability of SI units by the engineering profession is recognized. However, in this edition of the *Piping Handbook,* the English units of the Engineering Dimensional System are used. For those who must perform calculations in other units, useful conversion factors are presented in App. E1 and in Tables B8.1, B8.2, and B8.3.

THEORETICAL BACKGROUND

Some basic fluid properties used in this chapter are discussed below, but for more detailed information, see Refs. 1 and 2, listed at the end of this chapter.

Viscosity

There is an experimental fact that the fluid in immediate contact with a solid boundary has the same velocity as the boundary itself. For the case of Fig. B8.2

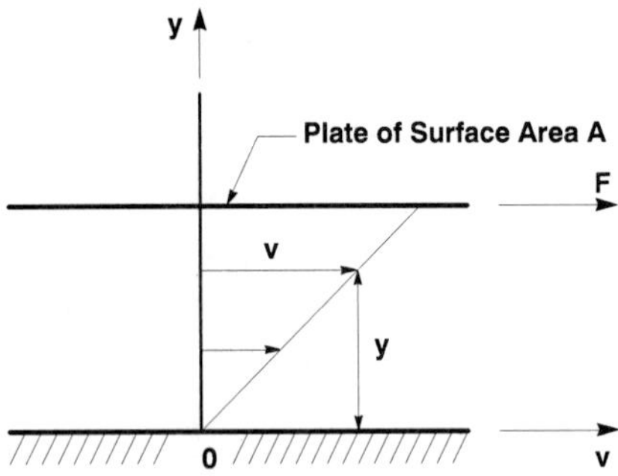

FIGURE B8.2 Fluid flow adjacent to solid boundary.

in which a fluid separates closely spaced parallel plates, F is directly proportional to A and v, and is inversely proportional to the distance y between the plates. The last statement is expressed in the form of Newton's law of viscosity:

$$F = \frac{\mu A v}{y} \tag{B8.2}$$

in which μ is the proportionality factor and is called the *dynamic* (or *absolute*) viscosity of the fluid.

The CGS unit of dynamic viscosity, called the *poise* (P), is gram/(centimeter · second) [g/(cm · s)], and:

$$\text{centipoise} = \frac{\text{poise}}{100}$$

Water at 68°F (20°C) has a viscosity of 1.002 cP. Conversion factors for viscosity are presented in Tables B8.1, B8.2, and B8.3.

Since the shear stress is $\tau = F/A$, Eq. (B8.2) could be written as $\tau = \mu v/y$, or in differential form as $\tau = \mu dv/dy$. The ratio dv/dy is called the *rate of angular deformation of the fluid for one-dimensional flow.* Viscosity is that property of a fluid by virtue of which it offers resistance to shear. In newtonian fluids, there is a linear relation between the applied shear stress and the resulting rate of deformation [μ = constant in Eq. (B8.2)]. In nonnewtonian fluids, this relationship is not linear. Gases and thin liquids tend to be newtonian fluids, while thick hydrocarbons may be nonnewtonian.

The kinematic viscosity ν, which is the ratio of dynamic viscosity to mass density ρ, expressed in units under consideration, is:

$$\nu = \frac{\mu}{\rho} g_c \tag{B8.3}$$

The use of kinematic viscosity tends to prevent errors when performing calculations in different dimensional systems.

Pressure Variation in a Static Fluid

The hydrostatic pressure existing at a point within a fluid body due to the weight of the fluid above that point is known as the *hydrostatic pressure.* In the case of

TABLE B8.1 Conversion Factors for Viscosity
(Force × time/area ≈ mass/length × time)

To obtain → multiply ↓, by ↘	Pa × s*	$\frac{\text{lbf} \times \text{s}}{\text{ft}^2}$	$\frac{\text{lbm}}{\text{ft} \times \text{s}}$	$\frac{\text{lbm}}{\text{h} \times \text{ft}}$	$\frac{\text{g}}{\text{cm} \times \text{s}}$ (poise)	$\frac{\text{kg}}{\text{m} \times \text{s}}$
Pa × s*	1	$\frac{(0.3048)^2}{9.80665 \times 0.45359237}$ = 0.020 885 4342	$\frac{0.3048}{0.45359237}$ = 0.671968 975	$\frac{0.3048 \times 3600}{0.45359237}$ = 2419.08831	10	1
$\frac{\text{lbf} \cdot \text{s}}{\text{ft}^2}$	$\frac{9.80665 \times 0.45359237}{(0.3048)^2}$ = 47.8802590	1	$\frac{980.665}{30.48}$ = 32.1740486	$\frac{980.665 \times 3600}{30.48}$ = 115826.575	$\frac{980.665 \times 453.59237}{30.48^2}$ = 478.802590	$\frac{980.665 \times 453.59237}{10 \times 30.48^2}$ = 47.8802590
$\frac{\text{lbm}}{\text{ft} \times \text{s}}$	$\frac{0.45359237}{0.3048}$ = 1.48816394	$\frac{30.48}{980.665}$ = 0.031080 9502	1	3600	$\frac{453.59237}{30.48}$ = 14.8816394	$\frac{453.59237}{304.8}$ = 1.48816394
$\frac{\text{lbm}}{\text{h} \times \text{ft}}$	$\frac{0.45359237}{0.3048 \times 3600}$ = 0.413378 873 × 10^{-3}	$\frac{30.48}{980.665 \times 3600}$ = 8.633597 27 × 10^{-6}	$\frac{1.0}{3600}$ = 0.000277 777778	1	$\frac{453.59237}{30.48 \times 3600}$ = 0.004133 78873	$\frac{453.59237}{304.8 \times 3600}$ = 0.000413 378873
$\frac{\text{g}}{\text{cm} \times \text{s}}$ (poise)	0.1	$\frac{30.48^2}{980.665 \times 453.59237}$ = 0.002088 54342	$\frac{30.48}{453.59237}$ = 0.067196 8975	$\frac{3600 \times 30.48}{453.59237}$ = 241.908831	1	0.1
$\frac{\text{kg}}{\text{m} \times \text{s}}$	1	$\frac{10 \times 30.48^2}{980.665 \times 453.59237}$ = 0.020885 4342	$\frac{304.8}{453.59237}$ = 0.671968 975	$\frac{10 \times 3600 \times 30.48}{453.59237}$ = 2419.08831	10	1

*SI units for ASME use.
Example: 1 lbf × s/ft^2 = 32.174 048 6 lbm/(ft × s).
From definitions: 1 poise = 1 P = 1 g/cm × s = 1 dyn × s/cm^2 = 0.1 kg/m × s = 0.1 N × s/m^2 = 0.1 Pa × s = 0.1 Pa × s.
Source: Ref. 6.

TABLE B8.2 Conversion Factors for Kinematic Viscosity (area/time)

To obtain → multiply ↓, by ↘	$\frac{m^{2*}}{s}$	$\frac{ft^2}{s}$	$\frac{cm^2}{s}$ (stoke)	$\frac{cm^2}{h}$	$\frac{m^2}{h}$
$\frac{m^{2*}}{s}$	1	$\frac{1.0}{(0.3048)^2}$ $= 10.7639104$	$(100)^2 = 10{,}000$	$(100)^2 \times 3600 = 36 \times 10^6$	3600
$\frac{ft^2}{s}$	$(0.3048)^2$ $= 0.092\ 903\ 04$	1	$(30.48)^2 = 929.030\ 4$	$(30.48)^2 \times 3600$ $= 3{,}344{,}509.440\ 000$	$(0.304\ 8)^2 \times 3600$ $= 334.450\ 944\ 000$
$\frac{cm^2}{s}$ (stoke)	$\frac{1.0}{(100)^2} = 10^{-4}$	$\frac{1.0}{(30.48)^2}$ $= 0.001\ 076\ 391\ 04$	1	3600	$\frac{3600}{(100)^2} = 0.36$
$\frac{cm^2}{h}$	$\frac{1{:}0}{(100)^2 \times 3600}$ $= 27.777\ 777\ 8 \times 10^{-9}$	$\frac{1.0}{(30.48)^2 \times 3600}$ $= 0.2\ 989\ 975\ 12 \times 10^{-6}$	$\frac{1.0}{3600}$ $= 0.000\ 277\ 777\ 778$	1	$\frac{1.0}{(100)^2} = 0.0001$
$\frac{m^2}{h}$	$\frac{1.0}{3600}$ $= 277.777\ 778 \times 10^{-6}$	$\frac{1.0}{(0.3048)^2 \times 3600}$ $= 2.989\ 975\ 12 \times 10^{-3}$	$\frac{(100)^2}{3600}$ $= 2.777\ 777\ 78$	$(100)^2 = 10{,}000$	1

*SI units for ASME use.
Example: 1 ft^2/s = 3600 ft^2/h.
Source: Ref. 6.

TABLE B8.3 Conversion Factors for Thermal Conductivity. (Energy/Time × Length × Temp. Diff. ~ Power Length × Tem. Diff.)

To obtain → Multiply ↓, by ↘	$\frac{Btu}{hr \times ft \times F}$	$\frac{ft \times lbf}{hr \times ft \times F}$ $\frac{lfb}{hr \times F}$	$\frac{Watt}{ft \times F}$	$\frac{Watt}{m \times C}$	$\frac{kp \times m}{hr \times m \times C} = \frac{kp}{hr \times C}$	$\frac{cal}{sec \times cm \times C}$	$\frac{kcal}{hr \times m \times C}$
$\frac{Btu}{hr \times ft \times F}$	1	$\frac{2.326 \times 10^7}{980.665 \times 30.48}$ = 778.169 262	$\frac{2.326 \times 453.592\ 37}{3\ 600}$ =0.293 071 070	$\frac{4.186\ 8 \times 453.592\ 37}{3600 \times 0.304\ 8}$ = 1.730 734 67	$\frac{4.186\ 8 \times 453.592\ 37}{980.665 \times 30.48 \times 10^{-4}}$ = 635.348 952	$\frac{453.592\ 37}{3\ 600 \times 30.48}$ =0.004 133 788 73	$\frac{453.592\ 37}{10 \times 30.48}$ = 1.488 163 94
$\frac{ft \times lbf}{hr \times ft \times F}$	$\frac{980.665 \times 30.48}{2.326 \times 10^7}$ = 0.001 285 067 46	1	$\frac{980.665 \times 453.592\ 37}{3\ 600 \times 10^7/30.48}$ = 0.000 376 616 097	$\frac{980.665 \times 453.592\ 37}{3\ 600 \times 10^5 \times 5/9}$ = 0.002 224 110 81	$\frac{453.592\ 37 \times 9/5}{1\ 000}$ = 0.816 466 266 000	$\frac{980.665 \times 453.592\ 37}{3\ 600 \times 2.326 \times 10^7}$ = 5.312 197 40 × 10^{-6}	$\frac{980.665 \times 453.592\ 37}{2.326 \times 10^8}$ = 0.001 912 391 06
$\frac{Watt}{ft \times F}$	$\frac{3\ 600}{2.326 \times 453.592\ 37}$ = 3.412 141 63	$\frac{3\ 600 \times 10^7/30.48}{980.665 \times 453.592\ 37}$ = 2.655.223 74	1	$\frac{9/5}{0.304\ 8}$ = 5.905 511 81	$\frac{3\ 600 \times 10^4 \times 9/5}{980.655 \times 30.48}$ = 2 167.900 61	$\frac{1.0}{30.48 \times 2.326}$ = 0.0014 105 072 6	$\frac{360}{30.48 \times 2.326}$ = 5.077 826 15
$\frac{Watt}{m \times C}$	$\frac{3\ 600 \times 0.304\ 8}{4.186\ 8 \times 153.592\ 37}$ = 0.577 789 316	$\frac{3\ 600 \times 10^5 \times 5/9}{980.665 \times 453.592\ 37}$ = 449.617 886	0.304 8 × 5/9 = 0.169 333 333	1	$\frac{3\ 600 \times 100}{980.665}$ = 367.097 839	$\frac{1.0}{418.68}$ = 0.002 388 458 97	$\frac{360}{418.68}$ = 0.859 845 228
$\frac{kp \times m}{hr \times m \times C}$	$\frac{980.665 \times 30.48 \times 10^{-4}}{4.186\ 8 \times 453.592\ 37}$ = 0.001 573 938 22	$\frac{10^3 \times 5/9}{453.592\ 37}$ = 1.22479 035	$\frac{980.665 \times 30.48 \times 5/9}{3\ 600 \times 10^4}$ = 0.000 461 275 759	$\frac{980.665}{3\ 600 \times 100}$ = 0.002 724 069 44	1	$\frac{980.665}{3\ 600 \times 4.186\ 8 \times 10^4}$ = 6.506 328 09 × 10^{-6}	$\frac{980.665}{4.186\ 8 \times 10^{-5}}$ = 0.002 342 278 11
$\frac{cal}{sec \times cm \times C}$	$\frac{3\ 600 \times 30.48}{453.592\ 37}$ = 241.908 831	$\frac{3\ 600 \times 2.326 \times 10^7}{980.665 \times 453.592\ 37}$ = 188 246.017	2.326 × 30.48 = 70.896 48	418.68	$\frac{3\ 600 \times 4.186\ 8 \times 10^4}{980.665}$ = 153 696.522	1	360
$\frac{kcal}{hr \times m \times C}$	$\frac{10 \times 30.48}{153.592\ 37}$ 0.671 968 975	$\frac{2.326 \times 10^8}{980.665 \times 453.592\ 37}$ 522.905 602	$\frac{2.326 \times 30.48}{360}$ = 0.196 934 667	$\frac{418.68}{360}$ = 1.163	$\frac{4.186\ 8 \times 10^5}{980.665}$ = 426.934 784	$\frac{1.0}{360}$ = 0.002 777 777 78	1

All values given in the rational fractions are exact and are taken from Appendix 4.

Example, $1 \frac{Watt}{ft \times F} - 1 \frac{Watt}{ft \times F} \times 3.412\ 141\ 632 \frac{Btu/(hr \times ft \times F)}{Watt/(ft \times F)} = 3.412\ 141\ 632 \frac{Btu}{hr \times ft \times F}$

Source: Reference 6.

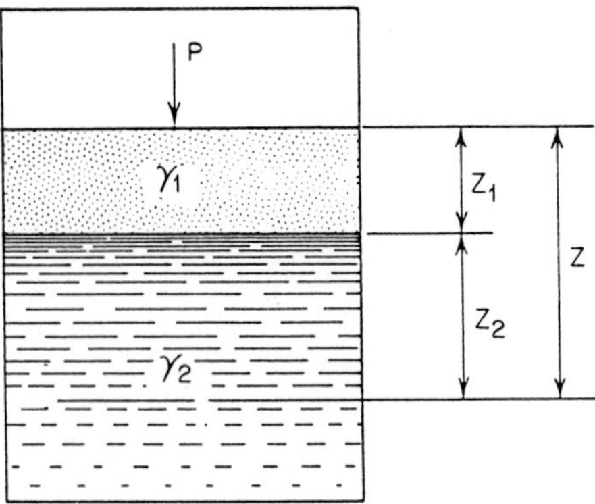

FIGURE B8.3 Hydrostatic pressure in two-fluid mixture.

gaseous fluids, the weight of the fluid column is relatively small unless great vertical heights are involved. With denser fluids such as liquids, the increase in pressure due to depth within the liquid can be of great significance. When applied to the two-fluid mixture in a container as shown in Fig. B8.3, the expression for hydrostatic pressure at a horizontal plane located a distance z below the free surface is:

$$p = p_1 + \gamma_1 z_1 + \gamma_2 z_2 \tag{B8.4}$$

Standard atmospheric pressure is defined as that pressure produced by a column of mercury of 760 mm length at a mercury density of 13.5950889 g/cm^3 (at 0°C or 32°F) and at an acceleration due to gravity of 32.1740 ft/s^2. On this basis, then:

$$1 \text{ standard atmosphere} = 14.6959488 \text{ psia}$$
$$= 1.01325 \text{ bar}$$

$$(1 \text{ bar} = 10^5 \text{ Pascals})$$

For many engineering calculations, it is sufficiently accurate to use 14.7 psia as being equivalent to 1 standard atmosphere.

Continuity

Applying the law of conservation of mass to a flow process yields a mass balance, or continuity, equation. The mass balance in physical processes is that the mass of substance added to the system is equal to the mass subtracted from the system, plus accumulation of any mass in the system.

Conservation of Energy

The first law of thermodynamics can be formulated as the principle of conservation of energy.[2] The change in the total energy within a control boundary, as shown in Fig. B8.4, is the difference between all types of added energy and all types of subtracted energy during a specified time interval. The change in energy

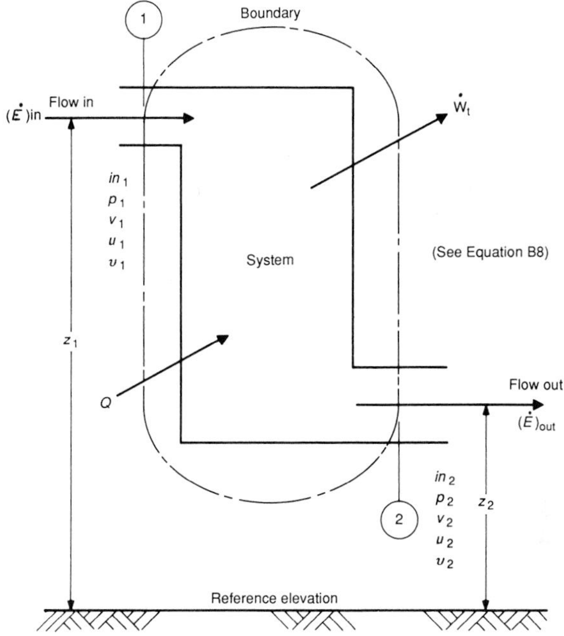

FIGURE B8.4 Energy balance in an open system.

of the system can be calculated by knowing the initial and final states of the system. The difference in states of the system at rest, being independent of the path from initial to final state, must be a property of the system or a parameter of state. It is called the *internal energy* and is denoted by U.

If $E_{in} = Q$ is the net heat flowing into the system during the process, and $E_{out} = W$ is the net work done by the system during the process (note that heat Q and work W are the energies in transit), then energy E_{in} or Q may be expressed as:

$$Q = (U_2 - U_1) + \frac{W}{J} \tag{B8.5}$$

where $J = 778.169\ (\text{ft} \cdot \text{lb}_f)/\text{Btu}$ (dimensional conversion factor)

The sum of the internal energy $u = U/m$ and the external pumping work ($p\upsilon$) which is required to maintain the flow:

$$h = u + \frac{p\upsilon}{J} \tag{B8.6}$$

is called the *enthalpy*. Since, u, p, and υ are all properties, then also h must be a state function or a property. The sum:

$$h_o = h + \frac{\text{v}^2}{2g_c J} \tag{B8.7}$$

is called the *stagnation enthalpy*.

Combining all variables from Fig. B8.4 with enthalpy yields an equation of the following form for an open system undergoing a steady flow with a mass flow rate of m:

$$\dot{Q} - \frac{\dot{W}_t}{J} = \dot{m}\left[(h_2 - h_1) + \frac{v_2^2 - v_1^2}{2J\,g_c} + \frac{g}{J\,g_c}(z_2 - z_1)\right] \tag{B8.8}$$

In a differential form, per unit mass, this equation may be regrouped to yield the following equation which is known from literature as the *mechanical energy balance:*

$$d\left(\frac{v^2}{2g_c}\right) + \frac{g}{g_c}\,dz + v\,dp = d'w_t - d'w_f \tag{B8.9}$$

where $d'w_t$ = "technical work" transmitted across the boundaries
$d'w_f$ = work done against fluid friction

STEADY SINGLE-PHASE INCOMPRESSIBLE FLOW IN PIPING

Characteristics of Incompressible Flow

Although there is no such thing in reality as an incompressible fluid, this term is applied to liquids. Yet sound waves, which are really pressure waves, travel through liquids. This is an evidence of the elasticity of liquids. In problems involving water hammer, it is necessary to consider the compressibility of the liquid. The compressibility of a liquid is expressed by its bulk modulus of elasticity which influences the wave speed in the liquid.

It should be explained here that when density changes of compressible fluids (gases or steam) are gradual and do not change by more than about 10 percent, the flow may be treated as incompressible with the use of an average density.

Bernoulli's equation, which is a special case of Eq. (B8.8), when supplemented by a frictional head loss H_f (expressed in feet of a column of the fluid) and a pump head term H_P (total dynamic head TDH, expressed in feet of a column of fluid) yields:

$$\frac{v_1^2}{2g} + \frac{p_1}{\gamma} + z_1 + H_p = \frac{v_2^2}{2g} + \frac{p_2}{\gamma} + z_2 + H_f \tag{B8.10}$$

It can be shown that the frictional head loss is:

$$H_f = \frac{\Delta p}{\gamma} = f\frac{v^2}{2g}\frac{L}{D} \tag{B8.11}$$

and the corresponding pressure drop (expressed in lb_f/ft^2) is:

$$\Delta p = f\rho\frac{v^2}{2g_c}\frac{L}{D} \tag{B8.12}$$

where f is the Darcy-Weisbach friction factor which is four times the Fanning friction factor.

For laminar flow:[1]

$$f = \frac{64}{Re} \tag{B8.13}$$

It applies to all roughnesses, as the head loss in laminar flow is independent of wall roughness. The Reynolds number is:

$$Re = \frac{Dv}{\nu} \tag{B8.14}$$

Expressions for calculating the loss of pressure in turbulent flow are based upon experimental data. Typical sources of engineering values are Refs. 7 and 8.

Applications: Water Systems

Water system design requirements and parameters are developed in the system descriptions, which include piping & instrument diagrams (P&IDs), and descriptions of the system individual components (pumps, valves, heat exchangers, etc.), design flow rates, valve alignments, control valve operation, and pipe sizes. Pipe sizes are determined by evaluation of water velocities in the system. Selection of velocities must consider many factors including:

- Pipe material
- Water quality
- System flow balance requirements
- Economic evaluation of pipe cost versus pumping costs
- Available pump head
- Water hammer prevention

Typical water flow velocities (feet per second) for various applications based on general industry practices are shown in Table B8.4.

A steady-state analysis of a water system is generally performed twice. The first calculation is needed to determine data required for the purchase of the sys-

TABLE B8.4 Typical Water Flow Velocities

Condensate pump suction	3 ft/s
Condensate pump discharge	10 ft/s
Booster pump discharge	10 ft/s
Feed pump suction (no deaerator)	10 ft/s
For systems with deaerators, the downcomer sizes should be optimized with a pump NPSH decay in mind.	
Feed pump discharge	20–25 ft/s
Circulating water system	≤ 9 ft/s
General service	5–10 ft/s

Source: Stone & Webster.

tem components. This initial analysis is performed after the system description and general arrangement drawings have been issued. The initial analysis proceeds as follows:

1. Issue the system description and general arrangement drawings.
2. Develop a preliminary piping layout based on P&IDs and general arrangement drawings.
3. Determine the system operating mode or modes (valve lineup, number of pumps, etc.) which are critical for equipment purchase.
4. Calculate the pressure drop in the piping for design flow rates.
5. Develop a system resistance curve.
6. Calculate the flow balance in branching systems.
7. Revise the system description (if required) to achieve desired system flow balance.
8. Develop a revised system resistance curve (if required).
9. Calculate values of parameters required for the purchase of system components (pump head, system pressures, control valve pressure drop, etc.).

The final steady-state hydraulic analysis of a system is usually performed after major equipment has been purchased and piping drawings have been prepared. This analysis is to confirm acceptable operation for all operating modes by reconciling the as-built piping and installed equipment with the initial analysis.

The final analysis proceeds as follows:

1. Determine the pressure drop in the system based on piping drawings and equipment suppliers' data for design flow rates (if required).
2. Develop system resistance curves (if required).
3. Plot pump curves on the system resistance curve to determine the system flow rates.
4. Compare design flow rates with calculated flow rates for all operating modes.

Due to problems which occur during construction, the as-built drawings often differ from the piping drawings. In most instances, these changes are minor and do not affect the results of the analysis. Should any major changes occur during construction, the effect on the analysis results would need to be evaluated.

Piping Configurations. Piping systems may be arranged in series, in parallel, and as branching pipes. For pipes in series, the total head loss is the sum of head losses in pipe sections, and the flow rate is the same at all sections. For pipes in parallel, the head loss is the same for each flow path, and the total flow rate is the sum of flows in parallel sections. For branching pipes, the analyst must combine the above two principles. The total energy at a junction must be the same for all branches of the pipe.

Analysis of a piping system begins with a determination of the head loss (pressure drop) due to pipe friction, form losses and equipment for the design flow rate. For ease in evaluating the total system performance, the head loss should be calculated for each pipe branch separately. The manual head loss calculations often are done on pressure drop sheets. Calculations of head loss should include a schematic sketch of the piping layout. The calculation should be a clear record of the design basis for head loss calculations so that as the design evolves, the significance of pipe changes is apparent. A second head loss calculation may be required if the pipeline is expected to operate at flow rates considerably different from the design flow rate.

Having calculated the head loss in each branch of the system for the design flow rate, the performance of the entire system can be calculated. System resistance curves, flow balances, pump design points, and system pressures are some of the products of the system analysis.

The complexity of the system analysis depends on the number of pipes in series, pipes in parallel, and pipe branches. Obviously the greater the number of pipes, the more complex the analysis becomes. The analysis is done either by manual calculation or by computer analysis. Computerized calculations are preferable.

Manual calculations are suited for relatively simple piping systems. Manual analysis with the use of graphs provides an easy method for analyzing systems with a number of pipes in parallel, pipes in series, and branching pipes. This technique is based on the assumption that head loss varies with the square of velocity. Table B8.5 presents flow data for water through various sizes of Schedule 40 steel pipe.

The hydraulic grade line (HGL) and energy grade line (EGL) are two useful engineering tools in the hydraulic design of any system in which a liquid is in a dynamic state.[1] The graphical representation, with respect to any selected datum, of the piezometric (static pressure + elevation) head for each point along the pipe under consideration as ordinate, plotted against distance along the pipe as the abscissa, is called the *hydraulic grade line.* Similar graphical representation of the total Bernoulli head (piezometric + kinetic), as described by Eq. (B8.10) is called the *energy grade line.* Figure B8.5 illustrates the HGL and EGL for a sample system.

Economic Optimization of Line Sizes. The objective of this subject is balancing the savings in cost obtained by reduction in the size of the piping diameter against the increased cost of pumping equipment and power necessary to provide for increased pressure drop.

For given flows, as the pipe diameter decreases, the flow velocity increases, which decreases the cost of erected piping, including fittings, hangers, supports, and labor as represented by cost *a* in Fig. B8.6. However, the piping pressure drop ratio increases to the 5th power with reduction in the pipe internal diameter ratio, as shown below. The same is true with the pumping power, which is proportional to the pressure drop in the line, as represented by the cost *b*. In addition, there is the cost of the pump and the pump drive, which increases as the larger pumping power is required, as represented by cost *c*.

d_2/d_1	$(d_1/d_2)^5$
0.99	1.050
0.98	1.106
0.97	1.165
0.96	1.230
0.95	1.292
0.90	1.694
0.80	3.052
0.70	5.950

TABLE B8.5 Flow of Water through Schedule 40 Steel Pipe

Discharge		Pressure drop per 100 ft and velocity in Schedule 40 pipe for water at 60°F															
gal/min	ft^3/s	Velocity, ft/s	Press. drop, lb/in^2	Velocity, ft/s	Press. drop, lb/in^2	Velocity, ft/s	Press. drop, lb/in^2	Velocity, ft/s	Press. drop, lb/in^2	Velocity, ft/s	Press. drop, lb/in^2	Velocity, ft/s	Press. drop, lb/in^2	Velocity, ft/s	Press. drop, lb/in^2	Velocity, ft/s	Press. drop, lb/in^2
		1/8 in		1/4 in		3/8 in		1/2 in									
0.2	0.000446	1.13	1.86	0.616	0.359												
0.3	0.000668	1.69	4.22	0.924	0.903	0.504	0.159	0.317	0.061	3/4 in							
0.4	0.000891	2.26	6.98	1.23	1.61	0.672	0.345	0.422	0.086								
0.5	0.00111	2.82	10.5	1.54	2.39	0.840	0.539	0.528	0.167	0.301	0.033						
0.6	0.00134	3.39	14.7	1.85	3.29	1.01	0.751	0.633	0.240	0.361	0.041						
0.8	0.00178	4.52	25.0	2.46	5.44	1.34	1.25	0.844	0.408	0.481	0.102	1 in		1¼ in			
1	0.00223	5.65	37.2	3.08	8.28	1.68	1.85	1.06	0.600	0.602	0.155	0.371	0.048			1½ in	
2	0.00446	11.29	134.4	6.16	30.1	3.36	6.58	2.11	2.10	1.20	0.526	0.743	0.164	0.429	0.044		
3	0.00668			9.25	64.1	5.04	13.9	3.17	4.33	1.81	1.09	1.114	0.336	0.644	0.090	0.473	0.043
4	0.00891			12.33	111.2	6.72	23.9	4.22	7.42	2.41	1.83	1.49	0.565	0.858	0.150	0.630	0.071
5	0.01114	2 in				8.40	36.7	5.28	11.2	3.01	2.75	1.86	0.835	1.073	0.223	0.788	0.104
6	0.01337	0.574	0.044	2½ in		10.08	51.9	6.33	15.8	3.61	3.84	2.23	1.17	1.29	0.309	0.946	0.145
8	0.01782	0.765	0.073			13.44	91.1	8.45	27.7	4.81	6.60	2.97	1.99	1.72	0.518	1.26	0.241
10	0.02228	0.956	0.108	0.670	0.046			10.56	42.4	6.02	9.99	3.71	2.99	2.15	0.774	1.58	0.361
15	0.03342	1.43	0.224	1.01	0.094	3 in				9.03	21.6	5.57	6.36	3.22	1.63	2.37	0.755
20	0.04456	1.91	0.375	1.34	0.158	0.868	0.056	3½ in		12.03	37.8	7.43	10.9	4.29	2.78	3.16	1.28
25	0.05570	2.39	0.561	1.68	0.234	1.09	0.083	0.812	0.041	4 in		9.28	16.7	5.37	4.22	3.94	1.93
30	0.06684	2.87	0.786	2.01	0.327	1.30	0.114	0.974	0.056			11.14	23.8	6.44	5.92	4.73	2.72
35	0.07798	3.35	1.05	2.35	0.436	1.52	0.151	1.14	0.704	0.882	0.041	12.99	32.2	7.51	7.90	5.52	3.64
40	0.08912	3.83	1.35	2.68	0.556	1.74	0.192	1.30	0.095	1.01	0.052	14.85	41.5	8.59	10.24	6.30	4.65
45	0.1003	4.30	1.67	3.02	0.668	1.95	0.239	1.46	0.117	1.13	0.064			9.67	12.80	7.09	5.85
50	0.1114	4.78	2.03	3.35	0.839	2.17	0.288	1.62	0.142	1.26	0.076	5 in		10.74	15.66	7.88	7.15
60	0.1337	5.74	2.87	4.02	1.18	2.60	0.406	1.95	0.204	1.51	0.107			12.89	22.2	9.47	10.21
70	0.1560	6.70	3.84	4.69	1.59	3.04	0.540	2.27	0.261	1.76	0.143	1.12	0.047			11.05	13.71
80	0.1782	7.65	4.97	5.36	2.03	3.47	0.687	2.60	0.334	2.02	0.180	1.28	0.060			12.62	17.59
90	0.2005	8.60	6.20	6.03	2.53	3.91	0.861	2.92	0.416	2.27	0.224	1.44	0.074	6 in		14.20	22.0

100	0.2228	9.56	7.59	6.70	3.09	4.34	1.05	3.25	0.509	2.52	0.272	1.60	0.090	1.11	0.036	15.78	26.9
125	0.2785	11.97	11.76	8.38	4.71	5.43	1.61	4.06	0.769	3.15	0.415	2.01	0.135	1.39	0.055	19.72	41.4
150	0.3342	14.36	16.70	10.05	6.69	6.51	2.24	4.87	1.08	3.78	0.580	2.41	0.190	1.67	0.077		
175	0.3899	16.75	22.3	11.73	8.97	7.60	3.00	5.68	1.44	4.41	0.774	2.81	0.253	1.94	0.102		
200	0.4456	19.14	28.8	13.42	11.68	8.68	3.87	6.49	1.85	5.04	0.985	3.21	0.323	2.22	0.130	8 in	
225	0.5013	...	...	15.09	14.63	9.77	4.83	7.30	2.32	5.67	1.23	3.61	0.401	2.50	0.162	1.44	0.043
250	0.557	...	...	...	...	10.85	5.93	8.12	2.84	6.30	1.46	4.01	0.495	2.78	0.195	1.60	0.051
275	0.6127	...	...	...	...	11.94	7.14	8.93	3.40	6.93	1.79	4.41	0.583	3.05	0.234	1.76	0.061
300	0.6684	...	...	...	...	13.00	8.36	9.74	4.02	7.56	2.11	4.81	0.683	3.33	0.275	1.92	0.072
325	0.7241	...	...	...	...	14.12	9.89	10.53	4.09	8.19	2.47	5.21	0.797	3.61	0.320	2.08	0.083
350	0.7798			...	...	...	...	11.36	5.41	8.82	2.84	5.62	0.919	3.89	0.367	2.24	0.095
375	0.8355			...	...	...	...	12.17	6.18	9.45	3.25	6.02	1.05	4.16	0.416	2.40	0.108
400	0.8912			...	...	...	...	12.98	7.03	10.08	3.68	6.42	1.19	4.44	0.471	2.56	0.121
425	0.9469			...	...	...	...	13.80	7.89	10.71	4.12	6.82	1.33	4.72	0.529	2.73	0.136
450	1.003	10 in		...	...	...	...	14.61	8.80	11.34	4.60	7.22	1.48	5.00	0.590	2.89	0.151
475	1.059	1.93	0.054			...	...	...	...	11.97	5.12	7.62	1.64	5.27	0.653	3.04	0.166
500	1.114	2.03	0.059			...	...	...	...	12.60	5.65	8.02	1.81	5.55	0.720	3.21	0.182
550	1.225	2.24	0.071			...	...	...	...	13.85	6.79	8.82	2.17	6.11	0.861	3.53	0.219
600	1.337	2.44	0.083			...	...	...	...	15.12	8.04	9.63	2.55	6.66	1.02	3.85	0.258
650	1.448	2.64	0.097	12 in		...	...	...	...	...	...	10.43	2.98	7.22	1.18	4.17	0.301
700	1.560	2.85	0.112	2.01	0.047			...	...	...	...	11.23	3.43	7.78	1.35	4.49	0.343
750	1.671	3.05	0.127	2.15	0.054	14 in		...	...	...	...	12.03	3.92	8.33	1.55	4.81	0.392
800	1.782	3.25	0.143	2.29	0.061			...	...	...	...	12.83	4.43	8.88	1.75	5.13	0.443
850	1.894	3.46	0.160	2.44	0.068	2.02	0.042	...	...	...	...	13.64	5.00	9.44	1.96	5.45	0.497
900	2.005	3.66	0.179	2.58	0.075	2.13	0.047	...	...	...	...	14.44	5.58	9.99	2.18	5.77	0.554
950	2.117	3.86	0.198	2.72	0.083	2.25	0.052			...	...	15.24	6.21	10.55	2.42	6.09	0.613
1000	2.228	4.07	0.218	2.87	0.091	2.37	0.057	16 in		...	...	16.04	6.84	11.10	2.68	6.41	0.675
1100	2.451	4.48	0.260	3.15	0.110	2.61	0.068			...	...	17.65	8.23	12.22	3.22	7.05	0.807
1200	2.674	4.88	0.306	3.44	0.128	2.85	0.080	2.18	0.042	...	...	...	...	13.33	3.81	7.70	0.948
1300	2.896	5.29	0.355	3.73	0.150	3.08	0.093	2.36	0.048	...	...	...	...	14.43	4.45	8.33	1.11

TABLE B8.5 Flow of Water through Schedule 40 Steel Pipe (*Continued*)

Discharge		Pressure drop per 100 ft and velocity in Schedule 40 pipe for water at 60°F															
gal/min	ft^3/s	Velocity, ft/s	Press. drop, lb/in^2	Velocity, ft/s	Press. drop, lb/in^2	Velocity, ft/s	Press. drop, lb/in^2	Velocity, ft/s	Press. drop, lb/in^2	Velocity, ft/s	Press. drop, lb/in^2	Velocity, ft/s	Press. drop, lb/in^2	Velocity, ft/s	Press. drop, lb/in^2	Velocity, ft/s	Press. drop, lb/in^2
1400	3.119	5.70	0.409	4.01	0.171	3.32	0.107	2.54	0.055					15.55	5.13	8.98	1.28
1500	3.342	6.10	0.466	4.30	0.195	3.56	0.122	2.72	0.063	18 in				16.66	5.85	9.62	1.46
1600	3.565	6.51	0.527	4.59	0.219	3.79	0.138	2.90	0.071					17.77	6.61	10.26	1.65
1800	4.010	7.32	0.663	5.16	0.276	4.27	0.172	3.27	0.088	2.58	0.050			19.99	8.37	11.54	2.08
2000	4.456	8.14	0.808	5.73	0.339	4.74	0.209	3.63	0.107	2.87	0.060	20 in		22.21	10.3	12.82	2.55
2500	5.570	10.17	1.24	7.17	0.515	5.93	0.321	4.54	0.163	3.59	0.091					16.03	3.94
3000	6.684	12.20	1.76	8.60	0.731	7.11	0.451	5.45	0.232	4.30	0.129	3.46	0.075	24 in		19.24	5.59
3500	7.798	14.24	2.38	10.03	0.982	8.30	0.607	6.35	0.312	5.02	0.173	4.04	0.101			22.44	7.56
4000	8.912	16.27	3.08	11.47	1.27	9.48	0.787	7.26	0.401	5.74	0.222	4.62	0.129	3.19	0.052	25.65	9.80
4500	10.03	18.31	3.87	12.90	1.60	10.67	0.990	8.17	0.503	6.46	0.280	5.20	0.162	3.59	0.065	28.87	12.2
5000	11.14	20.35	4.71	14.33	1.95	11.85	1.21	9.08	0.617	7.17	0.340	5.77	0.199	3.99	0.079	...	
6000	13.37	24.41	6.74	17.20	2.77	14.23	1.71	10.89	0.877	8.61	0.483	6.93	0.280	4.79	0.111	...	
7000	15.60	28.49	9.11	20.07	3.74	16.60	2.31	12.71	1.18	10.04	0.652	8.08	0.376	5.59	0.150	...	
8000	17.82	...	...	22.93	4.84	18.96	2.99	14.52	1.51	11.47	0.839	9.23	0.488	6.38	0.192	...	
9000	20.05	...	...	25.79	6.09	21.34	3.76	16.34	1.90	12.91	1.05	10.39	0.608	7.18	0.242	...	
10000	22.28	...	...	28.66	7.46	23.71	4.61	18.15	2.34	14.34	1.28	11.54	0.739	7.98	0.294	...	
12000	26.74	...	...	34.40	10.7	28.45	6.59	21.79	3.33	17.21	1.83	13.85	1.06	9.58	0.416	...	
14000	31.19	...	...	...	...	33.19	8.89	25.42	4.49	20.08	2.45	16.16	1.43	11.17	0.562	...	
16000	35.65	...	...	...	...	...	...	29.05	5.83	22.95	3.18	18.47	1.85	12.77	0.723	...	
18000	40.10	...	...	...	...	...	...	32.68	7.31	25.82	4.03	20.77	2.32	14.16	0.907	...	
20000	44.56	...	...	...	...	...	...	36.31	9.03	28.69	4.93	23.08	2.86	15.96	1.12	...	

Note: For pipe lengths other than 100 ft, the pressure drop is proportional to the length. Thus, for 50 ft of pipe, the pressure drop is approximately one-half the value given in the table . . . for 300 ft, three times the given value, etc. Velocity is a function of the cross-sectional flow area; thus, it is constant for a given flow rate and is independent of pipe length.

Source: Ref. 7.

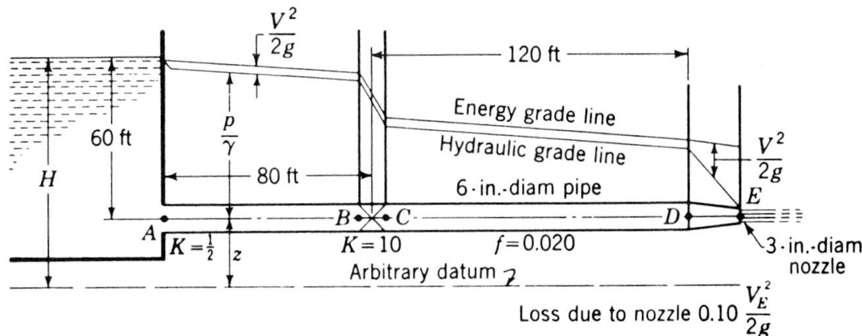

FIGURE B8.5 Hydraulic and energy grade lines. (*From Ref. 1.*)

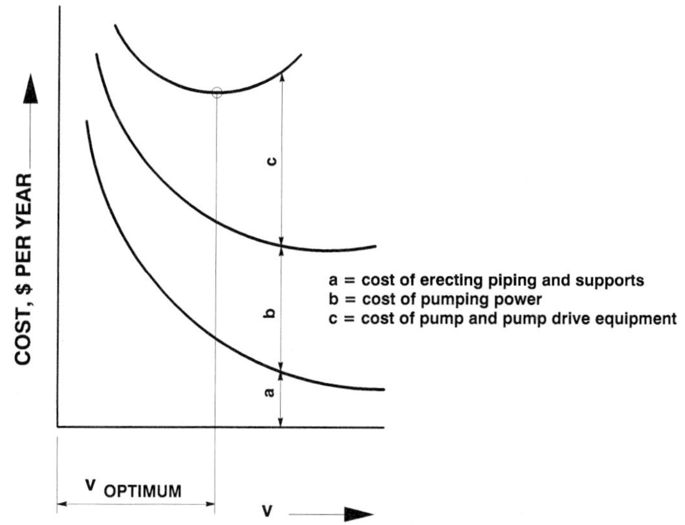

FIGURE B8.6 Flow velocity optimization plot.

The total cost, which is the sum of these three costs, $T_{COST} = a + b + c$ \$/year, is shown in Fig. B8.6 as reaching a minimum value at the optimum flow velocity. Each piping system should consider the optimization of flow velocity, optimum internal diameter, of the pipe under consideration. It is important to note that cost *b* depends strongly on the plant operating mode or the load factor, and on other economic indicators for a particular project. Industry data containing updated prices of equipment, piping, and labor are needed to implement this optimization procedure.

Sample Problem B8.1: Water System. What is the pressure drop $p_A - p_B$ (see Fig. B8.7), when water at 200°F flows in a piping system at the rate of $\dot{m} = 450{,}200\ \text{lb}_m/\text{h}$, $p_A = 500$ psia?

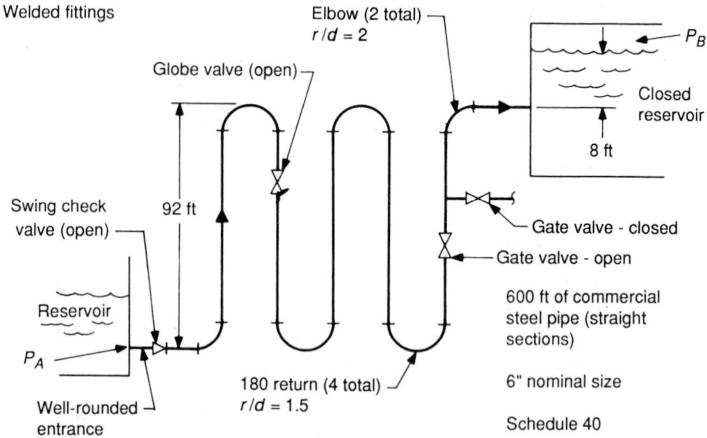

FIGURE B8.7 Sample problem for section "Applications: Water Systems."

Pipe Data. For 6-in nominal size, Schedule 40 pipe, the internal diameter $d = 6.065$ in, $D = 0.5054$ ft, and $A = 0.200$ ft^2.

Properties of Fluid. From App. I for $T = 200°$F and $P = 500$ psia, the dynamic viscosity of water is:

$$\mu = 63.5 * 10^{-7} \quad [(\text{lb}_f \cdot \text{s})/\text{ft}^2]$$

From 1979 ASME Steam Tables,[6] the specific volume of water at $T = 200°$F and $P = 500$ psia is:

$$\upsilon = 0.01661 \quad [\text{ft}^3/\text{lb}_m]$$

Then, the kinematic viscosity of water is [see Eq. (B8.3)]:

$$\nu = \frac{\mu}{\rho} g_c = (63.5 * 10^{-7})(0.01661)(32.174)$$

$$= 3.394 * 10^{-6} \ [\text{ft}^2/\text{s}]$$

The flow velocity is calculated from the continuity equation:

$$\text{v} = \frac{\dot{m}}{\rho A}$$

where $\dot{m}$

$$= \frac{450{,}200}{3600} = 125.06 \quad [\text{lb}_m/\text{s}]$$

$$\text{v} = \frac{125.06}{0.02006} 0.01661 = 10.35 \quad [\text{ft/s}]$$

Using Eq. (B8.14), the Reynolds number is:

$$Re = \frac{D\text{v}}{\nu} = \frac{(0.5054)(10.35)}{3.394 * 10^{-6}} = 1.542 * 10^6$$

Flow Resistance Coefficients.[7]

Local disturbance	K (f = 0.015)	
Entrance	K	= 0.04
4, 180° turns	(4)(22.18f)	= 1.33
1 globe valve (open)	340f	= 5.10
1 tee flow through run	20f	= 0.33
1 swing check valve (open)	100f	= 1.50
2 long radius elbows	(2)(14f)	= 0.42
1 gate valve (open)	8f	= 0.12
Exit	K	= 1.00
	ΣK	= 9.77

Darcy-Weisbach Friction Factor. For commercial steel pipe, from Ref. 7, the relative roughness is:

$$\frac{\epsilon}{D} = 0.0003$$

and from Ref. 7, for calculated Re and ϵ/D = 0.0003, the friction factor is:

$$f = 0.0153$$

Total Resistance

$$K_{\text{tot}} = \Sigma K + f\frac{L}{D}$$

$$= 9.77 + 0.0153\,\frac{600}{0.5054}$$

$$= 27.93$$

"Friction" Pressure Drop in Pipe. From Eq. (B8.12):

$$\Delta p = K_{\text{tot}}\rho\,\frac{\mathrm{v}^2}{2g_c}$$

$$= 27.93\,\frac{(10.35)^2}{(0.01661)(2)(32.174)}$$

$$= 2799 \quad [\text{lb}_f/\text{ft}^2]$$

$$= \frac{2799}{144} = 19.44 \quad [\text{psi}]$$

Pressure Difference $p_A - p_B$. From Bernoulli's Eq. (B8.10):

$$\frac{\rho_A \mathrm{v}_A^2}{2g} + p_A + \frac{g}{g_c} z_A\rho_A = \frac{\rho_B \mathrm{v}_B^2}{2g} + p_B + \frac{g}{g_c} z_B\rho_B + \Delta p$$

assuming $\rho_A = \rho_B = \rho$ and g = 32.174 [ft/s^2]:

$$p_A - p_B = \frac{g}{g_c}(z_B - z_A)\frac{\rho}{144} + \Delta p = \frac{100}{(144)(0.01661)} + 19.44 = 61.25 \quad \text{[psi]}$$

Applications: Oil and Other Liquid Systems

The calculation of oil flow through pipes is a much more complicated process than a similar calculation for water flow. Water properties are well defined and many, including viscosity, are nearly constant within liquid temperature ranges, whereas oil is quite different. No two oils have the same physical properties, and any given oil is subject to important physical changes at expected variable temperatures. When considering oil flow in pipes, the most important variable physical property is its viscosity.

Ordinary crude oil is not a homogeneous liquid; it is a very complex fluid composed of compounds of carbon and hydrogen which exist in petroleum in a wide variety of combinations. Physical properties change accordingly, and they also change as a result of temperature variations. Therefore, this handbook assumes that the user will research the fluid of interest in other source books such as the *Petroleum Processing Handbook*[9] to determine important properties including density and viscosity.

The subsection "Viscosity" in this handbook refers to Table B8.2 for conversion among several common units. Kinematic viscosity is defined by Eq. (B8.3), which in the CGS system is expressed in stokes or centistokes. To obtain centistokes, multiply stokes by 100. Table B8.6 lists values for conversion between centistokes and the corresponding indications of various viscometers. As noted above, viscosity is a key variable in determining the pressure drop or flow rate of oil through pipes.

Typical pressure drops for oil flowing through pipes ranging from ½ to 12 in in diameter are given in Table B8.7 for fluids with a density of approximately 7.78 lb/gal.

Figures B8.8 through B8.14 provide pressure drops versus flow rates for fluids ranging from gasoline to those with 10,000 s Saybolt Universal viscosity through 2- to 12-in Schedule 40 pipe.

Usually engineers in oil burning plants will know only two things regarding the fuel oil supplied to them, namely, the gravity and the viscosity. The gravity can be used to convert volume measurements of fuel oil into weight or vice versa. It can also be used to estimate the heating value of the fuel by substitution in any one of a number of formulas of which the following is probably the best known: Btu per pound = 18,250 + 40 × API gravity*. The API gravity of a mixture can be readily calculated proportionately as follows from the gravity of the components: What is the API gravity of a 40 percent mixture of a 10° API fuel and 60 percent of a 15° API fuel? Answer: 0.40 × 10° API + 0.60 × 15° API = 13° API.

A knowledge of the viscosity and how it varies with temperature is essential for the easy handling and efficient utilization of fuel oil. Heavier fuel oils must be heated to the proper temperature in order to obtain a viscosity of 150 Saybolt Universal at the burners for the most efficient atomization. It is also important to determine to what temperature the oil should be heated in the bunkers so it can be readily pumped through the fuel oil system. Generally the maximum temper-

*Significance of Tests of Petroleum Products—American Society for Testing and Materials.

TABLE B8.6 Kinematic Viscosity Conversion Table (Centistokes to Engler, Saybolt, and Redwood units)

Centi-stokes	Engler degrees	Saybolt seconds at 130 F	Redwood seconds at 140 F	Centi-stokes	Engler degrees	Saybolt seconds at 130 F	Redwood seconds at 140 F	Centi-stokes	Engler degrees	Saybolt seconds at 130 F	Redwood seconds at 140 F
2.0	1.140	32.66	30.95	18.0	2.644	89.37	78.45	41.0	5.465	190.6	168.3
2.5	1.182	34.46	32.20	19.0	2.755	93.48	82.10	42.0	5.590	195.1	172.3
3.0	1.224	36.07	33.45	20.0	2.870	97.69	85.75	43.0	5.720	199.6	176.4
3.5	1.266	37.67	34.70	21.0	2.984	101.9	89.50	44.0	5.845	204.2	180.4
4.0	1.308	39.17	35.95	22.0	3.100	106.2	93.25	45.0	5.975	208.8	184.5
4.5	1.350	40.78	37.20	23.0	3.215	110.5	97.05	46.0	6.105	213.4	188.5
5.0	1.400	42.38	38.45	24.0	3.335	114.8	100.9	47.0	6.235	218.0	192.6
5.5	1.441	43.98	39.80	25.0	3.455	119.1	104.7	48.0	6.365	222.6	196.6
6.0	1.481	45.59	41.05	26.0	3.575	123.5	108.6	49.0	6.495	227.2	200.7
6.5	1.521	47.19	42.40	27.0	3.695	127.9	112.5	50.0	6.630	231.8	204.7
7.0	1.563	48.79	43.70	28.0	3.820	132.4	116.5	52.0	6.890	241.1	212.8
7.5	1.605	50.44	45.00	29.0	3.945	136.8	120.4	54.0	7.106	250.3	221.0
8.0	1.653	52.10	46.35	30.0	4.070	141.2	124.4	56.0	7.370	259.5	229.1
8.5	1.700	53.80	47.75	31.0	4.195	145.6	128.3	58.0	7.633	268.7	237.2
9.0	1.746	55.51	49.10	32.0	4.320	150.0	132.3	60.0	7.896	277.9	245.3
9.5	1.791	57.21	50.55	33.0	4.445	154.5	136.3	62.0	8.159	287.2	253.5
10.0	1.837	58.91	52.00	34.0	4.570	159.0	140.2	64.0	8.422	296.4	261.6
11.0	1.928	62.42	55.00	35.0	4.695	163.5	144.2	66.0	8.686	305.6	269.8
12.0	2.020	66.03	58.10	36.0	4.825	168.0	148.2	68.0	8.949	314.8	277.9
13.0	2.120	69.73	61.30	37.0	4.955	172.5	152.2	70.0	9.212	324.0	286.0
14.0	2.219	73.54	64.55	38.0	5.080	177.0	156.2	72.0	9.475	333.3	294.1
15.0	2.323	77.35	67.95	39.0	5.205	181.5	160.3	74.0	9.738	342.5	302.2
16.0	2.434	81.25	71.40	40.0	5.335	186·0	164.3	75.0	9.870	347.2	306.3
17.0	2.540	85.26	74.85								

Supplementary Kinematic Viscosity Conversion Table

Centistokes................	2	6	10	20	30	40	50	60	70
Saybolt at 100 F	32.60	45.50	58.80	97.50	40.9	185.7	231.4	277.4	323.4
Saybolt at 210 F	32.83	45.82	59.21	98.18	141.9	187.0	233.0	279.3	325.7
Redwood at 70 F	30.20	40.50	51.70	85.40	123.7	163.2	203.3	243.5	283.9
Redwood at 200 F	31.20	41.50	52.55	86.90	126.0	166.7	208.3	250.0	291.7

ature that oil in the bunkers be heated to is 130°F in an effort to reduce the viscosity for easier pumping.

When two or more oils are blended, it is necessary to determine the properties of the mixture. Several methods are available to achieve the critical properties such as viscosity and most use some type of viscosity blending chart. The *Petroleum Processing Handbook*[9] is one source of such a blending chart which is a semi-logarithmic plot of kinematic viscosity versus either volume percent lower or higher viscosity material. A linear interpolation is used on this type of plot.

TABLE B8.7 Flow of Oils through Commercial Pipes

Size of pipe and average inside diameter, inches	Capacity, gallons per minute	Pressure drop in pounds per square inch per 100 ft. of pipe based on oils of 20° Bé. gravity — Viscosity in Saybolt Universal seconds					
		100	200	300	400	500	600
½ 0.622	2	6.59	14.3	21.7	29.1	36.9	43.6
	5	21.1	34.8	53.4	71.5	89.5	107.0
	7	37.7	49.2	74.6	101.0	129.0	152.0
	10	69.8	85.8	107.0	144.0	177.0	217.0
	15	143.0	174.0	194.0	217.0	267.0	325.0
¾ 0.824	2	2.16	4.74	7.04	9.48	11.6	14.2
	5	5.6	11.5	17.1	23.1	28.9	35.2
	7	9.96	16.1	24.5	32.8	40.5	48.9
	10	18.4	22.3	34.9	46.0	58.5	70.9
	15	37.9	45.7	51.5	70.4	88.3	106.0
1 1.05	5	2.06	4.32	6.68	8.85	11.0	13.4
	10	5.9	8.74	13.4	17.9	22.0	26.6
	15	11.9	14.5	19.7	26.6	33.2	40.3
	20	19.7	23.8	26.8	35.4	44.1	53.5
	25	29.0	35.1	39.3	44.2	54.5	66.8
1½ 1.61	10	0.783	1.57	2.40	3.22	3.98	4.86
	20	2.61	3.17	4.85	6.52	8.02	9.70
	30	5.30	6.43	7.15	9.68	12.1	14.7
	40	8.72	10.6	11.9	12.8	16.0	19.4
	50	12.7	15.7	17.4	18.8	19.8	24.2
2 2.067	10	0.266	0.578	0.875	1.17	1.47	1.79
	20	0.79	1.15	1.75	2.34	2.92	3.52
	30	1.60	1.93	2.62	3.50	4.40	5.30
	40	2.63	3.18	3.56	4.66	5.85	7.10
	50	3.86	4.68	5.28	5.88	7.26	8.85
4 4.026	50	0.174	0.198	0.307	0.412	0.494	0.612
	100	0.550	0.668	0.744	0.810	1.01	1.22
	150	1.09	1.36	1.50	1.62	1.74	1.84
	200	1.81	2.22	2.49	2.68	2.85	2.98
	250	2.67	3.26	3.67	3.97	4.20	4.41
6 6.065	100	0.0788	0.0956	0.115	0.158	0.194	0.236
	200	0.259	0.315	0.359	0.388	0.408	0.480
	500	1.27	1.55	1.74	1.88	2.01	2.20
	700	2.29	2.80	3.15	3.36	3.60	3.78
	1000	4.24	5.21	5.82	6.32	6.68	7.05
8 8.03	200	0.0696	0.0834	0.0947	0.102	0.128	0.155
	500	0.340	0.418	0.459	0.500	0.530	0.556
	1000	1.13	1.37	1.55	1.67	1.78	1.87
	1500	2.27	2.74	3.07	3.30	3.59	3.75
	2000	3.88	4.55	5.15	5.56	5.92	6.22
12 12.05	1000	0.165	0.203	0.228	0.242	0.258	0.272
	2000	0.552	0.670	0.750	0.810	0.866	0.906
	3000	1.15	1.34	1.50	1.63	1.73	1.83
	4000	1.91	2.22	2.50	2.69	2.84	3.00
	5000	2.90	3.34	3.63	3.97	4.22	4.41

To change capacities from gallons per minute to barrels (42 gal) per hour, multiply pressure drops by 1.43. To change from pressure drops per square inch per 100 ft to pressure drops per square inch per mile, multiply by 52.8.

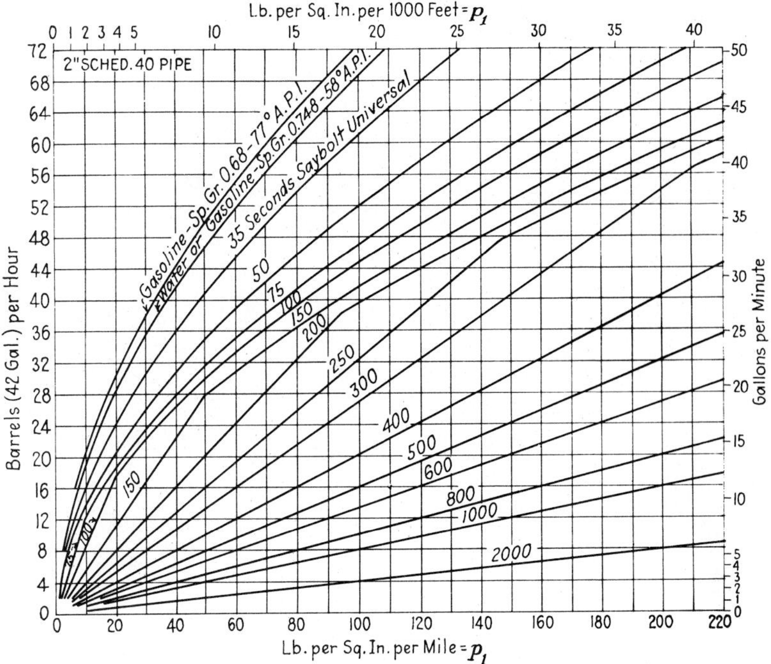

FIGURE B8.8 Pipe friction based on average Saybolt Universal viscosity and specific gravity of fluid in pipeline. Pressure drop $\Delta p = p_1 \times$ sp gr.

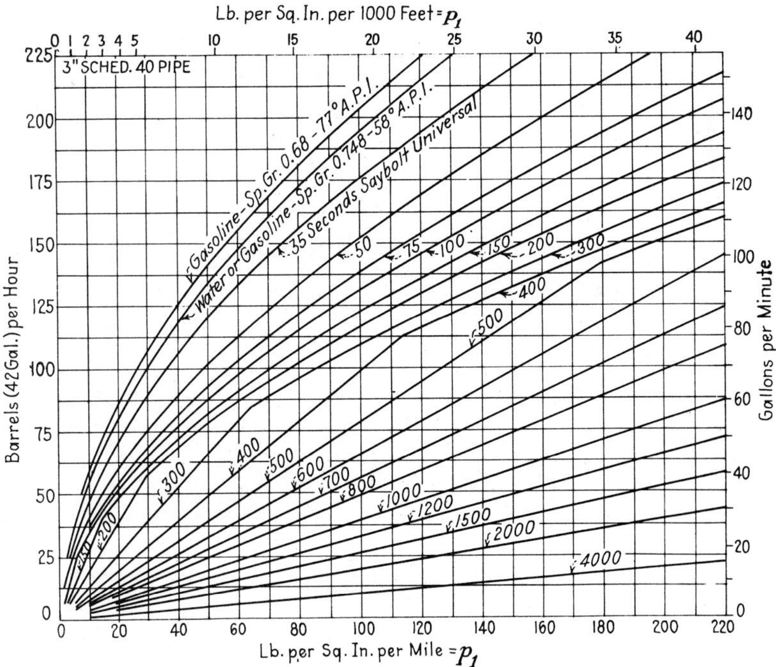

FIGURE B8.9 Pipe friction based on average Saybolt Universal viscosity and specific gravity of fluid in pipeline. Pressure drop $\Delta p = p_1 \times$ sp gr.

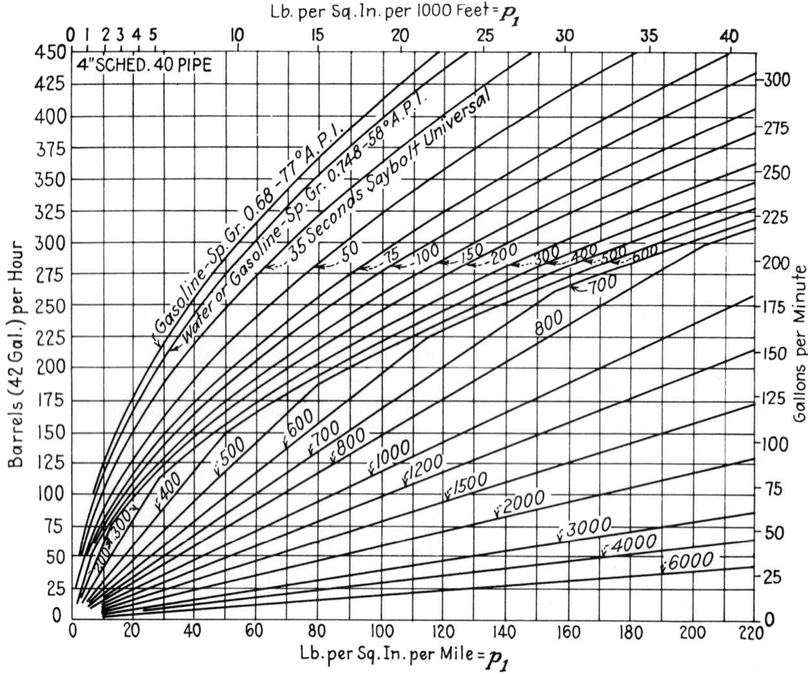

FIGURE B8.10 Pipe friction based on average Saybolt Universal viscosity and specific gravity of fluid in pipeline. Pressure drop $\Delta p = p_1 \times$ sp gr.

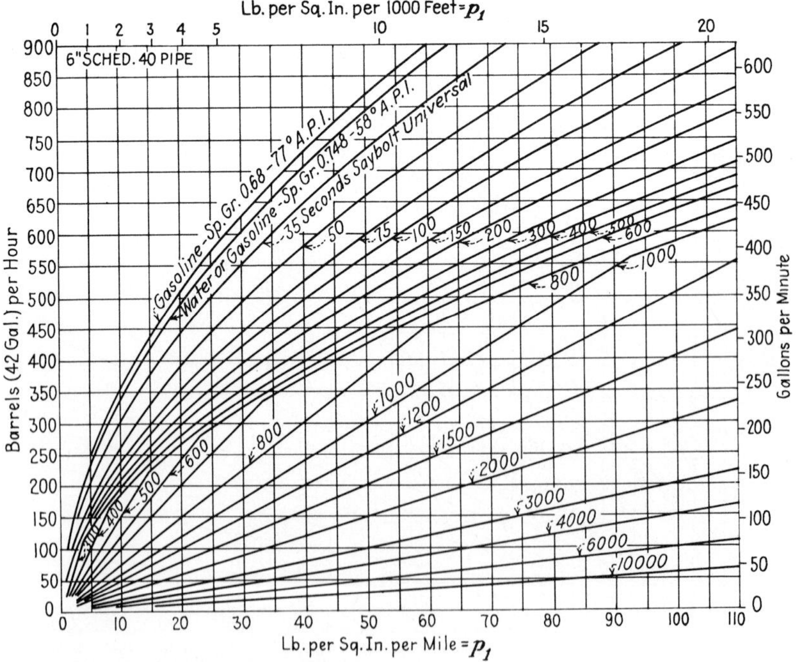

FIGURE B8.11 Pipe friction based on average Saybolt Universal viscosity and specific gravity of fluid in pipeline. Pressure drop $\Delta p = p_1 \times$ sp gr.

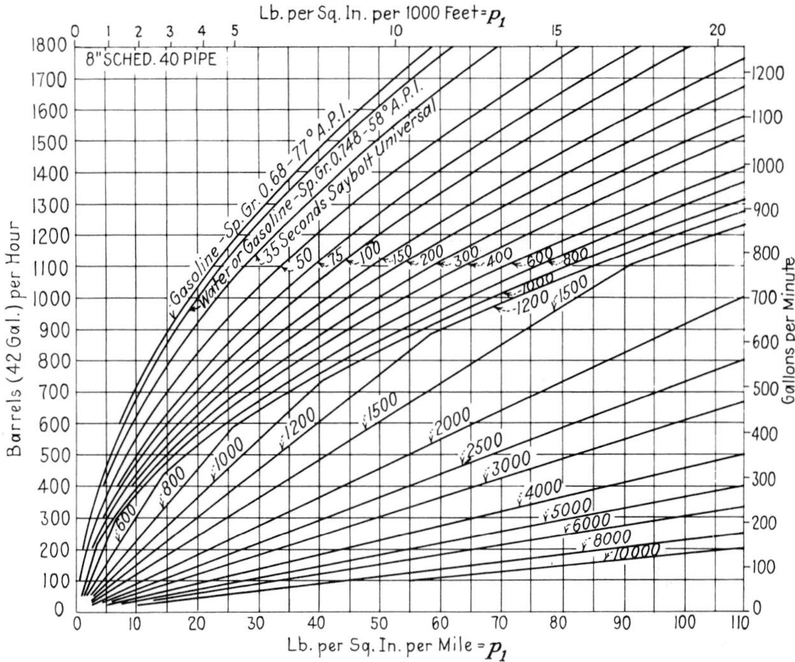

FIGURE B8.12 Pipe friction based on average Saybolt Universal viscosity and specific gravity of fluid in pipeline. Pressure drop $\Delta p = p_1 \times$ sp gr.

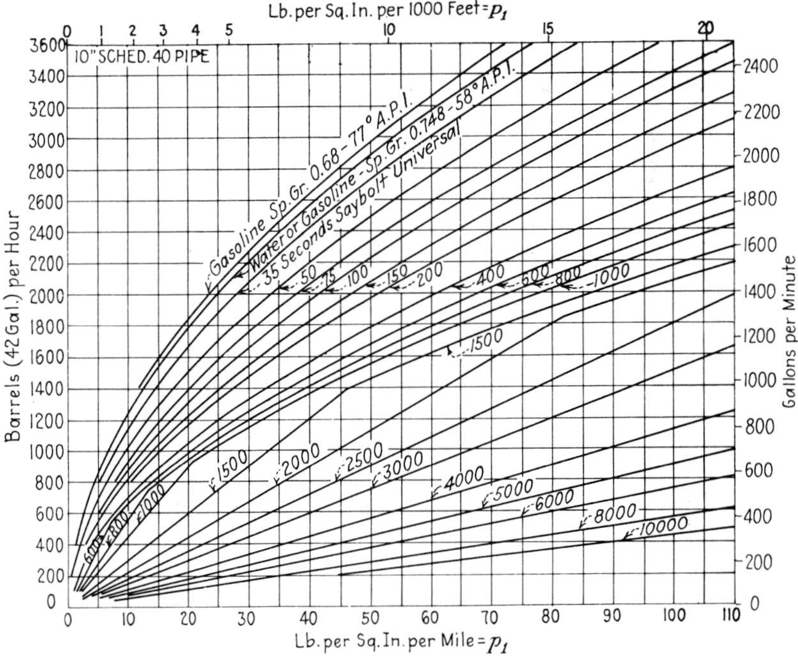

FIGURE B8.13 Pipe friction based on average Saybolt Universal viscosity and specific gravity of fluid in pipeline. Pressure drop $\Delta p = p_1 \times$ sp gr.

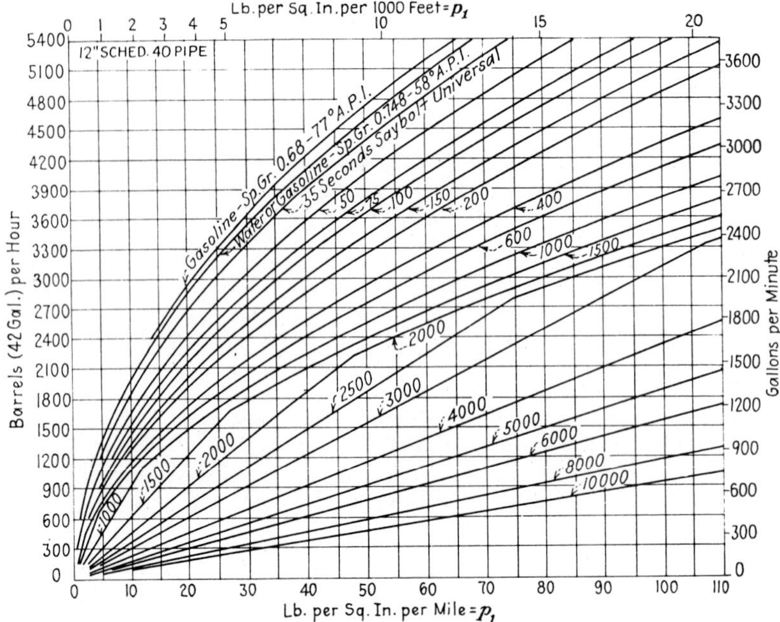

FIGURE B8.14 Pipe friction based on average Saybolt Universal viscosity and specific gravity of fluid in pipeline. Pressure drop $\Delta p = p_1 \times$ sp gr.

STEADY SINGLE-PHASE COMPRESSIBLE FLOW IN PIPING

Characteristics of Compressible Flow in Pipes

The term *compressible flow* implies variations in density of a fluid. These variations are the results of pressure and temperature changes from one point to another. The rate of change of density with respect to pressure is, therefore, an important parameter in the analysis of compressible flow, and it is closely connected with the velocity of sound.

Adiabatic, Constant-Area Flow with Friction. Adiabatic flow with friction is of interest in sizing safety valve discharge lines (vent lines) and other pipes where heat transfer may be neglected, and the flow may be restricted by sonic condition at the pipe exit (choked flow). Supersonic velocities will not be discussed here.

In this section, the flow of compressible gases in ducts of constant cross-sectional area will be discussed. For adiabatic conditions ($dQ = 0$), the energy equation for steady flow between two points, 1 and x, in a pipe may be written as follows [see Eq. (B8.7)]:

$$h_x + \frac{1}{J}\frac{v_x^2}{2g_c} = h_1 + \frac{1}{J}\frac{v_1^2}{2g_c} = h_o \tag{B8.15}$$

where h_o is the stagnation enthalpy.

Utilizing the continuity equation:

$$\dot{G} = \frac{\dot{m}}{A} = \frac{\mathrm{v}_x}{\upsilon_x} = \frac{\mathrm{v}_1}{\upsilon_1} \tag{B8.16}$$

For constant values of $\dot{m}$ and A, the following is found:

$$h_x + \frac{1}{J}\left(\frac{\dot{m}}{A}\right)^2 \frac{\upsilon_x^2}{2g_c} = h_1 + \frac{1}{J}\frac{\mathrm{v}_1^2}{2g_c} = h_o \tag{B8.17}$$

Equation (B8.17) represents the Fanno line (Ref. 10) which is the locus of the conditions in a cylindrical pipe of constant diameter. Having fixed v_1 and h_1 at the starting point, the stagnation enthalpy h_o is calculated. This enthalpy stays constant along the length L of the pipe. Choosing an arbitrary specific volume υ_x, the corresponding value of h_x can be calculated from Eq. (B8.17). The intersection of υ_x and h_x on the h–s diagram represents a point on the Fanno line. The sonic (choked) condition at the pipe exit shown at point 2 in Fig. B8.15 is defined by:

$$\frac{ds}{dh} = 0 \tag{B8.18}$$

Considering the work done against friction and that the heat generated within the fluid by internal friction for adiabatic flow $dq = 0$, the following expression is derived:

$$L_x = \frac{2DJg_c}{f}\int_{s_1}^{s_x} \frac{T}{\mathrm{v}^2}\,ds \tag{B8.19}$$

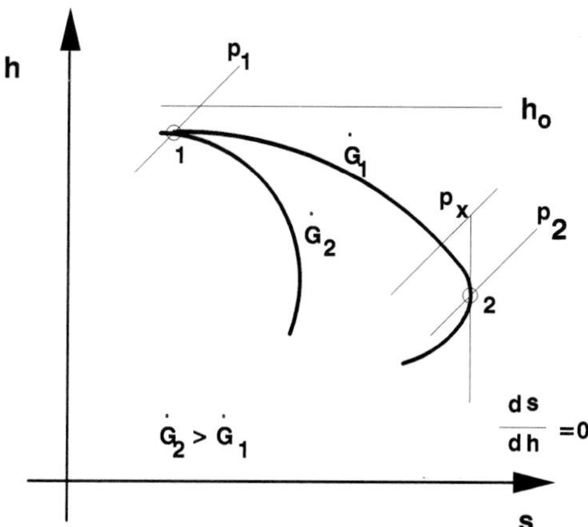

FIGURE B8.15 Enthalpy versus entropy for gas flow in pipe.

Equation (B8.19) represents the length L_x of the pipeline along which the pressure drops from p_1 to p_x. The described procedure yields a number of points on a Fanno line and also the coordinates of the function $T/v^2 = f(s)$. Integrating the function $T/v^2 = f(s)$ with fixed upper boundary (sonic conditions) and floating lower boundary of integration, the fluid parameters (static conditions) at any distance L upstream of the sonic (choked) plane can be found (convergence on the given length L). The procedure is explained in Fig. B8.16.

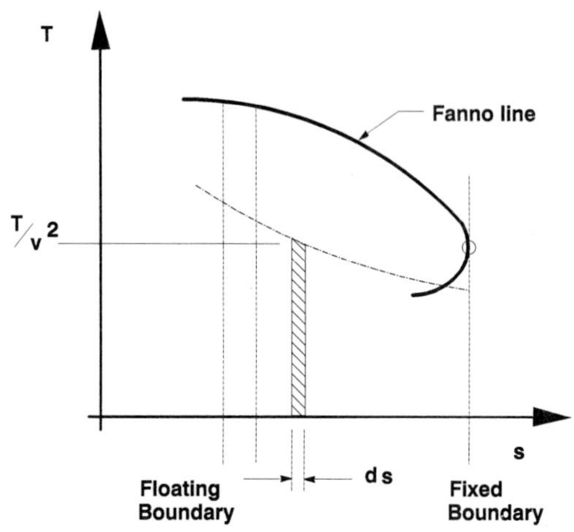

FIGURE B8.16 Fanno line plot.

If the calculated pressure at sonic exit is lower than the receiver pressure, the additional length of the pipe required for choked conditions at the exit should be calculated. Adding this fictitious pipe length to the actual length of the pipe, the "extended length" is found. This new length is now used for calculating the parameters at the start of the line (or after a valve).

A simplified approach for adiabatic, constant-area flow (Fanno line analysis) of a perfect gas (that has constant specific heat) is presented in Ref. 11 Following are a few of the working formulas suitable for practical computations developed in this reference book.

Choked Pressure

$$p^* = \frac{\dot{m}}{A}\sqrt{\frac{2RT_o}{g_c k(k+1)}} \tag{B8.20}$$

Static Pressure

$$\frac{p}{p^*} = \frac{1}{Ma}\sqrt{\frac{k+1}{2\left(1 + \frac{k-1}{2}Ma^2\right)}} \tag{B8.21}$$

Flow Velocity

$$\frac{\mathrm{v}}{\mathrm{v}^*} = \mathrm{Ma}\sqrt{\frac{\mathrm{k}+1}{2\left(1+\frac{\mathrm{k}-1}{2}\mathrm{Ma}^2\right)}} \tag{B8.22}$$

Maximum Duct Length

$$f\frac{L_{\max}}{D} = \frac{1-Ma^2}{k\,Ma^2} + \frac{k+1}{2\,k}\ln\frac{(k+1)Ma^2}{2\left(1+\frac{k-1}{2}Ma^2\right)} \tag{B8.23}$$

The quantities marked with an asterisk in these expressions, such as p^* and v*, represent the values of the stream properties at the section in the pipe where $Ma = 1$ (sonic plane). Since they are constants for a given adiabatic, constant-area flow, they may be regarded as convenient reference values for normalizing the equations. In order to find the change in some stream property, say, the pressure, between the sections where the Mach numbers are Ma_1 and Ma_2, respectively, the following expression is used:

$$\frac{p_2}{p_1} = \frac{(p/p^*)_{Ma_2}}{(p/p^*)_{Ma_1}} \tag{B8.24}$$

where $(p/p^*)_{Ma_2}$ is the value corresponding to Ma_2, and so forth.

In order to facilitate hand calculations, the important dimensionless ratios are tabulated in tables of Appendix B of Ref. 11 with Mach number as the argument.

Isothermal Flow in Long Pipelines. Isothermal flow with friction is of interest in connection with pipelines for transporting gas over long distances. When the lines are extremely long, there is sufficient pipe surface area for heat transfer to make flow nonadiabatic and approximately isothermal, especially when the gas temperature is not much different from the temperature of the surroundings. In such cases, the flow may not be treated as incompressible and the Darcy equation cannot be used for calculating the pressure drop.

Using the assumptions of constant elevation, no work being done by the fluid, continuity, isothermal flow, and that the friction coefficient does not change with the length of the pipe yields the following application equation:

$$\dot{m} = \frac{\pi}{4}\sqrt{\frac{2g_c}{fRT_1}}\sqrt{\frac{D^5}{L}p_{\text{ave}}\Delta p} \tag{B8.25}$$

For the same mass flow rate $\dot{m}$, the pipe diameter D may be decreased by increasing the average pressure p_{ave} in the pipeline and/or permitting a larger pressure drop Δp. One of the applications of Eq. (B8.25) is the economic analysis of long gasline systems requiring intermediate compression stations.

It has to be mentioned here that the Mach number:

$$Ma = \frac{1}{\sqrt{k}} \tag{B8.26}$$

represents a limit for continuous isothermal flow, in the same way that $Ma = 1$

represents a limit for continuous adiabatic flow (Ref. 11). The working formula for determining the maximum pipe length beyond which the continuous isothermal flow may not proceed (ideal gas approximation) is given by the following expression:

$$f\frac{L_{\max}}{D} = \frac{1 - k\,Ma^2}{k\,Ma^2} + \ln\,(k\,Ma^2) \tag{B8.27}$$

where Ma is the Mach number at the pipe inlet. For greater lengths, choking occurs and the mass flow rate decreases.

Sample problems for this subsection are presented under Applications: Sample Problems.

Applications: Steam Systems

In a case of compressible fluids such as air or steam, when density changes are small, the fluid may be considered as incompressible. Therefore, all design rules described in earlier sections are applicable to this case. For steam and gas systems, where the fluid density is small, static head is negligible and may be omitted in pressure drop calculations.

The velocities (feet per minute) shown in Table B8.8 may be considered for preliminary line sizing, but economic size should be determined as a function of pressure drop, pipe cost, and so on.

Special attention should be given to main steam lines and to hot and cold reheat piping since pressure drop in those systems affects heat rate and plant capability. Cost of piping must be compared with these effects for the most economical piping arrangement. Normally, total pressure drop in reheat piping and the boiler reheater should be 7 to 9 percent of high-pressure turbine exhaust pressure. It is desirable to use a smaller-diameter hot reheat line and a larger-diameter cold reheat line, taking a greater pressure drop in the more expensive, alloy, hot reheat line.

Extraction steam piping also affects heat rate and output, and normally this piping should be sized so that the pressure drop does not exceed about 5 percent of turbine stage pressure for the low-pressure and 3 to 4 percent for the higher-pressure lines.

Extraction steamlines should be designed for the pressure shown on full-load heat balance diagrams at 5 percent overpressure and valves wide open.

Continuously operating steamlines in process projects shall be designed on the basis of reasonable total pressure drop and, except for short leads such as to turbines and pumps, shall not generally exceed the conditions noted in Table B8.8.

TABLE B8.8 Typical Steam Line Velocity in FPS on Industrial Process Projects*

Steam pressure	Velocity	Pressure drop
50 psig and lower	175 ft/s	0.4 psi/100 ft
Over 50 psig (saturated steam)	120 ft/s	1.0 psi/100 ft
Over 50 psig (superheated steam)	175 ft/s	1.0 psi/100 ft

*Velocities are based on typical industrial process industry practice.
Source: Stone & Webster.

Applications: Air and Other Gas Systems

As indicated in the last section, "Applications: Steam Systems," in a case of compressible fluids such as air or steam when density changes are small, the fluid may be considered as incompressible. Therefore, all design rules described previously in the subsections "Characteristics of Incompressible Flow," "Applications: Water Systems," and "Applications: Steam Systems" are applicable to this section. For steam and gas systems, where the fluid density is small, static head is negligible and may be omitted in pressure drop calculations.

Table B8.9 (from Ref. 7) presents pressure drop for some typical values of air flow rates through piping from 1/8- to 12-in diameter.

Applications: Sample Problems

The most frequent application of single-phase compressible flow steamline analysis, normally encountered by engineers, is the sizing of safety valve vent lines. This analysis can be done either by using a computer program which is based on procedures discussed or a hand calculation similar to that discussed in this section. Users should compare their results with those obtained from procedures based on tables in Appendix B of Ref. 11.

The primary consideration in these analyses is to ensure that the vent line will pass the required flow without exceeding recommended backpressure limitations on valves with solidly connected vents or without blowing back in the case of open vent stacks.

Problem B8.2. During abnormal operation of a system, 72,000 lb/h of air must be released from a high-pressure air tank through a 4-in (Schedule 40) bypass line (vent line) into the atmosphere. Stagnation conditions in the vessel during this operation are: p_o = 600 psig (kept at a constant level by compressors): T_o = 120°F. Equivalent length of the vent line is 90 ft. Calculate the pressure p_1 that exists at the valve discharge (air may be treated as a perfect gas):

$$\dot{m} = 72{,}000 \text{ lb/h} = 20 \text{ lb/s}$$

$$T_o = 120°\text{F} = (460 + 120) = 580°\text{R}$$

$$p_o = 600 \text{ psig} = 614.7 \text{ psia (assumed atmospheric pressure of 14.7 psia)}$$

Internal pipe diameter D = 4.026 in (A = 0.0884 ft²).(See Fig. B8.17.)

Critical pressure at the pipe discharge is found by using Eq. (B8.20):

$$p^* = \frac{\dot{m}}{A}\sqrt{\frac{2RT_o}{g_c k(k+1)}} \quad \left[\frac{\text{lb}_f}{\text{ft}^2}\right]$$

$$= \frac{20}{0.0884}\sqrt{\frac{(2)(53.3)(580)}{(32.174)(1.4)(2.4)}}$$

$$= 5410.63 \quad \left[\frac{\text{lb}_f}{\text{ft}^2}\right] = 37.57 \quad [\text{psia}]$$

Pipe $f(L/D)$, from choked exit to the valve discharge, is calculated for complete turbulence friction factor f = 0.017:

TABLE B8.9 Flow of Air through Schedule 40 Steel Pipe

Free air q'_m ft³/min at 60°F and 14.7 psia	Compressed air ft³/min at 60°F and 100 psig	Pressure drop of air in pounds per square inch per 100 ft of Schedule 40 pipe For air at 100 lb/in² gauge pressure and 60°F temperature								
		⅛ in	¼ in	⅜ in	½ in					
1	0.128	0.361	0.083	0.018						
2	0.256	1.31	0.285	0.064	0.020					
3	0.384	3.06	0.605	0.133	0.042	¾ in				
4	0.513	4.83	1.04	0.226	0.071					
5	0.641	7.45	1.58	0.343	0.106	0.027				
							1 in			
6	0.769	10.6	2.23	0.408	0.148	0.037				
8	1.025	18.6	3.89	0.848	0.255	0.062	0.019			
10	1.282	28.7	5.96	1.26	0.356	0.094	0.029	1¼ in	1½ in	
15	1.922	...	13.0	2.73	0.834	0.201	0.062			
20	2.563	...	22.8	4.76	1.43	0.345	0.102	0.026		
25	3.204	...	35.6	7.34	2.21	0.526	0.156	0.039	0.019	
30	3.845	...		10.5	3.15	0.748	0.219	0.055	0.026	
35	4.486	...		14.2	4.24	1.00	0.293	0.073	0.035	
40	5.126	...		18.4	5.49	1.30	0.379	0.095	0.044	
45	5.767	...		23.1	6.90	1.62	0.474	0.116	0.055	2 in
50	6.408			28.5	8.49	1.99	0.578	0.149	0.067	0.019
60	7.690	2½ in		40.7	12.2	2.85	0.819	0.200	0.094	0.027
70	8.971			...	16.5	3.83	1.10	0.270	0.126	0.036
80	10.25	0.019		...	21.4	4.96	1.43	0.350	0.162	0.046
90	11.53	0.023		...	27.0	6.25	1.80	0.437	0.203	0.058
100	12.82	0.029	3 in		33.2	7.69	2.21	0.534	0.247	0.070
125	16.02	0.044			...	11.9	3.39	0.825	0.380	0.107
150	19.22	0.062	0.021		...	17.0	4.87	1.17	0.537	0.151
175	22.43	0.083	0.028		...	23.1	6.60	1.58	0.727	0.205
200	25.63	0.107	0.036	3½ in	...	30.0	8.54	2.05	0.937	0.264
225	28.84	0.134	0.045	0.022		37.9	10.8	2.59	1.19	0.331
250	32.04	0.164	0.055	0.027		...	13.3	3.18	1.45	0.404
275	35.24	0.191	0.066	0.032		...	16.0	3.83	1.75	0.484
300	38.45	0.232	0.078	0.037		...	19.0	4.56	2.07	0.573
325	41.65	0.270	0.090	0.043		...	22.3	5.32	2.42	0.673
					4 in					
350	44.87	0.313	0.104	0.050		...	25.8	6.17	2.80	0.776
375	48.06	0.356	0.119	0.057	0.030	...	29.6	7.05	3.20	0.887
400	51.26	0.402	0.134	0.064	0.034	...	33.6	8.02	3.64	1.00
425	54.47	0.452	0.151	0.072	0.038	...	37.9	9.01	4.09	1.13
450	57.67	0.507	0.168	0.081	0.042	...		10.2	4.59	1.26
475	60.88	0.562	0.187	0.089	0.047		...	11.3	5.09	1.40
500	64.08	0.623	0.206	0.099	0.052		...	12.5	5.61	1.55
550	70.49	0.749	0.248	0.118	0.062		...	15.1	6.79	1.87
600	76.90	0.887	0.293	0.139	0.073		...	18.0	8.04	2.21
650	83.30	1.04	0.342	0.163	0.086	5 in	...	21.1	9.43	2.60
700	89.71	1.19	0.395	0.188	0.099	0.032		24.3	10.9	3.00
750	96.12	1.36	0.451	0.214	0.113	0.036		27.9	12.6	3.44
800	102.5	1.55	0.513	0.244	0.127	0.041		31.8	14.2	3.90
850	108.9	1.74	0.576	0.274	0.144	0.046		35.9	16.0	4.40
900	115.3	1.95	0.642	0.305	0.160	0.051	6 in	40.2	18.0	4.91

TABLE B8.9 Flow of Air through Schedule 40 Steel Pipe (*Continued*)

Free air q'_m ft³/min at 60°F and 14.7 psia	Compressed air ft³/min at 60°F and 100 psig	Pressure drop of air in pounds per square inch per 100 ft of Schedule 40 pipe For air at 100 lb/in² gauge pressure and 60°F temperature								
950	121.8	2.18	0.715	0.340	0.178	0.057	0.023	...	20.0	5.47
1 000	128.2	2.40	0.788	0.375	0.197	0.063	0.025	...	22.1	6.06
1 100	141.0	2.89	0.948	0.451	0.236	0.075	0.030	...	26.7	7.29
1 200	153.8	3.44	1.13	0.533	0.279	0.089	0.035	...	31.8	8.63
1 300	166.6	4.01	1.32	0.626	0.327	0.103	0.041	...	37.3	10.1
1 400	179.4	4.65	1.52	0.718	0.377	0.119	0.047			11.8
1 500	192.2	5.31	1.74	0.824	0.431	0.136	0.054			13.5
1 600	205.1	6.04	1.97	0.932	0.490	0.154	0.061	8 in		15.3
1 800	230.7	7.65	2.50	1.18	0.616	0.193	0.075			19.3
2 000	256.3	9.44	3.06	1.45	0.757	0.237	0.094	0.023		23.9
									10 in	
2 500	320.4	14.7	4.76	2.25	1.17	0.366	0.143	0.035		37.3
3 000	384.5	21.1	6.82	3.20	1.67	0.524	0.204	0.051	0.016	
3 500	448.6	28.8	9.23	4.33	2.26	0.709	0.276	0.068	0.022	
4 000	512.6	37.6	12.1	5.66	2.94	0.919	0.358	0.088	0.028	
4 500	576.7	47.6	15.3	7.16	3.69	1.16	0.450	0.111	0.035	12 in
5 000	640.8	...	18.8	8.85	4.56	1.42	0.552	0.136	0.043	0.018
6 000	769.0	...	27.1	12.7	6.57	2.03	0.794	0.195	0.061	0.025
7 000	897.1	...	36.9	17.2	8.94	2.76	1.07	0.262	0.082	0.034
8 000	1025	...		22.5	11.7	3.59	1.39	0.339	0.107	0.044
9 000	1153	...		28.5	14.9	4.54	1.76	0.427	0.134	0.055
10 000	1282	...		35.2	18.4	5.60	2.16	0.526	0.164	0.067
11 000	1410	...			22.2	6.78	2.62	0.633	0.197	0.081
12 000	1538	...			26.4	8.07	3.09	0.753	0.234	0.096
13 000	1666	...			31.0	9.47	3.63	0.884	0.273	0.112
14 000	1794	...			36.0	11.0	4.21	1.02	0.316	0.129
15 000	1922	...				12.6	4.84	1.17	0.364	0.148
16 000	2051	...				14.3	5.50	1.33	0.411	0.167
18 000	2307	...				18.2	6.96	1.68	0.520	0.213
20 000	2563	...				22.4	8.60	2.01	0.642	0.260
22 000	2820	...				27.1	10.4	2.50	0.771	0.314
24 000	3076	...				32.3	12.4	2.97	0.918	0.371
26 000	3332	...				37.9	14.5	3.49	1.12	0.435
28 000	3588	...					16.9	4.04	1.25	0.505
30 000	3845	...					19.3	4.64	1.42	0.520

For lengths of pipe other than 100 ft, the pressure drop is proportional to the length. Thus, for 50 ft of pipe, the pressure drop is approximately one-half the value given in the table... for 300 ft, three times the given value, etc.

The pressure drop is also inversely proportional to the absolute pressure and directly proportional to the absolute temperature.

Therefore, to determine the pressure drop for inlet or average pressures other than 100 psi and at temperatures other than 60°F, multiply the values given in the table by the ratio $[(100 + 14.7)/(P + 14.7)]\,[(460 + t)/520]$ where P is the inlet or average gauge pressure in pounds per square inch, and t is the temperature in degrees Fahrenheit under consideration.

The cubic feet per minute of compressed air at any pressure is inversely proportional to the absolute pressure and directly proportional to the absolute temperature.

To determine the cubic feet per minute of compressed air at any temperature and pressure other than standard conditions, multiply the value of cubic feet per minute of free air by the ratio, $[14.7/(14.7 + P)]\,[(460 + t)/(520)]$

Calculations for Pipe Other than Schedule 40

To determine the velocity of water, or the pressure drop of water or air, through pipe other than Schedule 40, use the following formulas, $v_a = v_{40}(d_{40} / d_a)^2$, $\Delta P_a = \Delta P_{40}(d_{40} / d_a)^5$, where v = velocity, ft/s; d = internal diameter of pipe, in; ΔP = pressure drop, lb/in²

Subscript a refers to the schedule of pipe through which velocity or pressure drop is desired. Subscript 40 refers to the velocity or pressure drop through Schedule 40 pipe.

Source: Ref. 7.

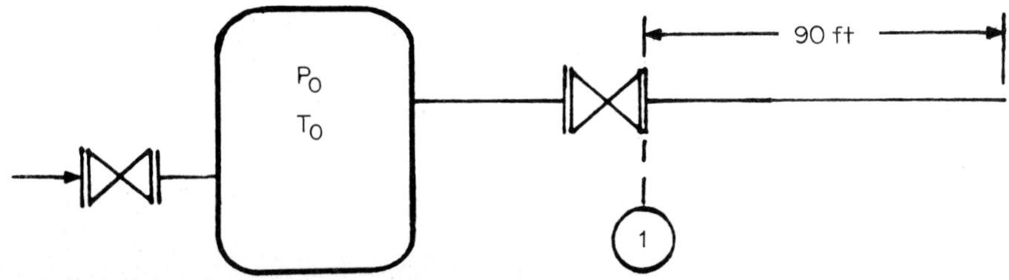

FIGURE B8.17 Sample problem: Compressible flow.

$$f\frac{L}{D} = 0.017\,\frac{(90)(12)}{4.026} = 4.5604$$

Using Fanno line values for $k = 1.4$, at $f(L/D)$, the following values are found:

$$\left(\frac{p}{p^*}\right) = 3.4193$$

$$Ma = 0.317$$

Therefore, from Eq. (B8.24):

$$p_1 = p^*\left(\frac{p}{p^*}\right) = (37.57)(3.4193) = 128.46 \quad [\text{psia}]$$

Problem B8.3. Double flow rate must be released from the tank. Check if the piping described in Problem B8.2 is adequate:

$$p^* = (2)(37.57) = 75.14 \quad [\text{psia}]$$

Then:

$$p_1 = (75.14)(3.4193) = 256.93 \quad [\text{psia}]$$

The piping is still adequate.

Problem B8.4. Keep the mass flow as in Problem B8.2 but double the length of the pipe. Check the pressure p_1:

$$f\left(\frac{L}{D}\right) = (2)(4.5604) = 9.1208$$

at this $f(L/D)$

$$\left(\frac{p}{p^*}\right) = 4.4854$$

$$Ma = 0.243$$

and

$$p_1 = (37.57)(4.4854) = 168.52 \quad [\text{psia}]$$

Problem B8.5. For the same mass flow and the same pipe as in Problem B8.2, assume that the air temperature in the vessel $T_o = 500°\text{F}$:

$$p^* = \frac{20}{0.0884}\sqrt{\frac{(2)(53.3)(460 + 500)}{(32.174)(1.4)(2.4)}}$$

$$= 6960.97 \left[\frac{\text{lb}_f}{\text{ft}^2}\right] = 48.34 \quad [\text{psia}]$$

$$p_1 = (3.4193)(48.34) = 165.29 \quad [\text{psia}]$$

Problem B8.6. For the same mass flow and the same pipe as in Problem B8.2, use the pipe diameter of 2 in (Schedule 40) and calculate p_1:

$$D = 2.067 \text{ in} \qquad (A = 0.0233 \text{ ft}^2)$$

$$p^* = 37.57 \frac{0.0884}{0.0233} = 142.54 \quad [\text{psia}]$$

$$f\left(\frac{L}{D}\right) = (0.019)\frac{(90)(12)}{2.067} = 9.9274$$

Then:

$$\left(\frac{p}{p^*}\right) = 4.644$$

$$Ma = 0.235$$

and

$$p_1 = (142.54)(4.644) = 661.96 \quad [\text{psia}]$$

The calculated pressure p_1 is too high. This line cannot be used. The pipe diameter is too small.

Problem B8.7. For the same flow and the same pipe length as in Problem B8.2, use the pipe diameter of 8 in (Schedule 40), and calculate p_1:

$$D = 7.981 \text{ in} \qquad (A = 0.3474 \text{ ft}^2)$$

$$p^* = 37.57 \frac{0.0884}{0.3474} = 9.56 \quad [\text{psia}]$$

Because $p^* < p_{atm}$, the line is not choked. To calculate the pressure p_1 after the valve, first the additional length of the pipe required for choked conditions at the exit should be calculated.

Assuming $p_{atm} = 14.7$ psia:

$$\frac{p_{atm}}{p^*} = \frac{14.7}{9.56} = 1.5376$$

From Fanno line values for $(p/p^*) = 1.5376$, the following is found:

$$f\left(\frac{L}{D}\right) = 0.2467$$

$$Ma = 0.6814$$

Therefore:

$$L_{add} = \frac{f(L/D)}{f/D} = \frac{(0.2467)(7.981)}{(0.014)(12)} = 11.72 \quad [\text{ft}]$$

and the fictitious pipe length from the choked pipe exit to the valve outlet:

$$L_{fict} = 90 + 11.72 = 101.72 \quad [\text{ft}]$$

The new $f(L/D)$ value is:

$$\left(f\frac{L}{D}\right)_{\text{fict}} = \frac{(0.014)(101.72)(12)}{7.981}$$
$$= 2.1412$$

For this $[f(L/D)]_{\text{fict}}$, the following is obtained from the Fanno line values:

$$\frac{p}{p^*} = 2.6306$$
$$Ma = 0.4096$$

Therefore: $$p_1 = (9.56)(2.6306) = 25.15 \quad \text{[psia]}$$

SINGLE-PHASE FLOW IN NOZZLES, VENTURI TUBES, AND ORIFICES

Theoretical Background and Sample Problems

Liquid Service. A nozzle or an orifice in a tank or reservoir may be installed in the wall or in the bottom (Fig. B8.18). In the case of a nozzle, the fluid emerges in the form of a cylindrical jet of the same diameter as the throat of the nozzle, but in the case of a sharp-edged orifice, the jet contracts after passing through it, attaining its smallest diameter (vena contracta) and greatest velocity some distance (about one-half of a diameter) downstream from the opening. The distance z from the opening in the bottom of the tank to the liquid free surface includes the length of a nozzle or the distance of the vena contracta from the bottom of the tank. The ratio of jet area A_2 at vena contracta to the area of an orifice A is called the *coefficient of contraction* C_c:

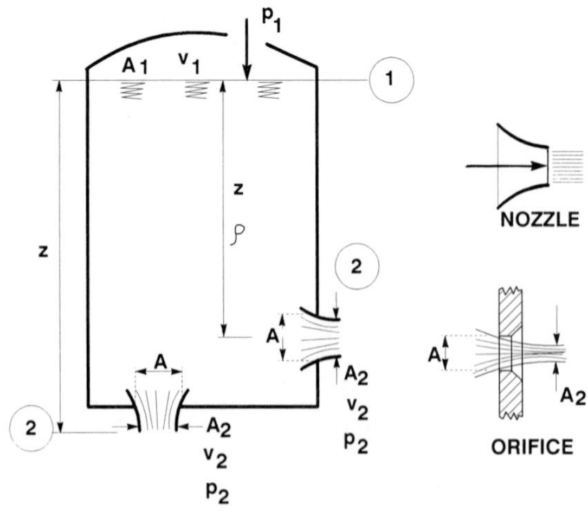

FIGURE B8.18 Typical nozzle or orifice.

$$C_c = \frac{A_2}{A} \tag{B8.28}$$

For a nozzle, $C_c = 1$.

Bernoulli's equation (B8.10) applied from a point 1 on the free surface to the center of the vena contracta, point 2, yields the following expression for the theoretical discharge velocity of fluid at the vena contracta ($v_1 = 0$):

$$v_2 = \sqrt{\frac{2g_c}{\rho}\left(p_1 + \rho z \frac{g}{g_c} - p_2\right)} \quad [\text{ft/s}] \tag{B8.29}$$

where p_1 and p_2 are absolute static (lb_f/ft^2) pressures at the free surface in the tank and in the vena contracta, respectively. If the fluid is discharged to the atmosphere, p_2 is the atmospheric pressure. For open vessels $p_1 = p_2$.

Hydraulic engineers sometimes prefer to use the following version of Eq. (B8.29):

$$v_2 = \sqrt{2g\left(\frac{p_1 - p_2}{\gamma} + z\right)} \tag{B8.30}$$

Equation (B8.29) or (B8.30) describes the theoretical velocity because the losses between the points 1 and 2 were neglected. The ratio of actual velocity v_a to the theoretical velocity v_t is called the *velocity coefficient* C_v.

The actual volumetric discharge rate from the orifice is the product of the actual velocity at the vena contracta and the area of the jet at the vena contracta. Using the above-described coefficient of contraction C_c, the mass flow rate from the opening may be calculated from the following expression:

$$\dot{m} = C_v C_c A \rho v_2 \tag{B8.31}$$

where

$$C_d = C_v C_c \tag{B8.32}$$

is called the *discharge coefficient*.

The velocities and quantities of liquid discharged from a submerged orifice or nozzle (Fig. B8.19) are determined by the same formulas as presented above.

The different symbols in Eqs. (B8.29) and (B8.30) are understood in this case as follows:

z $= z_A$ Immersion depth of the center of the opening relative to the free liquid level in reservoir A, ft

p_1 $= p_A$ Static pressure at the free surface in reservoir A, lb_f/ft^2

p_2 $= p_B + (g/g_c)\rho z_B$ Static pressure in the vena contracta jet, lb_f/ft^2

It is preferable that the actual coefficient of discharge C_d of an orifice or a nozzle be determined by calibration. Such calibration should encompass the entire range of flow rates to be experienced. If an orifice or a nozzle is manufactured according to the ASME specification (Ref. 12), then the appropriate values of C_d, as described in Ref. 12, may be used for flow rate calculations.

Consider now a flow nozzle installed in a pipe (Fig. B8.20). Because with an incompressible fluid, that is, a liquid, the temperature does not change, the density is constant. Thus, the continuity equation becomes:

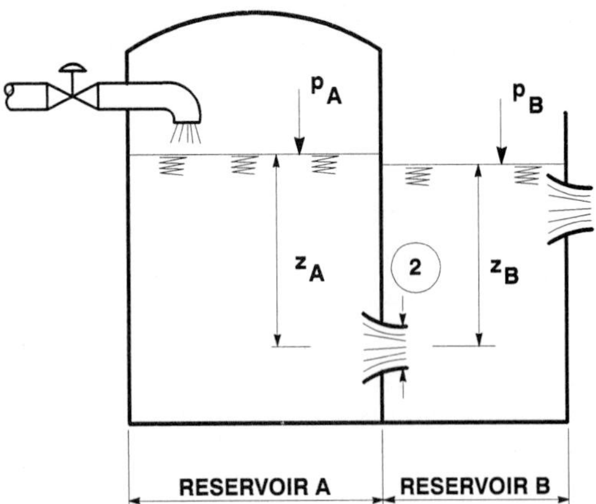

FIGURE B8.19 Typical submerged orifice or nozzle.

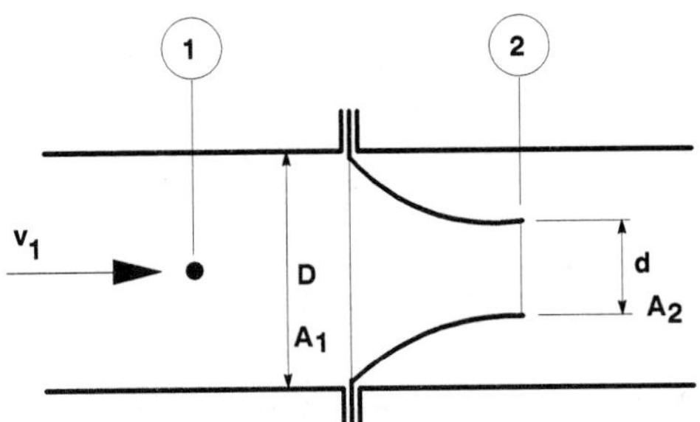

FIGURE B8.20 Flow nozzle in a pipe.

$$v_1 \frac{\pi D^2}{4} = v_2 \frac{\pi d^2}{4} \tag{B8.33}$$

and the Bernoulli equation, applied to section 1 in the pipe to section 2 at the nozzle outlet, is:

$$\frac{v_1^2}{2g_c} + \frac{p_1}{\rho} = \frac{v_2^2}{2g_c} + \frac{p_2}{\rho} \tag{B8.34}$$

Substituting v_1 from Eq. (B8.33) into Eq. (B8.34), the following expression for the theoretical velocity v_2 is obtained (no friction):

$$v_2 = \sqrt{2g_c \frac{p_1 - p_2}{\rho} \frac{1}{1 - \beta^4}} \quad \text{[ft/s]} \tag{B8.35}$$

where $\beta = d/D$, and the approach velocity factor, F_{va}, is:

$$F_{va} = \frac{1}{1 - \beta^4} \tag{B8.36}$$

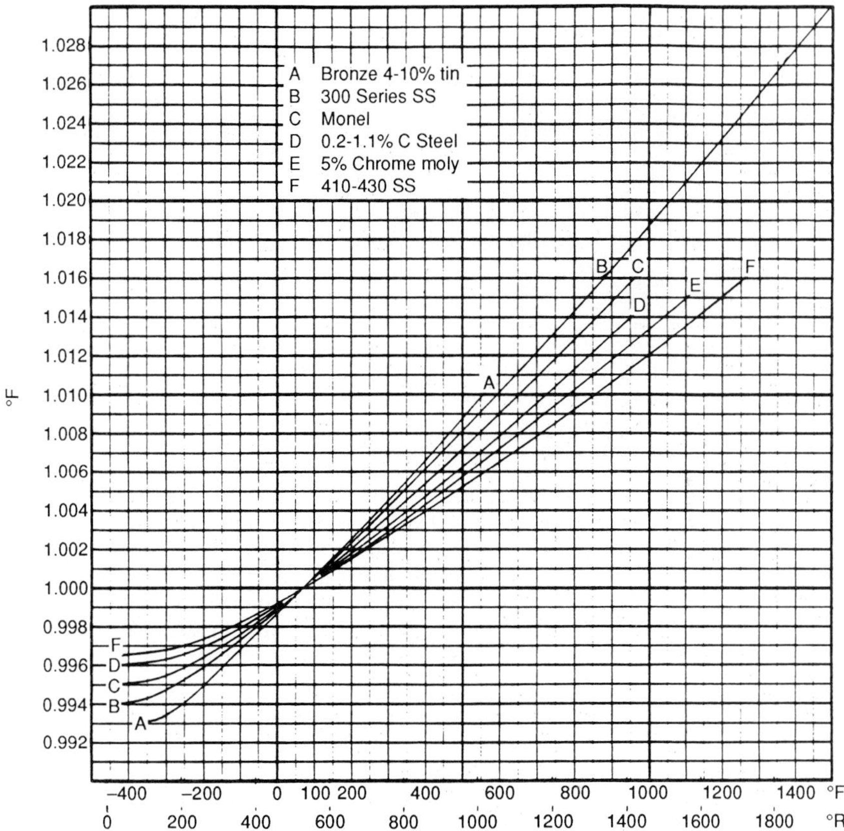

FIGURE B8.21 Area factors F_a for the thermal expansion of primary elements. (*From Ref. 12.*)

Because most materials expand or contract as their temperature changes, an area factor F_a (Fig. B8.21) for the thermal expansion of the primary element (nozzle or orifice) must be introduced in order to find an actual flow area. In general, the actual flow rate is less than the theoretical flow. Hence, to obtain the actual flow, the discharge coefficient C_d must be introduced (empirical value) into the theoretical Eq.(B8.35):

$$\dot{C}_d = \frac{\text{actual flow rate}}{\text{theoretical flow rate}} \tag{B8.37}$$

Thus, the actual mass flow rate through a flow nozzle is:

$$\dot{m} = F_a A_2 C_d v_2 \rho \tag{B8.38}$$

Equation (B8.38), which was derived for a flow nozzle, holds equally well for horizontal venturi tubes and for orifices. The following factor is frequently used in calculations and is called the *flow coefficient:*

$$C = \frac{C_d}{\sqrt{1 - \beta^4}} \tag{B8.39}$$

The values of C_d and C are different for each different type of primary element: venturi tube, flow nozzle, and orifice. Also, with flow nozzles and orifices, the values depend upon the locations of the pressure taps; and, with the orifice, the values differ with the type of inlet edge, whether square and thin or rounded. Values of C_d for long radius nozzles and square-edged orifices are available in Ref. 7. For a classical venturi tube, the discharge coefficients C_d are:[12]

	C_d
Rough-cast entrance cone	0.984
Machined entrance cone	0.995
Rough-welded sheet metal entrance cone	0.985

Each obstacle in a flow path generates a permanent pressure drop Δp_p. The pressure profile in a pipe due to an orifice insertion is shown in Fig. B8.22.

The permanent pressure loss through a primary element of a flowmeter, with an orifice, a nozzle, or a venturi tube, is shown in Fig. B8.23. The venturi tube has a low overall loss, due to the gradually expanding conical section, which aids in efficient reconversion of the high kinetic energy at the throat into pressure.

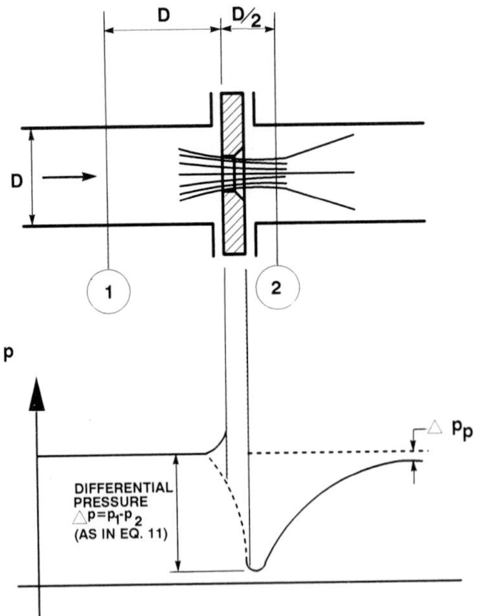

FIGURE B8.22 Pressure profile in a pipe.

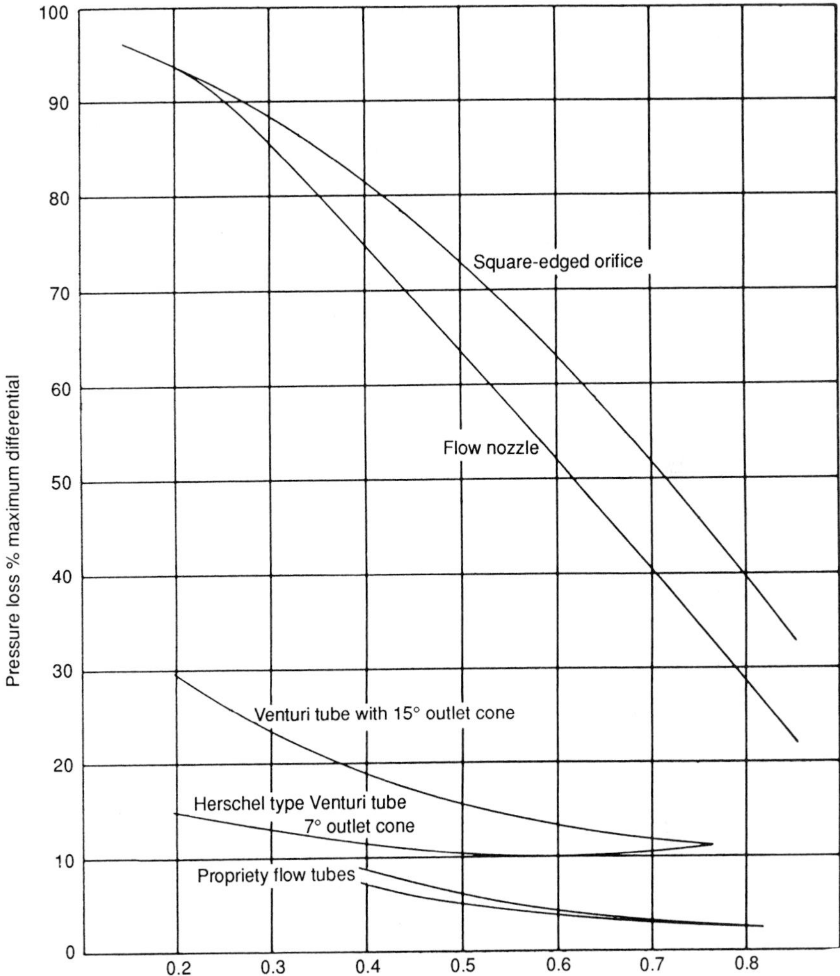

FIGURE B8.23 Overall pressure loss through several primary elements. (ASME, 1971) (*From Ref. 12.*)

Steam and Gas Service. Because the static head is not considered in gas systems, there is no need to distinguish the case of outflow from a reservoir from that of flow in a pipe.

Consider the flow of a perfect gas through a converging nozzle (Fig. B8.24). Neglecting potential energy and technical work, the following result is found for an isentropic flow:

$$v_{2s} = \sqrt{2g_c J(h_1 - h_{2s}) + v_1^2} \tag{B8.40}$$

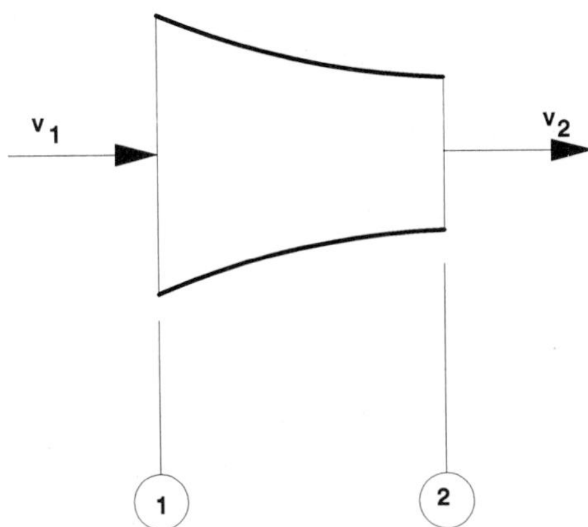

FIGURE B8.24 Fluid flow in a converging nozzle.

If there are no losses due to friction or heat transfer, the change of state of a gas is isentropic (s = const), and using the gas properties relationships (Ref. 2, typical) yields:

$$v_{2s} = \sqrt{2g_c J c_p (T_o - T_{2s})} \tag{B8.41}$$

which was first derived by De St. Venant and Wantzel in 1839. Equation (B8.41) is written in terms of stagnation conditions in a vessel, where:

$$T_o = T_1 + \frac{v_1^2}{2g_c J c_p} \tag{B8.42}$$

$$p_o = p_1 \left(\frac{T_o}{T_1}\right)^{k/(k-1)} \tag{B8.43}$$

The mass flow rate of a gas through the cross-section area A_2 can be calculated from:

$$\dot{m} = A_2 \psi_s \sqrt{\frac{p_o}{v_o}} \tag{B8.44}$$

where:

$$\psi_s = \sqrt{2g_c \frac{k}{k-1}\left[\left(\frac{p_2}{p_o}\right)^{2/k} - \left(\frac{p_2}{p_o}\right)^{(k+1)/k}\right]} \tag{B8.45}$$

Figure B8.25 shows the mass flow rate $\dot{m}$ plotted against p_2/p_o for a specified gas which is characterized by its $k = c_p/c_v$ value. The flow rate becomes zero when $(p_2/p_o) = 1$. As the pressure p_2 decreases, the flow rate increases to its maximum value at p^*/p_o (critical pressure ratio).

At $(p_2/p_o) < (p^*/p_o)$, Eq. (B8.44) predicts decreasing flow to zero at $(p_2/p_o) = 0$. However, experiments show that in a flow through a convergent

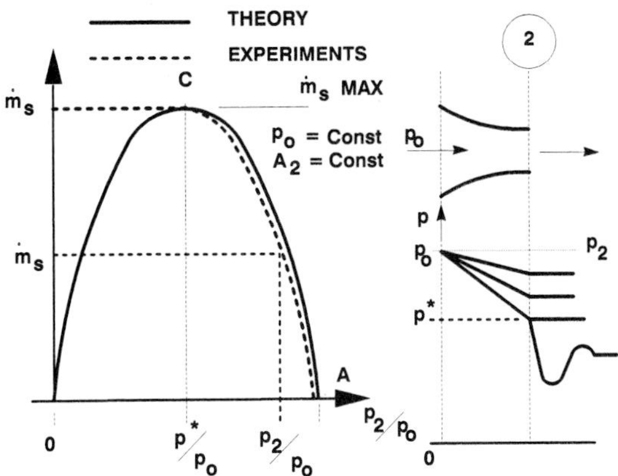

nozzle, the pressure at the exit cross-section A_2 cannot fall below the value of p^* and the flow rate does not change, no matter how low the back pressure p_2. At the critical pressure ratio the mass flow rate reaches its maximum value and the flow is said to be "choked." The maximum flow rate at the choke is called *critical flow rate.* Critical or mass limiting flow is characteristic of compressible fluid systems.

The value of the critical pressure ratio (p^*/p_o) may be found from the following conditions:

$$\frac{d\dot{m}_s}{d(p_2/p_o)} = 0 \quad \text{or} \quad \frac{d\psi_s}{d(p_2/p_o)} = 0 \tag{B8.46}$$

Applying the above condition to Eq. (B8.45) leads to:

$$\beta_s = \frac{p^*}{p_o} = \left(\frac{2}{k+1}\right)^{k/(k-1)} \tag{B8.47}$$

Substituting this in Eq.(B8.45) gives:

$$\psi_{s\max} = \sqrt{g_c k \left(\frac{2}{k+1}\right)^{(k+1)/(k-1)}} \tag{B8.48}$$

and the corresponding maximum flow rate of:

$$m_{s\max} = A_2 \psi_{s\max} \sqrt{\frac{p_o}{v_o}} \tag{B8.49}$$

For $(p_2/p_o) > p^*/p_o$, the mass flow rate must be calculated from Eq. (B8.49) or by the simplified equation presented by Bendemann in 1907.[10] It was proved by Bendemann that the arc AC in Fig. B8.25 can be replaced with a high degree of accuracy by a quadrant of an ellipse leading to:

$$\left(\frac{\dot{m}}{\dot{m}_{s\max}}\right)^2 + \frac{\left(\frac{p_2}{p_o} - \frac{p^*}{p_o}\right)^2}{\left(1 - \frac{p^*}{p_o}\right)^2} = 1 \qquad \text{(B8.50)}$$

The remaining thermodynamic parameters (in addition to the critical pressure p^*) at the nozzle throat, corresponding to the maximum flow rate of an ideal gas, are the following:

1. Temperature may be determined by using the relation for an isentropic expansion and Eq. (B8.47):

$$T^* = T_o \frac{2}{k + 1} \qquad \text{(B8.51)}$$

   For T_o see Eq. (B8.42).

2. Specific volume may be calculated from another relation for an isentropic expansion:

$$p_o \upsilon_o^k = (\beta_s p_o)(\upsilon^*)^k \qquad \text{(B8.52)}$$

3. The gas velocity may be calculated from the following expression:

$$\mathrm{v}_{2\max} = \alpha_s \sqrt{p_o \upsilon_o} \qquad \text{(B8.53)}$$

   where:

$$\alpha_s = \sqrt{2g_c \frac{k}{k + 1}} \qquad \text{(B8.54)}$$

   Because $p_o \upsilon_o = RT_o$:

$$\mathrm{v}_{2\max} = \alpha_s \sqrt{RT_o} \qquad \text{(B8.55)}$$

It will be observed that for the same perfect gas (α_s, R), the maximum velocity depends only upon its initial stagnation temperature (before a nozzle). It can be shown that Eq. (B8.53) or (B8.55) represents a velocity of sound in the choked exit plane of a nozzle:

$$\mathrm{v}_{2\max} = \sqrt{g_c k R T^*} = c \qquad \text{(B8.56)}$$

The ASME report on fluid meters[12] presents slightly modified equations which are more suitable for precise differential-pressure flow-metering techniques. See this document for procedures and more information about flow measurement using orifices, nozzles, or venturi tubes.

Sample problems concerning ASME fluid meters may be found in Refs. 12 and 13.

Flow Restriction Applications

Flow-restricting orifices are used where a continuous small flow of fluid is required. Flow-restricting orifices are usually not designed according to requirements of the ASME specification for fluid metering devices. In most cases, they are round openings drilled in an orifice plate. Their fluid entrance may be sharp-edged, beveled, or well-rounded.

It is suggested that reasonably accurate approximations in flow rate estimations may be obtained by using equations presented in the last subsection. If the entrance is well-rounded, C values would tend to approach those for ASME nozzles, whereas openings with square entrances would have characteristics similar to those for square-edged ASME orifices. The minimum allowable orifice size should be 1/16 in to prevent clogging. The recommended thicknesses of orifice plates are given in Table B8.10 for carbon steel or chrome-moly steel.

Application for Fluid Flow Metering

Recommended conditions, procedures, and data for measuring the flow of fluids, particularly with the three principal differential-pressure meters—the orifice, the flow nozzle, and the venturi tube—are presented in the ASME test codes.[12,13]

STEADY TWO-PHASE FLOW

A phase is simply one of the states of matter and can be either a gas, a liquid, or a solid. The general subject of two-phase flow includes gas-liquid, gas-solid, and solid-liquid flow.

The term *multicomponent* is used to describe flows in which the phases do not consist of the same chemical substance. In the petrochemical industries many processes involve the evaporation (and condensation) of binary ($n = 2$) and multicomponent mixtures. Pure single-component, two-phase flows are those during evaporation and condensation of the same chemical substance. For example, steam-water flow is a single-component, two-phase flow, while air-water is a two-phase, two-component flow.

The main emphasis of the following presentation is on the two-phase flow of water.

Regimes of Gas-Liquid Flow

Description. Cocurrent, simultaneous flows of gases and liquids occur in numerous components of plant equipment such as steam generators, drain lines, and oil and natural gas pipelines.

Ever since the earliest visual observations of two-phase flow, it has been recognized that there are natural varieties of flow patterns. In addition to the random character of each flow configuration, two-phase flows are never fully developed. In fact, the gas phase expands due to the pressure drop along a pipe leading to a modification of the flow structure. The flow pattern depends also upon the geometry changes of a flow channel (bends, valves, etc.). Flow patterns will be classified according to pipe geometry and flow direction (upward, downward, cocurrent, countercurrent), and several shown in Refs. 14 and 15 are discussed below.

TABLE B8.10 Restriction Orifice Plate, Pressure, Temperature Ratings
(Plate material—carbon steel or chrome-moly—allowable differential pressure across the orifice plate)

Pipe size, NPS	Orifice plate OD, in	Temp. °F	Pressure, psi — Plate thickness, in						
			⅛ in	3⁄16 in	¼ in	5⁄16 in	⅜ in	7⁄16 in	½ in
		300	333	748	1330	2078	—	—	—
1½	2⅞	600	283	637	1133	1770	2549	—	—
		900	149	335	596	932	1342	1826	2385
		300	214	482	857	1339	1928	2624	—
2	3⅝	600	182	411	730	1140	1642	2235	—
		900	96	216	384	600	864	1176	1536
		300	146	329	585	914	1316	1791	2339
2½	4⅛	600	125	280	498	778	1121	1525	1992
		900	66	147	262	410	590	803	1049
		300	99	222	395	616	888	1208	1578
3	5	600	84	189	336	525	756	1029	1344
		900	44	99	177	276	398	542	707
		300	59	134	239	373	537	731	955
4	6 3⁄16	600	50	114	203	317	457	623	813
		900	26	60	107	167	241	328	428
		300	28	62	110	172	248	337	440
6	8½	600	23	53	94	147	211	287	375
		900	12	28	49	77	111	151	197
		300	16	37	65	102	146	199	260
8	10⅝	600	14	31	55	86	125	169	221
		900	7	16	29	46	66	89	117
		300	10	24	42	65	94	128	167
10	12¾	600	9	20	36	56	80	109	142
		900	5	11	19	29	42	57	75
		300	7	17	30	46	67	91	119
12	15	600	6	14	25	40	57	78	101
		900	3	7	13	21	30	41	53
		300	6	14	25	39	55	76	99
14	16¼	600	5	12	21	33	47	64	84
		900	3	6	11	17	25	34	44
		300	5	11	19	30	42	58	76
16	18½	600	4	9	16	25	36	49	64
		900	2	5	9	13	19	26	34
		300	4	8	15	23	34	46	60
18	21	600	3	7	13	20	29	39	51
		900	2	4	7	10	15	20	27
		300	3	7	12	19	27	37	48
20	23	600	3	6	10	16	23	32	41
		900	1	3	5	8	12	17	22
		300	2	6	10	16	22	31	40
22	25¼	600	2	5	9	13	19	26	34
		900	1	3	4	7	10	14	18
		300	2	5	8	13	19	26	34
24	27¼	600	2	4	7	11	16	22	29
		900	1	2	4	6	8	12	15

Source: Stone & Webster data.

Upward Cocurrent Flow in Vertical Pipes. The main flow patterns encountered in a vertical pipe are shown in Fig. B8.26.[16] Bubbly flow is certainly the most widely known configuration, although at high velocity its milky appearance prevents it from being easily recognized. Bubbles are spherical only if their diameters do not exceed 1 mm; whereas beyond 1 mm, their shape is variable. Roumy distinguishes two bubbly flow patterns.[16] In the independent bubble configuration, bubbles are spaced and do not interact with each other. On the other hand, in the packed configuration, bubbles are crowded together and strongly interact with each other.

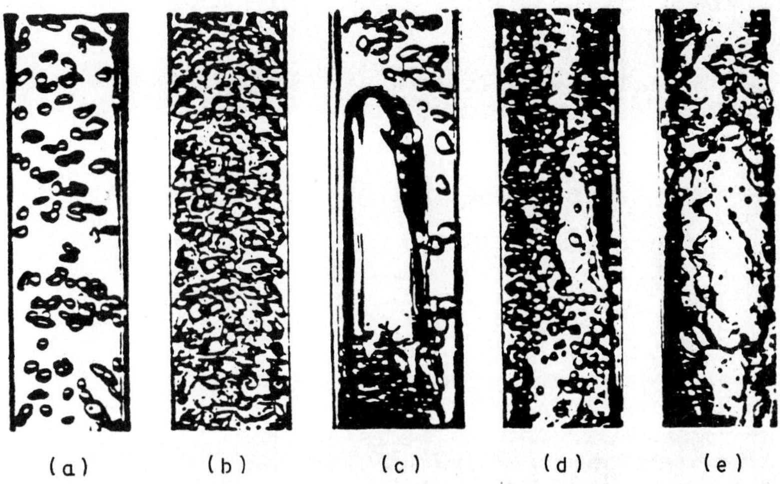

FIGURE B8.26 Upward cocurrent flow in a vertical pipe. Air-water flow patterns (Roumy, 1969); (*a*) Independent bubbles; (*b*) packed bubbles; (*c*) slug flow; (*d*) churn flow; (*e*) annular flow. Pipe diameter: 32 mm. (*From Ref. 16.*)

Slug flow is composed of a series of gas plugs. The head of a gas plug is generally blunt whereas its end is flat with a bubbly wake. A simple visual observation reveals that the liquid film which surrounds a gas plug moves downward with respect to the pipe wall.

Given a constant liquid flow rate, an increase of the gas flow rate leads to a lengthening and a breaking of the gas plugs. The flow pattern evolves toward an annular flow in a chaotic way. This transition configuration is called a *churn flow*.

Dispersed annular flow is characterized by a central gas core loaded with liquid droplets and flowing at a higher velocity than the liquid film which clings to the wall. Droplets are torn off from the crest of the waves which propagate on the surface of the liquid film. They diffuse in the gas core and can eventually impinge onto the film surface. Hewitt and Roberts evidenced the existence of a wispy annular flow where the liquid droplets gather into clouds within the central gas core (Fig. B8.27).[17]

Finally, if the wall temperature is high enough to vaporize the film, the droplets will constitute a mist flow. Figure B8.28 shows the configurations of a liquid-vapor flow in a heated pipe as a function of the wall heat flux. The liquid enters the pipe at a constant flow rate and at a temperature lower than the saturation temperature. When the heat flux increases, the vapor appears closer and closer to

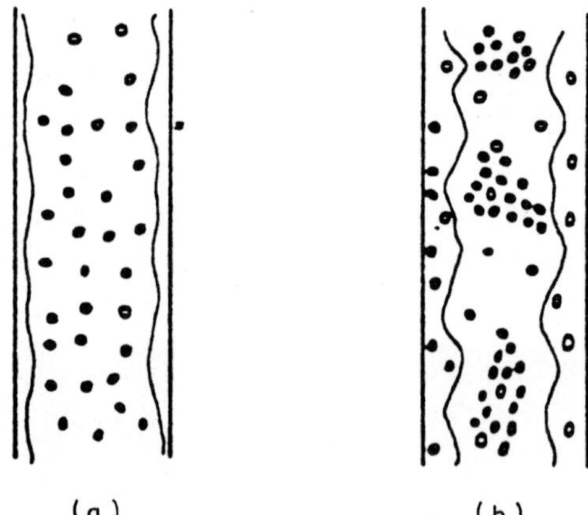

FIGURE B8.27 Annular flows. (*a*) dispersed; (*b*) wispy annular. (*From Ref. 17.*)

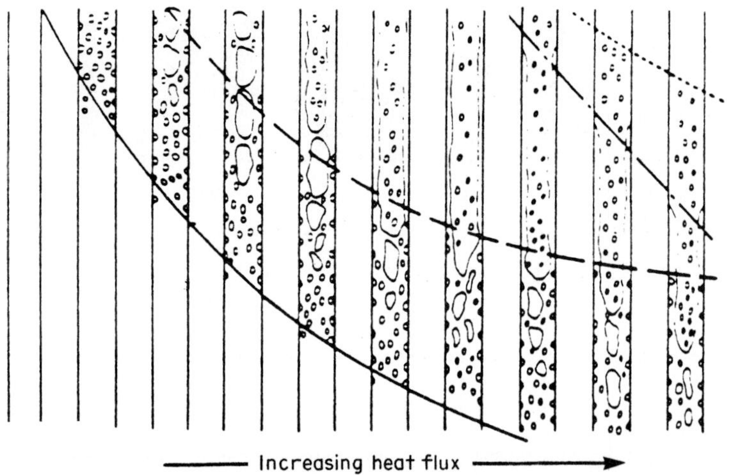

FIGURE B8.28 Convective boiling in a heated channel at a constant liquid flow rate. (_____) Onset of nucleate boiling. (– – –) End of nucleate boiling. (— . — . —) Dryout. (-----) Limit of the super-heated vapor region (Hewitt and Hall-Taylor, 1970). (*From Ref. 18.*)

the pipe inlet. The local boiling length is the extent of pipe where the bubbles are generated at the wall and condense in the liquid core where the liquid temperature is still lower than the saturation temperature.

The usual way of presenting results of observations of flow patterns is to plot them on a flow regime map. A flow map is a two-dimensional representation of the flow pattern existence domains. The coordinate systems are different according to different authors, and, so far, there is no agreement on the best coordinate systems. Selecting a series of flow maps is not easy, and no recommendation can be made since no method has proven entirely adequate so far.

Hewitt and Roberts' map (1969) is the most widely used chart for air-water and steam-water flows and is described in Refs. 14 and 15. The steam-water results of Bennett, et al.,[15] are also well represented in the Hewitt and Roberts' diagram.

Downward Cocurrent Flows in Vertical Pipes. So far the most comprehensive studies of downward cocurrent flow patterns are due to Oshinowo and Charles.[15] These authors distinguish six different flow configurations as shown in Fig. B8.29.

The downward bubbly flow structure is quite different from the upward bubbly flow configuration. In the latter case, bubbles are spread over the entire cross section, whereas in the downward flow, bubbles gather near the pipe axis. This coring effect is similar to the phenomenon observed in a flow of liquid loaded with solid particles whose density is smaller than the liquid density.

When the gas flow rate is increased, the liquid flow rate being held constant, the bubbles agglomerate into large gas pockets. The top of these gas plugs is dome shaped whereas the lower extremity is flat with a bubbly zone underneath. This slug flow is generally more stable than in the upward case.

The annular configuration can exhibit several aspects. For small liquid and gas flow rates, a liquid film flows down the wall (falling film flow). If the liquid flow rate is higher, bubbles are entrained within the film (bubbly falling film). When liquid and gas flow rates are increased a churn flow appears and can evolve into a dispersed annular flow for very high gas flow rates.

Oshinowo and Charles proposed a flow map, described in Ref. 15, which was obtained from their own experimental data. They studied two-component mixtures of air and different liquids flowing in a pipe 25.4 mm in diameter at a pressure of around 1.7 bar.

Cocurrent Flows in Horizontal and Inclined Pipes. The number of possible flow patterns in a horizontal pipe is higher than in a vertical pipe. This is due to the

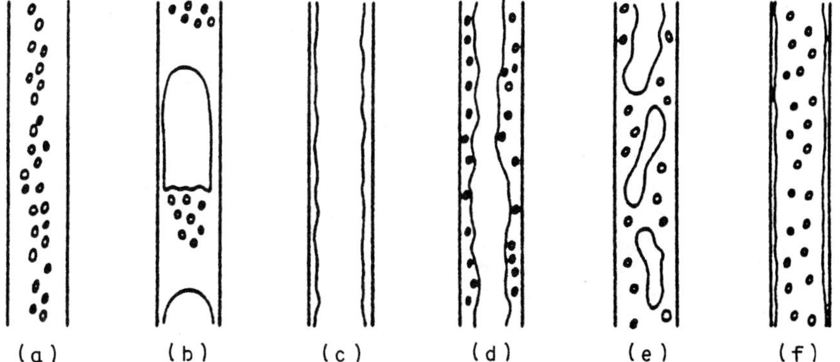

FIGURE B8.29 Downward cocurrent flow in a vertical pipe. Air-water patterns (Oshinowo and Charles, 1974); (*a*) bubbles, (*b*) slugs; (*c*) falling film; (*d*) bubbly falling film; (*e*) churn; (*f*) dispersed annular. (*From Ref. 19.*)

effect of gravity, which tends to separate the phases and to create a horizontal stratification.

Alves proposed the classification shown in Fig. B8.30.[20] In the bubbly flow configuration bubbles are moving in the upper part of the pipe. When the gas flow rate is increased, bubbles coalesce and a plug flow takes place. For low liquid and gas flow rates a stratified flow appears with a smooth interface. At higher gas rates waves propagate along the interface (wavy flow) and can reach the top wall of the pipe giving rise to a slug flow. Finally at high gas flow rates and low liquid flow rates, an annular flow can exist with a thicker film in the lower part of the pipe. These flow pattern names differ depending on the author. As an example, Taitel and Dukler class plug flows and slug flows under the same denomination (intermittent flow).[21]

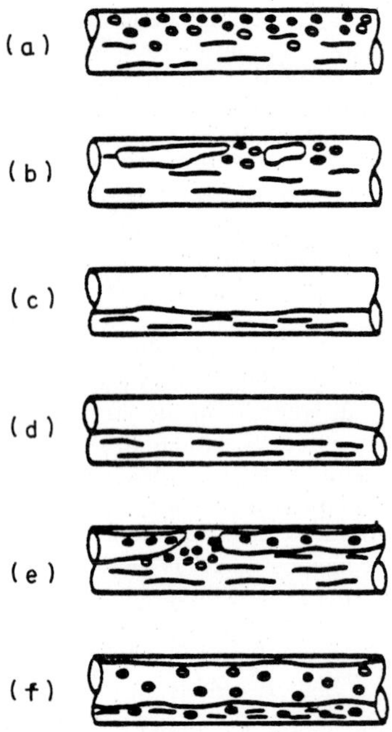

FIGURE B8.30 Flow patterns in a horizontal pipe. (*a*) bubbly flow; (*b*) plug flow; (*c*) stratified flow; (*d*) Wavy flow; (*e*) slug flow; (*f*) annular flow. (*From Ref. 20.*)

Figure B8.31 shows the evolution of a vaporizing flow in a horizontal pipe. The liquid enters the heated pipe with a low flow rate and at a temperature slightly lower than the saturation temperature. An important point concerns the upper part of the tube which can dry out periodically and then suffer a sudden increase in wall temperature. If the wall temperature is high enough, the wall dries out completely and the liquid droplets form a mist flow.

The earliest and the most durable of regime maps for two-phase gas-liquid flow in horizontal channels was proposed by Baker, as described in Ref. 22. He developed a very useful and informative representation of flow patterns for steam-water mixtures presented in Fig. B8.32.

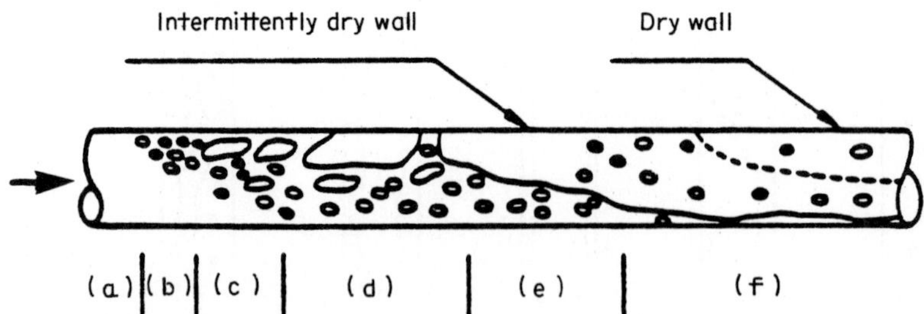

FIGURE B8.31 Flow pattern evaluation in a horizontal evaporator tube. (*a*) Liquid single-phase flow; (*b*) bubbly flow; (*c*) plug flow; (*d*) slug flow; (*e*) Wavy flow; (*f*) annual flow. (*From Ref. 15.*)

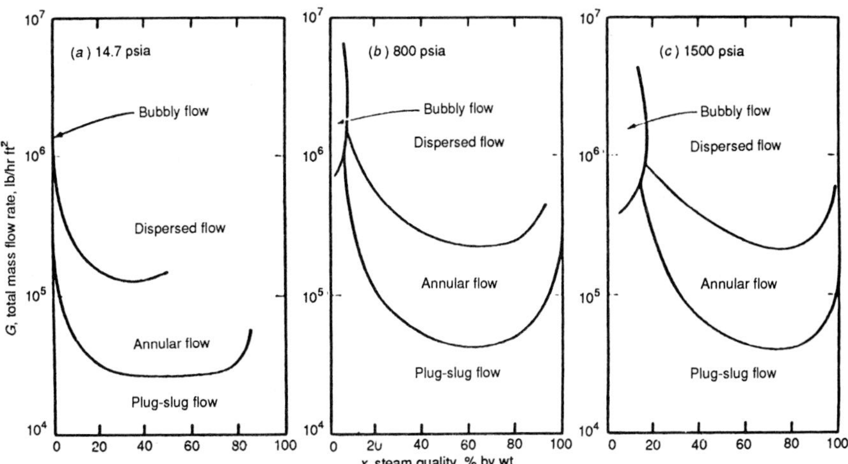

FIGURE B8.32 Flow patterns of two-phase flow. (*From Ref. 22.*)

Mandhan et al., constructed a flow-regime map based on 5935 data points, 1178 of which concern air-water flows.[15] Its coordinates are the superficial velocities j_f and j_g calculated at the test section pressure and temperature.

The general trends of Mandhane's map were utilized by Taitel and Dukler.[21] They developed a theoretical approach for the prediction of transition between flow regimes. This approach also provides considerable insight into the mechanisms of the transitions. The theoretical chart proposed by Taitel and Dukler (Fig. B8.33) uses different coordinate systems for different transitions considered.[21,23] It is, of course, not necessary to draw a flow regime map at all. Given any one set of flow conditions (flow rate, fluid properties, line inclinations, and line size), the flow pattern that exists for that condition can be determined by using a computerized calculating procedure. Figure B8.33 is valid only for strictly horizontal two-phase flow with both phases being turbulent. The Taitel-Dukler theory covers more general applications.

Flow regime coordinates in Fig. B8.33 are the following:

***Coordinate* X**

Defining

$$XA = (j_f D/\nu_f)^{-0.2} \tag{B8.57}$$

$$XB = 0.5\rho_f j_f^{\,2} \tag{B8.58}$$

$$XC = (j_g D/\nu_g)^{-0.2} \tag{B8.59}$$

$$X = 0.5\rho_g j_g^{\,2} \tag{B8.60}$$

the value of X is:

$$X = \sqrt{\frac{(XA)(XB)}{(XC)(XD)}} \tag{B8.61}$$

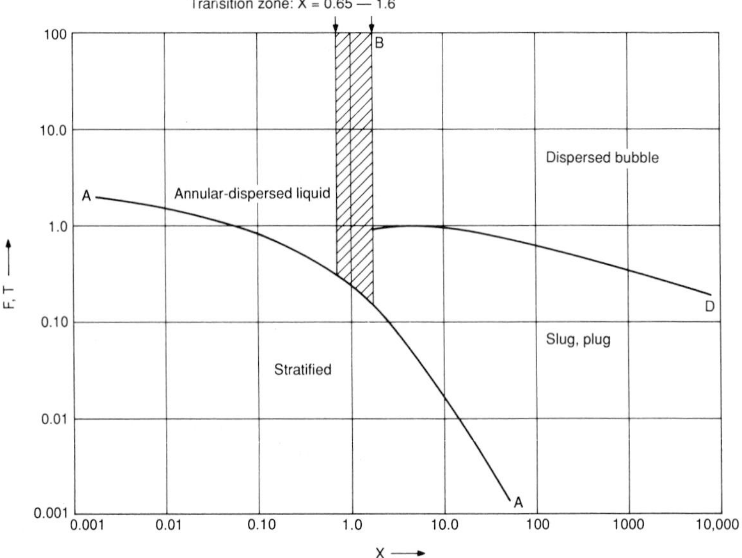

FIGURE B8.33 Generalized flow regime map for horizontal two-phase flow (Taitel-Dukler, 1976). (*From Ref. 21.*)

Coordinate* F *(Curve* A*)

$$F = \sqrt{\frac{\rho_g}{\rho_f - \rho_g}} \frac{j_g}{\sqrt{gD}} \tag{B8.62}$$

Coordinate* T *(Curve* D*)

Defining

$$TA = XA \tag{B8.63}$$

$$TB = XB \tag{B8.64}$$

$$TC = \frac{0.184 g_c}{D(\rho_f - \rho_g)g} \tag{B8.65}$$

the value of T is:

$$T = \sqrt{(TA)(TB)(TC)} \tag{B8.66}$$

Countercurrent Flow of Steam and Water. Countercurrent flow of steam and water may exist only when a pipe does not run full. There are two very important areas in a plant design where this kind of flow is carefully studied. One of these concerns the design of self-venting lines where the vapor is not carried down the drain pipe with the liquid but can rise counter to the liquid flow, continuously venting the pipe. The second concern deals with initiation of a condensation-induced water hammer in horizontal or nearly horizontal pipes containing steam and subcooled water. Designing self-venting lines is discussed later in this section of this chapter, while the condensation-induced water hammer is described in a later section, "Steam Condensation Induced Water Hammer."

The stratified countercurrent flow of a gaseous phase and a liquid is governed by the open channel flow criterion expressed as a liquid Froude number:

$$Fr = \frac{j_f}{\sqrt{gD}} \tag{B8.67}$$

The stratified void fraction α in a circular pipe versus dimensionless vapor gas is presented in Fig. B8.34.

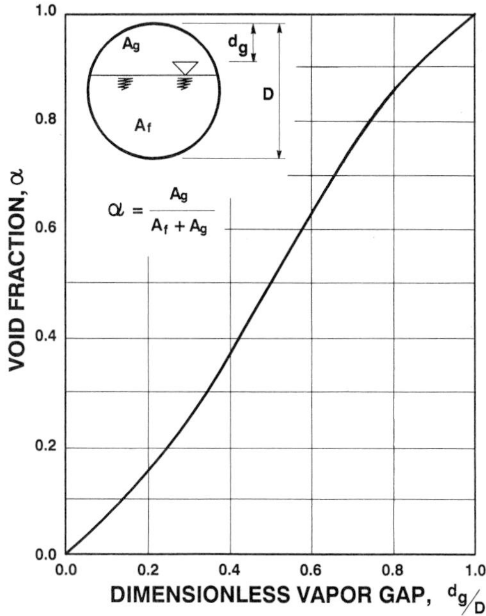

FIGURE B8.34 Void fraction in circular pipe versus dimensionless gap.

Pressure Drop in Gas-Liquid Flow ("Nonflashing" Flow)

A two-phase pressure drop consists of three components: a frictional term, an acceleration term (momentum change), and a static head (elevation) term.

The elevation pressure drop is similar to that for single-phase flow.

The acceleration pressure drop Δp_A is important in a case of heated channels (Ref. 24 should be consulted for further information) and is calculated as:

$$\Delta p_A = \frac{\dot{G}^2}{g_c}\left(\frac{d\bar{v}}{dL}\right)\Delta L \tag{B8.68}$$

where $(d\bar{v}/DL)$ is an increase of average specific volume of the mixture at the location considered.

The difficulty in calculating the pressure drop in two-phase flow systems is that detailed information on local mass flux, velocity, and density is practically never available.

By introducing a two-phase multiplier, the frictional pressure drop may be calculated from:

$$\left(\frac{\Delta p_F}{\Delta L}\right)_{TP} = \phi_{\mathrm{lo}}^2\left(\frac{\Delta p_F}{\Delta L}\right)_{\mathrm{lo}} \tag{B8.69}$$

where ϕ_{lo}^2 is the two-phase frictional pressure drop multiplier, $(\Delta p_F/\Delta L)_{TP}$ is two-phase frictional pressure drop, and $(\Delta p_F/\Delta L)_{\mathrm{lo}}$ is single-phase pressure drop obtained at the same mass flux when the fluid is entirely in the liquid phase.

A number of workers have carried out systematic comparison between various two-phase pressure drop correlations and data banks containing large numbers of experimental pressure drop measurements for steam-water mixtures. A summary of these various comparisons is given in Ref. 14.

Baroczy's correlation for ϕ_{lo}^2 is widely used in the U.S.[25] This correlation, which may be considered an extension of that by Lockhart and Martinelli, and Martinelli and Nelson, is given in two sets of curves:[26]

1. A plot of the two-phase pressure drop multiplier ϕ_{lo}^2 as a function of the property index PI is shown in Fig. B8.35 where:

$$PI = \frac{(\mu_f/\mu_g)^{0.2}}{\rho_f/\rho_g} \tag{B8.70}$$

2. Plots of a two-phase multiplier ratio as a function of the property index, and mass flux are shown in Fig. B8.36. This ratio multiplies ϕ_{lo}^2 whenever G deviates from 10^6 lb/(ft^2 · h). Baroczy's two-phase multiplier ratio varies with mass flux in a rather peculiar manner. Since several investigators have observed changes in pressure drop behavior characteristics with flow regime, these variations may be related to flow pattern changes.

Baroczy compared his correlation method with experimental data for water-air, water-steam, and other liquids including liquid metals, organic oils, and so on, over a wide range of conditions and found the agreement good.

A summary of the better-known correlations for the slip ratio S or the local void fraction α is given in Ref. 15. Systematic comparisons have been carried out between these various void fraction correlations and data banks containing large numbers of experimental measurements of either void fraction or fluid density measurements for steam-water mixtures. The various studies agreed that the most accurate void fraction correlations are those by Smith, CISE, and Chisholm,[14] with the latter having the added advantage of great simplicity. The standard deviation of error on the mean density is about 20 to 30 percent.

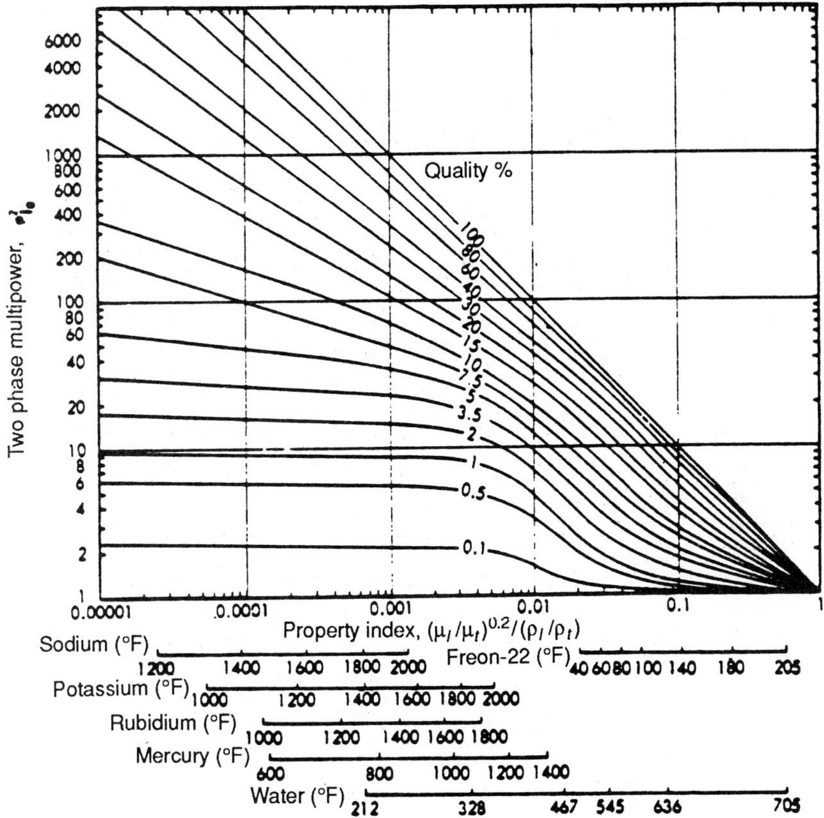

FIGURE B8.35 Two-phase friction pressure drop correlation for $G = 1 \times 10^6$ lb/(h ft^2). (*From Ref. 26.*)

Critical Gas-Liquid Flow (Flashing Flow)

General Remarks. Critical or mass limiting flow is characteristic of compressible fluid systems. In two-phase flow, a maximum or critical flow rate is also observed. When fluid travels in a horizontal pipeline, its pressure decreases until it reaches pipe exit. If a liquid in the upstream vessel is near the boiling point, its pressure in the pipeline may drop below the saturation point, resulting in vapor formation (flashing). Therefore, two-phase flow may occur starting from a certain cross section in the line. There are cases where fluid entering the pipe may already be a liquid-vapor mixture. If the pressure in the downstream receiver is reduced sufficiently to cause the flow velocity at the pipe exit to equal that of the speed of sound at local fluid conditions, the mass flow rate reaches its maximum value and the flow is said to be *choked.* The maximum flow rate at the choke is called *critical flow rate.*

At the present, there is no general model or correlation for critical two-phase flow which is valid for a wide range of pipe lengths, pipe diameters, and upstream conditions including subcooled liquid. For these reasons different models may be

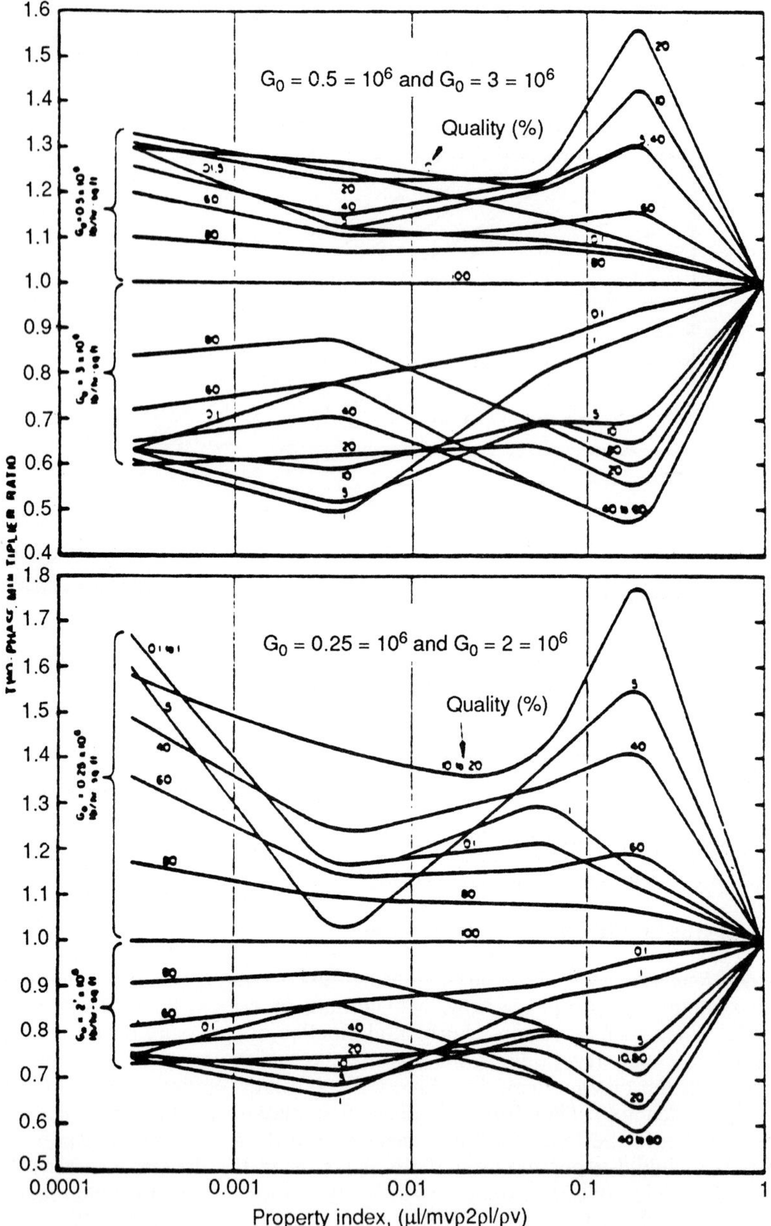

FIGURE B8.36 Mass-velocity correction versus property index. (*From Ref. 26.*)

more appropriate for designing specific two-phase flow discharge systems. The most frequently used models and calculation procedures are described in the following subsections.

Fanno Model[27] (Thermal Equilibrium, Homogeneous). All homogeneous-equilibrium models are based on three assumptions:

1. There is no slip between phases.
2. There is thermodynamic equilibrium between the liquid and vapor at all times.
3. The specific volume is calculated from the equation:

$$v = v_f + x(v_g - v_f) \tag{B8.71}$$

Equation (B8.17) represents the Fanno line as a locus of conditions in a pipe of constant diameter.

For the two-phase (wet steam) region, the following expressions hold:

$$h = h_f + xh_{fg} \tag{B8.72}$$

$$v = v_f + xv_{fg} \tag{B8.73}$$

$$s = s_f + xs_{fg} \tag{B8.74}$$

From Eqs. (B8.17), (B8.72), and (B8.73), the following is found:

$$x^2 + 2x\left[\frac{v_f}{v_{fg}} + Jg_c\left(\frac{A}{\dot{m}}\right)^2 \frac{h_{fg}}{v_{fg}^2}\right]_x + \frac{(v_f^2)_x - v_1^2 + 2Jg_c(A/\dot{m})[(h_f)_x - h_1]}{(v_{fg}^2)_x} = 0 \tag{B8.75}$$

Choosing a set of temperatures (or corresponding saturation pressures) below T_1, the appropriate values of h_f, h_g, v_f, and v_g are calculated by using steam tables.[8] Then, solving Eq. (B8.75) for each pressure, the corresponding steam quality x is computed. Each intersection of corresponding x and p values represents the Fanno line point on the h–s diagram. The choked condition at the pipe exit is defined by maximum entropy:

$$\frac{ds}{dh} = 0 \tag{B8.76}$$

The detailed calculation procedures for finding the static pressure p_L at the distance L from the choked exit is the same as described earlier in the subsection "Adiabatic, Constant-Area Flow with Friction."

Allen Model[28] (Thermal Equilibrium, Homogeneous). Basic equations are presented in the section "Characteristics of Compressible Flow in Pipes." By substituting the differential form of the continuity equation:

$$d\mathrm{v} = \left(\frac{\dot{m}}{A}\right) dv \tag{B8.77}$$

into basic equations discussed in the earlier section "Conservation of Energy," the following result is found:

$$v\,dp + \left(\frac{\dot{m}}{A}\right)^2 \left(\frac{v\,dv}{g_c} + \frac{fv^2}{2g_cD}\,dL\right) = 0 \qquad \text{(B8.78)}$$

Dividing Eq. (B8.78) by v^2 and integrating between locations 1 and x along the pipe, it follows that:

$$\left(\frac{\dot{m}}{A}\right)^2 = g_c \frac{\int_{p_1}^{p_x} -(dp/v_x)}{\ln(v_x/v_1) + (fL/2D)} \qquad \text{(B8.79)}$$

A similar equation is presented in the paper by Benjamin-Miller.[29]

The function $(\dot{m}/A)^2$ has its maximum value at choked pressure $p_x = p_c$ as illustrated in Fig. B8.37. For a given pipe diameter and piping layout (equivalent length L), the choked pressure may be calculated by assuming a friction factor and evaluating the integral numerically by taking small pressure increments and summing the values $\Delta p/\bar{v}$ where $\bar{v}$ is the average specific volume within the Δp limits.

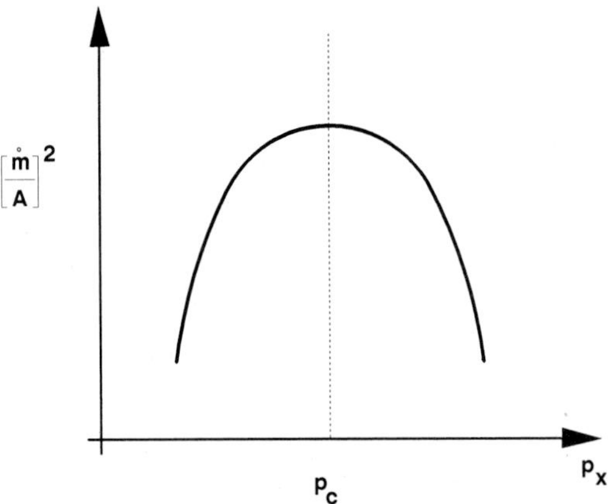

FIGURE B8.37 Maximum mass flux versus pressure relationship.

To evaluate the average specific volume, the Benjamin-Miller method assumes isentropic expansion between pressures p_{x-1} and p_x, while the Allen method assumes isenthalpic expansion between those pressures. Allen explains in his paper that the difference in specific volume, whether isenthalpic or isentropic expansion is used, will not be large.

Allen found a simple relationship between critical pressure, initial pressure, and rate of flow:

$$p_c = \frac{\dot{m}}{A} C_A p_1 \qquad \text{(B8.80)}$$

where:

$$C_A = \left[\left(\frac{h_{f1} - h_{f2}}{p_1 - p_2}\right)\left(\frac{2v_{fg1}}{h_{fg1} + h_{fg2}}\right)\frac{1}{g_c}\right]^{1/2} \tag{B8.81}$$

In computing the constant C_A, the subscript 1 refers to saturation conditions at the source (say, drain cooler exit) temperature and the subscript 2 refers to saturation conditions an infinitely small amount below this temperature. Practically, a finite interval of 1° less than the temperature at drain cooler exit is recommended for use.

For a given stagnation enthalpy in the upstream source, and for a given mass flux $\dot{m}/A$, the corresponding pipe exit critical pressure may be found using Fig. B8.38.

In order to apply Eq. (B8.79) for finding the pressure at the outlet of the control valve, or at any distance L from choked pipe exit, the value p_x should be replaced by the calculated critical pressure p_c, and the equation must be solved by trial for $p_1 = p_L$ (convergence on the lower limit of the integral in Eq. (B8.79).

Allen succeeded in writing Eq. (B8.79) in terms of a single variable by expressing the specific volume of a mixture in terms of pressure. See Ref. 28 for more details.

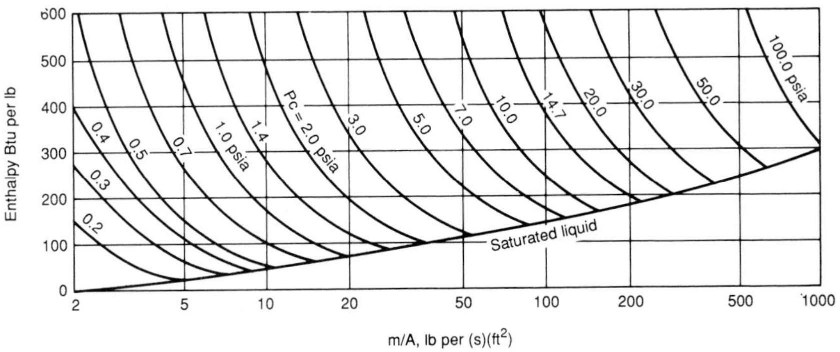

FIGURE B8.38 Pipe exit critical pressures for water-steam mixtures. (*Courtesy of Stone & Webster; method from Ref. 28.*)

Moody Model*[30,31] *(Thermal Equilibrium, Nonhomogeneous). Moody considers isentropic expansion of a homogeneous fluid in a converging nozzle (Fig. B8.24). The nozzle may be considered as a pipe of the diameter D_2 and the length $L = 0$.[30] The two-phase flow pattern at the exit is annular with no entrainment, but the exit velocities of each phase may differ, resulting in an exit slip ratio.

The Moody model is based on the same logic as described for the above homogeneous models, the only difference being that average vapor velocity and average liquid velocity are not equal (nonhomogeneous annular flow). He developed the Eq. (B8.82) for the mass flux G, using properties at nozzle exit (subscript 2).

Equation (B8.82) shows that $\dot{G}$ is a function of S_2 and p_2 when h_o and s_o are known (upstream stagnation properties). The expression for slip ratio S_2 at maximum flow rate was found to be as given by Eq. (B8.83).

$$\dot{G} = \sqrt{\frac{2g_cJ\left[h_o - h_f - \dfrac{h_{fg}}{s_{fg}}(s_o - s_f)\right]_2}{\left[\dfrac{S_2(s_g - s_o)v_f}{s_{fg}} + \dfrac{(s_o - s_f)v_g}{s_{fg}}\right]_2^2 \left[\dfrac{s_o - s_f}{s_{fg}} + \dfrac{s_g - s_o}{S^2 s_{fg}}\right]_2^2}} \qquad \text{(B8.82)}$$

$$S_2 = \left[\frac{v_g}{v_f}\right]_2^{1/3} \qquad \text{(B8.83)}$$

which indicates that for a maximum two-phase flow, S_2 depends only on p_2.

For known h_o and s_o, a maximum (choked) value of G_c is derived from the relation:

$$\frac{d\dot{G}}{dp_2} = 0 \qquad \text{(B8.84)}$$

The thermal parameters of the fluid at a distance L from the choked pipe exits are found in a similar way as it is described earlier in the section "Adiabatic, Constant-Area Flow with Friction."

References 30 and 31 should be consulted for additional information concerning the development of the Moody model. Exit properties for maximum steam-water discharges G_c, and stagnation enthalpies h_o, are shown in Fig. B8.39.

Computer solution of numerical integration of equations developed for choked lines lead to graphs presented in Fig. B8.40, taken from Ref. 24.

Henry-Fauske Model*[32] *(Nonequilibrium, Nonhomogeneous). Henry-Fauske model is of importance for choked two-phase flow through short pipes, nozzles, and orifices, where the fluid transit is short and the thermodynamic equilibrium between phases cannot be reached.[32] Since the phases have different densities, the pressure gradient will also tend to accelerate the lighter vapor phase more than the liquid. It is assumed in this model that the actual fluid quality may be determined by introducing a nonequilibrium parameter N.

For a given stagnation condition of p_o and x_o, the model predicts both the critical (choked) pressure and the flow rate. The theory is extended to subcooled inlet conditions. The calculated mass fluxes G_{max} corresponding to initial stagnation conditions p_o, h_o, in the wet-steam and subcooled liquid regions are presented in Figs. B8.41 and B8.42 (based on Ref. 25). The figures indicate how the max flux G_{max} increases with increased pressure p_o, holding the stagnation enthalpy at the same level (say, 200 Btu/lb). Steep stagnation pressure lines, especially those below 100 psia in Fig. B8.41, indicate extreme sensitivity to the fluid enthalpy in determining the mass fluxes at low qualities, near the saturation line $x = 0$. Therefore, a very careful estimation of stagnation conditions ahead of a valve, as well as the computerized Henry-Fauske model, are helpful in predicting reasonable values of mass fluxes in this wet-steam region.

In order to find the choked pressure p_c for a given $G_{max} = (\dot{m}/A)_{max}$, the calculated values $p_c = f(G_{max})$, corresponding to the same inlet stagnation enthalpy h_o and different stagnation pressures p_o, may be plotted in a similar way as p_c in Fig. B8.43. From such graphs, for a given value of G_{max} which has to be discharged through a nozzle, the appropriate choked pressure p_c, as well as the required stagnation pressure p_o needed for evacuating assigned flow, can be found.

The Henry-Fauske theory was developed for two-phase critical discharge of

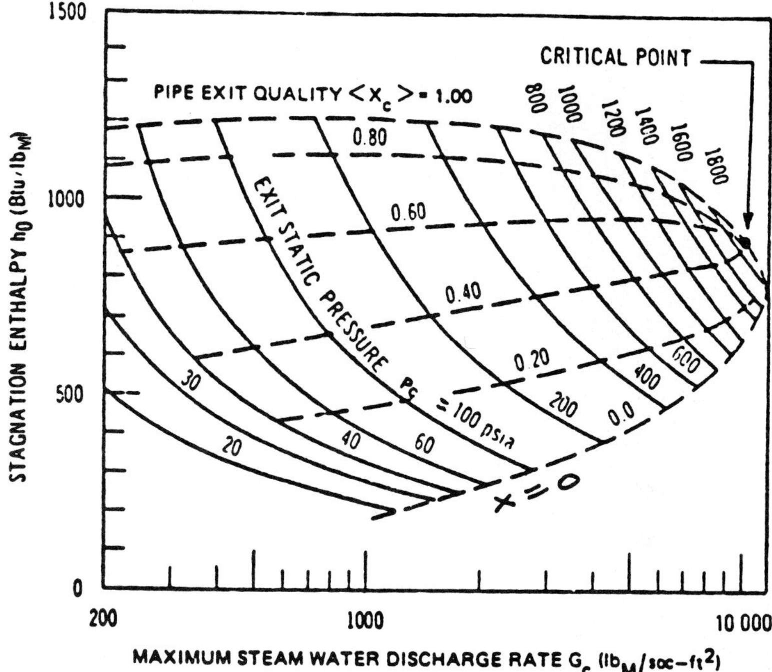

FIGURE B8.39 Exit properties for maximum steam-water discharge (Moody model). (*From Ref. 24.*)

single-component mixture through convergent nozzles. Because the orifices (e.g., valve orifices) cannot be considered as nozzles, a discharge coefficient of 0.84, as recommended in Ref. 32 should be applied. More information on this subject may be found in Ref. 33.

Comparison of Results Generated by Different Models. As mentioned in the section "General Remarks," there is no general model for two-phase critical flow which is valid for all cases of interest. Each model is only applicable for a limited range of parameters. The number of critical flow models available may cause some apprehension when the task of selecting one is approached.

In order to facilitate the selection of an appropriate model for a given problem, some useful graphs are presented below. Figures B8.44 and B8.45 show the critical (choked) pressure versus stagnation enthalpy for the mass fluxes of 50 and 600 lb/(ft^2 · s). The curves marked by *F, A, M,* and *H-F* represent the values corresponding to the following models, respectively: Fanno, Allen, Moody, and Henry-Fauske. Figures B8.46 to B8.49[34] represent a comparison of critical flow rates predicted by various models and their relations to experimental data. In view of the above, the following conclusions can be made:

1. A homogeneous equilibrium theory (Fanno model, Benjamin-Miller model, Allen model) may be used to predict critical flow rates from long pipes ($L/D \geq 40$). This theory provides a lower bound to all presented experimental

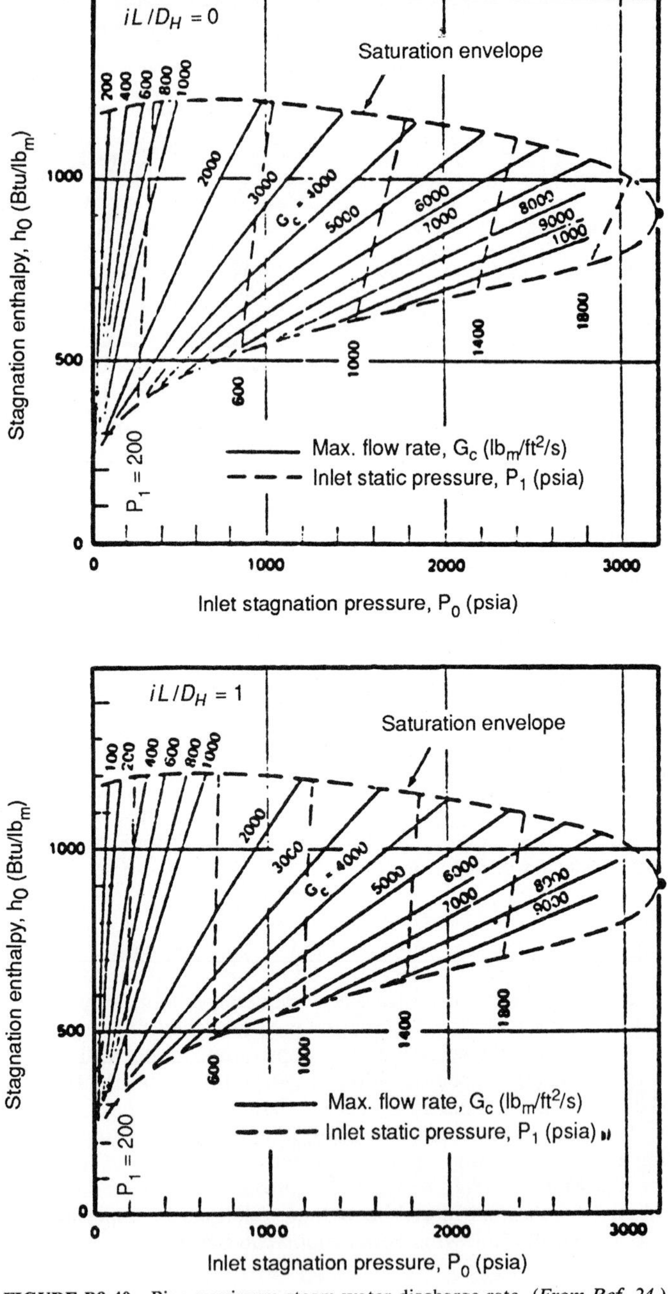

FIGURE B8.40 Pipe maximum steam-water discharge rate. (*From Ref. 24.*)

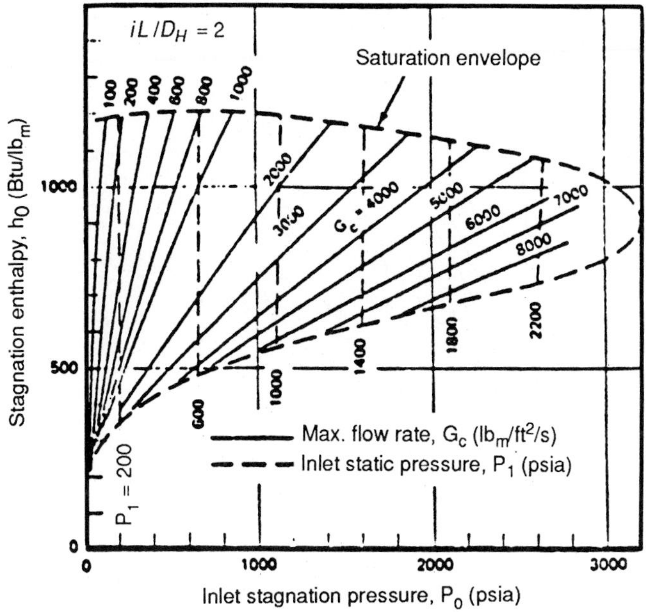

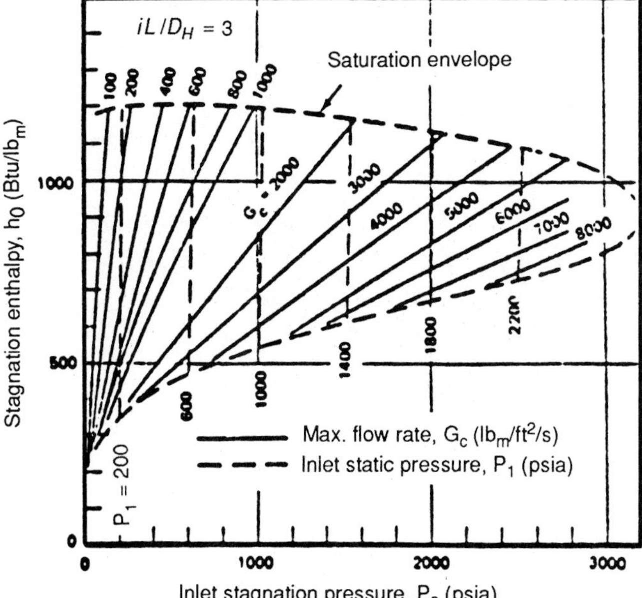

FIGURE B8.40 (*Continued*) Pipe maximum steam-water discharge rate. (*From Ref. 24.*)

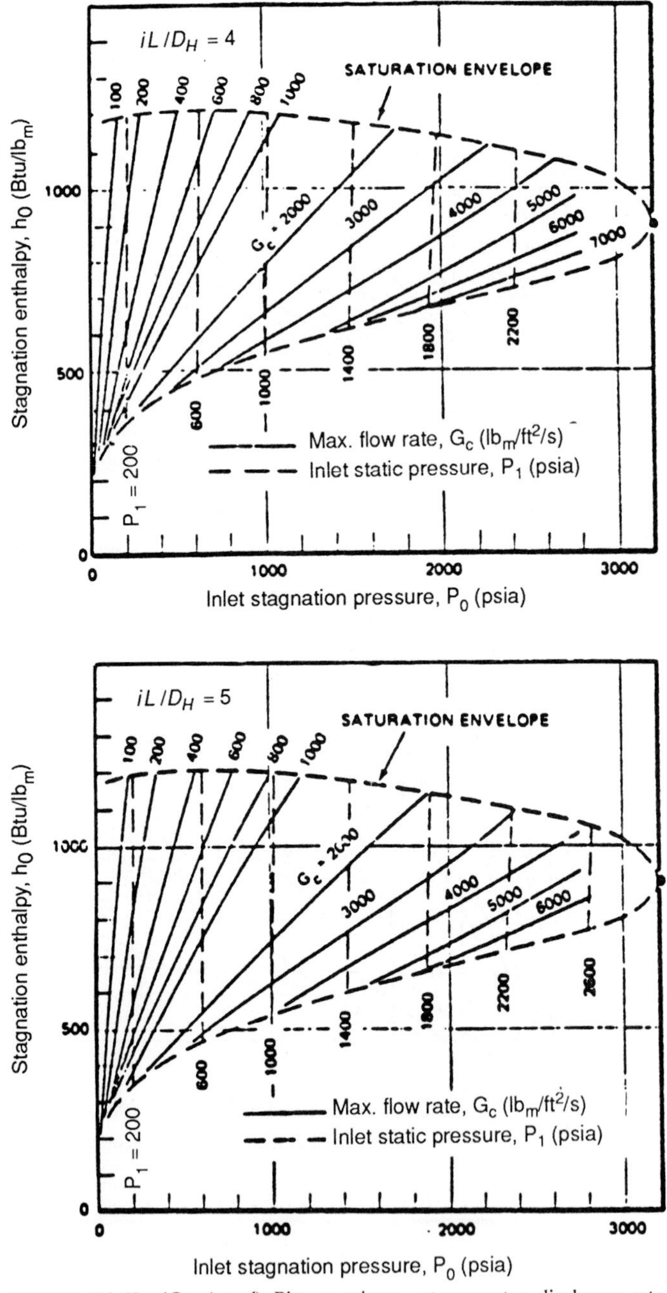

FIGURE B8.40 (*Continued*) Pipe maximum steam-water discharge rate. (*From Ref. 24.*)

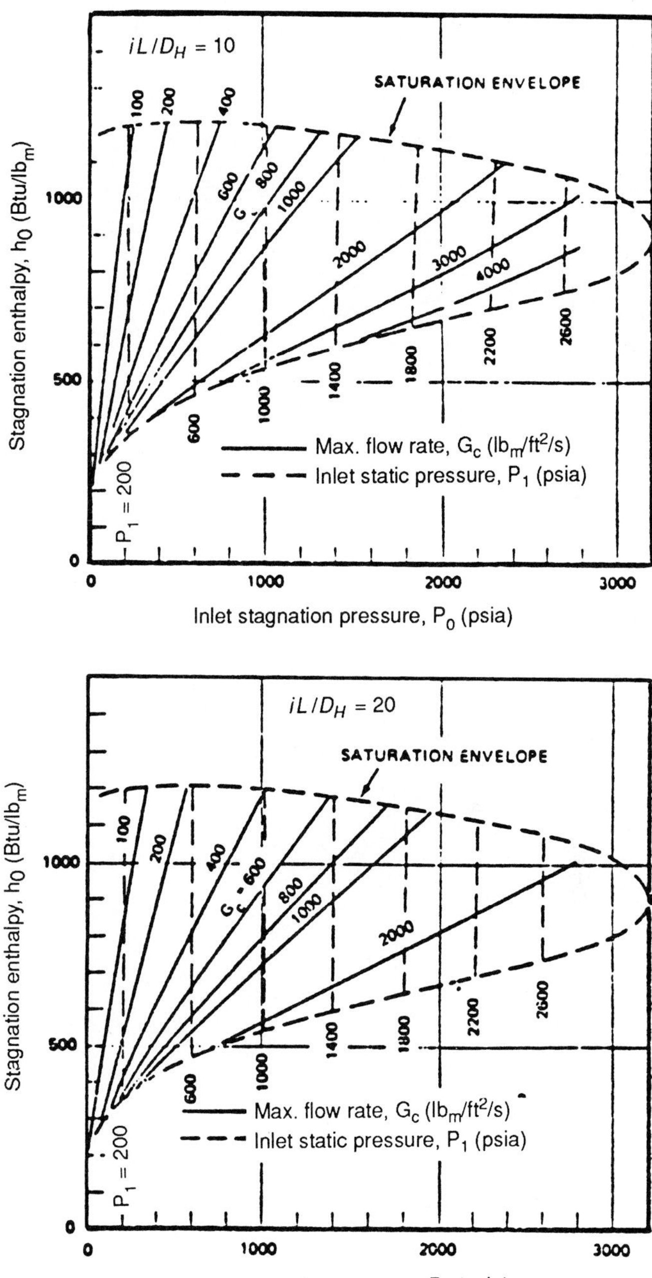

FIGURE B8.40 (*Continued*) Pipe maximum steam-water discharge rate. (*From Ref. 24.*)

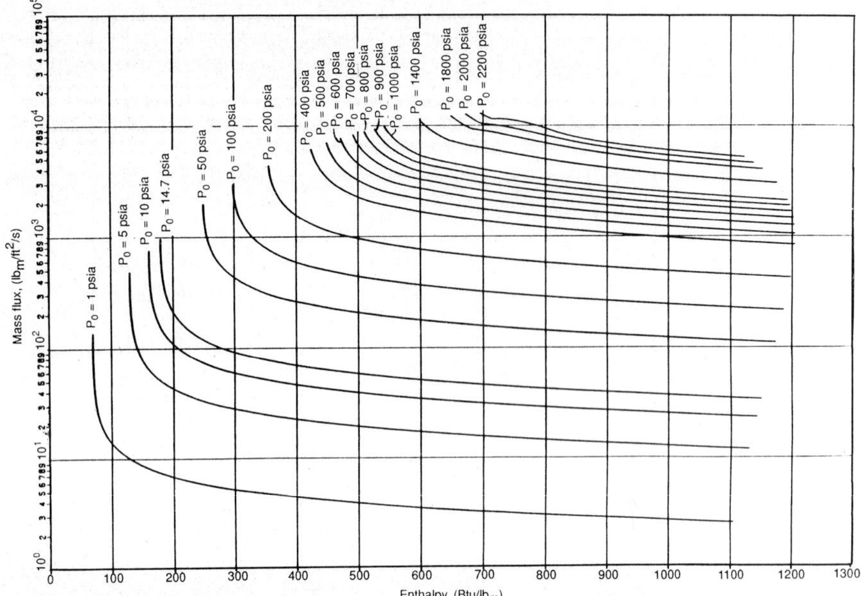

FIGURE B8.41 Two-phase critical flow (Henry-Fauske, Ref. 32). (*Ref. 25 data.*)

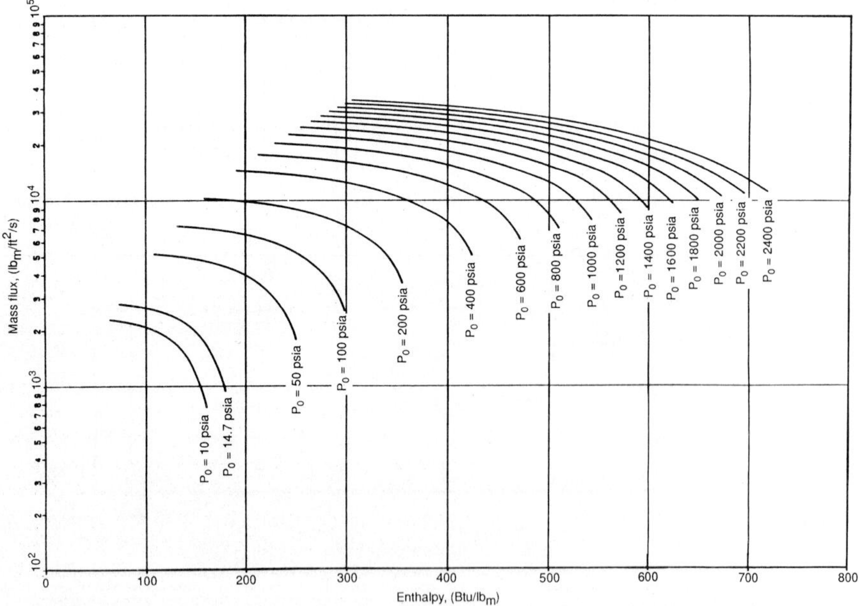

FIGURE B8.42 Subcooled critical flow (Henry-Fauske, Ref. 32). (*Ref. 25 data.*)

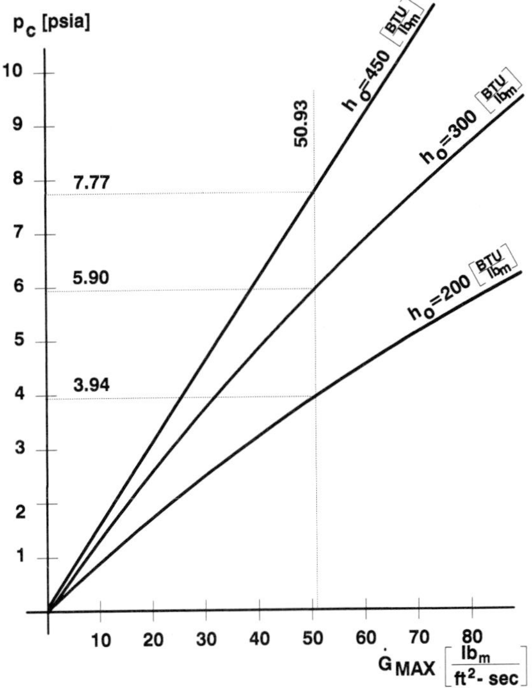

FIGURE B8.43 Critical two-phase flow by Henry-Fauske. (*Stone & Webster.*)

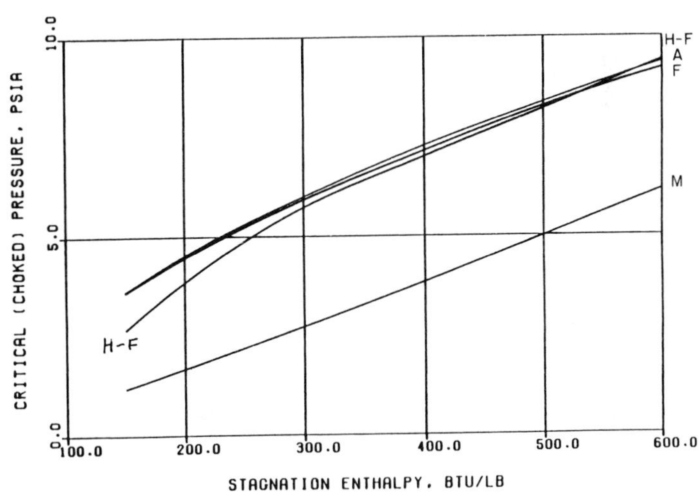

FIGURE B8.44 Stagnation enthalpy, Btu/lb (0 = 50 lb/ft^2 · s). (*Courtesy of Stone & Webster.*)

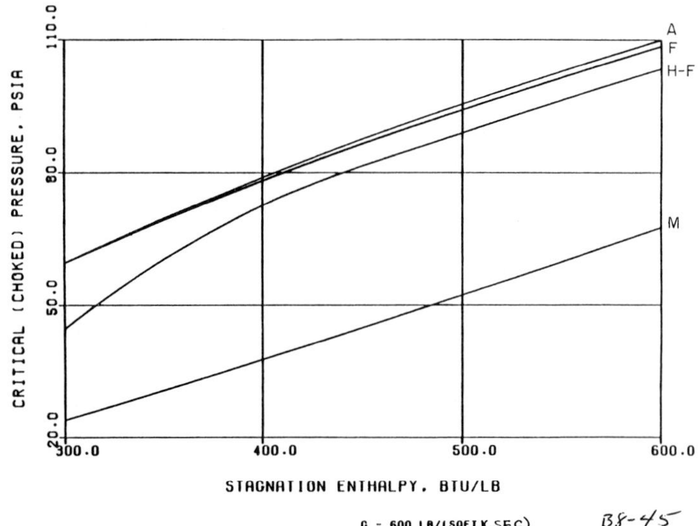

FIGURE B8.45 Stagnation enthalpy, Btu/lb (0 = 600 lb/ft² · s). (*Courtesy of Stone & Webster.*)

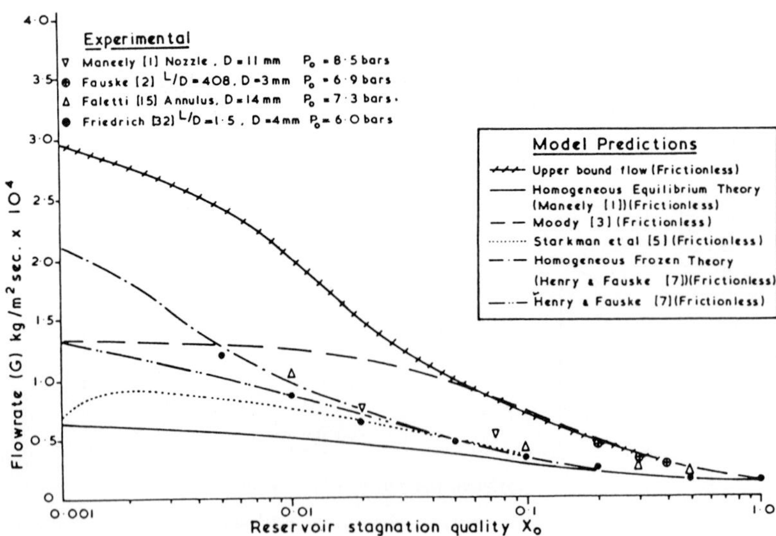

FIGURE B8.46 Comparison of critical flow models with data for P_o = 6.9 bar (100 psia). (*From Ref. 34.*)

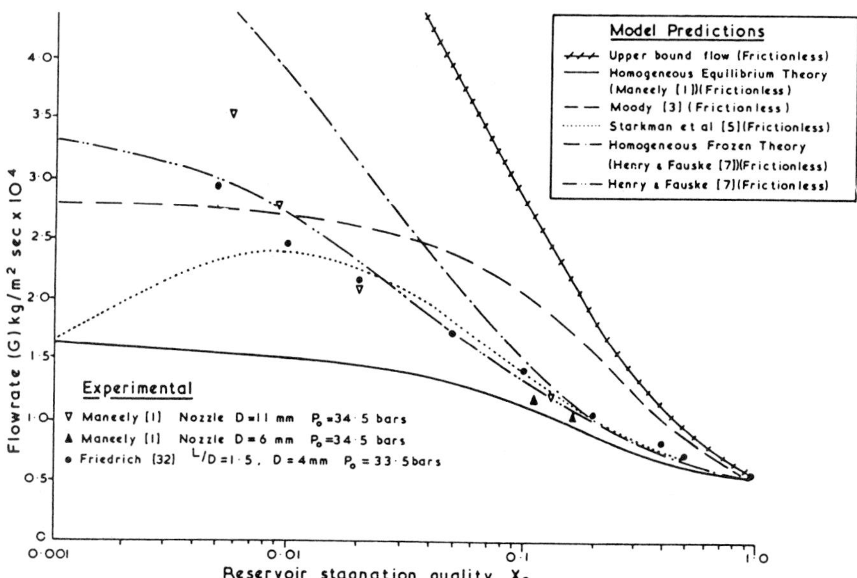

FIGURE B8.47 Comparison of critical flow models with data for P_o = 34.5 bar (500 psia). (*From Ref. 34.*)

data. It is generally recognized, however, that this theory underestimates the choked mass flux at low pressures and qualities. As the pressures and qualities increase, the answer improves.[35]

2. The Moody model predicts flow rates higher than those observed for stagnation qualities above 0.01; thus being conservative, it is recommended for U.S. water-reactor licensing calculations (e.g., LOCA). It should be explained here that the computer program RELAP4/MOD5[25] is a part of the Nuclear Regulatory Commission's (NRC) Water Reactor Evaluation Model (WREM). The critical flow model criterion of the WREM specifies use of the Henry-Fauske model in the subcooled region and the Moody critical flow model in the saturated region. To effect a smooth transition, therefore, the first point in the Moody critical flow tables has been placed in the last portion of the table for the Henry-Fauske critical flow model extended into the subcooled region. This limitation of the subcooled Henry-Fauske table should be realized during RELAP4 calculations.
3. The Henry-Fauske model is in reasonable agreement with short pipe (L/D < 5) and nozzle data for qualities greater than 0.001. Below this limit, however, agreement is less satisfactory and the model underestimates the flow of saturated water through orifices by about 50 percent.

Applications and Practical Considerations

Design Considerations. Very often, severe problems can occur if slug flow is present in the pipeline. A number of circumstances can occur which will lead to the generation of slug flow. For instance, if an oil-gas mixture flows along a pipe

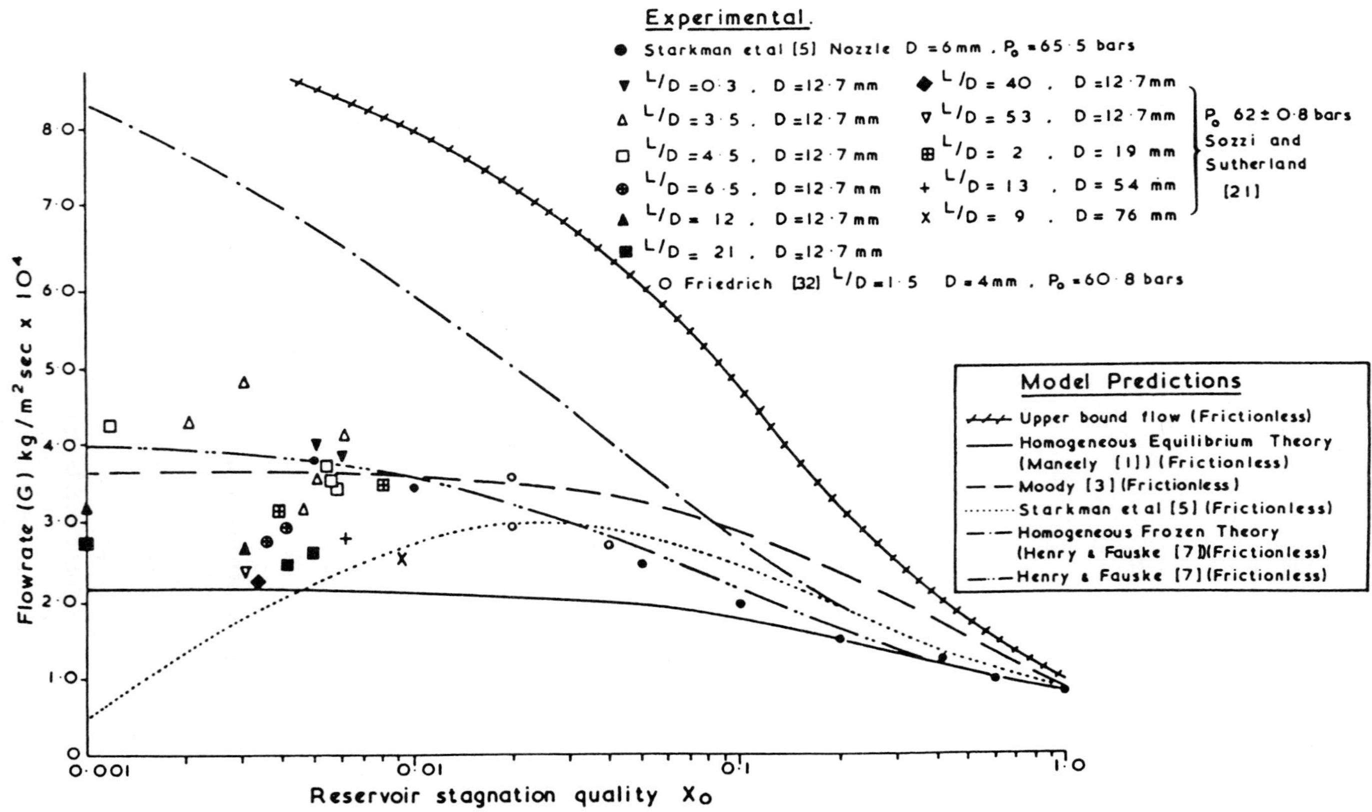

FIGURE B8.48 Comparison of critical flow models with data for P_o = 62 bar (900 psia). (*From Ref. 34.*)

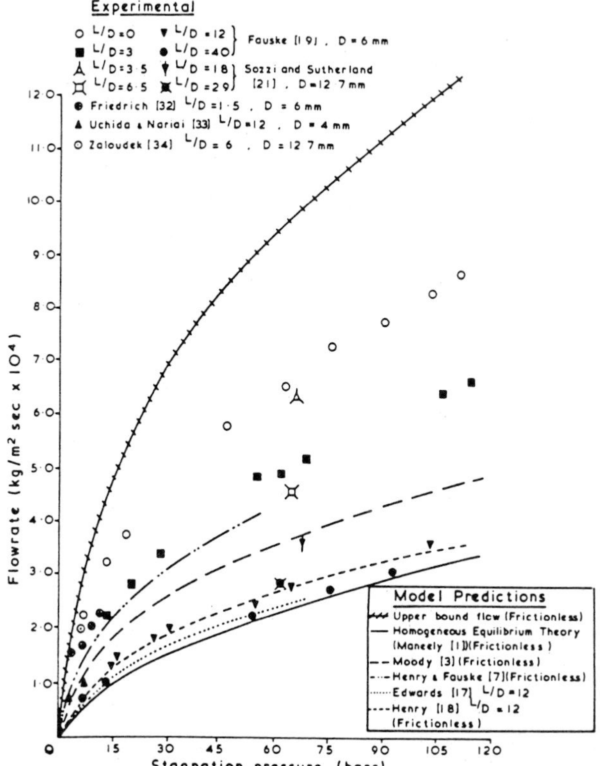

FIGURE B8.49 Discharge of saturated water through orifices, nozzles, and pipes. (*From Ref. 34.*)

on the sea-bed and then rises to a platform level up a vertical pipe, the liquid may collect upstream on the bend until it reaches a given level, at which point it is swept up the vertical leg, giving rise to mechanical problems in the platform equipment. In power plants, a similar phenomenon may occur in a pocketed line. Also, slug flow causes pressure fluctuations in piping, which can upset process conditions and cause inconsistent instrument sensing.

Existence of two-phase flow in a pipeline may also cause more severe fluid transient problems than in a single-phase condition. Should a two-phase flow condition be unavoidable in the system operation, the impact of two-phase flow transient condition should be taken into consideration in system design and piping support evaluation.

Slug flow can be avoided in piping by (a) reducing line sizes to the minimum permitted by available pressure differentials to achieve a safe mass flow rate as indicated in Figs. B8.32 or B8.33, (b) designing for parallel pipe runs that will permit increasing mass flux per pipe at low load conditions by removing one pipe from service, or (c) arranging the pipe configurations to protect against slug flow (for example, in a pocketed line where liquid can collect, slug flow might develop).

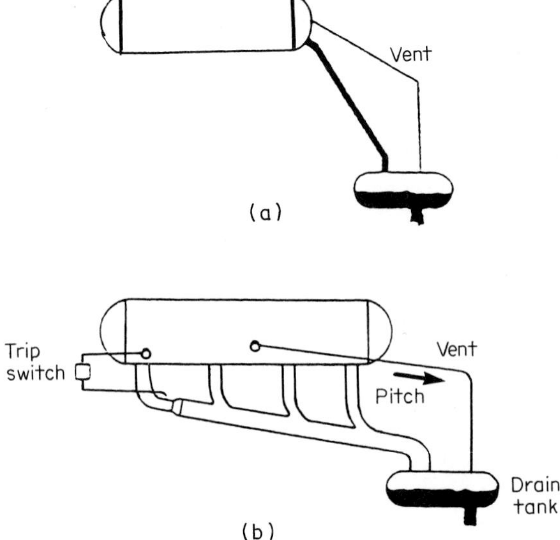

FIGURE B8.50 Drain system requirements. (*a*) Design of non-self-venting gravity drains; (*b*) design of self-venting gravity drains.

Slug flow will not occur in gravity-flow lines when appropriate venting is provided. Figure B8.50*a* shows a sketch of a non-self-venting gravity drain from a tank where the drain runs full and hence requires a separate vent line. Figure B8.50*b* shows the typical design of a self-venting gravity drain which does not require a vent but which usually has a separate vent line to maintain continuous drainage during the plant transients. A hard-tee connection (i.e., flow through the branch) at a low point can provide sufficient turbulence for more effective liquid carryover. A diameter adjustment coupled with gas injection can also alter a slug-flow pattern to bubble or dispersed flow. Gas addition, however, when used solely to avoid slug flows can be expensive.

Sample Problems

Problem B8.8. To illustrate the design of a self-venting line, an example of calculating the diameter and minimum slope required to discharge 362,000 lb_m/h of 200°F water (approximately 750 gpm) into a tank at ambient pressure is presented.

Diameter of the Self-Venting Line. A properly designed drain line with gravity flow will be self-venting if the liquid Froude number, Eq. (B8.67) is less than approximately 0.3:[36]

$$Fr = \frac{v}{\sqrt{gD}} \leq 0.3$$

where v = the liquid velocity if full, ft/s

g = 32.2 ft/s^2

D = inside pipe diameter, ft

$$v = \frac{(362{,}000\ lb_m/h)(0.016637\ ft^3/lb_m)}{(3600\ s/h)(\pi D^2/4)}$$

$$v = \frac{2.13}{D^2}$$

Substituting into the first equation of this problem yields:

$$0.3 \geq \frac{v}{\sqrt{gD}} \geq \frac{2.13}{D^2(32.2D)^{1/2}}$$

and $D^{2.5} \geq 1.251$
or $D \geq 1.094$ ft = 13.12 in

Use 14-in standard pipe (inside diameter = 13.25 in).

Minimum Slope of a Self-Venting Line. The minimum slope of a self-venting line may be calculated on the basis of an assumption of constant depth of water in line. Assuming this depth to be $D/3$, the flow in a pipe may be considered as an open channel flow. The Chezy formula describes velocity in such a flow[1] as:

$$v = C_1\sqrt{r_h(SL)} \qquad \text{ft/s}$$

where SL = slope of the pipe (channel), ft/ft
r_h = hydraulic radius, ft (see Table B8.11)
C_1 = Chezy coefficient

The Chezy coefficient C_1 is described by the widely accepted Manning formula[1] as:

$$C_1 = \frac{1.486}{n}(r_h)^{1/6}$$

where $n \approx 0.014$ is the Manning roughness factor for a metal conduit.

TABLE B8.11 Hydraulic Radius of the Cross Section of a Circular Conduit Flowing Part Full as a Ratio of the Pipe Diameter

$\frac{d}{D}$	0.00	0.01	0.02	0.03	0.04	0.05	0.06	0.07	0.08	0.09
0.0	0.000	0.007	0.013	0.020	0.026	0.033	0.039	0.045	0.051	0.057
0.1	0.063	0.070	0.075	0.081	0.087	0.093	0.099	0.104	0.110	0.115
0.2	0.121	0.126	0.131	0.136	0.142	0.147	0.152	0.157	0.161	0.166
0.3	0.171	0.176	0.180	0.185	0.189	0.193	0.198	0.202	0.206	0.210
0.4	0.214	0.218	0.222	0.226	0.229	0.233	0.236	0.240	0.243	0.247
0.5	0.250	0.253	0.256	0.259	0.262	0.265	0.268	0.270	0.273	0.275
0.6	0.278	0.280	0.282	0.284	0.286	0.288	0.290	0.292	0.293	0.295
0.7	0.296	0.298	0.299	0.300	0.301	0.302	0.302	0.303	0.304	0.304
0.8	0.304	0.304	0.304	0.304	0.304	0.303	0.303	0.302	0.301	0.299
0.9	0.298	0.296	0.294	0.292	0.289	0.286	0.283	0.279	0.274	0.267

Notes: (1) Read d/D down and increment by 0.01 horizontally. (2) d/D is the depth of liquid divided by conduit inside diameter.

For a pipe running one-third full, $d/D = 0.33$, and $r_h/D = 0.185$ from Table B8.11:

$$r_h = 0.185\ D = (0.185)(13.25) = 2.45\ \text{in} = 0.204\ \text{ft}$$

$$C_1 = \frac{1.486}{n}(r_h)^{1/6} = \frac{1.486(0.204)^{1/6}}{0.014}$$

$$= 81.4$$

The cross section of fluid flow for $d/D = 0.33$ is 0.29 as read from Fig. B8.34.

The actual fluid flow velocity v is:

$$\text{v} = \frac{2.13}{D^2(0.29)}$$

$$= \frac{2.13}{(1.094)^2(0.29)} = 6.14\ \text{ft/s}$$

Solving the Chezy equation for SL yields:

$$SL = \frac{\text{v}^2}{C_1{}^2 r_h}$$

$$= \frac{(6.14)^2}{(81.54)^2(0.2024)} = 0.028\ \text{ft/ft or } 0.33\ \text{in/ft}$$

Industrial experience requires the minimum slope for a self-venting line to be not less than 0.5 in/ft.[36] Therefore, in this case, the minimum slope should be 0.5 in/ft.

Problem B8.8: Fanno Model. Consider a refrigeration system filled with Freon-12. Boiling Freon at initial conditions $p_1 = 139.33$ psia and $T_1 = 104°\text{F}$ is relieved through a capillary tube of 1.5 mm internal diameter at the rate of 22.05 lb/hr. Calculate the choke pressure and the velocity of the fluid at choked exit conditions. The properties of refrigerant 12 are listed in the appendixes to this handbook.

Equations (B8.72) through (B8.75) are used to generate the table below. The procedure is explained by calculating results for a chosen temperature of 86°F.

For $T = 86°\text{F}$, from Freon tables:

$$p_{\text{sat}} = 108.04\ \text{psia}$$

$$v_f = 0.012396\ \text{ft}^3/\text{lb}_\text{m}$$

$$v_g = 0.37657\ \text{ft}^3/\text{lb}_\text{m}$$

$$h_f = 27.769\ \text{Btu/lb}_\text{m}$$

$$h_g = 85.821\ \text{Btu/lb}_\text{m}$$

$$s_f = 0.057301\ \text{Btu/(lb}_\text{m} \cdot °\text{R)}$$

$$s_g = 0.16368\ \text{Btu/(lb}_\text{m} \cdot °\text{R)}$$

For initial conditions (saturated liquid at 104°F):

$$h_1 = 32.067 \text{ Btu/lb}_m$$

$$v_1 = 0.012783 \text{ ft}^3/\text{lb}_m$$

Pipe cross-section area:

$$A = \frac{\pi}{4}\frac{(0.059)^2}{(12)^2} = 0.0000190 \quad [\text{ft}^2]$$

Mass flux:

$$\dot{G} = \frac{\dot{m}}{A} = \frac{22.05}{(3600)(0.0000190)} = 322.6 \frac{\text{lb}_m}{\text{ft}^2 \cdot \text{s}}$$

Equation (B8.75) may be rewritten as:

$$x^2 + Bx + C = 0$$

where:

$$B = 2\left[\frac{v_f}{v_g - v_f} - \left(\frac{1}{\dot{G}}\right)^2 \frac{h_g - h_f}{(v_g - v_f)^2} g_c J\right]$$

$$= 2[0.03404 + 105.300] = 210.7$$

$$C = \frac{v_f^2 - v_1^2 + 2(1/\dot{G})^2(h_f - h_1)g_c J}{(v_g - v_f)^2}$$

$$= \frac{(0.012396)^2 - (0.012783)^2 - (2)(0.103394)}{0.132623}$$

$$= -0.155923$$

Then, the quality x is equal to:

$$x = \frac{-B \mp \sqrt{B^2 - 4C}}{2} = 0.0740$$

and, from Eqs. (B8.72) to (B8.74):

$$h = h_f + x(h_g - h_f) = 32.06$$

$$v = v_f + x(v_g - v_f) = 0.03934$$

$$s = s_f + x(s_g - s_f) = 0.06517$$

Similar procedures for other pressures (or temperatures) result in values tabulated below. Underlined data in Table B8.12 show that the maximum entropy is 0.06946 and the corresponding pressure (critical pressure) is 17.94 psia.

The exit velocity is calculated from the continuity equation:

$$\mathrm{v} = c = \upsilon \frac{\dot{m}}{A} = (0.760557)(322.608) = 245.362 \quad [\text{ft/s}]$$

TABLE B8.12 Sample Data for Problem B8.9: Fanno Model (Boiling Freon-12, initial pressure = 139.33 psia, initial temperature = 104°F)

Pressure, psia	Temp., °F	Enthalpy, Btu/lb	Entropy, Btu/°F/lb	Spec. vol., ft^3/lb	Steam quality
139.33	104.00	32.07	0.06492	0.012783	0.00000
122.95	95.00	32.07	0.06502	0.024723	0.03818
108.04	86.00	32.06	0.06517	0.039340	0.07399
94.51	77.00	32.06	0.06539	0.057256	0.10774
82.28	68.00	32.05	0.06566	0.079250	0.13970
71.27	59.00	32.04	0.06600	0.106360	0.17006
61.39	50.00	32.03	0.06640	0.139801	0.19896
52.58	41.00	32.00	0.06686	0.181284	0.22648
44.76	32.00	31.95	0.06737	0.232750	0.25267
37.85	23.00	31.88	0.06791	0.297063	0.27750
31.78	14.00	31.77	0.06897	0.477890	0.32244
26.48	5.00	31.59	0.06846	0.377216	0.30084
21.89	−4.00	31.31	0.06935	0.603310	0.34187
17.94	−13.00	30.87	0.06946	0.760557	0.35839
14.56	−22.00	30.18	0.06910	0.953742	0.37094
11.71	−31.00	29.13	0.06797	1.189298	0.37809

TRANSIENT FLOW ANALYSIS

General Background

Hydraulic transients in piping systems result from rapid changes in operational functioning of components, such as pumps and valves. They may produce rapid momentum changes as the column of water in the system is suddenly stopped or started. Large momentum changes can subject system components to severe force transients, and these must be considered when designing the system to ensure its safe, reliable operation.

Piping transients may result in the formation of vapor cavities when the system pressure is reduced to the vapor pressure of the fluid. Vapor cavities can easily form in a piping system that has a wide elevation range, if permitted to drain after the system is shut down. This behavior is known as *column separation.* In a large cooling-water system, reclosure of water columns after separation can result in destructive water hammer loads if appropriate design consideration is not given to alleviate these effects.

Analysis of hydraulic transients should be performed for all major plant systems. In addition to the maximum and minimum design pressures for components and the forcing functions for pipe supports, the analysis will help appropriate procedures for system startup and shutdown and operator responses to power failure.

Transients that exhibit single-phase flow can be analyzed by one-dimensional wave theory. During a transient condition, pressure waves are generated throughout the system. In complex systems, these waves may be analyzed with the aid of a computer.

When simulating transient waves with a computer program, the unbalanced forces acting on a pipe segment can be determined by using the momentum theorem. In a water system, fluid-induced forces are exerted in a pipe segment by a pressure force acting normal to the inner surface of the pipe and a frictional force between fluid and pipe wall. The combination of these two elements is called a *segment force,* and it is collinear with the pipe axis.

Applying the momentum theorem to a straight pipe segment of length L that is bounded at both ends, the segment force is expressed as follows:[1]

$$F(t) = \frac{\partial}{\partial t} \int_{o}^{L} L\rho A\text{v}\, dx \tag{B8.85}$$

where the integration is done over the length of the segment. Note that a positive segment force acts along the pipe axis opposite to the direction of flow. Equation (B8.85) is valid for bounded pipe regardless of the bend angle at either end of the segment. (During steady-state conditions, there is no unbalanced force in the segment.)

When fluid density is not the same at both ends of a given segment, the unbalanced segment force may include a very significant momentum force equal to $\rho A\text{v}^2$. For example, for a slug of cold water upstream of a relief valve that discharges into a closed system with a downstream slug-mixture density of 30 lb/ft^3, velocities can become very large (approaching 500 ft/s): This would produce a segment force of 180,000 lb in a 12-in pipe. In the case of column separation or column rejoining (reclosure), large flow velocities are not necessary for significant peak pressures and segment forces. A modest fluid-closure velocity may produce a substantial peak pressure, which can be estimated from the following:[1]

$$\Delta p = 0.5\rho c \Delta\text{v} \tag{B8.86}$$

where Δv is the relative closure velocity of the fluid before impact. This peak pressure would create significant segment forces, which may be calculated with Eq. (B8.85). For example, for cold water with 10-ft/s closure velocity in a 10-in pipe, a peak pressure differential of 278 psi could produce segment forces as high as 5000 lb.

Typical Water Hammer Transients

Typical transients for a water-filled system include rapid valve closure, pump startup or pump trip within the circulating water system, feedwater system, or service water system of a power plant, or the cooling water system of a process plant. The water hammer phenomena in a piping system may result in the formation of vapor pockets at locations where the pressures are reduced to or below vapor pressure. This phenomenon is normally called *column separation.* The subsequent collapse of these vapor pockets may develop significant pressure spikes which should be taken into consideration in system design and analysis. Typical transients for a steam-filled system include turbine trip valve or isolation valve closure or steam bypass valve opening in the main steam system of a power plant.

NUREG-0582[37] has identified a number of high-energy systems in nuclear power plants that have had a history of transient-related problems. It also discussed the major causes of transient problems which are attributable to events such as pump startup, stopping, or seizure; pump startup with inadvertently voided discharge lines; valve opening, closing, and instability; check valve closure and delayed opening; water entrainment in steam lines; steam bubble collapse and mixing of subcooled water and steam from interconnected systems; slug impact due to rapid condensation; and column separation and subsequent rejoining. Some of these transient events can be prevented by implementing appropriate operating procedures. However, there are other events that may occur during normal-upset, emergency or faulted plant operating conditions. These flow transients in the piping system may cause significant dynamic loads and large reactions on the piping, piping supports, and connected equipment. These transient loads must be analyzed and incorporated into the pipe stress analysis and pipe support evaluation.

The service water system of a nuclear station illustrates the application. The system is designed to supply cooling water to remove heat from component heat exchangers throughout the station. Hydraulically, such a system is far more complex than a condenser circulating water system because there are numerous heat exchangers and considerable variations in piping elevations.

Figure B8.51 depicts the major components of a typical service water system in the case of a pressurized water reactor (PWR) unit; Fig. B8.52 shows the profile (vertical) of a portion of the system, including the control building air-conditioning water cooler (HVK) and emergency diesel engine cooling water (EGS) piping. Because of the high elevation of the HVK piping, a vapor gap will form on either side of the HVK heat exchanger when the service water pumps trip out. Also, because of the U-shaped piping configuration, a water column will be retained in the central portion of the HVK piping, as shown.

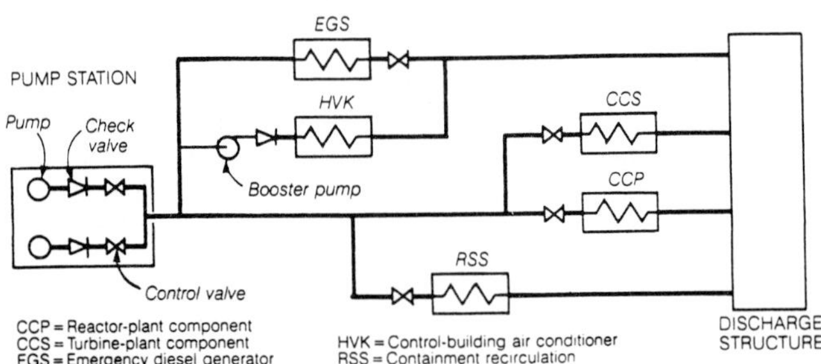

FIGURE B8.51 Major components of a service-water system for a PWR-type nuclear plant.

After the pump has restarted, the vapor gap at the inlet piping may close rapidly—resulting in severe pressure and force transients—unless proper care is taken in the design and operation of the system. Analysis by computer indicates a vapor-gap closure rate of 24 ft/s (fluid-closure velocity) before cavity collapse, resulting in pipe-segment forces up to 50,000 lb. By throttling the pump-discharge valve to reduce the vapor-gap closure of 7 ft/s, it was possible to reduce peak segment forces to 8000 lb and calculated peak impact pressure to 110 psig—close to the computer-calculated value of 105 psig. Thus verified for the calculation of

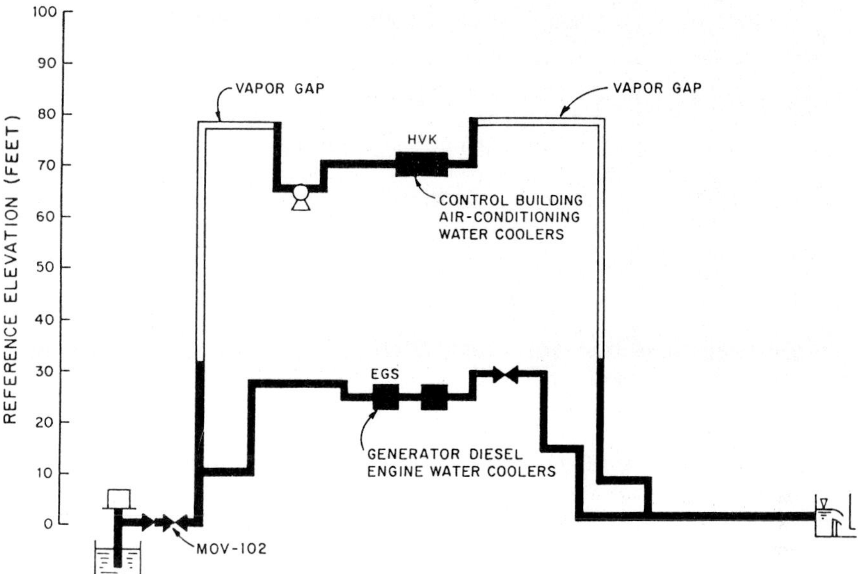

FIGURE B8.52 Partial service water system profile showing vapor gap.

peak pressures, the computer model has provided piping-segment forces needed for dynamic analysis.

Typical Steam Hammer Transients

Typical applications of a steam hammer program to perform analysis of compressible flow include turbine generator main steam and bypass system, safety relief valve blowdown, reactor core isolation cooling system, and high-pressure coolant injection system, and so on. The program could apply the method of characteristics with finite difference approximations for solutions of unsteady one-dimensional compressible fluid flows. The choking phenomena of steam flow are included in the program. In analyzing transients in a safety-relief valve discharge system, the program has an option to handle the flowdown forces at the interface of steam discharging into the suppression pool.

Steam-Condensation-Induced Water Hammer

Several flow conditions may lead to severe condensation-induced water hammers, with steam and water counterflow in a horizontal pipe events being potentially the most severe.[38] The water cannon events (subcooled water with condensing steam in a vertical pipe) may occur if the right geometry and system conditions exist. Pressurized water entering a vertical steam-filled pipe can cause a severe transient event over a wide range of piping geometry conditions. Hot water entering a lower-pressure line is also a very common condition for many nuclear power plant piping systems, especially for those heater drain dump lines which lead to the condenser.

Condensation-induced water hammers of the steam and water counterflow type can occur in a horizontal pipe with a steam and water stratified environment. If the water is highly subcooled, violent condensation may occur. This rapid condensation process will generate a significant steam flow above the water surface. The shearing forces at the interface between the steam and the subcooled water can create enough turbulence to generate a water slug which, in turn, will entrap an isolated steam pocket. Continued rapid condensation of the entrapped steam will accelerate the water slug into the void, causing a damaging water hammer. Two conditions must coexist in the system to initiate this condensation-induced water hammer, namely:

- A steam and subcooled water stratified flow exists in a significant length of a horizontal or near horizontal pipe.
- Substantial turbulence exists at the steam-water interface.

Depending on the void fraction and subcooling in the two-phase flow environment, the magnitude of the pressure wave generated by steam condensation and bubble collapse can be devastating. These events should be prevented from occurring by all practical means.

In summary, the significant condensation-induced water hammer conditions are:

- Subcooled water with condensing steam in a vertical pipe
- Steam and water counterflow in a horizontal pipe
- Pressurized water entering a vertical steam-filled pipe
- Hot water entering a lower pressure line

Practical Recommendations to Minimize Water Hammers

Protection Devices to Minimize Water Hammers. Various devices may be used to protect piping systems from damage by severe hydraulic transients. Check valves, standpipes, and accumulators prevent the occurrence of transients or reduce their effects.

Vacuum-breaker valves allow air to enter a piping system when a vacuum develops. Sometimes installed on heat exchangers, these relief valves may be located in the system where water-column separation may occur during transient conditions. Air in the system will significantly attenuate pressure spikes when water columns rejoin.

Check valves are sometimes used in service water systems to prevent column separation. Following a power loss, water in the high portion of a riser may drain out because of the large change in elevation, creating a vapor cavity. In this situation, a check valve may be installed at the bottom of the upstream riser to prevent water drainage and column separation.

A standpipe may be used in a cooling-water system where relief of vacuum pressure is necessary. Serving as a simple surge tank, the standpipe admits water and possibly air into the system. This either eliminates column separation or reduces the severity of column reclosure because of air injection.

An accumulator of proper size may effectively eliminate column separation after a power failure. The device, a closed container partially filled with water topped with compressed air, is usually installed immediately downstream from a pump station along with a check valve. Following a power failure, the pump coasts down and the pump head decreases rapidly, causing the check valve to close. Compressed air in the accumulator will supply fluid flow pneumatically into the main line to minimize downsurge and will provide a cushion on the return surge.

Field Tests for Hydraulic Transients. Frequently, field tests for hydraulic transient loading are conducted to accomplish the following:

- Verify that system design is adequate.
- Determine the best way to operate a system to avoid transients.
- Discover scenarios of unanticipated transients.
- Validate results obtained from analytical approaches.

The first three functions are often included as part of the normal startup and checkout of a new system.

Field tests serve to measure key parameters related to hydraulic transients while operating a system in a manner that could produce plausible transient events. Validation of analytical results involves either benchmarking the computation method for known operating parameters or benchmarking combined with checking of input variables—pump characteristics, valve opening or closing time, and so on. In many cases, transients introduced by valve opening or closing and by pump trip or startup are overpredicted and can be improved with the help of field test data.

System Design Considerations. There are three ways to prevent or minimize fluid transients through effective system design: (1) general design practices that either reduce or eliminate a possible transient; (2) adding special systems to either control or prevent a transient; and (3) performing time-history analysis of loads and designing a pipe support system to accommodate the transient loads.

General design practices should include:

- Designing all steamline piping to provide continuous downward slopes with provisions for low-point drainage to preclude the formation of water pockets (Upstream of a potentially isolable valve is considered to be a low point.)
- Where allowed by system design function, providing slow opening-closing valves
- Providing high-point vents or air release valves in water-filled lines to allow system venting to eliminate the formation of air pockets
- Using vacuum breakers to minimize potential dynamic effects by providing an air cushion in fluid-carrying pipelines that could otherwise be under an occasional vacuum

Special systems that preclude or diminish the effects of fluid transients include:

- Loop fill systems that continuously maintain the primary piping system filled and pressurized during operating modes when the primary system is on standby (e.g., fire protection system and emergency core cooling systems)
- Steamline preheating systems that continuously maintain or allow gradual

heatup of steam supply lines to reduce steam condensation and the development of water slugs that generate substantial loads on the piping system

The overall piping system stress qualifications should consider as appropriate all loading conditions, including the dynamic effects of fluid transients. Piping codes prior to 1970 were rather vague in this regard; however, the more recent ASME III codes are quite specific and require this consideration. The forcing functions generated by manual or computer modeling are incorporated into the stress analyses which are used as the basis for designing a support system that can accept the loads and maintain them within the code allowable limits for safe, continuous system operation.

Pump NPSH Transients

Net positive suction head (NPSH) is the often recognized, but just so often misunderstood abbreviation for NPSH. There are adjectives that modify NPSH: one is "available"; the other is "required." The available NPSH must always exceed the required NPSH. The difference between available and required NPSH is called *margin* and must be positive even during transient conditions. See Ref. 39 for more information on NPSH.

A turbine-generator load rejection event can create a significant NPSH decay, and this transient event must be evaluated.

The design of fluid systems must clearly establish criteria that address steady-state and transient conditions from the view of system or component protection. Certain types and severity levels of transients can be accommodated by the pump, but may not be by the system.

NPSH transients at a pump's suction may produce a reduction in total pressure below the minimum required to satisfy a pump's inlet requirements, or even below the minimum pressure required by the temperature of the fluid.

This reduction in pressure produces voids in the suction of the pump. It is the production of these vapor pockets, or bubbles, and the consequent collapse as the head, or pressure, is developed in the impeller, that generates the cavitation which produces the noise, vibration, and the pitting attack of the impeller's surfaces. The measurable consequences of cavitation on a pump's performance are a loss in flow and head resulting from voiding of vane passages, turbulence, and two-phase flow.

Short-term transients can be considered momentary events where the available NPSH falls below the required NPSH at a particular moment. These events will probably have no measurable effects on the pump mechanically and, because of their duration of fractions of a second to seconds, should not measurably affect performance which would be restored with restored suction conditions. An example of a short-term transient occurs during normal startup of a centrifugal where, in establishing a stable operating point, the pump would briefly run out past its design flow as the system fills. The NPSH available at run out flows can be less than the requirements of the pump, initiating a transient condition that is corrected as the system fills and flow stabilizes.

Transients which occur for periods that range from seconds to a few minutes will have a measurable impact on performance. Sustained suction conditions that are below minimum NPSH requirements will cause cavitation in the eye of the pump's suction impeller that will reduce flow and developed head. Long-term, low NPSH transients, by design, must be avoided.

Pump NPSH transients have been modeled and photographed under test conditions, and certain chronic transients have resulted in verifiable documented cases where NPSH conditions were the cause of pump problems. However, in spite of our sophistication and diligence in this area, over many decades NPSH transients and their impact remain qualitative rather than quantitative.

NPSH transients must be understood as well as the pump's capabilities to withstand such events; then margins can be applied initially in the design of the system.

Pipe Rupture

Jet forces F_j from postulated pipe rupture are expressed customarily in terms of the initial fluid pressure $p_{o'}$ and the break area A and take the form:

$$F_j = C_j p_o A \tag{B8.87}$$

where C_j is the jet coefficient. The value of jet coefficient depends on the fluid properties, temperature, pipe diameter, distance between the pressure reservoir and the pipe break locations, flow restrictions, and so on.

Upper Bound of Jet Force. To evaluate the upper bound of jet force, it is assumed that the pipe break location is very close to the pressure reservoir and the pressure drop, due to pipe flow friction and restrictions, is negligible.

1. *Nonflashing water:* For nonflashing water, the incompressible form of Bernoulli's equation neglecting the elevation changes and combining with the upper bound jet force conservatively yields:

$$F_j = 2p_o A \tag{B8.88}$$

2. *Saturated water:* For saturated water at a pressure of 1050 psia, Moody calculated the jet force to be approximately:[24.40]

$$F_j = 1.26 p_o A \tag{B8.89}$$

3. For breaks in pipelines containing gaseous fluid such as saturated steam, superheated steam, and helium gas, the critical flow condition exists is most cases. The gaseous flow is choked at the break exit.

For saturated steam and assuming that the ambient pressure is small in comparison with $p_{o'}$, then Eq. (B8.89) applies.

In the case of helium gas:

$$F_j = 1.3 p_o A \tag{B8.90}$$

Jet Force Reductions. Friction effects will tend to reduce jet forces, as will other restrictions to flow. These effects may be considered, if desired.

Cross-Sectional Area of Jet. For longitudinal breaks and circumferential breaks with full separation, the initial jet area is:

$$A = \frac{\pi D^2}{4} \tag{B8.91}$$

For circumferential breaks with limited separation, the initial jet area is:

$$A = \pi D\,dx \tag{B8.92}$$

where dx is the axial separation of the two circumferential breaks.

For numerical example problems, see Ref. 24.

REFERENCES

1. V. L. Streeter and E. B. Wylie, *Fluid Mechanics,* 8th ed., McGraw-Hill, New York, 1985.
2. E. F. Obert, *Thermodynamics,* McGraw-Hill, New York, 1962.
3. H. Schlichting, *Boundary Layer Theory,* McGraw-Hill, New York, 1955.
4. W. M. Rohsenow and H. Choi, *Heat, Mass and Momentum Transfer,* Prentice-Hall, Englewood Cliffs, New Jersey, 1961.
5. International Organization for Standardization, *Units of Measurements,* ISO Standards Handbook 2, 2d ed., Geneva, Switzerland, 1982.
6. The American Society of Mechanical Engineers, "1967 ASME Steam Tables," 4th ed., New York, 1979.
7. Crane Company, "Flow of Fluids Through Valves, Fittings and Pipe," Technical Paper No. 410 (or 410 M in metric units).
8. I. E. Idelchik, G. R. Malyavskaya, O. G. Martynenko, and E. Fried, *Handbook of Hydraulic Resistance,* Hemisphere Publishing, 1986.
9. W. F. Bland and R. L. Davidson, (eds.), *Petroleum Processing Handbook,* McGraw-Hill, New York, 1967.
10. A. Stodola, *Steam and Gas Turbines* (translated from 6th German ed.), McGraw-Hill, New York, 1927.
11. A. H. Shapiro, *The Dynamics and Thermodynamics of Compressible Fluid Flow,* Ronald Press, 1953.
12. ASME, *Fluid Meters, Their Theory and Applications,* ASME, 6th ed., 1971.
13. ASME, *Flow Measurements,* ASME Power Test Codes, The American Society of Mechanical Engineers, 1959.
14. A. E. Bergles, J. G. Collier, J. M. Delhaye, G. F. Hewitt, and F. Mayinger, *Two-Phase Flow and Heat Transfer in the Power and Process Industries,* Hemisphere Publishing, 1981.
15. J. M. Delhaye, M. Giot, and M. L. Riethmuller, *Thermohydraulics of Two-Phase Systems for Industrial Design and Nuclear Engineering,* Hemisphere Publishing and McGraw-Hill, New York, 1981.
16. R. Roumy, "Structure des Ecoulements Diphasiques Eau-air. Etude del la Fraction de Vide Moyenne et des Configurations d'Ecoulement," CEA-R-3892, 1969.
17. G. F. Hewitt and D. N. Roberts, "Studies of Two-Phase Flow Patterns by Simultaneous X-ray and Flash Photography," AERE-M 2159, 1969.
18. G. F. Hewitt and N. S. Hall-Taylor, *Annular Two-Phase Flow*, Pergamon, New York, 1970.
19. T. Oshinowo and M. E. Charles, "Vertical Two-Phase Flow. Part 1. Flow Pattern Correlations," *Can. J. Chem. Eng.*, vol. 52, 1974, pp. 25–35.
20. G. E. Alves, "Cocurrent Liquid-Gas Flow in a Pipeline Contactor," *Chem. Eng. Prog.*, vol. 50, no. 9, 1954, pp. 449–456.
21. Y. Taitel and A. E. Dukler, "A Model for Predicting Flow Regime Transitions in Hor-

izontal and Near-Horizontal Gas-Liquid Flow," *AIChE Journal,* 22, January 1976, pp. 47–55.

22. O. Baker, "Simultaneous Flow of Oil and Gas," *Oil and Gas Journal,* p. 185, vol. 53, 1954.
23. A. E. Dukler and Y. Taitel, "Flow Pattern Transitions in Gas-Liquid Systems: Measurement and Modeling," *Multiphase Science and Technology,* vol. 2, 1987.
24. R. T. Lahey, Jr., and F. J. Moody, *The Thermal-Hydraulics of Boiling Water Nuclear Reactor,* published by ANS, 1977.
25. Aerojet Nuclear Company, "RELAP4/MOD5, A Computer Program for Transient Thermal-Hydraulic Analysis of Nuclear Reactors and Related Systems," *User's Manual,* ANCR-NUREG-1335, 1976.
26. C. J. Baroczy, "A Systematic Correlation for Two-Phase Pressure Drop," NAA-SR-MEMO-11858, 1966.
27. H. D. Baehr, "Thermodynamik," Springer Verlag, 1962 (in German).
28. W. F. Allen, Jr., "Flow of Flashing Mixture of Water and Steam Through Pipes and Valves," *Trans. ASME,* April 1951.
29. M. W. Benjamin and J. G. Miller, "The Flow of a Flashing Mixture of Water and Steam Through Pipes," *Trans. ASME,* vol. 64, 1942.
30. F. J. Moody, "Maximum Flow Rate of a Single Component, Two-Phase Mixture," APED-4378, October 1963.
31. F. J. Moody, "Maximum Two-Phase Vessel Blowdown from Pipes," APED-4827, April 1965.
32. R. E. Henry and H. K. Fauske, "The Two-Phase Critical Flow of One-Component Mixtures in Nozzles, Orifices, and Short Tubes," *ASME Journal of Heat Transfer,* May 1971, pp. 179–187.
33. D. Abdollahian and A. Singh, "Prediction of Critical Flow Rates Through Power-Operated Relief Valves," ANS/ASME/AIChE, *Proceedings of the Second International Topical Meeting on Nuclear Reactor Thermal-Hydraulics,* vol. 2, Santa Barbara, Calif., 1983, pp. 912–918.
34. K. H. Ardron and R. A. Furness, "A Study of the Critical Flow Models Used in Reactor Blowdown Analysis," *Nuclear Engineering and Design,* 1976, p. 39.
35. P. Griffith, "Notes for the MIT Spring Course on Two-Phase Flow," 1976.
36. General Electric Company, "Moisture Separator and Reheater Drain Systems," GEK-37949A, 1977.
37. U.S. Nuclear Regulatory Commission NUREG-0582, "Water Hammer in Nuclear Power Plants," July 1979.
38. R. W. Bjorge, "Initiation of Water Hammer in Horizontal or Nearly-Horizontal Pipes Containing Steam and Subcooled Water," Ph.D. thesis, Massachusetts Institute of Technology, 1982.
39. I. J. Karassik et al., *Pump Handbook,* McGraw-Hill, New York, 1976.
40. F. J. Moody, "Prediction of Blowdown Thrust and Jet Forces," ASME Paper 69-HP-31.

BIBLIOGRAPHY

ANSI/ASME Code B31.1, 1980 edition, Appendix II, 1.0–1.2, "Nonmandatory Rules for the Design of Safety Valve Installations."

K. Goldmann, H. Firstenberg, and C. Lombardi, "Burnout in Turbulent Flow—A Droplet Diffusion Model," *Trans. ASME, Ser. C., J. Heat Transfer,* 83, 1961, pp. 158–162.

P • A • R • T • C

PIPING SYSTEMS

CHAPTER C1
WATER SYSTEMS PIPING

Michael G. Gagliardi
Manager, Mechanical Nuclear Engineering
Ebasco Services Incorporated
New York, N.Y.

Louis J. Liberatore
Staff Engineer
Ebasco Services Incorporated
New York, N.Y.

INTRODUCTION

General Description

Water-distribution systems which serve populated areas and industrial complexes, including offices, and light and heavy industry are classified broadly as being of the loop, gridiron, or tree types. Figure C1.1 describes these three types. Within the broad concept, all three types may be combined to be used as the building blocks for the overall system.

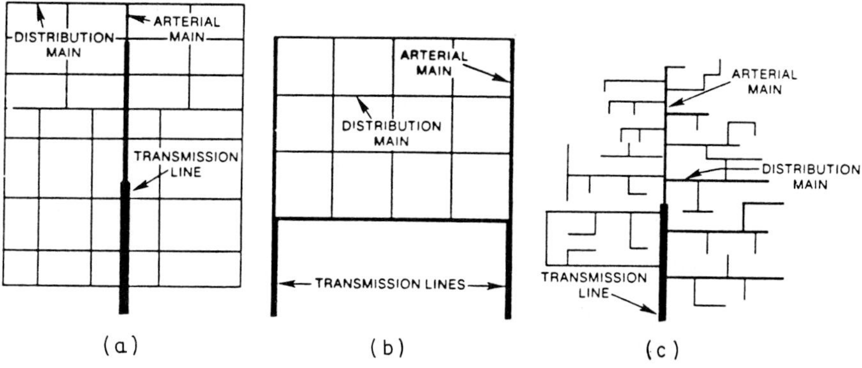

FIGURE C1.1 (*a*) Grid system; (*b*) arterial-loop system; (*c*) tree system. (*AWWA, Principles and Practices of Water Supply Operations,* Volume 3, Introduction to Water Distribution.)

In the *loop system,* large feeder mains that surround areas many city blocks square serve smaller crossfeed lines connected at each end into the main loop. In the *gridiron* (or *grid*) *system,* the piping is laid out in checkerboard fashion, with piping usually decreasing in size as the distance increases from the source of supply. In the *tree system,* there is a single trunk main, reducing in size with increasing distance from its source of supply; branch lines are supplied from the trunk.

The grid and loop systems provide better reliability because of their multiple paths. Grid and loop systems are often backed up with feeder pipes leading directly from the pumping station to remote distribution centers serving to bolster the supply to meet increased demands with growth of population.

Water-distribution systems are made up of pipes, valves, and pumps through which water is moved from the source to homes, offices, and industries that use the water. The distribution system may include facilities to store treated and untreated water for use during periods when demand is greater than the source can supply and when special service requirements must be satisfied.

The requirements of distribution systems are subject to the requirements of local ordinances and state laws and health regulations. Two important requirements of any water-distribution system are that it supply each user with a sufficient volume of water at adequate pressure. For treated water systems, it is also important that the quality of the water be maintained by the treatment facility and distribution system.

Types of Water Piping Systems

There are four general types of piping systems in water-distribution utilities: transmission lines, in-plant piping systems, distribution mains, and service lines. *Transmission lines* carry water from a source of supply to the distribution system. *Distribution mains* are the pipelines that carry water from transmission lines and distribute it throughout a service area such as a community or industrial complex. *Service lines* are small diameter pipes that run from the distribution mains to the user.

The prime objective of a distribution network is to supply a sufficient quantity of water to all parts of the system at pressures adequate for the requirements of the users, at all times and under all conditions of their demands. Therefore, the selection of pipe sizes, material, geometry, and configuration in distribution networks is influenced more by the necessity of maintaining adequate water pressure than by the economics of pumping costs.

Common industrial or power applications of water systems are condenser circulating water and service cooling water systems. A condenser uses circulating water to condense steam exhausted from the plant's turbines. In a large steam power plant, this requires a considerable amount of water to be continuously circulated. Consequently, since the circulating water directly affects the plant's efficiency and reliability, an efficient, reliable, and economical circulating water system is required.

Service water systems provide cooling water to a plant's components and heat exchangers and to other services required by the plant. Due to current environmental regulations, recirculation-type systems in which the same water is used repeatedly must be applied in most cases. Means of cooling the water is provided in the form of cooling towers, spray ponds, or cooling ponds. To compensate for evaporation, leakage, and blowdown, initial fill and makeup water has to be provided from a river, lake, or other large natural body of water. Some service water systems and in nuclear plants emergency service water systems may be essentially once through type systems.

A siphon system is one in which the siphon principle is used to carry the water through elevated parts of the system, such as the condenser, in order to reduce the pumping power required. These elevated portions of the water system operate under a partial vacuum. A pressure system is one in which the water flows under a positive head throughout. This system is generally used with recirculating systems, such as with cooling tower installations.

Vertical pumps set in an intake basin are usually the most suitable for circulating water and service water applications. The complexity of the intake structure is naturally affected by the number of pumps necessary for the system. Reliability points to the use of at least two pumps. The particular conditions for the plant will dictate the final choice—two pumps at two-thirds capacity each, at one-half capacity each, or at some other more favorable percentage each. The capacity selection has to be the subject of a careful analysis, taking into account such items as water requirement, the variation of pumping head, and the best efficiency range of the pumps.

The intake chambers for vertical pumps require careful design for good pump operation. The design must bring about a uniform and undistributed flow of water to the pump without whirl. Most pump manufacturers have design suggestions for intake chambers for their particular pumps. Each vertical-pump installation should be studied individually. There are no standard solutions to vertical-pump intake problems.

In-service water systems booster pumps may be required to ensure pressure to most distant higher elevation points without overpressurizing the lowest components. Horizontal pumps are generally suited for this application.

The intake piping to the suction of horizontal pumps should be designed so as to avoid air pockets. Also, the water-flow velocity should be made uniform over the suction inlet area by placing bends as far as possible from the pump inlet.

Discharge Structure. On the discharge end of once-through systems an underwater (or sealed) discharge must be provided to prevent entry of air into the piping, which would otherwise break the siphon action at the condenser. One means of providing this seal is through the use of a seal well, that is, a basin with a water level controlled by an overflow weir. The seal-well water level regulates the height of the siphon recovery and is the final elevation to which the circulating pump delivers the water.

Beyond the seal well, the discharge into the river or other body of water must be done in such a way that the discharge velocity is dissipated without washing away banks, tearing up the bottom, undermining the discharge piping, or permitting uncontrolled recirculation to the intake. These systems require attention to problems such as air binding and water hammer as discussed in this chapter.

A concern to maintaining reliability in plants using raw water for cooling is microbiological growth and sediment accumulation (silting). A critical concern in nuclear plants, keeping piping and components free of clogging or bacterial attack, requires careful prevention and maintenance programs.

The design of high-temperature, high-pressure piping such as boiler feedwater (FW) systems require considerable experience and study. Besides those typical hydraulic problems inherent in lower pressure, lower temperature systems, concern for flashing cavitation and the problems associated with handling two-phase flow and large system transients are encountered.

Velocities ranging from 10 to 25 ft/s are normal in these systems since the fluid is usually high-quality, low-solids water. Piping material can range from carbon steel, such as A53, ASTM A106, chrome and molybdenum alloy steels, such as ASTM A335, and other alloy steels.

FW piping is usually seamless and uses welded joints. Flanged connections where required must use a temperature-resistant gasket.

Network Analysis of Distribution Systems

The complexity of the analysis required for a well-designed water distribution system is comparable to that of utility electric power networks. Several procedures may be used for the analysis of flow in complex piping networks, such as the Hardy-Cross method. All of the methods involve the solution of a flow problem considering head losses of a complex distribution network resulting in extremely tedious and time-consuming trial and error calculations. With the development of state-of-the-art computer hardware and software, complex network problems involving hundreds of branches can be solved in a relatively short time. Illustrative Example C1.1 presents a sample problem using the Hardy-Cross method of flow network solution.

Illustrative Example C1.1

1. Make a skeleton drawing of the network. Indicate by appropriate arrows the points of constant flow input or output, constant head input or output (see Fig. C1.2).
2. Number all loops in the system in arbitrary sequence. Do not include "loops around loops." For example, in Fig. C1.3 there are two loops, not three. The large loop (*abcdefg*) is not numbered. The two basic loops (*abfg* and *bcdef*) are numbered.
3. Number each line. A line has two ends. An end may be a point at which water is drawn from or added to the system, one at which pipe characteristics change, or a tee joint. For example, in Fig. C1.4, the point x is the meeting of three lines, not two; point y is the meeting of two lines where an 8-in pipe joins a 10-in pipe; point z is simply a bend in the single pipe and is not the end of any line, although it could have been specified as one, if desired. Figure C1.4 shows the complete numbering of the system shown in Fig. C1.2. Note that each line is numbered once and only once, even though it may be in more than

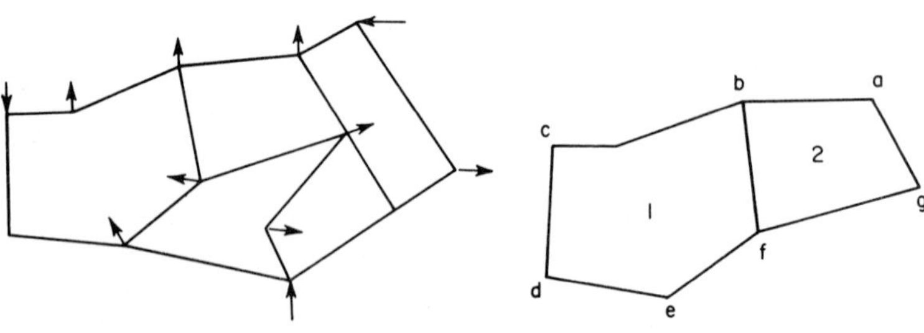

FIGURE C1.2

FIGURE C1.3

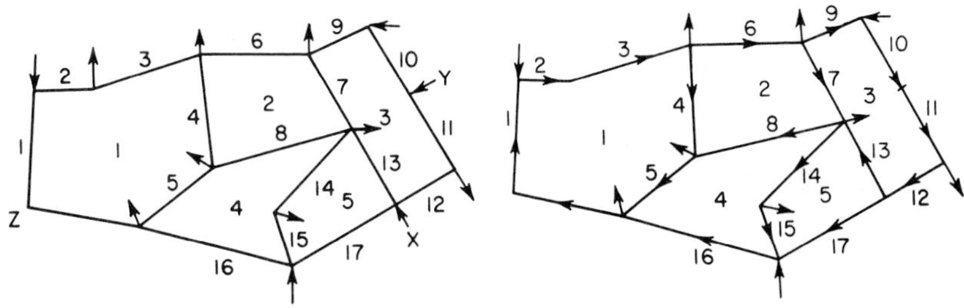

FIGURE C1.4

FIGURE C1.5

one loop. Also note that the numbering is serial; that is, if there are n branches, each of the numbers from 1 to n must be used in the numbering.

4. Assign a base direction. Put an arrow on each line in loop 1, indicating the clockwise direction (as shown in Fig. C1.5). Then put an arrow on each line in loop 2, indicating clockwise direction, except where a line which previously has been assigned a direction is encountered. Then the original assignment is not changed. In Fig. C1.5, line 4 is a member of loop 1 and also of loop 2 but has been given a base direction of loop 1. The line 4 assignment is not changed. This process is continued for every loop in the network, an arrow being assigned in a clockwise direction whenever it has not been assigned previously.
5. In water-distribution systems, the situation often is encountered where system pressure must be raised by the use of booster pumps in series with the supply pipeline. If the higher pressure area is connected to the remainder of the system at one point only, the two pressure-zone networks are hydraulically independent problems. If the pressure zones are connected at two or more points, the booster pumps must be included in the appropriate loops.

   For all loops containing booster pumps, an unbalanced or residual head H_0 must be determined. This is done by algebraically summing the assumed constant head changes at the boosters in a clockwise direction.

   Note that head *losses* are considered as positive in sign, so proceeding from the suction side of a pump to the discharge side gives a negative head loss.

   Following the hydraulic analysis, a check should be made to assure that the pumping head assumptions are sufficiently accurate. The resulting flow-rate values should allow optimum hydraulic design of the booster-station installations.
6. Additional "pseudo-loops" must now be added to the list if there is more than one constant head input (see Fig. C1.6). If the number of such inputs is m, trace $(m - 1)$ paths between inputs in the same manner in which the loops were traced, making sure that each constant head input is used at the end of at least one of these "loops." If the direction of procedure is from the lower to the higher input in each path, H_0 will be the positive difference in the head loss between the two inputs. If booster pumps are encountered, the head change across such pumps must be algebraically added to the head difference between the inputs in order to obtain the H_0 for the "pseudo loops."

   When the listing of all the loops has been completed (including the consideration of booster pumps), the work should be carefully checked, preferably by a second person, since any errors will completely upset the calculations.

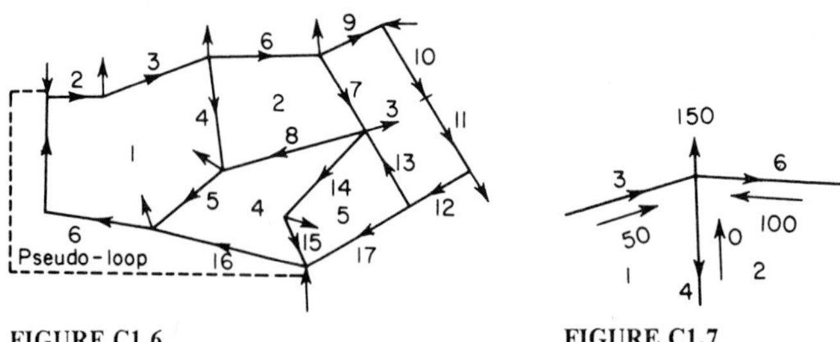

FIGURE C1.6

FIGURE C1.7

Note that pseudo-loops do not introduce any new lines. Note also that each pseudo-loop must be assigned its own number.

7. The only remaining task is to supply initial flow values and pipe characteristics which the computer can use as starting values for the calculations. The only restriction on these values is that they satisfy the mass balance condition at each junction. That is, the sum of the flow *into* a junction must equal the sum of the flows *out* of the junction. For example, Fig. C1.7 shows the junction of lines 3, 4, and 6; flows of 50 gpm in line 3 and 100 gpm in line 6 would satisfy the condition.

   Proceeding in this manner, balance every junction in the network, working toward the variable-flow (constant-head) inputs which can take up the slack. When all flows are specified, check the accuracy of the work by summing the inputs and outputs. If these sums are unequal, some computational error has been made and must be corrected. The complete schematic for this system is shown in Fig. C1.8. This schematic includes the assumed starting values of the flows.

Several personal computer (PC) based and main frame computer software programs have been developed to handle steady-state and transient flow analysis in piping systems with any configuration and including a variety of components such as storage tanks, pumps, check valves, pressure regulating valves, and variable pressure supplies. These programs can also consider extended period simulations which are simulations of a piping system over a long period of time. These simulations result in varying water levels in tanks or reservoirs and can take into account such occurrences as pump operation controlled by water levels or pressure levels. Demand patterns can be varied throughout the simulation providing modeling flexibility. These programs can handle any liquid using a variety of English and Standard International (SI) units.

In all flow network problems the following conditions must be satisfied:

1. The algebraic sum of the pressure drops around each circuit must be equal to zero.
2. The flow into each junction must equal the flow out of the junction.
3. The proper relationship between head loss h and discharge flow Q must be maintained for each pipe.

Parallel and Series Piping. A combination of two or more pipes connected so that the flow is divided among the pipes then joined again is a *parallel-pipe system.* In *series pipes* the same fluid flows through all the pipes, and the head losses

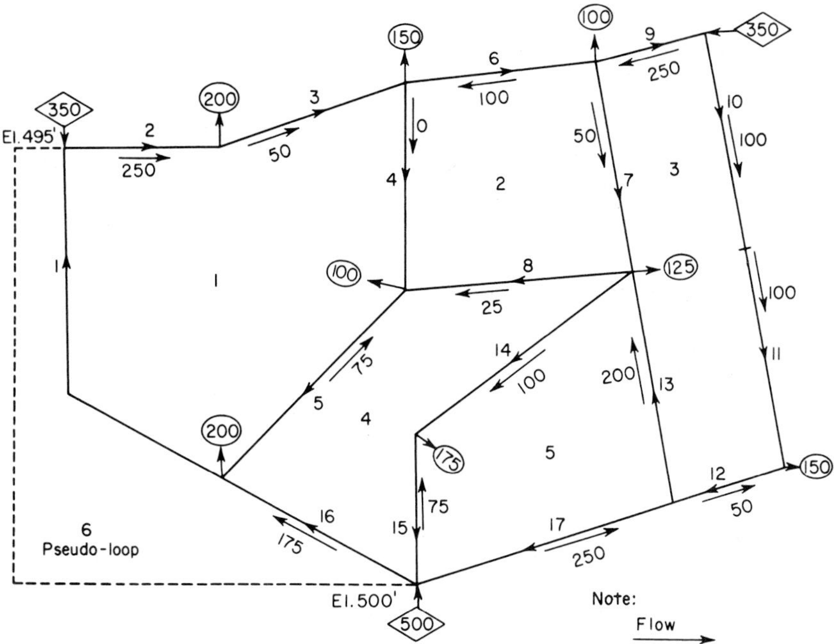

FIGURE C1.8

are cumulative; in a parallel piping system the head losses (h_{f1}, h_{f2}, h_{f3}) are the same in each parallel branch associated with that system, and the discharge flows (Q_1, Q_2, Q_3) are cumulative. In analyzing parallel-pipe systems, it is assumed that the minor losses are added into the lengths of each pipe as equivalent lengths.

From Fig. C1.9 the conditions to be satisfied are as follows:

$$h_{f1} = h_{f2} = h_{f3} = \left(\frac{P_A}{\gamma} + Z_A\right) - \left(\frac{P_B}{\gamma} + Z_B\right) \tag{C1.1}$$

in which Z_A, Z_B are elevations of points A and B, γ is the density of the fluid, and Q is the discharge flow through the approach pipe or the exit pipe.

Illustrative Example C1.2 applies to the parallel-pipe system shown in Fig. C1.9, where D_n, L_n, and ε_n are the diameter, length, and absolute roughness of the pipe in branch n.

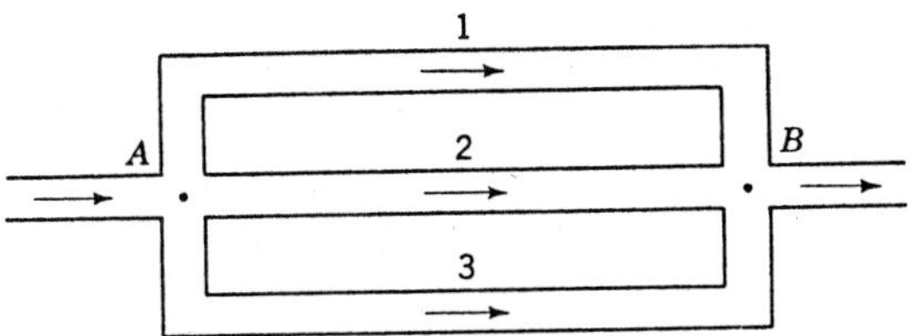

FIGURE C1.9 Parallel-pipe system.

Illustrative Example C1.2

$L_1 = 3000$ ft $\quad D_1 = 1$ ft $\quad \epsilon_1 = 0.001$ ft

$L_2 = 2000$ ft $\quad D_2 = 8$ in $\quad \epsilon_2 = 0.0001$ ft

$L_3 = 4000$ ft $\quad D_3 = 16$ in $\quad \epsilon_3 = 0.0008$ ft

$\rho = 2.00$ slugs/ft^3 $\quad \nu = 0.00003$ ft^2/s $\quad p_A = 80$ psi $\quad z_A = 100$ ft $\quad z_B = 80$ ft

For a total flow of 12 ft^3/s, determine flow through each pipe and the pressure at B.

Assume $Q'_1 = 3$ ft^3/s; then $V'_1 = 3.82$, $\mathbf{R}'_1 = 3.82 \times 1/0.00003 = 127{,}000$, $\epsilon_1/D_1 = 0.001$, $f'_1 = 0.022$, and

$$h'_{f1} = 0.022 \times \frac{3000}{1.0}\frac{(3.82)^2}{64.4} = 14.97 \text{ ft}$$

For pipe 2

$$14.97 = f'_2 \frac{2000}{0.667}\frac{V'^2_2}{2g}$$

Then $\epsilon_2/D_2 = 0.00015$. Assume $f'_2 = 0.020$; then $V'_2 = 4.01$ ft/s, $\mathbf{R}'_2 = 4.01 \times ⅔ \times 1/0.00003 = 89{,}000$, $f'_2 = 0.019$, $V'_2 = 4.11$ ft/s, $Q'_2 = 1.44$ ft^3/s.

For pipe 3

$$14.97 = f'_3 \frac{4000}{1.333}\frac{V'^2_3}{2g}$$

Then $\epsilon_3/D_3 = 0.0006$. Assume $f'_3 = 0.020$; then $V'_3 = 4.01$ ft/s, $\mathbf{R}'_3 = 4.01 \times 1.333/0.00003 = 178{,}000$, $f'_3 = 0.020$, $Q'_3 = 5.60$ ft^3/s.

The total discharge for the assumed conditions is

$$\Sigma Q' = 3.00 + 1.44 + 5.60 = 10.04 \text{ ft}^3\text{/s}$$

Hence

$$Q_1 = \frac{3.00}{10.04} \times 12 = 3.58 \text{ ft}^3\text{/s} \qquad Q_2 = \frac{1.44}{10.04} \times 12 = 1.72 \text{ ft}^3\text{/s}$$

$$Q_3 = \frac{5.60}{10.04} \times 12 = 6.70 \text{ ft}^3\text{/s}$$

Checking the values of h_1, h_2, h_3,

$$V_1 = \frac{3.58}{\pi/4} = 4.46 \qquad \mathbf{R}_1 = 152{,}000 \qquad f_1 = 0.021 \qquad h_{f1} = 20.4 \text{ ft}$$

$$V_2 = \frac{1.72}{\pi/9} = 4.93 \qquad \mathbf{R}_2 = 109{,}200 \qquad f_2 = 0.019 \qquad h_{f2} = 21.6 \text{ ft}$$

$$V_3 = \frac{6.70}{4\pi/9} = 4.80 \qquad \mathbf{R}_3 = 213{,}000 \qquad f_3 = 0.019 \qquad h_{f3} = 20.4 \text{ ft}$$

f_2 is about midway between 0.018 and 0.019. If 0.018 had been selected, h_2 would be 20.4 ft.

To find p_B,

$$\frac{p_A}{\gamma} + z_A = \frac{p_B}{\gamma} + z_B + h_f$$

or

$$\frac{p_B}{\gamma} = \frac{80 \times 144}{62.4} + 100 - 80 - 20.8 = 183.5$$

in which the average head loss was taken. Then

$$p_B = \frac{183.5 \times 2 \times 32.2}{144} = 81.8 \text{ psi}$$

Two types of problems occur: (1) with the elevation of the hydraulic grade line at A and B known, to find the discharge Q; (2) with Q known, to find the distribution of flow and the head loss. Sizes of pipes, fluid properties, and pipe wall roughnesses are assumed to be known.

The first type is the solution of a traditional pipe discharge problem as the head loss is the drop in the hydraulic grade line. The individual discharges are then added to determine the total discharge. The second type is more complex, as neither the head loss nor the discharge for any one pipe is known. This type of problem can be solved by

1. Assuming a discharge Q'_1, through pipe 1
2. Solving for h_f, using the assumed discharge
3. Using h'_{f1}, find Q'_2, Q'_3
4. With the three discharges flowing with a common head loss, now assume that the given Q is split up among the pipes in the same proportion as $Q1'_1$, Q'_2, Q'_3; thus,

$$Q_1 = \frac{Q'_1}{Q'} Q \qquad \text{(C1.2)}$$

$$Q_2 = \frac{Q'_2}{Q'} \qquad \text{(C1.3)}$$

$$Q_3 = \frac{Q'_3}{Q'} \qquad \text{(C1.4)}$$

5. Check the correctness of these discharges by computing h_{f1}, h_{f2}, h_{f3} for the computed Q_1, Q_2, and Q_3 flows.

Some other important relationships to consider in complex flow problems include the following: In the turbulent region, Reynolds number above 2,000 (see Moody diagram, Fig. C1.10, to determine region and establish relationship for head loss), the pressure drop h varies as the 1.85 power of the flow rate Q, that is:

$$\frac{h_1}{h_2} = \left(\frac{Q_1}{Q_2}\right)^{1.85} \qquad \text{(C1.5)}$$

In the viscous flow region (low flow, Reynolds number below 2,000), the pressure drop varies directly as the flow or

$$\frac{h_1}{h_2} = \frac{Q_1}{Q_2} \qquad \text{(C1.6)}$$

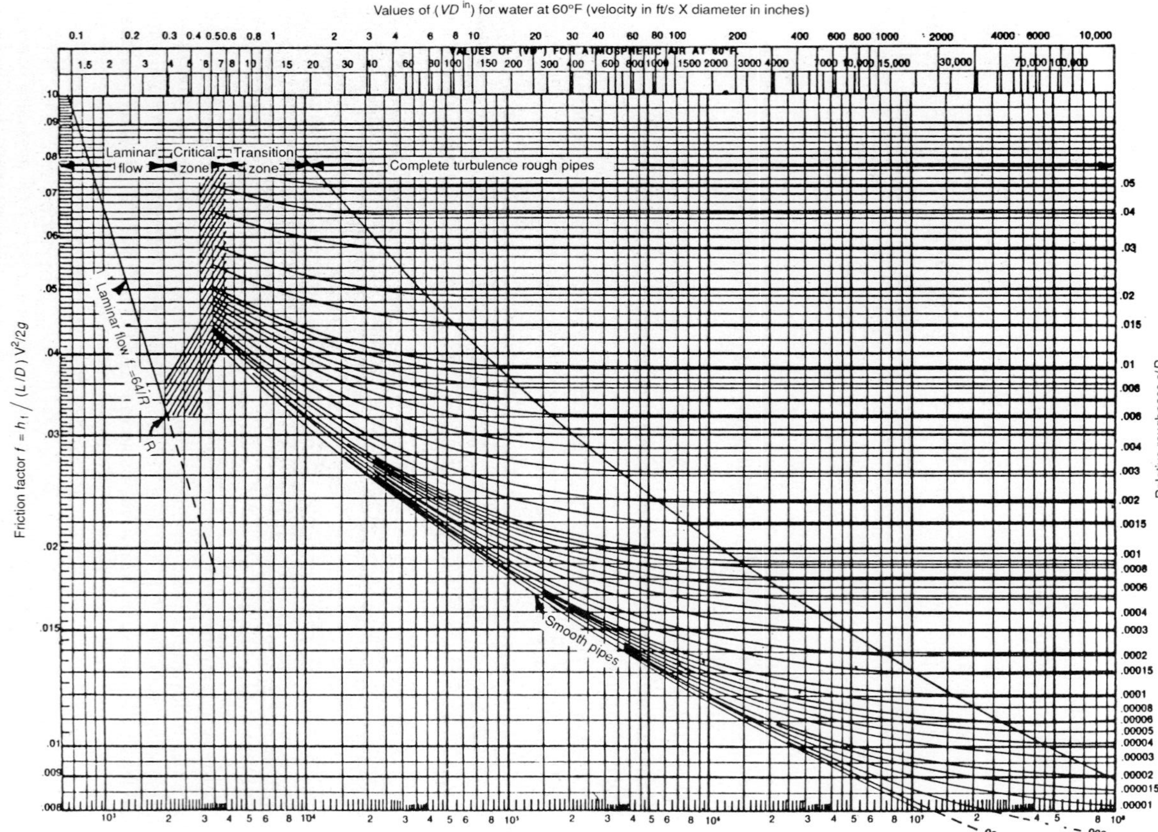

FIGURE C1.10 Moody diagram. Friction factors for any kind and size of pipe. (*Hydraulic Institute.*)

For the same flow, pressure drop varies approximately as the fifth power of the inside diameter D, so that

$$\frac{h_1}{h_2} = \left(\frac{D_1}{D_2}\right)^5 \tag{C1.7}$$

These relationships have been widely used and have resulted in handy pressure drop tables. Appendix E4 provides friction loss data for water and commonly used steel pipe and cast iron pipe in some representative sizes. Table C1.1 provides properties of water at various temperatures. Figure C1.10 provides friction factors for any kind and size of pipe.

HYDRAULIC AND ENERGY GRADE LINES

The concepts of hydraulic and energy grade lines are useful in analyzing more complex water flow problems (see Fig. C1.11). If at each point along a pipe system, the term p/γ is determined and plotted as a vertical distance above the center of the pipe, the connection of these points is the hydraulic grade line. More generally the plot of the two terms,

$$\frac{p}{\gamma} + Z$$

for the flow, as the ordinate against length along the pipe as abscissa, produces the hydraulic grade line.

The hydraulic grade line is the focus of heights to which liquid would rise in vertical glass tubes connected to piezometer openings in the line. When the pressure in the line is less than atmospheric, p/γ is negative and the hydraulic grade line is below the pipeline.

The energy grade line is a line joining a series of points marking the available energy in foot pounds per pound for each point along the pipe as ordinate, plotted against distance along the pipe as the abscissa. It consists of the plot of

$$\frac{V^2}{2g} + \frac{p}{\gamma} + Z$$

for each point along the line, where g = acceleration due to gravity (32.2 ft/s^2). The energy grade line is always vertically above the hydraulic grade line by a distance of $V^2/2g$, neglecting the kinetic-energy correction factor.

Pipes in Series

When two pipes of different sizes or roughness are connected so that fluid flows through one pipe then through the other pipe, they are said to be connected in series. By applying the energy equation, including all losses, for the system in Fig. C1.12 for known lengths and sizes of pipes, the relationship between head loss and flow can be expressed as

TABLE C1.1 Properties of Water at Various Temperatures from 32 to 705.4° F

Temp. (°F)	Temp. (°C)	Specific volume (Cu·ft/lb)	Specific gravity 39.2 F Reference	60 F Reference	68 F Reference	Wt in lb/cu ft	Vapor pressure psi abs
32	0	.01602	1.000	1.001	1.002	62.42	0.088
35	1.7	.01602	1.000	1.001	1.002	62.42	0.100
40	4.4	.01602	1.000	1.001	1.002	62.42	0.1217
50	10.0	.01603	.999	1.001	1.002	62.38	0.1781
60	15.6	.01604	.999	1.000	1.001	62.34	0.2563
70	21.1	.01606	.998	.999	1.000	62.27	0.3631
80	26.7	.01608	.996	.998	.999	62.19	0.5069
90	32.2	.01610	.995	.996	.997	62.11	0.6982
100	37.8	.01613	.993	.994	.995	62.00	0.9492
120	48.9	.01620	.989	.990	.991	61.73	1.692
140	60.0	.01629	.983	.985	.986	61.39	2.889
160	71.1	.01639	.977	.979	.979	61.01	4.741
180	82.2	.01651	.970	.972	.973	60.57	7.510
200	93.3	.01663	.963	.964	.966	60.13	11.526
212	100.0	.01672	.958	.959	.960	59.81	14.696
220	104.4	.01677	.955	.956	.957	59.63	17.186
240	115.6	.01692	.947	.948	.949	59.10	24.97
260	126.7	.01709	.938	.939	.940	58.51	35.43
280	137.8	.01726	.928	.929	.930	58.00	49.20
300	148.9	.01745	.918	.919	.920	57.31	67.01
320	160.0	.01765	.908	.909	.910	56.66	89.66
340	171.1	.01787	.896	.898	.899	55.96	118.01
360	182.2	.01811	.885	.886	.887	55.22	153.04
380	193.3	.01836	.873	.874	.875	54.47	195.77
400	204.4	.01864	.859	.860	.862	53.65	247.31
420	215.6	.01894	.846	.847	.848	52.80	308.83
440	226.7	.01926	.832	.833	.834	51.92	381.59
460	237.8	.0196	.817	.818	.819	51.02	466.9
480	248.9	.0200	.801	.802	.803	50.00	566.1
500	260.0	.0204	.785	.786	.787	49.02	680.8
520	271.1	.0209	.765	.766	.767	47.85	812.4
540	282.2	.0215	.746	.747	.748	46.51	962.5
560	293.3	.0221	.720	.727	.728	45.8	1133.1
580	304.4	.0228	.703	.704	.704	43.9	1325.8
600	315.6	.0236	.678	.679	.680	42.3	1542.9
620	326.7	.0247	.649	.650	.650	40.5	1786.6
640	337.8	.0260	.617	.618	.618	38.5	2059.7
660	348.9	.0278	.577	.577	.578	36.0	2365.4
680	360.0	.0305	.525	.526	.527	32.8	2708.1
700	371.1	.0369	.434	.435	.435	27.1	3093.7
720	374.1	.0503	.319	.319	.320	19.9	3206.2

Computed from Keenan & Keyes' Steam Table.

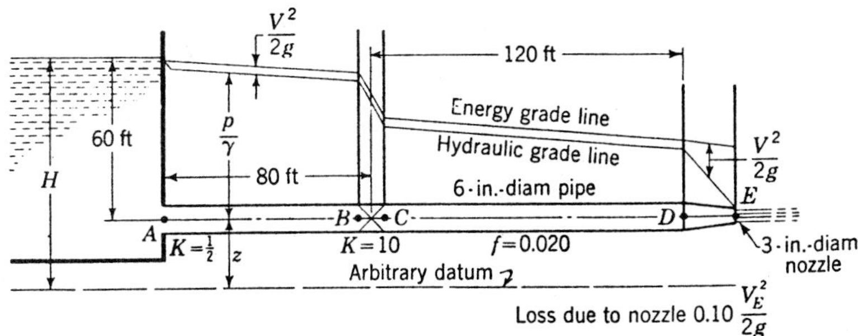

FIGURE C1.11 Hydraulic and energy grade lines.

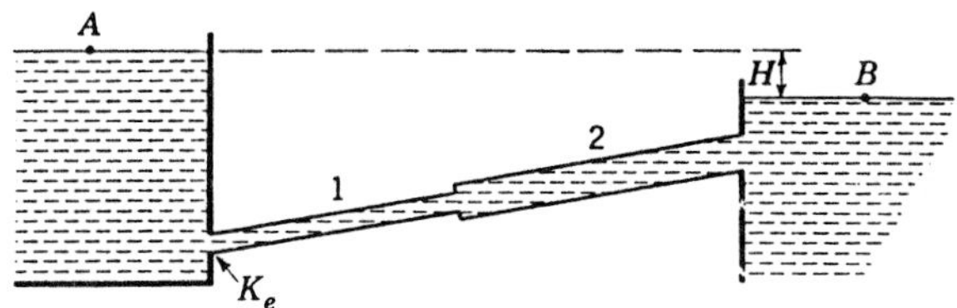

FIGURE C1.12 Pipes connected in series.

$$H = \frac{V_1^2}{2g}(C_1 = C_1 f_1 + C_2 f_2) \tag{C1.8}$$

in which C_1, C_2, and C_3 are known constants and defined as follows:

C_0 = resistance coefficient for valves, fittings, entrance, and exit losses. In this case, it is for exit loss. It is generally known as the k factor

C_1 = L_1/D_1 for one section of pipe of diameter D_2 and length L_2

C_2 = L_n/D_n for second section of pipe in series of diameter D_3 and length L_3

The equation can be expanded up to the term $(C_n f_{n-1})$,

where $C_n = L_n/D_n$ for an nth section of pipe in series of diameter D_n and length L_n

Series pipes may be solved by the method of equivalent lengths. Two pipe systems are said to be equivalent when the same head loss produces the same discharge in both systems. This is expressed as

$$\frac{f_1 L_1}{D_1^5} = \frac{f_2 L_2}{D_2^5} \tag{C1.9}$$

Solving for L_2,

$$L_2 = L_1 \frac{f_1}{f_2}\left(\frac{D_2}{D_1}\right)^5 \tag{C1.10}$$

which determines the length of a second pipe to be equivalent to that of the first pipe.

REFERENCE DOCUMENTS

Codes and Standards

American Society of Mechanical Engineers, Boiler and Pressure Vessel Code. The American Society of Mechanical Engineers (ASME) Code consists of 11 sections of which the following contain requirements for piping:

Section I Power Boilers

Section II Material Specifications

Section III Nuclear Power Plant Components

Section XI Rules for Inservice Inspection of Nuclear Power Plant Components

ASME B31, Code for Pressure Piping. The following sections of ASME B31, Code for Pressure Piping, specify requirements for water systems piping within their scope:

B31.1 Power Piping

B31.3 Chemical Plant and Petroleum Refinery Piping

B31.9 Building Services Piping

Standards, specifications, and so on, published by the following, when referenced in the governing code for design and construction of water systems piping, shall apply:

American Welding Society

American Petroleum Institute

American Water Works Association

Pipe Fabrication Institute

National Fire Protection Association (NFPA)

Table C1.2 provides a listing of the most commonly used industry codes and standards in water systems design, fabrication, assembly, and testing.

DESIGN BASIS CONSIDERATIONS

Design Criteria

Design Pressure. The following are guidelines for determining and specifying system design pressure requirements:

The design pressure shall be based on the maximum expected operating pressure of a particular system which is determined from the maximum operating pressure of the connected pump, pressure vessels, relief valve settings, and so on, depending on the type of system and equipment used. Reasonable margin shall be added to cover variations in expected maximum performance, transients, and control tolerances.

The internal design pressure including the effect of the static head and allow-

TABLE C1.2 AWWA/ANSI Standards for Water Systems Piping

AWWA designation	ANSI designation	Title
C-100		AWWA Standards for cast-iron pipe and fittings
C-104-85	A21.4	Standard for Cement-Mortar Lining for Ductile-Iron Pipe and Fittings for Water
C-105-88	A21.5	Standard for Polyethylene Encasement for Ductile-Iron Piping for Water and Other Liquids
C-110-87	A21.10	Standard for Ductile-Iron and Gray-Iron Fittings, 3 in through 48 in for Water and Other Liquids
C111-85	A21.11	Standard for Rubber Gasket Joints for Ductile-Iron and Gray-Iron Pressure Pipe and Fittings
C115-88	A21.15	Standard for Flanged Ductile-Iron Pipe with Threaded Flanges
C150-81	A21.50	Standard for Thickness Design of Ductile-Iron Pipe
C151-86	A21.51	Standard for Ductile-Iron Pipe, Centrifugally Cast in Metal Molds or Sand-Lined Molds, for Water or Other Liquids
C153-88	A21.53	Standards for Ductile-Iron Compact Fittings 3 in through 16 in for Water and Other Liquids
C-200	—	AWWA Standards for steel pipe
C200-86	—	Standard for Steel Water Pipe 6 in and Larger
C203-86	—	Standard for Coal-Tar Protective Coatings and Lining for Steel Water Pipelines Enamel and Tape—Hot Applied
C205-89	—	Standard for Cement-Mortar Protective Lining and Coating for Steel Water Pipe—4 in and Larger—Shop Applied
C206-88	—	Standard for Field Welding of Steel Water Pipe
C207-86	—	Standard for Steel Pipe Flanges for Waterworks Service Sizes 4 in through 144 in
C208-83	—	Standard for Dimensions for Fabricated Steel Water Pipe Fittings (Addendum C208A-1984)
C209-84	—	Standard for Cold-Applied Tape Coatings for the Exterior of Special Sections, Connections, and Fittings for Steel Water Pipelines
C210-84	—	Standard for Liquid Epoxy Coating Systems for the Interior and Exterior of Steel Water Pipelines
C213-85	—	Standard for Fusion Bonded Epoxy Coating for the Interior and Exterior of Steel Water Pipelines
C214-83	—	Standard for Tape Coating Systems for the Exterior of Steel Water Pipelines
C215-88	—	Standard for Extruded Polyolefin Coatings for the Exterior of Steel Water Pipelines
C216-89	—	Heat-Shrinkable Cross-Linked Polyolefin Coatings for the Exterior of Special Sections, Connections, and Fittings for Steel Water Pipelines
C300	—	AWWA Standard for Concrete Pipe
C300-89	—	Standard for Reinforced Concrete Pressure Pipe, Steel Cylinder Type for Water and Other Liquids
C301-84	—	Standard for Prestressed Concrete Pressure Pipe, Steel Cylinder Type for Water and Other Liquids

TABLE C1.2 AWWA/ANSI Standards for Water Systems Piping (*Continued*)

AWWA designation	ANSI designation	Title
C302-87	—	Standard for Reinforced Concrete Pressure Pipe, Noncylinder Type for Water and Other Liquids
C303-87	—	Standard for Reinforced Concrete Pressure Pipe, Steel Cylinder Type, Pretensioned for Water and Other Liquids
C400	—	AWWA Standards for asbestos-cement pipe
C400-80	—	Standard for Asbestos-Cement Distribution Pipe 4 in through 16 in (100 mm through 400 mm) NPS for Water and Other Liquids (Addendum—C400A-83, R 1986)
C401-83	—	Standard Practice for the Selection of Asbestos-Cement Distribution Pipe 4 in through 16 in (100 mm through 400 mm) for Water and Other Liquids (R 1986)
C402-89	—	Standard for Asbestos-Cement Transmission Pipe 18 in through 42 in (450 mm through 1050 mm) for Potable Water and Other Liquids
C403-89	—	Standard Practice for the Selection of Asbestos-Cement Transmission and Feeders Main Pipe Sizes 18 in through 42 in (450 mm through 1050 mm)
C500	—	AWWA Standards for valves and hydrants
C500-86	—	Gate Valves for Water and Sewerage System
C501-87	—	Cast-Iron Sluice Gates
C502-85	—	Dry-Barrel Fire Hydrants
C503-88	—	Wet-Barrel Fire Hydrants
C504-87	—	Standard for Rubber-Seated Butterfly Valves
C506-78	—	Standard for Backflow Prevention Devices, Reduced Pressure Principles, and Double Check Valves Types (R 1983)
C507-85	—	Ball Valves 6 in through 48 in
C508-82	—	Standard for Swing-Check Valves for Waterworks Service, 2 in through 24 in NPS
C509-87	—	Resilient-Seated Gate Valves for Water and Sewerage Systems
C540-87	—	Power Actuating Devices for Valves and Sluice Gates
C550-81	—	Protective Interior Coatings for Valves and Hydrants
C600	—	AWWA Standards for pipe laying
C600-87	—	Standard for Installation of Ductile-Iron Water Mains and Their Appurtenances
C602-83		Cement-Mortar Lining of Water Pipelines—4 in (100 mm) and Larger in Place
C603-78	—	Standard for Installation of Asbestos-Cement Pressure Pipe
C606-87	—	Grooved and Shouldered Joints
C651-86	—	Standard for Disinfecting Water Mains
C652-86	—	Disinfection of Water Storage Facilities

TABLE C1.2 AWWA/ANSI Standards for Water Systems Piping (*Continued*)

AWWA designation	ANSI designation	Title
C653-87	—	Disinfection for Water Treatment Plants
C654-87	—	Disinfection of Wells
C700	—	AWWA Standard for meters
C700-77	—	Cold Water Meters—Displacement Type
C701-88	—	Cold Water Meters—Turbine Type for Customer Service
C702-86	—	Cold Water Meters—Compound Type
C703-86	—	Cold Water Meters—Fire-Service Type
C704-70	—	Cold Water Meters—Propeller Type for Main Line Applications (R 1984)
C706-86	—	Direct Reading Remote—Registration Systems for Cold Water Meters
C707-82	—	Encoder-Type Remote—Registration Systems for Cold Water Meters
C708-82	—	Cold Water Meters—Multi-Jet Types
C800	—	AWWA Standard for service lines
C800-84	—	Underground Service Line Valves and Fittings
C900	—	AWWA Standards for plastic pipe
C900-89	—	Polyvinyl Chloride (PVC) Pipe, 4 in through 12 in for Water Distribution
C901-88	—	Polyethylene (PE) Pressure Pipe and Tubing ½ in through 3 in for Water Service
C902-88	—	Polybutylene (PB) Pressure Pipe and Tubing ½ in through 3 in for Water
C905-88	—	Polyvinyl Chloride (PVC) Water Transmission Pipe, Nominal Diameters 14 in through 36 in
C950-88	—	Fiberglass Pressure Pipe
D100	—	AWWA Standards for water storage
D100-84	—	Welded Steel Tanks for Water Storage
D101-53	—	Inspecting and Repairing Steel Water Tanks, Standpipes, Reservoirs, and Elevated Tanks for Water Storage (R 1986)
D102-78	—	Painting Steel Water Storage Tanks
D103-87	—	Factory Coated Bolted Steel Tanks for Water Storage
D110-86	—	Wire-Wound Circular Prestressed-Concrete Water Tanks
D120-84	—	Thermosetting Fiberglass-Reinforced Plastic Tanks
D130-87	—	Flexible-Membrane-Lining and Floating Cover Material for Potable-Water Storage
E100	—	AWWA Standard for vertical turbine pumps
E101-88	—	Vertical Turbine Pumps—Line Shaft and Submersible Types

ance for pressure surges shall not be less than the maximum sustained fluid operating pressure. Consideration shall also be given to pump shut-off pressure.

Piping subject to external pressure shall be designed for the maximum differential pressure anticipated during operating, shutdown, or test conditions excluding pressure tests. For buried piping this includes loadings due to earth cover and/or traffic.

In accordance with ASME B31.1 Paragraph 102.2.4, the piping system shall be considered safe for occasional short operating periods at higher than design pressure or temperature if the calculated stress value resulting from such a variation in pressure and/or temperature does not exceed the code allowable by more than 15 percent during less than 10 percent of any 24-h operating period or by more than 20 percent during less than 1 percent of any 24-h operating period.

A piping system is considered safe for operation if the maximum sustained pressure and temperature which may act at any part or component of the system does not exceed the maximum pressure and temperature determined in accordance with Code rules by the Power Piping Code ASME B31.1. Allowable stress values and pressure and temperature ratings are provided by the piping codes and the standards referenced therein.

Design Temperature. The following are guidelines for determining and specifying system design temperature requirements:

The design temperature shall be determined on the basis of the maximum expected operating temperature. The effects of pumping, throttling, heating, cooling, and so on, must be considered in the determination of the design temperature of the piping system.

Pipe Sizing Criteria. Typically, total piping system cost is approximately 7 to 8 percent of the total plant investment. These values range upward to 30 percent for municipal water systems. Besides affecting initial cost, selection of pipe sizes will affect operating costs due to their sensitivity to changes in pressure drop, heat losses, and maintenance requirements.

Selection of a pipe line size involves determination of an optimum size. For instance, if extra pumping is needed to boost the fluid pressure or if the heat rate will be affected adversely, the cost of the extra energy required becomes a significant factor in the evaluations. The optimum pipe size is obtained when the sum of installed and operating costs is at the minimum.

Piping optimization is not widely used in preliminary calculations. Where the pressure drop is defined by other considerations, the minimum pipe size compatible with good engineering practice will be selected.

Other related considerations which have an important impact on pipe size selection include the following:

1. Noise, which can result from high velocity flow, cavitation, or two-phase flow.
2. Vibration, which can result from noise, excessive velocities at changes in the direction of the fluid flow, or the causes of cavitation.
3. Erosion or corrosion due to chemical action of the fluid, excessive velocities, cavitation, and excessive turbulence at fittings, valves, branch connections, and so on.
4. Flow distribution, as the more uniform the cross-sectional velocity profile, the more likely the above factors will be reduced. This can be achieved by using

reasonable velocities along with a piping layout that will produce a smooth flow pattern.

5. Cavitation, which can result from the collapse of bubbles close to a metallic surface at a high enough velocity to cause erosion, and two-phase flow fluids.

Effects of Velocity. Higher allowable velocities will lead to smaller pipe sizes and higher pressure drops. Excessively high velocities can cause noise, vibration, and erosion. Velocities in pump suction lines shall be kept sufficiently low in order to maintain the pump's required net positive suction head (NPSH).

The pressure drop in a system can be decreased by selecting a larger pipe size or sometimes by using more than one pipe for the total flow.

For water piping systems, a velocity in the range of 4–15 ft/s is acceptable. Depending upon the material selected, piping design and size are either on the low or high side of this range. For example, for brass pipe a velocity of 4–15 ft/s would be recommended, whereas for steel pipe, a velocity of 7–10 ft/s is the recommended range although velocities to 30 ft/s may be acceptable. Higher velocities are acceptable if materials less susceptible to erosion, such as stainless steel, are selected. Concurrently reducing vibration and meeting system hydraulic requirements will reduce the piping's susceptibility to erosion. In all cases, it should be recognized that these ranges are recommended only if system operating requirements are also satisfied.

Pipe-Wall Thickness Selection. After determining the internal diameter of the pipe, the designer must select materials; their strength needs to be considered and a pipe-wall thickness or schedule as a function of temperature, pressure, corrosion, erosion, and vibration should be selected.

Pipe-wall thickness determination begins with the basic hoop stress in the pipe wall. This stress calculation ignores longitudinal wall stress that exists if the pipe has closed ends. An example of this is a flask or short header.

Advanced analysis shows that for thin-wall pipe, the outside diameter should be used in the hoop stress equation:

$$S = \frac{PD_o}{2t_{\min}} \tag{C1.11}$$

where P = internal design pressure (psig)
D_o = outside diameter of pipe (in)
$t_{\min}$ = minimum required pipe wall thickness (in)
S = allowable stress (psi)

This equation, called the Barlow formula, is the basis for most code stress pipe-wall thickness equations such as those provided in ASME B31.1 and B31.3. The formula also applies to thick-walled pipe.

The allowable stress in a pipe wall is not determined solely by the material's strength at the service temperature. In pipe with a longitudinal welded seam, the efficiency of the welded joint must also be considered.

ASME Power Piping Code Section B31.1 provides efficiency factors of 0.60 for a furnace butt weld, 0.85 for an electric resistance weld, and 1.00 for single or double butt weld with 100 percent radiography. Allowable stress for welded pipe is determined by multiplying the allowable stress value for seamless pipe at the given temperature by the efficiency factor. Tables of allowable stress in Codes, such as B31.1, include the joint efficiency. The B31.1 Code further specifies the following equations for determining the minimum required thickness of pipe wall at design temperatures:

$$t_m = \frac{PD_o}{2\,(SE + Py)} + A \tag{C1.12}$$

or

$$t_m = \frac{Pd + 2SEA + 2yPA}{2\,(SE + Py - P)} \tag{C1.13}$$

and the design pressure shall not exceed

$$P = \frac{2SE\,(t_m - A)}{D_o - 2y\,(t_m - A)} \tag{C1.14}$$

$$P = \frac{2SE\,(t_m - A)}{d - 2y\,(t_m - A) + 2\,t_m} \tag{C1.15}$$

where t_m = minimum, pipe wall thickness required, in (mm).

P = internal design pressure, psig [kPa (gauge)].

D_o = outside diameter of pipe, in (mm). For design calculations, the outside diameter of pipe as given in tables of standards and specifications shall be used in obtaining the value of t_m. When calculating the allowable working pressure of pipe on hand or in stock, the actual measured outside diameter and actual measured minimum wall thickness at the thinner end of the pipe may be used to calculate this pressure.

d = inside diameter of pipe, in (mm). For design calculations, the inside diameter of pipe is the maximum possible value allowable under the purchase specification. When calculating the allowable working pressure of pipe on hand or in stock, the actual measured inside diameter and actual measured minimum wall thickness at the thinner end of the pipe may be used to calculate this pressure.

SE = allowable stress for the material due to internal pressure and joint efficiency, at the design temperature, psi (kPa).

A = allowance for threading, grooving, mechanical strength, and the effects of erosion and/or corrosion, in (mm).

y = 0.4 for ferritic steels and austenitic steels for temperatures up to 900°F. For pipe with a D_o/t_m ratio less than 6, the value of y for ferritic and austenitic steels designed for temperature of 900°F (480°C) and below shall be taken as

$$Y = \frac{d}{d + D_o} \tag{C1.16}$$

If pipe is ordered by its nominal wall thickness, the manufacturing tolerance on wall thickness must be taken into account. After the minimum required pipe wall thickness t_m is determined by Eq. (1.12) or (1.13), this minimum thickness shall be increased by an amount sufficient to provide the manufacturing tolerance allowed in the applicable pipe specification or required by the process. The next heavier commercial wall thickness shall then be selected from thickness schedules such as contained in ANSI B36.10, ANSI B36.19, or from manufacturers' schedules for other than standard thickness. Refer to Chap. B2 for details on calculating the pipe wall thickness. For cast piping components or to compensate for thinning in bends, refer to the applicable code.

Where ends are subject to forming or machining for jointing, the wall thickness of the pipe, tube, or component shall not be less than t_m, after such forming and machining.

The thickness of gray and ductile iron pipe conveying liquids may be determined by selection from ANSI/AWWA C110/A21.10, ANSI/AWWA C115/A21.15, ANSI/AWWA C150/A21.50, ANSI/AWWA C151/A21.51, and Federal Specification WW-P-421.

While the thickness of steel pipe determined from Eq. (C1.12) or (C1.13) is theoretically ample for both bursting pressure and material removed in threading, the following minimum requirements are mandatory to provide added mechanical strength:

1. Where steel pipe is threaded and used for water service above 100 psi (700 kPa) with water temperature above 220°F (105°C), the pipe shall be seamless having the minimum ultimate tensile strength of 48,000 psi (330 MPa) and a weight at least equal to Schedule 80 of ANSI B36.10.
2. Where threaded brass or copper pipe is used for the services described above, it shall comply with pressure and temperature classifications permitted for these materials by other paragraphs of the applicable Code and shall have a wall thickness at least equal to that specified above for steel pipe of corresponding size.
3. Plain end nonferrous pipe or tube shall have minimum thicknesses as follows:
   a. For nominal sizes smaller than NPS 3/4, the thickness shall not be less than that specified for Type K of ASTM B 88 for copper tubing.
   b. For nominal sizes NPS 3/4 and larger, the wall thickness shall not be less than 0.049 in (1.25 mm). The wall thickness shall be further increased, as required by determining the minimum wall.
4. After calculating the required minimum wall thickness, a choice is usually made from standard thicknesses or schedules. Tables of pipe sizes and wall thicknesses (schedules) are presented in App. E2.
5. For piping systems other than steel or cast iron, refer to the applicable Code.

Design Features

Materials. This section discusses the most commonly used materials for water distribution system piping and piping components. Pipe description, material specifications characteristics, available sizes, normal use, and advantages or disadvantages of different materials are briefly noted in Table C1.3. It should also be noted that state and local regulations may govern the preferred types of pipe to be used for water distribution systems.

The selection of water distribution system piping is based upon the following considerations: strength, ductility, modulus of elasticity, toughness, corrosion, erosion and abrasion resistance, weldability, workability, and surface smoothness. The most common piping materials are steel, either low carbon, low alloy, or stainless; plastic, either thermoplastic or thermosetting with fiberglass rein-

TABLE C1.3 Piping Material Specifications, Standards, and Applications Material

Material	Industry standards	Common sizes (diam. in)	Normal maximum working pressure (psi)	Advantages	Disadvantages
Cast-iron pipe Ductile iron centrifugally cast in metal molds or sand-lined molds for water and other liquids	ANSI/AWWA/C151/A21.51 ASTM A377	4-30	350	Durable, strong, flexural strength, lighter weight than cast iron, greater carrying capacity, same external diameter, fracture resistant, easily tapped	Subject to electrolysis and attack from acid and alkali soils, heavy to handle; may require thrust blocks at changes in direction
Concrete Reinforced concrete pressure pipe, noncylinder type for water and other liquids	AWWA C302	12-168	50	Durable with low maintenance, good corrosion resistance, good flow characteristics, generally suited for gravity systems with low gradient	May deteriorate in alkaline soil if cement type is improper or in acid soil if not protected; may require thrust blocks at changes in direction
Prestressed concrete pressure pipe steel cylinder type for water and other liquids	ANSI/AWWA C301	16-120	250	Durable, low maintenance, good corrosion resistance, good flow characteristics, resists backfill, and external loads	Same as above

Steel		4-120	High	Easy to install, tensile strength, low cost, good hydraulically when lined, adapted to locations where some movement may occur	Subject to electrolysis, external corrosion in acid or alkali soil, poor corrosion resistance welded unless properly lined, coated, and wrapped; low resistance to external pressure in large sizes; air-vacuum valves imperative in large sizes; subject to tuberculation when unlined.
Seamless pipe for high temperature service	ASTM A106, A335., A376				
Seamless and seam welded pipe	ASTM A53, A312, A333, A714				
Forged and bored pipe for high-temperature service	ASTM A426, A451, A452				
POLYVINYL CHLORIDE(PVC)		4-36	200	Light weight, easy to install, excellent resistance to corrosion, good flow characteristics, high tensile and impact strength	Difficult to locate underground, requires special care when tapping, susceptible to damage during handling; above ground use may require more supports
PVC Plastic Pipe	ASTM D1785				
Schedule 40, 80 pressure rated (SDR) series	ASTM D2241				
PVC pressure pipe	ANSI/AWWA C900				
Copper		1/8-12	<700	Excellent resistance to corrosion; ideal for use in water supply and plumbing, including DWV, compressed air instrumentation lines, and heat transfer equipment	Copper deteriorates rapidly under high temperatures and repeated stresses
Seamless copper Standard sizes	ASTM B42				
Seamless copper Water tube for general plumbing purposes	ASTM B88				

TABLE C1.3 Piping Material Specifications, Standards, and Applications Material (*Continued*)

Material	Industry standards	Common sizes (diam. in)	Normal maximum working pressure (psi)	Advantages	Disadvantages
Copper (*continued*)					
Seamless threadless copper pipe	ASTM B302				
Seamless copper tube	ASTM B75				
Copper nickel					
Seamless Copper Nickel pipe and tube	ASTM B466			High level of corrosion resistance. Excellent resistance of copper (90%) nickel (10%) to corrosion and biofouling make ideal for use in piping systems for brackish or sea water. Copper (70%) nickel (30%) provides excellent service under the most adverse conditions; it offers excellent corrosion resistance to high velocity seawater, immune to stress corrosion cracking and resistant to the action of ammonia	May undergo a general attack or uniform thinning in aggressive environment containing high concentration of oxygen and carbon dioxide
Welded Copper Nickel pipe	ASTM B467				
Welded copper and copper Alloy heat Exchanger tube, 90-10 and 70-30	ASTM B543				

Aluminum		1/8-12	<300	Light weight, durable, strong, ideal for use in cryogenic systems, heat transpressure lines, process systems where performance requirements outweigh cost	Requires special techniques and skills for brazing or welding joints depending on type of alloy
Aluminum and aluminum-alloy seamless pipe extruded tube	ASTM B241				
Drawn seamless tube for condensers and heat exchangers	ASTM B234				
Aluminum-alloy drawn seamless tube	ASTM B210				
Titanium		1/8-12	High	Excellent for general corrosion resistance and elevated temperature service; lighter than steel while having strength comparable with alloy steel. Excellent for use cooling water, fresh brackish or salt, polluted, etc.; essentially immune to all forms of corrosion in condenser environments.	Resistance to biofouling is relatively poor when compared to copper alloy for similar applications; coupling of titanium tubes with copper alloy tube sheets in condensers may result in galvanic attack on the tube sheets in all types and make it essential to provide some form of cathodic protection
Seamless and welded titanium alloy pipe	ASTM B337				
Seamless and welded titanium alloy tube for condensers and heat exchangers	ASTM B338				

General note: Materials used for piping within the jurisdiction of ASME I and ASME III are designated by SA/SB numbers and are derived from ASTM "A" specifications. For material specifications referenced in ASME Boiler and Pressure Vessel Code, Sections I and III Applications, refer to Section II of that Code.

forcement; or concrete, lined and unlined. Aluminum, copper, brass, titanium, and high-nickel alloys are also used for some applications.

The metallurgy of the various types of metallic pipe material is somewhat complicated; however, an understanding of the basic physical properties and their effects is necessary for design engineers concerned with selecting pipe material, pipe fabrication process, welding process, and corrosion resistance properties. The most commonly accepted piping standards are those of the American Society for Testing and Materials (ASTM), American Water Works Association (AWWA), American National Standards Institute (ANSI), the American Society of Mechanical Engineers (ASME), and the American Petroleum Institute (API).

These standards have been developed by experimentation, testing, and experience. Most regulatory agencies use these guidelines to set their own requirements. Pipe manufacturers also publish product literature that is useful in pipe selection and installation. Table C1.4 summarizes the advantages and disadvantages of piping materials and related data of pipe joints and their applications.

Pipe Toughness. Pipe toughness is the ability of piping to absorb impact without brittle rupture. Service metal temperature is important in design since most pipe metals have a transition range over which ductile behavior changes to brittle behavior on impact as the temperature drops.

Gray and Ductile Iron Pipe. Cast-iron pipe has been the standard for water distribution systems worldwide for many years. There are more miles of this pipe in use today than any other pipe. Two types of cast iron commonly found in dis-

TABLE C1.4 Pipe Joints and Their Applications

Pipe material	Type of joints	Applications
Steel	Mechanical-type couplings	Pipes are less than 24 in od, especially linings
Steel	Welded joints	Pipes 24 in od and larger with inside coatings ideal for soft soils where settlement may be excessive
Steel	Flanged joints	Where valves or fittings are to be attached; ideal for soft soils where settlement may be excessive
Copper	Welded, flanged, threaded, screwed, or brazed, flared fittings	Generally used in underground water service, interior water, steam, gas, fuel, oil and for some underground drainage
Copper-nickel	Flanged, soldered, or brazed fittings used with hard drain temperatures; flared compression fittings used with annealed temper tubes	Used in condenser tubing, systems for sea or brackish water
Aluminum	Brazed, welded	Used in cryogenic systems, heat transfer, process systems and pressure lines
Titanium	Welded	Used with any kind of cooling water in large power plant condenser tubing

tribution systems are gray cast-iron pipe (CIP) and ductile-iron pipe (DIP). (In this text, CIP is used as the abbreviation for gray cast-iron pipe; it does not stand for cast-iron pipe in general.)

Gray cast-iron pipe is strong but brittle, usually offers a long service life, and is reasonably maintenance free. Ductile-iron pipe resembles CIP in appearance and has many of the same characteristics. It differs from CIP in that the graphite is distributed in the metal in spheroidal or nodular form—that is, in ball-shape form—rather than in flake form. This is achieved by adding a material called an inoculant, usually magnesium, to the molten iron. Ductile-iron pipe is much stronger, tougher, and more ductile than CIP. Gray cast-iron pipe has not been produced since 1980 due to the increased reliability of DIP, but it is still used in the manufacture of valves and fittings.

Although unlined cast iron has a certain resistance to corrosion, aggressive waters can cause the pipe to lose carrying capacity through corrosion and tuberculation. The process for lining pipe with a thin coating of cement mortar made it possible to minimize tuberculation and maintain the carrying capacity of the pipe. The cement-mortar lining is approximately 1/8 in (3 mm) thick and adheres to the pipe. The lined pipe may be cut or tapped without damage to the lining. Ductile-iron pipe internal lining can be cement mortar. Various thermoplastic and other epoxy-lined pipe are also available but are expensive. Generally these would be used to handle fluids considerably more corrosive than water. Bituminous external coating and polyethylene wraps are methods commonly used to reduce external corrosion.

Ductile-iron pipe is available with standard wall thickness (referred to as thickness classes) in diameters of 4 in (100 mm) and larger. The standard lengths are 18 and 20 ft (5.5 and 6.1 m). Ductile-iron pipe is strong and can withstand the working pressures found in distribution systems. It is also durable and can be cut and tapped in the field.

CIP and DIP Joints. Gray cast-iron and ductile-iron joints of the following types have been used to join pipe lengths together (listed in order of development):

- Flanged joints
- Bell-and-spigot joints
- Mechanical joints
- Ball-and-socket or submarine joints
- Push-on joints
- Bolted retainer-gland joints, to prevent pullout

Flanged joints are easy to make and require no special tools. They are used above ground in water plants, pump houses, and other places where rigidity, self-restraint, and tightness are required. Flanged joints will not flex and are not normally used underground.

Mechanical joint is made by bolting a movable follower ring on the spigot to a flange on the bell and compressing a rubber gasket to form a tight seal. The mechanical joint is less economical than previously mentioned joints but is easily made and requires no special skill. Since the bell-and-spigot ends need not fit tightly, each joint can be made to deflect slightly.

Ball-and-socket joints are special-purpose joints most commonly used for submerged installations. Their great advantage is that they can accommodate large deflections (up to 15°). This makes them very useful for pipe lines laid across mountainous terrain or under rivers. Boltless flexible pipe joints, designed on the brass-and-socket principle, are also available.

Push-on joints are the most popular joints in water distribution system installation today. The joint consists of a bell, with a specially designed recess to accept a rubber ring gasket, and a beveled-end spigot. The joint offers ease of installation, and when made up, the rubber ring gasket is compressed to produce water tightness and is locked in place against further displacement. Push-on joints are available in several designs. In addition to ease of installation and water tightness, the joint permits deflections of 3° to 5°, depending on the design, permitting installation on a curve or irregular terrain without the use of additional fittings. Small diameters may be assembled by hand; larger sizes usually require mechanical aids. For detailed discussion on the above joints, refer to Chap. A2 of this handbook.

Steel Pipe and Reinforced Concrete Pipe. Steel pipe and reinforced concrete pipe are sometimes used as large feeder mains in water distribution systems.

Plastic Pipe. Plastic pipe is also used in water distribution applications. Plastic pipe was first introduced in the United States around 1940.

Plastic pipe materials include polyvinyl chloride (PVC), chlorinated polyvinyl chloride (CPVC), polyethylene (PE), polybutylene (PB), and acrylonitrile-butadiene-styrene (ABS). ASTM and AWWA standards cover PVC pipe in sizes from 4 to 12 in NPS (100–300 mm) in diameter. Polyvinyl chloride pipe is a rigid pipe manufactured by an extrusion process. Fittings are made by a mold process. It is available in diameters up to 36 in (0.9 m); lengths of 20–40 ft (6–12 m); and various types, grades, and pressure ratings. Within a given nominal pipe diameter, there are several equivalent systems for specifying internal and external diameters. Larger than 12 NPS plastic pipe is normally specified by a sizing system other than the iron pipe size (IPS) system.

The AWWA standard is based on outside diameter, the same as DIP. A new sizing system, termed the standard dimension ratio-pressure rated (SDR-PR) system, is a ratio of the outside pipe diameter to wall thickness; see Table C1.5.

The SDR-PR system recognizes the strength properties of plastic and allows pipe of one pressure rating to be available in various sizes. State and local regulations should be checked to determine what types and sizes of plastic pipe are approved for a particular application. Manufacturers' recommendations should be consulted when tap diameters exceed 2 in.

PVC pipe may be joined by a bell-and-spigot push-on joint or by a solvent-weld joint. See Part D of this handbook for more detail.

Steel Pipe. Application may dictate the need for steel piping to be lined and/or coated with a plastic, tar, plastic-encapsulated cement, or rubber. ASME B31.1 Appendix IV, Corrosion Control for ASME B31.1—Power Piping Systems provides guidelines for the control of corrosion of steel piping.

Special applications might require the use of stainless steel, ductile-iron, copper, copper alloys, plastic, fiberglass, and aluminum alloy pipe. Use of these materials is generally based on their corrosion-resistance properties and ease of installation.

Copper is widely used in service connections for potable water. Heat exchanger tubing is a common use for nonferrous metals, including copper, copper-nickel alloys, titanium, and aluminum. Copper and its alloys, such as brass, are useful in this application because of their good thermal conductivity, favorable cold or hot working properties, and corrosion resistance.

Aluminum's low density and relatively high strength of its alloys provide good corrosion resistance and good working properties. Titanium and its alloys have

TABLE C1.5 PVC Type I Pressure Rated Pipe

Nominal pipe size (in)	o.d.	Min wall.	Average i.d.	Nominal weight per ft
		SDR 26 NSF—W.P. 160 PSI (Water @ 73.4°F.)		
½		See SDR 13.5		
¾		See SDR 21		
1	1.315	0.060	1.175	0.164
1¼	1.660	0.064	1.512	0.221
1½	1.900	0.073	1.734	0.284
2*	2.375	0.091	2.173	0.432
2½*	2.875	0.110	2.635	0.622
3*	3.500	0.135	3.210	0.915
3½	4.000	0.154	3.672	1.183
4*	4.500	0.173	4.134	1.494
5	5.563	0.214	5.109	2.288
6*	6.625	0.255	6.085	3.228
8*	8.625	0.332	7.921	5.468
10*	10.750	0.413	9.874	8.492
12*	12.750	0.490	11.710	11.956
14	14.000	0.538	12.860	14.430
16	16.000	0.615	14.696	18.810
18	18.000	0.692	16.534	23.860
20	20.000	0.769	18.370	29.470
24	24.000	0.923	22.043	42.520
		SDR 41—W.P. 100 PSI (Water @ 73.4°F.)		
18	18.000	0.439	17.070	15.370
20	20.000	0.488	18.970	18.920
24	24.000	0.585	22.748	27.320
		SDR 21 NSF—W.P. 200 PSI (Water @ 73.4°F.)		
½		See SDR 13.5		
¾	1.050	0.060	0.910	0.129
1	1.315	0.063	1.169	0.170
1¼	1.660	0.079	1.482	0.263
1½	1.900	0.090	1.700	0.339
2	2.375	0.113	2.129	0.521
2½	2.875	0.137	2.581	0.754
3	3.500	0.167	3.146	1.106
3½	4.000	0.190	3.596	1.443
4	4.500	0.214	4.046	1.825
5	5.563	0.265	5.001	2.792
6	6.625	0.316	5.955	3.964
8	8.625	0.410	7.755	6.679
		SDR 13.5 NSF—W.P. 315 PSI (Water @ 73.4°F.)		
½	0.840	0.062	0.696	0.104

Source: Harvel Plastics Inc. Product Bulletin 112/401, Effective 7.1.87, Revised 1/1/89, Harvel Plastics, Inc.

strength comparable to alloy steels at 60 percent of its weight. Corrosion resistance is superior to that of aluminum and even stainless steel.

Expansion, Flexibility, and Support

Expansion and Flexibility. Water systems piping shall have adequate flexibility to account for thermal expansion. Water systems operating under low temperature (less than 250°F) and low pressure (less than 300 psig) conditions are considered to have adequate flexibility if the following conditions are satisfied:

$$\frac{Dy}{(L - U)^2} \leq 0.03 \tag{C1.17}$$

where D = nominal pipe diameter (in)
y = thermal growth of the pipe (in)
L = length of pipe in the system (ft)
U = distance between rigid supports (ft)

If the above conditions are not met, a detailed analysis of the piping system may be required. Refer to Chap. B4 for more information.

In piping systems without adequate flexibility, thermal expansion may lead to failure of piping or anchors. It may also lead to joint leakage and excessive loads on nozzles. The thermal expansion of piping can be controlled by use of proper locations of anchors, guides, and snubbers. Where expansion cannot be controlled, flexibility shall be provided by use of bends, loops, or expansion joints (bellows or slip joints).

Detailed calculations for underground water piping systems are not needed unless significant settlement, seismic, or temperature changes are expected. Buried piping, although supported throughout by proper bedding and backfilling procedures, also requires thrust restraint for unrestrained joints at changes of direction.

Pipe Support Systems. Standard component supports are normally used to carry dead weight and thermal expansion loads. The dead weight and thermal supports typically used are defined in MSS-SP-58. Chap. B5 of this handbook discusses pipe support design and selection.

Valve Selection. In water piping systems, valves are generally used for isolating a section of a water main, draining the water line, throttling fluid flow, regulating water storage levels, controlling water hammer, bleed off of air, or preventing backflow.

Isolation. Gate valves are used to isolate specific areas of the system during repair work or to reroute water flow throughout the distribution system. An open gate valve allows water to flow through in a straight line. The valve may be closed during an emergency, such as a water main break, or during routine maintenance. Sizes range from ½ to 72 in in diameter.

Gate valves commonly used in water distribution systems are the iron-body bronze-mounted (IBBM) nonrising-stem (NRS) gate valves. Buried gate valves are usually NRS valves. In situations where an operator will need to know by observation whether a valve is open or closed, a rising-stem valve with an outside screw and yoke (OS&Y) is often used.

Service stops are valves used to shut off service lines to individual homes or businesses. Specific types of service stops include the corporation stop, which is tapped into the main, and the curb stop, which is located near the property line. Access to the curb stop is through a curb box. Small plug valves are used as curb stops and corporation stops.

Butterfly valves are also used for isolation purposes. Because the disk remains in the water path in all positions, the butterfly valve creates a higher head loss than the gate valve. The position of the disk also makes it difficult to clean scale from a pipeline because the pig or swab is blocked by the valve disk. However, butterfly valves open easily since the water pressures acting on one-half of the upstream side of the disk tends to force it open, balancing the pressure on the other half, which tends to force it closed. The cost of large butterfly valves (16 in and larger) is less than the cost of large gate valves, since large gate valves normally require reduction gears, a bypass valve, rollers, tracks, and scrapers. Butterfly valves should be located away from sources of turbulence to preclude damaging effects of turbulence to the disk.

Draining the Water Line. Drain or blow-off valves may be installed at low points to flush sediment from the main or to drain the entire main. Gate valves are commonly used as drain valves. Where rapid draining is not important, globe valves may be used to drain lines if sediment accumulation is not a problem. Although quicker to operate and less costly to repair than gate valves, 3 in and larger globe valves are less economical.

Throttling Flow and Regulating Water Storage Levels. In addition to on-off control of flow, globe valves may be used to regulate or throttle flow. This can be done manually, although it is usually done automatically. For throttling under low-flow and low-pressure conditions, butterfly valves are also used. Large plug valves may be used for throttling.

An altitude valve is a type of control valve, diaphragm, or piston type used to control the level of water in a tank supplied from a pressure system. There are two general types of altitude valves—single acting and double acting.

A single-acting altitude valve is used for filling the tank. A bypass line with check valves around the altitude valve is needed to permit back flow out of the tank and into the distribution system when the inlet pressure is lower than the tank. The tank discharges through a separate line or through a check valve in a bypass line around the altitude valve.

A double-acting altitude valve allows water to flow both to and from the tank. When the tank becomes full, the valve closes to prevent overflow. When the distribution pressure drops below the pressure exerted by the full tank, the valve opens to discharge water into the distribution system.

Controlling Water Hammer. In a water distribution system, opening or closing a valve too fast can cause water hammer; this phenomena is discussed in Chap. B8. Pressure relief valves are used to help control water hammer by releasing some of the energy that is created by a sudden stop in flow.

Bleed Off of Air. Air tends to collect in water lines. Under the pressure of the distribution system, air dissolves and can reappear as microscopic air bubbles, which gives water a cloudy appearance. A more common operating problem occurs when air collects in high places in the distribution system, producing air pockets. Air pockets effectively reduce the area of pipe through which water flows, causing an affect known as air binding. The result is pressure loss and increased pumping costs.

Air relief valves can be installed to eliminate these problems in pumping stations where air can enter the system and at high points where it can collect. Air

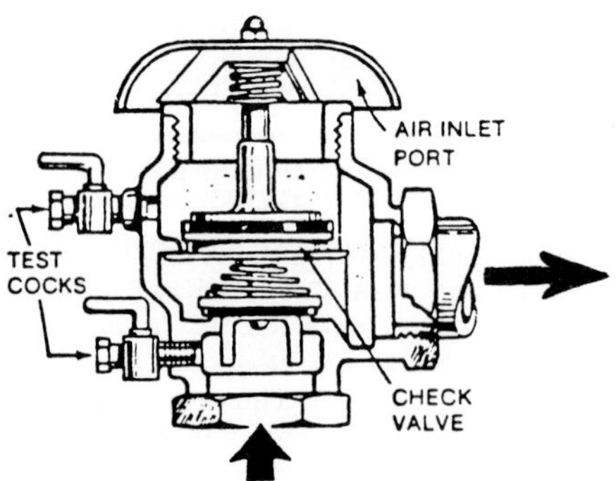

FIGURE C1.13 Pressure and vacuum breaker.

relief valves solve the problem by automatically venting any air that accumulates. Conversely, vacuum caused by column separation can be broken with vacuum relief valves. A combination of air and vacuum relief valves are also available.

Figure C1.13 shows a spring-loaded check valve that opens during forward flow and is closed by the spring when flow stops. When pressure drops to a low value, a second valve opens and allows air to enter this breaker. With this arrangement, the breaker can remain under supply pressure for long periods without sticking and can be installed upstream from the last shut-off valve.

Figure C1.14 shows an atmospheric vacuum breaker consisting of a check valve operated by water flow and a vent to the atmosphere. When flow is in the forward direction, the valve lifts and shuts off the air vent; when flow stops or reverses, the valve drops to close the water supply entry and open an air vent.

Preventing Back Flow. Back flow in a pump discharge line can be prevented by installing a check valve to allow flow in one direction only—away from the pump. However, check valves can contribute to water hammer problems, especially in the case of pump failure. Proper precautions to avoid or relieve these surges must be taken.

Back flow, or reversed flow, could result in contaminated or polluted water entering the water system. Back flow can occur through a cross connection under two conditions—back pressure and back siphonage (see Fig. C1.15). If a pressur-

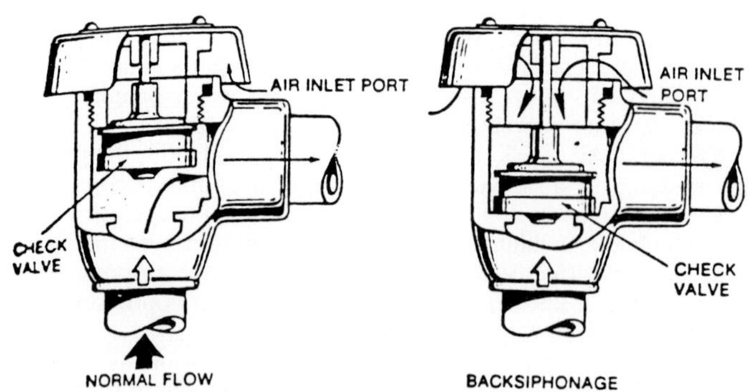

FIGURE C1.14 Atmospheric vacuum breaker.

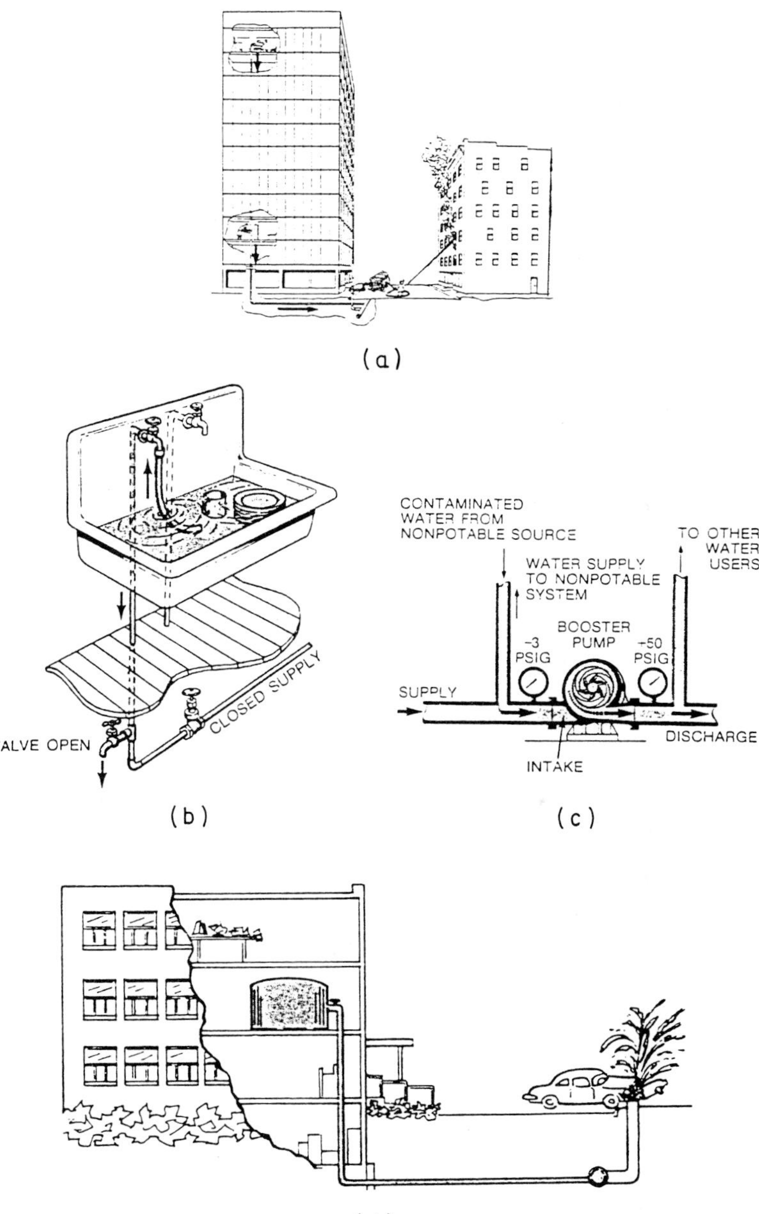

FIGURE C1.15 Examples of back flow due to back siphonage. (*a*) Back siphonage due to pressure loss; (*b*) Back siphonage—hose forms cross connection; (*c*) Back siphonage from a booster pump; (*d*) Back siphonage due to a broken main. (Introduction to Water Distribution Principles and Practices of Water Supply Operations, *American Water Works Association, 1986.*)

ized nontreated system is cross connected with a lower-pressure treated water system, then the pressure in the nontreated system can force nontreated fluid into the treated supply. This situation is referred to as back flow due to back pressure.

Valves are used as back flow prevention devices. A device that can be used in every cross-connection situation and with every degree of risk is the reduced-pressure-zone back flow. This device consists of two spring-loaded check valves with a pressure regulated relief valve located between them. Two check valves, even though well designed and constructed, are not considered sufficient protection because all valves leak from wear or obstruction. For this reason, a relief valve is positioned between the two checks. Typical back flow conditions are illustrated in Fig. C1.16.

If a treated water distribution system is cross connected to a nontreated source that is open to the atmosphere and if the pressure in the treated system falls below atmospheric pressure, then the pressure of the atmosphere can

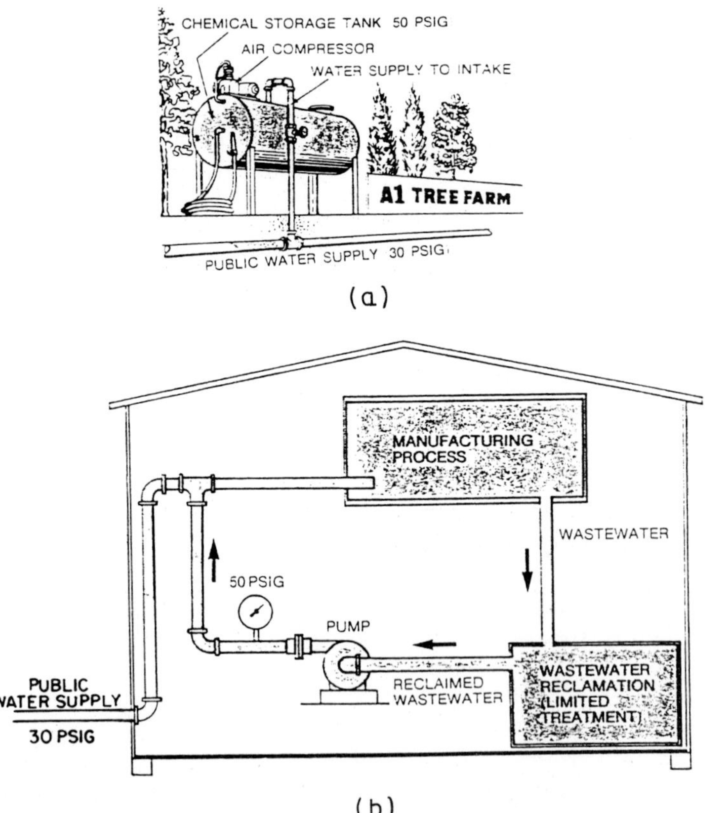

FIGURE C1.16 Examples of backflow due to back pressure. (*a*) Cross connection between pressurized nonpotable system and lower pressure potable system; (*b*) Backflow from recirculated system. (Introduction to Water Distribution Principles and Practices of Water Supply Operations, *American Water Works Association, 1986.*)

force the nontreated fluid into the treated supply. This situation is called back flow due to back siphonage. Examples of back siphonage conditions are overpumping by fire or booster pumps, undersized distribution piping, and a broken main. When back siphonage occurs, the partial vacuum pulls liquid back into the supply line. If air enters the line between a cross connection and the source of the vacuum, the vacuum will be broken and back siphonage will be prevented.

Water Hammer and Surge Control. The problem of water hammer in water piping systems consists of containing the pressure and dissipating the water flow energy. For example, the energy necessary to move the water through the piping is supplied by the pump. If a valve is suddenly closed at the end of the discharge line, the moving column of water is brought to a stop at the valve. The kinetic energy contained in the column of water, originally given to the water by the pump, is still present and must be dissipated. The column of water compresses, the pressure rises, and some of the kinetic energy is transformed to internal energy. The higher water pressure acts upon the pipe wall and does work in stretching it, but only a small percentage of energy will be lost in this. The pipe will obey the laws of vibration and return most of the energy to the water.

The water-hammer effects are obtained from equations that define relations between head and flow in the discharge line during the transient flow condition which results from water-hammer wave action.

Water-column separation might occur at high points near the hydraulic gradient on long discharge lines. This condition can create high-pressure conditions at the moment of the rejoining of the separated water columns.

Table C1.6 gives the water-hammer wave velocity as a function of diameter-to-thickness ratios for three different piping materials encountered frequently in water supply or distribution systems. In this tabulation, a is the wave velocity in feet per second, D/t is the dimensionless ratio of diameter to thickness, and E is the modulus of elasticity.

If a valve is closed in the time of one wave cycle (in the time a pressure wave travels to the other end of the pipeline and returns to the closing valve) or less, the water hammer should be calculated on basis of instant valve closure.

TABLE C1.6 Waterhammer Velocity in Piping Systems
(a, wave velocity in feet per second)

D/t	Steel ($E = 28 \times 10^6$ psi)	Cast iron ($E = 16 \times 10^6$ psi)	Transite ($E = 3.4 \times 10^4$ psi)
20	4,300	4,100	3,000
40	4,000	3,600	2,300
60	3,800	3,350	2,000
80	3,600	3,100	1,750
100	3,400	2,900	1,600
150	3,100	2,500	1,300
200	2,800	2,250	1,150
250	2,600	2,050	
300	2,400		

To determine time for wave cycle and water-hammer head for instantaneous valve closing, use the following equations:

$$T = \frac{2L}{a} \tag{C1.18}$$

$$h = \frac{aV}{g} \tag{C1.19}$$

where T = time for one wave cycle (s)
L = pipeline length (ft)
h = water-hammer head above static head (ft)
a = velocity of pressure wave (ft/s)
V = water velocity at instant before valve closure (ft/s)
g = 32.2 ft/s^2

To determine water hammer for slower valve closing, use

$$h_2 - h_1 = \frac{a(V_1 - V_2)}{g} \tag{C1.20}$$

where h_2 = pressure after partial closing of valve (ft)
h_1 = pressure before start of valve closing (ft)
$h_2 - h_1$ = pressure rise due to water hammer (ft)
V_1 = water velocity before start of valve closing (ft/s)
V_2 = water velocity after partial closing of valve (ft/s)

The phenomenon and damaging effects of water hammer is discussed in Chap. B8 of this handbook.

A means of eliminating water hammer is to permit the liquid to surge into a tank or be discharged to atmosphere. To suppress all the momentum in a long pipe system quickly would require high-pressure piping which is very costly. With a surge tank or relief valve as near the valve as feasible, the development of high pressure in this region is prevented.

Surge tanks may be classified as simple, orifice, and differential. The simple surge tank has an unrestricted opening into it and must be of adequate size so it will not overflow and will not be emptied to permit air to enter the pipeline. An orifice surge tank has a restricted opening between pipeline and tank and allows more rapid pressure changes in the pipeline. A more rapid pressure change causes a more rapid adjustment of flow. A differential surge tank is a combination of an orifice surge tank and a simple surge tank of small cross-sectional area.

Surge tanks under air pressure are used in certain circumstances such as after a reciprocating pump. They are generally uneconomical for large pipelines. Relief valves are available in various types from spring loaded to control blowdown diaphragm types.

Air Binding. Air which accumulates in water piping will reduce the effective cross-sectional area for water flow and thus increase pumping cost through the resulting extra head loss. Air enters the piping system from several sources, such as release of air from the water, air carried in through vortices into the pump suction, air leaking in through joints that may be under negative pressure, and the air originally present in the piping system before filling.

The water from the water source may be nearly completely saturated with air. If the temperature of this water is raised and the pressure is lowered by the siphon action, the water will release most of its air. However, this air release is not instantaneous but proceeds on a time-rate release and is therefore dependent upon the length of time the water remains in the piping. Experience indicates that the actual air release in a circulating water system of a conventional power-generating plant is probably in the order of 10 percent of calculated theoretical release.

On gentle downward slopes, a continuous air pocket may form along the top of the pipe for the entire slope. In a sharper downward slope, several air pockets may form, each air pocket terminating in a hydraulic jump. Slopes may require a water velocity in excess of 10 ft/s to assure that the piping remains free of air. In a 90° drop, an air pocket may form in the upper portion of the bend and a velocity in excess of 7 ft/s may be required for air elimination. Connections for air vents should be provided at all high points in the piping system and along sloping piping of considerable length. Air release valves at the high points of these systems' mains eliminate air pockets.

Corrosion and Erosion Effects

Corrosion. All raw water coming from wells, rivers, lakes, and ocean is an extremely dilute water solution of mineral salts and gases. The salts are mineral matter dissolved by water flowing over and through the earth layers. The salts are mainly sulfates, bicarbonates, chlorides of calcium, sodium, and magnesium. These minerals give water its hardness (destroying soap and preventing lather) and precipitate as a white limetype scale. The dissolved gases are atmospheric oxygen and carbon dioxide, picked up by water-atmosphere contact, such as spray, raindrops, and ammonia from decaying vegetable matter.

The dissolved gases are the prime agents of chemical corrosion which act on the metals of piping systems. The oxygen attacks the iron or steel, and the process is accelerated by the carbon dioxide. The rate and extent of the chemical corrosion are influenced by the amounts of mineral salts dissolved in the water.

The calcium content of water is used to measure the tendency of water to corrode or form scale. For this purpose values are assigned to the calcium content and the alkalinity of the water. Adjustments are made for temperature and the effect of total dissolved solids. The resultant value is compared with the observed pH of the water to determine whether it is corrosive or scaling. This is basically the Langelier saturation index as reported by Larson-Buswell in the *Journal of the American Water Works Association,* vol. 28, p. 1500, 1931.[4]

The water is corrosive when the Langelier index (calcium carbonate saturation index) is minus (−). The water analysis will generally, but not always, show a pH value below 7 (acidic). The scaling in water lines occurs when the Langelier index is plus (+). The water analysis will generally show a pH value above 7 (basic).

The precipitation of calcium carbonate as a scale or film thickness may be desirable as a means of protection against corrosion if the rate of buildup is sufficiently low. Calcium carbonate is undesirable on heat-transfer surfaces. Since temperature lowers the solubility of calcium carbonate and calcium sulfate, the Langelier index will vary for cold water and for warm water.

The exterior of unprotected buried metallic pipe is subject to similar chemical action due to exposure to water. In addition, the pipe exterior is susceptible to attack by aerobic and anaerobic bacteria, galvanic action, and stray electric currents. The chemical action on the pipe exterior may be more intense because of concentration of oxygen, salts, and other chemicals leached out of the surrounding earth by ground water.

Some forms of anaerobic bacteria that thrive only in the absence of free oxygen obtain their oxygen by the chemical breakdown of oxygen compounds in the earth with the resultant production of substances, such as hydrogen sulfide, that will corrode the base metallic buried pipeline. There are also many types of aerobic bacteria that produce sulfuric acid, sulfate, and ferric hydroxide—compounds that are all corrosive to steel or iron. Organic soil should be kept away from the vicinity of the pipeline to minimize possibility of this corrosive action.

Also, when iron or steel is in contact with a more cathodic material, for example copper or brass, a galvanic cell is formed, electrolysis results, and the corrosion rate of steel or iron increases. If iron or steel is in contact with a more anodic material, for example zinc, the zinc will be the affected material and the corrosion rate of the steel or iron will decrease.

There is some natural resistance to chemical corrosion of the base metallic pipe materials. The chemical-corrosion product, an oxide film, may build up sufficiently to slow down or prevent further corrosion.

The natural coating characteristics of the most commonly used piping materials are mentioned briefly. On cast iron, the rust (iron oxide) builds up into a strong adhesive coating that finally forms a barrier sufficient to stop or slow down further corrosion. The higher silicon cast iron has the best characteristics in this respect.

On steel, the rust powders and flakes off easily and does not build up into an adhesive, sufficiently protective coating. For concrete and cement piping, the corrosion is of a different form. These materials are subject to leaching of the free lime from the cement, deterioration in alkali soils, and attack by organic growth.

Cathodic Protection. If no protective coating is used, or if a low-cost, limited-life coating has been selected, cathodic protection may be considered as a means of limiting the main agent of corrosion, which is the electrochemical process. In this process, the moist earth is the electrolyte, two dissimilar materials are the anode and the cathode, and the pipe wall between them completes the electric circuit. This process may be set in motion in a number of different ways, among which are dissimilar metals, galvanic action of a single metal due to dissimilar soils, variation in moisture and chemical content of soil, nonuniformity of metal caused by mill scale, surface scarring, welding, and even temperature differentials.

The current flows from the anode to the cathode and causes corrosion at a rate greater than that which would occur by normal chemical means. Corrosion rate increases at the anode end and decreases at the cathode end. The anode is the point or area at which the current leaves the metal, and the cathode is the point at which the current enters the metal.

The electrochemical galvanic series (Table C1.7) gives the relation between metals. The metal listed nearer the top of the table is the anode that will waste away. The metal nearer the bottom of the table is the cathode and will be protected. The farther apart from each other the metals are located in the table, the greater the potential difference will be between them and the greater the corrosion rate of the anode end.

A typical example of the galvanic action of dissimilar metals is represented by a steam condenser having a steel shell, steel tube sheets, and copper-alloy tubes. The steel is nearer the anode end than is the copper alloy, and as a consequence the corrosion of the steel tube sheets and shell is accelerated. Always, that metal which is higher in the galvanic series will waste away.

TABLE C1.7 Galvanic Series

Anode end (least noble, the wasting end)
Magnesium
Magnesium alloys
Zinc
Aluminum
Aluminum alloys
Cadmium
Carbon steel
Cast iron
Stainless steel (active)
Soft solder
Tin
Lead
Nickel (active)
Brasses
Copper
Bronzes
Nickel-copper alloys
Nickel (passive)
Stainless steel (passive)
Titanium
Silver solder
Silver
Graphite
Gold
Platinum
Cathode end (most noble, the protected end)

Cathodic protection is a means of diverting the electrochemical corrosion from the pipeline to wasting anodes. There are two methods of providing cathodic protection. The least costly installation is the galvanic method based on natural battery action between position of metals in the electrochemical table. An anode or wasting piece is deliberately used. This approach requires very careful analysis of all the varying conditions involved.

The second and more costly cathodic protection is the impressed current method that requires an external source of electricity. The impressed current renders the piping cathodic to the surrounding soil by a controlled difference of potential.

In locations where there may be stray currents, the installation of removal wires at designated points so that the current may leave the pipeline should be considered. In other words, stray currents are used to provide cathodic protection for the pipeline.

Protective Coatings. Since corrosion of metal is a surface reaction, if a protective coating which is continuous, impervious, chemically inert, and electrically insulating can be bonded to the interior and/or exterior of the piping, corrosion cannot take place on the pipe surface as long as the protective coating remains in place undamaged and without cracks or pinholes.

The basis of selection for the best coating differs for the interior and exterior of the pipe. To perform its function properly, the coating on the interior of the pipe would be selected for its chemical inertness, imperviousness, adhesiveness, adjustment to pipe deformation, and resistance to erosion caused by the flowing water. The coating on the exterior of the pipe would be selected for its chemical inertness, electrical resistance, imperviousness, adhesiveness, adjustment to pipe deformation, and resistance to shear and compression due to varying earth conditions.

Galvanizing. The zinc used for galvanizing pipe is on the anodic (wasting) or electrochemical protective side of the steel and is wasted or changed to zinc compounds before the steel pipe will be attacked.

Coal-Tar Enamel. Specification AWWA C203 covers the coal-tar enamel protective coatings for steel water pipe. This standard delineates the specifications for the materials involved, method of application to the inside and/or outside of the piping, the thickness required, protection of the coatings, testing, and so on. The type of enamel is specified as AWWA coal-tar enamel and is described in this standard with full characteristics and the ASTM tests required.

Erosion. Erosion effects in water piping systems is fundamentally an accelerated form of corrosion and as such is distinguishable from mechanical processes such as erosion abrasion, and cavitation. Single-phase and two-phase erosion-corrosion differ in that a second damage process (droplet impingement) is available under two-phase conditions, and this can accelerate the overall rate of attack. In the most severe cases of two-phase erosion-corrosion, rapid rates of metal loss (approximately 40 mils/year) have been reported.

Erosion is observed only when specific combinations of material, water chemistry, and hydrodynamic conditions coexist. Most problems have been associated with plain carbon or very highly alloyed steels having a low content of dissolved oxygen and a pH less than about 9.3. Detrimental erosion occurs at temperatures within the range of 212–525°F (but most rapidly between 260 and 400°F) and is restricted to locations where the mass transfer coefficient is high. This can be either as a result of a high fluid flow rate or because of the presence of a geometric flow discontinuity such as an impingement orifice, bend, or tee.

Erosion under both single-phase and two-phase conditions can lead to a characteristic scalloped surface appearance, and in carbon steels the pearlite is preferentially attacked. Corrosion films are typically very thin and the surface sometimes appears polished, in marked contrast to adjacent regions which generally are more heavily oxidized.

Several laboratory investigations of erosion have been conducted (almost exclusively in single-phase water only flow), and major variables affecting the process have been identified and their effects documented. Quantitative mechanistic understanding is not yet complete, but empirical predictive models are available. Laboratory studies have shown the feasibility of a variety of mitigating actions and remedies, and several of these have been applied successfully. The physical picture of the erosion process that has emerged from this work is illustrated in Fig. C1.17. As can be seen, erosion can be viewed as a flow-accelerated corrosion process characterized by the presence of a poorly adherent magnetite film. The rate of metal removal depends on interactions between the rates of a number of subprocesses.

The laboratory work shows that three main groups of variables affect the rate of metal loss by erosion under single-phase conditions:

- Material variables (chiefly the chemical composition)
- Water chemistry variables (temperature, pH, oxygen concentration, impurity content)
- Hydrodynamic variables (flow rate, geometry)

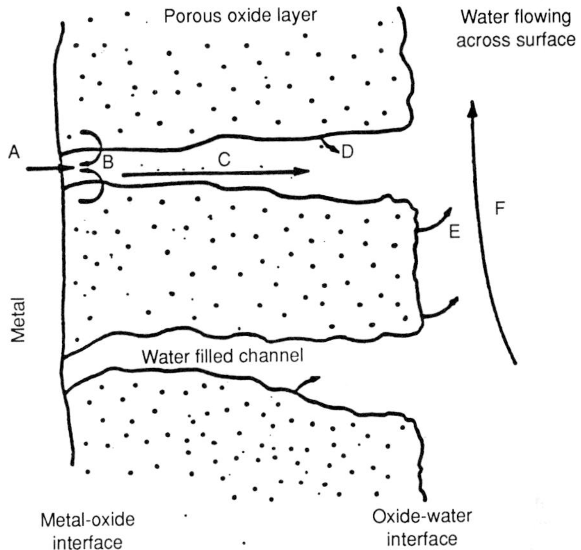

FIGURE C1.17 Phenomena occurring during erosion-corrosion.

Some largely empirical models have been developed[5] and are capable of metal loss predictions within the range of the data on which they are based.

Erosion occurs most readily in plain carbon steel. Austenitic stainless steels essentially are immune to erosion-corrosion. In ferritic steels, alloying elements such as molybdenum, copper, and particularly chromium (even when present at levels of 1 percent or less) can greatly improve the erosion-corrosion resistance.

The effect of steel composition depends on the severity of hydrodynamic conditions in that as conditions become more demanding, higher alloy contents are required to confer the same resistance to erosion-corrosion. However, in the relatively mild situation typical of feedwater piping, significant effects of small changes in material composition would be anticipated based on the laboratory data.

The effects of water temperature, pH, and dissolved oxygen content on erosion rate have been studied by a number of investigators. For the orifice configuration tested, the temperature dependence at pH 9.05 results in a flow-rate-dependent erosion rate peak at about 284°F, suggesting that the erosion-corrosion rate is controlled by oxide dissolution kinetics at low temperatures and by mass transfer limitations at higher temperatures. A marked decrease in erosion rate accompanies increases in pH.

Dissolved oxygen also has a marked effect in neutral water. Iron release rates from carbon steel in 100–400°F water at a flow rate of 6 ft/s have been shown to decrease by up to two orders of magnitude as oxygen concentration was increased over the range 1–200 ppb. Tests indicated that it is important to keep oxygen levels above 15–20 ppb, and oxygen dosing is a fairly common practice.

The effects on erosion of ionic impurities such as chlorides and sulfates have not been widely studied. In particular, there is little information which bears upon the question of whether a long prior period of adverse chemistry could have any irreversible effects.

Plant experience indicates that geometry and flow rate are important factors in erosion. The physical picture of the process shown in Fig. C1.17 suggests that the importance of geometry and flow rate rises through their influence on the rate of mass transfer of oxide dissolution products away from the oxide and water interface. Laboratory studies have confirmed that the mass transfer coefficient is the controlling parameter. However, the exponent of the power function relating erosion-corrosion rate and mass transfer coefficient is dependent on temperature. Since the mass transfer coefficients for simple flow geometries can be calculated, the existence of these empirical relationships allows the erosion rate to be estimated for a variety of situations. However, local mass transfer coefficients are sensitive to local geometrical discontinuities and at present can only be derived empirically.

Component redesign or flow path geometry improvements, aimed at reducing the mass transfer coefficient, can sometimes be used to remedy erosion-corrosion problems. Referring to Fig. C1.17, the phenomena occurring during erosion is as follows:

A. Iron hydroxides are generated:

$$Fe + 2H_2O \rightarrow Fe(OH)_2 + H_2$$

B. Magnetite is formed according to the Schikorr reaction:

$$3Fe(OH_2) \rightarrow Fe_3O_4 + H_2 + 2H_2O$$

C. A fraction of the hydroxides formed in step 2 and hydrogen generated in steps 1 and 2 diffuse along pores in the oxide.

D. Magnetite can dissolve in the pore.

E. Magnetite dissolves at the oxide and water interface.

F. Water flow removes the dissolved species by a convection mass transfer mechanism.

The following presents a sample calculation of the rate of erosion-corrosion under single phase flow conditions. The methodology used is based upon the work performed by Coney.[6,7] Physical and flow conditions for this example are:

pipe outside diameter	= 18.00 in
pipe inside diameter	= 17.00 in
wall thickness (nom.)	= 0.50 in
pH	= 9
pressure	= 367 psig
temperature	= 380°F
flow rate	= 5,000,000 lbm/h

For these conditions the Reynolds number, Re, is

$$\frac{VD_H\rho}{\mu} \qquad \text{(C1.21)}$$

where V = velocity (ft/s)
D_H = inside diameter (ft)
μ = dynamic viscosity (lbm/s ft)
ρ = density (lbm/ft^3)

Since the flow rate is known, the velocity can be calculated as follows:

$$W = \rho A V \tag{C1.22}$$

or

$$V = \frac{W}{A\rho} \tag{C1.23}$$

where W = flow rate (lbm/h)
A = flow area (ft^2)

and

$$Re = \frac{[55(\text{lbm/ft}^3)]\ [17(\text{ft/s})]\ [17/(12)\ \text{ft}]}{9.66 \times 10^{-5}\ (\text{lbm/s} \cdot \text{ft})} = 1.37 \times 10^7$$

$$V = \frac{(5 \times 10^6)\ (\text{lbm/h}) \times (1\ \text{h/3600s})}{(\pi/4)\ (17/12)^2\ \text{ft}^2 \times 55\ (\text{lbm/ft}^3)} = 17\ \text{ft}$$

The mass transfer rate is now calculated.

In order to calculate the mass transfer coefficient, an empirical correlation must be used. The Berger and Hau correlation is used to predict this coefficient. The recommended range of this correlation is $10^4 < Re < 10^6$. This correlation will be used for two reasons:

1. The turbulent flow, mass transfer correlations are based, by analogy, on heat transfer correlations. And heat transfer correlations of the same form as Berger and Hau typically are valid up to $Re = 10^7$.
2. In view of the limited mass transfer data at very high Reynolds number, this was judged to be the most suitable correlation for this application.

The Berger and Hau[8] correlation is

$$Sh = 0.0165\ Re^{0.86} Sc^{0.33} \tag{C1.24}$$

where the Sherwood number and the Schmidt number are dimensionless numbers defined as

$$Sh = \frac{KD_H}{d} \tag{C1.25}$$

$$Sc = \frac{\mu}{\rho d} \tag{C1.26}$$

where K = mass transfer coefficient (ft/s)
d = diffusivity (ft^2/s)

The diffusivity of dissolved species in water, d, is equal to about 135×10^{-9} ft^2/s for the temperature of interest. Thus,

$$Sc = \mu/\rho d = \frac{9.66 \times 10^{-5}\ (\text{lbm/s ft})}{55\ (\text{lbm/ft}^3) \times 135 \times 10^{-9}\ (\text{ft}^2/\text{s})} = 13$$

Using the Berger and Hau correlation and rewriting to solve for K,

$$K = \left(\frac{d}{D_H}\right) (0.0165)\,(\mathrm{Re})^{0.86}\,(\mathrm{Sc})^{0.33} \tag{C1.27}$$

Since everything is now known, K is calculated as follows:

$$K = \frac{[135 \times 10^{-9}\ (\mathrm{ft}^2/\mathrm{s})]}{(17/12)\ \mathrm{ft}} (0.0165)\,(1.37 \times 10^7)^{0.86}(13)^{0.33}$$

$$K = 0.005\ \mathrm{ft/s}\ \text{or}\ 1.53\ \mathrm{mm/s}$$

This mass transfer coefficient is for a straight pipe. For a rupture occurring near a change in flow direction, the straight pipe value should be increased. Figure C1.18 presents a wide range of data showing the increase of heat and mass transfer with bends of various geometries. At the value corresponding to the elbow (0.5), the data range from about 0.4 to almost 2.0 with most of the data less than 0.7. To bound the problem, three calculations have been made—a low case at a value of 0.4, a mid case at a value of 0.55, and a high case at a value of 0.7. For a straight pipe mass transfer coefficient of 1.53 mm/s, the low, mid, and high mass transfer coefficients in the elbow are 2.14, 2.35, and 2.60 mm/s, respectively.

There is a large amount of experimental data correlating mass transfer coefficient and erosion rate. The data are presented in Fig. C1.19. Using this figure, at the above values and a pH of 9.0, the predicted erosion-corrosion rates are about 0.4, 0.5, 0.8 mm/year, respectively.

How long would it take at the *n d* erosion rate to erode the pipe to one-half its original thickness?

$$\text{Time} = \frac{\text{thickness}}{\text{erosion rate}} \tag{C1.28}$$

$$\text{Time} = \frac{0.25\ \text{in} \times 25.4\ (\text{mm/in})}{0.5\ \text{mm/year}} = 12.7\ \text{years}$$

At the low rate, the time to thin the wall to half the original thickness is about 15.9 years. At the high erosion rate it is about 7.9 years.

Start Up, Operation, and Maintenance Considerations

Start Up and Operation. Before start up of water systems, the piping system should be cleaned as required. In some cases, chemical solvents are used followed by flushing. Care should be taken that components not compatible with the chemical being used are removed during the cleaning and flushing process. Temporary strainers are sometimes included in the system for preoperational purposes. This practice protects pumps and valve seating surfaces from large particles left in the pipe from the installation and testing periods. Temporary strainers are replaced by spool pieces when the system is ready for start up. The strainers should be kept on hand and installed in the system during major work or repair to the system and replaced by the spool pieces after the system is certified to be clean.

Before starting any system the following precautions should be taken and included in plant procedures:

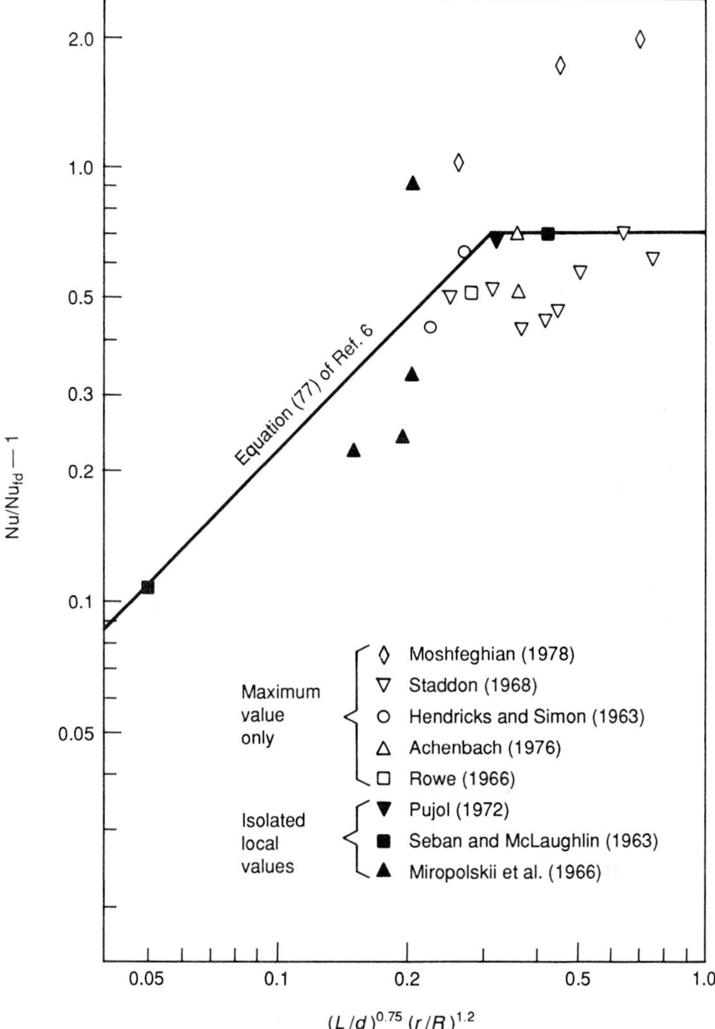

FIGURE C1.18 Comparison of Eq. (77) of Ref. 6 with the data of various authors on heat and mass transfer in bends.

1. Verify that the system is free of debris and has been cleaned and flushed thoroughly and that all testing and inspections are complete.
2. Verify that all components function as required. Rotate pumps by hand to ensure that impellers rotate freely, exercise valves several times, and, if necessary, adjust packing.
3. Verify that all instrumentation is in place and root valves are open.
4. Verify that the system piping and components are properly vented and drained; for high-pressure or high-temperature systems, ensure that vents and

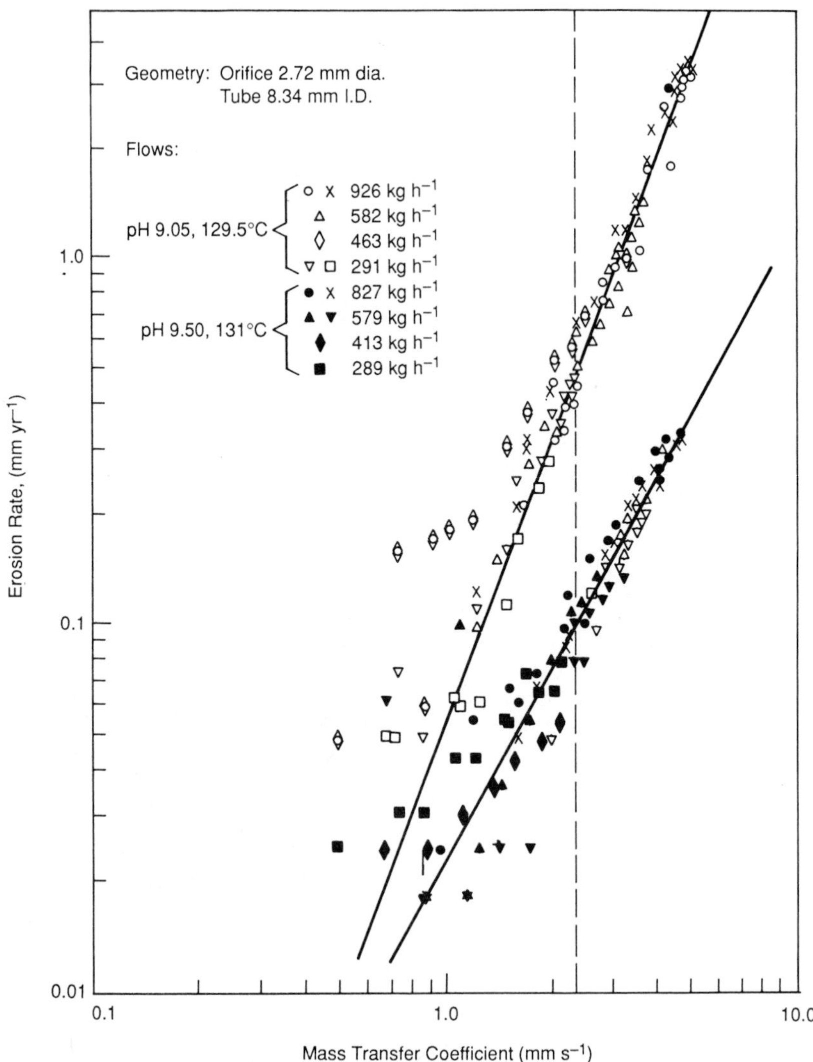

FIGURE C1.19 Mass transfer coefficient dependencies at different pHs. Correlations from loss profiles of specimens at four flow velocities.

drains are routed away from personnel or sensitive equipment. Close the valves as evidence of complete venting and drainage occur.

5. Verify valve line up intended for start up, especially the required positions of pump discharge valves and minimum flow recirculation valves.
6. Monitor the system's initial operation for anomalies, such as surges, spikes, leaks, and control failures, and make necessary adjustments or shut down the system and correct any problems.

7. Monitor system pressures, flow rates, and component response as a regularly planned routine. Maintain logs to develop a baseline condition for comparison for all modes of operation.
8. System shut down should generally be performed in reverse of the start-up procedure. It is important, however, that specific instructions for the shut down and securing of a system be included in operational procedures. In most cases drains should be left opened.
9. High-energy systems require extra precautionary measures such as tagging or locking valve positions and active overrides to controls and interlocks.

Flow Balancing. The system through which water flows offers resistance to flow (as discussed), such as friction, static head, and back pressure. Flow rates in a system vary depending on several factors such as

1. Variations in static head
2. Valve alignment
3. System demands or loads
4. Flow control valve operation

All these factors affect system resistance.

To design and operate a system, it would be ideal to assume steady-state operation. If steady-state operation is assumed and the system contains branch lines of flow, it is important for economy and system stability to proportion flow in accordance with the specific demands of each branch or service. This is achieved by flow balancing. On a theoretical basis, this is done by plotting the system resistance curves for each branch for a given system flow. Since the total flow must equal the sum of all branch flows and the total pressure drop across each branch between junction points must be identical, the flow divides to produce these identical losses. The sum of the flow rates through each branch at a particular head provides the total flow rate for the system at that head. The total system flow required can be determined from load data, and resistances can be added to the branch lines by adjusting throttle valve positions. In some cases orifices may be warranted.

This practice in the field can be tedious but may be made easier with the proper flow and pressure measuring instrumentation and favorable throttle valve characteristics.

Maintenance. A major factor in developing a maintenance program is the accessibility of piping and components. Although accessibility should be designed into a system, it is not always practical, such as in intake or municipal water systems where most of the piping is buried. In such cases, it is cost-effective to maintain piping upon failure. Generally, the buried portion of a system is limited to passive components. Piping and normally opened or closed valves not usually requiring normal maintenance fall into this category. Other active components such as pressure regulators, relief valves, pumps, and metering devices should be located in buildings, modules, or manholes and be readily accessible. For this latter category of components, a rigorous maintenance schedule should be developed.

Depending on the degree of importance for safety and operation of a system, an inspection program should be instituted. Inspection during operation is mostly visual; however, some systems may be fitted with taps or probes for on-line inspection.

Beyond inspection, there are components which must be maintained on a regular basis regardless of need. The component's manufacturer provides instruction manuals which should recommend whatever maintenance and replacement of parts it deems prudent. The station operator must also use his or her own judgment and operating experience in the development and implementation of maintenance procedures. Generally speaking, bolt tensions, gasket condition, and packing leakage can be monitored daily. Pressure and temperature monitoring can show evidence of component deterioration or malfunction. Problems such as clogged strainers, hung up valves, and leaks or breaks can be determined by reading instrumentation when the problem is not normally visible. Instrument calibration for both system components and testing instruments should be more tightly controlled although frequency varies. Inspection of coatings and painting on a scheduled basis will help eliminate corrosion problems to exposed components.

Accordingly, a sufficient supply of spare parts should be on hand. The system can be surveyed to determine which parts should be ordered. Operating experience can also help predict expected life of wearing parts.

Water sampling should be performed routinely. Depending on the service, it can determine the safety of the supply if it is for human consumption or prevent corrosive or abrasive attack on internal components of the system. Sampling procedures should be developed to locate sampling points and to facilitate storage or transportation to the testing laboratory. Procedures should also contain acceptance criteria for each system application and associated water treatment solutions for unacceptable water quality.

Occasionally water systems require flushing or mechanical cleaning. These operations remove bacteriological growth, silt deposits, and scale buildup on pipe walls. Again, strict procedures must be developed and adhered to in order to control work.

Installation

Designer's piping arrangement drawings and isometrics, if provided, are the governing documents for installation of piping. Unless existing site conditions warrant otherwise, piping shall be installed in accordance with the piping arrangement drawings.

Buried Piping. The trench for installation of buried piping shall be of sufficient width and depth to provide necessary bedding and cover, depending on traffic volume and depth of frost line and to facilitate joining, tapping, and future maintenance considerations. Pipe should be laid flat on the bedding and not be supported on the end by the bell. The trench should be filled in accordance with the specified requirements for fill material, rock size, and compaction. When specified, an insulating type of initial fill may be used.

For support of pipe, there are two generally accepted procedures. One is to support the piping temporarily during fit up, then to install the supports. The second procedure is to install the pipe support, then the piping. In either case, location tolerances should be provided and adhered to during design so as not to invalidate stress analysis calculations. Cold springing of pipe should also be avoided but, if required, kept within tolerances.

REFERENCES

1. *AWWA Principles and Practices of Water Supply Operations,* Volume 3, Introduction to Water Distribution, 1986, American Water Works Assocation, Denver, CO.

2. *Hydraulic Institute Engineering Data Book,* First Edition, Hydraulic Institute, Cleveland, OH.

3. Havel Plastics, Inc., Products Bulletin 112/401, Effective 7/1/87, Revised 1/1/89, Havel Plastics, Inc.

4. *Journal of the American Water Works Association,* 28, 1931 (Larson-Buswell), American Water Works Association, Denver, CO.

5. *EPRI Final Report—Single Phase Erosion-Corrosion of Carbon Steel Piping,* February 19, 1987, Nuclear Power Division, Electric Power Research Institute, Palo Alto, CA.

6. Coney, M. W. E., *Erosion-Corrosion: The Calculation of Mass-transfer Coefficients,* Central Electricity Generating Board, RD/L/N 197/80, Job No. VE 280, May 1981.

7. Coney, M. W. E., Wilkin, S. J., and Oates, H. S., *Thermal-Hydraulic Effects on Mass-Transfer Behaviour and on Erosion-Corrosion Metal Loss Rates,* Central Electricity Generating Board, TPRD/L/2349/N82, January 1983.

8. Berger, F. P. and Hau, K.-F. F-L., Mass Transfer in Turbulent Pipe Flow Measured by the Electrochemical Method. *Int. J. Heat Mass-Transfer,* 20:1185–1194, 1977.

CHAPTER C2

FIRE PROTECTION SYSTEMS PIPING

Daniel L. Arnold
Engineering Manager
Rolf Jensen & Associates, Inc.
Atlanta, Georgia

INTRODUCTION

Fire protection systems are unique since the majority of their service life may be spent in a static, no-flow condition. However, when required to operate in an emergency, fire protection systems can be critically important to property preservation, the continued operation of a facility, and the safety of building occupants. For this reason, the piping associated with fire protection systems must be designed to minimize service interruptions and be capable of operating reliably over an extended time.

To assure that fire protection systems are readily available, associated piping systems must be constructed in accordance with nationally recognized standards and applicable codes. Applicable codes may include international standards, federal regulations, and locally adopted requirements.

Fire protection piping, as treated in this chapter, deals with the network of distribution piping that delivers fire extinguishing agents. The chapter presents general information on piping materials, available references, and design considerations.

Types of Fire Suppression Systems

There are numerous types of fire suppression systems. Each uses piping or tubing to convey fire suppression agents throughout a protected area or to a specific fire hazard. The type of fire suppression system selected for a particular building or location depends on the character of the fire hazard, the value of the building and contents, the code and regulatory requirements, and the physical considerations such as environment and aesthetics. The common fire suppression systems are described below.

Automatic Sprinkler System. An automatic sprinkler system for fire protection purposes is a network of piping to which automatic sprinklers or spray nozzles are attached. The system is connected to an automatic water supply. The piping network and connected sprinklers or nozzles are distributed throughout the protected area in accordance with fire protection engineering standards.

Automatic sprinklers are sealed by a heat-responsive element such as a eutectic metal or frangible glass bulb. When the element of an automatic sprinkler reaches its predetermined operating temperature during a fire, the sprinkler opens, discharging water to control the fire. System operation is stopped manually by closing the system control valve. Control valves must remain open until the fire is completely extinguished and should be constantly attended while overhauling the fire area to guard against rekindling.

The wet-pipe system is the most common automatic sprinkler system due to its high reliability and fewer moving parts as compared to other more specialized suppression systems. There are other types available for special conditions such as preaction, dry pipe, and deluge systems.

Automatic wet pipe sprinkler systems are provided for general fire suppression throughout an area where fixed suppression is required and where there are no special considerations that restrict their use.

Located in the main supply header of each wet pipe system is an alarm check valve or water-flow switch; see Fig. C2.1. An alarm check valve is a free-swinging, hinged clapper that opens when water flows through the system and automatically resets when the flow stops. A water-flow switch consists of a paddle in the riser which lies motionless until water flows through the system. Alarm

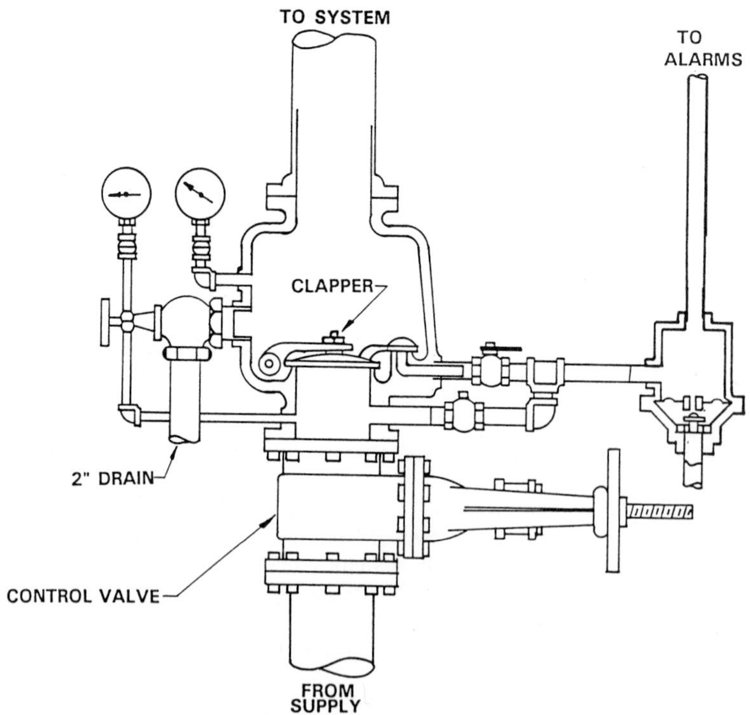

FIGURE C2.1 Typical wet pipe system alarm check valve.

check valves and water-flow switches are used to initiate sprinkler flow alarm signals.

Automatic preaction sprinkler systems are provided for fixed fire suppression where it is particularly important to prevent the accidental discharge of water into an area; see Fig. C2.2. The piping network of preaction sprinkler systems is maintained dry until water is needed for fire suppression. Sealed automatic sprinklers are installed on the piping network.

A deluge valve is provided for each preaction system. A deluge valve is a normally closed, automatic control valve located in the system riser, which prevents water from entering system piping until required. Operation of the deluge valve is achieved by an electrical, hydraulic, or pneumatic signal initiated by fire detectors located throughout the protected space. The deluge valve can also be operated manually.

Dry pipe automatic sprinkler systems are used primarily in unheated occupancies and structures. The dry pipe valve is located in the main supply header. All piping downstream of the dry pipe valve is charged with supervisory air or nitrogen to keep the dry pipe valve closed. When the sprinkler opens, the air is released, tripping the dry pipe valve and introducing water to the system piping; see Fig. C2.3.

Deluge systems are used to provide fire protection specifically for high hazard equipment or areas such as transformer areas and ammunition magazines. A deluge valve, similar to that described for the preaction system, is provided in the main supply header; see Fig. C2.4. Open spray nozzles are mounted to the piping network of deluge systems in lieu of sealed automatic sprinklers. When the deluge valve is opened by an electrical signal, water simultaneously sprays from all spray nozzles on the system.

Gaseous Fire Suppression Systems. Gaseous fire suppression systems are provided in areas where water is not the extinguishing agent of choice or where early

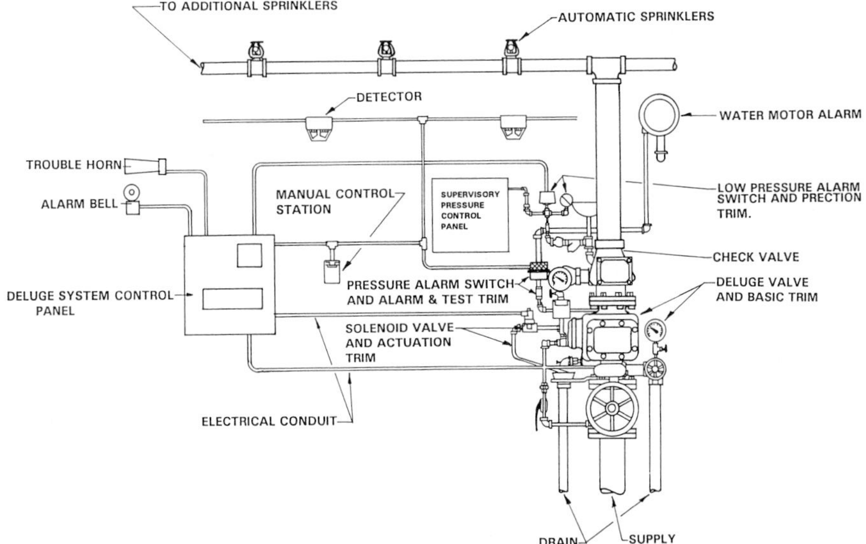

FIGURE C2.2 Typical preaction sprinkler system diagram.

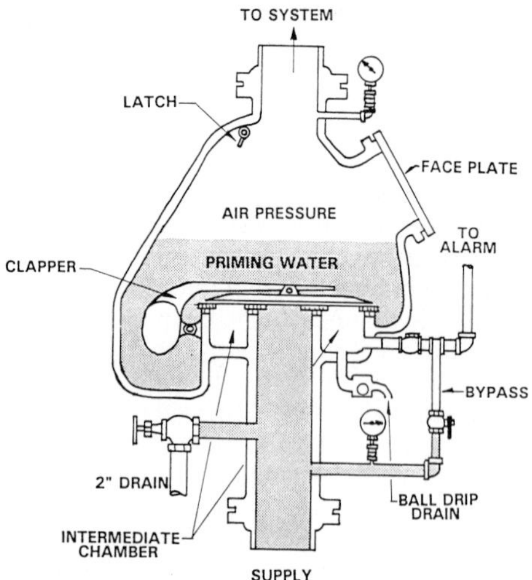

FIGURE C2.3 Schematic dry pipe valve diagram.

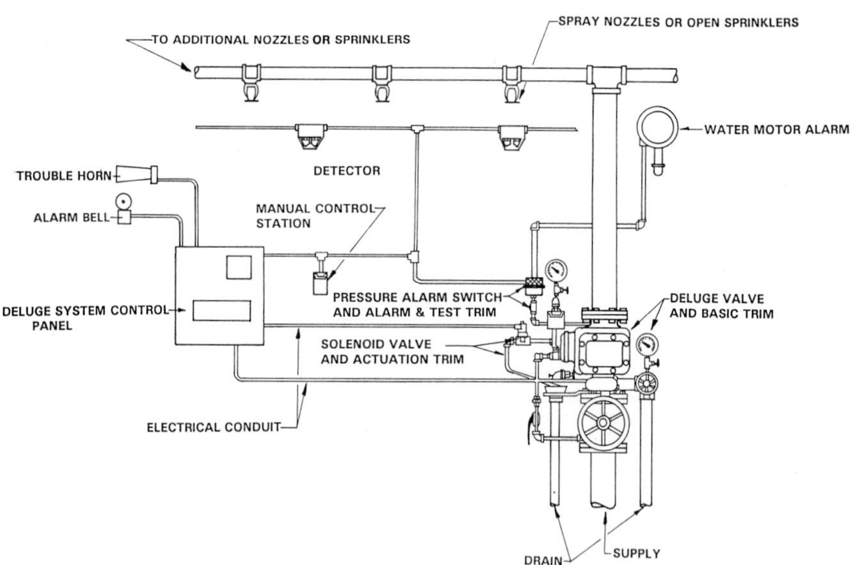

FIGURE C2.4 Typical deluge system diagram.

intervention into a fire condition is desired, such as electronic equipment rooms or water sensitive storage areas. It is often recommended that when gaseous fire suppression systems are provided in an area, they be supplemented with an automatic sprinkler system protecting the same area.

Gaseous fire suppression systems include carbon dioxide and halogenated gas (Halon) suppression systems. The extinguishing agent is generally stored in pressurized cylinders or tanks. The number of cylinders and quantity of extinguishing agent required for a particular area is dependent on the volume of the protected space, the design density required to achieve fire control, and the physical arrangement of system piping.

Automatic activation of these systems is initiated by an electrical signal from a detection system installed in the protected space. When the detection system senses the fire condition, the extinguishing agent is released into the protected space. Manual means for discharging the systems are also provided.

To be effective, the extinguishing agent must be maintained in the protected area at design concentrations for a sufficient length of time, often 10 min or more. To maintain the concentration level, the protected space should have boundary penetrations appropriately sealed to prevent leakage of the agent.

Foam Fire Suppression Systems. Foam fire suppression systems are used primarily to protect hazards or areas with flammable liquid fire hazards. The fire extinguishing foam is made from mixing foam concentrate and water within the piping network. The foam extinguishes fire by smothering the fire, suppressing flammable vapor production, and cooling the fuel in adjacent areas.

The specifications for piping material that carries foam concentrate and solution must be closely coordinated with the recommendations of the foam manufacturer. Often stainless-steel piping is recommended. Selecting an inappropriate piping material may increase maintenance needs on the system and decrease the piping's usable service life.

Standpipe Systems. A standpipe system consists of a network of piping that supplies water for manual fire suppression throughout a building or facility. The network is supplied by a fire protection water supply system, and it distributes water to normally closed hose stations or standpipe valve outlets. Standpipe systems are also often used to provide water supply to automatic fire suppression systems.

Water Supply System. Fire protection water supply systems distribute water from the source of supply to fire protection systems. The distribution system may be designed by the public water works company. In other instances, such as a large industrial plant, the water supply system may be privately owned and operated. Regardless of ownership, the design of the water supply to fire protection systems is a critical factor in assuring reliable fire protection.

A fire protection water supply system may consist of storage tanks, pumps, underground and aboveground pipe, and associated control valves.

The capacity and pressure required of a particular water supply system is generally related to the design demands of the water-based fire suppression system in conjunction with the normal water consumption for a facility. Additionally, local authorities and insurance organizations often prescribe fire flow requirements for individual properties.

In determining the required water supply, the following should be considered:

- Hazard classification of occupancy
- Building construction type
- Adjacent exposures or buildings
- Sprinkler protection
- Manual fire fighting
- Domestic and/or process water demands

REFERENCE DOCUMENTS

Codes and Standards

The nationally recognized standards which cover fire protection systems piping are those developed by the National Fire Protection Association (NFPA). Federal regulations may reference applicable NFPA standards but often include special requirements for unique fire protection systems such as those protecting marine, military, and nuclear facilities. Local, municipal, state, or other governmental regulations may also be followed. Insurance organizations publish standards which may be applicable to a particular project.

Fire protection codes often require specific equipment or materials to be listed. The term *listed* means that the equipment or material has been evaluated by a recognized product evaluation organization and has been found to meet appropriate standards or has been tested and evaluated for use in a particular fire protection application. Recognized organizations concerned with product evaluation publish lists of evaluated products. Additionally, the listing organization performs periodic inspections of the production of listed equipment and material. It is important that all fire protection piping products be used in applications consistent with their listings. Some generally recognized standards which relate to fire protection piping systems are given in Table C2.1.

There are additional NFPA codes, standards, and recommended practices and guides which may contain specific design criteria for particular hazards or facilities. A complete listing of available documents developed by NFPA technical committees is available by contacting the National Fire Protection Association, Batterymarch Park, Quincy, MA 02269.

Other Reference Documents

Other references related to the design of fire protection piping systems include the following:

1. NFPA Fire Protection Handbook, National Fire Protection Association
2. Fire Protection Equipment List, Underwriters Laboratories, Inc.
3. Factory Mutual System Approval Guide, A Guide to Equipment, Materials, and Services Approved by Factory Mutual Research for Property Conservation
4. Automatic Sprinkler Systems Handbook, National Fire Protection Association
5. The SFPE Handbook of Fire Protection Engineering, Society of Fire Protection Engineers
6. AWWA Handbooks, Manuals and Standards, American Water Works Association
7. Federal regulations, standards, and specifications
8. State and local codes and standards
9. Insurance organization standards and manuals
10. International standard organizations

TABLE C2.1 National Fire Protection Association (NFPA) Standards

Standard No.	Title
NFPA 13	Standard for the Installation of Sprinklers
NFPA 13D	Standard for the Installation of Sprinkler Systems in One- and Two-Family Dwellings and Mobile Homes
NFPA 13R	Standard for the Installation of Sprinkler Systems in Residential Occupancies Up to Four Stories in Height
NFPA 14	Standard for the Installation of Standpipe and Hose Systems
NFPA 20	Standard for the Installation of Centrifugal Fire Pumps
NFPA 22	Standard for the Installation of Water Tanks for Private Fire Protection
NFPA 24	Standard for Private Fire Service Mains and Their Appurtenances
NFPA 11	Standard on Foam Extinguishing Systems
NFPA 15	Standard for Water Spray Fixed Systems
NFPA 16	Standard for Deluge Foam-Water Sprinkler and Spray Systems
NFPA 16A	Recommended Practice for the Installation of Closed Head Foam-Water Sprinkler Systems
NFPA 231	Standard for General Storage, Indoor
NFPA 231C	Standard for Sprinkler Protection for Rack Storage of Materials
NFPA 12	Standard for Carbon Dioxide Extinguishing Systems
NFPA 12A, 12B	Standard for Halon Fire Extinguishing Systems
NFPA 17	Standard for Dry Chemical Extinguishing Systems
NFPA 17A	Standard on Wet Chemical Extinguishing Systems

Federal agencies and departments which regulate fire protection related matters reference NFPA standards extensively. However, specific federal regulations may be issued as authorized by the U.S. Congress. When issued for enforcement, these federal regulations are published in the Code of Federal Regulations, available from the General Services Administration. Departments of the U.S. government which issue regulations related to fire protection include the General Services Administration and Departments of Defense, Energy, Labor, Veterans Affairs, and Housing and Urban Development.

Several insurance organizations publish engineering standards used in the assessment of property insurance risks and that provide fire protection guidance to their insureds. These organizations include the Factory Mutual System, Industrial Risk Insurers, and Insurance Services office. Insurance associations exist for special underwriting risks such as nuclear, marine, textile, and food industries.

Most standards related to fire protection piping have been developed based on the needs and experience of a particular region or industry. In the United States and Canada, these standards are generally based on NFPA codes and standards.

With the increasing global community and economy, the need for awareness of standards from other countries and communities is becoming imperative. Examples include DIN (German Standards Institute), AFNOR (French Standards Association), and BSI (British Standards Institute). It should be recognized that many European standards are being harmonized by the European Center for Standardization.

DESIGN CONSIDERATIONS

To ensure that all necessary aspects that could have an impact on fire protection piping systems are considered, design goals should be established based on present and projected system needs and conditions. These design goals must consider expected fire suppression system demands, building locations, economic constraints, reliability, water supply source availability, design pressures, and environmental conditions.

Design Conditions

Working Pressure. The maximum working pressure of a particular installation must be considered when selecting the type and class of pipe to be used. To determine the maximum working pressure for water systems, designers should consider fire pump shut-off head, elevation changes, and the expected range of source pressures.

Pipe in underground fire service mains should withstand a working pressure of not less than 150 psig. Pipe in standpipe systems should be designed for a working pressure of at least 175 psi. When working pressures exceeding 175 psi are expected, all valves, fittings, and pipe should be rated for the appropriate pressure.

Sprinkler system devices such as automatic sprinklers, water-flow switches, and alarm valves are typically approved for installation in systems with working pressures up to 175 psi. Since sprinkler system components are generally intended for use up to 175 psi, higher pressures should be avoided. However, in some design situations such as high-rise buildings and large multibuilding complexes, higher pressures may be necessary. When these situations occur, sprinkler system components should be protected from excessive pressures. This is often accomplished using pressure control valves located strategically in the piping system.

Pressure control valves used in a fire protection system must be listed by a nationally recognized testing laboratory. Additionally, the setting of the valve should be closely controlled and its operability regularly maintained to assure that sufficient pressure is available to operate the fire protection systems properly.

When evaluating the impact of fire pump operating pressures on the maximum working pressure conditions, the total discharge pressure at pump shut-off should be considered. Centrifugal fire pumps may have a shut-off pressure of up to 140 percent of rated pressure.

The minimum operating pressure at the discharge point of sprinklers must be maintained to assure proper discharge flow and water spray distribution. Operating below these minimums can affect the ability of the sprinkler to control or suppress a fire.

NFPA 13 requires a minimum discharge operating pressure of 7 psi for any sprinkler. Higher minimums are required for devices with special applications such as nonstandard coverage areas and flows. For example, an extended coverage sidewall sprinkler can require up to 25 psi minimum operating pressure to achieve its rated coverage. The minimum pressure required for special application sprinklers are specific to manufacturer, model, and intended application. These higher minimums are described in the sprinkler's approval listing.

The working pressure for gaseous suppression systems can be substantially higher than those normally found in fire protection water systems. Storage cylinders with pressures of 600 to 850 psi are typical. These pressures result in substantial thrust forces and piping stress which are critical factors in the design of gaseous fire suppression system piping systems.

Water Source of Supply. Water for fire protection purposes can be obtained from public water systems, storage tanks, and raw sources such as rivers, lakes, and reservoirs. Whatever the source, it must be reliable and of sufficient flow and capacity to operate the connected fire protection systems.

The required flow capacity and residual pressure for a particular building is dependent on automatic sprinkler system demand, hose stream demand for interior fire-fighting purposes, and exterior hose stream demand from fire hydrants. The required flow capacity and residual pressure is often referred to as system demand. System demand is the minimum water supply required to operate the number of sprinklers contemplated to open from a particular fire hazard plus an allowance for manual hose stream operations. NFPA 14 provides requirements for the water supply necessary for various classes of standpipe systems. NFPA 13 covers water supply requirements for automatic sprinkler systems.

For fully sprinklered buildings, the water supply for sprinklers need not be added to the standpipe system demand required by NFPA 14. This is in recognition of the reduced likelihood of a large uncontrolled fire in a fully sprinklered building. For partially sprinklered buildings, the sum of the sprinkler system demand and standpipe system demand must be supplied by the water source.

The water demand requirements for a particular building may also be set forth by the loss prevention department of the property's insurance carrier.

When raw water is the supply source for automatic sprinkler systems, special provisions are required to reduce the accumulation of sediment in piping drop nipples. Return bends are required to prevent the accumulation of sediments in drop nipples which can eventually plug sprinklers, obstructing the flow of water; see Fig. C2.5.

All sprinkler system cross mains must be arranged for flushing with easily removable fittings. Internal inspections of fire protection piping systems which are supplied by raw water should be conducted regularly for conditions which would impair system effectiveness.

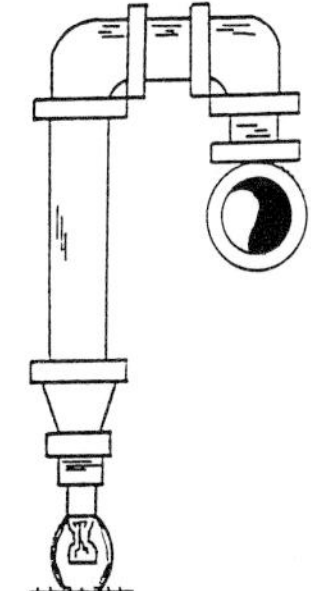

FIGURE C2.5 Typical return bend arrangement.

Temperature. Fire protection piping systems must be protected from freezing when temperatures cannot be constantly maintained above 40°F (4°C). For underground mains, burying pipes

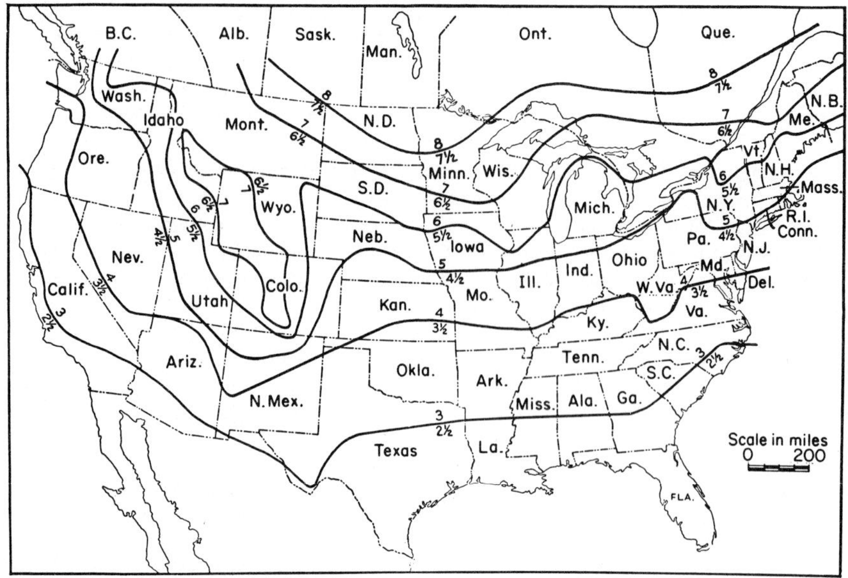

FIGURE C2.6 Depth of cover map.

below the frost line is the usual method to protect against freezing. The depth of cover for underground water mains to avoid freezing in different regions of the United States and Canada is shown in Fig. C2.6.

In fire protection system piping, there may be no circulation of water through the piping such as exists in a public waterworks system; therefore, the use of exterior insulation on exposed pipes is likely to be ineffective. Exposed short sections of pipe should be boxed or wrapped and heated.

Aboveground fire protection systems must also be protected from low temperatures. Where required, this is usually accomplished by providing special suppression systems such as a dry pipe sprinkler system.

Corrosive Environment. Corrosive conditions, both in soil and in air, require special precautions for fire protection system piping, fittings, and hangers. The precautions may involve the use of corrosion-resistant material or the application of protective coatings or wraps. The method of protection depends on the expected severity of the corrosive conditions.

External corrosion may cause a rapid deterioration and weakening of buried ferrous piping. Severely corrosive soil should be avoided if possible, but if not, special precautions are required to prevent corrosion-induced failure.

Protection against moderately corrosive conditions can be provided by using noncorrosive fill, coal-tar enamel coatings and wraps, and polyethylene encasement. NFPA 24 requires all ferrous fire service mains to be cement mortar lined, coated, and wrapped. Joints and fittings must be field coated and wrapped after assembly. Cement mortar lining should be in accordance with the requirements of AWWA C104 for cast iron pipe and fittings, AWWA C205 for steel water pipe, or AWWA C602 when lining pipe is in place. Polyethylene encasement in accordance with AWWA C105 is recognized for protecting ductile iron pipe. When coal-tar enamel protective coatings are used for steel water pipe, the requirements of AWWA C203 should be met.

Damage to pipe due to corrosive atmospheric conditions can be a problem in areas with high moisture, salt air, or fumes from corrosive chemicals. Where these conditions exist, corrosion-resistant pipe, fittings, and hangers should be used, or protective coating should be applied to exposed surfaces. In high moisture areas, consideration should be given to the use of galvanized pipe or copper tube for aboveground fire protection piping. Commercial grade corrosion-resistant paint can be used in a corrosive environment. As with any coating, maintenance is important for maximum effectiveness.

When a corrosive water supply exists, threaded or cut grooved thin wall pipe should be avoided. When these conditions exist, pipe with wall thickness in accordance with Schedule 40 should be used. Schedule 30 may be used in 8-in pipe and larger.

Seismic Integrity. Fire protection piping systems may require protection from damage when installed in areas subject to earthquakes. Areas where the potential for earthquake damage exists and special protection is required are determined by building codes, insurance requirements, and regulatory agencies.

Where required for aboveground piping, NFPA 13 provides earthquake protection design criteria for automatic sprinkler systems and standpipe systems. NFPA 13 does not require seismic protection but rather provides criteria where such protection is required by other codes and standards.

Flexible pipe couplings, lateral and longitudinal sway bracing, and prescribed pipe clearances are strategically used to minimize pipe damage and resulting system impairment. Flexible pipe couplings are used to allow differential movement between pipe sections and building parts. These couplings are provided for pipe 3½ in or larger. They should be installed at locations such as risers greater than 3 ft in length, where pipe passes through masonry floors and walls, and at or near building expansion joints. Swing joints must be used for seismic joints.

Annular clearances up to 2 in are necessary where pipe penetrates rigid barriers such as walls and floors, depending on pipe size, wall construction, and the location of flexible couplings and swing joints. Sway bracing is used to limit excessive lateral and longitudinal movement.

NFPA 13 provides additional details on the seismic protection of sprinkler systems. Alternate means of protecting automatic sprinkler systems from earthquake damage to those found in NFPA 13 may be used when based on an analysis that demonstrates system performance under expected dynamic seismic forces equivalent to the seismic performance of the building. For additional information on piping system stress analysis and supports, refer to Chaps. B.4 and B.5.

Aboveground Piping

Materials. Fire protection piping systems can convey many different suppression agents at a wide range of operating conditions. These agents include water and aqueous solutions, chemicals, and gases, each with different piping material requirements. Therefore, aboveground piping for fire protection systems must be designed and installed in accordance with the applicable NFPA standards relating to the system type being installed. The appropriate type of pipe is determined by design considerations, economics, ease of installation, troublefree service, and environmental factors.

Ferrous piping for water-based fire suppression systems must meet or exceed the requirements of ASTM A795, A53, or A135. Wrought steel pipe that meets the dimensional requirements of ANSI B36.10 may also be used.

NFPA 13 provides wall thickness requirements for steel pipe in sprinkler systems. The wall thicknesses are dependent on pipe size and the method of joining. Heavier wall pipe is required when threaded fittings or cut groove mechanical fittings are used; refer to Table C2.2.

Gaseous suppression systems are often subject to higher operating pressures than water-based suppression systems. Additionally, thrust forces are a major consideration in the acceptability of piping and fittings. For gaseous systems, ferrous piping must be black or galvanized steel pipe conforming to either ASTM A53, Grade A or B or ASTM A106, Grade A, B, or C. Other piping materials appropriate for high pressures such as stainless steel and copper may be used for gaseous systems. When used, pipe wall thickness must be calculated per the Power Piping Code (ASME B31.1). (Refer to Chap. B2 of this handbook.) NFPA Standards do not permit the use of ordinary cast iron, steel pipe conforming to ASTM A120 and nonmetallic pipe for gaseous suppression systems.

Copper tube for water-based fire suppression systems must meet or exceed the requirements of ASTM B75 or ASTM B88. Copper tube used in sprinkler systems shall have a wall thickness of type K, L, or M. Bending of copper tube (types K and L) is acceptable when performed in accordance with NFPA 13 which specifies minimum bend radius of six pipe diameters for pipe 2 in and smaller, and five pipe diameters for 2½ in and larger.

Many NFPA standards permit the use of other types of pipe in aboveground piping systems if they are investigated and listed by recognized testing organizations for fire protection service. When used, the pipes must be installed in accordance with the limitations of the organization's listing, including installation instructions.

When other pipe materials are investigated for use by a testing laboratory, many factors are considered, including

1. Working pressure rating
2. Hanger requirements
3. Unsupported vertical stability
4. Corrosion resistance
5. Joining methods
6. Operability at elevated temperatures

TABLE C2.2 Minimum Nominal Wall Thickness, Steel Pipe (A53), for Pressures up to 300 psi*

Fitting Type	Pipe Diameter		
	Up to 5 in	6 in	8 in and larger
Welded	Schedule 10	0.134 in	0.188 in
Roll grooved	Schedule 10	0.134 in	0.188 in
Threaded	Schedule 30	Schedule 30	Schedule 40
Cut grooved	Schedule 30	Schedule 30	Schedule 40

*Other pipe type and thickness may be used when the product has been evaluated and listed for fire protection service and installed per manufacturer's instructions.

Special thermoplastic piping systems have been investigated and approved for use in specific fire suppression system installations. The thermoplastic piping materials currently listed are chlorinated polyvinyl chloride (CPVC) and polybutylene (PB). Not all CPVC and PB pipe that is manufactured is acceptable for use in fire suppression systems. Only pipe that is specifically approved for fire protection service and that carries the listing mark of a nationally recognized independent testing laboratory such as Underwriters Laboratories (UL) may be used. The standards used to evaluate thermoplastic sprinkler pipe are Subject 1821 and UL 1887.

Special installation and design criteria exist for plastic fire suppression piping systems. These criteria are contained in the listing information for the material and relate to

1. Limitations on use based on hazard classification
2. Physical protection and pipe location
3. Hanger spacing
4. Piping restraint and deflection
5. Maximum ambient temperature

Piping Joints. There are several acceptable means of joining aboveground fire protection piping. Steel pipe with sufficient wall thickness may be joined using threaded connections; see Table C2.2. Threads must be cut to the requirements of ANSI/ASTM B1.20.1, Pipe Treads, General Purpose.

In normal applications, threaded fittings which have been listed for use in fire suppression systems by a recognized testing laboratory may be used with steel pipe with wall thicknesses less than Schedule 40 for pipe sizes less than 8 in and Schedule 30 for sizes 8 in or greater.

Sections of aboveground fire protection piping may be shop welded. However, field welding of fire protection piping should be avoided. When it is necessary to cut or weld inside a building, strict fire prevention precautions as described in NFPA 51B must be taken.

Welding methods for joining fire protection pipe are described in the NFPA standard for the type of suppression system being installed. Welding methods that comply with AWS D10.9, Level AR-3 or Section IX of the ASME Boiler and Pressure Vessel Code are often acceptable.

Mechanical couplings have become a popular method of joining aboveground fire suppression system piping. Mechanical couplings are an assembly of clamps, gaskets, and bolts. There are mechanical couplings appropriate for use on rolled groove, cut groove, and plain pipe ends.

Couplings for rolled groove pipe may be used with pipe having a minimum wall thickness of Schedule 10 unless specifically evaluated and approved for thinner wall pipe. Since cut grooved pipe reduces wall thickness similar to threads, fittings for cut grooved pipe are limited to Schedule 40 pipe in sizes less than 8 in. Schedule 30 may be used in pipe sizes larger than 8 in with cut grooved mechanical fittings. Only mechanical couplings which have been investigated and approved by a recognized testing laboratory specifically for fire protection service may be used. Working pressures, temperatures, system rigidity needs, and external and internal loads should always be investigated when considering mechanical joining methods. Many styles of mechanically joined piping components are available, including couplings, fittings, and valves.

Plain end pipe fittings are popular for sprinkler system installation due to the lack of needed pipe end treatment. When used, they must be listed and be used with pipe having a minimum wall thickness specified by the manufacturer's installation instructions.

In general, copper tubing used in fire protection systems should be joined by brazing in accordance with the requirements of ANSI B16.22. However, soldered joints may be permitted for wet-pipe systems protecting light and ordinary hazard occupancies when the maximum ceiling temperature is less than 150°F and the tube is concealed. Where soldering is used, the fittings shall conform to the requirements of ANSI B15.18.

Nonmetallic fire protection piping system components are joined using methods and materials that have been evaluated by the listing organization. Heat fusion is used to join PB pipe components, and a solvent cement is used for CPVC pipe. The methods and restrictions for joining these pipe materials are detailed in the product's approved installation guidelines published by pipe and fitting manufacturers. These guidelines are included as part of the project's special listing by the testing organization that evaluated and approved the product.

Due to operating pressures and thrust forces, fittings capable of withstanding higher pressures are required for gaseous suppression systems. Class 150 and cast iron fittings are not permitted. For example, fittings for 600 psig stored pressure Halon 1301 systems are required to have a working pressure of 1000 psi. Due to the cooling effect on the piping system of gaseous system discharges, the relationship between material temperature and pressure rating must be considered. For additional guidance, refer to ASME B31.1 and manufacturer information.

Hangers and Supports. The adequate support of aboveground fire protection piping systems is important. NFPA standards provide detailed information on methods and rules of proper support of sprinkler system piping. Other standards commonly reference NFPA 13 regarding hangers and supports.

Aboveground piping must be independently supported from building structure. Hangers may not be suspended from ceilings or nonstructural partitions. The points of connection to the structure must be capable of supporting the sprinkler system. Hanger components for aboveground piping systems are typically made of ferrous materials. Nonferrous hanger components may be used only when they have been evaluated as acceptable through fire testing and are listed by a recognized testing organization.

Hanger components that attach to the pipe or to the building structure, such as clamps, concrete inserts, and hangar rings, must be listed. If the hanger components are not listed, the hanger and installation methods should be designed to support five times the weight of the water-filled pipe plus 250 lb at each point of connection, in addition to any other applied loads at the point of hanging.

The maximum spacing between hangers is related to the piping's rigidity based on the piping material and the pipe size in accordance with the requirements of the installation standards or the piping's listing.

In general, hangers for steel pipe 1½ in and larger can be spaced no more than 15 ft apart. For steel pipe less than 1½ in, the maximum spacing is reduced to 12 ft.

Nonmetallic piping systems typically require closer hanger and support spacing. The maximum support spacings are detailed in the products approved installation guidelines published by the nonmetallic pipe manufacturer; for example, maximum hanger spacing for 1-in, UL-listed CPVC sprinkler piping is 6 ft, increasing to 10 ft for 3-in diameter CPVC pipe. For discussion of nonmetallic piping, refer to Part D of this handbook.

Many different styles and types of hangers are available for use with the various types of ceiling construction. Careful evaluation of hanger alternatives will result in a well-supported and reliable piping system. For additional details on hangers and supports, refer to Chap. B5.

Piping Layout and Design. The piping layout for most aboveground fire protection piping systems consists of risers, feed and cross mains, and branch lines. Branch lines are the pipe sections to which the sprinklers or nozzles are attached. The feed mains and cross mains are the pipes which supply the branch lines. Risers are essentially vertical feed mains.

The piping layout should be carefully considered to ensure that the resulting arrangement conforms to applicable standards in a manner that provides maximum efficiency and flexibility. Considerations should include the following:

1. Proper location and size of risers
2. Available water supply location and its characteristics
3. Building construction, height, and area
4. Hazard classification
5. Area's use or function such as the existence of cranes and large duct
6. Interferences to piping layout
7. Architectural considerations such as aesthetics

There are three basic piping layout configurations for sprinkler systems: tree, loop, and grid; see Fig. C2.7. The tree configuration is the traditional piping layout. Loop and grid layouts have become popular with the use of hydraulically designed systems. Additional information on determining pipe size is provided later.

Piping layout of aboveground fire protection piping associated with automatic sprinkler systems involves the following principles:

1. Provide automatic sprinklers throughout.
2. Do not exceed the maximum permitted area of protection per sprinkler.
3. Position sprinklers to optimize activation and water distribution.

The most effective automatic sprinkler system provides full area coverage throughout the protected premises. Partial systems are sometimes used to protect hazardous areas in an otherwise nonsprinklered facility. The use of partial systems should be carefully considered since it is difficult to predict with accuracy where a fire will occur. Additionally, a partial sprinkler system may not be capable of controlling a fully developed fire spreading from a nonsprinklered area.

To provide full area coverage, sprinklers should be provided for the following areas in accordance with the guidance of related installation standards:

1. All spaces and rooms
2. Combustible concealed spaces
3. Building shafts and service chutes
4. Underneath large ducts and platforms
5. Exterior canopies

The maximum protection area allowed for sprinklers is related to the hazard classification of the protected space and ceiling construction. The higher the hazard

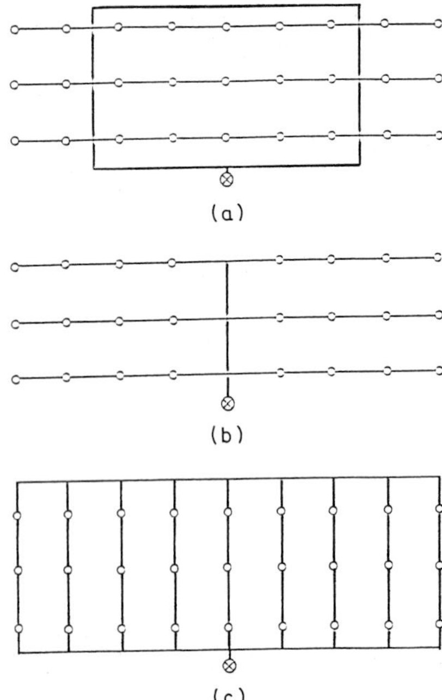

FIGURE C2.7 Typical sprinkler system piping layouts. (*a*) Grid; (*b*) tree; (*c*) loop.

classification, the smaller the maximum protected area allowed. Certain wood ceiling construction also requires a reduced maximum protection area. The maximum protection areas per standard sprinkler are shown in Table C2.3. Systems with nonstandard sprinklers may exceed these areas when designed and installed in accordance with the listing limitations of the special application sprinklers.

Sprinklers should be positioned to optimize activation and distribution. The spacing of branch lines and the location of sprinklers relative to walls, ceilings, structural members, and other obstructions are important factors.

The layout of aboveground fire protection piping for gaseous systems differs significantly from sprinkler systems. Rather than sprinklers located at prescribed spacing throughout an area, the objective of gas system nozzle placement is to achieve design gas concentrations by strategic nozzle placement to assure proper distribution and mixing of the gas when discharged. Also, due to working pressures up to 850 psig and discharge velocity, the support of the piping layout is critical. For additional information, refer to NFPA 12A or 12B for Halon suppression systems and NFPA 13 for carbon dioxide suppression systems.

Inspection and Testing. Before making connections to aboveground piping risers, underground piping must be flushed to remove any accumulated sediment or material which could affect suppression system operation. To assure adequate

TABLE C2.3 Maximum Protection Area per Standard Sprinkler

Hazard classification	Ceiling construction	Maximum protection (Area in ft^2)
Light	Smooth, beam, and girder	200; 225 if hydraulically calculated
	Open wood joist	130
	All other ceiling types	168
Ordinary	All ceiling types except high-piled storage	130
	High-piled storage	100
Extra hazard	All ceiling types	90; 100 if hydraulically calculated

cleaning, the minimum flow rate for the flushing should achieve a velocity of 10 fps in the underground piping system.

Newly installed pipe should be hydrostatically tested for leaks and to detect faulty pipe, fittings, or joints. No visible leakage is acceptable for interior fire protection piping. Aboveground fire protection piping should be tested at a minimum hydrostatic pressure of 200 psi for 2 h. If the maximum system pressures exceed 150 psi, the hydrostatic test pressure should be 50 psi greater than the maximum pressure. For example, if the maximum system pressure was 175 psi, the hydrostatic test pressure should be 225 psi minimum.

Underground Piping

Materials. The selection of appropriate underground pipe and fittings for fire service mains should consider economical installation, troublefree service, and easy maintenance. The review of these factors should include the required methods of trenching and laying, the characteristics of the water to be handled, and the corrosiveness of the soil. The pipe material should be recognized by applicable codes and standards and suitable for the conditions under which it will be installed.

The majority of pipe for underground fire service mains currently being installed is ductile iron. Ductile iron is a cast iron material where the primary graphite content occurs as nodules or spheroids. The graphite in this form maximizes impact resistance and ductility.

Ductile iron pipe that is manufactured in accordance with ANSI A21.51 may be used for fire protection service. The pipe class, or wall thickness, required for a particular condition should be determined on the basis of expected internal pressure, laying conditions (i.e., trench load, vehicle loads), and soil characteristics.

Design methods for determining the minimum required thickness of ductile iron pipe are provided in ANSI 21.50 (AWWA C150). These methods consider the conditions of trench load and internal pressure. The net thickness required to withstand the larger of bending stress or deflection caused by the trench load is added to the thickness required for the hoop stress due to internal pressure. Additional thickness is provided to the sum of these net thicknesses for service allowance and casting tolerance.

Steel pipe is seldom used for general underground fire mains. However, due to its high strength, it is suitable for locations subject to vibrations or shock as from railroad tracks, truck crossings, or highways. When used, steel pipe should conform to AWWA C200, be standard weight for the size (Schedule 40 minimum), and be lined and coated for corrosion protection.

Nonmetallic pipe is available for use in underground fire service mains. Nonmetallic materials used include polyvinyl chloride, reinforced concrete, and polyethylene and glass fiber reinforced composites. The benefits of these materials may include their light weight and corrosion resistance. When used for underground fire protection service, the pipe should be listed, comply with applicable AWWA standards, and be installed according to the manufacturer's installation instructions.

Joints. Joints for underground fire protection piping vary with piping material. The joint must be suitable for the pipe material, working pressures, and the particular installation conditions. Manufacturer's specifications should be followed.

The majority of cast iron or ductile iron joints are push-on or mechanical. Bell-and-spigot joints that use jute and molten lead are no longer used but are found in many existing installations.

Push-on joints use a special rubber gasket. The rubber gasket is placed in the bell end of the pipe. When the spigot end of the pipe is pushed into the socket past the gasket, the joint is formed. Push-on joints do not require packing or caulking. For more details on push-on type joints, refer to Chap. A2 of this handbook.

Mechanical joints are those which use a bolted follower ring or gland to hold the gasket in place. The ring is placed over spigot end, and bolts are used to compress the gasket and the bell end of the pipe. The bolts must be tightened per manufacturer's recommendations. They should also be coated to minimize corrosion. Refer to Chap. A2 of this handbook for additional information regarding this type of joints.

Joints for steel pipe may be welded, threaded, or mechanical. Welding should conform to AWWA C206, Standard for Field Welding of Steel Water Pipe Joints. Field welding can damage pipe coatings and linings. Any damage should be repaired.

Nonmetallic pipe is joined using approved butt fusion, push-on joints, mechanical fittings, or solvent cement and couplings. The limits of the approval and the manufacturer's recommendations should be followed.

Installation. There are many factors to be considered in laying underground pipe, including preparation of the trench, placing and aligning the pipe, making the joints, and anchoring, leak testing, and backfilling the installation.

Trenches should be carefully excavated to minimize crumbling walls or cave in. In sandy or lose soils, sheeting or bracing may be necessary. The bottom of the trench should conform to the grade of the pipeline. It may be necessary to excavate to an extra depth and prepare a stable pipe bed with a layer of firmly compacted soil. Pipes should not be laid in contact with rocks or boulders.

The trench should be wide enough to allow careful alignment of the pipe and convenient making up of the joint. Space should be provided below joints so the joint can be properly made and there will be no localized bearing on the joint.

Do not allow foreign material and water to enter the pipe during installation. Unattended open pipe ends should be closed by installing a watertight plug or by other means. Additionally, the flotation of pipe being installed due to excessive trench water should be avoided using necessary backfilling.

Most types of underground pipe joints are not designed to resist significant forces which would tend to pull them apart. The friction between the pipe and the ground resists this lateral movement. However, where such forces are significant such as bends, tees, plugs, and pipe laid with a steel stope, special consideration is required. A typical location where lateral restraint is required is at the connection of a fire service main to a fire protection system riser; see Fig. C2.8.

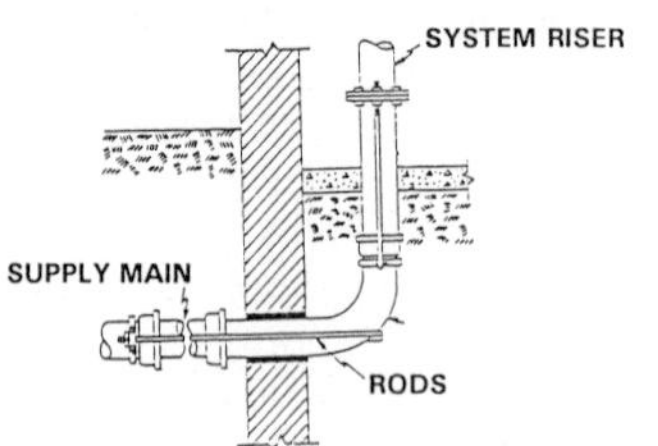

FIGURE C2.8 Typical fire protection riser restraint.

To restrict this lateral movement, joints at bends, tees, and plugs should be anchored by clamps, rods, bolts, or concrete thrust blocks. Mechanical fittings can be obtained with lugs for anchorage by tie rods. Pipe clamps and anchor straps are also available.

When determining anchorage methods, consideration must be given to the maximum forces produced by water pressure, the direction in which the

TABLE C2.4 Forces (in pounds) to be Resisted by Anchorage at Pipe Joints

Size, in.	Outside diam, in.	Separating force produced by 200 psi at tee and plug	Resultant of forces at bends		
			90 deg (1.41)	45 deg (0.765)	22½ deg (0.385)
6	6.90	7,500	10,600	5,740	2.890
8	9.05	12,900	18,200	9,880	4,970
10	11.10	19,400	27,400	14,850	7,480
12	13.20	27,300	38,200	20,850	10,520
14	16.65	38,400	54,200	29,400	14,800
16	17.80	49,700	70,100	38,000	19,150
20	22.96	76,400	108,000	58,400	29,450

forces act, and allowances for favorable conditions such as pipe to earth friction and the resistance that can be provided by various anchorage methods. Table C2.4 gives

1. Forces tending to separate single joints, such as at plugs, at tees, at the base of hydrants, and at bends having only one joint that needs anchorage
2. The resultant effect of the two forces acting at the ends of an elbow or bend

As indicated above, thrust blocks may also be used to anchor horizontal pipes at fittings; see Fig. C2.9. For maximum effectiveness, thrust blocks should bear against undisturbed soil leaving the pipe joint accessible for inspection and repair. Thrust blocks should only be considered satisfactory where the bearing soil is suitably supporting.

Thrust blocks should be made of concrete mixed not leaner than 1 part cement, 2½ parts sand, and 5 parts stone. The bearing area should ensure adequate resistance to the thrust force anticipated. NFPA 24 provides guidance on the required bearing area of concrete thrust blocks. The required bearing face area of concrete thrust blocks is dependent on the type of fitting being restrained such as ¼ bend or tee, the system's water pressure, and the undisturbed soil resistance.

Coatings and linings are used for underground piping installations to resist the corrosiveness of water and soil. Protective coal tar enamel per AWWA C203 and cement-mortar coatings and linings (AWWA C205) are used extensively.

Polyethylene encasement, in accordance with ANSI/AWWA C105, provides good protection to cast and ductile iron pipe exposed to corrosive soil conditions.

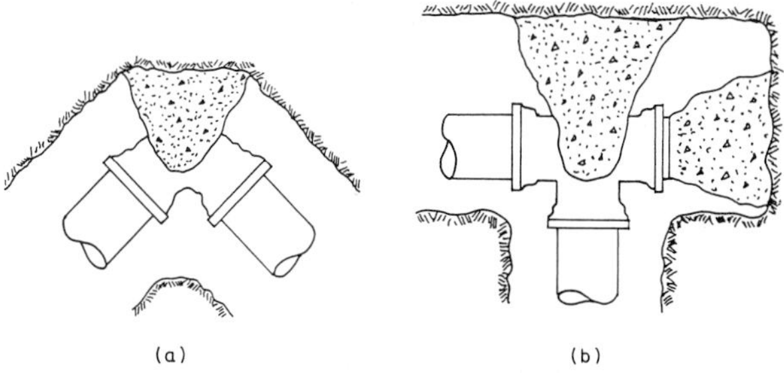

FIGURE C2.9 Typical thrust block configurations.

Cathodic protection systems may also be used for the external protection of iron and steel water mains.

The backfill for underground fire protection piping should be free of cinders, refuse, plant material, and rocks and of a type that will compact firmly. The top of the pipe shall be buried not less than 1 ft below the frost line for the locality but in no case less than 2½ ft to prevent mechanical injury. The depth of cover for underground fire protection piping to avoid freezing in different regions of the United States and Canada is shown in Fig. C2.6. Due to variations in pipe material performance under varying temperature conditions such as pipe contraction, manufacturer's instructions should always be followed.

Inspections and Testing. Newly installed fire protection mains should be inspected for faulty pipe or fittings. The piping should be hydrostatically tested for tightness. Before hydrostatically testing newly installed underground pipe, the installation should be firmly backfilled to above the center line of the pipe for cast and ductile iron pipe. For nonmetallic and pipe, the backfill should extend 1–2 ft over the pipe, except at joints. All joints, regardless of pipe type, should be left exposed until tests are satisfactorily completed. If used, thrust blocking should be adequately cured to restrict pipe movement.

Similar to aboveground fire protection piping, all new underground piping should be hydostatically tested at not less than 200 psi for 2 h. If the service pressure and rating of the pipe installed exceeds 150 psi, the hydrostatic test pressure should be 50 psi above the service pressure or rating.

An entire installation may be tested at once, or smaller sections may be tested using installed isolation valves. Ensure that all entrapped air is released from the test section using hydrants or connected aboveground fire suppression systems.

During the hydrostatic testing, all joints should be inspected for leakage. A small amount of leakage is permitted for new underground fire protection piping installations. The quantity of leakage is determined by pumping from a calibrated container. The maximum amount of leakage permitted by NFPA 24 for new pipe is 2 qt/h/100 gaskets or joints. The pipe diameter and the length of pipe gaskets are not considered. When a metal seated valve is used to isolate a test section, the allowable leakage is increased to compensate for valve leakage at a rate of 1 fl oz/in of valve diameter per hour.

Pipe Sizing

Most modern fire protection piping systems are hydraulically designed. That is, appropriate pipe sizes are determined considering system demand, available water supply, and pipe network considerations such as friction loss, elevation changes, and pipe type. The generally accepted formula for estimating friction loss in fire protection piping practice is the Hazen-Williams formula; its form is

$$f = \frac{4.52\,Q^{1.85}}{C^{1.85}\,d^{4.87}}$$

where f = friction loss (psi per ft)

Q = rate of flow (gpm)

C = Hazen-Williams pipe coefficient

d = internal pipe diameter (in)

TABLE C2.5 Values for C in Hazen-Williams Formula

Type of pipe	Value of C
New or in condition of new, pipe	
Cast iron—unlined	120
Cast iron—cement lined	130
Cast iron—bituminous lined	140
Nonmetallic	140
Steel	140

The value for coefficient C varies with pipe type and internal pipe conditions. The lower the capacity-carrying characteristics of the pipe, the lower the value of C. Table C2.5 lists C values for various pipe internal surfaces.

Due to the effects of water corrosiveness, increasing age affects the capacity-carrying characteristics of unlined cast iron pipe. The variations of C with age from mildly corrosive water is shown in Table C2.6. Table C2.7 gives the friction

TABLE C2.6 Variation in Hazen-Williams Coefficient C with Age of Unlined Cast Iron Pipe

Age of pipe, years	Value of c
New	120
10	105
15	100
20	95
30	87
50	75

loss of nominal pipe in psi per 1000 ft. While using Table C2.7, note the references to Tables C2.8 and C2.9. Appendix E4 of this handbook contains additional pressure drop tables for different sizes of pipe.

The degradation of C accelerates with moderately or severely corrosive water. In these cases, the coefficient may be reduced to 60–75 for 15-year-old pipe. A coefficient of $C = 100$ is commonly used for average conditions. In sections of fire protection pipe in which there is normally no flow, such as sprinkler systems, deterioration is less rapid. For underground piping, lined, nonmetallic pipe typically exhibits little decrease in capacity over time.

The determination of proper pipe size for gaseous suppression systems is by complex flow calculations that consider the two-phase (i.e., liquid and vapor) characteristics of the agent flowing through the pipe network. These calculation methods consider storage pressure, rate of pressure reduction, elevation, line pressure, heat transfer of the piping network, density, and other factors. Details on acceptable calculation methods can be found in NFPA 12 for carbon dioxide systems and NFPA 12A and 12B for Halon systems.

TABLE C2.7 Friction Loss in Pipes (*f* in Hazen-Williams Formula)
Loss in pressure in psi per 1,000 ft of pipe $C = 100$*

Flow (gpm)†	Inside diameter of pipe (in) (nominal)‡						
	4	6	8	10	12	14	16
50	1.47	0.204					
60	2.06	0.285					
75	3.11	0.430					
100	5.29	0.735	0.181				
125	8.0	1.11	0.273				
150	12.1	1.55	0.381				
200	19.1	2.65	0.652	.0220			
250	28.8	4.00	0.985	0.332			
300	40.4	5.62	1.38	0.466	0.192		
400	68.8	9.55	2.35	0.793	0.326	0.154	
500	104.0	14.4	3.53	1.20	0.493	0.233	0.122
750	220.0	30.5	7.52	2.54	1.04	0.493	0.257
1,000	375.0	52.0	12.8	4.32	1.78	0.839	0.438
1,500	—	110.0	27.1	9.14	3.76	1.78	0.928
2,000	—	—	46.2	15.6	6.41	3.30	1.58
2,500	—	—	—	23.5	9.69	4.57	2.39
3,000	—	—	—	33.0	13.6	6.41	3.34
4,000	—	—	—	—	23.1	10.9	5.69
5,000	—	—	—	—	34.9	16.5	8.60

*For values of *c* different from 100, multiply the flow rates listed in Table C2.7 by the conversion factors listed in Table C2.8.

†For approximate interpolation between the *tabulated flows,* calculate the actual flow as a percent of the next lower tabulated flow, convert this percent to the given value in Table C2.9 and take this percent of the friction loss corresponding to the lower tabulated flow.

‡Nominal pipe diameters were used in calculating the tabulated friction losses. The inside diameter of pipe conforming to American National Standards Institute or American Water Works Association Standards will vary with the process of manufacture, material, and pressure classification and can be found in the Standards. Some published tables of friction losses are based upon actual internal diameters. The errors caused by minor differences in internal diameter are usually not significant when considered in relation to the uncertainties in assumed coefficients and rates of flow. If desired, adjustment for differences in diameter can be made by applying a factor $d(\text{nominal})^{1.25}/d(\text{actual})^{1.25}$ to the values given in Table C2.9.

TABLE C2.8 Conversion for Different Values of *c* in Williams and Hazen Friction-loss Table

Value of *c*	Conversion factor
140	0.537
130	0.615
120	0.714
110	0.836
100	1.000
90	1.22
80	1.51
70	1.93
60	2.57

TABLE C2.9 Interpolation for Water Flows between Values Given in Table C2.7

Actual flow as % of next lower tabulated flow	Converted % to be applied to friction loss at next lower tabulated flow
105	110
110	120
115	130
120	140
125	151
130	162
135	174
140	186
145	198
150	211
160	238
180	296
200	361

Example: 8-in. pipe, flow 650 gpm.
Next lower tabulated flow 500 gpm.
Per cent of lower flow = 650 ÷ 500 = 130 per cent.
Converted per cent = 162.
Tabulated lower friction loss at 500 gpm = 3.53.
Approximate actual friction loss = 3.53 × 130 per cent = 4.58 psi/1,000 ft.

CHAPTER C3
STEAM SYSTEMS PIPING

Tadeusz J. Swierzawski
Consultant, Mechanical Division
Stone & Webster Engineering Corporation
Boston, Massachusetts

Daniel A. Van Duyne
Assistant Chief Engineer, Mechanical Division
Stone & Webster Engineering Corporation
Boston, Massachusetts

INTRODUCTION

General

In this chapter we will first review the basics of piping system design for steam systems applications; we will then consider specifics for underground steam piping and steam piping used in power plants.

Definitions

Condensate: Condensed steam.

Trunk line distribution system: Distribution system with large diameter line leaving the boiler plant. As lateral branches are installed off it for service, the diameter of the trunk line is gradually reduced.

Main and feeder network distribution system: Distribution system that receives its supply of steam through a high-pressure feeder main leading from the plant through the network; the size of the feeder main required in this case is not as large as in a trunk-line system with the same boiler-plant steam pressure.

Protective conduits (typically for underground steam lines): Enclosures for underground steam mains and services to protect the pipe and insulation from damage due to earth pressure and impact loadings, to allow free longitudinal expansion and contraction while held in proper alignment, and to prevent groundwater seepage or flooding by providing either drains or a completely waterproof structure.

Light water reactors (LWR): Nuclear power reactors of either the pressurized water reactor (PWR) or boiling water reactor (BWR) type.

ASME Class 1[1] piping: Includes main steam piping up to and including the first stop valve outside the reactor containment for BWRs and is designated as ASME Class 1.

ASME Class 2[1] piping: Includes main steam piping up to and including the first stop valve outside the reactor containment for PWRs and BWR main steam piping after the first isolation valve outside the reactor containment and is designated as ASME Class 2.

ASME B31.1[2] piping: Includes main steam piping downstream of the first stop valve outside the reactor containment in PWRs and piping external to boilers in fossil power plants and is constructed to the ASME B31.1 Code for Pressure Piping.

ASME B31.3[3] piping: Includes steam process piping in industrial plants and is designed to the ASME B31.3 Code for Process Piping.

DNBR: Departure from nucleate boiling ratio.

FFWT: Final feedwater temperature.

HARP: Heater above the reheat point.

IP: Intermediate pressure (turbine).

IV: (governor-operated) intercept valve.

LP: Low pressure (turbine).

NSSS: Nuclear steam supply system.

psi: Pressure drop in pounds (force) per square inch (lb/in^2).

psia: Pressure in pounds (force) per square inch, absolute.

psig: Pressure in pounds (force) per square inch, gauge.

PWHT: Postweld heat treatment

VWO: Valves wide open (steam turbine).

Nomenclature

A: An additional thickness to provide for material removed in threading, corrosion, or erosion allowance and material required for structural strength of the pipe, as appropriate (in). See Table C3.1 for selected values of A.

d: Inside diameter of pipe (in). In using Eq. (C3.4), the value of d is for the maximum possible inside diameter allowable under the purchase specifications.

D_o: Outside diameter of pipe (in). For design calculations, the outside diameter of pipe as given in ANSI B36.10M and ANSI B36.19M[4,5] and specifications shall be used in obtaining the value of t_m.

L: Length of pipe (ft).

P: Internal design pressure (psig).

P_a: Calculated maximum allowable internal pressure (psi) for straight pipe which shall at least equal the design pressure. P_a may be used for piping products with pressure ratings equal to that of straight pipe.[6] For pipe products where the pressure rating may be less than that of the pipe (e.g., flanged joints and reinforced branch connections where part of the required reinforcement is in the run pipe), the design pressure shall be used instead of P_a.

TABLE C3.1 Selected Values of $A^{1,2}$

Type of pipe	A (in)
Cast-iron pipe centrifugally cast or cast horizontally in green sand molds	0.14
Cast-iron pipe, pit cast	0.18
Threaded steel, wrought iron or nonferrous ¾ in nominal and smaller	0.065
Threaded steel, wrought iron or nonferrous 1 in nominal and larger	Depth of thread
Grooved steel, wrought iron, or nonferrous	Depth of groove plus 1/64 in
Plain end steel or wrought-iron pipe or tube for sizes 1 in and smaller	0.05
Plain end steel or wrought-iron pipe or tube for sizes over 1 in	0.065
Plain end nonferrous pipe or tube	0.00

TABLE C3.2 Minimum Thickness for Bending^{1,2}

Radius of bends	Minimum thickness recommended prior to bending*
6 pipe diameters or greater	$1.06t_m$
5 pipe diameters	$1.08t_m$
4 pipe diameters	$1.14t_m$
3 pipe diameters	$1.25t_m$

*t_m is determined by Eq. (C3.3) or (C3.4).

ΔP: Pressure drop (psi).

S: Maximum allowable stress for the material at the design temperature (psi).

t: Specified or actual wall thickness minus, as appropriate, material removed in threading, corrosion, or erosion allowance, material manufacturing tolerances, bending allowance (see Table C3.2), and material to be removed by counterboring (in).

t_m: Minimum required wall thickness (in). If pipe is ordered by its nominal wall thickness, the manufacturing tolerance on wall thickness must be taken into account.

W: Steam flow (lb_m/min).

Y: Density of steam (lb_m/ft^3) [used in Eq. (C3.2)].

y: A coefficient. For pipe with a D_o/t_m ratio less than 6, refer to Table C3.3. For pipe with a D_o/t_m ratio greater than 6, the value of y can be calculated from Eq. (C3.1), as follows:

$$y = \frac{d}{d + D_o} \tag{C3.1}$$

TABLE C3.3 Values of Coefficient $y^{1,2}$

Temperature (°F)	900 and below	950	1000	1050	1100	1150 and above
Ferritic steels	0.4	0.5	0.7	0.7	0.7	0.7
Austenitic steels	0.4	0.4	0.4	0.4	0.5	0.7

Types of Systems

Steam Distribution Systems. While there are no fixed standards for the design of steam-distribution piping systems, most of the systems fall into one of two general classes: (1) a trunk-line distribution network system and (2) a main and feeder distribution network system.

In case 1, the diameter of the trunk line leaving the boiler plant is large; as lateral branches are installed off it for service, the diameter of the trunk line is gradually reduced as the needs for carrying capacity are diminished.

In Case 2, the main and feeder network distribution system receives its supply of steam through a high-pressure feeder main leading from the plant through the network. Advantage is taken of the pressure drop available for the transportation of large volumes of steam to the low-pressure network.

Since piping is the largest individual factor in the selection and design of a steam distribution system, the major items that must be resolved are as follows: (1) pipe size, (2) wall thickness, (3) materials selection, (4) types of joints, (5) proper insulation, (6) a protective conduit for the pipe and insulation from water and mechanical damage, (7) drainage of condensate, (8) provision for thermal expansion with controlling anchorage, and (9) safety provisions.

Underground Steam Piping. Underground steam piping with its purpose of steam distribution has been a highly specialized field of engineering peculiar to the district-heating industry; however, now underground steam piping is also used to carry process steam. With the advent of the construction of groups of buildings such as housing developments, institutions, and industrial plants, central-heating systems and steam-distribution problems are no longer restricted to the district-heating industry. Steam piping systems may be buried and not readily available for enlargement, replacement, and repair. Such piping must be protected from ground elements and excessive heat losses; thus it is important that every phase of its design and operation be understood before such a system is installed. Figures C3.1 and C3.2 show typical steam distribution systems which would be fed by underground steam line.

See "Protective Conduits for Underground Lines" for more information about protective enclosures and "Drainage of Condensate" for condensate drainage considerations for underground piping.

Fossil Fueled Power Plants. The main steam system conducts superheated steam from the steam generator (economizer plus evaporator and superheater) to the turbines as shown in Figure B1.1 in Part B of this handbook. Figure C3.3 shows a simplified schematic of a main steam system. The main steam system may also provide steam for various auxiliary services.

Steam power-plant piping layouts, with due respect to economic factors, should consider its mechanical design or ability to function properly (and effi-

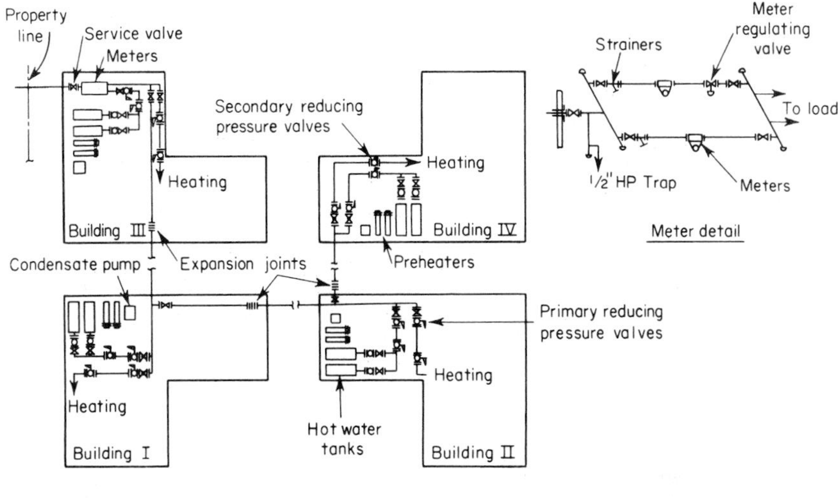

FIGURE C3.1 Typical steam distribution system for a housing development.

ciently) with respect to the mechanical equipment which it serves—its convenience from an operating standpoint and its appearance as a coordinated part of the plant layout. Although the relative importance of these basic points falls in the order named, each has an important bearing on the acceptability of any arrangement, as will be discussed in the following sections.

If two or more boilers are interconnected, the connections for main superheated steam piping are taken from the superheater outlet header through stop and check (combination) valves which are provided to prevent backflow of steam from the mains into a boiler in case of tube rupture or other boiler trouble.

Availability of steam generators has increased to the extent that single units are used almost universally to supply the steam to turbines of sizes up to approximately 1000 MW. In fossil-fueled power plants exhaust steam from the high-pressure turbine is piped back to the steam generator, where it is reheated in a special section before being returned to the inlet of the intermediate pressure or reheat turbine.

The cold reheat system (CRS) and the hot reheat steam (HRS) system shown in Fig. C3.4 are treated here in one single section because they are parts of the same overall system used for reheating steam. The CRS conducts steam from the outlet of the high-pressure turbine to the reheater inlet and provides steam for the auxiliary steam and extraction steam systems. The CRS is the normal auxiliary steam source, but at low loads auxiliary steam is usually augmented from the main steam system via a pressure reducing station.

The exhaust steam from the high-pressure turbine is transported in a single or multiple leads into the reheater inlet. Each reheater may have a desuperheater for reheater outlet temperature control and safety valves at the reheater inlet. A cross connection (cross tie) may be used between the two cold reheat leads to provide steam to the associated feedwater heater and the auxiliary steam system without creating a pressure imbalance at the reheater inlet. This system has provisions for isolation of the reheater for hydrostatic testing.

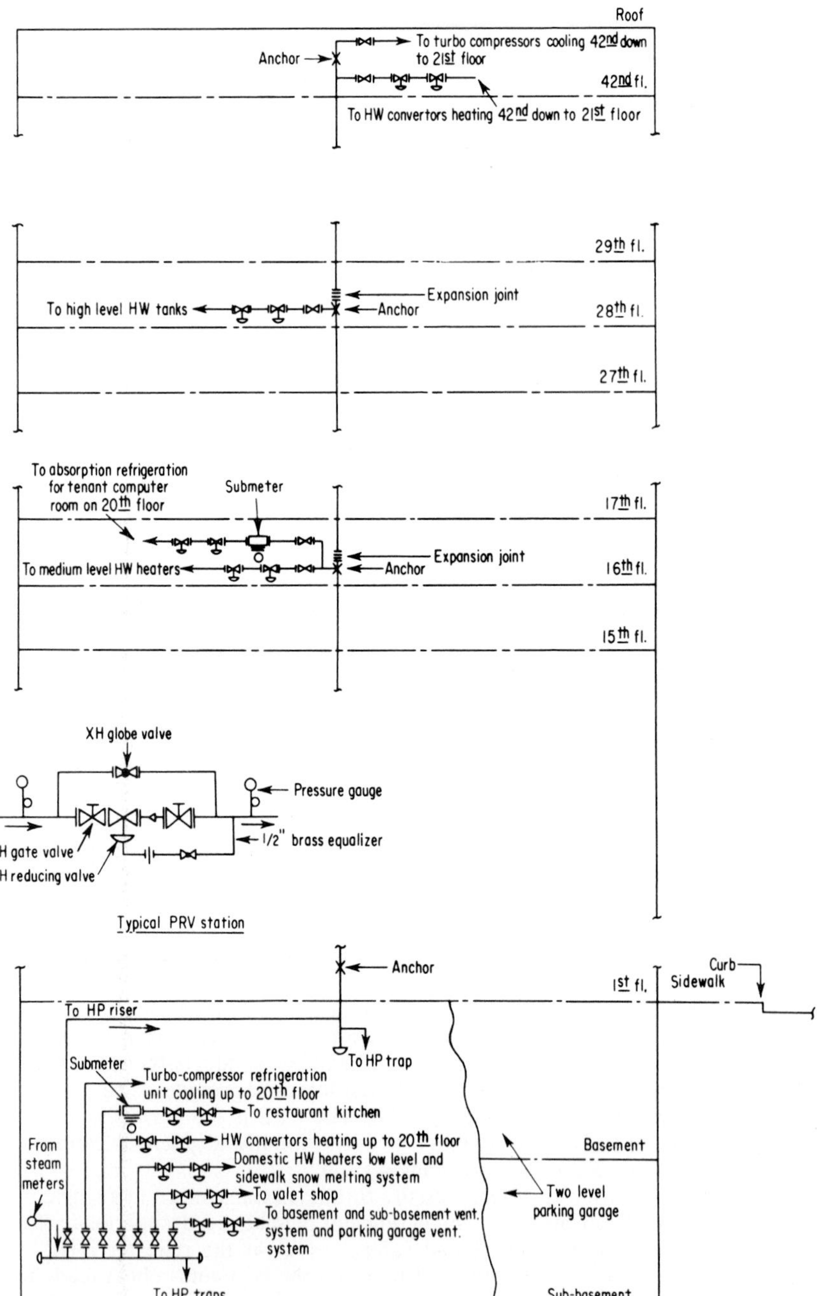

FIGURE C3.2 Typical high-pressure steam-distribution system in a tower office building.

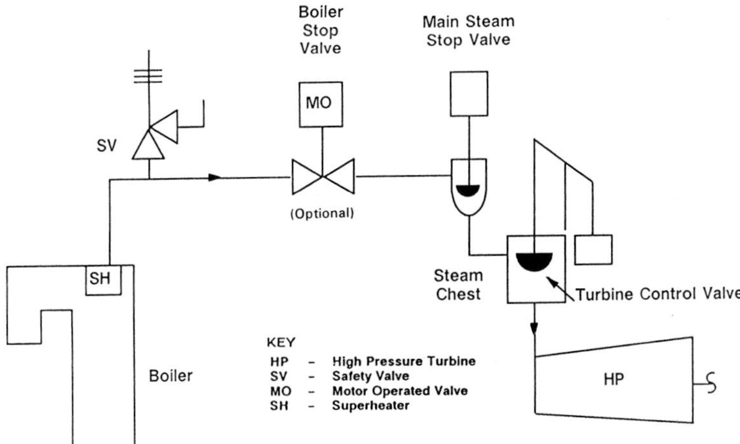

FIGURE C3.3 Main steam system.

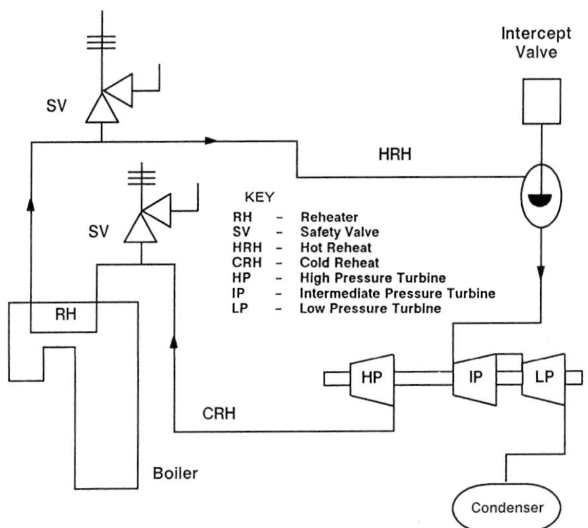

FIGURE C3.4 Reheat steam system.

The extraction steam system conducts steam from the high-pressure turbine, cold reheat line, intermediate pressure turbine, and low-pressure turbines to the feedwater heaters. This extraction steam is required for feedwater heating. Feedwater heating increases cycle efficiency; in addition, the extraction steam system may provide steam for the feedwater pump turbine. In nuclear units, the system also removes moisture from the turbine to provide protection for the lower pressure turbine blades and to increase turbine efficiency.

The hot reheat system conducts superheated steam from the reheater outlet header to the intermediate pressure turbine or to the inlet of the low-pressure turbine. One or more hot reheat leads may be furnished. Each lead has a safety valve for reheater protection. A cross connection between the two hot reheat leads is often furnished to ensure equal pressure at each reheat stop and intercept valve before entering the turbine.

Nuclear Fueled Power Plants. Only the stationary LWR, either of the PWR or BWR type, are discussed in this chapter as being typical of nuclear practice. In the typical BWR, cycle water is circulated through the reactor core, producing steam which is separated from recirculation water, dried in the top of the reactor vessel, and directed to the steam turbine. The reactor is the steam generator.

The PWR uses two closed systems—a primary system including the reactor and its cooling system and a secondary system including turbine-generator, condensate system, and feedwater system. The two systems communicate with each other at the steam-generator tube interface where pressurized water of the primary system transfers fission-reaction heat to the steam-generator feedwater in the secondary system, producing steam to drive the turbine generators. PWR power plants typically use two, three, or four primary loops, each containing a steam generator which transfers the energy from primary coolant to the feedwater on the secondary side.

The NSSS customarily consists of those components in contact with the reactor coolant, specialized auxiliary machinery.

The main steam system in LWRs transports steam from the outlet of the reactor and steam generators to the turbine stop valves. The main steam system also provides steam for various auxiliary services. Additionally, it provides a means of controlled heat release from the NSSS during periods of station electrical load rejection.

In PWRs, each steam generator supplies steam to a line which connects to a common main steam manifold. Valves in each line permit isolation of individual steam generators. Safety valves on each line provide pressure relief protection for the steam generators. The main steam manifold supplies steam to the high-pressure turbine throttle valves, the moisture separators/reheaters, and auxiliary steam loads.

The main steam system includes a turbine bypass system that provides a direct steam path to the condenser and to the atmosphere in order to prevent unnecessary reactor trips during load rejections up to 100 percent of full electrical load.

The turbine bypass system also permits testing NSSS at power levels up to 55 percent without having the turbine loaded. It is also used to maintain reactor coolant temperature during hot standby and shutdown operation and is used to conduct controlled cooldown of the plant to the point where the residual heat removal system can be placed in operation.

For the PWR and BWR, the main steam system design pressure is about 1200 psia (see Fig. C3.7) and 600°F. Actual design conditions depend upon the specific reactor design. Main steam line sizing is determined by performing an economic analysis of pressure drop, pipe cost, and so on, with a maximum velocity of 15,000 fpm (250 fps); see "Preventing Turbine Overspeed" for more flow velocity data. Preliminary pipe sizing is based on a 3 percent pressure drop from the steam generator outlet to the turbine stop valves. Wall thickness is calculated using the equations presented in "Design Pressure."

For PWR and BWR main steam lines, turbine water induction problems are of great concern as the main steam is saturated or contains some moisture. Prevention criteria are to be as specified in ASME Standard No. TDP-2.[7]

Each type of nuclear plant has isolation valves outside of the reactor containment. BWR plants also have isolation valves in main steam lines inboard of the reactor containment. Since the BWR main steam is radioactive, it must be shielded and special considerations must be made in its design to minimize crud traps (small deposits of radioactive contaminants in flow passages). Avoid the use of backing rings when welding pipes together.

In LWR nuclear plants, reheating is achieved in a combined moisture separator and reheater unit. High-pressure turbine exhaust steam passes through the moisture separator portion of the unit, where most of the moisture is removed mechanically. The steam is then reheated in one or two stages by passing over the bundles of tubes containing high-temperature heating steam. For single-stage reheat and for the second stage of two-stage reheat, the heating steam is supplied from the main steam system at a point ahead of the turbine stop valves. Pipe lines carrying live steam to reheaters belong to the main steam system. For two-stage reheat, the first-stage heating steam supply is from an intermediate stage of the high-pressure turbine. More advanced liquid metal cooled fast breeder reactor (LMFBR) nuclear power plants may use the liquid metal for steam reheating.

Industrial (Process) Power Plants. In industrial power plants, power generation may be considered as a by-product, depending upon the process steam requirements. Process steam is the main product of the cogenerating plant. Industrial turbines, with controlled exhaust and extraction pressures, are very efficient throttling devices (replacing valves), supplying steam at desired pressures to a process plant (e.g., oil refinery) while generating electricity. Variations in process heat and power demands usually do not coincide. In a case where electrical power generation requirements are specified, separate condensing steam turbines or gas turbines must be installed in the power plant to make up the difference between electric power demand and the process-heat-dependent power generated by industrial steam turbines.

A typical steam header in a modern industrial plant is shown in Fig. C3.5, which is a photograph of an installation in the Wabash, Indiana, plant of the General Tire Company. As may be noted, the boiler leads and connections supplying steam to process, heating, pumps, and so on, are brought down to this header, which is located near the floor so that all valves are readily accessible.

REFERENCE DOCUMENTS

Various codes and standards and other reference documents used for design of steam systems piping are listed under References at the end of this chapter. The reader may find other documents suitable for use in the area of steam systems piping.

Extensive safety requirements for power-piping systems are contained in the ASME Boiler and Pressure Vessel Code, Section I, Power Boilers, and ASME Section III, Nuclear Power Plant Components, and the ASME B31.1, Power Piping Code. In designing the component parts of piping systems within the jurisdiction of these codes, reference should be made to specific provisions as representing standards for minimum safety requirements, but this is not intended to indicate necessarily the best practice known to the art. Requirements of the Code for Pressure Piping are not compulsory in any state until they have been adopted as law by that state. They are in common use, however, and frequently are referred to in contract specifications and similar documents. Sometimes even though the state has not adopted codes, the insurance carriers make it a requirement to comply with certain codes.

FIGURE C3.5 Steam header in industrial plant. (*Courtesy of Valve World.*)

The selection of suitable dimensional standards for flanges, fittings, valves, pipe, and bolting for ordinary service conditions can be made from the appropriate publications of the American National Standards Institute (ANSI), American Standards for Testing and Materials (ASTM), and American Society of Mechanical Engineers (ASME) which are referenced in the governing Codes such as B31.1, ASME I, ASME III, and ASME B31.3.

DESIGN CONSIDERATIONS

Design Conditions

Design conditions or loadings include design pressure, design temperature, and design mechanical loads[1] which are generally defined as follows:

Design pressure is the most severe pressure expected during any normal or upset operating condition. Allowances are made for pressure surges, control system error, and system configuration effects such as static pressure heads.

Design temperature shall not be less than the expected maximum mean metal temperature through the thickness of the part being considered for the most severe normal or upset operating condition.

Design mechanical loads are selected so that when combined with the effects of design pressure, they produce the highest primary stresses of any coincident combination of loadings for expected normal or upset operating conditions.

Design Pressure. The factors which determine the size of a steam pipe for a specific installation are as follows: the initial steam pressure and steam temperature, the minimum permissible discharge pressure, allowable velocity, the quantity of steam, and the length of line including equivalent lengths for fittings. By knowing or assuming any one or all of these factors, the pipe size may be calculated by means of one of the pressure-drop formulas of Chap. B8 in Part B of this handbook.

Because of its ease of solution by means of a graphic chart, the Unwin formula[8,9,10a] has been widely used in the district-heating industry for many years:

$$\Delta P = \frac{0.0001306W^2L(1 + 3.6/d)}{Yd^5} \tag{C3.2}$$

Solutions of problems by use of this formula can be shortened by means of the Unwin chart shown in Fig. C3.6. This chart shows the pressure drop for 100 ft of pipe with various initial pressures and quantities of steam. The diagonal velocity lines are equivalent velocities and must be multiplied by the factor corresponding to the pipe diameter to obtain the actual velocity. Even though many engineers would prefer to use the calculated solution of Eq. (C3.2) or a small computer program, illustration of the chart method is included to provide a clear understanding of the variables being considered.

Example. Find the pressure drop per 100 ft in an 8-in straight pipe with a flow of 40,000 lb/h of saturated steam and an initial pressure of 150 psia.

Solution. On the horizontal scale for 8-in pipe, find 40,000 lb/h; draw an imaginary line vertically from this point to intersect the line marked 150 psia initial pressure. Read the equivalent velocity, 6700 fpm. Multiply this by 0.8638, the velocity factor given in Fig. C3.6*b* for 8-in pipe, thus obtaining the actual velocity, 5790 fpm. To find the pressure drop for 100 ft of pipe, follow right from the intersection (previously obtained) of the vertical line with the 150 psia pressure line and read the pressure drop of 0.80 from the scale at the right.

The pressure drops through fittings such as elbows, tees, and valves vary in proportion to the pressure drop through straight pipe. Because of this fact, it is possible to express the resistance of fittings as equivalent feet of straight pipe and to compute the pressure drop for the whole line as if it consisted only of straight pipe as discussed in Chap. B8 in Part B.

At elevated velocities, Unwin's formula gives pressure drops known to be higher than actual. In solving for pipe size, it gives sizes slightly larger than the more exact method.

Where grid or network systems are involved, the computation of pipe size, pressure drops, and quantities of steam flowing presents a more complex problem. The use of one of the many computer programs available is recommended for these applications.

Minimum pipe wall thickness for either heavy-wall or thin-wall pipe may be calculated from the following formulas from NC-3641 of Section III of the ASME Code[1] or equivalent, ASME B31.1 or B31.3 Code.[2,3] See Chap. B2 in Part B of this handbook for more information on wall thickness. The y factor adjusts for elastic or plastic properties of the respective materials over the expected range of operating temperatures:

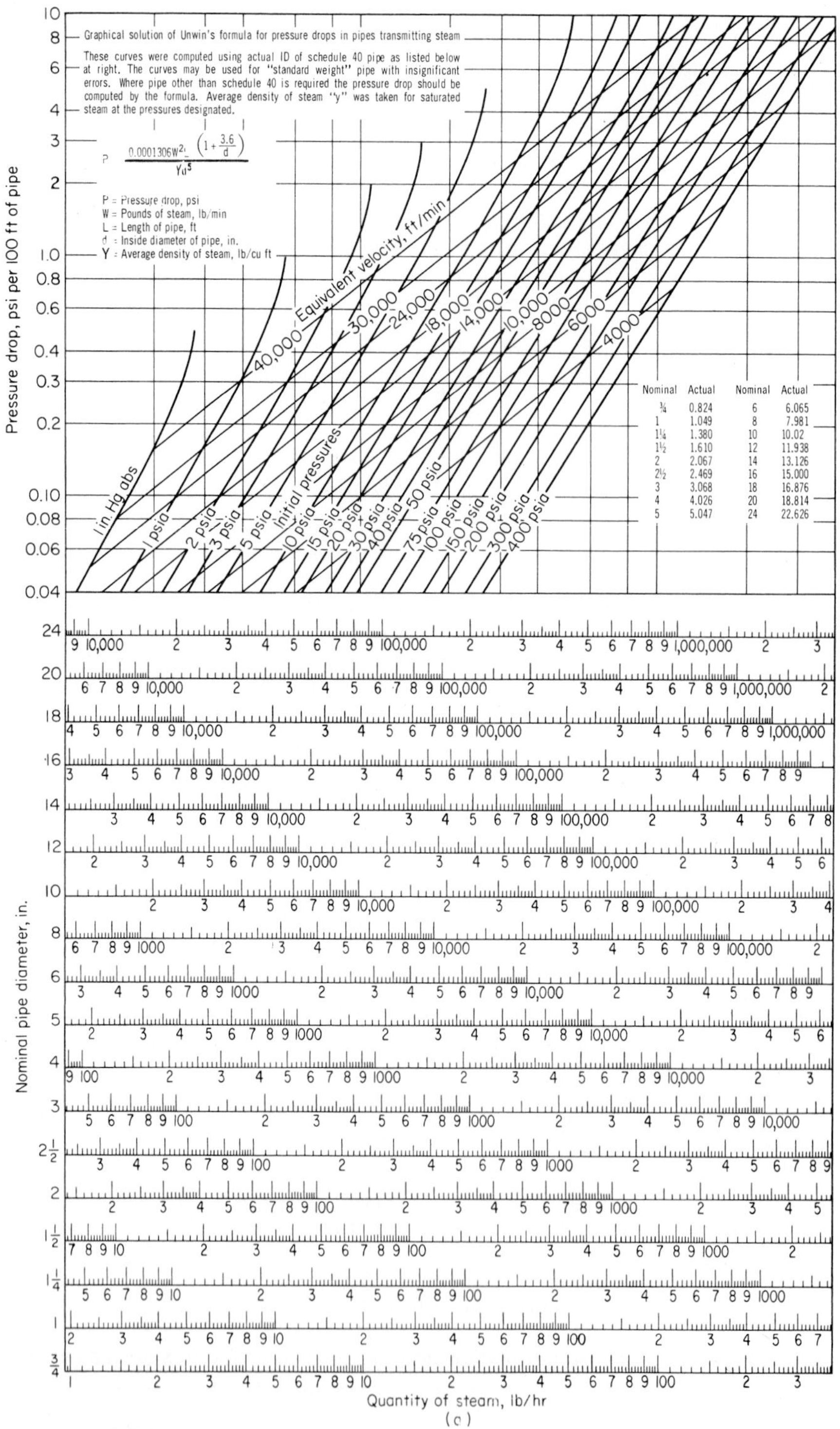

FIGURE C3.6*a* Unwin chart: Approximate solution for sizing pipe in district heating steam systems.

Velocity factor

To obtain the actual velocity from the curves, multiply the equivalent velocity found on the graph by the factor corresponding to the pipe diameter

Pipe diameter	Factor
$\frac{3}{4}$	0.1443
1	0.1792
$1\frac{1}{4}$	0.2278
$1\frac{1}{2}$	0.2598
2	0.3198
$2\frac{1}{2}$	0.3691
3	0.4376
4	0.5370
5	0.6322
6	0.7185
8	0.8638
10	1.000
12	1.116
14	1.182
16	1.281
18	1.374
20	1.465
24	1.627

(b)

Equivalent length of fittings and valves in feet of schedule 40 straight pipe

Nominal pipe size, in.	90° ell screwed	Tee screwed	90° ell welded $R/d=1\frac{1}{2}$	Miter ells 1-45°	Miter ells 1-90°	Miter ells 2-90°	Gate valve	Globe valve	Angle valve
$\frac{3}{4}$	2.06	4.12	0.82	1.03	4.12	1.37	0.48	22.9	11.4
1	2.62	5.24	1.05	1.31	5.24	1.75	0.61	29.1	14.6
$1\frac{1}{4}$	3.45	6.90	1.38	1.72	6.90	2.30	0.81	38.3	19.1
$1\frac{1}{2}$	4.02	8.04	1.61	2.01	8.04	2.68	0.94	44.7	22.4
2	5.17	10.3	2.07	2.58	10.3	3.45	1.21	57.4	28.7
$2\frac{1}{2}$	6.16	12.3	2.47	3.08	12.3	4.11	1.44	68.5	34.3
3	7.67	15.3	3.07	3.84	15.3	5.11	1.79	85.2	42.6
4	10.1	20.2	4.03	5.04	20.2	6.71	2.35	112	56
5	12.6	25.2	5.05	6.30	25.2	8.40	2.94	140	70
6	15.2	30.4	6.07	7.58	30.4	10.1	3.54	168	84.1
8	20.0	40.0	7.98	9.97	40.0	13.3	4.65	222	111
10	25.0	50.0	10.0	12.5	50.0	16.7	5.85	278	139
12	29.8	59.6	11.9	14.9	59.6	19.9	6.96	332	166
14	32.8	65.6	13.1	16.4	65.6	21.9	7.65	364	182
16	37.5	75.0	15.0	18.8	75.0	25.0	8.75	417	208
18	42.1	84.2	16.9	21.1	84.2	28.1	9.85	469	234
20	47.0	94.0	18.8	23.5	94.0	31.4	11.0	522	261
24	56.6	113	22.6	28.3	113	37.8	13.2	629	314

(c)

FIGURE C3.6 *b* and *c* (*Continued*)

$$t_m = \frac{PD_o}{2(S + Py)} + A \tag{C3.3}$$

$$t_m = \frac{Pd + 2SA + 2yPA}{2(S + Py - P)} \tag{C3.4}$$

Equation (C3.3) is used when outside pipe diameter D_o is considered for calculating required minimum wall thickness t_m, whereas Eq. (C3.4) is used with inside pipe diameter, d.

The allowable working pressure of pipe may be determined from the following equation:

$$P_a = \frac{2St}{D_o - 2yt} \tag{C3.5}$$

Large fossil power plants generally use a 2400 psig drumtype steam generator or 3500 psig once-through, supercritical steam generator. Small and medium-sized power plants of less than 200 MW may use lower standard pressures. For the large fossil power plants, the main steam design pressure is typically 2650 or 3860 psig (lowest superheater safety valve setting) and design temperature is 1015 or 1050°F (maximum expected superheater outlet temperature). The steam at these conditions is superheated so there is no moisture, but the lines must still have moisture removal capabilities.

ASME B31.1[2] Paragraph 122.1.2 states, in part, the following for boiler external piping:

> The minimum pressure and temperature and other special requirements to be used in the design for steam, feedwater, blowoff, and drain piping from the boiler to the valve or valves required.
>
> It is intended that the design pressure and temperature be selected sufficiently in excess of any expected operating conditions, not necessarily continuous, to permit satisfactory operation without operation of the overpressure protection devices. Also, since the operating temperatures of fired equipment can vary, the expected temperature at the connection to the fired equipment shall include the manufacturer's maximum temperature tolerance.

For steam piping the value of design pressure, *P,* to be used in Eqs. (C3.3), (C3.4), and (C3.5) shall according to ASME B31.1 Paragraph 122.1.2 be as follows:

> *(A.1)* for steam piping connected to the steam drum or to the superheater inlet header up to the first stop valve in each connection, the value of *P* shall be not less than the lowest pressure at which any drum safety valve is set to blow, and the *S* value shall not exceed that permitted for the corresponding saturated steam temperature.
>
> *(A.2)* For steam piping connected to the superheater outlet header up to the first stop valve in each connection, the design pressure, except as otherwise provided in (A.4) below shall be not less than the lowest pressure at which any safety valve on the superheater is set to blow, or not less than 85 percent of the lowest pressure at which any drum safety valve is set to blow, whichever is greater, and the *S* value for the material used shall not exceed that permitted for the expected steam temperature.
>
> *(A.3)* For steam piping between the first stop valve and the second valve, when one is required by Para. 122.1.7, the design pressure shall be not less than the expected maximum sustained operating pressure or 85 percent of the lowest pressure at which any drum safety valve is set to blow, whichever is greater, and the *S* value for the material used shall not exceed that permitted for the expected steam temperature.
>
> *(A.4)* For boilers installed on the unit system (i.e., one boiler and one turbine or other prime mover) and provided with automatic combustion control equipment responsive to steam header pressure, the design pressure for the steam piping shall be not less than the design pressure at the throttle inlet plus 5 percent, or not less than 85 percent of the lowest pressure at which any drum safety valve is set to blow, or not less than the expected maximum sustained operating pressure at any point in the piping system, whichever is greater, and the *S* value for the material used shall not exceed that permitted for the expected steam temperature at the superheater outlet. For forced-flow steam generators with no fixed steam and waterline, the design pressure shall also be no less than the expected maximum sustained operating pressure.
>
> *(A.5)* The design pressure shall not be taken as less than 100 psig [700 kPa (gauge)] for any condition of service or material.

For example, for a fossil power plant with the lowest pressure at which any drum safety valve is set to blow at 2400 psig, the design pressure shall not be less than maximum sustained operating pressure for all steam piping connected to the steam drum or to the superheater outlet header up to the first stop valve or 2040 psig (85 percent of 2400), whichever is greater, for piping as noted in A.2, A.3, or A.4 above.

Initial steam conditions at the steam-generator outlet depend upon the plant under consideration. For process plants, steam parameters are dictated by the process requirements. In nuclear power plants, steam parameters depend upon the characteristics of a nuclear reactor which is the source of heat for generating steam. The maximum permissible heat flux for fuel, reactor vessel pressure, and the DNBR are limiting factors influencing turbine initial steam conditions of nearly saturated steam at about 1100 psia (see Fig. C3.7) in a NSSS with a PWR. In BWR, the steam moisture problem assumes a major role in optimizing steam pressure at about 1000 psia.

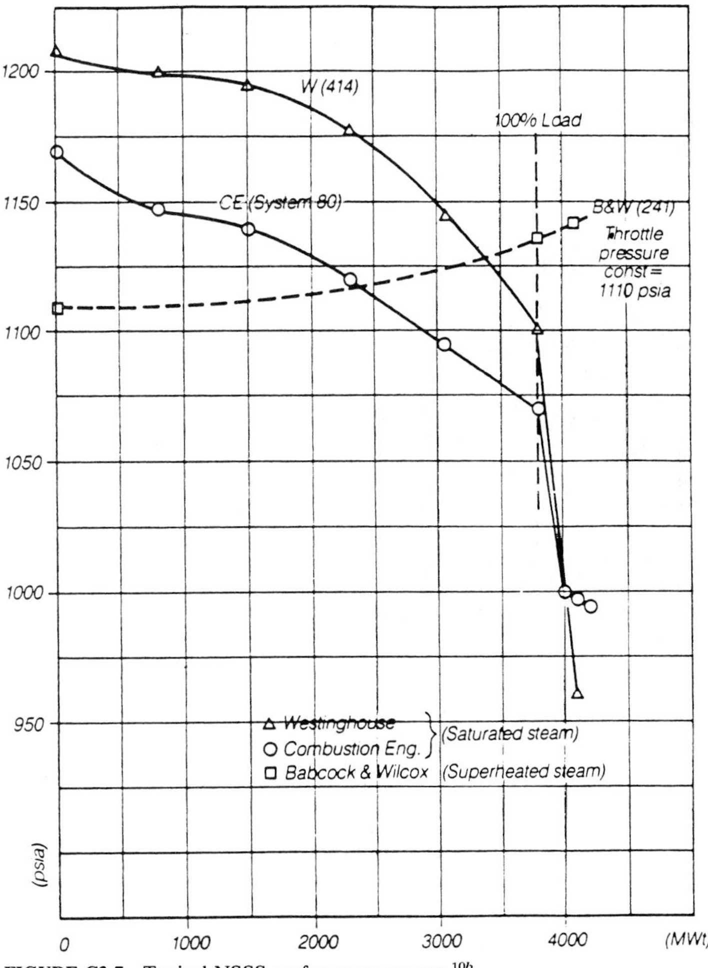

FIGURE C3.7 Typical NSSS performance curves.[10b]

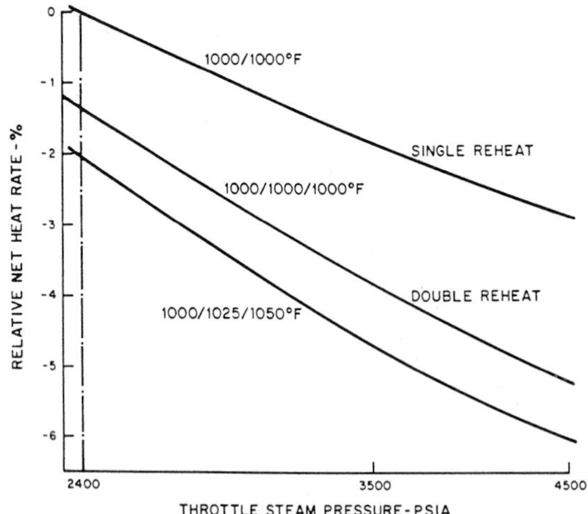

FIGURE C3.8 Effect of steam conditions on heat rate. (*R. C. Spencer, Design of Double Reheat Turbines for Supercritical Pressures, Proceedings of the American Power Conference, vol. 42, 1980.*)

The most freedom in selecting initial steam conditions is found in fossil-fueled power plants which may be designed as subcritical or supercritical units. The question then arises: What are the most economical initial steam conditions for the present technological state of the art? Generally, from a thermodynamic point of view, an increase in initial steam temperature or pressure increases the power plant cycle efficiency (see Fig. C3.8).

The cold reheat and hot reheat steam systems for fossil power plants are designed to ASME B31.1. For PWR and BWR equipped plants, the reheat lines are part of the turbine manufacturer's area of responsibility and are designed to turbine manufacturer's criteria which must equal or exceed the rules and requirements of ASME B31.1. Design temperatures and pressures are the maximum ones expected in the system (consult appropriate heat balance diagrams and system descriptions).

The extraction steam system is designed to ASME B31.1. The extraction line design pressure and temperature depend upon the specific extraction point from the turbine as shown on the full-load heat balance diagram at valve wide open (VWO) and 5 percent overpressure. Extraction steam is typically superheated steam except for the lowest extraction points which are in the wet steam region.

Extraction line sizing is determined by performing an economic analysis of pressure drop, pipe costs, and so on, with a maximum velocity limit of 15,000 fpm (250 fps) for superheated steam. See "Determining Reasonable Flow Velocity" for more flow velocity data. Since extraction steam piping affects heat rate and output of a plant, this piping is sized so pressure drop does not exceed about 5 percent of turbine stage pressure for the low-pressure lines and 3 percent for the higher pressure lines. However, these pressure drops should be as low as practical, especially those related to the higher pressure heaters. Also, keep them low to avoid erosion and corrosion problems.

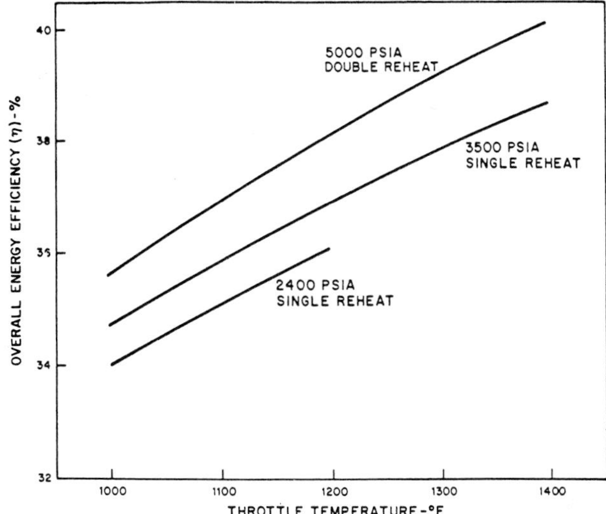

FIGURE C3.9 Effect of steam turbine throttle pressure and temperature on overall efficiency of a power plant. (*D. T. Beecher et al., Energy Conversion Alternatives Study, NASA CR-134941, Vol. I, Feb. 1976.*)

Design Temperature. Historical development of throttle parameters reveals that in the field of throttle temperatures certain stabilization was reached, while the throttle pressures are continuously rising and the condenser back pressure is steady at approximately 1 psia. The effect of steam-turbine throttle conditions on overall power plant efficiency is shown in Fig. C3.9. The results shown in Fig. C3.9 indicate that the efficiency increases by 1.5–2.5 points as pressure increases from 2400 to 5000 psia at constant throttle and reheat temperature of 1000–1400°F. It is also seen that the 3500 psia plant efficiency increases by nearly 4 points as the throttle temperature is increased from 1000 to 1400°F.

Some utilities have installed units using supercritical steam pressure; that is, pressures higher than 3206 psia at which the specific volume of steam and water are equal. Thus, main steam pressure for larger units lies between say 1450 psia and the supercritical region. Main steam temperature may range up to and beyond 1050°F resulting in upgrading of main steam line material as the temperature increases as noted below:

Up to 775°F—carbon steel

Up to 950°F—use ¼ Cr

From more than 950 to 1050°F—use 2¼ Cr

Beyond 1050°F—use austenitic stainless steel

Dynamic Effects. Consider transient effects for turbine trip and safety and relief valve discharge early in the design schedule so that pipe supports and structural steel are adequate to withstand dynamic loads. Also, nozzle design loads for safety and relief valve discharge should be considered. See Chap. B8 in Part B of this handbook for more information about transient flow analysis.

Weight Effects. See Section B2.4.3 in Part B of this handbook for consideration of weight effects.

Thermal Expansion and Contraction Loads. Thermal expansion of pipelines can be provided for by use of pipe bends, offsets, expansion loops, or changes in direction of the pipeline itself. Where pipe bends or offsets can be used or where the pipeline direction is changed to provide for expansion, the provisions of the appropriate codes must be followed as required to assure that all expansion stresses are within the applicable code limits.

Design Features

Pipe Size and Materials. Specific materials for components used in steam piping systems must satisfy the requirements of the applicable codes. Any standards or specifications quoted in this chapter are minimum requirements. Construction at least equal to that required by the applicable Codes is mandatory. The following requirements for piping and fittings are generally acceptable in the industry and may be used as a guide.

Steel pipe, either seamless or welded, is generally used, although the piping Codes permit a variety of materials and several types of welded pipe. For welding and bending, a carbon seamless steel pipe is recommended. Seamless or electric-resistance welded steel pipe A53, Grade B and seamless steel pipe A106, Grade B are popular selections.

Adjusted ratings at temperatures above and below 750°F for carbon and alloy steels and other alloys are given in the standards to govern their use under pressure or temperature other than the primary service ratings.

Selection of materials for temperatures above 750°F from the various grades of alloys described in ASTM specifications for high-temperature service is facilitated by reference to specific standards. The multiplicity of services in a large plant and the variety of dimensional standards and materials, possible joints, and different types of welding available make it desirable to provide proper standards, design details, and materials selection guidelines.

Main steam line sizing is determined by performing an economic analysis of pressure drop, pipe cost, and so on, with a maximum velocity limit of 15,000 fpm (250 fps); see "Determining Reasonable Flow Velocity" for more flow velocity data. Preliminary pipe sizing is often based on approximately a 3 percent pressure drop from the superheater outlet to the turbine stop valves. Wall thickness is calculated using the equations presented in "Design Pressure." Thus, the main steam piping between the turbine and boiler may consist of one or more lines with metallurgy varying from pressure-temperature rating requirements.

Because of the high cost of alloy piping, the selection of its size is usually the subject of an economic study where the increased pressure drop and its effect on turbine output are weighted against unit pipe cost including installation and hangers. For systems operating at or below 1050°F, chrome-moly alloy steel is often used in order to bring the pipe thickness down to an economically acceptable value for the steam pressures used.

Selection of the operating steam pressure and temperature of a turbine-generator installation is based on an economic study or on experience derived from previous installations. Studies have shown that higher pressures and temperatures are associated with larger units or with higher expected fuel costs.

Main steam piping at 1015°F is typically 2¼ percent chrome, 1 percent molybdenum steel to ASTM A335, Grade P22 in U.S. fossil power plants (see "Design Temperature"). Some European plants use higher levels of chrome—up to 9 percent.

Main steam piping for the PWR and BWR is typically carbon steel to ASTM A106, Grade B or C.

For the given initial steam conditions, power plant capacity, pipe material, and economic factors, the number of main steam leads should be optimized considering the steam generator manufacturer's header arrangement and the turbine valve arrangement.

In practical applications, steam-generator and main steam line material limitations, plant reliability, and economic considerations influence the initial steam conditions. For a given pipe, an increase of initial steam temperature above certain values requires lower initial pressure. The maximum allowable stress values in tension for pipe materials A335 Grade P22 low alloy steel and for A376 Grade TP316N steel versus metal temperature are shown in Fig. C3.10.

If better materials are used for main steam (MS) pipes, steam generators, and turbines, additional steam reheat may be beneficial since this will allow higher throttle pressure at fixed throttle steam temperature. This additional steam reheat not only guards a turbine against operation in a high moisture region, but it also increases fuel savings due to the carnotization of a Clausius-Rankine cycle (see "Understanding the Extraction Steam System").

The high-pressure main steam piping must be designed with attention to the need for adequate size (pipe diameter), thermal expansion, adequate piping sup-

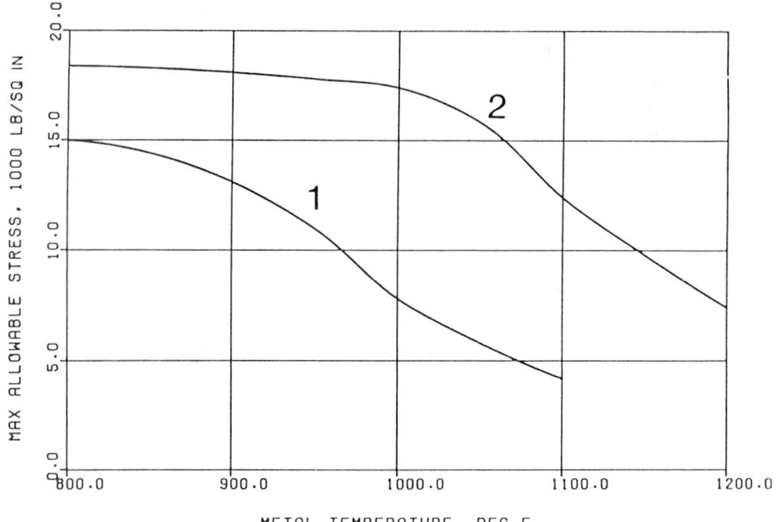

FIGURE C3.10 Maximum allowable stress values versus pipe temperature. (1) Steel A335 grade P22; (2) steel A376 grade TP316N. (*Based on ASME B31.1-1989 ed. Power Piping, Appendix A.*)

ports, and the drain of condensate, the last being especially important when warming up the system preparatory to a start.

Pressure drop in steam lines to turbines and heat losses to the surroundings have an important influence on power plant performance. Long runs of pipe or shorter ones of expensive, thick-walled, alloy steel high-pressure pipe create the need for an economic determination of optimum pipe size. The higher the steam velocity, the smaller the required size of pipe, but unfortunately the frictional pressure loss increases as the square of the velocity. In an economic analysis, the higher capital cost of larger pipe diameters, including insulation and support, must be weighed against the lower operating costs resulting from better plant performance. For more information on the economic optimization of line sizes, see Chap. B8 in Part B of this handbook. The amount of insulation to apply is a separate economic consideration and is also governed by the maximum allowable surface temperature for personnel protection.

Piping Joints. Pressure piping construction to a large extent is welded. Welded joints have generally replaced most of the screwed or flanged joints in new construction of steam distribution system mains and customers' service lines. Joint construction must satisfy the requirements of the code applicable for construction.

Although welding has largely displaced flanged joints, some flanged connections are required, particularly in making connections to flanged valves, expansion joints, or fittings where space limitations do not permit welding or where easy removal of a fitting or valve is desired. Malleable, cast-iron, bronze, or brass fittings may be used for pressures not exceeding 250 psi and temperatures not in excess of 450°F. Cast or forged carbon steel fittings are required for pressures above 250 psi. For temperatures above 775°F, carbon steel is not recommended. Welded fittings must comply with the American Standard for Factory-Made Wrought Steel Buttwelding Fittings (ANSI B16.9,[6]) or American Standard for Forged Steel Fittings, Socket-Welding and Threaded (ANSI B16.11,[11]) where applicable, and the material shall conform to ASTM Specification A216, A234, or A105. Special fittings or welded assemblies fabricated in either the shop or field are required to conform to the requirements of the appropriate codes.

The following conditions are outlined as a guide for design:

1. Pipe lighter than standard or Schedule 40 of Wrought Iron and Wrought Steel Pipe shall not be threaded.
2. Schedule 40 low-carbon-steel pipe may be used for steam pressures up to approximately 400 psia, where welded joints are used and corrosion is not a problem.
3. For lower pressures under the same conditions, Schedule 30 or even 20 may be acceptable.
4. Schedule 80 low-carbon-steel pipe either screwed or welded may be used for saturated steam pressures up to about 600 psi.
5. When steel pipe is threaded and used for steam pressures of 250 psi or greater or for water pressures in excess of 100 psig at temperatures of 220°F or higher, it shall be seamless of a quality at least equal to ASTM Specification A53 or A106 and of a weight at least equal to Schedule 80 in order to furnish added mechanical strength. Refer to Paragraph 10.4.1.2(b) of ASME B31.1.

Gaskets may be made of metal or other material which will not burn, char, or change in character so as not to perform the service intended. Asbestos is prohibited. Several alternatives for asbestos, including spiral-wound gasket applications, have recently been developed by the gasket manufacturers. A summary of the principal selections for a 125-MW fossil power plant is given in Table C3.4.

It is advisable to express a word of caution regarding the connection of cast-iron flanges and fittings to steel flanges and fittings. The Class 25 and Class 125 cast-iron flanges have plain faces, whereas the Class 250 flanges have raised faces which come out practically to the inner edge of the bolt holes. Steel flanges are narrower and stop some distance inside the bolt holes. The plain face or wide raised face on cast-iron fittings and flanges is a necessary precaution to prevent cracking the flange in drawing up the bolts. Numerous instances have been observed where cast-iron flanges have cracked when being bolted to raised-face steel flanges. In cases where it is necessary to bolt cast-iron and steel flanges together, the raised face of the steel flange should be machined down flush with the flange edge. For the same reason, when lap-joint pipe is made up with cast-iron flanges, the lapped end should be brought out to the inner edge of the bolt holes. These types of flange joints should only be used where permitted by the governing code.

For reasons similar to those stated above, it is inadvisable to use alloy steel bolts in cast-iron flanges. Commercial carbon-steel bolts (A370, Grade B) are amply strong for use in cast-iron flanges, and there is no occasion to risk cracking such flanges through the use of high-strength bolts. Steel flanges are properly designed, both as regards to dimensions and material, for use with a narrow raised face and alloy-steel bolts.

Valves. Valves must be of a design at least compatible with the service conditions and constructed of the materials allowed by the appropriate codes for design pressure and design temperature.

ANSI pressure class determination should be based on the design and operating conditions of the system (i.e., temperature and pressure). It should be noted that for valves with elastomeric or plastic gaskets, packing, or seating elements, the valves may not meet the entire range of pressure and temperature conditions for their designated ANSI or ASME pressure class.

For power work, the following codes give pressure temperature tables for valves:

ANSI B16.34, Valves-Flanges, Threaded, and Welding End[12]

MSS-SP-84, Valves-Socket Welding and Threaded Ends[13]

API-602, Compact Steel Gate Valves, Fifth Edition[14]

When selecting a valve, consider the following:

- The system safety category
- The design and operating conditions of the system or portion of the system in which the valve will be installed
- Valve function; that is, isolation, throttling, or modulating
- Radioactivity level, if any
- Whether or not pressure drop is critical
- ,Preferred end connection; that is, socket weld, butt weld, threaded, flanged, or mechanical
- Special features, including minimum operating time, stem leakoff required, exceptional tightness required for seat and/or seals, body taps
- Material requirements; that is alloy, carbon steel, hard facing, and so on

TABLE C3.4 Material Specification for Selected Lines (150 MW Circulating Fluidized Bed Power Plant)

	System					
Item	Main steam	Hot reheat	Cold reheat 1st point extraction	2nd point extraction	Condensate pump discharge; 4th, 5th, 6th point heater outlet	3rd–6th point extraction, raw water, service water, other condensate piping
Design pressure and temperature	1985 psig, 1015°F	590 psig, 1015°F	615 psig, 710°F	210 psig, 785°F	330 psig, 285°F	110 psig, 650°F or less
Pipe size (OD/NPS), thickness	16.1 t_{min} = 210 in. 11.7, t_{min} = 1.5 in.	22, Sch 80	24, Sch 60 8 in XS	8 Std	Over 2 Std Up to 2 in, Sch 80	Over 2 Std Up to 2 in, Sch 80
Pipe material	A335 P22	A335 P22	A106 Gr B	A335 P22	A106 Gr B	A53 Gr B
Pipe construction	Seamless	Seamless	Seamless	Seamless	Seamless	Seamless
Flange type	Not allowed	Welding neck	Welding neck	Welding neck	Welding neck	Slip-on or welding neck
Flange material	N/A	A182 F22	A105	A182 F11	A-105	A105
ANSI standard	N/A	B16.5	B16.5	B16.5	B16.5	B16.5 or MSS-SP-44
Class	N/A	900 RF	600 RF	300 RF	300 RF	150 RF
Gaskets	Not allowed	Spiral wound	Spiral wound	Spiral wound	Spiral metal	*
Fittings over 2 in						
ASTM Spec.	A234 WP22	A234 WP22	A234 WBP	A234 WP11	SA234 WPB	A234 WPB, or A216 WCB
ANSI Std.	B16.9, B16.28	B16.9, B16.28	B16.9, B16.28	B16.9, B16.28	B16.9, B16.28	B16.9, B16.28
Type	Butt weld	Butt weld	Butt weld	Butt weld	Butt weld	Butt weld or flanged
2 in and smaller						
ASTM Spec	A182 F22	A182 F22	A105	A182 F11	SA-105	A-105
ANSI	B16.11	B16.11, MSS-SP-79	B16.11, MSS-SP-79	B16.11, MSS-SP-79	B16.11, MSS-SP-79	B16.11, MSS-SP-79
Rating	9,000 class	3,000 class	3,000 class	3,000 class	3,000 class	3,000 class
Type	Socket weld	Socket weld	Socket weld	Socket weld	Socket weld	Socket weld

*Spiral wound if over 180°F; red rubber J-M Style 107, or equal if below 180°F; Vellumoid if oil.

Isolation and Control Valves. A selection of the proper valve for a particular purpose depends on the operating pressure and temperature and on the type of valve best suited for the use to which it will be put. In general, it is customary to use gate valves in locations where pressure drop through the valve is a consideration and where the valve will be either wide open or fully closed. Guard valves and shutoffs for boiler and turbine leads, and so on, are almost always of the gate type. Globe valves are commonly used in water, steam, and air lines for control or throttling purposes, as the globe type permits closer regulation of flow. Throttling often involves some steam cutting of the seat and disk, and these parts of globe valves are more easily repaired or replaced than in gates. Among such uses for globe valves are turbine and engine throttles, bypasses around traps or reducing valves, and hand-feed regulation on boilers.

Check valves are required in feed lines close to a boiler to prevent water or steam blowing back from the boiler if, for any reason, the feed line ruptures or its pressure fails. It is advisable to use check valves in individual pump or trap discharges before they join a common header and where different lines are joined together to discharge into a common header. A check valve cannot be counted on for closing a line off completely against pressure working back through, but it will stop a considerable flow. In pump discharges where the header remains under pressure after the pump is shut down, a gate valve should be provided in addition to the check. It is also desirable to provide a small relief valve on the pump suction to prevent pressure backing up through the pump and overpressurizing the suction side of the system pump.

With pressure-reducing or other control valves it is desirable to select a size that is loaded somewhere near capacity under normal operation, as such valves are then more stable in their operation. If there is considerable seasonal variation in the load on a reducing or control station, it is good practice to install a large and a small valve in parallel and use the one best fitting the load at any particular time. It is frequently desirable to install a hand-operated bypass around a control valve so that service can be maintained while the special valve is being repaired. Such an arrangement with provision for shutoff to repair the special valve while the line is in operation is shown in Fig. C3.11, which includes flanged valves. Advantage can be taken of improved valve design, wherein weld ends are used while still permitting the internals to be removed for repairs. Main steam lines usually do not use flanges. Filler pieces should be provided between a flanged reducing valve and the adjacent gates, as reducing valves usually are constructed so it is impossible to remove all the end-flange bolts when they are bolted directly to other valves or fittings, because of close clearance at the valve bonnets or fitting necks. For convenience in removing these valves from the line, flanged connections are used frequently at such points, even though the rest of the line is made with screwed or welded joints. When reducing from a pressure that requires the use of a heavy standard for flanges and fittings to a pressure with which a lighter standard is used or to which low-pressure equipment is con-

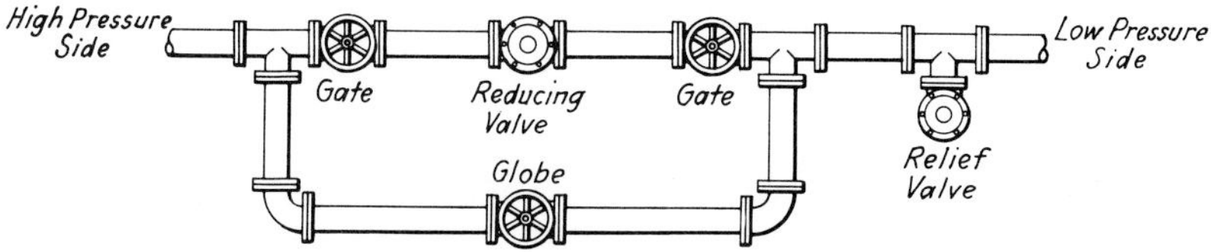

FIGURE C3.11 Pressure-reducing station.

nected, relief valves should be provided on the low-pressure side. The use of the heavier standard should be continued to the last valve ahead of the relief valve, as it is possible to have full pressure up to that point. A reducing valve seldom has a tight shutoff, and at times of negligible steam consumption, leakage through the valve may be enough to build up full line pressure on the low-pressure side. The ASME Code for Power Piping requires that the combined discharge capacity of the safety or relief valves shall be such that the pressure rating of the lower pressure piping will not be exceeded in case the reducing valve sticks open.

Safety and Relief Valves. Safety and relief valves are defined in Section I of the ASME Code[15] as follows:

> *Safety Valve:* An automatic pressure relieving device actuated by the static pressure upstream of the valve and characterized by full-opening pop action. It is used for gas or vapor service.
>
> *Relief Valve:* An automatic pressure relieving device actuated by the static pressure upstream of the valve which opens further with the increase in pressure over the opening pressure. It is used primarily for liquid service.
>
> *Safety Relief Valve:* An automatic pressure-actuated relieving device suitable for use either as a safety valve or relief valve, depending on application.

The construction and method of installing safety valves for power boilers are explained in Section I of the ASME Boiler and Pressure Vessel Code.[15] Part of the boiler safety valve requirements are excerpted as follows:

> Each boiler shall have at least one safety valve or safety relief valve and if it has more than 500 sq ft of bare tube water-heating surface, or if an electric boiler has a power input more than 1100 kW, it shall have two or more safety valves or safety relief valves. For a boiler with combined bare tube and extended water-heating surface exceeding 500 sq ft, two or more safety valves or safety relief valves are required only if the design steam generating capacity of the boiler exceeds 4000 lb/h.
>
> The safety valve or safety relief valve capacity for each boiler shall be such that the safety valve, or valves will discharge all the steam that can be generated by the boiler without allowing the pressure to rise more than 6 percent above the highest pressure at which any valve is set and in no case to more than 6 percent above the maximum allowable working pressure...
>
> One or more safety valves on the boiler proper shall be set at or below the maximum allowable working pressure. If additional valves are used the highest pressure setting shall not exceed the maximum allowable working pressure by more than 3 percent. The complete range of pressure settings of all the saturated-steam safety valves on a boiler shall not exceed 10 percent of the highest pressure to which any valve is set. Pressure setting of safety relief valves on high-temperature water boilers may exceed this 10 percent range.

Additional requirements and requirements for alternate protection against overpressure are included in Section I of the ASME Boiler and Pressure Vessel Code,[15] Section PG-67.

Where more than one safety or relief valve is used on a boiler or other pressure vessel, it is desirable to set one or more of the valves to relieve at a lower pressure than the rest. This serves as a warning before too much steam is lost through all the valves opening at once and also tends to facilitate repairs by confining any cutting action to the one or more valves that open first. In some cases an extra safety valve, known as the power-control valve, is set to blow before the others and is downstream of a gate valve so that it can be removed for repairs while the boiler or steam line is in service. The capacity of this valve cannot be considered in meeting Code or other safety requirements because it might be shut off. Where the hazard involved does not require the installation of a full-size relief valve, it is sometimes desirable to install a small-size pop valve as a telltale to give warning when the usual working pressure is exceeded. The operator can then attend to restoring it to normal conditions. A safety valve for use with a compressible fluid, such as steam or air, is distinguished from a relief valve in that a safety valve has an adjusting, or huddling, ring and chamber to control the amount the pressure blows down before the valve reseats.

Bolting. In bolting cast-iron flanges or steel flanges to cast-iron flanges, valves, fittings, and so on, bolts must be of carbon steel equivalent to ASTM A307, Grade B without heat treatment other than stress relief. Otherwise, they may be too strong for the cast-iron flanges. Threads in accordance with the coarse-thread series of the standard for screw threads, ANSI B1.1 and B18.2.1,[16,17] are recommended for carbon-steel bolts. Carbon-steel bolts conforming to A307 may be the standard regular or heavy hexagonal-head bolts and must be used with standard heavy semifinished hexagonal nuts, ANSI B18.2.2,[18] which conform to ASTM A194.

For high-temperature service or for insurance of a tight joint in the case of steel flanges, bolts or stud bolts should be of alloy steel, conforming to ASTM A193, typically Grade B7. Nuts must be of steel according to ASTM A194, Grade 2H.

Bolt and nut selection for use with flanges in power piping may often be made from the alloy-steel bolts listed in Table 112 and Section 108.5 of ASME B31.1.[2]

Following is a summary of the principal bolting selections for a 150-MW typical circulating fluidized bed power plant (see Table C3.4):

- *Bolt studs:* Eight threads per inch (coarse thread for less than 1 in) with two nuts.
- No bolted joints are allowed for main steam.
- *Stud material:* A193, B16 for hot reheat, if present; A193, B7 for all other systems noted.
- *Nuts:* Hexagon semifinished Heavy Series per B18.2.2.[18]
- *Nut material:* A194, Grade 4 for hot reheat; A194, Grade 2H nut standard for all other systems noted.

Protective Conduits for Underground Lines. Protective conduits for underground steam mains and services are necessary to protect the pipe and insulation from damage due to earth pressure and impact loadings, to allow free longitudinal expansion and contraction while held in proper alignment, and to prevent groundwater seepage or flooding by providing either drains or a completely waterproof structure.

There are many types of conduits used by the industry today. They can be categorized into the following general classifications: prefabricated, boxtype, solid pour, granular fused, and tunnels.

Prefabricated conduits are popular with utilities located in large cities where conditions exist such as congested subsurface, heavy surface traffic, tidewater,

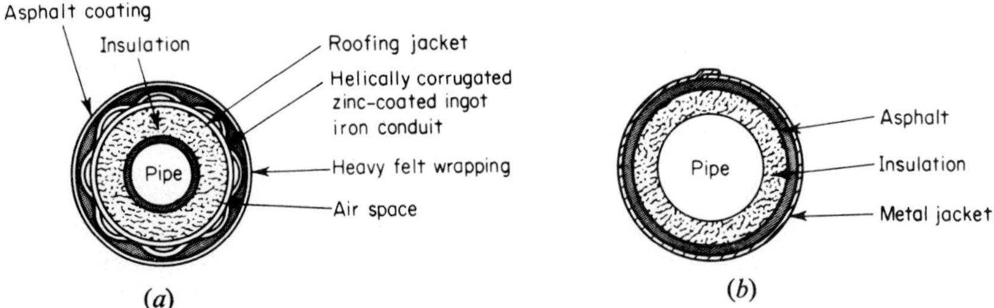

FIGURE C3.12 Prefabricated conduits for underground steam pipes. (*a*) Coated and wrapped corrugated conduit; (*b*) poured asphalt protects thermal insulation.

and rock areas. The ideal design for these conditions is one with the smallest cross-sectional area, consistent with thermal requirements, to fit into limited subsurface space. Two prefabricated designs are shown in Fig. C3.12.

In Fig. C3.12*a,* the insulated pipe is placed within a corrosion-resistant helically corrugated metal jacket which is protected with a heavy asphaltic coating. In Fig. C3.12*b,* the insulated pipe is placed concentrically within the corrosion-resistant metal jacket and the intervening space between jacket and insulation is poured full of high-melting-point asphalt, which is a protective medium for the insulation. Some of these prefabricated designs are also available with cast-iron casings with fittings designed for joints, bends, and so on.

Boxtype conduits have many variations. Depending on loading conditions, reinforcement may or may not be necessary. Drainage is provided by filling drainage pockets with crushed stone or gravel; these pockets may be installed on either or both sides of the conduit. They conduct water from the top of the conduit to the lower drain, which is connected to a sewer.

Solid-pour construction, such as is shown in Fig. C3.13*a* consists of poured structural concrete which is vibrated or tamped around a conventionally insulated steam line. If the insulation has a high compressive strength, such as is characteristic of diatomaceous earth or silica, it can support the pipe; otherwise, it should be supported independently. Reinforcing is sometimes used to prevent settlement. An eccentric space may be left around the pipe by using insulation larger than the outside diameter of the pipe. This permits the pipe to rise during the warming-up period without crushing the insulation.

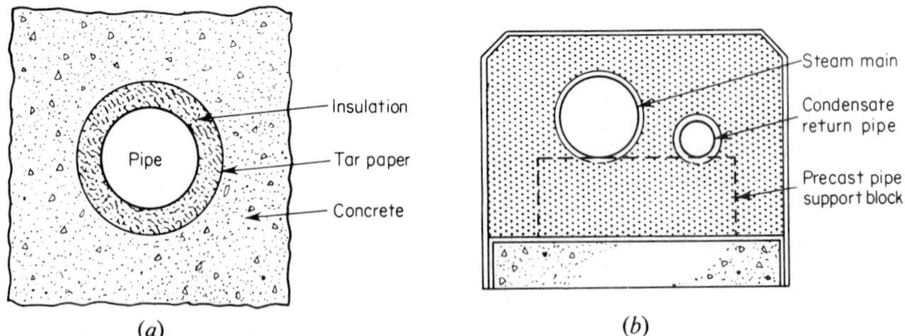

FIGURE C3.13 Solid-pour protection for underground steam pipe. (*a*) Insulated and wrapped pipe in structural concrete; (*b*) pipe surrounded by insulating concrete.

The design of Fig. C3.13*b* uses insulating concrete as a conduit. The concrete is poured around the steam pipe, which is supported on precast blocks of the same material. The insulating concrete may consist of any mixture of insulation or other cellular materials and portland cement or a mixture of a special foaming material mixed with portland cement; the aim is to create a cellular mass composed of minute air cells in the concrete so as to develop insulating qualities. The piping (which could be a number of pipes) is wrapped with corrugated paper before the insulation is poured, and a heavy-asphalt-coated waterproofing membrane, or equivalent waterproof sheeting, is installed to protect the top and side of the structure before backfilling.

Granular-fused types of conduits as shown in Fig. C3.14 consist of a granular bitumen, selected as to the required temperature range and poured and tamped around the pipe or pipes in a trench. Heat passing through the pipe forms three concentric zones: (1) a dense semiplastic core fused on by the pipe's own heat, (2) a sintered zone providing thermal insulation and moisture proofness, and (3) an outer layer of granules providing a final zone of thermal insulation and the load-bearing portion of the structure. Tunnels are very costly and are constructed only out of necessity.

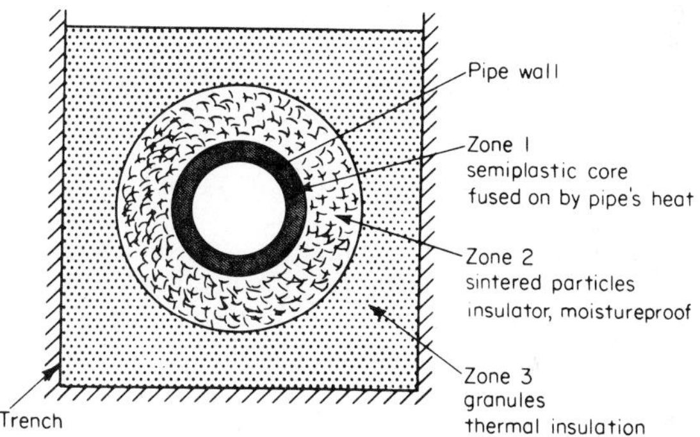

FIGURE C3.14 Granular-fused conduit for underground steam pipe.

Walk-through tunnels are usually not provided for steam mains unless they are required for underground passage between buildings, as in institutions, and then, they are usually constructed to accommodate other utilities services, such as electricity, water, and gas.

Provision for drainage and ventilation with special insulating methods for steam mains to prevent heat loss should be considered.

No attempt has been made to recommend the type of conduit to be used for a certain type of condition. The illustrations shown here are only a few of the many designs used in practice today. Choice and application depend upon the local conditions. In the economic evaluation, factors to be considered are life expectancy, operation and maintenance, foundations to prevent settlement, external and internal drainage, corrosive action of soil conditions, loads to be imposed, field installation, and necessary insulating properties.

Drainage of Condensate. To prevent a water slug type of water hammer and a possible rupture of the steam main, condensed steam or condensate within the steam main must be removed. Provisions must be made during construction to grade the steam main carefully so as to pitch the pipe down not less than 1 in within 50 ft in the direction of steam flow. Since it is impractical to assume that a continuous slope can be maintained in a long run of main, removal of condensate (drainage) must be provided at all low points where a water pocket will exist. In any event, recommended lengths of steam main for draining off condensate should not exceed 300–400 ft. Usually, the contour of the ground or subsurface structures will dictate the pitch of the steam main.

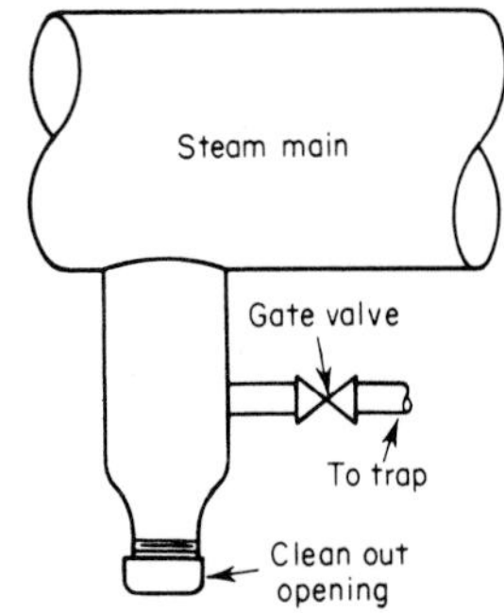

FIGURE C3.15 Drain pocket for steam-trap connection to low-velocity steam main.

When condensate is drained from a steam main, a drain pocket is welded to the bottom of the pipe to be drained. The diameter of the pocket should be about one-third the diameter of the line up to a maximum of 6 in for 18-in and larger mains. The pocket not only provides for condensate removal but also allows for sediment removal. Figure C3.15 shows this design. Figure C3.16 is a design which is in the form of a separator and is more prevalent in underground steam mains.

Sometimes, it is not economically feasible or practical to return the condensate from district-heating systems because of the high cost of installation and maintenance of pumping equipment, manholes, and return piping. For systems which supply high flow rates to concentrated loads, it probably will be economical to return the condensate in order to recover the energy in the condensate and to save the chemical treating and quantity of makeup water in the boiler plant.

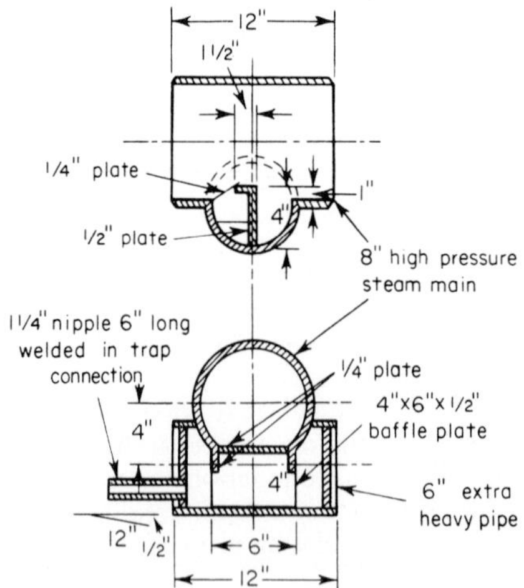

FIGURE C3.16 Drain pocket for steam-trap connection to high-velocity steam main.

In district-heating systems, traps off the drain pocket can be discharged through cooling coils to the city sewer system. City regulations may require the condensate discharge to be cooled before entering the sewer. In some cases, condensate discharge lines are piped to sump pits in manholes where the water is removed by float-controlled electric pumps which discharge it into the sewer.

In process and power system piping, condensate removal is an important consideration to prevent water slug formation in steam lines. Paragraph 122.11 of ASME B31.1[2] contains the following provisions for drains, drips, and steam traps:

> Drip lines from piping or equipment operating at different pressures shall not be connected to discharge through the same trap. If two or more traps discharge into the same header which is, or may be, under pressure, a stop valve and a check valve shall be provided in the discharge line from each trap.
>
> Trap discharge piping shall be designed for the same pressure as the inlet piping unless the discharge is vented to atmosphere, or is operated under low pressure and has no stop valves. In no case shall the design pressure of trap discharge piping be less than the maximum discharge pressure to which it may be subjected.

Steam traps are automatic devices used to trap or hold steam until it has condensed and to allow condensate and air to pass as soon as they accumulate. In general, a trap consists of a vessel in which the condensate accumulates, an orifice through which the condensate is discharged, a valve to close the orifice port, mechanisms to operate the valve, and inlet and outlet openings for the entrance and discharge of the condensate from the trap vessel. Steam traps are discussed in detail in Chap. A2.

All steam lines must be adequately drained of condensate. Even superheated steam lines need drainage since condensation forms during the warming up period or while the line is hot but without flow. The points to be drained are the low points in the line, moisture separators, drip pockets, and valves, especially in vertical lines. Horizontal portions of the steam lines should be pitched downward approximately ⅛ in/ft in the direction of flow or ¼ in/ft for lines that contain a steam and water mixture or that require draining periodically. Condensate flow against the steam should be avoided if possible.

Determining Reasonable Flow Velocity. Before proceeding beyond a preliminary layout of piping systems, it is necessary to determine pipe sizes which allow reasonable velocities and friction losses. The maximum allowable velocity of the fluid in a pipeline is that which corresponds to the permissible pressure drop from the point of supply to the point of consumption or that which does not result in excessive pipeline erosion. The economic optimization of line sizes is discussed in Chap. B8 in Part B of this handbook. The values of velocity listed in Table C3.5 are reasonable for use in such cases. The lower velocities should be used for small pipes and the upper ones for large ones. These values represent good average practice and may be used as a guide in many cases where actual pressure drops are not computed. Additional steam line velocity information is given in Chap. B8 in Part B of this handbook.

Erosive action on valve seats and similar exposed parts also affects permissible velocity. This action is much more pronounced in the case of wet steam than with superheated steam, and velocities should be correspondingly lower when there is much moisture in steam.

High velocities are sometimes used in dry steam lines where excess pressure

TABLE C3.5 Reasonable Design Velocities for Flow of Fluids in Pipes

Fluid	Pressure (psig)	Use	Reasonable Velocity	
			Feet per minute	Feet per second
Water	25–40	City water	120–300	2–5
Water	50–150	General service	300–600	5–10
Water	150 up	Boiler feed	600–1200	10–20
Saturated steam	0–15	Heating	4000–6000	67–100
Saturated steam	50 up	Miscellaneous	6000–10,000	100–167
Superheated steam	200 up	Large turbine and boiler leads	10,000–15,000	167–250

Source: Courtesy, Stone & Webster.

exists to absorb the higher pressure drop. The high velocity is not in itself objectionable. There is no appreciable erosion of the pipe walls and no undesirable effects except that high velocities are accompanied by considerable noise, which would be objectionable in heating systems in office buildings and dwellings but may not be in power plants.

In the selection of pipe sizes for exhaust lines from auxiliary turbines and similar services where high velocities may be used to advantage, consideration should be given to the limiting velocities which can be obtained. This limiting or sonic velocity has a definite value for each combination of steam enthalpy and pressure (see Chap. B8 in Part B).

It is often less expensive to use a larger standard pipe size than to use a unique pipe size. Table C3.4 shows a special size for main steam piping and standard sizes recommended for the other systems. Where space is at a premium, it may be necessary to use a smaller size than would otherwise be good practice. The calculation of velocity or pressure drop is a valuable check, but in the last analysis judgment is the deciding factor, and the blind use of generalized criteria is impracticable.

Special attention should be given to sizing hot and cold reheat piping since pressure drop in reheat systems affects heat rate and capability of the power plant. Cost of piping must be compared with these effects to arrive at the most economical piping arrangement. Normally, total pressure drop in reheat piping and the reheating section of the steam generator should be 7–9 percent of high-pressure turbine exhaust pressure. It is desirable to use a smaller diameter hot reheat line and larger diameter cold reheat line, taking a greater pressure drop in the more expensive (alloy) hot reheat line.

Loss Due to Steam Leaks. Modern power plants with almost all welded joints are subject to little leakage, only through the few remaining nonwelded joints such as in the valves or other equipment connections. The hazard of high-pressure, high-temperature steam leaks, together with the cost of chemicals required in most makeup systems, makes early repairs mandatory.

Existing low-pressure plants using flanged or screwed joints with considerable leakage and where maintenance is deferred are wasting both water and energy at a rate which depends on the size of the opening and steam pressure.

The value of steam which can be lost through a comparatively small leak becomes appreciable when considered over a period of time. The amounts of steam

TABLE C3.6 Loss Due to Steam Leaks*

Size of orifice (in)	Steam wasted per month (lb)	Water wasted (gal)
250-lb gauge		
½	1,780,000	213,600
⅜	1,001,000	120,100
¼	445,000	53,400
⅛	111,000	13,300
1/16	27,800	3,300
1/32	7,000	800
300-lb gauge		
½	2,125,000	255,000
⅜	1,195,000	143,400
¼	531,000	63,700
⅛	132,800	15,900
1/16	33,200	4,000
1/32	8,300	1,000
400-lb gauge		
½	2,804,000	336,500
⅜	1,577,000	189,200
¼	701,000	84,100
⅛	175,200	21,000
1/16	43,800	5,300
1/32	11,000	1,300
600-lb gauge		
½	4,157,000	498,800
⅜	2,338,000	280,600
¼	1,039,000	124,700
⅛	259,000	31,200
1/16	65,000	7,800
1/32	16,300	2,000

*Values in the table are based on the use of Grashof's formula.

which will escape through various size orifices at different pressures can be computed by an appropriate flow formula. Table C3.6, which was computed by Grashof's formula,[19] gives the pounds of steam lost per month and water wasted.

Expansion, Flexibility, and Supporting. Chapters B4 and B5 of Part B of this handbook provide basic information on the thermal expansion, the needed flexibility, and the supporting of piping systems. In underground steam piping, expansion joint fittings are more commonly used for thermal expansion owing to the limitations of space and costly trenching. Expansion joints are often used for application in underground steam mains.

The basic design of a sliptype joint consists of a cast-iron or steel body with a stuffing box and a sliding sleeve. The advantage of sliptype joints is the longer traverses that can be obtained to absorb pipe expansion; the disadvantage is the necessity of maintenance of packing. By the very nature of its construction, the slip joint is capable of absorbing only axial movement.

The metallic bellows expansion joint is composed of the following components: (1) the flexible element proper (corrugated metal tube), the rings enabling reinforced elements to support higher pressures, and the end collars or other structures serving to increase pressure capacity of cylindrical ends and transfer spring force and hydrostatic end thrust to connected piping and (2) the end nipples or flanges, and, for other than anchored axial joints, the various hardware items used to cross connect the ends, such as tie rods, hinges, and gimbals.

The advantages of the bellows type of joint which permit it for use in buried steam piping are that it requires no maintenance and can absorb a combination of axial, lateral, and angular movement within a large range of pressures and temperatures. Its disadvantage is the limitation of traverse, generally 7–8 in of axial movement.

Adequate anchoring and guiding are essential for the proper functioning of all expansion joints. Main anchors for end thrusts must be designed for the sum of the pressure thrust, the force required to deflect the joint, and the force due to friction in the piping guides. Intermediate anchors are used between balanced or doubletype joints, where the pressure thrust is balanced and the anchor need be designed only to restrict the expansion movement of the pipe. Guides are essential to assure proper pipe alignment into the expansion joint so that undesirable torques are not imposed on the expansion element.

Cold springing of expansion joints used only for lateral deflection can provide several advantages. The most significant of these is the optimization of the position of the expansion joint. In addition, the joint is more stable at high pressures, since the maximum angular displacement of the corrugations is reduced. Joints with internal sleeves and external covers must have adequate clearance to permit the lateral deflection of the expansion element. If the deflection can be reduced by 50 percent, these clearances can also be reduced by 50 percent. Internal sleeves can then be of maximum diameter, and external sleeves are held to a minimum diameter.

All piping systems expand with an increase in temperature in the system. Long runs of high-temperature piping generally use bends, fittings, and offsets to assist in keeping the stresses within the Code's allowable values and reactions (forces and moments) on the turbine or boiler connections within the manufacturer's allowable values. Rigid hangers, constant and variable support hangers, guides, and anchors are also useful to control the thermal growth but must be carefully engineered.

For piping such as main steam and hot reheat systems operating in the creep range (900°F and above), cold springing may prove useful in controlling the reactions on the equipment within the manufacturer's specified limits, although this technique is not universally recommended. (Cold spring is the process of cutting short the piping by a certain percentage of the expected thermal growth so that equipment loading in the cold, erected position is that percentage of the calculated hot reactions from the thermal analysis.) Cold springing does not reduce the stresses in the piping as allowed by the Code, but it may be useful in meeting the vendors' limits on allowable forces and moments on their equipment, provided the piping does not creep significantly at operating temperature. As the piping heats up and cycles through a number of thermal cycles in the creep range, the cold reactions will gradually be self-relieved to a certain extent. See ASME B31.1 Code, Paragraph 119.9[2] for a more complete discussion of this subject.

The piping designer in conjunction with the flexibility analyst needs to work closely with the vendors of the equipment in laying out the high-temperature piping and support systems and so these systems will meet all requirements over the life of the plant. Especially critical are the loads imposed on rotating equipment such as steam turbines since clearances are critical and distortion due to piping loads can cause real problems to this equipment.

The true economics of the system design must include all the above factors plus the normal cost factors (piping, insulation, hangers, etc.).

Fabrication, Assembly, and Erection. For underground steam piping, manholes are required for sectionalizing valves and bypass valve piping, trap piping and traps, some types of expansion joints, and convenience of location of other expansion joints and anchorage. Modern manholes are constructed of reinforced concrete, cast iron, or steel. In field-pour concrete construction, waterproofness should be assured by pouring the walls and floor monolithically. Prefabricated enclosures are pretested for waterproofness before installation. Other provisions that should be considered in manhole design are (1) adequate working space for maintenance, (2) clearance for removal of equipment, (3) ventilation, and (4) drainage. If the manhole floor elevation is below the sewer or if a sewer is not accessible for drainage piping from the manhole, sump pump manholes with automatic pumps or water ejectors must be provided. An example of such an installation is shown in Fig. C3.17. In prefabricated steam-main construction, prefabricated manholes with the necessary valves and piping installed may be delivered to the job site as a unit. This type of manhole is shown in Fig. C3.18.

With prefabricated installation, expansion joints of the bellows design may be delivered to the job site and welded into the steel piping and conduit of the steam main. This provides for directly buried expansion joints in a fully encapsulated system. Where convenient or necessary, expansion joints are installed in manholes with sectionalizing valves and thrusttype anchors. Traps are usually installed in their own manholes, that is, in a manhole separate from that which houses the sectionalizing valve, for ease of maintenance.

Steam piping in power plants is fabricated and installed in accordance with the appropriate Code. The following discussion highlights several of the requirements for ASME Class 2 piping,[1] but specific Codes should be reviewed as ap-

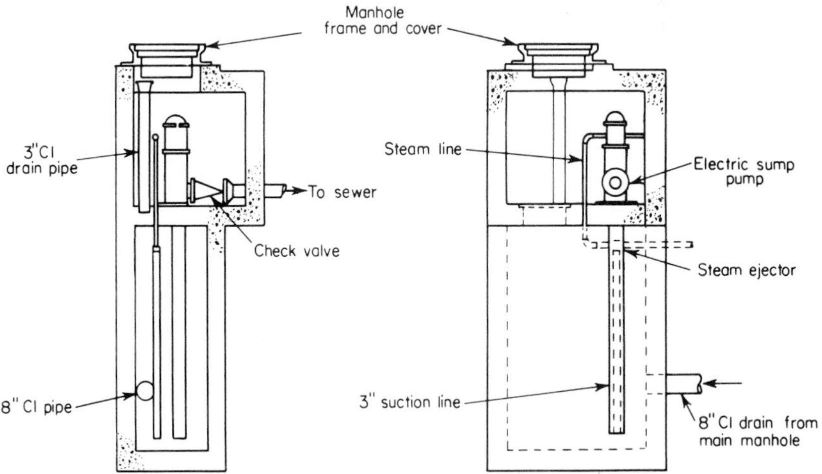

FIGURE C3.17 Manhole for sump pump.

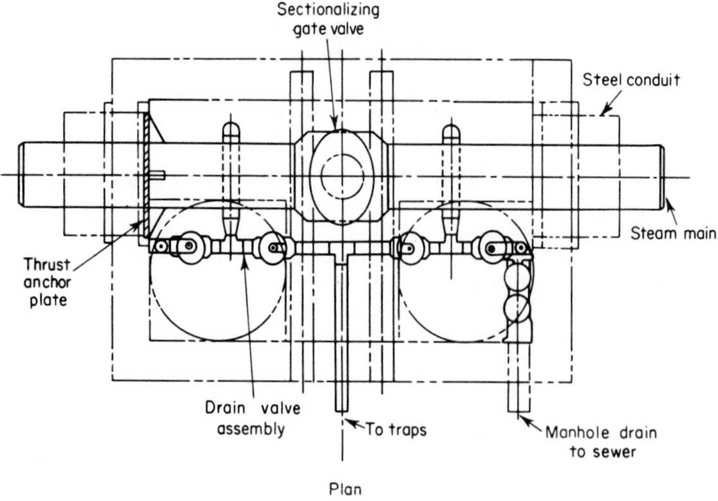

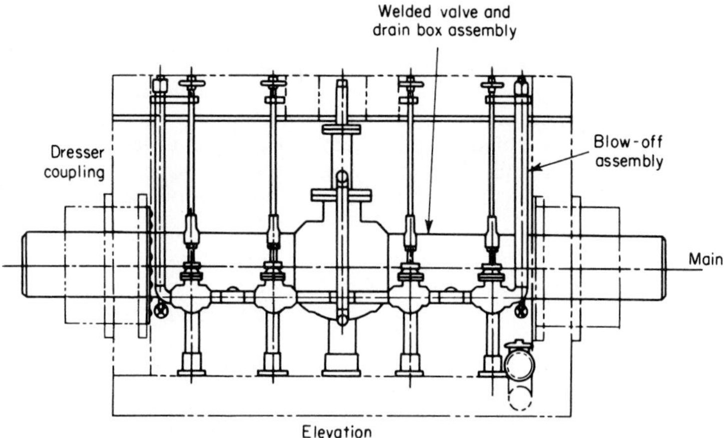

FIGURE C3.18 Manhole for sectionalizing valve and drain valve assembly.

plicable. Components, parts, and appurtenances shall be fabricated and installed in accordance with the rules in the Code and shall be manufactured from materials which meet the requirements of the Code.

Material for pressure-retaining parts shall carry identification markings which will remain distinguishable until the component is assembled or installed. If the original identification markings are cut off or the material is divided, either the marks shall be transferred to the parts cut or a coded marking shall be used to ensure identification of each piece of material during subsequent fabrication or installation. In either case, an as-built sketch or a tabulation of materials shall be made identifying each piece of material with the Certified Material Test Report, where applicable, and the coded marking. For studs, bolts, nuts, and heat exchanger tubes, it is permissible to identify the Certified Material Test Reports for material in each component in lieu of identifying each piece of material with the Certified material Test Report and the coded marking. Material supplied with a Certificate of Compliance and welding and brazing materials shall be identified and controlled so they can be traced to each component or installation of a piping system, or else a control procedure shall be used which ensures that the specified materials are used.

Material originally accepted on delivery in which defects exceeding the limits of acceptance are known or discovered during the process of fabrication or installation is unacceptable. However, the material may be used, provided the condition is corrected in accordance with the requirements for the applicable product form.

Materials may be cut to shape and size by mechanical means such as machining, shearing, chipping, or grinding or by thermal cutting. When thermal cutting is performed to prepare weld joints or edges, to remove attachments or defective material, or for any other purpose, consideration shall be given to preheating the material, using preheat schedules such as suggested in the Code.

Any process may be used to hot or cold form or bend pressure-retaining materials, including weld metal, provided the impact properties of the materials, when required, are not reduced below the minimum specified values or are effectively restored by heat treatment following the forming operation. Hot forming is defined as forming with the material temperature higher than 100°F below the lower critical temperature of the material.

When impact testing is required by the design specifications, a procedure qualification test shall be conducted using specimens taken from materials of the same specification, grade or class, and heat treatment and with similar impact properties as required for the material in the component. These specimens shall be subjected to the equivalent forming or bending process and heat treatment as the material in the component. Applicable tests shall be conducted to determine that the required impact properties of the Code are met after straining.

The tolerances for formed or bent piping include minimum wall thickness and ovality requirements. Bending processes shall be selected and qualified to maintain a wall thickness for bent piping sufficient to satisfy the requirements of the design calculations at the resultant section thickness. Unless otherwise justified by the design calculations, the ovality of piping after bending shall not exceed 8 percent as noted in Eq. (C3.6):

$$100 \times \frac{(D_{max} - D_{min})}{D_o} \leq 0.08 \qquad \text{(C3.6)}$$

where D_o = the nominal pipe outside diameter (in)

D_{min} = the minimum outside diameter after bending or forming (in)

D_{max} = the maximum outside diameter after bending or forming (in)

Parts that are to be joined by welding may be fitted, aligned, and retained in position during the welding operation by the use of bars, jacks, clamps, tack welds, or temporary attachments. When the inside surfaces of items are inaccessible for welding or fairing, alignment of sections shall meet the following requirements:

For circumferential joints, the inside diameters shall match each other within 1/16 in. When the items are aligned concentrically, a uniform mismatch of 1/32 in all around the joint can result. However, other variables not associated with the diameter of the item often result in alignments that are offset rather than concentric. In these cases, the maximum misalignment at any one point around the joint shall not exceed 3/32 in. Should tolerances on diameter, wall thickness, out of roundness, and so on, result in inside diameter variation which does not meet these limits, the inside diameters shall be counterbored, sized, or ground to produce a bore within these limits. Offset of outside surfaces shall be faired to at least a 3:1 taper over the width of the finished weld or, if necessary, by adding additional weld metal.

For longitudinal joints the misalignment of inside surfaces shall not exceed 3/32 in, and the offset of outside surfaces shall be faired to at least a 3:1 taper over the width of the finished weld or, if necessary, by adding additional weld metal.

No welding shall be undertaken until after the welding procedures which are to be used have been qualified. Only welders and welding operators who are qualified in accordance with Code requirements shall be used. All welding procedure qualification tests shall be in accordance with the requirements of the Code.

The method used to prepare the base metal shall leave the weld preparation with reasonably smooth surfaces. The surfaces for welding shall be free of scale, rust, oil, grease, and other deleterious material. The work shall be protected from deleterious contamination and from rain, snow, and wind during welding. Welding shall not be performed on wet surfaces. Rules for making welded joints in piping are described in the following paragraphs.

Backing rings which remain in place may be used for piping in accordance with the requirements of the Code. The materials for backing rings shall be compatible with the base metal, but spacer pins shall not be incorporated into the weld.

In double-welded joints, before applying weld metal on the second side to be welded, the root of full penetration double-welded joints shall be prepared by suitable methods such as chipping, grinding, or thermal gouging, except for those processes of welding by which proper fusion and penetrations are otherwise obtained and demonstrated to be satisfactory by welding procedure qualifications.

Where single-welded joints are used, particular care shall be taken in aligning and separating the components to be jointed so there will be complete penetration and fusion at the bottom of the joint for its full length.

When components of different diameters are welded together, there shall be a gradual transition between the two surfaces. The slope of the transition shall be such that the length–offset ratio shall not be less than 3:1, unless greater slopes are shown to be acceptable by analysis. The length of the transition may include the weld.

Thickness of weld reinforcements for piping is specified in Table C3.7 for double-welded butt joints. The limitation on the reinforcement given in Column 1 shall apply separately to both inside and outside surfaces of the joint. For single-welded butt joints, the reinforcement given in Column 2 shall apply to the inside surface, and the reinforcement given in Column 1 shall apply to the outside surface. The reinforcements shall be determined from the higher of the abutting surfaces involved.

Fillet welds may vary from convex to concave. A fillet weld in any single continuous weld may be less than the specified fillet weld dimension by not more than 1/16 in, provided the total undersize portion of the weld does not exceed 10

TABLE C3.7 Thickness of Weld Reinforcements for Piping[1]

Material nominal thickness (in)	Column 1: Both double-welded butt joints and outside single-welded butt joints	Column 2: Inside single-welded butt joints
Up to ⅛, incl.	3⁄32	3⁄32
Over ⅛–3⁄16, incl.	⅛	3⁄32
Over 3⁄16–½, incl.	5⁄32	⅛
Over ½–1, incl.	3⁄16	5⁄32
Over 1–2, incl.	¼	5⁄32
Over 2	The greater of ¼ in. or ⅛ times the width of the weld in inches	3⁄32

percent of the length of the weld. Individual undersize weld portions shall not exceed 2 in in length. In making socket welds, a gap of approximately 1⁄16 in at the end of the pipe shall be provided before welding; the gap need not be present nor be verified after welding.

Structural attachments shall conform reasonably to the curvature of the surface to which they are to be attached and shall be attached by full penetration, fillet, or partial penetration continuous or intermittent welds.

Attachments may be welded to the piping system after performance of the pressure test provided the welds do not require postweld heat treatment (PWHT); welds shall be restricted to fillet welds not exceeding ⅜ in throat thickness and to full penetration welds attaching materials not exceeding ½ in thickness; welds shall not exceed a total length of 24 in for fillet welds or 12 in for full penetration welds; and welds shall be examined as required by the Code.

The need for and temperature of preheat are dependent on a number of factors such as the chemical analysis, degree of restraint of the parts being joined, elevated temperature, physical properties, and material thicknesses. Preheat for welding or thermal cutting, when used, may be applied by any method which does not harm the base material or any weld metal already applied or which does not introduce deleterious material into the welding area which is harmful to the weld. PWHT may be accomplished by any suitable methods of heating and cooling, provided the required heating and cooling rates, metal temperature, metal temperature uniformity, and temperature control are maintained.

The threads of all bolts or studs shall be engaged in accordance with the design. Any lubricant or compound used in threaded joints shall be suitable for the service conditions and shall not react unfavorably with either the service fluids or any component material in the system. All threading lubricants or compounds shall be removed from surfaces which are to be seal welded. In bolting gasketed flanged joints, the contact faces of the flanges shall bear uniformly on the gasket and the gasket shall be properly compressed in accordance with the design principles applicable to the type of gasket used. All flanged joints shall be made up with relatively uniform bolt stress.

Examination, Inspection, and Testing. The inside of all pipes, valves, fittings, traps, and other apparatus shall be smooth, clean, and free from all blisters, loose mill scale, sand, and dirt when erected. Main steam lines must be steam blown and others should be before placing in service, if practical.

Before installation, all valves, fittings, and so on, shall be capable of withstanding an appropriate hydrostatic shell test, and piping shall be capable of meeting the hydrostatic test requirements contained in the respective material specifications under which it was purchased.

After installation, all piping systems shall be capable of withstanding a hydrostatic test pressure of one and one-half times the design pressure, except that the test pressure shall in no case exceed the adjusted pressure and temperature rating for 100°F as given in ANSI B16.5 or B16.34[20,12] for the material and pressure standard involved. After a boiler has been completed, it shall be subjected to pressure tests using water at no less than ambient temperature but in no case less than 70°F.[15] At no time during the hydrostatic test shall any part of the boiler be subjected to a stress greater than 90 percent of its yield strength at test temperature. For systems joined wholly with welded joints, the adjusted pressure rating shall be that for ring joint facing. For systems joined wholly or partly with flanged joints, the adjusted pressure shall be that for the facing type used.

When hydrostatic testing is impractical, it shall be required that the piping be tested with steam at a pressure at least equal to the pressure at which the piping is to be operated. These tests shall be made on sections or on the whole of the piping system, but the connections between the sections must be similarly tested.

Steam piping in power plants must be examined, inspected, and tested in accordance with the appropriate Code. The following discussion summarizes several general requirements for ASME Class 2 piping,[1] but specific Codes should be revised as applicable.

The examinations required by the Code shall be performed by personnel who have been qualified as required by Code, and results of the examinations shall be evaluated in accordance with the specified acceptance standards.

Following any nondestructive examination in which examination materials are applied to the piece, the piece shall be thoroughly cleaned in accordance with applicable materials or procedure specifications.

Acceptance examinations of welds and weld metal cladding shall be performed at the times stipulated below during fabrication and installation:

1. Radiographic examination of welds will usually be performed after an intermediate or final PWHT, when required.
2. Magnetic particle or liquid penetrant examinations of welds shall be performed after any required PWHT, except that welds in P-No. 1 materials may be examined either before or after PWHT.
3. All dissimilar metal weld joints such as in austenitic or high nickel to ferritic material or using austenitic or high nickel alloy filler metal to join ferritic materials which penetrate the wall shall be examined after final PWHT.
4. The magnetic particle or liquid penetrant examination of weld surfaces that are to be covered with weld metal cladding shall be performed before the weld metal cladding is deposited. The magnetic particle or liquid penetrant examination of weld surfaces that are not accessible after a PWHT shall be performed before the operation which caused this inaccessibility. These examinations may be performed before PWHT.
5. Ultrasonic examination of electroslag welds in ferritic materials shall be performed after a grain refining heat treatment, when performed, or after final PWHT.

The Codes provide specific acceptance standards for radiographic, ultrasonic, magnetic particle, liquid penetrant, and metallographic examination of all components. Unique examination requirements are usually specified for bellows expansion joints.

Pressure Testing. All pressure retaining components, appurtenances, and completed systems shall be pressure tested. Bolts, studs, nuts, washers, and gaskets are exempted from the pressure test. Pressure testing required by the applicable Code often must be performed in the presence of the inspector.

The installed system shall be pressure tested before initial operation. The pressure test may be performed progressively on erected portions of the system. Components and appurtenances shall be pressure tested before installation in a system except that the system pressure test may be substituted for a component or appurtenance pressure test provided the following is true: The component can be repaired by welding, if required, as a result of the system pressure test; the component repair weld can be postweld heat treated, if required, and nondestructively examined, as applicable; and the component is resubjected to the required system pressure test following the completion of repair and examination if the repair is required to be radiographed. Valves also require pressure testing before installation in a system.

All joints, including welded joints, shall be left uninsulated and exposed for examination during the test. Components designed to contain vapor or gas may be provided with additional temporary supports, if necessary, to support the weight of the test liquid. Expansion joints shall be provided with temporary restraints, if required, for the additional pressure load under test. Equipment that is not to be subjected to the pressure test shall be either disconnected from the component or system or isolated during the test by a blind flange or similar means. Valves may be used if the valves with their closures are suitable for the proposed test pressure. Flanged joints at which blanks are inserted to isolate other equipment during the test need not be retested. If a pressure test is to be maintained for a period of time and the test medium in the system is subject to thermal expansion, precautions shall be taken to avoid excessive pressure. The test equipment shall be examined before pressure is applied to ensure that it is tight and that all low-pressure filling lines and other items that should not be subjected to the test have been disconnected or isolated.

The component or system in which the test is to be conducted shall be vented during the filling operation to minimize air pocketing. Water or an alternative liquid, as permitted by the design specification, shall be used for the hydrostatic test. It is recommended that the test be made at a temperature that will minimize the possibility of brittle fracture. The test pressure shall not be applied until the component, appurtenance, or system and the pressurizing fluid are at approximately the same temperature.

The installed system shall be hydrostatically tested at not less than 1¼ times the lowest design pressure of any component within the boundary protected by the overpressure protection devices which have been provided. Following the application of the hydrostatic test pressure for the required time, all joints, connections, and regions of high stress such as regions around openings and thickness transition sections shall be examined for leakage, except in the case of pumps and valves, which shall be examined while at test pressure. Leakage of temporary gaskets and seals, installed for the purpose of conducting the hydrostatic test and which will be replaced later, may be permitted unless the leakage exceeds the capacity to maintain system test pressure for the required amount of time. Other leaks, such as from permanent seals, seats, and gasketed joints in components, may be permitted when specifically allowed by the design specifications.

The hydrostatic test requirements for bellows expansion joints require that the completed expansion joint shall be subjected to a hydrostatic test in accordance with the applicable code. This test may be performed with the bellows fixed in the straight position, at its neutral length, or, in some cases, with the bellows fixed at the maximum design rotation angle or offset movement.

Pressure test gauges used in pressure testing shall be indicating pressure gauges and shall be connected directly to the component. If the indicating gauge is not readily visible to the operator controlling the pressure applied, an additional indicating gauge shall be provided where it will be visible to the operator throughout the duration of the test. For systems with a large volumetric content, it is recommended that a recording gauge be used in addition to the indicating gauge. All test gauges shall be calibrated against a standard dead-weight tester or a calibrated master gauge. The test gauges shall be calibrated before each test or series of tests.

Special Design Criteria

Overpressure Protection. Construction of underground steam mains must adhere to the requirements as provided for in Section 122.14 of ASME B31.1[2] when the piping is within the jurisdiction of ASME B31.1.

Each main steam lead generally has safety valves at the superheater outlet and often a motor operated block valve which is used for isolation. A main steam stop valve at the boiler outlet facilitates boiler hydro testing before completion of turbine erection; whether or not to provide it is a matter of economics. In the case of two leads, the cross connection between the two main steam leads shall permit full closed testing of one turbine stop valve at a time while the turbine is operating. The cross connection line may have connections to the steam-generator feed pump turbines, auxiliary steam loads, turbine-generator gland seal system, and steam-generator restart system, depending on system design requirements.

Preventing Turbine Overspeed. Turbine water induction problems from main steam lines would normally only occur during start up or shortly after shutdown, so prevention criteria are limited to low point drain design per ASME Standard No. TDP-1.[21]

Steam storage in the turbine, reheater, and reheat piping presents a problem in turbine speed control because surplus energy can continue to be generated in the IP and the LP sections for some time after the main governor has reacted to a load decrease. This condition could result in dangerous overspeeding. To prevent this, governor-operated IVs are installed ahead of the point of reintroduction of the reheated steam to the turbine. The IVs are adjusted to begin closing when the normal speed is exceeded but before turbine speed has increased to the point (approximately 10 percent over normal RPM) where an emergency overspeed trip would shut down the whole unit.

Understanding Reheat Steam Systems. The example of power plant cycle efficiency improvement by increased initial steam conditions in "Design Temperature" suggests that maintaining the initial temperature constant over a range of pressures (as in the Carnot cycle) might also prove fruitful. This would require reheating the main steam to its initial temperature again after each infinitesimal amount of expansion in the turbine. Such a cycle might be said to have an infinite number of reheats.

There is also another important advantage derived from using steam reheat in a power cycle. Reheating is a feature associated with high steam pressures where there is insufficiently high initial steam temperature to yield an expansion that will end with an acceptable moisture content at the condenser pressure. This limit of moisture at the last stage of a turbine is roughly about 12–14 percent. Moisture in the steam flow causes erosion of turbine buckets, shortening their useful life, and also results in an efficiency loss of a turbine. The purpose of reheat is to take full advantage of higher initial steam parameters discussed in "Design Temperature" as they become commercially available.

The number of reheats to be used is both an engineering and an economic problem. The high-pressure, high-temperature steam generator, together with extra cost of a turbine, piping, and controls, makes plants with reheat more expensive than those without. Economically, single reheating has proved desirable on most machines above 100 MW. In actual practice the gain from single reheating is about 4–7 percent in thermal efficiency over an equivalent nonreheat cycle, depending on the size and construction of the turbine (mainly the governing stage). The gain in performance which can be realized from a second reheating is smaller than from the first. For large units operating at 3500 psig, the gain in performance due to a second reheat is in the neighborhood of 2 percent of the net heat rate (see Fig. C3.8). The performance gains for reheats beyond the second diminish rapidly, and the justification for greater expenditures for equipment depends strongly upon the price of fuel.

For every set of initial steam conditions there exists an optimum reheat pressure (or pressures in a double-reheat cycle) which yields the maximum efficiency (minimum heat rate) of the power plant. Studies made of the performance and economic application of single-reheat cycles indicate that for a seven (or less) feedwater heater cycle the cold reheat pressure is adjusted by the turbine manufacturer to provide either the optimum or desired final feedwater temperature. If eight (or more) feedwater heaters are used, it proves economical to design the cycle with the first point heater located above the reheat point (HARP extraction). Figure C3.19[22] indicates typical optimum reheat pressures corresponding to 2400 psig/1000°F/1000°F steam conditions for units taking extraction for the top heater at the reheat point (curve marked "Top Heater at Reheat Point") or above the reheat point where the optimum reheat pressure is found at the low point of a curve corresponding to required final feedwater temperature.

It is seen from this figure that minor deviations from the optimum reheat pressure have a small effect on performance; however, the performance penalties should always be estimated. Size of the turbine, throttle temperature, and reheat temperature have a relatively minor effect on the indicated optimum conditions, and Fig. C3.19 may therefore be considered as reasonably accurate for any single-reheat unit at indicated throttle pressure. However, if more accurate information is needed, it should be obtained from the turbine manufacturer. It should be noted that the relative performance indicated applies only at rated output.

The selection of optimum reheat pressures for a double-reheat turbine is an involved problem. It is affected not only by thermodynamic optimization but also by design limitations (e.g., second-reheat intercept-valve pressure must be fixed at about 300–400 psia at rated conditions due to the very large volumetric steam flows which would result at lower pressures). Figures C3.20 and C3.21[22] show the heat rate penalty which results from the use of reheat pressures other than optimum when the throttle pressure is 3500 and 4500 psig, respectively, and the second cold reheat pressure is fixed at 300 psia.

Table C3.8[23] compares the basic cycle parameters for the 738-MW, 3500-psig

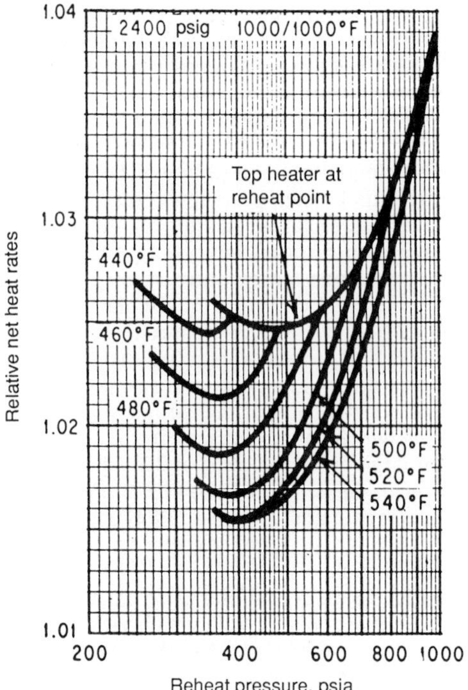

FIGURE C3.19 Relative performance for typical single-reheat units with various combinations of first-reheat intercept-valve pressure and final feedwater temperature.[22] (*General Electric Co.*)

unit and the 900-MW, 4500-psig design. The relatively higher first reheat pressure of the 4500-psig cycle makes possible an additional attractive thermodynamic gain when an eighth feedwater heater, extracting steam from the first reheat section, is added to the cycle. The use of a heater above the reheat point in the 3500-psig cycle will result in an improvement in the heat rate. In this case the required final feedwater temperature would set the limit on the HARP extraction pressure.

Understanding the Extraction Steam System. Regenerative feedwater heating improves the Rankine cycle efficiency. It is achieved by withdrawing a portion of steam flowing through the turbine at a number of points along the flow path (extraction, bleeding). The steam so extracted is condensed in a series of heat exchangers (feedwater heaters) through which flows the feedwater from the condenser on its return to the steam generator. Bleeding is also advantageous for the turbine in that it reduces the mass flow rate of steam in lower pressure stages, which is a desirable feature from the turbine designer's point of view. In most cases it reduces the exhaust losses which occur between the last turbine stage and the condenser. The regenerative feedwater heating arrangement is an integral part of all modern power plants.

One of the most frequent cycle changes which is considered in designing a power plant is whether or not to increase the number of feedwater heating ex-

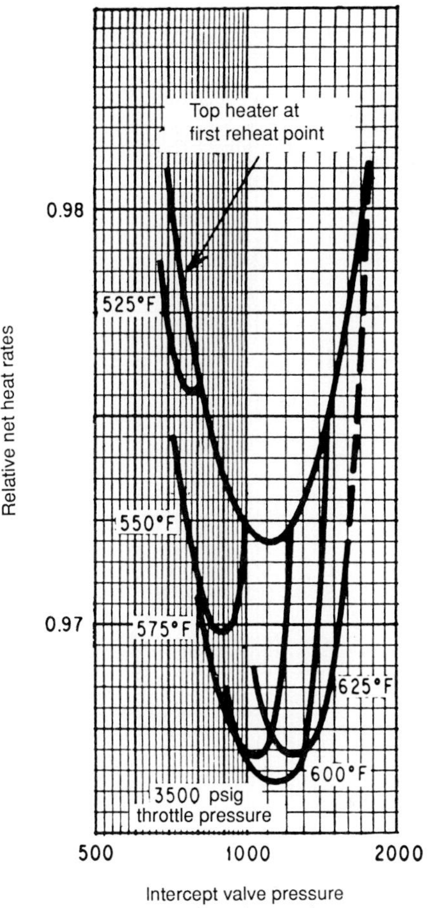

FIGURE C3.20 Relative performance for typical double-reheat units with various combinations of first-reheat intercept-valve pressure and final feedwater temperature.[22] (*General Electric Co.*)

traction points. This problem does not have a simple answer that is applicable in every case. Theoretical prediction of a change in station heat rate, as published by many authors, is based on the assumption that the enthalpy (or temperature) rise is divided equally among the heaters. This is difficult to achieve in practice since the actual feedwater enthalpy rises in the heaters are determined by the extraction location available in the turbine. As a result, the heat-rate gains due to additional heaters typically may be only about one-half the theoretical value. Normally for optimum application, the heater above the reheat point has a feedwater temperature rise of 50–100°F. The number of feedwater heaters to be used is also an economic consideration which must be determined on the basis of power plant cycle optimization. Recently, seven or eight heaters have been used in both fossil-fueled and nuclear power plants.

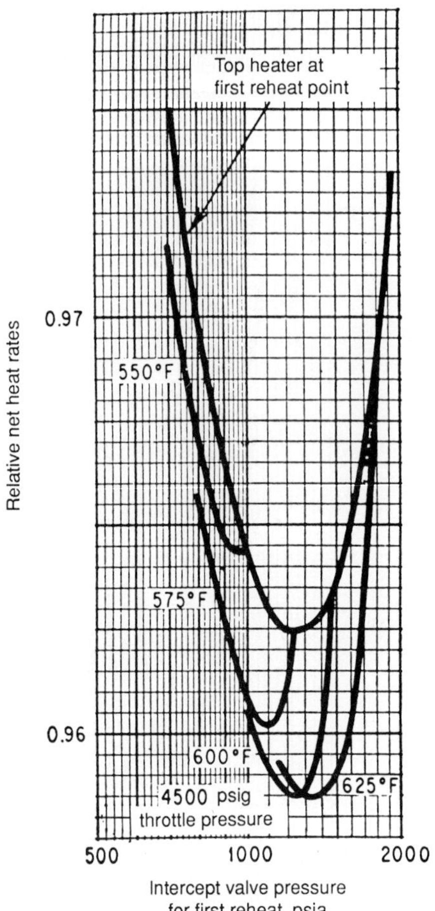

FIGURE C3.21 Relative performance for typical double-reheat units with various combinations of first-reheat intercept-valve pressure and final feedwater temperature.[22] (*General Electric Co.*)

It must be emphasized that the stage pressures at extraction points change with turbine load and are roughly proportional to the mass flow rate of steam to the following stage.

In a multistring cycle when a heater is removed from service, the extraction cross-tie arrangement will be one of the deciding factors in the turbine load limitation. Extraction cross ties are the means by which extraction pipes at the same pressure level can be connected.

TABLE C3.8 Basic Cycle Parameters[23]

	738 MW, 3500 PSIG	900 MW, 4500 PSIG
Maximum capability (MW)	826	1000
Maximum throttle pressures (psia)	3500	4500
First reheat intercept valve pressure (psia)	980	1360
Second reheat intercept valve pressure (psia)	340	390
Final feedwater temperature (°F)	548	590
Number of feedwater heaters	7	8

EXPERIENCE FEEDBACK

Use appropriate materials and welding processes and procedures for all piping. Be aware of erosion and corrosion problems in high moisture content steam lines (>5 percent moisture steam extraction lines) and in drain lines (feedwater heater, moisture separator, and reheat drain lines).

Preventing Turbine Water Induction

Turbine water induction prevention criteria for hot reheat lines in fossil-fired power plants are similar to those for main steam, that is, low point drain design per ASME TDP-1.[21] Turbine water induction prevention criteria for cold reheat lines are more of a concern due to reheater atemperator (regulating the hot reheat steam temperature) spray or feedwater heaters which extract steam from the cold reheat line. The special prevention criteria for cold reheat lines are specified in the ASME Standard No. TDP-1. For nuclear power plants with light water cooled reactors (PWR, BWR), cold reheat lines carry steam containing moisture. Therefore, the special prevention criteria specified in ASME Standard No. TDP-2 must be applied.

REPAIRS, REPLACEMENTS, AND MODIFICATIONS

Use of flanges in high-pressure and high-temperature steam may cause leakage problems. Defects in weld metal detected by examinations or by tests shall be eliminated and repaired when necessary or the indication reduced to an acceptable limit. Weld metal surface defects may be removed by grinding or machining and need not be repaired by welding, provided the remaining thickness of the section is not reduced below that required; the depression, after defect elimination, is blended uniformly into the surrounding surface; and the area is examined by a magnetic particle or liquid penetrant method in accordance with the Code after blending and now meets the acceptance standards. Defects detected by visual or volumetric method and located on an interior surface need only be reexamined by the method which initially detected the defect when the interior surface is inaccessible for surface examination. Excavations in weld metal, when repaired by welding, shall meet the requirements of the Code. The repaired area shall be heat treated, if required.

It is important for the convenience of the operating and maintenance personnel and to minimize the possibility of making mistakes that piping be arranged so as to make the purpose of each line and valve as obvious as possible. This is especially important when making repairs, replacements, and modifications to piping systems and components. Valves in general, and bypass valves in particular, should be installed so their purpose is evident at a glance. It is desirable that bypass valves be grouped together where all can be seen at once rather than placed in scattered locations close to separate pieces of equipment. The possibility of an error in operation or maintenance procedures can be further reduced by stenciling the purpose of each line on the pipe close to its valve. A simple illustration of this point is shown in Fig. C3.22, a photograph of the piping for a small deaerator.

FIGURE C3.22 How to make the purpose of valves obvious.

References

1. ASME Boiler and Pressure Vessel Code, Section III, Rules for Construction of Nuclear Power Plant Components, 1989 Edition, New York.
2. ASME Code for Pressure Piping, B31, an American National Standard, Power Piping, ASME B31.1, 1989 edition, ASME, New York.
3. ASME Code for Pressure Piping, B31, an American National Standard, Chemical Plant and Petroleum Refinery Piping, ANSI/ASME B31.3, 1990 edition, ASME, New York.
4. American Standard, Welded and Seamless Wrought Steel Pipe, B36.10M, 1985.
5. American Standard, Stainless Steel Pipe, B36.19M, 1985.

6. American Standard, Factory-Made Wrought Steel Buttwelding Fittings, ANSI B16.9, 1986.
7. ASME Standard TDP-2, Recommended Practices for the Prevention of Water Damage to Steam Turbines Used for Electric Power Generation, Nuclear Fueled Plants, 1985.
8. W. C. Unwin, "Flow of Gas in Mains and Distribution at High Pressure," *Proceedings of the Institute of Gas Engineers, in J. Gas Lighting, Water Supply,* etc., June 21, 1904, pp. 852–867.
9. National District Heating Association, *District Heating Handbook,* 3d ed.

10*a*. Graphical Solution of Unwin's Formula. Chart published by the National District Heating Association.

10*b*. T. J. Swierzawski, "Selected Considerations Related to Improvements in Thermal Performance of Nuclear Power Plants," *The 2nd International Topical Meeting on Nuclear Power Plant Thermal Hydraulics and Operations*, Tokyo, Japan, April 1986.

11. American Standard, Forged Steel Fittings, Socket-Welding and Threaded, ANSI B16.11, 1980.
12. American Standard, Valves—Flanges, Threaded, and Welding End, ANSI B16.34, 1988.
13. Manufacturers Standardization Society Standard, MSS-SP-84, Valves—Socket Welding and Threaded Ends, 1990.
14. American Petroleum Institute Standard, Compact Steel Gate Valves, Fifth Edition, API 602, 1985.
15. ASME Boiler and Pressure Vessel Code, Section I, Rules for Construction of Power Boilers, 1989 Edition, New York.
16. American Standard, Unified Inch Screw Threads (UN or UNR Thread Form), ANSI B1.1, 1989.
17. American Standard, Square and Hex Bolts and Screws Inch Series Including Hex Cap Screws and Lag Screws, B18.2.1, 1981.
18. American Standard, Square and Hex Nuts Inch Series, ANSI B18.2.2, 1987.
19. S. Crocker in *Piping Handbook, Fifth Edition,* edited by R. C. King, McGraw-Hill Book Company, New York, 1967.
20. American Standard, Pipe Flanges and Flanged Fittings, ANSI B16.5, 1988.
21. ASME Standard, TDP-1, Recommended Practices for the Prevention of Water Damage to Steam Turbines Used for Electric Power Generation, Fossil Fueled Plants, 1985.
22. R. L. Bartlett, *Steam Turbine Performance and Economics*, McGraw-Hill, New York, 1958.
23. R. C. Spencer, "Design of Double Reheat Turbines for Supercritical Pressures", *Proceedings of the American Power Conference*, vol. 42, 1980.

BIBLIOGRAPHY

American Standard, Wrought Steel Buttwelding Short Radius Elbows and Returns, ANSI B16.28, 1986.

Manufacturers Standardization Society Standard, MSS-SP-79, Socket-Welding Reducer Inserts, 1989.

Manufacturers Standardization Society Standard, MSS-SP-44, Steel Pipe Line Flanges, 1990.

CHAPTER C4
BUILDING SERVICES PIPING

Paul A. Bourquin, P.E.
Senior Vice President
Wolff & Munier, Inc.
Hawthorne, New York

Buildings contain many different piping systems to provide heating, cooling, and plumbing for human comfort; pneumatic controls to make the other building systems function; and sprinkler, fire standpipe, and roof drainage systems for safety. This chapter discusses the piping for heating, cooling, and compressed air. The piping for plumbing, roof drainage, refrigerant, and fire protection systems are treated in other chapters.

Building services piping operates at relatively low pressures and temperatures, usually below 400 psig and 400°F, thereby relieving engineers of many of the complexities they are faced with when designing pipelines in utility plants or refineries. The requirements of the ASME/ANSI Pressure Piping Code B31.9, Building Services Piping will be followed where applicable to give the designer a basic familiarity with code design.

BUILDING TYPES

The buildings that are to be serviced by the systems discussed in this chapter can be classified into several groups. These groups are listed below with examples of the types of structures which fall into each category:

Residential: Multistory apartment houses or condominiums

Commercial: Stores, shopping malls, showrooms, theaters, restaurants, and office buildings

Institutional: Schools, universities, and hospitals

Industrial: Factories, assembly plants, and, broadly, buildings where machines and people make or assemble products

Public: Libraries, police stations, municipal buildings, and other places of public usage or assembly

CONSTRUCTION CODES

Before construction of a building is started it is necessary to obtain a building permit from the authorities having jurisdiction over the property on which the building is to be placed. To obtain the permit, the plans must show that the building is to be built in conformance with the building code which has been adopted by the authority.

States, provinces, counties, cities, and towns have their own building codes or have adopted one of the several national codes written by groups organized for the purpose of writing building codes. Every plot of land is under the authority of a jurisdiction, and inquiry will lead to the proper construction codes.

Building codes are omnibustype documents which treat all aspects of construction and specify safety requirements, as well as reference safety codes written by other code-writing organizations. Building codes, safety codes, and standards are updated periodically by the issuing organizations, and the latest revision should be used. The revision date is usually found as a suffix to the code or standard number.

A listing of code-writing bodies and the specific sections of the codes which control piping follows:

Building Codes

1. BOCA—Building Officials & Code Administrators, International, Inc.: National Mechanical Code
2. ICBO—International Conference of Building Officials: Uniform Mechanical Code
3. BCCI—Southern Building Code Conference International, Inc.: Standard Mechanical Code

Requirements for piping are specified in the mechanical sections of the building codes and in the referenced safety sections. When referenced, the applicable section of the ASME/ANSI B31 Pressure Piping Code should be used.

Safety Codes

Safety codes give rules for the construction of piping systems. These codes are often listed as reference documents in building codes. They have also been adopted as law by some states. Applicable codes are listed in Table C4.1.

Component Standards

Component standards list dimensions, chemical composition, and tensile strength of materials. They sometimes contain pressure-temperature ratings. These standards are referenced in both the building codes and the safety codes. Component standards used for building piping are listed in Table C4.2.

Responsibility

The ultimate responsibility for the proper design and construction of a building and its various systems remains with the owner. The owner can use professionals

TABLE C4.1 Safety Codes

ASME—American Society of Mechanical Engineers
BPVC—Boiler and Pressure Vessel Code
Section I—Power Boilers
Section IV—Heating Boilers
Section VIII—Pressure Vessels
Section IX—Welding and Brazing Qualifications
ASME/ANSI B31—Code for Pressure Piping
B31.1—Power Piping
B31.3—Chemical Plant and Petroleum Refinery Piping
B31.5—Refrigeration Piping
B31.9—Building Services Piping
NFPA—National Fire Protection Association
National Fire Codes
NFPA 30—Flammable and Combustible Liquids Code
NFPA 30A—Automotive and Marine Service Station Code
NFPA 31—Installation of Oil Burning Equipment
NFPA 32—Drycleaning Plants
NFPA 50—Bulk Oxygen Systems at Consumer Sites
NFPA 51—Design and Installation of Oxygen-Fuel Gas Systems for Welding, Cutting, and Allied Processes
NFPA 51B—Cutting and Welding Processes
NFPA 54 (ANSI Z223.1)—National Fuel Gas Code
NFPA 58—Storage and Handling of Liquified Petroleum Gases
NFPA 88B—Repair Garages
NFPA 99—Health Care Facilities
NFPA 407—Aircraft Fuel Servicing

and experts to do the actual work and perform the inspections, but in the event of a failure, the owner is the first person who will be sought out for satisfaction. Based on the requirements of the local law, however, the division of responsibility may differ from one jurisdiction to another.

BASIC SYSTEMS

Each piping system in a building is used to convey a fluid to heat, cool, or perform some other service. Energy must be used to move the fluid: For a liquid, the energy is provided by a pump; for steam systems, a boiler; and for compressed air, a compressor. Systems fall into two broad categories:

1. Recirculating
2. Distributing

Steam

Steam systems are usually of a recirculating type to return the condensate to the boiler. If the condensate is not recoverable or may be contaminated, it is wasted

TABLE C4.2 Component Standards

Ferrous pipe and tubing	
ANSI/ASME	
B36.10	Welded and Seamless Wrought Steel Pipe Dimensions
B36.19	Stainless Steel Pipe Dimensions
ASTM	
A53	Pipe, Steel-Black and Galvanized, Welded and Seamless
A106	Seamless Carbon Steel Pipe for High Temperature Service
A135	Electric-Resistance-Welded Steel Pipe
A139	Electric Fusion Welded Steel Pipe
A211	Spiral-Welded Steel or Iron Pipe
A312	Seamless & Welded Austenitic Stainlêss Steel Pipe
A377	Ductile Iron Pressure Pipe
A539	Electric-Resistance-Welded Coiled Steel Tubing for Gas and Fuel Oil Lines
ANSI/AWWA	
A21.51/C151	Ductile Iron Pipe Centrifugally Cast in Metal Molds or Sand-Lined Molds for Water or Other Liquids
AGA/ANSI	
A21.52	Ductile Iron Pipe, Centrifugally Cast in Metal Molds or Sand-Lined Molds for Gas
Ferrous fittings and valves	
ASME/ANSI	
B16.1	Cast Iron Pipe Flanges and Flanged Fittings
B16.3	Malleable Iron Fittings, Classes 150 and 300
B16.4	Cast Iron Threaded Fittings, Classes 125 and 250
B16.5	Pipe Flanges and Flanged Fittings
B16.9	Factory-Made Wrought Steel Butt-Weld Fittings
B16.10	Face-to-Face and End-to-End Dimensions of Valves
B16.11	Forged Steel Fittings, Socket-Weld, and Threaded
B16.14	Ferrous Pipe Plugs, Bushings and Lock Nuts with Pipe Threads
B16.28	Wrought Steel Butt-Weld Short Radius Elbows and Returns
B16.33	Manually Operated Metallic Gas Valves for Use in Gas Piping Systems up to 125 psig
B16.34	Valves—Flanged, Threaded, and Welding End
B16.39	Malleable Iron Threaded Pipe Unions, Classes 150, 250, and 300
B16.42	Ductile Iron Pipe Flanges and Flanged Fittings, Classes 150 and 300
B16.47	Large Diameter Steel Flanges
ASTM	
A403	Wrought Austenitic Stainless Steel Piping Fittings
MSS	
SP-43	Wrought Stainless Steel Butt-Weld Fittings
SP-76	Butterfly Valves
SP-70	Cast Iron Gate Valves, Flanged and Threaded Ends
SP-71	Cast Iron Swing Check Valves, Flanged and Threaded Ends
SP-72	Ball Valves with Flanged or Butt-Weld Ends for General Service
SP-78	Cast Iron Plug Valves, Flanged or Threaded Ends
SP-83	Carbon Steel Pipe Unions, Socket-Weld or Threaded
SP-84	Steel Valves, Socket-Weld and Threaded Ends
SP-85	Cast Iron Globe and Angle Valves, Flanged and Threaded Ends

TABLE C4.2 Component Standards (*Continued*)

Copper and aluminum pipe and tubing	
ASTM	
B42	Seamless Copper Pipe, Standard Sizes
B43	Seamless Red Brass Pipe, Standard Sizes
B68	Seamless Copper Tube, Bright Annealed
B75	Seamless Copper Tube
B88	Seamless Copper Water Tube
B210	Aluminum-Alloy Drawn Seamless Tube
B241	Aluminum-Alloy Seamless Pipe and Seamless Extruded Tube
B251	General Requirements for Wrought Seamless Copper and Copper-Alloy Tube
B302	Threadless Copper Pipe
B547	Aluminum-Alloy Formed and Arc Welded Round Tube
Copper and aluminum fittings and valves	
ANSI/ASME	
B16.15	Cast Bronze Threaded Fittings, Classes 125 and 250
B16.18	Cast Copper-Alloy Solder Joint Pressure Fittings
B16.22	Wrought Copper and Copper-Alloy Solder Joint Pressure Fittings
B16.24	Bronze Pipe Flanges and Flanged Fittings, Classes 150 and 300
B16.26	Cast Copper-Alloy Fittings for Flared Copper Tubes
B361	Factory-Made Wrought Aluminum and Aluminum-Alloy Welding Fittings
MSS	
SP-80	Bronze Gate, Globe, Angle, and Check Valves
Other components and standards	
ANSI/ASME	
B1.20.1	Pipe Threads
B16.21	Nonmetallic Flat Gaskets for Pipe Flanges
B16.25	Butt-Weld Ends for Pipe, Valves, Flanges and Fittings
B18.2.1	Square and Hex Bolts and Screws
B18.2.2	Square and Hex Nuts
ASTM	
A36	Structural Steel
A183	Carbon Steel Track Bolts and Nuts
A193	Alloy-Steel and Stainless Steel Bolting Materials for High Temperature Service
A194	Carbon and Alloy-Steel Nuts for Bolts for High Pressure and High Temperature Service
A307	Carbon Steel Bolts and Studs, 60,000 psi Tensile Strength
B32	Solder Metal
ANSI/AWWA	
A21.11/C111	Rubber-Gasket Joints for Ductile Iron and Gray-Iron Pressure Pipe and Fittings
ANSI/AWS	
A5.1	Covered Carbon Steel Arc Welding Electrodes
A5.2	Iron and Steel Oxyfuel Gas Welding Rods
A5.3	Aluminum and Aluminum-Alloy Covered Arc Welding Electrodes
A5.6	Covered Copper and Copper-Alloy Welding Rods and Electrodes

TABLE C4.2 Component Standards (*Continued*)

Other components and standards (*continued*)	
ANSI/AWS (*continued*)	
A5.7	Copper and Copper-Alloy Bare Welding Rods and Electrodes
A5.8	Brazing Filler Metal
A5.9	Corrosion-Resisting Chromium and Chromium-Nickel Steel Bare and Composite Metal Cored and Stranded Welding Electrodes and Welding Rods
A5.10	Aluminum and Aluminum-Alloy Bare Welding Rods and Electrodes
A5.12	Tungsten Arc Welding Electrodes (Non-Consumable)
A5.18	Carbon Steel Filler Metals for Gas Shielded Arc Welding
A5.20	Carbon Steel Electrodes for Flux Cored Arc Welding
A5.22	Flux Cored Corrosion-Resisting Chromium and Chromium-Nickel Steel Electrodes
MSS	
SP-6	Standard Finishes for Contact Faces of Pipe Flanges and Connecting-End Flanges of Valves and Fittings
SP-25	Standard Marking System for Valves, Fittings, Flanges, and Unions
SP-58	Pipe Hangers and Supports—Materials, Design, and Manufacture
SP-69	Pipe Hangers and Supports—Selection and Application
SP-89	Pipe Hangers and Supports—Fabrication and Installation Practices
ASTM	
E488	Test Method for Strength of Anchors in Concrete and Masonry Elements
ANSI/AWWA	
A21.50/C150	Thickness Design of Ductile Iron Pipe
Nonmetallic pipe and fittings*	
ASTM	
F412	Definition of Terms Relating to Plastic Piping Systems
D1788	Specifications for Acrylonitrile-Butadiene-Styrene (ABS) Plastics, Rigid
D1527	ABS Plastic Pipe, Schs 40 and 80
D2235	Solvent Cement for ABS Plastic Pipe and Fittings
D2468	ABS Plastic Pipe Fittings, Sch 40
D3965	Rigid ABS Compounds for Pipe and Fittings
D2581	Specifications for Polybutylene (PB) Plastics Molding and Extrusion Materials
D2657	Standard Practice for Heat-Joining Polyolefin Pipe and Fittings
D2666	PB Plastic Tubing
D3000	PB Plastic Pipe (SDR-PR) Based on Outside Diameter
D1248	Specifications for Polyethylene (PE) Plastics Molding and Extrusion Materials
D2104	PE Plastic Pipe, Sch 40
D2447	PE Plastic Pipe. Schs 40 and 80, Based on Outside Diameter
D2683	Socket-Type PE Fittings for Outside Diameter Controlled PE Pipe and Tubing
D2737	PE Plastic Tubing
D3350	Specifications for PE Plastic Pipe and Fittings Materials
D1784	Specifications for Rigid Poly (Vinyl Chloride) (PVC) Compounds and Chlorinated Poly (Vinyl Chloride) (CPVC) Compounds

TABLE C4.2 Component Standards (*Continued*)

Nonmetallic pipe and fittings (*continued*)*	
ASTM (*continued*)	
F437	Threaded CPVC Plastic Pipe Fittings, Sch 80
F438	Socket-Type CPVC Plastic Pipe Fittings, Sch 40
F439	Socket-Type CPVC Plastic Pipe Fittings, Sch 80
F441	CPVC Plastic Pipe, Schs 40 and 80
F442	CPVC Plastic Pipe (SDR-PR)
F493	Specification for Solvent Cements for CPVC Plastic Pipe and Fittings
D2846	CPVC Plastic Hot- and Cold-Water Distribution Systems
D1785	PVC Plastic Pipe, Schs 40 and 80
D2241	PVC Pressure-Rated Pipe (SDR Series)
D2464	Threaded PVC Plastic Pipe Fittings, Sch 80
D2466	PVC Plastic Pipe Fittings, Sch 40
D2467	Socket-Type PVC Plastic Pipe Fittings, Sch 80
D2564	Solvent Cements for PVC Plastic Pipe and Fittings
D2855	Practice for Making Solvent-Cemented Joints with PVC Pipe and Fittings
D2310	Standard Classification for Machine-Made Reinforced Thermosetting-Resin Pipe (RTRP)
D2517	Reinforced Epoxy Resin Gas Pressure Pipe and Fittings
D2996	Filament-Wound Reinforced Thermosetting-Resin Pipe
D3517	Fiberglass (RTRP) Pressure Pipe
D3754	Fiberglass (RTRP) Sewer and Industrial Pressure Pipe

*This list is limited to some of the basic piping standards. There are more than 200 ASTM standards relating to plastic pipe.

and the system becomes, in effect, distributive. The return of a steam recirculating system is in the form of liquid condensate which is recovered through steam traps which pass the condensed steam in the form of a liquid but do not pass steam. Steam piping should be pitched downward in the direction of steam flow. Steam takeoffs should always be from the top of the main. As condensation is constantly occurring due to heat loss through the pipe walls, a trap must be installed at any rise in pipe elevation, at the end of horizontal mains, and at intermediate points of long runs. Traps are also used at the discharge of heating devices such as radiators, unit heaters, and steam coils.

The condensate system should be pitched downward in the direction of flow to the boiler or recovery tank. If it is not feasible to have continuous pitch to the boiler, a condensate pump can be used and the pump discharge line can change elevation as needed.

Some small low-pressure systems do not use traps and the condensate returns by gravity to the boiler through the supply or return pipes. All piping must pitch downward toward the boiler. Supply pipe steam velocities should be kept low to prevent slugs of condensate from being carried along with the steam which produce loud knocks at changes in direction. Noise is apt to occur during warm-up when cold pipes result in heavy condensation.

Condenser Water

A typical condenser water system recirculates the water from the refrigeration machine condenser, through the pump, up to the sprays on the top of the cooling tower, and from the basin of the tower back to the condenser. Since this is an open system, expansion of the water is not a factor; the water level in the basin compensates.

Because it is sprayed through the air, the water contains oxygen and other contaminants and becomes extremely corrosive. For many years chromates were used to inhibit steel pipe corrosion with good success. Recent environmental rules have prohibited chromate water treatment, and new methods which require closer control are being used. It is important to clean condenser water systems carefully and monitor treatment closely to prevent the destruction of the piping system.

Chilled Water

Chilled water systems are closed recirculating systems and therefore require expansion tanks. An open expansion tank can be used at the highest point of the system with makeup water provided through a tank level control device. Expansion can also be provided for by use of a closed hydropneumatic tank at any point in the system. For this type of tank the water level is maintained by a pressure pump controlled by a tank level sensor. As sufficient air pressure must be maintained above the water in the tank to support the water to the highest point of the system, it is sometimes necessary to provide a method for increasing the air pressure in the tank. The tank must be constructed for the proper pressure, and a relief valve must be provided. The corrosion problem is not as severe as in condenser water systems, but water treatment should be used.

Compressed Air

A compressed air system is an example of a distributive system. The air is piped to its point of use for a control or tool system, at which point energy is recovered, and it is reduced to atmospheric pressure and released.

PIPE DESIGN

Piping systems are designed to withstand the combined effects of the internal pressure and temperature of the contained fluid and other external stress imposing forces such as expansion or contraction, support spans, earthquake, and wind.

The internal pressure produces two types of stresses in the pipe:

1. Hoop stress, also called the tangential stress, which is the resultant of the radial pressure in the pipe. This force tends to split the pipe open along a seam.
2. Longitudinal stress, which is caused by the force of the pressure on the end of the pipe. This force is exerted equally on a closed end such as a cap or on an open component such as an elbow. It acts along the axis of the pipe and tends to pull the pipe apart around its circumference or at a joint. Other longitudinal stresses are caused by the pipe acting as a beam between supports and by expansion or contraction where the pipe ends are fixed by anchors.

Pressure

The internal pressure of steam systems is the gauge pressure of the steam as determined by the boiler control or the setting of a pressure-reducing valve. For compressed air or other gasses the gauge pressure is determined by the compressor control or the setting of a regulator. This pressure can be considered constant throughout the system.

For liquid piping systems the pressure at any point is determined by adding the pressure produced by the circulating pump to the static pressure. The static pressure is calculated in a state of no flow (pump off) and is the result of the weight of the liquid in the pipe from the highest point of the system to the lowest. This pressure varies with the elevation of each point in the piping.

The pressure in liquid piping systems is also referred to as *head.* This is the height of a column of the liquid in the pipe which would be supported by the pressure.

For water 1 ft^3 = 62.5 lb.

The footprint of a cubic ft is 144 in^2.

The weight per square inch of a column of water 1 ft high is

$$\frac{62.5}{144} = 0.43 \text{ psi or } 1 \text{ psi} = 2.3 \text{ ft of water}$$

For circulating liquid piping systems, the datum pressure is at the expansion tank. With an open tank the pressure at the surface is atmospheric (0 psig). A hydropneumatic tank has a regulator or relief valve which determines the pressure on the surface of the water in the tank. The pressure at other points in the system is found by adding the pressure produced by the circulating pump to the static pressure of the system. For a flowing closed loop system the pump pressure is highest at the pump discharge and will decrease, due to friction, as the fluid flows through the pipe and other elements of the system back to the pump suction. For a system which may have no flow with the pump running, the pump shutoff pressure may be imposed on the system depending on the location of the expansion tank. See Fig. C4.1 for examples.

In Fig. C4.1*b* a 30-story building with two subgrade levels could have a water piping elevation difference of 400 ft from the open expansion tank water surface in the penthouse down to the pump in the basement. This is called a static head of 400 ft, and the design static pressure would be calculated as follows:

$$400 \text{ ft} \times 0.43 \text{ psi/ft} = 172 \text{ psi}$$

Horizontal offsets in the piping system do not affect the static pressure.

The total pump head is 70 ft, which is required to overcome the resistance of the system. The pump discharge pressure in this example is 48 ft, or 21 psi, which when added to the static pressure would provide the system design pressure:

$$\text{Design pressure} = 172 \text{ psi} + 21 \text{ psi} = 193 \text{ psi}$$

All pipe, fittings, valves, pumps, and heat exchangers on the pump discharge part of the system must be designed to withstand this pressure at the lowest elevation. If it is possible to block the flow in the system, a study should be made to see if the pump shutoff pressure may cause a higher design pressure. On the suction side of the pump, the design pressure will be 172 psi, which will be the pressure when the pump is not operating. It is advantageous to locate the heat

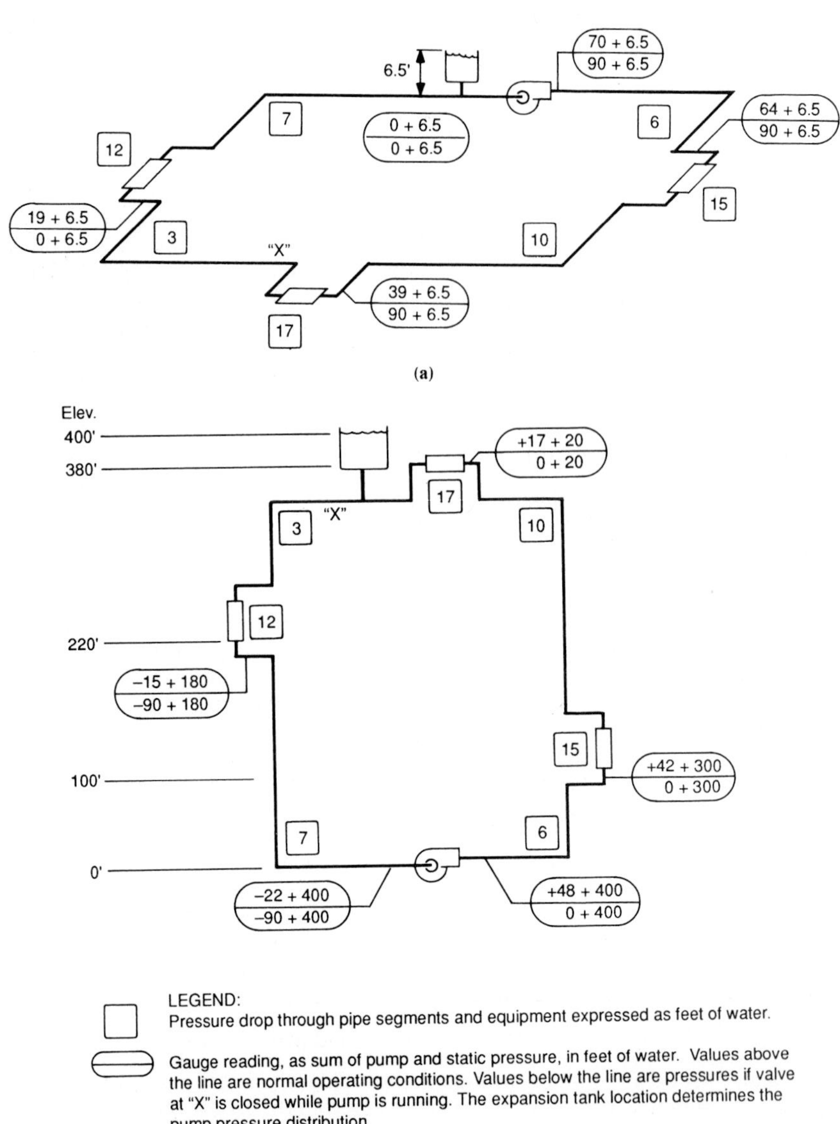
6.5'
70 + 6.5
90 + 6.5
7
0 + 6.5
0 + 6.5
6
64 + 6.5
90 + 6.5
12
15
19 + 6.5
0 + 6.5
3
"X"
10
39 + 6.5
90 + 6.5
17
(a)
Elev.
400'
380'
+17 + 20
0 + 20
17
"X"
3
10
12
220'
−15 + 180
−90 + 180
15
+42 + 300
0 + 300
100'
7
6
0'
−22 + 400
−90 + 400
+48 + 400
0 + 400
LEGEND:
Pressure drop through pipe segments and equipment expressed as feet of water.
Gauge reading, as sum of pump and static pressure, in feet of water. Values above the line are normal operating conditions. Values below the line are pressures if valve at "X" is closed while pump is running. The expansion tank location determines the pump pressure distribution.
Pump delivers 70 ft of head to overcome resistance of system. Shut-off head 90 ft.
(b)

FIGURE C4.1 (*a*) Horizontal piping system; (*b*) vertical piping system.

exchangers and associated valves in the lower pressure portion of the piping system. At the higher elevations it is economical to use valves and fittings of a lower pressure class rating.

Temperature

The pipe material temperature is considered to be the same as the temperature of the fluid in the pipe. For saturated steam systems it is the saturation temperature corresponding to the gauge pressure. The temperature is important since the strength of materials used in the systems decreases as the temperature rises. For ferrous materials in building systems fluid temperatures are not usually high enough to be a factor. However, copper and aluminum and the joints in these materials can be affected by the temperatures encountered. Thermoplastics, as a group, lose strength rapidly as the temperatures rise above 70°F. At 100°F they have lost approximately 25 percent of their strength and at 140°F less than 50 percent remains.

Allowable Stress

Allowable stresses for pipe materials are tabulated in the safety codes listed in Table C4.1. As this chapter concerns piping in buildings, Appendix A of ASME/ANSI B31.9, Building Services Piping, will be used for allowable stress values.

The published allowable stresses are determined for various materials by using the lowest of the following criteria for each material:

1. Cast iron or ductile iron
   a. One-tenth of the specified minimum yield strength at room temperature
   b. One-tenth of the tensile strength at the listed temperature
2. Malleable iron
   a. One-fifth of the specified minimum tensile strength at room temperature
   b. One-fifth of the tensile strength at the listed temperature
3. Other metals
   a. One-fourth of the specified minimum tensile strength at room temperature
   b. One-fourth of the tensile strength at the listed temperature
   c. Two-thirds of the specified minimum yield strength at room temperature
   d. Two-thirds of the yield strength at the listed temperature
4. Thermoplastics
   a. One-half of the hydrostatic design basis at design temperature as determined by the methods in ASTM D2837
5. Reinforced thermosetting resins
   a. One-half of the hydrostatic design basis at design temperature as determined by the methods in ASTM D2992

The tensile and yield strengths are found in the standards listed in Table C4.2 for each material specification. The shear and bearing stress values of pipe materials can be determined by applying multipliers:

$$\text{Shear stress} = 0.8 \times \text{basic allowable stress}$$

$$\text{Bearing stress} = 1.6 \times \text{basic allowable stress}$$

When pipe is produced by a method of manufacture which results in a welded longitudinal seam, the basic allowable stress for the pipe is modified by a factor E, called the joint efficiency factor. Table C4.3 lists some piping materials used

TABLE C4.3 Allowable Stresses

Material & ASTM Spec. No.	Grade	Available sizes (in)	*E* factor	Strengths		Max. allowable stress value SE (ksi) for metal temperature not over			
				Min. tensile (ksi)	Min. yield (ksi)	0–100°F	200°F	300°F	400°F
Carbon steel									
Seamless pipe and tube									
A 53	B	¼ to 26	1.00	60.0	35.0	15.0	15.0	15.0	15.0
A 106	B	¼ to 30	1.00	60.0	35.0	15.0	15.0	15.0	15.0
Electric resistance welded pipe and tube									
A 53	B	2 to 20	0.85	60.0	35.0	12.8	12.8	12.8	12.8
A 135	B	2 to 30	0.85	60.0	35.0	12.8	12.8	12.8	12.8
Furnace butt welded (continuous weld) pipe and tube									
A 53	F	½ to 4	0.60	45.0	25.0	6.8	6.8	6.8	6.8
Wrought welding fittings									
A 234	WPB	½ to 42	1.00	60.0	35.0	15.0	15.0	15.0	15.0
Stainless steel; welded pipe and tube									
A 312	TP304	¼ to 30	0.85	75.0	30.0	18.8	15.7	14.1	13.0
A 312	TP316	¼ to 30	0.85	75.0	30.0	18.8	16.2	14.6	13.4
Ductile iron									
C151/A21.51		4 to 48	—	60.0	42.0	4.2	4.2	4.2	4.2

Material & ASTM Spec. No.	Alloy	Available sizes (in)	Temper	Strengths		Max. allowable stress value SE (ksi) for metal temperature not over			
				Min. tensile (ksi)	Min. yield (ksi)	0–100°F	200°F	300°F	400°F
Aluminum Alloy; seamless pipe and tube									
B 241	3003	¾ to 24	0	14.0	—	3.4	3.4	2.4	1.4
B 241	6061	¾ to 24	T4	26.0	—	6.5	6.5	6.0	4.5
Copper and copper alloys; seamless pipe and tube									
B 75	102–142	¼–12	annealed	30.0	9.0	6.0	4.8	4.7	3.0
B 75	102–142	¼–12	light drawn	36.0	30.0	9.0	9.0	8.7	8.2
B 75	102–142	¼–12	hard drawn	45.0	40.0	11.3	11.3	11.0	4.3
B 88	102–122	¼–12	annealed	30.0	9.0	6.0	4.8	4.7	3.0
B 88	102–122	¼–12	drawn	36.0	30.0	9.0	9.0	8.7	8.2
Structural shapes for pipe supports									
A 36	—	—	—	58.0	36.0	11.6	11.6	11.6	11.6
Bolts, nuts, and studs for flanges and supports									
A 307	B	—	—	55.0	—	7.0	7.0	7.0	7.0

for building piping and the allowable stresses and E factors for each material. The listed values are SE, which is the product of the allowable stress S and the longitudinal joint efficiency factor E.

Pipe Wall Thickness

The dimensions of steel pipe are governed by ANSI/ASME B36.10M. This standard lists nominal pipe sizes (NPS) from NPS ⅛ to NPS 80 with a range of wall thicknesses for each size. There are 23 different standardized wall thicknesses listed for NPS 8 pipe. Not all of the diameters and thicknesses are manufactured by the mills on a regular basis. Before making a selection check with a pipe supply warehouse for availability. Mills will make large quantities on order.

Pipe sizes up to and including NPS 12 have a standardized outside diameter which is greater than the nominal size. For sizes NPS 14 and larger the outside diameter is equal to the nominal size.

Two different systems for describing commonly used pipe wall thicknesses have developed. The first, which has been used commercially for many years, uses these designations:

STD Standard

XS Extra strong, also called extra heavy

XXS Double extra strong

The second system developed from an attempt to establish a set of schedule numbers. Each schedule was to designate a wall thickness for each pipe size so the pressure-carrying capacity would be the same regardless of size. Some schedules in use are

Sch 40 Schedule 40

Sch 80 Schedule 80

Sch 160 Schedule 160

Standard and Sch 40 are the same thickness through NPS 10. Standard wall thickness is ⅜ in for all larger sizes, whereas Sch 40 becomes increasingly thicker. Extra strong and Sch 80 are the same thickness for all sizes through NPS 8. For all larger sizes extra strong pipe remains at ½ in wall thickness whereas Sch 80 becomes increasingly thicker.

Even though pipe stress calculations are seldom required for building piping, a brief review of the basics is given so that the principles can be understood. To find the required pipe wall thickness to contain the fluid at the design pressure use the following formula:

$$t_m = \frac{PD}{2SE} + A \qquad \text{(C4.1)}$$

Conversely, to find the pressure a pipe of known wall thickness will withstand the formula is

$$P = \frac{2SE(t_m - A)}{D} \qquad \text{(C4.2)}$$

where t_m = minimum required wall thickness (in)
P = internal design pressure (psig)
D = outside pipe diameter (in)
S = basic allowable stress (psi)
E = longitudinal joint efficiency factor
A = allowance for corrosion, mill tolerance, joint preparation, mechanical strength, and so on

For example, this is the procedure to find the proper wall for a 2-in 125-psig steam line with threaded fittings using A53 Grade B ERW pipe:

$$t_m = \frac{PD}{2SE} + A$$

$$P = 125 \text{ psig}$$

$$D = 2.375 \text{ in}$$

$$SE = 12{,}800 \text{ psi}$$

$$= 125 \times 2.375/2 \times 12{,}800 + A$$

$$= 0.012 \text{ in} + A$$

The factor A requires some consideration. The corrosion allowance depends on the expected service and the maintenance of a chemical water treatment system. For closed systems a value of 0.025 in may be sufficient. For open systems a value of 0.065 in or higher may be needed. These values can be varied for the expected service conditions in the system being designed. The allowable mill minus tolerance for A53 pipe is given in the ASTM specification as 12.5 percent of the nominal wall thickness. If the pipe is to be threaded or grooved, the depth of the groove or the depth of the thread, as listed in Table C4.4, plus a tolerance of 0.015 in should be allowed. For threaded NPS 2 standard weight pipe which has a nominal wall thickness of 0.154 in, the factor A will equal the sum of:

Corrosion allowance =	0.025
Mill tolerance = 0.125 × 0.154 =	0.019
Thread depth (from Table C4.4) = 0.070 + 0.015 =	0.085
Total A =	0.129

The minimum required wall thickness plus allowance is

$$t_m = 0.012 \text{ in} + 0.129 \text{ in} = 0.141 \text{ in}$$

Select the next larger commercially available wall thickness for use. In this case standard weight pipe with a wall thickness of 0.154 in is suitable for the service.

Mechanical strength should be considered as the overriding wall thickness needed to withstand the shear loads at hangers, the rough handling during installation, and such unanticipated use of the pipe as for scaffolding or hoisting after installation.

If in the above example the pipe was used for 5-psig steam, the wall thickness needed to withstand the pressure would be 0.0005 in. The allowances do not change so the minimum pipe wall thickness required would be 0.0005 in + 0.129 in, which is 0.1295 in. This points out that the prime consideration for low-pressure piping systems is the mechanical strength rather than pressure containment.

TABLE C4.4 Thread and Groove Depths (tolerances not included)

Nominal pipe size (NPS)	Thread depth (in)*	Groove depth (in)†
½	0.057	—
¾	0.057	0.056
1	0.070	0.063
1¼	0.070	0.063
1½	0.070	0.063
2	0.070	0.063
2½	0.100	0.078
3	0.100	0.078
4	0.100	0.083
5	0.100	0.084
6	0.100	0.085
8		0.092
10		0.094
12		0.109
14		0.109
16		0.109
18		0.109
20		0.109
24		0.172

*Dimension *h,* the height of thread, from ANSI/ASME B1.20.1, Pipe Threads—General Purpose. Pipe more than 6 in is rarely threaded.

†Reference groove depth as listed by Victaulic Co.

General Note: As the plus tolerance on the pipe's outside diameter increases, the fixed thread or groove diameter cuts away more of the pipe wall thickness. Victaulic limits the plus tolerance to 1/32 in. This is less than that allowed by ASTM A53 or A106. If the pipe is 1/32 in oversize, the groove or thread depth is increased by about 30 percent. If large pipe is to be used for grooving the plus tolerance should be stated on the pipe purchase order.

Elbows, Bends, and Miters

A change in pipe direction for building piping is done by using an elbow fitting. Bends and miters may also be used but are generally not economical when labor costs are high.

Additional wall thickness must be provided when a bend is made since the heel of the bend will stretch and thin. Table C4.5 gives the added thickness needed to provide the required minimum thickness in a 90° finished bend. The

TABLE C4.5 Pipe Thickness for Bends

Radius of bend pipe diameters*	Thickness increase†
6 or greater	$1.06t_m$
5	$1.08t_m$
4	$1.14t_m$
3	$1.24t_m$

*The pipe diameter is the nominal diameter.

†t_m is the minimum required wall thickness.

flattening of pipe during a bend should not result in a reduction of the original diameter by more than 8 percent.

For miter joints, an angular offset of 3° or less is considered to be the same as a girth weld and does not require an increase in the calculated pipe wall thickness. The angular offset is the change in direction of the axis of the pipe. A 3° offset will result in a miter angle of 1½° to be cut on the end of each of the pipe segments to be joined. Miters can be used for offsets from 3° to 45° at pressures of 50 psig or less. For miters at higher pressures the formulas in ASME/ANSI B31.9 or B31.3 should be used.

Pipe Branch Design

Branch connections for pipe sizes under NPS 3 are usually made with threaded or grooved tees. For larger pipe sizes with welded joints, branches are made by using a welding tee, an integrally reinforced outlet fitting, or by welding the branch pipe directly to the main using a full penetration weld.

When welding the branch directly to the main, the opening in the main for the branch pipe weakens the main and the need for added reinforcement to replace the removed material must be evaluated. Reinforcement is *not* needed when

1. The branch connection is made using a fitting made to an ANSI B16.9 listed standard.
2. The branch connection is made using a threaded or socket weld coupling not exceeding
   a. NPS 2 or
   b. One-fourth the nominal diameter of the main
   The coupling wall thickness should be not less than that of the branch pipe and welded to the main with a full penetration weld.
3. An integrally reinforced outlet fitting is used.

Table C4.6 lists the maximum internal pressure permitted without reinforcement for various combinations of branch and main sizes for ASTM A53 Grade B and A106 Grade B pipe at 90° and 45° branch angles. For pressures higher than those listed, use a fitting or use the rules in ASME/ANSI B31.1 for determining the required reinforcement and the method of application.

Closures and Blanks

Closures are made at the end of pipe runs using standard fittings such as caps, plugs, or blind flanges. A flat plate can also be used as a closure when it is welded to the pipe with a full penetration weld.

The minimum thickness of the closure plate, *tc,* can be calculated by

$$t_c = d\sqrt{CP/S} + A \tag{C4.3}$$

where d = inside pipe diameter (in)
C = $0.5t_m/T$, but not less than 0.3
t_m = minimum required pipe wall thickness (in)
T = nominal wall thickness of pipe (in)
P = internal design pressure (psi)
S = maximum allowable stress of plate material (psi)

TABLE C4.6 Maximum Operating Pressures for Unreinforced Welded Branch Connections—Valid to 650°F—for ASTM A53, A135, and A106 Grade B Pipe. S = 15,000 psi

NPS		Standard weight pipe Pressure (psig)		Extra heavy pipe Pressure (psig)	
Main	Branch	90° branch	45° branch	90° branch	45° branch
3	2	842	713	1252	1076
4	3	720	620	1105	963
	2	723	613	1096	942
6	4	567	490	934	816
	3	583	502	955	831
	2	582	494	945	813
8	6	489	426	826	724
	4	503	435	830	724
	3	515	444	846	735
	2	513	437	835	719
10	8	443	387	667	586
	6	448	389	674	589
	4	458	396	674	587
	3	468	403	686	595
	2	465	398	674	580
12	10	385	336	544	477
	8	387	337	570	499
	6	390	339	574	501
	4	356	345	573	498
	3	406	350	582	504
	2	404	345	570	491
14	12	346	302	483	424
	10	352	307	498	436
	8	353	308	522	457
	6	356	309	525	459
	4	363	314	523	455
	3	371	319	531	461
	2	368	314	520	448
16	14	301	263	420	368
	12	304	266	426	373
	10	310	270	439	384
	8	311	270	460	403
	6	313	272	463	404
	4	319	276	460	400
	3	325	280	467	404
	2	322	275	456	393
18	16	265	231	369	323
	14	269	234	376	329
	12	272	237	381	333
	10	276	241	393	343
	8	277	241	412	360
	6	279	242	413	360
	4	284	246	411	357
	3	290	249	416	360
	2	287	245	406	350

TABLE C4.6 Maximum Operating Pressures for Unreinforced Welded Branch Connections—Valid to 650°F—for ASTM A53, A135, and A106 Grade B Pipe. S = 15,000 psi (*Continued*)

NPS		Standard weight pipe Pressure (psig)		Extra heavy pipe Pressure (psig)	
Main	Branch	90° branch	45° branch	90° branch	45° branch
20	18	236	206	329	288
	16	239	209	333	292
	14	243	212	340	297
	12	245	214	345	301
	10	249	217	355	310
	8	250	218	372	325
	6	251	218	373	326
	4	256	221	371	322
	3	261	225	376	325
	2	258	221	366	315
24	20	196	171	273	238
	18	198	173	276	241
	16	200	175	280	245
	14	203	177	285	249
	12	205	179	289	253
	10	209	182	298	260
	8	209	182	312	273
	6	210	182	313	273
	4	214	185	310	269
	3	218	188	314	272
	2	215	184	306	263

Notes: Pressures change directly with the ratio of SE to S.

1. For A53 Type F Butt Weld Pipe use a multiplier of 0.75.
2. If the branch opening in the main cuts the seam of pipe with a longitudinal weld, multiply table values by
   Electric resistance weld pipe 0.85
   Type F butt weld pipe 0.45
3. Based on the rules in B31.1.
4. A 12½ percent mill tolerance and a 1/32-in corrosion allowance have been used to calculate the pipe wall thickness for this table.

A blank can be installed between two flanges to close off a portion of a system. The minimum thickness of the blank, t_b, can be calculated by

$$t_b = d_g\sqrt{3P/16S} + A \tag{C4.4}$$

where d_g = inside diameter of the gasket (in)

When the blank is to be left in place during operation, use the same nomenclature as above. If the blank is to be used only during hydrostatic testing, P is the test pressure and S is 0.95 times the specified minimum yield strength of the blank material.

PIPE MATERIALS

Pipe, fittings, and valves suitable for building service piping systems are available in many different materials. A brief discussion of the advantages and disadvantages of commonly used materials follows.

Steel

Steel (ASTM A53, A106) is the most widely used material for pipe. Its advantages are strength, availability, and economy. Because of its wide use there is a large pool of skilled labor knowledgeable in its installation. The disadvantages of steel pipe are its weight and low resistance to corrosion.

Copper

Copper tube (ASTM B88) is widely used as a material for sizes up to 3 or 4 in. It is available up to 8 in and on order up to 12 in, but it is not usually economical in the larger sizes. ASTM B88 gives dimensions for copper tube up to NPS 12, but it is not commonly made.

The nominal sizes of copper tubing approximate the inside diameter. The outside diameter is always ⅛ in larger than the nominal size. Copper tube for refrigerant service (ACR tube) is referred to by its actual outside diameter.

Copper tube is furnished in two tempers—drawn and annealed. The pressure rating of annealed tube is about 60 percent of that of drawn tube. When joints are made by brazing, the tube in the vicinity of the joint becomes annealed and the annealed pressure rating should be used. Drawn tubing is used for most building applications. Drawn tubing has been work hardened and is furnished in straight lengths. In the annealed form the tube is soft and very easily bent. It is furnished in coils or straight lengths.

Copper pressure tube is furnished in three standard thicknesses which vary with the size. The heaviest is Type *K*, followed by Type *L* and Type *M*. All three are suitable, but Type *M* is thin walled and does not have mechanical strength and is therefore easily damaged.

The advantages of copper are its resistance to corrosion and the fact that it is light weight. It is not as strong as steel and not available in sizes over 12 in. Copper can readily be bent using bending tools which keep the tube from flattening. The disadvantages of copper as a material are its low strength compared to steel and its high cost. The high cost is offset in the smaller sizes by the ease of installation.

Red Brass Pipe

Red brass pipe (ASTM B43) is made up to NPS 12 for standard weight and NPS 10 for extra strong weight. Copper tube has largely replaced brass pipe for reasons of economy. It is made to steel pipe dimensions and is sometimes used where the added mechanical strength is needed.

Thermoplastics

Thermoplastics, which are discussed in detail in Part D of this handbook, are widely used for liquids with temperatures up to 100°F for PVC and 180°F for CPVC. These plastics have exceptional resistance to corrosion, are light weight, easy to install, and readily available and economical. The disadvantages of thermoplastics are lower strength, which necessitates short hanger spans, high coefficients of expansion, which must be compensated for when using long straight runs of pipe, and the emission of toxic fumes under fire conditions. Fire codes in some locations do not permit the use of some plastic materials.

PVC pipe can be used as underground pressure pipe as well as for above ground service.

Thermosetting Resins

Thermosetting resins are used for reinforced thermosetting resin pipe (RTRP), also called fiberglass reinforced plastic (FRP) pipe. This pipe is made in larger sizes and can be used at higher pressures and temperatures than thermoplastics. It is suitable for aboveground or underground service. Refer to Part D, Chap. D2 of this handbook for more information. See manufacturers' data for specific ratings.

Ductile Iron

Ductile iron has replaced cast iron as water main pipe and is used for corrosion-resistant buried pipe. It is stronger and less brittle than cast iron. Ductile iron pipe is made in standard thickness classes, the use of which is determined by the depth of bury and laying condition. This pipe is made in standard sizes from NPS 3 to NPS 54.

Stainless Steel and Aluminum

Stainless steel and aluminum are expensive and therefore used only when corrosion resistance and strength are needed. Schedule 5S and Schedule 10S light-wall stainless pipe are commonly used to reduce cost.

Discontinued Pipe

Discontinued pipe materials, which are no longer made, are A72 wrought iron pipe, low-alloy (Yoloy) pipe, and ASTM A120 steel pipe.

PIPE FITTINGS AND JOINTS

Pipe joints fall into two basic categories—restrained and unrestrained. Restrained joints have inherent mechanical properties that prevent the joint from separating axially as well as containing the fluid. Unrestrained joints rely on packing to contain the fluid and the friction of the packing to keep the joints from separating. Systems with unrestrained joints must be provided with external restraint if they are to withstand pressure.

Bell and spigot piping is unrestrained. Mechanical joint and sleevetype couplings, which rely on a bolted gland retainer, are also unrestrained.

Threaded, flanged, and grooved joints are restrained. Welded, soldered, brazed, flared, and compression joints are considered to be restrained but are dependent on the skill of the assembler.

Some of the fittings discussed here are also made for higher pressures and temperatures than those mentioned.

Steel Pipe

Steel pipe is joined by threading, welding, or grooving. Threaded fittings are made of cast iron, malleable iron, ductile iron, and forged steel for joining pipe up to NPS 8. It is an industry standard, however, to change to welding or grooving at sizes above NPS 2. Threaded joints are made using ANSI/ASME B1.20.1 standard taper pipe threads. Pipe lighter than standard weight must not be threaded.

Cast iron fittings (ANSI/ASME B16.4) are made in Class 125 (standard) and Class 250 (extra heavy) and are used in most threaded applications. Cast iron is brittle, so if ductility or higher strength is required malleable or ductile iron fittings should be used.

Malleable iron threaded fittings (ANSI/ASME B16.3) are made in Classes 150 and 300 and are stronger and less brittle than cast iron. Care must be taken when threading steel pipe into Class 150 malleable iron fittings. The joint can be overtightened since the malleable fitting will stretch. Two threads should be exposed on the pipe when the joint is properly made up.

Ductile iron threaded fittings are made in Class 300 and fall between the Class 150 and 300 malleable iron fittings in strength. There is no listed ASTM standard for ductile iron threaded fittings.

Forged steel threaded fittings (ANSI B16.11) are made in smaller sizes and have high pressure-temperature ratings. They are used primarily for high-pressure steam.

Welding is normally used to join steel pipe from NPS 2½ and up. Welding fittings (ANSI B16.9) are made to match steel pipe diameters and wall thicknesses. They are furnished with a standard bevel end for butt welding to the pipe. Weld fittings have the same pressure-temperature ratings as the equivalent thickness seamless pipe.

For smaller sizes, forged steel socket weld fittings (ANSI B16.11) can be used for high-pressure service.

Properly made welded joints are as strong as the pipe, do not deteriorate, and have a smooth inside contour to minimize friction losses.

Flanges are used in large piping to connect to valves and equipment and to provide a means of disassembly in welded piping systems. Flanges are made of cast iron, ductile iron, bronze, and steel.

Cast-iron threaded flanges and flanged fittings (ANSI B16.1) are available in Classes 125 and 250. Ductile iron threaded flanges and flanged fittings (ANSI B16.42) are made in Classes 150 and 300. Threaded flanges are used for mounting equipment and valves in larger sized threaded pipe systems.

Class 125 cast-iron flanges have a flat face and are often mated to Class 150 steel flanges which have a raised face. Great care must be taken in tightening the bolts to avoid cracking the cast iron flange. The steel flanges can be ordered with the raised face removed by machining.

Steel flanges (ANSI B16.5) are made in Classes 150 and 300 for the services discussed in this chapter. They are available for higher pressures up to Class 2500. Steel flanges are generally used in welded piping systems. These flanges are made in weld neck, slip-on, and socket weld configurations. Classes 150 and 300 have a raised face.

Gasket material (ANSI B16.21) should be suitable for the fluid and pressure to be contained. Full face gaskets are used for flat faced flanged joints and ring gaskets are used in raised face joints. The flange bolts should be suitable for the pressure and flange facings.

Grooved pipe joining systems are made by several manufacturers, but as yet there is no ANSI standard for dimensions or pressure ratings. All the leading manufacturers use the same "standard" cut groove and roll groove dimensions. The joint consists of a circumferential groove cut into the outside of each pipe, close to the end to be joined. A coupling with continuous internal ridges engages the grooves in the pipe, and a gasket inside the coupling is expanded by the pressure in the pipe and seals against the pipe and coupling.

Grooves can be rolled into standard weight and lighter steel pipe. Cut grooves can be used on standard weight and thicker pipe. Gaskets are available for a great variety of liquids and gases.

These systems are extremely versatile and offer many advantages. They are easy to install, have inherent expansion compensation and flexibility, and are easy to disassemble. Grooved systems are not recommended for steam.

Copper Tube

Copper tube is joined by soldering or brazing using solder joint fittings (ANSI B16.18, B16.22) or using flare (ANSI B16.26) or compression fittings. Solder joints are also referred to as sweat joints.

Solder (ASTM B32) is defined as a filler metal whose melting point does not exceed 800°F. Brazing filler metals (ANSI/AWS A5.8) are specified as those whose melting points are 1000°F or higher. Soldered and brazed joints rely on capillary attraction to draw molten filler metal into the gap between the socket on the fitting end and the tube.

The strength of the joint depends upon the composition of the solder used. Brazing provides much stronger joints than soldering but anneals drawn tubing in the vicinity of the joint. It is against the law to use solders containing lead for drinking water piping systems, but they can be used for heating, cooling, and other building service applications. Table C4.7 gives working pressures for copper tube joints made by soldering and brazing.

Flared and compression-type fittings are made by many manufacturers for copper, steel, stainless steel, and aluminum tubing. Fittings are available for sizes NPS 2 and under but are used primarily in sizes NPS 1 and under. These joints can be taken apart and reassembled with ease. The manufacturers' catalogs give pressure ratings which can be as strong as the tube being joined.

Grooved joint couplings and fittings are made to match copper tube sizes. Since tube walls are thin, roll grooving is used.

TABLE C4.7 Pressure Ratings of Copper Tube Joints*

Solder or brazing alloy used in joints	Service temperature (°F max.)	Tube size, Types *K*, *L*, and *M* (in)					
		Water, noncorrosive liquids and gases					Steam
		¼–1	1¼–2	2½–4	5–8†	10–12†	All
50–50	100	200	175	150	130	100	—
Tin-lead	150	150	125	100	90	70	—
Solder	200	100	90	75	70	50	—
ASTM B32 Gr 50A‡	250	85	75	50	45	40	15
95–5	100	500	400	300	270	150	—
Tin-antimony	150	400	350	275	250	150	—
Solder	200	300	250	200	180	140	—
ASTM B32 GR 95TA	250	200	175	150	135	110	15
Brazing alloys	200	§	§	§	§	§	—
ANSI/AWS A5.8	250	300	210	170	150	150	—
(melting above 1000°F)	350	270	190	150	150	150	120

*Ratings for solder joints are from ASME/ANSI B16.22 and B16.18. Ratings for brazed joints and steam are from *The Copper Tube Handbook* published by the Copper Development Association, Inc.

†B31.9 prohibits compressed air or other gases above 20 psig in these sizes.

‡It is prohibited by law to use lead solder for potable water.

§The pressure rating of the tube determines the strength of this joint.

Bronze pipe flanges and flanged fittings (ANSI B16.24) are made in Classes 150 and 300 and can be used to join copper tube to flanged valves or equipment.

Red Brass Pipe

Red brass pipe is joined in the smaller sizes using cast bronze threaded fittings (ANSI/ASME B16.15) which are made in Classes 125 and 250. ANSI B16.24 flanges as mentioned above can be used for sizes above NPS 2.

Plastics

Plastic piping is the subject of Part D of this handbook. To aid those reading this chapter, a brief summary of joining methods and fittings is given here.

Solvent cementing is the most common method used for joining PVC (ASTM D2564, D2855), CPVC (F493) pipe, and ABS (D2235). Socket-type fittings are available in Sch 40 (PVC-ASTM D2466, CPVC-ASTM F438) and Sch 80 (PVC-ASTM D2467, CPVC-F439) for pressure systems to at least NPS 8. Check on availability before deciding to use larger sizes. Solvent cement joints are best made when the ambient temperature is between 40 and 100°F.

Threaded fittings up to NPS 4 are available for PVC (ASTM D2464) and CPVC (ASTM F437) piping systems. Only Sch 80 pipe should be threaded, and its pressure rating should be reduced by 50 percent. This type of joint can be taken apart and reassembled.

Flanges are available with solvent cement sockets up to NPS 8 and are made to Class 125 dimensions. Threaded flanges are made up to NPS 4. Flanges can be installed in cemented systems for ease in disassembly. An ASTM standard for thermoplastic flanges has yet to be developed.

Heat joining methods are also available for certain polyolefin joints (ASTM D2657), particularly for polypropylene. Pipe and fittings of this plastic are used for acid-resistant drainage systems. A wire filament is embedded in the socket of the joint then electrically heated with a special timing device to fuse the joint.

Bell and spigot PVC pressure pipe (AWWA C900) is made for underground water mains. The bell is made to contain an elastomer gasket (ASTM F477), and the joint is assembled by pushing the spigot into the bell. RTRP can be joined using a special taper on the end of the pipe which can be cemented into a matching bell on a coupling or fitting. Refer to Part D of this handbook.

Mechanical joint fittings (AWWA C110/ANSI A21.10) of ductile or cast iron can be used with PVC or RTRP pipe made to AWWA ductile pipe dimensions (AWWA C151/ANSI A21.51).

Ductile Iron

Ductile iron pipe is used primarily for underground systems. Fittings are rated in pressure classes (AWWA C110/ANSI A21.10).

Bell and spigot fittings are available and made of cast or ductile iron with bell ends. Pipe is made with one end belled. An elastomer gasket fits into a special groove in the bell, and the plain end of the pipe to be joined is pushed into the bell. This can be difficult with large pipe.

Mechanical joint fittings made from cast or ductile iron are available and are easier to work with in larger pipe sizes. The pipe is assembled into the bell before the gasket is in place, which reduces the friction. The gasket is then slid into the bell and a retainer ring is bolted into place to secure the gasket.

Grooved couplings are available and made to fit AWWA ductile iron pipe diameters (AWWA C151/ANSI A21.51) from NPS 3 to NPS 36. Special cut grooves are needed on the pipe. Grooved fittings are not made. Bell and spigot or mechanical joint fittings can be used.

Stainless Steel and Aluminum

Stainless steel and aluminum pipe and fittings are joined using the same methods as those for steel. The availability of fitting sizes is limited and should be investigated before using them in design.

MATERIAL SELECTION

The selection of the material to be used for each system is based on an evaluation of the following factors:

1. Requirements and limitations of the building and piping codes
2. The fluid in the pipe
3. The pressure and temperature of the fluid in the pipe
4. The location and external environment of the pipe
5. Availability of the material
6. The expected life of the facility where the system is to be installed
7. The installed cost of the system

The first four factors relate to safety and are of primary importance. The last three are related to the economics of the project and are weighted to suit.

Steam and Condensate Systems (to 150 psig)

Steel pipe is used for steam and condensate systems. Fittings for sizes under NPS 2½ are threaded cast iron. Larger fittings are steel with welding ends. Bronze valves are used for small sizes and cast or ductile iron, as required by the pressure, for valves NPS 2½ and over. Large gate valves in high-pressure systems should be provided with a bypass to allow warm-up of the downstream pipe before full opening. Equalizing the pressure with a bypass also allows the valve to open.

Ball valves and special steam butterfly valves may be used, but only if the valve has a gear operator so it cannot be opened rapidly. Ball valves can be used for small valves on low-pressure systems.

Copper tubing with brazed joints may be used up to 120 psig. The high energy content of steam makes mechanically restrained joints preferred. However, when space is limited, copper tubing with wrought brazed fittings, which is a much less bulky system than steel, can be used. Welding and brazing should be performed by qualified operators.

Steam condensate piping requires additional attention since condensate corrodes steel pipe. To lengthen the life of small size pipe, extra heavy weight pipe is often specified. For larger pipe sizes, the increased wall thickness of standard weight pipe provides added material to prolong pipe life. For small low-pressure systems, copper tube with 95-5 solder joints may be used. Fittings and valves for condensate are the same as those used for steam.

Table C4.8 shows some of the materials which can be used for steam and condensate service. Code pressure ratings for the various components or joints are listed. The table should not be read line by line. For example, any of the fittings and valves listed on lines D(1)–(4) may be used with any of the pipe types listed on lines D(1)–(3).

Recirculating Water Piping Systems

Closed recirculating water systems such as chilled water, hot water, and dual temperature water are used for heating and cooling. System temperatures can range from 50 to 200°F and the pressure can be above 200 psig on the lower floors of tall buildings.

For pipe NPS 2½ and larger the preferred material is steel with welded or grooved joints. The selection of materials for water piping systems NPS 2 and under is usually based on economics. The fluctuating price of materials and the skills of the labor to be used for installation make a number of combinations possible. For pressures above 150 psig and temperatures above 200°F, seamless or extra strong steel pipe with the proper class of threaded fittings or grooved joints can be used. Copper tube with solder or brazed joints is also widely used. For low temperatures, plastic piping systems can be used.

Bronze gate and ball valves are used on small pipe systems and cast-iron gate, plug, and butterfly valves are used with large pipe.

Table C4.9 lists materials for pipe, fittings, and valves and gives pressure ratings at 75, 150, and 220°F. The rating at 75°F can be used down to 0°F. Materials within any letter group can be mixed when a suitable joint can be made.

Condenser Water Systems

Refrigeration condenser water is usually cooled in a cooling tower where the water temperature is reduced by evaporation to the atmosphere. Lake, river, and well water are also used for condensing refrigerant. As oxygen and atmospheric contaminants are regularly induced into the water, corrosion becomes a major problem.

When a cooling tower is used, the same water plus makeup water are recirculated and chemicals must be added to the system to protect the pipe against corrosion. With river water or well water, corrosion-resistant materials must be used for the system or the interior of the pipe must be lined with a corrosion-resistant material.

Steel pipe is most often used for large condenser water piping systems aboveground. Where chemical treatment is possible it does not have to be lined, but the use of extra heavy pipe should be considered. Where chemicals cannot be used steel pipe can be cement lined using welded or grooved joints. Plastic lining

TABLE C4.8 Material Application Chart for Saturated Steam and Condensate*†

	Pipe						Fitting					Valve				
Line	Material	ASTM standard	Mfr. process	Weight	Joint	Pressure rating (psig)	Class	Material	Joint	ASTM standard	Pressure rating (psig)	Class	Material	Joint	Type	Pressure rating (psig)
	Low pressure to 15 psig; Medium pressure 16–90 psig															
	NPS 2 & smaller															
A. (1)	Steel	A 53 B	Type F(CW)	Std.	Thread	230	125	Cast iron	Thread	B 16.4	125	125	Bronze	Thread	Gate	140
(2)	Steel	A 53 B	Seamless	Std.	Thread	510	300	Ductile iron	Thread	—	430		Bronze	Thread	Ball	150
(3)	Copper	B 88	Drawn	Type L	95-5 Solder	15		Wrought copper	95-5 Solder	B 16.22	15					
(4)	Copper	B 88	Drawn	Type K	Braze	120		Wrought copper	Braze	B 16.22	120					
	NPS 2½–12															
B. (1)	Steel	A 53 B	ERW	Std.	Weld	530	Std	Wrought steel	Weld	B 16.9	580	125	Cast iron	Flange	Gate	150
(2)							150	Wrought steel	Flange	B 16.5	150		Steam trim	Wafer	Butterfly	150
(3)								Std ERW steel	90° weld branch		375					
	NPS 14–20															
C. (1)	Steel	A 53 B	ERW	Std.	Weld	335	Std	Wrought steel	Weld	B 16.9	395	125	Cast iron	Flange	Gate	100
(2)							150	Wrought steel	Flange	B 16.5	150	150	Ductile iron	Flange	Gate	208
(3)								Std ERW steel	90° weld branch		230		Steam trim	Lug wafer	Butterfly	150

	High pressure 91 to 150 psig at 366°F															
	NPS 2 & smaller															
D. (1)	Steel	A 53 B	ERW	Std.	Thread	645	250	Cast iron	Thread	B 16.4	250	200	Bronze	Thread	Gate	275
(2)		A 53 B	Seamless	XS	Thread	1215	300	Ductile iron	Thread	—	415	300	Bronze	Thread	Gate	410
(3)		A 106 B	Seamless	Std	Socket weld	1385	3000	Forged steel	Socket weld	B 16.11	2600					
(4)	Copper	B 88	Drawn	Type K	Braze	120		Wrought copper	Braze	B 16.22	120	800	Forged steel	Socket weld	Gate	1710
	NPS 2½–12															
E. (1)	Steel	A 53 B	ERW	Std	Weld	530	Std	Wrought steel	Weld	B 16.9	580	250	Cast iron	Flange	Gate	315
(2)		A 53 B	Seamless	Std	Weld	620	300	Wrought steel	Flange	B 16.5	300	150	Ductile iron	Flange	Gate	204
								Std SML steel	90° weld branch		440		Steam trim	Lug wafer	But-terfly	200
	NPS 14–20															
F. (1)	Steel	A 53 B	ERW	Std	Weld	335	Std	Wrought steel	Weld	B 16.9	395	150	Ductile iron	Flange	Gate	204
(2)		A 53 B	Seamless	Std	Weld	395	300	Wrought steel	Flange	B 16.5	300		Steam trim	Lug wafer	But-terfly	200
(3)		A 53 B	Seamless	XS	Weld	560		Std ERW steel	90° weld branch		230					
(4)								XH SML steel	90° weld branch		360					

*Pressure ratings for steel pipe are calculated using a mill tolerance of −12.5 percent of the wall thickness, the thread or groove depth, a corrosion allowance of 0.025 in for pipe to NPS 2, and 0.065 in for 2½ in and larger pipe. No pipe reinforcement value is applied for the strength of threaded fittings or grooved couplings.

†Pressure ratings are for the highest temperature and largest pipe size in each group. Higher ratings can be found for lower temperatures and smaller pipe sizes.

TABLE C4.9 Material Application Chart for Water Systems*,†,‡

	Pipe								Fitting		
						Pressure rating					
Line	Material	ASTM standard	Mfr. process	Wall thickness	Joint	@ 75°F (psig)	@ 150°F (psig)	@ 220°F (psig)	Material	Class	Joint
	In the building or above ground										
	NPS 2 and smaller										
G. (1)	Steel	A 53 B	Type F(CW)	Std.	Thread	230	230	230	Cast iron	125	Thread
(2)	Steel	A 53 B	Type F(CW)	Std.	Groove	275	275	275	D.I. or M.I.		Groove
(3)	Steel	A 53 B	Type F(CW)	Sch 10	Roll groove	400	400	400			
(4)	Copper	B 88	Drawn	Type L	95-5 solder	400	350	220	Wrought copper		95-5 solder
(5)	Copper	B 88	Drawn	Type K	Braze	380	380	300	Wrought copper		Brazed
(6)	Steel	A 53 B	Seamless	Std	Thread	510	510	510	Cast iron	250	Thread
(7)	Steel	A 53 B	Seamless	Std	Groove	605	605	605	Malleable iron	150	Thread
(8)	CPVC	F441	Seamless	Sch 40	Solvent	280	125	NR	CPVC	Sch 40	Solvent
	NPS 2½–12										
H. (1)	Steel	A 53 B	ERW	Std	Weld	530	530	530	Wrought steel	Std	Weld
(2)	Steel	A 53 B	ERW	Std	Groove	310	310	310	D.I. or M.I.		Groove
(3)	Steel	A 53 B	Seamless	Std	Weld	620	620	620	Cast iron	125	Flange
(4)	Steel	A 53 B	Seamless	Std	Groove	365	365	365	Wrought steel	150	Flange
(5)									Std ERW steel unreinforced 90° weld branch		
(6)									Wrought steel	300	Flange
(7)	Copper	B 88	Drawn	Type *L*	95-5 solder	150	150	130	Wrought copper		95-5 solder
(8)	Copper	B 88	Drawn	Type *L*	Braze	260	220	190	Wrought copper		Braze
(9)	Copper	B 88	Drawn	Type *K*	Braze	380	320	285			
(10)	Copper (to 6 in)	B 88	Drawn	Type *L*	Roll groove	300	300	300	Wrought copper	(to 6 in)	Roll groove
(11)	CPVC	F 441	Seamless	Sch 40	Solvent	130	70	NR	CPVC (to 6 in)	Sch 40	Solvent
(12)	CPVC	F 441	Seamless	Sch 80	Solvent	230	105	NR	CPVC (to 6 in)	Sch 80	Solvent
	NPS 14–20										
I. (1)	Steel	A 53 B	ERW	Std	Weld	335	335	335	Wrought steel	Std	Weld
(2)	Steel	A 53 B	ERW	Std	Groove	195	195	195	D.I. or M.I.		Groove
(3)	Steel	A 53 B	Seamless	Std	Weld	395	395	395	Cast iron	125	Flange
(4)	Steel	A 53 B	Seamless	Std	Groove	230	230	230	Wrought steel	150	Flange
(5)									Std ERW steel unreinforced 90° weld branch		
(6)									Wrought steel	300	Flange

Fitting				Valve						
	Pressure rating							Pressure rating		
ASTM standard	@ 75°F (psig)	@ 150°F (psig)	@ 220°F (psig)	Class	Material	Type	Joint	@ 75°F (psig)	@ 150°F (psig)	@ 220°F (psig)
B 16.4	175	175	160	125	Bronze	Gate	Thread	200	200	180
—	500	500	500		Bronze	Ball	Thread	400	400	400
B 16.22	375	320	220		Bronze	Ball	95-5 solder	400	350	220
B 16.22	445	375	310	200	Bronze	Gate	Thread	400	400	365
B 16.4	400	400	360							
B 16.3	300	300	255							
F 438	280	125	NR		CPVC	Ball	Socket	150	100	NR
B 16.9	510	510	510	125	Cast iron	Gate	Flange	150	150	130
—	500	500	500	150	Ductile iron	Gate	Flange	250	250	230
B 16.1	200	200	180	250	Cast iron	Gate	Flange	500	500	445
B 16.5	285	270	250		Buna-N liner	Butterfly	Wafer	150	150	NR
	375	375	375		EPDM liner	Butterfly	Wafer	150	150	150
B 16.5	720	710	695		Hi-service	Butterfly	Lug	250	250	250
B 16.22	150	150	130							
B 16.22	380	320	285	200	Cast iron	Plug	Flange	200	200	180
				500	Cast iron	Plug	Flange	500	500	445
—	300	300	300							
F 438	180	80	NR							
F 439	280	125	NR							
B 16.9	400	400	400	125	Cast iron	Gate	Flange	150	150	130
—	300	300	300	150	Ductile iron	Gate	Flange	250	250	230
B 16.1	150	150	130	250	Cast iron	Gate	Flange	300	300	275
B 16.5	275	255	235		Buna-N liner	Butterfly	Wafer	150	150	NR
	230	230	230		EPDM liner	Butterfly	Wafer	150	150	150
B 16.5	720	710	695		Hi-service	Butterfly	Lug	250	250	250

TABLE C4.9 Material Application Chart for Water Systems*,†,‡ (*Continued*)

Line	Pipe								Fitting		
						Pressure rating					
Line	Material	ASTM standard	Mfr. process	Wall thickness	Joint	@ 75°F (psig)	@ 150°F (psig)	@ 220°F (psig)	Material	Class	Joint
	Below ground (Corrosion protected materials from lines G, H, I may be used underground.)										
	NPS 4–20										
J. (1)	Ductile iron	A 21.51	Cast	Class 50	Mech. jt.	300	300	—	Ductile iron	Class B	Mech. joint
(2)	Ductile iron	A 21.15	Cast	Class 50	Flange	250	250	—	Ductile iron	Class B	Flange
	NPS 2–16							@ 210°F			
K. (1)	RTRP-11AF	D 2996	Fil. wound	—	B&S adhesive	150	150	150	RTRP-11AF		B&S or T.A.B.
	NPS 4–24						@ 100°F				
L. (1)	PVC	AWWA C900	Seamless	Class 100	B&S O-ring	100	60	NR	Ductile iron	Class B	Mech. joint
(2)	PVC	AWWA C900	Seamless	Class 150	B&S O-ring	150	90	NR			
(3)	PVC	AWWA C900	Seamless	Class 200	B&S O-ring	200	120	NR			

*Pressure ratings for steel pipe are calculated using (a) a mill tolerance or −12.5% of the wall thickness, (b) the thread or groove depth where used, and (c) a corrosion allowance of 0.025 to NPS 2 and 0.065 for NPS 2½ and larger. No pipe wall reinforcement value is applied for the strength of threaded fittings or grooved couplings.

†Copper tube pressures are based on the joint strength when soldered or brazed.

‡Pressure ratings are for the largest pipe size in each group. Smaller pipe sizes have higher ratings.

can also be used with grooved joints. Linings do not add to the strength of the pipe and may reduce the flow capacity.

Condenser water lines are often run below ground. Ductile iron or plastic water main pipe can be used since both the interior and exterior are corrosion resistant. Copper is a good alternative for both aboveground and underground condenser water piping. Table C4.9 lists some materials suitable for condenser water and the pressures at which they can be used.

Compressed Air

High temperatures are not usually involved, but compressed air contains stored energy and system failure can result in explosive reactions. Welding or a restrained joint should be used for large sizes. Do not use PVC or CPVC or any material subject to brittle failure for compressed air. Other thermoplastics which are not subject to brittle failure may be used. Check with the manufacturer of the material to be sure it is recommended for compressed air service.

Steel and copper pipe with malleable iron, cast-iron, or copper fittings are suit-

Fitting				Valve						
	Pressure rating							Pressure rating		
ASTM standard	@ 75°F (psig)	@ 150°F (psig)	@ 220°F (psig)	Class	Material	Type	Joint	@ 75°F (psig)	@ 150°F (psig)	@ 220°F (psig)
A 21.10	350	350	350	Same as listed in H and I above.						
A 21.10	250	250	—							
	150	150	@ 210°F 150	Same as listed in H and I above.						
A 21.10	350	350	350	Same as listed in H and I above.						

able materials for compressed air. Ball, butterfly, or gate valves can be used. For large sizes use welded, flanged, or grooved systems. ASME/ANSI B31.9, Building Services Piping, does not permit the use of copper with soldered joints for sizes above NPS 4 size at pressures exceeding 20 psig. Table C4.10 lists some suitable materials.

Fuel Gas

Pipe material for use in buildings should be standard weight, electric resistance weld, or seamless steel or copper tube Type *K* or *L*. Copper is not to be used if the gas contains more than an average of 0.3 grain of hydrogen sulfide per 100 standard cubic feet of gas. Fittings for steel pipe may be threaded malleable iron or ductile iron for sizes up to and including NPS 3. Some jurisdictions require cast fittings to be galvanized to seal casting pin holes. For sizes NPS 2½ and over, welded joints should be used with steel fittings and flanges.

Copper tube should be joined by brazing, using a brazing alloy that does not contain phosphorus, or by the use of flared connections. Table C4.10 lists some suitable materials.

TABLE C4.10 Material Application Chart for Air, Gas, and Oil Systems*,†,‡

Line	Pipe					Pressure rating			Fitting		
	Material	ASTM standard	Mfr. process	Wall thickness	Joint	@ 75°F (psig)	@ 150°F (psig)	@ 220°F (psig)	Material	Class	Joint
	Compressed air to 20 psig										
	NPS 2 and smaller										
O. (1)	Same materials as listed in Table C4.9 lines G (1) through (7).										
	NPS 2½–12										
M. (1)	Same materials as listed in Table C4.9 lines H (1) through (6), (8), and (9).										
	Compressed air to 100 psig										
	NPS 2 and smaller										
N. (1)	Same materials as listed in Table C4.9 lines G (1) through (3) and (5) through (7).										
(2)	Line (7), copper with solder joints may be used at or below NPS 4.										
	NPS 2½–12										
O. (1)	Same materials as listed in Table C4.9 lines H (1) through (6) and (8) through (10).										
	Fuel gas to 5 psig										
P. (1)	NPS 2 and smaller										
(2)	Steel	A 53 B	ERW	Std	Thread	430	430	430	Malleable iron	150	Thread
(3)	Steel	A 106	Seamless	Std	Thread	510	510	510	Ductile iron	300	Thread
(4)	Steel tube	A539,A254			Braze	Use mfrs. rating					
(5)	Copper tube	B 88		Type L, K	Braze, flare	315	315	260	Wrought copper		Brazed
(6)	Corrugated S.S. Conduit	AGA 1-87			System	Use mfrs. rating					
	NPS 2½–6										
Q. (1)	Steel	A 53	ERW	Std	Weld	695	695	695	Wrought steel	Std	Weld
(2)	Steel	A 106	Seamless	Std	Weld	815	815	815	Wrought steel	150	Flange
	Fuel oil										
	NPS 2 and smaller										
R. (1)	Steel	A 53	Seamless	XH	Thread	1215	1215	1215	Malleable iron	150	Thread
(2)	Steel	A 106	Seamless	XH	Thread	1215	1215	1215	Ductile iron	300	Thread
(3)	Steel tube	A539, A254			Braze				Forged steel	3000	Socket weld
	NPS 2½–12										
S. (1)	Steel	A 53	Seamless	Std	Weld	620	620	620	Wrought steel	Std	Weld
(2)	Steel	A 106	Seamless	Std	Weld	620	620	620	Wrought steel	150	Flange

*Pressure ratings for steel pipe are calculated using (a) a mill tolerance or −12.5% of the wall thickness, (b) the thread or groove depth where used, and (c) a corrosion allowance of 0.025 in to NPS 2 and 0.065 for NPS 2½ and larger. No pipe wall reinforcement value is applied for the strength of threaded fittings or grooved couplings.

†Copper tube pressures are based on the joint strength when soldered or brazed.

‡Pressure ratings are for the largest pipe size in each group. Smaller pipe sizes have higher ratings.

Fitting				Valve							
	Pressure rating							Pressure rating			
ASTM standard	@ 75°F (psig)	@ 150°F (psig)	@ 220°F (psig)	Class	Material	Type	Joint	@ 75°F (psig)	@ 150°F (psig)	@ 220°F (psig)
B 16.3	300	300	255	125	Bronze	Gate	Thread	200	200	180
A 395	500	500	460		Bronze	Ball	Thread	400	400	400
B 16.22										
B 16.9	510	510	510	125	Cast iron	Gate	Flange	150	150	130
B 16.5	285	270	250	150	Ductile iron	Gate	Flange	250	250	230
				200	Cast Iron	Plug	Flange	200	200	180
B 16.3	300	300	255	125	Bronze	Gate	Thread	200	200	180
A 395	500	500	460		Bronze	Ball	Thread	400	400	400
B 16.11	2600	2600	2600							
B 16.9	510	510	510	150	Ductile iron	Gate	Flange	250	250	230
B 16.5	285	270	250							

Fuel Oil

For piping to boilers operating at or under 15 psig steam or 30 psig water, use electric-resistance welded, seamless steel pipe, or copper tube, Type *K* or *L*. When threaded joints are used, malleable iron, ductile iron, or forged steel fittings may be used. A thread sealing compound suitable for oil should be used when making the joints. Joints in copper tube may be brazed or made by using flared or compression fittings.

Oil piping to boilers operating at steam pressures from 16 to 150 psig should be seamless steel, standard weight, when welded or extra strong if threaded. Threaded joints using ductile iron or forged steel fittings may be used if unavoidable. Valves should be made of ductile iron or brass.

Outside of the boiler room malleable iron threaded fittings may be used. Type *K* or *L* copper tube with brazed joints may also be used outside of the boiler room. Table C4.10 lists some suitable materials.

WELDING

Welding Qualification

To comply with the B31 Pressure Piping Code sections, all welding procedures and welders must be qualified as required by Section IX of the ASME Boiler and Pressure Vessel Code. B31.9, Building Services Piping, also permits qualification to AWS D10.9, Qualification of Welding Procedures and Welders for Piping and Tubing.

A welding procedure specification is written to outline a method for making a weld. It lists the conditions under which the weld must be made. The procedure is qualified by making a weld following the specification. If the weld passes the prescribed tests, the procedure is qualified as being able to make a sound weld.

Welders and welding operators are tested to see that they have the necessary skills by making a weld using a qualified weld procedure specification. If the weld passes the required test, the welder can be termed qualified by his employer.

Boiler External Piping

For boilers operating at pressures higher than 15 psig, the ASME Boiler and Pressure Vessel Code has administrative control over the welding on boiler external piping (BEP), which is broadly defined as the piping between the boiler and (1) the first steam stop valve, (2) the second feedwater valve, and (3) the second blowdown valve. When multiple boilers are connected to a header, the BEP extends to the second steam valve.

The boiler external piping must be fabricated and installed by an organization holding an ASME certificate and must be provided with data reports, inspection, and stamping as required by the boiler code.

The ASME Pressure Piping Code B31.1, Power Piping, does not require nondestructive examination of welds unless the pressure is above 1025 psig with a temperature above 350°F or the temperature is above 750°F at all pressures. If radiography or other nondestructive examination is desired in excess of code requirements, it should be clearly spelled out in the contract or specification.

HANGERS AND SUPPORTS

This section gives a brief outline of the use of hangers, supports, anchors, and guides. See Part B of this handbook for a detailed discussion of hangers and hanger loads.

Hangers support the pipe from above; supports bear the load of the pipe from below; anchors restrain pipe movement; guides allow only axial movement of the pipe and direct expansion forces.

Buildings are built to provide usable floor space; therefore, piping is usually hung from the structure above. Hangers are concentrated loads; an allowance of extra distributed dead load in the structure above mechanical rooms will not necessarily be sufficient to support the point loads of large pipelines or suspended equipment. Piping should be arranged to allow the use of major structural elements for hanging pipelines. Major loads that pipe hangers impose on a structure should be reviewed by a structural designer.

Hanger loads are calculated according to B31.9 by taking the following elements into account:

1. The dead weight of the pipe, fittings, valves, insulation, and the hanger itself.
2. The weight of liquid in the pipe.
3. Occasional loads, such as ice, wind, earthquake, a test liquid, and water hammer. Occasional loads need not be considered as acting concurrently.

The forces acting on pipe anchors are

1. The hanger loads listed above.
2. Forces caused by expansion and contraction of the pipe.
3. Expansion joint forces needed to overcome joint friction and separation. The separation force is the product of the line pressure and the largest inside cross-section area of the joint.

When supporting a riser, the entire weight of the riser pipe and the fluid or test liquid in that vertical section of the pipe must be carried by the base elbow or anchor.

Hanger design is based on using one-fifth of the minimum tensile stress of the hanger material. If the minimum tensile strength of a material is not known, a value of 9500 psi may be used. During hydrostatic testing this value may be increased to 24,000 psi. Allowable stress values for shear are 80 percent and for bearing are 160 percent of the values determined above.

The carrying capacity of threaded rod is based on the rod area at the root of the thread. The allowable stress is reduced by 25 percent for cut threads. See Table C4.11 for rod-carrying capacities.

Hangers should be designed to permit vertical adjustment. When this is done with threaded elements, double nuts or other locking devices should be used to prevent vibration from working the nut loose.

Hangers are attached to the structure by welding, beam clamps, concrete inserts, and metal deck inserts and by drilling expansion anchors directly into the concrete. Loads on inserts in, or attachments to, concrete should be limited to one-fifth of the ultimate strength of the attachment as determined by the manufacturer's tests. If the compressive strength of the concrete is unknown, it can be assumed to be 2500 psi and the results of the manufacturer's test derated propor-

TABLE C4.11 Capacities of Threaded Steel Rods

Rod diameter (in)	Root area of thread (in^2)	Unknown steel rolled threads, S = 9.5 ksi (lb)
¼	0.027	260
⅜	0.068	650
½	0.126	1200
⅝	0.202	1900
¾	0.302	2900
⅞	0.419	4000
1	0.551	5200
1⅛	0.693	6600
1¼	0.889	8400
1⅜	1.054	10000
1½	1.294	12300

tionally. Explosively actuated fasteners should not be used where a group of pins is needed to support a load.

Spring hangers should be used where expansion can cause vertical movement of a pipeline. For example, if a branch line is connected to a vertical heating-cooling riser on the twelfth floor of a high-rise building, the connection point could be 1 in above its ambient elevation during the heating season and 1 in below in the summer. Using spring hangers on the branch near the riser connection will provide support under both conditions.

Hanger spacing is often determined by the building steel available for suspending the pipe. Tables C4.12, C4.13, C4.14, and C4.15 show maximum spans for steel, copper, aluminum, and plastic pipe. These tables are calculated using a simple beam formula to determine stress and deflection. The basic allowable stress of the pipe will not be exceeded at double the span shown in the tables. The usable stress for bending is that which remains after the stresses due to longitudinal pressure have been deducted from the basic allowable stress of the material. The tables also have the deflection limited to the smaller of 0.2 in or 10 percent of the nominal pipe diameter at the listed span.

The tables show the effect of internal pressure on pipe spans. In general the deflection is the governing factor in the smaller sizes. The weights and spans have been calculated based on mineral fiber insulation.

EXPANSION AND FLEXIBILITY

Expansion in building service piping must be recognized and allowed for. The temperature extremes may run from 40°F for chilled water, which will result in the contraction of a pipe installed at 70°F, to 360°F for 150 psig steam, which will give rise to a much greater expansion of a pipe. Part B of this handbook gives a detailed discussion on expansion and flexibility as well as tables giving coefficients of expansion and the actual expansion for various materials between given temperatures.

Uncontrolled expansion forces can be harmful at equipment connections by imposing loads on the equipment. Manufacturers will provide the allowable

TABLE C4.12 Pipe Spans for Standard Weight Steel Pipe for Straight Runs with No Valves or Components

	Pipe and insulation						Pipe, water, and insulation					
	ASTM A53 type F butt weld S = 11,250			ASTM A53B Smls or ERW S = 15,000			ASTM A53 type F butt weld S = 11,250			ASTM A53B Smls or ERW S = 15,000		
	Pressure in pipe			Pressure in pipe			Pressure in pipe			Pressure in pipe		
Pipe size (NPS)	300 psig (ft)	15 psig (ft)	Hanger load (lb)	400 psig (ft)	15 psig (ft)	Hanger load (lb)	300 psig (ft)	15 psig (ft)	Hanger load (lb)	400 psig (ft)	15 psig (ft)	Hanger load (lb)
½	6	6	5	6	6	5	6	6	6	6	6	6
¾	7	7	9	7	7	9	7	7	11	7	7	11
1	9	9	16	9	9	16	9	9	19	9	9	19
1¼	11	11	26	11	11	26	11	11	32	11	11	32
1½	12	12	34	12	12	34	12	12	45	12	12	45
2	14	14	56	14	14	56	13	14	75	14	14	77
2½	16	16	96	16	16	96	15	16	128	16	16	129
3	17	17	139	17	17	139	16	17	189	17	17	195
4	20	20	224	20	20	224	18	19	317	20	20	334
5				22	22	338				22	22	530
6				24	24	478				23	24	782
8				28	28	829				25	28	1431
10				31	31	1309				28	31	2369
12				34	34	1746				28	33	3301
14				36	36	2038				29	34	3929
16				38	38	2500				29	35	5015
18				40	40	2993				28	36	6223
20				43	43	3515				28	36	7561
24				46	47	4638				27	37	10603

1. Spans based on lesser of (1) half that permitted by allowable stress in simple beam formula or (2) deflection of 0.1 × NPS to 0.2 in max. for empty pipe and fiberglass insulation.
2. Formulas used: Deflection span, ft = $[(384EIG/W)^{0.25}]/12$. Simple beam span, (ft) = $[\{8Z(S - SL)/W\}^{0.5}]/12$, where E = modulus of elasticity, I = moment of inertia, G = permitted deflection, Z = section modulus, S = allowable stress, SL = longitudinal stress caused by pressure, W = weight.
3. Hanger loads listed are the full weight of the longer span in each category.

TABLE C4.13 Pipe Spans for Copper Tube Type L for Straight Runs with No Valves or Components

Nominal pipe size (NPS)	Pipe and insulation: ASTM B88 type L hard drawn S = 11,300 to 250°F — Pressure in tube: 300 psig (ft)	15 psig (ft)	Hanger load (lb)	Pipe, water, and insulation: ASTM B88 type L hard drawn S = 11,300 to 250°F — Pressure in tube: 300 psig (ft)	15 psig (ft)	Hanger load (lb)
½	4	4	1	4	4	2
⅝	5	5	2	5	5	3
¾	5	5	3	5	5	4
1	7	7	6	7	7	8
1¼	8	8	8	8	8	13
1½	9	9	12	9	9	19
2	11	11	22	10	11	37
2½	12	12	33	10	12	56
4	14	14	52	12	13	88
3½	15	15	70	12	14	124
4	16	16	93	13	15	167
5	18	18	147	14	17	269
6	20	20	216	15	18	400
8	23	23	473	18	21	854
10	25	25	791	20	23	1480
12	28	28	1175	21	25	2214

1. Spans based on (1) lesser of half that permitted by allowable stress in simple beam formula or (2) deflection of 0.1 × NPS to 0.2 in max.
2. Formulas used: Simple beam span (ft) = $[\{8Z(S - SL)/W\}^{0.5}]/12$; Deflection span (ft) = $[(384EIG/W)^{0.25}]/12$, where E = modulus of elasticity, I = moment of inertia, G = permitted deflection, Z = section modulus, S = allowable stress, SL = longitudinal stress due to pressure, W = weight (lb/in).
3. Hanger loads listed are the full weight of the longest span in each category.

forces and moments. When necessary these forces can be controlled by providing an anchor or restraint on the piping at or near the connection.

Long straight runs of pipe with no offsets, or very short offsets, need to be checked for total linear expansion. These are found most often as risers in vertical buildings and mains in horizontal buildings. If the amount of expansion encountered cannot be absorbed by the flexibility in the pipe configuration at takeoffs from a main, or at the end of a main, an expansion joint or pipe loop must be used.

There are three categories of expansion joints: corrugated bellows, packed slip joints, and rotary or ball joints. All expansion joints must be placed between anchors to direct the forces and in some cases to keep the joint from separating due to the internal pressure.

1. Corrugated joints have no packing, which is an advantage since they do not develop leaks. The pipe leading to this type of joint must be provided with adequate guiding to keep the pipe axially aligned and direct the expansion forces into the joint. If the pipe is improperly guided the joint may squirm, which results in misalignment and possible catastrophic failure.
2. Packed slip joints have the disadvantage of packing which must be main-

TABLE C4.14 Pipe Spans for Standard Weight Aluminum Pipe for Straight Runs with No Valves or Components

	Pipe and insulation			Pipe, water, and insulation		
	ASTM B241 A93003-0 S = 3,400 to 200°F			ATM B241 A93003-0 S = 3,400 to 200°F		
	Pressure in pipe			Pressure in pipe		
Nominal pipe size (NPS)	180 psig (ft)	15 psig (ft)	Hanger load (lb)	180 psig (ft)	15 psig (ft)	Hanger load (lb)
½	6	6	2	6	6	3
¾	7	7	4	7	7	5
1	8	8	6	8	8	9
1¼	10	10	10	8	9	14
1½	12	12	13	9	10	19
2	13	14	19	9	10	31
2½	14	15	33	10	12	51
3	16	17	47	11	12	77
4	18	20	74	12	14	132
5	19	22	108	12	14	207
6	21	24	149	13	15	304
8	22	27	252	13	17	550
10	25	31	383	14	18	901
12	26	33	534	14	19	1329

1. Spans based on (1) lesser of half that permitted by allowable stress in simple beam formula or (2) deflection of 0.1 × NPS to 0.2 in max.

2. Formulas used: simple beam span (ft) = $[\{8Z(S - SL)/W\}^{0.5}]/12$; deflection span (ft) = $[(384EIG/W)^{0.25}]/12$, where E = modulus of elasticity, I = moment of inertia, G = permitted deflection, Z = section modulus, S = allowable stress, SL = longitudinal stress due to pressure, W = weight (lb/in).

3. Hanger loads listed are the full weight of the longest span in each category.

tained. They must also be guided, but catastrophic failure is unlikely unless an anchor fails.

3. Ball joints are used in pairs in branch pipes, or in the main after an elbow, to take up the expansion offset. These are packed swivel joints which need maintenance but will not separate.

Flexibility is inherent in a building system due to the elbows and offsets normally found necessary to get from the source to the destination. An unrestrained piping system will find its own point of least stress and have the lowest end forces. If an inspection of the pipe route does not reveal any places where the expected expansion will interfere with walls, columns, ducts, or other pipes, it is advisable not to introduce anchors. Detailed flexibility analysis methods can be found in Part B of this handbook.

TESTING

Before putting a system in service it should be tested for leakage or proof tested to demonstrate its ability to withstand the design pressure. This can be done by several methods.

TABLE C4.15 Pipe Spans for Sch 40 CPVC, PVC, ABS, and Sch 80 PP Thermoplastic Pipe, Including Water and Insulation

Nominal pipe size (NPS)	CPVC ASTM F441 #4120 Cell no. 23447						PVC ASTM D1785 #1120 Cell no. 12454				ABS ASTM D1517 Des. no. 1210				Polypropylene Schedule 80 pipe					
	Pipe pressure (psig)						Pipe pressure (psig)				Pipe pressure (psig)				Pipe pressure (psig)					
	73°F		100°F		180°F		73°F		100°F		73°F		100°F		73°F		100°F		180°F	
	100 (ft)	15 (ft)	100 (ft)	15 (ft)	100 (ft)	15 (ft)	100 (ft)	15 (ft)	100 (ft)	15 (ft)	100 (ft)	15 (ft)	100 (ft)	15 (ft)	100 (ft)	15 (ft)	100 (ft)	15 (ft)	100 (ft)	15 (ft)
½	3	3	3	3	3	3	3	3	3	3	3	3	3	3	3	3	3	3	2	3
¾	4	4	3	3	2	3	4	4	4	4	3	3	3	3	3	3	3	3	EX	3
1	4	4	4	4	3	3	4	4	4	4	4	4	4	4	4	4	4	4		3
1¼	5	5	5	5	EX	4	5	5	3	5	5	5	4	5	5	5	5	5		3
1½	6	6	6	6		4	6	6	6	6	5	6	4	5	6	6	5	6		3
2	7	7	7	7		4	7	7	7	7	5	6	4	5	6	7	5	6		4
2½	8	8	8	8		5	8	8	8	8	6	7	5	6	7	7	6	7		4
3	9	9	8	9		5	9	9	8	9	6	8	5	7	8	8	6	7		4
4	10	10	8	10		5	10	10	9	10	6	8	EX	7	8	9	6	8		5
5	10	11	9	10		5	10	12	9	11	EX	8		7						
6	11	12	9	11		6	11	12	9	11		9		8	9	11	7	9		5
8	11	13	9	12		6	11	13	9	12		9		8	9	12	EX	10		6
10	12	14	10	13		7	12	14	10	13		10		9						
12	12	15	10	13		7	12	15	10	14		11		9						

1. Spans based on lesser of (1) half that permitted by allowable stress in simple beam formula or (2) deflection of 0.1 × NPS to 0.2 in max. for empty pipe and glass fiber insulation.
2. Formulas used: Deflection span (ft) = $[(384EIG/W)^{0.25}]/12$; simple beam span (ft) = $[\{8Z(S\text{-}SL)/W\}^{0.5}]/12$, where E = modulus of elasticity, I = moment of inertia, G = permitted deflection, Z = section modulus, S = allowable stress, SL = longitudinal stress caused by pressure, W = weight.

EX—At this pressure the hoop stress in the pipe exceeds the allowable stress for this size and larger.

Hydrostatic Testing

Hydrostatic testing is the preferred method of testing since leaks are evident and the stored energy in the pipe is low. All joints should be exposed and uninsulated. Before initial filling, the following precautions should be observed:

1. A survey of the entire pipeline should be made to ensure that there are no open ends, all hangers, anchors, and guides are in place, and all joints are properly made.
2. All equipment that will be damaged by the test pressure should be isolated from the system.
3. Expansion joints and anchors in the pipeline should be checked for their ability to withstand the test pressure.
4. If the test pressure is to be left on the system for an extended period, a relief valve should be installed to prevent overpressurization due to expansion of the fluid caused by an increase in temperature.
5. Vents should be used to release air from high points of the system. Drains should be provided to remove the test liquid.

Test Medium. The test medium should be water unless there is a risk of freezing or water will damage the system. A glycol solution can be used if provisions are made to flush the system and properly dispose of the glycol at the end of the test. The system should be filled gradually and examined to detect leaks as the filling progresses.

Test Pressure. The test pressure required by B31.9 is 1.5 times the design pressure at all points of the system. For vertical systems the pressure at the bottom should not exceed the lower of

1. 90 percent of the minimum yield strength of the material or
2. 1.7 times the SE value of brittle materials and
3. The rated pressure of valves or equipment which will be subjected to the pressure.

Examination. Examination of the entire system for leaks should be made after the test pressure has been on the system for at least 10 min. If leaks are found they should be repaired and the test repeated until no leakage is detected.

Pneumatic Testing

Pneumatic testing is dangerous and should only be used within the following conditions:

1. There are no soldered joints in the system over NPS 4 size.
2. If brittle plastic, no pipe is larger than NPS 2 size.
3. Water will be detrimental to the system.
4. The test pressure does not exceed 150 psig.

The test medium should should be air, nitrogen, or any other innocuous gas.

Test Pressure. The test pressure for a pneumatic test is limited by B31.9 to a maximum of 1.25 times the design pressure. The pressure should be introduced gradually with the first increment not more than 10 psig, at which time an inspection can be made for major leaks. The pressure can then be raised in increments of 25 percent, pausing at each increment to allow system equalization.

Examination. Examination for leaks should be made after the full test pressure has been on the system for 10 min or more. Leaks can be detected by soap bubbles, special testing fluids, ultrasonic means, or test gauge monitoring.

Service Testing

Service testing can be used in place of the above tests for low-pressure systems that operate below 15 psig steam pressure or 30 psig water pressure. Proof testing is not needed, and an examination for leaks can be made when the system is placed in initial operation for the service intended.

PROBLEMS AND SOLUTIONS

Some of the most often occurring problems with piping systems are discussed below.

The cost of the installation is too high. If this comment is heard check the following:

1. Has seamless pipe been used when furnace butt weld or electric resistance weld pipe will be adequate?
2. Has Type *K* copper tubing been called for when Type *L* is adequate?
3. Have steel valves been used when ductile iron or cast-iron valves are strong enough?
4. Has radiography of welds been called for when it is not required by the piping code?
5. Are too many anchors and expansion joints called for when the system has inherent flexibility?
6. Have factory made tees or reinforced outlet fittings been used when welded branch connections are suitable?
7. Has the use of groovedtype joints been permitted where they are economical?
8. Has the use of plastic materials been considered?
9. Are pipe sizes too large? This requires analysis of the long-term pumping costs since smaller pipe sizes will result in greater friction.
10. Has the hanger spacing been checked to make sure the most economical spacing has been allowed?
11. Has the system configuration been reviewed to be sure the most direct route possible has been selected?

The system does not circulate the proper GPM. Look for the following:

1. Are all of the valves open?
2. Is the system fully vented?
3. Are all the strainers clean?
4. Is the pump providing the proper pressure differential?
5. Is there an obstruction in the piping? Check pressures along the system to find an unpredicted drop.

The system is noisy. Look for the following causes:

1. Has the system been fully vented? Air in the system is noisy.
2. Is the pump circulating too much fluid? If the required system head has been predicted higher than the actual one, the pump is circulating too much fluid and causing velocity noise. Throttle a pump balancing valve to introduce more pressure drop in the system. If this solves the problem, the pump impeller should be changed for long-term economy.

The chilled water expansion tank overflows. Investigate the following:

1. The makeup level in the tank is set too high and water is added when the system is operating cold.
2. The tank does not have the capacity for the expansion volume.
3. There is a cross connection to another system.

The steam pipes make knocking noises. Investigate the following:

1. The steam pipe has not been pitched in the direction of condensate flow.
2. The distance between drip legs on long runs is too great. If the knocking only occurs on warmup, look for a slower warmup method.

The water pipes hammer or shake. Water hammer can be caused by the following:

1. The sudden opening or closing of a valve. Automatic valves can be made to operate slower to prevent water hammer. Manual valves should always be operated slowly.
2. The starting or stopping of a pump. A spring loaded or hydraulically operated check valve should always be installed at the discharge of large pumps.
3. Air can be sucked into a piping system at a cooling tower basin and cause violent pipe movement. Install a vortex breaker in the basin.

Is the system ready to be placed in operation?

1. The pressure test should be performed and all blanks removed.
2. The system should be flushed, and the water treatment system should be operable.
3. A visual inspection should be made to be sure all open ends have been secured, all block valves are in the proper position for operation, all safety and relief valves are installed, and if a water system, the system is full and vented.
4. A visual inspection should be made to be sure adequate pipe expansion room is available. Some approximations that can be used for visual inspection are as

follows per 100 ft of straight pipe per 100°F from ambient temperature (do not use for exact calculations):

Ductile iron	Expands	1 in
Steel	Expands	1 in
Stainless steel	Do	1¼ in
Copper or brass	Do	1½ in
Aluminum	Do	2 in
RTRP	Do	2 in
PVC & CPVC	Do	5 in
Polypropylene	Do	6 in
ABS	Do	7 in
PB	Do	9½ in

5. Check that all anchors and guides are in place.

Life expectancy. Piping in power plants, refineries, industrial plants, and so on, where the piping is a part of the process is subjected to extreme service conditions and is therefore closely monitored. Building piping, while necessary to the operation of a building, is not subject to severe service and therefore not watched as closely. Because the pipe is built into walls and shafts, it is expensive to replace and is generally designed to last as long as the building.

The corrosion of steel pipe is the biggest concern. Closed recirculating systems such as chilled water or heating water can circulate for years with very little corrosion since no new oxygen is introduced into the system. Cooling tower piping systems require very close monitoring, and alternative materials should be considered.

BIBLIOGRAPHY

ASHRAE Handbook, American Society of Heating, Refrigerating and Air Conditioning Engineers, Atlanta, GA.

ASME Boiler & Pressure Vessel Code, ASME, New York.

Section I, Power Boilers

Section IV, Heating Boilers

Section VI, Recommended Rules for Care and Operation of Heating Boilers

Section IX, Welding and Brazing Qualifications

ASME/ANSI Code for Pressure Piping B31, ASME, New York.

B31.1, Power Piping

B31.3, Chemical Plant and Petroleum Refinery Piping

B31.5, Refrigeration Piping

B31.9, Building Services Piping

ASTM Annual Book of ASTM Standards, ASTM, Philadelphia, PA.

Volume 01.01, Steel-Piping, Tubing, Fittings

Volume 02.01, Copper and Copper Alloys

Volume 08.04, Plastic Pipe and Building Products

Piping Engineering, Tube Turns Division of Chemetron Corporation, Louisville, KY.

Copper Tube Handbook, Copper Development Association Inc., Greenwich, CT.

Victaulic Mechanical Piping Systems General Catalog, Victaulic Company of America, Easton, PA.

MSS Standard Practices, Manufacturers Standardization Society, Inc., Vienna, VA.

SP-58, Pipe Hangers and Supports—Materials, Design and Manufacture

SP-69, Pipe Hangers and Supports—Selection and Application

PFI Technical Bulletin, Pipe Fabrication Institute, Springdale, PA.

TB2, Reinforcement Tables for Branch Connections

CHAPTER C5
OIL PIPELINE SYSTEMS

Charles L. Arnold
Principal Pipeline Consultant
Bechtel Corporation
San Francisco, California

Lucy A. Gebhart
Pipeline Engineer
Bechtel Corporation
San Francisco, California

INTRODUCTION

This chapter has been prepared as a basic guide to the design of cross-country pipelines for liquid petroleum and related products. It focuses on the fundamentals of pipeline design, emphasizing practical guidelines for real systems. It provides a general overview of the system approach to design, which integrates the hydraulic, mechanical, and operations and maintenance aspects in the design of a system, along with project economic analysis, in determining the preferred pipeline system.

This chapter also includes discussion of design topics for related pipeline system components such as pump station location and sizing, material selection for pipe, metering, leak detection, and system control. Aspects of petroleum system design related to the special characteristics of some petroleum commodities are also addressed, in particular topics related to high-vapor-pressure systems, multiproduct systems, and systems requiring consideration of variable thermal properties of the fluid (i.e., hot oil systems). Finally, design considerations for seismic and underwater design of pipeline systems are outlined.

Scope

There are three basic codes developed by American Society of Mechanical Engineers (ASME) and the American National Standards Institute (ANSI) which govern the design of piping systems in chemical, petroleum liquid, and gas usage. Piping inside the boundaries of a chemical plant, refinery, or gas processing plant falls under the scope of ASME B31.3 and is covered separately in this handbook (see Chap. C7). Likewise, ASME B31.8, which covers gas transmission and distribution piping systems, is specifically addressed in Chap. C6 of this handbook.

This chapter specifically addresses oil transportation systems as covered by the ASME B31.4 Code—"Liquid Transportation Systems for Hydrocarbons, Liquid Petroleum Gas, Anhydrous Ammonia, and Alcohols."[1] In this chapter, references to specific sections of the ASME B31.4 Code will be cited as Section *n* of the Code, or in some cases as *the Code.*

Figure C5.1 is a schematic diagram illustrating the scope of ASME B31.4 for liquid petroleum piping systems.

Code Compliance

The ASME/ANSI Codes set forth the practices required for design and operation of safe pipeline systems. Section 400.1.1 of the Code states its purpose as follows:

> This code prescribes minimum requirements for the design, materials, construction, assembly, inspection, and testing of piping transporting liquids such as crude oil, condensate, natural gasoline, natural gas liquids, liquified petroleum gas, liquid alcohol, liquid anhydrous ammonia, and liquid petroleum products between producers' lease facilities, tank farms, natural gas processing plants, refineries, stations, ammonia plants, terminals (marine, rail, and truck), and other delivery and receiving points.

While the Code gives guidelines for pipeline system design, it is not intended to provide complete specifications for all phases of design and operation. Furthermore, there may be additional federal (or country), state, and local regulations governing pipeline design, construction, and operation.

It is the intent of this chapter to supplement the Code with discussion of the principles governing the design of a petroleum transportation system, identifying the analytical and design tools and procedures an engineer might use. This chapter in itself may not address all the design problems which will arise in the real world, and there are some cases where even the best design guide or reference cannot replace experience and judgment. Specific engineering and operating companies may also have guidelines which they require an engineer to follow in the course of designing a pipeline transportation system.

The responsibility rests with the engineer or designer to identify the specific requirements and applicable codes for a given system with regard to design and operational conditions, as well as to follow specific guidelines mandated by the engineering or operating company and any relevant additions to the Code and other regulations which may govern the design.

Codes, Standards, Specifications, and Recommended Practices

In general, pipelines which are designed in accordance with the Code will meet the requirements in the United States for liquid petroleum pipelines and associated facilities. In addition to the ASME B31.4 Code, the following codes, specifications, standards, regulations, and recommended practices may be applicable

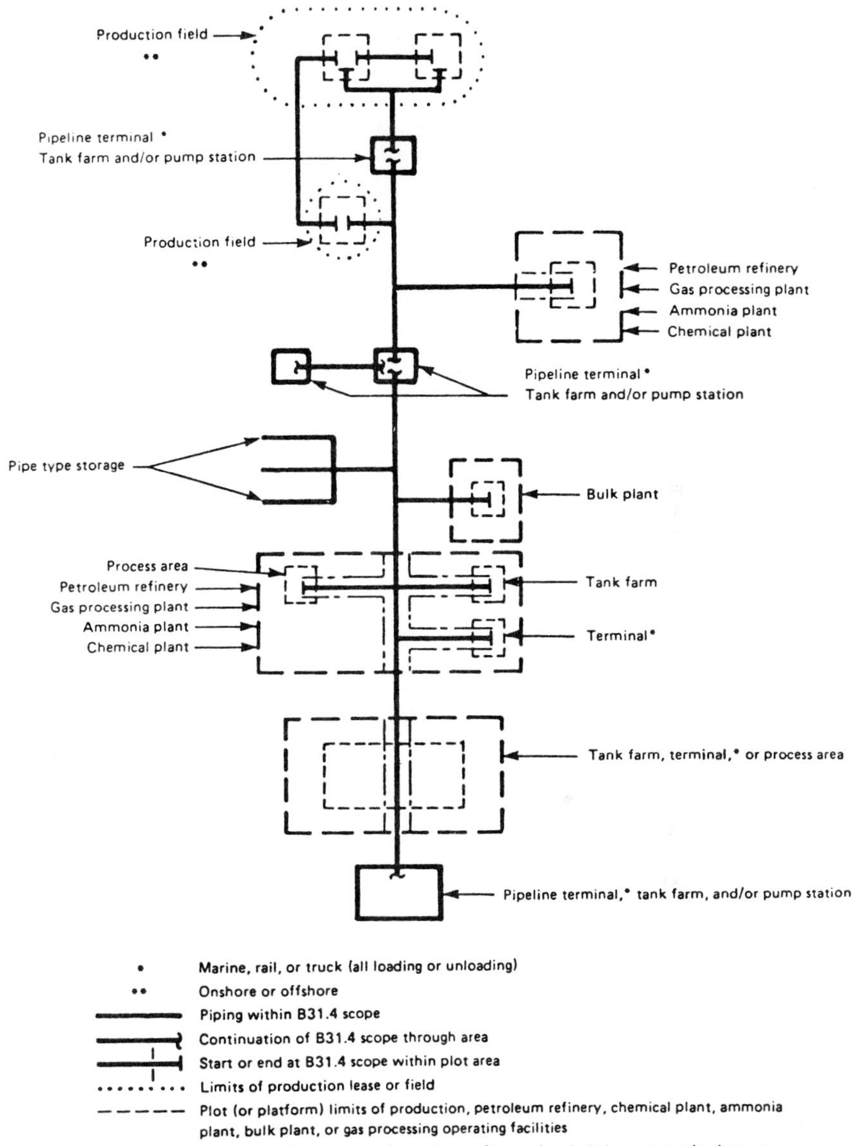

FIGURE C5.1 Scope of ASME B31.4.

to a proposed pipeline system or component thereof (this list is representative but is not comprehensive):

DESIGN

U.S. Code of Federal Regulations, Title 49, Part 195—Transportation of Hazardous Liquids by Pipeline (known as 49 CFR 195)

American Petroleum Institute (API) RP 1102—Recommended Practice for Liquid Petroleum Pipelines Crossing Railroads and Highways

American Petroleum Institute (API) RP 1111—Recommended Practice for Design, Construction, Operation and Maintenance of Offshore Hydrocarbon Pipelines

CAN/CSA-Z183—Canadian Oil Pipeline System Code

MATERIAL

API 5L—Specifications for Line Pipe

ANSI B16.5—Pipe Flanges and Flanged Fittings

ANSI B16.34—Valves—Flanged, Threaded, and Welding End

These codes, specifications, etc., also cross-reference additional standards and recommended practices that apply to various aspects of petroleum pipeline system components.

References

General references and supplements for the material covered in this chapter are supplied at the end of the chapter. The majority of the material covered in this chapter is also available in numerous petroleum reference books and technical papers through trade publications. No attempt has been made to reference all sources of relevant data available. Only those most pertinent to the discussions or those considered most useful are included here.

LIQUID PETROLEUM PIPELINE SYSTEMS

It is interesting to note that liquid petroleum pipeline systems have been in operation since the late 1800s, and, as illustrated in the title of the Code, a number of different commodities are transported via pipeline, with widely varying properties. Therefore, to begin the discussion of liquid petroleum pipeline systems, it is useful to illustrate some of the characteristics of fluids covered by the B31.4 Code, based on examples of real pipeline systems. Then, since this chapter is intended to be an aid in the selection of the preferred pipeline design, the concept of the "system approach" to design is introduced wherein the hydraulic, mechanical, and operations and maintenance aspects of design are integrated and evaluated with economic analysis to select the most economically attractive system.

Characteristics of Transported Commodities

ASME B31.4 covers a wide range of petroleum liquid commodities, including crude oils, residuals, products of refining such as diesel oil, jet fuel, gasoline, natural gas liquids, oil-water emulsions, anhydrous ammonia, and others. The physical properties of these commodities are also variable, and each pipeline system design is based on specific properties of an identified commodity or group of commodities in the case of multiproduct pipeline systems.

Table C5.1 identifies general characteristics of fluid commodities covered under the Code. The API gravity, viscosity, and temperature values shown (where available) are specific examples of operating pipeline systems and should only be considered as examples. The table also identifies specific considerations for different commodities which can have an effect on the design and operation of a pipeline system.

System Approach to Design

In the design of an oil transportation system, it is necessary to consider many aspects of design and operation as well as project economics in determining the optimum pipeline system to transport a commodity, or commodities, from a source to a destination. On a technical or engineering level, there are three aspects of design which are interrelated in the system approach to design:

- Hydraulic
- Mechanical
- Operations and maintenance

Decisions in one area of design directly affect, or limit, the options in another area. For example, it may be necessary to locate a pump station such that it is accessible, for example, on a main road, near an electrical power source. Thus the pipeline route will have an intermediate location point set in addition to the origin and terminal points. Likewise, preliminary design and cost estimating are not separate and independent procedures but are instead closely related and proceed concurrently.

The hydraulic design is the process of evaluating the physical characteristics of the commodity or commodities to be transported, the quantities to be transported, the pipeline route and topography, and the range of pressures, temperatures, and environmental conditions along the route. Identifying the number and location of pump stations with respect to the hydraulic characteristics of the system is also part of the hydraulic design. There may be several viable hydraulic designs for any given pipeline design basis and route. The most feasible is identified in conjunction with the owner or operator of the system, giving consideration to early use requirements and future capacity plans for the system.

For any one hydraulic design there are a number of mechanical system designs that can be developed to meet the criteria of the design basis and deliver the commodity from origin to destination. The mechanical design is governed by the codes and standards developed from experience in operating petroleum pipeline systems, and it focuses on selection of pipe material and the specification of physical line pipe properties such as pipe diameter and wall thickness as required by the stresses imposed on the system by the hydraulic and thermal conditions,

TABLE C5.1 Petroleum Commodity Characteristics for Pipelines

Commodity	Temp range (°F)	Specific gravity	API gravity	Viscosities		Pour point (°F)	Vapor press (psia)	Remarks: flow regime, rheology, general considerations
				cS at °F	cS at °F			
Residuals:	150–250	1.02	7.2	50,000 at 130	330 at 250	130	—	Common design temperature in the range of 150–225°F (65–105°C). [One design for 800°F (427°C) is operating.] Thermal design required. Laminar flow
	150–250	0.96	1.6	1000 at 100	45.7 at 210	90		
Crudes:								
General	40–160	0.84	12–40	11 at 68	4.1 at 122	55	15	Generally newtonian in range of operating temperature. May require thermal design considerations. Transition to fully turbulent flow
High wax content	70–140	0.81	35–45	7.4 at 122	3.3 at 140	95	15	Generally newtonian above cloud point; develops yield stress and nonnewtonian flow characteristics after static cooling. Transition to fully turbulent flow
Shale oils	40–120	Data still emerging on shale oil systems—current state of the art						Fluid properties vary with method of extraction/prep. for transport. Require considerations for vapor pressure, interfacial mixing and fluid properties when setting operations requirements, selecting pumps.
Products:								
No. 2 furnace	30–80	0.82–0.84	39	5.7 at 30	2.6 at 100	—		Newtonian fluids. Fully turbulent. May require thermal design.
Diesel—No. 1	30–80	0.83		2.8 at 30	1.4 at 100	—		Multiproduct hydraulic design is based on the combination of fluid properties in the system that produces the maximum system stress and required pumping horsepower at stations.
Diesel—No. 3	30–80	0.88		10 at 30	3.6 at 100	—		
Jet fuel	30–80	0.78		2.2 at 30	1.3 at 100	—		
Gasoline	30–80	0.71–0.73	65	0.8 at 30		—	30 at 100°F	Require considerations for vapor pressure, interfacial mixing, and fluid properties when setting operations requirements, selecting pumps.

Natural gas liquids (NGL) (and other high-vapor pressure petroleum liquids	30–130	0.5		0.23 at 30	0.2 at 100	–	200 at 60°F to 700 at 120°F	Pressure is maintained above the critical pressure to avoid two-phase flow. Special considerations for loading pumps and blowing down pipeline sections are required. Thermanl design is required. Fully turbulent flow is desired.
Other:								
Oil-water emulsions	There is no specific example pipeline system for this commodity, or the information is proprietary/confidential.							May be either an oil or water suspension; either extremely variable in fluid properties.
Alcohols	There is no specific example pipeline system for this commodity, or the information is proprietary/confidential.							Toxic, and flame may be invisible in daylight.
Anhydrous ammonia	There is no specific example pipeline system for this commodity, or the information is proprietary/confidential.							Toxic and corrosive—Code requires >0.2 percent water by wt. to inhibit stress corrosion cracking.
Carbon dioxide	There is no specific example pipeline system for this commodity, or the information is proprietary/confidential.							(See high-vapor-pressure liquids.) Dehydration is required for pipeline quality. Heavier than air, toxic in elevated concentrations.

yet within the limits set by the Code. Other aspects of the mechanical design include the type, size, and horsepower of pumps and other equipment or ancillary facilities required to meet the hydraulic-thermal design, such as heating stations, and the support or burial requirements for the pipeline.

The final aspect of design takes into consideration the day-to-day tasks of operating and maintaining the functional integrity of the system. These include the necessary control systems to operate the system within its design parameters and to promote safe and continuous operation.

The preferred pipeline system for a given set of conditions is selected through an economic comparison of several systems, seeking to identify the system that yields the best economic return on investment dollar depending on the initial and subsequent capital costs, the method of financing, and the operating and maintenance costs for the economic life of the investment. If alternatives require capital investments (e.g., for pumping stations) at different future dates, these costs should be compared on a present value basis and discounted at a real (after inflation) interest rate to ensure a valid, unbiased selection of the optimum system. The details of economic analysis are addressed in financial analysis references.[2]

Programs are available for hand-held calculators,[3,4] personal computers, and mainframe computers for separately performing the hydraulic, thermal, mechanical, and economic analyses and design of petroleum pipeline systems.

Concurrent with the hydraulic, mechanical, and operations and maintenance designs, the pipeline project team will also be performing many tasks related to the construction of the pipeline. These include technical and environmental surveys of the pipeline route and surrounding areas; preparation of environmental impact reports; acquisition of permits and rights-of-way, procurement of construction materials; development of construction costs considering pipe diameter, wall thickness, grade of steel, and welding procedures; and preparation of contract specifications and bidding papers. These and other topics related to construction of pipelines are covered in other handbooks.[5,6]

The investigation of any pipeline system begins by establishing the design basis for the commodity, then making a preliminary selection of pipe diameters and cost estimates for comparative economic attractiveness. If the preliminary estimates indicate further consideration is desirable, preliminary feasibility considerations begin by selecting possible routes and developing a preliminary design.

The preliminary cost estimates developed in the section "System Cost Analysis" and Table C5.7 are examples of initial order-of-magnitude cost estimates for selecting alternative pipe diameters and overall cost feasibility. The pipe diameters used in Table C5.7 are the basis for the discussion of the hydraulic designs in the section that follows. The pipe diameters used have been selected to illustrate considerations in the hydraulic analysis and the effect on mechanical design and cost rather than to select a preferred design for the example system discussed.

HYDRAULIC DESIGN

The hydraulic design integrates the physical characteristics of the transported commodity along a given pipeline route, within specified operating conditions as established in the design basis. The result of the hydraulic design is identification of the total system energy required to meet the design criteria. In addition, the hydraulic calculations indicate a range of feasible pipe diameters and preliminary spacing of pump stations along the route.

When the design is finalized (i.e., the route selected, pipe line size determined, and type of pipe selected), the hydraulic calculations are refined to determine the conditions for overpressure control during line shut-off and surges during operation. Hydraulic calculations can also be made for the variables in the operating conditions (temperature, ranges of viscosities for products pipelines, etc.) and for future expansion of system capacity.

Route Selection

Given the task of transporting a liquid commodity from one point to another, whether it is from the point of production or storage to a processing plant or from the process plant to distribution facilities, the first "selection" of route will logically be the shortest course, or a straight line.

While a straight-line route is a reasonable first approximation of the pipeline route, there are several common-sense reasons for deviation, including:

- Significant natural obstacles such as mountain ranges, rivers, swamps, etc.
- Minimizing of "control points" in the hydraulic profile (discussed later)
- Access for construction equipment and materials
- Permitting restrictions

A preliminary route is determined using suitable maps of the area which need to show geographic features such as contour lines, as well as towns, roads, rivers, railroads, existing pipelines and utility corridors, etc. World Aeronautical Charts are available for most parts of the world, on different scales, for this purpose. U.S. Geological Survey maps are particularly useful for pipeline routing in the United States. Aerial photographs are also useful. Several of the factors which will influence the selection of the design route may not be readily identifiable or resolved until later phases of design, in particular environmental and permit requirements and land acquisition. However, a preliminary route can be selected and later modified when more information on the specific and final route is available. Once an initial route is identified, the ground profile is plotted for use in the hydraulic design.

For example, Fig. C5.2 is a potential pipeline route profile for a crude oil pipeline design illustrated in this chapter. This example is a simplified profile. Real pipeline systems typically have much more detail, showing river crossings, mountain ranges, etc. It should be emphasized that this is not a cross section of a straight-line course from origin to terminal but represents the "real" route, avoiding major obstacles mentioned earlier. The selected route results in a 240-mile length. The elevation of points along the route are shown.

If more than one reasonable alternative route has been identified, all routes are plotted and designed and analyzed simultaneously. Following sufficient identification, the alternatives will be evaluated on the basis of construction, operation, and maintenance cost, and a comparative economic analysis spanning the effective project or system "life" is performed to identify the best alternative. This topic is covered in more detail in the section "System Cost Analysis."

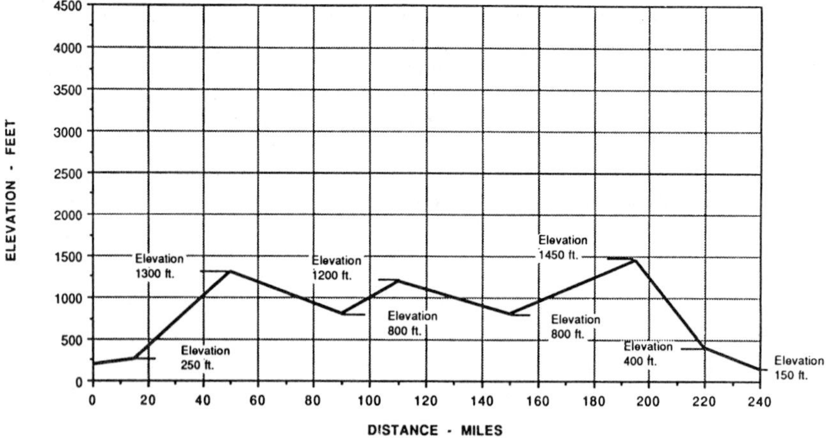

FIGURE C5.2 Example pipeline route profile.

Design Basis

When beginning the design of a pipeline system, it is necessary to define the basis of the design as completely as possible. The general parameters which are required for the design of the system include:

- System operating parameters, such as design "throughput," or flow rate, and operating temperature of the system
- Environmental conditions, such as ambient ground and air temperature (average and extremes)
- Properties of the transported fluid(s), or commodity, such as viscosity, specific gravity, vapor pressure, pour point temperature

System Parameters. There are a number of system parameters which are typically defined by the operating company or owner of the system. It is helpful for the design flow rate for the system to be defined as closely as possible. Maximum, minimum, and forecast future annual throughput of the pipeline system are required for good design, resulting in selection of the most economic line size as early as possible. This limits the iterations of the design as well as the range of alternatives.

The design throughput of an oil pipeline may vary by year and is usually expressed as the average annual pumping rate in barrels per calendar day (BPCD), also referred to as barrels per day (BPD), where the 42-gal barrel is the common unit used for transportation of most petroleum products such as crude oils, residuals, etc. The actual flow that a system must be capable of attaining to compensate for lost capacity from shutdowns and reduced flow conditions is the barrels per operating day (BPOD); therefore BPOD is greater than BPCD.

The ratio of BPCD to BPOD is the load, or service, factor:

$$\text{Load factor} = \frac{\text{BPCD}}{\text{BPOD}} \tag{C5.1}$$

A well-operated pipeline can be expected to have a load factor of 92 to 95 percent. For domestic pipelines, this may be used in the design procedure unless special circumstances dictate a lower factor. Pipelines to be located in remote areas, more complex systems with many pump stations, or pipelines operated with expected flow variations would be more reasonably designed to a lower load factor, 85 to 90 percent, to account for greater system "down time" (i.e., from interruptions in service as a result of operations and/or maintenance).

Environmental Parameters. The critical environmental parameter for the hydraulic design is the ambient temperature of the ground for buried pipelines and the air temperature for above-ground systems. Most locations will have seasonal variations, and long pipeline systems may have variations over the length of the system. It is important to identify the mean or average ambient temperature as well as the seasonal and local extremes.

Properties of the Commodity. Specification of the commodity to be transported includes identification of viscosity, density, vapor pressure, and pour point temperature. Some of these properties will have to be determined from laboratory tests on specific commodity samples. However, design may proceed on the basis of a "typical" commodity and include flexibility for a specified range of variation.

Viscosity is the physical property of fluids which resists flow and varies inversely with temperature. Besides density, viscosity is a key characteristic of the fluid to be considered in the design of liquid pipelines, having a significant effect on determining line size, station spacing, and pumping horsepower requirements. A discussion of viscosity, including the definitions of kinematic and absolute viscosity, has been included in Chap. B8. Several of the references may also be consulted for viscosity data on specific hydrocarbons and other fluids.[7,8,9,10]

For example, Fig. C5.3 shows the approximate viscosity-temperature (for a limited range of temperatures) relationship for a crude oil sample, having a viscosity of 7.1 centistoke (cS) at 130°F, and 23.7 cS at 60°F. This will be used for

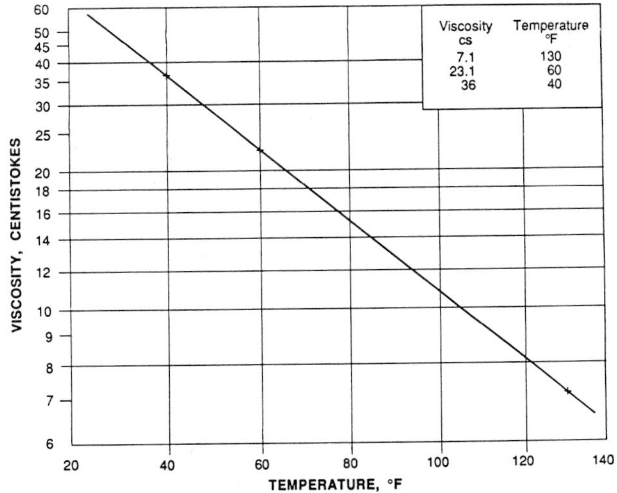

FIGURE C5.3 Example viscosity versus temperature crude oil.

the basis of the example. The viscosity-temperature is not truly linear, unless plotted on a special ASTM graph paper designed for this purpose. Figure C5.3 is plotted to the coordinates of ASTM D-341 for the temperature range of 40 to 130°F. See ASTM D-341 for further discussion.

For crude oils, the pour point of the oil (i.e., the temperature at which viscosity of a cooling oil abruptly increases) needs to be considered to determine if special measures are required to move the oil when ambient ground and air temperatures are below this temperature. An oil with a pour point temperature above the ambient condition will require dilution with a lighter stock oil, addition of a pour point depressant, or a heated pipeline system.

Isothermal Systems

Oil transportation pipelines typically will have some variation in temperature over the entire system. Since this then affects viscosity and has other design impacts, the discussion of the hydraulic design will continue based on isothermal—constant temperature—systems with the assumption that temperature variations are limited. Special considerations for nonisothermal systems are discussed later, in the section "Nonisothermal Systems."

Take, for example, a crude oil pipeline system with isothermal flow, 240 ground miles over a route with a maximum elevation of 1450 ft. Figure C5.2 shows the route profile for this example. The initial elevation is 200 ft at the origin and a liquid head level of 250 ft with respect to the datum is maintained at the terminal. The design basis is:

API gravity at 60°F	32.6
Specific gravity (based on water at 60°F)	0.8623
Ambient design temp	40.0°F
Viscosity at 60°F	23.7 cS
Viscosity at 130°F	7.1 cS
Flow rate (average)	180,000 BPCD
Service, or load, factor	0.9

It is necessary to make adjustments from the values given here at standard conditions (60°F) to the operating conditions[11,12]:

$$\text{Specific gravity}^{7,8} \text{ at } 40°\text{F} = 0.8623 \times 1.008 = 0.8692 \quad \text{(C5.2)}$$

$$\text{Viscosity at } 40°\text{F} = 36 \text{ cS} \quad \text{see Fig. C5.3}$$

$$\text{Design flow rate, adjusted to } 40°\text{F} = \frac{(180{,}000/0.9)}{1.0080} = 198{,}500 \text{ BPOD (rounded)}$$

System Energy

Both head H and pressure P are used in discussing system energy. The conversions from head to pressure, and vice versa, are given by the following formulas, for standard conditions (60°F), with correction for the specific gravity (sg) of the commodity, which adjusts the head or pressure to operating conditions[8,11,12]:

$$H\text{, ft} = 2.31 \times P, \frac{\text{psi}}{\text{sg}} \tag{C5.3a}$$

or

$$H\text{, m} = 9.807 \times P, \frac{\text{kPa}}{\text{sg}}$$

$$P\text{, psi} = H\text{, ft} \times 0.433 \times \text{sg} \tag{C5.3b}$$

or

$$P\text{, kPa} = H\text{, m} \times 0.102 \times \text{sg}$$

The total pressure drop in a pipeline system comprises three components:

1. Static pressure drop, due to changes in elevation
2. Acceleration pressure drop, due to changes in pipe geometry or phase
3. Friction pressure drop, due to flow rate, fluid properties, and pipe characteristics

When using the terminology *head* of fluid to denote system energy, the three components of head are typically referred to as static or elevation head H_s, velocity head H_v, and friction head H_f. These are the terms of the Bernoulli equation discussed in Chap. B8. Head loss expressed per unit of pipe length uses the lowercase h, with the same subscripts.

The elevation head H_s is the difference between the inlet and outlet elevations between points on a system. The velocity head component for long pipeline systems, and generally for systems with high head requirements, is a small percentage of the total head and is often assumed to be negligible in energy calculations.

Friction head loss, on the other hand, is the dominant effect in most liquid pipeline systems and can be calculated using one of the following equations previously discussed in Chap. B8:

- Darcy-Weisbach
- Fanning
- Hazen-Williams

The first two equations require a determination of the friction factor as a function of the Reynolds number. The Hazen-Williams equation accounts for friction in a coefficient "C," which is an empirically determined factor for the specific commodity.

Friction loss tables based on the Darcy-Weisbach formula are available in many engineering reference books for water and viscous liquids.[9,12] Generally these tables, as illustrated in App. E, give friction loss in units of head loss, feet per 100 ft of pipe length, or in pressure loss, psi per 1000 ft of pipe length. (Similar tables are available for SI units.)

Hydraulic calculations are not complex for pipelines with a single commodity having little variation in viscosity. The first task is to specify the nominal pipeline diameter, or range of diameters, based on the design flow rate.

An approximation of line size can be made for the design flow rate using the tables available in handbooks, standard fluid design manuals, or engineering reference books. The selection of line size is based on a preliminary economic analysis of alternatives discussed in the section "System Cost Analysis" and Table C5.7. Depending upon commodity, location, distance, and energy costs, velocities for preliminary estimates of line size are typically in the range of 4 to 7 ft/s. Figure C5.4 shows daily flow rate in MBPD versus nominal pipe diameter in inches for velocities between 2 to 10 ft/s.

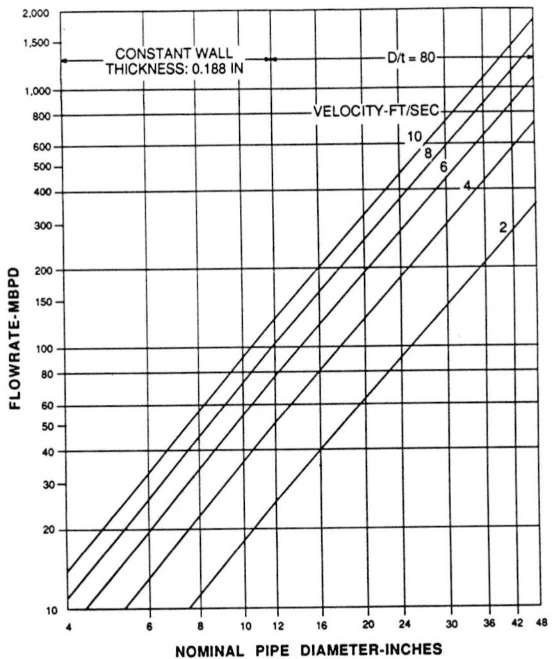

FIGURE C5.4 Flow rate versus nominal pipe diameter for velocities of 2, 4, 6, 8, and 10 ft/s.

At this point a preliminary choice of one or more wall thicknesses is necessary since eventually a cost of pipe will be found and because wall thickness will also be a component of the internal pressure and stress calculations. For a given diameter *D,* also referred to as the nominal pipe size (NPS), a range of wall thickness *t* is available.[13] Standard pipe wall thickness, depending on diameter, such as 0.250 or 0.375 in, may be used for the early design and later adjusted if special commodity or system requirements are identified, such as high pressure or limitation on the allowable pipe stress in a section of the system.

Returning to the example, for the flow rate of 198,500 BPOD (also written as 198.5 MBPD), assume that there are three diameters of pipe which are viable to illustrate the effect of pipe diameter on design. Table C5.2 can be constructed using a tabulation of the Darcy-Weisbach formula and a minimum wall thickness of 0.250 in for the three diameters.

Figure C5.5 is graphed using the data from Table C5.2. The use of log/log scales results in an approximately linear relationship between flow rate and unit

TABLE C5.2 Example: Pipeline System Parameters

Parameter	Pipeline diameter × wall thickness					
	16 in × 0.250 in		20 in × 0.250 in		24 in × 0.250 in	
Flow rate, MBPD	100	200	100	200	100	200
Roughness, in	0.0018	0.0018	0.0018	0.0018	0.0018	0.0018
Reynolds number	16,523	33,047	13,134	26,268	10,898	21,797
Darcy-Weisbach friction factor	0.0274	0.0233	0.0289	0.0245	0.0303	0.0255
Unit friction loss, ft/mile	42.76	145.57	14.34	48.52	5.91	19.91
Velocity, ft/s	4.96	9.92	3.14	6.29	2.16	4.31

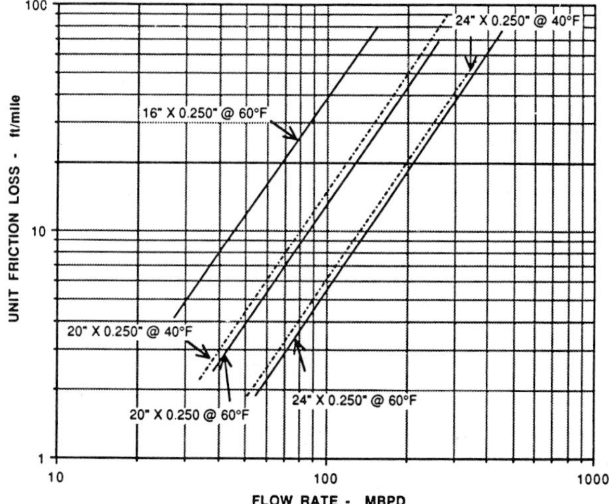

FIGURE C5.5 Unit friction loss versus flow rate.

friction loss h_f. Two of the diameters also show the effect of adjusted viscosity to the operating condition of 40°F.

It is important to point out the adjustments made for the viscosity at operating temperature. If a reference table based on water (sg = 1.0) to determine unit friction loss is used or if the friction loss chart is based on viscosity at 60°F, corrections would need to be made for the design temperature and viscosity of the specific commodity at operating conditions.

Summing the friction losses for a specific set of pipeline diameters along the system results in the net friction head that must be overcome by the pump stations of the system. In addition, the positive or negative static head due to elevation difference between the inlet and outlet must be considered in determining the pumping requirements of the system. Minor losses at valves and fittings between stations are normally ignored initially, or an allowance of an added length of pipe may be added to the scaled length to determine the estimated total system friction head loss.

Hydraulic Gradient

The hydraulic gradient is a profile representing the static head at any point in the pipeline system relative to a common datum elevation, which is usually mean sea level. Ground elevation is represented by the route elevation profile to the same datum. Energy added to the system through a pump station is plotted above the elevation profile. Head losses from friction, etc., are also shown graphically. For a pipeline system with constant parameters along the system, such as viscosity, specific gravity, and diameter, the hydraulic gradient will be a straight line with a slope equal to the friction loss per unit of length h_f for a specific flow rate. Therefore, the actual pressure in the pipeline at any point along the route is the difference between the hydraulic gradient and the ground elevation.

Figure C5.6 illustrates the slope of hydraulic gradients based on the 200-MBPD flow rate and the three diameters tabulated in Table C5.2, where an assumed pump station discharge head of 3500 ft is shown at the origin. The hydraulic gradient, being the available head at any point along the pipeline route, cannot be less than the elevation head (i.e., the pipeline route profile). Therefore, where the gradient intersects the pipeline elevation, a pump station is required. It is obvious from Fig. C5.6 that unrealistically high station discharge heads would be required for only one pump station using 16- or 20-in pipe. However, one station could be used on the 24-in pipeline by elevating the hydraulic gradient at the station discharge sufficiently to clear the ground profile at milepost 195. The station discharge head is usually selected on the basis of the allowable pressure rating for valves and fittings or the maximum allowable operating pressure (MAOP) of the selected pipe, whichever provides the required operating flexibility and is the most attractive economically.

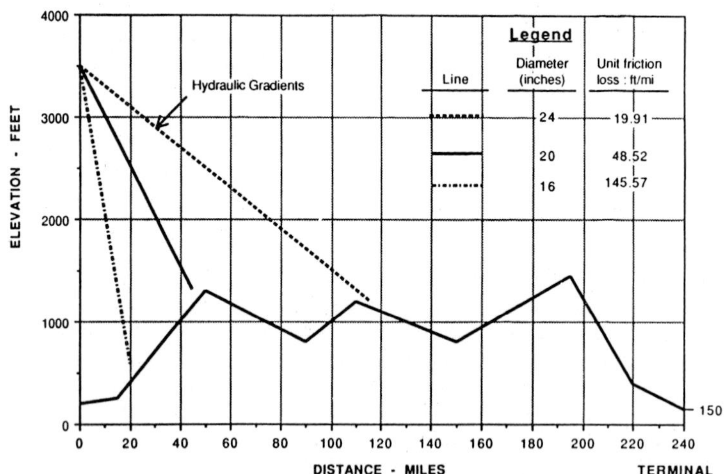

FIGURE C5.6 Example hydraulic gradient for 200 MBPD 16-, 20-, and 24-in nominal pipe diameters.

The example above for the 24-in pipeline illustrates an important use of the hydraulic gradient for identifying hydraulic control points on a pipeline route. A hydraulic control point is a high elevation point that will govern the inlet head for a section of pipeline as discussed above. In the example, the maximum elevation of 1450 ft is a control point. It is logical that the hydraulic gradient must clear the ground elevation control point but in doing so, two situations may result downstream, as shown in Fig. C5.7, for the hydraulic gradient of the 24-in line:

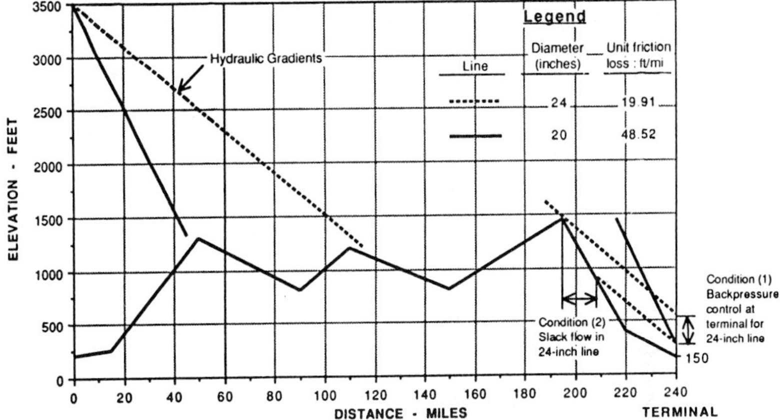

FIGURE C5.7 Example for clearance of a control point illustrating backpressure control and slack flow for 200 MBPD.

1. The line is designed to flow full requiring that backpressure control is provided at the terminal.
2. Without backpressure control, the length of line downstream of the control point will flow partially full in a cascade or slack-line condition. Slack flow is generally not a desirable condition in pipeline systems; however, it is possible to design a system to operate successfully with slack flow.

In this example, the hydraulic gradient of the 20-in line clears the elevation of the control point, and the considerations for backpressure control or slack flow only occur at lower flow rates.

Throughout this section, the example which will be discussed is a system with constant flow rate and one delivery point. In general, pipeline systems may have intermediate injection or delivery points along the route such that flow in the main line is increased or decreased. Hydraulic design of this type of system requires that the gradients for each section of the system with different flow rates be calculated and plotted in succession along the pipeline.

Maximum Allowable Operating Pressure

The MAOP is the limit of internal pressure allowed for straight pipe by Section 404.1.2 of the Code:

$$P_i = 2 \times S_A \times \frac{t}{D} \qquad \text{(C5.4)}$$

where P_i = maximum allowable operating pressure, psi (MPa)

D = outside diameter, in (m)
S_A = allowable stress, psi (MPa)
t = wall thickness, in (m)

Wall thickness t for calculation of the MAOP excludes additional thickness for corrosion allowance or imposed stresses such as concentrated loads at supports, thermal expansion on contraction, and bending. Determination of additional required wall thickness for these considerations is discussed in the section "Allowable Pipe Stress."

At this point, the reader is referred to the tabulations for specified minimum yield strength and allowable stress which can also be found in API Specification 5L—Specification for Line Pipe,[13] which is based on the restrictions on allowable stress per the Code.

The following example uses the tabulated allowable stress values of API Specification 5L and solves Eq. (C5.4) for P_i based on an assumed diameter, wall thickness, and material.

For the example pipeline system, assume the steel pipe grade is API 5L-X60, having a specified minimum yield strength of 60,000 psi. Table C5.3 identifies the MAOP for X60 pipe for three diameters and two wall thicknesses. [Calculations are based on Eq. (C5.4), with 43,200 psi as the allowable stress S_A used in place of S_H per the Code and based on tables in API Spec 5L.]

TABLE C5.3 Maximum Allowable Operating Pressure

Pipe nominal D., in	16		20		24	
Wall thickness, in	0.250	0.312	0.250	0.312	0.250	0.312
MAOP, psi	1350	1684	1080	1347	900	1123
MAOH, ft	3588	4475	2870	3580	2392	2985

The MAOP is used in the development of the system hydraulic design as a limit on the internal pressure component of the hydraulic gradient. When plotted above the route profile, the hydraulic gradient may not exceed this limit and still be designed to the B31.4 Code. MAOP is also used in determining the approximate number of pump stations.

Pump Stations

The calculated total system head required to achieve a given flow rate through a pipeline system with a selected diameter, which includes the total friction component and the static elevation difference between inlet and outlet, determines the pumping requirements of the system. As seen in Fig. C5.6, at least one intermediate pump station is required for the example system and pipe diameters illustrated. This section discusses a method of determining how many pump stations are required and locating them on the basis of hydraulic balance and a graphical method.

Number of Pump Stations. A rough number of pumping stations is found by dividing the total system pressure, or head required to overcome elevation changes and friction, by the maximum allowable operating pressure or head, MAOP or MAOH, for a specific diameter, wall thickness, and pipe material, using consistent units. For example:

$$\text{Number of pump stations} = \frac{\text{total system head}}{\text{MAOH}} \tag{C5.5}$$

For example, the first step to determining the number of pump stations is to determine the pressure or head required to overcome the frictional resistance caused by the flow of oil. Using a 20-in diameter pipe size with 0.250-in wall thickness (this is usually written 20 in × 0.250 in) and a hand-held calculator programmed with the Darcy-Weisbach equation, the unit friction head loss h_f for the design flow of 198,500 MBPD is 47.9 ft per mile.

Summing the friction head loss over the 240-mile measured length of the system yields 11,500 ft of total head loss due to friction H_f. There is also a static head H_s of 50 ft to be pumped against due to a change in elevation between the initial point and the head required at the terminal.

The total head H which is required from all pump stations on the system is the sum of the static head and the friction head:

$$\begin{aligned} H, \text{ft} &= H_f, \text{ft} + H_s, \text{ft} \\ &= 11{,}500 + 50 \\ &= 11{,}550 \end{aligned}$$

The MAOP of X60 pipe with dimensions 20 in × 0.250 in, expressed in head units, is 2870 ft. Therefore the number of pump stations required for the system is:

$$\begin{aligned} \text{Number of pump stations} &= \frac{11{,}550 \text{ ft}}{2870 \text{ ft}} \\ &= 4.02 \end{aligned}$$

Fractional stations may be accounted for by using heavier wall pipe for a limited length or by installing a booster station. These alternatives are evaluated on the basis of economics.

Location of Pump Stations. One pump station is located at the initial point of the pipeline. Downstream stations are initially located such that each of the pipeline sections will be in hydraulic balance (i.e., with each station having approximately the same differential head or pressure), thereby distributing the energy load equally.

$$\text{Pump station differential head} = \frac{\text{total system head}}{\text{number of pump stations}} \tag{C5.6}$$

In level terrain, this procedure for determining the number of pump stations would space stations equally along the route. However, in uneven terrain such as in the example, the stations can be balanced using a graphical technique.

Figures C5.8 and C5.9 illustrate stations located graphically for the 20- and 24-in pipelines by beginning at the elevation required at the terminal, then drawing the hydraulic gradient upstream to the intersection of the MAOH of the pipe, which is plotted above the ground elevation profile. This location establishes the actual discharge

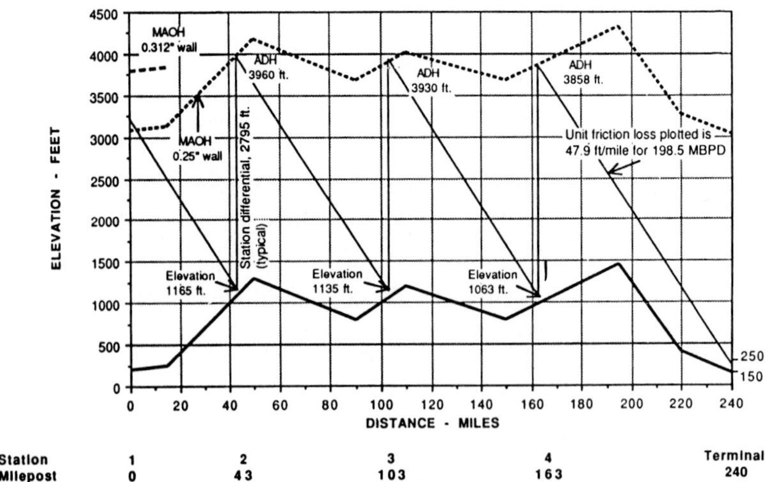

FIGURE C5.8 Hydraulic gradient for 20-in pipeline example, 198.5 MBPD.

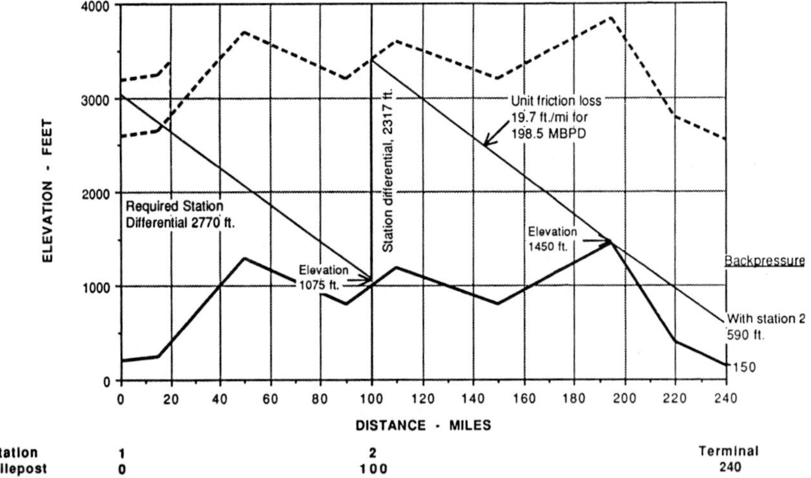

FIGURE C5.9 Hydraulic gradient for 24-in pipeline example, 198.5 MBPD with profile control point.

head (ADH) at the discharge valve of the pump station. The ADH is used here to avoid confusion with the term *total discharge head* (TDH), which refers to pump performance relative to the centerline elevation of a pump.[12]

The elevation of the hydraulic gradient into the station is plotted above the ground elevation by the allowance to provide station losses to the pumps, and the required net positive suction head (NPSH) at the pumps. In this example the allowance is 75 ft. The difference between the ADH and the elevation of the incoming head is the difference between the head at the station discharge valve and the station suction valve and is the station differential. In this example, the station differential is the MAOH of the pipe minus the allowance for station suction losses and NPSH. The example below shows the calculation for the station differentials for the 20- and 24-in systems:

Pipe O.D. and wall, in	20 × 0.250	24 × 0.250
MAOH, ft	2870	2392
Station loss, in, and NPSH, ft	75	75
Station differential, ft	2795	2317

The actual differential head required of the pumps, however, is greater than the station differential by adding the total of all station losses including a discharge pressure control valve if one is included in the station discharge piping.

Generally, an allowance of 50 to 100 ft of head at the intake of the pump station will account for the suction requirements of the pumps, station valves, and fittings losses. The estimate may be revised once specific pumps at a given station, station piping and control discharge head, and the associated suction head requirements are determined. Subsequent pump stations are located similarly with the use of a parallel-rule drafting tool to draw the gradients.

Alternatively, in more complicated terrain, pump station locations may be determined beginning at the initial point and the MAOH, then progressing downstream using the same graphical procedure. Using this procedure, the station is located at the elevation required for the incoming hydraulic gradient above the ground profile.

While it is not required that the pump stations along a pipeline system be equal in energy or discharge head, doing so results in a better operating environment. Other considerations may preclude a pure balance between the stations. For example, it may be necessary to fix or shift the location of intermediate pump stations due to unsuitable terrain, population and infrastructure criteria, or operations and maintenance considerations. Many designs, especially in rugged terrain or populated regions, may involve fixing certain key stations in accessible locations (i.e., near power lines), then locating other stations by the trial and error method. The graphical technique described here allows leeway to make these adjustments. A considerable amount of adjustment can be made with little effect on the total hydraulic design, with respect to pipe diameter or total pump station horsepower. However, moving stations to accommodate unequal station differentials may require using heavier wall pipe on the discharge of some stations. Figures C5.8 and C5.9 illustrate using 0.312 in of wall pipe for the 20- and 24-in systems at the discharge of station 1 to accommodate the hydraulic gradients that exceed the MAOH of 0.250-in wall pipe used farther downstream. In this case, the total requirement for heavier wall pipe to avoid a fractional station is installed at the initial station rather than distributing shorter sections to the discharge of the remaining stations.

In the cases illustrated, an increase in pipe wall thickness from 0.250 to 0.312 in results in increased friction loss between Stations 1 and 2 and the pumping head at Station 1. As shown below, the incremental friction loss for a 5-mile-long NPS 20 pipe is 7.5 ft, whereas for a 20-mile-long NPS 24 pipe it is 10 ft.

	Pipe size, NPS			
	20		24	
Wall thickness, in	0.250	0.312	0.250	0.312
Friction loss per mile, ft	47.9	49.4	19.7	20.2
Incremental friction, ft per mile due to increase in thickness, ft	49.4 − 47.9 = 1.5		20.2 − 19.7 = 0.5	
Length of pipe, mile	5		20	
Total incremental friction loss, ft	1.5 × 5 = 7.5		0.5 × 20 = 10.0	

In cases where the hydraulic gradient crosses deep defiles in the terrain, heavier wall pipe may be required to prevent exceeding the MAOH at these locations. Increased friction loss is determined for these locations that are similar to this example or by accumulating the distance and friction loss for each of the selected wall thicknesses.

System Growth and Station Bypass. Many pipeline systems are designed to initially operate at a reduced flow rate, with increments added in subsequent years. The design basis for these systems is usually the economic maximum future flow rate, considering oil characteristics, pipe diameter, wall thickness, and grade of steel.

During the early years of operation, only the stations necessary to transport the initial flow rate are installed. Stations are then added as flow rate is increased until the pipeline is fully developed. Intermediate pumping stations are installed at locations determined from the design for the maximum flow rate. As part of the system design, each station, except the first, is omitted from the system to establish the maximum flow rate possible through the system for that configuration. In other words, each omitted station creates a "bottleneck" in the system at some maximum flow rate.

For example, the bypass operating conditions for the 20- and 24-in pipeline systems are shown in Figs. C5.10 and C5.11. The hydraulic gradient bypassing a single station is drawn, and the unit head loss determined graphically or numerically. Then, using the unit friction loss graph constructed earlier for the specific crude oil transported in the system, Fig. C5.5, the limiting flow rate is found.

Designing for bypassing of station 2 in the 20- and 24-in systems would use 20 and 26 miles of 0.312-in wall pipe, respectively, as illustrated in Figs. C5.10 and C5.11. Pipe tonnage would be adjusted upward as shown in Table C5.8 to reflect the effect of ground elevation profile on the mechanical design and cost. Additional heavier wall pipe would be inserted in other locations where the MAOH would lie below the hydraulic gradient to accumulate the appropriate pipe tonnage. Additionally, the increasing wall thickness adds to the friction loss. In this case, the added friction loss would appear at station 1 and increase the operating horsepower.

Figures C5.9 and C5.11 illustrate, for the 24-in pipeline, a hydraulic control point establishing the elevation of the hydraulic gradient out of a station. The elevation added to clear a hydraulic control point adds to the static head difference for pumping. In Fig. C5.9, the added head for the design flow rate shifted the intermediate pump station closer to the hydraulic control point than for a level line. The added head was then moved to station 1 by limiting the discharge head at station 2 to the MAOP of 0.250-in wall pipe.

In this case, 0.312-in wall placed at station 1 for the design flow rate is available for increasing flow rate when bypassing station 2. The incremental length of

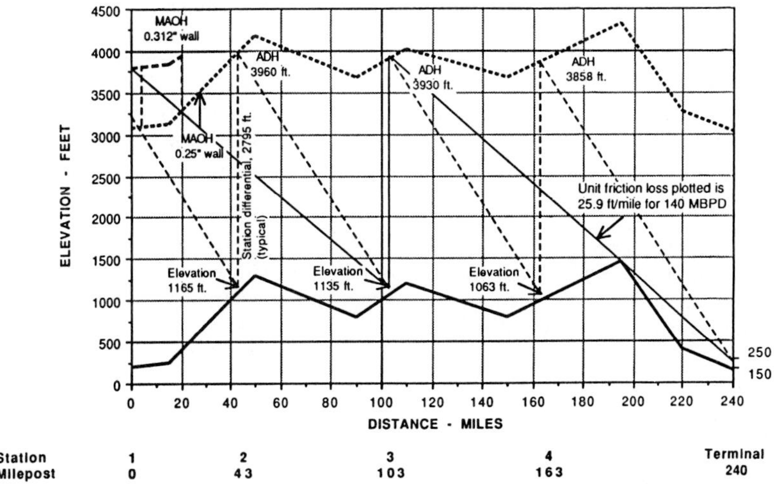

FIGURE C5.10 Hydraulic gradient for 20-in pipeline example, 140 MBPD with bypass at stations 2 and 4.

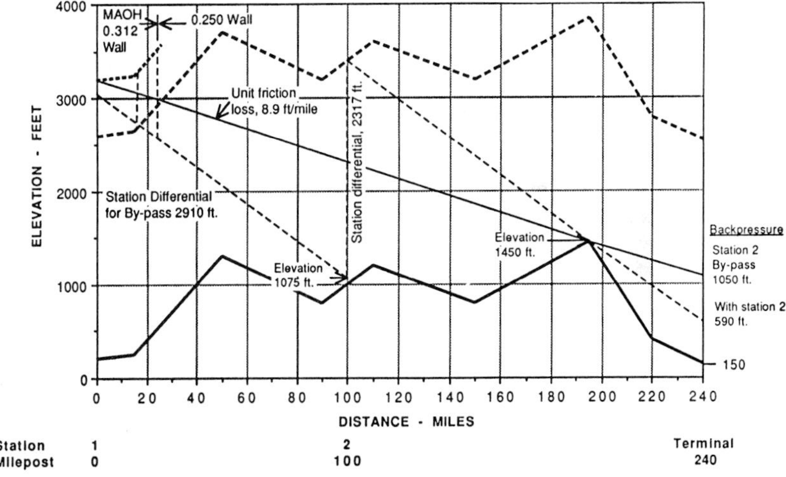

FIGURE C5.11 Example hydraulic gradient for 24-in pipeline, 125 MBPD with bypass at station 2 and profile control point.

0.312-in wall pipe at station 1 to bypass station 2 is 6 miles to use the available MAOH of 0.312-in wall pipe at station 1, which is illustrated in Fig. C5.11. Placing any length of 0.312-in wall pipe at station 2 would not be available for improving the flow rate to bypass station 2.

Depending upon the ground profile and the length of the system, bypassing one station may also allow bypassing other stations and redistributing units of pumping energy among the remaining stations as illustrated for the 20-in system in Fig. C5.10. This same analysis can also be used to establish the logical growth pattern for installation of intermediate stations.

Once the logical growth pattern has been established, and the hydraulic gradients plotted for each "stage" of growth, it is possible to identify sections of the pipeline where different wall thicknesses, material with higher yield strength, or reductions in line diameter may be economical. When the pipeline system is fully commissioned and operating, this design becomes the operating design for the system when any station is bypassed.

System Curves. The basis for selecting the actual type and size of pump unit(s) at a station is facilitated by construction of the resistance curve for the section the station must deliver energy to and superimposing it on the characteristic operating, or performance, curve of the pump(s) being considered.

The system resistance curve is constructed using the unit head loss curve, such as Fig. C5.5. For several flow rates, h_f is multiplied by the distance to the next station, resulting in the friction loss in that section of pipeline for a given flow rate. A graph can be constructed this way for a range of operating flow rates. Superimposed on this is the head required to overcome elevation changes between two stations, or the elevation of a critical profile point, plus an allowance for station losses resulting from flow through fittings, control valves, etc. The resulting curve is the system resistance curve and is concave upward (i.e., total friction head increases at a faster rate than flow rate increases). For example, Fig. C5.12 is the system resistance curve for station 1 of the 20-in-diameter example.

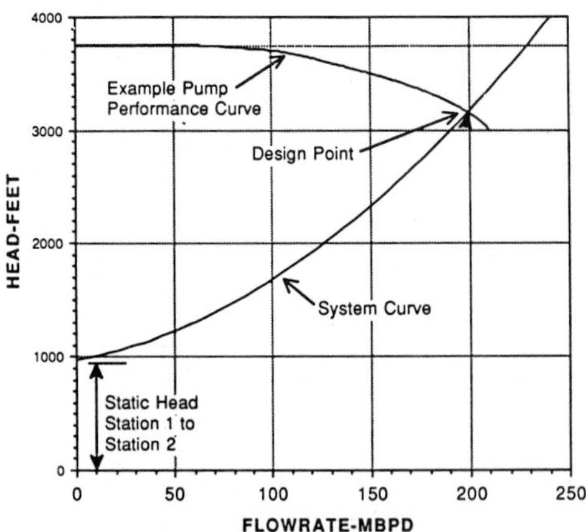

FIGURE C5.12 System curve for station 1 with example performance curve.

A pump unit of a given model and size can deliver a specific quantity of flow at a given discharge pressure or head. Pump manufactures publish this information in the form of head capacity curves, otherwise known as pump performance curves. Performance curves for centrifugal pumps are characteristically concave downward (i.e., delivered head decreases as capacity increases for a given model).

Superimposing a pump performance curve on the system resistance curve results in an intersection of the two curves, which is the operating point of the pump units at the station. The configuration of multiple pumps at a station is also considered. For example, pumps in parallel result in increased capacity without a similar increase in head delivered. Pumps in series yield increased head at a given capacity, which is roughly additive.[12,14] Therefore, the "net" performance curve of the combination of pumps at the station and the combination of pump units that satisfies the required pumping head and power requirements most efficiently are selected for that station.

Nonisothermal Systems[15,16]

Up to this point, the discussion of hydraulic design has focused on isothermal pipeline systems. However, in general the physical properties of petroleum commodities are temperature dependent, and the significance of temperature change varies with the commodity. Therefore, the true nature of liquid petroleum systems is nonisothermal.

Temperature Profiles. The basis for nonisothermal hydraulic design is an analysis of the thermal nature of the system, considering variations of external, or environmental, temperature as well as systemic factors (i.e., flow rate and its effects on friction, including friction heat generated). The analysis of a particular system's hydraulic design is based on an analysis of the temperature profile for the range of operating conditions.

The temperature profile can be calculated by the following equation (holds for any set of consistent units):

$$T(x) = T(a) + [T(0) - T(a)] \times e^{-uAx/wcv} \tag{C5.7}$$

where $T(x)$ = temperature at distance x along the pipeline, °F
$T(a)$ = ambient temperature, °F
$T(0)$ = temperature at the initial point, °F
u = overall heat transfer coefficient per unit of area, Btu(h · ft^2 · °F)
A = pipe surface area per unit of length, ft^2/ft
x = distance from initial point, ft
w = weight of fluid per unit of length, lb/ft
c = specific heat of fluid, Btu/(lb · °F)
v = velocity of fluid, ft/h

Hydraulic designs for heavier commodities, such as heavy crude oils and residual fuel oils, are influenced largely by the effect of temperature on viscosity and related friction losses. For lighter commodities, such as gasoline and natural gas liquids (NGL), hydraulic friction losses diminish continuously with declining flow rate, but for some heavy commodities, continuing reduction of flow rate may cause an increase in pump station discharge pressure to overcome the fluid's resistance to motion by increasing viscosity. Hydraulic design for the lighter commodities, although relatively independent of viscosity, is more dependent on vapor pressure, which is also a function of temperature.

In the discussion of isothermal oil systems, Fig. C5.5 illustrated that at constant flow rate, unit friction loss increases with declining temperature and that at constant temperature, unit friction loss reduces with declining flow rate. In non-

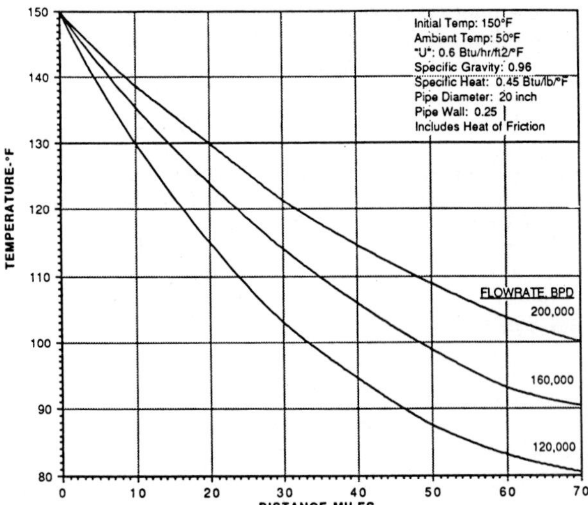

FIGURE C5.13 Example temperature profiles for 20-in hot oil system, 120, 160, and 200 MBPD.

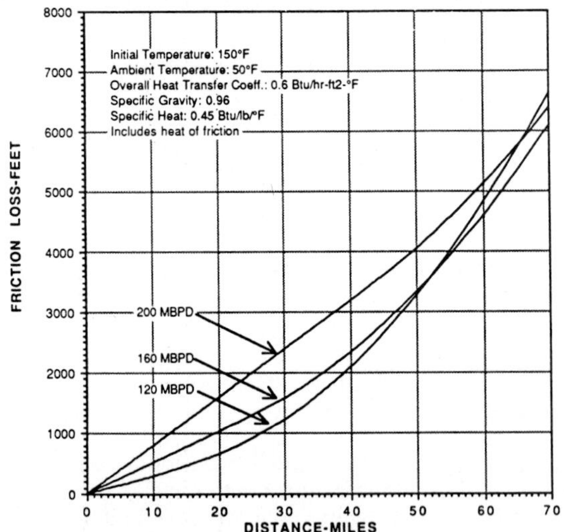

FIGURE C5.14 Example 20-in pipeline friction versus temperature, 120, 160, and 200 MBPD.

isothermal systems, temperature typically decreases at increasing distance from a heating station, increasing the unit friction in the direction of flow, which is illustrated in Fig. C5.13. Generally, total friction loss in a hot oil pipeline decreases with declining flow rate as in isothermal systems. However, for some commodities, friction loss reaches a minimum, then increases at lower flow rates. Figure C5.14 shows that pipeline friction for 120 MBPD lies above the pipeline friction for 160 MBPD at distances greater than 52 miles and above 200 MBPD at distances greater than 64 miles. Figure C5.14 includes the beneficial effect of heat of friction. Friction loss would be greater if heat of friction were ignored.

Heat of Friction. In a flowing fluid, the pressure dissipated by friction becomes heat. The effect of friction heating generally increases with flow rate, viscosity, insulation, and line length. Heat of friction should be considered at high flow rates to assure that overheating will not occur. Pipeline insulation to reduce heat loss during cold weather may contribute to overheating in summer.

When heat of friction is considered, the limit of cooling while the line is operating can be obtained by adding the temperature offset by heating to the ambient temperature. The temperature offset is calculated by solving the heat flow equation for temperature difference where the heat flow equals the frictional heat generated by a unit length of fluid flowing in the pipe:

$$(T - T_A) = \frac{Q}{u \times A} \tag{C5.8}$$

where Q = unit heat of friction, Btu/(h · ft)
u = heat transfer coefficient per unit of area, Btu/(h · ft^2 · °F)
A = pipe surface area per unit of length, ft^2/ft
T = flowing temperature, °F
T_A = ambient temperature, °F

The heat of friction Q can be calculated by converting the unit friction loss to heat and multiplying by the weight of fluid per unit length and velocity in feet per hour.

For large-diameter pipelines and high flow rates, heat generated by friction loss should also be included in the temperature profile, and the following equation results:

$$T(x) = \left[T(a) + \frac{Q}{uA}\right] + \left\{T(a) - \left[T(a) + \frac{Q}{uA}\right]\right\}e^{-uAx/wcv} \tag{C5.9}$$

where Q equals heat generated per linear unit of flowing liquid in a unit of time by converting the unit friction loss to thermal units as described above.

As an example, Fig. C5.13 shows calculated temperature profiles versus distance, including heat of friction, for a 13.5 API gravity crude oil, flowing between 120 and 200 MBPD in a 20-in-diameter pipeline. Figure C5.14 shows friction loss as a function of the distance from a pumping/heating station, with an initial temperature of 150°F to correspond with the temperature profiles of Fig. C5.13.

Thermal-Hydraulic Gradients. For initial hydraulic designs, once the temperature profile is determined, the system can be divided into discrete sections, and the thermal-hydraulic gradient calculated for appropriate sections using the average viscosity of the successive sections. Station locations and sizes are then determined in the manner described for the isothermal case.

For a more detailed hydraulic design, it is recognized that the hydraulic gradients of nonisothermal systems are curves of increasing friction loss as a result of cooling, as the distance from the pumping and heating station increases, unless heat of friction is a significant factor, such as in a large-diameter line at high flow rates and with a low initial temperature, wherein temperature may actually increase with distance from the station.

For example, Table C5.4 and Fig. C5.15 illustrate the hydraulic gradients of a pipeline system for a crude oil with 13.5 API gravity along the given route, for three flow rates, 120, 160, and 200 MBPD. The discharge temperature at each station is 150°F and the ground temperature is 50°F. Table C5.4 illustrates the effect of increasing distance between heating stations by comparing the difference of friction loss between station 1 and the remaining stations as flow rate increases.

This example illustrates, by comparison with Fig. C5.8, the effect of characteristics between a 32.6 and 13.5 API gravity crude on the hydraulic, mechanical, and operating designs that differ between an isothermal and nonisothermal system for a given pipeline route.

The hydraulic and mechanical designs of nonisothermal systems are more detailed than systems within the isothermal range at usual ambient temperatures by including the additional effects of temperature on stresses and of materials and components for the operating conditions. Mechanical design is discussed in the section "Mechanical Design."

Intermediate pumping and pumping and heating stations would be investigated in this example to determine the effect increasing the average flowing temperature would have on reducing average viscosity and the total friction loss and on establishing the required wall thickness between pumping stations. Further analysis would also include increasing the temperature of the heated crude leaving the stations and insulating the pipeline and analyzing other pipe diameters.

With some commodities such as heavy and waxy crudes, the effect of friction heating can be significant and results in a decrease in the energy required for pumping. The decrease in pumping energy is a factor of how much the viscosity decreases by increasing temperature and how sensitive the flow regime is to changes in viscosity (i.e., pressure drop in laminar flow is a stronger function of viscosity than in transition flow and is not a function of viscosity in the fully turbulent flow regime).

Shutdown and Restart. During periods when flow ceases, fluid in the pipeline cools statically without the heat of friction until flow resumes or the system reaches the ambient temperature. During static cooling, the temperature may be calculated by

$$T(t) = T(a) + [T(0) - T(a)] \times e^{-uAt/wc} \tag{C5.10}$$

where $T(t)$ = temperature at time t, °F

$T(a)$ = ambient temperature, °F

$T(0)$ = temperature at the initial point, °F

t = time from start of static cooling, h

On restart, the viscosity, determined by the local temperature, determines the local friction loss. The restart flow rate and pressure may be determined by dividing the pipeline into segments and then summing the friction losses of the segments. Repeating this process stepwise as sequential segments are displaced with heated oil determines the calculated startup pumping rate and restart time. Waxy crudes may develop a yield stress on cooling, which may require additional pressure to reestablish flow.

Figure C5.14 shows that a flow rate can be reached where friction loss can increase for decreasing flow rates. At lower flow rates, where friction loss increases, the friction loss may increase beyond the head capacity of the pumps or

TABLE C5.4 Example: Hot Oil System Station Discharge Heads

	Unit	Station 1 at origin			Station 2 at milepost 43			Station 3 at milepost 103			Station 4 at milepost 163		
Flow rate	MBPD	120	160	200	120	160	200	120	160	200	120	160	200
Friction loss	feet	2430	2630	3440	4820	4580	5130	4820	4580	5130	8050	7280	7390
Head at inlet downstream	feet	1190	1190	1190	1160	1160	1160	1090	1090	1090	250	250	250
Head out of station	feet	3620	3820	4630	5980	5740	6290	5910	5670	6220	8300	7530	7640
Station elevation	feet	200	200	200	1090	1090	1090	1060	1060	1060	990	990	990
Suction loss + NPSH	feet	100	100	100	100	100	100	100	100	100	100	100	100
Station differential	feet	3320	3520	4330	4790	4550	5100	4750	4510	5060	7210	6440	6550
Station loss	feet	100	100	100	100	100	100	100	100	100	100	100	100
Pumping head	feet	3420	3620	4430	4890	4650	5200	4850	4610	5160	7380	6540	6650

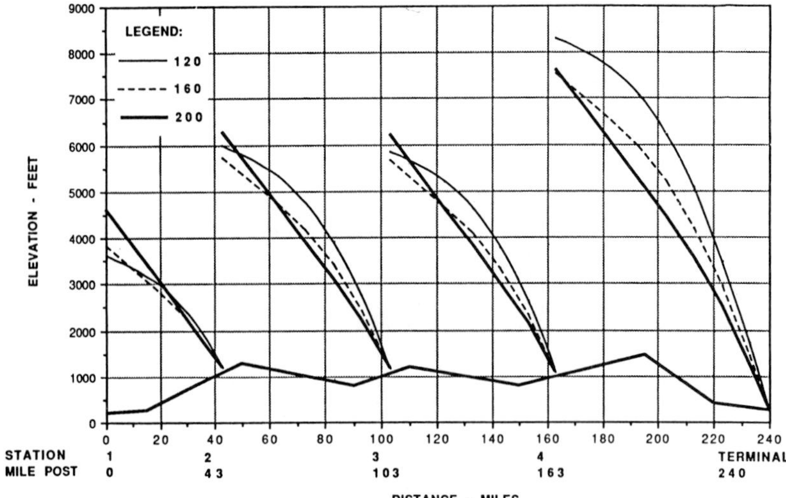

FIGURE C5.15 Example plot of hot oil pipeline hydraulic gradients for 120, 160, and 200 MBPD.

the pressure rating of the pipeline leaving the station. This condition would plug the pipeline and must be avoided. Methods to prevent the plugging of a pipeline that cools below the limit of pumpability, either by low flow rate or static cooling include:

- Maintaining the flow rate above the temperature at which the viscosity effect increases friction loss
- Blending with a cutter stock to reduce the viscosity at lower temperatures
- Displacing with a displacement fluid before static cooling
- Insulating and heat tracing

Pipelines using the displacement method generally require preheating before refilling the line with the commodity. These systems are preheated by circulating heated displacement fluid until the system is sufficiently warm to accept the heated commodity, and then allow time to redisplace if a false start should occur during the refill and restart operation.

MECHANICAL DESIGN

Mechanical design of a pipeline system is the selection of materials, including type of steel, diameter and wall thickness of pipe, as well as methods of support and/or restraint for the system in response to the loadings and stresses imposed on the pipeline system by physical pressures and forces such as the internal and external design pressures; static loadings and weight effects of the pipe, fluid, and soil; dynamic loadings from wind, waves, earthquake, etc.; and relative motion of connected components. These factors impose loadings on the pipe and result in longitudinal, hoop, and radial stresses which must be evaluated in the mechanical design of the piping system.

In addition to the mechanical factors that affect the allowable stress levels for design, the grade of steel and wall thickness determine welding procedures and affect construction cost. Section 434.8 of the Code prescribes the requirements for welding. Section 434.8.3 specifies welding qualifications with reference to API 1104 and Section IX of the ASME Boiler and Pressure Vessel Code. The engineer should refer to these references for welding requirements that may determine the maximum diameter, wall thickness, or grade steel for the system on the basis of cost.

The mechanical design of the pipeline, with respect to restraint against longitudinal, or axial, and radial motions, considers the pipeline as a unit and must provide that sufficient flexibility is designed into the system to ensure that expansion or contraction as a result of the internal or external loadings does not cause excessive stresses in the piping material, bending moments at joints, or excessive forces or moments at points of connection to equipment or at supports.

Mechanical piping system design primarily uses computer programs, many of which operate on personal computers. In most cases, the code requirements are built into the programs, so for a set of internal pressures and external loadings, the program will give the optimum wall thickness, pump locations, and maximum stress values on the basis of parameters input by the engineer. However, this does not preclude the possibility of error. The engineer must be able to accurately determine the required loadings and pressure and to analyze the computer results, verifying their validity to the overall system.

Line Pipe, Fittings, and Valves

Specifications for line pipe and for fittings, valves, and flanges are given in various API and ANSI standards and specifications including:

- ANSI/ASME B36.10M (Welded and Seamless Steel Pipe)
- ANSI/ASME B36.19M (Stainless Steel Pipe)
- API 5L, 5LU (Line Pipe)
- ANSI B16.5, 16.9, 16.10, 16.25, 16.28 (Flanges, Fittings, Valves)
- API 6D (Valves)
- API 600, 602, 603 (Valves)

Additional information on components of piping can be found in Chap. A2 of this handbook.

Line pipe is manufactured by several methods, the most common being seamless (SMLS), electric resistance welded (ERW), and submerged arc welded (SAW) in the form of longitudinal and helical (spiral) welds. Table C5.5 summarizes some of the characteristics of these four types of pipe manufacture. Each has advantages and disadvantages for different uses, and there are also economic and availability considerations which enter into the decision of the type that is specified and supplied on a particular project.

With respect to the mechanical design of a pipeline, the characteristic of line pipe which is of critical interest is the specified minimum yield strength (SMYS) of the material. API 5L Specification for Line Pipe is available in various strength

TABLE C5.5 Availability and Usage for Types of Line Pipe

	Seamless	ERW	SAW, longitudinal welds	SAW, helical (spiral) welds
Minimum diameter	$2\frac{3}{8}$ in or less.	$2\frac{3}{8}$ in or less.	16 to 20 in.	16 in.
Maximum diameter	16 in (possible for 26 in).	24 to 26 in.	64 to 84 in.	80 to 100+ in.
Maximum wall thickness	0.750 to 2.000 in.	0.312 to 0.750 in.	0.625 to 1.500 in.	0.500 to 1.500 in.
Grades	B through X-80. Grades X-60 and higher are heat treated.	B through X-70. X-52 and higher are made from controlled rolled skelp.	B through X-80. X-52 and higher are made from controlled rolled plate.	B through X-70. X-52 and higher are made from controlled rolled skelp.
Service	All services; on/offshore.	All services; generally not used for offshore.	All services; on/offshore.	Experience limited to less critical service; used as equivalent to SAW in other countries.
Relative cost	More expensive than ERW. Cost premium for larger diameters/wall thickness, higher grades.	Less expensive than seamless.	Less expensive than seamless, more than ERW (within size overlap range).	May be less expensive than SAW (depends on manufacturer).

grades, ranging from Grade B, rated at 35,000 psi (241 MPa) to X80, where the number refers to the SMYS in kips per square inch (ksi), a kip being 1000 lb.

There is some advantage to the higher strength grades, principally in that wall thickness may be reduced. In some cases this may have an economic impact on the project, since thinner walls translate into lower steel tonnage for the entire pipeline system, and this may be a significant factor, even though higher grades of pipe cost more per ton. Cost savings can also result from reduced time required to field weld the thinner wall sections. There are other considerations which will affect the decision to use higher strength, thinner wall pipe. These include aspects of construction which are the result of experience in the field, such as the way the pipe handles with regard to field bending, laying stresses, tendency to go "out of round," etc. In addition, there may be limitations placed on the grade of pipe and wall thickness used for a particular project, particularly for a system which will be in "sour" or corrosive service.

Pipe flanges and fittings are described by Class rating, which is not the same as the rated pressure for a particular class (i.e., Class 300 fittings and flanges have a nominal pressure rating of 720 psig, not 300 psig). Section 402.2.1 of the Code specifies for pressure-temperature ratings of piping components as follows: "Within the metal temperature limits of −20°F (−30°C) to 250°F (120°C), pressure ratings for components shall conform to those stated for 100°F (40°C) in material standards listed in Table 423.1."

Few pipelines now in operation use fittings heavier than Class 600, rated at 1400 psig, although the trend is toward higher pressure ratings. Deep water subsea pipelines, which have heavier wall thickness due to laying stresses and high external pressures, may have fittings and flanges rated at Class 1500.

Valves are rated similarly to flanges and fittings. Table C5.6 lists valves by class and temperature from API-6D, Table 2.1.

TABLE C5.6 Pressure Ratings of Valves (psig)

Class	−20 to 100°F	150°F	200°F	250°F
150	275	270	260	255
300	720	705	675	665
400	960	940	900	885
600	1440	1415	1350	1330
900	2160	2120	2025	1995
1500	3600	3540	3375	3325

Interpolation is permitted for intermediate temperatures.

Allowable Pipe Stress

ASME B31.4 establishes the allowable stress value S_A in psi (MPa) to be used in the temperature range −20 to 250°F (−30 to 120°C) for design calculations as:

$$S_A = \text{design factor} \times \text{SMYS} \times \text{joint weld factor}^1 \qquad \text{(C5.11)}$$

Table 402.3.1(a) of the Code tabulates allowable stress values for various grades of materials, specifications, and manufacturing processes.

The design factor varies for different conditions and is defined by Section 402.3.1(a) of the Code for new pipe as 0.72. The design factor is based on nominal wall thickness and includes consideration and allowance for manufacturing tolerances as provided for in the Code-approved material specifications (API 5L, etc.). Sections 402.3.1(b), (c), and (d) of the Code cover the design factors used for reclaimed or used pipe, pipe of unknown origin, and other special cases.

The SMYS of a pipe material is used as the basis of design because it is a property which can be determined for a specific material. Furthermore, steel generally behaves elastically below this stress level. Ultimate yield strength has also been used as the basis for design and may still be used in other countries. A particular batch of steel line pipe may be tested, and it may be determined that the elastic yield strength is higher than the nominal value; however, Section 402.3.1 (g) of the Code specifies that "In no case where the Code refers to the specified minimum value of a physical property shall a higher value of the property be used in establishing the allowable stress value."

The joint weld factor is included in calculating allowable stress as a consideration to the way the line pipe is manufactured. In most cases, the joint weld factor is 1.00, but it may be 0.80 or 0.60 for specific grades of steel and welding method of the manufacturing process. (Table 402.3.1) (a).[1]

It is useful at this point to clarify a question that arises periodically, as to whether the mechanical stress calculations are based on a particular design temperature (i.e., if S_A is temperature dependent). Section 401.3.1 of the Code states "it is not necessary to vary allowable design stress for metal temperatures between −20°F (−30°C) and 250°F (120°C)."[1] However, for applications where ground or air temperature is expected to be extremely low, seasonally or locally, the properties of pipe component materials at low temperatures should be considered to verify that the design will be adequate.

Allowable stress limits for shear and bearing are given in Section 402.3.1(e) of the Code. For shear, S_A should not exceed 45 percent of SMYS; for bearing, S_A is limited to 90 percent of SMYS. Limits on calculated stresses due to sustained loads and thermal expansion and due to occasional loads in operation and test conditions are specified in Sections 402.3.2 and 402.3.3 of the Code. In general, these limits fall within the definition of allowable stress given above; however, special circumstances may apply, and the engineer is encouraged to verify the stresses and relevant limitations.

Pipe Diameter

In the hydraulic design of a pipeline system, line size is initially based on a preliminary choice of diameter and wall thickness from experience and from simplified charts. Further calculations are needed to verify the selection and finalize the system design based on the Code requirements as well as considerations for project cost and material availability.

For most pipeline systems, the pipe cost, which is a function of the diameter and wall thickness, will be the highest material cost in the system. In addition, the size of pipe will have a direct effect on the cost of installation. Therefore, total project cost is affected by the selection of pipe size. For this reason, it is important to optimize the pipe diameter, wall thickness, and grade of steel to be used so that the overall project cost is contained.

As discussed in the hydraulic design section of this chapter, the diameter of pipe is a function of the design flow rate, and mechanical design considerations have little effect on diameter selection. However, internal and external pressure, allowable stress, etc., do affect the final design of the wall thickness for the selected diameter.

Wall Thickness

In the hydraulic design, a preliminary determination of wall thickness is based on experience for the preliminary selection of pipe diameter and grade of steel. The actual design of a system must reflect code requirements for the wall thickness, which is a function of internal design pressure and additional loads at the design temperature.

Internal Design Pressure. By definition, Section 401.2.2 of the Code prescribes:

> The piping component at any point in the piping system shall be designed for an internal design pressure which shall not be less than the maximum steady state operating pressure at that point, or less than the static head pressure at that point with the line in a static condition. The maximum steady state operating pressure shall be the sum of the static head pressure, pressure required to overcome friction losses, and any required back pressure.

The earlier discussion of MAOP in the hydraulic design, defined it as a function of diameter, wall thickness, and allowable stress for a material per Code restrictions. Section 402.2.4 of the Code provides for an additional allowance of 10 percent over the internal design pressure for surges and other variations from the normal operation.

Additional Loads. Additional loads for determining wall thickness include loadings applied on a pipeline system from pipeline material and commodity weight, wind, hydrostatic, and other external forces such as impact loads. In some applications, external pressure may be a significant factor in pipeline wall thickness determination. One such example is a subsea pipeline. Section 401.2.3 of the Code specifies that a component of the pipeline system shall be designed to withstand the maximum differential between external and internal design pressures.

Wall Thickness Calculation. Minimum wall thickness t is a function of internal pressure P_i, nominal diameter D, and the allowable stress S_A, as specified by Section 404.1.2 of the Code:

$$t = \frac{P_i D}{2\, S_A} \qquad \text{(C5.12)}$$

Nominal wall thickness t_n includes an allowance for manufacturing tolerance:

$$t_n = t + \text{allowance(s)} \qquad \text{(C5.13)}$$

The actual wall thickness used in the system will be equal to or greater than this calculated value. API 5L, Specifications for Line Pipe, lists commonly manufactured wall thickness for various grades of materials (Table 6.2).[13]

With respect to construction and installation of a pipeline, there is also a practical minimum wall thickness, based on handling during installation, since the pipe wall must be able to resist damage and maintain roundness during construction. Good judgment should be used in balancing higher strength steel materials versus heavier wall thickness.

It is not a requirement that wall thickness be constant for the entire length of the system. Specific sections of the system may have different wall thickness requirements as determined by the internal pressure and other imposed stresses and making use of the hydraulic gradient developed in the hydraulic design. Thinner wall may be installed at some distance downstream of pumping stations as the operating pressure in the system declines. There are economic benefits to using the minimum wall thickness allowed under Code design, however, there are other considerations, such as complications in construction (i.e., field welding), of a system with frequent variations in wall thickness. Practically speaking, changes in wall thickness should be limited. Furthermore, anticipated growth or expansion of the system capacity should be considered carefully.

Design of Restrained and Unrestrained Pipelines

A pipeline system is subjected to static and dynamic loads due to local environmental and operating conditions, and provision must be made for the system to have flexibility and expansion capability to prevent excessive stresses in the pipe or components, excessive bending or unusual loads at joints, or undesirable forces or moments at points of connection to equipment. The types of loadings which will affect the flexibility and expansion of the pipeline as a system include the following:

- Thermal expansion and contraction
- Internal pressure
- Fluid expansion
- Pipe and soil friction interaction
- External pressure
- Bending (sag or uplift) due to
  - Dead loads, including weight of the pipe, coatings, backfill, and unsupported pipe appurtenances
  - Live loads, such as liquid transported, wind, snow, earthquake, waves/currents

The stresses that may develop in the pipe are functions of both the loadings on the pipeline system and the degree of restraint against motion of or in the pipeline system or a section of the system. Stresses may be reduced to acceptable levels by a combination of anchors, extra depth of burial of piping, use of bends, loops or offsets in the piping, heavier wall piping components, etc. The reader is referred to Chaps. B4 and B5 for additional information on relevant topics of stress and supports.

In most cases, long, cross-country pipelines are buried for the following reasons:

1. Surface use of pipeline corridor
2. Protection from intentional and accidental damage

3. Protection against expansion and contraction from ambient temperature changes and radiant energy gains and losses
4. Minimizes variations of ambient temperature and resultant effects on fluid viscosity
5. Provides restraint longitudinally along pipeline length
6. Regulations against aboveground installation

Although a buried pipeline can be considered restrained against expansion or contraction radially and longitudinally as a result of the overburden of soil and soil-pipe friction, expansion calculations are necessary, per Section 419.6.4(b) of the Code, if significant temperature changes are expected, as in a heated oil system. Furthermore, thermal expansion of buried lines may cause movement at a transition to an aboveground section and at the termination points of the pipeline or a section of line. There may also be movement where the buried line changes direction (i.e., when there is insufficient soil restraint); therefore either the system is designed with adequate flexibility in these areas or anchors against motion should be provided.

Pipeline systems may be partially or wholly installed aboveground for reasons of economy of construction, maintenance, etc. Likewise, installation on the surface is practical for pipelines requiring insulation and/or heat tracing. An above-ground pipeline can be designed with longitudinal restraints at certain support locations such that expansion or contraction due to temperature or pressure changes is absorbed by axial, or longitudinal, compression or tension stress in addition to radial expansion. The additional consideration in aboveground pipeline systems is that beam bending stresses in spans between and at supports must be evaluated.

Thermal Expansion and Contraction. Thermal expansion and contraction calculations are necessary for buried and aboveground systems if a substantial temperature change is expected between installation and operation, such as when the line is to carry a hot oil or where significant variations in local environmental temperatures will occur. Thermal expansion may cause movement where the line changes direction or terminates or where a discontinuity occurs, such as a change in size. In some pipeline systems, these motions may not be restrained by anchors or supports (i.e., absorbed by direct axial stress of the pipe), in which case flexibility must be provided using loops, bends, or expansion joints.

The temperature change from the installation temperature to both the maximum and minimum design temperature is used to determine thermal expansion, with the linear coefficient of thermal expansion α for carbon and low-alloy high-tensile steel being 6.5×10^{-6} in/in °F (11.7×10^{-6} mm/mm °C) up to 250°F (120°C). For a section of pipe restrained at both ends, longitudinal stress $S_{L/T}$, due to a temperature change ΔT, is given by

$$S_{L/T}, \text{psi} = E\alpha\Delta T \tag{C5.14}$$

where E is the modulus of elasticity for steel 30×10^6 psi (2×10^5 MPa). An increase in temperature results in a compressive longitudinal stress, and vice versa. The general convention is that tensile stress is positive, and compressive stress is negative.

Internal Pressure. Internal pressure imposes a radial and longitudinal component of stress in the pipe material. The hoop stress, which is in the radial direction and is caused by internal pressure, was presented in Eq. (C5.4). In a free, or unrestrained, section of pipe material or pipeline system, as internal pressure increases, the length of a section will increase. Due to the Poisson effect, restrained systems will develop a component of longitudinal stress as a result of the internal pressure $S_{L/\mathrm{IP}}$, given by

$$S_{L/\mathrm{IP}}, \text{ psi (MPa)} = \upsilon S_H$$

$$= \upsilon P_i \frac{D}{2t} \tag{C5.15}$$

where υ equals Poisson's ratio, 0.30 for carbon steel. This is generally a tensile (positive) stress in the longitudinal direction. For most liquid lines maximum internal pressure will most likely occur during the hydrostatic test prior to operation.

Fluid Expansion. The expansion of the fluid in a liquid pipeline due to increase in temperature will impose an additional component of pressure which will have both a radial and longitudinal component, resulting in stress in the pipe material. This stress, $S_{L/E}$, is calculated in the same manner as the stresses from internal pressure, after conversion of the expansion of the fluid to equivalent pressure terms.

Pipe and Soil Friction Interaction. The combined longitudinal effects of temperature, pressure, and expansion of the fluid can cause significant stresses in long sections of pipe. For buried pipelines, the interaction of the pipe material and the soil will have an opposing friction effect, with the maximum stress due to soil friction occurring at the midlength of the section. The friction stress is calculated as follows, with any set of consistent units[17]:

$$S_{L/F} = L \times H \times \rho \times \frac{\sigma}{2t} \tag{C5.16}$$

where L = length of pipe section
H = depth of cover above the pipe
ρ = unit weight of the soil cover
σ = coefficient of friction between pipe and soil
t = wall thickness

External Pressure. For sections of pipe laid underwater, external hydrostatic pressure needs to be considered in the stress analysis. The hoop and longitudinal stresses resulting from hydrostatic pressure are calculated in the same way as the internal pressure stresses are calculated and are opposite in sign. A similar condition arises when thin walled pipe is drained and a vacuum is produced inside the pipe.

The critical collapse pressure P_{cr} for round thin walled pipe with a diameter-to-wall thickness is calculated by the equation[18]:

$$P_{\mathrm{cr}} = 2 \times E \times \frac{(t/D)^3}{1 - \upsilon^2} \tag{C5.17}$$

Equation (C5.17) assumes the stress is within the elastic range and the ratio of length to radius is greater than 20.

If the pipe section is out of round, the resistance to collapse is reduced by flattening. Additional detail is beyond the purpose here, and the reader should refer to appropriate references or specialists where collapse by external pressure is a concern.

Stiffening rings or heavier wall pipe may be used to prevent or limit collapse due to external pressure. Collapse of buried pipe can also be controlled by careful preparation of the bedding and compaction of the backfill around the pipe.

Considering all the longitudinal components of stress discussed to this point, the net longitudinal stress S_L imposed on a restrained pipeline due to the combined effects of internal pressure, external pressure, temperature, fluid expansion, and soil and pipe friction is the algebraic sum of Eqs. (C5.14), (C5.15), and (C5.16), where tensile stress is the positive convention:

$$S_L = S_{L/\mathrm{IP}} + S_{L/\mathrm{EP}} + S_{L/T} + S_{L/E} + S_{L/F} \tag{C5.18}$$

where S_L = net longitudinal stress, psi (MPa)
$S_{L/\mathrm{IP}}$ = internal pressure, psi (MPa)
$S_{L/\mathrm{EP}}$ = external pressure, psi (MPa)
$S_{L/T}$ = thermal expansion/contraction, psi (MPa)
$S_{L/E}$ = fluid expansion, psi (MPa)
$S_{L/F}$ = pipe and soil friction, psi (MPa)

The net longitudinal stress is generally tensile since the internal pressure and fluid expansion components are tensile, countered by the compressive temperature component and by the soil and friction interaction in buried pipeline systems, and the external hydrostatic pressure component is only applicable to deepwater crossings. Concentrated loadings at highway and railroad crossings require special consideration and may require casing. Refer to Section 434.13.4 of the Code.

It is important to note that in an axially restrained line, if the increase in temperature between placement of the line and operation is great enough, the compressive stress from the restraint to pipe "growth" by thermal expansion from Eq. (C5.14) will exceed the tensile stress due to internal pressure.

The numerical sum of the longitudinal stress and the hoop stress is called the equivalent tensile stress. In the case where net S_L is negative (i.e., compressive), the absolute values are used for pipe stresses, and the equivalent tensile stress is the sum of the absolute value of S_H and S_L. Section 419.6.4(b) of the Code specifies that the equivalent tensile stress for restrained lines is not to exceed 90 percent of the SMYS calculated for the nominal pipe wall thickness. Stresses may be reduced by burial, anchors, heavier pipe wall components, or expansion provisions such as loops, offsets, or bends. The following section discusses the beam bending stresses which are included in longitudinal stress calculations for aboveground portions of restrained lines, as well as unrestrained.

Bending or Beam Stresses. For most cases, welded pipelines are very stiff longitudinally and make good beams, capable of sustaining typical loadings from the weight of the pipe itself, the commodity within, and any coatings applied. However, a bending stress analysis is a standard part of pipeline system design, in particular aboveground sections, and is generally carried out by an expert, using one of several computer programs that are available.

When considering the structural nature of a pipeline system, one of two types of "beam" or structural models can be assumed. A buried pipeline, or section of a longer pipeline system, can be considered as continuously supported as long as the bedding it is laid on is uniform. Even if it is not perfectly uniform, the stiffness of the pipe will allow reasonable spans across low or soft areas in the soil bedding, or rock protrusions.[14]

An aboveground pipeline system and a buried pipeline system with nonuniform bedding can be modeled as a continuous beam system with multiple supports. At each support there will be reactions to the loadings on the pipeline, caused by pipe material, commodity, and coating weights, as well as loadings from environmental factors such as snow, earthquake, and wind.

Beam bending stresses resulting from sagging (deflection) are included in the analysis of maximum stresses. This means evaluating the pipeline for different loading cases and combinations of loadings. Considering the possibility for bending moments in the plane of the pipe (M_i), transverse to the plane (M_o), and torsionally (i.e., about the axis of the pipe; M_t), the resultant stress due to expansion from bending for unrestrained pipelines S_E is given by[1]

$$S_E = \sqrt{S_b^2 + 4 \times S_t^2} \tag{C5.19}$$

where S_b = equivalent bending stress, psi (MPa)
$= \sqrt{(i_i \times M_i)^2 + (i_o \times M_o)^2}/Z$
S_t = torsional stress, psi (MPa)
$= M_t/2 \times Z$
i_i = stress intensification factor (in-plane)
i_o = stress intensification factor (transverse)
Z = section modulus of pipe, in^3 (mm^3)

For straight pipe, the stress intensification factors (in-plane and transverse) have a value of 1; therefore the S_b component reduces to the flexure formula ($S_b = M/Z$). Stress intensification factors for elbows, miter bends, and tee sections of pipe can be found in Fig. 419.6.4(c) in the ASME B31.4 Code. This figure also includes a correction of pressure (Note 7) to be applied to large diameter, thin walled pipe fittings, since pressure can significantly affect the stress intensification and flexibility of these components.[1]

Code Stress Limits.[1] Sections 402.3.2 and 402.3.3 of the Code specify that the calculated stresses due to sustained loads, thermal expansion, and occasional loads are to be limited by:

- Internal pressure stresses—not to exceed S_A as discussed in the Code paragraph 402.3 (see earlier section of this chapter)
- External pressure stresses—considered within bounds where the minimum wall thickness has been calculated based on the code formulas
- Expansion stresses—allowable stress S_A values for the equivalent tensile stress for restrained lines is not to exceed 90 percent of SMYS; for unrestrained lines, it shall not exceed 72 percent of SMYS
- Additive longitudinal stresses—the sum of longitudinal stresses due to pressure, weight, and other external sustained loadings shall not exceed 75 percent of S_A calculated for the expansion cases stated above
- Additive circumferential stresses—the sum of circumferential stresses due to internal design pressure and external loads is not to exceed allowable stress calculated for internal pressure
- Longitudinal stresses from occasional loads—the sum of the longitudinal stresses produced by pressure, live and dead loads, and occasional loads such

as wind or earthquake (not considered concurrently), shall not exceed 80 percent of SMYS

Pump Selection

The concept of pump curves and preliminary pump selection was discussed earlier. Expanding on that discussion, it is assumed that the hydraulic and pipe design are essentially complete. The next step is to choose the pumps and drivers. The first decision is what type of driver—diesel, turbine, or electric. This is primarily an economic decision influenced by the availability of fuel and cost of electric power. When electric power is readily available, electric motors provide simple operation, low cost, and low maintenance; where natural gas or other unrefined fuels are available, gas turbines may give low fuel costs; and occasionally a steam turbine or some combination of machines proves the best choice. It may even be possible to draw off commodity from the system to fuel the pump drivers themselves, say with a diesel transport system.

Pumping horsepower requirements for the entire system, known as brake horsepower (bhp), can be calculated using the following formula, where H_T and P_T refer to the total system head and pressure requirements:

$$\text{bhp, hp} = \text{Flow, BPOD} \times H_T, \frac{\text{ft} \times \text{sg}}{136{,}000 \times \text{eff}} \qquad \text{(C5.20a)}$$

$$= \text{Flow, GPM} \times P_T, \frac{\text{psi}}{1714 \times \text{eff}} \qquad \text{(C5.20b)}$$

The percentage of pump efficiency typically ranges from 70 to 80 percent for centrifugal pumps to 90 percent for reciprocating pumps. Determination of brake horsepower for individual stations and pumps is similar, using the head or pressure required for the downstream section between one station and the next. As motor horsepower is provided in standardized increments, the provided horsepower will exceed the brake horsepower requirement. For example, in the example for the 20-in pipeline, assuming 82 percent efficiency for the pumps, the required system brake horsepower at the operating temperature is:

$$\text{bhp} = \frac{198{,}500 \times 11{,}550 \times 0.8692}{136{,}000 \times 0.82}$$

$$= 17{,}900 \text{ (rounded)}$$

Specification for the main line pumps should stress performance, efficiency, and ease of maintenance because over the life of a pipeline, the cost of fuel and power will be the major operating, or annual, expense. Therefore, 1 or 2 percent in pump or driver efficiency has a major impact. Consideration should also be given to variable speed centrifugal pumps, on the basis of economics, in order to satisfy varying flow and pressure or head requirements over the life of a pipeline system. It may be that the added cost of a variable speed pumping unit is favorable versus controlling station discharge pressure by throttling.

Valve Spacing

The location and spacing of section block valves along an oil pipeline is a matter of design procedure and may be dependent on factors such as the terrain that the pipeline is crossing. In general, valves should be installed at locations where they will contribute to the safe operation of the line and enhance the safety of the system. Section 434.15.2 of the Code provides the details for these considerations.

Typically, valves are installed at the origin and termination of a pipeline, at branch points, to provide isolation of a section and to facilitate hydrotesting (i.e., anywhere that the test pressure is differentiated such as at sections of higher operating pressure or a change in wall thickness). Section block valves should be located in easily accessible positions (e.g., above ground on a buried pipeline or in an impervious pit where the ground water level is high).

OPERATIONS AND MAINTENANCE DESIGN

Operating Conditions

The operating conditions for a pipeline system are defined by the operator of the system and are part of the design basis. Many of the design decisions are directly related to the operating philosophy of the system, both in an intermediate stage, as with a phased or growth system, and in the final system. For example, if the pipeline will be transporting a hot oil, it is critical that the system not be underdesigned (i.e., that the operating temperature be reasonably set, therefore leading to a viable design of the pumping and heating station requirements). Maintenance aspects of the system will also influence the design with regard to location of facilities (near fuel or power sources) and spacing of stations for cleaning facilities (pigging). The aspect of operator training and required level of supervision to monitor and verify the function of the pipeline system may dictate the location of controls and instrumentation readout panels, spacing of head operator and control locations, and required personnel to operate the facilities.

The following sections discuss some of the specific operation and maintenance considerations for liquid pipeline design.

Surge[11]

An important consideration in the design of liquid pipelines is surge, also known as water hammer. It is the pressure wave, and reflected wave, which travel through the fluid up and down a length of pipe when the velocity of the flowing column of fluid is altered or stopped suddenly. For example, water flowing at 10 ft/s can generate a surge pressure of 500 psi.

If a valve is closed against a flowing stream or if the velocity of the stream is slowed, as when a pump station is stopped, the kinetic energy of flow is converted to pressure energy, and a positive wave is sent upstream, at the velocity of sound in the medium:

$$v_s = \sqrt{\frac{144\, K \cdot g/\rho}{1 + [(K \cdot d \cdot C) / (E \cdot t)]}} \qquad \text{(C5.21)}$$

where v_s = speed of sound through commodity, ft/s
K = bulk modulus of liquid,[10] psi
g = 32.2 ft/s^2

ρ = density of liquid, lbm/ft^3
d = inside diameter of pipe, in
E = modulus of elasticity, psi
t = wall thickness, in
C = constant of pipe fixity (0.91 for an axially restrained line, 0.95 for unrestrained), dimensionless

In the case of a main line block valve closing instantaneously some distance downstream of a pump station, flow is stopped in the vicinity of the valve but continues to flow from the pump station. The continued pumping packs the line section between the pump and the valve. The surge wave beginning at the valve travels up the hydraulic gradient, reaching the pump several seconds later, raising the discharge pressure high enough to reach the shutoff head or setpoint of the pump, causing the discharge check valve to close. This closed valve reflects the wave back toward the valve, reinforcing the incoming wave and resulting in line pressure much higher than normal. This case is exaggerated as "instantaneous" while in reality valve closure is not. However, in a real situation such as described, the effect is measurable and can be dramatic.

The surge wave travels upstream and is reflected downstream, oscillating back and forth until its energy is dissipated in pipe wall friction. The amplitude of the surge wave, or the magnitude of the pressure surge P_{surge}, is a function of the change in velocity and the steepness of the wave front and is the inverse of the time it took to generate the wave:

$$P_{\text{surge}}, \text{psi} = v_s \times \rho \times \frac{\Delta V}{144g} \qquad \text{for } T < \frac{2L}{v_s} \tag{C5.22}$$

where v_s = speed of sound through liquid, ft/s (m/s)
ρ = density of liquid, lb/ft^3 (kg/m^3)
ΔV = total change in velocity, ft/s (m/s)
$2L/v_s$ = propagation time, s
T = valve closing time, s

This is only an approximation of the surge pressure magnitude for cases limited by the stated time of closure criteria.

Computer programs are used to give specific and accurate analysis of the maximum surge pressure, location of critical points in the system, etc.

The surges are attenuated by friction, and the surge arriving at any point on the line is less than at the origin of the surge wave. Nevertheless, when flow velocity is high and stoppage is complete, or when a pump station is bypassed suddenly, the surge energy generated can produce pressures high enough to burst pipe, sometimes at points distant from the point of origin of the event.

Another case of surge is where a pump station is shut down suddenly, as in a power failure at an electrically powered station, causing the station to be bypassed automatically. This produces a drastic change in velocity of the flow but not a complete stoppage. As in any other surge, a pressure wave is sent upstream and a refraction wave is sent back downstream from some point in the system where there is a sufficient discontinuity to permit total or partial reflection of the wave (i.e., a valve or major bend in the pipe). The positive surge wave, traveling upstream, reaches the next upstream station, raising both its discharge and suction pressure (with the wave effectively passing counter to flow through the pump since the check valve does not close). It is possible that this situation may shut down the upstream station on high discharge pressure, thereby starting a new surge wave which effectively adds to the first. This pyramiding of surges can travel hundreds of miles, knocking down pump station after station in domino style and producing higher and higher pressures upstream. In the downstream direction, the negative wave may shut down stations downstream of the event by starving their suction pressure.

Severe surge problems can be mitigated through the use of quick-acting relief valves, tanks, and gas-filled surge bottles. These facilities tend to be expensive single-purpose devices which are seldom needed and are often inadequately maintained by operators. An inexpensive feature of modern computer-based control systems is the "permissive" circuit which can be rigged so that the system cannot be operated at a rate above an intrinsically safe level unless all outside parameters are satisfied. The loss of a signal saying that a given station is operating, for example, will cause other stations' set points to be backed off to a level consistent with any surge situation.

Design for the control of surges requires a thorough understanding of the particular pipe, pump, valve, and tank systems and their instrumentation, and therefore the total evaluation of the surge problems on a new system must await a fairly complete design (pipe size, wall thickness, taper, profile, flow velocity, tankage requirements, block valve spacing, number, size and arrangement of pumping units, type of devices, and static relief systems must be defined), but certain rudimentary surge considerations can be observed from the beginning which will help "surge-proof" the system:

- Provide interlocks such that all pumping stops before main line block valves can be closed.
- Low flow velocities ensure that changes in velocity cannot be too great.
- Long station spacing ensures maximum surge attenuation.
- Multiple pumping units at each pump station minimize the opportunity for a complete station failure.
- Use looped feeders for electric stations.
- Leave a margin of error between the hydraulic gradient and the MAOP based on wall thickness.
- SCADA communications to prevent upstream pump stations from overpressuring the downstream section.

Section 402.2.4 of the Code addresses the topic of surge pressure, stating that "...the level of pressure rise due to surges and other variations from normal operations shall not exceed the internal design pressure at any point in the piping system and equipment by more than 10%."

Corrosion Protection[20,21]

Corrosion of a pipeline can be both external and internal. Internal corrosion, apart from exceptional cases of corrosive fluid components such as H_2S, is usually a gradual process resulting in a lowering of pipeline efficiency and is characterized by indentations and pits. Regular line cleaning with scraper "pigs," discussed later, can be used to care for the internal surface of most installations. Internal corrosion can also be controlled by injecting a corrosion inhibitor into the transported fluid. Another method of reducing internal corrosion is by internally lining the pipe. While there are few examples of existing pipeline systems using epoxy coatings, they give good protection, long life, and have a low friction factor.

External corrosion is a major factor in the design and operation of a pipeline system in that external corrosion can reduce the life of a pipeline and impair its safety. External corrosion is mitigated by application of a pipe coating and the installation of a cathodic protection system. The external coating increases the pipe-soil electrochemical resistance, and the cathodic protection (impressed current or galvanic anodes) system makes the pipe cathodic with respect to the surrounding soil.

There are a number of materials and methods available for external coatings, each having different benefits and a range of cost; they include fusion bonded epoxy (FBE), asphalt mastic and coal tar, and asphalt enamels. Selection of a coating should consider installation cost balanced with reliability and other concerns such as shipping cost, application site, chemical resistance, maximum service temperature, soil conditions, storage and handling, etc.

Further discussion of internal and external coatings and of cathodic protection systems can be found in Chap. C6. The same material is applicable to liquid petroleum systems.

ASME B31.4 does not specify that an allowance be made for corrosion in determining nominal wall thickness for fluids and services covered by the Code, with the provision that internal and external corrosion control is provided as directed by Chap. VIII of the Code.

Metering

Early methods for monitoring the volume of flow through a liquid pipeline relied on tankage gauge readings at different points along the system. With the development of computer monitoring systems, liquid pipeline systems are now monitored continuously. Metering devices used in the pipeline business today fall into four groups:

- Pressure drop
- Positive displacement
- Turbines
- Miscellaneous, including sonic and vortex meters

The first group measures the pressure drop created when flow is restricted, either across an orifice plate or a venturi tube. This type of meter is unable to identify variations in density and viscosity which may be variable in liquid petroleum services and therefore is not used for fine measurements except in conjunction with a viscosity or density meter.

The positive displacement group of meters includes a wide variety of mechanical devices which entrap a discrete quantity of fluid and move it physically from one side of the meter to the other. Screws, pistons, buckets, gears, and sliding vanes have been used successfully especially with viscous fluids, but they are subject to wear in continuous, high-capacity low-lubricity situations.

Turbine meters consist of a turbine wheel with tiny magnets mounted axially between sleeve bearings inside a short length of pipe. As the fluid turns the turbine wheel, the rotations are recorded and counted electronically by the passage of the magnets. These meters have been successful because they are simple in design, essentially having only one moving part. The turbine wheel, when in motion, tends to center itself in the pipe so there is almost no bearing friction.

All three of these meter types have their uses. The orifice, because it is rugged and low in cost, is used where exact measurement is not required. Positive displacement meters are used where the fluid is viscous and/or a high range of viscosity is expected and close measurements are needed for custody transfer. Turbine meters are used for the continuous, high-capacity bulk movement of fluids because they are reliable and accurate over a wide range of flow rates.

The fourth group consists of more specialized meters, and the reader is directed to manufacturers to obtain information on a particular type of meter or installation.

Leak Detection

Detection of major leaks is a major concern of pipeline operating companies. Large leaks can be detected with relatively simple instrumentation; however, detecting small leaks requires computer systems that can also monitor and account for variations in temperature, pressure, density, and composition. To illustrate the problem of leak detection, with an order of magnitude comparison, consider that for an 8-in line transporting 24,000 BPD, a detectable deficiency of ±2 percent of flow is a 14-gal/min leak. A 0.2 percent deficiency, detectable only with a sophisticated leak detection system, on a 48-in line pumping 2.4 million barrels per day (MMBPD) is a 140-gal/min leak. The hole in the pipe which would leak 140 gal/min at 500 psig is more than ½ in in diameter. The hole which would leak 14 gal/min at the same pressure is less than 1/16 in in diameter.

For years, basic leak detection has been based on the principle of line balance, based on the continuity of flow in the pipeline (i.e., flow in equals flow out). In its simplest form, leak detection can rely solely on readings of flow meters or tank gauges at periodic intervals, with a recurring discrepancy indicating a leak. For line balance to be viable, accurate flow measurements at both ends of the pipeline, or section of pipeline, must be made and reconciled regularly. In early pipeline systems, minor differences in reading time, temperature, line pack, etc., made the results somewhat erratic, but a continuous shortage in delivery for several hours was sufficient cause for a line patrol to be sent out or for the system to be shut down and pressure tested. Large leaks or line breaks were detected by comparing suction and discharge pressures with flow at pump stations (e.g., falling discharge pressure combined with increased flow meant a leak downstream; falling suction pressure combined with decreased flow meant a leak upstream).

In modern pipeline systems, computer-based monitoring systems compare flow rates and total flow and record the variables of the flow (temperature, viscosity, etc.). The determination of whether there is a leak in the system is based on the readout differentials. For reliable pipeline surveillance, both short-term (minutes) and long-term (days) calculations should be maintained and compared against respective threshold values.

Perhaps the more important problem of leak detection is location of the leak once it has been detected. While line balance will indicate the magnitude of the leak, other means are required to pinpoint the location between two monitoring positions.

A general method for locating a leak is to compare the hydraulic gradient for the flow measured upstream with that of the flow measured downstream, with the intersection of the hydraulic gradients being the approximate leak location. Methods for location of rapidly formed leaks include monitoring the rate of change of pressure and flow (dp/dt and dq/dt, respectively) and the pressure wave differential which occurs as the result of a rapid leak, propagating from the leak in both directions, with the velocity of sound in the liquid (see surge discussion). Similarly, the transient pressure wave can be monitored with transducers which will detect the transient pressure wave, its magnitude being the generated pressure wave at the leak location delayed exponentially as a function of the distance to the leak and the velocity of sound in the liquid.

More discussion on the theory of these leak detection and location methods can be found in standard hydraulic texts, as well as several of the references.[11]

Pipeline Pigs

A "smart" pig is an instrumented device which travels internally along a pipeline, monitoring the operating parameters (flow, temperature, etc.) and the physical condition of the pipe (wall thickness, corrosion, out-of-roundness, etc.). Pigs are also used to "listen" for the acoustical traces of leaks. Simple pigs, or spheres, are sometimes used to mark the transition between two commodities in a multiproduct pipeline.

The information the pig collects must be relayed to the master control center, and the pumping and heating stations along the route must have facilities designed for the handling of pigs, including launching and receiving traps.

Supervisory Control and Data Acquisition

Maintaining the integrity of a pipeline system which spans hundreds of miles is a complicated task, involving the monitoring, measurement, and analysis of a continuous flow of data from meters, pipeline pigs, transducers, etc. Just as critical as collecting the data from source points along the length of the system is the transmission of the data to the facilities of the pipeline operating company, which may be remote from the pipeline itself. There is also the aspect of control and/or coordination of multiple stations along the system so that they operate in conjunction with each other, rather than opposition.

Early pipeline systems relied on analysis of data at the local station and communication of information by voice over telephone or wireless. The operating philosophy of many pipeline systems today is to limit the number of operator-attended control stations. Today's technology relies instead on supervisory control and data acquisition (SCADA) systems to collect data from monitor points by fiber optics, microwaves, and satellite communications technology and transmit it to control stations. Here, high-speed computers analyze the data and perform on-line control functions required to maintain system parameters within their operating limits. Application of these technologies requires consideration early in the mechanical design to include the operating components for the intended modes of operation and control.

The result of this technology has been to maximize safety and operating efficiency of the system. Additionally, the collection and storage of operational information over the life of a pipeline system facilitates new pipeline design and operation through the verification of computer models which are used in the design phase to:

- Identify out-of-range operating variables
- Select preferred operations at and among pumping stations to control use of pumping units and cost of energy
- Identify limits for pumping rates and fluid properties of commodities transported
- Predict effects of modifying system facilities or operations or changing characteristics of commodities transported
- Schedule tenders, or contracts, for transporting commodities by various fluid properties, ownership, source, and destination

SYSTEM COST ANALYSIS

The first several sections of this chapter have shown that for a given proposed pipeline, there are many possible pipeline systems which can be designed to transport the commodity. Selection of the optimal pipeline route, diameter, material, wall thickness, pump station location, pump units, and operational equipment or facilities is typically the result of economic analysis and investment capital evaluation of the most reasonable scenarios developed through the design phase.

Typically, even before the detailed design of a pipeline system has begun, an order-of-magnitude cost study will be performed, with the goal of determining the feasibility of continuing to invest time and capital in the design phase of the project. At this point, a preliminary route may have been selected; however, other possibilities may still exist, and further developments may indicate a change in the assumptions or information available during the preliminary analysis. In conjunction with the hydraulic, mechanical, and operational and maintenance designs, the economic analysis progresses, at times steering the decisions made in the design.

For a typical cross-country pipeline project, the cost of pipe and its associated construction and installation costs can be as much as 80 percent of the capital investment; therefore, the selection of the pipe, with regard to type of material, size, etc., is very important. Trade-offs can be made during the design process with regard to diameter and wall thickness, grade of steel, and method of manufacture, as discussed in various earlier sections of this chapter. The cost of the pipe itself generally represents 25 to 50 percent of the total line cost, and the use of a reliable cost, based on current industry information, is significant. Annual cost indexes are published in a number of trade periodicals. Another source of costs for line pipe, as well as component fittings, valves, and installation factors, are manufacturing associations and the construction industry publications.

Using the information developed through the earlier example, a comparison of system designs can be made, as shown in Tables C5.7 and C5.8. These tables illustrate the general procedure for selecting an appropriate pipeline system for a specific route. The same comparisons can be made for different routes. These comparisons also illustrate that consideration should be given to the method of

TABLE C5.7 Example: Isothermal Crude Oil System

	Pipeline O.D. × Wall thickness		
	16 in × 0.250 in	20 in × 0.250 in	24 in × 0.250 in
Comparison of alternative diameters, 198,500 BPOD			
Unit friction loss, ft/mile	143.7	47.9	19.7
Length, miles	240	240	240
Pipeline friction, ft	34,488	11,496	4,728
Static head, ft	50	50	50
Total head, ft	34,538	11,546	4,778
Total pressure, psi	12,996	4,344	1,798
MAOH, ft	3,588	2,870	2,392
MAOP, psi	1,350	1080	900
Number of pump stations	9.63	4.02	2.00
Operating HP at 82 percent efficiency, HP	53,526	17,894	7,405
Preliminary cost comparison; capital cost ($1000)			
Pipeline			
Pipe unit weight, ton/mile	110.0	139.2	161.4
Pipe weight, ton	26,400	33,408	38,736
Total pipe cost at $700/ton	18,480	23,386	27,115
Unit installation cost, $/ft	35	40	45
Pipeline installation	44,352	50,688	57,024
Total pipeline cost	62,832	74,074	84,139
Pump stations			
Installed HP per station (includes 10% excess)	6,120	4,890	4,080
Base cost per station	1,000	1,000	1,000
Installed HP at $750/HP	4,590	3,668	3,060
Cost per station	5,590	4,668	4,060
Total pump station cost	55,900	18,670	8,120
Total system capital cost	118,732	92,744	92,259
Annual costs ($1000)			
Insurance at 1% of installed cost	1,187	927	923
Operating cost			
Pipeline at 1.5% of installed cost	942	1,111	1,262
Stations at 3% of installed cost	1,677	560	244
Financing cost (assuming 100% at 7% of total cost)	8,311	6,492	6,458
Power at $0.08/kW	26,535	8,871	3,671
Total annual operating cost	38,652	17,961	12,558

TABLE C5.8 Example: Isothermal Crude Oil System

Comparison of revised systems for bypass of station 2, 198,500 BPOD		
	Pipeline O.D. (in)	
	20	24
Length of pipe		
0.250 wall, miles	220	214
0.312 wall, miles	20	26
Unit friction loss		
0.250 wall, ft/mile	47.9	19.7
0.312 wall, ft/mile	49.4	20.2
Pipeline friction		
0.250 wall, ft	10,538	4,216
0.312 wall, ft	988	525
Total Pipeline friction, ft	11,526	4,741
Static head, ft	50	390
Number of pump stations, ft	4	2
Station losses at 75 ft/station, psi	300	150
Total pumping head, ft	11,876	5,281
Total pumping pressure, psi	4,469	1,987
Operating HP at 82% efficiency, HP	18,405	8,184
Cost comparison of revised systems; capital cost ($1000)		
Pipeline		
Pipe unit weight		
0.250 wall, ton/mile	139.2	161.4
0.312 wall, ton/mile	173.2	208.4
Pipe weight		
0.250 wall, ton	30,624	34,540
0.312 wall, ton	3,464	5,418
Total pipe weight, ton	34,088	39,958
Pipe cost at $700/ton	23,862	27,971
Unit installation cost, $/ft	40	45
Pipeline installation	50,688	57,024
Total pipeline cost	74,550	84,995
Pump stations		
Installed HP per station (includes 10% excess)	5,060	4,500
Base cost per station	1,000	1,000
Installed HP at $750/HP	3,795	3,375
Cost per station	4,795	4,375
Total pump station cost	19,180	8,750
Total system capital cost	93,730	93,745
Annual costs ($1000)		
Insurance at 1% of installed cost	937	937
Operating cost		
Pipeline at 1.5% of installed cost	1,118	1,275
Stations at 3% of installed cost	575	263
Financing cost (assuming 100% at 7% of total cost)	6,561	6,562
Power at $0.08/kW	9,124	4,057
Total annual operating cost	18,315	13,094

financing and appropriate economic factors, such as system growth and annual operating costs.

The reader should refer to a handbook on financial decision-making to learn more about the process of optimum system selection on economic grounds, including the processes of discounted cash flow analysis.[2]

SPECIAL OIL PIPELINE DESIGN TOPICS

Special Hydraulic Conditions

Pipeline systems which fall under the scope of the Code have a wide variety of physical properties, as shown in Table C5.1. Furthermore, two of the most important properties—viscosity and vapor pressure—vary with temperature. Earlier sections discussed pipeline systems that have no fluctuation in temperature; therefore, the hydraulic system design was simplified. Another section discussed hydraulic design of nonisothermal systems for heavier commodities, such as heavy crude and residual fuel oils, influenced by the effect of temperature on viscosity and related friction losses. Hydraulic design for the lighter commodities, such as gasoline and natural gas liquids, is relatively independent of viscosity but more dependent on vapor pressure considerations.

This section summarizes the special considerations which will arise in the design of multiproduct, high-vapor pressure, hot oil, and nonnewtonian fluid pipeline systems without attempting to be comprehensive. The reader is encouraged to consult specialists and/or specific references for more detailed discussions of the design of these systems than can be described here.

Multiproduct Pipelines. If the pipeline system is to transport fluids of differing properties such as in a batch-type operation of a multiproduct system, the design fluid should be considered as the fluid producing the greatest friction loss (i.e., with the greatest viscosity at the design operating temperature). Using this fluid assures that all pumping stations have adequate power and all sections between stations have adequate wall thickness to sustain the design pressure and flow rate. Furthermore, multiproduct systems may have intermediate deliveries to the system and discharges from the system (i.e., different flow rates for each pipeline section may need to be considered).

To avoid excessive mixing of products, the system should be designed for flow in the turbulent region. Batching pigs can be used to minimize interface mixing at low flow rates.

A special pipeline application uses a batch flow technique to transport unlike petroleum products sequentially in a single pipe without significant deterioration of the quality of the products by contamination. An explanation of the principle involves a discussion of laminar and turbulent flow regimes, as given in Chap. B8.

In laminar flow molecules in the flow travel forward in parallel, except near the pipe wall, where the velocity of the molecules is slowed by friction with the pipe material. The standard velocity profile is parabolic over the cross section of the pipe. Therefore, in the laminar flow region, an interfacial plane between adjacent products would rapidly deteriorate into a bulge, and the trailing product would tend to push through the leading product near the pipe center.

In turbulent flow, however, the molecules are in random motion, bouncing off other molecules as well as the pipe wall, and they tend to remain in relative position. Therefore the velocity-flow profile is almost constant over the cross section of the pipe and the fluid flows more like a "plug." The interfacial plane between adjacent batches in the pipeline tends to remain in place as the products proceed down the pipeline. Some mixing occurs, principally by diffusion, and is more a function of length (time) of transport than velocity. The amount of contamination of the leading product by the trailing one is the same as the contamination of the following product by the leading one. The transition from one product to another, as the interface zone passes a point, takes the shape of a sine wave. At some point as the interface wave passes the switching manifold, the flow is diverted to separate storage or distribution facilities. The contamination of one product in another is controlled by timing the "cut" or diversion. One way this is done is to place a gravitometer some distance upstream of the receiving terminal and to record its signal at the terminal so that the receiving operator has a preview of exactly what the interface will look like on arrival. From a design point, a batch flow pipeline has several hydraulic gradients, at least one for each product, as well as for the "mixing" zone.

High-Vapor-Pressure Pipelines. High-vapor-pressure systems are characterized by low density, low viscosity, and the necessity to operate the system at elevated pressure to maintain the fluid as a single phase in the pipeline. Single phase is maintained throughout the pipeline by maintaining the elevation of the hydraulic gradient above the ground profile by more than the head equivalent of the local vapor pressure. Backpressure regulators may be installed in terminals to maintain the required elevated gradient. In this sense, design of high-vapor-pressure lines differs from crude oil pipelines in that the parameter governing design is the vapor pressure, directly related to temperature, rather than viscosity, which is inversely related to temperature. In other words, the design is based on maximum temperature where maximum vapor pressure occurs versus the maximum viscosity or minimum temperature point, which is the design basis of viscous fluid systems.

Waxy and Heavy Crude Oils. One treatment for waxy and heavy crude oils is to heat the commodity as part of the transport operation. This concept was discussed earlier under nonisothermal hydraulics.

Chemical additives have also been used in oil pipeline systems to improve flow (i.e., reduce drag of heavy and waxy crudes). Initially developed to enhance oil well fracturing, these additives are hydrocarbon-based polymers. The additives are solutions of a high molecular weight copolymer or polyolefin, in a hydrocarbon solvent. Being a hydrocarbon, it distills according to the volatility of its fractions and is not distinguishable from the hydrocarbons originally present in crude oil. The additive reduces the turbulent flow, effectively expanding the transition flow region. It is injected into the flowing pipeline stream by a compact, skid-mounted chemical injection module consisting of injection pumps, positive displacement flowmeters with totalizers, and miscellaneous instrumentation. The polymer itself is so viscous that an inert gas, such as nitrogen, is used to help transfer the material from tank to pumps.

Injection of the additive must be at every station on a system because the polymer is degraded by the action of centrifugal pumps. Initially it was thought that only very large-diameter pipelines could benefit from use of drag reducers. Widespread tests have demonstrated that smaller crude and product lines can also benefit.

Multiphased Flow. The presence of gas and liquid phases of a product in sections or the entire pipeline length has been mentioned earlier in the discussion of slack flow. Multiphased flow is particularly a factor in flow lines used to gather the produced gas, oil, and water mixture from a well in a production field and transport it to separation facilities. Detailed knowledge of phase equilibrium conditions and related product properties as well as specific multiphased flow energy and hydraulic conditions is required for the design of these systems. Computer programs have been developed to include the possibility of multiphased flow and are commonly employed for the solution of the hydraulic design of these types of systems.

Seismic Considerations

The entire topic of designing facilities and systems, including pipeline systems for earthquakes, is of major importance in many parts of the world. With consideration to pipeline design, three major seismic hazards for buried pipelines are landslides, liquefaction of soil under and/or around the pipe, and differential fault movement and ground rupture. Ground motion (shaking) itself is a major consideration in the design of stations and terminal facilities and aboveground supports of a pipeline, but it has less effect on buried pipelines.[22,23] Differential settlement and faultline shifts with vertical and horizontal displacement will impose additional longitudinal stresses on the pipe material.

In route selection, a survey of alternate routes should consider evidence of past landslides (i.e., movement of the ground) triggered by a seismic occurrence. Slopes showing signs of recent instability or movement may be candidates for further disturbance in event of an earthquake. If slope instability involves deep translations and rotational displacement, the potential ground movements in the area may be large, and relocation of the pipeline may be more cost beneficial than expensive stabilization measures. If instability involves slumps and shallow slides, slope stabilization may be effective. In any event, seismic and geologic specialists should be consulted.

In saturated, cohesionless soils, for example nonplastic silts and medium-dense sands, liquefaction, wherein the soil temporarily is transformed into a liquid state, loosing shear strength and bearing capacity, may be the major hazard. Liquefaction leads to lateral spreading, loss of bearing capacity, and uplift of buried pipe due to increased buoyancy. Several measures can be taken to design a pipeline system crossing areas susceptible to liquefaction, including designing for moderate deformations as a result of uplift during an event, limited burial in the area, and adding weight coatings, thereby limiting uplift. Operationally, additional line block valves for shutoff in case of a failure following an earthquake may need to be provided.

Fault movement is the most dramatic earthquake occurrence and is potentially catastrophic for a pipeline. Pipeline alignment in fault zones should be such that the expected fault movement will produce tensile stresses in the pipe. If compressive stresses result, buckling may occur. The line should be laid in relatively straight sections, crossing the fault at an angle of 60 to 90° and without abrupt changes in direction or elevation, which might serve as an anchor during an earthquake. Depth of cover should be minimal, thereby reducing soil restraint, and the backfill material should be granular, medium-range soft sand, without large stones, placed well around the pipe to allow relatively free movement.

Two of the references[21,22] have additional material on the location and operation of pipelines across earthquake fault zones.

Underwater Pipeline Design

The subject of underwater pipeline design, in particular ocean or deepwater pipelines, is worthy of a separate chapter unto itself. Yet it is important to reference some of the related topics here, particularly with regard to the section of a pipeline system that may cross a river or lake, requiring placement underwater. There are generally two types of underwater pipelines—those that cross a relatively short distance and those that represent a minimal portion of the entire system design (e.g., a river crossing and the submarine pipeline which is primarily underwater). River crossings of a pipeline can be made by pulling the pipe string that was previously welded onshore or by laying the pipeline into a trench beneath the depth of scour and then covering it with protective backfill. Directional drilling techniques to "tunnel" under the riverbed are available where soil conditions, distance, construction space, and pipe properties are amenable. However, the design of an underwater, or submarine, pipeline has several special considerations, which may or may not also be major considerations in river crossings of terrain pipeline systems. The reader is referred to specific handbooks on the design of ocean pipelines for more specific information on this topic.

If a body of water is to be crossed as part of a terrain pipeline, and the decision is made to place the line underwater, a hydrographic survey of the area should be initiated. In addition to as much information as possible about the possible route, other hydrographic data are required such as current and tidal data, weather records, wave height and patterns, possible underwater obstructions along the route, and sources of impact, such as dragging anchors. Selection of the landfall location is also an important element in the route selection.

External pressure was discussed earlier and will be greatest on the empty pipeline after placement. The tendency of the pipe to buckle under external pressure should always be checked, and buckle arrestors (i.e., stiffer sections) should be included in the design to limit buckle propagation.

Selection of the material for the pipeline should consider that the stresses during the laying operation may govern the mechanical design. Note that these stresses do not occur in conjunction with internal pressure. The laying stresses can be categorized as direct pulling stresses from welding onshore and pulling into location or by tension when laying from a laybarge; bending stresses as the pipe is lowered to the riverbed or seabed; torsional stresses induced during the laying operation; and current and wave stresses during tow and placement.

Pipeline stability, once it is in place, is another consideration to ensure security of the pipeline. If additional weight is required, it can be provided by weights or anchors placed at intervals or by continuous weight coating. Placement of the pipeline in a trench and covering it with protective backfill is another alternative.

REFERENCES

1. ASME B31.4—1989 ed., Liquid Transportation Systems for Hydrocarbons, Liquid Petroleum Gas, Anhydrous Ammonia and Alcohols. American Society of Mechanical Engineers, New York, 1989.
2. Ikoku, C. U. *Economic Analysis and Investment Decisions.* John Wiley, New York. 1985.
3. Hein, M. A. *HP41 Pipeline Hydraulics and Heat-Transfer Programs.* PennWell Publishing Company, Tulsa, OK. 1984.

4. "Software Manual of Pipeline Programs for Hand-Held Calculators and Other Computers," (selected authors). *Oil and Gas Journal,* 1979 to 1987.

5. Hosmanek, M. *Pipeline Construction.* Petroleum Extension Service, Division of Continuing Education, University of Texas, Austin. 1984.

6. Schurr, B. *Manual of Practical Pipeline Construction.* Gulf Publishing, Houston. 1982.

7. ASTM D1250—Petroleum Measurements Tables (also designated API Standard 2540). American Standards of Testing and Materials, 1980.

8. Bell, H. S., ed., *Petroleum Transportation Handbook.* McGraw-Hill, New York. 1963.

9. "Flow of Fluids through Valves, Fittings and Pipe," Technical Paper no. 410 by Crane Company, New York. 1981.

10. *GPSA Engineering Data Book,* 10th ed. Gas Processors Suppliers Association. 1987.

11. Brater, E. F., and King, H. W. *Handbook of Hydraulics,* 6th ed. McGraw-Hill, New York. 1976.

12. *Cameron Hydraulic Data,* 17th ed. (C. C. Heald, ed.). Ingersoll-Rand, Woodcliff Lake, N.J., 1988.

13. API SPEC 5L—Specifications for Line Pipe, American Petroleum Institute, Washington, D.C., 1985.

14. Warring, R. H. *Pumps: Selection, Systems and Applications,* 2nd ed. Gulf Publishing, Houston. 1984.

15. Arnold, C. L. "Temperature Effects on Hydraulics and Fluid Mechanics," presented at 1981 Annual Pipeline Design and Construction Symposium, Dallas, TX. April 30, 1981.

16. Ford, P. E., Ellis, J. W., and Russell, R. J. "Pipelines for Viscous Fluids," Proceedings of the Fourth World Petroleum Congress—Section VIII/B. 1964.

17. Vincent-Genod, J. *Fundamentals of Pipeline Engineering.* Gulf Publishing, Houston. 1984.

18. Timoshenko, S., *Strength of Materials, Part II, Advanced Theory and Problems,* 3d ed., Robert E. Krieger Publishing, New York. 1956, p. 189.

19. Watkins, R. K. "Longitudinal Stresses in Buried Pipes." *Advances in Underground Pipeline Engineering.* American Society of Civil Engineers, New York, 1985.

20. Galka, R., and Yates, A. P. J. *Pipe Protection: A Review of Current Practice, 2nd Edition.* British Hydromechanics Research Association, Bedford, England. 1984.

21. Parker, M. E., and Peattie, E. G. *Pipeline Corrosion and Cathodic Protection,* 3d ed. Gulf Publishing, Houston. 1984.

22. "Seismic Response of Buried Pipes and Structural Components," a report by the ASCE Committee on Seismic Analysis. American Society of Civil Engineers, New York. 1983.

23. "Guidelines for the Seismic Design of Oil and Gas Pipeline Systems," prepared by the ASCE Committee on Gas and Liquid Fuel Lifelines, American Society of Civil Engineers, New York. 1984.

BIBLIOGRAPHY

API-Bul 5C3—Bulletin on Formulas and Calculations for Casing, Tubing, Drill Pipe and Line Pipe Properties, 5th ed. American Petroleum Institute, Washington, D.C., 1989.

API-RP 1102—Recommended Practice for Liquid Petroleum Pipelines Crossing Railroads and Highways, 5th ed. American Petroleum Institute, Washington, D.C., 1981.

API-RP 1110—Recommended Practice for the Pressure Testing of Liquid Petroleum Pipelines, 2nd ed. American Petroleum Institute, Washington, D.C., 1981.

API-RP 1111—Recommended Practice for Design, Construction, Operation and Maintenance of Offshore Hydrocarbon Pipelines, American Petroleum Institute, Washington, D.C.

CSA Z183—Oil Pipeline Transportation Systems. Canadian Standards Association. 1982.

CFR 49, Part 195—Transportation of Hazardous Liquids by Pipeline. Codes of Federal Regulation, 1990, U.S. Department of Transportation, Washington, D.C.

Fletcher, L. (ed.). "Design and Operation of Pipeline Control Systems." Proceedings of a session sponsored by the Pipeline Division of ASCE (San Francisco, Oct. 5, 1984). American Society of Civil Engineers, New York. 1984.

Herbich, J. B. *Offshore Pipeline Design Elements.* Sponsored by Marine Technology Society. Marcel Dekker, Inc., New York. 1981.

Kiefner, J. F., and Wall, T. A. "Calculations Help Control Stresses." *Oil and Gas Journal,* October 14, 1985.

"Lifeline Earthquake Engineering—Buried Pipelines, Seismic Risk and Instrumentation," ASME Third National Congress on Pressure Vessel and Piping, San Francisco, 1979.

Mendel, O. *Practical Piping Handbook.* PennWell Publishing Company, Tulsa, OK. 1981.

Mousselli, A. H. *Offshore Pipeline Design, Analysis, and Methods.* PennWell Publishing Company, Tulsa, OK. 1981.

"Pipeline Design for Hydrocarbon Gases and Liquids," report of the ASCE Committee on Pipeline Planning. American Society of Civil Engineers, New York. 1975.

Pipeline Safety Code, 4th ed. Institute of Petroleum, Model Code of Safe Practice in the Petroleum Industry, Part 6. John Wiley, London. 1982.

"Rules for Submarine Pipeline Systems," Det Norske Veritas, Oslo, Norway. 1981.

Seiders, E. J. (ed.). Pipeline Safety and Leak Detection, presented at the Petroleum Division Fall Workshop, Houston, TX, 1988. American Society of Mechanical Engineers.

Whitelaw, J. A., and Reppond, D. W. "Designs for Buried Pipeline Can Reduce Seismic Hazards." *Oil and Gas Journal,* October 17, 1988.

CHAPTER C6
GAS SYSTEMS PIPING

Peter H. O. Fischer
Manager, Pipeline Operations
Bechtel Corporation
Los Angeles, California

German R. Mayer, P.E.
Manager, Pipeline Projects
Bechtel Corporation
San Francisco, California

INTRODUCTION

General

Plentiful supply, moderate costs, environmental concerns, and national programs to reduce dependence on unstable oil supplies support an upward trend in the consumption of natural gas on a worldwide scale. In addition to the industrial and residential uses of natural gas, nontraditional markets for gas energy include use as a motor fuel for fleet and private vehicles, as a supply for gas-fired cogeneration and/or fuel cells, and as a component of a gas-coal mix to allow coal to meet clean air requirements.[1]

The American Gas Association (AGA) Gas Supply Committee in a report issued in 1987 estimated that gas supplies are expected to meet current consumption levels under all likely gas price scenarios throughout the period 1987 to 2010. Total supply includes conventional gas production, gas from tight formations, coal seam gas, and gas from coproduction of gas and brine.[1]

Definitions

Section B31.8 of the ASME Code for Pressure Piping, Gas Transmission and Distribution Piping Systems defines *gas* as follows:

> Gas, as used in this Code, is any gas or mixture of gases suitable for domestic or industrial fuel and transmitted or distributed to the user through a piping system. The common types are natural gas, manufactured gas, and liquefied petroleum gas distributed as a vapor, with or without the admixture of air.[2]

A comprehensive listing of the definitions of general terms used in the gas transmission industry is provided in Sections 803 through 806 of ASME B31.8.

Types of Systems

Figure C6.1, abstracted from ASME B31.8, gives an overview of the piping and facilities (indicated by solid lines) considered to be part of gas transmission and distribution piping systems.

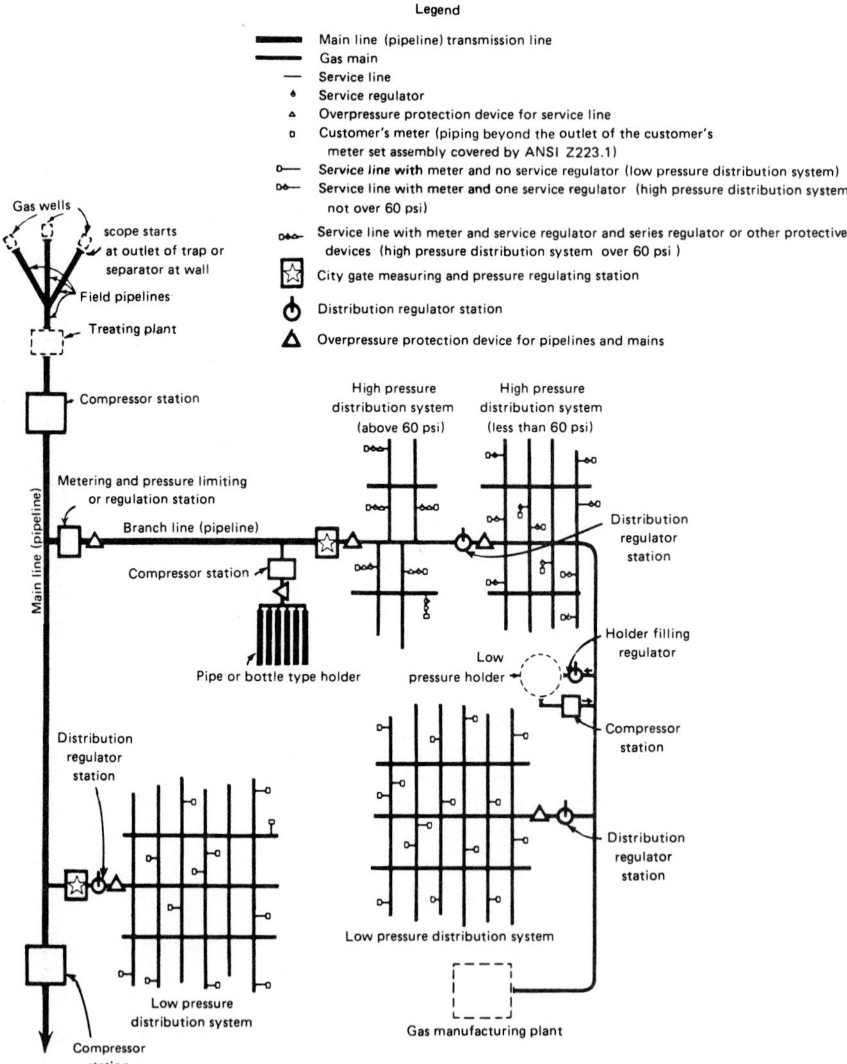

FIGURE C6.1 Gas transmission and distribution piping systems.[2]

Gas Gathering Systems. The gas gathering system consists of field pipelines transporting dry or wet gas from the wellheads to a central treating facility where initial separation of gas and liquids takes place.

Main Line (Pipeline) Transmission Lines. Main transmission pipelines transport gas from a source or sources of supply to one or more distribution centers or to one or more large volume customers or may interconnect sources of supply. Main transmission pipelines are usually characterized by larger-diameter pipe installed over longer distances with intermediate compressor stations.

Gas Distribution Systems. The gas distribution system is a piping system installed within a community to convey gas to individual service lines or other gas mains.

REFERENCE DOCUMENTS

Codes and Standards

In the United States the need for a national code for pressure piping became increasingly evident from 1915 to 1925. To meet this need, the American Engineering Standards Committee (later the American Standards Association) initiated Project B31 in March 1926 at the request of the American Society of Mechanical Engineers (ASME). A first edition was published in 1935 as an American Tentative Standard Code for Pressure Piping.[2]

A number of revisions culminated in the first edition of a separate integrated document known as the American Standard Code for Pressure Piping, Section 8, Gas Transmission and Distribution Piping Systems, published in 1952. In 1966, the name of the association was changed to United States of America Standards Institute and in 1969 changed again to American National Standards Institute (ANSI).[2]

The Pipeline Safety Act became effective in the United States in August 1968. As a result, Title 49 Parts 192 and 195 of the Code of Federal Regulations, "Transportation of Natural and Other Gas by Pipeline" and "Transportation of Liquids by Pipeline" became effective in November 1970.

Other Reference Documents

Countries other than the United States also have pipeline regulations of their own and some of these are more stringent than the U.S. codes, especially where licensing and leak testing are concerned. One must check before starting design intended for installation in other countries. Countries known to have codes are:

Algeria Germany
Austria Great Britain
Australia Japan
Belgium Italy
Canada USSR
France

Table C6.1 compares some of the major design parameters of the gas codes of a selection of countries.

TABLE C6.1 Gas Pipeline Code Comparison for Various Countries

Country and code	Location	Safety factor	% of yield	Design on min. or nom. wall thickness	Allowance for excess pressure	Main line valve frequency	Normal temperature range	Special remarks
ASME B31.8 1989	Class 1 Class 2 Class 3 Class 4	1.4 1.7 2.0 2.5	72 60 50 40	Nominal	Lesser of 10% overpressure or 75% of yield	Class 1, 20 miles Class 2, 15 miles Class 3, 10 miles Class 4, 5 miles	−20 to 450°F (−29 to 232°C) Derating above 250°F (121°C)	Class 1: Desert or farmland Class 2: Town outskirts Class 3: Residential and commercial Class 4: Town center, high rises
Canada—CSA Z184	Class 1 Class 2 Class 3 Class 4	1.4 1.7 2.0 2.5	72 60 50 42	Nominal	Lesser of 10% overpressure or 75% of yield	Class 1, 20 miles Class 2, 15 miles Class 3, 8 miles Class 4, 5 miles	−100 to 450°F (−74 to 232°C) Derating above 250°F (121°C)	
Australia—AS	Class 1 Class 2 Class 3	1.4 1.7 2.0	72 60 50	Nominal	10% overpressure	Class 1, 32 km Class 2, 24 km Class 3, 13 km Class 4, 8 km	−22 to 446°F (−30 to 230°C) Derating above 248°F (120°C)	Submarine lines S.F. 1.4 except at piers, risers, scraper traps S.F.: 1.7
Japan—petroleum code		2.0	50	Minimum	Lesser of 2 kg/cm^2 or 5%	Rivers, roads, railroads 4 km in city areas; 10 km in other areas		Additional safety factors for wind, snow, thermal changes, waves, currents, and earthquakes
Great Britain—BS-CP2010 - Part 2 1970 (also Institute of Petroleum and Institution of Gas Engineers)	Open country Near buildings	1.4 1.67	72 60	Minimum	10% overpressure	Every 10 miles in open country and as needed	−13 to 250°F (−25 to 120°C)	IGE uses a factor of 2.2 near habitations, also does not allow overpressure. Safety factors follow the U.S.S.R. National Gas Code classification based on types of fluid, line size, and population density

Belgium—Moniteur Arrete royal	Class 1 Class 2 Class 3 Class 4	1.4 1.7 2.0 2.5	72 60 50 40	Nominal	Test pressure	30 km maximum	up to 248°F (120°C)	
Germany—DIN 2470 Blutt 2	Class 1 Class 2 Class 3 Class 4	1.6 1.7 1.87 1.87 w/additional require-ments	62.5 60 53.5 53.5	Minimum	10% overpressure	Class 2, 10 km Class 3, 6 km		TUV or TUA specialists are to be consulted and will rule on matters not explicitly covered in DIN 2470 or referenced DIN standards
France—Journal Official	Class 1 Class 2	2.78 1.82	36 55	Minimum	0.9 test pressure	30 km maximum		Class 1, built-up area Class 2, all other areas
International Gas Union country code	1.4 to 2.5 restricted on ratio of yield to ultimate		72 to 40	Nominal	None	30 km maximum except desert		Adopted by EEC. Based on U.S. & U.S.S.R. codes, which differ in area classification

DESIGN

The specific identification of types of pipe generally used for high-pressure gas lines is given in the latest edition of the ASME Code for Pressure Piping, ASME B31.8. Special fittings for gas systems, together with typical details and design recommendations, are included in this chapter.

Basic Flow Equations

Rational Gas Flow Formula. Many equations for calculations involving isothermal gas flow in horizontal gas pipelines have been used by the pipeline industry with varying degrees of success over the years. The latest is the rational gas flow formula:

$$P_1^2 - P_2^2 = Bf\left(\frac{ZTGQ^2}{D^5}\right)L$$

where B = dimensional constant = 76.86
D = internal diameter, in
f = friction factor, dimensionless
L = length, miles
G = gas gravity (air = 1)
P_1 = initial line pressure, psia
P_2 = final line pressure, psia
Q = flow rate, thousands of cubic feet per hour (MCF/HR)
T = absolute temperature at which gas exists, °R
Z = compressibility factor, at average flow conditions, dimensionless

For flow in gas pipelines, to take into account differences in elevation, the pressure profile is determined using the rational gas flow formula with J. William Ferguson's elevation correction method:

$$P_1^2 - e^s P_2^2 = 76.86\,f\left(\frac{ZTGQ^2}{D^5}\right)L_e$$

where P_1 = upstream pressure, psia
P_2 = downstream pressure, psia
e = natural logarithmic base, 2.71828
e^s = elevation correction factor
f = friction factor
G = gas specific gravity (air = 1)
T = absolute gas flowing temperature, °R (°F + 460)
Q = gas flow rate, thousands of cubic feet per hour
Z = compressibility factor
L = length of pipe segment, miles
H = elevation difference over the segment, feet (positive uphill, negative downhill)
s = $GH/(26.647TZ)$
L_e = effective length, $(e^s - 1)L/s$
D = pipe internal diameter, in

Pipe Design Formula[2]**.** The design pressure for steel gas piping systems or the nominal wall thickness for a given design pressure is determined by the following formula:

$$P = \frac{2StFET}{D}$$

where P = permissible design pressure, psig
S = yield strength, psig
D = nominal outside pipe diameter, in
t = nominal pipe wall thickness, in
F = design factor. Value depends on location class. (See Table C6.2.) Exceptions are given in Table C6.3.
E = longitudinal pipe joint factor. This is a function of the type of pipe manufacture (see Table C6.4).
T = temperature derating factor (see Table C6.5).

Location Classes for Design. The possibility of damage to a gas pipeline increases with larger concentrations of buildings. One method of providing added protection is to lower the pipe stress level as a function of public activity. ASME B31.8 quantifies this activity by determining Location Class and relating the design of the pipeline to the appropriate design factor.

ASME B31.8 has defined Location Classes as follows:[2]

Location Class 1. A Location Class 1 is any 1 mile section that has 10 or fewer buildings intended for human occupancy. A Location Class 1 is intended to reflect areas such as wasteland, deserts, mountains, grazing land, farmland, and sparsely populated areas.

Location Class 2. A Location Class 2 is any 1 mile section that has more than 10 but fewer than 46 buildings intended for human occupancy. A Location Class 2 is intended to reflect areas where the degree of population is intermediate between Location Class 1 and Location Class 3 such as fringe areas around cities and towns, industrial areas, ranch or country estates, etc.

Location Class 3. A Location Class 3 is any 1 mile section that has 46 or more buildings intended for human occupancy except when a Location Class 4 prevails. A Location Class 3 is intended to reflect areas such as suburban housing developments, shopping centers, residential areas, industrial areas, and other populated areas not meeting Location Class 4 requirements.

Location Class 4. Location Class 4 includes areas where multistory buildings are prevalent, and where traffic is heavy or dense and where there may be numerous other utilities underground. Multistory means 4 or more floors above ground including the first or ground floor.

TABLE C6.2 Basic Design Factor F[2]

Location class	Design factor F
Location Class 1, Division 1	0.80
Location Class 1, Division 2	0.72
Location Class 2	0.60
Location Class 3	0.50
Location Class 4	0.40

TABLE C6.3 Design Factors for Steel Pipe Construction[2]

Facility	Location Class				
	1				
	Div. 1	Div. 2	2	3	4
Pipelines, mains, and service lines	0.80	0.72	0.60	0.50	0.40
Crossings of roads, railroads without casing:					
1. Private roads	0.80	0.72	0.60	0.50	0.40
2. Unimproved public roads	0.60	0.60	0.60	0.50	0.40
3. Roads, highways, or public streets, with hard surface and railroads	0.60	0.60	0.50	0.50	0.40
Crossings of roads, railroads with casing:					
1. Private roads	0.80	0.72	0.60	0.50	0.40
2. Unimproved public roads	0.72	0.72	0.60	0.50	0.40
3. Roads, highways, or public streets, with hard surface and railroads	0.70	0.72	0.60	0.50	0.40
Parallel encroachment of pipelines and mains on roads and railroads:					
1. Private roads	0.80	0.72	0.60	0.50	0.40
2. Unimproved public roads	0.80	0.72	0.60	0.50	0.40
3. Roads, highways, or public streets with hard surface and railroads	0.60	0.60	0.60	0.50	0.40
Fabricated assemblies	0.60	0.60	0.60	0.50	0.40
Pipelines on bridges	0.60	0.60	0.60	0.50	0.40
Compressor station piping	0.50	0.50	0.50	0.50	0.40
Near concentration of people in Location Classes 1 and 2	0.50	0.50	0.50	0.50	0.40

Design factor for pipelines or mains supported by railroad, vehicular, pedestrian, or pipeline bridges must be determined in accordance with the Location Class prescribed for the area in which the bridge is located, except that in Location Class 1, a design factor 0.6 must be used.

Flowing Temperature. The original equation developed to predict with reasonable accuracy the temperature of gas at any point along a transmission line was derived by Charles E. Schorre and presented in 1954.[3] However, with Schorre's equation, flowing gas temperatures continuously decrease with downstream distance, never reaching an equilibrium value as would be expected when the Joule-Thompson cooling effect is offset by heat gain from warmer surrounding soil.

In 1979, D. M. Coulter and M. F. Bardon[4] developed the following equation, which gives a logarithmically decreasing flowing gas temperature which asymptotically approaches a value below that of the ground temperature. Using the same nomenclature as Schorre (and P = pressure):

$$T_2 = \left[T_1 - \left(T_g + \frac{\mu}{a}\frac{dP}{dX}\right)\right]e^{-aX} + \left(T_g + \frac{\mu}{a}\frac{dP}{dX}\right)$$

TABLE C6.4 Longitudinal Joint Factor E[2]

Spec. no.	Pipe class	E factor
ASTM A 53	Seamless	1.00
	Electrical Resistance Welded	1.00
	Furnace Butt Welded—Continuous Weld	0.60
ASTM A 106	Seamless	1.00
ASTM A 134	Electric Fusion Arc Welded	0.80
ASTM A 135	Electric Resistance Welded	1.00
ASTM A 139	Electric Fusion Welded	0.80
ASTM A 211	Spiral Welded Steel Pipe	0.80
ASTM A 333	Seamless	1.00
	Electric Resistance Welded	1.00
ASTM A 381	Double Submerged-Arc-Welded	1.00
ASTM A 671	Electrical Fusion Welded	
	Classes 13, 23, 33, 43, 53	0.80
	Classes 12, 22, 32, 42, 52	1.00
ASTM A 672	Electric Fusion Welded	
	Classes 13, 23, 33, 43, 53	0.80
	Classes 12, 22, 32, 42, 52	1.00
API 5L	Seamless	1.00
	Electric Resistance Welded	1.00
	Electric Flash Welded	1.00
	Furnace Butt Welded	0.60
	Submerged Arc Welded	1.00

General note: Definitions for the various classes of welded pipe are given in ASME B31.8.

TABLE C6.5 Temperature Derating Factor T for Steel Pipe[2]

Temperature, °F	Temperature derating factor T
250 or less	1.000
300	0.967
350	0.933
400	0.900
450	0.867

General note: For intermediate temperatures, interpolate for derating factor.

where μ = Joule-Thompson coeffecient, °F/psi
X = distance, ft
P = pressure, psi
T_1 = initial gas temperature, °F
T = gas temperature at point X_2, °F
T_g = average ground temperature, °F
$a = 2\pi RU/(qC_p)$ where $\pi = 3.1416$
R = pipe radius, ft

U = heat-transfer coefficient, Btu/(h · °F · ft^2)
q = gas flowing, thousand cubic feet per hour (MCF/HR)
C_p = specific heat of gas at constant pressure, Btu/°F · MCF

Transmission Factor. The transmission factor, $(1/f)^{0.5}$, is one of the most difficult values to determine. The following equations have proven to be reasonably reliable. For laminar flow,

$$(1/f)^{0.5} = 4\ \ln(f^{0.5}R_e) - 0.6$$

For fully turbulent flow,

$$(1/f)^{0.5} = 4\ \ln\left(3.7\frac{D}{k}\right)$$

where f = friction coefficient
D = inside diameter of pipe, in
k = effective roughness, in
R_e = Reynolds number

The Reynolds number for flowing gas is determined by the following:

$$R_e = 0.71\frac{Q_b G P_b}{u D T_b}$$

where Q_b = flow rate ft^3 per day at P_b, T_b
G = gas specific gravity (air = 1.0)
P_b = base pressure, psia
u = viscosity, centipoise
D = internal diameter, in
T_b = base temperature, °R (°F + 460)

Compressibility Factor. The compressibility factor Z may be obtained from the standard gas compressibility factor chart, Fig. C6.2.

$$\text{Pseudo-reduced temperature } T_r = \frac{T}{T_c}$$

$$\text{Pseudo-reduced pressure } P_r = \frac{P}{P_c}$$

where T_c = absolute pseudo-critical temperature, °R
P_c = absolute pseudo-critical pressure, psia
P = absolute pressure at which the gas exists, psia
T = absolute temperature at which the gas exists, °R

Figure C6.3 provides convenient approximations for determining the pseudo-critical pressure and pseudo-critical temperature of gases when only the specific gravity of the gas is known. Otherwise, these values should be calculated based on actual gas composition.

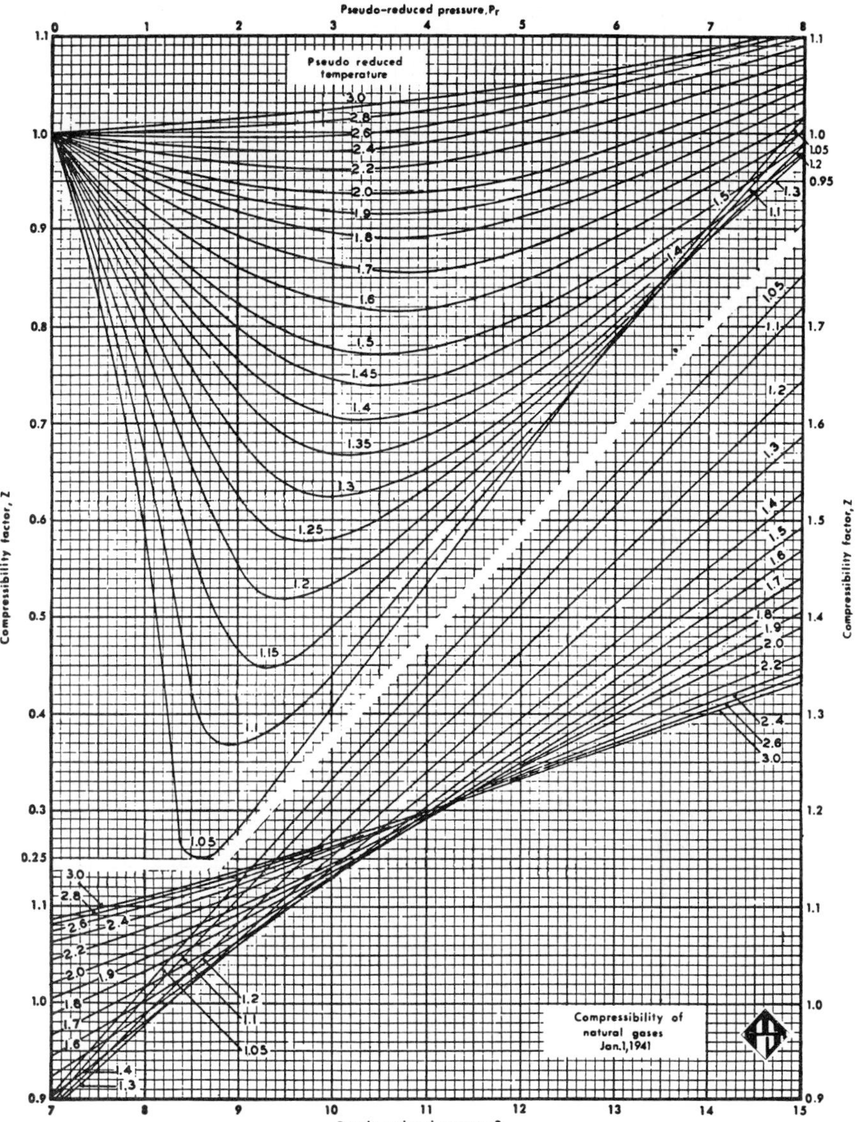

FIGURE C6.2 Compressibility factors for natural gas.[5]

Gas Gathering Systems

Condensate Formation. Liquids are created in gas pipelines when the temperature of the gas flowing through the pipeline drops below the dewpoint of the gas. Since the gas has both a hydrocarbon dewpoint and a water dewpoint, both hydrocarbons and water may condense and accumulate at low points in the pipeline

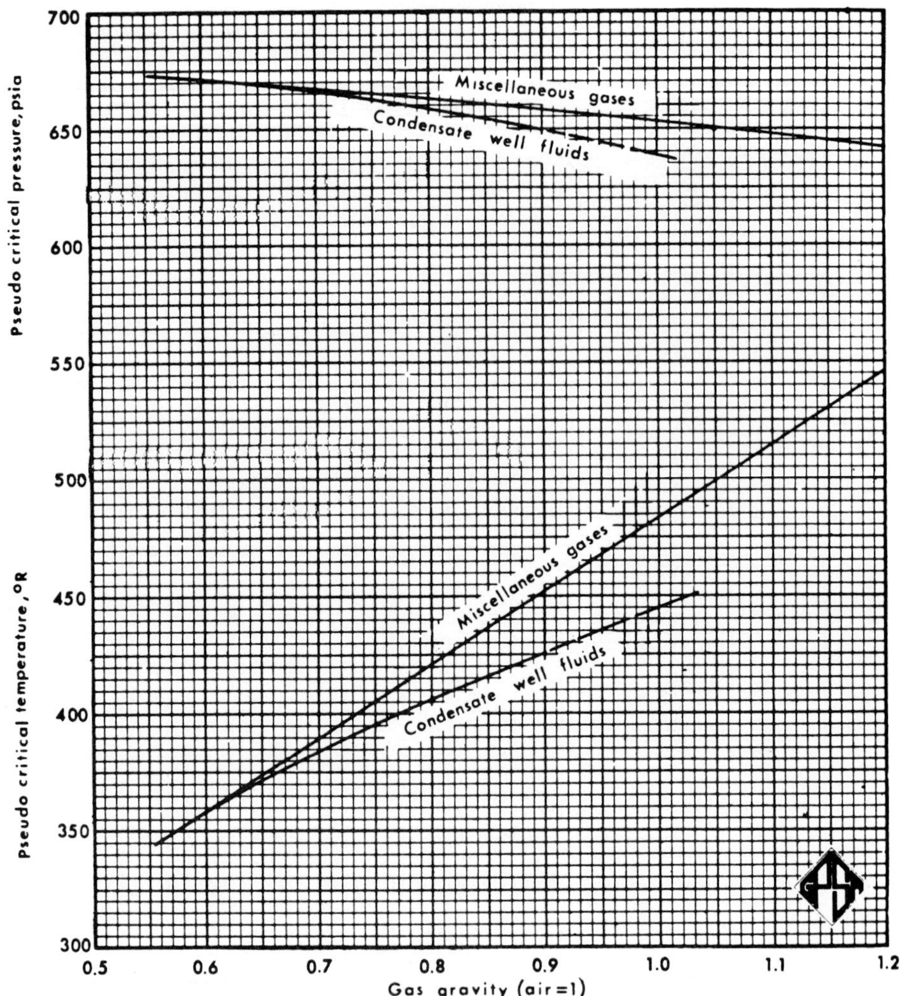

FIGURE C6.3 Pseudocritical properties of natural gas.[5]

to obstruct the flow of gas. In addition to the reduction in the capacity of the pipeline, the arrival of large volumes of liquid at downstream gas treatment facilities can result in severe damage to equipment or at the least cause erratic plant operation.

Two-Phase Flow. Accurate determination of pressure along the route of a wet gas pipeline operating in two-phase flow is difficult because the flow regimes associated with two-phase flow are numerous, complex, and difficult to define. Consideration must be given to gas and liquid fluid properties, flow regimes, pressure drop, and liquid holdup.

The major complication with two-phase flow is the variety of flow patterns that can be produced in a gas-liquid system. The type of flow pattern encountered depends upon the flow rates and properties (density and viscosity) of each of the gas and liquid components, surface tension between the two phases, pipe size, and pipe configuration. These factors have rendered the derivation of general flow correlations difficult.

New correlations for pressure loss and liquid holdup in gas-liquid pipelines are continuously being developed and are showing consistently reliable predictions for both holdup and pressure loss when compared to actual field operating data. The very number, complexity, and application-specific limitations of many of these correlations preclude the presentation and discussion of any particular correlation in this chapter. The reader is referred to current literature to obtain the latest information and formulas and to proprietary computer programs available on the market such as PIPEFLO by Neotechnology for multiphase pipeline network analysis.

Hydrate Considerations. Under unfavorable conditions, the low molecular weight hydrocarbons such as methane, propane, or butane form insoluble hydrate deposits in conjunction with water molecules. These hydrate crystals, which resemble ice or wet snow, have been known to virtually plug and stop gas transmission lines. Hydrates may also occur in equipment as a result of cooling due to pressure reduction. This can be a problem, particularly in pressure control valves and pressure regulators, which can literally freeze up.

The methods for determining the hydrate temperature and pressure are covered in detail in Section 15 of Ref. 5.

Hydrate formation can be prevented by maintaining the gas at a higher temperature, by dehydrating the gas, or by the injection of glycol or methanol. Where hydrate problems occur at pressure control valves or other equipment, localized heating of the equipment may provide an efficient solution.

Gas Transmission Systems

Route Selection. The most logical pipeline route is a straight line between the point of supply and the point of delivery. However, this tends to be the exception and not the rule. Physical terrain, soil conditions, built-up areas, population densities, and both natural and artificial obstacles together with the requirements of codes and standards will force deviations from a straight line route. Normally, the best route from a number of alternates will be that which results in the lowest overall project cost for the life of the project, considering material and equipment, construction, operation, and maintenance costs.

In evaluating possible pipeline routes, consideration must be given to the following:

1. *Location of facilities:* Although the locations of supply and delivery points are usually fixed, the siting of these facilities should take into account the availability of suitable pipeline routes whenever possible. On long transmission lines, the availability of suitable sites for intermediate compressor stations will also have an impact on the selection of the final route. And finally, the location of possible future supply and delivery connections to the main line may play a role in the selection of the main line route.

2. *Terrain:* The terrain which a pipeline traverses has a significant impact on construction costs and in some cases maintenance costs. For instance, the amount of rock along the pipeline route has a major impact on the cost of installation but little or none on operating or maintenance costs. However, areas of soil erosion, landslide areas, swamps, and river crossings increase both the cost of installation and the cost of maintenance. Such area usually require special construction methods and in many cases require stabilization measures after installation, which become a continuing maintenance item.

3. *Artificial infrastructure:* The pipeline route must normally also cross roads, highways, railroads, canals, and irrigation structures. Such crossings increase the installation costs since additional depth of burial and, in some instances, the installation of casing is required. The costs of crossing such obstacles must be weighed against the costs of alternate routes, if available.

4. *Populated areas:* The pipeline route should avoid populated areas to reduce the exposure of the population to hazards associated with the pipeline. This also works in reverse. By avoiding populated areas, the possibility of damage to or interference with the pipeline is reduced.

There is also an economic penalty in crossing populated areas. Most codes will require the use of a lower pipe design factor resulting in the installation of heavier wall pipe. More expensive construction methods may also be required and access to the pipeline for operation and maintenance will be reduced.

The area defined by Section 840 of ASME B31.8 in which the design of the pipeline is affected by the density of the population extends one-eighth of a mile on either side of the pipeline. Since future increases in population density can force the upgrading of an existing pipeline, the rate of development and probable direction of future population growth must be considered in selecting the pipeline route.

Diameter Selection and Station Spacing. Investment in pipe invariably represents the largest single expenditure for a transmission line. It is therefore imperative to find the size of pipe that will most cost effectively handle the required gas volumes. To do so, horsepower and compressor station spacing must be considered. While pipe tonnage represents a high first cost in investment compared to the installation of horsepower, the operating charges for pipe are relatively negligible and fuel and maintenance costs of compressor facilities are high. The savings realized by decreasing pipe tonnage can eventually be offset by the costs associated with the increased horsepower required.

An estimate of the pipe size required can be made from charts similar to Fig. C6.4 which plot gas transmission costs against throughput volume for varying pipe diameters. With the approximate pipe size(s) determined, a preliminary economic analysis can be performed using computer programs developed for economic studies of gas pipeline systems. These programs allow for the rapid comparison of a number of alternatives and the selection of the one apparently most economic.

With an optimum pipe size determined a more precise design of the selected system can be undertaken using the flow equations presented in the section "Basic Flow Equations."

Main Line Valves. Main line valves are installed in gas transmission lines for both safety and economic reasons. The basic concern is the loss of gas and associated hazard in the event of a break in the pipe. Factors which must be considered in determining the spacing of main line valves include the availability of continuous access to the valves, the conservation of gas, the time required to

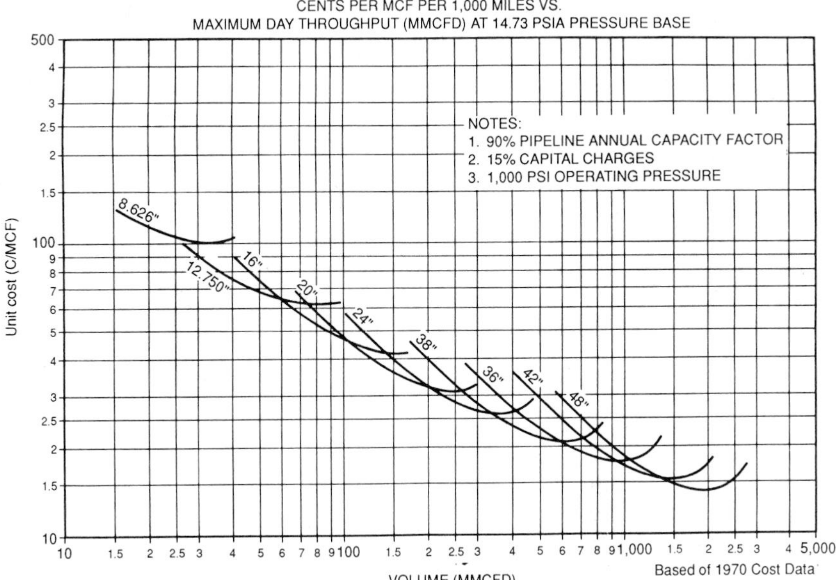

FIGURE C6.4 Effect of pipeline diameter on gas transmission costs.

blow down any given section of pipe in case of emergency or maintenance, continuity of gas service, operating flexibility, future development within the pipe section, and any significant natural conditions which could adversely affect the operation and security of the pipeline. The maximum spacing between valves is specified in Section 846 of ASME B31.8 and varies with the population density along the pipeline. Spacing on new transmission lines may not exceed the following:[2]

Location	Spacing (miles)
Class 1	20
Class 2	15
Class 3	10
Class 4	5

The main line valves should be full opening, through conduit type to allow for the passage of scrapers and inspection pigs. Both ball and gate valves are suitable, preferably weld end (avoids flange leaks) with shop welded transition pieces. Valves for underground service should have stem extensions to elevate the valve operators above grade and should also have lubricant and bleed lines extended for ease of access.

Blowdown Assemblies. Blowdown assemblies allow for the evacuation of gas from sections of pipeline under emergency conditions or for scheduled maintenance operations. A typical main line valve and blowdown assembly is shown in

Fig. C6.5. The primary consideration in sizing the piping for the assembly is the time required to blow down the section between two main line valves.

The following formula, found in the American Gas Association Manual, provides a means of determining venting time and blowdown valve size:

$$T_m = 0.0588 \frac{P_1^{1/3} G^{1/2} D^2 L F_c}{d^2}$$

where T_m = blowdown time, min
P_1 = initial line pressure, psia
G = specific gravity (air = 1.0)
D = inside diameter of pipeline, in
L = length of pipeline section, miles
d = inside diameter of blowdown, in

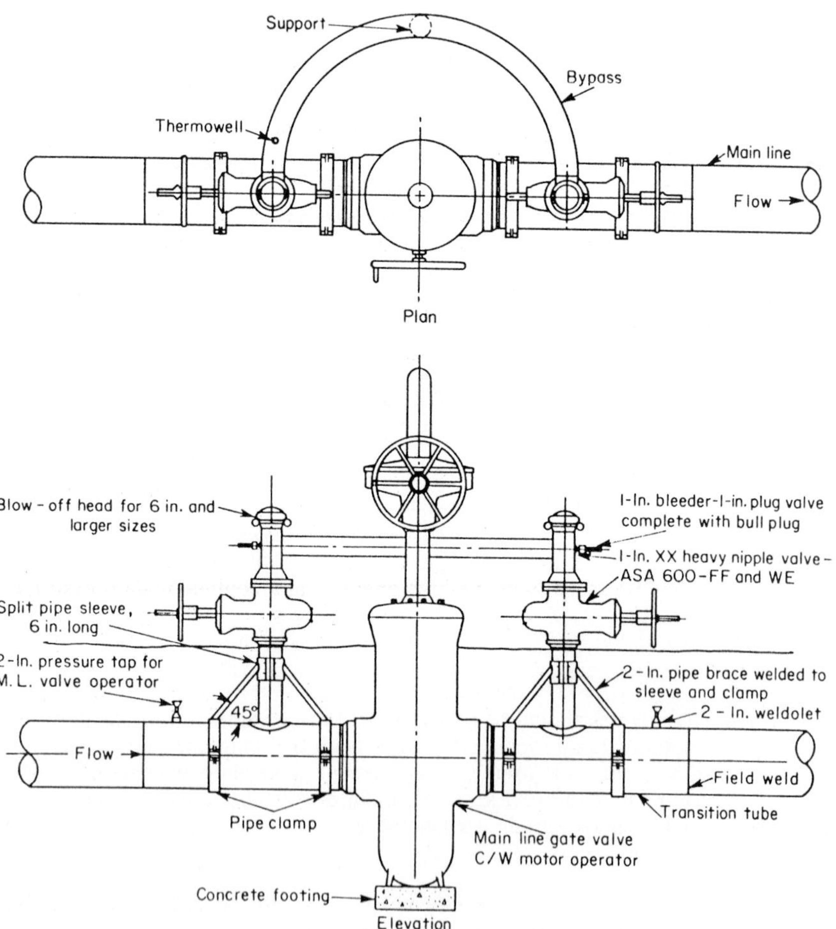

FIGURE C6.5 Main line gate valve equipped with bypass and blowoff arrangement.

F_c = choke factor:

Ideal nozzle	= 1.0
Through gate	= 1.6
Regular gate	= 1.8
Regular lube plug	= 2.0
Venturi lube plug	= 3.2

Supports for the blowdown assembly must be designed to not only carry the weight of valves and piping but also the thrust loads which will occur during venting.

Gas Distribution Systems

Although the service pipe that connects the street main to the customer's meter is usually installed by the gas supplying company, the actual ownership (and hence, responsibility) of the pipe varies. In some areas, the gas company owns the pipe all the way to the meter set assembly, but in others the customer owns the pipe from the property line.

Sizing of services may be facilitated by reference to published tables and charts.[6]

Many different bases are available and used in selecting the proper size. Some use a maximum pressure drop; some use a maximum size for all services; others make a detailed pressure-drop calculation for each installation. A brief description of the methods followed by a majority of the companies will be of interest. Consideration of connected load, length of service, and main pressure are made in selecting service-pipe size. Generally this combination of variables is handled by means of selecting an allowable pressure drop for a particular main pressure and then selecting the pipe size based upon length of service and anticipated load, which will result in a calculated pressure drop of less than the allowable. Since it is usually the easiest to determine the connected load on a gas service, this flow is used to size the service pipe. Some companies refine the principle by introducing a diversity factor based upon the assumption that all the appliances will not be in use at the same time. Others assume that the service will ultimately supply a piece of gas space heating equipment and size the service for this load even if no house heating load is installed at present.

The pressure drop allowed on a gas service is primarily a function of the pressure being carried on the mains during the time the mains are supplying the peak demand. In a low-pressure system which operates at a nominal main pressure of from 6 to 8 in of water column, most of the companies consider the maximum allowable pressure drop on a gas service to be either 0.3 or 0.5 in of water column (see Figs. C6.6 and C6.7). An intermediate-pressure system operating at a main pressure of from 1 to 15 psig will generally permit an allowable pressure drop of about 0.5 psig. High-pressure distribution systems permit allowable pressure drops of 0.5 to 3 psig. Another method of selecting the allowable pressure drop is to specify that it be some percentage of the main pressure during the maximum hour. One company selects a 20 percent drop, and another a 10 percent drop. Still another method is to specify a value of 10 for the difference in squares of the absolute inlet and outlet pressures.

Flow Characteristics of Low-Pressure Services.[7] Experiments in 1960 showed that the widely used Spitzglass equation for low-pressure gas flow required new resistance values. This flow is a function of inside diameter rather than surface smooth-

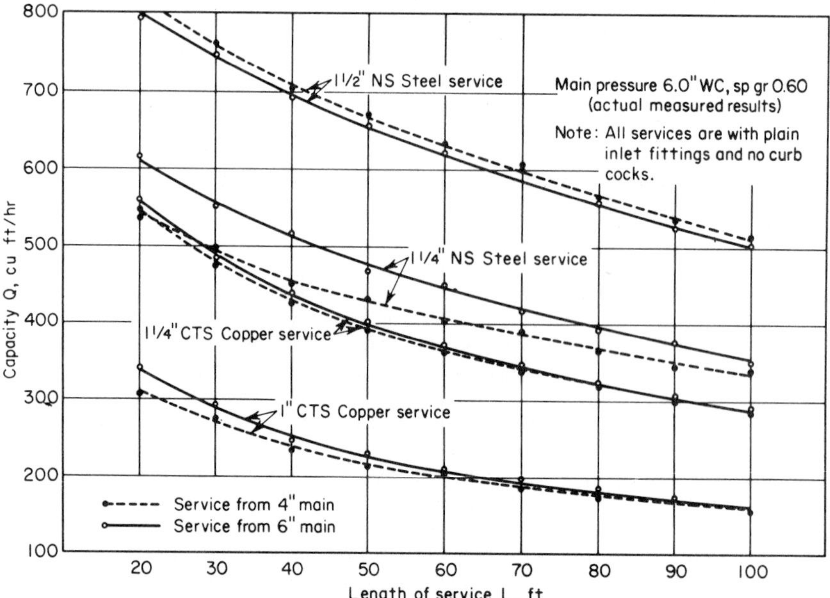

FIGURE C6.6 Total gas flow through LP services, 0.5-in water column total drop.

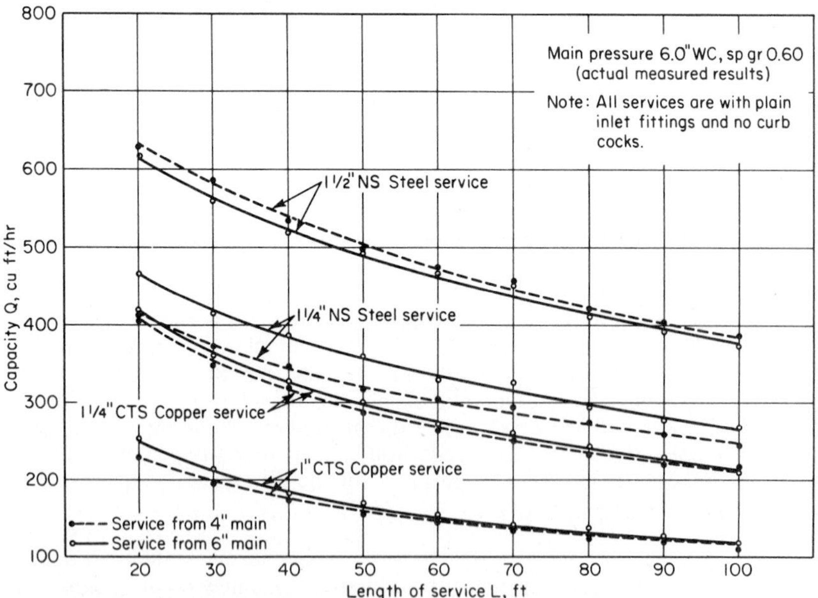

FIGURE C6.7 Total gas flow through LP services, 0.3-in water column total drop.

ness. Thus, tubing of smaller diameter does not have a capacity equal to that of a larger size steel pipe through which it may be drawn for replacement purposes.

The flow equation for either copper or steel services from ¾-in CTS copper to 1½-in NS steel was found to follow the form:

$$\text{ft}^3/\text{h} = \left[\frac{\text{total pressure drop in service, in } H_2O}{(K_p)(S/S')(L + L_{ef})}\right]^{0.54}$$

where K_p = pipe constant given in Table C6.6
S = sp gr of gas
S' = sp gr 0.60
L = length of service, ft
L_{ef} = equivalent length of fittings given in Table C6.7

TABLE C6.6 Values of K_p

¾-in CTS copper	1.622×10^{-6}
1-in ID plastic	0.279×10^{-6}
1-in CTS copper	0.383×10^{-6}
1¼-in CTS copper	0.124×10^{-6}
1¼-in NS steel	0.080×10^{-6}
1½-in NS steel	0.037×10^{-6}

TABLE C6.7 Equivalent Length of Fittings (ft)

1-in or 1¼-in curb cock for copper service	3.5
1¼-in curb cock for 1¼-in steel service	13.5
1½-in curb cock for 1½-in steel service	12.0
1½-in street elbow for 1¼-in steel service	7.5
1½-in street elbow for 1½-in steel service	7.5
1¼-in street tee for 1¼-in steel service	10.5
1½-in street tee on sleeve or 1¼-in hole in main	15.0
1¼ × 1 × 1¼-in street tee	23.0
1½ × 1¼ × 1½-in street tee	19.0
Combined outlet fittings:	
¾-in copper	2.0
1-in copper or plastic	6.0
1¼-in steel	8.0
1½-in steel	22.0

PIPE AND FITTINGS

Material

Most pipe for transmission pipelines worldwide is made to American Petroleum Institute (API) Specification 5L, Specification for Line Pipe, and virtually all the major pipe mills in the world qualify for the API stamp. The API specification covers pipe made from mild steel, Grades A and B (specified minimum yield strengths up to 35000 psi); pipe made from high strength steels (i.e., 5LX42, X46, X52 etc., the number following the X being the specified minimum yield strength expressed in thousands); and pipe made by welding coiled skelp into a helix (spiral) and welding the abutted edges.

Line pipe comes in a variety of sizes from 3.5 to 60 in in outside diameter. Special production runs of pipe up to 100 in in diameter have been made. Wall thicknesses are usually specified in $\frac{1}{32}$-in intervals, from $\frac{3}{16}$ in in small pipe sizes to $1\frac{1}{4}$ in in the larger diameters. American engineers use the decimal equivalent of $\frac{1}{32}$ in to express wall thickness (0.219 for $\frac{7}{32}$ in, 0.250 for $\frac{1}{4}$ in, etc.). Engineers in metric countries express the same wall thicknesses in millimeters (0.219 in = 5.56 mm, 0.250 in = 6.35 mm, etc.). Pipe sizes in the metric system tend to be expressed in multiples of 25 mm nominal (14 in = 350 mm, 24 in = 600 mm, etc.).

Pipe flanges and flanged fittings are manufactured according to ANSI B16.5, Steel Pipe Flanges and Flanged Fittings, which establishes pressure ratings for eight classes of flanges and fittings. ANSI B16.5 details the standard physical dimensions, number and size of bolt holes, etc., for each size fitting. The exact pressure rating of each "class" of fitting varies with type of steel and the design temperature.

Other Materials

Other materials which may be used in gas service include cast iron, ductile iron, plastic, and copper. These are generally limited to use in mains and service lines in distributions systems. A brief summary of the limitations of and restrictions to their use is given below.

Cast Iron. The use of cast-iron pipe is declining rapidly in favor of plastic pipe. When used, cast-iron pipe is limited to sizes 6 in in diameter or larger. Due to its brittleness, it may not extend through building walls and must not be installed in unstable soils or under buildings.

Ductile Iron. Ductile iron pipe must be manufactured in accordance with ANSI A21.52 Ductile Iron Pipe, Centrifugally Cast, in Metal Molds or Sand Lined Molds for Gas.

Plastic. Plastic pipe and components must be manufactured in accordance with the following American Society for Testing and Materials (ASTM) standards:

ASTM D 2513—Thermoplastic Gas Pressure Pipe, Tubing, and Fittings
ASTM D 2517—Reinforced Epoxy Resin Gas Pressure Pipe and Fittings

Copper. Copper tubing or pipe for use in gas mains is limited to pressures of 100 psi or less, must have a minimum wall thickness of 0.065 in, and must be hard drawn. Where the gas being transported contains more than an average of 0.3 grain of hydrogen sulfide per 100 standard cubic feet of gas, copper may not be used.

COMPRESSOR STATIONS

Types and Function

Whenever a gas has insufficient energy for transport, a compressor station must be used. The following types of compressor stations are presently in general use:

Field or gathering stations: These stations gather gas from wells in which pressure is not sufficient to produce a desired flow rate into transmission or distribution systems. Such stations may handle suction pressures from below atmospheric pressure up to 750 psig and volumes from a few thousand to many million cubic feet per day.

Repressurizing or recycling stations: This type of station is an integral part of a processing or secondary recovery facility, not involving transportation of natural gas to a consumer. Discharge pressures can exceed 6000 psig.

Storage field stations: These stations compress trunk line gas for injection into storage wells. Discharge pressures may range up to 4000 psig with compression ratios as high as 1 to 4. Some storage stations are designed to also permit the withdrawal of gas from storage and injecting it into high-pressure pipe lines. These field stations require precise design engineering because of the wide range of pressure-volume operating conditions encountered.

Distribution plant stations: Distribution plant stations compress gas from a holder supply to medium- or high-pressure distribution lines at about 20 to 100 psig or compress into bottle storage at pressures up to 2500 psig.

Pipeline booster stations: Stations of this type are used in gas transmission line service. The volume through these stations is usually quite large with compression ratios below 2. The pressure range is generally between 200 and 1000 psig, sometimes as high as 1200 psig.

Compressor Station Layout

A typical compressor station arrangement can be broken down into three systems:

- Main gas system
- Unit gas system
- Auxiliary gas system

Main Gas System. The main gas system includes main gas piping and equipment between the transmission line and the compressor unit suction and discharge leads. Equipment encompasses:

Station block valves
Station bypass valve
Station purge valve
Gas scrubbers
Orifice fitting
Station surge valve
Station relief valves
Station blowdown valve

Unit Gas System. The unit gas system includes the suction and discharge leads from the main gas header to the compressor and back to the main gas header. Equipment consists of:

Unit block valves
Unit bypass valve

Unit purge valve
Unit vent valve

Auxiliary Gas System. The following are auxiliary gas systems:

Fuel gas system
Starting gas system
Utility gas system

Gas enters the station through the station block valve, passes through scrubber, orifice meter, and unit suction block valve to compressors where it is compressed and discharged through a unit discharge block valve to the station discharge piping and block valve back to transmission line.

Piping

Compressor station piping design is governed by the minimum federal safety standard "DOT Title 49 Part 192 Transportation of Natural and Other Gas by Pipeline." For high-pressure transmission service the law specifies certain types of pipe materials.

According to ASME B31.8, Compressor Station Piping must be Class 3 construction, [i.e., a design factor of 0.5 (Table C6.2) has to be used in the Steel Pipe Design Formula]. After possible pipe sizes are determined and preliminary station layouts are prepared, pressure drop studies and thermal piping stress studies will follow.

The next step is a thermal piping stress analysis to determine that stresses in station piping stay within allowable limits and forces and moments on compressor flanges do not exceed those set by the compressor manufacturer. Pipe stresses are caused by temperature increase when gas is being compressed. Temperature rise across a compressor can range from 0.2 to 3°F per 1 psi increase in pressure, depending on the specific gas and pressure involved.

In order to keep noise levels in station piping within acceptable limits, the following maximum allowable velocities in plant piping should be considered.

Station main gas piping	2500 ft/min
Unit main gas piping	1500 ft/min
Fuel gas piping	1500 ft/min

Another consideration in designing compressor station piping is the installation of piping either above or below ground. Aboveground piping is easier to monitor for gas leaks and to maintain. However, yard access may be impeded. For buried piping, the reverse would apply.

A further point to look into in piping design is the location of drains in piping required to properly drain the water out of station piping after hydrostatic testing.

Components and Equipment

The following are additional components and equipment required:

1. Sheltered or outdoor installations: *a.* buildings and, *b.* packaged components—smaller units.

2. Station valves—operators: Ball, plug and gate, gas or electric operated.
3. Scrubbers
   a. Horizontal Inline-Type Scrubber. Because of the effect of the helicoidal tuyere, the gas after entering the vessel is subjected to an extended centrifugal motion, the heavier particles being thrown to the periphery of the vessel. Here the particles are forced into the annular space, at which point all solid particles are trapped and ejected through the drain. The inlet stream, freed from entrainment, continues through the outlet of the separator. The secondary vortex breaker prevents reentrainment, thereby extending flow range. The scrubber has the following characteristics:

Efficiency	99.5 percent of all solids or liquids
Pressure drop	Low
Cost	Medium
Simple installation	Wide flow range

   b. Vertical Tube-Type Scrubber. Dust-laden gas enters the tube tangentially, creating a high centrifugal force that projects solids and liquid droplets to the walls of the tube. Clean gas reverses flow at the vortex and passes through the concentric outlet tube to the outlet plenum. Impurities continue downward to a storage area.
4. Flowmeters
5. Purge, relief, surge, and blowdown systems

OTHER FACILITIES

Pressure Relief

The following discussion on pressure relief valves is quoted from Ref. 8. A *safety-relief* valve is an essential and important piece of equipment on virtually any pressured system. Required by the ASME-UPV Code, among others, it must be carefully sized to pass the maximum flow produced by emergency conditions.

Sizing. The sizing formulas for vapors and gases fall into two general categories, based on the flowing pressure with respect to the general categories and based on the flowing pressure with respect to the discharge pressure. When the ratio of P_1 (set pressure plus allowable accumulation) to P_2 (outlet pressure) is greater than 2, the flow through the relief valve is sonic; the flow reaches the speed of sound for the particular flowing medium. Once the flow becomes sonic, the velocity remains constant; it cannot go supersonic. No decrease of P_2 will increase the flow rate.

Sonic Flow. In accordance with API RP 520, Part 1, Section 3.3, the formulas used for calculating orifice areas for sonic flow are:

$$A = W \frac{\sqrt{TZ}}{CKP_1 \sqrt{M}}$$

or

$$A = Q \frac{\sqrt{Z}}{18.3KP_1(C/356)\sqrt{520/T}\sqrt{29/M}}$$

where A = calculated orifice area, in^2
W = flow capacity, lb/h
Q = flow capacity, standard ft^3/min
M = molecular weight of flowing media
T = inlet temperature, absolute (°F + 460)
Z = compressibility factor
C = gas constant based on ratio of specific heats (Table C6.8)
K = valve coefficient of discharge
P_1 = inlet pressure, psia (set pressure + accumulation + atmospheric pressure)

Subsonic Flow. The second general category for vapor or gas sizing is generally when P_2 is greater than half of P_1 (backpressure greater than half of inlet pressure). Using k (ratio of specific heats) and P_1/P_2 (absolute), confirm from Fig. C6.8 that the subsonic (low pressure) flow formula is required. If so, determine the F factor. If it is not, use above sonic flow formula:

$$A = Q \frac{\sqrt{GI}}{863\text{KF}\sqrt{(P_1 - P_2)P_2}}$$

where G = specific gravity
F = factor obtained from Fig. C6.8
P_2 = outlet pressure, psia (backpressure + atmospheric pressure)

After determining the calculated orifice area, select the next largest standard orifice size from the relief valve manufacturer's catalog.

Selection. The fundamental selection involves the consideration of the two basic types of relief valves more commonly used; conventional spring-loaded relief valves have:

- Competitive price at lower pressures and in smaller sizes
- Wide range of chemical compatibility
- Wide range of temperature compatibility, particularly at higher temperatures

The disadvantages are:

- Metal-to-metal seat not tight near set pressure and usually after valve relieves.
- Sensitive to conditions that can cause chatter and/or rapid cycling.
- Protection against effects of backpressure is expensive, pressure limited, and can create additional maintenance problems.
- Testing of set pressure not easily accomplished.

Advantages of pilot-operated-type relief valves are:

- Seat tight to set pressure (Fig. C6.9)
- Ease of setting and changing set and blowdown pressures
- Can achieve short blowdown without chatter
- Pop or modulating action available
- Easy maintenance

TABLE C6.8 Gas Constant Based on Ratio of Heats

Gas	Mol. wt.	C_p/C_v	C
Acetylene	26	1.26	343
Air	29	1.40	356
Ammonia	17	1.31	348
Argon	40	1.67	378
Benzene	78	1.12	329
Butadiene	54	1.12	324
Carbon dioxide	44	1.28	345
Carbon monoxide	28	1.40	356
Ethane	30	1.19	336
Ethylene	28	1.24	341
Freon 22	86.47	1.18	335
Helium	4	1.66	377
Hexane	86	1.06	322
Hydrogen	2	1.41	357
Hydrogen sulfide	34	1.32	349
Methane	16	1.31	348
Methyl mercapton	48.11	1.20	337
N-Butane	58	1.09	326
Natural gas (0.60)	17.4	1.27	344
Nitrogen	28	1.40	356
Oxygen	32	1.40	356
Penetane	72	1.07	323
Propane	44	1.13	330
Propylene	42	1.15	332
Sulfur dioxide	64	1.29	346
VCM	62.5	1.18	335

- Easy verification of set pressure without removing relief valve from service
- Flexible; options for remote operation, backpressure protection, valve position indication

The disadvantages are:

- Maximum temperature limitations
- Should not be used in extremely dirty or polymerizing-type service

Inlet Piping. The proper design of inlet piping is extremely important. Very often, safety relief valves are added to an installation at the most physically convenient location, with little regard to flow considerations. Pressure loss during flow in a pipe always occurs. Depending upon the size, geometry, and inside surface condition of the pipe, the pressure loss may be large (20, 30, or 40 percent) or small (less than 5 percent).

API RP 520, Part 2, recommends a maximum inlet pipe pressure loss to a safety relief valve of 3 percent. This pressure loss shall be the sum total of the inlet loss, line loss, and when a block valve is used, the loss through it. The loss should be calculated using the maximum rated flow through the pressure relief valve.

The 3 percent maximum inlet loss is a commendable recommendation but often very difficult to achieve. If it cannot be achieved, the effects of excessive

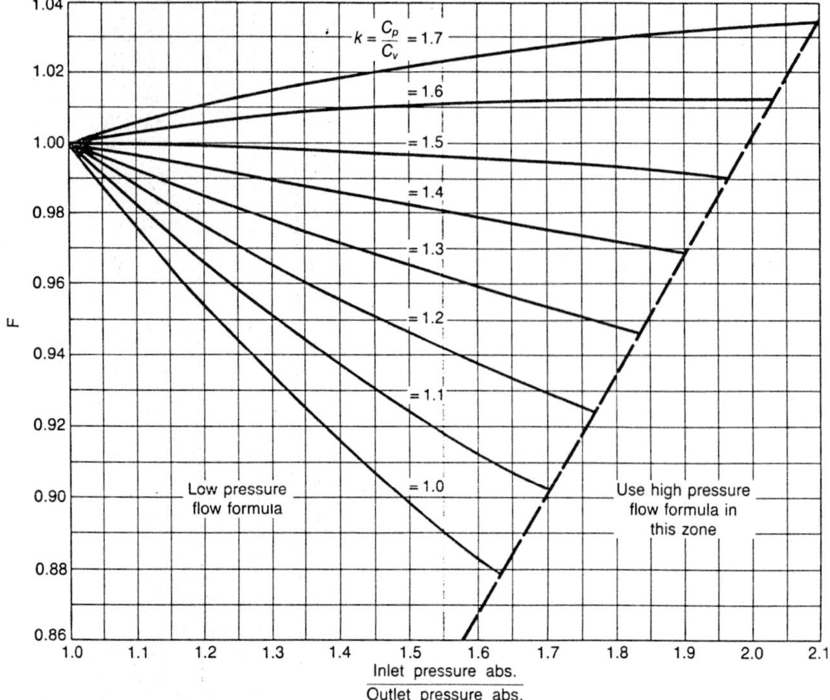

FIGURE C6.8 Low-pressure and high-pressure flow formulas.

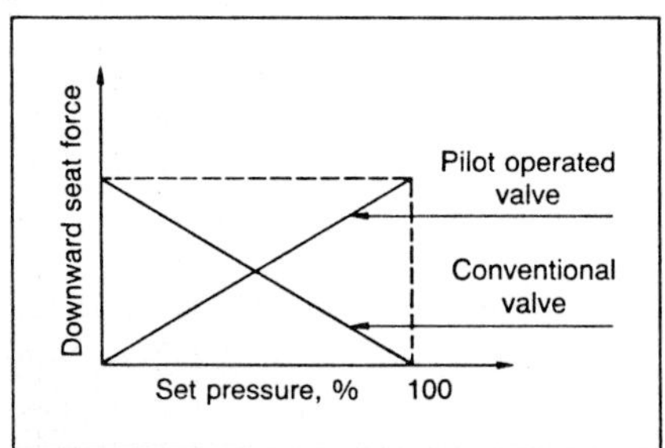

FIGURE C6.9 Setting seat pressure to set flow pressure.

inlet pressure should be known. These effects are rapid or short cycling with direct spring-operated valves or resonant chatter with pilot-operated relief valves. In addition, on pilot-operated relief valves, rapid or short cycling may occur when the pilot pressure-sensing line is connected to the main valve inlet. Each of these conditions results in a loss of capacity.

Pilot-operated valves can tolerate higher inlet losses when the pilot senses the system pressure at a point not affected by inlet pipe pressure drop. However, even though the valve operates satisfactorily, reduced capacity will still occur because of inlet pipe pressure losses. The sizing procedure should consider the reduced flowing inlet pressure when required orifice area A is calculated.

A conservative guideline to follow is to keep the equivalent L/D ratio (length/diameter) of the inlet piping to the relief valve inlet to 5 or less.

Discharge Piping. Discharge piping for direct spring-operated valves is more critical than for pilot-operated valves. As with inlet piping, pressure losses occur in discharge headers with large equivalent L/D ratios. Excessive backpressure will reduce the lift of a direct spring-operated valve and enough backpressure (15 to 25 percent of set plus overpressure) will cause the valve to reclose. As soon as the valve closes, the backpressure in the discharge header decreases, and the valve opens again with the result that rapid cycling can occur.

Pilot-operated relief valves with the pilot vented to the atmosphere or with a pilot balanced for backpressure are not affected by backpressure. However, if the discharge pressure can ever exceed the inlet pressure, a backflow preventer must be used. The valve-relieving capacity for either direct or pilot-operated relief valves can be affected by backpressure if the flowing pressure with respect to the discharge pressure is below critical (subsonic flow).

Balanced bellows valves (direct spring-operated) have limitations on maximum permissible backpressure due to the collapse pressure rating of the bellows element. This limitation will in some cases be less than the backpressure limit of a conventional valve. Manufacturer's literature should be consulted in every case. If the bellows valve is used for systems with superimposed backpressure, the additional built-up backpressure under relieving conditions must be added to arrive at maximum backpressure.

The performance of pressure relief valves, both as to operation and flow capacity, can be achieved through the following good discharge piping practices:

- Discharge piping must be at least the same size as the valve outlet connection and may be increased when necessary to larger sizes.
- Flow direction changes should be minimized, but when necessary use long radius elbows and gradual transitions.
- If the valve has a drain port on the outlet side, it should be vented to a safe area. Avoid low spots in the discharge piping; preferably pitch piping away from valve outlet to avoid liquid trap at valve outlet.
- Proper pipe supports to overcome thermal effects, static loads due to pipe weight, and stresses that may be imposed due to reactive thrusts forces must be considered.

Reactive Force. On large-orifice, high-pressure valves, the reactive forces during valve relief are substantial and external bracing may be required (Fig. C6.10). API RP 520, Part 2 gives the following formula for calculating this force:

$$F = \frac{Q_h}{366}\sqrt{\frac{kT}{(k+1)M}} \tag{C6.1}$$

where F = reactive force at valve outlet centerline, lb

Q_h = flow capacity, lb/h

k = ratio of specific heats (C_p/C_v)

T = inlet temperature, absolute (°F + 460)

M = molecular weight of flowing media

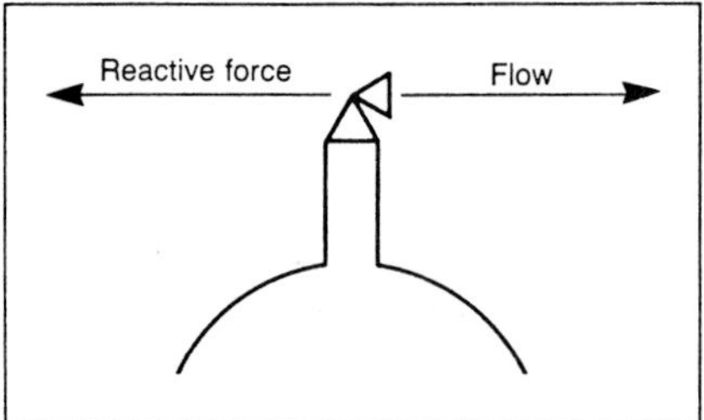

FIGURE C6.10 Reactive forces during valve relief.

If bracing is not feasible, a dual outlet valve (available in some pilot-operated safety relief valves) can be used. The reactive forces are equal but opposite, resulting in zero force on the valve or outlet but if redirected, can still impose loads that must be reacted in some manner.

An approximate method for calculating the reactive force is to multiply the following pressure (psig) by the orifice area (in^2); the reactive force obtained will be on the high side.

Example	A-G Type 22312R68 "R" Orifice (16.00 in^2)
Set pressure	1000 psi, 10 percent overpressure
Flowing media	Natural gas, 60°F $M = 17.4$ ($G = 0.060$) $C = 344$ ($k = 1.27$)
Reactive force	= 10,143 lb, using Eq. (C6.1)
Reactive force	= 17,600 lb, using approximate method

Testing. Using DOT Title 49, Part 192.739 as a guideline, each valve should be inspected at least once a year to determine that it:

- Is in good mechanical condition
- Will operate at the correct set pressure
- Has adequate capacity and is operationally reliable
- Is properly installed and protected from dirt, liquids, or other conditions that might prevent proper operation.

The effectiveness of a safety relief valve installation depends greatly on proper sizing and selection of the valve type, suitable installation conditions, and proper and timely testing of the valve.

Pressure Regulation

Good regulation of high-pressure gas is important for two reasons:

- To protect low-pressure equipment from becoming overpressured and becoming a hazard.
- If located near an orifice meter station, a regulator station must operate smoothly or it will result in a chart record which is impossible to interpret accurately.

Types of Regulator. The type of load to be controlled usually dictates the type of regulator to be selected:

- *Globe bodied:* The globe-bodied control valve has been the standard for the gas industry for many years and performs well in a wide range of applications. When properly sized and mated with a good controller, it is very well suited for difficult applications such as pressure control on a plant where load changes rapidly. The globe-bodied regulator is inherently noisy with the exception of some relatively new designs. It becomes very large and difficult to maintain when sized to control large loads with low-pressure drops.
- *Expansible tube:* The expansible tube regulator is extremely simple in its construction and principle of operation. On many applications it does not require a controller but operates with a pilot regulator. The expansible tube is by nature quieter than many regulators and it has good low-flow characteristics and tight shut-off. It does have a minimum differential pressure requirement of 0.68 to 3.45 bars (10 to 50 psi), depending on size, which precludes its use in some applications.
- *Ball valve:* The ball valve is best suited to large loads where small pressure drops are required. This is often the case in flow control applications or where pipeline pressure varies from time to time. The ball valve can perform a double duty when used as a monitor regulator and as an isolating valve on one side of a primary regulator. Ball valves are inherently noisy and can give erratic control if not maintained properly. Because of their large capacity, it is often important to consider pressure losses through adjacent piping when sizing this type of regulator.

Influence on Station Design. Whichever type regulator is selected, the way it is installed is important for successful operation. For most applications, it is best to have at least two parallel units. If the load varies widely, "split range" control is used, which requires two regulators, one to control low flows and a large-capacity unit to come in when required. For critical applications, it is desirable to have a standby regulator which would not normally be used but would open in an emergency.

When it is necessary to make a very large pressure reduction, it is best handled in two stages with each regulator making about half the cut. Two-stage regulation can enhance measurement if one stage is placed on each side of the measuring station.

The upstream regulators should be used to control pressure in order to maintain a constant pressure at the measuring station. Downstream regulators can be used to control flow or to control customer pressure. As much volume as possible should be provided between the two regulator gauges; otherwise stable control may be difficult to achieve.

Overpressure Protection. The most important part of designing a regulator station is being as sure as possible that the system downstream cannot be overpressured. The first step in protection is strict compliance with all safety regulations and, in particular, with Part 192 of the Department of Transportation Safety Regulations (DOT 192). Basically DOT requires secondary protection if it is possible for the pressure upstream of a regulator to reach a level which will be dangerous to the downstream system. Secondary protection may be provided either by a monitor regulator installed in series with the primary regulator or by a relief valve.

There are systems, however, where a safety or relief valve may be the best selection. If a regulator feeding a small capacity system fails suddenly, a monitor may be too slow to prevent the system from being overpressured. A relief valve should be provided on small downstream systems to take off valve leakage if the load is cut off completely.

Gas Metering

Efficient measurement and control of high-pressure natural gas is vital. A measurement error of only 1 percent can cost thousands of dollars a day at a single large-volume station.

Measurement. Measurement requirements in the gas industry can be broken into two broad categories:

1. *Custody transfer:* Most important and demanding is custody transfer at a station which measures gas flowing from one company to another. The orifice meter dominates the field in measurement of large volumes of gas at high pressure. However, turbine and rotary meters are being used more and more where better accuracy is justified.
2. *Control:* Control, or check measurement, is used for routing gas toward a customer, control of compressor stations, etc. Repeatability is the most important feature of such measurement, and the complexity and expense of custody transfer-type metering are not usually justified.

Gas industry meters operate on simple physical principals and have a long history of reliable and accurate performance. The commonly used meters fall into two categories: displacement types such as diaphragm, rotary lobed impeller, and rotary vane shown in Figs. C6.11 and C6.12; and rate-of-flow (velocity) meters such as orifice meter and turbine meters shown in Fig. C6.13.

Tables C6.9 and C6.10 list factors which affect selection of the five meter types used in the gas industry. The tables are intended as a guide and should not be used without further study when selecting the best meter for a particular application. Capacity ranges given in Table C6.9 are shown graphically in Fig. C6.14. The ranges are representative and are intended to give preliminary information for specific problems. Since the chart reflects information from many manufacturers, individual model lines may not have the exact capacity shown.

Leak Detection

Leak detection for both gas and liquid pipelines, and especially for those fluids such as ethane and propane that are sometimes liquid and sometimes gas, has been an industry problem for many years, but public attention to the subject has not been widespread until recently.

The safety record of the pipeline industry is quite good. Measured in terms of accidents (defined by the U.S. Department of Transportation regulations in terms of dollars and fluid loss) per 1000 miles of pipeline, deaths per ton mile of cargo moved or any other scale, the pipeline record is better than other means of transportation by at least an order of magnitude.

The causes of leaks are many and varied but over 70 percent are accounted for by accidental damage from external sources, corrosion, and defective pipe. Historically, the number 1 source of leaks has been external corrosion, but modern

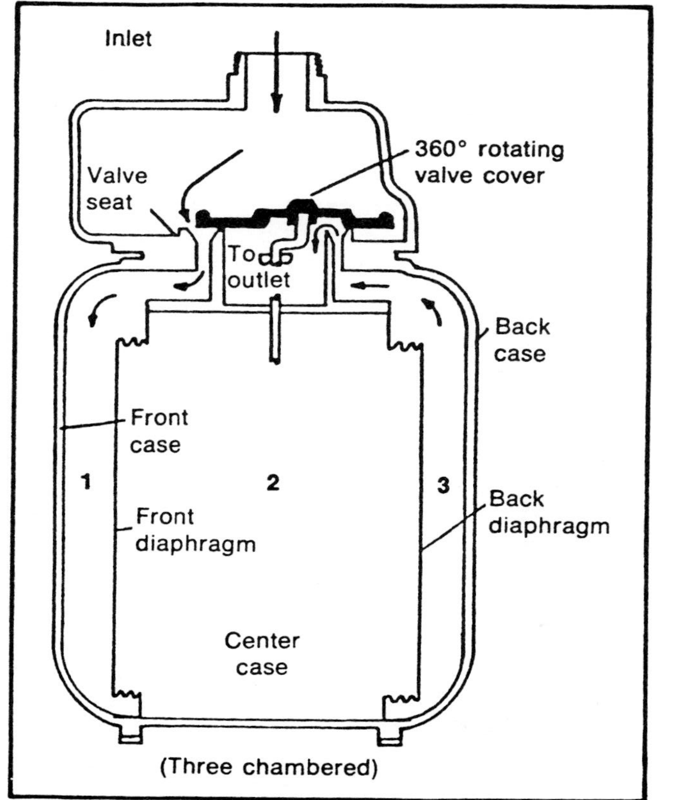

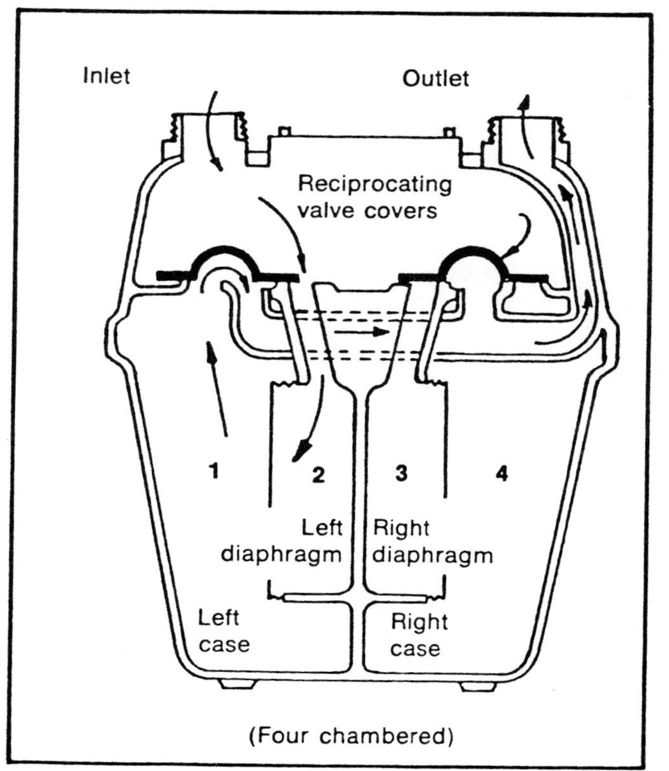

FIGURE C6.11 Diaphragm displacement meters operate by alternately filling chambers of known volume.[9]

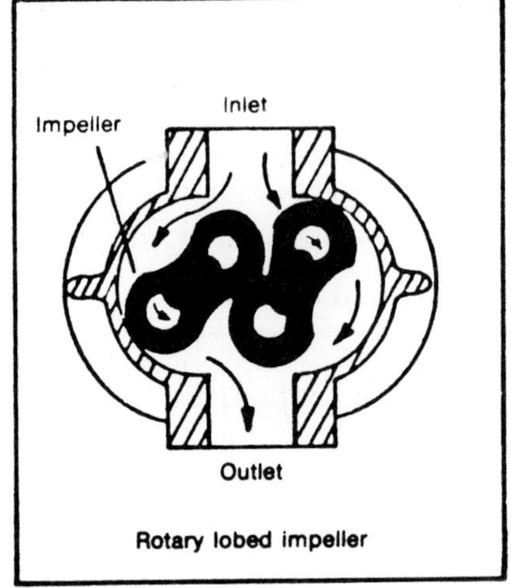

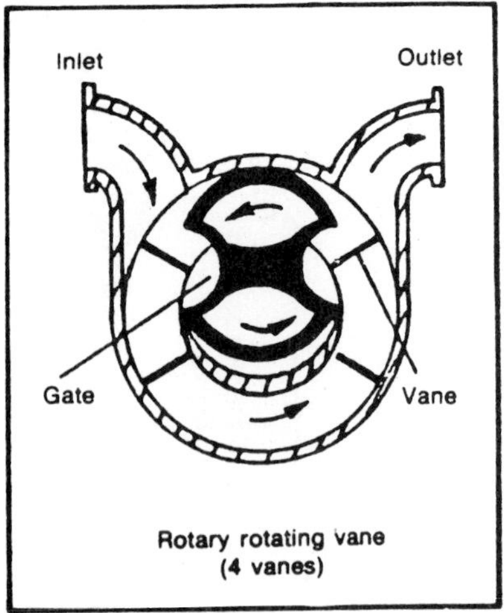

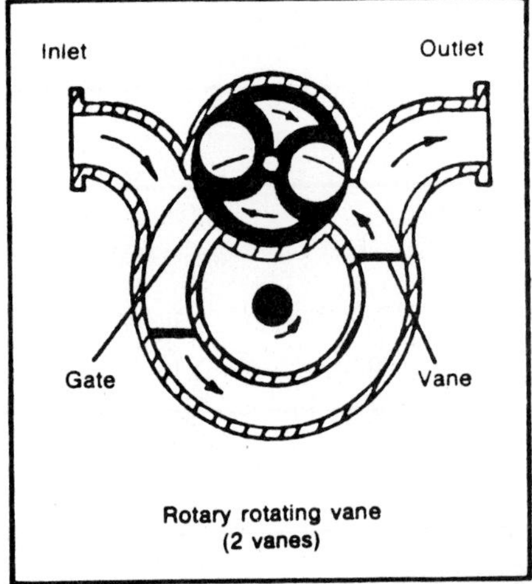

FIGURE C6.12 Rotary diaphragm meters, like diaphragm meters, measure gas flow by alternate filling and emptying of fixed volume chambers. A counter or dial registers the total gas amount.[9]

pipe coatings and the almost universal application of cathodic protection system in the last 30 years have all but eliminated corrosion leaks on the newer pipelines.

Modern mill practices and more conscientious hydrostatic testing have almost eliminated defective pipe as a cause of leaks on new lines. Of ever increasing importance is the problem of damage to pipelines by construction machinery. The problem is made worse by the fact that the very machine that does the damage comes equipped with a spark or hot exhaust to ignite the fuel. It is recommended that such protective measures as extra depth burial, concrete slabs, and/or warning tapes buried above the pipe and extra line markers be considered at major road crossings and in city growth areas.

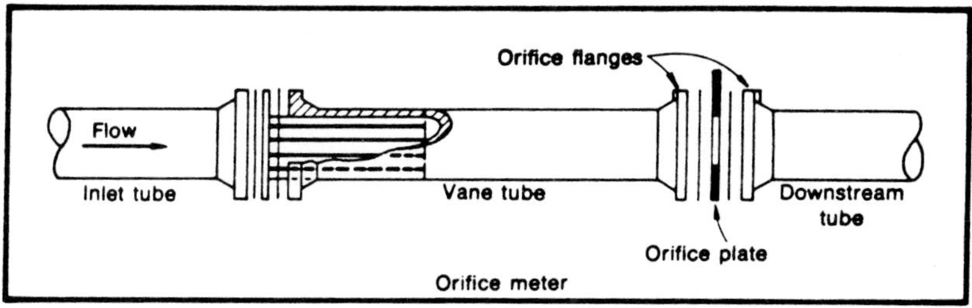

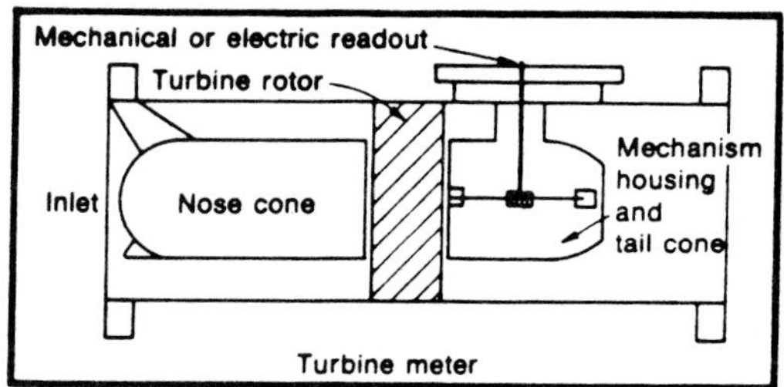

FIGURE C6.13 Velocity flow meters pass a steady gas stream. Flow is detected as a differential pressure in the orifice meter and as rotor movement in the turbine meter.[9]

One of the problems with leak detection is the frustration of knowing that when a leak has been detected, a large part of the damage has already been done. It is well and good to say that it is needed to ensure early remedial action. However, one must know what action to take—for example to immediately slam a main line valve shut would almost invariably be the wrong thing to do on a liquid line. Any automatic action following a leak signal must be thought out very carefully. The other frustrating problem with leak detection is that it may be impossible to do it at any reasonable cost.

Rudimentary leak detection has been carried out for years by reporting hourly meter readings. Minor differences in reading time, temperature, line pack, etc., made the result somewhat erratic but a continuous shortage in delivery on a small pipeline was sufficient cause to assume a leak. Large leaks or line breaks were detected (sometimes automatically) by the combinations of suction and discharge pressure with flow at pump stations (i.e., falling discharge pressure combined with increased flow meant a leak downstream; falling suction combined with decreased flow meant a leak upstream). The difference today is in the magnitude of the problem. The ±2 percent deficiency detectable on an 8-in line transporting 25 million cubic feet per day (25 MMCFD) was a 0.5-MMCFD leak; the ±0.2 percent deficiency (maybe) detectable by a sophisticated leak detection system on a 42-in line transporting 1000 MMCFD is a 2-MMCFD leak.

Some new, patented devices which "listen" for the pressure wave which is generated by a suddenly occurring leak are state-of-the-art leak detection. They define a leak as "the loss that can be measured with existing instrumentation" (probably about 0.1 percent of steady state flow).

Real time computer modeling techniques have been developed which have already improved the sensitivity of old fashioned "meter in meter out" techniques

TABLE C6.9 Summary of Gas Meter Selection Factors[9]

	Gas properties				Meter characteristics					
Meter type	Maximum pressure (psig)*	Flowing fluid temperature limits (°F)	Suitability for corrosive gas	Influence of condensate	Base maximum capacity range (Mcfh)	Accuracy % of reading 10 50 90	Base rangeability	Type of scale	Common construction materials†	Pressure loss at base maximum capacity (psi)
Diaphragm displacement	1000	−30 to +140	No	Potential	0.2 to 12	±1	200:1	Uniform	A1/CI/Brass/ Plastic/Zinc alloy	0.02 to 0.07
Rotary rotating vane	1440	−40 to +145	Yes (with special bearings & materials)	None	3 to 38	±1	25:1	Uniform	Anod. A1/300/SS/ Steel/Bronze	0.04
Rotary lobed impeller	1440	−40 to +140	Yes (with special bearings & materials)	None	1.5 to 102	±1	20:1	Uniform	A1/CI/Steel	0.07
Gas turbine	1440	−40 to +145	Yes (with special materials)	None	4 to 150	±1	5:1 to 25:1	Uniform	A1/DU/CS	0.04
Orifice	5000	−65 to +500	Yes	Potential	22 to 1500	±1½	3:1 to 4½:1	Square root	FS/SS/CS	2.5

*Refer to manufacturer's published literature for the maximum rating of a specific meter size.
†Common construction materials

AL = aluminum
CI = cast iron
CS = cast steel
DU = ductile iron
FS = forged steel
SS = stainless steel

TABLE C6.10 Summary of Gas Meter Selection Factors[9]

	Installation factors				Economic factors				
						Life expectancy			
Meter type	Normal line size (in)	Straight pipe reqmts (no. of pipe diameter)	Ambient temperature range (°F)	Limitations	Approx. first cost*	Between repairs (years)	Total (years)	Maintenance cost†	Installation cost*
Diaphragm displacement	¼ to 4	None	−30 to +40	Horizontal	220	8 to 10	30 to 40	L	M
Rotary rotating vane	2 to 6	None	−40 to +145	Not critical	130	3 to 6	10 to 25	M	L
Rotary lobed impeller	1.5 to 10	None	−40 to +140	Horizontal & leveled	150	3 to 6	10 to 25	M	L
Gas turbine	2 to 12	4 to 10	−40 to +145	Horizontal	125	3 to 6	10 to 25	M	L
Orifice	2 to 16	3 to 40	−40 to +170	Horizontal	100	1 to 3	10 to 15	H	M

*First cost ratio is based upon a 2-in orifice meter sized to measure approximately 5 Mcfh at atmospheric pressure.

†Maintenance and installation costs.
H = high
M = medium
L = low

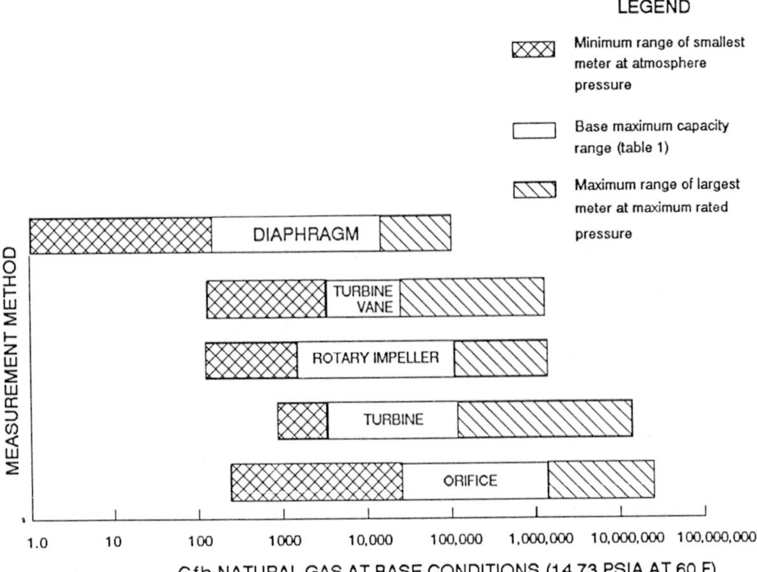

FIGURE C6.14 Gas measurement meter capacities cover a wide flow range.[9]

even under transient flow conditions. Accuracy is achieved by the use of remote terminal units (RTUs), high-precision instrumentation, and an on-line computer.

Scrapers and Scraper Handling

Purpose of Scrapers and Spheres. At various times throughout the life of a pipeline a need is found to send a scraper (pig) or sphere through the line. This may be for any one or combination of the following:

- *Initial cleaning:* On completion of a section of pipeline, at least one scraper is run to push out air as well as trash, rocks, weld slag, etc., preparatory to hydrostatic testing.
- *Sizing run:* Before a line is put in service, usually before hydrostatic testing, a scraper with a steel plate, or a "caliper" pig, is run in order to make sure the pipe was not damaged during construction and that all future pigs will pass unrestricted.
- *Dewatering scrapers and spheres:* During commissioning of the pipeline, a number of scrapers are used to displace the water left after hydrostatic testing.
- *In-service cleaning:* During normal pipeline operation pipelines accumulate water, corrosion products (rust, scale), condensate, and compressor oil, sand, and dust which must be cleaned out on a regular basis to maintain efficiency of operation.
- *Deslugging:* On some two-phase pipelines spheres are run on a scheduled basis to push condensate into the slug catchers.
- *Inspection:* Special instrumented smart pigs may be run from time to time to check corrosion, listen for leaks, or check damage after an accident.

Some of these uses require special scrapers with knives (for cutting waxy deposits) or brushes (for cleaning rust and scale) or noise-making devices, etc.

Scraper Handling Devices. Because of the variety of uses and types of scraper, a variety of scraper-handling equipment is required. The basic horizontal scraper trap has proved to be convenient for general services (see Fig. C6.15). There are two types of scraper traps, one for launching and one for receiving. The differ-

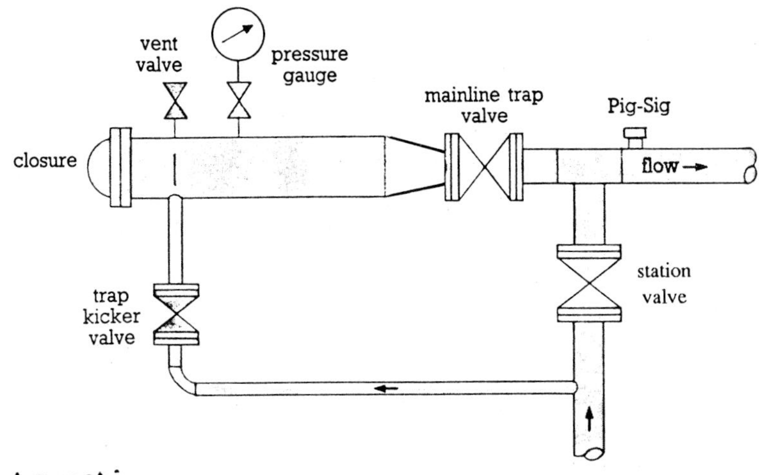

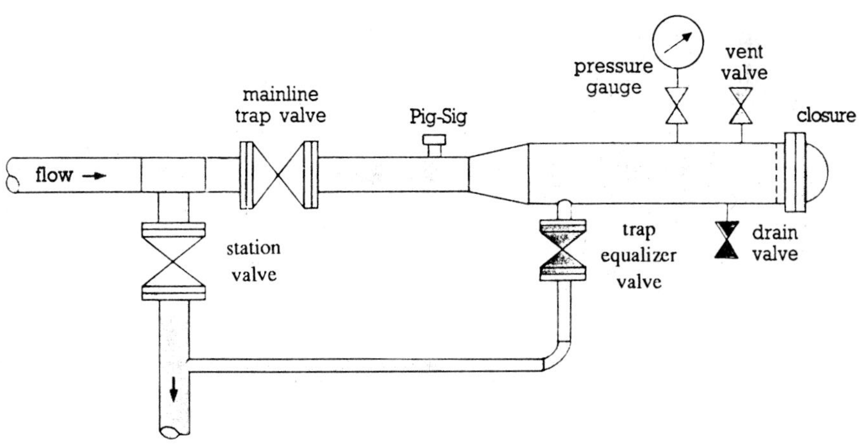

FIGURE C6.15 Launching and receiving scraper traps for gas service.[10] (*Used by permission of T. D. Williamson, Inc.*)

ence between the two is the point of entry of the bypass (kicker/equalizer) lines into the barrel and sometimes the barrel length.

The traps are equipped with a barrel one to two sizes larger than the main line, a hinged closure, a full opening (through conduit) trap valve, side entering barred tee, and a bypass line with a valve. The trap operates to launch or receive a scraper by pinching the station valve with the trap and the bypass valves both open. This controls the movement of the pig out of the launcher or into the receiver.

Scraper traps are placed at the initial and terminal points of the pipeline and at any other point where it is required to start or receive a scraper. It is not always necessary to put traps at a change in pipe size since multisized scrapers are available (within limits) nor is it necessary to put scraper traps at every compressor station.

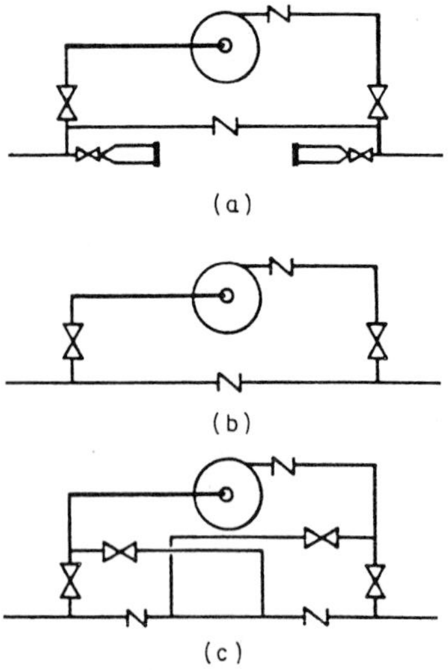

(a)

(b)

(c)

FIGURE C6.16 (*a*) Piping arrangement for station with scraper traps, (*b*) for passing scrapers with station shutdown, and (*c*) for passing scrapers without station shutdown.

Figure C6.16 is a simple piping diagram of the essential pipe and valves in a booster station with scraper traps (*a*), without traps (*b*) where a scraper is passed by shutting down, and with pipe and valving for bypass without shutdowns (*c*).

Pin Ball Machines. On some pipelines where spheres are used frequently for cleaning condensate or for separating product batches, a special adaptation of the basic scraper traps allows for loading of several spheres in the launcher and releasing them on signal or by timer. The receiver is long enough to receive as many spheres as can be launched. Both launcher and receiver are sloped so that gravity aids in the operation.

Layout. Scrapers and spheres are heavy, especially in the larger sizes and may require davits or cranes to maneuver them into and out of the trap. Traps should also be located with the ingress and egress of trucks and machinery in mind. Gravity drainage to a sump tank should also be provided.

CORROSION PROTECTION

Pipelines are subject to both external and internal corrosion. In general, external corrosion is the more serious in that it can seriously reduce the life of the pipeline and impair its safety. Internal corrosion, apart from exceptional cases of corrosive fluid components such as H_2S and CO_2, is usually a much more gradual process resulting in a lowering of pipeline efficiency. Regular line cleaning as discussed in the previous section can be used to take care of the pipe internally for most installations. However, with an unusually corrosive fluid, internal coatings or inhibition might also be needed.

The corrosion rate of steel is higher in low-resistance soils or in sea water than it is in high-resistance soils such as dry sands. Also corrosion rates are higher in turbulent waters than when the water is stagnant.

The two principal parts of corrosion protection are pipe coatings (or insulation) to increase the pipe to soil resistance and cathodic protection (impressed current or galvanic anodes) to make the pipe always cathodic.

Coatings

For many years pipelines were coated with processed natural substances such as asphalt or coal-tar enamels, usually reinforced with fiberglass mesh and overwrapped with felt sheeting. Relatively recently synthetic organic plastic materials have been developed extensively for external coatings and are now in common use.

Coal-Tar and Asphalt Enamels. Coal-tar and asphalt enamel coatings are applied hot over a cold primer and are usually reinforced with a fiberglass inner wrap and protected with a felt outer wrap. They are often applied "over the ditch" by special "coat and wrap" machines.

Asphalt Mastics. Asphalt mastic coating is commonly applied in a shop. It forms a thick (½ to ⅝ in) layer around the pipe and has mineral fillers which provide built-in reinforcement. Generally, it does not require an outer wrap. Mastics have been extensively used for submarine lines, swamp crossings, and some land lines. They are not in common use today.

Extruded Plastic Coatings. Extruded plastic coatings are usually applied in a shop, after sandblast cleaning of the pipe and priming with a thin layer of mastic. They afford good protection because the mastic exudes through a nick or a scratch and heals the break in the coating. Other systems include a bonded polyethylene jacket with a nonbonded PVC outer rockguard. Extruded coatings are often used where long-term dependability is required such as for city or swamp service. They are also used on submarine lines.

Tape Coatings. Tape coatings usually have an adhesive backing and may be applied in a shop or yard or over the ditch. Polyethylene and polyvinyl chloride tapes are readily available. Polyethylene may be preferred for its chemical, thermal, and physical stability.

Fusion-Bonded Epoxies. Fusion-bonded epoxies are very thin coatings applied in a shop. The pipe is degreased, sandblasted, and heated to the fusion temperature. Epoxy powder is applied electrostatically and fuses to a uniform coat on the heated surface. Pinholes and other defects may be repaired with hot stock or with a liquid epoxy resin and catalyst compound. The coating is tough and easy to repair. However, it must not be too thick or it may crack during bending, and it is subject to cathodic disbonding.

Field Joint Coatings. When each length of pipe has been coated individually, the field joint must still be protected after it is welded. Generally, the field joint coating material should be similar to the main pipe coating and suitable for application in relatively uncontrolled field conditions.

Asphalt and coal-tar enamels are usually applied by hand, along with the appropriate inner and outer wraps. Plastic tapes are also applied by hand. For extruded plastic coatings, plastic shrink sleeves with mastic backing may be used. They are slipped over the pipe before the joint is welded and are moved to cover the joint afterward. Applications of heat then causes the plastic to shrink tightly down on the joint and on the coating to either side.

Mastic is usually made up by surrounding the joint with a thin metal shield and pouring hot mastic into the shielded space. This usually bonds well with the mastic to either side and makes a satisfactory tight joint.

Fusion-bonded epoxy coatings may be made up by heating the joint with a preheat coil and applying the resin powder. The pipe surfaces must be sandblasted before the powder is applied.

Liquid epoxy resin and catalyst compound, applied by hand, may also be used for field joint protection.

Internal Coatings. Where corrosive components are encountered in the gas being transported, the pipe can be protected internally by adding inhibitors to the fluid or coating the pipe internal wall. An economic comparison can assist in making the choice. Continued use of inhibitors tends to be costly and introduces complications both in injection equipment and in the resultant effect on delivered gas analysis.

Coal-tar and asphaltic components for internal coating are not difficult to apply, but where deterioration occurs during the life of the pipeline, they can seriously affect the pipeline efficiency.

Epoxy coatings give good protection, long life, and a low friction factor. The main problem is to get a satisfactory bond between the shop-applied internal coating and the machine- or hand-applied coating on the joints after welding. This involves special applications which must be conscientiously applied in the field.

Cathodic Protection

Cathodic protection is a method of inhibiting corrosion that has been used for over a century. It is applicable to all types of pipe metals, although it is used mainly for steel pipe. The technique is to connect the pipeline to an anode bed and to make the pipe behave as a cathode by impressing a direct current voltage so that the anode bed corrodes and not the pipe. Where this is not practical, as on long water crossings, passive protection may be attained by attaching sacrificial anodes directly to the pipe (see Fig. C6.17).

Cathodic protection is particularly important for coated pipe; it overcomes the effect of pinholes or accidental breaks in the coating, which would otherwise permit local corrosion cells to form. Being localized, these cells are highly active and can rapidly penetrate the pipe wall.

Insulating flanges are necessary to ensure the integrity of the section of pipeline to be protected. Test leads installed at intervals along the length of the line during construction are used to monitor the voltage levels during the life of the pipeline. Regular measurements and maintenance, as required, ensure the continued effectiveness of the cathodic protection system.

Impressed Current Systems. Impressed current systems are generally installed where alternating current is available for the rectifier units (see Fig. C6.18). Since

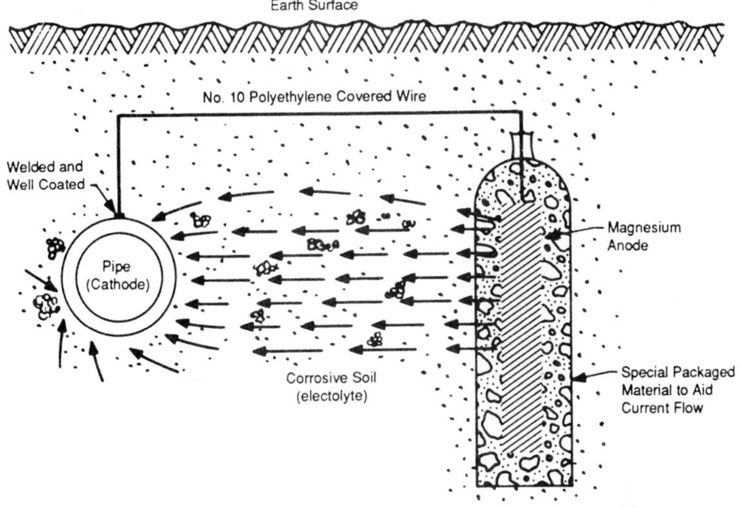

FIGURE C6.17 Cathodic protection for underground pipe.[11]

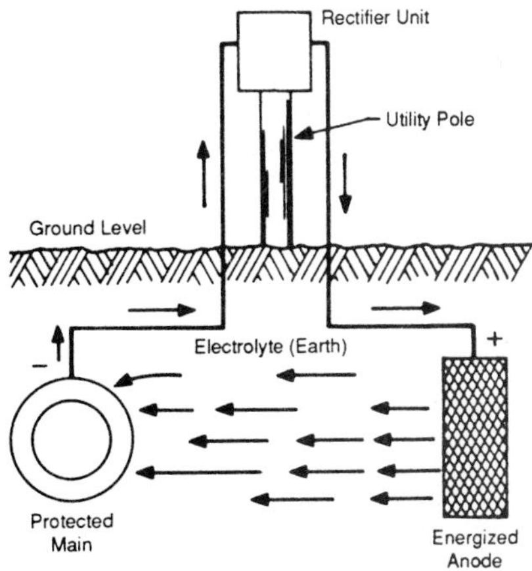

FIGURE C6.18 Rectifier and ground bed installations.[11]

the units can be obtained in a wide range of output voltage and current combinations, they are used (1) where soil or water electrical resistivities are high and (2) where total current requirements are large.

Direct current for cathodic protection is provided by a transformer-rectifier combination. Where alternating current is not available for standard rectifier units, alternative direct current sources that can be used are thermoelectric generators, diesel generators, and solar power systems.

Typical anode materials for impressed current systems are graphite, high-silicon iron, lead-silver alloys, and platinum on a titanium or columbium substrate. On occasion scrap steel is used but usually only because of its availability. As an anode material, steel is consumed at a rate of about 20 lb per ampere year. Consumption for graphite and silicon iron is approximately 1 lb per ampere year, for lead-silver, about 0.1 lb per ampere year, and for platinum about 6 to 20 mg per ampere year.

Galvanic Protection. Galvanic protection is a passive system which depends on the principle that coupling two dissimilar metals in the same electrolytic environment causes accelerated corrosion of the more active and protection of the less active metal. The more active metal is preferentially consumed and is, therefore, called a sacrificial anode.

Galvanic protection is often used in preference to an impressed current installation as a temporary measure during construction or when the current requirements are low and the electrolyte has relatively low resistivity (less than 10,000 Ω · cm). Clearly, it has an advantage when there is no source of electrical power or when a completely underground system is desired. Capital investment will generally be lower and is often the most economical method for short-life protection.

Insulating Joints

Insulating joints are required for electrical isolation of the cathodically protected section of the pipe from those sections not protected. In their simplest form, the joints may be flange connections with special arrangements of insulating gaskets, washers, and sleeves to prevent electrical contact between the bolts and the flanges. Preferably, they may be special unitized couplings in which the insulating features are factory assembled, and the entire assembly can be welded into the pipeline (see Figs. C6.19 and C6.20).

Insulating flange connections are made aboveground whenever possible. This simplifies installation, makes periodic inspection easier, and minimizes the possibility of a short circuit across the joint.

For underground installation an external application of bitumen or compound molding, covering the whole insulated flange connection, protects it from corrosion and dirt and is absolutely necessary for maintaining the insulating properties of the joint.

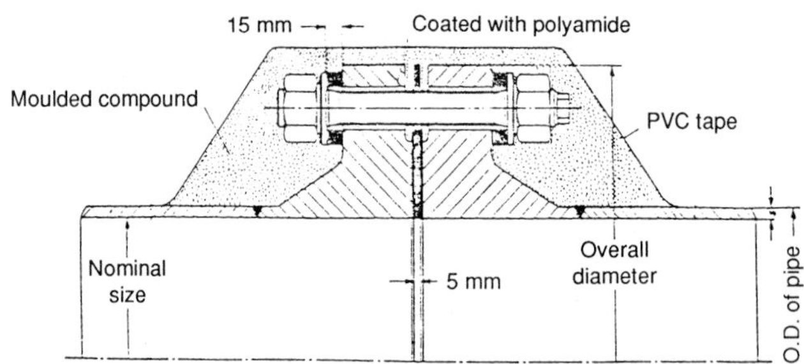

FIGURE C6.19 Insulating flange connection for laying in the ground.

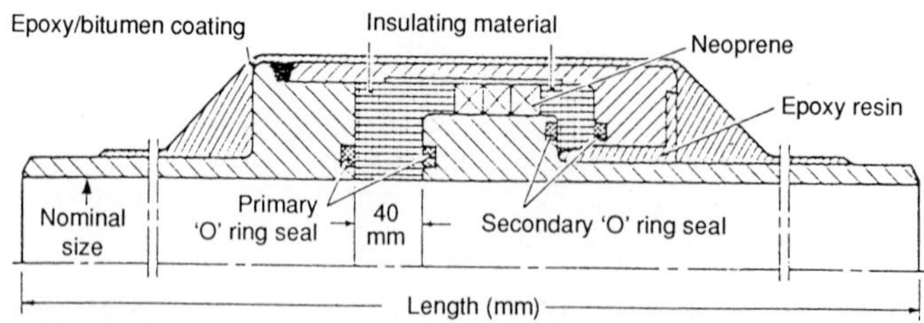

FIGURE C6.20 Monobloc insulating coupling ready for erection by welding in.

INSPECTION AND TESTING

Inspection

ASME B31.8 specifies that for piping systems that will operate at 20 percent or more of the specified minimum yield strength of the pipe, the quality of welding must be checked by nondestructive inspection. Such inspection may consist of radiographic examination, magnetic particle testing, or other acceptable methods. The trepanning method of nondestructive testing is prohibited.

The minimum number of field butt welds to be selected on a random basis from each day's construction for examination are given below. Each weld so selected must be examined over its entire circumference, or as an alternative, the equivalent length of weld must be examined if only a section of the circumference of each weld is examined.

1. 10 percent of welds in Class 1 locations.
2. 15 percent of welds in Class 2 locations.
3. 40 percent of welds in Class 3 locations.
4. 75 percent of welds in Class 4 locations.
5. 100 percent of welds in offshore pipelines, in compressor stations, and at major or navigable river crossings, major highway crossings, and railroad crossings, if practical, but in no case less than 90 percent. All tie-in welds not subjected to a pressure proof test must be examined.

All welds which are inspected must meet the standards of acceptability of the American Petroleum Institute Standard API 1104. If a weld does not, it must be repaired and reinspected.

If the pipe is intended to operate at 40 percent or less of the specified minimum yield strength, and pipe size is less than 6-in nominal pipe size (NPS 6), or the number of welds to be inspected is small enough to make nondestructive testing impractical, welding may be inspected visually and approved by a qualified welding inspector.

Testing

ASME B31.8 requires that all pipelines and mains which will be operated at a hoop stress of 30 percent or more of the specified minimum yield strength of the pipe must be given a test for at least 2 h to prove strength after construction and before being placed in operation. The test requirements are summarized in Table C6.11.

Pipelines and mains operating at less than 30 percent of the specified minimum yield strength of the pipe, but greater than 100 psi, must be tested as follows:

- In Class 1 locations, a leak test must be made at a pressure in the range from 100 psi to that pressure required to produce a hoop stress of 20 percent of the specified minimum yield strength in all cases where the line is to be stressed to 20 percent or more of the specified minimum yield strength of the pipe and gas or air is the test medium.

TABLE C6.11 Test Requirements for Pipelines and Mains to Operate at Hoop Stresses of 30 Percent or More of the Specified Minimum Yield Strength of the Pipe[2]

1	2	3	4	5
		Pressure test prescribed		Maximum allowable operating pressure, the lesser of
Location class	Permissible test fluid	Minimum	Maximum	
1, Division 1	Water	1.25 × m.o.p.	None	t.p. ÷ 1.25
1, Division 2	Water	1.1 × m.o.p.	None	t.p. ÷ 1.1 or d.p.
	Air	1.1 × m.o.p.	1.1 × d.p.	
	Gas	1.1 × m.o.p.	1.1 × d.p.	
2	Water	1.25 × m.o.p.	None	t.p. ÷ 1.25 or d.p.
	Air	1.25 × m.o.p.	1.25 × d.p.	
3&4*	Water	1.40 × m.o.p.	None or d.p.	t.p. ÷ 1.40 or d.p.

m.o.p. = maximum operating pressure (not necessarily the maximum allowable operating pressure)
d.p. = design pressure
t.p. = test pressure
*For exceptions, see para. 841.322(d) of ASME B31.8.

General note: This table brings out the relationship between test pressures and maximum allowable operating pressures subsequent to the test. If an operating company decides that the maximum operating pressure will be less than the design pressure, a corresponding reduction in prescribed test pressure may be made as indicated in the Pressure test prescribed, minimum, column. However, if this reduced test pressure is used, the maximum operating pressure cannot later be raised to the design pressure without retesting the line to the test pressure prescribed in the Pressure test prescribed, maximum, column. See paras. 805.214, 845.213, and 845.214 of ASME B31.8.

TABLE C6.12 Maximum Hoop Stress Permissible During Test[2]

	Class location, % of specified minimum yield strength		
Test medium	2	3	4
Air	75	50	40
Gas	30	30	30

In Class 2, 3, and 4 locations, at the least in accordance with the requirements listed in Table C6.11, except that gas or air may be used as the test medium within the maximum limits set in Table C6.12.

For pipelines and mains that will operate at less than 100 psi, gas may be used as the test medium at the maximum pressure available in the distribution system at the time of the test.

OPERATING AND MAINTENANCE

The ASME B31.8 Code has established comprehensive guidelines for the operation and maintenance of gas transmission systems. Pertinent sections have been paraphrased below.

Pipeline Maintenance

Continuing Surveillance of Pipelines. To maintain the integrity of a pipeline system, procedures for the continuing surveillance of its facilities must be established and implemented. Where unusual operating and maintenance conditions occur, such as failures, leakage history, drop in flow efficiency due to internal corrosion, or substantial changes in cathodic protection requirements, appropriate action must be taken.

When a facility is in unsatisfactory condition, a planned program must be initiated to abandon, replace, or recondition and proof test the facility. If the facility cannot be reconditioned or phased out, the maximum allowable operating pressure must be reduced in accordance with code requirements.

Pipeline Patrolling. A periodic pipeline program must be implemented to observe surface conditions on and adjacent to the pipeline right of way, indications of leaks, construction activity other than that performed by the company, and any other factors affecting the safety and operation of the pipeline. Patrols must be performed:

- At least once each year in Class 1 and 2 locations
- At least once each 6 months in Class 3 locations
- At least once each 3 months in Class 4 locations

Weather, terrain, size of line, operating pressures, and other conditions will be factors in determining the need for more frequent patrol. Main highways and railroad crossings must be inspected with greater frequency and more closely than pipelines in open country.

Maintenance of Cover at Road Crossings and Drainage Ditches. If the cover over the pipeline at road crossings and drainage ditches has been reduced below the requirements of the original design and is found to be unacceptable, additional protection such as barriers, culverts, concrete pads, casing, lowering of the line, or other suitable means must be provided.

Maintenance of Cover in Cross-Country Terrain. Where the cover over the pipeline in cross-country terrain does not meet the original design, and it is determined to be at an unacceptable level, additional protection must be provided by replacing the cover, lowering the line, or other suitable means.

Leakage Surveys. Periodic leakage surveys on the line must be a part of any company's operating and maintenance plan. The type of surveys selected must be effective for determining if potentially hazardous leakage exists. The extent and frequency of such surveys must be determined by the operating pressure, piping age, class location, and whether the transmission line transports gas without an odorant.

Pipeline Markers. Signs or markers must installed where necessary to indicate the presence of a pipeline at road, highway, railroad, and stream crossings. Additional signs and markers must be installed along the remainder of the pipeline at locations where there is a probability of damage or interference. The signs or markers must include the words *Gas* (or name of gas transported) *Pipeline,* the name of the operating company, and the telephone number (including area code) where the operating company can be contacted.

Distribution Piping Maintenance

Patrolling. Distribution mains must be patrolled in areas where construction activity, physical deterioration of exposed piping and supports, or any natural causes could result in damage to the pipe. The frequency of the patrolling must be a function of the severity of the conditions which could cause failure or leakage and the subsequent hazards to public safety.

Leakage Surveys. The type or types of surveys selected for the operating and maintenance plan of a company's gas distribution system must be effective for determining if potentially hazardous leakage exists. Some procedures which may be employed are:

- Surface gas detection surveys
- Subsurface gas detector survey (including bar hole surveys)
- Vegetation surveys
- Pressure drop test
- Bubble leakage test
- Ultrasonic leakage test

Miscellaneous Facilities Maintenance

Compressor Station Maintenance

Compressor and Prime Movers. The starting, operating, and shutdown procedures for all gas compressor units must be established by the operating company and appropriate steps taken to see that the approved practices are followed.

Inspection and Testing of Relief Valves. All pressure-relieving devices in compressor stations must be inspected and/or tested, and all devices except rupture disks must be operated periodically to determine that they open at the correct set pressure. Any defective or inadequate equipment found must be promptly repaired or replaced. All remote control shutdown devices must be inspected and tested at least annually to make sure that they function properly.

Isolation of Equipment for Maintenance or Alterations. Procedures must be established and followed for the isolation of units or sections of piping for maintenance, and for purging prior to returning units to service.

Storage of Combustible Materials. All flammable or combustible materials in quantities beyond those required for everyday use or other than those normally used in compressor buildings must be stored in a separate structure built of noncombustible material located a suitable distance from the compressor building. All aboveground oil or gasoline storage tanks must be protected in accordance with NFPA 30.

Maintenance of Pressure-Limiting and Pressure-Regulating Stations. All pressure-limiting stations, relief devices, and other pressure-regulating stations and equipment must be subject to systematic, periodic inspections and suitable tests, or reviewed to determine that they are:

1. In good mechanical condition. Visual inspections must be made to determine that equipment is properly installed and protected from dirt, liquids, or other conditions that might prevent proper operation. The following must be included in the inspection where appropriate:
   a. Station piping supports, pits, and vaults for general condition and indications of ground settlement.
   b. Station doors and gates and pit vault covers to ensure that they are functioning properly and that access is adequate and free from obstructions.
   c. Ventilating equipment installed in station buildings or vaults for proper operation and for evidence of accumulation of water, ice, snow, or other obstructions.
   d. Control, sensing, and supply lines for conditions which could result in a failure.
   e. All locking devices for proper operation.
   f. Station schematics for correctness.
2. Adequate from the standpoint of capacity and reliability of operation for the service in which they are employed and set to function at the correct pressure.
   a. If acceptable operation is not obtained during the operational check, the cause of the malfunction must be determined and the appropriate components must be adjusted, repaired, or replaced as required. After repair, the component must again be checked for proper operation.
   b. At least once each calendar year, relief valves must be reviewed for sufficient capacity. If it is determined that the relieving equipment is of insufficient capacity, new or additional equipment must be installed to provide adequate capacity.

Whenever abnormal conditions are imposed on pressure or flow control devices, the incident must be investigated and the device inspected and/or repaired. Abnormal conditions may include regulator bodies which are subjected to erosive service conditions or contaminants from upstream construction and hydrostatic testing. Necessary inspections include:

1. An inspection or test, or both, of stop valves must be made to ensure that the valves will operate and are correctly positioned. The following must be included:
   a. Station inlet, outlet, and bypass valves
   b. Relief device isolating valves
   c. Control, sensing, and supply line valves
2. The final inspection procedure must include the following:
   a. A check for proper position of all valves with special attention given to regulator station bypass valves, relief device isolating valves, and valves in control, sensing, and supply lines.
   b. Restoration of all locking and security devices to proper position.

Valve Maintenance

Pipeline Valves. Pipeline valves required to operate during an emergency must be inspected periodically and partially operated at least once a year to provide safe and proper operating conditions. Inspection must include:

1. Routine valve maintenance procedures must at the least include the following:
   a. Servicing in accordance with written procedures by adequately trained personnel

   b. Accurate system maps for use during routine or emergency conditions
   c. Valve security to prevent service interruptions, tampering, etc., as required
   d. Employee training programs to familiarize personnel with the correct valve maintenance procedures
2. Emergency valve maintenance procedures include:
   a. Written contingency plans to be followed during any type emergency
   b. Training personnel to anticipate all potential hazards
   c. Furnishing tools and equipment as required, including auxiliary breathing equipment, to meet anticipated emergency valve servicing and/or maintenance requirements.

Distribution System Valves. Valves, used for the safe operation of a gas distribution system, must be checked and serviced, including lubrication where necessary, at sufficiently frequent intervals to assure their satisfactory operation. Inspection must include checking of alignment to permit use of a key or wrench and clearing from the valve box or vault any debris which would interfere with or delay the operation of the valve. System maps showing valve location should be available.

Service Line Valves. Outside shut-off valves installed in service lines supplying places of public assembly, such as theaters, churches, schools, and hospitals, must be inspected and lubricated where required at sufficiently frequent intervals to assure their satisfactory operation. The inspection must determine if the valve is accessible, if the alignment is satisfactory, and if the valve box or vault, if used, contains debris which would interfere with or delay the operation of the valve. Unsatisfactory conditions encountered must be corrected.

Valve Records. A record must be maintained for locating all pipeline valves and distribution system valves which are needed for safe system operation or may need to be operated under emergency conditions. These records may be maintained on operating maps, separate files, or summary sheets and must be readily accessible to personnel required to respond to emergencies.

Prevention of Accidental Operation. To prevent accidental operation of any of the valves covered above, certain precautions must be taken. Some recommended actions are as follows:

1. Lock valves in aboveground settings readily accessible to the general public that are not enclosed by a building or fence
2. Lock valves located in vaults if accessible to the general public
3. Identify the valve by tagging, color coding, or any other suitable means of identification

DECOMMISSIONING AND ABANDONMENT

Guidelines for decommissioning or abandoning gas transmission facilities are given in the ASME B31.8 Code. Pertinent sections are paraphrased below for reference.

Abandoning of Transmission Facilities

Any plan for abandoning transmission facilities must include the following provisions:

1. Facilities to be abandoned must be disconnected from all sources and supplies of gas such as other pipelines, mains, crossover piping, meter stations, control lines, and other appurtenances
2. Facilities to be abandoned in place must be purged of gas with an inert material and the ends sealed (except as in 3 below).
3. If precautions are taken to ensure that no liquid hydrocarbons remain in the facilities to be abandoned, the facilities may be purged with air. If the facilities are purged with air, precautions must be taken to ensure that a combustible mixture is not present after purging.

Abandoning of Distribution Facilities

To abandon inactive facilities, such as service lines, mains, control lines, equipment, and appurtenances for which there is no further planned use, the following provisions must be addressed:

1. If the facilities are abandoned in place, they must be physically disconnected from the piping system. The open ends of all abandoned facilities must be capped, plugged, or otherwise effectively sealed. The need for purging the abandoned facility to prevent the development of a potential combustion hazard must be considered and appropriate measures taken. Abandonment is not complete unless the volume of gas or liquid hydrocarbons contained within the abandoned section has been determined to pose no potential hazard. Air or inert gas may be used for purging, or the facility may be filled with water or other inert material. If air is used for purging, the facility must be checked to ensure that a combustible mixture is not present after purging. Consideration must also be given to any effects the abandonment may have on an active cathode protection system and appropriate action taken.
2. In cases where a main is abandoned, together with the service lines connected to it, insofar as service lines are concerned, only the customer's end of such service lines need be sealed as stipulated above.
3. Service lines abandoned from the active mains should be disconnected as close to the main as practical.
4. All valves left in the abandoned segment should be closed. If the segment is long and there are few line valves, consideration should be given to plugging the segment at intervals.
5. All above-grade valves, risers, and vault and valve box covers must be removed. Vault and valve box voids must be filled with suitable compacted backfill material.

Temporarily Disconnected Facilities

Whenever service to a customer is disconnected, one of the following actions must be taken:

1. The valve that is closed to prevent the flow of gas to the customer must be provided with a locking device or other means designed to prevent the opening of the valve by unauthorized persons.

2. A mechanical service or fitting that will prevent the flow of gas must be installed in the service line or in the meter assembly.
3. The customer's piping must be physically disconnected from the gas supply and the open pipe ends sealed.

REFERENCES

1. William R. Quarles, "Capital Spending to Increase in the Natural Gas Industry," *Pipeline Industry,* February 1989, pp. 13–14.
2. The American Society of Mechanical Engineers, ASME Code for Pressure Piping, B31, "Gas Transmission and Distribution Piping Systems," ASME B31.8—1989 ed. and ASME B31.8a–1990 Addenda.
3. Charles E. Schorre, "Flow Temperature in a Gas Pipeline," *The Oil and Gas Journal,* September 27, 1954, pp. 26–28.
4. D. M. Coulter and M. F. Bardon, "Revised Equation Improves Flowing Gas Temperature Prediction," *The Oil and Gas Journal,* February 26, 1979, pp. 107–108.
5. Gas Processors Suppliers Association, *Engineering Data Book,* 9th ed., 1972.
6. B. E. Hunt et al., "Gas Service Design—Final Report of Task Group" (DMC-56-14), *Proc. AGA,* 1956, pp. 272–287.
7. D. Menegakis and E. H. Luntey, "Experimental Investigation of Flow Characteristics of Low-Pressure Service" (DMC-61-15), *Proc. AGA,* 1961.
8. Gary B. Emerson, "Relief Valve Sizing, Selection, Installation and Testing," *Pipeline Industry,* February 1987, pp. 25–32.
9. Richard A. Sutton, "Part 1—How Columbia Gas Selects Distribution Meters," *Pipeline Industry,* December 1978, pp. 33–38., "Part 2—How Columbia Gas Selects Distribution Meters," *Pipeline Industry,* January 1979, pp. 51–53.
10. T. D. Williamson, Inc., Piping Systems Brochure, Feb. 1986, Tulsa Oklahoma.
11. Richard C. Berger, "How Construction Forces Can Control Corrosion," *Pipeline & Gas Journal,* February 1984, pp. 29–32.

CHAPTER C7
CHEMICAL AND REFINERY PIPING SYSTEMS

Richard C. Getz, P.E.
Chief Piping Engineer
United Engineers and Constructors
Philadelphia, Pennsylvania

Before the publication of the current Pressure Piping Code ASME/ANSI B31.3, the Code was titled Petroleum Refinery Piping, with direct application to the petroleum refinery industry and, by inference, the guiding document for the chemical process industry. With the production and retitling of B31.3 to Chemical Plant and Petroleum Refinery Piping, this Code has expanded its scope into areas where it previously had only been used as reference.

As applied to this section, *process* piping refers, except as excluded below, to all piping within the property limits of a process installation (chemical or petroleum), loading terminals, natural gas processing plants, bulk, specialty and commodity chemical plants, compounding plants, or tank farms. Included in this scope are piping systems which convey all fluids such as fluidized solids, oil, gas, steam, air, water, chemicals, and refrigerants. Figure C7.1 is a typical example of process piping. Excluded from code jurisdiction are various low-pressure piping systems, plumbing, fired heater piping, and other special applications.

Petroleum refinery piping is generally characterized as large-diameter piping, with metallic materials of construction, operating at or near elevated temperatures and pressures. Most refinery piping is conveying hydrocarbons or is in close proximity of hydrocarbon containing equipment. Consequently, the potential for a fire is ever present.

Typical chemical plant piping applications are characterized by relatively small (2 in and smaller) diameter piping, lower operating pressures and temperatures, and corrosive fluids. The use of exotic alloy materials, thermoplastics, and thermoset resin materials of construction is common. Many chemical plant piping systems convey flammable and/or toxic substances.

Modern process facilities handle a myriad of chemical substances and compounds at various process temperatures and pressures. The piping systems specified to transport these fluids must be compatible with the intended service conditions. Since most of these facilities are one of a kind, the process piping materials of construction selection will be specific application driven. Today,

FIGURE C7.1 Typical process plant piping.

there are numerous construction materials available, both metallic and nonmetallic. Installed cost variations may be as much as tenfold, depending upon the materials suitable for the application. Life-cycle cost considerations must also be taken into account during the material selection phase; in addition, the availability of various components must enter any selection criteria.

Whereas process piping material selection is specifically application driven and will vary dramatically on a case-by-case basis, utility piping systems are basically the same across various process plants. Basic materials of construction and mode of construction will differ little from one plant to another for services such as air, water, and steam.

Table C7.1 is a hypothetical illustration of services and piping material classes for a typical process facility. Note the material variations on the process side of the plant relative to the relatively constant material types in the utility area of the facility.

Most process piping is welded construction except where flanged disassembly is required. Due to the high-value-added cost of the process fluid and its potential for damage to personnel or the environment upon release, integrally welded joints for all size range is common.

Energy costs have made integrally welded joints for steam and other utility services more attractive when evaluated against conventional threaded joints and their potential for leakage. Threaded joint construction is still used for low severity service but is typically limited to smaller (2 in and less) sizes and services that offer little or no threat to personnel safety or environmental harm.

TABLE C7.1 Hypothetical Process Plant *X, Y, Z* Piping Service Index

Service material and class	Service conditions psig	°F	Piping
Reactor feed	60	1250	Class 600 stainless steel Type 310
Liquid sulfur	130	310	Class 150 carbon steel jacketed
Molten TiO_2	3500	520	Class 2500 stainless steel Type 316
Wet chlorine	3	90	Reinforced thermoset resin
Acetic acid	150	70	Class 150 stainless steel Type 304
20% caustic	50	Amb	Polyvinyl chloride
Hydrogen	180	1500	Class 2500 Ni-iron-chrome
			Refractory lined alloy and carbon steel
Process vapors	100	400	Glass-lined carbon steel
Process liquid	100	100	TFE-lined carbon steel
Methane feed	250	925	Class 300 1¼ Cr - ½ mo
	Utilities		
Instrument air	100	100	Class 125 galvanized carbon steel
Potable water	60	Amb	Class 125 galvanized carbon steel
Low-pressure, steam and condensate	60	310	Class 150 carbon steel
High-pressure, steam and condensate	325	429	Class 300 carbon steel
Deionized water	150	275	Class 150 stainless steel Type 316L
Steam tracing	60	310	Copper
Firewater	175	100	Class 125 carbon steel
Boiler feed water	350	430	Class 300 carbon steel
Eyewash-safety shower	70	40	Class 125 galvanized carbon steel
Boiler chemical feed	400	Amb	Stainless steel Type 316 tubing
Heat transfer fluid	75	600	Class 300 carbon steel

REFERENCE CODES AND STANDARDS

Code application, jurisdiction, and specific scopes are covered in Chap. A4, Part A, Piping Codes and Standards, of this handbook. This chapter closely parallels the requirements outlined in ASME B31.3, Chemical Plant and Petroleum Refining Piping, 1990 edition.

Codes and standards referenced in this chapter are as follows:

ASME B31.3, Chemical Plant and Petroleum Refinery Piping

ASME/ANSI B16.5, Pipe Flanges and Flanged Fittings

ASME/ANSI B16.11, Forged Steel Fittings, Socket-Welding and Threaded
ASME/ANSI B16.47, Large Diameter Steel Flanges
ASME VIII, Pressure Vessels
ASME IX, Welding and Brazing Qualifications
API 5L, Specification for Line Pipe
API 601, Metallic Gaskets for Raised Face Pipe Flanges
API 605, Large Diameter Carbon Steel Flanges
MSS SP-44, Steel Pipe Line Flanges
MSS SP-45, Bypass and Drain Connection Standard
ASME/ANSI B16.1, Cast Iron Pipe Flanges and Flanged Fittings, Classes 25, 125, 250 and 800
ASME/ANSI B16.24, Bronze Pipe Flanges and Flanged Fittings, Classes 150 and 300
ASME/ANSI B16.42, Ductile Iron Pipe Flanges and Flanged Fittings, Classes 150 and 300
ASME/ANSI B16.31, Non-Ferrous Pipe Flanges
ASME/ANSI B36.10M, Welded and Seamless Wrought Steel Pipe
ASME/ANSI B36.19M, Stainless Steel Pipe
ASME/ANSI B16.9, Factory Made, Wrought Steel Butt Welding Fittings
ASME/ANSI B16.25, Butt Welding Ends

DESIGN CONSIDERATIONS

Design Conditions

The definitions below are quoted from the Chemical Plant and Petroleum Refinery Piping Code, ASME/ANSI B31.3.

Internal Design Pressure. The piping component shall be designed for an internal pressure representing the most severe condition of coincident pressure and temperature expected in normal operation (including fluid head). The most severe condition of coincident pressure and temperature under normal operation shall be that condition which results in the greatest required pipe thickness and the highest flange rating.

External Design Pressure. The piping component shall be designed for the maximum differential pressure (including fluid head) at the coincidental temperature that can act externally on the component in the piping system, taking into consideration the failure of external or internal pressure.

Design Temperature. The design temperature is the temperature representing the most severe condition of coincident pressure and temperature as explained above and shall be determined as follows:

1. For fluid temperatures below 32°F, the metal temperature shall be taken as the fluid temperature.

2. For fluid temperatures 32°F and above, the metal temperature for uninsulated components shall be no less than the following values:
   a. Threaded and welding end valves, pipe, welding fittings, and other components having wall thicknesses comparable to that of pipe—95 percent of the fluid temperature.
   b. Flanged valves, flanged fittings, and flanges (except lap joints)—90 percent of the fluid temperature.
   c. Lap joint flanges—85 percent of the fluid temperature.
   d. Bolting—80 percent of the fluid temperature.
3. *Externally insulated piping:* The fluid temperature shall be used unless calculations, previous tests, or service experience based on measurements support the use of other temperatures. Where piping is heated by heat tracing or jacketing, the effect of such heating shall be incorporated into the establishment of the design temperature.
4. *Internally insulated piping:* The design metal temperature shall be based on heat transfer calculations or tests.

Discussion of Design Pressure and Temperature. The designer is cautioned that regardless of the combination of operating pressure and temperature which results in the most severe condition from a stress standpoint, the selection of materials will often be governed by the extremes of operating temperature and pressure.

In designing piping systems, it is simpler to use fluid temperature instead of metal temperature. Substantial savings, however, can be realized by making the calculations or tests in items 3 or 4 above to determine the metal temperature, especially if the fluid temperatures are high. Not only can flange (and valve) ratings and pipe thicknesses often be reduced, but occasionally less expensive materials can be used.

In the application of item 2 above, any savings in the cost of piping components must be balanced against the present worth of the additional heat to be lost from the piping components if the insulation is omitted. Flanges and valves which are to be left bare should be clearly marked on the piping drawings so they are not inadvertently insulated.

Cooling Effects on Pressure. The cooling of a gas or vapor in a piping component may reduce the pressure sufficiently to create an internal vacuum. In such a case, the piping component must be capable of withstanding the external pressure at the lower temperature, or provisions must be made to break the vacuum.

Fluid Expansion Effects. Although the Code states that provision shall be made in the design either to withstand or to relieve increased pressure caused by the heating of static fluid in a piping component, most operators have found it unnecessary to provide a relief valve on piping components which may be blocked in. The reason for this is that flanged joints, valve packing glands, or valve seating surfaces usually will leak before the pressure buildup becomes excessive as the result of the heating of the blocked-in fluid. Valves using resilient seating materials have made this more of a problem because of their ability to seal even more tightly as the pressure increases. Consequently, when such valves are used, greater consideration should be given to the problem of possible pressure buildup in the valve body cavity. A similar problem can be created in a blocked-in section of chemical piping by the increase in pressure which can result from the evolution of gas caused either by an increase in fluid temperature or by a chemical reaction (e.g., hydrogen is a product of corrosion).

Dynamic Effects. The Code requires that the piping designer take into account wind and earthquake forces, although not concurrently, in design of the piping. It also requires that the designer consider impact forces (including hydraulic shock) and vibration. Vibration and impact forces create complex design problems. This is especially true in the case of vibration because difficult-to-detect low-amplitude high-frequency vibrations often produce the most dangerous stresses. Refer to Chap. B4, Part B of this handbook for stress analysis of piping systems.

Most piping systems will vibrate to some extent. The exciting forces causing vibration in piping may be (1) mechanical vibration of connected equipment, such as compressors, pumps, and vessels; (2) wind-produced vortices that form alternately on opposite sides of cylindrical surfaces; or (3) internal pulsations in flowing fluids, such as those set up by reciprocating pumps and compressors. Vibration from this latter cause can generally be kept within controllable limits by limiting pressure pulsations to 1–3 percent of line pressure.

Where vibration is expected, good design should include the following:

- Adequate foundations, especially for reciprocating pumps and compressors.
- Strategic location of pipe guides and supports to reduce vibration. They should be installed so as to cause minimum restraint to normal thermal movements. The use of sway braces of the energy-absorbing or instant-counterforce-acting type is recommended for control of undesirable pipeline movement. Rigid braces are also effective in controlling movement provided they do not restrict the flexibility of the piping and their effect is taken into account in the design of the piping for flexibility. Where pulsating flow exists, piping should be supported at all changes in direction, and cantilever sections must be avoided.
- Avoidance of small branch connections. Additional supports, such as gusset plates, may help reduce vibration problems.
- An acoustical study to determine if dampening equipment is needed. Failure can be caused by resonance of some part of the system with the pulsation frequency. Pressure pulsations can be minimized by the use of hydropneumatic accumulators, snubbers, or surge drums. These pulsation-reducing devices should be installed as close as possible to the pulsation-producing equipment.
- Any sudden change in the flow velocity or pressure in a liquid line will produce hydraulic shock (water hammer). The common cause of water hammer is the rapid closure of a valve.

Water hammer requires careful consideration because it can damage equipment associated with piping and instrumentation even though the permissible pressures for piping components may not be exceeded. In addition to the use of slow-closing valves, the installation of surge tanks, pneumatic chambers, spring-operated relief valves, or shock absorbers is sometimes used to help control this phenomenon. Refer to Chap. B8, Part B of this handbook for more details.

Weight Effects. In the design of piping and its supports, it is required that live loads, dead loads, and loads and forces from other causes be taken into account. Refer to Chap. B2, Part B of this handbook.

Pressure-Temperature Ratings for Piping Components. Pressure-temperature ratings have been established for certain piping components. Those that have been accepted by the Code are maintained in the standards listed in the Code. These established ratings should not be exceeded by the expected normal operating conditions; however, during shutdown, start-up, or an interruption in the normal operation of a process unit, conditions more severe than normal may be characteristic of a service. Depending on the frequency and duration of these more severe conditions, the Code permits adjustment of the pressure ratings as follows:

- When the increased operating condition will not exceed 10 h at any one time or 100 h/yr, it is permissible to increase the pressure rating at the temperature existing during the increased operating condition by a maximum of 33 percent.
- When the increased operating condition will not exceed 50 h at any one time or 500 h/yr, it is permissible to increase the pressure rating at the temperature existing during the increased operating condition by a maximum of 20 percent.

Some of the conditions which should be investigated for the short time period are the centrifugal-pump shutoff pressure or the pressure at the maximum point of the pump characteristic curve, centrifugal-compressors surge-point pressure, stalling pressure of reciprocating pumps and reciprocating compressors, and the set pressure of relief valves which limit pressure in the piping.

The Code permits extrapolation of accepted pressure-temperature ratings where these ratings do not extend to the upper material temperature limits permitted by the Code provided this is done in accordance with the applicable Code rules.

Allowable Stresses. The stress values indicated in the Code are grouped according to the material temperature. Except in the case of welded pipe, welded fittings, castings, and structural-grade material, these values are the basic allowable stresses for materials. For welded pipe and welded fittings, the tables show the product of the basic allowable stresses and the applicable joint factor (SE),

where S = allowable stress
E = joint efficiency factor

The Code describes the various supplementary inspections which permit the allowable stresses and casting-quality factors to be increased. The casting-quality factors, however, do not apply to valves, flanges, and fittings which conform to the standards listed in the Code.

The bases for establishing the basic allowable stress values for ferrous, nonferrous, and nonmetallic materials are described in the Code. The Code user will need to refer to this only in those cases where he desires to establish allowable stresses for an unlisted material. The allowable stress values in shear and bearing are 0.80 and 1.60, respectively, times the values contained in the Code.

For other than normal operations, the allowable adjustments in pressure-temperature ratings, set forth above, are applicable to allowable stress values in calculations concerning components, such as pipe, which do not have established pressure-temperature ratings. The designer should remember that the more severe operating condition may be caused by increased contents load rather than by increased temperature or pressure. For example, a pipeline may have been designed for gaseous service. Subsequent process changes required the pipeline to see liquid service at nominally the same pressure and temperature as the previous gas service. The basic pipe spans would need to be reevaluated due to the increase in the dead weight bending loads realized by the change from gas to liquid service.

Limits on Calculated Stresses Due to Sustained Loads and Thermal Expansion. This subject is covered in detail in Chap. B8, Part B of this handbook.

Corrosion and Erosion Allowances. A commonly used nominal value of corrosion allowances is $\frac{1}{10}$ in for carbon-steel and alloy-steel piping in hydrocarbon service. When carbon steel is used for utility piping, that is, steam, air, and water, a commonly used corrosion allowance is 0.05 in. The allowance for corrosion or erosion should be added to all surfaces exposed to the process fluid. Threading and grooving allowances, weld joint factors, and other mechanical strength factors shall be as required by the applicable code.

Although lines under railroad tracks and roadways outside process plant limits are usually installed in steel pipe sleeves in conformance with the requirements of the railroad or highway departments involved, it seldom is necessary to perform this protection for underground steel lines in a process plant. For practical considerations, however, at least 24 in of cover should be provided for unsleeved lines under roads and railroads (for railroads, the 24 in are measured from bottom of ties to top of pipe). This applies especially to railroads, because a derailment could cause the failure of any line closer to the surface; see Fig. C7.2.

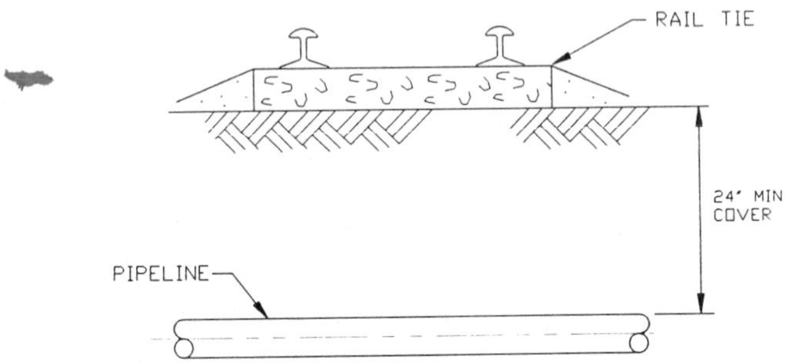

FIGURE C7.2 Pipeline under railroad.

Pressure Design of Piping Components

Straight Pipe. The equations given in the Code consider pressure and mechanical, corrosion, and erosion allowances. In addition to these factors, the Code requires that all designs, not only those for straight pipe, be checked for adequacy of mechanical strength under the applicable loadings discussed previously. The Code gives equations for determining the thickness of straight pipe for the outside diameter to thickness ratios Do/t greater than 6. The pressure design of piping having a diameter to thickness ratio of 6 or less requires special considerations which encompass design and material factors, such as theory of failure, fatigue, and thermal stress and is addressed in App. X of the Code. Refer to Chap. B2, Part B of this handbook for calculation of the required minimum wall thickness of the pipe.

Another practical rule used by some operating companies is to make pipe nipples for tapped and socket-welding openings in valves, piping accessories, and process equipment one schedule heavier than the remainder of the line. It is also fairly common practice to make the piping for ¾-in and smaller connections from a header to the first valve or fitting one schedule heavier than the remainder of the branch line unless the header is smaller than 2-in pipe size.

Bends and Elbows. The minimum required thickness of a pipe bend, after bending, should be determined as for straight pipe, provided the bending operation does not produce a difference between the maximum and minimum diameters greater than 8 percent for internal pressure service and 3 percent for external pressure.

Elbows that are manufactured in accordance with any of the standards listed in the Code are considered to be suitable for use at the pressure-temperature rating specified in the listed standard. In the case of standards under which elbows are made to a nominal pipe thickness (such as ANSI B16.9), the elbows are considered suitable for use with pipe of the same nominal thickness.

Branch Connections. Branch connections may be made by the use of integrally reinforced fittings or welding outlet fittings, by welding the branch pipe directly to the run pipe, or by threading under greatly restricted conditions.

Branches at angles other than 90° should not be used except when flow and pressure-drop considerations greatly outweigh the difficulty in obtaining satisfactory welds and the greater difficulty in adequately reinforcing such connections.

The Code gives rules governing the design of the pipe-to-pipe branch connections to sustain internal and external pressure. Branch connections in which the smaller angle between the axis of the branch and the run is less than 45°, impose special design and fabrication problems. The rules given in the Code for angles between 45° and 90° may be used as a guide, but sufficient additional strength must be provided to ensure safe and satisfactory service, and these branch connections shall be designed to meet the requirements of the Code.

For welded pipe to pipe branch connections, the stress concentration at the junction increases rapidly as the size of the branch approaches the size of the run. This is also true with most welding outlet fittings. Consequently, in services which involve considerable cycling due to pressure or temperature or both, it is usually good practice to make the branch connections with butt-welding tee fittings. The use of butt-welding tees is also considered good practice for full-sized branches in many flammable and hazardous services. If there is at least one size reduction in the branch, pipe-to-pipe branch connections should, as a rule, be acceptable for all services except those with severe cycling. Also, the cost of butt-welding reducing tees is such that savings will result if pipe-to-pipe branches or welding outlet fittings are used.

Branch connections made by either socket-welding or threading the branch pipe directly to the run pipe are limited by the Code to nonflammable, nontoxic services below 150 psig and 400°F. Such connections are seldom used in process piping applications.

Strength of Branch Connections. A pipe having a branch connection is weakened by the opening that must be made in it, and unless the wall thickness of the pipe is sufficiently in excess of that required to sustain the pressure, it may be necessary to provide reinforcement. Certain branch connections may be made without the necessity of their use being supported by engineering calculations. In other cases, the amounts of reinforcement required to sustain the pressure in welded pipe-to-pipe branches are outlined clearly in ASME/ANSI B31.3.

Openings in Closures. Code rules govern the design of openings in closures when the size of the opening is not greater than one-half of the inside diameter of the closure. The basis for the design of these openings in closures is the same as it is for branch connections. Closures with larger openings must be designed as reducers, except that flat closures must be designed as flanges.

Miters. The thickness of each segment of the miter must be designed in the same manner as straight pipe. This thickness, however, does not allow for the discontinuity stresses which are present at junctions between miter segments. The discontinuity stresses are reduced for a given miter as the number of segments is increased. The discontinuity stresses may be neglected for miters in nonflammable, nontoxic, noncyclic services at pressures less than 100 psig. Miter elbows are commonly used for large-diameter, low-pressure (50 psig or less) water, drainage, and vent piping. Even for these relatively noncritical services, 90° miter elbows should have three or more segments.

Attachments. Attachments to piping, both external and internal, are required to be designed so they will not cause flattening of the pipe, excessive localized bending stresses, or detrimental thermal gradients in the pipe wall. It is important that such attachments be designed to minimize stress concentrations, particularly in cyclic services.

Closures. The design of welded flat closures and ellipsoidal, spherically dished, hemispherical, and conical closures is encountered relatively infrequently. Closure fittings manufactured in accordance with ASME/ANSI B16.9 and B16.11 are considered suitable for use at the pressure-temperature ratings specified by such standards. In the case of standards under which closure fittings are made to nominal pipe thickness, the closure fittings shall be considered suitable for use with pipe of the same nominal thickness.

In process plant piping the most commonly used closure is probably the butt-welding cap. Bolted flanged covers are used only where access is needed.

Flanges. Flanges manufactured in accordance with ASME/ANSI standards B16.5, B16.31, B16.1, B16.24, and B16.42 shall be considered suitable for use at the pressure-temperature ratings specified by such standards. Flanges not made in accordance with the standards listed above shall be designed in accordance with Section VIII Division 1 of the ASME Boiler and Pressure Vessel Code, except that the requirements for fabrication, assembly, inspection, and testing and the pressure and temperature limits for materials of ASME/ANSI B31.3 shall govern.

Blanks. The pressure design thickness for permanent blanks must be calculated in accordance with formula 15 of B31.3. Corrosion or erosion allowances and manufacturer's minus tolerance must be added to the thickness thus determined.

In the design of blanks which are not to be used during operation of the piping (e.g., during shutdowns or during testing), an allowable stress equal to 90 percent of the yield strength is sometimes used. It is suggested, however, that ¼ in is a practical minimum thickness for carbon-steel blanks.

Pressure Design of Other Components. In general, pressure-containing components are to satisfy Code requirements. However, if the design of similarly shaped or proportioned components has been proven successful by performance under comparable service conditions, provision for the use of such components is allowed by the Code. Alternatively, if pressure design of the particular component is proven either by an experimental stress analysis or by tests made in accordance with the ASME Boiler and Pressure Vessel Code, such component will be deemed to satisfy the requirements of ASME/ANSI B31.3.

Selection and Limitations of Piping Components

Pipe and Fittings. Furnace lap-weld ferrous pipe, furnace butt-weld ferrous pipe, spiral-weld ferrous pipe, and fusion-welded steel pipe (made to ASTM A134, A53, Type *F,* API 5L furnace butt-welded and A211) are not permitted for hydrocarbons or other flammable fluids within process unit limits or for hazardous fluids in any location. Nonferrous pipe made by similar manufacturing process is similarly restricted.

The use of bell-and-spigot fittings is limited to water and drainage service. Also, pipe couplings made of cast, malleable, or wrought iron are not permitted for flammable fluids within process limits or hazardous fluids in any area. In addition, they cannot be used for flammable fluids outside process unit limits at design temperatures above 300°F or design pressures above 400 psig.

Comments on Selection and Limitations of Components. There are many Code mandated restrictions for the use of fittings, bends, intersections, and valves which are not universally applicable to all process plant services, and such components are not easily codified. Some guides, however, are of value to the designer and are listed:

- Welding fittings are usually preferred to flanged fittings, not only for economic reasons but because the potential for leakage is reduced.
- Pipe bends are preferred to butt-welding elbows for reciprocating compressor suction and discharge piping, vapor relief-valve discharge piping, and piping conveying corrosive fluids (such as acid where turbulence in a fitting may cause excessive corrosion).
- Bends or dead-end tees should be used for piping which conveys pulverized abrasive solids suspended in gas in the dilute phase. Dead-end tees (so arranged that the flow will impinge against the dead end) have a longer life than bends in abrasive service and should be used if the system can be designed to accommodate the resulting increase in pressure drop.
- Bends should be used for dense-phase flow of pulverized abrasive solids and for all piping which handles either pulverized or granular solids suspended in liquids or granular solids suspended in gases.
- If the flow is through a branch into a header (or run pipe) in a piping system which transports pulverized abrasive solids suspended in gas in the dilute phase, a dead-end cross (so arranged that the flow will impinge against the dead end) should be used.
- In services with very high corrosion rates, butt-welding fittings with the same inside diameter (ID) as the attached pipe (if not the same, consider taper boring the component with the smaller ID) are preferred to threaded and socket-welding fittings.
- Threaded cast-iron fittings should not be used in pressurized process and utility piping.
- Threaded plugs are preferred to pipe caps for threaded end closures to reduce dead-end corrosion problems.
- In most process plants, internal corrosion is a greater problem than external corrosion. Consequently, it is common practice that all ¾ in and larger steel and cast-iron (and all 2½ in and larger brass) gate, globe, and angle valves (located above grade) be of the outside screw and yoke type.
- Although valves with seal-welded or pressure seal bonnets and welding ends are commonly used in steam service, 2½ in and larger valves in process service are usually equipped with flanged ends.
- Valves operated in full-open or block service are generally gate valves. Butterfly valves, ball valves, nonlubricated plug valves, and lubricated plug valves may be considered as possible alternates to gate valves.
- As a general rule, hand-operated throttling valves for services where fine control is not required, and those for control valve bypasses should be globe valves (integral stem and plug preferred) for sizes 2 in and smaller and gate valves for sizes 2½ in and larger. For severe throttling service and where close control is required, a conventional control valve with a hand operator should be used.

The only other common application for globe valves in process service is for mixing purposes.

- Solid-wedge and flexible-wedge gate valves are generally preferred to split and double-disk valves. Split-wedge and double-disk valves are generally used for clean liquids and noncondensing gases only.
- Gate valves with Teflon inserts in the seat rings are very satisfactory in liquid butane and propane services.
- Where coking may occur in blocked connections, a flushing connection should be added between the valve and the process line or equipment. The flushing medium may be oil, gas, or steam (if water can be tolerated in the system).
- If the pressure differential across a closed gate valve is approximately equal to the pressure rating of the valve, consideration should be given to providing a pressure-equalizing bypass around the valve. Consideration should also be given to bypasses for valves in steam lines for warm-up purposes. When bypasses are provided, they should be sized in accordance with the Manufacturers Standardization Society (MSS) of the Valve and Fitting Industry's Standard Practice (SP) MSS SP-45 (bypass and drain connection standard). A gate valve should be provided in the bypass line.
- Drain or bypass connections may be tapped (or socket welded) into a valve body where necessary to simplify piping or to assure complete drainage.
- Do not use check valves in vertical lines in which the flow is downward.
- If a valve is installed with the stem lower than horizontal, the valve bonnet should be provided with a drain.
- The designer should consider providing gear operators for all 6-in and larger lubricated plug valves and all 14-in and larger gate valves.
- The use of double block valves should be kept to a minimum. Double block valves, however, are required for sample connections and for drains which are connected to a closed drain system. Double block valves or their equivalent should be used where contamination must be prevented.

Under certain conditions double block valves are also needed where it is necessary to remove essential equipment from service for cleaning or repairs while the unit continues in operation. However, even in these cases often a single block valve with provisions for blinding will suffice. Such equipment must be provided with a spare, or it must be possible to bypass it temporarily without shutting down the unit. The nature of the fluid, its pressure and temperature, and many other factors must be considered when determining the need for double block valves. Generally, the need for double block valves at equipment should be considered if the fluid is hazardous or very corrosive or if the fluid is above 500°F. Where double block valves are used, a ¾-in valve should be installed between the block valves; see Fig. C7.3.

Some ball valves and nonlubricated valves, when equipped with a bleeder between the seats, have been satisfactory substitutes for double block valves. Both conventional gate valves with Teflon inserts in the seat rings and flexible-wedge gate valves have also been equipped with a bleeder connection and used in place of double block valves.

Selection and Limitations of Flanges. The Code restricts the use of screwed flanges as it does any threaded pipe joint. Slip-on flanges must not be used in installations where many large temperature cycles are expected or if the flanges

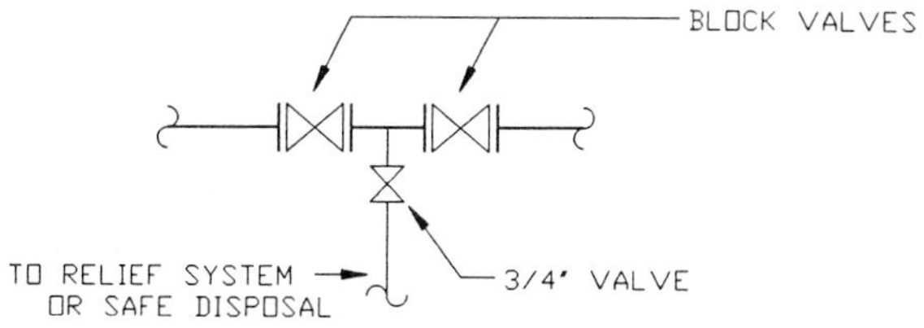

BLOCK VALVES

TO RELIEF SYSTEM OR SAFE DISPOSAL

3/4' VALVE

FIGURE C7.3 Double block and bleed.

are not insulated. The following statements on flanges are included for the designer's guidance.

- The use of cast-, nodular-, wrought-, and malleable-iron screwed flanges should be avoided.
- In services with very high corrosion rates, the bore of weld neck flanges should be the same inside diameter as the attached piping (if not the same, consider taper boring the component with the smaller inside diameter).
- The bore of weld neck orifice flanges should match the inside diameter of the attached pipe.
- ANSI Standard B16.47, Large Diameter Steel Flanges, governs steel flanges in sizes larger than 24 in. However, the designer must ensure that the flange drilling on such flanges will match that of the equipment to which it is to be attached. MSS-SP-44, Steel Pipe Line Flanges, and API 605, Large-Diameter Carbon Steel Flanges, are also useful standards when using flanges larger than 24 in.

Selection and Limitations of Blanks. In a process plant, blanks are usually required to isolate individual pieces of equipment at shutdown and to block off selected process lines positively at the process unit limits. They are also needed during operation wherever positive shutoff is required to prevent leakage of one fluid into another.

Blanks should be located in horizontal lines if possible. Blanks should not be used in vertical water and steam lines in climates where danger of freezing exists.

Circular-handletype blanks can be used for raised face joints in locations where the lines can be sprung easily to permit installation of the blanks. As a rule, this is easily accomplished only in 4-in and smaller lines. Figure-eight-type blanks are used for larger lines, and even then, jackscrews may be needed to install the blank. Blanks should be made from a plate specification, approved for use in ASME/ANSI B31.3, of substantially the same chemical composition as the pipe.

Selection and Limitations of Gaskets. Gaskets must be made of materials which are not injuriously affected by the nature of the fluid or its temperature under anticipated operating conditions. Nonmetallic gaskets are usually not permitted above 750°F gasket design temperature. Various elastomeric gasketing materials have very low (200°F) maximum service temperatures. Also, nonmetallic gaskets should not be used in nonconfining (flat or raised face) flanged joints at gasket design pressure-temperature conditions above the ratings of Class 600 flanges per ANSI B16.5, except that for nonflammable, nontoxic service fluids, the limiting ratings can be those of Class 900 flanges.

The use of metal or metal nonmetallic filled gaskets is not limited as to pressure provided the gasket materials are suitable for maximum fluid temperatures.

The American Petroleum Institute has developed API 601, a standard covering the dimensions of spiral-wound and double-jacketed gaskets. The spiral-wound gasket covered by API 601 has been widely used with great success with raised face flanges as a replacement for the ringtype joint.

Selection and Limitations of Flange Facings. When Class 125 cast-iron or flat face nonmetallic flanges are bolted to Class 150 steel flanges, the 1/16-in raised face on the steel flanges should be removed. If the raised face is removed and a full-face gasket is used, either high-strength carbon-steel bolting or alloy-steel bolting may be used. However, if the face is not removed (or if the face is removed and a ring gasket extending only to the inner edge of the bolt holes is used), the bolting material may not be of higher strength than carbon steel ASTM A307 Grade B.

Selection and Limitations of Bolting. Carbon-steel machine bolts may be used to make flange connections for bolt metal temperature from −20 to 400°F inclusive. This restriction is quite conservative with regard to the pressure limit. Also, these bolts can be used quite safely to the limits of the Class 300 pressure class as permitted in ANSI B16.5 with the use of appropriate gasketing material.

The most widely used bolting materials in process plant are ASTM A193 Grade B7 stud bolts with ASTM A194 Grade 2H heavy semifinished hexagonal nuts. These materials are acceptable from −50 to 1000°F. A number of operating companies use these materials almost exclusively to simplify inventories and reduce the possibility of misapplication of carbon-steel bolting. Carbon-steel bolting may be used with nonmetallic gaskets with flanged joints rated Class 300 and lower for bolt metal temperatures at −20–400°F.

Selection and Limitations of Strainers. The material for the strainer body (including bolting) should be equal to material specified for the valves in the same service. The screen material generally should be the same as the valve trim (e.g., 11–13 percent chrome or Type 316 stainless steel for most services). The location of permanent strainers (as contrasted to the temporary cone type which is installed at a flanged joint) also merits attention.

Centrifugal and reciprocating pumps handling material containing solids should have permanent strainers provided in the suction lines to the pump or in the vessel from which the pump takes suction. The free area of such strainers should be not less than three times the cross-sectional area of the suction line.

In addition, permanent strainers or filters should be provided in the piping for the protection of the equipment indicated in Table C7.2. The maximum clear opening for screens in these strainers varies with the application, but it should not exceed the value recommended for the particular type of equipment. The available pressure differential usually determines the minimum clear opening for screens.

Permanent strainers should have baskets which can be flushed clean during operation or easily removed for cleaning. If considerable clogging of strainers is anticipated, the strainers should be of the self-cleaning or the duplex type to permit continual flow of clean liquid.

Piping Joints. The type of piping joint used must be suitable for the pressure-temperature conditions and should be selected by giving consideration to joint tightness and mechanical strength under the service conditions (including thermal expansion) and to the nature of the fluid handled with respect to corrosion, ero-

TABLE C7.2 Recommended Strainer Screen Openings

Type of equipment	Recommended maximum clear opening
Screw, gear, or cam type rotary pumps	1/4 in
Seals, gland, flushing-oil systems to pumps and compressors	1/16 in
Fuel-oil lines to burners	1/8 in
Steam turbines (if not integral)	1/8 in
Air supply to pneumatically actuated equipment	35 mesh
Centrifugal pump (suction side)	40 mesh conical
Upstream of restriction orifices in bleed services	1/16 in
Reciprocating pump (suction side)	40 mesh conical
Energizing fluids to ejectors	1/4 in
Tubular coolers and condensers using unfiltered cooling water	1/4 in

sion, flammability, and toxicity. In general, the number of disassembly joints is minimized; most joints are welded if the material is weldable.

Welded piping is used almost exclusively for hydrocarbons and other flammable fluids. Bypass piping, alternate process connections, and auxiliary piping systems such as gland oil, seal oil, lubricating oil, fuel gas, fuel oil, heating or cooling oil, flushing oil, flue gas, and blowdown piping, are included. Welded construction is also used for all piping outside process unit limits which is used for the transfer of hydrocarbons or most other process fluids. Piping which is threaded or welded, depending primarily on economic considerations, includes piping (other than specifically mentioned above) in services such as drains, vents, pumpouts, sample connections, and certain instrument leads, which contain process fluids only upon intermittent or occasional use of the piping involved and which are not an integral or essential part of the process system.

Welded Joints. The Code permits welded joints in all instances in which it is possible to qualify welding procedures, welders, and welding operators in conformance with the rules of the Code. There are, however, a few minor additional considerations for seal-welded and socket-welded joints. For example, the Code cautions against the use of socket-welded construction in cases where severe crevice corrosion or erosion occurs. The Code also states that seal welds may be used to avoid joint leakage but that they shall not be considered as contributing any strength to the joint.

Flanged Joints. The number of flanged joints in a piping system is usually determined by maintenance and erection considerations, including sufficient flanged joints for insertion of blanks during shutdown.

Expanded Joints. This type of joint is more commonly used on the piping and tubes for refinery heaters. They are excluded for use in hazardous and toxic services.

Threaded Joints. The threading of pipe with a wall thickness less than ANSI B36.10 standard wall is not permitted, and the use of threaded joints where severe crevice corrosion or erosion may occur should be avoided. Economics will limit the use of threaded piping to small pipe sizes (2 in and smaller) for most services; however, it is used for most galvanized piping.

All pipe threads on piping components must be taper pipe threads in accordance with ANSI B1.20.1 except for the following:

- Pipe threads other than taper pipe threads may be used for piping components, where tightness of the joint depends upon a sealing surface other than the threads and where experience or tests have demonstrated that such threads are suitable for the condition.
- Couplings, 2 in and smaller, with straight-tapped pipe threads may be used on piping components with taper pipe threads if the design conditions do not exceed 150 psig or 400°F and if the fluid handled is nonflammable and nontoxic.

Flared, Flareless, and Compression Joints. Piping joints using flared, flareless, or compression-type tubing fittings may be used within the limitations of applicable standards or specifications. In the absence of such standards or specifications, the engineer shall determine that the type of fitting selected is adequate and safe for the design conditions in accordance with the following requirements:

- The pressure design shall meet the requirements of the code.
- A suitable quantity of the type and size of fitting to be used shall meet successful performance tests to determine the safety of the joint under simulated service conditions.
- Fittings and their joints shall be suitable for the tubing with which they are to be used.
- Fitting shall not be used in services which exceed the manufacturer's maximum pressure-temperature recommendations.

Caulked Joints. The term *caulked joints* applies to joints of the bell-and-spigot type which are permitted only for water service at pressures suitable for the pipe to which they are applied. Provisions must be made to prevent disengagement of the joints at bends and dead ends and to support lateral reactions produced by branch connections or other causes.

Brazed and Soldered Joints. Fillet-brazed or fillet-soldered joints may not be used, but soldered-type and silver brazed socket-type joints are permitted in nonflammable nontoxic service. The low melting point of brazing alloys shall be considered where possible exposure to fire is involved.

Flexibility and Support of Process Plant Piping. The general challenge of flexibility and expansion of piping systems is treated in Chap. B4, Part B of this handbook, and the support of piping systems is covered in Chap. B5, Part B of this handbook. Techniques, methods, and procedures developed in those chapters are applicable to process plant piping.

Fabrication, Assembly, and Erection of Process Plant Piping. Fabrication, assembly, and erection of process plant piping are covered at length in Chap. A6, Part A of this handbook. Complete detail and exact requirements which relate to fabrication, assembly, and erection of the process piping systems are given in ASME/ANSI B31.3. Important phases of the Code treatment of these topics are covered briefly below.

Materials for Welding. All filler materials must comply with the requirements in Section IX ASME, Boiler and Pressure Vessel Code. If backing rings are used in services where their presence will result in severe corrosion or erosion, it is required that the backing ring be removed after welding and the internal joint ground smooth.

End Preparation. ASME/ANSI B16.25 provides dimensional standards for weld end bevels. Preferably, the ends of pipe and the edges of plate which are to be formed into pipe should be shaped by machine. Other methods of shaping may be used provided a reasonably smooth surface suitable for welding and free from tears, slag, scale, and grease is attained. Oxygen or arc cutting is acceptable only if the cut is reasonably smooth and true and all slag is cleaned from the flame cut surfaces.

If piping component ends are machined on the inside for backing rings, such machining must not result in a finished wall thickness, after welding, less than the minimum design thickness plus corrosion and erosion allowances. Generally a root gap of ⅛ in is used for joints (including branch connections) without backing rings, except that where the pipe wall thickness is less than 3/16 in, a 1/16-in root gap is generally used.

Other Alignment Considerations. Flange bolt holes should straddle the established center lines unless other orientation is required to match the flange connections on equipment. Slip-on flanges should be positioned so the distance from the face of the flange to the pipe end is about equal to the nominal pipe-wall thickness plus ⅛ in. Welding neck orifice flanges should be the same bore as the pipe to which they are attached and must be aligned accurately. Longitudinal seams in adjoining lengths of welded pipe should be staggered and located to clear openings and external attachments.

The following restrictions, limitations, or guidelines apply to welding of process piping:

- Projection of weld metal into the pipe bore at welded butt joints should not exceed 1/16 in for pipe 8 in and smaller or ⅛ in for larger pipe. Excessive projection on accessible joints should be removed. Welds attaching welding neck orifice flanges to pipe should be ground smooth on the inside.
- On ferritic materials to be used below 20°F, the welder's identification mark should preferably be made with ink stencil. Steel stamping should be avoided.
- Welding procedure for fillet welds and seal welds. The Code does not permit cracks in fillet or seal welds and limits undercutting to 1/32 in for these welds. Fillets welds may vary convex to concave. If seal welding of threaded joints is performed, the Code requires that all exposed threads be covered by the seal weld and that the welding be done by qualified welders. In addition to the Code requirements, it is strongly recommended that (1) threaded joints be made dry (without thread compound), (2) seal welds be at least two-pass (preferably three-pass) welds using a 3/32- or ⅛-in electrode (5/32-in electrode is acceptable for 2½-in and larger pipe size), and (3) valve and union ends be welded by the electric arc process to minimize distortion and ensure that valves be closed during welding.

Welding Procedures. Before welding, all surfaces must be cleaned and free from paint, oil, rust, scale, and other detrimental material. Furthermore, welding is prohibited if there is impingement of any rain, snow, sleet, or high wind on the weld area.

The following Code requirements apply to girth butt welds and any longitudinal butt weld in a piping component which is not made in accordance with a standard or specification:

- If the external surfaces of the two components are not aligned, the girth butt weld must be tapered between the two surfaces.
- Tack welds, if not made by a qualified welder using the same procedure as the completed weld, must be removed. Tack welds which are not removed should be made with an electrode which is the same as or equivalent to the electrode to be used for the first pass. Tack welds which have cracked must be removed.
- The types and limitations of imperfection required to be evaluated with various types of examinations are shown in the Code.

Preparation and Welding Procedure for Welded Branch Connections. Branch connections (including specially made integrally reinforced branch connection fittings) which abut the outside surface of the run wall or which are inserted through an opening cut in the run wall must be so arranged as to provide a good fit and be attached by means of full-penetration groove welds.

The recommendations for spacing and location of branch connections contained in Pipe Fabrication Institute (PFI) Standard ES7 should be followed. A good fit must be provided between reinforcing rings and saddles and the parts to which they are attached. When rings or saddles are used, a vent hole is provided (at the side and not at the crotch) in the ring or saddle to reveal leakage in the weld between branch and main run and to provide venting during welding and heat-treating operations. Reinforcing pads should be proportioned so the diameter of a vent hole is not greater than one-third of the pad width.

Tolerances for Welded Piping. A widely accepted tolerance on face-to-face and center-to-face dimensions of welding piping is ±⅛ in. As for the location of the flanges, their lateral translation in any direction from the specified position should not exceed 1⁄16 in. Also, the alignment of flanges should not deviate from the specified position, measured across any diameter, by more than 1⁄32 in.

Qualification. Qualification of the welding procedures to be used and of the performance of welders and welding operators is required to comply with the ASME Boiler and Pressure Vessel Code, Section IX.

Defect Repairs. Weld defects which require repair must be removed. All repair welds must be made with the same welding procedure initially used.

Bending and Forming. Pipe may be bent by any hot or cold method consistent with material characteristics of the pipe being bent and the intended service. It may be bent to any radius which will result in a bend arc surface which is free of cracks and buckles.

Hot bending and forming must be done within a temperature range consistent with material characteristics, end use, or postweld heat treatment. It is recommended that hot bends in pipe sizes 1½ in and larger be packed with high-temperature silica sand and that the pipe be uniformly heated before the hot bending operation. When pipe must be threaded before bending, forging, or heat treating, all exposed threaded surfaces should be protected during heat treatment.

Cleaning after Fabrication. Following fabrication, all loose scale, weld spatter, slag, sand, and other foreign material should be removed from the piping.

PFI Standard ES5 is an acceptable standard for cleaning fabricated piping. Piping should not be painted in the fabricating shop (i.e., before it has been erected and tested).

Heat Treatment. The welding procedure qualification establishes the necessity for preheating and postheating welds (and the temperatures and soaking period to be used) in order to restore or obtain the physical properties of the materials (such as strength, ductility, and corrosion resistance) needed to satisfy end-use requirements.

Bolting Procedure for Flanged Joints. It is a good practice to apply an antiseize thread compound to the bolts before the nuts are installed. A mixture of graphite and oil is one of the best substances for this purpose.

In bolting joints using spiral-wound gaskets, the gasket should be compressed to about 25 percent of the original thickness. Spiral wound gaskets conforming to API Standard 601 have a 0.125-in thick outside gauge and centering ring. The gasket is seated when it is compressed until the flange faces touch the gauge ring.

Steel-to-cast-iron flanged joints must be assembled carefully in order to prevent damage to the cast-iron flange. Both flanges in steel-to-cast-iron flanged joints should be flat faced. These joints should be made up with extreme care, taking up on bolts uniformly after fitting flanges into close parallel and lateral alignment. Flanges which connect piping to mechanical equipment, such as pumps, turbines, or compressors, should be fitted up in close parallel and lateral alignment before tightening the bolting.

Cast-Iron Bell-and-Spigot Piping. Bell-and-spigot joints in cast-iron piping must be assembled using poured lead or other joint compound suitable for the service. Usually each cast-iron bell-and-spigot joint is packed with hemp, poured full of lead (with a minimum number of pours), then caulked. The depression of lead below the face of the bell, after joint caulking, should not exceed ¼ in. (Lead wool can be used where it is not permissible to pour lead.)

Threaded Piping. Any compound or lubricant used on threads must be suitable for the service conditions and compatible with both the service fluid and the piping materials. Threaded joints which are to be seal welded should be made up without any thread compound.

Erection of Corrugated Expansion Joints. Corrugated expansion joints should be installed as-shipped from the manufacturer or compressed for the cold condition at erection, depending on anticipated direction and magnitude of movement after the pipeline reaches operating temperature. The manufacturer's recommended total travel should preferably straddle the calculated travel.

Erection of Valves. Valve packing glands should be checked for the quality and quantity of packing, and lubricated plug valves should be provided with proper lubricant.

Erection of Pipe Supports. In addition to the major supports specified by the design drawings, minor supports as found necessary in the field should also be installed to prevent undesirable vibration, sag, lateral movement, and stresses. Spring hangers, including constant-support type, should be checked for proper adjustment of travel and be correctly positioned for the cold condition of erection.

Cleaning of Lines after Assembly and Erection. After completion of erection, scale, dirt, welding electrodes, slag, and other foreign material should be removed from the lines. Particular attention should be given to the cleaning of air lines and compressors and blower, pump, and turbine inlet piping. Before the initial operation, steam lines to turbines and to the steam ends of reciprocating compressors and pumps should be blown down with steam, 100 psig or higher.

All practical precautions should be taken to prevent the introduction of foreign matter into pumps, instruments, and other equipment. Cleaning may be accomplished by flushing out the lines. Temporary strainers should be used at pumps during the flushing operation unless spools or valves can be conveniently dropped out and suitable deflectors provided to prevent refuse from entering the pumps. Consideration should be given in dismantling those lines which cannot be adequately cleaned by flushing.

Examination Inspection and Testing. Before initial operation, a piping installation should be inspected to the extent necessary to assure compliance with the engineering design and with the material, fabrication, assembly, and test requirements of the Code. An employee representative of the owner should be responsible for this inspection. This examiner may delegate performance of any part of the inspection to inspectors who may be employees of his own organization, of an engineering or scientific organization, or of a recognized insurance or inspection company.

A nondestructive examination (NDE) plan should be consistent with service severity, incorporating process and mechanical factors. The NDE plan focus should be on those pipelines where failure would produce the most harm to personnel or property. Some issues the designer should consider when developing the NDE plan are as follows:

- *Service factors:* Corrosivity, toxicity, and flammability—hazardous nature of flowing media. The designer would want to apply a sound NDE program to detect flaws in these dangerous streams.
- *Mechanical factors:* Temperature, pressure, cyclic conditions, thermal bending stresses, vibration. Fatigued and highly stressed lines are more likely to fail than low-stressed lines. Detection and removal of flaws can provide additional service life.

Visual Examination. All welds are required to be capable of compliance with the limitations on imperfections specified in the Code for visual examination. Visual examination consists of observation by the inspector of whatever portions of a component or weld are exposed to such observation, either during or after manufacture, fabrication, assembly, or test.

Supplementary Types of Examination. The following supplementary types of examination are not required unless specified by the engineering design because of special service conditions requiring a high degree of freedom from imperfections. If such examination is specified for a weld, it is only required that the weld examined be repaired, if necessary, so the weld imperfections comply with the limitations in the Code for the type of examination used. If supplementary types of examination are specified, they should be performed after completion of any postheat treatment where required. If any of the following types of examination is specified by the engineering design, it should be performed to the extent as follows:

Magnetic Particle: An area to be examined by magnetic particle examination can be completely examined or examined on a random sampling basis, as specified.

Random Radiography: X-Ray or gamma-ray radiography may be used. The selection of the method should be dependent upon its adaptability to work being radiographed. When random radiography of welds is specified by the engineering design, it should be done on the number of welds designated. The engineering design shall specify the extent to which each examined weld should be radiographed. Random radiography may also be used for examination of piping components such as a valve or fitting to any extent specified by the engineering design.

100 Percent Radiography: If 100 percent radiography is specified for welds in piping, each weld in the piping shall be completely radiographed.

Hardness Tests: The extent of hardness testing required shall be specified by the engineering design, considering the severity of the service, type of material, and other pertinent factors.

Pressure Tests. Before initial operation, piping must be pressure tested to assure leak tightness. If repairs or additions are made following the pressure tests, the affected piping is retested, except that in the case of minor repairs or additions the owner may waive retest requirements. The pressure test is maintained for a sufficient time, but not less than 10 min, to determine if there are any leaks.

Water is commonly used as the test fluid except if there is a possibility of damage due to freezing or if the operating fluid or piping material would be adversely affected by water. If hydrostatic testing is not considered practical, a pneumatic test using air or another nonflammable gas may be substituted.

A preliminary air test at not more than 25 psig is made before hydrostatic test in order to locate major leaks. If pressure tests are conducted at low metal temperatures, the possibility of brittle fracture must be considered. Hydrostatic pressure tests are conducted at 1.5 times nominal design pressure, adjusted for temperature. (See applicable portions of Code for formulas.) Pneumatic tests are conducted at 1.1 times nominal design pressure.

Test Preparation. All joints, including welds, are to be left uninsulated and exposed for examination during the test. If a joint has been previously tested in accordance with the Code, it may be insulated or covered. Piping designed for vapor or gas shall be provided with additional temporary supports, if necessary, to support the weight of the test liquid. Expansion joints shall be provided with temporary restraint, if required, for the additional pressure load under test or shall be isolated from the test.

Equipment which is not to be included in the test shall be either disconnected from the piping or isolated by valves or blanks.

All pressure gauges, gauge glasses, flowmeter pots, liquid level float gauges, and all other pressure parts of instruments, together with the piping connecting the instruments to the main piping, should be included in the hydrostatic test. Relief valves and rupture disks should not be subjected to the pressure test. If a pressure test is to be maintained for a period of time and the test liquid in the system is subject to thermal expansion, precautions must be taken to avoid excessive pressure buildup.

Pneumatic Testing. If piping is tested pneumatically, the test pressure is set at 110 percent of the design pressure. Pneumatic tests include a preliminary check at not more than 25 psig; the pressure is then increased gradually in steps providing sufficient time to allow the piping to equalize strains during test and to check for leaks.

Test Records. Records must be made of the tests, including date of test, identification of piping tested, test fluids, test pressure, and approval by inspector.

Special Design Piping Systems

Jacketed Pipe. Jacketed pipelines (Fig. C7.4) are commonly used to convey certain fluids in process facilities. Process fluids that require stringent temperature control (i.e., molten sulfur) are good candidates for jacketed pipe applications. Molten materials (i.e., polymers) where high-temperature maintenance is required are also candidates for jacketed pipe construction.

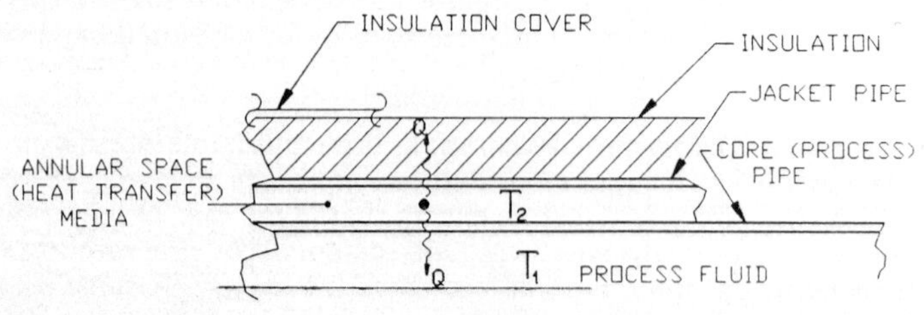

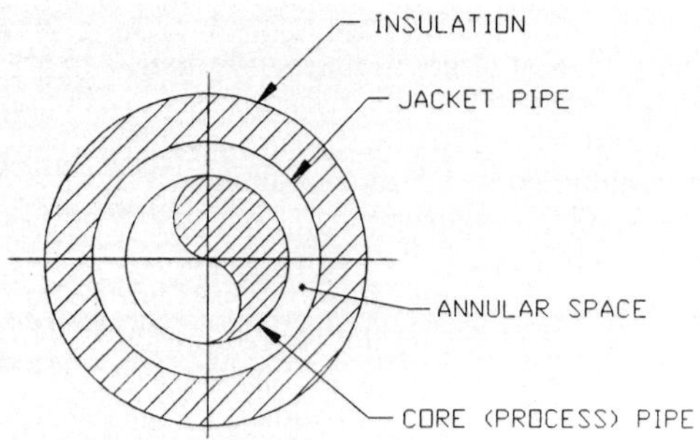

FIGURE C7.4 Heat transfer jacketed pipe.

The advantages afforded by jacketed pipeline construction over other (i.e., tracing) heat transfer methods can be briefly described as follows:

- Uniformity of heat input around circumference of process pipe
- Tighter temperature control over entire pipeline length
- Elimination of cold spots that may cause process fluid degradation or localized freezing

Various heating media are used for temperature control of the process fluid. Liquid and vapor phase fluids are used, each with their own specific advantages and design requirements.

Liquid Phase. Jackets are considered as circuits each having its own valved supply and return connections from the pipe header. The number of jackets included in a circuit is a function of heating medium heat loss, pressure drop through the jacket circuit, and position and location of the piping. As indicated in Fig. C7.5, fluid supply introduced at the lowest tapping of a jacketed part pass

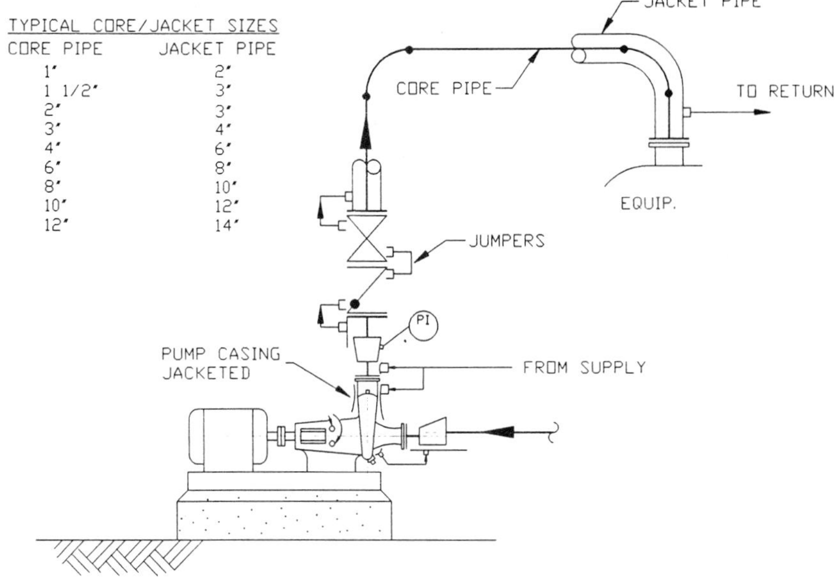

FIGURE C7.5 Jacket liquid phase—circuitry.

through the jacket and exits at the highest jacket tapping. This method of piping continues for the length of the circuit. Jumpers (Fig. C7.6) are used to carry the liquid across flanged connections. The fluid is then returned to the heater via the return header, and a new connection from the supply main feeds the next circuit. Temperature of any circuit is controlled by throttling the quantity of fluid flowing to the jackets.

Vapor Phase (Fig. C7.7). The number of jacket sections heated with a condensing vapor in any one circuit has the same considerations as liquid heating mediums. Unlike liquid phase heat transfer fluids, a jacketed pipeline heated with a condensing vapor requires the vapor inlet pipe be connected to the uppermost jacket tap rather than the lowest tap. Jumpovers (Fig. C7.8) carry vapor to the top of each section or flanged joint. The condensate is drained from each section or flanged joint, collected, and piped to a trapped common return header.

Jacketed pipeline construction details vary depending upon process factors. When maximum heat transfer is desired, a full jacket is used (where the jacket pipe is welded to the back of oversize flanges). This technique will minimize any potential cold spots. Partial jackets are used for those services where product contamination or danger of hazardous conditions could arise if product and heating media could mix or where temperature control is not critical and localized cold spots would not be detrimental to pipeline performance.

Nonintegral, or strap on, jackets are usually used to provide a means of heat transfer to a pipeline component that may not be adaptable to integral jacketing (i.e, valve bodies). In all cases, the heat transfer distribution circuitry must be properly designed and installed for satisfactory performance.

Vacuum jacketed piping systems are used to transfer cryogenic temperature process fluids. The vacuum is established to minimize heat gain from the atmo-

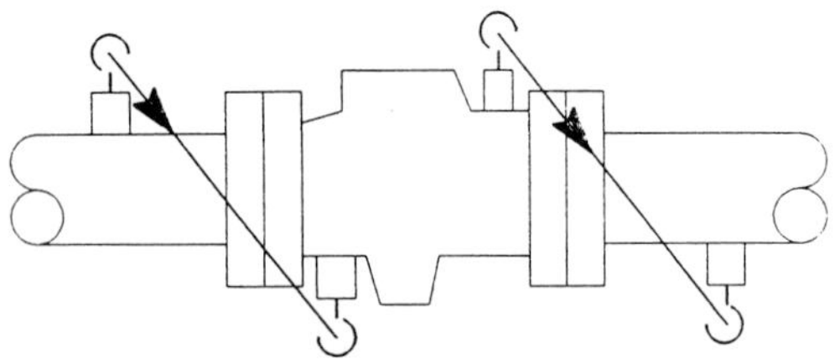

FIGURE C7.6 Jumper-liquid phase jacketed pipe.

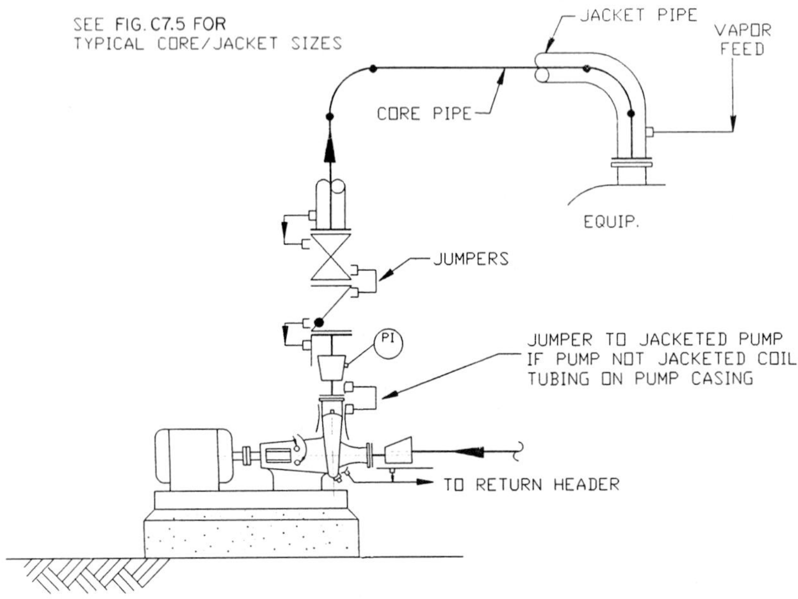

FIGURE C7.7 Jacket vapor phase—circuitry.

sphere to the cryogenic fluid. With the addition of external insulation, little heat gain and resulting fluid vaporization will be realized. Cryogenic systems' piping is covered in depth in Chap. C8, Part C of this handbook.

Refractory-Lined Pipe. Often elevated temperature pipelines can be candidates for internally lined systems. Economic material advantages can be gained by applying thermal insulation on the inside of the pipeline rather than on the outside. Some of the immediate advantages are as follows:

- Pipe wall (pressure boundary) temperature is considerably lower, thereby allowing the use of lower alloy or carbon steel material at higher allowable stresses.
- Decreased thermal expansion allowing tighter piping layout.

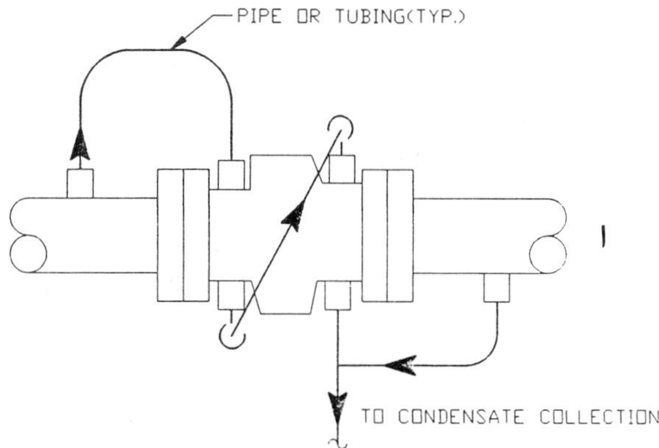

FIGURE C7.8 Jumper vapor phase jacketed pipe.

These factors affect lower installed cost. The economic advantage becomes increasingly significant as temperatures, pressures, and pipe sizes increase.

The design of internally lined pipe entails a number of unique considerations. One consideration is the effect of the process fluid on the thermal conductivity of the refractory. Low molecular weight gases, such as hydrogen, tend to permeate refractory linings. Influencing parameters include pressure, temperature, refractory density and composition, application techniques, and refractory cell structure. Empirical correction factors based on field experience have been developed for various processes and combinations of refractory types.

Selection of lining materials and design details are important to ensure a properly functioning system. Depending on service conditions, either a single-layer or dual-layer lining can be used. An advantage of the dual layer is that the internal layer can be made of a relatively dense refractory with good strength and erosion-resistance properties without concern for insulating values. This thickness should be sufficient to prevent through-wall cracking.

The outer layer next to the pressure boundary pipe wall can be a lightweight refractory with good insulating properties to provide maximum temperature reduction. Various schemes for refractory installation have been used for dual layer linings—some designs are shown in Figs. C7.9, C7.10, and C7.11.

Typical application for refractory-lined steel-shell transfer lines are found in process applications when high temperatures (greater than 1000°F) and significant flow rates (and corresponding large diameter pipelines) are concurrent design parameters. For example, such lines with diameters of 60 in and larger often carry flue gases from regeneration vessels, cokers, and crackers in refinery application. Smaller diameter lines are often used to transport liquid metals.

Plastic-Lined Pipe. Internal plastic liners are available in numerous materials that can be selected for use for a specific media. Some commonly used lining materials are polypropylene, tetrafluoroethylene (TFE) and polytetrafluoroethylene (PTFE), Teflon, vinyldiene fluoride (PVDF-Kynar), and polyvinyl chloride and chlorinated polyvinyl chloride (PVC, CPVC). The properties of plastic-lined piping components permit these materials to be considered for the transportation of a variety of combinations and concentrations of fluids.

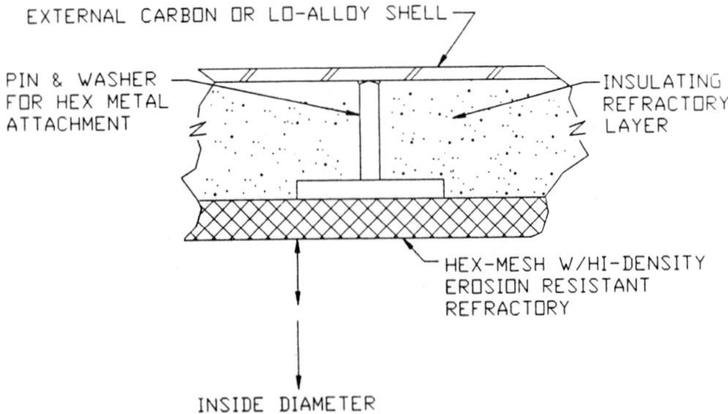

FIGURE C7.9 Dual layer refractory lining for erosive service.

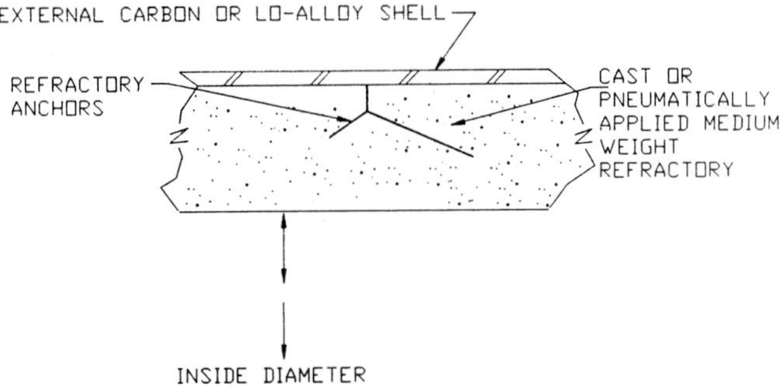

FIGURE C7.10 Single refractory lining for clean service.

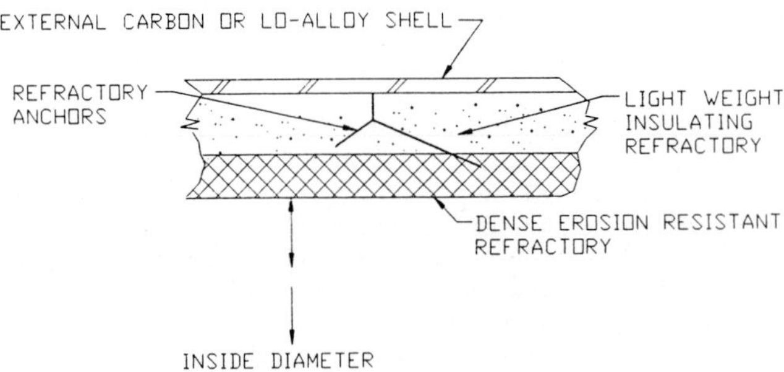

FIGURE C7.11 Dual layer refractory lining for hydrogen bearing streams.

Corrosion resistance is one consideration for use of plastic-lined piping; however, other applications can also be candidates for plastic-lined piping, some of which are as follows:

- Maintenance of process fluid purity, such as the processing of food products.

- Water treatment facilities and laboratory waste disposal systems.
- For secondary containment of dangerous, flammable or environmentally damaging liquids. Because of its steel outer shell construction, the lined pipe will remain structurally self-supporting at higher temperatures and is less likely to rupture than solid plastic pipe.

Thermoplastic and thermoset resin piping systems are described further in Part D of this handbook.

Glass and Glass-Lined Pipe. Borosilicate glass is used for specific piping services in the chemical and pharmaceutical industries. Some advantages afforded by glass piping are as follows:

- Outstanding corrosion resistance for a variety of aggressive chemicals
- Smooth porefree surface
- Transparency
- Lack of effect on taste and odor
- Inertness

Glass-lined piping components offer much of the same advantages afforded by solid glass piping and offer secondary containment, increased resistance to shock loads, and installed economies compared to other corrosion-resistant systems.

System Layout

General Piping Arrangement. Piping arrangement and layout is discussed in depth in Chap. B3, Part B of this handbook. The following are some important considerations in the design and layout of the various piping systems.

A general rule in piping layout is that lines should be located in as neat and orderly manner (in groups or banks whenever practicable) as is consistent with economical design, pressure loss considerations, and satisfactory supporting arrangements. With the exceptions of water, drainage, and pumpout lines, the accepted practice on a process unit is to run the piping overhead, providing 7 ft or more of clear headroom over walkways and platforms. Piping in a process unit should not be located at grade especially in areas where frequent personnel traffic is likely.

All piping and equipment requiring regular attention of the operating and maintenance personnel should be readily accessible. Also, adequate, clear working spaces (minimum width 3 ft) should be maintained around equipment such as pumps, heat exchanger, control valves, instruments, and tower manways, which require frequent servicing. Consideration should be given to providing lateral (and vertical) clearance for the use of motorized materials-handling equipment in maintenance work. If practical, all valves should be located so they may be replaced and operated from grade, permanent platforms or small portable platforms. If the bottoms of the handwheels are more than 6 ft above a platform or grade level or if otherwise inaccessible, the valve should be equipped with extension stems or chain operators.

Chain operators should be used on threaded end valves only if the ends of the valves are seal welded or if the valve is in a vertical line. Also, the use of chain operators on 1-in and smaller valves is undesirable because of the possibility of bending the valve yoke or stem.

High temperature lines in a process plant pose problems for other piping and equipment. For example, hot lines, with temperature higher than 100°F, should be routed so as to avoid electrical conduits. Steam and condensate should not be discharged into the ground in the vicinity of electrical conduits. Lines containing corrosive chemicals should not be located near hot lines or other sources of heat.

Rack Piping. A pipeway is the space allocated for routing several parallel adjacent pipelines. A pipe rack (Fig. C7.12) is the structure used for carrying the pipelines and electrical and instrument trays. The pipe rack is usually constructed of steel or concrete frames called bents on top of which the pipeline rests.

Pipe racks are expensive; therefore their lengths should be minimized. Pipe racks are necessary for arranging the process and service pipelines throughout the plant and are used in secondary ways, principally to provide a protected location for auxiliary equipment, pumps, utility stations, manifolds, fire fighting, and first aid stations. Lighting and other fixtures can be fitted to the pipe rack columns. Air-cooled heat exchanger can be supported above the pipe rack.

Some other considerations when arranging piping on pipe racks are as follows:

- Place utility and service piping on the upper level of double-deck pipe racks.
- Do not run piping over columns as this will prevent adding another level.
- Locate large liquid-filled pipelines near columns to reduce bending stresses on pipe rack beams.
- Allow for future space approximately 25 percent of final width.

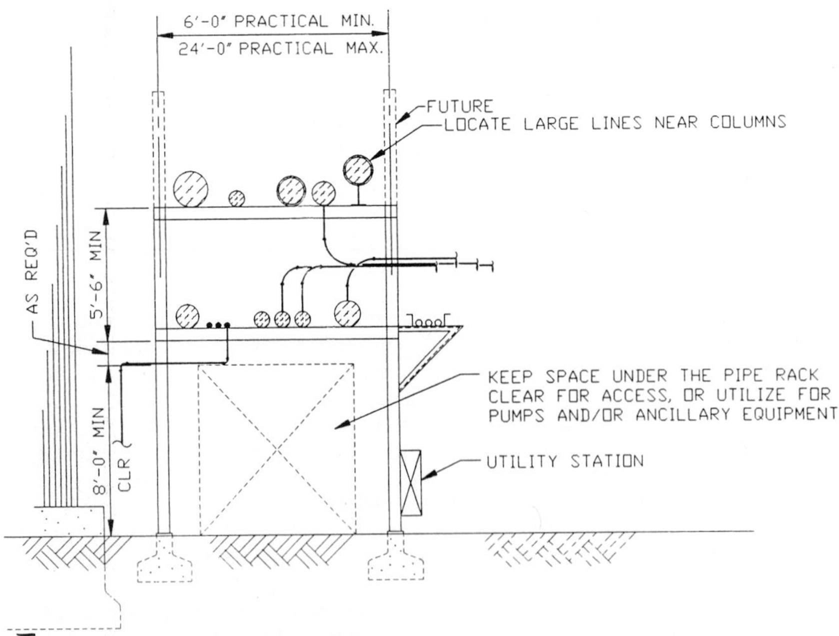

FIGURE C7.12 Pipe rack cross section.

- When possible, place electrical and instrument trays on outriggers or brackets to prevent interference with pipes leaving the pipe rack.
- Adjust elevation (up or down) of horizontal lines when making a change in direction. This will avoid blocking space for future lines.
- Support piping on sleepers at grade if roads or walkways will not be required over the pipeway at a later date.
- Minimum clearance under the pipe rack is a function of the available mobile lifting equipment requiring access and the minimum vertical clearance determined by the basic plant design parameters.
- Group hot pipeline at one side of the pipe rack for ease of support.

Vent and Drain Piping. Valved drain and vent connections should be provided for most types of process equipment. These drain and vents should be located on the equipment, if practical, but may be located in connected piping where there are no valves or blocks between the drain or vent connections and the equipment. Piping to these connections should be arranged to drain the equipment and the connected piping to the appropriate process drainage system. The alternate to complete drainage is a start-up procedure for water removable, such as displacement by circulation, gradual heating during start-up, dry gas purging, or high-velocity gas purging. Multistage pumps, furnace headers, control valves, and horizontal pipe that deflects between supports are typical locations where it is usually impractical to provide complete drainage.

Generally, drain connections to closed drainage systems should be provided with double block valves and with a bleed connection or sight drain between the block valves. The sight drain should be drawn from the spool piece between the double blocks, and the sight drain should be provided with a single block valve blinded or plugged.

Drains, vents, and pump outs for piping and equipment in vacuum service should be blinded or plugged during operation of the unit to prevent the entrance of air. There are, however, drain and vent connections which need not be connected to a drainage system. Examples are connections which are not hazardous if left open, connections for checking water accumulations, and vessel vents which are not needed during operation. These drains and vents should be provided with a block valve and a pipe plug.

In lines containing hazardous fluids, a drain should be provided between block and check valves where fluids could be trapped. Where check and block valves separate a hazardous fluid from process piping or other process equipment, the block valve should be located between the check valve and the process piping or equipment.

Water drainage from vessels in light-ends service can be complicated by the refrigeration effect of light hydrocarbons that vaporize at atmospheric pressure. An ice plug formed by this refrigeration effect can prevent proper valve closure, and hazardous vapors will be released when the ice melts. In most cases heat tracing or other means of heating drain lines and valves will prevent freezing.

Means should be available for removing the operating liquid contents from all vessels and heat exchanger units and the connected piping. Although process lines and pumps should be used for this purpose, an auxiliary pumping out system may be needed. (A permissible alternate is to use steam to remove the contents of the equipment by pressure.) On pressure vessels, pumpout connections should be provided for side drawoffs as well as at the bottom of the vessel.

The recommended minimum size for pipeline drains and vents is ½ in. The recommended minimum size of drains and vents is 1 in for vessels and ¾ in for all other equipment. However, the size of vent and drain connections should be such that the water used for hydrostatic test or flushing may be drained off without pulling a vacuum. On some small pumps, compressors, and turbine and steam engines drives, ¾-in or larger drain and vent connections are not economical. In such cases, ½-in drains and vents are acceptable.

Instrument Piping and Sample Connections

Sample Connections. On a process unit, sample connections should be provided on all feed and product streams and on such intermediate streams as are necessary for control and testing. Sample piping should be as short as possible and be adequately braced to enable it to resist unexpected external loads and to protect it from damage when valves are operated. If the piping is carefully supported and anchored, it is permissible to use an equipment drain for sample purposes. It is preferred that connections directly on pumps, compressors, and other equipment subject to vibration are not used for sample connections if other connections where samples might be taken are available. Sample connections may be used for the installation of pressure gauges. It is suggested that the minimum size of the first nipple attached to the piping or equipment from which the sample is taken be ¾ in. A block valve of the same size as the nipple should be installed at the end of the nipple. A second valve should be installed in the sample line as close to the sampling point as practicable. A sample cooler will sometimes be necessary to assure safe handling of the stream being sampled.

Instrument Piping. Instrument as used here includes all piping and piping components used to connect air or hydraulically operated control apparatus. It does not include instruments or permanently sealed fluid-filled tubing systems furnished with instruments as temperature pressure responsive devices. Instrument piping must meet all the applicable requirements of the associated principal piping systems and the following:

- The design pressure and temperature for instrument piping are to be determined with consideration of short-time conditions. If it presents a more severe condition, the temperature of the piping during periodic operation of the blowdown valve should be considered a short-time condition.
- Consideration must be given to the mechanical strength (including fatigue) of small instrument connections or apparatus.
- Instrument piping containing fluids which are normally static and subject to freezing must be protected by heat tracing or other heating methods.
- When it is necessary to blow down or bleed instrument piping systems containing hazardous fluids, consideration must be given to the safe disposal of such fluids.

Pump Piping. Permanent strainers should be provided upstream to pumps handling streams which are likely to contain foreign material such as sand and scale. Temporary strainers, preferably of the cone type, should be provided for initial unit start-up where permanent strainers are not provided and should be located as close as possible to the pump suction nozzle. A block valve should be provided in the suction line of each pump and located upstream of the strainer. For dirty streams where two or more pumps take suction from a single header, the block valves should be located as close as is practical to the header to minimize the collection of dirt upstream of the valve.

A block valve should be provided in the discharge line of each pump. A check valve should be installed in the discharge line of each centrifugal or rotary pump unless there is no possibility of a reversal of flow or pressure surge under any condition. The check valve should be located between the pump and the block valve with a drain between the block and the check valve.

When the discharge line contains a quick-closing valve, the necessity for shock-absorbing equipment should be investigated if the closing time of the valve cannot be increased to a safe level. Where a remotely located valve can be closed against the pump and the pump does not shut off automatically or cannot be shut off immediately, a recirculating line should be provided from the pump discharge back to the point of suction. The purpose of recirculating the fluid is to prevent damage to the pump due to overheating. The minimum size of the recirculating line should be ¾ in. The line should be equipped with at least one gate valve and an orifice sized to restrict flow to the minimum pumping rate of the pump.

Standby pumps, which may be idle during plant operations and which have to start quickly, should be provided with warm-up lines if pump design temperature exceeds 450°F or if the process fluid will solidify at atmospheric temperature. The purpose of these warm-up lines is to eliminate undesirable thermal effects on lines and equipment and plugging of idle pump and piping materials. A warm-up line should consist of a ¾-in-valved bypass around the pump discharge block and check valves. The standby pump should be kept at operating temperature by opening the warm-up lines and cracking the suction block valve(s) to permit a small flow back through the idle pumps.

If the process fluid will solidify at atmospheric temperature and the suction and discharge lines are not heated and insulated, an additional ¾-in-valved bypass should be provided from the discharge lateral to the suction lateral and the header side of the valves. When the pump is removed from service, these laterals should be kept at operating temperature by opening this bypass valve to permit a small flow. The above warm-up lines should be heat traced if the process fluid will solidify at atmospheric temperature. Warm-up lines should be checked for adequate flexibility for the differential expansion between the pump discharge line and the warm-up line.

Exchanger Piping. Generally, bypass piping around exchanger should be provided only where required for temperature control. There may be cases when the increase in operating efficiency resulting from cleaning or repair during the operation of the rest of the process unit, would justify the cost of installing a bypass. All streams which are to be heated should enter at the bottom of the exchanger, and all streams to be cooled should enter at the top of the exchanger.

Block valves need not be provided on the process side of the exchanger, except where the valve is needed for flow control or where the exchanger may be bypassed while the unit is running. A pipe spool, elbow, or some such removable piece (other than the block valve) should be provided adjacent to the channel section of any exchanger which will be opened while the unit is in operation. Lines to condensers should be sized to provide sufficient velocity to carry condensed liquids along with the vapors. Pockets must be avoided in these lines.

Pressure Vessel Piping. The piping designer is cautioned that often limitations concerning piping connections at vessels are detailed in Section VIII of the ASME Boiler and Pressure Vessel Code. Piping between the vessels protected by the same relief valve and piping between the vessel and the vessel's relief valve may be in that category. The piping designer should coordinate piping requirements with the vessel designer to achieve the optimum nozzle locations. For economy and ease of support, piping at a tower should drop or rise immediately upon leaving the tower nozzle and run parallel and as close as practical to the vessel itself.

Process requirements usually govern the location of valves in vessel piping. However, block valves should generally be provided at vessel nozzles for the following:

- Vapors and reboiler lines.
- Safety and relief valves.
- Side-stream draw-off lines.
- Cracking unit transfer line containing quench valves and furnace transfer lines to vacuum vessels.
- Lines containing check valves located outside of building and within 30 ft in a horizontal direction from the vessel nozzle.
- When vessel nozzles are located inside the vessel skirt and a valve is needed, a connecting pipe may be attached to the nozzle and a valve bolted to the end of the pipe outside the skirt.

Another consideration in deciding where valves are needed is that the rupture of a line connected below the vessel liquid level (or to the dense phase in a fluid-solids vessel) would drain the vessel unless there was a valve for this purpose. In deciding whether these valves are needed, the likelihood of mechanical damage to the line will be the prime consideration. Small lines (2 in and smaller) are more susceptible to damage than large lines. If a line connecting two pressure vessels below the normal liquid level is short (e.g., less than 20 ft long), no valves need be provided unless required for process reasons.

Compressor Piping. Special precaution is necessary in the design and fabrication of the piping at or near compressors to reduce fatigue failures. This piping should be designed to have the minimum of overhanging weight. (This is mostly a problem with high-pressure compressors where valves are very heavy.) Butt-welding fittings should be used wherever practical, and fit-up should be accurate. Braces should be provided as needed to reduce vibration, and consideration should be given to grinding all welds to remove surface discontinuities.

A check valve should be installed in the discharge line from any centrifugal or rotary compressor discharge into a system from which liquid or gas may flow backward through the compressor. The check valve should be located as close as possible to the compressor.

When a compressor takes suction from a header, the suction lateral should preferably be connected to the top of the header. However, if the lateral is at least one pipe size smaller than the header, it is permissible to make a center-line connection to the side of the header. Temporary screens should be provided for initial compressor start-up and should be located as close as possible to the compressor unless permanent screens or filters are installed immediately adjacent to the compressor.

Means to reduce excessive surge and vibration should be provided as necessary in the suction and discharge lines of all reciprocating and rotary compressors and located as close as practical to the compressor. Where surge chambers are provided, the connecting pipe should extend into the bottom of the chamber. The flow characteristics of centrifugal compressors should be investigated to determine if devices are required to prevent surging during start-up. Knockout drums should be provided upstream of all compressors except those which handle gases with no possibility of condensate being formed. That is, most air and nitrogen compressors do not require knockdown drums. Compressor suction lines between the knockdown drum and the compressor should be as short as possible, without pockets, and horizontal and sloped toward the compressor. Also, for wet gas compressors, this portion of the suction line should be insulated. It may require auxiliary heating in the form of heat tracing to prevent condensation. Where the line between the knockdown drum and the compressor is long, low points in compressor suction lines should be provided with drains to remove any possible accumulation of liquid. If the suction line normally operates under vacuum conditions, all drains between the knockdown drum and the compressor should discharge into the knockdown drum. Compressor discharge piping should be analyzed for flexibility under thermal load resulting from the heat of compression.

Storage Tank Piping. Many storage tanks do not require separate discharge and suction headers, connections, and piping. (Separate discharge and suction connections are required if it is necessary to have facilities for recirculation or blending and there is no mixer in the tank.) Valves adjacent to tank nozzles should be of steel to assure adequate fire resistance when required. Filling lines for tanks containing flammable fluids should discharge near the bottom of the tank without free-fall because of the danger of static electricity being created. Where it is necessary to provide top connections on tanks containing liquid with flash points of less than 100°F, floating swing lines should be installed to avoid free-fall.

Expansion bends should be used only in those cases where anticipated tank settlement or thermal expansion will cause the line or tank to be overstressed. Unless a complete water draw-off valve or an arrangement to prevent freezing is used, a pumpout connection should be provided and may be installed as a crossover from the water draw-off line to the tank suction line.

Fired Heater Piping. A permanent steam-air decoking method for cleaning coke buildup from heater tube internal surface connections should be made on heaters requiring frequent decoking and where installation of a temporary steam air header would necessitate considerable dismantling of the process piping. On heaters with parallel coils (passes), blanks are required to separate the coils for decoking connections. Dropout spools or blanks should be provided for all decoking operations which are also installed for steam out. The alternate to steam air decoking is to provide sufficient flanged fittings for mechanical decoking.

A test connection on the heater charging line, provided with a block valve and a blank, should be installed. This should be permanently connected to the test pump if such a pump is provided with the unit.

Generally, every heater should be provided with a blowdown valve (or valves) operable from a safe location. Such valves should be installed on charge, transfer, or outlet lines as required by the heater design and service characteristics and at an elevation equal to or below that of the lowest coil. Blowdown valves should be sized for a flow area approximately equal to that of the largest tube in the coil but should not be less than a 2-in pipe size. Where there is a likelihood of coke formation, manually operated valves should be of the globe type with the pressure against the bottom of the disk.

Steam-out valves to be used in conjunction with the blowdown system may be installed on transfer lines. Valves should not be less than a 2-in pipe size and be operable from a location remote from the heater. If the process fluid pressure is at all times greater than the steam pressure, two block valves and one check valve are required between the steam and the heater coil, one block valve next to the heater coil followed immediately by the check valve. A bleeder should be installed between the block valve nearest the heater coil and the check valve. If the process fluid pressure is at all times less than the steam pressure, one block valve and one check valve with a bleeder between them will normally be sufficient. The check valve should be located between the block valve at the process equipment.

All fuel gas supplied to heaters should pass through a dry drum located as close as practical to the heaters. The supply main, the branch lines, and the distribution headers between the dry drum and the heater should be pitched downward in the direction of flow and be without pockets. If this is not possible, a condensate leg with valved and plugged drain connections should be provided at the low point. Branch lines should be connected to the top of the header. A remotely controlled or a remotely located block valve should be provided in the supply line of each heater. Wherever heater outlet temperatures are controlled by regulation of the fuel supply, automatic fuel-regulator valves should be provided upstream of the distribution header. Each gas burner should be supplied with a steel shutoff valve installed in a position such that a person operating the valve will not be in close contact with the aspirator.

For heaters using fuel oil, the burner oil piping system generally consists of a burner oil storage tank, a burner oil pump and oil supply main with a strainer, a branch line to each heater, a distribution header at the heater, and an oil return main back to the storage tank. Circulating fuel oil lines should be sized to carry 200 percent of the maximum design fuel load of the heater. A remotely controlled or a remotely located block valve should be provided in the supply main or in the branch lines to each heater, and shutoff valves should be provided between the firing valves and the headers. Fuel oil lines should be sloped from the burner shutoff valves toward the burners to provide natural drainage. A pressure-reducing valve should be installed as close to the heater as practical and upstream of the burners to regulate the pressure at the burners. A recirculating bypass should be provided between the branch line to the distribution header and the return branch line from the distribution header.

The atomizing piping system used in conjunction with the burner oil consists of a branch line from a live steam main and a distribution header around each burner. Branch lines should have a capacity of approximately twice the combined steam requirements of all burners supplied by the branch. Steam traps should be provided on the atomizing steam header where necessary to prevent water from reaching the burners.

Relief Valve and Flare Header Piping. Relief valve piping in a process plant should be in accordance with API RP 520, Recommended Practice for the Design and Construction of Pressure Relieving Systems. Piping for relief valves protecting unfired pressure vessels should be in accordance with the applicable requirements of Section VIII of the ASME Boiler and Pressure Vessel Code.

The discharge of all pressure relief valves should be piped to a safe place for disposal. Liquid and readily condensible hydrocarbons are usually discharged to a closed system. Pressure relief valves discharging light hydrocarbons which are not likely to condense or accumulate at grade can frequently be safely vented to the atmosphere from the tops of tall towers. Discharging to the atmosphere reduces the size and cost of closed piping systems otherwise required and is the preferred method where it does not create a hazard and where recovery facilities are not necessary. The term *closed system* refers to the typical pressure relief valve collecting system at a process unit, wherein the discharge of pressure relief valves is collected in a piping system for disposal at a safe location. A blowdown drum, which may be integral with a blowdown stack, is usually provided for separating the vapors and collecting liquids. Vapors are vented to the atmosphere through a flare or blowdown stack. Frequently this system is combined with any required facilities for emergency blowdown or depressurizing of equipment.

Block valves should not be used before or after pressure relief valves except where necessary to permit continuous operation of the process unit or equipment. Where block valves are used, the installation should conform to the requirements of Section VIII of the ASME Code when protecting an unfired pressure vessel.

On certain vessels, pressure relief valve leakage and consequent premature shutting down of the process unit can be anticipated. These vessels should be provided with a sufficient number of pressure relief valves (and accompanying block valves) so that in the event of pressure relief valve leakage it will be possible to shut off any one defective valve and replace it while the vessel is in service and still retain full calculated relieving capacity.

Pressure relief valves should be located so that the inlet piping is short and direct and self-draining with no pockets. However, on installations where pressure pulsations or turbulence are likely to effect the pressure relief valve (e.g., discharge side of reciprocating compressors and pumps), it may be desirable to locate the valve farther from the source in a more stable pressure region. The differential between operating and valve set pressures is also important when the operating pressure is not steady. A large differential will tend to reduce valve maintenance costs.

For gas, vapor, or flashing liquid service, the inlet piping pressure drop at design flow should not exceed 3 percent of the safety relief valve set pressure. Nor should the inlet piping to a pressure valve be smaller than the valve inlet nominal pipe size. The inlet piping includes all piping between the protected equipment and the inlet flange of the valve. Excessive pressure drop in the inlet piping will cause valve chatter (extremely rapid opening and closing of the valve), which may lower the valve capacity and damage valve seating surfaces.

Outlet piping for pressure relief valves discharging flammable vapors directly to the atmosphere should normally be equipped with steam and drain connections controlled from grade. Outlet piping from pressure relief valves should be equipped with drains or be otherwise suitably piped to prevent accumulation of liquids at the valve outlet. Pressure relief valve outlet piping for water or other liquids should be self-draining.

Separate pressure relief valve lines should be provided for each valve discharging directly to atmosphere. On towers, the pressure relief valve vent piping should be extended at least 10 ft above the nearest working platform within a radius of 40 ft. Outlet piping should be arranged so the pressure relief valve discharge will not impinge on any equipment.

Pressure relief valve discharge piping connecting to a closed system should be self-draining to the blowdown drum, vent stack, or other means for liquid vapor separation and disposal. The main headers are frequently sloped to assure drainage. A continuous purging connection should be considered for closed system piping to prevent flammable mixtures resulting from possible pressure relief valve leakage. Where necessary or desirable to detect leaking pressure relief valves, a

¾-in valved and plugged drain connection should be provided at the outlet of each valve.

The sudden initiation of relief valve outflow can cause severe stresses in attached equipment and structures. Consequently, such factors as the high- and low-temperature properties of material, thermal expansion, vibration, and fatigue must be considered in designing pressure relief valve discharge piping.

Generally, the most difficult and important feature associated with sizing relief valve discharge lines and headers is the determination of the maximum probable flow. The flow is based on the number of valves which may discharge simultaneously owing to a fire or abnormal process conditions. To do this, the layout of the unit must be considered along with many possible abnormal operation conditions.

The permissible back pressure must also be determined. Generally, the back pressure should not exceed 10 percent of the set pressure for unbalanced safety valves. Balanced pressure relief valves will operate satisfactorily at higher back pressures (approximately 30 percent of the set pressure), and, consequently, their use will sometimes result in a more economical relieving system.

One method for increasing the permissible back pressure when using unbalanced valves is to set the valve at some pressure below the vessel design pressure.

If the permissible back pressures vary widely and if the suggestions made above are not feasible, it might be economical to provide one high-pressure and one low-pressure relieving system.

Pressure relief valve discharge piping should be sized so that any back pressure that may exist or develop will not reduce the capacity of the pressure relief valve below that required to protect the equipment. Regardless, the discharge piping for each pressure relief valve should not be smaller than the nominal pipe size of the pressure relief valve outlet.

Utility Piping. Air, steam, or water connections to process piping or process equipment should be temporary unless they serve as part of the process. Temporary connections should consist of a block valve, a check valve, and a blind flange. The block valve should be located between the check valve and the process piping or equipment. Both valves should conform to the specification of the more severe service.

When a permanent air, steam, or water connection to process piping or process equipment other than exchanger is needed, a check valve, a ¾-in bleed, and a blank should be provided in addition to the block valve. The block valve should be located between the check valve and the process piping or equipment, and the bleed should be located between the two valves. There must be a block valve in the utility line, upstream of the check valve, to permit installation of the blank. All valves downstream of the blank should conform to the specification of the more severe service. A block and check valve should be considered in the steam supply line to an exchanger if the utility side of the exchanger is at a lower pressure than the process side. The block valve should be located between the exchanger and the check valve.

Utility and drain connections at the bottom of the equipment may be manifolded into a single header in order to simplify piping connections to the vessel, except that steam connections should not be in the same manifold as the drain and pump-out connections.

Service outlets for steam, water, and air should be 1 in. Outlets should be located so working areas and process equipment can be reached with a single 50-ft length of hose.

Water Piping. Many process plants have three water systems—high pressure (for fire fighting), low pressure (for cooling and use in the process), and drinking water. One of the most severe design problems in many parts of the United States is to locate or protect water systems from freezing. One way to do this is to place the piping underground and below the frost lines. However, in a process unit much of the water piping must be above grade. If the piping is out of doors and in intermittent or standby service, it should be heat traced and insulated. In cold climates, heat tracing and insulating should be considered for water lines with low continuous flow rates. (An alternative to heat tracing and insulating is to provide a bypass to a drain so that flow in the water line is continuous and at a high enough rate to prevent freezing.) On water mains, the high-point vent between block valves should be protected from mechanical damage as well as from freezing.

Drains should be provided on any water line located above the frost line so that it can be drained when it is shut down. Such drain connections and valves should generally be located underground, and drains should, where practicable, connect to a sewer. Drainage facilities should also be provided for the water side of heat exchanger.

Water injected into a process stream normally is taken from the low-pressure water system. However, where salt water is used for the cooling-water and firewater systems, water for process purposes should be taken from the potable water system.

On units where cooling-water failure could create a hazard, the firewater main should be connected to the cooling-water main for emergency cooling. If cooling-water booster pumps are used, the connections should be made downstream of the pumps.

Each heat exchanger used as a cooler that is essential for the operation of the unit should be provided with a single block valve in the water line located either upstream or downstream of the exchanger. Each exchanger that may be removed from service during operation of the unit should have a block valve in both the inlet and outlet piping. Multiple shells or exchanger in series, which cannot operate independently of each other, should be considered as a single exchanger.

The water supplied to shell and tube coolers and condensers should pass through a strainer. If a strainer is not provided at the water pump or in the supply main, individual strainers should be provided in the branch line. The necessity for installing an oil separator drum, a gas disengaging drum, or a bypass filter in the cooling-tower water return system should be considered.

Water from exchangers should generally be sent to a clear water sewer or cooling-water return system. Sample connections should be provided for detection of process leaks. However, a separate connection need not be provided for this purpose if other connections (e.g., drains and vents) can be used. A ¾-in valved and plugged vent should be provided on top of the first horizontal section of the water line downstream of the exchanger. The vent should be plugged during operation of the unit.

For chemical cleaning of exchanger using cooling water, connections should be provided to the inlet and outer nozzles on the water side of each exchanger. The connections should be between any block valve and the exchanger. If there is no block valve, a pair of flanges must be provided nearby so the piping can be blanked off during cleaning. It is suggested that the chemical cleaning connections be NPS 1½ in and equipped with a blind flange.

Sufficient connections to the water system should be provided so water can be supplied to the pressure vessels on the process unit for washing out or hydrostatic testing. These connections should be from the cooling-water system if the pressure in the system is adequate to supply water to the top of the tallest tower on the unit; otherwise, the connections should be to the firewater system.

Normally vessels need not be permanently connected to a source of water. If a permanent connection is made, it should be at the bottom of the vessel and should be blanked off when the vessel is in operation.

Air Piping. Most process plants have a plant air system not only for use in the processes but to operate tools, equipment, and instruments.

Where necessary, the intakes of air compressors should be designed to minimize the noise level. Filters should be provided in the intake piping to reciprocating and rotary air compressors when they take suction from the atmosphere. Filters will sometimes be necessary for centrifugal air compressors. When a filter is not provided for a centrifugal air compressor taking suction from the atmosphere, the intake piping should be provided with a bird screen. Filters preferably should be a dry, replaceable-cartridge type. Such filters should have an open area not less than three times the area of the intake pipe. The oil bathtype filter should not be used with centrifugal air compressors.

Low points in the discharge line from an air compressor should be avoided because it is possible for lube oil to be trapped and subsequently ignited. If low points are unavoidable, they should be provided with drains.

When condensed moisture in air lines is undesirable from a process standpoint or the possibility of moisture freezing exists, consideration should be given to providing a dry drum in the supply line near the process unit. The drum should be located where it will not be exposed to heat from other equipment. The drum's diameter should not be less than 24 in. Based on estimated future air requirements, the size of the drum should be such that the velocity in the drum does not exceed 15 fpm during shutdown periods when maintenance equipment is being used and the capacity be equal to at least 6 percent of the free air requirements per minute during normal operation.

In climates where freezing is possible, the bottom 18 in of the dry drum should be insulated and heat traced. The drum drain (or blow off) should also be traced or insulated. All blow-off connections should be installed pointing downward so any rust or scale blown out will not endanger personnel.

Air piping should slope downward to dry drums or moisture traps or be horizontal. Branch connections to air headers should be to the top of the pipe. Block valves should be provided in all branch lines.

When an air line is connected to process piping, two block valves, a check valve, and a bleeder should be provided. Consideration should also be given to providing a removable section of line or hose in order to guard against inadvertent operation.

Air for operating instruments is normally taken from the plant air system. For process units, a steam-driven compressor should be furnished to supply instruments in case of failure of the main supply. Where plant air is the primary source and the possibility of a power failure is remote, electrically driven compressors may be used.

If an air dryer is not provided, it may be necessary to insulate and heat trace instruments and lines which are out of doors or in unheated buildings to prevent freezing.

In extensive instrument air systems, the piping should be arranged with header and subheaders, such that groups of instruments may be isolated from the systems without affecting the air supply to all instruments. Block valves should be provided at the instrument air headers in all branch lines to instruments. Leads to individual instruments should be ½-in pipe size but may terminate with

¼-in valves. It is suggested that headers serving from 1 to 25 instruments be 1-in pipe size and headers serving from 26 to 75 instruments be 2-in pipe size.

Steam Condensate Piping. Process plants usually have two live steam systems and an exhaust steam system. One of the live steam systems generally operates in the range of 100–150 psig. The exhaust steam system normally operates at a pressure less than 30 psig. The design problems associated with these systems are not all similar to those encountered in a central power station; consequently, a brief discussion on process plant steam piping requirements follows.

The principal concern is to supply clean dry steam to the equipment using it. In accomplishing this, it is desirable to connect all branch lines (except condensate collection points) to the top of horizontal steam mains. However, if the line to steam driver is at least one size smaller than the main and the steam has a considerable amount of superheat, it is permissible to make a center-line connection to the side of the steam main. With other steam conditions, it probably will be necessary to install a knockout pot or drum or a steam separator in addition to making the connection to the top of the main. Pockets should be avoided in the line to the turbine.

Connections to exhaust headers should preferably be made to the top of the header so the condensate in the header does not run back into the driver.

In the steam line to a steam driver a block valve(s) should be located at the driver and be easily accessible for operating purposes. A single gate valve is needed in the exhaust line from each steam driver which does not exhaust directly to the atmosphere or directly into an individual condenser. (Valves need not be provided where two or more drivers, which will never be shut down separately, exhaust to the same condenser.) This gate valve should be installed at the driver so the position of the gate (i.e., open or closed) will be obvious to the operator whenever he is required to operate the inlet valve.

Wherever steam is exhausted to the atmosphere and could create such hazards as burns, freezing of condensate on walkways, or blanketing of working area with a heavy fog, the line should be fitted with an exhaust head and a drain to a sewer. The use of a silencer should be considered where noise nuisance is likely.

The flexibility of steam piping should be attained through the use of expansion bends and spring hangers. The use of expansion joints is discouraged. However, where the size and arrangement of exhaust lines prevent the use of expansion bends, a joint may be provided, preferably adjacent to the flange on the steam driver. Particular attention should be given to the anchorage and support of the connecting piping.

A check valve should be considered for the steam supply line to an exchanger if the steam side of the exchanger operates at a lower pressure than the process side of the exchanger. The steam supply for smothering, snuffing, service hoses, space heating, and auxiliary or protective heating should be connected to a source that will not be shut off during unit shutdowns or to a source that will not be shut off when the steam to a piece of equipment, such as a turbine, is shut off.

For fire-protection purposes, smothering (or snuffing) steam usually is required for fired heaters and for relief valve discharge lines. When required by the service, means should be available for purging process equipment with steam or inert gas. For example, each pressure vessel in hydrocarbon service should be provided with a steam-hose connection near the bottom if not permanently connected to the source of steam. Where a permanent connection is made, it should be blinded during operation of the unit.

Condensate Removal and Steam Traps. Condensate should preferably be discharged into an oilfree drain system, but under no circumstances should it be discharged into a sanitary sewer. Consideration should be given to a condensate collection system in installations which involve a large number of steam traps. When condensate is to be discharged to a cast-iron or concrete sewer or a concrete sewer box, the hazard of vaporizing hydrocarbons which may exist in the sewer should be considered. Also, to avoid damage to the concrete, the connection should be below the water level. If there is an insufficient quantity of water for quench, the condensate should be first led to an atmosphere-pressure drain tank.

Steam traps should be provided for the removal of condensate from collection points in live and exhaust steam systems, in particular from condensate drip legs, drains on steam turbines, steam separators, connectors, unit heaters, and terminal ends of companion piping. All low points in steam lines (except steam companion lines and the ends of long headers) should be provided with drip legs. It may also be necessary to install drip legs at intermediate points on headers with long sections at one elevation (i.e., in addition to those low points at the end).

When a valve is installed in steam piping in such a manner that condensate can collect above the valve, a trapped drain should be provided above the valve seat. Whenever possible, a steam trap should be installed below and close to the equipment pipeline being drained, but the trap should be easily accessible for periodic inspection. Each trap should serve only one collection point. Where large quantities of condensate are expected, either condensate pots or condensate drains should be provided.

Drains from turbine shaft packing glands and from governor valve stem packing glands should preferably be connected to an open drain system. The drain lines and headers should be of sufficient size to prevent a back-pressure buildup. Also, untrapped drains should be provided at the lowest point of the steam end of each reciprocating pump and compressor.

Drains not discharging into a closed drainage system should discharge downward and should be arranged so rising steam does not create a hazard or condense on equipment, such as a turbine or pump. The condensation of rising steam on such equipment can create lube oil contamination. One thing that can be done to help eliminate this problem is to quench the condensate.

Probably the principal cause of freezing of steam traps is improperly designed discharge lines. Steam trap discharge lines should be sloped for drainage where possible. In cases where freezing is likely, no part of the trap discharge header should be at an elevation above that of the trap discharge. Pockets in the discharge lines should be avoided. Long trap-discharge lines, if not in heated enclosures, should be insulated. Trap-discharge lines in heated enclosures need to be insulated only if necessary for burn protection. To decrease further the possibility of freezing, steam trap bodies should not be insulated except if the following circumstances make it advisable:

- The trap is installed downstream of automatic steam controls that could shut the steam off for long periods of time.
- The trap is installed in a location where operators might be burned by the bare metal surfaces.
- The trap is part of a heat recovery system where retention of heat is important.
- The trap is installed to handle exhaust steam condensate that contains quantities of cylinder oil.

Inverted-bucket and thermodynamic steam traps, which are commonly used in process plant, are generally installed without strainers. Steam traps should be selected for a continuous discharge rate which is the actual condensate rate multiplied by a safety factor. A safety factor of at least 3 should be used for inverted bucket type traps and thermodynamic traps. (A larger safety factor is needed for traps draining jacketed equipment, and trap manufacturers should be consulted.) In borderline cases offering a choice between two trap sizes, the smaller trap is usually preferred.

Air should be used as the actuating medium for vacuum steam traps. The use of live steam is not recommended because live steam is likely to vaporize the lower temperature vacuum system condensate. If live steam must be used as the actuating medium, it may be necessary to spray water into the trap interior to keep the trap operating properly.

Steam Companion Piping for Auxiliary Heating. The most commonly encountered situations requiring auxiliary heating are for

1. Piping in which the fluid temperature could drop below the pour point or freezing point and for piping in which the fluid is subject to coagulation, excessive viscosity, or salting out
2. Hydrocarbon vapor and gas piping where condensate formation and icing will affect the safety and operation of the equipment such as might be caused by the reduction in pressure that takes place through a control, throttle, or relief valve
3. Lube-and-seal-oil systems for compressors and turbines.

Auxiliary heating is normally not needed for freeze prevention and viscosity maintenance on equipment in intermittent service if the equipment is drained, flushed, blown, or steamed out when there is no flowing stream or if the equipment is far enough underground to prevent freezing. Auxiliary heating normally is furnished by external steam companion piping (steam tracing). Other acceptable methods of heating piping and other equipment are internal steam tracing, steam jacketing, hot water tracing and jacketing, and electric tracing. Details of various heat tracing techniques are covered in detail in Chap. 6, Part B of this handbook.

It is desirable that each steam companion line be continuous from the header to a trap at the end of the line without any vents, drains, branches, or dead-end extensions at intermediate points. Each companion line should have a block valve at the upstream end and be arranged so flow is generally downward, avoiding pockets as much as possible and leaving no section of the companion line at a greater elevation than the companion header.

In the design of the companion piping system, provisions should be made for the differential expansion between the traced line and the tracer. Live steam is preferred for steam companion piping in colder climates unless a lower temperature is required. When the piece of equipment which is to be kept hot is irregular in shape such as traps, strainers, valves, and pumps, tubing must be used. The item should be spirally wrapped starting at the top and working toward the bottom. Several lines to be traced may be grouped inside a single covering of insulation if they are to be maintained at the same temperature.

Companion piping on specific process pipe lines and on instrument differential pressure leads handling volatile streams should be separated from these lines by a 1-in wide asbestos tape wrapped around the steam liens at 2-ft intervals to give a total thickness of about ½ in. Separation may also be effected by using small blocks of insulation securely wired to the lines.

CASE HISTORIES—CHALLENGES AND SOLUTIONS

Process plants offer the piping designer some unique challenges not found elsewhere. The combinations of demanding service requirements and mechanical needs will necessitate innovative designs and solutions. Included here are a few practical approaches to problem resolution:

Challenge: A thermoplastic pipeline carrying demineralized water was failing from a water-hammer effect.

Solution: The problem source was diagnosed as a fast-acting, lever-operated quarter-turn valve that was opening and closing too fast. A gear operator was installed to slow down the valve motion, thereby eliminating the water hammer.

Challenge: A steel compressor discharge line was suffering high-frequency and high-amplitude vibrations.

Solution: Additional pipe supports were installed to change the natural frequency of the piping geometry and reduce the response to the compressor perturbing forces.

Challenge: An erosive slurry was causing material loss at changes of direction in a conventionally constructed piping system of elbows and tees. Space constraints did not allow for long sweep turns.

Solution: Dead-end tees were installed where the solids filled the impact area of the tee. The abrasive solids then wore on themselves, thereby protecting the pressure boundary.

Challenge: A heater outlet line was expected to operate at about 1500°F. The attendant expansion and stress analysis difficulties were magnified since the process piping material was well into the creep range. Premature failure was expected.

Solution: The hot metal heater line was transitioned into an internally refractory-lined system near the heater outlet. The lower shell temperature eliminated the probability of creep rupture failure, simplified the expansion and stress analysis problems, and reduced system maintenance.

Challenge: Transport of raw seawater through a large diameter (98-in) pressure pipe was to be buried in a corrosive soil in a remote locale.

Solution: A filament-wound reinforced thermoset resin piping system was used. A custom designed laminate wall structure afford internal and external corrosion resistance, adequate strength to accommodate the expected mechanical loads, and a relatively lightweight pipe for ease of installation.

Challenge: Fugitive emissions of a highly flammable, potentially explosive vapor were found escaping from numerous valve bonnets.

Solution: Conventional-style valves were replaced with extended bonnet valves using primary and secondary packing and purge ports between packing rings. The secondary packing virtually eliminated all emissions.

The previous discussion only touches the volume of challenges and solutions encountered in the area of process piping. Reference literature has documented the many variables encountered in this extensive subject.

BIBLIOGRAPHY

1. ASME B31.3, *Chemical Plant and Petroleum Refinery Piping,* 1990 Edition, published by ASME, New York, N.Y.
2. M. W. Kellogg, *Design of Piping Systems,* John Wiley & Sons, N.Y.
3. Sherwood and Whitestance, *The Piping Guide,* Syentek Books Co.
4. Crocker and King, *Piping Handbook,* Fifth Edition, McGraw-Hill, N.Y.
5. R. E. Johnson, *Specifying Plastic-Lined Piping,* Chemical Engineering, May 1982.
6. H. Thielsch, *Defects and Failures in Pressure Vessels and Piping,* Reinhold Publishing Corp.
7. P. A. Schweitzer, *Handbook of Corrosion Resistant Piping,* Industrial Press Inc.

CHAPTER C8
CRYOGENIC PIPING SYSTEMS

INTRODUCTION

Nicholas P. Theophilos, Design Engineer
Union Carbide Industrial Gases, Linde Division

Cryogenics* is the science and technology associated with very low temperatures. Depending on one's point of view, any temperature below −20°F can be set to establish such a demarcation. Here the −20°F point has been selected because it normally represents the onset of embrittlement for ordinary carbon steels in typical structural applications.

Cryogenics is not a separate branch of physics, since it obeys all laws of ordinary physics. In fact, cryogenics is low-temperature physics. The reasons for its special treatment, therefore, are not because of its uniqueness as a science but rather because of the very special problems it creates as a technology. These problems relate to embrittlement of materials, large displacements (expansion and contraction), rapid change of phase due to large heat fluxes (big delta T), and small latent heats of the fluids involved.

In order to obtain a better appreciation of the special considerations involved in cryogenic piping system applications, it was felt that it would be necessary to review the behavior of materials at cryogenic temperatures and the physical and thermodynamic properties of cryogenic fluids. These considerations are covered in the sections "Properties of Cryogenic Fluids" and "Materials Used in Cryogenic Piping Systems." Additionally, cryogenic piping systems design is discussed in the sections "Piping Systems Design—Fluids" and "Piping Systems Design—Mechanical."

From the strictly heuristic point of view of fundamental applications of scientific principles there are hardly any differences between cold box piping and all other types, Nevertheless, we are making a special topic of cold box piping because of the confined spaces involved and the conceptual arrangements required to satisfy logistically workable and economically feasible process considerations. Such piping is discussed in the section "Cold Box Piping."

The coverage on cryogenic distribution systems, as provided under the sections "Liquid Storage and Conversion Systems" and "Mobile Equipment System," considers more than just piping; it covers the functional design philosophy of cryogenic fluid storage and distribution and provides quite an insight into the logistics of the entire operation.

With the advent of chip making, the need for ultra-high-purity inert gases has come into clear focus and industry has responded to this need by developing suitable storage and distribution systems. These aspects of cryogenic piping systems are discussed in the last section.

*From the Greek "kryo-genikos" meaning cold generation.

Naturally, the drive behind most technologies is economic in nature and in this respect cryogenics is no exception. This is certainly much more so when it comes to liquid distribution because there is no other motive. Industrial gases can certainly be distributed in the compressed gaseous form, even in bulk quantities, if costs are not a consideration. Such economic aspects are discussed in the next section.

References for each section are at the end of each section.

ECONOMIC PARAMETERS OF CRYOGENIC FLUID DISTRIBUTION SYSTEMS

Norman H. White, Applications Engineer
Union Carbide Industrial Gases, Linde Division

Cryogenic processes are typically applied to commodity chemicals that exist as gases at normal ambients. Such gases are liquified at reduced temperatures and are normally maintained at saturated conditions. The processing techniques generally involve both the liquid and gaseous phases and exploit the dramatic change in physical and thermodynamic properties that occur with changes of state. Cryogenic temperatures are applied at each step in the process of bringing these gases to the final consumer, including production, distribution, and storage.

Typically, industrial gases are found in mixtures in which some of the components have commercial value. Two of the most important examples of such mixtures are gases from certain hydrocarbon wells and the earth's atmosphere. The feed streams are separated by liquification and subsequent fractional distillation with the produced streams delivered either in the gaseous or liquid state. Cryogenic distillation allows a wide choice in the degree of product separation ranging from crude to extremely fine. It is a highly efficient process with power consumed chiefly in refrigeration lost to the environment and pressure lost in the product streams. Economic considerations in the liquification and distillation processes involve trade-offs between operating efficiency and capital expenditure. No other method is as versatile or as effective as cryogenic separation of industrial gas mixtures for commodity usage.

The key to industrial gas distribution and storage operations is the use of a cost-effective method that increases product density. In this manner, the transport or storage vessel is reduced to manageable dimensions. The historical solution to this problem is elevated pressures at ambient temperatures. Indeed, high-pressure cylinders and receivers made to a variety of DOT and ASME specifications are widely used today to store gases at pressures exceeding 2000 psig. These vessels are typically of single piece forged construction involving fabrication and inspection procedures that enable them to safely operate at ultimate to design stress ratios of 3:1. In spite of these measures to achieve an efficient package, the ratio of vessel weight to product weight is extremely high, and therefore, the cost of storing and transporting gases in such vessels is very high relative to the value of the product they contain. Some of the characteristics of typical high-pressure receivers and cylinders are summarized in Table C8.1.

Liquification is another approach that has been widely used to achieve the high product density necessary for efficient storage and transportation of industrial gases. Gases typically emerge from the liquification process saturated at approximately atmospheric pressure and from this point are transported and stored in cryogenic vessels. These are typically double-walled vessels with an inner container designed for the working pressure and temperature of the product and an outer casing designed for ambient temperatures and external pressure. In between the two vessels is a high-performance insulation system which is usually evacuated for the purpose of further enhancing thermal protection. Structural members to support the inner container and piping to provide access to it are also located in the insulation space. Heat is continuously entering the vessel through the insulation, supports, and internal piping. This heat will make the liquid contents boil, and the resultant gas must be removed from the tank if the pressure and temperature of the contents are to be held steady. For this reason heat leak must be minimized.

The thermal efficiency of the tank design is expressed in terms of the percentage of full capacity of a specific product that will be lost in a day when the tank pressure is held at 0 psig. This parameter is termed the normal evaporation rate (NER). Table C8.2 summarizes specifications of typical tankage applied in various production and distribution functions.

Comparison of the economics of gaseous state storage at high pressures to liquid state storage at cryogenic temperatures is important since both methods have their place in industry today. The primary costs to be compared are those for storage vessel construction, transportation, power to achieve the storage state, and product loss. The construction methods used in forged high-pressure receivers are very different from those used in high thermal efficiency cryogenic temperature vessels. However, for small product quantities, the resultant cost per pound, in both these cases, is sufficiently close to be considered equal for rough comparison. Storage vessel and transportation costs for alternative systems can, therefore, be compared on the basis of the ratio of vessel weight to product weight. Table C8.3 compares this ratio for several products stored in typical gas and liquid storage vessels. Generally, the total cost (capital plus operating) to liquify product is lower than that required to compress it as a gas to receiver pressure. Indeed, receivers are most often charged with gas pumped to pressure in the liquid state and subsequently heated to ambient temperature rather than by gas state compression. Product losses to be expected in various liquid systems can be estimated from the NER specifications given in Table C8.2. From these considerations, it can be concluded that the reduced vessel weights possible with cryogenic liquid phase storage makes this the economically desirable approach in most cases. Gas storage should be considered where the requirement involves small quantities or long periods of nonuse or difficulty in disposing of the gas boiloff expected in a cryogenic system.

The piping used in cryogenic systems obviously must meet the structural demands imposed by low temperatures. From an economic point of view, the thermal efficiency of the piping system must be carefully considered since the heat addition to the system will ordinarily result in loss of product. There are two important factors of product loss involved in piping systems that must be considered: refrigeration required to bring the line to operating temperature (cool down) and steady state heat addition. Table C8.4 gives these parameters for

TABLE C8.1*a* Typical Specifications for Seamless Forged Pressure Vessels—ASME Pressure Vessel Specification. Sizes, Capacities, and Design Pressures

Material						Max. length					Design pressures		
Type	Grade	Class	O.D.	Min wall	Nominal wall*	ft	ft	in	Water vol.	Vessel weight	Appendix 14-70 SF = 3	SF = 4 No CA†	SF = 4 1/16 CA†
IV	—	—	8⅝	1.313	1.501	114.2	19	0	3.09	2170	12134	9118	8626
IV	—	—	12¾	1.531	1.750	205.6	28	0	12.6	5757	9299	6987	6673
IV	—	—	16	0.902	1.031	165.0	37	4	38.3	6160	4133	3100	2880
IV	—	—	16	1.177	1.345	210.5	28	10	26.6	6070	5471	4103	3873
V	5	B	16	1.193	1.364	213.2	28	10	26.5	6147	6343	4726	4463
IV	—	—	16	1.416	1.619	248.7	23	6	19.7	5844	6668	5001	4763
IV	—	—	16	1.531	1.750	266.3	23	0	18.5	6125	7255	5442	5202
IV	—	—	18	0.860	0.983	178.7	30	0	40.3	5360	3477	2608	2412
IV	—	—	18	1.255	1.434	253.7	21	0	24.6	5330	5169	3877	3673
IV	—	—	20	1.093	1.250	250.3	22	4	35.0	5600	4000	3000	2821
IV	—	—	24	0.817	0.934	230.1	25	0	62.9	5755	2450	1837	1692
IV	—	—	24	1.121	1.282	311.1	20	0	46.4	6221	3397	2548	2400

*Nominal Wall = Min. Wall ÷ .875
†Internal Corrosion Allowance
Source: Christy Park Industries, Inc.

TABLE C8.1*b* Typical Specifications for Seamless Forged Pressure Vessels—Dot Pressure Vessel Specification. Dimensions and Data for Typical Vessel Sizes

DOT	O.D.	Min. wall	Length*		Weight (lb)	Vol. (ft^3)	Hydrogen (scf)	Oxygen (scf)†	Nitrogen (scf)†	Helium (scf)†	HCL no.	Nat. gas (scf)
			ft	in								
DOT-3A-1800	24	0.558	6	11	1101	15.7					628	Not approved
DOT-3AX-1800	24	0.558	30	0	4800	80.53					3261	Not approved
DOT-3AAX-2400	22	0.536	17	6	2474	38	5,660	7,274	6,595	6,316	—	8,129
	22	0.536	34	4	4800	77.65	11,567	14,865	13,477	12,905	3141	16,570
	22	0.536	36	0	5041	81.6	12,158	15,621	14,162	13,562	—	17,413
DOT-3AAX-3855	18	0.705	34	4	5095	48.5	10,913	14,278	12,382	12,346		15,005
	18	0.705	36	0	5342	50.9	11,543	14,985	12,995	12,957		15,747
	18	0.705	40	0	5934	56.8	13,280	16,722	14,501	14,458		17,572
DOT-3T-2400	22	0.415	18	6	2220	41.5	Not approved	7,944	7,206	6,897		Not approved
	22	0.415	34	4	3890	79.6	Not approved	15,276	13,850	13,263		Not approved
	22	0.415	36	0	3996	83.7	Not approved	16,023	14,527	13,911		Not approved
	22	0.415	40	0	4440	93.3	Not approved	17,899	16,228	15,540		Not approved
DOT-3T-2740	22	0.472	18	6	2290	41.0	Not approved	8,939	7,979	7,690		Not approved
	22	0.472	34	4	4263	78.8	Not approved	17,180	15,335	14,780		Not approved
	22	0.472	36	0	4469	82.8	Not approved	18,052	16,113	15,531		Not approved
	22	0.472	38	6	4779	88.7	Not approved	19,338	17,262	16,637		Not approved
	22	0.472	40	0	4964	92.3	Not approved	20,123	17,962	17,312		Not approved

*Length can be varied to meet specific requirements.
†Includes 10 percent overfill.
Source: Christy Park Industries, Inc.

TABLE C8.2 Typical Specification for Cryogenic Storage Vessels

Model	Nominal capacity (gal)	Diameter (ft-in)	Height (ft-in)	Working pressure (psig)	Tare weight (lb)	Normal O_2 evaporation rate (%)
TM-500	530	5-0	15-6	250	5,450	0.50
TM-900	904	6-6	15-9	250	9,700	0.4
TL-1500	1,523	6-6	15-9	125	8,940	0.4
TM-1500	1,517	6-6	15-9	250	10,300	0.4
TL-3000	3,016	8-0	15-11	83	11,400	0.5
TM-3000	3,000	8-0	15-11	250	15,500	0.5
TL-6000	5,903	8-0	25-9	77	20,700	0.3
TM-6000	5,889	8-0	25-9	250	27,800	0.3
TL-9000	8,900	9-6	29-9	65	26,500	0.26
TM-9000	8,900	9-6	29-9	250	37,000	0.26
TL-11000	11,000	10-2	31-7	65	34,200	0.25
TM-11000	11,000	10-2	31-7	250	47,000	0.25
TM-11000-HP	11,000	10-2	31-7	375	56,500	0.25
TL-13000	13,000	10-2	36-2	62	38,000	0.23
TM-13000	13,000	10-2	36-2	250	50,000	0.23
TM-13000-HP	13,000	10-2	36-2	375	63,600	0.23

Source: Union Carbide Industrial Gases.

TABLE C8.3 Weight Ratio Contents/Vessel

Vessel parameters			Weight ratio: contents/vessel			
Water volume (ft^3)	Working pressure (psig)	Tare weight (lb)	O_2	N_2	Ar	H_2
		Cryogenic vessels				
200	250	10300	1.41	1.00	1.72	—
802	250	27800	2.02	1.43	2.46	—
1738	250	50000	2.48	1.76	3.29	—
1176	150	46400				.1123
		High-pressure receivers				
62.9	2450	5735	.161	.13	.199	.008
26.5	6343	6147	.147	.104	.175	.007
3.1	12134	2170	—	.044	.078	—

Source: Ref. 3; courtesy, Union Carbide Industrial Gases.

TABLE C8.4 Heat Addition[a] Steady State and Cooldown

Insulation	Nominal pipe size (NPS) 1	2	3	4
Steady state heat addition (Btu/ft)				
Uninsulated[b]	283	535	787	1037
Polyurethane foam[c] thickness				
1 in	16.0	25.3	32.9	41.3
2 in	10.8	15.4	19.8	24.1
3 in	8.8	12.2	15.2	18.2
Vacuum insulation[d]	0.5	0.8	1.0	1.3
Cool-down (Btu/ft)				
CU tube[e]	20.2	54.0	103.0	166.0
SS pipe[f]	27.4	50.6	95.7	157.5
Polyurethane foam[g] thickness				
1 in	2.5	3.7	4.9	7.9
2 in	7.3	9.6	12.2	16.5
3 in	14.5	18.0	22.0	28.0
Vacuum insulation[g]	2.1	3.8	7.2	11.8

[a]Liquid nitrogen service.
[b]8-mph wind over frosted uninsulated line.
[c]Closed-cell polyurethane foam with PVC cover.
[d]Evacuated laminar radiation shields.
[e]ASTM B-88 Type L.
[f]Schedule 5
[g]Heat addition due to insulation only.
Source: Ref. 4; Courtesy, Union Carbide Industrial Gases.

uninsulated lines, lines insulated with closed cell polyurethane foam, and lines insulated with radiation shields in high vacuum (super insulation).

PROPERTIES OF CRYOGENIC FLUIDS

Theodore F. Fisher, Process Engineer
Union Carbide Industrial Gases, Linde Division

Physical and thermodynamic properties of cryogenic fluids constitute important data that are needed for the design of cryogenic piping systems. The following discussion, tables, and figures are furnished with this need in mind.

Transport property data are readily available for the more common pure cryogenic fluids. The bibliography at the end of this section lists sources of physical properties and thermodynamic charts, which supply the detailed data required for the design of piping systems. Table C8.5 summarizes some of the more important properties for a number of cryogenic fluids.

TABLE C8.5 Physical and Thermodynamic Properties of Cryogenic Fluids

	Helium	Hydrogen (equil.)	Hydrogen (normal)	Neon	Nitrogen	Carbon monoxide	Air	Fluorine	Argon	Oxygen	Methane	Krypton	Nitric oxide	R-14
Formula	He	e-H2	n-H_2	Ne	N_2	CO	Air	F_2	Ar	O_2	CH_4	Kr	NO	CF_4
Molecular weight	4.0026	2.0159	2.0159	20.183	28.013	28.011	28.96	37.997	39.948	31.999	16.043	83.8	30.006	88.005
Triple point														
Temperature, °F		−434.8	434.6	−415.5	−346.0	−337.1	−357. −351.0	−363.3	−308.8	−361.8	−296.5	−251.0	−262.6	−298.9
Pressure, psia		1.02	1.04	6.26	1.82	2.23		0.037	10.0	0.021	1.7	10.6	3.18	0.022
heat of fusion, BTU/lb		25.0	25.0	7.1	11.1	12.8		5.78	12.7	6.0	25.2	8.4	33.0	3.42
Normal boiling point														
Temperature, °F	−452.1	−423.2	−423.0	−410.9	−320.4	−312.6	−317. −312.4	−306.7	−302.5	−297.3	−258.7	−244.0	−241.2	−198.3
Density, lb/ft^3														
Liquid	7.798	4.418	4.428	75.35	50.46	49.3	54.56	94.1	86.98	71.27	26.5	150.6	79.3	100.6
Vapor	1.04	0.0835	0.0831	0.596	0.287	0.275	0.280	0.33	0.363	0.279	0.114	0.53	0.19	0.476
Heat of vaporization, BTU/lb	8.72	193	192	37.0	85.7	92.8	88.2	74.0	70.2	91.7	219	46.4	198	61.6
Specific heat, BTU/lb · °F														
Liquid	1.08	2.32	2.33	0.44	0.488	0.516	0.47	0.366	0.478	0.406	0.82	0.126		
Vapor		2.90	2.83	0.28	0.266	0.31	0.27	0.195	0.235	0.24	0.50			
Viscosity, lb/ft · h														
Liquid	0.0087	0.0322	0.0322		0.399	0.411	0.431	0.605	0.653	0.457	0.281	1.0		
Vapor	0.0022	0.0027	0.0027	0.0102	0.0136	0.0136	0.0153	0.0102	0.0179	0.0167	0.0106	0.0259	0.0206	
Thermal conductivity, BTU/(h · ft°F)														
Liquid	0.015	0.069	0.069	0.066	0.079	0.081	0.081	0.0915	0.071	0.0868	0.108	0.052		
Vapor	0.006	0.0089	0.0089	0.0046	0.0043	0.0040	0.0044	0.0042	0.0033	0.0044	0.0069	0.0024	0.0063	
Critical point														
Temperature, °F	−450.3	−400.3	−399.9	−380.0	−232.5	−220.5	−221.3	−200.2	−188.1	−181.1	−115.8	−82.8	−135.1	−50.2
Pressure, psia	33.23	187.7	188.1	395	492.9	507.4	547	808	710	736.9	673.1	796	949	543
Compressibility	0.305	0.309	0.305	0.307	0.292	0.294	0.292	0.28	0.292	0.300	0.289	0.291	0.256	0.274
Density, lb/ft^3	4.33	1.92	1.94	30.2	19.4	18.8	20.5	39	33.4	26.2	10.1	56.7	32.5	39.1
Gas at 1 atm, 70°F														
Density, lb/ft^3	0.0103	0.0052	0.0052	0.0521	0.0724	0.0724	0.0724	0.0983	0.1034	0.0821	0.0416	0.2172	0.0777	0.228
Specific heat, BTU/lb°F	1.25	3.56	3.42	0.246	0.248	0.249	0.240	0.197	0.125	0.220	0.533	0.0598	0.24	0.166
Specific heat ratio	1.66	1.39	1.41	1.66	1.40	1.40	1.40	1.358	1.67	1.40	1.31	1.68	1.40	
Viscosity, lb/ft · h	0.0477	0.0215	0.0215	0.0755	0.0424	0.0426	0.0443	0.0571	0.0542	0.0494	0.0269	0.0600	0.0462	
Thermal conductivity, BTU/(h · ft°F)	0.086	0.109	0.106	0.0280	0.0149	0.0143	0.0150	0.0153	0.0100	0.0204	0.0191	0.0053	0.0147	

	Ozone	Xenon	Ethylene	Ethane	Nitrous oxide	Hydrogen chloride	Acetylene	Carbon dioxide	Hydrogen sulfide	Propylene	Propane	Ammonia
Formula	O_3	Xe	C_2H_4	C_2H_6	N_2O	HCl	C_2H_2	CO_2	H_2S	C_3H_6	C_3H_8	NH_3
Molecular weight	47.998	131.30	26.038	30.070	44.013	36.461	26.038	44.010	34.080	42.081	44.097	17.031
Triple point												
Temperature, °F	−315.3	−169.2	272.5	297.9	−131.5	−226.4	−114	−69.8	−122.0	−301.5	−305.8	−107.9
Pressure, psia	0.00017	11.8	0.014	0.00014	12.7	2.0	17.4	75.1	3.8	10^{-8}	10^{-10}	0.88
heat of fusion, BTU/lb		7.5	51.4	40.9	63.9	23.6	62	81.4	30.0	30.7	34.3	143
Normal boiling point												
Temperature, °F	−168.2	−162.6	−154.7	−128.0	−127.2	−121.0	Subl. −119.2	Subl −109.4	−76.5	−53.9	−43.8	−28.1
Density, lb/ft^3												
Liquid	84.5	190.8	35.5	34.1	77	74.3	S: 38.7	S: 97.6	60.0	38.0	36.4	42.6
Vapor	0.231	0.60	0.130	0.127	0.194		0.0131	0.176		0;14	0.151	0.0556
Heat of vaporization, BTU/lb	127	41.4	208	210	162	191	S: 352	S: 247	236	188	183	589
Specific heat, BTU/lb · °F												
Liquid			0.57	0.581						0.52	0.531	1.066
Vapor			0.27	0.32						0.31	0.33	0.559
Viscosity, lb/ft · h		1.186	0.387	0.402								
Liquid		0.0315	0.0140	0.0143							0.501	0.617
Vapor					0.0227	0.0211	0.0165		0.02		0.0150	
Thermal conductivity, BTU/(h · ft°F)												
Liquid	0.13	0.0032	0.111	0.080	0.10					0.12	0.079	0.32
Vapor		0.0019	0.005	0.0050	0.0051		0.0065	0.005	0.0052		0.0061	0.011
Critical point												
Temperature, °F	10.3	61.9	49.7	90.4	97.6	124.5	95.3	87.9	212.7	197.5	206.3	270.3
Pressure, psia	802	847	742	710	1052	1198	890	1070	1306	667	617.4	1636
Compressibility	0.281	0.288	0.269	0.285	0.274	0.266	0.274	0.274	0.283	0.274	0.277	0.242
Density, lb/ft^3	27.2	69.0	14.2	12.7	28.2	26.3	14.4	29.2	21.8	14.5	13.7	14.7
Gas at 1 atm, 70°F												
Density, lb/ft^3	0.124	0.3416	0.0729	0.0783	0.1146	0.0949	0.0678	0.1146	0.0890	0.107	0.116	0.0445
Specific heat, BTU/lb°F		0.0383	0.368	0.419	0.21	0.189	0.403	0.202	0.241	0.364	0.395	0.516
Specific heat ratio		1.68	1.25	1.20	1.30	1.41	1.24	1.30	1.32		1.14	1.32
Viscosity, lb/ft · h		0.0552	0.0242	0.0223	0.0348	0.0346	0.0247	0.0358	0.0312	0.0206	0.0196	0.0244
Thermal conductivity, BTU/(h · ft°F)		0.0032	0.0119	0.0121	0.0098	0.0079	0.0122	0.0092	0.0082		0.0101	0.0135

Source: Union Carbide Industrial Gases

Thermodynamic charts show pressure-temperature-phase-specific volume (or density) and enthalpy (heat content, H) relationships for a specific fluid under a variety of formats. Entropy data are often included but are not generally required for the purposes under consideration in this chapter.

Figure C8.1 is a pressure-enthalpy chart for nitrogen, which will be used to illustrate the behavior of fluids in transport systems, and various uses of pressure-temperature-volume-enthalpy (P-T-V-H) data. The most obvious use may be to determine the density (the reciprocal of specific volume) of a fluid which is being transported under constant pressure and temperature conditions. Referring to the chart, for example, the specific volume of nitrogen at 100 psia and 60°F is found to be 2.0 ft^3/lb, corresponding to a density of 0.50 lb/ft^3. Other uses of the chart are illustrated in the examples which follow.

The dome-shaped curve at the bottom left of the chart encloses the two-phase (vapor-liquid) region. Nitrogen at pressure and temperature conditions to the left of the dome is a saturated or subcooled liquid. Conditions to the right of the dome correspond to saturated or superheated vapor. Vapor and liquid phases coexist within the dome at a unique pressure for a specific temperature. This pressure is designated as the vapor pressure of the fluid at that temperature. The length of the isobar (constant pressure) line between the sides of the dome is proportional to the heat input which is required to completely vaporize the fluid at the pressure and corresponding temperature. The fraction of the fluid which is vapor at a condition corresponding to a specific point along the isobar is equivalent to the fraction of its distance from the saturated liquid line, divided by the total length of the isobar within the dome.

The point at the very peak of the dome represents a unique condition of pressure and temperature which is designated the critical point of the fluid. As nitrogen approaches 493 psia and −232.5°F from any direction, all distinctions between the characteristics of vapor and liquid phases disappear. Liquid being warmed at a pressure above the critical pressure behaves as a dense fluid, which gradually approaches the characteristics of a high-pressure gas at higher temperatures, without passing through any observable phase change.

The vertical lines on a pressure-enthalpy chart correspond to a constant heat content condition. Movement along these lines designates an isenthalpic or adiabatic (no heat input or output) process. Flow through reasonably short or insulated runs of piping or fittings, or at a temperature close to that of the environment, approaches isenthalpic behavior.

When a fluid flowing within a pipe is throttled through a valve, the change in its pressure-temperature-phase and density relationships is essentially isenthalpic. If a fluid is initially a gas which is reduced in pressure, its decrease in pressure and density may be accompanied by a change in temperature (Joule-Thomson effect). In most cases this will be a decrease in temperature, as shown, for example, by nitrogen at 1000 psia and −140°F following the 60 Btu/lb isenthalpy to about −216°F when throttled to atmospheric pressure. At conditions more remote from the critical point, the temperature change is less dramatic. A rise in temperature may occur when the fluid is well above its critical temperature. This is shown by nitrogen at 500°F when throttled from an initial pressure above about 3000 psia, but it is more typically encountered with hydrogen or helium under normal processing conditions.

When a saturated liquid is throttled to lower pressure, an isenthalpic line is followed into the dome, indicating partial vaporization of the downstream fluid. For example, if liquid nitrogen which is initially saturated at 400 psia and −240°F is throttled to atmospheric pressure, the downstream condition will be about 55

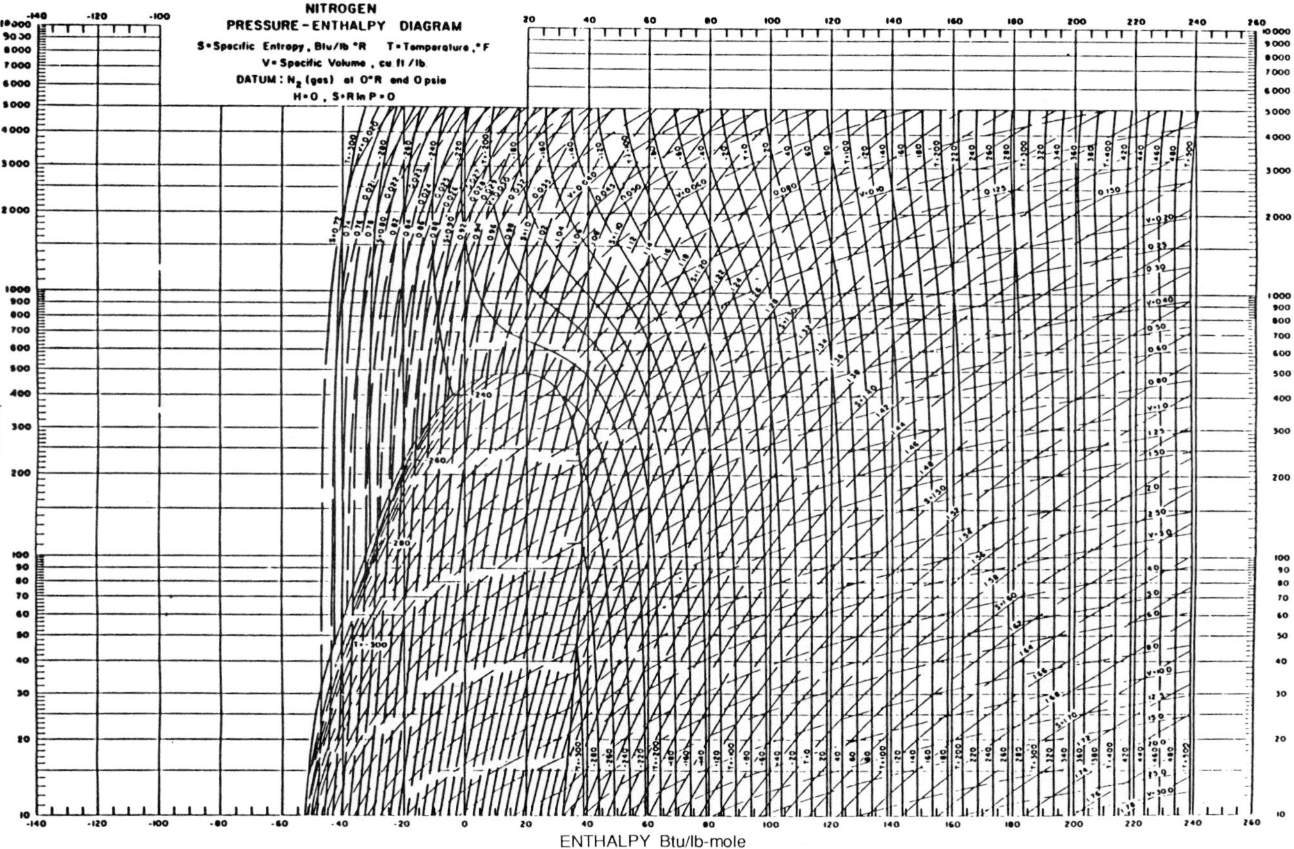

FIGURE C8.1 Nitrogen pressure-enthalpy diagram. (*Courtesy, Union Carbide Industrial Gases.*)

percent vapor. This will be accompanied by change in specific volume from about 0.032 to 1.92 ft^3/lb, or a factor of 60 decrease in density. If nitrogen vapor is throttled from a point above the dome, it may partially liquify (retrograde condensation) or, within a very narrow range of conditions, pass through the two-phase region before ending up as all vapor at a final low-pressure condition.

The large change in specific volume which may occur with throttling—and particularly vaporization—may necessitate a substantial increase in the diameter of downstream piping in order to maintain reasonable velocities of the fluid. If the process leads to a two-phase downstream condition, there may be some slippage between the phases (i.e., the relative fraction of the vapor and liquid inventoried in the piping may differ from that of the net throughput). Two-phase flow characteristics can be extremely complex, and determination of this behavior is beyond the scope of this discussion.

An important consideration in the design of a cryogenic piping system arises from recognition that a confined fluid cannot increase in specific volume when heated. Extreme overpressurization can result when a cryogenic liquid which is trapped between valves is warmed, for example by heat leak. Confined nitrogen which is initially liquid at 40 psia will (following the 0.021-ft^3/lb specific volume line) exceed 650 psia when warmed to −300°F and approach 5000 psia at −260°F. This makes overpressure protection mandatory wherever entrapment of a cryogenic liquid or (initially) high-pressure cold gas is possible.

Pressure-enthalpy charts for other cryogenic fluids are similar to that for nitrogen. Charts for oxygen and argon are provided as Figs. C8.2 and C8.3, respectively. Sources of charts for additional fluids are given in the Bibliography at the end of this section. The thermodynamic property interrelationships of these fluids are similar to those which have been illustrated for nitrogen.

Rapid exposure of a partially confined cryogenic liquid to heat may result in overpressurization even when some outlet is provided. The possibility of a cryogenic liquid spill into a warmer environment must be considered in designing any enclosure around a piping system.

Certain fluids require special consideration. Liquid carbon dioxide, for example, is not stable at atmospheric pressure. (This is indicated on Table C8.5 by a triple point pressure above 1 atmosphere.) Loss of pressure on a liquid carbon dioxide system (or a high-pressure gas system at sufficiently low temperature) will result in dry ice formation.

Physical and thermodynamic property data on cryogenic mixtures are not readily available in easily applied chart or tabular format (as they are for pure fluids). Required design data are usually computer generated from complex correlations for specific cases. Knowledge of pure fluid behavior will, however, provide an awareness of many factors which require consideration in the design of a cryogenic piping system.

Among the specific differences between the behavior of pure fluids and that of mixtures is a potentially much broader range of temperature and pressure conditions within the two-phase region for the latter case and differences in composition between the individual phases and that of the overall mixture. There is also a possibility of one of the components freezing under conditions wherein other components may exist as gas and/or liquid mixtures. Although a component which is below its normal freezing point may have considerable solubility in the liquid or gas mixture, the possibility of solid formation should be anticipated when any of the components is present in a mixture at a temperature below its triple point. This situation may result from the mixing of two streams, neither of which contains a frozen component.

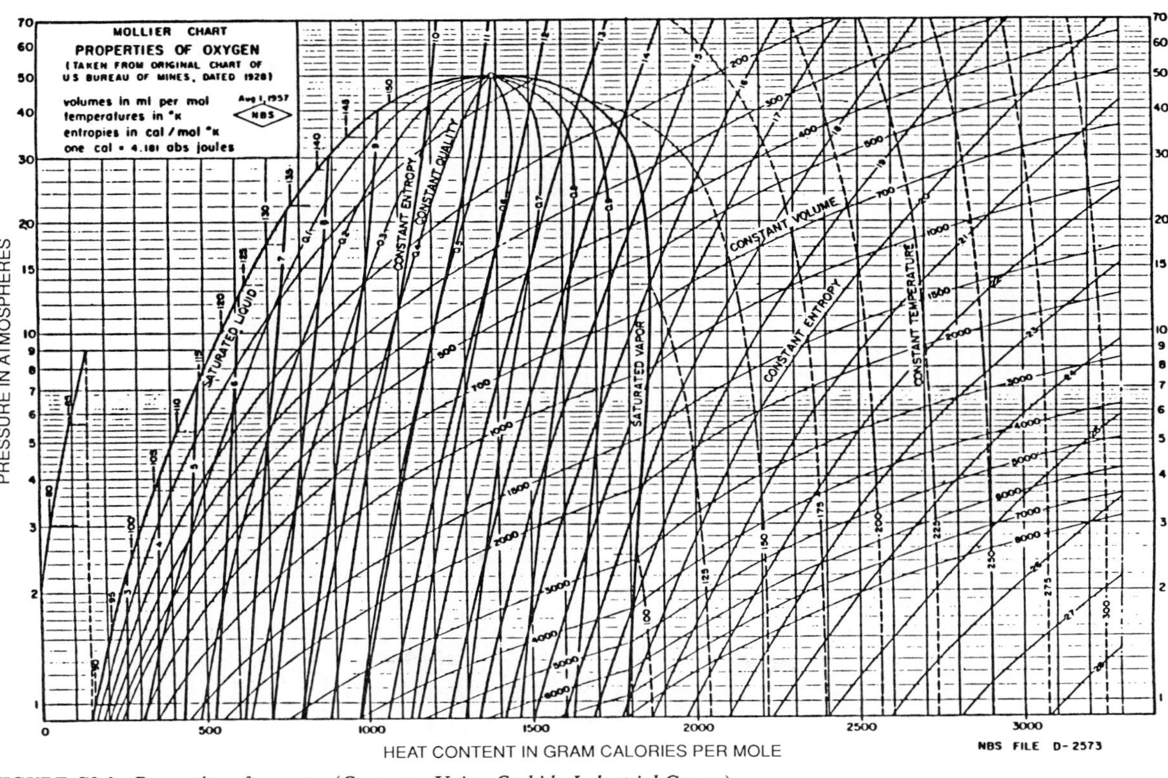

FIGURE C8.2 Properties of oxygen. (*Courtesy, Union Carbide Industrial Gases.*)

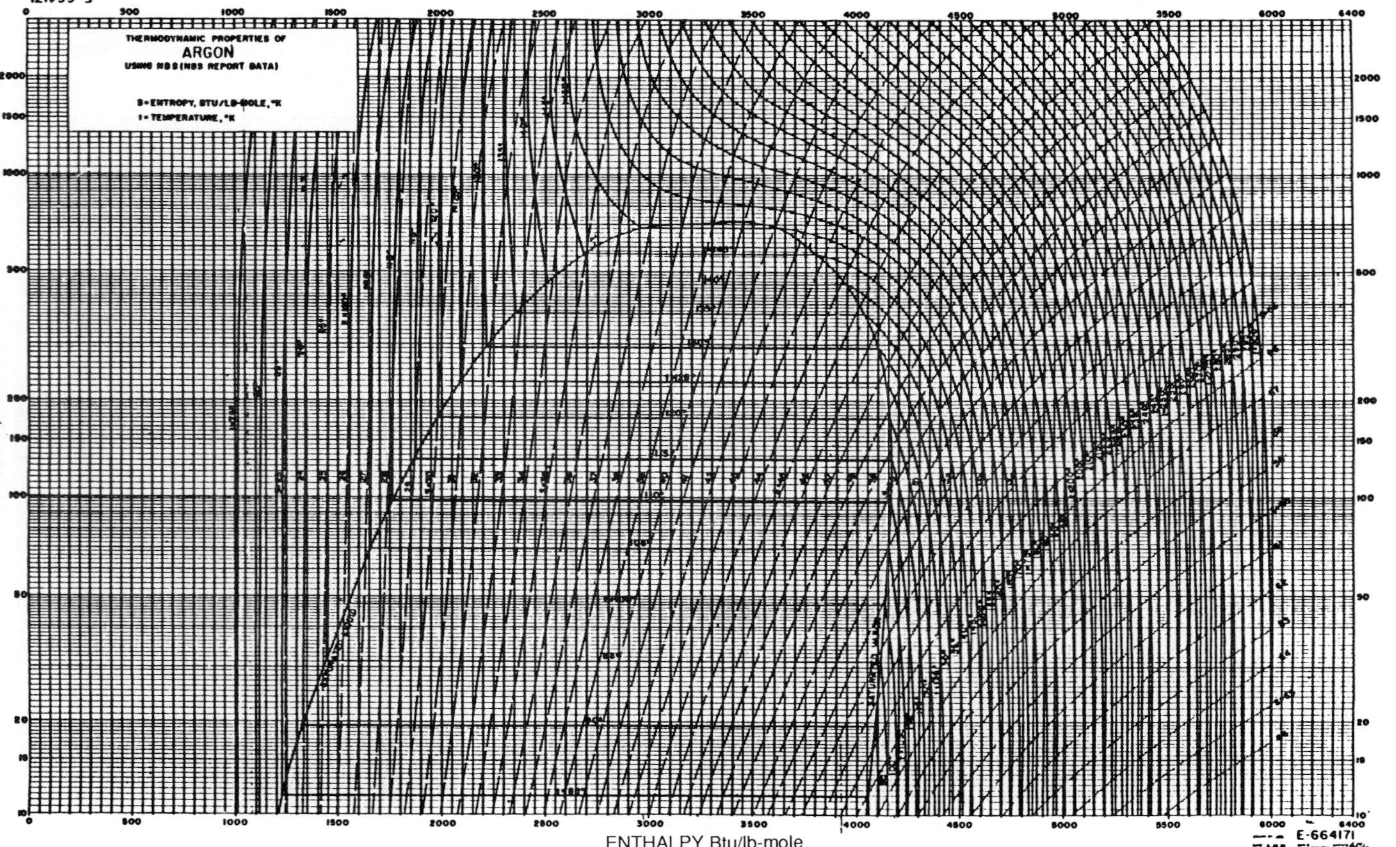

FIGURE C8.3 Thermodynamic properties of argon. (*Courtesy, Union Carbide Industrial Gases.*)

The designer should have knowledge of the combustion and physiological properties of the fluids being handled. Air and other oxidants must be excluded from piping transporting flammable fluids and from any enclosure (such as a cold box) into which leakage may occur. Conditions which may concentrate flammable contaminants present a danger in air separation. A number of cryogenic fluids are toxic, and all can present an asphyxiation hazard, particularly within confined areas. Venting of these fluids must take into account the possibility that temperature differences from the ambient air may lead to unexpected localized concentration buildups which threaten the safety of personnel and equipment.

BIBLIOGRAPHY

Sources of Physical and Thermodynamic Properties of Cryogenic Fluids

Physical Properties

CRC Handbook of Chemistry and Physics, CRC Press, Boca Raton, FL.

Lange, *Lange's Handbook of Chemistry,* McGraw-Hill, New York.

Perry and Chilton, *Chemical Engineers' Handbook,* McGraw-Hill, New York.

Reid, R. C., J. M. Prausnitz and T. K. Sherwood, *The Properties of Gases and Liquids,* 3d ed., McGraw-Hill, New York, 1977. Contains estimating methods for physical properties of pure fluids and mixtures.

Thermodynamic Properties

Matheson Gas Data Book, Matheson, Lyndhurst, NJ, 1980. *The Matheson Unabridged Gas Data Book,* Matheson, East Rutherford, NJ, 1975. Contains tabular data for most cryogens.

Maxwell, J. B., *Data Book on Hydrocarbons,* D. Van Nostrand, Princeton, NJ, 1950. Contains tabular and chart data for light hydrocarbons.

Perry, R. H., and Green, D. W., *Chemical Engineers' Handbook,* 6th ed., McGraw-Hill, New York, 1984. Contains tabular data and thermodynamic charts for many cryogens.

Raznjevic, Kuzman, *Handbook of Thermodynamic Tables and Charts,* Hemisphere Publishing, Washington, 1976.

Selected Values of Properties of Hydrocarbons and Related Compounds, American Petroleum Institute Research Project 44, Thermodynamics Research Center, Texas A&M University, College Station, TX, 1977.

Tables of Thermodynamic and Transport Properties of Air, Argon, Carbon Dioxide, Carbon Monoxide, Hydrogen, Nitrogen, Oxygen, and Steam (originally issued as NBS Circular 564), Pergamon Press, New York, 1960.

MATERIALS USED IN CRYOGENIC PIPING SYSTEMS

Robert Zawierucha, Materials Engineer
Union Carbide Industrial Gases, Linde Division

Important considerations in the selection of structural materials for cryogenic piping systems include suitable mechanical and physical properties, compatibility with process fluids, fabricability, cost, and compliance with regulatory codes.

The subject of materials for cryogenic applications has been generously treated in the technical literature. A few selected works in this area appear in the References and Bibliography at the end of this section for those interested in additional background information. The prime focus of this section, however, is cryogenic piping for the chemical process industry and commercial cryogenic distribution applications. Consequently, this overview will be deliberately limited in its scope and coverage, and many materials, such as those used in aerospace applications, will not be covered. The materials that will be covered include ferrous alloys, nonferrous alloys, and nonmetallic materials.

FERROUS MATERIALS

Ferrous alloys most often encountered in cryogenic piping applications are usually classified as ferritic or austenitic types. The terms *austenitic* and *ferritic* refer to the predominant crystallographic phases ferrite or austenite which are body centered cubic (BCC) and face centered cubic (FCC), respectively.

Ferritic Alloys

Most of the steels in common use are ferritic. This classification also covers steels which are martensitic as a result of heat treatment. Low cost, ease of fabricability, and high strength via heat treatment are major reasons for their popularity. Ferritic alloys, however, can exhibit a ductile to brittle toughness transition. The Charpy Impact Test by which measurements of energy absorption, lateral expansion, and ductile fracture appearance are made is the *most* common method of measuring this transition.

However, the ductile to brittle transition temperature in ferritic steels can be influenced by a number of variables involved in the steel-making process which control the levels of residual elements, inclusion shape, and heat treatment techniques which in turn control grain size and crystallographic morphology. A comprehensive treatment of these approaches is beyond the scope of this document. However, alloying with nickel, reduction of sulphur levels, and reductions in grain size may be stated as the most popular approaches with ferritic alloys.

Table C8.6 contains a listing of some of the most common ferritic alloy steels

TABLE C8.6 Typical Ferrous Alloys Used in Cryogenic Piping

Alloy	Minimum temperature (°F)*	ASME specification	Comment
C-Mn steel†	-50	SA-333 Grade 1	Aluminum killed, fine grain practice
2-1/4% Ni steel†	-100	SA-333 Grade 7	Aluminum killed, fine grain practice
3-1/2% Ni steel†	-150	SA-333 Grade 3	Aluminum killed, fine grain practice
9% Ni steel†	-320	SA-333 Grade 8	Aluminum killed, fine grain practice
304 stainless steel‡	-425	SA-312	
304L stainless steel‡	-425	SA-312	
316 stainless steel‡	-325	SA-312	
316L stainless steel‡	-325	SA-312	
347 stainless steel‡	-425	SA-312	

*Design minimum temperature for which material is normally suited without impact testing other than that required by material specification.

†Ferritic steels

‡Austenitic steels

Source: Union Carbide Industrial Gases

used in cryogenic piping. Minimum service temperatures are included as well as the applicable ASME specification.

Table C8.7 contains a listing of the same materials as Table C8.6 with typical data, including some data not found in the ASME specifications. Included in Table C8.7 are mechanical properties such as strength, impact, and elongation, as well as thermodynamic data such as thermal expansion and thermal conductivity at cryogenic temperatures (boiling point at 1 atmosphere).

Austenitic Alloy Steels

In addition to the ferritic alloy steels, Tables C8.6 and C8.7 also contain a listing of austenitic stainless steels which are likely to be encountered in cryogenic piping. Most of the austenitic alloy steels used in cryogenic piping are chromium-nickel stainless steels of the AISI 300 type, such as 304, 304L, 316, and 316L. Other stainless steels classified as martensitic, duplex, and precipitation hardening also exist; however, the preceding alloys are most commonly used in cryogenic piping for chemical process and distribution applications.

A major consideration in the use of the 300 series stainless steels is the improvement in toughness properties and elimination of the sharp ductile to brittle transition found in ferritic steels. These alloys have been used to contain and distribute liquid hydrogen and helium. Consequently, although the austenitic stainless steels were originally developed for corrosion resistance, it is their toughness and excellent fracture properties which have resulted in their selection for cryogenic piping applications.

TABLE C8.7 Typical Mechanical and Physical Properties of Ferrous Alloys

Alloy	ASME spec.	Ultimate temp. (°F)	Tensile str. (ksi)	0.2% offset yield strength (ksi)	% elong. in 2 in (%)	Charpy impact strength (ft · lb)	Thermal expansion (in/in/°F × 10^{-6})	Thermal conductivity [Btu/(h · ft · °F)]
C-Mn steel†	SA 333	RT	55	30	21[1]	70	6.5	30
	Grade 1	−50°F				50	3.0	—
2-1/4% Ni steel‡	SA 333	RT	65	35	18[1]	58	—	—
	Grade 7	−100°F	75	40		20	—	—
3-1/2% Ni steel‡	SA 333	RT	100	75	18[1]	96	5.8	21
	Grade 3	−150°F	120	85		22	4.8	—
9% Ni steel‡	SA 333	RT	115	90	25	47	5.8	15.7
	Grade 8	−320°F	170	135	27	25	4.8	7.6
304 stainless steel	SA 312	RT	85	38	45	115	9	9
	TP 304	−425°F	250	70	3	75	1.2	0
304L stainless steel	SA 312	RT	80	37	45	60	Same as 304	Same as 304
	TP 304L	−425°F	225	65	31	60	Same as 304	Same as 304
316 stainless steel	SA 312	RT	87	38	45	—	9	9
	TP 316	−325°F	197	65	56	—	7.8	0
316L stainless steel	SA 312	RT	85	38	45	—	Same as 316	Same as 316
	TP 316L	−325°F					Same as 316	Same as 316
347 stainless steel	SA 312	RT	90	65	50	60	8.7	8.5
	TP 347	−425°F	230	70	38	45	7.2	0

*Minimum value as stated in ASME Specification SA-333.
†Normalized, aluminum killed, fine grain practice.
‡Quenched and tempered.
Source: Union Carbide Industrial Gases

Of the 300 series alloys, the AISI 304 composition is the most popular as measured by tonnage. It should be noted that there is a preference for AISI 316L in the electronics industry for the distribution of high-purity gases that are free of particulates. Furthermore, piping and tubing used in this application are frequently electro-polished and the trend is for an increase in electro-polishing for ultra-high-purity applications.

Cast versions of the common austenitic stainless steels which may be used if valve applications exist. CF-3, CF-3M, CF-8, and CF-8M are cast equivalents of 304L, 316L, 304, and 316, respectively. Note that the cast stainless steels have been optimized for castability and increased delta ferrite levels may adversely affect toughness at cryogenic temperatures. Similar effects may be observed in austenitic weld filler metals, and both castings and weld fillers should be well characterized and qualified prior to use in cryogenic piping systems.

NONFERROUS ALLOYS

Nonferrous alloys encountered in cryogenic piping are usually of the aluminum, cuprous, or nickel families. All three alloy families exhibit no ductile to brittle toughness transition due to the FCC crystal lattice. Common nonferrous alloys used in cryogenic piping appear in Tables C8.8 and C8.9.

Aluminum Alloys

Common aluminum alloys used in cryogenic process piping do not represent the gamut of what is available in aluminum alloys. The high-strength aluminum alloys commonly used in aerospace applications are not used in the chemical process

TABLE C8.8 Typical Nonferrous Alloys Used in Cryogenic Piping

Alloy	Tempers	Minimum temperature (°F)*	ASME spec.
1100 aluminum	O, H11	−452	SB 210
3003 aluminum	O, H112	−452	SB 210
5052 aluminum	O, H32	−452	SB 210
5083 aluminum	O, H112	−452	SB 210
5086 aluminum	O, H112	−452	SB 210
6061 aluminum	T6	−452	SB 210
Copper (C10200, C12200)	Annealed	−325	SB 75
Copper-nickel (70600)	Annealed	−325	SB 467
Copper-nickel (C71500)	Annealed	−325	SB 467
Monel 400, Ah-Cu alloy	Annealed	−325	SB 165

*Design minimum temperature for which material is normally suitable without impact testing other than that required by material specification.

Source: Union Carbide Industrial Gases

TABLE C8.9 Typical Mechanical and Physical Properties of Nonferrous Alloys*

Alloy	ASME Spec.	Temp. (°F)	Ultimate tensile str. (ksi)	0.2% offset yield strength (ksi)	% Elong. in 2 in (%)	Charpy impact strength (ft · lb)	Thermal expansion (in/in/°F × 10^{-6})	Thermal conductivity ([Btu/(h · ft · °F)]
100 aluminum	SB-210	RT	13S 24H	5S 22H	40S 10H	16S 70H	13	125
0, H112		−452	24S	8S	56S		5	160
3003 aluminum	SB-210	RT	16S 29H	6S 27H	35S 7H	16S 70H	12	92
0, H112		−452	32S	9S	48S		5	85
5052 aluminum	SB-210	RT	43S 46H	37S 42H	12S 8H		13.2	75
0, H32		−452	76S 86H	47S 55H	42S 30H			7
5082 aluminum	SB-210	RT	42S 44H	21S 28H	22S 16H		13	68
0, H1112		−452	63S	23S	32S		5	4
5086 aluminum	SB-210	RT	38S 42H	17S 17H	22S 12H		13.2	73
0, H112		−452	78S 96H	20S 26H	38S 30H			
6061 aluminum	SB-210	RT	45	40	12	10	10	99
T6		−452	70	55	25	12	2	
Copper (C10200, C12200)	SB-75	RT	33	10	45	56	9.5	150
		−325	52	14	69	75	5.0	75
90/10 Cu-Ni	SB-467	RT	44	16	42		9.5	26
CDA 706		−325						
70/30 Cu-Ni	SB-467	RT	44	20	40	79	9	17
CDA 715		−325	85	31	60	87		
Monel 400	SB-165	RT	80	25	42	50	7.5	15
		−325	115	50	64	50	2.5	9

*The letters S and H designate soft and hardened conditions respectively.
Source: Union Carbide Industrial Gases

industry because of the lack of code coverage or other considerations such as corrosion.

Aluminum compositions of the 5083 and 6061 types constitute the highest-strength alloys used in cryogenic applications and tempers of these alloys are suitable at temperatures as low as −452°F. From the toughness standpoint 5083 would be preferred. However, prolonged exposure to temperatures as high as 150°F during thaws can result in both a reduction in corrosion resistance and toughness. For low-strength applications, the 3000 series are also used.

Copper and Cuprous Alloys

Most of the early work in the cryogenic industry was accomplished through the use of copper process vessels, piping, and tubing. Aluminum has extensively replaced copper and cuprous alloys in the fabrication of air separation plants. However, copper and cuprous piping are still extensively used in piping and tubing runs from cryogenic tankage for several reasons: ease of fabrication (e.g., brazing, soldering), resistance to ignition, and combustion resistance in oxygen-enriched systems. It may be noted that orbital arc welded stainless piping systems are increasingly replacing cuprous piping for the delivery of particulate- and contaminant-free electronic grade gases.

Nickel Alloys

While nickel alloys could be used in cryogenic piping systems on the basis of their mechanical properties, their cost generally precludes their use. One major exception is the use of Monel, a nickel-copper alloy, which may be used in oxygen systems to minimize ignition tendencies where there is concern of impingement or the potential for high velocities.

NONMETALLIC MATERIALS

Although the bulk of the materials used in cryogenic pipelines are metals, nonmetallic materials have critical functions in cryogenic pipeline components such as valves and insulation. Numerous nonmetallic components have been used in cryogenic pipeline applications, and it would be beyond the scope of this section to cite them all. A brief listing of some of the more common nonmetallic materials which are used in pipeline components is found in Tables C8.10 and C8.11.

Again, compatibility and mechanical and physical properties must be considered in the selection of nonmetallic materials. With respect to compatibility, liquid oxygen is the commercial cryogen of greatest concern because of its large usage. Typical tests or experimental parameters covered in the selection of nonmetallic materials for oxygen service include autoignition temperature, heat of combustion, impact test, and oxygen index. Materials that are compatible with oxygen generally have high autoignition temperatures, low heats of combustion, high energy absorption in impact tests, and high oxygen index values. Reference 1 gives advice about the specific criteria levels required for different applications. It is recommended that nonmetallic materials be qualified for oxygen service on a batch by batch basis.

Liquid fluorine is of greater concern from the compatibility standpoint with

TABLE C8.10 Typical Examples of Nonmetallic Materials Used in Cryogenic Piping

Application	Material
Gaskets	Durabla (bonded or compressed asbestos) Grafoil (flexible graphite)
Insulation (fiber)	Mineral wool Fiberglass
Insulation (foam)	Polyurethane Styrofoam Foamglass(R)
Insulation (powder)	Perlite Vermiculite
Insulation (sheet)	Mylar Aluminum opacified paper
Insulation (support brock)	Transite Micarta
Valve packing, seals, and lubricants	Viton KEL-F Glass reinforced teflon Fluorolube

Source: Union Carbide Industrial Gases

respect to nonmetallic materials. However, the industrial usage of liquid fluorine is very low in comparison to the other industrial gases. Reference 2 should be consulted for information on liquid fluorine systems.

The mechanical and physical properties of significance in the use of nonmetallic materials relate to their application in the form of insulation, gaskets, seals, and lubricants. It may be noted that they are not currently used as structural materials. Development of composites is expected to significantly increase the use of nonmetallic materials as structural components in cryogenic piping.

JOINING

Welding is the most common joining technique used in cryogenic piping. Brazing and soldering may be encountered in cuprous piping. With the recent trends toward high purity, there is a greater tendency to use orbital arc welded stainless steel piping in lieu of brazed copper alloy piping.

Joint selection, filler metals, process qualification, and welder qualification are covered in detail by the ASME Code and the ANSI/ASME B31.3 Piping Code. Generally, areas which have been welded, brazed, or soldered experience thermal effects which may exhibit reduced strength, ductility, or toughness if not properly controlled. Appropriate mechanical tests as specified by applicable codes must be performed to verify suitability.

Within components such as cold boxes there may be occasions when transitions must be made between metals such as stainless steel and aluminum forming

TABLE C8.11 Typical Mechanical and Physical Properties of Nonmetallic Materials

Material	Tensile strength (ksi)	Modulus of elasticity (ksi)	Specific gravity	Thermal expansion (in/in/°F × 10^{-5})	Thermal conductivity [Btu/(h · ft · °F)]	Specific heat (Btu · lb · °F)
Durabla (asbestos gasket material)	4-11					
Grafoil (flexible graphite)	.75–1.00	200	1.1	0.02	432*	0.7
Mineral wool fiber insulation	.27		0.15–0.2	0		
Fiberglas fiber insulation	50–500	10,000		0.5–0.8		
Perlite powder insulation						
Vermiculite powder insulation			0.14			
Mylar sheet insulation	17–18	700	1.39		0.022	
	36	1600				
Aluminum opacified paper sheet insulation	No longer available					
Transite insulation block						

*Through thickness conductivity.
Source: Union Carbide Industrial Gases

joints which are difficult to weld directly or are unweldable. Transition joints are commercially available which may be classified as mechanical, brazed, diffusion bonded, or explosion bonded. Such joints have been successfully used in cryogenic applications. However, caution is advised to avoid the in-service problems such as leaks, embrittlement, or actual joint separation. Careful vendor evaluation, joint design evaluation, and attention to fabrication details are required when dissimilar metal joints are required.

ENVIRONMENTAL AND SAFETY CONCERNS

A number of materials both metallic and nonmetallic may be toxic, carcinogenic, tetragenic, or have other properties which are considered environmentally undesirable (i.e., effects on the ozone layer). For example, the use of asbestos-containing materials such as Durabla or certain vermiculite ores may be further restricted or phased out. Similar concerns exist with cadmium-bearing brazing alloys. The recent concern about ozone depletion caused by chlorofluorocarbons could affect polyurethane foam production. Consequently, many common materials currently used in cryogenic piping systems are likely to be phased out over time.

Recent legislation mandates that material safety data sheets (MSDS) be obtained for all industrial substances that are in industrial use. This obviously includes the metallic and nonmetallic materials used in cryogenic piping. Failure to comply with the requirements and implementation provisions of the original legislation can have serious consequences for fabricators.

Equipment used in oxygen services have additional requirements such as cleaning and velocity limitations which must be considered. Oxygen equipment must be cleaned to eliminate contaminants such as hydrocarbons and metal particulates which could serve as ignition sources. Reference 3 provides a general discussion of this issue. In addition, velocity limitations may be placed on certain classes of material to preclude ignition by particle impingement. See Reference 4 for velocity limitations and design considerations pertinent to oxygen systems.

Other industrial gases that might be encountered in cryogenic pipelines have specific hazards associated with them (i.e., flammability, toxicity, etc.). These are summarized in MSDS forms available from industrial gas suppliers.

REFERENCES

1. "Guide for Evaluating Nonmetallic Materials for Oxygen Service," ASTM Specification G63-83a, American Society for Testing and Materials, Philadelphia, PA.
2. "Design Handbook for Liquid Fluorine Ground Handling Equipment," Technical Report No. AFRPL-TR-65-133, Air Force Rocket Propulsion Laboratory, Edwards Air Force Base, 1965.
3. Hust, J. G., and Clark, A. F., "A Survey of Compatibility of Materials, with High Pressure Oxygen Service," *Cryogenics,* vol. 13, 1973, pp. 325–335.
4. "Cleaning Equipment for Oxygen Service," CGA Pamphlet G.4.1-1985, Compressed Gas Association, Arlington, VA.

BIBLIOGRAPHY

"Handbook on Materials for Superconducting Machinery," Technical Report MCIC-HB-04, Metal & Ceramics Information Center, Battelle, Columbus, OH, November 1974.

Metals Handbook, Desk Edition, American Society for Metals, Cleveland, OH, 1985, pp. 20.24–20.34.

Reed, R. P., and Clark, A. F., "Materials at Low Temperatures," American Society for Metals, Metals Park, OH, 1983.

Schwartzberg, F. R., "Cryogenic Materials Data Handbook," vols. I and II, Technical Report AFML-TR-64-280, Wright Patterson Air Force Base, OH, July 1970.

PIPING SYSTEM DESIGN—FLUIDS

M. F. Melchioris, Piping Design Engineer
Union Carbide Industrial Gases, Linde Division

PIPE SIZING CRITERIA

For all cryogenic fluids, except oxygen gas, pipe sizing is based on pressure drop considerations (for oxygen gas, see the section "Oxygen Gas Piping"). The pressure drop criteria presented in Chap. B8 is applicable to cryogenic fluids when the fluid is either in the liquid phase or gas flow regime. A convenient form for the pressure drop equation is:

$$P = 1.078 \times 10^{-4} K\rho V^2$$

where P = pressure drop, pounds per square inch
K = total resistance coefficient for the pipe segment
ρ = weight density of fluid, pounds per cubic foot
V = mean velocity of fluid, feet per second

Heat transfer into a saturated cryogenic liquid or a drop in pressure, such as across a throttling valve, can cause a portion of the liquid to flash into a gas (see the section "Properties of Cryogenic Fluids"). A two-phase (gas-liquid) flow condition must then be considered when determining the required pipe size and pressure head requirements (see the next section). The two-phase flow condition will result in larger pressure drop losses or larger pipe size requirement than for the liquid phase flow condition. Therefore, whenever possible, the fluid should be maintained in a single-phase flow condition. The liquid can be subcooled to prevent flashoff due to pressure decreases such as across throttling valves. Use of pipe insulation can reduce the heat leak into the piping, thereby reducing flashoff.

TWO-PHASE FLOW

Many empirical correlations have been developed to predict the pressure drop due to two-phase flow. One approach is described below and has an expected accuracy of ±30 percent.

The two-phase flow of cryogenic fluids is complicated by the fact that because of heat infiltration liquid is continuously being vaporized. The maximum velocity of the vapor-liquid mixture is limited to that of the velocity of sound in the fluid mixture. This will be much lower than for liquids because of the high adiabatic compressibility resulting from the vapor.

The method of calculating the critical velocity is as follows:

1. Plot a curve of P in pounds per square inch absolute as abscissa versus specific volume of the mixture in cubic feet per pound as ordinate.
2. On the P-V curve, plot a line whose slope $= -G^2/144g$ [where G = lb/(s · ft^2) and $g = 32.2$].
3. Where the slope is tangent to the P-V curve read the critical pressure.

If the critical pressure is greater than the downstream terminal pressure, a shock wave will result, causing noisy, unstable operation. In the design of any line, the size should be so chosen as to eliminate this critical-flow condition.

The various flow patterns for two-phase flow were established by Baker and are shown in Fig. C8.4,

where G = gas-phase mass velocity, lb/(h · ft^2) of total pipe area
L = liquid-phase mass velocity, lb/(h · ft^2) of total pipe area
ρ_G = gas-phase density, lb/ft^3
ρ_L = liquid-phase density, lb/ft^3
$\lambda = [(\rho_G/0.075)(\rho_L/62.3)]^{1/2}$
$\psi = (73/\nu)[\mu_L(62.3/\rho_L)^2]^{1/3}$

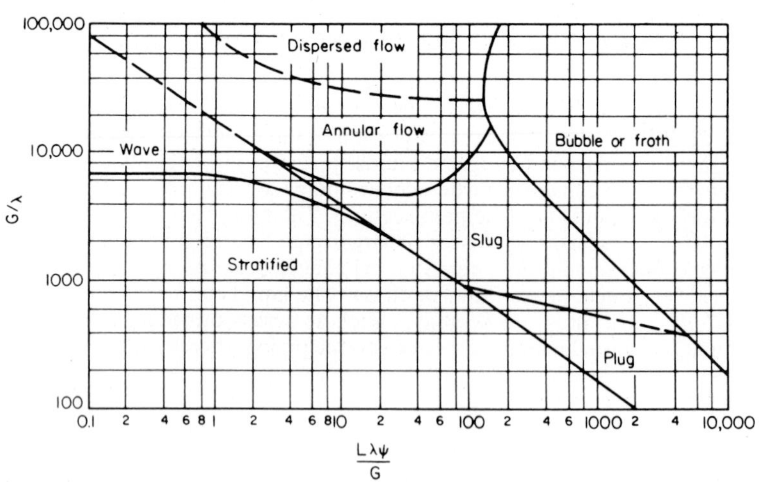

FIGURE C8.4 Two-phase flow patterns. (*O. Baker, Oil Gas J., vol. 53, p. 185, 1954.*)

ν = surface tension of liquid phase, dynes/cm
μ_L = liquid-phase viscosity, centipoises

These patterns may be described as:

1. *Bubble Flow:* Characterized by the formation of individual bubbles along the upper surface of the tube.
2. *Froth Flow:* Foamlike mixture of small bubbles and liquids intimately mixed during turbulent flow.
3. *Plug Flow:* Type of flow which is likely to exist during the transition period from bubble to stratified flow. It is characterized by the flow of large plugs of liquid. Large pressure fluctuations occur during this type of flow.
4. *Stratified Flow:* Flow of liquid phase along the bottom of the pipe, while the gas phase occupies the upper portion. The gas phase flows at a much higher velocity than the liquid at the bottom.
5. *Wavy Flow:* Wavy flow is similar to stratified flow. Waves are formed at the gas-liquid interface as the fluid flows.
6. *Slug Flow:* Consists of alternate sections of gas and liquid. It occurs as the proportion of gas increases by heat addition. Bubbles formed by heat added to the fluid enlarge and rise in the pipe to increase the size of the gas slug.
7. *Annular Flow:* The flow of a continuous liquid layer along the tube wall. The gas flows in the central core at a much higher velocity.
8. *Dispersed, Spray, or Mist Flow:* Flow in which fine liquid drops are suspended by surface tension. There is no significant relative velocity between the liquid and gas phase.

Because of heat transfer to the flowing fluid the relative distribution of liquid and vapor is altered progressively and the pressure drop must be calculated in finite steps. The final result is a summation of the finite pressure drops.

For those cases in which the pressure drop across the pipe is small compared with the absolute pressure, the pressure drop resulting from the flow of a boiling mixture is made up of two parts: (1) the pressure drop due to two-phase flow and (2) the pressure drop resulting from the rate of increase of momentum of the mixture as it flows.

The two-phase pressure drop for all types of flow except slug flow is correlated by the parameters given in Fig. C8.5.

The value of the abscissa for various flow regimes is given by the following expressions:

Liquid	Vapor	
t	t	$X^2 = (W_t/W_g)^{1.8}(\rho_g/\rho_t)(\mu_t/\mu_g)^{0.2}$
ν	t	$X^2 = K(\text{Re}_g)^{-0.8}(W_t/W_g)(\rho_g/\rho_t)(\mu_t/\mu_g)$
t	ν	$X^2 = 1/K(\text{Re}_t)^{0.8}(W_t/W_g)(\rho_g/\rho_t)(\mu_t/\mu_g)$
ν	ν	$X^2 = (W_t/W_g)(\rho_g/\rho_t)(\mu_t/\mu_g)$

K = 296 for commercial pipe

= 348 for smooth tubes

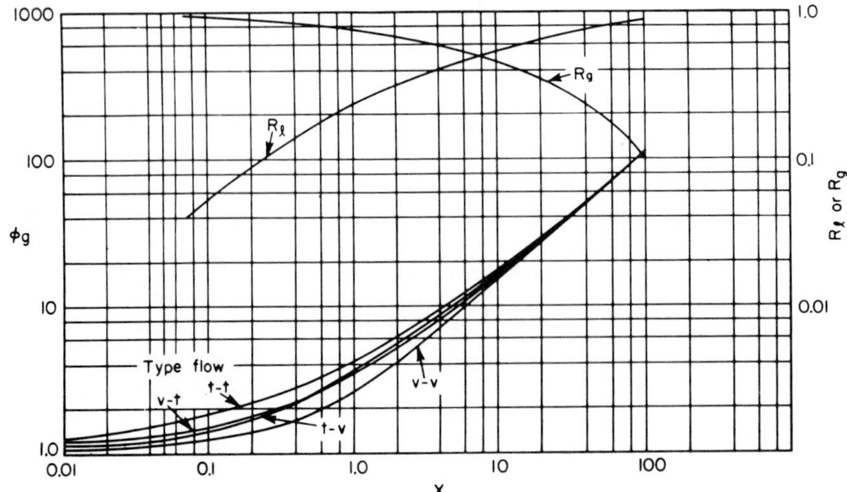

FIGURE C8.5 Pressure drop in two-phase flow (*R. W. Lockhart and R. C. Martinelli, Chem. Eng. Progr., vol. 45, p. 39, 1949.*)

$$\left(\frac{\Delta P}{\Delta L}\right)_{TP} = (\phi)^2 \left(\frac{\Delta P}{\Delta L}\right)_g$$

where flow t = turbulent flow, Reynolds number >2000
flow v = viscous flow, Reynolds number <1000
W = weight flow per unit time
ρ = density lb/ft^2
μ = viscosity
$(\Delta P/\Delta L)_{TP}$ = two-phase flow pressure drop
$(\Delta P/\Delta L)_g$ = pressure drop due to gas-phase flow only in pipe

Subscript l denotes the liquid phase; subscript g denotes the gas phase and ϕ is read from Fig. C8.5.

Method of Calculation

1. Calculate the Reynolds number Re_t and Re_g based on each phase flowing separately through pipe.
2. From Re_t and Re_g determine the type of flow.
3. Calculate X using the appropriate equation and read ϕ from the curve.
4. Calculate $(\Delta P/\Delta L)_g$ for the gas phase only using single-phase pressure-drop correlation.
5. $(\Delta P/\Delta L)_{TP} = \phi^2(\Delta P/\Delta L)_g$.

The correlation giving the X versus ϕ relationship is accurate in the range of about ±30 percent, and for design safety the pressure drop should be increased by at least 30 percent.

For the calculation of pressure drop in a line having slug flow, see Ref. 1 at the end of this section.

The pressure drop due to momentum change is

$$A_p \Delta Pa = \frac{W_l v_l}{g} + \frac{W_g v_g}{g} - \frac{W_T v_0}{g}$$

where A_p = pipe area, ft^2
ΔPa = pressure drop due to momentum change, lb/ft^2
W = weight flow, lb/s
v = velocity, ft/s
g = 32.2 ft/s^2
sub l = liquid phase at exit of line
sub g = gas phase at exit of line
sub T = at inlet of line where only liquid phase exists
sub 0 = velocity at inlet of line

Two extreme cases of exit conditions can exist, the actual condition probably lying between the two:

1. Liquid and vapor completely mixed (fog)
2. Liquid and vapor completely separated

For the first case, $v_g = v_t$ and

$$\Delta Pa = \frac{G^2}{g\rho_l}\left[\left(1 - \frac{W_g}{W_T}\right) + \left(\frac{W_g}{W_T}\right)\left(\frac{\rho_l}{\rho_g}\right) - 1\right]$$

For the second case,

$$\Delta Pa = \frac{G^2}{g\rho_l}\left[\frac{(1 - W_g/W_T)^2}{R_l} + \frac{(W_g/W_T)^2 \rho_l}{R_g \rho_g} - 1\right]$$

For flow through any other type of fitting or valve, use the single-phase pressure-drop correlation multiplied by the ϕ factor for two-phase flow.

OXYGEN GAS PIPING

For oxygen gas, the fluid velocity must also be considered when determining the pipe size. The allowable velocity is a function of the oxygen gas pressure and temperature and the material of the piping and its components. For systems with pressures up to 1000 psi and a maximum temperature of 200°F, carbon steel and stainless steel piping are acceptable provided that the maximum allowable gas velocity as shown in Fig. C8.6 is not exceeded.

When the velocity is exceeded, copper or a copper base material such as brass or monel is required. The velocity criteria should also be considered at valves. For example, sonic velocity can occur at throttling and safety valves. Copper base materials are required for the parts of the valves where high velocity can occur.

When oxygen gas impinges directly on ferrous piping, such as from a side back feed line, the allowable velocity must be reduced to one-half the values given in Fig. C8.6 or the impingement surface must be a copper base alloy. When

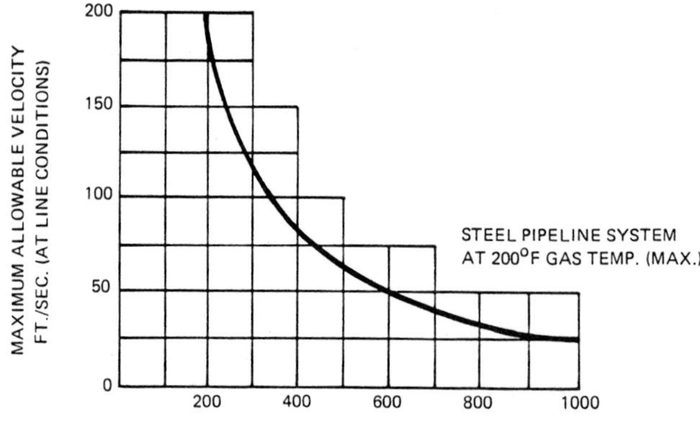

FIGURE C8.6 Maximum velocity versus internal pressure for steel pipelines (from Ref. 1).

the velocity returns to acceptable levels, the copper base alloy is extended for eight diameters in pipe length before returning to ferrous piping. For a more thorough discussion of material requirements for oxygen service, see Ref. 2 at the end of this section.

PIPING ARRANGEMENT

Consideration of the cryogenic fluid properties has an effect on the piping arrangement. Because the cryogenic fluid is colder than ambient air, the continuous heat leak from ambient air to the piping system is a design consideration.

Whenever cryogenic liquid can be trapped between two valves, a line block safety valve must be provided to prevent overpressurization caused by heat transfer from the ambient air into the cryogenic fluid. As noted earlier in the discussion of cryogenic fluid, containment of the pressure increase caused by vaporization of the trapped cryogenic liquid is not practical and use of a line block safety valve is mandatory. When cryogenic gas can be trapped between two valves, the pressure rise should be calculated to ensure that the piping system design pressure is not exceeded when the trapped gas warms up to ambient temperature. When the design pressure can be exceeded, a line block safety valve should be used.

Traps are normally designed into liquid piping systems to prevent undesirable heat leak from a branch of the piping system when it is not in use. To form a trap at the branch and main run connection, a vertical rise equal to a minimum of two pipe diameters is provided in the branch. Heat transfer into the nonflowing branch will cause the liquid to vaporize and a gas-liquid interface will form in the trap. The gas-liquid interface in the vertical rise will prevent flow from the main run into the nonflowing branch. Undesirable heat transfer and liquid loss is therefore reduced. When liquid flow is required through the branch, the liquid will flow through the trap.

When a cryogenic liquid line is initially put in service, the warm piping will cause liquid flashoff which could restrict the flow during the two-phase flow transient period. When it is possible to precool the lines, the piping can be sized for liquid phase flow, which will result in smaller piping. If rapid cooldown is required, the piping must be sized for two-phase flow.

As a good approximation, the amount of liquid required to cool down a line is

$$W = \frac{M_m\, C_{pm}\, \Delta T_m}{(H_v - H_l)}$$

where W = liquid, lb
M_m = line to be cooled, lb
C_{pm} = mean specific heat of line, Btu/(lb · °F)
ΔT_m = temperature change through which line is cooled, °F
H_v = enthalpy of fluid as a vapor at ambient conditions, Btu/lb
H_l = initial enthalpy of liquid when entering line, Btu/lb

REFERENCES

1. Griffith and Wallis, *Journal of Heat Transfer,* August 1961, pp. 307–320.
2. CGA Pamphlet G-4.4, "Industrial Practices for Gaseous Oxygen Transmission and Distribution Piping Systems," Compressed Gas Association, Inc., Arlington, VA.

PIPING SYSTEM DESIGN—MECHANICAL

M. F. Melchioris, Piping Design Engineer
Union Carbide Industrial Gases, Linde Division

APPLICABLE CODES AND MATERIALS

The applicable code for the design of cryogenic piping systems is the ASME/ANSI B31.3 piping code.[1] B31.3 presents a design procedure to determine the stresses in the pipe due to fluid pressure, pipe and fluid weight, and thermal expansion and contraction of the pipe. The code also lists piping materials by ASTM specification number. The allowable stress as a function of design temperature is listed for each material. The listed minimum design temperature is used to determine which materials are suitable for cryogenic service.

Some of the materials can be used in lower-temperature service provided the specified impact test is performed to ensure adequate ductility at the lower temperature. The B31.3 code also lists pipe fittings and bolting by ASTM number so that materials compatible with the pipe can be selected. For more discussion of materials, see the section "Materials Used in Cryogenic Piping Systems."

Other piping design considerations such as corrosion resistance, chemical compatibility, and material melting point can affect the material selection. Some examples of these conditions include:

1. Copper base materials are not recommended for ammonia or acetylene service because the copper alloys react with the fluid.
2. Stainless steel piping is recommended for flammable fluids such as hydrogen rather than lower melting point materials such as copper or aluminum unless the piping is protected to prevent exposure to heat in the event of a fire.
3. As discussed earlier, copper base materials are recommended for high-velocity oxygen service.

See Chap. B2 for a discussion of formulas which can be used to calculate the pipe wall thickness as required to contain the fluid pressure.

ECONOMIC PIPE SIZING

When a piping system is being designed, an initial pipe size must be selected before the piping configuration is developed. This subsection provides recommendations for selecting the initial pipe size. However, after the piping system is developed, the pipe size selection is finalized by calculating the pressure drop based on the actual piping configuration and comparing it to the pressure head available.

For cryogenic liquid lines, the initial pipe size should be selected so that the liquid velocity is in the range of 5 ft/s.

For cryogenic gas lines, the initial pipe size is selected based on the available pressure head (see the section "Pipe Sizing Criteria"). When estimating pipe length and number of fittings to calculate the initial pipe size, add 50 percent to the total resistance coefficient, K, to allow for pipe loops and extra fittings as may be required for thermal flexibility. To ensure good process control, one-third of the total system pressure drop should be allotted to the control valves.

Except for oxygen gas, as discussed in the section "Oxygen Gas Piping," there are no specific gas velocity limitations, However, the gas velocity is normally less than 400 ft/s to avoid noise problems and excessive pressure drop. For systems with pressure up to 400 psi, gas velocities in the range of 50 to 200 ft/s are common.

When the pipe size of the gas line affects the plant power consumption, the initial pipe size can be selected by using the following equation. The equation calculates the pipe diameter which provides the minimum total cost, that is pipe capital cost plus operating power cost.

$$D = (0.0275)\left(\frac{MTUCY}{EA}\right)^{0.16}\frac{Q^{0.48}}{P^{0.32}}$$

where M = molecular weight
T = absolute temperature, K
U = average compression temperature, K
C = cost of power, \$/kW · h
Y = evaluation period, years
E = compressor overall efficiency
A = pipe cost fraction, \$ per 100-ft/in pipe diameter
Q = flow rate, ft^3/h at normal temperature and pressure
P = absolute pressure, psia

D = inside pipe diameter, in

PIPING COMPONENTS

The types of valves used in cryogenic service are similar to those used for conventional fluids except that the valves may require unique design features due to the cryogenic fluids. Metallic and nonmetallic materials must be suitable for the low-temperature service. See the discussions on materials in the section "Materials Used in Cryogenic Service." Valve types used include gate, globe, butterfly, ball, check, and safety.

The stems of gate, globe, butterfly, and ball valves are extended so that the valve packing and operator remain at ambient temperature when cryogenic fluid is in the valve body. The valves are oriented so that the packing is at a higher elevation than the valve body so that a gas pocket can form inside of the extension tube, thereby making the extension tube more effective at insulating the packing from the cold temperatures. Figure C8.7 shows an extended stem gate valve.

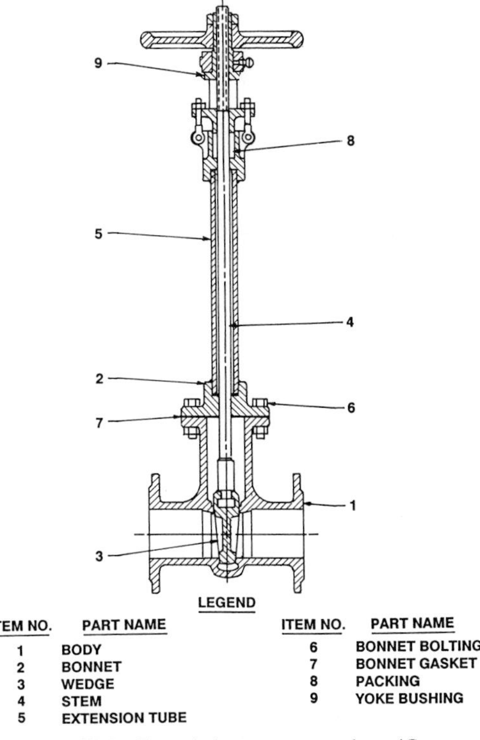

FIGURE C8.7 Extended stem gate valve. (*Courtesy, Union Carbide Industrial Gases.*)

Valves must be designed so that liquid cannot be trapped in a portion of it when the valves are cycled from open to closed. If liquid is trapped, heat input from the atmosphere will cause the liquid to vaporize and overpressurize the valve. When a conventional gate valve is closed, it is possible for liquid to be trapped in the connect area. When a gate valve is used in cryogenic service, the bonnet area is vented to one side of the valve so that liquid cannot be trapped in the bonnet area. Figure C8.8 shows one method that is used to vent the bonnet area of a gate valve.

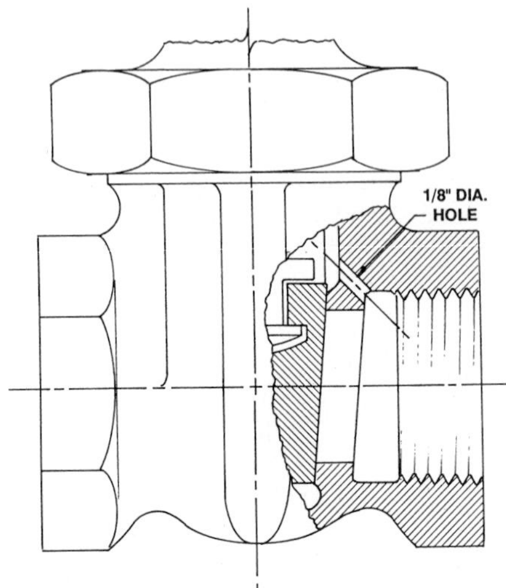

FIGURE C8.8 Gate-valve—vented bonnet. (*Courtesy, Union Carbide Industrial Gases.*)

With a conventional ball valve, liquid can be trapped between the ball and the seat. Therefore, for ball valves in cryogenic service, the ball area is vented to one side of the valve.

During the cooldown or thaw of a cryogenic piping system, different parts of the valves may cool down or warm up at different rates, resulting in varying rates of contraction and expansion. The valve design must consider that the valves are required to be operational during these conditions.

FLEXIBILITY ANALYSIS AND SUPPORTS FOR CRYOGENIC PIPING SYSTEMS

Piping flexibility analysis is an important design consideration because the large difference between ambient and cryogenic temperature will result in significant pipe shrinkage. The analysis methods used are similar to those required for conventional piping systems as discussed in Chap. B4. The one exception is that the piping contracts in cryogenic service rather than expands as experienced with high-temperature service. However, since the commercially available flexibility analysis computer programs have the temperature range as an input value and the program calculates the resulting pipe shrinkage, the analysis methods become identical to those used for conventional piping systems.

When possible, cryogenic piping is routed so that the piping configuration provides adequate flexibility. This means that the pipe is routed so that there are Z, L, and U bends provided to take up the pipe movement while keeping the stresses within the allowable range. When additional flexibility is required, flexibility can be increased by the addition of expansion loops (U bends) or the addition of flexible metal hose. The flexible hose has the advantages that larger pipe movements can be accommodated and a more compact design is possible. The flexible hose has the disadvantage that more pipe supports may be required to guide the piping so that the pipe movement is taken up by the hose.

The flexible hose is located so that pipe movement is lateral to it. The braided cover on the flexible hose prevents any significant pipe movement which is axial to the flexible hose. A flexible hose acts similarly to an expansion joint. The discussion in Chap. B4 on pipe support and guides at expansion joints is applicable to a system using flexible hose. The recommended upper pressure limit for flexible hose and expansion joints is 1000 psi.

Hanging-style pipe supports can be used to accommodate significant pipe movement in both the lateral and axial directions. Roller-style pipe supports can be used to accommodate large axial pipe movement. When the amount of pipe movement exceeds the capability of a hanger or roller pipe support system, a fixed support located in the center of the pipe span can be effective in reducing the amount of movement.

When an uninsulated cryogenic line is supported, a portion of the pipe support will be at cryogenic temperature. The lower temperature should be considered when selecting the materials for the pipe support and its hardware.

INSTALLATION OF PIPING COMPONENTS

To minimize potential leaks, welding and brazing are the most common assembly methods for cryogenic piping systems. In accordance with the B31.3 Code, solder joints are not acceptable for cryogenic service. However, to allow removal of valves and other piping components for maintenance, other assembly methods are used at piping components.

Piping components with threaded fittings are used in small sizes, usually NPS 1½ pipe size and smaller. Flange joints are used for larger sizes, with flanges provided in accordance with ANSI B16.5. For raised face and flat face flanges, gaskets are usually compressed asbestos sheet, spiral-wound stainless steel with asbestos or teflon filler, and flexible graphite such as Grafoil.

INSULATION SYSTEMS

Most piping in liquid cryogenic service is insulated. The only reasons a line would not be insulated are that (1) its use is very infrequent and brief, (2) it is a temporary installation, or (3) the flow rates are continuous and at a very high rate. The following table shows the heat leak from a frosted uninsulated line containing liquid nitrogen and subjected to an 8-mile per hour wind.

Pipe size	Heat leak per foot of pipe (Btu/h)
1	280
2	530
4	1000

The type of insulation used for cryogenic piping includes (1) expanded foams such as polyurethane and foamglass, (2) powder insulations such as perlite, and (3) vacuum insulated pipe. For an insulation system to remain effective, the vapor barrier system must keep atmospheric moisture from entering the insulation space and freezing against the cryogenic line. When this occurs, the ice that is formed will destroy the insulation system.

When the cryogenic liquid is colder than the boiling point of oxygen (−297°F), oxygen can condense out of the air and collect in the insulation space. For this situation, the insulation system should be noncombustible in the presence of oxygen.

The typical values for thermal conductivity are shown in Table C8.12. The expanded foam insulations use a plastic covering, such as PVC or neoprene sheeting, to provide the vapor barrier protection. The initial capital cost is usually lower than the other systems, but more frequent maintenance is required to maintain a tight vapor barrier.

TABLE C8.12 Thermal Conductivity of Pipe Insulation Materials at an Insulation Mean Temperature of −100°F

Insulation	Thermal conductivity [Btu/(h · ft · °F)]
Urethane foam	0.012
Foamglass	0.024
Perlite (at atmospheric pressure)	0.018
Perlite (vacuum at 1 μm)	7.9×10^{-4}
Laminar radiation shielding (vacuum at 0.1 μm)	2.1×10^{-5}

Powder insulation is generally used when several piping segments and equipment can be grouped in one area. A metal jacket or casing is used to contain the perlite around the piping and equipment. When the insulation space is maintained at atmospheric pressure, it must be purged with a dry gas, such as nitrogen, to keep atmospheric moisture out of the casing. For improved heat transfer performance, the casing can be made vacuum tight and the insulation space evacuated to high vacuum. For power insulation, high vacuum is a pressure level less than 1 μm of mercury when the line is at the normal cryogenic temperature operating condition (1 μm of mercury is equal to 10^{-3} torr). See Table C8.12 for typical thermal conductivity values. When considering this insulation option, the difficulty in maintaining a vacuum-tight casing must be evaluated against the difficulty of maintaining a dry gas purge with consideration of the difference in heat transfer performance.

Vacuum insulated piping (VIP) is constructed of a stainless steel inner pipe that contains the fluid, a stainless steel outer jacket to form the vacuum space, and insulation in the vacuum space. The insulation is normally laminar radiation shielding that consists of alternate layers of a reflective material, such as aluminum foil, and an insulation material, such as glass paper. For vacuum insulated pipe, the required vacuum is a pressure level less than 0.1 μm, when the line is at the normal cryogenic operating condition. Heat leak by conduction and radiation is reduced by the laminar radiation shielding. The heat leak by convection is reduced by the vacuum.

Nonmetallic spacers are required in the vacuum space to support the inner pipe within the outer pipe. Bellows are required in one pipe to account for the differential expansion between the inner and outer pipe. Due to the level of fabrication required, the VIP is normally factory fabricated. To accommodate field insulation, a mechanical joint is required between the pipe segments. A bayonet assembly is one type of joint that has a long heat leak path between the inner pipe and the flange connection which is in the outer pipe. Figure C8.9 illustrates the bayonet assembly.

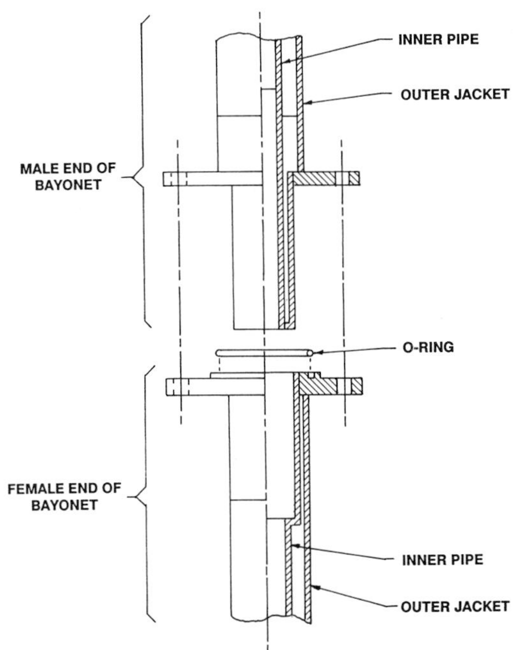

FIGURE C8.9 Vacuum insulated pipe bayonet assembly. (*Courtesy, Union Carbide Industrial Gases.*)

Table C8.12 shows typical values of insulation thermal performance for laminar radiation shielding. For a vacuum insulated piping system, the thermal performance of the straight pipe is affected by the number and type of spacers. The thermal performance of fittings and other components depend on fabrication techniques. Table C8.13 shows typical heat leak performance values for commercially available vacuum insulated pipe.

TABLE C8.13 Typical Heat Leak Values for Vacuum Insulated Pipe When the Fluid Temperature Is −320°F.*

Pipe size (in)†	Pipe	Elbow	Tee	Flex hose	Bayonet and field welded joints	Valve
¾	0.47	1.50	2.00	1.41	15	14
1	0.50	2.30	3.40	1.50	21	15
1½	0.58	2.60	4.00	1.74	23	19
2	0.64	3.00	4.50	1.92	24	25
3	0.79	3.80	5.70	2.37	48	64
4	0.92	4.40	6.50	2.76	84	79
6	1.20	6.00	9.00	3.60	110	120
8	1.46	7.40	11.00	4.38	140	180

*Heat leak values are BTU/h/ft of pipe and flex hose and are BTU/h for each fitting and valve.
†For the inner pipe.

The total cost including the piping system and the refrigeration cost due to heat leak should be considered when selecting an insulation system. For many permanent piping systems, the vacuum insulated pipe provides the cost effective insulation system.

CLEANING

All materials used in oxygen piping systems or connected with oxygen systems should be cleaned before the system is put into service. The cleaning must remove mill scale, rust, dirt, weld slag, oil, grease, and other organic materials. The purpose of the cleaning is to remove hazardous hydrocarbons and particulate contaminants which could ignite and cause a fire in an oxygen atmosphere. To obtain additional information on cleaning requirements, see Ref. 2. For a discussion of the factors affecting ignition in an oxygen system, see Ref. 3.

All materials used in cryogenic systems, other than oxygen, are also cleaned before the system is put in service. The purpose of the cleaning is to reduce contaminants to the point where they will not migrate, seize up moving parts at low temperatures, or prevent the attainment of high-product purity or vacuum levels.

VENTS

When a cryogenic fluid is vented, such as from safety valves, safe disposal of the fluid must be considered. The fluid should be directed so that it will not contact personnel, because the low temperature can cause burns.

When cold, most cryogenic fluids are heavier than air. A heavier-than-air gas can displace the air and create a potential for asphyxiation. Some cryogenic fluids such as hydrogen are flammable. An oxygen-enriched atmosphere can promote flammability of other materials. Therefore, all vents should be located outdoors and directed so that high concentrations will not collect in confined areas.

REFERENCES

1. ASME/ANSI B31.3, Chemical Plant and Petroleum Refinery Piping, ASME Code for Pressure Piping, American Society of Mechanical Engineers, New York, 1990.
2. CGA Pamphlet G-4.1, *Cleaning Equipment for Oxygen Service,* Compressed Gas Association, Arlington, VA.
3. ASTM G88, Standard Guide for Designing Systems for Oxygen Service, American Society for Testing and Materials, Philadelphia, PA, 1984.

COLD BOX PIPING

J. K. Howell, Cold Box Engineer
Union Carbide Industrial Gases, Linde Division

A cold box is a mechanical system that insulates an entire low-temperature process mechanical embodiment in a single assembly instead of insulating each pipe and vessel individually. The cold equipment and piping are installed in an airtight steel insulation casing. The void space inside the casing is filled with insulation powder or fibers.

The design of cold box piping requires knowledge of some basic engineering disciplines such as strength of materials, fluid mechanics, and heat transfer. Also required are some piping design skills such as flexibility analysis, pressure drop calculations, and material selection. These skills are discussed elsewhere in this handbook.

Cold box piping is usually designed to meet the engineering requirements of ASME Piping Code B31.3[1] for safe design. Further, the designer should also be aware of the design considerations for safe operation and maintenance of cold boxes discussed in Section 8 of Ref. 2 of this article.

What follows is a description of problems unique to the design of cold box piping, and, where appropriate, suggested methods of solution.

DESIGN REQUIREMENTS FOR COLD BOX PIPING

General System Architecture

Cold boxes tend to be tall vertical structures. They often contain tall distillation columns and brazed aluminum heat exchangers which are oriented for vertical flow (see Fig. C8.10). A minimum of 12 in of space is usually provided between the casing and the piping and equipment inside. This space provides adequate in-

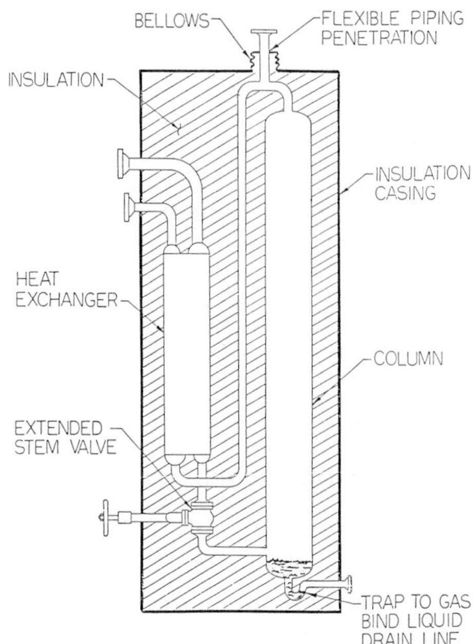

FIGURE C8.10 Schematic cold box diagram. (*Courtesy, Union Carbide Industrial Gases.*)

sulation for the cold equipment and access to the piping for fabrication, testing, and maintenance. Valves and controls above grade level are located so as to be easily accessible from operating platforms. The piping arrangement within the cold box is very compact to minimize the amount of insulation and to fit shop fabricated cold boxes within the maximum outline which can be shipped by rail or by truck. Components requiring frequent operator attention are located at ground level. Heat exchangers operating in parallel are located at the same elevation to avoid flow unbalance due to density differences between the warm and cold streams.

Insulation System

Before start-up, the insulation is purged with a dry inert gas, usually nitrogen. A small flow of purge gas is continued during operation to maintain a slight positive pressure in the casing and to prevent inflow of oxygen and moisture. Pipe and valve stem penetrations through the casing must be airtight.

Equipment and pipes operating at different temperatures must be separated to minimize unwanted heat flows. The piping and pipe supports must withstand loads from the insulation material. These loads occur when fibrous insulation is being packed into the casing, when powder insulation is being loaded or removed from the casing (especially if it is damp and frozen), and when pipes are moving through the insulation due to thermal expansion and contraction.

Reliability

Piping inside the cold box is not readily accessible for maintenance. Repair of a simple leak requires that the cold box be out of service for several days. Erosion caused by insulation powder propelled by the leaking fluid can enlarge a small leak or cause another leak in nearby equipment. For these reasons, leak-tight integrity of cold box piping is extremely important. The number of flanged joints inside a cold box is kept to a minimum. Welded or brazed pipe joints are used wherever possible. Transition joints are often used to join stainless valves to aluminum pipe. Weld-end, top-entry, extended-stem globe valves afford maximum reliability. These valves do not have any flanges inside the casing and the plug and seat can be replaced from outside the cold box. When flanges are used inside the casing, proper torquing of flange bolts must be assured. Threaded joints and tubing compression fittings are usually not used inside the casing. Flanges are sometimes installed in a separate, small, insulated compartment where they can be accessed without removing the insulation from the entire cold box. A careful leak test is done on all piping inside the casing before the insulation is installed.

Thermal Expansion and Contraction

Typically, cold box piping is heated to 250°F before start-up to remove any moisture and then cooled to operating temperature, as low as −320°F in an air separation plant. In that case, the total temperature range is 570°F. The total thermal expansion and contraction of aluminum pipe in this range is 7.5 in/100 ft. The flexibility analysis of the piping must consider the full temperature range as well as any more severe differential temperature conditions which may occur during upset, thaw, or cool-down. Clearance must be allowed for the pipes to move through their full expansion and contraction range. This clearance is especially critical for pipes connected near the top of tall vessels. Pipe penetrations through the casing will restrict the movement of the pipe unless the casing is made flexible by use of a metal bellows or a rubber boot.

Piping Installation Details

Pipes running from liquid pipes or vessels to the casing must have thermal traps (upward rise of at least two pipe diameters) to gas bind the connection and keep the cold liquid away from the warm casing. Thermal traps are also used in gas pipes to prevent unwanted convection currents.

Extended stem valves used in cold pipes are installed with the body located at least 12 in inside the casing and with the packing gland and the operator or handwheel located outside. Liquid valve stems are tilted upward at least 15° to gas bind the stem extension tube and keep the liquid away from the packing.

Supports for cold pipes are commonly made from austinitic stainless steel because it is strong, ductile at cryogenic temperature, and has a relatively low thermal conductivity. Blocks of insulation material can be included in the supports to reduce heat leak from the casing.

Low-point drains are provided where necessary to remove process liquids for shutdown and to remove liquid water formed during thaw of the cold box.

Cleaning

Cold box piping is cleaned to remove contaminants (see the earlier discussion of cleaning in this chapter). The cleaning is usually done before the pipe is installed. Pipe ends and other openings are covered after completion of fabrication to maintain cleanliness.

DESIGN REQUIREMENTS FOR SPECIAL SITUATIONS

Reversing Exchangers

Reversing heat exchangers and regenerators are often used in air separation cold boxes. The piping in the reversing streams is subjected to cyclic pressure variations. Fatigue analysis should be performed on this piping. Paragraph K304.8 of Ref. 1 can be used as a guide for the fatigue analysis. Liquid water may be present in the warm end piping from these heat exchangers. Possible corrosion and freeze-up problems should be considered in the design of this piping.

Liquid Oxygen Evaporation

Most liquid oxygen contains some traces of hydrocarbon contaminants. If this liquid vaporizes in a dead ended pipe or crevice, the hydrocarbons will concentrate. Explosive mixtures can result. Crevices should be avoided in liquid oxygen piping where a source of heat is present. Drains and other dead ended pipes connected to a liquid oxygen source should be trapped as close as possible to the liquid source. Liquid oxygen piping should be well insulated from the casing and from warmer process equipment.

Vacuum Insulated Cold Boxes

In some cold boxes, a vacuum is pumped on the insulation casing to reduce the heat leak. In such cases, the piping and vessels are tested with helium and a mass spectrometer to find and repair tiny leaks which would greatly reduce the effectiveness of the vacuum insulation. The design pressure for such piping is based on the absolute rather than the gauge pressure.

Instrument Lines

Instrument pipes in the cold box must be designed with care. These pipes have a small diameter and are easily damaged. The design of instrument lines should provide adequate flexibility, support, and protection from insulation and human loads. Equation (16), paragraph 319.4.1(c), of Ref. 1 can be used for assuring adequate flexibility of instrument lines.

Flammable Fluids

Some cold boxes process flammable fluids such as hydrogen, hydrocarbons, carbon monoxide, etc. Any aluminum or other low melting temperature piping outside the casing should be protected by fireproof insulation. The insulation should not contain any chemicals which could corrode or otherwise degrade the piping material. Discharges from drains, vents, and relief valves must be piped to a flare stack or some other safe means of disposal. Connections are provided on dead ended pipes for purging air from the lines before admitting the flammable process fluids.

REFERENCES

1. ASME/ANSI B31.3, "Chemical Plant and Petroleum Refinery Piping," ASME Code for Pressure Piping, B31 An American National Standard, latest ed.
2. CGA Pamphlet C-8, "Safe Practices Guide for Air Separation Plants," Compressed Gas Association, Arlington, VA, latest ed.

LIQUID STORAGE AND CONVERSION SYSTEMS

N. H. White, Applications Engineer
Union Carbide Industrial Gases, Linde Division

Bulk liquid storage systems are often used at or near the point of final consumption. They typically supply product in three basic forms: as a gas at less than the storage tank working pressure, as a gas at high pressure, and as a liquid. The equipment and piping required may vary depending on which of these delivery forms is used. However, several general considerations apply to the piping design in all cases.

The working pressure of all lines in the system should be at least 10 percent higher than that of the tank. Relief valves required to protect the piping should be set at this pressure. This configuration of relief valve set points will ensure that if the tank is overpressured, it will be relieved in the gas phase from its safety valve and not in the liquid phase through one of the pipeline relief valves.

Vaporizers are commonly used in storage and conversion systems to gasify liquid product and bring it to ambient temperatures. The piping downstream of these units should be designed for liquid product temperature up to the point that a safety device is installed to automatically shut off flow in the event that low temperature is sensed. The response time of this device must be compatible with the failure modes of the vaporization system used. The set point chosen must consider the design temperature of all components downstream.

Relief valves should be installed between any set of valves that can trap liquid or low-temperature gas if they are simultaneously closed. These "blocked line relief valves" are required to relieve pressure build-up due to heat addition. They should be set at the line working pressure and sized to relieve gas at conditions of maximum heat addition. Particular attention should be paid to situations where the source of heat is due to natural convection associated with bare lines and ambient vaporizers. They are of special concern because even moderate winds can greatly increase the rate of heating and subsequent pressure buildup.

Bulk liquid units are typically employed in intermittent flow situations. For this reason, liquid lines will frequently have to be cooled down to liquid temperature when product is demanded. Therefore, such lines should in general be sized to pass the required flow in the gaseous state to avoid excessive pressure drop at the start of demand.

Design for cryogenic temperatures requires that particular attention be paid to line flexibility. Although large displacements are involved due to the extreme temperature changes, they can be accommodated by conventional techniques including flexible hose sections and expansion loops. In addition, most piping materials experience reduced impact strength at cryogenic temperatures. The materials section in this chapter deals in depth with this topic.

Thermal trapping is a piping technique that can be used in any liquid line, whether or not it is insulated, to minimize heat input to an idle piping section or branch. Where the branch is oriented vertically, a gas-liquid interface will be established near the connection, and the majority of this section will be filled with gas approaching ambient temperature. This configuration greatly reduces heat input to the branch when it is idle.

LOW-PRESSURE BULK CONVERSION SYSTEMS

The most common bulk liquid units deliver gas at a pressure less than the storage tank working pressure. The working pressure of such systems is typically less than 250 psig. However, systems with as high as 600 psig working pressure tankage have been used commercially. The delivery pressure must be limited to approximately 80 percent of the tank working pressure to allow for both tolerance on the safety relief valve setting and buildup in product pressure during periods of nonuse. Figure C8.11 illustrates a typical piping configuration, and item numbers in the following discussion refer to this figure.

The lines connecting the vessel to the safety devices (1 and 2) are sized to meet ASME pressure vessel construction requirements. Normally both a relief valve and a burst disk are included. The relief valve is sized to handle the vapor generated by the heat load that results from the loss of insulation (loss of vacuum), and the bursting disc is designed to handle both loss of vacuum and fire conditions. The assumptions used to size these devices are usually per guidelines given by the Compressed Gas Association in its pamphlet CGA S-1.3 "Pressure Relief Device Standards." Where it is critical that the system not be taken out of service for periodic inspection of the relief devices, it is common to use a dual set of safety devices with a diverter valve.

The filling and refilling characteristics of both low- and high-pressure systems

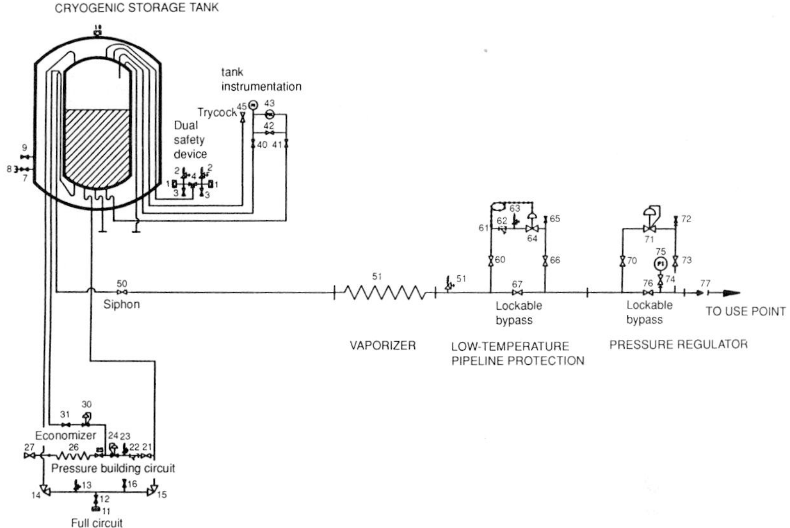

FIGURE C8.11 Low-pressure bulk conversion system. (*Courtesy, Union Carbide Industrial Gases.*)

are the same. In either case, both gas and liquid phase lines are provided. Tank pressure is controlled during the fill operation by adjusting the portion of refill liquid entering the gas phase relative to the liquid phase with valves 14 and 15. Filling in the gas phase reduces tank pressure by condensing the vapor in the gas space of the storage tank. Filling in the liquid phase increases tank pressure because the liquid acts as a piston compressing the gas space. These systems are filled from bulk transports either by a centrifugal pump or by pressurizing the supply vessel. The centrifugal pumps used for transfer applications are typically one- or two-stage designs matched for the discharge rate and pressure by adjusting the speed and impeller cut as in conventional service. Special piping considerations for the fill lines involve the line size which should be adequate to permit refill to occur in a reasonable time period and yet should not be oversized to limit heat input. Insulation is not typically used on these lines because of their short length, intermittent use, and high transfer rate. Helium and hydrogen fill operations are an exception to this practice because of their extremely low temperature and low heat of vaporization. Other products may justify the use of insulation based on their value.

The tank instrument piping includes a gas phase line called the full trycock that extends into the gas space to the full liquid level. This line is used in manual fill operations by the operator to determine when the tank is full. The operator opens valve 45 when the fill is nearly complete and observes when liquid is discharged, indicating the full level has been reached. Also included in the instrument piping are a liquid and gas phase line used to measure contents and pressure.

Control of pressure between refills is accomplished by the combined actions of the pressure building, economizer, and relief systems. The pressure building circuit consists of components 21 through 26 and is driven by the difference in densities of the liquid and gas phases. It functions when product flows from the liquid phase line, is vaporized in the pressure building vaporizer 26, and returns to the gas space. Attention must be paid to the very low driving pressure in this circuit.

In the case of low or intermittent demand, the product vapor pressure can increase above the delivery pressure. At this point pressure building is no longer necessary, and it is desirable to withdraw product in the gas phase. Product delivered in the gas phase will remove approximately 5 times the heat from the tank as the same mass withdrawn in the liquid phase. A scheme to automatically switch to gas phase supply at high tank pressure is called an economizer. A siphon cycle economizer is illustrated in items 30 and 31 together with the siphon line inside the tank. When tank pressure is high, regulator 24 is closed and back pressure valve 31 is open. The siphon is broken in this way and gaseous product is withdrawn through valve 50. The siphon is reestablished at reduced tank pressure when the positions of valves 24 and 31 are reversed. There are alternative economizer designs and care must be taken in selecting one because most are limited to low instantaneous flows.

The product is next vaporized and superheated in vaporizer 51. Vaporizers are classified as ambient where the required heat is derived from the atmosphere or as powered where the heat is supplied from another source such as steam, electricity, or hot water. Relief valve 52 provides protection in the event of a block line condition, and special attention should be paid to the vaporizer power source in selecting this valve. A low-temperature pipeline protection device (LTPP) is normally included downstream of the vaporizer to automatically shut down flow in the event of vaporizer failure. This shutdown device generally is the transition point between cryogenic and noncryogenic pipe design. Blocked line relief devices may also be required downstream of the LTPP to allow for warm up of gas from its setpoint to ambient temperature. The LTPP is usually piped in a bypass arrangement to allow calibration and servicing with the system on-line. A pipeline regulator is often included at the system discharge point and should also be installed in a bypass arrangement to provide for servicing.

HIGH-PRESSURE BULK CONVERSION SYSTEMS

In some applications, gas is required at greater than the working pressure of the tank. In such cases, it is pumped in the liquid state, vaporized, super-heated to ambient temperature, and regulated to the required pipeline pressure. A bank of high-pressure receivers is usually included to match the demand to the pump discharge rate. The pump is typically a reciprocating single-acting design ranging in configuration from simplex to triplex. Figure C8.12 gives a typical flow diagram of a high-pressure conversion system. Item numbers referred to in the following discussion are given in this figure. Piping practice for much of the system remains unchanged from the low-pressure gas system discussed above. However, special considerations must be made for the piping both up- and downstream of the pump to ensure proper system operation.

The fluid at the suction flange must be subcooled in order to meet the net positive suction pressure (NPSP) requirements. The degree of subcooling required ranges from 2 to 10 psi depending on pump design, condition, and discharge requirements. The liquid at the bottom of the storage tank is normally subcooled by

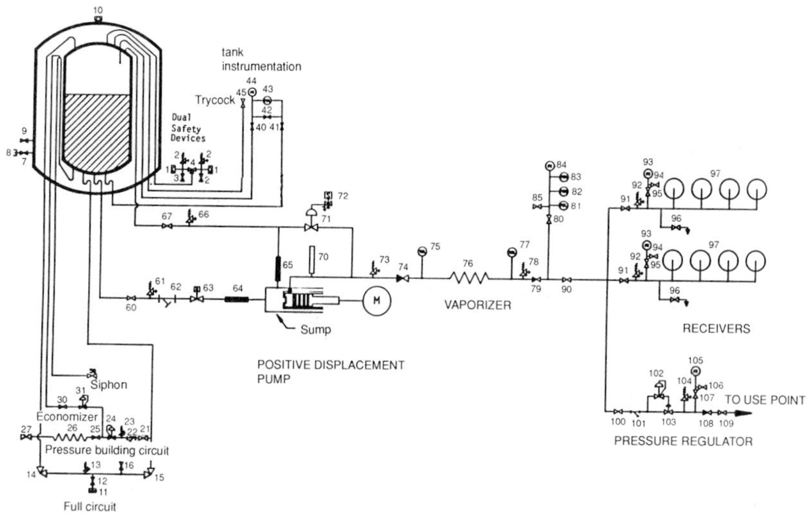

FIGURE C8.12 High-pressure bulk conversion system. (*Courtesy, Union Carbide Industrial Gases.*)

approximately the static head of the liquid phase above it. The suction piping must minimize both heat input and pressure drop to preserve the available subcooling. Pressure losses are primarily due to the fluid acceleration required by the reciprocating pump and may be calculated using classical analysis. Heat input generally increases with pipe diameter but may be essentially eliminated by the use of high-performance insulation. Figure C8.13 summarizes the acceleration and heat loss expected for a typical simplex pump. The total NPSP would apply in the case of bare lines, and the graph shows that an optimum suction line diameter can be selected for any discharge rate. A vacuum insulated line essentially eliminates the heat component and only the acceleration component would apply. Insulation should always be used where the tank working pressure is low, the demand is intermittent, or the pump discharge rate is low. It is used almost universally in hydrogen pumping systems.

The pump is usually contained in a small vacuum insulated jacket called a sump which is connected to the storage tank gas and liquid phases. As the first step in priming the pump, components 60 through 67 are used to fill the sump with liquid. The product vaporized in cooling these components is returned to the tank gas phase, increasing system pressure and thereby available subcooling. The second step is to run the pump with valve 71 open, which circulates liquid through the pump and discharges it to the tank gas phase, cooling the pump compression chamber. Finally, valve 71 is closed, forcing the pump to discharge against receiver pressure which is being held by check valves 74 and 77. Temperature sensor 75 is used to check for pump prime.

The piping downstream from the pump must be designed for cryogenic temperatures and high pressures as well as the pulsations and vibration generated by the reciprocating pump. Temperature switch 77 is used to shut off the pump in

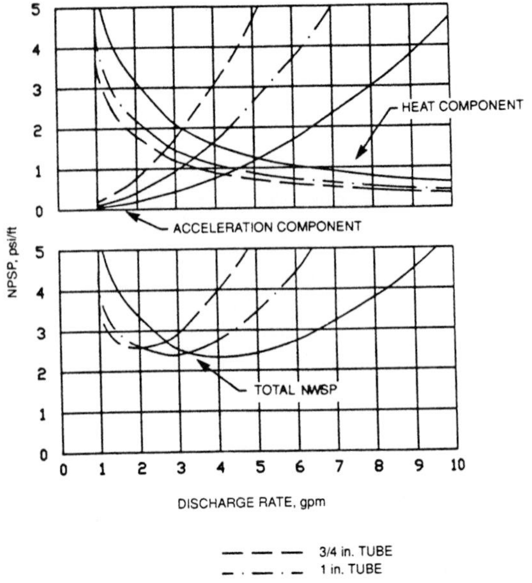

FIGURE C8.13 NPSP components simplex cryogenic pump (LN2 service). (*Courtesy, Union Carbide Industrial Gases.*)

the event of low pipeline temperature. Although noncryogenic materials may be used downstream of this sensor, it is common practice to use cryogenic materials for all high-pressure piping in the bulk delivery system.

BULK LIQUID DELIVERY SYSTEMS

Figure C8.14 gives a flow diagram for a unit that delivers liquid product. Minimizing heat addition is the primary design consideration in these systems, particularly since it is usually refrigeration rather than the fluid itself that is required at the use point. Therefore, some degree of pipe insulation can almost always be justified. However, it should be pointed out that the storage tank insulation performance is often more critical to satisfactory system operation than that of the piping insulation, and care should be taken to be sure that both insulation systems are properly maintained.

Ideally, the system should be configured so that the entire pressure requirement can be provided by the static head of the tank contents. This eliminates the need for pressure building in the storage tank, which can easily represent several times the heat load from all other sources. Ordinarily the tank must be elevated to provide the required pressure. A horizontal tank is preferred to a vertical tank in this case, both because the elevated foundation is less expensive and because the variation in static head, as the contents are withdrawn, is reduced. Where pressure building is required and the tank will not be elevated, a vertical tank is pre-

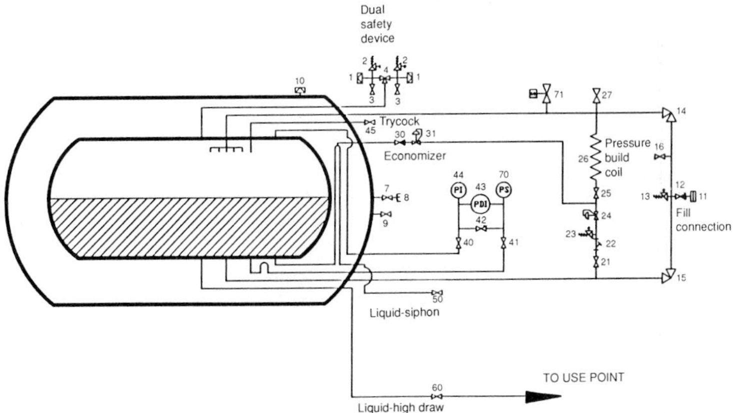

FIGURE C8.14 Bulk liquid system. (*Courtesy, Union Carbide Industrial Gases.*)

ferred to minimize heat transfer through the gas to liquid interface area of the vessel.

It is frequently necessary to maintain the product delivery temperature in the tank either within a narrow range or below a threshold. Since the liquid is always nearly saturated, this can be accomplished by venting the product in the gas phase to hold tank pressure constant. An automatic system consisting of pressure switch 70 and solenoid valve 71 can be used for this purpose. It is the temperature of the product at the bottom of the tank that is of interest. For this reason the pressure switch is connected to the liquid phase instrument line rather than to the gas phase line from which tank pressure is normally read.

MOBILE EQUIPMENT SYSTEM

R. C. Cipolla, Cryogenic Equipment Engineer
Union Carbide Industrial Gases, Linde Division

Bulk atmospheric gases (i.e., oxygen, nitrogen, and argon), hydrogen, and helium are transported as liquids in various styles of double-walled tank trucks. This section presents the transport vehicles used for the distribution of atmospheric gases. The most common road transport used for the distribution of cryogenic liquids is the tractor semitrailer system.

A double-walled cylindrical tank truck has a product tank, or inner vessel, enclosed in an outer casing or jacket. The inner container is designed, manufactured, and tested to Section VIII of the ASME Boiler and Pressure Vessel Code. The casing is designed to the requirements of CGA-341, Standard for Insulated Cargo Tank Specification for Cryogenic Liquid, Compressed Gas Association. The annular space between the inner container and casing is insulated and evacuated in order to minimize the amount of heat leak from the casing shell to the cryogenic liquid. Heat transferred to the product will result in a pressure rise and the eventual venting of product. The inner container is normally supported or attached to the casing at or near the liquid container heads. The fifth wheel king-pin coupler, landing gear support legs, and tandem axle assembly for the cryogenic semitrailer are standard automotive components available to the trucking industry.

Atmospheric gases are typically transported in low-pressure semitrailers equipped with an on-board pumping system for off-loading the product to a customer storage tank. This pumper allows for a maximum payload at a gross combination weight limit of 80,000 lb (i.e., limit for a five-axle tractor trailer combination). Inner containers are rated for a working pressure of 25 to 45 psig and constructed from type 304 stainless steel or AA-5083 aluminum. Casings are constructed from either carbon steel or aluminum. Carbon steel casings are used along with stainless inner vessels, and aluminum casings are matched to aluminum inner vessels. A lightweight casing design can be achieved by providing closely spaced circumferential ring stiffeners. These ribs reinforce the vessel against the external pressure which results from the evacuation of the annular space.

Atmospheric gases have different product densities which can result in product optimized equipment. When optimized for an 80,000-lb road limit, a nitrogen-sized trailer has a volume of about 8000 gal, an oxygen trailer about 6000 gal, and an argon trailer about 4500 gal.

Liquid oxygen, nitrogen, and argon are exempt from Department of Transportation (DOT) hazardous material regulations in regard to design requirements for the semitrailer when the pressure in the container is maintained at less than 25.3 psig during transportation. However, DOT markings and shipping preparation along with regulations pertaining to the trucking components such as braking and lighting systems are all considerations when designing a cryogenic transport. Basic requirements for vessel, structural supports, structural protection, and external piping are covered by CGA-341 specifications.

Piping systems for cryogenic transports can be separated into two categories: internal piping, which includes piping internal to the inner vessel as well as annular space piping, and external piping, which includes all piping exterior to the casing envelope. Both internal and external piping systems are normally fabricated with butt welded joints.

INTERNAL PIPING

An internal piping system is constructed from stainless steel for stainless inner vessels and aluminum for aluminum inner vessels. A typical arrangement of internal piping has the piping penetrating the vessel walls through the rear head as compared to a location on the cylindrical portion of the vessel (see Fig. C8.15 for a typical piping schematic). Piping sections inside the inner vessel are not subject to internal pressure. Design considerations for piping supports and attachments include loadings imposed by road vibration and thermal contraction due to temperature as low as −320°F.

Annular space piping configurations must have consideration given to heat leakage, flexibility, and trapping of liquid or gas. Since heat leakage to the cryo-

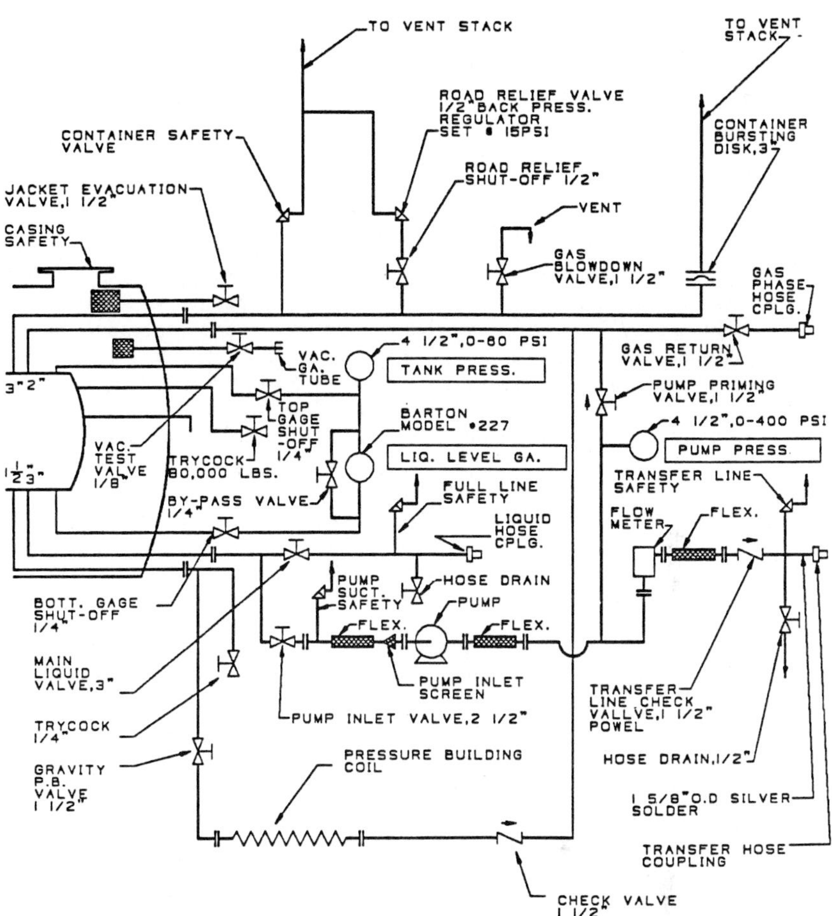

FIGURE C8.15 Flow diagram, cryogenic trailer specification. (*Courtesy, Union Carbide Industrial Gases.*)

genic liquid is primarily by conduction down the pipe wall, the length of the piping run is of importance. Stresses induced by thermal contraction of the line or movement of the inner container relative to the casing are typically accommodated by providing adequate flexibility in the form of loops and bends. Stress levels due to flexibility are generally kept to within the allowable limits imposed by ANSI/ASME B31.3, Chemical Plant and Petroleum Refinery Piping. Piping bellows to accommodate thermal displacements are not generally considered an acceptable way of providing flexibility to annular space piping.

Testing of internal piping systems includes pressure, leak, and other examinations when required by codes. Pressure testing procedures follow ANSI B31.3 requirements. Nozzle penetrations at the inner vessel wall may require examination of the welds using dye penetrant techniques. Annular space piping is thoroughly leak tested since any leaks would deteriorate the vacuum pressure within the insulation spaces. A helium mass spectrometer technique is the most common method used for vessel and piping leak testing.

EXTERNAL PIPING

The external piping system is commonly located inside an operating compartment at the rear of the trailer; however, the pressure-building coil is located under the trailer forward of the tandem axle assembly. The rear grouping has accident protection features provided by the tandem axle subframe, rear bumper, and the piping compartment.

Stainless steel external piping systems are used on all transports regardless of inner vessel construction with the exception of the pressure-building coil. The coil is normally constructed of aluminum extruded fin tubing which is flanged to stainless steel inlet and outlet piping. The function of the coil will be discussed later. Valves, gauges, and other instruments and controls are not necessarily constructed from stainless steel materials. Bronzes, copper alloys, and aluminum alloys are used for various component parts. Internal and external piping runs are joined at the casing penetrations with piping flanges. For aluminum vessel designs these flanges provide a simple way of connecting aluminum internal lines to stainless steel external lines.

As Fig. C8.15 schematically illustrates, cryogenic transports typically have two liquid phase lines. The first piping run is used to fill the trailer at the production plant and to unload the trailer using the off-loading pump. The second liquid line is used for the pressure building system. Pressure in the inner vessel can be increased by vaporizing a small amount of liquid. The coil with a large heat transfer surface can readily vaporize liquid and return the warmer gas to the ullage space of the trailer. This pressure buildup is performed prior to and during the off-loading process in order to maintain adequate suction pressure for the centrifugal transfer pump.

Gas phase piping includes a pressure relief device line which directly communicates with the vapor space near the midpoint of the top centerline. A spring-loaded pressure relief valve and a rupture disk device are normally provided on the cryogenic transport. These devices are designed to maintain pressure at a safe level under emergency conditions, including exposure of the vessel to a fire. Pressure relief devices are designed to the requirements of Section VIII of the ASME Code as well as CGA pamphlet Sl.2, "Pressure Relief Device Standards—Cargo and Portable Tanks for Compressed Gases." The same piping run is also used to provide for a manual vent system and a "road relief" circuit. The road relief valve, typically set at about 15 psig, controls the inner vessel pressure during transportation. Venting of product when the pressure reaches the road relief valve setting can occur. A shutoff valve upstream of the road relief allows isolation during off-loading and, therefore, permits the buildup of pressure to an operating level near the working pressure of the inner vessel.

A second gas phase circuit has multiple use as a gas phase outlet, pressure-building coil return, and a transfer pump recirculation line.

An optional sparger, or gas phase fill header, has various uses. A typical design has a pipe running along the top of the inner vessel with a series of holes which act as spray nozzles for the liquid. With this line teed from the main fill line, transfer of liquid to the trailer can occur in an all liquid phase, all gas phase or any combination. Gas phase filling of an empty trailer through the sparger will cause a partial collapse of the gas pressure. This pressure reduction by collapse eliminates product losses that would occur by manually venting the vessel. A sparger also provides a means of uniformly cooling down or shrinking a warm inner vessel. Uneven shrinking of a inner vessel from near ambient to liquid temperatures can cause structural damage to the vessel, supports, or piping.

Inner vessel pressure, transfer pump discharge pressure, and liquid contents are monitored with trailer-mounted gauges. A differential pressure indicator is the most common device used for contents measurement. Liquid and gas taps are provided for this gauge.

Sampling of product in the inner vessel is sometimes necessary in order to determine product purity. A liquid tap typically teed externally to a liquid line can be used for sampling.

The liquid fill level can be detected by a small-diameter fixed-length dip tube. Flow of liquid through the line indicates the liquid level in the tank. If the semitrailer is used to transport various cryogenic liquids, dip tubes positioned at different levels would be provided.

As previously stated, the annular space is evacuated. Pressure levels below 100 μm of mercury are achieved by pumping down the space through a line equipped with a shutoff valve. A filter on this pipe eliminates problems with the insulation material being carried to the valve or pump. The vacuum level can be monitored using the trailer-equipped thermocouple gauge tube. Vacuum level is determined by a portable vacuum meter.

Cryogenic piping must be protected from overpressurization in any piping section which can be isolated. After liquid or gas has been transferred, the cold product warms up and builds pressure which can exceed the bursting pressure of the pipe or fittings. For this reason "block line safety" devices are a mandatory feature in cryogenic piping circuits.

Thermal contraction of external pipes can cause high forces when constrained. Typically, centrifugal transfer pumps and meter systems are protected from these forces by expansion bellows in the inlet and discharge piping. Piping bellows are made from corrugated stainless steel hose.

HIGH-PURITY PIPING

John W. Burr, Development Engineer
Union Carbide Industrial Gases, Linde Division

Ultra-high-purity piping systems are encountered in gas and liquid supply systems used in the fabrication of electronic semiconductor chips and optical disk systems. The requirements for purity in these cases involve contamination levels of only a few parts per billion (ppb) of foreign material.

Semiconductor fabrication involves the forming of alternate layers of very fine lines of semiconducting material on the surface of a nonconducting material such as silicon. The lines form miniature ultra-large-scale-integrated (ULSI) electronic circuits on a small chip of silicon. A large number of these small chips are simultaneously fabricated on a single silicon wafer. The line or feature size of ULSI chips is approaching 0.5 μm or less. Any particle one-tenth of the line size can cause a defect in the circuit, resulting in the chip being destroyed. To keep the percentage output yield of defect-free chips high, the gases and liquids used in the fabrication of ULSI circuits must be very pure and free of particulate contamination.

The present requirements for gas purity are that trace gas contamination must be held to the parts-per-billion level. For particulates, contamination with particles greater than about one-tenth the semiconductor feature size must be held to less than five particles per cubic foot of gas.

Fabrication of optical discs for both audio-visual and computer application involves the same small feature size and hence ultra-clean fabrication and operational technology. The requirements for purity in optical disc fabrication are generally the same as for semiconductor fabrication. The only difference is the application and device medium.

GENERAL GAS SUPPLY SYSTEM DESCRIPTION

Gas supply systems can range from small liquid storage pad packages to large air separation plants. The specialty gas requirements can usually be met with high-pressure gas cylinders. However, regardless of the gas supply system size or complexity, it must be able to supply to the customer gases in ultra-high-purity condition.

The typical system consists of a bulk liquid storage tank, a vaporizer to convert the liquid to gas, control systems to regulate the flow and pressure delivered to the customer, gas purifiers to remove any residual trace gas impurities, filtration systems to remove particulate contamination, and a piping system to conduct the gas to the customer's point of use. A schematic representation of a typical supply system is shown in Fig. C8.16.

Alternate supplies may involve complex air separation plants for customers requiring large quantities of gases or high-pressure cylinders for customers whose gas requirements are minimal. Many of the specialty gases required for semiconductor fabrication are also supplied in special high-purity cylinders. Cylinder installations typically do not include purifiers.

Semiconductor fabricators require ultra-high-purity gases at the point of fabrication inside their facility. Normally a gas supplier would only be required to supply gas to some interface point near the customer's plant boundary. However, for ultra-clean systems, the gas supplier is required to provide the ultra-high-purity gas to the final point of use. This means that the gas supplier is responsible for ultra-high-purity piping system within the customer's plant, in addition to the gas storage and conditioning system at the plant boundary. This is to assure that the gas meets the customer's requirements at the point of use.

The key characteristic of an ultra-high-purity gas supply system is that it must be free of trace gas impurities, particulates, metal ions, and other contaminants.

Trace gas impurities are typically trapped within parts of the system during component manufacture and system assembly and later released into the process stream. Joining materials, laps within joints, screw threads, and other cavities can trap moisture, welding gases, and other contaminants and can become long-

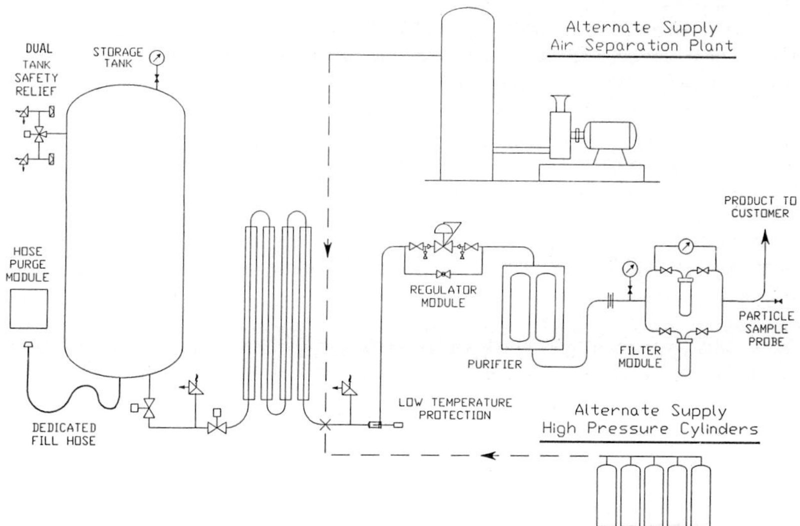

FIGURE C8.16 Ultra-high-purity gas supply system. (*Courtesy, Union Carbide Industrial Gases.*)

term outgassing sources. For this reason such geometric discontinuities must be eliminated from the system by design.

Additionally, gas-absorbing materials, especially nonmetallics, must be avoided. Contaminant gases dissolve into many materials of construction. This occurs primarily in nonmetallic components such as valve seats and filter cartridges, but it can also occur in the metallic parts. Since there is so much steel existing in the tanks, component bodies, and piping, dissolved gases within these components could present a serious contamination problem. Special manufacturing procedures as well as final cleaning and conditioning procedures must be considered to reduce dissolved gases within steels and other metals used in ultraclean systems.

Many nonmetallic materials, particularly those used in the filter cartridges, tend to dissolve gases, primarily water vapor. This is one of the more serious contaminants in semiconductor fabrication. Special materials or special procedures to remove the contaminants are necessary to prevent these contaminants from entering the system.

Particulate contamination is found in many of the same places as the trace gas contaminants (i.e., crevices, joints between mating pieces, dead end legs off the process stream, etc.). In addition, particulates become trapped within indentations and small voids which exist within the natural roughness of most materials. For this reason all pipe and component surfaces exposed to process streams must be polished to as smooth a surface as possible to eliminate contamination from this source.

Particles can also be generated by erosion of the process stream over sharp edges which are left in the piping and flow path of system components. Also, contact between moving parts will scrape off metal particulates from contacting surfaces. Therefore, sharp edges and rough surfaces must be eliminated from the design.

Stray metallic ions are also a serious contamination in semiconductor fabrication systems. Unusual metals used in the fabrication process of other metallic parts could leave undesirable metal ions on or within the parts. These must also be eliminated as much as possible.

Many process conditions promote the generation of particulates or promote the release of particulates from component surfaces. High process gas velocities will erode particulates from component surfaces. Vibration or mechanical shocks will shake loose particulates resting on process flow surfaces. These actions can be caused by either external conditions, such as adjacent machinery which is running roughly, or they can result from valve or regulator operation within the system. Valve or regulator operation may also generate particles resulting from contact and erosion of moving surfaces.

Finally, transient process conditions can also generate contamination. Variations in system temperatures may enhance release of trace gas impurities or cause thermal expansion and contraction which could cause particulate release. Outgassing and particulate shedding upon system start-up can also release contamination into the process stream.

SYSTEM DESIGN AND FABRICATION REQUIREMENTS

Ultra-high-purity gas supply systems must meet extremely stringent design, fabrication, installation, checkout, and commissioning requirements. A brief description of such requirements is presented below.

Design Requirements

Traditional industrial gas systems are considered clean when contamination levels meet parts-per-million requirements. In the semiconductor industry, however, contamination levels must meet purity requirements on the order of only a few parts per billion. For instance, particulate cleanliness may require the following specifications; fewer than

5 particles/ft^3 >0.1 μm in size
1 particle/ft^3 >0.3 μm in size
0.5 particle/ft^3 >0.5 μm in size

Since very small particulate contamination can be trapped in small voids, pits, or imperfections on metal surfaces, it is important that such surface imperfections be virtually eliminated by polishing the surface to a very smooth finish. Surface finish roughness may be required to be less than 10-μin average roughness.

Materials of Construction

The obvious starting point in meeting requirements of this type is to use the right materials. The only material generally acceptable for fabrication of piping and system components is 316L (low carbon) stainless steel. In addition, the steel must have low sulfur content in order to obtain better weld qualities. Where 0.035 percent sulfur content is acceptable for normal applications, the sulfur content for ultra-high-purity systems should be in the 0.008 percent range. Also, all steel for one piping run should be of a single heat. This is to ensure that chemical properties of the two pieces at a weld joint are similar. Further, tubing fittings such as elbows should be prefabricated from the same heat of steel as the straight tubing stock.

Standard elastomeric seal materials such as Teflon, Kel-f, etc., are acceptable for seals. However, if metallic seals can be designed into the system, greater leak-tight systems will be possible. For this reason valves with bellows seals on the stems should be used to totally isolate the product stream from external contamination. Some good diaphragm sealed valves are also available.

Purity and cleanliness are important. Outgassing of dissolved or adsorbed gases must be reduced to very low levels. This is sometimes accomplished by special material processing, such as passivating steels in acid mixtures, or by subjecting steels to a double melt process. It may be necessary to "bake out" some of the components, or the entire system if the contamination is serious, in order to drive off the adsorbed gases.

Special material treatments to improve particulate or ionic emission in metals involves surface polishing. The methods usually employed to accomplish this are mechanical polishing, chemical polishing, and electro-polishing.

Any material not used in the particular semiconductor fabrication step could cause unacceptable contamination. Heavy metal ions will cause severe problems any time they enter the semiconductor fabrication process.

Finally, all materials used in the construction of any system component must be mutually compatible in order to avoid chemical or electrical reactions between different materials and to prevent corrosion problems.

Welding

All process line tube joining must be done by autogenous orbital butt welding only, performed by special automatic process welding machines. The joint to be welded is purged both internally and externally with inert argon to avoid oxidation of the hot weld surface. A good orbital weld must be free of stain on the product surface, can have no undercut or concavity of the weld area, and must have a generally smooth process surface finish. Weld buildup and metal puddling are not acceptable.

Weld samples must be made at the beginning of each day or whenever the weld machine variables are changed. These weld samples must be marked with the welder's name, and a log of all sample welds must be maintained.

Lap and socket welding is unacceptable since it does not permit full penetration and leaves overlaps and crevices which are very difficult to clean. Hand welding, which leaves rough surfaces on the interior, is also unacceptable. Rough surfaces promote particulate generation.

Weld requirements for field welds are the same as for shop welds. That is, all welds must be autogenous orbital welds wherever possible. Field welders must pass the same certification requirements as shop welders.

Fabrication and Installation

Fabrication and installation procedures for ultra-high-purity systems are not dramatically different from normal industrial practice, except for cleanliness. Because of such similarity, cleanliness becomes analogous to quality or safety in that it involves a culture rather than a procedure. Procedures are easy to establish; cultures are not. A culture requires a completely new approach to the fabrication and installation process.

Fabrication of ultra-high-purity gas supply systems must be conducted in as clean a manner as possible. As much of the work as possible should be done in an enclosed shop where clean conditions can be controlled more easily. Generally, it is less expensive to build as much as possible in the shop. In order to provide the cleanest conditions possible, all component fabrication must be conducted in a clean room, preferably of at least Class 100 quality.

Since cleanliness is a prime requirement, the cleaning process is one of the most critical steps in all stages of fabrication and assembly. As stated earlier, in addition to general cleaning, cleanliness involves polishing (i.e., mechanical, chemical, or electropolishing). All parts must be cleaned at critical steps in the manufacturing process and must be kept clean between fabrication steps and during shipping.

The best method of cleaning is to perform an electro-polishing procedure to the component. This procedure actually removes some of the surface metal, along with any contaminants on the surface.

All gas supply system components such as valves and regulators should be fabricated in a shop where cleanliness can be controlled. Also, some subassemblies and piping "spools" should be fabricated in a cleanliness controlled shop area. However, fabrication of the gas distribution piping to interconnect the various components is usually done in the field at the job site. This will expose the components and piping pieces to contamination from the atmosphere. Therefore, careful consideration must be made of division of work between the shop and the field.

Components must be packaged in such a manner as to ensure the cleanliness level is maintained. Double wrapping in plastic bags, preferably Saran, is a minimum requirement. Piping must have the ends capped and the tubes must be pressurized with pure nitrogen or argon, or the tubes must be sealed in a pressurized container. Consideration must be given to both the final destination and the route of transit. Changes in altitude, and hence pressure, could damage packaging.

Once the clean components arrive at the plant site, they must be stored in a manner such that the cleanliness of the component will be maintained. While waiting for installation, all components must be stored in their protective wraps and only opened when ready to be finally installed.

Special care and procedures must be used to install the component without introducing unnecessary contamination. It may be necessary to provide a portable clean room at the installation site to perform some of the required on-site fabrication in as clean an environment as possible. The clean room should be at least Class 100 quality. As much of the subcomponent assembly as possible should be done in this portable clean room. After subcomponent fabrication, all open ends of "spool pieces" must be covered with plastic wrap, again Saran wrap, and maintained sealed until just prior to welding into the final assembly.

It may also be necessary to construct a clean environment around some parts of the system to ensure that minimum contamination is introduced as final component installation is performed.

In order to reduce the possibility of atmospheric contamination entering the partially fabricated piping system, the portion of the system which has been completed must be continually purged with a pure, inert gas filtered to sub-micrometer levels. Since inert gases are used for purging and maintaining clean conditions, safety provisions must be made to protect personnel against asphyxiation.

In addition, all tools which are used on the system which might come in contact with process gas surfaces must be maintained in as clean a condition as possible. These tools must be specially cleaned, used only on the clean system, and stored in the clean room or clean environment.

QUALITY ASSURANCE

An independent quality assurance representative should be provided to establish, review, and approve all of the quality assurance procedures to be used in the design and installation of a gas supply system. These procedures should include the following:

1. Inspection of the precleaned, ultra-high-purity components
2. Written procedures for using the automatic tube welder in fabrication of the ultra-high-purity piping system
3. Qualifying personnel to work on the ultra-high-purity piping system, including welder qualifications
4. Leak testing the ultra-high-purity piping system
5. Sampling and analyzing samples of the gas for analytic tests for oxygen, hydrocarbons, and particulates

The quality assurance representative should also oversee the final analysis of the high-purity system to verify that it meets all construction and purity specifications.

CHECKOUT AND STARTUP

Verification of the cleanliness of the system may be required at several points in the project installation schedule in addition to certification at completion. This may involve gas purity certification, particulate cleanliness certification, and certification of the leak tightness of the system.

To ensure that minimal contamination will migrate into the ultra-high-purity gas supply system during operation, the gas supply piping system must be as leak tight as possible. Extensive helium leak testing using a helium-argon test mixture must be conducted in which each component, piping system weld, and mechanical joint is separately tested. Specification of leak rates of less than 10^{-8} (cubic centimeter/second at one atmosphere) of helium may be required.

Since the output yield of semiconductor fabrication plants is so dependent upon the purity and cleanliness level of the gas supplies, knowledge of system operation and cleanliness level is necessary at all times to warn of poor operation or impending degradation of performance. Continuous, automatic artificial intelligence-expert system computer controlled monitors are being developed which will very rapidly survey all critical operating parameters, gas purities, and particulate cleanliness. These expert systems will make an analysis of system maladies and either send out alarms with suggested corrections or make correcting system adjustments automatically.

CHAPTER C9
REFRIGERATION SYSTEMS PIPING

Sam P. Soling
St. Onge Ruff Associates
York, Pennsylvania

The broad term *refrigeration* refers to a general science concerned with the use of producing temperatures below normal for commercial or other useful purposes. Refrigeration piping is used in conjunction with refrigeration equipment. Refrigerants are fluids which absorb heat by evaporating at a lower temperature and pressure and transfer heat out when they condense at a higher pressure and temperature. The increase in pressure necessary to elevate the temperature level is produced by a compressor of the reciprocating, rotary, or centrifugal types or in the case of an absorption system, by the transfer of heat, thereby boiling the more volatile refrigerant out of a solution.

Fluids which do not change state are sometimes used to transfer heat in an indirect system. Such fluids are called secondary coolants. To be classified as a secondary coolant, the fluid must be used for the transfer of heat without a change in its state and must have no flash point or a flash point above 150°F as determined by the American Society of Testing Materials method in ASTM Specification D93. Brine is a secondary coolant which is a solution of salt and water.

Many fluids have been used as volatile refrigerants in the evaporation, compression, condensing, and expansion cycle. This chapter will deal with application and structural design of piping for the more commonly used volatile refrigerants such as ammonia and some of the halogenated hydrocarbons. It will also cover general methods for other refrigerants where specific tables are not presented. Since volatile refrigerants are used in the liquid, vapor, and mixture phases, each of these will be treated separately.

Many fluids have been used for brines. Originally, the term *brines* applied to salt solutions such as calcium chloride or sodium chloride. The use of such salt brines permitted the transfer of heat at lower temperature levels without introducing refrigerant of the volatile type into refrigerated spaces. These brines were commonly used for cold storage plants, ice plants, or commercial and process refrigeration.

Solutions of glycols are also used as secondary coolants. Ethylene glycol and propylene glycol are most commonly used for this purpose. Several other compounds or mixtures have been developed specifically for the purpose of heat-transfer media. These compounds are specifically designed to have high thermal capacities, low viscosities, and other desirable properties for high-heat-transfer and low-pressure losses.

Two major Codes relate to refrigeration piping. The first of these was the American Standard Safety Code for Mechanical Refrigeration, ASA B9.1. This Code is reviewed and revised periodically. The most recent edition at the time of this writing was issued in 1989. This Code is sponsored by the American Society of Heating, Refrigerating, and Air Conditioning Engineers and has been adopted by many states and municipalities to be the existing law in these localities. This Code will be referred to frequently in this section and will be designated as ANSI/ASHRAE 15.

Another important Code on piping is the American Standard B31 Code for Pressure Piping.[2] Section B31.5 covers refrigeration piping and relates to structural design rules, fabrication, construction, and testing. The Code in this section will be referred to as the ANSI/ASME B31.5 Code.

This section recognizes and uses the definitions included in Section 2 of ASHRAE 15. The definitions included in Section 500.2 of the ASME B31.5 Code are also used. In general, the definitions in these two Codes coincide with definitions in the ASME Boiler Code, Section VIII usually called the Unfired Pressure Vessel Code.[3] In addition, the ASME B31.5 Code recognizes and refers to the basic definitions of the American Welding Society.

Other basic definitions accepted by the refrigeration and air-conditioning industry are given in *ASHRAE Terminology of Heating, Ventilating, Air-Conditioning and Refrigerating*, published by the American Society of Heating, Refrigerating, and Air Conditioning Engineers.

REFRIGERATION CYCLES

Compression System

Figure C9.1*a* shows a typical single-stage refrigeration cycle plotted on a pressure-enthalpy chart. Figure C9.1*b* is a typical diagram of a single-stage compression system and shows a compressor, a condenser, an optional receiver, an expansion device, and an evaporator. The state points on the line diagram of Fig. C9.1*b* are numbered to correspond to the same points on the pressure-enthalpy chart of Fig. C9.1*a*.

In a typical system, P_1 represents the pressure in the evaporator corresponding to the temperature at which the refrigerant is evaporating. From point 1 to point 2, the refrigerant vapor is carried to the compressor in a suction line. The pressure-enthalpy chart indicates a small pressure drop in this line. From point 2 to point 3, the refrigerant vapor is compressed. The connection between points 3 and 4 represents the discharge, or hot-gas, line, and the pressure-enthalpy chart also indicates the pressure drop due to friction in this line. Desuperheating and condensation of the refrigerant at constant pressure in the condenser occur between points 4 and 5. The liquid line is represented by the section between point 5 and point 6. From point 6 to point 7, there is a representation of the pressure reduction or "expansion" through the expansion device. From point 5 to point 8, the refrigerant is a mixture of liquid and vapor because of the expansion at constant enthalpy. At point 8, the liquid and vapor mixture enters the evaporator; heat transferred from the evaporator results in evaporation of the liquid in the

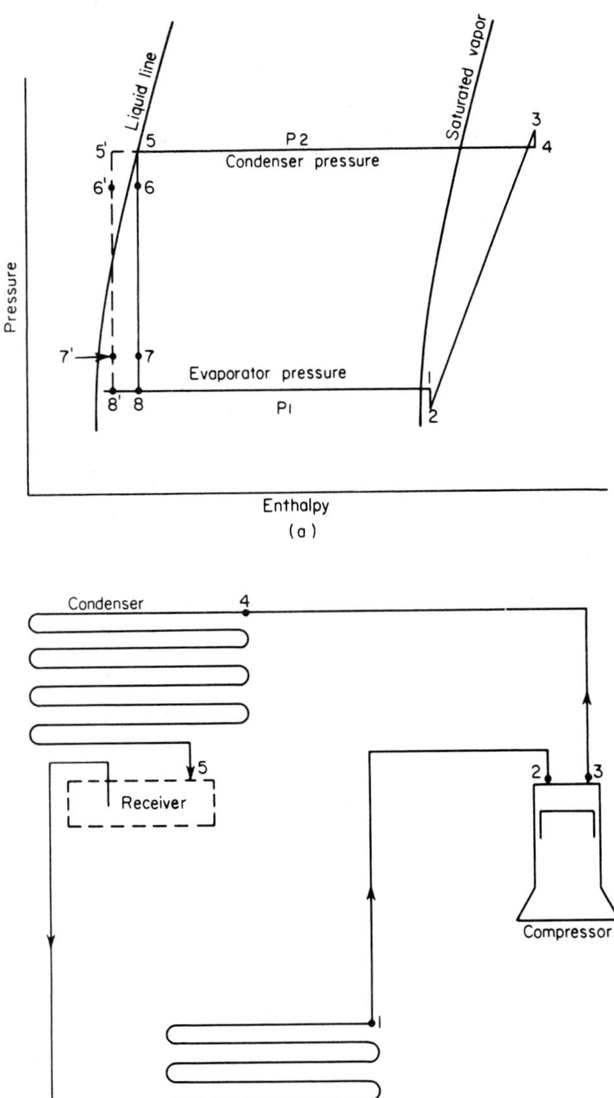

FIGURE C9.1 Compression refrigeration cycle. (*Courtesy, Carrier Air Conditioning Co.*)

liquid-vapor mixture and, as indicated, a slight superheating of the resultant vapor. At point 1, the cycle is completed.

In this chapter, the line between points 1 and 2 will be referred to as the suction line. The pipe or tubing between points 3 and 4 will be designated as the hot-gas line. The piping between points 5 and 6 will be termed the liquid line before expansion, and that between points 7 and 8 will be called the liquid line after expansion.

Absorption System

Figure C9.2 shows a typical piping arrangement for an absorption-refrigeration cycle in which a lithium salt solution is used with water as the refrigerant. In such a system, the absorber and generator serve the same purpose as the compressor in a compression system. The refrigerant is absorbed in solution at low pressure in the absorber and is pumped to the generator, where the refrigerant is boiled out of the solution at high pressure. The state points are numbered in Fig. C9.2 to correspond to the similar points of the compression system cycle in Fig. C9.1*a*.

For other types of absorption systems (such as ammonia and water), a rectifier or a fractionating-column type of purifier would be installed between the generator and condenser. Also, the evaporator might be remote or of some other type,

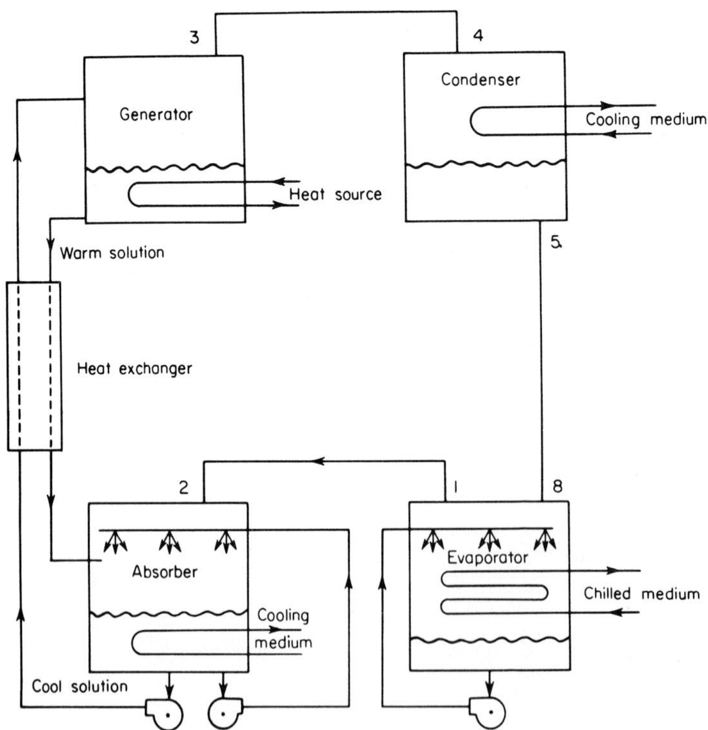

FIGURE C9.2 Absorption refrigeration cycle, lithium bromide type. (*Courtesy, Carrier Air Conditioning Co.*)

such as a flooded cooler or a direct expansion type in which the refrigerant passes through coils.

In a consideration of the piping of an absorption system, the piping between points 3 and 4, 5 and 8, and 1 and 2 may be handled in a manner similar to those of a compression system when the different fluids and flow rates which might be encountered are taken into account. The solution lines between the absorber and generator and through the heat exchanger would be treated as brine lines in the case of lithium bromide or lithium chloride systems. For ammonia-water systems, properties of aqueous solutions of ammonia are available, and these lines would be designed in the same manner as would be brine or water lines.

Present-day absorption-refrigeration systems are integrally designed as complete units, and in many instances the entire assembly including the piping is made at the factory. For these reasons it is seldom necessary to consider piping for an absorption system as a separate design problem. However, the principles involved in this chapter could be applied to the corresponding sections of an absorption system.

Flow Rates

Refrigeration is usually measured in tons of refrigeration or in Btu's per hour. The relationship between these is

$$1 \text{ ton of refrigeration} = 200 \text{ Btu/min or } 12{,}000 \text{ Btu/h}$$

For selection of piping, it is necessary to relate refrigeration rate to flow rate of the refrigerant. Since the refrigerant changes state in the process, it is customary to calculate refrigerant flow in pounds per hour or per minute. This weight flow rate for a constant rate of refrigeration will be constant throughout the system (for a single-stage system). The volume of refrigerant handled at the various points of the cycle can be determined if the density at these various points is known. Since the volume handled in a suction line is important not only for the piping selection but also for the compressor selection, charts are frequently made showing the cubic feet per minute per ton of refrigerant gas or vapor at this point.

To calculate the flow rate of refrigerant for a given rate of refrigeration the following procedure is used:

$$\frac{\text{lb}}{(\text{min ton})} = \frac{200}{(h_g - h_f)} \tag{C9.1}$$

where h_g is the enthalpy of dry saturated vapor at the evaporator outlet pressure or temperature and h_f is the enthalpy of the liquid refrigerant at the expansion device inlet.

To calculate the flow rate it is necessary to have available thermodynamic properties of the refrigerant. After the mass-flow rate has been determined as shown in the above equation, it is possible to determine the volume flowing at various points in the system.

If the specific volume of the refrigerant is known at the state point leaving the evaporator corresponding to point 1 in Fig. C9.1*a*, the volume at that point can be determined as follows:

$$\frac{\text{cfm}}{\text{ton}} = \frac{200\ V}{(h_g - h_f)} \tag{C9.2}$$

where V is the specific volume of the vapor at the evaporator suction temperature. For purposes of design, this value may be taken as the specific volume of dry, saturated vapor.

In the above equations, h_g corresponds to the enthalpy of the refrigerant vapor at point 1 leaving the evaporator. In the flooded type of cooler, this state point is close to saturation, but for a direct-expansion coiltype evaporator, the vapor may be slightly superheated. For low amounts of superheat, it is permissible to ignore the superheat for purposes of selecting the piping, although this should not be done for compressor selection or for very accurate determination of the velocity.

h_f corresponds to the enthalpy of the liquid entering the expansion device. Some subcooling of liquid may exist at this point, in which case point 6′ would result in a more accurate figure. However, for selection of piping, it is customary to use the enthalpy of the saturated liquid at the condensing temperature (point 5) for determining the flow rate. Later considerations will show when the actual state of the liquid may have an effect on the liquid line size. A later example will illustrate methods of calculation for flow rate by weight and by volume in the suction line for a typical refrigerant.

Table C9.1 shows flow rates in pounds per minute per ton and cubic feet per minute per ton at the suction condition for three common refrigerants: ammonia, refrigerant 12 (dichlorodifluoromethane), and refrigerant 22 (monochlorodi-

TABLE C9.1 Refrigerant Flow Rate

Refrigerant	717 Ammonia			12 Dichlorodifluoro-methane			22 Monochlorodifluoro-methane		
Condensing temp, F	90	100	110	90	100	110	90	100	110
Evaporating temp, F	Lbs/(min ton)								
40	0.418	0.428	0.439	3.8	3.98	4.17	2.81	2.94	3.09
20	0.422	0.432	0.444	3.96	4.15	4.36	2.88	3.03	3.19
0	0.426	0.438	0.450	4.12	4.34	4.57	2.97	3.13	3.30
−20	0.434	0.445	0.457	4.32	4.55	4.82	3.07	3.24	3.42
	Cfm/ton (suction line)*								
40	1.71	1.75	1.8	3.02	3.17	3.32	1.91	1.99	2.1
20	2.57	2.63	2.70	4.47	4.68	4.92	2.75	2.90	3.07
0	3.96	4.07	4.18	6.81	7.16	7.55	4.15	4.38	4.62
−20	6.55	6.72	6.90	10.84	11.42	12.1	6.54	6.90	7.28
	Approximate cfm/ton (discharge line)†								
40	0.878	0.805	0.715	1.55	1.42	1.31	1.07	0.94	0.87
20	0.950	0.860	0.776	1.64	1.50	1.38	1.12	1.00	0.93
0	1.023	0.935	0.850	1.72	1.58	1.46	1.19	1.10	0.99
−20	1.128	1.023	0.929	1.84	1.68	1.57	1.29	1.17	1.10

*Based on saturated enthalpies and specific volume at 10°F superheat.

†Based on discharge specific volume at 30° above isentropic discharge temperature and 10°F suction superheat.

fluoromethane). This table is based on enthalpies read at saturated conditions of refrigerant vapor and liquid as mentioned above, but the suction cubic feet per minute per ton is based on a specific volume corresponding to 10° superheat.

The cubic feet per minute per ton in the discharge line cannot be calculated readily because the actual temperature at the end of the compression is a function of the compressor design and efficiency and these will vary among various manufacturers. The discharge cubic feet per minute per ton for single-stage systems can be approximated by the following formula:

$$\text{Discharge cfm/ton} = \text{suction cfm/ton} \times \frac{P_1}{P_2} \times 1.2 \qquad \text{(C9.3)}$$

where P_1 = absolute pressure at suction
P_2 = absolute pressure at discharge

This formula is not exact, but it will serve as an approximation when it is necessary to determine approximate velocities in discharge lines.

Table C9.2 shows the temperature-pressure relationship of some of the common refrigerants. It is customary to evaluate pressure losses in refrigerant lines in terms of the number of degrees change in saturation temperature as a result of this loss. Many refrigeration manufacturers show refrigerant pressure-drop curves in terms of equivalent degrees. Table C9.2 shows the operating pressures which may be expected in refrigerant lines under certain temperature conditions.

PRESSURE DROP IN REFRIGERATION PIPING

Suction Lines

Figure C9.1*a* shows that the compressor in a refrigeration system must pump the refrigerant from point 2 to point 3. The evaporator pressure is established by the temperature level in the evaporator. Any loss in pressure in the suction line will require the compressor to operate at a lower suction pressure. The greater the loss in the suction line, the lower will be the suction pressure. Since the vapor expands at a lower pressure and since the compressor is essentially a volume device, a reduction in suction pressure causes a reduction in the capacity of the compressor because less weight of refrigerant will be handled and the refrigerant must be pumped through a greater pressure differential. An increased pressure differential also will require more power per ton of refrigeration to drive the compressor. On the other hand, if the suction line from the evaporator is made too large, the cost of installation of the line and the stop valves or controls which are placed in it may be so great as to outweigh the economic advantage of improved performance of the compressor. Also, it will be pointed out later than an excessively large suction line can result in difficulty owing to inadequate oil return to the compressor.

Pressure drop and costs are not the only considerations in the selection of suction lines for refrigeration systems. The halogenated hydrocarbons which are commonly used as refrigerants are miscible with oil and tend to carry oil in circulation with the refrigerant. Suction lines for refrigerants of this type must be sized so that the oil in circulation is carried through the system properly and is

TABLE C9.2 Temperature-Pressure Chart (Pressure: Pounds per Square Inch Gauge)†

Temp. (°F)	Refrigerants 12	22	114	500	717
−140	29.401*	—	—	—	—
−130	29.084*	28.519*	—	—	—
−120	28.617*	27.738*	—	—	—
−110	27.950*	26.623*	—	—	—
−100	27.019*	25.075*	29.523*	—	27.410*
−90	25.753*	22.976*	29.259*	24.869*	26.136*
−80	24.065*	20.191*	28.994*	22.884*	24.356*
−70	21.858*	16.564*	28.474*	20.301*	21.921*
−60	19.023*	11.926*	27.955*	16.990*	18.656*
−50	15.440*	6.0851*	27.014*	12.811*	14.358*

Temp. (°F)	Refrigerants 12	22	114	500	717
−40	10.977*	0.5717	26.073*	7.6082*	8.7956*
−38	9.9666*	1.3763		6.4300*	
−36	8.9139*	2.2138		5.2028*	
−34	7.8180*	3.0852		3.9250*	
−32	6.6777*	3.9914		2.5951*	
−30	5.4916*	4.9333		1.2118*	1.7050*
−28	4.2586*	5.9119		0.1113	0.0784*
−26	2.9773*	6.9283		0.8455	0.7974
−24	1.6466*	7.9832		1.6082	1.6715
−22	0.2651*	9.0778		2.4002	2.5850
−20	0.5739	10.213	22.893*	3.2222	3.5392
−18	1.3043	11.390		4.0751	4.5355
−16	2.0615	12.610		4.9596	5.5752
−14	2.8463	13.873		5.8766	6.6596
−12	3.6592	15.181		6.8267	7.7902

−10	4.5011	16.535	20.638*	7.8110	8.9683	50	46.736	84.062	3.9226	57.523	74.254
−8	5.3725	17.936		8.8301	10.195	60	57.766	101.64	7.9967	70.587	92.622
−6	6.2742	19.385		9.8849	11.473	70	70.204	121.45	12.722	85.346	113.78
−4	7.2067	20.883		10.976	12.802	80	84.154	143.66	18.553	101.93	137.99
−2	8.1710	22.431		12.105	14.185	90	99.725	168.44	24.384	120.47	165.54
0	9.1675	24.030	17.829*	13.272	15.622	100	117.03	195.97	31.452	141.10	196.70
2	10.197	25.682		14.478	17.116	110	136.18	226.44	39.436	163.96	231.77
4	11.260	27.387		15.724	18.668	120	157.29	260.03	48.407	189.20	271.06
6	12.358	29.147		17.011	20.280	130	180.50	296.96	58.438	216.96	314.88
8	13.491	30.962		18.340	21.953	140	205.93	337.45	69.603	247.41	363.54
10	14.660	32.834	14.374*	19.712	23.688	150	233.73	381.72	81.982	280.71	417.39
12	15.865	34.765	13.597*	21.127	25.488	160	264.05	430.06	95.656	317.06	476.76
14	17.108	36.754	12.789*	22.586	27.354	170	297.04	482.76	110.71	356.66	542.04
16	18.389	38.805	11.950*	24.091	29.288	180	332.91	540.19	127.24	399.75	613.58
18	19.709	40.916	11.078*	25.642	31.291	190	371.86	602.82	145.33	446.64	695.37
20	21.069	43.090	10.174*	27.240	33.366	200	414.14	671.32	165.09	497.70	777.16
25	24.645	48.809	7.7564	31.6065	39.1165	210	460.06	—	186.62	553.47	874.14
30	28.486	59.945	5.1219*	35.973	44.867	220	510.01	—	210.05	614.80	971.11
35	32.604	61.519	2.2371*	40.9995	51.6425	230	564.63	—	235.49	—	1085.51
40	37.009	68.550	0.4384	46.026	58.418						

*In Hg vacuum.
†Contributed by B. C. Seam, Engineering Supervisor, Bechtel Corporation, Gaithersburg, Maryland.
Source: *ASHRAE Handbook*,[4] 1989.

returned to the compressor. The sizing of suction risers is extremely important and will be discussed in more detail later in this chapter.

Discharge Lines

Figure C9.1*a* shows that the condenser pressure will be established by the temperature of the cooling medium of the refrigeration system and that a loss in the discharge line will result in a higher pressure at point 3 than at point 4. Excessive loss in the discharge line between the compressor and condenser will result in an elevated discharge pressure. When the discharge pressure rises, the compressor must pump against a higher pressure differential, and this will increase the power requirement and reduce the volumetric efficiency and capacity of the compressor. For economy, however, the discharge line must not be too large and the cost of installation investment must be weighed against the operating penalty.

Liquid Lines

Pressure losses in refrigerant liquid lines are not so critical as those in the suction and discharge lines. The pressure reduction between points 5 and 8 of Fig. C9.1*a* is inherent in the cycle. For good control and economic sizing of control valves, most of this pressure reduction is obtained in the expansion device between points 6 and 7. If the pressure loss in the liquid line before the expansion device is excessive, the reduction in pressure in the liquid line may cause vapor formation, called "flashing." Excessive flashing of the refrigerant liquid will penalize the performance of the control valve or expansion device. It is possible to avoid flashing of refrigerant for some distance in the liquid line if the liquid refrigerant is deliberately subcooled below the condensing temperature. This, however, entails extra condenser surface or auxiliary equipment and is not always economically sound. Good practice dictates that the liquid-line pressure drop should be limited to result in a reasonable selection of the expansion device and to minimize the flashing of the vapor before it flows through the expansion device.

For that portion of the liquid line after the expansion device shown between points 7 and 8 in Fig. C9.1*a*, there will be a considerable amount of vaporized refrigerant, or "flash gas," in the refrigerant mixture. Excessive pressure drop in this line will again penalize the expansion device, and it is customary to size this line generously.

Selection of Line Sizes

Suction Lines. It is customary to size suction lines so that the total loss in pressure is approximately equivalent to about 2°F drop in saturation temperatures for halogenated hydrocarbon refrigerants and not over 1°F for ammonia. It is also customary to allow less pressure drop for lower temperature installations because of the increased change in saturation temperature per pound change in pressure and the added penalties on compressors and compressor selection. Table C9.2, which relates pressure and temperature for various common refrigerants, can be used to decide what pressure drops are allowable to stay within the equivalent temperature loss indicated above. Since pressure loss is a function of length of line and number of fittings as well as flow rate and diameter, it will be obvious that short runs of lines may be sized somewhat smaller and that longer runs will require larger lines for the same loss.

Table C9.3 shows the ammonia-carrying capacity of ½–12-in steel suction pipelines. The main table headings are in saturated suction temperature. The subheadings show allowable pressure drops varying from ½ psi per 100 ft to 3 psi per 100 ft at higher temperature levels and up to 2 psi per 100 ft at lower temperature levels. The main body of the table shows the tons of refrigeration which will result in the pressure drop listed.

The variation in mass-flow rate per ton of refrigeration is not significant over a broad range of condensing temperature, and it is satisfactory to use Table C9.3 over a range of condensing temperatures from 80 to 110°F without correction.

Table C9.4 shows the carrying capacity for refrigerant 12 (dichlorodifluoromethane) suction lines over a range of saturated suction temperatures and for various allowable pressure drops from ½ psi per 100 ft to 3 psi per 100 ft. Again, the body of the table shows tons of refrigeration which will result in the pressure drop shown. Since the mass-flow rate per ton of refrigeration varies only slightly with condensing temperature, the table is usable over a range of 90–120°F without correction. Since steel pipe or copper tubing may be used with refrigerant 12, both of these are shown in the table. Table C9.5 shows similar information for the carrying capacity in suction lines for refrigerant 22 (monochlorodifluoromethane).

Discharge Lines. Pressure loss in hot gas or discharge lines should be minimized because of the adverse effects on compressor volumetric efficiency and power requirements. It is not desirable to exceed 1° or 2° equivalent loss in the discharge line. Because of the change in pressure-temperature relationships, this will result in somewhat higher permissible pressure drops in discharge lines. Discharge lines should be selected on the basis of 2 or 3 psi per 100 ft.

Table C9.6 shows the carrying capacity of steel pipe sizes for discharge lines for ammonia for various pressure drops. The table is based on an average condensing temperature of 90°F saturation and a flowing temperature of 250°F. The mass-flow rate of refrigerant is based on 20°F evaporating temperature, but the table is usable over a fairly wide range of condensing temperatures, say from 80 to 110°F, and for evaporating temperatures from 10 up to 30°F without correction.

Table C9.7 shows the carrying capacity of steel pipe or copper tubing for refrigerant 12 in discharge lines for various pressure drops. This table is based on an average condensing temperature of 105 and a 40°F saturated evaporating temperature with a flowing temperature in the discharge line of 175°F. These conditions represent a good average for air conditioning for comfort cooling conditions and for the purposes of line sizing may be used over a fairly wide range of conditions. Table C9.8 shows similar information for the carrying capacity of discharge lines for refrigerant 22.

Liquid Lines. When a system is equipped with a refrigerant receiver downstream of the condenser as a storage place to maintain a seal of liquid refrigerant on the control devices, the liquid line entering the receiver is usually sized generously to assure free flow from the condenser to the receiver. An allowable velocity of 100 fpm is typical in the selection of liquid-receiver inlet piping. To assure pressure equalization between condensers and receivers and to prevent vapor binding, lower velocities should be used in smaller size liquid lines. If 100 fpm is used as

TABLE C9.3 Suction-Line Capacities (Tons), Refrigerant 717 (Ammonia)*,†

Line size, in.‡	Saturated suction temperature, F																
	−30			−20			0			20				40			
	Pressure drop, psi/100 ft			Pressure drop, psi/100 ft			Pressure drop, psi/100 ft			Pressure drop, psi/100 ft				Pressure drop, psi/100 ft			
IPS	½	1	2	½	1	2	½	1	2	½	1	2	3	½	1	2	3
½	0.44	0.62	0.88	0.50	0.72	1.02	0.65	0.92	1.31	0.82	1.18	1.70	2.40	1.02	1.45	2.06	2.92
¾	0.96	1.37	1.96	1.11	1.58	2.24	1.45	2.06	2.93	1.81	2.60	3.70	5.23	2.25	3.22	4.61	6.52
1	1.92	2.72	3.85	2.13	3.01	4.26	2.74	3.9	5.61	3.5	4.98	7.06	8.70	4.33	6.14	8.84	10.8
1¼	4.8	6.95	9.85	5.43	7.80	11.1	7.07	10.1	14.6	8.99	12.95	18.5	22.8	11.18	16.15	23.1	28.3
1½	7.3	10.5	14.9	8.25	11.9	16.8	10.7	15.5	22.0	14.6	19.7	27.8	34.2	17.1	24.2	34.5	42.6
2	14.1	20.5	29.0	15.9	23.9	32.5	20.9	29.6	42.7	26.4	38.0	53.7	67.1	32.8	46.8	66.7	82.0
2½	22.8	32.6	46.1	25.3	36.1	52.0	33.3	47.7	68.2	42.3	60.2	85.6	105.0	52.5	75.0	106.5	131.0
3	40.1	57.5	81.4	45.1	64.6	91.5	59.1	84.2	121	74.5	106.5	151	187.5	92.5	132	190	233
4	83.5	119	169	93.0	132	186	121	172	244	153	218	305	378	190	269	382	469
5	150	214	303	168	238	341	218	312	443	276	394	555	683	342	485	690	849
6	244	344	487	274	388	550	354	505	715	447	637	900	1,110	558	789	1,125	1,380
8	500	710	1,000	560	796	1,128	726	1,039	1,468	920	1,308	1,850	2,270	1,135	1,615	2,295	2,810
10	900	1,280	1,810	1,010	1,435	2,020	1,305	1,860	2,645	1,645	2,350	3,310	4,100	2,040	2,900	4,140	5,035
12	1,450	2,050	2,900	1,625	2,310	3,280	2,100	2,780	4,280	2,675	3,820	5,410	6,600	3,325	4,685	6,670	8,200

*Courtesy of Air Conditioning Refrigeration Institute (ARI).
†Based on fluid flow at 90°F saturated condensing temperature.
‡Data based on Sch 40 steel pipe, except that 1 in and smaller are based on Sch 80.

TABLE C9.4 Suction-Line Capacities (Tons), Refrigerant 12*,†

Line size, in.‡		Saturated suction temperature, F																			
		−40				−20				0				20				40			
		Pressure drop, psi/100 ft				Pressure drop, psi/100 ft				Pressure drop, psi/100 ft				Pressure drop, psi/100 ft				Pressure drop, psi/100 ft			
IPS	OD	½	1	2	3	½	1	2	3	½	1	2	3	½	1	2	3	½	1	2	3
	½	....	....	....	....	....	....	....	....	....	....	....	0.24	....	....	0.25	0.31	....	.....	0.31	0.39
	⅝	....	....	0.20	0.25	....	....	0.28	0.35	....	0.25	0.36	0.45	....	0.32	0.47	0.58	0.28	0.40	0.59	0.74
½	...	....	....	0.24	0.29	....	0.22	0.32	0.39	....	0.29	0.41	0.51	0.26	0.37	0.53	0.66	0.32	0.46	0.66	0.81
	¾	....	0.23	0.35	0.44	0.22	0.32	0.47	0.59	0.29	0.42	0.62	0.77	0.37	0.54	0.79	0.99	0.47	0.68	1.00	1.23
	⅞	0.25	0.37	0.53	0.67	0.34	0.50	0.73	0.91	0.45	0.65	0.95	1.19	0.57	0.84	1.22	1.51	0.72	1.05	1.54	1.89
¾	...	0.24	0.35	0.51	0.63	0.33	0.47	0.68	0.83	0.43	0.61	0.87	1.08	0.55	0.78	1.13	1.38	0.69	0.98	1.38	1.70
	1⅛	0.51	0.75	1.10	1.36	0.69	1.00	1.47	1.85	0.91	1.32	1.94	2.42	1.17	1.69	2.47	3.10	1.46	2.14	3.14	3.91
1	...	0.47	0.67	0.96	1.20	0.63	0.90	1.29	1.58	0.82	1.17	1.67	2.06	1.05	1.51	2.14	2.64	1.30	1.87	2.65	3.27
	1⅜	0.90	1.32	1.92	2.40	1.22	1.77	2.58	3.22	1.61	2.34	3.39	4.22	2.04	2.94	4.30	5.34	2.58	3.74	5.52	6.75
1¼	...	0.97	1.38	1.99	2.43	1.29	1.85	2.62	3.24	1.68	2.40	3.42	4.22	2.16	3.06	4.35	5.40	2.68	3.81	5.41	6.73
	1⅝	1.44	2.09	3.00	3.78	1.92	2.82	4.10	5.10	2.54	3.68	5.37	6.62	3.25	4.70	6.90	8.46	4.05	6.02	8.65	10.7
1½	...	1.47	2.09	2.98	3.65	1.94	2.78	3.95	4.87	2.52	3.62	5.12	6.29	3.25	4.59	6.54	8.00	4.00	5.70	8.10	9.83
	2⅛	2.94	4.35	6.30	7.80	4.01	5.80	8.44	10.6	5.28	7.70	11.1	13.8	6.70	9.75	14.2	17.6	8.47	12.3	17.9	22.1
2		2.88	4.05	5.78	7.10	3.80	5.40	7.70	9.5	4.94	7.04	10.0	12.2	6.35	8.95	12.7	15.6	7.87	11.1	15.7	19.3
	2⅝	5.27	7.71	11.2	13.9	7.10	10.4	15.1	18.7	9.44	13.7	19.6	24.5	12.0	17.4	25.4	31.0	15.0	21.9	31.5	39.3
2½	...	4.52	6.50	9.24	11.4	6.02	8.55	12.2	14.9	7.86	11.2	15.7	19.3	10.1	14.1	20.2	24.7	12.4	17.6	25.0	30.7
	3⅛	8.53	12.4	17.9	22.4	11.3	16.6	23.9	29.9	15.0	21.8	31.4	39.3	19.0	27.8	40.5	49.9	23.8	34.9	50.5	62.6
3	...	8.02	11.4	16.2	19.9	10.6	15.0	21.6	26.5	13.8	19.6	27.8	34.4	17.6	24.9	35.2	43.5	21.9	30.7	44.3	54.0
	3⅝	12.7	18.4	26.8	32.9	16.9	24.5	35.8	44.4	22.2	32.1	46.7	57.9	28.5	41.2	59.3	74.1	35.6	51.6	75.0	92.0
	4⅛	17.7	25.8	37.4	46.3	23.7	34.6	50.1	62.3	31.4	45.8	65.5	82.0	40.0	57.7	83.5	105	50.3	72.5	106	130
4	...	16.6	23.8	32.8	41.2	21.8	31.4	44.4	55.5	28.6	40.6	57.2	70.7	36.2	51.8	73.4	90.4	45.0	64.4	91.0	111
	5⅛	32.1	46.3	67.0	83.2	42.7	62.3	90.7	113	56.1	81.3	119	146	71.7	103	150	186	89.7	130	189	233
5	...	30.0	42.6	60.9	74.6	39.9	56.5	80.3	99.3	51.7	73.6	105	128	65.9	93.0	134	162	81.5	117	165	202
	6⅛	51.2	74.5	108	134	69.4	100	145	180	91.0	131	190	236	115	165	240	300	145	209	300	371
6	...	48.7	69.0	97.5	121	64.5	91.2	130	159	83.5	118	167	207	106	151	217	264	132	187	265	325
8	...	99.5	140	199	245	130	186	262	322	170	242	339	422	216	306	435	540	268	381	540	669
10	...	181	258	361	446	239	338	480	588	310	438	618	759	395	553	788	965	489	689	970	1,205
12	...	290	406	574	706	382	538	765	936	490	695	984	1,210	623	882	1,260	1,550	774	1,093	1,548	1,890

*Courtesy of ARI.

†Based on fluid flow at 105°F saturated condensing temperature.

‡IPS data based on Sch 40 steel piping. OD data based on Type L copper tubing.

TABLE C9.5 Suction-Line Capacities (Tons), Refrigerant 22*,†

Line size, in.‡		Saturated suction temperature, F																			
		−40				−20				0				20				40			
		Pressure drop, psi/100 ft				Pressure drop, psi/100 ft				Pressure drop, psi/100 ft				Pressure drop, psi/100 ft				Pressure drop, psi/100 ft			
IPS	OD	½	1	2	3	½	1	2	3	½	1	2	3	½	1	2	3	½	1	2	3
	½	....	....	....	....	....	....	.....	.....	....	.....	.....	0.38	....	.....	0.39	0.47	.....	0.33	0.48	0.59
	⅝	....	....	0.34	0.40	....	....	0.43	0.52	....	0.38	0.55	0.69	0.34	0.48	0.71	0.87	0.41	0.60	0.87	1.09
½	...	....	....	0.40	0.48	....	0.36	0.49	0.61	0.31	0.45	0.63	0.78	0.40	0.56	0.80	0.96	0.49	0.70	0.99	1.21
	¾	....	0.39	0.56	0.69	0.37	0.52	0.74	0.90	0.46	0.67	0.94	1.17	0.57	0.84	1.21	1.53	0.73	1.16	1.54	1.90
	⅞	0.42	0.61	0.87	1.07	0.53	0.79	1.12	1.41	0.70	1.02	1.46	1.84	0.88	1.28	1.90	2.33	1.13	1.64	2.35	2.93
¾	...	0.39	0.56	0.79	0.98	0.51	0.73	1.03	1.26	0.65	0.93	1.31	1.62	0.81	1.16	1.64	2.02	0.99	1.42	2.03	2.58
	1⅛	0.83	1.19	1.73	2.16	1.09	1.56	2.29	2.81	1.40	2.02	2.90	3.77	1.78	2.61	3.86	4.72	2.28	3.28	4.80	5.83
1	...	0.74	1.05	1.50	1.81	0.98	1.38	1.95	2.42	1.25	1.78	2.51	3.12	1.56	2.25	3.18	3.92	1.96	2.72	4.04	4.90
	1⅜	1.45	2.11	3.05	3.89	1.90	2.76	4.14	5.03	2.50	3.61	5.20	6.40	3.16	4.61	6.76	8.20	4.04	5.69	8.35	10.3
1¼	...	1.53	2.20	3.18	3.86	1.98	2.84	4.05	5.00	2.53	3.74	5.15	6.37	3.21	4.50	6.43	7.97	4.04	5.69	8.05	9.94
	1⅝	2.30	3.38	4.76	5.91	2.96	4.35	6.23	7.77	3.93	5.60	8.10	9.96	4.95	7.17	10.8	12.9	6.23	8.94	13.1	16.2
1½	...	2.33	3.35	4.76	6.05	3.00	4.35	6.15	7.54	3.93	5.49	7.79	9.52	4.81	6.91	9.65	12.0	6.03	8.47	12.0	14.9
	2⅛	4.79	6.9	10.1	12.4	6.24	9.05	13.2	16.3	8.11	11.7	17.0	21.5	10.3	14.9	22.2	27.0	13.1	19.0	27.1	34.4
2	...	4.47	6.4	8.95	10.8	5.70	8.19	11.7	14.1	7.37	10.3	14.9	18.4	9.31	13.0	18.7	22.8	11.5	16.4	23.2	28.5
	2⅝	8.30	12.1	17.3	21.6	10.9	15.8	23.1	28.4	14.0	20.6	29.0	38.4	18.0	26.2	39.2	47.0	22.8	32.8	47.0	59.0
2½	...	7.15	10.1	14.1	17.3	9.1	12.9	18.6	23.0	11.7	16.5	24.0	29.3	14.7	20.9	29.4	38.0	18.2	26.2	37.7	46.3
	3⅛	13.2	19.4	27.7	35.2	17.7	25.2	36.8	45.8	22.6	33.0	47.7	58.3	28.9	41.8	60.0	76.6	35.7	52.9	76.2	94.4
3	...	12.4	17.9	25.4	31.8	16.1	23.4	34.2	40.4	21.0	29.6	42.0	52.3	26.1	38.0	53.0	64.3	32.5	45.7	65.6	81.2
	3⅝	20.0	28.9	41.3	50.8	25.8	38.1	53.8	67.7	33.7	49.2	70.4	86.2	43.2	62.7	89.0	113	54.6	78.2	114	139
	4⅛	27.4	40.0	57.6	73.6	36.6	52.7	76.3	96.4	47.7	68.5	99.5	125	59.5	87.5	128	160	77.2	112	161	199
4	...	25.6	36.4	50.5	63.5	32.0	47.1	65.9	81.7	42.7	59.1	84.5	107	53.7	75.5	105	129	64.7	91.7	132	162
	5⅛	50.0	73.7	107	132	66.0	96.5	138	174	89.3	124	180	222	110	159	230	286	135	199	286	357
5	...	46.4	65.7	92	114	59.3	85.0	121	149	77.4	109	154	189	96.5	133	191	234	118	167	234	291
	6⅛	79.6	117	169	207	105	151	215	272	136	193	284	359	171	251	373	450	218	314	457	573
6	...	75.0	105	149	180	96	138	195	239	124	176	249	309	156	222	312	396	192	270	397	477
8	...	156	222	317	392	206	291	410	515	262	373	530	656	331	470	660	820	407	579	827	1,010
10	...	274	396	533	678	365	519	719	890	458	670	920	1,120	570	817	1,140	1,400	718	1,020	1,420	1,740
12	...	442	606	882	1,060	555	810	1,140	1,420	717	1,020	1,490	1,810	901	1,290	1,830	2,250	1,130	1,600	2,280	2,810

*Courtesy of ARI.
†Based on fluid flow at 105°F saturated condensing temperature.
‡IPS data based on Sch 40 steel piping. OD data based on Type L copper tubing.

TABLE C9.6 Discharge and Liquid-Line Capacities (Tons), Refrigerant 717 (Ammonia)*,†

Line size,‡,§,¶ in.	Discharge lines				Liquid lines	
	Temperature, 250 F				To receiver	To system
	Pressure drop, psi/100 ft				Velocity, fpm	Pressure drop, psi/100 ft
IPS	½	1	2	3	100	2
⅜	....	....	....	....	8.5	11.6
½	1.28	1.85	2.65	3.25	13.6	23.5
¾	2.84	4.03	5.83	7.15	25.2	53.2
1	5.68	8.06	11.6	14.2	42.1	105
1¼	14.7	21.1	30.4	37.2	75.3	225
1½	22.2	31.5	45.0	55.0	103	351
2	43.0	61.4	87.6	107	197	805
2½	68.6	98.5	140	171	280	1,280
3	122	174	246	300	432	2,270
4	244	351	497	608	745	4,630
5	450	638	900	1,100		
6	734	1,030	1,470	1,800		
8	1,480	2,110	3,010	3,650		

*Courtesy of ARI.

†Based on fluid flow at 90°F saturated condensing temperature and 20°F saturated evaporating temperature.

‡Data on sizes 2 in and over based on Sch 40 steel pipe.

§Data on sizes 1 in and below based on Sch 80 steel pipe.

¶Data for discharge line sizes 1¼ and 1½ in based on Sch 40 steel pipe; for liquid-line sizes 1¼ and 1½ in based on Sch 80 steel pipe.

the design velocity for liquid to the receiver, vapor equalizing lines should be provided from the top of the receiver to the top of the condenser.

It is possible to design the liquid drain lines from the condenser so as to assure equalization without necessity of adding separate equalizer lines. A later table will show design information for this condition.

Based on a design velocity of 100 fpm in the liquid line to the receiver, Tables C9.6, C9.7, and C9.8 show the carrying capacity of liquid lines to the receiver for refrigerants 717 (ammonia), 12 (dichlorodifluoromethane), and 22 (monochlorodifluoromethane).

Liquid lines between the receiver and the expansion device should be sized on the basis of pressure loss, and 2 psi per 100 ft of run is recommended for normal conditions. Additional consideration due to vertical rise of liquid lines, liquid lines through heated spaces, or other conditions causing excessive flashing will be discussed in the next section.

Tables C9.6, C9.7, and C9.8 show the carrying capacity of various sized liquid lines to the system for refrigerants 717, 12, and 22 based on a pressure drop of 2 psi per 100 ft.

In the sizing of any refrigerant lines, the valves and fittings must be taken into account. Since their line size will not be known before the pressure drop is evaluated, it is recommended that the length of run be estimated and that an addi-

TABLE C9.7 Discharge and Liquid-Line Capacities (Tons), Refrigerant 12*,†

Line size,‡ in.		Discharge lines				Liquid lines	
		Temperature, 175 F				To receiver	To system
		Pressure drop, psi/100 ft				Velocity, fpm	Pressure drop, psi/100 ft
IPS	OD	½	1	2	3	100	2
	½	....	0.33	0.48	0.60		
	⅝	0.42	0.62	0.90	1.13	3.18	4.23
½	...	0.48	0.70	0.98	1.21	3.20	3.62
	¾	0.73	1.06	1.54	1.92	4.77	7.25
	⅞	1.11	1.62	2.36	2.92	6.61	11.2
¾	...	1.02	1.46	2.06	2.54	5.90	8.17
	1⅛	2.26	3.30	4.80	6.02	11.2	23.1
1	...	1.94	2.78	3.96	4.80	9.85	16.1
	1⅜	3.96	5.72	8.25	10.3	17.1	40.0
1¼	...	3.98	5.72	8.15	9.95	17.5	34.4
	1⅝	6.27	9.10	13.4	16.5	24.3	64.0
1½	...	5.97	8.45	12.1	14.8	24.1	52.6
	2⅛	13.0	18.8	27.3	34.0	42.3	133
2	...	11.6	16.6	23.4	29.0	45.7	123
	2⅝	23.1	33.7	48.0	60.2	65.1	236
2½	...	18.4	26.6	37.4	45.5	65.5	197
	3⅛	36.9	53.6	77.5	95.5	93.0	376
3	...	32.4	46.2	65.1	80.0	101	350
	3⅝	54.6	79.2	113	140	126	565
	4⅛	76.7	111	160	198	163	795
4	...	67.1	94.7	135	165	174	712
	5⅛	138	199	288	357		
5	...	122	172	244	298		
	6⅛	222	320	455	570		
6	...	195	280	394	480		
8	...	398	573	810	985		
10	...	725	1,030	1,450	1,770		
12	...	1,145	1,625	2,310	2,830		

*Courtesy of ARI.

†Based on fluid flow at 105°F saturated condensing temperature and 40°F saturated evaporating temperature.

‡IPS data based on Sch 40 steel piping except that liquid lines 1½ in and smaller are Sch 80. OD data based on Type *L* copper tubing.

tional 50 percent allowance be made for preliminary consideration to allow for valves and fittings. After selection of a line size, the pressure losses in the valves and fittings may be evaluated and the overall pressure drop for the line with its valves and fittings should be determined. If the total pressure drop exceeds the recommended allowable drop for reasons of inadequate selection, excessive length of run, or other conditions peculiar to the installation, the line size should be reselected and reevaluated.

TABLE C9.8 Discharge and Liquid-Line Capacities (Tons), Refrigerant 22*,†

Line size in.‡		Discharge lines				Liquid lines	
		Temperature, 200 F				To receiver	To system
		Pressure drop, psi/100 ft				Velocity, fpm	Pressure drop, psi/100 ft
IPS	OD	½	1	2	3	100	2
	½	0.33	0.48	0.69	0.86	2.34	2.89
	⅝	0.59	0.88	1.27	1.63	3.78	5.48
½	...	0.71	1.00	1.40	1.71	3.81	4.65
	¾	1.05	1.53	2.22	2.74	5.55	9.20
	⅞	1.64	2.36	3.42	4.32	7.85	14.3
¾	...	1.50	2.09	3.00	3.82	7.05	10.3
	1⅛	3.29	4.71	6.91	8.64	13.4	29.2
1	...	2.82	4.09	5.75	6.98	11.7	20.2
	1⅜	5.71	8.37	12.1	15.1	20.4	51.5
1¼	...	5.75	8.21	11.6	13.8	20.9	44.1
	1⅝	8.97	13.1	19.0	23.6	28.9	83.0
1½	...	8.64	12.4	17.2	21.6	28.8	66.4
	2⅛	19.3	27.2	40.5	49.8	50.4	168
2	...	16.6	23.6	33.1	41.9	54.6	159
	2⅝	32.9	48.2	68.8	87.0	77.6	296
2½	...	26.6	39.2	53.2	65.8	77.9	248
	3⅛	53.2	77.1	111	136	111	475
3	...	47.2	66.4	93.7	116	120	459
	3⅝	79.0	115	165	203	150	742
	4⅛	111	163	232	291	194	984
4	...	95	133	189	232	207	911
	5⅛	199	292	419	522	303	
5	...	171	239	346	425	325	
	6⅛	316	459	658	823	434	
6	...	281	409	572	681	471	
8	...	588	844	1,180	1,440	815	
10	...	1,020	1,430	2,040	2,490	1,280	
12	...	1,640	2,320	3,300	4,080	1,840	

*Courtesy of ARI.

†Based on fluid flow at 105°F saturated condensing temperature and 40°F saturated evaporating temperature.

‡IPS data based on Sch 40 steel piping except that liquid lines 1½ in and smaller are Sch 80. OD data based on Type *L* copper tubing.

Fitting Equivalent. Two methods are used for determining the pressure losses in valves and fittings. The first of these is the *equivalent-length method,* where each valve or fitting is assumed to have a pressure drop equal to an equivalent length of pipe of the same size as the valve or fitting. In order to use the equivalent-length method, a line size is first estimated both as to diameter and length. Table C9.9 shows equivalent lengths for valves and fittings which should be added to the estimated actual length of the line in order to evaluate the total pressure loss. The equivalent lengths of the valves and fittings should be added to the actual net

TABLE C9.9 Equivalent Lengths of Valves and Fittings*,†

Ferrous valves and fittings‡,§

Line size, in. IPS	Globe valve		Angle valve		Short-radius ell			Long-radius ell			Tee, line-flow			Tee, branch-flow		
	Screwed	Flanged	Screwed	Flanged	Screwed	Flanged	Welded	Screwed	Flanged	Welded	Screwed	Flanged	Welded	Screwed	Flanged	Welded
½	29	...	16	...	4.1	....	....	...	....	....	1.8	...	....	4.7		
¾	31	...	16	...	4.7	....	....	2.5	....	....	2.5	...	....	5.6		
1	35	57	16	19	5.3	1.6	1.8	2.8	1.5	1.2	3.4	1.0	1.6	6.8	3.8	2.5
1¼	46	69	19	22	7.1	2.2	2.3	3.3	2.0	1.6	4.9	1.3	2.0	9.2	4.9	7.1
1½	51	76	19	22	7.9	2.6	2.6	3.4	2.2	1.8	5.9	1.4	2.0	9.9	5.8	8.4
2	63	89	20	25	9.0	3.2	3.4	3.6	2.7	2.3	8.1	1.7	2.5	12.6	7.2	10.5
2½	..	101	..	28	...	3.8	4.2	...	3.0	2.7	...	1.9	2.9	....	8.4	13
3	..	123	..	36	...	4.9	5.3	...	3.7	3.4	...	2.4	3.6	....	11	16
4	..	155	..	48	...	6.2	7.2	...	4.5	4.5	...	2.9	4.5	....	14	22
5	..	190	..	63	...	8.1	9.2	...	5.4	5.7	...	3.5	5.1	....	17	27
6	..	227	..	78	...	9.5	11	...	6.1	6.8	...	4.1	6.1	...	20	33
8	..	295	..	110	...	13	15	...	7.1	9.0	...	4.7	7.1	....	27	44
10	..	370	..	142	...	16	18	...	8.7	11	...	5.6	8.7	....	32	56
12	..	465	..	173	...	19	22	...	10	14	...	6.2	10	....	39	68

Non-ferrous valves and fittings‡

Line size, in. OD	Globe valve		Angle valve		Short-radius ell		Long-radius ell		Tee, line-flow		Tee, branch-flow	
	Screwed	Other¶	Screwed	Other¶	Screwed	Other¶	Screwed	Other¶	Screwed	Other¶	Screwed	Other¶
½	40	70	21	24	4.7	4.7	...	3.2	1.9	1.7	5.1	6.6
⅝	39	72	22	25	5.4	5.7	...	3.9	2.3	2.3	6.2	8.2
¾	39	75	23	25	6.2	6.5	2.9	4.5	2.9	2.9	7.1	97
⅞	45	78	23	28	7.0	7.8	3.7	5.3	3.7	3.7	8.2	12
1⅛	54	87	25	29	8.1	2.7	4.2	1.9	5.2	2.5	11	8.0
1⅜	64	102	27	33	9.9	3.2	4.6	2.2	6.9	2.7	13	10
1⅝	75	115	28	34	12	3.8	5.0	2.6	8.7	3.0	14	12
2⅛	95	141	30	39	14	5.2	5.4	3.4	12	3.8	19	16
2⅝	..	159	..	44	....	6.5	...	4.2	....	4.6	....	20
3⅛	..	185	..	53	....	8.0	...	5.1	....	5.4	....	25
3⅝	..	216	..	66	....	10	...	6.3	....	6.6	....	30
4⅛	..	248	..	76	....	12	...	7.3	....	7.3	....	35
5⅛	..	292	..	96	....	14	...	8.8	....	7.9	....	42
6⅛	..	346	..	119	....	17	...	10	....	9.3	....	50

*Courtesy of ARI.

†$L_e = K(D/f)$.

‡Friction factors f determined at "practical" Reynolds numbers based on 40°F suction lines having pressure drop of 1.8 psi per 100 ft.

§Based on Sch 40 pipe.

¶Flare, sweat, flanged, and so on, and based on Type L copper tubing.

run length of the pipe in order to arrive at the total length which may be used in the calculation of pressure loss or with the tables.

Another method for estimating pressure losses in valves and fittings consists of relating the loss to the velocity head. This method is considered to be somewhat more accurate but is more awkward to use. In order to determine the pressure drop in the valve or fitting, the velocity in the corresponding line size for the flow rate is determined, the velocity head is calculated and is multiplied by a K factor to result in the number of velocity heads lost in the valve or fitting. Since this loss is expressed in feet, it must be converted to pounds per square inch lost by use of the density of the refrigerant at the flowing condition. The refinement of this calculation is usually not justified, but where pressure drops must be determined with a great degree of accuracy, the K factors are useful. Table C9.10 shows K factors for various types of valves and fittings.

General Method. The tables and charts presented in this chapter provide a means for quick selection or checking of suction, discharge, and liquid lines for a few of the most commonly used refrigerants in applied refrigeration systems. In Chap. B8, general methods are outlined for calculating the frictional losses and pressure drop for fluids flowing in pipes or tubing. These methods can be applied directly to refrigerants.

Pressure losses for refrigerants flowing without a change of state should be calculated using the Darcy-Weisbach formula:

$$H = f\left(\frac{L}{D}\right)\left(\frac{V^2}{2g}\right) \qquad \text{(C9.4)}$$

where f = friction factor
L = length of pipe (ft)
D = diameter of pipe (ft)
V = velocity (fps)
g = acceleration of gravity
= 32.17 ft/s^2
H = head loss (ft)

Friction factors depend on the roughness of the pipe and the Reynolds number of the fluid flowing. Since the Reynolds number depends on pipe diameter, velocity, density, and viscosity of the fluid, it is necessary to have the physical properties and the thermodynamic properties of a refrigerant and the conditions of operation before a detailed analysis can be made of pressure drop in the system. For friction factors in suction or discharge lines of refrigerants flowing at normal velocities and within the band of pressure drops previously recommended, the friction factor on a dimensionless basis will be 0.0025–0.003. For conservative determinations, it is recommended that the latter figure be used. The friction factor for normal liquid-line velocities will be from 0.003 to 0.004. The latter figure is recommended.

Example. The following example will give a detailed determination of flow rates, volume flow, line selection, and pressure-drop determination for a typical case of a refrigerant not listed in the basic common tables in this chapter.

Assume a refrigeration system using refrigerant 500 with a capacity of 100 tons at an evaporating temperature of 20°F and a condensing temperature of 100°F. Further assume a suction line of 50-ft net length with one globe valve and two elbows. Assume the discharge line to be 30-ft net length with one globe valve and

TABLE C9.10 *K*-Factors (Velocity Heads) for Valves and Fittings[a,b]

Ferrous valves and fittings[c]

Line size, in. IPS	Globe valve		Angle valve		Short-radius ell			Long-radius ell			Tee, line-flow			Tee, branch-flow		
	Screwed	Flanged	Screwed	Flanged	Screwed	Flanged	Welded	Screwed	Flanged	Welded	Screwed	Flanged	Welded	Screwed	Flanged	Welded
½	15	....	8.4	...	2.1	....	....	....	....	....	...	....	....	2.4		
¾	11	....	5.7	...	1.7	....	....	0.9	....	....	0.9	....	....	2.0		
1	9.3	15.5	4.3	5.0	1.4	0.43	0.46	0.73	0.40	0.32	0.9	0.26	0.43	1.8	1.0	1.37
1¼	8.4	12.8	3.5	4.0	1.3	0.40	0.42	0.60	0.37	0.29	0.9	0.24	0.36	1.7	0.90	1.31
1½	7.8	11.5	2.9	3.4	1.2	0.39	0.40	0.52	0.34	0.27	0.9	0.22	0.31	1.5	0.88	1.27
2	7.0	9.9	2.2	2.8	1.0	0.36	0.38	0.40	0.30	0.25	0.9	0.19	0.28	1.4	0.80	1.17
2½	....	9.0	...	2.5	...	0.34	0.37	....	0.27	0.24	....	0.17	0.26	...	0.75	1.13
3	....	8.3	...	2.4	...	0.33	0.36	....	0.25	0.23	...	0.16	0.24	...	0.72	1.10
4	....	7.5	...	2.3	...	0.31	0.35	....	0.22	0.22	...	0.14	0.22	...	0.68	1.05
5	....	7.0	...	2.3	...	0.30	0.34	....	0.20	0.21	...	0.13	0.19	...	0.64	1.01
6	....	6.7	...	2.3	...	0.28	0.32	....	0.18	0.20	...	0.12	0.18	...	0.60	0.98
8	....	6.2	...	2.3	...	0.27	0.31	....	0.15	0.19	...	0.10	0.15	...	0.57	0.93
10	....	6.0	...	2.3	...	0.25	0.30	....	0.14	0.18	...	0.09	0.14	...	0.52	0.90
12	....	6.0	...	2.3	...	0.25	0.29	....	0.13	0.18	...	0.08	0.13	...	0.50	0.88

Nonferrous valves and fittings[d,e,f]

Line size, in. OD	Globe valve, flare or sweat	Angle valve flare or sweat	Short-radius ell, flare or sweat	Long-radius ell, flare or sweat	Tee, line-flow, flare or sweat	Tee, branch-flow, flare or sweat
½	37	12.8	2.5	1.7	0.9	3.5
⅝	28	9.9	2.2	1.5	0.9	3.2
¾	23	7.8	2.0	1.4	0.9	3.0
⅞	19	6.7	1.9	1.3	0.9	2.8
1⅛	15.0	5.0	0.46	0.32	0.43	1.37
1⅜	13.4	4.4	0.42	0.29	0.36	1.33
1⅝	12.0	3.5	0.40	0.27	0.31	1.29
2⅛	10.4	2.9	0.38	0.25	0.28	1.19

[a]Courtesy of ARI.
[b]$K = 2gh/V^2$.
[c]Based on Sch 40 pipe.
[d]Based on Type *L* copper tubing.
[e]For screwed valves and fittings, use ferrous *K* factors.
[f]For OD sizes above 2⅛ in, use welded ferrous *K* factors.

three elbows. Assume the liquid line to be 15 ft with two elbows and two angle valves. The following detailed analysis will determine the flow rates, line sizes, and pressure drops together with the total equipment temperature loss.

Referring to Fig. C9.1 and Eq. (C9.1), the weight of refrigerant flow is

$$\text{lb/(min ton)} = \frac{200}{(h_g - h_f)}$$

From thermodynamic properties of refrigerant 500,

$$h_g \text{ (at 20°F saturated)} = 94.924 \text{ Btu/lb}$$

$$h_f \text{ (at 110°F saturated)} = 39.699 \text{ Btu/lb}$$

$$\frac{\text{lb}}{\text{(min ton)}} = \frac{200}{(94.924 - 39.699)} = 3.62$$

For 100 tons,

$$\text{lb/min} = 100 \times 3.62 = 362$$

To determine volume flowing in the suction line, from thermodynamic properties

$$V_g \text{ (at 20°F)} = 1.1294$$

Using Eq. (C9.2),

$$\text{cfm/ton} = 200 \times \frac{V_g}{(h_g - h_f)}$$

$$= \frac{(200 \times 1.1294)}{(94.924 - 39.699)} = 4.09$$

$$\text{cfm (in suction line)} = 4.09 \times 100 = 409$$

The approximate cubic feet per minute in the discharge line may be determined by Eq. (C9.3) and the pressure-temperature relationship of the refrigerant. Absolute pressures must be used in this equation.

From Table C9.2,

Refrigerant 500 at 20°F has a saturation pressure of 27.24 psig or 41.94 psia.

Refrigerant 500 at 110°F has a saturation pressure of 163.96 psig or 178.66 psia.

$$\text{Approximate discharge cfm/ton} = \text{suction cfm/ton} \times \frac{P_1}{P_2} \times 1.2$$

$$= \frac{(4.09 \times 41.94)}{(178.66 \times 1.2)} = 1.15$$

$$\text{Discharge cfm} = 1.15 \times 100 \text{ tons} = 115$$

The liquid volume may be determined from the weight flow rate and the density of the liquid at the condensing temperature.

$$\text{Density of refrigerant 500 at 110°F} = 67.962 \text{ lb/ft}^3$$

$$\text{cfm of liquid refrigerant} = \frac{362}{67.962} = 5.33$$

To illustrate the general method, the fitting resistance will be evaluated by the equivalent-length method.

For the suction line, assume a copper tube of 4⅛ in OD. From tables of copper tube sizes, 4⅛-in-OD Type *L* copper tubing has an ID of 3.905 in and an internal area of 11.92 in^2. The total equivalent length of suction piping may be determined by adding the equivalent length of fittings to the net length of run.

From Table C9.9,

Equivalent length of one 4⅛-in globe valve	=	248 ft
Equivalent length of two 4⅛-in elbows	=	24 ft
Length of net run	=	50 ft
Total equivalent length	=	322 ft

Velocity of suction gas in 4⅛-in OD lines is

$$V = \frac{(\text{cfm} \times 144)}{(60 \times \text{area})} = \frac{(409 \times 144)}{(60 \times 11.92)} = 82.3 \text{ fps}$$

$$\text{Diameter of tube} = \frac{3.905}{12} = 0.326 \text{ ft}$$

From Eq. (C9.4)

$$H = \frac{fLV^2}{2Dg} = \frac{0.003 \times 322 \times 82.7 \times 82.3}{2 \times 0.326 \times 32.2} = 313 \text{ ft}$$

At suction conditions, specific volume of vapor was 1.1294 ft^3/lb; therefore, the density of vapor is

$$\frac{1}{1.1294} = 0.885 \text{ lb/ft}^3$$

$$\text{Pressure drop (psi)} = \frac{313 \times 0.885}{144} = 1.92$$

From the pressure-temperature equivalent (Table C9.2) a pressure drop of 1.92 psi at 20°F is equivalent to approximately 2.4°F penalty.

The use of an angle valve instead of a globe valve would reduce the total equivalent length to 150 instead of 322 ft. The pressure drop would then be

$$\left(\frac{150}{322}\right) \times 313 \text{ ft} = 146 \text{ ft or } 0.897 \text{ psi}$$

which would incur a penalty of approximately 1.1°F. Hence the selection of a 4⅛-in-OD suction line is satisfactory from a pressure-drop standpoint. From Table C9.21, a 4⅛-in Type *L* tube is satisfactory for a working pressure of 250 psi which (see Table C9.18) exceeds the required design pressure of 120 psi.

For the *discharge line,* try a 2⅝-in-OD line. From tables of copper tube sizes, the 2⅝-in-OD Type *L* tubing has an ID of 2.465 in and an internal area of 4.76 in^2.

From Table C9.9, and the fittings specified, the equivalent length is determined as follows:

$$\text{Equivalent length of } 2\tfrac{5}{8}\text{-in globe valve} = 159 \text{ ft}$$

$$\text{Equivalent length of three } 2\tfrac{5}{8}\text{-in elbows} = 19.5 \text{ ft}$$

$$\text{Net length of } 2\tfrac{5}{8}\text{-in line} = \underline{30.0 \text{ ft}}$$

$$\text{Total equivalent length} = 208.5 \text{ ft}$$

Velocity of discharge gas in a 2⅝-in-OD line is calculated and found to be 57.9 fps

$$\text{Tube diameter} = \frac{2.465}{12} = 0.206 \text{ ft}$$

The friction factor may again be assumed to be 0.003.
From Eq. (C9.4) the friction pressure drop is

$$H = \frac{fLV^2}{2Dg} = \frac{0.003 \times 208.5 \times 57.9 \times 57.9}{2 \times 0.206 \times 32.2} = 158 \text{ ft}$$

Approximate specific volume of discharge gas is

$$\frac{\text{cfm discharge}}{\text{cfm suction}} \times V_g$$

or

$$\left(\frac{115}{409}\right) \times 1.1294 = 0.32 \text{ ft}^3/\text{lb}$$

$$\text{Density of discharge gas} = \frac{1}{0.32} = 3.12 \text{ lb/ft}^3$$

$$\text{Pressure drop (psi)} = 158 \times \frac{3.12}{144} = 3.42 \text{ psi}$$

From Table C9.2, at 110°F, a pressure drop of 3.42 psi for refrigerant 500 is equivalent to approximately 1.4°F.

For the *liquid line to the receiver,* a velocity of 100 fpm is recommended for general use. The required area of line is

$$\text{Area} = \text{cfm (liquid)/velocity} = \frac{5.33}{100} = 0.0533 \text{ ft}^2 = 7.7 \text{ in}^2$$

which requires a 3⅝-in liquid line to the receiver.

For the liquid line to the system expansion valve, assume a 2⅛-in-OD copper tube.

For Type *L,* 2⅛ in OD, the inside diameter is 1.985 in and the inside area is 3.09 in^2.

The equivalent length of liquid line is

$$\text{Equivalent of two } 2\tfrac{1}{8}\text{-in angle valves} = 78 \text{ ft}$$

$$\text{Equivalent of two } 2\tfrac{1}{8}\text{-in elbows} = 10.4 \text{ ft}$$

$$\text{Net run of } 2\tfrac{1}{8}\text{-in line} = \underline{15 \text{ ft}}$$

$$\text{Total equivalent length} = 103.4 \text{ ft}$$

Velocity of liquid in a 2⅛-in copper tube is

$$\text{Velocity} = \frac{\text{cfm} \times 144}{60 \times \text{area}} = \frac{5.33 \times 144}{60 \times 3.09} = 4.1 \text{ fps}$$

For an assumed friction factor of 0.004, friction drop is calculated from Eq. (C9.4):

$$H = \frac{fLV^2}{2Dg} = \frac{0.004 \times 103.4 \times 4.1 \times 4.1}{2 \times 0.1655 \times 32.2} = 0.65 \text{ ft}$$

$$\text{Pressure drop} = 0.65 \times \frac{68.05}{144} = 0.308 \text{ psi}$$

This pressure drop is too low for an economical selection, so try a 1⅝-in line:

$$\text{Diameter} = \frac{1.505}{12} = 0.1255 \text{ ft}$$

$$\text{Area} = 1.771 \text{ in}^2$$

Equivalent lengths are:

Equivalent lengths of two 1⅝-in valves =	68 ft
Equivalent lengths of two 1⅝-in elbows =	7.6 ft
Net run =	15.0 ft
Total equivalent length =	90.6 ft

$$\text{Velocity} = \frac{\text{cfm} \times 144}{60 \times \text{area}} = \frac{5.33 \times 144}{60 \times 1.771} = 7.22 \text{ fps}$$

$$H = \frac{fLV^2}{2Dg} = \frac{0.004 \times 90.6 \times 7.22 \times 7.22}{2 \times 0.1255 \times 32.2} = 2.34 \text{ ft}$$

Pressure drop = 2.34 × 68.05/144 = 1.106 psi, and this is satisfactory. The liquid line after expansion is usually sized one size larger than the main liquid line. In this case, a 2⅛-in OD line could be used.

For the illustration, the selected line sizes are as follows:

Suction line	4⅛ in OD
Discharge line	2⅝ in OD
Liquid line to receiver	3⅝ in OD
Liquid line to expansion valve	1⅝ in OD
Liquid line after expansion valve	2⅛ in OD

The compressor selection must be adjusted for the suction and discharge line loss, so the compressor should be selected for conditions equivalent to 18.9°F saturated suction (1.1°F line loss) and 111.4°F saturated discharge (1.4°F line loss).

Secondary Coolant. Piping for secondary coolant service whether of salt brine, glycols, or other compounds which may be used for transfer of heat without a change of state, may be designed by the rules and principles which apply to water piping. As in the design of water piping, consideration must be given to pressure losses, pumping head, and cost of investment. Additional considerations for brine piping involve compatibility of materials with the brine to be circulated and evaluation of pressure drop, taking into account the density and viscosity of the flowing liquid.

When choosing a secondary coolant at moderately low temperature levels, it is customary to select a coolant which has a freezing point approximately 20°F below the operating temperature expected in the system. To use a more concentrated coolant (having a lower freezing point) may be safer in operation but is uneconomical from the standpoint of reduced heat transfer and increased pumping costs. The principles in Chap. B8 may be applied for sizing of secondary coolant piping. To assist in the analysis of such piping, Table C9.11 shows the concentration, specific gravity, freezing point, and the viscosity at a temperature 20°F above the freezing point for four common secondary coolants. If the desired freezing point is known, this table will show the viscosity at a temperature 20°F above the freezing point and the density of the brine solution. This information will enable determination of the friction factor and, subsequently, calculations of the pressure loss. Standard equivalents may be used for valves and fittings in the brine circuit.

Oil Return. The halogenated hydrocarbon refrigerants are miscible with oil. These refrigerants in the liquid phase will carry oil without separation. In a compressor, some oil will be carried out with the refrigerant because of the gas velocities and oil coatings of the refrigerant compressor parts. The oil will be carried into the condenser, where it will go into solution with the liquid refrigerant and will be carried to the evaporator in the liquid phase. Since the evaporator acts as a still, it will tend to boil off refrigerant vapor and leave the oil in the evaporator. In the case of shell-and-tube evaporators where the refrigerant is in the shell, there will be a tendency for the oil concentration in the evaporator to increase. One common way of returning oil from such an evaporator is to provide a small line from the bottom which bleeds a controlled amount of refrigerant liquid and oil mixture into a point in the suction line which must loop down below the evaporator. Another method is to maintain a slight amount of refrigerant liquid carryover from the evaporator which will carry oil into the suction line, after which a suitable heat exchanger may be used to complete the vaporization of the suction gas by heat exchange with warm liquid coming from the condenser to the evaporator. This method will not result in thermodynamic loss as long as the energy used to vaporize the refrigerant in the suction line comes from the liquid refrigerant being fed to the evaporator.

After the oil is introduced into the suction line by either of the means described above or is automatically returned to the suction line from a direct expansion evaporator where the refrigerant is boiling inside tubes, it is necessary to maintain velocities in the suction line which will assure that the oil will be returned to the compressor. For suction lines chosen in the conventional manner and within the limits of the pressure drops and velocities recommended earlier, oil will return in horizontal runs or in lines running downward. However, for vertical risers, there is a possibility of oil collecting on the sides of the pipe wall and running back toward the evaporator unless the velocities in the vertical risers are maintained.

The necessary velocities expressed in terms of minimum tons capacity for oil entrainment in vertical suction risers have been determined, and tables are included in this chapter to guide in the selection of such risers. Table C9.12 shows

TABLE C9.11 Secondary Coolant Properties

Concentration %	Calcium chloride			Sodium chloride			Ethylene glycol			Propylene glycol		
	Specific gravity*	Freezing point, F	Viscosity†	Specific gravity*	Freezing point, F	Viscosity†	Specific gravity*	Freezing point, F	Viscosity†	Specific gravity*	Freezing point, F	Viscosity†
0	1.00	32°	1.27	1.00	32°	1.27	1.00	32°	1.27	1.00	32°	1.27
5	1.044	27.7	1.45	1.035	27.0	1.45	1.005	30	1.50	1.004	29.5	1.60
10	1.087	22.3	1.80	1.072	20.4	1.75	1.012	25.2	1.80	1.009	26.0	2.0
15	1.133	13.5	2.5	1.111	12.0	2.3	1.019	21	2.03	1.014	24.5	2.6
20	1.182	−0.4	4.1	1.15	1.8	3.4	1.026	15.8	2.9	1.018	20.5	3.7
23.3‡	.....	.....	.....	1.175	−6.0	4.7	....	....	....	....	....	...
25	1.233	−21.0	9.7	1.191	16.0	3.0	1.033	10.5	3.7	1.022	15.6	5.2
29.87‡	1.290	−67.0	38.0	.....	....	....	....	....	....	....	....	...
30	1.295	−50.8	25.7	.....	....	....	1.039	3.2	5.0	1.027	10.0	7.7
35	.....	.....	.....	.....	....	....	1.046	−1.0	6.3	1.031	3.5	12.6
40	.....	.....	.....	.....	....	....	1.053	−12.0	11.0	1.035	−4.0	24.0
45	.....	.....	.....	.....	....	....	1.059	−17.0	15.0	1.038	−15.0	45.0
50	.....	.....	.....	.....	....	....	1.066	−33.0	29.0	1.041	−24.0	85.0
55	.....	.....	.....	.....	....	....	1.072	−43.0	55.0	....	....	...

*Specific gravities compared with water at 60°F.
†Viscosities in centipoises at 20°F above freezing point.
‡Eutectic points.

TABLE C9.12 Minimum Tonnage for Oil Entrainment Up Vertical Suction Risers, Refrigerant 12

Copper tubing, Type ACR

Tube OD	7/8		1 1/8		1 3/8		1 5/8		2 1/8		2 5/8		3 1/8		3 5/8		4 1/8		5 1/8		6 1/8		8 1/8	
Area (in^2)	0.484		0.825		1.256		1.780		3.094		4.770		6.812		9.213		11.97		18.67		26.83		46.85	
Suct temp.*	*T*	*F*	*T*	*F*	*T*	*F*	*T*	*F*	*T*	*F*	*T*	*F*	*T*	*F*	*T*	*F*	*T*	*F*	*T*	*F*	*T*	*F*	*T*	*F*
−40	0.31	3.1	0.61	2.9	1.0	2.7	1.6	2.6	3.2	2.4	5.5	2.2	8.6	2.1	12.5	2.0	17.3	1.9	30.2	1.8	47.4	1.7	95.0	1.6
−20	0.40	1.8	0.77	1.7	1.3	1.5	2.0	1.5	4.0	1.4	6.9	1.3	10.8	1.2	15.6	1.1	21.8	1.1	37.8	1.0	59.6	1.0	119.1	0.9
0	0.47	1.1	0.93	1.0	1.6	0.9	2.4	0.9	4.8	0.8	8.2	0.8	12.8	0.7	18.8	0.7	26.0	0.6	45.4	0.6	71.4	0.6	143.0	0.5
+20	0.55	0.65	1.1	0.6	1.8	0.6	2.8	0.5	5.6	0.5	9.7	0.5	14.9	0.4	22.0	0.4	30.6	0.4	53.0	0.4	83.1	0.4	167.0	0.3
+40	0.63	0.40	1.2	0.4	2.1	0.4	3.2	0.3	6.4	0.3	11.1	0.3	17.1	0.3	25.0	0.3	34.8	0.2	61.0	0.2	95.6	0.2	191.5	0.2

Steel pipe, standard weight (Sch 40)

IPS.	3/4		1		1 1/4		1 1/2		2		2 1/2		3		3 1/2		4		5		6		8	
Area (in^2)	0.533		0.864		1.495		2.036		3.355		4.788		7.393		9.89		12.73		20.01		28.99		50.0	
Suct temp.*	*T*	*F*	*T*	*F*	*T*	*F*	*T*	*F*	*T*	*F*	*T*	*F*	*T*	*F*	*T*	*F*	*T*	*F*	*T*	*F*	*T*	*F*	*T*	*F*
−40	0.35	3.9	0.65	3.6	1.3	3.2	1.9	3.1	3.5	2.9	5.5	2.8	9.5	2.6	13.6	2.5	18.7	2.4	33.0	2.3	52.0	2.2	103.8	2.0
−20	0.44	2.2	0.81	2.1	1.6	1.8	2.4	1.7	4.4	1.7	6.9	1.6	11.8	1.5	17.0	1.4	23.4	1.4	41.3	1.3	65.4	1.3	129.6	1.1
0	0.53	1.3	0.97	1.2	1.9	1.1	2.9	1.0	5.3	1.0	8.3	1.0	14.2	0.9	20.4	0.9	28.0	0.8	49.0	0.8	78.4	0.8	152.0	0.7
+20	0.62	0.8	1.1	0.7	2.3	0.7	3.3	0.6	6.2	0.6	9.7	0.6	16.6	0.5	23.9	0.5	32.7	0.5	57.4	0.5	92.0	0.4	181.0	0.4
+40	0.71	0.5	1.3	0.5	2.6	0.4	3.8	0.4	7.1	0.4	11.2	0.4	19.1	0.3	27.5	0.3	37.6	0.3	66.2	0.3	105.5	0.3	208.0	0.3

*T = tons of refrigeration, F = friction drop, degrees F per 100 ft equivalent length at tons shown.
Source: Courtesy Carrier Air Conditioning Co.

the minimum tonnage for oil entrainment in vertical suction risers of refrigerant 12 both in copper tubing and in steel pipe. The actual velocities required to return oil in these vertical risers will vary with the operating temperature and the size of the pipe. The minimum velocity for oil return in vertical risers varies from approximately 1500 fpm in 5⅛-in tubing at 40°F to as high as 5500 fpm at −40°F in the same size tubing. Correspondingly, there is a diameter effect, and velocities 750 fpm at 40°F to 2500 fpm at −40°F are sufficient to return oil in vertical risers for tubing of approximately 1⅛ in diameter.

Table C9.13 shows similar minimum tonnage capacity for refrigerant 22 for oil return in vertical suction risers.

For other refrigerants of the halogenated hydrocarbon type, approximately the same velocities should be maintained under similar operating conditions and with the same size pipe. For such cases, it is recommended that the corresponding velocities be calculated from the basic table shown for refrigerant 12 or refrigerant 22 and that the same velocities be applied to the other refrigerants.

Since vapor in the line between the compressor and the condenser may also be subject to the same problem of oil carryover from the compressor, it is necessary to check the discharge lines also to be sure that any oil that gets into a vertical riser will be carried into the condenser and continue in circulation. Table C9.14 shows the minimum tonnages required for oil entrainment in vertical hot-gas risers for refrigerant 12 and refrigerant 22.

Where a refrigeration system operates over a range of capacities, it is evident that a suction riser which is properly sized for the full-load condition may not handle adequately a partial-load condition. In such a case, a double-suction-riser construction can be used. Figure C9.3 shows a typical double-suction riser. Lines *A* and *B* are sized so the two together will handle the full load with proper oil return in accordance with Table C9.12 or C9.13 or with equivalent velocities for other refrigerants. If the minimum operating tonnage is determined, line *A* should be sized to assure oil return for this minimum tonnage. At the low tonnage, the trap at the bottom of the loop on line *B* will seal with oil and line *A* will handle the reduced load with suitable oil return. Care must be observed that the oil-holding capacity of the trap is such that it will not hold enough oil to deplete the compressor of necessary oil charge during the partial-load operation. Similar arrangements can be made on discharge lines when necessary.

Suction lines should be designed so that oil from an active evaporator does not drain into an idle one.

Figure C9.4*a* shows multiple evaporators on different floor levels and the compressor above. Each suction line is brought upward and looped into the top of the common suction line to prevent oil from draining into inactive coils.

Figure C9.4*b* shows multiple evaporators stacked with the compressor above. Oil cannot drain into the lowest evaporator because the common suction line drops below the outlet of the lowest evaporator before entering the suction riser.

Figure C9.4*c* shows multiple evaporators on the same level with the compressor located below. The suction line from each evaporator drops down into the common suction line so oil cannot drain into an idle evaporator. An alternate arrangement is shown in Figure C9.4*d* for cases where the compressor is above.

Figure C9.5 illustrates typical piping for evaporators above and below a common suction line. All horizontal runs should be level or pitched toward the compressor to assure oil return.

The traps shown in the suction lines after the coil suction outlet are recommended by various thermal expansion valve manufacturers to prevent erratic operation of the thermal expansion valve. The expansion valve bulbs are located in

TABLE C9.13 Minimum Tonnage for Oil Entrainment Up Vertical Suction Risers, Refrigerant 22*

Copper tubing, Type ACR																								
Tube OD	7/8		1 1/8		1 3/8		1 5/8		2 1/8		2 5/8		3 1/8		3 5/8		4 1/8		5 1/8		6 1/8		8 1/8	
Area (in^2)	0.484		0.825		1.256		1.780		3.094		4.770		6.812		9.213		11.97		18.67		26.83		46.85	
Suct temp.*	*T*	*F*	*T*	*F*	*T*	*F*	*T*	*F*	*T*	*F*	*T*	*F*	*T*	*F*	*T*	*F*	*T*	*F*	*T*	*F*	*T*	*F*	*T*	*F*
−40	0.45	1.6	0.88	1.5	1.5	1.4	2.3	1.3	4.6	1.2	8.0	1.2	12.3	1.1	18.1	1.1	25.0	1.0	43.6	0.9	68.5	0.9	137.1	0.8
−20	0.56	1.0	1.1	0.9	1.8	0.8	2.8	0.8	5.6	0.7	9.7	0.7	15.1	0.6	21.9	0.6	30.6	0.6	53.0	0.6	83.6	0.5	167.0	0.5
0	0.66	0.6	1.3	0.5	2.2	0.5	3.3	0.5	6.7	0.4	11.4	0.4	17.9	0.4	26.2	0.4	36.3	0.4	63.3	0.3	99.5	0.3	199.3	0.3
+20	0.77	0.4	1.5	0.3	2.5	0.3	3.9	0.3	7.8	0.3	13.5	0.3	20.8	0.2	30.8	0.2	42.7	0.2	74.1	0.2	116.2	0.2	234.0	0.2
+40	0.89	0.2	1.8	0.2	3.0	0.2	4.6	0.2	9.1	0.2	15.8	0.2	24.3	0.2	35.4	0.1	49.4	0.1	86.5	0.1	135.6	0.1	272.0	0.1
Steel pipe, standard weight (Sch 40)																								
IPS	3/4		1		1 1/4		1 1/2		2		2 1/2		3		3 1/2		4		5		6		8	
Area (in^2)	0.533		0.864		1.425		2.036		3.355		4.788		7.393		9.89		12.73		20.01		28.99		50.0	
Suct temp.*	*T*	*F*	*T*	*F*	*T*	*F*	*T*	*F*	*T*	*F*	*T*	*F*	*T*	*F*	*T*	*F*	*T*	*F*	*T*	*F*	*T*	*F*	*T*	*F*
−40	0.51	2.0	0.93	1.9	1.8	1.6	2.7	1.6	5.0	1.5	7.9	1.4	13.5	1.3	19.4	1.3	26.7	1.2	47.2	1.2	74.3	1.1	148.1	1.0
−20	0.63	1.2	1.1	1.1	2.3	1.0	3.3	0.9	6.2	0.9	9.7	0.9	16.6	0.8	23.9	0.7	33.0	0.7	58.2	0.7	92.1	0.7	182.4	0.6
0	0.74	0.7	1.4	0.7	2.7	0.6	4.0	0.6	7.4	0.5	11.5	0.5	19.9	0.5	28.5	0.5	39.2	0.4	68.7	0.4	110.0	0.4	212.5	0.4
+20	0.87	0.5	1.6	0.4	3.2	0.4	4.7	0.3	8.7	0.3	13.6	0.3	23.4	0.3	33.6	0.3	46.0	0.3	80.6	0.3	129.1	0.3	254.5	0.2
+40	1.0	0.3	1.8	0.3	3.6	0.2	5.3	0.2	10.0	0.2	15.7	0.2	26.4	0.2	38.6	0.2	53.0	0.2	93.0	0.2	148.0	0.2	292.0	0.1

**T* = tons of refrigeration, *F* = friction drop, degrees F per 100 ft equivalent length at tons shown.
Source: Courtesy Carrier Air Conditioning Co.

TABLE C9.14 Minimum Tonnage for Oil Entrainment Up Vertical Hot-Gas Risers*

Refrigerant 12

Copper tubing, Type ACR

Tube OD	7/8		1 1/8		1 3/8		1 5/8		2 1/8		2 5/8		3 1/8		3 5/8		4 1/8		5 1/8		6 1/8	
Area (in²)	0.484		0.825		1.256		1.78		3.094		4.77		6.812		9.213		11.97		18.67		26.83	
Disch sat temp.*	*T*	*F*	*T*	*F*	*T*	*F*	*T*	*F*	*T*	*F*	*T*	*F*	*T*	*F*	*T*	*F*	*T*	*F*	*T*	*F*	*T*	*F*
80	0.71	0.2	1.4	0.4	2.3	0.4	3.6	0.3	7.2	0.3	12.3	0.3	19.2	0.3	28.0	0.3	38.6	0.2	68.0	0.2	106.0	0.2
90	0.75	0.3	1.5	0.4	2.5	0.4	3.9	0.4	7.7	0.3	13.3	0.3	20.6	0.3	30.4	0.3	42.0	0.3	73.0	0.3	114.0	0.2
100	0.80	0.3	1.6	0.4	2.6	0.4	4.0	0.4	8.0	0.4	14.0	0.3	21.7	0.3	31.8	0.3	43.7	0.3	76.6	0.3	120.0	0.3
110	0.87	0.4	1.7	0.5	2.9	0.5	4.4	0.4	8.9	0.4	15.0	0.4	23.7	0.3	34.8	0.3	48.0	0.3	83.0	0.3	131.0	0.3
120	0.95	0.4	1.9	0.5	3.1	0.5	5.0	0.5	9.7	0.4	16.9	0.4	26.0	0.4	38.2	0.4	52.5	0.3	91.5	0.3	143.0	0.3

Refrigerant 22

Copper tubing, Type ACR

Tube OD	7/8		1 1/8		1 3/8		1 5/8		2 1/8		2 5/8		3 1/8		3 5/8		4 1/8		5 1/8		6 1/8	
Area (in²)	0.484		0.825		1.256		1.78		3.094		4.77		6.812		9.213		11.97		18.67		26.83	
Disch sat temp.*	*T*	*F*	*T*	*F*	*T*	*F*	*T*	*F*	*T*	*F*	*T*	*F*	*T*	*F*	*T*	*F*	*T*	*F*	*T*	*F*	*T*	*F*
80	1.1	0.5	2.1	0.6	3.5	0.6	5.4	0.5	10.6	0.5	18.1	0.4	28.4	0.4	42.0	0.4	57.6	0.4	101.0	0.4	146.0	0.3
90	1.2	0.5	2.3	0.7	3.8	0.6	5.8	0.6	11.5	0.5	19.9	0.5	31.2	0.5	45.6	0.5	62.5	0.4	110.0	0.4	158.0	0.3
100	1.25	0.5	2.5	0.7	4.1	0.7	6.3	0.6	12.5	0.6	21.6	0.6	33.9	0.5	49.6	0.5	68.5	0.5	120.0	0.5	173.0	0.4
110	1.4	0.6	2.7	0.8	4.4	0.7	6.9	0.7	13.7	0.6	23.5	0.6	38.1	0.6	53.7	0.5	74.0	0.5	129.0	0.5	187.0	0.4

*T = tons of refrigeration, F = friction drop, degrees F per 100 ft equivalent length at tons shown.
Source: Courtesy Carrier Air Conditioning Co.

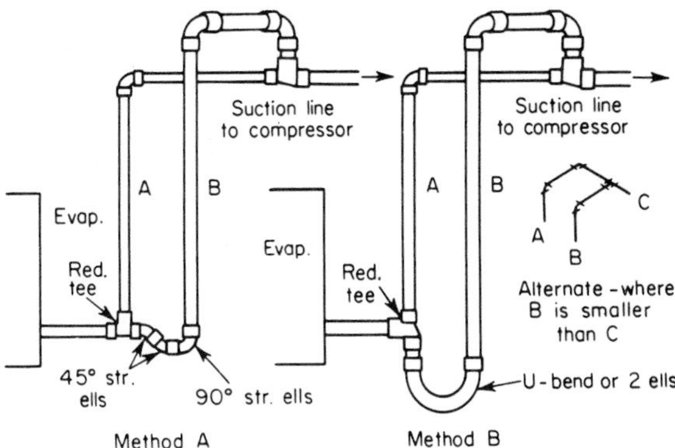

FIGURE C9.3 Double suction riser construction. (*Courtesy, Carrier Air Conditioning Co.*)

the suction lines between the coils and these traps. The traps serve as drains and help prevent liquid from accumulating under the expansion valve bulbs during compressor off-cycles. They are useful only where straight runs or risers are encountered in the suction line leaving the coil outlet.

The illustrations and descriptions of piping cover some of the more common arrangements. There are many arrangements of multiple evaporators, multiple compressors, or multiple condensers with variations which are beyond the scope of this chapter. However, the principles outlined herein may be applied to many types of installations.

Flashing in Liquid Lines. As previously discussed, excessive pressure drop in refrigerant liquid lines leaving the condenser or receiver may cause excessive formation of vapor or "flash gas." Since the vapor occupies a much greater volume than the liquid, the remaining liquid will be forced along the pipe at a much higher velocity but with an average density approaching that of the mixture of liquid and vapor. In a pipe that conveys a fluid which vaporizes as it flows, the pressure drop will gradually increase. The pressure drop in a line which carries a flashing liquid or in a liquid line after expansion can be considerably greater than that in the line before expansion or vaporization.

In liquid refrigeration lines which are vertical or inclined even in the case of low velocities or low frictional pressure drops, the static-head effect of the liquid in the line may cause vaporization or flashing of the liquid near the top of the vertical run. In such cases, the liquid lines should be sized generously to minimize frictional losses, and liquid subcooling may be used if possible in the plant design. Provision for subcooling of liquid is usually beyond the scope of the piping designer and must be obtained in conjunction with the designer of the condenser or the base equipment. Excessive flash gas in long runs of liquid lines may be partially avoided by insulation if the line is to run through a warm ambient temperature or by the installation of a liquid subcooler.

The pressure-temperature relation of various refrigerants can be used as a guide in determining the amount of subcooling required to overcome pressure drop in the liquid line and to overcome the effect of static head in vertical liquid lines. Since liquid subcooling cannot always be obtained, generous sizing of liquid lines may be required.

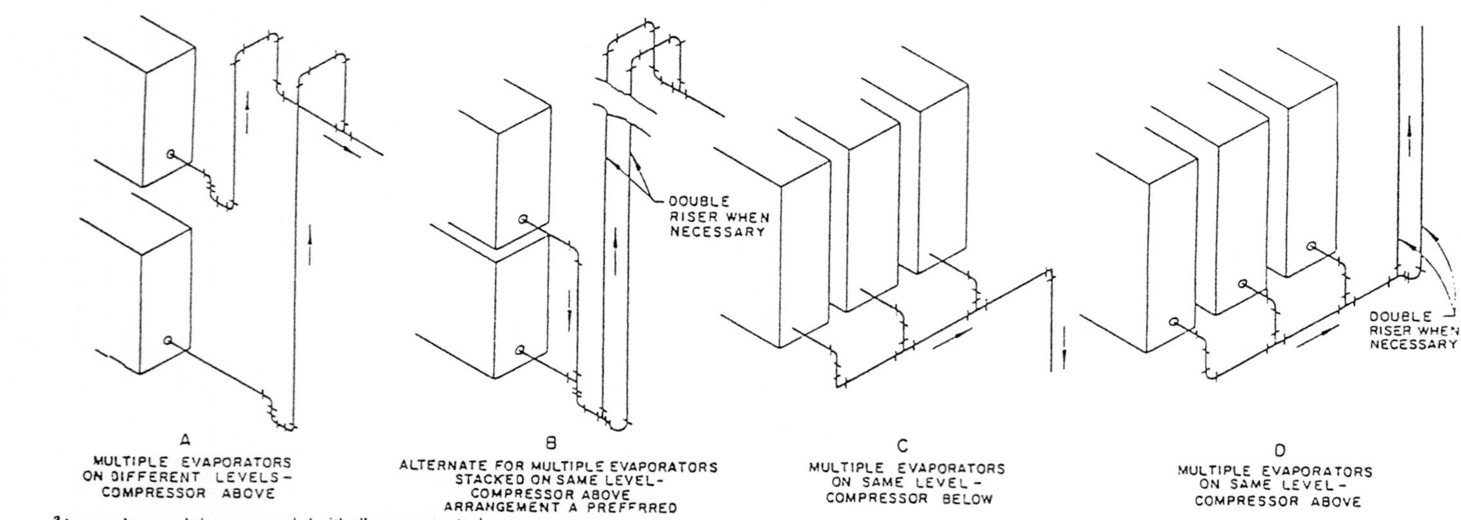

[a] A pumpdown cycle is recommended with all arrangements shown.

FIGURE C9.4 Arrangements of suction line loops.

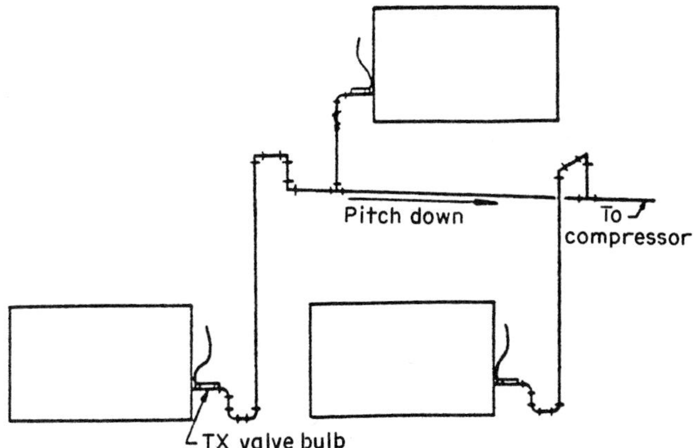

FIGURE C9.5 Typical piping from evaporators located above and below common suction line.

For liquid lines after the expansion device, it is usually customary to size these lines one size larger than the liquid line which leads to the expansion device. Commercially obtainable thermal-expansion valves which operate on superheat from the evaporator are usually made with the outlet one size larger than the inlet. In all cases, close coupling of the expansion device to the evaporator is recommended so that the liquid line after expansion will be as short as possible.

Condenser Drains. In the section on selection of line sizes, liquid lines were sized on the basis of about 100 fpm. When liquid lines run vertically downward from the bottom of a horizontal shell-and-tube or shell-and-coil condenser to a receiver and where full equalization of the condenser and receiver is required through the condensate drain line, it may be necessary to maintain velocities less than 100 fpm. Table C9.15 shows the maximum velocity based on full pipe area which should be allowed in vertical outlets from condensers to assure full equalization and to prevent gas binding. This table is equally applicable to pipe or copper tubing and is applicable to any refrigerant compatible with the materials, since it is based on the hydraulic effect of liquid flowing into vertical outlets. If the velocities exceed those shown in this table, separate vapor-equalizing lines should be used.

Marine Condensers. Refrigerant condensers for use on land usually have one single liquid outlet. Condensers intended for use on shipboard are usually mounted fore and aft to take advantage of the lower pitch angle. Outlets are provided at each end of the condenser. These are piped into a common header having a single outlet in the center. Each condenser outlet should be capable of handling the entire full capacity of the condenser.

Insulation. Since refrigerating systems are designed basically to produce temperatures below normal, many of the pipelines in a refrigeration system will be at temperatures below the dew point of the surrounding air. In addition to a heat gain from the surroundings, condensation, commonly called sweating, will form

TABLE C9.15 Maximum Velocity in Condenser Vertical Drains to Assure Equalization

Inside diameter, in.	Max. velocity, fpm
1	44.4
2	62.8
3	76.8
4	88.8
5	99.5
6	108.3
7	117.0
8	125.0
9	133.2
10	140.0
11	147.0
12	154.0

Table shows maximum velocity permitted in round vertical drains, based on full pipe area, to assure full equalization with receiver.

Based on the formula: Velocity (fpm) $= 44.4\sqrt{d}$ where d is in inches.

on these pipes. The condensation may be objectionable and even harmful. It is customary to insulate all refrigerant lines where there is a possibility of condensation on the refrigerant line.

The amount and type of insulation depend upon the operating-temperature level. Chapter B7 covers insulation in detail with recommendations for economical thicknesses, types, and application.

An important consideration in the application of any insulation to refrigerant piping is to assure that there is a vapor seal on the outside of the insulation. If the insulation type is such that moisture can enter it, the natural difference in vapor pressure between the surrounding atmosphere and the surface of the pipe will result in moisture migration into the insulation and eventually to the surface of the pipe. Vapor seals at all joints and on the outside of the insulation are essential to assure efficient performance and the avoidance of future trouble.

It is also necessary to avoid thermal bridges between the cold piping and the outside ambient air or atmosphere. Pipe hangers for supports should not contact the cold piping and should be arranged so that the supports bear against saddles of adequate area which are outside the insulation. Any metal in contact with the cold pipe wall, because of the higher conductivity of the metal hanger or rod, will cause condensation on the hangers.

MECHANICAL PIPING DESIGN

Code Requirements

The Safety Code for Mechanical Refrigeration, ANSI/ASHRAE 15, was written to assure the safe design, construction, installation, operation, and inspection of every refrigerating system using a volatile refrigerant. ANSI/ASHRAE 15 Standard covers refrigerant piping, valves, fittings, and related parts.

The ASME Code for Pressure Piping, Section B31.5,[2] is a standard covering the minimum requirements for materials, design, fabrication, assembly, test, and inspection of refrigeration piping, including limitations.

The ANSI/ASHRAE 15 and ANSI/ASME B31.5 Codes recognize that refrigeration equipment of the self-contained or unit type which has been designed in accordance with good practice and which has been submitted to an approved, nationally recognized testing laboratory which provides uniform testing and examination procedures and which has a follow-up inspection service of current production of such units will be construed as meeting the requirements of either of these codes.

The ANSI/ASME B31.5 Code also excludes water piping from consideration but does include secondary coolant piping.

Materials. Since volatile refrigerants have different chemical compositions, some materials are incompatible with certain refrigerants. Table C9.16 shows that several common materials are not compatible with certain of the refrigerants; it also indicates certain other limitations from a Code standpoint. It will be noted that cast-iron pipe is not permitted by the Pressure Piping Code for any volatile refrigerant; cast-iron valves and fittings of approved types are permitted.

Copper and brass are not compatible with ammonia. Aluminum, zinc, and magnesium are not suitable for use with methyl chloride. Zinc and magnesium are not suitable for use with any of the halogenated hydrocarbon refrigerants or ammonia.

In addition to the limitations and qualifications shown in Table C9.16, each of the eight listed refrigerants is classified according to the Underwriters' Laboratories classification and according to a group classification system covered in the ANSI/ASHRAE 15 Safety Code.

The ANSI/ASHRAE 15 classification of Group 1 includes refrigerants which are not considered toxic or inflammable in the ordinary sense (remember that many refrigerants can smother if present in heavy concentrations).

Group 2 in the ASHRAE classification covers toxic or flammable refrigerants, and Group 3 covers highly flammable or explosive refrigerants.

TABLE C9.16 Material Compatibility

	Refrigerant number							
Material	11	12	113	22	114	500	717	40
Carbon steel	S*	S	S	S	S	S	S	S
Wrought iron	S	S	S	S	S	S	S	S
Cast-iron pipe	NP*	NP	NP	NP	NP	NP	NP	NP
Copper or brass	S	S	S	S	S	S	NS	S
Aluminum	Q*	Q	Q	Q	Q	Q	Q	NS
Zinc	NS*	NS	NS	NS	NS	NS	NS	NS
Magnesium	NS	NS	NS	NS	NS	NS	NS	NS
ASHRAE 15-78 group	1	1	1	1	1	1	2	2
Underwriters Laboratory class	5a	6	4–5	5a	6	5a	2	4

*NP, not permitted by ASME B31.5 Code; NS, not satisfactory; Q, qualified—moist refrigerant may corrode—consult supplier; S, permissible.

Limitations. Table C9.17 shows the limitations of various types of materials and group classifications with respect to the service with refrigerants. It will be noted that steel pipe must be Sch 40 or heavier for use with volatile refrigerants with certain limitations with respect to size.

ANSI/ASME B31.5 no longer permits the use of butt-welded carbon steel. However, it does permit the use of listed electric-resistance welded pipe and tube.

Cast-iron pipe is not allowed for any volatile, flammable, or toxic refrigerant but may be used for water or nonvolatile brines. Cast iron is not allowed for temperatures below −150°F.

Copper or brass tubing may be used with any refrigerant with which it is compatible and in any size or pressure when selected by the design rules. If copper tubing is erected on the premises, it must be Type *K, L,* or ACR.

The ANSI/ASHRAE 15 Safety Code has certain requirements for institutional, public assembly, residential, and commercial occupancies. These rules prohibit the carrying of refrigerant piping through floors except that it may be carried from the basement to the first floor or from the top floor to a machinery penthouse to the roof.

Refrigerant piping may be connected to a condenser on the roof if it is carried through an approved rigid and tight, continuous, fire-resistant pipe duct or shaft having no openings on intermediate floors, or it may be carried on the outer wall of the building provided it is not located in an air shaft, closed court, or any other similar opening enclosed within the outer walls of the building.

For Group 1 refrigerants, the refrigerant piping may be carried through floors intermediate between the first floor and the top floor provided it is enclosed in an approved, rigid and tight, continuous fire-resisting pipe or shaft where it passes through intermediate spaces not served by the system. The piping of direct systems need not be enclosed where it passes through space served by that system. The pipe duct or shaft must be vented to the outside or to a space served by the system.

The ANSI/ASHRAE 15 Code further requires rigid or flexible metal enclosures for soft, annealed copper tubing used for refrigerant piping which contains other than Group 1 refrigerants.

Schedule 40 wrought-steel or wrought-iron pipe may be used for design working pressure not exceeding 300 psig provided lap-welded, electric-resistance-welded, or seamless pipe is used for nominal pipe sizes over 4 in.

For limitations on copper tubing, refer to ANSI/ASHRAE 15.

Joints on copper tubing containing Group 2 or Group 3 refrigerants as classified by the ANSI/ASHRAE 15 Code must be brazed. Solder joints are prohibited in such systems.

A brazed joint is obtained by the joining of metal parts with alloys which melt at temperatures higher than 800°F but less than the melting temperatures of the joined parts.

A soldered joint is obtained by the joining of metal parts with metallic mixtures or alloys which melt at temperatures below 800 and above 400°F.

Systems containing 100 lb of refrigerant with positive-displacement compressors should have stop valves on each inlet of each liquid receiver and on each branch liquid and suction line except on receivers which are in a condensing unit or which are an integral part of the condenser.

Refrigerant piping crossing an open space which affords a passage way in any building must be not less than 7½ ft above the floor unless against a ceiling of such space.

TABLE C9.17 Piping Limitations

Type or material	ASHRAE 15 group number	Line size (in)	Service	Limitation
Carbon steel or wrought iron	2 or 3	1½ or smaller	Refrigerant liquid	Sch 80 or heavier
	1	6 or smaller	Refrigerant liquid	Sch 40 or heavier
	2 or 3	2 through 6	Refrigerant liquid	Sch 40 or heavier
	1, 2, 3	6 or smaller	Refrigerant vapor	Sch 40 or heavier
Butt-welded carbon steel or wrought iron	Any	Any	Refrigerant liquid	Not permitted by Code
Cast-iron pipe	Any	Any	Refrigerant	Not permitted for refrigerants
Cast-iron pipe	—	—	Brine	Permitted above −150°F
Cast-iron pipe	—	—	Any	Not permitted below −150°F
Copper or brass	Any (except ammonia)	Any	Any (except ammonia)	Type K or L if erected on premises; soft annealed may not exceed 1⅜ in OD
			Ammonia	Not compatible
Aluminum, zinc, magnesium	—	—	Methyl chloride	Not compatible
Magnesium	—	—	Halogenated hydrocarbons	Not compatible
Threaded joints	3	1 and smaller	Any	Seal welded or braze
		See limitation	Group 3 fluids	Not allowed over 1 in
			Salt brines	Not allowed over 6 in
			Any	Not lighter than Sch 40 up through 6 in; not lighter than Sch 30 on 8, 10, or 12 in

Free passageway must not be obstructed by refrigerant piping. Refrigerant piping must not be placed in any elevator, dumb-waiter, or other shaft containing a moving object or any shaft which has openings to living quarters or to main exit hallways. Refrigerant piping is not to be placed in public hallways, lobbies, or stairways except that such refrigerant piping may pass along a public hallway if there are no joints in the section in the public hallway and provided that nonferrous tubing of 1⅛ in OD and smaller is contained in a rigid metal pipe.

Limitations on Threaded Joints. Threaded joints must be seal welded or brazed for Group 3 refrigerants. Threaded joints larger than 1 in in size should not be used for Group 3 fluids and must be no larger than 6 in in size for salt brine. Threaded joints must not be used on lighter than Sch 40 pipe up through 6 in in diameter or on lighter than Sch 30 pipe for 8-, 10-, or 12-in pipe.

Design Pressures

Minimum design pressures for piping for many refrigerants are included in the ANSI/ASME B31.5 Pressure Piping Code. These design pressures are based on good practice and are correlated with certain minimum field test pressures which are established on the basis of both ambient temperature considerations and existing practice on some of the older refrigerants.

Table C9.18 gives industry practice for minimum design pressures for field erected systems, originally based on ANSI/ASME B31.5. The refrigerant numbering system established by ANSI/ASHRAE 34 has been referred to earlier in this chapter. Table C9.18 also lists the refrigerant number and chemical formula for many of these refrigerants.

For refrigerants not listed in Table C9.18, the general rule is that the components shall be designed for an internal pressure representing the most severe conditions of coincident pressure and temperature expected in normal operation, including shutdown. Consideration must also be given to external design pressure. Refrigeration piping systems must be designed to resist collapse when the internal pressure is zero absolute and the external pressure is atmospheric. This is to permit drying the pipe by evacuation.

For secondary coolant service, the maximum pressure which may be encountered in operation should also be used as a guide in designing the system, considering the possibilities of shutoff head of any pumps, static head due to elevation or long vertical runs, water hammer, and thermal expansion of fluid when the lines are valved off. The ASHRAE and ANSI/ASME B31.5 Codes require that the refrigerant piping must be capable of withstanding certain leak field test pressures. Table C9.19 shows the minimum refrigerant leak field test pressures for systems designed in accordance with Table C9.18 and is based on ASA B31.5, but higher than those given in the current ANSI/ASME B31.5. The notes indicate the basis for determining certain of these test pressures together with a basis for determining test pressures for refrigerants other than those listed.

For limited-charge systems equipped with a pressure-relief device, the piping must be designed for a pressure not less than the setting of the pressure-relief device.

A limited-charge system in one in which, with a compressor idle, the internal volume and the total refrigerant charge are such that the design pressure will not be exceeded by the complete evaporation of the refrigerant charge.

TABLE C9.18 Minimum Design Gauge Pressures psig *,†,‡

Group†	Refrigerant	Name	Chemical formula	Low-pressure side	High side: Water or evaporation cooled	High side: Air cooled
1	R-11	Trichlorofluoromethane	CCl_3F	15	15	21
1	R-12	Dichlorodifluoromethane	CCl_2F_2	85	127	169
1	R-13	Chlorotrifluoromethane	$CClF_3$	521	547	547
1	R-13B1	Bromotrifluoromethane	$CBrF_3$	230	321	410
1	R-14	Tetrafluoromethane	CF_4	529	529	529
1	R-21	Dichlorofluoromethane	$CHCl_2F$	15	29	46
1	R-22	Chlorodifluoromethane	$CHClF_2$	144	211	278
1	R-30	Methylene chloride	Ch_2Cl_2	15	15	15
2	R-40	Methyl chloride	CH_3Cl	72	112	151
1	R-113	Trichlorotrifluoroethane	CCl_2FCClF_2	15	15	15
1	R-114	Dichlorotetrafluoroethane	$CClF_2CClF_2$	18	35	53
1	R-115	Chloropentafluoroethane	$CClF_2CF_3$	123	181	238
3	R-170	Ethane	C_2H_6	618	695	695
3	R-290	Propane	C_3H_8	129	188	244
1	R-C318	Octafluorocyclobutane	C_4F_8	34	59	85
1	R-500	Dichlorodifluoromethane, 73.8%, and ethylidene fluoride 26.2%	CCl_2F_2/CH_3CHF_2	102	153	203
1	R-502	Chlorodifluoromethane, 48.8%, and chloropentafluoroethane, 51.2%	$CHClF_2/CClF_2CF_3$	162	232	302
3	R-600	N-butane	C_4H_{10}	23	42	61
3	R-601	Isobutane	$CH(CH_3)_3$	39	63	88
2	R-611	Methyl formate	$HCOOCH_3$	15	15	15
2	R-717	Ammonia	NH_3	139	215	293
1	R-744	Carbon dioxide	CO_2	955	1058	1058
2	R-764	Sulfur dioxide	SO_2	45	78	115
3	R-1150	Ethylene	C_2H_4	732	732	732

*Selection of higher design pressure may be required to satisfy actual shipping, operating, or standby conditions.

†It shall be the responsibility of the owner to establish the refrigerant group for refrigerants used which are not listed above according to ANSI/ASHRAE 34.

‡To convert psig to kPa gauge, multiply psig by 6.895. Group Designation from ANSI/ASHRAE 34.

TABLE C9.19 Refrigerant Piping—Minimum Refrigerant Leak Field Test Pressures

ASHRAE group	Number	Name	Chemical formula	Minimum field refrigeration leak test pressure (psig) High side	Low side
2	717	Ammonia	NH_3	300	150
3	600	Butane	C_4H_{10}	95	50
1	744	Carbon dioxide	CO_2	1,500	1,000
1	12	Dichlorodifluoromethane	CCl_2F_2	235	140
1	500	Dichlorodifluoromethane 73.8% and ethylidene fluoride 26.2%	CCl_2F_2 / $CH_3—CHF_2$	285	150
2	1,130	Dichloroethylene	$C_2H_2Cl_2$	30	30
1	30	Dichloromethane (methylene chloride)	CH_2Cl_2	30	30
1	21	Dichloromonofluoromethane	$CHCl_2F$	70	40
1	114	Dichlorotetrafluoroethane	$C_2Cl_2F_4$	50	50
3	170	Ethane	C_2H_6	1,200	700
2	160	Ethyl chloride	C_2H_5Cl	60	50
3	1,150	Ethylene	C_2H_4	1,600	1,200
3	601	Isobutane	$(CH_3)_3CH$	130	70
2	40	Methyl chloride	CH_3Cl	210	120
2	611	Methyl formate	$HCOOCH_3$	50	50
1	22	Monochlorodifluoromethane	$CHClF_2$	300	150
1	13	Monochlorotrifluoromethane	$CClF_3$	685	685
3	290	Propane	C_3H_8	300	150
2	764	Sulfur dioxide	SO_2	170	85
1	11	Trichloromonofluoromethane	CCl_3F	20	20
1	113	Trichlorofluoroethane	$C_2Cl_3F_3$	20	20

Note 1: For refrigerants not listed in the table, the test pressure for the high-pressure side shall not be less than the saturated vapor pressure of the refrigerant at 150°F. The test pressure for the low-pressure side shall not be less than the saturated vapor pressure of the refrigerant at 110°F. However, the test pressure for either the high or low side need not exceed 125 percent of the critical pressure of the refrigerant. In no case shall the test pressure be less than 20 psig.

Note 2: When a compressor is used as a booster to obtain a low pressure and discharges into the suction line of another system, the booster compressor is considered a part of the low side, and values listed under the low-side column in the table shall be used for both high and low sides of the booster compressor provided that a low-pressure stage compressor of the positive-displacement type shall have a pressure-relief valve.

Note 3: In field testing systems using nonpositive-displacement compressors, the entire system shall be considered for field test purposes as the low-side pressure.

The *required thickness* of pipe or tubing is determined from the following equations and nomenclature:

$$t_m = t + c \tag{C9.5}$$

$$t = \frac{PD_o}{2(S + Py)} = \frac{Pd}{2(S + Py - P)} \tag{C9.6}$$

$$P = \frac{2St}{(D_o - 2yt)} \tag{C9.7}$$

where t_m = minimum required thickness (in)
c = allowance for grooves, threads, tolerances, corrosion, erosion
P = internal design pressure (psig)

D_o = outside diameter of pipe (in)
d = inside diameter of pipe (in)
S = allowable stress (psi)
t = calculated thickness

y is a material coefficient for which

Ductile nonferrous materials = 0.4

Ferritic steels = 0.4

Austenitic steels = 0.4

The design of piping for external pressure involves the use of charts to determine factors which are used to calculate the thickness or allowable working pressure. The method and charts referred to in paragraphs UG 28 and UG 31 of Section VIII of the ASME Boiler and Pressure Vessel Code[3] are acceptable for design of pipes and tubes subject to external pressure.

The Pressure Piping Code recognizes and permits the use of the design rules of the ASME Unfired Pressure Vessel Code for closures, flanges, and blind flanges. For blanks, the following equation should be used:

$$t = d_g\sqrt{3P/16S}$$

where t = required thickness (in)
d_g = inside diameter of gasket for raised or flat-face flanges or the pitch diameter of retained gasket flanges (in)
P = internal or external design pressure (psig)
S = allowable stress (psi)

Since the Pressure Piping Code[2] and the Safety Code for Mechanical Refrigeration both limit the minimum thickness of steel pipe and the minimum thickness for copper tubing for erection on the premises, it is possible to calculate the maximum working pressure for these commonly used weights of pipe or tubing.

Table C9.20 shows the maximum allowable internal working pressure for seamless steel pipe in the permitted schedule numbers. The maximum allowable external pressure has also been calculated for a length-over-diameter ratio in excess of 15 which will be common in most piping systems. It will be evident that the allowable working pressure for the permitted thicknesses of pipe is usually far in excess of that required by the design working pressure requirements for the various refrigerants. In most cases except in unusual circumstances, in cases where shock may be anticipated, or in the larger sizes, no further checking of allowable working pressure will be necessary.

Table C9.21 shows rated internal working pressures for type ACR copper tubing normally used for air conditioning and refrigeration service in the field. Ratings are given for both drawn (hard temper) and annealed tubing.

The ANSI/ASME B31.5 Pressure Piping Code establishes certain allowable working stresses for many grades of pipe and tubing in various materials. These allowable stresses are to be used in conjunction with the design equations listed above for special calculations.

Low-Temperature Design Criteria. It is recognized that certain materials tend to become brittle at low temperatures and may be subject to failure which would not occur normally at usual temperatures or at elevated temperatures. The transition

TABLE C9.20 Allowable Working Pressures for Carbon Steel Refrigerant Piping

Nominal size, in.	Schedule no.	Allowable internal working pressure, psig	Allowable external working pressure, psig
1/8	40	1,890	2,070
	80	3,510	2,860
1/4	40	1,490	2,000
	80	2,880	2,700
3/8	40	1,300	1,660
	80	2,500	2,280
1/2	40	1,126	1,580
	80	2,210	2,140
3/4	40	994	1,320
	80	1,890	1,800
1	40	866	1,210
	80	1,680	1,670
1 1/4	40	773	980
	80	1,470	1,410
1 1/2	40	740	890
	80	1,390	1,270
2	40	670	750
2 1/2	40	665	810
3	40	624	700
3 1/2	40	600	640
4	40	580	580
6	40	534	450
8	40	515	390
10	40	496	340
12	STD.	436	260

For internal pressure:
Based on minimum wall thickness; no corrosion or erosion allowance: thread allowance factor from ANSI/ASME B.1.201
Allowable stress = 12,000 psi
y (material coefficient) = 0.4
$P = 2St/(D_0 - 2yt)$.
For external pressure:
Based on minimum wall thickness: no corrosion, erosion, threading or grooving allowance
Yield: 24,000 psi to 30,000 psi
L/D_0 = 15 or greater where D_0 = outside pipe diameter in inches; L = maximum straight length of run between flanges, elbows, caps or stiffening rings (in)

temperature at which certain materials become brittle is not well defined. Some ferrous materials may pass through the transition range at normal temperatures, while others may not become brittle until quite low temperatures are attained. The ASME Unfired Pressure Vessel Code[3] arbitrarily establishes a temperature of −20°F as a point below which all vessels constructed of carbon or low-alloy steels should be impact tested, with certain exemptions.

Refrigeration piping is frequently subject to temperatures below normal atmospheric temperatures to the degree that embrittlement may occur, and the ASME B31.5 Piping Code also requires impact tests on certain materials subject to temperature below −20°F. There are certain materials and certain conditions under which impact tests are not required. The exemptions are as follows[2]:

1. No impact tests are required for aluminum, austenitic stainless steel in Grades 304, CF8, 304L, CF3, 316, CF8M or 321 or copper, red brass, copper nickel alloys, or nickel copper alloys.

TABLE C9.21 Rated Internal Working Pressures (psi) for Copper Tube Type ACR*

Size and wall thickness (in)	Rated internal working pressure (psi)						
	100°F		200°F		300°F		400°F
	Annealed	Drawn	Annealed	Drawn	Annealed	Drawn	Annealed or drawn
1/8 (.030)	3130	—	3090	—	2620	—	1310
3/16 (.030)	1990	—	1950	—	1650	—	820
1/4 (.030)	1450	—	1420	—	1200	—	600
5/16 (.032)	1230	—	1200	—	1020	—	510
3/8 (.030)	900	1350	880	1300	740	1180	370
3/8 (.032)	1010	—	990	—	840	—	420
1/2 (.032)	740	—	730	—	610	—	300
1/2 (.035)	800	1200	780	1150	660	1060	330
5/8 (.035)	640	—	630	—	530	—	260
5/8 (.040)	740	1110	720	1060	610	980	300
3/4 (.042)	650	980	630	930	530	850	260
7/8 (.045)	590	890	570	840	480	770	240
1 1/8 (.050)	510	770	490	720	420	670	210
1 3/8 (.055)	460	690	440	650	370	590	180
1 5/8 (.060)	430	650	410	600	350	560	170
2 1/8 (.070)	370	560	360	530	300	480	150
2 5/8 (.080)	350	530	340	500	280	450	140
3 1/8 (.090)	330	500	320	470	270	430	130
3 5/8 (.100)	320	480	300	440	260	420	130
4 1/8 (.110)	300	450	290	430	240	380	120

*Based on *S* values as follows: 100°F—6,000 psi, annealed; 9,000 psi, drawn; 200°F—5,900 psi; annealed, 8,700 psi drawn; 300°F—5,000 psi, annealed, 8,000 psi, drawn; 400°F—2,500 psi, annealed or drawn, according to American National Standard Code for Pressure Piping, Refrigeration Piping, ANSI B31.5.

Source: Copper Development Association, Inc.

2. No impact tests are required for bolting material conforming with A193, Grade B7 for use at temperatures above −50°F.
3. No impact tests are required for bolting materials conforming with A320, Grades L7, L10, and L43 at temperatures above −150°F or above −225°F for A320, Grade L9.
4. No impact test is required for material used in fabricating a piping system for metal temperatures between −20 and −150°F when the most severe condition of pressure (internal if above atmospheric and external if below atmospheric) does not produce a stress exceeding 40 percent of the allowable material stress.

For low-temperature application, the use of nonferrous materials or the stainless steel mentioned will normally be satisfactory. The use of nickel-steel pipe in conjunction with the use of nickel-steel pressure vessels has long been an acceptable material for low-temperature when these materials are subjected to and pass the impact-testing requirements.

Impact tests, when conducted, shall follow the requirements of ANSI/ASME B31.5, Paragraph 523.2.2.

The standard 10- by 10-mm specimen is used if the thickness of the material being tested is 7⁄16 in or greater. For material that is not of sufficient thickness to permit preparation of full-size specimens, tests may be made on the largest possible of the subsize specimens listed below.

The impact properties for each size specimen are as listed below.

Size of specimen (mm)	Minimum impact value required (ft · lb)
10 × 10	15
10 × 7.5	12.5
10 × 5	10
10 × 2.5	5

In welded fabrication, the weld also is required to meet the impact-test requirement.

Expansion and Contraction. Since refrigeration piping systems are subject to changes in temperature, some precautions must be taken to assure that these changes in temperature during operation or during shutdown are considered in the design of the piping and in the design of supports and flexibility.

Piping systems must be designed to have sufficient flexibility to prevent thermal expansion from causing

1. Failure of piping or anchors from overstress or overstrain
2. Leakage at joints
3. Detrimental distortion of connected equipment resulting from excessive thrusts or moments

Expansion strains are usually taken up by bending or torsion or by compression and tension. The concentration of stresses will be different in each case.

Bending or torsional flexibility may be provided by the use of bends, loops, or offsets. While swivel joints, ball joints, and corrugated expansion joints are recognized by the Pressure Piping Code, some of these are not considered desirable for volatile refrigerant piping. Bends, loops, and offsets are generally used to provide flexibility. Loops and cold springing also may be used in the design of piping. Chapter B4 covers in detail the general design considerations involved in expansion and flexibility of piping. These same principles must be applied to refrigeration piping, and the maximum temperature cycle involved in the installation should be taken into account in determining the nature and direction of the stresses which may be caused by temperature effects. As mentioned above under "Insulation," pipe hangers or supports for low-temperature piping normally will not be in direct contact with the metal portion of the piping. Consideration must be given to the insulation in types of support which are peculiar to refrigeration piping when the problems of expansion and flexibility are considered.

Miter Joints. The ANSI/ASME B31.5 Pressure Piping Code gives details for design of branch connections where the angle between the axes of the branch and of the run is between 45 and 90°. Branch connections less than 45° impose special design and fabrication problems. The Code permits connections to be made by the use of tees and welding outlet fittings such as cast or forged nozzles or couplings, or by attaching a branch pipe directly to the run pipe by welding.

Normally the use of standard forged fittings of the butt-welding or socket types will provide sufficient strength in the case of a branch connection to permit application of these fittings without additional reinforcement. However, when a branch connection is welded into a hole cut into the main-run pipe, it is recognized that certain reinforcement may be required. The analysis of the extent of reinforcement, if required, is similar to that used on unfired pressure vessels for nozzle connections. The complete detailed analysis for such determinations and the means of determining the amount of reinforcement required are shown in the ANSI/ASME B31.5 Pressure Piping Code or in the ASME Unfired Pressure Vessel Code. The general method is to calculate the amount of metal cut out of the pipe and to calculate the amount of metal which is added by extra thickness of the pipe wall over that required for strength, the extra metal in the nozzle or branch connection, and the extra metal which would be added within certain limiting geometric zones by welding. If the added metal provides the equivalent of the amount of metal which had been cut out, no additional reinforcement is necessary. Reinforcement metal usually is provided in the form of a ring or a saddle which is welded to the run pipe. This reinforcement material must be added within certain limiting dimensions as defined by the Code. The use of ribs, gussets, or clamps is permissible to stiffen the branch connection, but their areas cannot be counted as contributing to the reinforcement area.

Welding, Brazing, and Soldering. Joints in piping which is to convey volatile refrigerants are usually made by welding, brazing, or soldering. This does not exclude the use of flanged connections, which are commonly used to connect valves or control devices in refrigeration piping. Flanged connections are commonly used to connect to pressure vessels and to compressors. Couplings of the friction type with seal rings may be used for refrigerants when the materials are compatible with the refrigerant and when the pressures permit. Use of such fittings is normally confined to low-pressure refrigerants. Limitations on threaded connections have already been listed and frequently threaded connections, where used and where permissible, will be seal welded.

Welded joints may be used in any materials for which it is possible to qualify the welding procedures, the welders, and the welding operators.

Butt welds are permitted. Usually, backing rings are used in butt-welded joints, but where it is necessary to have a smooth interior surface or where the backing ring may result in severe corrosion or erosion, the joint may be welded without backing rings provided the piping is suitably cleaned. Socket welds are permitted under the Pressure Piping Code. The Pressure Piping Code defines in detail the required weld sizes and joint arrangements which are recommended for use in welded-part construction.

The ASME Boiler and Pressure Vessel Code, Section IX, defines in detail the qualification of welding procedures and welders' performance requirements for unfired pressure vessel construction. These rules have been adopted by the Pressure Piping Code ANSI/ASME B31.5 and form part of the requirements of that Code. The welders and the procedures should be qualified to assure that their quality is in conformance with these codes.

The ANSI/ASHRAE 15 Safety Code for Mechanical Refrigeration requires brazing of certain joints. Also certain joints in restricted areas may require high-melting-point filler material. The ANSI/ASME B31.5 Pressure Piping Code defines the filler metal used in brazing to be nonferrous metal or alloy having a melting point above 800°F but below that of the metal being joined. Good practice in cleanliness of joints and the use of proper fluxes is required for brazed joints.

Brazing procedures and operators, except for sockettype capillary joints, should be qualified in accordance with the requirements of Section IX of the ASME Boiler and Pressure Vessel Code.

For soldered joints, the ANSI/ASME B31.5 Code defines the solder metal to be a nonferrous metal or alloy having a melting point below 800°F and below that of the metal being joined. Good soldering technique requires proper cleanliness and preparation of the joints, proper joint clearances, and proper heating. Procedures to be used on soldering or brazing sockettype joints are outlined in a publication of the Copper Development Association, Publication 25[5].

Miscellaneous Considerations

Corrosion. The ANSI/ASME B31.5 Pressure Piping Code recognizes that corrosion or erosion may be factors to be considered in piping design. When corrosion or erosion is expected, an increase in wall thickness of the components above that dictated by other design requirements is to be provided consistent with the expected life of the particular piping involved. In the basic equation for calculating the pressure ratings of pipe or in determining the required wall thickness of pipe, the Pressure Piping Code requires the addition of a factor to the calculated wall thickness to result in the actual thickness required. The factor includes allowance for threading, groove depth, and manufacturers' minus tolerance plus corrosion and erosion allowances.

Corrosion allowance on the inside of piping for volatile refrigerants is not mandatory. The refrigerant is recirculating and is usually charged into the system in a commercially pure state after thorough cleaning and evacuation of the entire system. When installed in accordance with good practice, a leaktight refrigerating system will not tend to corrode and it is not customary to add corrosion allowances. It is possible, with certain of the halogenated hydrocarbon refrigerants when contaminated with noncondensible gases or with water which may leak into a system under vacuum, to have corrosive products form. On some occasions, in hermetically sealed refrigeration systems, compressor motor burnouts have resulted in formation of contaminants which also may be damaging to the inside of the system. However, these considerations are not properly part of the piping design and are usually the result of carelessness or misapplication. It is ordinarily not necessary to add corrosion allowance to volatile-refrigerant piping.

For secondary coolant piping, especially with salt brines, the consideration of possible corrosion should be kept in mind in the design of the piping system. Ordinarily standard-weight pipe for either volatile-refrigerant use or for brine piping inherently has sufficient strength so that the normal wall thickness of pipes which are used are much heavier than are required for the actual pressure service, and therefore it may not be necessary to add additional allowances for corrosion.

Fittings. The Pressure Piping Code permits the use of standard fittings provided they are compatible with the refrigerant or fluid. The standard ratings of forged steel flanges, fittings, and similar parts may be used for refrigerant service.

Bell-and-spigot fittings may be used only for water and drainage service.

Couplings made of cast, malleable, or wrought iron may not be used on pipe containing flammable or toxic fluids. Wrought-iron couplings are subject to the same limitations in temperature, stress, and service which apply to cast-iron screwed fittings.

Valves. Cast-iron gate valves and plug cocks must not be used in liquid-refrigerant lines unless consideration is given to the expansion of liquid trapped in a space when the valve is closed.

Several manufacturers make standard lines of refrigeration fittings which do not fall into the classification of ANSI Standards for forged-steel valves. These valves and fittings over long years of usage have gained acceptability and are widely used and acceptable for refrigeration service to the degree recommended by the manufacturer.

Other Factors. The Pressure Piping Code lists the following dynamic effects which should be taken into account in the design of refrigerant piping.

1. Impact forces (including hydraulic shock) caused by either external or internal conditions.
2. The effect of wind loading on exposed piping.
3. Piping systems located in regions where earthquakes are a factor are to be designated for a horizontal force in conformity with the good engineering practice using governmental data as a guide in determining the earthquake force. However, this force is not to be considered as acting concurrently with lateral wind force.
4. Piping shall be arranged and supported with consideration for vibration.

The Pressure Piping Code also calls attention to the following weight effects which should be taken into account in the design of piping:

1. Live loads such as the weight of the fluid transported and snow and ice loads if the latter will be encountered. If low-temperature piping is not insulated, there can be a buildup of ice on the pipe even in high ambient temperatures.
2. Dead loads, consisting of the weight of the piping components and insulation and other superimposed loads.
3. Test loads which consist of the weight of the test fluid in the pipe.

REFERENCES

1. ASHRAE Standard Safety Code for Mechanical Refrigeration, ANSI/ASHRAE 15, American Society of Heating, Refrigerating and Air Conditioning Engineers, Inc., 1791 Tullie Circle NE, Atlanta, GA 30329.
2. Refrigeration Piping, ANSI/ASME B31.5, American Society of Mechanical Engineers, New York.
3. ASME Boiler and Pressure Vessel Code, Section VIII, Unfired Pressure Vessels, American Society of Mechanical Engineers, New York.
4. ASHRAE Handbook—1989. Fundamentals. American Society of Heating, Refrigerating and Air Conditioning Engineers, Inc., Atlanta, Ga.
5. "Soldered and Brazed Joints in Copper Tube," Publication No. 25, Copper Development Association, Inc., Greenwich, Conn.

CHAPTER C10
TOXIC AND HAZARDOUS SYSTEMS PIPING

Richard C. Getz, P.E.
Chief Piping Engineer
United Engineers & Constructors, Inc.
Philadelphia, Pennsylvania

Modern processing, manufacturing, and energy-generation facilities contain extensive quantities of pressurized pipelines. The fluids conveyed may be flammable, toxic, reactive, or corrosive (or a combination of these) to the degree that would cause harm to people and/or property if released and therefore are considered hazardous even if they are innocuous in their ambient state.

The selection of materials, the types of components, and the construction and support methods are extremely important in the design of these piping systems. Additionally, all of the piping design criteria are important for the establishment of maintenance and preventative maintenance programs. To these ends, identification of the hazards associated with the fluids to be handled is the most important step in the design process.

Various methods may be used to identify these hazards. The Right-to-Know Act extends to all people who have a need to know the properties of any chemicals; hence there can be no reasons for ignorance of potential hazards during the design phase. It is incumbent upon the owner to ensure that all knowledge of the expected contents of each piping system is conveyed to the designer, and it is the designer's responsibility to be sure that all of the pertinent information has been received and used.

During this analytical phase other factors can be brought into the study, such as

How has this material been handled in the past?

Has it been handled successfully?

Although yesterday's successes may not guarantee tomorrow's suitability, some useful history of the substance can provide meaningful information. Bad experiences, as well as positive experiences, must be considered to achieve the best possible design.

DEFINITIONS

Some of the terms or phrases used in this chapter are defined below. Other common piping related terms are defined in Chap. A1 of this handbook.

Flammability is the degree of susceptibility to ignition, combustion, or explosion of a material under specific environmental conditions.

Hazard level is an evaluation of a particular material with respect to its inherent dangerous properties, which include but are not limited to toxicity, flammability, and reactivity.

Hazard review is a comprehensive review designed to identify potential hazards and to produce specific recommendations that will reduce the probability of their occurrence and/or its consequences.

Lethal is used to describe a material or substance that is life hazardous or having a very high hazard level.

Life hazardous is when a life-threatening condition exists due to the toxicity, flammability, or reactivity of a material.

Potentially lethal is when life hazardous conditions exist under abnormal or emergency conditions only.

Reactivity is the susceptibility of materials to release energy either by themselves or in combination with other materials.

Safeguarding is the provision of protective and/or preventative measures that reduce the probability of life hazardous and/or potentially lethal conditions.

Toxicity is the ability of a chemical molecule or compound to produce injury once it reaches a susceptible site in or on the body.

Included herein are definitions from various industry consensus standards:

- From ANSI/ASME B31.3, Chemical Plant and Petroleum Refinery Piping. *Category M fluid service* is a fluid service in which the potential for personnel exposure is judged to be significant and in which a single exposure to a very small quantity of a toxic fluid, caused by leakage, can produce serious irreversible harm to people on breathing or bodily contact, even when prompt restorative measures are taken.
- *Fluid service* is a general term concerning the application of a piping system, considering the combination of fluid properties, operating conditions, and other factors which establish the basis for design of the piping system.
- From ANSI/ASME Boiler and Pressure Vessel Code, Section VIII, Division 1, Rules for Construction of Pressure Vessels. *Lethal substances* mean poisonous gases or liquids such that a very small amount of the gas or vapor of the liquid mixed or unmixed with air is dangerous to life when inhaled.

CODES AND STANDARDS

Construction codes mandate minimum requirements necessary to provide for public safety and property protection. The Codes use cautionary verbiage to alert the owner and designer that not all combinations of events can be predicted and the owner and designer must use their skills and knowledge when confronted with potentially hazardous circumstances.

The piping codes do not list hazardous substances; therefore, the owner and designer must identify these potentially harmful streams. This task may appear difficult to accomplish; however, there is sufficient information available, and consensus codes and standards provide assistance to one confronted with the challenges of designing and constructing a piping system for handling hazardous or toxic fluids.

One such code is ANSI/ASME B31.3, Chemical Plant and Petroleum Refinery Piping. This document provides the user with both a method for determination of hazardous and toxic fluid services and rules for the selection, design, fabrication, inspection, and testing of these unique systems. (Codes: The jurisdiction and the scope are discussed in Chap. A4.)

The following is a list of pertinent codes, standards, and regulations relevant to the construction of hazardous and toxic piping systems:

- U.S. Nuclear Regulatory Commission, Regulatory Guide 1.143

  Design Guidance for Radioactive Waste Management Systems, Structures and Components Installed in Light-Water-Cooled Nuclear Power Plants

- American National Standards Institute (ANSI)

  ANSI/ASME B31.1, Code for Pressure Piping-Power Piping

  ANSI/ASME B31.3, Chemical Plant and Petroleum Refinery Piping

  ANSI Z9.2, Fundamentals Governing the Design and Operation of Local Exhaust Systems

  ANSI Z37.1-Z37.8, Z37.19, Z37.2-Z37.29, Z37.31-Z37.39 (acceptable concentration levels of various toxic dusts and gases)

- American Society of Mechanical Engineers (ASME)

  ANSI/ASME Boiler and Pressure Vessel Code, Section VIII, Division 1, Rules for Construction of Pressure Vessels

- Code of Federal Regulations (CFR)

  Title 29, CFR, Chapter XVII, Occupational Safety and Health Administration (OSHA)

  Title 40, CFR, Chapter I, Environmental Protection Agency (EPA)

  Title 49, CFR, Chapter I, Research and Special Programs Administration Department of Transportation

- Chlorine Institute Pamphlet 6, Piping Systems for Dry Chlorine
- NIOSH, Registry of Toxic Effects of Chemical Substances
- National Fire Protection Association

  ANSI/NFPA, Standard on Explosion Prevention Systems

  ANSI/NFPA 91, Standard for the Installation of Blower and Exhaust Systems for Dusts, Stock, and Vapor Removal or Conveying

  NFPA 491M, Manual of Hazardous Chemical Reactions

  NFPA 101, Life Safety Code

  NFPA 30, Flammable and Combustible Liquids Code

NFPA 49, Hazardous Chemical Data

NFPA 325M, Fire Hazard Properties of Flammable Liquids, Gases, and Volatile Solids

- Industrial Ventilation Manual, American Conference of Governmental Hygienists

The following are miscellaneous publications. Although they do not have the same jurisdiction as codes and consensus standards, they can provide the user and designer with a valuable tool in their efforts toward identification of hazardous and toxic fluids.

- Accident Prevention Manual for Industrial Operations, National Safety Council
- American Insurance Association (Chemical Hazards Bulletins, Research Reports)
- Fire Protection Handbook, National Fire Protection Association
- Dangerous Properties of Industrial Materials, N. Irving Sax, Reinhold Publishing Corporation
- Handbook of Dangerous Materials, N. Irving Sax, Reinhold Publishing Corporation
- Loss Prevention Data Books, Factory Mutual System
- Industrial Hygiene and Toxicology, F. A. Patty, Interscience Publishers
- Corrosion Data Survey, National Association of Corrosion Engineers

DESIGN CONSIDERATIONS

Identification of Toxic and Hazardous Systems

Various methods are used to identify piping systems that may be conveying hazardous and toxic fluids. Although it is beyond the scope of this chapter to describe in detail the different hazard review techniques, we can present some basic objectives common to these techniques:

- Identify the hazards.
- Evaluate the frequency of occurrence and consequences of a postulated failure.
- Develop specific recommendations to eliminate or control the identified hazards.

Some of the common industry hazard analysis techniques are referred to as

- What if
- Fault-tree analysis
- Hazop
- How can

Typical hazard reviews are conducted at various stages of a project:

- A preliminary hazard review is conducted early in the project. Its purpose is to highlight major potential hazards and define those areas where further study is needed.

- The design hazard review is performed after the process design has been completed. Its purpose is to ensure the process hazards have been identified and necessary corrective recommendations have been implemented.
- A design verification review is the final review during the design phase. Its purpose is to ensure that all items relating to previous hazard reviews have been satisfactorily resolved.
- Operational readiness and safety audits are conducted prior to plant start-up. Its purpose is to ensure that the design measures established as a result of the various hazards' reviews have been incorporated into the construction and operation of the facility.

Whatever technique is chosen, it should be used by experts in that field, and their recommendations should be followed to reduce probabilities of occurrence of hazardous conditions and situations, thereby increasing the reliability of the piping system.

After the identification of those systems that represent the potential for hazardous and toxic service conditions, precautions should be taken to segregate them adequately from the less severe systems. The early identification and segregation will ensure that the unique aspects of the hazardous and toxic system will be addressed throughout the design, construction, and operation.

Factors other than steady-state pressure-temperature parameters must also be taken into account. These factors should include but not be limited to the following:

- Temperature and/or pressure excursions, whether as part of the normal operating mode or as the result of an operating upset condition. Even short-term temperature excursions for materials operating in the creep range can seriously reduce expected service life.
- Anticipated number of cycles, either full or partial, and length of cycle duration. A continuous process would have less cycles than a batchtype process during the life-cycle of the operating plant.
- Corrosiveness of the hazardous and toxic fluid to the containment materials of construction.
- Determine whether there is anything in the expected design that may induce thermal gradients. An example is a heavy wall pipe that is partially insulated or an uninsulated warm and hot line subject to environmental elements such as a heavy rain.
- Effects of a water hammer condition should be addressed; two-phase flow can induce hydraulic loads.
- Dynamic loading can be expected when piping systems are joined to reciprocating and/or rotating machinery. High-amplitude and high-cycle vibration are detrimental to material fatigue life and serviceability of mechanical joints.
- Seismic loadings and their dynamic effect on piping and components must be addressed where applicable.
- Thermal stresses from system expansion or contraction must be considered.

Although it is difficult to anticipate all events that could occur singularly or in combination, most of the previously mentioned factors can be determined early in the design stage via a hazards review. ANSI/ASME B31.3, Chemical Plant and Petroleum Refinery Piping, code provides a diagram to help determine whether a

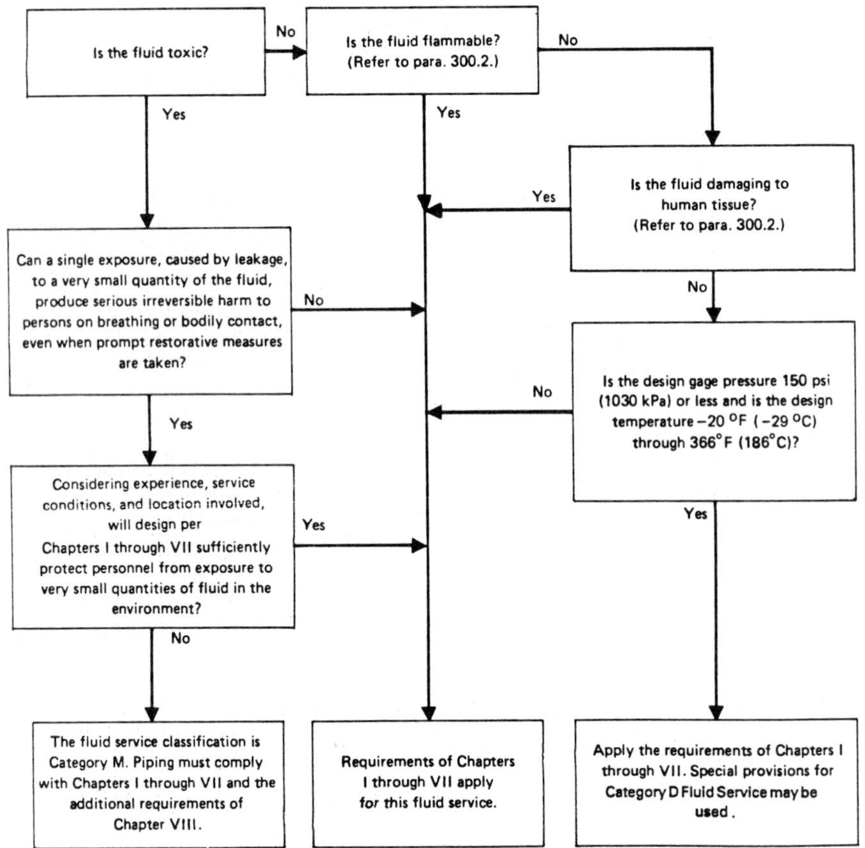

FIGURE C10.1 Guide to classifying fluid services. (Appendix M of ASME/ANSI B31.3.)

particular fluid service should be classified category M fluid service (see Fig. C10.1). When in doubt, a conservative approach would be to adopt the provisions of category M fluid service without referring to it as such.

Materials Selection and Limitation

The selection process for the piping system materials of construction will necessarily become more rigorous for a system conveying hazardous and toxic fluids. The selection process will entail the usual criteria of flowing media characteristics and temperature and additional considerations of some other load cases as the result of a hazards analysis.

Perhaps the worst-case scenario for a piping system would be the catastrophic type failure of a sudden rupture. The designer should ensure that the basic materials of construction are not susceptible to that mode of failure.

The use of low-ductility materials, such as cast iron and glass, should be avoided; their demonstrated brittle behavior and sensitivity to thermal and mechanical shock loadings limit their serviceability and range of use. Low-ductility materials are susceptible to brittle failure from thermal shock when exposed to fire or fire-fighting measures.[1]

Material selection should take into account the suitability of the piping to resist deterioration in service. Information can be obtained for determining material performance in various corrosive environments.[2,3] This information usually provides either qualitative (acceptable or not acceptable) or quantitative data. If stated quantitatively, the data will be expressed as a uniform corrosive rate (mils per year). This information can be used to establish a realistic corrosion allowance or to select a construction material that will not suffer corrosion penetration during the expected service duration. The latter approach is preferable, since data provided are expressed as a uniform rate and cannot account for localized points of accelerated corrosion and, consequently, reduced service life.

The designer should be cautioned that in actual service corrosion will probably not occur uniformly throughout the piping system and all its components. Corrosion is usually accelerated at crevices, under backing rings, in threaded joints, in socket-welded joints, at weld metal areas and heat-affected zones, or in other stagnant, low flow areas of the piping system. Because a corroding pipeline may have large areas of advanced stages of local corrosion, the designer would be well advised to select a material that will not corrode in the anticipated service. For example, liquid fluorine can be contained in a number of material types. However, data[3] suggest that carbon steel will suffer significant yearly corrosive penetration. This can be managed by providing substantial allowances for corrosion in the pipe wall thickness. However, as previously discussed, a material can be selected that will not suffer the same service degradation of carbon steel and can provide a greater assurance of service reliability.

Material selection for all combinations of hazard and toxic fluid piping systems is beyond the scope of this chapter. Numerous sources can provide meaningful data for use in the materials selection process. However, prudent practice suggests that a material choice for a hazardous and toxic fluid piping system would clearly be a choice where the chosen materials of construction would not be subjected to the deleterious effects of corrosion.

Material selection should also consider the potential adverse effects of dissimilar materials in contact. Galvanic action at dissimilar joints can promote accelerated corrosion where none had been anticipated.

Time-dependent material properties must be addressed during the material selection process. Under certain operating conditions and over a period of time, various construction materials lose ductility, decrease in strength, continue to deform (creep), or otherwise deteriorate. The designer must be conscious of the time-dependent behavior of a candidate material being considered for use in a hazardous and toxic fluid piping system.

The user of thermoplastic materials for pipeline components (i.e., gasket and packing) must recognize the possibility of material loss during a fire. An otherwise serviceable component may become a significant leak path in the aftermath of a local fire that may have destroyed the thermoplastic item.

Joints and Joining

Piping joints must be suitable for the pressure-temperature conditions, and the designer must be ultrasensitive to the nature of the fluid when it is hazardous and toxic. Joints can be categorized into two broad classes: welded joints and mechanical joints.

Welded Joints. Probably one of the most common methods of pipe joining is the use of a butt weld (Fig. C10.2). In hazardous and toxic systems this type of joint offers the following advantages:

- A butt-weld joint does not produce a local stress riser as do some other types of welded joints. A butt-weld joint has good fatigue resistance.
- A butt-weld joint can be readily examined by most conventional nondestructive techniques.
- Butt-weld joints should be made without the use of backing rings and inserts, eliminating a crevice that would be detrimental in corrosive, dynamic, or cyclic operation.
- Field welding of butt joints is simpler than other, more configurated joints (i.e., nozzle welds). This consideration alone can improve the resulting pipeline serviceability in hazardous and toxic service.
- The resulting profile of the properly produced butt weld lends itself more readily to a uniform distribution of any external loadings, and the uniformity of thickness through the joint reduces or eliminates peak stresses produced by thermal gradients.

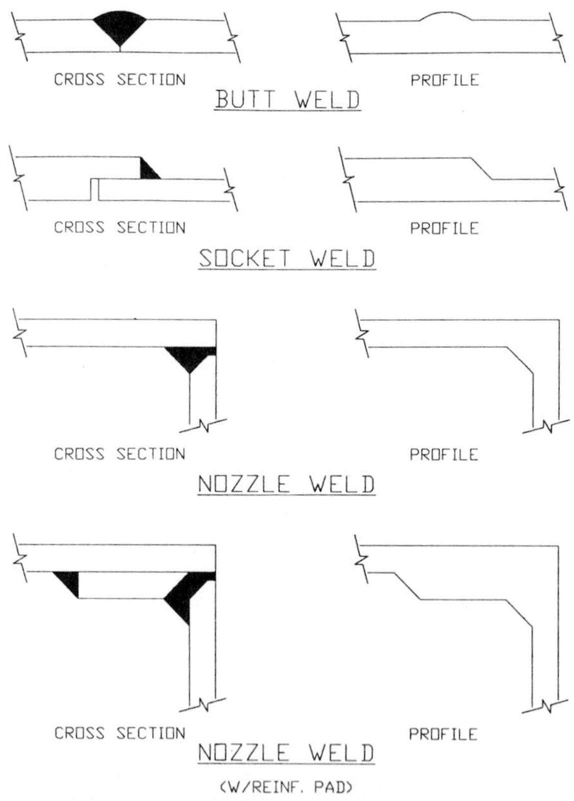

FIGURE C10.2 Welded joints.

Socket-weld joints are used extensively in smaller (2 in and less) pipe connections. Although they have performed satisfactorily in demanding service, they have some disadvantages.

- The geometry of the joint produces a crevice where a corrodent may accelerate attack (see Fig. C10.2).
- Socket-weld joints are a stress riser, which must be considered in view of external loadings (i.e., thermal) and other loadings that would adversely affect cyclic life of the system.
- Profile of a completed socket-weld joint illustrates the varying cross-sectional thicknesses of material, leading to a nonuniformity of stress distribution that will intensify peak stresses produced during heatup and cooldown cycles.
- Examination of a socket-weld joint is limited to visual and surfacetype examination. The varying cross section through the joint does not lend itself to subsurface examination.

Nozzle type (stub-in) welds should be avoided in hazardous toxic piping systems. When making a branch in a header pipe, the preferred methods are butt-weld fittings (i.e., tees), integrally reinforced outlet fittings, or similar items that produce a smooth geometrical transition between branch and run.

The nozzle weld contains the following disadvantages when used in hazardous and toxic service:

- The joint geometry produces a very high stress riser which in turn leads to a significant lowering of expected cycle life.
- Any subsurface nondestructive examination of a nozzletype weld is difficult due to joint geometry and varying thicknesses, thereby limiting examination to detection of surface defects only.
- Field welding of nozzles can be complicated by the need to shape pipe ends and to customize fit up. This arduous task can only detract from quality workmanship, potentially leaving inclusions, fissures, or lack of penetration, all initiators of fatigue failures.

Flat reinforcing pads, although an improvement over straight stub-in connections, have many of the same disadvantages evident in straight nozzle connections:

- Due to the joint geometry, external stresses will be concentrated in areas of weld metal and the heat-affected zones of the parent material. Because weld metal has low ductility, elastic and plastic distribution of loadings will be transmitted through the weld metal with the potential of initiating surface or subsurface incipient cracks.

Mechanical Joints. The following must be considered in regard to the use and selection of mechanical joints.

- Tapered, threaded joints have more disadvantages than socket-weld joints previously discussed (i.e., crevice corrosion, stress riser). Because these joints are sensitive to service failures, they are to be avoided.
- Threaded components of a specialty nature should be located where external loadings are minimal or nonexistent.

- Expanded joints must not be used in hazardous and toxic piping systems. Expanded joints tend to lose tightness when subjected to vibration, thermal expansion, and contraction forces or externally applied mechanical loads. The lesser reliability of this type of construction precludes its use for handling hazardous and toxic fluids.

Flanges conforming to ANSI standard B16.5, Pipe Flanges and Flanged Fittings, and ANSI B16.47, Large Diameter Steel Flanges, have been used successfully under all service conditions. The use of flanged joints at terminal equipment, valves and instruments and for assembly and disassembly is common industry practice. Flanges manufactured to other consensus standards (e.g., MSS SP-44, API-605) although entirely suitable for many applications, do not have the same degree of interchangeability and compatibility with flanged piping components that flanges conforming to ANSI standards provide. Therefore, their use may be limited and discussion herein will be limited to ANSI standard flanges.

Selection of the flangetype is an important choice. Two general categories are used:

- Bolt-torque energized
- Internal pressure energized

The first type includes raised face, tongue and groove, ringtype joint male and female, and lap joint flanges. Gasket materials used for this category include various sheet stocks of rubber, cork, asbestos, elastometric composition, metallic-elastomer mixes as spiral wound and laminated, and solid metal rings made as flat washers.

These flanges offer the advantages of ease of joint assembly, standardized dimensions, and general availability. A disadvantage is that the assembly requires high bolt loading during initial seating to retain sufficient gasket pressure in service, a consideration if the joint is subject to cyclic service or is operating near or in creep range of the various assembly materials.

In the second type, initial sealing is achieved by the gasket and facing geometry with relatively low bolt load. The gasket seating force increases with pressure. O-rings made of elastomeric materials are most commonly used in this style of seal. Metallic, proprietary seals such as delta ring, double cone lens joint, and metallic O-rings are also used. These joints are more responsive to fluctuating pressure levels. However, all of these joints require fine surface finishes, tight tolerances, and great care in assembly. Metal-to-metal gasketing creates potential crevice corrosion.

Virtually all of the aforementioned gasket and facing combination have been applied in hazardous and toxic piping systems successfully. Prudent selection, application, and location within the piping geometry will ensure safe, reliable service.

Although consensus standard flanges have enjoyed success over the years, the user must be attentive to the proper application of a mechanical joint in a hazardous and toxic piping system. Some design considerations the user of a flange in a hazardous and toxic piping system should be aware of are as follows:

- Published flange ratings (pressure-temperature material class) presumes the proper selection of gasketing and bolting materials. The flange joint is an assembly of flange, gasket, and bolting. In most cases the gasket seating surface is critical to proper joint sealing. Should the user be inattentive to the interdependency of the flange assembly components, he may have a joint that will not withstand the mechanical or service-induced loadings.

- The designer should locate flanged joints within the piping geometry where external loads from expansion and contraction and wind are low. Total loading across a flanged joint is important to evaluate. The total load would include internal pressure and temperature of the contained fluid and various external loads as previously discussed.[1]
- Flange ends (i.e., weld neck and socket weld) should be consistent with the overall system joining methods. Discontinuities, crevices, and localized stress risers are all present at any particular type of flange, just as they are at other connective points. Threaded and slip-on-type flanges should be avoided in hazardous or toxic piping systems due to their susceptibility to localized crevice corrosion and the stress riser effect of their construction.
- Some designers build in additional flange performance reliability by upgrading the nominal pressure class rating for the expected service conditions. For example, he or she might use a Class 300 flange where a Class 150 flange would meet or exceed the pressure-temperature conditions of the fluid stream. By doing this, the designer is providing additional safety margin for carrying extraneous loads and/or a more reliable seal.
- Assembly of the joint in the field cannot be overlooked. Fit up, gasket alignment, bolting preload, and other factors can affect joint performance throughout service life.
- Flanged joints in hazardous and toxic fluid service usually can be safeguarded with flange shields for use in confining any escaping fluid.
- Proprietary pipe connector mechanical joints (i.e., clamped and hub) or other designs have performed well in demanding services. The designer should apply the same parameters for the selection of any proprietary joint as he would for choosing class and type of flange.

COMPONENT SELECTION

Piping components are discussed at great length in this and other books (see Chap. A2). Discussion herein will be limited to concerns of specific component choices in toxic and hazardous fluid piping systems.

Piping (pipe, valves, and fittings) should conform to applicable codes and consensus standards. The selection and specification of components that meet accepted consensus standards will assure the specifier that certain chemistry, mechanical, and dimensional requirements will be met. Use of other than standard components will require additional qualifications as outlined in the applicable code. Component selection should adhere to commercially available items as much as possible. Special fabrications, forgings, castings, and other specialties complicate the design and may cause difficulties later when replacements are required.

Choice of isolation or control devices (valves) shall be made using the same criteria as in other services. However, due to the potential consequences associated with leakage of a hazardous and toxic fluid, subtle design features of various valves must be considered.

The body material of the valve must be compatible with the fluid under normal and upset conditions. The selection of basic material form (i.e., cast or forged) should be made consistent with the selection of other piping elements in the system. The result from a hazards analysis scenario should be used to determine whether a specific material mechanical property would be advantageous under abnormal conditions. For example, the use of low-ductility castings may not be prudent if shockingtype dynamic loadings are anticipated.

The choice of generic valve type would be made with the usual considerations of function, flow media, pressure drop, and pressure-temperature parameters. Additional considerations for any candidate valve for hazardous and toxic fluid service would include the following:

- Stem leakage to the environment
- Leakage through the shutoff element (i.e., gate, ball, plug)
- Other possible leak paths within the confines of the valve (i.e., body joints)(see Fig. C10.3)
- Ease of maintenance
- Service or test case histories

Service failures have resulted from cracking in a valve body originating from shrinkage cavities in the cast valve wall.[4] However, through-wall cracking of valves is rare; but leakage through stem and bonnet is common.[5]

Recent design enhancements have produced valves with sophisticated devices

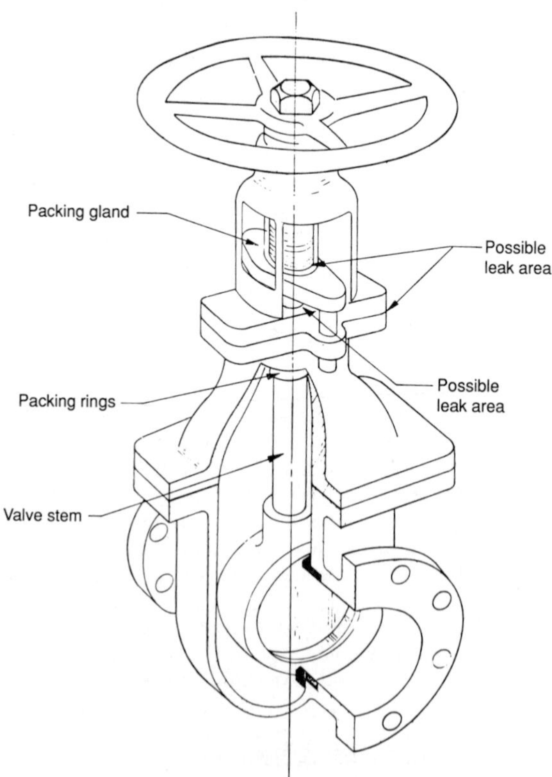

FIGURE C10.3 Conventional gate valve.

to eliminate valve stem leakage. Most of these designs use a proprietary bellowslike device to effect a hermetic seal. These devices along with conventional backup seals have provided a solution to the problem of fugitive emissions emanating from valve stem areas. Some designs of fire-safe valves have incorporated devices to reduce or eliminate stem leakage.

Metal-seated gate valves should be of the flexible-disk, split-wedge, or double-disk design to provide tighter shutoff than solid-wedge designs and to prevent jamming of the gate if the valve cools down in the closed position. The wedge and/or body seat should be hardfaced to prevent galling.

Valve connecting joints are subject to the same limitations as other piping components. Larger size (3 in and larger) valves typically are equipped with flanged ends for use in conventional services. The user should be cautious in the use of flanged end valves without considering the possible leakage at the flange joint; use of a weld end valve in hazardous and toxic service should be a primary choice.

Valving arrangements, such as double block and bleed, and valve shield devices will be discussed later in this chapter (refer to Fig. C10.4).

- Pipe-supporting elements are an integral part of the piping system and must be considered as such. Integral attachments must provide the same degree of metallurgical, geometrical, and fabrication consistency as other components. Improper support design and attachment have caused accelerated failures in piping systems.[4]

In an overall sense, the pipe-support design must be integral with the analytical evaluation of the piping geometry to assure the piping system will respond as anticipated. Changes in the *installed* piping system need to be evaluated relative to the *designed* piping system including supports.

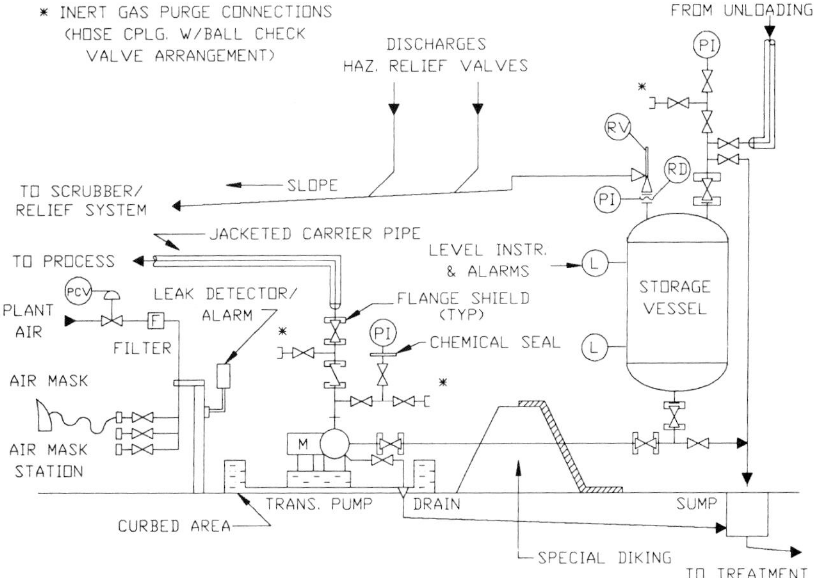

FIGURE C10.4 Composite system diagram.

FABRICATION, ASSEMBLY, AND ERECTION

Without proper care during the fabrication and erection phase, hazardous and toxic piping system integrity can be compromised. The literature is dotted with service failures directly attributable to poor fabrication techniques. Lack of weld penetration and poor fitup in circumferential butt welds have been the cause of numerous service failures. A myriad of other fabrication-induced failures have been experienced for a variety of reasons. The reader need only consult related literature for identification and explanation of the numerous cases.[4]

The user can use various nondestructive examination (NDE) methods to minimize any failure potential as a result of substandard workmanship. A successful application of a nondestructive examination plan entails that the test and procedure be suited both to inspection objectives and flaws to be detected. For instance, with an inappropriate technique, gross errors are possible in detecting and characterizing flaws. For example, liquid penetrant examination will only detect surface flaws; subsurface inclusions would go undetected if only surface examination is used.

Of particular concern is the failure to detect flaws that may seriously impair performance. Conversely, the detection of flaws having little or no bearing on performance may result in unnecessary and sometimes counterproductive repairs.[4] It is necessary that the designer understand the type of flaws that may be detrimental to serviceability and the appropriate NDE techniques be used to detect those particular flaws.

Identification and segregation of those piping systems handling toxic and hazardous materials during the assembly and erection phase (including testing) of the project will ensure that the prescribed NDE plan will be thoroughly implemented. Without proper segregation, the toxic and hazardous system may take on the appearance of the other conventional piping systems also present in the plant. This mixing in the field could lead to an inadequately installed system.

The acceptable criteria for any discovered defects would be as outlined in the applicable code.[6] The user and designer may want to restrict certain defect limits to less than allowed by the applicable code. For example, he or she may want to restrict the Code-allowable amount of porosity in a butt weld in a highly corrosive, cyclic service subjected to high externally applied mechanical loads. The presence of even a minor flaw in this severe service could lead to premature failure.

It must be cautioned that pressure tests will be carried out in static conditions and will not duplicate all service loadings. It is therefore important that mechanical joints either be minimized or located where extraneous loadings are of little consequence.

As can be seen, the NDE program outlined herein has been applied to welded joints. Mechanical joints should be tested hydrostatically, pneumatically, or with a sensitive gas and bubble formation test.

SYSTEM DESIGN

All of the previous discussions focused on increasing piping system safety and reliability by providing superiority of materials of construction, mode of construction, and the selection of components appropriate to the risks associated with handling toxic and hazardous fluid services.

The designer can use other aspects of plant layout and operation that can be expected to produce a safer installation. Plant layout should reflect the hazardous and toxic nature of the fluid conveyed. The design should provide for isolation, buffer areas, or protective devices to contain or recover inadvertent discharge of hazardous streams.[7]

The following are some considerations the designer should implement when arranging a hazardous and toxic piping system and related equipment:

- Open air process units to lessen the potential for concentrating hazardous vapors.
- Containment dikes for collecting spills of hazardous liquids. The diked area would be equipped with a collection sump and means for safe removal. Hazardous and toxic fluid spills would be collected in a dedicated system to eliminate any cross-contamination with other streams.
- Physical containment of the entire process area, with instrumentation for remote monitoring and control.
- Ventilation to remove hazardous vapors for safe disposal during emergency conditions. Ventilation may be the most important technique for controlling toxic air contaminants. General ventilation continually exchanges a supply of fresh air while exhausting air within the entire workplace. Local ventilation removes vapors, mists, and dusts continually from around equipment where hazardous and toxic fluids are contained. Either type of ventilation will require a scrubber to strip the vented air before its release to atmosphere.
- Use the inherent piping geometry, proper location of pipe anchors, pipe loops, and other integral techniques to compensate for thermal expansion and contraction. Avoid the use of mechanical devices to eliminate the effect of expansion and contraction. Bellows and other types of expansion joints should be used only with the utmost care and adequate safeguarding.
- Pressure relief system for safe discharge during upset conditions, blowdown, or cleanout. The relief system would be piped to the hazardous fluid treatment system.
- Double-block and bleed valve arrangements on all hard piped connections where personnel may be required to enter a vessel.
- Engineered barriers and shields at mechanical joints to protect personnel from leakage.
- Guards or barricades to protect the piping from accidental mechanical abuse.
- Plant arrangement that would control access to hazardous areas and provide a safe distance between the hazard and plant and/or public populated areas.
- The system must limit the quantity of hazardous fluid that can escape in the event of a pipe rupture. Minimizing the quantity of hazardous fluid present at any time is a means of protecting people and property in the event of a piping failure.
- Various process controls need to be used to protect the system from excursions of temperature, pressure, or flow rates.
- A systematic monitoring and leak detection program should be implemented to determine whether harmful releases are being experienced.

Table C10.1 summarizes many recommendations for use in identification, design, installation, operation, and maintenance of piping systems used for conveying toxic and hazardous materials. Figures C10.5, C10.6, and C10.7 illustrate some practical solutions to the challenges presented by toxic and hazardous piping systems.

TABLE C10.1 Recommended Practices for Toxic and Hazardous Systems Piping

Do	Do not
Identify and segregate hazardous and toxic piping systems.	View hazardous and toxic piping systems as conventional systems.
Determine consequence of piping failure, for example, quantities released, personnel exposure, harm to the environment, and loss of property.	Assume piping system cannot fail.
Understand operating mode of system, including service variations and upset conditions.	Expect operating conditions to be without variation.
Consider dynamic effects such as water hammer, vibrations, and seismic.	Overlook potential dynamic effects.
Perform stress analysis incorporating total loadings expected.	Disregard dynamic effects working in combination with steady-static loads.
Select materials that will not deteriorate in service.	Choose materials subject to corrosive attack.
Use ductile materials only.	Use low-ductility materials such as cast iron and glass.
Eliminate and minimize the use of mechanical joints; provide safeguarding.	Use mechanical joints indiscriminately without safeguarding.
Provide smooth transitions at welded joints.	Have abrupt changes in joint geometry.
Distribute discontinuities (i.e., shape and materials).	Concentrate discontinuities (local stress risers) (see Fig. C10-8).
Choose valving consistent with hazardous and toxic service. Provide designs to minimize fugitive emissions.	Use conventional stem packing arrangement and mechanical ends.
Use appropriate NDE techniques to assure quality fabrication and construction.	Rely totally on hydrostatic pressure tests.
Arrange plant considering the toxic and hazardous character of fluid conveyed.	Arrange plant without considerations of prevailing wind, isolation and buffer areas, need for barriers, and engineered safeguards.
Use piping geometry to compensate for thermal expansion and contraction.	Use expansion joints.
Provide a collection and disposal system for pressure relief of toxic and hazardous systems during upset conditions.	Vent directly to atmosphere without proper treatment.
Instrument the plant to minimize worker and public exposure to potential accidental releases of toxic and hazardous fluids.	Provide instrumentation systems that require local control or are a maintenance item.
Isolate toxic and hazardous piping systems during fabrication, erection, testing, and commissioning to ensure proper NDE requirements are met.	Comingle with other conventional piping materials and fabrication.

TABLE C10.1 Recommended Practices for Toxic and Hazardous Systems Piping (*Continued*)

Do	Do not
Provide an in-service monitoring program for early detection of leaks.	Wait for a catastrophic event.
Maintain service records through the life of system. Periodically monitor critical elements.	Install the system and forget about it.
Design and maintain supports as an integral part of the piping system.	Treat piping supports as independent components, not an integral part of the piping system.
Provide details of critical elements to construction.	Leave critical fabrication and assembly details to be provided by field.
Provide mechanism for positive identification of piping materials of construction (e.g., material certifications).	Rely on specifying materials with no follow-up.

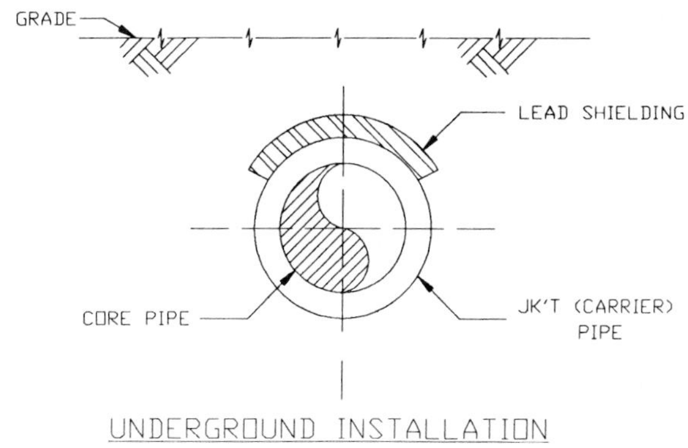

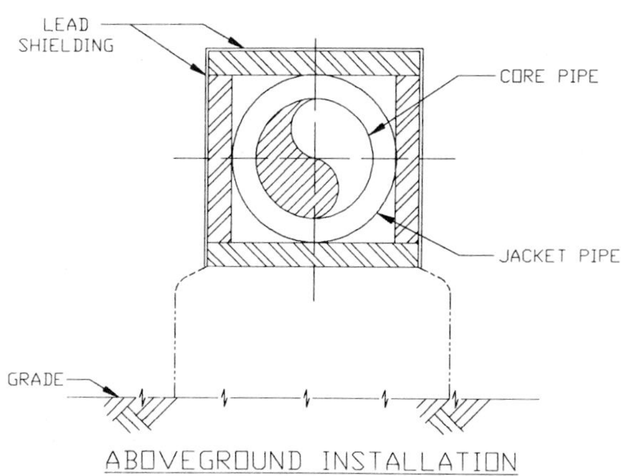

FIGURE C10.5 Piping systems handling radioactive wastes.

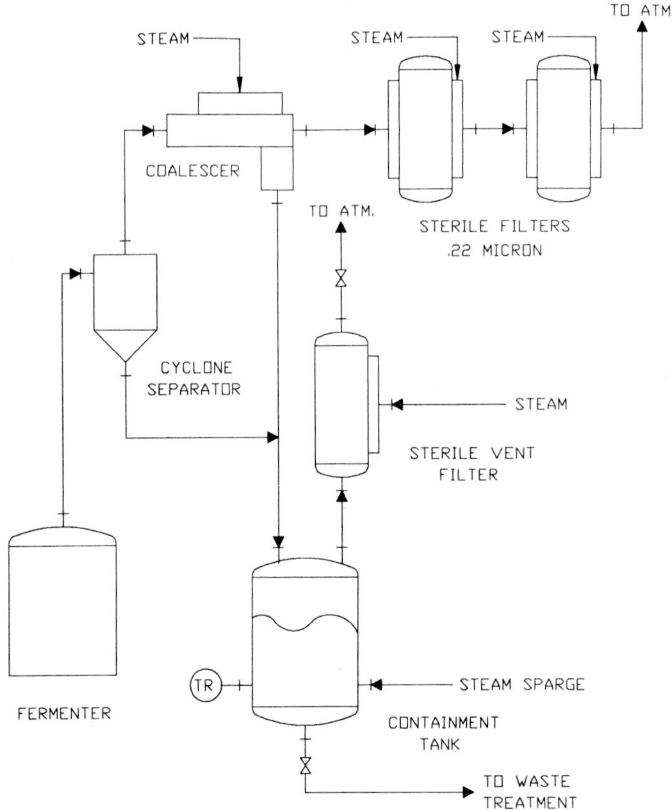

FIGURE C10.6 Sterile barriers at fermenter.

SYSTEM MONITORING PROGRAM

After the hazardous and toxic piping system has been carefully designed, fabricated, erected, tested, and placed in service, it is important that a comprehensive system be used to monitor the serviceability of the various components. In-service monitoring will detect harmful releases so that proper action can then be implemented. However, a program for anticipation and detection of potential problematic events before their appearance, offers a greater advantage over a system that detects release of hazardous fluids only. To that end, it is important that the system conveying hazardous and toxic fluids retain their unique identification all through the design and construction phase and into the operating phase of the plant. In-service monitoring methods must be known before the design phase so special requirements can be included in the design.

The operator should maintain service records of the system and its critical components.[8] These records should be of sufficient detail to provide information of any significant change to the physical structure of the pipeline components and

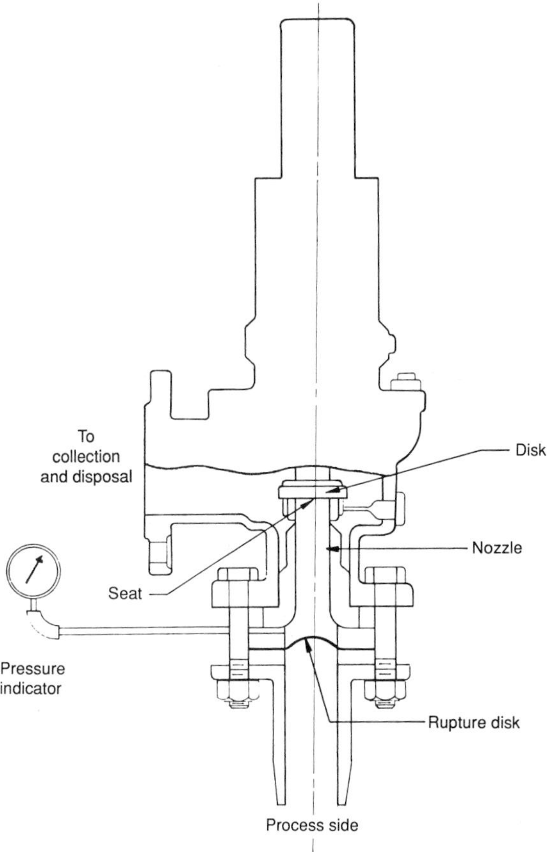

FIGURE C10.7 Rupture disk and relief valve assembly.

be able to identify any shift in the various operating parameters. Even slight changes in temperature, pressure, or the chemical composition of flowing media can manifest in increased corrosion, decreased serviceability of elastomer components, or a decrease in life expectancy of materials operating at or near the creep range.

The operator should know the hazardous and toxic system in sufficient detail to focus any monitoring plan on those areas where early signs of distress may appear. To attain this important knowledge of the hazardous and toxic piping systems, the operator must know the essential design parameters and other possible events that could cause a catastrophic failure or conditions that have a time-dependency component (i.e., accelerated corrosion due to operating at a higher than design temperature). A way of gaining this important information is to be a member of a hazards-analysis team or to have access to the information contained within a hazards analysis study.

One technique that may be used is a composite drawing that illustrates the entire hazardous and toxic system, including all equipment, pipelines, supports,

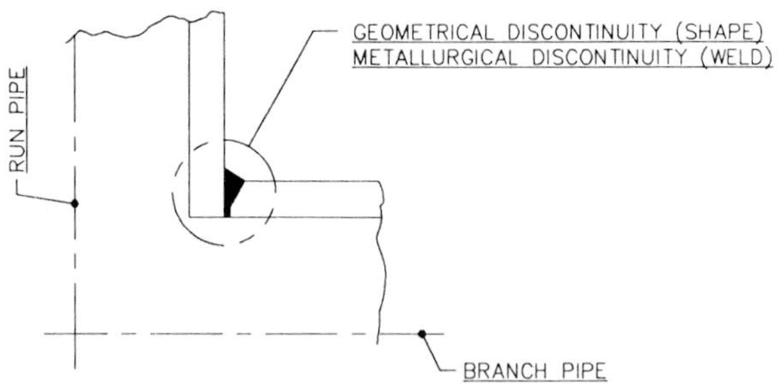

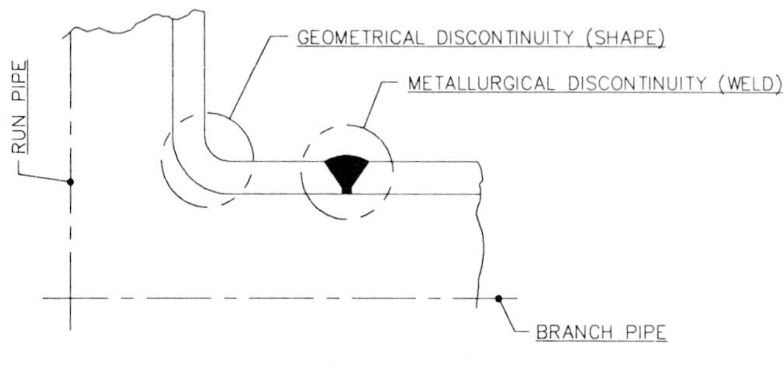

FIGURE C10.8 Treatment of discontinuities.

valves, and components within the pressure boundary of the system (see Fig. C10.4). This drawing can be used for developing a plan for monitoring the integrity of the system. Various parts can be identified where expected discontinuities, potential leak paths at valve stems, or other possible areas of concern are envisioned. The plan can then establish a datum point and measure any change from that point.

Conventional NDE methods can be used for detection of material thickness loss or cracking. If more sophistication need be used, other techniques can be used for in-service monitoring, such as acoustic emission.

Acoustic emission techniques are occasionally used to locate areas of flaw growth in pressure vessels and piping during operation. The advantage of acoustic emission monitoring is that it is in real time and offers the operator a tool to monitor the hazardous and toxic system remotely, thereby reducing the potential for personnel harm and property damage. This technique can be used for homogenous materials (i.e., steels) and for fiber composite materials (i.e., reinforced thermoset resin). Acoustic emission can also be used for gaseous leak detection.

The following is a partial list of acoustic emission monitoring applications. If the expected failure mode is anticipated, proper implementation of an appropriate acoustic emission program can provide the early warning needed by the operator to initiate remedial action.

Partial List of Reported Acoustic Emission Monitoring Application(s)

Incipient fatigue cracking
Incipient stress corrosion cracking
Incipient hydrogen embrittlement cracking
Fiber composite cracking
Leak detection
Detection of creep
Thermal shock cracking
Cavitation detection
Unbonds in honeycombs and laminates

Advantages of Acoustic Emission Monitoring

Nonintrusive technique
Remote detection
Identification of growing, on-stream defects

Any plan the operator develops should be sensitive to the probable cause of system failure, be cognizant of those components and subcomponents prone to premature failure, and be aware of the various techniques and methods apt to detect potential points of failure.

Monitoring of materials subjected to the combination of aggressive fluids and radioactive elements or compounds is understandably difficult. Failed components in such an environment may be so contaminated by radioactive materials that a complete examination is impossible. Predictability of a material's performance in such a high-radiation environment is complicated. A careful, comprehensive monitoring plan should be implemented taking advantage of the remote aspects associated with various NDE techniques.

SELECTED CASE HISTORIES

It is true that in most cases of major failures, multiple causative factors were in place, acting in combination, thereby producing the catastrophe. It is also true that many failures could have been averted with proper design, installation, operation, and maintenance.

A catastrophe in Bhopal, India, in 1984 is illustrative of such a case. A massive leak of highly toxic methyl isocyanate (MIC) caused more than 2500 deaths. Complete details have not been released, but available information suggests that the loss of containment of the MIC resulted from overheating that may have been caused by water contamination of the MIC. It also appears that several of the engineering features included in the plant design to prevent such a release had not been properly maintained and were not operative at the time of incident.[8] This tragedy illustrates the point that merely good design practice and proper installation will not ensure system safety through the plant life if adequate maintenance is not practiced. Perhaps if a system of periodic system performance monitoring had been in place and practiced, the system would have been operable when needed, avoiding the catastrophe.

Failure of glass-level gauges and sight ports have caused many serious incidents.[9] Glass components should not be used on vessels containing flashing flammable or toxic liquids without adequate protection measures. Proper safeguarding for glasstype components would entail the use of an armorlike substance (i.e., clear resin) covering the glass so that in the event of glass fracture, the armor material will contain the toxic and hazardous substance for a time until a controlled safe shutdown could be accomplished. When level glasses are used, they should be fitted with ball and check valves which will prevent a massive leak if the glass breaks. As with any scheme, only proper operation, maintenance, and periodic monitoring will ensure proper system operation.

Wrong materials have inadvertently been installed with disastrous results.[4,9] The most spectacular failure of this sort occurred when the exit pipe from a high-pressure ammonia converter was constructed from carbon steel instead of the 1¼ percent chrome, ½ percent molybdenum as specified. Hydrogen attack created a hole leak path, and the reaction force pushed the converter over. This failure would have been avoided if the specified metal had been used. Although it may appear negligent to mix-up materials, it may be more understandable if one realizes that visual identification is not adequate to differentiate between the carbon steel and specified low alloy. Additionally, external manufacturer markings can become obliterated in the field or may be missing altogether. Positive metallurgical identification is imperative.

Many small-bore pipes (1½ in and less) have failed by fatigue caused by vibration. Supports for these pipes are usually field run, and inadequacies may not be apparent.[4,9] These failures can be avoided by the development of a programmatic checklist that identifies *all* pipelines, equipment pieces, and related items forming the scope of the toxic and hazardous system. An inspection and monitoring plan then addresses design, construction, operation, and maintenance concerns throughout the plant life.

The previous case histories depict some events that could have been avoided with proper forethought. It must be emphasized that during the service life of a plant, perhaps as high as 30 years, only diligent stewardship will ensure safe, reliable operation of toxic and hazardous piping systems. This stewardship begins at the early design phase and continues through the detail design into the construction, erection, and test phase and into operation. However, this is only the beginning; a comprehensive monitoring and preventative maintenance program must be developed and implemented to ensure a hazardfree system.

SUMMARY

Although many factors affect the reliability of a piping system conveying a toxic and hazardous fluid, careful stream identification, prudent material and component selection, quality fabrication and installation, thoughtful system design, and a comprehensive monitoring program will provide the user a safe, troublefree piping system.

REFERENCES

1. M. W. Kellogg, *Design of Piping Systems*, John Wiley & Sons, NY, 1967.
2. P. A. Schweitzer, *Handbook of Corrosion Resistant Piping*, Industrial Press, NY, 1969.
3. National Association of Corrosion Engineers, *Corrosion Data Survey*, Houston, TX, 1985.
4. H. Thielsch, *Defects and Failures in Pressure Vessels & Piping*, Reinhold Publishing Corp., NY, 1965.
5. M. J. Wallace, "Controlling Fugitive Emissions," *Chemical Engineering*, August 1979.
6. ASME B31.3, *Chemical Plant and Petroleum Refinery Piping*, 1990 edition.
7. W. R. Payne, "Toxicology and Process Design," *Chemical Engineering*, April 1978.
8. A. S. Krisher, "Plant Integrity Programs," *Chemical Engineering Progress*, 1986.
9. T. A. Kletz, *What Went Wrong?—Case Histories of Process Plant Disasters*, Gulf Publishing Company, 1985.

CHAPTER C11
SLURRY AND SLUDGE PIPING

Ramesh L. Gandhi
Chief Slurry Engineer
Bechtel Corporation
San Francisco, California

DEFINITION AND BACKGROUND

Slurry is a mixture of solids and liquid. A sludge denotes a mud or a concentrated slurry having considerable amount of fine material that imparts high viscosity. Typical examples of slurry are the solid-liquid mixtures encountered in mineral processing plants and dredged material from waterways and dams. Most of the slurries are made up with water. However, industrial paints, rocket fuel, coal-oil mixture, and coal-methanol slurries are made up with liquids other than water.

River sediment in the form of slurry appears to have been handled since ancient times.[1] All ancient civilizations appeared on river banks. Maintenance of waterways require periodic dredging which results in a sand and silt water slurry. Today, dredging represents the largest volume of solids handled in slurry form. Slurry transport is also used for dam construction.

Blatch[2] was the first one to report hydraulic test results for a sand-water slurry flowing through a 1-in diameter pipe. Gregory,[3] O'Brien and Folsom,[4] and Howard[5] reported results of tests of clay, sand, and gravel slurries. The flow of muds and sludges through pipes was first examined by Caldwell and Babbit.[6] The first large-scale experimental program on flow of slurries through pipes was reported by Durand.[7] The correlations proposed by Durand and his co-workers serve as a basis for the present-day design methods. Design of a slurry piping system involves

- Selection of pipe diameter
- Estimate of friction loss and pumping requirements
- Selection of pipe material, valves, and fittings
- Selection of pumps
- Selection of instruments and control system for safe and reliable operation

Pipelines transporting liquids such as oil and water can be operated at any velocity up to the pipeline's design limit. In most slurry applications, a certain minimum velocity needs to be maintained in order to keep solids from settling out in horizontal sections of the pipe. The velocity below which particles tend to settle out and form a deposit in the pipe is called deposition velocity. The pipe diameter should be selected such that the velocity in the pipeline is maintained above the deposition velocity over the operating range of flow rates.

The operating flow rate range is determined by the expected range of solids throughput and slurry concentration. Solids throughput is defined as the weight of solids to be transported per unit time. It is normally expressed as tons per hour (tph). The slurry concentration is expressed as weight of solids per unit weight of slurry or volume of solids per unit volume of slurry.

The slurry concentration may be established by the requirements of the upstream or downstream processing plants. This is normally the case with inplant piping. In the case of long-distance pipelines, it becomes advantageous to adjust the slurry characteristics and concentration in order to reduce the cost of the pipeline system.

The deposition velocity and friction loss in a given size pipe at a given concentration depends upon slurry flow behavior. The selection of pipe material, valves, fittings, and pumps depends upon the velocity of flow, abrasivity of the slurry, and pumping pressures, which are in turn governed by the slurry flow behavior.

SLURRY FLOW BEHAVIOR

Flow of slurry in pipes depends upon the interaction between the solids and liquid as well as between the slurry and the pipe. Depending upon the velocity of flow, pipe diameter, solids size distribution, fluid properties, and solid characteristics, four different flow conditions can be encountered in a horizontal or nearly horizontal pipeline.[8] These are homogeneous flow, heterogeneous flow, intermediate regime, and saltation regime.

Homogeneous Flow

Homogeneous flow implies that the solid particles are uniformly distributed across the pipeline cross section. Homogeneous flow, or a close approximation to it, is encountered in slurries of high concentrations and fine particle sizes. Slurries exhibiting homogeneous flow properties do not tend to settle and form a deposit under flowing conditions. Typical examples of homogeneous slurries are sewage sludge, coal water fuel, clays, drilling mud, paper pulp, titania, fine limestone (cement-kiln-feed slurry), thorium oxide, and many other finely ground materials.

Heterogeneous Flow

In heterogeneous flow condition, there is a pronounced concentration gradient across the pipeline cross section. Slurries at low concentration with fast-settling (coarse particles) solids generally exhibit heterogeneous flow. Typical examples are sand and gravel slurries, coarse coal slurries, and coarse tailings slurries.

Intermediate Regime

Intermediate regime of flow occurs when some of the particles are homogeneously distributed while others are heterogeneously distributed. Most industrial applications involve particles having a wide range of particle sizes. Intermediate regime of flow is expected with transportation of tailings slurry from mineral processing plants and transportation of coal-water slurries.

Saltation Regime

With fast-settling particles, the fluid turbulence may not be sufficient to keep the particles in suspension. The particles travel by discontinuous jumps or rolls along a sliding or stationary bed on the pipe bottom. This type of flow will occur with coarse sand and gravel slurries.

IN-PLANT SYSTEMS

In-plant systems generally involve horizontal, vertical, and inclined sections of pipe. The pipe lengths are generally short. A large number of bends, valves, and fittings may be present in such systems. The pressure losses due to bends, valves, and fittings may be a significant part of the total friction loss. Static head due to change in pipe elevation may be a significant part of the total pumping head requirements for in-plant systems. Typical examples of in-plant systems are the slurry preparation plants, mineral beneficiation plants, and municipal and industrial waste treatment plants.

In a mineral beneficiation plant, different types of slurries may be handled in the same plant. The slurry concentration as well as the particle size distribution of the slurry may change as the mineral passes through various grinding, separation, and settling stages. Large variations in slurry characteristics as well as flow rate may be encountered in the same section of pipe due to changes in ore characteristics or plant operations. The pipes should be sized for these anticipated variations.

LONG-DISTANCE PIPELINES

A number of slurry pipelines has been built to transport solid particles. The materials transported include coal, limestone, Kaolin clay, China clay, iron concentrate, copper and nickel concentrates, phosphate concentrates, gold ore, fly ash, sludges, and mineral tailings. Because of the relatively long length of these pipelines, pressure losses through bends and fittings are not a significant part of the total friction loss. Pumping requirements should include changes in pipeline elevation which could be substantial in long-distance pipelines traversing rugged terrain.

Because of the relatively large investment in a long-distance pipeline, it is generally advantageous to adjust the characteristics of the slurry to suit the pipeline requirements. The slurry concentration, particle size distribution, and throughput are generally controlled within relatively narrow operating limits. The material is generally finely ground to obtain pseudohomogeneous flow condition in the pipeline.

Slurry pipeline systems range from single-station, low-pressure centrifugal pump installations to multistation high-pressure reciprocating pump systems. In all cases, the basic requirement for successful slurry pumping is to maintain pipeline flow above a minimum operating velocity. The minimum operating velocity is set at a desired margin of safety above the critical velocity. The critical velocity in turn is determined by the solids screen analysis, solids density, and concentration, as well as the specific system characteristics—pipe diameter, slurry temperature, and so on.

In a positive displacement system, the flow is controlled by varying the pump speed. This can be accomplished by the use of a fluid coupling, eddy current coupling, ac or dc variable speed drive, or a hydraulic clutch system. For optimum system efficiency, most of the pumps should be operated at their maximum design speed.

A large number of long-distance slurry pipelines are in operation throughout the world. Table C11.1 presents a list of selected long-distance slurry pipelines.

SLURRY CHARACTERISTICS

Slurries can be classified as settling suspensions and nonsettling suspensions. Settling suspensions require turbulence to maintain individual particles in motion or in suspension. With finely divided solids, homogeneous flow could also be achieved for settling suspensions in turbulent flow. Nonsettling suspensions, as the name implies, do not settle, even under no-flow condition.

TABLE C11.1 List of Selected Long-Distance Slurry Pipelines

Material	System and location	Throughput (mmtpy)	Length (mile)	Start of operation
Coal	Consolidation Coal, Ohio	1.2	105	1957
	Russia	4.0	7	1966
	France (Merlebach)	1.5	6	1954
	Black Mesa, Arizona	5.0	273	1970
	Japan	0.3	16	1965
Limestone	Trinidad	0.5	6	1959
	Rugby Cement, England	1.5	57	1964
	Calaveras, California	1.4	17	1971
	Gladstone, Australia	2.0	15	1981
Iron concentrate	Savage River, Tasmania	2.3	53	1967
	Pena Colorado, Mexico	1.6	30	1974
	Las Truchas, Mexico	1.4	17	1975
	Sierra Grande, Mexico	1.9	20	1977
	Samarco, Brazil	12.0	247	1978
	Kudremukh, India	7.5	42	1980
	La Perla, Argentina	4.5	237	1983
Iron sands	Waipipi, New Zealand	1.0	4	1971
Copper concentrate	Bougainville, Papua New Guinea	1.0	17	1972
	West Irian, Indonesia	0.3	69	1973
	Pinto Valley, Arizona	0.4	11	1974
	Kennecott Chino, New Mexico	0.7	7	1982
	KBI, Turkey	0.9	40	1973
	Kennecott, Utah	0.7	17	1987
Phosphate concentrate	Valep, Brazil	2.0	70	1979
	Chevron, Utah	1.8	94	1986
	Maton, India	0.1	7	1983

Mineral concentrates, tailings, and coal-water slurries require turbulence to maintain particles in suspension. Digested sludge and coal-water-fuel slurries do not settle under static conditions.

The flow characteristics of a settling suspension are largely governed by the settling velocity of solids in it. The flow characteristics of nonsettling suspension are governed by its rheological characteristics and density. Most commercial slurries contain appreciable amounts of finely divided solids that change the rheological properties of the suspending fluid. For these slurries, both the settling characteristics of solids and rheological properties and the density of the slurry become important.

Slurry Density

Density of a slurry is given by the following equation:

$$\rho_m = \frac{100}{(C_w/\rho_s) + [(100 - C_w)/\rho_l]} \tag{C11.1}$$

where ρ_l = the density of suspending liquid (lb/ft^3) (kg/m^3)
ρ_m = the density of mixture (lb/ft^3) (kg/m^3)
ρ_s = the density of the solids (lb/ft^3) (kg/m^3)
C_w = the solids concentration by weight in slurry (%)

Measurements of slurry concentration, solids density, and liquid density are straightforward. The slurry concentration is obtained by evaporating the liquid component from a known weight of slurry and measuring the weight of the dried solids.

The density of slurry in a pipe may be measured by using a nuclear density meter or by measuring head loss per unit length in up and down flowing vertical pipes arranged in the form of an inverted U. The slurry density may also be measured by collecting a sample in a suitably designed specific gravity bottle or by a Marcy balance. It should be noted that these devices measure specific gravity of slurry. The density of slurry is computed by multiplying its specific gravity by the density of water.

Example C11.1. Slurry concentration is determined by drying to constant weight a sample of slurry in an oven maintained at 220°F. Determine the slurry concentration based on the following data:

Weight of empty dry container	0.1 lb (0.0454 kg)
Weight of container plus slurry	0.32 lb (0.1454 kg)
Weight of container plus dry solids	0.21 lb (0.0954 kg)

Solution

Weight of dry solids = 0.21 − 0.1
= 0.11 lb (0.05 kg)

Weight of slurry = 0.32 − 0.1
= 0.22 lb (0.1 kg)

Weight % solids, C_w = weight of solids × 100/weight of slurry
= 50%

Example C11.2. Determine the density of the slurry considered in Example C11.1 if the solids and liquid specific gravities are 3.0 and 1.0, respectively.

Solution

$$\text{The density of solids } \rho_s = 3.0 \times 62.4 \text{ lb/ft}^3 \ (3000 \text{ kg/m}^3)$$

$$\text{The density of liquid } \rho_1 = 1.0 \times 62.4 \text{ lb/ft}^3 \ (1000 \text{ kg/m}^3)$$

$$\text{Solids concentration } C_w = 50\%$$

Substituting in Eq. (C11.1) we get

$$\text{Density of slurry } \rho_m = 93.6 \text{ lb/ft}^3 \ (1500 \text{ kg/m}^3)$$

Example C11.3. A nuclear density meter gives a slurry specific gravity of 1.167 for a coal-water slurry. If the specific gravity of coal is 1.4, find the weight percent coal in slurry.

Solution

$$\text{Density of the coal slurry } \rho_m = 1.167 \times 62.4 \text{ lb/ft}^3$$

$$= 72.82 \text{ lb/ft}^3 \ (1167 \text{ kg/m}^3)$$

$$\text{Density of water } \rho_1 = 62.4 \text{ lb/ft}^3 \ (1000 \text{ kg/m}^3)$$

$$\text{Density of coal, } \rho_s = 1.4 \times 62.4$$

$$= 87.36 \text{ lb/ft}^3 \ (1400 \text{ kg/m}^3)$$

Rearranging Eq. (C11.1) we get

$$C_w = 100 \frac{[\rho_s(\rho_m - \rho_w)]}{[\rho_m(\rho_s - \rho_w)]}$$

$$= 50\%$$

Slurry Rheology

In the presence of subsieve particles (smaller than 35 μ) and at relatively high concentrations, the slurry flow properties are governed by its rheology. Slurries that do not contain particles smaller than 35 μ or that are at low concentration exhibit heterogeneous flow behavior. Heterogeneous flow properties are not governed by slurry rheology.

Rheology is the relationship between the shear stress and the corresponding rate of shear in a slurry under laminar flow condition. The friction loss in a pipeline depends upon the rheology of the slurry in homogeneous and intermediate flow regimes. In the case of pure liquids, the shear stress is directly proportional to the rate of shear in laminar flow. The proportionality constant is called viscosity of the liquid. This type of flow behavior is called newtonian. Liquids containing long-chain polymers and finely ground solids exhibit a nonlinear relationship between shear stress and the rate of shear under laminar flow conditions. Such slurries are said to exhibit nonnewtonian flow properties.

Depending upon the size distribution of solids, slurry concentration, and interaction between solids and liquid, the slurry may have newtonian or nonnewtonian flow properties.

Slurries containing nonflocculated particles generally exhibit newtonian flow behavior. Nonnewtonian flow behavior is generally encountered with flocculated suspensions.

Some slurries require a certain minimum stress before flow starts. For example, fresh concrete does not flow over a chute until a certain slope is exceeded. The slurry is said to possess a yield stress which must be exceeded in order to initiate flow.

The rheology of a newtonian fluid is expressed by its viscosity, which is the ratio of shear stress to the corresponding rate of shear. Two or more parameters are needed to describe the rheological properties of a nonnewtonian liquid. Bingham plastic, pseudoplastic and yield pseudoplastic models are generally used to describe the flow behavior of slurries. The relationships between the shear stress and shear rate for these rheological models are as follows:

Newtonian:

$$\tau = \mu\gamma \tag{C11.2}$$

Bingham plastic:

$$\tau = \tau_y + \eta\gamma \tag{C11.3}$$

Pseudoplastic:

$$\tau = K\gamma^n \tag{C11.4}$$

Yield pseudoplastic

$$\tau = \tau_y + K\gamma^n \tag{C11.5}$$

where τ = the shear stress (lbf/ft^2) (Pa)
τ_y = the yield stress (lbf/ft^2) (Pa)
γ = the rate of shear (velocity gradient) (1/s)
μ = newtonian viscosity (lbfs/ft^2) (Pa · s)
η = plastic viscosity (coefficient of rigidity) (lbfs/ft^2) (Pa · s)
n = flow behavior index
K = consistency index (lbfs^n/ft^2) (Pa · s^n)

Example C11.4. The following rheology test results were obtained for a sample of a mineral tailings slurry containing 50 percent solids by weight:

Rate of shear (1/s)	Shear stress (lbf/ft^2)	$\tau - \tau_y$ (lb/ft^2)
0	0.125	0
0.1	0.1256	0.0006
1.0	0.1280	0.0030
5	0.1343	0.0093
10	0.1400	0.0150
20	0.1494	0.0244
40	0.1647	0.0397
80	0.1895	0.0645
100	0.2004	0.0754
150	0.2250	0.1000
200	0.2474	0.1224
300	0.2876	0.1626
400	0.3240	0.1990

Rate of shear (1/s)	Shear stress (lbf/ft^2)	$\tau - \tau_y$ (lb/ft^2)
500	0.3575	0.2325
600	0.3890	0.2640
700	0.4190	0.2940
800	0.4480	0.3230

Solution. The shear stress at zero shear rate is 0.125 lbf/ft^2. The slurry yield stress is therefore 0.125 lbf/ft^2 (2 · 61 Pa).

Find the difference between the observed shear stress, τ, and the yield stress, τ_y, at each shear rate as shown in the third column of the above table.

A plot of ($\tau - \tau_y$) versus shear rate on an arithmetic scale (Fig. C11.1) shows a nonlinear relationship. A similar plot on log-log scale (Fig. C11.2) shows a linear relationship (note that the date for zero shear rate is excluded). The value of $\tau - \tau_y$ at a shear rate of 1 gives the value of K in $lbfs^n/ft^2$. The slope of the line gives the value of the flow behavior index, n. The results from the graph are

$$K = 0.003 \text{ lbfs}^n/\text{ft}^2$$

$$n = 0.7$$

Estimate of Slurry Rheology. Correlations between slurry concentration and rheology of the slurry for newtonian and Bingham plastic slurries have been proposed by various investigators. These relationships may be used for preliminary estimates when rheology test results are not available.

The viscosity of a slurry depends upon the volume fraction of solids in slurry. The volume fraction of solids is determined using the following:

$$C_v = C_w \frac{\rho_m}{100\rho_s} \tag{C11.6}$$

where C_v is the volume fraction solids in slurry.

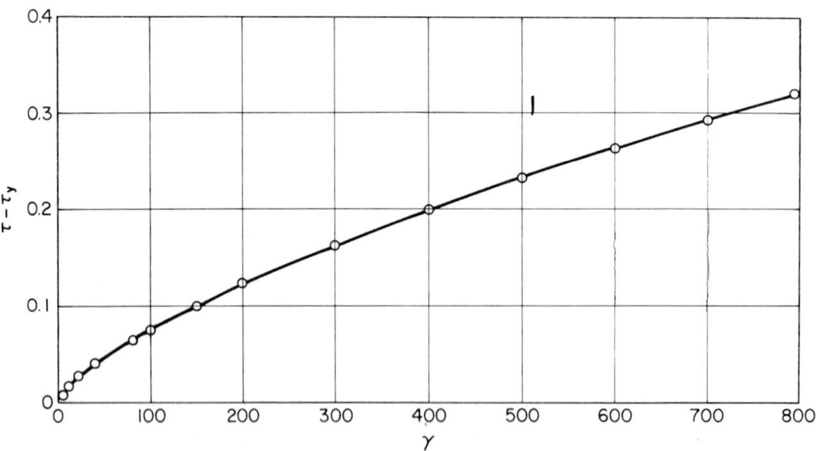

FIGURE C11.1 Shear stress ($\tau - \tau_y$) as a function of rate of shear γ plot for Example C11.4.

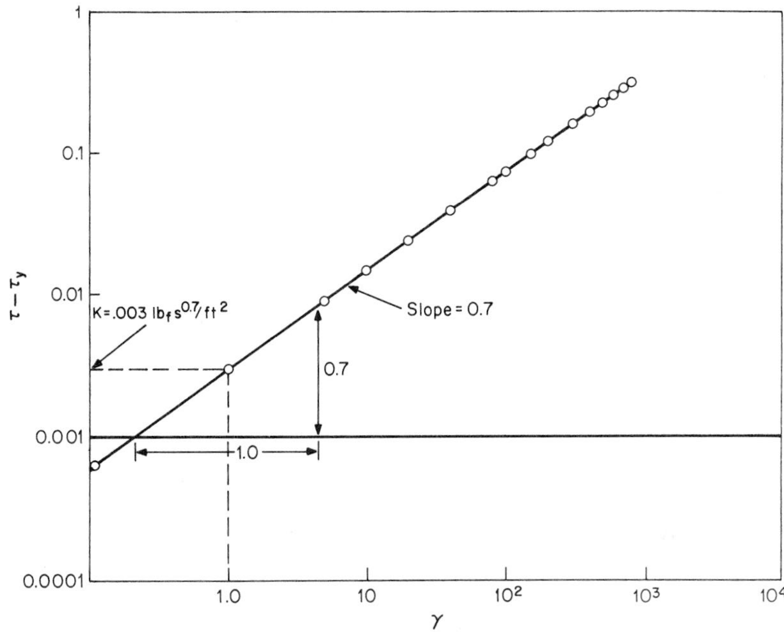

FIGURE C11.2 Shear stress ($\tau - \tau_y$) as a function of rate of shear plot for Example C11.4.

The viscosity of slurries exhibiting newtonian behavior can be estimated by the correlation proposed by Thomas[9]:

$$\frac{\eta_m}{\eta_o} = 1 + 2.5C_v + 10.05C_v^2 + .00273e^{16.6C_v} \tag{C11.7}$$

where η_m = the viscosity of the mixture (lbf · s/ft²) (Pa · s)
η_o = the viscosity of the suspending liquid (lbf · s/ft²) (Pa · s)

Chong et al.[10] proposed the following equation for concentrated suspensions of spherical particles:

$$\frac{\eta_m}{\eta_o} = \left[1 + 0.75\left(\frac{C_v/C_{voo}}{1 - C_v/C_{voo}}\right)\right] \tag{C11.8}$$

where C_{voo} is the maximum packing concentration of the solids in slurry. Equation (C11.8) should be used for values of C_v greater than 0.4.

Gay et al.[11] proposed the following correlations for estimating the Bingham plastic viscosity and yield stress based on their experimental data:

$$\eta_m = \mu\exp\left\{\left[2.5 + \left(\frac{C_v}{C_{voo} - C_v}\right)^{0.48}\right]\frac{C_v}{C_{voo}}\right\} \tag{C11.9}$$

$$\tau_y = 200\left(\frac{d}{C_{voo} - C_v}\right)\left(\frac{C_{voo}}{1 - C_{voo}}\right)^2 / (\xi^{1.5}\sigma_g^{\ 2}) \qquad \text{(C11.10)}$$

where μ = the viscosity of the suspending medium (lbf · s/ft^2)
η_m = the Bingham plastic viscosity of the slurry (lbf · s/ft^2)
τ_y = the yield stress of the slurry (lbf/ft^2)
d = the geometric mean particle diameter (ft)
ξ = the particle shape factor defined as the ratio of the surface area of a sphere of equivalent volume to the surface area of the particle
σ_g = the geometric standard deviation of the particle diameter

SLURRY HYDRAULICS

Homogeneous Flow

The friction loss for a homogeneous slurry depends upon the rheological characteristics of the slurry. The flow through a pipeline can be laminar or turbulent depending upon the velocity of flow. For a nonsettling suspension such as sewage sludge or a highly concentrated coal-water-fuel slurry, laminar flow may be encountered. Turbulent flow should be maintained when the suspension exhibits settling tendency. It is, therefore, necessary to estimate the velocity at which transition from laminar to turbulent flow occurs.

Transition Velocity. The transition velocity is defined as the velocity below which laminar flow is encountered. For a newtonian slurry, the transition velocity corresponds to a Reynolds number of 2000. The Reynolds number should be based on the viscosity of the slurry.

For slurries exhibiting Bingham plastic rheology, the transition velocity is governed by the Reynolds number as well as the Hedstrom number. The Reynolds number should be defined using the plastic viscosity. The critical Reynolds number corresponding to the transition velocity can be estimated from a knowledge of the physical properties of the slurry and the pipe system from Fig. C11.3 proposed by Hanks and Pratt.[12] In this figure the Reynolds number, Re, and the Hedstrom number, He, are defined as follows:

$$\mathrm{Re} = \frac{VD\rho}{\eta} \qquad \text{(C11.11)}$$

$$\mathrm{He} = \frac{D^2\rho\tau_y}{\eta^2 g_c} \qquad \text{(C11.12)}$$

where D = the pipe inside diameter (ft) (m)
V = the average flow velocity (ft/s) (m/s)
g_c = the dimension conversion factor (32.2 lbm ft/lbf·s^2) (1 for SI units)
η = the Bingham plastic viscosity (lbf · s/ft^2)
ρ = the fluid density (lb/ft^3) (kg/m^3)
τ_y = the yield stress (lbf/ft^2) (Pa)

For slurries exhibiting pseudoplastic rheology, the transition velocity is governed by the flow behavior index of the slurry. Figure C11.4 shows the variation of transition

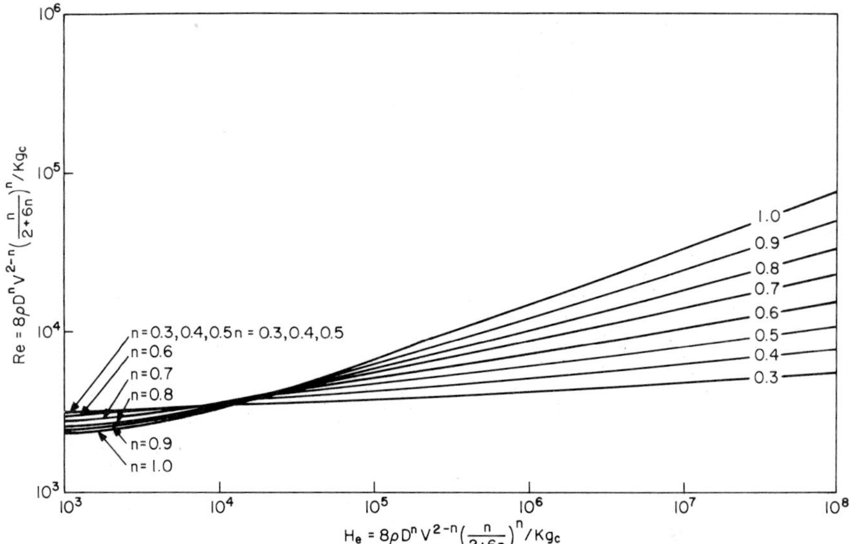

FIGURE C11.3 Laminar-turbulent transition Reynolds number (Re_c) as a function of Hedstrom number (He) for Bingham plastic slurries, where D = pipe ID (ft); V = velocity (ft/s), ρ = density (lb/ft^3), K = consistency (lbf · s/ft^2), τ_y = yield stress (lb/ft^2), n = flow behavior index, g_c = 32.2 (lb · s/lbf · ft).

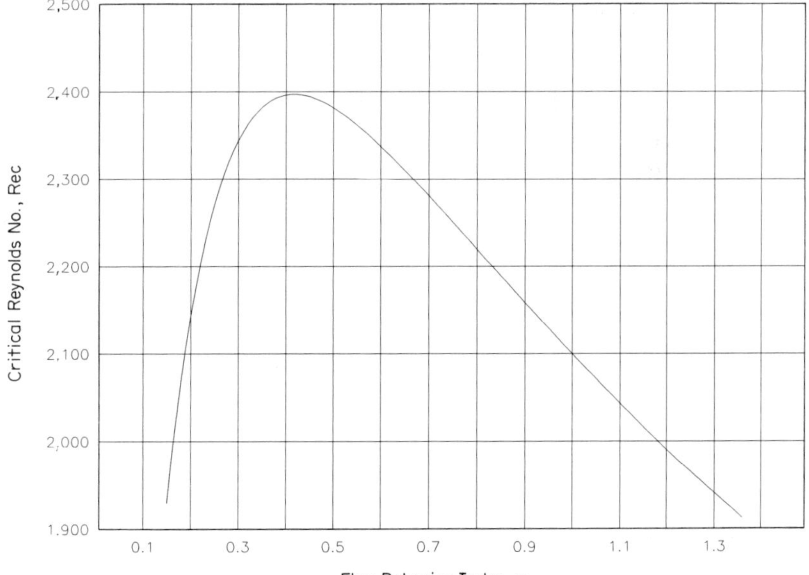

FIGURE C11.4 Laminar-turbulent transition Reynolds number (Re_c) as a function of flow behavior index, n for pseudoplastic slurries, where D = pipe ID (ft), V = velocity (ft/s), ρ = density (lb/ft^3), K = consistency (lbf · s^n/ft^2), g = 32.2 (lb · s/lbf · ft^2).

critical Reynolds number with the flow behavior index n. It should be noted that the Reynolds number for a pseudoplastic slurry is given by the following:

$$\mathrm{Re}_p = 8\rho D^n V^{2-n} \left(\frac{n}{2 + 6n}\right)^n / K g_c \tag{C11.13}$$

where K = the consistency index, lbf s^n/ft^2 (Pa · s^2)
Re_p = the Reynolds number for a pseudoplastic slurry
n = the flow behavior index

For a yield pseudoplastic slurry, the generalized Reynolds number corresponding to the transition critical velocity can be estimated using the following equations proposed by Hanks and Ricks[13]:

$$\mathrm{Re}_{cp} = \frac{6464n}{(1 + 3n)^n} (2 + n)^{\frac{2+n}{1+n}} \frac{((1 - x)^2/(1 + 3n) + 2x(1 - x)/(1 + 2n) + x^2/(1 + n))^{2-n}}{(1 - x)^n} \tag{C11.14}$$

The value of x, which is the ratio of the yield stress to wall shear stress, at the critical Reynolds number is obtained from the following equation:

$$\mathrm{He}_{yp} = \frac{3232}{n} (2 + n)^{\frac{2+n}{1+n}} \left(\frac{x}{(1 - x)^{1+n}}\right)^{\frac{2-n}{n}} \left(\frac{1}{1 - x}\right)^n \tag{C11.15}$$

The Reynolds number, Re_{cp} in Eq. (C11.14) is the same as that for a pseudoplastic Eq. (C11.13). The Hedstrom number He_{yp} is defined as follows:

$$\mathrm{He}_{yp} = \frac{D^2 \rho \tau_y}{K^2 g_c} \left(\frac{\tau_y}{K}\right)^{(2/n)-2} \tag{C11.16}$$

For given slurry characteristics and pipe diameter, the Hedstrom number is computed using Eq. (C11.16). This value of Hedstrom number is used in Eq. (C11.15) to compute the value of x. This value is then used in Eq. (C11.14) to compute the critical Reynolds number. The corresponding transition velocity is calculated using Eq. (C11.13). Figure C11.3 also shows the variation in critical Reynolds number with Hedstrom number for the yield pseudoplastic slurries.

Example C11.5. Estimate the laminar-turbulent transition critical velocity for the slurry considered in Example C11.4 if the pipe ID is 12 in and solids specific gravity is 3.0.

Solution

Given Data

Pipe ID, D = 12 in
= 1 ft

Yield stress of slurry = 0.125 lbf/ft^2

Slurry consistency (K) = 0.0003 lbf s^n/ft^2

Solids specific gravity = 3.0

Solids concentration (Cw) = 50

Computed Results

Solids density $\rho_s = 3 \times 62.4$ lbm/ft^3

Liquid density $\rho_l = 62.4$ lbm/ft^3

Substituting in Eq. (C11.1), we get

Slurry density, $\rho_m = 93.6$ lbm/ft^3

Compute the Hedstrom number He_{yp} using Eq. (C11.16):

$$\mathrm{He}_{yp} = 9.9 \times 10^5$$

From Fig. C11.3 find the critical Reynolds number corresponding to $n = 0.7$ and the computed value of the Hedstrom number:

$$\mathrm{Re}_{yp} = 8850$$

Next, compute the laminar-turbulent transition velocity using Eq. (C11.13):

$$V = 3.58 \text{ ft/s}$$

Friction Loss in Laminar Flow. For homogeneous flow condition, the friction loss in a pipeline is estimated using the following equation:

$$h = 4f\left(\frac{L}{D}\right)\frac{V^2}{2g} \tag{C11.17}$$

where D = pipe ID (ft) (m)
L = pipe length (ft) (m)
V = velocity of flow (ft/s) (m/s)
f = friction factor
h = friction loss (ft) (m) of slurry head
g = acceleration due to gravity (ft/s^2) (m/s^2)

The friction factor for a newtonian slurry is given by the following:

$$f = \frac{16}{\mathrm{Re}} \tag{C11.18}$$

$$\mathrm{Re} = \frac{VD\rho}{\mu_m g_c} \tag{C11.19}$$

where ρ = the slurry density (lbm/ft^3) (kg/m^3)
μ_m = the slurry viscosity (lbf s/ft^2) (Pa · s)
g_c = 32.2 lbm (ft/lbf s^2) (1 for SI units)

The laminar flow friction factor for the Bingham plastic slurries is given by the following:

$$\frac{f}{16} = \frac{1}{\text{Re}} + \frac{\text{He}}{6\text{Re}^2} - \frac{\text{He}^4}{3f^3\text{Re}^8} \qquad \text{(C11.20)}$$

The Reynolds number, Re, and the Hedstrom number, He, are given by Eqs. (C11.11) and (C11.12), respectively.

The laminar flow friction factor for a pseudoplastic slurry is given by Dodge and Metzner[14] as follows:

$$f = \frac{16}{\text{Re}_p} \qquad \text{(C11.21)}$$

where Re_p is the generalized Reynolds number given by Eq. (C11.13).

The laminar flow friction factor for a slurry exhibiting yield-pseudoplastic flow behavior is given by the following:

$$f = \frac{16}{Re_p \Psi} \qquad \text{(C11.22)}$$

where the Reynolds number, Re_p, is given by Eq. (C11.13) and the laminar flow function, Ψ, is derived from the following formula:

$$\Psi = (1 + 3n)^n (1 - x)^{1+n} \left(\frac{(1 - x)^2}{1 + 3n} + \frac{2x(1 - x)}{1 + 2n} + \frac{x^2}{1 + n} \right)^n \qquad \text{(C11.23)}$$

The value of the ratio of the yield stress to wall shear stress, x, is obtained from the following equation:

$$\text{Re}_p = 2\text{He}_{yp} \left(\frac{n}{1 + 3n} \right)^2 \left(\frac{\Psi}{x} \right)^{\frac{2-n}{n}} \qquad \text{(C11.24)}$$

Figure C11.5 shows a plot of Ψ as a function of Hedstrom number for various values of the flow behavior index n.

Example C11.6. Estimate the friction loss in a 12-in ID pipe at a velocity of 3 ft/s for the slurry considered in Example C11.4.

Given Data

Pipe ID (D) = 1 ft

Slurry density = 93.6 lb/ft^3

Velocity (V) = 3 ft/s

K (from Example C11.5) = 0.003 lbf s^n/ft^2

τ_y (from Example C11.5) = 0.125 lbf/ft^2

n (from Example C11.5) = 0.7

= 32.2 lbm ft/lbf s^2

He_{yp} (from Example C11.5) = 9.9×10^5

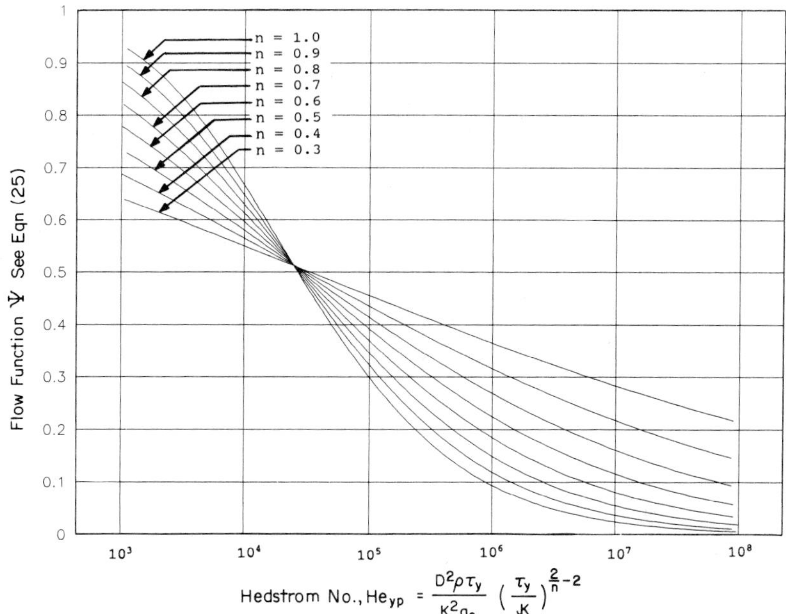

FIGURE C11.5 Laminar flow function Ψ as a function of Hedstrom number for yield pseudoplastic slurries He = $D^2\rho\tau_y(\tau_y/K)^{2/n_s - 2}/K^2 g_c$, where D = pipe ID (ft), V = velocity (ft/s), ρ = density (lb/ft^3), K = consistency (lbf · s^n/ft^2), τ_y = yield stress (lb/ft^2), n = flow behavior index, g_c = 32.2 (lb · s/lbf · ft^2).

Computations. Using Eq. (C11.13), find Re_p = 7023. The Reynolds number is less than the critical Reynolds number. The flow is laminar.

From Fig. C11.5, the value of Ψ is found to be 0.184.

$$\text{The friction factor } f = 16/Re_{yp}\Psi$$
$$= 0.0124$$

$$\text{Friction loss per unit length} = 4fv^2/2gD$$
$$= 6.92 \times 10^{-3} \text{ ft/ft}$$

Friction Loss in Turbulent Flow. The friction factor for newtonian liquids in turbulent flow regime is given by the Colebrook equation as follows:

$$\frac{1}{\sqrt{f}} = 4\log\left(\frac{D}{2\varepsilon}\right) + 3.48 - 4\log\left(1 + \frac{9.35D}{2\varepsilon Re\sqrt{f}}\right) \qquad \text{(C11.25)}$$

where ε is the roughness of the pipe in feet.

Figure C11.6 shows friction factor as a function of Reynolds number with the relative roughness ε/D as a parameter.

Hanks and Dadia[15] developed friction-factor–Reynolds-number relationship for Bingham plastic liquids. This relationship was later modified by Hanks[16]

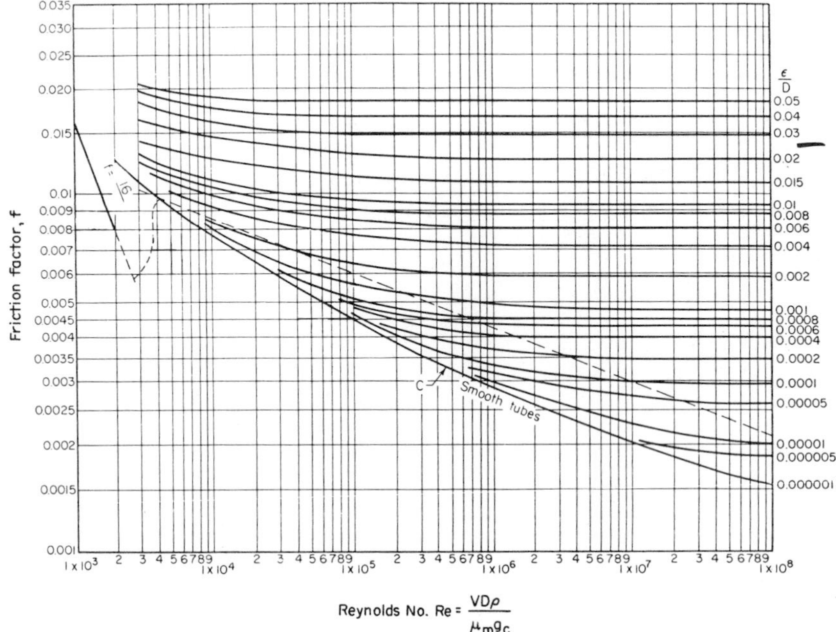

FIGURE C11.6 Friction factor, f, as a function of Reynolds number for newtonian slurries, where D = pipe ID (ft), V = velocity (ft/s), ρ = density (lb/ft^3), μ = viscosity (lbf · s/ft^2), g_c = 32.2 (lb · s/lbf · ft^2).

based on further analysis of their data. Figure C11.7 shows the friction-factor–Reynolds-number relationship for Bingham plastics. It should be noted that for Hedstrom numbers greater than 500,000, there is a long transition region between the laminar turbulent transition critical Reynolds number and the Reynolds number at which the friction factor curve for the Bingham plastics intersects the newtonian curve. The friction factor for Bingham plastic in this transition region are significantly greater than the newtonian friction factor.

Dodge and Metzner[14] carried out a semitheoretical analysis of the turbulent flow of pseudoplastic liquids in smooth pipe. They proposed the following equation for friction factor in turbulent flow:

$$\frac{1}{\sqrt{f}} = \frac{4}{n^{0.75}} \log(\mathrm{Re}_p f^{1-n/2}) - \frac{0.4}{n^{1.2}} \tag{C11.26}$$

Figure C11.8 shows a plot of friction factor as a function of Reynolds number for various values of n based on Eq. (C11.26).

Hanks and Ricks[16] used the concept of mixing length in developing a semitheoretical relationship between friction factor and Reynolds number for turbulent flow of pseudoplastics. This method was extended by Hanks[17] for estimating turbulent flow friction factors for yield pseudoplastic fluids. Interested readers should see the referenced articles for further details.

Figure C11.9 presents friction-factor–Reynolds-number relationship with Hedstrom number as a parameter for selected values of flow behavior index n.

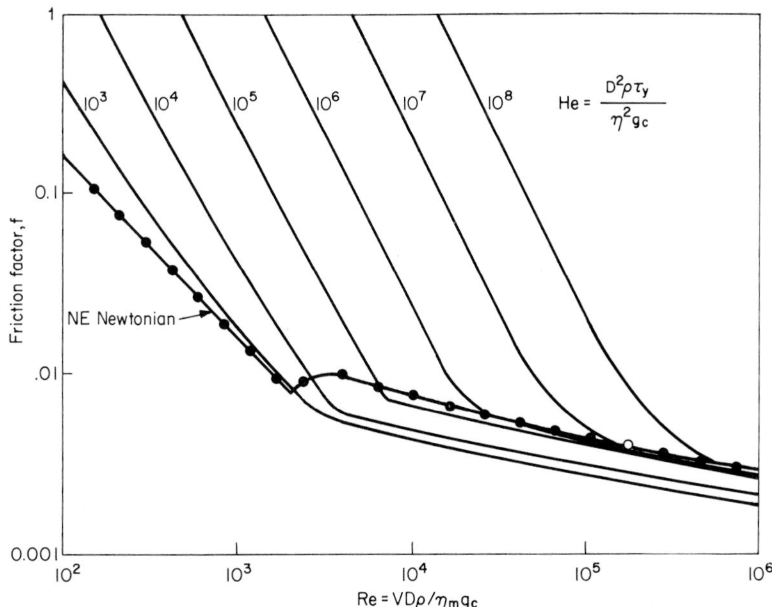

FIGURE C11.7 Friction factor, *f*, as a function of Reynolds number for Bingham plastic slurries. Re = $VD\rho/\mu g_c$, where D = pipe ID (ft), V = velocity (ft/s), ρ = density (lb/ft^3), μ = viscosity (lbf · s/ft^2), g_c = 32.2 (lb · s/lbf · ft^2).

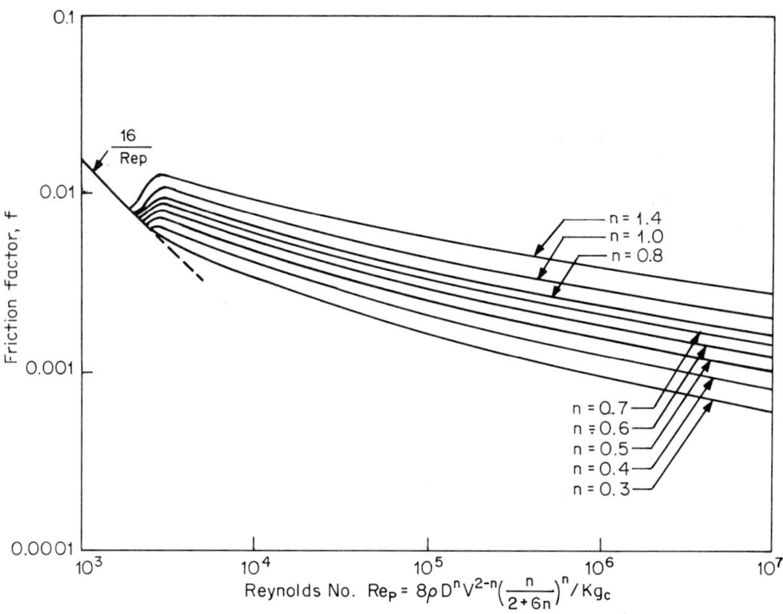

FIGURE C11.8 Friction factor, *f*, as a function of Reynolds number for pseudoplastic slurries. Re = $8\rho D^n V^{2-n}[n/(2 + 6n)]^n/Kg_c$, where D = pipe ID (ft), V = velocity (ft/s), ρ = density (lb/ft^3), K = consistency (lbf · s^n/ft^2), g_c = 32.2 (lb · s/lbf · ft^2).

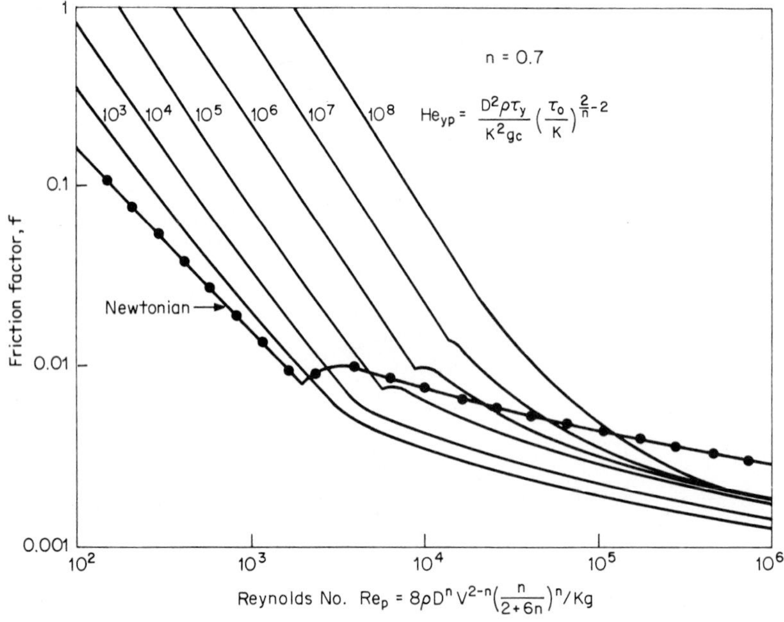

(a)

FIGURE C11.9 (*a*) Friction factor, *f*, as a function of Reynolds number for yield pseudoplastic slurries for $n = 0.7$. $He = D^2\rho\tau_y(\tau_y/K)^{2/n,-2}]/K^2g_c$, $Re = 8\rho D^n V^{2-n}[n/(2+6n)]^n/Kg_c$, where D = pipe ID (ft), V = velocity (ft/s), ρ = density (lb/ft^3), K = consistency (lbf · s^n/ft^2), τ_y = yield stress (lb/ft^2), g_c = 32.2 lb · s/lbf · ft^2.

The following friction-factor–Reynolds-number relationship developed by Torrance[18] is applicable to turbulent flow of newtonian, pseudoplastic, and Bingham plastic as well as yield pseudoplastic fluids in smooth pipes:

$$\frac{1}{\sqrt{f}} = \frac{2.69}{n} - 2.95 + \frac{4.53}{n}\log(1-x) + \frac{4.53}{n}\log\left(\mathrm{Re}_T\sqrt{f^{2-n}}\right) + \frac{0.68}{n} \tag{C11.27}$$

$$\mathrm{Re}_T = \frac{8\rho D^n V^{2-n}}{K8^n} \tag{C11.28}$$

Torrance also extended his analysis to fully rough wall turbulent friction factor for nonnewtonian fluids and obtained the following relationship:

$$\frac{1}{\sqrt{f}} = 4.07\log\left(\frac{D}{2\varepsilon}\right) + 6.0 - \frac{2.65}{n} \tag{C11.29}$$

Example C11.7. Estimate the friction loss in a 12-in ID pipe at a velocity of 5 ft/s for the slurry considered in Example C11.5.

Since the velocity of flow is greater than the transition velocity (see Example C11.5), the flow will be turbulent. Find the Reynolds number corresponding to the given velocity using the slurry rheological data given in Example C11.5:

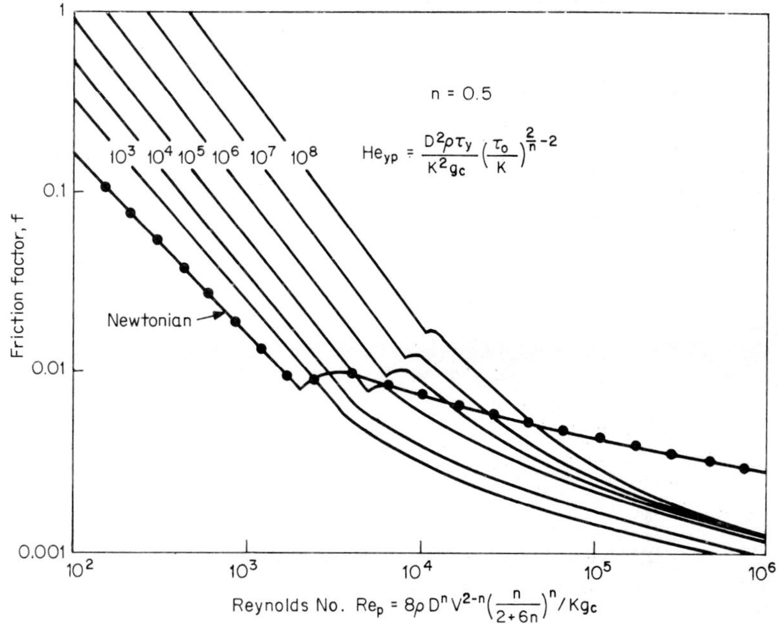

(b)

FIGURE C11.9 (*b*) friction factor, *f*, as a function of Reynolds number for yield pseudoplastic slurries for $n = 0.5$. $He = [D^2\rho\tau_y(\tau_y/K)^{2/n,-2}]/K^2g_c$, $Re = 8\rho D^nV^{2-n}[n/(2+6n)]^n/Kg_c$, where D = pipe ID (ft), V = velocity (ft/s), ρ = density (lb/ft^3), K = consistency (lbf · s^n/ft^2), τ_y = yield stress (lb/ft^2), g_c = 32.2 lb · s/lbf · ft^2.

$$Re_p = 13{,}644$$

The Hedstrom number from Example C11.5 = 9.9×10^5. Using Fig. C11.9*c* find the friction factor, $f = 0.008$. Substituting the values of *D, V, f,* and *g* in Eq. (C11.17), find the head loss per unit length of pipe:

$$h = 0.0124 \text{ ft slurry/ft pipe}$$

Heterogeneous Flow

Fluid turbulence is needed to maintain particles in suspension or motion in a horizontal pipe carrying settling suspension. At low velocities, particles settle down and form a stationary bed on the bottom of a horizontal pipe.

Deposition Velocity. The velocity below which bed deposits form is called deposition velocity. Operating the pipeline at or below the deposition velocity for a prolonged time could result in a pipeline blockage. The minimum operating velocity in a slurry pipeline should be kept greater than the deposition velocity in order to prevent pipeline blockages.

A number of empirical correlations have been proposed for estimating the

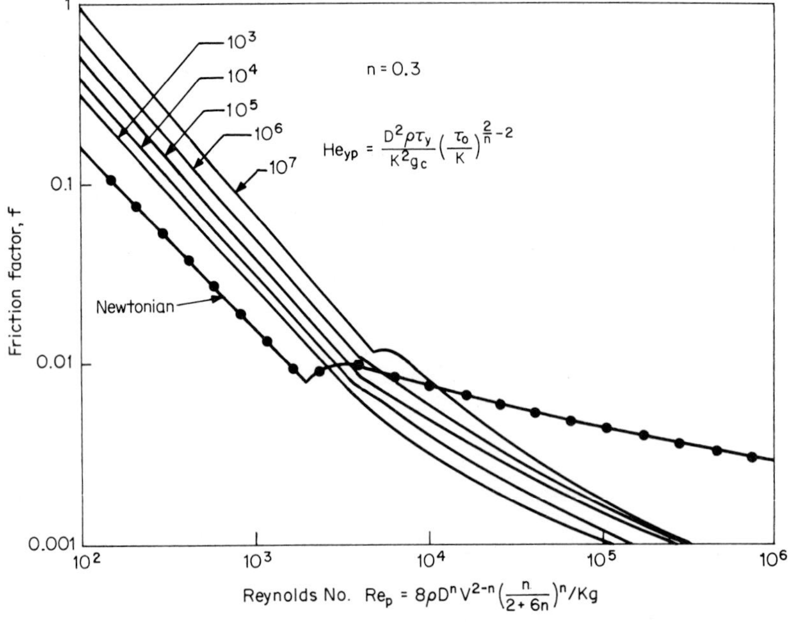

(c)

FIGURE C11.9 (*c*) Friction factor, *f*, as a function of Reynolds number for yield pseudoplastic slurries for $n = 0.3$. He = $[D^2\rho\tau_y(\tau_y/K)^{2/n,-2}]/K^2g_c$, $Re = 8\rho D^n V^{2-n}[n/(2+6n)]^n/Kg_c$, where D = pipe ID (ft), V = velocity (ft/s), ρ = density (lb/ft^3), K = consistency (lbf · s^n/ft^2), τ_y = yield stress (lb/ft^2), g_c = 32.2 lb · s/lbf · ft^2.

deposition velocity. For uniform size particles, the Durand[19] correlation given as follows is widely used:

$$V_D = F_L\sqrt{2gD(s-1)} \tag{C11.30}$$

The value of F_L in Durands correlation can be obtained from Fig. C11.10.
In Eq. (C11.30):

D = the pipe ID (ft) (m)
V_D = the deposition velocity (ft/s) (m/s)
g = acceleration due to gravity (ft/s^2) (m/s^2)
s = the solids specific gravity

Zandi and Govatos,[20] Wasp et al.,[21] and Graf et al.[22] proposed slight modifications of Durand correlation incorporating additional data.

Most industrial applications involve nonuniform size particles. Pilot plant test results or prior experience with similar material is generally used for estimating deposition velocity for nonuniform size particles.

Oroskar and Turian[23] developed a semiempirical correlation which can be used for nonuniform size particles. Their correlation is as follows:

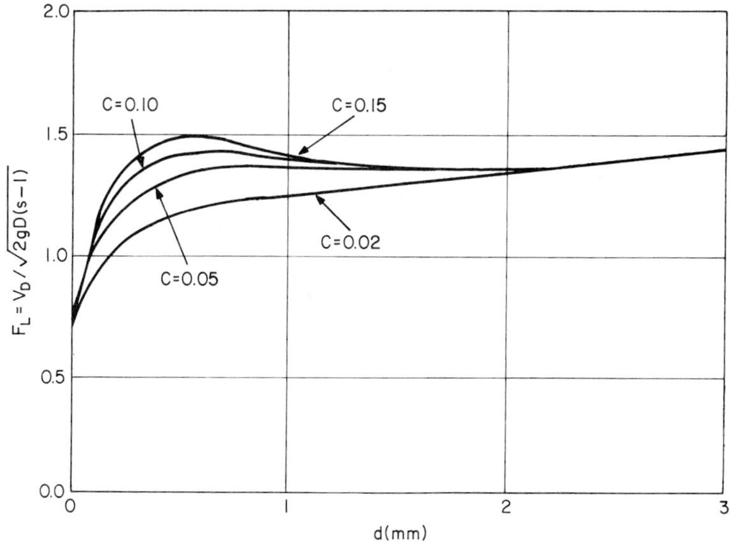

FIGURE C11.10 Durand coefficient F_L, as a function of particle size. $F_L = V_D/\sqrt{[2gD(s - 1)]}$, where V_D = deposition velocity (ft/s), D = pipe ID (ft), g = 32.2 ft/s, s = solids specific gravity.

$$V_D = \left\{5C_v(1 - C_v)^{2m-1}\left(\frac{D}{d}\right)\left(\frac{D\rho\sqrt{gd(s - 1)}}{\mu g_c}\right)^{\frac{1}{8}}/Z\right\}^{\frac{8}{15}}\sqrt{gd(s - 1)} \qquad \text{(C11.31)}$$

where V_D = the deposition velocity (ft/s) (m/s)
Z = a function of V/V_D as shown in Fig. C11.11
d = the mean diameter of particles (ft) (m)
w = the settling velocity of solid particle in slurry (ft/s) (m/s)
m = the hindered settling velocity exponent as a function of particle Reynolds number shown in Fig. C11.12
w_o = the settling velocity of solid particle in clear water of infinite extent (ft/s) (m/s)
μ = the viscosity of water (lbf · s/ft^2)

The settling velocity w_o of a single particle in a fluid is given as follows:

$$w_o^2 = \frac{4gd(\rho_s - \rho_l)}{3C_D\rho_l} \qquad \text{(C11.32)}$$

where C_D = drag coefficient
d = particle diameter (ft) (m)
g = acceleration due to gravity, 32.2 ft/s^2 (9.81 m/s^2)
ρ_s = density of solid particle (lb/ft^3) (kg/m^3)
ρ_l = density of liquid (lb/ft^3) (kg/m^3)
w_o = settling velocity (ft/s) (m/s)

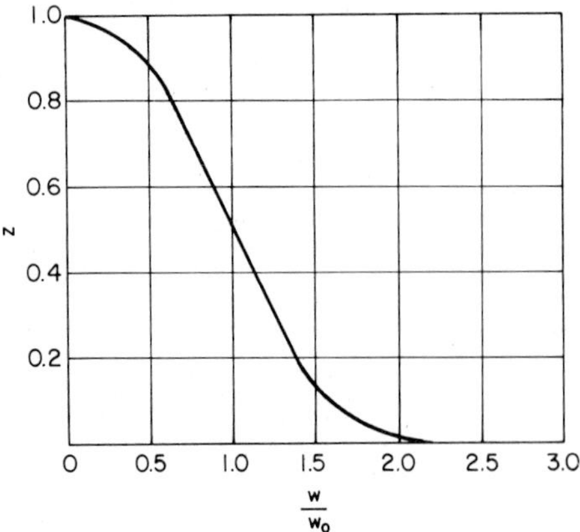

FIGURE C11.11 z as a function of w/w_o for use in Eq. C11.31, where w = hindered settling velocity of solid particle in slurry (ft/s), w_o = settling velocity of single particle in suspending liquid (ft/s).

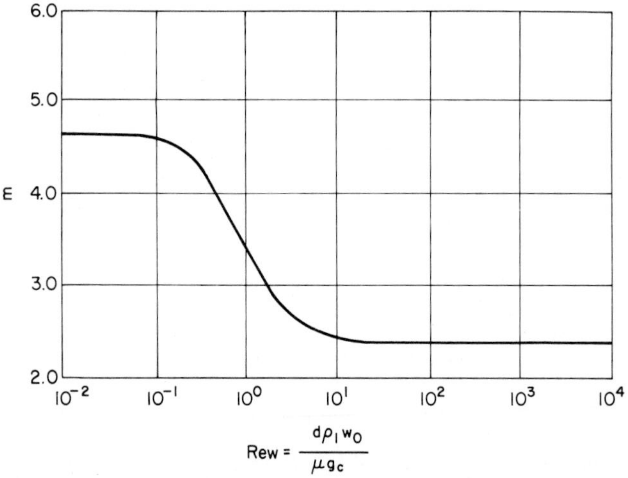

FIGURE C11.12 Hindered settling velocity exponent, m, as a function of particle Reynolds number, where d = particle diameter (ft), w_o = particle settling velocity (ft/s), ρ_l = liquid density (lb/ft^3), μ = viscosity of fluid (lbf · s/ft), g_c = 32.2 lb · s/lbf · ft.

The drag coefficient of spherical particles depends upon particle Reynolds number, Re_w.

For $Re_w < 0.1$, the drag coefficient, C_D, is given by the following equation:

$$C_D = \frac{24}{Re_w} \tag{C11.33}$$

where

$$\mathrm{Re}_w = \frac{d\rho_l w_o}{\mu g_c} \tag{C11.34}$$

μ = viscosity of liquid (lbf · s/ft^2) (Pa · s)

The value of C_D equals 0.4 for particle Reynolds number greater than 1000. Figure C11.13 shows variation in drag coefficient with particle Reynolds number for spherical particles.

Use of $C_D\mathrm{Re}_w^2$ instead of C_D allows determination of the particle Reynolds number from a plot of $C_D\mathrm{Re}_w^2$ versus Re_w shown in Fig. C11.13:

$$C_D\mathrm{Re}_w^2 = \frac{4gd^3\rho_l(\rho_s - \rho_l)}{(3\mu^2 g_c^2)} \tag{C11.35}$$

The drag coefficient is larger for a nonspherical particle compared to a spherical particle of the same diameter. Experimental data on settling velocity as a function of particle diameter can be used to establish the relationship between C_D and Re_w.

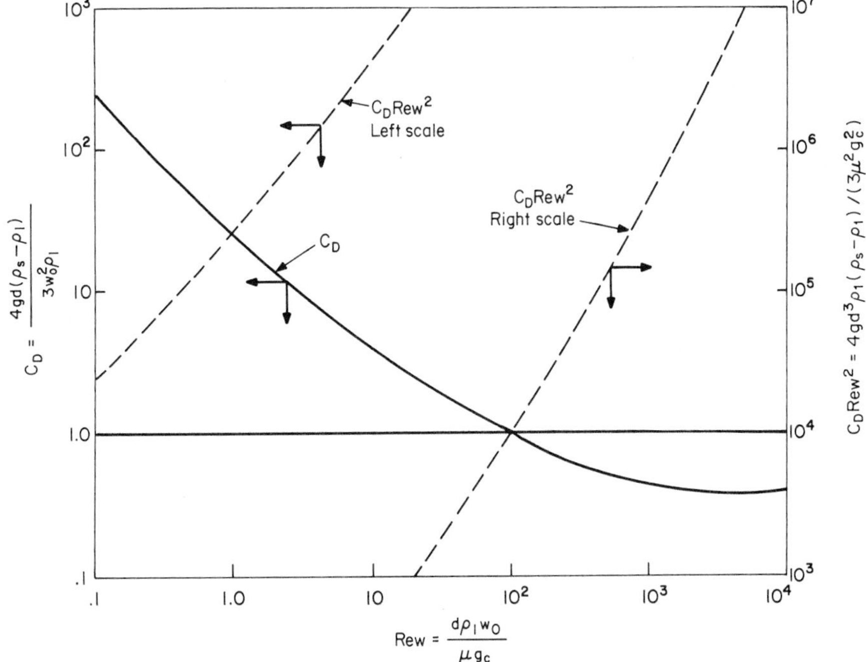

FIGURE C11.13 Drag coefficient, C_D, of spherical particles settling in liquid. $C_D = 4gd(\rho_s - \rho_l)/[3w^2\ \rho_l]$, where d = particle diameter (ft), w_o = settling velocity (ft/s), g = acceleration due to gravity (32.2 ft/s), ρ_s = particle density (lb/ft^3), ρ_l = liquid density (lb/ft^3).

The settling velocity of a solid particle in slurry (hindered settling velocity) is given by the following:

$$w = w_o(1 - C_v)^m \tag{C11.36}$$

where the exponent m varies with the particle Reynolds number as shown in Fig. C11.12.

Wilson[24,25] developed a semitheoretical analysis for heterogeneous slurries. He has presented nomographs for estimating deposition velocity for particles larger than 0.15 mm in diameter.

Example C11.8. Estimate deposition velocity for a sand-water slurry in a 12-in ID pipeline. The diameter of sand particles is 0.2 mm and its specific gravity is 2.65. The slurry concentration is 31.9 percent solids.

Solution. Using Eq. (C11.1), find the slurry density, $\rho_m = 77.9$ lb/ft³. Using Eq. (C11.6) find the volume fraction solids $C_v = 0.15$. From Fig. C11.10, find the value of F_L equal to 1.3 based particle diameter equal to 0.2 mm and C_v equal to 0.15. In Eq. (C11.30) substitute the following values to compute the deposition velocity V_D in ft/s:

$$F_L = 1.3;\ D = 1\text{ ft};\ g = 32.2\text{ ft/s}^2;\ S = 2.65;\ V_D = 13.4\text{ ft/s}$$

Example C11.9. Estimate deposition velocity for a coal-water slurry having a mean particle diameter equal to 0.01 in in a 12-in pipe. The volume fraction coal in slurry is 0.4 and the specific gravity of coal is 1.4. The viscosity of water is 2.1×10^{-5} lbf · s/ft².

Given

$$d = 0.01\text{ in}$$

$$= 8.33 \times 10^{-4}\text{ ft}$$

$$\rho_l = 62.4\text{ lb/ft}^3$$

$$\rho_s = 1.4 \times 62.4 = 87.36\text{ lb/ft}^3$$

$$\mu = 2.1 \times 10^{-5}\text{ lbf} \cdot \text{s/ft}^2$$

$$g_c = 32.2\text{ lbm} \cdot \text{ft/lbf} \cdot \text{s}^2$$

$$C_D\text{Re}w^2 = 84.5$$

From Fig. C11.13, $\text{Re}_w = 2.7$. Using Eq. (C11.33) find w_o equal to 0.035 ft/s. Next, find the hindered settling velocity exponent m from Fig. C11.12:

$$m = 2.8$$

Now, compute the hindered settling velocity using Eq. (C11.34):

$$w = 0.0084\text{ ft/s}$$

The ratio $w/w_o = 0.24$.

From Fig. C11.11 find Z equal to 0.98.

Next compute the deposition velocity using Eq. (C11.31).

$$V_D = 3.5\text{ ft/s}$$

Friction Loss for Heterogeneous Flow-Horizontal Pipes. The formula proposed by Durand[7] is widely used for graded solids. The formula is based on sand and gravel slurries with particle sizes ranging from 0.2 to 25 mm, pipe diameters from 38 to 580 mm, and solids concentrations up to 60 percent by volume. The Durand formula is as follows:

$$\frac{i - i_w}{i_w C_v} = 81\left(\frac{V^2\sqrt{C_D}}{(s - 1)gD}\right)^{-1.5} \tag{C11.37}$$

where i = the friction loss for slurry in ft (m) water per ft (m)
i_w = the friction loss for water in ft (m) water per ft (m) at the same velocity
C_D = the drag coefficient of suspended solid particle settling in fluid of infinite extent
S = the specific gravity of solid particles

Zandi and Govatos[20] concluded from an examination of 2549 data points that Durand's formula predicted the observed head losses fairly well once the saltation data were separated from the heterogeneous flow data. Turian and Yuan[26] divided the available data into heterogeneous flow, saltation flow, and flow with stationary bed and developed correlations applicable to each individual type of flow. Their correlations fit the available data better than the Durand's correlation.

In Eq. (C11.37), the difference $(i - i_w)$ represents an increase in pressure drop due to the presence of solids in the slurry. The effect of particle size on slurry pressure drop is accounted for by the inclusion of drag coefficient, C_D.

In most industrial applications, the particle size is not uniform. A mean value of C_D will have to be used in Eq. (C11.37). Equation (C11.37) may be rewritten as follows:

$$\frac{i - i_w}{i_w} = 81\left(\frac{V^2}{(s - 1)gD}\right)^{-1.5} C_v C_D{}^{-0.75} \tag{C11.38}$$

$$\frac{i - i_w}{i_w} = 81\left(\frac{V^2}{(s - 1)gD}\right)^{-1.5} \sum_{i=1}^{N} C_{vi} C_{Di}^{-0.75} \tag{C11.39}$$

where C_{vi} = the volume fraction of solids having size d_i
d_i = the particle size of ith fraction
C_{Di} = the drag coefficient of particle having size d_i
N = the total number of size fractions into which the given particle size distribution is divided.

It should be noted that the particle size distribution is determined by screen analysis. Tyler mesh screens are widely used. The diameters of openings for Tyler screens are given in Table C11.2.

Example C11.10. Estimate friction loss for a coal water slurry in a 12-in ID pipe at a velocity of 8 ft/s based on the following data:

Coal specific gravity	= 1.4
Volume fraction coal in slurry	= 0.2
Pipe roughness	= 0.002 in
Viscosity of water	= 2.1×10^{-5} lbf · s/ft^2

TABLE C11.2 Tyler Screen Sizes

Sieve designation		Sieve opening		Nominal wire diam.		Tyler equivalent designation
Standard	Alternate	Millimeters	Inch (approx. equivalents)	Millimeters	Inch (approx. equivalents)	
107.6 mm	4.24 in	107.6	4.24	6.40	0.2520	
101.6 mm	4 in	101.6	4.00	6.30	0.2480	
90.5 mm	3½ in	90.5	3.50	6.08	0.2394	
76.1 mm	3 in	76.1	3.00	5.80	0.2283	
64.0 mm	2½ in	64.0	2.50	5.50	0.2165	
53.8 mm	2.12 in	53.8	2.12	5.15	0.2028	
50.8 mm	2 in	50.8	2.00	5.05	0.1988	
45.3 mm	1¾ in	45.3	1.75	4.85	0.1909	
38.1 mm	1½ in	38.1	1.50	4.59	0.1807	
32.0 mm	1¼ in	32.0	1.25	4.23	0.1665	
26.9 mm	1.06 in	26.9	1.06	3.90	0.1535	1.050 in
25.4 mm	1 in	25.4	1.00	3.80	0.1496	
22.6 mm	⅞ in	22.6	0.875	3.50	0.1378	0.883 in
19.0 mm	¾ in	19.0	0.750	3.30	0.1299	0.742 in
16.0 mm	⅝ in	16.0	0.625	3.00	0.1181	0.624 in
13.5 mm	0.530 in	13.5	0.530	2.75	0.1083	0.525 in
12.7 mm	½ in	12.7	0.500	2.67	0.1051	
11.2 mm	7/16 in	11.2	0.438	2.45	0.0965	0.441 in
9.51 mm	⅜ in	9.51	0.375	2.27	0.0894	0.371 in
8.00 mm	5/16 in	8.00	0.312	2.07	0.0815	2½ mesh
6.73 mm	0.265 in	6.73	0.265	1.87	0.0736	3 mesh
6.35 mm	¼ in	6.35	0.250	1.82	0.0717	
5.66 mm	No. 3½ in	5.66	0.223	1.68	0.0661	3½ mesh
4.76 mm	No. 4	4.76	0.187	1.54	0.0606	4 mesh
4.00 mm	No. 5	4.00	0.157	1.37	0.0539	5 mesh
3.36 mm	No. 6	3.36	0.132	1.23	0.0484	6 mesh
2.83 mm	No. 7	2.83	0.111	1.10	0.0430	7 mesh
2.38 mm	No. 8	2.38	0.0937	1.00	0.0394	8 mesh
2.00 mm	No. 10	2.00	0.0787	0.900	0.0354	9 mesh
1.68 mm	No. 12	1.68	0.0661	0.810	0.0319	10 mesh
1.14 mm	No. 14	1.41	0.0555	0.725	0.0285	12 mesh
1.19 mm	No. 16	1.19	0.0469	0.650	0.0256	14 mesh
1.00 mm	No. 18	1.00	0.0394	0.580	0.0228	16 mesh
841 μ	No. 20	0.841	0.0331	0.510	0.0201	20 mesh
707 μ	No. 25	0.707	0.0278	0.450	0.0177	24 mesh
595 μ	No. 30	0.595	0.0234	0.390	0.0154	28 mesh
500 μ	No. 35	0.500	0.0197	0.340	0.0134	32 mesh
420 μ	No. 40	0.420	0.0165	0.290	0.0114	35 mesh
354 μ	No. 45	0.354	0.0139	0.247	0.0097	42 mesh
297 μ	No. 50	0.297	0.0117	0.215	0.0085	48 mesh
250 μ	No. 60	0.250	0.0098	0.180	0.0071	60 mesh
210 μ	No. 70	0.210	0.0083	0.152	0.0060	65 mesh
177 μ	No. 80	0.177	0.0070	0.131	0.0052	80 mesh
149 μ	No. 100	0.149	0.0059	0.110	0.0043	100 mesh
125 μ	No. 120	0.125	0.0049	0.091	0.0036	115 mesh
105 μ	No. 140	0.105	0.0041	0.076	0.0030	150 mesh
88 μ	No. 170	0.088	0.0035	0.064	0.0025	170 mesh
74 μ	No. 200	0.074	0.0029	0.053	0.0021	200 mesh
63 μ	No. 230	0.063	0.0025	0.044	0.0017	250 mesh
53 μ	No. 270	0.053	0.0021	0.037	0.0015	270 mesh
44 μ	No. 325	0.044	0.0017	0.030	0.0012	325 mesh
37 μ	No. 400	0.037	0.0015	0.025	0.0010	400 mesh

Particle size distribution

Diameter (in)	Weight %
0.24	10
0.12	40
0.06	40
0.03	10

Solution. Find the drag coefficient of particles of individual size fraction.

Particle diameter (ft)	$C_{Di}\text{Re}_w{}^2 i$	Re_i	C_{Di}
0.02	1.17×10^6	1710	0.4
0.01	1.46×10^5	520	0.54
0.005	1.83×10^4	145	0.87
0.0025	2.29×10^3	36	1.76

Now,

$$\sum_{i=1}^{N} C_{vi} C_{Di}^{-0.75} = 0.269$$

Next, find the friction loss, i_w, for water:

The Reynolds number for water = 2.38×10^7

Relative roughness of pipe = 0.0002/12

= 0.000167

From Fig. C11.6, find $f = 0.00325$.

Next find

$$i_w = 4f \frac{V'^2}{2gD}$$

$$= 0.0129 \text{ ft/ft}$$

Using Eq. (C11.39), get

$$\frac{i - i_w}{i_w} = 1.97$$

Thus,the friction loss for slurry,

$$i = 2.97\, i_w$$

$$= 0.0383 \text{ ft water/ft}$$

Friction Loss for Saltation Flow-Horizontal Pipes. At low velocities or with particles having large settling velocities, flow with a moving bed or saltating particles may arise.

The saltation flow regime is encountered when N_I is less than 40:

$$N_I = \frac{V^2\sqrt{C_D}}{[C_v gD(s - 1)]} \tag{C11.40}$$

Newitt et al.[27] have developed the following formula based on their experiments:

$$\frac{i - i_w}{i_w C_v} = 66(s - 1)\frac{gD}{V^2} \tag{C11.41}$$

Babcock[28] has proposed the following formula based on his tests:

$$\frac{i - i_w}{i_w C_v} = 60.6(s - 1)\frac{gD}{V^2} \tag{C11.42}$$

Friction Loss for Intermediate Regime. Most of the formulas for heterogeneous flow and saltation regime are applicable to uniform particles. For a mixture of two or more size fractions, it is necessary to determine an average particle diameter for use in various formulas. Use of a weighted average diameter, or drag coefficient, was illustrated in the previous section.

For slurries containing finely ground particles, a better approach is to divide solid particles into a fraction that is carried in homogeneous flow and a fraction that is carried in heterogeneous or saltation regimes. The friction losses for each fraction are computed separately using the appropriate formula, then added together to obtain the total slurry friction loss. This approach also allows the use of nonnewtonian flow properties of the slurry in computing homogeneous friction loss. Wasp et al.[29] have successfully used this approach for correlating coal slurry data.

The method of Wasp is an iterative procedure which works as follows[30]:

1. Divide the total size fraction into a heterogeneous part and a homogeneous part using the following formula:

$$\log\left(\frac{C}{C_A}\right) = -1.8\left(\frac{w}{\beta\kappa u^*}\right) \tag{C11.43}$$

where C = the volume fraction solids at $0.98D$ from bottom
C_A = the volume fraction solids at pipe axis
$U^* = \sqrt{\tau_w g_c/\rho}$ is the friction velocity (ft/s) (m/s)
w = the settling velocity of particles (ft/s) (m/s)
β = the ratio of the mass transfer coefficient to the momentum transfer coefficient (about 1)
κ = the Von Karman constant = 0.4

It is assumed that for each size fraction, the fraction C/C_A is homogeneously distributed and the remainder is heterogeneously distributed. At the start of the iteration, the slurry may be assumed to be homogenous to compute the initial value of wall shear stress.

2. Compute the friction losses for the homogeneous part using the rheological properties of the slurry. Compute the friction loss for the heterogeneous part

using Durand's formula. The sum of the two parts gives an initial estimate of the slurry friction losses.

3. Determine the C/C_A values of each size fraction based on the value of friction loss estimated in step 2.
4. Based on these new values of C/C_A, determine the fraction of solids in the homogeneous phase and the heterogeneous phase.
5. Recompute the friction loss for slurry as in step 2. This provides a new estimate of slurry friction losses. The iteration is continued until the new estimate closely agrees with the previous estimate. The computational procedure is suitable for analysis using a digital computer.

Friction Loss in Vertical Pipes. In vertical pipe flow, there is an absence of a concentration gradient. The slurry flow may be treated as homogeneous flow. For coarse particles, the friction loss for slurry has been found to be the same as that for water at the same velocity. With fine particles, the viscosity of the slurry should be considered in computing the friction losses. The friction factor for slurry is estimated using the equations presented under "Friction Loss in Turbulent Flow."

Friction Losses in Inclined Pipes. Worster and Denny[31] proposed the following equation relating the friction loss for inclined pipes to that for horizontal pipe for heterogeneous slurries:

$$i_\theta = i_w + (i - i_w) \cos \theta \tag{C11.44}$$

where i_θ = friction loss in an inclined pipe
θ = angle of inclination of the pipe from horizontal

Equation (C11.44) suggests that the friction losses in an inclined pipe are the same with both up and down flow and are smaller than that in a horizontal pipe.

Experimental evidence presented by Kao and Hwang[32] shows that the friction losses in an inclined pipe with upflow first increases then decreases after the angle reaches a certain magnitude. In the case of downflow, the friction losses are less than that for horizontal pipe flow.

Friction Losses in Pipe Fittings. Turian et al.[33] have shown that the friction loss in fittings can be approximated using the relations for single-phase newtonian fluid provided the density of the liquid is set equal to that of the slurry.

DESIGN FEATURES

System Components

Figure C11.14 shows a sketch of a slurry pipeline system. It includes a slurry preparation plant, pipeline, pump stations, and slurry receiving terminal.

Slurry handling operation in a preparation plant may involve pipelining of slurry from one processing unit to another. Example C11.11 illustrates the design of such a piping system.

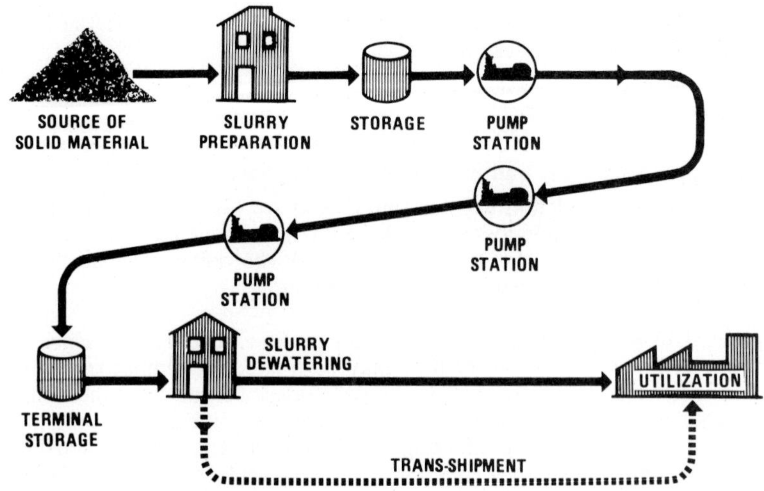

FIGURE C11.14 Sketch of a slurry pipeline system.

Example C11.11. Determine the pipe size and pumping requirements for the slurry piping system schematically shown in Fig. C11.15. The design basis for the system is as follows:

Maximum flow rate (gpm)	2000
Minimum flow rate (gpm)	1500
Slurry concentration (wt%)	50
Specific gravity of solids	3.0
Slurry specific gravity	1.5

Estimated deposition velocity in 12-in pipe is 7.5 ft/s.

Average value of C_D for use in Eq. (C11.37) is 50

Viscosity of water (lbf · s/ft^2 is 2×10^{-5}

Use steel pipe with a roughness of 0.00015 ft.

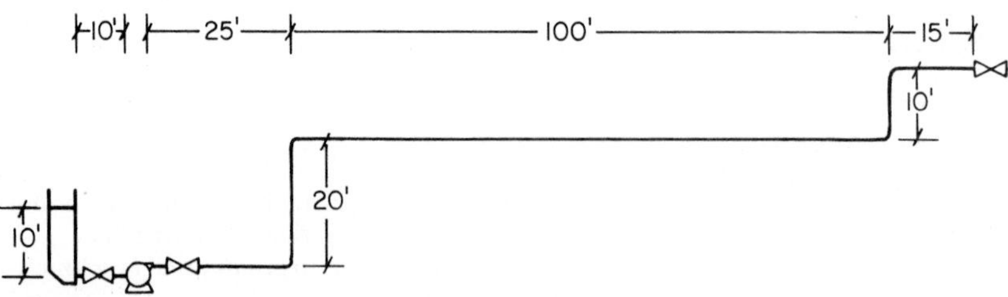

FIGURE C11.15 Sketch of a slurry piping system for Example C11.10.

Solution. Assuming that the deposition velocity varies as the square root of pipe inside diameter [refer to Eq. (C11.30)], estimate the pipe diameter that will give a velocity greater than deposition velocity at the minimum flow rate of 1500 gpm.

$$\text{Minimum flow} = 1500 \text{ gpm}$$
$$= 3.34 \text{ ft}^3/\text{s}$$
$$\text{Let } D = \text{pipe ID (ft)}$$
$$\text{Deposition velocity} = 7.5\sqrt{D} \text{ ft/s}$$

Let the velocity at the minimum flow rate be 1 ft/s above the estimated deposition velocity. Thus,

$$D = 0.79 \text{ ft}$$

Try 10-in pipe (10.75 OD) with 0.25-in wall thickness and 0.375-in thick rubber lining:

$$\text{Pipe ID} = 10.75 - 2 \times (0.25 + 0.375)$$
$$= 9.5 \text{ in}$$
$$= 0.79 \text{ ft}$$

The computations of friction loss and pumping head requirement are summarized in Table C11.3.

TABLE C11.3 Summary of Hydraulic Computations

Item	Min flow	Max flow	Remarks
1. Flow rate (cfs)	3.34	4.45	Given
2. Velocity (ft/s)	6.8	9.1	Flow/area
3. Solids (Sp. Gr.)	3.00	3.00	Given
4. Volume fractions solids (Cv)	0.25	0.25	Use Eq. (C11.6)
5.	6.45	11.46	
6. $(i - i_w)/i_w$	1.24	0.52	Use Eq. (C11.37)
7. Viscosity of water (lbf · s/ft)	2.00*E* − 05	2.00*E* − 05	Given
8. Density of water (lb/ft)	62.4	62.4	Given
9. Reynolds No. for water	5.22*E* + 05	6.95*E* + 05	
10. Relative roughness of pipe	0.00019	0.00019	Roughness/pipe ID
11. Friction factor for water	0.00384	0.00375	From Fig. C11.6
12. Friction loss for water (ft/ft)	0.0140	0.0243	Use Eq. (C11.17)
13. Friction loss for slurry:			
a. Horizontal pipe: ft water/ft	0.0313	0.0370	Use Eq. (C11.37)
ft slurry/ft	0.0209	0.0247	ft slurry = ft water/slurry sp.gr.
b. Vertical pipe: ft slurry/ft	0.0140	0.0243	Same as water
14. Length of horizontal pipe (ft)	150	150	
15. Length of vertical pipe (ft)	30	30	
16. Friction loss, ft slurry:			
a. Horizontal pipe, ft slurry	3.13	3.70	Given
b. Vertical pipe, ft slurry	0.42	0.73	Given
17. Head loss through 3 valves	0.37	0.66	$3 \times V/2g \times K$ valve
18. Head loss through 4 bends	1.31	2.33	$4 \times V/2g \times K$ bends
19. Total head loss, ft slurry	5.23	7.42	Sum of 16(a), 16(b), 17, and 18
20. Static head, ft slurry	20	20	Elevation (outlet-inlet)
21. Total pumping head, ft slurry	25.23	27.42	Sum of 19 and 20

Pipe Wall Thickness

The pipe wall thickness should be sufficient to withstand the expected maximum pressure in the pipe. Most pipelines are designed to have a service life of 10 years or more. Allowance for corrosion and/or erosion should be included in determining the pipe wall thickness.

The design of slurry pipelines should follow ANSI B31.11 Code. It provides methods for determining the allowable stresses in the pipeline.

The pipe wall thickness is computed using the following equation:

$$t = \frac{pD}{2S} + c \tag{C11.45}$$

where t = pipe wall thickness (in) (mm)
p = maximum design pressure in the pipe (psi) (Pa)
S = maximum allowable design stress (psi) (Pa)
c = allowance for corrosion and erosion (in) (mm)

The maximum allowable pipe stress is given as follows per ANSI B31.11:

$$S = 0.8E \times \text{specified minimum yield strength of the pipe (psi)} \tag{C11.46}$$

where E = weld joint factor as summarized in Table C11.4.

Example C11.12. Determine the wall thickness required at point A in the pipeline shown in Fig. C11.16. A 12-in steel pipe is used. The pipe is manufactured in accordance with API 5LX. The minimum yield strength of the pipe is 52,000 psi. The design hydraulic gradient is shown in Fig. C11.15. The design head at point A is 800 ft. The slurry specific gravity is 1.5. The metal loss due to corrosion and erosion is estimated to be 4 mils per year (mpy) and the design life of the pipeline is 25 years.

Solution.

$$\begin{aligned}\text{The maximum design pressure at } A &= 800 \text{ ft slurry}\\ &= 800 \times 1.5 \times 62.4/144 \text{ psi}\\ &= 520 \text{ psi}\end{aligned}$$

$$\begin{aligned}\text{The maximum allowable stress} &= 0.8 \times \text{minimum yield strength of pipe steel}\\ &= 0.8 \times 52{,}000\\ &= 41{,}600 \text{ psi}\end{aligned}$$

$$\begin{aligned}\text{Corrosion allowance } c &= 0.004 \times 25\\ &= 0.1 \text{ in}\end{aligned}$$

$$\begin{aligned}\text{Wall thickness } t &= \frac{pD}{2S} + c\\ &= \frac{520 \times 12.75}{2 \times 41{,}600} + 0.1\\ &= 0.18 \text{ in}\end{aligned}$$

The next higher commercially available pipe wall thickness is 0.188 in.

TABLE C11.4 Weld Joint Factor

Spec. No.	Pipe type*	Weld joint factor E	
		Pipe mfd. before 1959	Pipe mfd. after 1958
ASTM A 53	Seamless	1.00	1.00
	Electric resistance welded	0.85†	1.00
	Furnace lap welded	0.80	0.80
	Furnace butt welded	0.60	0.60
ASTM A 106	Seamless	1.00	1.00
ASTM A 134	Electric fusion (arc) welded, single or double pass	0.80	0.80
ASTM A 135	Electric resistance welded	0.85†	1.00
ASTM A 139	Electric fusion welded, single or double pass	0.80	0.80
ASTM A 155	Electric fusion welded	0.90	1.00
ASTM A 381	Electric fusion welded, double submerged arc welded	—	1.00
ASTM A 672	Electric fusion welded	—	1.00‡
API 5L	Seamless	1.00	1.00
	Electric resistance welded	0.85 (2)	1.00
	Electric flash welded	0.85 (2)	1.00
	Electric induction welded	—	1.00
	Submerged arc welded	—	1.00
	Furnace lap welded	0.80	0.80§
	Furnace butt welded	0.60	0.60

*Definitions for the various pipe types (weld joints) are given in para. 1100.2.

†A weld joint factor of 1.0 may be used for electric resistance welded or electric flash welded pipe manufactured prior to 1959, where (*a*) pipe furnished under this classification has been subjected to supplemental tests and/or heat treatments as agreed to by the supplier and the purchaser, and such supplemental tests and/or heat treatment demonstrates the strength characteristics of the weld to be equal to the minimum tensile strength specified for the pipe or (*b*) pipe has been tested as required for a new pipeline in accordance with para. 1137.4.1.

‡For classes and grades that have been hydrostatically and nondestructively tested to specification requirements.

§Manufacture was discontinued and process deleted from API 5L in 1962.

Corrosion and Erosion Control. In a slurry pipeline, metal loss is expected to be a result of corrosion with possible erosion of the corrosion products taking place simultaneously. Under some conditions mechanical abrasion also will play a part in producing the metal loss.

Erosive wear (abrasive) is governed by the size, shape, and angularity of the solids, slurry concentration, and velocity of flow. In a slurry pipeline, these parameters are interdependent to some extent. For example, use of large size solids requires an increase in minimum transportation velocity. It has been found that above some critical velocity, the abrasive wear increases as the cube of slurry velocity. Wear also increases as the size of solid particles increases. Thus, by reducing the size of the solids, the abrasive wear can be substantially reduced due to the combination of lower required velocity and reduction in wear due to

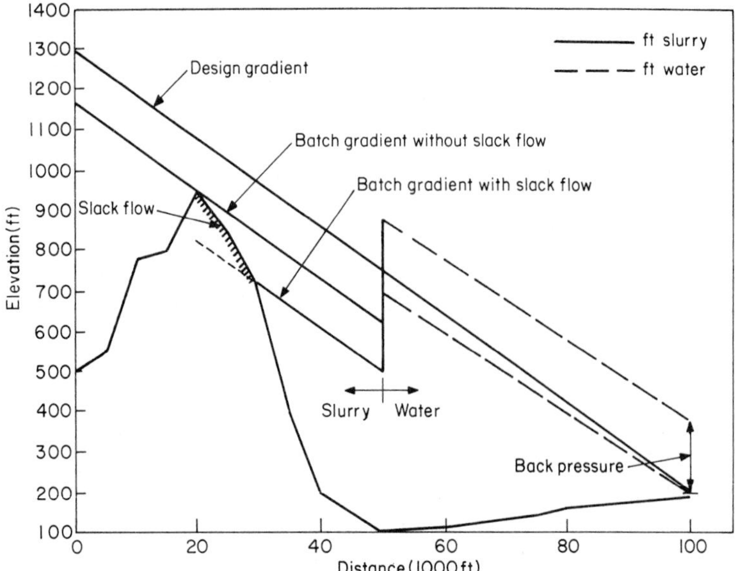

FIGURE C11.16 Pipeline profile and hydraulic gradient showing design gradient, batch gradient, and slack-flow area.

smaller particle size. The effect of slurry concentration on the abrasive wear is more complicated.

From experience, it has been found that the metal loss due to abrasion is insignificant if the velocity of flow is less than about 10 ft/s (3 m/s). For long-distance slurry pipelines, velocities in the range of 4–6 ft/s (1.2–2.0 m/s) results in an optimum design from the standpoint of economics. Thus, when possible a particle size should be selected so the slurry is nearly homogeneously suspended at velocities in the 4–6 ft/s range.

Corrosion can be controlled by passivating either the anodic or the cathodic reactions at the pipe wall. Elimination of dissolved oxygen and the adjustment of slurry pH can reduce the corrosion rate substantially.[34] In most long-distance slurry pipelines carrying mineral concentrate, the slurry pH is adjusted to 9.0 or higher using lime to reduce corrosion rate.

Slack Flow and Its Control

Slack flow occurs when the available static head between the discharge point of the pipeline and a given point in the pipeline exceeds the friction loss between those points at a given flow rate. It usually occurs just downstream of peaks where the hydraulic gradient intersects the ground profile. For example, Fig. C11.15 shows slack flow area.

In slack flow areas, the velocity of flow is governed by the pipe slope. The velocity in the slack flow area is higher than that in the fully packed pipe section. The section of pipe which flows full and is under pressure is called fully packed. In slack flow sections, the pipe flows partly full and pressure generally drops below atmospheric. Erosion takes place in the slack flow area due to higher flow velocity as well as possible cavitation at the point where the pipeline changes from slack flow to a packed flow condition.

Due to the higher density of slurry compared to that of water, slack flow can occur when water is displaced by slurry. For example, Derammelaere and Chapman[35] have described the development of slack flow in the Samarco iron concentrate pipeline during operation of the pipeline in batch mode. This is illustrated in Fig. C11.15 at point S where the hydraulic gradients for pipe full of slurry as well as for batch operation are shown.

Slack flow can be avoided by using a smaller diameter pipe section or orifice chokes. The velocity of flow in the smaller diameter pipe should not exceed about 3 m/s, otherwise erosion may occur. Either orifice chokes or a combination of orifice chokes and smaller diameter pipe may be used to achieve flexibility and economy.

Storage

It is seldom that the flow of solids in a slurry pipeline matches its rate of production at the mine site or its rate of use at the terminal facility. To provide means by which the pipeline can operate efficiently by transporting solids in a nearly continuous fashion, storage facilities are required at both ends of the line. The type and amount of storage are determined by the specific system operating parameters as well as the characteristics of the material being transported.

Tanks

For live slurry storage at the head and tail end of a long-distance pipeline system, agitated tanks are typically used. The number and size to be used are determined by comparing the plant and pipeline availabilities in addition to normal engineering criteria such as allowable soil loadings and an economic analysis of available sizes. If the pipeline is to be operated in batch mode for the first few years of operation, due to low throughput requirements, the size of the tanks may be determined by the minimum batch length chosen for the system. This, in turn, is a function of pipeline length.

Agitated storage may also be required at intermediate pump stations. Its purposes would be to facilitate pipeline section reconnection during restart operations as well as hold slurry which has been flushed from station piping during pump change out.

Ponds

For storage of high volumes of solids over intermediate- to long-term periods, storage ponds are used. Ponds are generally used when the storage requirement exceeds 24 h of pipeline flow. A prime advantage of this method of storage is that no energy is required to maintain suspension of the solids.

Slurry storage ponds can be classified into two basic types: semiactive and dead storage. Semiactive ponds are equipped for recovery and reslurrying on short notice. Depending on design, it may be necessary to remove all the stored material before refilling the pond.

The major consideration in pond design is recovery of a uniform solids particle size distribution. Segregation of the solids can occur during the filling or recovery operation. The resulting coarse and fine slurry slugs can be difficult to handle in the downstream processes. Three major recovery methods have been used: mechanical, dredge, and the Marconaflo technique.

Mechanical recovery uses conventional earth-moving equipment to remove the settled solids. Before the start of recovery operations, the bulk of the free water must be removed from the pond by evaporation, natural drainage, sand points, or an underdrain system.

Dredge recovery requires the maintenance of a water layer over the settled solids to float the dredge. The recovery dredge may be maintained on site or moved in for the recovery operation, depending on whether semiactive or dead storage is required.

The Marconaflo technique[36] was originally developed to remove settled slurries from the hold of ore ships. A high-pressure water jet on a rotating head undercuts and reslurries the solids and the material flows into an underdrain system for pumping away.

Slurry Pumps

Centrifugal as well as positive displacement pumps are available for pumping slurry. The maximum pressure capability and maximum particle size limits for different types of pumps is shown in Fig. C11.17.[37]

The positive displacement pumps can be divided into piston, diaphragm, and plunger pumps. Piston pumps can be used for relatively less abrasive materials, while the diaphragm and plunger pumps are used for handling abrasive slurries at high pressures. The initial capital costs and maintenance costs of positive dis-

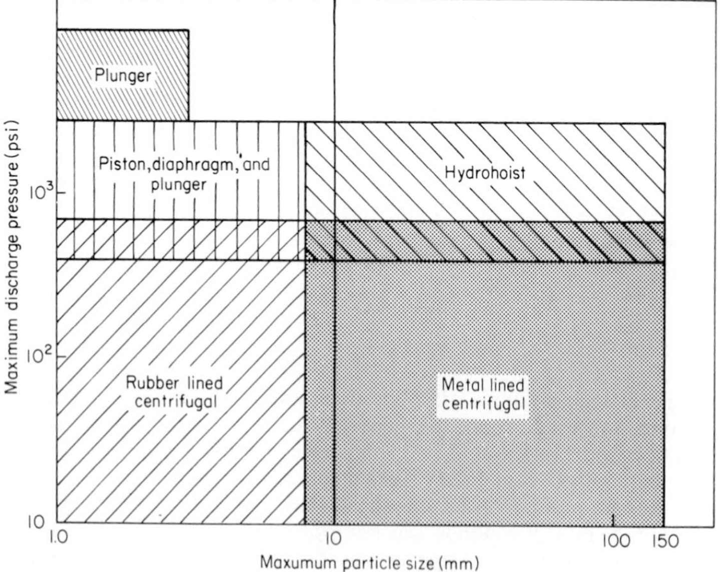

FIGURE C11.17 Maximum particle size and pressure capabilities of different types of pumps.

placement pumps are higher than those of centrifugal pumps, but their hydraulic efficiency is 85 percent as compared to about 60–70 percent for the centrifugal slurry pumps. The flow rate per pump is limited in the case of positive displacement pump, while the head developed per pump is limited to about 30 m in the case of centrifugal slurry pumps.

Centrifugal Slurry Pumps. Centrifugal pumps are extensively used for pumping slurry under relatively low pressure. The main advantages of these pumps are as follows:

1. High flow rates can be achieved with a single unit at a relatively low initial cost. Pumps capable of 25,000 gallons per minute are available.
2. Very few moving and wearing parts are involved.
3. They are simple to operate as well as maintain.
4. There is no practical restriction on the maximum size of solids that can be handled.
5. The flow through the pump is pulse free.
6. They require little space.
7. Valves are not required for the operation of the pump.

Centrifugal pumps have the following shortcomings:

1. The maximum discharge pressure is limited to less than 130 psi for a single-stage pump. With several pumps in series, a maximum pressure of about 750 psi can be achieved.
2. The flow rate of the pump is governed by the system pressure.
3. Attrition of friable solids may occur due to the high velocity of flow through the pump. Attrition can become important if the slurry has to pass through a number of pumps.
4. Seal water is required for good packing life. The seal water dilutes the slurry. The amount of dilution could become significant when the slurry passes through a number of pumps.
5. Centrifugal slurry pumps are made robust because of the abrasion of parts coming in contact with slurry, resulting in a low pump efficiency.

The centrifugal pump parts exposed to wear from slurry are the casing, impeller, and gland seal. Mechanical gland seals have not proven to be effective in slurry installations. A seal that incorporates a water flush to keep solids from entering the gland is used in most cases.

To obtain good service, the casing and impeller should be lined with abrasion-resistant material. Both rubber-lined and Ni hard (metal-lined) units are used extensively. The size of the solids to be pumped determines the type of pump to be selected. Rubber-lined pumps are generally used with particles up to about 0.375 in, and Ni-hard pumps are used for coarser slurries. However, if material with sharp cutting edges, such as crushed glass, is being pumped, Ni-hard pumps can be used even for relatively fine solids. Rubber-lined pump parts usually have longer service life with fine materials. Coal is one material for which Ni-hard-lined pumps have shown better parts life than rubber-lined units.

To obtain good pump parts life, it is good practice to limit the impeller tip speed to less than 4000 ft/min. The pump parts' life on units running faster than this speed drops in proportion to the square of the impeller tip speed.

Reciprocating Pumps. Reciprocating pumps have several desirable features:

1. The flow rate of the pump is independent of the system pressure.
2. They can meet any reasonable system discharge pressure requirement. Pumps capable of producing 2300 psi pressure have been used in magnetite pipelines. Units capable of discharge pressures over 5000 psi are available.
3. The overall efficiency of the pump, including drive train, is relatively high—on the order of 85 percent.
4. Pipeline flow rate can be determined without the use of a flow meter.

The following disadvantages are associated with this type of unit:

1. The maximum flow rate per pump is limited to less than about 3900 gallons per minute. Furthermore, this capacity is only available at relatively low discharge pressures. Therefore, a large number of pumps operating in parallel is needed to handle the high flow rates and working pressures found in large long-distance systems. For example, seven 1250 hp units are used at each pump station of the Samarco system.
2. Initial capital costs and maintenance costs are usually high. Skilled labor is required for operation and maintenance.
3. Variable speed drives are needed to vary flow rates.
4. The flow through the pump is pulsating, which requires greater attention to station piping design to avoid vibration and fatigue problems.
5. The maximum size of particles that can be pumped is restricted by the check valve seal requirements.

For material with a maximum particle size of less than 0.1 in and with discharge pressures up to about 2000 psi, either piston or plunger pumps can be used. Slurry with a maximum particle size of 0.1–0.25 in may also be handled with these types of pumps if special design pump valves are used.

The decision to use piston or plunger pumps is usually based on the results of a Miller abrasivity test. Material with a Miller number below 30 can be handled using piston pumps; material with a Miller number above 60 should be pumped with plunger units. Between these values, the type of pump to use is based upon other considerations, such as

- Piston pumps can be of double-acting design so about twice the flow rate can be obtained for the same physical pump size.
- Plunger pumps are more adaptable to flushing and lubrication. A flushed stuffing box can prolong part's life. However, a flush fluid free of solids must be provided, and some dilution of the slurry will result.

Piston Pumps. Figure C11.18 shows the fluid end of a conventional piston pump. With such a unit, major wear occurs on the pistons, valves, cylinder liners, piston rods, and packing.

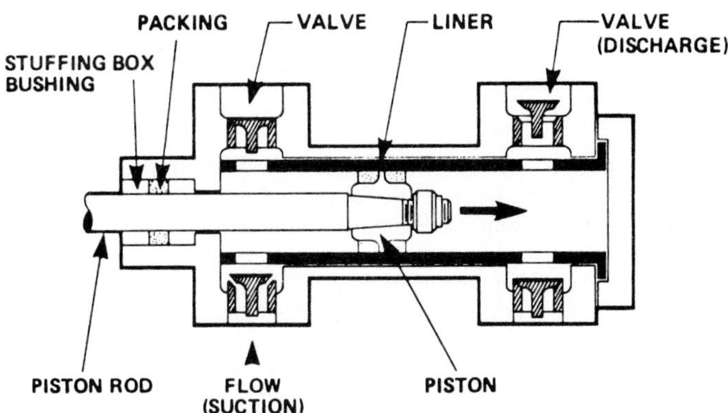

FIGURE C11.18 Piston pump fluid end.

To reduce maintenance, several variations of piston pumps have been developed that limit the number of working parts that are in contact with the slurry. One such type is the diaphragm pump, the fluid end of which is sketched in Fig. C11.19. By pumping hydraulic fluid with a piston pump, the diaphragm is alternately squeezed and expanded. With this type of pump, the valves and diaphragm are the only parts that experience major wear. One disadvantage is the possibility

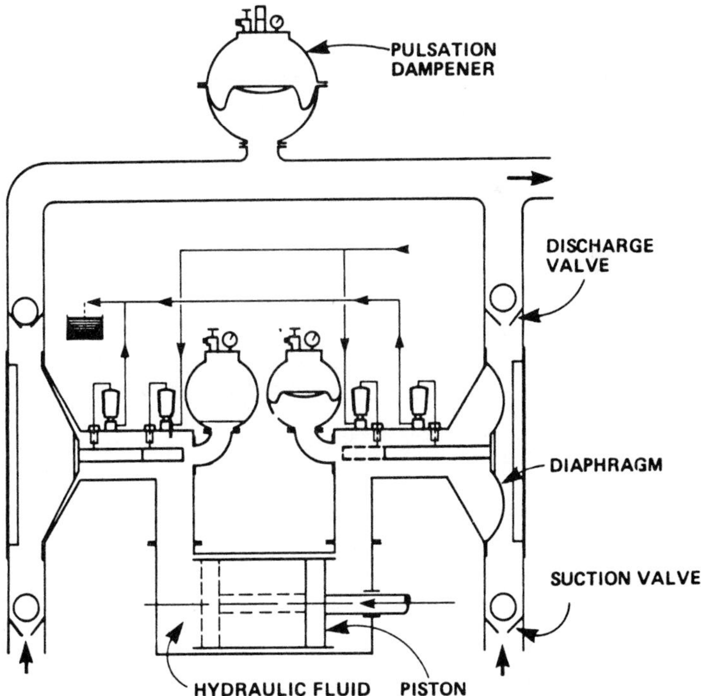

FIGURE C11.19 Diaphragm pump fluid end.

of diaphragm failure. However, this can be minimized through careful pump design by limiting the amount of diaphragm flexing and selecting the membrane elastomer carefully. In addition, suppliers offer visual and audible monitoring systems to detect diaphragm failure. Units are available that will limit the amount of slurry entering the propelling fluid chamber.

Plunger Pumps. Because abrasive slurries can greatly reduce the life of the pistons and cylinder liners of conventional piston units, plunger pumps are often used. This type of pump maintains a clear liquid barrier between the plunger and packing by means of a flushing system, as shown in Fig. C11.20. Major wear is limited to the plunger, valves, and packing.

Hydrohoists. The limitation of maximum particle size for positive displacement pumps and the limitation of pump discharge pressure for centrifugal pumps can be eliminated by using a hydrohoist system. Figure C11.21 shows a sketch of a hydrohoist pumping system. The hydrohoist consists of one or more chambers that can be filled with solids either in dry form (lock hopper) or as a slurry. Once the chamber is filled with solids, water under high pressure is admitted to the chamber to push the solids into the pipeline. A major application is in vertical transportation of coarse ore from deep mines.

The advantages of these devices are as follows:

1. Large-size solids can be pumped under high pressure without having them pass through a number of centrifugal pumps. The attrition of solids is thereby reduced.
2. A high-pressure water pump, having relatively high efficiency compared to a centrifugal slurry pump, is used.
3. The number of moving parts exposed to slurry is considerably less than with reciprocating units.
4. The life of slurry check valves is extended because of the reduced frequency of operation compared to reciprocating pumps; normally the valves open and close in the presence of water.
5. A larger capacity can be obtained with a hydrohoist than with a reciprocating pump operating at the same discharge pressure.

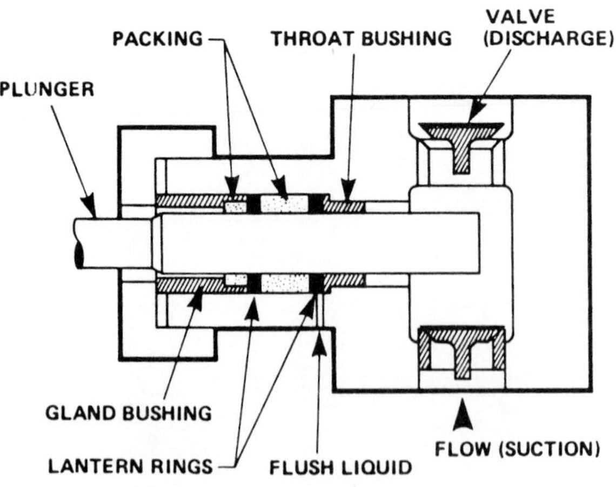

FIGURE C11.20 Plunger pump fluid end.

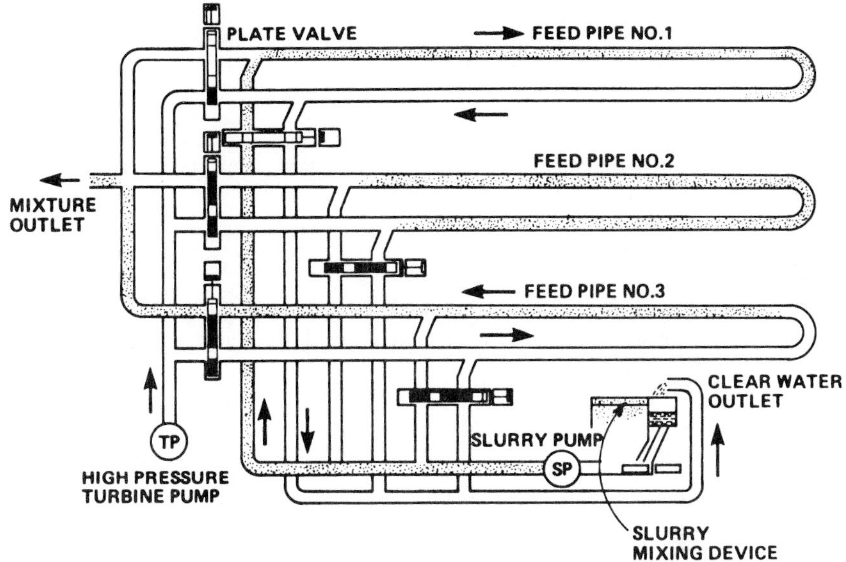

FIGURE C11.21 Sketch of a hydrohoist system.

The disadvantages of the hydrohoist system are as follows:

1. A sophisticated control system is needed to open and close the slurry valves.
2. The flow rate of this device depends upon the system discharge pressure since a high-pressure centrifugal water pump is used.

Pump Horsepower. The pump horsepower is computed using the following equation:

$$\text{HP} = \frac{QP}{F \times \text{Eff}}$$

where HP = Pumping power (hp) (kW)
Q = Flow rate (gpm) (m^3/h)
P = Increase in pressure across the pump (psi) (Pa)
F = Conversion factor, 1714 HP/(gpm · psi) [3.6×10^6 kW/(m^3/h · Pa]

Example C11.13. Select type, number, and power of pumps based on the following:

Flow rate = 300 gpm

Rise in head across the pump(s) = 260 ft

Specific gravity of slurry = 1.5

Solution

$$\text{The pressure rise across the pumps} = 260 \text{ ft} \times 1.5 \times 62.4/144 \text{ psi}$$
$$= 169 \text{ psi}$$

Select centrifugal slurry pumps. Assume three pumps operating in series to get a total head rise of 260 ft. Assuming a pump efficiency is 70 percent, the horsepower per pump is obtained as follows:

$$\text{HP} = \frac{3000 \times 169/3}{1714 \times 0.7}$$

$$= 141 \text{ use 150 hp motors}$$

Life of Pump Parts

The typical wear parts' life for the different types of slurry pumps is given in Table C11.5. The actual life depends upon pump speed, discharge pressure, as well as abrasivity of the slurry.

Selection of Valves

There are many manufacturers and types of valves offered for low-pressure slurry service. These designs are backed by years of experience in the minerals and dredging industries. Selection of valves for use in positive displacement slurry pump stations introduces some new parameters besides those that must be considered in selecting low-pressure valves. These include vibration of the piping and packing of solids into the void spaces in the valves.

Generally speaking, in high-pressure duty where a full opening is not required, lubricated plug valves have been applied. Depending on the service, the wetted parts of these valves should be hardfaced to minimize abrasive wear during the opening or closing cycle. Gate or ball valves have been used where a full round opening is required. Both of these types introduce the problem of removing solids from the body cavity after the valve has been operated in slurry. Again, it may be necessary to hardface some wetted parts to extend the valve life.

Slurry valves will require regular lubrication and maintenance in order to retain their usefulness. An important feature in selecting valves is the ease of disassembly for maintenance. Some ball valves, for instance, have to be sent to the factory to renew the seats or seals. Some maintenance problems may be a matter of solids finding their way into the valve workings, with the solution being disas-

TABLE C11.5 Pump Wear Parts Life

	Part life (h)			
Pump part	Piston	Plunger pump	Diaphragm pump	Centrifugal pump
Piston rod	3000	—	—	—
Plunger sleeve	—	700	—	—
Piston liner	4000	—	—	—
Packing	6000	400	—	1500
Diaphragm	—	—	8000	—
Valve seat	1000	800	5000	—
Valve body	1000	800	3000	—
Valve insert	1000	500	3000	—
Impeller	—	—	—	6000
Casing	—	—	—	6000

sembly and clean up of the valve. Some slurries (such as limestone) may have some cementation properties. Therefore, it is good practice to lubricate and exercise valves that are otherwise infrequently used regularly.

Abrasion Control

Pipe abrasion is a major concern in coarse solids pipelines. In temporary systems, the pipeline may be replaced periodically. For example, the phosphate mining industry in Florida replaces steel pipes after about 2 years of operation.[38] Use of special pipes, nonferrous pipes, and steel pipes lined with abrasion-resistant linings can significantlv increase the life of pipe carrying coarse solids.

Nonferrous Pipes. For low-pressure applications, a number of nonferrous materials can be considered for transporting coarse solids.[39]

Wood. Wood has some unique advantages. It can handle a wide pH range and acts as a naturally good insulator. Furthermore, wood pipe undergoes minimal expansion or contraction with temperature change. However, high-grade fir and redwood suitable for making into pipe are becoming relatively scarce and expensive. Therefore, there has been a tendency to turn to other materials, in particular, plastic pipes.

Polyethylene. Ultrahigh molecular weight polyethylene has been extensively used for tailings line construction. For sizes up to 14-in diameter, it is available with pressure ratings up to 250 psi. This material is a major improvement over unlined carbon steel or wood and under good conditions lasts many times longer than the material it replaces.

Polybutylene. Polybutylene pipe is a flexible thermoplastic pipe that has been developed relatively recently. Depending upon the diameter, it is normally available with pressure ratings up to 200 psi, although higher pressure pipes have been made. Polybutylene has two major advantages as compared to polyethylene; namely, the tensile strength is higher and it is less affected by extremes of temperature.

Concrete. Concrete pipe is available as nonreinforced and reinforced pressure pipe. The latter type is available in various forms with diameters exceeding 40 in. For tailings service, special hard gravel can improve the life. When wear is found in the pipe, repairs are made by using a mix of hard ceramic chips in epoxy. Evidence to date indicates that concrete in general has only a fair abrasion resistance.

Polyurethane. Polyurethane pipe has been used for some in-plant systems, in particular, coal cleaning plants. The polyurethane is spun cast into flanged spools up to 12 in in diameter.

Other Nonferrous Materials. Polyvinylchloride (PCV) polypropylene and acrylonitrile-butadiene-styrene (ABS) have been made into pipe but have poor wear resistance compared to polyethylene and polybutylene. They should not be considered for long life slurry applications.

Fiberglass pipe is available with ceramic chips or tile embedded into the inner surface. This material is expensive, and once the chips are broken out, the epoxy matrix that holds the chips has poor wear resistance and the pipe quickly fails. Also, joint systems designed to take high pressure are prone to failure. Abrasive particles attack the groove left at the joint, and early failure may take place.

Internally Lined Steel

Many of the materials discussed in the previous section can also be used as a lining within a steel outer shell. By using such a system, the inherent pressure limitations of nonferrous materials can be overcome. Table C11.5 lists three major lined steel alternatives. In most cases, however, the use of a composite pipe leads to difficulty in joining the pipe sections. Most systems require the use of steel flanges with the lining material serving as the gasket material.

In Situ Lining of Steel Pipe

Methods of in situ internal lining of welded steel pipe have been developed. A plastic pipe is pulled inside a steel pipe, and the annular space between the two pipes is grouted with cement. This method reduces the number of mechanical joints in the pipe considerably.

Instrumentation

The instrumentation used and the major parameters measured and controlled are as follows.

Pressure. Both the suction and discharge pressures at a pump station are normally measured and controlled. Suction pressure is controlled to prevent cavitation. The discharge pressure is controlled so as not to exceed the maximum allowable pressure for the pump unit or the pipeline system. Bourdontype pressure gauges with an isolation diagram are generally used.

Density. The pumping head requirement varies with the changes in slurry density. If the slurry density is not controlled within the pipeline design limits, the pipeline system can be overpressurized or the flow velocity could drop below deposition critical velocity.

Slurry density is normally measured by radiation gauges. Density meters may be needed at each pump station in a multiple station system. The density meters are normally installed on the suction side of a pump station.

Density is controlled by adding dilution water to high-density slurry. If the density is too low, the slurry can be returned to the preparation plant for concentrating.

Flow Rate. Because slurries are often abrasive by nature, normal in-line flow-measuring devices such as orifice plates and pitot tubes are not applicable. For primary slurry flow metering, magnetic flow meters are considered the best choice. However, these meters are not useful with oil slurries. Ultrasonic flow meters may be required in applications where magnetic flow meters are unsuitable. However, these meters are not accurate and reliable.

Slurry Tank Level Measurement. A number of different systems have been used for recording the level in slurry storage tanks. Conductance probes for level alarms and a pressure sensor mounted in the wall of the tank are used to obtain a rough indication of level of slurry in the tank. This procedure has proven satisfactory, although the accuracy of the level measurement with a pressure sensor is affected by changes in slurry density. The way to compensate for this effect is available but not commonly used. Sonic devices provide direct measurement of tank level.

REFERENCES

1. Linssen, J. G. Th., The Performance and the Future Development of Dredging Equipment, *First International Symposium on Dredging Technology,* University of Kent at Canterbury, England BHRA Fluid Engineering, Paper A1, Sept. 17–19 (1975).
2. Blatch, N. S., Discussion of Works for the Purification of the Water Supply of Washington, D.C., by Hazem, A., and Hardy E. D., *Trans. ASCE,* vol. 57, pp.400–409 (1906).
3. Gregory, Pumping Clay Slurry Through a Four-Inch Pipe, *Mech. Eng.* vol. 49, no. 6 (1927).
4. O'Brien, M. P., and Folsom, R. G., The Transportation of Sand in Pipelines, University of California (Nov. 12, 1937).
5. Howard, G. W., Transportation of Sand and Gravel in a Four-Inch Pipe, ASCE, vol. 104 (1939).
6. Caldwell, D. H., and Babbit, H. E., The Flow of Muds, Sludges and Suspensions in Circular Pipes, *Trans. A.I.Ch.E.* vol. 37, pp. 237–266, (Feb. 1941).
7. Durand, R., Basic Relationships of the Transportation of Solids in Pipes—Experimental Research, *Proc. International Assoc. for Hydraulic Research,* Minneapolis, Minn., Sept. 1–4 (1953).
8. Zandi, I., Hydraulic Transport of Bulky Materials, Paper 1: Advances in Solid-Liquid Flow in Pipes and Its Application, Edited by Iraj Zandi, Pergamon Press (1971).
9. Thomas, D. G., Transport Characteristics of Suspension: VIII. A Note on the Viscosity of Newtonian Suspensions of Uniform Spherical Particles, *J. Colloid Science,* vol. 20, no. 3, pp. 267–277 (March 1967).
10. Chong, J. S., Christiansen, E. B., and Baer, A. D., Rheology of Concentrated Suspensions, *Journal of Applied Polymer Science,* vol. 15, pp. 2007–2021 (1971).
11. Gay, E. D., Nelson, P. A., and Armstrong, W. P., Flow Properties of Suspensions with High Solids Concentration, *AIChE Journal,* vol. 15, no. 6, pp. 815–822 (Nov. 1969).
12. Hanks, R. W., and Pratt, D. R., On The Flow of Bingham Plastic Slurries in Pipes and Between Parallel Plates, *Society of Petroleum Engineers Journal,* pp. 342–346 (Dec. 1967).
13. Hanks, R. W., and Ricks, B. L., Laminar-Turbulent Transition in Flow of Pseudoplastic Fluids with Yield Stress, *J. Hydronautics,* vol. 8, no. 4., pp. 163–166 (Oct. 1974).
14. Dodge, D. W., and Metzner, A. B., Turbulent Flow of Non-Newtonian Systems, *AIChE Journal,* vol. 5, no. 2, pp. 189–204 (June 1959).
15. Hanks, R. W., and Dadia, B. H., Theoretical Analysis of the Turbulent Flow of Non-Newtonian Slurries in Pipes, *AIChE Journal,* vol. 17, pp. 554–557 (May 1971).
16. Hanks, R. W., and Ricks, B. L., Transitional and Turbulent Pipeflow of Pseudoplastic Fluids, *J. Hydronautics,* vol. 9, no. 1, pp. 18–23, (Jan. 1975).
17. Hanks, R. W., Low Reynolds Number Turbulent Pipeline Flow of Pseudohomogeneous Slurries, *Hydrotransport* 5, pp. C2.23–C2.34. BHRA Fluid Engineering, Cranfield, Bedford, England (May 1978).
18. Torrance, B Mck., *South African Mechanical Engineer,* vol. 13, p. 89 (1963).
19. Durand, R., The Hydraulic Transportation of Coal and Other Materials in Pipes, *Colloq. of National Coal Board,* London (Nov. 1952).
20. Zandi, I., and Govatos, G., Heterogeneous Flow of Solids in Pipeline, *Proc. Hydraulics Division, ASCE,* vol. 93, pp. 145–159 (May 1967).
21. Wasp, E. J., et al., Deposition Velocities and Spatial Distribution of Solids in Slurry Pipelines, *Hydrotransport* 1, BHRA Fluid Engineering, Cranfield, Bedford, U.K., (Sept. 1970).

22. Graf, W. H., Robinson, M., and Ucel, O., The Critical Deposit Velocity for Solid-Liquid Mixtures, *Hydrotransport* 1, BHRA Fluid Engineering, Cranfield, Bedford, U.K. (Sept. 1970).

23. Oroskar, A. R. , and Turian, R. M., The Critical Velocity in Pipeline Flow of Slurries, *AIChE Journal,* vol. 26, no. 4, pp. 550–558 (July 1980).

24. Wilson, K. C., and Judge, D. G., Analytically-Based Nomographic Charts for Sand-Water Flow, *Hydrotransport* 5, BHRA Fluid Engineering, Cranfield, Bedford, U.K. (May 1978).

25. Wilson, K. C., Deposition-Limit Nomograms for Particles of Various Densities in Pipeline Flow, *Hydrotransport* 6, BHRA Fluid Engineering Cranfield, Bedford, U.K., (Sept. 1979).

26. Turian, R. M., and Yuan, T., Flow of Slurries in Pipelines, *AIChE Journal* vol. 23, no. 3, pp. 232–243 (May 1977).

27. Newitt, D. M., Richardson, J. F., Abbott, M., and Turtle, R. B., Hydraulic Conveying of Solids in Horizontal Pipes, *Trans. Inst. Chem. Engrs.,* vol. 33, pp. 93–110 (1955).

28. Babcock, H. A., Heterogeneous Flow of Heterogeneous Solids, in *Advances in Solid-Liquid Flow in Pipes and Its Application,* I. Zandi, Editor., pp. 125–148, Pergamon Press, N.Y. (1971).

29. Wasp, E. J., Regan, T. J., Withers, J. G., Cook, P. A. C., and Clancey, J. T., Cross Country Pipeline Hydraulics, *Pipeline News,* vol. 35, pp. 20–28 (July 1963).

30. Wasp, E. J., Kenney, J. P., and Gandhi, R. L., Solid Liquid Flow Slurry Pipeline Transportation, *Trans Tech Publications* (1977).

31. Worster, R. L., and Denny, D. F., Hydraulic Transport of Solid Materials in Pipes, *Proc. Institute of Mechanical Engineering,* vol. 169 (1955).

32. Kao, D. T. Y., and Hwang, L. Y., Critical Slope for Slurry Pipeline Transporting Coal and Other Solid Particles, *Hydrotransport* 6, BHRA Fluid Engineering, Cranfield, Bedford, England (Sept. 1979).

33. Turian, R. M., Hsu, F. L., and Selim S., Friction Losses for Flow of Slurries in Pipeline Bends, Fittings and Valves, *Particulate Science and Technology,* vol. 1, no. 4, pp. 365–392 (1983).

34. Gandhi, R. L., Ricks, B. L., and Aude, T. C., Control of Corrosion-Erosion in Slurry Pipelines, *First International Conference on Internal and External Protection of Pipes,* BHRA Fluid Engineering, Cranfield, Bedford, England, Paper G-2, (Sept. 1972).

35. Derammelaere, R. H., and Chapman, J. P., Slack Flow in the World's Largest Iron Concentrate Slurry Pipeline, *Proc. 4th International Technical Conference on Slurry Transportation,* Las Vegas, Nevada, (March 28–30, 1979).

36. Lutjen, G. P., Marconaflo: The System and the Concept. *Engineering and Mining Journal,* pp. 67–75 (May 1970).

37. Gandhi, R. L., Snoek, P. E., and Carney, J. C., An Evaluation of Slurry Pumps, *Proc. 5th International Technical Conference on Slurry Transportation,* Lake Tahoe, Nevada (March 26–28, 1980).

38. Faddick, R. R., and Staman, O. D., Pipeline Transportation of Phosphate Slurries: A Survey, Colorado School of Mines Research Institute, *Mineral Industries Bulletin,* vol. 20, no. 6 (Nov. 1977).

39. Snoek, P. E., and Carney, J. C., Pipeline Material Selection for Transport of Abrasive Tailings, *Proc. 6th International Technical Conference on Slurry Transportation,* Las Vegas, Nevada, (March 24–27, 1981).

CHAPTER C12
WASTEWATER AND STORMWATER PIPING SYSTEMS

Dr. Ashok L. Lagvankar, P.E., DEE
Vice President and Chief Engineer
Harza Environmental Services, Inc.
Chicago, Illinois

John P. Velon, P.E., DEE
Vice President
Harza Environmental Services, Inc.
Chicago, Illinois

This chapter presents information on the analysis and design of wastewater and stormwater piping systems. Wastewater piping systems convey water-borne domestic, commercial, or industrial wastes to a point of discharge and/or treatment. Such systems are also known as *sanitary sewer collection systems.* Stormwater piping systems convey captured stormwater runoff to points of discharge. Such systems are also known as *storm sewer systems.* There are systems, chiefly in older cities, that convey both domestic, commercial, and industrial wastewater and stormwater runoff in a single piping system. Such systems are called *combined sewer systems.*

The design of both wastewater and stormwater collection systems must comply with standards of city, county, and state, and federal regulation agencies. Permits must be obtained for the disposal of domestic, commercial, and industrial waste into receiving streams. Stormwater discharges to receiving streams may also require permits.

This chapter focuses on public and/or private wastewater and stormwater collection systems generally serving a number of buildings. For information about in-building plumbing, which is not covered in this chapter, the reader is referred to Chap. 13 of this handbook. The scope of this chapter does not permit examination of appropriate piping materials for handling exotic wastes such as oils, acids, and nuclear materials. See Chap. 1 of this handbook for specific information on piping systems to convey such wastes.

DEFINITIONS

A selection of terms used in wastewater and stormwater technology follows. Additional terms are defined at the place of use.

Sewer: A pipe or conduit that carries wastewater or drainage water.

Wastewater: The spent water of a community. From the standpoint of source, it may be a combination of liquid and water-carried wastes from residences, commercial buildings, industrial plants, and institutions, together with any ground water, surface water, and stormwater that may be present. The term *wastewater* is often used instead of the less inclusive *sewage.*

Stormwater: The portion of water that runs off the ground during and immediately following a rainstorm, snow melt, or other flooding event.

Sanitary sewer: A sewer that carries liquid and water-carried wastes from residences, commercial buildings, industrial plants, and institutions, together with minor quantities of storm, surface, and ground waters that are not admitted intentionally.

Storm sewer: A sewer that carries stormwater and surface water, street wash, and other waters or drainage but excludes domestic or industrial wastewater. It is also called a *storm drain.*

Combined sewer: A sewer intended to receive both wastewater and storm or surface water.

Building sewer: In plumbing, the extension from the building drain to the public sewer or other place of disposal; also called *house connection.*

Lateral sewer: A sewer that discharges into a branch or other sewer and has no other common sewer tributary to it.

Branch sewer: A sewer that receives wastewater from a relatively small area and discharges into a main sewer serving more than one branch-sewer area.

Main sewer: The principal sewers to which branch sewers are tributary; also called *trunk sewer.*

Intercepting sewer: A sewer that receives dry-weather flow from a number of branch or main sewers or outlets and frequently additional predetermined quantities of stormwater (in combined sewer systems) and conducts such waters to a point for treatment or disposal.

Outfall sewer: A sewer that receives wastewater from a collection system or from a treatment plant and carries it to a point of final discharge.

Sanitary system: A sewer system composed exclusively of sanitary sewers which carry only wastewater and to which stormwater, surface water, and ground water are not intentionally admitted; also referred to as *separate system* or *separate sanitary system.*

Storm sewer system: A system composed only of sewers carrying stormwater, surface water, street wash, and other wash waters or drainage from which domestic wastewater and industrial wastes are excluded.

Combined sewer system: A system of sewers receiving both surface runoff and wastewater.

QUANTITY OF FLOW

Quantity of Wastewater Flow

One of the first critical steps in the design of sanitary sewers is the estimation of design flows of wastewater tributary to them. Sanitary sewers must be designed to provide capacity for the present and estimated future quantities of domestic sewage, commercial and industrial wastes, and ground-water infiltration.

Design Period and Tributary Flow. Lateral and branch sewers should be designed for the ultimate population density to be expected in the area served. Larger sewers are commonly designed to handle the flows to be expected from 25 to 50 years in the future. The estimation of future flows should be arrived at only after a detailed study of the land usage, population growth trends, water-consumption rates, commercial and industrial growth, etc. Forecasts of population, commercial, industrial, and institutional development expected to be in place at the end of the design period should be obtained from local and regional planning organizations adjusted when appropriate by site-specific knowledge.

Quantities of Flow—Overview. Wastewater tributary to sanitary sewers emanates from residential, commercial, and industrial land uses. The most important single index in estimating flows to the sanitary sewer from each of these land-use sources is the rate of water consumption. Flows entering the sanitary sewers generally can be estimated as the water consumption, less an allowance for uses outside the building that typically do not return to the sewer (landscaping, car washing, etc.), plus an allowance for infiltration of clear water into the sewers. It should be noted that water consumption means water delivered to the user. The quantity of water actually pumped into the water distribution system includes water that will be lost as leakage en route to users. If the designer has pumping records available, as opposed to consumption (or water billing) records, an allowance of 5 to as much as 20 percent should be made for leakage in the potable water system.

The quantities of water consumption depend to some degree on the type of plumbing fixtures in residential and commercial buildings. Low-flow plumbing fixtures, if prevalent in an area, will decrease water consumption and consequent sewage flows. The presence of such fixtures should be accounted for in estimates.

When estimates of sanitary sewage flow from small areas are being prepared, the most accurate procedure is to make separate estimates of the various classifications of flow which make up the total. The classifications which are commonly used are domestic, commercial, and industrial sewage flows and ground-water infiltration.

Quantity of Domestic Sewage. Domestic sewage flow, excluding allowances for infiltration from residential land uses, is generally equivalent to water consumption minus an accounting for outside the home water use that does not flow back to the sanitary sewer (landscaping, car washing, etc.). The amount of water used for such outside uses varies depending on climate of the area and landscaping practices. In temperate climates, such as the Midwest and eastern United States, there is a distinct seasonal fluctuation of domestic water consumption due to significant outside water uses for landscaping, and so on. Peak water consumption occurs during summer seasons and minimum consumption occurs during winter months when outside uses are negligible. In such areas the wastewater flow rate (excluding infiltration) is best gauged by observing water-consumption records during winter months. In warmer, drier climates where landscaping water use continues year-round, an estimated allowance can be made during lowest water-consumption months. The ratio of annual domestic sewage flow, excluding infiltration, to annual water consumption (water billing, if available) ranges from 70 percent in arid areas with extensive landscaping water use to 95 percent in areas where little or no landscaping water use is practiced.

The average per capita domestic water consumption varies from about 40 to 120 gpd, depending upon the character of the area and the economic status of the population. If water is supplied through meters, accurate estimates of average per capita consumption can be made. If water is supplied unmetered, estimates have to be based on the water-consumption rates or measured sewer flow rates which are known to prevail in other areas of similar character. Table C12.1 presents typical wastewater flow rates from residential sources.[1]

Quantity of Commercial Sewage. The quantity of sewage flow from commercial areas varies widely depending upon the nature of the commercial activity. Allowances made for the quantity of sewage from commercial areas in large sewer districts are commonly in terms of either: gallons per day per acre, gallons per day per square foot of floor space, or gallons per day per capita or per employee. Allowances made by wastewater utilities for average wastewater flow from commercial areas vary from 2000 to 60,500 gpd per acre and from 15 to 500 gpd per capita. Table C12.2 shows typical average commercial flows for various commercial categories.[2] It is evident from the wide range of estimates shown in Table C12.2 that, for any area in which commercial activity is an important factor, the estimate of sewage flow should be based on a special study of the area.

Quantity of Industrial Sewage. The flow of sewage from industrial establishments may be purely sanitary sewage, or it may also include water-borne industrial wastes. Estimates of the sanitary sewage are made by the procedures already considered. The quantity of industrial wastes can be determined only by special studies of the individual industrial activities. When large industrial waste flows are involved, the problem of collection and disposal of these wastes usually requires special engineering studies.

Quantity of Infiltration. The rate of infiltration of ground water into sewers is influenced by the size, age, and condition of the sewers; the position of the sewers with respect to the ground-water table; the character of the soil; and the amount

TABLE C12.1 Typical Wastewater Flow Rates from Residential Sources

Source	Unit	Flow, gal/unit · d	
		Range	Typical
Apartment			
High rise	Person	35–75	50
Low rise	Person	50–80	65
Hotel	Guest	30–55	45
Individual residence			
Typical home	Person	45–90	70
Better home	Person	60–100	80
Luxury home	Person	75–150	95
Older home	Person	30–60	45
Summer cottage	Person	25–50	40
Motel			
With kitchen	Unit	90–180	100
Without kitchen	Unit	75–150	95
Trailer park	Person	30–50	40

Source: Adapted in part from Ref 1.

TABLE C12.2 Average Commercial Wastewater Flow

Type of establishment	Avg. flow, gpd/cap
Stores, offices, and small businesses	12–25
Hotels	50–150
Motels	50–125
Drive-in theaters (3 persons per car)	8–10
Schools (no showers), 8-h period	8–35
Schools (with showers), 8-h period	17–25
Tourist and trailer camps	80–120
Recreational and summer camps	20–25

Source: Adapted in part from Ref. 2.

of precipitation. The infiltration rate for any one system will vary from season to season. There has been extensive work in quantifying and removing excessive infiltration in existing systems in the United States during the period of 1975 to 1990. Consequently, sewer monitoring results of infiltration are often available for existing sewer systems and should be consulted.

Specifications for installation of new sewers limit infiltration to be between 200 and 500 gpd per inch-diameter mile of sewer at the time of testing. Allowances for infiltration at the end of the design period are higher than the initial test allowances. Design allowances for infiltration for new sanitary sewers at the end of the design period range from 10,000 to 40,000 gpd/mi of sewer.

Quantity of Exfiltration. Exfiltration is the process of wastewater inside the sewer flowing out of the sewer through joints, pipe cracks, and so on, into the surrounding bedding. Exfiltration is usually not of concern with respect to estimating wastewater flows because sanitary sewers generally do not function under pressurized conditions. However, exfiltration is important to consider in conjunction with protection of ground water in pressurized tunnels. In near-surface open-channel sewers, significant exfiltration can occur only when the sewer is in gross disrepair in granular soils or is cross-connected to a storm sewer or other outlet. In cases where flow monitoring indicates exfiltration, investigation of the situation is warranted to determine the flow path of the sewage.

Flow Variations. Sewers must be designed to handle the peak flow rates that are expected to occur at the end of the design period. It is also desirable to design them so as to minimize the problem of solids deposition during the early years of use when the flows may be much lower than the future flows. The flows vary from day to day and from hour to hour within each day. The ratio of the absolute maximum future flow rate to the initial minimum rate may vary from about 3 to 1 for large sewers serving highly developed areas to more than 20 to 1 for small sewers serving areas still under development. Figure C12.1 presents commonly used formulas for predicting ratios of maximum to average flows and minimum to average flows as a function of connected population.[2] These formulas assume dry-weather conditions (without excessive inflow) and no unusual industrial-use patterns. Where nondomestic flows make up significant amount of flow, the flow patterns should be considered separately.

Fixture-Unit Basis of Design. For very small tributary populations and for institutions such as schools, hospitals, hotels, and factories, the required capacities of

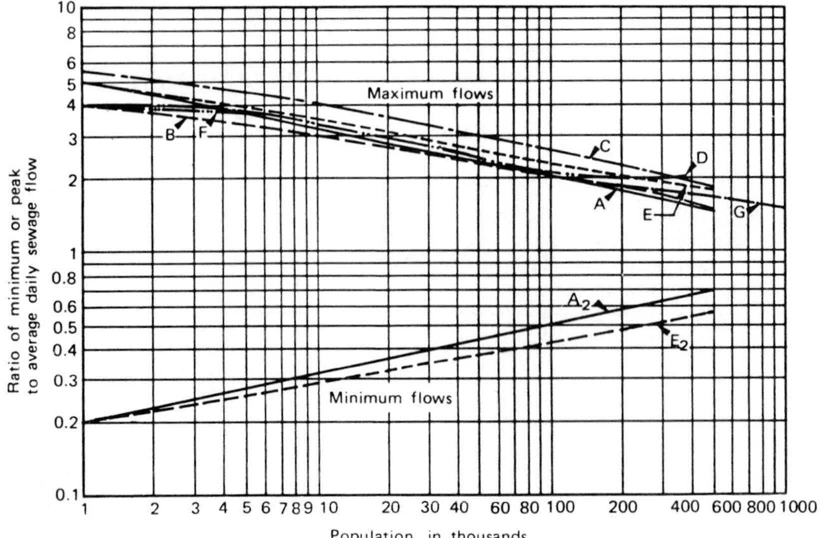

Curve A source: Babbitt, H.E., "Sewerage and SewageTreatment." 7th Ed., John Wiley & Sons, Inc. New York (1953).

Curve A_2 source: Babbitt, H.E., and Baumann, E.R., "Sewerage and Sewage Treatment." 8th Ed., John Wiley & Sons, Inc., New York (1958).

Curve B source: Harman. W.G., "Forecasting Sewage at Toledo under Dry Weather Conditions," Eng. News-Rec. 80, 1233 (1918)

Curve C Source: Youngstown Ohio, report

Curve D source: Maryland State Department of Health curve prepared in 1914. In "Handbook of Applied Hydraulics." 2nd Ed., Mcgraw-Hill Book Co., New York (1952)

Curve E source: Gifft, H.M., "Estimating Variations in Domestic Sewage Flows." Waterworks and Sewerage, 92,175 (1945).

Curve F source: "Manual of Military Construction." Corps of Engineers, United States Army, Washington, D.C.

Curve G source: Fair, G.M. and Geyer, J.C., "Water Supply and Waste-Water Disposal." 1st Ed., John Wiley & Sons, Inc., New York (1954)

Curves A_2, B, and G were constructed as follows:

$$\text{Curve } A_2, \quad \frac{5}{P^{0.167}}$$

$$\text{Curve B}, \quad \frac{14}{4 + \sqrt{P}} + 1$$

$$\text{Curve G}, \quad \frac{18 + \sqrt{P}}{4 + \sqrt{P}}$$

in which P equals population in thousands.

FIGURE C12.1 Ratio of Extreme Flows to Average Daily Flow Compiled from Various Sources. (*Adapted from Ref 2.*)

sanitary sewers may be estimated from the "fixture-unit flow rates" defined by various local and national plumbing codes. The designer should consult local applicable plumbing codes for estimating such flows.

Requirements of Regulatory Agencies. Some state regulatory agencies have established definite per capita flow rates to be used when detailed studies and estimates of expected flows have not been made. The designer should look into municipal and state regulations to obtain such regulatory requirements.

Typical of such regulations are the recommended standards of the Great Lakes–Upper Mississippi River Board of State Sanitary Engineers. These standards, used widely by state agencies in the United States in regard to the design of sanitary sewers, are as follows[3]:

Design Period. In general, sewer systems should be designed for the estimated ultimate tributary population, except in considering parts of the systems that can be readily increased in capacity. Similarly, consideration should be given to the maximum anticipated capacity of institutions.

Design Factors. In determining the required capacities of sanitary sewers, the following factors should be considered:

1. Maximum hourly quantity of sewage
2. Additional maximum sewage or waste from industrial plants
3. Ground-water infiltration

Design Basis. There are two methods accepted. These are:

- *Per capita flow:* New sewer systems should be designed on the basis of an average daily per capita flow of sewage of not less than 100 gallons per capita per day (gpcd). This figure is assumed to cover normal infiltration, but an additional allowance should be made where conditions are unfavorable. Generally the sewers should be designed to carry, when running full, not less than the following daily per capita contributions of sewage, exclusive of sewage or other waste from industrial plants:

  Laterals and submain sewers: 400 gpcd.

  Main, trunk, and outfall sewers: 250 gpcd.

  Interceptors: Intercepting sewers, in the case of combined sewer systems, should fulfill the above requirements for trunk sewers and have sufficient additional capacity to care for the necessary increment of stormwater. Normally no interceptor should be designed for less than 350 percent of the gauged or estimated dry-weather flow.

- *Alternate method:* When deviations from the foregoing per capita rates are demonstrated, a brief description of the procedure used for sewer design must be included in the design report.

Summary of Considerations for Determination of Sanitary Sewer Capacity. The following summary of the considerations necessary for the determination of sanitary sewer capacity is given in the *American Society of Civil Engineers (ASCE) Manual of Engineering Practice, No. 37* [*Water Pollution Control Federation* (*WPCF*) *Manual of Practice, No. 9*], "Design and Construction of Sanitary and Storm Sewers":

The quantity of wastewater which must be transported is based on full consideration of the following:

1. The design period during which the predicted maximum flow will not be exceeded, usually 25 to 50 years in the future.
2. Domestic sewage contributions based on future population and future per capita water consumption, or if a more satisfactory parameter than water consumption is available, that parameter should be used. Careful analysis should be made of population distributions and the relationship of maximum and minimum to average per capita sewage flows. The fixture-unit method of estimat-

ing peak rates should be employed for small populations, giving due care to the forecasting of the probable number of fixture units and water use per capita. When large areas are to be considered, the peak rate of flow per capita or per acre sometimes is decreased as area and population increase.

3. In some instances, maximum flow rates may be determined almost entirely by extraneous flows, the source of which may be foundation, basement, roof, or areaway drains, storm runoff entering through manhole covers, or infiltration. Foundation, roof, and areaway drain connections to sanitary sewers should be prohibited. Proper construction and yard grading practices should be mandatory. Nevertheless, there may be times when strict prohibition may not be feasible or even practicable. In any event, some storm and surface water will get into separate sanitary sewers, and a judgment allowance, therefore, must be made.
4. Commercial-area contributions are sometimes assumed to be adequately provided for in the peak allowance for per capita sewage flows in small communities. A per acre allowance for comparable commercial areas based on records is the most reasonable approach for larger communities.
5. Industrial waste flows should include the estimated employee contribution, estimated or gauged allowances per acre for industry as a whole, and estimated or actual flow rates from plants with process wastes which may be permitted to enter the sanitary sewer.
6. Institutional wastes are usually domestic in nature although some industrial wastes may be generated by manufacturing at prisons, rehabilitation centers, and so on. Peak and minimum design flow rates from persons in the institution are discussed elsewhere.
7. Air-conditioning and industrial cooling waters, if permitted to enter sewers, may amount to 1.5 to 2.0 gpm per ton of non-water-conserving cooling units. Unpolluted cooling water should be kept out of separate sanitary sewers.
8. Infiltration may occur through defective pipe, pipe joints, and structures. The probable amount should be evaluated carefully. Design allowances should be larger (under some circumstances very much larger) than those stipulated in construction specifications for which acceptance tests are made very soon after construction. Underevaluation of infiltration is one reason why some sewers have become overloaded.
9. The relative emphasis given to each of the foregoing factors varies among engineers. Some have set up single values of peak design flow rates for the various classifications of tributary area, thereby integrating all contributory items. It is recommended, however, that maximum and minimum peak flows used for design purposes be developed step by step, giving appropriate consideration to each factor which may influence design.

Design Example for Calculating Wastewater Flows

Example. Determine the average and peak wastewater design flows for the downstream end of a sanitary sewer designed to serve an area of 340 acres. Current residential population is 1000 (primarily single-family homes). Projected ultimate population in the future is 3500 (3000 people in single-family homes and 500 people in apartments). Additionally, commercial and institutional employment in the service area currently is 150 employees, primarily in office and non-water-intensive businesses. Future commercial-institutional employment in the service area is projected to be 900 again in similar types of businesses. There is negligible existing industrial development, but an 80-acre industrial park is planned.

Solution. The solution proceeds stepwise through the parameters to be considered:

1. *Select the design period of the sewer:* For this sewer the design condition is the ultimate future condition.

2. *Estimate average unit rate of flow of domestic sewage from residential areas:* Based on water billing records of existing residential single-family accounts, average water consumption during winter months (November to January) is 80 gpcd, and during summer months (June to August) it is 110 gpcd.

Apartments in adjacent communities have winter water unit consumption rates of 65 gpcd. Average wastewater flow from the residential sector is, therefore, 80 gpcd times 1000 people, or 80,000 gpd for existing conditions. For future conditions, the average flow is 3000 times 80 gpcd for single-family residential plus 500 times 65 gpcd for apartments, which totals to 272,500 gpd.

3. *Estimate commercial-institutional wastewater flows:* Using flow rates from similar establishments in adjacent communities, it was found that an average per employee flow rate of 25 gpd was appropriate. Current flow from this sector is estimated to be 3800 gpd, and future average flow would be 900 employees times 25 gpcd, or 22,500 gpd.

4. *Estimate the industrial wastewater:* Current industrial flow is negligible. Projection of future flow requires a thorough analysis of the planned industrial park. It was determined in consultation with the developer, municipal officials, and the zoning board that the ultimate future industrial park would have 800 employees and one significant water-intensive industry. The water-intensive industry will discharge process water to the sewer at the rates of 40,000 gpd (average) and 80,000 gpd (peak hour). Total estimated average future flow is estimated to be 800 employees times an allowance of 25 gpd per employee (20,000) plus 40,000 gpd for process water. These two factors total 60,000 gpd of average future flow.

5. *Estimate an allowance for the peak quantity of future infiltration:* Because the sewers are of new construction, the future allowance for infiltration will be 15,000 gpd per mile of sewer. Estimate sewer length roughly at 1 mi of sewer per 25 acres of service area. Estimated sewer length for the 340-acre area is 13.6 mi. Allowance for infiltration will be 15,000 gpd per mile times 13.6 mi of sewer, or 204,000 gpd.

6. *Determine peaking factor:* Peaking factor is taken from Fig. C12.1 using curve *G*. Where significant nonresidential sources are present, it is common to substitute population equivalent for population in the equation. Population equivalents of commercial and industrial flow are estimated by dividing the per employee water rate by the per resident water rate and multiplying by the number of employees. For the future condition the population equivalent of the commercial industrial component is 25 gpcd divided by 78 gpcd (weighted-average rate of single-family and apartment usage) times 1700 employees (900 commercial and 800 industrial). The total population equivalents in the future would be 3500 population plus 545 population equivalents for commercial-industrial sector, or 4045. The peaking factor from curve *G* would be 3.33.

7. *Estimate peak design flows:* Peak design flows are calculated by first assessing the average flow rate from residential, commercial, and industrial sectors excluding infiltration and specialized industrial waste. This would be 272,500 gpd (residential) plus 22,500 gpd (commercial) plus 20,000 gpd (industrial), or 315,000 gpd. Multiply this sum by the peaking factor of 3.33 to obtain a peak hourly flow of 1,050,000 gpd. Add to this the peak allowance for infiltration of 204,000 gpd and peak flow of process water from the water intensive industry of 80,000 gpd. The total peak hour design flow would be 1,334,000 gpd.

Quantity of Stormwater Flows

Stormwater runoff is the portion of precipitation which flows over the ground surface during and after a precipitation event. The design of storm sewers and combined sewers requires the estimation of design stormwater flows. There are a number of methods to develop peak stormwater flows. Among these are the rational formula, the hydrograph method, the inlet method, and the unit hydrograph method. These basic methodologies have been incorporated into a number of proprietary and public domain computer programs. Among the public domain programs, Illinois Urban Drainage Area Simulator (ILLUDAS) and Stormwater Management Model (SWMM), are in common usage in the profession.

Discussion of each of the methods is beyond the scope of this chapter. As such, discussion herein will be limited to a description of the rational method, the simplest and most basic analytic procedure in common use.

Level of Protection. The development of peak design flows for storm sewers is based on the area-specific political, economic, regulatory, and meteorological conditions. Typical practice is to design the system to accommodate the peak flows of a rainfall event that has a probability of recurring every 3 to 10 years. The duration and intensity of that rainfall, established from area meteorological history, are used to estimate the peak design flows.

The Rational Method. The rational method is the most basic procedure in common usage for the computation of the rates of stormwater runoff for storm sewer design. The rate of runoff Q in cubic feet per second is given by the equation:

$$Q = CiA \tag{C12.1}$$

in which A is the size of the drainage area in acres, i is the average intensity of rainfall during the design event in inches per hour for a duration equal to the time of concentration of the area, and C is a runoff coefficient whose value depends principally upon the character of the area.

The assumption behind the rational method is that the runoff rate for a given rainfall intensity will increase and reach its maximum when the duration of the rainfall reaches the time of concentration of the area (the time required for the runoff to flow from the remotest point of the area to the point where Q is being measured). By this assumption, the maximum runoff rate Q which can be expected to occur with any given frequency will be produced by a storm having the maximum average rainfall intensity corresponding to the given frequency and duration. The application of the method requires knowledge of the rainfall intensity-duration-frequency characteristics for the locality. Although the assumptions are not strictly in accord with the mechanics of the runoff process, the rational method has proven to be a practical procedure for storm-drain design because the accumulated experience has resulted in practicable values for the runoff coefficient. Reported practice generally limits the use of the rational method to urban areas less than 5 square miles. For larger areas application of hydrograph methods is usually warranted.

Time of Concentration. The time of concentration is the time required for the runoff to flow from the remotest point of the drainage area to the point under design. This is the minimum time necessary to permit the entire area to contribute to the flow at the point. The time of concentration consists of the inlet time, or time required for the runoff at the upper end of the area to reach the nearest inlet, plus the time of flow in the sewer from this inlet to the point being considered.

Inlet time will vary with the nature of the surface, the slopes, the nature of the established drainage channels such as street gutters, and the antecedent conditions. Because the inlet time is small, it is commonly chosen on the basis of experience. In densely developed areas with a high percentage of paved surfaces and closely spaced inlets, an inlet time as low as 5 min may be assumed. In moderately developed urban areas with flat slopes, the inlet time may vary from 10 to 20 min. In flat residential areas having a relatively low percentage of paved surface, the inlet time may be as high as 30 min. It is possible to make estimates of the inlet time by calculating the time of flow over the various types of surfaces, but such estimates can rarely be made with a high degree of accuracy.

The time of flow in the sewer is computed from the hydraulic properties of the sewer, the common practice being to use the average flowing-full velocity computed for the prevailing slope. The use of full-flow velocity in the calculation of the time of flow in the sewer is a convenient method of making the initial determination of time of concentration for hand calculations. The design can be checked to account for variance in travel times for partially filled or surcharged sewers and rainfall pattern during the design storm.

The calculation of the time of concentration is illustrated by the following example:

Example. Determine the time of concentration for the drainage system shown in Fig. C12.2.

Solution. First, estimate the time for conveyance of runoff from the remotest catchment area to the remotest inlet. In a moderately developed urban area, use

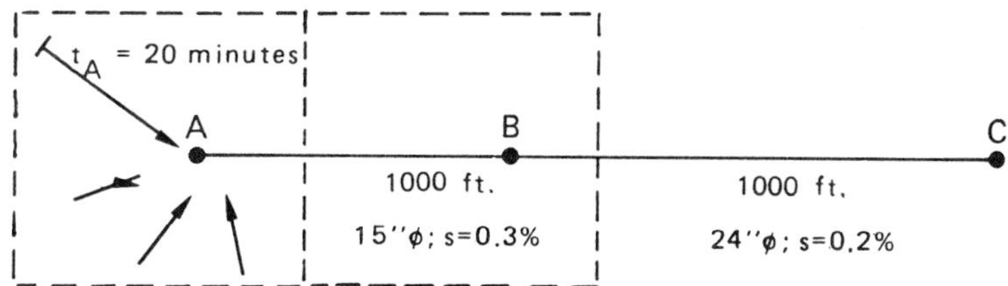

Time of Concentration for Design Point C =

$$
\begin{aligned}
t_A &= 15.0 \text{ minutes} \\
+\ t_{A\text{-}B} &= 5.7 \text{ minutes} \\
+\ t_{B\text{-}C} &= 5.2 \text{ minutes} \\
\hline
t_C &= 25.9 \text{ minutes}
\end{aligned}
$$

FIGURE C12.2 Calculation of time of concentration.

15 min for flow from the catchment area to travel to point *A*. Next calculate time for flow to travel from *A* to *B* using Manning equation [Eq. (C12.2)]. Velocity in the *A* to *B* section at full flow is 2.9 ft/s. The time to travel from *A* to *B* is 1000 ft divided by 2.9 ft/s, or 5.7 min. Similarly, the time to travel from *B* to *C* at a full-flow velocity of 3.2 ft/s is 5.2 min. The time of concentration, therefore, to point *A* is 15 min, to point *B* is 20.7 min, and to point *C* is 25.9 min.

Rainfall Frequency. It is usually prohibitive, on the basis of cost, to construct storm sewers capable of handling the largest conceivable storms. Current practice is to use storm rainfalls having an average expected frequency of once every 3 to 10 years for the design of storm sewers in residential areas and storms of 10 to 30 years for commercial and high-value districts.

Rainfall Intensity-Duration-Frequency Relationships. The rainfall characteristics which must be known for storm sewer design are presented in a very concise manner by the intensity-duration-frequency curves, which can be prepared from a long record of precipitation at a given station. Figure C12.3 shows such a set of curves prepared by the Illinois State Water Survey for the northeastern area of Illinois for the years 1901 to 1983.[4] This figure shows, for example, that for a system to be designed with a time of concentration of 30 min and a design recurrence

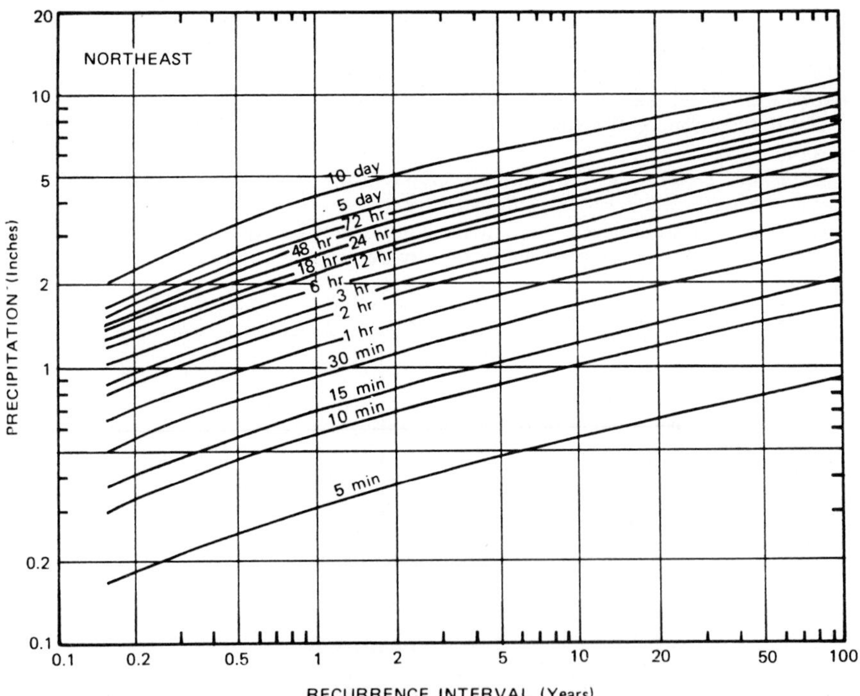

FIGURE C12.3 Frequency distributions of rainfall for Northeast Illinois climatic section for storm periods of 5 min to 10 days and recurrence intervals of 2 months to 100 years. (*From F. A. Huff and J. R. Angel, "Frequency Distributions and Hydroclimatic Characteristics of Heavy Rainstorms in Illinois," Illinois State Water Survey, 1989.*)

interval of 5 years that the total precipitation is 1.3 in. This means that the total rainfall of 1.3 in during a 30-min period will be equaled or exceeded in the area for which the curves apply, on the average, once every 5 years. The rainfall-intensity term i in the rational method equation [Eq. (C12.1)] is expressed in units of inches per hour. For the 30-min duration, 5-year recurrent storm, the rainfall intensity would be 1.3 in divided by 0.5 h, or 2.6 in/h. Similar data to that shown on Fig. C12.3 for many other localities have been compiled and published by the U.S. Weather Bureau. State agencies in charge of stormwater management should be consulted to obtain rainfall intensity-duration-frequency data accepted for use in local areas as such data vary widely by geographic region.

Runoff Coefficient. The runoff coefficient C in the rational method formula is the variable which is least susceptible to precise determination. Whereas the use of the runoff coefficient implies that there is a constant ratio of runoff to rainfall, the actual ratio for a given area will depend upon the condition of the area at the time of occurrence of the storm and will increase with the duration of the storm. A more logical procedure than the rational method for storm-drain design would be to subtract the rainfall losses due to infiltration and retention in surface depressions and to distribute the remainder as an actual hydrograph of runoff. Computer programs attempt to account for such parameters. However, because of the great variability in the time distribution of the rainfall itself as well as the difficulty in estimating the quantities of infiltration and surface depression storage, the use of the rational formula simply involves an estimate of the value of the runoff coefficient C. A common practice is the use of average coefficients for various types of districts, the coefficients being assumed to be constant throughout the storm duration. The range of values reported to be in common use is shown in Table C12.3.

For specific small areas it is more logical to relate the value of C to the actual type of surface. Coefficients commonly used are shown in Table C12.4. When an

TABLE C12.3 Runoff Coefficients by Land Use

Type of area	Runoff coefficient, C
Business	
Downtown areas	0.70–0.95
Neighborhood areas	0.50–0.70
Residential	
Single-family areas	0.30–0.50
Multiunits, detached	0.40–0.60
Multiunits, attached	0.60–0.75
Residential (suburban)	0.25–0.40
Apartment dwelling areas	0.50–0.70
Industrial	
Light areas	0.50–0.80
Heavy areas	0.60–0.90
Parks, cemeteries	0.10–0.25
Playgrounds	0.20–0.35
Railroad yard areas	0.20–0.40
Unimproved areas	0.10–0.30

Source: Adapted from Ref. 2.

TABLE C12.4 Runoff Coefficients by Surface Characteristics

Character of surface	Runoff coefficient
Streets:	
Asphaltic	0.70–0.95
Concrete	0.80–0.95
Brick	0.70–0.85
Drives and walks	0.75–0.85
Roofs	0.75–0.95
Lawns, sandy soil:	
Flat, 2%	0.05–0.10
Average, 2 to 7%	0.10–0.15
Steep, 7%	0.15–0.20
Lawns, heavy soil:	
Flat, 2%	0.13–0.17
Average, 2 to 7%	0.18–0.22
Steep, 7%	0.25–0.35

Note: Percentages given are average ground surface slopes in the catchment area.
Source: Adapted from Ref. 2.

area is made up of different types of surfaces, a common procedure is to use a weighted-average coefficient.

The coefficients in Tables C12.3 and C12.4 are designed for use for storms of 5- to 10-year frequencies. For less frequent, higher-intensity storms, the coefficients should be higher, because the infiltration and surface retention will be smaller proportions of the total precipitation. Likewise, for more frequent, lower-intensity storms, the coefficients should be lower than indicated in the tables.

HYDRAULICS OF SEWERS

Where possible it is common practice to design sewers to flow with a free water surface. Hydraulically, this condition is termed *open-channel flow.* The advantages of open-channel flow are twofold: (1) the free water surface will allow ventilation of the sewers, and (2) the velocities at lower flows can be kept reasonably high to facilitate self-cleansing of the sewers. In areas where the depth of excavation for a sewer becomes uneconomically large, a lift (pumping) station and force main are commonly installed to convey wastewater to a location where gravity open-channel flow can resume. The force main is termed such because sewage is "forced" by pumping, and the flow is pressurized flow as opposed to open-channel (gravity) flow.

The capacity of a sewer pipe should be sufficient to carry the peak flow rate to be anticipated at the end of the design period, and the slope should be sufficient to provide for self-cleansing velocities during the early years of use. It is common practice to design sanitary sewers with slopes sufficient to provide for velocities of 2 ft/s when flowing full. Experience shows that with such slopes trouble from deposits is seldom encountered. Storm sewers are commonly designed for a minimum full-flow velocity of 3 ft/s in order to resuspend sediment deposited from intermittent storm events.

Although the flow in sewers is seldom steady or uniform, it is impracticable in most cases to take this into account, and each section of the sewer is usually designed with the assumption that the flow is steady and uniform. In certain conditions it is important to check the impacts of nonuniform flow. These conditions include the drawdown, backwater, and hydraulic jump conditions. These specialized conditions are briefly discussed later in this chapter. Specialized computer programs that dynamically route flows can analyze such conditions. Such programs, currently proprietary, should be entering the public domain in coming years.

Uniform Flow Formulas and Calculations

The most widely used formula for calculating open-channel uniform flow in sewers is the Manning equation. The Manning equation is:

$$V = \left(\frac{1.486}{n}\right) r^{2/3}\, S^{1/2} \tag{C12.2}$$

where V = mean velocity, ft/s
n = dimensionless roughness coefficient
r = hydraulic radius, ft, which is the wetted cross-sectional area divided by the wetted perimeter
S = the slope of the energy gradient

The roughness coefficient n varies from 0.010 for smooth surfaces to as high as 0.10 for rough natural channels. It is common practice to use values of Manning's n of 0.013 for sewer design. This value makes some allowance for the future condition of the pipe as well as disturbances in the flow resulting from rough joints, interior coatings of grease or other matter, and flow disturbances due to changes in pipe size, junctions, and so on. Figure C12.4 is a diagram for the solution of the Manning equation applied to circular pipes flowing full, with n equal to 0.013.

Pipes Flowing Full. The Manning equation can be transformed to conveniently determine the quantity of flow of a circular pipe *running full* at a particular slope as follows:

$$Q = \frac{0.463}{n} D^{8/3} S^{1/2} \tag{C12.3}$$

where Q is quantity of flow at full flow, ft^3/s; D is diameter of pipe, ft; and n and S are defined above.

Similarly, other convenient transformations of the full pipe Manning formula are as follows:

$$V = \frac{0.590}{n} D^{2/3} S^{1/2} \tag{C12.4}$$

$$S = \left(\frac{4.66}{D^{16/3}}\right) n^2 Q^2 \tag{C12.5}$$

$$D = \left(\frac{1.33}{S^{3/16}}\right) Q^{3/8} n^{3/8} \tag{C12.6}$$

Pipes Flowing Partly Full. Manning's equation and Fig. C12.4 are convenient for determining flows and velocities for pipes running full. When determining velocities and depths of flow when pipes are flowing under partially full conditions, diagrams

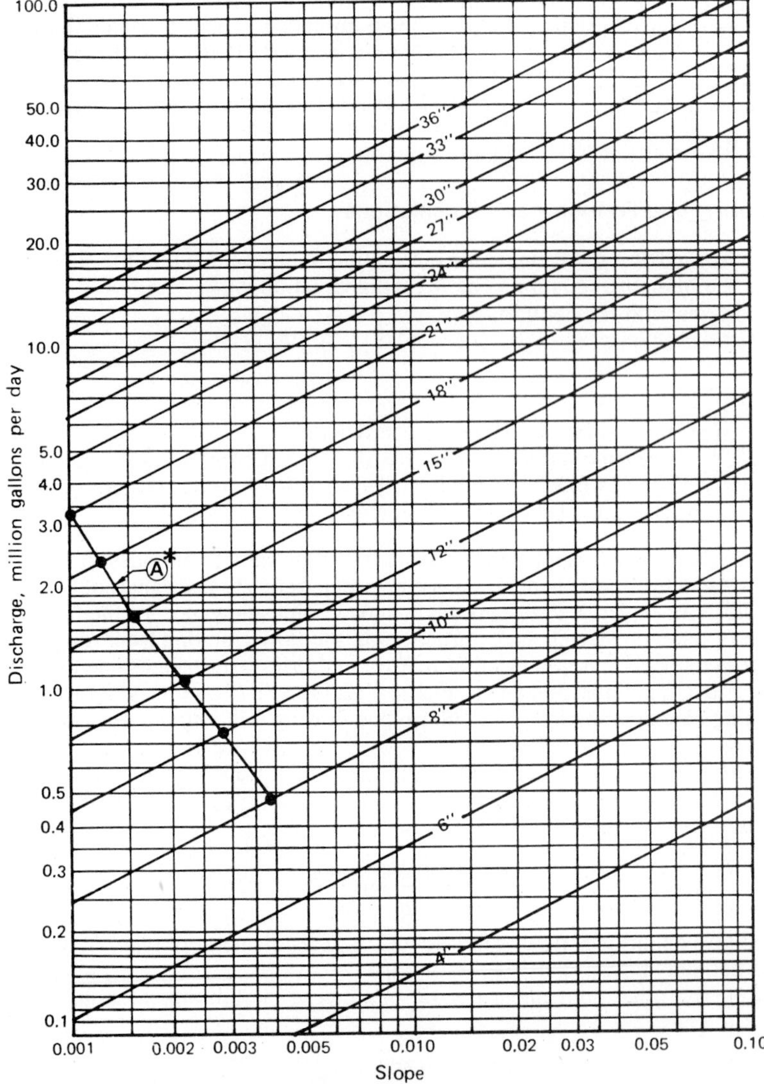

FIGURE C12.4 Discharge of circular pipes (running full) based on Manning formula:

$$Q = A\left(\frac{1.486}{n}\right)r^{2/3}S^{1/2}$$

where Q = flow rate, A = cross-sectional area, r = hydraulic radius, S = slope of energy gradient, and n = coefficient of roughness = 0.013. *Curve Ⓐ—Minimum Slopes per "Recommended Standards for Sewage Works," Great Lakes—Upper Mississippi River Board of State Sanitary Engineers. (*Adapted from Ref. 3.*)

such as Fig. C12.5 are used. Figure C12.5 gives the hydraulic elements of circular pipes flowing partly full. This figure shows the ratios of the values of the various elements to the values for the flowing-full condition. The cross-sectional area and the hydraulic radius are purely geometric functions and hence independent of n. The velocity and discharge for any particular ratio of depth to diameter depend upon whether n is assumed to be constant or variable with the depth. Velocity and discharge curves computed from both assumptions are shown.

The use of Manning equation, Fig. C12.4, and of Fig. C12.5 are illustrated by the following example:

Example. A 36-in sewer pipe (n = 0.013) is laid on a slope of 2.00 ft per 1000 ft. Find the discharge capacity and velocity of flow of the sewer flowing full, and find the depth of flow and the velocity when the flow rate is 20.0 ft³/s. Assume that n varies with the depth of flow.

Solution. Using the transformed Manning equation for full flow:

$$Q = \frac{0.463}{n} D^{8/3} S^{1/2}$$

the full-flow discharge capacity is 29.8 ft³/s (n = 0.013, D = 3.0 ft, S = 0.002). This checks against Fig. C12.4 which yields a reading of 20.0 mgd (1.0

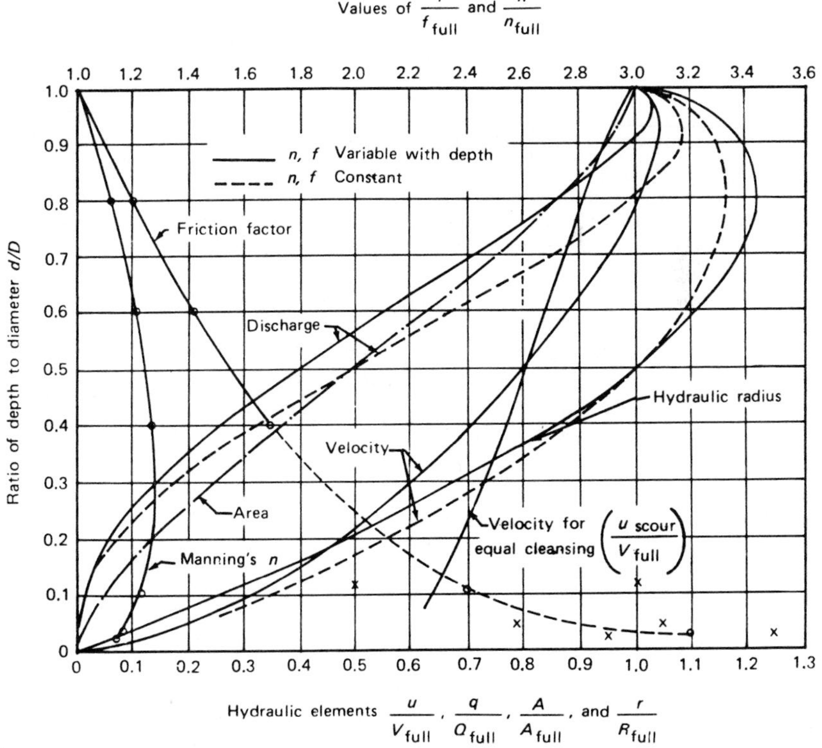

FIGURE C12.5 Hydraulic elements of circular sewers. (*Adapted from Ref. 8.*)

mgd ≈ 1.5 ft^3/s). Velocity at full flow from Manning's equation is 4.22 ft/s as calculated from Eq. (C12.4).

At 20 ft^3/s the sewer is partially full. Referring to Fig. C12.5, q/Q full = 20.0/29.8 = 0.67. The term q is the flow in cubic feet per second for the partially full condition being analyzed. Start at the bottom axis at q/Q = 0.67 and read up to a point on the solid-line discharge curve (n, f variable with depth). Refer to the left axis to determine that the ratio of the actual depth to the full depth (d/D full) is 0.67. Then, reading across to the velocity curve, the mean velocity at d/D = 0.67 is 0.94 of the full-flow velocity. Consequently, for a flow of 20 ft^3/s the depth of flow will be 0.67 times 36 in, or 24 in, and the mean velocity will be 0.94 times 4.22 ft/s, or 3.97 ft/s.

Figure C12.5 also shows the relative velocities required to obtain equal cleansing of the pipe at all depths of flow. This is based on T. R. Camp's analyses of the movement of granular materials in open channels.[5] The diagram indicates that, if a sewer has self-cleansing velocities under flowing-full conditions, the velocity will also be self-cleansing for all flow conditions at depths greater than one-half the diameter. This concept is important for designing sewers that initially will experience relatively small flows but are installed to accommodate large flow in the future when population growth occurs. In the early years the sewer may never experience full-flow conditions. The use of the velocity for equal cleansing curve on Fig. C12.5 is best illustrated by example.

Example. A 36-in-diameter sewer is being installed to accommodate a future flow of 14.0 ft^3/s. Design flow in the first 10 years of installation will be 2.5 ft^3/s. The target full-flow velocity for self-cleansing is 2.0 ft/s. Determine the necessary velocity for equivalent scouring at 2.5-ft^3/s flow.

Solution. First, determine d/D for 2.5 ft^3/s. This is done by first calculating q/Q (2.5 ft^3/s/14.0 ft^3/s = 0.18). Then read up from the bottom axis to the discharge curve, and then read d/D from the left axis as 0.33. Then read to the right to the velocity for equal cleansing curve to determine the v scour/V full ratio. The term v scour is that velocity required at a given depth d that will provide equivalent scouring to that provided at full depth (d/D = 1.0). The v scour/V full ratio for d/D = 0.33 is 0.74. If the target self-cleansing velocity at full flow is 2.0 ft/s, then an equivalent cleansing velocity of d/D = 0.33 is 0.74 × 2.0 ft/s, or 1.48 ft/s. Determine what slope is needed to provide 1.48-ft/s velocity at d/D = 0.33. Velocity at d/D = 0.33 is determined by reading from the left axis to the velocity curve and then down to determine v/V full. The v/V full ratio is 0.64. If the self-cleansing velocity of full flow is targeted at 2.0 ft/s, the actual velocity at d/D = 0.33 would be 2.0 × 0.64 = 1.28 ft/s. Because the actual velocity at d/D = 0.33 of 1.28 ft/s is less than the required self-cleansing velocity of 1.48 ft/s, there is a probability of deposition of solids in the first 10 years. To remedy this situation, it is advised to steepen the slope of the sewer to be installed to provide a full-flow velocity greater than 2.0 ft/s.

Surcharged Flow. Although sewers are not usually designed to flow in a pressurized condition, often the analysis of existing sewer systems requires examination of conditions where the quantity of flow exceeds the full-flow design of the sewer. Additionally, sometimes storm sewers are designed to surcharge slightly under peak flow. The analysis of flow under such pressurized conditions in the waterworks area has traditionally been carried out using the Hazen-Williams formula. The Manning equation can also be used.

Example. A 36-in sewer pipe (n = 0.013) is laid on a slope of 2.00 ft per 1000 ft. The length of the sewer is 1000 feet. Find the level of water in the upstream manhole with respect to the sewer invert when 50 ft^3/s is conveyed. Assume that the downstream discharge water level is equal to the crown of the sewer at the downstream end.

Solution. Solving for S, the Manning equation under full-flow conditions becomes:

$$S = \frac{4.66Q^2n^2}{D^{5.33}}$$

At $Q = 50$ ft^3/s, $n = 0.013$, $D = 3.0$ ft, S becomes 0.0056. It should be noted that under surcharged conditions, S in the Manning formula is taken to be the slope of the hydraulic gradient rather than the actual slope of the physical pipe. Therefore, for 1000 ft the required gradient would be 5.6 ft to convey 50 ft^3/s of flow. The drop in elevation of the pipe over the 1000 ft at 0.002 slope is 2.0 ft. Therefore, the surface of the water at the upstream manhole would be surcharged 3.6 ft above the crown of the pipe.

Hydraulic Impact of Bends and Manholes. In extensive sewer systems, most of the pipes will have mild slopes, and the flows will be subcritical. Extra energy losses occur at all transitions where changes occur in size, slope, or direction of the pipe and at junctions where several pipes come together. If the transitions are properly designed to allow for these energy losses, the condition of uniform flow may be approximated in the individual lines, and the flows will never be at depths greater than the depths computed on the assumption of uniform flow. However, if the transitions are not properly designed, pipes may at times flow at depths greater than the computed depths and surcharge may occur under peak flow conditions. The hydraulic principles involved in transition design are illustrated in Fig. C12.6, which shows a transition where a pipe flows into a larger pipe laid on a flatter grade. Due to the turbulence created by the flow expansion there will be a head loss h_t. In order to prevent the upstream pipe from flowing at a depth greater than its normal depth d_1, the relative elevations of the pipes must be such that the energy gradient of the downstream pipe is lower than that of the upstream pipe by the amount of h_t. This requires that the invert of the downstream

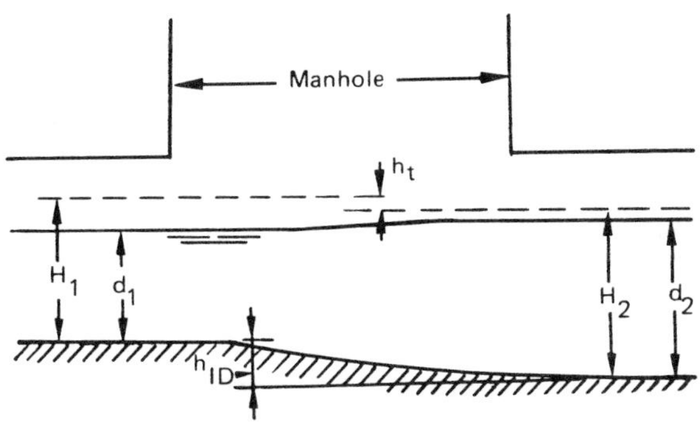

FIGURE C12.6 Flow profile at junction.

be placed below that of the upstream pipe by the amount h_{ID}. The relationship between the various vertical dimensions is given by the equation

$$h_{ID} = H_2 - H_1 + h_t \qquad \text{(C12.7)}$$

The equation assumes that the loss is concentrated at the center of the transition. As actually constructed, the transition will take place within a manhole, and the channel section within the manhole is made to provide for a gradual transition between the two pipes. If the computed invert drop h_{ID} is negative, it is usually taken as zero, and the pipe inverts are placed at the same elevation.

The energy loss h_t is usually small, but it can assume fairly high values when high velocities are involved. Data on the magnitudes of h_t are scarce, but such data as are available indicate that h_t can be represented as a fraction of the change in velocity heads in accordance with the equation

$$h_t = K\Delta\left(\frac{v^2}{2g}\right) \qquad \text{(C12.8)}$$

where h_t is the energy loss in feet caused by the transition, K is a constant, characterizing the degree of the hydraulic disturbance of the transition, v is the velocity in feet per second, and g is 32.2 ft/s^2.

Values of K for smooth transitions might be taken as low as 0.10 for increasing-velocity transitions and 0.20 for decreasing-velocity transitions. Increased transition losses occur when a sewer line changes direction and at junctions where one or more branch sewers join a main sewer. Reliable information on the transition head losses in such cases is almost entirely lacking. The hydraulic design of junctions may be considered as the design of two or more transitions, one for each path of flow. The exit sewer is common to all paths, and its invert must be placed at the lowest computed elevation. Because of the lack of information on the transition losses, allowances are usually made in accordance with the judgment of the designer. An arbitrary procedure which is commonly adopted is to allow about twice as much loss along flow paths in junctions as compared with the allowances for simple transitions involving the same velocities.

Aspects of Nonuniform Flow

As stated previously, routine design procedures for sewers assume uniform flow conditions. This assumption, although not precisely correct, is adequate in the majority of situations. It is noted that some situations require a more thorough analysis of non-uniform-flow conditions.

The significance of some non-uniform-flow conditions are illustrated by reference to Fig. C12.7. The flow conditions for each of the cases shown are explained briefly below:

1. A channel with a mild slope discharges freely into the atmosphere (Fig. C12.7*a*). Critical depth will occur at the outlet. The depth will increase at successive sections upstream until the normal depth is reached, beyond which the flow will be uniform. In many cases the length of the channel will be much less than the distance required to develop normal depth.

   The significance of this is that for a short stretch of pipe, its actual capacity to carry flow without surcharge may be much greater than would be calculated by assuming uniform flow.

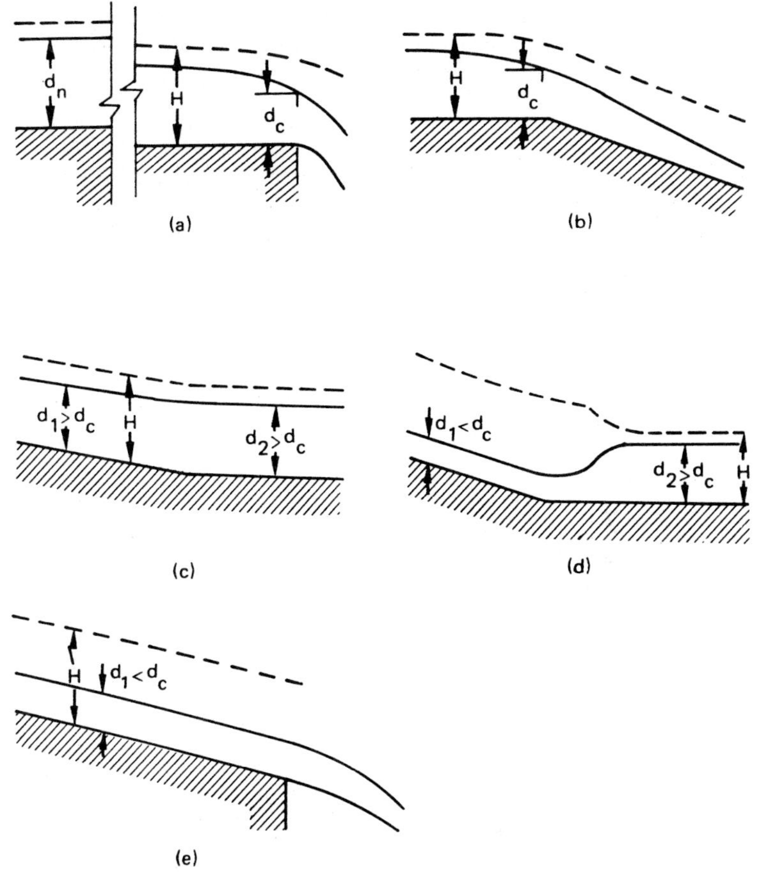

Note:
d_n = normal depth, d_c = critical depth, $d_{1,2}$ = actual flow depth, H = energy head

FIGURE C12.7 Examples of nonuniform flow profiles.

2. A channel changes slope from mild to steep (Fig. C12.7*b*). The conditions upstream of the junction will be the same as for paragraph 1. The depth downstream of the junction will decrease and approach normal depth.
3. A channel changes slope from a mild slope to another mild slope (Fig. C12.7*c*). Upstream of the junction there will be a gradual decrease in the depth from section to section.
4. A channel changes slope from steep to mild (Fig. C12.7*d*). In this case, the change from the upstream supercritical flow to the downstream subcritical flow will take place suddenly in a "hydraulic jump." The position of the jump may be either upstream or downstream of the junction, depending upon the relative values of the various parameters which control the flow pattern.

5. A steep slope discharges into the atmosphere (Fig. C12.7*e*). In this case the flow will be at the normal supercritical flow, provided that the upstream control has permitted normal depth to be developed.

The foregoing discussion on nonuniform flow is presented to show the reader the importance of understanding these principles. For detailed presentation of nonuniform flow, the reader is referred to textbooks on the subject.[6,7]

DESIGN ASPECTS OF SEWERS

Sewer designs are governed by minimum design standards in many areas. In the United States various city, county, or state standards may apply. Various countries also have national standards. Many of these standards repeat similar design criteria.

Commonly Used Design Criteria. Commonly used design criteria include the following:

- *Minimum size for gravity sewers:* 8-in diameter.
- *Minimum depth:* Sufficient to receive sewage from basements and prevent freezing.
- *Slope:* As required to produce velocities no less than 2.0 ft/s when flowing full at $n = 0.013$.

  Uniform slope between manholes.

  Velocities greater than 15 ft/s must be protected.
- *Alignment:* Sewers 24 in or less should be straight between manholes.
- *Changes in pipe size:* When a smaller sewer joins a larger pipe, the invert of larger sewer should be lowered to maintain energy gradient. This criteria is approximately met by matching 0.8 depth point of both sewers, or by matching crowns in smaller pipes.
- *Manholes:* To be placed at the end of each line; at all changes in grade, size, or alignment; at all intersections of sewers and at intervals of not more than approximately 400 ft in smaller sewers and 600 ft in larger sewers.
- *Relation to water mains:* Sewers shall be at least 10 ft horizontally from water mains, and where they cross, an 18-in minimum separation shall be maintained.
- *Design pressure:* Sewers convey wastewater under atmospheric pressure. In some instances, where the wastewater must be pumped, the pipe must be designed for maximum expected internal pressure. For additional information, see Chap. C1.
- *Design temperature:* Wastewater and stormwater are generally at ambient temperatures. Temperature of the stormwater may vary significantly depending on the season, such as winter and summer; however, no special consideration is warranted for the design of underground sewer pipes.

Design Examples

Design examples are presented for a sanitary sewer system and a storm sewer system in the following section. These examples are adapted from Ref. 8.

Design Example of Sanitary Sewer System

Example. Design a sanitary sewer system for the residential district shown in Fig. C12.8. The district is two-thirds developed; therefore, the probable future population density can be estimated without making a detailed population study. It is estimated that the future average saturation population density will be 65 persons per acre. The maximum hourly rate of flow of sewage is estimated at 250 gpcd. The maximum rate of ground-water infiltration to the sewers, to be provided for, is 2000 gpd per acre.

The minimum size of sewer is to be 8 in. The minimum velocity of flow in the sewer when full is to be 2.0 ft/s. The capacity of the sewers will be determined using Manning equation with a recommended n value of 0.013.

Since the homes in this area have basements, the minimum depth below the street surface to the top of the sewers will be 7.0 ft. (In areas where basements are not normally constructed, the depth of cover to the top of the sewer may be as little as 3.0 ft.)

Solution

1. Draw a line to represent the proposed sewer in each street or alley to be served. Near the line indicate by an arrow the direction in which the sewage is to

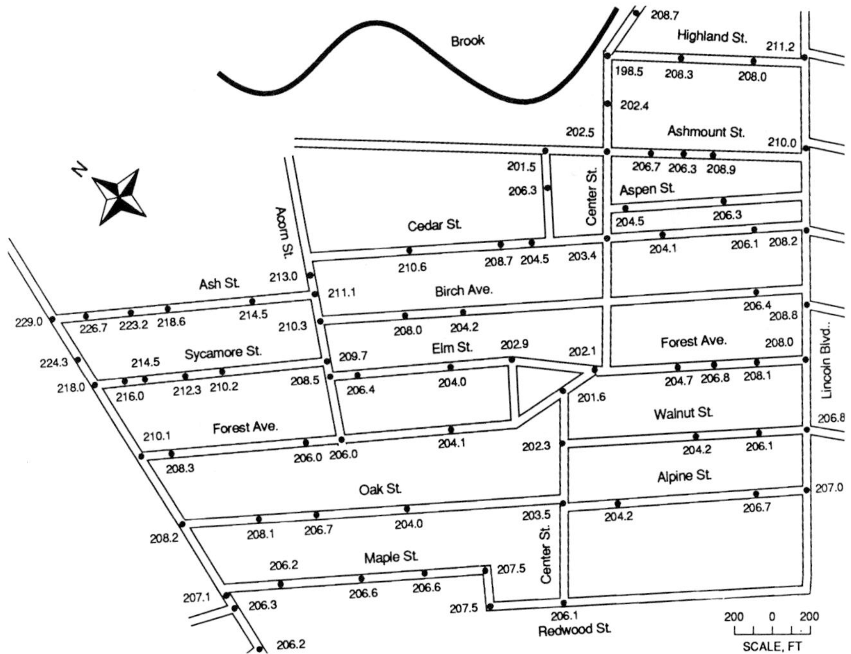

FIGURE C12.8 Typical map for the design of sanitary sewers. (*Adapted from Ref. 8.*)

flow. Except in special cases, the sewer should slope with the surface of the street. It is usually more economical to plan the system so that the sewage from any street will flow to the point of disposal by the most direct (and consequently the shortest) route.

The lines representing the system will often resemble a tree and its branches. In general, the laterals connect with the submains; and these, in turn, connect with the main or trunk sewer, which leads to the point of discharge.

2. Locate the manholes, giving each an identification number.

3. Sketch the limits of the service areas for each lateral, unless a single lateral will be required to accommodate an area larger than can be served by the minimum size of sewer with the minimum slope, in which case a further subdivision may be required. Where the streets are laid out, the limits may be assumed as being midway between them. If the street layout is not shown on the plan, the limits of the different services areas cannot be determined as closely and the topography may serve as a guide.

4. Measure the acreages of the several service-areas. For this, a planimeter will give results with sufficient accuracy. At this point, the design may be represented as shown in the plan view in Fig. C12.9.

5. Prepare a tabulation, such as that shown in Table C12.5, with columns for the different steps in the computation and a line for each section of sewer between manholes. This tabulation is a concise, time-saving method and shows both the data and the results in orderly sequence for subsequent use.

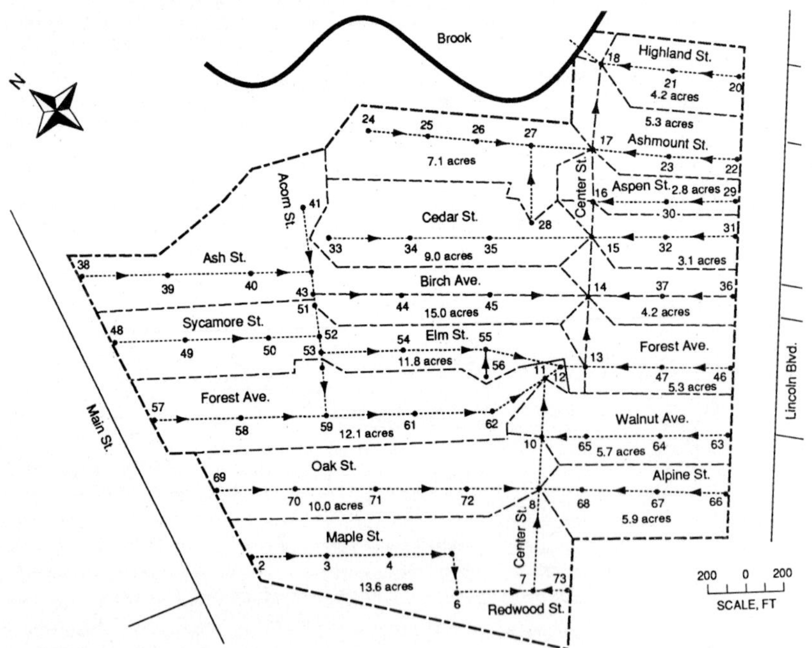

FIGURE C12.9 Map showing manholes, sewer lines, and subareas for sanitary sewer design example. (*Adapted from Ref. 8.*)

TABLE C12.5 Computations for a Sanitary Sewer

Line (1)	From manhole no. (2)	To manhole no. (3)	Location (4)	Length, ft. (5)	Area, acres: Incre-ment (6)	Area, acres: Total (7)	Sewage, mgd* (8)	Ground water at 2000 gpad (9)	Total max. flow, sewage and ground water*: mgd (10)	Total max. flow, sewage and ground water*: ft^3/s (11)	Size of sewer, in (12)	Slope, ft/ft (13)	Veloc-ity, ft/s (14)	Capac-ity, ft^3/s (15)	Surface eleva-tion, upper end (16)	Invert elevation: Upper end (17)	Invert elevation: Lower end (18)
1	57	58	Forest Ave.	380	...		...		...		8	0.004	2.2	0.77	208.2	200.40	198.89
2	58	59	Forest Ave.	370	...		...		...		8	0.004	2.2	0.77	206.4	198.89	197.40
3	59	61	Forest Ave.	365	...		...		...		8	0.004	2.2	0.77	205.2	197.40	195.94
4	61	62	Forest Ave.	370	...		...		...		8	0.004	2.2	0.77	204.3	195.94	194.46
5	62	11	Forest Ave.	240	...	12.1	0.196	0.024	0.220	0.34	8	0.004	2.2	0.77	202.0	194.46	193.50
6	11	12	Forest Ave.	130	35.2	47.3	0.767	0.095	0.862	1.34	12	0.0023	2.2	1.70	201.6	189.30	189.00
7	12	13	Forest Ave.	82	11.8	59.1	0.960	0.118	1.078	1.67	12	0.0023	2.2	1.70	202.1	189.00	188.81
8	13	14	Center St.	280	5.3	64.4	1.046	0.129	1.175	1.82	15	0.0017	2.3	2.70	202.8	188.56	188.08
9	14	15	Center St.	275	19.2	83.6	1.360	0.167	1.527	2.37	15	0.0017	2.3	2.70	203.2	188.08	187.61
10	15	16	Center St.	113	12.1	95.7	1.555	0.191	1.746	2.70	15	0.0020	2.4	2.90	203.6	187.61	187.38
11	16	17	Center St.	245	2.8	98.5	1.600	0.197	1.797	2.78	15	0.0024	2.6	3.20	203.7	187.38	186.79
12	17	18	Center St.	375	12.4	110.9	1.800	0.222	2.022	3.13	15	0.0024	2.6	3.20	202.3	186.79	185.89
13	18	19	Right of way	130	4.2	115.1	1.871	0.230	2.101	3.26	15	0.0025	2.7	3.27	196.0	185.89	185.56

*Based upon a maximum rate of 250 gpcd and 65 persons per acre. Since the capacity of the minimum size of sewer with a minimum velocity of 2.0 ft/s is 0.7 ft^3/s, equivalent to the maximum rate of discharge from 24.6 acres, and since 24.6(65 × 250 + 2,000)1.55/1,000,000 = 0.7 all laterals will be 8 in in diameter (the minimum), as no lateral is to serve an area exceeding 24.6 acres.

Source: Adapted from Ref. 8.

Use col. 1 for numbering the lines of the table, for ready reference. Determine by inspection the manhole that is farthest from the point of discharge, and enter its identification number in the first line of col. 2, and the number corresponding to the manhole next on the line toward the trunk sewer in col. 3. Enter the name of the street or alley in col. 4, the length between manholes in col. 5, and the area in acres to be served by the sewer at a point just upstream up the lower manhole in col. 6.

On the next line enter the corresponding data for the next stretch of sewer, and in col. 7 enter the sum of the areas listed in col. 6. The area in col. 7 is the basis for computing the required capacity of the sewer. Enter the data for each section of sewer in the above manner, following the line to the point of discharge, including the trunk or main sewer.

Enter in col. 8 the rate of flow in the sewer, which is equal to the maximum per capita rate of sewage flow multiplied by the assumed future density multiplied by the area shown in col. 7.

Enter in col. 9 the rate of allowance for ground-water infiltration, which is equal to the rate per acre to be provided for, multiplied by the area in col. 7.

Column 10 contains the sums of the figures in cols. 8 and 9, in millions of gallons per day. In col. 11 this rate is converted to cubic feet per second, which is the more convenient way of expressing the capacity of sewers, since most diagrams and tables indicate the capacity of circular pipes in cubic feet per second (1.0 mgd = 1.54 ft^3/s).

Column 12 contains the required sewer sizes; col. 13, the slope; col. 14, the velocity when the sewer is full; and col. 15, the capacity. Column 16 contains the elevations of the street surface at the manhole corresponding to the identification number in col. 2. Columns 17 and 18 contain the invert elevations of the upper and lower ends, respectively, of each reach of sewer.

In selecting the sewer sizes and slopes, the designer makes use of profiles, such as the one shown in Fig. C12.10. This allows the designer to select a minimum sewer size and slope that will carry the computed flow and that also will meet minimum depth criteria.

Design of Storm Sewer System

Example. Design a storm sewer system for the area shown in Fig. C12.11. The location of the proposed main storm sewer that is to receive the stormwater from the district is shown on the map, and the invert elevation is known at the point where the proposed branch storm sewer is to be connected and for which provision has been made in the design of the main storm sewer. The required lowest elevation of the invert of the branch storm sewer is therefore known at the proposed point of discharge into the main storm sewer.

A careful study of local conditions, including the present and probable future development of the district, indicates that 70 percent of the surfaces in the district are expected to be impervious. The inlet time has been assumed to be 20 min.

The rate of rainfall is to be taken from the assumed curve of intensity of precipitation represented by the formula $i = 20.4/t^{0.61}$, in which i is rainfall intensity in inches per hour, and t is rainfall duration in minutes. This formula represents the average rate of rainfall for a duration of t min which may be expected to be equaled or exceeded on the average once in a 5-year period. The rainfall and runoff curves are shown on Fig. C12.12.

Note that the intensity of precipitation curve in Fig. C12.12 is site specific for the geographic area under consideration. For different areas different curves can

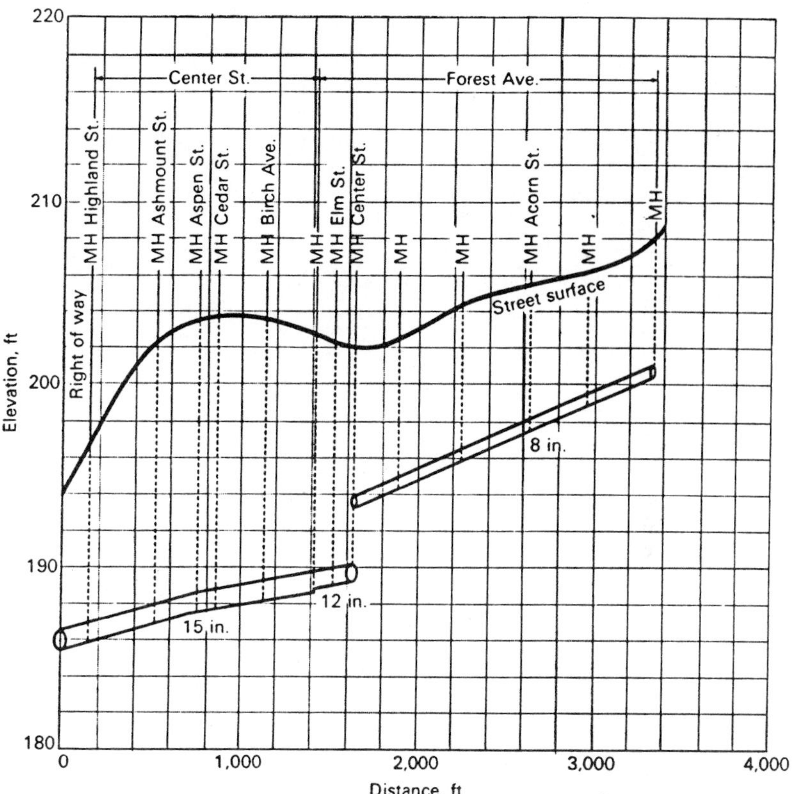

FIGURE C12.10 Typical profile for sanitary sewer design example. (*Adapted from Ref. 8.*)

be obtained normally from state agencies. Note also that the intensity of precipitation is for the 5-year recurrent interval storm. For design of systems for greater or lesser levels of protection different curves would be used. The runoff curves are provided for various coefficients of runoff C. The curves plot out the C times i terms in the rational method equation $Q = CiA$. Units for runoff are in inches of runoff per hour. To illustrate the chart use, assume that a particular area has a time of concentration of 60 min and the area is 70 percent impervious. The runoff would be read directly from the Ci curve for 70 percent imperviousness at 1.02 in of runoff per hour. Multiplying this number by the drainage area in acres yields the runoff flow in cubic feet per minute (one cubic foot per second is almost exactly equivalent to one acre-inch per hour). Note that 70 percent impervious area yields a coefficient of runoff C of about 0.60 for a 60-min duration storm. The weighted-average coefficient of runoff C of an area composed of 70 percent impervious area at a runoff coefficient of 0.80 and 30 percent pervious area with a runoff coefficient of 0.15 is 0.605. Runoff from a 60-acre area with a composite C of 0.605 for a storm with an intensity of precipitation of 1.68 in per hour would be 61 ft^3/s. This result can also be obtained by reading 1.02 in/h off the 70 percent impervious curve on Fig. C12.12 for a 60-min duration and multiplying by the area of 60 acres to yield 61 ft^3/s.

FIGURE C12.11 Map for storm sewer design example. (*Adapted from Ref. 8.*)

While it was recognized that storm sewers designed on this basis might be overtaxed on the average of once in about 6 years, it was not considered reasonable to provide for storms of greater intensity, because of the greater cost. During the earlier years of the life of the storm sewers, they will be able to carry the runoff from higher rates of rainfall than they will be able to carry later, because the assumed coefficients of runoff are based upon future rather than present conditions. A progressive increase in impervious surface and runoff will be caused by the gradual substitution of roofs and paved areas for presently unimproved areas. In the future, when the district is more densely built up and funds are available, relief sewers can be constructed to provide for higher runoff rates, if flooding has become sufficiently serious to warrant the expenditure.

Figure C12.11 shows the drainage area. Street elevations are shown in Fig. C12.8. The limits of this area are influenced not only by the surface contours but also by the service areas of existing storm sewers. In a district where the surface slopes are moderate and generally uniform, contour maps may not be required. Instead, surface elevations may be adequate if they are obtained for street intersections, for high and low points, and at locations of change of surface slope.

The storm sewers are to be designed, in general, with the crown at a depth of at least 5 ft below the surface of the street. The minimum size of drain is to be 12 in. The assumed minimum mean velocity is 3.0 ft/s when flow is at full depth.

The capacities of the sewers are to be determined using a value of $n = 0.013$. Velocities for flow at design conditions may be higher for storm sewers than for sanitary sewers for the reason that the design storm flow is many times greater than the peak sewage flow in sanitary sewers. Because of high velocities, it is

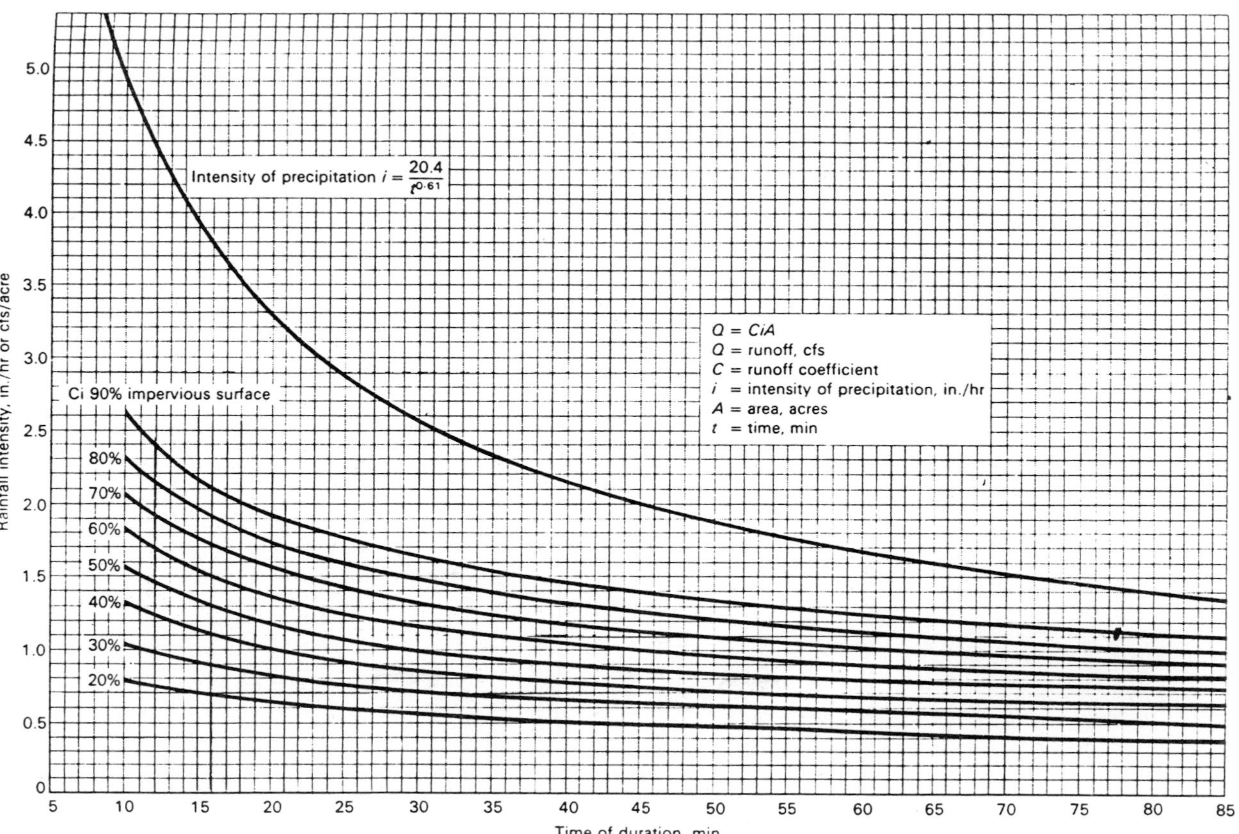

FIGURE C12.12 Precipitation-intensity curves for various degrees of imperviousness. (*Adapted from Ref. 8.*)

important to provide for additional head to compensate for losses, such as those due to bends, manholes, transitions, and velocity changes.

Solution

1. Draw a line to represent the storm sewer in each street or alley to be served. Place an arrow near each sewer to show the direction of flow. The sewers should, in general, slope with the street surface. It will usually prove to be more economical, however, to lay out the system so that the water will reach the main storm sewer by the most direct route.

2. Locate the manholes tentatively, giving to each an identification number. In this example, a manhole is to be placed at each bend or angle, at all junctions of storm sewers, at all points of change in size or slope, and at intermediate points where the distance exceeds 350 ft on 12- to 24-in sections and 400 ft for larger sections.

Where a good velocity would be maintained during practically all conditions of flow, and the sewer is large enough for workers to walk without stooping, intervals between manholes up to 600 ft may be used. Sufficient manholes should be built to allow access for inspection and cleaning. Later, when the profiles are drawn and the final slopes are fixed, it may be desirable to change the locations for some manholes so that the sewers would be at the most advantageous depth, particularly where the slope of the street surface is not substantially uniform. Other considerations, such as obstacles underground, may require the installation of additional manholes, due to change in alignment or special forms of construction involved in junctions or connections with other sewers.

3. Sketch the limits of the drainage area tributary at each manhole. The assumed character of future development and the topography will determine the proper limits.

4. Measure each individual area by planimeter or other method that will give equally satisfactory results.

5. Prepare a tabulation to record the data and steps in the computations of each section of sewer between manholes.

The computations for a selected line of this section are shown in Table C12.6. Computations are carried out as follows: Columns 1, 2, 3, 4, 5, and 6 are identified from the layout map of the system in Fig. C12.11. Column 7 is the travel time to the upstream manhole considered. For line 1, 20.0 min is the time for overland flow to reach the farthest inlet. For line 2 it is 20.0 min plus the travel time (1.7 min) in the sewer from points 1 to 2. The travel time in section (col. 8) is calculated by dividing col. 4 (length) by col. 13 (velocity) and then dividing the result by 60 to convert the time to minutes. Column 9 (runoff rate) is read from Fig. C12.12 using the 70 percent impervious curve for the time listed in col. 7. Column 10 (required capacity in cubic feet per second) is calculated by multiplying col. 9 by col. 6.

Columns 11 and 12 involve designing the pipe segment. A particular size and slope is determined from the Manning equation. Slope is determined by the constraints of: (1) maintaining adequate cover of 5.0 ft over the crown of the pipe at upstream and downstream manhole, (b) maintaining a minimum full-flow capacity of 3.0 ft/s; and (c) avoiding as much as possible excessive excavation depths. Column 13 (velocity) is determined from the Manning equation and from Fig. C12.5 for determining velocities in partially filled sewers. Column 14 is deter-

TABLE C12.6 Computations for Storm Sewers*

	From (1)	To (2)	Location (3)	Length, ft (4)	Increment, acres (5)	Total acres (6)	To upper end, min. (7)	In section, min. (8)	Rate of runoff, cfs per acre (9)	Required capacity, ft³/s (10)	Pipe size, in (11)	Slope, ft/ft (12)	Velocity, ft/s (13)	Capacity, ft³/s (14)	Surface elevation (15)	Fall, ft (16)	Upper end (17)	Lower end (18)
1	1	2	Maple St.	300	2.3	2.3	20.0	1.7	1.56	3.6	15	0.003	2.9	3.6	206.2	0.90	199.90	199.00
2	2	3	Maple St.	300	2.4	4.7	21.7	1.7	1.51	7.1	21	0.002	3.0	7.2	206.5	0.60	198.50	197.90
3	3	4	Maple St.	300	2.2	6.9	23.4	1.6	1.47	10.1	24	0.002	3.2	10.2	206.6	0.60	197.65	197.05
4	4	5	Maple St.	165	1.5	8.4	25.0	0.7	1.42	11.9	24	0.0027	3.8	11.9	207.1	0.45	197.05	196.60
5	5	6	Redwood St.	325	2.2	10.6	25.7	1.4	1.40	14.9	27	0.0023	3.8	15.0	207.4	0.75	196.35	195.60
6	6	7	Center St.	400	3.1	13.7	27.1	1.7	1.38	18.9	30	0.0021	3.9	18.8	206.1	0.84	195.35	194.51
7	7	8	Center St.	35	6.0	19.7	28.8	0.2	1.34	26.4	30	0.004	5.2	26.0	203.2	0.14	194.51	194.37
8	8	9	Center St.	230	10.2	29.9	29.0	1.0	1.34	40.0	42	0.0016	4.1	41.0	203.2	0.37	193.37	193.00
9	9	10	Center St.	240	5.7	35.6	30.0	0.9	1.32	47.0	42	0.0022	4.9	47.0	201.9	0.53	193.00	192.47
10	10	11	Forest Ave.	110	11.9	47.5	30.9	0.4	1.31	62.3	48	0.0018	4.9	62.0	201.6	0.20	191.97	191.77
11	11	12	Forest Ave.	95	11.1	58.6	31.3	0.3	1.30	76.3	54	0.0015	4.8	76.0	202.1	0.14	191.27	191.13
12	12	13	Center St.	295	5.6	64.2	31.6	0.9	1.30	83.3	54	0.0018	5.2	84.0	202.8	0.53	191.13	190.60
13	13	14	Center St.	260	17.2	81.4	32.5	0.8	1.28	104.0	60	0.0015	5.3	103.0	203.2	0.39	190.10	189.71
14	14	15	Center St.	145	13.7	95.1	33.3	0.5	1.27	121.0	66	0.0013	5.1	121.0	203.6	0.19	189.21	189.02
15	15	16	Center St.	225	2.9	98.0	33.8	0.7	1.26	124.0	66	0.0014	5.2	126.0	203.7	0.32	189.02	188.70
16	16	17	Center St.	380	13.2	111.2	34.5	1.1	1.25	139.0	66	0.0017	5.8	129.0	202.5	0.65	188.70	188.05†
17	17	18	Private land	165	4.1	115.3	35.6	—	1.24	143.0	48	0.0096	11.4	143.0	196.0	1.59	187.32	185.73

*Figures in col. 8 are obtained by dividing those in col. 4 by 60 and by the figures in col. 13. Figures in col. 10 are obtained by multiplying those in col. 6 by the figures in col. 9. Figures in col. 16 are obtained by multiplying those in col. 4 by the figures in col. 12.

†The difference in the elevations of the 66-in sewer at point 17 and the 48-in outlet sewer would allow for velocity head increase and bend losses.

Source: Adapted from Ref. 8.

mined as full-flow capacity of the pipe as determined from the size and slope. The capacity should equal or exceed the required capacity listed in col. 10.

Column 15 (surface elevation) is taken from street maps, survey information, or in preliminary design from topographical maps. Column 16 (fall) is calculated by multiplying col. 12 (slope) by col. 4 (length). Column 17 (invert at upstream end) presents the invert at the upstream manhole. At the farthest manhole the required depth of cover of 5.0 ft normally governs as a starting point. Note that in line 1 the invert of 199.90 plus the pipe size of 1.30 ft (15-in pipe with allowance for pipe thickness) gives an elevation of the top of the pipe as 201.20. This is 5.0 ft below the surface.

Column 18 (invert elevation at downstream end) is calculated as col. 17 (upstream invert) minus col. 16 (fall). The lower or downstream elevation should be checked to ensure that a minimum of 5.0 ft of cover is maintained. It should be noted that when pipe sizes change, the crowns of the sewers are generally matched. Therefore col. 17 on line 2 is 0.50 ft lower than col. 18 for line 1. This is accounted for by the increase in pipe size from 15 to 21 in.

Each lateral is then designed in a similar way. If necessary, the first design of the submain is subsequently modified so as to serve the laterals properly. It is possible in some cases to omit some manholes on lateral storm sewers, using the inlet substructures at which junctions, changes in size, direction, or slope may be made.

In relatively flat areas the design is limited by the total available head (elevation difference) between the outlet of the system and the upstream area. In such cases, the designer needs to be cognizant of the overall limitation when starting the analysis. In such areas larger pipe sizes at flatter slopes will be required.

Finally, the outlet conditions at the downstream end of the system may be variable especially if the storm sewer discharges to a stream or river. In such cases, the designer must determine an outlet design condition (level of the river) to form the basis for the hydraulic computations. The outlet condition, which may be the 5-year recurrent stage level in the river, is determined in conjunction with policymakers. The outlet condition selected will, especially in flat areas, have a large impact on the design and its cost.

SEWER PIPE

Available Pipe Materials

The materials of which street sewer pipes are most commonly constructed are vitrified clay pipe, concrete, plastic, asbestos cement, and ductile iron pipe. Suggested specifications and other pertinent information about sewer pipes is presented in Table C12.7.

Vitrified Clay Pipe. Vitrified clay pipe is manufactured in standard and extra-strength classifications. It is widely used especially in smaller sizes because of its resistance to corrosive wastewaters. It is available in sizes from NPS 4 to 36 in laying length from 1 to 10 ft.

Concrete Pipe. Concrete pipe is available either in plain or reinforced classifications in various strength categories. Concrete pipe is widely used especially in larger sizes. Unreinforced concrete pipe is available in sizes from NPS 4 to 36 in, and reinforced concrete pipe is available in sizes from NPS 12 to 144 in five

TABLE C12.7 General Information and Suggested Specifications for Piping Material and Joints

General Information and Suggested Specifications for Piping Material and Joints						
PIPE MATERIALS SPECIFICATIONS	COMMERCIAL SIZE (in.)	COMMERCIAL LENGTH (ft.)	JOINTS TYPES & MATERIAL SPECIFICATIONS	MAX. BURIED DEPTH (ft.)	INSTALLATION SPECIFICATIONS	REMARKS
Vitrified Clay Pipe						
ASTM C700 ASTM C700	4-12 15-24	2 5	ASTM C425 ASTM C425 Bell and spigot joint & rubber gaskets	Buried depth shall be calculated based on bedding conditions. For calculation refer to clay pipe engineering manual.	ASTM C12 for Installation ASTM C828 for Air testing ASTM C1091 for Infiltration and exfiltration testing	
Concrete Pipe						
ASTM C14	6-24	3½, 4, 6, 7½	Mortar ASTM C443 Bell & spigot joint, rubber gaskets	Buried depth for 6" Sewer up to 25: for 24" sewer up to 10 ft.	ASTM C924 for Air testing ASTM C969 for Infiltration and Exfiltration testing	Classification, 1-3 for gravity sewer only
Reinforced Concrete Pipe						
ASTM C76	12-120	6, 7½, 12	ASTM C443 Bell & spigot joint, rubber gaskets	35	ASTM C924 for Air testing ASTM C969 for Infiltration and Exfiltration testing ASTM C1103 for Joint acceptance testing	Classification, I-V for both pressure and gravity. For the strength of each class, see latest revision of ASTM C76
ASTM C361	12-42 48-108	8 7½	ASTM C433 Bell & spigot joint, rubber gaskets	20	ASTM C924 for Air testing ASTM C969 for Infiltration and Exfiltration testing ASTM C1103 for Joint acceptance testing	Classification, A, B, C, D. For low head 25, 50, 75, 100, 125 feet only

TABLE C12.7 General Information and Suggested Specifications for Piping Material and Joints (*Continued*)

PIPE MATERIALS SPECIFICATIONS	COMMERCIAL SIZE (in.)	COMMERCIAL LENGTH (ft.)	JOINTS TYPES & MATERIAL SPECIFICATIONS	MAX. BURIED DEPTH (ft.)	INSTALLATION SPECIFICATIONS	REMARKS
Reinforced Concrete Elliptical Culvert						
ASTM C507	14 x 23 to 63 x 93	7½	ASTM C877 External sealing bands, bell & spigot	15 feet for horizontal elliptical pipe	ASTM C924 for Air testing ASTM C969 for Infiltration and Exfiltration testing ASTM C1103 for joint acceptance testing	Classification, HE-A HE-I to HE-IV, VE-II to VE-VI
Reinforced Concrete Box Culvert						
ASTM C789	36 x 36 to 144 x 144	7½	ASTMC877 Bell & spigot	60		For buried depth for specific culvert, consult with concrete products manufacturer
ASTM C850	36 x 36 to 144 x 144	7½	ASTMC877 Bell and spigot	1		
PVC Pipe						
ASTM D1785	1/8 - 12 15-27	20 20	ASTM D2672 socket joint ASTM D2564 Solvent cement, bell and spigot	Buried depth shall be calculated based on bedding conditions, and sewer strength	ASTM D2321 for Installation ASTM D2774 for Installation ASTM C969 for Infiltration and Exfiltration testing ASTM C924 for Air testing	Schedule 40, 80, 120. For up to 6" pipe For both pressure and gravity sewers

ASTM D2241			ASTM D2672 socket joint ASTM D2564 solvent cement ASTM D3139 for pipe joints ASTM F477 rubber gasket Bell & spigot	ASTM D 2321 for Installation ASTM C924 for Air testing ASTM C969 for Infiltration and Exfiltration testing	Classification SDR from 13.5 - 64 For both pressure and gravity sewer
ASTM D2729	2-6	20	ASTM D2672 socket joint	ASTM D2321 for Installation ASTM C924 for Air testing ASTM C969 for Infiltration and Exfiltration testing	for gravity sewer
ASTM F679	18-36	20	ASTM D3139 for pipe joint ASTM F477 rubber gasket Bell & spigot	ASTM D2321 for Installation ASTM C924 for Air testing ASTM C1103 for Infiltration and Exfiltration	for gravity sewer
ASTM F794	4-48	20	ASTM D3139 ASTM F477 Bell & spigot	ASTM D2321 for Installation ASTM C924 for Infiltration and Exfiltration testing ASTM C1103 for joint acceptance testing	for gravity sewer

TABLE C12.7 General Information and Suggested Specifications for Piping Material and Joints

PIPE MATERIALS SPECIFICATIONS	COMMERCIAL SIZE (in.)	COMMERCIAL LENGTH (ft.)	JOINTS TYPES & MATERIAL SPECIFICATIONS	MAX. BURIED DEPTH (ft.)	INSTALLATION SPECIFICATIONS	REMARKS
<u>Asbestos Cement Pipe</u>						
ASTM C296	4-36	10-16	ASTM D1869 Coupling Rubber rings	Buried depth shall be calculated based on bedding conditions and sewer strength	ASTM C966 for Installation ASTM C924 for Air testing ASTM C969 for Infiltration and Exfiltration testing ASTM C1103 for Joint acceptance testing	For both pressure and gravity sewer classification 100, 150, 200. Type I and Type II
ASTM C428	4-42	10-16	ASTM D1869 Coupling Rubber rings	Buried depth shall be calculated based on bedding conditions and sewer strength	ASTM C966 for Installation ASTM C924 for Air testing ASTM C969 for Infiltration and Exfiltration testing ASTM C1103 for joint acceptance testing	For gravity sewer Classification 1500, 2400, 3300, 4000, 5000, 6000, 7000, type I and type II
<u>Ductile Iron Pipe</u>						
AWWAC 151/ ANSI A21.51	4-36	18	AWWAC111 ANSI A21.11 Bell & spigot mechanical, Push on, flange, rubber gaskets	32		Classification, 50-56 laying condition and buried depth governs the pipe classifications

strength classifications. For details of classification refer to ASTM Standard C-76. Joints are normally made with rubber gaskets in grooves formed in the tongue. Other jointing systems are also available.

Advantages of concrete pipe are: wide range of sizes, laying lengths, and strengths. A disadvantage of concrete pipe for sewers is that it is subject to corrosion under acid conditions. If flow velocities are insufficient to prevent the deposition of organic solids, septic conditions may result. Hydrogen sulfide gas produced by anaerobic decomposition of organic matter becomes oxidized to produce sulfuric acid, which damages the pipe. This condition can usually be prevented by designing the sewers so that self-cleansing velocities will occur most of the time. Protective linings, including coal-tar, coal-tar epoxy, vinyl, and epoxy mortars, can be used to prevent corrosion where wastewater is expected to be highly acidic or where deposition of solids is anticipated. Additional information can be obtained from the American Concrete Pipe Association.

Plastic Pipe. Plastic pipe used in sewage systems include PVC (polyvinyl chloride), ABS (acrylonitrile-butadiene-styrene), and PE (polyethylene). All offer advantages of corrosion resistance, low-friction characteristics, and light weight. Plastic pipes are generally less rigid. They require proper bedding and lateral support. Polyethylene, ABS pipes are also available in a variety of standard and proprietary products. The reader is referred to manufacturers for more information on these pipes.

Asbestos Cement Pipe. Asbestos cement pipe (ACP) has been widely used in sewer design. Advantages claimed for ACP is corrosion resistance and light weight. Note that ACP is one of the asbestos products that may be banned by the federal government in the future.

ACP is available in sizes up to 42 in. Joints are commonly with grooves and rubber rings.

Bituminized fiber pipes are obsolete. They may be found in old installations. The pipe sections may be replaced by essentially any other pipe material.

Ductile Iron Pipe. Ductile iron pipe (DIP) is employed in sewerage primarily for force mains and for piping in and around buildings. It is generally not used for gravity sewer applications.

APPURTENANCES AND SPECIAL STRUCTURES

Essential to all sewerage systems are the appurtenant structures such as service connections, manholes, junction chambers, stormwater inlets, and diversion chambers. The design of such structures is not covered in detail in this chapter, but typical designs for the most commonly used appurtenances will be presented briefly.

Building Service Connections

Figure C12.13 shows typical details of service connections to a sanitary sewer laid in a relatively shallow trench; Fig. C12.14 shows a typical connection to a

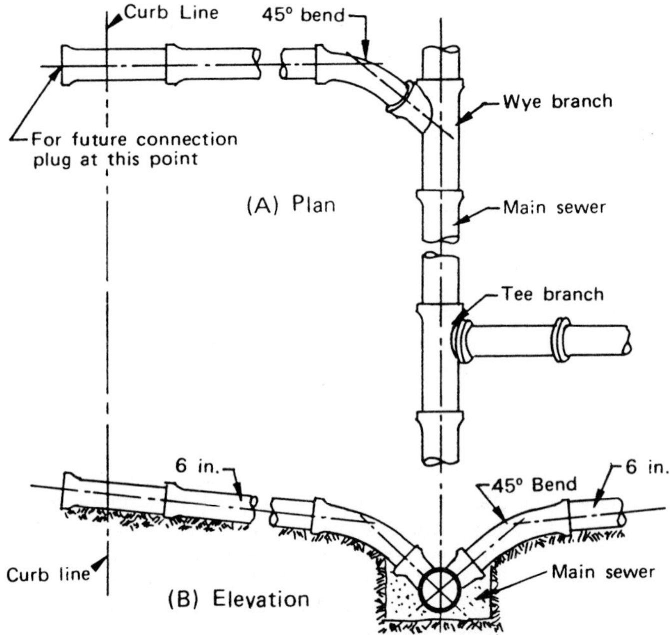

FIGURE C12.13 Typical service connections to a shallow sewer.

deep sewer. It is noted that the connection shown in Fig. C12.13 makes use of either a wye branch or a tee branch in the main sewer line.

Junction Chambers and Manholes

Figure C12.15 shows a typical design for a junction chamber and manhole for relatively small sewers. For junctions of large sewers, a special underground structure will ordinarily be required, and the entrance to it will be provided for by a manhole located at one side. Such chambers and manholes are required at every sewer junction and at every point where the sewer changes in size, slope, direction, or elevation. It is general practice to install sewers in straight lines between manholes, except that for the larger sizes (36 in and above) which may be laid on curves. Manholes are usually installed at the upper end of every lateral sewer and in straight-line sewers so that the spacing will not exceed about 400 to 600 ft. Figure C12.16 shows typical details of a "drop manhole" at a point where a sewer takes an abrupt drop in grade.

Stormwater Inlets

Stormwater inlets, which carry stormwater from the streets to the storm sewers, are located upstream of the crosswalks at street intersections and at low points. The designs vary considerably, and most cities have adopted their own standard design details. There are three general types of inlets: (1) curb inlets, which have

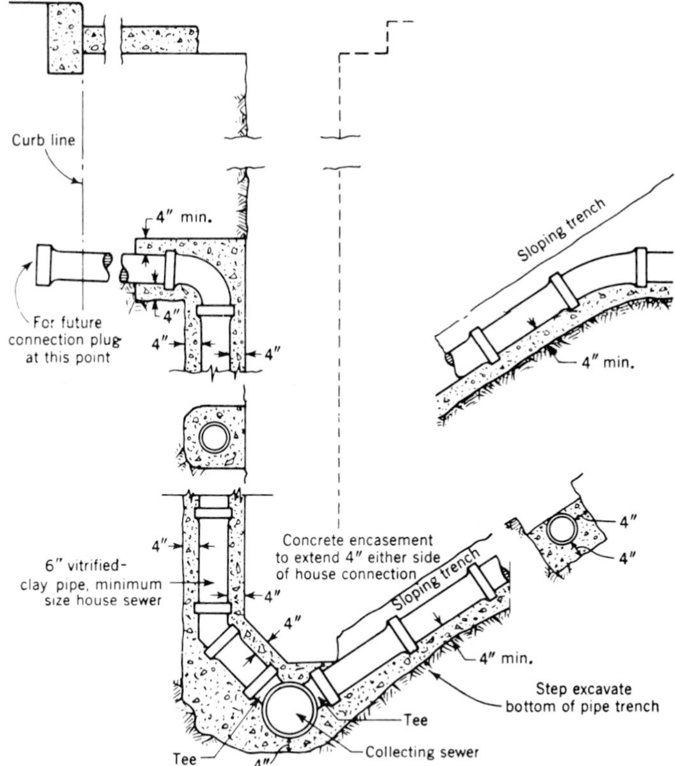

FIGURE C12.14 Typical service connection to a deep sewer.

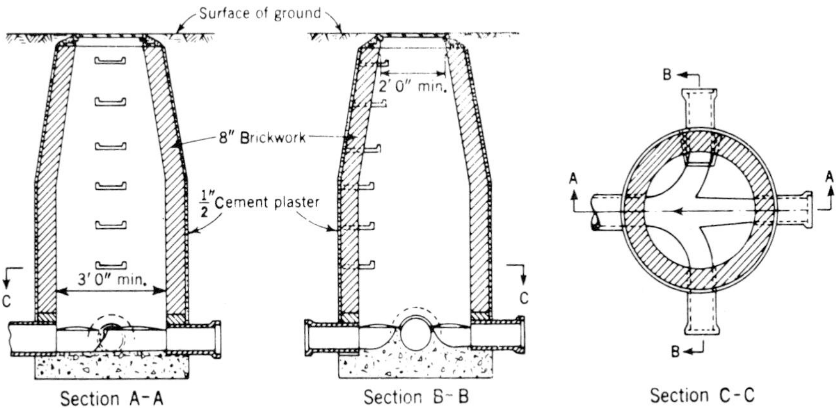

FIGURE C12.15 Junction chamber and manhole for small sewers.

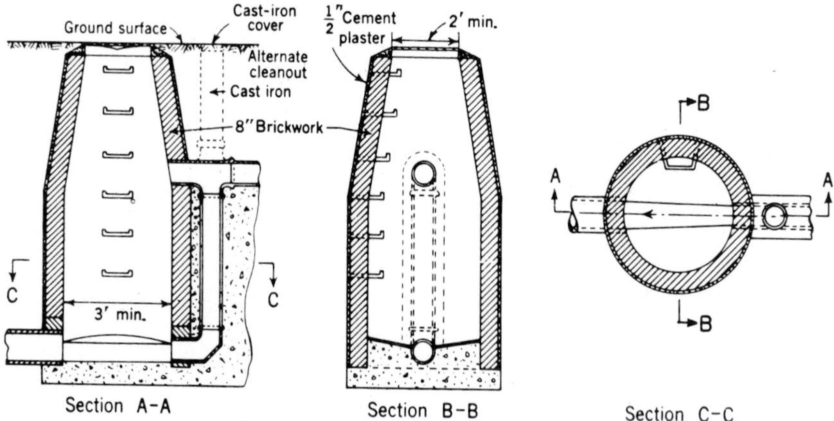

FIGURE C12.16 Drop manhole.

a vertical opening in the curb; (2) gutter inlets, in which a horizontal opening in the gutter is covered by a cast-iron grating; and (3) combination inlets, which combine both the above features. Many types and sizes of standard castings are available for the construction of inlets.

Sewage Pumping

In many sewerage-system layouts it is necessary to provide for pumping at one or more points. The required pumping capacity will vary from a few gallons per minute for stations serving only a few laterals to many millions of gallons per day for stations serving large districts. The smaller stations are frequently built underground, either as built-in-place or as complete factory-assembled units. Pumping is usually done with nonclog centrifugal pumps, although pneumatic ejectors are sometimes used for the smaller installations. For detailed discussion of the design of sewage pumping stations, the reader is referred to the ASCE-WPCF manual.[2]

STRUCTURAL REQUIREMENTS

Sewers must be installed so as to be able to withstand the loads imposed upon them by the weight of the earth and any superimposed loads. The supporting strength of a buried pipe depends upon the installation conditions as well as the structural properties of the pipe itself. Sewer pipes are classed as rigid pipes, which cannot deform materially without cracking. For rigid pipes in trenches, the load can be represented by the equation:

$$W = CwB^2 \tag{C12.9}$$

where W = load, lb/ft of length
w = weight of the soil, lb/ft^3

B = width of trench, ft
C = coefficient whose value depends upon type of soil and ratio of depth of cover to trench width

Table C12.8 gives values for C, and Table C12.9 gives values of w to be used in the equation.

If a pipe is placed on undisturbed ground and covered with a fill, the load can be estimated from the equation:

$$W = CwD^2 \tag{C12.10}$$

where D is the pipe diameter.

Table C12.10 gives values of C for the latter condition, which is known as the "projection condition." The load on a pipe placed in a trench will increase with the trench width until it equals the load for the project condition. If there is a doubt as to whether the "ditch condition" or the "projection condition" controls, the load should be calculated by both formulas and the minimum value used.

In addition to the load of the backfill, some allowance should be made for the superimposed loads caused by vehicles. It is usually safe to assume that H-20 wheel loads will be the greatest live loads to be supported. *H-20 loads* refer to

TABLE C12.8 Values of C for Use in Formula $W = CwB^2$

Ratio of depth to trench width	Sand and damp topsoil	Saturated topsoil	Damp clay	Saturated clay
0.5	0.46	0.46	0.47	0.47
1.0	0.85	0.86	0.88	0.90
1.5	1.18	1.21	1.24	1.28
2.0	1.46	1.50	1.56	1.62
2.5	1.70	1.76	1.84	1.92
3.0	1.90	1.98	2.08	2.20
3.5	2.08	2.17	2.30	2.44
4.0	2.22	2.33	2.49	2.66
4.5	2.34	2.47	2.65	2.87
5.0	2.45	2.59	2.80	3.03
5.5	2.54	2.69	2.93	3.19
6.0	2.61	2.78	3.04	3.33
6.5	2.68	2.86	3.14	3.46
7.0	2.73	2.93	3.22	3.57
7.5	2.78	2.98	3.30	3.67
8.0	2.81	3.03	3.37	3.76
8.5	2.85	3.07	3.42	3.85
9.0	2.88	3.11	3.48	3.92
9.5	2.90	3.14	3.52	3.98
10.0	2.92	3.17	3.56	4.04
11.0	2.95	3.21	3.63	4.14
12.0	2.97	3.24	3.68	4.22
13.0	2.99	3.27	3.72	4.29
14.0	3.00	3.28	3.75	4.34
15.0	3.01	3.30	3.77	4.38
Very great	3.03	3.33	3.85	4.55

Source: Iowa State Univ. Eng. Expt. Sta. Bull. 47.

TABLE C12.9 Weights of Ditch-Filling Materials

Material	Lb/cu ft
Dry sand	100
Ordinary (damp) sand	115
Wet sand	120
Damp clay	120
Saturated clay	130
Saturated topsoil	115
Sand and damp topsoil	100

TABLE C12.10 Values of C for Projection Conditions (values of C for use in formula $W = CwD^2$)

Ratio, depth of cover/pipe diam	0.5	1.0	1.5	2.0	2.5	3.0	3.5	4.0
C	0.6	1.2	2.0	3.0	4.2	5.6	7.5	10.0

trucks having a gross weight of 20 tons, 80 percent of which is on the rear axle, each rear wheel carrying 8 tons. Table C12.11 gives the percentage of wheel loads that can be assumed to be transmitted to buried pipe.[9]

Pipe Bedding Conditions

The supporting strength of a rigid pipe depends upon the type of bedding used in the installation of the pipe. Four general types of bedding conditions have been defined for ditch conduits:

Type 1, impermissible bedding: Little or no care is taken to shape the foundation to fit the lower part of the pipe or to fill and tamp around the pipe.

Type 2, ordinary bedding: The soil at the bottom of the trench is shaped to fit the lower part of the pipe with reasonable closeness for a width of at least 50 percent of the pipe diameter; and the remainder of the pipe is covered to a height of at least 6 in above its top by granular material which is hand placed and tamped.

Type 3, first-class bedding: The pipe is carefully bedded on fine granular material in an earth foundation carefully shaped to fit the bottom part of the pipe for a width at least 60 percent of the diameter; the remainder of the pipe is entirely surrounded to a height at least 1.0 ft above the top by granular materials placed by hand in layers not exceeding 6 in and thoroughly tamped.

Type 4, concrete cradle bedding: The lower part of the pipe is embedded in concrete.

Load Factors

The load factors, or the ratios of the supporting strength to the crushing load as determined by the three-edge bearing method (ASTM Methods C497) for the various types of bedding are generally taken as follows:

TABLE C12.11 Percentage of Wheel Loads Transmitted to Underground Pipes for Unpaved Roadway of Berm Areas*
(tabulated figures show percentage of wheel load applied to 1 lin ft of pipe)

Depth of backfill over top of pipe, ft	Trench width at top of pipe, ft						
	1	2	3	4	5	6	7
1	17.0	26.0	28.6	29.7	29.9	30.2	30.3
2	8.3	14.2	18.3	20.7	21.8	22.7	23.0
3	4.3	8.3	11.3	13.5	14.8	15.8	16.7
4	2.5	5.2	7.2	9.0	10.3	11.5	12.3
5	1.7	3.3	5.0	6.3	7.3	8.3	9.0
6	1.0	2.3	3.7	4.7	5.5	6.2	7.0

Live loads transmitted are practically negligible below 6 ft.
* These percentages include both live load and impact transmitted to the pipe.

Impermissible bedding	1.1
Ordinary bedding	1.5
First-class bedding	1.9
Concrete cradle bedding	2.2–3.4

The factors for the concrete cradle bedding depend upon the amount and quality of the concrete that is used. The value of 2.2 will generally apply when the concrete extends from about one-quarter of the pipe diameter (with minimum of 6 in) below the pipe to the height where the lower 120° sector radii intersect the outside of the pipe. If the concrete is carried up to cover the entire bottom half of the pipe, the load factor may be as high as 3.4. If the entire pipe is encased in concrete with a minimum of $0.25D$ (or 4 in) both above and below, the load factor may be as high as 4.5.

Safety Factor

The specified minimum strength by the three-edge bearing method for a rigid pipe should be divided by an appropriate safety factor in order to obtain the working strength. Some engineers use safety factors as low as 1.0 to 1.2 for reinforced-concrete pipe culverts. For street sewers a safety factor of 1.5 is recommended by the ASCE-WPCF manual.[2]

OPERATION AND MAINTENANCE

Protecting the investment in sewer systems requires attention to installation, testing, operations, maintenance, and cleaning.

Start-up, Operation, and Maintenance

Sanitary and storm sewer systems should be cleaned and tested prior to being put into service. For both types of sewer systems, tests should be carried out to confirm the following items:

- Alignment
- Elevations and pipe grades
- Watertightness of joints
- Location and accessibility of manholes

Relevant data should be used to prepare a sewer atlas which will include locations and elevations of all sewers. This atlas is generally crucial to good operation and maintenance of the system.

After the sewer system is put into use, adequate care and maintenance must be provided to ensure continued good operation. Preventative maintenance will include:

- Spot inspections of sewers for damage at manholes
- Identification and correction of system misuse such as illegal connections
- Identification of missing or damaged manhole covers
- Ensuring that electromechanical equipment is in good working order
- Inspections to identify accumulations of grit and grease and the intrusion of roots into the system

Testing

Pipe joints shall be tested for watertightness using an air test or a water test. Air tests shall be carried out in accordance with ASTM C924. Water tests shall be infiltration tests where the depth of ground water is sufficient or exfiltration tests where ground water is insufficient to submerge the section to be tested. No standard testing methods are available for infiltration-exfiltration tests, but local agencies generally have maximum allowable rates.

Inspections

Periodically, the insides of sewers should also be inspected. For sewers too small to enter, inspections are done using TV cameras which are pulled through the sewers between manholes. For larger sewers, inspections are carried out by wading or floating through them in a boat. The frequency of such inspections will depend on the type of system (sanitary or storm); the pipe material and joint types; and the local experience with the system.

In all sewer inspections, safety is a major concern especially with sanitary sewers where there is a danger of septicity. Precautions should be taken to avoid exposure to dangerous gases or to the absence of oxygen. Adequate ventilation must be provided, and rules must be established to indicate the appropriate course of action in emergencies. In addition, all local safety requirements must be satisfied, and personnel should be given adequate training and equipment.

Cleaning

Based on results of inspections, sewers and appurtenances such as manholes and catch basins must be cleaned. Large sewers which can be entered can be cleaned by hand. Smaller sewers can be cleaned by pulling, thrusting, or dragging some form of instrument through them. Cleaning can also be accomplished by flushing the sewer using a sudden rush of water down the sewer at high velocity. Turbine sewer cleaners utilize revolving blades and are effective in cutting through roots and greases.

On storm sewer systems, it is important to clean catch basins which are designed to intercept settleable solids before they enter the sewers. Catch basins are cleaned by hand, suction pumps, or grab buckets.

DISPOSAL OF WASTEWATER AND STORMWATER

Stormwater can ordinarily be disposed of by discharge into any natural drainage channel. Sanitary sewage and industrial wastewaters containing objectionable constituents must be disposed of in accordance with the requirements of the local health and environmental authorities. The most satisfactory method of disposal of sanitary sewage is to convey it to an adequate public sewerage system. In areas which do not have public sewerage systems, individual disposal systems must be provided. These will vary in size from septic-tank systems for private residences to large treatment plants handling the wastes from large institutions and industries. The design of such systems is beyond the scope of this handbook. See Ref. 1 for further information.

REFERENCES

1. Metcalf & Eddy, Inc., *Wastewater Engineering: Treatment, Disposal and Reuse,* 3rd ed., McGraw-Hill, New York, 1991.
2. Design and Construction of Sanitary and Storm Sewers, "ASCE Manuals of Engineering Practice—No. 37" ("WPCF Manual of Practice No. 9").
3. Recommended Standards for Sewage Works, adopted by Great Lakes—Upper Mississippi River Board of State Sanitary Engineers, 1978 edition.
4. *Frequency Distributions and Hydroclimatic Characteristics of Heavy Rainstorms in Illinois, Bulletin 70,* Illinois State Water Survey, 1989.
5. T. R. Camp, "Design of Sewers to Facilitate Flow," *Sewage Ind. Wastes,* vol. 18, no. 1, January 1946, p. 3.
6. Ven Te Chow, *Open-Channel Hydraulics,* McGraw-Hill, New York, 1959.
7. E. F. Brater, *Handbook of Hydraulics,* 6th ed., McGraw-Hill, New York, 1976.
8. Metcalf & Eddy, Inc., *Wastewater Engineering: Collection, Treatment and Disposal,* McGraw-Hill, New York, 1972.
9. Clay Sewer Pipe Association, Inc., "Clay Pipe Engineering Manual."

CHAPTER C13
PLUMBING PIPING SYSTEMS

Michael Frankel, CIPE
Chief Plumbing Engineer
Jacobs Engineering Group, Inc.
Somerset, New Jersey

This chapter provides the necessary criteria to enable accurate pipe sizing for various plumbing systems and cost-effective selection of appropriate plumbing piping material.

Plumbing systems directly affect the health and safety of the public, and thus they are distinguished from other piping systems by the following general requirements:

1. The design, materials, and installation of the systems are directly regulated by a plumbing code.
2. System design must be approved by an authorized code official charged with the responsibility of assuring plumbing code compliance.
3. A permit for installation of the systems must be obtained from the authority having jurisdiction.
4. The systems shall be installed by an individual duly licensed by the authority having jurisdiction for determining the competence of an individual to obtain a Plumbing installation license. This may not be required in some jurisdictions.
5. The installed systems are required to be inspected and approved by an authorized code official charged with the responsibility of code enforcement.

The basic plumbing systems are:

1. Sanitary drainage systems
2. Vent systems
3. Stormwater drainage systems
4. Potable water systems
5. Fuel gas systems

CODES AND STANDARDS

Plumbing codes establish minimum acceptable standards for the design and installation of the various plumbing systems and for the components that comprise them. There are five building codes, that along with their associated plumbing codes, have found general acceptance over large areas of the country. They are:

1. Building Officials Code Authority (BOCA). Plumbing code: BOCA National Plumbing Code.
2. International Association of Plumbing and Mechanical Officials (IAPMO). Plumbing code: Uniform Plumbing Code.
3. National Association of Plumbing-Heating-Cooling Contractors (PHCC). Plumbing code: National Standard Plumbing Code.
4. Southern Building Code Congress International (SBCC). Plumbing code: National Standard Plumbing Code.
5. Council of American Building Officials (CABO).

Some states and large cities have adapted codes separate from these building codes. In addition, some local authorities using specific regional building codes have adapted plumbing codes other than the one usually associated with that building code. Because of this nonstandardization, the plumbing code used for each specific project must be obtained from a responsible code official.

Many of the tables and charts appearing in this chapter are used only to illustrate and augment discussions of the system sizing procedures and design methods and should not be used for actual design purposes. The information pertaining to those systems is included in the approved plumbing code and should be the primary criteria for accepted methods and sizes for use on any project.

There are many nationally recognized standards that establish dimensions, manufacturing methods, material composition, tests, and numerous other details specific to individual components of the plumbing system. A partial list of organizations originating such standards adapted by various plumbing codes are:

American National Standards Institute (ANSI)
American Society of Mechanical Engineers (ASME)
American Society of Sanitary Engineers (ASSE)
American Society of Testing Materials (ASTM)
America Water Works Association (AWWA)
American Welding Society (AWS)
Cast Iron Soil Pipe Institute (CISPI)
National Fire Protection Association (NFPA)
National Sanitation Foundation (NSF)
Plumbing and Drainage Institute (PDI)
Underwriters Laboratories (UL)

FIXTURE UNITS

The fixture unit (FU) is an arbitrary, comparative value assigned to a specific plumbing fixture, device, or piece of equipment. FU values represent the probable flow that fixture will discharge into a drainage system or use (demand) from a potable water supply system, compared to other fixtures.

The use of fixture units for plumbing systems was expanded upon by the late Roy B. Hunter, of the former Bureau of Standards. In the years since the development of the "Hunter method," new fixtures, changes in the patterns of water use, and water conservation would now result in oversized water and drainage piping systems if the original criteria were used. Long-term data and modern statistical methods and analyses have resulted in revised figures, which were used in calculating the tables and charts provided for this revision of the handbook.

Since sanitary discharge and water demand FUs are different, the designation DFU for drainage fixture units and WFU for potable water fixture units will be used to differentiate between them.

PLUMBING FIXTURES

A plumbing fixture is any approved receptacle specifically designed to receive human and other waterborne waste and discharge that waste directly into the sanitary drainage system, usually with the addition of water. Ideal fixture materials should be nonabsorbent, nonporous, nonoxidizing, smooth, and easily cleaned.

Plumbing codes usually mandate the number and type of fixtures that must be provided for specific building usage, based on the proposed population. Provisions for the handicapped have been made an integral part of code requirements, mandating the number, layout, and barrier-free access to those fixtures.

Potable water discharged from specific plumbing fixtures may be restricted to a maximum flow rate mandated by water conservation requirements. Refer to specific code provisions for these restrictions.

Table C13.1 lists average drainage and vent DFUs, hot and cold water WFUs, gallons-per-minute (gpm) flow, and branch size information for typical fixtures.

EQUIVALENT LENGTH OF PIPING

When calculating the pressure loss through a pressurized piping system, one of the factors to be considered is the equivalent length of pipe. This is the actual pipe run plus an additional length, expressed as a number of feet of straight pipe, that would have the same friction loss as that occurring through various fittings, valves, and so on. Figure C13.1 gives the straight run of pipe for both water and gas systems equal to various valve types and fittings for different pipe sizes. An often used and generally conservative method of quickly finding the equivalent run is to add 50 percent to the actual measured pipe run.

PIPING MATERIALS: GENERAL

When selecting materials and jointing methods for use in a project, the plumbing code usually specifies pipe materials permitted to be used for the various systems along with any restrictions on their use. The code also may stipulate accepted standards that govern the manufacture of the components, their tolerances, and their installation. The piping materials to be discussed are all accepted for use in various national, regional, and most local codes. Some materials may not be acceptable for use in specific local codes.

TABLE C13.1 Typical Plumbing Fixture Schedule

	Drainage			Water			
Fixture type	DFU	Size trap	Size vent	WFU	Size cold	Size hot	Flow, gpm
Automatic clothes washer	3	2	1½	2	½	½	5
Bathroom group (WC, LAV, SH / BT) FV	8			8			
Bathroom group (WC, LAV, SH / BT) tank	6			6			
Bathtub (BT), with or without SH	2	1½	1½	2	½	½	5
Bidet	1	1¼	1¼	1	½		2
Clinic sink	6	3	1½	2	½	½	3
Dishwasher, domestic	2	1½	1½	1	½	½	3
Dental lavatory, cuspidor and unit	1	1½	1½	1	½	½	1
Drinking fountain	½	1¼	1¼	½	½		½
Floor drain	5	3	1½				
Kit. sink & tray, with food grinder	4	2	1½	2	½	½	3
Kit. sink & tray, single 1½-in trap	2	1½	1½	2	½	½	3
Kit. sink & tray, multiple 1½-in traps	3	1½	1½	2	½	½	3
Lavatory, private	1	1¼	1¼	1	⅜	⅜	2
Lavatory (LAV), public	2	1¼	1¼	2	⅜	⅜	2
Laundry tray, 1 or 2 compartments	2	1½	1½	2	½	½	5
Shower (SH) per head or stall	2	2	1½	2	½	½	3
Sink service (SS), trap standard	3	3	1½	3	¾	¾	4
Sink service, P trap	2	1½	1½	2	½	½	4
Sink pot & scullery	2	1½	2	2	½	½	4.5
Sink bar	1½	1½	1½	1	½	½	2
Sink flushing rim	6	3	1½	5	1		15–30
Sink surgeon's	3	2	1½	2	½	½	2.5
Sink wash fountain, per faucet	2	1½	1½	2	½	½	2.5
Urinal pedestal, blowout	6	3	1½	10	1		15–40
Urinal washout	4	2	1½	5	¾		10–20
Water closet (WC), private flush valve (FV)	6	3	1½	10	1		15–40
Water closet private tank type	4	3	1½	5	½		3–5
Water closet private pressure tank	4	3	1½	4	½		3–5
Water closet public flush valve	6	3	1½	10	1		15–40
Water closet public tank type	4	3	1½	5	½		3–5
Water closet public pressure tank	4	3	1½	4	½		3–5
Fixture not listed	1	1¼	1¼				
Fixture not listed	2	1½	1½				
Fixture not listed	3	2	1½				
Fixture not listed	5	3	1½				
Fixture not listed	6	4	2				
Hose bibb or sill cock, public				5	¾		5
Hose bibb or sill cock, private				3	½		3
Water supply not listed				1	⅜		
Water supply not listed				2	½		
Water supply not listed				3	¾		
Water supply not listed				10	1		

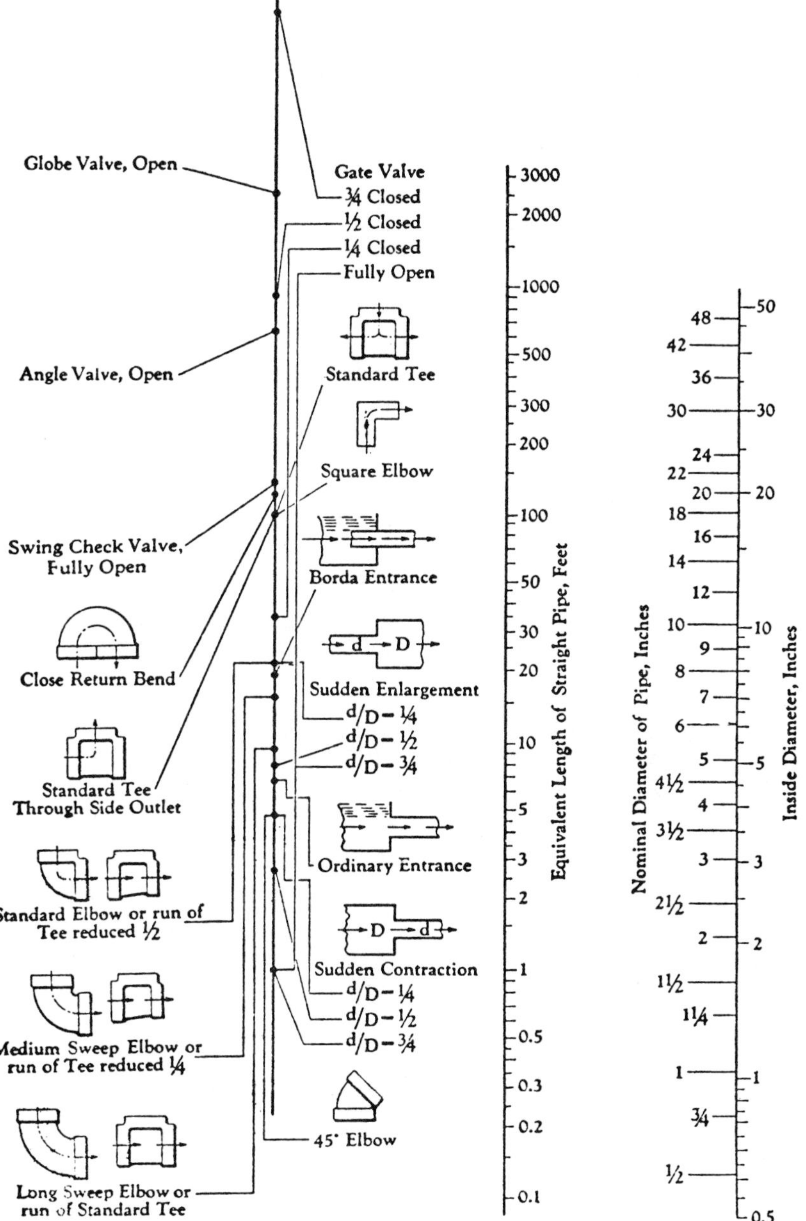

FIGURE C13.1 Resistance of valves and fittings to flow of fluids.

When renovating a plumbing system or when circumstances require a unique design, it may be necessary to request deviation from the accepted list of materials in order to match existing piping or to obtain special design characteristics. To review such requests, the authorities require enough information to determine if the intent of the applicable code provisions are adhered to in terms of safety and suitability of the materials for the intended purpose.

The design engineer is ultimately responsible for selecting and specifying the most suitable pipe, fittings, and jointing methods used for any project. The following characteristics of pipe and fittings are important considerations in making that selection:

1. Corrosion resistance of the pipe and fittings. This is a measure of their ability to resist both the internal corrosive effects of the fluid likely to flow through the pipe, and the effects of soils or ambient conditions on its exterior. Corrosion can be reduced or eliminated by the application of a suitable coating, encasement, lining, and cathodic protection.
2. Total installed cost, which includes the cost of the pipe and fittings, assembly of the joints, handling, and the cost of the support system for the piping.
3. Physical strength of the pipe and fittings, which is the ability to withstand the internal pressure of the liquid and external physical damage that may occur either during installation or after being placed in service.

Metallic Pipe and Fittings

Cast Iron Soil Pipe and Fittings (CI). Cast iron soil pipe is acceptable for any nonpressure, noncorrosive sanitary and stormwater drainage service. Three types of pipe are manufactured: service (or standard) weight, extra heavy, and hubless. Two types of pipe ends are available: hub and spigot and hubless. Three types of joints are used: caulked and compression gaskets used with hub and spigot pipe, and compression couplings used only for hubless pipe.

Hub and spigot cast iron soil pipe and fittings must conform to the following standards:

1. ASTM A-74, Cast-Iron Soil Pipe and Fittings
2. ASTM C-564, Rubber Gaskets for Cast-Iron Soil Pipe and Fittings
3. ASTM HSN, Neoprene Rubber Gaskets for Hub and Spigot Cast Iron Soil Pipe and Fittings
4. CISPI 301, Hubless Cast-Iron Soil Pipe and Fittings for Sanitary and Storm Drain, Waste, and Vent Piping Applications
5. CISPI 310, Patented Joint for Use in Connection with Hubless Cast-Iron Sanitary System

Acid-Resistant Cast Iron Pipe and Fittings (AR). This pipe material is used for nonpressure drainage service for corrosive liquids whose corrosion potential is too severe for CI pipe. AR pipe is CI pipe containing between 14.25 and 15 percent silicon and small amounts of manganese, sulphur, and carbon. It is manufactured only in extra-heavy grade. It is available with two types of pipe ends; hub and spigot, or hubless. The hub and spigot ends can be joined by caulking. Hubless pipe is joined by the use of compression coupling.

Acid-resistant cast iron pipe and fittings must conform to the following standards:

1. ASTM A-518, Corrosion Resistant High Silicon Iron Castings
2. ASTM A-861, High Silicon Iron Pipe and Fittings

Ductile Iron (DI) Pipe and Fittings. DI pipe is suitable for any noncorrosive plumbing service. Ductile iron pipe is fabricated of a cast iron alloy in which graphite replaces the carbon that is present in cast iron soil pipe. It is available for use either as a nonpressure gravity sewer pipe or pressure pipe. Eight pressure ratings are available: Class 50 (25 psi), Class 51 (50 psi), Class 52 (100 psi), Class 53 (125 psi), Class 54 (175 psi), Class 55 (200 psi), Class 56 (250 psi), Class 57 (350 psi), and gravity sewer pipe. Three types of joints are used: mechanical, gasketed, or flanged.

Ductile iron pipe and fittings must conform to the following standards:

1. ASTM/C151, Ductile-Iron Pipe, Centrifugally Cast in Metal Molds or Sand-Lined Molds, for Water and Other Liquids
2. ASTM C 115, Flanged Ductile Iron and Gray Iron Pipe with Threaded Flange
3. ASTM C 111, Rubber-Gasket Joints for Ductile-Iron and Gray-Iron Pressure Pipe and Fittings
4. ANSI A21.10, Gray and Ductile Iron Fittings, 2 through 48 in for water and other liquids

Steel Pipe (ST). Steel pipe is commonly used in vent systems, drainage systems where human waste is not discharged, for indirect waste lines, potable water systems, and for fuel gas piping. Steel pipe can be obtained with threaded ends, plain ends, and beveled ends. Four types of joints are used: screwed, grooved, flanged, and welded. Fitting materials commonly used are steel (for welded joints), malleable iron (either screwed or grooved joints), and cast iron (either screwed or flanged joints). Steel pipe for plumbing systems should be galvanized to retard corrosion.

Steel pipe and fittings used for plumbing systems must conform to the following standards:

1. ASTM A-53, Pipe, Steel, Black, and Hot-Dipped, Zinc-Coated Welded and Seamless
2. ASTM B16.1, Cast-Iron Pipe Flanges and Flanged Fittings, Classes 25, 125, 250, and 800
3. ASTM B16.3, Malleable-Iron Threaded Fittings
4. ASTM B16.5, Steel Pipe Flanges and Flanged Fittings
5. ASTM B16.9, Wrought Steel Buttweld Fittings
6. ASTM B16.12, Cast-Iron Threaded Drainage Fittings

Copper Water Tube. Copper tube is used for domestic water service. Copper tube is fabricated of 99.9 percent copper with plain ends, and in three types: K, L, and M. Each type has the same outside diameter, with K tube having the greatest wall thickness and pressure rating, and M the least. Each of the three types of copper tube is also available in either drawn (hard) and annealed (soft) forms. Fittings can be either wrought copper or cast bronze. Copper tube can be joined by either flared, soldered, or brazed joints.

Copper tube and fittings must conform to the following ASTM standards:

1. ASTM B-88, Seamless Copper Water Tube
2. ANSI B-16.22, Wrought Copper and Copper Alloy Solder Joint Pressure Fittings
3. ANSI B-16.18, Cast Bronze Cast Copper Alloy Solder Joint Pressure Fittings
4. ANSI B-16.26, Cast Copper Alloy Fittings for Flared Copper Tubes

Copper Tube, Type DWV (Drainage, Waste, and Vent). It is a nonpressure, thin wall drainage pipe, primarily used in residential buildings, and in commercial buildings for indirect waste lines or local branch lines where human waste is not discharged. It is a seamless tube, made from almost pure copper (99.9 percent) and is available only in drawn form with plain ends. Joints can be either soldered or brazed.

Type DWV copper tube for drainage systems must conform to the following standards:

1. ASTM B-306, Copper Drainage Tube (DWV)
2. ASTM B16.29, Wrought Copper and Wrought Copper Alloy Solder Joint Drainage Fittings, DWV
3. ASTM B16.23, Cast Copper Alloy Solder Joint Drainage Fittings, DWV

Brass Pipe. It is generally used in local branch drainage lines where this alloy resists specific corrosive drainage effluent, in alterations to match existing work and as a pressure pipe for potable water in sizes larger than 4 in, where soldering or brazing is impractical. Brass pipe is manufactured from an alloy containing 85 percent copper and 15 percent zinc with plain ends. Joints can be either screwed, soldered, or flanged.

Brass pipe and fittings must conform to the following standards:

1. ASTM B43, Seamless Red Brass Pipe
2. ANSI B16.24, Bronze Flanged Fittings
3. ANSI B16.15, Bronze Pipe Flanges and Flanged Fittings, Classes 150 and 300
4. ANSI B16.18, Cast Bronze Solder-Joint Pressure Fittings

Plastic Pipe and Fittings

General. Plastic pipe is manufactured in a great variety of compositions, many of which are suitable for plumbing systems. The applicable code is the most important factor in selecting the type of plastic pipe for any specific purpose. All plastic pipe, components, and jointing methods used in potable water systems must be approved by the NSF. Plastic pipe must be closely integrated with the selection of hangers and the entire pipe support system.

The advantages of plastic pipe include: excellent resistance to a wide variety of chemical and waste effluents, resistance to aggressive soils, availability in long lengths, low resistance to fluid flow, and low initial cost. Disadvantages include: poor structural stability (requiring additional supports), lower pressure ratings at elevated temperatures, susceptibility of some types of plastics to physical changes due to exposure to sunlight, low resistance to solvents, and production of toxic gases released upon combustion of some types of plastics.

Three designations are used to express pressure rating and wall thickness: schedule, (dimensions are outside diameter controlled, matching iron pipe size); standard dimensional ratio (a pressure rating only); and dimensional ratio (a pressure rating only using nonstandard dimensional ratios).

Polyvinyl Chloride Pipe (PVC). PVC is used for potable water and drainage systems. It is one of the most widely used of the plastic pipes. It has a low pressure and temperature rating and very poor resistance to solvents.

PVC pipe and fittings must conform to the following standards:

1. ASTM D-1785, PVC Plastic Pipe, Schedules 40, 80, and 120
2. ASTM D-2241, PVC Plastic Pipe (SDR-PR)
3. ASTM D-2466, PVC Plastic Pipe Fittings, Schedule 40
4. ASTM D-2665, PVC Drain, Waste and Vent Pipe and Fittings

Chlorinated Polyvinyl Chloride Pipe (CPVC). CPVC is used for potable water and drainage systems. It has the same characteristics as PVC and is used where a stronger piping system with higher pressure and temperature ratings are required.

CPVC pipe and fittings must conform to the following standards:

1. ASTM F-441 CPVC Plastic Pipe, Schedules 40 and 80
2. ASTM D-2846 CPVC Pipe, Fittings, Solvent Cements and Adhesives for Potable Hot and Cold Water Systems
3. ASTM F-439, Socket-Type CPVC Plastic Pipe Fittings, Schedule 80

Polypropylene Pipe (PP). This material is widely used for chemical drainage piping systems. PP pipe and fittings are manufactured from flame-retardant material and is available in Schedules 40 or 80. Joining methods include solvent cement joints, threaded joints, or mechanical-type joints. (Only Schedule 80 can be threaded.)

It must conform to ASTM Standard D-2146, PP Plastic Pipe and Fittings Schedules 40 and 80.

Polyethylene Pipe (PE). It is widely used for underground fuel gas and foundation drainage piping. It is joined by socket and butt heat fusion.

It must conform to the following standards:

1. ASTM D-2239, PE Plastic Pipe SDR
2. ASTM D-2447, PE Plastic Pipe, Schedules 40 and 80, DR

Acrylonitrile Butadiene Styrene (ABS). ABS is widely used as drainage pipe and is available in Schedules 40 and 80 with plain or socket ends. Joints are made by either solvent cement or threaded connections. Only Schedule 80 can be threaded.

ABS pipe and fittings must conform to ASTM Standard D-2661, ABS plastic drain, waste, and vent pipe and fittings.

Other Pipe

Glass Pipe. This pipe is primarily used for gravity drainage of various corrosive liquids. Glass pipe is fabricated from a low-expansion borosilicate glass having a low alkali content. Glass pipe is joined by compression couplings.

Vitrified Clay Pipe (VC). This pipe is suitable for use in underground gravity drainage systems where resistance to a wide variety of corrosive effluent and aggressive soils is required. It is manufactured in standard and extra-strong grades with hub and spigot ends. Clay pipe is joined by mortar joints.

MISCELLANEOUS FITTINGS

Adapters are used to join two different pipe materials or piping with dissimilar joint ends. Most plumbing codes require the use of approved adapters.

Dielectric fittings are used to connect dissimilar metallic pipes together to avoid galvanic corrosion that would quickly weaken the pipe at the joint. The principal of the joint is a gasket that isolates one pipe from the other.

Unions are fittings used to connect two fixed pipes together, neither of which are capable of being turned. The union consists of three interconnected pieces; two internally threaded ends and a center piece that draws the ends together when rotated.

JOINTS

A joint is a connection between one pipe and either another pipe or a fitting. It must be able to withstand the greatest pressure capable of being exerted upon it. Most plumbing codes refer to standards that govern the methods and materials used in forming joints. The selection of the joining methods is determined by the type of pipe and fittings used, the maximum pressure expected in the system, and the need for disassembly.

Caulked Joints

A caulked joint, illustrated in Fig. C13.2 is a rigid, nonpressure-type joint. This joint consists of a rope of oakum or hemp that is packed into the annular space

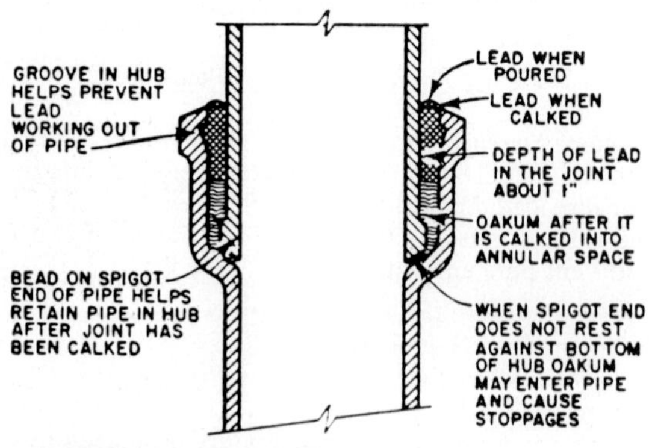

FIGURE C13.2 Caulked joint.

around spigot end. For acid-resistant cast iron pipe, hydras magnesium aluminum silicate, reinforced with Fiberglas, is used as a packing material instead of oakum. Molten lead 1 in in depth is then poured into the annular space on top of the rope. The lead is then driven (caulked) farther into the joint. In use, the hemp or oakum swells when it absorbs water and further increases the joint's ability to resist leaking.

Because caulked joints are labor intensive, they have generally been replaced by either compression coupling or gasketed joints for most cast iron joint applications where permitted by code.

Compression Couplings

The compression coupling illustrated in Fig. C13.3, is a rigid, nonpressure type of joint that can be easily disassembled. The coupling consists of an inner elastomeric gasket and an outer metallic sleeve with an integral bolt used for tightening and compressing the gasket.

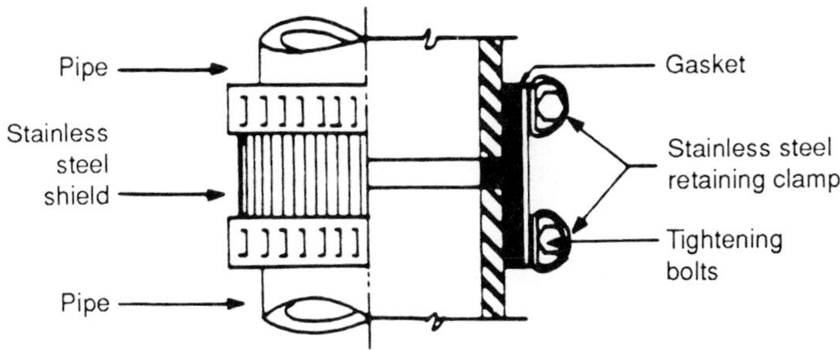

FIGURE C13.3 Compression coupling.

This joint is preferred for above-ground installations because of its ease of assembly and strength. Underground, the metallic sleeve often fails after years of service due to corrosion by surrounding soil or fill. The standard governing the fabrication of this type of joint is Cast Iron Soil Pipe Institute Standard 310.

Screwed Joints

The screwed joint illustrated in Fig. C13.4, is a rigid, pressure-type joint that can be easily disassembled. Such a joint can be used with any plain-end pipe that has the necessary wall strength and thickness to have threads cut into it. The joint uses threads on two pipe ends (or on a pipe and a fitting) to draw the two pieces together and form a leak-proof seal. The threads used for pipe are known as American tapered pipe threads (APT). This type of joint is generally limited to pipe 3 in and smaller in diameter, because of the great effort required to turn a pipe of larger size in making a joint.

Applicable plumbing codes usually specify the type of pipe that can be threaded. The standard governing the fabrication of this type of joint is ANSI B2.1.

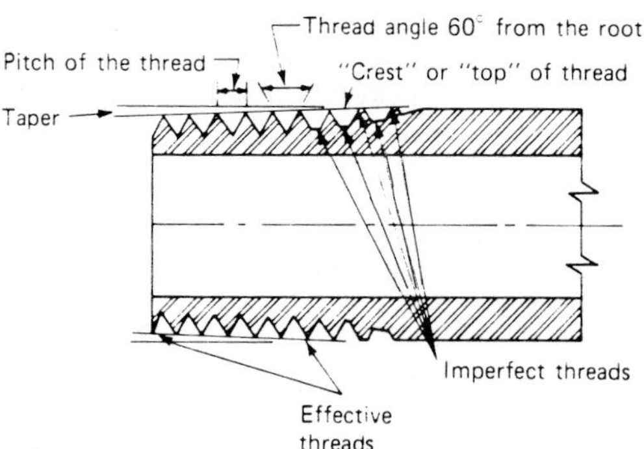

FIGURE C13.4 Screwed joint.

Soldered Joints

A soldered joint, illustrated in Fig. C13.5, is a rigid, pressure-type joint made with a filler metal called *solder,* that when heated to its melting point, is drawn by capillary action into the annular space between the pipe and fitting. When the solder cools, it adheres to the walls of both pipe and fitting, creating a joint that is suitable for any installation for which the piping itself is acceptable. This type of joint is generally limited to pipe having diameters no larger than 4 in because of the difficulty of applying heat evenly to larger joints.

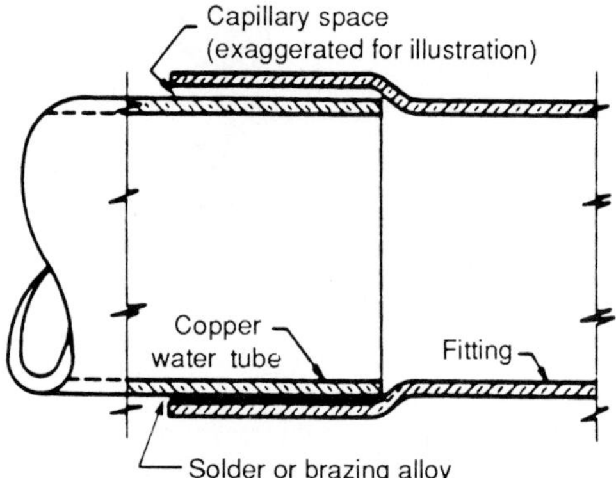

FIGURE C13.5 Soldered joint.

Solder melts at a temperature of 900°F or lower. When used in drainage and vent systems, an alloy of tin and lead or tin and antimony is used. Solder and flux used for potable water systems must contain no lead.

The solder material should conform to ASTM Standard B-32, Solder Metal.

Brazed Joints

A brazed joint is similar to the soldered joint, except for the melting point of the filler metal, which is higher than 900°F. This joint is used where higher pressure ratings are required than that used for solder joints. Various compositions of the filler metal are available for various applications.

Brazing metal should conform to AWS Standard A5.8, Filler Metals for Brazing.

Gasketed Joints

A gasketed joint, illustrated in Fig. C13.6, is a flexible, pressure-type joint using an elastomeric gasket under compression. The joint is capable of being easily disassembled. It is well suited for both above-ground and underground installations.

Various manufacturers produce pipe ends and gasket configurations for different applications that are not compatible with each other.

Standards governing the fabrication of gasketed joints are: ASTM C-564 (Metallic Pipe), ASTM D-3212 (Plastic Pipe), and CISPI HSN (Cast Iron Soil Pipe).

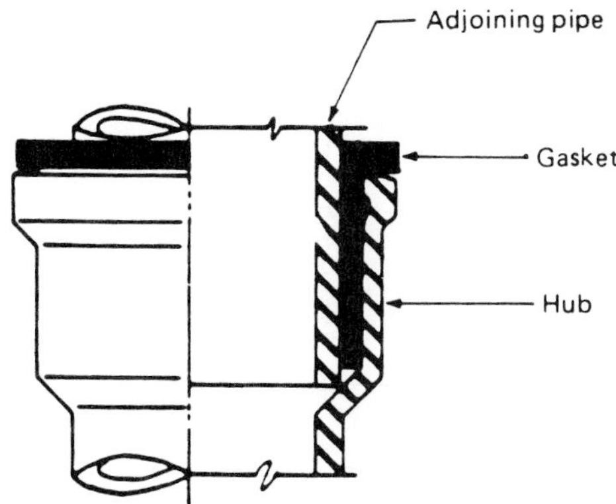

FIGURE C13.6 Gasketed joint.

Mechanical Joints

A mechanical joint is a pressure-type joint that can easily be disassembled and uses nuts and bolts to draw together a pipe and gland to compress a gasket around the pipe, forming a leak-proof seal. A typical joint is illustrated in Fig. C13.7. Many different kinds of proprietary mechanical joints are available to achieve varying degrees of flexibility and pressure rating. It is highly resistant to being pulled apart.

Grooved Joints

A grooved joint, illustrated in Fig. C13.8, is a pressure-type joint for metallic pipe that can easily be disassembled. The joint consists of an inner elastomeric gasket and an outer split metallic sleeve with an integral bolt used for tightening. The outer sleeve has extensions at each end that fit into grooves cut in or rolled into the pipe near the ends to be joined. Roll grooves are used to form joints when the pipe wall is not thick enough to have a groove cut in it. Different styles are avail-

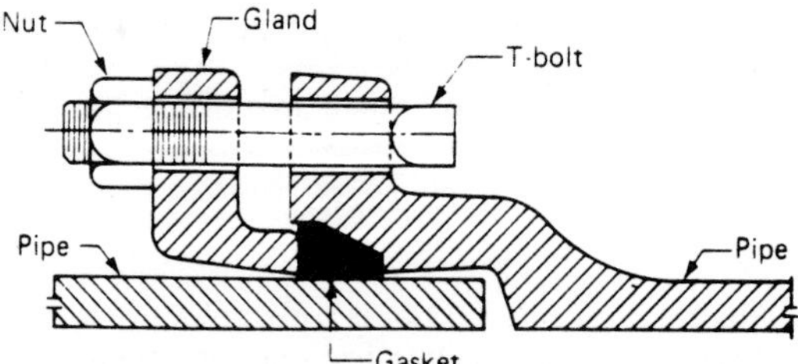

FIGURE C13.7 Mechanical joint.

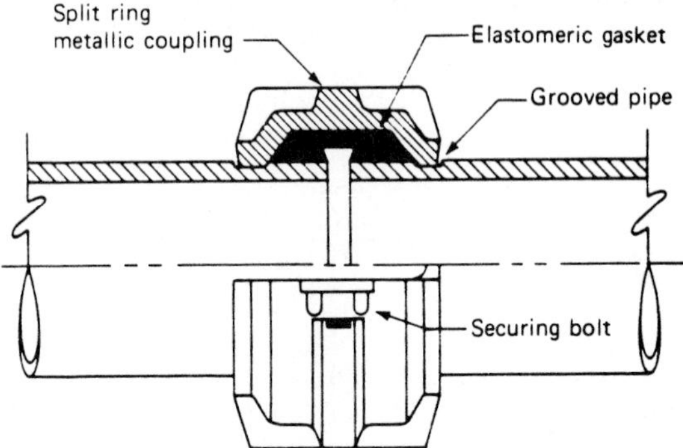

FIGURE C13.8 Grooved joint.

able to achieve varying degrees of flexibility and pressure rating. It is highly resistant to being pulled apart.

Standards governing the fabrication of grooved joints are: AWWA C606, Couplings; ASTM D-735, Gasket; and ASTM D-183, Bolts.

Flanged Joints

A flanged joint, illustrated in Fig. C13.9, is a rigid, pressure-type joint that uses nuts and bolts through a raised projection on the end of a pipe to draw the ends of the pipe together against a gasket to form a leak-proof joint. It can easily be

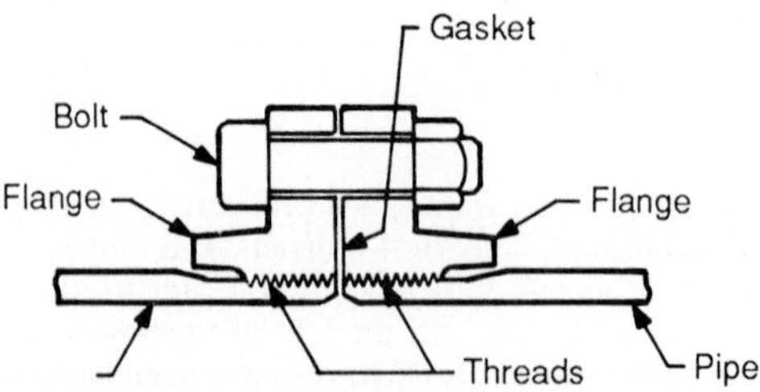

FIGURE C13.9 Flanged joint.

disassembled. Flanges can be cast integrally, or they can be welded or screwed onto a plain-end pipe. The face of the mating flanges can be raised or flat face. A variety of proprietary methods of flange attachment to plain end pipes are also available.

Welded Joints

The welded joint, illustrated in Fig. C13.10, is a pressure-type joint most often used for steel pipe. Welding is accomplished by bringing both pipe walls, at the joint, to the melting point and fusing them together with the addition of metal to allow for correct wall thickness and strength. The necessary amount of heat for welding is produced either by a high-temperature flame or an electric arc formed between the welding electrode and the pipe. In order to properly buttweld pipe, the pipe ends must be specially prepared depending on the pipe thickness, pipe metal composition, and welding method. Two types of joints are possible: butt, illustrated in Fig. C13.10*a*, and socket, illustrated in Fig. C13.10*b*. The four methods of welding are: gas tungsten arc (GTAW) or TIG welding; gas metal arc welding (GMAW) or MIG welding; shielded metal arc welding (SMAW); and oxy-fuel torch welding.

Flared Joints

This is a rigid, pressure-type joint used only with annealed (soft) copper tubing. A flared joint, illustrated in Fig. C13.11, is made by first placing a loose, threaded coupling nut on one end of the pipe, then cold forming that end with a mandrel

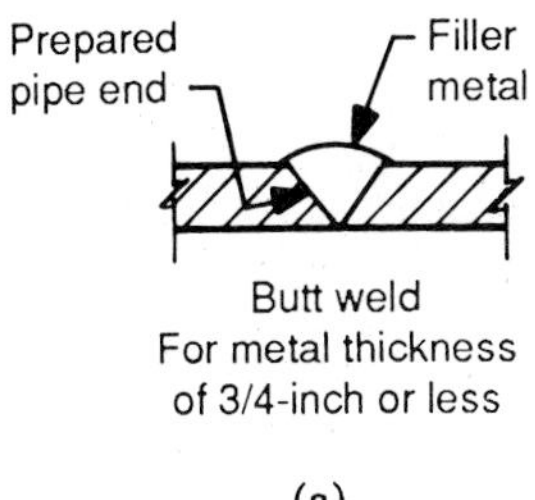

(a)

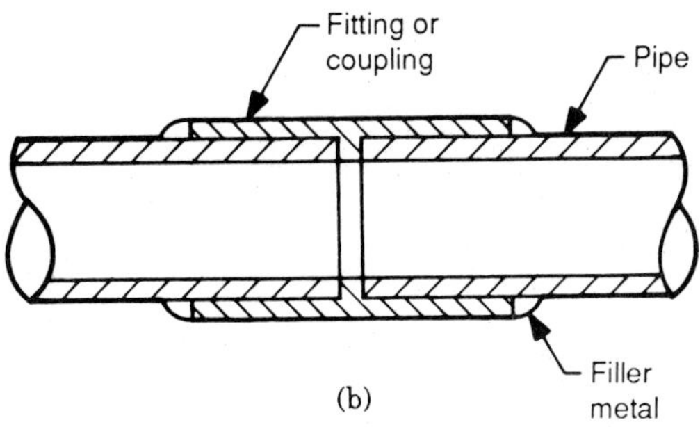

(b)

FIGURE C13.10 (*a*) Buttweld joint. (*b*) Socket weld joint.

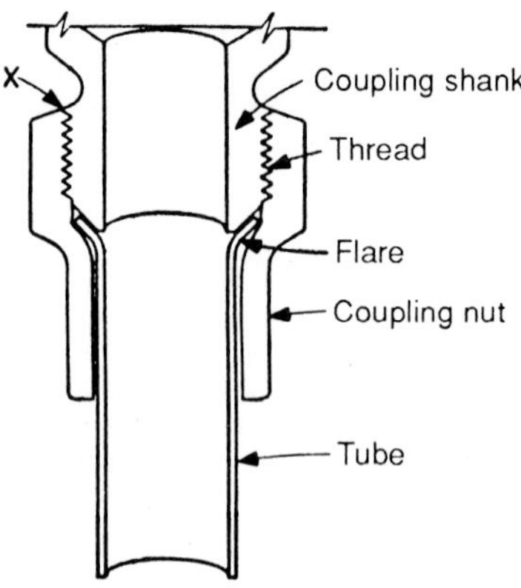

that enlarges the pipe end to fit a mating end on a threaded coupling shank. The screwed coupling nut and shank are then turned, drawing the pipe ends together to form a leak-proof seal.

JOINTS FOR PLASTIC PIPE

Solvent Cement Joints

This is a rigid, pressure-type joint that is suitable for any type of installation for which the piping itself is acceptable. A solvent cement joint, illustrated in Fig. C13.12, is made by spreading a combination of solvent and cement on the surfaces to be joined. They react chemically by dissolving the surface of the pipe and fitting it comes in contact with. The two components are put together while wet. After drying, the two components are fused into a homogeneous mass, producing a leak-proof joint. Solvent welded joints can be used only with specific matching types of plastic pipe and fittings with plain and socket ends.

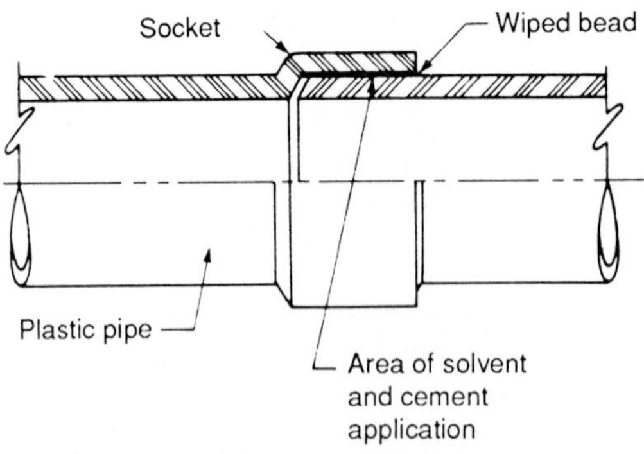

FIGURE C13.12 Solvent cement joint for plastic pipe.

The following standards govern the use of solvent cement—depending on the type of pipe for which the cement will be used. ASTM D-2235, Solvent Cement for ABS Plastic Pipe and Fittings; ASTM D-2541, Solvent Cement for PP Plastic Pipe and Fittings; and ASTM D-2564, Solvent Cements for PVC Plastic Pipe and Fittings.

Heat Fusion Joints

A heat fusion joint is a rigid, pressure-type joint that is only suitable for thermoplastic pipe materials. Heat is used to melt the plastic pipe surfaces and fuse them together into a homogeneous mass. Two types of joints are available: butt, illustrated in Fig. C13.13*a,* and socket, illustrated in Fig. C13.13*b.* Such a joint can be used only with two plain-end plastic pipes (butt joints) or a plain end pipe and a socket fitting with resistance wire inside the socket manufactured specifically for this purpose. For butt heat fusion, the ends of the pipe are heated to the melting point with an outside source of heat, usually a flat, electrically heated plate. The

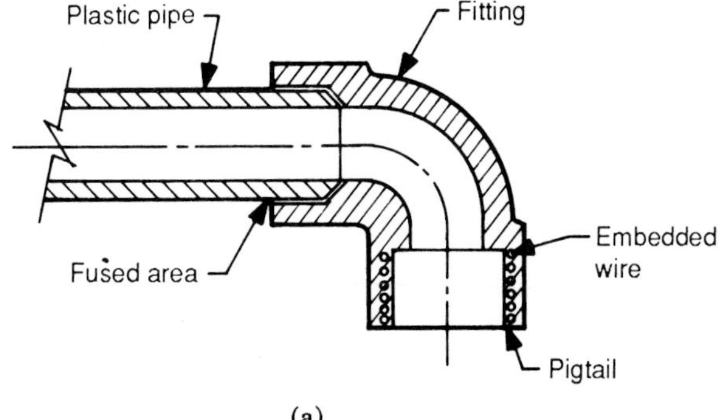

(a)

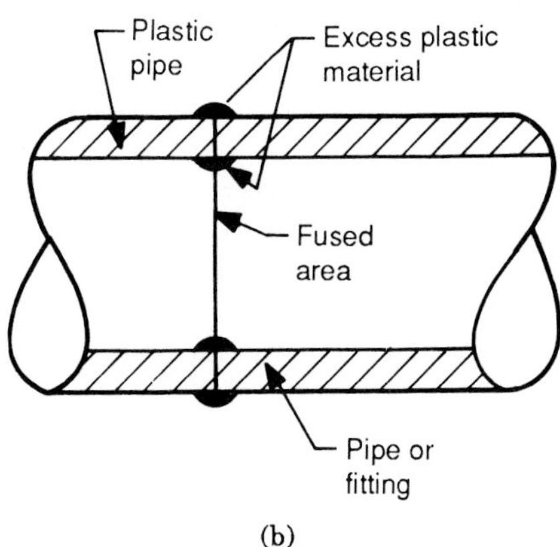

(b)

FIGURE C13.13 (*a*) Socket heat fused joint for plastic pipe. (*b*) Buttheat fused joint for plastic pipe.

heated plate is removed, and the two ends of the pipe are brought together. The socket joint is made by placing the plain end of the pipe into the socket, connecting leads from wire embedded inside the fitting to a proprietary electric source and heating the socket to the melting point to fuse the pipe to the inside of the socket. The manufacturer's instructions must be followed carefully throughout each phase of the process. It is accepted practice to have the mechanics making up the joint certified by the manufacturer as being properly qualified in the correct procedures.

The standard governing the fabrication of heat fused joints is ASTM D 2657, Heat Joining of Thermoplastic Pipe and Fittings.

For more details on nonmetallic piping systems, refer to Part D of this handbook.

SANITARY DRAINAGE SYSTEM PIPING

System Description

The sanitary drainage system conveys waterborne effluent discharged from plumbing fixtures and other equipment to an approved point of disposal. The sanitary system receives all liquid waste except stormwater or unacceptably treated process or chemical drainage.

System Components

Major components of the sanitary drainage system are pipe and fittings, joints, valves, traps, cleanouts, drains, interceptors, sewage ejectors, and sump pumps.

Nomenclature

Approved: Accepted for the intended purpose, as an appropriate design or for installation into a plumbing system by a responsible code official or other agency exerting jurisdiction for a specific project.

Backwater valve: A commonly used term for a type of check valve used in a drainage system.

Branch: A horizontal run of pipe not considered a house drain or stack.

Branch interval: A branch interval is the distance measured along the stack, within which horizontal drainage branches are connected to a drain stack. This distance is usually one story height, but never less than 8 ft, 0 in.

Building drain: The lowest horizontal part of the drainage piping system; considered the principal pipe conveying sanitary effluent by gravity to a point outside the building.

Building sewer: The continuation of the house drain from a point outside the building wall to the actual connection to an adequate and approved point of disposal, such as a public sewer or private sewage disposal system.

Building trap: A trap installed on the house sewer to prevent the circulation of sewer gas between the building sewer and the building drain.

Chemical waste: Any substance that may cause harm to the sanitary piping system, treatment facility, or environment without being treated or neutralized prior to discharge into the sanitary drainage system.

Cleanout: A gas-tight, water-tight pipe fitting with a removable plug that is used to obtain access to the inside of a drainage pipe for cleaning or maintenance.

Combined drainage system: A drainage system that combines sanitary effluent and stormwater runoff in a single piped system.

Fitting: A device used to connect one or more pipes together and/or to change the direction of a straight run of pipe.

Floor drain: A plumbing fixture that removes liquid effluent from the surface of floors and other areas.

House drain: A commonly used term for *building drain.*

House sewer: A commonly used term for *building sewer.*

House trap: A commonly used term for *building trap.*

Indirect waste: Any waste pipe not connected directly into the drainage system, that discharges through an air gap into a fixture, interceptor, trap, or drain.

Interceptor: A device that separates, retains, and allows removal of specific harmful material suspended in the waste stream, while permitting the remaining acceptable liquid effluent to be discharged into the drainage system.

Invert: The elevation of the inside, bottom of a drainage pipe.

Offset: Any change in direction of a stack from vertical or any change in direction of a horizontal drainage line.

Pitch: The distance that one end of a pipe is lower than the other end, expressed as a percent of the total length of run or as a dimension, in inches or feet per foot of run.

Plumbing fixture: Any approved receptacle or device specifically designed to receive human or other waterborne waste and discharge that waste directly into the sanitary drainage system, often with the addition of water.

Runout: A commonly used term for the horizontal drainage line from a stack at its lowest level.

Slope: A commonly used term for *pitch.*

Soil line: Any pipe that conveys human waste.

Stack: A vertical drainage line, usually more than three floors in height.

Trap: A device that maintains a water seal preventing the passage of sewer gas, vermin, air, and odors originating from inside the drainage system, while permitting the unrestricted passage of liquid waste into the drainage system.

Waste line: A drainage pipe not conveying human waste, chemical waste, or stormwater drainage.

Major System Components

Cleanouts. Codes mandate that cleanouts generally be provided at the base of stacks before the pipe changes direction from vertical to horizontal, at changes in horizontal pipe direction greater than 45° and along horizontal runs of pipe every 50 ft. Typical cleanouts are illustrated in Fig. C13.14*a* and cleanout components in Fig. C13.14*b*.

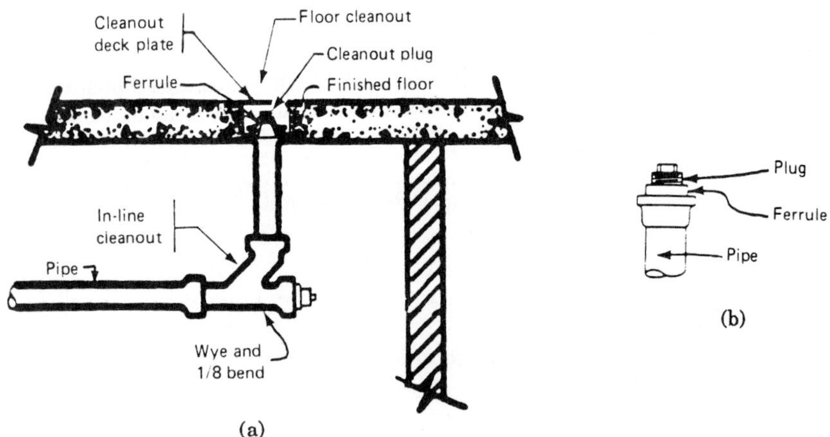

FIGURE C13.14 (*a*) Typical cleanouts. (*b*) Cleanout components.

Floor Drains. A floor drain is a receptacle used to remove liquid effluent from building interior floor areas and other locations. A typical floor drain is illustrated in Fig. C13.15. It provides a receptacle for spills, washdown, and effluent to be collected and routed directly into the sanitary drainage piping system. Code provisions do not specify where a drain should be located. However, most codes regulate the minimum seal requirements for drain traps, the minimum open area of grates and strainers, and the mandatory inclusion of certain individual components (such as removable secondary strainers or sediment buckets) for drains in some locations. A standard commonly cited in the selection of floor drains is ANSI A112.21-1, Floor Drains. Drains consist of the following components:

1. Drain body.
2. Grates located at the top of a drain permit liquid effluent to enter the drain body, while excluding larger solids and foreign matter. Grates are classified as follows:
   a. Light duty: foot traffic only
   b. Medium duty: live wheel loads up to 2000 lb

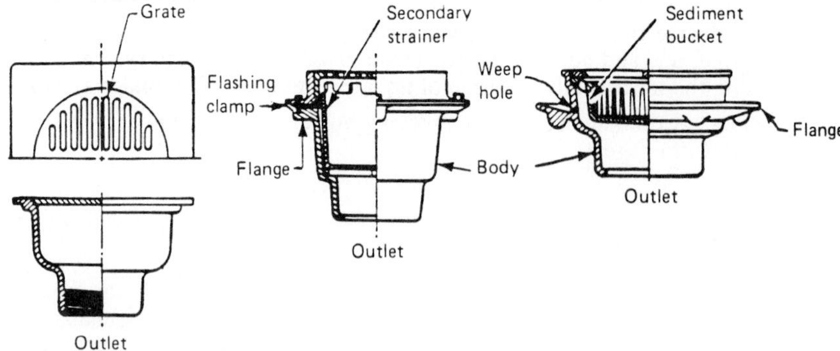

FIGURE C13.15 Typical floor drains.

c. Heavy duty: live wheel loads up to 5000 lb
d. Extra-heavy duty: live wheel loads 5000 lb or more
3. A secondary strainer may be installed below the grate in a drain that does not have a sediment trap.
4. A sediment trap (or bucket) is a removable device inside the drain body, that may be installed to trap and retain small solids that pass through the grate.
5. A flashing ring or clamp is a device used to secure flashing directly to the body of the drain.

Interceptors. Plumbing codes require that any substance harmful to either the building drainage system, the public sewer, or the municipal sewage treatment process be prevented from being discharged into the public sewer system. Among such materials are grease, flammable liquids, sand, or other substances objectionable to the local authorities.

Traps. A fixture trap, illustrated in Fig. C13.16 is a U-shaped section of pipe of the necessary depth to retain sufficient liquid required by code. All fixtures and equipment directly connected to the sanitary drainage system are required to have traps.

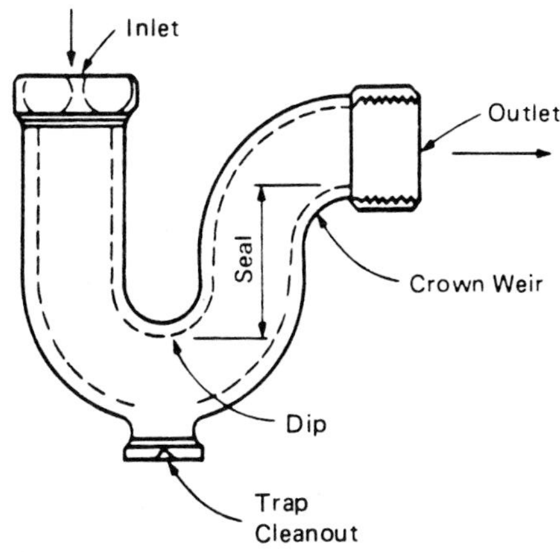

FIGURE C13.16 Typical fixture trap.

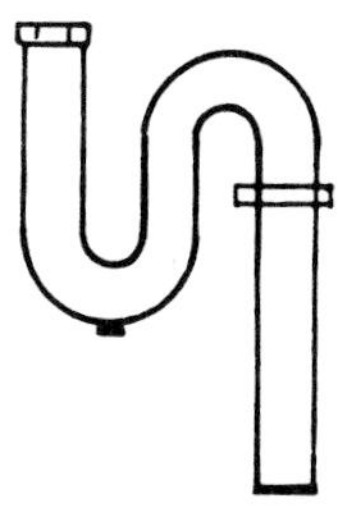

FIGURE C13.17 S trap.

In general, traps must: (a) be self-cleaning; (b) provide a liquid seal of at least 2 in with larger seals where required; (c) conform to local code requirements in terms of minimum size; (d) provide an accessible cleanout; (e) be capable of rapidly draining a fixture. All traps must be vented in an approved manner, except for specific conditions waived by local code requirements or authorities.

Traps that are prohibited by code include: traps requiring moving parts to maintain the seal; full S-type traps, illustrated in Fig. C13.17; crown vented traps, illustrated in Fig. C13.18; and drum traps, illustrated in Fig. C13.19. Drum traps may be permit-

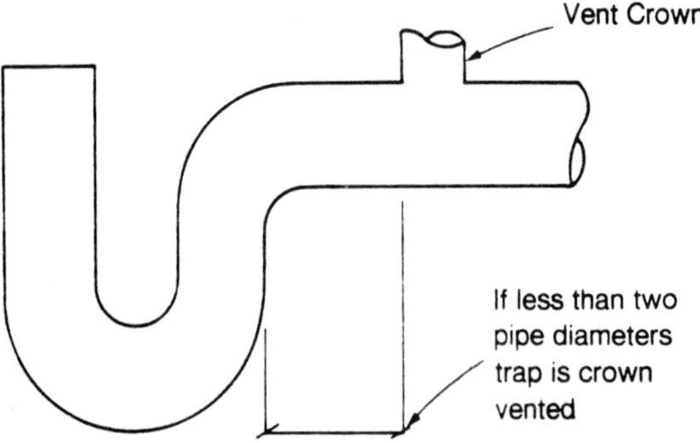

FIGURE C13.18 Crown vented trap.

ted by some codes for use on special-use sinks, such as in laboratories.

The branch drainage line extending from the trap to the vent is called the *trap arm* and is illustrated in Fig. C13.20. The maximum length of the trap arm is shown in Table C13.2.

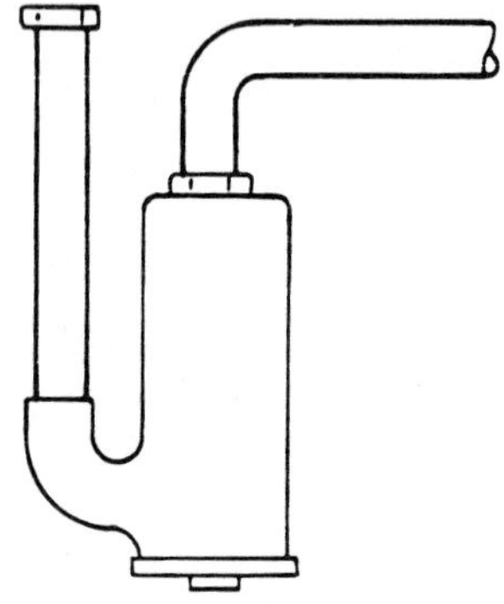

FIGURE C13.19 Drum trap.

Sanitary System Design

The design of the gravity drainage piping system is strictly regulated by the applicable plumbing code. All codes include charts, similar to those presented here, that permit the design engineer to properly size all horizontal and vertical pipes based on the accumulated drainage fixture unit discharge and slope of the pipe.

The pitch of the drainage system must provide sufficient velocity to produce a "scouring action" that will convey all solids along with the liquid stream. The recommended minimum velocity for ordinary sewage is 2 feet per second (fps) to prevent the settlement of solids out of the effluent stream. When grease is in suspension, the velocity should be at least 4 fps.

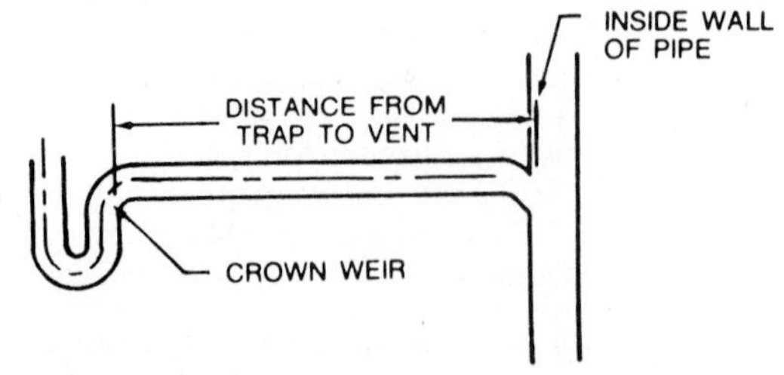

FIGURE C13.20 Trap arm.

TABLE C13.2 Maximum Length of Trap Arm

Diameter of trap arm, in	Distance—trap to vent
1¼	3 ft, 6 in
1½	5 ft
2	8 ft
3	10 ft
4	12 ft

For use with Fig. C13.20.

Accepted practice for low-rise buildings with relatively uniform discharge is to size horizontal drainage lines to flow half full under average design conditions. High-rise buildings produce higher velocities and turbulence in building drains that can fill portions of the piping system as much as three-quarters full, with completely full pipes expected for short distances at stack runouts. This is acceptable, providing that pipe size ultimately allowed for half-full pipes.

The following is a simplified method of sizing the drainage piping system:

1. Establish the location, size, and invert of the point of ultimate disposal of sanitary effluent. Determine if sump or ejector systems will be required and locate them.
2. Locate and layout drainage branch lines, stacks, and the house sewer.
3. Start with the individual device or fixture at the farthest and most remote point of the system or branch, for which the code specifies two drainage values. The first is the drainage fixture unit (DFU) value that will be used to size the drainage piping system. The second is a minimum size of the trap, which is the minimum individual branch pipe size. For the minimum trap size, refer to Table C13.1 for typical values. If a fixture or device is not listed, use either the unlisted value based on the size of the discharge, or ask the local code official for the accepted value.
4. When more than one fixture or waste line discharges into a horizontal branch pipe, the size of the horizontal drainage line is determined by both the pitch of the drainage line and the accumulated total number of DFUs discharging into it. Refer to Table C13.3 for sizes of branch lines. Refer to Table C13.4 for the size of building drains and sewers.
5. To determine the size of a horizontal drainage line based on flow given in gpm rather than DFUs, refer to Table C13.5. Use the appropriate pitch and velocity combinations as necessary to select a size. When there is a combination flow of both DFUs and gpm into a horizontal line or stack, generally accepted practice is to assign 2 FUs for each gpm to allow sizing based on DFUs.
6. The size of a stack is governed by the total DFU discharge into it, and its height. Refer to Table C13.3 using the applicable column and the total DFUs for the stack to find the stack size.
7. To size a stack based solely on gpm, refer to Table C13.6. Two generally accepted recommendations regarding the maximum cross-sectional area which may be occupied with water flowing down a stack are ¼ and 7⁄24, depending on the code used and the requirements of the local authority having jurisdiction. Separate columns are provided for each of these two values. Accepted prac-

TABLE C13.3 Horizontal Fixture Branches and Stacks
(Maximum number of fixture units that may be connected to:)

Diameter of pipe	Any horizontal fixture branch*	One stack of three branch intervals or less	Stacks with more than three branch intervals	
			Total for stack‡	Total at one branch interval
in	dfu	dfu	dfu	dfu
1½	3	4	8	2
2	6	10	24	6
2½	12	20	42	9
3	20†	48†	72†	20†
4	160	240	500	90
5	360	540	1100	200
6	620	960	1900	350
8	1400	2200	3600	600
10	2500	3800	5600	1000
12	3900	6000	8400	1500
15	7000			

*Does not include branches of the building drain.

†Not more than 2 water closets or bathroom groups within each branch interval nor more than 6 water closets or bathroom groups on the stack.

‡Stacks shall be sized according to the total accumulated connected load at each story or branch interval and may be reduced in size as this load decreases to a minimum diameter of ½ of the largest size required.

TABLE C13.4 Building Drains and Sewers*

Diameter of pipe, in	Maximum number of fixture units that may be connected to any portion of the building drain or the building sewer — Slope per foot			
	1/16-in	⅛-in	¼-in	½-in
2			21	26
2½			24	31
3			42†	50†
4		180	216	250
5		390	480	575
6		700	840	1,000
8	1,400	1,600	1,920	2,300
10	2,500	2,900	3,500	4,200
12	2,900	4,600	5,600	6,700
15	7,000	8,300	10,000	12,000

*On-site sewers that serve more than one building may be sized according to the current standards and specifications of the administrative authority for public sewers.

†Not over two water closets or two bathroom groups, except that in single-family dwellings, not over three water closets or three bathroom groups may be installed.

TABLE C13.5 Approximate Discharge Rates in gpm and Velocities in Sloping Drains (flowing half full*)
(discharge rate and velocity†)

Actual inside diameter of pipe, in	1/16-in/ft slope		1/8-in/ft slope		1/4-in/ft slope		1/2-in/ft slope	
	Disch., gpm	Vel., fps	Disch., gpm	Vel., fps	Disch., gpm	Vel., fps	Disch., gpm	Vel., fps
1¼							3.40	1.78
1⅜					3.13	1.34	4.44	1.90
1½					3.91	1.42	5.53	2.01
1⅝					4.81	1.50	6.80	2.12
2					8.42	1.72	11.9	2.43
2½			10.8	1.41	15.3	1.99	21.6	2.82
3			17.6	1.59	24.8	2.25	35.1	3.19
4	26.70	1.36	37.8	1.93	53.4	2.73	75.5	3.86
5	48.3	1.58	68.3	2.23	96.6	3.16	137.0	4.47
6	78.5	1.78	111.0	2.52	157.0	3.57	222.0	5.04
8	170.0	2.17	240.0	3.07	340.0	4.34	480.0	6.13
10	308.0	2.52	436.0	3.56	616.0	5.04	872.0	7.12
12	500.0	2.83	707.0	4.01	999.0	5.67	1413.0	8.02

*Half full means filled to a depth equal to one-half of the inside diameter.
†Computed from the Manning formula for ½-full pipe, $n = 0.015$.
For ¼ full: Multiply discharge by 0.274.
Multiply velocity by 0.701.
For ¾ full: Multiply discharge by 1.82.
Multiply velocity by 1.13.
For full: Multiply discharge by 2.00.
Multiply velocity by 1.00.
For smoother pipe: Multiply discharge and velocity by 0.015 and divide by n value of smoother pipe.

TABLE C13.6 Drainage Stack Capacity, gpm

Pipe diameter, in	¼ full	7/24 full
1¼	5	6.5
1½	8.1	10.5
2	17.5	22.6
2½	31.8	41
3	52.1	67.2
4	111	143
5	202	261
6	336	423
8	709	915

tice is to use the ¼-full criteria, which closely approximates the allowable flow from a horizontal pipe flowing full at ¼-in pitch.

8. If a stack should offset more than 45° from the vertical, the horizontal offset portion of the stack must be sized as a house drain. If the offset size is larger than that portion of the stack higher than the offset, the larger size must be carried down from the offset to the lowest level. The portion of the stack above the offset may remain unchanged.

9. The purpose in differentiating between branch intervals and the actual number of horizontal soil or waste branch lines entering the stack is to prevent overloading the stack in a short distance. Many codes limit the number of SFUs allowed in a branch interval.

Flow conditions in the offset portion of a stack create severe turbulence. Because of the resulting pneumatic effects, all branch connections that normally would be made at the level of the offset should be carried down 10 pipe diameters of the stack below the level of the offset.

Suds Pressure Areas. Appliances and fixtures normally using detergents, such as kitchen sinks, bathtubs, showers, dishwashers, and clothes washers, could discharge a large quantity of detergents into the drainage system. When flowing through the drainage piping, turbulence causes large amounts of suds to be generated. The suds accumulate in the lower portions of the drainage system and can remain there for a considerable period of time. When additional liquids flow into these sections of the system, the suds are displaced and will follow the path of least resistance. Enough suds pressure can be built up to force the suds through a fixture trap. Suds pressure areas exist in the following parts of the drainage system, as illustrated in Fig. C13.21:

1. For an upper-level stack offset serving fixtures on two or more floors above the offset, there are two suds areas. The first area, C13.21*a,* extends 40 pipe diameters of the stack upward from the base of the offset. The second,

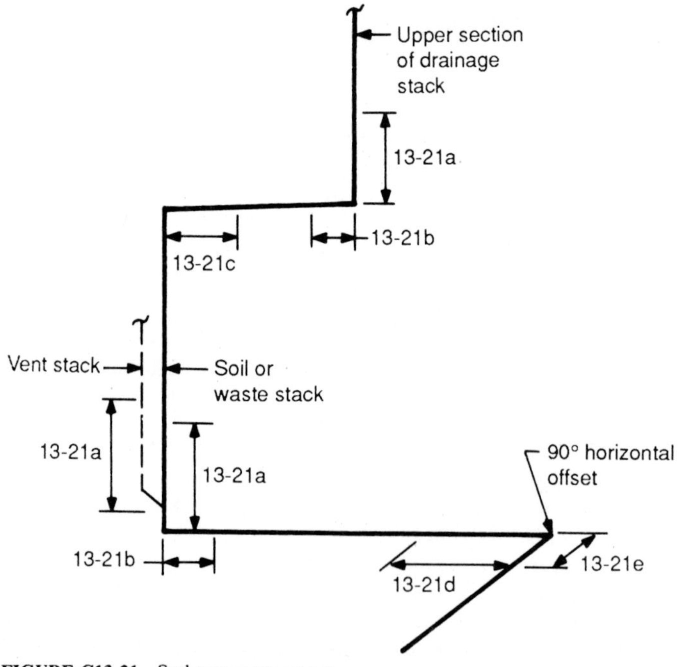

FIGURE C13.21 Suds pressure areas.

C13.21*b,* extends 10 pipe diameters horizontally downstream from the point of change in direction.

2. For an upper-level stack offset turning from horizontal back to vertical, there is one area, C13.21*c,* extending 40 pipe diameters of the stack upstream from the fitting changing direction from horizontal to vertical.
3. In the horizontal runout from a stack when the pipe changes direction horizontally with a fitting greater than 45°, there are two areas. The first, C13.21*d,* extends 40 pipe diameters of the horizontal pipe upstream from the change in direction. The second, C13.21*e,* is 10 pipe diameters downstream.

When suds pressure areas are anticipated, no pipe shall connect to any of the areas indicated as C13.21*a* to C13.21*e*. Refer to Table C13.7 for actual distances based on pipe size.

TABLE C13.7 Suds Pressure Area Distance Determination

Pipe size, in	40 diameters	10 diameters
1½	5 ft 0 in	1 ft 6 in
2	7 ft 0 in	1 ft 6 in
2½	8 ft 0 in	2 ft 0 in
3	10 ft 0 in	2 ft 6 in
4	13 ft 0 in	3 ft 6 in
5	17 ft 0 in	4 ft 0 in
6	20 ft 0 in	5 ft 0 in

For use with Fig. C13.21.

SANITARY VENT SYSTEM PIPING

System Description

The sanitary vent system is a network of pipes directly connected to the sanitary drainage piping system for the purpose of limiting air pressure fluctuations within the sanitary drainage piping to ±1 in of water column.

There are two primary reasons why the vent system is an integral and necessary adjunct to any drainage piping network:

1. It prevents the loss of fixture trap seals.
2. It permits the smooth flow of water in the drainage system.

Other lesser problems will be prevented if air pressure in the drainage system is excessive, such as:

1. Unsightly movement of water levels in water closet bowls
2. The possibility of sewer gases discharging through a fixture trap
3. Noise in the drainage system due to the "gurgling" of water

System Components

The components of the vent system consist of pipes directly connected to the drainage piping network. The piping materials are the same as that of the sanitary drainage system, except that the fittings may be of the short-turn type in lieu of the required long-turn fittings of the drainage system.

Code Considerations

Most problems occurring in the drainage system not resulting from blockages are caused by fluctuations in air pressure. These problems can be either eliminated or reduced to a level where they are no longer objectionable by designing a vent system that limits these variations in the drainage piping network to a generally accepted figure of ±1 in of water column. This basic design criteria has been used to determine vent sizing and allowable lengths that appear in modern codes.

Nomenclature

The following definitions are presented to avoid any differences in terminology between this handbook and any local, regional, and national codes.

Branch vent: A branch vent is a vent that connects one or more individual or common vents to a vent stack or a stack vent.

Circuit vent: A circuit vent is a branch vent that serves two or more traps and extends from a connection to a drainage line, in front of the last fixture connection, to a vent stack.

Common vent: A single vent line serving two fixtures.

Continuous vent: A vertical vent that is a continuation of the waste line from a fixture to which it is connected.

Developed length: Developed length is the total length of a vent pipe measured along the centerline of that pipe, from point to point.

Fixture battery: Any group of two or more fixtures that discharge into a common horizontal waste or soil branch.

Individual vent: A vent that connects directly to only one fixture and extends to either a branch vent or vent stack.

Loop vent: A loop vent is a branch vent that serves two or more traps, and extends from a point in front of the last fixture connection to a stack vent.

Main vent: A main vent is the principal vent of a building, remaining undiminished in size from the connection with the drainage system to its terminal.

Relief vent: A relief vent is an auxiliary vent that connects the vent stack to the soil or waste stack in multistory buildings; it is used to equalize pressure between them. This connection will occur at offsets and at set vertical intervals determined by code.

Revent: Another name for an individual vent.

Stack vent: A stack vent is the extension of a soil or waste stack above the highest horizontal drainage connection to that stack. It is also the name of a method of venting using the stack as a branch vent connection.

Suds venting: A method of venting where there is a suds pressure zone.

Trap arm: The trap arm is that portion of the drain pipe between the trap and the vent.

Vent extension: The height of the vent above the roof at its terminal.

Vent header: A single pipe at the highest level of a building connecting the top of vent stacks in order to penetrate the roof only once.

Vent terminal: The open-air location where the end of the vent stack is placed, generally above the roof.

Vent stack: A vertical pipe, extending one or more stories and terminating in the outside air.

Wet vent: A vent line that may also serve as a drain pipe.

System Design

General Vent System Design Considerations. Differences in pressure within drainage piping are caused by the flow of water. When water is flowing under design conditions in a horizontal drain (approximately one-half full), the air above the liquid will be forced into movement by the friction between the flowing water and the air. In a stack, the water flows around the perimeter of the pipe leaving a central core of air except when overloaded.

The following principles govern the design of the vent system:

1. When design flow is exceeded, the pipes are completely filled with water. This compresses the air ahead of, and creates a vacuum behind, the solid front of water.
2. The air moving in a vent pipe has friction losses similar to that of flowing water. This is why the longer the pipe, the larger the size.
3. The amount of air displaced is proportional to the amount of water flowing in the drainage pipe. The flow is determined by using drainage fixture units (DFU).
4. The size of a vent stack should be a minimum of one-half of the size of a drainage stack. The size of a branch vent should be a minimum of one-half the size of the branch drainage line it serves.
5. In a plumbing code, where the heading for soil or waste size refers to "stack size," it should also be used for horizontal branch soil and waste stacks. Since the venting requirements for a stack are more severe than that for a horizontal drainage line, there is a small safety factor.
6. All fixture vents must rise above the flood level of the fixture served so as not to act as a waste line in the event the drain line becomes blocked.

Developed Length Measurement. The developed length of an individual or common vent is measured from its point of connection with the fixture trap arm to where it connects with the branch vent or vent stack. The developed length of a branch vent is taken from the farthest connection with a waste branch from the point being sized. The developed length of a vent stack is taken from its connection with the soil or waste stack to its terminal above the roof.

Sizing of Vent Stacks, Vent Branches, and Fixture Vents. Plumbing codes contain the information necessary to size a vent system. A typical vent sizing chart is presented in Table C13.8.

In order to enter Table C13.8, there are three items that must be known: (1) the total DFU count of the soil or waste line associated with the vent being sized,

TABLE C13.8 Size and Length of Vents

Size of soil or waste stack, in	Fixture units connected	Diameter of vent required, in								
		1¼	1½	2	2½	3	4	5	6	8
		Maximum length of vent, ft								
1½	8	50	150							
2	12	30	75	200						
2	20	26	50	150						
2½	42		30	100	300					
3	10		30	100	100	600				
3	30			60	200	500				
3	60			50	80	400				
4	100			35	100	260	1000			
4	200			30	90	250	900			
4	500			20	70	180	700			
5	200				35	80	350	1000		
5	500				30	70	300	900		
5	1100				20	50	200	700		
6	350				25	50	200	400	1300	
6	620				15	30	125	300	1100	
6	960					24	100	250	1000	
6	1900					20	70	200	700	
8	600						50	150	500	1300
8	1400						40	100	400	1200
8	2200						30	80	350	1100
8	3600						25	60	250	800
10	1000							75	125	1000
10	2500							50	100	500
10	3800							30	80	350
10	5600							25	60	250

(2) the developed length of the vent being sized, and (3) the size of the soil or waste branch or stack.

Having calculated these items, enter the table with the most severe condition of soil pipe size or DFUs. Then read horizontally across until you come to the figure that meets or exceeds the developed length that you calculated. Read up to find the correct size of the vent. Use the following as a guide to sizing:

1. For vent stacks, use the total DFU load for the drainage stack and the full developed length of the vent to find the size. Vent stacks must be undiminished in size for its entire length.
2. For branch vents, use the longest developed length from the point where the size is being determined to the farthest connection to the waste line.
3. For individual fixture vent size, refer to Table C13.1.
4. For building trap vents and fresh air inlets, the size should be a minimum of one-half the size of the building drain.

Vent Terminals. The vent pipe passing through the roof must remain open under all circumstances. The two conditions that would cause the exposed pipe to become blocked are frost closure and snow closure. Local codes and authorities will provide the minimum extension to avoid closure by accumulated snow on a roof.

In the absence of specific code requirements, the following can be used as a guide to locating vent extensions. The vent extension shall not be located under, or within 10 ft, 0 in, of any window, door or ventilating opening unless it is 2 ft, 0 in, above that opening. If the terminal is through a building wall, it shall be located a minimum of 10 ft, 0 in from the property line, a minimum of 10 ft, 0 in, above grade, and not under any overhang.

Relief Vents. Soil or waste stacks with no offsets, in buildings having more than 10 branch intervals, shall be provided with a relief vent at each tenth interval starting at the top floor. Offsets in the drainage stacks may also be required to have relief vents.

There are several acceptable configurations allowed by the various codes. In general, the lower end of the relief vent shall connect to the soil or waste stack below the horizontal branch serving the floor required to have the relief vent. The upper end of the relief vent shall connect to the vent stack no less than 3 ft, 0 in above that same floor level. The size shall be equal to that of the vent stack to which it is connected or the drainage stack, whichever is smaller.

Circuit and Loop Vents. These venting schemes are intended to provide a more economical means of venting than the individual vent. It is allowed only for venting of floor-mounted fixtures such as water closets, shower stalls, and floor drains and may not be acceptable in all code jurisdictions. Circuit venting is illustrated in Fig. C13.22.

Circuit venting requires a uniformly sized drainage line with at least two, but not exceeding eight, fixtures connected in a battery arrangement. The circuit vent

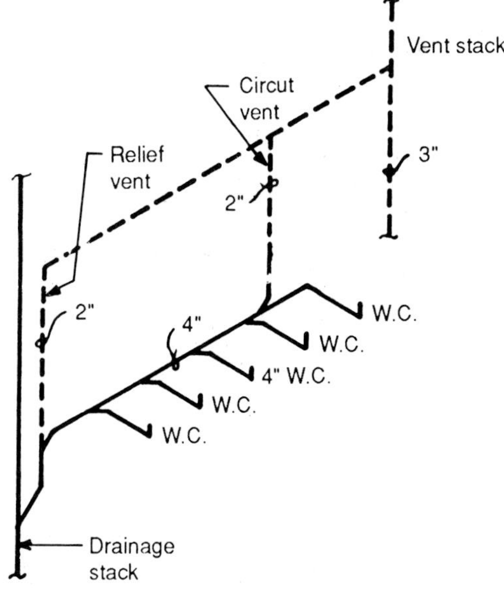

FIGURE C13.22 Detail of circuit vent.

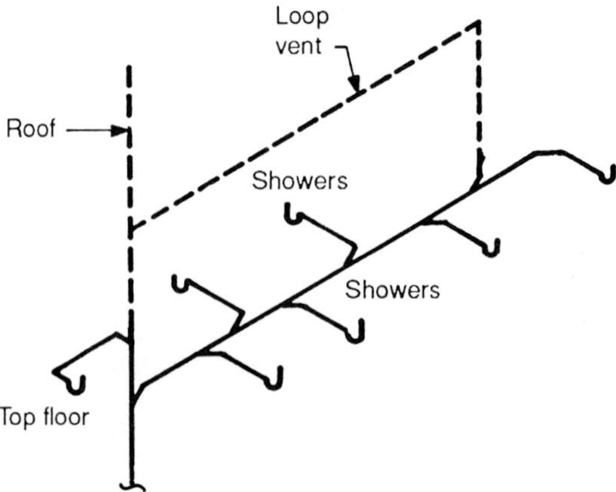

FIGURE C13.23 Detail of loop vent.

connects to the vent stack from the horizontal drain line from a point between the two most remote fixtures. In addition to the circuit vent, a relief vent is required to be connected to the horizontal drain line at the end of the battery, or every eight fixtures. The sizes of each shall be one-half the size of the horizontal drainage line or the full size of the vent stack, whichever is smaller.

Loop venting is the same as circuit venting except for the connection of the branch vent to the building system. The loop vent "loops" back to the stack vent instead of the vent stack. This is illustrated in Fig. C13.23.

Wet Vents. A wet vent is a combined vent-drain line that receives drainage from fixtures in addition to serving as a vent pipe. Wet vents are primarily used in residential-type projects, and permitted only in a limited number of codes. Wet vents shall conform to the guidelines provided in each respective code.

Suds Relief Vents. Suds pressure zones are illustrated in Fig. C13.21. If a drainage connection to these zones is made, a relief vent must be installed from the base of the suds pressure zone of the drainage stack to a nonpressure zone. Typical suds pressure relief vent sizes are shown in Table C13.9.

TABLE C13.9 Suds Pressure Relief Vents

Waste size	Relief vent size
1½	2
2	2
2½	2
3	2
4	3
5	4
6	5
8	6

TABLE C13.10 Size and Length* of Sump Vents

Discharge capacity of sump pump, gpm	Diameter of vent, in					
	1¼	1½	2	2½	3	4
	(Maximum equivalent length of vent, in feet, given below)					
10	NL†	NL	NL	NL	NL	NL
20	270	NL	NL	NL	NL	NL
40	72	160	NL	NL	NL	NL
60	31	75	270	NL	NL	NL
80	16	41	150	380	NL	NL
100	10‡	25	97	250	NL	NL
150	NP§	10^3	44	110	370	NL
200	NP	NP	20	60	210	NL
250	NP	NP	10	36	132	NL
300	NP	NP	10‡	22	88	380
400	NP	NP	NP	10^3	44	210
500	NP	NP	NP	NP	24	130

*Developed length plus an appropriate allowance for effects of entrance losses and friction due to fittings, changes in direction, and changes in diameter. Suggested allowances may be obtained from NBS Monograph 31 or other acceptable sources. An allowance of 50 percent of the developed length may be assumed if a more precise value is not available.

†No limit; actual values greater than 500 ft.

‡Less than 10 ft.

§Not permitted.

Sump and Ejector Vents. All codes require the venting of ejector pits, since they are gasketed and air tight. Many codes may also require the venting of sump pits. The vent pipe size is determined from the gpm discharge of the pump and the developed length from the pit to its connection with the building vent system or vent terminal. Table C13.10 is a typical method of sizing such vents.

Vent Headers. When combining several vent stacks into a common header at the highest level to penetrate the roof only once, various codes require that the combined vent pipe be sized using the combined DFUs of all the connected vents and the single longest developed length of all the vent stacks being combined. Other codes require that the combined vent stacks have a minimum cross-sectional area of all the separate vent pipes being combined. Table C13.11 lists the cross-sectional area for Schedule 40 steel pipe in square inches.

TABLE C13.11 Nominal Pipe Cross-Sectional Area, in^2 (Schedule 40 pipe)

1½ in	1.767
2 in	3.1416
2½ in	4.908
3 in	7.068
4 in	12.566
5 in	19.635
6 in	28.274

INTERIOR STORMWATER DRAINAGE SYSTEM PIPING

Purpose

The purpose of the stormwater drainage system is to collect stormwater runoff from building roofs and ancillary areas exposed to the weather, and convey the runoff to an approved point of disposal. Ancillary areas of a building include areaways, walkways, canopies, and balconies.

Components

Major components of the stormwater drainage system are drains and the pipe itself. Drain manufacturers make various types of roof and specialty drains as well as accessories used for the installation of drains in all areas of a project. Any pipe material suitable for the sanitary drainage system is also acceptable for use in the stormwater drainage system.

Nomenclature

Areaway: An enclosed excavated area below grade level open to the weather.

Canopy: A small roof protecting a window or entrance.

Conductor: Stormwater piping inside of a building.

Design point: The specific point in the piping network where pipe size is calculated.

Downspout: A vertical pipe attached to gutters installed on the outside of a building.

Drain: A receptacle for the collection and removal of stormwater from surfaces exposed to the weather into the stormwater drainage piping network.

Duration: A commonly used term for time of concentration.

Flow rate: The measurement of a volume of water over time, such as cubic feet per second.

Frequency: The estimated number of years that elapse between the reoccurrence of storms with a specific intensity.

Gutter: An open horizontal channel used to collect stormwater, usually made of sheet metal or wood, attached to the lowest point of a pitched roof.

Imperviousness factor: A number indicating the percent of rainfall available as runoff and not absorbed into the ground, absorbed by plants, left as puddles, or lost to evaporation during the rainstorm, expressed as a decimal.

Inlet time: A frequently used term for overland flow time.

Intensity: The rate at which rain falls for design purposes, measured in inches per hour.

Leader: A vertical pipe carrying stormwater either inside or outside the building.

Overflow: A positive and fail-safe outlet for removal of stormwater that has reached a predetermined height above the roof level.

Overland flow time: The time rainwater takes to travel on the ground from the farthest point of an outside area to a drain, measured in minutes.

Piping network: The entire stormwater drainage system, including all drains and pipe, to the point of disposal.

Return period: A commonly used term for frequency.

Rate of rainfall: A commonly used term for intensity.

Runoff: The actual flow rate of stormwater discharging into the piping network.

Scupper: A penetration through a parapet above the roof level serving as an overflow.

Sidewall area: Vertical surfaces that contribute runoff to the stormwater system.

Stormwater: Liquid effluent resulting from any form of precipitation, such as rain, snow, hail, or sleet.

Time in pipe: The length of time stormwater will take to reach one design point from another design point while inside the piping network.

Time of concentration: The length of time a rainstorm will persist for design purposes, usually calculated by adding the overland flow time to the time in pipe.

Tributary area: The entire area contributing runoff into any portion of the piping network.

General System Criteria

General Information. The design of the stormwater drainage system requires that the following information be obtained in order to establish design criteria:

1. Local climatic conditions
2. Local building and plumbing code restrictions
3. Building use
4. Building construction and pitch of roof
5. Location, size, depth, type, and availability of the ultimate point of disposal
6. Total size of roof and ancillary tributary areas
7. Method of connecting to public sewers
8. Allowable methods of disposal and permits required if public sewers are not available
9. Client standards and preferences

Design Storm. The stormwater system is designed to remove the maximum expected runoff in a given period of time. The ability to calculate the flow rate is complicated by not being able to accurately predict many of the factors affecting the actual amount of runoff resulting from any given storm. In order to calculate the estimated maximum runoff, an artificial "design storm" must be created. This storm will serve as a simulation capable of predicting runoff volume accurately enough to provide a basis for the design of the piping network.

The design storm is based on actual rainfall records and is plotted in convenient form by the National Oceanic and Atmospheric Administration/National Weather Service (NOAA).

Special consideration should be given to the degree of protection provided for the building and its contents by the stormwater drainage system. This depends on the importance of the facility or the space below the roof, the importance of uninterrupted service and the value of the equipment or material installed or stored. Since a severe thunderstorm or hurricane may produce rainfall rates larger than the code has been based on, the property value may require the design engineer to select a rainfall rate greater than the minimum provided for by the code.

Overflows. A fail-safe method must be provided to immediately remove excess runoff from a roof before the water level rises to a point where damage would result. The most commonly used method are scuppers that will allow the excess runoff to be discharged directly off the roof and down the side of the building. Codes often stipulate the height that scuppers shall be installed above the roof level, and the size of the opening. One disadvantage of scuppers is that wave action may allow water to spill out of the scuppers before the design depth is reached.

Various codes mandate the use of a separate overflow piping network in addition to the regular stormwater drainage system. This network could connect either to the regular system independently outside a building or directly to the point of disposal. Another type of overflow protection is the use of separate overflow pipes or drains, in addition to the regular drains, connecting to the regular stormwater drainage piping. Applicable codes and local authorities will provide the information required for design.

Sidewall Area Calculation. Precipitation falling on vertical walls located on roofs and other areas, such as penthouse walls and stair towers, will add to the total runoff calculated for the horizontal area. Therefore, It is necessary for these vertical areas to be added to the horizontal tributary area. To calculate the amount of sidewall area that shall be added to the horizontal roof segment, determine the square foot area of the single or two largest adjacent walls that would contribute runoff to any one roof drain. Divide this area in half, because a lesser amount of rain falls on vertical surfaces than on horizontal surfaces. Add the calculated vertical sidewall area to the horizontal tributary area for each roof drain to obtain the total tributary area.

Roof Drainage Systems. There are two types of systems used to remove stormwater runoff: conventional and limited discharge. The conventional system removes runoff as quickly as it accumulates. The limited discharge system removes only a portion of the runoff, storing the remainder temporarily on the roof or elsewhere on the site. Selection of the appropriate method depends on the capacity and/or availability of disposal facilities and acceptance of the proposed method by local code and authorities.

Roof Drainage Design Procedures

Conventional Roof Drainage Procedure. Design of the conventional roof drainage system consists of the following general steps:

1. Locate drains on roofs and ancillary areas throughout all areas of the project discharging into the system.
2. Determine the code overflow requirements.
3. Route the stormwater piping and overflow systems.

4. Select the rainfall intensity.
5. Size the piping network by first calculating the total tributary area for each individual drain located in each specific section of the roof and other areas. For horizontal pipe, the sizing procedure starts at the most remote part of the system. Determine the pipe pitch and total tributary area for each horizontal pipe section from design point to design point.
6. Use the appropriate charts to size the roof drains and the piping network. Enter the charts provided in the applicable code with the intensity, pitch, and total tributary area. Select the figure at the intersection of the criteria used for that specific chart and choose a pipe size corresponding to a figure equal to or greater than the calculated tributary area. Notes associated with different chart types allow conversion to another rainfall rate if different from that of the chart. For vertical piping, use the tributary area discharging into the vertical leader, increasing the size as additional areas are added. Refer to Table C13.12 for a typical table to size both roof drains and the vertical leaders. Table C13.13 will allow sizing of horizontal pipes depending upon pitch and tributary area. Table C13.14 allows the sizing of gutters. These tables are from the National Standard Plumbing Code which uses 4 in/h as the basis for the tables. Local rates shall always take precedence over a regional value.
7. When a very large building, such as a warehouse or factory, has a total tributary area in excess of that appearing in the code, other methods must be used to calculate the volume of runoff and size the pipe. The most commonly used method for the calculation of runoff is the rational formula:

$$Q = A \times I \times R \tag{C13.1}$$

where Q = quantity (flow rate) of stormwater runoff, ft^3/s (cfs) (1 ft^3/s = 448 gpm)
A = tributary area, acres (1 acre = 43,560 ft^2 or sf)
I = imperviousness factor: use a value of 1.0
R = rate (intensity) of rainfall, in/h (Select the rainfall rate based on information contained in the previous discussion.)

TABLE C13.12 Size of Vertical Conductors and Leaders*

Diameter of leader or conductor†, in	Maximum projected roof area	
	ft^2	gpm
2	544	23
2½	987	41
3	1,610	67
4	3,460	144
5	6,280	261
6	10,200	424
8	22,000	913

*Based upon a maximum rate of rainfall of 4 in/h and on the hydraulic capacities of vertical circular pipes flowing between one-third and one-half full at terminal velocity, computed by the method of NBS Mono. 31. Where maximum rates are more or less than 4 in/h, the figures for drainage area shall be adjusted by multiplying by 4 and dividing by the local rate in inches per hour.

†The area of rectangular leaders shall be equivalent to that of the circular leader or conductor required. The ratio of width to depth of rectangular leaders shall not exceed 3 to 1.

TABLE C13.13 Size of Horizontal Storm Drains*

Diameter of drain, in	Maximum projected area for drains of various slopes					
	⅛-in slope		¼-in slope		½-in slope	
	ft^2	gpm	ft^2	gpm	ft^2	gpm
3	822	34	1,160	48	1,644	68
4	1,880	78	2,650	110	3,760	156
5	3,340	139	4,720	196	6,680	278
6	5,350	222	7,550	314	10,700	445
8	11,500	478	16,300	677	23,000	956
10	20,700	860	29,200	1,214	41,400	1,721
12	33,300	1,384	47,000	1,953	66,600	2,768
15	59,500	2,473	84,000	3,491	119,000	4,946

*Based upon a maximum rate of rainfall of 4 in/h. Where maximum rates are more or less than 4 in/h, the figures for drainage area shall be adjusted by multiplying by 4 and dividing by the local rate in inches per hour.

TABLE C13.14 Size of Roof Gutters*

Diameter of gutter†, in	Maximum projected roof area for gutters	
	1/16-in slope†	
	ft^2	gpm
3	170	7
4	360	15
5	625	26
6	960	40
7	1380	57
8	1990	83
10	3600	150

*Based upon a maximum rate of rainfall of 4 in/h. Where maximum rates are more or less than 4 in/h, the figures for drainage area shall be adjusted by multiplying by 4 and dividing by the local rate in inches per hour.

†Gutters other than semicircular may be used provided they have an equivalent cross-sectional area.

‡Capacities given for slope of 1/16 in/ft shall be used when designing for greater slopes.

The design procedure is the same as described for conventional drainage system, except that Eq. (C13.1) is used to calculate the flow rate in horizontal system piping. Determine the slope of the pipe, and use Table C13.13 to find the pipe size up to the largest areas in the chart. When a larger area is reached, use Table C13.15, entering the chart with the gpm and the pitch. Table C13.16 provides a direct conversion in both gph and gpm per square foot of tributary area for various rainfall rates. For rates not shown, add any lesser rate together to find the new rate.

In some areas of the country, the public sewer is a combined sanitary and stormwater system. This requires the sanitary and stormwater systems to be combined either inside or outside the building wall. The combined system is usually sized on the basis of DFUs. In order to size the combined system, the square

TABLE C13.15 Flow Capacity for Stormwater Piping Systems in gpm, Flowing Full (n = 0.013)

Pipe diameter, in	Roof drain and vertical downspouts	Horizontal storm drainage		
		Slope, in/ft		
		1/8	1/4	1/2
2	30	—	—	—
2½	54	—	—	—
3	92	36	51	80
4	192	77	110	174
6	563	220	315	449
8	1208	494	696	987
10	2600	943	1302	1800
12	6000	1526	2154	2800
15		2873	3500	4950

TABLE C13.16 Rainfall Conversion Data

Rainfall, in/h	gph per 1 ft^2	gpm per 1 ft^2
3.0	1.870	0.0312
2.9	1.808	0.0302
2.8	1.745	0.0291
2.7	1.683	0.0281
2.6	1.621	0.0270
2.5	1.558	0.0260
2.4	1.496	0.0250
2.3	1.434	0.0239
2.2	1.371	0.0229
2.1	1.309	0.0218
2.0	1.247	0.0208
1.9	1.184	0.0198
1.8	1.122	0.0187
1.7	1.060	0.0177
1.6	0.997	0.0166
1.5	0.935	0.0156
1.4	0.873	0.0146
1.3	0.810	0.0135
1.2	0.748	0.0125
1.1	0.686	0.0114
1.0	0.623	0.0104

feet of tributary area of the stormwater system is converted to DFUs in order to calculate the size of the combined building drainage network. Table C13.17 can be used as a guide for the conversion of square feet into DFUs.

Limited Discharge Roof Drainage Systems

There are two reasons to select a limited discharge roof drainage system: voluntarily for economic reasons or by necessity. The voluntary use of the limited dis-

TABLE C13.17 Fixture Unit-Drainage Square-Footage Equivalent

Drainage area, ft^2	Fixture unit equivalent
180	6
260	10
400	20
490	30
1,000	105
2,000	271
3,000	437
4,000	604
5,000	771
7,500	1,188
10,000	1,500
15,000	2,500
20,000	3,500
28,000	5,500
Each additional 3 ft^2	1 fixture unit

Note: IGPM = 19 ft^2.

charge system will allow smaller piping throughout the entire network, resulting in a saving for the total installed cost of the piping. Job conditions may mandate the use of a limited discharge system. Urban areas may have overloaded sewers, allowing only a small portion of the runoff to be discharged into the public sewer in order to prevent additional overloading.

In order to limit the discharge, install roof drains with factory preset grate openings that allow only a predetermined amount of stormwater to enter the drain. The allowable amount to be discharged must be given to the manufacturer in order to have the proper drain openings set correctly at the factory. The runoff not discharged must be temporarily stored on the roof or on site.

A simplified design procedure for the limited discharge roof drainage system consists of the following general steps:

1. Determine if the code and local authorities will permit limited discharge roof drainage for the project.
2. Establish the maximum allowable flow rate, in gpm, permitted to discharge into the ultimate point of disposal from all sources of water for the entire site. If the system is voluntary, some iterations will be necessary after the piping is run to find the optimum pipe size compared to the amount of stormwater stored on the roof. If discharge is into a sewer, the authorities having jurisdiction will provide the allowable discharge amount. If discharge is into a waterway, the existing site conditions should be calculated. The difference between the existing and proposed discharge volume is the volume of water that should be stored on the roof. If drains in areas other than the roof are present, it is not generally possible to store water from those areas. The discharge from these drains will have to be separated from the total roof discharge to find the actual discharge allowed from the roof alone.
3. Decide on the length of time water is to remain on the roof. Generally accepted practice is to allow between 12 and 18 h, starting with as short a time

as possible. Lengthen the time as necessary to obtain the calculated rate of discharge.

4. Since the structural engineer must design the roof to support the additional load of stored water, a mutual agreement between the plumbing and structural engineers must be made to determine the depth. If the roof is flat, the generally accepted depth is 3 in. If the roof is pitched, add 3 in water depth above the high point of the roof. Table C13.18 gives the weight for each inch of water.

TABLE C13.18 Weight of Rainfall

Amount of rain, in	Weight of water, lb/ft^2
6	31.21
5	26.01
4	20.81
3	15.61
2	10.40
1	5.20

5. Find the total amount of rainfall, in inches, that will fall on the roof for the time established in Step 3. The amount is obtained from Fig. C13.24, which is a 24-h, 10-year return rainfall. A 10 percent reduction in the 24-h figure approximates an 18-h rainfall.
6. Divide the figure found in Step 5 with the time found in Step 3 (e.g., 8 in divided by 12 h equals 0.67 in/h). Then determine the gpm discharge using Table C13.16. Compare this figure with the maximum allowable discharge from the roof found in Step 2. Adjust the retention time as required to match the allowable discharge rate.
7. Consideration must be given for a heavy rainfall occurring after the design storm ends, while some water remains on the roof during the draindown time. This rainfall should be the actual 1-h duration, with the same return period selected for the project. This 1-h rainfall will deposit a number of inches of rain. The actual drainage rate will be the figure selected in Step 2. By subtracting the drainage rate figure (in inches per hour) from the 1-h rainfall figure, the actual total allowable rainfall depth will be found. This depth must not exceed the depth established in Step 4. If it does, additional depth of storage is required, or the client must be willing to accept rainwater spilling out of the emergency overflow once during the return period selected.
8. Locate drains on roof, and find the total number of drains required.
9. Route the stormwater piping and overflow systems.
10. Divide the number of drains into the total roof discharge allowed to find the gpm flowing out of each drain. Using Table C13.15, size the individual drains and branches using the gpm and pitch of the pipe. Limited discharge roof drains are set at the factory for the specified maximum flow rate established by the engineer.
11. Size the piping network based on accumulated gpm flow at each design point and the pitch of the pipe.

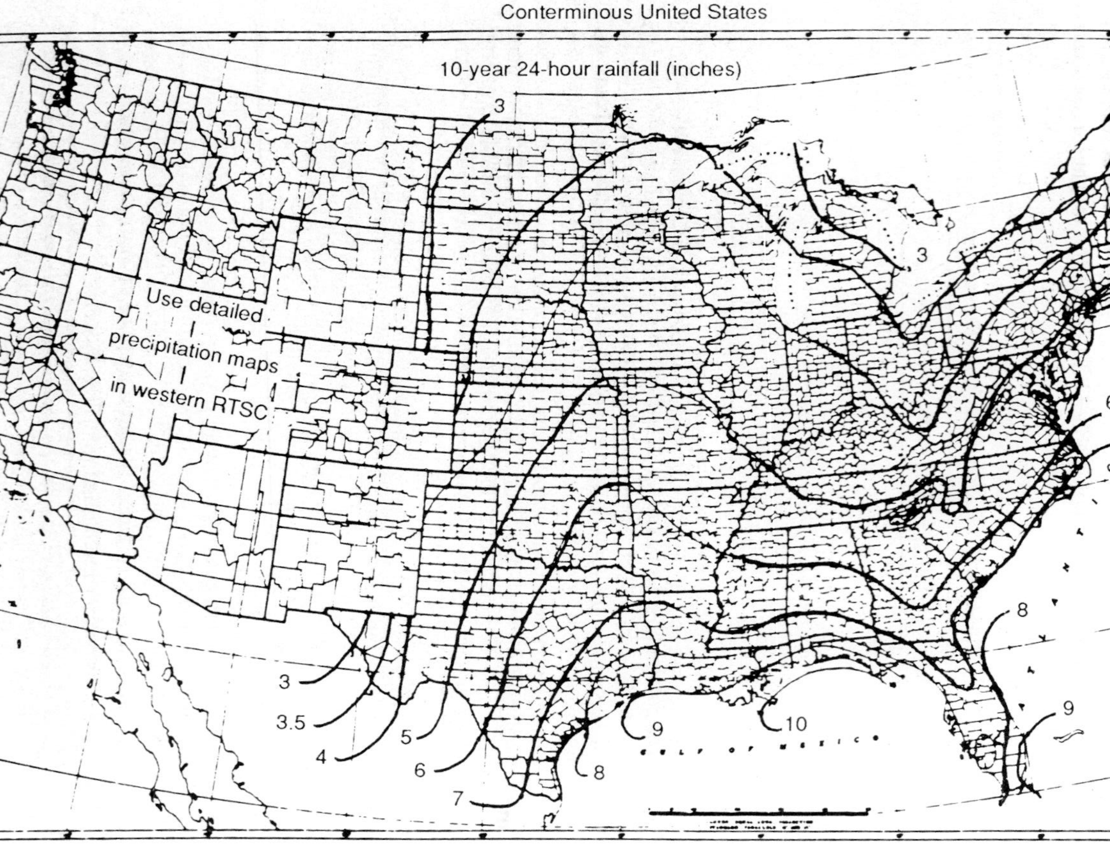

FIGURE C13.24 10-year, 24-h rainfall.

Conventional Roof Drainage System Considerations

1. It is important that the local codes and authorities be consulted for the method used to connect to the public sewers.
2. It is good practice to place two roof drains in any area (except very small areas) to provide at least one drain to accept flow if the other is blocked by debris.
3. Limit the square footage of any individual roof drain to a maximum of 5000 ft^2 of tributary area.

Limited Discharge Roof Drainage System Considerations

1. Drains should be no greater than 200 ft apart.
2. Drains should be no greater than 50 ft from the end of a tributary area.

POTABLE WATER SUPPLY SYSTEMS AND PIPING

Purpose

The purpose of the potable water system is to provide water suitable for human consumption that has adequate purity, pressure, temperature, and volume to satisfy all applications for a specific project or purpose.

Components

The total potable water system consists of cold water, hot water, and hot water temperature maintenance subsystems. Major components of the various subsystems are water meters, treatment facilities, backflow preventers, storage tanks, pumps, pressure reducing devices, water heating equipment, temperature maintenance wiring, hot water circulating pumps, system valves, and piping.

Nomenclature

Air gap: An unobstructed separation between a source of potable water and any source of contamination.

Backflow: Any reversal of the flow of water from its intended direction.

Back pressure: Backflow caused by an increase of normal pressure.

Back siphonage: Backflow caused by a lowering of normal pressure.

Booster hot water system: A secondary water heating system used to heat water to a higher temperature than that of the primary water heating system.

Cavitation: A phenomenon of flowing water caused by the rapid formation and collapse of air cavities, which results in the pitting of surfaces on which they occur.

City water: A commonly used term for potable water.

Contaminant: Any impurity or toxic substance which, when introduced into a potable water supply, will create a health hazard or threaten the well being of a consumer.

Cross connection: Any physical connection between a potable water system and any potential source of contamination not protected by an approved device specifically designed to prevent flow between the two.

Demand: Estimated flow rate expected under normal operating conditions, most often expressed in gallons per minute (gpm).

Direct fired: A water heater whose primary heat source is an integral part of the water heater assembly.

Domestic water: Potable water primarily intended for direct human use, such as that supplied to plumbing fixtures.

Evacuation-type plumbing fixtures: Plumbing fixtures, such as water closets and urinals, used to receive and discharge waterborne human bodily waste.

Hot water: Water at a temperature higher than ambient, established by generally accepted practice or code as being suitable for a specific application.

Hydraulically remote: Farthest from the source of supply in terms of total pressure lost through the entire water supply piping system.

Impurity: Any physical, chemical, or biological substance found in water, making it undesirable for a specific use or degrading the purity of potable water.

Indirect fired: A water heater whose primary heat source is generated remotely from the water heater.

Ion: Atoms, either singly or in groups, that have an electric charge.

Maximum acceptable pressure: The highest pressure that will not cause a nuisance or produce premature and accelerated damage to any component.

Maximum building demand: The estimated flow of water from the maximum fixture demand plus highest water demand from various equipment throughout a building.

Maximum fixture demand: The greatest estimated flow of water resulting from the probable maximum simultaneous use of intermittently operated plumbing fixtures.

Minimum acceptable pressure: The lowest water pressure permitting safe, efficient, and satisfactory operation of the most hydraulically remote fixture or component.

Normal pressure: The design or expected force per unit area at any point in a water system, usually expressed as pounds per square inch.

Point-of-use heater: Locations immediately adjacent to the fixtures and/or equipment requiring hot water, as compared to a remote, centralized location serving an entire building, area, or project.

Pollutant: Any nontoxic impurity that may create a moderate or minor hazard to the water supply.

Potable water: Water of sufficient purity to meet various standards established as being fit for human consumption.

Pressure zone: A water distribution system within any area of a building having a common source of water supply or pressure origin.

Raw water: Water received directly from the supply source.

Recovery rate: The amount of water capable of being heated to the design temperature per unit of time in a water heater.

Residual water pressure: The pressure of water available in a piping system when water is flowing.

Service hot water: Hot water intended for commercial, industrial, or domestic use within a building.

Static water pressure: The pressure of water in a piping system during the time no water is flowing.

Street pressure system: A pressure zone supplied from a public water main, using only the pressure available in that main.

Toxic substance: A commonly used term for pollutant.

Ultra low flush(ULF): Used to describe water closets requiring low volume of water (1.6 gpm) for discharge into the sanitary drainage system in lieu of the 3.5 gpm which is currently the standard.

Usable storage volume: The gallons of hot water available for use before the introduction of water into the tank faster than it can be heated lowers the water temperature below acceptable limits. A generally accepted value is 70 percent.

Water hammer: A sharp, instantaneous rise in pressure of water in a pipe and the resulting shock wave caused by a sudden stoppage of moving water.

Water Meters

When water is provided by a utility company, meters are required for billing purposes. The utility company will usually require a specific location for the meter, such as inside the building or in a pit adjacent to the property or building, clearance around the meter needed for reading the meter, piping arrangement, and other regulations to discourage any attempt to bypass the meter.

The water meter is usually a part of a water meter assembly that could include shut-off valves, strainers, a test tee for testing the accuracy of the meter by the local authorities, and a meter bypass.

The selection of a meter type, if not mandated by the utility company or local authorities, is based on accuracy and pressure loss through the meter at the intended flow rates. Water meter types include disc meter, turbine meter, and compound meter.

Contamination Prevention

Impurities found in potable water are divided into three general classifications: severe, moderate, and minor. It is necessary to evaluate each facility as a whole and also at specific points of use to protect against potential hazards. Local health, building codes, and ordinances must be consulted for specific requirements, and for the suitability of any device for a specific application.

There are five basic methods of preventing the contamination of a potable water system due to cross connections with, and backflow from, a potentially contaminated source:

1. *Air gap:* Suitable for severe hazards, the air gap is passive, and is the only fail-safe method of preventing backflow. The allowable air gap dimension is provided in local codes and is usually 2 in.

2. *Pressure-type vacuum breaker:* Suitable for minor hazards, the pressure-type vacuum breaker is a mechanical device designed to prevent backflow caused only by back pressure conditions. It is designed to operate under continuous pressure on both sides of the device.
3. *Atmospheric-type vacuum breaker:* Suitable for minor hazards, the atmospheric-type vacuum breaker is a mechanical device designed to prevent backflow caused only by back siphonage conditions. It is designed to operate with pressure on only one side of the device.
4. *Double check valve:* Suitable for moderate hazards, a double check valve, illustrated in Fig. C13.25, is a mechanical device consisting of two independently operating, soft seat, swing check valves. It prevents backflow from back siphonage and back pressure conditions.

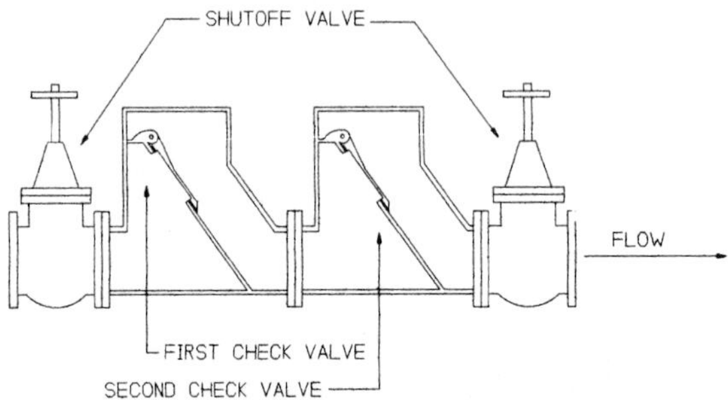

FIGURE C13.25 Double check valve backflow preventer.

5. *Reduced pressure zone:* Suitable for severe hazards, a reduced pressure zone backflow preventer, illustrated in Fig. C13.26, is a mechanical device consisting of two independently operating, soft seat, spring loaded, check valves with the addition of an independent, differential pressure relief valve installed between the check valves, in the region of the assembly called the *reduced pressure zone.* The relief valve will open under backflow conditions and discharge all the upstream water under pressure until the condition is corrected. With the potential for large flows, it is accepted practice to direct the discharge outside a facility.

The supplier of public potable water is responsible to protect the public distribution system from contamination by a nearby contaminated water source, such as a pond or stream, that may be used by a fire department to fight a fire at the project location. Depending on local regulations, it may be necessary to provide backflow protection on the site main before it connects to the public source of water.

Another aspect of contamination protection is the disinfection of the entire interior potable water system. A commonly referenced standard is AWWA C651, Standard for Disinfecting Water Mains.

Water Velocity

The velocity of water flowing through the piping system should be kept low enough to prevent objectionable noise, water hammer, and accelerated compo-

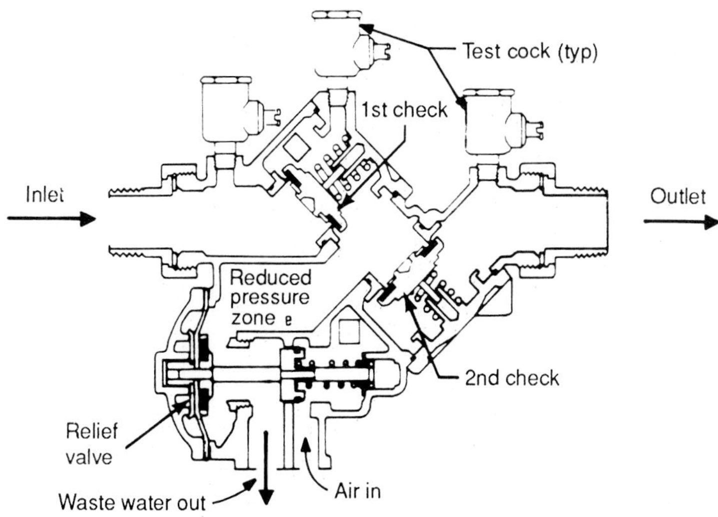

FIGURE C13.26 Reduced pressure zone backflow preventer.

nent wear due to erosion or nuisance splashing of water from fixtures. In order to avoid excessive noise, generally accepted practice for commercial buildings is to limit water velocity to between 6 and 8 fps. For industrial projects, 10 fps is acceptable in work areas where the noise is not noticeable. If the water supply is controlled by a quick closing valve, the velocity should be limited to approximately 4 fps to avoid water hammer.

Water Distribution Systems

The purpose of a water distribution system is to deliver both hot and cold water, in an acceptable range of pressure and temperature, to fixtures and devices in all parts of a building. All distribution systems can be divided into either upfeed or downfeed systems.

In the downfeed system, illustrated in Fig. C13.27, water is supplied down through vertical pipes to the lowest point in the pressure zone. In an upfeed system, illustrated in Fig. C13.28, the water is supplied upward through vertical pipes to the highest point in the pressure zone.

The decision as to which system to use is based on several factors, such as where enough room exists to run the hot and cold water distribution mains, origin of the source of supply, and economics. The downfeed system generally has smaller pipe sizes and is usually more economical than an upfeed system.

Estimating Water Demand

The calculation of maximum demand is approximate because it is not possible to predict exactly how many fixtures or various pieces of equipment may be in use at the same time. This estimate should allow for the probability that the calcu-

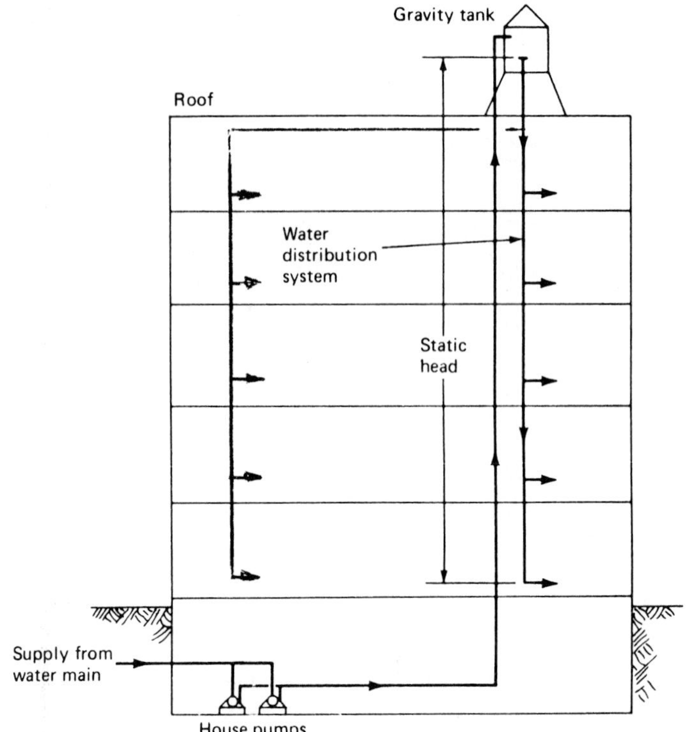

FIGURE C13.27 House tank and downfeed water distribution system.

lated flow will be exceeded occasionally but not permit wasteful oversizing because the estimate will be rarely exceeded.

The method used to calculate the maximum probable demand is based on water fixture units (WFU). Refer to Table C13.1 for WFU values for typical fixtures. Table C13.19 permits determination of the estimated flow rate in gpm for accumulated WFUs at any point in the system. Interpolate to find intermediate values.

Table C13.19 is divided into two columns, one for systems containing flush valve operated water closets in addition to other fixtures, and the other for flush tank water closets and any other fixtures. Use the appropriate column for the specific branch or system under design, with a value of 75 percent of the WFU value for fixtures using both hot and cold water.

Design of the Water Supply Distribution System

The water supply system must achieve the following basic objectives:

1. Deliver an adequate volume of water to the most hydraulically remote fixture during minimum pressure and maximum flow conditions

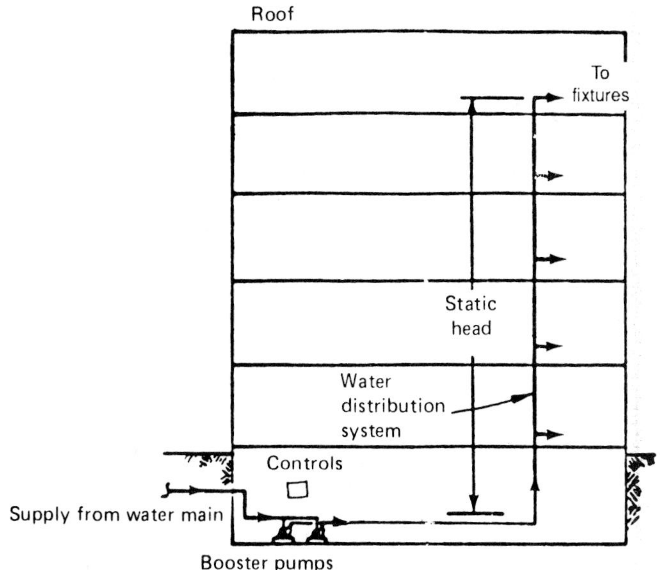

FIGURE C13.28 Booster pump system and upfeed water distribution system.

2. Provide adequate water pressure to the most hydraulically remote fixture during minimum pressure and maximum flow conditions
3. Prevent excessive water velocity during maximum flow conditions

The process of pipe sizing and component selection is an iterative one, requiring the design engineer to first assume initial values, and recalculate if necessary using new values if the initial assumptions prove wrong. Use the following simplified method as a guide to sizing. Additional criteria regarding system components are presented later in this chapter.

The basic method of system design is to first establish values that are fixed, such as fixture operating pressure and the difference in static height of that fixture from the pressure source. The pipe size, which is adjustable, would then be selected so that the remaining system pressure, in the form of friction loss of the water flowing through the pipe, will be used while not exceeding recommended velocity figures. For a pumped system, the design engineer has the ability to increase the total dynamic head of the pump in order to provide additional pressure to the piping network if desired. Simplified design of a street pressure system is as follows:

1. Find the static and residual source water pressure and the elevation at which the pressures are obtained. The residual pressure is the basis of the sizing procedure.
2. Determine by rough calculation if water pressure booster or reducing systems are necessary. If pressure adjustment is required, select the appropriate system.
3. Locate main runs and route the water distribution system piping.
4. Calculate pressure losses in the distribution system as follows:

TABLE C13.19 Maximum Probable Demand, gpm

Water supply fixture units	Maximum probable flow, gpm: Tank-type water closets	Maximum probable flow, gpm: Flushometer-type water closets	Water supply fixture units	Maximum probable flow, gpm: Tank-type water closets	Maximum probable flow, gpm: Flushometer-type water closets
1	1		120	25.9	75.7
2	3		125	26.5	76.5
3	5		130	27.1	77.3
4	6		135	27.7	78.1
5	7	27.2	140	28.3	78.8
6	8	29.1	145	29.0	79.6
7	9	30.8	150	29.6	80.3
8	10	32.3	160	30.8	81.6
9	11	33.7	170	32.0	82.9
10	12.2	35	180	33.3	84.2
12	12.4	37.3	190	34.5	85.3
14	12.7	39.3	200	35.7	86.5
16	12.9	41.2	220	38.1	88.6
18	13.2	42.8	240	40.5	90.5
20	13.4	44.3	260	43.0	92.3
22	13.7	45.8	280	45.4	94.0
24	13.9	47.1	300	47.7	95.6
26	14.2	48.3	400	59.6	102.0
28	14.4	49.4	500	71.2	108.0
30	14.7	50.5	600	82.6	113.0
35	15.3	53.0	700	93.7	117.0
40	15.9	55.2	800	105.0	120.0
45	16.6	57.2	900	115.0	123.0
50	17.2	59.1	1000	126.0	126.0
55	17.8	60.8	1500	175.0	175.0
60	18.4	62.3	2000	220.0	220.0
65	19.0	63.8	2500	259.0	259.0
70	19.7	65.2	3000	294.0	294.0
75	20.3	66.4	3500	325.0	325.0
80	20.9	67.7	4000	352.0	352.0
85	21.5	68.8	4500	375.0	375.0
90	22.2	69.9	5000	395.0	395.0
95	22.8	71.0	6000	425.0	425.0
100	23.4	72.0	7000	445.0	445.0
105	24.0	73.0	8000	456.0	456.0
110	24.6	73.9	9000	461.0	461.0
115	25.3	74.8	10000	462.0	462.0

a. Estimate the maximum flow in the building water service. This is done by adding all WFUs and converting them to gpm using Table C13.19.

b. Calculate the loss of pressure in the building water service from the source into the building. Add (or subtract) the height difference between source and height of main distribution piping inside building, friction loss of the service line, meter, BFP, valves, and all equipment contributing to the loss of pressure. Allow 5 to 10 psi for future losses in water supply source pressure, if applicable.

c. Find the height of the most hydraulically remote fixture from the height of main distribution piping.

d. Find the pressure required to operate the most hydraulically remote fixture from Table C13.20.

5. Add the result of Steps b, c, and d together, and subtract from the figure obtained from Step 1.
6. Calculate the total equivalent run of water piping to the farthest fixture.
7. Divide the pressure calculated in Step 3 into the pipe run calculated in Step 4 to find the friction loss allowable for the piping system.
8. Using appropriate pipe friction loss charts or tables and the estimated water demand, size the piping at each design point.

For a pumped system, Steps 1, 3, 4*a,* and 4*b* will determine the suction pressure at the inlet to the pump. Steps 4*c* and 4*d* will establish the fixed pressure requirements, and the design engineer would then select a pump with enough pressure to allow a reasonable friction loss in the piping system.

Adjusting Water Pressure

If the pressure in the water source is sufficient to supply the most hydraulically remote fixture in a building, a street pressure system is the most economical selection.

When the pressure is not adequate, it must be increased. Systems used are elevated water tank, booster pumps, or hydropneumatic tank systems. Sizing should be based on accepted practices, such as those published by the ASPE. If the pressure is excessive, it must be reduced to an acceptable level.

TABLE C13.20 Minimum Acceptable Operating Pressures for Various Plumbing Fixtures

Fixture	Pressure, psi
Basin faucet	8
Basin faucet, self-closing	12
Sink faucet, ⅜ in (0.95 cm)	10
Sink faucet, ½ in (1.3 cm)	5
Dishwasher	15–25
Bathtub faucet	5
Laundry tub cock, ¼ in (0.64 cm)	5
Shower	12
Water closet ball cock	15
Water closet flush valve	15–20
Urinal flush valve	15
Garden hose, 50 ft (15 m), and sill cock	30
Water closet, blowout type	25
Urinal, blowout type	25
Water closet, low-silhouette tank type	30–40
Water closet, pressure tank	20–30

Excessive Water Pressure. Water pressure in a water distribution system is considered excessive if it will damage, or create conditions that will damage, components of the water distribution system or create a nuisance such as by having water splash out of a fixture during use.

There is no precise value of water pressure below which the pressure will never damage a water distribution system, and above which will always damage the system. A widely accepted value is 70 psi. Often, the maximum permissible water pressure is stipulated by code.

A *pressure-regulating valve (PRV)* is a device used to lower and automatically maintain the pressure of water within predetermined design parameters for both dynamic flow and static conditions. This is done by a closure device that opens and closes an orifice in response to fluctuations in outlet (regulated) pressure. The degree of closure depends upon the ability of a sensing mechanism to detect changes in water pressure at the outlet side of the valve.

Pressure-regulating valves fall into two general categories: (1) direct operated and (2) pilot operated.

The direct-operated PRV has the closure member controller in direct contact with water pressure in the outlet (regulated) side of the valve. When the outlet pressure varies, the differing pressure causes the closure member to open or close by an amount necessary to achieve the desired outlet pressure. Direct-operated valves are lowest in initial cost and provide the least accuracy in regulating the outlet pressure. They produce a pressure reduction in proportion to the flow—the larger the flow, the less the pressure in the discharge line.

A pilot-operated valve is a combination of two pressure-regulating valves in a single housing. It consists of a primary valve (or pilot) that is in direct contact with water pressure in the outlet (regulated) side of the valve, and the main valve that contains the closure member. The pilot valve senses variations in the outlet pressure and magnifies closure member travel to achieve the desired outlet pressure.

The pilot-operated valves are the highest in initial cost but provide the greatest degree of accuracy over a wider range of pressure and flow conditions than a direct-operated PRV.

Selecting a Pressure-Regulating Valve

Various manufacturers offer different types of PRVs. The different valves represent a compromise between price, capacity, accuracy, and speed of response. Such information is provided by the manufacturers for use in valve selection.

The following considerations affect the selection of a pressure-regulating valve:

1. *Minimum flow rate:* The minimum rate of flow (other than zero) expected in the piping section under design.
2. *Maximum flow rate:* The maximum rate of flow expected in the piping section under design.
3. *Nature of flow:* Whether the flow rate is reasonably constant or intermittent.
4. *Location of the pressure regulating valve:* Is the valve located at the beginning or end of a branch?
5. *Maximum inlet pressure:* The highest pressure expected at the inlet of the pressure-regulating valve.
6. *Outlet pressure:* The pressure that the regulating valve must maintain.

7. *Fall-off:* The difference between the design pressure at which the system has been set and the actual outlet pressure found in the piping—a difference usually limited to approximately 15 psi.
8. *Pressure differential:* The difference in pressure between the inlet and outlet of the pressure-regulating valve. If this difference is excessive, cavitation may result.
9. *Accuracy of pressure regulation:* The degree of accuracy desired to be maintained within the regulated water distribution system.
10. *Speed of response to changes in pressure:* If this response is too rapid, a noise called *chatter* may result. If the response is too slow, an unacceptably wide variation of pressure may occur at the outlet.
11. If flow requirements are beyond the recommended capacity of a single PRV, multiple PRVs in parallel are commonly used to allow for low and high flows in a single supply branch.

Pipe Size Selection

Pipe sizing is accomplished using maximum allowable pressure loss, demand, and velocity at the design point. Two methods of determining these values are available. The first uses prepared tables for each pipe size, with the velocity and pressure loss for various flow rates. The second is the use of nomographs, where on a single graph, all the information is displayed. The tables are regarded as more accurate, and the nomographs are more convenient. The nomographs in Figs. C13.29, C13.30, and C13.31 are provided for smooth, fairly rough, and rough pipes. Charts have been made for specific plastic pipe materials, but space limitations prevent reproduction in this handbook.

SERVICE HOT WATER SYSTEM

Purpose

The purpose of the service hot water system is to heat raw water from ambient temperature to a desired higher temperature and to deliver hot water to terminal points with a delay of less than 20 s. A booster hot water system heats primary service hot water to a higher temperature for a specific purpose.

Water Heating Methods and Equipment

Water heaters are categorized as instantaneous, semi-instantaneous, or storage. They can be either directly or indirectly fired. Fuels most frequently used to heat water are electricity, fuel oil, fuel gas, and solar energy.

Instantaneous-type heaters consist of a unit that heats water as quickly as the demand flow rate requires. They have no storage and a high recovery rate. Advantages include little floor space required, low initial cost, and factory preassembly into a package ready to install. Disadvantages are difficult control of outlet water temperature and high Btu requirements for the heating medium. For

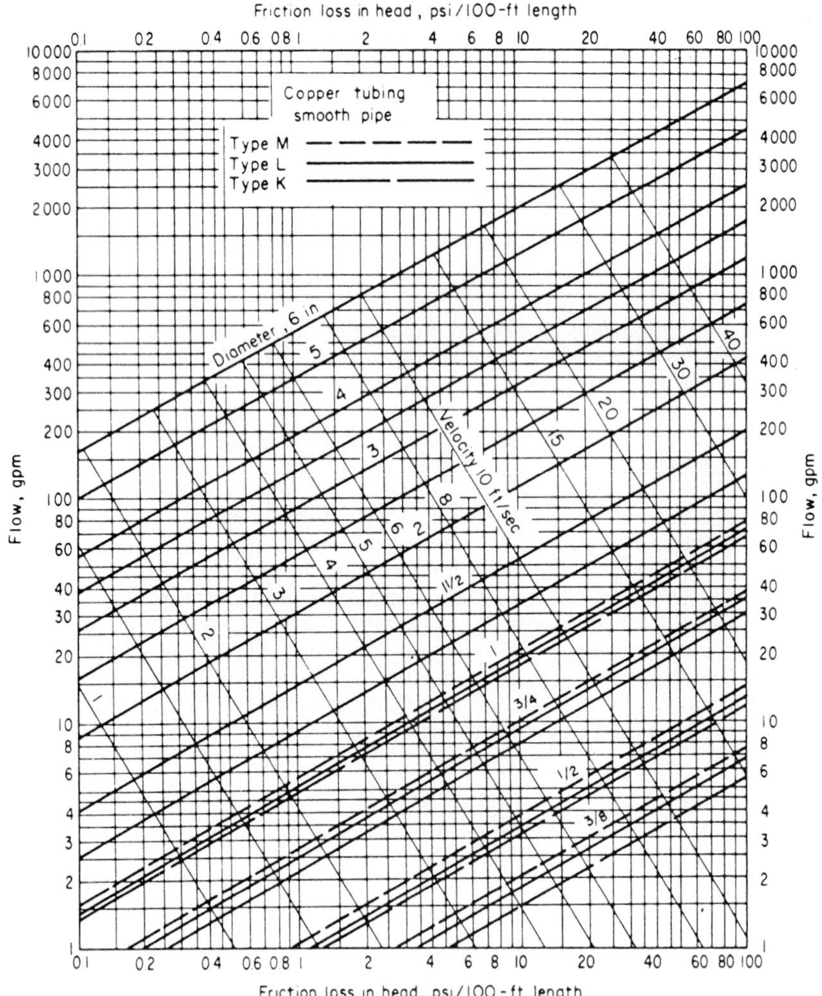

FIGURE C13.29 Friction loss for smooth pipe.

instantaneous-type heaters, an accurate approximation for steam can be calculated by multiplying the gpm requirements by 50 lb/h. This type of heater is almost always indirectly fired using either steam or hot water supplied from a central heating plant, steam utility system, or a boiler.

Semi-instantaneous-type heaters are similar to the instantaneous type except for having a limited water storage capacity, which permits easier control of outlet water temperature. This type of heater can be either directly or indirectly fired and is preferred over the instantaneous type. The far greater majority of installations are of the indirect fired type.

Storage-type heaters have a large storage capacity and lower recovery rate. This system consists of either a combination storage tank and a direct- or

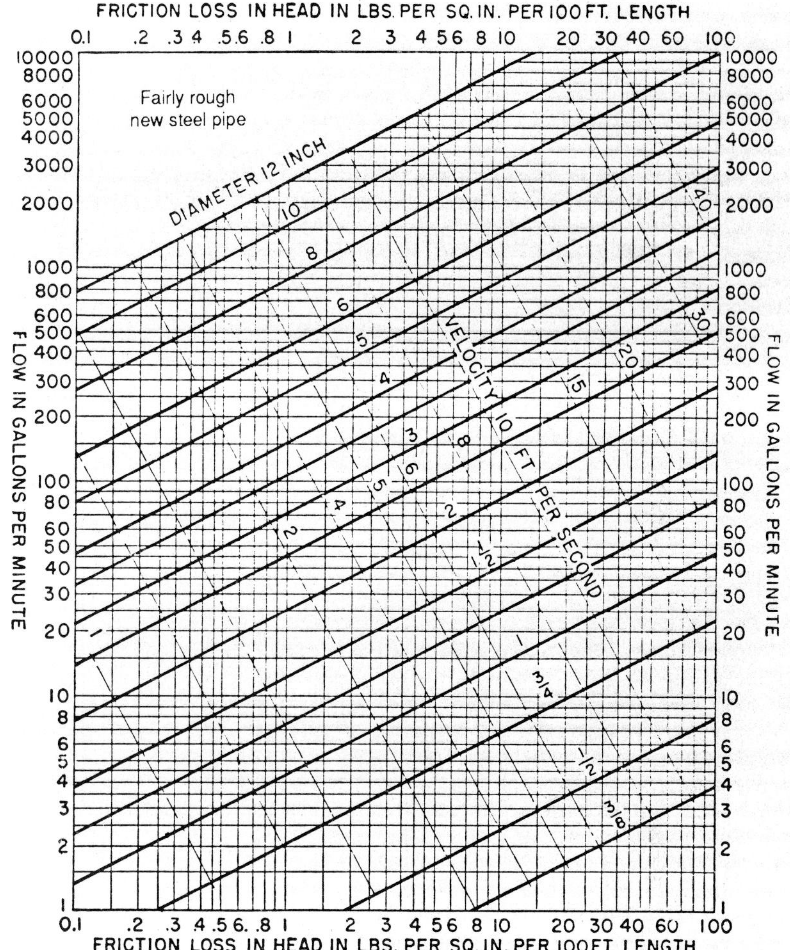

FIGURE C13.30 Friction loss for fairly rough pipe.

indirect-type immersion heater inside the tank or separate water heater and storage tank. This system should be considered when high peak surge loads are encountered for short periods of time and when a limited source of heat energy exists. Disadvantages include large amount of floor space and high initial cost. Advantages include a low instantaneous heat energy demand rate.

Point-of-use heaters are used for isolated and remote locations where it is not economical to run piping from the primary service hot water system.

The choice of a primary fuel for heating water depends on the following considerations:

1. Availability of fuel
2. Cost

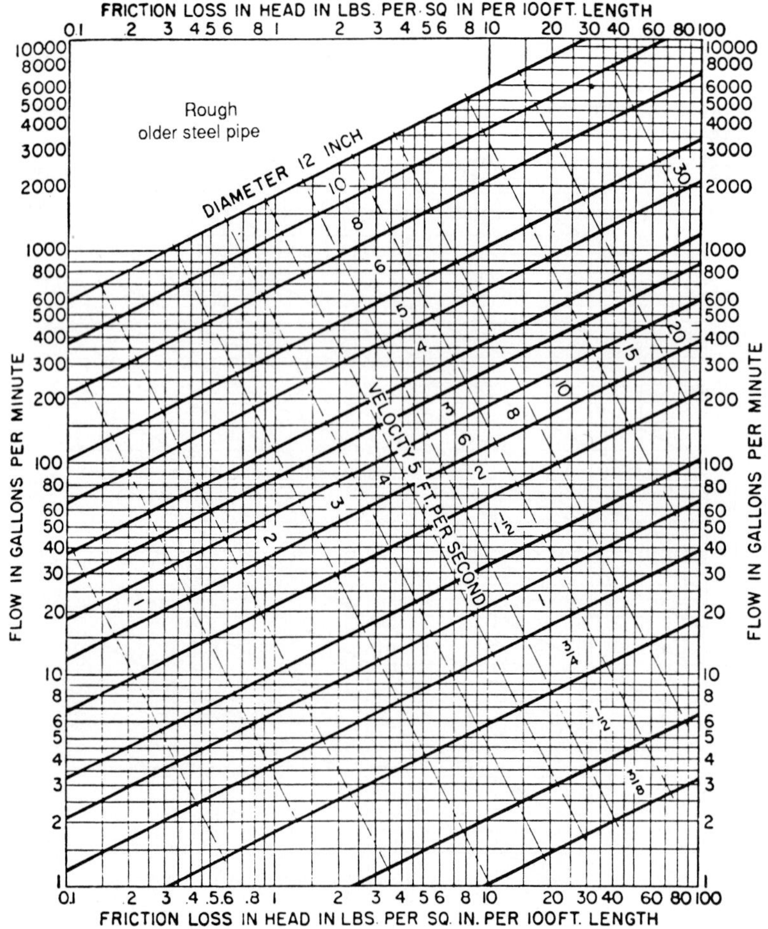

FIGURE C13.31 Friction loss for rough pipe.

3. Availability of heating equipment using the desired fuel
4. Space requirements and cost for vents or flues
5. Client preferences

Acceptable Hot Water Temperature

Generally accepted practice limits hot water temperature to a maximum of 140°F to various plumbing fixtures and other pieces of equipment. Other codes and standards, particularly for hospitals and health care facilities, may further restrict the allowable temperature of the hot water supplied for various uses.

Water Heater Sizing

Hot water usage and flow rates are characterized by intermittent periods of peak, sustained, and low to no-flow conditions. The pattern and usage requirements vary depending on building type, population, time of day, and commercial process requirements.

When designing a project for a client or agency with a preference for specific heater types and sizing criteria, such as cities and states, the federal government and Department of Defense, the design engineer must follow the methods established by the client. When no criteria exists, sizing should be based on accepted practices, such as those published by the ASPE and ASHRE.

Safety Devices

The heating of water produces pressures in the hot water piping network and equipment higher than that in the cold water system. This pressure may exceed the rating of the various components if not relieved. To accomplish this, three devices are provided. Temperature relief valves discharge water from the system in the event that water temperature is excessive, pressure relief valves discharge water from the system if the design pressure is exceeded, and energy cutoff devices are provided to stop the flow of fuel or heating medium to the heater if the temperature or pressure value is exceeded anywhere in the system. It is common practice to combine the temperature and pressure relief into a single safety device.

Hot Water Temperature Maintenance

Generally accepted practice requires that hot water delivered to terminal points of the system reach utilization temperature in 20 s or less. To accomplish this, two methods are used to maintain the water temperature. The first uses a separate water circulation pipe to return the cooler hot water from the end of the system back to the heater for reheating, thereby keeping the primary water hot. The second method uses a self-regulating electric heater wire attached directly to the hot water pipe. This heater wire becomes hot only in response to a drop in temperature of the water pipe to a predetermined point.

The electric heater strip has become the preferred method of temperature maintenance because of its lower initial and operating costs compared to a recirculated system.

FUEL GAS SYSTEMS PIPING

System Description

The fuel gas system delivers gas used to provide light or heat energy in sufficient volume and pressure required for the satisfactory operation of all connected devices.

Description of Fuel Gases

There are two major fuel gases; natural gas (NG) and liquefied petroleum gas (LPG). Manufactured gas and mixed gas are no longer used because of the abandonment of the obsolete processes that created them and the widespread availability of natural gas.

NG and LPG are hydrocarbon compounds obtained from the separation of gas mixtures occurring naturally at the wellhead of crude oil or gas-producing wells, or as a by-product of the oil refining process. *LPG* is a term applied to a group of hydrocarbons such as propane, butane, isobutane, and pentane, or various mixtures of each. Propane and butane are the principal constituents of LPG. The major component of NG is methane. Since NG and LPG are odorless, an odorant is added to make detection of leaking gas possible. Refer to Table C13.21 for the properties of fuel gases.

TABLE C13.21 Physical Properties of Fuel Gases

	Propane	Natural gas
Formula	C_3H_8	CH_4
Molecular weight	44.907	16.402
Melting or freezing point, °F	−44	−296.5
Specific gravity of gas	1.52	0.6
Specific gravity of liquid, 60°F	0.588	
Latent heat of vaporization, Btu/lb	183	
Vapor pressure, psi at 60°F	92	
lb/gal of liquid, 60°F	4.24	
Gal/lb of liquid, 60°F	0.237	
Btu/lb of gas	21591	23891
Btu/ft^3 of gas	2516	1050
Btu/gal of gas	91547	
ft^3 of gas/gal of liquid	36.39	59
ft^3 of gas/lb of liquid	8.58	23.6
ft^3 of air reqd. to burn 1 ft^3 of gas	23.87	9.53
Flame temperature	3595	3419
Upper flammability limit in air, %	9.5	15
Lower flammability in air, %	2.37	5

There is usually a variation in the reported hydrocarbon mixtures that make up LPG from various suppliers. The values presented here are average values and sufficient for engineering design of systems. It is recommended that an analysis of the LPG product actually furnished by the supplier of LPG for each specific project be obtained if extreme precision is required.

System Components

NG and LPG system components include: storage tanks, meters, regulators, mixers, vaporizers, and pumps (if required), in addition to the pipe.

Nomenclature

Btu: An abbreviation for British thermal unit, which is the quantity of heat required to raise the temperature of one pound of water one degree Fahrenheit.

Connected load: The sum of the rated input of every device connected to the entire fuel gas system, expressed as either Btu or cfh.

cfh: The abbreviation for cubic feet per hour (or ft^3/h).

cfm: The abbreviation for cubic feet per minute (or ft^3/min).

Device: Any appliance or piece of equipment utilizing fuel gas to produce light, heat, or heat energy.

Diversity factor: An estimate of the maximum probable simultaneous use of the connected devices, expressed either as a decimal or percentage.

Input: The total amount of fuel gas required for proper operation at the inlet to a device.

Latent heat of vaporization: The amount of heat required to change state from solid to liquid, liquid to vapor, or vice versa.

Maximum probable demand: The estimated maximum amount of fuel gas per unit of time that is expected to be in simultaneous use, expressed as either Btu or cubic feet per hour. This is the connected load multiplied by the diversity factor.

Maximum LPG liquid tank capacity: To allow space for propane vaporization, 85 percent is the maximum permitted filling level.

Minimum LPG liquid tank capacity: To allow time for resupply, 10 to 15 percent is recommended. Absolute low level is 5 percent.

Output: Actual Btus available to perform the intended function of the device, usually expressed as a percent of the input and taking into consideration the efficiency of the device.

Pressure measurement: Pressures less than 1 psi are measured in inches of water column (in WC). One psi equals 26.35 in WC.

Required pressure: The minimum pressure necessary for satisfactory operation of any device.

Regulator: A device used to reduce a variable inlet pressure to a constant outlet pressure under variable flow conditions.

Specific gravity: As applied to fuel gas, it is the ratio of the weight of a given volume of gas to the same volume of air under the same conditions.

Codes and Standards

There are a number of nationally recognized codes and standards governing the manufacture, design, installation, and testing of LPG and NG systems, piping, and components. The principle codes are NFPA 54, National Fuel Gas Code, and NFPA 58, Standard for Storage and Handling Liquified Petroleum Gases. Some insurance carriers, such as Factory Mutual, have standards that in many aspects may be more stringent than those listed.

System Operating Pressures

The maximum allowable system operating pressure of fuel gas is governed by NFPA 54, unless local codes or insurance carriers have more stringent requirements. NG systems are not permitted to exceed 5 psig unless all of the following conditions are met:

1. Local authorities permit a higher pressure.
2. All piping is welded.
3. The pipe runs are enclosed for protection or located in a well-ventilated space that will not permit gas to accumulate.
4. The pipe is run inside buildings or areas used only for industrial processes, research, warehouse, or boiler and/or mechanical equipment rooms.

LPG pressures up to 20 psig are permitted only if all of the following conditions are met:

1. Buildings are used exclusively for industrial, experimental laboratories or other buildings with similar use requirements.
2. The building is constructed in accordance with NFPA 58, Chapter 7.

Equivalent Length of Piping

Refer to Fig. C13.1 for equivalent lengths of fittings. It is common practice not to use the vertical length of piping when calculating the total run for NG systems. Since NG is lighter than air, it expands as the gas rises at the rate of 0.1 in WC for every 15 ft of elevation. This additional pressure created as the gas rises closely approximates the pressure lost to friction inside the pipe. Since LNG is heavier than air, the entire length of run is used for design of this system.

Maximum Probable Demand

For some types of buildings, such as multiple dwellings and laboratories, the total connected load is not used to size the piping system since all of the connected devices will not be used at the same time. For design purposes, it will be necessary to apply a diversity factor to reduce the total connected load when calculating the maximum probable demand.

This calculation first requires the listing of every device using gas in the building and the demand in Btu/h for each. The manufacturer of each device should be consulted to find its actual input gas consumption. If this information is not available, refer to Table C13.22 for gas demand of typical appliances.

For multiple dwellings, refer to Figs. C13.32 and C13.33 for a direct reading of the quantity of gas based on the number of apartments. A diversity factor has been used to create the chart. For individual risers, refer to Fig. C13.34 for a direct reading of the pipe size. For laboratories, refer to Table C13.23 for the diversity factor. Where laboratories are part of a school, use no diversity for entire rooms, and consult with the school authorities to find the total number of rooms that might be in use at once. If there is no conclusive answer, use no diversity.

For industrial or process installations, and for major gas using equipment such as boilers and water heaters in all building types, a diversity factor is generally not used because it is possible for all connected equipment to be in use at the same time.

System Design Procedures for Natural Gas

General. Natural gas is usually obtained from a franchised public utility obligated to provide gas to every customer that requests this service. As part of this

TABLE C13.22 Average Gas Demand, Btu

Appliance or device	Demand
Residential equipment	
Clothes dryer	35,000
Range	65,000
Oven	40,000
Top burners	25,000
Water heater, instantaneous, each gpm	70,000
Water heater, storage, each 10 gal	10,000
Barbecue	50,000
Commercial kitchen equipment	
Broiler, small	30,000
Broiler, large	60,000
Broiler and roaster combination	66,000
Coffee maker, per burner	6,000
Coffee urn, 5 gal	28,000
Coffee urn, 10 gal	56,000
Coffee urn, 15 gal	84,000
Deep fat fryer, 45 lb fat	50,000
Deep fat fryer, 75 lb fat	75,000
Doughnut fryer, 200 lb fat	72,000
Oven, baking and roasting, two deck	100,000
Oven, baking, three deck	96,000
Oven, revolving	210,000
Range, hot top	45,000
Range, fry top	50,000
Range, hot top and oven	90,000
Range, fry top and oven	100,000
Laboratory equipment	
Bunsen burner, range	1,000 to 10,000
Bunsen burner, commonly used	3,000
Biological culture cabinet	5,000
Miscellaneous equipment	
Gas engine, per hp	10,000
Boiler, per boiler hp	50,000
Commercial log lighter	50,000

service, the utility company supplies and installs the service line from the utility main, in addition to providing a regulator-meter assembly in or adjacent to the building.

Requirements differ regarding the placement of the meter assembly. It could be installed either in an underground, exterior meter pit, at an above-ground exterior location exposed to the weather, or inside the building in a well-ventilated area or mechanical equipment room.

The following criteria and information should be obtained in writing from the public utility company:

1. Btu content of the gas provided
2. Minimum pressure of the gas at the outlet of the meter
3. Extent of the installation work done by the utility company and the point of connection by the contractor

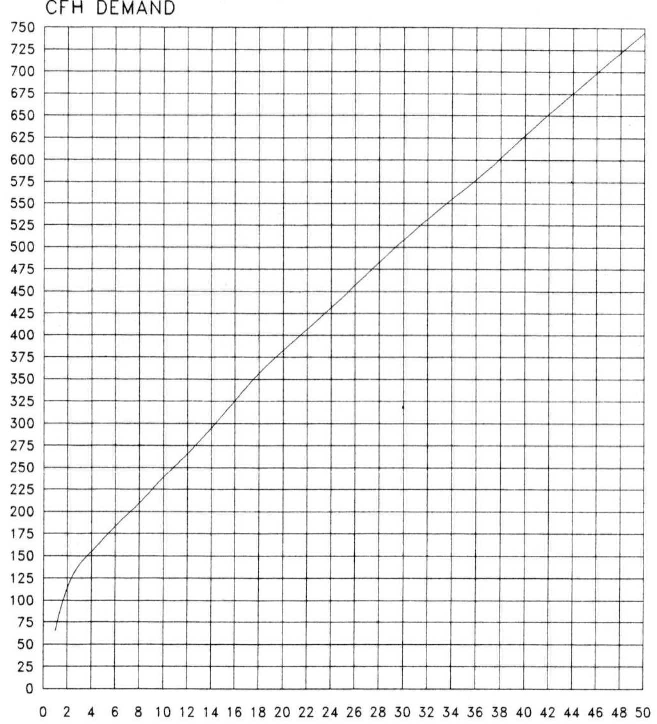

FIGURE C13.32 Gas demand for multiple dwellings, to 50 apartments.

4. The location of the utility supply main and the proposed run of pipe on the site by the utility company
5. Acceptable location of the meter and/or regulator assembly, and any work required by the owner to allow the assembly to be installed (such as a meter pit or slab on grade)
6. Types of service available and the cost of each

In order for the utility company to provide this data, they require the following information from the design engineer:

1. The total connected load. The utility company will use their own diversity factor to calculate the maximum probable demand and size of the service line.
2. Minimum and maximum pressure requirements for the most demanding device.
3. Site plan indicating the location of the proposed building on the site and the specific area of the building where the proposed NG service will enter the building.
4. Preferred location of the meter/regulator assembly.
5. Expected date of the start of construction.

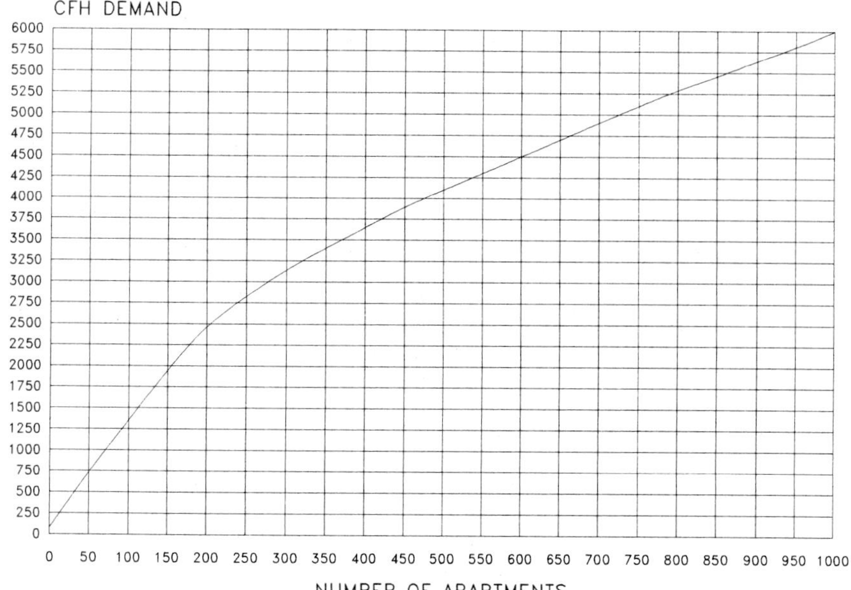

FIGURE C13.33 Gas demand for multiple dwellings, over 50 apartments.

Pipe Sizing Procedure. In order to size the piping system, the following information must be calculated or established:

1. Pressure available from the utility company
2. Allowable friction loss for gas flowing through the piping system
3. Equivalent length of the piping system
4. Maximum probable demand
5. The pipe sizing method acceptable to local codes

Pressure Available from the Utility Company. The minimum pressure that the utility company will guarantee for the project will be provided upon request to the company.

Allowable Friction Loss in the Piping System. The minimum guaranteed pressure supplied by the utility company after the meter-regulator assembly could be as low as 4 to 7 in of water column (WC). Because of this, the friction loss of NG through the piping system must be kept low in order to have sufficient pressure to properly operate the terminal appliance or equipment. A range of between 0.2 to 0.5 in WC are considered generally acceptable friction loss values, depending on the actual pressure available. When the available pressure is higher than expected, a higher allowable loss figure can be used for economy of pipe sizing.

NG Pipe Sizing Methods. The most common and conservative method for sizing NG piping systems is by the use of tables, such as the series of charts prepared

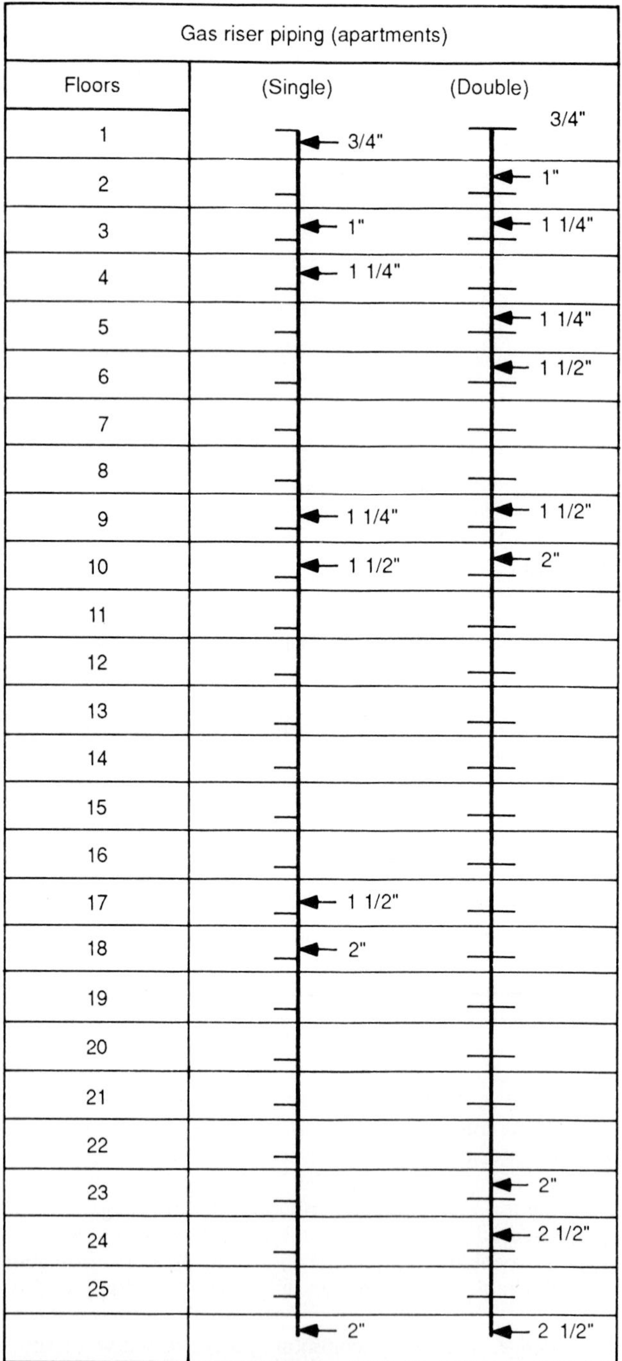

FIGURE C13.34 Gas riser sizing for apartments.

TABLE C13.23 Laboratory Diversity Factors

No. of outlets	Average % use	Max.
1–5	100	
6–10	75	90
11–20	60	75
21–70	40	60
71–150	30	50
Over	20	40

by the American Gas Association and included in Appendix C of the National Fuel Gas code, NFPA 54. One of them is Table C13.24 for 0.3 in WC pressure drop, 1 psig initial pressure of gas or less, and a specific gravity of 0.60. Other methods using proprietary calculators are available. They are considered more accurate for complex projects than the prepared tables.

Table C13.24 is used by first calculating the equivalent run of pipe from the outlet of the meter or regulator (or point of connection) to the farthest point of use. *This is the only distance used.* Find the pipe size at the intersection of the above distance column with a cfh figure that equals or exceeds the calculated cfh figure at each design point. Branches are sized using the same distance column and cfh figure as the main.

Pipe Materials. Allowable piping materials are listed in NFPA 54 and the applicable codes. The most often used material for underground lines are high-density polyethylene (PE). For above-ground lines, black steel, ASTM A53 pipe is used with cast or malleable iron screwed fittings in sizes 4 in and smaller and buttwelded joints 5 in and larger.

Joints for PE should be butt-type, heat fused joints. Socket-type joints have been found to introduce a stiffness in the joint area that is undesirable. Joints for steel pipe shall be screwed for pipe sizes 4 in and smaller. Sizes 5 in and larger should be welded.

Code does not permit plastic pipe to be run above ground. In order to make the transition from plastic pipe underground to steel pipe above ground, a transition fitting must be used. A typical fitting is illustrated in Fig. C13.35.

Liquefied Petroleum Gas Design Considerations

System Components

Storage Tanks. Propane storage tanks are constructed of steel. Because of the relatively high pressure developed by the propane vapor, all tanks must be designed in conformance with the ASME Pressure Vessel Code.

Tanks are usually referred to as either Department of Transportation (DOT) cylinders or ASME tanks. DOT cylinders generally range in size from 1- to 420-lb capacity. ASME tanks generally range in size from 500- to 60,000-gal capacity.

In the following discussions, all capacities referring to *gallons* are gallons of water, and all weights are of liquid propane, unless indicated otherwise.

TABLE C13.24 Pipe Sizing Table for Pressures under 1 lb Approximate Capacity of Pipes of Different Diameters and Lengths in Cubic Feet per Hour with Pressure Drop of 0.3 in Water Column and 0.6 Specific Gravity

Pipe size of Schedule 40 standard pipe, in	Internal diameter, in	Total equivalent length of pipe, ft										
		50	100	150	200	250	300	400	500	1000	1500	2000
1.00	1.049	215	148	119	102	90	82	70	62	43	34	29
1.25	1.380	442	304	244	209	185	168	143	127	87	70	60
1.50	1.610	662	455	366	313	277	251	215	191	131	105	90
2.00	2.067	1,275	877	704	602	534	484	414	367	252	203	173
2.50	2.469	2,033	1,397	1,122	960	851	771	660	585	402	323	276
3.00	3.068	3,594	2,470	1,983	1,698	1,505	1,363	1,167	1,034	711	571	488
3.50	3.548	5,262	3,616	2,904	2,485	2,203	1,996	1,708	1,514	1,041	836	715
4.00	4.026	7,330	5,038	4,046	3,462	3,069	2,780	2,380	2,109	1,450	1,164	996
5.00	5.047	13,261	9,114	7,319	6,264	5,552	5,030	4,305	3,816	2,623	2,106	1,802
6.00	6.065	21,472	14,758	11,851	10,143	8,990	8,145	6,971	6,178	4,246	3,410	2,919
8.00	7.981	44,118	30,322	24,350	20,840	18,470	16,735	14,323	12,694	8,725	7,006	5,997
10.00	10.020	80,130	55,073	44,225	37,851	33,547	30,396	26,015	23,056	15,847	12,725	10,891
12.00	11.938	126,855	87,187	70,014	59,923	53,109	48,120	41,185	36,501	25,087	20,146	17,242

NFPA 54.

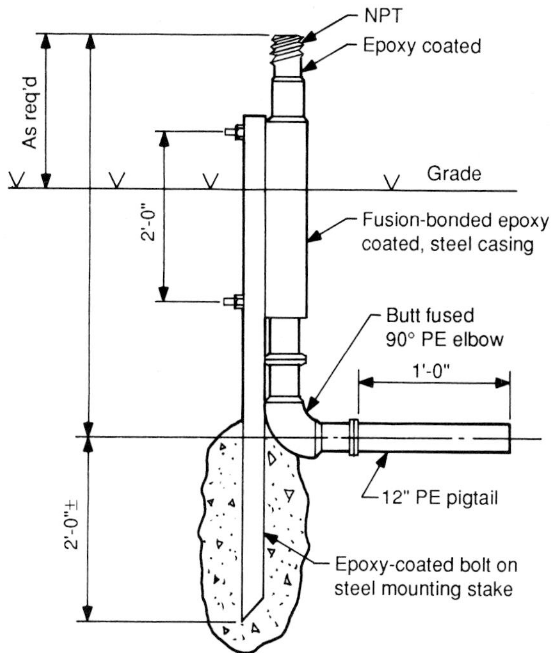

FIGURE C13.35 Transition riser detail.

Tanks can be located either above ground or underground. The advantages of an underground tank are:

1. The tank is not visible if aesthetics are a consideration.
2. There is greater vaporization of liquid in the winter due to the constant temperature, which is about 50°F.

The disadvantages are:

1. Anchoring may be necessary to prevent floating in areas of high water tables.
2. The initial cost is higher.
3. Inspection and maintenance are more difficult.

If the underground tank is subject to traffic or potential damage, the top of a tank should be a minimum of 2 ft below grade. In remote locations, 6 in of cover is considered adequate. A manhole giving access to the valves, gauges, connections, and so on must be provided for maintenance and inspection.

Spacing of Equipment. The factors to be considered in locating equipment are:

1. Code required clearance from other buildings, roads, property lines, and any other equipment. These clearances also depend on insurance carrier requirements and client preferences. Factory Mutual requirements, indicated in Fig. C13.36, are considered conservative. Distances recommended by NFPA and other carriers may vary.

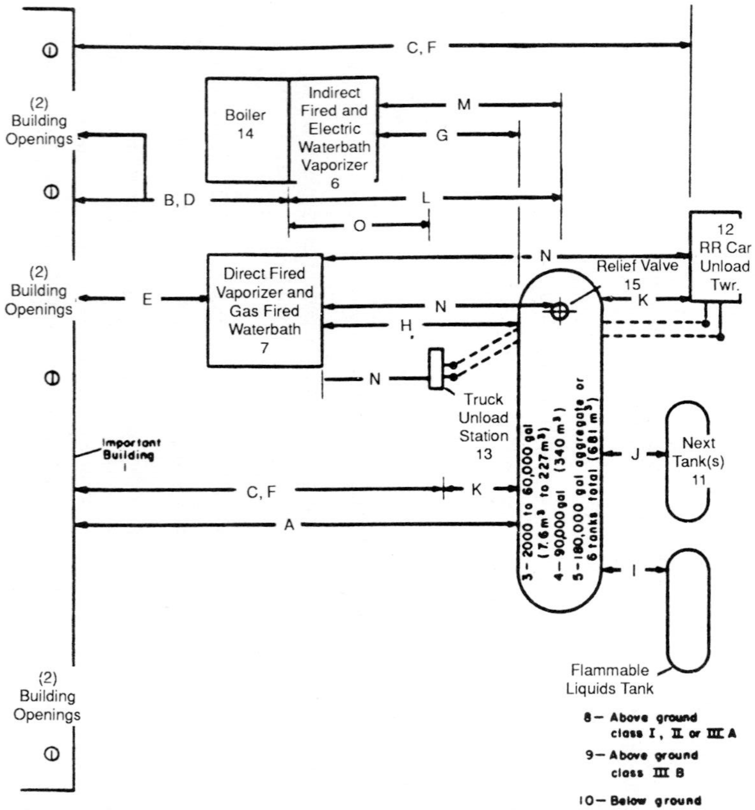

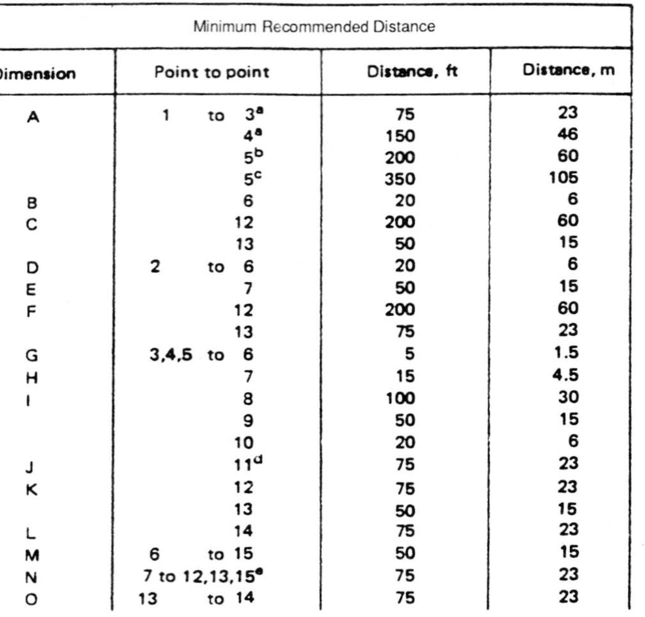

FIGURE C13.36 Minimum spacing between LPG system components.

Dimension	Point to point	Distance, ft	Distance, m
	Minimum Recommended Distance		
A	1 to 3[a]	75	23
	4[a]	150	46
	5[b]	200	60
	5[c]	350	105
B	6	20	6
C	12	200	60
	13	50	15
D	2 to 6	20	6
E	7	50	15
F	12	200	60
	13	75	23
G	3,4,5 to 6	5	1.5
H	7	15	4.5
I	8	100	30
	9	50	15
	10	20	6
J	11[d]	75	23
K	12	75	23
	13	50	15
L	14	75	23
M	6 to 15	50	15
N	7 to 12,13,15[e]	75	23
O	13 to 14	75	23

Notes:

a. For single tanks only. Treat multiple tanks as No. 5.
b. For buildings with hydrant protection.
c. For buildings without hydrant protection.
d. 5 ft (1.5 m) for tanks within a group.
e. For tanks smaller than 2000 gal (7.6 m^3), 25 ft (7.6 m).

2. Accessibility of storage tank for fuel delivery.
3. Location of underground utilities.
4. Location of low points of a site, which should be avoided in placements.
5. Client preferences.

Tank Foundations and Support. Tank foundations must provide a stable means of support. Uneven settlement will lead to errors in fuel gauging and stresses leading to broken lines.

The weight of the tank should be calculated using water, since there may be a requirement for hydrostatic testing during the life of the system. An approximation of the gross weight of a steel tank can be obtained by multiplying the capacity in gallons by 11 for water and 6 for propane. Some larger tanks may require a temporary, intermediate center support during the time it is filled with water for hydrostatic testing.

Regulators. The purpose of a regulator is to reduce a variable inlet pressure to a constant outlet pressure under variable flow conditions. The inlet pressure is the vapor pressure developed inside the storage tank. Figure C13.37 can be used to determine the vapor pressure for various mixtures of propane and butane that might be provided by suppliers. The following information is required to size a regulator:

1. The minimum significant and maximum flow rate possible
2. Maximum and minimum container pressure
3. Required outlet pressure desired
4. Manufacturer's rating curves for the regulator

Regulators are manufactured in single- and double-stage models and are selected using the capacity and rating curves supplied by the manufacturer.

Pressure Relief Devices. The purpose of the pressure relief is to automatically vent propane vapor to the atmosphere upon reaching a predetermined high pressure. It can be an integral part of a pressure regulator or it can be separately installed on the storage tank, as part of vaporizers and mixers or in the piping system itself. Sizing is based on vapor flow through the device.

Excess Flow Valve. An excess flow valve permits the flow of vapor or liquid in both directions, but shuts off the flow of liquid or vapor in only one direction when the flow exceeds a predetermined limit. It is recommended that an excess flow device be placed on all connections to a larger tank except for the safety relief valve and filler connection, which should have an integral one.

The capacity is calculated from the largest expected flow and its mounting position (horizontal or vertical). Valves are selected to close at between 150 and 200 percent of expected maximum flow.

Service Line Valves. A service line valve controls the flow of propane from a DOT cylinder. This is a multipurpose valve, that could contain a shut-off valve, relief valve, pressure gauge, and filler valve all in one body.

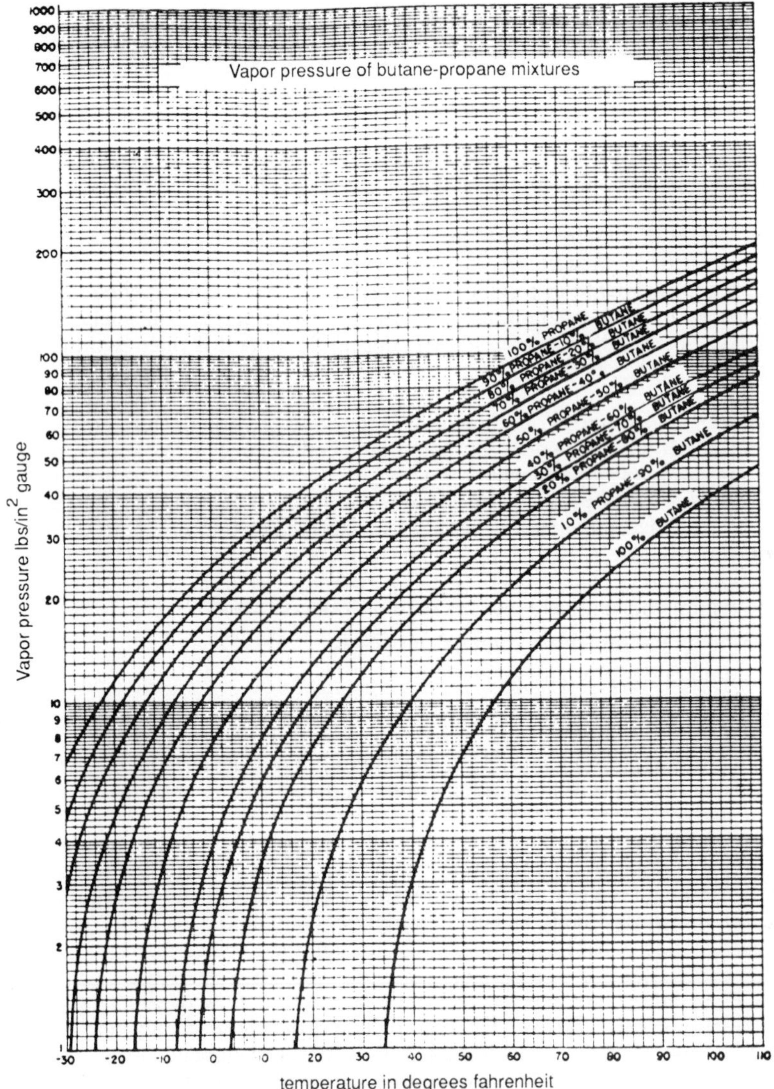

FIGURE C13.37 Vapor pressure of LPG mixtures.

Filler Valves. A filler valve permits the flow of liquid in one direction only, and is used for filling larger tanks with liquid. This valve contains an integral back pressure check valve that will prevent the loss of vapor or liquid from the tank if the fill hose or a fitting ruptures. A preferred location is at the filling connection used by the delivery truck hose.

Vapor Equalizing Valves. When large tanks are filled, the liquid propane added to the storage tank compresses the vapor in the tank. A vapor equalizing line, connected between the vapor area of the receiving tank and the delivery truck during propane deliveries, will equalize pressure in both containers and to lessen the head requirements of the supply pump. The vapor equalizing valve is installed in the equalizing line tank truck hose connection and permits the flow of vapor both ways.

Liquid-Level Gauges. When a larger tank is filled on the basis of volume, it is important that a liquid-level gauge be installed on the tank to indicate the exact liquid level inside.

Miscellaneous Equipment. In addition to the devices indicated, it is often desirable to install a liquid temperature gauge and a pressure gauge on a tank to aid in diagnosing any potential problem.

Vaporizer. A vaporizer is a device that uses an additional source of heat to convert liquid propane into a vapor. It is required when sufficient quantities of liquid propane cannot be vaporized quickly enough inside the storage tank to satisfy the maximum demand.

Propane Mixers. A mixer, also called a *blender* or *proportioner,* is a device used to combine pure propane and air together into a mixture of both gases. It is required when propane will be used as a direct substitute for natural gas, because the much higher Btu content of 100 percent propane will not allow the use of the same burner orifices and gas train as natural gas. Experience has shown that a mixed gas content of 1450 Btu/ft^3 provides the best burning characteristics.

LPG System Design Considerations

Prior to sizing components, initially decide what equipment may be required, if the tank is to be above or below ground, the proposed location of the tank and other required equipment on the site, and if the propane system will be in constant use or used only periodically, such as for emergency operation or as an occasional substitute for interrupted natural gas.

Calculate the maximum hourly and daily demand, and determine if this demand is continuous, such as that for a process used all day, or intermittent, such as that used for heating purposes. If only a small part of the demand is continuous, the entire load should be considered continuous.

Storage Tank Sizing and Selection. The storage tank volume is based on one of two factors: a reasonable return schedule for the local supplier, or the amount of liquid propane that has to be vaporized by the ambient air in order to satisfy the maximum demand.

Using the return schedule, a preliminary starting point for determining actual capacity is a 10-day usable supply for continuous demand and between 3 to 5 days usable supply for intermittent, or standby purposes. With the maximum propane level of 85 percent and the minimum level between 10 and 15 percent of the capacity, the usable tank capacity is about 75 percent of the total tank capacity.

Determining the amount of propane capable of being vaporized in a storage tank is presented in Ref. 8.

LPG Pipe Sizing Methods. The same charts used to size NG are used to size propane. Propane is often mixed with air to provide a direct substitute for natural gas. Field experience has found that if mixed gas with the same Btu value (1050) as natural gas is used, the mixture will not burn properly. As a result, it is common practice to provide a mixed gas with a value of 1450 Btu/ft^3. This mixture has a specific gravity of 1.3. Since gas flow varies inversely as the square root of the specific gravity, the use of a conversion factor can be used with readily available natural gas charts to size the mixed gas–natural gas piping system. The conversion factor for 1450 Btu/ft^3 mixed gas is 0.68, which is used to multiply the capacity found in NG tables in order to convert the chart figure to the actual mixed gas value.

The most widely accepted pipe sizing method for 100 percent propane systems is based on the "Clifford" method, available in Ref. 6. To use the charts in NFPA 54, a conversion factor of 0.63 should be used.

High-pressure charts are provided in the NFPA 54 for LPG vapor pressures up to 50 psi. Chart usage is similar to NG sizing procedure. Proprietary calculators are available that are considered more accurate for larger and more complex projects.

Piping Materials. The piping materials are the same as for the NG system.

REFERENCES

1. C. M. Harris, *Handbook of Utilities and Services for Buildings,* McGraw-Hill, New York, 1990.
2. American Society of Plumbing Engineers, *ASPE Databooks,* Chaps. 1–22 inclusive, Sherman Oaks, Calif.
3. Crane Co., *Flow of Fluids Through Valves, Fittings and Pipe,* Technical Paper No. 410, New York, 1976.
4. American Society of Heating, Refrigeration and Air Conditioning Engineers, Inc., *ASHRE Syystems Handbook,* Atlanta, Ga. 1987.
5. New York City Housing Authority, *Plumbing Design Criteria,* New York.
6. Earle A. Clifford, *A Practical Guide to LP-Gas Utilization, LP-Gas Magazine,* Duluth, Minn. 1973.
7. Department of Commerce, U.S. Weather Bureau, National Oceanic and Atmospheric Administration, *Technical Paper No. 40,* Washington, D.C., 1961.
8. L. Denny, L. Luxon, and B. E. Hall, *Handbook Butane-Propane Gases,* Chilton Co., Los Angeles, 1962.

P • A • R • T • D

NONMETALLIC PIPING

CHAPTER D1
THERMOPLASTIC PIPING

Stanley A. Mruk
Executive Director
Plastics Pipe Institute
Wayne, New Jersey

SCOPE

This chapter is a summary of the properties and uses of thermoplastics piping. Included is a review of advantages and limitations and general and basic information on materials, properties, standardization and design, and installation. This information is intended to guide the reader in evaluating the applicability of thermoplastics piping for an intended application; in choosing the appropriate material and product; and in the proper design and installation of the piping. References are provided for more detailed information and further guidance on these and related subjects.

INTRODUCTION

Plastics piping is made from either of two basic groups of synthetic materials: thermoplastic and thermosetting. Thermoplastics can be softened and reshaped repeatedly by the application of heat in contrast to thermosetting materials, which are "set," or "cured," or hardened into a permanent shape after a single heating, normally during factory manufacture. Thermoplastic materials include minimal reinforcements, whereas thermosetting resins are almost always combined with reinforcements, such as Fiberglas, and sometimes fillers, such as sand, to produce structurally integrated composite constructions. Chapter D2 in this handbook is devoted to Fiberglas piping.

Principal Materials

Thermoplastics account for the lion's share of plastics used for piping. During 1989, over 95 percent of the 4200 million pounds of plastics that went into pipe, conduit, and fittings consisted of thermoplastics.[1] Polyvinyl chloride (PVC) accounted for about three-quarters of all thermoplastic pipe. The second most widely used thermoplastic is polyethylene (PE), accounting for about a 15 percent share, followed by acrylonitrile-butadiene-styrene (ABS), representing about a 4 percent share. The balance, about 6 percent, consisted of special-purpose materials, such as chlorinated polyvinyl chloride (CPVC), polybutylene (PB), polypropylene (PP), and various fluorinated polymers, principally polyvinylidene fluoride (PVDF).

Somewhat more footage of thermoplastic pipe is now being installed than that of all other types of piping materials combined.[2] However, total dollar value of installed thermoplastic pipe is second to and only about one-quarter of that of the leading material, steel.[3] This is because thermoplastics' current principal use is in the smaller sizes. But the very successful track record in these sizes has led to the increased acceptance and use of the larger diameters. As of this writing, thermoplastic pipe is available through 60 in for pressure uses and 108 in for sewer and drain.

The first thermoplastic tubes were made in Germany during the 1930s from a PVC copolymer. Thermoplastics pipe was first manufactured commercially in the United States in 1940 from cellulose acetate butyrate (CAB) and was used by the Southern California Gas Company for distributing natural gas. Volume production commenced in 1948 when polyethylene was adopted for water services. ABS and PVC pipe were first commercially made in the United States in 1949 and 1950, respectively. However, the serious start of the evolution of the U.S. thermoplastic pipe industry into its current status took place during the mid-1950s when the first American Society for Testing and Materials (ASTM) group for plastics pipe standardization was organized and soon after started issuing standards which led to the establishment of thermoplastic piping as an engineering material. Also significant during this period were the many advances that were introduced in polymer chemistry, materials, formulation, and fabrication technology.

In 1955, total U.S. shipments of thermoplastic pipe were estimated at under 40 million pounds. By 1989, within one generation, output has increased a hundredfold and still continues to grow.

Available Products

Plastics pipe and fittings are available in a vast array of materials, diameters, wall thicknesses, and designs. For nonpressure applications special wall constructions are offered, such as double wall, ribbed, and foamed core, which are designed to more economically achieve the required longitudinal and diametrical pipe stiffness. Most of these products are covered by national standards. The more popular of these are listed in Table D1.1 together with a brief notation of size range and primary application. As the updating of existing and the writing of new product standards is a dynamic ongoing process, the reader is advised to contact standards-issuing organizations for the latest status. The Plastics Pipe Institute* publishes a periodically updated report, PPI TR-5, which includes a comprehensive listing of North American and International Standards Organization standards on thermoplastics piping. There are also many commercially available piping appurtenances, such as shown in Table D1.2, that are made from plastics but which are not covered by any national standard. In addition, some pipe and fitting manufacturers and their distributors can custom fabricate components that may or may not be shown in product catalogs. These specials include manholes

*See the listings of trade associations at the end of this chapter.

TABLE D1.1 Principal Standards for Thermoplastics Piping

Product standard*	Product description or abbreviated title of standard	Nominal sizes, range (in)	Principal applications
ABS, PP, PVC			
ASTM D3311	DWV plastic-fittings patterns	1¼–8	Drain, waste, vent
ABS, PVC			
ASTM D2680	ABS and PVC composite sewer pipe	4–15	Sewer, drain
ASTM F409	Accessible and replaceable tube and fittings	1¼–1½	Drain, waste, vent
ASTM F480	Thermoplastic water-well casing	2–16	Water-well casing
ABS			
ASTM D1527	ABS pipe, Schedules 40 and 80	⅛–12	Cold water; industrial
ASTM D2282	ABS pipe, dimension-ratio series	⅛–12	Cold water; industrial
ASTM D2468	ABS fittings, Schedule 40, socket	⅛–8	Cold water; industrial
ASTM D2661	ABS DWV pipe and fittings	1¼–6	Drain, waste, vent
ASTM D2750	ABS utility conduit and fittings	1–6	Electrical duct
ASTM D2751	ABS sewer pipe and fittings	3–12	Sewer; drain
ASTM F628	ABS foam-core DWV	1¼–6	Drain, waste, vent
PB			
ASTM D2662	PB pipe, dimension-ratio series (ID-based)	½–6	Hot and cold water; industrial
ASTM D2666	PB tubing	½–2	Hot and cold water; industrial
ASTM D3000	PB pipe, dimension ratio	½–6	Hot and cold water; industrial
ASTM D3309	PB, hot and cold water distributing	⅛–2	Hot and cold water; water service
ASTM F809	PB, larger-diameter pipe	3–42	Hot and cold water; industrial
ASTM F845	Plastic insert fittings for PB pipe	⅜–¾	Hot and cold water; water service
AWWA C902	PB pipe and tubing for water service	½–3	Cold-water service
PE, PB, PVC			
ASTM D2513	Thermoplastic gas-pressure pipe and fittings	¼–24	Natural gas distribution
PE			
AASHTO M252 (interim)	Corrugated PE drainage tubing	3–15	Subsurface drainage
AASHTO M294 (interim)	Corrugated PE pipe	12–24	Drainage
ASTM D2104	PE pipe, Schedule 40 (ID-based)	½–6	Cold water; industrial

TABLE D1.1 Principal Standards for Thermoplastics Piping (*Continued*)

Product standard*	Product description or abbreviated title of standard	Nominal sizes, range (in)	Principal applications
PE (*continued*)			
ASTM D2239	PE pipe, dimension-ratio series (ID-based)	½–6	Cold water; industrial
ASTM D2447	PE pipe, Schedules 40 and 80	½–12	Cold water; industrial
ASTM D2609	Plastic insert fittings for PE pipe	½–6	Cold water
ASTM D2683	PE fittings, socket-fusion type	½–4	Cold water; natural gas; industrial
ASTM D2737	PE tubing	½–2	Cold water
ASTM D3035	PE pipe, dimension ratio	½–6	Water; industrial
ASTM D3261	PE fittings, butt-fusion type	½–48	Cold water; natural gas; industrial
ASTM F405	PE corrugated tubing and fittings	3–6	Drain; leaching fields
ASTM F667	PE corrugated tubing, larger diameter	8–24	Drainage; leaching fields
ASTM F714	PE pipe, larger diameter	3–48	Cold water; sewer; drain; industrial
ASTM F771	PE pipe for irrigation	½–6	Irrigation
ASTM F810	PE pipe for drainage and waste disposal	3–6	Drain; leaching fields
ASTM F876	Cross-linked PE (PEX) tubing	¼–2	Hydronic heating
ASTM F877	Cross-linked PE (PEX) tubing	¼–2	Hot and cold water
ASTM F892	PE corrugated pipe with smooth interior	4	Sewer; drain; conduit
ASTM F894	PE profile wall pipe, large diameter	18–120	Sewer; drain; industrial
ASTM F1055	PE electrofusion fittings	½–12	Natural gas; water; industrial
AWWA C901	PE pipe and tubing for water service	½–3	Cold water service
AWWA C906	PE pipe for water distribution, large diameter	4–63	Water distribution
API 15LE	PE line pipe	½–12	Oil and gas production; cold water
PVC			
AASHTO M278 (interim)	PVC drainage pipe	—	Drainage
ASTM D1785	PVC pipe, Schedules 40, 80, and 120	⅛–12	Cold water; industrial
ASTM D2241	PVC pipe, dimension-ratio series	⅛–36	Cold water; industrial
ASTM D2466	PVC fittings, Schedule 40, socket	⅛–8	Cold water; industrial
ASTM D2467	PVC fittings, Schedule 80, socket	⅛–8	Cold water; industrial
ASTM D2665	PVC DWV pipe and fittings	1¼–12	Drain; waste; vent

TABLE D1.1 Principal Standards for Thermoplastics Piping (*Continued*)

Product standard*	Product description or abbreviated title of standard	Nominal sizes, range (in)	Principal applications
PVC *continued*)			
ASTM D2672	PVC pipe, bell-end	⅛–12	Cold water; industrial
ASTM D2729	PVC drain pipe and fittings	2–6	Drainage
ASTM D2949	PVC thin-wall DWV pipe 3 in	3	Drain, waste, vent
ASTM D3034	PVC sewer pipe and fittings	4–15	Sewer; drainage
ASTM F512	PVC conduit for buried application	1–6	Electrical conduit
ASTM F679	PVC large-diameter sewer pipe and fittings	18–36	Sewer; drainage
ASTM F758	PVC underdrain piping	4–8	Drain
ASTM F789	PVC sewer pipe, Type 46	4–15	Sewer; drain
ASTM F794	PVC sewer pipe, ribbed wall	4–48	Sewer; drain
ASTM F949	PVC sewer pipe, corrugated-wall, smooth ID	4–10	Sewer; drain
AWWA C900	PVC for water distribution	4–12	Water distribution
AWWA C905	PVC pipe for water distribution, large diameter	14–36	Water distribution
UL 514	PVC electrical outlet boxes and fittings	½–6	Electrical-conduit
UL 651	PVC rigid nonmetallic conduit	½–6	Electrical-conduit
NEMA TC-2	PVC electrical plastic conduit	½–6	Electrical-conduit
NEMA TC-3	PVC fittings for conduit	½–6	Electrical-conduit
CPVC			
ASTM D2846	CPVC hot and cold water distributing	⅜–2	Hot and cold water; distribution
ASTM F437	CPVC fittings, Schedule 80, threaded	¼–6	Hot and cold water; industrial
ASTM F438	CPVC fittings, Schedule 40, socket	¼–6	Hot and cold water; industrial
ASTM F439	CPVC fittings, Schedule 80, socket	¼–6	Hot and cold water; industrial
ASTM F441	CPVC pipe, Schedules 40 and 80	¼–12	Hot and cold water; industrial
ASTM F442	CPVC pipe, dimension-ratio series	¼–12	Hot and cold water; industrial

*Issuing organizations:

(AASHTO) American Association of State Highway and Transportation Officials, Room 341 National Pressure Building, Washington, D.C. 20045.

(API) American Petroleum Institute, Publications and Distribution Section, 1220 L Street, N.W., Washington, D.C. 20005.

(ASTM) American Society for Testing & Materials, 1916 Race Street, Philadelphia, PA 19103.

(AWWA) American Water Works Association, 6666 West Quincy Avenue, Denver, CO 80235.

(NEMA) National Electrical Manufacturer's Association, 2101 L Street, N.W., Suite 300, Washington, D.C. 20038-1581.

(UL) Underwriters Laboratories, Inc., 333 Pfingsten Road, Northbrook, IL 60062-2096.

A more complete listing of North American Standards, including those of the Canadian Standards Association (CSA), is available in Technical Report *TR-5,* "Standards for Plastics Piping," issued by the Plastics Pipe Institute, Wayne Interchange Plaza II, 155 Route 46 West, Wayne, NJ 07470.

TABLE D1.2 Size Range of Readily Available Thermoplastic Valves and Other Appurtenances (in Inches)

Product	PVC	CPVC	PE	PP	PVDF
Valves					
Needle	¼–½	¼–½	—	—	—
Ball	¼–6	¼–4	—	½–4	½–6
Gate	½–6	½–6	—	—	—
Butterfly	4–24	—	—	—	—
Check	¼–16	½–4	—	½–4	½–4
Globe	⅛–4	—	—	—	—
Diaphragm	½–10	½–10	—	—	½–6
Foot	½–4	½–4	—	½–4	—
Solenoid	⅛–3	⅛–3	—	—	—
Gas service	—	—	¼–6	—	—
Strainers	⅛–3	⅛–3	—	—	—
Saddles, tapping	2–8	—	2–12	—	—
Expansion joints	½–12	½–6	—	½–4	½–4
Flange adapters	2–12	2–6	2–42	2–6	2–6

for both infrastructure and industrial applications. Fabricated fittings intended for pressure service are often reinforced by an overwrap with a glass-fiber–thermosetting resin composite.

Principal Uses

Thermoplastics piping is routinely used for many common pressure and nonpressure applications. Approximately 80 percent of the newly installed mains and 90 percent of the services for gas distribution are made of PE. Over 90 percent of rural water distribution mains and about 40 percent of municipal mains are made of PVC. Most of the smaller-diameter piping installed for agricultural and turf irrigation is made primarily from PE and PVC. In oil and gas production, significant quantities of PE pipe are used to convey water and well gases. Thermoplastics piping is also frequently used for commercial and industrial applications such as for conveying chilled and process waters, aqueous solutions of corrosive chemicals, and slurries.

More than half of all thermoplastic pipe made goes into nonpressure uses. Over 85 percent of the newly installed underground building-sewer connections are made of PVC. PVC also accounts for a similar share of the sewer collection mains in sizes 4 through 18 in. About 80 percent of new single-family dwellings include either PVC or ABS drain, waste, and vent (DWV) piping. Most drainage systems, including those for building foundations, leaching fields, agriculture, and road construction now consist of thermoplastics piping, mostly PE and PVC. And both PVC and PE are increasingly used for larger-diameter sewers and culverts.

Advantages and Limitations

There are a number of general advantages that have sparked the widespread adoption of thermoplastics piping for so many pressure and nonpressure uses. The most universally recognized one is its virtual freedom from attack by ambient water and moisture. Thermoplastics piping is not subject to surface attacks in any way comparable to the rusting or environmental corrosion of metals. Thermoplastics, being nonconductors, are immune to the electrochemical-based corrosion process induced by electrolytes such as acids, bases, and salts. In addition, plastics pipe materials are not vulnerable to biological attack. In sum, thermoplastics are not subject to corrosion in most environments in both above- and underground service. This has resulted in negligible costs for maintenance and external protection such as painting, plastic coating, galvanizing, electroplating, wrapping, and cathodic protection.

Another general advantage offered by thermoplastics is their lower specific gravity, which results in ease of handling, storage, and installation, as well as in lower transportation costs. The very smooth pipe surfaces yield low friction factors and very low tendency to fouling. And they also offer very good abrasion resistance, even when conveying slurries that rapidly abrade harder materials.

High deformation capacity without fracture—or "strainability" (Janson)[4]—is another important performance feature, particularly for underground service. In carrying a load, buried flexible pipes deform (deflect) to develop support from the surrounding embedment. This interaction of pipe and soil provides a pipe-soil structure which is capable of supporting earth fills and surface live loads of considerable magnitude and which could fracture stronger but more brittle materials.

Thermoplastics piping, particularly in the sizes below around 18 in, can be competitive to piping of other materials. In the larger sizes, thermoplastics will oftentimes overcome a first-cost disadvantage due to their lower operating and maintenance costs and longer life.

The principal limitations of thermoplastics arise from their relatively low strength and stiffness and greater sensitivity of mechanical properties to temperature. As a result, their primary use is for gravity and lower-pressure applications with relatively low upper temperature limits. Despite these restrictions, thermoplastics piping satisfies the design requirements for a very broad range of applications, as already described.

Successful design with thermoplastics requires recognition of their viscoelastic nature. These materials do not exhibit the relatively simple stress-strain relationship that is characteristic of metals. Duration of loading, as well as temperature and environment, can have a profound effect on their stress-strain response and rupture strength. The influence of duration of loading, temperature, and environment vary not only from one class of thermoplastic material to another, for example, between PVC and PE, but can also significantly differ within the same generic material depending on the specific nature of the polymer (molecular weight, molecular weight distribution, degree of branching, extent of copolymerization with other monomers, and so on), the type and quantity of polymer additives and modifiers, and the processing conditions. These factors must be recognized when characterizing the engineering properties of thermoplastic piping, including allowable stress, strain, and upper temperature limits; and it goes without saying that for effective design and installation, they must also be considered by the user and specifier.

Compared to traditional piping materials, thermoplastics have high coefficients of thermal expansion-contraction. For example, thermal expansion rate can be from 6 to 10 times greater than that of metal pipe. This must be recognized both in design and installation, particularly for above-ground applications where resultant piping reaction may require frequent use of expansion loops or pipe supports. For above-ground piping, more attention may also need to be given to proper pipe restraint because the low mass of thermoplastics provides less inertia against pipe movement induced by sudden changes in the velocity of the fluid flowing through the pipe. Additionally, above-ground thermoplastics piping should be positioned or protected against possible accidental mechanical damage.

Since thermoplastics are combustible, their use in certain locations may be delimited by fire safety concerns and regulations. Construction and building codes address these concerns through various requirements including the placing of the pipe inside suitable fire-resistant walls and chases and the use of fire stops around pipe penetrations through such structures.

THERMOPLASTIC PIPING MATERIALS

The term *polymer* (from the Greek *poly,* meaning "many," and *mer* meaning "unit") is used to denote the long-chain or network structure of macromolecules that are produced either by nature or humans. The latter are referred to as *synthetic polymers.* Polymers which are the base material for plastics are also termed *resins.* Plastics are compounds composed of resins and additives. Since thermoplastics are capable of being softened by heating and hardened by cooling, they can be shaped into articles by operations, such as by molding or extrusion, which take advantage of this capability. Additives are incorporated in a plastics composition to achieve specific effects during fabrication or service. The precise nature and amounts of these additives depend on the plastic and its inherent properties; the processing method used to convert it to a finished article; and any desired modification of properties to achieve certain aesthetic, performance, or economic objectives. The main classes of additive that may be used in thermoplastic piping compositions include the following:

Heat stabilizers: To improve thermal stability, particularly during processing.

Antioxidants: To protect against oxidation during processing and in service.

Ultraviolet screens, or stabilizers: To protect against ultraviolet radiation from the sun during outdoor storage or weather exposed service.

Lubricants: To facilitate and improve fabrication by reducing viscosity and lessening frictional drag through dies and other surfaces.

Pigments: To color the product.

Processing aids: To improve fusion during processing and, thereby, enhance the homogeneity of material and properties.

Property modifiers: To enhance a particular property such as impact strength or flexibility.

Fillers: Most often used to reduce volume cost; however, fillers may also be used to increase stiffness or to modify processing characteristics.

Additives are essential components of most thermoplastic piping compositions. They facilitate processing, enhance certain properties, give a product a distinctive appearance and color, and provide required protection during fabrication and service. There are only a few thermoplastics [e.g., certain fluorinated polymers such as polyvinylidene fluoride (PVDF)] that do not require the incorporation of some type of additive because they have sufficient natural thermal stability and aging and weathering resistance.

The precise nature and quantities of additives that can be used for piping compositions are delimited by their effect on engineering properties, such as rigidity, impact strength, chemical resistance, creep resistance, and rupture strength under long-term loading. For example, the use of an inorganic filler can compromise the natural resistance of polymers to very strong acids or bases. Also too much filler, or use of a filler of a coarser grade, or its improper dispersion can introduce physical discontinuities, or faults, that can compromise long-term strength and ductility. Another example is the excessive use of liquid stabilizers or lubricants which tend to plasticize the plastic and, thereby, make it less creep resistant.

Polymer architecture exerts profound influence on material properties. A most important parameter is the length of the molecular chain. The longer the chain, the larger and heavier the molecule. Molecule size is denoted by its molecular weight, which is the sum of the atomic masses of all the elements in the molecule. Since all the molecules in a polymer are not of the same size, the degree of polymerization is usually expressed by the polymer's average molecular weight. The nature of the distribution of molecular sizes also bears a significant influence on a number of physical and mechanical properties. Thermoplastics used for piping applications tend to be of relatively high molecular weight (generally over 100,000) and of relatively narrow molecular weight distribution so as to impart the optimum properties for long-term service under load. However, the molecular weight cannot be so high as to hinder fabrication of the end product.

Another molecular architecture parameter is the length and frequency of branches along the main polymer chain. Branching may be controlled through copolymerization. For example, polyethylene pipe compositions are copolymers of ethylene with small amounts of other olefins such as propylene, butene, pentene, and hexene. Although the amount of comonomers used is low, and, thereby, the polymer still falls under the classification of polyethylene, this results in enough variation in the molecular structure, principally in the number of short branches along the linear molecular chains, to exert significant influence on engineering properties. Many commercial polymers, including polypropylene (PP) and polybutylene (PB), are also partial copolymers.

Polymer architecture also affects the relative physical arrangement of molecules to one another and, thereby, the polymer's physical properties. Generally, the long molecules in polymers tend to align themselves near each other in a random symmetry analogous to spaghetti in a bowl. This unprecise physical order is referred to as an *amorphous state.* PVC and ABS are amorphous materials. Certain polymers, such as PE, PP, and PB, are partly crystalline materials because portions of their polymer chains organize themselves in close and very well-ordered arrangements called *crystallites*; the other portions lie in the amorphous regions. The stronger physical bonds in the well-ordered closely packed crystalline regions have significant influence over mechanical properties such as strength, stiffness, and toughness. The extent of crystallization and the size and nature of crystalline regions, as well as the nature of the interconnective network of molecules running between the crystalline regions, can all be influenced by tailoring molecular structure.

The many possible variations in polymer structure combined with the different types and amounts of additives result in a great diversity of plastic compositions, even within a particular polymer group, such as polyvinyl chloride (PVC) or polyethylene (PE). The defining and classifying of such compositions is, understandably, not a simple task. The primary standard plastic material specifications are issued by the American Society for Testing and Materials (ASTM). The first ASTM standards classified plastic materials by a "Type, Grade, and Class" system in accordance with three key properties. However, with the growing need to better define plastic materials by more than just three properties, a number of ASTM material standards have gone to a cell classification system whereby each of a number of primary properties is defined by a property cell number depending on the property value. All the resultant cell numbers—there can be as many as needed—are then listed in a specified order. For example, referring to Table D1.3, in accordance with the cell classification system of ASTM D 3350, "Standard Specification for Polyethylene Plastics Pipe and Fittings Materials," Class 234424 polyethylene designates a material with the following range of properties:

Density: 0.926 to 0.940 g/cm^3

Melt index: <0.4 to 0.15

Flexural modulus: 80,000 to <110,000 psi

Tensile strength at yield: 3000 to <3500 psi

Environmental stress crack resistance condition B: 50 percent max. failure in 24 h

Hydrostatic design basis at 23°C: 1600 psi

Although this newer cell-type format is a major improvement in classifying and specifying piping materials by a broader array of significant property and performance characteristics, it may not always be sufficiently definitive. The manufacturer may have to be consulted for further information. For example, two PE materials with the same ASTM material cell classification may have strength under long-term loading that responds somewhat differently to increasing temperature, or to fatigue loading, or to chemical environments.

A brief description of the major materials used for thermoplastics piping follows.

Polyvinyl Chloride (PVC)

In its virgin state PVC is a translucent colorless rigid polymer. When PVC was first commercialized, it was softened by the addition of plasticizers and primarily used in the manufacture of such items as luggage, upholstery, garden hose, wire coating, floor tiles, and laboratory tubing. Subsequent advances in extrusion and molding equipment and in stabilizer and lubrication additives allowed for the extrusion of the much more viscous, rigid compositions which are the only ones suitable for piping. PVC used for piping is, therefore, sometimes referred to as *rigid PVC* (or *uPVC* in Europe) to denote it is an unplasticized material.

Of the commonly available thermoplastics, rigid PVC offers the highest strength and stiffness at least cost, which accounts for its having become the leading plastic material for both pressure and nonpressure piping. Major uses include: water mains; irrigation; drain, waste, and vent (DWV); sewage and drainage; well casing; electric conduit; and power and communications duct. A much broader range of pipe sizes and wall thicknesses, fittings, valves, and appurtenances is available in PVC than in any other plastic.

PVC piping is joined primarily by two techniques: solvent cementing and elastomeric seals. Although it can be joined by thermal fusion, its melt viscosity is too high for making reliably strong joints under field conditions.

PVC piping is made only from rigid compounds containing no plasticizers and relatively small quantities of other ingredients. To minimize adverse effects on

TABLE D1.3 Primary Properties: Cell Classification Limits for PE Materials in Accordance with ASTM D3350

Destination order no.	Property	Test method	Cell limits 0	1	2	3	4	5	6
1	Density, g/cm³	D 1505	. . .	0.910–0.925	0.926–0.940	0.941–0.955	>0.955	. . .	
2	Melt index	D 1238	. . .	>1.0	1.0 to 0.4	<0.4 to 0.15	<0.15	*	†
3	Flexural modulus, psi	D 790	. . .	<20,000	20,000 to <40,000	40,000 to 80,000	80,000 to 110,000	110,000 to <160,000	>160,000
4	Tensile strength at yield, psi	D 638	. . .	<2,200	2,200–<2,600	2,600–<3,000	3,000–<3,500	3,500–<4,000	>4,000
5	Environmental stress crack resistance	D 1693	. . .				. . .		
	a. Test condition			A	B	C			
	b. Test duration, h			48	24	192			
	c. Failure, max, %			50	50	20			
6	Hydrostatic design basis, psi (23°C)	D 2837	NPR‡	800	1,000	1,250	1,600	. . .	

*Materials with melt index less than cell 4 but which have flow rate <4.0 g/10 min when tested in accordance with D 1238, Condition F.

†Materials with melt index less than cell 4 but which have flow rate <0.30 g/10 min when tested in accordance with D 1238 but at 310°C with total load of 12,480 g.

‡Nonpressure rated.

long-term strength and chemical resistance, minimal quantities of additives are used in pressure pipe compounds. To improve impact strength for conduit and other applications that may be subject to mechanical abuse, small quantities of solid polymeric impact modifiers (not plasticizers, which are generally liquids) may be incorporated into the composition. If improved stiffness is desired, filler, generally very finely divided calcium carbonate, is added. Combinations of these and other additives can be used to optimize a rigid PVC composition for its intended applications. The enhancement of a property by the use of additives may often require a trade-off with some other property.

To allow a specifier to define rigid PVC requirements, ASTM has established two material specifications based on the property cell classification system. One of these is ASTM D 1784, "Standard Specification for Rigid Poly(Vinyl Chloride) and Chlorinated Poly(Vinyl Chloride) Compounds," which classifies PVC materials in accordance with the nature of the polymer and five primary properties. The cell class limits that describe the polymer and four of the primary properties are presented in Table D1.4. The requirements for the fifth property, chemical resistance, are presented in Table D1.5. The manner in which a rigid PVC material is identified by this classification system is illustrated by a Class 12454-B rigid PVC material which, according to Tables D1.4 and D1.5, would have to meet the following minimum requirements:

Base resin: poly(vinyl chloride) homopolymer

Property and minimum value:
- Impact strength (Izod), 0.65 ft·lbf/in
- Tensile strength, 7000 psi
- Modulus of elasticity in tension, 400,000 psi
- Deflection temperature under load, 158°F
- Chemical resistance, meet the requirements of Suffix B in Table D1.5

Most PVC pressure pipe is made from materials that meet the minimum requirements of cell 12454-C which, to maximize long-term strength, are compounded with minimal quantities of processing additives and stabilizers and weathering protectants. An older "Type and Grade" system found in earlier versions of D 1784, although technically obsolete, has become rather well established and continues to be referenced by a number of piping standards. New issues of D 1784 include a table (see Table D1.6) that cross-references the former with the new designations.

For pressure pipe applications, the cell classification system of ASTM D 1784 is often complemented by one additional material requirement. All PVC pressure pipe standards require that the pipe be made from a formulation with a specified minimum long-term strength that has been established in accordance with ASTM D 2837, "Standard Method for Obtaining the Hydrostatic Design Basis for Thermoplastic Pipe Materials." Standards for products intended for the transport of potable water also require that the material meet certain minimum chemical extraction requirements designed to protect water quality.

Most ASTM and a number of other PVC pressure pipe standards identify PVC stress-rated materials by a four-digit number, the first two of which designate its type and grade in accordance with Table D1.6, and the last two identify, in hundreds of pounds per square inch, the material's maximum recommended hydrostatic design stress (HDS) for water at 73°F. According to ASTM convention, the

TABLE D1.4 Primary Properties: Cell Classification Limits for PVC Materials in Accordance with ASTM D 1784

Designation order no.	Property and unit	Cell limits								
		0	1	2	3	4	5	6	7	8
1	Base resin	Unspecified	Poly(vinyl chloride) homo-polymer	Chlori-nated poly(vinyl chloride)	Vinyl copol-ymer					
2	Impact strength (Izod) min.:									
	ft·lb/in of notch	Unspecified	<0.65	0.65	1.5	5.0	10.0	15.0		
3	Tensile strength, min.:									
	MPa									
	psi	Unspecified	<5,000	5,000	6,000	7,000	8,000			
4	Modulus of elasticity in tension, min.:									
	psi	Unspecified	<280,000	280,000	320,000	360,000	400,000	440,000		
5	Deflection temperature under load, min.:									
	264 psi									
	°F	Unspecified	<131	131	140	158	176	194	212	230
	Flammability	*	*	*	*	*	*	*	*	*

Note: The minimum property value will determine the cell number although the maximum expected value may fall within a higher cell.

*All compounds covered by this specification when tested in accordance with Method D635 shall yield the following results: average extent of burning of <25 mm; average time of burning of <10 s.

TABLE D1.5 Suffix Designation for Chemical Resistance

Solution	Suffix A	Suffix B	Suffix C	Suffix D
H_2SO_4 (93%)—14 days immersion at 55 ± 2°C				
Change in weight:				
Increase, max, %	1.0*	5.0*	25.0	NA†
Decrease, max, %	0.1*	0.1*	0.1	NA
Change in flexural yield strength:				
Increase, max, %	5.0*	5.0*	5.0	NA
Decrease, max, %	5.0*	25.0*	50.0	NA
H_2SO_4 (80%)—30 days immersion at 60 ± 2°C				
Change in weight:				
Increase, max, %	NA	NA	5.0	15.0
Decrease, max, %	NA	NA	5.0	0.1
Change in flexural yield strength:				
Increase, max, %	NA	NA	15.0	25.0
Decrease, max, %	NA	NA	15.0	25.0
ASTM Oil No. 3—30 days immersion at 23°C				
Change in weight:				
Increase, max, %	0.5	1.0	1.0	10.0
Decrease, max, %	0.5	1.0	1.0	0.1

*Specimens washed in running water and dried by an air blast or other mechanical means shall show no sweating within 2 h after removal from the acid bath.

†NA = not applicable.

TABLE D1.6 Classification of Commercial Types and Grades of Rigid Poly(vinyl Chloride) Piping Materials: Comparison of Former and New Designations

Former commercial type and grade from former specification D 1784–65T	Cell classification class from tables D1.4 and D1.5
Rigid PVC materials:	
Type I, Grade 1	12454-B
Type I, Grade 2	12454-C
Type I, Grade 3	11443-B
Type II, Grade 1	14333-D
Type III, Grade 1	13233
CPVC:	
Type IV, Grade 1	23447-B

maximum HDS is one-half the material's hydrostatic design basis (HDB), which refers to the material's long-term hydrostatic strength (LTHS) category when established in accordance with ASTM D 2837. The following describes the most common PVC stress-rated materials defined by this designation system:

- PVC 1120 is a Type 1, Grade 1, PVC (minimum cell class 12454-B) with a maximum recommended HDS of 2000 psi for water at 73°F.
- PVC 2110 is a Type 2, Grade 1, PVC (minimum cell class 14333-D) with a maximum recommended HDS of 1000 psi.
- PVC 2116 is a Type 2, Grade 1, PVC (same minimum cell class as above) with a maximum recommended HDS of 1600 psi.

Since by the ASTM convention the maximum recommended HDS is one-half the material's HDB, the HDBs for these materials are 4000 psi, 2000 psi, and 3200 psi, respectively.

The Plastics Pipe Institute has defined a generic PVC 1120 formulation that provides for certain specified alternative choices of ingredients and formulation quantities that allow the final compound to satisfy both the short- and long-term requirements established for this classification. This formulation, which is published in PPI TR-3, "Policies and Procedures for Developing Recommended Hydrostatic Strengths and Design Stresses for Thermoplastic Pipe Materials," is periodically updated to include any new alternate choices of ingredients that have been validated by both short-term and long-term tests.

The other PVC material specification is ASTM D 4396, "Standard Specification for Rigid Polyvinyl Chloride (PVC) and Related Plastic Compounds for Non-Pressure Piping Products." As indicated by its title, this specification covers compounds intended only for nonpressure uses. It is similar to D 1784 in that it is also based on the cell format and most of the same primary classification properties.

Chlorinated Polyvinyl Chloride (CPVC)

As implied by its name, CPVC is a chemical modification of PVC. It is very similar to PVC in many properties, including strength and stiffness at ambient temperature. But the extra chlorine in CPVCs structure increases the maximum operating temperature limit about 50°F above that for PVC. Thus, CPVC can be used up to nearly 200°F for pressure uses and up to about 210°F for nonpressure applications. Principal uses for CPVC are domestic hot and cold water piping, residential fire-sprinkling piping, and for many industrial applications which take advantage of its elevated-temperature capabilities and superior chemical resistance.

CPVC materials are also classified by the previously discussed ASTM D 1784. Similar to PVC, most CPVC standards covering pressure-rated products identify stress-rated materials by a four-digit number combining the older type and grade designation with the material's maximum recommended HDS for water at 73.4°F. Currently, the only recognized stress-rated CPVC designation is CPVC 4120, signifying a Type IV, Grade 1, material in accordance with D 1784 (see Table D1.6), with a maximum recommended hydrostatic design stress of 2000 psi for water at 73°F (23°C) in accordance with ASTM D 2837. In addition, most CPVC pipe standards covering products intended for elevated-temperature service require that the CPVC material have no less than a recommended HDS of 500 psi (equivalent to an HDB of 1000 psi) for water at 180°F.

Polyethylene (PE)

Polyethylene is possibly the best-known member of the polyolefin family (materials derived from the polymerization of olefin gases including ethylene, propylene, and butylene) because it has penetrated so widely into everyday household uses. PE in its virgin form is a translucent and tough substance with a waxylike feel and appearance. As is the case for the other polyolefins, PE is a partly crystalline and partly amorphous material. The properties of a PE are determined by its molecular structure. PE consists of a backbone of a long molecular chain from which short chain branches occasionally project. The length, type, and frequency of distribution of these branches, as well as other parameters such as molecular weight and molecular weight distribution, determine the degree of crystallinity and the network of molecules that anchor the crystal-like regions to one another. These structural characteristics greatly influence the short- and long-term mechanical properties of PE.

The extent of crystallinity of a PE polymer is reflected by its density—the higher-density materials have more crystalline regions, which results in greater stiffness and tensile strength. However, as the crystallinity increases, there is some compromise in ductility and toughness. Polyethylene polymers used for piping are classified into three types: a low-density, relatively flexible form; a medium-density, somewhat stiffer and less-flexible form; and a high-density form, which is more rigid and stronger. Most pressure pipe is made of materials of densities lying around the high end of the medium-density PEs and the lower end of high density. This range has established itself as offering the best balance of toughness, flexibility, and strength. Nonpressure pipe is primarily made from the more rigid, higher-density materials.

PE, which is somewhat less strong and rigid than PVC at ambient temperature, is the second most used plastic pipe material primarily because of its outstanding toughness and ductility, and flexibility, even at low temperatures. PE pipes do not fracture under the expansive action of freezing water. PE gas pipes can be safely "squeezed-off" (clamped tightly) by suitable procedures to shut down the flow of gas in an emergency. Also, PE gas pipe is very resistant to rapid crack propagation or brittle fracture. These two last-named characteristics are important reasons why PE pipe is used in over 85 percent of all current new installations of piping for gas distribution.

PE pipes have also good fatigue endurance. This, plus their ability to dampen water hammer shock, has led to their use in situations such as in sewer force mains where repeated cyclic pressure changes occur.

The high strainability and fracture resistance of PE has led to its selection for unstable soils and situations where axial bending and diametrical deflection are anticipated. Example installations that utilize this feature are methane collection systems for solid waste sites, pipes installed by directional trenchless boring techniques, lake and river crossings, and outfall pipes.

PE pipe is also used for the rehabilitation of old pipelines. Lengths of PE pipe which have been joined to the required length are pulled, or sometimes pushed, inside the old line. New rehabilitation procedures have evolved by which, for ease of insertion, the diameter of the liner PE pipe is reduced by a squeeze-down procedure or by folding. Once inside the old pipe, the strain memory in the material is relieved by heating, allowing the PE pipe to reround so it fits snugly inside the existing pipe.

The low stiffness of PE permits the coiling of smaller-diameter pipe, generally up to about 4 in, although pipe up to 6-in diameter has been coiled for special jobs. The coil length can be hundreds, and sometimes more than a thousand, feet, depending on material, wall thickness, and diameter. PE pipe is readily heat fusible and can be joined to itself or to fittings by the butt fusion process. PE fittings are also available for joining to pipe by the socket fusion and electrofusion processes.

To protect the polymer during processing, storage, and service, PE piping compounds contain small quantities of heat stabilizers, anti-oxidants, and ultraviolet (UV) screens or stabilizers. Black PE pipe materials incorporate very finely divided carbon black as both a coloring pigment and a screen to protect the polymer from the potentially damaging UV radiation in sunlight. Nonblack piping compositions, which are primarily used for gas distribution, include a UV stabilizer in addition to a coloring pigment, usually tan, orange, or yellow.

The primary specification for identifying and classifying PE piping materials is ASTM D 3350, "Standard Specification for Polyethylene Pipe and Fittings Materials." Standard D 3350 employs the cell class format to cover the diversity of materials suitable for piping. As shown by Table D1.3, this specification classifies PE materials by a matrix of six primary properties and a specified range of cell values for each of these properties. In addition, an ending code letter is used to designate the incorporation of a colorant and UV stabilizer. An example of how PE materials are identified by this system was illustrated previously.

Prior to issuance of ASTM D 3350, most PE piping standards referred to ASTM D 1248, "Standard Specification for Polyethylene Plastics Molding and Extrusion Materials," for material requirements. ASTM D 1248 classified PE by type, representing the material's density category, and grade, reflecting a combination of properties, primarily the melt flow or processing characteristics. Similar to PVC, PE piping standards classify PE stress-rated materials by means of a four-digit number, the first two of which refer to the older type and grade designation and the last two represent, in hundreds of pounds per square inch, the material's maximum recommended hydrostatic design stress (HDS) for water at 73°F. To relate this older designation system to the newer cell system of D 3350, the latter standard includes a cross reference. The crossovers recognized by the 1990 Edition of D 3350 are presented in Table D1.7.

Polybutylene (PB)

Polybutylene is a polyolefin with a stiffness resembling that of low-density polyethylene, while its long-term strength is greater than that of high-density polyeth-

TABLE D1.7 Classification of Commercial Polyethylene (PE) Piping Materials: Comparison of Two Designation Systems

Material designation per traditional coding based on ASTM D 1248 and maximum recommended HDS for water at 73°F	Minimum cell property value required to meet traditional coding, per cell classification system of ASTM D 3350*
PE 1404	PE 122111
PE 2406	PE 213333
PE 3406	PE 324433
PE 3408	PE 334434

*For property cells 1 and 6 material values must fall within the values shown for the indicated cell. For the other properties material values may fall within the given, or higher numbered cells. Refer to Table D1.3 for definition of property ranges for each cell.

ylene (PE). Its most distinctive feature, however, is that its long-term strength is less affected by increasing temperature than that of PE. While most PEs have an upper temperature limit of around 140°F, the limit for PB is nearly 200°F.

Major applications for PB pipe and tubing are for water service lines and for applications that take advantage of its improved elevated-temperature strength. They include piping for residential hot and cold water distribution, residential fire sprinklers, and industrial uses such as hot effluent lines.

PB is similar to PE in its chemical resistance and heat fusibility. PB piping materials are covered by ASTM D 2581, "Standard Specification for Polybutylene (PB) Plastics Molding and Extrusion Materials." Materials for pressure applications are designated as PB 2110, signifying a Type 2, Grade 1, material in accordance with D 2581 with a maximum recommended HDS of 1000 psi for water at 73°F. Hot water piping standards generally require that the PB material also have an established recommended maximum HDS of 500 psi for water at 180°F.

Cross-Linked Polyethylene (PEX)

Cross-linked polyethylene is, as its name implies, actually a thermoset. It is covered in this chapter because PEX pipe and tubing is made from PE—a thermoplastic—by essentially the same extrusion process used to manufacture the other thermoplastic pipes. The difference is that in the case of PEX there is a cross-linking of the polymer that occurs during or soon after the extrusion. The cross-linking may be effected in two principal ways: by the addition of cross-linking compounds to the PE material that are triggered into action by either the extrusion, or higher postextrusion temperatures; or by subsequent electron or ultra-high-frequency radiation. There are various chemical cross-linking agents used, including peroxides, azo compounds, and silanes.

The final properties of the PEX depend on the degree and the temperature of cross-linking as well as on the particular methods used. Generally, PEX is not much stronger than PE at ambient temperature, but it is much less affected by increasing temperature. PEX can be used for pressure applications up to about 200°F.

Currently, PEX pipe and tubing made in North America is furnished in diameters through 2 in. Principal applications are for under-floor heating systems, for melting of ice and snow on roads and sidewalks, and for hot-cold water piping.

Polypropylene (PP)

Polypropylene is a polyolefin similar in properties to high-density PE but somewhat harder, more temperature resistant, and lighter in weight, but less tough. It is also similar to PE in its chemical resistance and its heat fusibility. As in the case of PE, PP can be joined to itself by socket fusion, butt fusion, and electrofusion.

Because of its greater stiffness and better tolerance to elevated temperatures, PP is sometimes chosen over PE where these qualities are advantageous—i.e., for above-ground piping and for conveying hotter liquids. One principal application is in corrosive drainage piping, for which PP offers better solvent resistance than either ABS or PVC. A line of PP corrosive drainage piping made from a flame-retardant grade of material is available for use in laboratories, hospitals, and by industry. Another principal application for PP is for conveying corrosive chemicals under pressure. For this application socket fusion systems of pressure-rated PP pipe and fittings are available through 6 in. As of this writing, a standard covering solid wall PP piping for above- and below-ground drainage is under development at ASTM. There are no standards for PP pressure pipe; all available products are proprietary.

Polypropylene materials are classified by ASTM D 4101, "Standard Specification for Polypropylene Molding and Extrusion Materials," into two types. Type I covers materials that have the highest rigidity and strength but offer only moderate toughness. Type II covers materials (copolymers of propylene with ethylene or other olefins) which tend to be less rigid and strong but have improved toughness, particularly at lower temperatures. Both types are used for pipe.

Acrylonitrile-Butadiene-Styrene (ABS)

ABS plastics are made by combining styrene-acrylonitrile copolymers with copolymers formed by reacting styrene-acrylonitrile with butadiene. The butadiene copolymers impart toughness, while the acrylonitrile copolymers contribute strength, rigidity, and hardness. The result is a tough, relatively strong plastic that is easy to mold and extrude.

The ABS family covers a wide range of materials. The proportions of the basic components and the way in which they are combined can be varied to produce a wide range of end properties. Major uses of ABS for pipe is in the manufacture of drain-waste-vent (DWV) piping for which it offers good rigidity, temperature resistance, low-temperature toughness, and the ability to make fast-setting solvent-cemented joints. ABS is also used in the manufacture of a composite pipe consisting of two concentric thermoplastic tubes integrally braced, with the resultant annular space filled with portland cement–perlite concrete. ABS was used for pressure piping, but it has been largely displaced by the stronger, more chemically resistant and more economical PVC. However, compressed air piping made from a proprietary extra-tough, shatter-resistant composition is currently marketed in Europe and the United States.

ABS materials are classified by ASTM D 1788, "Standard Specification for Rigid Acrylonitrile-Butadiene-Styrene (ABS) Plastics," in accordance with a cell class system by which each of three properties—impact strength (toughness), tensile stress at yield (short-term strength), and deflection temperature under load (temperature resistance)—is accorded a cell number depending on the property value. The ABS ASTM specification for DWV pipe requires that the material have a minimum cell classification of ABS 2-2-2 which signifies the following minimum properties: notch impact strength of 2 ft · lb/in of notch, 180°F deflection temperature, and 4000-psi tensile strength.

Fluoroplastics

Fluoroplastics designate a broad family of paraffinic polymers that have some or all of the hydrogen replaced by fluorine. Included among these materials are the following polymers which are used for pipe or pipe liners: Polytetrafluoroethylene (PTFE), a completely fluorinated polymer with the predominant repeating unit consisting of tetrafluoroethylene; polychlorotri-fluoroethylene (PCTFE), similar to PTFE; and polyvinylidene fluoride (PVDF) with the dominant repeating unit consisting of difluoroethylene. This family also includes copolymers such as perfluoroalkoloxies (PFAs) which are similar to PTFE in chemical structure except they include occasional side chains of fully fluorinated olefins that are bound to the main chain via an oxygen atom. Other members include ethylene-chlorotrifluoroethylene (ECTFE) copolymer which is primarily a 1:1 copolymer of ethylene and chlorotrifluoroethylene, and fluorinated ethylene propylene copolymer which is produced by copolymerization of tetrafluorothylene with hexafluoropropylene.

Fluorinated polymers have outstanding resistance to chemicals and excellent resistance to solvents. They also offer improved elevated-temperature properties and are very stable and durable. Most members of this family require little or no addition of processing or thermal stabilizers. For this reason they are often specified when exceptional purity of water or other liquid must be maintained. These materials are also very fire resistant.

Pipe, tubing, and socket fusion fittings through about 6 in are commercially manufactured from PVDF, PFA, and ECTFE copolymer. PVDF has good strength, wear resistance, and creep resistance and can be used over a temperature range from about −100 to 300°F. PFA has somewhat less strength and creep resistance but has greater toughness and can be used up to over 400°F. ECTFE copolymer is somewhere between these materials.

Fluorinated plastics have also outstanding resistance to weathering and radiation. These materials do not require the use of additives to achieve weathering and ultraviolet resistance. Because of their immunity to radiation, they are used in the reprocessing of nuclear wastes and similar radiation-intensive exposure.

JOINING METHODS

Plastics piping may be joined by different methods (see Table D1.8), depending on the characteristics of the material. For example, ABS, PVC, and CPVC can be solvent cemented. However, the polyolefins (PE, PB, and PP) and fluoropolymers cannot be joined by this method because of their high solvent resistance, but they can readily be heat fused. Both heat fusion and solvent cementing yield monolithic joints of maximum strength and chemical resistance.

In solvent cementing, the mating spigot-socket or saddle-pipe surfaces are

TABLE D1.8 Methods for Joining Thermoplastic Pipe

Joining method	ABS	PVC	CPVC	PE	PP	PB	PVDF
Solvent cementing	x	x	x	—	—	—	—
Heat fusion	—	—	—	x	x	x	x
Threading*	x	x	x	x	x	—	x
Flanged connectors†	x	x	x	x	x	x	x
Grooved joints‡	x	x	x	x	x	—	x
Mechanical compression§	x	x	x	x	x	x	x
Elastomeric seal	x	x	x	x	x	x	x
Flaring	—	—	—	x	—	x	—

*Thermoplastics pipes thinner than Schedule 80 should not be threaded.
†Flanged adapters are fastened to pipe by heat fusion, solvent cementing, or threading.
‡Grooving requires a minimum pipe wall thickness, which depends on the material.
§In most cases, internal stiffeners are required to support the pipe against the compressive forces.

readily fused when softened by the action of the solvent. Procedures for solvent cementing PVC pipe and fittings are included in ASTM Standard Practice D 2855.

In heat fusion, softening of the surfaces is achieved by melting. There are two techniques used for melting the surfaces. One is to heat them with a specially designed heating iron just prior to joining. In the other, the surfaces are first mated and then brought to the proper melt temperature by means of heating wires embedded in the socket. This latter technique, called *electrofusion,* is used with PE gas piping and PP industrial drainage piping.

There are two heat fusion jointing systems. One of them is the socket-spigot system, similar to that used for solvent cementing. In the other the butt ends of pipe and/or fittings are squared off precisely, heated, and then brought together under pressure. Socket fusion is limited to the smaller sizes, generally not above 4 in. Larger-diameter pipe is heat fused by the butt fusion process. Automatic, portable equipment for field joining by butt fusion is available to join all available pressure pipe sizes. ASTM Recommended Practice D 2657 covers socket, butt, and saddle heat fusion. Electrofusion joining is covered by ASTM F 1290.

Flanged connections are often used for industrial applications, particularly when making transitions to nonplastic components such as to a metal valve or to a tank outlet, or when it is advantageous to provide for easy removal of a pipe section or other component from the system for cleaning, maintenance, or other purpose. Flange connectors can be applied on the pipe by heat fusion (socket or butt), or solvent cementing, depending on the material.

Much of the pipe used for buried water and sewer lines and drains is made with bell-and-spigot connections that include an elastomeric gasket to seal the joint. The bell, including the gasket cavity, is usually formed during pipe or fitting manufacture and is an integral part of the product. Rubber-gasketed connectors facilitate construction and produce tight joints, even when made under foul weather and poor field conditions when solvent cementing and heat fusion joining may be adversely affected. ASTM D 3112 prescribes requirements for joints for nonpressure applications and D 3139 covers those for pressure uses.

Threading is also sometimes used. Molded threads with reduced roots are preferred because cut threads are more notch sensitive. Molded threaded adapters are available for solvent cementing to PVC and CPVC pipe. If pipe is to be threaded, it is generally recommended that its wall thickness be not less than that corresponding to Schedule 80. Threaded connections of any type are prohibited for gas distribution.

Mechanical compression fittings are also used, particularly when making transition connections to dissimilar materials. Mechanical connectors are the only option for connecting PEX since it cannot be solvent cemented or heat fused. Most, but not all, mechanical fittings designed for plastics use compressed elastomeric gaskets for sealing; and to ensure this seal is not lost through pipe deformation, the design incorporates a metal sleeve that fits inside the plastic pipe to stiffen it against the compression forces. Many compression fittings are also designed to prevent pipe pullout by service-induced forces.

DIMENSIONING SYSTEMS

Thermoplastic pipe is made to a number of dimensioning systems based on controlled outside diameter. Pipe from all thermoplastic materials is manufactured to the standard outside diameters of iron pipe sizes (IPS) of commercial wrought steel pipe (ANSI B36.10). In this diameter system most plastic pipes are also offered with wall thicknesses corresponding to those of Schedule 40 and 80 IPS pipe. More common, however, is plastic pipe with IPS dimensioned outside diameters but with walls sized in accordance with the standard dimension ratio (SDR) principle whereby all pipe sizes in an SDR series have the same ratio of outside diameter to minimum wall thickness. The SDRs adopted by ASTM and other organizations are from the following series: 11, 13.5, 17, 21, 26, 32.5, 41, and so on. However, other dimension ratios are also used, in which case they are identified by the prefix DR. The broad acceptance of the SDR dimensioning system arises from the fact that certain performance ratings are directly proportional to the ratio of diameter to wall thickness. For example, all pipe sizes made from the same material and to a constant SDR series have the same pressure rating and pipe stiffness.

Other diameter systems to which plastic pipe is made include:

- *Cast-iron (CI) pipe sizes:* PVC pipes are available in this sizing system to facilitate connections to valves and hydrants and water works fittings which, generally, are made to cast-iron sizes.
- *Copper tubing sizes (CTS):* CTS-sized pipe can be joined using compression and flare fittings designed for copper. Plastic pipe in CTS sizes is used for gas and water services and for hot-cold water piping.
- *Plastic pipe sizes:* Certain products, particularly nonpressure pipes (for example, PE and PVC sewer), are available in sizes unique to plastic.
- *International Standards Organization (ISO) sizes:* Some of the larger-diameter PE pipes are made to an internationally established outside-diameter series. There is also a series of PE pressure-rated pipes, up to 6 in diameter, that is made with inside diameters equal to those of Schedule 40 iron pipe size pipes. In this series, which is designed to be joined by insert fittings, the outside diameter is determined by the wall thickness required to satisfy the pipe pressure rating.

Many manufacturers are also prepared to manufacture nonstandard sized pipe on special order. Such pipe may be required for a particular situation, such as for sliplining inside an existing corroded line with the largest possible plastic pipe.

PROPERTIES

Approximate values of some of the physical properties for the more common generic thermoplastic piping materials are given in Table D1.9. The actual property values for a particular commercial material depend not only on its ASTM classification but also on its specific composition, material state (i.e., some anisotropy may result because of processing conditions), and the combination and history of previously applied stresses.

Compared to traditional piping materials, thermoplastics are lighter in weight, have lower heat capacities, are poorer conductors of heat, and have significantly larger coefficients of expansion-contraction. They are also less strong and rigid; however, they offer sufficient rigidity and strength to satisfy the performance requirements of a great many applications, and they are not subject to the gradual degradation by rusting and other corrosive processes that commonly afflict traditional pipe materials. In selecting appropriate long-term values of strength and

TABLE D1.9 Approximate Physical Properties of Principal Thermoplastic Piping Materials

Property at 75°F	ASTM test no.	ABS	PVC	CPVC	PE	PB	PVDF
Specific gravity	D792	1.08	1.40	1.54	0.95	0.92	1.76
Tensile strength, psi (10^3)	D638	7.0	8.0	8.0	3.2	4.2	7.0
Tensile modulus, psi (10^3)	D638	340	410	420	120	55	220
Impact strength, Izod ft·lbf/in notch	D256	4	1	1.5	>10	>10	3.8
Coeff. of linear expansion, in/in · °F (10^{-6})	D696	60	30	35	90	72	70
Thermal conductivity, Btu · in/h · ft · °F	C177	1.35	1.1	1.0	3.2	1.5	1.5
Specific heat, Btu/lb · °F	—	0.34	0.25	0.20	0.55	0.45	0.29
Approx. operating limit,							
°F, nonpressure	—	180	150	210	160	210	300
°F, pressure	—	160	130	180	140	180	280

stiffness for design purposes, the special consideration with plastics is over their unique load-deformation response.

Plastics are viscoelastic materials. Their deformation and strength properties vary with temperature and duration of loading and can also be affected by certain environments. Tensile strength and stiffness values, such as given in Table D1.9 and which have been obtained by means of short-term mechanical tests adapted from metal testing, are not appropriate for design of piping systems that are subjected to long-term loading. In the case of metals the conventional tensile test is used to define basic properties such as elastic modulus, proportional limit, and yield strength. These are important not only for defining and specifying a metal but they are also basic constants for use in design equations based on elastic theory, whereby strain is assumed to be proportional to stress. Although very few materials are perfectly elastic, the assumption of such behavior in metals is usually sufficiently close for purposes of engineering design.

The stress-strain response for plastics is curvilinear and can depart greatly from an assumption of proportionality. Furthermore, the viscoelastic nature of plastics results in a relationship between stress and strain that is greatly influenced by time (e.g., rate of straining in a tensile test), temperature, and environment. As depicted schematically by Fig. D1.1, the stress-strain response and fracture strength much depend on the test conditions. For example, with some thermoplastics a reduction in but two decades in tensile strain rate, or an increase in temperature of around 50°F can result in nearly a doubling of the strain response. Also, as depicted by this figure, the stress-strain response is nonlinear, although near the origin the behavior is fairly linear. Accordingly, plastics have no true elastic constants, such as elastic modulus or proportional limit, nor do they have sharply defined yield points. The reported tensile elastic modulus for plastics represents a tangent modulus that is calculated by extending the initial portion of the load-extension curve, as obtained under a specified set of test conditions. Since plastics have no true elastic properties, the propriety of applying the term "elastic modulus," or even "modulus," in describing the stiffness or rigidity of a plastic has been questioned. However, such a constant has proven

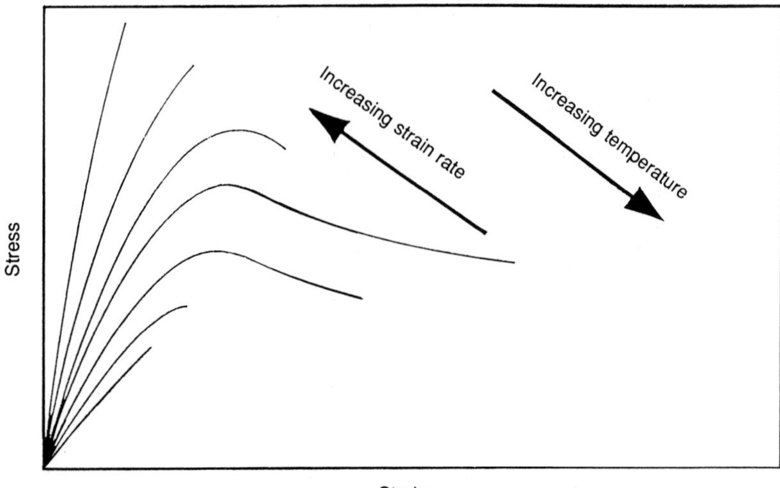

FIGURE D1.1 Tensile stress-strain response of a thermoplastic exhibiting ductile behavior at intermediate strain rates. Possible effects of strain rate and temperature are shown.

useful both for specification and design when its arbitrary nature and its dependence on duration of loading, temperature, and similar factors is recognized.

Even though plastic's behavior is inelastic, most of the equations for stress analysis such as for pipe, beams, and pressure vessels, which have been derived on the assumption of elastic behavior, can still be used provided values for strength and stiffness are appropriately established. Thus, property values obtained through short-term tests should be used only for predicting response under loads of short duration (short-term tests, of course, also have important value for defining and classifying plastic materials). To be able to use elastic equations to forecast response under longer-term load requires the use of "effective" values of strength and modulus that account for long-term loading effects—the development of such values usually involves some form of extrapolation protocol. Similar approaches exist with other inelastic engineering materials. As discussed later on, the protocols for plastics generally rely on information from longer-duration testing for defining effects of time and temperature on design limits. In fact, many of the tests and procedures have evolved from methods used to define structural behavior of metals at very high temperatures, under which condition they also behave as viscoelastic materials.

Viscoelasticity

As the name implies, viscoelastic materials respond to stress as complex aggregates of many different elastic and viscous (fluid) elements. The springs in the highly simplified model of Fig. D1.2 represent the elastic elements of a polymer (e.g., chain rigidity, chemical bonds, and crystallinity), each spring having a different constant that represents a time-independent modulus of elasticity. The dashpots represent the fluid elements (e.g., molecules slipping past each other), each one having a different viscosity or time-dependent stress-strain response.

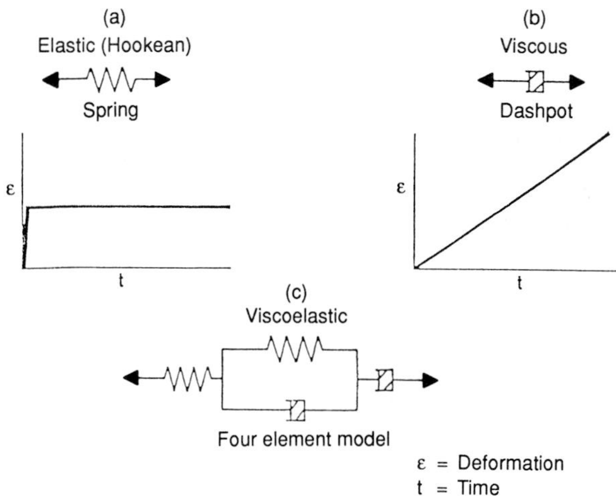

FIGURE D1.2 Model of viscoelastic behavior.

When a constant load is applied and sustained on this model, it results in an initial deformation which continues to increase indefinitely (Fig. D1.3). This phenomenon of continuing deformation, which also occurs in concrete, soft metals, wood, and structural metals at very high temperatures, is called *creep.* If the load is removed after a certain time (say, at point t_i in Fig. D1.3), there occurs a rapid initial strain recovery followed by a continuing recovery that occurs at a steadily decreasing rate; in this model recovery is never complete. However, if the creep

Creep (constant load)

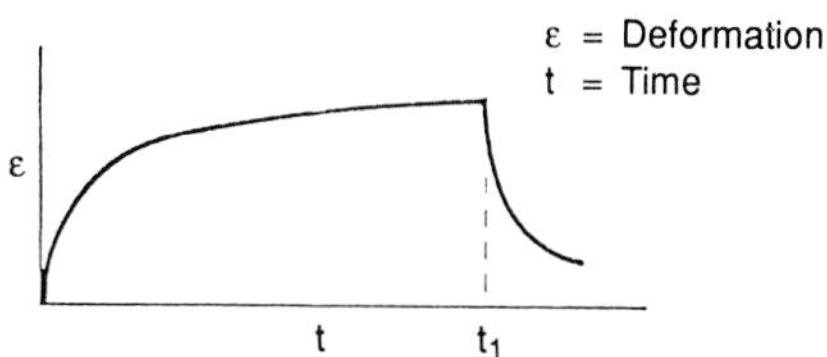

Stress relaxation (constant deformation)

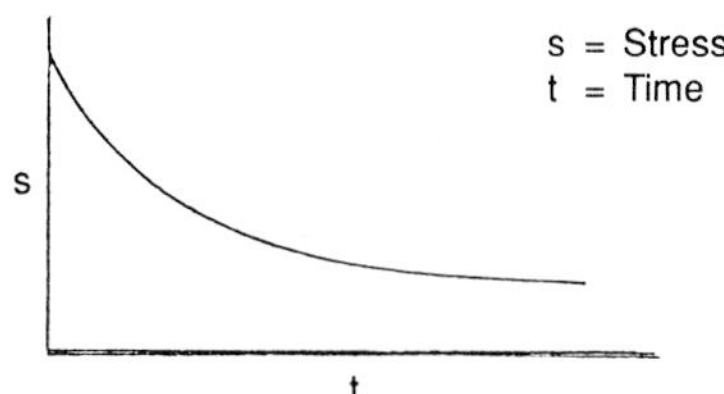

FIGURE D1.3 Viscoelastic responses.

strain does not cause irreversible structural changes and sufficient time is allowed, the strain recovery will be almost complete. The rate and extent of deformation and recovery are sensitive to temperature and can also be influenced by environmental effects such as by the absorption of solvents or other materials with which the plastics may have come in contact with while under stress.

An analogous response of viscoelastic materials is stress-relaxation. The initial load required to achieve a certain deformation will tend to gradually relax when that deformation is kept constant (see Fig. D1.3). Initially, stress-relaxation occurs rapidly and then steadily decreases with increasing time.

Tensile Creep

Each material has a characteristic stress-strain–time-temperature function. The primary form for characterizing this function is a family of tensile creep curves. Creep curves plotted on cartesian coordinates (Fig. D1.4) generally show three continuous stages: a first stage marked by large and initial deformation; a second stage where deformation continues at a relatively slow and constant rate; and a third stage during which rupture occurs. In ductile plastics third-stage creep usually includes a distinct elongation or yielding just prior to rupture. In nonductile plastics, rupture occurs abruptly during second-stage creep. As illustrated by Fig. D1.5, which has been obtained on a certain pipe grade PE material, tensile creep curves are frequently plotted with log time as the abscissa. This is a more practical way of representing the information over the time range of engineering interest, and it also facilitates extrapolation of data to longer times. Many mathematical methods have been proposed to describe the creep behavior of plastic materials in terms of stress, strain, and time. One such method, which is contained in an American Society of Civil Engineers standard design practice, presents constants for PVC and PE that have been verified by tests lasting nearly 20 years.[5]

Any point on any creep curve gives a stress-strain ratio. This ratio is usually designated as the "creep modulus," or sometimes as the "apparent modulus," and is used for design calculations where the stress is prescribed but the strain is free to vary. Creep modulus curves derived from the creep data on Fig. D1.5 are presented in Fig. D1.6.

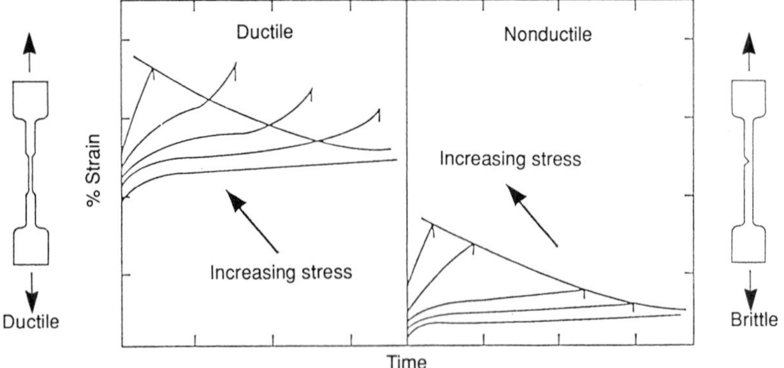

FIGURE D1.4 Schematic of creep rupture strength of thermoplastics in tension.

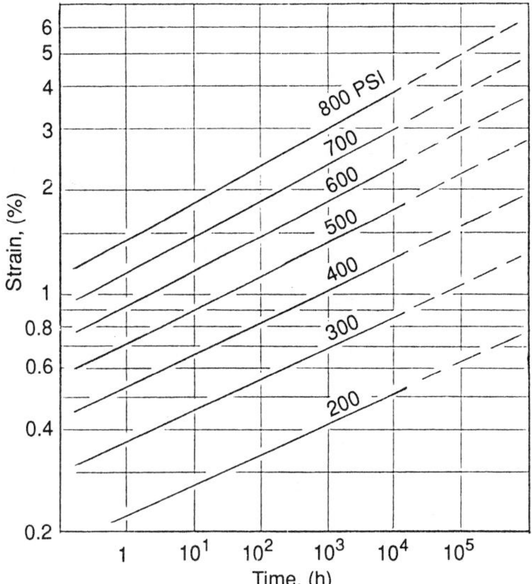

FIGURE D1.5 Tensile creep response for high-density polyethylene pipe material.

Analogous "stress-relaxation moduli" can be derived from stress-relaxation data. The stress-relaxation modulus is the properly defined quantity which is required for design calculations in which the strain is prescribed and the stress is free to vary. However, the numerical difference between the relaxation modulus and the creep modulus is often small when the strain is small and elapsed times are matched. Accordingly, the two can often be used interchangeably for engineering design. A procedure has been proposed to define the range over which the above assumption provides a reasonable approximation.[5] This reference includes a detailed discussion on how viscoelastic properties can be used to model stress-strain response under a variety of time-dependent loading conditions.

The creep modulus for a certain duration of loading is often expressed as a fraction of the modulus of elasticity measured by a particular short-term tensile test. Such a simplified representation (see Table D1.10) assumes that over the range of engineering stresses for which the approximation will be used, creep modulus is independent of stress intensity. The consequence of this simplification is usually small and acceptable for most pipe design. The reciprocal of this fraction, called the "creep factor," was introduced by Ref. 5 and is used by certain design practices, that is, those of the American Association of State Highway and Transportation Officials (AASHTO) Design Specification for Culverts.[6]

Even though the deformation behavior of plastics is rather complicated, successful design can be simplified by using a creep or an apparent modulus that reasonably reflects response under the anticipated loads. For example, a buried pipe may be subjected to relatively short periods of externally induced high loads, followed by longer periods at lower loading. For such conditions, Boltzman superposition theory can be used to estimate an appropriate creep modulus.[5] More often though, a simpler design check assuming a worse-case condition is all that is required. For example, the wall thickness of a buried pipe

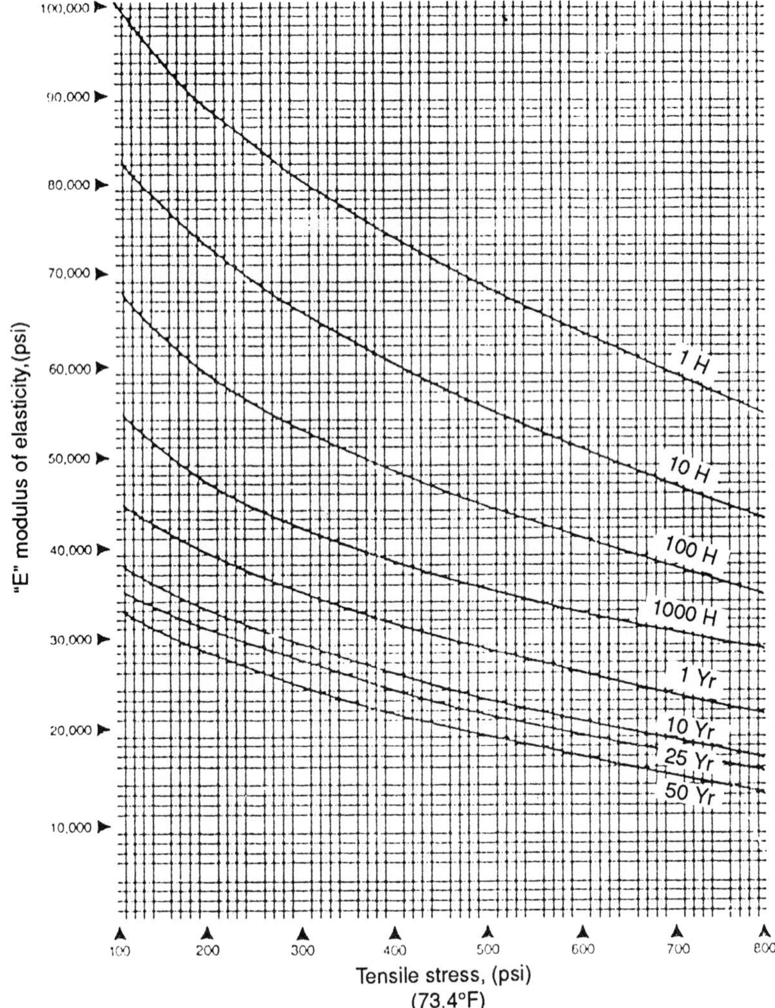

FIGURE D1.6 Creep modulus of elasticity versus stress intensity for a high-density polyethylene.

is first selected based on pressure rating considerations and then a check is made to ensure this wall is sufficient to withstand a given combination of traffic (short-term) and soil (long-term) loads. To conduct the design check, it can be assumed that the two loads are continuously present; if the pipe is adequate to this task, then no further calculation need be made. While procedures such as this are often used to simplify design, care must be exercised not to make unduly conservative assumptions which could rule out a fully acceptable construction that would be justified by a more refined evaluation.

TABLE D1.10 Approximate Ratio of Creep Modulus to Short-Term Modulus as a Function of Duration of Loading, for 73°F

Duration of uninterrupted loading, h	Approximate ratio of creep to short-term modulus for 73°F*	
	High-density PE	Type I PVC
1	0.80	0.84
100	0.52	0.60
10,000	0.28	0.40
438,000 (50 yr)	0.22	0.34

*Exact ratio depends on the specific material, stress field, residual fabrication stresses, and other factors.

Allowable Strength When Load Is Constant

The relationship between tensile load and time-to-fracture is described by the creep-rupture envelope of tensile creep curves (see Fig. D1.4). Each material has a characteristic envelope depending on its specific composition.

The stress versus lifetime envelope (more commonly referred to as the *stress-rupture envelope*) of thermoplastic materials intended for pressure piping is determined by means of long-term pressure tests conducted on pipe specimens made from the material under evaluation. The use of pipe specimens more closely replicates the combined hoop and axial stresses generated by pressure in actual service. Pipe testing is performed in accordance with ASTM D 1598, "Time to Failure of Plastic Pipe Under Constant Internal Pressure." Sufficient pressure versus time-to-fail points are obtained for a set of temperature and test environment conditions to define a hoop stress-rupture envelope through at least 10,000 h. The following relationship, commonly known as the *ISO equation* (denoting its adoption by the International Standards Organization), is used to calculate the pipe hoop stress generated by internal test pressure:

$$S = \frac{P}{2} \times \frac{D_m}{t} \tag{D1.1}$$

where S = hoop stress, psi
P = internal pressure, psi
D_m = mean pipe diameter, in
t = minimum pipe wall thickness, in

The stress-rupture data obtained by D 1598 testing are plotted on log stress versus log time-to-fail coordinates. If, as is generally the case for materials that qualify for pressure piping, the data plot along a straight line, the best least-squares line is determined mathematically and extrapolated to the 100,000-h intercept to forecast that material's long-term hydrostatic strength (LTHS). The extrapolation procedure used is that of ASTM D 2837, "Obtaining Hydrostatic Design Basis for Thermoplastic Pipe Materials." A material's hydrostatic design basis (HDB) is established by categorizing its LTHS into one of a series, as follows, of standard long-term strength values in which the numbers ascend in increments of 25 percent: 1000 psi, 1250 psi, 1600 psi, 2000 psi, and so on. The purpose of classifying LTHSs into a limited number and prescribed value of HDBs is to simplify

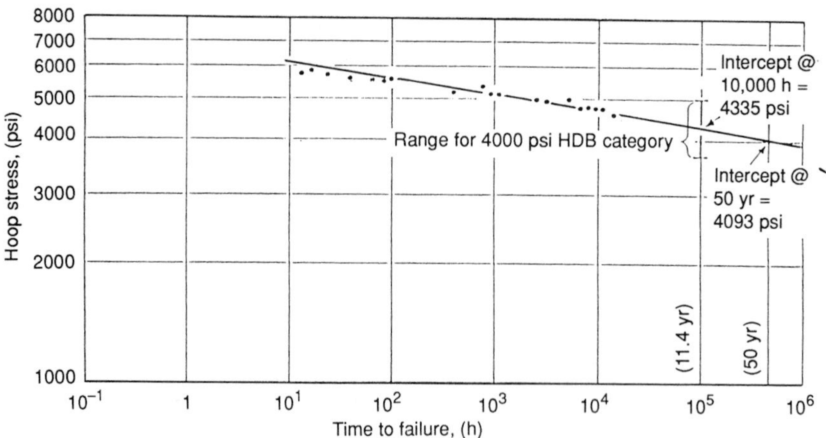

FIGURE D1.7 Application of method ASTM D 2837 to establish the hydrostatic design basis (HDB) for 73°F for a typical PVC 1120 material.

material standardization and pipe design. Figure D1.7 illustrates the application of the ASTM D 2837 procedure on a PVC 1120 material.

Nearly every standard specification for thermoplastic pressure pipe requires that the material from which the pipe is made have an established HDB for the standard condition of water at 73°F. Standards covering natural gas piping require evaluation of the HDB with natural gas as the test medium. For most materials, testing with water or gas make little difference on the resultant HDB. Piping materials intended for hot water applications are generally required to also have an established HDB for water at 180°F. The Plastics Pipe Institute (PPI) publishes a periodically updated list of thermoplastic pipe materials with HDBs that have been established in accordance with ASTM D 2837 and the additional requirements included in PPI TR-3, "Policies and Procedures for Developing Recommended Hydrostatic Design Stresses for Thermoplastic Pipe Materials." Materials carrying a PPI recommended HDB are listed in PPI TR-4, "Recommended Hydrostatic Strengths and Design Stresses for Thermoplastic Pipe and Fittings Compounds." A number of pressure pipe standards require that the material have a PPI listed HDB.

The HDB forecasted by method ASTM D 2837 is predicated on a fundamental assumption: The log stress versus log time-to-fail line defined by the data obtained through 10,000 h (1.14 years) will continue its straight course through at least the 100,000-h (11.4-year) intercept. To lend confidence to this assumption, certain requirements and procedures, as follows, have been established:

PE. To ensure that the inherent ductility of PE is not compromised by potential chemical or physical aging mechanisms, the only PE materials that qualify for pressure service are those that have been suitably protected by addition of stabilizers and those that have an HDB that has been independently validated by certain elevated-temperature sustained-pressure requirements that have been added to ASTM D 2837 and PPI TR-3.

It has been established that PE pipe under long-term load can fail by one of three distinct failure modes[7] (see Fig. D1.8):

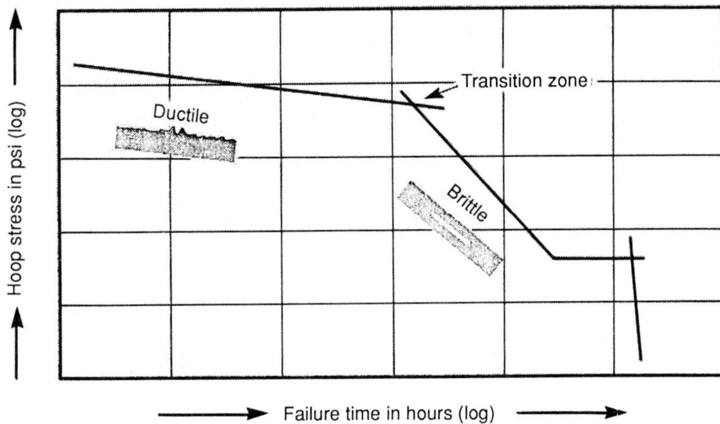

FIGURE D1.8 Schematic of the stress-rupture behavior of a polyethylene material subject to slow crack growth (first downturn) and chemical degradation (second downturn). Such materials are excluded from pressure service.

- The first is a ductile failure whereby the specimen ruptures as a consequence of gross yielding at a location subjected to maximum tensile stress. The lower the stress, the longer it takes for yielding to occur. No irreversible structural damage occurs in the material prior to yielding.
- The second is a slit failure which is the end result of a slow crack growth mechanism, whereby localized stress concentrations caused by minute defects or inclusions in the material spawn small cracks which then grow as long as sufficient stress is present. As shown by Fig. D1.8, strength regresses faster in the brittle slow crack zone than projected by the ductile failures.
- The third is the result of a breakdown of the polymer molecule through oxidative or other degradation. When this occurs, the polymer strength can be severely compromised.

PE pipe materials can qualify for pressure service only when independent tests validate the continuance of the ductile failure zone through at least 100,000 h. The procedure for validation is based on the observation that the onset of the slow crack growth mechanism is greatly accelerated by increasing temperature. This is illustrated by Fig. D1.9, in which the downturn portions of the stress-rupture lines (denoting the slow crack growth regions) shift to significantly shorter times as the temperature increases. By correlating the test temperature with the time required to reach the brittlelike failure zone, one may validate the assumption inherent to ASTM D 2837, that is, that there will be no downturn prior to the time of extrapolation. Materials that do not validate are rejected for pressure service[7].

The fracture mechanics of the slow crack growth process are being extensively investigated.[8,9] The results promise the development of simpler accelerated tests by which a material's potential to resist failure by this mechanism may be accurately assessed in relatively short periods of time.

For protection against oxidative degradation, current PE piping standards require that the material include appropriate amounts and types of stabilizer packages. Accelerated aging studies, as well as those conducted on pipe that has been

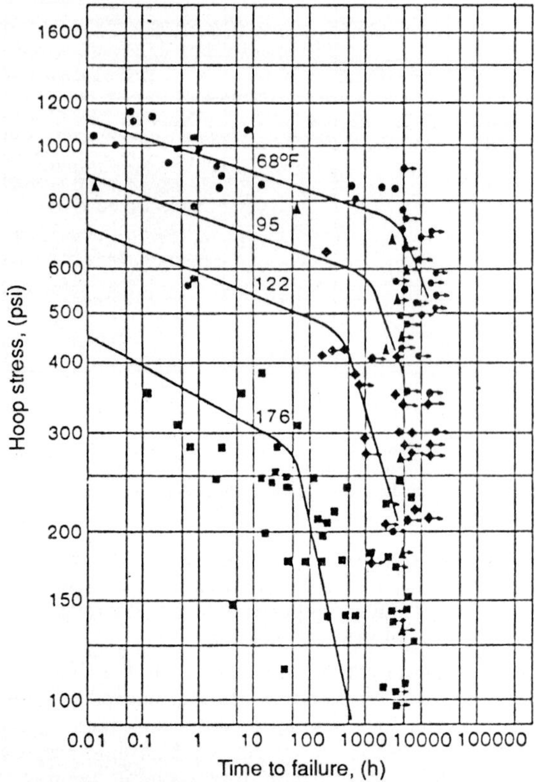

Arrows indicate test in progress.

FIGURE D1.9 Effect of temperature on the time required to reach downturns caused by slow crack growth. This material does not qualify for pressure service.

in actual service for many years, confirm that PE can be suitably protected against oxidative degradation over its long anticipated service life when conveying water, gas, and similar fluids.[7]

Other Materials Downturns in the stress-rupture behavior have also been observed with other thermoplastics. A material's susceptibility to physical and chemical changes that can lead to the downturn in long-term strength depends not only on its chemical nature (that is, PVC, PE, PB, and so on) but also on its polymer architecture (molecular weight, molecular weight distribution, branching, and so on), and quality of additive system. For example, resistance to downturning in PVC depends on molecular weight. For this reason PPI TR-3 imposes a minimum solution viscosity (a measure of molecular weight) requirement on PVC materials for pressure pipe service.

There is no formal general protocol for ensuring that a forecast of long-term strength by method ASTM D 2837 is not subject to a future downturning. The reliability of each forecast has to be judged with consideration of the material's potential physical and chemical aging processes under the anticipated service conditions. A positive source of confidence is the lack of downturns in stress-rupture data that extend much beyond 10,000 h—certain materials have been on test for over 10 years. Another is the absence of a downturn during at least the first 10,000 h of test data obtained at higher test temperatures, say, at least 20°C, than for the intended service. The listings of HDBs issued by PPI have been arrived at in consideration of these various means of enhancing the confidence in forecasting longer-term strength from shorter-time data.

Failure Strain under Constant Load

ASTM method D 2837 imposes a limitation on the HDB: It cannot result in more than 5 percent diametrical pipe expansion at 100,000 h. For all the presently used stress-rated materials, the 5 percent expansion strengths are not the limiting factor. It should be noted, however, that for materials that qualify for pressure service, measurements of diametrical expansion show that at hoop stresses equal to a material's LTHS, the pipe's diametrical expansion at 100,000 h is at least 3 percent. The capacity to safely tolerate such strain under conditions of continuous loading greatly accounts for thermoplastics ductilelike behavior. At the maximum allowable hoop stress resulting from internal working pressure—which by North American practice cannot exceed one-half the material's HDB—there is, generally, ample cushion to safely tolerate the additional strains generated by other loads. And, as discussed later on, even greater latitude is offered by the fact that under conditions where stresses are diminished by stress relaxation, strain limits are generally significantly larger.

However, when pipe failure occurs in the downturn (i.e., the steeper-sloped) regions denoted by stress-rupture diagrams (see Fig. D1.9), it is under much less diametrical expansion than that which occurs in the flatter-sloped regions. In the steeper regions, which denote a shift to a brittlelike failure mechanism, there is not only a more rapid regression of rupture strength with time but also of strain capacity. This has been borne out by performance problems experienced in buried pipe applications with some of the early PE pipe materials which, after some time of successful service, failed even though they were subjected to relatively low strains. Assurance of good long-term strain capacity is another reason for the exclusion from pressure service those materials which have not been stress-rated in accordance with current protocols.

The mechanism that results in downturns yields failures of "brittle" appearance, is one of slow crack development and growth. With plastics the term *brittle* or *brittlelike* often refers to failure by this mechanism. It does not necessarily mean that the material has become brittle and shatters on impact, in the common sense of the word.

Allowable Strain

As explained earlier, when deformation is kept constant during loading time, the load gradually relaxes, reducing the stress intensity which works to generate and propagate cracks. As a result, the long-term strain capacity of a material maintained at constant deformation is higher than if deformation had been achieved over the same time of loading by free creeping. Consequently, allowable strains for constant strain loading conditions can be higher than for constant load service. Also, it is easier for a plastic material to qualify for buried nonpressure service where the principal stresses tend to relax because of pipe deformation and stress relaxation. The relaxing stress gives more leeway in formulating nonpressure pipe compositions—for example, polymers of lower crack resistance can be used, and certain additives, such as fillers, which increase stiffness, can also be incorporated into the composition. However, even though there is more leeway, care has to still be exercised in the selection of a material with sufficient resistance to cracking, particularly when used for underground service where it will be subjected to multiaxial stressing which encourages crack development and growth. Some compositions that seemed suitably crack resistant based on limited shorter-time testing and field experience proved otherwise after longer times—in some cases even after a number of years of successful service. It has proven more difficult to develop a relatively short time test to qualify a material for constant strain loading conditions. Consequently, material requirements in most nonpressure pipe standards have been established by a combination of experience, certain specific long-term evaluations,[10,11] and judgment. These have worked very well. For new materials and for materials not covered by consensus standards, the designer should inquire how recommended limits have been established. Design limits for constant strain loading conditions are referenced by some product specifications and design manuals.

There are a number of standards for nonpressure PE pipe that require pipe to be made only from materials with established HDBs. This is to ensure a sufficient level of resistance against brittlelike failure by slow crack growth to preclude the possibility of premature failure. As noted earlier, slow crack growth can occur in some PEs after long times at relatively low stresses. Even though stress relaxes with time, it may not relax sufficiently to preclude failure by slow crack growth in materials subject to this long-term failure mechanism. Furthermore, there are many sources of persistent tensile stresses in buried pipe such as by pipe ovalization, beam bending, point loading, residual manufacturing stress, expansion-contraction forces, and longitudinal restraints, which could be sufficient to initiate crack development and growth. Sometimes these stresses also have a dynamic component. In view of this, these standards take the prudent course of requiring an HDB to ensure high resistance to slow crack growth and, thereby, maximum strain capacity. The establishment of more easy to measure requirements for slow crack growth resistance for nonpressure PE pipe materials is under evaluation by industry and ASTM.

Fatigue Resistance

As with any other engineering material, repeated cyclic stressing can cause a reduction of strength through fatigue—that is, plastics too have an endurance limit. Also, brittlelike fracture is more likely under cyclic loads than under a similar static load. There is a similarity between the shape of fatigue or *S–N* (stress amplitude versus number of cycles to failure) curves for plastics and those obtained for metals. *S–N* data[12,13] for PVC and PE pipes indicate fatigue sensitivity varies not only from one plastic to another but can be different within the same plastics family depending on molecular structure, formulation, and fabrication quality. Also, stress risers caused by fitting or joint geometry can magnify the effects of fatigue. By suppressing the amplitude of the stress change to the "safe" values indicated by *S–N* curves, long service life can be achieved. A number of design practices prescribe such limits.[14,15,17]

Consequence of Failure Mode

The potential failure mode of a plastic pipe may determine the choice of material and/or the limiting service conditions. Some thermoplastics are subject to failure by shattering, which could present a hazard to personnel when the pipe is either used for the above-ground conveying of pressurized gases or pressure tested with compressed air. For this reason, the Plastics Pipe Institute recommends that no thermoplastic pipe be tested with compressed air or used for above-ground compressed gas service unless it has been evaluated and is specifically recommended by the manufacturer for that service.

Thermoplastic pipe failure can also occur by a rapid crack propagation (RCP) process, whereby axial cracks may travel long distances and at high speeds in a pressurized pipeline. Resistance to RCP has been linked to the absorbed energy in the Charpy impact test, a property highly influenced by temperature. The possibility for the occurrence of RCP increases with increasing pressure and pipe diameter. In certain applications, such as gas distribution, a design objective is to ensure that the selected pipe material in the largest pipe diameter when operating at maximum pressure and minimum temperature will not lead to RCP should the pipe be accidentally damaged. Calculation procedures for evaluating compliance with this objective have been proposed.[16]

CHEMICAL RESISTANCE

A major reason for the broad acceptance of thermoplastics piping is its virtual freedom from attack by ambient moisture and other common corrosives. This applies not only to external surroundings, both below as well as above ground, but also to the materials being conveyed. Being nonconductors, plastics are immune to galvanic or electrochemical corrosion. The electrical activity of electrolytes such as acids, bases, and salts is of no consequence. Plastics piping is generally durable and not subject to a gradual degradation of material and properties comparable to the rusting of steel, or atmospheric corrosion of copper, or pitting of concrete. Painting, wrapping, lining cathodic, and other forms of corrosion protection are not required. Maintenance costs are minimal.

However, as indicated by the abbreviated chemical-resistance guide shown in Table D1.11, for each plastic there are certain substances that can be hostile. Identifying the suitability of a plastic to a specific chemical is inherently more complex than for metals for several reasons. One of these, as previously elaborated, is that even within the same family of plastics the individual members may differ in molecular composition and in the nature and quantity of compounding additives. All of these can affect chemical resistance. Chemical-resistance charts, such as the abbreviated Table D1.11, are not to be considered more than a preliminary guide.

Another reason is that plastics can interact with chemical environments by a number of different mechanisms that can vary in rate, and degree of impairment of performance properties, depending on concentration, temperature, and stress level. The primary mechanisms are as follows:

Chemical attack: A chemical environment can attack certain active sites on the polymer chain, leading to chain scission and, ultimately, to degradation of the polymer. The rate of attack can determine suitability of the plastic—if very slow, or if arrested by the addition of an appropriate stabilizer, long service

TABLE D1.11 Thermoplastic Piping Materials: Chemical-Resistance Guide for Ambient Temperatures*

Attacking chemicals	ABS	PVC I	PVC II	CPVC	PE	PB	PP	PVDF
Inorganic compounds								
Acids, dilute	G	G	L	G	G	G	G	G
Acids, concentrated 80%	L	L	L	G	L	L	L	G
Acids, oxidizing	L	P	P	L	P	P	P	G
Alkalies, dilute	G	G	G	G	G	G	G	G
Alkalies, concentrated 80%	L	G	L	G	G	G	G	G
Gases, acid (HCl and HF), dry	L	L	L	L	G	G	G	G
Gases, acid (HCl and HF), wet	L	G	L	G	G	G	G	G
Gases, ammonia, dry	L	G	L	G	G	G	G	G
Gases, halogens, dry	L	L	L	L	L	L	P	G
Gases, sulfur gases, dry	P	G	L	G	G	L	P	G
Salts, acidic	G	G	G	G	G	G	G	G
Salts, basic	G	G	G	G	G	G	G	G
Salts, neutral	G	G	G	G	G	G	G	G
Salts, oxidizing	L	L	L	L	G	G	G	G
Organic compounds								
Acids	G	G	G	L	G	G	G	G
Acid anhydrides	L	L	L	P	L	L	L	L
Alcohols, glycols	L	G	L	G	L†	G	G	G
Esters, ethers, ketones	P	P	P	P	L	L	L	L
Hydrocarbons, aliphatic	L	L	L	G	L	L	L	G
Hydrocarbons, aromatic	P	P	P	L	P	P	P	G
Hydrocarbons, halogenated	L	L	L	L	P	P	P	L
Natural gas (fuel)	G	G	G	G	G	G	G	G
Mineral oil	G†	G	G	G	L†	G	G	G
Oils, animals and vegetable	G†	G	G	G	L†	G	G	G
Synthetic gas (fuel)	L	L	L	L	L	L	L	G

*G, good; P, poor; L, limited knowledge. These ratings are only for general guidance. For determination of actual chemical resistance under anticipated conditions, more detailed information should be consulted.

†Stress crack resistant grade should be used.

life can be realized. Strong oxidizers will attack many plastics. This is why antioxidants are generally included in the stabilization "package" of many thermoplastic compounds.

Solvation: The absorption by a plastic of an organic solvent is called *solvation.* Its effect may range from a slight swelling and softening, with minor effect on properties, to a complete solution. For example, the solvent cementing of ABS and PVC is based on solvation.

Plasticization: When a liquid hydrocarbon is an imperfect solvent, that is when it is miscible with a polymer but unable to dissolve it, plasticization may result. The plasticizing effect can vary over a wide range, depending on the relative miscibilities. Plasticization can significantly compromise mechanical properties, including creep-rupture behavior.

Environmental Stress-Cracking: Under this mechanism, fracture of the plastic will occur after some time in response to a synergistic action of stress and environment. Stress-cracking agents tend to be strong surface wetting liquids, such as detergents, surfactants, glycols, and alcohols, which, by some mechanism not yet fully identified, facilitate the development and growth of microcracks under stress. Susceptibility to stress-cracking can be mitigated by selecting stress-crack-resistant grades of material.

In all the above mechanisms, stress, reagent concentration, and time are important variables. In addition, temperature has a profound effect on the results. Raising the temperature raises the reaction rates rapidly. It also causes a polymer to expand and become more penetrable, more permeable, and more soluble. Forecasting performance at higher temperatures from lower-temperature results can, therefore, be risky.

Because chemical resistance can be limited by any of the above mechanisms and affected by many variables, shorter-time "soak" tests cannot reliably predict actual performance. Most chemical-resistance charts are based on such information. They should be considered only as an initial guide. A more effective method of evaluating the effects of chemicals, particularly for pressure piping, is creep rupture with the specimen immersed in or in contact with the test environment. A creep-rupture test method specifically designed for evaluating the chemical resistance of plastics pipe is the subject of ISO Standard ISO 8584, "Thermoplastic Pipes for Industrial Applications Under Pressure—Determination of the Chemical Resistance Factor and of the Basic Stress." Chemical resistance obtained under this or similar methods is reported by some manufacturers. Successful previews in the same or similar service are also reliable indicators. Lacking this, the actual chemical resistance will be best established by actual service testing. Manufacturers and trade associations can assist in obtaining this information. An advantage of service testing is that it is sure to include some minor chemical component, such as a surfactant, which could influence the final result.

Effect on Fluids Being Conveyed

Plastic pipe materials are generally inert to the materials being transported. Because their composition includes minimal or no quantities of materials which can produce dissolved ions, plastics pipes are often used for the transport of pure materials, including deionized water. For food service there are pipes available that have been made from materials approved by the Federal Food and Drug Administration.

Plastic pipe resins are also neutral to potable water. However, because it is possible to formulate pipe with certain ingredients, such as stabilizers, catalysts, or modifiers, that could leach substances adversely impacting on water quality, most standards and codes require that pipe intended for this service meet the requirements of National Sanitation Foundation Standard 61, "Drinking Water System Components—Health Effects (October 1988)."

DESIGN AND INSTALLATION

As discussed earlier, thermoplastics are viscoelastic materials; thus they exhibit a profoundly different stress-strain and stress-rupture response than do elastic materials. Nonetheless, elastic equations used for other types of piping are frequently applicable to thermoplastics piping, provided that the latter's engineering behavior is represented through appropriately derived values of apparent modulus and strength. Since these properties are greatly influenced by the history of the material's exposure to stress as well as by temperature and environment, proper use of traditional elastic equations requires appropriately established, or estimated, property values. Long-term strength values for certain conditions (such as for water at 73°F) are available, and in some cases (i.e., maximum recommended hydrostatic design stress), they may even be part of the product standard. In addition, some codes and suggested design protocols either list or give procedures for arriving at appropriate values of strength, stiffness, and allowable strain. As this is a still developing technology, not all properties for all materials are yet available in a standardized basis. However, for the more frequently used materials, sufficient information for the majority of applications is either available or may be adequately estimated.

While the viscoelasticity of thermoplastics somewhat complicates the process of selecting appropriate material constants, materials with high strain capacity help to facilitate design. Under a great many conditions thermoplastics display a ductilelike behavior: They are able to deform significantly before fracture. This behavior helps to redistribute stresses and preclude failure by localized stress intensification which could initiate cracking in brittlelike materials. Within certain limits, design can be based on average stress; localized stress concentrations may be ignored. Also, as previously pointed out, the strain capacity of thermoplastics under constant strain (where stresses can gradually decrease through stress relaxation) is often significantly greater than that under constant load (where stresses intensify as the material deforms). Allowable strain limits under constant strain can, therefore, often be greater than fracture strains for sustained loading. For example, several investigations of PVC and PE pipes subjected to constant deflections over long periods of time show that the materials did not fail at strains as high as 5 to 10 percent.[10,11] These same strains correspond to relatively short lifetimes when the materials were subject to constant load.

One simplification commonly employed in the North American design of flexible (i.e., can undergo significant permanent deformation without cracking) buried plastic pipes assumes that internal pressure (constant load) and external loading (resulting largely in pipe bending stresses relieved by both pipe deformation and stress relaxation) are acting independently—that is, the pipe wall thickness is chosen on the basis of internal fluid pressure and then a separate analysis is made to make sure that the pipe is sufficiently strong and structurally stable under the external loads acting alone.[14,15] In effect, localized fiber stresses due to bending are neglected provided that the pipe is installed in accordance with recommended practice. For cases where localized strain may limit design, there are available more rigorous design protocols based on a combined loading analysis.[5]

There are many potential factors that could cause a material with apparent high strain capacity to shift from a ductilelike to a brittlelike state and fail at lower than expected strains. These can include: temperature, environment, duration of loading, nature of loading (unrelieved or relieved by stress relaxation), stress triaxiality, fatigue, scale factors (such as wall thickness), damage (cuts, gouges), stored energy in system, material imperfections (voids, contaminants), polymer aging, and chemical attack. Judging from the good track record, these influences, although difficult to quantify, appear to have been adequately considered for typical uses by standards, codes, and installation practices. In addition, material requirements and improvements have been evolving to provide more durability in pipe products. Furthermore, design practices give recognition to the effects on strength and stiffness and other long-term engineering properties by time, temperature, and environment. Finally, installation recommendations address many of the unique characteristic of plastics that can affect structural integrity and durability. To be sure, the state of the art of thermoplastics piping is still evolving, and considerable work yet remains to be accomplished to further define materials properties and performance limits. Such work is ongoing, and for the latest information the reader is encouraged to refer to the growing literature, particularly to the papers presented at the various conferences that address plastics piping technology.

As demonstrated by the successful record of experience, sufficient information is available to successfully support proper application of thermoplastics over a very broad range of engineering uses. To best realize the performance potential of thermoplastics piping, the user should base materials selection and utilize design and installation practices on information, such as standards and recommended practices, which has been developed under the technical scrutiny of the consensus process of an established professional society or technical association, as well as upon the recommendations and reports issued by industry and independent sources. A number of such references are provided at the end of this chapter.

The following material is an introduction to some of the more basic aspects of the design and installation of thermoplastics piping. More detailed recommendations suitable to a particular product and situation should be sought and followed.

COMMON CONSIDERATIONS

Internal Hydraulic Pressure

Thermoplastic pipe is pressure rated by means of the ISO equation (D1.1). When rewritten to solve for pressure, this equation takes the form:

$$PR = 2(HDS) \times \frac{t}{D_m} \qquad \text{(D1.2)}$$

where PR = pipe pressure rating, psi
t = minimum wall thickness, in
D_m = mean diameter, in
HDS = $HDB \times DF$, where DF is the pipe design factor

Values of *HDS* for water at 73°F are specified by most ASTM and other standards covering water and gas applications. The *HDS* for water is generally established by multiplying the material's *HDB* by a design factor (*DF*) of 0.5. For gas pipe a *DF* of 0.32 is prescribed by the federal code. A *DF* smaller than 0.5 may also be used in applications where greater compensation for other factors (e.g., surges or temperature) is advisable, or where the fluid conveyed may have some effect on the pipe material properties.

By assuming that a pipe's outside diameter is equal to $D_m + t$, the above equation takes the following form:

$$PR = \frac{2(HDS)}{DR - 1} \tag{D1.3}$$

where DR = ratio of average outside pipe diameter to minimum wall thickness

This equation is used to compute the pressure rating of DR dimensioned pipe.

There is no equivalent design system for establishing the pressure rating of fittings. Some fitting standards require that fittings have a short-term burst strength, and in some cases a 1000-h strength, that is not less than that of the pipe for which the fitting is intended. This is not always sufficient to ensure that fittings will have long-term strengths that are comparable to that of the pipe. Because of their geometry, the stresses generated by pressure on a fitting body are more complex and not amenable to easy calculations. Furthermore, such complex stress fields tend to result in a faster regression of strength with duration of loading than occurs in the simple cylindrical pipe shape. For these reasons, even if a fitting matches the short-term strength of a pipe, it may become the weak link over the long term. This is more likely the case where there are cyclic pressures that can weaken the material by fatigue, usually at fitting stress risers.

Fittings are also sometimes made from a lower molecular weight or better lubricated compound in order to facilitate molding. Such compounds may have somewhat weaker creep-rupture and fatigue characteristics which may need to be compensated by thicker walls. For these reasons the designer should not presume that a fitting labeled with a certain schedule (say, Schedule 40 or 80) will have the same pressure rating as the pipe of the same designation. Suggested maximum pressure ratings for PVC fittings Schedule 40 and 80 have been published[17] (see Table D1.12), and these are significantly lower than for pipe. Because of differences in mold design and other factors, it is best for the designer to consult with the fitting manufacturer for fitting pressure-rating recommendations.

Surge Pressure

Transient and regularly recurring surge pressures may cause damage to pipe and fittings by either of two possible effects: The surge exceeds the short-term fracture strength of pipe or one of the components; or—and this is the more likely possibility—the repetitive changes in pressure exceed the fatigue endurance limit. Transient, or water hammer, surges result from sudden changes in velocity. The pressure rise caused by the velocity change can be calculated by the same equations used for calculating the effects of water hammer in other pipes. The only difference is that with plastics the appropriate material modulus is that for the condition of dynamic, instantaneous response (about 150,000 psi and 450,000 psi for high-density PE and PVC, respectively, at 73°F).

Because of the lower stiffness of plastics, the surge pressure rise that occurs from water hammer is significantly lower than for metallic piping. For example, the surge pressure rise for each foot-per-second change in flow velocity is from about 16 to 20 psi and 8 to 12 psi for PVC and PE, respectively, at 73°F. The exact value depends on pipe wall thickness—the thicker the wall, the larger the pressure rise. (It should be noted that thermoplastic pipes have a certain ability to withstand short-term pressures exceeding their pressure class. This is due to the

TABLE D1.12 Suggested Maximum Working Pressures for Water for 73°F for Schedule 40 and Schedule 80 PVC Fittings*

	Schedule 40		Schedule 80	
Nominal size	Burst pressure	Working pressure	Burst pressure	Working pressure
½	1910	358	2720	509
¾	1540	289	2200	413
1	1440	270	2020	378
1¼	1180	221	1660	312
1½	1060	198	1510	282
2	890	166	1290	243
2½	970	182	1360	255
3	840	158	1200	225
3½	770	144	1110	207
4	710	133	1040	194
5	620	117	930	173
6	560	106	890	167
8	500	93	790	148
10	450	84	750	140
12	420	79	730	137

*To be used only as a general guide. Actual allowable working pressures may vary widely with field conditions, particularly with surge pressures. Manufacturer should be consulted for recommendations.

Source: See Ref. 17.

stress versus time-to-failure characteristics of the pipe material. However, if the short-term pressure rise exceeds short-term burst strength, failure will result.)

Entrapped air in a pipeline can lead to the development of sudden shifts of water column at very high velocity, sometimes high enough to fracture the pipe. For this reason, when plastic piping systems are first filled with water for operation or testing, they should be filled carefully, and relatively slowly, to minimize air entrapment. Air should be vented from the high points before the system is pressurized. In addition, other precautions should be taken—such as carefully laying pipe to grade or using air vent–vacuum relief lines—to minimize the entrapment of air in operating pipelines.

When there exists a frequently recurring pressure surge of significant amplitude—say, over 25 percent of operating pressure—the designer should evaluate the piping for adequacy of fatigue endurance. The resistance to fatigue varies from material to material. PE pipe is quite tolerant; for example, modern materials can withstand frequent surging up to one-half of the pipe pressure rating even when the pipe is operating at its full rating based on only static pressure. In the case of PVC pipe, the following equation has been proposed[12] for estimating the maximum total hoop stress, due to both static and cyclic pressure, that PVC pipe can tolerate as a function of the total number of cycles:

$$S' = \left(\frac{5.05 \times 10^{21}}{C'}\right)^{0.204} \tag{D1.4}$$

where S' = maximum allowable total hoop stress, psi (no safety factor)
C' = total number of cycles

Resistance to External Pressure

The performance of flexible pipe with thin walls that is made from materials with low modulus of elasticity can sometimes be limited by buckling under external pressure which can result from external hydrostatic loading, from internal negative pressure, from the temporary vacuum that may accompany pressure surging, or from a combination of these. The buckling resistance of plastic pipes may be estimated using the following adaptation of the elastic buckling equation for thin tubes:

$$P_c = \left(\frac{24EI}{1 - \nu^2}\right) \times D_m^3 \times C \tag{D1.5}$$

where P_c = critical buckling pressure, psi
E = apparent modulus of elasticity (for short-term loading conditions, use the values of E and ν as obtained from short-term tensile tests; for long-term loading, appropriate values as determined from long-term loading tests should be employed.)
I = pipe wall moment of inertia
ν = Poisson's ratio (approximately 0.35 to 0.45 for long-term loading)
D_m = mean diameter, in = inside diameter + 2 (wall centroid)
C = ovality correction factor, $(r_o/r_i)^3$, where r_i is the major radius of curvature of the ovalized pipe, and r_o is the radius assuming no ovalization

For pipe of solid wall construction, for which $I = t^3/12$, the above equation is usually expressed as follows:

$$P_c = \left(\frac{2E}{1 - \nu^2}\right) \times \left(\frac{t}{D_m}\right)^3 \times C \tag{D1.6}$$

where t = pipe wall thickness, in

According to this equation, pipe made to a constant ratio of diameter to wall thickness has the same resistance to hydraulic collapse independent of pipe diameter.

Temperature Effects

As the system temperature increases, thermoplastics piping becomes more ductile, increases in impact strength, and decreases in short- and long-term strength. The opposite effects take place with decreasing temperature. The exact effect depends not only on material class but the specific composition. For example, there are PEs suitable for service up to a high of 160°F, whereas other PEs might only have sufficient strength through about 120°F. In the case of fittings, wall thickness and product design also bear influence on the effects of temperature on strength.

The best way to determine the effect of temperature on long-term strength is through stress-rupture testing. PPI TR-4, "Recommended Hydrostatic Strengths and Design Stresses for Thermoplastic Pipe and Fittings Compounds," lists recommended *HDB*s for various commercial grade thermoplastics for temperatures up to 200°F. Table D1.13 lists approximate temperature derating factors for some of the more commonly used materials.

Because of its effect on stiffness, increasing temperature also lowers the col-

TABLE D1.13 Thermoplastics Pipe: Approximate Pipe Derating Factors (F_T) for Elevated Temperatures*

Temperature	PE	PB	PVC, Type I	CPVC	PVDF
70	1.0	1.0	1.0	1.0	1.0
80	0.95	0.97	0.88	—	0.93
90	0.90	0.92	0.75	—	0.87
100	0.80	0.86	0.62	0.78	0.82
110	0.75	0.82	0.50	—	0.76
120	0.70	0.77	0.40	0.65	0.71
130	0.50	0.72	0.30	—	0.65
140	0.40	0.68	0.22	0.50	0.61
150	0.20	0.62	NR	—	0.57
160	NR	0.58	NR	0.40	0.54
180	NR	0.48	NR	0.25	0.47
200	NR	0.40	NR	0.20	0.47
210	NR	NR	NR	NR	0.41
220	NR	NR	NR	NR	0.38
250	NR	NR	NR	NR	0.35
280	NR	NR	NR	NR	0.28

*Check with manufacturer for recommendations for the specific pipe material under consideration. Fitting derating may occur at different rate.

lapse resistance of plastic pipe. As evident from inspection of Eq. (D1.5), this effect is in direct proportion to the change in the apparent modulus of elasticity. This modulus changes with temperature at a rate roughly parallel to the strength derating factors given in Table D1.13.

Other principal effects to be considered in piping design and installation are those resulting from thermoplastics' high coefficient of expansion-contraction. Some potential consequences to be considered include:

1. Piping that is installed hot may cool sufficiently after installation to generate substantial tensile forces—the final connection should be made after the pipe has equilibrated to ambient, or desired, temperature.
2. Unrestrained pipe may shrink enough to pull out from elastomeric gasket or compression joints. The pipe should be adequately restrained by the use of anchors, or the fitting should be designed to either resist pull-out forces or to tolerate the maximum anticipated pipe movement.
3. Piping exposed to cyclic temperature changes may be susceptible to fatigue damage at points subject to repetitive bending.
4. Pipe installed at depressed temperatures may buckle if the compression forces developed on subsequent heating are not adequately relieved.

CONSIDERATIONS FOR ABOVE-GROUND USES

Thermoplastic piping systems in above-ground service must be properly supported to avoid excessive stresses and sagging. Valves and other heavy piping components should be individually supported. Piping should be located, or protected, to avoid mechanical damage. The piping layout should have sufficient flexibility or other means of mitigating excessive bending and axial stresses and fatigue effects induced by repetitive expansion-contraction.

Supports and Anchors

Horizontal runs require the use of hangers that are carefully aligned and are free of rough and sharp edges. Many hangers designed for metal pipe are suitable for thermoplastic pipe as well. These include the shoe, clamp clevis, sling, and roller types. To preclude high localized support pressures, it is generally advisable to modify the hangers by increasing the bearing area by inserting a protective sleeve of plastic between the pipe and the hanger.

Vertical lines must also be supported at intervals to reduce loads on the lower fittings. This can be accomplished by using riser clamps or double bolt clamps located just below a coupling or other fitting to support the pipe. When so located, these provide the necessary support without excessive compression of the pipe.

Anchors are used in thermoplastic piping systems as fixed points from which to direct expansion-contraction and other movements in a defined direction. Their placement is selected to prevent overloading of the piping, particularly at changes of direction where pipe movement could generate excessive bending and axial stresses. Anchors should be placed as close to elbows and tees as possible. Guides are used to allow axial motion while preventing transverse movement. Both anchors and guides may be used in the control of expansion and contraction of pipelines and should be located to prevent overstressing of the pipe. A "flexibility" analysis is used to determine suitable arrangements for anchors and guides.

Support Spacing

Support spacing requirements are computed using the same beam deflection equations used for metal piping but on the basis of either the material's long-term maximum allowable stress (where stress may be the limiting factor) or the long-term apparent modulus (where beam deflection may be the limitation) for the maximum anticipated service temperature. Maximum beam deflection, or sag, is frequently the controlling factor. Typical support spacing recommendations are presented in Table D1.14.

Expansion-Contraction

There are several methods used for controlling or compensating for axial and bending stresses caused by thermal expansion. Piping runs may include changes in direction which will allow the thermally induced length changes to be taken up safely. Where this method is employed, the pipe must be able to float except at anchor points.

When the piping layout does not include sufficient changes in direction, appropriate expansion loops or offsets have to be provided. The size of the loops and offsets depend on the design (see Fig. D1.10), and the change in length of pipe that has

TABLE D1.14 Typical Recommended Maximum Support Spacing, in Feet, for Thermoplastic Pipe for Continuous Spans and for Uninsulated Lines Conveying Fluids of Specific Gravity up to 1.35

Pipe dimension	PVC			CPVC				PVDF				PP			
Nominal diam., in	60°F	100°F	140°F	60°F	100°F	140°F	180°F	80°F	100°F	140°F	160°F*	60°F	100°F	140°F	180°F
Wall Schedule 40															
½	4½	4	2½	5	4½	4	2½	3¾	3½	2		1¾	1¾	1½	1¼
¾	5	4	2½	5½	5	4	2½	4	3¾	2½		2	2	1¾	1¾
1	5½	4½	2½	6	5½	4½	2½	4¼	4	2½		2	2	2	1¾
1¼	5½	5	3	6	5½	5	3	—	—	—		2½	2¼	2	2
1½	6	5	3	6½	6	5	3	4½	4¼	2½		2½	2½	2¼	2
2	6	5	3	6½	6	5	3	4½	4½	2¾		3	2¾	2½	2¼
3	7	6	3½	8	7	6	3½					3½	2¾	3	2¾
4	7½	6½	4	8½	7½	6½	4					4	3¾	3½	3
6	8½	7½	4½	9½	8½	7½	4½								
8	9	8	4½												
Wall Schedule 80															
½	5	4½	2½	5½	5	4½	2½	4½	4½	2½		2	2	2	1½
¾	5½	4½	2½	6	5½	4½	2½	4½	4½	3		2½	2½	2¼	2
1	6	5	3	6½	6	5	3	5	4¾	3		2½	2½	2¼	2
1¼	—	—	—	—	—	—	—	—	—	—		3	2¾	2½	2½
1½	6½	5½	3½	7	6½	5½	3½	5½	5	3		3	3	2¾	2½
2	7	6	3½	7½	7	6	3½	5½	5¼	3		3½	3¼	3	2¾
3	8	7	4	9	8	7	4					4	4	3½	3½
4	9	7½	4½	10	9	7½	4½					4½	4½	4	3½
6	10	9	5	11	10	9	5								
8	11	9½	5½												

*Continuous support recommended.

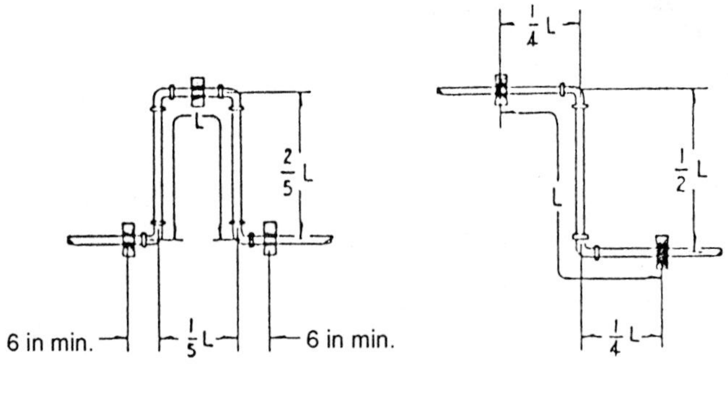

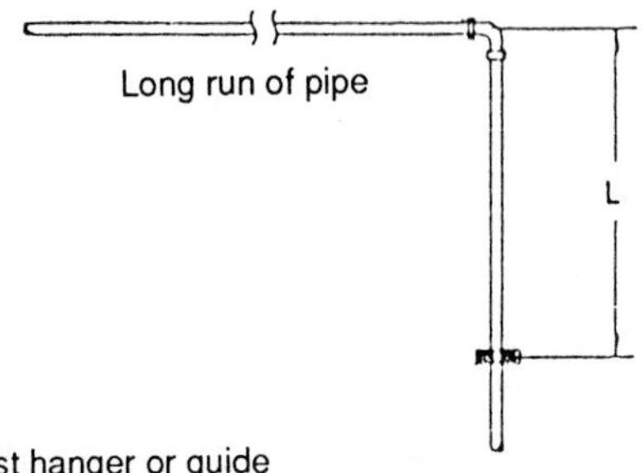

FIGURE D1.10 Expansion loops relieve thermal stresses by transforming them to bending stresses. The minimum loop strength L is generally determined by the maximum allowable bending stress or strain.

to be accommodated. The dimensions of loops and offsets are calculated using the equation, as follows, for cantilevered beams loaded at one end[18]:

$$L = 1.5 \times \left(\frac{E}{S}\right)^{1/2} \times D_o(\Delta L)^{1/2} \tag{D1.7}$$

where L = loop length, in
E = modulus of elasticity at the working temperature, psi
S = maximum allowable stress at the working temperature, psi
D_o = outside pipe diameter, in
ΔL = change in length due to temperature change, in

The following, somewhat more conservative equation, which has been developed under the assumption that a pipe fitting joining to cantilevered sections of pipe restrains rotation of the pipe ends, is also sometimes used[19]:

$$L = 3 \times \left(\frac{E}{S}\right)^{1/2} \times D_o(\Delta L)^{1/2} \tag{D1.8}$$

For both equations, the convention is to use the short-term apparent modulus in combination with the maximum allowable long-term working stress (often the same value is used as the material's hydrostatic design stress). Since expansion-contraction does not occur instantaneously and since the working stress applies to a condition of constant load where, unlike this situation, there is no stress reduction due to stress relaxation, this combination is conservative. This helps to compensate for other factors, such as fatigue effects caused by repetitive expansion-contraction. Since for most plastics E and S vary with temperature at approximately the same rate, calculations based on ambient temperature values are generally also appropriate for a fairly wide operating temperature range.

Because the fitting restrains the pipe, a separate check should be made with the manufacturer regarding the fittings' capacity to absorb expansion-contraction stresses and bending moments.

Expansion joints of bellows and piston designs are available and sometimes used. However, piston expansion joints for pressure applications are generally expensive. Proper alignment of piston joints is critical to prevent binding. Bellows-type joints can accept some lateral movement.

CONSIDERATIONS FOR BELOW-GROUND USES

Design and installation of thermoplastic pipe recognizes its "flexible" conduit behavior. As noted earlier, the word *flexible* primarily signifies that the pipe has the capacity to sustain significant deflection without failure. Conduits that are strong and stiff and that fail at low deformations are classified as "rigid." Both systems must be designed and installed to avoid deformations at any point along the system which could result in cracking; however, flexible pipe gives greater leeway to this requirement.

Buried flexible conduits have the virtue of being able to deflect downward vertically and outward horizontally under vertical earth loads thereby mobilizing passive lateral support for the pipe, in turn precluding significant further downward deflection. Flexibility also works to mobilize soil arching and so reduces the net vertical load acting on the conduit; that is, as the pipe moves downward in reaction to the load, a greater portion of this load is shed off to the soil columns on the side of the pipe. Thus, flexible pipe and surrounding soil interact and behave as a structural system. In this system, pipe deflection is controlled more by soil stiffness than by pipe flexural stiffness, and the soil arching characteristics bear great influence on the system's load carrying capacity. Flexible pipe properly installed in stable soils can resist very substantial loads.

Since the pipe and the soil system interact, design and installation of buried flexible pipe must always consider both the pipe and the soil around it. If a designer allows different pipes in a specification, he or she should consider, for each type of pipe, both the backfill specification and installation quality requirements. One of the advantages of flexible pipe is that the quality of installation can readily be checked via a deflection test after installation is complete. A particular benefit of most thermoplastic pipe is its high strainability which allows it to deform considerably in generating soil support. To take economic advantage of this benefit, many of the newer larger-diameter thermoplastic pipes are offered with relatively low wall stiffness, which requires that careful attention be given to proper design and installation in order to ensure durable and stable performance. Since the soil and the pipe must always work together to constitute a system, the designer has to consider both when evaluating alternate pipe materials.

Wall Stiffness

Flexible conduits need to have sufficient wall stiffness to resist excessive deformation during handling and installation as well as to maintain shape and structural stability in the final pipe-soil system under the imposed loads anticipated during the installation design life. The ultimate shape (i.e., deflection) of a flexible pipe in service is the probabilistic sum of any original pipe out-of-roundness, deformations caused during handling-installation, and deflections resulting from the response of the installed soil-pipe system to external loads. Many equations relating the postinstallation deformation response of flexible conduits [for example, see Eqs. (D1.5) and (D1.13)] include the term EI/r_m^3. This term is commonly referred to as *ring stiffness.* A standard test method, ASTM D 2412, "External Loading Properties of Plastic Pipe by Parallel Plate Loading," is used to determine a pipe's short-term ring stiffness. The method is based on the following relationship, which has been derived for elastic conduits:

$$\frac{F}{\Delta y} = \frac{EI}{0.149 r_m^3} \tag{D1.9}$$

where F = the parallel plate, short-term load applied to the pipe to produce a given percentage deflection, generally 5 percent, lb

Δy = measured change, or deflection, of the inside diameter in the direction of load application, in

r_m = the midwall radius of the pipe (determined by subtracting the average wall thickness from the outside diameter and dividing the difference by 2), in

E = the apparent modulus of elasticity under condition of test, psi

I = the moment of inertia of the pipe wall, in^4/in

The quantity F/y is called the *pipe stiffness.* Minimum short-term pipe stiffness requirements are included in most standards for thermoplastic pipe intended for buried applications. Many such documents offer more than one series of different standard pipe stiffness categories. Categorizing pipes in accordance with pipe stiffness has proven useful to designers and installers. This parameter is an important indicator of a flexible conduit's deformation response, particularly under service loads. However, anticipated performance comparisons based on the pipe stiffness alone cannot always be made for a variety of reasons including the following:

- Pipe stiffness is a short-term property, which, because of differences in viscoelastic behavior from one material to the next, does not always bear the same relationship to a pipe's long-term stiffness.
- Service deflections are controlled more by soil stiffness than by pipe flexural stiffness, which, among other things, means that quality of installation and nature of soil are often more important to controlling ultimate deflection than is pipe stiffness.
- The same pipe stiffness for two pipes of widely different diameters does not mean that the pipes have an equivalent resistance to deformation by handling and installation (including compaction) loads; in a constant pipe stiffness series, as pipe diameter increases, so does the pipe's relative resistance to deformation.

To illustrate the last point, consider the example of two pipes, one 6-in and the other 60-in, each with a pipe stiffness of 60 lb/in · in. A 60-lb/in of pipe length parallel plate load applied on either of the two pipes will produce the same decrease in vertical diameter, 1 in. Clearly, this deformation is much more significant to the 6-in than to the 60-in pipe. Because of this, certain larger-diameter standards categorize pipe in accordance with a ring stiffness constant (*RSC*). The *RSC* is defined as the parallel plate load in pounds per foot of pipe length, which causes a 1 percent reduction in diameter when measured at 3 percent deflection. Essentially the same test, ASTM D 2412, as used to measure pipe stiffness, is used to measure *RSC*. The main difference is the form in which the results are expressed. Pipe with equal *RSC* will undergo equal percent deflection under equal load. *RSC* is related to pipe properties by the following relationship:

$$RSC = \frac{6.44\,EI}{D_m^2} \tag{D1.10}$$

where D_m = pipe mean diameter and all other symbols as previously designed

The *RSC* concept is an adaptation of the flexibility factor (*FF*), as follows, which is used by the American Association of State Highway and Transportation Officials (AASHTO) to classify the handling and installation flexibility of another flexible conduit, corrugated metal pipe and culvert:

$$FF = \frac{D_m^2}{EI} \tag{D1.11}$$

where D_m = the mean pipe diameter

The *FF* is now used by AASHTO to classify thermoplastic pipes and culverts.

The classification of larger-diameter thermoplastic pipe based on the handling and installation flexibility concept recognizes that the ultimate deformation of installed conduit is greatly influenced by the conduit's response to all the short-term loads experienced during handling, placement of bedding, compaction, and other installation operations. This generally means that as pipe diameter increases, less pipe stiffness (as measured by ASTM D 2412) is required for satisfactory handling and installation. However, after selecting a conduit of adequate wall stiffness based on handling and installation concerns, the designer still has to check for structural adequacy under the anticipated lifetime service loads.

Thermoplastic pipes and conduit for nonpressure buried applications is available in a broad range of wall stiffnesses. As discussed previously, the larger the pipe size, generally the lower the pipe stiffness. Up to about 4-in-diameter pipe stiffness is generally above 50 and can be over 200 lb/in · in. In the midsizes, through about 18 in, pipe stiffness ranges from about 30 to 60 lb/in · in. Above this size, values tend to decrease, down to 8 psi for the largest sizes.

AASHTO recommends a maximum flexibility factor (*FF*) of 9.5×10^{-2} in/lb for both PVC and PE; this value was originally established for corrugated aluminum conduit. This is equivalent to an *RSC* of about 65 lb/ft × percent deflection. *RSC*s of commercially available larger diameter (18 through 96 in) PE pipes range from 40 to 120.

Loads on Buried Flexible Pipes

The load acting on a buried pipe consists of dead load and surcharge load. The *dead load* is the permanent load from the weight of soil and pavement above the pipe and sometimes, from any surcharge loads applied at the ground surface. *Surcharge loads* may, or may not, be permanent. Surface-applied wheel loads are called *live loads.*

As flexible pipe undergoes vertical deflection, it encourages arching over the pipe. Part of the weight of the backfill soil is transferred onto the trench walls, thereby reducing the load directly acting on the pipe. In 1930 Marston published a design method for determining loads on buried pipe that accounts for arching. His method is widely accepted and can be found in *ASCE Manual No. 60.*[20] It has been shown that the Marston load when used with an appropriate coefficient of earth pressure results in a conservative design load that accounts for load reduction due to arching.[21]

A more conservative approach, which is frequently used for the typical trench burial depths, is to assume no arching. Under this assumption, the pipe sees the full weight of the prism of soil directly above. The prism load can be simply calculated by the following equation:

$$P_s = WH \tag{D1.12}$$

where P_s = vertical soil pressure, lb/in^2
W = unit weight of soil, lb/in^3
H = height of soil mass above the top of the pipe, in

Surcharge, as well as traffic, railway, aircraft, and other live loads, are generally estimated using the same methods and equations employed for other conduits, such as presented in *ASCE Manual No. 60*[20] and in the AASHTO manual.[6]

The loadings to which a pipe-soil system is subjected can neither exceed the performance limitations of the pipe nor the design limitations established for the system. The primary performance limits for thermoplastic pipe are wall cracking, excessive ring deflection, wall crushing, and wall buckling.

Wall Cracking. When the bending strains exceed the strain capacity of the pipe material, wall cracking will occur. However, since high strain capacity is a major attribute of thermoplastics, this limitation is seldom a design consideration. Work is underway by industry and ASTM to develop test methods to evaluate limits of wall cracking under long-term loading.

Excessive Ring Deflection. While a measure of deflection is essential to activate the soil-pipe system, excessive deflection is clearly undesirable. Too large a deflection could lead to a reversal in curvature over the top of the pipe, imperiling structural stability with possible pipe buckling. However, reversal of curvature is typically not a risk unless there is a reduction of at least 20 percent in vertical diameter. A more typical concern is the impairing of pipe flow capacity, ability to clean the system with mechanical devices, and retention of sufficient pipe roundness to allow future tap-ins by the application of saddles or similar devices.

To avoid all of these problems, buried installations of thermoplastic pipe are generally designed and installed to achieve a long-term maximum deflection of from 7.5 to 10 percent. Usually, this is translated to a short-term, or installation deflection, requirement of from 5 to 7.5 percent. Since flexible pipe and the surrounding soil work as a system, compliance to maximum deflection requirements requires proper attention be also given to quality of soil embedment and construction.

The extent to which a flexible pipe will deflect when embedded in a given quality of soil may be estimated by a variety of methods. One of the better-known relationships, sometimes called the *Iowa formula,* was developed for flexible metal conduits at Iowa State University. A modification of this equation is:

$$\frac{\Delta x}{D_i} = \frac{L_D K P}{EI/r_m^3 + 0.061E'} \tag{D1.13}$$

where Δx = horizontal deflection of the pipes, in (For relatively small deflections, the change Δy in vertical diameter of a circular section deforming elliptically is equal to $1.10x$. As an approximation, it is often assumed $\Delta y = \Delta x$.)

D_i = pipe inside diameter prior to loading, in

L_D = deflection lag factor compensating for the time dependence of soil deformation, dimensionless

K = bedding constant which varies with the angle of bedding (i.e., bedding support), dimensionless (The bedding constant ranges from 0.110 for a point support on the bottom of a pipe to 0.083 for full support. For plastic pipe, the typical value is taken as 0.10.)

P = total vertical pressure acting on the pipe, lb/in^2

r_m = pipe radius, in

E = modulus of elasticity of pipe material, lb/in^2

I = moment of inertia of pipe wall per unit of length, in^4/in (For round pipe, $I = t_a^3/12$, in which t_a is the average wall thickness.)

E' = modulus of passive soil resistance, lb/in^2

The modulus of passive soil resistance E' is an empirical measure of the stiffness of the soil in resisting pipe deflection and is dependent on both soil stiffness and pipe radius. Because of this, E' is not a true soil property that can be evaluated by laboratory tests. E' must be obtained empirically by back-calculating it from measured pipe deflections. The most extensive field study to determine E' was conducted by A. Howard of the U.S. Bureau of Reclamation.[22] In this study data were collected from 113 field installations on different types of flexible pipe buried up to 50 ft deep. For this work Howard assumed: A prism load, $K = 0.1$ and $L_D = 1.0$; and, in the case of plastic pipes he computed the term EI/r_m^3 based on the material's short-term apparent modulus. The resultant back-calculated values of E' are tabulated in Table D1.15 in accordance with soil type. This table also identifies soils by the classification system of ASTM D 2321, "Standard Practice for Underground Installation of Thermoplastic Pipe for Sewers and Other Gravity-Flow Applications."

For the ordinary situation, it is seldom necessary to go through the Iowa equation calculation, for when the recommended installation practices in ASTM D 2321 are followed, initial installed deflections can quite readily be held to below 7.5 percent.

Various other methods have been proposed for estimating deflection, for example by Watkins,[23] Gaube,[24] and Brown and Lytton.[25] Recently, more precise approaches to forecasting deflection have been made possible by the application of computer-run finite element analysis programs that involve rational and measurable soil properties as opposed to the empirical E'.[26]

Wall Crushing. Wall crushing can occur when the maximum compressive stress, which generally develops at the 3 and 9 o'clock positions on a pipe, exceeds the material's short- or long-term compressive strength. This situation is generally only of concern with thinner walled pipes under deep burial. A material's limiting

TABLE D1.15 Bureau of Reclamation Values of E' for Iowa Formula for Initial Average Deflection of Pipe

Soil type for pipe embedment material per ASTM D 2321	Soil type description (United Classification System, ASTM D 2487)	E', lb/in^2 for degree of compaction of embedment (proctor density, %)*			
		Dumped	Slight (>85%)	Moderate (85–95%)	High (>95%)
I	Manufactured angular, granular materials (crushed stone or rock, broken coral, cinders, etc.)	1000 (+4%)	3000 (+4%)	3000 (+3%)	3000 (+2%)
II	Coarse-grained soils with little or no fines	NR†	1000 (+4%)	2000 (+3%)	3000 (+2%)
III	Coarse-grained soils with fines	NR	NR	100 (+3%)	2000 (+2%)
IV	Fine-grained soils	NR	NR	NR	NR
V	Organic soils (peats, mulches, clays, etc.)	NR	NR	NR	NR

*Values in parentheses give the approximate limit of deflection beyond the average deflection that is computed by using the given E' values. These limits are for pipe of relatively low stiffness. As pipe stiffness increases, the limit is narrowed.

†NR indicates use not recommended by ASTM D 2321.

long-term compressive strength should be determined by long-term tests. Generally, the compressive strength is significantly higher than the long-term tensile strength. As a conservative approximation, some designers assume that long-term compressive strength is equal to long-term tensile strength.

Wall Buckling. When pipe deformation is restrained by stiffer soil around the pipe, its resistance to buckling [Eq. (D1.5)] is significantly increased. Various equations have been proposed[27] to account for this increase. AWWA Standard C 950, "Standard for Fiberglass Pressure Pipe," introduced an equation which is also used for design of thermoplastic pipe.

INSTALLATION

Concurrent with the development of structural design methods for thermoplastics, there also have developed installation practices dedicated to these materials. Noteworthy among these are ASTM D 2774, "Underground Installation of Thermoplastic Pressure Piping," and ASTM D 2321, "Underground Installation of Thermoplastic Pipe for Sewers and Other Gravity-Flow Applications." A commentary on the installation issues that are critical to the long-term performance of flexible nonpressure plastics pipe has been offered by T. J. McGrath.[28] Installation, as well as design, recommendations are also issued by various professional and trade associations. A number of these references are identified in the following section, "Sources of Additional Information."

NEW DEVELOPMENTS

The inroads that thermoplastics piping has made in fuel gas distribution, sewer, water, agricultural and highway drainage, and in various industrial uses has generated many studies regarding the durability and engineering performance of these materials. Topics of particular recent interest relate to the use of these materials for larger-diameter applications for which certain limits of performance, such as maximum depth of burial, buckling resistance, compressive wall strength, are often design limiting. A reader interested in these topics, as well as in the general state of the art, should consult the proceedings of the following periodically held symposia and conferences:

Proceedings of International Conferences on Pipeline Design and Installation, American Society of Civil Engineers, 345 East 47th Street, New York, NY 10017.

Proceedings of the Symposium on Buried Plastic Pipe Technology, American Society for Testing & Materials, 1916 Race Street, Philadelphia, PA 19103.

Proceedings of the Fuel Gas Plastic Pipe Symposium, American Gas Association, 1515 Wilson Boulevard, Arlington, VA 22209.

Proceedings of the National Conference on Flexible Pipes, Center for Geotechnical and Groundwater Research, Ohio University, Athens, OH 45701.

Proceedings of the International Conferences on Plastics Pipe, Plastics and Rubber Institute, 11 Hobart Place, London, England SW1W OHL.

SOURCES OF ADDITIONAL INFORMATION

Publications Related to Standards

The following publications contain much information that is useful for all applications of plastics piping, particularly with respect to design and installation:

ASME Guide for Gas Transmission and Distribution Piping Systems. Available from American Society of Mechanical Engineers, United Engineering Center, 345 East 47th Street, New York, NY 10017, (212) 705–7722.

AGA Plastic Pipe Manual for Gas Service. Available from American Gas Association, 1515 Wilson Boulevard, Arlington, VA 22209, (703) 841–8454.

Maintenance of Operation of Gas Systems, November, 1970, Army TM5-654; NAVFAC-MO-220; Air Force AFM 91-6. Available from Superintendent of Documents, U.S. Government Printing Office, Washington, D.C. 20402.

Trade Associations

Various trade associations issue reports, manuals, and lists of references on properties and design and installation of plastics piping. A listing of current literature offerings may be obtained by contacting these organizations at the following addresses:

Thermoplastic Pipe (Industrial, Gas Distribution, Sewerage, Water, and General Uses)

The Plastics Pipe Institute, a Division of The Society of the Plastics Industry, Inc., Wayne Interchange Plaza II, 155 Route 46 West, Wayne, NJ, 07470.

Thermoplastics Pipe (Plumbing Applications)

Plastics Pipe & Fittings Association, 800 Roosevelt Road, Building C, Suite 20, Glen Ellyn, IL 60137.

PVC Piping (Water Distribution, Sewerage, and Irrigation)

Uni-Bell PVC Pipe Association, 2655 Villa Creek Drive, Suite 155, Dallas, TX 75234.

Codes

Thermoplastics piping for plumbing, heating, cooling and ventilating, sewer, water, fire protection, gas distribution, and other hazardous materials may be subject to the provisions of a code or other regulation. Nearly all plumbing codes allow plastics piping for certain applications. The major model building and plumbing codes from which most such codes are derived are issued by the following organizations:

BOCA: *National Building Code, BOCA National Mechanical Code, and BOCA National Plumbing Code.* Building Officials and Code Administrators, International, Inc., 4051 West Flossmoor Road, Country Club Hills, IL 60478.

CABO: *One and Two Family Dwelling Code.* Council of American Building Officials, 5203 Leesburg Pike, Falls Church, VA 22041.

IAPMO: *Uniform Plumbing Code.* International Association of Building and Mechanical Officials, 20001 Walnut Drive South, Walnut, CA 91789-2855.

ICBO: *Uniform Building Code* and *Uniform Mechanical Code.* International Conference of Building Officials, 5360 South Workman Mill Road, Whittier, CA 90601.

PHCC: *National Standard Plumbing Code.* National Association of Plumbing-Heating-Cooling Contractors, Post Office Box 6808, Falls Church, VA 22040-6808.

SBCI: *SBCCI Southern Building Code, SBCCI Southern Standard Plumbing Code,* and *SBCCI Southern Standard Mechanical Code.* Southern Building Code Congress International, 900 Montclair Road, Birmingham, AL 35213.

Plastics piping for other applications may also be covered by other codes, such as the following:

American National Standards Institute

ANSI B31.3 Chemical Plant and Petroleum Refinery Piping.

ANSI B31.8 *Gas Transmission and Distribution Piping Systems.*

ANSI Z223 *National Fuel Gas Code.*

Department of Transportation, Hazardous Materials Board, Office of Pipeline Safety Operations

Code of Federal Regulations (CFR), Title 49, Part 192, Transportation of Natural Gas and Other Gas by Pipeline: Minimum Federal Safety Standards.

Code of Federal Regulations (CFR), Title 49, Part 195, Transportation of Liquids by Pipeline, Minimum Federal Safety Standards.

The National Fire Protection Association (Quincy, MA) Model Codes

NFPA 30 *Flammable and Combustible Liquids Codes.*

NFPA 54 *National Fuel Gas Code.*

NFPA 70 *National Electrical Code.* *

NFPA 70A *Electrical Code for One and Two Family Dwellings.*

NFPA 34 *Outdoor Piping.*

**National Electrical Code* is a registered trademark of the National Fire Protection Association, Quincy, MA 02269.

Some standards and various jurisdictions and authorities require that, before a pipe may be used for certain applications, it first must be approved for that use by a recognized, or specifically designated, organization. Organizations and approval programs for plastic pipe include the following:

For Potable Water

NSF International, NSF Building, Post Office Box 1468, Ann Arbor, MI, 48106.

Canadian Standards Association, 178 Rexdale Boulevard, Rexdale, Ontario, Canada, M9W 1R3

For Drain, Waste, and Vent

NSF International and Canadian Standards Association (see above).

For Meat- and Food-Processing Plants

U.S. Department of Agriculture, 14th and Independence S.W., Room 0717 South, Washington, D.C. 20250.

For Underground Fire Protection Systems

Underwriters Laboratories Inc., 333 Pfingsten Road, Northbrook, IL 60062.

Factory Mutual Research Corporation, 1151 Boston-Providence Turnpike, Post Office Box 688, Norwood, MA 02062.

For Underground Gasoline and Petroleum Lines

Underwriters Laboratories Inc. (see above).

REFERENCES

1. Source: *Facts & Figures of the U.S. Plastics Industry,* 1990 Edition, The Society of the Plastics Industry, Inc., Washington, D.C.
2. Bill Bregar, "Special Report: Plastic to Remain Leading Pipe Material," *Plastics News,* February 19, 1990, p. 10.
3. Source: BCC Report P-043N, "The Competitive Pipe Market: Materials, Applications, Directions," Business Communications Co., Inc., Stamford, Connecticut.
4. L. E. Janson, *Plastic Pipes for Water Supply and Sewage Disposal,* published by Neste Chemicals, Stenungsund, Sweden, Stockholm, 1989.
5. F. J. Heger, R. E. Chambers, and A. G. H. Dietz, "Structural Plastics Design Manual," *ASCE Manual of Engineering Practice No. 63,* American Society of Civil Engineers, New York, May 1984.
6. Section 18, Soil-Thermoplastic Pipe Interaction Systems, *Standard Specifications for Highway Bridges,* American Association of State Highway and Transportation Officials (AASHTO), Washington, D.C.
7. Stanley A. Mruk, "The Durability of Polyethylene Piping," *STP 1093, Buried Plastic Pipe Technology,* American Society for Testing and Materials, Philadelphia, October 1990.
8. P. E. O'Donoghue, M. F. Kanninen, C. H. Poplar, and M. M. Mamoun, "A Fracture Mechanic's Assessment of the Battelle Slow Crack Growth Test for Polyethylene Gas Pipe Materials," *Proceedings of the Eleventh Plastic Fuel Gas Pipe Symposium,* October 3–5, 1989, San Francisco, published by the American Gas Association, Arlington, Virginia.
9. X. Lu, R. Qian, and N. Brown, "Notchology—The Effect of the Notching Method on the Slow Crack Growth Failures in a Tough Polyethylene," *Journal of Materials Science,* **26**, 1991, p. 26.
10. L. E. Janson, "Plastic Gravity Sewer Pipes Subject to Constant Strain by Deflection," *Proceedings of the International Conference on Underground Plastic Pipe,* American Society of Civil Engineers, New York, March 1981.
11. A. P. Moser, O. K. Shupe, and R. R. Bishop, "Is PVC Pipe Strain Limited After All These Years?" *STP 1093, Buried Plastic Pipe Technology,* American Society for Testing and Materials, October 1990.
12. H. W. Vinson, "Response of PVC Pipe to Large, Repetitive Pressure Surges," *Proceedings of the International Conference on Underground Plastic Pipe,* American Society of Civil Engineers, New York, March 1981.
13. J. A. Bowman, "The Fatigue Response of Polyvinyl Chloride and Polyethylene Piping Systems," *STP 1093, Buried Plastic Pipe Technology,* American Society for Testing and Materials, October 1990.
14. "PVC Pipe—Design and Installation," *AWWA Manual M23,* American Water Works Association, 1980.
15. "AWWA Committee Report, Design and Installation of Polyethylene Pipe Made in Accordance with C 906," American Water Works Association (AWWA), 1992.
16. M. F. Kanninen, P. E. O'Donoghue, J. W. Cardinal, S. T. Green, R. Curr, and J. G. Williams, "A Fracture Mechanic's Analysis of Rapid Crack Propagation and Arrest in Polyethylene Pipes," *Proceedings of the Eleventh Plastic Fuel Gas Pipe Symposium,* October 3–5, 1989, American Gas Association, Arlington, Virginia.
17. R. D. Bliesner, *Designing, Operating and Maintaining Piping Systems Using PVC Fittings,* published by the PVC Fittings Division of the Irrigation Association, Arlington, Virginia, 1987.
18. R. Hall, "Design and Installation of Above Ground Thermoplastic Piping Systems,"

Managing Corrosion with Plastics, vol. V, National Association of Corrosion Engineers, Houston, 1983.

19. M. W. Kellog Company, *Design of Piping Systems,* Wiley, New York, 1956.
20. *ASCE Manual No. 60,* "Gravity Sanitary Sewer Design and Construction," American Society of Civil Engineers, New York, 1982.
21. L. J. Petroff, "Review of Relationship Between Internal Shear Resistance and Arching in Plastic Pipe Installations, *STP 1093, Buried Plastic Pipe Technology,* American Society for Testing and Materials, October 1990.
22. A. K. Howard, "The USBR Equation for Predicting Flexible Pipe Deflections," *Proceedings of the International Conference on Underground Plastic Pipe,* American Society of Civil Engineers, New York, March 1981.
23. R. K. Watkins, E. Szpak, and W. B. Allman, *Structural Design of PE Pipes Subjected to External Loads,* Engineering Experimental Station, Utah State University, Logan, Utah, 1974.
24. E. Gaube, *Bemessen von Kanalrohren aus PE Hart und PVC Hart,* Kunststoffe, 1977, pp. 353–356.
25. F. A. Brown and R. L. Lytton, "Design Criteria for Buried Flexible Pipe," *Proceedings of Pipeline Materials and Design,* American Society of Civil Engineers, San Francisco, 1984.
26. K. M. Chua and L. J. Petroff, "Predicting Performance of Large Diameter Profile Wall HDPE Pipe," *Proceedings of the Second International Conference on Case Histories in Geotechnical Engineering,* St. Louis, Missouri, 1988.
27. I. D. Moore, and E. T. Selig, "Use of Continuum Buckling Theory for Evaluation of Buried Plastic Pipe Stability," *STP 1093, Buried Plastic Pipe Technology,* American Society for Testing and Materials, October 1990.
28. T. J. McGrath, R. E. Chambers, and P. A. Sharff, "Recent Trends in Installation Standards for Plastic Pipe," *STP 1093, Buried Plastic Pipe Technology,* American Society for Testing and Materials, October 1990.

CHAPTER D2
FIBERGLASS PIPING SYSTEMS

Kenneth J. Oswald
Smith Fiberglass Products Inc.
Little Rock, Arkansas

Fiberglass was at one time considered a new and exotic material to be used only when metals could not be used to handle a particularly corrosive service. Today, however, fiberglass is just one of many materials considered by knowledgeable designers when faced with a piping project. Piping engineers must be able to design fiberglass piping systems with the same confidence as when designing a metallic system. The material in this chapter is chosen to give the engineer an insight into the important design criteria for both above-ground and underground fiberglass piping systems.

BENEFITS AND LIMITATIONS OF FIBERGLASS PIPING SYSTEMS

Piping systems made from fiberglass offer many benefits to the designer, but the limitations of fiberglass must also be considered. The choice of the proper material for a given piping system always involves a comparison of the positive and negative aspects of each material being considered, and each piping system may present a new set of conditions by which a given material must be judged.

Benefits

Corrosion Resistance. Fiberglass pipe is corrosion resistant inside and out. The external corrosion resistance is not necessarily equal to the internal corrosion resistance; but, if required, pipe can be properly constructed to have the same corrosion resistance on both internal and external surfaces.

The corrosion resistance of most fiberglass pipe is excellent and broad ranged.[19,22] This broad range of chemical resistance is useful on new construction where one type of fiberglass pipe will often handle several chemical services which could require a different type of metallic pipe for each service. However, care must be taken to stay within the chemical-resistance ability of the specified piping system; no material is the answer to all corrosion problems.

Light Weight. The high strength-to-weight ratio of filament wound fiberglass pipe produces many of the other benefits listed in this section. Perhaps the greatest benefit of this property is the reduction of installation costs. Two workers can easily carry 20 ft of 8-in filament wound epoxy fiberglass pipe rated for 225 psig service at 225°F. Even on simple installations the light weight of this type of pipe contributes to installed cost savings. The larger the pipe size, the greater the installation cost savings with the light weight of fiberglass pipe. In many installations the light weight of fiberglass pipe allows the designer to specify lighter hangers and supports. Increased safety during installation is an additional benefit of lightweight pipe since the pipe is easier to handle and the inertia of moving pipe is less.

Low Installed Cost. The low installed cost of fiberglass pipe is the major reason for the trend toward increased fiberglass pipe usage[6] in services as diverse as seawater filtration plants above the arctic circle and phosphoric acid plants in Florida. A recent study has shown that on an installed-cost basis, fiberglass pipe is less expensive than unpainted Schedule 40 black steel in sizes larger than 2-in nominal diameter, and it is much less expensive than corrosion-resistant alloy piping.[23]

The installed-cost advantage of fiberglass pipe has been the reason for using this type of pipe in some applications, such as potable water piping, where corrosion is not a problem.

Ease of Installation. The low installed cost of fiberglass pipe is for the most part the result of the ease of installing this type of pipe. The light weight of fiberglass pipe allows for the use of lighter lifting equipment and sometimes eliminates the need for lifting equipment. And in many instances the flexibility of this type of pipe eliminates the need for small-angle fittings or pipe bending equipment.[21] The pipe is easily cut with carbide tipped blades, and bonded joints are quickly and easily assembled in the field. *A word of caution:* Bonded fiberglass joints are easily assembled by trained crews, but untrained crews can make a disaster of the joints in an otherwise well-engineered and specified fiberglass piping system. Always specify that the installation crews must be trained by a representative of the manufacturer and that only those who have been trained may make joints. More on this subject later.

Availability. Mass-produced pipe in 1- to 16-in nominal diameter are available from distributors that represent the major manufacturers of fiberglass pipe in the United States. Additionally, these major manufacturers and many smaller manufacturers produce pipe made on a custom basis for a particular market or for specific projects. By specifying pipe made according to ASTM or other nationally recognized codes and standards, the designer can easily find several sources for the pipe needed for a given project.

Ease of Repair. Minor damage to fiberglass pipe or leaking joints can be repaired in the field using simple tools, and no flame is required. This allows for repairs to be made in many plant areas where a plant shutdown would be required for a metallic piping system or where only a temporary repair could be made.

Hydraulically Smooth. The interior surface of fiberglass pipe is extremely smooth and tends to resist buildup so the inner surface usually remains in this condition throughout the service life of the pipe. A Hazen-Williams flow coefficient of 150, an absolute roughness of 0.0002 in, is common for most fiberglass pipe, and flow tests run on pipe after many years in service show that buildup or increased friction losses are the exception rather than the rule for this type of pipe.

Inherent Flexibility. Most filament wound pipe is rather flexible, having a modulus of elasticity in the axial direction of 1 to 3 million psi. As mentioned previously, this flexibility is a definite advantage during installation. Because of the relatively low axial modulus of this type of pipe, the axial forces generated during temperature changes are only about one-twentieth of those developed by Schedule 40 steel. This allows the designer to use relatively lightweight anchors to restrain the pipe and in most instances eliminates the need for expansion loops or expansion joints.

Abrasion Resistance. Fiberglass pipe is not the solution to abrasion problems encountered in pneumatic conveying of materials, but the addition of a liquid medium has a tremendous effect on the performance of fiberglass pipe conveying abrasive materials. Calcium carbonate slurries and fly ash slurries in coal fired power plants are often handled by fiberglass pipe with a standard wall construction. If increased abrasion resistance is required in slurry applications, abrasion-resistant materials such as alumina can be added to the pipe liner. Pipes with a ceramic alumina bead liner have been used successfully for several years to handle severe bottom ash slurry abrasion problems in coal fired power plants.[5,18]

Because it does not depend upon a protective oxide film for corrosion resistance, fiberglass pipe is not subject to the combination of corrosion and abrasion which occurs with metals when handling high-velocity water. Tests with water conveyed at 100 ft/s have failed to show any wear in a 355-h study.[3] And fiberglass pipes have been operated in liquid services at velocities as high as 25 ft/s. Liquid handling piping systems are normally designed to operate in the range of 5 to 10 ft/s. This usually makes most efficient use of a given size of pipe and helps to eliminate the possibility of severe water hammer in the system.

Low Maintenance. The exterior corrosion resistance of fiberglass pipe is usually enough to protect the pipe from any plant atmosphere to which it will be exposed. When fiberglass pipe is painted, the paint usually has a longer life because tiny nicks and scratches do not cause corrosion products to develop and lift the coating as on a metal pipe. If fiberglass pipe is to be installed outdoors in strong sunlight, the pipe should be painted to protect it from ultraviolet surface degradation. Such degradation, while giving the appearance of severe attack, is actually a very mild surface phenomenon. Several studies have shown that the physical properties of pipe with severe surface attack from exposure to ultraviolet radiation show no measurable difference from the physical properties of similar unexposed pipe.[7] But the appearance of such pipe can become unacceptable in a year or less in extreme southern climates where ultraviolet exposure is intense. As with indoor piping, paint life on fiberglass pipe installed outdoors is excellent.

Low Thermal and Electrical Conductivity. The low electrical conductivity of most fiberglass pipe (some pipe is made to be electrically conductive for special applications) is often used to isolate stray currents from metallic equipment for corrosion protection. The low thermal conductivity of this type of pipe is a definite energy-saving advantage in most applications. However, the thermal resistance in the pipe wall is usually insignificant in determining whether insulation will be required in above-ground applications because the greatest resistance to heat transfer usually occurs at the air-pipe wall interface under bare pipe conditions.

Fungal, Bacterial, and Rodent Resistance. Some plastics (thermoplastics) are subject to fungal, bacterial, and/or rodent attack, but fiberglass pipe offers no nourishment or attraction to these annoyances. Under stagnant conditions some marine growths will attach to fiberglass surfaces, but they do not attack or penetrate the pipe and are usually easily removed.

Limitations

Temperature Rating. Some pipe manufacturers advertise a 300°F temperature rating, but in reality very little fiberglass pipe is used at temperatures above 250°F. The strength and chemical resistance of most fiberglass pipe drop rapidly at temperatures above 225°F. This does not mean that the pipe disintegrates at higher temperatures. At least two manufacturers currently produce pipe, fittings, and adhesive bonded joints which meet the requirements of Military Specification P28584A for 125-psig condensate return service at temperatures up to 250°F. This specification requires that these components hold 187 psig for seven days at 300°F during qualification testing and also pass cyclic bending and pressure tests at temperatures to 300°F.

Pressure. Some 2- through 8-in nominal diameter oil field fiberglass pipe is commercially available in pressure ratings from 2000 to 4000 psig. Pipe larger than 8 in is generally available in pressure ratings as high as 600 psig.

Susceptibility to Mechanical Damage. Fiberglass pipe, especially smaller-diameter, thin wall pipe, can be damaged by sharp impact blows to the point of leaking. Fiberglass pipe should be protected if it is to be installed in an area where it can be physically abused.

Fiberglass pipelines that are painted should be identified so they are not mistaken for steel lines. In some instances painted fiberglass lines have been damaged when they were mistaken for steel lines and were used to support a chain hoist used for heavy lifting.

Low Modulus of Elasticity. Compared to steel pipe, fiberglass pipe requires closer support spacing. However, the support spacing for fiberglass pipe is not as close as that required for thermoplastic pipe.

Most commercially available fiberglass pipe 16-in nominal diameter and smaller can normally be buried with only minimal precautions (smooth trench bottom and clean backfill) unless it must bear the weight of full legal highway loads. Most road crossings with this type of pipe are accomplished by placing the fiberglass pipe in a conduit which bears the traffic load. Direct burial of fiberglass pipe which must bear full legal traffic loads requires special consideration of burial depth, soil characteristics, and degree of backfill compaction (see section, "Buried Pipe Installations," in this chapter).

Fiberglass pipes 18-in diameter and larger are usually not stiff enough to be buried without special compaction procedures. It is most important that the manufacturer's burial instructions be followed explicitly when installing this type of pipe. Ditch preparation, type of backfill, placement of backfill, ground water level, and backfill compaction are extremely important considerations when burying large-diameter fiberglass pipe.

Lack of Standard Joint. All fiberglass pipe can be joined by the butt-n-wrap method which is the equivalent of welding. However, most commercially available 16-in-diameter and smaller pipe and fittings are made to be joined using adhesive bonded joints. Three adhesive bonded joint designs are in general use: fully tapered bell and spigot, semitapered joint, and straight socket-spigot joint (see section, "Joining Systems" in this chapter). Pipe, fittings, and adhesive from one manufacturer are generally not interchangeable with products from another, even if two manufacturers use the same basic type of bonded joint.

Movement under Severe Water Hammer. Water hammer movement is a problem only when the water hammer is severe. This is more of a factor in fiberglass piping systems because of the light weight and the relatively low bending modulus of this type of pipe. Severe water hammer problems are usually best solved by eliminating the source of the water hammer, such as quick closing valves or pump start-up surges. If this cannot be done, the magnitude of the pressure peak can be lowered considerably by the use of accumulators and other devices. As with pipe made from any material, the peak pressure should not exceed the pressure rating of the piping system. If movement occurs in the system after making changes to minimize water hammer, the system should be anchored to prevent damage due to excessive movement. Calculation of the peak pressure under quick valve closing conditions can be used as the maximum possible pressure for design conditions in piping systems with minor elevation changes. If the piping system has elevation changes of great magnitude, other methods of calculation are required.

RESINS AND CORROSION RESISTANCE

As with the metals used to make pipe, each resin system used in the manufacture of fiberglass pipe has particular strengths and weaknesses.

Custom fiberglass equipment and fiberglass pipe larger than 16 in in diameter are generally made from polyester resins because of the ease of handling large quantities of this type of resin. Some small-diameter pipe is made from polyester or vinyl ester resins, but most 1- through 16-in-diameter pipe are manufactured from epoxy resin systems which are easier to handle in mass-production processes. For special corrosion applications outside the capabilities of polyester or epoxy resins, pipe made from furan resins or phenolic resins are available.

Epoxy Resins

The chemical resistance and physical properties of an epoxy resin system depend upon both components of the system: the basic resin and the curing or crosslinking agent. Two general types of epoxy resins are in common usage today: bisphenol-A epoxies and epoxy novolacs. The bisphenol epoxies are much more widely used because they are more economical and easier to handle during fabrication. The epoxy novolacs are employed where increased temperature resistance and/or better solvent resistance are required. Both types of epoxies can be cured with a great variety of curing agents, and the choice of curing agent has much influence on the properties of the final product. The two most common resin systems used in the manufacture of epoxy fiberglass pipe are bisphenol epoxies cured with aromatic amines and bisphenol epoxies cured with aromatic anhydrides.

In the 1- through 16-in-diameter range, bisphenol epoxies cured with aromatic amines produce pipe with the balance of physical, chemical, and economic properties which are needed for most fiberglass piping applications. Pipe made from these resin systems have an upper temperature limit of 250°F and are resistant to salt solutions and rather severe alkaline and solvent exposures. Dilute acids are also handled with this type of pipe. If increased solvent resistance is required, an epoxy novolac resin system is recommended.

In the 2- through 16-in-diameter range, bisphenol epoxies cured with aromatic anhydrides are used to manufacture pipe for use in oil field and water handling applications where the chemical resistance of an aromatic amine cured epoxy resin is not required. When used within the temperature and chemical limits of the resin system, these pipes give excellent service. These pipes have an upper temperature limit of approximately 150°F and are less chemical resistant than pipe made from aromatic amine cured epoxy resin systems. Anhydride cured resin systems have no resistance to alkaline solutions and are rapidly attacked by water at temperatures above the rated temperature.

Neither of these epoxy resin systems is resistant to strong mineral acids or strong oxidizers.

Polyester Resins

The chemical resistance and physical properties of commercially available polyesters and vinyl esters—a special class of chemical resistant polyesters—are uniform for a given resin because all of these resins are cured using styrene as the crosslinking agent. When one knows of a successful application of a particular polyester or vinyl ester resin, one need not be concerned about the curing agent. During the manufacture of fiberglass pipe, an initiator or "catalyst," is added to the styrene-polyester mixture to cause the two components to react and solidify. In almost all chemical services the choice of initiator system is of no consequence. However, the choice of initiator system has been found to affect the chemical resistance of fiberglass pipe in some extremely aggressive chemical services such as hot, wet chlorine, or sodium hypochlorite.

Fiberglass pipe are generally manufactured from any of four types of chemical-resistant polyester resins: vinyl esters and isophthalic, chlorendic, and bisphenol-A fumarate polyesters. Each particular resin has different chemical, physical, and economic properties, and the choice of resin is critical.

Vinyl Ester Resins. Until the development of vinyl ester resins, it was not practical to mass produce small-diameter polyester pipe. Now, however, 1- through 16-in pipe manufactured from vinyl ester resins are commercially available from several manufacturers. Pipe made from these resins have good physical strength and, in general, better impact strength than other chemical-resistant polyesters. These resins have excellent resistance to oxidizers and strong mineral acids and good resistance to alkaline environments. Standard vinyl esters are limited to 200 to 225°F in most applications, while more costly high-performance vinyl esters are suitable for general use up to 250°F and in some special applications have been used in temperatures as high as 350°F.

Bisphenol-A Fumarate Polyester Resins. The bisphenol-A fumarate polyester is the original high-volume, commercially available, chemical-resistant polyester resin. This type of resin has been produced for over 35 years and, until the advent of vinyl ester resins, this type of polyester was the resin most widely used in the manufacture of chemical-resistant fiberglass products. The chemical resistance of this type of resin is roughly equivalent to that of vinyl ester resins at temperatures up to 250°F, but bisphenol-A fumarate resins are more rigid than vinyl esters, and this makes them unsuitable for the manufacture of small-diameter pipe on mass-production equipment. The rigidity of this resin is the major reason for its displacement by vinyl esters from its former position as the most widely used chemical-resistant resin. In large-diameter pipe and large reaction vessels, resin rigidity is not a disadvantage, and this resin is still used in the manufacture of this type of equipment, especially when it is to be used in services where this resin has proven successful in past applications.

Chlorendic Polyester Resins. The chlorendic polyester resins have a chlorinated backbone in their molecular structure which makes them particularly well suited for elevated-temperature applications, up to 350°F. In most resin classifications there is an increase in rigidity when chemical structure changes are made to give increased temperature performance to the resin. Chlorendic resins follow this general rule and are more rigid than bisphenol-A fumarate resins. The molecular structure of chlorendic resins gives them excellent resistance to concentrated mineral acids and highly oxidizing environments, but poor resistance to alkaline solutions. The solvent resistance of these resins is very good when compared to other polyesters.

Isophthalic Polyester Resins. The isophthalic polyesters are the least expensive and least chemical resistant of the corrosion-resistant polyester resins. For service temperatures up to 180°F, these resins generally have good resistance to water, dilute acids, and very weak alkaline solutions and good resistance to petroleum solvents such as gasoline and oil. There are many grades of isophthalic polyester resins. It is important to choose an isophthalic polyester which is compatible with the service being handled.

Other Resins

Furan resins are difficult to work with, and this presently limits their application to systems which require the unique blend of superior acid, alkali, and solvent resistance at temperatures up to 300°F offered by these resins. However, furan resins are not suitable for handling oxidizing services.

Phenolic resins also require special processing techniques which limit their economical application to systems that require superior acid and solvent resistance at temperatures to 300°F.

MANUFACTURING PROCESSES

Filament Winding

Most machine-made fiberglass pipe (Fig. D2.1) is manufactured using the filament winding process. This method of manufacture allows for the best utilization

FIGURE D2.1 Typical installation of filament wound epoxy pipe with spray-up fittings.

of the tensile strength of the glass. In this process a band of continuous glass fiber strands are pulled through a resin bath and are then mechanically laid down on a rotating mandrel at a precisely controlled winding angle. When the desired wall thickness is reached, the resin is solidified by applying heat or by allowing the pipe to cure for a longer time at room temperature. The fibers are accurately orientated and uniformly tensioned to best absorb the stresses developed in pressure applications. Another advantage of filament winding is that the bundles of strands can be opened and more thoroughly wetted by the resin. Probably most important, however, is the fact that the fibers are not crimped, such as when braiding or weaving a fabric. When a fiber is crimped, it is abraded and loses up to 40 percent of its strength. The efficient use of glass in filament winding produces fiberglass parts with a strength-to-weight ratio which cannot be surpassed by other production methods. An example is a 2-in-diameter epoxy filament wound fiberglass pipe with 0.070-in wall. This pipe bursts at over 3000 psig and weighs only 0.4 lb/ft. This type of performance accounts for the popularity of the filament winding process in fiberglass pipe manufacture.

Centrifugal Casting

In the centrifugal casting method of fiberglass pipe manufacture, a sleeve of fiberglass cloth and/or mat is placed inside a mandrel which is then rotated at high speed. A nozzle is then inserted into the mandrel, and resin is injected into the mandrel where the centrifugal force moves the resin through the glass sleeve to wet out the glass fibers. When the correct ratio of glass and resin has been achieved, the mandrel is heated to solidify the resin so the pipe can be removed from the mandrel. Because centrifugally cast pipe is formed inside of a mandrel, it has a smooth, uniform outside surface. This method of manufacture requires thicker walls for a given pressure class because it makes less efficient use of the tensile strength of the reinforcing glass, but it has the advantage of being able to use some highly reactive resin systems which are difficult to use in the filament winding process.

FITTINGS MANUFACTURING PROCESSES

Compression Molding

Most high-volume fittings, such as small-diameter flanges, elbows, and tees (Fig. D2.2), are manufactured by the compression molding process. In this method of manufacture, the molding compound, a mixture of resin and reinforcements, is placed in a heated mold which solidifies the resin and shapes the part under high pressure. For a given resin system, the strength of molded fittings is determined by wall thickness and the type and amount of reinforcement in the molding compound.

Filament Winding

Filament wound fittings are manufactured by winding continuous glass filaments over the outside of a mandrel or liner which gives the fitting its shape. The winding patterns for fittings, other than flanges, are very complicated, and until the development of computer controlled winders, all filament wound fittings were handmade, that is, the glass pattern was laid down by hand. Most filament wound fittings are

FIGURE D2.2 Compression molded elbows, tee, and flange. Grooved adapter, threaded adapter, and sleeve coupling machined from filament wound stock.

still handmade, but many are now made on computer controlled winders. This method of manufacture is especially suited for producing 8- through 16-in fittings for normal service pressures and for 2- through 16-in high-pressure fittings.

Spray-up

Custom fittings and many fittings 6 in in diameter and larger are manufactured by the spray-up method. Elbows of this type are shown in Fig. D2.1. In the spray-up process chopped glass fibers and resin are simultaneously applied to the outside of a mold from special spraying equipment which chops the glass fibers and mixes resin components just before spraying. The glass-resin mix is then processed with a hand roller to remove trapped air and produce a dense laminate. This process is especially suited for use with polyester resins, but is very difficult to use with epoxy resin systems. This method of manufacture can be used to produce complex shapes with relatively low cost tooling.

Hand Lay-up or Contact Molding

In this process resin and glass fibers in the form of chopped strand mat, woven cloth, or unidirectional mat are combined by hand and applied to the outside of a mold. Trapped air is removed by using brushes, squeegees, and/or rollers. This method of manufacture has the advantage of allowing for the application of unidirectional glass or some other special reinforcement in areas where a fitting may require additional strength. Hand lay-up procedures are often combined with spray-up and filament winding to produce a fitting.

Mitered Fittings

Elbows and tees are often made by cutting pipe so that adjoining pieces are properly aligned and then wrapping or applying spray-up material over the joint (Fig. D2.3). This method is especially useful for making custom elbows or for producing elbows at a construction site.

PIPE WALL CONSTRUCTION

One of the advantages of using fiberglass materials to manufacture a corrosion-resistant pipe is that the wall construction can be varied to meet the requirements of specific applications. These various wall constructions are a part of the manufacturing operation and not the result of adding a liner or coating after the pipe is manufactured.

The simplest pipe wall construction is found in unlined pressure pipe designed for use in light corrosion applications. These are filament wound pipe with 20 to 30 percent resin in the structural wall. When made from vinyl ester resins or aromatic amine cured epoxy resins, this type of pipe is highly resistant to fresh and salt water in the pH range from 2 to 13, solutions of neutral or nearly neutral

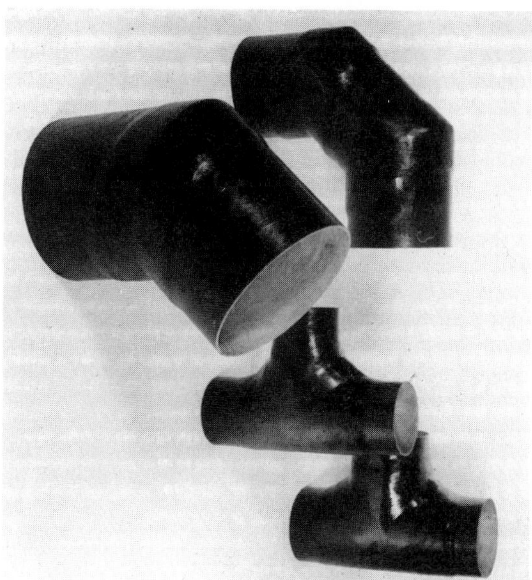

FIGURE D2.3 Mitered tees and elbows.

salts, dilute acid or alkaline waste streams, sour crude oil, and other "light" corrosion services. The uses for unlined filament wound pipe are numerous and well documented,[15,16,17] but the uses are limited to areas where the glass is unaffected by the service conditions. This type of pipe is often used in applications where the fluid being handled is not particularly corrosive but where external corrosion by acid soils or a corrosive atmosphere could be a problem.

Several manufacturers produce filament wound pipe with resin-rich, reinforced liners 0.01- to 0.05-in thick to protect the glass fibers in the structural wall. The liners contain 80 to 90 percent resin reinforced with a chemical-resistant glass, such as C-glass, or an organic fiber. These liners improve the chemical resistance of fiberglass pipe and are necessary for chemical services which would attack the glass used in the filament wound structural wall. Pipes with this liner construction have been used to handle many severe corrosion problems in the chemical process industries, pulp mill bleach plants, phosphoric acid production, and pollution control.[15,16,17,24]

Centrifugally cast pipes are produced with unreinforced resin liners 0.05 to 0.08 in thick. The thicker liner and absence of liner reinforcement make this type of pipe particularly useful for handling some severe acid and alkali services which can attack the reinforcement in reinforced liners.

Most who deal with corrosion problems are familiar with the 0.110-in-thick reinforced liner specified in National Bureau of Standards Voluntary Standard PS15-69 for Custom Contact Molded Reinforced Polyester Chemical Resistant Process Equipment. As stated in the title, this specification was developed for handmade products. However, the liner construction specified in this document is sometimes employed in filament wound piping systems, especially custom manufactured systems. This liner construction consists of an inner liner of from

0.010 to 0.020 in of resin-rich reinforced liner followed by a minimum of 0.100 in of laminate reinforced with not less than 20 nor more than 30 percent by weight noncontinuous glass strands having fiber lengths from 0.5 to 2.0 in.

The authors of PS15-69 specified this thick liner system because the liners were to be handmade, and in 1969 when this specification was written, this meant that the liners would contain a certain amount of voids and imperfections. Modern machine-made filament wound pipes with thinner liners now handle most of the chemical services that once required the thick PS15-69 liner. This liner specification is still used for some extremely corrosive applications where the added cost of the thick liner is justified. And a double-thickness PS15-69 liner is specified for applications such as hot, wet chlorine where the resin is attacked by the service and a thick sacrificial liner is required for economical service life.

Special abrasion-resistant liners are built into some fiberglass pipe for use in slurry applications. An abrasion-resistant fiberglass piping system lined with ceramic alumina beads is used to handle the severe corrosion and erosion problems encountered in bottom ash slurries from coal fueled power plants.[5,18]

CODES, STANDARDS, AND SPECIFICATIONS

Most fiberglass pipe is manufactured to meet the physical, chemical, and dimensional requirements of some nationally recognized pipe standard. There are at least nine different ASTM specifications covering 1- through 144-in-diameter fiberglass pipe, three U.S. military specifications, two API specifications, and an AWWA specification covering fiberglass pipe. For some applications this type of pipe is tested by an independent laboratory for certification or to meet the exacting requirements of some specific applications. Additionally, fiberglass piping has been installed in compliance with many safety codes such as are listed in the U.S. Code of Federal Regulations or the B31 pressure piping codes of the American Society of Mechanical Engineers.[14]

Most of these specifications use standard test methods developed by ASTM to determine the properties of fiberglass pipe. A summary of the test methods, specifications, and recommended practices for fiberglass pipe developed by ASTM and other standards organizations are listed in Tables D2.1 through D2.5. Knowledgeable piping engineers should become familiar with these and refer to the proper documents when specifying fiberglass pipe.

JOINING SYSTEMS

Fiberglass pipes are joined by two basic methods: adhesive bonded joints and overwrapped joints. Three basic adhesive bonded joints—fully tapered, semitapered, and straight socket-spigot joints—are used on mass-produced piping systems. These joints are shown in Figs. D2.4, D2.5, and D2.6. Overwrapped joints, Figure D2.7, are generally used on custom-made pipe and large-diameter pipe. Additionally, several manufacturers of fiberglass pipe offer proprietary mechanical joining systems which utilize threads, o-rings, and/or lock rings. These proprietary joining systems have gained acceptance in oil field applications where long runs of pipe must sometimes be installed under adverse conditions.

TABLE D2.1 Test Methods

Tensile properties
ASTM D1599. Short-time Hydraulic Failure Pressure of Plastic Pipe, Tubing, and Fittings.
ASTM D 2105. Longitudinal Tensile Properties of Reinforced Thermosetting Plastic Pipe and Tube.
ASTM D2290. Apparent Tensile Strength of Ring or Tubular Plastics by Split Disk Method.
Compressive properties
ASTM D695. Compressive Properties of Rigid Plastics.
Bending properties
ASTM D2925. Measuring Beam Deflection of Reinforced Thermosetting Plastic Pipe under Full Bore Flow.
Internal pressure strength, long term
ASTM D1598. Time to Failure of Plastic Pipe under Constant Internal Pressure.
ASTM D2143. Cyclic Pressure Strength of Reinforced Thermosetting Plastic Pipe.
ASTM D 2992. Obtaining Hydrostatic Pressure Design Basis for "Fiberglass" (Glass-Fiber-Reinforced Thermosetting Resin) Pipe and Fittings. (Procedure A for cyclic pressures. Procedure B for steady pressures.)
Pipe stiffness
ASTM D2412. External Loading Properties of Plastic Pipe by Parallel Plate Loading.
ASTM D2924. External Pressure Resistance of Reinforced Thermosetting Resin Pipe.

TABLE D2.2 Recommended Installation Practices

ASTM D3839. Standard Practice for Underground Installation of Flexible Reinforced Thermosetting Resin Pipe and Reinforced Plastic Mortar Pipe.

AWWA C950, Appendix A. Design Requirements and Criteria for RTRP and RPM for Water Service.

API RP5L4. Recommended Practice for the Care and Use of Reinforced Thermosetting Resin Casing and Tubing.

API 1615. Installation of Underground Petroleum Storage Systems.

The fully tapered bell-and-spigot joint uses matching tapers machined on the outside of the pipe and the inside of the fitting bell. The tapered surfaces can be pushed together to produce an extremely thin glue line for good adhesion. The matching tapers are locked by the wedging action, and this holds the joint in position while the adhesive is curing. This type of joint is used extensively in all types of fiberglass piping systems. It was developed for the relatively high pressure applications encountered in oil field applications and is used on nearly all bonded fiberglass pipe joints designed for high-pressure service. The thin glue line in this type of joint enables it to handle thermal stresses encountered in lines which experience large temperature changes.

TABLE D2.3 Specifications and Classifications for Process Piping

ASTM D2310. Standard Classification for Machine-Made Reinforced Thermosetting Resin Pipe.

ASTM D2517. Standard Specification for Reinforced Epoxy Resin Gas Pressure Pipe and Fittings. (Covers filament wound epoxy resin pipe and reinforced epoxy resin fittings in sizes 2 through 12 in.)

ASTM D2996. Standard Specifications for Filament Wound "Fiberglass" (Glass-Fiber-Reinforced Thermosetting Resin) Pipe. (Covers 1-in through 16-in pipe made from epoxy, polyester, or furan resins with glass fiber reinforcement.)

ASTM D2997. Standard Specification for Centrifugally Cast Reinforced Thermosetting Resin Pipe. (Covers 1½-in through 14-in pipe made from epoxy or polyester resin with glass fiber reinforcement.)

ASTM 4024. Standard for Reinforced Thermosetting Resin Flanges. (Covers ½-in through 24-in flanges with ANSI B16.5, 150-lb bolt hole circles. Classifies flanges by pressure rating, burst pressure, sealing test pressure, and bolt torque limits.)

ANSI/AWWA C950. Standard for Glass-Reinforced Thermosetting Resin Pressure Pipe. (Covers 1- through 144-in pipe made from polyester or epoxy resins, with or without siliceous sand aggregate, for pressurized water systems operating at pressures to 250 psig. This standard contains information on quality-control tests, design, shipping, handling, storage, and installation.)

API 15LR, Sixth Edition. Specification for Low Pressure Fiberglass Line Pipe. (Covers 2-in through 16-in pipe and fittings made from thermosetting resins for cyclic operating pressures to 1000 psig. This standard contains information on quality-control tests, hydrostatic mill tests, dimensions, weights, material properties, physical properties, and minimum performance requirements.)

API 15 AR. Specification for Reinforced Thermosetting Resin Casing and Tubing. (Covers 1½-in through 10¾-in tubing or casing used for drilling, producing, and disposal operations.)

MIL-P-24608A. For 1- through 12-in epoxy resin pipe and fittings for U.S. Navy shipboard piping systems operating at 200 psig at temperatures up to 150°F.

MIL-P-28584A. For 2- through 12-in pipe and fittings for use in continuous service at 125 psig and 250°F in condensate return systems.

MIL-P-29206. For 2- through 12-in epoxy or polyester resin pipe intended for an operating pressure of 150 psig at temperatures up to 150°F and pressure surges to 275 psig in liquid petroleum products lines installed below ground.

The semitapered joining system consists of a straight spigot which is machined on the end of the pipe and a tapered bell with an internal shoulder which acts as a pipe stop. This stop is an advantage in close tolerance plumbing. The end of the pipe interferes with the bottom of the tapered bell over a short distance before hitting the stop. This produces a thin glue line with good adhesion at the end of the bonding area near the stop.

The straight socket-spigot joint has no tapered surfaces, and the void between the pipe and fitting surfaces is filled with a rather thick adhesive.

The fully tapered and semitapered joining systems use a field tool to machine the ends of the pipe in the field. The straight socket-spigot joining system requires that the bonding area on the outside of the pipe be sanded before joining.

Overwrapped joints, known as *butt-n-wrap joints,* are made by wrapping fiberglass laminate materials around the pipe joint area. The pipe is prepared by

TABLE D2.4 Specifications and Classifications for Water and Sewer Piping

ASTM D3262. Standard Specification for "Fiberglass" (Glass- Fiber-Reinforced Thermosetting Resin) Sewer Pipe. (Covers 8- through 144-in pipe made from polyester resin and siliceous sand with glass fiber reinforcement for conveying sanitary sewage, stormwater, and some industrial wastes.)

ASTM D3517. Standard Specification for "Fiberglass" (Glass-Fiber-Reinforced Thermosetting Resin) Pressure Pipe. (Covers 8- through 144-in pipe made from polyester or epoxy resin, with or without siliceous sand, with glass fiber reinforcement for conveying water in systems which operate under internal head pressure of 500 ft or less.)

ASTM D3754. Standard Specification for "Fiberglass" (Glass-Fiber-Reinforced Thermosetting Resin) Sewer and Industrial Pressure Pipe. (Covers 8- through 144-in pipe made from polyester or epoxy resin, with or without siliceous sand, with glass fiber reinforcement for conveying sanitary sewage, stormwater, and many industrial wastes and corrosive fluids at pressures to 250 psig.)

ASTM D3840. Standard Specification for Reinforced Plastic Mortar Pipe Fittings for Nonpressure Applications. (Covers pipe fittings with gasketed joints intended for gravity flow systems conveying water, sewage, and some industrial wastes.)

ASTM D4161. Standard Specification for "Fiberglass" (Glass-Fiber-Reinforced Thermosetting Resin) Pipe Joints Using Flexible Elastomeric Seals. (Covers gasketed joints intended for pressures to 250 psig.)

ASTM F1173. Specification for Epoxy Resin Fiberglass Pipe and Fittings to be used for Marine Applications. (Covers 1- through 36-in marine piping systems in which resistance to seawater, chemicals, and sea environment is required.)

sanding the outside of the pipe ends in the area to be overwrapped. The outside diameters of the two pipes must be reasonably close. If they do not match, fiberglass putty is used to smooth the transition between the two diameters. The joint is made by applying several layers of fiberglass mat and/or woven roving that have been saturated with resin. This method essentially builds a pipe coupling over the ends of the two pipes.

PHYSICAL PROPERTIES

A search of the literature will often yield tables of "fiberglass design properties" or something similar. These tables are usually of little use since they seldom give enough information to design a piping system using a specific type of pipe. The best source of piping system design information is the manufacturer of the specific type of pipe to be used on a given project, and it is often necessary to obtain physical properties information from several manufacturers so the installed cost of their products can be compared. Design properties determined by ASTM test methods should be used whenever possible.

Selecting the Proper Pipe for a Given Application

When selecting a piping system for a particular application, the three primary considerations are chemical compatibility, pressure rating, and temperature rating.

TABLE D2.5 Listings, Approvals, and Piping Codes Applicable to Fiberglass Piping

National Sanitation Foundation, Standard No. 14. Tests and lists pipe, fittings, and adhesives used for conveying drinking water. Components are tested extractable toxic substances, and the manufacturer's published physical properties are certified.

Underwriters' Laboratories, Inc.
- a. Standard for testing and listing fiberglass piping for use as underground fire water mains.
- b. Standards for testing and listing fiberglass piping for conveying petroleum fuels and alcohol-petroleum fuel blends underground.

Factory Mutual Research. Approval standard for plastic pipe and fittings for underground fire protection service.

ANSI/ASME B31.3, Chemical Plant and Petroleum Refinery Piping Code. Lists specifications for fiberglass pipe acceptable for use within the code and establishes criteria for the use and installation of these piping systems.

ANSI/ASME B31.1, Power Piping Code. Lists specifications for fiberglass pipe acceptable for use within the code and establishes criteria for the use and installation of these piping systems.

ANSI/ASME B31.8, Gas Transmission and Distribution Piping Code. Lists fiberglass pipe manufactured in compliance with ASTM D2517 as being acceptable for use within the code. Also establishes criteria for installation and uses of these piping systems.

U.S. Department of Transportation, Title 49, Part 192, Code of Federal Regulations Covering Transportation of Natural and Other Gas by Pipeline, Minimum Federal Safety Standards. Lists fiberglass pipe manufactured in compliance with ASTM D2517 as being acceptable for use within the code.

ASME Boiler and Pressure Vessel Code Case N155. Lists the design criteria for fiberglass piping systems in Section III, Division I, Class 3, applications in nuclear power plants.

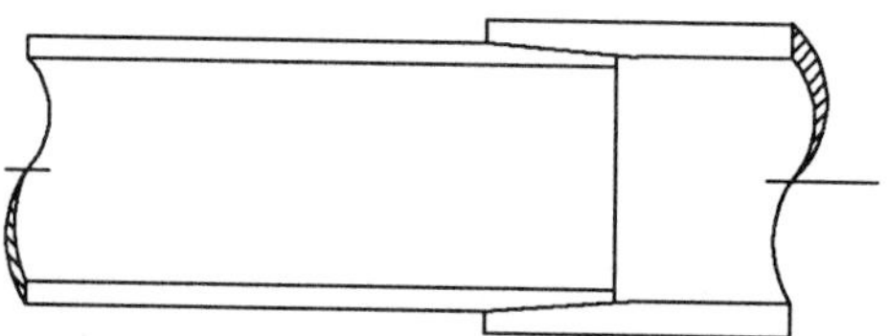

FIGURE D2.4 Fully tapered joint.

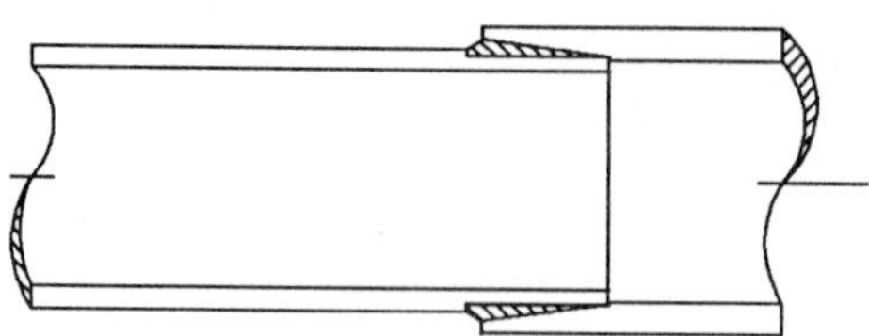

FIGURE D2.5 Semitapered joint.

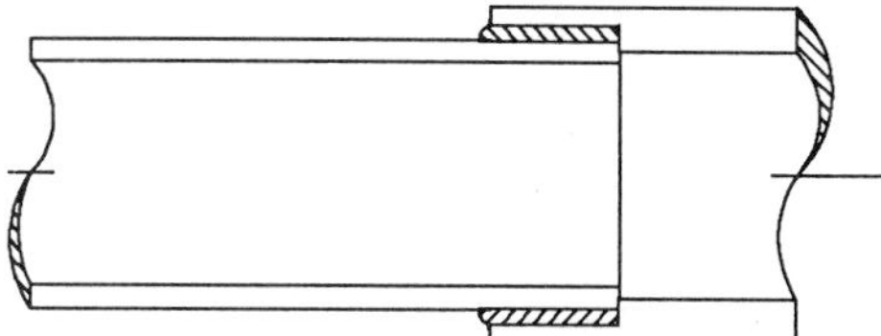

FIGURE D2.6 Straight socket-spigot joint.

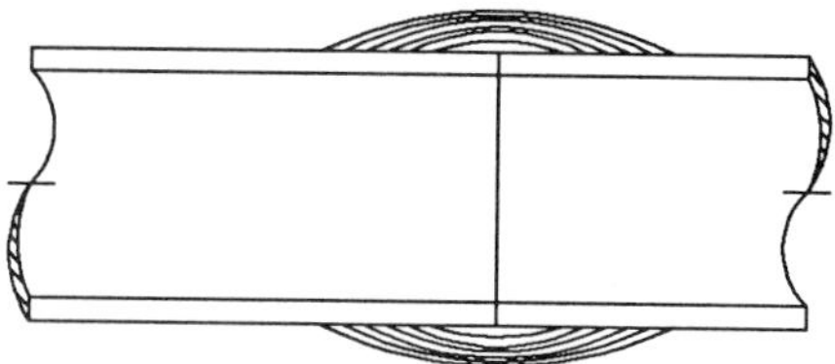

FIGURE D2.7 Butt-wrap joint.

Chemical compatibility information is available from various sources. Most commercial pipe manufacturers publish a chemical-resistance chart developed from laboratory testing, field case histories, and/or resin manufacturers data; and the National Association of Corrosion Engineers publishes a summary of nonmetals chemical-resistance data which can be consulted.[10] As with any material, when mixtures of chemicals are to be handled, it is best to consult with the pipe or resin manufacturer to determine whether a particular mixture will produce more attack than would be anticipated from information in the published tables.

Pressure rating and temperature rating are closely related. The design hoop stress at a given temperature for a given pipe is usually determined by using the procedures described in ASTM D2992. This method of determining design stress is referenced in nearly all national standards (with the exception of some U.S. military specifications). ASTM D2992 references two ASTM test methods: ASTM D1598, "Standard Test Method for Time-to-Failure of Plastic Pipe Under Steady Pressure," and ASTM D2143, "Standard Test Method for Cyclic Pressure Strength of Reinforced Thermosetting Resin Pipe." Both of these test methods are based on time to failure versus pressure regression testing, and both require that 18 or more specimens be tested to failure and that at least 1 pipe specimen be on test for 10,000 h before failing. In the case of the cyclic pressure test, the pipe sample must go through 15,000,000 full pressure cycles during the 10,000-h period. The data from each test are accumulated, and a linear regression line plotting hoop tensile stress versus cycles to failure, or hours to failure, is established (see Fig. D2.8). The steady pressure regression line is extrapolated to an estimated stress at 100,000 h, and the cyclic data are extrapolated to an estimated stress at 150,000,000 cycles. The data must meet specified statistical quality standards before being considered valid. The design stress for steady pressure rating calculations is typically taken as one-half of the extrapolated 100,000-h hoop stress. Cyclic pressure ratings are based on the extrapolated 150,000,000-cycle hoop stress.

Two important points regarding both test methods should be noted:

1. In both test methods, pipe samples can be tested with either free-end conditions (unrestrained ends) where the test specimen has both hoop and axial

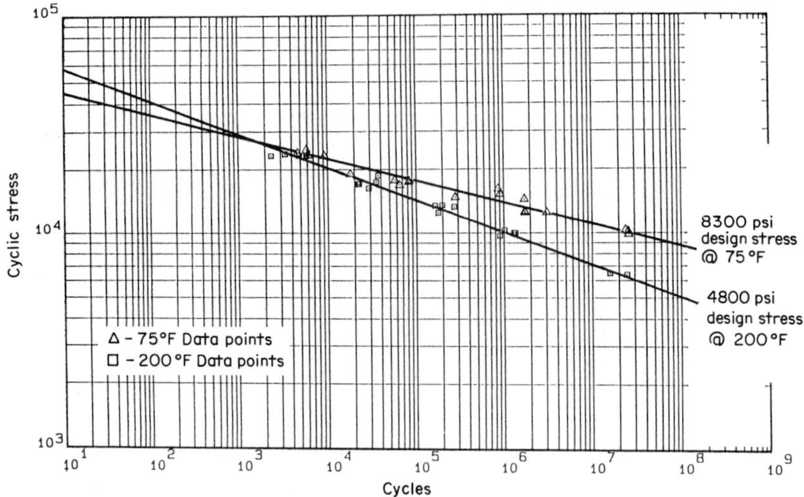

FIGURE D2.8 Cyclic regression per ASTM D2992. Aromatic amine cured epoxy filament wound pipe.

forces acting on it, or with restrained ends where only hoop forces are acting on the sample. It is important to know the end conditions used when pipe design data are evaluated using these test methods.

2. Both tests can be conducted at elevated temperatures. Some manufacturers have conducted these tests at room temperature and at a higher temperature, usually 150 or 200°F. The design stresses at elevated temperatures are lower than those obtained at room temperature.

Typical regression curves showing cyclic pressure data at two different temperatures for filament wound aromatic amine cured epoxy pipe are shown in Fig. D2.8. The design hoop stresses for each temperature are shown at the far right of each regression curve. These stresses are related to the cyclic pressure rating of a fiberglass pipe by the relationship:

$$S = \frac{P(D - t)}{2t}$$

where S = design hoop stress, psi
P = internal pressure rating, psig
D = average reinforced outside diameter, in
t = minimum reinforced wall thickness, in

ASTM D2992 contains further design information.

Piping System Design Properties

To properly evaluate a given piping system design, the engineer must know the physical properties which govern the reaction of the piping system to various operating conditions. Numerous computer programs are available to guide one through this exercise, but the calculations are only as good as the input data supplied to the program. It is important to note that many of the physical properties of fiberglass pipe have different values when measured in different directions. For example, the modulus of elasticity of filament wound pipe measured in the hoop direction is usually much different from the same property measured in the axial direction. This difference is caused by the alignment of the reinforcing glass fibers in relation to the direction of the force being applied when determining a particular physical property.

The physical properties of different types of fiberglass pipe vary considerably within the 75 to 250°F temperature range. And, as with most materials, different properties vary at different rates. Most physical properties decrease with temperature, but the short-term burst strength of some epoxy fiberglass pipe increases with an increase in temperature. A comparison of some of the 75 and 210°F physical properties of a particular epoxy fiberglass pipe serves as an example. The axial or longitudinal tensile strength and compressive strength vary at the same rate; the 210°F values are about one-half of the 75°F values. However, the hydrostatic design stresses determined by ASTM D2992 at these temperatures differ by a factor of 0.76, and the burst strength actually increases with temperature by a factor of 1.48. A comparison of the temperature effect on the burst strength and long-term design strength of this pipe shows that a burst test is not a good basis for determining a long-term design stress and illustrates the necessity of obtaining proper design information from the pipe manufacturer.

Since the modulus of elasticity of fiberglass pipe varies with the direction of measurement relative to reinforcement, the Poisson's ratio also varies. Most computer programs for calculating piping system stresses call for Poisson's ratio as a required input. It is usually best to use an "average" value of 0.3 for Poisson's ratio in these programs, because for a given type of pipe, it is almost impossible to utilize the exact Poisson's ratio for each design situation. And, piping systems designed using the average value of 0.3 have performed well for many years.

PIPING SYSTEM DESIGN

Supports

When designing and specifying the support system components of a fiberglass piping system, the two most important considerations are support design and support spacing.[4,8]

When designing the pipe support system, avoid the use of supports which provide point contact or narrow supporting areas. A good general rule for pipe diameters less than 16 in is to use supports which have at least 120° of contact and which are wide enough so that the bearing stress on the pipe wall does not exceed 85 psi. Many standard supports used with steel pipe meet these requirements. Note that roller supports do not meet these requirements unless a support sleeve is installed between pipe and roller. Pipe greater than 18-in diameter usually require more support area because of the lower pipe wall-to-diameter ratios encountered in these sizes. The minimum support requirements for this type of pipe are 150° of contact and a hanger width equal to one-third the nominal diameter of the pipe.

Because the outside surface of most fiberglass pipe is slightly irregular, it is good practice to install ⅛-in-thick rubber sheet between pipe and support to prevent point loading on the irregular pipe surface. This is particularly important when supporting pipe is larger than 12-in diameter.

If the pipe is in a system which could impart vibration or pulsing action into the pipe, areas of contact with supports should be protected to prevent abrasion of the pipe wall by the support. This can be accomplished by bonding a metal or fiberglass sleeve to the pipe in the anticipated wear areas. Alternatively, the vibration can be eliminated by installing a flexible coupling between the pipe and the vibrating machinery.

Support Spacing

Fiberglass piping systems generally require closer support spacing than the same size of metallic pipe, but wider spacing than is required for the same size of thermoplastic pipe. The major reason for this is the difference in the axial modulus of elasticity of the different materials. Also, the support spacing of fiberglass pipe decreases with an increase in temperature, because the modulus of elasticity varies significantly within the operating temperature range of this type of pipe. For example, the support spacing of a filament wound vinyl ester pipe at 200°F is about 65 percent of the 75°F value. For an aromatic amine cured epoxy filament wound pipe, the 225°F support spacing is 85 percent of the 75°F value.

Most manufacturers publish support spacing information based on one-half inch of midspan deflection. This information is usually based on a bending modulus of elasticity determined in a long-term support spacing test such as ASTM D2925.

When a section of a fiberglass pipe line is installed in a vertical run, it should be restrained with riser clamps at about the same spacing as is required for supports in horizontal runs.

Valves and other heavy equipment must be supported independently of the pipe in both vertical and horizontal installations.

If pipe is to be installed directly on the ground, avoid excessive bending. Refer to the manufacturer's design bending stress or published minimum bending radius.

Anchors

Piping system design conditions often call for restraint of fiberglass pipe, primarily to handle thermal expansion and to prevent line movement due to water hammer.

If a fiberglass line is connected to significant runs of metallic pipe or to other pipe which can develop greater thermal forces (such as a larger-diameter fiberglass line), the connecting pipe must be anchored to prevent the fiberglass pipe from absorbing all thermal movement generated by both piping systems.

Anchors used to restrain thermal forces in fiberglass pipe are normally much lighter than those required for metallic pipe. During a temperature change, the low axial modulus and thin pipe walls of fiberglass pipe produce forces which are about $\frac{1}{30}$ of those generated in Schedule 40 steel pipe under the same conditions. When calculating anchor loads for fiberglass pipe, it is important to use the axial or longitudinal modulus of elasticity.

Typical anchors for fiberglass pipe are shown in Figs. D2.9, D2.10, and D2.11.

The most common anchor, Fig. D2.9, is made from a loose-fitting clamp installed between two sets of bonded anchor saddles or between one set of saddles and a fitting to prevent movement in either direction. Anchor saddles should cover the full circumference of the pipe on both sides of the anchor clamp (unless a fitting is used on one side of the clamp) and should have a minimum length equal to the nominal diameter of the pipe. These saddles can be adhesive bonded to the pipe or built up from glass cloth or mat on the outside of the pipe. This type of anchor also acts as a support, so the clamp should have the same minimum width of the supports designed for the system.

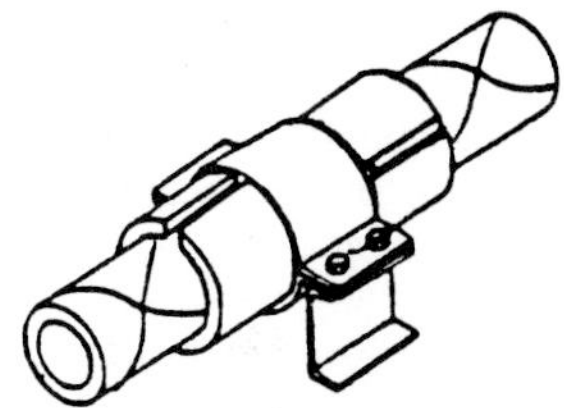

FIGURE D2.9 Pipe anchor made from loose-fitting clamp and two sets of bearing saddles.

Operating experience with fiberglass piping systems has shown that it is good practice to anchor long straight runs of above-ground piping at approximately 300-ft intervals. These anchors prevent accumulated pipe movement due to vibration, water hammer, thermal expansion, and other sources of minor cyclic movement.

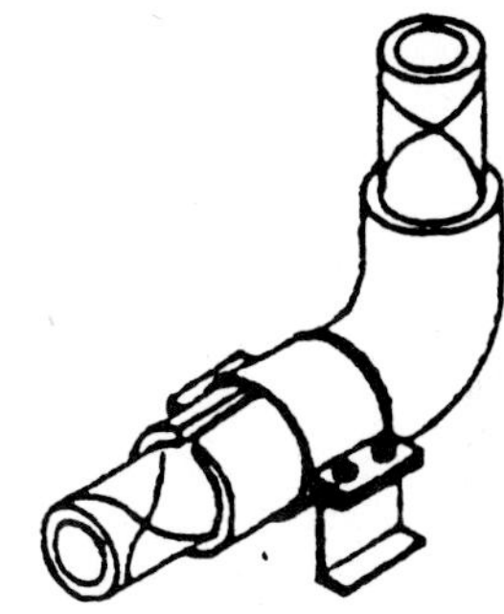

FIGURE D2.10 Pipe fitting anchor made from loose-fitting clamp and one set of bearing saddles. Fitting upset acts as second bearing saddle.

Guides

Guides are used to control thermal forces and prevent all but axial movement of the pipe. The guiding mechanism must be loose so as to allow free axial movement but must be rigidly attached to the supporting structure so the pipe can move only in the axial direction. The lateral forces being restrained by the guide are only about 1/10 of the axial forces. Guides normally act as supports and should have the same minimum bearing surface as the supports designed for the system. A typical guide is shown in Fig. D2.12.

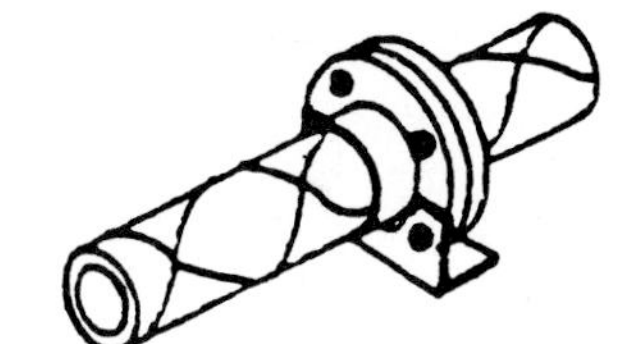

FIGURE D2.11 Anchoring pipe at a flange.

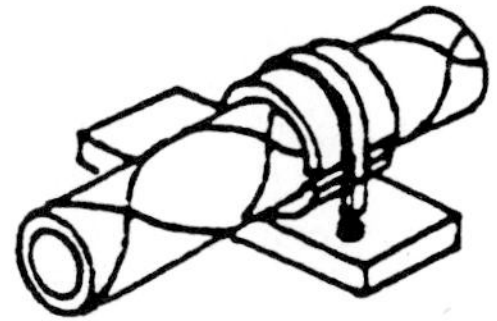

FIGURE D2.12 Pipe guide made from a U bolt and one set of bearing saddles.

Thermal Expansion

The axial thermal expansion coefficient of fiberglass pipe depends on the type of resin and the orientation of the glass fibers in the pipe wall. Most filament wound pipe is manufactured from vinyl ester, polyester, or epoxy resin with a winding angle of 54.75°. This type of pipe has an axial thermal expansion coefficient of about 1×10^{-5} in/in · °F, about two times that of steel, or about the same as aluminum. Pipe manufactured with a different glass orientation can have a different expansion coefficient. Since the thermal forces developed in fiberglass pipe are small, restraining equipment (guides, anchors, etc.) need not be as strong or heavy as for steel piping.

In fiberglass piping system design, thermal expansion is handled using the same principals that are used when designing metallic piping systems. But the different physical properties of fiberglass pipe sometimes direct the designer to use some methods which are not practical with metallic pipe.

Guide Spacing

In most instances the most efficient method of handling thermal expansion in a fiberglass piping system is to anchor both ends of each straight run and guide the pipe at intervals which prevent buckling from columnar instability.[4,20] Most manufacturers publish guide spacing information based on the difference between the installation temperature and the anticipated maximum operating temperature. If the pipe manufacturer's guide spacing information is not available, the proper guide spacing can be calculated using the physical properties of the pipe, and the end load developed during expansion.[9] The variables are related as follows by the Euler load formula for a column pin-ended at both ends:

$$E_cACT = \frac{0.6854E_bI}{L^2}$$

where E_c = compressive modulus of elasticity, psi
A = cross-sectional area, in^2
C = coefficient of thermal expansion, in/in · °F
T = temperature increase, °F
E_b = bending modulus of elasticity, psi
I = moment of inertia, in^4
L = maximum distance between guides, ft

The guide spacing should be compared with the required support spacing so the most economical spacing of supports and guides can be used.

Changes in Direction

As with most piping materials, if the temperature change is small and the piping layout is such that there are no long straight pipe runs, it is usually not necessary to make any special provisions for thermal expansion, because changes in direction in the pipeline can perform the function of expansion loops. However, it is necessary to analyze the design in these instances to determine whether changes in direction in the pipeline are capable of absorbing length changes. Most fiberglass fittings are designed with walls much thicker than the pipe. This allows the designer to consider these fittings as rigid connections in stress analysis of these systems. Saddles and laterals are exceptions. These fittings are usually not designed to withstand bending, so the pipe must be anchored to protect these fittings from thermal forces if installed in a system which will operate with a significant temperature change.

In some systems with changes in direction, it is practical to install hangers or supports that are capable of allowing enough movement in the pipeline to absorb thermal expansion from moderately long runs of pipe. In these instances the bending stress in the pipe at a given direction change depends on the total change in length to be absorbed and the distance to the first secure hanger or guide past the direction change. The bending stress under these conditions can be calculated by considering the pipe beyond the change in direction as a cantilevered beam anchored by the first secure hanger beyond the change in direction. The design bending stress of the pipe must not be exceeded. All other supports in the pipe run must be rigid enough to prevent lateral movement or buckling.

When large thermal movements must be absorbed at a change in direction, a short length of chemical and pressure compatible hose can be installed at the change in direction to absorb line movement. In many installations, this method is the most economical means of handling large thermal movements when guide spacing cannot be employed. Specification sheets are available from hose manufacturers. Minimum bending radius, chemical resistance, and temperature and pressure ratings should be checked to determine the suitability of a particular hose for a given application.

Expansion Joints

Various types of expansion joints have been successfully used with fiberglass pipe. Because the forces developed in fiberglass pipe during a temperature change are relatively small as compared with metallic pipe, it is essential to specify expansion joints which are activated by low forces. Specification sheets are available from expansion joint manufacturers.[11] Chemical resistance, temperature rating, and pressure rating must be checked to determine the compatibility of a particular expansion joint for a given application.

Four important considerations which apply to piping systems containing expansion joints are as follows:

1. The expansion joint must be selected and installed (Fig. D2.13) so that it can accommodate the maximum axial motion which can occur in either expansion or contraction in the system. In most cases this requires that some preset (compression from maximum elongation) be accomplished during installation. The amount of preset can be calculated using the following relationship:

$$\text{Length of preset} = \frac{R(T_i - T_{\min})}{T_{\max} - T_{\min}}$$

where R = rated movement of expansion joint, in
T_i = installation temperature, °F
$T_{\min}$ = minimum system temperature, °F
$T_{\max}$ = maximum system temperature, °F

2. Suitable anchors must be provided to restrain the expansion joint.
3. Guides must be installed to assure that the pipe will move directly into the expansion joint. These guides should be installed at 4 and 14 nominal pipe diameters from the expansion joint.

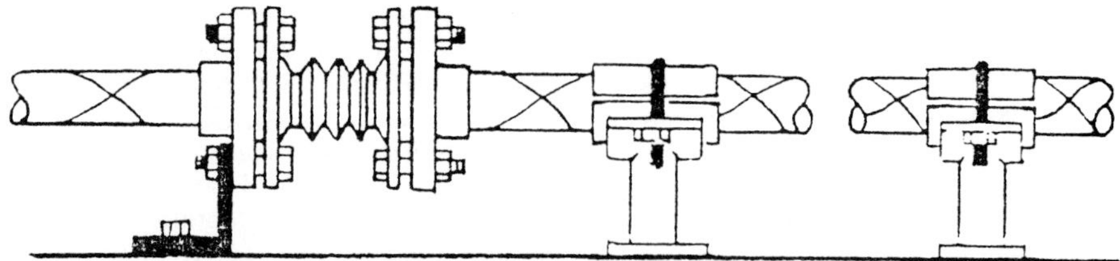

FIGURE D2.13 Expansion joint. First guide, 4 diameters distance from expansion joint. Second guide, 14 diameters distance from expansion joint.

4. Supports must be a type that prevents lateral movement and must be installed at a spacing which will prevent pipe buckling at the activation force of the expansion joint.

Expansion Loops

The cost of expansion loops usually makes this the least desirable method of handling thermal expansion in fiberglass piping systems. Expansion loops for fiberglass piping are designed in the same manner as for pipe made from other materials. The design bending stress for the pipe must not be exceeded and, as with expansion joints, proper supports and guides must be used to direct pipe movement into the expansion loop. The loop should be installed with the proper preset to handle design conditions.

Fiberglass expansion loops must be fabricated from pipe and fittings since the pipe cannot be bent in the field. Fiberglass fittings are usually made with much thicker walls than the pipe and in most designs can be considered as rigid connections during stress analysis of the loop design. Most manufacturers supply expansion loop design information to simplify design of this type of system.

Thermal Expansion in Buried Lines

Soil restraint will prevent movement of most buried fiberglass pipelines because this type of pipe develops relatively small forces during a temperature change. Special precautions are not necessary if the pipe is buried at least 3 ft deep and the soil is a type capable of restraining the line. Sand, loam, clay, and silt are suitable backfill for restraining the pipe. However, some manufacturers recommend thrust blocking or anchoring buried pipelines which will be subjected to large temperature changes to protect fittings and joints from bending stresses. This appears to be a function of joint and fitting strength and should be discussed with any prospective supplier.

Qualification of Installation Crews

Because there is no standard fiberglass pipe joining system, it is essential that installation crews be properly trained in the joining procedures to be used on a specific job. Installation crews should be trained by representatives of the pipe manufacturer, and test assemblies should be pressure tested before general pipe fabrication and installation are started. The specified test pressure for the installed piping system is the minimum test pressure requirement for assemblies fabricated during the training session. A more rigorous qualification procedure for installation crews is described in Section A328.2 of ASME B31.3. This procedure calls for pressurizing the qualification assemblies to four times the design pressure. If the design pressure is close to the rated pressure of the piping components, this procedure is equivalent to a destructive test, even if a leak is not produced, because the test pressure will overstress the test assembly.

An alternate qualification test for installation crews is to subject the fabricated assemblies to 10 pressure cycles at 1½ times the rated pressure of the piping system. This test does not meet the requirements of ASME B31.3 but is rigorous enough to assure bonding crew qualification. This type of test will not overstress the qualification assemblies. See the next section, "Pressure Test," for more information on this type of testing.

Pressure Test

The last specification, and perhaps one of the most important, in the design of a piping system, is the pressure test performed after installation. The design engineer should specify a test which is severe enough to eliminate bad joints or components which can cause problems when the line is put into operation. However, the test must not be so severe that pipeline components are damaged during testing. The recommended test procedure is a cyclic pressure test in which the piping system is subjected to 10 pressurization cycles with water at 1.5 times the anticipated or design operating pressure. After the last pressure cycle, the test pressure is maintained for 1 to 8 h while the line is inspected for leaks. When higher test pressures are desired, the test pressure should not exceed 1.5 times the rated pressure of the lowest rated component in the system.

For low-pressure drain lines, the cyclic pressure test may be replaced by a steady pressure test.

Lines which can be subjected to extreme temperature cycles—such as steam condensate lines, hot water lines, and chilled water lines—should be tested using the cyclic test procedure at 1.5 times the rated pressure, or some other pressure which will approximate the thermal stresses which will be encountered during operation. This type of test should be specified even if the system is to operate at a relatively low pressure.

Large or complicated installations should be tested in subsections as they are completed. Pressure tests should be performed on a small section of any installation as early as possible in the construction schedule to assure that proper bonding techniques are being used. This is particularly important when the installation personnel have not previously used a particular bonding system.

Testing any type of pipe with air or gas can be dangerous, and the light weight, flexibility, and elasticity of fiberglass pipe create different conditions than are present with steel pipe. If a failure should occur while testing fiberglass pipe with air or gas, the system can be subjected to considerable whipping and other shock-induced conditions due to the sudden release of stored energy. If a line must be tested with air or gas, the pipe and fittings should be secured to protect against possible injury to personnel and/or damage to the pipe and adjoining equipment.

BURIED PIPE INSTALLATIONS

When being considered for buried applications fiberglass pipe falls into two categories: (1) pipe with a stiffness of 40 or greater, and (2) pipe with a stiffness less than 40 psi. Experience has shown that pipe with a stiffness of 40 psi or greater, measured according to ASTM D2412, can be buried with fewer restrictions than are required for lower-stiffness pipe.

Pipe with Stiffness Greater Than 40 psi

Most 1- to 16-in-diameter pipe have pipe wall-to-diameter ratios which give the pipe a stiffness greater than 40 psi. Pipe in this size range installed in areas where there will be no heavy vehicular traffic can usually be buried to a maximum depth of 7 ft with normal compaction of backfill. With more compaction, these pipes can be buried to a maximum depth of 23 ft. If greater burial depth is required, a stiffer pipe or special backfill material can be used.

If this type of pipe is to be buried in an area with heavy vehicular traffic, it must normally be buried at least 3 ft deep. Special care must be given to placement and compaction of backfill material to ensure that the pipe is properly supported by the soil. Road crossings and railroad crossings are usually made by installing the pipe in a conduit.

When burying fiberglass pipe, it is important to have the trench bottom as uniform and continuous as possible. High spots in the trench bottom cause uneven backfill loading on the pipe. Sharp bends and changes in elevation in the line should also be avoided (do not exceed the recommended bending radius of the pipe). If the trench is excavated through rock or shale ledges, the trench should be dug slightly deeper, and a layer of sand or clean backfill at least 6 in deep should be used in the bottom of the ditch to assure that the pipe does not bear directly on rock. During installation, pipe should be covered as soon as possible to eliminate the possibility that the pipe might be damaged by floating if the ditch is flooded or by shifting of the line due to cave-in of the ditch wall. Clean backfill material should be used until the pipe is covered to a depth of at least 1 ft. This material should be free of sharp rocks, heavy boulders, large frozen lumps, or other objects which could damage the pipe. When installing pipe in cold weather, it is extremely important that frozen backfill be avoided in the region around the pipe. Frozen backfill will eventually thaw, leaving the pipe with insufficient support. If vibratory or similar tamping equipment is used to compact the backfill, clean backfill must be placed 6 in under, over, and around the pipe to eliminate stones which could be driven into the pipe wall during compaction.

Pipe with Stiffness Less than 40 psi

Most pipes larger than 16 in diameter have pipe wall-to-diameter ratios which yield an ASTM D2412 pipe stiffness of less than 40 psi. Particular care must be given to the selection, installation, and compaction of backfill when installing this type of pipe. Experience has shown that a minimum stiffness of 9 psi is required if thin wall pipe is to be buried. Do not attempt to bury pipe with a stiffness of less than 9 psi. In the early days of the fiberglass industry, many large-diameter lines with a pipe stiffness of less than 9 psi were damaged by earth loads immediately after burial. These failures caused the industry to form a study group which developed a method for designing pipe for buried applications. This method was then published in ANSI/AWWA C950-81, AWWA Standard for Glass Fiber Reinforced Thermosetting Resin Pressure Pipe.[1] This standard is an excellent reference when considering burial of large-diameter, thin wall fiberglass pipe.

Site Selection

The most important variable in the design of a buried thin wall fiberglass pipe system is the modulus of soil reaction of the backfill. This property of soil depends on the nature of the material: gravel, sand, clay, loam, and so on, and the amount of compaction applied to the material during installation. It is important, therefore, that a preliminary soil exploration of the site be conducted before contacting a pipe manufacturer. Borings should be made at least 2 ft below the bottom of the trench. The preliminary examination should determine the following:

1. Soil classification per ASTM D423 and the amount of fines which pass through a No. 200 sieve per ASTM D1140
2. Location of discontinuities in the soil
3. Water table level
4. Soil stability (depends on type of soil, density, and moisture content)

Excavation of Trench

In stable soils the trench should be excavated to a depth of 8 to 12 in below the grade line. Trench width should be kept to a minimum, while allowing adequate room for compaction equipment to pass between the pipe and the trench wall (Fig. D2.14).

In unstable soils it is necessary to either overexcavate the trench or put permanent shoring in the trench. When overexcavating, the trench width should be three times the pipe diameter (Fig. D2.15). If shoring is used (Fig. D2.16), the shoring material must have soil corrosion resistance as good as or better than the pipe. In either type of unstable soil installation, it is likely that the trench bottom will be unstable. If the trench bottom is unstable, it will require a minimum of 18 in of excavation below the grade line so a satisfactory foundation can be built below the pipe bedding material. The trench bottom must be kept dry during installation.

The pipeline must be protected from uneven settlement if the line runs through an area where the material beneath the trench changes from rock to soil, or where the pipe leaves a rigid support, such as a building, and enters soil. Any settlement occurring on the soil side of the change can induce excessive shear loading and bending in the pipe. The effects of the transition can be reduced by allowing for an even transition from soil to rock. The soil must be overexcavated as in the case of a trench in unstable soils for at least three diameters from the change in strata and then returned to grade in three additional diameters of ditch length. The trench can be excavated per the specifications for a stable trench throughout the rock.

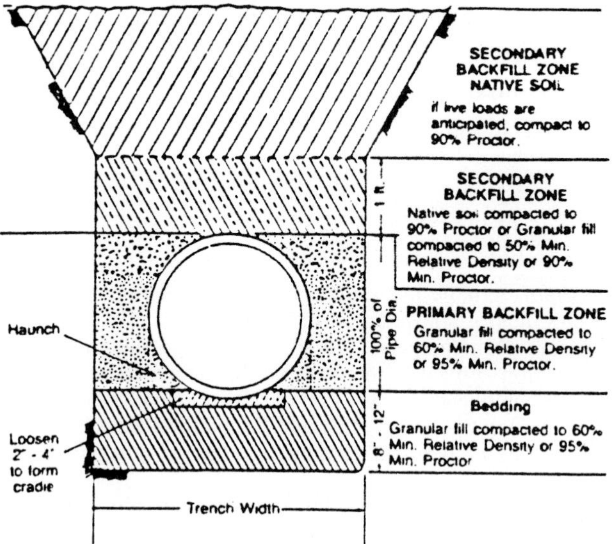

FIGURE D2.14 Burial in stable soils. (*Smith Fiberglass Products Inc.*)

Placement of Bedding

Bedding must be placed on a firm, dry trench bottom. Bedding material must be granular with little or no fines and no particles exceeding ¾ in. Clear the trench bottom of all rocks or clumps larger than 2 in. Add bedding material in 2- to 4-in lifts until the bed is 4 in above pipe grade. Compact each lift to at least 60 percent relative density per ASTM D2049 or at least 95 percent Procter density per ASTM D698. After compaction is complete, loosen an area of the bedding approximately 1 ft wide and 3 in deep, forming a cradle or trough for the pipe. The pipe can then be lowered into the ditch and set into the loosened area. The pipe must set level—eliminate any high or low spots which can cause uneven bearing and pipe wall failure.

If the bedding is below the water table, or if water will flow through the trench any time during the service life of the line, bedding must be either pea gravel or crushed stone with a maximum particle size of 3/16 in or well-graded gravel with a maximum particle size of ¾ in. Sand should be used under these circumstances only after consultation with a geotechnical engineer.

Placement of Backfill

There are two backfill zones. The primary zone extends from the bedding to the top of the pipe. The secondary zone extends from the top of the pipe to the surface.

Placement and compaction of backfill material in the primary zone are critical to successful installation of thin wall, large-diameter pipe, because the material in this area supports the flexible pipe. Backfill in the primary zone must be granular material with a maximum particle size of ¾ in containing little or no fines. If the trench is below the water table or water will flow through the trench, use only the

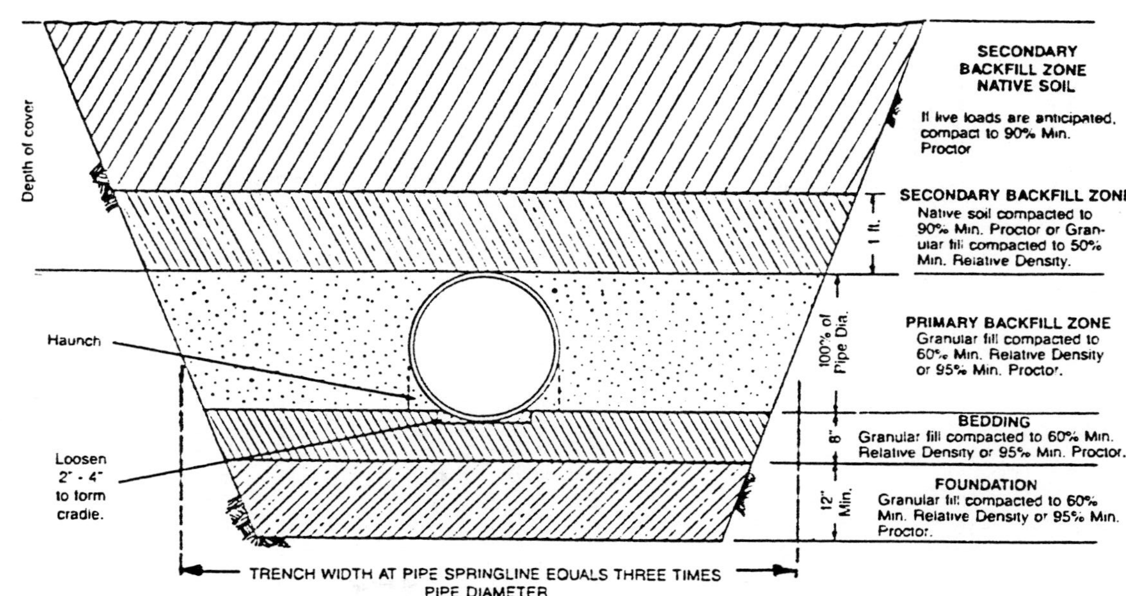

FIGURE D2.15 Burial in unstable soils by overexcavation. (*Smith Fiberglass Products Inc.*)

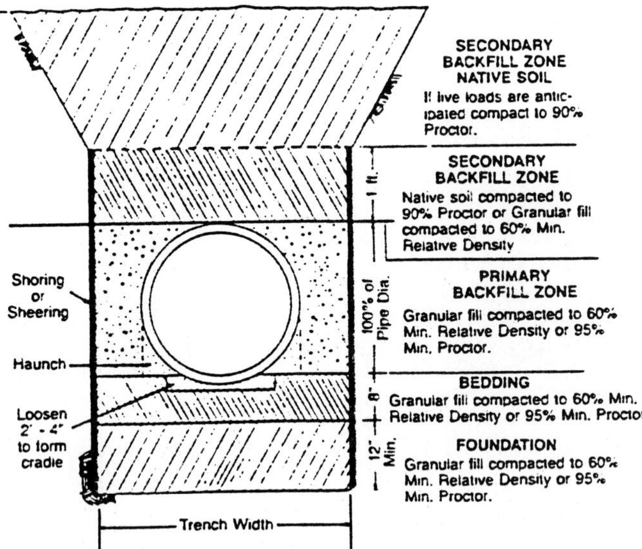

FIGURE D2.16 Burial in unstable soils by shoring. (*Smith Fiberglass Products Inc.*)

materials specified above for bedding materials under these conditions. Initial backfill material should be placed under the pipe haunches on both sides of the pipe and thoroughly hand tamped into place using a 2 × 4. The remainder of the backfill in the primary zone should be placed in 6-in lifts using a vibratory compactor between lifts to obtain at least 60 percent relative density per ASTM D2049 or at least 95 percent Procter density per ASTM D698. When placing the primary backfill, it is advantageous to allow the pipe to assume an elliptical shape with an increase in vertical diameter not to exceed 4 percent.

Backfill in the secondary zone can be native soil. In this zone, if the pipe is installed in an area where no live loads are anticipated, only the first foot of backfill above the pipe need be compacted to 50 percent relative density if less than 12 percent fines, or to 90 percent Procter density if greater than 12 percent fines. If live loads are anticipated, compaction should be carried to the surface. The total depth of burial required is dependent upon pipe design and operating conditions. A minimum burial depth of 4 ft with good compaction is required for most pipe to handle H20 wheel loading. Construction vehicles and roller compactors should not be allowed until the minimum burial depth is achieved.

Quality Checks during Installation

Compaction tests should be performed during installation to determine that the specified compaction is being achieved. The inspector should also regularly determine the amount of increase in the vertical diameter of the pipe.

Positive Projecting Conduits

If pipe is installed in a levee or embankment in such a way that the pipe is located above the surface of the natural ground and then covered, the embankment width must be at least five pipe diameters and must be compacted to the density recommended for that material if it were trench backfill.

OTHER CONSIDERATIONS

Static Electricity

In most industrial applications the generation of static electricity by fluids flowing through fiberglass pipes is not a problem. The amount of static electricity generated in a fiberglass pipe depends upon the flow rate, electrical conductivity of the fluid, turbulence, and surface area at the interface between the fluid and pipe. The effects of static electricity usually become a problem only if a dry, electrically nonconductive gas or liquid flows at a high velocity through an ungrounded system.

Relatively nonconductive liquids such as JP-4 jet fuel are handled in buried fiberglass lines with no static electricity hazard if normal grounding procedures are employed.[2]

Dry gases flowing at high velocity in above-ground pipe lines are the most common source of static electricity problems with fiberglass pipe or ducts, especially if nonconductive particles are entrained in the gas stream. Under these conditions a charge may build up on an ungrounded valve. If the charge is not drained off by humid air, personnel who come in contact with the valve may experience an electric shock. This situation can be easily remedied by grounding the valve.

Under the unique conditions where an electrically nonconductive fluid must be handled in an above-ground application, the pipe can be grounded by wrapping a wire helically around the outside of the pipe. If this method of grounding cannot be used, premium fiberglass pipe with a conductive wall may be specified.[13] However, this type of pipe is more costly than standard fiberglass pipe and is not readily available.

Water Hammer

Water hammer can cause severe movement in any piping system. But, because of the relatively light weight and low tensile modulus of fiberglass, a given water hammer force can produce more movement in a fiberglass piping system than in a steel system. Quick closing valves or rapid pump start-up or shutdown are the most common causes of water hammer. Good piping system design will minimize water hammer by the use of reasonable flow rates, slow closing valves, and controlled pump start-up. In some applications surge tanks and pump bypass valves can be used to temper the shock of water hammer that can occur for reasons which are beyond the control of the designer, such as power outage in a system with a large elevation change.

If a system in operation has pressure surges severe enough to move the pipeline, the system must be anchored at each change in direction. Laterals and saddles are particularly susceptible to damage by this type of movement. Buried lines are normally restrained by good soil compaction at changes in direction, but lines buried in unstable soil or less than 3 ft deep can be worked out of the ground by water hammer pulses.

Water hammer pressure surges can be calculated using the Talbot formula as follows:

$$P = \frac{aWV}{144g} = \frac{a}{g}\frac{SV}{2.3}$$

in which

$$a = \frac{12}{\left[\frac{w}{g}\left(\frac{1}{K} + \frac{d}{Et}\right)\right]^{1/2}}$$

where a = wave velocity, ft/s
P = pressure above normal, psig
V = flow velocity, ft/s
W = specific weight of fluid, lb/ft^3
S = specific gravity of fluid
K = bulk modulus of compressibility of liquid, psi
E = modulus of elasticity of pipe wall, psi
d = inside diameter of pipe, in
g = acceleration due to gravity, 32.2 ft/s · s
t = pipe wall thickness, in

Weathering

Exposure to sunlight produces appearance changes in all types of fiberglass pipe. The rate of appearance change depends on the type of resin in the pipe wall, the length of exposure, and the intensity of the sunlight and is caused by degradation of surface resin by ultraviolet radiation. Under the severe conditions encountered in the southwestern United States, the surface resin can be rapidly degraded to the extent that glass fibers will be exposed on the outside surface of the pipe. Although surface resin degradation may occur rather rapidly, further degradation is prevented by the ability of glass to absorb ultraviolet radiation. Studies have shown that surface degradation has little effect on the performance characteristics of a fiberglass piping system.[7] In some applications the appearance of glass fibers on the surface of a piping system is aesthetically unpleasing. In these instances the exterior of the pipe can be painted with any solvent-based paint. Urethane paints have been found to be particularly effective in withstanding ultraviolet degradation. Pigments and ultraviolet inhibitors have been added to resins to slow the effects of sunlight. These methods slow the effects of sunlight, but are not as effective as painting the pipe. Pigments are gradually weathered away as the resin is degraded, and ultraviolet inhibitors inhibit, but do not prevent, the effects of sunlight.

Surface degradation of fiberglass bonding surfaces by sunlight can be a serious problem during installation of a piping system, because bonding to a weathered surface always produces a weak joint. Studies by a major fiberglass pipe manufacturer have shown that ultraviolet radiation causes a change in the chemical composition of surface resin. This results in a very thin layer of material which has relatively weak adhesion to the substrate. Bonding to this weak layer produces a weak joint. If pipe or fittings joints have been exposed to direct sunlight for more than 2 days, the bonding surfaces should be lightly sanded to remove any weak layer on the surface. The sanding should be kept to a minimum, removing only enough material to expose a fresh surface.

The effects of ultraviolet degradation on bonding surfaces are not easily seen, so care should be taken to prevent degradation. If surface fibers are visible and easily removed by scraping, the degradation is obvious. However, in less severe cases the only evidence of weathering may be a slight color change on the surface. Always follow the manufacturer's instructions concerning storage of pipe and fittings, and keep end caps and other weathering protection intact during storage.

Abrasion Resistance

Test installations have shown that fiberglass pipe with standard wall constructions have very little resistance to the abrasion encountered in pneumatic conveying applications.[5] However, the addition of a liquid medium has a tremendous effect on the performance of fiberglass pipe conveying abrasive materials. Fiberglass piping systems are used routinely to handle fly ash slurries, finely ground limestone slurries, and wet scrubber discharge slurries in flue gas desulfurization systems at coal fired power plants. In some of these applications, fiberglass pipes have lasted 6 to 8 years handling streams which wore through steel pipe in 6 months. The good performance of fiberglass pipe in these applications is probably the result of the corrosion resistance of fiberglass. What appears to be severe abrasion in metallic pipe is probably accelerated corrosion caused by the removal of protective oxides by the abrasive action of the fluid. The abrasive action need not be too great to be able to accelerate the corrosion rate of metals. Only a thin protective film need be wiped from the corroding metal surface by the abrasive fluid. The appearance of this type of wear is similar to that of pure abrasion, since the most severe material removal occurs in the areas of highest turbulence. If abrasive action is not extreme, and corrosion is the true culprit, fiberglass pipe may be the answer to this type of problem. As with any material selection for an abrasive application, it is best to install a test spool in abrasive services unless the manufacturer can supply case history information for a similar application. Laboratory abrasion tests can give guidance, but they do not always give good indications of the abrasion resistance of a given type of pipe in a slurry application.

Fiberglass pipe with a ceramic bead liner is available for handling severely abrasive slurries. This type of pipe has been found to be particularly effective in handling bottom ash slurries from coal fired power plants and ore slurries.[18] This type of pipe has also handled pneumatic conveyed, shredded municipal refuse in a coal fired power plant,[12] showing only superficial wear after handling 25 times as much material as wore through Schedule 40 steel pipe. Moisture in the refuse apparently prevented excessive heat buildup because heat buildup prevents this type of pipe from being used to pneumatically convey dry materials.

High-Pressure Piping Systems

High-pressure fiberglass pipe, tubing, and casing are available from several manufacturers in sizes from 1½- to 8-in diameter with pressure ratings of 1000 to 4000 psig. These pipes are designed for oil field applications and are joined with threaded connections. Fittings selection is limited in these pressure ranges, and complicated piping configurations are difficult to achieve with these products.

Reclaiming an Existing Line

In some situations the light weight and flexibility of fiberglass pipe can be used to the advantage of the designer who is faced with the possibility of replacing an existing buried pipeline. A fiberglass line can be assembled and inserted into the existing line as an economical alternative to the installation of a new pipeline.[25] This method has been used to reclaim chemical sewers, water mains, natural gas mains and distribution lines, and oil field flow lines.

REFERENCES

1. ANSI/AWWA C950, AWWA Standard for Fiberglass Pressure Pipe, American Water Works Association, Denver, Colo., October 1, 1989.
2. K. C. Bachman, and J. C. Munday, "Evaluation of the Hazard of Static Electricity in Nonmetallic POL Systems—Static Effects in Handling Jet Fuel in Fiberglass Reinforced Plastic Pipe," *Technical Report No. AFWL-TR-72-90,* U.S. Air Force Weapons Laboratory, Kirkland Air Force Base, New Mexico, June 1973.
3. J. W. Ball, "Report of Flow Tests—Fiberglass Nozzle," Colorado State University, Fort Collins, Colorado, November 1971.
4. W. F. Britt, "Providing Proper Supports for Reinforced Thermoset and Non-Reinforced Thermoplastic Process Pipe," Paper No. 92, Corrosion '83, National Association of Corrosion Engineers, Houston, Texas, 1983.
5. D. Daavettilla and K. J. Oswald, "The Use of Fiberglass Pipe with a Ceramic Bead Lining to Handle Bottom Ash Slurry Applications," 7th International Coal and Lignite Utilization Conference, *Industrial Publications,* Houston, 1984, pp. 400–407.
6. J. S. Dorsey, "Using Reinforced Plastics for Process Equipment," *Chemical Engineering,* September 15, 1985, pp. 104–114.
7. *Derakane News,* Dow Chemical U.S.A., Midland, Mich. November 1983, p. 4.
8. H. A. Ershig, "Above Ground FRP Pipe Design and Support Systems," TAPPI 1987 Engineering Conference, Technical Association of the Pulp and Paper Industry, Atlanta, 1987, pp. 61–71.
9. *Fiberglass Pipe Handbook,* SPI Fiberglass Pipe Institute, New York, 1989.
10. Norman Hamner, *Nonmetals Section—Corrosion Data Survey,* 5th ed., National Association of Corrosion Engineers, Houston, 1975.
11. *Technical Handbook,* 5th ed., Rubber Expansion Joint Division, Fluid Sealing Association, Philadelphia, 1980, Chap. IV, pp. 10–11.
12. D. L. Klumb, "Union Electric Company's Solid Waste Utilization System," Union Electric Company, St. Louis, Mo., March 20, 1975.
13. J. H. Mallinson, *Corrosion-Resistant Plastic Composites in Chemical Plant Design,* Marcel Dekker, Inc., New York, 1988, p. 447.
14. A. M. May, "The Influence of Codes and Standards on Uses of Fiberglass Pipe," Chem Show Technical Presentation, New York, 1985.
15. K. J. Oswald, "Evaluation of Epoxy and Vinyl Ester Filament Wound Pipe Removed from Chemical and Petroleum Services," *Managing Corrosion with Plastics, vols. I, II, and III,* NACE, Houston, 1977, pp. 240–244.
16. K. J. Oswald, "The Effect of 25 Years of Oil Field Flow Line Service on Epoxy Fiber Glass Pipe," *Materials Performance,* vol. 27, no. 8, NACE, Houston, 1988, pp. 81–83.
17. K. J. Oswald, "The Effects of Chemical and Petroleum Service on Epoxy and Vinyl

Ester Filament Wound Pipe," K. J. Oswald, Corrosion '84, Paper No. 18, NACE, Houston, 1984.

18. K. J. Oswald, "The Performance of Fiberglass Pipe with an Abrasion Resistant Ceramic Bead Liner in Bottom Ash Slurry Applications," Paper No. 529, Corrosion '89, NACE, Houston, 1989.
19. J. A. Rolston, "When and How to Select Plastics," *Chemical Engineering,* October 29, 1984, pp. 70–75.
20. H. Summerhill, "Using Fiberglass Pipe in Condensate Return Systems," *Plant Engineering,* August 17, 1978, pp. 133–135.
21. W. A. Szymanski, "The Use and Performance of RTP in Waste Water Treatment Applications," Paper No. 320, Corrosion '83, NACE, Houston, 1983.
22. R. C. Talbot, "Using Fiberglass Reinforced Plastics," *Chemical Engineering,* October 29, 1984, pp. 76–82.
23. A. T. Tuthill, "Installed Cost of Corrosion Resistant Piping," Parts 1, 2, 3, 4, and 5, *Chemical Engineering,* March 3, pp. 113–115; March 31, pp. 125–128; May 26, pp. 99–100; and June 23, pp. 131–133, 1986.
24. P. E. Woodward and A. M. May "Designing and Specifying of Fiberglass Piping Systems," XVIII, *Journee d'Etude Internationales,* Center de Documentation du Verre Textile et des Plastics Renforces, Paris, France, March 23, 1983.
25. D. E. Wyman, "Pipeline Insertion Under the Sea," *Journal of New England Water Works Association,* vol. 89, no. 3, 1975.

P • A • R • T • E

APPENDIXES

APPENDIX E1
CONVERSION TABLES

Ervin L. Geiger, P.E.
Engineering Supervisor
Bechtel Corporation
Gaithersburg, Maryland

TABLE E1.1 Conversion Factors—Frequently Used U.S. Customary Units to SI Standard Units

To convert	To	Multiply by	To convert	To	Multiply by
Acceleration			Power		
Feet per sq second	Meters per sq second	0.3048	Btu per second	Watt	1054.350
Area					
Square feet	Square meters	0.09290304	Foot pounds per second	Watt	1.355818
Energy			Horsepower	Watt	746.
Btu	Joule	1055.06	Pressure		
Calorie	Joule	4.19002	Atmosphere	Newtons per sq meter	101325.0
Foot pound	Joule	1.355818	Bar	Newtons per sq meter	100000.
Watthour	Joule	3600.355818	Kilogram per sq cm	Newtons per sq meter	98066.50
Force			Pounds per sq in.	Newtons per sq meter	6894.757
Dyne	Newton	0.00001	Torr (mm Hg 0°C)	Newtons per sq meter	133.322
Kilogram	Newton	9.80665			
Pound	Newton	4.448222	Viscosity		
Length			Centipoise	Newton second per meter	0.001
Foot	Meter	0.3048000			
Mile (U..S statute)	Meter	1609.344	Pounds per foot second	Newton second per meter	1.488164
Mass			Volume		
Pound	Kilogram	0.4535924	Cubic foot	Cubic meter	0.02831685
Slug	Kilogram	14.59390	Gallon (U.S. liquid)	Cubic meter	0.003785412
Ton (2000 lb)	Kilogram	907.1847			

TABLE E1.2 Mass Equivalents

To obtain →, multiply ↓, by ↘	Pound (avdp)	Ounce (avdp)	Gram	Kilogram	Slug	Ton (short)*
Pound (avdp)	1	16	453.6	0.4536	0.0310	0.0005
Ounce (avdp)	0.0625	1	28.3495		0.0019	
Gram	0.0022	0.0353	1	0.001	68.5 E−5	
Kilogram	2.2046	35.274	1000	1	0.0685	0.0011
Slug	32.174	514.784	1.459 E+4	14.5939	1	
Ton (short)*	2000			907.185		1

*1 short ton = 0.8926 long tons.
1 short ton = 0.9072 metric tons.

TABLE E1.3 Length Equivalents

To obtain →, multiply ↓, by ↘	Inch	Foot	Miles (statute)	Milli-meter	Centi-meter	Meter	Kilometer
Inch	1	0.08333		25.4	2.54	0.0254	
Foot	12	1		304.8	30.48	0.0348	
Mile (statute)	63360	5280	1			1609.344	1.609344
Millimeter	0.03937	0.003281		1	0.1	0.001	
Centimeter	0.3937	0.032808		10	1	0.01	
Meter	39.3701	3.28084		1000	100	1	0.001
Kilometer	39,370	3280.8	0.62137		100,000	1000	1

TABLE E1.4 Area Equivalents

To obtain →, multiply ↓, by ↘	Square inch	Square foot	Acre	Square millimeter	Square centimeter	Square meter
Square inch	1	0.006944		645.16	6.4516	0.00064516
Square foot	144	1	2.2956 E−5	92903.04	929.0304	0.09290
Acre		43560	1			4046.8564
Square millimeter	0.00155			1	0.01	1 E−6
Square centimeter	0.1550	0.001076	2.5 E−8	100	1	0.0001
Square meter	1550.0031	10.76391	0.000247	1 E+6	10,000	1

TABLE E1.5 Volume Equivalents

To obtain →, multiply ↓, by ↘	U.S. gallon	Imperial gallon	Cubic inch	Cubic foot	Liter	Cubic meter	Barrel (oil)
U.S. gallon	1	0.83267	231	0.13368	3.7853	0.00378	0.02381
Imperial gallon	1.2009	1	277.42	0.16054	4.5459	0.00455	0.02859
Cubic inch	0.004329	0.003604	1	0.000579	0.0164	0.000016	0.00010
Cubic foot	7.4805	6.2288	1728	1	28.316	0.02832	0.17813
Liter	0.26418	0.21997	61.025	0.0353	1	0.001	0.00629
Cubic meter	264.17	219.97	61023.74	35.3147	1000	1	6.2899
Barrel (oil)*	42	34.977	9702.00	5.614	158.983	0.15876	1

*The capacity of a barrel varies with industries as follows:
1 barrel of beer = 31 U.S. gallons
1 barrel of wine = 31.5 U.S. gallons
1 barrel of oil = 42 U.S. gallons
1 barrel of whiskey = 45 U.S. gallons

TABLE E1.6 Volumetric Flow Rate Equivalents

To obtain →, multiply ↓, by ↘	U.S. gallons per minute	Imperial gallons per minute	U.S. million gallons per day	Cubic feet per second	Cubic meter per hour	Liter per second	Barrel (42 gallons) per minute	Barrel (42 gallons) per day
U.S. gallons per minute	1	0.8327	0.00144	0.00223	0.2271	0.0631	0.0238	34.286
Imperial gallons per minute	1.201	1	0.00173	0.002676	0.2727	0.0758	0.02859	41.176
U.S. million gallons per day	694.4	578.25	1	1.547	157.7	43.8	16.53	23810
Cubic feet per second	448.83	373.7	0.646	1	101.9	28.32	10.686	15388
Cubic meters per second	15850	13199	22.83	35.315	3600	1000	377.4	543447
Cubic meters per minute	264.2	220	0.3804	0.5886	60.0	16.667	6.290	9058
Cubic meters per hour	4.403	3.67	0.00634	0.00982	1	0.2778	0.1048	151
Liter per seconds	15.85	13.20	0.0228	0.0353	3.60	1	0.3773	543.3
Liter per minute	0.2642	0.220	0.000380	0.000589	0.060	0.0167	0.00629	9.055
Barrel (42 gallons) per minute	42	34.97	0.0605	0.09357	9.5256	2.65	1	1440
Barrel (42 gallons) per day	0.0292	0.0243	0.000042	0.000065	0.00662	0.00184	0.00069	1

TABLE E1.7 Density Equivalents

To obtain → multiply ↓, by ↘	Grams per cubic centimeter	Pounds per cubic inch	Pounds per cubic foot	Slugs per cubic foot	Kilograms per cubic meter
Grams per cubic centimeter	1	0.03613	62.42806	1.9403	1000
Pounds per cubic inch	27.67991	1	1728	53.708	27679.905
Pounds per cubic foot	0.01602	0.0005787	1	0.031081	16.01846
Slugs per cubic foot	0.51538	0.0186	32.17	1	515.379
Kilograms per cubic meter	0.001	3.6128E-5	0.06243	0.00194	1

TABLE E1.8 Pressure Equivalents

To obtain → multiply ↓, by ↘	Pounds per square inch	Pounds per square foot	Atmosphere	Kilograms per square centimeter	Kilograms per square meter	Inches water (68°F)	Foot water (68°F)	Inches mercury (32°F)	Millimeters mercury (32°F)	Bar	Mega-Pascal (MPa)
Pounds per square inch	1	144	0.068046	0.070307	703.070	27.7276	2.310636	2.03602	51.71497	0.068947	0.006895
Pounds per square foot	0.0069444	1	0.000473	0.000488	4.88242	0.1926	0.01605	0.014139	0.35913	0.000479	0.0000479
Atmosphere	14.696	2116.22	1	1.03323	10332.27	407.484	33.9570	29.9213	760	1.01325	0.101325
Kilograms per square centimeter	14.2233	2048.155	0.96784	1	10000.	394.38	32.8650	28.959	735.559	0.98067	0.098067
Kilograms per square meter	0.001422	0.204768	0.0000968	0.0001	1	0.03944	0.003287	0.002896	0.073556	0.000098	0.0000098
Inches water (68°F)	0.036065	5.1972	0.002454	0.00253	25.375	1	0.08333	0.073430	1.8651	0.002466	0.000249
Foot water (68°F)	0.432781	62.3205	0.029449	0.03043	304.275	12	1	0.88115	22.3813	0.029839	0.0029839
Inches mercury (32°F)	0.491154	70.7262	0.033421	0.03453	345.316	13.6185	1.1349	1	25.4	0.033864	0.0033864
Millimeters mercury (32°F)	0.0193368	2.78450	0.0013158	0.0013595	13.59509	0.53616	0.044680	0.03937	1	0.001333	0.0001333
Bar	14.5038	2088.55	0.98692	1.01972	10197.2	402.156	33.5130	29.5300	750.062	1	0.10
MPa	145.038	20885.5	9.8692	10.1972	101972.0	4021.56	335.130	295.300	7500.62	10.0	1

TABLE E1.9 Energy Equivalents

To obtain → multiply ↓, by ↘	Btu*	Calorie (gram)†	Foot pound	Horsepower hour	Joule	Kilowatt hour	Kilogram meter
Btu	1	251.8	778.169	0.000393	1055.06	0.000293	107.586
Calorie (gram)†	0.00397	1	3.08596	1.56 E−6	4.184	1.16 E−6	0.426649
Foot pound	0.00129	0.32405	1	5.05 E−7	1.35582	3.77 E−7	0.13825
Horsepower hour	2544.5	641616	1.98 E+6	1	2.68 E+6	0.7457	273745
Joule	0.00095	0.2390	0.73756	3.72 E−7	1	2.77 E−7	0.102
Kilowatt hour	3412.97	860421	2.65 E+6	1.34102	3.6 E+6	1	367098
Kilogram meter	0.00929	2.344	7.233	3.65 E−6	9.807	2.72 E−6	1

*Based on 1 Btu = 778.169 ft · lb.
†Based on 1 Cal = 4.1840 Joules

TABLE E1.10 Power Equivalents*

To obtain →, multiply ↓, by ↘	Horsepower	Foot pound per second	Watt	Btu per hour
Horsepower	1	550	745.7	2544
Foot pound per second	0.00182	1	1.3558	4.626
Watt	0.00134	0.7376	1	3.412
Btu per hour	0.00039	0.2161	0.2931	1

*Based on 1 Btu = 778.169 ft · lb.

TABLE E1.11 Conversion Factors for Thermal Conductivity, k

To obtain →, multiply ↓, by ↘	Watt/ m · °K	Cal/ (s · cm · °K)	Btu/ (h · ft · °R)	Btu · in/ (h · ft² · °R)
W/m · °K	1	0.0022901	0.57818	6.9381
W/cm · °K	100	0.23901	57.818	693.81
W/ft · °R	5.9055	0.014114	3.4144	40.973
W/in · °R	70.866	0.16937	40.973	491.68
cal/(s · cm · °K)	418.4	1	241.91	2902.9
cal/(s · in · °R)	296.50	0.70866	171.43	2057.2
kcal/(h · m · °K)	1.1622	0.0027778	0.67197	8.0636
Btu/(s · ft · °R)	6226.5	14.882	3600	43200
Btu/(s · in · °R)	74717 E+4	178.58	43200	5.184 E+5
Btu · in/(s · ft² · °R)	518.87	1.2401	300	3600
Btu/(h · ft · °R)	1.7296	0.0041338	1	12
Btu/(h · in · °R)	20.755	0.049605	12	144
Btu in/(h · ft² · °R)	0.14413	0.00034448	0.083333	1
lb · ft/(h · ft · °R)	0.0022241	5.3157 E−6	0.0012859	0.015431

TABLE E1.12 Prefix Names of Multiples and Submultiples of Units

Prefix	Measure	Decimal equivalent	Exponential expression
Alto	one-quintillionth	0.000 000 000 000 000 001	E−18
Femto	one-quadrillionth	0.000 000 000 000 001	E−15
Pico	one-trillionth	0.000 000 000 001	E−12
Nano	one-billionth	0.000 000 001	E−9
Micro	one-millionth	0.000 001	E−6
Milli	one-thousandth	0.001	E−3
Centi	one-hundredth	0.01	E−2
Deci	one-tenth	0.1	E−1
Uni	one	1.0	E0
Deka	ten	10.0	E+1
Hecto	one hundred	100.0	E+2
Kilo	one thousand	1 000.0	E+3
Mega	one million	1 000 000.0	E+6
Giga	one billion	1 000 000 000.0	E+9
Tera	one trillion	1 000 000 000 000.0	E+12

APPENDIX E2
PIPE PROPERTIES

Dr. Chakrapani Basavaraju
Engineering Specialist
Bechtel Corporation
Gaithersburg, Maryland

TABLE E2.1 Principal Properties of Commercial Pipe

Nominal pipe size, outside diam., (in)	Schedule number*			Wall thickness (in)	Inside diam. (in)	Inside area, (in^2)	Metal area (in^2)	Ft^2 outside surface, per ft	Ft^2 inside surface per ft	Weight per ft (lb)†	Weight of water per ft (lb)	Moment of inertia, (in^4)	Elastic section modulus (in^3)	Radius gyration (in)	Plastic section modulus (in^3)
	a	b	c												
	...	...	5S	0.065	0.710	0.396	0.158	0.220	0.186	0.538	0.171	0.012	0.029	0.275	0.039
	...	...	10S	0.083	0.674	0.357	0.197	0.220	0.177	0.671	0.155	0.014	0.034	0.269	0.048
½	40	Std	40S	0.109	0.622	0.304	0.250	0.220	0.163	0.851	0.132	0.017	0.041	0.261	0.059
0.840	80	XS	80S	0.147	0.546	0.234	0.320	0.220	0.143	1.088	0.101	0.020	0.048	0.251	0.072
	160	...	...	0.187	0.466	0.171	0.383	0.220	0.122	1.304	0.074	0.022	0.053	0.240	0.082
	...	XXS	...	0.294	0.252	0.050	0.504	0.220	0.066	1.714	0.022	0.024	0.058	0.220	0.096
	...	...	5S	0.065	0.920	0.665	0.201	0.275	0.241	0.684	0.288	0.025	0.047	0.349	0.063
	...	...	10S	0.083	0.884	0.614	0.252	0.275	0.231	0.857	0.266	0.030	0.057	0.343	0.078
¾	40	Std	40S	0.113	0.824	0.533	0.333	0.275	0.216	1.131	0.230	0.037	0.071	0.334	0.100
1.050	80	XS	80S	0.154	0.742	0.432	0.435	0.275	0.194	1.474	0.188	0.045	0.085	0.321	0.125
	160	...	...	0.218	0.614	0.296	0.570	0.275	0.161	1.937	0.128	0.053	0.100	0.304	0.154
	...	XXS	...	0.308	0.434	0.148	0.718	0.275	0.114	2.441	0.064	0.058	0.110	0.284	0.179
	...	...	5S	0.065	1.185	1.103	0.2553	0.344	0.310	0.868	0.478	0.0500	0.0760	0.443	0.102
	...	...	10S	0.109	1.097	0.945	0.413	0.344	0.2872	1.404	0.409	0.0757	0.1151	0.428	0.159
1	40	Std	40S	0.133	1.049	0.864	0.494	0.344	0.2746	1.679	0.374	0.0874	0.1329	0.421	0.187
1.315	80	XS	80S	0.179	0.957	0.719	0.639	0.344	0.2520	2.172	0.311	0.1056	0.1606	0.407	0.233
	160	...	...	0.250	0.815	0.522	0.836	0.344	0.2134	2.844	0.2261	0.1252	0.1903	0.387	0.289
	...	XXS	...	0.358	0.599	0.2818	1.076	0.344	0.1570	3.659	0.1221	0.1405	0.2137	0.361	0.343
	...	...	5S	0.065	1.770	2.461	0.375	0.497	0.463	1.274	1.067	0.1580	0.1663	0.649	0.219
	...	...	10S	0.109	1.682	2.222	0.613	0.497	0.440	2.085	0.962	0.2469	0.2599	0.634	0.350
1½	40	Std	40S	0.145	1.610	2.036	0.799	0.497	0.421	2.718	0.882	0.310	0.326	0.623	0.448
1.900	80	XS	80S	0.200	1.500	1.767	1.068	0.497	0.393	3.631	0.765	0.391	0.412	0.605	0.581
	160	...	...	0.281	1.338	1.406	1.429	0.497	0.350	4.859	0.608	0.483	0.508	0.581	0.744
	...	XXS	...	0.400	1.100	0.950	1.885	0.497	0.288	6.408	0.412	0.568	0.598	0.549	0.921
	...	...	5S	0.065	2.245	3.96	0.472	0.622	0.588	1.604	1.716	0.315	0.2652	0.817	0.347
	...	...	10S	0.109	2.157	3.65	0.776	0.622	0.565	2.638	1.582	0.499	0.420	0.802	0.560
2	40	Std	40S	0.154	2.067	3.36	1.075	0.622	0.541	3.653	1.455	0.666	0.561	0.787	0.761
2.375	80	XS	80S	0.218	1.939	2.953	1.477	0.622	0.508	5.022	1.280	0.868	0.731	0.766	1.018
	160	...	...	0.343	1.689	2.240	2.190	0.622	0.422	7.444	0.971	1.163	0.979	0.729	1.430
	...	XXS	...	0.436	1.503	1.774	2.656	0.622	0.393	9.029	0.769	1.312	1.104	0.703	1.667

	...	...	5S	0.083	3.334	8.73	0.891	0.916	0.873	3.03	3.78	1.301	0.744	1.208	0.969
	...	...	10S	0.120	3.260	8.35	1.274	0.916	0.853	4.33	3.61	1.822	1.041	1.196	1.372
3	40	Std	40S	0.216	3.068	7.39	2.228	0.916	0.803	7.58	3.20	3.02	1.724	1.164	2.333
3.500	80	XS	80S	0.300	2.900	6.61	3.02	0.916	0.759	10.25	2.864	3.90	2.226	1.136	3.081
	160	...	...	0.437	2.626	5.42	4.21	0.916	0.687	14.32	2.348	5.03	2.876	1.094	4.128
	...	XXS	...	0.600	2.300	4.15	5.47	0.916	0.602	18.58	1.801	5.99	3.43	1.047	5.118
	...	...	5S	0.083	4.334	14.75	1.152	1.178	1.135	3.92	6.40	2.811	1.249	1.562	1.620
	...	...	10S	0.120	4.260	14.25	1.651	1.178	1.115	5.61	6.17	3.96	1.762	1.549	2.303
4	40	Std	40S	0.237	4.026	12.73	3.17	1.178	1.054	10.79	5.51	7.23	3.21	1.510	4.312
4.500	80	XS	80S	0.337	3.826	11.50	4.41	1.178	1.002	14.98	4.98	9.61	4.27	1.477	5.853
	120	...	...	0.437	3.626	10.33	5.58	1.178	0.949	18.96	4.48	11.65	5.18	1.445	7.242
	160	...	...	0.531	3.438	9.28	6.62	1.178	0.900	22.51	4.02	13.27	5.90	1.416	8.415
	...	XXS	...	0.674	3.152	7.80	8.10	1.178	0.825	27.54	3.38	15.29	6.79	1.374	9.968
	...	...	5S	0.109	6.407	32.2	2.231	1.734	1.677	5.37	13.98	11.85	3.58	2.304	4.628
	...	...	10S	0.134	6.357	31.7	2.733	1.734	1.664	9.29	13.74	14.40	4.35	2.295	5.647
6	40	Std	40S	0.280	6.065	28.89	5.58	1.734	1.588	18.97	12.51	28.14	8.50	2.245	11.280
6.625	80	XS	80S	0.432	5.761	26.07	8.40	1.734	1.508	28.57	11.29	40.5	12.23	2.195	16.600
	120	...	...	0.562	5.501	23.77	10.70	1.734	1.440	36.39	10.30	49.6	14.98	2.153	20.718
	160	...	...	0.718	5.189	21.15	13.33	1.734	1.358	45.30	9.16	59.0	17.81	2.104	25.176
	...	XXS	...	0.864	4.897	18.83	15.64	1.734	1.282	53.16	8.17	66.3	20.03	2.060	28.890
	...	...	5S	0.109	8.407	55.5	2.916	2.258	2.201	9.91	24.07	26.45	6.13	3.01	7.905
	...	...	10S	0.148	8.329	54.5	3.94	2.258	2.180	13.40	23.59	35.4	8.21	3.00	10.636
	20	...	...	0.250	8.125	51.8	6.58	2.258	2.127	22.36	22.48	57.7	13.39	2.962	17.540
	30	...	...	0.277	8.071	51.2	7.26	2.258	2.113	24.70	22.18	63.4	14.69	2.953	19.311
	40	Std	40S	0.322	7.981	50.0	8.40	2.258	2.089	28.55	21.69	72.5	16.81	2.938	22.210
8	60	...	...	0.406	7.813	47.9	10.48	2.258	2.045	35.64	20.79	88.8	20.58	2.909	27.448
8.625	80	XS	80S	0.500	7.625	45.7	12.76	2.258	1.996	43.39	19.80	105.7	24.52	2.878	33.050
	100	...	...	0.593	7.439	43.5	14.96	2.258	1.948	50.87	18.84	121.4	28.14	2.847	38.326
	120	...	...	0.718	7.189	40.6	17.84	2.258	1.882	60.63	17.60	140.6	32.6	2.807	45.013
	140	...	...	0.812	7.001	38.5	19.93	2.258	1.833	67.76	16.69	153.8	35.7	2.777	49.745
	...	XXS	...	0.875	6.875	37.1	21.30	2.258	1.800	72.42	16.09	162.0	37.6	2.757	52.778
	160	...	...	0.906	6.813	36.5	21.97	2.258	1.784	74.69	15.80	165.9	38.5	2.748	54.230

TABLE E2.1 Principal Properties of Commercial Pipe (*Continued*)

Nominal pipe size, outside diam., (in)	Schedule number*			Wall thickness (in)	Inside diam. (in)	Inside area, (in^2)	Metal area (in^2)	Ft^2 outside surface, per ft	Ft^2 inside surface per ft	Weight per ft (lb)†	Weight of water per ft (lb)	Moment of inertia, (in^4)	Elastic section modulus (in^3)	Radius gyration (in)	Plastic section modulus (in^3)
	a	b	c												
10 *10.750*	...	...	5S	0.134	10.482	86.3	4.52	2.815	2.744	15.15	37.4	63.7	11.85	3.75	15.103
	...	...	10S	0.165	10.420	85.3	5.49	2.815	2.728	18.70	36.9	76.9	14.30	3.74	18.489
	20	...	...	0.250	10.250	82.5	8.26	2.815	2.683	28.04	35.8	113.7	21.16	3.71	27.568
	...	...	...	0.279	10.192	81.6	9.18	2.815	2.668	31.20	35.3	125.9	23.42	3.70	30.597
	30	...	...	0.307	10.136	80.7	10.07	2.815	2.654	34.24	35.0	137.5	25.57	3.69	33.490
	40	Std	40S	0.365	10.020	78.9	11.91	2.815	2.623	40.48	34.1	160.8	29.90	3.67	39.381
	60	XS	80S	0.500	9.750	74.7	16.10	2.815	2.553	54.74	32.3	212.0	39.4	3.63	52.573
	80	...	...	0.593	9.564	71.8	18.92	2.815	2.504	64.33	31.1	244.9	45.6	3.60	61.246
	100	...	...	0.718	9.314	68.1	22.63	2.815	2.438	76.93	29.5	286.2	53.2	3.56	72.384
	120	...	...	0.843	9.064	64.5	26.24	2.815	2.373	89.20	28.0	324	60.3	3.52	82.939
	140	...	...	1.000	8.750	60.1	30.6	2.815	2.291	104.13	26.1	368	68.4	3.47	95.396
	160	...	...	1.125	8.500	56.7	34.0	2.815	2.225	115.65	24.6	399	74.3	3.43	104.695
12 *12.750*	...	...	5S	0.156	12.438	121.4	6.17	3.34	3.26	20.99	52.7	122.2	19.20	4.45	24.744
	...	...	10S	0.180	12.390	120.6	7.11	3.34	3.24	24.20	52.2	140.5	22.03	4.44	28.443
	20	...	...	0.250	12.250	117.9	9.84	3.34	3.21	33.38	51.1	191.9	30.1	4.42	39.068
	30	...	...	0.330	12.090	114.8	12.88	3.34	3.17	43.77	49.7	248.5	39.0	4.39	50.917
	...	Std	40S	0.375	12.000	113.1	14.58	3.34	3.14	49.56	49.0	279.3	43.8	4.38	57.445
	40	...	...	0.406	11.938	111.9	15.74	3.34	3.13	53.53	48.5	300	47.1	4.37	61.886
	...	XS	80S	0.500	11.750	108.4	19.24	3.34	3.08	65.42	47.0	362	56.7	4.33	75.073
	60	...	...	0.562	11.626	106.2	21.52	3.34	3.04	73.16	46.0	401	62.8	4.31	83.543
	80	...	...	0.687	11.376	101.6	26.04	3.34	2.978	88.51	44.0	475	74.5	4.27	100.078
	100	...	...	0.843	11.064	96.1	31.5	3.34	2.897	107.20	41.6	562	88.1	4.22	119.717
	120	...	...	1.000	10.750	90.8	36.9	3.34	2.814	125.49	39.3	642	100.7	4.17	138.396
	140	...	...	1.125	10.500	86.6	41.1	3.34	2.749	139.68	37.5	701	109.9	4.13	152.508
	160	...	...	1.312	10.126	80.5	47.1	3.34	2.651	160.27	34.9	781	122.6	4.07	172.399

14 *14.000*	...	...	5S	0.156	13.688	147.20	6.78	3.67	3.58	23.0	63.7	162.6	23.2	4.90	29.900
	...	...	10S	0.188	13.624	145.80	8.16	3.67	3.57	27.7	63.1	194.6	27.8	4.88	35.867
	10	...	...	0.250	13.500	143.1	10.80	3.67	3.53	36.71	62.1	255.4	36.5	4.86	47.271
	20	...	...	0.312	13.376	140.5	13.42	3.67	3.50	45.68	60.9	314	44.9	4.84	58.467
	30	Std	...	0.375	13.250	137.9	16.05	3.67	3.47	54.57	59.7	373	53.3	4.82	69.633
	40	...	...	0.437	13.126	135.3	18.62	3.67	3.44	63.37	58.7	429	61.2	4.80	80.416
	...	XS	...	0.500	13.000	132.7	21.21	3.67	3.40	72.09	57.5	484	69.1	4.78	91.167
	...	...	...	0.562	12.876	130.2	23.73	3.67	3.37	80.66	56.5	537	76.7	4.76	101.545
	60	...	...	0.593	12.814	129.0	24.98	3.67	3.35	84.91	55.9	562	80.3	4.74	106.660
	...	...	...	0.625	12.750	127.7	26.26	3.67	3.34	89.28	55.3	589	84.1	4.73	111.889
	...	...	...	0.687	12.626	125.2	28.73	3.67	3.31	97.68	54.3	638	91.2	4.71	121.869
	80	...	...	0.750	12.500	122.7	31.2	3.67	3.27	106.13	53.2	687	98.2	4.69	131.813
	...	...	...	0.875	12.250	117.9	36.1	3.67	3.21	122.66	51.1	781	111.5	4.65	150.956
	100	...	...	0.937	12.126	115.5	38.5	3.67	3.17	130.73	50.0	825	117.8	4.63	160.166
	120	...	...	1.093	11.814	109.6	44.3	3.67	3.09	150.67	47.5	930	132.8	4.58	182.519
	140	...	...	1.250	11.500	103.9	50.1	3.67	3.01	170.22	45.0	1127	146.8	4.53	203.854
	160	...	...	1.406	11.188	98.3	55.6	3.67	2.929	189.12	42.6	1017	159.6	4.48	223.931
16 *16.000*	...	...	5S	0.165	15.670	192.90	8.21	4.19	4.10	28.00	83.5	257	32.2	5.60	41.375
	...	...	10S	0.188	15.624	191.7	9.34	4.19	4.09	32.00	83.0	292	36.5	5.59	47.006
	10	...	...	0.250	15.500	188.7	12.37	4.19	4.06	42.05	81.8	384	48.0	5.57	62.021
	20	...	...	0.312	15.376	185.7	15.38	4.19	4.03	52.36	80.5	473	59.2	5.55	76.798
	30	Std	...	0.375	15.250	182.6	18.41	4.19	3.99	62.58	79.1	562	70.3	5.53	91.570
	...	...	...	0.437	15.126	179.7	21.37	4.19	3.96	72.64	77.9	648	80.9	5.50	105.872
	40	XS	...	0.500	15.000	176.7	24.35	4.19	3.93	82.77	76.5	732	91.5	5.48	120.167
	...	...	...	0.562	14.876	173.8	27.26	4.19	3.89	92.66	75.4	813	106.6	5.46	134.002
	...	...	...	0.625	14.750	170.9	30.2	4.19	3.86	102.63	74.1	894	112.2	5.44	147.826
	60	...	...	0.656	14.688	169.4	31.6	4.19	3.85	107.50	73.4	933	116.6	5.43	154.542
	...	...	...	0.687	14.626	168.0	33.0	4.19	3.83	112.36	72.7	971	121.4	5.42	161.201
	...	...	...	0.750	14.500	165.1	35.9	4.19	3.80	122.15	71.5	1047	130.9	5.40	174.563
	80	...	...	0.843	14.314	160.9	40.1	4.19	3.75	136.46	69.7	1157	144.6	5.37	193.866
	...	...	...	0.875	14.250	159.5	41.6	4.19	3.73	141.35	69.1	1193	154.1	5.36	200.393
	100	...	...	1.031	13.938	152.5	48.5	4.19	3.65	164.83	66.1	1365	170.6	5.30	231.383
	120	...	...	1.218	13.564	144.5	56.6	4.19	3.55	192.29	62.6	1556	194.5	5.24	266.745
	140	...	...	1.437	13.126	135.3	65.7	4.19	3.44	223.64	58.6	1760	220.0	5.17	305.750
	160	...	...	1.593	12.814	129.0	72.1	4.19	3.35	245.11	55.9	1894	236.7	5.12	331.993

TABLE E2.1 Principal Properties of Commercial Pipe (*Continued*)

Nominal pipe size, outside diam., (in)	Schedule number*			Wall thickness (in)	Inside diam. (in)	Inside area, (in^2)	Metal area (in^2)	Ft^2 outside surface, per ft	Ft^2 inside surface per ft	Weight per ft (lb)†	Weight of water per ft (lb)	Moment of inertia, (in^4)	Elastic section modulus (in^3)	Radius gyration (in)	Plastic section modulus (in^3)
	a	b	c												
18 *18.000*	...	...	5S	0.165	17.670	245.20	9.24	4.71	4.63	31.00	106.2	368	40.8	6.31	52.486
	...	...	10S	0.188	17.624	243.90	10.52	4.71	4.61	36.00	105.7	417	46.4	6.30	59.649
	10	...	...	0.250	17.500	240.5	13.94	4.71	4.58	47.39	104.3	549	61.0	6.28	78.771
	20	...	...	0.312	17.376	237.1	17.34	4.71	4.55	59.03	102.8	678	75.5	6.25	97.624
	...	Std	...	0.375	17.250	233.7	20.76	4.71	4.52	70.59	101.2	807	89.6	6.23	116.508
	30	...	...	0.437	17.126	230.4	24.11	4.71	4.48	82.06	99.9	931	103.4	6.21	134.824
	...	XS	...	0.500	17.000	227.0	27.49	4.71	4.45	93.45	98.4	1053	117.0	6.19	153.167
	40	...	...	0.562	16.876	223.7	30.8	4.71	4.42	104.75	97.0	1172	130.2	6.17	170.954
	...	...	...	0.625	16.750	220.5	34.1	4.71	4.39	115.98	95.5	1289	143.3	6.15	188.763
	...	...	...	0.687	16.626	217.1	37.4	4.71	4.35	127.03	94.1	1403	156.3	6.13	206.029
	60	...	...	0.750	16.500	213.8	40.6	4.71	4.32	138.17	92.7	1515	168.3	6.10	223.313
	...	...	...	0.875	16.250	207.4	47.1	4.71	4.25	160.04	89.9	1731	192.8	6.06	256.831
	80	...	...	0.937	16.126	204.2	50.2	4.71	4.22	170.75	88.5	1834	203.8	6.04	273.078
	100	...	...	1.156	15.688	193.3	61.2	4.71	4.11	207.96	83.7	2180	242.2	5.97	328.496
	120	...	...	1.375	15.250	182.6	71.8	4.71	3.99	244.14	79.2	2499	277.6	5.90	380.904
	140	...	...	1.562	14.876	173.8	80.7	4.71	3.89	274.23	75.3	2750	306	5.84	423.335
	160	...	...	1.781	14.438	163.7	90.7	4.71	3.78	308.51	71.0	3020	336	5.77	470.386
20 *20.000*	...	...	5S	0.188	19.634	302.40	11.70	5.24	5.14	40	131.0	574	57.4	7.00	71.869
	...	...	10S	0.218	19.564	300.6	13.55	5.24	5.12	46	130.2	663	66.3	6.99	85.313
	10	...	...	0.250	19.500	298.6	15.51	5.24	5.11	52.73	129.5	757	75.7	6.98	97.521
	...	...	...	0.312	19.376	294.9	19.30	5.24	5.07	65.40	128.1	935	93.5	6.96	120.947
	20	Std	...	0.375	19.250	291.0	23.12	5.24	5.04	78.60	126.0	1114	111.4	6.94	144.445
	...	...	...	0.437	19.126	287.3	26.86	5.24	5.01	91.31	124.6	1286	128.6	6.92	167.273
	30	XS	...	0.500	19.000	283.5	30.6	5.24	4.97	104.13	122.8	1457	145.7	6.90	190.167
	...	...	...	0.562	18.876	279.8	34.3	5.24	4.94	116.67	121.3	1624	162.4	6.88	212.403
	40	...	...	0.593	18.814	278.0	36.2	5.24	4.93	122.91	120.4	1704	170.4	6.86	223.412
	...	...	...	0.625	18.750	276.1	38.0	5.24	4.91	129.33	119.7	1787	178.7	6.85	234.701
	...	...	...	0.687	18.626	272.5	41.7	5.24	4.88	141.71	118.1	1946	194.6	6.83	256.354
	...	...	...	0.750	18.500	268.8	45.4	5.24	4.84	154.20	116.5	2105	210.5	6.81	278.063
	60	...	...	0.812	18.376	265.2	48.9	5.24	4.81	166.40	115.0	2257	225.7	6.79	299.140

	...	...	...	0.875	18.250	261.6	52.6	5.24	4.78	178.73	113.4	2409	240.9	6.77	320.268
	80	...	...	1.031	17.938	252.7	61.4	5.24	4.70	208.87	109.4	2772	277.2	6.72	371.343
	100	...	...	1.281	17.438	238.8	75.3	5.24	4.57	256.10	103.4	3320	332	6.63	449.564
	120	...	...	1.500	17.000	227.0	87.2	5.24	4.45	296.37	98.3	3760	376	6.56	514.500
	140	...	...	1.750	16.500	213.8	100.3	5.24	4.32	341.10	92.6	4220	422	6.48	584.646
	160	...	...	1.968	16.064	202.7	111.5	5.24	4.21	379.01	87.9	4590	459	6.41	642.442
	...	...	5S	0.188	21.624	367.3	12.88	5.76	5.66	44	159.1	766	69.7	7.71	89.446
	...	...	10S	0.218	21.564	365.2	14.92	5.76	5.65	51	158.2	885	80.4	7.70	103.435
	10	...	...	0.250	21.500	363.1	17.16	5.76	5.63	58	157.4	1010	91.8	7.69	118.271
	20	Std	...	0.375	21.250	354.7	25.48	5.76	5.56	87	153.7	1490	135.4	7.65	175.383
	30	XS	...	0.500	21.000	346.4	33.77	5.76	5.50	115	150.2	1953	177.5	7.61	231.167
	...	...	...	0.625	20.750	338.2	41.97	5.76	5.43	143	146.6	2400	218.2	7.56	285.638
22	...	...	...	0.750	20.500	330.1	50.07	5.76	5.37	170	143.1	2829	257.2	7.52	338.813
22.000	60	...	...	0.875	20.250	322.1	58.07	5.76	5.30	197	139.6	3245	295.0	7.47	390.706
	80	...	...	1.125	19.750	306.4	73.78	5.76	5.17	251	132.8	4029	366.3	7.39	490.711
	100	...	...	1.375	19.250	291.0	89.09	5.76	5.04	303	126.2	4758	432.6	7.31	585.779
	120	...	...	1.625	18.750	276.1	104.02	5.76	4.91	354	119.6	5432	493.8	7.23	676.034
	140	...	...	1.875	18.250	261.6	118.55	5.76	4.78	403	113.3	6054	550.3	7.15	761.602
	160	...	...	2.125	17.750	247.4	132.68	5.76	4.65	451	107.2	6626	602.4	7.07	842.607
	...	...	5S	0.218	23.564	436.1	16.29	6.28	6.17	55	188.9	1152	96.0	8.41	123.301
	10	...	10S	0.250	23.500	434	18.65	6.28	6.15	63.41	188.0	1316	109.6	8.40	141.021
	...	...	...	0.312	23.376	430	23.20	6.28	6.12	78.93	186.1	1629	135.8	8.38	175.080
	20	Std	...	0.375	23.250	425	27.83	6.28	6.09	94.62	183.8	1943	161.9	8.35	209.320
	...	...	...	0.437	23.126	420	32.4	6.28	6.05	109.97	182.1	2246	187.4	8.33	242.657
	...	XS	...	0.500	23.000	415	36.9	6.28	6.02	125.49	180.1	2550	212.5	8.31	276.167
	30	...	...	0.562	22.876	411	41.4	6.28	5.99	140.80	178.1	2840	237.0	8.29	308.788
24	...	...	...	0.625	22.750	406	45.9	6.28	5.96	156.03	176.2	3140	261.4	8.27	341.576
24.000	40	...	...	0.687	22.626	402	50.3	6.28	5.92	171.17	174.3	3420	285.2	8.25	373.490
	...	...	...	0.750	22.500	398	54.8	6.28	5.89	186.24	172.4	3710	309	8.22	405.563
	60	...	...	0.968	22.064	382	70.0	6.28	5.78	238.11	165.8	4650	388	8.15	513.800
	80	...	...	1.218	21.564	365	87.2	6.28	5.65	296.36	158.3	5670	473	8.07	632.768
	100	...	...	1.531	20.938	344	108.1	6.28	5.48	367.40	149.3	6850	571	7.96	774.131
	120	...	...	1.812	20.376	326	126.3	6.28	5.33	429.39	141.4	7830	652	7.87	894.044
	140	...	...	2.062	19.876	310	142.1	6.28	5.20	483.13	134.5	8630	719	7.79	995.313
	160	...	...	2.343	19.314	293	159.4	6.28	5.06	541.94	127.0	9460	788	7.70	1103.215

TABLE E2.1 Principal Properties of Commercial Pipe (*Continued*)

Nominal pipe size, outside diam., (in)	Schedule number*			Wall thickness (in)	Inside diam. (in)	Inside area, (in^2)	Metal area (in^2)	Ft^2 outside surface, per ft	Ft^2 inside surface per ft	Weight per ft (lb)†	Weight of water per ft (lb)	Moment of inertia, (in^4)	Elastic section modulus (in^3)	Radius gyration (in)	Plastic section modulus (in^3)
	a	b	c												
26 *26.000*	...	...	...	0.250	25.500	510.7	19.85	6.81	6.68	67	221.4	1646	126.6	9.10	165.771
	10	...	...	0.312	25.376	505.8	25.18	6.81	6.64	86	219.2	2076	159.7	9.08	205.891
	...	Std	...	0.375	25.250	500.7	30.19	6.81	6.61	103	217.1	2478	190.6	9.06	246.258
	20	XS	...	0.500	25.000	490.9	40.06	6.91	6.54	136	212.8	3259	250.7	9.02	325.167
	...	...	...	0.625	24.750	481.1	49.82	6.81	6.48	169	208.6	4013	308.7	8.98	402.513
	...	...	...	0.750	24.500	471.4	59.49	6.81	6.41	202	204.4	4744	364.9	8.93	478.313
	...	...	...	0.875	24.250	461.9	69.07	6.81	6.35	235	200.2	5458	419.9	8.89	552.581
	...	...	...	1.00	24.000	452.4	78.54	6.81	6.28	267	196.1	6149	473.0	8.85	625.333
	...	...	...	1.125	23.750	443.0	87.91	6.81	6.22	299	192.1	6813	524.1	8.80	696.586
28 *28.000*	...	...	...	0.250	27.500	594.0	21.80	7.33	7.20	74	257.3	2098	149.8	9.81	192.521
	10	...	...	0.312	27.376	588.6	27.14	7.33	7.17	92	255.0	2601	185.8	9.78	239.197
	...	Std	...	0.375	27.250	583.2	32.54	7.33	7.13	111	252.6	3105	221.8	9.77	286.195
	20	XS	...	0.500	27.000	572.6	43.20	7.33	7.07	147	248.0	4085	291.8	9.72	378.167
	30	...	...	0.625	26.750	562.0	53.75	7.33	7.00	183	243.4	5038	359.8	9.68	468.451
	...	...	...	0.750	26.500	551.6	64.21	7.33	6.94	218	238.9	5964	426.0	9.64	557.063
	...	...	...	0.875	26.250	541.2	74.56	7.36	6.87	253	234.4	6865	490.3	9.60	644.018
	...	...	...	1.000	26.000	530.9	84.82	7.33	6.81	288	230.0	7740	552.8	9.55	729.333
	...	...	...	1.125	25.750	520.8	94.98	7.33	6.74	323	225.6	8590	613.6	9.51	813.023
30 *30.000*	...	...	5S	0.250	29.500	683.4	23.37	7.85	7.72	79	296.3	2585	172.3	10.52	221.271
	10	...	10S	0.312	29.376	677.8	29.19	7.85	7.69	99	293.7	3201	213.4	10.50	275.000
	...	Std	...	0.375	29.250	672.0	34.90	7.85	7.66	119	291.2	3823	254.8	10.48	329.133
	20	XS	...	0.500	29.000	660.5	46.34	7.85	7.59	158	286.2	5033	335.5	10.43	435.167
	30	...	...	0.625	28.750	649.2	57.68	7.85	7.53	196	281.3	6213	414.2	10.39	539.388
	40	...	...	0.750	28.500	637.9	68.92	7.85	7.46	234	276.6	7371	491.4	10.34	641.813
	...	...	...	0.875	28.250	620.7	90.06	7.85	7.39	272	271.8	8494	566.2	10.30	742.456
	...	...	...	1.000	28.000	615.7	91.11	7.85	7.33	310	267.0	9591	639.4	10.26	841.333
	...	...	...	1.125	27.750	604.7	102.05	7.85	7.26	347	262.2	10653	710.2	10.22	938.461

32 *32.000*	...	...	...	0.250	31.500	779.2	24.93	8.38	8.25	85	337.8	3141	196.3	11.22	252.021
	10	...	...	0.312	31.376	773.2	31.02	8.38	8.21	106	335.2	3891	243.2	11.20	313.299
	...	Std	...	0.375	31.250	766.9	37.25	8.38	8.18	127	332.5	4656	291.0	11.18	375.070
	20	XS	...	0.500	31.000	754.7	49.48	8.38	8.11	168	327.2	6140	383.8	11.14	496.167
	30	...	...	0.625	30.750	742.5	61.59	8.38	8.05	209	321.9	7578	473.6	11.09	615.326
	40	...	...	0.688	30.624	736.6	67.68	8.38	8.02	230	319.0	8298	518.6	11.07	674.652
	...	...	...	0.750	30.500	730.5	73.63	8.38	7.98	250	316.7	8990	561.9	11.05	732.563
	...	...	...	0.875	30.250	718.3	85.52	8.38	7.92	291	311.6	10372	648.2	11.01	847.893
	...	...	...	1.000	30.000	706.8	97.38	8.38	7.85	331	306.4	11680	730.0	10.95	961.333
	...	...	...	1.125	29.750	694.7	109.0	8.38	7.79	371	301.3	13023	814.0	10.92	1072.898
34 *34.000*	...	...	...	0.250	33.500	881.2	26.50	8.90	8.77	90	382.0	3773	221.9	11.93	284.771
	10	...	...	0.312	33.376	874.9	32.99	8.90	8.74	112	379.3	4680	275.3	11.91	354.093
	...	Std	...	0.375	33.250	867.8	39.61	8.90	8.70	135	376.2	5597	329.2	11.89	424.008
	20	XS	...	0.500	33.000	855.3	52.62	8.90	8.64	179	370.8	7385	434.4	11.85	561.167
	30	...	...	0.625	32.750	841.9	65.53	8.90	8.57	223	365.0	9124	536.7	11.80	696.263
	40	...	...	0.688	32.624	835.9	72.00	8.90	8.54	245	362.1	9992	587.8	11.78	763.575
	...	...	...	0.750	32.500	829.3	78.34	8.90	8.51	266	359.5	10829	637.0	11.76	829.313
	...	...	...	0.875	32.250	816.4	91.01	8.90	8.44	310	354.1	12501	735.4	11.72	960.331
	...	...	...	1.000	32.000	804.2	103.67	8.90	8.38	353	348.6	14114	830.2	11.67	1089.333
	...	...	...	1.125	31.750	791.3	116.13	8.90	8.31	395	343.2	15719	924.7	11.63	1216.336
36 *36.000*	...	...	...	0.250	35.500	989.7	28.11	9.42	9.29	96	429.1	4491	249.5	12.64	319.521
	10	...	...	0.312	35.376	982.9	34.95	9.42	9.26	119	426.1	5565	309.1	12.62	397.384
	...	Std	...	0.375	35.250	975.8	42.01	9.42	9.23	143	423.1	6664	370.2	12.59	475.945
	20	XS	...	0.500	35.000	962.1	55.76	9.42	9.16	190	417.1	8785	488.1	12.55	630.167
	30	...	...	0.625	34.750	948.3	69.50	9.42	9.10	236	411.1	10872	604.0	12.51	782.201
	40	...	...	0.750	34.500	934.7	83.01	9.42	9.03	282	405.3	12898	716.5	12.46	932.063
	...	...	...	0.875	34.250	920.6	96.50	9.42	8.97	328	399.4	14903	827.9	12.42	1079.768
	...	...	...	1.000	34.000	907.9	109.96	9.42	8.90	374	393.6	16851	936.2	12.38	1225.333
	...	...	...	1.125	33.750	894.2	123.19	9.42	8.89	419	387.9	18763	1042.4	12.34	1368.773
42 *42.000*	...	...	...	0.250	41.500	1352.6	32.82	10.99	10.86	112	586.4	7126	339.3	14.73	435.771
	...	Std	...	0.375	41.250	1336.3	49.08	10.99	10.80	167	579.3	10627	506.1	14.71	649.758
	20	XS	...	0.500	41.000	1320.2	65.18	10.99	10.73	222	572.3	14037	668.4	14.67	861.167
	30	...	...	0.625	40.750	1304.1	81.28	10.99	10.67	276	565.4	17373	827.3	14.62	1070.013
	40	...	...	0.750	40.500	1288.2	97.23	10.99	10.60	330	558.4	20689	985.2	14.59	1276.313
	...	...	...	1.000	40.000	1256.6	128.81	10.99	10.47	438	544.9	27080	1289.5	14.50	1681.333
	...	...	...	1.250	39.500	1225.3	160.03	10.99	10.34	544	531.2	33233	1582.5	14.41	2076.554
	...	...	...	1.500	39.000	1194.5	190.85	10.99	10.21	649	517.9	39181	1865.7	14.33	2461.500

TABLE E2.1 Principal Properties of Commercial Pipe (*Continued*)

Notes: The following formulas were used in the computation of the values shown in the table:

Weight† of pipe per foot (pounds) $= 10.6802t(D - t)$

Weight of water per foot (pounds) $= 0.3405d^2$

Square feet outside surface per foot $= 0.2618D$

Square feet inside surface per foot $= 0.2618d$

Inside area (square inches) $= 0.785d^2$

Area of metal (square inches) $= 0.785(D^2 - d^2)$

Moment of inertia (inches^4) $= 0.0491(D^4 - d^4)$ $= A_M R_g^2$

Elastic section modulus (inches^3) $= \dfrac{0.0982(D^4 - d^4)}{D}$

Plastic section modulus $= \dfrac{(D^3 - d^3)}{6}$

Radius of gyration (inches) $= 0.25\sqrt{D^2 + d^2}$

A_M = area of metal (square inches)
d = inside diameter (inches)
D = outside diameter (inches)
R_g = radius of gyration (inches)
t = pipe wall thickness (inches)

*Schedule numbers: Standard weight pipe and Schedule 40 are the same in all sizes through 10 in; from 12 in through 24 in, standard weight pipe has a wall thickness of ⅜ in. Extra-strong weight pipe and Schedule 80 are the same in all sizes through 8 in; from 8 in through 24 in, extra-strong weight pipe has a wall thickness of ½ in. Double extra-strong weight pipe has no corresponding schedule number.

a: ANSI/ASME B36.10 Steel Pipe Schedule Numbers.

b: ANSI/ASME N36.10 Steel Pipe Nominal Wall Thickness Designations.

c: ANSI/ASME B36.19 Stainless Steel Pipe Schedule Numbers.

†The ferritic stainless steels may be about 5 percent less and the austenitic stainless steels about 2 percent greater than the values shown in this table which are based on weights for carbon steel.

APPENDIX E3
TUBE PROPERTIES

Ervin L. Geiger
Engineering Supervisor
Bechtel Power Corporation
Gaithersburg, Maryland

TABLE E3.1 Weights of Tubing, Pound per Foot Length

OD, in.	Wall thickness, in.															
	0.050	0.095	0.150	0.200	0.250	0.300	0.360	0.400	0.460	0.500	0.625	0.750	0.875	1.000	1.125	1.250
0.500	0.2403	0.4109	0.5607													
0.750	0.3738	0.6646	0.9612	1.175	1.335											
1.000	0.5073	0.9182	1.362	1.709	2.003	2.243	2.461									
1.250	0.6408	1.172	1.762	2.243	2.670	3.044	3.422	3.631								
1.500	0.7743	1.426	2.163	2.777	3.338	3.845	4.383	4.699	5.109	5.340						
1.750	0.9078	1.679	2.563	3.311	4.005	4.646	5.344	5.767	6.338	6.675	7.509					
2.000	1.041	1.933	2.964	3.845	4.673	5.447	6.305	6.835	7.566	8.010	9.178	10.01				
2.250	1.175	2.186	3.364	4.379	5.340	6.248	7.267	7.903	8.794	9.345	10.85	12.02				
2.500	1.308	2.440	3.765	4.913	6.008	7.049	8.228	8.971	10.02	10.68	12.52	14.02	15.19			
2.750	1.442	2.694	4.165	5.447	6.675	7.850	9.189	10.04	11.25	12.02	14.18	16.02	17.52	18.69		
3.000	1.575	2.947	4.566	5.981	7.343	8.651	10.15	11.11	12.48	13.35	15.85	18.02	19.86	21.36		
3.250	1.709	3.201	4.966	6.515	8.010	9.452	11.11	12.18	13.71	14.69	17.52	20.03	22.19	24.03		
3.500	1.842	3.455	5.367	7.049	8.678	10.25	12.07	13.24	14.93	16.02	19.19	22.03	24.53	26.70		
3.750	1.976	3.708	5.767	7.583	9.345	11.05	13.03	14.31	16.16	17.36	20.86	24.03	26.87	29.37	31.54	
4.000	2.109	3.962	6.168	8.117	10.01	11.85	14.00	15.38	17.39	18.69	22.53	26.03	29.20	32.04	34.54	
4.250	2.243	4.216	6.568	8.651	10.68	12.66	14.96	16.45	18.62	20.03	24.20	28.04	31.54	34.71	37.55	
4.500	2.376	4.469	6.969	9.185	11.35	13.46	15.92	17.52	19.85	21.36	25.87	30.04	33.88	37.38	40.55	
4.750	2.510	4.723	7.369	9.719	12.02	14.26	16.88	18.58	21.08	22.70	27.53	32.04	36.21	40.05	43.55	46.73

5.000	2.643	4.977	7.770	10.25	12.68	15.06	17.84	19.65	22.30	24.03	29.20	34.04	38.55	42.72	46.56	50.06
5.250	....	5.230	8.170	10.79	13.35	15.86	18.80	20.72	23.53	25.37	30.87	36.05	40.88	45.39	49.56	53.40
5.500	....	5.484	8.571	11.32	14.02	16.66	19.76	21.79	24.76	26.70	32.54	38.05	43.22	48.06	52.57	56.74
5.750	....	5.738	8.971	11.85	14.69	17.46	20.72	22.86	25.99	28.04	34.21	40.05	45.56	50.73	55.57	60.08
6.000	....	5.991	9.372	12.39	15.35	18.26	21.68	23.92	27.22	29.37	35.88	42.05	47.89	53.40	58.57	63.41
6.250	....	6.245	9.772	12.92	16.02	19.06	22.65	24.99	28.45	30.71	37.55	44.06	50.23	56.07	61.58	66.75
6.500	....	....	10.17	13.46	16.69	19.86	23.61	26.06	29.67	32.04	39.22	46.06	52.57	58.74	64.58	70.09
6.750	....	....	10.57	13.99	17.36	20.67	24.57	27.13	30.90	33.38	40.88	48.06	54.90	61.41	67.58	73.43
7.000	....	....	10.97	14.52	18.02	21.47	25.53	28.20	32.13	34.71	42.55	50.06	57.24	64.08	70.59	76.76
7.250	....	....	11.37	15.06	18.69	22.27	26.49	29.26	33.36	36.05	44.22	52.07	59.57	66.75	73.59	80.10
7.500	....	....	....	15.59	19.36	23.07	27.45	30.33	34.59	37.38	45.89	54.07	61.91	69.42	76.60	83.44
7.750	....	....	....	16.13	20.03	23.87	28.41	31.40	35.81	38.72	47.56	56.07	64.25	72.09	79.60	86.78
8.000	....	....	....	16.66	20.69	24.67	29.37	32.47	37.04	40.05	49.23	58.07	66.58	74.76	82.60	90.11
8.250	....	....	....	17.19	21.36	25.47	30.34	33.54	38.27	41.39	50.90	60.08	68.92	77.43	85.61	93.45
8.500	....	....	....	17.73	22.03	26.27	31.30	34.60	39.50	42.72	52.57	62.08	71.26	80.10	88.61	96.79
8.750	....	....	....	....	22.70	27.07	32.26	35.67	40.73	44.06	54.23	64.08	73.59	82.77	91.61	100.1
9.000	....	....	....	....	23.36	27.87	33.22	36.74	41.96	45.39	55.90	66.08	75.93	85.44	94.62	103.5
9.250	....	....	....	....	24.03	28.68	34.18	37.81	43.18	46.73	57.57	68.09	78.26	88.11	97.62	106.8
9.500	....	....	....	....	24.70	29.48	35.14	38.88	44.41	48.06	59.24	70.09	80.60	90.78	100.6	110.1
9.750	....	....	....	....	25.37	30.28	36.10	39.94	45.64	49.40	60.91	72.09	82.94	93.45	103.6	113.5
10.000	....	....	....	....	26.03	31.08	37.06	41.01	46.87	50.73	62.58	74.09	85.27	96.12	106.6	116.8

TABLE E3.2 Diameter and Wall-Thickness Tolerances for Seamless Hot-Finished Mechanical Tubing of Carbon and Alloy Steel (AISI)*

Specified size, OD, in.	Ratio of wall thickness to OD	OD tolerance		Wall thickness tolerance, %							
				0.109 in. and under		Over 0.109 to 0.172 in., incl.		Over 0.172 to 0.203 in., incl.		Over 0.203 in.	
		Over	Under	Over	Under	Over	Under	Over	Under	Over	Under
Under 3	All wall thicknesses	0.023	0.023	16.5	16.5	15	15	14	14	12.5	12.5
3–5½, excl.	All wall thicknesses	0.031	0.031	16.5	16.5	15	15	14	14	12.5	12.5
5½–8, excl.	All wall thicknesses	0.047	0.047	...	...	..	..	14	14	12.5	12.5
8–10¾, incl.	5% and over	0.047	0.047	...	...	..	..	..	..	12.5	12.5
8–10¾, incl.	Under 5%	0.063	0.063	...	...	..	..	..	..	12.5	12.5

*The common range of sizes of hot-finished tubes is 1½ to and including 10¾ in outside diameter with wall thickness not less than 0.095 in (no. 13 BWG) or 3 percent or more of the outside diameter. For sizes under 1½ or over 10¾ in outside diameter, the tolerances are commonly negotiated between the purchaser and producer.

Source: Steel Products Manual.

TABLE E3.3 Diameter and Wall-Thickness Tolerances for Seamless Cold-Worked Mechanical Tubing of Carbon and Alloy Steel (AISI)[1]

Size, OD, in.	Unannealed or finish-annealed				Soft annealed or normalized				Quenched and tempered				Wall thickness all conditions, %	
	OD, in.		ID, in.		OD, in.		ID, in.		OD, in.		ID, in.			
	Over	Under	Over	Under	Over	Under	Over	Under	Over	Under	Over	Under	Over	Under
3⁄16–1⁄2, excl.[2,3]	0.004	0	....	....	0.006	0.002	....	....	0.010	0.010	....	....	15	15
1⁄2–11⁄2, excl.[2,3,4,5]	0.005	0	0	0.005	0.008	0.002	0.002	0.008	0.015	0.015	0.015	0.015	10	10
11⁄2–31⁄2, excl.[2,3,4,5]	0.010	0	0	0.010	0.015	0.005	0.005	0.015	0.030	0.030	0.030	0.030	10	10
31⁄2–51⁄2, excl.[4,5]	0.015	0	0.005	0.015	0.023	0.007	0.015	0.025	0.045	0.045	0.045	0.045	10	10
51⁄2–8, excl.[5], wall less than 5% OD	0.030	0.030	0.035	0.035	0.060	0.060	0.070	0.070	....	....	....	....	10	10
51⁄2–8, excl, wall from 5 to 7.5% OD	0.020	0.020	0.025	0.025	0.040	0.040	0.050	0.050	....	....	....	....	10	10
51⁄2–8, excl.[4], wall over 7.5% OD	0.030	0	0.015	0.030	0.045	0.015	0.037	0.053	....	....	....	....	10	10
8–103⁄4, incl.[5], wall less than 5% OD	0.045	0.045	0.050	0.050	....	....	....	....	....	....	....	....	10	10
8–103⁄4, incl., wall from 5 to 75% OD	0.035	0.035	0.040	0.040	....	....	....	....	....	....	....	....	10	10
8–103⁄4, incl.[4], wall over 7.5% OD	0.045	0	0.015	0.040	....	....	....	....	....	....	....	....	10	10

[1]For tolerances closer than those indicated, availability and applicable tolerances for tubing less than 3⁄16 in OD or larger than 10¾-in. OD, the producer should be consulted.

[2]For those tubes with inside diameter less than ½ in (or less than ⅝ in when the wall thickness is more than 20 percent of the outside diameter), which are not commonly drawn over a mandrel, note[4] is not applicable. Unless otherwise agreed upon by the purchaser and producer, the wall thickness may vary 15 percent over and under that specified, and the inside diameter is governed by the outside diameter and wall-thickness tolerances shown.

[3]For tubes with inside diameter less than ½ in (or less than ⅝ in when the wall thickness is more than 20 percent of the outside diameter), which can be produced by the rod or bar mandrel process, the tolerances are as shown in the above table except that the wall-thickness tolerances are 10 percent over and under the specified wall thickness.

[4]Many tubes with inside diameter less than 50 percent of outside diameter, with wall thickness more than 25 percent of outside diameter, with wall thickness over 1¼ in or weighing more than 90 lb/ft are difficult to draw over a mandrel. Unless otherwise agreed upon by the purchaser and producer the inside diameter may vary over or under by an amount equal to 10 percent of the wall thickness and the wall thickness may vary 12½ percent over and under that specified.

[5]Tubing having a wall thickness less than 3 percent of the outside diameter cannot be straightened properly without a certain amount of distortion. Consequently, such tubes, while having an average outside diameter and inside diameter within the tolerances shown in the above table, require an ovality tolerance of 0.5 percent over and under nominal outside diameter, this being in addition to the tolerances indicated in the above table.

Source: AISI Steel Products Manual.

APPENDIX E4

FRICTION LOSS FOR WATER IN FEET PER 100 FT OF PIPE

(Excerpted from Hydraulic Institute Engineering Data Handbook with permission.)

TABLE E4.1 Friction Loss for Water in Feet per 100 Ft of Pipe

4 in nominal		Steel-schedule 40 ID = 4.026 in ε/D = 0.000447			Asphalt-dipped cast iron ID = 4.00 in ε/D = 0.00120		
Discharge							
ft^3/s	gal/min	V (ft/s)	$V^2/2g$ (ft)	h_f feet per 100 ft of pipe	V (ft/s)	$V^2/2g$ (ft)	h_f feet per 100 ft of pipe
0.0111	5	0.126	0.000247	0.00310	0.128	0.000253	0.00325
0.0223	10	0.252	0.000987	0.01017	0.255	0.00101	0.01080
0.0446	20	0.504	0.00395	0.0344	0.511	0.00405	0.03700
0.0668	30	0.756	0.00888	0.0702	0.766	0.00912	0.0770
0.0981	40	1.01	0.0158	0.118	1.02	0.0162	0.131
0.111	50	1.26	0.0247	0.176	1.28	0.0253	0.199
0.134	60	1.51	0.0355	0.245	1.53	0.0365	0.278
0.156	70	1.76	0.0484	0.325	1.79	0.0496	0.370
0.178	80	2.02	0.0632	0.415	2.04	0.0648	0.476
0.201	90	2.27	0.0800	0.515	2.30	0.0820	0.594
0.223	100	2.52	0.0987	0.624	2.55	0.101	0.725
0.245	110	2.77	0.119	0.744	2.81	0.123	0.869
0.267	120	3.02	0.142	0.877	3.06	0.146	1.03
0.290	130	3.28	0.167	1.017	3.32	0.171	1.19
0.312	140	3.53	0.193	1.165	3.57	0.199	1.38
0.334	150	3.78	0.222	1.32	3.83	0.228	1.58
0.356	160	4.03	0.253	1.49	4.08	0.259	1.78
0.379	170	4.28	0.285	1.67	4.34	0.293	2.00
0.401	180	4.54	0.320	1.86	4.60	0.328	2.24
0.423	190	4.79	0.356	2.06	4.85	0.366	2.49
0.446	200	5.04	0.395	2.27	5.11	0.406	2.74
0.490	220	5.54	0.478	2.72	5.62	0.490	3.28
0.535	240	6.05	0.569	3.21	6.13	0.583	3.88
0.579	260	6.55	0.667	3.74	6.64	0.685	4.54
0.624	280	7.06	0.774	4.30	7.15	0.794	5.25
0.668	300	7.56	0.888	4.89	7.66	0.912	6.03
0.713	320	8.06	1.01	5.51	8.17	1.04	6.87
0.758	340	8.57	1.14	6.19	8.68	1.17	7.75
0.802	360	9.07	1.28	6.92	9.19	1.31	8.68
0.847	380	9.58	1.43	7.68	9.70	1.46	9.66
0.891	400	10.1	1.58	8.47	10.2	1.62	10.7
0.936	420	10.6	1.74	9.30	10.7	1.79	11.7
0.980	440	11.1	1.91	10.2	11.2	1.96	12.8
1.025	460	11.6	2.09	11.1	11.7	2.14	14.0
1.069	480	12.1	2.27	12.0	12.3	2.33	15.3
1.114	500	12.6	2.47	13.0	12.8	2.53	16.6
1.225	550	13.9	2.99	15.7	14.0	3.06	19.9
1.337	600	15.1	3.55	18.6	15.3	3.65	23.6
1.448	650	16.4	4.17	21.7	16.6	4.28	27.7
1.560	700	17.6	4.84	25.0	17.9	4.96	32.1
1.671	750	18.9	5.55	28.6	19.1	5.70	36.7
1.782	800	20.2	6.32	32.4	20.4	6.48	41.6
1.894	850	21.4	7.13	36.5	21.7	7.32	46.8
2.005	900	22.7	8.00	40.8	23.0	8.20	52.3
2.117	950	23.9	8.91	45.3	24.3	9.14	58.1

TABLE E4.1 Friction Loss for Water in Feet per 100 Ft of Pipe (*Continued*)

4 in nominal		Steel-schedule 40 ID = 4.026 in ε/D = 0.000447			Asphalt-dipped cast iron ID = 4.00 in ε/D = 0.00120		
Discharge							
ft³/s	gal/min	V (ft/s)	$V^2/2g$ (ft)	h_f feet per 100 ft of pipe	V (ft/s)	$V^2/2g$ (ft)	h_f feet per 100 ft of pipe
2.228	1 000	25.2	9.87	50.2	25.5	10.1	64.2
2.451	1 100	27.7	11.9	60.5	28.1	12.3	78.2
2.674	1 200	30.2	14.2	72.0	30.6	14.6	92.8
2.896	1 300	32.8	16.7	84.3	33.2	17.1	108.2
3.119	1 400	35.3	19.3	97.6	35.7	19.9	126
3.342	1 500	37.8	22.2	112	38.3	22.8	144
3.565	1 600	40.3	25.3	127	40.8	25.9	164
3.788	1 700	42.8	28.5	143	43.4	29.3	185
4.010	1 800	45.4	32.0	160	46.0	32.8	207
4.233	1 900	47.9	35.6	178	48.5	36.6	230
4.456	2 000	50.4	39.5	196	51.1	40.5	255

6 in nominal		Steel-schedule 40 ID = 6.065 in ε/D = 0.000293			Asphalt-dipped cast iron ID = 6.00 in ε/D = 0.000800		
Discharge							
ft³/s	gal/min	V (ft/s)	$V^2/2g$ (ft)	h_f feet per 100 ft of pipe	V (ft/s)	$V^2/2g$ (ft)	h_f feet per 100 ft of pipe
0.0223	10	0.111	0.000192	0.00146	0.113	0.000200	0.00157
0.0446	20	0.222	0.000767	0.00487	0.227	0.000800	0.00523
0.0668	30	0.333	0.00172	0.00988	0.340	0.00180	0.01070
0.0891	40	0.444	0.00307	0.0164	0.454	0.00320	0.0179
0.111	50	0.555	0.00479	0.0244	0.567	0.00500	0.0268
0.134	60	0.666	0.00690	0.0337	0.681	0.00720	0.0374
0.156	70	0.777	0.00939	0.0445	0.794	0.00980	0.0496
0.178	80	0.888	0.0123	0.0564	0.908	0.0128	0.0635
0.201	90	0.999	0.0155	0.0698	1.02	0.0162	0.0789
0.223	100	1.11	0.0192	0.0843	1.13	0.0200	0.0958
0.267	120	1.33	0.0276	0.118	1.36	0.0288	0.130
0.312	140	1.55	0.0376	0.155	1.59	0.0392	0.178
0.356	160	1.78	0.0491	0.198	1.82	0.0512	0.229
0.401	180	2.00	0.0621	0.246	2.04	0.0648	0.282
0.446	200	2.22	0.0767	0.299	2.27	0.0800	0.346
0.490	220	2.44	0.0927	0.357	2.50	0.0968	0.415
0.535	240	2.66	0.110	0.419	2.72	0.115	0.490
0.579	260	2.89	0.130	0.487	2.95	0.135	0.570
0.624	280	3.11	0.150	0.560	3.18	0.157	0.655
0.668	300	3.33	0.172	0.637	3.40	0.180	0.745
0.713	320	3.55	0.196	0.719	3.63	0.205	0.846
0.758	340	3.78	0.222	0.806	3.86	0.231	0.952
0.802	360	4.00	0.240	0.898	4.08	0.259	1.06
0.847	380	4.22	0.277	0.993	4.31	0.289	1.18
0.891	400	4.44	0.307	1.09	4.54	0.320	1.30

TABLE E4.1 Friction Loss for Water in Feet per 100 Ft of Pipe (*Continued*)

6 in nominal		Steel-schedule 40 ID = 6.065 in ε/D = 0.000293			Asphalt-dipped cast iron ID = 6.00 in ε/D = 0.000800		
Discharge				h_f feet per 100 ft of pipe			h_f feet per 100 ft of pipe
ft^3/s	gal/min	V (ft/s)	$V^2/2g$ (ft)		V (ft/s)	$V^2/2g$ (ft)	
0.936	420	4.66	0.338	1.20	4.76	0.353	1.43
0.980	440	4.89	0.371	1.31	4.99	0.387	1.57
1.025	460	5.11	0.405	1.42	5.22	0.423	1.71
1.07	480	5.33	0.442	1.54	5.45	0.461	1.86
1.11	500	5.55	0.479	1.66	5.67	0.500	2.02
1.23	550	6.11	0.580	1.99	6.24	0.605	2.42
1.34	600	6.66	0.690	2.34	6.81	0.720	2.84
1.45	650	7.22	0.810	2.73	7.37	0.845	3.33
1.56	700	7.77	0.939	3.13	7.94	0.980	3.87
1.67	750	8.33	1.08	3.57	8.51	1.12	4.45
1.78	800	8.88	1.23	4.03	9.08	1.28	5.06
1.89	850	9.44	1.38	4.53	9.64	1.44	5.69
2.01	900	9.99	1.55	5.05	10.2	1.62	6.34
2.12	950	10.5	1.73	5.60	10.8	1.80	7.02
2.23	1 000	11.1	1.92	6.17	11.3	2.00	7.73
2.45	1 100	12.2	2.32	7.41	12.5	2.42	9.80
2.67	1 200	13.3	2.76	8.76	13.6	2.88	11.2
2.90	1 300	14.4	3.24	10.2	14.7	3.38	13.0
3.12	1 400	15.5	3.76	11.8	15.9	3.92	15.1
3.34	1 500	16.7	4.31	13.5	17.0	4.50	17.4
3.56	1 600	17.8	4.91	15.4	18.2	5.12	19.8
3.79	1 700	18.9	5.54	17.3	19.3	5.78	22.3
4.01	1 800	20.0	6.21	19.4	20.4	6.48	24.8
4.23	1 900	21.1	6.92	21.6	21.6	7.22	27.6
4.46	2 000	22.2	7.67	23.8	22.7	8.00	30.5
4.68	2 100	23.3	8.45	26.2	23.8	8.82	33.6
4.90	2 200	24.4	9.27	28.8	25.0	9.68	36.8
5.12	2 300	25.5	10.1	31.4	26.1	10.6	40.1
5.35	2 400	26.6	11.0	34.2	27.2	11.5	43.5
5.57	2 500	27.8	12.0	37.0	28.4	12.5	47.1
5.79	2 600	28.9	13.0	39.9	29.5	13.5	51.0
6.02	2 700	30.0	14.0	42.9	30.6	14.6	55.2
6.24	2 800	31.1	15.0	46.1	31.8	15.7	59.6
6.46	2 900	32.2	16.1	49.4	32.9	16.8	64.1
6.68	3 000	33.3	17.2	52.8	34.0	18.0	68.8
7.13	3 200	35.5	19.6	59.9	36.3	20.5	78.0
7.58	3 400	37.8	22.2	67.4	38.6	23.1	88.0
8.02	3 600	40.0	24.8	75.5	40.8	25.9	98.7
8.47	3 800	42.2	27.7	84.1	43.1	28.9	110
8.91	4 000	44.4	30.7	93.1	45.4	32.0	122

TABLE E4.1 Friction Loss for Water in Feet per 100 Ft of Pipe (*Continued*)

8 in nominal		Steel-schedule 40 ID = 7.981 in ε/D = 0.000226			Asphalt-dipped cast iron ID = 8.00 in ε/D = 0.00060		
Discharge							
ft^3/s	gal/min	V (ft/s)	$V^2/2g$ (ft)	h_f feet per 100 ft of pipe	V (ft/s)	$V^2/2g$ (ft)	h_f feet per 100 ft of pipe
0.0223	10	0.0641	0.0000639	0.000401	0.0638	0.0000633	0.000399
0.0446	20	0.128	0.000256	0.001320	0.128	0.000253	0.001320
0.0668	30	0.192	0.000575	0.00266	0.191	0.000570	0.00269
0.0891	40	0.257	0.00102	0.00442	0.255	0.00101	0.00447
0.111	50	0.321	0.00160	0.00652	0.319	0.00158	0.00664
0.134	60	0.385	0.00230	0.00904	0.382	0.00228	0.00920
0.156	70	0.449	0.00313	0.01190	0.447	0.00310	0.01210
0.178	80	0.513	0.00409	0.0151	0.511	0.00405	0.0154
0.201	90	0.577	0.00518	0.0186	0.574	0.00513	0.0191
0.223	100	0.641	0.00639	0.0224	0.638	0.00633	0.0232
0.267	120	0.770	0.00920	0.0311	0.766	0.00911	0.0323
0.312	140	0.898	0.0125	0.0410	0.893	0.0124	0.0428
0.356	160	1.03	0.0164	0.0521	1.02	0.0162	0.0548
0.401	180	1.15	0.0207	0.0644	1.15	0.0205	0.0681
0.446	200	1.28	0.0256	0.0780	1.28	0.0253	0.0828
0.490	220	1.41	0.0309	0.0928	1.40	0.0306	0.0989
0.535	240	1.54	0.0368	0.1088	1.53	0.0365	0.1163
0.579	260	1.67	0.0432	0.1260	1.66	0.0428	0.135
0.624	280	1.80	0.0501	0.144	1.79	0.0496	0.155
0.668	300	1.92	0.0575	0.163	1.91	0.0570	0.176
0.713	320	2.05	0.0655	0.184	2.04	0.0648	0.198
0.758	340	2.18	0.0739	0.206	2.17	0.0732	0.222
0.802	360	2.31	0.0828	0.229	2.30	0.0820	0.248
0.847	380	2.44	0.0923	0.253	2.43	0.0914	0.275
0.891	400	2.57	0.102	0.279	2.55	0.101	0.304
1.003	450	2.89	0.129	0.348	2.87	0.128	0.380
1.11	500	3.21	0.160	0.424	3.19	0.158	0.464
1.23	550	3.53	0.193	0.507	3.51	0.191	0.557
1.34	600	3.85	0.230	0.597	3.83	0.228	0.658
1.45	650	4.17	0.271	0.694	4.15	0.267	0.767
1.56	700	4.49	0.313	0.797	4.47	0.310	0.884
1.67	750	4.81	0.360	0.907	4.79	0.356	1.01
1.78	800	5.13	0.409	1.02	5.11	0.405	1.14
1.89	850	5.45	0.462	1.147	5.42	0.457	1.29
2.01	900	5.77	0.518	1.27	5.74	0.513	1.44
2.12	950	6.09	0.577	1.41	6.06	0.571	1.60
2.23	1 000	6.41	0.639	1.56	6.38	0.633	1.76
2.45	1 100	7.05	0.773	1.87	7.02	0.766	2.14
2.67	1 200	7.70	0.920	2.20	7.66	0.911	2.53
2.90	1 300	8.34	1.08	2.56	8.30	1.07	2.94

TABLE E4.1 Friction Loss for Water in Feet per 100 Ft of Pipe (*Continued*)

8 in nominal		Steel-schedule 40 ID = 7.981 in ε/D = 0.000226			Asphalt-dipped cast iron ID = 8.00 in ε/D = 0.00060		
Discharge							
ft^3/s	gal/min	V (ft/s)	$V^2/2g$ (ft)	h_f feet per 100 ft of pipe	V (ft/s)	$V^2/2g$ (ft)	h_f feet per 100 ft of pipe
3.12	1 400	8.98	1.25	2.95	8.93	1.24	3.40
3.34	1 500	9.62	1.44	3.37	9.57	1.42	3.91
3.56	1 600	10.3	1.64	3.82	10.2	1.62	4.45
3.79	1 700	10.9	1.85	4.29	10.8	1.83	5.00
4.01	1 800	11.5	2.07	4.79	11.5	2.05	5.58
4.23	1 900	12.2	2.31	5.31	12.1	2.29	6.19
4.46	2 000	12.8	2.56	5.86	12.8	2.53	6.84
4.90	2 200	14.1	3.09	7.02	14.0	3.06	8.26
5.35	2 400	15.4	3.68	8.31	15.3	3.65	9.80
5.79	2 600	16.7	4.32	9.70	16.6	4.28	11.47
6.24	2 800	18.0	5.01	11.20	17.9	4.96	13.3
6.68	3 000	19.2	5.75	12.8	19.1	5.70	15.2
7.13	3 200	20.5	6.55	14.5	20.4	6.48	17.3
7.58	3 400	21.8	7.39	16.4	21.7	7.32	19.5
8.02	3 600	23.1	8.28	18.4	23.0	8.20	21.9
8.47	3 800	24.4	9.23	20.5	24.3	9.14	24.4
8.91	4 000	25.7	10.2	22.6	25.5	10.1	27.0
10.03	4 500	28.9	12.9	28.5	28.7	12.8	34.0
11.1	5 000	32.1	16.0	35.1	31.9	15.8	42.0
12.3	5 500	35.3	19.3	42.5	35.1	19.1	51.0
13.4	6 000	38.5	23.0	50.5	38.3	22.8	60.5
14.5	6 500	41.7	27.0	59.1	41.5	26.7	71.0
15.6	7 000	44.9	31.3	68.3	44.7	31.0	82.0
16.7	7 500	48.1	36.0	78.1	47.9	35.6	94.0
17.8	8 000	51.3	40.9	88.6	51.1	40.5	107

10 in nominal		Steel-schedule 40 ID = 10.020 in ε/D = 0.000180			Asphalt-dipped cast iron ID = 10.00 in ε/D = 0.000480		
Discharge							
ft^3/s	gal/min	V (ft/s)	$V^2/2g$ (ft)	h_f feet per 100 ft of pipe	V (ft/s)	$V^2/2g$ (ft)	h_f feet per 100 ft of pipe
0.0223	10	0.0407	0.0000257	0.000138	0.0409	0.0000259	0.000140
0.0446	20	0.0814	0.000103	0.000451	0.0817	0.000104	0.000460
0.0891	40	0.163	0.000412	0.00149	0.163	0.000415	0.00154
0.134	60	0.244	0.000926	0.00304	0.245	0.000934	0.00315
0.178	80	0.325	0.00165	0.00505	0.327	0.00166	0.00525
0.223	100	0.407	0.00257	0.00747	0.409	0.00259	0.00783
0.267	120	0.488	0.00370	0.0103	0.490	0.00373	0.01085
0.312	140	0.570	0.00504	0.0136	0.572	0.00508	0.0144
0.356	160	0.651	0.00659	0.0174	0.654	0.00664	0.0183
0.401	180	0.732	0.00834	0.0215	0.735	0.00840	0.0227

TABLE E4.1 Friction Loss for Water in Feet per 100 Ft of Pipe (*Continued*)

10 in nominal		Steel-schedule 40 ID = 10.020 in ε/D = 0.000180			Asphalt-dipped cast iron ID = 10.00 in ε/D = 0.000480		
Discharge							
ft³/s	gal/min	V (ft/s)	$V^2/2g$ (ft)	h_f feet per 100 ft of pipe	V (ft/s)	$V^2/2g$ (ft)	h_f feet per 100 ft of pipe
0.446	200	0.814	0.0103	0.0260	0.817	0.0104	0.0276
0.490	220	0.895	0.0125	0.0309	0.899	0.0126	0.0329
0.535	240	0.976	0.0148	0.0362	0.980	0.0149	0.0387
0.579	260	1.06	0.0174	0.0417	1.06	0.0175	0.0449
0.624	280	1.14	0.0202	0.0478	1.14	0.0203	0.0514
0.668	300	1.22	0.0232	0.0542	1.23	0.0233	0.0583
0.780	350	1.42	0.0315	0.0719	1.43	0.0318	0.0778
0.891	400	1.63	0.0412	0.0917	1.63	0.0415	0.0990
1.003	450	1.83	0.0521	0.114	1.84	0.0525	0.1235
1.11	500	2.03	0.0643	0.138	2.04	0.0648	0.151
1.23	550	2.24	0.0778	0.164	2.25	0.0785	0.181
1.34	600	2.44	0.0926	0.192	2.45	0.0934	0.214
1.45	650	2.64	0.109	0.224	2.66	0.110	0.250
1.56	700	2.85	0.126	0.256	2.86	0.127	0.288
1.67	750	3.05	0.145	0.291	3.06	0.146	0.328
1.78	800	3.25	0.165	0.328	3.27	0.166	0.370
1.89	850	3.46	0.186	0.368	3.47	0.187	0.415
2.01	900	3.66	0.208	0.410	3.68	0.210	0.462
2.12	950	3.87	0.232	0.455	3.88	0.234	0.512
2.23	1 000	4.07	0.257	0.500	4.09	0.259	0.565
2.45	1 100	4.48	0.311	0.600	4.49	0.314	0.680
2.67	1 200	4.88	0.370	0.703	4.90	0.373	0.805
2.90	1 300	5.29	0.435	0.818	5.31	0.438	0.945
3.12	1 400	5.70	0.504	0.940	5.72	0.508	1.09
3.34	1 500	6.10	0.579	1.07	6.13	0.584	1.25
3.56	1 600	6.51	0.659	1.21	6.54	0.664	1.42
3.79	1 700	6.92	0.743	1.36	6.94	0.749	1.60
4.01	1 800	7.32	0.834	1.52	7.35	0.840	1.78
4.23	1 900	7.73	0.929	1.68	7.76	0.936	1.97
4.46	2 000	8.14	1.03	1.86	8.17	1.04	2.17
4.90	2 200	8.95	1.25	2.23	8.99	1.26	2.64
5.35	2 400	9.76	1.48	2.64	9.80	1.49	3.12
5.79	2 600	10.6	1.74	3.08	10.6	1.75	3.63
6.24	2 800	11.4	2.02	3.56	11.4	2.03	4.18
6.68	3 000	12.2	2.32	4.06	12.3	2.33	4.79
7.13	3 200	13.0	2.63	4.59	13.1	2.66	5.47
7.58	3 400	13.8	2.97	5.16	13.9	3.00	6.18
8.02	3 600	14.6	3.33	5.76	14.7	3.36	6.91
8.47	3 800	15.5	3.71	6.40	15.5	3.74	7.68
8.91	4 000	16.3	4.12	7.07	16.3	4.15	8.50

TABLE E4.1 Friction Loss for Water in Feet per 100 Ft of Pipe (*Continued*)

10 in nominal		Steel-schedule 40 ID = 10.020 in ε/D = 0.000180			Asphalt-dipped cast iron ID = 10.00 in ε/D = 0.000480		
Discharge							
ft³/s	gal/min	V (ft/s)	$V^2/2g$ (ft)	h_f feet per 100 ft of pipe	V (ft/s)	$V^2/2g$ (ft)	h_f feet per 100 ft of pipe
10.03	4 500	18.3	5.21	8.88	18.4	5.25	10.7
11.1	5 000	20.3	6.43	10.9	20.4	6.48	13.2
12.3	5 500	22.4	7.78	13.2	22.5	7.85	15.9
13.4	6 000	24.4	9.26	15.6	24.5	9.34	18.9
14.5	6 500	26.4	10.9	18.3	26.6	11.0	22.2
15.6	7 000	28.5	12.6	21.1	28.6	12.7	25.8
16.7	7 500	30.5	14.5	24.3	30.0	14.6	29.6
17.8	8 000	32.5	16.5	27.5	32.7	16.6	33.6
18.9	8 500	34.6	18.6	30.9	34.7	18.7	37.8
20.1	9 000	36.6	20.8	34.6	36.8	21.0	42.2
21.2	9 500	38.7	23.2	38.5	38.8	23.4	46.9
22.3	10 000	40.7	25.7	42.6	40.9	25.9	51.8

12 in nominal		Steel-schedule 40 D = 11.938 in ε/D = 0.000151			Asphalt-dipped cast iron ID = 12.00 in ε/D = 0.000400		
Discharge							
ft³/s	gal/min	V (ft/s)	$V^2/2g$ (ft)	h_f feet per 100 ft of pipe	V (ft/s)	$V^2/2g$ (ft)	h_f feet per 100 ft of pipe
0.223	100	0.287	0.00128	0.00325	0.284	0.00125	0.00320
0.267	120	0.344	0.00184	0.00448	0.340	0.00180	0.00445
0.312	140	0.401	0.00250	0.00590	0.397	0.00245	0.00589
0.356	160	0.459	0.00327	0.00747	0.454	0.00320	0.00752
0.401	180	0.516	0.00414	0.00920	0.511	0.00405	0.00932
0.446	200	0.573	0.00511	0.0111	0.567	0.00500	0.01129
0.490	220	0.631	0.00618	0.0132	0.624	0.00605	0.0135
0.535	240	0.688	0.00735	0.0155	0.681	0.00720	0.0158
0.579	260	0.745	0.00863	0.0180	0.738	0.00845	0.0182
0.624	280	0.802	0.0100	0.0206	0.794	0.00980	0.0208
0.668	300	0.860	0.0115	0.0233	0.851	0.0113	0.0236
0.780	350	1.00	0.0156	0.0306	0.993	0.0153	0.0316
0.891	400	1.15	0.0204	0.0391	1.13	0.0200	0.0404
1.00	450	1.29	0.0259	0.0485	1.28	0.0253	0.0500
1.11	500	1.43	0.0319	0.0587	1.42	0.0313	0.0604
1.23	550	1.58	0.0386	0.0698	1.56	0.0378	0.0718
1.34	600	1.72	0.0460	0.0820	1.70	0.0450	0.0845
1.45	650	1.86	0.0539	0.0950	1.84	0.0528	0.0990
1.56	700	2.01	0.0626	0.109	1.99	0.0613	0.115
1.67	750	2.15	0.0718	0.124	2.13	0.0703	0.131
1.78	800	2.29	0.0817	0.140	2.27	0.0800	0.148
1.89	850	2.44	0.0922	0.156	2.41	0.0903	0.166
2.01	900	2.58	0.103	0.173	2.55	0.101	0.184
2.12	950	2.72	0.115	0.191	2.69	0.113	0.203
2.23	1 000	2.87	0.128	0.210	2.84	0.125	0.224

TABLE E4.1 Friction Loss for Water in Feet per 100 Ft of Pipe (*Continued*)

12 in nominal		Steel-schedule 40 D = 11.938 in ε/D = 0.000151			Asphalt-dipped cast iron ID = 12.00 in ε/D = 0.000400		
Discharge							
ft^3/s	gal/min	V (ft/s)	$V^2/2g$ (ft)	h_f feet per 100 ft of pipe	V (ft/s)	$V^2/2g$ (ft)	h_f feet per 100 ft of pipe
2.45	1 100	3.15	0.154	0.251	3.12	0.151	0.272
2.67	1 200	3.44	0.184	0.296	3.40	0.180	0.321
2.90	1 300	3.73	0.216	0.344	3.69	0.211	0.372
3.12	1 400	4.01	0.250	0.395	3.97	0.245	0.428
3.34	1 500	4.30	0.287	0.450	4.26	0.281	0.488
3.56	1 600	4.59	0.327	0.509	4.54	0.320	0.552
3.79	1 700	4.87	0.369	0.572	4.82	0.361	0.621
4.01	1 800	5.16	0.414	0.636	5.11	0.405	0.695
4.23	1 900	5.45	0.461	0.704	5.39	0.451	0.774
4.46	2 000	5.73	0.511	0.776	5.67	0.500	0.858
4.90	2 200	6.31	0.618	0.930	6.24	0.605	1.03
5.35	2 400	6.88	0.735	1.093	6.81	0.720	1.22
5.79	2 600	7.45	0.863	1.28	7.38	0.845	1.43
6.24	2 800	8.03	1.00	1.47	7.94	0.980	1.65
6.68	3 000	8.60	1.15	1.68	8.51	1.13	1.88
7.13	3 200	9.17	1.31	1.90	9.08	1.28	2.13
7.58	3 400	9.75	1.48	2.13	9.65	1.45	2.41
8.02	3 600	10.3	1.65	2.37	10.2	1.62	2.70
8.47	3 800	10.9	1.84	2.63	10.8	1.81	3.00
8.91	4 000	11.5	2.04	2.92	11.3	2.00	3.31
10.03	4 500	12.9	2.59	3.65	12.8	2.53	4.18
11.1	5 000	14.3	3.19	4.47	14.2	3.13	5.13
12.3	5 500	15.8	3.86	5.38	15.6	3.78	6.17
13.4	6 000	17.2	4.60	6.39	17.0	4.50	7.30
14.5	6 500	18.6	5.39	7.47	18.4	5.28	8.55
15.6	7 000	20.1	6.26	8.63	19.9	6.13	9.92
16.7	7 500	21.5	7.18	9.88	21.3	7.03	11.4
17.8	8 000	22.9	8.17	11.20	22.7	8.00	13.0
18.9	8 500	24.4	9.22	12.6	24.1	9.04	14.7
20.1	9 000	25.8	10.3	14.1	25.5	10.1	16.4
21.2	9 500	27.2	11.5	15.7	26.9	11.3	18.2
22.3	10 000	28.7	12.8	17.4	28.4	12.5	20.2
24.5	11 000	31.5	15.4	21.0	31.2	15.1	24.2
26.7	12 000	34.4	18.3	24.8	34.0	18.0	28.8
29.0	13 000	37.3	21.6	28.9	36.9	21.1	34.0
31.2	14 000	40.1	25.0	33.5	39.7	24.5	39.7
33.4	15 000	43.0	28.7	38.4	42.6	28.1	45.7
35.6	16 000	45.9	32.7	43.7	45.4	32.0	51.8
37.9	17 000	48.7	36.9	49.2	48.2	36.1	58.2
40.1	18 000	51.6	41.4	55.2	51.1	40.5	65.0
42.3	19 000	54.5	46.1	61.5	53.9	45.1	72.1
44.6	20 000	57.3	51.1	68.1	56.7	50.0	79.8

TABLE E4.1 Friction Loss for Water in Feet per 100 Ft of Pipe (*Continued*)

16 in nominal		Steel-schedule 40 ID = 15.000 in ε/D = 0.000120			Asphalt-dipped cast iron ID = 16.00 in ε/D = 0.000300		
Discharge							
ft^3/s	gal/min	V (ft/s)	$V^2/2g$ (ft)	h_f feet per 100 ft of pipe	V (ft/s)	$V^2/2g$ (ft)	h_f feet per 100 ft of pipe
0.668	300	0.545	0.00461	0.00769	0.479	0.00356	0.00581
0.891	400	0.726	0.00820	0.0129	0.638	0.00633	0.00980
1.114	500	0.908	0.0128	0.0193	0.798	0.00989	0.0148
1.34	600	1.09	0.0184	0.0269	0.957	0.0142	0.0207
1.56	700	1.27	0.0251	0.0356	1.12	0.0194	0.0276
1.78	800	1.45	0.0328	0.0454	1.28	0.0253	0.0354
2.01	900	1.63	0.0415	0.0563	1.44	0.0320	0.0441
2.23	1 000	1.82	0.0512	0.0683	1.60	0.0396	0.0537
2.67	1 200	2.18	0.0738	0.0953	1.91	0.0570	0.0760
3.12	1 400	2.54	0.1004	0.127	2.23	0.0775	0.101
3.56	1 600	2.90	0.131	0.163	2.55	0.101	0.130
4.01	1 800	3.27	0.166	0.203	2.87	0.128	0.163
4.46	2 000	3.63	0.205	0.248	3.19	0.158	0.200
5.57	2 500	4.54	0.320	0.377	3.99	0.247	0.307
6.68	3 000	5.45	0.461	0.535	4.79	0.356	0.435
7.80	3 500	6.35	0.627	0.718	5.58	0.485	0.584
8.91	4 000	7.26	0.820	0.921	6.38	0.633	0.754
10.02	4 500	8.17	1.04	1.15	7.18	0.801	0.948
11.1	5 000	9.08	1.28	1.41	7.98	0.989	1.17
13.4	6 000	10.9	1.84	2.01	9.57	1.42	1.66
15.6	7 000	12.7	2.51	2.69	11.2	1.94	2.26
17.8	8 000	14.5	3.28	3.49	12.8	2.53	2.96
20.1	9 000	16.3	4.15	4.38	14.4	3.20	3.73
22.3	10 000	18.2	5.12	5.38	16.0	3.96	4.57
24.5	11 000	20.0	6.20	6.49	17.6	4.79	5.50
26.7	12 000	21.8	7.38	7.69	19.1	5.70	6.52
29.0	13 000	23.6	8.66	8.99	20.7	6.69	7.63
31.2	14 000	25.4	10.04	10.4	22.3	7.75	8.81
33.4	15 000	27.2	11.5	11.9	23.9	8.90	10.1
35.6	16 000	29.0	13.1	13.5	25.5	10.1	11.5
37.9	17 000	30.9	14.8	15.3	27.1	11.4	13.0
40.1	18 000	32.7	16.6	17.2	28.7	12.8	14.6
42.3	19 000	34.5	18.5	19.2	30.3	14.3	16.3
44.6	20 000	36.3	20.5	21.2	31.9	15.8	18.1
49.0	22 000	39.9	24.8	25.5	35.1	19.1	21.8
53.5	24 000	43.6	29.5	30.2	38.3	22.8	25.9
57.9	26 000	47.2	34.6	35.4	41.5	26.7	30.4
62.4	28 000	50.8	40.2	41.0	44.7	31.0	35.3
66.8	30 000	54.5	46.1	47.0	47.9	35.6	40.5
71.3	32 000	58.1	52.4	53.5	51.1	40.5	46.0
75.8	34 000	61.7	59.2	60.2	54.3	45.7	51.9
80.2	36 000	65.4	66.4	67.2	57.4	51.3	58.1
84.7	38 000	69.0	74.0	75.0	60.6	57.1	64.7
89.1	40 000	72.6	82.0	83.0	63.8	63.3	71.7
93.6	42 000	76.2	90.4	91.5	67.0	69.8	79.1
98.0	44 000	79.9	99.2	101	70.2	76.6	86.9
102.5	46 000	83.5	108.4	110	73.4	83.7	95.0
107	48 000	87.1	118	118	76.6	91.2	103
111	50 000	90.8	128	128	79.8	98.9	112

TABLE E4.1 Friction Loss for Water in Feet per 100 Ft of Pipe (*Continued*)

18 in nominal		Steel-schedule 40 ID = 16.876 in ε/D = 0.000107			Asphalt-dipped cast iron ID = 18.00 in ε/D = 0.000267		
Discharge				h_f feet per 100 ft of pipe			h_f feet per 100 ft of pipe
ft³/s	gal/min	V (ft/s)	V²/2g (ft)		V (ft/s)	V²/2g (ft)	
0.668	300	0.430	0.00288	0.00437	0.378	0.00222	0.00328
0.891	400	0.574	0.00512	0.00730	0.504	0.00359	0.00554
1.114	500	0.717	0.00799	0.0109	0.630	0.00618	0.00832
1.34	600	0.861	0.0115	0.0152	0.756	0.00889	0.01162
1.56	700	1.00	0.0157	0.0201	0.883	0.0121	0.0154
1.78	800	1.15	0.0205	0.0256	1.01	0.0158	0.0197
2.01	900	1.29	0.0259	0.0318	1.13	0.0200	0.0245
2.23	1 000	1.43	0.0320	0.0386	1.26	0.0247	0.0298
2.67	1 200	1.72	0.0460	0.0541	1.51	0.0356	0.0420
3.12	1 400	2.01	0.0627	0.0719	1.77	0.0484	0.0560
3.56	1 600	2.30	0.0819	0.092	2.02	0.0632	0.0728
4.01	1 800	2.58	0.1036	0.114	2.27	0.0800	0.0910
4.46	2 000	2.87	0.128	0.139	2.52	0.0988	0.110
5.57	2 500	3.59	0.200	0.211	3.15	0.154	0.170
6.68	3 000	4.30	0.288	0.297	3.78	0.222	0.240
7.80	3 500	5.02	0.392	0.397	4.41	0.303	0.320
8.91	4 000	5.74	0.512	0.511	5.04	0.395	0.415
10.02	4 500	6.45	0.647	0.639	5.67	0.500	0.525
11.1	5 000	7.17	0.799	0.781	6.30	0.618	0.645
13.4	6 000	8.61	1.15	1.11	7.56	0.889	0.920
15.6	7 000	10.0	1.57	1.49	8.83	1.21	1.24
17.8	8 000	11.5	2.05	1.93	10.09	1.58	1.61
20.1	9 000	12.9	2.59	2.42	11.3	2.00	2.02
22.3	10 000	14.3	3.20	2.97	12.6	2.47	2.48
26.7	12 000	17.2	4.60	4.21	15.1	3.56	3.56
31.2	14 000	20.1	6.27	5.69	17.7	4.84	4.85
35.6	16 000	22.9	8.19	7.41	20.2	6.32	6.34
40.1	18 000	25.8	10.36	9.33	22.7	8.00	8.02
44.6	20 000	28.7	12.8	11.5	25.2	9.88	9.88
49.0	22 000	31.6	15.5	13.9	27.7	12.0	11.90
53.5	24 000	34.4	18.4	16.5	30.3	14.2	14.10
57.9	26 000	37.3	21.6	19.2	32.8	16.7	16.50
62.4	28 000	40.2	25.1	22.2	35.3	19.4	19.1
66.8	30 000	43.0	28.8	25.5	37.8	22.9	21.9
71.3	32 000	45.9	32.7	29.0	40.3	25.3	24.9
75.8	34 000	48.8	37.0	32.8	42.9	28.6	28.1
80.2	36 000	51.6	41.4	36.8	45.4	32.0	31.5
84.7	38 000	54.5	46.2	40.8	47.9	35.7	35.1
89.1	40 000	57.4	51.2	45.0	50.4	39.5	38.9
93.6	42 000	60.2	56.4	49.7	53.0	43.6	42.9
98	44 000	63.1	61.9	54.5	55.5	47.8	47.0
102	46 000	66.0	67.7	59.5	58.0	52.3	51.3
107	48 000	68.9	73.7	64.8	60.5	56.9	55.8
111	50 000	71.7	79.9	70.2	63.0	61.8	60.5
123	55 000	78.9	96.7	84.8	69.3	74.7	73.0
134	60 000	86.1	115	101	75.6	88.9	86.7
145	65 000	93.2	135	118	82.0	104.4	101.8
156	70 000	100.4	157	136	88.3	121	118

TABLE E4.1 Friction Loss for Water in Feet per 100 Ft of Pipe (*Continued*)

20 in nominal		Steel-schedule 40 ID = 18.812 in ε/D = 0.0000957			Asphalt-dipped cast iron ID = 20.00 in ε/D = 0.000240		
Discharge							
ft^3/s	gal/min	V (ft/s)	$V^2/2g$ (ft)	h_f feet per 100 ft of pipe	V (ft/s)	$V^2/2g$ (ft)	h_f feet per 100 ft of pipe
0.668	300	0.346	0.00186	0.00258	0.306	0.00146	0.00197
0.891	400	0.462	0.00331	0.00432	0.408	0.00259	0.00332
1.114	500	0.577	0.00517	0.00645	0.511	0.00405	0.00496
1.34	600	0.692	0.00745	0.00897	0.613	0.00583	0.00691
1.56	700	0.808	0.0101	0.01186	0.715	0.00794	0.00918
1.78	800	0.923	0.0132	0.0152	0.817	0.0104	0.0117
2.01	900	1.039	0.0168	0.0188	0.919	0.0131	0.0146
2.23	1 000	1.15	0.0207	0.0227	1.02	0.0162	0.0177
2.67	1 200	1.38	0.0298	0.0318	1.23	0.0233	0.0249
3.12	1 400	1.62	0.0406	0.0422	1.43	0.0318	0.0332
3.56	1 600	1.85	0.0530	0.0538	1.63	0.0415	0.0427
4.01	1 800	2.08	0.0671	0.0669	1.84	0.0525	0.0533
4.46	2 000	2.31	0.0828	0.0812	2.04	0.0648	0.0650
5.57	2 500	2.89	0.129	0.123	2.55	0.1013	0.0998
6.68	3 000	3.46	0.186	0.174	3.06	0.146	0.140
7.80	3 500	4.04	0.254	0.232	3.57	0.198	0.188
8.91	4 000	4.62	0.331	0.298	4.08	0.259	0.243
10.02	4 500	5.19	0.419	0.372	4.59	0.328	0.306
11.1	5 000	5.77	0.517	0.455	5.11	0.405	0.376
13.4	6 000	6.92	0.745	0.645	6.13	0.583	0.533
15.6	7 000	8.08	1.014	0.862	7.15	0.794	0.721
17.8	8 000	9.23	1.32	1.11	8.17	1.04	0.935
20.1	9 000	10.39	1.68	1.39	9.19	1.31	1.18
22.3	10 000	11.5	2.07	1.70	10.2	1.62	1.45
26.7	12 000	13.8	2.98	2.44	12.3	2.33	2.07
31.2	14 000	16.2	4.06	3.29	14.3	3.18	2.80
35.6	16 000	18.5	5.30	4.26	16.3	4.15	3.66
40.1	18 000	20.8	6.71	5.35	18.4	5.25	4.62
44.6	20 000	23.1	8.28	6.56	20.4	6.48	5.67
49.0	22 000	25.4	10.02	7.91	22.5	7.84	6.85
53.5	24 000	27.7	11.9	9.39	24.5	9.33	8.13
57.9	26 000	30.0	14.0	11.0	26.5	10.95	9.54
62.4	28 000	32.3	16.2	12.7	28.6	12.7	11.1
66.8	30 000	34.6	18.6	14.6	30.6	14.6	12.7
71.3	32 000	36.9	21.2	16.6	32.7	16.6	14.4
75.8	34 000	39.2	23.9	18.7	34.7	18.7	16.3
80.2	36 000	41.5	26.8	20.9	36.8	21.0	18.2
84.7	38 000	43.9	29.9	23.2	38.8	23.4	20.2
89.1	40 000	46.2	33.1	25.7	40.8	25.9	22.4
93.6	42 000	48.5	36.5	28.4	42.9	28.6	24.7
98	44 000	50.8	40.1	31.3	44.9	31.4	27.1
102	46 000	53.1	43.8	34.2	47.0	34.3	29.7
107	48 000	55.4	47.7	37.1	49.0	37.3	32.4
111	50 000	57.7	51.7	40.0	51.1	40.5	35.2
123	55 000	63.5	62.6	48.3	56.2	49.0	4

TABLE E4.1 Friction Loss for Water in Feet per 100 Ft of Pipe (*Continued*)

20 in nominal		Steel-schedule 40 ID = 18.812 in ε/D = 0.0000957			Asphalt-dipped cast iron ID = 20.00 in ε/D = 0.000240		
Discharge							
ft³/s	gal/min	V (ft/s)	$V^2/2g$ (ft)	h_f feet per 100 ft of pipe	V (ft/s)	$V^2/2g$ (ft)	h_f feet per 100 ft of pipe
134	60 000	69.2	74.5	57.4	61.3	58.3	50.4
145	65 000	75.0	87.4	67.2	66.4	68.5	59.0
156	70 000	80.8	101.4	77.8	71.5	79.4	68.4
167	75 000	86.6	116	89.3	76.6	91.1	78.6
178	80 000	92.3	132	102	81.7	103.7	89.5
189	85 000	98.1	150	115	86.8	117	101
201	90 000	103.9	168	129	91.9	131	113
212	95 000	109.6	187	143	97.0	146	126
223	100 000	115.4	207	158	102.1	162	139

24 in nominal		Steel-schedule 40 ID = 22.624 in ε/D = 0.0000796			Asphalt-dipped cast iron ID = 24.00 in ε/D = 0.000200		
Discharge							
ft³/s	gal/min	V (ft/s)	$V^2/2g$ (ft)	h_f feet per 100 ft of pipe	V (ft/s)	$V^2/2g$ (ft)	h_f feet per 100 ft of pipe
0.668	300	0.239	0.000891	0.00107	0.213	0.000703	0.000821
0.891	400	0.319	0.00158	0.00178	0.284	0.00125	0.00137
1.114	500	0.399	0.00247	0.00267	0.355	0.00195	0.00205
1.34	600	0.479	0.00356	0.00371	0.426	0.00281	0.00284
1.56	700	0.559	0.00485	0.00490	0.496	0.00383	0.00376
1.78	800	0.638	0.00633	0.00621	0.567	0.00500	0.00480
2.01	900	0.716	0.00801	0.00767	0.638	0.00633	0.00597
2.23	1 000	0.798	0.00989	0.00928	0.709	0.00782	0.00724
2.67	1 200	0.958	0.0142	0.0129	0.851	0.01126	0.0102
3.12	1 400	1.12	0.0194	0.0171	0.993	0.0153	0.0135
3.56	1 600	1.28	0.0253	0.0219	1.135	0.0200	0.0173
4.01	1 800	1.44	0.0321	0.0272	1.276	0.0253	0.0216
4.46	2 000	1.60	0.0396	0.0330	1.42	0.0313	0.0262
5.57	2 500	1.99	0.0618	0.0499	1.77	0.0489	0.0398
6.68	3 000	2.39	0.0891	0.0700	2.13	0.0703	0.0563
7.80	3 500	2.79	0.121	0.0934	2.48	0.0957	0.0759
8.91	4 000	3.19	0.158	0.120	2.84	0.125	0.098
10.02	4 500	3.59	0.200	0.149	3.19	0.158	0.122
11.1	5 000	3.99	0.247	0.181	3.55	0.195	0.149
13.4	6 000	4.79	0.356	0.257	4.26	0.281	0.211
15.6	7 000	5.59	0.485	0.343	4.96	0.383	0.284
17.8	8 000	6.38	0.633	0.441	5.67	0.500	0.368
20.1	9 000	7.18	0.801	0.551	6.38	0.633	0.464
22.3	10 000	7.98	0.989	0.671	7.09	0.782	0.571
26.7	12 000	9.58	1.42	0.959	8.51	1.126	0.816
31.2	14 000	11.2	1.94	1.29	9.93	1.53	1.11
35.6	16 000	12.8	2.53	1.67	11.35	2.00	1.43
40.1	18 000	14.4	3.21	2.10	12.76	2.53	1.80
44.6	20 000	16.0	3.96	2.58	14.2	3.13	2.21
49.0	22 000	17.6	4.79	3.10	15.6	3.78	2.67

TABLE E4.1 Friction Loss for Water in Feet per 100 Ft of Pipe (*Continued*)

24 in nominal		Steel-schedule 40 ID = 22.624 in ε/D = 0.0000796			Asphalt-dipped cast iron ID = 24.00 in ε/D = 0.000200		
Discharge							
ft^3/s	gal/min	V (ft/s)	$V^2/2g$ (ft)	h_f feet per 100 ft of pipe	V (ft/s)	$V^2/2g$ (ft)	h_f feet per 100 ft of pipe
53.5	24 000	19.2	5.70	3.67	17.0	4.50	3.16
57.9	26 000	20.7	6.69	4.29	18.4	5.28	3.71
62.4	28 000	22.3	7.76	4.96	19.9	6.13	4.32
66.8	30 000	23.9	8.91	5.68	21.3	7.03	4.97
71.3	32 000	25.5	10.13	6.42	22.7	8.00	5.65
75.8	34 000	27.1	11.4	7.22	24.1	9.04	6.35
80.2	36 000	28.7	12.8	8.08	25.5	10.13	7.10
84.7	38 000	30.3	14.3	9.00	26.9	11.3	7.90
89.1	40 000	31.9	15.8	9.98	28.4	12.5	8.75
93.6	42 000	33.5	17.5	11.0	29.8	13.8	9.63
98.0	44 000	35.1	19.2	12.1	31.2	15.1	10.5
102	46 000	36.7	20.9	13.2	32.6	16.5	11.5
107	48 000	38.3	22.8	14.3	34.0	18.0	12.5
111	50 000	39.9	24.7	15.5	35.5	19.5	13.6
123	55 000	43.9	29.9	18.7	39.0	23.6	16.4
134	60 000	47.9	35.6	22.3	42.6	28.1	19.5
145	65 000	51.9	41.8	26.2	46.1	33.0	22.9
156	70 000	55.9	48.5	30.4	49.6	38.3	26.5
167	75 000	59.8	55.7	34.8	53.2	44.0	30.5
178	80 000	63.8	63.3	39.4	56.7	50.0	34.7
189	85 000	67.8	71.5	44.4	60.3	56.5	39.2
201	90 000	71.8	80.1	49.7	63.8	63.3	43.9
212	95 000	75.8	89.3	55.5	67.4	70.5	48.9
223	100 000	79.8	98.9	61.5	70.9	78.2	54.2
245	110 000	87.8	120	74.0	78.0	94.6	65.5
267	120 000	95.8	142	88.0	85.1	113	78.0
290	130 000	103.7	167	103	92.2	132	91.5
312	140 000	112	194	119	99.3	153	106
334	150 000	120	223	137	106	200	121

30 in nominal		Steel-schedule 40 ID = 29.000 in ε/D = 0.0000621			Asphalt-dipped cast iron ID = 30.00 in ε/D = 0.000160		
Discharge							
ft^3/s	gal/min	V (ft/s)	$V^2/2g$ (ft)	h_f feet per 100 ft of pipe	V (ft/s)	$V^2/2g$ (ft)	h_f feet per 100 ft of pipe
0.891	400	0.194	0.000587	0.000540	0.182	0.000512	0.000466
1.114	500	0.243	0.000917	0.000805	0.227	0.000800	0.000695
1.34	600	0.291	0.00132	0.001115	0.272	0.00115	0.000964
1.56	700	0.340	0.00180	0.00147	0.318	0.00157	0.00128
1.78	800	0.389	0.00235	0.00187	0.363	0.00205	0.00163
2.01	900	0.437	0.00297	0.00231	0.408	0.00259	0.00202
2.23	1 000	0.486	0.00367	0.00280	0.454	0.00320	0.00244
2.67	1 200	0.583	0.00528	0.00390	0.545	0.00461	0.00343
3.12	1 400	0.680	0.00719	0.00514	0.635	0.00627	0.00452
3.56	1 600	0.777	0.00939	0.00652	0.726	0.00819	0.00577

TABLE E4.1 Friction Loss for Water in Feet per 100 Ft of Pipe (*Continued*)

30 in nominal		Steel-schedule 40 ID = 29.000 in ε/D = 0.0000621			Asphalt-dipped cast iron ID = 30.00 in ε/D = 0.000160		
Discharge							
ft^3/s	gal/min	V (ft/s)	$V^2/2g$ (ft)	h_f feet per 100 ft of pipe	V (ft/s)	$V^2/2g$ (ft)	h_f feet per 100 ft of pipe
4.01	1 800	0.874	0.0119	0.00814	0.817	0.0104	0.00720
4.46	2 000	0.971	0.0147	0.00986	0.908	0.0128	0.00876
5.57	2 500	1.21	0.0229	0.0148	1.13	0.0200	0.0132
6.68	3 000	1.46	0.0330	0.0206	1.36	0.0288	0.0186
7.80	3 500	1.70	0.0449	0.0276	1.59	0.0392	0.0248
8.91	4 000	1.94	0.0587	0.0354	1.82	0.0512	0.0320
10.02	4 500	2.19	0.0742	0.0440	2.04	0.0648	0.0400
11.14	5 000	2.43	0.0917	0.0535	2.27	0.0800	0.0488
13.4	6 000	2.91	0.132	0.0750	2.72	0.115	0.0690
15.6	7 000	3.40	0.180	0.100	3.18	0.157	0.0923
17.8	8 000	3.89	0.235	0.129	3.63	0.205	0.119
20.1	9 000	4.37	0.297	0.161	4.08	0.259	0.149
22.3	10 000	4.86	0.367	0.196	4.54	0.320	0.183
26.7	12 000	5.83	0.528	0.277	5.45	0.461	0.260
31.2	14 000	6.80	0.719	0.371	6.35	0.627	0.351
35.6	16 000	7.77	0.939	0.478	7.26	0.819	0.455
40.1	18 000	8.74	1.19	0.598	8.17	1.04	0.572
44.6	20 000	9.71	1.47	0.732	9.08	1.28	0.703
55.7	25 000	12.1	2.29	1.13	11.3	2.00	1.09
66.8	30 000	14.6	3.30	1.61	13.6	2.88	1.57
78.0	35 000	17.0	4.49	2.17	15.9	3.92	2.13
89.1	40 000	19.4	5.87	2.83	18.2	5.12	2.77
100	45 000	21.9	7.42	3.56	20.4	6.48	3.50
111	50 000	24.3	9.17	4.38	22.7	8.00	4.30
134	60 000	29.1	13.2	6.23	27.2	11.5	6.19
156	70 000	34.0	18.0	8.43	31.8	15.7	8.39
178	80 000	38.9	23.5	11.0	36.3	20.5	10.9
201	90 000	43.7	29.7	13.8	40.8	25.9	13.8
223	100 000	48.6	36.7	17.0	45.4	32.0	17.0
245	110 000	53.4	44.4	20.6	49.9	38.7	20.5
267	120 000	58.3	52.8	24.5	54.5	46.1	24.4
290	130 000	63.1	62.0	28.7	59.0	54.1	28.6
312	140 000	68.0	71.9	33.3	63.5	62.7	33.1
334	150 000	72.9	82.5	38.2	68.1	72.0	38.0
356	160 000	77.7	93.9	43.3	72.6	81.9	43.2
379	170 000	82.6	106	48.8	77.2	92.5	48.7
401	180 000	87.4	119	54.7	81.7	104	54.7
423	190 000	92.3	132	60.8	86.2	116	61.0
446	200 000	97.1	147	67.1	90.8	128	67.6
468	210 000	102	162	73.8	95.3	141	74.5
490	220 000	107	177	81.0	99.8	155	81.7
512	230 000	112	194	88.6	104	169	89.2
535	240 000	117	211	96.7	109	184	97.0
557	250 000	121	229	106	113	200	105

TABLE E4.1 Friction Loss for Water in Feet per 100 Ft of Pipe (*Continued*)

36 in ID				Steel $\varepsilon/D = 0.0000500$	Cast iron $\varepsilon/D = 0.000133$
Discharge					
ft^3/s	gal/min	V (ft/s)	$V^2/2g$ (ft)	h_f feet per 100 ft of pipe	h_f feet per 100 ft of pipe
2.23	1 000	0.315	0.00154	0.000988	0.00101
2.67	1 200	0.378	0.00222	0.00137	0.00140
3.12	1 400	0.441	0.00303	0.00181	0.00186
3.56	1 600	0.504	0.00395	0.00231	0.00237
4.01	1 800	0.567	0.00500	0.00285	0.00295
4.46	2 000	0.630	0.00618	0.00344	0.00357
5.57	2 500	0.788	0.00965	0.00517	0.00538
6.68	3 000	0.946	0.0139	0.00721	0.00751
7.80	3 500	1.103	0.0189	0.00957	0.0101
8.91	4 000	1.26	0.0247	0.0122	0.0129
10.02	4 500	1.41	0.0313	0.0152	0.0161
11.14	5 000	1.58	0.0386	0.0185	0.0196
13.4	6 000	1.89	0.0556	0.0260	0.0276
15.6	7 000	2.21	0.0756	0.0345	0.0369
17.8	8 000	2.52	0.0988	0.0442	0.0475
20.1	9 000	2.84	0.125	0.0551	0.0593
22.3	10 000	3.15	0.154	0.0670	0.0724
26.7	12 000	3.78	0.222	0.0942	0.103
31.2	14 000	4.41	0.303	0.126	0.139
35.6	16 000	5.04	0.395	0.162	0.180
40.1	18 000	5.67	0.500	0.203	0.227
44.6	20 000	6.30	0.618	0.248	0.279
55.7	25 000	7.88	0.965	0.378	0.430
66.8	30 000	9.46	1.39	0.540	0.617
78.0	35 000	11.03	1.89	0.724	0.832
89.1	40 000	12.6	2.47	0.941	1.08
100	45 000	14.1	3.13	1.18	1.36
111	50 000	15.8	3.86	1.45	1.68
134	60 000	18.9	5.56	2.07	2.40
156	70 000	22.1	7.56	2.81	3.25
178	80 000	25.2	9.88	3.66	4.23
201	90 000	28.4	12.5	4.59	5.34
223	100 000	31.5	15.4	5.64	6.58
267	120 000	37.8	22.2	8.05	9.50
312	140 000	44.1	30.3	10.9	12.9
356	160 000	50.4	39.5	14.2	16.8
401	180 000	56.7	50.0	17.9	21.3
446	200 000	63.0	61.8	22.1	26.3
557	250 000	78.8	96.5	34.4	41.0
668	300 000	94.6	139	49.4	58.8
780	350 000	110	189	67.0	80.0
891	400 000	126	247	87.3	105

TABLE E4.1 Friction Loss for Water in Feet per 100 Ft of Pipe (*Continued*)

42 in ID				Steel ε/D = 0.0000429	Cast iron ε/D = 0.000114
Discharge					
ft³/s	gal/min	V (ft/s)	V²/2g (ft)	h_f feet per 100 ft of pipe	h_f feet per 100 ft of pipe
2.23	1 000	0.232	0.000833	0.000471	0.000481
3.34	1 500	0.347	0.00187	0.000977	0.000997
4.46	2 000	0.463	0.00333	0.00164	0.00168
5.57	2 500	0.579	0.00521	0.00246	0.00252
6.68	3 000	0.695	0.00750	0.00343	0.00353
7.80	3 500	0.811	0.0102	0.00454	0.00470
8.91	4 000	0.926	0.0133	0.00580	0.00602
10.02	4 500	1.042	0.0169	0.00720	0.00750
11.14	5 000	1.16	0.0208	0.00874	0.00915
13.4	6 000	1.39	0.0300	0.0122	0.0128
15.6	7 000	1.62	0.0408	0.0162	0.0172
17.8	8 000	1.85	0.0533	0.0208	0.0222
20.1	9 000	2.08	0.0675	0.0258	0.0276
22.3	10 000	2.32	0.0833	0.0314	0.0337
26.7	12 000	2.78	0.120	0.0441	0.0477
31.2	14 000	3.24	0.163	0.0591	0.0641
35.6	16 000	3.71	0.213	0.0758	0.0829
40.1	18 000	4.17	0.270	0.0944	0.104
44.6	20 000	4.63	0.333	0.115	0.127
55.7	25 000	5.79	0.521	0.176	0.196
66.8	30 000	6.95	0.750	0.250	0.279
78.0	35 000	8.11	1.02	0.334	0.377
89.1	40 000	9.26	1.33	0.433	0.490
100	45 000	10.42	1.69	0.545	0.619
111	50 000	11.6	2.08	0.668	0.760
134	60 000	13.9	3.00	0.946	1.09
156	70 000	16.2	4.08	1.27	1.48
178	80 000	18.5	5.33	1.66	1.92
201	90 000	20.8	6.75	2.08	2.42
223	100 000	23.2	8.33	2.57	2.98
267	120 000	27.8	12.0	3.67	4.30
312	140 000	32.4	16.3	4.98	5.82
356	160 000	37.1	21.3	6.46	7.58
401	180 000	41.7	27.0	8.12	9.58
446	200 000	46.3	33.3	10.00	11.8
557	250 000	57.9	52.1	15.6	18.4
668	300 000	69.5	75.0	22.3	26.5
780	350 000	81.1	102	30.4	36.1
891	400 000	92.6	133	39.6	47.2
1002	450 000	104.2	169	50.1	59.7
1114	500 000	116	208	67.7	73.6

TABLE E4.1 Friction Loss for Water in Feet per 100 Ft of Pipe (*Continued*)

48 in ID				Steel ε/D = 0.0000375	Cast iron ε/D = 0.000100
Discharge					
ft³/s	gal/min	V (ft/s)	$V^2/2g$ (ft)	h_f feet per 100 ft of pipe	h_f feet per 100 ft of pipe
3.34	1 500	0.266	0.00110	0.000508	0.000521
4.46	2 000	0.355	0.00195	0.000855	0.000883
5.57	2 500	0.443	0.00305	0.00129	0.00133
6.68	3 000	0.532	0.00440	0.00180	0.00185
7.80	3 500	0.621	0.00598	0.00238	0.00245
8.91	4 000	0.709	0.00782	0.00304	0.00314
10.02	4 500	0.798	0.00989	0.00378	0.00391
11.14	5 000	0.887	0.01221	0.00458	0.00474
13.4	6 000	1.064	0.0176	0.00636	0.00667
15.6	7 000	1.24	0.0239	0.00844	0.00890
17.8	8 000	1.42	0.0313	0.0108	0.0114
20.1	9 000	1.60	0.0396	0.0134	0.0142
22.3	10 000	1.77	0.0489	0.0163	0.0173
26.7	12 000	2.13	0.0703	0.0229	0.0244
31.2	14 000	2.48	0.0957	0.0305	0.0327
35.6	16 000	2.84	0.125	0.0391	0.0422
40.1	18 000	3.19	0.158	0.0488	0.0529
44.6	20 000	3.55	0.195	0.0598	0.0648
55.7	25 000	4.43	0.305	0.0910	0.0996
66.8	30 000	5.32	0.440	0.128	0.142
78.0	35 000	6.21	0.598	0.172	0.192
89.1	40 000	7.09	0.782	0.222	0.248
100.2	45 000	7.98	0.989	0.278	0.314
111.4	50 000	8.87	1.221	0.341	0.384
134	60 000	10.64	1.76	0.484	0.548
156	70 000	12.4	2.39	0.652	0.742
178	80 000	14.2	3.13	0.849	0.968
201	90 000	16.0	3.96	1.06	1.22
223	100 000	17.7	4.89	1.30	1.50
267	120 000	21.3	7.03	1.87	2.15
312	140 000	24.8	9.57	2.51	2.92
356	160 000	28.4	12.5	3.26	3.81
401	180 000	31.9	15.8	4.11	4.83
446	200 000	35.5	19.5	5.05	5.97
557	250 000	44.3	30.5	7.88	9.28
668	300 000	53.2	44.0	11.3	13.4
780	350 000	62.1	59.8	15.3	18.2
891	400 000	70.9	78.2	20.0	23.7
1002	450 000	79.8	98.9	25.2	29.9
1114	500 000	88.7	122.1	31.1	36.8
1225	550 000	97.5	148	37.6	44.5
1337	600 000	106.4	176	44.7	53.0

TABLE E4.1 Friction Loss for Water in Feet per 100 Ft of Pipe (*Continued*)

60 in ID				Steel $\varepsilon/D = 0.0000300$	Cast iron $\varepsilon/D = 0.0000800$
Discharge					
ft^3/s	gal/min	V (ft/s)	$V^2/2g$ (ft)	h_f feet per 100 ft of pipe	h_f feet per 100 ft of pipe
4.46	2 000	0.227	0.000800	0.000293	0.000298
5.57	2 500	0.284	0.00125	0.000440	0.000446
6.68	3 000	0.340	0.00180	0.000612	0.000621
7.80	3 500	0.397	0.00245	0.000810	0.000824
8.91	4 000	0.454	0.00320	0.00103	0.00105
10.02	4 500	0.511	0.00405	0.00128	0.00131
11.14	5 000	0.567	0.00500	0.00155	0.00159
13.4	6 000	0.681	0.00720	0.00216	0.00223
15.6	7 000	0.794	0.00980	0.00285	0.00297
17.8	8 000	0.908	0.0128	0.00365	0.00382
20.1	9 000	1.021	0.0162	0.00454	0.00476
22.3	10 000	1.13	0.0200	0.00550	0.00579
26.7	12 000	1.36	0.0288	0.00766	0.00815
31.2	14 000	1.59	0.0392	0.0102	0.0108
35.6	16 000	1.82	0.0512	0.0131	0.0140
40.1	18 000	2.04	0.0648	0.0163	0.0174
44.6	20 000	2.27	0.0800	0.0198	0.0212
55.7	25 000	2.84	0.125	0.0301	0.0325
66.8	30 000	3.40	0.180	0.0424	0.0460
78.0	35 000	3.97	0.245	0.0567	0.0618
89.1	40 000	4.54	0.320	0.0730	0.0800
100.2	45 000	5.11	0.405	0.0916	0.100
111.4	50 000	5.67	0.500	0.112	0.124
134	60 000	6.81	0.720	0.158	0.176
156	70 000	7.94	0.980	0.213	0.237
178	80 000	9.08	1.28	0.275	0.307
201	90 000	10.21	1.62	0.344	0.387
223	100 000	11.3	2.00	0.420	0.478
267	120 000	13.6	2.88	0.600	0.688
312	140 000	15.9	3.92	0.806	0.930
356	160 000	18.2	5.12	1.04	1.20
401	180 000	20.4	6.48	1.32	1.52
446	200 000	22.7	8.00	1.62	1.87
557	250 000	28.4	12.5	2.52	2.92
668	300 000	34.0	18.0	3.60	4.20
780	350 000	39.7	24.5	4.88	5.71
891	400 000	45.4	32.0	6.34	7.42
1002	450 000	51.1	40.5	8.01	9.40
1114	500 000	56.7	50.0	9.87	11.6
1225	550 000	62.4	60.5	11.9	14.0
1337	600 000	68.1	72.0	14.1	16.7
1448	650 000	73.8	84.5	16.6	19.6
1560	700 000	79.4	98.0	19.2	22.7
1671	750 000	85.1	112.6	22.0	26.0
1782	800 000	90.8	128	25.0	29.6

TABLE E4.1 Friction Loss for Water in Feet per 100 Ft of Pipe (*Continued*)

84 in ID				Steel $\varepsilon/D = 0.0000214$	Cast iron $\varepsilon/D = 0.0000571$
Discharge					
ft^3/s	gal/min	V (ft/s)	$V^2/2g$ (ft)	h_f feet per 100 ft of pipe	h_f feet per 100 ft of pipe
6.68	3 000	0.174	0.000469	0.000121	0.000122
8.91	4 000	0.232	0.000833	0.000203	0.000206
11.14	5 000	0.289	0.00130	0.000306	0.000309
13.4	6 000	0.347	0.00188	0.000425	0.000432
15.6	7 000	0.405	0.00255	0.000562	0.000573
17.8	8 000	0.463	0.00333	0.000717	0.000731
20.1	9 000	0.521	0.00422	0.000891	0.000910
22.3	10 000	0.579	0.00521	0.00108	0.00110
26.7	12 000	0.695	0.00750	0.00150	0.00154
31.2	14 000	0.811	0.01021	0.00199	0.00205
35.6	16 000	0.926	0.0133	0.00255	0.00262
40.1	18 000	1.042	0.0169	0.00316	0.00327
44.6	20 000	1.16	0.0208	0.00384	0.00400
55.7	25 000	1.45	0.0326	0.00579	0.00606
66.8	30 000	1.74	0.0469	0.00810	0.00858
78.0	35 000	2.03	0.0638	0.0108	0.0115
89.1	40 000	2.32	0.0833	0.0139	0.0148
100.2	45 000	2.61	0.105	0.0174	0.0185
111.4	50 000	2.89	0.130	0.0212	0.0226
134	60 000	3.47	0.180	0.0298	0.0321
156	70 000	4.05	0.255	0.0398	0.0431
178	80 000	4.63	0.333	0.0513	0.0558
201	90 000	5.21	0.422	0.0640	0.0700
223	100 000	5.79	0.521	0.0781	0.0866
267	120 000	6.95	0.750	0.111	0.122
312	140 000	8.11	1.021	0.149	0.166
356	160 000	9.26	1.33	0.193	0.216
401	180 000	10.42	1.69	0.242	0.272
446	200 000	11.6	2.08	0.297	0.334
557	250 000	14.5	3.26	0.458	0.516
668	300 000	17.4	4.69	0.649	0.740
780	350 000	20.3	6.38	0.880	1.00
891	400 000	23.2	8.33	1.14	1.30
1002	450 000	26.1	10.5	1.44	1.65
1114	500 000	28.9	13.0	1.78	2.04
1225	550 000	31.8	15.8	2.14	2.47
1337	600 000	34.7	18.0	2.54	2.94
1448	650 000	37.6	22.0	2.97	3.45
1560	700 000	40.5	25.5	3.43	4.00
1671	750 000	43.4	29.3	3.93	4.58
1782	800 000	46.3	33.3	4.47	5.20
1894	850 000	49.2	37.6	5.04	5.87
2005	900 000	52.1	42.2	5.64	6.58
2117	950 000	55.0	47.0	6.29	7.32
2228	1 000 000	57.9	52.1	6.95	8.10

Note: No allowance has been made for age, differences in diameter, or any abnormal condition of interior surface. Any factor of safety must be estimated from the local conditions and the requirements of each particular installation.

Source: Friction loss, h_f, is derived from Darcy-Weisbach equations. (*Hydraulic Institute Engineering Data Book,* Table IIIB-4).

INDEX

ABOUT THE EDITOR IN CHIEF

Mohinder L. Nayyar is the piping and valve specialist in the Mechanical Engineering Department of the Gaithersburg Regional Office of Bechtel Corporation. He is chairman of the Bechtel Inservice Inspection Task Force, and he is responsible for the development of the Mechanical Department training programs, design guides, and standards. Mr. Nayyar is a contributor to the seventh edition of the *McGraw-Hill Encyclopedia of Science & Technology* and has published technical papers.

Mr. Nayyar is a registered engineer in Virginia and a member of ASME. He is a member of ASME B31.1, Power Piping, Code Committee and participates in ASME Section XI Code Committee activities as an observer. He resides in Brookeville, Maryland.

ABOUT THE EDITOR IN CHIEF